The Circle

Figure	Sketch	Circumference (c)	Area (A)
Circle		$c = 2\pi r$ $c = \pi d$	$A = \pi r^2$ $A = \dfrac{\pi d^2}{4}$

Perimeters and Areas of Plane Figures

Figure	Sketch	Perimeter (P)	Area (A)
Triangle		$P = s_1 + s_2 + b$	$A = \dfrac{1}{2}\,ba$
Trapezoid		$P = s_1 + s_2 + b_1 + b_2$	$A = \dfrac{1}{2}(b_1 + b_2)a$
Parallelogram		$P = 2\ell + 2w$	$A = \ell a$
Rhombus		$P = 4s$	$A = sa$
Rectangle		$P = 2\ell + 2w$	$A = \ell w$
Square		$P = 4s$	$A = s^2$

Basic Technical Mathematics with Calculus

Second Edition

Stuart R. Porter
Monroe Community College

John F. Ernst
Monroe Community College

HarperCollins*CollegePublishers*

This book is dedicated to my wife, Joyce, without whose patience, love, and understanding this book would not have been possible.

Sponsoring Editor: Anne Kelly
Development Editor: Ann Shaffer
Project Editor: Lisa A. De Mol
Design Administrator: Jess Schaal
Text Design: Ellen Pettengell
Cover Design: Ellen Pettengell
Cover Photo: Superstock: Porsche 930 (911 Turbo)
Production Administrator: Randee Wire
Compositor: Interactive Composition Corporation
Printer and Binder: R. R. Donnelley and Sons

Architectural drawing in lower portion of designs on pages 115, 249, 529, 675, 756, 904, and 1113 is courtesy of Skidmore, Owings & Merrill.

Basic Technical Mathematics with Calculus, second edition

Library of Congress Cataloging-in-Publication Data

Porter, Stuart R., 1932–
  Basic technical mathematics with calculus / Stuart R. Porter, John F. Ernst. —
2nd ed.
    p.   cm.
  Includes index.
  ISBN 0-673-46176-9
  1. Mathematics.   I Ernst, John F.   II Title.

QA37.2.P665 1995
510—dc20   94-13873
                                                        CIP

95 96 97 98 9 8 7 6 5 4 3 2

Table of Contents

Appendices

Preface

This second edition of *Basic Technical Mathematics with Calculus* is designed to provide students with the necessary mathematical skills to pursue a course of study in scientific or engineering technology or to prepare students for calculus and provide an introduction to differential and integral calculus. To use this book effectively, students should have completed two years of high-school mathematics (algebra and geometry). For students requiring a quick review, Chapters 1 through 3 offer a concise summary of essential topics in algebra and geometry. The text can be used in a one-, two- or three-semester program.

A Note on Graphics Calculators

This book has been written to accommodate the full range of attitudes toward graphics calculators. Instructors who rely on the graphics calculator as an essential teaching tool will appreciate the special boxed keystrokes, examples, and exercises. This material is designed both to reinforce mathematical concepts and to familiarize students with the most popular models of graphics calculators.

Instructors who do not use the graphics calculators can easily skip the graphics calculator material, since it is all clearly marked. In some cases, concepts are covered more than once to make the book effective for both graphics calculator users and nonusers. For example, the "Introduction to the Graphics Calculator" in Chapter 1 introduces basic calculator techniques and concepts. It also includes a discussion of the coordinate plane, which is repeated and expanded upon in Chapter 5 for the benefit of all students.

Features

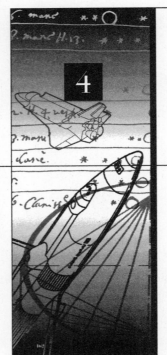

Chapter Openers

Chapter openers present a real-world problem drawn from principles taught in *Basic Technical Mathematics with Calculus.* Students solve these problems later on in the chapter.

4 Linear Equations and Dimensional Analysis

The Saturn rocket often moves objects from the surface of earth to outer space. Though you may have seen many fiery blast-offs, have you thought about the forces needed to lift these metal giants into orbit?

The weight of the Saturn 5 before takeoff is 6,391,120 pounds! The thrust of the engine must at least exceed the weight of the rocket, or it will never take off. The first stage thrust of the Saturn 5 rocket is about 8 million pounds, more than enough to insure a lift-off.

The resultant acceleration of the rocket is computed by determining the difference between the downward pull of its weight and the upward push of the thrust. The equation is $F_{net} = ma$ (m stands for the mass of the object, which is in this case is 198,000 slugs). What is the acceleration of the rocket lift-off? The solution is given in Section 4.4, Example 6.

Graphics Calculator Boxes, Examples, and Exercises

Throughout the text, boxes provide calculator solutions to section examples. These include keystrokes for the most popular models—the T1-82 and the Casio fx-7700G. Graphics calculator examples and exercises are indicated by an identifying symbol so they can be skipped if desired.

---

380   Chapter 8   Graphs of Trigonometric Functions

EXAMPLE 7

Use a graphics calculator to graph the functions in the interval $0° \le \theta \le 360°$.

a. $y = 4\cos 2\theta$ **b.** $y = 2\cos\dfrac{\theta}{2}$

c. $y = \dfrac{1}{2}\cos 3\theta$ **d.** $y = \dfrac{1}{4}\cos\dfrac{\theta}{3}$

For each of the functions determine the amplitude, the period, and the number of cycles in $0° \le \theta \le 360°$.

Solution   The graphs of the functions are given in Figure 8.26a, b, c, and d, respectively.

a. The amplitude is 4. The period $= \dfrac{360°}{2}$ or 180°. There are two complete cycles of the curve in the interval $0° \le \theta \le 360°$, as shown in Figure 8.26a.

b. The amplitude is 2. The period $= \dfrac{360°}{\frac{1}{2}}$ or 720°. There is one-half of a complete cycle of the curve in 360°, as shown in Figure 8.26b.

c. The amplitude is $\dfrac{1}{2}$. The period $= \dfrac{360°}{3}$ or 120°. There are three complete cycles of the curve in 360°, see Figure 8.26c.

d. The amplitude is $\dfrac{1}{4}$. The period $= \dfrac{360°}{\frac{1}{3}}$ or 1080°. There is one-third of a cycle in 360°, see Figure 8.26d.   ◆

FIGURE 8.26

8.3 Exercises

In Exercises 1–12, determine the following. **a.** The amplitude **b.** The period of each function **c.** The number of cycles in $0 \le \theta \le 2\pi$.

1. $y = 3\sin\theta$ **2.** $y = \dfrac{1}{2}\cos\theta$ **3.** $y = 0.75\cos\theta$ **4.** $y = 1.3\sin\theta$

5. $y = \sin 3\theta$ **6.** $y = \cos\dfrac{\theta}{4}$ **7.** $y = -6\cos\dfrac{\theta}{4}$ **8.** $y = 2\sin\pi\theta$

9. $y = \dfrac{1}{3}\sin\dfrac{\theta}{3}$ **10.** $y = -\pi\cos\dfrac{\pi}{2}\theta$ **11.** $y = -\dfrac{1}{3}\cos\dfrac{1}{3}\theta$ **12.** $y = \dfrac{5}{3}\sin\dfrac{5}{3}\theta$

Examples

Approximately 1300 examples include detailed, well-explained solutions. Approximately 220 of these examples illustrate technical applications. Where appropriate, examples include a brief descriptive title to help students understand the purpose of the example and to aid in studying for examinations.

Cautionary Remarks

Common student errors and difficulties are highlighted and identified with the heading "Caution."

8.1   Periodic Functions   **353**

EXAMPLE 1   **Determining the Period**

For the function in Figure 8.2, determine the period p.

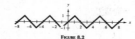

FIGURE 8.2

Solution   The function goes through one complete cycle for each of the intervals from -8 to -4, -4 to 0, 0 to 4, and 4 to 8. Thus, the period is four units. Using the definition, we see that:

$$f(-4) = f(-4 + 4) = f(0) = 0$$
$$f(1) = f(1 + 4) = f(5) = 1.$$
$$f(-1) = f(-1 + 4) = f(3) = -1 \quad \blacklozenge$$

CAUTION   We may be tempted to let $p = 2$ in Example 1. To see that this is incorrect, let $x = 1$ and substitute for x and p in the equality $f(x + p) = f(x)$. Thus, $f(1 + 2) \neq f(1)$ since $f(1 + 2) = -1$ and $f(1) = 1$. Clearly the function has only traveled through one-half of its full cycle. To determine the period p of the function, use the statement $f(x + p) = f(x)$ for all x. If we select $x = -4$, then $f(-4 + 2) = f(-2) = 0$ and $f(-4) = 0$, which gives the wrong impression–that 2 is the period. Therefore, always try more than one value to check. ■

As shown in Figure 8.2, the graph goes through one complete cycle from 0 to 4, completing four cycles from -8 to 8.

DEFINITION 8.2

A **cycle** of a periodic function is a portion of the graph of the function from any point on the graph to the first point at which the graph starts repeating itself. (See Figure 8.3.)

FIGURE 8.3

Key Terms

Essential terms are highlighted in boldface type within the explanations and examples.

Definitions, Formulas, Rules, and Procedures

These items are outlined in boxes to stress the importance of the material.

---

452   Chapter 10   Systems of Linear Equations and Determinants

10.1

Introduction to Linear Equations

In Chapter 4, we learned how to solve linear equations with one variable, and in Chapter 5, we graphed linear equations in two variables. In technical fields, people commonly work with linear equations with two variables. The general form of the linear equation in two variables is $ax + by + c = 0$. A specific example is:

$$3x + 4y - 12 = 0, \quad \text{where } a = 3, b = 4 \quad \text{and} \quad c = -12.$$

We can tell that the equation is linear since the exponents of both the variables x and y are 1. For example, $3x^2 - 5y + 2 = 0$ is not a linear equation because the exponent of the variable x is 2. Another way of recognizing a linear equation is by its graph, which is always a straight line (as illustrated in Chapter 5).

A linear equation may take the form $y = mx + b$, where m and b are constants and x and y are the independent and dependent variables, respectively. This form of the equation of a line is called the **slope-intercept form** because m is the slope of the line, and b is the ordinate of the y-intercept. The **slope** of the line is a ratio of two numbers that indicates the slant of the line with respect to the positive x-axis. The slope is also defined as the angle the line forms with respect to the positive x-axis. The **y-intercept** is the point where the line crosses the y-axis.

DEFINITION 10.1

The **slope-intercept form** of an equation of a line is:

$$y = mx + b.$$

The equation $3x + 4y - 12 = 0$ can be written in the form $y = mx + b$.

$$4y = -3x + 12 \qquad \text{Subtracting } 3x \text{ and adding 12 to both sides of the equation}$$
$$y = \frac{-3}{4}x + 3 \qquad \text{Dividing both sides of the equation by 4.}$$

By rewriting the equation in this form, we see that $m = \frac{-3}{4}$ and $b = 3$. The graph is sketched in Figure 10.1. The line crosses the y-axis at the point (0, 3), the y-intercept.

Now we will look at the slope. As shown in Figure 10.2, draw a line parallel to the y-axis through the point $A(-4, 6)$ and another line parallel to the x-axis through the point $B(8, -3)$. These lines intersect at the point $C(-4, -3)$. The resulting figure is right triangle ABC, with a right angle at $C(-4, -3)$. The length of the line segment BC is:

$$\overline{BC} = -4 - 8 = -12$$

(0, 3)

$y = -\frac{3}{4}x + 3$

(4, 0)

FIGURE 10.1

$A(-4, 6)$

$\Delta y = 9$

$C(-4, -3)$ $B(8, -3)$

$\Delta x = -12$

FIGURE 10.2

49. From a land-based radar station it is determined that the bearing of ship A is 70° and it is a distance of 3.5 km from the station, while ship B has a bearing of 110° and is a distance of 5.2 km from the station. Determine the distance between the two ships.

50.* One side of a triangle is 3.472 times as long as another side x. The angle between them is 32° 01'.
   a. Determine the other two angles.
   b. Determine the third side in terms of the side x.

51.* A local carnival has a ferris wheel. There are 18 seats equally spaced on the wheel. If the radius of the wheel is 18 ft, determine the distance between any two adjacent seats on the wheel. (See the illustration.)

Writing About Mathematics

1. Two landscape designers were asked to design a triangular-shaped park. The city council does not understand why each design is shaped differently. Each designer claims that, with the information given, the shape of the triangle is correct. The only information given the designers was "two sides and an angle opposite one of the sides." Write a letter explaining why two solutions are possible.

2. You and a friend, having chopped down a tree, now need to move its trunk 25 ft to the edge of the road so that it can be trucked to the sawmill. You have attached two ropes to the log, and each of you is pulling on a rope. Using the sum of two vectors, explain how anyone can determine the force exerted by pulling on the two ropes.

Chapter Test

In Exercises 1 and 2, sketch the vectors on a coordinate system.

1. $A = 4/30°$
2. $B = 6/150°$

In Exercises 3 and 4, use the fact that $A = 3.5/137°$ to determine the following.

3. The horizontal component
4. The vertical component

In Exercises 5 and 6, use the fact that $A = 5.0/45°$ and $B = 7.0/215°$ to determine the following.

5. $A + B$
6. $B - A$
7. Determine the resultant of $5.0/-45° - 7.0/115° + 3.0/210°$.

In Exercises 8–13, determine the measure of the three other parts of the triangle, if possible. See the diagram for proper labeling.

8. $A = 42°, B = 60°, b = 15$
9. $B = 35°, b = 12, a = 16$
10. $A = 75°, b = 10, c = 20$
11. $B = 35°, b = 16, a = 12$
12. $C = 160°, a = 8, b = 12$
13. $a = 12, b = 25, c = 24$

Writing about Mathematics

The writing exercises at the end of each chapter encourage a deeper understanding of concepts. They also can be used as topics for group discussion. Answers are not given for these exercises because they are open-ended and instructors may use them in different ways.

---

*35. Write a computer program to determine the mean of a set of data.

*36. Write a computer program to determine the standard deviation of a set of data.

*37. The following formulas can both be used to calculate the standard deviation of ungrouped data. Show that these equations are equal.

$$s = \sqrt{\frac{\sum (x_i - \bar{X})^2}{n - 1}} \qquad s = \sqrt{\frac{n(\sum x_i^2) - (\sum x_i)^2}{n(n - 1)}}$$

Writing About Mathematics

1. Your friend does not understand how your school determines grade point averages. You recognize that the grade point average is the same as a weighted mean. Write a few paragraphs, including examples, to explain how to calculate a grade point average so that your friend does understand.

2. In labor negotiations the union states that the average pay of company employees is $35,000 per year. On the other hand, management claims that the average pay of employees is $42,000 per year. Both claims are correct because one group is using the median and the other is using the mean. Discuss why the difference exists and why each used a different average to make their point.

3. A psychology instructor assumes that her test grades are normally distributed. That is, she assigns letter grades by the method shown in the diagram. On a particular exam a student scores 116 out of a possible 120. The letter grade the teacher assigns is a C. Explain how that grade is possible using her technique.

4. Using the grading technique shown in Exercise 3, discuss the following questions. Is it possible that no student will receive an A? Is is possible that no student will fail?

Chapter Test

In Exercises 1–5, use the data set 2, 2, 3, 4, 4, 5, 6, 8, 8, 9.

1. Determine the mean.
2. Determine the median.
3. Determine the mode.
4. Determine the range.
5. Determine the standard deviation.

Content Highlights

- Review material is collected in Chapters 1 through 3. Sections 1.5, 1.6, and 3.1 are important in understanding other parts of the text. The graphics calculator is introduced in Section 1.5, and the discussion provides basic rules for working with the calculator. The rules for expressing answers for problems containing approximate numbers are discussed in Section 1.6. These rules are used throughout the text. Section 3.1 is a review of plane figures and emphasizes angles, providing helpful background material for trigonometry in Chapter 6.

- In Chapter 4, Sections 4.1 and 4.2 provide the rules for solving simple linear equations. These rules are then applied in Sections 4.3 and 4.4 in working with dimensional units. Ratio, proportion, and variation are discussed early in the text because they are important tools in technical courses.

- Chapter 5 gives a brief introduction to the concept of function, providing the student with the basic terminology that can be used to build on the concept of function as we move throughout the text. Strong emphasis is given to graphing and to the use of graphs in understanding how functions behave.

- Chapters 6, 7, and 8 provide an early introduction to trigonometry as is required by some technical areas. In Chapter 6 the trigonometric functions are developed using only degree measurement; radian measure is then introduced. In Chapter 8 conventional graphing techniques for the trigonometric functions are discussed as well as graphics calculator techniques.

- Chapter 9 gives an introduction to imaginary and complex numbers. The applications in this chapter are directed primarily toward the field of electronics. Sections 9.1 and 9.2 are beneficial to the student in understanding imaginary roots of nonlinear equations.

- Chapters 10, 11, and 12 deal with the techniques of solving linear equations, quadratic equations, and equations of higher degree. The calculator is introduced as a tool in solving systems of equations using determinants and matrices. Additional graphs are used to help provide a better understanding of the techniques for determining roots of equations.

- In Chapter 13 logarithmic and exponential functions and the graphing of these functions is discussed. Techniques of solving equations, and logarithms to other bases, in addition to base 10 and base e are stressed in this chapter.

- In Chapter 14 the techniques of solving linear, quadratic, and higher-degree inequalities are discussed. The sign method is used to solve inequalities of higher degree.

- In Chapter 15 trigonometric identities, such as sum and difference of two angles, double angles, and half-angles, are developed. Graphing is introduced as a method of checking the equality of trigonometric statements. Methods of solving trigonometric equations and inverse trigonometric functions also are discussed.

- Chapter 16 provides the student with an introduction to arithmetic and geometric series. An important topic in this chapter is the binomial theorem. It is perhaps the only time the student will see how we calculate all the terms or how a specific term of the binomial expression is determined.

- Chapter 17 on statistics can be taught any time after Chapter 4.
- Chapter 18, analytic geometry, is a study of the conic sections starting with the center (vertex) at the origin. The concept of translation has been developed for those who want to go beyond the basics.
- Chapters 19 through 24 provide a development of differentiation and integration of algebraic and transcendental functions. The calculus is presented with many examples and graphs to help provide an intuitive as well as a formal understanding of limit, continuity, differentiation, and integration.
- Chapters 25 and 26 provide an introduction to first-order and high-order linear differential equations. Different methods of solving differential equations are discussed, including Laplace transforms.
- Appendix A contains a brief discussion of the International System of Units (SI) and the United States Customary System (USCS) of measurements.
- In Appendix B the techniques of using a scientific calculator in problem solving are discussed. This material is discussed in the appendix so it will not be confused with the material presented on the graphics calculator.
- Appendix C is a table of the areas of a standard normal distribution.
- Appendix D contains a short table of integrals.

Answers to Odd-numbered Exercises

Answers to half the exercises in the book allow students to check their work quickly.

Exercises

Section Exercises   More than 7000 exercises are provided, covering the full range of concepts. Approximately 2200 of these exercises are technical applications.

Chapter Review Exercises   These end-of-chapter exercises tie together chapter material, encouraging students to review concepts covered in the section exercises. More than 1900 are included.

Chapter Tests   The comprehensive chapter tests allow students to practice for in-class examinations.

Supplements

This edition is accompanied by an extensive supplemental package that includes answers, solutions, and testing materials for both students and instructors.

For the Instructor

Instructor's Manual   This manual includes answers to all even-numbered exercises, solutions to selected exercises, and two test forms per chapter. In addition, it includes "Tips for Using Graphics Calculators in Class" and copies of log and semi-log paper that can be reproduced for homework assignments.

HarperCollins Test Generator/Editor for Mathematics with QuizMaster is available in IBM and Macintosh versions and is fully networkable. The Test Generator enables instructors to select questions by objective, section, or chapter, or to use a ready-made test for each chapter. The Editor enables instructors to edit any preexisting data or to easily create their own questions. The software is algorithm driven, allowing the instructor to regenerate constants while maintaining problem type, providing a nearly unlimited number of available test or quiz items. Instructors may generate tests in multiple-choice or open-response formats, scramble the order of questions while printing, and produce up to 25 versions of each test. The system features printed graphics and accurate mathematical symbols. It also features a preview option that allows instructors to view questions before printing and to replace or skip questions if desired. QuizMaster enables instructors to create tests and quizzes using the Test Generator/Editor and save them to disk so that students can take the test or quiz on a stand-alone computer or network. QuizMaster then grades the test or quiz and allows the instructor to create reports on individual students or classes.

For the Student

Student's Solution Manual   Prepared by Bill Ferguson and Ken Seidel of Columbus State Community College, this manual contains solutions to every other odd-numbered section and review exercise. (ISBN 0-673-46376-1)

Interactive Tutorial Software with Management System   This innovative package is available in IBM or Macintosh versions and is fully networkable. As with the Test Generator/Editor, this software is algorithm driven, automatically regenerating constants so students will not see the numbers repeat in a problem type if they return to any particular section. The tutorial is self-paced and provides unlimited opportunities to review lessons and practice problem solving. When students give a wrong answer, they can request to see a problem worked out. The program is menu driven for ease of use, and on-screen help can be obtained at any time with a single keystroke. Students' scores are automatically recorded and can be printed out for a permanent record. The optional Management System lets instructors record student scores on disk and print diagnostic reports for individual students or classes. In addition, there is a specific set of computer programs, with examples, that students can use when working text exercises.

GraphExplorer   With this sophisticated software, available in IBM and Macintosh versions, students can graph rectangular, conic, polar, and parametric equations; zoom; transform functions; and experiment with families of equations quickly and easily.

College Algebra with Trigonometry: Graphing Calculator Investigations This supplemental text, written by Dennis Ebersole of Northampton County Area Community College, provides investigations that help students visualize and explore key concepts, generalize and apply concepts, and identify patterns. (ISBN 0-06-500888-X)

Acknowledgments

Special thanks are due to the reviewers who offered valuable thoughts and suggestions:

Haya Adner,
Queensborough Community College

A. David Allen,
Ricks College

Richard D. Armstrong,
St. Louis Community College at Florissant Valley

Larry Badois,
County College of Morris

Philip E. Buechner, Jr.,
Cowley County Community College

Jim Cordle,
Westmoreland County Community College

Marie A. Dupuis,
Milwaukee Area Technical College

David Durdin,
North Harris County College

W. R. Ellis,
St. Petersburg Junior College

William D. Ferguson,
Columbus State Community College

Rita Fischbach,
Illinois Central College

Dewey Furness,
Ricks College

Hubert D. Haefner,
Monroe Community College

Wanda Hebert,
San Jacinto College

Robert L. Kimball,
Wake Technical Community College

Robert E. Lawson,
San Antonio College

Lynn G. Mack,
Piedmont Technical College

Mark Manchester,
Morrisville College

Darlene Miller,
Vermont Technical College

Kurt Mobley,
Augusta Technical Institute

Parthasarathy Rajagopal,
Kent State University

Barbara A. Ries,
Chippewa Valley Technical College

Thomas M. Rourke,
North Shore Community College

Doris Schoonmaker,
Hudson Valley Community College

Arlene M. Shenburne,
Montgomery College

Sharon Smith,
Augusta Technical College

Thomas J. Stark,
Cincinnati Technical College

Gwen Huber Terwilliger,
University of Toledo

Jisheng Wang,
Community College of Southern Nevada

Margaret M. Womble,
Wayne Community College

Kelly Wyatt,
Umpqua Community College

Anne Zeigler,
Florence-Darlington Technical College

My thanks also go to the students of Monroe Community College—particularly Helen Fox and Joe Morrissey—for their help in class-testing the manuscript. Their comments have proven extremely useful. Bill Ferguson and Ken Seidel, both of Columbus State Community College, have my gratitude for writing the Student's Solution Manual, a difficult and time-consuming task. I wish to thank Priscilla Ware, of Franklin University, Sharon Smith, of Augusta Technical Institute, Kathleen Pirtle, of Franklin University, Jeff Hildebrand, the University of Wisconsin–Madison, and Angel Andrew, Mark Harris, Debbie McMahon, and Laura Titus of Monroe Community College, for doing an outstanding job of checking answers to all of the exercises in the book. Many thanks to my colleagues at Monroe Community College who made suggestions for this edition: Peter Collinge, Patricia Kuby, Rosemary Mahoney, Elizabeth Heston, and Hubert Haefner. A special thank-you goes to Rob Sells, who went the extra mile in developing technical exercises and in proofreading. He also served as an adviser to me when John Ernst decided, for personal reasons, that he could not contribute to this revision. The HarperCollins editorial staff have been a joy to work with. I wish to extend a special thank-you to Ann Shaffer, Linda Youngman, Lisa De Mol, and my Sponsoring Editor, Anne Kelly. I'd especially like to thank my wife, Joyce, and my immediate family, Adam, Andrew, Brian, Emily, Lisa, Matthew, Molly, Teri, and Tod. Without their support and great sacrifices, this revision could not have become a reality.

Stuart R. Porter

Number Systems

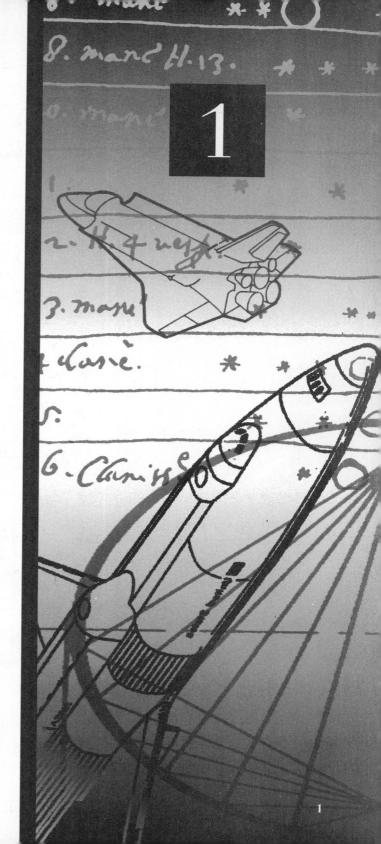

O n reentry the space shuttle endures extreme heat (with temperatures as high as 6000° F) generated on its surfaces by atmospheric friction. In fact, in early flights, the shuttle lost some of its protective tiles due to this intense heat. The tile loss did not present any immediate danger to the astronauts, but the situation had to be remedied. To solve the problem, engineers had to determine the temperature on the leading edge of the shuttle. This temperature, measured in degrees Kelvin, is known as *stagnation temperature*. The stagnation temperature can be approximated by using the temperature of the surrounding air, the specific heat ratio of air, and the number of times the speed of sound the body is moving. The calculations for stagnation temperature involve operations on real numbers (see Example 5 in Section 1.5).

This chapter will show how to perform operations on the real numbers and develop an understanding of their properties, and introduce the calculator as a tool that can be helpful in problem solving.

1.1

Numbers

The study of technical mathematics requires a knowledge of different kinds of numbers. These types range from the counting numbers to the complex numbers.

The set of **counting numbers** {1, 2, 3, 4, 5, 6, . . . } (also called **positive integers** or **natural numbers**) is the set of numbers to which we were introduced when we learned to count. Somewhere along the way a zero (0) was added to this set of counting numbers. The new set was called the set of **whole numbers** {0, 1, 2, 3, 4, 5, 6, . . . }. The three dots after the 6 is called an ellipses and indicates that the elements continue in the same manner. As mathematics evolved people quickly discovered that not every problem has a solution that is a whole number, so other kinds of numbers appeared.

The first written evidence of a zero was in India in 876.

EXAMPLE 1 Whole Number Solutions

Find a whole number that is a solution to the following problems.

a. $3 + 2$ **b.** $4 - 7$

Solution

a. For the problem $3 + 2$, we simple add, and the sum is the solution, the whole number 5.
b. For the problem $4 - 7$, there is no solution that is a whole number. It is not possible to subtract a large number from a smaller one and find a solution that is a whole number. (Remember that whole numbers do not include negative numbers.) ◆

German mathematician Leopold Kronecker (1823–1891): "God created the integers; all the rest is the work of man."

If we expand our set of numbers to include the **negative integers,** { . . . , −6, −5, −4, −3, −2, −1}, then $4 - 7$ has a solution: -3. When the set of counting numbers, zero, and the set of negative integers is combined into a single set, we have a new set, called the **integers** { . . . , −6, −5, −4, −3, −2, −1, 0, 1, 2, 3, 4, 5, 6, . . . }.

Decimals and fractions are commonly used in solving technical problems. Fractions and their decimal equivalents are members of the set of **rational numbers.**

DEFINITION 1.1
A **rational number** is a number of the form a/b, where a and b are integers and $b \neq 0$.

The following are examples of rational numbers:

$$\frac{1}{2}, \frac{3}{4}, \frac{-5}{4}, \frac{1}{3}, \frac{133}{99}, \frac{-7}{6}, \frac{2}{3}, \frac{0}{4}, \frac{11}{-2}, \frac{25}{1}, \frac{1}{1}, \frac{6}{2}.$$

Each of the rational numbers also can be expressed as a decimal. A rational number in fraction form easily can be changed to its equivalent decimal form using a calculator.

EXAMPLE 2   **Fractions to Decimals**

Change the following fractions to their decimal equivalents.

a. $\dfrac{1}{2}, \dfrac{3}{4}, \dfrac{-5}{4}$ **b.** $\dfrac{1}{3}, \dfrac{133}{99}, \dfrac{-7}{6}$

Solution   A calculator will display the following results.

a. $\dfrac{1}{2} = 1 \div 2 = 0.5 \qquad \dfrac{3}{4} = 3 \div 4 = 0.75 \qquad \dfrac{-5}{4} = -5 \div 4 = -1.25$

b. $\dfrac{1}{3} = 0.3333333 \qquad \dfrac{133}{99} = 1.3434343$

$\dfrac{-7}{6} = -1.6666666 \quad \text{or} \quad -1.6666667$

(For the final calculation in **b**, did the display on your calculator have a last digit of 6 or 7? If the last digit was a 6, your calculator cuts off, or truncates, the remaining series of 6s. If the last digit was a 7, your calculator automatically rounds off the answer.) ◆

The decimal equivalents of the fractions in Example 2a are examples of **terminating decimals.** The decimal equivalents of the fractions in 2b are examples of **repeating decimals.** Each repeating decimal can be written with a bar above the digit(s) to show which repeats, as follows.

$$0.333\ldots = 0.\overline{3} \qquad 1.3434\ldots = 1.\overline{34} \qquad -1.166\ldots = -1.1\overline{6}$$

Numbers such as $\sqrt{2}, \sqrt{3}, \sqrt{5},$ or π cannot be expressed in the form a/b or as a terminating or repeating decimal. This set of numbers is called the set of **irrational numbers.**

> **DEFINITION 1.2**
>
> A number that cannot be expressed in the form a/b, where a and b are integers and $b \neq 0$, is called an **irrational number.** Another way to express the definition is that an irrational number is a nonrepeating, nonterminating decimal expression.

Examples of irrational numbers are:

$$\frac{\sqrt{3}}{2}, \ -\sqrt{7}, \ \frac{3}{\sqrt{2}}, \ \frac{\pi}{5}, \ \sqrt{19}, \ \sqrt{\frac{38}{2}}.$$

When the sets of all the numbers discussed so far—counting, whole, integers, rational, and irrational—are placed in a single set, this collection is called the set of **real numbers.** Although it is impossible to list all the real numbers, they can be indicated by using a geometrical concept called the number line. A **number line** is a straight line that has an arbitrary point we call **zero,** or the **origin.** By convention, all the points to the right of zero are positive numbers, and those to the left of zero are negative (Figure 1.1). Arrows on the ends of the line indicate that the number line extends indefinitely to the right and the left of the origin. A point on the number line can be labeled with a real number, called the **coordinate** of the point. Also, each real number can be represented as a point on the number line, called the **graph** of the given number.

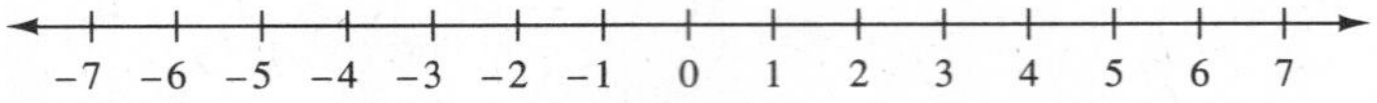

$$\begin{array}{cccccccccccccccc} & & & & & & & & & & & & & & & \\ \hline -7 & -6 & -5 & -4 & -3 & -2 & -1 & 0 & 1 & 2 & 3 & 4 & 5 & 6 & 7 \end{array}$$

FIGURE 1.1

EXAMPLE 3   Graphs of Real Numbers

Draw a number line and indicate the position of the following real numbers on the number line.

$$-4\frac{1}{2}, \sqrt{2}, \frac{9}{2}, -1.75, -\sqrt{7}, \pi, \frac{\sqrt{214}}{2}, \frac{-7.5}{1}$$

Solution   The solution is shown in Figure 1.2.   ◆

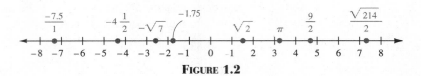

FIGURE 1.2

A number of the form $\sqrt[n]{-a}$, where n (called the *index*) is an *even* integer and a is positive, is called an **imaginary number.** For example, $\sqrt{-4}$ is an imaginary number. There is no real number that, when multiplied by itself, will have a product of -4. (The expression $\sqrt{-4}$ and $\sqrt[2]{-4}$ mean the same thing—the square root of -4. It is common practice to omit the index 2 when indicating a square root.) By definition $j = \sqrt{-1}$. So another way of writing $\sqrt{-4}$ is

$$\sqrt{-4} = \sqrt{4}\,\sqrt{-1} = 2j.$$

(In many mathematics texts, $i = \sqrt{-1}$. However, for electrical applications, i = current; so, to keep *i*s straight, we will let $j = \sqrt{-1}$.) Imaginary numbers will be discussed in more detail in Chapter 9.

A number of the form $\sqrt[n]{-a}$, where n is an *odd* integer, is a real number. The $\sqrt[3]{-8} = -2$, since $(-2)(-2)(-2) = -8$.

EXAMPLE 4 Real Numbers and Imaginary Numbers

Indicate whether the following numbers are real or imaginary.

a. $\sqrt{-9}$ b. $\sqrt[4]{-16}$ c. $\sqrt[3]{-27}$ d. $-\sqrt{4}$

Solution

a. $\sqrt{-9}$ is an imaginary number since the index, 2, is an even integer.
b. $\sqrt[4]{-16}$ is an imaginary number since the index, 4, is an even integer.
c. $\sqrt[3]{-27}$ is a real number since 3 is an odd integer. In fact, $\sqrt[3]{-27} = -3$, since $(-3)(-3)(-3) = -27$.
d. An equivalent way of writing $-\sqrt{4}$ is $-(\sqrt{4})$, indicating that the operation of square root is performed and then the minus sign is assigned to the answer. Thus $-\sqrt{4} = -2$, a real number. ◆

A number formed using a real number and an imaginary number is called a *complex number*. An example of a complex number is $3 + 4j$.

DEFINITION 1.3

A **complex number** is a number that can be written in the form $a + bj$, where a and b are real numbers and $j = \sqrt{-1}$. If $b = 0$, then $a + bj$ is a real number. If $a = 0$, then we have a number of the form bj, which is called a **pure imaginary number.**

Any real number can be expressed as a complex number. For example, $5 = 5 + 0j$. To get an overview of the number system and how the numbers fit in the system, study Figure 1.3.

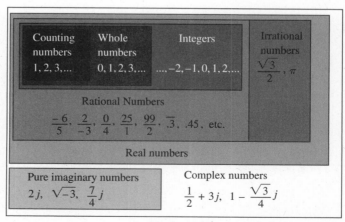

FIGURE 1.3

1.1 Exercises

In Exercises 1–8, determine which of the sums and differences represent a counting number, a negative integer, or neither.

1. $4 + 5$ **2.** $5 - 7$ **3.** $6 - 4$ **4.** $11 - 14$

5. $5 - 5$ **6.** $13 - 9$ **7.** $11 - 11$ **8.** $15 - 17$

In Exercises 9–20, determine if the numbers are rational or irrational.

9. $\dfrac{22}{7}$ **10.** $\sqrt{2}$ **11.** $3.14\overline{14}$ **12.** π

13. $3.141441444\ldots$ **14.** $\sqrt{16}$ **15.** $\dfrac{\sqrt{9}}{4}$ **16.** $\sqrt{3} + \sqrt{2}$

17. $3.14159\ldots$ **18.** $1.4142136\ldots$ **19.** $\sqrt{16} + \sqrt[5]{32}$ **20.** $\dfrac{0}{\sqrt{2}}$

In Exercises 21–34, draw a number line and indicate the approximate position of the real numbers on the number line.

21. 5 **22.** -8 **23.** $\dfrac{5}{4}$ **24.** -0.75

25. $\sqrt{9}$ **26.** $-\sqrt{2}$ **27.** $\dfrac{0}{14}$ **28.** $-5\dfrac{1}{2}$

29. $\sqrt{3}$ **30.** π **31.** $\dfrac{\sqrt{36}}{2}$ **32.** $\dfrac{18}{\sqrt{81}}$

33. $\dfrac{-\sqrt{121}}{\sqrt{121}}$ **34.** $\dfrac{\sqrt{49}}{\sqrt{7}}$

In Exercises 35–48, determine whether the numbers are real, pure imaginary, or complex.

35. $3 + 4j$ **36.** $7j$ **37.** $\dfrac{23}{3}$ **38.** $-17j$

39. $-11 + 9j$ **40.** $4\sqrt{3}$ **41.** $-\dfrac{3}{4}j$ **42.** $\sqrt[3]{-125}$

43. $\sqrt{7} + 5j$ **44.** $-\sqrt[3]{125}$ **45.** $-\sqrt{-36}$ **46.** $\sqrt[4]{-81}$

47. $\sqrt{\dfrac{0}{-4}}$ **48.** $-\sqrt{16} + \dfrac{\sqrt{-27}}{\sqrt{9}}$

49. Construct a chart, different from the one in Figure 1.3, to illustrate the relationship between the sets of numbers in the complex number system. (*Hint:* Use a tree-type diagram.)

1.2

Relationships Between Numbers

Not all quantities or expressions are equal. We can show the relationships between quantities by using appropriate symbols.

The symbols **<** **(less than)** and **>** **(greater than)** are called **signs of inequality.**

$a = b$	a equals b
$a < b$	a is less than b
$a > b$	a is greater than b
$a \leq b$	a is less than or equal to b
$a \geq b$	a is greater than or equal to b
$a < x < b$	a is less than x and x is less than b
$a \leq x \leq b$	a is less than or equal to x and x is less than or equal to b

The two inequalities $a < x < b$ and $a \leq x < b$ are called **double inequalities.**

Example 1 illustrates the use of inequality and equality signs.

EXAMPLE 1 Relationships Between Numbers

For the following pairs of numbers, insert the appropriate symbols $(<, >, \text{ or } =)$ between the numbers.

a. $3, 6$ **b.** $-3, -8$ **c.** $5, 0$

d. $-\pi, -3.14159$ **e.** $-\sqrt{3}, -\sqrt{2}$ **f.** $4\frac{1}{3}, -2\frac{1}{7}$

Solution

a. 3 is less than 6, therefore $3 < 6$.
b. -3 is greater than -8, therefore, $-3 > -8$. Using a number line like the one in Figure 1.4 will help show why.

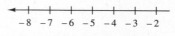

FIGURE 1.4

Note that -8 is to the left of -3. Any number to the left of another number on the number line is less than that number.

c. 5 is greater than 0, therefore $5 > 0$.
d. $-\pi$ is less than -3.14159, therefore $-\pi < -3.14159$.
e. $-\sqrt{3}$ is less than $-\sqrt{2}$, therefore, $-\sqrt{3} < -\sqrt{2}$.
f. $4\frac{1}{3}$ is greater than $-2\frac{1}{7}$, therefore, $4\frac{1}{3} > -2\frac{1}{7}$. ◆

Many real numbers satisfy the expression of the inequality $x < 2$. A few of these real numbers are $-1, 0, \frac{1}{2}, -\sqrt{2}$, and $-\pi$. The set of all the real numbers that satisfy such an expression is called the **solution set.** We can use a number line to show the set of numbers that makes this a true statement of inequality. The part of the number line that is marked to indicate the solution set is called the **graph of the solution set.**

For some problems we will need to show that 2 is also a member of the solution set. In this case we would write $x \le 2$, read: "x is less than or equal to 2."

EXAMPLE 2 **Graphs of Solution Sets**

Illustrate the solution set for each of the following on a number line.

a. $x < 2$ **b.** $x \ge -2$ **c.** $-\pi \le x \le \pi$

Solution

a. The expression $x < 2$ tells us that the solution set consists of all the real numbers less than 2, but not including 2. To illustrate this on a number line, draw an open circle at 2 (meaning 2 is not included) and an arrow to the left of 2, as shown in Figure 1.5.

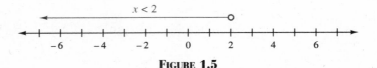

FIGURE 1.5

b. The statement $x \ge -2$ tells us that the solution contains all the real numbers greater than and equal to -2. To illustrate the solution set on the graph, draw a solid dot at -2 and an arrow pointing to the right at -2 as shown in Figure 1.6

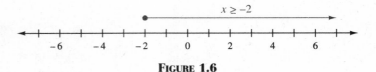

FIGURE 1.6

c. We read the double inequality, $-\pi \le x \le \pi$, "negative pi is less than or equal to x, and x is less than or equal to pi." The word *and* in the inequality statement means that the solution is the set of points common to both inequalities $-\pi \le x$ and $x \le \pi$. An equivalent way of writing $-\pi \le x$ is

$x \geq -\pi$. With this change, the statement is $x \geq -\pi$ and $x \leq \pi$, which may make it easier to see that the solution is the set of points between $-\pi$ and π, including the end points. To illustrate this solution, draw solid dots at $-\pi$ and π and connect the dots with a line, as shown in Figure 1.7. ◆

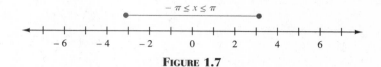

FIGURE 1.7

EXAMPLE 3 **Write a Statement Representing a Graph**

Write a statement using equality and/or inequality symbols that describes the set of points illustrated by the graphs in Figures 1.8, 1.9, and 1.10.

a.

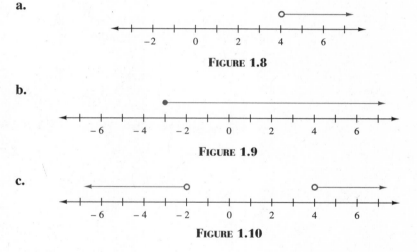

FIGURE 1.8

b.

FIGURE 1.9

c.

FIGURE 1.10

Solution

a. In Figure 1.8, the open circle at 4 and the arrow pointing to the right of 4 indicates the set of all real numbers greater than but not including 4. It is written symbolically as $x > 4$.

b. In Figure 1.9, the solid dot at -3 and the arrow pointing to the right of -3 indicates the set of all the real numbers greater than or equal to -3. It is written symbolically as $x \geq -3$ or $-3 \leq x$.

c. In Figure 1.10, there are two open circles at -2 and 4 and arrows pointing to the left of -2 and to the right of 4. The solution set consists of two sets of points, those defined by $x < -2$ or those defined by $x > 4$. The solution set is written $x < -2$ or $x > 4$. ◆

CAUTION   When writing a double inequality such as $2 < x < 4$, do not commit a common error by thinking that one can equivalently write $2 > x > 4$. This is incorrect since 2 is not greater than 4. Use the "less than" symbols and place the smaller number on the left. ■

Absolute Value

The *absolute value* of a real number is its distance or magnitude or its value disregarding its sign. For example, the absolute value of -4 is 4, and the absolute value of 5 is 5.

> **DEFINITION 1.4**
> The **absolute value** of the real number x, symbolized $|x|$, is
> $$|x| = \begin{cases} x \text{ if } x > 0 \\ 0 \text{ if } x = 0 \\ -x \text{ if } x < 0 \end{cases}$$

The absolute value of any real number is positive or zero. In fact, the absolute value of 0 is the only real number whose absolute value is zero. Look at the following.

$$\text{If } a = -7, \text{ then } |-7| = -(-7) = 7$$
$$\text{If } a = 0, \text{ then } |0| = 0$$

EXAMPLE 4   Absolute Value of Real Numbers

Determine the absolute value of the following real numbers.

a. $+\dfrac{3}{4}$ b. -3.427 c. $-\pi$ d. $-\sqrt{5}$ e. $(-4)^2 - 22$

Solution

a. $\left|\dfrac{+3}{4}\right| = \dfrac{3}{4}$ b. $|-3.427| = -(-3.427) = 3.427$

c. $|-\pi| = \pi$ d. $|-\sqrt{5}| = \sqrt{5}$

e. $|(-4)^2 - 22| = |16 - 22| = |-6| = 6$ ◆

From the definition we can see that an absolute value cannot be negative, that is when

$x > 0$, $|x| = x$,   which is non-negative ($|3| = 3$)
$x = 0$, $|x| = 0$,   which is non-negative, and
$x < 0$, $|x| = -x$,   which is non-negative since x itself is negative ($|-3| = -(-3) = 3$).

1.2 Exercises

In Exercises 1–9, insert the appropriate symbol ($<$, $>$, or $=$) between the pairs of numbers.

1. 3, 14

2. $-5, -21$

3. $-\sqrt{3}, -\sqrt{2}$

4. $-\sqrt{81}, -9$

5. $\sqrt{5}, 2.236068\ldots$

6. $\frac{22}{7}, \pi$

7. $\frac{14}{4}, 3\frac{1}{4}$

8. $\frac{1}{4} - \frac{1}{3}, \frac{1}{3} - \frac{1}{4}$

9. $\frac{1}{2} + \frac{1}{3}, \frac{3}{2} - \frac{2}{3}$

In Exercises 10–21, graph the statements on a number line and list three values of x that satisfy the inequality.

10. $x > 2$

11. $x < 1.4$

12. $x > -3$

13. $x < -2$

14. $-4 \leq x$

15. $4 \geq x$

16. $-3 < x < 5$

17. $3 \leq x < 11$

18. $-5.4 < x < 4.3$

19. $-10.4 \leq x < 12.2$

20. $x < -3$ or $x > 5$

21. $x \leq 1$ or $x \geq 7$

In Exercises 22–31, write a statement using equality and/or inequality symbols that describes the set of points illustrated by each of the graphs.

22.
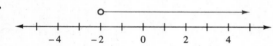

23.

24.
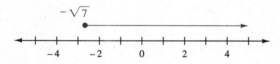

25.

26.

27.
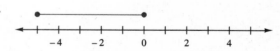

28.
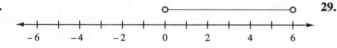

29.
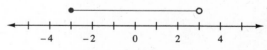

30.
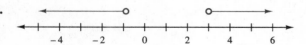

31.
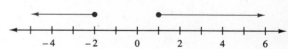

In Exercises 32–44, find the absolute value of the numbers.

32. $|-4|$

33. $|9|$

34. $|-\sqrt{11}|$

35. $|9 - 5|$

36. $\left|\frac{3}{7} - \frac{1}{5}\right|$

37. $|-\sqrt[3]{8}|$

38. $\left|3\frac{1}{7} - \pi\right|$

39. $|(-2)^2|$

40. $|(-2)^3|$

41. $|-2|^2$

42. $|(-2)^2 - (-2)^3|$

43. $|-4.361|$

44. $|5.73 - 8.16|$

1.3

Real Numbers: Properties and Operations

With the exception of one chapter, almost all the work in this text deals with the set of real numbers or expressions that represent real numbers. To use real numbers properly, we need to understand the set of properties for addition and multiplication of real numbers. Many of the properties may be familiar from arithmetic. Their names, however, may not.

PROPERTIES FOR ADDITION AND MULTIPLICATION OF REAL NUMBERS

(Let a, b, and c represent real numbers.)

Property	Addition	Multiplication
Closure Property	$a + b$ is a unique real number. $a + b$ is called the sum of a and b, and a and b are called the addends of the sum.	$a \cdot b$ is a unique real number. $a \cdot b$ is called the product of a and b, and a and b are called factors of the product.
Commutative Property	$a + b = b + a$. The sum of two numbers is unchanged when the order of the addends is reversed.	$a \cdot b = b \cdot a$. The product of two numbers is unchanged when the order of the factors is reversed.
Associative Property	$(a + b) + c = a + (b + c)$. The sum is the same regardless of the manner in which the addends are grouped.	$(a \cdot b) \cdot c = a \cdot (b \cdot c)$. The product is the same no matter how the factors are grouped.
Identity Property	There is a unique number, 0, such that $a + 0 = a$ and $0 + a = a$. 0 is called the **additive identity.**	There is a unique number, 1, such that $a \cdot 1 = a$ and $1 \cdot a = a$. 1 is called the **multiplicative identity.**

Inverse Property	For each number a, there is a unique number, $-a$, such that $a + (-a) = 0$ and $(-a) + a = 0$. $-a$ is called the **negative** of a or the **additive inverse** of a or the opposite of a.	For each nonzero number a, there is a unique number, $\frac{1}{a}$, such that $a \cdot \frac{1}{a} = 1$ and $\frac{1}{a} \cdot a = 1$. $\frac{1}{a}$ is called the **reciprocal** of a or the **multiplicative inverse of** a.
Distributive Property	$a \cdot (b + c) = (a \cdot b) + (a \cdot c)$ or $(b + c) \cdot a = (b \cdot a) + (c \cdot a)$. The product of a real number with a sum of real numbers is equal to the sum of the products of those numbers.	

EXAMPLE 1   Identify the Real Number Property

Match each mathematical statement in column A with the relevant property in column B.

Column A	**Column B**
a. $6 \cdot (4 + 3) = 6 \cdot 4 + 6 \cdot 3$	**1.** Multiplicative identity
b. $3 + 4 = 4 + 3$	**2.** Distributive property
c. $3 \cdot (4 \cdot 7) = (3 \cdot 4) \cdot 7$	**3.** Commutative property of addition
d. $(4 + 3) \cdot 6 = 4 \cdot 6 + 3 \cdot 6$	**4.** Associative property of multiplication
e. $(6 \cdot 4) \cdot 8 = 8 \cdot (6 \cdot 4)$	**5.** Additive identity
f. $4 + 0 = 4$ and $0 + 4 = 4$	**6.** Associative property of addition
g. $3 + (4 + 7) = (3 + 4) + 7$	**7.** Commutative property of multiplication
h. $4 \cdot (1/4) = 1$ and $(1/4) \cdot 4 = 1$	**8.** Additive inverse
i. $2 \cdot 1 = 2$ and $1 \cdot 2 = 2$	**9.** Multiplicative inverse
j. $8 + (-8) = 0$ and $(-8) + 8 = 0$	

Solution

a. 2   **b.** 3   **c.** 4   **d.** 2   **e.** 7   **f.** 5   **g.** 6   **h.** 9   **i.** 1   **j.** 8

The operations of subtraction and division may be expressed in terms of addition and multiplication respectively.

> **DEFINITION 1.5**
> The **subtraction** of b from a, written $a - b$, can be written $a - b = a + (-b)$.

In other words, subtraction may be thought of as adding a to the negative of b. The result of the subtraction of b from a is called the **difference.**

To add and subtract any real numbers we must understand the rules for addition of signed numbers.

RULES FOR ADDITION OF SIGNED NUMBERS

Rule I   If a and b have like signs (that is, either both are positive or both are negative), compute $|a| + |b|$. Then place the common sign of a and b in front of the sum. The result is the sum a and b.

Rule II   If a and b have unlike signs (one positive and the other negative) and $|a| = |b|$, than $a + b = 0$.

Rule III   If a and b have unlike signs and $|a| \neq |b|$, subtract the smaller absolute value from the larger absolute value. Then place the sign of the number having the larger absolute value in front of the difference. The resulting number is the sum of a and b.

CAUTION   The real number properties for addition and multiplication do not necessarily hold for subtraction and division. The commutative property, for example, does not hold for the operation of subtraction ($8 - 5 \neq 5 - 8$). ■

To show that a property does not hold for a particular operation, we must find one case that does not hold. An illustration that shows that a property does not hold is called a **counterexample.** A counterexample that shows division is not commutative is $4 \div 2 \neq 2 \div 4$.

EXAMPLE 2   Addition of Signed Numbers

Determine the sum of the signed numbers.

a. $(+4) + (+7)$ **b.** $(-4) + (-7)$ **c.** $(-4) + (+4)$
d. $(-7) + (+4)$

Solution

a. Since the signs of the numbers are alike, apply Rule I.

$$|+4| + |+7| = 4 + 7 = 11$$

Since both signs are positive, the sum is positive. Thus,

$$(+4) + (+7) = 11.$$

b. Since the signs of the numbers are alike, apply Rule I.

$$|-4| + |-7| = 4 + 7 = 11$$

Since both signs are negative the sum is negative. Thus,

$$(-4) + (-7) = -11.$$

c. Since $|-4| = |+4|$, apply Rule II.

$$(-4) + (+4) = 0$$

d. Since the signs of the numbers are not alike, and $|-7| \neq |+4|$, apply Rule III.

$$|-7| - |4| = 7 - 4 = 3$$

Since the larger number is negative, the result is negative. Thus,

$$(-7) + (+4) = -3. \quad \blacklozenge$$

EXAMPLE 3 **Subtraction of Signed Numbers**

Find the difference of the signed numbers.

a. $(+6) - (+4)$ **b.** $(-6) - (-4)$ **c.** $(-6) - (+6)$
d. $(-6) - (-6)$

Solution To find the difference in each of the problems, apply the definition of subtraction in the first step and then apply the rules of addition.

a. $(+6) - (+4) = (+6) + (-4)$ **Definition of subtraction**
$\qquad\qquad = +(|+6| - |-4|)$ **Rule III of addition**
$\qquad\qquad = +(6 - 4)$
$\qquad\qquad = 2$

b. $(-6) - (-4) = (-6) + (+4)$ **Definition of subtraction**
$\qquad\qquad = -(|-6| - |+4|)$ **Rule III of addition**
$\qquad\qquad = -(6 - 4)$
$\qquad\qquad = -2$

c. $(-6) - (+6) = (-6) + (-6)$ **Definition of subtraction**
$\qquad\qquad = -(|-6| + |-6|)$ **Rule I of addition**
$\qquad\qquad = -(6 + 6)$
$\qquad\qquad = -12$

d. $(-6) - (-6) = (-6) + (+6)$ **Definition of subtraction**
$\qquad\qquad = 0$ **Rule II of addition** $\blacklozenge$

EXAMPLE 4   Application: Addition of Signed Numbers

The balance in Lisa's checking account on May 1 was $85.97. During the month of May she made the following transactions.

May 8	deposited	$74.80
May 17	check for Sears	$34.75
May 23	check for garage	$28.95

Using the rules of signed numbers write an addition statement that represents her balance on May 23. Determine the balance.

Solution

$$\begin{aligned}
\text{Balance} &= \$85.97 + (+\$74.80) + (-\$34.75) + (-\$28.95) \\
&= \$160.77 + (-\$34.75) + (-\$28.95) \\
&= \$126.02 + (-\$28.95) \\
&= \$97.07
\end{aligned}$$

Lisa's balance on May 23 is $97.07.   ◆

Before looking at the rules for multiplication and division of signed numbers, we will need a definition for the operation of division.

> **DEFINITION 1.6**
>
> The **division** of a by a nonzero number b, written $a \div b$ or $\frac{a}{b}$ is defined as
>
> $$\frac{a}{b} = c, \text{ if and only if } a = b \cdot c.$$

We also define $a \div b$ as $a \cdot \frac{1}{b}$. The result of the division of a by b is called the **quotient.**

Just as the rules for addition and subtraction are similar, the rules for multiplication and division also are very similar and can be discussed together. To multiply (or divide) a and b, determine the absolute value of each number and compute $|a| \cdot |b|$ (or $|a| \div |b|$). $\left(\text{or } \frac{|a|}{|b|}\right)$.

If the signs of a and b are alike, the product (or quotient) is positive. If the signs are unlike, the product (or quotient) is negative.

RULES FOR MULTIPLICATION AND DIVISION OF SIGNED NUMBERS

Rule IV   If a and b have like signs, then their product (or quotient) is a positive number. In symbols, write $+(|a| \cdot |b|)$ (or $+(|a| \div |b|)$).

Rule V   If a and b have unlike signs, then their product (or quotient) is a negative real number. In symbols, write $-(|a| \cdot |b|)$ (or $-(|a| + |b|)$).

EXAMPLE 5   **Multiplying Signed Numbers**

Find the product or the quotient of the signed numbers.

a. $(+6) \cdot (+5)$ **b.** $(-6) \div (-2)$ **c.** $(+6) \cdot (-5)$ **d.** $(-6) \div (+2)$

Solution

a. Applying Rule IV,

$$
\begin{aligned}
(+6) \cdot (+5) &= +(|+6| \cdot |+5|) \\
&= +(6 \cdot 5) \\
&= +30.
\end{aligned}
$$

b. Applying Rule IV,

$$
\begin{aligned}
(-6) \div (-2) &= +(|-6| \div |-2|) \\
&= +(6 \div 2) \\
&= +3.
\end{aligned}
$$

c. Applying Rule V,

$$
\begin{aligned}
(+6) \cdot (-5) &= -(|+6| \cdot |-5|) \\
&= -(6 \cdot 5) \\
&= -30.
\end{aligned}
$$

d. Applying Rule V,

$$
\begin{aligned}
(-6) \div (+2) &= -(|-6| \div |+2|) \\
&= -(6 \div 2) \\
&= -3. \quad \blacklozenge
\end{aligned}
$$

The order in which we perform operations is determined by the operation and/or the use of parentheses. The parentheses in the expression $(35 - 15) \div 5$ indicate that the operation of subtraction is performed first, then the operation of division.

$$
\begin{aligned}
(35 - 15) \div 5 &= 20 \div 5 \\
&= 4
\end{aligned}
$$

On the other hand, the expression $35 - 15 \div 5$ has no parentheses to tell us what do first, so we look at the operations. First perform all multiplications and divisions, from *left* to *right* ($\rightarrow$) through the expression, then perform all additions and subtractions in the same order—from left to right.

$$35 - \boxed{15 \div 5} = 35 - 3$$
$$= 32$$

RULES FOR EVALUATION OF EXPRESSIONS

Rule VI   Work from the innermost grouping symbols (if any) out and perform operations in order from left to right. (Grouping symbols may be parentheses, braces, brackets, etc.)

Rule VII   Perform the operations of finding roots or raising numbers to a power.

Rule VIII   Perform multiplications and divisions from left to right.

Rule IX   Perform additions and subtractions from left to right.

Eventually we will study exponents, such as 3^4 or 6^2, and radicals, such as $\sqrt{25}$. Exponents and radicals are evaluated before multiplication and division.

EXAMPLE 6   **Order of Operations**

Find the value of the expression: $-16 \div 4 - 8(-2) + 7$.

Solution   The parentheses around the -2 in this expression indicate multiplication. Since no grouping is indicated with parentheses, first perform all the operations of multiplication and division. Then perform the operations of addition and subtraction.

$$\boxed{-16 \div 4} - \boxed{8 \cdot (-2)} + 7 =$$
$$\boxed{-4 - (-16)} + 7 =$$
$$+12 + 7 = 19 \quad \blacklozenge$$

CAUTION   Always work from left to right. For example,

$$14 \div 2 \cdot 7 \neq 1$$

The correct solution is 49:

$$\boxed{14 \div 2} \cdot 7 = 7 \cdot 7$$
$$= 49 \quad \blacksquare$$

EXAMPLE 7   Order of Operations

Find the value of the expression $\dfrac{4-7}{3} + \dfrac{6+8}{7}$.

Solution   To find the value of the expression, first combine the values in the numerator of the fractions.

$$\frac{4-7}{3} + \frac{6+8}{7} = \frac{-3}{3} + \frac{14}{7}$$

The fractions are grouping symbols that indicate the operation of division. Following the rules for evaluation of expressions, perform the division and then the addition.

$$= -1 + 2$$
$$= 1 \quad \blacklozenge$$

In mechanics, the laws of equilibrium state that if an object is at rest, the sum of the forces acting on that object must be zero. Assume that forces pushing an object down are negative and forces pushing an object up are positive. The force is measured in pounds (1b) or newtons (N).

EXAMPLE 8   Application: Sum of Forces

Determine whether the beam in Figure 1.11 is at rest. In other words, is the sum of the positive forces balanced by the sum of the negative forces acting on the plank?

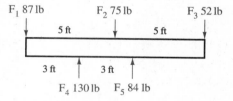

FIGURE 1.11

Solution   The sum of the forces must equal zero for an object to remain at rest. Using the convention that down forces are negative and up forces are positive we can write the following statement.

$$(-F_1) + (-F_2) + (-F_3) + F_4 + F_5 = 0 \quad \text{Is this true?}$$
$$(-87) + (-75) + (-52) + 130 + 84 = ? \quad \text{Substituting}$$
$$(-162) + (-52) + 130 + 84 = ?$$
$$(-214) + 130 + 84 = ?$$
$$(-84) + 84 = ?$$
$$0 = 0$$

Since the total sum of the forces is equal to zero, the object remains at rest. $\blacklozenge$

1.3 Exercises

In Exercises 1–8, determine the number indicated.

1. The additive inverse of 6

2. The reciprocal of $\dfrac{2}{3}$

3. The negative of -5

4. The multiplicative inverse of $\dfrac{1}{7}$

5. The additive inverse of -7

6. The reciprocal of -13

7. The additive inverse of $-\dfrac{1}{3}$

8. The negative of $\dfrac{3}{13}$

In Exercises 9–18 fill in the blank so that the resulting statement illustrates the indicated property of real numbers. (*a* and *b* represent real numbers.)

9. The commutative property of multiplication: $8 \cdot 9 =$ _____ .

10. The additive inverse of 4 is _____ .

11. The associative property of multiplication: $4 \cdot (6 \cdot 7) =$ _____ .

12. The multiplicative inverse of $\dfrac{-4}{7} =$ _____ .

13. The distributive property: $3 \cdot (a + b) =$ _____ .

14. The additive identity: $18 +$ _____ $= 18$.

15. The commutative property of addition: $3 + (a + b) =$ _____ .

16. The multiplicative identity: $4 \cdot$ _____ $= 4$.

17. The associative property of addition: $3 + (14 + 19) =$ _____ .

18. The distributive property: $(a + 4) \cdot b =$ _____ .

In Exercises 19–28, identify the real number property that is illustrated in each case. (*a* and *b* represent real numbers.)

19. $0 + a = a$

20. $\dfrac{1}{a} \cdot a = 19$

21. $3 + a = a + 3$

22. $a + (b + 5) = (a + b) + 5$

23. $3 \cdot (2 + b) = 3 \cdot 2 + 3 \cdot b$

24. $(a + b) + 8 = 8 + (a + b)$

25. $5 + -5 = 0$

26. $[(2 \cdot a) \cdot b] \cdot 4 = 2 \cdot [a \cdot (b \cdot 4)]$

27. $(5 + 13) \cdot b = 5 \cdot b + 13 \cdot b$

28. $-15 \cdot \dfrac{1}{-15} = 1$

In Exercises 29–46, evaluate each of the given expressions by performing the indicated operations.

29. $7 + 6$

30. $8 + 5$

31. $(-8) + (+5)$

32. $(-6) + (+7)$

33. $15 - 9$

34. $18 - 12$

35. $(-18) - (+27)$

36. $(-19) - (+15)$

37. $(4)(3)$

38. $(+17)(+17)$

39. $(-4)(+3)$

40. $(-6)(5)$

41. $(-1.5)(-0.3)$

42. $(+2.4)(-0.5)$

43. $(-0.18) \div (-0.3)$

44. $(-0.18) \div (+0.3)$

45. $(+2.1) \div (-0.3)$

46. $(-2.4) \div (-0.8)$

In Exercises 47–48, give examples using real numbers that show that the indicated property does not hold for **a.** subtraction and **b.** division.

47. commutative property

48. associative property

In Exercises 49–63, evaluate each of the given expressions using the rules for evaluating expressions.

49. $-13 + (-12) - 14 + (-5)$

50. $-6 + (-14) - 16 + (-8)$

51. $(-4)(+3)(-5)$

52. $(-6)(-2)(-7)$

53. $(+7)(-6)(+8)$

54. $\dfrac{(-6)(8)}{-12}$

55. $\dfrac{-96}{(-3)(4)}$

56. $\dfrac{(-4)(-6)}{-8}$

57. $11 - 3 \cdot (7)$

58. $-2 \cdot (7 - 4) + 8$

59. $(-2)(7) - 4 + 8$

60. $\dfrac{8}{5 - 3} + 2 - 7$

61. $\dfrac{3 - (-4)}{9 - 2}$

62. $\dfrac{13 + (-2)}{3} - \dfrac{9 - 5}{-2}$

63. $5 - (-9)(-5) \div (-3) + 11 - (-13)$

64. The amount of electric charge always stays the same. If you begin with two negative charges, you will end with two negative charges. Initially, your shoes and a rug are neutral electrically; the total charge is zero. When you scuff your shoes against the fluffy rug, the rug acquires a charge of -0.00030 coulombs. What must be the new charge of your shoes?

65. In Chicago, the temperature during a 24-hour period dropped from 55°F to -5°F. Find the change in temperature.

66. The difference between apparent magnitude and absolute magnitude is used to find the distance to a star. If the apparent magnitude is $+18$ and the absolute magnitude is -6, what is the difference?

67. The balance in Andrew's checking account on April 5 was $95.97. During the month of April he made the following transactions:

April 10	deposited	$444.80
April 17	check for car repairs	$ 54.75
April 25	check for rent	$350.00

Using the rules of signed numbers, write an addition statement that represents his balance on April 25. Determine the balance.

68. Two numbers are used to record an individual's blood pressure. For example, normal blood pressure for a 30-year-old is about 120/80. If an individual's blood pressure is recorded as **(a)** 140/90 and **(b)** 90/70, indicate the change in blood pressure using signed numbers.

In Exercises 69–70 determine whether the beam conforms to the laws of equilibrium (see Example 8).

69.

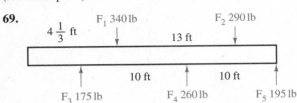

70.

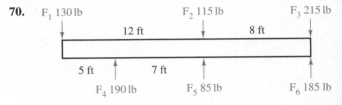

1.4

Properties of Zero

Although zero appears to be an innocent number, it frequently causes problems for students. To help avoid problems, study carefully the various properties of zero.

PROPERTIES OF ZERO

1. For any real number a, $a + 0 = a$. This is the identity property of addition, which was discussed in the previous section.
2. For any real number a, $a - 0 = a$. That is, $4 - 0 = 4$.
3. For any real number a, $a \cdot 0 = 0$. That is, $6 \cdot 0 = 0$. This is called the multiplicative property of zero.
4. If a and b are real numbers and $a \cdot b = 0$, then $a = 0$ or $b = 0$ or both equal zero. This will be an important rule in later chapters when we seek solutions to equations by the technique of factoring.
5. If $b \neq 0$, then $\dfrac{0}{b} = 0$. That is, $\dfrac{0}{5} = 0$. This property can be demonstrated by using the definition of division. $\dfrac{a}{b} = c$ if and only if $a = b \cdot c$. When $a = 0$ and $b \neq 0$, $a = b \cdot c$ becomes $0 = b \cdot c$. We can see that, for the statement to be true, c must equal zero. There is no other real number that, when multiplied times b, will give the product of 0.
6. If $a \neq 0$, then $\dfrac{a}{0}$ is undefined. That is, $\dfrac{-6}{0}$ is undefined. This property can also be demonstrated using the definition of division. Let $a \neq 0$ and $b = 0$, than $a = b \cdot c$ becomes $a = 0 \cdot c$. There is no real number c that can be multiplied by zero to obtain a nonzero real number a. Thus, we say that $\dfrac{a}{0}$ is undefined.

> **7.** If $a = 0$ and $b = 0$, than $\dfrac{a}{b} = \dfrac{0}{0}$ is indeterminate. Indeterminate expressions are studied in calculus. However, we will bend the rules slightly and use the definition of division to suggest why the quotient of $\dfrac{0}{0}$ is called indeterminate. With $a = 0$ and $b = 0$, $a = b \cdot c$ becomes $0 = 0 \cdot c$. The right side of the equation implies that c could be any real number. We cannot determine what c is. Thus, there is no unique solution to the expression $\dfrac{0}{0}$, and it is said to be indeterminate.

EXAMPLE 1 Illustrating Properties of Zero

Evaluate the following by performing the indicated operations.

a. $\dfrac{2 \cdot 4 - 8}{9}$ **b.** $\dfrac{-5 + 11(-2)}{6 + (-6)}$ **c.** $\dfrac{4\sqrt{7} - 4\sqrt{7}}{\sqrt{7} - \sqrt{7}}$

Solution

a. $\dfrac{2 \cdot 4 - 8}{9} = \dfrac{8 - 8}{9}$

$\qquad = \dfrac{0}{9}$

$\qquad = 0 \qquad$ **Property 5**

b. $\dfrac{-5 + 11(-2)}{6 + (-6)} = \dfrac{-5 + (-22)}{0}$

$\qquad = \dfrac{-27}{0}$

$\qquad = \text{Undefined} \qquad$ **Property 6**

c. $\dfrac{4\sqrt{7} - 4\sqrt{7}}{\sqrt{7} - \sqrt{7}} = \dfrac{0}{0}$

$\qquad = \text{Indeterminate} \qquad$ **Property 7** ◆

1.4 Exercises

In Exercises 1–12, evaluate each of the given expressions by performing the indicated operations.

1. $3 + (6 - 6)$ **2.** $7(\sqrt{2} - \sqrt{2})$ **3.** $\sqrt{6} - \sqrt{0}$ **4.** $\dfrac{0}{4}$

5. $\dfrac{3 - 3}{3}$ **6.** $\dfrac{2 \cdot 3 - 6}{12}$ **7.** $\dfrac{-2 \cdot 3 + 7 \cdot 5}{5 - 5}$ **8.** $\dfrac{-4 + 7(-3)}{4 + -4}$

9. $\dfrac{(-5)(2) + 7}{-8 + 5}$ **10.** $\dfrac{\sqrt{3} - \sqrt{3}}{\sqrt{5}}$ **11.** $\dfrac{7 - 7}{\pi - \pi}$ **12.** $\dfrac{2\sqrt{6} - 2\sqrt{6}}{\sqrt{6} - \sqrt{6}}$

In Exercises 13–16, let *a*, *b*, and *c* be any real numbers.

13. Is $\dfrac{a}{a}$ always equal to 1?

14. Is $\dfrac{a-4}{a-4}$ always equal to 1?

15. Is $\dfrac{a-b}{a-b}$ always equal to 1?

16. Is $\dfrac{a(b+c)}{a(b+c)}$ always equal to 1?

17. a. What is the result on your calculator when $4 \div 0$?
    b. Divide 4 by the smallest positive number you can find on the calculator. What is the quotient?

1.5

Introduction to the Graphics Calculator

Throughout the text are boxes referring to the graphics calculator. If you are using a graphics calculator, you should work through these sections to develop an even better understanding of concepts explained in each chapter. (If you are not using a graphics calculator, feel free to skip this section as well as any other sections on the graphics calculator.)

The calculators used in developing this text material are the CASIO fx-7700G and the Texas Instruments TI-82. This section will help you become familiar with and comfortable using each calculator.

In general, both calculator keyboards are divided into three regions. The lower region contains the numbers and operation symbols of

$$+ \qquad - \qquad \times \qquad \div .$$

Also in this region, each calculator has a key labeled $\boxed{\text{EXE}}$ or $\boxed{\text{ENTER}}$, which is the key used to display the answer. The middle region contains the scientific keys, such as

$$\sin \qquad \cos \qquad \sqrt{} \qquad x^2.$$

The top region contains the keys relating to graphing. In addition both calculators have menus that are used for special functions.

Calculators are actually small computers, and can perform many different functions if given the proper instructions. We instruct the calculator by pressing the correct keys and by placing the calculator in the proper modes. **Modes** are sets of instructions that control the operations of the calculator by setting certain parameters (boundaries or limits).

Mode CASIO fx-7700G

Run / COMP
G-type: REC/CON
angle: Deg
Display: Nrml

FIGURE 1.12

When we turn on the fx-7700G, we see a display like the one shown in Figure 1.12. It shows the current operating modes of the calculator. The display tells us that we can perform calculations and run programs which are stored in the calculator. The angular measure (angle) is in degrees, and the result will have a normal display, meaning that the results will be displayed up to ten digits. To change the mode display, press $\boxed{\text{MODE}}$. The calculator now offers several choices for each category. For example, say we want to write and store a program in the calculator. Simply press $\boxed{2}$; the system will change from computation mode to writing mode. To return to computational mode, press $\boxed{\text{MODE}}$; and then $\boxed{1}$. To leave the mode display, press any key. To review the current modes of the calculator, press and hold the $\boxed{\boxed{\text{M}}\text{Disp}}$ key.

Mode Texas Instruments TI-82

Norm Sci Eng
Float 0123456789
Rad Deg
Function Param
Connected Dot
Sequence Simul
Gird Off Gird On
Rect Polar

FIGURE 1.13

For the TI-82, press $\boxed{\text{MODE}}$ to see the mode menu, shown in Figure 1.13. The words or numerals that are highlighted indicate the current modes of the calculator. If Rad is highlighted, for example, the calculator is in the radian mode. To change a mode, use the arrows to move the cursor (or blinking light) to the desired symbol; then press $\boxed{\text{ENTER}}$. To change the angular measure to degrees move the cursor to Deg and then press $\boxed{\text{ENTER}}$. Degree will be highlighted, showing that the calculator is in degree mode. To leave the mode menu, press $\boxed{\text{CLEAR}}$.

CAUTION   You must have the calculator in the correct mode. Otherwise you may get a wrong answer or no answer at all. ■

Algebraic logic is used to perform computations on these graphics calculators. **Algebraic logic** dictates that numbers and operations are entered from left to right in the same manner we read a statement. This is illustrated in Example 1.

In this and all other calculator boxes, "fx-7700G" refers to the CASIO calculator; "TI-82" refers to the Texas Instrument calculator. The column called "Keystrokes" indicates the keys you will press on the calculator. The column labeled "Screen Display" duplicates what is shown on the screen after pressing the keys.

EXAMPLE 1   **Calculating with a Calculator**

Perform the following calculations.

a. $15 \times 8 + 64 \div 4$ **b.** $\dfrac{37 + 29}{16 - 13}$

Solution

Keystrokes	Screen Display	Answer
a. fx-7700G: 15 $\boxed{\times}$ 8 $\boxed{+}$ 64 $\boxed{\div}$ 4 $\boxed{\text{EXE}}$	$15 \times 8 + 64 \div 4$	136
b. TI-82: $\boxed{(}$ 37 $\boxed{+}$ 29 $\boxed{)}$	$(37 + 29)/(16 - 13)$	
$\boxed{\div}$ $\boxed{(}$ 16 $\boxed{-}$ 13 $\boxed{)}$ $\boxed{\text{ENTER}}$		22

In part b the numerator, $37 + 29$, and the denominator, $16 - 13$, must be placed in parentheses on either calculator. If parentheses are not used, either calculator will perform the operations in the following order.

$$37 + 29 \div 16 - 13$$

Notice that the general procedure for finding the answer is the same for both calculators. ◆

EXAMPLE 2   **Squares and Roots with a Calculator**

Perform the following calculations.

a. 12^2 b. $5^2 \cdot 6^2$ c. $\sqrt{7} \cdot \sqrt{343}$ d. $\dfrac{\sqrt{2}}{\sqrt{3}}$

Solution

	Keystrokes	Screen Display	Answer
a.	fx-7700G: 12 SHIFT x^2	12^2	144
b.	TI-82: 5 x^2 × 6 x^2	$5^2 \times 6^2$	
	ENTER		900
c.	fx-7700G: √ 7 × √ 343 EXE	$\sqrt{7} \times \sqrt{343}$	49
d.	TI-82: 2nd √ 2 ÷ 2nd √ 3	$\sqrt{2} / \sqrt{3}$	
	ENTER		.8164965809

◆

For Example 2 we used either the SHIFT key on the fx-7700G or the 2nd key on the TI-82 to access a function that appears above the key.

Calculators have a key for subtraction − and a key to indicate a negative value (−). To enter a −3, we would use the (−) key.

Keystrokes	Screen Display
fx-7700G: SHIFT (−) 3	−3
TI-82: (−) 3	−3

EXAMPLE 3   **Negative Numbers and Fractions**

Perform the following calculations.

a. $\dfrac{(-6)8}{14(-5)}$ b. $\dfrac{4(-8) - 5(-4)}{6 - 9}$

Solution

Keystrokes	Screen Display	Answer

a. fx-7700G: [SHIFT] [(−)] 6 [×] 8
[÷] [(] 14 [×] [SHIFT] [(−)] 5 [)] $-6 \times 8 \div (14 - 5)$
[EXE] 0.6857142857

b. TI-82: [(] 4 [×] [(−)] 8 [−] 5 [×]
[(−)] 4 [)] [÷] [(] 6 [−] 9 [)] $(4*^-8 - 5*^-4)/(6 - 9)$
[ENTER] 4 ◆

Additional Features: Storage and Decimal Places

When multiplying or dividing many different numbers by the same number, we can save keystrokes by using the storage (or memory) of the calculator. The storage locations correspond to letters of the alphabet. If we want to store 23 in location (cell) A, press:

fx-7700G: 23 [→] [ALPHA] [A] [EXE]

TI-82: 23 [STO] [ALPHA] [A] [ENTER]

To determine what is stored at location A, press

[ALPHA] [A] [EXE] (or [ENTER])

and the calculator will display 23, the value stored at location A.

When dealing with problems involving money, we want to express the answer in dollars and cents, that is, to two decimal places. The instructions for the calculators are as follows.

fx-7700G

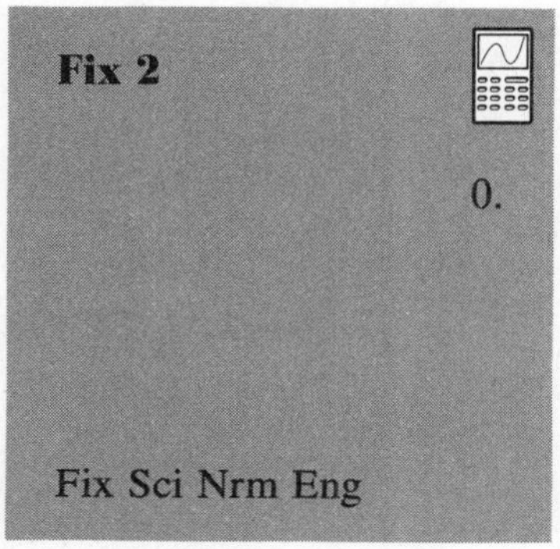

Press [SHIFT] [DISP], and a menu appears at the bottom of the screen, as shown in Figure 1.14. Press [F1], and "Fix" will appear at the top of the screen. Since we want 2 decimal places, enter 2 and press [EXE]. We can fix the number of decimal points, from 0 to 9 inclusive, by entering the desired number.

TI-82

Press [MODE]. Norm should be highlighted. Move cursor to "Float," shown in Figure 1.13 and highlight the 2. Press [CLEAR]. To select any number of decimal places between 0 and 9 inclusive, highlight the desired number.

When you place the calculator in this mode, all answers will be expressed to the number of decimal places you select. (Remember that you want two decimal places for money.)

Fix 2

0.

Fix Sci Nrm Eng

FIGURE 1.14

EXAMPLE 4 **Application: Pay Roll Problem**

The manager of a local grocery store is preparing the gross pay of six people who stock shelves. Calculate the gross pay for each person (a through f) using the storage and decimal features on your calculator and the information in the table.

	Hours Worked	Hourly Rate ($)
a.	24.5	5.75
b.	15.0	5.75
c.	17.5	5.75
d.	25.0	6.35
e.	32.0	6.35
f.	38.0	6.35

Solution   The fx-7700G is used to calculate the gross pay for a, b, and c. The TI-82 is used to calculate d, e, and f.

fx-7700G: All of the keystrokes and the complete screen display are given in the steps below. Begin by placing 5.75 into memory storage cell A. For a, press $\times$, and the screen display is 5.75 $\times$. Enter 24.5 and press EXE; the result is 140.88, the pay of worker a. To calculate b and c's pay use the keystrokes ALPHA A to bring the 5.75 out of memory.

	Keystrokes	Screen Display	Answer
	5.75 $\to$ ALPHA A EXE	5.75 $\to$ A	5.75
a.	$\times$ 24.5 EXE	5.75 $\times$ 24.5	140.88
b.	ALPHA A $\times$ 15 EXE	A $\times$ 15	86.25
c.	ALPHA A $\times$ 17.5 EXE	A $\times$ 17.5	100.63

TI-82: All of the keystrokes and the complete screen display are given in the steps below. Begin by placing 6.35 in storage cell A as shown on the first line. For d, press $\times$, and the screen display is ANS*. Insert the number 25 and press ENTER; the result is 158.75, the pay of worker d. To calculate the pay of workers e and f, the keystrokes ALPHA A bring the 6.35 out of memory.

	Keystrokes	Screen Display	Answer
	6.35 STO A ENTER	6.35 $\to$ A	6.35
d.	$\times$ 25 ENTER	ANS*25	158.75
e.	ALPHA A $\times$ 12 ENTER	A*32	203.20
f.	ALPHA A $\times$ 38 ENTER	A*38	241.30

Thus, the gross wages of the six people who stock shelves are $140.88, $86.25, $100.63, $158.75, $203.20 and $241.30. ◆

There is no need to clear the memory in either calculator; when a new value is entered, it replaces the old value.

EXAMPLE 5   **Application: Determining Stagnation Temperature**

The temperature in degrees Kelvin on the leading edge of a body, known as the stagnation temperature T_s, can be approximated by the formula

$$T_s = T_a \left[1 + \left(\frac{r-1}{2} M^2 \right) \right].$$

T_a is the temperature of the ambient, or surrounding air, r is the specific heat ratio for air, and M is a Mach number or the number of times the speed of sound the body is moving. Calculate the stagnation temperature when $T_a = 230°$ K, $r = 1.4$, and $M = 4$.

Solution

$$T_s = \boxed{230} \left[1 + \left(\frac{\boxed{1.4} - 1}{2} \right) \boxed{4^2} \right] \quad \textbf{Substituting}$$

$$= 230 \;\boxed{\times}\; \boxed{(}\; 1 \;\boxed{+}\; \boxed{(}\; 1.4 \;\boxed{-}\; 1 \;\boxed{)}\; \boxed{\div}\; 2 \;\boxed{\times}\; 4 \;\boxed{x^2}\; \boxed{)}\; \boxed{\text{ENTER}}$$

$$= 966$$

Thus $T_s = 966°$K. ◆

A BASIC computer program, which can be used to calculate T_s, is found on the tutorial disk.

Graphing on the fx-7700G

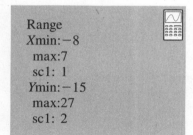

Range
Xmin:−8
 max:7
 sc1: 1
Ymin:−15
 max:27
 sc1: 2

FIGURE 1.15

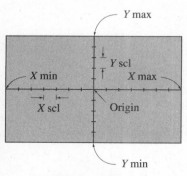

FIGURE 1.16

Even if you are using the TI-82, be sure to read this section carefully, since much of it applies to the TI-82 as well.

To draw a graph with a calculator, we must provide the calculator with specific instructions. Begin by placing the calculator in the proper mode. Since we will be graphing on a coordinate plane the mode display should read

G-type: REC/CON.

If it does not, press $\boxed{\text{MODE}}$ $\boxed{\text{SHIFT}}$, and Mode menu 2 will appear on the screen. The symbol for rectangular coordinates is

+ : Rec.

To change to rectangular mode, press $\boxed{\text{MODE}}$ $\boxed{\text{SHIFT}}$ $\boxed{+}$.

We now need to give the calculator instructions so that the graph is properly located on the screen. To accomplish this, press $\boxed{\text{RANGE}}$. The range parameter screen should now be displayed, as shown in Figure 1.15. The values for Xmin and max indicate the distances from the **origin** (intersection of the two axis) to the left and right of the screen respectively (Figure 1.16). The "sc1" is the scale on the **x-axis** (or horizontal axis). The number after sc1 indicates the number of units between the small lines, or **hash marks**, on the x-axis. The values for Ymin and max indicate the distance from the origin to the bottom and the top of the screen, respectively, along the **y-axis** (or vertical axis). The sc1 is the scale on the y-axis.

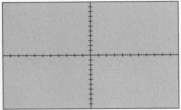

Range
Xmin: -10
 max: 10
 sc1: 1
Ymin: -10
 max: 10
 sc1: 1

FIGURE 1.17

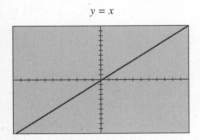

FIGURE 1.18

Now change the values for the range by moving the cursor to the right of Xmin and performing the following keystrokes.

| SHIFT | | $-$ | 10 | EXE | 10 | EXE | 1 | EXE | SHIFT |
| $-$ | 10 | EXE | 10 | EXE | 1 | EXE | .

The range display should now be the same as in Figure 1.17. To enter negative values for the range, either use $-$ or $(-)$. Now press RANGE, and a second range list will appear on the screen. This list relates to polar coordinates, which will be discussed later. Press RANGE again and then press G⇔T. The G⇔T key shifts the screen display from displaying a graph to displaying text, or vice versa. The screen display (see Figure 1.18) shows a horizontal and vertical axis system that intersects in the center of the screen. On each axis there are twenty small lines or hash marks. Since there is a scale of 1, the distance between each hash mark is 1 unit.

Now that the range is set, we will graph the straight line $y = x$. To "graph the straight line" means that we will give the calculator instructions to draw a picture of the line on the screen.

EXAMPLE 6   **Graph a Straight Line**

Graph the straight line $y = x$.

Solution   To graph the line we perform the following keystrokes.

Keystrokes	Screen Display
GRAPH	Graph $y =$
x,θ,T	Graph $y = x$
EXE	(See Figure 1.19)

The graph is a straight line that passes through the origin.   ◆

The line $y = x$ should be the same distance from the x-axis as the y-axis in Figure 1.19. This is not true. Why? Notice that the distance between the hash marks on the x-axis are farther apart than the hash marks on the y-axis. Why? The width of the screen is 96 dots, and the height is 64 dots. Therefore, if the same scale is used on both axes, the hash marks must be farther apart on the x-axis than on the y-axis. The ratio of the width to height is $96:64$ or $3:2$. To calculate the same length between hash marks on both scales, use the proportion

$$\frac{3}{2} = \frac{h}{v},$$

where h is the number of units on the x-axis, and v is the number of units on the y-axis. We can write the equation in two forms. One is for determining the number of units on the y-axis v, given the number of units on the x-axis, h.

$y = x$

FIGURE 1.19

The other equation is for determining the number of units on the x-axis h, given the number of units on the y-axis, v.

$$v = \frac{2 \cdot h}{3} \quad \text{or} \quad h = \frac{3 \cdot v}{2}$$

For example, if we want the total distance on the x-axis to be 18 units, to determine the total number of units on the y-axis, v, use the equation

$$v = \frac{2 \cdot h}{3}$$

$$= \frac{2 \cdot 18}{3} = 12 \qquad \textbf{Substituting 18 for } \textbf{\textit{h}}.$$

With the scale 18 units on the x-axis and 12 units on the y-axis, we will have the same distance between hash marks on both the x- and the y-axis.

Using the units 18 and 12, we must now reset the range. In order to have the origin in the center of the screen, divide the number of units by 2 and let Xmin $= -9$, max $= 9$, sc1 $= 1$, Ymin $= -6$, max $= 6$, sc1 $= 1$.

Now that the range is set, we again can graph the straight line $y = x$. Follow the key strokes as given in Example 6, and the result is as in Figure 1.20. Now the scales appear the same on both sets of axes, and the line $y = x$ is the same distance from both axes.

Now graph the straight line $y = 2x$. The keystrokes are $\boxed{\text{GRAPH}}$ 2 $\boxed{\text{x,}\Theta\text{,T}}$ $\boxed{\text{EXE}}$. On the calculator screen we have the graphs of both lines, $y = x$ and $y = 2x$. To display only one graph at a time, clear the screen before graphing the next line. To clear the screen use the following keystrokes: $\boxed{\text{SHIFT}}$ $\boxed{\text{Cls}}$ $\boxed{\text{EXE}}$. Note above the key $\boxed{\text{F5}}$ are the letters Cls, which means **clear screen.** To check whether the lines are erased, press $\boxed{\text{G}\Leftrightarrow\text{T}}$. The $\boxed{\text{G}\Leftrightarrow\text{T}}$ key shifts the screen display between the graph screen and the text screen. The lines $y = x$ and $y = 2x$ are now gone, and only the x- and y-axes remain on the screen. Now graph $y = 2x$, and the result will be as in Figure 1.21.

In Example 7 we will draw a graph that is not a straight line. Before getting into that example, reset the range. To save space, we list the range (Ymin, max, sc1, Ymin, max, sc1) in the following manner now and in the future: $[-12, 12, 2, -3, 12, 2]$.

EXAMPLE 7   Graph of a Curve

Graph $y = 3x^2 - 2$.

Solution   First clear the screen and then enter the new set of values for the range. To graph $y = 3x^2 - 2$, use the following keystrokes.

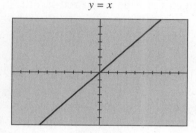

$y = x$

FIGURE 1.20

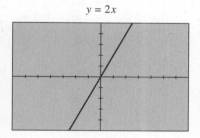

$y = 2x$

FIGURE 1.21

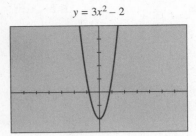

$y = 3x^2 - 2$

FIGURE 1.22

Keystrokes	Screen Display
GRAPH	Graph $y =$
3	Graph $y = 3$
x,Θ,T	Graph $y = 3x$
SHIFT x^2	Graph $y = 3x^2$
−	Graph $y = 3x^2 -$
2	Graph $y = 3x^2 - 2$
EXE	(See Figure 1.22)

The curve we have graphed is called a parabola and will be discussed in later chapters. ◆

The fx-7700G has one feature that makes it easy to graph a standard form of all the equations that are listed on the keyboard. These equations are log, ln, sin, cos, $\sqrt{}$, etc. For example, to graph sin, press GRAPH sin EXE , and the calculator will graph the sin curve. Note that we do not need to enter a variable nor set the range. When we ask the calculator to graph these standard equations, the calculator automatically sets the range (referred to as a **default setting** since it is done automatically).

Graphing on the TI-82

Normal Sci Eng
Float 0123456789
Radian Degree
Func Par Pol Seq
Connected Dot
Sequential Simul
Full Screen Split

FIGURE 1.23

WINDOW FORMAT
Xmin $= -10$
Xmax $= 10$
Xscl $= 1$
Ymin $= -10$
Ymax $= 10$
Yscl $= 1$

FIGURE 1.24

If you skipped over the material on graphing with the fx-7700G go back and read it carefully. Some material in that section applies to both calculators and it is not repeated here.

The size of the screen is the same for both calculators, 96 dots by 64 dots. This means that the ratio is $3 : 2$ and the formulas for determining the scale for the x- and y-axes are the same.

$$v = \frac{2 \cdot h}{3} \quad \text{or} \quad h = \frac{3 \cdot v}{2}.$$

For graphing on a coordinate plane, the TI-82 should be in the Function mode. To place the calculator in the proper mode, press MODE . If Func is not highlighted, highlight it. You can highlight Func by placing the cursor over Func and pressing ENTER (Figure 1.23).

To set the range, press WINDOW , and the window settings will be displayed as in Figure 1.24. (You can see that the settings for both calculators are the same.) To change a value of the window setting, move the cursor to the right of the equal sign of the value you wish to change by pressing ∇ . Now enter the new value and press ENTER . For example, to change Xmin to -47, move the cursor to the right of equal sign and perform the following keystrokes: (−) 47 ENTER . The line will read Xmin $= -47$.

One difference between the two calculators is in the procedure for entering the equation of a curve. In the top row of keys on the TI-82 there is a Y= key. Press Y= , and a menu, shown in Figure 1.25, will appear. The menu shows ten different Ys, so it is possible to enter ten different equations at one time. However, the calculator will graph only those functions for which the equal sign is highlighted.

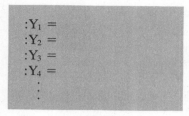

:$Y_1 =$
:$Y_2 =$
:$Y_3 =$
:$Y_4 =$
:
·
·

FIGURE 1.25

We will graph both a straight line and a curve as was done in Examples 6 and 7, using the same window settings $[-12, 12, 2, -8, 8, 2]$ for both examples.

EXAMPLE 8 **Graph of a Straight Line**

Graph the line $y = 2x + 3$.

Solution To graph the line $y = 2x + 3$, use the following keystrokes.

Keystrokes	Screen Display
Y=	:$Y_1 =$
2	:$Y_1 = 2$
x,T,Θ	:$Y_1 = 2x$
+	:$Y_1 = 2x +$
3	:$Y_1 = 2x + 3$
GRAPH	(See Figure 1.26)

Note that when we entered the two, the equal sign was automatically shaded. The result is a straight line as illustrated in Figure 1.26. ◆

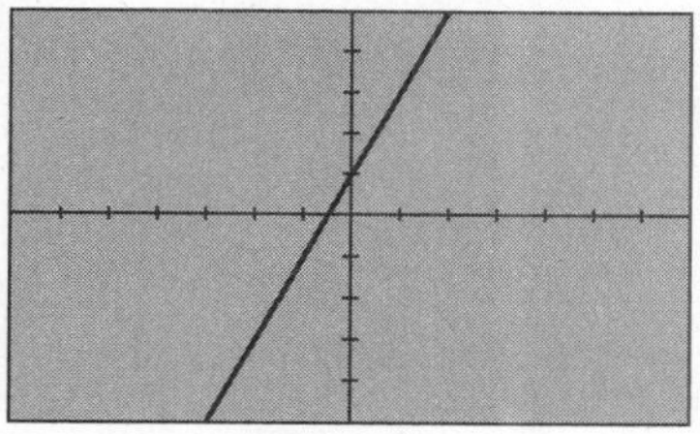

EXAMPLE 9 **Graph of a Curve**

Graph $y = x^2 - 5x + 6$.

Solution We will enter this equation opposite Y_2. If the equal sign opposite Y_1 is shaded, move the cursor to the equal sign opposite Y_1 and press ENTER. This tells the calculator not to graph Y_1 when you press GRAPH. To graph $y = x^2 - 5x + 6$, use the following keystrokes.

Keystrokes	Screen Display
Y=	:$Y_2 =$
x,T,Θ	:$Y_2 = x$
x²	:$Y_2 = x^2$
−	:$Y_2 = x^2 -$
5	:$Y_2 = x^2 - 5$
x,T,Θ	:$Y_2 = x^2 - 5x$
+	:$Y_2 = x^2 - 5x +$
6	:$Y_2 = x^2 - 5x + 6$
GRAPH	(See Figure 1.27)

$y = 2x + 3$

FIGURE 1.26

$y = x^2 - 5x + 6$

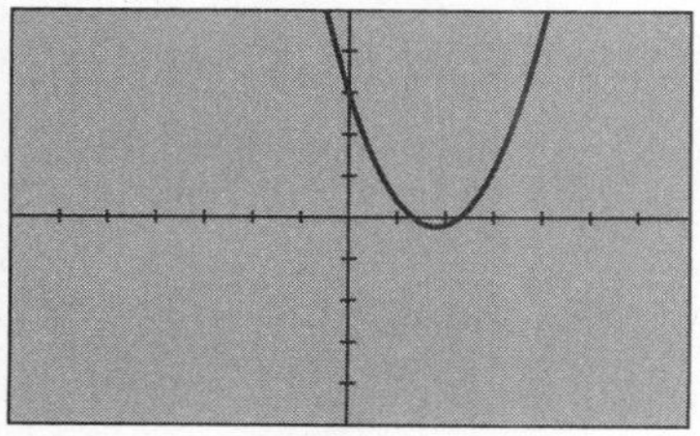

FIGURE 1.27

Figure 1.27 shows the graph of $y = x^2 - 5x + 6$. The curve is a parabola. ◆

$y = 2x + 3$ and $y = x^2 - 5x + 6$

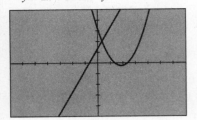

FIGURE 1.28

To graph equations Y_1 and Y_2 at the same time, press $\boxed{Y=}$. Now shade the equal sign to the right of Y_1 and Y_2 and then press $\boxed{\text{GRAPH}}$. The equations are graphed in the order in which they were written, the straight line first and the parabola second (see Figure 1.28). From the graph we can see that the line and the parabola intersect at one point, possibly at a second point. Later in the text, we will need to see if they intersect at two points. Change the range settings (try $[-8, 8, 2, -2, 17, 2]$) and see if there are two points of intersection. We will also need to change the window settings so that we can see the *complete graph,* in other words, where the graph crosses an axis, goes up, goes down, changes direction, and/or intersects another curve.

We now have learned some basic calculator operations. As we continue through the text, more special keys will be introduced as they are needed. The calculator is a tool, like a pen or a pencil. It must given instructions and will do only those things we tell it to do.

1.5 Exercises

In Exercises 1–14, use your calculator to determine the value of each of the given expressions.

1. $3 - (-2) + (-7) - 6$

2. $5 + (-6) - (-11) + 13$

3. $-13 + (-12) - 14 + (-5)$

4. $(-4)(+3)(-5)$

5. $(-6)(-2)(-7)$

6. $(+7)(-6)(+8)$

7. $\dfrac{(-6)(8)}{-12}$

8. $12 - 3(5 - 9) + 3$

9. $-2(7 - 4) + 8$

10. $(-2)(7) - 4 + 8$

11. $\dfrac{8}{5 - 3} + 2 - 7$

12. $\dfrac{13 + (-2)}{3} - \dfrac{9 - 5}{-2}$

13. $\dfrac{4(-8) - 5(-4)}{4 - 7}$

14. $7 + (-8)(-2) - (6) \div (-3) + 14 - (-7)$

In Exercises 15–39, use your calculator to determine the value of the given expressions.

15. $13^2 \cdot \pi^2$

16. $2 - (\sqrt{3})(3.1416)^2$

17. $\dfrac{\sqrt{11}}{\sqrt{47}}$

18. $\sqrt{\dfrac{11}{47}}$

19. $\dfrac{\sqrt{13} - \sqrt{61}}{\sqrt{83}}$

20. $\sqrt{\dfrac{81.7622 - 41.8765}{56.982}}$

21. $(-8.937)(13.7856)(-14.9762)$

22. $(\sqrt{-6.34})(-2.345)(-13.756)$

23. $(8.37)^2(-6.19)(13.79)$

24. $\dfrac{(-11.34)(18.7654)}{14.379}$

25. $\dfrac{-97.486}{(-35.75)(4.789)}$

26. $\dfrac{(-14.387)(14.372 - 14.372)}{-8.765}$

27. $\dfrac{\sqrt{14.79}\,(3.456 - 8.732)}{16.937 - 16.937}$

28. $(-34.86)(18.96 - 43.21) + 43.15$

29. $\dfrac{24.94}{106.42 - 376.41} + 67.39 - 79.46$

30. $\dfrac{\sqrt{34.89} - (-69.73)}{98.75 - 107.24}$

31. $\dfrac{136.785 + (-207.456)}{307.162} - \dfrac{346.751 - 961.88}{-43.617}$

32. $(-342.651)(876.477) + 909.706 \div (-3.04) + 8.569$

33. $\dfrac{(-4.76)(324.56) - (51.82)(61.487)}{6.852 - 13.796}$

34. $316.75 + (\sqrt{64}) \cdot (-314.86) - 6.817 \div (-5.147) + 143.782 \div (-714.86)$

35. If a soda can has a radius of 31 mm and a height of 116 mm, find the volume of the can in cubic millimeters. The formula for finding the volume of a cylinder is $V = \pi r^2 h$. (*Hint:* Use the calculator value for π.)

36. Find the amount of tin needed to construct a rectangular box whose length, width, and height are respectively $l = 12$ in., $w = 8$ in., and $h = 6$ in. Assume there is no waste. The formula for finding the surface area of a rectangular box is $S = 2lw + 2lh + 2wh$.

37. The natural frequency of a pendulum of length (L) is $f = \dfrac{1}{2\pi}\sqrt{\dfrac{g}{L}}$, where g is the acceleration due to gravity. Find f if $g = 32$ ft/s^2 and $L = 2$ ft. (*Hint:* Use the calculator value of π.)

38. A convenient way of computing the speed of an object at the end of an oscillator (spring) is the formula

$$v = \sqrt{\dfrac{k}{m}(x_0^2 - x^2)},$$

where k is the spring constant, m is the mass at the end of the spring, x_0 is the initial position of the mass at the end of the spring, and x is a position of the mass at the end of the spring after release. Find the speed of the mass if $k = 5.0$ N/m, $m = 0.20$ kg, $x_0 = 0.12$ m and $x = 0.04$ m.

39. One important part in the design of high-flying supersonic planes is the amount of lift created by the air currents against the wing. The amount of lift for a particular section, called an airfoil section, is calculated with the formula

$$C = \dfrac{2\pi A a}{2 + \sqrt{4 + (A\sqrt{M^2 - 1})^2}},$$

where a is the angle between the air stream and the airfoil section, A is the area of the section, and M is the Mach number of the aircraft's speed. If $a = 0.100$ radians, $A = 80.0$ ft^2, and $M = 2$, what is C?

In Exercises 40–44, set the following range or window settings on your calculator.

40. $(-13, 13, 2, -18, 18, 5)$

41. $(-6, 12, 2, -3, 8, 2)$

42. $(-12, 24, 2, -8, 16, 2)$

43. $(-2, 52, 5, -2, 34, 5)$

44. $(-100, 200, 50, 100, 100, 50)$

45. To have the same distance between hash marks on the x- and y-axes, what should be the scale on the y-axis if the scale on the x-axis is 72 units?

46. To have the same distance between hash marks on the x- and y-axes, what should be the scale on the y-axis if the scale on the x-axis is 600 units?

47. To have the same distance between hash marks on the x- and y-axes, what should be the scale on the x-axis if the scale on the y-axis is 72 units?

48. To have the same distance between hash marks on the x- and y-axes, what should be the scale on the x-axis if the scale on the y-axis is 18 units?

In Exercises 49–58, graph the following on your graphics calculator. Use the range or window settings $(-9, 9, 2, -6, 6, 2)$.

49. $y = 2x + 4$

50. $y = 2x - 4$

51. $y = 2x^2 - 5$

52. $y = 3x^2 + 2$

53. $y = -2x^2 + 6$

54. $y = -0.2x^2 - 1$

55. $y = x^2 - 2x - 5$

56. $y = 0.8x^2 + 4x + 3$

57. $y = -0.7x^2 - 5x - 6$

58. $y = -0.7x^2 + 5x - 6$

In Exercises 59–62, graph both equations on the same set of axes. Start with the range or window settings of $(-9, 9, 2, -6, 6, 2)$. You may have to readjust the range or window setting to see the complete graph.

59. $y = 2x + 5, y = -3x + 2$

60. $y = 4x - 3, y = 3x^2 - 4$

61. $y = -2x^2 + 5, y = 2x - 3$

62. $y = 0.5x^2 - 0.7x - 0.5, y = -0.8x^2 - 0.6x + 0.9$

1.6

Significant Digits

When the number of objects in a group, such as the number of bolts in a box, is counted, the result is always an exact number that can be represented by an integer. When an object, such as the length of a bolt, is measured, there is always a certain amount of doubt about whether the measurement provided the true length of the object. Since all measurements are uncertain, we can never really know what the true value is. Therefore, scientists will use an agreed-upon measurement referred to as the **accepted value.** The **deviation** of a measurement is the difference between the measurement and the accepted value. Deviations from the accepted value can be grouped into two categories: accuracy and precision.

The **accuracy** of a measurement indicates how closely the measurement agrees with its accepted value. The error of a measurement is directly related to the deviation, and is often expressed in the form of a percentage:

$$\text{Percentage of error} = \frac{|\text{deviation}|}{\text{accepted value}} \times 100\%.$$

Whereas a deviation can be large or small, its relative size is best ascertained by finding the percentage or error.

EXAMPLE 1 Percentage of Error

Determine the percentage of error for parts a and b.

a. A bolt is measured at 4.7 cm; the accepted value is 5.0 cm.
b. A piece of wood is measured to be 370 cm; the accepted value is 376 cm.

Solution

a. The deviation is 0.3 cm (5.0 cm − 4.7 cm = 0.3 cm).

$$\text{Percentage of error} = \frac{|0.3 \text{ cm}|}{5.0 \text{ cm}} \times 100\%$$
$$= 6\%$$

b. The deviation is 6 cm (376 cm − 370 cm = 6cm)

$$\text{Percentage of error} = \frac{|6 \text{ cm}|}{376 \text{ cm}} \times 100\%$$
$$= 1.6\% \quad \blacklozenge$$

The 0.3 cm in part a of Example 1 seems to be a small deviation compared to the 6 cm in b. However, the percentage errors are 6% and 1.6% respectively. Therefore, the measurement of the wood was more accurate then the measurement of the bolt even though the deviations were just the opposite. The size of an error depends upon the deviation and the relative size of the measured quantity.

The **precision** of a measurement is an indication of the repeatability of a measurement. If there is a very small difference in several measurements of the same object, we can conclude that the measurement is precise.

A way to distinguish between accuracy and precision is to consider the following tasks. When we put an extra shelf in a bookcase, we need not be concerned whether we measure with a bent ruler or a straight ruler as long as we use the same tool to measure both bookcase and shelf and we measure in the same manner. Certainly, the final length of the shelf will not be accurate if we use a bent ruler; that is, the shelf would not be near the true length of the bookcase. But both bookcase and shelf will be equally inaccurate. Only the precision of the measurements matter; the accuracy of the measurement, or nearness to the true value, is immaterial.

However, if we are ordering the shelf from the store, the accuracy of the measurement is just as important as the precision of the measurement if the new shelf is to fit perfectly. The ruler used at the store will have to be as accurate as the ruler we used at home to measure the bookshelf.

Since technology is based upon measurements, and measurements are represented with approximate numbers, we must learn how to work with approximate numbers. This involves determining the number of significant digits in a measurement, performing calculations with approximate numbers, and rounding approximate numbers.

CONVENTION FOR USING DECIMALS AND MEASUREMENTS IN CALCULATIONS

We will adopt the convention that any number written with a decimal point and/or a measurement unit will be assumed to be an approximate number. If a number has no decimal point or no measurement unit, assume it to be exact. All counting number are exact. The exception to this rule are definitions:

$$12 \text{ in.} = 1 \text{ ft}; \quad 1 \text{ mile} = 5{,}280 \text{ ft}; \quad 1 \text{ in.} = 2.54 \text{ cm.}$$

These numbers are all exact, even though they are measurements and/or decimals.

In an equation such as $V = \dfrac{4}{3}\pi r^3$, the fraction $\dfrac{4}{3}$ is exact and so is the exponent 3. π is an approximate number. In fact, all irrational numbers, such as π, $\sqrt{2}$, $\sqrt{3}$, $\sqrt{7}$, are approximate numbers.

In each approximate number, there is one digit that is a good guess—called the *doubtful digit*. If we measure the length of a wire, as shown in Figure 1.29, using a meter stick marked only every centimeter (the bottom one), we might record the value as 5.8 cm. The 5 is certain because the end of the wire clearly falls between the 5 and 6. The 8 is a guess, or doubtful digit, since it falls

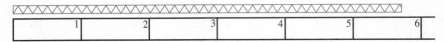

FIGURE 1.29

somewhere between the divisions of the measuring instrument. (The smallest division of space that the human eye can estimate with any precision is a tenth.) Though the 8 is considered an estimate, it is a reasonable guess, and we still consider it significant. Consequently, there are two significant digits in the measurement. Also, the precision of the unit of measurement is to the tenth. If we measure the length of the wire with a more precise ruler (see Figure 1.30), we will obtain a value of 5.76 cm. In this case, the 5.7 is certain, since the wire falls between 5.7 and 5.8. The 6 is doubtful, but it is, again, a good guess. Hence, 5.76 cm is a more accurate measurement, and it has three significant digits, making the precision of the unit of measurement to the hundredth.

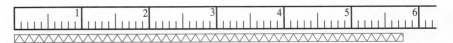

FIGURE 1.30

When the same measurement is obtained for repeated trials, the assumption is that the deviation, or error, of the measurements is half of the magnitude represented by the doubtful digit. Therefore, the true length of the wire would be assumed to lie between 5.75 cm and 5.85 cm for the first measurement. In the second case, the true value would be assumed to be between 5.755 cm and 5.765 cm.

The measurement 5.76 cm can also be expressed in meters as 0.0576 m, which also contains 3 significant digits. The zeros preceding the 5 in 0.0576 are not significant digits.

> **DEFINITION 1.7 Significant Digits**
>
> In an approximate number, a digit is significant if it is not a zero, if it is an underlined zero, if it is a zero between two nonzero digits, or if it is terminal zero on the right side of the decimal point.

One problem we have in defining significant digits is determining the number of significant digits in a number such as 7500 mm. If the length is 7500 mm to the nearest hundred millimeters, then there are two significant digits. On the other hand, if the length is 7500 mm to the nearest millimeter, then there are four significant digits. In order to determine whether the zeros are significant digits, we must know something about the number. Sometimes a line is placed under or over the last significant zero. This book uses the convention of underlining. For example, in the number 7500 the line under the zero indicates that there are three significant digits: 7, 5, and 0.

EXAMPLE 2   **Number of Significant Digits**

Determine the number of significant digits in each set of approximate numbers.

Solution

Approximate numbers	Number of significant digits
a. 3, 0.3, 0.003, 30, 300	1
b. 300, 30, 32, 3.2, 0.032, 3.0, 0.30	2
c. 5670, 500, 506, 5.06, 0.0650, 0.500, 0.0506	3
d. 3478, 3400, 3000, 3004, 0.03000, 0.3004	4

◆

Rounding Approximate Numbers

To calculate approximate numbers, we need to round approximate numbers to the desired decimal place. There are various ways to round. The rules this text uses follow.

RULES FOR ROUNDING

If the digit to be rounded is to the left of the decimal point

Rule X   If the digit to the right of the one to be rounded is less
than 5, leave the digit to be rounded unchanged; replace
all remaining digits up to the decimal point with zeros and
drop all digits to the right of the decimal point.

Rule XI   If the digit to the right of the one to be rounded is 5 or
greater, add 1 to the digit to be rounded; replace all
remaining digits up to the decimal point with zeros and
drop all digits to the right of the decimal point.

If the digit to be rounded is to the right of the decimal point

Rule XII   If the digit to the right of the one to be rounded is less
than 5, leave the digit to be rounded unchanged and drop
all the remaining digits.

Rule XIII   If the digit to the right of the one to be rounded is 5 or
greater, add 1 to the digit to be rounded and drop all the
remaining digits.

EXAMPLE 3   **Rounding Approximate Numbers**

Round the numbers to the nearest tenth.

a. 396.41   **b.** 396.48   **c.** 396.45

Solution

a. 396.4 1 to the nearest tenth is 396.4   **Rule XII**
b. 396.4 8 to the nearest tenth is 396.5   **Rule XIII**
c. 396.4 5 to the nearest tenth is 396.5   **Rule XIII** ◆

EXAMPLE 4   **Rounding Approximate Numbers**

Round the numbers to the decimal place indicated in parentheses.

a. 25,683.1 (hundreds)   **b.** 25,683.7 (tens)   **c.** 2.5683 (hundredths)

Solution

a. 25,6 83.1 to the nearest hundred is 25,700   **Rule XI**
b. 25,6 8 3.7 to the nearest ten is 25,680   **Rule X**
c. 2.5 6 83 to the nearest hundredth is 2.57   **Rule XIII** ◆

In answers **a** and **b** of Example 4, the 7 and 8 are significant digits. The zeros
to the right of the 7 and 8 are placeholders.

Calculating with Approximate Numbers

The answer when we calculate with approximate numbers cannot be more accurate than the least accurate number in the data. Several different systems of rules for calculating with approximate numbers exist. This text uses the following.

RULES FOR CALCULATING WITH APPROXIMATE NUMBERS

Rule XIV   Add or subtract the numbers, then round the result to the decimal position of the least precise number involved.

Rule XV   Multiply or divide the numbers, then round the result to the same number of significant digits as the least accurate number involved.

A consequence of using the rules for calculating with approximate numbers is that the result will contain one doubtful digit. In general, calculations in this text are performed as if "chained" on a calculator, and the result is rounded at the end. **Chained** means doing all the calculation on the calculator and then rounding the final result. If we must dsiplay and/or reenter subresults, carry two more digits than required in order to maintain the proper accuracy in the rounded result.

EXAMPLE 5   **Addition and Subtraction of Approximate Numbers**

Perform the indicated operations and round the results using the rules for approximate numbers.

a. $3.43 + 1.124 + 0.943$ **b.** $5.42 - 2.56789$ **c.** $6740 + 3525$

Solution

a.   The least precise number is 3.43. Therefore, the sum will be rounded to the hundredths place.

$$
\begin{array}{r}
3.4\boxed{3} \\
1.124 \\
\underline{0.943} \\
5.497 \quad \text{Adding} \\
5.50 \quad \text{Rounding to hundredths}
\end{array}
$$

b.   The least precise number is 5.42. Therefore, the difference will be rounded to the hundredths place.

$$
\begin{array}{r}
5.4\boxed{2} \\
-\ 2.56789 \\
\underline{} \\
2.85211 \quad \text{Subtracting} \\
2.85 \quad \text{Rounding to hundredths}
\end{array}
$$

c. The least precise number is 6740. The sum is rounded to the tens place.

$$
\begin{array}{r}
6,740 \\
+\ 3,525 \\
\hline
10,265 \quad \text{Adding} \\
10,270 \quad \text{Rounding to tens} \quad \blacklozenge
\end{array}
$$

EXAMPLE 6   Application: Multiplication of Approximate Numbers

Find the area of a desk top that is 0.75 m wide and 1.42 m long, using the formula $A = lw$.

Solution   The measurements 0.75 m and 1.42 m contain 2 and 3 significant digits, respectively. The least accurate number is 0.75. The product will be rounded to two significant digits.

$$
\begin{aligned}
A &= (1.42 \text{ m})(0.75 \text{ m}) \quad \text{Substituting}\\
&= 1.065 \text{ m}^2 \quad\quad\quad \text{Multiplying}\\
&= 1.1 \text{ m}^2 \quad\quad\quad\ \text{Rounding to two significant digits} \quad \blacklozenge
\end{aligned}
$$

EXAMPLE 7   Application: Multiplication and Division of Approximate Numbers

The formula for the volume of a cylinder is $V = \pi r^2 h$. If the volume, V, is 12 in^3, and the radius, r, is 2 in., what is the height, h?

Solution   The measurement 12 in.^{3} and 2 in. contain 2 and 1 significant digits, respectively. The answer will be rounded to one significant digit. Solve $V = \pi r^2 h$ for h.

$$
\frac{V}{\pi r^2} = \frac{\pi r^2 h}{\pi r^2} \quad \text{Dividing both sides of the equation by } \pi r^2.
$$

$$
\frac{V}{\pi r^2} = h
$$

$$
\begin{aligned}
h &= \frac{12 \text{ in.}^3}{(\pi)(2 \text{ in})^2} \quad \text{Substituting}\\
&= 0.9549 \quad\quad\ \text{Calculating}\\
&= 1 \text{ in.} \quad\quad\ \ \text{Rounding to one significant digit.} \quad \blacklozenge
\end{aligned}
$$

EXAMPLE 8   Application: Resistance

If two resistors R_1 and R_2, are connected in parallel, the equivalent resistance, R, is given by the formula $R = \dfrac{R_1 R_2}{R_1 + R_2}$.

a. If $R_1 = 2.3475$ and $R_2 = 1.3456$, find R.
b. If $R_1 = 0.08$ and $R_2 = 0.0456$, find R.

Solution

a.

$$R = \frac{(\;2.3475\;)(\;1.3456\;)}{2.3475 \;+\; 1.3456} \qquad \text{\textbf{Substituting}}$$

In the denominator, both numbers have the same precision, so their sum will have five significant digits. The numbers in the numerator have the same accuracy, so their product will have five significant digits. The quotient therefore will be accurate to five significant digits.

$$= 0.855324 \qquad \text{\textbf{Calculating}}$$
$$= 0.85532 \qquad \text{\textbf{Rounding to five significant digits}}$$

b.

$$R = \frac{(\;0.08\;)(\;0.0456\;)}{0.08 \;+\; 0.0456} \qquad \text{\textbf{Substituting}}$$

$$= \frac{0.003648}{0.1256} \qquad \text{\textbf{Showing all digits in calculation}}$$

The denominator of the fraction is accurate to the hundredths place, so it has two significant digits. The product in the numerator has one significant digit. Therefore the quotient will have only one significant digit.

$$= 0.029045 \qquad \text{\textbf{Dividing}}$$
$$= 0.03 \qquad \text{\textbf{Rounding to one significant digit}} \quad \blacklozenge$$

1.6 Exercises

In Exercises 1–6, determine the percentage of error. The measurement is followed by the accepted value, enclosed in parentheses.

1. 4.75 in. (5.00 in.)

2. 23.4 cm (23.5 cm)

3. 315 ft (325 ft)

4. 0.002 mm (0.005 mm)

5. 275,000 km (278,000 km)

6. 3,130,000 mi (3,150,000 mi)

In Exercises 7–16, round the numbers to the decimal place indicated in the parentheses. (*Hint*: See Examples 3 and 4.)

7. 476.1 (tens)

8. 56.7 (units)

9. 2.34 (tenths)

10. 345.2 (tens)

11. 86.5 (units)

12. 46.75 (tenths)

13. 0.04655 (ten thousandths)

14. 475,000 (ten thousands)

15. 4615 (hundreds)

16. 0.055 (hundredths)

In Exercises 17–26, perform the indicated operations. The numbers are approximate numbers.

17.
```
    234.5
   18.734
+  567.04
```

18.
```
   56.941
     0.81
+  176.4
```

19. $524.1 - 73.456$

20. $276.45 - 1.0476$

21. $34.75 - 6.187 + 87.942 - 605.703$

22. $106 - 90 + 764 - 375$

23. $(76.4)(3.25)$

24. $(2,\underline{4}00)(104)$

25. $264 \div 0.05$

26. $1001 \div 13$

In Exercises 27–39, the measures are approximate numbers.

27. When you perform calculations on approximate numbers with a calculator, the answer on the screen may contain the maximum number of digits the calculator can display. The display may be 8 or more digits, unless the number of digits is fixed. Divide 2 by 7. Round your answer to
 a. two significant digits
 b. three significant digits.
 c. five significant digits.

28. A water sample at 4°C has a density of 988 kg/m³. The commonly accepted value for pure water is 1000 kg/m³.
 a. How many significant digits does the experimental value contain?
 b. What is the percentage of error?

29. What is the total mass of four pieces of plate glass if the masses of the four pieces are 34.5 kg, 4.75 kg, 376.5 kg, and 10.04 kg?

30. What is the total length of three cells that are 0.034 mm, 0.0306 mm, and 0.0335 mm long?

31. A carpet is to be installed in a room whose length is measured to be 12.71 m and width is 3.46 m. Find the area of the room.

32. A rocket is traveling at the rate of 7.8642 km/hr. How far will the rocket travel in 8.00 hrs $(d = rt)$?

33. If the circumference of a bicycle wheel is 81.681 in., what is the approximate diameter of the wheel $(c = \pi d)$?

34. One solution to the quadratic equation $6.82x^2 + 8.72x - 7.13 = 0$ is

$$x = \frac{-(8.72) + \sqrt{(8.72)^2 - 4(6.82)(-7.13)}}{2(6.82)}.$$

Solve for x.

35. Find R when $R_1 = 1.0034$ and $R_2 = 3.47621$ (*Hint*: See Example 8).

36. Stress in a beam is given by $S = \dfrac{My}{I}$. If $M = 19,200$, $y = 3.25$, and $I = 36.0$, find S.

37. For $P = \dfrac{100.0R}{(0.500 + R)^2}$ and $R = 0.2514$, find P.

38. Given $I = \dfrac{400}{r^2}$. Find illuminance **a.** if $r = 2.071$ m and **b.** if $r = 2.831$ m.

39. The equation $\dfrac{1}{p} + \dfrac{1}{q} = \dfrac{1}{f}$ is known as the mirror equation. p and q are the distances, respectively, of the object and the image from the surface of the mirror. f is the focal length. Find q if $p = 10.0$ cm and $f = 15$ cm.

Review Exercises

In Exercises 1–9, determine whether the following numbers are rational, irrational, real, pure imaginary, or complex. There may be more than one answer.

1. $\sqrt{144}$

2. $8.464674677\ldots$

3. $\sqrt[3]{-1728}$

4. $21j$

5. $\sqrt[4]{-64}$

6. $3.727272\ldots$

7. $\sqrt{\dfrac{0}{-98}}$

8. $4 + 7j$

9. $\sqrt{4} - \sqrt{16j}$

In Exercises 10–18, insert the appropriate symbol ($<$, $>$, or $=$) between the pairs of numbers.

10. $75, 34$

11. $-141, -32$

12. $-\sqrt{64}, -\sqrt{625}$

13. $\sqrt{13}, 3.6055512\ldots$

14. $3\dfrac{1}{4}, 3 + \dfrac{1}{4}$

15. $\dfrac{133}{4}, 33\dfrac{3}{4}$

16. $\dfrac{7}{8} - \dfrac{8}{7}, \dfrac{8}{7} - \dfrac{7}{8}$

17. $1\dfrac{5}{8} - \dfrac{13}{8}, 1\dfrac{7}{8} - \dfrac{15}{8}$

18. $\sqrt{3} - \sqrt{2}, \sqrt{2} - \sqrt{3}$

In Exercises 19–25, graph the following on a number line and list three values of x that satisfy the inequality.

19. $x > 4$

20. $x < 5$

21. $x \geq -7$

22. $x \leq 9$

23. $-9 \leq x < 11$

24. $-\sqrt{7} \leq x \leq \sqrt{13}$

25. $-3\dfrac{1}{3} < x \leq 4\dfrac{1}{3}$

In Exercises 26–31, find the absolute value of the numbers.

26. $|-9|$

27. $|-\sqrt{9}|$

28. $\left|3\dfrac{1}{3} - 4\dfrac{1}{7}\right|$

29. $|-\sqrt[3]{27}|$

30. $|(-2)^3 + 2^3|$

31. $\left|2\dfrac{1}{3} - \dfrac{11}{4}\right|$

In Exercises 32–38, identify the property illustrated in each case. (x and y represent real numbers.)

32. $x + 3 = 3 + x$

33. $0 + x = x$

34. $x + (-x) = 0$

35. $\dfrac{1}{4} \cdot 4 = 1$

36. $(3 + x) + 7 = 3 + (x + 7)$

37. $4 \cdot (x + y) = 4 \cdot x + 4 \cdot y$

38. $(3x)(y) = 3(xy)$

In Exercises 39–58, evaluate each of the given expressions by performing the indicated operations.

39. $(-7) + (-14)$

40. $(-6)(-15)(+3)$

41. $(-75) \div (15)$

42. $-4(3 - 11) + 9$

43. $\dfrac{-13 + 7 + 6}{4 - 11}$

44. $\dfrac{11 - 18}{47 - 47}$

45. $\dfrac{-3(13 - 7) + 18}{11 - 3 \cdot 2 - 5}$

46. $5 + (-3)^2 + \dfrac{4 - 4}{6 - 3} + (-2)(-3)$

47. $\dfrac{(4 - 7)^2 + (-3)^2}{7 + (7 - 9)}$

48. $\dfrac{-13 - 6(-2) + 1}{11 + (9)(-2) + 7}$

49. $-43.765 + 32.1876$

50. $(23.14)(-316.71)(2176.31)$

51. $-3.47628 - (-1.7653)$

52. $\dfrac{(-3.782)(84.76)}{-3.472}$

53. $\dfrac{-34.76 + 18.95}{\pi - \pi}$

54. $\dfrac{2.345}{14.76 - 18.25} + 15.1 - 16.3$

55. $\dfrac{-6.415 + 6.415}{15.237} - 11.678$

56. $(42.35)(-13.78) \div 32.47$

57. $[\pi + (-6)(-7) \div 6] \div (3.25 + 8.76)$

58. $\dfrac{(3.12)^2 - (14.85)^2}{(2.31)^2 - (8.5)^2} - (15.95)^2$

In Exercises 59–70, perform the indicated operations. The measures are approximate numbers.

59. 137.5 lb + 13.73 lb

60. 22.45 in. − 107.3 in. + 67.05 in. − 10.01 in.

61. (34.85 yd) × (1.35 yd)

62. $(100.\ \text{km})(0.32\ \text{km})$

63. 3400. mm² ÷ 0.056 mm

64. 247.5 in³ ÷ 1728 in³

65. (432.7 in.)(4.03 in.) ÷ 144 in²

66. (3.4 ft)(4.75 ft)(11.2 ft) ÷ 27 ft³

67. (0.75 mm + 2.34 mm − 0.47 mm)(2.51 mm)

68. Mt. Whitney has an elevation of 14,494 ft, while Death Valley's elevation is −282 ft. What is the difference in altitudes?

69. At noon the temperature was −16°F. The weather bureau recorded a change of +13°F between 12 P.M. and 3 P.M. and a change of −14°F between 3 P.M. and 6 P.M. What was the temperature at 6 P.M.?

70. An airplane is flying at an altitude of 3800 ft. Air currents cause it to drop 1500 ft. The pilot is able to control the plane and regain 1100 ft. Write an expression that describes the plane's change in altitude in relation to the original altitude. (*Hint*: consider a drop in altitude as negative.)

71. a. Write a program in BASIC language to evaluate the expression $5x^3 + 6\sqrt{x} - 7\sqrt[4]{x} + 2$.
   b. Using your program, find the value of the expression when x equals 3, 11, 98, and 247.

✐ Writing About Mathematics

1. Write a brief report on Leopold Kronecker's contributions to mathematics.

2. Our system of numeration is called the Hindu-Arabic system of numeration. Write a brief report on the development of the Hindu-Arabic system, with special emphasis on the development of zero.

3. Our number system is referred to as the decimal system, base 10. There are other systems called the binary, octal, and hexadecimal. You can do calculations in each of these systems on your calculator. Describe the differences between these systems and give examples where they may be used.

4. Instructions for a problem read, "Determine the sum of $\dfrac{1}{3} + \dfrac{1}{3} + \dfrac{1}{3}$ and record your answer to the nearest hundredth." Your answer is 1.00, while your friend's answer is 0.99. Explain how and why there is this difference in your answers.

Chapter Test

Identify the following numbers as: counting number, whole number, integer, rational number, real number, or complex number. (There may be more than one answer for each question.)

1. -14

2. $\dfrac{3}{4}$

3. $3 + 4j$

4. $1.3131131113\ldots$

5. 45

6. $4.35353535\ldots$

Insert the proper symbol ($<$, $>$, $=$) between the numbers.

7. $0.7, \dfrac{7}{10}$

8. $-3.4, 4.5$

9. $-3\dfrac{1}{4}, -3.375$

Graph on a number line.

10. $x \geq 3$

11. $-5 < x \leq 4$

12. $-4 \geq x$

Find the absolute value.

13. $|-7|$

14. $\left|\dfrac{1}{5} - \dfrac{1}{7}\right|$

Name the property illustrated by the statement.

15. $3x + 4 = 4 + 3x$

16. $4(x + 7) = 4x + 28$

17. $5x + (-5x) = 0$

18. $(2x)y = 2(xy)$

Evaluate the expressions without the use of a calculator.

19. $(-11) + (-5) + (-4)$

20. $\dfrac{23 - 23}{19 - 14}$

21. $\dfrac{6 + (-8)}{7 - 7}$

Round to the indicated decimal place.

22. 0.04853 (thousandths)

23. 2149 (hundreds)

24. $21{,}742$ (thousands)

Evaluate the expressions using a graphics calculator. The numbers are approximate numbers.

25. $163.2 + 24.27 + 439$

26. $\dfrac{4.7}{5.3} + (6.7)(-4.8) + 1.3$

27. $\sqrt{\dfrac{3.45 + 6.79}{11.3}}$

28. $(3.4)(1.59 - 11.8)^2$

Review of Algebra

H ave you ever wondered about how far a police radar system is effective? The maximum range in miles of an experimental police radar unit can be approximated by the equation:

$$R_{max} = 32.4\sqrt[4]{PL^2TF^{1/3}}$$

where P is the transmission power, L is the wavelength of the radar emission, T is the pulse length of the radar signal, and F is the frequency of the pulses. Knowing how to solve this equation requires a solid knowledge of radicals and exponents. We will review both in this chapter. (The solution to this problem is presented as Example 10 in Section 2.4.)

2.1

Addition and Subtraction of Algebraic Expressions

A letter used to represent an unknown quantity is called a **variable.** Traditionally in algebra letters at the end of the alphabet are chosen as variables. However, since many formulas in technical areas use different English and/or Greek letters this text is not limited to letters at the end of the alphabet. The endsheets contain a list of Greek letters.

An **algebraic expression** may be a constant, a variable, or a collection of constants and/or variables that are united by operation and/or grouping symbols.

Examples of expressions are:

$$3 \qquad x \qquad x + y$$

$$\frac{3x^2 - 5xz}{11y} \qquad 3x^2 + 7x^2 \qquad 6\sqrt{3x} + 11\sqrt{3x}$$

The algebraic expression $3x - y$ consists of two terms: $3x$ and $-y$. A **term** may be either a single constant or variable or a product or quotient of one or more variables and constants. The term $3x$ has a **numerical** factor of 3 and a **literal** factor of x. (The word *literal* is derived from the same base as the word *letter*.) The numbers or variables multiplied in a multiplication problem are called **factors.** The numerical factor of $-y$ is -1 even though the 1 is not shown ($-y = -1y$). The algebraic expression $5y(y + 3y^2) - 3x$ consists of two terms $5y(y + 3y^2)$ and $-3x$. The term $5y(y + 3y^2)$ has **factors** of $5y$ and $(y + 3y^2)$. The factor $y + 3y^2$ is composed of the two terms y and $3y^2$.

The expression $\dfrac{3x^2 - 5xz}{11y}$ is a single term; the numerator of which consists of two terms, $3x^2$ and $-5xz$ and the denominator a single term, $11y$.

An algebraic expression that contains only one term, such as 5, $3x$, $4x^2$, is called a **monomial.** The algebraic expressions $2x + 3y$ and $2x^2 + 3x$ each have two terms and are called **binomials.** An algebraic expression with three terms is called a **trinomial,** and one with two or more terms is called a **multinomial.**

For the algebraic expression $2x^2 - 5x^2$, the literal coefficient of 2 and 5 is x^2. Since the literal coefficient is the same for both terms (the letters and exponents are the same), we say that $2x^2$ and $5x^2$ are **like terms.** In the expression $6\sqrt{3x} + 11\sqrt{3x}$, the product under the radical is $3x$ for both terms so we classify $6\sqrt{3x}$ and $11\sqrt{3x}$ as like terms.

DEFINITION 2.1

Two or more terms that have the same literal coefficients (the letters and the exponents are the same) are called **like terms.**

EXAMPLE 1 **Identifying Like Terms**

In each of the following, two of the three terms are like terms. Identify the pair of like terms and explain why the third term is not a like term.

a. $\frac{4}{3}\pi r^3$, πr^2, $4\pi r^2$ **b.** I^2R, $\frac{1}{2}RI^2$, IR

c. $\dfrac{Ea - Eg}{2R}$, $\dfrac{2(Ea - Eg)}{R}$, $\dfrac{Ea - E}{R}$

Solution

a. The term $\frac{4}{3}\pi r^3$ contains the factor r^3, which is not a factor of the terms πr^2 and $4\pi r^2$. Thus, πr^2 and $4\pi r^2$ are the like terms since the literal factor, r^2, is the same in both terms. (π is a numerical constant.)

b. The terms I^2R and $\frac{1}{2}RI^2$ contain identical literal factors. The term IR is different from the other terms since the factor I has an exponent of 1.

c. The terms $\dfrac{Ea - Eg}{R}$ and $\dfrac{2(Ea - Eg)}{R}$ are like terms since $\dfrac{Ea - Eg}{2R}$ can be written $\dfrac{1}{2}\dfrac{Ea - Eg}{R}$. The only difference in the two terms is the numerical coefficient. The term $\dfrac{Ea - E}{R}$ is different since it does not contain the factor g in the second term of the numerator. ◆

Now look at how we will use this information to add and subtract. In adding and subtracting algebraic expressions, we combine like **terms** as illustrated in Example 2.

EXAMPLE 2 **Combining Like Terms**

In each of the following algebraic expressions combine the like terms.

a. $3a + 4b + 9a$ **b.** $x^2y - 3x^2y + 7x^2y$ **c.** $6\sqrt{3x} + 11\sqrt{3x}$

Solution

a. In the expression $3a + 4b + 9a$, the terms $3a$ and $9a$ are like terms. Use the commutative property to rewrite the expression as $3a + 9a + 4b$. Add the like terms, and the result is $12a + 4b$.

b. The expression $x^2y - 3x^2y + 7x^2y$ consists of three like terms. To combine the like terms, add the numerical coefficients 1, (-3) and 7, whose sum is 5. The sum of the like terms is $5x^2y$.

c. In the expression $6\sqrt{3x} + 11\sqrt{3x}$, the terms are alike since the same product is under both radicals. The simplified form of the expression $6\sqrt{3x} + 11\sqrt{3x}$ is $17\sqrt{3x}$. ◆

In algebra, as in arithmetic, the order in which operations are performed is essential. The tools that are used to show the order of operations are **parentheses** ( ), **brackets** [ ], and **braces** { }. The general practice is that parentheses are used first and innermost, then brackets, and then braces.

The instruction "simplify an algebraic expression," means that we are to remove all parentheses, brackets, braces and to combine like terms. This will be illustrated in Example 3.

CAUTION

$$-(3x - 5) \neq 3x - 5$$

When removing a grouping symbol, change all the signs inside the grouping symbol to the opposite sign if the symbol is preceded by a minus sign. If the grouping symbol is preceded by a plus sign, do not change any of the signs inside the grouping symbol.

For example, $-(3x - 5) = -3x + 5$ and $+(3x - 5) = 3x - 5$. ■

EXAMPLE 3   **Simplify the Algebraic Expressions**

Simplify each expression.

a. $8 - (11 - 17)$

b. $5 + (3x - 4) - (2x + 5)$

c. $2x + 2 - [5x + (3 - 11x)]$

d. $3a - \{4c + [7a - 5c - 2(9a + 11c + 6)]\}$

Solution

a. $8 - (11 - 17) = 8 - (-6)$

$\qquad\qquad\qquad = 8 + 6$ **Definition of subtraction**

$\qquad\qquad\qquad = 14$

b. $5 + (3x - 4) - (2x + 5)$

$\quad = 5 + 1(3x - 4) + (-1)(2x + 5)$ **Definition of subtraction**

$\quad = 5 + 3x - 4 - 2x - 5$ **Distributive property**

$\quad = 3x - 2x + 5 - 4 - 5$ **Commutative property**

$\quad = x - 4$ **Combining like terms**

c.   In a problem such as this with more than one set of grouping symbols, remove the inner symbols first and work from the inside out until all grouping symbols are removed and the problem is simplified.

$$2x + 2 - [5x + (3 - 11x)]$$

$= 2x + 2 - [5x + 3 - 11x]$

$= 2x + 2 - [-6x + 3]$ **Combining like terms**

$= 2x + 2 + (-1)[-6x + 3]$ **Definition of subtraction**

$= 2x + 2 + 6x - 3$ **Distributive property**

$= 8x - 1$ **Combining like terms**

d. Again, start by removing the parentheses and working out until all "fences" are removed.

$$= 3a - \{4c + [7a - 5c + (-2)(9a + 11c + 6)]\}$$ **Definition of subtraction**

$$= 3a - \{4c + [7a - 5c - 18a - 22c - 12]\}$$ **Distributive property**

$$= 3a - \{4c + [-11a - 27c - 12]\}$$ **Combining like terms**

$$= 3a - \{4c - 11a - 27c - 12\}$$ **Removing brackets**

$$= 3a - \{-11a - 23c - 12\}$$ **Combining like terms**

$$= 3a + (-1)\{-11a - 23c - 12\}$$ **Definition of subtraction**

$$= 3a + 11a + 23c + 12$$ **Distributive property**

$$= 14a + 23c + 12$$ **Combining like terms**

◆

An essential part of algebraic problem solving is being able to translate verbal statements into mathematical statements or expressions.

EXAMPLE 4   From Verbal Statements to Mathematical Expressions

Translate the following verbal statements into mathematical expressions.

a.   6 less than 7 times a number.
b.   11 more than twice a number.
c.   One-half of three times a number.

Solution

a.   The first thing we must do is select a letter to represent the number. Arbitrarily select the letter n. Seven times the number is written as $7n$. "Less than" means that we subtract 6 from the product. Thus, the solution is $7n - 6$.

b.   Let x represent the unknown number. Twice the number is represented as $2x$. "More than" means that we add 11 to the product. The solution is $2x + 11$.

c.   Let x be the number. Three times the number is $3x$. One-half of three times the number is $\frac{1}{2}(3x)$ or $\frac{3x}{2}$ or $\frac{3}{2}x$. ◆

CAUTION   In Example 4**a**, an incorrect answer, $6 - 7n$, is frequently given. If the problem had read either "7 times a number less than 6" or "6 less 7 times a number," then the correct translation would be $6 - 7n$. ■

2.1 Exercises

In Exercises 1–6, identify each of the terms of the algebraic expression.

1. $6x + y$

2. $3x - 5yz$

3. $\dfrac{a + b - c}{3}$

4. $3x^2 + \dfrac{4x^2y}{z}$

5. $-4xy + (6x + 4y)^2$

6. $\dfrac{ab}{c} + 7bc - 6b^2$

In Exercises 7–38, simplify the following algebraic expressions.

7. $3a + 5a + 7b$

8. $5a - 11b + 14b$

9. $2x - 3y + 9y$

10. $8y - 11x - 9y$

11. $15r - 7s + 8r - 9s - 17r$

12. $3s + 5r - 7r + 2s + 6r$

13. $3\pi r^2 - 4\pi r + 7\pi r - 5\pi r^2$

14. $6a^2b + 7ab^2 - 5a^2b + 4ab^2$

15. $\dfrac{1}{2}x + \dfrac{1}{3}x + \dfrac{1}{4}y$

16. $\dfrac{1}{5}v - \dfrac{1}{7}v + \dfrac{2}{3}t$

17. $\dfrac{5}{8}x^2 + \dfrac{3}{4}x^2 - \dfrac{1}{2}x^2$

18. $\dfrac{1}{6}\pi r^3 - \dfrac{4}{3}\pi r^3 + \dfrac{1}{9}\pi r^3$

19. $0.03r^2 - 1.45r^2 + 0.06r^2$

20. $0.125s^2 - 3.041s^2 + 0.047s^2$

21. $0.314r^2 + 0.0314r^2 - 3.14r^2$

22. $8.14s^2 + 11.031s^2 - 31.008s^2$

23. $4 + (7 + 11)$

24. $8 - (6 - 13)$

25. $3x + (5x - 7)$

26. $11x - (7x - 6)$

27. $4 + (5a - 7) + (8 - 2a)$

28. $7a - 2(4a + 6) - (5a - 9)$

29. $2x + [5y + 6x + (3y + 7x)]$

30. $8y - [3y + 2x + (x - 6y)]$

31. $3a - 5[2b + 7a - 3(b - a)]$

32. $2a - [3b - (5a + 2b) + (6a - 4b)]$

33. $5r - [2s - (8r - 3s) - (4s - 7r)]$

34. $6r - \{5s + 2[2r - 3s + (13s - 11r)]\}$

35. $6r - \{5s - 2[2r - 3s + (13s - 11r)]\}$

36. $6a - \{5b - 2[2a - 3b - (13b - 11a)]\}$

37. $-3a - 5\{2b + (5a - b) + 2[2a - (3b + 5a)]\}$

38. $4x - \{3y - 9(2x + y) - [4x - (5y - 7x)]\}$

In Exercises 39–48, write an algebraic expression for each of the statements.

39. Seven times a number.

40. Six more than 5 times a number.

41. The sum of negative eight and one-half a number.

42. A number whose square is decreased by 7.

43. The sum of the square root of five times the number and the square of the number.

44. The sum of the cubes of two numbers x and y.

45. The cube of the sum of two numbers x and y.

46. Two terms with one variable.

47. Two terms with three variables.

48. Three terms, two variables, and all numerical coefficients are equal.

In Exercises 49–52, simplify the expressions.

49. In electronics problems involving Kirchhoff's second law the following expressions were encountered:
    a. $4 - 0.1I_1 + 6 - 0.2I_2 - 2.0I_1$
    b. $5 - 0.5(I_2 - I_1) - 3.0(I_2 - I_1) + 6 - 0.2I_2$
    c. $2.0I_1 + 5 - 0.5(I_2 - I_1) - 3.0(I_2 - I_1) + 0.1I_1 - 4$

50. In a motion problem, the following expressions for acceleration and average speed are given:
    a. $\dfrac{3\text{m/s} - 2\text{m/s}}{4s}$ b. $\dfrac{1}{2}(2\text{m/s} + 3\text{m/s})$

51. In strength of materials problems, the following expression is encountered:
    $$\left(\frac{\pi d^2}{4}\right)\left(\frac{15\sigma^2}{8}\right) + \frac{\pi}{4}(D^2 - d^2)\sigma^2$$

52. In computing the total energy of a ball in vertical flight, the following expression could be used:
    $$\frac{1}{2}mv^2 + mgh + \frac{1}{2}\left(\frac{2}{5}mR^2 \cdot \frac{v^2}{R^2}\right)$$

2.2

Exponents

In Chapter 1 you were asked to evaluate 9^2. From past experience you automatically would assume that $9^2 = 9 \cdot 9 = 81$. You may recall that the superscript 2 is referred to as the **exponent** of the number 9, which is called the **base.** The expression 9^2 is generally read as "the second power of 9." The exponent tells how many times the base appears as a factor. In general, the expression a^n is called the nth **power of a** where n is the exponent and a is the base.

In the expression $3 \cdot 9^2$, note that the exponent, 2, acts only on the 9 and not on the 3.

CAUTION   When n is an even counting number $-a^n \neq (-a)^n$. For example, $-3^2 \neq (-3)^2$, rather $-3^2 = -(3 \cdot 3) = -9$. Whereas, $(-3)^2 = (-3)(-3) = 9$. ■

> **DEFINITION 2.2**
>
> $a^n = \underbrace{a \cdot a \cdot a \cdot \ldots \cdot a}_{n \text{ factors}}$ (n is a counting number)

EXAMPLE 1 **Evaluating Numbers to a Power**

Evaluate

a. 3^4 **b.** $(-5)^3$ **c.** $\left(\dfrac{1}{7}\right)^3$ **d.** -5^3

Solution

a. 3^4 is read "the fourth power of 3." Here we write 3 four times as dictated by the exponent and multiply

$$3^4 = 3 \cdot 3 \cdot 3 \cdot 3 = 81$$

b. $(-5)^3$ is read "negative five cubed." Here we write -5 three times as dictated by the exponent and multiply

$$-5 \cdot -5 \cdot -5 = -125.$$

c. $\left(\dfrac{1}{7}\right)^3$ is read "one seventh cubed."

$$\left(\frac{1}{7}\right)^3 = \frac{1}{7} \cdot \frac{1}{7} \cdot \frac{1}{7} = \frac{1}{343}.$$

d. -5^3 is read "the negative of five cubed."

$$-5^3 = -(5 \cdot 5 \cdot 5) = -125. \quad \blacklozenge$$

Note in parts **b** and **d** of Example 1, the answers are the same, which may seem to be a contradiction to the caution. When n is an odd counting number

$$(-a)^n = -a^n$$

The calculator can be used to raise a number to a power, see Example 2.

EXAMPLE 2 **Evaluating Numbers to a Power with a Calculator**

Evaluate

a. $16 - x^2$, when $x = 3$ **b.** 1.5^6

Solution For any number that is raised to the second power (squared), enter the number and press $\boxed{x^2}$. Your calculator will have one of the following keys, which is used to raise a number to a power higher than 2, a $\boxed{y^x}$ key, an $\boxed{x^y}$ key, or a $\boxed{\wedge}$ key.

a. $16 - x^2 = 16 - \boxed{3}^2$ **Substituting 3 for x**

$$= 16 \boxed{-} 3 \boxed{x^2} \boxed{\text{ENTER}} 7$$

There is no need for parentheses because the calculator recognizes that we are not multiplying $(-3)(-3)$ but rather finding $-(3 \cdot 3)$.

b. fx–7700G: 1.5 $\boxed{x^y}$ 6 $\boxed{\text{EXE}}$ 11.390625, or

TI–82: 1.5 $\boxed{\wedge}$ 6 $\boxed{\text{ENTER}}$ 11.390625 $\blacklozenge$

Note the TI–82 has a cube and a cube root function under the MATH menu.

In addition to understanding the meaning of a^n, we must also know the laws for exponents that relate to multiplying and dividing. The first part of the discussion will be limited to positive exponents ($n > 0$ and $m > 0$). To find the product of $a^4 \cdot a^5$ we can use Definition 2.2

$$a^4 \cdot a^5 = \underbrace{\underbrace{a \cdot a \cdot a \cdot a}_{4 \text{ factors}} \cdot \underbrace{a \cdot a \cdot a \cdot a \cdot a}_{5 \text{ factors}}}_{9 \text{ factors}} = a^9$$

LAW 2.1

$$a^n \cdot a^m = a^{n+m}$$

Law 2.1, called the **product rule,** states that when multiplying two quantities with the same base, each of which is raised to a power, all we need do is add the exponents and keep the same base.

EXAMPLE 3   Product Rule

Determine the products using the product rule.

a. $a^3 \cdot a^2$ **b.** $3^8 \cdot 3^7$ **c.** $x^5 \cdot y^2$

Solution

a. Since the bases are alike, the product rule can be applied, and we add the exponents.

$$a^3 \cdot a^2 = a^{3+2} = a^5$$

b. Since the bases are alike, the product rule can be applied, and we add the exponents.

$$3^8 \cdot 3^7 = 3^{8+7} = 3^{15} \qquad 3^{15} = 14,348,907$$

(Use your calculator to check this answer.)

c. In $x^5 \cdot y^2$ the bases are not alike, therefore, we cannot apply the product rule. ◆

The product of $(a^2)^3$ can be found by using Definition 2.1 and Law 2.1.

$$(a^2)^3 = a^2 \cdot a^2 \cdot a^2 \quad \textbf{Definition 2.2}$$
$$= a^{2+2+2} \qquad \textbf{Product rule (Law 2.1)}$$
$$= a^6$$

> **LAW 2.2**
>
> $$(a^n)^m = a^{n \cdot m}$$

Law 2.2 is called the **power rule.**

EXAMPLE 4   Power Rule

Evaluate the following using the power rule. In part **b** use a calculator to determine the numerical value.

a. $(a^2)^5$ **b.** $(3^4)^5$

Solution

a. Since a^2 is being raised to the power of 5, we can use the power rule and multiply the exponents.

$$(a^2)^5 = a^{2 \cdot 5} = a^{10}$$

b. Since 3^4 is being raised to the power of 5, we can use the power rule and multiply the exponents,

$$(3^4)^5 = 3^{4 \cdot 5} = 3^{20} \quad \blacklozenge$$

Definition 2.2 and the commutative property of multiplication can be used to show that $(a \cdot b)^4 = a^4 \cdot b^4$.

$$
\begin{aligned}
(a \cdot b)^4 &= (a \cdot b) \cdot (a \cdot b) \cdot (a \cdot b) \cdot (a \cdot b) &&\text{Definition 2.2} \\
&= a \cdot a \cdot a \cdot a \cdot b \cdot b \cdot b \cdot b &&\text{Commutative property} \\
&= a^4 \cdot b^4
\end{aligned}
$$

> **LAW 2.3**
>
> $$(a \cdot b)^n = a^n \cdot b^n$$

Law 2.3 states that the power of a product is equal to the product of the powers.

EXAMPLE 5   Evaluating a Product Raised to a Power

Evaluate the following using Law 2.3. In part **b** use a calculator to determine the numerical value.

a. $(a \cdot b)^5$ **b.** $(4^2 \cdot 7)^3$

Keystrokes for evaluating Example 4b.
fx–7700G: 3 $\boxed{x^y}$ 4 $\boxed{x^y}$ 5 $\boxed{\text{EXE}}$
3486784401
TI–82: 3 $\boxed{\wedge}$ 4 $\boxed{\wedge}$ 5 $\boxed{\text{ENTER}}$
3486784401

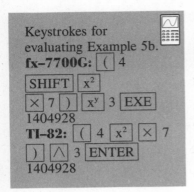

Keystrokes for
evaluating Example 5b.
fx–7700G: (4
SHIFT x²
× 7) xʸ 3 EXE
1404928
TI–82: (4 x² × 7
) ∧ 3 ENTER
1404928

Solution

a. $(a \cdot b)^5 = a^5 \cdot b^5$ **Power of a product (Law 2.3)**

b. $(4^2 \cdot 7)^3 = (4^2)^3 \cdot 7^3$ **Power of a product (Law 2.3)**

           $= 4^6 \cdot 7^3$ **Power rule (Law 2.2)**

           $= 4096 \cdot 343$, or $1,404,928$ ◆

Using Definition 2.2, we can write the quotient $\left(\dfrac{a}{b}\right)^3$ as follows:

$$\left(\frac{a}{b}\right)^3 = \frac{a}{b} \cdot \frac{a}{b} \cdot \frac{a}{b}$$

$$= \frac{a^3}{b^3}.$$

LAW 2.4

$$\left(\frac{a}{b}\right)^n = \frac{a^n}{b^n}, \qquad b \neq 0$$

Law 2.4 states that the quotient of two numbers to the same power is equal to the quotient of the numbers raised to the common power.

EXAMPLE 6   **Evaluating a Quotient Raised to a Power**

Evaluate the following using Law 2.4.

a. $\left(\dfrac{a}{b}\right)^6$ b. $\left(\dfrac{7}{8}\right)^4$ c. $\left(\dfrac{1}{4}\right)^5$

Solution

a. $\left(\dfrac{a}{b}\right)^6 = \dfrac{a^6}{b^6}$ **Power of a quotient (Law 2.4)**

b. $\left(\dfrac{7}{8}\right)^4 = \dfrac{7^4}{8^4}$ **Power of a quotient (Law 2.4)**

         $= \dfrac{2401}{4096}$

         ≈ 0.5862

c. $\left(\dfrac{1}{4}\right)^5 = \dfrac{1^5}{4^5}$ **This is a special case of Law 2.4, where $a = 1$**

         $= \dfrac{1}{1024}$ ◆

The quotient $\dfrac{a^7}{a^3}$ can be written as:

$$\frac{a^7}{a^3} = \frac{a \cdot a \cdot a \cdot a \cdot a \cdot a \cdot a}{a \cdot a \cdot a}$$

$$= \frac{\cancel{a}^1 \cdot \cancel{a}^1 \cdot \cancel{a}^1 \cdot a \cdot a \cdot a \cdot a}{\cancel{a}_1 \cdot \cancel{a}_1 \cdot \cancel{a}_1}$$

$$= a \cdot a \cdot a \cdot a$$

$$= a^4$$

Law 2.5, called the **quotient rule,** shows a shorter method of arriving at the same result, that is $\dfrac{a^7}{a^3} = a^{7-3} = a^4$.

LAW 2.5

$$\frac{a^m}{a^n} = \begin{cases} a^{m-n} & \text{if } m \geq n \quad \text{and} \quad a \neq 0 \\ \dfrac{1}{a^{n-m}} & \text{if } m < n \quad \text{and} \quad a \neq 0 \end{cases}$$

The quotient rule states that when two quantities with the same base are being divided, all we need do is subtract the exponents and keep the base.

EXAMPLE 7 **Quotient Rule**

Evaluate the following using the quotient rule

a. $\dfrac{a^7}{a^3}$ **b.** $\dfrac{a^4}{a^4}$ **c.** $\dfrac{a^5}{a^{12}}$

Solution

a. $\dfrac{a^7}{a^3} = \boxed{a^{7-3}} = \boxed{a^4}$ **Quotient rule (Law 2.5)** $\{m > n\}$

b. $\dfrac{a^4}{a^4} = \boxed{a^{4-4}} = \boxed{a^0}$ **Quotient rule (Law 2.5)** $\{m = n\}$

c. $\dfrac{a^5}{a^{12}} = \dfrac{1}{\boxed{a^{12-5}}} = \dfrac{1}{\boxed{a^7}}$ **Quotient rule (Law 2.5)** $\{m < n\}$ ◆

The result in part **b** of Example 7 leads to the following law.

LAW 2.6

$$a^0 = 1, \qquad a \neq 0$$

Law 2.6 tells us that any base, except 0, raised to the exponent zero is 1. Try this one on your calculator by determining 3452^0.

Consider the fraction, $\dfrac{2^3}{2^4}$. Applying the quotient rule we can write 2^{3-4} or 2^{-1}.

Now let's look at how to work with negative exponents. To do this return to the original expression $\dfrac{2^3}{2^4} = \dfrac{8}{16} = \dfrac{1}{2}$.

Apparently $2^{-1} = \dfrac{1}{2}$. The negative sign on the exponent tells us only about the size of a number, not the negativeness of its base. Law 2.7 formally states this result.

LAW 2.7

$$a^{-n} = \frac{1}{a^n} \quad \text{and} \quad \frac{1}{a^{-n}} = a^n \text{ for } a \neq 0$$

Law 2.7, the **negative exponent rule,** shows that we can move a factor from the numerator to the denominator of a fraction (or vice versa) by changing the sign of the exponent. Common practice dictates that results are expressed with positive exponents. However, some equations exist that commonly are written with negative exponents—for example, $PV = P(1 + i)^{-n}$.

EXAMPLE 8 **Negative Exponent Rule**

Write the following with positive exponents.

a. a^{-3} **b.** $\dfrac{1}{a^{-5}}$ **c.** $\dfrac{3^{-2}}{5^{-4}}$ **d.** $\dfrac{x^{-3}}{x^4 y^{-2}}$

Solution

a. $a^{-3} = \dfrac{1}{a^3}, a \neq 0$ Negative exponent rule (Law 2.7)

b. $\dfrac{1}{a^{-5}} = a^5, a \neq 0$ Negative exponent rule (Law 2.7)

c. $\dfrac{3^{-2}}{5^{-4}} = \dfrac{3^{-2}}{1} \cdot \dfrac{1}{5^{-4}}$

$= \dfrac{1}{3^2} \cdot \dfrac{5^4}{1}$ Negative exponent rule (Law 2.7)

$= \dfrac{5^4}{3^2}$

d. $\dfrac{x^{-3}}{x^4 y^{-2}} = \dfrac{x^{-3}}{x^4} \cdot \dfrac{1}{y^{-2}}$

$= \dfrac{1}{x^{4-(-3)}} \cdot \dfrac{y^2}{1}$ Negative exponent rule (Law 2.7) and quotient rule (Law 2.5)

$= \dfrac{y^2}{x^7}$ ◆

Without formal proof or demonstration, accept that the laws of exponents will hold when the bases and the exponents are real numbers. This is illustrated in Example 9.

EXAMPLE 9 Noninteger Bases and Exponents

Evaluate using the definitions and laws of exponents. In part b use a calculator to determine a numerical value.

a. $\dfrac{\left(\dfrac{1}{3}\right)^4}{\left(\dfrac{1}{2}\right)^2}$ **b.** $\dfrac{(0.3)^2}{(0.5)^3}$ **c.** $\dfrac{(3^2)^{\frac{1}{2}}}{20^0}$

Solution

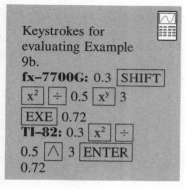

Keystrokes for evaluating Example 9b.

fx–7700G: 0.3 [SHIFT] [x^2] [÷] 0.5 [x^y] 3 [EXE] 0.72

TI–82: 0.3 [x^2] [÷] 0.5 [∧] 3 [ENTER] 0.72

a. $\dfrac{\left(\dfrac{1}{3}\right)^4}{\left(\dfrac{1}{2}\right)^2} = \dfrac{\dfrac{1}{3^4}}{\dfrac{1}{2^2}}$ **Power of a quotient (Law 2.4)**

$\phantom{\dfrac{\left(\dfrac{1}{3}\right)^4}{\left(\dfrac{1}{2}\right)^2}} = \dfrac{1}{3^4} \cdot \dfrac{2^2}{1}$ **Division of fractions**

$\phantom{\dfrac{\left(\dfrac{1}{3}\right)^4}{\left(\dfrac{1}{2}\right)^2}} = \dfrac{2^2}{3^4}$ or $\dfrac{4}{81}$

b. $\dfrac{(0.3^2)}{(0.5)^3} = \dfrac{(0.3)(0.3)}{(0.5)(0.5)(0.5)}$ **Definition 2.2**

$\phantom{\dfrac{(0.3^2)}{(0.5)^3}} = \dfrac{0.09}{0.125}$ or 0.72

c. $\dfrac{(3^2)^{\frac{1}{2}}}{20^0} = \dfrac{3^{2 \cdot \frac{1}{2}}}{1}$ **Power rule (Law 2.2) and Law 2.6**

$\phantom{\dfrac{(3^2)^{\frac{1}{2}}}{20^0}} = 3^1$ or 3 ◆

EXAMPLE 10 Applying the Definitions and Laws of Exponents

Evaluate the following using the definitions and laws of exponents. Write the answers with positive exponents.

a. $a^{\frac{1}{3}} \cdot a^{\frac{1}{2}}$ **b.** $\dfrac{a^{\frac{4}{5}}}{a^{\frac{2}{3}}}$ **c.** $(a^{\frac{1}{3}})^{\frac{3}{5}}$ **d.** $\left(\dfrac{a}{b}\right)^{-\left(\frac{1}{2}\right)}$

Solution

a. $a^{\frac{1}{3}} \cdot a^{\frac{1}{2}} = a^{\frac{1}{3}+\frac{1}{2}}$ **Product rule (Law 2.1)**

$\phantom{a^{\frac{1}{3}} \cdot a^{\frac{1}{2}}} = a^{\frac{5}{6}}$

b. $\dfrac{a^{\frac{4}{5}}}{a^{\frac{2}{3}}} = a^{\frac{4}{5}-\frac{2}{3}}$ **Quotient rule (Law 2.5)** $\{m > n\}$

$\qquad = a^{\frac{2}{15}}$ $\left(\dfrac{4}{5} - \dfrac{2}{3} = \dfrac{12 - 10}{15} = \dfrac{2}{15}\right)$

c. $\left(a^{\frac{1}{3}}\right)^{\frac{3}{5}} = a^{\frac{1}{3} \cdot \frac{3}{5}}$ **Power rule (Law 2.2)**

$\qquad = a^{\frac{1}{5}}$

d. $\left(\dfrac{a}{b}\right)^{-\frac{1}{2}} = \dfrac{a^{-\frac{1}{2}}}{b^{-\frac{1}{2}}}$ **Power of a quotient (Law 2.4)**

$\qquad = \dfrac{b^{\frac{1}{2}}}{a^{\frac{1}{2}}}$ **Negative exponent rule (Law 2.7)**

$\qquad = \left(\dfrac{b}{a}\right)^{\frac{1}{2}}$ ◆

By Law 2.7, $x^{-1} = \dfrac{1}{x}$. Recall that the reciprocal of x is $\dfrac{1}{x}$. Thus, the special case that x^{-1} is the reciprocal of x. Using the concept of reciprocal, Example **10 d** can be evaluated as $\left(\dfrac{a}{b}\right)^{-\frac{1}{2}} = \left(\dfrac{b}{a}\right)^{\frac{1}{2}}$.

EXAMPLE 11   Application: Physics

An expression used in physics is $\dfrac{MG}{rc^2}\left(1 + \dfrac{MG}{rc^2}\right)^{-1}$. Simplify the expression and write it with positive exponents.

Solution   To simplify the expression, first write the factor with a positive exponent.

$\dfrac{MG}{rc^2}\left(1 + \dfrac{MG}{rc^2}\right)^{-1} = \dfrac{MG}{rc^2} \, \dfrac{1}{\left(1 + \dfrac{MG}{rc^2}\right)}$ **Negative exponent rule (Law 2.7)**

$\qquad\qquad = \dfrac{MG}{rc^2 + MG}$ **Multiplication** ◆

EXAMPLE 12   Applying Definitions and Laws of Exponents

Simplify the following expressions by applying the definitions and the laws of exponents. Write the final results with positive exponents.

a. $\left(3^4 x^2 y^4 z^{\frac{1}{6}}\right)^3$ **b.** $\dfrac{6^4 x^3 y^7 z^{\frac{5}{3}}}{6^3 x^5 y^7 z^{\frac{2}{3}}}$

Solution

a. $\left(3^4 x^2 y^4 z^{\frac{1}{6}}\right)^3 = (3^4)^3 (x^2)^3 (y^4)^3 (z^{\frac{1}{6}})^3$ **Power of a product (Law 2.3)**

$\qquad = 3^{4 \cdot 3} x^{2 \cdot 3} y^{4 \cdot 3} z^{(\frac{1}{6})3}$ **Power rule (Law 2.2)**

$\qquad = 3^{12} x^6 y^{12} z^{\frac{1}{2}}$

$\qquad = 531{,}441 x^6 y^{12} z^{\frac{1}{2}}$

b. $\dfrac{6^4 x^3 y^7 z^{\frac{5}{3}}}{6^3 x^5 y^7 z^{\frac{2}{3}}} = \dfrac{6^{4-3}}{1} \cdot \dfrac{1}{x^{5-3}} \cdot \dfrac{y^{7-7}}{1} \cdot \dfrac{z^{\frac{5}{3}-\frac{2}{3}}}{1}$ **Quotient rule (Law 2.5)**

$\qquad\qquad = \dfrac{6}{1} \cdot \dfrac{1}{x^2} \cdot \dfrac{y^0}{1} \cdot \dfrac{z^{\frac{3}{3}}}{1}$

$\qquad\qquad = \dfrac{6 \cdot 1 \cdot 1 \cdot z}{1 \cdot x^2 \cdot 1 \cdot 1}$ **Law 2.6**

$\qquad\qquad = \dfrac{6z}{x^2}$ ◆

EXAMPLE 13   Applying Definitions and Laws of Exponents

Simplify the following expressions by applying the definitions and the laws of exponents. Then use a calculator to find the decimal value.

a. $\dfrac{a^3 b^{\frac{7}{3}}}{a^6 b^{\frac{2}{3}}}$ $a = 3,\ b = 4$ **b.** $(13^3 x^7 y^{\frac{1}{3}})^{\frac{2}{3}}$ $x = 0.5,\ y = 1.3$

Solution

a. $\dfrac{a^3 b^{\frac{7}{3}}}{a^6 b^{\frac{2}{3}}} = \dfrac{a^3}{a^6} \cdot \dfrac{b^{\frac{7}{3}}}{b^{\frac{2}{3}}}$

$\qquad\qquad = \dfrac{1}{a^{6-3}} \cdot \dfrac{b^{\frac{7}{3}-\frac{2}{3}}}{1}$ **Quotient rule (Law 2.5)**

$\qquad\qquad = \dfrac{1}{a^3} \cdot \dfrac{b^{\frac{5}{3}}}{1}$

$\qquad\qquad = \dfrac{b^{\frac{5}{3}}}{a^3}$

$\qquad\qquad = \dfrac{4^{\frac{5}{3}}}{3^3}$ **Substituting $a = 3$ and $b = 4$**

Some calculators do not express rational numbers in the form $\dfrac{a}{b}$. Therefore to evaluate $4^{\frac{5}{3}}$, the exponent must be expressed in decimal form, that is $(5 \div 3)$.

$$4\ \boxed{\wedge}\ \boxed{(}\ 5\ \boxed{\div}\ 3\ \boxed{)}\ \boxed{\div}\ 3\ \boxed{\wedge}\ 3\ \boxed{\text{ENTER}} = 0.3733099$$

or

$$4\ \boxed{\wedge}\ \boxed{(}\ 5\ \boxed{\div}\ 3\ \boxed{)}\ \boxed{\times}\ 3\ \boxed{\wedge}\ \boxed{(-)}\ 3\ \boxed{\text{ENTER}} = 0.3733099$$

The second method is possible since $\dfrac{4^{\frac{5}{3}}}{3^3} = 4^{\frac{5}{3}} \cdot 3^{-3}$. The fx–7700G has a fraction key labeled $\boxed{a^{\frac{b}{c}}}$. When we use the fraction key, the keystrokes are:

$$4\ \boxed{x^y}\ \boxed{(}\ 5\ \boxed{a^{\frac{b}{c}}}\ 3\ \boxed{)}\ \boxed{\div}\ 3\ \boxed{x^y}\ 3\ \boxed{\text{EXE}} = 0.3733099$$

b. $(13^3 x^7 y^{\frac{1}{3}})^{\frac{2}{3}} = (13^3)^{\frac{2}{3}}(x^7)^{\frac{2}{3}}(y^{\frac{1}{3}})^{\frac{2}{3}}$ **Power of a product (Law 2.3)**

$\qquad\qquad = 13^{\frac{3}{1} \cdot \frac{2}{3}} x^{\frac{7}{1} \cdot \frac{2}{3}} y^{\frac{1}{3} \cdot \frac{2}{3}}$ **Power rule (Law 2.2)**

$\qquad\qquad = 13^2 x^{\frac{14}{3}} y^{\frac{2}{9}}$

$\qquad\qquad = 13^2 (0.5)^{\frac{14}{3}}(1.3)^{\frac{2}{9}}$ **Substituting $x = 0.5$ and $y = 1.3$**

$\qquad\qquad = 7.0534372$ ◆

CAUTION Try evaluating $(-1.3)^{2/9}$ on your calculator, using the keystrokes
$\boxed{(}\ \boxed{(-)}\ 1.3\ \boxed{)}\ \boxed{\wedge}\ \boxed{(}\ 2\ \boxed{\div}\ 9\ \boxed{)}$. The result is an error message. Why? Now
evaluate the expression using $((-1.3)^2)^{(1 \div 9)}$. The answer is 1.0600363. ■

This section concludes with a summary of the definitions and laws of exponents.

DEFINITIONS AND LAWS OF EXPONENTS

Definition 2.2 $a^n = \underbrace{a \cdot a \cdot a \cdot \ \ldots \ \cdot a}_{n \text{ factors}}$ (n is counting number)

Law 2.1 $a^n \cdot a^m = a^{n+m}$	**Product rule**
Law 2.2 $(a^n)^m = a^{n \cdot m}$	**Power rule**
Law 2.3 $(a \cdot b)^n = a^n b^n$	**Power of a product**
Law 2.4 $\left(\dfrac{a}{b}\right)^n = \dfrac{a^n}{b^n}$	**Power of a quotient**
Law 2.5 $\dfrac{a^m}{a^n} = \begin{cases} a^{m-n} & \text{if } m \geq n \text{ and } a \neq 0 \\ \dfrac{1}{a^{n-m}} & \text{if } m < n \text{ and } a \neq 0 \end{cases}$	**Quotient rule**
Law 2.6 $a^0 = 1, a \neq 0$	
Law 2.7 $\left.\begin{array}{l} a^{-n} = \dfrac{1}{a^n} \\ \dfrac{1}{a^n} = a^n \end{array}\right\} \quad a \neq 0$	**Negative exponent rule**

2.2 Exercises

In Exercises 1–28 simplify each expression using the definitions and the laws of exponents. Write the answer with positive exponents.

1. $x^4 \cdot x^{13}$

2. $4^7 \cdot 4^{27}$

3. $(3^4 x^{11})^5$

4. $\left(\dfrac{x^2}{y^3}\right)^4$

5. $\left(\dfrac{1}{x^3}\right)^5$

6. $b^y \cdot b^z$

7. $(a \cdot b)^0$

8. $(-2x)^3$

9. $\dfrac{a^7}{a^4}$

10. $\dfrac{b^8}{b^{11}}$

11. a^{-4}

12. $\dfrac{1}{a^{-7}}$

13. $\dfrac{4^{-3}}{7^{-5}}$

14. $\dfrac{x^{-7}}{y^3 z^{-4}}$

15. $x^{\frac{1}{4}} x^{\frac{1}{3}}$

16. $\dfrac{x^{\frac{3}{4}}}{x^{\frac{2}{5}}}$

17. $\left(\dfrac{1}{z^{-3}}\right)^5$

18. $x^2 y^3 (x^4 y^5)^2$

19. $\dfrac{15a^3 b^2}{25a^2 b^2}$

20. $\dfrac{27x^{13} y^4}{(3x^2 y^5)^2}$

21. $(5^{\frac{3}{4}} a^{16} b^{\frac{5}{4}})^4$

22. $\dfrac{6x^{-4} y^7}{(2^6 x^3 y^2)^{\frac{1}{3}}}$

23. $\dfrac{(xy^2 z^3)^n}{(x^n y^n z^n)^2}$

24. $(3x^2 y^7 z^{\frac{1}{3}})^0$

25. $(25^0 a^2 b^3 c^{-1})^{-2}$

26. $\left(\dfrac{3a^2 b^3}{xy}\right)^{-3}$

27. $\left(\dfrac{5x^3 y^4}{3^{-2} x^5 y^3 z^2}\right)^{-4}$

28. $\dfrac{(27x^2 y^3)^{\frac{1}{3}}}{(4x^5 y^7)^{\frac{1}{2}}}$, $x = 2.1$, $y = 0.5$

In Exercises 29–32 use your calculator to show that the expressions are equal.

29. 3^4, 9^2

30. 0.25^{-3}, 4^3

31. 12^4, $(3^4)(4^4)$

32. $\dfrac{1}{3^3}$, $81^{-(\frac{3}{4})}$

In Exercises 33–54 use a calculator to find the decimal value of each expression. (*Hint:* In some exercises you can save time by first simplifying the expression using the definitions and laws of exponents.)

33. $(11.23)^3$

34. $\dfrac{\left(\frac{1}{4}\right)^2}{\left(\frac{1}{3}\right)^2}$

35. $\dfrac{(2^4)^{\frac{1}{2}}}{(4^{\frac{1}{2}})^2}$

36. $19 - x^3$, $x = 2$

37. $19 - x^4$, $x = 2$

38. $19 - x^{-3}$, $x = -2$

39. $19 - x^{-4}$, $x = -2$

40. $a^3 \cdot b^4$, $a = 1.2$, $b = 2$

41. $(3^3 x^4)^5$, $x = 0.3$

42. $\left(\dfrac{x^{1.4}}{y^{3.2}}\right)^3$, $x = 2$, $y = 4$

43. $\left(\dfrac{1}{x^3}\right)^6$, $x = 2.5$

44. $\left(\dfrac{x^4}{x^4}\right)^{13}$, $x = 5.6$

45. $\left(\dfrac{x^4}{y^4}\right)^0$, $x = 5.6$, $y = 3.6$

46. $\dfrac{a^8}{a^5}$, $a = 3.4$

47. $\dfrac{x^5}{x^9}$, $x = 4.5$

48. $a^4 b^5 (a^2 b^3)$, $a = 1.1$, $b = 1.2$

49. $a^2 b^3 (a^4 b^5)^2$, $a = 1.1$, $b = 1.2$

50. $\dfrac{8x^3 y^2}{12x^2 y^5}$, $x = 3$, $y = 4$

51. $\dfrac{2^6 a^3 b^2}{6a^{-4} b^7}$, $a = 1.5$, $b = 2.05$

52. $\dfrac{a^n b^n c^n}{(ab^3 c^4)^n}$, $a = 2$, $b = 3$, $c = 4$

53. $36^0 x^3 y^{-3} z^{-1}$, $x = 0.2$, $y = 1.1$, $z = 0.3$

54. For what value of x is $x^0 \ne 1$?

55. Explain the difference between the product rule and the power rule. Give an example of each.

56. Is the following true?

$$5x^{-4} = \frac{1}{5x^4}$$

If it is not true, write $5x^{-4}$ without a negative exponent.

57. Is the following true?

$$7x^{-2} = \frac{1}{7x^2}$$

If it is not true, write $7x^{-2}$ without a negative exponent.

58. The total resistance, R_T, of two conductors in parallel is given by

$$R_T = \frac{1}{R_2^{-1} + R_1^{-1}}.$$

Write the expression with positive exponents and simplify.

59. The focal length, F, of two lenses with focal lengths F_1 and F_2 separated by a distance d is given by

$$F = \frac{1}{F_2^{-1} + F_1^{-1} - d(F_1 F_2)^{-1}}$$

Write the expression with positive exponents and simplify. (The notation $d(F_1 F_2)$ is used to indicate the distance form F_1 to F_2.)

2.3

Scientific Notation

The scientist and the technician may work with extremely large and extremely small numbers. For example, the spaceship *Columbia* flew at the rate of 17,800 mph in space. The heat on the bottom of the ship as it reentered the earth's atmosphere was estimated to be 2600°F. The size of a certain virus is 0.0000005 mm.

Each of these numbers can be expressed in a more convenient form, called **scientific notation.**

> **DEFINITION 2.3**
> A number written in **scientific notation** is written as a product of a positive integer from 1 to 9 inclusive and a power of 10.

The definition can be represented symbolically as $p \times 10^k$, where p is a positive integer such that $1 \le p \le 9$, and k can be any integer.

EXAMPLE 1 **From Numbers in Ordinary Notation to Scientific Notation**

Use definition 2.4 to write the three numbers given in the opening paragraph of this section in scientific notation.

Solution

	Number	Scientific Notation
a.	17,800 mph	1.78×10^4 mph
b.	2600°F	2.6×10^3 °F
c.	0.0000005 mm	5.0×10^{-7} mm

Note that in scientific notation there is a single digit to the left of the decimal point in each case: 1, 2, and 5. Also note that the exponents of 10 are the integers 4, 3, and −7 and that the exponent is the same as the number of places the decimal point was moved in changing to scientific notation. In part a the decimal point was moved four places to the left (17,800.). In parts b and c the decimal point was moved 3 places to the left and seven places to the right, respectively. In part c the 5 is in the ten-millionths place, and we can write

$$0.0000005 = \frac{5}{10,000,000} = \frac{5}{10^7} = 5 \times 10^{-7}. \quad \blacklozenge$$

EXAMPLE 2 **From Scientific Notation to Ordinary Notation**

Express the following numbers in ordinary decimal notation.

a. 1.4×10^8 mi²
b. 6.0×10^{-6} mg
c. 9.46×10^{12} km

Solution To change numbers that are written in scientific notation into ordinary decimal notation, reverse the procedure used in Example 1. (Recall that $10^3 > 1$ and $10^{-3} < 1$. Therefore, 3.00×10^3 is a larger number in ordinary notation, whereas 3.00×10^{-3} is a smaller number in ordinary notation.)

	Scientific Notation	Ordinary Notation
a.	1.4×10^8 mi²	140,000,000 mi²
b.	6.0×10^{-6} mg	0.000006 mg
c.	9.46×10^{12} km	9,460,000,000,000 km

Note that when the exponent is positive, as in a and c, we move the decimal point to the right (1.40000000) to change to ordinary notation. When the exponent is negative, as in part b, we move the decimal point to the left (000006.). $\quad \blacklozenge$

EXAMPLE 3 **Adding and Subtracting Numbers in Scientific Notation**

Write the numbers in scientific notation, perform the indicated operations, and express the result in scientific notation.

a. 34,000,000,000 + 25,000,000,000

b. 0.000000019 − 0.0000000027

Solution

a. $34{,}000{,}000{,}000 + 25{,}000{,}000{,}000 = 3.4 \times 10^{10} + 2.5 \times 10^{10}$
Since the power of ten is the same for both numbers, add the 3.4 and the 2.5 and attach the power of ten to the sum. Thus,

$$3.4 \times 10^{10} + 2.5 \times 10^{10} = (3.4 + 2.5) \times 10^{10}$$
$$= 5.9 \times 10^{10}$$

b. $0.000000019 - 0.0000000027 = 1.9 \times 10^{-8} - 2.7 \times 10^{-9}$
In this case, the powers of ten are not the same. To be able to subtract, the numbers must be written so that they have the same power of ten. We can select any convenient power, our choice is 10^{-8}. We will write 2.7×10^{-9} as 0.27×10^{-8}. Now we can subtract; the answer is

$$0.000000019 - 0.0000000027 = 1.9 \times 10^{-8} - 0.27 \times 10^{-8}$$
$$= 1.63 \times 10^{-8} \quad \blacklozenge$$

EXAMPLE 4 **Multiplying and Dividing Numbers in Scientific Notation**

Write the numbers in scientific notation, perform the indicated operations, and express the result in scientific notation.

a. 0.0000075×0.00034

b. $3{,}800{,}000 \div 500$

Solution

a. $0.0000075 \times 0.00034 = (7.5 \times 10^{-6}) \times (3.4 \times 10^{-4})$
$= (7.5 \times 3.4) \times (10^{-6} \times 10^{-4})$
$= 25.5 \times 10^{-10}$
$= (2.55 \times 10^{1}) \times 10^{-10}$
$= 2.55 \times 10^{-9}$

b. $\dfrac{3,800,000}{500} = \dfrac{3.8 \times 10^6}{5 \times 10^2}$

$\qquad\qquad = \dfrac{3.8}{5} \times \dfrac{10^6}{10^2}$

$\qquad\qquad = 0.76 \times 10^{6-2}$

$\qquad\qquad = 0.76 \times 10^4$

$\qquad\qquad = (7.6 \times 10^{-1}) \times 10^4$

$\qquad\qquad = 7.6 \times 10^3$ ◆

Example 4 illustrates how scientific notation can be useful in performing calculations. Often a technician or scientist needs only an estimated answer. For example, in computing how much power a certain tool will draw, a rough answer usually is adequate. We, too, can use estimating techniques to get a ballpark answer to a problem. If the final answer and the estimate are not in the same ballpark (that is, within a power of ten of each other), then some minor computational error was probably made.

Making estimates involves only a few simple rules. These depend upon the retention of only a few simple arithmetic identities. The chart below contains these identities.

Arithmetic Identities		
$10,000 = 10^4$		$2 = 2^1$
$1000 \;\;= 10^3$		$4 = 2^2$
$100 \;\;\;= 10^2$		$8 = 2^3$
$10 \;\;\;\;= 10^1$		$16 = 2^4$
		$32 = 2^5$
$2^{10} \;\;\;\approx 10^3$		$64 = 2^6$

There are only two rules. The *first rule* is when a number is near a power of 10, round to that power of 10. The *second rule* is when a number is near a power of 2, round to that power of 2. Then we simply apply what we know about exponents.

Examples of rounding include $9300 \approx 10^4$, $17 \approx 2^4$, $89 \approx 10^2$. Sometimes we can construct a combination:

$$37,000 = 37 \times 1000 = 37 \times 10^3 \approx 2^5 \times 10^3.$$

Some numbers can be expressed more than one way—for example, $9 \approx 2^3$ or 10^1. The symbol $\approx$ means approximately equal to.

Remember we are only estimating, so we should be more concerned with the speed and ease of a calculation rather than with its accuracy.

EXAMPLE 5   Application: Estimating Mass of Sun to Mass of Earth

A certain calculation for the mass of the sun to the mass of the earth yields the following ratio:

$$\frac{93{,}600{,}000 \times 93{,}600{,}000 \times 93{,}600{,}000 \times 27.3 \times 27.3}{240{,}000 \times 240{,}000 \times 240{,}000 \times 365.2 \times 365.2}.$$

Estimate the mass of the sun to the mass of the earth.

Solution   Using the arithmetic identities and the two rules for estimating, we can replace each of the measurements with an estimate.

$$93{,}600{,}000 \approx 94 \times 1{,}000{,}000 \approx 10^2 \times 10^6 \approx 10^8 \quad \text{and} \quad 27.3 \approx 2^5,$$
$$240{,}000 = 24 \times 10{,}000 \approx 2^5 \times 10^4 \quad \text{and}$$
$$365.2 \approx 400 = 4 \times 100 = 2^2 \times 10^2$$

Substituting these estimates in the initial ratio we have:

$$\frac{10^8 \times 10^8 \times 10^8 \times 2^5 \times 2^5}{2^5 \times 10^4 \times 2^5 \times 10^4 \times 2^5 \times 10^4 \times 2^2 \times 10^2 \times 2^2 \times 10^2}$$

Using the rules of exponents we have: $\dfrac{10^{24} \times 2^{10}}{10^{16} \times 2^{19}} = \dfrac{10^8}{2^9}$ Recall that $2^{10} \approx 10^3$.

When we multiply the numerator and denominator by 2, we have:

$$\frac{10^8 \times 2}{2^9 \times 2} \approx \frac{10^8 \times 2}{10^3} = 10^5 \times 2.$$

Therefore, the mass of the sun to the mass of the earth is approximately 2×10^5 to 1. This is not a bad estimate since the actual answer is 330,000. Depending on how you estimate, your answer may vary. However, the estimate should be within at least a factor of 10 of the actual answer. ◆

EXAMPLE 6   Calculations in Scientific Notation with a Calculator

Evaluate the expression with a calculator using scientific notation.

$$\frac{(8 \times 10^6)(7 \times 10^3)}{2 \times 10^{-4}}$$

Solution   Before touching the calculator, find the answer using scientific notation. Rewriting the expression above we have:

$$\frac{8 \times 7}{2} \times \frac{10^6 \times 10^3}{10^{-4}} = 28 \times 10^{13}$$
$$= 2.8 \times 10^{14}$$

The solution using a calculator is:

fx–7700G: 8 EXP 6 × 7 EXP 3 ÷ 2 EXP SHIFT (−) 4 EXE
   2.8E + 14
TI–82:   8 2nd EE 6 × 7 2nd EE 3 ÷ 2 2nd EE (−) 4
   ENTER 2.8 E 14

Thus the solution is 2.8×10^1. ◆

EXAMPLE 7   **Calculations in Scientific Notation with a Calculator**

Use a calculator to evaluate the following expression.

$$\frac{(5.36 \times 10^{12})^3}{(1.82 \times 10^{-7})(4.63 \times 10^{-9})}$$

Solution

TI–82: 5.36 2nd EE 12 ∧ 3 ÷ (1.82 2nd EE (−) 7 × 4.63
2nd EE (−) 9) ENTER 1.8274352 E 53

Thus,

$$\frac{(5.36 \times 10^{12})^3}{(1.82 \times 10^{-7})(4.63 \times 10^{-9})} = 1.83 \times 10^{53}$$

Does this answer seem realistic? Estimate the answer using scientific notation. Express 5.36, 1.82 and 4.63 to the nearest integer. Thus,

$$\frac{(5.36 \times 10^{12})^3}{(1.82 \times 10^{-7})(4.63 \times 10^{-9})} \approx *\frac{(5 \times 10^{12})^3}{(2 \times 10^{-7})(5 \times 10^{-9})}$$

$$\approx \frac{5^3 \times (10^{12})^3}{2 \times 5 \times 10^{-7} \times 10^{-9}}$$

$$\approx \frac{125 \times 10^{36}}{10 \times 10^{-16}}$$

$$\approx 1.25 \times 10^{53}$$

The estimated result, 1.25×10^{53}, shows that the calculator result is realistic. The estimated result is smaller due to a rounding error and the fact that one number was cubed, which magnified the error. ◆

2.3 Exercises

In Exercises 1–12, write the number in scientific notation.

1. 5000   **2.** 367,000   **3.** 0.0004   **4.** 0.00345

5. 7   **6.** 27   **7.** 0.3   **8.** 0.33

9. 56,000,000   **10.** 632,400   **11.** 0.000000877   **12.** 7382.60

In Exercises 13–22, express the number in each statement in scientific notation.

13. One kilometer is approximately 39,370 in.

14. The mean distance from the earth to the sun is approximately 148,320,000,000 meters.

15. One degree is approximately 0.01745 rad.

16. One atomic-mass unit is approximately 0.00000000000000000000000166 g.

17. The mean distance from the earth to the moon is approximately 382,000 km.

18. One estimate of the size of the universe is 100,000,000,000,000,000,000,000,000 meters.

19. The mass of the smallest particle known to humans, the mass of the electron, is 0.00000000000000000016 kilogram.

20. A coulomb represents a grouping of 6,250,000,000,000,000,000 excess protons or electrons.

21. The average number of heartbeats per normal life-time is 2,838,000,000.

22. Determine the number of seconds in a year and write the result in scientific notation.

In Exercises 23–34, write the numbers in ordinary decimal notation.

23. 8.5×10^3

24. 4×10^3

25. 5.67×10^{-4}

26. 5×10^{-4}

27. 3×10^{-1}

28. 4.4×10^{-1}

29. 9×10^0

30. 8.76×10^0

31. 8.75×10^7

32. 6.53×10^{-7}

33. 3.4×10^{-9}

34. 9.873×10^{10}

In Exercises 35–44, write the numbers in ordinary decimal notation.

35. One yard is equal to 5.682×10^{-4} miles.

36. Absolute zero is -4.60×10^2 °F.

37. The half-life of uranium 235 is 7.5×10^8 years.

38. One millimeter is equal to 1×10^{-3} liters.

39. The mass of the uranium atom 235, which was the fuel pellet for the atomic bomb, is 1.67×10^{-27}.

40. One centimeter is equal to 1×10^{-5} kilometers.

41. The area of Russia is approximately 8.649×10^6 square miles.

42. The area of the United States is approximately 3.615×10^6 square miles.

43. The thickness of an oil film is approximately 5×10^{-7} centimeters.

44. Some sources estimate that the population of the world is 5 billion people.

In Exercises 45–52, use scientific notation to estimate the answer for each exercise and then use a calculator to perform the indicated operations. Express answers in scientific notation.

45. $(5 \times 10^6) + (7 \times 10^6)$

46. $(6.1 \times 10^{-2}) + (5.3 \times 10^{-2}) - (4.8 \times 10^{-2})$

47. $\dfrac{(4 \times 10^5)(3.5 \times 10^3)}{2 \times 10^4}$

48. $\dfrac{(5 \times 10^7)(4.6 \times 10^9)}{6.3 \times 10^5}$

49. $\dfrac{(2.3 \times 10^3)(7.4 \times 10^5)}{5 \times 10^{-4}}$

50. $\dfrac{(3.7 \times 10^6)(8.6 \times 10^8)}{6.7 \times 10^{-5}}$

51. $\dfrac{(6.34 \times 10^{-5})(8.75 \times 10^7)^3}{(5.4 \times 10^7)(3.4 \times 10^3)^2}$

52. $\left[\dfrac{(3.4 \times 10^8)(6.8 \times 10^{-5})(2.7 \times 10^9)^2}{(86{,}000)^3(0.00048)^3}\right]^2$

In Exercises 53–62 solve the problems using scientific notation with a calculator and express the answers in scientific notation.

53. Using the equation rate is equal to distance divided by time, determine how long it takes a spaceship traveling 1.78×10^4 mph to travel the 2.4×10^5 mi from the earth to the moon.

54. From fluid kinematics we have the formula:

$$v = \frac{V}{A},$$

where v = velocity of the fluid, V = volume of the fluid flow per unit time, and A = cross-sectional area. If $V = 5.0 \times 10^{-2}$ ft³/s and $A = 6.94 \times 10^{-3}$ ft², find v.

55. Using the formula from Exercise 54, find V if $A = 4.0 \times 10^{-2}$ m² and $v = 2.0$ m/s.

56. In chemistry the pH of a particular solution can be found by the expression

$$pH = \frac{E - (2.80 \times 10^{-1})}{5.9 \times 10^{-2}}$$

where E is the electric potential. Find the pH if $E = 4.75 \times 10^{-1}$.

57. If a computer can do a calculation in 0.000002 second, how long, in seconds, would it take the computer to do 8 trillion (8,000,000,000,000) calculations?

58. Calculate the volume of a box that has dimensions of 6000 by 9700 by 4700 millimeters.

59. The mass of one proton is 0.00000000000000000000000167248 gram. Find the mass of one billion protons.

60. If light travels at 3×10^{10} centimeters per second, find the magnitude of a light year in centimeters. (A light year is the distance traveled by light in one year.)

61. The lighting system for a recreational area consists of 70 fixtures at 1500 W each and 40 fixtures at 1650 W each. Yearly usage of the lighting system in the recreational area is 300 hours. At a rate of \$0.000045 per watt-hour (Wh), what is the yearly cost for operating the lights?

62. Under certain conditions the heat flow through a brick wall may be determined using the formula:

$$\text{Heat Flow} = 1.7 \times 10^{-4} \frac{\text{length of wall} \times \text{width of wall}}{\text{thickness of wall}}$$

Determine the heat flow through a wall that is 18 meters high by 40 meters long and 0.48 meter thick. The rate of heat flow is expressed in calories per second.

63. An electron volt is an atomic packet of energy equaling 1.6×10^{-19} joules. When one single nucleus of uranium splits apart (fission) about 200,000,000 electron volts are released. If 3.2×10^{26} atoms of uranium undergo such a fission process simultaneously, how many joules of energy will be released? If one pot of coffee requires about 680 joules, about how many pots can be heated by that amount of energy?

Problems 64–65 do not require exact answers. Estimate your answers and then compare your estimate with the estimate in the back of the book.

64. A politician insists that every taxpayer in the United States pay $10,342 to retire the national debt. Assuming that there are about 180 million taxpayers in the country and that the national debt is 1.24 trillion dollars, quickly determine if the politician is in the right ballpark.

65. If a spaceship travels at 1.27×10^6 m/s (just beyond the escape velocity for sun) about how long will it take to travel to a star with a planetary system approximately 20 light years (or 9.46×10^{15} meters)^2 away. Use the equation for speed equaling distance over time. Change your answer to years to decide if such a trip is feasible. One year is about π-teen million seconds or 3.14×10^7 seconds.

66. The following equation can be used to determine the force between two electrons. On the right side of the equation are appropriate numbers with units omitted.

$$F = \frac{kq_1q_2}{R^2} = \frac{9 \times 10^9 (1.6 \times 10^{-19})^2}{(0.4 \times 10^{-4})^2}$$

In the equation q is the symbol for charge, k is a constant, and R is the distance between the charges. Simplify the expression first and then solve for the force.

67. The following equation can be used to determine the acceleration due to gravity. On the right side of the equation are appropriate numbers with units omitted.

$$a = \frac{GM}{R^2} = \frac{\left(\frac{2}{3} \times 10^{-11}\right)(6 \times 10^{24})}{\left(\frac{19}{3} \times 10^6\right)^2}$$

In the equation M is the mass at the earth, a is acceleration, G is gravity, and R is the earth's radius. Simplify the expression first and then solve for the acceleration.

68. The kinetic energy of a space ship is given by

$$K = \frac{1}{2}mv^2 = \frac{1}{2}(2 \times 10^5)(12 \times 10^3)^2$$

where m is the mass of the ship, and v is the speed. Simplify the expression first and then solve for the kinetic energy.

2.4

Roots and Radicals

In Section 2.1 we explored definitions and laws of exponents. For example, we learned that in applying Definition 2.1 we found that $5^2 = 5 \cdot 5$, or 25. Using these same definitions and laws we can find the value of $25^{\frac{1}{2}}$. If we substitute 5^2 for 25, the expression is $(5^2)^{\frac{1}{2}}$. Now apply the power rule, and the result is 5^1 or 5. An equivalent way of writing $25^{\frac{1}{2}}$ is $\sqrt{25}$, read "the square root of 25." The symbol $\sqrt{25}$ asks the question, what number multiplied by itself is 25? As we already have discovered, the answer is 5, which is called the **root.** A second answer is -5 since $(-5)(-5)$ is 25. This means that the answer to the square root of 25 could be $+5$ or -5. Unless otherwise indicated, when we find the even root of a number the answer will be the positive root. This positive root is called the **principal square root,** or the **principal root.**

> **DEFINITION 2.4**
>
> $a^{\frac{1}{n}} = \sqrt[n]{a}$, where $\sqrt{}$ is the **radical sign,** a is the **radicand,** and n is the index of the radical. The answer is called the **root.** $(a \geq 0)$ When $n = 2$, we assume the index and write the square root of 27 as $\sqrt{27}$.

EXAMPLE 1   Radical Expressions

Evaluate and, if possible, find a real number solution.

a. $\sqrt{16}$ **b.** $\sqrt[3]{-8}$ **c.** $\sqrt[4]{81}$ **d.** $\sqrt[5]{32}$ **e.** $\sqrt{-9}$

Solution

a. $\sqrt{16}$ (square root is 16), $\sqrt{16} = 4$
b. $\sqrt[3]{-8}$ (cube root of -8), $\sqrt[3]{-8} = -2$, since $-2 \cdot -2 \cdot -2 = -8$. Note that the solution has only one possible answer, -2; $+2$ does not work.
c. $\sqrt[4]{81}$ (fourth root of 81), $\sqrt[4]{81} = 3$
d. $\sqrt[5]{32}$ (fifth root of 32), $\sqrt[5]{32} = 2$. Again there is only one solution; this time $+2$ works, but not -2. From this we may generalize that odd roots have only one solution, while even roots could have two solutions.
e. $\sqrt{-9}$ (square root of -9). There is no *real number* that when multiplied by itself will give a product of -9. ◆

To indicate the negative solution of an even root, a negative sign is placed in front of the radical. To include both the principal and negative roots, a $\pm$ sign (read "plus or minus") is placed in front of the radical.

EXAMPLE 2   Simplifying Radical Expressions

Find real number solutions for the radical expressions.

a. $-\sqrt{25}$ **b.** $\pm\sqrt{64}$ **c.** $\sqrt{8^2}$

Solution

a. The $\sqrt{25} = 5$, thus $-\sqrt{25} = -5$
b. The $\sqrt{64} = 8$, thus $\pm\sqrt{64} = \pm 8$
c. The $\sqrt{8^2} = 8^{\frac{2}{2}} = 8^1$ or 8. ◆

Since $a^{m/n} = (a^m)^{1/n} = (a^{1n})^m$, Law 2.8 follows.

LAW 2.8

If m and n are both integers with n a positive number, and if $\sqrt[n]{a}$ exists as a real number, then

$$a^{\frac{m}{n}} = \sqrt[n]{a^m} = (\sqrt[n]{a})^m.$$

EXAMPLE 3 **Change from Exponentional to Radical Form**

Write each of the following exponentional expressions in radical form.

a. $3^{\frac{3}{2}}$ b. $(-3)^{\frac{3}{2}}$ c. $4^{-\frac{3}{2}}$

Solution

a. $3^{\frac{3}{2}} = \sqrt{(3)^3} = \sqrt{3^2 \cdot 3} = 3\sqrt{3}$
b. $(-3)^{\frac{3}{2}} = \sqrt{(-3)^3} = \sqrt{-27}$
c. $4^{-\frac{3}{2}} = \dfrac{1}{4^{\frac{3}{2}}} = \dfrac{1}{\sqrt{4^3}} = \dfrac{1}{\sqrt{64}} = \dfrac{1}{8}$ ◆

The $\sqrt{a^2}$ is a special situation.

LAW 2.9

$$\sqrt{a^2} = |a|$$

The result of Law 2.9 follows from the fact that any nonzero number squared is a positive number; hence, the root can only be positive.

EXAMPLE 4 **Applying Law 2.9**

Simplify the following expressions.

a. $3^{\frac{2}{2}}$ b. $(-3)^{\frac{2}{2}}$

Solution

a. $3^{\frac{2}{2}} = \sqrt{(3)^2} = |3| = 3$ **Law 2.8, Law 2.9**
b. $(-3)^{\frac{2}{2}} = \sqrt{(-3)^2} = |-3| = 3$ **Law 2.8, Law 2.9** ◆

Using the information to this point on the relation between the definitions of exponents and the definitions of radicals, consider the following problem?

$$(9 \cdot 4)^{\frac{1}{2}} = \sqrt{36} = 6$$

Note that

$$\sqrt{9} \cdot \sqrt{4} = 3 \cdot 2 = 6$$

This result leads to the generalization in Law 2.10.

LAW 2.10

$$(a \cdot b)^{\frac{1}{n}} = \sqrt[n]{a \cdot b} = (\sqrt[n]{a})(\sqrt[n]{b})$$

If n is even, a and b must be non-negative.

Similarly,

$$\left(\frac{a}{b}\right)^{\frac{1}{n}} = \frac{a^{\frac{1}{n}}}{b^{\frac{1}{n}}} \text{ leads to Law 2.11.}$$

LAW 2.11

$$\sqrt[n]{\frac{a}{b}} = \frac{\sqrt[n]{a}}{\sqrt[n]{b}}$$

If n is even, a and b must be non-negative and $b \neq 0$.

Suppose we take a^n to the third power, $(a^n)^3$. This means

$$a^n \cdot a^n \cdot a^n = a^{n+n+n} = a^{3n}$$

To generalize, we have $(a^n)^m = a^{nm}$. This leads to Law 2.12.

LAW 2.12

$$\sqrt[m]{\sqrt[n]{a}} = \sqrt[m \cdot n]{a}$$

EXAMPLE 5 **Illustrating Laws 2.10 and 2.11**

Using the definitions for exponents and radicals, simplify the following expressions. (To simplify means to remove part or all of the expression from the radical, if possible.)

a. $\sqrt{54x^2}$ **b.** $3\sqrt[5]{64y^7}$ **c.** $\sqrt{\dfrac{625}{243}}$

Solution

a. $\sqrt{54x^2} = \sqrt{2 \cdot 3 \cdot \boxed{3^2} \cdot \boxed{x^2}}$

$\phantom{\sqrt{54x^2}} = \sqrt{6} \sqrt{3^2} \sqrt{x^2}$ Law 2.10

$\phantom{\sqrt{54x^2}} = \sqrt{6} \, (3)(x)$ Law 2.9

$\phantom{\sqrt{54x^2}} = 3x\sqrt{6}$

b. $3\sqrt[5]{64y^7} = 3\sqrt[5]{2 \cdot \boxed{2^5} \, \boxed{y^5} \, y^2}$

$\phantom{3\sqrt[5]{64y^7}} = 3\sqrt[5]{2} \, \sqrt[5]{2^5} \, \sqrt[5]{y^5} \, \sqrt[5]{y^2}$ Law 2.10

$\phantom{3\sqrt[5]{64y^7}} = 3 \cdot \sqrt[5]{2} \cdot 2 \cdot y \cdot \sqrt[5]{y^2}$

$\phantom{3\sqrt[5]{64y^7}} = 6y\sqrt[5]{2y^2}$

c. $\sqrt{\dfrac{625}{243}} = \dfrac{\sqrt{5^4}}{\sqrt{3^4 \cdot 3}}$ Law 2.11

$\phantom{\sqrt{\dfrac{625}{243}}} = \dfrac{(5^4)^{\frac{1}{2}}}{(3^4 \cdot 3)^{\frac{1}{2}}}$ Definition 2.4

$\phantom{\sqrt{\dfrac{625}{243}}} = \dfrac{5^2}{3^2 \cdot 3^{\frac{1}{2}}}$ Power rule (Law 2.2)

$\phantom{\sqrt{\dfrac{625}{243}}} = \dfrac{25}{9\sqrt{3}}$ Definition 2.4   ◆

EXAMPLE 6   Simplifying Radical Expressions

Using the definitions for exponents and radicals, simplify the following expressions.

a. $\sqrt[3]{\sqrt{128}}$ b. $\sqrt[n]{x^{2n}y^{n+1}}$

Solution

a. $\sqrt[3]{\sqrt{128}} = (\sqrt{128})^{\frac{1}{3}}$ Definition 2.4

$\phantom{\sqrt[3]{\sqrt{128}}} = (128^{\frac{1}{2}})^{\frac{1}{3}}$ Definition 2.4

$\phantom{\sqrt[3]{\sqrt{128}}} = 128^{\frac{1}{6}}$ Power rule (Law 2.2)

$\phantom{\sqrt[3]{\sqrt{128}}} = (2^6 \cdot 2)^{\frac{1}{6}}$ Substituting: $128 = 2^7 = 2^6 \cdot 2$

$\phantom{\sqrt[3]{\sqrt{128}}} = 2 \cdot 2^{\frac{1}{6}}$ Power rule (Law 2.2)

$\phantom{\sqrt[3]{\sqrt{128}}} = 2\sqrt[6]{2}$

b. $\sqrt[n]{x^{2n}\,y^{n+1}} = \sqrt[n]{x^{2n}} \, \sqrt[n]{y^n y}$ Law 2.10

$\phantom{\sqrt[n]{x^{2n}\,y^{n+1}}} = (x^{2n})^{\frac{1}{n}} \, (y^n y)^{\frac{1}{n}}$ Law 2.8

$\phantom{\sqrt[n]{x^{2n}\,y^{n+1}}} = x^{\frac{2n}{n}} y^{\frac{n}{n}} y^{\frac{1}{n}}$ Power rule (Law 2.2)

$\phantom{\sqrt[n]{x^{2n}\,y^{n+1}}} = x^2 y y^{\frac{1}{n}}$

$\phantom{\sqrt[n]{x^{2n}\,y^{n+1}}} = x^2 y \sqrt[n]{y}$ Law 2.8   ◆

In Example 5 **c,** the answer contains a radical in the denominator. When we work without a calculator, it is more convenient to evaluate the expression if there is no radical in the denominator. For example, $\dfrac{1}{\sqrt{3}} = \dfrac{1}{1.7320508}$ is difficult to evaluate without a calculator. However, if we multiply the numerator and denominator by $\sqrt{3}$, the expression is easier to evaluate.

$$\frac{1}{\sqrt{3}} \cdot \frac{\sqrt{3}}{\sqrt{3}} = \frac{1.7320508}{3}$$

The procedure for eliminating the radical in the denominator is called **rationalizing the denominator.** Essentially, we just multiply the final result by 1, where $1 = \dfrac{\sqrt{3}}{\sqrt{3}}$. To obtain a solution to Example 5 **c** by rationalizing the denominator we would do the following.

$$\frac{25}{9\sqrt{3}} = \frac{25}{9\sqrt{3}} \cdot \frac{\sqrt{3}}{\sqrt{3}} \quad \text{Rationalizing the denominator}$$

$$= \frac{25\sqrt{3}}{9 \cdot 3}$$

$$= \frac{25\sqrt{3}}{27}$$

Procedure to Rationalize Denominator

To rationalize a denominator with a radical having an index n, multiply the numerator and denominator of the fraction by the smallest quantity that will raise the exponent of the denominator to a number that is evenly divisible by n.

EXAMPLE 7 Rationalize the Denominator

Rationalize the denominator in each expression.

a. $\sqrt{\dfrac{2}{3}}$

b. $\dfrac{\sqrt{a}}{\sqrt{b^3}}$

c. $\sqrt[3]{\dfrac{3}{2x^3}}$

Solution

a.   Since $n = 2$, we want the exponent of 3 to be divisible by 2. Multiply both the numerator and denominator by 3.

$$\sqrt{\frac{2}{3}} = \sqrt{\frac{2}{3} \cdot \frac{3}{3}}$$

$$= \sqrt{\frac{6}{3^2}}$$

$$= \frac{\sqrt{6}}{\sqrt{3^2}} \qquad \text{Law 2.11}$$

$$= \frac{\sqrt{6}}{3} \qquad \text{Law 2.9}$$

b. $\dfrac{\sqrt{a}}{\sqrt{b^3}} = \sqrt{\dfrac{a}{b^3}}$ $\qquad$ **Law 2.11**

Since $n = 2$, multiply the numerator and denominator by b to make the exponent of b divisible by 2.

$$\sqrt{\frac{a}{b^3}} = \sqrt{\frac{a}{b^3} \cdot \frac{b}{b}}$$

$$= \sqrt{\frac{ab}{b^4}}$$

$$= \frac{\sqrt{ab}}{\sqrt{(b^2)^2}}$$

$$= \frac{\sqrt{ab}}{b^2}$$

c.   Since $n = 3$, the exponent of 2 and x must be divisible by 3. Therefore, multiply both numerator and denominator by $2^2 x$.

$$\sqrt[3]{\frac{3}{2x^2}} = \sqrt[3]{\frac{3}{2x^2} \cdot \frac{2^2 x}{2^2 x}}$$

$$= \sqrt[3]{\frac{12x}{2^3 \cdot x^3}}$$

$$= \frac{\sqrt[3]{12x}}{2x} \qquad \blacklozenge$$

The denominators of the fractions in Example 7 are monomials. So how do we rationalize the denominator when it is a binomial? To rationalize a denominator of the form $\sqrt{a} + b$ we multiply both numerator and denominator by the conjugate $\sqrt{a} - b$. The product of $\sqrt{a} + b$ and its conjugate $\sqrt{a} - b$ is $a - b^2$. The product is free of radicals. Look at $\dfrac{5}{\sqrt{x} + 1}$. In this case, multiply the numerator and denominator by $\sqrt{x} - 1$, the conjugate of $\sqrt{x} + 1$ to rationalize the denominator. The result is

$$\frac{5}{\sqrt{x} + 1} \cdot \frac{\sqrt{x} - 1}{\sqrt{x} - 1} = \frac{5\sqrt{x} - 5}{x - 1}.$$

To complete this discussion of radicals, let's look at the specific conditions under which we can add, subtract, multiply, and divide them.

Addition and Subtraction of Radicals

Expressions containing radicals can be added or subtracted if the radicals are the same—that is, the radicals have the same radicand and the same index.

EXAMPLE 8 Adding and Subtracting Radicals

Where possible simplify the radicals and then add or subtract.

a. $\sqrt{x} + 5\sqrt{x}$ b. $8\sqrt[3]{5} - 14\sqrt[3]{625}$ c. $\dfrac{1}{2}\sqrt{24} + 3\sqrt{\dfrac{2}{3}} - 2\sqrt{\dfrac{3}{2}}$

Solution

a. Since the radicands are the same, in $\sqrt{x} + 5\sqrt{x}$ we can combine the like radicals using the distributive law.

$$\sqrt{x} + 5\sqrt{x} = (1 + 5)\sqrt{x} = 6\sqrt{x}$$

b. To combine the radicals $8\sqrt[3]{5} - 14\sqrt[3]{625}$, the radicands must be the same. Can we express the 625 with a radicand of 5? Yes.

$$\sqrt[3]{625} = \sqrt[3]{5^3 \cdot 5} = \sqrt[3]{5^3} \cdot \sqrt[3]{5} = 5\sqrt[3]{5}$$

Substituting we have:

$$8\sqrt[3]{5} - 14 \cdot 5\sqrt[3]{5} = 8\sqrt[3]{5} - 70\sqrt[3]{5}$$
$$= (8 - 70)\sqrt[3]{5} \qquad \text{Distributive property}$$
$$= -62\sqrt[3]{5}$$

c. In order to combine the radicals $\dfrac{1}{2}\sqrt{24} + 3\sqrt{\dfrac{2}{3}} - 2\sqrt{\dfrac{3}{2}}$, the radicands must be the same.

$$\sqrt{24} = \sqrt{4 \cdot 6} = \sqrt{4} \cdot \sqrt{6} = 2\sqrt{6}$$

$$\sqrt{\frac{2}{3}} = \sqrt{\frac{2}{3} \cdot \frac{3}{3}} = \frac{\sqrt{6}}{\sqrt{9}} = \frac{\sqrt{6}}{3}$$

$$\sqrt{\frac{3}{2}} = \sqrt{\frac{3}{2} \cdot \frac{2}{2}} = \frac{\sqrt{6}}{\sqrt{4}} = \frac{\sqrt{6}}{2}$$

Substituting we have:

$$\frac{1}{\cancel{2}}(\cancel{2}\sqrt{6}) + \frac{1}{\cancel{3}}\left(\frac{\sqrt{6}}{\cancel{3}}\right) - \frac{1}{\cancel{2}}\left(\frac{\sqrt{6}}{\cancel{2}}\right) = \sqrt{6} + \sqrt{6} - \sqrt{6}$$

$$= \sqrt{6} \quad \blacklozenge$$

Now let's turn our attention to multiplying and dividing radicals.

Multiplication and Division of Radicands

The radicands of two radicals may be multiplied or divided if the indexes of the radicals are the same. Also, if the indexes are even, the radicands must be positive.

EXAMPLE 9   **Multiplying and Dividing Radicals**

If possible find the product or the quotient of the following radicals and simplify. Assume that x, y and z represent positive real numbers.

a. $\sqrt{3x} \cdot \sqrt{5yz}$ **b.** $\dfrac{\sqrt{21xy}}{\sqrt{14yz}}$ **c.** $\dfrac{\sqrt[3]{4}}{\sqrt{2}}$

Solution

a. Since the indexes of the radicals $\sqrt{3x} \cdot \sqrt{5yz}$ are the same, 2, we can multiply the radicands.

$$\sqrt{3x} \cdot \sqrt{5yz} = \sqrt{15xyz}$$

b. The indexes of the radicals of $\dfrac{\sqrt{21xy}}{\sqrt{14yz}}$ are the same, 2. Therefore, we can divide the radicands.

$$\frac{\sqrt{21xy}}{\sqrt{14yz}} = \sqrt{\frac{21xy}{14yz}}$$

$$= \sqrt{\frac{3x}{2z}}$$

Note that the answer may be written $\dfrac{\sqrt{3x}}{\sqrt{2z}}$. By multiplying by $\dfrac{\sqrt{2z}}{\sqrt{2z}}$ we have $\dfrac{\sqrt{6xz}}{2z}$, which is a form preferred by some instructors. However, all three forms are correct results.

c. The indexes of the radicals of $\sqrt[3]{4}$ and $\sqrt{2}$ are not the same, so we cannot divide the radicals in this form. However, by writing $\sqrt[3]{4}$ and $\sqrt{2}$ with fractional exponents, we can apply the quotient rule for exponents.

$$\frac{\sqrt[3]{4}}{\sqrt{2}} = \frac{\sqrt[3]{2^2}}{\sqrt{2}}$$

$$= \frac{2^{\frac{2}{3}}}{2^{\frac{1}{2}}} \qquad \textbf{Law 2.8}$$

$$= 2^{\frac{2}{3}-\frac{1}{2}} \qquad \textbf{Quotient rule (Law 2.5)}$$

$$= 2^{\left(\frac{4}{6}\right)-\left(\frac{3}{6}\right)} \qquad \textbf{Substituting}$$

$$= 2^{\frac{1}{6}} \quad \text{or} \quad \sqrt[6]{2} \quad \blacklozenge$$

EXAMPLE 10 **Application: Police Radar System**

Look back now to the effective range of a radar system (the introduction to this chapter). Use a calculator to evaluate

$$R_{\max} = 32.4 \sqrt[4]{PL^2TF^{\frac{1}{3}}}$$

when $P = 6.00 \times 10^1$ W, $L = 1.00 \times 10^{-2}$ m, $T = 1.00 \times 10^{-6}$ s, and $F = 1.00 \times 10^3$ Hz.

Solution

$$R_{\max} = 32.4 \sqrt[4]{PL^2TF^{\frac{1}{3}}}$$

$$= 32.4\sqrt[4]{(\,6.00 \times 10^1\,)(\,1.00 \times 10^{-2}\,)^2(\,1.00 \times 10^{-6}\,)(\,1.00 \times 10^3\,)^{\frac{1}{3}}}$$

Substituting

$$= 32.4 \; \boxed{\times} \; \boxed{(} \; 6.00 \; \boxed{\text{2nd}} \; \boxed{\text{EE}} \; 1 \; \boxed{\times} \; \boxed{(} \; 1.00 \; \boxed{\text{2nd}} \; \boxed{\text{EE}} \; \boxed{(-)} \; 2 \; \boxed{)}$$
$$\boxed{x^2} \; \boxed{\times} \; 1.00 \; \boxed{\text{2nd}} \; \boxed{\text{EE}} \; \boxed{(-)} \; 6 \; \boxed{\times} \; \boxed{(} \; 1.00 \; \boxed{\text{2nd}} \; \boxed{\text{EE}} \; 3$$
$$\boxed{)} \; \boxed{\wedge} \; \boxed{(} \; 1 \; \boxed{\div} \; 3 \; \boxed{)} \; \boxed{)} \; \boxed{\wedge} \; \boxed{(} \; 1 \; \boxed{\div} \; 4 \; \boxed{)} \; \boxed{\text{ENTER}} \; \text{(TI-82)}$$

$$= .507087 \quad \textbf{Calculator reading}$$

Therefore, the maximum range of the radar is 0.507 miles. $\blacklozenge$

A BASIC computer program which can be used to calculate $R_{\max}$ is found on the tutorial disk.

Section 2.4 concludes with a summary of the definitions and laws of radicals.

DEFINITIONS AND LAWS OF RADICALS

Definition 2.4 $a^{1/n} = \sqrt[n]{a}$

Law 2.8 $a^{m/n} = \sqrt[n]{a^m} = (\sqrt[n]{a})^m$

Law 2.9 $\sqrt{a^2} = |a|$

Law 2.10 $\sqrt[n]{ab} = (\sqrt[n]{a})(\sqrt[n]{b})$

Law 2.11 $\sqrt[n]{\dfrac{a}{b}} = \dfrac{\sqrt[n]{a}}{\sqrt[n]{b}}$

Law 2.12 $\sqrt[m]{\sqrt[n]{a}} = \sqrt[n \cdot m]{a}$

2.4 Exercises

Write Exercises 1–8 in radical form.

1. The square root of 15.
2. The fifth root of 11.
3. The cube root of 13.
4. The square root of 17.
5. The seventh root of 2.
6. The cube root of $\dfrac{3}{32}$.
7. The square root of $\dfrac{5}{13}$.
8. The cube root of -31.

For Exercises 9–26 determine the principal value of each of the following without using a calculator.

9. $\sqrt{36}$
10. $\sqrt{144}$
11. $\sqrt{121}$
12. $\sqrt[3]{125}$

13. $\sqrt[3]{64}$
14. $\sqrt[3]{-8}$
15. $\sqrt{0.01}$
16. $\sqrt{\dfrac{49}{81}}$

17. $\sqrt[5]{32}$
18. $\sqrt{9x^2y^4}$
19. $\sqrt{169x^4}$
20. $\sqrt[3]{8x^3y^6}$

21. $\sqrt[5]{a^{10}b^5z^{15}}$
22. $\sqrt[4]{81x^4y^8}$
23. $\sqrt[3]{-64x^9y^6}$
24. $\sqrt[5]{-32a^5b^{10}}$

25. $\sqrt[3]{8^2}$
26. $\sqrt[4]{9^2}$

In Exercises 27–44 use the definitions for exponents and radicals to simplify the expressions. Then, where possible, use your calculator to compute the numerical results.

27. $\sqrt{12}$
28. $\sqrt{125}$
29. $\sqrt[3]{81}$
30. $\sqrt[3]{625}$

31. $\sqrt[4]{\dfrac{2}{7}}$
32. $\sqrt{\dfrac{24x^2}{18y^3}}$
33. $\sqrt{\dfrac{14x^3}{2x}}$
34. $\sqrt{\dfrac{9a^0 - b^0}{a^0 + (3b)^0}}$

35. $\sqrt{50a^3b^4}$
36. $\sqrt[3]{-24x^3y^6}$
37. $\sqrt[3]{\dfrac{9x^6y^2}{3y^2}}$
38. $\sqrt[n]{a^{n+1}}$

39. $\sqrt[3k]{x^{6k}}$
40. $\sqrt{\sqrt{81x^4}}$
41. $\sqrt[3]{\sqrt{729}}$
42. $\sqrt[3]{\sqrt[3]{-x^9}}$

43. $\sqrt[3]{\sqrt[4]{\dfrac{(6x)^0 - y^0}{x^0 + (7y^2)^0}}}$
44. $\sqrt[3]{\sqrt[4]{x^{12n}}}$

In Exercises 45–60, perform the indicated operations and simplify the result. Do not use a calculator.

45. $\sqrt{3} + 4\sqrt{3}$

46. $5\sqrt{8} + 7\sqrt{18}$

47. $7\sqrt{75} - 6\sqrt{27} + 2\sqrt{48}$

48. $11\sqrt{28} + 6\sqrt{\dfrac{1}{7}} - 5\sqrt{112}$

49. $2\sqrt{24} + 7\sqrt{54} - 5\sqrt{150}$

50. $5\sqrt[3]{24} - \sqrt[3]{81} + 2\sqrt[3]{375}$

51. $\sqrt{y} + \sqrt{y^3} + \sqrt{y^7}$

52. $\sqrt{3}\sqrt{5}$

53. $\sqrt{11} \cdot \sqrt{22}$

54. $3\sqrt{7} \cdot \sqrt{14}$

55. $\sqrt{xyz} \cdot \sqrt{wxy}$

56. $\sqrt[3]{4} \cdot \sqrt[3]{10}$

57. $\sqrt[4]{125} \cdot \sqrt[4]{5}$

58. $\sqrt[3]{9} \cdot \sqrt[3]{3}$

59. $\sqrt[4]{xy^2} \cdot \sqrt[4]{x^3y^2}$

60. $\sqrt[3]{x^2yz^4} \cdot \sqrt[3]{xy^2z^5}$

In Exercises 61–74, rationalize the denominator and simplify. Do not use a calculator.

61. $\dfrac{\sqrt{21}}{\sqrt{7}}$

62. $\dfrac{3\sqrt{12}}{\sqrt{3}}$

63. $\dfrac{\sqrt{55}}{\sqrt{35}}$

64. $\dfrac{\sqrt{12ab}}{\sqrt{14ac}}$

65. $\dfrac{\sqrt[3]{9}}{\sqrt{3}}$

66. $\dfrac{\sqrt[4]{a^5}}{\sqrt[4]{a^3}}$

67. $\dfrac{\sqrt[3]{a^2b^4}}{\sqrt[4]{a^2b^3}}$

68. $\dfrac{\sqrt[3]{8a^4b^2}}{\sqrt{2ab}}$

69. $\dfrac{4}{\sqrt{x}+2}$

70. $\dfrac{5x}{3-\sqrt{x}}$

71. $\dfrac{\sqrt{x}+2}{\sqrt{x}-1}$

72. $\dfrac{3+2\sqrt{x}}{2\sqrt{x}-2}$

73. $\dfrac{3x}{\sqrt{2x}-1}$

74. $\dfrac{4x}{\sqrt{5x}+3}$

75. A business must determine the amount a tool or piece of equipment decreases in value each year for income tax purposes. This decrease in value is called depreciation. The formula $R = N\left[1 - \left(\dfrac{S}{C}\right)^{\frac{1}{3}}\right]$ can be used to calculate R, the percentage depreciation for each year. N is the useful life of the item, C is the original cost, and S is the salvage value. Calculate R to two decimal places if
a. $N = 6$, $C = \$12{,}500$, $S = \$2500$.
b. $N = 4$, $C = \$14{,}000$, $S = \$3800$.

76. To determine the time it would take for a wrench to fall from the roof of a building, the following equation may be used.

$$t = \sqrt{\dfrac{2d}{a}}$$

where $a = 9.8\ m/s^2$ and d is the distance it will fall. If the building is 20.0 meters high, how long will it take for the wrench to hit the ground?

77. An automatic hammer uses a spring to fire nails into wood. The energy stored in the spring is converted into the motion energy of the nail. The equation is

$$\text{energy} = \dfrac{1}{2}\,mv^2.$$

If the energy stored in the spring is $10.0\ \text{kg} \cdot m^2/s^2$, and the mass of the nail is $0.010\ \text{kg}$, what is the speed of the nail as it leaves the gun?

2.5

Multiplication of Algebraic Expressions

To multiply algebraic expressions we must be able to apply the definition of exponents discussed in Section 2.2. This is clearly illustrated in Example 1 where two monomials are multiplied.

EXAMPLE 1   Monomial Times a Monomial

Determine the following products.

a. $(3x^2y)(4xy^3)$ **b.** $(4x^2z^3)(-5x^3y^4z^5)$ **c.** $(-8x^3y^2)(-9x^{-5}y^4z^2)$

Solution

a. $(3x^2y)(4xy^3) = \boxed{3} \cdot \boxed{4} \cdot \boxed{x^2} \cdot \boxed{x} \cdot \boxed{y} \cdot \boxed{y^3}$ The commutative property of multiplication allows us to write the factors in this order.

$$= 12 \cdot x^{2+1} \cdot y^{1+3}$$ Definition 2.2
$$= 12x^3y^4$$

b. $(4x^2z^3)(-5x^3y^4z^5) = 4 \cdot (-5) \cdot x^2 \cdot x^3 \cdot y^4 \cdot z^3 \cdot z^5$
$$= -20x^5y^4z^8$$

c. $(-8x^3y^2)(-9x^{-5}y^4z^2) = (-8)(-9) \cdot x^3 \cdot x^{-5} \cdot y^2 \cdot y^4 \cdot z^2$
$$= 72x^{-2} \cdot y^6 \cdot z^2$$
$$= \frac{72y^6z^2}{x^2} \quad \blacklozenge$$

To multiply a monomial and a binomial expression, use the distributive property: $a(b + c) = ab + ac$. As we will soon observe, in some cases the distributive property will need to be applied more than once.

EXAMPLE 2   Monomial Times a Binomial

Determine the product of $7x^2$ and $4x + 3y^2$.

Solution $7x^2(4x + 3y^2)$ is of the same form as $a(b + c)$.

$$a(b + c) = ab + ac$$

$$\boxed{7x^2}(\boxed{4x} + \boxed{3y^2}) = \boxed{7x^2} \cdot \boxed{4x} + \boxed{7x^2} \cdot \boxed{3y^2}$$
$$= 7 \cdot 4 \cdot x^2 \cdot x + 7 \cdot 3 \cdot x^2 \cdot y^2$$
$$= 28x^3 + 21x^2y^2 \quad \blacklozenge$$

CAUTION When multiplying a negative monomial times a binomial, remember to multiply each term of the binomial by the negative monomial. You must once again remember to change signs when appropriate.

$$-3(x - y) = (-3)(x) + (-3)(-y) = -3x + 3y$$

Also be careful when removing parentheses.

$$-(x + y) = (-1)(x + y) = -x - y. \ \blacksquare$$

EXAMPLE 3   Binomial Times a Binomial

Determine the product of $(x + 3)(x - 5)$.

Solution $(x + 3) \cdot (x - 5)$ is of the same general form as $a \cdot (b + c)$.

$$a(b + c) = ab + ac$$

$$(x + 3) \cdot (x + -5) = (x + 3) \cdot x + (x + 3) \cdot -5$$
Distributive property (binomial × binomial)

$$= x \cdot x + 3 \cdot x + x \cdot (-5) + 3 \cdot (-5)$$
Distributive property (binomial × monomial)

$$= x^2 + 3x + (-5x) + (-15)$$

$$= x^2 - 2x - 15. \ \blacklozenge$$

Now let's develop a general rule for finding the product of two binomial expressions. In general

$$(x + a)(x + b) = (x + a) \cdot x + (x + a) \cdot b \quad \text{Distributive property}$$
$$= x \cdot x + a \cdot x + x \cdot b + a \cdot b \quad \text{Distributive property}$$
$$= x^2 + ax + bx + ab$$
$$= x^2 + (a + b)x + ab \quad \text{Applying distributive property in reverse}$$

Therefore

$$(x + a)(x + b) = x^2 + (a + b)x + ab$$

EXAMPLE 4   Binomial Times Binomial

Determine the product of

a. $(x - 3)(x + 11)$ **b.** $(3x^2 + 7y^2)(5x^2 - 6y^2)$

Solution

a. Using the general rule developed for finding the product of two binomial expressions we have the following result.

$$(x - 3)(x + 11) = x^2 + (-3 + 11)x + -3 \cdot 11$$
$$= x^2 + 8x + (-33)$$
$$= x^2 + 8x - 33.$$

b. The binomials $(3x^2 + 7y^2)$ and $(5x^2 - 6y^2)$ are different from the expressions in part **a** in two ways: (1) there are literal factors in both terms of each binomial, and (2) each literal factor has a numerical coefficient. Applying the distributive property twice the result is:

$$(3x^2 + 7y^2)(5x^2 - 6y^2) = \boxed{3x^2 \cdot 5x^2} + \boxed{7y^2 \cdot 5x^2} + \boxed{3x^2(-6y^2)}$$
$$+ \boxed{7y^2(-6y^2)}$$

$$= 15x^4 + \underbrace{35x^2y^2 + (-18x^2y^2)} + (-42y^4)$$
$$\text{like terms}$$

$$= 15x^4 + 17x^2y^2 - 42y^4.$$

In fact, when the literal factors are the same for both binomials as in part **b** the answer will be of this form. ◆

Another way of finding the product of two binomials is called the FOIL method. The acronym **FOIL** stands for: product of the **F**irst two terms, product of the **O**uter two terms, product of the **I**nner two terms, and the product of the **L**ast two terms, as illustrated in the following example.

$$
\overset{F}{} \quad \overset{L}{} \qquad \overset{F}{} \quad \overset{O}{} \quad \overset{I}{} \quad \overset{L}{}
$$
$$(x + 3)(x - 5) = x \cdot x + x \cdot (-5) + 3 \cdot x + 3 \cdot (-5)$$
$$= x^2 + (-5x) + 3x + (-15)$$
$$= x^2 - 2x - 15$$

Now let's consider two special cases. In the first case find the product of two identical binomials.

1. The product of

$$(x + a)^2 = (x + a)(x + a)$$
$$= x^2 + ax + ax + a^2$$
$$= x^2 + 2ax + a^2$$

When both factors are the same, the first term of the product is the square of the first term (x^2), the second term is twice the product of the two terms $(2ax)$, and the third term is the square of the second term (a^2). This product, $x^2 + 2ax + a^2$, is given a special name, **perfect square trinomial.**

2. Another special situation is the product of two binomials that differ only in the sign between the terms. For example:

$$(x + a)(x - a) = x^2 + ax - ax + a(-a)$$
$$= x^2 + 0 \cdot x + (-a^2)$$
$$= x^2 - a^2$$

This product, $x^2 - a^2$ is given the special name **difference of two squares.**

CAUTION $(x + a)^2 \neq x^2 + a^2$; for example, $(3 + 2)^2 \neq 3^2 + 2^2$ ■

This section concludes with examples of raising a binomial to a power and multiplying a trinomial times a trinomial.

EXAMPLE 5 **Binomial to a Power and Trinomial Times Trinomial**

Find the product of

a. $(x + 3)^3$ **b.** $(x - 3)^2(x^2 + 5x + 3)$

Solution

a. $(x + 3)^3$ can be written

$(x + 3)(x + 3)(x + 3)$

$\quad = (x^2 + 6x + 9)(x + 3)$ **Perfect square trinomial**

$\quad = (x^2 + 6x + 9) \cdot x + (x^2 + 6x + 9) \cdot 3$ **Distributive property**

$\quad = x^3 + 6x^2 + 9x + 3x^2 + 18x + 27$

$\quad = x^3 + 9x^2 + 27x + 27$ **Combining like terms**

As indicated in the solution, when a binomial is raised to a power it is better to write the expression as a product and then multiply.

b. To find the product of $(x - 3)^2$ and $(x^2 + 5x + 3)$ first find $(x - 3)^2$ and then apply the distributive property.

$(x - 3)^2(x^2 + 5x + 3) = (x - 3)(x - 3)(x^2 + 5x + 3)$

$\qquad\qquad\qquad = (x^2 - 6x + 9)(x^2 + 5x + 3)$

                                          Perfect Square Trinomial

$\qquad\qquad\qquad = (x^2 - 6x + 9) \cdot x^2 + (x^2 - 6x + 9) \cdot 5x$

$\qquad\qquad\qquad\quad + (x^2 - 6x + 9) \cdot 3$ **Distributive property**

$\qquad\qquad\qquad = x^4 - 6x^3 + 9x^2 + 5x^3 - 30x^2 + 45x$

$\qquad\qquad\qquad\quad + 3x^2 - 18x + 27$ **Distributive property**

$\qquad\qquad\qquad = x^4 - 6x^3 + 5x^3 + 9x^2 - 30x^2 + 3x^2$

$\qquad\qquad\qquad\quad + 45x - 18x + 27$

$\qquad\qquad\qquad = x^4 - x^3 - 18x^2 + 27x + 27$

                                          Combining like terms

                                                ◆

CAUTION $(x + 3)^3 \neq x^3 + 3^3$, for example $(4 + 2)^3 \neq 4^3 + 2^3$ ■

2.5 Exercises

In Exercises 1–46, find the product of the expressions.

1. $(3x^2)(5x^4)$
2. $(8x^3)(9x^5)$
3. $(4xy)(7x^2y^3)$
4. $(-5x^2y)(-8x^3y^2)$
5. $(-3x^3y^4)(6x^8y^9)$
6. $(5x^2y^7)(-7x^3y^9)$
7. $(3ab)^2(-5ab)$
8. $(6a^5b^4)(-3ab^3)^2$
9. $(4p^2q)^2(6pq^3)$
10. $(-7p^2q^2)(4p^2q^3)^2$
11. $(4a)(ab)(-a^2b)$
12. $(-3ab)(-4a^2b)(ab^2)$
13. $8a(a + 4)$
14. $5a(a^2 - 6)$
15. $5p^2(p^2 + 7p)$
16. $-6q^3(q^2 - 5q)$
17. $3ab(2a^2b - 5a)$
18. $-7ab(3abc + 8a^2)$
19. $(a + b)(a + b)$
20. $(x + 9)(x + 9)$
21. $(2x + 5)(3x + 7)$
22. $(4y + 8)(6y + 13)$
23. $(a + b)(a - b)$
24. $(x - 5)(x + 5)$
25. $(2x - 3)(2x + 3)$
26. $(4x + 8a)(4x - 8a)$
27. $(3x - 4y)(2x + 7y)$
28. $(3xy + 7z)(5xy - 8z)$
29. $(5a^2 + 2c)(7a^2b + 3c^2)$
30. $(3ab - 4c)(5a^2b^2 - 6c^2)$
31. $(3x + 4)(x^2 - 2x + 5)$
32. $(2x - 1)(x^2 - 3x + 7)$
33. $(2a + 3)(2a^2 + 7a + 8)$
34. $(5a - 4)(3a^2 + 11a - 9)$
35. $(2x^2 - 4x)(2x^2 + 5x - 1)$
36. $(3x^4 - 7x^2)(2x^2 - 8x + 3)$
37. $(2x + 3y)^2$
38. $(5x + 6y)^2$
39. $2a(3a - 7b)^2$
40. $2p(p - qx^2)^2$
41. $(x + 3)(x - 4)(x + 5)$
42. $(3 + x)(4 - x)(x + 5)$
43. $(2a - 3b)^3$
44. $3a(2a - 3b)^3$
45. $5x(2x + 1)^2(x - 3)$
46. $[(2x - 4)^2(x - 3)]^2$

In Exercises 47–49, find the indicated products and simplify.

47. In working with equations of uniformly accelerated motion, we arrive at the following expression for distance.

$$\frac{\frac{1}{2}(v + v_0)(v - v_0)}{a}$$

48. In a problem dealing with internal energy and the specific heat of a gas we arrive at the following expression. (The subscripts indicate different temperatures.)

$$\frac{3}{4}R(T_2 - T_1) + R(T_2 - T_1)$$

49. In determining the value of sales in a particular business the expression $(500 + 75n)(8 - 2n)$ is found.

50. a. By arbitrarily letting $x = 3$ and $y = 4$ show that $x^2 + y^2 \neq (x + y)^2$.
    b. Choose a different pair of numbers for x and y to show that $x^2 + y^2 \neq (x + y)^2$.
    c. Algebraically show that $x^2 + y^2 \neq (x + y)^2$.

51. a. By arbitrarily letting $x = 3$ and $y = 4$ show that $x^3 + y^3 \neq (x + y)^3$.
    b. Choose a different pair of numbers for x and y to show that $x^3 + y^3 \neq (x + y)^3$.
    c. Algebraically show that $x^3 + y^3 \neq (x + y)^3$.

2.6

Division of Algebraic Expressions

The expression $a^3 \div a^2$ can be expressed as the fraction $\dfrac{a^3}{a^2}$. The quotient can be found by applying the quotient rule for exponents discussed in Section 2.2. For example,

$$\overset{\text{dividend}\quad\text{quotient}}{\frac{a^3}{a^2}} = a^{3-2} = a^1 = a$$

$$\text{divisor}$$

EXAMPLE 1 **Quotients of Algebraic Expressions**

Find the quotient in each of the following.

a. $6x^4 \div 2x$ **b.** $15x^7y^5z^4 \div 3x^4y^2z^3$
c. $(4a^3 - 10a^2 + 2b) \div 2b$ **d.** $(8x^2y^3 - 24x^3y^2) \div 4xy^2$

Solution

a. $6x^4 \div 2x = \dfrac{6x^4}{2x}$

$$= \frac{6}{2} \cdot \frac{x^4}{x}$$

$$= 3x^{4-1}$$

$$= 3x^3$$

b. $15x^7y^5z^4 \div 3x^4y^2z^3 = \dfrac{15x^7y^5z^4}{3x^4y^2z^3}$

$$= \frac{15}{3} \cdot \frac{x^7}{x^4} \cdot \frac{y^5}{y^2} \cdot \frac{z^4}{z^3}$$

$$= 5 \cdot x^{7-4} \cdot y^{5-2} \cdot z^{4-3}$$

$$= 5x^3y^3z$$

c. $(4a^3 - 10a^2 + 2b) \div 2b = \dfrac{4a^3 - 10a^2 + 2b}{2b}$

Recall from arithmetic that we can find the sum or difference of two fractions when the denominators are alike.

$$\frac{1}{3} + \frac{4}{3} = \frac{1+4}{3} = \frac{5}{3}$$

Using this concept in reverse we can write

$$\frac{4a^3 - 10a^2 + 2b}{2b} = \frac{4a^3}{2b} - \frac{10a^2}{2b} + \frac{2b}{2b}$$

$$= \frac{2a^3}{b} - \frac{5a^2}{b} + 1$$

d. $(8x^2y^3 - 24x^3y^2) \div 4xy^2 = \dfrac{8x^2y^3 - 24x^3y^2}{4xy^2}$

$$= \frac{8x^2y^3}{4xy^2} - \frac{24x^3y^2}{4xy^2}$$

$$= 2x^{2-1}y^{3-2} - 6x^{3-1}y^{2-2}$$

$$= 2xy - 6x^2 \quad \blacklozenge$$

The algebraic expressions $4a^3 - 10a^2 + 6a$, $8x^2y^3 - 24x^3y^2$, and $x + 1$ are called *polynomial expressions* or *polynomials*. **Polynomials** are algebraic expressions whose coefficients are real numbers and in which the exponents of all the variables are positive integers. The expressions $3x^{\frac{1}{2}} + x^2$ and $\sqrt{-9x^{\frac{1}{4}}} + 6$ are not polynomial expressions since they have fractional exponents. Additionally, an expression with negative exponents is not a polynomial expression. The **degree of a term** in a polynomial expression is the sum of the exponents of the terms variables. The degree of $4a^3$ is 3, and the degree of $8x^2y^3$ is 5 $(2 + 3)$. The **degree of a polynomial expression** is the same as the degree of its highest degree term. Thus, for $4a^3 - 10a^2 + 6a + 8$, the degree is 3, and for $5x^4 - 24x^3$, the degree is 4. The constant term, 8, has a degree of Ø.

The process of dividing two polynomials is very much the same as the procedure used to find the quotient of two real numbers. Let us review by considering the problem 147 divided by 4.

$$
\begin{array}{r}
36 \\
4\overline{)147} \\
\end{array}
$$

$4 \times 3 \longrightarrow$	12
difference $\rightarrow$	27 $\leftarrow$ **bring down 7**
$4 \times 6 \longrightarrow$	24
difference $\rightarrow$	3 $\leftarrow$ **remainder**

Dividend = Divisor × Quotient + Remainder

Where $0 \le$ Remainder $<$ Divisor

$147 = 4 \times 36 + 3$

The result can also be expressed as $\dfrac{147}{4} = 36 + \dfrac{3}{4}$.

If the terms of both the divisor or the dividend are not in descending order (from highest degree to lowest degree) or in ascending order (from lowest degree to highest degree), rearrange them so that the terms of both are in the same order. Without proper ordering of the terms, it is difficult to perform the division.

EXAMPLE 2 **Dividing Two Polynomials**

Divide $x^2 + 5x + 6$ by $x + 2$

Solution   In order to perform the division of two polynomials, the degree of the divisor must be less than the degree of the dividend. In this example, the degree of the divisor is one, and the degree of the dividend is two.

Now we ask what should we multiply the divisor $(x + 2)$ by so that the first term of the product is x^2. In this case we multiply x by x to obtain x^2, the first term of the dividend. Thus, x is the first term of the quotient and is written above the corresponding term, in this case x^2.

$$x + 2 \overline{) x^2 + 5x + 6}$$

Now multiply the divisor, $x + 2$, by x and write the product $x^2 + 2x$

$$
\begin{array}{r}
x \phantom{{}+5x+6} \\
x + 2 \overline{) x^2 + 5x + 6} \\
\underline{x^2 + 2x} \phantom{{}+6}
\end{array}
$$

under the appropriate like terms of the dividend and then subtract (remember to change the signs of both x^2 and $2x$). As part of the subtraction bring down the 6. Since $3x$ is the same degree as the divisor, it is necessary to repeat the procedure.

$$
\begin{array}{r}
x \ + 3 \\
x + 2 \overline{) x^2 + 5x + 6}
\end{array}
$$

product $(x + 2) \cdot x \longrightarrow x^2 + 2x$

difference $\longrightarrow 3x + 6 \longleftarrow$ **bring down 6**

product $(x + 2) \cdot 3 \longrightarrow 3x + 6$

difference $\longrightarrow 0 \longleftarrow$ **remainder**

What multiplied by x will give a product of $3x$? The answer is 3. Multiply the divisor by 3 and the product is $3x + 6$. Place the product under the appropriate terms and subtract. The answer is zero and the remainder is zero. ◆

The quotient, $x + 3$, in Example 2 is a polynomial. When two polynomials are divided, the quotient is a polynomial. What happens when the quotient has a remainder? If a remainder exists, it will also be a polynomial as illustrated in the next example.

EXAMPLE 3 **Division of Polynomials with a Remainder**

Divide $2x^3 - 4x^2 + 5$ by $x^2 + x - 1$

Solution   Note that there is no first degree term in the dividend. In setting up the problem we must include it with a coefficient of zero.

$$
\begin{array}{r}
2x \;-\; 6 \\
x^2 + x - 1 \overline{)\,2x^3 - 4x^2 + 0x + 5\,} \\
\end{array}
$$

product $(x^2 + x - 1) \cdot 2x \longrightarrow \quad 2x^3 + 2x^2 - 2x \quad \downarrow$

difference $\longrightarrow \qquad\qquad\qquad -6x^2 + 2x + 5$

product $(x^2 + x - 1) \cdot (-6) \rightarrow \qquad -6x^2 - 6x + 6$

difference $\longrightarrow \qquad\qquad\qquad\qquad 8x - 1 \;\leftarrow$ remainder

Thus

$$\frac{2x^3 - 4x^2 + 5}{x^2 + x - 1} = 2x - 6 + \frac{8x - 1}{x^2 + x - 1}.$$

Check: dividend =   quotient × divisor   + remainder

$$2x^3 - 4x^2 + 5 = (2x - 6)(x^2 + x - 1) + (8x - 1) \quad \blacklozenge$$

2.6 Exercises

In Exercises 1–24, perform the division.

1. $2a^2 \div 3a^2$

2. $32b^3 \div 8b^2$

3. $-42a^3b^4 \div 6a^2b$

4. $-72c^2d^3 \div -9cd^2$

5. $-36b^{14} \div -9b^3$

6. $-52p^2q^8 \div (-13pq^5)$

7. $(4xy - 8xy^2) \div 2xy$

8. $(51x^2y - 34x^3y) \div (-17x^2)$

9. $(6\pi rh + 18\pi r^2h) \div 3\pi rh$

10. $(-26m^2x^2 + 39mx^2) \div (-13mx)$

11. $(35x^2y^2 - 28x^2y + 14xy^2) \div (-7xy)$

12. $(121a^3b^2 - 55a^2b^3 - 22a^3b^3) \div 11a^2b^2$

13. $(m^2 + 7m + 12) \div (m + 3)$

14. $(6p^2 + 7p - 3) \div (3p - 1)$

15. $(9a^2 + 12ab + 4b^2) \div (3a + 2b)$

16. $(49a^2 - 25b^2) \div (7a - 5b)$

17. $(8x^2 - 2xy - 21y^2) \div (2x + 3y)$

18. $(27a^3 - 125) \div (3a - 5)$

19. $(64x^3 + 1) \div (4x + 1)$

20. $(x^3 - 18x - 35) \div (x - 5)$

21. $(y^3 - 25y + 66) \div (y + 6)$

22. $(z^4 - 5z^2 + 4) \div (z^2 - 1)$

23. $(q^4 - q^2 + 16) \div (q^2 - 3q + 4)$

24. $(b^5 - b^4 - b^3 + b^2 + 16b - 16) \div (b^2 + 3b + 4)$

In Exercises 25–38, perform the division and write all remainders in rational form.

25. $(a^2 - 7a + 2) \div (a - 3)$

26. $(b^2 + 1) \div (b - 1)$

27. $(6x^2 + 7x + 5) \div (3x + 5)$

28. $(16x^2 - 1) \div (2x - 1)$

29. $(t^3 - 4t^2 + 7t + 4) \div (t + 2)$

30. $(8r^2 - r - 19r^3 + 15r^4 - 1) \div (5r^2 - 3r - 1)$

31. $(x^3 - 7x + 5x^2 + 2) \div (x^2 - x - 2)$

32. $(2x^3 - 9x^2y + 7xy^2 + 6y^3) \div (x - 3y)$

33. $(16x^4 - 96x^3y + 216x^2y^2 - 216xy^3 + 81y^4) \div (2x - 3y)$

34. $(r^5 - 13r^2 + 17r - 2r^4 + 5) \div (3r + 1 + r^3 - 4r^2)$

35. $(8.1\alpha^2\beta^2\gamma + 7.2\alpha^3\beta^2\gamma^2 - 3.6\alpha^4\beta^3\gamma^3) \div (0.09\alpha^2\beta^2\gamma)$

36. $[8(\theta + \phi)^2 - 16(\theta + \phi)^4 + 12(\theta + \phi)^6] \div [4(\theta + \phi)]$

37. $[8\pi(EI + P)^4 - 32\pi(EI + P)^2 + 96\pi(EI + P)] \div [16\pi(EI + P)^2]$

38. $\left(\dfrac{1}{8}\theta^3 - \dfrac{9}{4}\theta^2\phi + \dfrac{27}{2}\theta\phi^2 - 27\phi^3\right) \div \left(\dfrac{1}{2}\theta - \phi\right)$

39. The average velocity of a certain particle in feet per second is given by

$$\bar{v} = \frac{3t^3 - 5t^2 - 7t + 6}{3t - 2}$$

where t is in seconds. Perform the indicated division and find a simplified expression.

40. The displacement, s, of a particle was found to vary according to

$$s = \frac{2t^3 - 11t^2 + 21t - 18}{t - 3}$$

where t is in seconds. Perform the indicated division and find a simplified expression for s.

2.7

Factoring

In this section we will do the reverse of multiplying algebraic expressions, an operation called factoring.

The expression $3x^2 + 6y^2$ can be written as $3x^2 + 3 \cdot 2 \cdot y^2$. Now recall the distributive property, $a(b + c) = ab + ac$. By applying the distributive property in reverse, we see that the 3 corresponds to the a,

$$ab + ac = a(b + c)$$
$$3x^2 + 6y^2 = \boxed{3}x^2 + \boxed{3} \cdot 2 \cdot y^2 = \boxed{3}(x^2 + 2y^2)$$

and we can write the expression in the form on the right. The 3 is the largest factor that is common on both terms. Thus, 3 is called the **Greatest Common Factor** (or **GCF**). The product $3(x^2 + 2y^2)$ is the prime factorization of $3x^2 + 6y^2$. **Prime factorization** or **product of the prime factors** means that there is no other number, letter, or symbol, except 1 that is common to all the terms of the expression.

EXAMPLE 1 **Factoring Common Terms**

Express each of the following as a product of its prime factors.

a. $-4xy + 8y^2$ **b.** $7a^2b - 14ab^2 + 21a^2b^2$

Solution

a. The GCF to both terms is $4y$. Thus,

$$-4xy + 8y^2 = \boxed{-4y} \cdot x + \boxed{4y} \cdot 2y$$
$$= 4y(-x + 2y) \qquad \text{4y is the GCF}$$

b. In examining the expression $7a^2b - 14ab^2 + 21ab^2b^2$ we see that each term contains factors of 7, a, and b and that the GCF to all three terms is $7ab$. Thus we can write

$$7a^2b - 14ab^2 + 21a^2b^2 = \boxed{7ab} \cdot a - \boxed{7ab} \cdot 2b + \boxed{7ab} \cdot 3ab$$
$$= 7ab(a - 2b + 3ab) \quad \blacklozenge$$

In examining the terms of the expression $3x^2y + 6x + 5xy + 10$ we see that there is no common factor in all four terms. It is still useful to factor the expression into still smaller groups. The two terms $3x^2y$ and $6x$ have a common factor of $3x$, and the terms $5xy$ and 10 have a common factor of 5. Thus

$$3x^2y + 6x + 5xy + 10 = 3x(xy + 2) + 5(xy + 2).$$

Note that by pairing up the terms and factoring a common term from each of the pairs of terms, we have uncovered an expression with two terms each with a common factor of $xy + 2$. Factor again.

$$3x(xy + 2) + 5(xy + 2) = (xy + 2)(3x + 5).$$

Thus

$$3x^2y + 6x + 5xy + 10 = (xy + 2)(3x + 5).$$

This method of factoring is called **factoring by grouping terms,** and it will be helpful in factoring trinomials of the form $ax^2 + bx + c$. In Example 3 of Section 2.5 we found the product of $(x + 3)$ and $(x - 5)$ to be $x^2 - 2x - 15$. The second to last line of that example is the key to reversing the process of factoring $x^2 - 2x - 15$. In the expression $x^2 + 3x - 5x - 15$, the sum of the two terms $3x$ and $-5x$ is $-2x$. The product of the numerical coefficients of $3x$ and $-5x$ is -15. Note that by applying factoring by grouping terms we have the following result.

$$x^2 + 3x - 5x - 15 = x(x + 3) + -5(x + 3)$$
$$= (x + 3)(x - 5).$$

In general, the trinomial $ax^2 + bx + c$ can be factored if we can find a pair of integers such that their product is ac and their sum is b.

This technique is illustrated in Example 2.

EXAMPLE 2   Factoring by Grouping Terms

Factor completely $2x^2 - 9x + 9$.

Solution   We cannot factor the expression $2x^2 - 9x + 9$ in this form. However, it is possible to factor the expression using factoring by grouping and common-term factoring. Begin by searching for two factors whose sum is -9 and whose product is 18 ($2 \cdot 9$). Select -3 and -6; the sum is -9, and the product is 18. Thus we can rewrite

$$2x^2 - 9x + 9 \text{ as } 2x^2 - 6x - 3x + 9$$

and factor by grouping.

$$2x^2 - 6x - 3x + 9 = 2x(x - 3) + (-3)(x - 3).$$

Now we have like terms inside the parentheses and can apply common-term factoring:

$$2x^2 - 6x - 3x + 9 = (x - 3)[2x + (-3)]$$
$$= (x - 3)(2x - 3). \quad \blacklozenge$$

A summary of the technique for factoring a trinomial follows.

METHOD FOR FACTORING A TRINOMIAL

Assume that we want to factor $ax^2 + bx + c$ (where a, b, and c, are integers).

1. Multiply a and c (remember the signs).
2. Write down b (remember the sign).
3. Search for a pair of numbers whose product is ac and whose sum is b.
4. Write the bx term as the sum of two terms, the sum of whose coefficients is b.
5. Factor by grouping terms (factoring by grouping).

EXAMPLE 3   **Factoring a Trinomial**

Factor the quadratic trinomial expressions.

a. $x^2 + 8x + 15$ **b.** $x^2 - x - 20$
c. $3x^2 + 5x - 12$ **d.** $x^2 + 5x + 7$

Solution

a. To factor the expression $x^2 + 8x + 15$ we need to find two numbers whose sum is 8 and whose product is 15. Also, since 8 and 15 are both positive, the numbers we are looking for are both positive.

Factors of 15	Sum of factors
$1 \cdot 15$	$1 + 15 = 6$
$3 \cdot 5$	$3 + 5 = 8$

Since $3 \cdot 5 = 15$ and $3 + 5 = 8$ we can write:

$$x^2 + 8x + 15 = x^2 + 3x + 5x + 15$$
$$= x(x + 3) + 5(x + 3) \quad \text{Factoring by grouping terms}$$
$$= (x + 3)(x + 5) \quad \text{Common-term factoring}$$

b. In the expression $x^2 - x - 20$ the constant term is -20. Therefore, the numbers we are looking for must have opposite signs. In addition, since b is negative (-1), the larger number must be negative.

Factors of -20	Sum of factors
$-20 \cdot 1$	$-20 + 1 = -19$
$-10 \cdot 2$	$-10 + 2 = -8$
$-5 \cdot 4$	$-5 + 4 = -1$

Thus, the numbers we are looking for are -5 and 4.

$$x^2 - x - 20 = x^2 \boxed{-5x + 4x} - 20$$
$$= x(x - 5) + 4(x - 5) \quad \text{**Factoring by grouping terms**}$$
$$= (x + \boxed{4})(x - \boxed{5}) \quad \text{**Common-term factoring**}$$

c. For the expression $3x^2 + 5x - 12$ we are looking for two numbers whose product is -36 and whose sum is 5. Since the product is negative, the numbers must have opposite signs, and since b is positive (5), the larger number must be positive.

Factors of -36	Sum of factors
$36 \cdot (-1)$	$36 + (-1) = 35$
$18 \cdot (-2)$	$18 + (-2) = 16$
$12 \cdot (-3)$	$12 + (-3) = 9$
$9 \cdot -4$	$9 + (-4) = 5$

The numbers are 9 and -4, so we write the bx term as the sum $9x - 4x$.

$$3x^2 + 5x - 12 = 3x^2 \boxed{+ 9x} - \boxed{4x} - 12$$
$$= 3x(x + 3) - 4(x + 3) \quad \text{**Factoring by grouping terms**}$$
$$= (x + 3)(3x - 4) \quad \text{**Common-term factoring**}$$

d. For the expression $x^2 + 5x + 7$ we are looking for two numbers whose sum is 5 and whose product is 7. Since 5 and 7 are both positive, the factors must be positive.

Factors of 7	Sum of factors
$1 \cdot 7$	$1 + 7 = 8$

There are no other factors of 7, and the sum of these factors is not equal to 5. Therefore, the expression $x^2 + 5x + 7$ is **prime** (it cannot be factored using the integers). In a later chapter you will be shown how this expression can be factored using real numbers. ◆

The quadratic expression $4x^2 + 20x + 25$ is a special form known as a **perfect square trinomial.** An expression of this type can be identified by observing that $\frac{1}{2}(20) = \sqrt{4 \cdot 25}$ or that $\frac{1}{2}b = \pm\sqrt{a \cdot c}$. In factoring this trinomial the two numbers we are looking for, whose sum is b, will both be 10. Thus,

$$4x^2 + 20x + 25 = 4x^2 + 10x + 10x + 25$$
$$= 2x(2x + 5) + 5(2x + 5) \quad \text{**Factoring by grouping terms**}$$
$$= (2x + 5)(2x + 5) \quad \text{**Common-term factoring**}$$
$$= (2x + 5)^2$$

Note that the coefficient of the x term is the square root of a (4) and that the constant term (5) is the square root of c (25). Thus, when we recognize that a quadratic expression is a perfect square, in order to factor we find the square root of a and the square root of c and write the answer, $(\sqrt{ax} \pm \sqrt{c})^2 = (2x + 5)^2$, where $\pm$ is determined by the sign of b.

EXAMPLE 4 Factoring Perfect Square Trinomials

Factor the quadratic trinomial expressions.

a. $x^2 - 6x + 9$ **b.** $16a^2 + 56a + 49$

Solution

a. To factor the expression $x^2 - 6x + 9$ we are looking for two numbers whose product is 9 and whose sum is -6. Since $\frac{1}{2}(-6) = -3$ and $-\sqrt{9} = -3$, the numbers we are looking for are -3 and -3. Thus, the expression is a perfect-square trinomial and the factors are $(x - 3)$, $(x - 3)$, and the solution is $(x - 3)^2$.

$$x^2 - 6x + 9 = (x - 3)^2.$$

b. To factor $16a^2 + 56a + 49$ we need to look for two numbers whose product is 784 and whose sum is 56. Since $\frac{1}{2}(56) = 28$ and $\sqrt{784} = 28$, the two numbers we are looking for are 28 and 28. The expression is a perfect square trinomial.

$$16a^2 + 56a + 49 = (4a + 7)^2. \quad \blacklozenge$$

A quadratic expression of the form $x^2 - a^2$ is known as the **difference of two squares**. It is factored as follows.

$$x^2 - a^2 = (x - a)(x + a).$$

CAUTION $a^2 + b^2$ is prime. $a^2 + b^2 \neq (a + b)(a + b)$. The product of $(a + b)(a + b) = a^2 + 2ab + b^2$. $\blacksquare$

EXAMPLE 5 Factoring the Difference of Two Squares

Factor the quadratic expressions.

a. $x^2 - 25$ **b.** $25x^2 - 64y^2$

Solution

a. $x^2 - 25 = x^2 - 5^2$ **The difference of two squares**

Thus, $x^2 - 25 = (x - 5)(x + 5)$

b. $25x^2 - 64y^2 = (5x)^2 - (8y)^2$ **The difference of two squares**

Thus, $25x^2 - 64y^2 = (5x - 8y)(5x + 8y)$ $\quad \blacklozenge$

A cubic expression of the form $x^3 - a^3$ is known as the **difference of two cubes.** It is factored as follows.

$$x^3 - a^3 = (x - a)(x^2 + ax + a^2).$$

Now show that the product of $(x - a)(x^2 + ax + a^2) = x^3 - a^3$.

A cubic expression of the form $x^3 + a^3$ is known as the **sum of two cubes.** It is factored as follows.

$$x^3 + a^3 = (x + a)(x^2 - ax + a^2)$$

EXAMPLE 6   **Factoring the Difference of Two Cubes**

Express $8y^3 - 27$ as a product of prime factors.

Solution   The expression $8y^3 - 27$ can be written as $(2y)^3 - (3)^3$. In this form we can identify the expression as the difference of two cubes.

$$\begin{aligned} 8y^3 - 27 &= (2y)^3 - (3)^3 \\ &= (2y - 3)[(2y)^2 + (2y)(3) + (3)^2] \\ &= (2y - 3)(4y^2 + 6y + 9) \quad \blacklozenge \end{aligned}$$

One general remainder: When trying to factor a polynomial, always look for the greatest common factor first; then try other forms of factoring.

2.7 Exercises

In Exercises 1–70, express each expression as a product of prime factors.

1. $3a + 12b$
2. $6a - 15b$
3. $7r + 14s$
4. $9r - 18s$
5. $2x - 12xy$
6. $5xy + 20x$
7. $5xy - 15x^2z + 20xz$
8. $4x^2y - 8xy^2 + 14x^2y^2$
9. $3ab + 12bc + 5ad + 20cd$
10. $2xy + 14yz + 7wx + 49wz$
11. $15rs - 25rt - 18ks + 30kt$
12. $38RI_1 - 19RI_2 - 34II_1 + 17II_2$
13. $15R_1I_1 - 30R_1I_2 - 35R_2I_1 + 70R_2I_2$
14. $28AB - 56AC + 26BD - 52CD$
15. $6A^2BD - 14ADC - 15A^2B^2 + 35ABC$
16. $P^2QTU - P^2QVW + PRSTU - PRSVW$
17. $\frac{1}{3}AB - \frac{2}{3}AC + \frac{2}{5}PB - \frac{4}{5}PC$
18. $\frac{3}{7}\pi r_1^2 - \frac{15}{49}\pi r_1^2 h - \frac{2}{9}\pi r_2^2 + \frac{10}{63}\pi r_2^2 h$
19. $0.06x^2y + 0.08xy^2 + 0.09xyz + 0.12y^2z$
20. $0.15x^2y^2 - 0.21wxyz + 0.35uvxy - 0.49uvwz$
21. $x^2 + 5x + 4$
22. $x^2 - 3x - 4$
23. $3a^2 + 6a + 3$
24. $5b^2 - 35b + 60$
25. $t^2 + 9t + 14$
26. $r^2 + 8r + 15$
27. $r^2 - 49$
28. $25 - a^2$
29. $-2 - 5b + 3b^2$
30. $6z^2 - 11z - 10$
31. $p^2 + 3p - 6$
32. $2p^2 + 2p - 4$
33. $11z + 5z^2 - 12$
34. $s^2 + s + 1$
35. $6s^2 + 16s - 6$
36. $5x^2 + 16xy + 3y^2$
37. $17p^2 + 6pq - 11q^2$
38. $3r^2 + 10rs + 3s^2$
39. $4w^2 - 20w + 25$
40. $7w^2 - 25w - 12$

41. $5x^2 + 15x - 28$ **42.** $8x^2 + 18x + 9$ **43.** $3z^2 + 10z + 3$ **44.** $2w^2 + 7w - 15$

45. $7R^2 - 16R + 4$ **46.** $x^2 + 6x + 9$ **47.** $w^2 + 14w + 49$ **48.** $4p^2 - 20p + 25$

49. $9p^2 + 42p + 49$ **50.** $16t^2 + 24t + 9$ **51.** $4s^2 - 12s + 9$ **52.** $12s^2 - 60s + 75$

53. $64m^2 - 80mn + 25n^2$ **54.** $49x^2 + 112xy + 64y^2$ **55.** $81w^2 - 72wz + 16z^2$

56. $121a^2 + 154ab + 49b^2$ **57.** $x^2 - 9$ **58.** $a^2 - 36$

59. $81r^2 - 16$ **60.** $5 - 5a^2$ **61.** $x^2y^2 - 25$

62. $100a^2b^2 - 49x^2y^2$ **63.** $(a + b)^2 - d^2$ **64.** $(z^2 + 16)^2 - 25y^2$

65. $p^2(q + 3) - 16q - 48$ **66.** $3x^2 - 108$ **67.** $x^3 - 64$

68. $27x^3 + 125$ **69.** $-16x^3 - 54y^3$ **70.** $x^4 + 8xy^3$

71. The cross sectional area of a pipe with outside diameter D and inside diameter d is given by the expression $\dfrac{\pi D^2}{4} - \dfrac{\pi d^2}{4}$. Write the expression as a product of its prime factors.

72. The surface area of a cylinder is given by the expression $T = \pi rh + 2\pi r^2$. Write the expression as a product of its prime factors.

73. If the factors of a polynomial are $(2x - 5y)$ and $(3x + 6y)$, determine the polynomial.

In Exercises 74–76 determine a factored expression that represents the shaded area.

74.

75.

76.

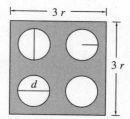

77. The expression for the difference in the volume of two spheres with radii of r_1 and r_2 is given by the expression $\dfrac{4}{3}\pi r_2{}^3 - \dfrac{4}{3}\pi r_1{}^3$. Factor the expression.

78. Given a second degree polynomial expression, explain the steps you would take to factor the expression.

2.8

Multiplication and Division of Fractions

To find the product of arithmetic fractions, we multiply the numerators and the denominators and then reduce the resulting fraction to lowest terms. We can use the same procedure to multiply algebraic fractions.

> **EXAMPLE 1** **Multiplying Fractions**
>
> Determine the product of $\dfrac{5}{12}$ and $\dfrac{8}{15}$.

Solution

$$\frac{5}{12} \cdot \frac{8}{15} = \frac{5 \cdot 8}{12 \cdot 15}$$

$$= \frac{40}{180}$$

$$= \frac{20 \cdot 2}{20 \cdot 9}$$

$$= \frac{2}{9} \qquad \text{\textbf{Reducing to lowest terms}}$$

This method will always work. However, a great deal of time may be spent reducing the product fraction to lowest terms. A more practical method may be to factor the numerators and denominators of the fractions, divide both numerator and denominator by the common factors (reduce), and then multiply.

$$\frac{5}{12} \cdot \frac{8}{15} = \frac{\overset{1}{\cancel{5}}}{\underset{1}{\cancel{2}} \cdot \underset{1}{\cancel{2}} \cdot 3} \cdot \frac{\overset{1}{\cancel{2}} \cdot \overset{1}{\cancel{2}} \cdot 2}{3 \cdot \underset{1}{\cancel{5}}}$$

$$= \frac{2}{9} \quad \blacklozenge$$

The second technique, reducing first and then multiplying, is by far the best method for multiplying algebraic fractions. The product of algebraic fractions should be expressed in simplest form. By simplest form we mean that the greatest common divisor of the numerator and denominator is 1. For example, in the fraction $\frac{5}{7}$ the greatest common divisor of 5 and 7 is 1. Therefore $\frac{5}{7}$ is in its simplest form.

EXAMPLE 2   **Multiplying Algebraic Fractions**

Reduce the fractions, multiply and express your answer in simplest form.

a. $\dfrac{2x + 6}{2x + 8} \cdot \dfrac{6x + 12}{3x + 9}$ **b.** $\dfrac{2x^2 - 7x - 15}{x^2 - 25} \cdot \dfrac{x^2 - 2x - 35}{7 - x}$

Solution

a. $\dfrac{2x + 6}{2x + 8} \cdot \dfrac{6x + 12}{3x + 9} = \dfrac{\overset{1}{\cancel{2}}(\overset{1}{\cancel{x + 3}})}{\underset{1}{\cancel{2}}(x + 4)} \dfrac{\overset{2}{\cancel{6}}(x + 2)}{\underset{1}{\cancel{3}}(\underset{1}{\cancel{x + 3}})}$ **Common-term factoring and reducing**

$$= \dfrac{2(x + 2)}{x + 4} \qquad \text{\textbf{You can leave the answer in this form,}}$$

$$= \dfrac{2x + 4}{x + 4} \qquad \text{\textbf{or you can multiply it out like this.}}$$

b. $\dfrac{2x^2 - 7x - 15}{x^2 - 25} \cdot \dfrac{x^2 - 2x - 35}{7 - x}$

$= \dfrac{(2x + 3)\overset{1}{\cancel{(x - 5)}}}{\underset{1}{\cancel{(x - 5)}}\underset{1}{\cancel{(x + 5)}}} \cdot \dfrac{(x - 7)\overset{1}{\cancel{(x + 5)}}}{7 - x}$ **Factoring and reducing**

$= \dfrac{(2x + 3)(x - 7)}{7 - x}$

Note that if $x - 7$ were in the denominator rather than $7 - x$, we could reduce. Recall that $(a - b) = -(b - a)$ or $(a - b) = (-1)(b - a)$. Thus:

$$\frac{(2x + 3)(x - 7)}{7 - x} = \frac{(2x + 3)(-1)\overset{1}{\cancel{(7 - x)}}}{\underset{1}{\cancel{7 - x}}}$$

$$= -(2x + 3) \quad \text{or} \quad -2x - 3 \quad \blacklozenge$$

CAUTION A common error made in reducing fractions when the operations of addition and/or subtraction is involved is:

$$\frac{2 + 8}{2} = \frac{\overset{1}{\cancel{2}} + 8}{\underset{1}{\cancel{2}}} = 9.$$

When we remember the correct order of operations, we can see that the correct result is

$$\frac{2 + 8}{2} = \frac{10}{2} = 5, \quad \text{not 9.}$$

By factoring the common terms, the error could have been avoided.

$$\frac{2 + 8}{2} = \frac{\overset{1}{\cancel{2}}(1 + 4)}{\underset{1}{\cancel{2}}} = 5 \quad \blacksquare$$

DEFINITION 2.5 Equality of Fractions

Let $\dfrac{a}{b}$ be any fraction and n a nonzero real number. Then

(1) $\dfrac{a}{b} = \dfrac{an}{bn} = \dfrac{na}{nb}$ and   (2) $\dfrac{a}{b} = \dfrac{c}{b}$ if and only if $a = c$

When we learned to divide fractions in arithmetic, we were told to "invert the divisor and multiply." Why does this work? The reason it works is illustrated in Example 3.

EXAMPLE 3   **Illustrating Why We Invert the Divisor**

Divide $\frac{2}{3}$ by $\frac{5}{7}$.

Solution

$$\frac{2}{3} \div \frac{5}{7} = \frac{\dfrac{2}{3}}{\dfrac{5}{7}}$$

$$= \frac{\dfrac{2}{3} \cdot \dfrac{7}{5}}{\dfrac{5}{7} \cdot \dfrac{7}{5}} \qquad \text{Definition 2.5 and } \frac{7}{5} \text{ is reciprocal of } \frac{5}{7}.$$

$$= \frac{\dfrac{2}{3} \cdot \dfrac{7}{5}}{1} \qquad \text{Product of a fraction and its reciprocal is 1.}$$

$$= \frac{14}{15} \quad \blacklozenge$$

The reciprocal of the algebraic fraction $\dfrac{x+2}{x+3}$ is the fraction $\dfrac{x+3}{x+2}$. The product of an algebraic fraction and its reciprocal is 1, the same as that of an arithmetic fraction and its reciprocal.

EXAMPLE 4   **Division of Algebraic Fractions**

Divide $\dfrac{x+4}{x+2}$ by $\dfrac{x+5}{3x}$.

Solution

$$\frac{x+4}{x+2} \div \frac{x+5}{3x} = \frac{x+4}{x+2} \cdot \frac{3x}{x+5} \qquad \text{Invert the divisor}$$

$$= \frac{(x+4)(3x)}{(x+2)(x+5)} \quad \text{or} \quad \frac{3x^2+12x}{x^2+7x+10} \quad \blacklozenge$$

EXAMPLE 5   **Factoring and Division of Algebraic Fractions**

Find the quotient and express it in simplest form.

$$\frac{x^2-25}{x^2+8x+15} \div \frac{x^2-2x-15}{3x+9}$$

Solution

$$\frac{x^2 - 25}{x^2 + 8x + 15} \div \frac{x^2 - 2x - 15}{3x + 9}$$

$$= \frac{x^2 - 25}{x^2 + 8x + 15} \cdot \frac{3x + 9}{x^2 - 2x - 15} \qquad \text{\textbf{Invert the}} \atop \text{\textbf{divisor}}$$

$$= \frac{\overset{1}{\cancel{(x + 5)}}\overset{1}{\cancel{(x - 5)}}}{\underset{1}{\cancel{(x + 5)}}\underset{1}{\cancel{(x + 3)}}} \cdot \frac{3\overset{1}{\cancel{(x + 3)}}}{\underset{1}{\cancel{(x - 5)}}(x + 3)} \qquad \text{\textbf{Factor completely}} \atop \text{\textbf{and reduce}}$$

$$= \frac{3}{x + 3} \qquad \text{\textbf{Multiply}} \;\blacklozenge$$

EXAMPLE 6 **Multiplication and Division of Algebraic Fractions**

Perform the indicated operations and express the result in simplest form.

$$\frac{2y}{3y} \div \frac{7y}{6y} \cdot \frac{14y}{9y}$$

Solution   Using the rules for priority of operations, perform the division first.

$$\frac{2x}{3y} \div \frac{7x}{6y} \cdot \frac{14x}{9x} = \frac{2\overset{1}{\cancel{x}}}{3y} \cdot \frac{\overset{2}{\cancel{6y}}}{\underset{1}{\cancel{7x}}} \cdot \frac{\overset{2}{\cancel{14x}}}{9y} \qquad \text{\textbf{Invert the divisor}} \atop \text{\textbf{and reduce}}$$

$$= \frac{8x}{9y} \quad\blacklozenge$$

2.8 Exercises

In Exercises 1–15, determine the product and express the product in simplest form.

1. $\dfrac{2}{3} \cdot \dfrac{6}{7}$

2. $\dfrac{8}{9} \cdot \dfrac{15}{20}$

3. $\dfrac{5p}{6m} \cdot \dfrac{12m^2}{10}$

4. $\dfrac{xy^2}{9xz} \cdot \dfrac{3x}{(xy)^2}$

5. $\dfrac{x}{6} \cdot \dfrac{x + 3}{x - 2}$

6. $\dfrac{x + 3}{x + 4} \cdot \dfrac{x + 5}{x - 6}$

7. $\dfrac{2x + 6}{3x + 12} \cdot \dfrac{x + 4}{x + 3}$

8. $\dfrac{x + 5}{x + 7} \cdot \dfrac{x^2 + 7x}{xy + 5y}$

9. $\dfrac{a^2 - b^2}{9} \cdot \dfrac{12}{a + b}$

10. $\dfrac{a + 3}{p - 4} \cdot \dfrac{p^2 - 16}{a^2 - 9}$

11. $\dfrac{x^2 + 6x + 9}{3xyz} \cdot \dfrac{6xy}{x^2 - 9}$

12. $\dfrac{m^2 - 9n^2}{8pq} \cdot \dfrac{2p}{m^2 - 6mn + 9n^2}$

13. $\dfrac{a^2 - b^2}{a^2 - 2ab + b^2} \cdot \dfrac{a^2 - b^2}{a^2 + 2ab + b^2}$

14. $\dfrac{3m^2 - m - 2}{9m^2 + 12m + 4} \cdot \dfrac{3m^2 + 2m}{m^3 - m^2}$

15. $\dfrac{p^2 + p - 12}{p^2 + 8p + 16} \cdot \dfrac{3p^2 + 12p}{9 - p^2}$

In Exercises 16–30, determine the quotient and express the quotient in simplest form.

16. $\dfrac{3}{4} \div \dfrac{9}{16}$

17. $\dfrac{6}{7} \div \dfrac{15}{28}$

18. $\dfrac{3x}{5y} \div \dfrac{6x^2}{15y^3}$

19. $\dfrac{ab^2}{9ac} \div \dfrac{(ab)^2}{3a}$

20. $\dfrac{a}{6} \div \dfrac{a-2}{a+3}$

21. $\dfrac{a+3}{a+4} \div \dfrac{a+5}{a-6}$

22. $\dfrac{x+3}{x+4} \div \dfrac{xy+3y}{x^2+7x}$

23. $\dfrac{3x+9}{4x+16} \div \dfrac{x+3}{x+4}$

24. $\dfrac{p^2-q^2}{5} \div \dfrac{p-q}{15}$

25. $\dfrac{p^2-p-6}{p^2-9} \div \dfrac{p^2-4}{p+3}$

26. $\dfrac{m^2+7m-8}{m^2-64} \div \dfrac{m^2+2m+1}{8m-64}$

27. $\dfrac{x^2+2xy+y^2}{x^2-2xy-3y^2} \div \dfrac{x^2-y^2}{x-3y}$

28. $\dfrac{x^2+5x+6}{x^2+6x+9} \div \dfrac{x^2-9}{3-x}$

29. $\dfrac{6m^2-11m-10}{3m^2-m-2} \div \dfrac{2m^2+5m-25}{m^2+4m-5}$

30. $\dfrac{4x^2-1}{6x^2+5x-4} \div \dfrac{2x^2+x-1}{3x^2+7x+4}$

In Exercises 31–36, find the indicated result and express the result in simplest form.

31. $\dfrac{3x}{4y} \div \dfrac{2y}{3x} \cdot \dfrac{6x}{9y}$

32. $\dfrac{3x}{4y} \cdot \dfrac{2y}{3x} \div \dfrac{6x}{9y}$

33. $\dfrac{5x^2}{7y} \div \left(\dfrac{5x}{3z} \cdot \dfrac{2x^2}{21y}\right)$

34. $\dfrac{2a^2}{3b} \cdot \left(\dfrac{6a}{7b} \div \dfrac{12a^2}{35b^3}\right)$

35. $\dfrac{R^2-r^2}{r^2+Rr} \cdot \dfrac{R(R-r)}{(R-r)^2} \div \dfrac{R^2-3Rr+2r^2}{Rr-2r^2}$

36. $\dfrac{16I^4R^2-9}{4\left(I^2R+\frac{3}{4}\right)} \cdot \dfrac{I^4R^2-3I^2R-28}{2I^4R^2-32} \div \dfrac{8I^4R^2-62I^2R+42}{8I^2R-32}$

37. A weight attached to a wire places a longitudinal strees and strain on the wire. The ratio of the longitudinal stress $\left(\dfrac{F}{A}\right)$ and the longitudinal strain $\left(\dfrac{\Delta l}{l}\right)$ is called Young's modulus, Y. $Y = \dfrac{F/A}{\Delta l/l}$ Simplify the right side of the equation.

38. In a mass spectrometer, the radius of curvature a charged particle possesses depends upon its mass. Hence, by properly calibrating the instrument and measuring the radius, the mass of small particles can be determined. The equation used by the mass spectrometer is $R = \dfrac{my}{qB}$. The frequency of oscillation is given by the expression $f = \dfrac{v}{2\pi R}$. Substitute the first equation into the second and simplify.

2.9

Addition and Subtraction of Fractions

The general procedure for adding or subtracting algebraic fractions is the same as the one used to add or subtract arithmetic fractions. The operations of addition or subtraction can be performed only when the numerators are written over a common denominator. If the denominators are alike (common denominators), the operations can be performed as in Example 1.

EXAMPLE 1   **Addition of Fractions with Common Denominators**

Perform the indicated operations and write the answer in simplest form.

a. $\dfrac{2}{7} + \dfrac{3}{7}$ **b.** $\dfrac{x}{3yz} - \dfrac{5x}{3yz}$

Solution

a. $\dfrac{2}{7} + \dfrac{3}{7} = \dfrac{1}{7}(2 + 3)$ **Factoring the common term**

$\qquad = \dfrac{5}{7}$

b. $\dfrac{x}{3yz} - \dfrac{5x}{3yz} = \dfrac{1}{3yz}(x - 5x)$ **Factoring the common term**

$\qquad = \dfrac{-4x}{3yz}$ ◆

When fractions do not have common denominators, we must find a common denominator before we perform the addition or subtraction. Generally, it is to our advantage to find the **lowest common denominator** (lcd), which is the smallest number or algebraic expression that is divisible by all the denominators. For example, the lcd of the fractions $\dfrac{3}{2x^2}$, $\dfrac{5}{6xy^2}$, and $\dfrac{7}{4yz}$ is $12x^2y^2z$. What method is used to determine that $12x^2y^2$ is the lcd? The first step is to write each denominator as the product of its prime factors. The factors of each denominator are: $2x^2 = 2x^2$, $6xy^2 = 2 \cdot 3xy^2$, and $4yz = 2^2yz$. In these three denominators, the unrelated factors are 2, x, y, 3 and z. These factors to the highest power in any one of the denominators are 2^2, x^2, y^2, 3 and z. The product of these factors is $12x^2y^2z$, which is the lcd.

EXAMPLE 2   **Adding and Subtracting Algebraic Fractions**

Perform the indicated operations and write the results in simplest form.

a. $\dfrac{3}{2x^2} + \dfrac{5}{6xy^2} + \dfrac{7}{4yz}$ **b.** $5(x^2 - 1)^{-1} - 3(x^2 + 2x + 1)^{-1}$

Solution

a. In the previous discussion the lcd of the fractions $\dfrac{3}{2x^2}$, $\dfrac{5}{6xy^2}$, and $\dfrac{7}{4yz}$ was found to be $12x^2y^2z$. Now ask the following questions:

$$2x^2 \cdot \underline{\ \ ?\ \ } = 12x^2y^2z \qquad 2x^2 \cdot \underline{6y^2z} = 12x^2y^2z$$
$$6xy^2 \cdot \underline{\ \ ?\ \ } = 12x^2y^2z \qquad 6xy^2 \cdot \underline{2xz} = 12x^2y^2z$$
$$4yz \cdot \underline{\ \ ?\ \ } = 12x^2y^2z \qquad 4yz \cdot \underline{3x^2y} = 12x^2y^2z$$

Now multiply the numerator and denominator of each fraction by the expression that replaced the question mark. Recall that multiplying the numerator and denominator by the same number (other than zero) is the same as multiplying the fraction by 1 and will not change the value of the fraction.

$$\frac{3}{2x^2} \cdot \frac{6y^2z}{6y^2z} = \frac{18y^2z}{12x^2y^2z}$$

$$\frac{5}{6xy^2} \cdot \frac{2xz}{2xz} = \frac{10xz}{12x^2y^2z}$$

$$\frac{7}{4yz} \cdot \frac{3x^2y}{3x^2y} = \frac{21x^2y}{12x^2y^2z}$$

Thus:

$$\frac{3}{2x^2} + \frac{5}{6xy^2} + \frac{7}{4yz} = \frac{18y^2z}{12x^2y^2z} + \frac{10xz}{12x^2y^2z} + \frac{21x^2y}{12x^2y^2z}$$

$$= \frac{18y^2z + 10xz + 21x^2y}{12x^2y^2z}.$$

b. Before we can subtract the two expressions we must rewrite the expression so that there are no negative exponents. Then we must find the lcd.

$$5(x^2 - 1)^{-1} - 3(x^2 + 2x + 1)^{-1} = \frac{5}{x^2 - 1} - \frac{3}{x^2 + 2x + 1}$$

Def. of negative exponents

To find the lcd, first factor the denominators.

$$x^2 - 1 = (x + 1)(x - 1) \quad \text{and} \quad x^2 + 2x + 1 = (x + 1)^2$$

Thus, the lcd is $(x - 1)(x + 1)^2$, which also can be written as

$$(x - 1)(x + 1)(x + 1) = (x^2 - 1)(x + 1).$$

Substituting we have:

$$\frac{5}{x^2 - 1} - \frac{3}{x^2 + 2x + 1} = \frac{5}{x^2 - 1} \cdot \frac{x + 1}{x + 1} - \frac{3}{(x + 1)^2} \cdot \frac{x - 1}{x - 1} \quad \textbf{(Step 1)}$$

$$= \frac{5x + 5}{(x^2 - 1)(x + 1)} - \frac{3(x - 1)}{(x + 1)^2(x - 1)}$$

$$= \frac{5x + 5 - (3x - 3)}{(x - 1)(x + 1)^2}$$

$$= \frac{5x + 5 - 3x + 3}{(x - 1)(x + 1)^2}$$

$$= \frac{2x + 8}{(x - 1)(x + 1)^2}$$

$$= \frac{2(x + 4)}{(x - 1)(x + 1)^2} \quad \blacklozenge$$

Note that this minus sign applies to all terms in the numerator of the second fraction, not just the first term.

In Step 1 of Example 2**b** why was the first fraction multiplied by $\dfrac{x+1}{x+1}$ and the second fraction by $\dfrac{x-1}{x-1}$? For the answer, look back at Example 2a.

Complex Fractions

Now that we understand the operations of addition, subtraction, multiplication, and division we can use them to simplify complex fractions. Working with complex fractions may cause some concern. The name is unfortunate because there is nothing complex or mysterious about working with complex fractions.

> **Simple Rule**
>
> To simplify a complex fraction, reduce both numerator and denominator to a single fraction, then perform the indicated division.

EXAMPLE 3 **Simplify a Complex Fraction**

Simplify $\dfrac{\dfrac{1}{x}+3}{\dfrac{5}{y^2}}$.

Solution Following the rule, the first thing we want to do is write the numerator as a single fraction. The denominator is already a single fraction. To do this we must find the sum of $\dfrac{1}{x}$ and 3.

$$\frac{1}{x}+3 = \frac{1}{x}+\frac{3x}{x}$$
$$= \frac{1+3x}{x}$$

Substituting we have:

$$\frac{\dfrac{1}{x}+3}{\dfrac{5}{y^2}} = \frac{\dfrac{1+3x}{x}}{\dfrac{5}{y^2}} \qquad \textbf{Dividing}$$

$$= \frac{1+3x}{x}\cdot\frac{y^2}{5} \qquad \textbf{Inverting and multiplying.}$$

$$= \frac{y^2+3xy^2}{5x} \qquad \blacklozenge$$

EXAMPLE 4   Simplify a Complex Fraction

Simplify: $\dfrac{\dfrac{x}{y} + \dfrac{x+y}{x-y}}{\dfrac{x}{y} - \dfrac{x-y}{x+y}}$.

Solution   The lcd for the fraction in the numerator is $y(x - y)$, and the lcd for the fractions in the denominator is $y(x + y)$. Multiplying the fractions in the numerator and the denominator by the appropriate factors we have:

$$\frac{\dfrac{x}{y} + \dfrac{x+y}{x-y}}{\dfrac{x}{y} - \dfrac{x-y}{x+y}} = \frac{\dfrac{x(x-y)}{y(x-y)} + \dfrac{y(x+y)}{y(x-y)}}{\dfrac{x(x+y)}{y(x+y)} - \dfrac{y(x-y)}{y(x+y)}}$$

$$= \frac{\dfrac{x^2 - xy + yx + y^2}{y(x-y)}}{\dfrac{x^2 + xy - yx + y^2}{y(x+y)}} \qquad \textbf{Expanding}$$

$$= \frac{\dfrac{x^2 + y^2}{y(x-y)}}{\dfrac{x^2 + y^2}{y(x+y)}} \qquad \textbf{Collecting like terms}$$

$$= \frac{x^2 + y^2}{y(x-y)} \div \frac{x^2 + y^2}{y(x+y)} \qquad \textbf{Dividing}$$

$$= \frac{\overset{1}{\cancel{x^2 + y^2}}}{\underset{1}{\cancel{y}(x-y)}} \cdot \frac{\overset{1}{\cancel{y}(x+y)}}{\underset{1}{\cancel{x^2 + y^2}}} \qquad \textbf{Inverting, multiplying, and reducing}$$

$$= \frac{x+y}{x-y} \quad \blacklozenge$$

2.9 Exercises

In Exercises 1–52, perform the indicated operations and write the result in simplest form.

1. $\dfrac{2}{x} + \dfrac{3}{2x}$

2. $\dfrac{1}{a} + \dfrac{1}{b}$

3. $1 + \dfrac{1}{p+q}$

4. $\dfrac{2}{m^2} + \dfrac{1}{m}$

5. $\dfrac{5}{ab^2} - \dfrac{1}{a^2 b}$

6. $\dfrac{3a}{c^2} - \dfrac{2c}{a^2}$

7. $\dfrac{1}{x-1} - \dfrac{1}{x+1}$

8. $\dfrac{4}{x^2 + x} + \dfrac{7}{4x+4}$

9. $\dfrac{1}{a+2} - \dfrac{2}{a+3}$

10. $\dfrac{3}{\theta - \pi} - \dfrac{4}{\theta + \pi}$

11. $\dfrac{x}{y-z} + \dfrac{x}{z-y}$

12. $\dfrac{1}{x} - \dfrac{1}{y} + \dfrac{1}{z}$

13. $\dfrac{a}{a-1} + \dfrac{2a}{1-a}$

14. $\dfrac{a}{R^2} - \dfrac{c}{R} + 2R$

15. $\dfrac{m-1}{m+1} - \dfrac{m+1}{m-1}$

16. $\dfrac{p-q+r}{m-n} + \dfrac{p-q-r}{n-m}$

17. $\dfrac{3}{x^5y^2} - \dfrac{5}{xy^4z}$

18. $\dfrac{1}{a^2(a+1)} - \dfrac{1}{a^5(a-1)}$

19. $\dfrac{a}{a^2-4} + \dfrac{a}{2+a}$

20. $\dfrac{1}{x^2-1} - \dfrac{1}{(x-1)^2}$

21. $\dfrac{a-b}{c-d} + \dfrac{a-b}{d-c}$

22. $\dfrac{x}{y+y^2} - \dfrac{1}{y(y+y^2)}$

23. $\dfrac{1}{4x^2+4x+1} + \dfrac{1-x}{1-2x}$

24. $\dfrac{3(a+1)}{a^2+6a-16} - \dfrac{1}{a-2}$

25. $\dfrac{a^2}{p^2q^2} + \dfrac{b^2}{p^2r^2} - \dfrac{c^2}{q^2r^2}$

26. $\dfrac{3\alpha}{1-\alpha^2} + \dfrac{2}{\alpha+1} - \dfrac{3}{1-\alpha}$

27. $\dfrac{x-1}{x+3} + \dfrac{x-2}{x-3} - \dfrac{x^2}{x^2-9}$

28. $\dfrac{R}{R^2-6R-7} + \dfrac{2}{R-7} - \dfrac{3}{R+1}$

29. $\dfrac{a^2+b^2}{a^2-b^2} - \dfrac{a}{a+b} + \dfrac{b}{a-b}$

30. $\dfrac{1}{2x+2} + \dfrac{5}{3x-3} - \dfrac{3x-1}{1-x^2}$

31. $\dfrac{1}{x+y} - \left(\dfrac{1}{x-y} - \dfrac{1}{x^2-y^2}\right)$

32. $2 - \left(\dfrac{3}{a} + \dfrac{4}{a-2}\right)$

33. $\dfrac{2}{t^2-12t+35} - \dfrac{1}{t^2-25} - \dfrac{1}{t^2-49}$

34. $\left(\dfrac{1}{a} + \dfrac{1}{a-1}\right) - \left(\dfrac{1}{x-1} - \dfrac{1}{x-2}\right)$

35. $5m - \dfrac{4m^2+7m-1}{2+m} - 6$

36. $\left(p + \dfrac{1}{p} + 2\right)\left(1 - \dfrac{2}{p+1}\right)$

37. $\left(\dfrac{a}{b} + \dfrac{b}{c} + \dfrac{c}{a}\right)\left(\dfrac{abc}{8}\right)$

38. $\left[1 - \left(\dfrac{x}{x-y}\right)^2\right]\left[1 - \dfrac{x}{y}\right]$

39. $\left(\dfrac{a}{a-b} - \dfrac{2ab}{a^2-b^2}\right)\left(\dfrac{1}{a} - \dfrac{2}{a-b}\right)$

40. $\left(\dfrac{1}{x^2-4x+3} - \dfrac{4}{x^2-9}\right)\left(x + \dfrac{3(3x-1)}{x-7}\right)$

41. $2(x-3)^{-1} + 3(x+3)^{-1}$

42. $(6x^2+1)^{\frac{1}{2}} - x^2(x^2+1)^{-\frac{1}{2}}$

43. $3(x-1)(x+3)^{-1} - (x-1)^2(x+3)^{-2}$

44. $(x+2)(x-3)^{-1} + (x-3)(x+2)^{-1}$

45. $\dfrac{3+\dfrac{1}{4}}{\dfrac{3}{4}-5}$

46. $\dfrac{5}{\dfrac{1}{3}+\dfrac{3}{7}}$

47. $\dfrac{m+\dfrac{1}{n}}{m-\dfrac{1}{n}}$

48. $\dfrac{\dfrac{i^2}{5}-5}{1+\dfrac{i}{5}}$

49. $\dfrac{I}{I-\dfrac{E}{R}}$

50. $\dfrac{\dfrac{E^2}{e^2}-1}{\dfrac{E^2+e^2}{2Ee}}$

51. $\dfrac{1-\dfrac{x-y}{x+y}}{1+\dfrac{x-y}{x+y}}$

52. $\dfrac{\dfrac{I-i}{I+i}+\dfrac{I+i}{I-i}}{\dfrac{I-i}{I+i}-\dfrac{I+i}{I-i}}$

Review Exercises

In Exercises 1–9, simplify each expression by using the rules of exponents.

1. $a^3 \cdot a^{13}$

2. $b^5 \cdot b^{15}$

3. $(x^4y^3)^7$

4. $\left(\dfrac{m^7}{n^5}\right)^3$

5. $\left(\dfrac{a^3 \cdot b^4}{c^5}\right)^0$

6. $\left(\dfrac{1}{p^{-4}}\right)^7$

7. $\dfrac{35x^4y^{11}}{7x^3y^6}$

8. $\dfrac{32a^4b^9}{(4a^2b^5)^2}$

9. $\left(\dfrac{2m^3n^4}{pq}\right)^{-3}$

In Exercises 10–16, write the numbers in scientific notation.

10. 60,000

11. 3250

12. 13.3

13. 0.4

14. 0.0027

15. 0.00000579

16. The human eye is normally capable of responding to light waves whose lengths are in the range of 0.000038 to 0.000076 cm.

In Exercises 17–23, write the numbers in ordinary decimal notation.

17. 7.052×10^6

18. 9.931×10^{-4}

19. 4.32×10^0

20. 5.3×10^{-4}

21. 4.62×10^{-5}

22. 3.45675×10^3

23. One gram is equal to 3.5×10^{-2} oz.

In Exercises 24–29, solve the problems using scientific notation. Write your answer in scientific notation.

24. $10^5 \times 10^9 \times 10^{-4}$

25. $\dfrac{10^{13} \times 10^{15}}{10^7 \times 10^9}$

26. $25{,}000 \times 0.002 \times 500 \times 0.00004$

27. The coefficient of linear expansion indicates the fractional change in the length of a rod per degree change in temperature. For tungsten the coefficient is 4.4×10^{-6} per degree Celsius. What is the fractional change in the length of a tungsten rod when the temperature rises from 10°C to 30°C?

28. A light year is the distance light can travel in a year's time. One light year is approximately 6×10^{12} mi. If a star is 375 light years away from earth, how far is it in miles?

29. One electron volt is the kinetic energy an electron acquires when it is accelerated in an electric field produced by a difference of potential of 1 volt. 1 electron volt = 1.60×10^{-19} J. How many joules equal five million electron volts?

In Exercises 30–35, use the definitions for exponents and radicals to simplify the expressions.

30. $\sqrt{27}$

31. $\sqrt[3]{32}$

32. $\sqrt[3]{a^{12}}$

33. $\sqrt[3]{\dfrac{129x^0 - y^0}{x^0 + (7b)^0}}$

34. $\sqrt[n]{a^{3n}}$

35. $\sqrt{\sqrt[3]{y^{6n}}}$

In Exercises 36–43, perform the indicated operations and simplify the results.

36. $\sqrt{7} + 2\sqrt{7}$

37. $4\sqrt{72} - 8\sqrt{50} + 3\sqrt{32}$

38. $\sqrt{m} + \sqrt{m^5} + \sqrt{m^7}$

39. $\sqrt{11} \cdot \sqrt{13}$

40. $\sqrt{13} \cdot \sqrt{26}$

41. $\sqrt{abc} \cdot \sqrt{abd}$

42. $\sqrt[3]{a^2bc^4} \cdot \sqrt[3]{ab^2c^5}$

43. $\dfrac{\sqrt{18x^2y^3}}{\sqrt{2xy}}$

In Exercises 44–53, simplify the following algebraic expressions.

44. $5x - 8x + 73x$

45. $3xy^2 + 7xy^2 - 18xy^2$

46. $6r - 7s + 14s - 8r$

47. $-4\phi\pi + 11RI - 2\phi\pi + 5RI$

48. $\dfrac{2}{3}a^2 - \dfrac{1}{7}a^2 + \dfrac{5}{6}a^2$

49. $\dfrac{2}{5}b^2 + \dfrac{7}{15}b^2 - \dfrac{11}{35}b^2$

50. $3x - (4x + 7)$

51. $-7x - (11x - 13)$

52. $4m - 6[3n + 8m - 5(n - m)]$

53. $8p - \{6q + 3[2p - 7q + (13q - 11p)]\}$

In Exercises 54–81, perform the indicated operations and write the answer in simplest form.

54. $(4a^2)(7a^3)$

55. $(11r^2s^2)(-13r^3s^4)$

56. $72m^2n^3 \div 12mn^2$

57. $48p^4q^2 \div (-16p^3q)$

58. $\dfrac{8r}{9s} \cdot \dfrac{27s^3}{56r^2}$

59. $\dfrac{m+2}{m+3} \cdot \dfrac{m+4}{m-5}$

60. $\dfrac{3m}{5n} \div \dfrac{6m^2}{15n^3}$

61. $\dfrac{xy}{9xy} \div \dfrac{(xy)^2}{3x}$

62. $\dfrac{1}{x+2} + \dfrac{3}{x+3}$

63. $\dfrac{2x}{y^2} - \dfrac{2y}{x^2}$

64. $(3m-4n)(2m+7n)$

65. $(5R-4)(3R^2+11R-9)$

66. $(5x-11y)^2$

67. $(x+4)(x-3)(x+5)$

68. $(4pq-8pq^2) \div 2pq$

69. $(102a^2y - 51a^3y) \div (-17a^2)$

70. $(8m^2 - 2mn - 21n^2) \div (2m+3n)$

71. $(a^3 - 8a + 5a^2 - 12) \div (a^2 - a - 2)$

72. $\dfrac{m^2 - 36n^2}{6mn} \cdot \dfrac{3m}{m^2 - 12mn + 36n^2}$

73. $\dfrac{x^2 + 4x - 21}{x^2 + 2x - 15} \cdot \dfrac{x^2 + 6x + 5}{x^2 + 8x + 7}$

74. $\dfrac{x+7}{x+9} \div \dfrac{xy+7y}{x^2 + 18x + 81}$

75. $\dfrac{8x^2 - 2}{6x^2 + x - 1} \div \dfrac{2x^2 + 7x - 4}{3x^2 + 11x - 4}$

76. $\dfrac{4x^2}{7y} \div \left(\dfrac{4x}{3z} \cdot \dfrac{2x^2}{21y}\right)$

77. $\dfrac{2R^2}{3I} \cdot \left(\dfrac{6R}{7I} \div \dfrac{12R^2}{35I^3}\right)$

78. $\dfrac{x-2}{x-4} + \dfrac{x-3}{x+4} - \dfrac{x^2}{x^2 - 16}$

79. $\dfrac{1}{2m+2} + \dfrac{5}{3m-3} - \dfrac{3m-1}{1-m^2}$

80. $\left(m + \dfrac{2}{m} + 3\right)\left(1 - \dfrac{2}{m+1}\right)$

81. $\dfrac{\dfrac{x^2+1}{x} - 2}{\dfrac{x^2+3x+5}{x} + 3}$

In Exercises 82–90, write each expression as a product of prime factors.

82. $3xy + 12yz + 5xw + 20zw$

83. $6x^2yw - 14xwz - 15x^2y^2 + 35xyz$

84. $6w^2 + 31w + 35$

85. $6r^2 - 6s^2$

86. $4x^2 + 3x - 10$

87. $64x^2 - 8x - 2$

88. $(x+y)^2 - z^2$

89. $10a^2 - 3ab - 18b^2$

90. $(x^2 + 25)^2 - 25y^2$

91. Is $(10^4)^3 = 10^{4^3}$? Explain your answer using rules of exponents and order of operations.

92. Prove: $\sqrt[m]{\sqrt[n]{a}} = \sqrt[mn]{a}$

93. Given: $\sqrt{5 + 2\sqrt{6}} = \sqrt{2} + \sqrt{3}$
   a. Verify on a calculator.
   b. Prove the relation.

94. Factor $x^2 + 2xy - 8y^2$ by writing $-8y^2$ in the form $y^2 - 9y^2$.

95. Consider two consecutive positive integers. Which is larger, the average of their squares or the square of their average?

✎ Writing About Mathematics

1. A friend has difficulty understanding that $125^{\frac{1}{3}}$ is equal to $625^{\frac{1}{4}}$. Write a short paper with illustrations explaining why they are equal.

2. A friend has difficulty simplifying the expression $\dfrac{\sqrt{5}}{\sqrt[3]{5}}$. Write a set of instructions explaining how the expression can be simplified.

3. Find an article in a newspaper or magazine that uses scientific notation. Write a brief review of the article with emphasis on how scientific notation is used.

4. Write a short paper discussing how, when, where, and why you need to be able to work with fractions in your chosen career field.

Chapter Test

Simplify each expression.

1. $\dfrac{54x^5y^8}{9x^3y^6}$

2. $\dfrac{(r^2 - s^2)^0}{r - s}$

Write the number in scientific notation.

3. 123,000,000

4. 0.000214

Write the number in ordinary notation.

5. 3.4×10^5

6. 2.87×10^{-8}

Perform the calculation in scientific notation on a calculator and leave the answer in scientific notation.

7. $3.24 \times 10^5 + 5.16 \times 10^4$

8. $\dfrac{(2.4 \times 10^7)(4.0 \times 10^{-5})}{8.0 \times 10^{-9}}$

Simplify the expressions.

9. $\sqrt{8x^3} + \sqrt{18x^3}$

10. $5x^3y + 2yx^3 + 4xy - 7yx + 9$

11. $\dfrac{\sqrt{x^3y^5}}{\sqrt{x^5y^3}}$

12. $\dfrac{2 + \dfrac{x}{3}}{\dfrac{x + 2}{3}}$

Write each expression as a product of prime factors.

13. $2rs + 14st + 7ru + 49tu$

14. $9x^2 - 25$

15. $x^3 + 8x^2 - 32x$

16. $25x^2 + 20xy + 4y^2$

Perform the indicated operations and simplify.

17. $(x^2 - 2x - 14) \div (x + 3)$

18. $\dfrac{x + 5}{x - 3} \div \dfrac{xy + 5y}{x^2 - 4x + 3}$

19. $\dfrac{x + 2}{x - 7} + \dfrac{x - 1}{x^2 - 49} - \dfrac{x}{x + 7}$

20. $\dfrac{5r^3}{13s^2} \div \dfrac{5r}{13s} \cdot \dfrac{26r^2}{15s}$

Review of Geometry

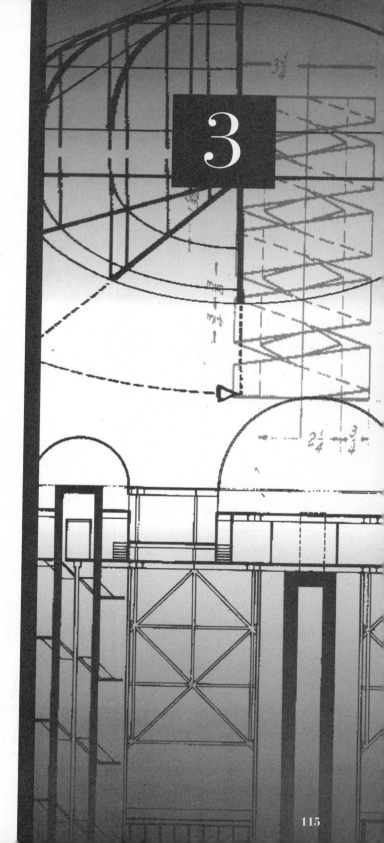

3

I n Rochester, New York, the pier at the outlet of the Genesee River into Lake Ontario is a favorite place for walks and fishing. The pier consists of a recently repaired section and an older section that is crumbling and in need of repair. The newer section is a rectangular block of concrete that is 26 ft wide and has a height of 20 ft. The older section can be divided into two shapes. The bottom part is a rectangular block; the top is a trapezoidal-shaped block. Since the basic structure of the old pier is solid, a repair solution is to widen and build up the older section so that it is the same width and height as the new part. The engineers must calculate the amount of concrete needed to add the required width and height to the 1250-ft older section. To estimate the amount of concrete, the engineers need to find the area of the rectangle and the trapezoid. When they know these areas, they can use them to find the volumes of rectangular prisms and trapezoidal prisms. Their calculations are illustrated in Section 3.4, Example 4.

3.1

Plane Figures

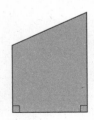

FIGURE 3.1

Understanding the basic principles of geometry is essential in many technical fields. For instance, the design engineer for a new aircraft may decide that the tail section should have the shape of a trapezoid. A surveyor may find that the shape of a parcel of land is a parallelogram joined with an isosceles triangle. And a scientist might describe the base of a crystal in a residue as a dodecagon. Do you recognize the shapes of these geometric figures (see Figure 3.1)? Can you list their properties and determine the perimeter and area of each figure? By the end of this chapter, which provides a brief review of the essentials of geometry, you should have these and other important skills.

The line (straight line) and the point are two elements of many figures constructed on a plane. A **plane** is an infinitely large flat surface. Look at Figure 3.2 and think of **line** AB, symbolized as $\overleftrightarrow{AB}$, as extending in either direction without bounds. **Line segment** AB, symbolized as $\overline{AB}$, is a measurable piece of line AB. The notation $m(AB)$ or $|AB|$ may be used to symbolize the length of the line segment. Now turn to Figure 3.3, which shows the **ray** AB, symbolized $\overrightarrow{AB}$. Ray AB starts at point A and moves in the direction of point B. The ray BA $(\overrightarrow{BA})$ starts at point B and moves in the direction of point A.

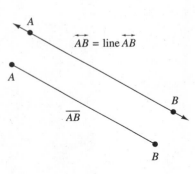

FIGURE 3.2

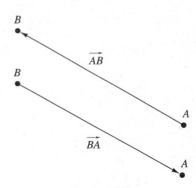

FIGURE 3.3

Angles

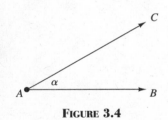

FIGURE 3.4

The union of rays $\overrightarrow{AB}$ and $\overrightarrow{AC}$ forms angle BAC, symbolized as $\angle BAC$, depicted in Figure 3.4. Sides $\overrightarrow{AB}$ and $\overrightarrow{AC}$ of $\angle BAC$ intersect at point A. Point A is called the **vertex** of $\angle BAC$. You can also designate $\angle BAC$ using only the vertex letter, $\angle A$, or by using a character such as the Greek letter inside the angle, $\angle \alpha$. See the inside cover of the text for a list of Greek letters.

Take two pencils, place one on top of the other on the positive x-axis, (with the erasers on the left). Put a pin through the erasers at the origin. Rotate the top pencil in a counterclockwise direction, "upward" (see Figure 3.5). The opening between the two pencils is called the measure of the angle, and in this case, the measure is positive. Now return your pencils to the positive x-axis and rotate the

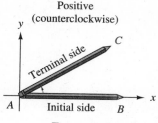

FIGURE 3.5

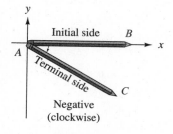

FIGURE 3.6

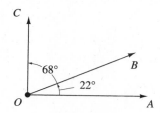

FIGURE 3.9

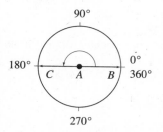

FIGURE 3.10

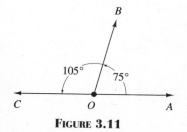

FIGURE 3.11

top pencil in a clockwise direction, "downward" (see Figure 3.6). The measure of this angle is negative.

For $\angle ABC$, ray $\vec{AB}$ is called the **initial ray,** and ray $\vec{AC}$ the **terminal ray.** If the initial ray visually coincides with the positive x-axis and the rotation to the terminal ray is counterclockwise, the angle is in **standard position.**

The basic measure of an angle is the degree: $1° = \dfrac{1}{360}$ of a complete revolution. In Figure 3.7, $\vec{AC}$ initially coincides with $\vec{AB}$. When $\vec{AC}$ is rotated counterclockwise one complete revolution, the measure of the angle is 360°. If $\vec{AC}$ is rotated one-fourth of a revolution (Figure 3.8), then the measure of $\angle BAC$ is 90°. Any angle whose measure is 90° is called a **right angle.** The measure of angle BAC is often indicated by the symbol $m(\angle BAC)$. This text uses the more common notation, $\angle BAC$.

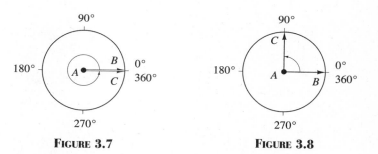

FIGURE 3.7   **FIGURE 3.8**

In Figure 3.9, $\angle AOB$ and $\angle BOC$ are adjacent angles. Two distinct angles are **adjacent angles** if, and only if, they have a common vertex and a common side between them. Angles AOB and BOC are acute angles. If the measure of an angle is greater than 0° but less than 90° it is an **acute angle.** Two acute angles are called **complementary angles** if the sum of their measures is 90°. Thus, $\angle AOB$ and $\angle BOC$ are complementary angles since $\angle AOB + \angle BOC = 22° + 68° = 90°$. Each angle is a complement of the other. Thus, $\angle AOB$ is the complement of $\angle BOC$ and $\angle BOC$ is the complement of $\angle AOB$.

A rotation of $\vec{AC}$ halfway around the circle, shown in Figure 3.10, describes $\angle BAC$ with a measure of 180°. An angle with a measure of 180° is called a **straight angle.**

Two distinct angles are called **supplementary angles** if the sum of their measures is 180°. Thus, $\angle AOB$ and $\angle BOC$ in Figure 3.11 are supplementary angles, since $\angle AOB + \angle BOC = 75° + 105° = 180°$. Each angle is a **supplement** of the other. Thus, $\angle AOB$ is the supplement of $\angle BOC$ and $\angle BOC$ is the supplement of $\angle AOB$.

Degrees may be divided into minutes, and minutes may be divided into seconds, just as the yard is divided into feet, and feet into inches. Similar to the usual measures for time, 1 degree is equal to 60 minutes ($1° = 60'$), and 1 minute is equal to 60 seconds ($1' = 60''$), or 1 degree is equal to 3600 seconds ($1° = 3600''$).

We can use these facts to change $90°$ to a measure in degrees, minutes, and seconds.

$$90° = 89° + 1°$$
$$= 89° + 60' \qquad 1° = 60'$$
$$= 89° + 59' + 1'$$
$$= 89° + 59' + 60''$$

This result typically is written without the plus signs: $89° \ 59' \ 60''$.

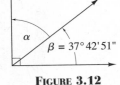

FIGURE 3.12

EXAMPLE 1   **Finding the Measure of the Complement of an Angle**

Determine the measure of $\angle\alpha$ in Figure 3.12.

Solution   From Figure 3.12 we see that $\angle\alpha$ and $\angle\beta$ are complementary angles. Thus, $\angle\alpha + \angle\beta = 90°$, or

$$\angle\alpha = 90° - \angle\beta$$
$$= 90° - \boxed{37° \ 42' \ 51''}$$

To subtract, change $90°$ to degrees, minutes, and seconds.

$$\begin{array}{r} 89° \ 59' \ 60'' \\ - \ 37° \ 42' \ 51'' \\ \hline 52° \ 17' \ \ 9'' \end{array}$$

Therefore, $\angle\alpha = 52° \ 17' \ 9''$.   ◆

EXAMPLE 2   **Application: Surveying Problem**

When surveying a lot, the surveyor determines that the measure of $\angle\alpha$ is $132° \ 14' \ 24''$, shown in Figure 3.13. What is the measure of $\angle\beta$?

FIGURE 3.13

Solution   From the surveyor's drawing we can see that $\angle\alpha$ and $\angle\beta$ are supplementary angles, that is, $\angle\alpha + \angle\beta = 180°$.

$$\angle\beta = 180° - \angle\alpha$$
$$= 180° - \boxed{132° \ 14' \ 24''}$$

To subtract, change $180°$ to degrees, minutes, and seconds.

$$\begin{array}{r} 179° \ 59' \ 60'' \\ - \ 132° \ 14' \ 24'' \\ \hline 47° \ 45' \ 36'' \end{array}$$

The angle β, inside the lot, measures $47° \ 45' \ 36''$.   ◆

Assume that the measure of $\angle BOC$ is 31° 29′ 32″. Some calculators are not designed to conveniently perform calculations with the measure of the angle expressed in this form. However, if the measure is expressed in decimal form, the process becomes simpler. So let's look at how to convert from degrees, minutes, and seconds to decimal degrees and vice versa.

Since angle measures are approximate numbers, you must decide the number of places you want to retain to the right of the decimal point. To determine the number of places when you convert from degrees, minutes, seconds to decimal degrees, read Table 3.1 from left to right. For example, the measure 43° 20′ is given to the nearest ten minutes, so the decimal equivalent is correctly recorded to the tenths place, one place to the right of the decimal point. The decimal equivalent of 43° 20′ is 43.333333 . . . , which, rounded properly, is 43.3°. If you know the angle in decimal degrees and are converting to degrees, minutes, and seconds, count the number of places to the right of the decimal point and read Table 3.1 from right to left. The measure 43.42° is given to the nearest hundredth of a degree. Therefore, when you convert to degrees, minutes, and seconds you would record the answer to the nearest minute. 43.42°, to the nearest minute, is 43° 25′.

TABLE 3.1   Angle In

Degrees, Minutes, Seconds	Decimal Degrees
Nearest ten minutes 43° 20′	1 place to right of decimal point 43.3°
Nearest minute 43° 25′	2 places to right of decimal point 43.42°
Nearest ten seconds 43° 25′ 40″	3 places to right of decimal point 43.428°
Nearest second 43° 25′ 47″	4 places to right of decimal point 43.4297°

Now lets turn to a procedure to change an angle measure from degrees, minutes, and seconds to degrees in decimal form.

Procedure to Change Degrees, Minutes, Seconds to Degrees in Decimal Form

If the measure is in degrees, minutes, and seconds:

1. Divide the seconds by 60 and add this result to the number of minutes. This result is the number of minutes in decimal form.
2. Divide the result in step 1, minutes, by 60 and add this to the number of degrees. This result is the number of degrees in decimal form.
3. Use the rules in Table 3.1 to round to the correct number of decimal places.

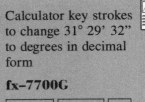

Calculator key strokes to change 31° 29' 32" to degrees in decimal form

fx–7700G

SHIFT MATH F4 31
F1 29 F1 32 F1
EXE
31.492222

TI–82

(32 ÷ 60 + 29)
÷ 60 + 31 ENTER
31.492222

EXAMPLE 3   Changing from Degrees, Minutes, Seconds To Degrees in Decimal Form

Change 31° 29' 32" to degrees in decimal form.

Solution   Follow the procedure for changing the measure to decimal form.

a. $32'' \div 60 + 29 = 0.533333' + 29' = 29.533333'$
b. $29.53333' \div 60 + 31° = 31.492222°$.
c. Rounding the answers using Table 3.1 we see that $31° 29' 32'' = 31.4922°$.   ◆

Next is a procedure to change degrees in decimal form to degrees, minutes, and seconds.

Procedure to Change Degrees in Decimal Form to Degrees, Minutes, and Seconds

1. Multiply the decimal part of the number by 60. The product is a number of minutes.
2. Multiply the decimal part of the answer in step 1 by 60. The product is the number of seconds.
3. Use Table 3.1 to round the answer.

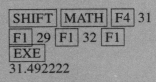

Calculator key strokes to change 27.5425° to degrees, minutes, and seconds.

fx–7700G

SHIFT MATH
F4 27.5425 EXE
F2 27° 32' 33"

TI–82

27.5425 2nd ANGLE
4 ENTER Answer is
27° 32' 33"

EXAMPLE 4   Change from Decimal Degrees to Degrees, Minutes, and Seconds

Change 28.64° to degrees, minutes, and seconds.

Solution   Follow the procedure to change from degrees in decimal form to degrees, minutes, and seconds.

a. $0.64° \times 60 = 38.40'$.
b. $0.40' \times 60 = 24.0''$
c. The rules in Table 3.1 tell us that the answer should be expressed to the nearest minute since there are two places to the right of the decimal point in 28.64°. Therefore, $28.64° = 28° 38'$.   ◆

The measure of angles in degrees is practical in some areas, but a more convenient unit of measure in applied areas is the radian. Radian measure will be discussed in detail in Chapter 6.

Lines

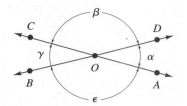

FIGURE 3.14

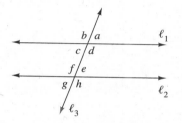

FIGURE 3.15

Two lines on a plane either: (1) never intersect, in which case they are considered to be parallel to one another; (2) coincide, in which case they may be considered to be essentially the same line; or (3) intersect at a single point. Lines $\overrightarrow{AC}$ and $\overrightarrow{BD}$ in Figure 3.14 intersect at point O. The four angles formed by the intersection are α (alpha), β (beta), γ (gamma), and ϵ (epsilon). Angles β and ϵ are on opposite sides of the vertex and are called **vertical angles,** as are α and γ. By careful examination of Figure 3.14, we should be able to deduce that all vertical angles are equal to one another. Angles α and β are supplementary angles as are α and ϵ. Observe that

$$\angle\alpha + \angle\beta = 180° \quad \text{and} \quad \angle\alpha + \angle\epsilon = 180°,$$

therefore, $\angle\beta = \angle\epsilon$ since they are supplements of the same angle, α. Using similar reasoning can you conclude that $\angle\alpha = \angle\gamma$?

Two parallel lines (ℓ_1 and ℓ_2) that are crossed by a third line (ℓ_3), called a **transversal**, form eight distinct angles (Figure 3.15). The five properties that follow illustrate important relations between the measures of the angles. a, b, g, and h are called *exterior angles*. c, d, e, and f are called *interior angles*. In Figure 3.15 angles a and e are called *corresponding angles*. **Corresponding angles** have different vertices, they are on the same side of the transversal, and one is an interior angle, and the other is an exterior angle. Two angles, such as d and f, on opposite sides of the transversal are called **alternating angles.**

Properties of the Angles Formed by Two Parallel Lines Cut by a Transversal

1. Exterior angles on the same side of the transversal are supplementary.

   $$\angle a + \angle b = 180° \quad \text{and} \quad \angle b + \angle g = 180°$$

2. Interior angles on the same side of the transversal are supplementary.

   $$\angle c + \angle f = 180° \quad \text{and} \quad \angle d + \angle e = 180°$$

3. Corresponding angles are equal.

   $$\angle a = \angle e, \angle b = \angle f, \angle c = \angle g, \quad \text{and} \quad \angle d = \angle b.$$

4. Alternate exterior angles are equal.

   $$\angle a = \angle g \quad \text{and} \quad \angle b = \angle h$$

5. Alternate interior angles are equal.

   $$\angle c = \angle e \quad \text{and} \quad \angle d = \angle f$$

EXAMPLE 5   **Determining the Measures of Angles Given Two Parallel Lines and A Transversal**

In Figure 3.16 $\overline{AD}$ is parallel to $\overline{BC}$ and $\overline{BF}$ is parallel to $\overline{ED}$. Determine the measure of the following.

1. $\angle a$ **2.** $\angle b$ **3.** $\angle c$ **4.** $\angle d$ **5.** $\angle f$ **6.** $\angle g$

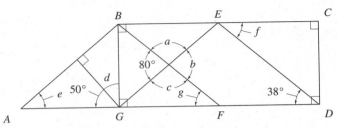

FIGURE 3.16

Solution

1. $\angle a$ is the supplement of an 80° angle. $\angle a + 80° = 180°$. Therefore, $\angle a = 100°$.
2. $\angle b = 80°$, since $\angle b$ and the 80° angle are vertical angles.
3. $\angle c = 100°$, since $\angle c$ and $\angle a$ are vertical angles.
4. $\angle d$ is the complement of a 50° angle; therefore, $\angle d + 50° = 90°$ and $\angle d = 40°$.
5. $\angle f = 38°$. If two parallel lines are cut by a transversal, then the alternate interior angles are equal.
6. $\angle g = 38°$. If two parallel lines are cut by a transversal, then the corresponding angles are equal. ◆

Polygons

A **polygon** is a closed plane figure with three or more sides. A polygon with three sides is called a **triangle.** Triangles may be classified according to the measure of their angles or the length of their sides. *The sum of the measures of the angles of a triangle is 180°.* A triangle that contains three **acute angles** (angles whose measures lie between 0° and 90°) is called an **acute triangle** (Figure 3.17a). A triangle that contains an **obtuse angle** (an angle whose measure lies between 90° and 180°) is called an **obtuse triangle** (Figure 3.17b). A triangle that contains a **right angle** (an angle measuring 90°) is called a **right triangle** (Figure 3.17c). Since right triangles are often used in solving problems, the sides have special names. The side opposite the right angle is called the hypotenuse, and the other two sides are called legs; sometimes the legs are referred to as the base and the altitude.

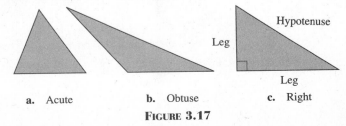

a. Acute   **b.** Obtuse   **c.** Right

FIGURE 3.17

A triangle in which no two sides are equal is called a **scalene triangle** (Figure 3.18a). A triangle with two sides the same length is called an **isosceles triangle** (Figure 3.18b). The angles opposite the equal sides have equal measure. A triangle with three equal sides is called an **equilateral triangle** (Figure 3.18c). The three angles have equal measure (60° each) since they are opposite equal sides. The equilateral triangle is also called an **equiangular triangle.**

a. Scalene   **b.** Isosceles   **c.** Equilateral

FIGURE 3.18

TABLE 3.2   Common Polygons

Number of Sides	Name
3	Triangle
4	Quadrilateral
5	Pentagon
6	Hexagon
7	Heptagon
8	Octagon
9	Nonagon
10	Decagon
12	Dodecagon
20	Icosagon

A polygon formed by the union of four line segments is called a **quadrilateral.** Quadrilaterals may be classified as to whether the figure has zero, one, or two pairs of opposite sides that are parallel. A quadrilateral with no parallel sides is called a **trapezium.** A quadrilateral with only one pair of parallel sides is called a **trapezoid** (Figure 3.19a). The other two sides may or may not be of equal length. If the two nonparallel sides are of equal length, it is called an **isosceles trapezoid** (Figure 3.19b). A quadrilateral with both pairs of opposite sides parallel is called a **parallelogram** (Figure 3.19c). A parallelogram that has four equal sides is called a **rhombus** (Figure 3.19d). A parallelogram that contains a right angle is called a **rectangle** (Figure 3.19e). A rectangle that contains four equal sides is called a **square** (Figure 3.19f). A square is an example of a regular polygon. A **regular polygon** is a polygon that is both equilateral and equiangular. A list of common polygons is found in Table 3.2. The polygons are named using Greek prefixes that correspond to the number of sides.

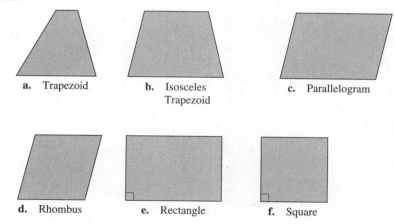

a. Trapezoid   **b.** Isosceles Trapezoid   **c.** Parallelogram

d. Rhombus   **e.** Rectangle   **f.** Square

FIGURE 3.19

Circles

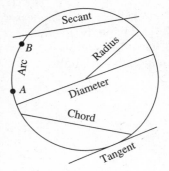

FIGURE 3.20

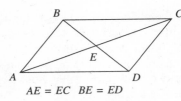

$AE = EC$ $BE = ED$

FIGURE 3.23

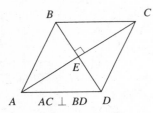

$AC \perp BD$

FIGURE 3.24

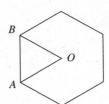

FIGURE 3.25

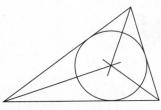

FIGURE 3.26

Another commonly used plane figure is a circle (Figure 3.20). A **circle** is a path of all the points that are a given distance from a fixed point called the **center.** The distance from the center to the path of points is called the **radius.** A **chord** is a line segment whose end points are points of a circle. A chord that passes through the center of a circle is called a **diameter** of the circle. The portion of the circles connecting points A and B is called **arc AB,** symbolized $\overset{\frown}{AB}$. $\overset{\frown}{AB}$ is called a *minor arc*, and $\overset{\frown}{BA}$ is called a *major arc*. A straight line that intersects a circle in two points is called a **secant.** A straight line that intersects a circle at one and only one point is called a **tangent.**

The distance around a circle is called the **circumference.** The ratio of the circumference to the diameter is the irrational number π. Thus, $\pi = c/d$, or equivalently, $c = \pi d$. In other words, circumference is a multiple of the diameter.

Two other regions of the circle that are important in technical applications are the sector and the segment of a circle. A **sector** of a circle is that part of the interior of a circle bounded by two radii and an arc. See Figure 3.21. A **segment** of a circle is that part of the interior of a circle bounded by a chord and its arc as shown in Figure 3.22.

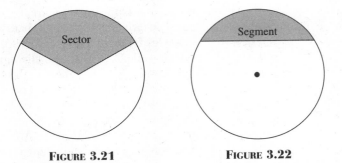

FIGURE 3.21　　　　　　　**FIGURE 3.22**

To supplement the ideas presented in this section, a list of geometrical proven theorems and formulas follows.

1. The diagonals of a parallelogram bisect each other (Figure 3.23).
2. The diagonals of a rhombus are perpendicular (Figure 3.24).
3. The sum of the measures of the interior angles of a polygon of n sides is equal to $(n - 2)\,180°$.
4. Each interior angle of a regular polygon of n sides is equal to $[(n - 2)/n]180°$.
5. The central angle of a regular polygon of n sides contains $\dfrac{360}{n}$ degrees. In Figure 3.25 $\angle AOB$ is an example of a central angle.
6. The intersection of the bisectors of the angles of a triangle determine the center of the inscribed circle (Figure 3.26).

A circle is *inscribed* in a polygon when all the sides of the polygon are tangent to the circle. The center of the circle is called the *incenter*.

7. The intersection of the perpendicular bisectors of the sides of a triangle determines the center of the circumscribed circle (Figure 3.27). A circle is said to be *circumscribed* about a polygon when all the vertices of the polygon lie on the circle. The center of the circle is called the *circumcenter* of the polygon.

8. If a radius of a circle bisects a chord that is not a diameter, then that radius is perpendicular to the chord (Figure 3.28). (If $\overline{RS} = \overline{ST}$, then $\overline{OU} \perp \overline{RT}$.)

9. A tangent to a circle is perpendicular to the radius drawn to the point of contact (Figure 3.28). Tangent at $Q \perp \overline{QO}$.

10. A diameter perpendicular to a chord bisects the chord and the arc determined by the chord (Figure 3.28). ($\overline{VU} \perp \overline{RT}$, so $\overline{RS} = \overline{ST}$, and $\overarc{RU} = \overarc{UT}$.)

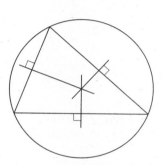

FIGURE 3.27

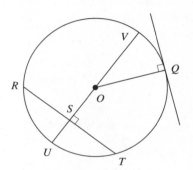

FIGURE 3.28

3.1 Exercises

In Exercises 1–8, change each angle measure to degrees in decimal form. Apply the rules in Table 3.1.

1. 15° 20′

2. 94° 40′

3. 104° 38′

4. 115° 49′

5. 72° 59′ 20″

6. 87° 42′ 30″

7. 73° 59′ 42″

8. 103° 35′ 49″

In Exercises 9–16, change each angle measure to degrees, minutes, and seconds. Apply the rules in Table 3.1.

9. 73.4°

10. 84.7°

11. 105.41°

12. 142.72°

13. 121.456°

14. 94.731°

15. 27.4586°

16. 94.7314°

17. Find the measure of angle α in the figure.

18. Find the measure of angle β in the figure.

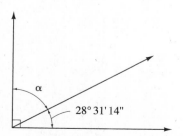

28° 31′ 14″

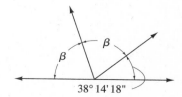

38° 14′ 18″

19. Find the measures of angles a, b, c, and d in the figure, given that ℓ_1 and ℓ_2 are parallel.

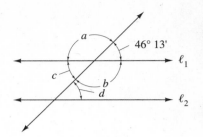

20. A surveyor is asked to divide the plot of ground $ABCD$ into lots by constructing lines parallel to AD (shown). If $\angle ADC = 115°$, how many degrees must he make angles DEF and DGH? If $\overline{AD}$ and $\overline{BC}$ are perpendicular to $\overline{AB}$, how many degrees are in $\angle DCB$?

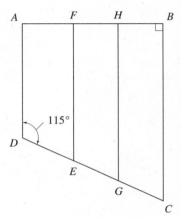

21. In the figure, the timber $\overline{BF}$ supports a stairway and makes an angle of 150° with the floor, $\overline{AB}$. What size must the carpenter make the angles CBD and CDB so that $\overline{CB} \perp \overline{AB}$ and $\overline{CD} \parallel \overline{AB}$?

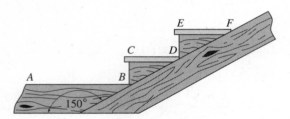

22. In making the legs for a picnic table, Jane nails two pieces of wood together as shown so that $\overline{AO} = \overline{OC}$, $\overline{OD} = \overline{OB}$, and $\angle AOC = 50°$. At what exterior angle must she cut the pieces of wood at A and C so that they will stand flat on the floor? How many degrees must angles D and B contain in order that the top of the table will be parallel to the floor (line AC)?

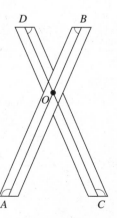

In Exercises 23–28, determine the measure of the missing angles and name the figure.

23.

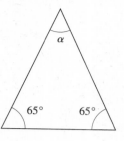

24.

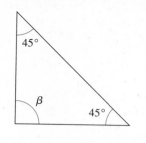

25.

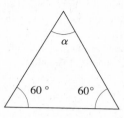

26.

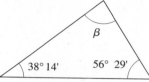

27.

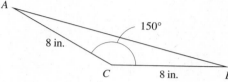

28.

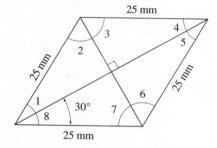

In Exercises 29–30, find the measures of the angles indicated on the figure.

29.

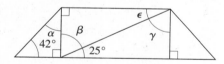

30.

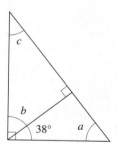

31. Take three straws and connect them with three pins to form a triangle. Since a triangle can be identified by its sides, this operation defines a unique triangle. Push on any one of the vertices and note that the shape of the triangle does not change. Now connect four straws with four pins to form a parallelogram. Push on any one of the vertices and note that the model is altered.

    a. Given three pins, three straws, 3 in., 4 in., and 6 in. long, construct a triangle. Do these three lengths determine a unique triangle?

    b. Given four pins, four straws, two 3 in. long and two 4 in. long, construct a parallelogram. Do these four lengths determine a unique parallelogram?

    c. Use the information from **a** and **b** to explain why carpenters always brace walls with triangular shapes.

32. When analyzing the behavior of light as it travels from air to water, it is useful to investigate the angle of incidence of the light ray in the air and the angle of refraction of the light ray as it penetrates water. In the figure, two angles are given. However, these are not the angles that are customarily used. Instead the complements with respect to the perpendicular line (called the normal) are used. Compute the angle of incidence and the angle of refraction.

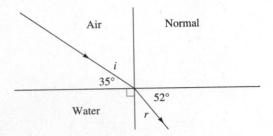

3.2

Similar Polygons

In order to be able to solve problems involving similar figures one must be able to set up a ratio and a proportion. A **ratio** is a comparison of two quantities that may be expressed as a fraction. For example, the ratio of the height of a tree to its shadow may be expressed as $\dfrac{\text{height}}{\text{shadow}}$, or height : shadow. A **proportion** is a statement that two ratios are equal. The following equalities are examples of proportions.

$$\frac{3}{4} = \frac{6}{8} \qquad \frac{x}{3} = \frac{4}{7} \qquad \frac{a}{b} = \frac{c}{d}$$

A more extensive discussion of proportions is given in Section 4.5.

A project engineer has the task of constructing the trapezoid-shaped tail section mentioned in Section 3.1. When working from the reduced drawings of the design engineer, how can the project engineer determine the actual size? One method of determining the actual size is by using the properties of similar polygons.

> **DEFINITION 3.1**
> **Similar polygons** are polygons that have corresponding angles with equal measure and corresponding sides that are proportional.

Corresponding sides and corresponding angles are those sides and angles that can be matched up when the polygons are in the same relative position. Similar polygons have the same shape, but may be of different sizes. In Figure 3.29 the corresponding parts of the trapezoids have the same letters.

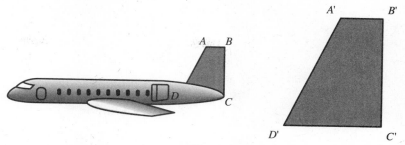

FIGURE 3.29

Now consider a special definition for similar triangles.

DEFINITION 3.2

If the measure of the angles of $\triangle ABC$ are equal to the measures of the corresponding angles of $\triangle A'B'C'$, then $\triangle ABC$ is **similar** to $\triangle A'B'C'$, symbolized by $\triangle ABC \sim \triangle A'B'C'$.

The symbol, $\sim$, is read "is similar to."

In Figure 3.30 $\angle A = \angle A'$, $\angle B = \angle B'$, and $\angle C = \angle C'$. Therefore, by Definition 3.2, $\triangle ABC \sim \triangle A'B'C'$. By Definition 3.2, the triangles in Figure 3.30 have corresponding sides that are proportional. Thus, we can set up the following proportions:

$$\frac{\overline{A'B'}}{\overline{A'C'}} = \frac{\overline{AB}}{\overline{AC}}, \frac{\overline{A'B'}}{\overline{B'C'}} = \frac{\overline{AB}}{\overline{BC}}, \frac{\overline{A'C'}}{\overline{B'C'}} = \frac{\overline{AC}}{\overline{BC}}$$

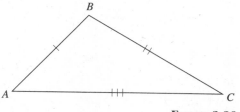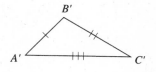

FIGURE 3.30

DEFINITION 3.3

If the corresponding sides of $\triangle ABC$ and $\triangle DEF$ are proportional, then $\triangle ABC$ **is similar to** $\triangle DEF$ ($\triangle ABC \sim \triangle DEF$).

Note that in order for polygons with more than three sides to be similar, the corresponding sides must be proportional, and the measures of the corresponding angles must be equal. For two triangles to be similar they only need to have the corresponding angles equal or the corresponding sides proportional.

EXAMPLE 1   Application: Given Right Triangles Show Triangles Are Similar

In the sketch of the truss (Figure 3.31) $\overline{BE} \parallel \overline{GD}$. (Read "*BE* is parallel to *GD*.") Show that $\triangle BCE$ is similar to $\triangle GCD$.

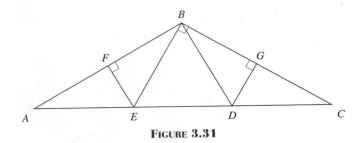

FIGURE 3.31

Solution   Given $\overline{BE} \parallel \overline{GD}$, show that $\triangle BCE \sim \triangle GCD$. To show that the triangles are similar, we must show that the corresponding angles are equal or that the corresponding sides are proportional. Since $\angle BCE$ is common to both triangles, we have one angle in each triangle with equal measure. Using the given fact that $\overline{BE} \parallel \overline{GD}$, we can state that

$$\angle CDG = \angle CEB \quad \text{and} \quad \angle DGC = \angle EBC.$$

Why? If two parallel lines are cut by a transversal, then the corresponding angles are equal. Therefore, the measures of the angles of $\triangle BCE$ are equal to the measures of the corresponding angles of $\triangle GCD$. Thus, by Definition 3.2, $\triangle BCE \sim \triangle GCD$. ◆

EXAMPLE 2   Given Right Triangles Show Triangles Are Similar

In right $\triangle ABC$ (Figure 3.32), $\angle C$ is a right angle, and $\overline{CD} \perp \overline{AB}$. Show that $\triangle ABC \sim \triangle BCD$.

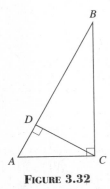

FIGURE 3.32

Solution   We can show that the triangles are similar by showing that the measure of the corresponding angles are equal. $\angle CDB$ is a right angle since $\overline{CD} \perp \overline{AB}$. The $\angle CDB = \angle BCA$ since all right angles are equal. Since $\angle DBC$ is common to both triangles, we can conclude that $\angle DBC = \angle ABC$. The sum of the measures of the angles of a triangle equals 180°. Thus ∠CDB + ∠DBC + ∠BCD = 180° and

$$\angle BCA + \angle ABC + \angle CAB = 180°.$$

Since both statements are equal to 180°, they are equal to each other

$$\angle CDB + \angle DBC + \angle BCD = \angle BCA + \angle ABC + \angle CAB.$$

Substituting those equal quantities we have

$$90° + \angle ABC + \angle BCD = 90° + \angle ABC + \angle CAB.$$

Subtracting 90° and ∠ABC from both sides we have ∠BCD = ∠CAB. Therefore, the corresponding angles are equal, and by Definition 3.2, ΔABC ~ ΔBCD. ◆

Using arguments similar to the ones given in Example 2, show that ΔABC ~ ΔACD.

EXAMPLE 3   **Application: Measuring the Width of a Swamp**

A surveyor has the task of determining the width of a swamp (see Figure 3.33). From point A she can see clearly both ends of the swamp, points B and C. She knows that the distance from point A to point C is 2.4 km and that ∠BAC = 50°.

Solution   The surveyor can find the distance from point B to point C using similar triangles (see Figure 3.34). Standing at point A she sights to the right edge of the swamp, point C. At a distance of 800 m from the point along the line of sight she drives a stake into the ground, point E. She then constructs an angle of 50° with its vertex at point E, forming ∠FEA. She then sights to the left edge of the swamp, point B, and names the point D where $\overleftrightarrow{AB}$ and $\overleftrightarrow{FE}$ intersect. She measures the length of $\overline{DE}$, which is 640 m. The lines $\overleftrightarrow{DE}$ and $\overleftrightarrow{BC}$ are parallel since they form equal corresponding angles with the transversal, $\overleftrightarrow{AC}$. From the work in Example 2, we can conclude that when two angles of one triangle have the same measures as the corresponding angles of the other triangle, the third angles also have equal measures. From this we can conclude that ∠ADE = ∠ABC and ΔABC ~ ΔADE. Since the triangles are similar, the corresponding sides are proportional, and we can set up the following proportion:

$$\frac{\overline{BC}}{\overline{DE}} = \frac{\overline{AC}}{\overline{AE}}.$$

Substituting:

$$\frac{\overline{BC}}{640 \text{ m}} = \frac{2.4 \text{ km}}{800 \text{ m}}$$

$$\overline{BC} = \frac{(2.4 \text{ km})(640 \text{ m})^{1}}{800 \text{ m}_{1}}$$

$$\overline{BC} = 1.92 \text{ km}.$$

The surveyor found the length of the swamp to be 1.9 km. ◆

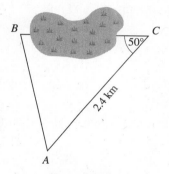

FIGURE 3.33

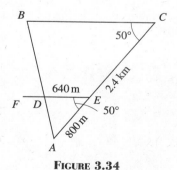

FIGURE 3.34

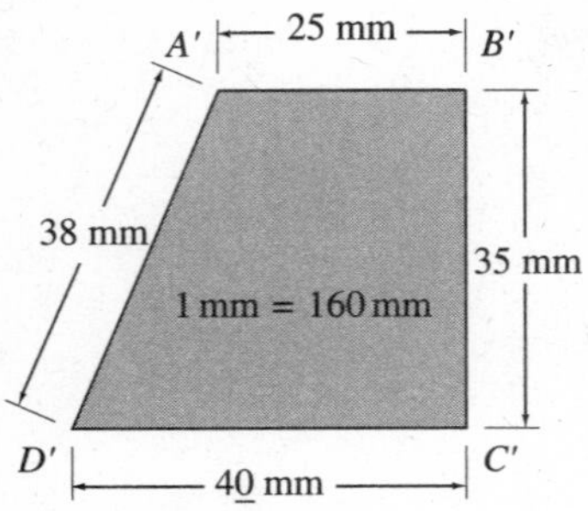

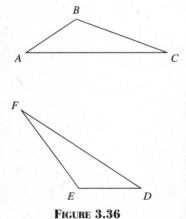

A' |←— 25 mm —→| B'

38 mm

35 mm

1 mm = 160 mm

D' |←———— 40 mm ————→| C'

FIGURE 3.35

EXAMPLE 4   Application: Similar Polygons

A sheet-metal mechanic must determine how large a piece of material needs to be to cover one side of a trapezoidal tailpiece. A scale drawing is given in Figure 3.35. Determine the dimensions of the actual tailpiece.

Solution   The scale drawing and the actual tailpiece are similar polygons. Thus, to calculate the actual length of each side, the mechanic sets up a proportion for each side using the lengths given in the diagram and the scale (1 mm = 160 mm).

$$\frac{\text{Actual length of side}}{\text{Actual length in scale}} = \frac{\text{Length of side on scale drawing}}{\text{Scale (in mm)}}$$

$$\frac{\overline{AB}}{160 \text{ mm}} = \frac{25 \text{ mm}}{1 \text{ mm}} \qquad \overline{AB} = \frac{25 \text{ mm}}{1 \text{ mm}}(160 \text{ mm}) = 4000 \text{ mm}$$

$$\frac{\overline{BC}}{160 \text{ mm}} = \frac{35 \text{ mm}}{1 \text{ mm}} \qquad \overline{BC} = \frac{35 \text{ mm}}{1 \text{ mm}}(160 \text{ mm}) = 5600 \text{ mm}$$

$$\frac{\overline{CD}}{160 \text{ mm}} = \frac{40 \text{ mm}}{1 \text{ mm}} \qquad \overline{CD} = \frac{40 \text{ mm}}{1 \text{ mm}}(160 \text{ mm}) = 6400 \text{ mm}$$

$$\frac{\overline{DA}}{160 \text{ mm}} = \frac{38 \text{ mm}}{1 \text{ mm}} \qquad \overline{DA} = \frac{38 \text{ mm}}{1 \text{ mm}}(160 \text{ mm}) = 6080 \text{ mm}$$

Thus, the mechanic needs a piece of material at least 6.4 m wide (6400 mm) and 5.6 m long (5600 mm).   ◆

Congruent triangles are a special case of similar triangles for which the corresponding sides are equal. (Congruent triangles have equal corresponding angles and equal corresponding sides.) The symbol used to indicate that two triangles are congruent is ≅. $\triangle ABC \cong \triangle DEF$, is read "triangle *ABC* is congruent to triangle *DEF*." The following list contains important statements about congruent triangles.

B

A   C

F

E   D

FIGURE 3.36

1. If two sides and the included angle (SAS) of one triangle are equal to two sides and the included angle of another triangle, then the two triangles are congruent. In Figure 3.36, $\angle A = \angle D$, $\overline{AB} = \overline{DE}$, and $\overline{AC} = \overline{DF}$.
2. If three sides of one triangle are equal to the three sides (SSS) of another triangle, then the two triangles are congruent. In Figure 3.36, $\overline{AB} = \overline{DE}$, $\overline{AC} = \overline{DF}$, and $\overline{BC} = \overline{EF}$.
3. If two angles and the included side (ASA) of one triangle are equal to two angles and the included side of another triangle, then the two triangles are congruent. In Figure 3.36, $\angle A = \angle D$, $\angle B = \angle E$, and $\overline{AB} = \overline{DE}$.

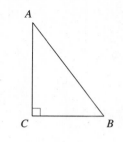

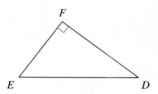

FIGURE 3.37

4. If two angles and a side of one triangle are equal to two angles and the corresponding side of another triangle, then the two triangles are congruent. In Figure 3.36, $\angle A = \angle D$, $\angle B = \angle E$, and $\overline{BC} = \overline{EF}$.

5. If the hypotenuse and an acute angle of one right triangle are equal to the hypotenuse and the acute angle of another right triangle, then the two triangles are congruent. In Figure 3.37, $\overline{AB} = \overline{DE}$ and $\angle A = \angle D$.

6. If the hypotenuse and a leg of one right triangle are equal to the hypotenuse and a leg of another right triangle, then the two triangles are congruent. In Figure 3.37, $\overline{AB} = \overline{DE}$ and $\overline{AC} = \overline{DF}$.

In statements 1, 2, and 3 the acronyms SAS, SSS, and ASA, respectively, are a way of remembering the theorems.

EXAMPLE 5 **Congruent Triangles Formed by the Diagonal of a Parallelogram**

Show that a diagonal of a parallelogram divides the parallelogram into two congruent triangles.

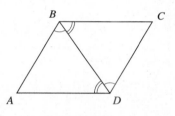

FIGURE 3.38

Solution Draw any parallelogram $ABCD$ and a diagonal $\overline{BD}$ as shown in Figure 3.38. $\angle DBC = \angle BDA$ and $\angle CDB = \angle ABD$ since they are alternate interior angles of the pairs of parallel lines $\overline{AB}$, $\overline{DC}$ and $\overline{BC}$, $\overline{AD}$. $\overline{BD} = \overline{BD}$, since any quantity is equal to itself. Therefore, $\triangle ABD \cong \triangle BCD$ since they have two angles and an included side, respectively, equal (Statement 3). ◆

3.2 Exercises

In Exercises 1–4, determine whether the figures are similar. State a definition that supports your answer.

1.

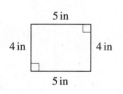

2.

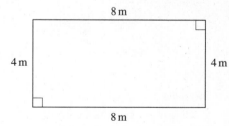

3.

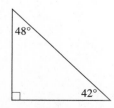

4.

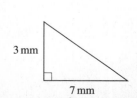

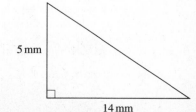

In Exercises 5–8, state the definition that explains why the triangles are similar, and then determine the values of the missing parts α, β, γ, ϕ, θ, a, b, t, x, y as indicated.

5.

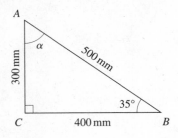

6.

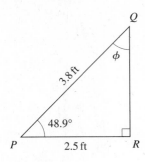

7.

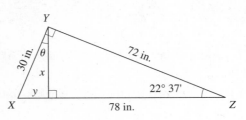

8.

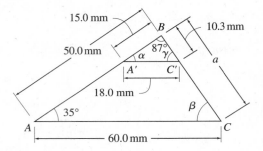

9. A tree 12 ft tall stands on the ground 542 ft from a vertical cliff. A boy whose eyes are 5′5″ above the ground level and who is standing 125 ft from the tree as shown in the figure, sees the top of the cliff in line with the top of the tree. How high is the cliff?

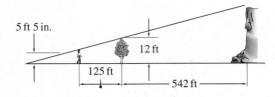

10. An aerial photograph of a city is made to the scale 1 in. = 1600 ft. If the focal length (distance from the lens to the image) of the camera in the figure was $6\frac{3}{4}$ in., how high was the photographer when he took the picture?

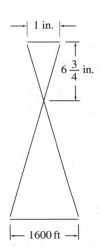

11. A film 3 in. high is being projected on a screen. How high will the projection be if the light is 7 in. behind the film, and the light is 6 ft 5 in. from the screen?

12. Two similar polygons have a ratio of similitude (ratio between corresponding sides) of 5 : 12. If the sum of their perimeters is 360 in., find the perimeter of each in feet.

13. A flagpole casts a shadow of 65 m. A stick $1\frac{1}{2}$ m in length casts a shadow of 5 m. How tall is the flagpole?

14. In constructing the triangular truss, *RST*, as depicted in the figure, the engineer found it necessary to add a brace, $\overline{XY}$, which is parallel to $\overline{RS}$. How long must the brace be if $\overline{ST} = 15$ m, $\overline{SY} = 8$ m, and $\overline{RS} = 11$ m?

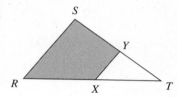

15. The blueprint for a solar panel has a scale of 20 mm $= 1\frac{1}{2}$ m. The width of the panel is $4\frac{1}{5}$ m. What distance on the blueprint represents this width?

16. The blueprint for an airplane has a scale of 5 mm $= 3\frac{1}{2}$ m. The length of the wing on the blueprint is 36 mm. What is the actual length of the wing?

17. On a map the scale is $\frac{1}{2}$ in. $= 75$ mi. If the actual distance from Rochester, New York, to Philadelphia, Pennsylvania, is 330 mi, what is the distance between the two cities on the map?

18. On a map the scale is 1 mm $= 8$ km. If the distance on the map from Buffalo to San Francisco is 560 mm, what is the actual distance in kilometers?

19. In the figure, $\angle ACB = \angle CAB = \angle ABC$ and $\overline{AB} = 275$ km. What is the length of $\overline{AC}$? Explain how you arrived at the answer.

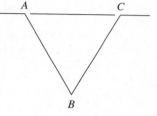

20. Use the relationships of similar triangles to determine the distance across the lake, $\overline{DE}$ as shown in the figure given $\angle CED = \angle ABC$. Explain how you determined the length of $\overline{DE}$.

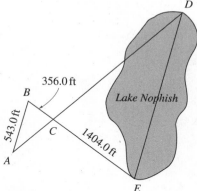

21. As shown, the person holds a stick 12 in. long at a distance of 2.0 ft from her eye; her eye is 192 ft from the tree. The stick is held so it is parallel to the tree as the person sights along the top of the stick to point *A*, the top of the tree, and along the bottom of the stick to point *B*, the base of the tree.
    a. Explain why triangles *CDE* and *CAB* are similar triangles.
    b. Determine the height of the tree.

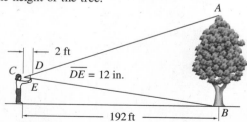

22. In the figure shown, $\overline{AD} \parallel \overline{EF}$, $\overline{AC} = \overline{BD} = 15$ m, $\overline{AD} = 24$ m, and $\overline{FG} = 2\,\overline{GB}$. Determine $\overline{EF}$.

23. Explain why $\triangle ABD \cong \triangle BDC$ in the figure. Also determine the missing parts α, β, and x.

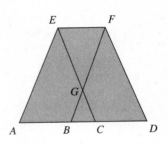

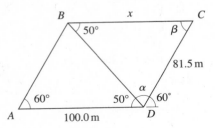

24. $\overline{MB}$ is an edge view of a mirror (see figure). Light from an object at A strikes the mirror at C and is reflected to the eye at E. The person's mind projects the mirror to A', forming the image at A'. It is known that $\angle MCE = \angle ACB$ and that $\overline{CB} \perp \overline{AA'}$. Prove that $\overline{AB} = \overline{A'B}$.

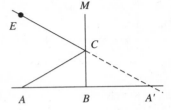

25. A bridge is to be constructed across a river. An engineer wearing a baseball cap estimates the width of the river by standing on one bank, looking at the opposite bank, and tipping the visor of the cap until the line of sight is just barely cut off (a). The engineer then looks down the river bank on this side and notes where the line of sight is allowed by the visor of the cap (b). The engineer paces off the second distance, discovering that it is 120 yards. Sketch the two triangles formed by the engineer's view and estimate the distance across the river.

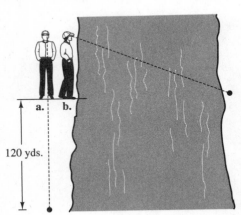

26. You are asked to measure the height of an inside wall of a warehouse. Since you cannot find a ladder tall enough, you borrow a pocket mirror from a salesclerk and place the mirror on the floor. You then move away from the mirror until you can see the reflection of the top of the wall in the mirror (see figure).
    a. Explain why $\triangle HFM \sim \triangle TBM$.
    b. If your eyes are $5\frac{1}{2}$ feet above the floor, $(\overline{HF})$, you are $2\frac{1}{2}$ feet from the mirror $(\overline{FM})$, and the mirror is 20 feet from the wall, how high is the wall $(\overline{BT})$?

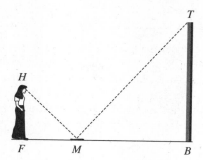

27. You are floating in a balloon $\frac{1}{2}$ mile above Geneseo, New York. Geneseo is 60 miles from Lake Ontario. Can you see Lake Ontario from Geneseo? Use the figure and assume that it is a very clear day! (Recall that the radius of the Earth is 4000 miles.)

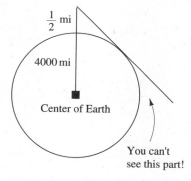

28. As shown, the stress on a member (a brace) of a bridge assembly is found by determining the distance $\overline{OC}$. O is the intersection of $\overleftrightarrow{CA}$ and $\overleftrightarrow{DE}$. What is the distance $\overline{OC}$?

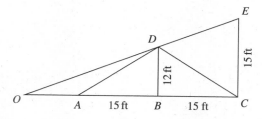

29. Often two similar triangles can have different units of measurement. For example, when a block rests on the incline plane ABC, as shown, the forces can be represented by sides TR, TS, and RS of $\triangle TRS$. Given that TR is parallel to BC and that the right angles are given as shown:
   a. Prove that triangles ABC and TRS are similar.
   b. List the corresponding angles and sides of the similar triangles.

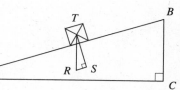

3.3

Perimeter and Area of Plane Figures

The **perimeter** of a polygon is the sum of the lengths of the sides. For the tail section in Example 4 in Section 3.2, the perimeter is the sum of the lengths of the line segments $\overline{AB}$, $\overline{BC}$, $\overline{CD}$, and $\overline{DA}$. These lengths, respectively, are 4.0 m, 5.6 m, 6.4 m, and 6.1 m. The perimeter of the trapezoid is 22.1 m. The **area** of a polygon is the region enclosed by its sides. The measure of area is given in square units. The formulas used to find the perimeters and areas of plane figures are given in Table 3.3.

TABLE 3.3   Perimeters and Areas of Plane Figures

Figure	Sketch	Perimeter (P)	Area (A)
Triangle		$P = s_1 + s_2 + b$	$A = \dfrac{1}{2}ba$
Trapezoid		$P = s_1 + s_2 + b_1 + b_2$	$A = \dfrac{1}{2}(b_1 + b_2)a$
Parallelogram		$P = 2\ell + 2w$	$A = \ell a$
Rhombus		$P = 4s$	$A = sa$
Rectangle		$P = 2\ell + 2w$	$A = \ell w$
Square		$P = 4s$	$A = s^2$

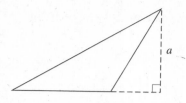

FIGURE 3.39

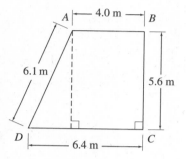

FIGURE 3.40

The Egyptians used the Pythagorean relation as a tool to survey land and build pyramids thousands of years ago. When the Greek mathematician, Pythagoras, proved the relationship to be true, he was so impressed with its power that he sacrificed 40 oxen to the gods.

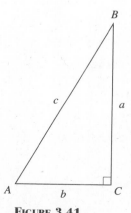

FIGURE 3.41

When discussing the area of a plane figure, such as a trapezoid, we will be asked to find the area of the trapezoid. But what we really are being asked to do is to find the area *inside* the trapezoid.

The letter a in the area formula for the triangle represents the altitude. The altitude is the perpendicular distance from a vertex to the opposite side (or side extended) of the triangle (Figure 3.39). In a formula for the area of a quadrilateral, the letter a again represents altitude, which is the perpendicular distance between two parallel sides.

EXAMPLE 1 **Area Of A Trapezoid**

Find the area of the trapezoid-shaped tail section in Figure 3.40.

Solution We will show two different methods of solving the problem. *Method 1:* Draw a line from point A that is perpendicular to line DC (see Figure 3.40). This line divides the trapezoid into a right triangle and a rectangle. By finding the area of each separately and then adding the results, we determine the area of the trapezoid.

Triangle:
$$A = \frac{1}{2}ba$$
$$= \frac{1}{2}(2.4)(5.6)$$
$$= 6.72 \text{ m}^2$$

Rectangle:
$$A = lw$$
$$= (5.6)(4.0)$$
$$= 22.4 \text{ m}^2$$

Trapezoid:
$$A = 6.72 + 22.4$$
$$= 29.12 \text{ or } 29 \text{ m}^2$$

Method 2: Use the formula for finding the area of a trapezoid given in Table 3.3.

Trapezoid:
$$A = \frac{1}{2}(b_1 + b_2)a$$
$$= \frac{1}{2}(6.4 + 4.0)(5.6)$$
$$= 29.12 \text{ or } 29 \text{ m}^2 \quad \blacklozenge$$

The results in Example 1, are the same by both methods. In some situations, as this example illustrates, we may be we able to break up the figure into recognizable polygons to find the area of a complex figure.

When we try to determine the altitude of a right triangle, we encounter a special case: The altitude, a, is actually one of the sides. Additionally, a special relationship exists between the lengths of the sides of the right triangle. This is known as the **Pythagorean relation:** $a^2 + b^2 = c^2$. Figure 3.41 shows a conventional labeling of a right triangle. Note that the right angle is $\angle C$ and that the side opposite the right angle is side c. A similar situation holds for the other two angles and sides: Side a is opposite $\angle A$, and side b is opposite $\angle B$. Under these conditions the Pythagorean relation would state that $c^2 = a^2 + b^2$. Since this relationship can be applied in countless situations, it is an important equation in geometry.

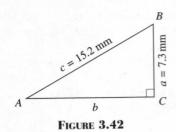

FIGURE 3.42

EXAMPLE 2 **Pythagorean Relation: Perimeter and Area of Right Triangle**

Find the perimeter and the area of the right triangle in Figure 3.42.

Solution In order to find the perimeter of the triangle, we need to know the length of side b. The length of side b can be found by using the Pythagorean relation.

$$c^2 = a^2 + b^2.$$

Solving for b

$$b^2 = c^2 - a^2$$
$$b = \sqrt{c^2 - a^2}.$$

Substituting:

$$b = \sqrt{(\,15.2 \text{ mm}\,)^2 - (\,7.3 \text{ mm}\,)^2}$$
$$= \sqrt{231.04 \text{ mm}^2 - 53.29 \text{ mm}^2}$$
$$= \sqrt{177.75 \text{ mm}^2}$$
$$= 13.3 \text{ mm}$$

Now we can find the perimeter using a variation of the formula given in Table 3.3, where $a = s_1$ and $c = s_2$.

$$P = a + b + c$$
$$= 7.3 \text{ mm} + 13.3 \text{ mm} + 15.2 \text{ mm} \qquad \text{Substituting}$$
$$= 35.8 \text{ mm}.$$

We find the area by substituting in the formula

$$A = \frac{1}{2}ba$$

$$= \frac{1}{2}(\,13.3 \text{ mm}\,)(\,7.3 \text{ mm}\,) \qquad \text{Substituting}$$

$$= 49. \text{ mm}^2. \quad \blacklozenge$$

EXAMPLE 3 **Perimeter and Area of a Parallelogram**

Find the perimeter and area of the parallelogram in Figure 3.43.

Solution To find the perimeter, use the formula

$$P = 2l + 2w$$
$$= 2(\,15 \text{ m}\,) + 2(\,10 \text{ m}\,) \qquad \text{Substituting}$$
$$= 30 \text{ m} + 20 \text{ m}$$
$$= 50 \text{ m}.$$

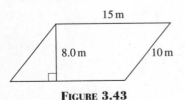

FIGURE 3.43

The area is found by using the formula

$$A = la$$

$$A = (\,\boxed{15 \text{ m}}\,)(\,\boxed{8.0 \text{ m}}\,) \quad \textbf{Substituting}$$

$$= 120 \text{ m}^2. \quad \blacklozenge$$

The perimeter of a **regular polygon** (a polygon in which all sides are equal) can be determined using the formula

$$P = ns$$

where n is the number of sides, and s is the length of each side. The area of a regular polygon can be determined by the formula

$$A = n\left[\frac{1}{2}(as)\right].$$

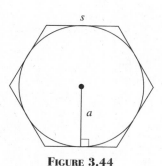

FIGURE 3.44

In the formula, a is called the **apothem,** which is the radius of the inscribed circle, or the perpendicular distance from the center of the polygon to one side (Figure 3.44).

If the regular polygon is divided into triangles as in Figure 3.45, then $\frac{1}{2}(as)$ is the area of one of the triangles of the regular polygon. Multiplying the area of one triangle by n, the number of triangles, which is the same as the number of sides, gives the area of the regular polygon.

If we rearrange the factors in the formula, we have

$$A = \left(\frac{1}{2}a\right)(ns),$$

where

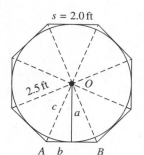

FIGURE 3.45

$$ns = P \qquad \textbf{\textit{P} is the perimeter}$$

$$= \frac{1}{2}aP \qquad \textbf{Substituting for \textit{ns}}$$

EXAMPLE 4 **Finding Perimeter and Area of a Regular Polygon**

Find the perimeter and area of the octagonally shaped regular polygon in Figure 3.45.

Solution To find the perimeter, use the formula

$$P = ns$$

$$= \boxed{8}(\,\boxed{2.0 \text{ ft}}\,) \quad \textbf{Substituting}$$

$$= 16 \text{ ft.}$$

The formula for finding the area is $A = \frac{1}{2}aP$. Before we can find the area, we need to know the length of the apothem, a. The lines joining the vertices of the octagon and the center form eight congruent isosceles triangles. Thus, $\overline{AO} = \overline{BO} = 2\frac{1}{2}$ ft. The altitude of an isosceles triangle bisects the base. Thus the apothem will bisect $\overline{AB}$ and can be found by using the Pythagorean relation.

$$c^2 = a^2 + b^2$$
$$a = \sqrt{c^2 - b^2} \qquad \textbf{Solving for } a$$
$$= \sqrt{(\boxed{2.5 \text{ ft}})^2 - (\boxed{1 \text{ ft}})^2} \qquad \textbf{Substituting}$$
$$= \sqrt{6.25 \text{ ft}^2 - 1 \text{ ft}^2}$$
$$= \sqrt{5.25} \text{ ft.}$$

Substituting in the formula for the area:

$$A = \frac{1}{2}aP$$
$$= \frac{1}{2}(\sqrt{\boxed{5.25}} \text{ ft})(\boxed{16 \text{ ft}})$$
$$= 8\sqrt{5.25} \text{ ft}^2 \quad \text{or} \quad 18. \text{ ft}^2. \quad \blacklozenge$$

This technique will work for finding the area of any **regular polygon.**

The formulas for finding the **circumference,** c (distance around), and the **area,** A, of a circle are found in Table 3.4.

TABLE 3.4

Figure	Sketch	Circumference (c)	Area (A)
Circle		$c = 2\pi r$	$A = \pi r^2$
		$c = \pi d$	$A = \dfrac{\pi d^2}{4}$

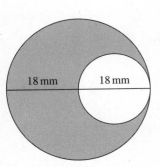

FIGURE 3.46

EXAMPLE 5   Application: Area of a Circular Region

In performing a pressure test, an engineer needs to know the area of the shaded region in Figure 3.46. Knowing the area will help determine the amount of pressure on a tube, the unshaded region. Determine the area of the shaded region.

Solution The area of the shaded region can be found by subtracting the area, A_1, of the smaller circle from the area, A_2, of the larger circle.

$$A = A_2 - A_1$$
$$= \pi r_2^2 - \pi r_1^2 \qquad \text{r_2 and r_1 are the radii of the respective circles}$$
$$= \pi(r_2^2 - r_1^2) \qquad \text{Common-term factoring}$$
$$= \pi[(\;18\text{ mm}\;)^2 - (\;9\text{ mm}\;)^2] \qquad \text{Substitute for r_2 and r_1}$$
$$= 243\text{ mm}^2 \qquad \text{Square and then subtract}$$
$$= 760\text{ mm}^2. \quad \blacklozenge$$

CAUTION In Example 5, the radii must be squared before subtracting. That is $r_2^2 - r_1^2 \neq (r_2 - r_1)^2$. Substitute numbers for r_1 and r_2 and show that the statement is not true. ■

3.3 Exercises

In Exercises 1–10, determine the perimeter and area of the figures.

1.
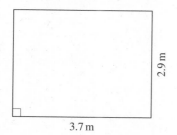
2.9 m
3.7 m

2.
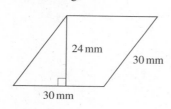
24 mm
30 mm
30 mm

3.
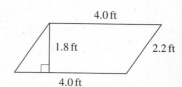
4.0 ft
1.8 ft
2.2 ft
4.0 ft

4.
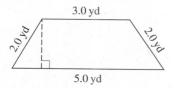
3.0 yd
2.0 yd
2.0 yd
5.0 yd

5.
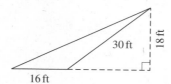
30 ft
18 ft
16 ft

6.
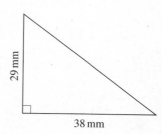
29 mm
38 mm

7.
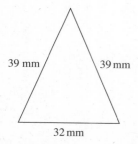
39 mm
39 mm
32 mm

8.
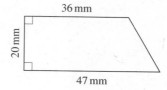
36 mm
20 mm
47 mm

9.

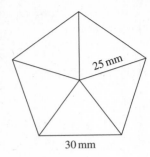

25 mm

30 mm

10.

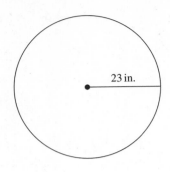

23 in.

11. To form a bolt with a square head as shown in the figure, four equal pieces are cut from a round bronze rod.
- **a.** What is the size of the central angle drawn in the figure?
- **b.** If $r = 1.000$ in., determine the total area of the pieces that are cut away to make the head of the bolt.

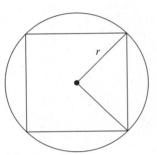

r

12. A bolt with a hexagonal head is cut from a bronze rod as shown.
- **a.** What is the measure of $\angle AOB$? Knowing that side $\overline{AO} = \overline{BO} = r$, what must be the measure of $\angle OAB$ and $\angle OBA$?
- **b.** How is the length of AB related to r?
- **c.** If $r = 1.000$ in., find the length of the apothem from O to $\overline{AB}$.
- **d.** Determine the total area of the pieces that are cut away to make the head of the bolt.

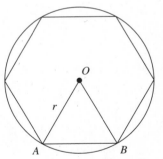

O

r

A B

13. A real estate developer bought a parcel of land that was subdivided as indicated in the figure.
- **a.** Find the area of Lot 1.
- **b.** Find the area of Lot 2.
- **c.** Find the area of Lot 3.
- **d.** Find the area of Lot 4. (*Hint:* Lot 5 contains the missing data.)
- **e.** In Lot 5, the measurement 114.86 ft was taken perpendicular to the lot line. What is the area of the parallelogram?
- **f.** What kind of quadrilateral is lot 6? Find its area.
- **g.** Using the results of **a–f**, find the total area of the parcel of land.

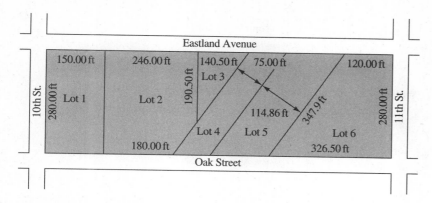

14. How many square feet of concrete walk will it take to go around a circular swimming pool 32. ft in diameter if the width of the walk is to be 4.0 ft?

15. If a circular disk with a diameter of 4.00 ft is cut from a piece of metal 4.00 ft square, what percent of the metal is wasted?

16. How many 3-in. diameter pipes will be needed to carry the flow of a 15-in. diameter pipe? The flow of water depends on the area of the cross-section of the pipe.

17. How do the areas of the bases of a cone compare if the radii are 3 in. and 9 in., respectively?

18. If you wish to double a square in area, by how much must you multiply each side?

19. Without changing the rate of flow, the amount of water flow through a pipe was doubled. Does this mean that the radius of the pipe delivering the water was doubled? By what factor should the diameter be changed?

20. If a map is drawn to a scale of 1 in. : 2000 ft, and a square region on the map has an area of 3.6 in^2, what is the area of the region in the real world in square inches? In square feet?

21. The length, width, and height of the classroom in the figure are 9.0 m, 11 m, and 3.5 m, respectively. What is the length of the diagonal from one corner on the floor to the opposite corner on the ceiling.

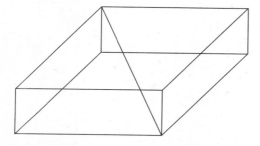

22. A farmer wants to build a pen. He has enough fencing material to build a pen with a perimeter of 200 ft. He can build the pen in any one of the following shapes: equilateral triangle, square, hexagon, or circle. Which shape should the farmer select if his pen is to be the maximum area? What is the area of the largest pen?

23. A baseball diamond measures 90 ft from home plate to first base and 90 ft from first to second base. Using these measurements, how far is it from home plate to second base?

24. A single steel rail is constructed to be a mile long (5280 ft) and placed down during the winter; it is wedged between the rest of the railroad tracks so that it cannot move along the plane of the Earth. Due to the summer heat, it expands a total of $\frac{1}{4}$ ft, thus buckling in the center as shown.
    a. Guess how high the midpoint is forced up from ground level.
    b. Calculate how high the midpoint is forced up from ground level.

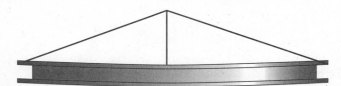

25. The gravitational force always points straight down to the center of the Earth. However, this force can be broken down into two parts, or components, as shown in the figure. The two components are situated so that they are always perpendicular to one another. If the component f_1 is 25 Newtons and if the weight of an object is 75 Newtons, what is the force of the f_2 component?

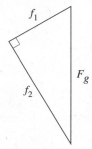

26. Determine the hypotenuse in the following examples:

Leg 1	Leg 2	Hypotenuse
12.0	8.0	
12.0	2.0	
12.00	0.50	
12.0	0.10	

Now complete the following statement. If one leg is much greater than the other leg, then the hypotenuse is _____ as the larger leg.

27. A goldfish starts at a point on the edge of a circular goldfish pond and swims due north 60.0 ft and hits the edge of the pond. The fish then turns and swims due east 80.0 ft and hits the edge of the pond again. What is the area of the pond? (*Hint:* Sketch the path of the fish inside a circle.) What is the geometrical name for the line joining the two points where the fish hit the edge of the pond?

3.4

Geometric Solids, Surface Area, Volume

The proof that Archimedes (287–212 B.C.) was most proud of was proof that the volume of a sphere is two-thirds that of the smallest cylinder enclosing it. In fact, he requested that a diagram of a sphere in a cylinder be engraved on his tombstone. When the Romans broke through the defenses of Syracuse where he lived, a Roman soldier found him drawing circles in the sand. Archimedes exclaimed, "Do not disturb my circles." The soldier then killed him. The location of his tomb was forgotten for over a century. A century after his death, Cicero, the Roman governor of Sicily, discovered a tombstone with a figure of a sphere inside a cylinder, which marked the burial place of Archimedes.

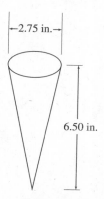

FIGURE 3.47

Familiar shapes that suggest volume are the cube, the rectangular prism (a box), the cone (ice cream cone), the pyramid, and the sphere (a ball). The solids are illustrated in Table 3.5 along with the formulas used to find their surface areas and volumes. One of the most common figures for which you will need to calculate the volume is the right rectangular prism. A **prism** is a geometric figure whose bases are parallel and equal, and whose surfaces are polygons. The **lateral area,** L, of a prism is the total area of the faces other than the bases. The total area, T, of a prism is the lateral area plus the area of the bases. The volume of any prism can be found by multiplying the area of the base times the height, or $V = Bh$. The formula commonly used for finding the volume of a rectangular prism is $V = lwh$, where l times w is the area of the base. To find the volume of a cone, use the following formula:

$$V = \frac{1}{3}\pi r^2 h$$

were πr^2 is the area of the base and h is the height.

EXAMPLE 1   Volume of a Cone

Jose is designing a mold to make a plastic glass in the shape of a cone (Figure 3.47). The inside diameter of the glass is 2.75 in., and the inside height is 6.50 in. How many ounces of liquid can the glass hold if it is filled within 0.25 in. of the top?

Solution   The volume of the cone can be determined using the formula

$$V = \frac{1}{3}\pi r^2 h$$

were r is the radius, and h is the height. Given that the diameter is 2.75 in., the radius is 1.375 in. Since we are asked to determine the amount of liquid within 0.25 in. of the top, the height is 6.50 in. − 0.25 in., or 6.25 in.

$$V = \frac{1}{3}\pi (\boxed{1.375})^2 (\boxed{6.25}) \quad \text{Substituting}$$
$$= 12.374 \text{ in}^3$$

TABLE 3.5

Cube

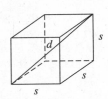

$d = s\sqrt{3}$
$L = 4s^2$
$T = 6s^2$
$V = s^3$

Right rectangular prism

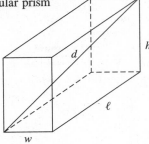

$d = \sqrt{\ell^2 + w^2 + h^2}$
$L = 2(wh + \ell h)$
$T = 2(\ell w + wh + \ell h)$
$V = \ell wh$ or $V = Bh$

Right triangular prism

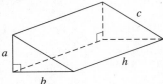

$L = ah + bh + ch$
$T = ah + bh + ch + ah$
$V = \dfrac{1}{2}\, abh$ or $V = Bh$

Pyramid

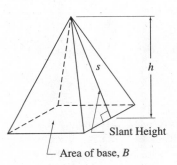

Slant Height

Area of base, B

$L = \dfrac{1}{2}\, Ps$ (P = Perimeter of base)

$T = \dfrac{1}{2}\, Ps + B$

$V = \dfrac{1}{3}\, Bh$

Frustum of a square pyramid

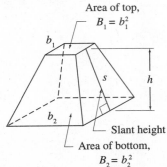

Area of top,
$B_1 = b_1^2$

Slant height

Area of bottom,
$B_2 = b_2^2$

$L = 2(b_1 + b_2)s$
$T = 2(b_1 + b_2)s + B_1 + B_2$

$V = \dfrac{h}{3}(B_1 + B_2 + \sqrt{B_1 B_2})$

Cylinder

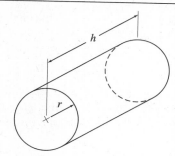

$$L = 2\pi rh$$
$$T = 2\pi rh + 2\pi r^2$$
$$V = \pi r^2 h$$

Portion of a cylinder

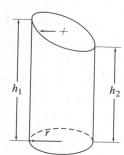

$$L = \pi r(h_1 + h_2)$$
$$V = \frac{\pi r^2}{2}(h_1 + h_2)$$

Cone

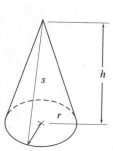

$$L = \pi r \sqrt{r^2 + h^2} \quad \text{or} \quad L = \pi rs$$
$$T = \pi r^2 + \pi rs$$
$$V = \frac{1}{3}\pi r^2 h$$

Frustum of a cone

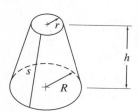

$$L = \pi(R + r)\sqrt{h^2 + (R - r)^2} \quad \text{or}$$
$$L = \pi(R + r)s$$
$$T = \pi(R + r)s + \pi R^2 + \pi r^2$$
$$V = \frac{\pi}{3}(R^2 + Rr + r^2)h$$

Sphere

$$T = 4\pi r^2 \quad \text{or} \quad T = \pi d^2$$
$$V = \frac{4}{3}\pi r^3$$

Since 1 in.^{3} = 0.5541 oz, we can divide both sides by 1 in.^{3}, and the result is the fraction $\dfrac{0.5541 \text{ oz}}{1 \text{ in 3}}$. Multiplying the volume by the fraction, we have

$$\frac{12.374 \text{ in.}^3}{1} \cdot \frac{0.5541 \text{ oz}}{1 \text{ in.}^3} = 6.86 \text{ oz.}$$

The glass will hold 6.86 oz of liquid if it is filled to 0.25 in. from the top. However, if the glass is filled to the top, it holds more than 7 oz; in fact, it holds 7.13 oz. With this capacity Jose would call it a 7-oz glass.   ◆

EXAMPLE 2   Height and Surface Area of a Cylinder

An engineer is asked to design a cylindrical container that will hold 8.0 oz of liquid. A second condition is that the diameter of the container be 2.25 in. so that it is easy to pick it up with one hand.

a.   Determine the height of the cylinder.
b.   Determine the surface area of the cylinder (see Figure 3.48)

h

—2.75 in.—

FIGURE 3.48

Solution

a.   To determine the height of the cylinder, we must first change the volume, 8.0 oz, to cubic inches. Then use the formula to find the volume of a cylinder, $V = \pi r^2 h$. Using the fact that 1 in.^{3} = 0.5541 oz, perform the following multiplication to convert ounces to cubic inches.

$$(8.0 \text{ oz})\left(\frac{1 \text{ in.}^3}{0.5541 \text{ oz}}\right) = 14.4 \text{ in.}^3$$

$$V = \pi r^2 h$$

$$h = \frac{V}{\pi r^2} \qquad \textbf{Solving for } h$$

$$= \frac{14.4 \text{ in.}^3}{\pi(\,1.125 \text{ in.}\,)^2} \qquad \textbf{Substituting for } V \textbf{ and } r$$

$$= 3.62 \text{ in.}$$

b.   To determine the surface area, use the formula

$$T = 2\pi r h + 2\pi r^2$$

$$= 2\pi(\,1.125 \text{ in.}\,)(\,3.62 \text{ in.}\,) + 2\pi(\,1.125\,)^2 \quad \textbf{Substituting}$$

$$= 33.5 \text{ in.}^2$$

A cylindrical container that holds 8.0 oz and has a diameter of 2.25 in. will have a height of 3.6 in. and a total surface area of 34 in^2.   ◆

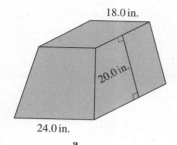

18.0 in.

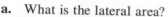

20.0 in.

24.0 in.

a.

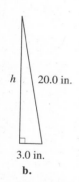

h 20.0 in.

3.0 in.

b.

FIGURE 3.49

If the top of a cone or pyramid is sliced off parallel to the base, the new figure is called a **frustum of a cone** or a **frustum of a pyramid.**

EXAMPLE 3   **Application: Volume of a Frustum of a Pyramid**

The frustum of a right regular pyramid, Figure 3.49a, is used as the base of a large post. Each edge of the lower base is 24.0 in., each edge of the upper base is 18.0 in., and the slant height is 20.0 in.

a. What is the lateral area?
b. What is the volume of the base?

Solution   To find the lateral area, use the formula

$$L = 2(b_1 + b_2)S \text{ from Table 3.5}$$
$$= 2(24.0 \text{ in. } + 18.0 \text{ in.})(20.0 \text{ in.}) \quad \textbf{Substituting}$$
$$= 1680 \text{ in.}^2$$

To find the volume, use the formula

$$V = \frac{h}{3}(B_1 + B_2 + \sqrt{B_1 B_2}).$$

B_1 is the area of the upper base; thus

$$B_1 = (18.0 \text{ in.})^2 = 324.0 \text{ in.}^2$$

B_2 is the area of the lower base; thus,

$$B_2 = (24.0 \text{ in.})^2 = 576.0 \text{ in.}^2$$

h is the perpendicular distance between the two bases. Taking a slice along the edge, we see that h is the height of a right triangle as shown in Figure 3.49b. Using the Pythagorean relation, we have

$$h = \sqrt{c^2 - b^2}$$
$$= \sqrt{(20.0 \text{ in.})^2 - (3.0 \text{ in.})^2}$$
$$= 19.8 \text{ in.}$$

Substituting in the formula we have

$$V = \frac{19.8 \text{ in.}}{3}(324.0 \text{ in.}^2 + 576.0 \text{ in.}^2 + \sqrt{(324.0 \text{ in.}^2)(576.0 \text{ in.}^2)}$$
$$= 8791 \text{ in.}^3 \quad \text{or} \quad 8800 \text{ in.}^3 \quad \blacklozenge$$

What happens when we do not have a regular geometric solid? The best way to handle this situation is to divide the solid into regular geometric solids.

EXAMPLE 4   **Application: Pier Problem**

For the situation posed at the beginning of the chapter, calculate the amount of concrete needed to build up the old section of pier so that it is the same size as the new part. A common measure of concrete volume is the yard, which is actually 1 cubic yard, or 27 cubic feet. Find the amount of concrete needed in yards.

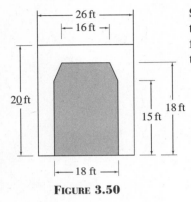

FIGURE 3.50

Solution   Looking at the sketch in Figure 3.50, we can see that the cross-sectional area of the concrete needed can be found by subtracting the shaded area from the area of the outer rectangle. If this resulting area is then multiplied by the length of the pier to be rebuilt, we will obtain the volume of concrete needed.

$$A = \text{outer rectangle} - \text{shaded area}$$
$$= \text{outer rectangle} - (\text{rectangle} + \text{trapezoid})$$
$$= (26 \text{ ft})(20 \text{ ft}) - [(18 \text{ ft})(15 \text{ ft}) + \frac{1}{2}(16 \text{ ft} + 18 \text{ ft})(3 \text{ ft})]$$
$$= 520 \text{ ft}^2 - (270 \text{ ft}^2 + 51 \text{ ft}^2)$$
$$= 199 \text{ ft}^2$$
$$V = (199 \text{ ft}^2)(1250 \text{ ft})$$
$$= 248{,}750 \text{ ft}^3$$
$$= (248{,}750 \text{ ft}^3)\left(\frac{1 \text{ yd}}{27 \text{ ft}^3}\right) \quad \textbf{Converting to yards}$$
$$= 9213 \text{ yd}$$

Thus, the amount of concrete needed is approximately 9200 yd.   ◆

A BASIC computer program which can be used to calculate the volume of concrete needed is found on the tutorial diskette.

3.4 Exercises

1.   The radius of the earth is approximately 3958.6 mi. If three-fourths of the earth's surface is water, how many square miles of land are there?

2.   A silo is in the form of a right prism. The side of the octagonal base is 9.2 ft, and the silo is 35.0 ft high. How many gallons of paint will be needed to paint the lateral surface of the silo if one gallon of paint will cover 250 ft² of surface?

3.   How many cubic feet are there in a tank 8.00 ft long, 5.00 ft wide, and 3.00 ft deep?

4.   If the total area of a cube is 78.60 in.², what is its volume?

5.   A farmer has decided to build a cylindrical silo 28 ft high. He wants the silo large enough to provide 1 ft³ of silage for each cow each day. If 25 cows are to be fed for 150 days, what should the diameter of the silo be?

6.   In a classroom 35 ft long, 25 ft wide, and 10 ft high, is there sufficient air space for 36 pupils? Assume 300 ft³ is the recommended allowance for each pupil.

7. The grain bin on a combine is in the shape of an inverted cone as shown in the figure. The diameter of the base of the cone is 58.0 in. and the altitude is 64.0 in. How many square inches of sheet metal are in this open-topped bin?

8. The rectangular swimming pool shown is 75 ft long and 25 ft wide. It is 12 ft deep at one end and 3 ft deep at the other end. The bottom is flat at the deep end for 10 ft and then has a uniform slope to the shallow end.
  a. How many gallons of water does the pool hold?
  b. What is the weight of the water in the pool when the pool is full? (*Hint:* 1 ft³ = 7.5 gal; fresh water weighs 62.4 lb per ft³.)

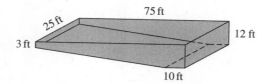

9. A concrete pier for a bridge is shaped as shown in the figure. The bases are rectangles with semicircles added at the ends. Determine the number of cubic yards of concrete needed to build one pier.

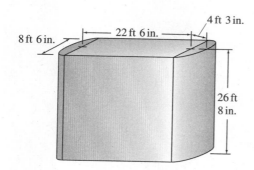

10. Oranges with a diameter of 2.0 in. sell for $1.80 per dozen, while those with a 3.2-in. diameter sell for $3.60 per dozen. Which are more economical to buy?

11. A spherical shell of cast iron has an external diameter of 6.34 in., and the thickness of the shell is 0.523 in. Find its weight if 1 in.³ of cast iron weighs 0.2604 lb.

12. A concrete form used in making bridge supports has the shape of a frustum of a cone with base radii of 12 ft and 18 ft. If the frustum is 36 ft tall and has a 3-ft-diameter hollow cylindrical core, determine the volume of the concrete needed to make one support.

13. Determine the slant height and the lateral surface area of the frustum of the cone in Exercise 12.

14. Compute the weight of a steel washer, as shown, if steel weighs 0.283 lb per cubic inch.

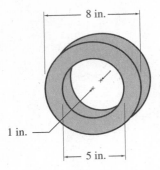

15. As shown, a water tank is shaped like a cylinder with a hemisphere on the bottom and a cone on the top. The diameter of the cylinder is 16.0 ft, its height is 20.5 ft, and the height of the cone is 3.4 ft. Determine the number of gallons of paint required to paint the entire tank inside and outside (cone, cylinder, and hemisphere) with two coats. A gallon of paint will cover 300 ft^2.

16. A motorist traveling in Europe ran out of gas. She walked to a station and was given a circular container 240 mm across the top, 300 mm across the bottom, and 380 mm high. She was told that it contained 30.0 liters of gasoline. Did it? How many liters did it contain?

17. The amount of paint needed to paint a 1 ft × 1 ft × 1 ft cube costs about $1. Suppose a 12 ft × 12 ft × 12 ft cube is painted in the same manner. What is the cost of the paint for the larger cube?

18. How many one in. cubes can you fit into a one ft cube?

19. How many one-in.-diameter spheres can you fit into a one-ft cube? Actually, the amount you can fit into the cube will be less than the amount you compute. Do you know why?

20. An engineer for a highway department has designed a divider to be placed between two lanes of a highway as shown in part a of the figure. A cross-section of the divider is a trapezoid with a semicircle on its top as shown in part b. The upper base, which is also the base of the semicircle, is 18.0 in., and the distance between the upper and lower bases is 36.0 in. (See part c.) The lower base has a rectangular shape that extends 40.0 in. below the finished road surface, and the lower base of the trapezoid is 24.0 in. (See part d.)

a. Find the number of cubic yds of concrete required per linear ft of construction.

b. Find the number of square ft of surface area for each linear ft of the exposed part of the divider.

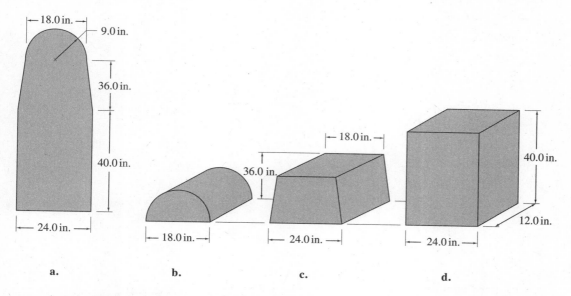

a.   b.   c.   d.

21. The Sells family is thinking about purchasing a circular swimming pool 12 ft in diameter and 4 ft deep. They would like to place the pool on their deck, which is built to support 28,000 lbs. Could they safely place the pool on the deck? (*Hint:* 1 ft³ = 7.5 gal; fresh water weighs 62.4 lb per cubic ft.)

Review Exercises

In Exercises 1–8, use a calculator to change each angle measure to degrees in decimal form. Apply the rules in Table 3.1.

1. 17° 23′

2. 39° 40′

3. 123° 35′

4. 125° 32′ 50″

5. 254° 12′ 20″

6. 147° 27′ 40″

7. 38° 29″ 31″

8. 142° 29′ 43″

In Exercises 9–16, use a calculator to change each angle measure to degrees, minutes, and seconds. Apply the rules in Table 3.1.

9. 33.7°

10. 79.8°

11. 176.14°

12. 8.23°

13. 1.234°

14. 124.789°

15. 25.4321°

16. 276.4587°

In Exercises 17–22, determine for each figure the values of the missing parts (angles and sides) indicated by a letter.

17.

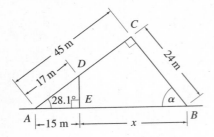

18.

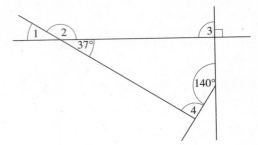

19.

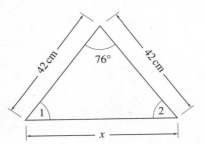

20.

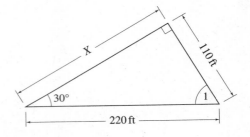

21.

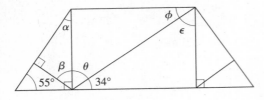

22.

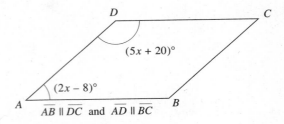

23. Too often mathematics is not associated with romance. To correct that impression, consider the following problem. A ladder used for an elopement is 20 ft long and just reaches a window sill 16 ft above the ground. How far from the house is the foot of the ladder?

24. Two balls have diameters of 30 mm and 90 mm. How much greater is the volume of the second ball than the volume of the first ball?

25. Find the volume of a frustum of a cone if the radius of the lower base is 18.5 in., the radius of the upper base is 11.2 in., and the height is 10.6 in.

26. As shown, a piece of tin 4.0 ft square was bent in half on a line parallel to one edge so that it formed a 90° angle. Other pieces of tin were soldered across the ends to form a drinking trough. The ends of the trough are isosceles triangles, and the open top of the trough is a rectangle.

    a. What is the depth of the trough?

    b. How much tin will it take to make the trough, including the ends?

    c. How many gallons of water are in the trough when it is filled to $\frac{2}{3}$ of the total depth?

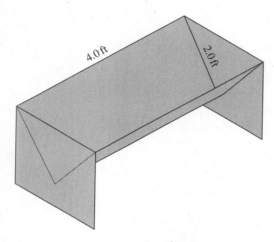

27. If the edges of a cubical box are multiplied by $1\frac{1}{2}$, what effect does this have on the volume?

28. A farmer is designing a bin with a rectangular base to hold 1500 bushels of wheat. He knows that 1 bushel occupies approximately 1.244 ft^3. What should be the height of the bin if the base must be 12.5 ft by 15.4 ft?

29. Find the area of an equilateral triangle whose side is 29.75 mm.

30. A cylindrical fruit can is 4.25 in. high; the diameter of its base is 3.40 in. Find the amount of metal needed to make the can. Do not consider seams and overlapping areas.

31. A manufacturer plans to make lamp shades in the shape of a frustum of a square pyramid. For one shade, each side of the lower base is $6\frac{1}{8}$ in., each side of the upper base is $3\frac{3}{8}$ in., and the distance between bases is $7\frac{3}{4}$ in. Determine the number of square inches of material needed to cover the four sides of the lamp shade.

32. The rain that falls on a horizontal flat roof of a home is carried to a cylindrical cistern 7.00 ft in diameter. The roof is a rectangle that measures 22.0 ft by 28.0 ft. How many inches of rain must fall in order to fill the cistern to a depth of 8.00 ft?

33. A strawberry box, shaped as shown, is 5.0 in. square at the top and 4.5 in. square at the bottom. The box is 3.0 in. deep. Will it hold 1 qt of berries when it is level full, if 1 dry quart equals approximately 67 in³?

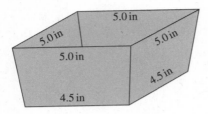

34. A spherical storage tank for gas is 50.0 ft in diameter. How many gallons will it hold? In this case, assume that each cubic foot contains 7.50 gal.

35. A water tank is made up of two geometric solids, a cylinder placed on top of a hemisphere. The diameter of the tank is 22.0 m, and the height of the cylindrical part is 25.0 m. Compute the capacity of the tank in liters.

36. Richard is replacing an oak board in the bed of his trailer. The bed of the trailer has a steel channel on both sides with a 1.50-in. lip as shown in the figure. The width of the trailer bed and the board is 50.0 in. When he places the left end of the board in the channel, how large a block must he place in the center of the trailer bed so that he can bend the board and get it past the lip on the right side?

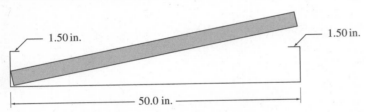

37. How many cubic feet of sand are there in a conical pile 9.5 ft high if the diameter of the base is 13.0 ft?

38. A road bed is cut through a hill as shown. Find the cross-sectional area of the cut-out region.

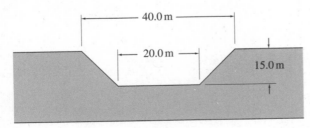

39. A contractor is building a concrete walk with a width of 5.00 ft around the outside of a circular swimming pool. How many cubic yards of concrete are needed for the walk if it is to be 4.00 in. thick?

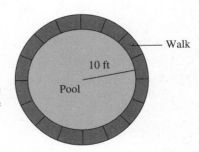

40. For the equilateral triangle in the figure assume that the side is s, the altitude is h, and the area is A. Show that $h = \dfrac{s\sqrt{3}}{2}$ and $A = \dfrac{s^2\sqrt{3}}{4}$.

41. The area of a circle is 36 m². Two circles concentric with the original one are drawn within the circle so that they divide the original area into three equal parts. Find the radii of the two smaller circles.

42. The diagonals of a rhombus intersect at right angles, and one-half the product of these diagonals gives the area. The figure shows the diagonals to be 21.56 ft and 12.45 ft.
   a. Use the diagonals to find the area of the rhombus.
   b. Find the area of the rhombus using the formulas given in Table 3.3 and compare your answers.
   c. The formula for **a** would be $A = \dfrac{1}{2}d_1 d_2$. Show why this formula works.

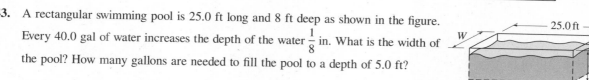
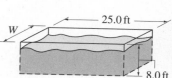

43. A rectangular swimming pool is 25.0 ft long and 8 ft deep as shown in the figure. Every 40.0 gal of water increases the depth of the water $\dfrac{1}{8}$ in. What is the width of the pool? How many gallons are needed to fill the pool to a depth of 5.0 ft?

44. The cylinder in the figure is 28.5 m long and 6.0 m in diameter. How many liters of oil will be needed to fill it to a depth of 5.5 m? (*Hint:* Difficult without trigonometry)

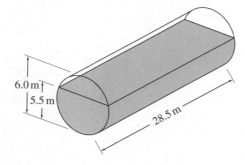

45. The following poem by A. C. Orr pays tribute to Archimedes, the "Immortal Syracusan." It gives the decimal expansion of π, correct to 30 decimal places! Can you see how?

   Now I, even, I, would celebrate
   In rhymes unapt, the great
   Immortal Syracusan, rivaled nevermore,
   Who in his wondrous lore,
   Passed on before,
   Left men his guidance
   How to circles mensurate.
Adam C. Orr, "Immortal Syracusan," *The Literary Digest*, Jan. 20, 1906.

✍ Writing About Mathematics

1. Use the following directions to draw a figure on paper. Draw a line four units north and stop. Continue by drawing three units to the east and stop. Now return to the starting point. What kind of shape is it? If these directions are always strictly followed, will the shape always be a triangle? What is the the total number of internal degrees of the triangle? Discuss the results.

2. Now try this experiment. Sketch the following path on a globe. From a point on the equator, travel north until you get to the North Pole. Turn 90° and travel south until you come to the equator again. Now turn 90° and travel along the equator until you come to your starting point. How many sides are there to the shape you have drawn? What kind of shape do you have? What is the total number of degrees enclosed in the figure? Why don't the rules of triangles you studied in this chapter hold in this case? Go to the library and investigate "other geometries," geometries that are not Euclidean. Write a short paper to explain how this strange thing—a triangle with more than 180°—could happen.

Chapter Test

1. Write 28° 17' in decimal form.

2. Write 29.347° in degrees, minutes, seconds form.

3. Given ∠A = 35.8°, determine the measure of the complement of ∠A.

4. Determine the measure of angle A.

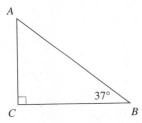

5. Determine the measure of angle A.

6. Determine the measure of angles B and C.

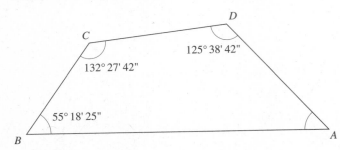

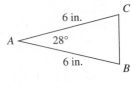

7. Determine the measure of angles 1, 2, 3 and 4.

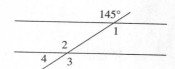

In Exercises 8 and 9, determine a value for x.

8.

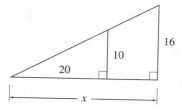

9.

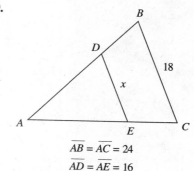

$\overline{AB} = \overline{AC} = 24$

$\overline{AD} = \overline{AE} = 16$

10. Determine the perimeter and the area of the figure.

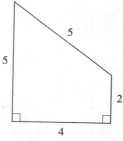

In Exercises 11–13, use the fact that the diameter of a penny is $\frac{3}{4}$ in. and the thickness is $\frac{1}{16}$ in.

11. Determine the circumference of a penny.

12. Determine the area of the face of a penny.

13. Determine the number of cubic inches of metal needed to make a penny.

14. Determine the volume and the surface area (the inside, the outside, and the ends) of a hollow cylinder 100 mm long, with 40 mm outside diameter and 20 mm inside diameter.

15. A spherical shell of cast iron has an external diameter of 7.38 in. The thickness of the shell is 0.625 in. Find its weight if 1 in³. of cast iron weighs 0.2604 lb.

Linear Equations and Dimensional Analysis

The Saturn rocket often moves objects from the surface of earth to outer space. Though you may have seen many fiery blast-offs, have you thought about the forces needed to lift these metal giants into orbit?

The weight of the Saturn 5 before takeoff is 6,391,120 pounds! The thrust of the engine must at least exceed the weight of the rocket, or it will never take off. The first stage thrust of the Saturn 5 rocket is about 8 million pounds, more than enough to insure a lift-off.

The resultant acceleration of the rocket is computed by determining the difference between the downward pull of its weight and the upward push of the thrust. The equation is $F_{net} = ma$ (m stands for the mass of the object, which in this case is 198,000 slugs). What is the acceleration of the rocket lift-off? The solution is given in Section 4.4, Example 6.

4.1

Solving Linear Equations

In the review of algebraic concepts in Chapter 2 we worked with algebraic expressions. If two expressions have an equal sign between them, the two expressions form an **equation.** Equations are called **linear equations** when the exponent of the variable is 1. Examples of linear equations in one variable are listed below.

$$2x + 3 = x + 4$$
$$4x - 12 = 4(x - 3)$$
$$\frac{x + 4}{3} = \frac{x + 5}{2}$$
$$3x + 2 = 3(x + 1)$$

An equation is a tool used in solving problems. The question asked by the equation is "What value (or values) may be substituted for the variable that will make the equation a true statement of equality?" For the equation, $2x + 3 = x + 4$, we can show that, if $x = 1$, the expression on the right side of the equal sign and the expression on the left side of the equal sign will have the same value, 5.

$$2x + 3 = x + 4$$
$$2(1) + 3 = 1 + 4$$
$$5 = 5$$

Thus, we say that 1 is the solution (or root) of the equation. The **solution (or root)** of an equation is the number or numbers that make the equation a true statement of equality.

Rules for Solving Equations

Addition property:	If $a = b$, then $a + c = b + c$ for all real numbers a, b, and c.
Subtraction property:	If $a = b$, then $a - c = b - c$ for all real numbers a, b, and c.
Multiplication property:	If $a = b$, then $a \cdot c = b \cdot c$ for all real numbers a, b, and c.
Division property:	If $a = b$, then $a/c = b/c$ for all real numbers a, b, and c, $c \neq 0$.

Example 1 illustrates the properties of addition, subtraction, multiplication, and division that are used in solving linear equations.

EXAMPLE 1   Solving Equations

Solve:

a. $x - 5 = 7$ **b.** $x + 3 = 2$ **c.** $\dfrac{x}{3} = 4$ **d.** $7x = 35$

Solution   To solve the equations we must isolate the variable on one side of the equal sign.

a. To isolate x in the equation $x - 5 = 7$, we need to add 5 to the algebraic expressions on each side of the equal sign.

$$x - 5 \boxed{+ 5} = 7 \boxed{+ 5} \quad \text{Addition property}$$
$$x = 12$$

b. In the equation $x + 3 = 2$, we need to subtract 3 from the algebraic expressions on each side of the equal sign to isolate x.

$$x + 3 \boxed{- 3} = 2 \boxed{- 3} \quad \text{Subtraction property}$$
$$x = -1$$

c. For the equation $\dfrac{x}{3} = 4$, we need to multiply the algebraic expressions on each side of the equal sign by 3 $\left(\text{the reciprocal of } \dfrac{1}{3}\right)$ to isolate x.

$$\dfrac{x}{3} \boxed{\cdot 3} = 4 \boxed{\cdot 3} \quad \text{Multiplication property}$$
$$x = 12$$

d. For the equation $7x = 35$, we need to divide the algebraic expression on each side of the equal sign by 7 to isolate x.

$$\dfrac{7x}{\boxed{7}} = \dfrac{35}{\boxed{7}} \quad \text{Division property}$$
$$x = 5 \quad \blacklozenge$$

Is the solution correct? To check a solution to a problem, substitute the root back into the original equation. If the equation balances (the right side equals the left side), then the solution is correct.

> **Procedure to Solve Linear Equations in One Variable**
>
> 1. Remove the parentheses using the distributive property.
> 2. Combine the like terms on both sides of the equal signs.
> 3. Use the addition and/or subtraction properties to rewrite the equation with all the terms not containing the variable on one side of the equal sign and the terms containing the variable on the other side. The result will be an equation of the form $ax = b$.
> 4. Use the mutiplication or division property to isolate the variable. The result will be of the form $x = c$, where c is a real number.
> 5. Check the solution in the original equation.

EXAMPLE 2 **Solving an Equation and Checking the Solution**

Solve the equations and check the solution.

a. $2x - 5(x - 4) = 14$ **b.** $2 = 5 + 7(p + 1)$

Solution

a.

$$2x - 5(x - 4) = 14$$
$$2x - 5x + 20 = 14 \qquad \textit{Step 1 } \textbf{Distributive property}$$
$$-3x + 20 = 14 \qquad \textit{Step 2 } \textbf{Combining like terms}$$
$$-3x + 20 - 20 = 14 - 20 \qquad \textit{Step 3 } \textbf{Subtraction property}$$
$$-3x = -6$$
$$\frac{-3x}{-3} = \frac{-6}{-3} \qquad \textit{Step 4 } \textbf{Division property}$$
$$x = 2$$

Check:

$$2x - 5(x - 4) = 14, \text{ for } x = 2$$
$$2(2) - 5(2 - 4) = 14$$
$$4 - 5(-2) = 14$$
$$4 + 10 = 14$$
$$14 = 14 \ \checkmark$$

b.

$$2 = 5 + 7(p + 1)$$
$$2 = 5 + 7p + 7 \qquad \textit{Step 1 } \textbf{Distributive property}$$
$$2 = 12 + 7p \qquad \textit{Step 2 } \textbf{Combining like terms}$$
$$2 - 12 = 12 - 12 + 7p \qquad \textit{Step 3 } \textbf{Subtraction property}$$
$$-10 = 7p$$
$$\frac{-10}{7} = \frac{7p}{7} \qquad \textit{Step 4 } \textbf{Division property}$$
$$\frac{-10}{7} = p$$

Check:

$$2 = 5 + 7(p + 1), \text{ for } p = \frac{-10}{7}$$

$$2 = 5 + 7\left(\frac{-10}{7} + 1\right)$$

$$2 = 5 - 10 + 7$$

$$2 = 2 \checkmark \quad \blacklozenge$$

CAUTION   Before performing any operations, be certain that the results of the operation will isolate the variable. Solve

$$3x + 4 = 5$$

$$3x + 4 - 5 = 5 - 5$$

$$3x - 1 = 0$$

Subtracting 5 from both sides of the equal sign did not isolate the variable by itself on one side of the equal sign.   ■

EXAMPLE 3   **Solving an Equation with Decimals**

Solve the equation and check the solution.

$$4.0x - 0.48 = 0.80x + 4.0$$

Solution   This equation may be solved either with the decimals or by multiplying each term by 100 to eliminate the decimals. We will solve the problem with the decimals.

$$4.0x - 0.48 = 0.80x + 4.0$$

$$4.0x - 0.48 + 0.48 = 0.80x + 4.0 + 0.48 \qquad \textit{Step 3} \text{ Addition property}$$

$$4.0x = 0.80x + 4.48$$

$$4.0x - 0.80x = 0.80x - 0.80x + 4.48 \qquad \textit{Step 3} \text{ Subtraction property}$$

$$3.20x = 4.48$$

$$\frac{3.20x}{3.20} = \frac{4.48}{3.20} \qquad \textit{Step 4} \text{ Division property}$$

$$x = 1.4$$

Check:

$$4.0x - 0.48 = 0.80x + 4.0, \text{ for } x = 1.4$$

$$4.0(1.4) - 0.48 = 0.80(1.4) + 4.0$$

$$5.6 - 0.48 = 1.12 + 4.0$$

$$5.12 = 5.12 \checkmark \quad \blacklozenge$$

Now work Example 3 again, but this time multiply each term of the equation by 100. Your answer should be the same, $x = 1.4$.

4.1 Solving Linear Equations 167

EXAMPLE 4 Solving an Equation That Is True for All **x.**

Solve: $4x - 12 = 4(x - 3)$.

Solution

$$4x - 12 = 4(x - 3)$$
$$4x - 12 = 4x - 12 \quad \textit{Step 1 } \textbf{Distributive property}$$
$$4x - 12 + 12 = 4x - 12 + 12 \quad \textit{Step 3 } \textbf{Addition Property}$$
$$4x = 4x$$
$$4x - 4x = 4x - 4x \quad \textit{Step 3 } \textbf{Subtraction property}$$
$$0 = 0$$

The result tells us that we can put any number in place of x, so the solution is all real numbers. ◆

The equation in Example 4 belongs to a special group of equations called **identities.** It is called an identity because the equation is true for all permissible values of the variable.

EXAMPLE 5 Solving an Equation

Solve: $3(x - 4) - 5 = -3(2x + 7) + 5x$.

Solution

$$3(x - 4) - 5 = -3(2x + 7) + 5x$$
$$3x - 12 - 5 = -6x - 21 + 5x \quad \textit{Step 1 } \textbf{Distributive property}$$
$$3x - 17 = -x - 21 \quad \textit{Step 2 } \textbf{Combining like terms}$$
$$3x - 17 + 17 = -x - 21 + 17 \quad \textit{Step 3 } \textbf{Subtraction property}$$
$$3x = -x - 4$$
$$3x + x = -x + x - 4 \quad \textit{Step 3 } \textbf{Addition property}$$
$$4x = -4$$
$$\frac{4x}{4} = \frac{-4}{4} \quad \textit{Step 4 } \textbf{Division property}$$
$$x = -1 \quad ◆$$

The check in Example 5 is left to the student.

4.1 Exercises

In Exercises 1–25 solve each linear equation and check the solution.

1. $r + 5 = 11$ **2.** $3x - 6 = 24$ **3.** $16 = 6n + 4$ **4.** $11 = 3 - 4R$

5. $12 = 14R + 21$ **6.** $15m - 7 = -37$ **7.** $3 + 5r = 7 + 4r$ **8.** $5 - 3r = 7 - 4r$

9. $27 - 6N = 4N - 7N$ **10.** $3r + 6 = 2r + 15$ **11.** $5x + 6 = 2x + 1$ **12.** $12x - 1.2 = 3x + 1.5$

13. $3(s + 2) = 2s + 12$

14. $5s - 7 = 4(11 - s)$

15. $2x + 4 - 5x = 6 - 13x + 11$

16. $2x + 3 - 5 = 7x - 2 + 5x$

17. $0.3a + 4.2 = 3.6$

18. $-4(x + 3) + 2x = 5(3x - 4)$

19. $6S + 3(4 + S) = 8$

20. $4.08 - 6m = 5.2 - 4m$

21. $2.5(2x - 3.2) = -4.1(2.3x + 5.1)$

22. $-13(3x - 5) = 12x - 4(3x + 7)$

23. $23x - 14(3x + 5) = 11(12 - 7x) + 5$

24. $0.34(0.01x - 0.08x) + 0.03 = -0.08(0.07x - 0.06)$

25. $3.4(2x - 2) - 0.8x = 2.4(3x - 4) - 1.2x + 2.8$

4.2

Equations with Fractions and Literal Equations

The same general procedure used to solve the linear equations in the previous section can be used for solving an equation with fractions. However, since many people prefer not to work with fractions we will learn how to simplify the work by using the least common denominator (lcd). We will determine the lcd for the fractions of the equation and multiply both sides of the equation by the lcd. This is illustrated in Example 1.

EXAMPLE 1   **Solving An Equation with Fractions**

Solve the equation and check the solution.

$$\frac{2t}{3} + \frac{1}{2} = \frac{3}{4}$$

Solution   To simplify the equation, find lcd and multiply both sides of the equation by the lcd. The lcd for 2, 3, and 4 is 12.

$$\frac{2t}{3} + \frac{1}{2} = \frac{3}{4}$$

$$12\left(\frac{2t}{3} + \frac{1}{2}\right) = 12\left(\frac{3}{4}\right) \qquad \textbf{Multiplying both sides of equation by the lcd}$$

$$12\left(\frac{2t}{3}\right) + 12\left(\frac{1}{2}\right) = 12\left(\frac{3}{4}\right) \qquad \textbf{Distributive property and reducing}$$

$$8t + 6 = 9$$

Now we have an equation without fractions and can proceed as in the previous section.

$$8t + 6 - 6 = 9 - 6 \qquad \textit{Step 3} \textbf{ Subtraction property}$$

$$8t = 3$$

$$\frac{8t}{8} = \frac{3}{8} \qquad \textit{Step 4} \textbf{ Division property}$$

$$t = \frac{3}{8}$$

Check:

$$\frac{2t}{3} + \frac{1}{2} = \frac{3}{4}, \text{ for } t = \frac{3}{8}$$

$$\frac{2}{3}\left(\frac{3}{8}\right) + \frac{1}{2} = \frac{3}{4}$$

$$\frac{1}{4} + \frac{1}{2} = \frac{3}{4} \checkmark \quad \blacklozenge$$

When solving an equation of the type, $\frac{1}{7}(x + 8) = \frac{2}{5}(x - 4)$, it is generally better to multiply both sides of the equation by the lcd before applying the distributive property. This is illustrated in Example 2.

EXAMPLE 2   **Solving an Equation with Fractions**

Solve the equation

$$\frac{1}{7}(x + 8) = \frac{2}{5}(x - 4)$$

and check the solution.

Solution   The lcd of 5 and 7 is 35, therefore we will multiply both sides of the equation by 35.

$$35\left(\frac{1}{7}\right)(x + 8) = 35\left(\frac{2}{5}\right)(x - 4)$$

$$5(x + 8) = 14(x - 4)$$

$$5x + 40 = 14x - 56 \qquad \textit{Step 1} \text{ Distributive property}$$

$$5x + 40 + 56 = 14x - 56 + 56 \qquad \textit{Step 2} \text{ Addition property}$$

$$5x + 96 = 14x$$

$$5x - 5x + 96 = 14x - 5x \qquad \textit{Step 3} \text{ Subtraction property}$$

$$96 = 9x$$

$$\frac{96}{9} = \frac{9x}{9} \qquad \textit{Step 4} \text{ Division property}$$

$$\frac{32}{3} = x$$

Check:

$$\frac{1}{7}(x + 8) = \frac{2}{5}(x - 4), \text{ for } x = \frac{32}{3}$$

$$\frac{1}{7}\left(\frac{32}{2} + 8\right) = \frac{2}{5}\left(\frac{32}{3} - 4\right)$$

$$\frac{1}{7}\left(\frac{32 + 24}{3}\right) = \frac{2}{5}\left(\frac{32 - 12}{3}\right)$$

$$\frac{1}{7} \cdot \frac{56}{3} = \frac{2}{5} \cdot \frac{20}{3}$$

$$\frac{8}{3} = \frac{8}{3} \checkmark \quad \blacklozenge$$

Sometimes we will encounter equations in which the variable is in the denominator of a fraction. Example 3 illustrates how to handle that situation.

EXAMPLE 3 Variables in the Denominator

Solve the equation and check the solution.

$$\frac{3}{x} + 5 = \frac{2}{x}$$

Solution In this problem x cannot be equal to zero. Why? If $x \neq 0$, we can treat x as a nonzero real number and multiply both sides of the equation by the lcd, x.

$$\frac{3}{x} + 5 = \frac{2}{x}$$

$$x\left(\frac{3}{x} + 5\right) = x\left(\frac{2}{x}\right) \qquad \text{Multiply both sides of the equation by } x$$

$$x\left(\frac{3}{x}\right) + x(5) = x\left(\frac{2}{x}\right)$$

$$3 + 5x = 2$$

$$3 \boxed{-3} + 5x = 2 \boxed{-3} \qquad \textit{Step 3} \text{ Subtraction property}$$

$$5x = -1$$

$$\frac{5x}{\boxed{5}} = -\frac{1}{\boxed{5}} \qquad \textit{Step 4} \text{ Division property}$$

$$x = -\frac{1}{5}$$

Check:

$$\frac{3}{x} + 5 = \frac{2}{x}$$

$$\frac{3}{\dfrac{-1}{5}} + 5 = \frac{2}{\dfrac{-1}{5}}$$

$$\frac{15}{-1} + 5 = \frac{10}{-1}$$

$$-15 + 5 = -10$$

$$-10 = -10 \; \checkmark \; \blacklozenge$$

EXAMPLE 4   **An Equation with No Solution**

Solve the equation and check the solution.

$$\frac{x - 4}{x^2 + x - 12} = \frac{24}{21x + 84}$$

Solution

$$\frac{x - 4}{x^2 + x - 12} = \frac{24}{21x + 84}$$

$$\frac{x - 4}{(x + 4)(x - 3)} = \frac{24}{21(x + 4)} \qquad \text{Factoring the denominator}$$

The lcd is $(x + 4)(x - 3)(21)$. Multiply the left side of the equation by $\dfrac{21}{21}$ and the right side of the equation by $x - 3$.

$$\frac{x - 4}{(x + 4)(x - 3)}\left(\frac{21}{21}\right) = \frac{24}{(21)(x + 4)}\left(\frac{(x - 3)}{(x - 3)}\right)$$

Since the denominators are equal if you multiply both sides by $(x + 4)(x - 3)(21)$, the denominator will divide out, and you are left with

$$21x - 84 = 24x - 72$$

$$21x - 24x - 84 = 24x - 24x - 72 \qquad \textit{Step 3} \text{ Subtraction property}$$

$$-3x - 84 + 84 = -72 + 84 \qquad \textit{Step 3} \text{ Addition property}$$

$$-3x = 12$$

$$\frac{-3x}{-3} = \frac{12}{-3} \qquad \textit{Step 4} \text{ Division property}$$

$$x = -4$$

Check:

$$\frac{x-4}{x^2+x-12}=\frac{24}{21x+84}, \text{ for } x=-4.$$

$$\frac{-4-4}{(-4)^2+(-4)-12}=\frac{24}{21(-4)+84}$$

$$\frac{-8}{0}=\frac{24}{0}$$

What appeared to be the solution, -4, when substituted back into the original equation results in fractions with zeros in the denominators. The value -4 cannot be a solution, and since it is the only possibility, the equation has no solution. ◆

A value that appears to be a solution (as in Example 4) but cannot be due to restrictions on the equation is called an **extraneous solution.**

EXAMPLE 5   **Solving an Equation with Fractions**

Solve: $\dfrac{3}{t-2}=\dfrac{-2}{t-4}.$

Solution   The lcd for the two fractions is $(t-2)(t-4)$. Multiply both sides of the equation by the lcd.

$$(t-2)(t-4)\frac{3}{t-2}=(t-2)(t-4)\frac{-2}{t-4}$$

$$(t-4)(3)=(t-2)(-2)$$

$$3t-12=-2t+4$$

$$3t+2t-12=-2t+2t+4$$

$$5t-12=4$$

$$5t-12+12=4+12$$

$$5t=16$$

$$\frac{5t}{5}=\frac{16}{5}$$

$$t=\frac{16}{5} \quad ◆$$

Often in mathematics and science courses we are given an equation expressed in terms of several variables and are asked to express it in terms of one of those variables. To do this, treat each of the variables, except the one for which you are solving, as if they were constants. Then solve for the variable desired using the properties previously discussed. (Such equations are called **literal equations.**

EXAMPLE 6 **Solving a Literal Equation**

The perimeter of a rectangle is given by the formula, $P = 2l + 2w$. Solve for the length, l.

Solution To solve for l, isolate l on one side of the equal sign and move everything else to the other side of the equal sign.

$$P = 2l + 2w$$

$$P - 2w = 2l + 2w - 2w \qquad \textit{Step 3 } \textbf{Subtraction property}$$

$$P - 2w = 2l$$

$$\frac{P - 2w}{2} = \frac{2l}{2} \qquad \textit{Step 4 } \textbf{Division property}$$

$$\frac{P - 2w}{2} = l, \quad \text{or} \quad \frac{P}{2} - w = l \quad \blacklozenge$$

For the equality $\dfrac{a}{b} = \dfrac{c}{d}$, an equivalent way of writing the equality is $\dfrac{b}{a} = \dfrac{d}{c}$.

$$(bd)\frac{a}{b} = (bd)\frac{c}{d} \qquad \textbf{Multiplying bothsides by } bd$$

$$da = bc$$

$$\frac{da}{ac} = \frac{bc}{ac} \qquad \textbf{Dividing both sides by } ac$$

$$\frac{d}{c} = \frac{b}{a}.$$

this result is used in Example 7.

EXAMPLE 7 **Application: Solving for Total Resistance**

Solve the equation $\dfrac{1}{R_T} = \dfrac{1}{R_1} + \dfrac{1}{R_2}$ for R_T.

Solution Start by adding the fractions on the right side of the equal sign. The lcd of the two fractions is $R_1 R_2$.

$$\frac{1}{R_t} = \left(\frac{1}{R_1}\right)\left(\frac{R_2}{R_2}\right) + \left(\frac{1}{R_2}\right)\left(\frac{R_1}{R_1}\right)$$

$$\frac{1}{R_T} = \frac{R_2 + R_1}{R_1 R_2}$$

Since the two fractions are equal to each other, we can take the reciprocal of both sides of the equation and write

$$R_T = \frac{R_1 R_2}{R_1 + R_2} \quad \blacklozenge$$

4.2 Exercises

In Exercises 1–20 solve the equations and check the solution.

1. $\frac{1}{2}x + \frac{1}{3} = \frac{2}{3}$

2. $\frac{1}{2}y + \frac{1}{3} = \frac{1}{4}$

3. $\frac{R}{4} + 2R = -\frac{2}{3}$

4. $\frac{2}{3}(I + 4) = 5 + 3I$

5. $\frac{15}{Y} + 4 = 9$

6. $\frac{4}{3}(R + 3) = 36$

7. $\frac{1}{R} + \frac{2}{R} + \frac{3}{R} = 2$

8. $\frac{5}{y} + 1 = \frac{6}{y}$

9. $\frac{2t - 1}{2t} = \frac{3}{t} + \frac{1}{2}$

10. $\frac{3}{5x} + \frac{2x - 3}{3x} = 4$

11. $\frac{1}{a - 1} = \frac{2}{a - 2}$

12. $\frac{3}{x - 2} = \frac{2}{x - 3}$

13. $\frac{2}{x - 5} = \frac{10}{x - 5}$

14. $\frac{6}{x - 3} = \frac{18}{x - 3}$

15. $\frac{3}{x^2 - 25} = \frac{4}{x^2 - 7x + 10}$

16. $\frac{2}{x^2 - 5x + 4} = \frac{3}{x^2 - 16}$

17. $\frac{3}{x^2 - 3x} = \frac{4}{x^2 - 4x}$

18. $\frac{5}{x^2 - 5x} = \frac{7}{x^2 - 7x}$

19. $\frac{9}{x^2 - 4} = \frac{8}{x^2 - 2x}$

20. $\frac{5}{x^2 + 6x + 8} = \frac{8}{x^2 + 2x}$

In Exercises 21–41, solve each of the formulas for the indicated letter.

21. $PV = nRT$, for P (ideal gas law)

22. $i = prt$, for t (simple interest formula)

23. $A = \frac{1}{2}bh$, for h (area of a right triangle)

24. $A + B + C = 180°$, for C (sum of the angles of a triangle)

25. $P = 2l + 2w$, for w (perimeter of rectangle)

26. $s = \frac{1}{2}gt^2$, for g (position as a function of time of a falling body)

27. $C = \frac{5}{9}(F - 32)$ for F (Fahrenheit to Celsius)

28. $A = \frac{1}{2}(b_1 + b_2)h$, for b_1 (area of a trapezoid)

29. $V = \dfrac{1}{3}\pi r^2 h$, for h (volume of a right circular cone)

30. $P = \dfrac{A}{1 + ni}$, for n (present value)

31. $F = \dfrac{Gm_1 m_2}{r^2}$, for m_2 (Newton's law of gravitation)

32. $s = \dfrac{a - rl}{1 - r}$, for r (geometric series)

33. $\dfrac{1}{R_T} = \dfrac{1}{R_1} + \dfrac{1}{R_2}$, for R_2 (electronics)

34. It is necessary to reinforce concrete with steel mesh or rebars in order to have a floor or slab without joints. A formula that can be used to determine the type of steel mesh is

$$A = \frac{12tP}{100}.$$

A is the steel mesh cross-sectional area in square inches, *t* is the concrete's actual thickness in inches, and *P* is the percentage of steel required (15% for crack containment, 5% for reinforcement). Solve for *t*.

35. A formula used to determine the spacing for rebars in concrete is

$$S = \frac{100A}{Pt}.$$

S is the spacing between bars in inches, *A* is the rebar's cross-sectional area in square inches, *t* is the thickness of concrete in inches, and *P* is the percentage of steel required. Solve for *P*.

36. In working with fluid power, a basic measurement for overall heat loss is to subtract useful energy from input energy.

$$E_L = E_1 - E_2.$$

The relationship may also be written:

$$E_L = 1.48QP(1 - \mu)$$

where μ is system efficiency, Q is pump flow (gpm), and P is pump discharge pressure (psig). Solve for P.

37. The formula $I_1 + \dfrac{I_2}{2} + \dfrac{S_1}{\Delta t} - \dfrac{Q_1}{2} = \dfrac{S_2}{\Delta t} + \dfrac{Q_2}{2}$ was developed to indicate the water-storage levels in a flood-control area. I_1 and I_2 are inflow rates at times 1 and 2, Q_1 and Q_2 are outflow rates at times 1 and 2, S_1 and S_2 are storage values at times 1 and 2, and Δt is the time interval (time 2 − time 1).
  a. Solve for S_1.
  b. Solve for S_2 when I_1, Q_1, and S_1 are equal to zero.

38. When two parallel wires are close together, magnetic forces can be a factor in bringing them closer together. The formula that relates this force to the current of one wire and the magnetic field of the other wire is $F = ILB$. Here the B represents the magnetic field, I is the current, and L is the length of the wire that is parallel. Solve for the current.

39. The equation that relates the image distance to object difference and focus is given in the equation

$$\frac{1}{D_0} + \frac{1}{D_i} = \frac{1}{f}.$$

When designing a camera it is useful to know how for away to put the film from the lens. Solve for D_i.

40. In the basic force equation, $F = ma$, the force should be expressed as F_{net}, the sum of all forces acting on an object. Suppose that a submerged can is filled with air, eventually its buoyant force will exceed its weight. The next force can be expressed as

$$F_{net} = F_b - F_g.$$

Write an expression for the subsequent acceleration of the can.

41. The rotational energy of an object is given by the equation

$$E = \frac{1}{2}I\omega^2,$$

where I is the rotational "mass" of an object, and ω is the angular velocity. Another way to think of ω is to relate it to linear velocity, $v = \omega r$. The equation for the rotational mass of a cylinder is, $I = \frac{1}{2}mr^2$. Solve for the linear velocity of a cylinder rotating about its axis in terms of E and m only. This would be the speed of the outer edge of the cylinder.

4.3

Mathematical Operations with Dimensional Units

Common Metric Prefixes

giga	10^9	5 G—earth's population
mega	10^6	$1 M—one million dollars
kilo	10^3	40 km/hr—speed limit
centi	10^{-2}	1.5 cm—width of a thumbnail
milli	10^{-3}	35 mm—camera lens
micro	10^{-6}	1 μm—diameter of a bacterium
nano	10^{-9}	1 ns—time between operations in a computer

The answer of a technical problem has two components: the numerical value and the dimensional units of measurement. Few students forget to include the numerical value in an answer, but many forget the unit of measurement. The number 125.6 is worthless to indicate the amount of water. The answer could be units of gallons, quarts, pints, or even cups. The amount of liquid is dramatically different depending on which unit of measurement we are calculating.

Two systems of measurement are used today. They are the **United States Customary System (USCS)** and the **International System of Units (SI)** (Systeme International d'Unites), sometimes referred to as the metric system. Today the United States is the only industrial country in the world not exclusively using the SI system. Most major industries use the SI system for the purpose of trade. (The metric system was made official by Congress in 1866, but the USCS, which has never been official, has not been replaced.) Because more industries are using the SI system the United States is finding it necessary to adopt its use. Appendix A contains a discussion of the SI system and has tables displaying the SI units. There are also tables for converting from SI units to USCS units and tables for converting from the USCS units to the SI units.

Both systems of measurement have basic dimensional units from which all other units are derived. For example, the SI base unit of length is the meter, the USCS base unit of length is the foot. Some of the basic units for both systems are given in Table 4.1

TABLE 4.1   Quantities and Their Associated Base Units

Quantity	Quantity Symbol	USCS Units		SI (Metric) Units	
		Name	Symbol	Name	Symbol
Length	L	foot	ft	meter	m
Mass (SI base unit)	M, m	slug		kilogram	kg
Force (British base unit)	F	pound	lb	newton	N
Time	t	second	s	second	s
Electric current	I, i	ampere	A	ampere	A
Temperature	T	Fahrenheit	°F	Celsius	°C
Amount of substance	n	mole	mol	mole	mol
Luminous intensity	I	candle power	cp	candela	cd

If the dimensional unit in Table 4.1 is treated as an algebraic quantity, we can see how additional dimensional units can be derived (see Table 4.2).

TABLE 4.2

Quantity	Dimensional Unit	USCS Unit	SI Unit
Area = (length)^2	L^2	ft^2	m^2
Volume = (length)^3	L^3	ft^3	m^3
Weight density = $\dfrac{\text{weight}}{\text{volume}}$	$\dfrac{F}{L^3}$	$\dfrac{lb}{ft^3}$	$\dfrac{N}{m^3}$
Mass density = $\dfrac{\text{mass}}{\text{volume}}$	$\dfrac{M}{L_3}$	$\dfrac{slug}{ft^3}$	$\dfrac{kg}{m^3}$
Velocity = $\dfrac{\text{change in distance}}{\text{time}}$	$\dfrac{L}{t}$	$\dfrac{ft}{s}$	$\dfrac{m}{s}$
Acceleration = $\dfrac{\text{change in velocity}}{\text{time}}$	$\dfrac{L}{t^2}$	$\dfrac{ft}{s^2}$	$\dfrac{m}{s^2}$
Pressure = $\dfrac{\text{force}}{\text{area}}$	$\dfrac{F}{L^2}$	$\dfrac{lb}{ft^2}$	Pa
Work = distance × force	$L \cdot F$	ft · lb	J
Power = $\dfrac{\text{work}}{\text{time}}$	$\dfrac{L \cdot F}{t}$	$\dfrac{ft \cdot lb}{s}$	W
Volt = $\dfrac{\text{work}}{\text{charge}}$	$\dfrac{L \cdot F}{t \cdot I}$	V	V

When we work with dimensional units, we frequently need to change from one unit to another in order to perform the calculations. For example, we may need to change 45 miles per hour to kilometers per hour or to change cubic inches to cubic feet. To perform these conversions, we must identify unit fractions that are called **conversion factors.** Knowing that 5280 feet make up 1 mile, we can write the statement in equation form.

$$5280 \text{ ft} = 1 \text{ mile}$$

If we divide both sides by 1 mile, the result is the conversion factor.

$$\frac{5280 \text{ ft}}{1 \text{ mile}} = 1$$

We could also divide both sides by 5280 ft to arrive at the conversion factor

$$\frac{1 \text{ mile}}{5280 \text{ ft}} = 1.$$

Other conversion factors are $\frac{36 \text{ in.}}{1 \text{ yd}} = 1, \frac{1 \text{ yd}}{36 \text{ in.}} = 1, \frac{1 \text{ km}}{1000 \text{ m}} = 1, \frac{1000 \text{ m}}{1 \text{ km}} = 1,$

$\frac{1 \text{ m}^3}{1 \text{ kl}} = 1, \frac{1 \text{ kl}}{1 \text{ m}^3} = 1, \frac{1 \text{ ft}}{12 \text{ in}} = 1, \frac{12 \text{ in}}{1 \text{ ft}} = 1.$

When we multiply an expression by one of these conversion factors, we are multiplying by the multiplicative identity 1. Therefore, even though we have changed the form of the expression, we have not changed its value.

CAUTION   Note that conversion factors derived from exact units are exact. Those derived from approximate units are approximate. For example, 1 mile is commonly stated as 1.6 km, or more precisely, as 1.609 km. Both values are approximate. Thus, a conversion factor could limit the number of significant digits in the answer. On the other hand, 1 hour is equal to 60 minutes by definition. A conversion factor derived from these units is exact, and it does not affect the number of significant digits. ■

EXAMPLE 1   Conversion Factor: USCS to SI Units

Convert 55 mi/hr (mph) to km/hr.

Solution   To do the conversion, we need the equality 1.0 mi = 1.6 km, which we can convert to a unit fraction by dividing both sides by 1.0 mi.

$$\frac{1.0 \text{ mi}}{1.0 \text{ mi}} = \frac{1.6 \text{ km}}{1.0 \text{ mi}}$$

$$1 = \frac{1.6 \text{ km}}{1.0 \text{ mi}}.$$ **This conversion factor is not exact, since it is derived from approximate units.**

Why divide by 1.0 mi? We need miles in the denominator for our unit-conversion fraction so that the numerator and denominator have a common factor of mi. Multiply

$$55. \frac{\text{mi}}{\text{hr}} = 55. \frac{\overset{1}{\cancel{\text{mi}}}}{\text{hr}} \times \frac{1.6 \text{ km}}{\underset{1}{1.0 \cancel{\text{mi}}}}$$

$$= 88 \frac{\text{km}}{\text{hr}}.$$

We started with average speed and ended with average speed. ◆

EXAMPLE 2   Converting Linear Unit to Linear Unit

Convert 55 mi/hr to feet per second.

Solution   To convert mi/hr to ft/s we need three conversion factors.

$$1 \text{ mi} = 5280 \text{ ft} \qquad 1 \text{ hr} = 60 \text{ min} \qquad 1 \text{ min} = 60 \text{ s}$$

$$\frac{1 \text{ mi}}{1 \text{ mi}} = \frac{5280 \text{ ft}}{1 \text{ mi}} \qquad \frac{1 \text{ hr}}{60 \text{ min}} = \frac{60 \text{ min}}{60 \text{ min}} \qquad \frac{1 \text{ min}}{60 \text{ s}} = \frac{60 \text{ s}}{60 \text{ s}}$$

$$1 = \frac{5280 \text{ ft}}{1 \text{ mi}} \qquad \frac{1 \text{ hr}}{60 \text{ min}} = 1 \qquad \frac{1 \text{ min}}{60 \text{ s}} = 1 \qquad \textbf{Conversion factors}$$

$$55 \frac{\text{mi}}{\text{hr}} = 55 \frac{\cancel{\text{mi}}}{\cancel{\text{hr}}} \times \frac{5280 \text{ ft}}{1 \cancel{\text{mi}}} \times \frac{1 \cancel{\text{hr}}}{60 \cancel{\text{min}}} \times \frac{1 \cancel{\text{min}}}{60 \text{ s}}$$

$$= \frac{(55)(5280) \text{ ft}}{(60)(60) \text{ s}} \qquad \textbf{Note that the conversion factors are exact}$$

$$= 81 \text{ ft/s.} \quad ◆$$

EXAMPLE 3   Converting Cubic Units to Cubic Units

Convert 45 cubic feet to cubic yards.

Solution   To convert ft³ to yd³, we need a unit fraction expressing yards in terms of feet. Given 1 yd = 3 ft, divide by 3 ft.

$$\frac{1 \text{ yd}}{3 \text{ ft}} = \frac{3 \text{ ft}}{3 \text{ ft}}$$

$$\frac{1 \text{ yd}}{3 \text{ ft}} = 1.$$

The unit fraction we need is $\frac{1 \text{ yd}}{3 \text{ ft}}$. To obtain a conversion factor in units cubed, multiply

$$\frac{1 \text{ yd}}{3 \text{ ft}} \times \frac{1 \text{ yd}}{3 \text{ ft}} \times \frac{1 \text{ yd}}{3 \text{ ft}} = \frac{1 \text{ yd}^3}{27 \text{ ft}^3}$$

Now convert 45 ft³ to yd³ by multiplying by the new conversion factor.

$$45 \text{ ft}^3 = 45 \ \overset{1}{\cancel{\text{ft}^3}} \times \frac{1 \text{ yd}^3}{27 \ \underset{1}{\cancel{\text{ft}^3}}}.$$

$$= \frac{45}{27} \text{ yd}^3$$

$$= 1.7 \text{ yd}^3. \quad \blacklozenge$$

In Example 3 we started with a measure of volume and ended with a measure of volume. Only the units were changed. The answer 1.7 yd³ contains two significant digits, which is the same as the number of significant digits in 45. ft³. The number of digits in 1 yd and 3 ft do not affect the answer since they are, by definition, exact numbers.

EXAMPLE 4   Converting British Thermal Units to Joules

Convert 30̲0 British thermal units per minute (Btu/min) to joules per second (J/s).

Solution   We need two conversion factors—one that expresses joules in terms of Btu and one that expresses minutes in terms of seconds.

$$1.00 \text{ J} = 9.48 \times 10^{-4} \text{ Btu} \qquad 1 \text{ min} = 60 \text{ s}$$

$$\frac{1.00 \text{ J}}{9.48 \times 10^{-4} \text{ Btu}} = \frac{9.48 \times 10^{-4} \text{ Btu}}{9.48 \times 10^{-4} \text{ Btu}} \qquad \frac{1 \text{ min}}{60 \text{ s}} = \frac{60 \text{ s}}{60 \text{ s}}$$

$$\frac{1.00 \text{ J}}{9.48 \times 10^{-4} \text{ Btu}} = 1 \qquad\qquad \frac{1 \text{ min}}{60 \text{ s}} = 1 \qquad \textbf{Conversion factors}$$

Multiplying by the conversion factors, we can divide numerator and denominator by Btus and minutes, and the solution is in joules/second.

$$300 \ \frac{\text{Btu}}{\text{min}} = 300 \ \frac{\overset{1}{\cancel{\text{Btu}}}}{\cancel{\text{min}}} \times \frac{1.00 \text{ J}}{9.48 \times 10^{-4} \ \underset{1}{\cancel{\text{Btu}}}} \times \frac{1 \ \overset{1}{\cancel{\text{min}}}}{60 \text{ s}}$$

$$= \frac{30\underline{0} \text{ J}}{(9.48 \times 10^{-4}) \, 60 \text{ s}}$$

$$= 5.27 \times 10^3 \text{ J/s.} \quad \blacklozenge$$

EXAMPLE 5   **Converting Inches to Centimeters**

Convert 8.00 in. to centimeters.

Solution   From Table A.4 we see that 1 in. = 25.4 mm. To make the conversion, we can convert inches to millimeters and then use 1 cm = 10 mm to convert millimeters to centimeters.

$$8.00 \text{ in.} \times \frac{25.4 \text{ mm}}{1 \text{ in.}} \times \frac{1 \text{ cm}}{10 \text{ mm}} = 20.32 \text{ cm}$$

$$= 20.3 \text{ cm}$$

Rounding to three significant digits. The conversion factors in this case are exact.) ◆

4.3 Exercises

In Exercises 1–12, perform the indicated conversion and then select the conversion **a-l** that matches your answer. The tables in Appendix A may be helpful.

1. 34 m = _____ cm   **a.** 6.84 miles

2. 2340 m = _____ km   **b.** 0.667 ft^3

3. 25.3 m = _____ yd   **c.** 3400 cm

4. 15 in. = _____ cm   **d.** 73.0 kg

5. 6.0 yd^2 = _____ m^2 **e.** 2.340 km

6. 5.00 gal = _____ ℓ   **f.** 37.0° C

7. 11.0 km = _____ miles **g.** 27.7 yds

8. 5.00 slug = _____ kg   **h.** 18.9 ℓ

9. 13.4 ℓ = _____ m^3 **i.** 4.25 yd^3

10. 3.25 m^3 = _____ yd^3 **j.** 38 cm

11. 98.6° F = _____ C   **k.** 0.0134 m^3

12. 5.00 gal = _____ ft^3 **l.** 5.0 m^2

In Exercises 13–45, perform the indicated conversions. The tables in Appendix A may be helpful.

13. 8.5 m to millimeters   **14.** 360 in^2 to square feet

15. 24 km to miles   **16.** 125 kg to pounds

17. 440 yd to meters   **18.** 88 km/hr to miles per hour

19. 60 mph to feet per second   **20.** 15 milliseconds to seconds

21. 1300 A to milliamperes   **22.** 17 mm^3 to cubic centimeters

23. 32 ft^2 to square meters   **24.** 88 ft/s to kilometers per hour

25. 5.6 MΩ (megohms) to ohms

26. 6.3 kcal to calories

27. 145 Btu/min to joules per second

28. 4.7 pF (picofarads) to farads

29. 70,000 Ω to megohms

30. 6500 W to kilowatts

31. 40 mpg to kilometers per liter

32. 55 Btu to joules

33. 175 hps (horsepower) to watts

34. 15 cal to joules

35. 6 hr 17 min 13 sec to seconds

36. 4.36 lb to grams

37. 3.76 qt to liters

38. 90 hp to kilowatts

39. 120 kilowatts to horsepower

40. 2.00 atm pressure to megapascal

41. 24.0 yd^2 to square feet

42. 24.0 yd^3 to cubic feet

43. A room is 4.00 yd 2.00 ft 7.00 in. long. What is its length in meters?

44. A cube is 6 ft on a side.
   a. What is the volume in ft^3?
   b. What is the volume in yd^3?

45. A liter of wine occupies how many cubic inches? (A liter is approximately the volume of a cube—10.0 cm on a side.)

In Exercises 46–48 use the fact that sometimes a conversion factor can be derived from the knowledge of a constant in two different system of units.

46. The speed of light is 186,000 mi/s, or 2.98×10^8 m/s. Use these facts to find the number of kilometers in a mile.

47. The weight of 1 ft^3 of water is 62.4 lb. Also, 1 cc (cm^3) of water has a weight of 1 gram. Use these facts to find the weight in pounds of a 1 kg mass. Recall that 1000 g is 1 kg.

48. A furlong is equal to 220 yards; a fortnight is two weeks. A good speed for the 100 yard dash is 10 yd/s. Change 10 yd/s into furlongs per fortnight.

4.4

Formulas and Dimensional Analysis

In this section the methods of working with equations developed in Section 4.1 and the methods of working with dimensional units developed in Section 4.3 are tied together. The technician frequently works with special equations called formulas. A **formula** is an equation with a special set of symbols that relate to a particular area of study. For example, $d = vt$, $V = IR$, $i = prt$, and $V = lwh$ are all formulas.

EXAMPLE 1   **Evaluating a Formula**

The formula $d = \frac{1}{2}gt^2$ expresses the relationship describing the distance, d, that a freely falling object (starting at rest) will fall in t seconds. Determine the distance, d, when $g = 9.81\frac{m}{s^2}$ and $t = 3.0$ s.

Solution

$$d = \frac{1}{2}gt^2$$

$$= \frac{1}{2}\left(9.81\,\frac{m}{s^2}\right)(3.0\text{ s})^2 \quad \text{Substituting numbers and dimensional units}$$

$$= \frac{1}{2}\left(9.81\,\frac{m}{\cancel{s^2}_1}\right)(9.0\,\cancel{s^2}^1) \quad \text{Squaring and reducing}$$

$$= 44.145\text{ m}$$

$$= 44\text{ m} \quad \text{Rounding to two significant digits}$$

The answer is expressed with the correct number of significant digits and the correct unit of measurement. ◆

Acceleration of Gravity

When a value for the acceleration of gravity, g, is required, $g = 9.81\,\frac{m}{s^2}$ will be used in the SI system and $g = 32.2\,\frac{ft}{s^2}$ will be used in the USCS system.

EXAMPLE 2   Formula: Freely Falling Object

Find the distance that a freely falling object will fall in 3.0 s if we take $g = 32.2\,\frac{ft}{s^2}$

Solution

$$d = \frac{1}{2}gt^2$$

$$= \frac{1}{2}\left(32.2\,\frac{ft}{s^2}\right)(3.0\text{ s})^2 \quad \text{Substituting numbers and dimensional units}$$

$$= \frac{1}{2}\left(32.2\,\frac{ft}{\cancel{s^2}_1}\right)(9.0\,\cancel{s^2}^1) \quad \text{Squaring and reducing}$$

$$= 144.9\text{ ft}$$

$$= 140\text{ ft} \quad \text{Rounding to two significant digits} \quad ◆$$

Should the results for Examples 1 and 2 be the same? Can you determine whether the results are the same? What would cause a slight difference in the answers?

EXAMPLE 3   Volume of a Solid: Dimensional Units

The width, w, of a rectangular solid can be calculated from the formula $w = \dfrac{V}{lh}$, where $V =$ volume, $l =$ length, and $h =$ height of the solid. Find the width of the rectangular solid whose volume is 5616 in^3, whose length is 24 in., and whose height is 18 in.

Solution

$$w = \frac{V}{lh}$$

$$= \frac{5616 \text{ in}^3}{(24 \text{ in})(18 \text{ in})} \quad \textbf{Substituting}$$

$$= \frac{5616 \text{ in}^3}{432 \text{ in}^2}$$

$$= 13 \text{ in}^{3-2}$$

$$= 13 \text{ in.}$$

Note that the rules of exponents apply equally well to dimensional units.   ◆

We must understand the difference between *weight* (gravitational force) and *mass*. The relationship between weight and mass is best understood by knowing the equation which relates them: $F_g =$ weight $= mg$ where g is the local value for the acceleration due to gravity (32.2 ft/s^2 or 9.81 m/s^2) and $m =$ mass. The weight of an object varies in value from place to place since the acceleration due to gravity varies; the mass stays the same. In the English, or USCS, system, weight is expressed in pounds, and mass is expressed in slugs. In the SI system, weight is expressed in newtons, while mass is expressed in kilograms. Often cans and other containers misrepresent weight with units of mass; for example, 134 grams could represent the mass of a can, but not its weight.

EXAMPLE 4   Application: Kinetic Energy

If the weight of a moving truck is 6400 lb, and it travels at 20 ft/s, determine its kinetic energy.

Solution   The kinetic energy of a moving body is given by

$$KE = \frac{1}{2}mv^2,$$

where m is the mass and v is the magnitude of the velocity. If the weight, rather than the mass is given, you must find the mass by dividing the weight by g. (Be sure to use a value of g consistent with your units of weight.)

$$KE = \frac{1}{2}mv^2$$

$$= \frac{1}{2} \frac{6400.\text{lb}}{\left(32.2\frac{\text{ft}}{s^2}\right)} \left(\frac{20.\text{ft}}{s}\right)^2 \quad \text{Substituting (since weight is in lb, we use 32.2 ft/s}^2 \text{ for } g)$$

$$= \frac{1}{2} \left(\frac{198.8 \text{ lb}}{\text{ft s}^{-2}}\right) \left(\frac{400. \text{ ft}^2}{s^2}\right)$$

$$= \frac{39{,}760. \text{ lb ft}^{2-1}}{s^{-2+2}}$$

$$= 40{,}000 \text{ ft lb} \quad \text{Rounding to two significant digits, since 20. ft has only 2 significant digits.}$$

$$\text{or } 4.0 \times 10^4 \text{ ft lb} \quad \text{(ft lb is a common measurement for work or energy.)} \quad \blacklozenge$$

EXAMPLE 5 **Application: Ohm's Law**

Consider the formula known as Ohm's law, $I = \dfrac{E}{R}$. Compute I, where $E = 880$ V and $R = 22$ Ω.

Solution

$$I = \frac{E}{R}$$

$$= \frac{880 \text{ V}}{22 \text{ } \Omega}$$

This is a case where we cannot cancel the units $\dfrac{\text{V}}{\Omega}$ because they do not have a common factor. However, $\dfrac{1 \text{ V}}{1 \text{ } \Omega} = 1$ A, so

$$I = \frac{880 \text{ V}}{22 \text{ } \Omega}$$

$$= 40\frac{\text{V}}{\Omega}$$

$$= 40 \text{ A.} \quad \blacklozenge$$

EXAMPLE 6 **Application: Saturn 5 Rocket Problem**

For the situation posed at the beginning of the chapter, evaluate $a = \dfrac{F_{net}}{m}$, where

$m = \dfrac{w}{g}$ when $F = 7{,}680{,}000$ lb, $w = 6{,}391{,}120$ lb and $g = 32.2$ ft/s². Express the acceleration in gs.

Solution   To determine a, we must first calculate a value for m. The net force (F_{net}) is $F_{\text{net}} = F - w$.

$$m = \dfrac{w}{g}$$

$$= \dfrac{6{,}391{,}120 \text{ lb}}{32.2 \text{ ft/s}^2} \qquad \left(\text{slugs} = \dfrac{\text{lb}}{\text{ft/s}^2}\right)$$

$$= 198{,}500 \text{ slugs} \qquad \textbf{Carrying one extra digit}$$

$$a = \dfrac{F_{\text{net}}}{m}$$

$$= \dfrac{7{,}680{,}000 \text{ lb} - 6{,}391{,}120 \text{ lb}}{198{,}500 \text{ slugs}}$$

$$= 6.493 \text{ ft/s}^2$$

The acceleration at ignition is 6.49 ft/s². The value shown in the calculation above was left at one extra significant digit since we have one more calculation to perform.

$$\text{acceleration in } g\text{s} = \dfrac{a}{g}$$

$$= \dfrac{6.493 \text{ ft/s}^2}{32.2 \text{ ft/s}^2}$$

$$= 0.202$$

The acceleration in gs is 0.202 to three significant digits, and is a dimensionless quantity. ◆

A BASIC computer program that can be used to calculate is found on the tutorial disk.

4.4 Exercises

In Exercises 1–23, solve for the indicated quantity. In finding the solution, pay strict attention to dimensional units and to significant digits.

1. Ohm's law, a physics equation that determines the flow of charge in a circuit, can be expressed as $V = IR$. The voltage is given by V, the current or charge flow is I, and the resistance to the flow is R. The unit for voltage is a volt that can be expressed as an amp · Ω. Find the current if the voltage is 120 volts and the resistance is 260 Ω.

2. For a uniformly accelerating object starting from rest, the velocity is given by $v = at$. For Example 6, which pertains to the Saturn 5 rocket, determine the velocity of the rocket after 4.50 seconds have passed.

3. The distance that an object moves (starting from rest) when accelerated constantly is given by the formula $d = \frac{1}{2}at^2$. How far had the Saturn 5 rocket moved in 4.50 s?

4. A rock is dropped into a well to determine the depth of the well. Since the rock is dropped from rest, use the formula $d = \frac{1}{2}at^2$. In this case, the acceleration is the acceleration due to gravity—32.2 ft/s². If the rock takes 4.5 s to fall, how deep is the well?

5. The formula for finding the volume of a regular cone is $V = \frac{1}{3}\pi r^2 h$. In this formula V = volume, r = radius, and h = height. Determine the height of the cone whose volume is 78.28 in.³ and whose radius is 3.25 in.

6. The length, l, of a rectangular solid can be calculated from the formula $l = \frac{V}{wh}$. Determine the length of the rectangular solid given: volume of 5411.52 mm³, width of 13.0 mm, and height of 18.5 mm.

7. If the mass of a moving vehicle is 10,000 kg and it travels at 9.2 m/s, determine its kinetic energy. Recall that the kinetic energy is determined by the formula $K = \frac{1}{2}mv^2$.

8. If the weight of a vehicle is 10,000 N and it travels at 9.20 m/s, determine its kinetic energy. Recall that the weight is not the mass. Weight is the mass of an object times the acceleration of gravity.

9. If the mass of a moving truck is 2880 kg and it travels at 6.00 m/s determine its kinetic energy.

10. The formula $v = v_0 + at$ enables us to determine the velocity, v, at any instant, t, given the initial velocity v_0, and constant acceleration, a. A bottle-nosed dolphin is observed to increase its speed uniformly from 1.2 m/s to 8.3 m/s in a time of 5.0 s. Determine its average acceleration.

11. An equation that can be developed from the formula in Exercise 10 is

$$s = v_0 t + \frac{1}{2}at^2$$

where s is distance. Using the information from Exercise 10, determine how far the dolphin travels while accelerating.

12. A water beetle accelerates from rest at 0.8 m/s². How far will it have traveled in 1.5 s? (See Exercises 10 and 11.)

13. The time, t (called the period), for one complete swing of a simple pendulum is given by the formula

$$t = 2\pi \sqrt{\frac{L}{G}}$$

where t is time in seconds, L is the length of the pendulum in feet, and g is the acceleration constant 32.2 ft/s². Compute t for a pendulum whose length is 24 in.

14. The area of a trapezoid is given by $A = \dfrac{1}{2}(b_1 + b_2)a$. Determine the area of the trapezoid with bases 1362 mm and 1586 mm and altitude 749 mm.

In Exercises 15 and 16 use the formula for the equivalent resistance of three parallel resistors: to determine the equivalent resistance, R, given the three resistances in each exercise.

$$R = \cfrac{1}{\dfrac{1}{R_1} + \dfrac{1}{R_2} + \dfrac{1}{R_3}}$$

15. 100 Ω, 150 Ω, 180 Ω

16. 2.7×10^3 Ω, 1.6×10^3 Ω, 3.1×10^3 Ω

17. What is the maximum load that can be hung from a steel wire $\dfrac{1}{4}$ in. in diameter if its elastic limit $\left(\dfrac{F}{A}\right)$ is not to be exceeded? The formula for finding the maximum load is $F = A$ (elastic limit), where F is the maximum load, A is the cross-sectional area of the wire $\left(\dfrac{\pi D^2}{4}\right)$, and the elastic limit for for steel is 36,000 lb/in.^{2}.

18. A golf shoe has 10 cleats, and each cleat has an area of 0.10 in.^{2} in contact with the floor. Assume that when a person walks there is one instant when all 10 cleats support the entire weight of the 150 lb person. What is the pressure exerted by the cleats on the floor? Express the answer in SI units. The pressure, P, is determined with the formula $P = \dfrac{F}{A}$, where F is the force, and A is the area of contact with the floor.

19. The fundamental equation for gravitational force—the inherent force between any pair of objects possessing mass—is given below.

$$F_g = \frac{Gm_1m_2}{R^2},$$

where $G = \dfrac{2}{3} \times 10^{-10}\dfrac{\text{Nm}^2}{\text{kg}^2}$ and m_1 = mass of object #1, m_2 = mass of object # 2, and R = the distance between the centers of the objects.

   a. Let the earth be object #1 with a mass of 6.0×10^{24} kg; let the ball be object #2. The distance between the centers is basically the radius of the earth, which is 6.4×10^6 m. Determine the weight of the ball.

   b. The weight of an object on the surface of the earth is given by the following equation: F_g = weight = mg, where $g = 9.81$ m/s^2. Find the weight of a 4.2 kg ball.

20. When the temperature in an object rises, its heat content increases. The equation that ties together heat and temperature change is $Q = cm \, \Delta T$, where Q is heat gain or loss, m is the mass of the object, c is the specific heat capacity, and ΔT is the change in temperature.

   a. An inventor decides to use 100 kg of alcohol in a car radiator instead of water. The specific heat of the alcohol is 2400 joules/kg°C. If the temperature rises 40°C, what is the heat gain in the alcohol radiator?

   b. If the same amount of water as alcohol is used and the same change in temperature occurs, what is the heat gain in the water radiator? For water, $c = 4800$ joules/kg°C. Water possesses the highest specific heat capacity of all readily abundant materials.

21. We exist at the bottom of an ocean of air and, thus, are under pressure—15 lb of air lies on every square inch of our bodies. If we go under water, we are under additional pressure. The added pressure due to water is given by the formula $P = \rho g \, \Delta h$. Where ρ is the density of water, 1.94 slugs/ft³; g is 32.2 ft/s²; and Δh is how far you are beneath the surface. What is the additional pressure on you if

   a. you are 12 ft below the surface?

   b. you are 120 ft below the surface?

   c. you are 1 m. below the surface?

22. The formula for the surface area of a sphere is $A = 4\pi r^2$. What is the surface area for a balloon that has a radius of 0.25 meters? If the radius is doubled, first estimate the new surface area of the balloon. Now calculate the surface area for a balloon that has a radius of 0.50 meters.

23. The formula for the volume of a sphere is $V = \dfrac{4}{3}\pi r^3$. What is the volume of a balloon that has a radius of 0.25 meter? If the radius is doubled, estimate the new volume of the balloon. Now calculate the volume of a balloon that has a radius of 0.50 meter. Estimate the volume of a balloon that has a radius of 0.75 meter.

4.5

Ratio and Proportion

In any technology size comparisons are often necessary. For example, in electronics many TVs, radios, tape recorders, and CDs have been reduced to pocket-sized instruments, although some TV screens have dramatically increased in size. A useful method of comparing things is a **ratio**—a comparison of two quantities with the same units, which may be expressed as a fraction. For example, suppose one metal shed design requires 300 bolts, while a different one requires only 100 bolts. The following ratio can be set up.

$$\frac{300 \text{ bolts}}{100 \text{ bolts}} = \frac{3}{1}.$$

This ratio also may be expressed in the form 3 : 1, which is read "3 to 1." The ratio tells us that there are three times as many bolts in the first design as there are in the second design.

Because fractional computations are sometimes cumbersome, a ratio may be converted into a decimal fraction where the numerator is a decimal and the denominator is 1. This type of conversion is a common practice in technical work. Examples include

$$\frac{1}{4} \text{ as } \frac{0.25}{1}, \frac{3}{8} \text{ as } \frac{0.375}{1}, \quad \text{and} \quad \frac{7}{4} \text{ as } \frac{1.75}{1}.$$

The ratio of two unlike quantities (different units) is a **rate.**

EXAMPLE 1   Application: Ratio

Emily walks 36 miles in 3 hours. Determine the ratio of the distance to the time.

Solution   Emily's average velocity is the ratio of the distance to the time. The ratio is $\dfrac{36 \text{ mi}}{3 \text{ h}} = \dfrac{12 \text{ mi}}{1 \text{ h}}$, or we could write 12mi : 1h. ◆

EXAMPLE 2   Ratio of Circumference to Diameter

A wheel of a boat trailer is 14 in. in diameter and covers a distance of about 44 in. as its rolls along a flat surface (see Figure 4.1). Determine the ratio of the distance around the wheel to the diameter of the wheel.

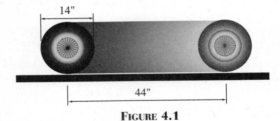

14"

44"

FIGURE 4.1

Solution   The ratio of the distance around the wheel to the diameter is approximately $\dfrac{44 \text{ in.}}{14 \text{ in.}} = \dfrac{22}{7}$, or in decimal to 1 form, $\dfrac{3.1}{1}$. By taking a more precise measurement to two more significant digits, the decimal to one ratio takes on an approximate value of π. If the new measurements are diameter of wheel 14.00 in. and the wheel rolls 43.98 in., then the ratio is $\dfrac{43.98 \text{ in.}}{14.00 \text{ in.}} = \dfrac{3.141}{1}$. ◆

If both numbers in a ratio are measured in the same unit of measurement, as in Example 2, then the ratio is a dimensionless quantity. However if the two terms are measured in different units, then the units are part of the ratio as in Example 1.

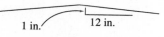

FIGURE 4.2

1 in.   12 in.

EXAMPLE 3   Application: Slope of the Roof

The roof of the Porter's family room rises 1 in. vertically for each 12 in. measured horizontally (see Figure 4.2). These measurements are called **rise** and **run,** respectively. Determine the ratio of the rise to the run.

Solution   The ratio that expresses the comparison of the rise to the run is called the **pitch** or **slope** of the roof. The ratio or slope in this case is $1 : 12$.   ◆

DEFINITION 4.2

A **proportion** is a statement that two ratios are equal, $\dfrac{a}{b} = \dfrac{c}{d}$ is a proportion provided $b \neq 0$ and $d \neq 0$.

The proportion $\dfrac{a}{b} = \dfrac{c}{d}$ is read "a is to b as c is to d," and it is sometimes expressed $a : b = c : d$. A proportion is a statement of equality. Therefore, if we know three parts of a proportion, we can solve for the fourth part using the rules for solving equations.

EXAMPLE 4   Solving for One Part of a Proportion

Given the proportion $\dfrac{x}{13} = \dfrac{4}{19}$, solve for x.

Solution

$$\frac{x}{13} = \frac{4}{19} \qquad \textbf{Given}$$

$$(13)\left(\frac{x}{13}\right) = (13)\left(\frac{4}{19}\right) \qquad \textbf{Multiply both sides by 13}$$

$$x = \frac{52}{19}, \text{ or } 2.7 \quad ◆$$

EXAMPLE 5   Proportion with Fractions

Given the proportion $\dfrac{x}{11} = \dfrac{\frac{2}{7}}{\frac{3}{4}}$, solve for x.

Solution

$$\frac{x}{11} = \frac{\dfrac{2}{7}}{\dfrac{3}{4}} \qquad \text{Given}$$

$$\frac{x}{11} = \left(\frac{2}{7}\right)\left(\frac{4}{3}\right) \qquad \textbf{Performing the division on the right side of the equal sign}$$

$$\frac{x}{11} = \frac{8}{21} \qquad \textbf{Substituting}$$

$$(11)\left(\frac{x}{11}\right) = (11)\left(\frac{8}{21}\right) \qquad \textbf{Multiplying both sides by 11}$$

$$x = \frac{88}{21}, \text{ or } 4.2 \quad \blacklozenge$$

There is more than one way of expressing a proportion. It is helpful to set up each ratio with the same units of measurement. By setting the ratios up in this form, the measurements will divide out, they will be unitless, and thus, the calculations will be simplified. This is illustrated in Example 6.

EXAMPLE 6   Application: Scale to Actual Distance

Determine the distance in kilometers from Buffalo, New York, to Chicago via Detroit. The scale distance on a map of the United States is measured at 15.6 cm. The scale of the map is 1 cm = 56 km.

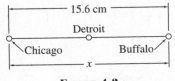

FIGURE 4.3

Solution   The scale on the map from Buffalo to Chicago is 15.6 cm. The scale of the map is 1 cm = 56 km. Let x = the actual distance in kilometers (see Figure 4.3). With this information we can set up two ratios.

a. The ratio of actual distance between the cities, x, to the actual distance of the scale, 56 km, is $\dfrac{x}{56 \text{ km}}$

b. The ratio of the distance on the map, 15.6 cm, to the scale, 1 cm, is $\dfrac{15.6 \text{ cm}}{1 \text{ cm}}$.

These two ratios are equal, so we can write the proportion:

$$\frac{\text{actual distance}}{\text{actual scale}} = \frac{\text{map distance}}{\text{map scale}}$$

$$\frac{x}{56 \text{ km}} = \frac{\overset{1}{\cancel{156 \text{ cm}}}}{\underset{1}{\cancel{1 \text{ cm}}}} \qquad \textbf{Reduce}$$

$$\overset{1}{\cancel{(56 \text{ km})}}\left(\frac{x}{\cancel{56 \text{ km}}}\right) = \left(\frac{15.6}{1}\right)(56 \text{ km}) \qquad \textbf{Multiply both sides by 56 km}$$

$$x = 873.6 \text{ km}$$

Thus, the distance from Buffalo to Chicago is 874 km. $\blacklozenge$

EXAMPLE 7 **Application: Spaceship Problem**

If a spaceship travels at a rate of 17,800 mph, use a proportion to determine how many hours are required to fly the 6,300,000 mi between a space station and Earth.

Solution Since the spaceship travels at 17,800 mph, it travels 17,800 mi in one hour. Distance from Earth to the space station is 6,300,000 mi; x = number of hours it takes the spaceship to make the trip (see Figure 4.4). With this information we can write the following proportion.

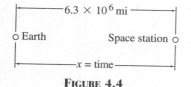

FIGURE 4.4

$$\frac{\text{total number of hours}}{1 \text{ hr}} = \frac{\text{total number of miles}}{\text{miles in one hour}}$$

$$\frac{x}{1 \text{ hr}} = \frac{6.3 \times 10^6 \text{ mi}}{1.78 \times 10^4 \text{ mi}}$$

$$(1 \text{ hr})\left(\frac{x}{1 \text{ hr}}\right) = \frac{6.3 \times 10^6}{1.78 \times 10^4}(1 \text{ hr}) \qquad \textbf{Multiply both sides by 1 hr}$$

$$x = 3.5393 \times 10^2 \text{ hr}$$

The time required for the spaceship to make the trip is 3.5×10^2 hr. Use the following keystrokes on the calculator to find the value of x.

fx–7700G: 6.3 ⏞EXP⏞ 6 ⏞÷⏞ 1.78 ⏞EXP⏞ 4 ⏞EXE⏞ 353.932584

TI–82: 6.3 ⏞2ⁿᵈ⏞ ⏞EE⏞ 6 ⏞÷⏞ 1.78 ⏞2ⁿᵈ⏞ ⏞EE⏞ 4 ⏞ENTER⏞ 353.932584

Therefore $x = 3.5 \times 10^2$ hr. ◆

4.5 Exercises

In Exercises 1–16, express each as a ratio in simplest form.

1. 15 km to 48 km

2. 63 min to 1001 min

3. 42 days to 8 wk

4. 250 ℓ to 2 kℓ

5. 24 m² to 96 m²

6. 15 ft to 15 yd

7. $3\frac{1}{4}$ cm to $4\frac{1}{3}$ cm

8. 6 g to $14\frac{1}{3}$ g

9. The ratio of the height of a tree 75 ft tall and the length of its shadow, 105 ft long.

10. The slope of a roof that has a rise of 12 ft and a run of 4 ft.

11. The slope of a roof that has a rise of $3\frac{1}{2}$ m and a run of $2\frac{1}{2}$ m.

12. The ratio of the height of a woman 5′6″ tall and the length of her shadow, 12′8″ long.

13. $6xy - 4y^2$ to $9x^2 - 6xy$.

14. $4xy - 8x^2$ to $10x^2 - 5xy$

15. $x^2 - 4y^2$ to $x^2 + 4xy + 4y^2$.

16. $9x^2 - 64y^2$ to $9x^2 - 48xy + 64y^2$

17. There are two types of cholesterol: low-density lipoprotein (LDL—considered the bad type of cholesterol) and high-density lipoprotein (HDL—considered the good type of cholesterol). Some doctors recommended that the ratio of low-density to high-density cholesterol be less than or equal to 4 : 1. Mr. Rivera's cholesterol test showed that his low-density cholesterol measured 225 mg per dℓ and his high-density cholesterol measured 54 mg per dℓ. Is Mr. Rivera's low- to high-density level greater than, less than, or equal to the recommended 4 : 1 ratio?

18. Mrs. Rivera had her cholesterol checked. The low-density cholesterol measured 167 mg per dl, and her high-density cholesterol measured 52 mg per dl. Is Mrs. Rivera's low- to high-density level less than or equal to the recommended $4:1$ ratio? (*Hint:* See Exercise 17.)

19. The wheel of a truck used in an open copper mine is 3.00 m in diameter and has a circumference of 9.42 m. Determine the ratio of the circumference to the diameter.

In Exercises 20–27 solve each proportion for the unknown.

20. $\dfrac{x}{15} = \dfrac{2}{3}$

21. $\dfrac{t}{9} = \dfrac{5}{6}$

22. $\dfrac{11}{3} = \dfrac{y}{9}$

23. $\dfrac{7}{4} = \dfrac{x}{6}$

24. $\dfrac{x}{5} = \dfrac{\frac{3}{10}}{\frac{7}{4}}$

25. $\dfrac{y}{3} = \dfrac{\frac{2}{3}}{\frac{4}{5}}$

26. $\dfrac{x}{4} = \dfrac{0.25}{1}$

27. $\dfrac{x}{8} = \dfrac{0.375}{1}$

In Exercises 28–47, set up a proportion and solve.

28. The distance between two cities measures 2.5 cm on a map. The scale on the map is 1.5 cm = 24.7 km. How far apart are the cities?

29. The distance between two mountaintops on a map is 85 cm. The scale of the map is $1:4372$. How far apart are the mountaintops?

30. Two cities that are exactly 15 mi apart appear as 3.456 in. apart on the map. What is the scale on the map?

31. A blueprint is made to a scale of $3:5$. If a part appears to be 20.15 mm long on the blueprint, how long is the actual part?

32. An 8-ft-long board appears as 3 in. long on a blueprint. What is the scale of the blueprint?

33. If the ratio of the width of a rectangle to its length is $\dfrac{5}{7}$ and its width is 62.5 m, how long is it?

34. If the ratio of the length to the width of a rectangle is $\dfrac{7}{4}$ and its length is 91 mm, how wide is it?

35. If a truck can travel 85 km on 14 liters of gas, how far can the truck travel on 18 liters?

36. If a pleasure boat can travel 48 mi in $1\frac{5}{6}$ hr, how far can it travel in $2\frac{3}{4}$ hr?

37. If there are 12 g of nitric acid in 80 g of solution, how many grams of nitric acid would be in 110 g of solution?

38. How much alcohol is requried to make 100 cc of a $3:5$ alcohol/water solution?

39. A 35-mm color slide measures 23 mm by 34 mm. If you print a picture from a slide that has a length of 120 mm, what is its width?

40. If you enlarge a 35-mm slide so that the length of the enlargement is 5 in., what is its width? (See Exercise 39.)

41. The weight of a 3.0-ft piece of I-beam is 137 lb. A contractor needs to lift a 38-ft piece. How much weight will he be lifting?

42. If you were charged a commission of $1450 on the sale of a $25,000 house, what would be the commission (at the same rate) on a house selling for $73,850?

43. The mixture of gasoline to oil for a two-cycle engine is $16:1$. If Adam has 16 oz of oil, how many gallons of gasoline must he add to the oil to have the proper mixture?

44. The color paint Heather picked for her room is mixed in a ratio of $5:1$; that is, five parts of white to one part of blue.
   a. What is the ratio of white to the total amount of paint needed?
   b. What is the ratio of blue to the total amount of paint needed?
   c. If she needs a total $1\frac{1}{2}$ gal of paint for the room, how many gal of white will she need?
   d. If she needs a total of $1\frac{1}{2}$ gal of paint for the room, how many gal of blue will she need?

45. The clarity of resolution goes up as the area of a television screen increases. The diagonals of a pair of TV sets is 15 in. for a portable and 36 in. for a console model.
   a. Write the decimal form of the ratio of the diagonals of the TVs.
   b. The length of one side of the portable set is 10.6 in. Using the ratio from part **a** determine the length of one side of the console.
   c. Recalling that the area of a square screen is the length squared, find the ratio of the areas. How much clearer would the resolution of the larger set be than the smaller set?

46. Ratios can be used to determine the heights of tall objects. The sun casts its rays along the same line of sight for a flagpole as it does for a meter stick. The shadow of the flagpole is 12.3 m long, while the shadow of the meter stick (when held straight up) is 0.45 m. What is the height of the flagpole?

47. Two houses are made into the shape of a cube (the roofs have not yet been built). The length of the smaller house is 20 ft; the length of the larger house is 30 ft.
   a. What is the ratio of the lengths?
   b. What is the ratio of the areas?
   c. What is the ratio of the volumes?
   d. If the heating cost is estimated by volume, how much more heat will be required to heat the larger house than the smaller house?

4.6

Variation

When two related quantities change in the same manner, we say that there is a **direct proportion** (or **direct variation**) between the quantities. Consider the speed of a car and the distance it travels along the same straight stretch of highway (Table 4.3). When the speed is doubled, the distance is doubled because of the speed. The distance is proportional to the speed; they vary directly with one another. This relationship can be expressed in different ways.

TABLE 4.3

Constant Speed for 2 hr	Distance Traveled
45 mph	90 mi
55 mph	110 mi
65 mph	130 mi
90 mph	180 mi

Expression	Method
Distance is proportional to speed	verbal
Distance $\propto$ speed	symbolical
$d = kv$	formula

TABLE 4.4

Constant Speed for 2 hr	Gal Gas Used
45 mph	4.00
55 mph	5.98
65 mph	8.35
90 mph	16.00

TABLE 4.5

Constant Speed for 2 hr	Time to Cover 30 mi
45 mph	40 min
55 mph	33 min
65 mph	28 min
90 mph	20 min

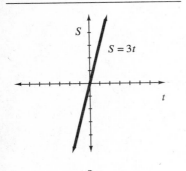

a.

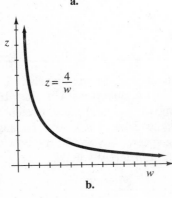

b.

FIGURE 4.5

The bottom expression is a formula that expresses the distance traveled, d, with respect to the speed, v, and k is a constant, called the **constant of proportionality (or constant of variation).** The formula shows that doubling the speed will cause a doubling of the distance.

Not all cause and effect relationships vary in a direct manner. Consider gas consumptions at various speeds (Table 4.4). Note that when the speed is doubled from 45 mph to 90 mph, the gas consumption is not doubled; rather it is quadrupled. Speed and gas consumption do not vary in a direct manner.

Suppose we investigate the time required to drive a given distance at a set speed (Table 4.5). Now the second quantity decreases as the first quantity increases. In fact, as the speed doubles, the time is decreased by one-half. This is an example of an **inverse variation (or an inverse proportion).** A simple formula that expresses this relation is $v = \dfrac{k}{t}$, where k is again some constant of proportionality.

EXAMPLE 1   Algebraic Statements of Variation

Write an algebraic statement of equality that represents each of the statements.

a. S varies directly with t.　　**b.** z varies inversely as w.
c. z is directly proportional to y.　　**d.** A is inversely proportional to B.
e. x varies as y.　　**f.** x varies inversely as y.

Solution

a. $S = kt$　**b.** $z = \dfrac{k}{w}$　**c.** $z = ky$　**d.** $A = \dfrac{k}{B}$　**e.** $x = ky$

f. $x = \dfrac{k}{y}$　◆

In Example 1, parts a, c, and e are examples of direct variation. A picture of a direct variation is a straight line as is illustrated in Figure 4.5a. Parts b, c, and f of Example 1 are examples of an inverse variation. A picture of an inverse variation is a curve as is illustrated in Figure 4.5b.

Procedure for Solving a Variation Problem

1. Determine the type of variation.
2. Write an algebraic statement of equality.
3. Solve for k.
4. Substitute the value, found for k in Step 3, for k in the algebraic equation in Step 2. The result is the statement of variation.

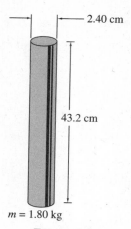

2.40 cm

43.2 cm

$m = 1.80$ kg

FIGURE 4.6

EXAMPLE 2 Solving for the Constant of Variation

For each of the problems write a statement of variation following the suggested procedure.

a. The mass of a steel rod is directly proportional to its length. A steel rod with a length of 0.432 m has a mass of 1.80 kg. See Figure 4.6.

b. The strength of a horizontal wooden beam (rectangular cross-section), with a concentrated load applied in the center, varies inversely with the distance between supports. A beam that is 3.6 m long and supported at each end has a safe load-capacity of 378 N. What is the safe load-capacity when the length is 12.5 m?

Solution

a. Let m = mass of steel rod ($m = 1.80$ kg);
  l = length of steel rod ($l = 0.432$ m); and
  k = constant of proportionality.

  Recognizing that this is a direct-variation problem, we can write the algebraic statement of equality, $m = kl$.

  Since we want to solve for k, divide both sides of the equation by l.

$$\frac{m}{l} = \frac{kl}{l}$$

or

$$k = \frac{m}{l}$$

$$k = \frac{1.80 \text{ kg}}{0.43 \text{ m}} \quad \text{Substituting for } l \text{ and } m$$

$$= 4.17 \frac{\text{kg}}{\text{m}}$$

  Now substituting the value of k in our original equation, we obtain a statement of direct variation for a particular value of k.

$$m = \left(4.17 \frac{\text{kg}}{\text{m}}\right) l$$

  We can now substitute any value for l and obtain the corresponding value for m.

b. Refer to Figure 4.7.
  Let l = length of the beam ($l = 3.6$ m);
    S = safe load-capacity ($S = 378$ N); and
    k = constant of variation.

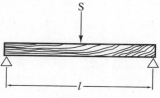

S

l

Figure 4.7

Recognizing that it is an inverse-variation problem, we can write the algebraic statement $S = \dfrac{k}{l}$. We want to solve for k; therefore, multiply both sides by l.

$$l(S) = \overset{1}{\cancel{l}}\left(\dfrac{k}{\underset{1}{\cancel{l}}}\right)$$

or

$$k = lS$$
$$k = (\ \boxed{3.6\ \text{m}}\ )(\ \boxed{378\ \text{N}}\ ) \qquad \textbf{Substituting for } l \textbf{ and } S.$$
$$= 1400\ \text{Nm}.$$

Now substituting the value of k in our original equation, we obtain a statement of inverse variation for a particular value of k.

$$S = \dfrac{1400\ \text{NM}}{l}.$$

Thus, for a beam of length 12.5 m, the safe load-capacity is found by substituting in the equation.

$$S = \dfrac{1400\ \text{Nm}}{\boxed{12.5\ \text{m}}}$$
$$= 110\ \text{N} \quad \blacklozenge$$

A technical problem may contain a variable that is dependent on two or more variables. For example, the volume of a cylinder is dependent upon the radius, r, of the base and the height, h. The algebraic statement that expresses this is $V = \pi r^2 h$, where π is the constant of variation. One way of reading this statement is, "The volume varies jointly with the square of the radius and the height." The type of variation referred to in this statement is called **joint variation.**

If we were working in mechanical technology, a variation problem we might face would be to determine the speed of a gear on a shaft driven by a gear on a motor. The ratio of the number of teeth, T_1, of the gear on the shaft to the number of teeth, T_2, of the gear on the motor is the same as the inverted ratio of the speed, S_1, of the gear on the shaft to the speed, S_2, of the gear on the motor. This statement expressed in algebraic form is $\dfrac{T_1}{T_2} = \dfrac{S_2}{S_1}$. Solving for T_1 we have

$$\overset{1}{\cancel{T_2}}\left(\dfrac{T_1}{\underset{1}{\cancel{T_2}}}\right) = T_2\left(\dfrac{S_2}{S_1}\right) \qquad \textbf{Multiply both sides by } T_2$$

$$T_1 = \dfrac{S_2}{S_1}T_2.$$

The gear problem can also be stated as a combined joint- and inverse-variation problem with $k = 1$. The speed, S_1, on a gear on a shaft driven by a gear on a motor varies jointly as the number of teeth, T_2, on the gear on the motor and the speed, S_2, of the gear on the motor and inversely as the number of teeth, T_1, on the gear on the shaft. This statement of the problem is stated algebraically as $S_1 = \dfrac{T_2 S_2}{T_1}$. The equation is another way of expressing the relationship between the speed of the gears and the number of teeth on a gear in this mechanical example. Using our algebraic skills, we can show that both ways of expressing the gear problem are equivalent.

EXAMPLE 3   **Joint and Inverse Variation**

For each of the following statements write an equivalent algebraic statement.

a. T varies jointly as S and P.
b. R varies directly as T and inversely as P.
c. Z varies jointly as T and the square of L.
d. Q varies jointly as R and the square root of T and inversely as the square of L.

Solution

a. $T = kSP$ **b.** $R = \dfrac{kT}{P}$ **c.** $Z = kTL^2$ **d.** $Q = k\dfrac{R\sqrt{T}}{L^2}$ ◆

CAUTION   In a variation problem, the word *jointly* indicates a product and the word *inversely* indicates a quotient. For joint variation, the constant of proportionality is part of the product. For an inverse variation, the k is always in the numerator. ■

EXAMPLE 4   **Application: Snow Removal Problem**

The time required to clear the snow in a driveway varies jointly as the length of the driveway, the width of the driveway, and the square of the average depth of the snow and, inversely, with the number of snow blowers used. When two snow blowers are used, it takes $\dfrac{3}{4}$ hr to clean a 600 ft driveway that is 14 ft wide when the snow has an average depth of $2\dfrac{1}{2}$ ft. How long will it take to clean 0.25 mi of driveway that is 15 ft wide using five snow blowers, when the average depth of the snow is 3.5 ft?

Solution This problem is a combination of joint and inverse variation. The first thing to do is to write an algebraic statement. Second, solve for k, and third, find the specific value for the time.

$$\text{Let } t = \text{time to clean the driveway,}$$
$$w = \text{width of the driveway,}$$
$$l = \text{length of the driveway,}$$
$$d = \text{depth of the snow,}$$
$$d^2 = \text{square of the depth of the snow, and}$$
$$n = \text{number of snow blowers.}$$

The algebraic statement then is

$$t = \frac{klwd^2}{n}.$$

Solve the equation for k.

$$nt = \left(\frac{klwd^2}{\cancel{n}_1}\right)\cancel{n}^1 \qquad \textbf{Multiply both sides by } n$$

$$\frac{nt}{lwd^2} = \frac{klwd^2}{lwd^2} \qquad \textbf{Divide both sides of the equation by } lwd^2$$

$$k = \frac{nt}{lwd^2}.$$

To find the value of k substitute the values $t = \dfrac{3}{4}$ hr, $l = 600$ ft, $w = 14$ ft, $d = 2\dfrac{1}{2}$ ft, $d^2 = 6.25$ ft^2, and $n = 2$ into the equation.

$$k = \frac{(2)\left(\dfrac{3}{4}\text{ hr}\right)}{(600\text{ ft})(14\text{ ft})(6.25\text{ ft}^2)}$$

$$= 0.0000286\,\frac{\text{hr}}{\text{ft}^2}$$

or

$$k = 2.86 \times 10^{-5}\,\frac{\text{hr}}{\text{ft}^4}.$$

Substituting the value for k in the equation.

$$t = \left(2.86 \times 10^{-5} \frac{\text{hr}}{\text{ft}^4}\right)\frac{lwd^2}{n}.$$

Using this equation we can determine the time required to clean the 0.25 mi driveway. Before we substitute we must change 0.25 mi to ft. 1 mi = 5280 ft, so 0.25 mi = 1320 ft.

$$t = \left(2.86 \times 10^{-5} \frac{\text{hr}}{\text{ft}^4}\right)\frac{(1320 \text{ ft})(15 \text{ ft})(3.5 \text{ ft})^2}{5}$$

$$= \left(2.86 \times 10^{-5} \frac{\text{hr}}{\text{ft}^4}\right)\frac{(1320 \text{ ft})(15 \text{ ft})(12.25 \text{ ft}^2)}{5}$$

$$= 1.39 \text{ hr}.$$

Thus, the time required to clear the snow would be about 1.4 hr.   ◆

The most difficult part of working with variation is recognizing the type: direct, inverse, joint, or combinations of two or more. Carefully examine the wording in the examples so that you recognize the key words.

4.6 Exercises

In Exercises 1–20, write an algebraic statement that represents each of the statements.

1. D varies directly as t.

2. A varies inversely as d.

3. C varies as d.

4. T varies inversely as w.

5. R varies jointly as V and W.

6. L varies directly as P and inversely as Q^2.

7. Z varies as the square of D and inversely as T.

8. G varies jointly as R and S^2 and inversely as the cube of P.

9. T varies jointly as R and the square root of x and inversely as the square of y.

10. The force, F, exerted by the wind against the wing of a plane increases with the area, A, of the wing.

11. The life, L, of a tire is less at higher speeds, V.

12. The area, A, of a circle varies as the square of the radius, r.

13. The area, A, of an equilateral triangle varies as the square of its altitude, a.

14. The length of time, t, required to empty a tank varies inversely as the square of the diameter, d, of the drain pipe.

15. Under certain conditions, the thrust, T, of a propeller varies jointly as the fourth power of its diameter, d, and the square of the number, n, of revolutions per second.

16. Newton's Law of Gravitation states that the force, F, exerted between two bodies varies jointly as their masses, m_1 and m_2, and inversely as the square of the distance, d, between them.

17. The resistance, R, of a wire to the flow of electricity varies directly as its length, l, and inversely as the square of its diameter, d.

18. The strength, s, of a rectangular beam with a concentrated load in the center and supported at both ends varies jointly as its width, w, and the square of its depth, d, and inversely as its length, l.

19. The time, t, in hours that it takes x men to build y houses varies directly as the number of houses and inversely as the number of men.

20. The number, n, of vibrations per second of a stretched cord varies directly as the square root of the stretching force, f, and inversely as the product of its length, l, and diameter, d.

In Exercises 21–36,

a. Write an algebraic statement that represents the variation problem.
b. Solve for k and find a numerical value for k.
c. Substitute the value found for k in b in the algebraic statement written in part a.
d. Using the algebraic statement from part c, solve for the desired quantity.

21. If y varies directly as the square of x and $y = 3$ when $x = 4$, find the value of x that makes $y = 12$.

22. If the square of y varies directly as the cube of x and $y = 4$ when $x = 2$, find the value of x when $y = 54$.

23. A photographer found that the exposure time needed for making an enlargement from a negative varies as the area of the enlargement. If she takes 10 s for a print $2\frac{1}{2}$ in. by $3\frac{1}{2}$ in., how long will she take for an enlargement 6 in. by 8 in.?

24. The volume, V, of an ideal gas varies directly as its temperature, T, as long as the temperature is expressed in absolute units like °K.
    a. A hot air balloon has an initial volume of 2000 m³ at 293°K. Suppose that the air is heated with a torch so that the temperature is raised to 393°K. What will be the new volume of the balloon?
    b. A balloon can be used as a thermometer. At room temperature (293°K), the volume of a balloon is 200 m³. When the balloon is carried outside, its volume increases to 220 m³. What is the temperature of the outside air?

25. The pressure, P, varies directly as the temperature. The water vapor in a pressure cooker is at room temperature (293°K) and at atmospheric pressure (14.7 lbs/in²). If the temperature of the pressure cooker is raised to 473°K, what is the new pressure inside the cooker?

26. The pressure varies inversely with the volume. A balloon maintains the same temperature as it rises 3000 ft. The volume of the balloon initially is 2000 m³ while its pressure is just one atmosphere (1 atm). Suppose the outside pressure at this high altitude is just 0.63 atm. What is the new volume of the balloon?

27. The current, I, in amperes in an electric circuit varies inversely as the resistance, R, in ohms when the electromotive force is constant. If in a certain circuit I is 15 A when R is 2 Ω, find I when R is 0.2 Ω.

28. The force, P, of the wind on a plane surface at right angles to the direction of the wind varies jointly as the area, A, of the surface and the square of the speed, v, of the wind. If the force on 16 ft² is 8 lb when the speed of the wind is 10 mph, determine the force on a surface 6 ft by 12 ft when the wind's speed is 50 mph.

29. If a 2-in. × 8-in. timber 16 ft in length will hold 500 lb when placed with the 2-in. side down, how much will a 1-in. × 8-in. timber hold when placed with the 1-in. side down? Use the formula developed in Exercise 18.

30. If a 2-in. × 6-in. timber will hold 1200 lb, how much will a 6-in. × 6-in. timber of the same length hold? (See Exercise 29.)

31. If a cord 4 m long and 2 mm in diameter vibrates n_1 times per second under a stretching force of 25 N, what stretching force would cause this cord to vibrate $2n_1$ times per second? Use the formula developed in Exercise 20.

32. The time required to dig a river channel varies jointly as its length, width, and square of its depth and inversely with the number of machines used. If it takes 1600 hr to dig 1 km of a channel that is 30 m wide and 6 m deep using 40 earth-moving machines, how long will it take to dig a channel $4\frac{1}{2}$ km long that is 32 m wide and 10 m deep using 90 earth-moving machines?

33. The loss in pressure of a liquid flowing through a pipe varies as the length and inversely as the square of the diameter of the pipe. If a water pipe 89 m long and 0.29 m in diameter results in a drop in water pressure of 600 N/m², determine the drop in pressure for a water pipe 264 m long whose diameter is 0.55 m.

34. The speed of light in air is 2.98×10^8 m/s. The index of refraction of air is 1.000. The speed of light in water is 2.24×10^8 m/s while its index of refraction is 1.33.
    a. Is the relationship between the index of refraction and the speed of light a direct or inverse relationship?
    b. The speed of light is very difficult to measure in solids. The index of refraction in a diamond is 2.48. What is the speed of light in that solid?

35. Subatomic particles that possess charge have their paths deflected into circular arcs. The mass of the particle is proportional to the radius of the arc. The standard arc for measuring the mass of unknown particles is that for the proton. A proton has a mass of 1.67×10^{-27} kg and its arc has a radius of 2.1300 m. If the arc of a particle has radius of 0.00116 m, what is its mass? (Incidentally, the unknown particle is the familiar electron.)

36. Currents generate magnetic fields. The amount of rotation of a compass is proportional to the size of the current. Thus, magnets can be used to estimate the size of currents in unknown wires without risking shock when testing the wires. If one amp of current causes a compass to deflect 23°, then how much current exists in a wire that causes the compass to deflect 34°?

Review Exercises

In Exercises 1–10, solve the equation and check the solution.

1. $x + 4 = 13$
2. $3R - 7 = 20$
3. $5a + 9 = 7a - 13$
4. $6y - 7 + 5y = 4y + 21$
5. $5z + 13 = 26 - 8z$
6. $34x + 30 = 12x - 40(20 - 23x)$
7. $\frac{1}{5}R + 4 = \frac{1}{3}R$
8. $\frac{4}{x} + 3 = \frac{7}{x}$
9. $\frac{2}{x + 3} = \frac{5}{x - 4}$
10. $y(y + 5) - 7 = (y + 5)(y - 3)$

In Exercises 11–20, perform the indicated conversions.

11. 11.5 m to millimeters

12. 17.5 ft^2 to square inches

13. 75.4 mi to kilometers

14. 950 mA to amperes

15. 27.5 m to yards

16. 45 mph to feet per second

17. 65 mph to kilometers per hour

18. 75 km/liter to miles per gallon

19. 40 hp (horsepower) to watts

20. 15 j to calories

In Exercises 21–23, use this statement: The teacher stated that two-thirds of her class were girls.

21. Express the meaning of the ratio $\frac{2}{3}$ as a comparison.

22. What is the ratio of the number of boys to the entire class?

23. If there are 27 pupils in the class, how many are girls?

24. An advertisement for raisins states, "In 10 out of 15 homes you will find Iron-Rich Raisins." If you visited 120 homes, how many homes probably would have Iron-Rich Raisins according to the information in the advertisement?

In Exercises 25–29, express each as a ratio in simplest form.

25. $r^2 - 9s^2$ to $r^2 + 6rs + 9s^2$

26. $3x^3y - 12xy^3$ to $6x^3y + 12x^2y^2 + 6xy^3$

27. $x^2 + 5x + 6$ to $x^2 - 2x - 8$

28. $x - 3$ to $x^3 - 27$

29. $x^2 - 2x + 4$ to $x^3 + 8$

30. How much would 6 oz of spinach seed cost if the price of the seed is 78 cents per quarter pound?

31. If 1 in. on the road map represents a distance of 20 mi, what is the actual distance between Robesonia and Harrisburg, which are $1\frac{13}{16}$ in. apart?

32. A blueprint is made to a scale of 4 : 5. If a part appears as 30.17 mm long on a blueprint, how long is the actual part?

33. The Smith family traveled 315 mi in $7\frac{1}{2}$ hr. At the same average rate, how much farther could they have traveled in three more hours?

34. If a car traveling at an average rate of 45 mph reaches its destination in 6 hr, 20 min, how fast must a car travel to cover the same distance in 4 hr, 15 min?

In Exercises 35–40, write an algebraic statement that represents each of the statements.

35. c varies directly as t.

36. p varies jointly as t and r.

37. z varies jointly as R and t^2 and inversely as the square of w.

38. The area, A, of a sphere varies as the square of the radius, r.

39. An hourly employee's pay varies directly with the number of hours worked.

40. The rate of consumption of wood in a fireplace varies inversely as the density of the wood.

In Exercises 41–62, solve for the indicated quantity, paying strict attention to dimensional units and to significant digits.

41. For a uniformly accelerating object the velocity, v, is given by $v = at$. Find a when $v = 24.0$ ft/s and $t = 7.5$ s.

42. If the weight of a moving car is 3500. lb and its kinetic energy is 49,000. ft lb, determine the rate at which the car is traveling. $\left(KE = \dfrac{1}{2} mv^2. \right)$

43. The area of a square is given by $A = s^2$. If the area is 36.9 cm², how many significant digits should be in the value for s? Why? Find s in mm.

44. The volume of a cone is given by the formula $V = \dfrac{1}{3}\pi r^2 h$. If $V = 36.44$ in³ and $h = 2.03$ in., what value of π should be used in the computation of r? Find r.

45. If 312 bricks were used to edge a rectangular grassy patch whose length was twice its width, how many bricks were used on each side?

46. A car averaged 26. mpg in city driving and 40. mpg in highway driving. If the car used 15. gal of gas on a 460.-mi trip, how much of the trip was in the city?

47. A landlady earned $2190 in one year from the rental of two rooms. Determine the rent she charged per month per room if the rent on the first room was $10 per month more than the rent on the second room and the first room was vacant for three months.

48. A student rode a bicycle for 3 hr then parked it and hiked up a mountain trail for 3 more hours. She rode 6 mph faster than she hiked and traveled 36 mi in all. How fast did she hike?

49. A man can exert a force of 100 lb on the end of a 20-ft lever. Where must he place the fulcrum under the lever to raise a 400-lb weight? Use the formula $d = \dfrac{fl}{f + w}$, where f is the force, l is the length of the lever, and w is the weight.

50. At a distance of 6 in. from the pivot, Jack grips a pair of pliers with a force of 56 lb (see the illustration). The forces are inversely proportional to the distances from the pivot. What is the pressure on the wire if the wire is $\dfrac{1}{2}$ in. from the pivot?

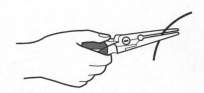

51. If y varies directly as the square of z, and $y = 3$ when $z = 5$, find the value of z that makes $y = 18$.

52. If the square of w varies directly as the cube of z and $w = 8$ when $z = 4$, determine the value of z when $w = 108$.

53. The compression ratio of an engine is the volume above the piston when it is at the bottom of the cylinder compared with the volume above the piston when the piston is at the top of the cylinder. In the gasoline engine (see the illustration), the compression ratio varies from 7 : 1 to 10 : 1, and the higher the compression ratio, the greater the efficiency. In a diesel engine the compression ratio is around 16 : 1. What is the compression ratio of an engine when the volumes mentioned above are 84 in³ and 12 in³, respectively?

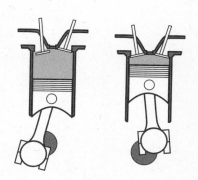

54. If the water pressure on a square inch of surface is 4.33 lb/in² when submerged 10 ft, what will be the pressure per square inch at a depth of 80 ft?

55. Determine the volume of the frustum of a cone if the radius of the lower base is 18.5 in., the radius of the upper base is 11.2 in., and the height is 10.6 in.

56. The volume of gas in a closed vessel is inversely proportional to the pressure on it. The volume of a gas is 800 ft^3 when it is under a pressure of 15 lb/in^2. What is the volume of the gas when it is under a pressure of 25 lb/in^2?

57. Two balls have diameters of 30.0 mm and 90.0 mm. How many times greater is the volume of the second than that of the first?

58. What is the ratio of the surface areas of two similar solids if their volumes are represented by 160π and 540π?

59. Two motor boats start at the same point and travel in opposite directions. One boat travels at 40 kilometers per hour; the other boat travels at 35 kilometers per hour. In how many hours will the boats be 262.5 kilometers apart?

60. Matthew stored his power boat on a trailer in the backyard. To take the pressure off the tires, he wants to raise the trailer and place blocks under the axle to each wheel. The boat and trailer weigh 2800 lb. If he can place the fulcrum 1 ft from the axle and he has a lever that is 12 ft long, as shown, how much force must he exert on the end of the lever in order to lift one wheel (half the total weight) off the ground?

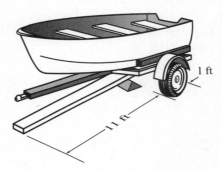

61. The equation $E = 8lwh^2$ is an algebraic model for energy, E (in foot-pounds), that can be expected from sea waves, where l is the length (in feet) of the wave (distance between successive crests), h is the height (in feet) of the wave (vertical distance from trough to crest), and w is the width (in feet) of the wave. (See the illustration.)
  a. What are the units associated with the constant 8?
  b. Solve this equation for w.
  c. If a wave is 50 ft long and 8 ft high and has 400,000 ft lb of energy, how wide is the wave?

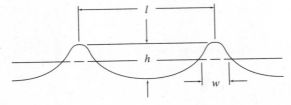

62. Solid-state digital watches use a vibrating quartz crystal to regulate the time. The formula $s = v/32{,}768$ is an algebraic model that gives the number of seconds elapsed, s, after v vibrations of the quartz crystal.
  a. What are the units associated with the constant 32,768?
  b. After the quartz crystal vibrates 1,048,567 times, how many seconds have elapsed?
  c. Solve the formula for v.
  d. How many vibrations does the quartz crystal make in 1 s?
  e. How many vibrations does the quartz crystal make in 5 s?

✎ Writing About Mathematics

1. Write a report for employees to help them understand the following statement. "The resistance of a conductor is directly proportional to its length." Use examples to illustrate how a technican can determine the resistance of an electric wire over a distance of many miles.

2. You are designing a deck and want to determine whether the deck will support a circular hot tub that is four ft deep and has a diameter of six ft. The following statement is a key to the solution of the problem. "The maximum safe load of a rectangular beam varies jointly as the width and the square of its depth and inversely as the length of the beam." Interpret the statement and do the calculations to determine the proper-sized beams needed to support the hot tub.

3. Write a report that illustrates the amount of exercise needed to burn calories that you consume from different types of snack foods. For example, you could write about the number of miles you must run to burn the calories from a milk shake.

Chapter Test

In Exercises 1–3, express the statements as a ratio in simplest form.

1. 3 ft to 12 ft

2. 1 day to 1 week

3. $x^2 - 16$ to $x^2 + 2x - 8$

In Exercises 4 and 5, solve each proportion for the unknown.

4. $\dfrac{x}{4} = \dfrac{1}{3}$

5. $\dfrac{\frac{1}{2}}{3} = \dfrac{4}{y}$

In Exercises 6–9, determine the root of the equation.

6. $4x - 7 = x + 5$

7. $2x - 3(5x - 7) + 4x = 2x - 4(3x - 6)$

8. $\dfrac{x + 5}{3} - \dfrac{x}{5} = \dfrac{7}{12}$

9. $1.34x + 2.71 = 3.45x - 4.75 + 6.41x$

In Exercises 10 and 11, solve for the indicated letter.

10. $V = \pi r^2 h$, for h

11. $V = \dfrac{\pi r^2}{2}(h_1 + h_2)$, for h_2.

In Exercises 12–15, perform the conversions.

12. 15.71 cm to mm   **13.** 5327 cm² to m²   **14.** 32.6 j/s to BTU/min   **15.** 1500 kW to hp

16. If a car can travel 120 mi on 6.0 gal of gas, how far can it travel on 7.5 gal?
   a. Write an algebraic statement.
   b. Solve for the constant of variation.
   c. Solve for the required quantity.

17. y varies directly as the square of x. Given $y = 1.5$, $x = 9$, determine the value for x when $y = 15$.

18. y varies inversely as the square of x. Given $y = 2$, $x = 2$, determine the value for x when $y = 20$.

19. A bolt holds together two pieces of material with a combined thickness of 46 mm. The length of the head of the bolt is $\frac{1}{7}$ of its total length. The length of the nut (which is flush with the end of the bolt) is $\frac{1}{5}$ of the total length of the bolt. There is no washer. What is the height of the nut?

20. If the ratio of the length to the width of a rectangle is $5:2$ and its length is 75 cm, how wide is the rectangle?

Functions and
Graphs

A sporting goods manufacturer is designing a new high-tech ski boot. Part of the design problem involves understanding the different angles the lower leg assumes as the individual glides down a slope. The engineers use a movie camera to record the actual motion of the leg, making it possible to measure the various angles made by the leg at specific time intervals. These measurements give the engineers a set of data that can be graphed. The graph provides the designer with a picture of the conditions that must be met by the new boot. Another piece of information needed by the designer is the location of the midpoint between the maximum and the minimum amount the leg is able to rotate. Solving this problem involves being able to graph a function and determine the midpoint between two points. These and other skills relating to functions are developed in this chapter. In Example 4 of Section 5.2 we will use actual data and derive the information needed to assist the designer in developing the high-tech ski boot.

5.1

Functions

Whether you realize it or not, you use the concept of a function frequently in your everyday life. The cost of a car trip is a function of the cost of gasoline. The acceleration of an airplane on takeoff is a function of the thrust of its engines. The depth of field in a photograph is a function of the lense opening.

In each of these situations we are saying that the first thing is dependent on the second. For example, as the price of gasoline goes up (or down), it costs more (or less) to run a car. However, the total cost of running a car is affected by other variables, not just by the price of fuel.

In mathematics the word *function* is used in much the same way, but more restrictively. Mathematicians will agree that "The area of a square is a function of the length of its side." Thus, the mathematical function demands that one quantity is uniquely determined by one or more quantities.

In our discussion of function we will be using the language of sets. Recall that a **set** is a well-defined collection of objects. A set indicated by listing the elements in a set of braces is said to be in **roster form.** For example $A = \{1, 2, 3, 4\}$. The numbers 1, 2, 3, and 4 are called members or elements of set A. Set A may also be indicated by describing the set: set A is the set of the first four natural numbers. The set may also be indicated using **set builder notation.**

$$A = \{x \mid x \in N \ \text{ and } \ x < 5\}$$

Set A is the set of all the elements x such that (the conditions x must meet to be a member of the set). For set A the conditions are that x is an element of the set of counting numbers and that x is less than 5.

A function can be expressed in different ways. One is by definition, another by algebraic equation or formula, and still another by using sets of ordered pairs. For example, the area of a square is determined by the length of the side: $A = s^2$. In this formula, s is the **independent variable,** and A is the **dependent variable.** A is the dependent variable because its value depends upon our choice of the value for s. Since we can select any positive value for s, s is the independent variable. The volume of a sphere is dependent on the length of its radius, $V = \frac{4}{3}\pi r^3$. In the formula of the sphere, r is the independent variable, and V is the dependent variable.

DEFINITION 5.1
A **function** is a rule that assigns for each value of the independent variable a unique value of the dependent variable.

The relationship between quantities doesn't necessarily need to be expressed as a formula; sometimes a table is useful. Table 5.1 gives the relationship between the temperature and the number of chirps a cricket makes per minute.

TABLE 5.1

Temperature	Number of Chirps per Minute
78°F	272
75°F	260
70°F	240
65°F	220
71°F	244
75°F	260
81°F	284
82°F	288
85°F	300
91°F	324

The temperature and the chirps were recorded on ten consecutive days. As the temperature increased so did the number of chirps. Mathematically then, we say that the number of chirps is a function of the temperature. In other words, there is a distinct number of chirps per minute for each temperature. The temperature cannot be a function of the number of chirps since there is more than one temperature that corresponds to a certain number of chirps. For example, there were 260 chirps recorded for two temperature readings of 75 degrees. Thus, the temperature is the independent variable, and the number of chirps per minute is the dependent variable.

Remember, a function also can be expressed as a **set of ordered pairs.** Using the data from Table 5.1, the number of chirps per minute can be expressed as a function of the temperature by writing the set of ordered pairs: {(78°F, 272), (75°F, 260), (70°F, 240), (65°F, 220), (71°F, 244), (75°F, 260), (81°F, 284), (82°F, 288), (85°F, 300), (91°F, 324)}. The first number in each ordered pair represents the independent variable, T, and the second number represents the dependent variable, C. In general, the ordered pair is written (T, C). A formula that may be used to express the number of cricket chirps as a function of temperature is $C = 4T - 40$.

EXAMPLE 1   Function: Independent and Dependent Variables

Each of the following is an example of a function. In each case identify both the independent variable(s) and the dependent variable.

a. $y = 4x + 7$ **b.** $s = 32t$ **c.**

A	−3	2	5	−5
B	4	9	14	4

d. $V = \dfrac{1}{3}\pi r^2 h$ **e.** $F = ma$ **f.** $2A + 3B - 4 = 0$

Solution

a. For the function $y = 4x + 7$, the independent variable is x , and the dependent variable is y . By convention x is used to designate an independent variable, and y is used to designate a dependent variable.

b. For the function $S = 32t$, the distance, S , is dependent upon the time, t , for its value. Thus, t is the independent variable, and s is the dependent variable. By convention, time is the independent variable.

c. Unless other information is provided in the table, by convention, take the first element to be the independent variable. Thus, A is the independent variable, and B is the dependent variable.

d. The function $V = \dfrac{1}{3}\pi r^2 h$ has two independent variables, r and h. V depends on r and h for its values; therefore V is the dependent variable. There is no restriction on the number of independent variables. However, a function will have one, and only one, dependent variable. (Remember, the symbol π represents a constant.)

e. For the function $F = ma$, F is dependent upon m and a for its value. Thus, the independent variables are m and a, and the dependent variable is F.

f. For the function $2A + 3B - 4 = 0$, we could select A or B as the independent variable. However, by convention, the independent variable is selected by alphabetical order. So, in this case, A is the independent variable, and B is the dependent variable. ◆

The dilemma demonstrated in Example 1, can be eliminated if we use **functional notation.** In functional notation, $y = \sqrt{x} + 3$ would be written as $f(x) = \sqrt{x} + 3$. You simply replace y with $f(x)$. In fact, $y = f(x)$. The symbol $f(x)$ is then read as the **value of x** or **f of x,** or for a specific case, **function of x equals** $\sqrt{x} + 3$.

It is important to remember that f is the name of the function and $f(x)$ is the value of the function. The algebraic expression f tells what operations to perform on the independent variable x. The final result after the operations have been performed on the independent variable x is represented by $f(x)$. Examples of functional values for the function $f(x) = 3x - 5$ are $f(-3), f(0)$, and $f(2)$. To determine the values, substitute the values in the given equation.

$$\text{For } x = -3, f(\,\boxed{-3}\,) = 3(\,\boxed{-3}\,) - 5 = -9 - 5 = -14$$
$$\text{For } x = 0, f(\,\boxed{0}\,) \quad = 3(\,\boxed{0}\,) - 5 \quad = 0 - 5 \quad = -5$$
$$\text{For } x = 3, f(\,\boxed{3}\,) \quad = 3(\,\boxed{3}\,) - 5 \quad = 9 - 5 \quad = 4$$

The most common name for a function is f. For the independent variable, the most common name is x. However, there are many other names that can be used. For example, $f(x) = 4x^2 + 2x - 1$, $g(t) = 4t^2 + 2t - 1$, $h(s) = 4s^2 + 2s - 1$ and $\phi(s) = 4s^2 + 2s - 1$ are all names for the same function.

The independent variable is essentially a place holder. The function could just as well be written as $f(\quad) = 4(\quad)^2 + 2(\quad) - 1$, with parentheses replacing the letters. To evaluate $f(-2)$, place the -2 in the parentheses:

$$f(-2) = 4(-2)^2 + 2(-2) - 1$$
$$= 11$$

EXAMPLE 2 Evaluating a Function

Determine the value of the function $f(x) = 3x^2 + 2x - 4$ for the values given.

a. $f(2)$ b. $f(2t)$ c. $f(t + 2)$

Solution

a. Replacing x with 2, the result is:

$$f(2) = 3(\,\boxed{2}\,)^2 + 2(\,\boxed{2}\,) - 4$$
$$= 12$$

b. Replacing x with $2t$, the result is:

$$f(2t) = 3(\;2t\;)^2 + 2(\;2t\;) - 4$$
$$= 3(4t^2) + 4t - 4$$
$$= 12t^2 + 4t - 4$$

c. Replacing x with $t + 2$, the result is:

$$f(t + 2) = 3(\;t + 2\;)^2 + 2(\;t + 2\;) - 4$$
$$= 3(t^2 + 4t + 4) + 2t + 4 - 4$$
$$= 3t^2 + 12t + 12 + 2t$$
$$= 3t^2 + 14t + 12 \quad \blacklozenge$$

CAUTION The results of parts a and c of Example 3 show that $f(t + 2) \neq f(t) + f(2)$.
Since $3t^2 + 14t + 12 \neq 3t^2 + 2t - 4 + 12$.
In general, $f(a + b) \neq f(a) + f(b)$. ∎

It may be easier to evaluate a function with the use of a calculator. Your calculator has either a $\boxed{y^x}$ key or a $\boxed{\wedge}$ key for raising a number to a power.

EXAMPLE 3 **Determining Functional Values with a Calculator**

For the function $f(x) = 5x^3 - 4x^2 + 2x - 3$, determine $f(3.7)$ using a calculator.

Solution To find the value of the function using a calculator use the following keystrokes.

$5 \boxed{\times} 3.7 \boxed{\wedge} 3 - 4 \boxed{\times} 3.7 \boxed{x^2} + 2 \boxed{\times} 3.7 \boxed{-} 3 \boxed{\text{ENTER}}\ 202.905$

This method works well if we want to evaluate the function for a few values. If we want to evaluate the function for many values, we can save time by writing a calculator or computer program. ◆

Writing a Program for the Graphics Calculator

To expedite the process of determining functional values, here is a demonstration of how to write a program for the graphics calculator.

For fx–7700G

To write a program, the calculator must be in the write mode. To enter the write mode press:

Keystrokes	Screen Display
$\boxed{\text{MODE}}$ 2	Wrt/Comp
	4164 Bytes Free
	P0 empty
	P1 empty Storage
	P2 empty Locations

Move the blinking cursor to the location where we want to store the program. Select P0, press $\boxed{\text{EXE}}$, and a blank screen will appear. Now write the program.

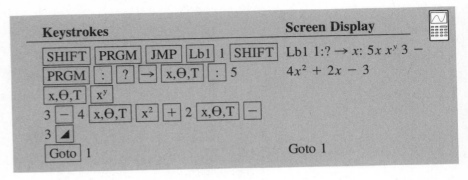

Keystrokes

$\boxed{\text{SHIFT}}$ $\boxed{\text{PRGM}}$ $\boxed{\text{JMP}}$ $\boxed{\text{Lb1}}$ 1 $\boxed{\text{SHIFT}}$
$\boxed{\text{PRGM}}$ $\boxed{:}$ $\boxed{?}$ $\boxed{\rightarrow}$ $\boxed{\text{x,}\Theta\text{,T}}$ $\boxed{:}$ 5
$\boxed{\text{x,}\Theta\text{,T}}$ $\boxed{x^y}$
3 $\boxed{-}$ 4 $\boxed{\text{x,}\Theta\text{,T}}$ $\boxed{x^2}$ $\boxed{+}$ 2 $\boxed{\text{x,}\Theta\text{,T}}$ $\boxed{-}$
3 $\boxed{\blacktriangleleft}$
$\boxed{\text{Goto}}$ 1

Screen Display

Lb1 1:? $\rightarrow$ x: $5x\ x^y$ 3 $-$
$4x^2 + 2x - 3$

Goto 1

The program is finished. To return to the computional mode press $\boxed{\text{MODE}}$ 1. To run the program press:

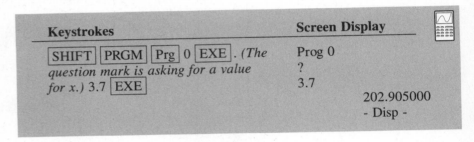

Keystrokes

$\boxed{\text{SHIFT}}$ $\boxed{\text{PRGM}}$ $\boxed{\text{Prg}}$ 0 $\boxed{\text{EXE}}$. *(The question mark is asking for a value for x.)* 3.7 $\boxed{\text{EXE}}$

Screen Display

Prog 0
?
3.7
                    202.905000
- Disp -

To determine additional values press $\boxed{\text{EXE}}$, the question mark will appear, and then follow the steps used previously. To quit the program press $\boxed{\text{EXE}}$ twice. To determine values for other functions, enter the write mode, and replace the old function with a new function.

For TI–82

The three choices at the top of the screen are EXEC, EDIT, and NEW, which stand for executing a program, editing a program, and entering a new program, respectively. To execute or edit a program, use the down arrow to highlight the desired program number, and then use the right arrow to highlight either EXEC or EDIT. By pressing the right arrow twice, we move the cursor to NEW, which indicates that we want to enter a new program.

Keystrokes

$\boxed{\text{PRGM}}$ $\boxed{\blacktriangleright}$ $\boxed{\blacktriangleright}$

Screen Display

EXEC EDIT NEW
1:

When we press NEW, the calculator asks us to enter the name of the program. (By default the calculator is in the ALPHA mode, and we do not have to press the ALPHA key.) Call the program FUNCTION.

After pressing the PRGM key, use the down arrow to see the number 9, which is Lbl. To leave the programming mode, press 2nd QUIT. We have entered the program and have called the function Y_1 in the program. But we have not entered the function.

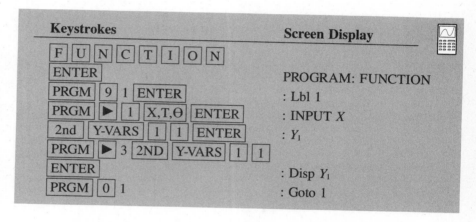

To enter the function, press Y = key. Since we labeled the function Y_1 in the program, enter the function after $Y_1 =$. We could have called the function any Y_i in the program as long as we use the same Y_i when entering the function. To run the program press:

Use the down arrow to shade the 1, and if EXEC is not highlighted, highlight EXEC.

Keystrokes	Screen Display	
ENTER	Prgm FUNCTION	
ENTER	?3.7	
ENTER	202.905000	
	?	

To quit the program press $\boxed{\text{2nd}}$ $\boxed{\text{QUIT}}$. An advantage of the TI–82 is that we can enter any function in the Y = register and graph it and/or determine functional values without changing the program.

CAUTION Does your calculator give an error message when you try to find $(-4)^3$? If so, the calculator does not distinguish between integral and nonintegral powers, and therefore, cannot perform the operation using the $\boxed{y^x}$. You can write a sequence of keystrokes using the $\boxed{x^2}$ key to avoid this problem. For example, using the function in Example 3, determine $f(-3.7)$.

$5 \boxed{\times} \boxed{(} 3.7 \boxed{+/-} \boxed{)} \boxed{x^2} \boxed{\times} 3.7 \boxed{+/-} \boxed{-} 4 \boxed{\times} 3.7$
$\boxed{+/-} \boxed{x^2} \boxed{+} 2 \boxed{\times} 3.7 \boxed{+/-} \boxed{-} 3 \boxed{=} -318.425.$ ■

EXAMPLE 4   Two Parallel Resistors

The total resistance of two resistors in parallel is a function of the individual resistance, R_1 and R_2 (see Figure 5.1).

$$R = g(R_1, R_2) = \frac{R_1 R_2}{R_1 + R_2}$$

Find:

a. $g(3, 4)$ **b.** $g\left(\frac{1}{3}, \frac{2}{7}\right)$

Solution

a. $R = g(R_1, R_2) = \dfrac{R_1 R_2}{R_1 + R_2}$

$$g(3, 4) = \frac{3 \cdot 4}{3 + 4}$$
$$= \frac{12}{7}$$
$$\approx 2$$

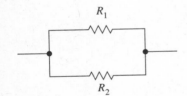

R_1

R_2

FIGURE 5.1

b. $g\left(\dfrac{1}{3},\dfrac{2}{7}\right) = \dfrac{\dfrac{1}{3}\cdot\dfrac{2}{7}}{\dfrac{1}{3}+\dfrac{2}{7}}$

$= \dfrac{\dfrac{2}{21}}{\dfrac{7+6}{21}}$

$= \dfrac{2}{21}\times\dfrac{21}{13}$

$= \dfrac{2}{13}$

≈ 0.15 ◆

In Example 4a given the point (3, 4) we have one significant digit, thus the answer 2. In Example 4b the fractions $\left(\dfrac{1}{3},\dfrac{2}{7}\right)$ are taken as exact numbers, thus we can take as many or as few digits as we please. Thus the answer can be written as 0.2, 0.15, 0.154, or with more digits.

EXAMPLE 5   **Given a Statement, Write a Function**

For each statement write a function in functional notation.

a.   The number of miles traveled as a function of time and the average rate of speed, given that the average rate of speed is 55 mph.
b.   The cost of renting a car is $18 per day plus $0.16 per mi. Write a function that expresses the cost of renting the car as a function of the daily fee and the number of miles driven.

Solution

a.   Start by assigning letters to represent the quantities.

Let n = number of miles traveled

t = time in hours

r = average rate of speed

Given that the number of miles traveled is a function of time and the average rate of speed, we can write

$$n = r\cdot t$$

Knowing that the average speed is 55 mph, we can substitute 55 for r.

$$n = 55t$$

Expressed in function notation, with an average rate of 55 mph, the function is:

$$f(t) = 55t.$$

b.   Assign letters to represent the quantities.

$$\text{Let } c = \text{cost of rental}$$
$$d = \text{number of days}$$
$$\$18d = \text{cost for } d \text{ days}$$
$$\$0.16n = \text{cost for } n \text{ miles}$$

The cost is a function of the cost for number of days and the cost for number of miles driven. Thus,

$$c = \$18d + \$0.16n$$

Expressed in functional notation the cost is

$$f(d, n) = \$8d + \$0.16n \quad \blacklozenge$$

Compound Function

A function may be defined by more than one equation. Such a function is called a **compound function.** For example, a single equation may define the rate of pay for the day shift (7 A.M.–3 P.M.). Assume that the equation is $P(r) = 8r$, where r is the hourly rate. Another equation may define the rate of pay for the evening shift (3:30 P.M.–11:30 P.M.). Assume the equation for this shift is $P(r) = 8(r + \$0.45)$. We can combine these two equations to create a compound function.

$$P(r) = \begin{cases} 8r & \text{Day shift} \\ 8(r + \$0.45) & \text{Evening shift} \end{cases}$$

EXAMPLE 6   **Compound Function**

Evaluate the function given by

$$f(x) = \begin{cases} 3x + 2, \text{ for } x \leq 1 \\ x^2 - 1, \text{ for } x > 1 \end{cases}$$

for the following.

a. $x = -1$ **b.** $x = 1$ **c.** $x = 2$

Solution

a.   Since $x = -1$, which is less than 1, use the function $f(x) = 3x + 2$.
$$f(-1) = 3(-1) + 2 \quad \text{Substituting}$$
$$= -1$$

b.   Since $x = 1$, use the function $f(x) = 3x + 2$.
$$f(1) = 3(1) + 2 \quad \text{Substituting}$$
$$= 5$$

c.   Since $x = 2$, which is greater than 1, use the function $f(x) = x^2 - 1$.
$$f(2) = 2^2 - 1 \quad \text{Substituting}$$
$$= 3 \quad \blacklozenge$$

EXAMPLE 7   **Application: Gross Pay**

Rob is paid $8.50 for each hour he works in a 40-hour week. If he works more than 40 hours in a week, he is paid time and a half.

a.   Write a function that will determine his gross pay (the amount received before taxes).

b.   Determine Rob's gross pay if he worked 46 hours last week.

Solution   Assign letters for the quantities.

a.   Let

G = gross pay

h = number of hours worked for the week

$12.75 = $ hourly rate for time and a half (1.5 · $8.50)

$8.50 · h = $ gross pay for $0 \leq h \leq 40$

$8.50 · 40 + $12.75(h - 40) = $ gross pay when $h > 40$

Since two expressions define the gross pay, we can combine them and write a compound function.

$$G(h) = \begin{cases} \$8.50 \cdot h, \text{ if } 0 \leq h \leq 40 \\ \$340 + \$12.75(h - 40), \text{ if } h > 40 \end{cases}$$

b.   Since Rob worked 46 hours last week, he has 6 hours of overtime. Thus, use the function for $h > 40$.

$$G(46) = \$340 + \$12.75(46 - 40)$$
$$= \$416.50$$

Thus, Rob's gross pay for 46 hours of work is $416.50.   ◆

Domain and Range

The set of real numbers that can be substituted for the independent variable and give real numbers for the dependent variable is called the **domain** of the function. The set of real numbers obtained for the dependent variable is called the **range** of the function.

EXAMPLE 8   **Pressure Function: Domain and Range**

The formula $P = 64d$ is used to determine the pressure, P, on objects that are immersed in saltwater, d feet below a free surface. Determine the domain, D, and the range, R.

Solution   For the function $P = 64d$, the depth, d, is measured as a positive number of feet. Thus, the domain is $D = \{d \mid d \geq 0\}$. The symbol $\{d \mid d \geq 0\}$ is read "The set of all elements d, such that d is greater than or equal to zero." Since d is nonnegative, P must be nonnegative, and the range then is $R = \{P \mid P \geq 0\}$.   ◆

EXAMPLE 9   Domain and Range

Determine the domain and range for the function $f(x) = x^2 + 5$.

Solution   We can replace the independent variable x with any real number and obtain real values for $f(x)$. Thus, the domain is $D = \{x \mid x \in R\}$. ($x \in R$ means x is an element of the set of real numbers.) If we let $x = -3$, then

$$f(-3) = (-3)^2 + 5$$
$$= 9 + 5$$
$$= 14$$

We know that squaring a negative number will yield a positive product. Thus, the functional value will always be five greater than the number squared. The least possible value of the function is 5, which occurs when $x = 0$. Thus, $R = \{f(x) \mid f(x) \geq 5\}$. ◆

EXAMPLE 10   Domain and Range

Determine the domain and range for the function $y = \dfrac{1}{x - 2}$.

Solution   To determine the domain of the function $y = \dfrac{1}{x - 2}$, ask the following question. "What real numbers may be substituted for x and not give real values for y?" An obvious answer is $x = 2$. For $x = 2$, the denominator would be zero, and we know that division by zero is undefined. Since 2 is the only restriction, state the domain in a negative sense, $D = \{x \mid x \neq 2\}$. This statement implies that all real numbers, except 2, may be substituted for x to obtain real values for y.

Are there any restrictions on the range? Can y be equal to all real numbers? By just looking at the function, it is difficult to tell. However, by solving for x, the solution becomes apparent.

$$y = \frac{1}{x - 2}$$

$$(x - 2)y = (x - 2)\frac{1}{x - 2} \quad \textbf{Multiply both sides by } x - 2$$

$$xy - 2y = 1$$
$$xy = 2y + 1$$
$$x = \frac{2y + 1}{y}$$

From this result, we can see that we could substitute any real number, except zero, for y and obtain a real number for x. Thus $R = \{y \mid y \neq 0\}$. ◆

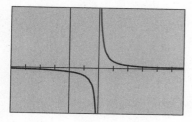

FIGURE 5.2

A graph may help to determine the domain and range of a function. For example, if we use a calculator to graph the function in Example 10 (shown in Figure 5.2), using the range settings of $[-3, 7, 1, -5, 5, 1]$, we can see that when $x = 2$, the function is not defined. Also by using the zoom or trace function, we can see that y cannot be equal to zero.

EXAMPLE 11   Domain and Range

Determine the domain and range for the function $y = \sqrt{x - 3}$.

Solution   To determine the domain of the function $y = \sqrt{x - 3}$, ask the following question. "What real numbers may be substituted for x and not result in real values for y?" Remember that we cannot find the square root of a negative number. If x is less than 3, then $x - 3$ is less than zero, and the number under the radical is negative. Thus, to obtain a real value for y, x must be greater than or equal to 3. Thus, $D = \{x \mid x \geq 3\}$. Recall that $\sqrt{x}$ means only the positive, or principal, root, and that means that y can never be a negative value. Thus for $y = \sqrt{x - 3}$, y cannot be negative. The following set of ordered pairs is obtained by substituting values for x, starting with 3: $\{(3, 0), (4, 1), (5, \sqrt{2}), \ldots, (12, 3), y \ldots, (28, 5), \ldots\}$. Examining the ordered pairs shows that the smallest value that y will attain is zero. As we substitute larger values for x, y becomes larger and larger. Thus, $R = \{y \mid y \geq 0\}$.   ◆

For the function in Example 11, use a calculator to graph the function, using the range settings $[0, 28, 2, -1, 7, 1]$. Is the path of the curve continuously increasing or decreasing as we move from left to right?

5.1 Exercises

In Exercises 1–8, identify the independent variable(s) and the dependent variable.

1. $P = 64d$

2. $F = \dfrac{9}{5}C + 32$

3. $C = \dfrac{5}{9}(F - 32)$

4. $y = f(x) = \sqrt{4x - 1}$

5. $s = f(r, \theta) = r \cdot \theta$

6. $V = f(r, h) = \dfrac{1}{3}\pi r^2 h$

7. $F = \theta(m, a) = ma$

8. $SA = \theta(\ell, w, h) = 2\ell w + 2\ell h + 2wh$

In Exercises 9–26, determine the value of the function.

9. $f(x) = 3x^2 + 7x - 5$:   **a.** $f(3)$ **b.** $f(-4)$ **c.** $f(a + h)$

10. $f(x) = 2x^2 - 7x + 11$:   **a.** $f(-2)$ **b.** $f(5)$ **c.** $f(a + h)$

11. $f(t) = \dfrac{t + 5}{t - 3}$:   **a.** $f(2)$ **b.** $f(7.4)$ **c.** $f(-3.7)$

12. $g(R) = \dfrac{R^2 - R + 6}{R - 3}$:   **a.** $g(2)$ **b.** $g(3)$ **c.** $g\left(\dfrac{3}{8}\right)$

13. $f(t) = 3t^2 + 2t - \sqrt{t}$:   **a.** $f(3.217)$ **b.** $f(5.613)$ **c.** $f(\pi)$

14. $f(z) = z^2 - \sqrt{z}$:   **a.** $f(5)$ **b.** $f(0.375)$ **c.** $f\left(\dfrac{2}{3}\right)$

15. The circumference of a circle is given by $C(r) = 2\pi r$, where r is the length of the radius (see the drawing). Find:

   a. $C(2.34$ in.$)$ b. $C(6.41$ m$)$ c. $C\left(\dfrac{5}{11}$ in.$\right)$

16. The area of a circle is given by $A(r) = \pi r^2$, where r is the length of the radius. Find:

   a. $A(2.34$ in.$)$ b. $A(6.41$ m$)$ c. $A\left(\dfrac{5}{11}$ in.$\right)$

17. The total surface area of a cube is given by the function $f(s)$, where s is the length of the side of the cube (see the illustration) and $f(s) = 6s^2$. Find:

   a. $f(3.75$ m$)$ b. $f(6.05$ in.$)$ c. $f(13.42$ mm$)$

18. If distance $s = f(t) = -16t^2 + 64t - 48$, where t is in seconds and s is in feet, find:

   a. $f(0)$ b. $f(1)$ c. $f(2.05)$ d. $f(3)$

19. Resistance, R, is a function of voltage, V, given by $R = \phi(V) = \dfrac{V}{0.21}$. Find:

   a. $\phi(1.57)$ b. $\phi(3.75)$

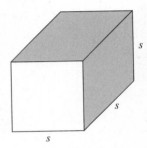

20. The measure of the angle θ in radians is given by $\theta = f(s, r) = \dfrac{s}{r}$, where s is the length of the arc determined by $\angle\theta$ and r is the length of the radius of the circle (see the drawing). Find:

   a. $f(4.71, 3)$ b. $f(15.71, 5)$

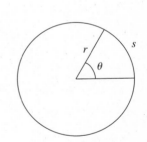

21. The formula for finding the volume of a torus (a perfectly shaped doughnut), shown in the illustration, is $V(r, R) = 2\pi^2 R r^2$. Find:

   a. $v(0.50, 1.0)$ b. $v(1.2, 3.4)$

22. Horsepower, HP, is given by $\phi(F, d, t) = \dfrac{Fd}{550t}$, where F is the force in pounds, d is the distance in feet, and t is the time in seconds. Find:

   a. $\phi(340, 8.7, 4.8)$ b. $\phi(650, 7.6, 3.7)$

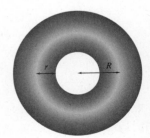

23. $f(x) = \begin{cases} x - 3, & \text{for } x < 0 \\ 2x + 5, & \text{for } x \geq 0 \end{cases}$ Find:

   a. $f(-1)$ b. $f(0)$ c. $f(1)$

24. $f(x) = \begin{cases} \dfrac{x^2 - 9}{x - 3} & \text{for } x \neq 3 \\ 6, & \text{for } x = 3 \end{cases}$ Find:

   a. $f(2)$ b. $f(3)$ c. $f(4)$

25. $f(x) = \begin{cases} \dfrac{x^2 - 4}{x - 2}, & \text{for } x \neq 2 \\ 4, & \text{for } x = 2 \end{cases}$ Find:

    a. $f(0)$ **b.** $f(2)$ **c.** $f(4)$

26. Resistance, R, in ohms, is given by $\theta(P, L, A) = \dfrac{PL}{A}$, where P is the resistivity in ohm meters, L is the length in meters, and A is the cross-section of the wire in square meters. Find:

    a. $\theta(17, 7.5, 1 \times 10^{-5})$

    b. $\theta(32, 12.5, 1.5 \times 10^{-5})$

In Exercises 27–40, determine the domain and range for each function. Use the graphics calculator to help verify the solution.

27. $y = 3x + 4$

28. $R = 6S - 7$

29. $f(t) = t^2 + 5$

30. $\phi(s) = s^2 - 3$

31. $SA = f(r) = 4\pi r^2$

32. $V = f(r) = \dfrac{4}{3}\pi r^3$

33. $y = f(x) = \dfrac{3}{x}$

34. $y = f(x) = \dfrac{4}{x - 1}$

35. $y = f(x) = \dfrac{x + 1}{x - 1}$

36. $y = f(x) = \dfrac{x - 1}{1 - x}$

37. $s = f(t) = \sqrt{t + 2}$

38. $s = f(t) = \sqrt{t - 2}$

39. $p = f(\ell) = \sqrt{3 + \ell}$

40. $p = f(\ell) = \sqrt{3 - \ell}$

In Exercises 41–45, write a function in functional notation for each statement, using the given letters.

41. The number of seconds, s, in h hours.

42. The Celsius reading corresponding to $x°$ Fahrenheit.

43. The distance, d, traveled at 55 mph in t hours.

44. The number of millimeters, y, in x kilometers.

45. The perimeter of a rectangle is 40 m. Express the area $A(w)$ as a function of the width, w.

46. **a.** Given a constant density, express the pressure of a liquid as a function of the height of the surface above the bottom.
    b. Given a set depth, express the pressure of a liquid as a function of the density.
    c. How could you express the pressure as a function of both height and density?

47. A printer agrees to print a brochure for a sum of $300 plus $0.15 for each copy. Express the cost as a function of the number of copies.

48. An employee of Company B receives $7.00 for each hour she works in a regular 8-hour day. If she works more than 8 hours in any one day, she receives time and a half for the next 2 hours and double time for the 2 hours that follow the first overtime period. Write a function that will determine her exact wage for a single day at any time between 0 and 12 hours.

49. The fixed costs per day for a doughnut shop are $400, and the variable costs are $1.80 per dozen doughnuts produced. If x dozen are produced daily, express the daily cost $C(x)$ as a function of x.

50. A farmer purchases 10 km of fencing to enclose a rectangular piece of grazing land along a straight river. If no fencing is required along the river and the sides perpendicular to the river are x km long, find a formula for the area $A(x)$ of the rectangle in terms of x. What is the domain of the function?

5.2

Rectangular Coordinates

We have heard that a picture is worth a thousand words. This is especially true for the technician who, without a graph, may not be able to interpret the data collected. In this section we will develop the basic tools for constructing graphs of functions.

In the discussion of functions, such as $f(x) = x + 6$, we found that many ordered pairs can satisfy a particular function. Some of the ordered pairs that satisfy $f(x) = x + 6$ are $(0, 6)$, $(-6, 0)$, $(3, 9)$, $(-3, 3)$, and $(-5, 1)$. One way to gain an idea of the many ordered pairs that satisfy the equation is to draw a graph of the equation on the coordinate plane. The pattern of the data points may lead to a deeper understanding of the function. To graph an equation, we must understand the rectangular coordinate system.

The **rectangular coordinate system,** also called the **Cartesian coordinate system,** consists of lines that are perpendicular to each other. In Figure 5.3, both lines have the same scale. The **scale** is the number of units between the hash marks on the axes. In some cases, we may wish to make one scale larger than the other. Convention says that the horizontal line is called the axis of the independent variable, in this case, the x-axis. The vertical line conventionally is considered the axis of the dependent variable, in this case, the y-axis. The point of intersection of the x- and y-axes is called the **origin.** The scales above and to the right of the origin on the x- and y-axes are positive, and the scales below and to the left of the origin on the x- and y-axes are negative. The axes divide the plane into four parts, which are named the first, second, third, and fourth **quadrants.**

Every point on the plane can be represented by an **ordered pair** (x, y), called the **set of coordinates of the point.** The point is called the **graph** of the ordered pair. The first coordinate, x, is called the **x-coordinate** (or **abscissa**), and the second coordinate, y, is called the **y-coordinate** (or **ordinate**). For example, the ordered pair $(5, 3)$ represents a specific point on the plane. The point $(5, 3)$ has an x value of 5 and a y value of 3 so that the point is 5 units to the right of the y-axis and 3 units above the x-axis. The origin is represented by the ordered pair $(0, 0)$.

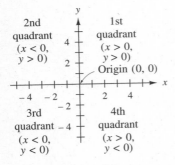

FIGURE 5.3

CAUTION The ordered pairs $(15, 4)$ and $(4, 15)$ are not equal. If you live at number 15 on 4th Street, but give the address as number 4 on 15th Street, will your guest arrive for dinner? ■

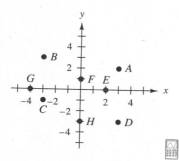

FIGURE 5.4

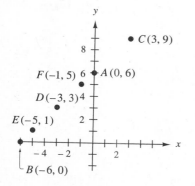

FIGURE 5.5

EXAMPLE 1 Determine the Coordinates of the Graph

Determine the coordinates of points A, B, C, D, E, F, G, and H, in Figure 5.4.

Solution Point A has an x value of 3 and a y value of 2; therefore, its ordered pair is (3, 2). Similarly, the other ordered pairs are: $B(-3, 3)$, $C(-3, -1)$, $D(3, -3)$, $E(2, 0)$, $F(0, 1)$, $G(-4, 0)$, and $H(0, -3)$. The two points, E and G, that are on the x-axis have zero for the ordinate, and the two points, F and H, that are on the y-axis have a zero for the abscissa. ◆

EXAMPLE 2 Graph the Ordered Pairs

Graph the points $A(0, 6)$, $B(-6, 0)$, $C(3, 9)$, $D(-3, 3)$, $E(-5, 1)$, and $F(-1, 5)$, using your calculator.

Solution Follow the instructions given in the boxes. The points graphed in Figure 5.5 all fulfill the conditions imposed by the function $f(x) = x + 6$. ◆

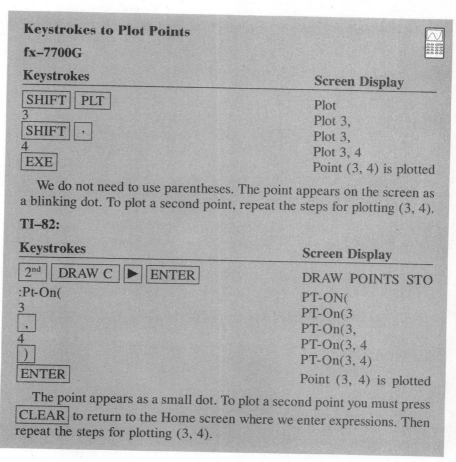

Keystrokes to Plot Points

fx–7700G

Keystrokes	Screen Display
SHIFT PLT 3	Plot Plot 3,
SHIFT , 4	Plot 3, Plot 3, 4
EXE	Point (3, 4) is plotted

We do not need to use parentheses. The point appears on the screen as a blinking dot. To plot a second point, repeat the steps for plotting (3, 4).

TI–82:

Keystrokes	Screen Display
2nd DRAW C ▶ ENTER :Pt-On(	DRAW POINTS STO PT-ON(
3	PT-On(3
,	PT-On(3,
4	PT-On(3, 4
)	PT-On(3, 4)
ENTER	Point (3, 4) is plotted

The point appears as a small dot. To plot a second point you must press CLEAR to return to the Home screen where we enter expressions. Then repeat the steps for plotting (3, 4).

In applied problems that deal with mechanics, civil technology, and navigation, we must be able to determine the distance between two points. The

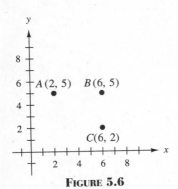

FIGURE 5.6

following discussion develops the distance formula, which allows us to find the distance between two points quickly and accurately.

To begin, the distance between two points $P_1(x_1, y_1)$ and $P_2(x_2, y_2)$ is denoted by $\overline{P_1 P_2}$. There are two preliminary comments: $\overline{P_1 P_2} \geq 0$ and $\overline{P_1 P_2} = \overline{P_1 P_2}$. In other words, distance is always considered to be positive, and the order of the labeling of a distance does not change the distance.

Let us take a look at a specific example before getting to the general formula. Consider the three points $A(2, 5)$, $B(6, 5)$, and $C(6, 2)$, shown in Figure 5.6. The points A and B lie on the same horizontal line; thus we find $\overline{AB} = 4$ by counting (on the x-axis). Also, $\overline{BA} = 4$ by counting (on the x-axis). $\overline{AB}$ can also be found by taking the absolute value of the difference of the abscissas, thus, $\overline{AB} = |6 - 2|$ or $\overline{BA} = |2 - 6|$. From previous experience we know that $|6 - 2| = |2 - 6| = 4$. Since B and C lie on a line parallel to the y-axis, we can use similar arguments to show that $\overline{CB} = 3$.

$\overline{AC}$ cannot be found by the techniques used to find $\overline{AB}$ and $\overline{CB}$ since the line segment AC is not parallel to either one of the axes. However, since $\triangle ABC$ is a right triangle, we can use the Pythagorean relation to find $\overline{AC}$.

$$(\overline{AC})^2 = (\overline{AB})^2 + (\overline{BC})^2$$
$$= 4^2 + 3^2$$
$$= 16 + 9$$
$$= 25$$

Taking the square root of both sides we have $\overline{AC} = 5$.

Now let's perform comparable operations for the general case. Let P_1, P_2, and P be points with coordinates (x_1, y_1), (x_2, y_2), and (x_1, y_2), respectively (see Figure 5.7). From the figure, we see that $PP_2 = |x_2 - x_1|$ and $P_1 P = |y_2 - y_1|$. Using the Pythagorean relation, we have the following.

$$\overline{P_1 P_2}^2 = \overline{PP_2}^2 + \overline{P_1 P}^2$$
$$= |x_2 - x_1|^2 + |y_2 - y_1|^2$$
$$= (x_2 - x_1)^2 + (y_2 - y_1)^2 \qquad \text{For any real number } |z|^2 = z^2$$
$$\overline{P_1 P_2} = \sqrt{(x_2 - x_1)^2 + (y_2 - y_1)^2} \qquad \begin{array}{l}\text{Take the principal square} \\ \text{root of both sides}\end{array}$$

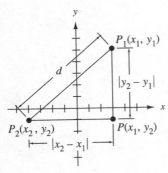

FIGURE 5.7

This demonstration serves as a proof to the following theorem.

THEOREM 5.1 Distance Formula

If $P_1(x_1, y_1)$ and $P_2(x_2, y_2)$ are any two points in the plane, then the distance, $\overline{P_1 P_2}$, between them is given by

$$\overline{P_1 P_2} = \sqrt{(x_2 - x_1)^2 + (y_2 - y_1)^2}.$$

EXAMPLE 3   Distance Formula: Isosceles Triangle

Show that the triangle with vertices $A(-2, 6)$, $B(4, 8)$, and $C(2, 4)$ is isosceles.

Solution To show that triangle ABC is isosceles, we must show that two of its sides are equal in length. Using the distance formula, we find that

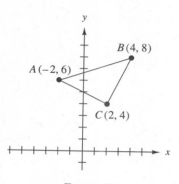

FIGURE 5.8

$$\overline{AB} = \sqrt{[4 - (-2)]^2 + (8 - 6)^2}$$
$$= \sqrt{6^2 + 2^2}$$
$$= \sqrt{40}.$$
$$\overline{AC} = \sqrt{[2 - (-2)]^2 + (4 - 6)^2}$$
$$= \sqrt{4^2 + (-2)^2}$$
$$= \sqrt{20}.$$
$$\overline{BC} = \sqrt{(2 - 4)^2 + (4 - 8)^2}$$
$$= \sqrt{(-2)^2 + (-4)^2}$$
$$= \sqrt{20}.$$

Therefore, $\overline{AC} = \overline{BC}$ and $\triangle ABC$ is isosceles. The triangle is shown in Figure 5.8. ◆

The midpoint of a line segment can be found by using the midpoint formula.

THEOREM 5.2 Midpoint Formula

If $P_1(x_1, y_1)$ and $P_2(x_2, y_2)$ are any two points in the plane, then the coordinates of the midpoint of the line segment from P_1 to P_2 are given by

$$\left(\frac{x_1 + x_2}{2}, \frac{y_1 + y_2}{2} \right).$$

CAUTION When using the distance formula and the midpoint formula, the order in which we select the points is not important. For the points $(1, 3)$ and $(-4, 5)$, we can select $(1, 3)$ as P_1 and $(-4, 5)$ as P_2, or $(-4, 5)$ as P_1 and $(1, 3)$ as P_2. The results in the formulas will be the same. However, we cannot mix the x and y values of P_1 and P_2. We cannot let $x_1 = 1$ and $y_1 = 5$, as this will give an incorrect result in both formulas. ■

EXAMPLE 4 Application: Ski Boot Problem

To help the ski boot designer mentioned at the beginning of the chapter, graph the function and give the coordinates of the point midway between $(0.10, 16)$ and $(0.15, -3)$. The data are given in the table.

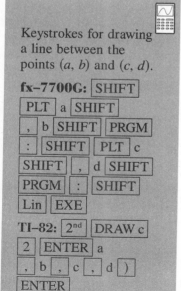

Keystrokes for drawing a line between the points (a, b) and (c, d).

fx–7700G: SHIFT
PLT a SHIFT
, b SHIFT PRGM
: SHIFT PLT c
SHIFT , d SHIFT
PRGM : SHIFT
Lin EXE

TI–82: 2nd DRAW c
2 ENTER a
, b , c , d)
ENTER

Time	°Rotation	Time	°Rotation	Time	°Rotation
0.00	0	0.40	−28	0.80	−5
0.05	17	0.45	−31	0.85	−5
0.10	16	0.50	−23	0.90	−3
0.15	−3	0.55	−12	0.95	−4
0.20	−10	0.60	12	1.00	−2
0.25	−14	0.65	17	1.05	−2
0.30	−17	0.70	18	1.10	0
0.35	−20	0.75	6		

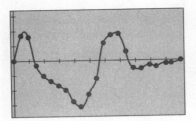

FIGURE 5.9

Solution   Use a graphics calculator to draw the graph of the data points. The keystrokes we will use to draw a line are shown in the box. The calculator will draw a line segment connecting the two points. Enter the second and third point and continue the process until the complete graph is drawn. After all the points are plotted and the lines are drawn, the result will be as shown in Figure 5.9. The graph could also be drawn with a computer using a graphing program.

The coordinates of the point midway between $(0.70, 18)$ and $(0.45, -31)$ are given by

$$\left(\frac{x_1 + x_2}{2}, \frac{y_1 + y_2}{2}\right) = \left(\frac{0.70 + 0.45}{2}, \frac{18 + (-31)}{2}\right)$$

$$= (0.58, -6.5). \quad \blacklozenge$$

A BASIC computer program which can be used to find the midpoint of a line segment is found on the tutorial disk.

5.2 Exercises

In Exercises 1 and 2, plot the given points on a Cartesian coordinate system.

1. $A(3, 5)$, $B(-3, -7)$, $C(-3, 4)$, $D(5, -4)$,
   $E(-2, 0)$ $F(3, 0)$, $G(0, -2)$, $H(0, 5)$, $I(0, \sqrt{2})$

2. $A(2, 5)$, $B(-3, 2)$, $C\left(\frac{7}{2}, \frac{3}{2}\right)$, $D(-4.5, -3.5)$,
   $E(-3, 0)$, $F(4, 0)$, $G(0, -5)$, $H(3, 6)$

In Exercises 3 and 4, plot the given points using a graphics calculator.

3. $A(4, 5)$, $B(-3, 7)$, $C(-4, -5)$, $D(6, -4)$,
   $E(7, 0)$, $F(0, -7)$, $G(-5, 0)$, $H(\sqrt{3}, -\sqrt{2})$

4. $A\left(\frac{2}{3}, \frac{3}{4}\right)$, $B\left(-\frac{1}{4}, \frac{6}{5}\right)$, $C\left(-\frac{7}{3}, -\frac{9}{5}\right)$,
   $D\left(\frac{5}{3}, -\frac{7}{9}\right)$, $E(-1.3, -2.4)$, $F(3.7, -5.6)$,
   $G(-6.4, 0)$, $H(4.3, 6.7)$

In Exercises 5 and 6, determine the coordinates of the given points.

5.

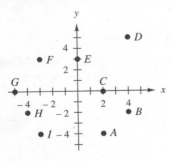

6.

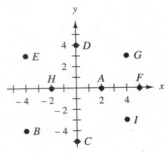

In Exercises 7–14, determine $\overline{P_1 P_2}$.

7. $P_1(3, 7)$; $P_2(-2, 5)$

8. $P_1(-5, 6)$; $P_2(4, 3)$

9. $P_1(-7, -11)$; $P_2(6, 3)$

10. $P_1(2, 3)$; $P_2(-4, -7)$

11. $P_1(3.4, 7.5)$; $P_2(6.3, 7.4)$

12. $P_1(-6.5, -7.4)$; $P_2(-5.4, 6.4)$

13. $P_1\left(3\frac{1}{2}, -7\frac{1}{2}\right)$; $P_2\left(-3\frac{1}{2}, 7\frac{1}{2}\right)$

14. $P_1(-0.54, 0.44)$; $P_2(7.5, -8.4)$

In Exercises 15–20, **a.** graph the points and **b.** determine the coordinates of the midpoint of the line segment joining them.

15. $P_1(2, 7)$; $P_2(5, 2)$

16. $P_1(9, 6)$; $P_2(2, 2)$

17. $P_1(2, 5)$; $P_2(-3, 4)$

18. $P_1(-3, -4)$; $P_2(2, 3)$

19. $P_1(-\pi, -6)$; $P_2(\sqrt{3}, -3)$

20. $P_1(\sqrt{2}, 0)$; $P_2(-\sqrt{2}, \pi)$

In Exercises 21–32, use reasoning similar to that used in Example 3.

21. **a.** Show that the triangle with vertices $A(0, 1)$, $B(2, 3)$, and $C(3, 0)$ is isosceles.
　　b. Find the perimeter of triangle ABC.

22. **a.** Show that the triangle with vertices $A(-4, 5)$, $B(4, 4)$, and $C(0, -2)$ is isosceles.
　　b. Determine the perimeter of triangle ABC.

23. Show that the points $A(-5, -1)$, $B(-1, 3)$, and $C(-1, -5)$ are the vertices of a right triangle.

24. Determine the area of the right triangle in Exercise 23.

25. Show that the points $A(-1, 1)$, $B(6, 0)$, and $C(3, -3)$ are the vertices of a right triangle.

26. Find the area of the right triangle in Exercise 25.

27. Show that the points $A(-5, -1)$, $B(-2, 2)$, $C(3, -3)$, and $D(0, -6)$ are vertices of a rectangle.

28. Determine all the points on the y-axis that are a distance of 5 from the point $(3, 5)$.

29. The ordinate of a point (or points) is 2, and its distance from $(7, 3)$ is $\sqrt{122}$. Find the abscissa of the point (or points).

30. What is the length of the longest straight line that can be drawn on a sheet of $8\frac{1}{2}$- by 11-in. typing paper?

31. Use a convenient coordinate system to show that the diagonals of a parallelogram bisect each other.

32. If guy wires are to run from the top of a 100-m tall antenna to the ground 20 m from the base of the antenna, how long must the wires be?

33. Using the information given in Example 3 and theorem 5.2 determine the area of the isosceles triangle ABC in Figure 5.8.

In Exercises 34 and 35, use your graphics calculator to draw the triangle. Then determine whether the triangle is isosceles, equilateral, or right.

34. $A(-4, 3)$, $B(3, 5)$, $C(3, 1)$

35. $A(1, 3)$, $B(6, -2)$, $C(-4, -2)$

5.3

Graph of a Function

Why do we draw graphs of functions? In problem solving, we may be able to write a function that is difficult to solve. As we have seen, a graph of the function can provide us with information that we may not be able to obtain from the function. For example, is the function a straight line or a curve? Does it have a high point, low point, move up or down as we move from left to right, or are there any holes in the curve? In general, how does the function behave? Also, as

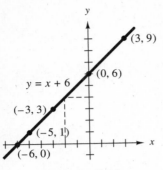

FIGURE 5.10

we become more familiar with graphs, we will be able to examine a function and have an idea of what the graph looks like.

In Section 5.2, we saw that the ordered pairs $(0, 6)$, $(-6, 0)$, $(3, 9)$, $(-3, 3)$, and $(-5, 1)$ all satisfy the function $y = x + 6$. Then in Figure 5.5, we plotted those points. Now with a straight edge we can draw one line that passes through all of these points as shown in Figure 5.10. As part of the graph, label the points and write the function on one side of the line. With proper labeling, the figure can be identified for future reference. If we add another function to the graph, we will find it easier to identify each function. When extended indefinitely in either direction, this straight line passes through all the points on the plane that satisfy the equation $y = x + 6$. For example, locate the point $(-2, 4)$ on the graph; we see that the line passes through this point. Thus, the ordered pair $(-2, 4)$ is a solution to the equation. Since there are an infinite number of ordered pairs of points that satisfy the equation, the graph is a powerful way to illustrate the solution.

Now examine the graph in Figure 5.10. The figure is a straight line. As we move from left to right the line goes up, or increases, and does not change its direction. The line crosses the x-axis at $(-6, 0)$ and the y-axis at $(0, 6)$. The points at which the line crosses the x- and y-axes are called the **x-intercept** and the **y-intercept**, respectively. In general, the x-intercept has coordinates $(x, 0)$, and the y-intercept has coordinates $(0, y)$. There are no discontinuities (holes in the graph); that is, for each value of x there is a unique value for y. The domain of the function is $D = \{x \mid x \in R\}$. The range of the function is $R = \{y \mid y \in R\}$.

The equation $y = x + 6$ is called a linear equation or linear function. We can tell that it is a **linear equation or linear function** for two reasons: (1) the graph of the equation is a straight line, and (2) the exponents of the variables x and y are 1. The general form of a linear equation as $ax + by = c$, where $a \neq 0$ and $b \neq 0$.

For example, in the equation $x^2 + 3y = 4$, the exponent of the variable x is not 1; therefore, it is not a linear equation. The function $y = \dfrac{1}{x}$ can also be written in the form $y = x^{-1}$, or $xy = 1$; it, therefore, is not a linear function as a later graph will show.

When we graph a linear function we will find it useful to solve the equation for the dependent variable y. With the equation in this form, we can use the graphics calculator to graph the function. Also, with this form, it may be easier to determine ordered pairs, which we use to establish a table of values.

EXAMPLE 1   Graphing a Linear Equation

Graph the equation $4x + 3y = 12$.

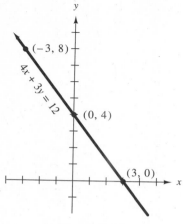

FIGURE 5.11

x	y
−3	8
0	4
3	0

Graph $y = \dfrac{-4}{3}x + 4$

on the graphics calculator. Use the range settings (−12, 12, 2, −8, 8, 2)

fx–7700G:

| SHIFT | Cls | EXE |

| GRAPH | (| SHIFT |

| (−) | 4 | ÷ | 3 |) |

| x,θ,T | + | 4 | EXE |

TI–82:

| y = | (| (−) | 4 | ÷ | 3 |

|) | x,T,θ | + | 4 |

| GRAPH |

Solution The equation $4x + 3y = 12$ is a linear equation. Thus, the graph is a straight line. To help determine ordered pairs, solve the equation for y.

$$4x + 3y = 12$$
$$3y = -4x + 12$$
$$y = \frac{-4}{3}x + 4$$

We can determine the ordered pairs by substituting any real number for x and obtaining a corresponding real number for y.

Let $x = -3$	Let $x = 0$	Let $x = 3$
$y = \dfrac{-4}{3}(-3) + 4$	$y = \dfrac{-4}{3}(0) + 4$	$y = \dfrac{-4}{3}(3) + 4$
$= 8$	$= 4$	$= 0$

The linear equation $4x + 3y = 12$ is graphed in Figure 5.11. ◆

The keystrokes for graphing the equation $4x + 3y = 12$ on the graphics calculators are given in the box. The resulting graph is the same as in Figure 5.11.

In examining these we see that the line falls or decreases as we move from left to right on the graph. This observation can be verified; using the trace and/or zoom feature of the graphics calculator. The x and y intercepts are (3, 0) and (0, 4), respectively. We also can see that the domain and range are $D = \{x \mid x \in R\}$ and $R = \{y \mid y \in R\}$.

Generally, the domain of all linear functions is the set of all real numbers unless a problem dictates restrictions on the domain. A problem that deals with the length of a fence would impose two restrictions: the length could not be negative nor could the fence have an indefinite length. In general, the range of all linear functions is all real numbers with two exceptions. First, restrictions imposed by problems limit the values of the domain. Secondly, when we are dealing with a constant function, the range is just one value. In the case of the constant function, $f(x) = c$, the range is $R = \{f(x) \mid f(x) = c\}$. This particular limitation is investigated in Example 2.

EXAMPLE 2 Graph of a Constant Function

Graph the function $f(x) = 4$.

Solution For the function $f(x) = 4$, x is the independent variable, and $f(x)$ is the dependent variable. Thus, name the vertical axis the $f(x)$-axis and the horizontal axis the x-axis. The function $f(x) = 4$ is a special function called a constant function. This means that the graph will always be 4 units above the x-axis for all values of x. If we write the function as $f(x) = 0 \cdot x + 4$, we can see that any real number can be substituted for x and that the functional value always will be 4. The domain is $\{x \mid x \in R\}$ and the range is $\{f(x) \mid f(x) = 4\}$. To determine a table of values, substitute any value for x and obtain the value of f for $f(x)$.

$$\text{Let } x = -5 \quad \text{Let } x = 0 \quad \text{Let } x = 6$$
$$f(-5) = 4 \qquad f(0) = 4 \quad f(6) = 4$$

The function is graphed in Figure 5.12 ◆

x	$f(x)$
-5	4
0	4
6	4

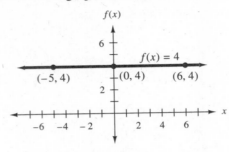

FIGURE 5.12

 EXAMPLE 3   **Graph the Equation with a Graphics Calculator**

Graph $2y - 64x + 7 = 0$.

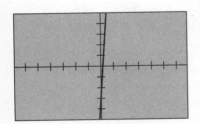

FIGURE 5.13

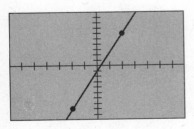

FIGURE 5.14

Solution   Since the equation is a linear equation, the graph will be a straight line. Solve the equation for y.

$$2y - 64x + 7 = 0$$
$$2y - 64x + 64x + 7 - 7 = 64x - 7$$
$$\frac{2y}{2} = \frac{64x}{2} - \frac{7}{2}$$
$$y = 32x - 3.5$$

Use the range settings $(-12, 12, 2, -8, 8, 2)$ to graph the equation. The graph on the calculator screen is given in Figure 5.13. The line is so close to the y-axis that it is difficult to tell anything about the graph. To get a better picture of the equation, change the range to $(-5, 5, 1, -70, 70, 10)$. The graph with the new range settings is given in Figure 5.14. The graph is easier to read; We can see that the curve is increasing and that x- and y-intercepts are very close to the origin. We also can read values from the graph. For example, when $x = 1$, we can estimate that $y \approx 29$. Another way to estimate the value of y when $x = 1$ is to use the trace feature on the calculator. ◆

In Example 3, the second range setting changed the scale on the x-axis to 1 and on the y-axis to 10. As we saw, the different scales produced a graph that was easier to interpret.

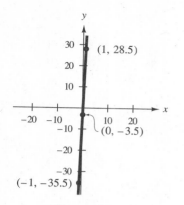

FIGURE 5.15

However, when we use two different scales be careful because we may misinterpret the results if the graph is not read carefully. The different scales will produce a line that is tilted differently from a graph in which the scales are identical. Compare Figures 5.14 and 5.15. In Figure 5.15 the line appears to be almost straight up and down, while in Figure 5.14, the line is slanted more to the right. Thus, a different visual effect is produced.

EXAMPLE 4   Graph of a Square Root

Graph $y = \sqrt{x - 4}$

Solution   The dependent variable is y, and the independent variable is x. To obtain real values for y, $x - 4$ must be greater than or equal to 0, or $x \geq 4$. The domain $D = \{x \mid x \geq 4\}$. Therefore y must always be greater than 0; thus, $R = \{y \mid y \geq 0\}$. From these restrictions we know that there will be no points to the left of 4 and below the y-axis.

The table of values is found by selecting values for x and solving for y. (See Figure 5.16 for the graph.) ◆

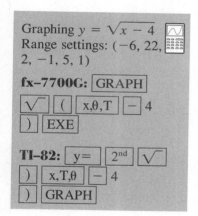

Graphing $y = \sqrt{x - 4}$
Range settings: $(-6, 22, 2, -1, 5, 1)$

fx-7700G: | GRAPH |

| √ | | (| | x,θ,T | | − | 4

|) | | EXE |

TI-82: | y= | | 2nd | | √ |

|) | | x,T,θ | | − | 4

|) | | GRAPH |

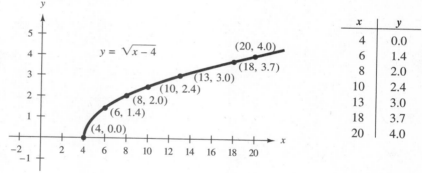

x	y
4	0.0
6	1.4
8	2.0
10	2.4
13	3.0
18	3.7
20	4.0

FIGURE 5.16

Now with the graphics calculator draw the graphs of $y = \sqrt{x + 4}$, $y = \sqrt{x}$, and $y = \sqrt{x - 4}$ on the same coordinate axes system. Your result should look like Figure 5.17. Examining the graph we see that:

1. All three functions continually increase (just keep going up) as we move from left to right.
2. The graphs of all three functions start on the x-axis; thus, the smallest value for y for all three functions is zero. The range is the same for all three functions: $R = \{y \mid y \geq 0\}$.
3. The graph of $y = \sqrt{x + 4}$ starts four units to the left of the origin. The domain is $D = \{x \mid x \geq -4\}$.
4. The graph of $y = \sqrt{x}$ starts at the origin. The domain is $D = \{x \mid x \geq 0\}$.
5. The graph of $y = \sqrt{x - 4}$ starts four units to the right of the origin. The domain is $D = \{x \mid x \geq 4\}$. We can see that, for a function of the form $y = \sqrt{x \pm a}$, the $+a$ will shift the curve to the left of the origin a units, and the $-a$ will shift the curve to the right of the origin a units.

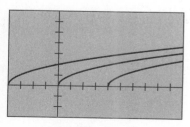

FIGURE 5.13

Quadratic Function

A function of the form $f(x) = ax^2 + bx + c$, where a, b, and c are real numbers and $a \neq 0$, is called a **quadratic function.** The graph of a quadratic function is called a *parabola.* We will look at two quadratic functions: $y = -2x^2 + 4$ and $y = 2x^2 - 4$. These two functions are sketched in Figure 5.18. Notice that both graphs have a point called the *vertex.*

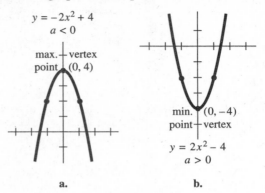

FIGURE 5.18

The **vertex** is the highest point or the lowest point on a parabola. The vertex is a maximum point if the parabola opens down, Figure 5.18a, and it is a minimum point if the parabola opens up, Figure 5.18b. Some of the important facts about these two parabolas are listed in the box.

FACTS ABOUT PARABOLAS

Function	Left of Vertex	Vertex	Right of Vertex
$f(x) = -2x^2 + 4$	Function increasing (rising)	Maximum value $(0, 4)$	Function decreasing (falling)
		Curve opens down $a = -2 < 0$	
$f(x) = 2x^2 - 4$	Function decreasing (falling)	Minimum value $(0, -4)$	Function increasing (rising)
		Curve opens up $a = 2 > 0$	

The coordinates of the vertex of the parabola can be determined using the formulas $x = \dfrac{-b}{2a}$ and $y = f\left(\dfrac{-b}{2a}\right)$, where a and b are the coefficients of the x^2

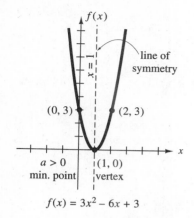

(0, 3)   (2, 3)

$a > 0$ (1, 0)
min. point   vertex

$f(x) = 3x^2 - 6x + 3$

FIGURE 5.19

x	y
-5	7
-4	0
-3	-5
-2	-8
-1	-9
0	-8
1	-5
2	0
3	7

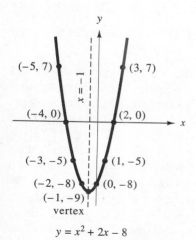

$(-5, 7)$ $(3, 7)$

$(-4, 0)$ $(2, 0)$

$(-3, -5)$ $(1, -5)$
$(-2, -8)$ $(0, -8)$
$(-1, -9)$
vertex

$y = x^2 + 2x - 8$

FIGURE 5.20

and the x term, respectively. For the function $f(x) = 3x^2 - 6x + 3$, $a = 3$, $b = -6$, and $c = 3$. Since $a > 0$, the vertex of the parabola is a minimum point. To determine the x-coordinate of the vertex, substitute for a and b.

$$\frac{-b}{2a} = \frac{-(-6)}{2(3)} = 1$$

Since $\dfrac{-b}{2a} = 1$, to determine the y-coordinate of the vertex, determine $f(1)$.

$$f(1) = 3(1)^2 - 6(1) + 3 = 0.$$

The vertex of the function $f(x) = 3x^2 - 6x + 3$ is (1, 0), and is graphed in Figure 5.19.

For all quadratic functions in x there is a line that passes though the vertex and is parallel to the y-axis. This special line is called a **line of symmetry.** As the name suggests, the parabola can be folded symmetrically along this line such that the two halves of the curve fit exactly on top of each other. For all quadratic functions, the equation $x = \dfrac{-b}{2a}$ is the line of symmetry. The value $\dfrac{-b}{2a}$ is the x-coordinate of the vertex. Thus, the line of symmetry will always pass through the vertex of the parabola.

For the function $f(x) = 3x^2 - 6x + 3$, the line of symmetry is $x = \dfrac{-(-6)}{2(3)}$, or $x = 1$.

In Figure 5.19, we can see the points (0, 3) and (2, 3) are symmetric points on the curve. The two points are the same distance, 1 unit, from the line of symmetry, and they are the same distance 3 units, above the vertex.

EXAMPLE 5   Graphing a Parabola

Graph $y = x^2 + 2x - 8$.

Solution   The function $y = x^2 + 2x - 8$ is a parabola, with $a = 1$, $b = 2$, and $c = 8$. Since $a > 0$, the vertex is a minimum point and the curve opens upward. The vertex is the point $(-1, -9)$; that is, $\dfrac{-b}{2a} = \dfrac{-2}{2} = -1$, and $f(-1) = -9$. Knowing the vertex, we also know that the line of symmetry is $x = -1$. Since the vertex is a minimum point and the y-coordinate of the minimum point is -9, the range is $R = \{y \mid y \geq -9\}$. Recall that the y-intercept is found by setting $x = 0$. When $x = 0$, $y = -8$; thus, the y-intercept is $(0, -8)$. Knowing that the curve is symmetric to the line $x = -1$, a point symmetric with the point $(0, -8)$ is $(-2, -8)$, see Figure 5.20. To get additional points to sketch a smooth curve, pick the values 3, 2, and 1 for x. Calculating the values for y, we have the ordered pairs (3, 7), (2, 0), and $(1, -5)$. The points symmetric with these three points are $(-5, 7)$, $(-4, 0)$ and $(-3, -5)$, respectively. Plotting these points, we can sketch the graph of $y = x^2 + 2x - 8$ as in Figure 5.20. ◆

Rational Function

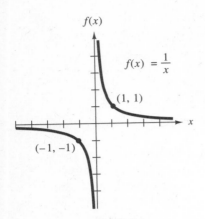

FIGURE 5.21

The function $h(x) = \dfrac{f(x)}{g(x)}$, where $f(x)$ and $g(x)$ are polynomials is called a **rational function.** Examples of rational functions are $f(x) = \dfrac{1}{x}, f(x) = \dfrac{x+1}{x-1}$, and $f(x) = \dfrac{x^2 + 3x + 1}{x^2 - 1}$. For the rational function $f(x) = \dfrac{1}{x}, x \neq 0$; thus, the domain is $\{x \mid x \neq 0\}$. If we solve for x, we would have $x = \dfrac{1}{f(x)}$; thus, $f(x) \neq 0$. The range is $\{f(x) \mid f(x) \neq 0\}$. To create a table of values, we can find values for $f(x)$ by selecting any value for x except $x = 0$. Looking at the table of values

x	$f(x)$	x	$f(x)$
0.01	100	-0.01	-100
0.1	10	-0.1	-10
1	1	-1	-1
$\vdots$	$\vdots$	$\vdots$	$\vdots$
10	0.1	-10	-0.1
100	0.01	-100	-0.01

and the graph in Figure 5.21, we see that for small values of $x, f(x)$ becomes very large and for large values of $x, f(x)$ becomes very small. For small values of x, the curve gets very close to the line $x = 0$ (the y-axis), but never touches or crosses the line. Also, for large values of x, the curve gets very close to the line $f(x) = 0$ (the x-axis), but never touches or crosses the line. These two lines, $x = 0$ and $y = 0$ are special lines for the function $f(x) = \dfrac{1}{x}$, called *asymptotes*.

A straight line that the curve gets closer and closer to, but never touches, for increasing and decreasing values of x is an **asymptote**. The line $x = 0$ is called a **vertical asymptote**, and the line $f(x) = 0$ is called a **horizontal asymptote.** A vertical asymptote coincides with, or is parallel to, the y-axis, and a horizontal asymptote coincides with, or is parallel to, the x-axis. We say that the curve $f(x) = \dfrac{1}{x}$ is asymptotic to the lines $x = 0$ and $f(x) = 0$. Asymptotes will exist when we graph a rational function.

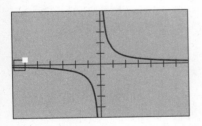

FIGURE 5.22

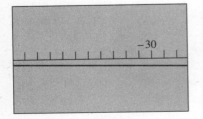

FIGURE 5.23

An asymptote may be illustrated with a graphics calculator. The fx–7700G and the TI-82 have a ZOOM key. When pressing this key a menu appears that contains several different choices, one of which is called BOX. When we activate this function, a blinking cursor appears on the screen. We move the cursor oo an area we wish to enlarge, such as the one in Figure 5.22. Press EXE or ENTER, which will locate one vertex of a rectangle. Now move the cursor up or down and to the left or right so that you form a rectangle around the area you wish to enlarge. When the rectangle is properly located, press EXE or ENTER. The area inside the rectangle will cover the complete screen as shown in Figure 5.23. Now we can see that the curve does not touch the x-axis as appears will happen in Figure 5.22 if the curve is extended. Figure 5.23 illustrates this for values of x between -30.9 and -29.8.

x	y
-4	$-\dfrac{1}{6}$
-3	$-\dfrac{1}{5}$
-2	$-\dfrac{1}{4}$
-1	$-\dfrac{1}{3}$
0	$-\dfrac{1}{2}$
1	-1
$-1\dfrac{1}{2}$	-2
$1\dfrac{3}{4}$	-4
$1\dfrac{7}{8}$	-8
$2\dfrac{1}{8}$	8
$2\dfrac{1}{4}$	4
$2\dfrac{1}{2}$	2
3	1
4	$\dfrac{1}{2}$
7	$\dfrac{1}{5}$

EXAMPLE 6 **Graph of a Rational Function**

Graph $f(x) = \dfrac{1}{x - 2}$.

Solution In Section 5.1, we found the domain and range for the function $y = \dfrac{1}{x - 2}$ to be $D = \{x \mid x \neq 2\}$ and $R = \{y \mid y \neq 0\}$. From this, we know that the curve will not exist when $x = 2$ and $y = 0$. The lines $x = 2$ and $y = 0$, then, are asymptotes for this function. Substituting values for x and solving for y, we arrive at the table of values. Additional points are needed to show both branches of the curve and the effect of the asymptotes. See Figure 5.24 for the graph. ◆

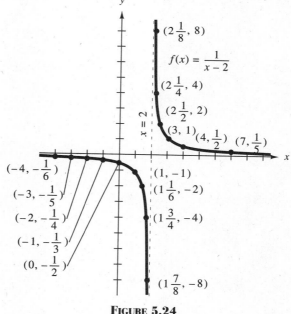

FIGURE 5.24

5.3 Exercises

1. Determine the asymptotes for the functions.

    a. $y = \dfrac{1}{x - 1}$

    b. $y = \dfrac{1}{x - 2}$

    c. $y = \dfrac{1}{x - 3}$

2. For the function $y = \dfrac{a}{x - b}$ (a and b are constants) determine the asymptotes.

In Exercises 3–12, match the function with the domain, range, and asymptotes.

3. $f(x) = x + 3$

    a. $D = \{t \mid t \neq 3\}; R = \{f(t) \mid f(t) \neq 1\};$
        asymptotes: $t = 3, f(t) = 1$

4. $f(x) = \dfrac{x^2}{4} - 3$

    b. $D = \{x \mid x \in R)\}; R = \{f(x) \mid f(x) \in R\};$
        no asymptotes

5. $f(s) = \dfrac{1}{s + 2}$

    c. $D = \{x \mid x \neq 5\}; R = \{f(x) \mid f(x) \neq 0\};$
        asymptotes: $x = 5, f(x) = 0$

6. $f(x) = \dfrac{1}{x - 5}$

    d. $D = \{x \mid x \in R\}; R = \{f(x) \geq -3\};$
        no asymptotes

7. $f(t) = \dfrac{t + 5}{t - 3}$

    e. $D = \{x \mid x \geq -2\}, R = \{f(x) \mid f(x) \geq 0\};$
        no asymptotes

8. $f(t) = \sqrt{t - 1}$

    f. $D = \{s \mid s \neq -2\}; R = \{f(s) \mid f(s) \neq 0\};$
        asymptotes: $s = -2, f(s) = 0$

9. $f(x) = \sqrt{2 - x}$

    g. $D = \{t \mid t \geq 1\}; R = \{f(t) \mid f(t) \geq 0\};$
        no asymptotes

10. $f(x) = \sqrt{x + 2}$

    h. $D = \{x \mid x \leq 2\}; R = \{f(x) \mid f(x) \geq 0\};$
        no asymptotes

11. $f(x) = x^3 - 1$

12. $f(x) = x^3 - 3x + 1$

In Exercises 13–22, use the information you obtained in Exercises 3–12 to help you graph the functions. Where appropriate, write a program for the calculator to help create the table of values.

13. $f(x) = x + 3$

14. $f(x) = \dfrac{x^2}{4 - 3}$

15. $f(s) = \dfrac{1}{s + 2}$

16. $f(x) = \dfrac{1}{x - 5}$

17. $f(t) = \dfrac{t + 5}{t - 3}$

18. $f(t) = \sqrt{t - 1}$

19. $f(x) = \sqrt{2 - x}$

20. $f(x) = \sqrt{x + 2}$

21. $f(x) = x^3 - 1$

22. $f(x) = x^3 - 3x + 1$

In Exercises 23–30, graph the function. To help graph the functions, determine the domain, range, and asymptotes. Where appropriate, write a program for the calculator to help create a table of values.

23. $f(x) = 2x - 1$

24. $f(x) = \dfrac{1}{2}x^2 + 4$

25. $f(x) = 3x^2 - 5x - 7$

26. $f(s) = \dfrac{1}{s}$

27. $f(s) = \dfrac{s - 3}{s - 4}$

28. $f(x) = \sqrt{x - 3}$

29.

A	0	1	−1	2	−2	3	4	5
B	−5	−4	−6	−3	−7	−2	−1	0

30.

L	1	2	3	4	5	6
V	98	102	99	101	102	98
L	7	8	9	10	11	12
V	99	103	100	97	101	104

31. In biology there is an approximate rule for temperate climates, called the bioclimatic rule, that states that in spring and early summer periodic phenomena, such as blossoming for a given species, appearance of certain insects, and ripening of fruit, usually come about four days later for each 500 ft of altitude. Stated as a formula: $d = 4(a/500)$, where $d =$ change in days and $a =$ change in altitude in feet. Graph the equation for $0 \leq a \leq 3500$.

32. The conversion of kilograms into pounds on the earth's surface is approximately $W_{lb} = 2.2\, m_{kg}$. Graph this equation letting W_{lb} be the dependent variable and m_{kg} the independent variable in kilograms. Can you use this graph to estimate weights?

33. The classic relationship between current, I, and voltage, V, is given by Ohm's law: $V = IR$, where R is the resistance.
  a. Graph this equation when $R = 6.0\ \Omega$.
  b. What voltage yields a current of 1.3 A?
  c. Resistance is sometimes referred to as a current regulator. The voltage from the sockets is a constant 120 V. By varying the resistance, the current changes. Plot a graph of I versus R. Note that this is not a linear function.
  d. What resistance yields a current of 20 A?

34. The formula used to convert kilometers to miles is $m = 0.62\, d$, where d is the distance in km.
  a. Graph this equation.
  b. Using the graph, convert 40 mi to km.

35. Newton's law of mutual attraction (also known as the law of gravity) is $F = \dfrac{km_1m_2}{r^2}$.
  Use $km_1m_2 = 10^{-9}$ to perform the following.
  a. Graph this equation.
  b. What value of r produces a force of 4×10^{-11}?

36. When an object travels closer and closer to the speed of light, the mass of the object (as viewed by someone at rest) dramatically increases. The equation is given by:

$$m = \frac{m_0}{\sqrt{1 - \left(\dfrac{v}{c}\right)^2}},$$

where v is the speed of the object and c is the speed of light. Recall that $c = 3 \times 10^8$ m/s. Sketch a graph of the mass as a function of speed when $m_0 = 1$ (m_0 is an initial mass).
  a. Use the graph to explain why the speed of light is often called the ultimate speed of the universe.
  b. In the graph, the curve is asymptotic to a line. What is the equation of the line?

In Exercises 37–47, use a graphics calculator to graph the functions on the same coordinate axes. Then examine the graphs and discuss how and why they are similar and/or different. Is there something about the function that causes the graphs to be similar and/or different?

37. $y = 2x + 5$, $y = -2x + 5$

38. $y = 2x + 5$, $y = 2x - 5$, $y = 2x$

39. $y = \sqrt{x - 6}$, $y = \sqrt{x}$, $y = \sqrt{x + 6}$

40. $y = \sqrt{5 - x}$, $y = \sqrt{x}$, $y = \sqrt{5 + x}$

41. $y = x^2$, $y = 2x^2$, $y = 4x^2$

42. $y = x^2 + 7$, $y = x^2$, $y = x^2 - 7$

43. $y = x^2 + 3x - 5$, $y = x^2 - 3x - 5$

44. $y = -x^2 + 3x + 5$, $y = -x^2 - 3x + 5$

45. $y = \dfrac{1}{x - 3}$, $y = \dfrac{4}{x - 3}$

46. $y = \dfrac{1}{x - 3}$, $y = \dfrac{1}{x + 3}$

47. $y = \dfrac{x + 2}{x - 3}$, $y = \dfrac{x - 2}{x - 3}$

5.4

Inverse Functions

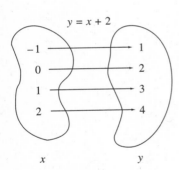

FIGURE 5.25

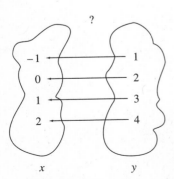

FIGURE 5.26

In Section 5.1 we learned that a statement of equality, such as $y = x + 2$, is a function. In fact, this statement is called a **one-to-one function.** That is, for each value substituted for x, we obtain a different value for y. For a function that is not one-to-one, each value substituted for x yields only one value for y, but not necessarily a different value. An example of a function that is not one-to-one is $y = x^2$. When $x = 1$, $y = 1$, and when $x = -1$, $y = 1$. In this case, each value of x did not yield a different value for y.

An example of a one-to-one function is given in Figure 5.25. Figure 5.25 gives a set of values, set X, which is called the domain, and a function $y = x + 2$, which is referred to as a rule. When you take each element in set X and apply the rule, you obtain a new set of values, which we shall call set Y. In Figure 5.25 the chosen domain is $\{-1, 0, 1, 2\}$, and the resultant range is $\{1, 2, 3, 4\}$. Using the domain and the range, we can write the ordered pairs $(-1, 1)$, $(0, 2)$, $(1, 3)$ and $(2, 4)$ that represent the function.

A question now arises: Is it possible to take the information given in Figure 5.25 and write an equation that would take the elements in set Y and give us the elements in set X? If $f(x) = x + 2$ represents the original function then the question mark in Figure 5.26 represents the equation such that for each value of y, we will get a corresponding value of x. This equation is called the **inverse function** of f. The inverse function is indicated by the symbol f^{-1}. (Here -1 is not a negative exponent, but rather a part of the symbol that means inverse function.)

To find the inverse of a function, we use the following procedure.

PROCEDURE FOR FINDING THE INVERSE OF A FUNCTION

1. Interchange the role of the x and y.
2. Solve for y.
3. Replace the y with the notation $f^{-1}(x)$; this notation indicates the inverse of the function.

EXAMPLE 1 **Determine the Inverse Function**

Determine the inverse function of $y = x + 2$.

Solution Applying the procedure for finding the inverse of a function, do the following.

a. Interchange the role of the x and the y.

$$x = y + 2$$

b. Solve for y.

$$x - 2 = y + 2 - 2$$
$$x - 2 = y \quad \text{or} \quad y = x - 2$$

c. $f^{-1}(x) = x - 2$

Thus $f^{-1}(x) = x - 2$ is the inverse function of $f(x) = x + 2$, or stated more formally, if $f(x) = x + 2$, then $f^{-1}(x) = x - 2$. ◆

CAUTION Finding the inverse of a function is not accomplished by adding or subtracting a constant as it appears to be in Example 1. For example, $f^{-1}(x) = 3x + 5$ is not the inverse function of $f(x) = 3x - 5$. This is illustrated in Example 2. ∎

EXAMPLE 2 **Determine the Inverse Function; Sketch the Function and the Inverse Function**

Determine the inverse function of $y = 3x - 5$. Sketch the function and the inverse function on the same set of axes.

Solution Applying the procedure for finding the inverse, interchange x and y. Thus, $x = 3y - 5$. Now solve for y.

$$x + 5 = 3y - 5 + 5$$
$$x + 5 = 3y$$
$$\frac{x + 5}{3} = \frac{3y}{3}.$$

Therefore,

$$y = \frac{x + 5}{3}.$$

Thus $y = \dfrac{x + 5}{3}$ is the inverse function of $y = 3x - 5$ or, more formally, if $f(x) = 3x - 5$, then $f^{-1}(x) = \dfrac{x + 5}{3}$.

The graphs of $f(x)$ and $f^{-1}(x)$ are mirror images of each other with respect to the line $y = x$. This is illustrated in Figure 5.27. ◆

$f(x) = 3x - 5$

x	$f(x)$
0	-5
1	-2
2	1

$f^{-1}(x) = \dfrac{x + 5}{3}$

x	$f^{-1}(x)$
-5	0
-2	1
1	2

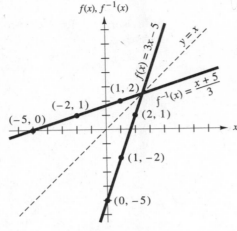

FIGURE 5.27

EXAMPLE 3   **Inverse of a Rational Function**

a. Determine the inverse of $y = \dfrac{2}{x - 3}$.

b. Determine the domain and range of the function and of its inverse.

Solution

a. Apply the procedure, by interchanging the x and y.

$$x = \frac{2}{y - 3}.$$

Solve for y.

$$(y - 3)x = (y - 3)\frac{2}{y - 3} \quad y \neq 3$$

$$xy - 3x = 2$$

$$xy - 3x + 3x = 2 + 3x$$

$$\frac{xy}{x} = \frac{2 + 3x}{x}$$

$$y = \frac{2 + 3x}{x} \quad \text{or} \quad f^{-1}(x) = \frac{2 + 3x}{x}.$$

b. For $y = \dfrac{2}{x - 3}$, we can see that $x \neq 3$. Thus, $D = \{x \mid x \neq 3\}$. If we solve the equation for x, we would have $x = \dfrac{2 + 3y}{y}$. From this equation, we can see that $y \neq 0$; thus, $R = \{y \mid y \neq 0\}$. For the inverse function $y = \dfrac{2 + 3x}{x}$, $x \neq 0$; therefore, $D = \{x \mid x \neq 0\}$. Solving the inverse function for x, we would have $x = \dfrac{2}{y - 3}$ and $y \neq 3$; therefore, $R = \{y \mid y \neq 3\}$. ◆

> Note that the range of the inverse of the function is the same as the domain of the original function. Also, the domain of the inverse of the function is the same as the range of the original function.

EXAMPLE 4 Inverse of the Inverse of a Function

For the function $y = x + 2$, determine
a. the inverse of the function;
b. the inverse of the inverse of the function.
Before reading the solution, try to predict the answers.

Solution

a. Apply the procedure by interchanging the x and the y.

$$x = y + 2$$
$$x - 2 = y + 2 - 2 \qquad \text{Solving for } y$$
$$y = x - 2 \quad \text{or} \quad f^{-1}(x) = x - 2.$$

b. We want to find the inverse of $f^{-1}(x) = x - 2$. For convenience, use the form $y = x - 2$. Apply the procedure by interchanging the x and the y.

$$x = y - 2$$
$$x + 2 = y - 2 + 2 \qquad \text{Solving for } y$$
$$y = x + 2 \quad \text{or} \quad [f^{-1}(x)] = x + 2$$

The result for part b is the original function. ◆

In Example 4 we started with a one-to-one function. Finding the inverse of the inverse of the function resulted in the original function. This result will occur only when the original function is a one-to-one function and the inverse of the original function is one-to-one.

5.4 Exercises

In Exercises 1–18, determine the inverse function of each of the functions.

1. $y = f(x) = x + 5$

2. $y = f(x) = x - 4$

3. $y = f(x) = 2x + 7$

4. $y = f(x) = 3x - 5$

5. $y = f(x) = 2(x - 4)$

6. $y = f(x) = 3(2 - x)$

7. $y = f(x) = \dfrac{x + 4}{2}$

8. $y = f(x) = \dfrac{6 + x}{5}$

9. $y = f(x) = \dfrac{9 - 2x}{3}$

10. $y = f(x) = \dfrac{2x - 5}{4}$

11. $y = f(x) = \dfrac{1}{2}(x + 7)$

12. $y = f(x) = \dfrac{1}{5}(2x + 3)$

13. $y = f(x) = \dfrac{1}{x + 2}$

14. $y = f(x) = \dfrac{1}{x - 5}$

15. $y = f(x) = \dfrac{2}{x + 1}$

16. $y = f(x) = \dfrac{4}{2x + 5}$

17. $y = f(x) = \dfrac{5}{3x + 2}$

18. $y = f(x) = \dfrac{4}{4x - 1}$

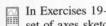 In Exercises 19–30, sketch the graph of the functions in Exercises 1–12. On the same set of axes sketch the inverse of each of the functions.

 In Exercises 31–36, determine the domain and range of each of the functions in Exercises 13–18 and then determine the domain and range of its inverse function.

In Exercises 37–38,
a. determine the inverse of the function;
b. determine the inverse of the inverse of the function.

37. $f(x) = 3x + 5$

38. $f(x) = \dfrac{2x}{3} - 4$

39. Given the function $y = f(x) = \dfrac{2}{11x - 9}$, determine the domain and range. Now find a function that has the same range as the original function's domain and same domain as the original function's range.

40. Given the function $y = f(x) = \dfrac{2x}{x - 3}$, determine the domain and range. Now find a function that has the same range as the original function's domain and same domain as the original function's range.

41. By examining the sketches for Exercises 19–30, can you determine the geometric relation between the function and its inverse?

Review Exercises

In Exercises 1–10 determine the value of the functions.

1. $f(x) = 3x + 5$ **a.** $f(-3)$ **b.** $f(2)$ **c.** $f(h + 3)$

2. $g(t) = 5t - 4$ **a.** $g(-4)$ **b.** $g(3)$ **c.** $g(k - 5)$

3. $h(t) = t^2 + 4$ **a.** $h(-3)$ **b.** $h(3)$ **c.** $h(z + 2)$

4. $k(a) = \dfrac{5}{a}$ **a.** $k(-2)$ **b.** $k(25)$ **c.** $k(m - 3)$

5. The volume of a cube is given by $V(s) = s^3$, where s is the length of a side. Determine:
    a. $V(2.35 \text{ mm})$
    b. $V(7.34 \text{ m})$
    c. $V\left(\dfrac{6}{7} \text{ in}\right)$

6. The total surface area of a rectangular solid is given by $T(l, w, h) = 2(lw + wh + lh)$. Determine
    a. $T(3 \text{ cm}, 4 \text{ cm}, 5 \text{ cm})$
    b. $T(4.3 \text{ m}, 7.5 \text{ m}, 8.3 \text{ m})$
    c. $T(0.060 \text{ mm}, 0.040 \text{ mm}, 0.030 \text{ mm})$

7. Force, F, is a function of mass, m, and acceleration, a, given by $F(m, a) = ma$. Determine
    a. $F(0.300 \text{ kg}, 10 \text{ m/s}^2)$
    b. $F(21.000 \text{ kg}, 75,000 \text{ m/s}^2)$

8. $f(x) = \begin{cases} 4x - 5, & \text{for } x \leq 0 \\ x^2 - 5, & \text{for } x > 0 \end{cases}$
    Determine: **a.** $f(-3)$ **b.** $f(0)$ **c.** $f(5)$

9. $s(t) = \begin{cases} 5t - 4, & \text{for } t < 4 \\ 7t + 11, & \text{for } t \geq 4 \end{cases}$
    Determine: **a.** $s(2)$ **b.** $s(4)$ **c.** $s(6)$

10. $s(t) = \begin{cases} \dfrac{t^2 - 49}{t - 7}, & \text{for } t \neq 7 \\ 4, & \text{for } t = 7 \end{cases}$
    Determine: **a.** $s(5)$ **b.** $s(7)$ **c.** $s(9)$

In Exercises 11–20, determine the domain and range for each function.

11. $y = x + 13$ 12. $2B = 6A - 5$ 13. $f(t) = t^2 - 5$ 14. $f(t) = t^2 + 5$

15. $f(x) = \dfrac{4}{x - 1}$ 16. $f(R) = \dfrac{2}{R + 3}$ 17. $f(L) = \dfrac{L + 1}{L - 1}$ 18. $f(L) = \dfrac{L - 3}{L + 3}$

19. $g(s) = \sqrt{s + 5}$ 20. $g(s) = \sqrt{5 - s}$

In Exercises 21–26, write a function that describes the statement.

21. The number of liters in a gallon.

22. The area of a square as a function of its side, s.

23. The cost of water used in a home as a function of the number of gallons used.

24. The distance a marble rolls down an inclined plane varies directly as the square of the time during which it rolls.

25. The circumference, C, of a circle as a function of its diameter, d.

26. The area, A, of a circle as a function of its diameter, d.

27. The length of a rectangle is 5 mm more than 3 times its width, x. Express the area, A, of the rectangle as a function of its width.

28. **a.** An amusement park charges an admission fee of $5.00 plus an additional charge of $0.25 for each ride. Express the total cost, $C(x)$, as a function of the number, x, of rides taken.
    b. Determine the total cost if you take 16 rides.

In Exercise 29–34, determine $\overline{P_1 P_2}$ and the midpoint of $\overline{P_1 P_2}$.

29. $P_1(4, 2)$; $P_2(7, 5)$
30. $P_1(-3, 5)$; $P_2(6, 7)$
31. $P_1(-4, -3)$; $P_2(4, 3)$
32. $P_1(4, -7)$; $P_2(-6, 3)$
33. $P_1(-2.4, 3.5)$; $P_2(-7.1, -6.4)$
34. $P_1(0, -3.4)$; $P_2(3.5, -4.1)$

 In Exercises 35–44, graph the function.

35. $y = f(x) = x + 7$
36. $y = g(x) = x - 3$
37. $B = 3A^2 - 4A + 5$
38. $B = 4A^2 - 5A + 7$
39. $f(t) = \dfrac{t + 3}{t - 2}$
40. $f(t) = \dfrac{t - 5}{t + 7}$
41. $y = 12x - x^3$
42. $y = x^4 - 4x^2 + 2$
43. $S = \sqrt{5 - t}$
44. $S = \sqrt{t - 5}$

In Exercises 45–53, **a.** determine the inverse function of each function; **b.** determine the domain and range of each of the functions and then determine the domain and range of its inverse function; **c.** sketch the graph of each of the functions and its inverse function on the same set of axes.

45. $y = f(x) = x - 7$
46. $y = f(x) = x + 6$
47. $y = f(x) = 5x - 4$
48. $y = f(x) = 6x + 5$
49. $y = f(x) = \dfrac{2x + 4}{3}$
50. $y = f(x) = \dfrac{1}{3}(2x + 6)$
51. $y = f(x) = \dfrac{3}{x}$
52. $y = f(x) = \dfrac{1}{x + 2}$
53. $y = f(x) = \dfrac{-1}{x + 2}$

54. An automobile and a truck leave the same point at the same time traveling due south at 48 mph and due east at 30 mph, respectively. Find an expression for their distance apart, d, measured in miles, at the end of t hours.

55. A sea plane flying 300 statute miles per hour at an altitude of 600 ft passes over a buoy. Find an expression for its distance, d, measured in miles from the buoy t hours later.

56. Find the lengths of the sides and thereby prove that the points are vertices of an isosceles triangle: $A(2, 3)$, $B(6, 7)$, $C(6, -1)$.

57. **a.** Show that the points $(1, 5)$, $(2, 4)$, and $(-3, 5)$ are on a circle with center $(-1, 2)$.
    b. What is the radius of the circle?

58. An average driver traveling r miles per hour requires f feet to stop his car, where r and f are related as in the following table.

r	20	30	40	50	60	70
f	40	80	125	180	250	340

a. Represent the data graphically.
b. From the graph estimate the number of feet needed to stop when traveling at the rate of 55 mph.

59. A taxi charges $1.15 as soon as you enter the car, plus $0.25 after each $\frac{1}{6}$ mile.

    a. Write a function $f(x)$ expressing the cost of riding x miles in this cab.
    b. Construct a table of values.
    c. Graph the function.

60. Show that the midpoints of the sides of a rectangle are the vertices of a rhombus. [*Hint:* Place the rectangle so that its vertices are at the points (0, 0), (a, 0), (0, b), and (a, b)].

61. Use a convenient coordinate system to show that the midpoint of the hypotenuse of a right triangle is equidistant from its vertices.

62. Prove Theorem 5.2. The proof requires two parts:
    a. Show that $\overline{P_1 M} = \overline{MP_2}$ and **b.** Show that $\overline{P_1 M} + \overline{MP_2} = \overline{P_1 P_2}$; that is, that the points are collinear.

63. The perpendicular bisector of the line segment $\overline{PQ}$ is the line that is perpendicular to $\overline{PQ}$ and that divides $\overline{PQ}$ into two line segments of the same length. If P and Q are $(-2, 5$ and $(4, 9)$, respectively, determine the equation of the perpendicular bisector of $\overline{PQ}$.

64. A rectangular box with a square base and open top is to be made from 48 ft² of lumber. Express the volume, V, of the box as a function of a side x, measured in feet, of the square base. Determine, by means of a graph, the approximate dimensions and volume of the largest box that can be made.

65. Write a BASIC program that will find the distance between any two points in a plane.

66. Write a program for your calculator that will find the distance between any two points in a plane.

✎ Writing About Mathematics

1. Your supervisor asks you to design a rectangular corral for a client's horse. The pen must meet the following conditions: the area must be a maximum; one side of the enclosure will be the barn, and the other three sides will be wire fence. There is 75 ft of wire fence available for the three sides. By letting l represent the length of the rectangle that is the same length as the side of the barn and x represent the length of the other side of the rectangle, you write the quadratic function $f(x) = 75x - 2x^2$, which is a general formula for the area of the rectangle. Use the given information to write a plan explaining how you arrived at the dimensions that will determine a corral of maximum area.

2. Write a paper that explains how to determine the inverse of a function.

3. Are the inverses of all functions, functions? Explain how someone can determine whether the inverse of a function will be a function.

Chapter Test

In Exercises 1 and 2, identify the independent variable(s) and the dependent variable for each function.

1. $y = 3x - 12$

                                        2. $C = 2\pi r$

In Exercises 3 and 4, determine the values of the function $f(x) = 3x^2 - 7x + 5$ for the values given.

3. $f(-2)$ **4.** $f(a + h)$

In Exercises 5 and 6, determine the domain and range of the function.

5. $f(x) = x^2 + 2$ **6.** $f(x) = \sqrt{x + 5}$

In Exercises 7 and 8, $P_1(5, -4)$ and $P_2(-3, 2)$ are points in the plane

7. Determine $\overline{P_1 P_2}$.   **8.** Determine the midpoint of $\overline{P_1 P_2}$.

In Exercises 9 and 10, graph the function. Determine whether or not asymptotes exist. If they do exist, use the asymptotes to help graph the function.

9. $f(x) = 3x - 4$ **10.** $f(x) = \dfrac{1}{x + 1}$

In Exercises 11 and 12, determine the inverse function of the function.

11. $f(x) = 3x - 4$ **12.** $f(x) = \dfrac{2}{x + 3}$

In Exercises 13 and 14, determine a function that describes the statement.

13. The number of inches in a foot.

14. The Old Car Rental Agency rents cars one day at a time. The agency charges a daily fee of \$125 plus \$0.50 per mile. Express the total cost, $C(x)$, of a day's rental as a function of the number, x, of miles driven.

15. Use the information from Problem 14 to find the total cost of renting a car from Old Car to drive 321 mi.

For Exercises 16 and 17, sketch the graphs of the functions. Discuss their similarities and differences and what may cause their similarities and differences.

16. $y = -5x + 6, y = 5x + 6$ **17.** $y = 3x^2 - 4, y = -3x^2 - 4$

Introduction to Trigonometry

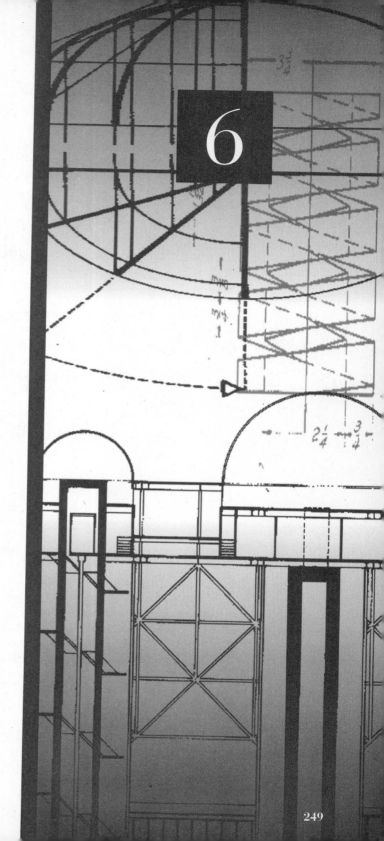

6

The pull starter on your lawn mower consists of a cord wrapped around a pulley that is attached to the crankshaft of the motor. When you pull the cord, the motor is turned and frequently starts. The greater the distance the cord is extended in a given time, the greater the speed at which the crankshaft turns and the better your chances of starting the motor. The ability to start the motor depends upon the angular motion of the crankshaft. (See Section 6.7, Example 5.)

6.1

The Trigonometric Functions

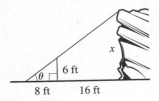

FIGURE 6.1

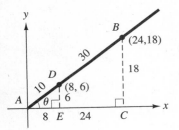

FIGURE 6.2

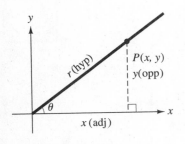

FIGURE 6.3

The word *trigonometry* means triangle measurement. In this section, the right triangle is used to introduce the six trigonometric functions. These functions are basic tools for measuring the various parts of a triangle. If your understanding of angles and angle measurement is a little rusty, review Section 3.1 before beginning this chapter.

In Section 3.2 we used similar triangles to find the height of the cliff, shown in Figure 6.1, to be 18 ft. A surveyor, using an instrument called a transit, would measure the angle θ and then use right triangle trigonometry to determine the height of the cliff. θ is the acute angle of a right triangle. To understand how the surveyor makes the determination, we must understand the six trigonometric functions.

The measure of angle θ in Figure 6.2 determines the position of the terminal side of the angle. Thus, the terminal side (dependent variable) is dependent on the measure of the angle (independent variable). If we pick two points $(8, 6)$ and $(24, 18)$ on the terminal side of the angle θ, we can draw perpendiculars to the initial side, creating right triangles ABC and ADE as shown in Figure 6.2. These right triangles are similar since the corresponding angles are equal. Therefore, the corresponding sides are proportional, and we can determine the lengths of the other sides. With this information we can write the following proportions.

$$\frac{\overline{DE}}{\overline{AE}} = \frac{\overline{BC}}{\overline{AC}} \qquad \frac{\overline{AE}}{\overline{AD}} = \frac{\overline{AC}}{\overline{AB}} \qquad \frac{\overline{DE}}{\overline{AD}} = \frac{\overline{BC}}{\overline{AB}}$$

$$\frac{6}{8} = \frac{18}{24} \qquad \frac{8}{10} = \frac{24}{30} \qquad \frac{6}{10} = \frac{18}{30}$$

$$\frac{3}{4} = \frac{3}{4} \qquad \frac{4}{5} = \frac{4}{5} \qquad \frac{3}{5} = \frac{3}{5}$$

These results show that the measure of angle θ determines the ratio of two sides and not the position of the point on the terminal ray or the size of the triangle. By selecting a point on the terminal side of the angle, as shown in Figure 6.3, three distances—x, y, and r—are determined. The distances are the **x-coordinate,** the **y-coordinate** and the **radius vector** (the distance from the origin to the point P). The side r, opposite the right angle, is called the hypotenuse (hyp). The side y, opposite the angle θ, is called the opposite side (opp). The side x, adjacent to angle θ, is called the adjacent side (adj). Using these three distances, we can create six distinct ratios of the sides of the triangle. Each of the the six ratios is used to define one of the six trigonometric functions given in Table 6.1.

CAUTION   When given a point on the terminal side of an angle, we always must construct the angle in standard position. When drawing the perpendicular to form the right triangle, *always* draw it perpendicular to the x-axis (horizontal axis). ■

TABLE 6.1   Definitions of Trigonometric Functions

Function	Abbreviation for Function Value	Defining Ratio for Function Value	Defining Ratio for Right Triangle
Sine	$\sin\theta$	$\sin\theta = \dfrac{y}{r}$	$\sin\theta = \dfrac{\text{opp}}{\text{hyp}}$
Cosine	$\cos\theta$	$\cos\theta = \dfrac{x}{r}$	$\cos\theta = \dfrac{\text{adj}}{\text{hyp}}$
Tangent	$\tan\theta$	$\tan\theta = \dfrac{y}{x}$	$\tan\theta = \dfrac{\text{opp}}{\text{adj}}$
Cotangent	$\cot\theta$	$\cot\theta = \dfrac{x}{y}$	$\cot\theta = \dfrac{\text{adj}}{\text{opp}}$
Secant	$\sec\theta$	$\sec\theta = \dfrac{r}{x}$	$\sec\theta = \dfrac{\text{hyp}}{\text{adj}}$
Cosecant	$\csc\theta$	$\csc\theta = \dfrac{r}{y}$	$\csc\theta = \dfrac{\text{hyp}}{\text{opp}}$

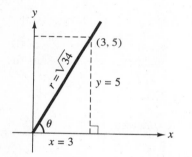

FIGURE 6.4

EXAMPLE 1   **Evaluating Trigonometric Functions**

Determine the six trigonometric functions of θ given that the terminal side of the angle in standard position passes through (3, 5). Give the answers in exact form.

Solution   Figure 6.4 shows a sketch of θ in standard position. The point (3, 5) is on the terminal side of the angle. From this point draw a perpendicular to the x-axis, creating a right triangle. Thus $x = 3$ and $y = 5$. We can find r using the Pythagorean relation.

$$r = \sqrt{(3)^2 + (5)^2}$$
$$= \sqrt{34}.$$

The functional values are:

$$\sin\theta = \frac{y}{r} = \frac{5}{\sqrt{34}}, \qquad \csc\theta = \frac{r}{y} = \frac{\sqrt{34}}{5}, \qquad \tan\theta = \frac{y}{x} = \frac{5}{3},$$

$$\cos\theta = \frac{x}{r} = \frac{3}{\sqrt{34}}, \qquad \sec\theta = \frac{r}{x} = \frac{\sqrt{34}}{3}, \qquad \cot\theta = \frac{x}{y} = \frac{3}{5}.$$

The denominators of the fractions were not rationalized so that the sides of the triangle could be identified easily. These answers are in exact form since the fractions and square roots were not changed to decimal form.   ◆

By examining Table 6.1 and Example 1 we can see that $\csc\theta$ is the reciprocal of $\sin\theta$, $\sec\theta$ is the reciprocal of $\cos\theta$, and $\cot\theta$ is the reciprocal of $\tan\theta$. The reciprocal trigonometric functions are listed in Table 6.2

TABLE 6.2   Reciprocal Trigonometric Functions

$\csc \theta = \dfrac{1}{\sin \theta},$	$\sin \theta \neq 0$	or	$\sin \theta = \dfrac{1}{\csc \theta},$	$\csc \theta \neq 0$
$\sec \theta = \dfrac{1}{\cos \theta},$	$\cos \theta \neq 0$	or	$\cos \theta = \dfrac{1}{\sec \theta},$	$\sec \theta \neq 0$
$\cot \theta = \dfrac{1}{\tan \theta},$	$\tan \theta \neq 0$	or	$\tan \theta = \dfrac{1}{\cot \theta},$	$\cot \theta \neq 0$

These reciprocal trigonometric functions are important when you use a calculator to find functional values of angles.

When dealing with right triangle trigonometry, θ is a first quadrant angle and x and y are always positive.

EXAMPLE 2   Applying Trigonometric Ratios

Given that $\tan \theta = \dfrac{7.0}{9.0}$ and that θ is a first quadrant angle, determine the five remaining trigonometric functions.

Solution   Given that $\tan \theta = \dfrac{7.0}{9.0}$ and knowing that $\tan \theta = \dfrac{y}{x}$ we can see that $x = 9.0$ and $y = 7.0$. The point (9.0, 7.0) is a point on the terminal side of the angle, as shown in Figure 6.5. The value of r can be determined using the Pythagorean relation:

$$r = \sqrt{(x)^2 + (y)^2} = \sqrt{(9.0)^2 + (7.0)^2} = \sqrt{130}.$$

We can now determine the other five trigonometric functions knowing that $x = 9.0$, $y = 7.0$, and $r = \sqrt{130}$.

$$\sin \theta = \frac{7.0}{\sqrt{130}} = 0.61 \quad \cos \theta = \frac{9.0}{\sqrt{130}} = 0.79 \quad \cot \theta = \frac{9.0}{7.0} = 1.3$$

$$\sec \theta = \frac{\sqrt{130}}{9.0} = 1.3 \qquad \csc \theta = \frac{\sqrt{130}}{7.0} = 1.6 \quad \blacklozenge$$

The functional values can be expressed in exact form or in decimal form as illustrated in Example 2. Remember when expressing a result in decimal form to use the rules for significant digits.

EXAMPLE 3   Applying Trigonometric Ratios

Given that $\sin \theta = \dfrac{3.0}{10.0}$ and that θ is a first quadrant angle, determine $\cos \theta$ and $\cot \theta$.

Solution   Given $\sin \theta = \dfrac{3.0}{10.0}$ and the definition $\sin \theta = \dfrac{y}{r}$, we can see that $y = 3.0$ and $r = 10.0$. To determine the value of $\cos \theta$, we must determine a

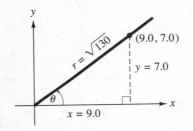

FIGURE 6.5

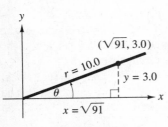

FIGURE 6.6

$$\sin \theta = \frac{1}{2}$$

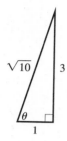

$$\tan \theta = \frac{3}{1}$$

FIGURE 6.7

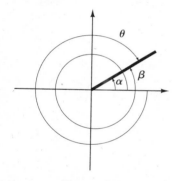

FIGURE 6.8

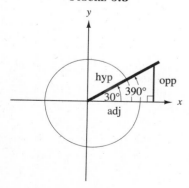

FIGURE 6.9

value for x, since $\cos \theta = \frac{x}{r}$. Use the Pythagorean relation to determine x.

$$x = \sqrt{(r)^2 - (y)^2} = \sqrt{(10.0)^2 - (3.0)^2} = \sqrt{91} = 9.5.$$

Knowing the values for x and y (9.5, 3.0), which is a point on the terminal side of the angle, it is possible to construct the right triangle as shown in Figure 6.6. The figure helps us determine that $\cos \theta = \frac{9.5}{10.0} = 0.95$ and that $\cot \theta = \frac{9.5}{3.0} = 3.2.$ ◆

We can determine the other five trigonometric functions given $\sin \theta = 0.5$ or $\tan \theta = 3$. In each case, the functional values can be expressed as fractions— $0.5 = \frac{1}{2}$ and $3 = \frac{3}{1}$. Using the definitions $\sin \theta = \frac{\text{opp}}{\text{hyp}}$ and $\tan \theta = \frac{\text{opp}}{\text{adj}}$ as well as the Pythagorean relation, we can label the right triangles as shown in Figure 6.7.

Having labeled the right triangles, we can now determine the ratios for all the triangles.

When two angles are placed in standard position and their terminal sides coincide, they are said to be **coterminal**. For example, in Figure 6.8, $\alpha = 30°$ and $\beta = 390°$ are coterminal since $390° = 360° + 30°$. Note that there are many coterminal angles for each angle. Another example using Figure 6.8 is that if $\theta = -330°$, then α and θ are coterminal angles.

EXAMPLE 4 Determining Coterminal Angles

Given $\theta = 25°$, determine four coterminal angles coterminal to θ.

Solution Four coterminal angles can be found by adding multiples of 360° to 25° and/or subtracting multiples of 360° from 25°.

a. $25° + 360° = 385°$ **b.** $25° + 720° = 745°$
c. $25° - 360° = -335°$ **d.** $25° - 720° = 695°$ ◆

For any two or more coterminal angles the ratios $\frac{\text{opp}}{\text{hyp}}$, $\frac{\text{opp}}{\text{adj}}$, etc., are the same for each angle. Therefore, the values of the trigonometric functions of coterminal angles are the same. For example the $\sin 30° = \sin 390°$. Look at Figure 6.9 to see that the same right triangle applies to both angles, and therefore, the sine of the angles have the same ratio.

The definitions can be used to show that certain ratios are equal.

EXAMPLE 5 Applying Trigonometric Ratios

Using the definitions in Table 6.1, show that the statement $\sin \theta = \frac{1}{\csc \theta}$ is true.

Solution Given that $\sin \theta = \dfrac{y}{r}$ and $\csc \theta = \dfrac{r}{y}$, we can substitute for $\csc \theta$ in the statement:

$$\sin \theta = \frac{1}{\csc \theta}$$

$$= \frac{1}{\dfrac{r}{y}}$$

$$= \frac{y}{r}$$

$$= \sin \theta.$$

This shows that $\sin \theta = \dfrac{1}{\csc \theta}$ for all θ, except when $\csc \theta = 0$. ◆

6.1 Exercises

In Exercises 1–12, match each of the trigonometric functions with one of the figures **a**, **b**, or **c**.

1. $\tan \theta = \dfrac{\sqrt{11}}{5}$

2. $\sin \theta = \dfrac{\sqrt{11}}{6}$

3. $\cos \theta = \dfrac{4}{\sqrt{65}}$

4. $\tan \theta = \dfrac{7}{4}$

5. $\cos \theta = \dfrac{5}{6}$

6. $\sin \theta = \dfrac{3}{4}$

7. $\sec \theta = \dfrac{4}{\sqrt{7}}$

8. $\csc \theta = \dfrac{6}{\sqrt{11}}$

9. $\cot \theta = \dfrac{4}{7}$

10. $\tan \theta = \dfrac{3}{\sqrt{7}}$

11. $\cos \theta = \dfrac{\sqrt{7}}{4}$

12. $\sin \theta = \dfrac{7}{\sqrt{65}}$

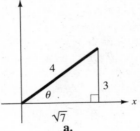
a.

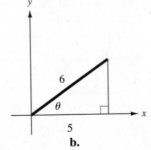
b.

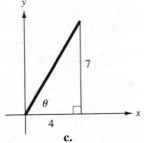
c.

In Exercises 13–22, determine the six trigonometric functions of θ (in standard position) given that the terminal side passes through the given point. Give the answers in exact form, see Example 1.

13. (3, 4) **14.** (8, 6) **15.** (4, 4) **16.** (2, 1)

17. (12, 15) **18.** (7, 5) **19.** (1, 2) **20.** (5, 3)

21. $(\sqrt{3}, 1)$ **22.** $(\sqrt{2}, \sqrt{2})$

In Exercises 23–28, determine the six trigonometric functions of θ (in standard position) given that the terminal side passes through the given point. Give the answers in decimal form.

23. (6.1, 7.4) **24.** (8.45, 7.63) **25.** (2.03, 3.40)

26. (5.0, 4.0) **27.** $(\sqrt{5.0}, \sqrt{11})$ **28.** $(\sqrt{3.0}, \sqrt{4.0})$

In Exercises 29–38, use the given trigonometric function (remember that θ is a first quadrant angle) to detemine the other trigonometric functions.

29. $\tan \theta = \dfrac{1}{4}$; determine $\sin \theta$ and $\sec \theta$.

30. $\sin \theta = \dfrac{3}{5}$; determine $\tan \theta$ and $\sec \theta$.

31. $\tan \theta = \dfrac{2}{3}$; determine $\cos \theta$ and $\csc \theta$.

32. $\cos \theta = \dfrac{1}{2}$; determine $\sin \theta$ and $\cot \theta$.

33. $\csc \theta = \dfrac{5}{2}$; determine $\sin \theta$ and $\tan \theta$.

34. $\sin \theta = \dfrac{\sqrt{3}}{2}$; determine $\cot \theta$ and $\cos \theta$.

35. $\cos \theta = \dfrac{\sqrt{2}}{2}$; determine $\csc \theta$ and $\tan \theta$.

36. $\tan \theta = \sqrt{3}$; determine $\sec \theta$ and $\cot \theta$.

37. $\cot \theta = 0.40$; determine $\sin \theta$ and $\sec \theta$.

38. $\sec \theta = 1.5$; determine $\csc \theta$ and $\cot \theta$.

In Exercises 39–44, determine two coterminal angles to the given angle and sketch the angles.

39. 35° **40.** 120° **41.** −45° **42.** 210°

43. −135° **44.** 150°

In Exercises 45–48, use the definitions in Table 6.1 to show that the statements are true.

45. $\tan \theta = \dfrac{\sin \theta}{\cos \theta}$ **46.** $\cot \theta = \dfrac{\cos \theta}{\sin \theta}$ **47.** $\sec \theta = \dfrac{1}{\cos \theta}$ **48.** $\csc \theta = \dfrac{1}{\sin \theta}$

6.2

Evaluating the Trigonometric Functions

The discussion in Section 6.1 provides us with a technique for finding a value for each of the six trigonometric functions for a given angle using the lengths of the sides of the triangle. However, in general, the angle is given and not the sides of the triangle. This section develops techniques for determining the trigonometric functional value for any acute angle. The discussion begins with a few special angles and then goes on to develop calculator techniques for finding any angle.

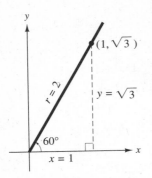

FIGURE 6.10

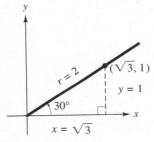

FIGURE 6.11

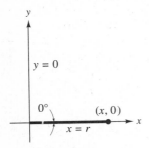

FIGURE 6.12

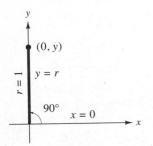

FIGURE 6.13

EXAMPLE 1   Evaluating the Trigonometric Functions for 60°

Determine the exact values for the six trigonometric functions of a 60° angle.

Solution   A 60° angle in standard position is shown in Figure 6.10. Recall from geometry that for a 30°, 60°, 90° right triangle, the sides opposite the respective angles are in the ratio $1 : \sqrt{3} : 2$. A convenient point to pick on the terminal side of the angle is $(1, \sqrt{3})$ as shown in Figure 6.10. A right triangle with $x = 1$, $y = \sqrt{3}$, and $r = 2$ is formed by drawing a perpendicular to the x-axis. With this figure, we can write the six trigonometric functions for the 60° angle.

$$\sin 60° = \frac{\sqrt{3}}{2} \quad \cos 60° = \frac{1}{2} \quad \tan 60° = \frac{\sqrt{3}}{1}$$

$$\csc 60° = \frac{2}{\sqrt{3}} \quad \sec 60° = \frac{2}{1} \quad \cot 60° = \frac{1}{\sqrt{3}} \quad \blacklozenge$$

EXAMPLE 2   Evaluating the Trigonometric Functions for 30°

Determine the exact values for the six trigonometric functions for a 30° angle.

Solution   In Figure 6.10, the other acute angle of the triangle is 30°. (Recall that the sum of the two acute angles of a right triangle is 90°.) With the 30° angle in standard position, the values of x and y are interchanged, and the point on the terminal side of the angle is $(\sqrt{3}, 1)$. Drawing a perpendicular to the x-axis and labeling the result gives us the triangle shown in Figure 6.11. Using this figure we now can write the six trigonometric functions of a 30° angle in exact form:

$$\sin 30° = \frac{1}{2} \quad \cos 30° = \frac{\sqrt{3}}{2} \quad \tan 30° = \frac{1}{\sqrt{3}}$$

$$\csc 30° = \frac{2}{1} \quad \sec 30° = \frac{2}{\sqrt{3}} \quad \cot 30° = \frac{\sqrt{3}}{1}. \quad \blacklozenge$$

Examine Fig. 6.12 and note that when there is no rotation of the terminal side of the angle, $x = r$ and $y = 0$. With these conditions $\theta = 0°$ and:

$$\sin 0° = \frac{0}{r} = 0, \quad \cos 0° = \frac{x}{r} = 1, \quad \tan 0° = \frac{0}{x} = 0.$$

Examine Fig. 6.13 and note that when there is a 90° rotation of the terminal side, $x = 0$ and $r = y$. With these conditions, $\theta = 90°$, and:

$$\sin 90° = \frac{y}{r} = 1, \quad \cos 90° = \frac{0}{r} = 0, \quad \tan 90° = \frac{y}{0} = \text{undefined}.$$

When a number is divided by zero, the result is a very large number that cannot be represented by a real number. Therefore, rather than saying the result is undefined, as we just did for $y \div 0$, the symbol for infinity (∞) is used to represent the result.

Using methods similar to Examples 1 and 2, we can determine the values for the six trigonometric functions of 45°. Using the information above and the reciprocal relationships of the six trigonometric functions, we can also determine the cotangent, secant, and cosecant for 0° and 90°. All of these values are given in Table 6.3.

TABLE 6.3 Values for Trigonometric Functions

Abbreviation			Exact Values			Decimal Values		
θ	0°	30°	45°	60°	90°	30°	45°	60°
sin	0	$\dfrac{1}{2}$	$\dfrac{1}{\sqrt{2}}$	$\dfrac{\sqrt{3}}{2}$	1	0.500	0.707	0.866
cos	1	$\dfrac{\sqrt{3}}{2}$	$\dfrac{1}{\sqrt{2}}$	$\dfrac{1}{2}$	0	0.866	0.707	0.500
tan	0	$\dfrac{1}{\sqrt{3}}$	1	$\sqrt{3}$	∞	0.577	1.000	1.732
cot	∞	$\sqrt{3}$	1	$\dfrac{1}{\sqrt{3}}$	0	1.732	1.000	0.577
sec	1	$\dfrac{2}{\sqrt{3}}$	$\sqrt{2}$	2	∞	1.155	1.414	2.000
csc	∞	2	$\sqrt{2}$	$\dfrac{2}{\sqrt{3}}$	1	2.000	1.414	1.155

The function values in Table 6.3 are used frequently. If you remember the function values, you will save a lot of time by not recalculating them or searching for the table. From Table 6.3 you see that the value of sin 30° = cos 60°, or sin 30° = cos(90° − 30°). That is, the sine of the angle is equal to the cosine of the complement of the angle. When this condition holds, the functions are called **cofunctions.** Thus, sine and cosine are cofunctions, tangent and cotangent are cofunctions, and so are secant and cosecant.

EXAMPLE 3 **Application Using Trigonometric Functions**

A 50.0 ft flagpole is placed on the center of a flat roof. Guy wires are used to support the pole as shown in Figure 6.14. If the guy wires are anchored 20.0 ft from the foot of the flagpole determine the following.
a. How long is each guy wire?
b. How high up the pole are the guy wires fastened?

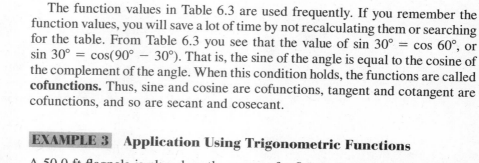

20.0 ft.   20.0 ft.

FIGURE 6.14

Solution

a. To determine the length of the guy wire, use the the cosine function. Let the length of the guy wire be ℓ.

$$\cos \theta = \frac{\text{adj}}{\text{hyp}}$$

$$\cos 30° = \frac{20.0}{\ell} \qquad \textbf{Substituting, Figure 6.3}$$

$$\frac{\sqrt{3}}{2} = \frac{20.0}{\ell} \qquad \textbf{Substituting, Table 6.3}$$

$$\sqrt{3}\,\ell = 40.0$$

$$\ell = \frac{40.0}{\sqrt{3}}, \text{ or } 23 \text{ ft}$$

Therefore the length of the guy wire is 23. ft.

b. To determine the distance up the pole to the point at which the wire is fastened, use the tangent function. Let the height up the pole be y ft.

$$\tan \theta = \frac{\text{opp}}{\text{adj}}$$

$$\tan 30° = \frac{y}{20.0} \qquad \textbf{Substituting, Figure 6.13}$$

$$\frac{1}{\sqrt{3}} = \frac{y}{20.0} \qquad \textbf{Substituting, Table 6.3}$$

$$y = \frac{20.0}{\sqrt{3}}, \text{ or } 12 \text{ ft}$$

Therefore, the guy wire is fastened 12 ft up the pole from the base. ◆

There is no exact value for $\sin 43°$ or $\cos 17°$. Values for the trigonometric functions of all angles can be calculated to as many decimal places as we choose, using infinite series. This technique is used internally in calculators and computers to give decimal approximations for trigonometric functions. In this text calculators are used to give the approximate values for the trigonometric functions of angles.

There are three different measures for angles, degrees (deg), radians (rad), and gradients (grad). For example, $90° = \frac{\pi}{2}$ rad $= 100$ grad. All three measures are usually on the calculator. Therefore, be certain that the calculator is in the correct mode when determining trigonometric functional values. Check your calculator instruction book to determine how to change from one mode to another. Radian measure will be discussed in Section 6.6. The gradient measure is used in some countries that use the SI System. Gradients are mentioned for information only; the unit will not be used in this text.

Procedure to Compute the Functional Values of sin, cos, and tan	Example
	Compute sin 60°.
	fx–7700G
1. Is calculator in correct mode?	1. Calculator is in degree mode.
2. Press the SIN , COS or TAN key.	2. SIN
3. Enter the angle measure.	3. SIN 60
4. Press the EXE or ENTER key.	4. SIN 60 EXE
	0.866025
	TI–82
	SIN 60 ENTER
	0.866025

CAUTION The sin 90° ≠ 0.894. The reason for the incorrect answer is that the calculator is in the wrong mode; the sin 90° = 1. ∎

When we solve problems dealing with degrees, the number of significant digits is dependent on the degree of accuracy of the angle measure, as shown in Table 6.4. Note that the number of digits in angular measure *does not* determine the number of significant digits of the trigonometric function of the angle. For example, angles of 1.3°, 123.4° and 45.5° all have 3 significant digits in their trigonometric functional values. For uniformity in recording and checking answers, follow the rules given in Table 6.4 unless other directions are given.

TABLE 6.4 Rules for Significant Digits in Angular Measure

Angles to the Nearest Degree	Significant Digits for Trigonometric Functions
One degree	2
Ten minutes, or tenth of a degree	3
One minute, or hundredth of a degree	4
Ten seconds, or thousandths of a degree	5
One second, or ten-thousandths of a degree	6

Any angle displayed as a whole number will be considered to be to the nearest degree. For example, 30°, 120°, 240°, 75°, and 3° are considered to be to the nearest degree, and any trigonometric functional value of these angles would be expressed with two significant digits.

EXAMPLE 4   Evaluating Trigonometric Functions: Calculator

Compute the values of the functions. Use the rules in Table 6.4 to round the answer.

a. sin 143°   **b.** tan 14.3°

Solution   Set the calculator in degree mode.

a.   SIN 143 EXE 0.601815. The measure of the angle is to the nearest degree; therefore, round the functional value to two significant digits: sin 143° = 0.60.

b.   TAN 14.3 ENTER 0.2548968. The measure of the angle is to the tenth of a degree; therefore, round the functional value to three significant digits: tan 14.3° = 0.255.   ◆

EXAMPLE 5   **Evaluating Trigonometric Functions: Calculator**

Compute cos [30.0° + 2(360.0°)].

Solution   First compute [30.0° + 2(360.0°)]; the value is 750.0°. The COS 750.0 = 0.8660254. The measure is to the nearest tenth of a degree, so round the functional value to three significant digits. Thus, cos 750.0° = 0.866. Recall from Table 6.3 that cos 30.0° = 0.866. The 30.0° angle and the 750.0° angle are coterminal angles. Recall from Section 6.1 that coterminal angles have the same functional value.   ◆

Having learned how to find the trigonometric functional values of sine, cosine, and tangent, we must now learn how to reverse the process to determine the angle given the functional value. In functional notation:

$$\sin^{-1}(\text{functional value}) = \text{angle},$$

where the -1 indicates the inverse operation and not a negative exponent. For example $\cos^{-1}(0.50) = 60°$. In section 5.4 we discussed the inverse concept for simple algebraic functions. The procedure for trigonometric functions is easier since we need learn only a few steps on the calculator or how to read a table. (A rigorous definition of inverse trigonometric functions is given in Chapter 15.)

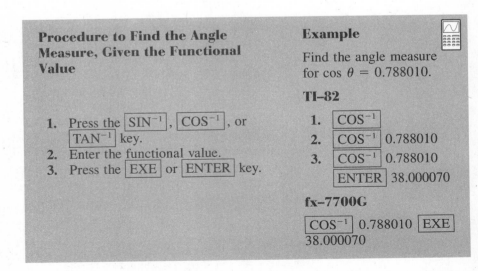

Procedure to Find the Angle Measure, Given the Functional Value

1. Press the SIN⁻¹ , COS⁻¹ , or TAN⁻¹ key.
2. Enter the functional value.
3. Press the EXE or ENTER key.

Example

Find the angle measure for cos θ = 0.788010.

TI–82

1. COS⁻¹
2. COS⁻¹ 0.788010
3. COS⁻¹ 0.788010
    ENTER 38.000070

fx–7700G

COS⁻¹ 0.788010 EXE
38.000070

EXAMPLE 6 **Determine the Angle Given the Functional Value**

Use a calculator to find the angle measure in decimal degrees and in degrees, minutes, seconds where appropriate.

a. $\sin \theta = 0.45879$ **b.** $\tan \theta = 11.32$

Solution

a. If $\sin \theta = 0.45879$, then $\sin^{-1}(0.45879) = \theta$.

$\boxed{\text{SIN}^{-1}}\, 0.45879 = 27.309056$. The functional value has five significant digits, so round the angle to thousandths of a degree, or ten seconds. Thus, $\theta = 27.309°$, $\theta = 27° \, 18' \, 30''$.

b. If $\tan \theta = 11.32$, then $\tan^{-1}(11.32) = \theta$.

$\boxed{\text{TAN}^{-1}}\, 11.32 = 84.95164$. The functional value has four significant digits, so round the angle to hundredths of a degree, or one minute. Thus, $\theta = 84.95°$, or $\theta = 84° \, 57'$.

We can check these results taking the value obtained for θ (not rounded off) and pressing the function key. For example, $\boxed{\text{TAN}}\, 84.95164 = 11.32$. ◆

The discussion to this point has been limited to the three functions: sine, cosine, and tangent. The discussion of the reciprocal functions, cosecant, secant, and cotangent, has been delayed because there are no keys for these functions on the calculator. The reciprocal functions are:

$$\csc \theta = \frac{1}{\sin \theta}, \quad \sec \theta = \frac{1}{\cos \theta}, \quad \cot \theta = \frac{1}{\tan \theta}.$$

To determine the functional values of $\csc \theta$, $\sec \theta$ and $\cot \theta$, and $\cot \theta$, first determine the functional values of $\sin \theta$, $\cos \theta$, and $\tan \theta$. Then determine the reciprocal of the functional values.

EXAMPLE 7 **Evaluating Reciprocal Trigonometric Functions**

Compute the functional value of the following.

a. $\csc 32.7°$ **b.** $\sec 9.27°$

Solution

a. Recall that $\csc \theta = \dfrac{1}{\sin \theta}$. Thus, the $\csc 32.7° = \dfrac{1}{\sin 32.7°}$. To determine the functional value, determine the sine of the angle and then take the reciprocal.

$\boxed{\text{SIN}}\, 32.7 = 0.5402403$ **Functional value of the reciprocal function**

$0.5402403 \,\boxed{x^{-1}} = 1.8510281$ **Reciprocal of functional value**

Rounding to three significant digits, $\csc 32.7° = 1.85$.

b. Recall that $\sec \theta = \dfrac{1}{\cos \theta}$. Thus, $\sec 9.27° = \dfrac{1}{\cos 9.27°}$. To determine the functional value, determine the cosine of the angle and then take the reciprocal.

$$\boxed{\text{COS}}\ 9.27 = 0.9869402 \qquad \textbf{Functional value of the reciprocal function}$$

$$0.9869402\ \boxed{x^{-1}} = 1.0132326 \qquad \textbf{Reciprocal of functional value}$$

Rounding to four significant digits, $\sec 9.27° = 1.013$. ◆

CAUTION   The procedure for determining the functional value for cosecant, secant, and cotangent is the same. Note that in determining the functional value, we are determining the reciprocal value of the function (sine, cosine, or tangent) and not the inverse of the function. ∎

If $\csc \theta = 1.83$, what is the measure of θ to the nearest tenth of a degree? The procedure for determining the angle given the functional value is just the reverse of determining the functional value. To determine θ, first find the reciprocal of the functional value and then determine the inverse of the reciprocal function. This procedure is illustrated in Example 8.

EXAMPLE 8   **Determining the Angle Given the Functional Value**

Given $\csc \theta = 1.83$, find θ.

Solution   Remember that $\sin \theta = \dfrac{1}{\csc \theta}$, or $\sin \theta = \dfrac{1}{1.83}$. To determine θ, take the reciprocal of the functional value and then the inverse of the sin function.

$$\sin \theta = \frac{1}{\csc \theta}$$

$$= \frac{1}{1.83}$$

$$\sin^{-1} \frac{1}{1.83} = \theta$$

$$33.123677 = \theta$$

A more direct approach for determining θ for $\csc \theta = 1.83$, using a calculator is: $\boxed{\text{SIN}^{-1}}\ 1.83\ \boxed{x^{-1}}\ \boxed{\text{EXE}}$ (or $\boxed{\text{ENTER}}$) $= 33.123677$.

Rounding to the nearest tenth of a degree, $\theta = 33.1°$. Therefore, $\csc 33.1° = 1.83$. ◆

EXAMPLE 9 **Determining the Angle Given the Functional Value**

Determine the value of θ given the following.

a. $\cot \theta = 2.41421$ **b.** $\sec \theta = 1.127$.

Solution

a. Given $\cot \theta = 2.41421$ and $\tan \theta = \dfrac{1}{\cot \theta}$. Thus,

$$\tan \theta = \frac{1}{2.41421}, \quad \textbf{Substituting}$$

or

$$\tan^{-1}\left(\frac{1}{2.41421}\right) = 22.50003 \quad \textbf{Inverse of the reciprocal of the functional value}$$

$$\theta = 22.5000° \quad \textbf{Rounding to ten-thousandths of a degree}$$

b. Given $\sec \theta = 1.127$ and $\cos \theta = \dfrac{1}{\sec \theta}$. Thus,

$$\cos \theta = \frac{1}{1.127}, \quad \textbf{Substituting}$$

or

$$\cos^{-1}\left(\frac{1}{1.127}\right) = 27.46268 \quad \textbf{Inverse of the reciprocal of the functional value}$$

$$\theta = 27.46° \quad \textbf{Rounding to hundredths of a degree}$$

We can check the results of each by finding the reciprocal functional value for θ (not rounded off) and then pressing $\boxed{x^{-1}}$. For example, ($\boxed{\text{TAN}}$ 22.50003) $\boxed{x^{-1}}$ 2.41421, which checks part **a.** ◆

6.2 Exercises

In Exercises 1–3, use the trigonometric functions and Table 6.3 to determine values for x, y, or r.

1.

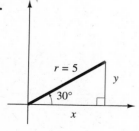

$r = 5$, $30°$, x, y

2.

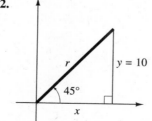

r, $y = 10$, $45°$, x

3.

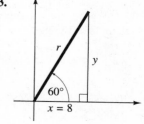

r, y, $60°$, $x = 8$

4. A painter's ladder, which is 40 ft long, is placed against a building. The angle formed between the ladder and the ground is 60°, as shown in the illustration.
   a. Determine the distance from the base of the building to the point where the ladder touches the building.
   b. Determine the distance from the foot of the ladder to the base of the building.

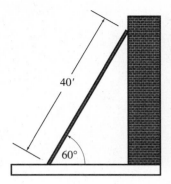

5. A screen for a radar device shows an aircraft 12.5 mi away at an elevation of 45° with the horizontal, as shown in the drawing.
   a. Determine the height of the aircraft at its present location.
   b. Determine the distances from the radar screen to a point on the ground directly below the aircraft.

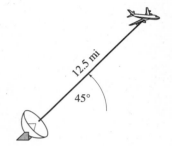

6. A technician on a battleship with a sonar device determines that it is 7.5 miles to a point where a submarine is below the surface. The angle formed between the horizontal and a line to the sumbarine is 30°, as shown in the illustration.
   a. Determine the distance between the submarine and the battleship.
   b. Determine the depth of the submarine.

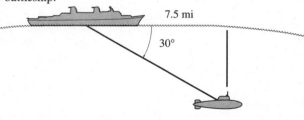

7. Surveyors use angular measurements to accurately determine a distance. The first measurement may be the length of the baseline shown in the sketch as AB. The point B is established such that the line AB is perpendicular to the line BX. A telescope with an angular scale on it is used to determine the precise angle made between AX and AB. Use the appropriate trigonometric relationship to determine the length, BX, if AB is 55.00 m and the angle θ is 32.34°.

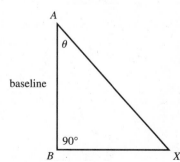

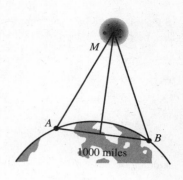

8. Measurements of long distances with measuring tapes may not be accurate due to temperature changes that alter the length of the tape. Angular measurements made with light rays do not change due to temperature changes. Therefore, angular measurements with light rays are preferred. Suppose the baseline shown in the figure in 7 is 84.3 m. Determine line BX if θ is 89.1°.

9. In measuring the distance between the earth and the moon, the baseline may be the distance between two observatories. Assume that the distance between observatories A and B shown in the drawing is 1000 miles. If the angle that both observatories measure with respect to the baseline is $\overline{89.925}°$, what is the approximate distance to the moon from point A on earth?

In Exercises 10–32, compute the functional value and round the answer using the rules in Table 6.4.

10. sin 12°

11. sin 78.4°

12. cos 78°

13. cos 11.6°

14. tan 19°

15. tan 34.3°

16. tan 2.05°

17. tan 23.6°

18. cos 45.7°

19. cos 25.45°

20. sin 1.01°

21. sin 0.29°

22. sin 26.93°

23. cos 73.15°

24. tan 56.42°

25. tan 15° 12′

26. cos 13° 28′

27. sin 21° 23′ 20″

28. sin 28° 25′ 34″

29. cos 68° 54′ 40″

30. cos 11° 38′ 47″

31. tan 37° 29′ 50″

32. tan 37° 35′ 43″

In Exercises 33–34, use your knowledge of geometry to determine the exact functional values for the six trigonometric functions for the given angle.

33. 45°

34. 90°

In exercises 35–40, compute the functional value and round the answer using the rules in Table 6.4.

35. sin [23° + 2(360°)]

36. cos [17° + 2(360°)]

37. tan [34° + 2(360°)]

38. sin [53.2° + 2(180°)]

39. cos [67.4° + 2(360°)]

40. tan [59.5° + 2(180°)]

In Exercises 41–52, determine θ in decimal degrees and in degrees, minutes, seconds. Use the rules in Table 6.4 to round the answer.

41. sin θ = 0.23

42. cos θ = 0.23

43. tan θ = 0.23

44. sin θ = 0.500

45. cos θ = 0.866

46. tan θ = 0.577

47. sin θ = 0.866

48. cos θ = 0.500

49. tan θ = 1.73

50. tan θ = 0.5773

51. cos θ = 0.7071

52. sin θ = 0.7431

In Exercises 53–64, compute the functional value and round the answer using the rules in Table 6.4.

53. csc 12°

54. csc 78.4°

55. sec 78°

56. sec 11.6°

57. cot 19°

58. cot 34.3°

59. cot 34° 10′

60. sec 42° 20′

61. csc 55° 30′

62. csc 55° 35′

63. sec 42° 25′

64. cot 34° 15′

In Exercises 65–76, determine θ in decimal degrees and in degrees, minutes, seconds. Use the rules in Table 6.4 to round the answer.

65. $\csc \theta = 19$ **66.** $\csc \theta = 1.00$ **67.** $\sec \theta = 1.0$ **68.** $\sec \theta = 52.1$

69. $\cot \theta = 28$ **70.** $\cot \theta = 0.017$ **71.** $\cot \theta = 1.732$ **72.** $\cot \theta = 0.5774$

73. $\sec \theta = 1.154$ **74.** $\sec \theta = 2.000$ **75.** $\csc \theta = 2.000$ **76.** $\csc \theta = 1.154$

A roadway with a radius of curvature, R (in meters), is banked at an angle θ. The maxium speed, v (in kilometers per hour), at which a vehicle can travel without skidding or leaving the road is given by $v = \sqrt{\dfrac{1800R \tan \theta}{121}}$. In Exercises 77–80, determine the maximum safe speed for each θ and R.

77. $\theta = 6.07°, R = 540$ m   **78.** $\theta = 2.1°, R = 930$ m

79. $\theta = 3.209°, R = 520$ m   **80.** $\theta = 7.2°, R = 510$ m

81. When a light ray passes from air to another transparent material, it is refracted (bent) according to Snell's law. The ratio of the sine of the angle of incidence to the sine of the angle of refraction is a constant for any two given materials. That is,   Light ray →

$$n = \frac{\sin \theta_i}{\sin \theta_r}.$$

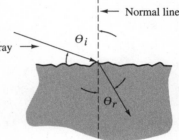

The angles θ_i and θ_r are formed by the ray of light and a line perpendicular to the surface of the material (see the illustration). The line perpendicular to the surface of the material is called the normal line. The constant, n, is called the index of refraction of the material the light ray is entering. The greater the value for the index of refraction, the more the light is bent in the refractive material. Determine the index of refraction, n, for **a.** Water: $\theta_1 = 70.0°$, $\theta_r = 35.6°$; and **b.** Oil: $\theta_i = 70.0°$, $\theta_r = 38.2°$.

82. A diamond has the highest index of refraction at 2.4365. If the angle of incidence of air stays at 70° (as in Exercise 81), what is the angle of refraction in the diamond?

83. A wagon is pulled by a parent with a force of 30 lbs, and the handle of the wagon makes an angle 15° with the ground as shown in the illustration. As a result of the pulling action, two forces are exerted on the wagon. One force, F_x, is parallel to the ground, and the other force, F_y, pulls the wagon into the air. Determine the following.
a. F_x **b.** F_y

84. In the design of a house, the architect placed the chimney so that it is centered at the peak of the roof, as shown in the sketch. If the chimney is 30.0 in. wide, determine the distance down from the peak of the roof at which the carpenter will have to cut the opening given the following angles that the roof peak forms. (*Hint:* Construct a right triangle.)
a. 90°   **b.** 60°

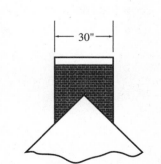

85. Where will the carpenter cut the opening for the chimney on each side of the peak if the chimney is not centered? The roof forms an angle of 60° at the peak, as shown in the illustration.

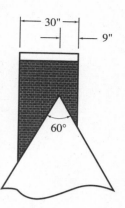

6.3

The Right Triangle and Applications

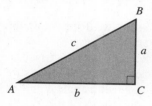

FIGURE 6.15

Now that we have learned the basic trigonometric ratios, let us look at the simplest application of these—the right triangle. For uniformity in our discussions, unless noted otherwise, we will label the right triangle as shown in Figure 6.15. Capital letters A, B, and C will indicate the angles, as well as the corresponding vertices of the triangle. C always names the right angle. The sides opposite the angles will be named with the corresponding lowercase letters, a, b, and c.

In the introduction to Section 6.1 we learned that a surveyor could find the height of a cliff by measuring angle and using right triangle trigonometry rather than similar triangles. Let's look at how it is done.

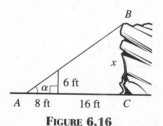

FIGURE 6.16

EXAMPLE 1 **Application: Solving a Right Triangle Given an Angle and a Side**

Determine the height of the cliff in Figure 6.16 given that $A = 36.9°$ and $b = 24$ ft.

Solution Use the given information to construct a right triangle as in Figure 6.17. The triangle helps us see that we need to find the length of side a, which is opposite angle A. We know that $A = 36.9°$ and the side adjacent to angle A, $b = 24$ ft. There are two trigonometric ratios in Table 6.1 that include angle A, side a, and side b. The trigonometric ratios are $\tan \theta = \dfrac{\text{opp}}{\text{adj}}$ and $\cot \theta = \dfrac{\text{adj}}{\text{opp}}$. Since the tangent key is on the calculator, it is more convenient to work with the tangent ratio. (The cotangent ratio would give the same result.) Translate the ratio using the symbols of the problem as follows:

FIGURE 6.17

$$\tan A = \frac{a}{b}$$

$a = b(\tan A)$ **Solving a**

$= 24(\tan 36.9°)$ **Substituting**

$= 18.$ ft   **Rounding to 2 significant digits**

Thus, the height of the cliff is 18. ft, which is the same as the result obtained using similar triangles. ◆

CAUTION   When rounding the answer to a problem containing a mixture of angle measurements and lengths of line segments, we must take into account all of the measurements. In Example 1, $A = 36.9°$ is a measure to the nearest tenth of a degree, which has three significant digits, and $b = 24$ ft is a measurement, which has two significant digits. The answer is rounded to the least number of significant digits, in this case 2. Therefore, the answer is 18. ft. ■

For the triangle in Figure 6.17, we can determine the length of side c and the measure of angle B. Since angle A and angle B are complementary angles, $B = 90° - A$, or $B = 53.1°$.

The length of side c in Figure 6.18 can be found using two different methods: right triangle trigonometry or the Pythagorean relation.

Method 1: To determine the length of the hypotenuse, we must use a trigonometric ratio that contains the hypotenuse. Limiting ourselves to the functions found on the calculator, we could select sine or cosine. Working with angle A and 18 ft, the opposite side, to determine the hypotenuse indicates the ratio opp/hyp, which means working with the sine function.

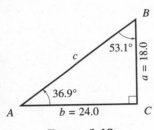

B

$53.1°$

$a = 18.0$

c

$36.9°$

A $b = 24.0$ C

FIGURE 6.18

$$\sin A = \frac{a}{c}.$$

$c = \dfrac{a}{\sin A}$ **Solving for c**

$= \dfrac{18}{\sin 36.9°}$ **Substituting**

$= 30.$ ft   **Rounding to 2 significant digits**

Knowing the measure of angle B, we could obtain the same result using the ratio $\sin B = \dfrac{b}{c}$.

Method 2: Using the Pythagorean relation, $c_2 = a^2 + b^2$:

$c = \sqrt{a^2 + b^2}$ **Solving for c**

$= \sqrt{18^2 + 24^2}$ **Substituting**

$= 30.$ ft.

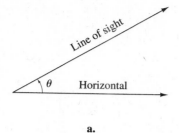

a.

When calculating the length of the third side of a right triangle the Pythagorean relation can be used as a check on the calculation by trigonometric ratios, or the calculation by the trigonometric ratios can be used as a check on the value calculated by the Pythagorean relation.

In navigation and surveying, names have been given to certain angles. The **angle of elevation** or **angle of inclination** is the angle between the horizontal and the line of sight to an object located above the horizontal (Figure 6.19a). The **angle of depression** is the angle between the horizontal and the line of sight to an object located below the horizontal (Figure 6.19b).

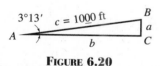

b.
FIGURE 6.19

EXAMPLE 2 Application: Solving a Right Triangle Given an Angle and a Side

Looking at a cross section of a piece of roadway, the angle of inclination of the road is 3° 13′. Measuring along the surface of the road from point A to point B, the distance is 1000 ft. Determine the rise of the roadway above point A at point B.

Solution First sketch and label a right triangle. From the information given, we know that $A = 3° 13′$ and the length of side c is 1000 ft (Figure 6.20). The rise for 1000 ft of roadway is the length of side a. The length of side a can be found using the sine ratio.

$$\sin A = \frac{a}{c}$$

$$a = c(\sin A) \qquad \text{Solving for } a$$
$$= 1000(\sin 3° 13′) \qquad \text{Substituting}$$
$$= 56.111937$$

FIGURE 6.20

The rise of the roadway above point A at point B is 56.1 ft to three significant digits. ◆

EXAMPLE 3 Application: Solving a Right Triangle Given Two Sides

The pilot's radar tells him that the distance from the plane to the end of the runway is 5.47 mi. His altimeter reads 2320 ft. Determine the glide angle (angle of depression) to the end of the runway.

Solution Label the triangle as in Figure 6.21. Using the fact that if two parallel lines are cut by a transversal then the alternate interior angles are equal, angle

A is the same as the angle of depression. To determine angle A, use the sine function since we know sides a and c $\left(\dfrac{\text{opp}}{\text{hyp}}\right)$.

$$\sin^{-1}\left(\frac{a}{c}\right) = A \quad \left(\sin^{-1}\frac{a}{c} = A\right)$$

$$\sin^{-1}\left(\frac{2320}{28{,}880}\right) = \qquad \textbf{Substituting}\ \left(5.47\ \text{mi} \times \frac{5280\ \text{ft}}{1\ \text{mi}} = 28{,}880\ \text{ft}\right)$$

$$4.607673° =$$

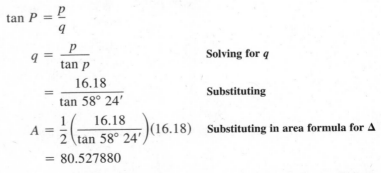

FIGURE 6.21

Therefore, under ideal conditions, the glide angle should be 4.6° for the plane to touch down at the end of the runway.   ◆

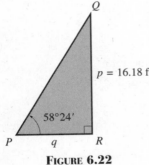

FIGURE 6.22

EXAMPLE 4   **Application: Using Right Triangle Trigonometry to Determine the Area of a Triangle**

Determine the area of a right triangle PQR if one of the perpendicular sides is 16.18 ft and the opposite angle is 58° 24′.

Solution   Sketch the triangle and label as in Figure 6.22. The formula for finding the area of a triangle is $A = \dfrac{1}{2}ba$. We know the altitude, $p = 16.18$ ft. In order to find the area, we must find the length of side q, which is the base of the triangle. We can do this using the tangent ratio.

$$\tan P = \frac{p}{q}$$

$$q = \frac{p}{\tan p} \qquad\qquad \textbf{Solving for } q$$

$$= \frac{16.18}{\tan 58°\ 24'} \qquad\qquad \textbf{Substituting}$$

$$A = \frac{1}{2}\left(\frac{16.18}{\tan 58°\ 24'}\right)(16.18) \qquad \textbf{Substituting in area formula for } \Delta$$

$$= 80.527880$$

The area of the triangle is 80.53 ft² (rounded to four significant digits).   ◆

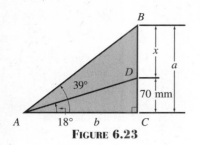

FIGURE 6.23

EXAMPLE 5   Application: Applying Trigonometric Ratios

In Figure 6.23, determine the length of side x (BD).

Solution   Since triangle ABD is not a right triangle, we cannot apply any of the trigonometric ratios directly to it. However, triangle ABC is a right triangle, and using the tangent ratio, we have the following.

$$\tan A = \frac{a}{b}$$

Also from the drawing we know that $a = x + 70$ and $A = 39°$. Substituting we have:

$$\tan 39° = \frac{x + 70}{b}$$

Solving for x, we have:

$$x = b(\tan 39°) - 70.$$

If we can determine the length of side b, we can find x. Using the tangent ratio and triangle ADC, we can determine the length of side b.

$$\tan 18° = \frac{70}{b} \qquad \text{or} \qquad b = \frac{70}{\tan 18°}$$

Substituting for b, we have:

$$x = \frac{70}{\tan 18°}(\tan 39°) - 70 = 104.458129.$$

Therefore, to one significant digit, $x = 100$ mm.   ◆

6.3 Exercises

In Exercises 1–12, given two parts of a right triangle, determine the other three parts, (see the sketch). Use the rules of significant digits in determining the final answer.

1. $A = 15.3°$, $c = 16.4$ in.
2. $B = 31.5°$, $a = 9.24$ m
3. $B = 64.2°$, $c = 31.4$ mm
4. $A = 47.3°$, $b = 82.3$ mm
5. $A = 45°$, $b = 1.82$ mm
6. $A = 73.4°$, $a = 2.06$ in.
7. $B = 18.2°$, $a = 3.04$ ft
8. $B = 30°$, $b = 17.4$ mm
9. $B = 45°$, $b = 22.3$ in.
10. $B = 52.4°$, $c = 0.384$ m
11. $A = 60°$, $a = 14.7$ m
12. $A = 63.7°$, $c = 15.9$ in.

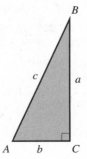

In Exercises 13–16, determine the area of each right triangle *PQR* (see the drawing). Use the rules for significant digits in determining the final answer.

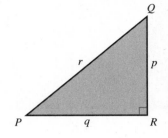

13. $P = 29.4°$, $p = 3.45$ m

14. $Q = 45.3°$, $p = 32.1$ in

15. $Q = 58.4°$, $p = 30.0$ mm

16. $P = 83.4°$, $p = 74.3$ m

17. The string attached to a kite is 175 ft long. The angle formed by the line of the string and the horizontal is 63.3°. How high is the kite when the string runs taut?

18. The Washington Monument is approximately 556 ft high. When the shadow of the monument is 159 ft in length, what is the angle of elevation of the sun?

19. Find the size of the acute angles of a right triangle whose sides are in the ratio 3 : 4 : 5.

20. Find the area of the parallelogram whose sides are 15.00 in. and 25.00 in. if the angle between these sides is 72° 45′.

21. At a distance of 14,265 ft from the foot of the Empire State Building the angle of elevation to its top is 5.42°. Find the height of the building.

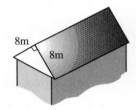

22. A steep gabled building roof has ends in the shape of an isosceles triangle (see the illustration). The two ends are to be painted. The contractor estimates the rafters to be 8 m long. Determine the area to be painted for the following.
  a. The rafters come together at an angle of 90°.
  b. The rafters come together at a 60° angle.

23. A surveyor at point *B* needs to measure the width of a river. Selecting a tree at point *P* directly across the river from point *B* (see the drawing), she turns and measures the distance *FB* to be 55.00 ft. At point *F*, she measures angle *θ* and finds it to be 82.23°. What is the width of the river?

24. An airplane pilot finds the angle of depression of a beacon light on the ground to be 14° 20′. His altimeter shows that he is 400 m high. How far does he estimate his ground distance to be from the beacon?

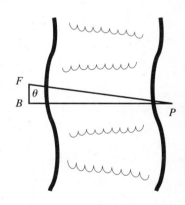

25. From a window 26 m above the ground, the angle of elevation to the top of a nearby building is 24° 20′, and the angle of depression to the bottom of the building is 14° 40′. Find the height of the building.

26. What is the angle of the roof with the attic floor if the ridgepole is 14.3 ft above the attic floor and the attic is 74.5 ft wide?

27. The angles of elevation to the top of a mountain from *A* and *B* on opposite sides of its base are 33° 10′ and 48° 15′ (see the drawing). The top of the mountain is known to be 1512 ft higher than the line made by connecting the two points, *A* and *B*. Determine the length of a tunnel through the mountain from *A* to *B*.

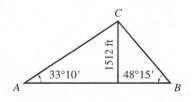

28. Determine the measure of angle A, which is formed by the diagonal of a cube and the diagonal of one of the faces as shown in the sketch. The two diagonals are drawn from the same vertex.

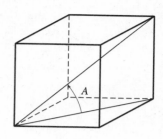

29. A pilot wishes to land at a point on the field that is estimated to be 3750 ft from the present location of the plane. Is it possible to descend to the field in a straight line if a safe glide angle is 10.4° and the altitude is 1540 ft? (See the drawing.)

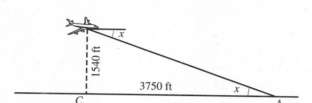

30. An AC electric circuit has an inductive reactance $X_L = 10.0$ Ω and a resistance $R = 22.3$ Ω. In such a circuit the **impedance** Z, in ohms, can be represented by the hypotenuse of a right triangle with X_L and R as its other sides, as shown in the illustration. The angle θ is called the **phase angle.** Determine the phase angle and the impedance.

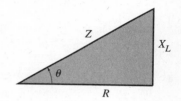

31. An AC circuit has an impedance $Z = 22.0$ Ω. If the circuit has a resistance $R = 18.9$ Ω, determine the inductive reactance X_L and the phase angle θ. (See Exercise 30.)

32. The right triangle shown in part a of the drawing is called a **power triangle,** and is used in solving AC-circuit problems. The hypotenuse represents the **apparent power** P_A, measured in volt-amperes (VA). The other sides represent **real power** P, measured in watts (W), and the **reactive power** P_R, measured in vars. The angle θ is called the **phase angle.** In a particular circuit the real power is $P = 23.0$ W and the reactive power is $P_R = 5.1$ vars. Determine the apparent power P_A and the phase angle (see part b in the sketch.)

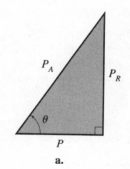

a.

33. If the real power in a circuit is 1.73 W and the phase angle θ is 66.5°, determine the apparent power and the reactive power. (See Exercise 32.)

34. An escalator inclined at an angle of 42.6° moves at a constant speed of 0.6 m/s. A person steps on the escalator and reaches the top in 42. s. Through what vertical height has the person been lifted?

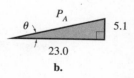

b.

35. A piston and rod assembly has the dimensions shown in the drawing. Determine the linear advance of the piston in the cylinder as the throw arm moves from the position shown to a position in line with the cylinder.

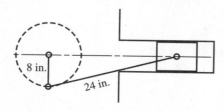

36. A block mason is setting a door frame. To ensure square corners, he measures the diagonals of the rectangular opening. What is the length of each diagonal if the doorjamb is square? (See the sketch.)

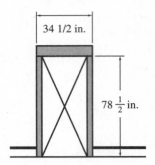

34 1/2 in.

$78\frac{1}{2}$ in.

37. A weight hangs on a 15.0 m chain from a high ceiling. If the weight is pulled sideways so that the chain forms an angle of 24.8° with its vertical position, how much is the weight raised vertically?

38. It has been found that the strongest beam that can be cut from a log (see the drawing) may be obtained by dividing the diameter into three equal parts, $\overline{AC}$, $\overline{CD}$, and $\overline{DB}$. Then, perpendiculars $\overline{CE}$ and $\overline{DF}$ are drawn. The rectangle $AEBF$ represents the cross section of such a beam. If the diameter $\overline{AB}$ of the log is 24″, find the width w and the depth d of the beam.

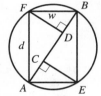

39. A right triangle is inscribed in a circle of radius 56 mm. One angle of the triangle is 64°. Determine the lengths of the two sides.

40. A beam of gamma rays is to be used to treat a tumor known to be 5.7 cm beneath the patient's skin. To avoid damaging a vital organ, the radiologist moves the source over 8.3 cm. (See the illustration.)
   a. At what angle to the patient's skin must the radiologist aim the gamma-ray source to hit the tumor?
   b. How far will the beam have to travel through the patient's body before reaching the tumor?

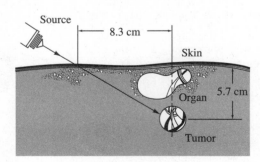

Source

8.3 cm

Skin

5.7 cm

Organ

Tumor

In Exercises 41 and 42, determine the values of x and and y.

41.

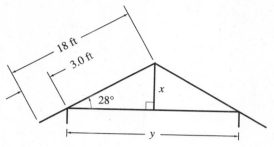

42.

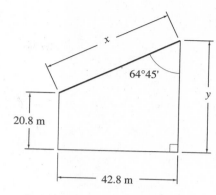

In Exercises 43 and 44, determine the values of α and β.

43.

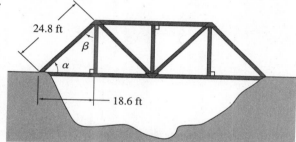

44.

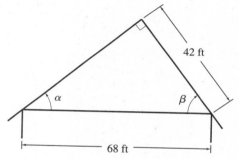

6.4

Signs of Trigonometric Functions

Until this section, the discussion of the trigonometric functions has been limited to angles in the first quadrant. An angle in standard position is named by the quadrant in which the terminal side is located. For example, a 127° angle in standard position is called a second-quadrant angle since its terminal side is in the second quadrant (90° < 127° < 180°).

Why should we be concerned if the angle is a second-, third-, or fourth-quadrant angle? First recall that the values of the six trigonometric functions are determined by x, y, and r. The ordered pair (x, y) is a point on the terminal side of the angle, and r is always positive. Therefore, the six trigonometric functions are positive or negative depending in which quadrant the terminal side of the angle lies. This is illustrated in Figure 6.24. In Section 6.5 we will be evaluating functions in all four quadrants. To do that we must know where a function is positive or negative.

If we drop a perpendicular to the x-axis from the point on the terminal side of the angle, we create a triangle called the **reference triangle.** We will label the acute angle of the reference triangle, whose adjacent side is on the x-axis, angle α. (see Figure 6.24). This angle is called the **reference angle.**

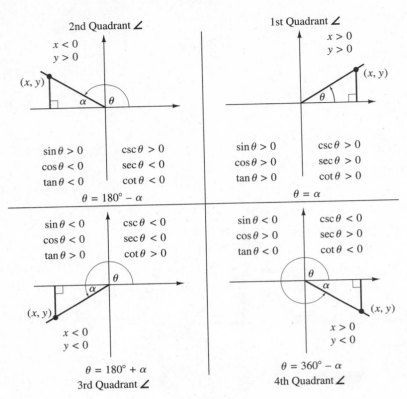

FIGURE 6.24

EXAMPLE 1   Signs of Trigonometric Functions

Determine the algebraic sign of the following trigonometric functions.

a. sin 195°   **b.** cos 135°   **c.** tan 315°

Solution

a. The terminal side of the 195° angle in standard position is in the third quadrant (see Figure 6.25). The algebraic sign of the y value of any third-quadrant angle is negative. Therefore, sin 195° is negative since $\sin \theta = \dfrac{y}{r}$.

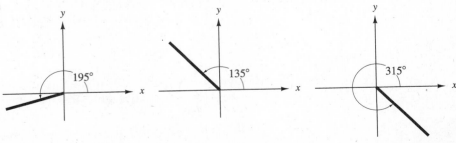

FIGURE 6.25   **FIGURE 6.26**   **FIGURE 6.27**

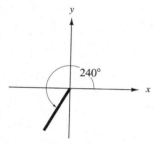

FIGURE 6.28

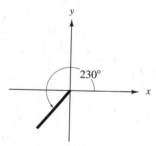

FIGURE 6.29

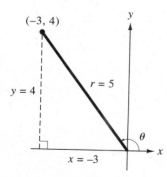

FIGURE 6.30

b. The terminal side of the 135° angle is in the second quadrant (Figure 6.26). The algebraic sign of the x value of any second-quadrant angle is negative. Therefore cos 135° is negative, since $\cos \theta = \dfrac{x}{r}$.

c. The terminal side of the 315° angle is in the fourth quadrant (Figure 6.27). The algebraic sign of the y value of any fourth-quadrant angle is negative, and the algebraic sign of the x value is positive. Therefore, tan 315° is negative, since $\tan \theta = \dfrac{y}{x}$. ◆

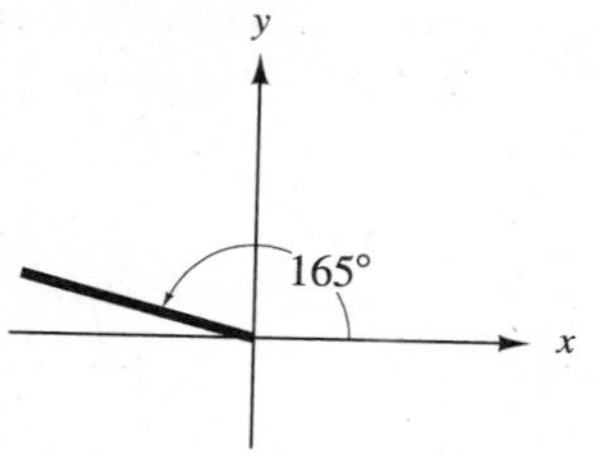

EXAMPLE 2   Signs of Trigonometric Functions

Determine the algebraic sign of the following trigonometric functions.

a. sin 165°   **b.**   cos 240°   **c.**   tan 23°

Solution

a. The terminal side of a 165° angle is in the second quadrant (Figure 6.28). The algebraic sign of the y value of a second quadrant angle is positive. Therefore, the sin 165° is positive.

b. The terminal side of a 240° angle is in the third quadrant (Figure 6.29). The algebraic sign of the x value of a third quadrant angle is negative. Therefore, cos 240° is negative.

c. The terminal side of a 230° angle is in the third quadrant (Figure 6.30). The algebraic sign of the x and y values are both negative in the third quadrant. Therefore, tan 230° is positive. ◆

To determine the six trigonometric functions for a specific point on the plane, it is necessary to draw a reference triangle. This is illustrated in Example 3.

EXAMPLE 3   Evaluating Trigonometric Functions

Find the six trigonometric functions of θ if the terminal side of the angle passes through $(-3, 4)$.

Solution   Draw the terminal side of the angle through the point $(-3, 4)$. Then construct the reference triangle as in Fig 6.31. Using the Pythagorean relation, we determine r is 5. From the reference triangle it is possible to determine the six trigonometric functions:

$$\sin \theta = \frac{4}{5} \quad \cos \theta = \frac{-3}{5} \quad \tan \theta = \frac{4}{-3}$$

$$\csc \theta = \frac{5}{4} \quad \sec \theta = \frac{5}{-3} \quad \cot \theta = \frac{-3}{4}.$$

θ is a second quadrant angle and the ratios show that the sine and cosecant are positive, while the other four functions are negative. ◆

FIGURE 6.31

Examining Figure 6.24 and the results of Examples 1–3, we can see that each of the six trigonometric functions is positive in two quadrants and negative in two quadrants. The signs of the functions are listed in Table 6.5.

TABLE 6.5   Signs of Trigonometric Functions

Quadrant	Function					
	$\sin \theta$	$\cos \theta$	$\tan \theta$	$\cot \theta$	$\sec \theta$	$\csc \theta$
1st Quadrant $0 < \theta < 90°$	+	+	+	+	+	+
2nd Quadrant $90° < \theta < 180°$	+	−	−	−	−	+
3rd Quadrant $180° < \theta < 270°$	−	−	+	+	−	−
4th Quadrant $270° < \theta < 360°$	−	+	−	−	+	−

EXAMPLE 4   **Evaluating Trigonometric Functions**

Determine the quadrant(s) in which the terminal side of θ lies, subject to the following conditions: $\cos \theta > 0$ and $\tan \theta < 0$.

Solution   Table 6.5 shows that $\cos \theta > 0$ in the first and fourth quadrants, while the $\tan \theta < 0$ in the second and fourth quadrants. The only quadrant in which cosine is positive and tangent is negative is the fourth quadrant. Therefore, the terminal side of θ lies in the fourth quadrant.   ◆

6.4 Exercises

In Exercises 1–36, determine the algebraic sign of the trigonometric functions.

1. $\sin 75°$ **2.** $\cos 124°$ **3.** $\tan 195°$ **4.** $\sin 295°$

5. $\cos 220°$ **6.** $\tan 250°$ **7.** $\sin 140°$ **8.** $\cos 82°$

9. $\tan 350°$ **10.** $\sin 225°$ **11.** $\cos 358°$ **12.** $\tan 22.5°$

13. $\sin 230°$ **14.** $\cos 150°$ **15.** $\tan 170°$ **16.** $\sin 290°$

17. $\cos 225°$ **18.** $\tan 95°$ **19.** $\sin 365°$ **20.** $\cos 0.79°$

21. $\tan 370°$ **22.** $\sin 192°$ **23.** $\cos 640°$ **24.** $\tan 136°$

25. $\csc 85°$ **26.** $\sec 135°$ **27.** $\cot 240°$ **28.** $\csc 97°$

29. $\sec 223°$ **30.** $\cot 285°$ **31.** $\csc 210°$ **32.** $\sec 300°$

33. $\cot 45°$ **34.** $\csc 331°$ **35.** $\sec 0.25°$ **36.** $\cot 150°$

In Exercises 37–44, determine the six trigonometric functions of θ for which the terminal side of θ passes through the given point. Determine exact values.

37. $(-6, 8)$ **38.** $(6, -8)$ **39.** $(-3, -4)$ **40.** $(-1, -2)$

41. $(4, -5)$ **42.** $(-3, 5)$ **43.** $(2, 5)$ **44.** $(-2, 3)$

In Exercises 45–56, determine the quadrant(s) in which the terminal side of θ lies, subject to the following conditions.

45. $\sin \theta > 0$, $\cos \theta > 0$

46. $\tan \theta < 0$, $\sin \theta > 0$

47. $\cos \theta < 0$, $\tan \theta < 0$

48. $\cos \theta < 0$, $\sin \theta > 0$

49. $\sin \theta < 0$, $\tan \theta > 0$

50. $\cos \theta > 0$, $\csc \theta > 0$

51. $\cot \theta > 0$, $\sin \theta > 0$

52. $\tan \theta > 0$, $\sec \theta < 0$

53. $\sec \theta > 0$, $\csc \theta < 0$

54. $\cot \theta > 0$, $\sin \theta < 0$

55. $\cot \theta < 0$, $\sec \theta < 0$

56. $\cot \theta < 0$, $\cos \theta < 0$

6.5

Trigonometric Functions of any Angle

This section teaches us two processes. First, it shows us how to determine the functional value of any of the six trigonometric functions with an angle in any quadrant, and second, how to determine the angle or angles when given a functional value for one of the six trigonometric functions. A technician's calculation shows that the cos θ is a negative value. Is this answer wrong? The answer could be correct and the information in this section will help support that conclusion.

The first objective is not too difficult with the use of a calculator. To find the values for sine, cosine, or tangent, enter the measure of the angle, press the appropriate keys, and read the result.

EXAMPLE 1   **Determining Functional Values**

Determine the values of the trigonometric functions using a calculator.

a. sin 98°   **b.** cos 195°   **c.** tan 285°

Solution

a. $\sin 98° = 0.990268$, or $\sin 98° = 0.99$.
b. $\cos 195° = -0.965926$, or $\cos 195° = -0.97$.
c. $\tan 285° = -3.732051$, or $\tan 285° = -3.7$.   ◆

In Example 1, the calculator result contains the functional value as well as the appropriate sign. The sign is determined by the quadrant in which the terminal side of the angle lies.

For example, determine the functional values for sin 35°, sin 145°, sin 215°, and sin 325°. The calculator values are listed below.

$$\sin 35° = 0.573576 \qquad \sin 145° = 0.573576$$
$$\sin 215° = -0.573576 \qquad \sin 325° = -0.573576$$

As we would expect, based on the discussion in Section 6.4, the functional values of the sine are positive in quadrants 1 and 2 and negative in 3 and 4. Look at the results more closely to note some interesting facts. The sine is positive in quadrants 1 and 2, and the functional values are identical. The sine is negative in quadrants 3 and 4, and the functional values are identical. In fact, the absolute value of the functional values are the same for all four angles. Why?

To see why, draw a 35°, 145°, 225°, and 325° angle, as shown in Figure 6.32. On the terminal side of each angle, select a point and draw a reference triangle. Each of these reference triangles cotains an acute angle whose vertex is at the origin and whose adjacent side is on the x-axis (horizontal axis); this is the reference angle named α. In each of the triangles in Figure 6.32, the reference angle, α = 35°.

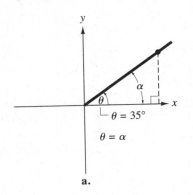

a.

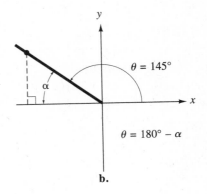

b.

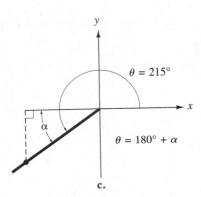

c.

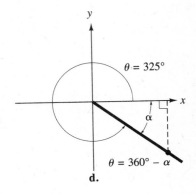

d.

FIGURE 6.32

Recall that the functional values are determined by a ratio of two of the three values x, y, or r. If we select a specific point on the terminal side of each angle, such as $(3.2, \sqrt{5.0})$, $(-3.2, \sqrt{5.0})$, $(-3.2, -\sqrt{5.0})$ and $(3.2, -\sqrt{5.0})$, in each case $r = 3.9$. Then, the functional values for the angles in Figure 6.32 are as listed below.

$$\sin 35° = \frac{\sqrt{5.0}}{3.9} = 0.57 \qquad \sin 145° = \frac{\sqrt{5.0}}{3.9} = 0.57$$

$$\sin 225° = \frac{-\sqrt{5.0}}{3.9} = -0.57 \qquad \sin 325° = \frac{-\sqrt{5.0}}{3.9} = -0.57.$$

In general, the absolute value of the functional value of angle θ in any quadrant is equal to the functional value of the reference angle α. That is, $|F(\theta)| = F(\alpha)$, where F is one of the six trigonometric functions.

CAUTION Why all the discussion and concern about the reference angle? All six trigonometric functions are positive in two quadrants and negative in two quadrants. A calculator will provide us with only one of these angles—the reference angle. Therefore, the reference angle is essential in determining the desired angles. ■

EXAMPLE 2 **Determining Angles Using Reference Angles**

Given $\tan \theta = 0.753$, find θ for $0° \le \theta < 360°$.

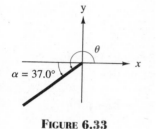

$\alpha = 37.0°$

FIGURE 6.33

Solution To solve the problem, we must determine all the angles θ such that $\tan \theta = 0.753$. There are two possible answers for θ. Since $\tan \theta$ is positive in the first and third quadrants, θ could be a first-or a third-quadrant angle.

Using the calculator, $\tan^{-1}(0.753) = 36.97975$, or to the nearest tenth of a degree, 37.0°. The reference angle is 37.0°, which is the same as the first-quadrant angle. How can we determine the third-quadrant angle? Construct a reference angle, as shown in Figure 6.33. From the diagram, we can see that:

$$\theta = 180° + \alpha$$
$$= 180° + 37.0°, \text{ or } \theta = 217.0°.$$

◆

RULES TO DETERMINE $\theta°$

In general to determine the angle θ, knowing the quadrant and the reference angle α, use the following rules.

First-quadrant angle $\theta = \alpha$
Second-quadrant angle $\theta = 180° - \alpha$
Third-quadrant angle $\theta = 180° + \alpha$
Fourth-quadrant angle $\theta = 360° - \alpha$

Why was the restriction $0° \le \theta < 360°$ placed on θ in Example 2? Because there is an unlimited number of coterminal angles that have the same functional values. For example, the angles 397.0° and 757.0° are, respectively, 37.0° + 360° and 37.0° + 720.0°.

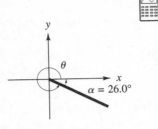

FIGURE 6.34

EXAMPLE 3   Determine Angles Using a Reference Angle

Given $\sin \theta = -0.438$, determine θ for $0° \leq \theta < 360°$.

Solution   For $\sin \theta = -0.438$, $\theta = $?, there are two possible values for θ since $\sin \theta$ is negative. The two values are third- and fourth-quadrant angles. The reference angle is:

$$\alpha = \boxed{\text{SIN}^{-1}}\,|-0.438| = 25.97634,$$

or to the nearest tenth of a degree, $26.0°$ (see Figure 6.34). The third- and fourth-quadrant angles are:

Third Quadrant	**Fourth Quadrant**
$\theta_3 = 180° + \alpha$	$\theta_4 = 360° - \alpha$
$\quad = 180° + 26.0°$	$\quad = 360° - 26.0°$
$\quad = 206.0°$	$\quad = 334.0°.$

The two values of θ that make $\sin \theta = -0.438$ a true statement are $\theta_3 = 206.0°$ and $\theta_4 = 206.0°$ and $334.0°$. ◆

CAUTION   The most frequent mistake made when determining the angle given the functional value is to answer with only one angle—the angle given by the calculator. Remember that, unless conditions dictate otherwise, for all six trigonometric functions, the answer contains two angles for a positive functional value or two angles for a negative functional value. ■

EXAMPLE 4   Determining Functional Values

Determine $\sin \theta$ when $\cos \theta = -0.4426$ and $\tan \theta < 0$ for $0° \leq \theta < 360°$.

Solution   Knowing θ, finding the $\sin \theta$ is easy using the calculator. The problem then is: for what angle θ will $\cos \theta = -0.4426$ and $\tan \theta < 0$? The only quadrant in which the cosine and tangent are both negative is the second quadrant. For a cosine negative functional value, the calculator will always give a second quadrant angle. Using the calculator:

$$\boxed{\text{COS}^{-1}}\,\boxed{(}\,-0.4426\,\boxed{)} = 116.26989.$$

The calculator result is a second quadrant angle, and $\theta = 116.27°$. Now we can find the answer by finding $\sin \theta$, $\sin 116.26989 = 0.8967192$. Earlier in the text we learned that calculator solutions are "chained." In this problem it means finding:

$$\boxed{\text{SIN}}\,\boxed{(}\,\boxed{\text{COS}^{-1}}\,\boxed{(}\,\boxed{(-)}\,0.4426\,\boxed{)}\,\boxed{)} = 0.8967192.$$

Rounding the answer to four significant digits, $\sin \theta = 0.8967$. ◆

EXAMPLE 5 Application: Determine the Angle

The approximate time it takes a tennis ball, thrown by a machine, to reach its maximum height can be measured by the formula $t = \dfrac{v \sin A}{g}$, where t = time, v = initial velocity, g = gravitational pull, and A = angle of the initial path (or velocity) with the horizontal. Determine A if $t = 3.0$ s, $v = 35$ m/s, and $g = 9.8$ m/s².

Solution The problem is determining angle A. To do this, solve the formula for $\sin A$, substitute the given values in the revised formula, and then find the angle.

$$t = \frac{v \sin A}{g}, \text{ for } \sin A$$

$$\frac{g}{v}\left(\frac{v \sin A}{g}\right) = \frac{g}{v}(t)$$

$$\sin A = \frac{(t)(g)}{v}$$

$$= \frac{(3.0 \text{ s})(9.8 \text{ m/s}^2)}{35 \text{ m/s}}$$

$$= 0.8400000$$

Therefore, $A = \sin^{-1}(0.8400000) = 57.14012$ and $A = 57.°$ or $123.°$ rounded to two significant digits. (Recall that the sine of an angle is positive in the first and second quadrants; see Figure 6.35). The $57.°$ solution represents the machine throwing the ball "forward," and the $123.°$ represents the machine throwing the ball "backward." Since the machine normally would be set for only one throwing direction, choose $A = 57.°$. ◆

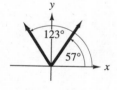

FIGURE 6.35

6.5 Exercises

In Exercises 1–10, match the function with the correct functional value.

1. $\sin 33°$ **a.** 57.29
2. $\cos 37°$ **b.** -0.4574
3. $\tan 89°$ **c.** 0.7965
4. $\sin 115° \, 56'$ **d.** 0.5446
5. $\cos 117° \, 13'$ **e.** -0.6652
6. $\tan 121° \, 52' \, 48''$ **f.** -0.7431
7. $\sin 127.2°$ **g.** 0.7986
8. $\cos 131.7°$ **h.** -0.9163
9. $\tan 137.5°$ **i.** 0.8993
10. $\sin 228°$ **j.** -1.6078

In Exercises 11–30, determine the value of the trigonometric function.

11. cos 231°   12. tan 237°   13. sin 243.5°   14. cos 245.5°

15. tan 247.5°   16. sin 273.4°   17. cos 275° 34′   18. tan 281° 15′

19. sin 562° 35′   20. cos 632.6°   21. tan 453.7°   22. sin (−156.3°)

23. cos (−48.3°)   24. tan (−100°)   25. csc 100°   26. sec 105°

27. tan 130° 45′ 32″   28. csc 192.5°   29. sec 225.7°   30. cot 315.°

In Exercises 31–48, determine θ for $0° \leq \theta < 360°$.

31. $\sin \theta = 0.74$ 32. $\cos \theta = 0.61$ 33. $\tan \theta = 0.75$ 34. $\sin \theta = -0.688$

35. $\cos \theta = -0.745$ 36. $\tan \theta = -1.21$ 37. $\sin \theta = 0.5446$ 38. $\cos \theta = 0.7660$

39. $\tan \theta = 1.428$ 40. $\sin \theta = -0.9396$ 41. $\cos \theta = -0.5074$ 42. $\tan \theta = -0.6249$

43. $\csc \theta = 1.82$ 44. $\sec \theta = 1.25$ 45. $\cot \theta = 1.43$ 46. $\csc \theta = -1.56$

47. $\sec \theta = -1.82$ 48. $\cot \theta = -0.625$

In Exercises 49–54, determine the functional value that satisfies the given conditions.

49. Find $\sin \theta$ when $\cos \theta = -0.58779$ and $\tan \theta > 0$.
    50. Find $\sin \theta$ when $\tan \theta = -0.70021$ and $\cos \theta < 0$.

51. Find $\cos \theta$ when $\sin \theta = 0.17365$ and $\tan \theta > 0$.
    52. Find $\tan \theta$ when $\cos \theta = -0.78801$ and $\sin \theta > 0$.

53. Find $\tan \theta$ when $\sin \theta = -0.57358$ and $\cos \theta > 0$.
    54. Find $\cos \theta$ when $\tan \theta = 1.23490$ and $\sin \theta < 0$.

55. The range x of a projectile fired with velocity v_0 at an angle α with the horizontal is given by $x = \dfrac{v_0^2 \sin 2\alpha}{g}$, where g is gravational acceleration. Determine x in terms of v_0 and g for the given values of α.

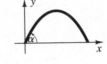

    a. $\alpha = 15°$ b. $\alpha = 45°$ c. $\alpha = 75°$

    d. Using the sketch, explain why x is the greatest when $\alpha = 45°$, which is the angle of maximum range.

56. Exercise 55 illustrates that two projectiles fired with the same initial velocity at angles of 15° and 75° have exactly the same range. In general, complementary angles determine the same range if all other factors are equal. Therefore something must be different; after all, the angles are drastically different. It turns out that the time of flight will be different for different-sized angles. The time of flight is twice the time it takes an object to reach its maximum height. From Example 5 we learned that the time for a tennis ball to reach its maximum height is given by $t = \dfrac{v \sin A}{g}$. If the speed of a tennis ball is 35 m/s and the acceleration due to gravity g is 9.8 m/s/s, determine the time of flight for the ball for the given angle of elevation.

    a. 75°   b. 15°.

    c. Which angle has the largest time of flight? How much larger is it?

57. When a punter kicks a football, he can obtain the greatest range if he kicks it at an angle of 45°.

    a. Most professional punters, however, try to kick the ball with an angle of about 55°. Why?

    b. Why would a kick angled at 35° not do as well? Recall the range for 35° would be the same as for 55°. (*Hint:* Refer to Exercise 56.)

58. A formula for determining the area of a triangle, knowing sides a and b and angle C, is $A = \frac{1}{2} ab \sin C$. Determine the area of a triangle for which $a = 34.5'$, $b = 60.4'$, and $C = 148.3°$.

59. In calculating the area of a triangular tract of land a surveyor uses the formula in Exercise 58. The measurements of the triangle are $a = 324$ m, $b = 196$ m, and $C = 118.7°$. Determine the area of the triangle.

6.6

Introduction to Radian Measure

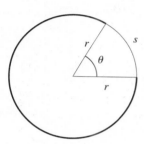

FIGURE 6.36

The degree is an arbitrary measure, which some believe was selected because there are approximately 360 days in a year. A second unit of angle measure that is commonly used in technology is the *radian*. The **radian measure** of an angle is determined by the ratio of the length of the arc subtended by a central angle and the radius of the circle. In Figure 6.36, the measure of angle θ in radians is $\theta = \frac{s}{r}$. In the case where $s = r$, the measure of θ is exactly one radian.

> **DEFINITION 6.1**
>
> One **radian** is the measure of a central angle of a circle that subtends an arc equal in length to the radius of the circle. (See Figure 6.36.)

Figure 6.37 shows several angles measured in radians.

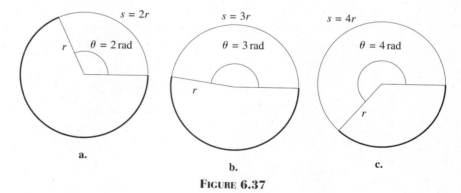

FIGURE 6.37

The length of the arc s may be determined by the formula:

$$s = r\theta,$$

where θ is the measure of the central angle in radians (abbreviated rad). The measure of the central angle can be determined by the formula:

$$\theta = \frac{s}{r} \text{ rad.}$$

The circumference of a circle ($s = 2\pi r$) subtends an angle of 360°. Thus 2π is the radian measure of an angle of 360°. Therefore,

$$2\pi \text{ rad} = 360°.$$

To arrive at the basic relationship between degree measure and radian measure, divide both sides of the equation by 2. That answer shows:

RADIANS = DEGREES

$$\pi \text{ rad} = 180°$$

Divide both sides of the equation $\pi \text{ rad} = 180°$ by π to obtain:

$$1 \text{ rad} = \frac{180°}{\pi}, \text{ or approximately, } 57.3°.$$

Divide both sides by the equation $\pi \text{ rad} = 180°$ by 180 to obtain:

$$1° = \frac{\pi \text{ rad}}{180},$$

or approximately, 0.01745 rad. These calculations provide angle conversion factors.

ANGLE CONVERSION FACTORS

Radians to degrees $\dfrac{180°}{\pi} = 1$

Degrees to radians $\dfrac{\pi}{180°} = 1$

EXAMPLE 1   Changing Angle Measure

Convert from degrees to radians.

a. 45°   **b.** 150°

Solution

a. Using the conversion factor to convert from degrees to radians, we have

$$45° = \overset{1}{\cancel{45°}}\left(\frac{\pi}{\underset{4}{\cancel{180°}}}\right) = \frac{\pi}{4}$$

The answer in this form, $\frac{\pi}{4}$, is called exact. However, if we change π to a decimal form and divide by 4, the answer would be approximate.

b. Using the conversion factor to convert from degrees to radians, we have the following:

$$150° = 150°\left(\frac{\pi}{180°}\right) = \frac{5\pi}{6}. \quad \blacklozenge$$

CAUTION   Note that a radian is a dimensionless quantity since a radian is the ratio of one distance to another. When radians are used, it is customary not to use a symbol to indicate the measure of the angle. The answer in Example 1 is $\frac{5\pi}{6}$, which means the angle is in radian measure. Angle measures such as 2, 1.4, etc., will mean 2 rad, 1.4 rad, etc. Angle measures that relate to a measure in degrees will be indicated by 2° or 2 degrees. ∎

EXAMPLE 2   **Changing Angle Measure**

Convert the following from radians to degrees.

a. $\frac{\pi}{6}$ **b.** $\frac{3\pi}{2}$

Solution

a. Multiply by the conversion factor to convert radians to degrees.

$$\frac{\pi}{6} = \frac{\pi}{6}\left(\frac{180°}{\pi}\right)$$
$$= 30°$$

b. Multiply by the conversion factor to convert radians to degrees.

$$\frac{3\pi}{2} = \frac{3\pi}{2}\left(\frac{180°}{\pi}\right)$$
$$= 270° \quad \blacklozenge$$

It is fairly easy to see a relationship develop, i.e., $0° = 0$, $90° = \frac{\pi}{2}$, $180° = \pi$, and $360° = 2\pi$. Now check these using conversion factors.

A sketch of the angles helps us gain a better understanding of the relationship between degrees and radians. Draw the angles $\theta = 60°$ and $\phi = \frac{\pi}{3}$ in standard position as shown in Figure 6.38.

When converting from radians to degrees and the radian measure contains a factor of π, then the answer will be exact. This was illustrated in Example 2. When the radian measure is in decimal form, the number of significant digits in the answer is determined by the number of significant digits in the radian measure.

When converting from degrees to radians and the angle to be converted is a whole number of degrees and π is not changed to a decimal form, then the answer is exact. This is illustrated in Example 1. If an exact answer is changed to decimal form, the number of significant digits is determined by the calculator value of π. For convenience in this text, exact answers changed to decimal form are expressed to the nearest hundredth. In the case where the angle is expressed in decimal form, the angle determines the number of significant digits.

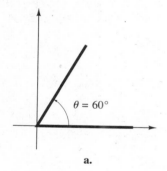

a.

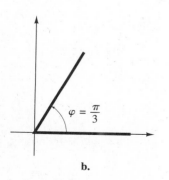

b.

FIGURE 6.38

EXAMPLE 3 Changing Angle Measure

Convert from radians to degrees, expressing the answer in approximate form.

a. 1.000   **b.** 2.3

Solution

a. $1.000 = 1.000\left(\dfrac{180°}{\pi}\right)$

$\qquad = 57.29578°$ **Using the calculator value for** π.

$\qquad = 57.30°$ **To 4 significant digits**

b. $2.3 = 2.3\left(\dfrac{180°}{\pi}\right)$

$\qquad = 131.78029$

$\qquad = 130.°$ **To 2 significant digits** ◆

EXAMPLE 4 Application: Determining the Length of an Arc

A flywheel has a radius of 11 ft (Figure 6.39). What is the length of an arc associated with a central angle that measures 1.6 rad?

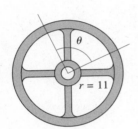

FIGURE 6.39

Solution To determine the length of the arc, use the formula $s = r\theta$, with $r = 11$ ft and $\theta = 1.6$ rad. (θ must be in radian measure.)

$$s = (11 \text{ ft})(1.6) \quad \textbf{Substituting}$$
$$= 18. \text{ ft}$$

Thus, the central angle of 1.6 rad subtends an arc of 18. ft on the flywheel. ◆

EXAMPLE 5 Determining Functional Values

Determine the values of the trigonometric functions using a calculator.

a. $\sin \dfrac{\pi}{5}$ **b.** $\cos 1.21$ **c.** $\tan 1.724$

Solution Make sure that the calculator is in radian mode.

a. $\sin(\pi \div 5) = 0.5877853$. Since the measure, $\dfrac{\pi}{5}$, is considered to be an

exact measure, $\sin \dfrac{\pi}{5} = 0.5877853$.

b. $\cos 1.21 = 0.3530194$. The answer to 3 significant digits is $\cos 1.21 = 0.353$.

c. $\tan 1.724 = -6.476111$. The answer to 4 significant digits is $\tan 1.724 = -6.476$. ◆

In part a of Example 5, the answer is $\sin \dfrac{\pi}{5} = 0.5877853$. Determine the angle θ in degrees ($\theta < 90°$) for $\sin \theta = 0.5877853$.

$$\sin^{-1} 0.5877853 = 36.00000°$$

Use the conversion formula to change $\dfrac{\pi}{5}$ to degrees. $\left(\dfrac{\pi}{5} \times \dfrac{180°}{\pi} \right) = 36°$. These results show that $\sin \dfrac{\pi}{5} = \sin 36°$, or the functional value of an angle in radian measure is the same as the functional of the same angle in degree measure.

Recall from Section 6.5 that in determining the angle when given the functional value two possible answers exist. We also learned that the reference angle is an important tool in determining the second angle. A problem frequently occurs when finding angles in radian measure, which are the dividing lines for each quadrant. That is what radian measures are equal to: $\dfrac{\pi}{2}, \dfrac{3\pi}{2}$, and 2π. These values are indicated in Table 6.6

TABLE 6.6   Angle Measures that Indicate Quadrants

	Degrees	Radians	Radians (Decimal)
	0°	0	0
Quadrant 1			
	90°	$\dfrac{\pi}{2}$	1.571
Quadrant 2			
	180°	π	3.142
Quadrant 3			
	270°	$\dfrac{3\pi}{2}$	4.712
Quadrant 4			
	360°	2π	6.283

EXAMPLE 6   **Determine Angle, Given Functional Value**

Given $\cos \theta = -0.342$, determine θ for $0 \le \theta < 2\pi$.

Solution   The interval $0 \le \theta < 2\pi$ indicates θ is in radians. For $\cos \theta = -0.342$, $\theta = ?$ Two possible values for θ exist since the cosine function is negative in the second and third quadrants. With the calculator in radian mode:

$$\cos^{-1}(-0.342) = 1.919841.$$

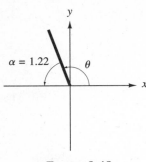

FIGURE 6.40

Rounding to three significant digits, $\theta = 1.92$. From Table 6.6 we can see that θ is a second-quadrant angle.

To determine the third-quadrant angle we must determine the reference angle α (Figure 6.40).

$$\alpha = \pi - \theta$$
$$= \pi - 1.92$$
$$= 1.22$$

A third-quadrant angle is found by finding the sum of π and the reference angle (Figure 6.41).

$$\theta = \pi + \alpha$$
$$\theta = \pi + 1.22$$
$$= 4.36$$

Therefore, the second- and third-quadrant angles that make $\cos \theta = -0.342$ a true statement are $\theta = 1.92$ and $\theta = 4.36$. ◆

Use Table 6.7 to determine an angle in radian measure using the reference angle.

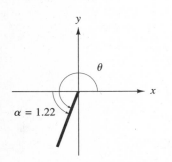

FIGURE 6.41

TABLE 6.7   Rules to Determine θ
(θ in radians)

Quadrant	Rule
1	$\theta = \alpha$
2	$\theta = \pi - \alpha$
3	$\theta = \pi + \alpha$
4	$\theta = 2\pi - \alpha$

6.6 Exercises

In Exercises 1–14, change the degree measures to radian measures. In each case give the answer in exact form. Draw each angle in standard position.

1. 15°	**2.** 30°	**3.** 45°	**4.** 60°
5. 75°	**6.** 120°	**7.** 135°	**8.** 150°
9. 210°	**10.** 240°	**11.** −225°	**12.** −60°
13. 300°	**14.** 315°		

In Exercises 15–24, change the degree measures to radian measures. In each case give the answer in approximate form to the nearest hundredth.

15. 330°	**16.** 22.5°	**17.** 212.5°	**18.** 4°
19. −120°	**20.** −60°	**21.** 39°	**22.** 190°
23. 290°	**24.** 100°		

In Exercises 25–44, change the radian measures to degree measures. If the answer is not exact, round the answer using the rules for significant digits. Draw each angle in standard position. (Use calculator only for Exercises 40–44.)

25. $\dfrac{\pi}{8}$

26. $\dfrac{\pi}{6}$

27. $\dfrac{\pi}{4}$

28. $\dfrac{\pi}{3}$

29. $\dfrac{5\pi}{12}$

30. $\dfrac{2\pi}{3}$

31. $\dfrac{3\pi}{4}$

32. $\dfrac{5\pi}{6}$

33. $\dfrac{7\pi}{6}$

34. $\dfrac{4\pi}{3}$

35. $\dfrac{5\pi}{4}$

36. $-\dfrac{\pi}{3}$

37. $\dfrac{5\pi}{3}$

38. $\dfrac{7\pi}{4}$

39. $\dfrac{11\pi}{6}$

40. 13

41. 1.74

42. 6.282

43. 1.00

44. 143

In Exercises 45–58, determine the value of the trigonometric function.

45. sin 1.32

46. cos 1.51

47. tan 1.05

48. sin 2.34

49. cos 2.44

50. tan 2.67

51. sin 3.45

52. cos 3.75

53. tan 4.25

54. sin 5.32

55. cos 6.00

56. tan 6.00

57. sin 3.42

58. cos 3.42

In Exercises 59–74, determine θ for $0 \le \theta < 2\pi$.

59. $\sin \theta = 0.969$

60. $\cos \theta = 0.61$

61. $\tan \theta = 1.74$

62. $\sin \theta = 0.718$

63. $\cos \theta = -0.764$

64. $\tan \theta = -0.510$

65. $\sin \theta = -0.304$

66. $\cos \theta = -0.821$

67. $\tan \theta = 2.01$

68. $\sin \theta = -0.821$

69. $\cos \theta = 0.960$

70. $\tan \theta = -0.291$

71. $\sin \theta = -0.275$

72. $\cos \theta = -0.961$

73. $\tan \theta = -1.732$

74. $\sin \theta = -0.866$

75. Through how many radians does the minute hand of a clock rotate in 15 minutes? Convert to degrees.

76. On a flywheel with a 70-mm radius, how long is an arc subtended by a central angle of 2.10 rad? See the sketch.

77. A telescope is designed so that it can distinguish two points 1.0 mm apart at a distance of 1.6 km. The angular separation of the two points at that distance is 6.25×10^{-6} rad. What is the angular separation in degrees?

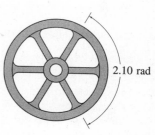

2.10 rad

78. Suppose that a pendulum of length 250 mm swings through an arc of length 150 mm. (See the drawing.) Through how many degrees does the pendulum swing?

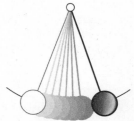

150 mm

79. An engineer is to design a cloverleaf for an exit from an interstate highway. A part of the cloverleaf can be thought of as an arc of a circle with a radius of 420 m. The central angle of this arc is 135°. Determine the length of this part of the cloverleaf. See the sketch.

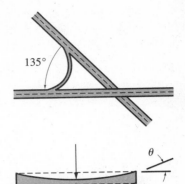

80. For a simple beam acted on by a concentrated load, the angle of deflection in radians in given by $\theta = \dfrac{P\ell^2}{16EI}$. Determine θ in (a) radians and (b) degrees when $P = 46{,}000$ N, $\ell = 0.800$ m, and $EI = 10^{14}$ Nm^2. See the drawing.

81. While rounding a curve, one front wheel of a truck rotates 286° 40′ and a second front wheel rotates 5.00 rad. Which wheel is closer to the inside of the curve? (*Hint:* Outside wheels rotate more.)

82. The area of a parallelogram with sides a and b and the included angle θ can be found using the formula $A = ab \sin \theta$. Determine the angle between the two sides if $A = 36.3$ in.^{2}, $a = 8.40$ in., and $b = 6.40$ in.

83. Rotational speeds are often given in rpm (revolutions per minute), but for many calculational purposes it is necessary to know the rotational speed in radians per second. Convert 6.000×10^3 rpm to radians per second (rad/s). Remember, one revolution is equivalent to 360° or 2π rad.

84. The earth makes one complete revolution in a day. Through how many radians does it rotate in a day? If average angular velocity is given by angular rotation divided by the time of the rotation, what is the earth's angular velocity in rad/s?

85. If the average diameter of the spool the string is wound around on a yo-yo is $\frac{1}{2}$ in. and the yo-yo winds up on 24 in. of string, how many revolutions does it turn?

86. The platen (paper roller) on a computer printer is $1\frac{1}{2}$ in. in diameter. How many rotations of the platen are necessary to exactly advance an 11 in. sheet of paper. Give the answer in revolutions, in radians, and in degrees.

87. The bicycle chain sprocket on the rear wheel has a radius of 75 mm and the one on the pedals has a radius of 185 mm. If the pedals turn through 4.5 rad, does more chain pass over one sprocket wheel than the other? Through how many radians does the rear sprocket wheel turn?

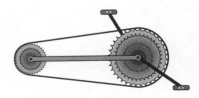

88. If a 32.00-in.-wide refrigerator door swings through a 3.00-in. arc before the light comes on, through how many radians does it swing? How far from the hinge does the light switch need to be located if it takes a 0.25-in. motion to turn on?

6.7

Industrial Applications

Technologists use trigonometry to solve many problems. We shall illustrate just a few from various fields.

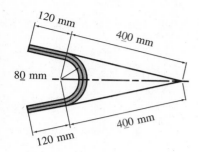

FIGURE 6.42

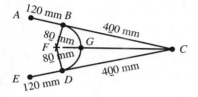

FIGURE 6.43

EXAMPLE 1 Application: Determining Distances Using Radian Measure

Determine the total length of steel rod needed to make the curved bracket which is sketched in Figure 6.42.

Solution

$$\text{Length of rod needed} = \overline{AB} + \overline{ED} + \overset{\frown}{BGD}$$

$$\overline{AB} = 120 \text{ mm} \qquad \textbf{Given}$$

$$\overline{ED} = 120 \text{ mm} \qquad \textbf{Given}$$

Need to determine the length of $\overset{\frown}{BGD}$. This can be done using the formula $s = r\theta$. $\overline{AC}$ and $\overline{EC}$ are tangent to $\overset{\frown}{BGD}$, as shown in Figure 6.43.

$$\overline{FB} \perp \overline{AC} \text{ and } \overline{FD} \perp \overline{EC} \qquad \textbf{Radii are} \perp \textbf{to tangent to curve}$$

$$\triangle CDF \cong \triangle CBF \qquad \textbf{SAS}$$

$$\angle BFD = 2\angle CFD \qquad \textbf{Congruent triangles}$$

Using right triangle trigonometry, we have:

$$\tan \angle CFD = \frac{400 \text{ mm}}{80 \text{ mm}},$$

or

$$\angle CFD = \tan^{-1} \frac{400 \text{ mm}}{80 \text{ mm}}$$

$$= 1.373$$

$$\angle BFD = 2(1.373) \quad \text{or} \quad 2.746.$$

Substituting in the formula $s = r\theta$, we have:

$$\text{length of } BGD = (80 \text{ mm})(2.746)$$

$$= 219.7 \text{ mm}$$

$$\text{length of rod} = 219.7 \text{ mm} + 120 \text{ mm} + 120 \text{ mm}$$

$$= 460 \text{ mm}. \quad \blacklozenge$$

It is sometimes necessary to calculate the area of circular sectors as in Figure 6.44. The area of a circle is $A = \pi r^2$. Knowing that the central angle of a circle measures 2π rad, we can express the area of the circle as $A = \frac{1}{2}(2\pi)r^2$.

The central angle of a semicircle measures π rad or half the central angle of a circle. The area of a semicircle is half the area of a circle or:

$$\frac{\frac{1}{2}(2\pi)r^2}{2} = \frac{1}{2}(\pi)r^2.$$

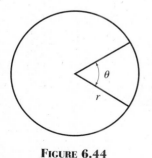

FIGURE 6.44

The central angle of a quarter circle measures $\dfrac{\pi}{2}$ or one-fourth the central angle of a circle. The area of a quarter circle is one-fourth the area of a circle or:

$$\frac{\dfrac{1}{2}(2\pi)r^2}{4} = \frac{1}{2}\left(\frac{\pi}{2}\right)r^2.$$

In each case, the area of the portion of the circle is one-half the central angle (expressed in radians) times r^2.

AREA OF CIRCULAR SECTOR

$$A = \frac{1}{2}\theta r^2, \text{ where } \theta \text{ is measured in radians.}$$

EXAMPLE 2   **Application: Determine the Area of a Circular Sector**

Determine the outside surface of a cone formed from a circle with a diameter of 6.00 m. The curve is formed by cutting a circular sector with a central angle of 1.50 rad from the circle and joining the cut edges of the remaining sector, as shown in Figure 6.45.

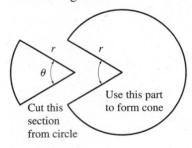

Solution

$$A = \frac{1}{2}\theta r^2 \qquad\qquad \textbf{Area of a sector of a circle}$$

$$r = 3.00 \text{ m} \qquad\qquad \tfrac{1}{2}\textbf{ diameter}$$

$$\theta = (2\pi - 1.50) \qquad\qquad \textbf{Central angle of circular sector}$$
$$\qquad\qquad\qquad\qquad\qquad\qquad \textbf{forming the cone}$$

$$A = \frac{1}{2}(2\pi - 1.50)(3.00 \text{ m})^2$$

$$= 21.5 \text{ m}^2.$$

The outside surface area of the cone is 21.5 m². ◆

Angular and Linear Velocity Applications

The time rate of change in angular displacement is called the **angular velocity.** Angular velocity is usually represented by the Greek letter ω (omega). Thus, if an object rotates through an angle θ in a time t, its average angular velocity is given by $\omega = \theta t$. Using the formula $s = r\theta$, we can determine the distance along a curve. If we divide each member by time t, we have:

$$\frac{s}{t} = \frac{r\theta}{t} \quad \text{or} \quad r\frac{\theta}{t},$$

where:

$$\frac{s}{t} \text{ is } \frac{\text{distance}}{\text{unit of time}} = \text{velocity } v,$$

and

$$\frac{\theta}{t} \text{ is } \frac{\text{angular rotation}}{\text{unit of time}} = \text{angular velocity } \omega.$$

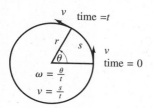

FIGURE 6.46

If we replace $\frac{s}{t}$ with v and $\frac{\theta}{t}$ with ω in the equation $\frac{s}{t} = \frac{r\theta}{t}$ we have $v = r\omega$. This equation states that the velocity of a point along a circular path is equal to the product of the radius of the path and the angular velocity ω (in radians per unit of time). When working with the equation $v = r\omega$, remember that θ must be in radians. Figure 6.46 illustrates the relation between angular and linear velocity.

VELOCITY ALONG CIRCULAR PATH

$v = r\omega$, where ω is measured in radians per unit of time.

EXAMPLE 3   **Application: Determining Linear Velocity**

A wheel is rotating at 7.0 rad/s. The wheel has a 14 in. radius. What is the velocity of a point on its rim in inches per second?

Solution   We know that $\omega = 7.0$ rad/s and $r = 14$ in. Substitute in the formula: $v = r\omega$.

$$v = (14 \text{ in.})(7.0 \text{ rad/s})$$
$$= 98 \text{ in./s.}$$

A point on the rim of a wheel is moving at the rate of 98 in./s. (Remember, a radian is a dimensionless quantity.) ◆

EXAMPLE 4   **Application: Determining Angular and Linear Velocity**

A pulley 120 mm in diameter turns at 500 revolutions per minute (rpm). Determine its angular velocity in radians per second as well as the velocity of the belt being driven by the pulley.

Solution   We know that the angular velocity is 500 rpm. To change from rpm to rad/s, use the conversion factors $\left(\dfrac{1\ \text{min}}{60\ \text{s}}\right)$ and $\left(\dfrac{2\pi\ \text{rad}}{1\ \text{rev}}\right)$.

Thus,

$$\omega = 500\ \frac{\text{rev}}{\text{min}}\left(\frac{1\ \text{min}}{60\ \text{s}}\right)\left(\frac{2\pi\ \text{rad}}{1\ \text{rev}}\right)$$

$$= 52.36\ \frac{\text{rad}}{\text{s}}.$$

The angular velocity is 52 rad/s.

The velocity of the belt is the same as the velocity of a point on the rim of the pulley. Therefore, use the equation from Example 3.

$$v = r\omega$$

$$= (60\ \text{mm})\left(52.36\ \frac{\text{rad}}{\text{s}}\right)$$

$$= 3142\ \frac{\text{rad mm}}{\text{s}}$$

Since radians are dimensionless, do not include rad in the answer. Therefore, the velocity of the belt is $v = 3100$ mm/s. ◆

EXAMPLE 5   Application: Determine Angle and Rotation

If the cord of the lawn mower in the chapter opener is wrapped around a 3.0 in. diameter pulley and 24 in. of cord are pulled out in 1.0s, determine the following.

a.   Through how many degrees does the crankshaft rotate?
b.   What is the average rpm? (See Figure 6.47.)

Figure 6.47

Solution

a.   The arc length of the circumference of the pulley that passes the point where the cord leaves the pulley is the same as the length of cord that is pulled out. Since $s = r\theta$,

$$\theta = \frac{s}{r}$$

$$= \frac{24\ \text{in.}}{1.5\ \text{in.}} \qquad \textbf{Substituting}$$

$$= 16.0\ \text{rad}$$

$$16.0\ \text{rad}\left(\frac{180°}{\pi\ \text{rad}}\right) = 917° \qquad \textbf{Converting radians to degrees}$$

The pulley and, therefore, the crankshaft is rotated 16.0 rad or 920.°.

b. To determine average rpm of the crankshaft, we must determine the average angular velocity.

$$\omega = \frac{\theta}{t}$$

$$\omega = \left(\frac{16.0 \text{ rad}}{1.0 \text{ s}}\right)\left(\frac{1 \text{ rev}}{2\pi \text{ rad}}\right)\left(\frac{60 \text{ s}}{1 \text{ min}}\right) = 150 \text{ rev/min}.$$

Therefore, when 24 in. of cord is pulled out, in 1.0 s the pulley and the crankshaft turn through an angle of 920° at an average angular velocity of 150 rpm. By increasing the length of the cord, we can increase ω and ensure that the motor turns. ◆

A BASIC computer program which may be used to solve this problem is found on the tutorial disk.

Compass Applications

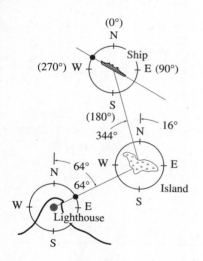

FIGURE 6.48

In navigation or surveying, direction may be specified by stating either the bearing or the heading. The **bearing** of one object to another object is the angle between a north-south line and a line through the center of the objects. The mariner's compass commonly is used to measure the angle between the two lines. The compass assumes a circle of 360° with 0° (also 360°) at north and the angular measure increasing in a clockwise direction making east lie at 90°, south lie at 180°, and west lie at 270°. Compare this with method of angular measure based on the x, y-axis system that is used conventionally for most other purposes. In Figure 6.48, the bearing of the ship from the island is found by drawing a north-south line through the center of the island and then drawing a line through the center of the island and the center of the ship. The angle between the two lines is approximately 16°. Thus, the bearing of the ship from the island is 344°.

Also in Figure 6.48, the angle between a north-south line through the lighthouse and a line through the island and the lighthouse is 64°. Thus, the bearing of the island from the lighthouse is 64°.

The **heading** is the direction of travel of a ship or aircraft according to the mariner's compass. In Figure 6.49, the heading of the ship is 300°.

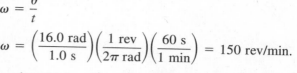

FIGURE 6.49

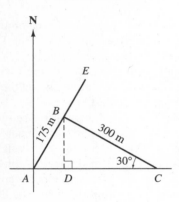

FIGURE 6.50

FIGURE 6.51

EXAMPLE 6   **Application: Determining Angles Using Trigonometric Functions**

A sailboat is sailing on the course shown in Figure 6.50.

a.   Determine the heading of the sailboat.
b.   Determine the bearing of the sailboat from point C.

Solution

a.   We can determine the heading of the sailboat by determining the measure of $\angle NAB$. Begin by finding the measure of $\angle BAC$ and subtracting it from 90°. To find $\angle BAC$, draw a perpendicular line from point B to AC, as shown in Figure 6.51. Since BD is the shorter leg of the triangle BDC, BD must be one-half the hypotenuse (a theorem in geometry). Thus $BD = 150$ m (or $\sin 30° = \dfrac{BD}{300}$, $BD = 300 \sin 30°$).

$$\sin \angle BAD = \frac{150 \text{ mm}}{175 \text{ mm}} \text{ or } 0.85714$$

$$\angle BAD = 59.0° \qquad \text{Nearest tenth}$$

$$\angle NAB = 90° - 59.0° \quad \text{or} \quad 31.0°.$$

Therefore, the heading of the sailboat is 31.0°.

b.   To determine the bearing of the sailboat from point C, draw a north-south line through point C and determine the clockwise angle of the north-south line with line BC. Since the north-south line is perpendicular to the line AC, the angle is 270° plus 30° or 300°. Thus, the bearing of the sailboat from point C is 300°.   ◆

6.7 Exercises

1.   The pendulum of a grandfather clock is 1 m long and oscillates through an angle that subtends an arc of 0.349 m. Determine the angle in degrees and radians that the pendulum makes in one complete swing.

2.   A swing for a baby is made with a windup motor on a frame. The seat of the swing is suspended from the top of the frame with rods that are 3.0 ft long.
    a.   Determine the length of the arc through which the swing moves if it moves though an angle of 55°.
    b.   If the windup motor runs for 15 min and the swing takes 5 s to go through one complete motion (back and forth), how many mi will the baby, seated in the swing, travel in 15 min?

3. When an object is held away from the eye, it subtends an angle. The angle may be determined by using the formula $\theta = \dfrac{s}{r}$. Determine the size of the subtended angle in radians and in degrees when a blood cell of diameter 7.5 μm is 25 cm away from the eye.

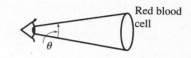

4. A nautical mile is the length of the arc on the surface of the earth subtended by an angle of 1 min whose vertex is at the center of the earth. Determine the number of ft in a nautical mi using 3960 statute mi as the radius of the earth. Recall that there are 5280 ft in a statute mi.

5. Resolving power (RP) refers to the ability of an optical instrument to distinguish between two bodies very close together. If the resolving power is low, small objects in a microscope are often blurred together to look like one. The resolving power equation for white light is:

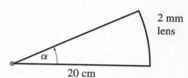

$$RP = \frac{2.0 \times 10^{-7}}{\sin (0.5\alpha)},$$

where α is the angle subtended by the lens and the object.
   a. Find the resolving power when the object is 20 cm away from a 2 mm lens.
   b. Find the resolving power when the object is 40 cm away from a 2 mm lens.
   c. Find the resolving power when the object is 20 cm away from a 4 mm lens.
   d. Comment on the design of a microscope that will enhance the clarity of the view.

6. For red light, the resolving power is given by:

$$RP = \frac{1.6 \times 10^{-7}}{\sin (0.5\alpha)}.$$

   For blue light the resolving power is given by:

$$RP = \frac{2.6 \times 10^{-7}}{\sin (0.5\alpha)}.$$

   a. Which color resolves better?
   b. Determine the resolving power for red and blue light when the 4 mm lense is 20 cm away from the object.

7. A highway department is designing a new exit for an expressway. The design is shown in the sketch. Determine the length of roadway needed to create the new exit.

8. For the road in Exercise 7, determine the number of cubic meters of concrete needed for the curved section. The thickness of the concrete is 350 mm and the width of the road is 9.60 m.

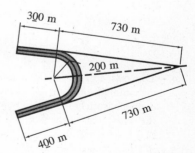

9. The Brighton Highway Department is replacing the curved part of the curbing on one side of the intersection of two streets. The streets intersect at an angle of 55° and the length of the radius describing the arc is 25 ft (see the drawing). Determine the cost of replacing the curb if the curb costs $45 per ft.

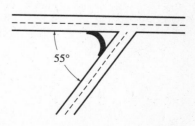

10. A circular sector whose central angle is 240° is cut from a circular piece of sheet metal of diameter 180 mm. A cone is then formed by bringing the two straight edges of the sector together. What is the lateral surface area of the cone?

11. A bullet is discovered in the wall of a room with an entry point 2.5 m above the floor. The path of the bullet in the wall is inclined upward with an 18° angle of elevation. Determine how far from the wall the gun was fired with the given conditions.
    a. The gun was fired from the floor.
    b. The gun was fired from 1.2 m above the floor.

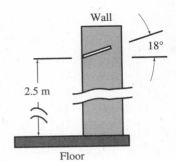

12. The diameter of the earth is approximately 7920 mi when measured across the equator.
    a. What is the linear velocity of the earth on the equator?
    b. What is the linear velocity of the earth at 42° latitude? (*Hint*: The diameter of rotation is less than 7920 miles! Look at the diagram and note that the equatorial radius is parallel to the radius at the 42° latitudinal line.)
    c. What is the linear velocity of the earth at the North Pole?

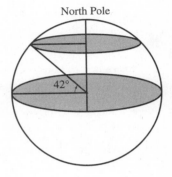

13. A satellite in a circular orbit 800 mi above the earth makes one complete orbit every 90.0 min. What is its linear speed? Use 4000 mi for the length of the radius of the earth.

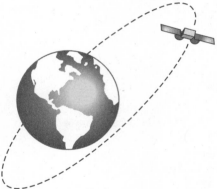

14. The town council is planning a baseball field. The two major expenses are seeding the outfield and erecting a fence around the curved portion of the outfield. The distance from home plate to the outer edge of the infield is 90 ft. The angle between the outer edge of the first and third baselines is 90°, and the radius of the sector is 350 ft. (See the diagram.)
    a. Determine the area of the outfield.
    b. Determine the number of feet of fencing needed for the curved portion of the outfield.

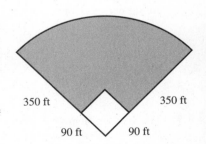

15. The 260-ft Penn Central Railroad Bridge crossing the Genesee River at Rochester Harbor is a swing bridge built on a pier in the middle of the river. The bridge is supported by six $2\frac{1}{2}$-ft-diameter vertical wheels that, as the bridge turns, roll around the circumference of a horizontal 40-ft-diameter plate mounted on the support pier. How many revolutions do the support wheels make as the bridge rotates 90° from full open to full close?

16. Engineers often need to know the tip speeds of rotating objects like propellers or the rim speeds of turbines. Usually these devices are rated by only their angular velocity. A certain cargo plane, for example, has propellers which are 16 ft in diameter. What must be the propeller-tip speed in feet per second when the shaft is turning at 1200 rpm?

17. A steam turbine rotor is 6.00 ft in radius and turns at 1200 rpm. How many miles does a point on its rim travel in a 24-hour day? (*Hint:* distance = linear speed × time.)

18. A wind-driven electrical generator has a 24.0-ft-diameter propeller.
   a. Find the angular velocity in radians per second at which the tip of the propeller will reach the speed of sound (approximately 1087 ft/s).
   b. Express the angular velocity in revolutions per second and rpm.

19. A satellite is orbiting the earth with a linear velocity of 15,742 mph. The orbit is approximately a circle with radius of 5265 mi. Determine the angular velocity of the satellite in revolutions per hour.

20. A spacecraft is circling the moon once every 1.45 hr. If the altitude of the craft is constant at 120 mi and the radius of the moon is 1080 mi, compute the linear velocity of the spacecraft.

21. The wheels of a scooter have a 200-mm diameter. If the wheels are turning at the rate of 28 rpm, compute the distance in meters the scooter will travel in 3 min.

22. A seesaw rotates through a 40° angle. If the end of the seesaw travels 1.0 m, how long is it?

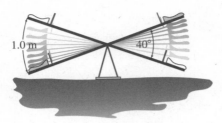

23. A 10.0 in. phonograph record plays at a constant rate of $33\frac{1}{3}$ rpm.

    a. What is the linear speed on the outer edge of the record? What is the angular velocity on the outer edge of the record?

    b. What is the linear speed at the distance of 3.0 in. from the center of the record? What is the angular velocity at this point?

24. A car is traveling at 40 mph. Its wheels have a 14-in. radius. Find the angle in radians through which a wheel rotates in 10 s.

25. A small pulley 60 mm in diameter is connected by a belt to a larger pulley 150 mm in diameter as shown in the drawing. The small pulley is turning at 120 rpm.

    a. Determine the angular velocity of the small pulley in radians per second.

    b. Determine the linear velocity of the rim of the small pulley.

    c. What is the linear velocity of the rim of the small pulley? Explain.

    d. Determine the angular velocity of the large pulley in radians per second.

    e. How many rpm is the large pulley turning?

26. The wheels of a bicycle have a 26-in. diameter. When the bicycle is being ridden so that the wheels rotate at 12 rpm, how far will the bike travel in 1 min?

27. To start an outboard motor, a rope, which is wound on an 8.0 in. diameter drum, is given a strong pull. At what linear speed is the rope pulled if the engine needs to be rotated at 600 rpm in order to start?

28. A shaft 3 in. in radius is being turned down on a lathe at 200 rpm. The cutting tool removes a continuous ribbon of metal. How many linear feet of the metal ribbon will be cut off in 5 min?

29. A ship is sailing on the course shown in the illustration.

    a. Determine the heading of the ship (∠NAB).

    b. Determine AC.

    c. Determine the measure of ∠ABC.

    d. Determine the bearing of line BC.

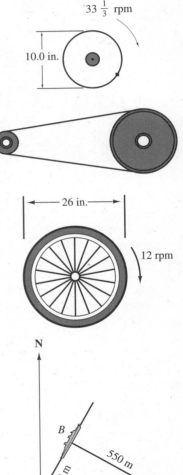

30. Orienteering is the competitive sport of finding one's way across country using a map and a compass. An orienteer finds herself on the bank of a stream that the map shows to run relatively straight and to make an angle of 30° with the river it intersects. A church steeple can be seen bearing 25° from the orienteer's position. (See the sketch.) If the church is 1500 m from the stream intersection, determine the following distances.
  a. From the orienteer to the church.
  b. From the orienteer to the stream intersection

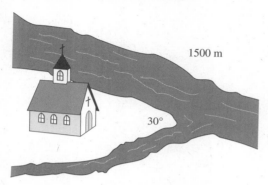

1500 m

30°

31. An orienteer finds himself on an east-west fireroad through the forest. He can see a fire spotter's tower at a bearing of 315°. He also can see a radio antenna that he knows is 2000 m from the tower on a bearing of 210° from the tower. (See the drawing.) How far is he from the following?
  a. The tower
  b. The antenna

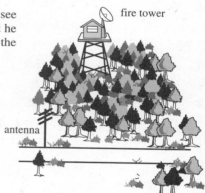

fire tower

antenna

Review Exercises

In Exercises 1–5, change the degree measures to radian measures. In each case give the answer in exact form and in approximate form using the rules for significant digits.

  1. 70°   **2.** 160°   **3.** 220°   **4.** 310°   **5.** −80°

In Exercises 6–10, change the radian measures to degree measures. If the answer is not exact, express the answer using the rules for significant digits. (Use calculator for 8 and 9 only)

  6. $\dfrac{\pi}{12}$ **7.** $\dfrac{11\pi}{12}$ **8.** 1.45   **9.** 2.0   **10.** $-\dfrac{\pi}{4}$

In Exercises 11–15, determine the six trigonometric functions of θ if the terminal side passes through the given point. Give the answer in exact form.

11. $(5, 7)$ **12.** $(-3, 2)$ **13.** $(-\sqrt{2}, -\sqrt{2})$ **14.** $(4, -3)$

15. (π, π)

In Exercises 16–20, use the given trigonometric function to determine the indicated trigonometric functions. Give each answer in exact form.

16. $\sin \theta = \dfrac{4}{5}$; determine $\cot \theta$ and $\cos \theta$.

17. $\tan \theta = -\dfrac{\sqrt{3}}{2}$; determine $\cos \theta$ and $\csc \theta$.

18. $\cos \theta = 0.5$; determine $\sin \theta$ and $\tan \theta$.

19. $\cot \theta = \dfrac{1}{7}$; determine $\csc \theta$ and $\cos \theta$.

20. $\csc \theta = \dfrac{\sqrt{5}}{-2}$; determine $\cos \theta$ and $\cot \theta$.

In Exercises 21–30, compute the functional value using the rules for significant digits.

21. $\sin 18.6°$ **22.** $\cos 96.4°$ **23.** $\tan 1.72$ **24.** $\csc 113°$

25. $\sec 195.4°$ **26.** $\cot 286°$ **27.** $\cos 3.45$ **28.** $\tan 142°$

29. $\sin 2.57$ **30.** $\cos 194.2°$

In Exercises 31–40, determine the angle measure in degrees and in radians $(0 \le \theta < 360°, 0 \le \theta < 2\pi)$.

31. $\sin \theta = 0.23456$ **32.** $\cos \theta = -0.39875$ **33.** $\tan \theta = 0.24008$ **34.** $\csc \theta = -4.93707$

35. $\sec \theta = -3.86370$ **36.** $\cot \theta = -1.45501$ **37.** $\cos \theta = 0.76604$ **38.** $\sin \theta = -0.90631$

39. $\sin \theta = -1.00000$ **40.** $\tan \theta = 1.00000$

In Exercises 41–45, solve each right triangle ABC for the missing parts. Angle C is the right angle.

41. $A = 38.5°$, $c = 19.4$ mm   **42.** $A = 48.7°$, $b = 3.92$ in.   **43.** $B = 24.7°$, $c = 24.2$ in.

44. $B = 68.9°$, $a = 3.42$ mm   **45.** $A = 0.75$, $a = 14.5$ m

46. The most efficient operating angle for a conveyor used to elevate crushed ore in a smelter is 30°. If the ore is to be elevated 215 ft, what length conveyor is needed?

47. A surveyor at point P is trying to determine his position between two buildings of known height. (See the drawing.)
   a. Determine his distance from building A.
   b. What is the distance between the two buildings?

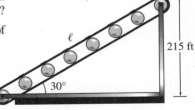

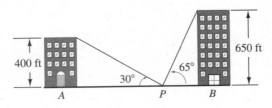

48. Determine the length of water pipe needed to run from point P to point Q in the illustration.

49. With respect to the harbor master's observation post, the resultant velocity v of a ship (in knots) is given by $v = \sqrt{67 - 28 \cos \theta}$. For $v = 9$, determine θ.

50. The large hand on a wall clock is 19 in. long.
  a. Through what angle (in both radians and degrees) does the hand move in 20 min?
  b. How many inches does its tip move in 20 min?

51. Suppose a bike wheel has a radius of 13.0 in. If the bike is rolling at 10.0 mph, through what angle does a spoke turn in 1.00 min?

52. If the bike wheel in Exercise 51 made 200 revolutions in a minute, how fast was the bike traveling?

53. One leg of a right triangle is 92 mm long. The tangent of the angle between the other leg and the hypotenuse is 1.56. How long is the other leg?

54. The angle of depression of a car from a balloon is 67.5°. If the balloon is 3000 ft from the car, how high is the balloon?

55. The angle of elevation to the top of a tower from the tip of its shadow (along level ground) is $\dfrac{\pi}{6}$. If the tower is 60 ft tall, how long is the shadow?

56. The angle of elevation to a mountain peak from point A is 65.0°. The angle of elevation to the same mountain peak from point B, 100 ft farther away from the peak, is 61.0°. How high is the mountain, to the nearest foot?

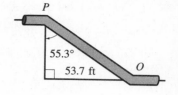

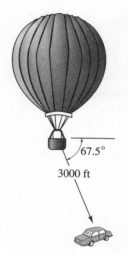

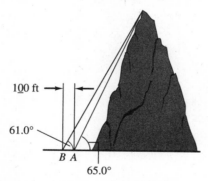

57. The distance a spring is displaced during simple harmonic motion is given by $d = 8 \cos (100\ t)°$. Determine the displacement, in centimeters, for the following.
  a. $t = 0\ s$ **b.** $t = 1.3\ s$ **c.** $t = 3.5\ s$

58. In a certain circuit containing a capacitance the voltage $V = 45 \sin \left(\omega t - \dfrac{\pi}{2} \right)$. Find V for the following.
  a. $\omega t = 3\dfrac{\pi}{4}$ **b.** $\omega t = 0$.

59. The cross-section of the rivet in the illustration is in the shape of a segment of a circle. The area of a circular segment is given by $A = \left(\dfrac{r^2}{2}\right)(\theta - \sin\theta)$, where θ is in radians. Determine the area of the bolt head if $r = 2$ cm and $\theta = 115°$.

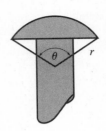

60. Three concentric circles have radii of 2.0, 3.5, and 4.0 m. What is the area of the region between the inner and the middle circles bounded by radii forming a central angle of 78.0°?

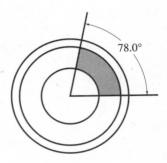

61. A pendulum 1.3 m long oscillates through an angle that subtends an arc length of 3.2 m. Determine the angle and express the results in degrees and radians.

62. Determine the length of an arc through which a pendulum 1.3 m long swings if it swings through a total angle of 10°.

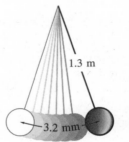

63. A conical tent as illustrated is made from a circular piece of canvas 18.0 ft in diameter, with a central angle of 150.0° removed. What is the surface area of the tent?

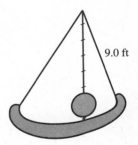

64. Determine the total length of steel rod needed to meet the conditions established in the illustration.

65. A formula for the area A of a parallelogram with sides a and b and included angle θ is $A = ab \sin\theta$. Find the area of a piece of plastic, where $a = 1.5$ m, $b = 0.90$ m, and $\theta = 118°$.

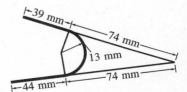

66. A surveyor measures the angle of elevation to the top of a building to be 49°. She then moves 75 ft closer to the building and measures the angle of elevation to be 65°. How high is the building? (See the drawing.)

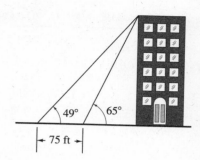

67. A ship is sailing on the course shown in the illustration.
   a. Determine the heading of a the ship (angle *NAB*).
   b. Determine the bearing of the line *BC*.

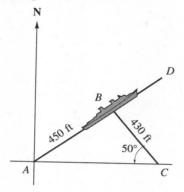

68. The area of a segment of a circle can be found using the formula $A = \frac{1}{2}r^2(\theta - \sin\theta)$ (θ must be in radians). A segment of a circle is the region bounded by a chord and its related arc (see the sketch). Use this formula to determine the number of gallons of oil in a tank of radius 1.5 m and length 3.0 m. The depth of oil in the tank is 1.2 m (see the diagram).

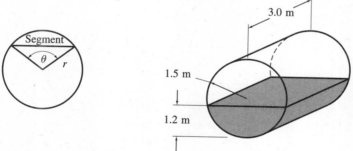

69. If a wheel with a 100-mm radius must have five holes equally spaced with their centers a distance of 40 mm from the center. Determine the *x* and *y* coordinates of each hole.

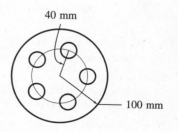

70. One of the formulas used to calculate the value of cos *x* comes from calculus:

$$\cos x = 1 - \frac{x^2}{2} + \frac{x^4}{2 \cdot 3 \cdot 4} - \frac{x^6}{2 \cdot 3 \cdot 4 \cdot 5 \cdot 6} + \cdots$$

The formula continues indefinitely, increasing in accuracy. The angle *x* must be in radians. Using your calculator, check the formula results with the four terms given against the calculator value for cos *x* for the following.

   a. $x = \dfrac{\pi}{3}$ **b.** $x = \dfrac{\pi}{4}$

✍ Writing About Mathematics

1. A classmate has difficulty understanding the reference triangle and the reference angle. Write a short explanation of each including examples of the proper use of the reference triangle and the reference angle.

2. Write a short paper explaining the difference between linear and angular velocity. Include in your discussion one example that illustrates the use of each type of velocity.

Chapter Test

In Exercises 1 and 2, change the degree measures to radian measures. In each case give the answer in exact form and in approximate form to the nearest hundredth.

1. $145°$

2. $55°$

In Exercises 3 and 4, change the radian measure to degree measure. If the answer is not exact, express it to the nearest hundredth.

3. $\dfrac{8\pi}{12}$

4. 1.5

In Exercises 5 and 6, use the given trigonometric function to determine the indicated trigonometric functions. Give each answer in exact form.

5. $\cos\theta = \dfrac{3}{5}$; determine $\tan\theta$ and $\csc\theta$.

6. $\cos\theta = \sqrt{3}$; determine $\sin\theta$ and $\sec\theta$.

In Exercises 7–12, compute the functional value.

7. $\sin 33.5°$

8. $\cos 1.59$

9. $\tan 128°$

10. $\cot 210°$

11. $\csc 2.500$

12. $\sec 310.0°$

In Exercises 13–15 determine, θ for $0° \le \theta < 360°$.

13. $\sin\theta = 0.57$

14. $\tan\theta = -0.839$

15. $\sec\theta = -1.414$

16. For right triangle ABC, determine the measure of angle A, given that $a = 14$ in. and $c = 18$ in.

17. For right triangle ABC, determine the measure of side a, given $\angle A = 28°$ and $b = 42$ mm.

18. The angle of elevation from point x to the top of a building is $45°$. If point x is 45 ft from the base of the building, how high is the building?

19. A circular piece of canvas 6.0 m in diameter is being used to make a conical tent. To make the tent, a pie-shaped piece with a central angle of $120°$ is cut from the piece of canvas. What is the surface area of the tent?

20. A spacecraft is circling Saturn once every 5.00 hours. If the altitude of the craft is constant at 125 mi and the radius of Saturn is 37,550 mi, compute the linear velocity of the spacecraft.

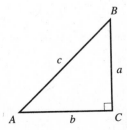

Vectors and Oblique Triangles

7

Your car is stuck in the mud. You have pushed and pulled, but it is still stuck. Now you attach one end of a rope to a tree about 110 feet ahead of the car and the other end to the car itself so that the rope is tightly stretched. You apply a relatively modest force at the center of the rope, pulling it two feet to the side. This action frees your car. How? (See Example 7 in Section 7.3.)

7.1

Introduction to Vectors

When studying quantities in technical areas, we soon discover that some kinds of quantities can be described by means of a single number. The number of screws in a bracket, the length of a bridge, the height of a building, and the value of a trigonometric ratio are all examples of this type of quantity. In fact, most quantities that we have encountered are examples of this type. Quantities that are undirected and can be fully described by a number are called **scalars.**

There is another type of quantity that possesses not only magnitude, but also direction. In other words, these quantities are *directed* magnitudes, which we called **vectors.** Airspeed coupled with a specific angular heading is an example of this second kind of quantity. Whenever the airspeed and the direction are specified, the specific vector term is called *velocity*.

> **DEFINITION 7.1**
> Quantities having both direction and magnitude are called **vector quantities.**

A vector quantity is represented in the text in boldface type—**V.** For the symbol **OP**, the letter **O** indicates the initial point, or tail, of the vector, and the letter **P** indicates the terminal point, or head, of the vector.

A vector quantity is represented in a diagram by means of an arrow whose length (to a suitable scale) indicates the **magnitude,** or size, of the vector quantity and whose head indicates the **direction** of the vector. Say the vector **OP** has a magnitude of 80 lb. This is indicated with the symbol $|\mathbf{OP}| = 80$ lb, or **OP** = 80 lb. Look at Figure 7.1, which shows that the direction of the vector is 125° with respect to the positive x-axis.

Vectors may be drawn from any point in the plane, as is shown in Figure 7.2b. However, for calculation purposes, they are most commonly drawn in standard

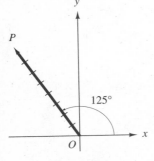

FIGURE 7.1

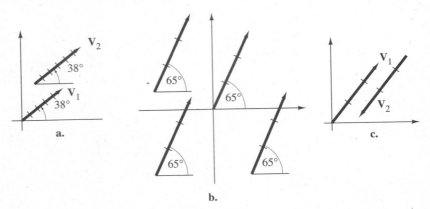

FIGURE 7.2

position; that is, with the initial point at the origin of the coordinate system. When a vector is in standard position, it is identified by its length and the angle it makes with the positive *x*-axis. In this position it is easy to use the definitions of the trigonometric functions in Chapter 6. This text uses the notation $\mathbf{V} = |\mathbf{V}|\angle\theta$ to fully describe a vector. $|\mathbf{V}|$ represents the magnitude and θ the direction of the vector. (For example, in Figure 7.1 $\mathbf{OP} = 80$ lb/$\underline{125°}$.)

DEFINITION 7.2

1. Two vectors $\mathbf{V}_1$ and $\mathbf{V}_2$ are **parallel** if $\mathbf{V}_1$ and $\mathbf{V}_2$ have the same direction (see Figure 7.2a).
2. Two or more vectors are **equal** if the vectors have the same direction and the same magnitude (see Figure 7.2b).
3. $\mathbf{V}_1 = -\mathbf{V}_2$, if and only if, $|\mathbf{V}_1| = |\mathbf{V}_2|$ and $\mathbf{V}_1$ and $\mathbf{V}_2$ are in opposite directions (see Figure 7.2c).

We will discover that addition of vectors is done in a special way that sets them apart from other quantities we have studied. In arithmetic $3 + 4$ is equal to 7. However, as shown in the following example, when vectors are added, $3 + 4$ may not equal 7!

A child rides her bicycle to the grocery store by traveling four blocks east and three blocks north. (Note that the vector is specified because both the magnitude and the direction are stated). Figure 7.3 illustrates this with two vectors $\mathbf{HA}$ and $\mathbf{AG}$. A third vector can be drawn from the starting point, home, to the ending point, the grocery store. This new vector is identified in Figure 7.3 as $\mathbf{HG}$. $\mathbf{HG}$ is called the **vector sum** of the two vectors $\mathbf{HA}$ and $\mathbf{AG}$. This vector sum is also referred to as the **resultant vector,** or simply, the **resultant.** How far did the two displacements, $\mathbf{HA}$ and $\mathbf{AG}$, move her from the starting point? The net displacement is the vector sum for the trip, represented by $\mathbf{HG}$. In Figure 7.3 we can see that the displacement is the hypotenuse of a right triangle. Using the Pythagorean relation, we have:

$$|\mathbf{HG}| = \sqrt{|\mathbf{HA}|^2 + |\mathbf{AG}|^2}$$
$$= \sqrt{(4)^2 + (3)^2}$$
$$= 5 \text{ blocks.}$$

CAUTION $|\mathbf{HA}| + |\mathbf{AG}| \neq |\mathbf{HG}|$. $|\mathbf{HA}| + |\mathbf{AG}| = 4 + 3 = 7$. We have just seen that $|\mathbf{HG}| = 5$. This demonstrates that the magnitude of the resultant $\mathbf{HG}$ is **not** the sum of the magnitudes of the vectors $\mathbf{HA}$ and $\mathbf{AG}$. ∎

The direction of the resultant can be determined by using right-triangle trigonometry. In many cases, we may find it helpful to place the figure in standard position with respect to the rectangular coordinate axes as shown in Figure 7.4. Using Figure 7.4, $\tan\theta = \dfrac{3}{4}$; hence $\theta = 36.9°$. The resultant vector is finally expressed as $\mathbf{HG} = 5$ blocks $\underline{/37°}$.

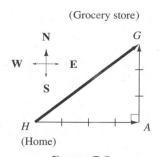

(Grocery store)

FIGURE 7.3

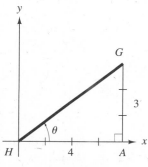

FIGURE 7.4

The vector **HG** is called the resultant, and the two vectors **HA** and **AG** are called **component vectors** of **HG**. When the figure is oriented to the coordinate axes as in Figure 7.4, it is convenient to refer to **HA** as the *x*-**component** and **AG** as the *y*-**component** since they are in the direction of the *x*-axis and *y*-axis, respectively.

EXAMPLE 1   Application: Determining *x*- and *y*-Components

A wagon is pulled with a force of 35 lb by a person holding the handle at an angle of 40° with the ground (Figure 7.5). Determine the components of the force in both the horizontal and the vertical directions.

FIGURE 7.5

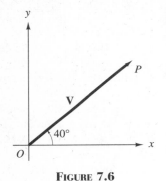

Solution   We are given the resultant vector **V**, which has a magnitude of 35 lb and a direction of 40°. Figure 7.6 illustrates this vector **V** in standard position with respect to *x*- and *y*-axis. We are asked to determine the component of the force in the *x*-direction (or *x*-component) and the component of the force in the *y*-direction (or *y*-component). We can determine the components geometrically by constructing perpendiculars to the *x*- and *y*-axes from point *P* (Figure 7.7). This process determines the component vectors $\mathbf{V}_x$ and $\mathbf{V}_y$. These components often are referred to as the **projection** of the vector **V** on to the *x*- and *y*-axes. Now we can construct a vector **SP** parallel to $\mathbf{V}_y$ and of equal magnitude as shown in Figure 7.7b. Recall from Definition 7.2 that these two vectors are equal.

FIGURE 7.6

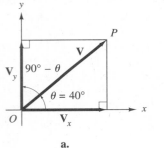

a.

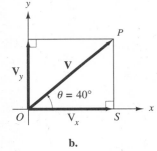

b.

FIGURE 7.7

Then use right-triangle trigonometry to determine $\mathbf{V}_x$ and $\mathbf{V}_y$.

$$\cos\theta = \frac{|\mathbf{V}_x|}{|\mathbf{V}|} \quad \text{or} \quad |\mathbf{V}_x| = 35\cos 40°$$
$$= 27 \text{ lb}$$

$$\sin\theta = \frac{|\mathbf{V}_y|}{|\mathbf{V}|} \quad \text{or} \quad |\mathbf{V}_y| = 35\sin 40°$$
$$|\mathbf{V}_y| = 22 \text{ lb}$$

Thus, $\mathbf{V}_x = 27$ lb $/0°$ and $\mathbf{V}_y = 22$ lb $/90°$. Since $\mathbf{V}_x$ and $\mathbf{V}_y$ always lie parallel to the *x*- and *y*-axes, respectively, identify the vectors with the subscripts. The wagon is moved horizontally with 27 lb force and pulled up with 22 lb force. We also can determine the magnitude of $\mathbf{V}_y$ by using the cosine function since $\cos(90° - \theta) = \sin\theta$; thus $|\mathbf{V}_y| = |\mathbf{V}|\cos(90° - \theta)$. ◆

DEFINITION 7.3

If vector **V** with magnitude $|\mathbf{V}|$ and acting at angle θ with the horizontal is composed of a horizontal component $\mathbf{V}_x$ and a vertical component $\mathbf{V}_y$ such that $\mathbf{V} = \mathbf{V}_x + \mathbf{V}_y$, then the magnitude of the horizontal component is:

$$|\mathbf{V}_x| = |\mathbf{V}|\cos\theta,$$

and the magnitude of the vertical component is:

$$|\mathbf{V}_y| = |\mathbf{V}|\sin\theta.$$

EXAMPLE 2 **Application: Determining x- and y-Components**

A child pushes on the handle of a toy with a force of 8.0 lb at an angle of 55° as shown in Figure 7.8. How much of his force acts horizontally (to push the toy) and how much acts vertically?

Solution The resultant vector **F** is illustrated in standard position in Figure 7.9. (Do not incorrectly jump to the conclusion that $|\mathbf{F}_x| = 8.0\cos 55°$. With the vector in standard position, the initial side of the angle is the positive x-axis.) The correct identification of the vector is a force of 8.0 lb at a direction of 325° or −35° (Figure 7.10). The magnitude of the horizontal component is:

$$|\mathbf{F}_x| = 8.0\cos 325° \quad \text{or} \quad |\mathbf{F}_x| = 8.0\cos(-35°)$$
$$= 6.6 \text{ lb} \qquad\qquad = 6.6 \text{ lb.}$$

The magnitude of the vertical component is:

$$|\mathbf{F}_y| = 8.0\sin 325° \quad \text{or} \quad |\mathbf{F}_y| = 8.0\sin(-35°)$$
$$= -4.6 \text{ lb} \qquad\qquad = -4.6 \text{ lb.}$$

(The negative sign indicates that this is a downward force as shown in the diagram.)

Thus $|\mathbf{F}_x| = 6.6$ lb and $|\mathbf{F}_y| = -4.6$ lb. ◆

In Examples 1 and 2 we determined the component vectors given the resultant vector and the angle. Is it possible to determine the resultant and the angle given the component vectors? Example 3 illustrates this process using Definition 7.4.

FIGURE 7.8

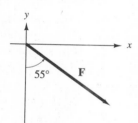

FIGURE 7.9

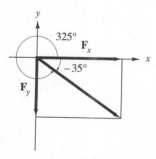

FIGURE 7.10

DEFINITION 7.4

Given the vector components $|\mathbf{V}_x|$ and $|\mathbf{V}_y|$, the magnitude of the vector **V** is given by $\mathbf{V} = \sqrt{|\mathbf{V}_x|^2 + |\mathbf{V}_y|^2}$. The angle that vector **V** makes with the positive x-axis is given by $\tan\theta = \dfrac{|\mathbf{V}_y|}{|\mathbf{V}_x|}$.

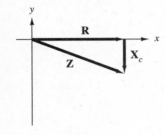

FIGURE 7.11

EXAMPLE 3   Application: Determing the Resultant Vector

An AC circuit has a resistance **R** of 30 Ω, acting at 0° and a capacitive reactance **X**$_c$ of 11 Ω acting at 270° (Figure 7.11). These two vectors form the components of a resultant vector **Z**, which is called the impedance of the circuit. What is the impedance **Z**?

Solution   The resultant **Z** is the hypotenuse of a right triangle. To construct the right triangle, move the vector **X**$_c$ so that its tail is against the head of **R**, and then draw the resultant **Z** as shown in Figure 7.12a. Determine the magnitude using the Pythagorean relation:

$$|\mathbf{Z}| = \sqrt{|\mathbf{R}|^2 + |\mathbf{X}_c|^2}$$
$$= \sqrt{30^2 + 11^2}$$
$$= 32.0 \ \Omega.$$

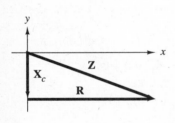

a.

Determine the direction of **Z** using the tangent function:

$$\tan \theta = \frac{-11}{30}$$
$$\theta = \tan^{-1}\left(\frac{-11}{30}\right)$$
$$= -20°$$

The reference angle $\alpha = 20°$. From the diagram of the vectors, we know that the angle is a fourth-quadrant angle; thus $\theta = 360° - 20°$, or 340°. The impedance of the circuit is, **Z** = 32 Ω /340°. Note that the same resultant vector **Z** could be obtained by moving **R** so its tail is against the head of **X**$_c$ as shown in Figure 7.12b. ◆

b.

FIGURE 7.12

CAUTION   In Example 3, when the resultant vector **Z** is constructed, the initial point of **Z** cannot be placed at the terminal point of **X**$_c$ or **R**. The initial point of **Z** must coincide with the initial points of **X**$_c$ and **R** as shown in Figure 7.13. ∎

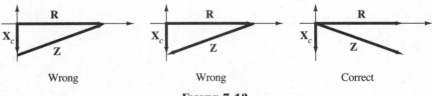

FIGURE 7.13

7.1 Exercises

In Exercises 1–10, indicate which are scalar quantities and which are vector quantities. Briefly state why you selected scalar or vector quantity.

1. The flow of electricity in a wire from switch to bulb.
2. The weight of a spaceship.
3. The flight of a spaceship in a lunar orbit.
4. The area of a computer chip.
5. The height of the CN tower in Toronto.
6. The force exerted by a spring.
7. The amount of frozen water in a pipe.
8. The number of seconds it takes to read this statement.
9. The force exerted by stretching a spring from point A to point B.
10. The sound from the television to your car.

In Exercises 11–14, draw the vectors on the same coordinate system, each having the origin as its initial point.

11. $\mathbf{V}_1 = 140\underline{/33°}$
12. $\mathbf{V}_2 = 220\underline{/115°}$
13. $\mathbf{V}_3 = 160\underline{/270°}$
14. $\mathbf{V}_4 = 130\underline{/220°}$

In Exercises 15–18, draw the vectors for Exercises 11–14 using the indicated point instead of the origin as the initial point.

15. $\mathbf{V}_1, P(5, 5)$
16. $\mathbf{V}_2, P(0, 10)$
17. $\mathbf{V}_3, P(-5, 0)$
18. $\mathbf{V}_4, P(5, -5)$

In Exercises 19–30, determine the horizontal and vertical components of the vector quantities. Sketch the vector and its components.

19. $\mathbf{V} = 75$ cm/s $\underline{/85°}$
20. $\mathbf{V} = 120$ cm/s $\underline{/130°}$
21. $\mathbf{F} = 25$ lb $\underline{/1.72}$
22. $\mathbf{F} = 38$ lb $\underline{/72°}$
23. $\mathbf{D} = 48$ km $\underline{/220°}$
24. $\mathbf{D} = 148$ mi $\underline{/35°}$
25. $\mathbf{F} = 145$ dynes $\underline{/-38°}$
26. $\mathbf{F} = 228$ dynes $\underline{/138.4°}$
27. $\mathbf{A} = 32.2$ cm/s^2 $\underline{/55°}$
28. $\mathbf{A} = 32.2$ cm/s^2 $\underline{/-42.5°}$
29. $\mathbf{F} = 0.25$ kg $\underline{/185°}$
30. $\mathbf{F} = 24.5$ kg $\underline{/0.750}$

In Exercises 31–40, determine the magnitude of the resultant vector and the angle it makes with the positive x-axis. Sketch the resultant and its component vectors.

31. $\mathbf{F}_x = 35$ lb, $\mathbf{F}_y = 47$ lb
32. $\mathbf{F}_x = -42$ lb, $\mathbf{F}_y = 55$ lb
33. $\mathbf{D}_x = 345$ km, $\mathbf{D}_y = -135$ km
34. $\mathbf{D}_x = 85$ mi, $\mathbf{D}_y = 55$ mi
35. $\mathbf{F}_x = -17$ dynes, $\mathbf{F}_y = 34$ dynes
36. $\mathbf{F}_x = -118$ dynes, $\mathbf{F}_y = -75$ dynes
37. $\mathbf{V}_x = -42$ cm/s, $\mathbf{V}_y = -78$ cm/s
38. $\mathbf{V}_x = 110$ ft/s, $\mathbf{V}_y = 115$ ft/s
39. $\mathbf{A}_x = -18$ ft/s^2, $\mathbf{A}_y = 24$ yd/s^2
40. $\mathbf{A}_x = 24$ cm/s^2, $\mathbf{A}_y = -12$ cm/s^2

41. A sailboat heads due east at a rate of 12.0 mph, and the current is flowing north at 4.00 mph. Add the vectors to determine the resultant velocity of the boat. (See the sketch).

current

42. Two boys are hauling a wagonload of wood. One boy is pushing with an effective force of 32 lb in a horizontal direction to the right, and the other boy is pulling with an effective force of 50 lb in the same direction. What single force would give the same result?

43. The lift of an airplane wing is 750 lb. The drag is 300 lb. What is the magnitude and the direction of the resulting force? See the sketch.

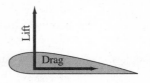

44. The force of 800 N is acting on the wing of an airplane at an angle of 25° with the vertical. Determine the vertical and horizontal components. See the drawing.

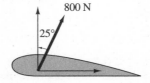

45. In moving a load of cement, Jan is exerting a force of 85 lb on the handles of her wheelbarrow. As indicated by **AD** in the drawing, part of this force is used to lift the legs of the wheelbarrow off the ground; the rest of the force, indicated by **AB**, is used in pushing the wheelbarrow forward. The 85-lb force is applied at an angle of 50° with the horizontal.
  a. Determine the force used to lift the legs of the wheelbarrow off the ground.
  b. Determine the force used to push the wheelbarrow forward.

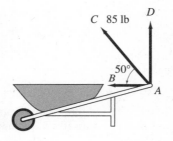

46. An AC circuit has two sources in series. Determine the total voltage if $V_1 = 15$ V at 0° and $V_2 = 26$ V at 90°.

47. An AC circuit has a resistance **R** of 25.0 Ω acting at 0.0° and a capacitive reactance X_c of 38.0 Ω acting at 270.0° Determine the impedance **Z** that is the resultant of these components.

7.2

Vector Addition

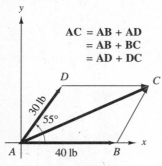

FIGURE 7.14

$$AC = AB + AD$$
$$= AB + BC$$
$$= AD + DC$$

Sometimes it is desirable to combine two or more vectors into one resultant vector; that is, to find one vector that represents (in magnitude and direction) the same effect as the combined original vectors.

Suppose that two forces, one of 40 lb and the other of 30 lb, act upon point A. The 40 lb vector has a direction of 0°, and the 30 lb vector has a direction of 55° (Figure 7.14). We can construct a parallelogram of forces $ABCD$. The diagonal of the parallelogram **AC** is the resultant vector.

The resultant **AC** also can be obtained by constructing vector **BC** at the end of the vector **AB**, **BC** being equal to **AD**. This completes a triangle of forces ABC, and as before, **AC** is the resultant. The resultant **AC** is the sum of **AB** and **BC**, or:

$$AC = AB + BC.$$

Another approach is to draw **DC** at the end of the vector **AD**; **DC** is equal to **AB**. This completes the triangle of forces ADC; again **AC** is the resultant vector, and:

$$AC = AD + DC.$$

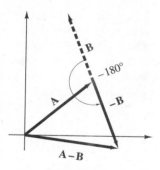

FIGURE 7.15

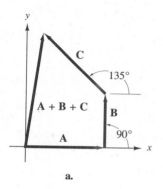

a.

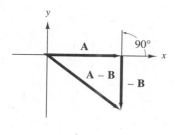

b.

FIGURE 7.16

Since **AD** = **BC** and **AB** = **DC**, we can substitute for **AB** and **BC** in the equality:

$$\mathbf{AC} = \mathbf{AB} + \mathbf{BC}.$$

We also know that:

$$\mathbf{AC} = \mathbf{AD} + \mathbf{DC}.$$

Therefore,

$$\mathbf{AD} + \mathbf{DC} = \mathbf{DC} + \mathbf{AD} \quad \text{Substituting}$$

In other words, the two vectors are independent of the order in which they are added, that is, vector addition is commutative.

Any number of vectors **A**, **B**, **C**, **D**, **E**, etc., can be added by placing them end-to-end in any order—the tail of the second at the head of the first, the tail of the third at the head of the second, and so on until all of the vectors are used. The vector drawn from the tail of the first to the head of the last is the resultant vector, indicating the sum of vectors.

We find the difference of the vectors **A** and **B** by adding the negative of vector **B** to vector **A**. That is, **A** − **B** = **A** + (−**B**). Figure 7.15 shows how to draw the vector **A** + (−**B**). Begin by drawing the vector **A**. At the head of **A** place the tail of vector **B**. Since you are adding a −**B**, measure the angle for **B**, rotate it 180°, and then draw vector **B**. The difference, **A** − **B** is the vector drawn from the tail of vector **A** to the head of vector **B**. Example 1 illustrates both processes.

EXAMPLE 1   **Determine the Sum of Vectors by Construction**

Given the vectors **A** = 5 mm $\underline{/0°}$, **B** = 4 mm $\underline{/90°}$, and **C** = 6 mm $\underline{/135°}$, estimate the vector sums and differences by drawing the given vectors.

a. **A** + **B** + **C**   **b.** **A** − **B**

Solution

a. Using an appropriate scale, construct the vectors **A**, **B**, and **C** as shown in Figure 7.16a. All vectors are drawn with the angle in standard position (with the initial side on the *x*-axis or on a line parallel to the *x*-axis). Connect the initial point of **A** with the terminal point of **C**. This vector is the sum of the vectors **A** + **B** + **C** (the resultant vector). Measuring the length of the resultant vector (**A** + **B** + **C**) shows that the magnitude of the sum of the three vectors is approximately 8 mm. The direction of the resultant vector with respect to the positive *x*-axis is approximately 85°. Therefore, the resultant vector is approximately 8 mm $\underline{/85°}$.

b. To detemine the difference of **A** − **B**, write the difference as a sum, **A** + (−**B**). Now draw the vectors using an appropriate scale as shown in Figure 7.16b. Since **B** is negative, rotate **B** 90° at the terminal point of vector **A**, and then move four units in the opposite direction. Connect the terminal point of −**B** with the initial point of **A**, this is the vector **A** − **B**. The direction of the resultant vector is approximately 315°, and the magnitude is approximately 6 mm. The difference of the two vectors, **A** − **B**, is approximately 6 mm $\underline{/315°}$. ◆

In Section 7.1 we learned how to determine the horizontal and vertical components of the individual vectors using trigonometric functions. Look at Figure 7.17. For two vectors **A** and **B**, the sum of the horizontal components $(|\mathbf{A}_x| + |\mathbf{B}_x| = |\mathbf{R}_x|)$ is the horizontal component of the resultant vector **R**. The sum of the vertical components $(|\mathbf{A}_y| + |\mathbf{B}_y| = |\mathbf{R}_y|)$ is the vertical components of the resultant vector **R**. The magnitude of the resultant vector is:

$$|\mathbf{R}| = \sqrt{|\mathbf{R}_x|^2 + |\mathbf{R}y|^2},$$

and the direction of the resultant vector is:

$$\theta = \tan^{-1}\frac{|\mathbf{R}_y|}{|\mathbf{R}_x|}.$$

This is illustrated in Example 2.

EXAMPLE 2   The Sum of Vectors Using the Component Method

Given the vectors $\mathbf{A} = 30 \text{ mm } /30°$ and $\mathbf{B} = 40 \text{ mm } /60°$, determine the resultant vector, indicating magnitude and direction.

Solution   Vectors **A** and **B** are sketched in Fig. 7.17. To determine the resultant vector **R**, we must first determine the horizontal and the vertical components of the vectors **A** and **B**.

Vector	Horizontal Component	Vertical Component
$\mathbf{A} = 30 \text{ mm } /30°$	$\begin{aligned}\|\mathbf{A}_x\| &= \|\mathbf{A}\|\cos\theta \\ &= 30\cos 30° \\ &= 26.0.\end{aligned}$	$\begin{aligned}\|\mathbf{A}_y\| &= \|\mathbf{A}\|\sin\theta \\ &= 30\sin 30° \\ &= 15.0.\end{aligned}$
$\mathbf{B} = 40 \text{ mm } /60°$	$\begin{aligned}\|\mathbf{B}_x\| &= \|\mathbf{B}\|\cos\theta \\ &= 40\cos 60° \\ &= 20.0.\end{aligned}$	$\begin{aligned}\|\mathbf{B}_y\| &= \|\mathbf{B}\|\sin\theta \\ &= 40\sin 60° \\ &= 34.6.\end{aligned}$

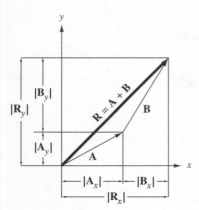

FIGURE 7.17

The horizontal and vertical components of the resultant vector are the respective sums of the horizontal and vertical components.

$$|\mathbf{R}_x| = |\mathbf{A}_x| + |\mathbf{B}_x| \quad \text{and} \quad |\mathbf{R}_y| = |\mathbf{A}_y| + |\mathbf{B}_y|$$
$$= 26.0 + 20.0 \qquad\qquad = 15.0 + 34.6$$
$$= 46.0 \text{ mm} \qquad\qquad = 49.6 \text{ mm}$$

To determine the magnitude of the resultant $\mathbf{R} = \mathbf{A} + \mathbf{B}$, use the Pythagorean relation since $|\mathbf{R}_x|$ and $|\mathbf{R}_y|$ are the legs of a right triangle.

$$|\mathbf{R}| = \sqrt{|\mathbf{R}_x|^2 + |\mathbf{R}_y|^2}$$
$$= \sqrt{(46.0)^2 + (49.6)^2}$$
$$= 68 \text{ mm}$$

To determine the direction of the resultant vector $\mathbf{R} = \mathbf{A} + \mathbf{B}$, use the tangent function.

$$\theta = \tan^{-1}\frac{|\mathbf{R}_y|}{|\mathbf{R}_x|}$$

$$= \tan^{-1}\left(\frac{49.6}{46.0}\right)$$

$$= 47.2$$

Therefore $\alpha = 47°$.

Since this value is positive, θ could be a first- or third-quadrant angle. From examining Figure 7.17, or knowing that both $\mathbf{R}_x$ and $\mathbf{R}_y$ are positive, θ must be a first-quadrant angle. Thus, $\theta = 47°$. Therefore, the resultant vector is $\mathbf{R} = 68$ mm $\underline{/47°}$. ◆

Calculator Keystrokes for Example 2

fx–7700G

Keystrokes	Display
$\|\mathbf{A}_x\|$: 30 [cos] 30 [EXE]	25.980762
[→] [ALPHA] A [EXE] *	25.980762
$\|\mathbf{B}_x\|$: 40 [cos] 60 [EXE]	20.000000
[→] [ALPHA] B [EXE]	20.000000
$\|\mathbf{A}_y\|$: 30 [sin] 30 [EXE]	15.000000
[→] [ALPHA] C [EXE]	15.000000
$\|\mathbf{B}_y\|$: 40 [sin] 60 [EXE]	34.641016
[→] [ALPHA] D [EXE]	34.641016
$\|\mathbf{R}_x\|$: [ALPHA] A [+] [ALPHA] B [EXE] †	45.980762
[→] [ALPHA] E [EXE]	45.980762
$\|\mathbf{R}_y\|$: [ALPHA] C [+] [ALPHA] D [EXE]	49.641016
[→] [ALPHA] F [EXE]	49.641016
$\|\mathbf{R}\|$: [√] [(] [ALPHA] E (SHIFT) [x^2] [+]	
[ALPHA] F [SHIFT] [x^2] [)] [EXE]	67.664326
θ: [SHIFT] [$\tan^{-1}$] [(] [ALPHA] F [÷]	
[ALPHA] E [)] [EXE]	47.192124

TI–82

Keystrokes	Display
$\lvert\mathbf{A}_x\rvert$: 30 [cos] 30 [ENTER]	25.980762
[STO] [ALPHA] A [ENTER] *	25.980762
$\lvert\mathbf{B}_x\rvert$: 40 [cos] 60 [ENTER]	20.000000
[STO] [ALPHA] B [ENTER]	20.000000
$\lvert\mathbf{A}_y\rvert$: 30 [sin] 30 [ENTER]	15.000000
[STO] [ALPHA] C [ENTER]	15.000000
$\lvert\mathbf{B}_y\rvert$: 40 [sin] 60 [ENTER]	34.641016
[STO] [ALPHA] D [ENTER]	34.641016
$\lvert\mathbf{R}_x\rvert$: [ALPHA] A [+] [ALPHA] B [ENTER] †	45.980762
[STO] [ALPHA] E [ENTER]	45.980762
$\lvert\mathbf{R}_y\rvert$: [ALPHA] C [+] [ALPHA] D [ENTER]	49.641016
[STO] [ALPHA] F [ENTER]	49.641016
$\lvert\mathbf{R}\rvert$: [2nd] [$\sqrt{}$] [(] [ALPHA] E [x^2] [+]	
[ALPHA] F [x^2] [)] [ENTER]	67.664326
θ: [2nd] [$\tan^{-1}$] [(] [ALPHA] F [÷]	
[ALPHA] E [)] [ENTER]	47.192124

EXAMPLE 3 **The Difference of Two Vectors Using Component Method**

Given the vectors $\mathbf{A} = 15$ cm $\underline{/35°}$ and $\mathbf{B} = 18$ cm $\underline{/64°}$, determine $\mathbf{A} - \mathbf{B}$. (See Figure 7.18.)

Solution To determine the difference of two vectors $\mathbf{A} - \mathbf{B}$ we need to apply the rule for subtraction of signed numbers; that is, $\mathbf{A} - \mathbf{B} = \mathbf{A} + (-\mathbf{B})$. Recall that $-\mathbf{B}$ has the same magnitude but the opposite direction of $\mathbf{B}$. Using the same technique as in Example 1, determine the horizontal and vertical components.

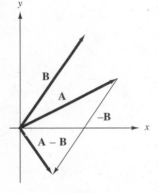

FIGURE 7.18

Vector	Horizontal Component	Vertical Component	Comments
$\mathbf{A} = 15$ cm $\underline{/35°}$	$\lvert\mathbf{A}_x\rvert = \lvert\mathbf{A}\rvert \cos\theta = 15\cos 35°$ $= 12.3$	$\lvert\mathbf{A}_y\rvert = \lvert\mathbf{A}\rvert \sin\theta = 15\sin 35°$ $= 8.60$	Components of $\mathbf{A}$
$\mathbf{B} = 18$ cm $\underline{/64°}$	$\lvert\mathbf{B}_x\rvert = \lvert\mathbf{B}\rvert \cos\theta = 18\cos 64°$ $= 7.89$	$\lvert\mathbf{B}_y\rvert = \lvert\mathbf{B}\rvert \sin\theta = 18\sin 64°$ $= 16.2$	Components of $\mathbf{B}$
	$\lvert\mathbf{A}_x\rvert + (-\lvert\mathbf{B}_x\rvert) = 12.3 + (-7.89)$ $\lvert\mathbf{R}_x\rvert = 4.4$	$\lvert\mathbf{A}_y\rvert + (-\lvert\mathbf{B}_y\rvert) = 8.6 + (-16.2)$ $\lvert\mathbf{R}_y\rvert = -7.6$	Add $\mathbf{A}$ and $-\mathbf{B}$

* Places answer in a storage cell, in this case in storage cell A.
† Recalls values stored in storage cells, in this case storage cells A and B.

To determine the magnitude of the resultant, **A** − **B**, use the Pythagorean relation:

$$|\mathbf{A} - \mathbf{B}| = \sqrt{(4.4)^2 + (-7.6)^2}$$
$$= 8.8$$

To determine the direction of the resultant vector **R** = **A** − **B**, use the tangent function.

$$\theta = \tan^{-1}\left(\frac{-7.6}{4.4}\right)$$
$$= -60°.$$

Therefore, $\alpha = 60°$.

Since the vertical component is negative and the horizontal component is positive, the resultant is in the fourth quadrant; thus, $\theta = 300°$. Therefore, **A** − **B** = 8.8 cm $/300°$. ◆

EXAMPLE 4 **Application: Resultant by Component Method**

Determine the resultant effect of the three people pulling the car shown in Figure 7.19.

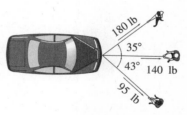

FIGURE 7.19

Solution We wish to determine the resultant effect of the three forces on the car. One of the most useful tenets of nature dictates that actions that are perpendicular to one another do not affect one another. Two forces that are perpendicular may be referred to as orthogonal forces. Hence, we can analyze a problem like this one by breaking the forces into orthogonal components— separating the x- and y-components. Begin by setting up x- and y-axes as shown in Figure 7.20.

Now find the resultant by adding the x-components together to find the x-component of the resultant and adding the y-components together to find the y-component of the resultant.

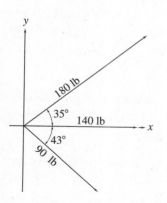

FIGURE 7.20

Vectors	x-Component	y-Component
180 lb $\underline{/35°}$	$180 \cos 35° = 147.4$ lb	$180 \sin 35° = 103.2$ lb
140 lb $\underline{/0°}$	$140 \cos 0° = 140.0$ lb	$140 \sin 0° = 0.0$ lb
95 lb $\underline{/43°}$	$95 \cos(-43°) = 69.5$ lb	$95 \sin(-43°) = -64.8$ lb
Resultant	$|\mathbf{R}_x| = 356.9$ lb	$|\mathbf{R}_y| = 38.4$ lb

The results show that the resultant force $\mathbf{R}$ has x- and y-components of 356.9 lb and 38.4 lb, respectively. The resultant is shown in Figure 7.21. To find the magnitude, use the Pythagorean relation:

$$|\mathbf{R}| = \sqrt{(356.9)^2 + (38.4)^2}$$
$$= 360 \text{ lb.}$$

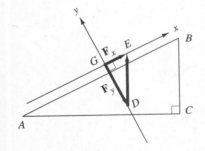

FIGURE 7.21

The direction of the resultant $\mathbf{R}$ is found by determining the angle θ.

$$\theta = \tan^{-1}\left(\frac{38.4}{356.9}\right)$$
$$\theta = 6°.$$

The pulling results in a force of 360 lb in a direction of 6° with respect to the positive x-axis. ◆

A smaller force is needed to push a wheelbarrow up a ramp than to lift the weight the same vertical distance. Vector components can explain this. There are two triangles—the triangle formed by the ramp and the **force triangle** formed by the weight of the wheelbarrow on the ramp (Figure 7.22). It can be shown that the two triangles are similar and that angles A and D are equal. To more easily analyze the force triangle, rotate the x, y-coordinate system into the position shown in Figure 7.22. The force triangle consist of two components: the force parallel to the incline (called $\mathbf{F}_x$) and the force acting perpendicular to the incline (called $\mathbf{F}_y$). The force parallel to the inclined plane is determined in Example 5.

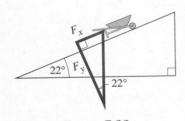

FIGURE 7.22

EXAMPLE 5   Application: Force on an Incline

If a wheelbarrow with a load of 210 lb is pushed up a ramp whose angle of inclination is 22°, determine the magnitude of the force parallel to the ramp (Figure 7.23).

Solution   The force parallel to the ramp is $\mathbf{F}_x$. Using similar triangles, we can determine that the angle opposite $\mathbf{F}_x$ is 22°. In the force triangle, the hypotenuse is 210 lb. Using the sine function:

$$\sin 22° = \frac{\mathbf{F}_x}{210}$$
$$\mathbf{F}_x = 210 \sin 22°$$
$$= 79 \text{ lb.}$$

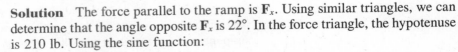

FIGURE 7.23

Using the Pythagorean relation or right triangle trigonometry, we can determine that the other leg of the force triangle is 190 lb. Thus, F_y = 190 lb. The force may be represented either as 210 lb or as two components, one acting along the incline (79 lb) and one acting perpendicular to the incline (190 lb). If the force is expressed as two components, additional information becomes available for understanding the problem. ◆

Recall that forces can be separated at right angles to one another. Hence, to move the weight up the incline, a counterforce greater than 79 lb is needed. This force is far less than the force needed to lift the weight the same vertical distance.

CAUTION   Although all vectors have magnitude and direction, the converse is not true. All quantities that have both magnitude and direction are not necessarily vectors. We learned that the sum of two vectors is independent of the order in which they are added:

$$A + B = B + A.$$

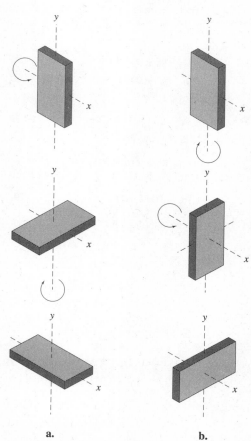

a.   b.

FIGURE 7.24

This is true for displacements; the order in which two successive displacements are made has no effect on the resultant displacement. This is also true for velocity, acceleration, momentum, angular velocity, etc.—all are quantities that are vectors. However, finite angular displacements do not obey the commutative law. The final position of an object after two successive angular displacements does depend on the order of the displacements. Angular displacement is probably the only quantity you will encounter that has magnitude and direction but is not a vector. For example, Figure 7.24a shows a 90° rotation of a rectangular block about the x-axis, followed by a 90° rotation about the y-axis. Figure 7.24b shows a 90° rotation about the y-axis and then a 90° rotation about the x-axis. The final position of the block is not the same. This proves that angular displacement is not a vector quantity since it does not obey the commutative property. ■

7.2 Exercises

In Exercises 1–14, estimate the vector sums and differences by drawing the given vectors. $\mathbf{A} = 4$ in. $/0°$, $\mathbf{B} = 3$ in. $/90°$, $\mathbf{C} = 5$ in. $/45°$, $\mathbf{D} = 6$ in. $/135°$, $\mathbf{E} = 4$ in. $/180°$

1. $\mathbf{A} + \mathbf{B}$
2. $\mathbf{A} + \mathbf{C}$
3. $\mathbf{C} + \mathbf{D}$
4. $\mathbf{D} + \mathbf{E}$
5. $\mathbf{A} + \mathbf{C} + \mathbf{E}$
6. $\mathbf{A} + \mathbf{D} + \mathbf{E}$
7. $\mathbf{B} + \mathbf{C} + \mathbf{E}$
8. $\mathbf{C} + \mathbf{D} + \mathbf{E}$
9. $\mathbf{A} - \mathbf{B}$
10. $\mathbf{B} - \mathbf{C}$
11. $\mathbf{E} - \mathbf{A}$
12. $\mathbf{A} - \mathbf{D}$
13. $\mathbf{A} + \mathbf{B} + \mathbf{C} + \mathbf{E}$
14. $\mathbf{A} + \mathbf{B} + \mathbf{C} + \mathbf{D}$
15. Show that $\mathbf{B} + \mathbf{B} + \mathbf{B} = 3\mathbf{B}$
16. Show that $\mathbf{A} + \mathbf{A} + \mathbf{A} + \mathbf{A} = 4\mathbf{A}$

In Exercises 17–20, draw the vectors to determine the sum or difference. Use the result of Exercise 16 and the definitions for $\mathbf{A}$, $\mathbf{B}$, $\mathbf{C}$, $\mathbf{D}$, and $\mathbf{E}$ as given for Exercises 1–14.

17. $2\mathbf{A} + 3\mathbf{B}$
18. $3\mathbf{C} - 2\mathbf{D}$
19. $4\mathbf{D} - 5\mathbf{E}$
20. $2\mathbf{A} + 3\mathbf{E}$

In Exercises 21–36, determine the resultant vector using the component method.

21. $\mathbf{R}_x = 3.45$ cm
    $\mathbf{R}_y = 2.75$ cm

22. $\mathbf{R}_x = 4.75$ in.
    $\mathbf{R}_y = 6.75$ in.

23. $\mathbf{A} = 3.4$ ft $/180°$
    $\mathbf{B} = 5.4$ ft $/74°$

24. $\mathbf{A} = 27.1$ yd $/125°$
    $\mathbf{B} = 11.4$ yd $/97°$

25. $\mathbf{V}_1 = 11$ m $/185°$
    $\mathbf{V}_2 = 27$ m $/118°$

26. $\mathbf{V}_1 = 22$ in. $/295°$
    $\mathbf{V}_2 = 17$ in. $/215°$

27. $\mathbf{A} = 0.71$ in. $/93.5°$
    $\mathbf{B} = 0.44$ in. $/192.4°$

28. $\mathbf{A} = 3.45$ ft $/138.4°$
    $\mathbf{B} = 8.75$ ft $/298.8°$

29. $\mathbf{A} = 17$ cm $/118°$
    $\mathbf{B} = 24$ cm $/159°$

30. $\mathbf{A} = 29$ m $/185°$
    $\mathbf{B} = 16$ m $/265°$

31. $\mathbf{A} = 0.42$ in. $/17.9°$
    $\mathbf{B} = 0.71$ in. $/93.5°$

32. $\mathbf{A} = 2.75$ ft $/2.2°$
    $\mathbf{B} = 3.45$ ft $/138.4°$

33. $\mathbf{V}_1 = 4.85$ m $/0.82$
    $\mathbf{V}_2 = 0.67$ m $/1.82$
    $\mathbf{V}_3 = 1.73$ m $/3.01$

34. $\mathbf{V}_1 = 2.14$ mm $/6.14$
    $\mathbf{V}_2 = 11.1$ mm $/3.45$
    $\mathbf{V}_3 = 7.4$ mm $/0.96$

35. $\mathbf{V}_1 = 5.0$ ft $/24°$
    $\mathbf{V}_2 = 24.0$ ft $/137°$
    $\mathbf{V}_3 = 4.0$ ft $/203°$
    $\mathbf{V}_4 = 9.0$ ft $/313°$

36. V_1 = 13.0 cm $\underline{/39.4°}$

   V_2 = 3.0 cm $\underline{/76.8°}$

   V_3 = 9.4 cm $\underline{/214.5°}$

   V_4 = 18.6 cm $\underline{/156.4°}$

37. A canoe leaves shore and travels as follows on a large, still lake. It goes 2.00 mi at 180°, then 0.50 mi at 150°, and finally, 1.00 mi at 30° with all angles measured counterclockwise from the east. How far must the canoe go to follow a straight-line path back to its starting point?

38. An airplane takes off from an airport with a velocity of 200 mph at an angle of 30° above horizontal.

   a. How fast is it rising?

   b. How fast is it moving parallel to the earth?

39. An airplane is flying due east at a velocity of 375 mph relative to the surrounding air. However, the air itself is part of a strong wind that is blowing straight west with a velocity of 75 mph relative to the ground. Determine the velocity of the airplane relative to the ground. Use the sketch.

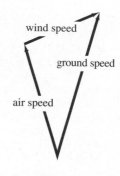

ground speed =
air speed + wind speed

40. Determine the resultant effect of three tugboats pulling a larger tanker as shown in the illustration.

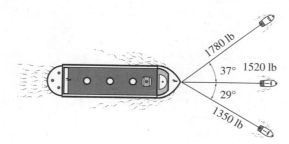

41. Determine the resultant effect of four people pulling on a car as shown in the drawing.

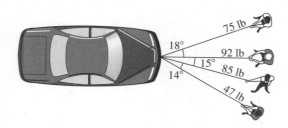

42. Philip is pushing a wheelbarrow with a 275 lb load up a ramp whose angle of inclination is 28°.
   a. Determine the magnitude of the force parallel to the ramp.
   b. Determine the magnitude of the force perpendicular to the incline.

43. Cheryl is pushing a wheelbarrow up a ramp whose angle of inclination is 14°. The magnitude of the force parallel to the ramp is 75 lb.
   a. Determine the weight of the wheelbarrow on the ramp.
   b. Determine the magnitude of the force perpendicular to the incline.

44. When something moves in a circle, its speed remains the same, but its direction changes. Therefore, there are two different vectors as shown in the diagram. The vector **E** points to the east, while the vector **S** points to the south. If the speed of both vectors is 20 m/s, what is the change in velocity as the object moves from **E** to **S**? In other words, what is **S** − **E**?

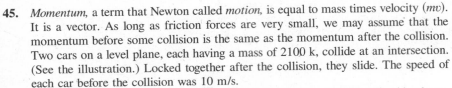

45. *Momentum,* a term that Newton called *motion,* is equal to mass times velocity (*mv*). It is a vector. As long as friction forces are very small, we may assume that the momentum before some collision is the same as the momentum after the collision. Two cars on a level plane, each having a mass of 2100 k, collide at an intersection. (See the illustration.) Locked together after the collision, they slide. The speed of each car before the collision was 10 m/s.
   a. What is the *y*-momentum for car *A* before the collision? What is the *y*-momentum for car *B* before the collision? Explain why the net momentum in the *y*-direction for the combination after the collision is zero.
   b. What is the *x*-momentum for car *A* before the collision? What is the *x*-momentum for car *B* before the collision? What is the total momentum for the combination after the collision?
   c. What is the speed of the combination after the collision?

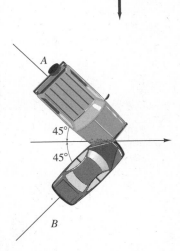

46. Car *A* and car *B* collide at an angle of 90° at a level intersection. Car *A* has a mass of 2100 kg and was traveling at 10 m/s. Car B was traveling at a rate of 3 m/s and has a mass of 5000 kg. (*Hint:* See Exercise 45.)
   a. What is the *y*-momentum for car *A* before the collision? What is the *y*-momentum for car *B* before the collision?
   b. What is the *x*-momentum for car *A* before the collision? What is the *x*-momentum for car *B* before the collision? What is the total momentum for the combination after the collision?
   c. What is the speed of the combination after collision?

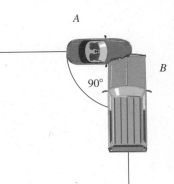

7.3

Applications of Vectors

EXAMPLE 1 **Application: Resultant by Component Method**

An individual in a motorboat is traveling from home to the store on the other side of the river (Figure 7.25). The driver sets a course of 15° with respect to a line joining points H and S to compensate for the rate of the river current, 3.0 mph and the rate of the boat, 12 mph. With the set course and a river width of 2.0 mi, will the boat meet the opposite shore at point S? If not, where will the boat meet the opposite shore?

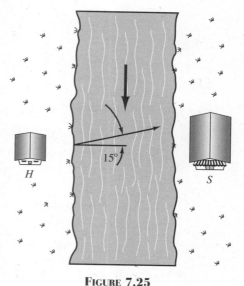

Figure 7.25

Solution   By finding the resultant vector of the vectors representing the velocity of the boat and the current of the river, we can answer both questions. Since point S is directly across the river from point H, the direction of the resultant vector must be zero if the boat is to meet the opposite shore at point S.

Vector	x-Component	y-Component
12 mph $/15°$	$12 \cos 15° = 11.6$	$12 \sin 15° = 3.1$
3.0 mph $/270°$	$3 \cos 270° = 0$	$3.0 \sin 270° = -3.0$
Resultant	$|R_x| = 11.6$	$|R_y| = 0.1$

$$\theta = \tan^{-1}\left(\frac{0.1}{11.6}\right) \quad \text{or} \quad \theta = 0.49°$$

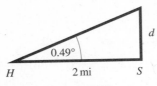

FIGURE 7.26

FIGURE 7.27

FIGURE 7.28

FIGURE 7.29

Since the direction of the resultant vector is not zero, but is positive, the boat will meet the opposite shore north of the store. To determine where, use right triangle trigonometry (Figure 7.26).

$$\tan 0.49° = \frac{d}{2 \text{ mi}}$$

or

$$d = (2 \text{ mi})(\tan 0.49°)$$
$$= 0.017 \text{ miles}$$

Therefore, the boat will meet the shore 0.02 mile, or 100 ft, north of the store. ◆

When an object moves at a constant velocity, the sum of all external forces is zero. Of course, an object at rest is "moving" at a very particular velocity—zero. If a body is moving at a constant velocity (or if a body is at rest), it is said to be in **equilibrium.**

EXAMPLE 2 **Application: Holding an Object**

Suppose Yuri, a weight lifter from Russia, holds a 325 lb sign straight up over his head (Figure 7.27). What force must he exert to keep the sign above his head?

Solution If Yuri is to keep the sign at rest and not let it crush him by falling, then the forces on the sign must balance out to zero. The weight of the sign is directed down, and Yuri is pushing straight up with a force of 325 lb. ◆

Even though an object is in equilibrium, forces may be exerted as was demonstrated by Yuri. For example, if a 10 lb hammer is resting on a table then it is in equilibrium. Its weight is balanced by the table pushing up 10 lb. This 10 lb force on the hammer due to the table pushing perpendicular to the surface is referred to as the **normal force** (Figure 7.28). The word *normal* in mathematics often refers to perpendicular. When dealing with forces, the normal force is perpendicular from the surface. However, the normal force does not always equal the weight of the object. For example, if someone leans on the hammer with an additional force of 30 lb and the hammer remains at rest, then the normal force must be 40 lb. Sometimes the normal force can be less than the weight of the object.

EXAMPLE 3 **Application: Forces Using Component Method**

A 325 lb sign is supported from a wall by a cable and a brace as shown in Figure 7.29. What are the magnitudes of the forces in the cable and in the brace to keep the sign in equilibrium?

Solution To help understand the problem, draw what is called a *free-body diagram,* which is used in all branches of technology. To form a **free-body diagram,** we first isolate the body we are interested in from its surroundings. Essentially we cut it away (Figure 7.30) and often idealize it as in Figure 7.31. We can then use vector algebra to determine the unknown forces.

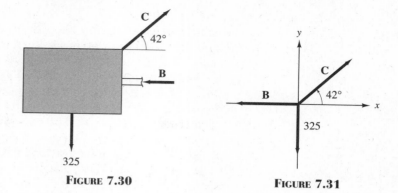

Figure 7.30   **Figure 7.31**

Since the sign is in equilibrium (it remains at rest), we know that the vector sum of the forces is zero. (It can be shown that if a vector is zero, each of its components must be zero.) We can find the x-component of the resultant force by adding the x-components of each of the individual forces; this sum must be equal to zero for the sign to be in equilibrium. Also, the sum of the y-components must be zero for equilibrium. Table 7.1 shows the components of all the forces acting on the free-body diagram of the sign.

TABLE 7.1

Force	x-Component	y-Component
B	$\lvert\mathbf{B}\rvert \cos 180° = -\lvert\mathbf{B}\rvert$	$\lvert\mathbf{B}\rvert \sin 180° = 0$
C	$\lvert\mathbf{C}\rvert \cos 42° = 0.743\,\lvert\mathbf{C}\rvert$	$\lvert\mathbf{C}\rvert \sin 42° = 0.669\,\lvert\mathbf{C}\rvert$
Weight of sign	$325 \cos(-90°) = 0$	$325 \sin(-90°) = -325$

Adding the x-components and setting the sum equal to zero gives:

$$-\lvert\mathbf{B}\rvert + 0.743\,\lvert\mathbf{C}\rvert + 0 = 0.$$

Adding the y-components and setting the sum equal to zero gives:

$$0 + 0.669\,\lvert\mathbf{C}\rvert - 325 = 0.$$

From this equation:

$$\lvert\mathbf{C}\rvert = \frac{325}{0.669}$$
$$= 485.8.$$

Substituting this result into the first equation gives:

$$\lvert\mathbf{B}\rvert = 0.743(485.8) = 360.9.$$

The force of the cable on the sign is 490. lb, and the force of the brace on the sign is 360. lb. ◆

Frictional and Normal Forces

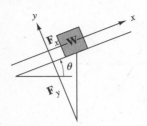

FIGURE 7.32

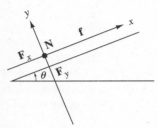

FIGURE 7.33

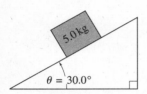

FIGURE 7.34

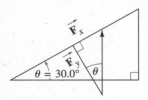

FIGURE 7.35

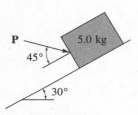

FIGURE 7.36

The force that holds an object on an inclined plane or the acceleration with which the object slides on the plane can be found using vector calculations. Figure 7.32 illustrates a typical situation. An object of weight W rests on an incline. The incline makes an angle θ with the horizontal. The angle θ is sometimes called the *angle of the incline* or the *angle of inclination.*

We can analyze the inclined plane problem by breaking up the weight ($\mathbf{W}$) of the object into two components: $\mathbf{F}_x$, the force parallel to it and $\mathbf{F}_y$, the force perpendicular to the weight. The vectors $\mathbf{F}_x$ and $\mathbf{F}_y$ are the two legs of the force triangle. Now idealize the object as a point and label the vectors on the diagram (Figure 7.33). Note that we also have the frictional force, $\mathbf{f}$, acting on the object parallel to the plane. There also must be a normal force acting on the object perpendicular to the surface. If the body is to stay in equilibrium, then the forces parallel to the incline (or along the orthogonal axes) must cancel to zero. Therefore,

$$|\mathbf{f}| = |\mathbf{F}_x| \quad \text{and} \quad |\mathbf{F}_y| = |\mathbf{N}|.$$

These equalities state that the forces parallel ($\mathbf{F}_x$) and perpendicular ($\mathbf{F}_y$) to the surface have the same magnitudes, but opposite directions, respectively to the frictional vector ($\mathbf{f}$) and the normal vector ($\mathbf{N}$). Note that the normal force is less than the weight of the object.

EXAMPLE 4   Application: Friction and Normal Forces

A 5.0 kg block rests on an inclined plane that makes an angle of 30.0° with the horizontal (Figure 7.34). **a.** What must be the frictional force between the block and the plane if it remains in equilibrium? **b.** What is the normal force?

Solution   Construct a force triangle and determine $\mathbf{F}_x$ and $\mathbf{F}_y$. Now use the statements of equality: $|\mathbf{f}| = |\mathbf{F}_x|$ and $|\mathbf{F}_y| = |\mathbf{N}|$ to determine $|\mathbf{f}|$ and $|\mathbf{N}|$.

a.   From the force triangle in Fig. 7.35, we can determine that:

$$|\mathbf{F}_x| = (\boxed{5.0})(\boxed{9.81})(\sin \boxed{30.0°})$$
$$= 24.525000.$$

Recall that to change mass to weight, multiply by 9.81.

Substituting for $\mathbf{F}_x$ in the equality $|\mathbf{f}| = |\mathbf{F}_x|$, $\mathbf{f} = 25.$ N.

b.   From the force triangle in Figure 7.35, we can determine that:

$$|\mathbf{F}_y| = (\boxed{5.0})(\boxed{9.8})(\cos \boxed{30.0°})$$
$$= 42.478546.$$

Substituting for $\mathbf{F}_y$ in the equality $|\mathbf{F}_y| = |\mathbf{N}|$, $\mathbf{N} = 42.$ N. ◆

EXAMPLE 5   Application: Friction and Normal Forces

If a force $\mathbf{P}$ of 25 N applied at an angle of 45° is required to keep the block in Example 4 in equilibrium, what is the new frictional force and the new normal force on the block? (See Figure 7.36.)

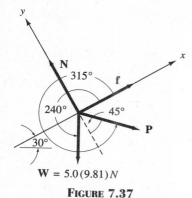

FIGURE 7.37

Solution If the block is in equilibrium, the sums of the horizontal and the vertical force components equal zero. A diagram like Figure 7.37, which shows the angles for the force vectors, helps us to identify and write down the component vectors, as shown in Table 7.2. The forces $\mathbf{F}_x$ and $\mathbf{F}_y$, calculated in Example 4, are negative due to the gravitational pull of the weight.

TABLE 7.2

Force	x-Component	y-Component
Weight	$\mathbf{F}_x = -24.5$	$\mathbf{F}_y = 42.5$
f	$|\mathbf{f}| \cos 0° = |\mathbf{f}|$	$|\mathbf{f}| \sin 0° = 0$
N	$|\mathbf{N}| \cos 90° = 0$	$|\mathbf{N}| \sin 90° = |\mathbf{N}|$
Weight	$25 \cos(315°) = 17.7$	$25 \sin(135°) = -17.7$

The sum of the x-components and the y-components are equal to zero, thus:

$$-24.5 + |\mathbf{f}| + 0 + 17.7 = 0 \qquad -42.5 + 0 + |\mathbf{N}| + (-17.7) = 0$$

Therefore, when $|\mathbf{P}| = 25$ N, the new frictional force $|\mathbf{f}|$ is 6.8 N, and the new normal force $|\mathbf{N}|$ is 60. N. ◆

EXAMPLE 6 Application: Friction and Normal Forces

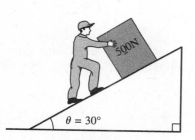

FIGURE 7.38

A 500 Newton block rests on an inclined plane that makes an angle of 30° with the horizontal. A worker maintains a constant velocity as he pushes the block up the incline (Figure 7.38).

a. If the worker pushes with a force of 400 N up the incline, what must be the frictional force on the incline?
b. What is the normal force?

Solution

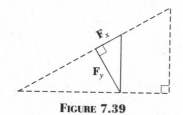

FIGURE 7.39

a. First construct the force triangle shown in Figure 7.39, and calculate the force parallel to the surface, $\mathbf{F}_x$.

$$|\mathbf{F}_x| = 500 (\sin 30)$$
$$= 250 \text{ N}$$

Now break up the weight into the orthogonal components, as shown in Figure 7.40. All of the forces along the axis parallel to the incline's surface must add up to zero. Thus $\mathbf{F}_x + \mathbf{f} +$ pushing force $= 0$. Since $\mathbf{F}_x$ is pushing down the incline, the force is negative.

$$-250 + \mathbf{f} + 400 = 0$$
$$\mathbf{f} = -150.$$

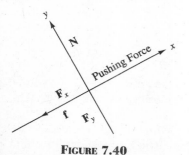

FIGURE 7.40

Therefore, the frictional force is −150 N since it is working against the worker pushing the block up the ramp.

b. In an orthogonal diagram, the sum of the forces perpendicular to the plane must equal zero. Thus $\mathbf{F}_y + \mathbf{N} = 0$. We know that $|\mathbf{F}_y| = |\mathbf{N}|$ and that $\mathbf{F}_y$ is pushing down and is a negative force. To determine $|\mathbf{F}_y|$, refer back to the force triangle in Figure 7.39.

$$|\mathbf{F}_y| = 500 (\cos 30)$$
$$= 430. \text{ N}$$

Therefore, the normal force, $\mathbf{N}$, is 430. N. ◆

Incidentally, another equation for the frictional force is $\mathbf{f} = \mu\mathbf{N}$ where μ is called the coefficient of friction and is related to the roughness between the two surfaces. For Example 4, determine the coefficient of friction between the block and the incline.

EXAMPLE 7   Application: Sum of Forces

The problem on the first page of Chapter 7 has a rope tightly stretched between a car and a tree 110. ft away. If a 45. lb sideways pull in the center of the rope pulls it 2° out of line, what tension is produced on the rope?

Solution   The distance between the tree and the car is 110. ft and the sideways pull at the center of the cable pulls the cable 2° out of line as shown (not to scale) in Figure 7.41. Summing force components in the y direction gives:

$$\mathbf{T} \sin 2° + \mathbf{T} \sin 2° - 45 = 0$$
$$2 \mathbf{T} \sin 2° = 45$$
$$\mathbf{T} = \frac{45.}{2 \sin 2°}.$$
$$= 640. \text{ lb}$$

The calculations show that a relatively small sideward force of 45. lb results in pulling the car with the much greater force of 640. lb—a far greater force than anyone can achieve by pulling and/or pushing the car. ◆

A BASIC computer program which can be used to calculate the tension in a cable which has been pulled sideways at the center may be found on the tutorial disk.

EXAMPLE 8   Application: Resultant Vector

A light aircraft is flying on a course of 135° at a speed of 165 mph. The wind is blowing east at 25 mph. What is the aircraft's resultant speed over the ground?

Solution   The resultant speed of the aircraft will be the sum of the velocity vector of the aircraft and the velocity vector of the wind. In standard position,

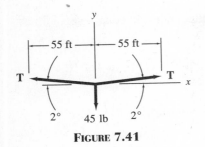

FIGURE 7.41

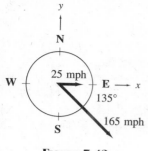

FIGURE 7.42

the angle for the aircraft's velocity vector is $-45°$ and the angle for the wind is $0°$ (Figure 7.42). Therefore, the sum of the x-components is $165\ \cos(\,-45°\,) + 25\ \cos(\,0°\,) = 142$. The sum of the y-components is $165\ \sin(\,-45°\,) + 25\ \sin(\,0°\,) = -117$. Therefore, the magnitude of the resultant is:

$$\sqrt{(\,142\,)^2 + (\,-117\,)^2} = 180.\ \text{mph}$$

and the direction of the resultant is:

$$\theta = \tan^{-1}\!\left(\frac{-117}{142}\right) \quad \text{or} \quad \theta = -39°.$$

So, the aircraft has an "over-the-ground" speed of 180 mph at a heading of $(90° + 39°)$, or $129°$. ◆

7.3 Exercises

1. Tod is swimming across the Rhine River, which he knows has a 6 km/h current. To compensate for the current he heads in a direction upstream at an angle of 120° with the current (see the illustration.) If he swims through the water at a velocity $v = 3$ km/h, what is the resultant velocity R (the magnitude of the total velocity and its angle with the current)?

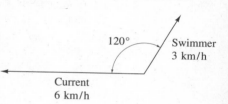

2. Teri is swimming across the Genesee River, which she knows has a current of 1.8 km/h. If she swims at the rate of 3.8 km/h, at what angle with respect to the shore must she swim to end up directly across the river from the starting point?

3. A rocket traveling at 1000 km/h ascends at an angle of 80.0° with the horizontal. What are the horizontal and vertical components of the velocity?

4. A Boston taxi driver (who should be fined for littering) is traveling at 25 km/h and throws a candy wrapper out the window perpendicular to his direction at a speed of 5 km/h. What is the velocity of the candy wrapper as it leaves his hand (the speed and direction with respect to the taxi's direction)? Use the sketch.

5. The heading of a boat traveling 15.0 mph upriver makes an angle of 115° with the current, which is 5.0 mph. What is the boat's resultant velocity (the speed and direction with respect to its heading)?

6. An orienteer leaves her starting point and walks 275 m 38° north of east, then 120 m east, and finally, 75 m 42° south of west.
   a. How far is she from her starting position?
   b. In what direction would she walk to return directly to the starting position?

7. Calculate the force (neglecting friction, i.e., $\mathbf{f} = 0$) needed to pull an object up an inclined plane if $W = 875$ lb for each of the following angles.
   a. $\theta = 30.0°$ b. $\theta = 13.5°$ c. $\theta = 43.5°$

8. Molly's Meat Market is placing a new sign outside the shop. The sign has a mass of 205 kg and is supported from the wall by a cable and a brace. The cable makes an angle of 32° with the wall, and the brace is perpendicular to the wall. What are the magnitudes of the forces in the cable and the brace?

9. The new roller rink sign weighs 365. lb and is mounted to the wall by a brace perpendicular to the wall and a cable making an angle of 37° with the wall. What are the magnitudes of the forces in the cable and the brace?

10. Rob has a sign that weighs 820 lb and a supporting wire with a tested strength of 1100 lb.
    a. If he hangs the sign as shown in the drawing, will the wire support the sign?
    b. At what minimum angle must the sign be hung for the wire to support the sign?
    c. What must be the tensile strength of the brace to keep the sign away from the wall?

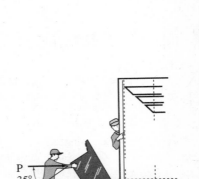

11. A truck weighing 6500 kg must be pulled up a ramp that is inclined 10° with the horizontal. If friction is neglected, what force in newtons must be exerted barely to move the truck?

12. A 50-kg box sits on a 37° incline. Find the normal force that acts on it. How large a force must friction supply if the box is not to slide?

13. A refrigerator rests on an adjustable ramp. The refrigerator has a mass of 455 kg, and it begins to slip on the ramp when the angle of inclination reaches 12°.
    a. What is the coefficient of friction between the refrigerator and the ramp?
    b. What is the normal force and the friction force on the refrigerator when $\theta = 8°$. (*Hint:* See Example 4.)

14. As shown in the illustration, movers have a washer that has a mass of 140 kg on the inclined plane. If a mover applies a force **P** at an angle of 35° to the inclined plane, what is the friction force and the normal force on the washer?

15. When a football player pushes a 200-lb blocking dummy on level ground, he must exert a force of 80 lb. What force parallel to the slope must he exert to push it up a 5° grade? See the sketch.

16. A 170-lb tightrope walker causes a 6-in. sag in a 50-ft cable, which initially was stretched tight. Assuming the walker is in the center of the cable and that we can neglect the weight of the cable, what is the tension produced in the cable? See the illustration.

17. A 1730-N block attached to a cable is loaded onto a truck by picking it up with a backhoe and driving it over the truck bed to be lowered. The operator misjudges his height, and the block hits the back of the truck. The lifting cable is 15.0° out of plumb before the backhoe stops. What is the tension in the cable? See the sketch.

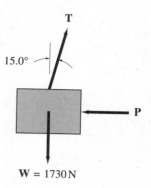

7.4

The Law of Sines

In our previous work with trigonometric ratios, we were able to solve for the unknown parts of a right triangle when only two parts were known and we knew that one angle was a right angle. In this and the next section we will develop formulas to determine the length of the sides and measures of the angles of an **oblique triangle** (no right angles).

In solving for parts of an oblique triangle, it is necessary to know three pieces of information, one of which must be the length of a side. The combinations are listed below.

1. One side and any two angles.
2. Two sides and an angle opposite one of them.
3. Two sides and an included angle.
4. Three sides.

The law of sines, the proof of which follows, can be used to solve oblique triangles for statements 1 and 2 of the list. The law of cosines, which will be discussed in the next section, can be used to solve oblique triangles for statements 3 and 4.

The *law of sines* states that *the sides of any triangle are in proportion to the sines of the opposite angles.*

To keep communication clear, side a is opposite angle A, side b is opposite angle B, and side c is opposite angle C, as shown in Figure 7.43. A line perpendicular to side c from vertex C forms two right triangles. The line perpendicular to side c is labeled h. For the right triangle ACD:

$$\sin A = \frac{h}{b},$$

and for the right triangle CBD:

$$\sin B = \frac{h}{a}.$$

Solving both of these equations for h, leads to:

$$h = b \sin A \quad \text{and} \quad h = a \sin B.$$

FIGURE 7.43

Since both are equal to h we can set them equal to each other:

$$b \sin A = a \sin B.$$

Multiplying both sides by $\dfrac{1}{\sin A \sin B}$,

$$\frac{a}{\sin A} = \frac{b}{\sin B}.$$

Drawing a line perpendicular to side b from vertex B and using a similar sequence of steps produces:

$$\frac{a}{\sin A} = \frac{c}{\sin C}.$$

Also, drawing a line perpendicular to side a from vertex A and using a similar sequence of steps produces:

$$\frac{b}{\sin B} = \frac{c}{\sin C}$$

These ratios may be developed from any triangle. However, in the case of an obtuse triangle, a side may need to be extended to form the right triangle. The combined results of the three proportions produce Theorem 7.1.

THEOREM 7.1 Law of Sines

In any triangle ABC:

$$\frac{a}{\sin A} = \frac{b}{\sin B} = \frac{c}{\sin C}.$$

Since the ratios are all equal we can write the law of sines in the form:

$$\frac{\sin A}{a} = \frac{\sin B}{b} = \frac{\sin C}{c}.$$

This form may be more convenient to use when we are solving for the angle.

CAUTION Regardless of which form we are using, remember that we can work with only two of the ratios at a time. For example:

$$\frac{a}{\sin A} = \frac{b}{\sin B} \quad \text{or} \quad \frac{a}{\sin A} = \frac{c}{\sin C} \quad \text{or} \quad \frac{b}{\sin B} = \frac{c}{\sin C} \quad \text{or}$$

$$\frac{\sin A}{a} = \frac{\sin B}{b} \quad \text{or} \quad \frac{\sin A}{a} = \frac{\sin C}{c} \quad \text{or} \quad \frac{\sin B}{b} = \frac{\sin C}{c}. \quad \blacksquare$$

EXAMPLE 1 Law of Sines: Given Two Angles and One Side

Given $A = 24°$, $B = 82°$, and $a = 17$ mm, determine the measures of the other three parts of the triangle: angle C, side b, and side c.

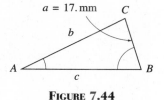

$a = 17.$ mm C

b

A c B

FIGURE 7.44

Solution First sketch and label the triangle as shown in Figure 7.44. To determine angle C, use the fact that $C = 180° - (A + B)$ so that, by substituting, $C = 180° - (\,24° + 82°\,)$, or $C = 74°$. Since we know two angles and the side opposite one angle, use the law of sines. Knowing A, a, and B suggests using $\dfrac{a}{\sin A} = \dfrac{b}{\sin B}$. Solving for b, we have:

$$b = \frac{a \sin B}{\sin A}.$$

Substituting,

$$b = \frac{17 \ \sin \ 82°}{\sin \ 24°} \quad \text{or} \quad b = 41. \text{ mm.}$$

To determine side c we can use $\dfrac{a}{\sin A} = \dfrac{c}{\sin C}$ since we know A, C and a. Solving for c we have:

$$c = \frac{a \sin C}{\sin A}.$$

Substituting,

$$c = \frac{17 \ \sin \ 74°}{\sin \ 24°} \quad \text{or} \quad 40. \text{ mm}$$

Therefore, the measure of the three other parts of the triangle are $C = 74°$, $b = 41$ mm, and $c = 40$ mm. ◆

If we are given two sides and the angle opposite one of the sides, there could be 0, 1, or 2 solutions. Example 2 illustrates, the case where there is no solution.

EXAMPLE 2 Law of Sines: Given Two Sides and One Angle

Given $a = 10$, $b = 3$, and $B = 50°$, determine the measures of the remaining three parts of the triangle, angles A and C and side c.

Solution Given two sides and the angle opposite one of the these sides, use the law of sines. Knowing a, b, and B suggests using $\dfrac{\sin A}{a} = \dfrac{\sin B}{b}$. Solving for $\sin A$ we have:

$$\sin A = \frac{a \sin B}{b}.$$

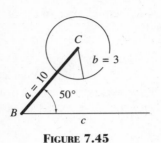

FIGURE 7.45

Substituting results in:

$$\sin A = \frac{10 \sin 50°}{3}$$

$$= 2.55.$$

The sine of any angle is a real number larger than or equal to -1 and smaller than or equal to 1 ($-1 \leq \sin A \leq 1$). The results show that $\sin A = 2.55$. Thus, there is no solution. If we draw a figure to scale, as shown in Figure 7.45, we can see that side b would never intersect side c. Therefore, there can be no triangle. The answer is that there is no solution for the given information. ◆

How do we know when a problem has one or two solutions? If we are given sides a, b, and angle A, it is usual to determine angle B using the law of sines.

$$\left(\sin B = \frac{b \sin A}{a} \right)$$

Recall that the sine function is positive in the first and second quadrants. When we evaluate for angle B, if B is an acute angle, there are two possible answers for B. If there are two possible answers for B, then the solution to the problem is two distinct triangles. This is illustrated in Example 3.

EXAMPLE 3 **Law of Sines: Given Two Sides and One Angle**

Given $a = 15$, $b = 18$, and $A = 27°$, determine the measure of angle B.

Solution With the given information we use $\dfrac{\sin A}{a} = \dfrac{\sin B}{b}$ and solve for $\sin B$. Thus, $\sin B = \dfrac{b \sin A}{a}$. Substituting results in:

$$\sin B = \frac{18 \sin 27°}{15}$$

$$B = \sin^{-1}\left(\frac{18 \sin 27°}{15} \right)$$

$$= 33°.$$

Since the sine function is positive in the first and second quadrants, there are two possible answers for angle B:

$$B_1 = 33°; \quad B_2 = 180° - 33°, \quad \text{or} \quad 147°.$$

Since there are two possible answers for B, there will also be two possible values for angle C and side c.

Case 1: With $B_1 = 33°$, then:

$$C_1 = 180° - (27° + 33°)$$

$$= 120°.$$

We can determine c_1 using the law of sines.

$$c_1 = \frac{a \sin C_1}{\sin A}$$

$$= \frac{15 \sin 120°}{\sin 27°}$$

$$= 29.$$

Case 2: With $B_2 = 147°$, then:

$$C_2 = 180° - (27° + 147°)$$
$$= 6°.$$

Determine C_2 using the law of sines.

$$c_2 = \frac{a \sin C_2}{\sin A}$$

$$= \frac{15 \sin 6°}{\sin 27°}$$

$$= 3.5$$

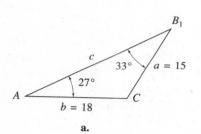

a.

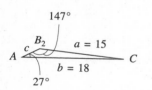

b.

FIGURE 7.46

Thus, the two possible triangles are:

Case 1			Case 2		
$A = 27°$	$a = 15$		$A = 27°$	$a = 15$	
$B_1 = 33°$	$b = 18$		$B_2 = 147°$	$b = 18$	
$C_1 = 120°$	$c_1 = 29$		$C_2 = 6°$	$c_2 = 3.5$	

Cases 1 and 2 are illustrated in Figures 7.46a and 7.46b, respectively.   ◆

Given two angles and a side, there is only one solution as illustrated in Example 4.

EXAMPLE 4   Application: Given Two Angles and a Side

A surveyor needs to know the three sides and the three angles of a triangular plot as shown in Figure 7.47. Determine the measures of the missing parts.

Solution   From the figure we know that $A = 43.8°$, $B = 63.7°$, and $c = 486.3$ m. Knowing A and B, we can find C using the fact that $C = 180° - (A + B)$.
Substituting results in:

$$C = 180° - (43.8°) + (63.7°)$$
$$= 72.5°.$$

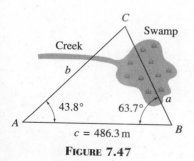

FIGURE 7.47

Now we can use the following to solve for a and b, respectively.

$$\frac{a}{\sin A} = \frac{c}{\sin C} \quad \text{and} \quad \frac{b}{\sin B} = \frac{c}{\sin C}$$

$$a = \frac{c \sin A}{\sin C} \quad \text{and} \quad b = \frac{c \sin B}{\sin C}$$

Substituting results in:

$$a = \frac{(486.3)(\sin 43.8°)}{\sin 72.5°} \quad \text{and} \quad b = \frac{(486.3)(\sin 63.7°)}{\sin 72.5°}$$

$$= 352.9 \text{ m} \qquad\qquad = 457.1 \text{ m.}$$

Thus, $C = 72.5°$, $a = 353.$ m, and $b = 457.$ m. ◆

In conclusion, let us summarize the situation when there may be two triangles. The possibility of having two triangles arises only when two sides, say a and b, and an angle A, opposite side a are given. If A is an acute angle there may be no triangle, one triangle, or two triangles. If A is an obtuse angle there may be no triangle or one triangle. If $a < b$ there may be no triangle, one triangle, or two triangles. If $a = b$, there may be no triangle or one triangle. If $a > b$, there is one triangle.

In Example 2, given angle B, sides a and b, and $b < a$, there could have been 0, 1, or 2 triangles. The result in that case was no triangle.

In Example 3, given angle A, sides a and b, and $a < b$, there could have been 0, 1, or 2 solutions. The result in that case was two triangles.

7.4 Exercises

In Exercises 1–22, determine the measures of the three other parts of the triangle, if possible.

1. $A = 68°$, $B = 79°$, $b = 8.0$ cm

2. $A = 48°$, $C = 100°$, $c = 18$ cm

3. $A = 60°$, $a = 9.0$ in., $b = 7.0$ in.

4. $A = 40°$, $C = 35°$, $a = 10$ in.

5. $a = 5.14$ mm, $b = 4.31$ mm, $A = 64.3°$

6. $b = 672$ km, $c = 630$ km, $B = 82.7°$

7. $C = 30.5°$, $a = 4.0$ in., $c = 10.5$ in.

8. $B = 40°$, $b = 18$ in., $c = 10$ in.

9. $b = 11.1$ cm, $c = 15.8$ cm, $B = 150.7°$

10. $a = 0.432$ cm, $c = 0.765$ cm, $C = 105.2°$

11. $B = 112° \ 30'$, $C = 27° \ 40'$, $b = 426$ ft

12. $C = 97°30'$, $c = 5.0 \times 10^{-4}$ m, $b = 6.0 \times 10^{-5}$ m

13. $C = 45°$, $b = 2.5$ in., $c = 2.5$ in.

14. $B = 30°$, $b = 5$ ft, $c = 10$ ft

15. $A = 1.7$, $B = 1.1$, $a = 53$ cm

16. $B = 0.48$, $b = 0.72$ cm, $c = 0.75$ cm

17. $c = 1.7 \times 10^4$ m, $b = 2.3 \times 10^5$ m, $B = 82.3°$

18. $C = 30.9°$, $b = 9.3$ in., $c = 4.0$ in.

19. $B = 119°$, $C = 47°$, $b = 8.0$ ft

20. $B = 50°$, $b = 8$ ft, $c = 8$ ft

21. $A = 60°$, $a = 72.0$ mm, $b = 78.0$ mm

22. $B = 37°$, $b = 0.475$ m, $c = 0.625$ m

23. Is it possible to use the law of sines, if you are given two sides and the angle opposite one of the sides is 90°? Explain! What is the result when $A = 90°$, $a = 31$ cm, and $b = 22$ cm?

24. If $A = 48°$, $a = 15$ cm, and $b = 18$ cm, determine the measure of the largest angle of the triangle correct to the nearest minute. (*Hint*: There are two answers.)

25. Two points, B and C, are on opposite sides of a swamp. To measure the distance across the swamp, point A, from which B and C are visible, is chosen. The distances $\overline{AB}$ and $\overline{AC}$ are measured to be 945.6 ft and 887.8 ft, respectively. Angle A was found to be 56° 44′. Determine the distance across the swamp. Use the drawing.

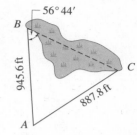

26. A distant tower on shore is sighted from points at opposite ends of a ship 650 ft long (see the sketch). If the sight lines make angles of 87.4° and 87.8° with the line joining the two points, how far is the tower from the nearest of the two points?

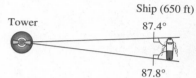

27. Two fire lookout towers P and Q are 28.5 km apart. Tower P spots a fire at R and measures angle QPR to be 49.3°, whereas tower Q measures angle PQR to be 43.7°. Which tower is closer to the fire and by how much?

28. An antenna 25 m high is on the top edge of a building. From a point on the ground, the angles of elevation to the top and bottom of the antenna are 58.7° and 46.3°. What is the distance from the point on the ground to the top of the antenna? (See the illustration.)

29. From point A on top of a building, the angle of depression to point C on the ground is observed to be 56°, while from a window at point B (18 m directly below A) the angle of depression to point C is 43°. Determine the distance from point B to point C.

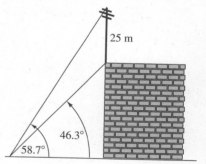

30. A surveyor sights to the top and bottom of an antenna from point A. Determine the height of the antenna $\overline{BC}$. (See the sketch.)

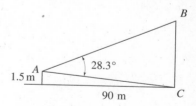

31. A radio operator on Coast Guard patrol boat A receives a distress call from a ship that he locates at 56.4° (nautical compass). At the same time, an operator on Coast Guard boat B picks up the call from a position that he locates at 111.5°. He also locates patrol boat A to be at 210.6°, a distance of 105 mi from his position.
  a. Explain why the law of sines is used to locate the ship.
  b. How far was the ship in distress from the closest of the two Coast Guard boats? (See the diagram.)

32. A pilot wants to fly to a city that is 35° south of east from the starting position. If the wind is blowing from the south at 48 mph and the plane flies at 275 mph with respect to the air, determine the heading of the plane.

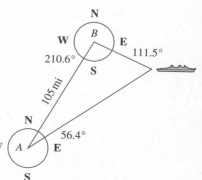

7.5

The Law of Cosines

In the previous section we learned that oblique triangles can be solved by the law of sines only when either (1) a side and any two angles or (2) two sides and an angle opposite one of them is known. We need another formula to solve oblique triangles when either two sides and an included angle are known or when three sides are known. This formula is called the **Law of Cosines.**

The *law of cosines* states that *the square of any side is equal to the sum of the squares of the other two sides decreased by twice their product times the cosine of the angle between them.*

To develop the formula $a^2 = b^2 + c^2 - 2bc \cos A$, which is one form of the law of cosines, use triangle ABC in Fig. 7.48. In the triangle, h is the altitude from vertex C to side c. There are now two right triangles so the cosine function and the Pythagorean relation can be used to develop a law of cosines. Begin by determining the value of a^2 in triangle BCD:

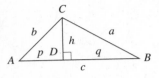

FIGURE 7.48

$$a^2 = h^2 + q^2 \tag{1}$$

and:

$$q = c - p. \tag{2}$$

Squaring both sides,

$$q^2 = c^2 - 2cp + p^2. \tag{3}$$

Substituting [3] into [1],

$$a^2 = h^2 + c^2 - 2cp + p^2. \tag{4}$$

In right triangle ADC:

$$h^2 + p^2 = b^2.$$

So:

$$h^2 = b^2 - p^2. \tag{5}$$

Substituting [5] into [4],

$$a^2 = b^2 - p^2 + c^2 - 2cp + p^2$$
$$= b^2 + c^2 - 2cp. \tag{6}$$

In right triangle ADC:

$$\cos A = \frac{p}{b} \quad \text{or} \quad p = b \cos A. \tag{7}$$

Substituting [7] into [6],

$$a^2 = b^2 + c^2 - 2c(b \cos A)$$
$$= b^2 + c^2 - 2bc \cos A. \tag{8}$$

If we go back and draw the altitude from point A or C and follow the same procedure, we can derive the following equations:

$$b^2 = a^2 + c^2 - 2ac \cos B \quad \text{and} \quad c^2 = a^2 + b^2 - 2ab \cos C.$$

The three equations are the law of cosines.

THEOREM 7.2   Law of Cosines

$$a^2 = b^2 + c^2 - 2bc \cos A \quad \text{or} \quad \cos A = \frac{b^2 + c^2 - a^2}{2bc}$$

$$b^2 = a^2 + c^2 - 2ac \cos B \quad \text{or} \quad \cos B = \frac{a^2 + c^2 - b^2}{2ac}$$

$$c^2 = a^2 + b^2 - 2ab \cos C \quad \text{or} \quad \cos C = \frac{a^2 + b^2 - c^2}{2ab}$$

EXAMPLE 1   Law of Cosines: Given Two Sides and Included Angle

Determine the measures of angle A, and B, and side c of the triangle in Figure 7.49.

Solution   Since we are given the two sides (a and b) and the included angle (C), we can use the law of cosines to solve for side c, the side opposite the given angle.

$$c^2 = a^2 + b^2 - 2ab \cos C$$
$$c = \sqrt{a^2 + b^2 - 2ab \cos C}$$

Substituting the given values:

$$c = \sqrt{(11)^2 + (8)^2 - 2(11)(8)(\cos 98°)}$$
$$= 14.5.$$

Now that we have determined side c, angle A is found most easily by using the law of sines.

$$\sin A = \frac{a \sin C}{c}$$
$$= \frac{11 \sin 98°}{14.5}$$
$$A = \sin^{-1}\left(\frac{11 \sin 98°}{14.47}\right)$$
$$= 49°$$

The sine function is positive in the first and second quadrants, but we use only the first-quadrant solution. (Why?)

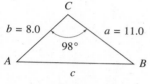

$b = 8.0$ $a = 11.0$
$98°$
A c B

FIGURE 7.49

$$B = 180° - (98° + 49°) = 33°.$$

Therefore, $c = 14$ units, angle $A = 49°$, and angle $B = 33°$. ◆

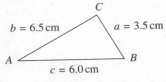

$b = 6.5\,\text{cm}$ $a = 3.5\,\text{cm}$ $c = 6.0\,\text{cm}$

FIGURE 7.50

EXAMPLE 2   Law of Cosines: Given Three Sides

Determine the measures of angle A, angle B, and angle C for the triangle in Figure 7.50.

Solution   Given the lengths of the three sides, we can determine the measure of an angle by using the law of cosines. To determine angle B, use the formula:

$$\cos B = \frac{a^2 + c^2 - b^2}{2ac}$$

Substituting the given values:

$$\cos B = \frac{(3.5)^2 + (6.0)^2 - (6.5)^2}{2(3.5)(6.0)}$$

$$B = \cos^{-1}\left[\frac{(3.5)^2 + (6.0)^2 - (6.5)^2}{2(30)(6.0)}\right]$$

$$= 82.°$$

At this point we can use either the law of cosines or the law of sines to find the measure of the other two angles.

To find the measure of angle A, use the formula:

$$\cos A = \frac{b^2 + c^2 - a^2}{2bc}$$

Substituting values:

$$\cos A = \frac{(6.5)^2 + (6.0)^2 - (3.5)^2}{2(6.5)(6.0)}$$

$$A = \sin^{-1}\left[\frac{(6.5)^2 + (6.0)^2 - (3.5)^2}{2(6.5)(6.0)}\right]$$

$$= 32.°$$

To find the measure of the third angle, use the fact that the sum of the angles of a triangle are 180°.

$$C = 180° - (32.° + 82.°) = 66.°$$ ◆

The best procedure to follow when finding the measure of the three angles given the three sides, is to find the measure of the largest angle first. In Example 2, if we had determined the measure of angle A first and then used the law of sines to find the measure of B, we would have had two possible choices: the ambiguous case $B = 82°$ or $B = 180° - 82° = 98°$. Which one would we select? With the fixed sides, only one measure can be the correct value, $B = 82.°$

By finding the largest angle first, B in this case, and then using the law of sines to find A, there still were two choices for A, $A = 32.°$ or $A = 180° - 32.° = 148°$. However, the $148°$ is an impossible answer since $148° + 82°$ is greater than $180°$. Therefore, it is to our advantage to work with the largest angle first when using the law of cosines.

EXAMPLE 3   **Application: Law of Cosines**

The attic walls of the A-frame house in Figure 7.51 are in the form of an isosceles triangle. The equal legs are each 10 m long, and the included angle is $50°$. Determine the width of the floor.

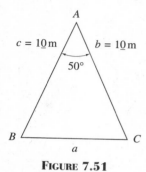

$c = 10$ m $b = 10$ m

$50°$

B _____ C

a

FIGURE 7.51

Solution   Since we know two sides and the included angle and want to find the third side, use the law of cosines.

$$a^2 = b^2 + c^2 - 2bc \cos A$$

Substituting the given values:

$$a^2 = 10^2 + 10^2 - 2(10)(10) \cos 50°$$
$$= 71.4$$
$$a = 8.5 \text{ m.}$$

The width of the of the floor is 8.5 m.   ◆

7.5 Exercises

In Exercises 1–12, determine the measures of the three other parts of the triangle.

1. $a = 12, b = 10, C = 78°$

2. $a = 9, b = 10, C = 47°$

3. $b = 45.0, c = 40.0, A = 59.9°$

4. $b = 38, c = 35, A = 42.1°$

5. $a = 52.0, c = 32.0, B = 104.3°$

6. $a = 17, c = 30, B = 23.5°$

7. $a = 20, b = 25, c = 28$

8. $a = 21, b = 22, c = 25$

9. $a = 17.8, b = 10.3, c = 14.2$

10. $a = 0.12, b = 0.08, c = 0.05$

11. $a = 6.743, b = 4.765, c = 8.512$

12. $a = 41.37, b = 59.73, c = 61.43$

13. If $a = 31.6$, $b = 53.4$, and $c = 34.9$, determine the measure of the smallest angle of the triangle correct to the nearest minute.

14. If $A = 48°$, $c = 15$ cm, and $b = 18$ cm, determine the measure of the largest angle of the triangle correct to the nearest minute.

15. Is it possible to use the law of cosines if we are given two sides and the included angle is $90°$? Explain! What is the result when $C = 90°$, $b = 31$ cm, and $a = 22$ cm?

16. A plot of land in the form of a parallelogram has sides of 43 m and 68 m. If one diagonal measures 72 m, determine the angles of the parallelogram.

17. The navigator of a reconnaissance plane determines that the distances from the plane to two radar stations, A and B, are 3592.7 m and 4591.2 m, respectively. If the angle between the two lines of sight to A and B is $123.4°$, determine the distance between the two stations (to the nearest tenth of a meter).

18. Two ships, as shown in the diagram, sail from the same dock P at the same time. One travels on a course of 035° at 22 knots. The other travels on a course of 158° at 20 knots. How far apart will they be in $2\frac{1}{2}$ hours? (*Hint:* One knot equals one nautical mph.)

19. To pull out a car stuck in the sand, forces of 415 lb and 325 lb are applied, acting at an angle of 49° 36′ with one another. What is the magnitude of the resultant force applied to the car?

20. Two magnetic forces of 60 N and 85 N act on an object to produce a resultant force of 95 N. Determine the angle that the resultant makes with each force.

21. Pat hikes into the wood in a northeast direction for 8.0 mi then turns due east and hikes 7.0 mi. How far is she from her starting point?

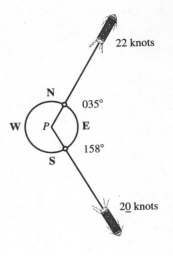

22. Two forces $F_1 = 675$ lb and $F_2 = 1250$ lb act on a point of a bridge girder at an angle of 85° to each other. Determine the resultant force $\mathbf{R}$ (the magnitude and the direction measured from F_1).

23. A guy wire runs 17.8 m from the top of a pole to the level ground. The original wire is replaced with a 25.7-m wire that runs from the top of the pole to a spot on the ground 10.8 m farther from the pole than the old wire. What angle does the new wire make with the ground? See the sketch.

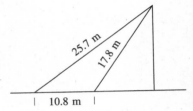

24. A radio antenna is to be placed on a sloping roof that makes an angle of 25° with the horizontal. The top of the antenna is 28 ft from the roof, and two of the guy wires are placed 14 ft from the base of the antenna as shown in the drawing. Determine the length of these two guy wires.

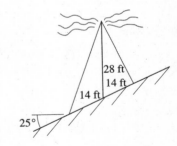

Review Exercises

In Exercises 1–8, determine the horizontal and vertical components of the vector quantities. Sketch the vector and its components.

1. $\mathbf{V} = 47$ cm/s /68°

2. $\mathbf{V} = 125$ cm/s /135°

3. $\mathbf{F} = 35$ lb /210°

4. $\mathbf{F} = 48$ lb /300°

5. $\mathbf{D} = 120$ km /−30°

6. $\mathbf{D} = 85$ km /−120°

7. $\mathbf{a} = 32.2$ cm/s² /1.72

8. $\mathbf{a} = 32.2$ cm/s² /0.52

In Exercises 9–16, determine the magnitude of the resultant vector and the angle it makes with the positive x-axis. Sketch, showing the resultant and the component vectors.

9. $\mathbf{F}_x = 241.0$ lb, $\mathbf{F}_y = 18.0$ lb

10. $\mathbf{F}_x = 85$ lb, $\mathbf{F}_y = 45$ lb

11. $\mathbf{D}_x = -14$ km, $\mathbf{D}_y = 15$ km

12. $\mathbf{D}_x = -21$ km, $\mathbf{D}_y = 24$ km

13. $\mathbf{F}_x = -0.472$ N, $\mathbf{F}_y = -0.965$ N

14. $\mathbf{F}_x = -98.76$ N, $\mathbf{F}_y = -45.72$ N

15. $\mathbf{V}_x = 0.007$ m/s, $\mathbf{V}_y = -0.028$ m/s

16. $\mathbf{V}_x = 4.2 \times 10^5$ m/s, $\mathbf{V}_y = 9.8 \times 10^3$ m/s

17. A swimmer heads due west at a rate of 3.0 mph, and the current of the river is flowing north at 2.0 mph. Determine the resultant vector. Sketch the resultant with respect to a point on the opposite bank of the river.

18. An AC circuit has two sources in series. Determine the total voltage if $V_1 = 22$ V at 0° and $V_2 = 33$ V at 90°.

19. An AC circuit has a resistance $\mathbf{R}$ of 33 Ω, acting at 0°, and a capacitive reactance $\mathbf{X}_c$ of 38 Ω, acting at 270°. Determine the impedance $\mathbf{Z}$ that is the resultant of these vectors.

20. A stone is thrown into the air with a speed of 22 m/s directed at an angle of 42° with the horizontal. Determine the magnitudes of the horizontal and vertical vector components of the stone's initial velocity.

21. A truck weighing 20,000 lb is placed on a 7° inclined plane. Determine the magnitudes of the vector components of the weight that are (a) parallel and (b) perpendicular to the inclined surface.

In Exercises 22–25, determine the resultant vector.

22. $\mathbf{A} = 8.7$ ft $/125°$
     $\mathbf{B} = 9.8$ ft $/65°$

23. $\mathbf{A} = 34.2$ cm $/47°$
     $\mathbf{B} = 67.4$ cm $/88°$

24. $\mathbf{V}_1 = 27.85/10°$
     $\mathbf{V}_2 = 37.21/170°$
     $\mathbf{V}_3 = 64.25/260°$

25. $\mathbf{V}_1 = 0.472/20°$
     $\mathbf{V}_2 = 0.365/40°$
     $\mathbf{V}_3 = 0.596/50°$

26. A 42-kg weight is on a ramp inclined at 31°. If there is no friction, what force, parallel to the ramp, is necessary to keep the weight at rest?

27. A 2400-lb car is on a hill inclined at 12°. How much force, pushing up the hill, is required to keep the car from rolling? (Assume no friction.)

28. A force of 45 lb is required to keep a weight of 80 lb from sliding down an inclined plane. What is the angle of inclination? (Assume no friction.)

29. A manual lawnmower requires a force of 50.0 N parallel to the ground in order to move it through the grass. If the handle is inclined at 38° to the ground, how much force is necessary along the handle? (See the illustration.)

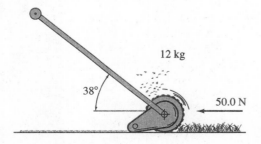

12 kg

38°

50.0 N

30. If the lawnmower in Exercise 29 weighs 12 kg and is pushed up a hill of 8°, how much force exerted along the handle is necessary to keep it moving through the grass? The handle is still inclined 38° to the slope, and you must take into account both the 50 N from Exercise 29 and the weight of the lawnmower. (See the illustration.)

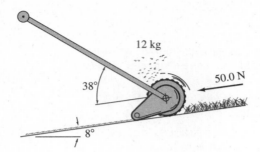

In Exercises 31–42, determine the measures of the three other parts of the triangle, if possible.

31. $A = 38.2°$, $B = 67.1°$, $b = 42.0$

32. $A = 72.3°$, $C = 84.5°$, $c = 72.0$

33. $a = 11.24$, $b = 14.72$, $A = 23.0°$

34. $a = 24.7$, $b = 28.65$, $A = 28°$

35. $a = 19$, $c = 15$, $C = 40°$

36. $b = 29.4$, $c = 26.7$, $C = 36.0°$

37. $A = 50.9°$, $b = 45.0$, $c = 40.0$

38. $B = 42.1°$, $a = 38.7$, $c = 35.6$

39. $a = 21.4$, $b = 22.8$, $c = 25.6$

40. $a = 33.7$, $b = 28.9$, $c = 51.9$

41. $C = 146.3°$, $a = 210$, $b = 120$

42. $a = 149$, $b = 233$ $c = 330$

43. A hiker leaves her camp and walks directly west 5.5 mi and then turns and walks 3.0 mi in a direction 40° south of east. She then decides to walk directly to her camp. How far and in what direction must she walk?

44. From town A a straight road leads to town B, a distance of 8.0 mi. From B another road leads directly to C, a distance of 7.0 miles. The angle between the two roads is 70°. A road is to be built directly from A to C. Determine the distance from A to C and the direction with respect to $\overline{AB}$. (See the diagram.)

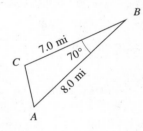

45. The maximum angle between the 18-in. handles of a pair of shrubbery clippers is 68°. When it is completely open, what is the distance between the ends of the handles? (See the diagram.)

46. A receiver on a football team gets the signal directing him to run to the right. He has to run 20 ft parallel to the line of scrimmage, turn at an angle of 120°, then run 40 ft to point A in order to be ready for the pass. If he starts 6 ft behind the position of the ball, and everything is in his favor for a successful pass, how many yards did he gain assuming that he was tackled at point A?

47. Two boys whose homes are 1.00 mi apart on a north-south line looked at a plane directly above their homes. The boy nearest to the plane read an elevation of 47° 10′; the boy farthest away read 38° 40′. How far away was the plane from the nearer boy?

48. A pilot wishes to fly northwest. If her airspeed is 180 km/h and there is a 60 km/h wind from due south, how many degrees west of north should she fly so that the resultant direction will be north-west?

49. From a land-based radar station it is determined that the bearing of ship A is $70°$ and it is a distance of 3.5 km from the station, while ship B has a bearing of $110°$ and is a distance of 5.2 km from the station. Determine the distance between the two ships.

50.* One side of a triangle is 3.472 times as long as another side x. The angle between them is $32°\ 01'$.
   a. Determine the other two angles.
   b. Determine the third side in terms of the side x.

51.* A local carnival has a ferris wheel. There are 18 seats equally spaced on the wheel. If the radius of the wheel is 18 ft, determine the distance between any two adjacent seats on the wheel. (See the illustration.)

✎ Writing About Mathematics

1. Two landscape designers were asked to design a triangular-shaped park. The city council does not understand why each design is shaped differently. Each designer claims that, with the information given, the shape of the triangle is correct. The only information given the designers was "two sides and an angle opposite one of the sides." Write a letter explaining why two solutions are possible.

2. You and a friend, having chopped down a tree, now need to move its trunk 25 ft to the edge of the road so that it can be trucked to the sawmill. You have attached two ropes to the log, and each of you is pulling on a rope. Using the sum of two vectors, explain how anyone can determine the force exerted by pulling on the two ropes.

Chapter Test

In Exercises 1 and 2, sketch the vectors on a coordinate system.

1. $\mathbf{A} = 4\underline{/30°}$

2. $\mathbf{B} = 6\underline{/150°}$

In Exercises 3 and 4, use the fact that $\mathbf{A} = 3.5\underline{/137°}$ to determine the following.

3. The horizontal component

4. The vertical component.

In Exercises 5 and 6, use the fact that $\mathbf{A} = 5.0\underline{/45°}$ and $\mathbf{B} = 7.0\underline{/215°}$ to determine the following.

5. $\mathbf{A} + \mathbf{B}$

6. $\mathbf{B} - \mathbf{A}$

7. Determine the resultant of $5.0\underline{/-45°} - 7.0\underline{/115°} + 3.0\underline{/210°}$.

In Exercises 8–13, determine the measure of the three other parts of the triangle, if possible. See the diagram for proper labeling.

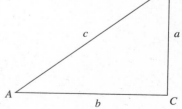

8. $A = 42°$, $B = 60°$, $b = 15$

9. $B = 35°$, $b = 12$, $a = 16$

10. $A = 75°$, $b = 10$, $c = 20$

11. $B = 35°$, $b = 16$, $a = 12$

12. $C = 160°$, $a = 8$, $b = 12$

13. $a = 12$, $b = 25$, $c = 24$

14. An aircraft is flying at 580 mph on a course of 75°. The surrounding wind is blowing west at 65 mph. What is the resultant speed of the plane over the ground?

15. A 20.0 kg crate rests on a 25° incline. Determine the normal force that acts on the crate. How large a force must friction supply if the box is not to slide?

16. A surveyor is measuring a large triangular plot of land. She determines that two angles and the included side of the triangle are 72°, 53°, and 2.3 mi. Determine the third angle and the other two sides.

17. From a point on the ground, the angles of elevation to the top of and to the bottom of an antenna on the top of a building are 43.2° and 28.7°, respectively. If the building is 145 ft high, how tall is the antenna?

18. Two forces whose magnitudes are 154 lb and 213 lb act on an object. The angle between the forces is 79.3°. Determine the magnitude of the resultant force.

Graphs of Trigonometric Functions

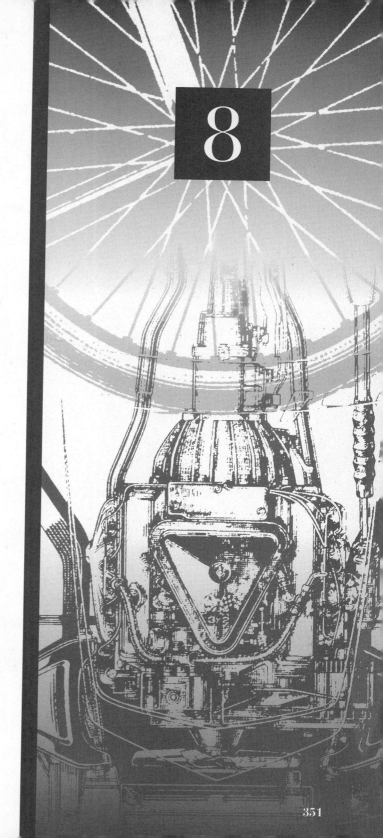

Automobile shock absorbers smooth the ride of the average car, creating a gentle up-and-down motion that buffers us from the many bumps in the road. This up-and-down motion causes the passengers to vary their position along an imaginary y-axis. The motion of the passenger is best illustrated by the graph of a trigonometric equation. A general equation for such a motion is:

$$y = A \sin Bt.$$

Changing the constants A or B dramatically alters the kind of ride we experience. For example, if A is increased (implying that the shock absorber is not working efficiently), we are more aware of the bumps. If B is allowed to increase, the up-and-down motion is more frequent. When a shock absorber is designed, the values of A and B are altered to produce the best ride. These ideas are explored in Example 6, Section 8.3.

8.1

Periodic Functions

Many natural functions are periodic functions. The seasons of the year, the alternation of night and day, and the phases of the moon are all periodic. Periodic means the time it takes to go through one complete cycle. The moon goes through one complete cycle in $27\frac{1}{3}$ days.

Mathematics encompasses many periodic functions. An example is the graph in Figure 8.1 that shows that the values of the function are repeated at regular intervals.

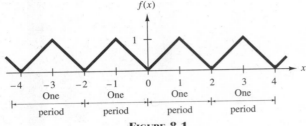

FIGURE 8.1

Examine Figure 8.1 closely to see that:

$$f(\boxed{-4}) = f(\boxed{-2}) = f(\boxed{0}) = f(\boxed{2}) = f(\boxed{4}) = 0,$$

and

$$f(\boxed{-3}) = f(\boxed{-1}) = f(\boxed{1}) = f(\boxed{3}) = 1.$$

This function goes through one complete **cycle** from $x = -4$ to $x = -2$, from $x = -2$ to $x = 0$, from $x = 0$ to $x = 2$, and from $x = 2$ to $x = 4$. This information tells us that the **period** p is two units. Another complete cycle of the function is from $x = -3$ to $x = -1$, a distance of two units, or one period.

DEFINITION 8.1

A function f is **periodic** if there is a positive number p such that for every real number x, whenever x is in the domain of f, so is $x + p$ and $f(x + p) = f(x)$. The smallest such positive number p is then called the **period** of the function.

EXAMPLE 1 **Determining the Period**

For the function in Figure 8.2, determine the period p.

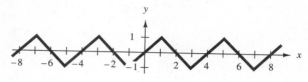

FIGURE 8.2

Solution The function goes through one complete cycle for each of the intervals from -8 to -4, -4 to 0, 0 to 4, and 4 to 8. Thus, the period is four units. Using the definition, we see that:

$$f(-4) = f(-4 + 4) = f(0) = 0$$
$$f(1) = f(1 + 4) = f(5) = 1.$$
$$f(-1) = f(-1 + 4) = f(3) = -1 \quad \blacklozenge$$

CAUTION We may be tempted to let $p = 2$ in Example 1. To see that this is incorrect, let $x = 1$ and substitute for x and p in the equality $f(x + p) = f(x)$. Thus, $f(1 + 2) \neq f(1)$ since $f(1 + 2) = -1$ and $f(1) = 1$. Clearly the function has only traveled through one-half of its full cycle. To determine the period p of the function, use the statement $f(x + p) = f(x)$ for all x. If we select $x = -4$, then $f(-4 + 2) = f(-2) = 0$ and $f(-4) = 0$, which gives the wrong impression—that 2 is the period. Therefore, always try more than one value to check. ∎

As shown in Figure 8.2, the graph goes through one complete cycle from 0 to 4, completing four cycles from -8 to 8.

DEFINITION 8.2

A **cycle** of a periodic function is a portion of the graph of the function from any point on the graph to the first point at which the graph starts repeating itself. (See Figure 8.3.)

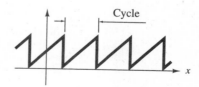

FIGURE 8.3

All six trigonometric functions are periodic, a fact that will become obvious to you as you learn to graph the functions. To illustrate that the sine function and the cosine function both have a period of 2π, let's use the unit circle. A **unit circle** is a circle with a radius of one. On the unit circle, the functional values repeat after each cycle, illustrating that the functions are periodic. Every central angle determines a point of the unit circle. The measure of the central angle of the entire circle is 2π. Central angles θ and $\theta + 2\pi$ determine the same point on the unit circle (Figure 8.4). Recall from right-triangle trigonometry that $x = r \cos \theta$ and $y = r \sin \theta$. Since we are dealing with a unit circle with $r = 1$, $x = \cos \theta$ and $y = \sin \theta$. Any point P on the unit circle will have coordinates $(\cos \theta, \sin \theta)$ or $(\cos (\theta + 2\pi), \sin (\theta + 2\pi))$. It follows that $\cos \theta = \cos (\theta + 2n\pi)$ and $\sin \theta = \sin (\theta + 2n\pi)$, where n is an integer. Key values for the first quadrant are given in Table 8.1. See Exercises 15 and 16 for problems on the unit circle.

TABLE 8.1

θ	$x = \cos \theta$	$y = \sin \theta$
0	1.000	0.000
$\dfrac{\pi}{6}$	0.866	0.500
$\dfrac{\pi}{4}$	0.707	0.707
$\dfrac{\pi}{3}$	0.500	0.866
$\dfrac{\pi}{2}$	0.000	1.000
$0 + 2\pi$	1.000	0.000
$\dfrac{\pi}{6} + 2\pi$	0.866	0.500
$\dfrac{\pi}{4} + 2\pi$	0.707	0.707
$\dfrac{\pi}{3} + 2\pi$	0.500	0.866
$\dfrac{\pi}{2} + 2\pi$	0.000	1.000

$$P(x, y) = (r \cos \theta, r \sin \theta)$$
$$= (r \cos (\theta + 2\pi), r \sin (\theta + 2\pi))$$
$$= (r \cos (\theta + 2n\pi), r \sin (\theta + 2n\pi))$$

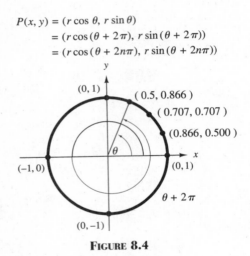

FIGURE 8.4

Is 2π the smallest possible positive value for p? The functional value, $\sin \theta = 1$, occurs only when $P = (0, 1)$. This happens only when $\theta = \dfrac{\pi}{2}$ or when $\theta = \dfrac{\pi}{2} + 2n\pi$. Therefore, the period p of the sine function is 2π.

By similar reasoning, $\cos \theta = -1$ only when $P = (-1, 0)$. This is true only when $\theta = \pi$ or $\theta = \pi + 2n\pi$. Thus, $\cos \pi = \cos (\pi + 2n\pi)$, and cosine has a period of 2π.

   Knowing that sine and cosine functions have periods of 2π is useful for graphing the functions. When graphing the sine and cosine functions, we only need to draw the graph for an interval of 2π. We can easily extend the graph for any values of θ by simply repeating what we have already drawn.

8.1 Exercises

In Exercises 1–10, examine the graphs. **a.** Determine whether the curve represents a periodic function. **b.** If it is periodic, determine a numerical value for the period p.

1.

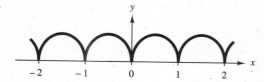

2.

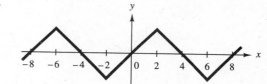

3.

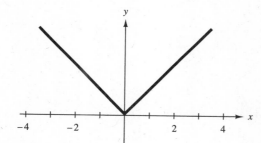

4.

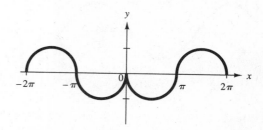

5.

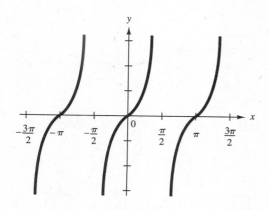

6.

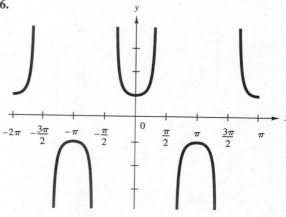

7.

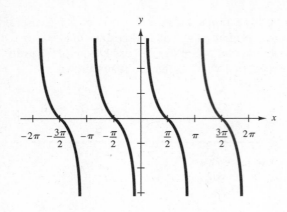

8.

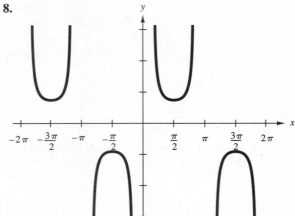

9.

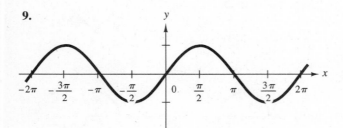

10.

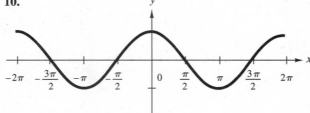

In Exercises 11–14, decide the following for each illustration. **a.** Determine whether the diagram is periodic. **b.** If it is periodic, determine the period.

11.

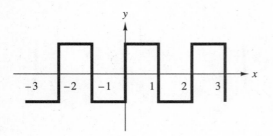

12.

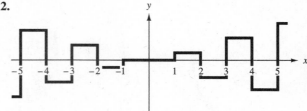

13.

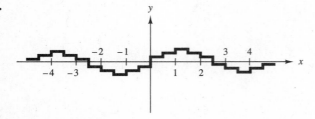

14.

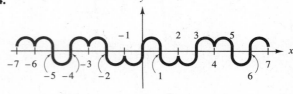

15. Calculate values for $x = \cos\theta$ and $y = \sin\theta$ in the 2nd, 3rd, and 4th quadrants. Let
$$\theta = \frac{2\pi}{3}, \frac{3\pi}{4}, \frac{5\pi}{6}, \pi, \frac{7\pi}{6}, \frac{5\pi}{4}, \frac{4\pi}{3}, \frac{3\pi}{2}, \frac{5\pi}{3}, \frac{7\pi}{4}, \frac{11\pi}{6}, 2\pi.$$

16. For the values of θ in Exercise 15 show that $\cos\theta = \cos(\theta + 2\pi)$ and $\sin\theta = \sin(\theta + 2\pi)$

8.2

Graphs of the Trigonometric Functions

This section deals with the basic graphs of the six trigonometric functions. These graphs will demonstrate some of the properties of the functions. Since all six functions are periodic, the concepts developed in Section 8.1 will help us construct these graphs.

Recall that in constructing a graph, we must identify the independent variable and the dependent variable. For the function $y = \sin\theta$, the angle θ (whether it is expressed in degrees or radians) is the independent variable, and y is the dependent variable.

The symbol for the angle may be any variable, x, t, θ, etc. However, in general discussions, θ will be used to represent the angle.

EXAMPLE 1   Graph Sine Function

Construct the graph of $y = \sin\theta$ for one cycle of the curve. Use the interval $0 \le \theta \le 2\pi$ ($0° \le \theta \le 360°$). Determine coordinates for intervals of $\frac{\pi}{4}$ rad (or 45°).

Solution   Because for the function $y = \sin\theta$, θ is the independent variable and y is the dependent variable, label the horizontal axis θ and the vertical axis y. The values for θ and $\sin\theta$ are given in Table 8.2. (To plot the point $\left(\frac{\pi}{4}, 0.7071\right)$, we move $\frac{\pi}{4}$ units to the right of the y-axis and 0.7071 units above the x-axis.) The graph of $y = \sin\theta$ is sketched in Figure 8.5.

TABLE 8.2

θ		$\sin\theta$
Radians	**Degrees**	
0	0°	0
$\frac{\pi}{4}$	45°	0.7071
$\frac{\pi}{2}$	90°	1.0000
$\frac{3\pi}{4}$	135°	0.7071
π	180°	0
$\frac{5\pi}{4}$	225°	−0.7071
$\frac{3\pi}{2}$	270°	−1.0000
$\frac{7\pi}{4}$	315°	−0.7071
2π	360°	0

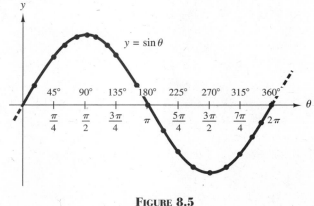

FIGURE 8.5

As we examine the graph of the curve $y = \sin \theta$, note that for the interval $0 < \theta < \pi \, (0 < \theta < 180°)$ the curve is above the horizontal axis. This is what we would expect, since the functional values are positive in the first and second quadrants, as we learned in Chapter 6. For the interval $\pi < \theta < 2\pi$ $(180° < \theta < 360°)$, the curve is below the horizontal axis, which corresponds to the fact that $\sin \theta$ is negative in the third and fourth quadrants.

Recall from the definition of periodic functions in Section 8.1 that as the curve reaches the value of 2π, it has gone through one complete cycle.

DEFINITION 8.3

The **period** of a trigonometric function is the number of degrees or radians needed to complete one cycle. (See Figure 8.6.)

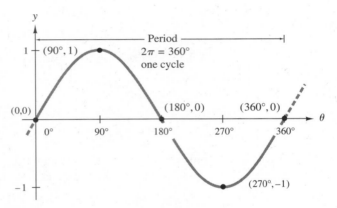

FIGURE 8.6

Since the sine function is periodic with period 360° (or 2π), we can identify five key points that help to sketch one cycle. The key points in degrees are (0°, 0), (90°, 1), (180°, 0), (270°, −1), (360°, 0), see Figure 8.6.

By studying Table 8.2 and Figure 8.5 we can determine the following properties of the sine function.

PROPERTIES OF THE SINE FUNCTION

1.   **The domain is the set of all real numbers.** The curve does not stop at 0 and 2π; it continues to the right and left indefinitely. (See Figure 8.7.)

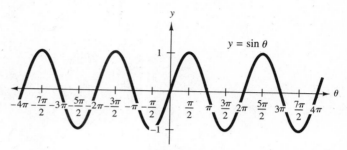

FIGURE 8.7

2.   **The range is the closed interval $-1 \le y \le 1$.** In every cycle of the sine function the function achieves a maximum and a minimum value. In the interval $0 \le \theta \le 2\pi$ the sine function has a maximum value of 1 at $\dfrac{\pi}{2}$ and a minimum value of -1 at $\dfrac{3\pi}{2}$.

3.   **The y-intercept is 0** (where the curve crosses the vertical axis).

4.   **The θ intercepts ($\sin \theta = 0$) are $0, \pm\pi, \pm2\pi, \pm3\pi \dots$** (where the curve crosses the horizontal axis, see Figure 8.7).

5.   **$\sin(\theta + \pi) = -\sin \theta$ for all θ.** Let $\theta = \dfrac{\pi}{4}$, then:

$$\sin\left(\frac{\pi}{4} + \pi\right) = \sin\left(\frac{5\pi}{4}\right) = -0.7071 = -\sin\frac{\pi}{4}.$$

6.   **$\sin(-\theta) = -\sin \theta$ for all θ.** (The sine is an odd function.) Let $\theta = \dfrac{3\pi}{4}$, then:

$$\sin\left(-\frac{3\pi}{4}\right) = -0.7071 = -\sin\frac{3\pi}{4}.$$

7.   **$\sin(\pi - \theta) = \sin \theta$ for all θ.** Let $\theta = \dfrac{7\pi}{4}$, then:

$$\sin\left(\pi - \frac{7\pi}{4}\right) = \sin\left(-\frac{3\pi}{4}\right) = -\sin\frac{3\pi}{4} - 0.7071 = \sin\frac{7\pi}{4}$$

8.   **The sine function is periodic with period 2π.** Figure 8.7 also illustrates that the curve goes through one complete cycle in 2π rad.

Some of the properties listed for the sine function are proven in Chapter 15 where trigonometric identities are discussed. To demonstrate that these statements are true, use the graphics calculator.

Graphing Trigonometric Functions with a Graphics Calculator

A first step in graphing any trigonometric function is to select the proper mode, degrees, or radians. The fx–7700G can be changed from the degree mode to the radian mode by pressing $\boxed{\text{SHIFT}}$ and $\boxed{\text{DRG}}$. When a menu appears at the bottom of the screen, press $\boxed{\text{F}_2}$ $\boxed{\text{EXE}}$, to put the calculator in the radian mode. To change the calculator to degree mode, press $\boxed{\text{SHIFT}}$ $\boxed{\text{DRG}}$ $\boxed{\text{F}_1}$ $\boxed{\text{EXE}}$.

To change the TI–82 to degrees or radians, press $\boxed{\text{MODE}}$ and then highlight the desired mode, Rad or Deg.

When we have selected the mode, we need to determine an appropriate range. For example, to graph four periods of $y = \sin \theta$, use the following range settings $(-4\pi, 4\pi, \dfrac{\pi}{2}, -2, 2, 0.5)$. Both fx–7700G and the TI–82 will accept π when setting the range.

If Property 5 for the sine function —that is, $\sin(\theta + \pi) = -\sin \theta$ for all θ— is true, then the graphs of $y = \sin(\theta + \pi)$ and $y = -\sin \theta$ should be identical. Graph both functions using the range $(-2\pi, 2\pi, \dfrac{\pi}{2}, -1.5, 1.5, 0.5)$. The graphs are given in Figure 8.8. Are the graphs identical? One way to check is to draw both graphs on the same coordinate plane.

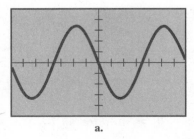   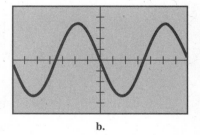

a.   b.

FIGURE 8.8

For Property 6, construct the graphs for $y = \sin(-\theta)$ and $y = -\sin \theta$. Are the graphs identical? For Property 7, construct the graphs for $y = \sin(\pi - \theta)$ and $y = \sin \theta$. Are the graphs identical?

TABLE 8.3

θ		$\cos \theta$
Radians	Degrees	
0	0°	1.0000
$\dfrac{\pi}{4}$	45°	0.7071
$\dfrac{\pi}{2}$	90°	0
$\dfrac{3\pi}{4}$	135°	−0.7071
π	180°	−1.0000
$\dfrac{5\pi}{4}$	225°	−0.7071
$\dfrac{3\pi}{2}$	270°	0
$\dfrac{7\pi}{4}$	315°	0.7071
2π	360°	1.0000

EXAMPLE 2   Graph Cosine Function

Construct the graph of $y = \cos \theta$ for the interval $0 \le \theta \le 2\pi$ ($0° \le \theta \le 360°$). Determine coordinates for intervals of $\dfrac{\pi}{4}$ rad (or 45°).

Solution   For the function $y = \cos \theta$, θ is the independent variable, and y is the dependent variable. Label the horizontal axis θ and the vertical axis y. The values for θ and $\cos \theta$ are given in Table 8.3. The graph of $y = \cos \theta$ is sketched in Figure 8.9.

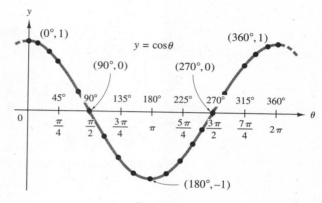

FIGURE 8.9

The graph of $\cos \theta$ illustrates the fact that the function is positive in the first quadrant $\left(0 < \theta < \dfrac{\pi}{2}\right)$ and the fourth quadrant $\left(\dfrac{3\pi}{2} < \theta < 2\pi\right)$. In these quadrants the curve is above the horizontal axis. It also shows that the function is negative in the second quadrant $\left(\dfrac{\pi}{2} < \theta < 2\pi\right)$ and the third quadrant $\left(2\pi < \theta < \dfrac{3\pi}{2}\right)$ since the curve lies below the horizontal axis.

Recall from the discussion of periodic functions in Section 8.1 that the cosine function is periodic, having a period of 360° (or 2π). The curve goes through one complete cycle in 360°. Using this information and the graph in Figure 8.9, we can identify five key points that help to sketch one cycle of the cosine curve. The key points in degrees are (0°, 1) , (90°, 0) , (180°, −1) , (270°, 0) , (360°, 1) , see Figure 8.9.

Using Table 8.3 and Figure 8.9, the following properties of the cosine functions may be determined.

PROPERTIES OF THE COSINE FUNCTION

1. **The domain is the set of all real numbers.** The curve does not stop at 0 and 2π; it continues to the right and left indefinitely. (See Figure 8.10.)
2. **The range is the closed interval $-1 \le y \le 1$.** In every cycle of the cosine function, the function achieves maximum and minimum values. The function achieves maximum values when $\theta = 0$ and 2π, and a minimum value when $\theta = \pi$.

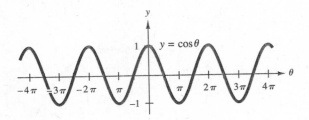

FIGURE 8.10

3. **The y-intercept is 1.**
4. **The θ intercepts ($\cos \theta = 0$) are $\pm\dfrac{\pi}{2}, \pm\dfrac{3\pi}{2} \pm\dfrac{5\pi}{2}, \ldots$** (See Figure 8.10.)
5. $\cos (\pi + \theta) = \cos (\pi - \theta) = -\cos \theta$ **for all θ.** Let $\theta = \dfrac{\pi}{4}$, then:

$$\cos\left(\pi + \frac{\pi}{4}\right) = \cos \frac{5\pi}{4} = -0.7071 = -\cos \frac{\pi}{4};$$

$$\cos\left(\pi - \frac{\pi}{4}\right) = \cos \frac{3\pi}{4} = -0.7071 = -\cos \frac{\pi}{4}.$$

6. $\cos (-\theta) = \cos \theta$ **for all θ.** Cosine is an even function. Let $\theta = \dfrac{\pi}{3}$, then:

$$\cos\left(\frac{-\pi}{3}\right) = 0.5000 = \cos \frac{\pi}{3}.$$

7. **The cosine function is periodic with period 2π.** Figure 8.10 also illustrates that the curve goes through one complete cycle in 2π rad.

If we shifted the cosine curve in Figure 8.10 $\dfrac{\pi}{2}$ rad to the right and placed it on the sine curve in Figure 8.7, it would fit exactly. Study Figure 8.7 to see that

TABLE 8.4 $y = \tan \theta$

θ		
Radian	**Degrees**	**tan θ**
$-\dfrac{\pi}{2}$	$-90°$	Undefined
$\dfrac{-5\pi}{12}$	$-75°$	-3.7321
$-\dfrac{\pi}{3}$	$-60°$	-1.7321
$-\dfrac{\pi}{4}$	$-45°$	-1.0000
$-\dfrac{\pi}{6}$	$-30°$	-0.5774
$-\dfrac{\pi}{12}$	$-15°$	-0.2679
0	$0°$	0
$\dfrac{\pi}{12}$	$15°$	0.2679
$\dfrac{\pi}{6}$	$30°$	0.5774
$\dfrac{\pi}{4}$	$45°$	1.0000
$\dfrac{\pi}{3}$	$60°$	1.7321
$\dfrac{5\pi}{12}$	$75°$	3.7321
$\dfrac{\pi}{2}$	$90°$	Undefined

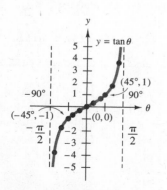

FIGURE 8.11

$\sin\left(\dfrac{\pi}{2} - \theta\right) = \cos \theta$. This result is proven in Chapter 15. We also can illustrate that this is a true equality by graphing $y = \sin\left(\dfrac{\pi}{2} - \theta\right)$ and $y = \cos \theta$ on the same graph.

To help us construct the graph of the tangent function, recall that $\tan \theta = \dfrac{\sin \theta}{\cos \theta}$, $\cos \theta \neq 0$. For those values of θ in which the cosine function is zero (see Property 4 for cosine), the tangent function is not defined. The tangent function is 0 for those values when $\sin \theta$ is 0 (see Property 4 for sine).

The value π is the smallest positive number p for which $\tan(\theta + p) = \tan \theta$ for all values of θ. Examples of this are $\tan 0 = \tan \pi = 0$, $\tan\left(\dfrac{\pi}{3}\right) = \tan\left(\dfrac{\pi}{3} + \pi\right) = 1.732$, and $\tan\left(-\dfrac{\pi}{2}\right) = \tan \dfrac{\pi}{2}$, which are undefined.

For $\dfrac{-\pi}{2} < \theta < \dfrac{\pi}{2}$, $\tan \theta$ is defined and takes on a unique value for each value of θ in the interval. The tangent function goes through one complete cycle in π radians (or 180°).

EXAMPLE 3   Graph of the Tangent Function

Construct the graph of $y = \tan \theta$ for the interval $-\dfrac{\pi}{2} < \theta < \dfrac{\pi}{2}$. Determine coordinates for intervals of $\dfrac{\pi}{12}$ rad (or 15°).

Solution   From Table 8.4 we can see that as we get closer to $\theta = -\dfrac{\pi}{2}$, the functional value, $\tan \theta$, becomes larger in the negative direction. In fact, if we calculate $\tan \theta$ for values of θ that are closer to $-\dfrac{\pi}{2}$, we would obtain negative values for the tangent function whose absolute values increase without bound. For example, if $\theta = -\dfrac{74\pi}{150}$, then $\tan \theta = -47.7$, and if $\theta = -\dfrac{149\pi}{300}$, then $\tan \theta = -95.5$. The same type of results can be shown for values of θ close to $\dfrac{\pi}{2}$. For values of θ that are closer and closer to $-\dfrac{\pi}{2}$ and $\dfrac{\pi}{2}$, the absolute values increase without bound, or in other words, they become undefined at $-\dfrac{\pi}{2}$ and $\dfrac{\pi}{2}$. Look at Figure 8.11, where the dashed lines show $\theta = -\pi 2$ and $\theta = \pi 2$; these lines are called asymptotes. The asymptotes will occur whenever the function is undefined; that is, $\theta = \pm\dfrac{\pi}{2}, \pm\dfrac{3\pi}{2}, \pm\dfrac{5\pi}{2}, \ldots$. The symbol $\pm\infty$ stands for "undefined" for $\tan \theta$ when $\theta = \pm\dfrac{\pi}{2}$, etc. ◆

Using Table 8.4 and Figure 8.11, the following properties of the tangent function can be determined.

PROPERTIES OF THE TANGENT FUNCTION

1. The domain is the set of real numbers except for $\pm\dfrac{\pi}{2}$, $\pm\dfrac{3\pi}{2}$, $\pm\dfrac{5\pi}{2}$, $\pm\dfrac{7\pi}{2}$, ...
2. The range is the set of all real numbers.
3. The y-intercepts are is 0.
4. The θ-intercepts are 0, $\pm\pi$, $\pm2\pi$, $\pm3\pi$, ...
5. Tan $(\theta + \pi)$ = tan θ for all θ in the domain of the tangent. That is, the tangent is a periodic function with period π. (See Fig. 8.12.)

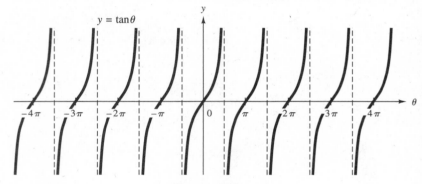

FIGURE 8.12

6. Tan $(-\theta) = \dfrac{\sin(-\theta)}{\cos(-\theta)}$ or   tan $(-\theta) = \dfrac{-\sin\theta}{\cos\theta}$;   therefore tan $(-\theta) = -$tan θ for all θ in the domain of the tangent. (The tangent is an odd function).
7. Tan θ is undefined (approaches infinity) as θ approaches $\pm\dfrac{\pi}{2}$, $\pm\dfrac{3\pi}{2}$, $\pm\dfrac{5\pi}{2}$, . . . from the left or the right. (See Figure 8.12.)

It is evident by now that $y = \tan\theta$ has a period of π (or 180°). Therefore, when graphing one period of the the function $y = \tan\theta$, we will generally find it easier and more convenient to graph the function about the origin as shown in Figure 8.11. This means that the origin is in the interval that contains one cycle of the function being graphed. To sketch one cycle of the tangent function, use the asymptotes $x = -\dfrac{\pi}{2}$ and $x = \dfrac{\pi}{2}$ and three key points, $(-45°, -1)$, $(0°, 0)$, and $(45°, 1)$, which are shown in Figure 8.11.

The graphs of the secant, cosecant, and cotangent functions are given in Figures 8.13, 8.14, and 8.15, respectively. As in the graph of the tangent, the asymptotes, shown by vertical dashed lines, are values of θ for which the secant, cosecant, and cotangent functions are undefined. A list of the properties of each of these functions is asked for in Exercises 10–12 at the end of this section.

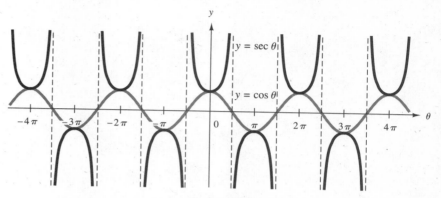

FIGURE 8.13

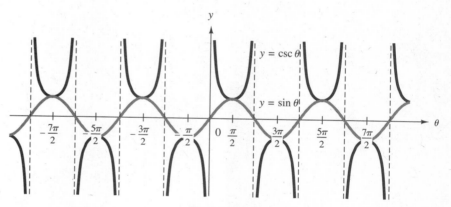

FIGURE 8.14

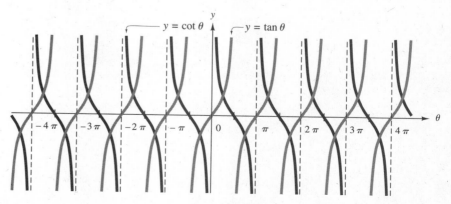

FIGURE 8.15

It is helpful in graphing the functions to recall that:

$$\sec \theta = \frac{1}{\cos \theta}, \cos \theta \neq 0$$ **The secant function is undefined where $\cos \theta = 0$.**

$$\csc \theta = \frac{1}{\sin \theta}, \sin \theta \neq 0$$ **The cosecant function is undefined where $\sin \theta = 0$.**

$$\cot \theta = \frac{1}{\tan \theta}, \tan \theta \neq 0$$ **The cotangent function is undefined where $\tan \theta = 0$.**

The cosecant and cotangent functions are undefined when $\theta = 0, \pm\pi, \pm 2\pi, \pm 3\pi, \ldots$, that is, when the sine and tangent functions are 0.

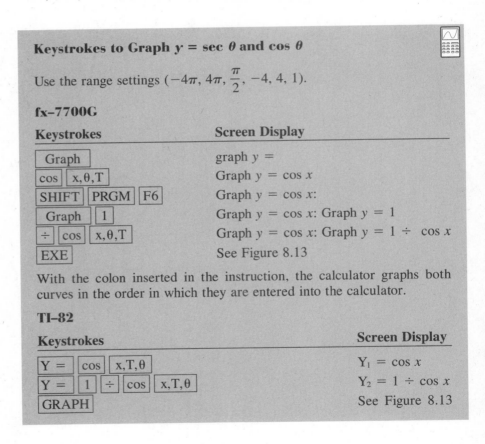

Keystrokes to Graph $y = \sec \theta$ and $\cos \theta$

Use the range settings $(-4\pi, 4\pi, \frac{\pi}{2}, -4, 4, 1)$.

fx–7700G

Keystrokes	Screen Display
Graph	graph $y =$
cos x,θ,T	Graph $y = \cos x$
SHIFT PRGM F6	Graph $y = \cos x$:
Graph 1	Graph $y = \cos x$: Graph $y = 1$
÷ cos x,θ,T	Graph $y = \cos x$: Graph $y = 1 \div \cos x$
EXE	See Figure 8.13

With the colon inserted in the instruction, the calculator graphs both curves in the order in which they are entered into the calculator.

TI–82

Keystrokes	Screen Display
Y = cos x,T,θ	$Y_1 = \cos x$
Y = 1 ÷ cos x,T,θ	$Y_2 = 1 \div \cos x$
GRAPH	See Figure 8.13

Key points for each of the six trigonometric functions that simplify graphing one cycle of the function are given in Table 8.5.

TABLE 8.5 Key Points for Graphing Trigonometric Functions

Functions			Key Points			
$\sin\theta$	radians	$(0, 0)$	$\left(\dfrac{\pi}{2}, 1\right)$	$(\pi, 0)$	$\left(\dfrac{3\pi}{2}, -1\right)$	$(2\pi, 0)$
	degrees	$(0°, 0)$	$(90°, 1)$	$(180°, 0)$	$(270°, -1)$	$(360°, 0)$
$\cos\theta$	radians	$(0, 1)$	$\left(\dfrac{\pi}{2}, 0\right)$	$(\pi, -1)$	$\left(\dfrac{3\pi}{2}, 0\right)$	$(2\pi, 1)$
	degrees	$(0°, 1)$	$(90°, 0)$	$(180°, -1)$	$(270°, 0)$	$(360°, 1)$
$\tan\theta$	radians	$\left(-\dfrac{\pi}{2}, -\infty\right)$	$\left(-\dfrac{\pi}{4}, -1\right)$	$(0, 0)$	$\left(\dfrac{\pi}{4}, 1\right)$	$\left(\dfrac{\pi}{2}, \infty\right)$
	degrees	$(-90°, -\infty)$	$(-45°, -1)$	$(0°, 0)$	$(45°, 1)$	$(90°, \infty)$
$\cot\theta$	radians	$(0, \infty)$	$\left(\dfrac{\pi}{4}, 1\right)$	$\left(\dfrac{\pi}{2}, 0\right)$	$\left(\dfrac{3\pi}{4}, -1\right)$	$(\pi, -\infty)$
	degrees	$(0°, \infty)$	$(45°, 1)$	$(90°, 0)$	$(135°, -1)$	$(180°, -\infty)$
$\sec\theta$	radians	$(0, 1)$	$\left(\dfrac{\pi}{2}, \infty\right)$	$(\pi, -1)$	$\left(\dfrac{3\pi}{2}, -\infty\right)$	$(2\pi, 1)$
	degrees	$(0°, 1)$	$(90°, \infty)$	$(180°, -1)$	$(270°, -\infty)$	$(360°, 1)$
$\csc\theta$	radians	$(0, \infty)$	$\left(\dfrac{\pi}{2}, 1\right)$	$(\pi, -\infty)$	$\left(\dfrac{3\pi}{2}, -1\right)$	$(2\pi, \infty)$
	degrees	$(0°, \infty)$	$(90°, 1)$	$(180°, -\infty)$	$(270°, -1)$	$(360°, \infty)$

8.2 Exercises

In Exercises 1–6, match the function with its graph.

1. $y = \sin \theta$

2. $y = \cos \theta$

3. $y = \tan \theta$

4. $y = \cot \theta$

5. $y = \sec \theta$

6. $y = \csc \theta$

a.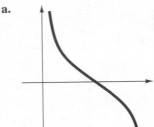

b.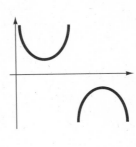

c.

d.

e.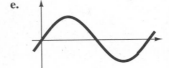

f.

In Exercises 7–12, make a list of the properties of the functions. The list should include domain, range, y-intercept, θ intercepts, period, and where the function is undefined.

7. Sine function
8. Cosine function
9. Tangent function
10. Cosecant function
11. Secant function
12. Cotangent function

In Exercises 13–24, graph each of the functions for the indicated interval.

13. $y = \sin \theta$ for values of θ at intervals of $15°$, $-180° \le \theta \le 180°$.
14. $y = \cos \theta$ for values of θ at intervals of $15°$, $-180° \le \theta \le 180°$.
15. $y = \tan \theta$ for values of θ at intervals of $45°$, $-270° \le \theta \le 270°$.

16. $y = \sin \theta$ for values of θ at intervals of $30°$, $0° \leq \theta \leq 360°$.

17. $y = \cos \theta$ for values of θ at intervals of $30°$, $0° \leq \theta \leq 360°$.

18. $y = \cos \theta$ for values of θ at intervals of $\dfrac{\pi}{4}$, $0° \leq \theta \leq 3\pi$.

19. $y = \sin \theta$ for values of θ at intervals of $\dfrac{\pi}{4}$, $0° \leq \theta \leq 3\pi$.

20. $y = \tan \theta$ for values of θ at intervals of $\dfrac{\pi}{4}$, $-\pi < \theta < \pi$.

21. $y = \csc \theta$ for values of θ at intervals of $45°$, $-180° \leq \theta \leq 360°$

22. $y = \sin \theta$ for values of θ at intervals of $\dfrac{\pi}{4}$ rad, $\dfrac{-5\pi}{2} \leq \theta \leq \dfrac{5\pi}{2}$.

23. $y = \cos \theta$ for values of θ at intervals of $60°$, $0° \leq \theta \leq 480°$.

24. $y = \tan \theta$ for values of θ at intervals of $\dfrac{\pi}{4}$ rad, $-\dfrac{5\pi}{2} < \theta < \dfrac{5\pi}{2}$.

In Exercises 25–28, graph the functions on the same set of axes for the indicated interval. Determine coordinates for intervals of $\dfrac{\pi}{6}$ or $30°$.

25. $y = \sin \theta$ and $y = \csc \theta$ for $-360° \leq \theta \leq 360°$.

26. $y = \sin \theta$ and $y = \csc \theta$ for $-270° \leq \theta \leq 270°$.

27. $y = \cos \theta$ and $y = \sec \theta$ for $\dfrac{-3\pi}{2} \leq \theta \leq \dfrac{3\pi}{2}$.

28. $y = \tan \theta$ and $y = \cot \theta$ for $-\pi \leq \theta \leq \pi$.

In Exercises 29–46, which trigonometric functions (sine, cosine, tangent, secant, cosecant, cotangent) have the following properties? There may be more than one answer.

29. The domain is the set of real numbers.

30. The domain is the set of real numbers excluding 0, $\pm\pi$, $\pm2\pi$, $\pm3\pi$, ...

31. The function has period of π.

32. The y-intercept is 1.

33. The range is the set of all real numbers.

34. There is no y-intercept.

35. The domain is the set of real numbers excluding $\pm\dfrac{\pi}{2}$, $\pm\dfrac{3\pi}{2}$, $\pm\dfrac{5\pi}{2}$, ...

36. The range is $-1 \leq y \leq 1$.

37. The range is $|y| \geq 1$.

38. The function has a period of 2π.

39. The θ-intercepts are 0, $\pm\pi$, $\pm2\pi$, $\pm3\pi$, ...

40. The θ-intercepts are $\pm\dfrac{\pi}{2}$, $\pm\dfrac{3\pi}{2}$, $\pm\dfrac{5\pi}{2}$, ...

41. The y-intercept is 0.

42. $f(-\theta) = f(\theta)$

43. $f(-\theta) = -f(\theta)$

44. $f(\theta + \pi) = f(\theta)$

45. $f(\theta + \pi) = -f(\theta)$

46. $f(\pi - \theta) = f(\theta)$

8.3

Sinusoids:
$$y = A \sin B\theta \text{ and}$$
$$y = A \cos B\theta$$

In the discussion of the cosine function in Section 8.2 we discovered that if the cosine curve is shifted $\frac{\pi}{2}$ rad to the right, it fits exactly on the sine curve. The family of curves that look like the sine curve are called **sinusoids.** The word *sinusoids* can be split into two parts. *Sinus-*, which comes from the same origin as *sine,* and *-oid,* a suffix that means *like.* Many "sine-like" curves are encountered in electronic testing devices, alternating currents, wave actions, and forces in rotating mechanisms. Look back at Figures 8.11 and 8.12. Note that the graph of tangent function is not a "sine-like" curve.

Before getting into specific applications, let's discuss the general forms of the sine and cosine curves. The most general form of the curves is expressed by the functions:

$$y = A \sin (B\theta + C) \text{ and } y = A \cos (B\theta + C).$$

For the functions, A, B, and C are constants, and θ is a variable that represents the angular measure.

Each of the constants A, B, and C has an individual effect on the curve. (For example, $y = A \sin \theta$, $y = \sin B\theta$, $y = \sin (\theta + C)$. Therefore, we will examine functions with each of the constants. Then we will consider the general form, $y = A \sin (B\theta + C)$ and $y = A \cos (B\theta + C)$. We will see that A affects the height of the curve, B affects the period, and C causes the curve to be shifted left or right.

Amplitude: $y = A \sin \theta$
and $y = A \cos \theta$

The maximum distance a sinusoid curve rises and falls above and below its horizontal axis is called the **amplitude.** The amplitude is shown by $|A|$. (Recall that $|A|$ means the absolute value of A.) The amplitudes of the functions $y = \boxed{1}$ $\sin \theta$, $y = \boxed{3} \cos \theta$, and $y = - \boxed{0.80} \cos \theta$ are $\boxed{1}$, $\boxed{3,}$ and $\boxed{0.80}$, respectively.

> **DEFINITION 8.4**
> The amplitude of $y = A \sin \theta$ and $y = A \cos \theta$ is the maximum value that maybe attained for y. Therefore, the amplitude $= |A|$.

EXAMPLE 1   Sine Graph: Effect of Amplitude

Sketch the graph of $y = 2 \sin \theta$ for $0 \le \theta \le 2\pi$, using intervals of $\frac{\pi}{4}$.

Solution   Use the values from the columns headed θ and $2 \sin \theta$ in Table 8.6. to sketch the graph of $y = 2 \sin \theta$, which is shown in Figure 8.16. This figure also shows $y = \sin \theta$ sketched on the same axes to illustrate the difference between $y = 2 \sin \theta$ and $y = \sin \theta$. Observe that the only difference between the two curves is their distances above and below the horizontal axis. For the

θ	$\sin \theta$	$2 \sin \theta$
0	0	0
$\dfrac{\pi}{4}$	0.707	1.41
$\dfrac{\pi}{2}$	1.000	2.00
$\dfrac{3\pi}{4}$	0.707	1.41
π	0	0
$\dfrac{5\pi}{4}$	−0.707	−1.41
$\dfrac{3\pi}{2}$	−1.000	−2.00
$\dfrac{7\pi}{4}$	−0.707	−1.41
2π	0	0

TABLE 8.6 $y = 2 \sin \theta$

function $y = 2 \sin \theta$, the maximum distance above and below the horizontal axis is two units. Two is the coefficient of $\sin \theta$ for the function $y = 2 \sin \theta$. Therefore, from the definition, the table of values, and the graph, we can determine that the amplitude of $y = 2 \sin \theta$ is 2. ◆

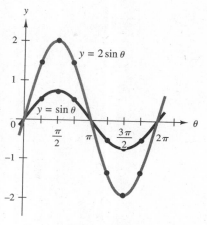

FIGURE 8.16

EXAMPLE 2   Cosine Graph: Effect of Amplitude

Use the graphics calculator to sketch the graphs of $y = -0.8 \cos \theta$, $y = 0.8 \cos \theta$, and $y = \cos \theta$ for $0 \le \theta \le 2\pi$.

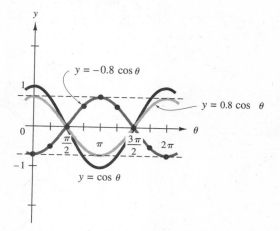

FIGURE 8.17

Solution   The three curves are sketched in Figure 8.17. Notice that the function $y = -0.8 \cos \theta$ is the mirror image of $y = 0.8 \cos \theta$. Therefore, when $y = 0.8 \cos \theta$ is positive, $y = -0.8 \cos \theta$ is negative; when $y = 0.8 \cos \theta$ is negative, $y = -0.8 \cos \theta$ is positive. The curves have the same shape and period. Another difference between the curves is the amplitude. The amplitude of $y = \boxed{-0.8}\cos\theta$ and $y = \boxed{0.8}\cos\theta$ is $\boxed{0.8}$, while the amplitude of $y = \cos\theta$ is $\boxed{1}$. ◆

fx–7700G

Keystrokes	Screen Display
GRAPH SHIFT (−) 0.8 cos x,θ,T	Graph $y = -0.8\cos x$
SHIFT PRGM F6	Graph $y = -0.8\cos x$:
GRAPH 0.8 cos x,θ,T	
SHIFT PRGM F6	Graph $y = -0.8\cos x$:
	Graph $y = 0.8\cos x$:
GRAPH cos x,θ,T	Graph $y = -0.8\cos x$:
	Graph $y = 0.8\cos x$:
	Graph $y = \cos x$
EXE	Figure 8.18

TI–82

Keystrokes	Screen Display
Y = (−) 0.8 cos x,T,θ ENTER	$Y_1 = -0.8\cos x$
0.8 cos x,T,θ ENTER	$Y_2 = 0.8\cos x$
cos x x,T,θ	$Y_3 = \cos x$
GRAPH	Figure 8.18

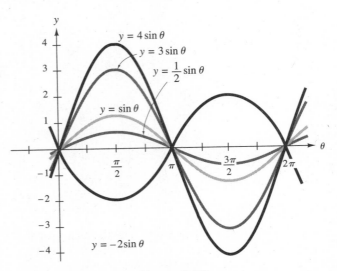

FIGURE 8.18

The amplitude of the function $y = \boxed{4} \sin \theta$ is $\boxed{4}$. Sketch the graph of $y = A \sin \theta$ when $A = -2, \frac{1}{2}, 1, 3,$ and 4 using your calculator. Note the effect of the change of the amplitude on the curves. Compare your results with Figure 8.18.

PROPERTIES OF $y = A \sin \theta$ AND $y = A \cos \theta$

1. The amplitude of $y = A \sin \theta$ and $y = A \cos \theta$ is given by the absolute value of A.
2. If the coefficient of the function is negative, the graph of the curve is a mirror image (with respect to the horizontal axis) of the same function with a positive coefficient.
3. The period of both of these functions is 2π rad or $360°$.

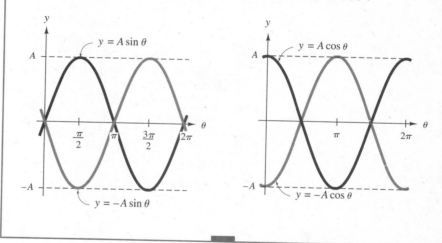

Period: $y = \sin B\theta$ and $y = \cos B\theta$

To determine the effect that B has on the curve, we will examine functions of the form $y = A \sin (B\theta + C)$ and $y = A \cos (B\theta + C)$ where $A = 1$ and $C = 0$.

Remember that the period of a function is the distance or time required to complete one cycle. For the functions $y = \sin B\theta$ and $y = \cos B\theta$, the period can be determined by dividing 2π or $360°$ by the absolute value of B ($|B|$).

DEFINITION 8.5

The period of $y = A \sin B\theta$ and $y = A \cos B\theta$ is:

$$\frac{2\pi}{|B|} \quad \text{or} \quad \frac{360°}{|B|}.$$

EXAMPLE 3   **Sine Graph: Effect of Period**

Sketch the following graphs.

a. $y = \sin 3\theta$ for $0 \leq \theta \leq 360°$

b. $y = \sin \dfrac{1}{3}\theta$ for $0 \leq \theta \leq 6\pi$

Solution

a. Use the definition of period to determine the period of $y = \sin \boxed{3}\,\theta$.

$$\frac{360°}{|B|} = \frac{360°}{|\boxed{3}|} = 120°$$

With a period of 120°, the curve goes through three cycles in 360°. For the function $y = \sin \theta$, $B = 1$, and the curve goes through one cycle in 360° degrees. Note that when the value of B is a counting number greater than 1, the length of the period is decreased. The first period of the function $y = \sin 3\theta$ is in the interval from $0° \leq \theta \leq 120°$. This interval can be divided into four equal regions, corresponding to the four quadrants, $0° \leq \theta \leq 30°$, $30° \leq \theta \leq 60°$, $60° \leq \theta \leq 90°$, and $90° \leq \theta \leq 120°$. The key values $(0°, 30°, 60°, 90°,$ and $120°)$ of this interval determine the key points used in sketching the graph (see Table 8.7 and Figure 8.19). The second and third periods are in the intervals from $120° \leq \theta \leq 240°$ and $240° \leq \theta \leq 360°$, respectively. The key values for these periods can be found by dividing each period into four equal regions. These values are given in Table 8.7.

TABLE 8.7 $y = \sin 3\theta$

θ	3θ	$\sin 3\theta$
0°	0°	0
30°	90°	1
60°	180°	0
90°	270°	−1
120°	360°	0
150°	450	1
180°	540	0
210°	630°	−1
240°	720°	0
270°	810°	1
300°	900°	0
330°	990°	−1
360°	1080°	0

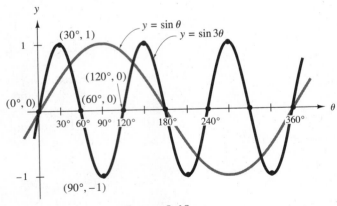

FIGURE 8.19

TABLE 8.8 $y = \sin \dfrac{1}{3}\theta$

θ	$\dfrac{1}{3}\theta$	$\sin \dfrac{1}{3}\theta$
0	0	0
$\dfrac{\pi}{2}$	$\dfrac{\pi}{6}$	0.50
π	$\dfrac{\pi}{3}$	0.87
$\dfrac{3\pi}{2}$	$\dfrac{\pi}{2}$	1
2π	$\dfrac{2\pi}{3}$	0.87
$\dfrac{5\pi}{2}$	$\dfrac{5\pi}{6}$	0.50
3π	π	0
$\dfrac{7\pi}{2}$	$\dfrac{7\pi}{6}$	-0.50
4π	$\dfrac{4\pi}{3}$	-0.87
$\dfrac{9\pi}{2}$	$\dfrac{3\pi}{2}$	-1
5π	$\dfrac{5\pi}{3}$	-0.87
$\dfrac{11\pi}{2}$	$\dfrac{11\pi}{6}$	-0.50
6π	2π	0

b. $y = \sin \dfrac{1}{3}\theta$.

Use the definition of period to determine the period of $y = \sin \dfrac{1}{3}\theta$.

$$\frac{2\pi}{|B|} = \frac{2\pi}{\dfrac{1}{3}} = 6\pi$$

For this function with $|B|$ less than 1, note that the length of the period is increased, becoming greater than 2π. (See Figure 8.20.) With a period of 6π, the curve $y = \sin \dfrac{1}{3}\theta$ goes through one cycle in 6π radians, whereas the curve of $y = \sin \theta$ goes through three cycles in 6π radians. Even though the period is 6π, the interval from $0 \le \theta \le 6\pi$ can be divided into four regions corresponding to the four quadrants $\left(0 \le \theta \le \dfrac{3\pi}{2}, \dfrac{3\pi}{2} \le \theta \le 3\pi, 3\pi \le \theta \le \dfrac{9\pi}{2}, \text{ and } \dfrac{9\pi}{2} \le \theta \le 6\pi\right)$. (See Table 8.8.) These intervals give us the key values $0, \dfrac{3\pi}{2}, 3\pi, \dfrac{9\pi}{2}$, and 6π that determine the key points for sketching the curve.

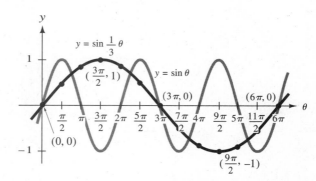

FIGURE 8.20   ◆

CAUTION   Does the period you calculated for the function make sense? A quick check for the function $y = \sin B\theta$ follows.

1. If $|B| > 1$, then the period is less than $360°$ or 2π.
2. If $|B| = 1$, then the period is equal to $360°$ or 2π.
3. If $|B| < 1$, then the period is greater than $360°$ or 2π. ■

GUIDELINES FOR SKETCHING $y = A \sin B\theta$ AND $y = A \cos B\theta$

Step 1. Determine the amplitude. (Amplitude $= |A|$)

Step 2. Determine the period. $\left(\text{Period} = \dfrac{360°}{|B|} \text{ or } \dfrac{2\pi}{|B|}.\right)$

Step 3. Determine the key points for the period. Determine the left and right end points of the period. Each period is divided into four equal intervals corresponding to the four quadrants. The end points of these intervals are the five key values, which are found by dividing the period by 4, multiplying the result by 0, 1, 2, 3, and 4, and adding each of the products to the left end point of the interval. Substitute the key values into the function to determine the functional values. With these results we can write the ordered pairs that we call the key points.

EXAMPLE 4   Graphing Using Amplitude, Period, and Key Points

Sketch one period of $y = 4 \cos \dfrac{3}{5}\theta$.

Solution   Use the guidelines to sketch the function $y = \boxed{4} \cos \dfrac{3}{5} \theta$.

a.   Amplitude $= \boxed{4.}$ **Step 1**

b.   Period $= \boxed{\dfrac{2\pi}{\dfrac{3}{5}}}$ or $\dfrac{10\pi}{3}$.   **Step 2**

c.   Find the key points for the period.   **Step 3**

1.   Let the left end point of the interval be 0. Then the right end point is $0 + \dfrac{10\pi}{3}$ or $\dfrac{10\pi}{3}$. We could select any other value for the left end point, and the width of an interval would still be $\dfrac{10\pi}{3}$.

2.   Divide the period by 4: $\dfrac{1}{4}\left(\dfrac{10\pi}{3}\right) = \dfrac{5\pi}{6}$.

3.   Multiply $\dfrac{5\pi}{6}$ by 0, 1, 2, 3, and 4 to obtain the key values $\boxed{0}$, $\boxed{\dfrac{5\pi}{6}}$, $\boxed{\dfrac{10\pi}{6}}$, $\boxed{\dfrac{15\pi}{6}}$, and $\boxed{\dfrac{20\pi}{6}}$. Note in this case that the fractions are not reduced to make the graphing easier. With a denominator of 6, divide the horizontal scale (θ axis) so each unit is comprised of six parts (Figure 8.21).

4.   Reduce the factional values found in part c and substitute 0, $\dfrac{5\pi}{6}$, $\dfrac{5\pi}{3}$, $\dfrac{5\pi}{2}$, and $\dfrac{10\pi}{3}$ for θ in the function $y = 4\cos\dfrac{3}{5}$ to calculate the corresponding values for y. The key points are $\boxed{(0, 4)}$, $\boxed{\left(\dfrac{5\pi}{6}, 0\right)}$, $\boxed{\left(\dfrac{5\pi}{3}, -4\right)}$, $\boxed{\left(\dfrac{5\pi}{2}, 0\right)}$, and $\boxed{\left(\dfrac{10\pi}{3}, 4\right)}$.

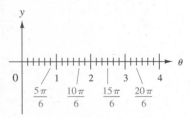

FIGURE 8.21

Now use these ordered pairs to sketch one period of $y = 4 \cos \dfrac{3}{5}\theta$. (See Figure 8.22.)

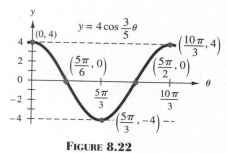

FIGURE 8.22

Example 5 requires using our graphing skills to interpret an application.

EXAMPLE 5   **Application: Respiratory Cycle; a Sine Function**

When a person is at rest, one respiratory cycle is comprised of inhaling and then exhaling. The rate of flow y (in liters per second) of air in and out of a person's lungs at time t (in seconds) is given by $y = 0.5 \sin \dfrac{2\pi t}{5}$. Sketch one cycle of the curve.

Solution   Use the suggested guidelines to sketch the function $y = 0.5 \sin \dfrac{2\pi t}{5}$

a. Amplitude $= |0.5| = 0.5$.   **Step 1**

b. Period $= \dfrac{2\pi}{|2\pi/5|} = 5$.   **Step 2**

c. Find the five key points for the period.   **Step 3**
   1. Let the left end point of the period be 0. Then the right end point is 0 + 5 or 5.

   2. Divide the period by 4: $\dfrac{5}{4} = 1.25$.

   3. Multiply 1.25 by 0, 1, 2, 3, and 4. The results are 0, 1.25, 2.50, 3.75, and 5.00. (For ease in graphing, divide the horizontal scale, the t-axis, so that each unit is comprised of four parts.)

   4. Substitute the key values 0, 1.25, 2.50, 3.75, and 5.00 for t in the function $y = 0.5 \sin \dfrac{2\pi t}{5}$. Then calculate the corresponding values for y. The key points are $(0, 0)$, $(1.25, 05)$, $(2.50, 0)$, $(3.75, -0.5)$, and $(5, 0)$.

Use the key points to sketch $y = 0.5 \sin \dfrac{2\pi t}{5}$ (See Figure 8.23.)   ◆

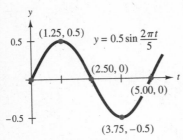

$$y = 0.5 \sin \frac{2\pi t}{5}$$

(1.25, 0.5)

(2.50, 0)

(5.00, 0)

(3.75, −0.5)

FIGURE 8.23

Examine Figure 8.23 to observe several things.

1. One complete cycle of inhaling and exhaling takes 5 units of time; the time could be in seconds.
2. In the intervals $0 < t < 1.25$ and $3.75 < t < 5.00$, the curve is increasing; a person would be inhaling.
3. In the interval $1.25 < t < 3.75$, the curve is decreasing; a person would be exhaling.
4. The point $(1.25, 0.5)$ is a maximum point on the curve, and the maximum amount of air would be in the lungs.
5. The point $(3.75, -0.5)$ is a minimum point on the curve, and a minimum amount of air would be in the lungs.

On the first page of this chapter questions were raised about designing shock absorbers. In Example 6, the graphs help determine which shock absorber gives the best ride.

EXAMPLE 6   **Application: Shock Absorber Problem**

In designing a shock absorber, an engineer takes into consideration a number of things so that the car rides smoothly. Looking at a very simplified form of the car, the ride is dependent upon the spring in the shock absorber and the mass of the car. The bounce in the ride can be expressed as a sinusoidal curve, $y = A \sin Bt$.

a. The engineer tests three shocks in the same car, obtaining the following results: Shock 1—$A = 3$, $B = 1$; Shock 2—$A = 5$, $B = 1$; Shock 3—$A = 7$, $B = 1$.
    Sketch the three curves and determine which value of A will give the best ride.
b. The oscillating motion caused by the shocks is affected by the mass of the car. Assume that a particular shock is in both a large and a small car and that the functions defining the oscillating motion are $y = 3 \sin 1.5t$ for the larger car and $y = 3 \sin 4.5t$ for the smaller car. Sketch the graph to determine which car gives the better ride for that shock.

Solution

a. With $B = 1$, the period of all of the functions is 2π. For $y = 3 \sin t$, the key points are $(0, 0)$, $\left(\frac{\pi}{2}, 3\right)$, $(\pi, 0)$, $\left(\frac{3\pi}{2}, -3\right)$, and $(2\pi, 0)$. For $y = 5 \sin t$, the key points are $(0, 0)$, $\left(\frac{\pi}{2}, 5\right)$, $(\pi, 0)$, $\left(\frac{3\pi}{2}, -5\right)$, and $(2\pi, 0)$. For $y = 7 \sin t$, the key points are $(0, 0)$, $\left(\frac{\pi}{2}, 7\right)$, $(\pi, 0)$, $\left(\frac{3\pi}{2}, -7\right)$, $(2\pi, 0)$. These three curves are sketched in Figure 8.24. Look at Figure 8.24c where you see that an amplitude of 7 causes a greater vertical displacement (a bigger bounce) and a less comfortable ride. Thus, the function in Figure 8.24a, with the smallest amplitude, 3 in this case, gives the smoothest ride.

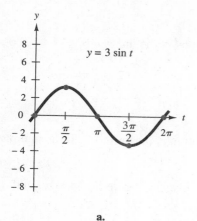

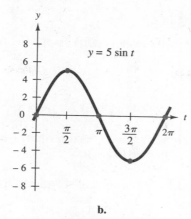

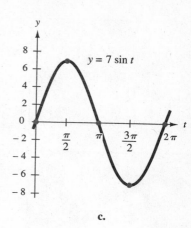

a.

b.

c.

FIGURE 8.24

b. The amplitude, $A = 3$, is the same for both functions. For $y = 3 \sin 1.5t$, the period is 4.2; the key points are $(0, 0)$, $(1.0, 3)$, $(2.1, 0)$, $(3.1, -3)$, and $(4.2, 0)$. For $y = 3 \sin 4.5t$, the period is 1.4; the key points are $(0, 0)$, $(0.4, 3)$, $(0.7, 0)$, $(1.1, -3)$, $(1.4, 0)$, $(1.8, 3)$, $(2.1, 0)$ $(2.5, -3)$, $(2.8, 0)$, $(3.2, 3)$, $(3.5, 0)$, $(3.9, -3)$, and $(4.2, 0)$. The curves are sketched in Figure 8.25.

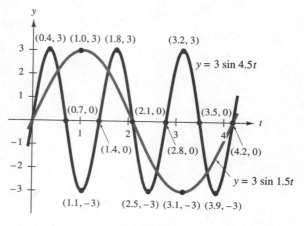

FIGURE 8.25

The graph shows that the heavier car has fewer vertical displacements (bounces) per unit of time; therefore, it gives a better ride with the same shock absorbers than the lighter car. For the sinusoid $y = A \sin Bt$, A is affected by the elasticity of the spring of the shock and B is affected by the spring of the shock and the mass of the car. This example demonstrates that a sinusoid is an effective tool for interpreting the motion. ◆

A BASIC computer program which can be used to find the key values is found on the tutorial disk.

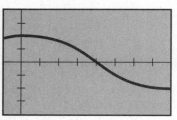

a.

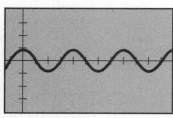

b.

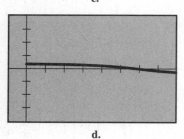

c.

EXAMPLE 7

Use a graphics calculator to graph the functions in the interval $0° \leq \theta \leq 360°$.

a. $y = 4 \cos 2\theta$ **b.** $y = 2 \cos \dfrac{\theta}{2}$

c. $y = \dfrac{1}{2} \cos 3\theta$ **d.** $y = \dfrac{1}{4} \cos \dfrac{\theta}{3}$

For each of the functions determine the amplitude, the period, and the number of cycles in $0° \leq \theta \leq 360°$.

Solution   The graphs of the functions are given in Figure 8.26a, b, c, and d, respectively.

a.   The amplitude is 4. The period $= \dfrac{360°}{2}$ or 180°. There are two complete cycles of the curve in the interval $0° \leq \theta \leq 360°$, as shown in Figure 8.26a.

b.   The amplitude is 2. The period $= \dfrac{360°}{\frac{1}{2}}$ or 720°. There is one-half of a complete cycle of the curve in 360°, as shown in Figure 8.26b.

c.   The amplitude is $\dfrac{1}{2}$. The period $= \dfrac{360°}{3}$ or 120°. There are three complete cycles of the curve in 360°, see Figure 8.26c.

d.   The amplitude is $\dfrac{1}{4}$. The period $= \dfrac{360°}{\frac{1}{3}}$ or 1080°. There is one-third of a cycle in 360°, see Figure 8.26d.   ◆

d.

FIGURE 8.26

8.3 Exercises

In Exercises 1–12, determine the following. **a.** The amplitude **b.** The period of each function **c.** The number of cycles in $0 \leq \theta \leq 2\pi$.

1. $y = 3 \sin \theta$

2. $y = \dfrac{1}{2} \cos \theta$

3. $y = 0.75 \cos \theta$

4. $y = 1.3 \sin \theta$

5. $y = \sin 3\theta$

6. $y = \cos \dfrac{\theta}{2}$

7. $y = -6 \cos \dfrac{\theta}{4}$

8. $y = 2 \sin \pi\theta$

9. $y = \dfrac{1}{3} \sin \dfrac{\theta}{3}$

10. $y = -\pi \cos \dfrac{\pi}{2}\theta$

11. $y = -\dfrac{1}{3} \cos \dfrac{1}{3}\theta$

12. $y = \dfrac{5}{3} \sin \dfrac{5}{3}\theta$

In Exercises 13–16, match the function with its graph.

13. $y = \cos \dfrac{1}{2}\theta$

a.
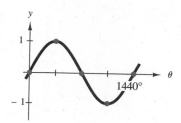

b.
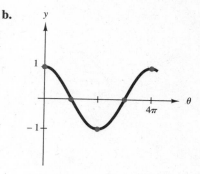

14. $y = 2 \cos 3\theta$

15. $y = \sin \dfrac{1}{4}\theta$

c.
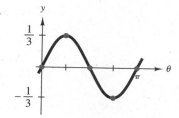

d.
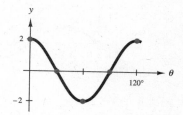

16. $y = \dfrac{1}{3} \sin 2\theta$

In Exercises 17–28, sketch the graph of each function for one complete cycle starting with the origin. Label the key points.

17. $y = 2 \sin \theta$ **18.** $y = \sin 2\theta$ **19.** $y = \cos \dfrac{\theta}{2}$ **20.** $y = \dfrac{1}{2} \cos \theta$

21. $y = 3 \cos 2\theta$ **22.** $y = 2 \cos 2\theta$ **23.** $y = 0.5 \sin \dfrac{\pi}{4}\theta$ **24.** $y = 4 \sin 0.5\theta$

25. $y = -\dfrac{1}{2} \sin 2\theta$ **26.** $y = -2 \sin \dfrac{\theta}{2}$ **27.** $y = -\dfrac{2}{3} \cos \dfrac{2}{3}\theta$ **28.** $y = -\pi \cos \dfrac{\pi}{2}\theta$

In Exercises 29–32, sketch the graphs of the functions and label the key points.

29. $y = 3 \sin \theta$ for $-\pi \le \theta \le \pi$ using values of θ at intervals of $\dfrac{\pi}{12}$.

30. $y = 4 \cos \theta$ for $-\pi \le \theta \le \pi$ using values of θ at intervals of $\dfrac{\pi}{12}$.

31. $y = \dfrac{1}{2} \cos \theta$ for two cycles starting with $-\dfrac{3\pi}{2}$.

32. $y = 0.75 \sin \theta$ for two and one-half cycles starting with $-270°$.

33. Find an interval on which $\sin 2\theta$ increases from -1 to 1.

34. Find an interval on which $\cos 4\theta$ decreases from 1 to -1.

In Exercises 35–38, write a sine or cosine function that describes the path of the curve.

35.

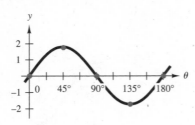

36.

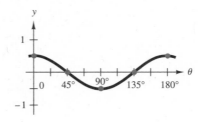

37.

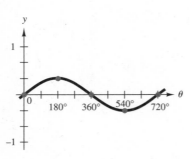

38.

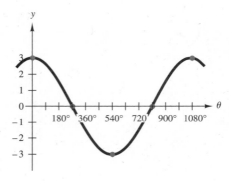

39. A weight at the end of a spring is pulled downward and then released. The position
    y (in centimeters) of the weight at time t (in seconds) is given by $y = \dfrac{5}{2} \cos 8t$.

    a. Determine the amplitude and period of the spring motion.
    b. Graph one cycle starting at $t = 0$.
    c. Compute y when $t = 1.5$ seconds.

40. After exercising for a few minutes, a person has a respiratory cycle for which the
    velocity of air flow is approximated by $v = 1.75 \sin \dfrac{\pi t}{2}$ (t in seconds).

    a. Determine the time for one full respiratory cycle (inhale and exhale).
    b. Determine the number of cycles per minute.
    c. Sketch one cycle of the function.

41. A 2-lb mass vibrates on a spring according to the equation $y = 0.25 \cos 2.5t$ where
    y is measured in feet and t in seconds.
    a. Determine the time required to go through one complete cycle.
    b. Determine the number of cycles in a minute.
    c. Sketch one cycle of the curve.

42. The wave theory of light suggests that the light has an electrical component associ-
    ated with it, which is referred to as an electric field. The electric field of a light wave
    varies in a periodic manner as shown by $E = E_0 \sin \omega t$. Here ω is related to the
    frequency of the light. In particular: $\omega = 2\pi f$. Recall that the general equation for
    a sinusoidal wave is $y = A \sin Bt$. Use the information in the chart to answer the
    following questions.

intense red light $E = 10 \sin (2\pi \cdot 4.3 \times 10^{14}t)$

dim red light $E = 1 \sin (2\pi \cdot 4.3 \times 10^{14}t)$

blue light $E = 10 \sin (2\pi \cdot 7.0 \times 10^{14}t)$

a. Does the brightness of light correspond to A or B in the equation?
b. Does the color of light correspond to A or B in the above equation?
c. What is the frequency of red light?
d. What is the frequency of blue light?

43. Sound waves may be represented by the equation $l = l_0 \sin \omega$ where l represents the loudness of the sound. Consider two different notes of the same intensity:

$$L = 10 \sin (2\pi \cdot 262t) = \text{C note}$$
$$L = 10 \sin (2\pi \cdot 330t) = \text{E note}$$

a. Which note, the C or the E, has the highest frequency?
b. If you wanted to make a C note twice as loud, how might you write the equation?
c. A G note has a frequency of 392 cps. Represent the intensity of a G note which has the same loudness as the original E note.

44. The resonant frequency of oscillations f_0 is dependent upon the mass and the spring constant.

$$f_0 = \frac{1}{2\pi}\sqrt{\frac{k}{m}},$$

where f_0 is the resonant frequency, k is the spring constant and m is the mass. (Resonance is the effect produced when the amplitude of a body is increased greatly by a periodic force at the same, or nearly the same, frequency.) The period of the oscillations, T, is given by the following.

$$T = 2\pi\sqrt{\frac{m}{k}}$$

In Example 6, if the heavier automobile is 2200 lb, the lighter automobile is 1500 lb, and the spring constant $k = 50$ lb/in^2 determine the following.
a. The resonant frequency for each car
b. The period of oscillation for each car
c. Whether the results in a and b agree with the conclusion reached in Example 6

8.4

General Sinusoids: $y = A \sin (B\theta + C)$ and $y = A \cos (B\theta + C)$

In Section 8.3, for the general curve $y = A \sin (B\theta + C)$ when $C = 0$, we learned that the constants A and B affect the amplitude and the period of the curve, respectively. We also observed that neither has an effect on the other. Now let's look at the effect that the constant C has on the curve.

The value of C has the effect of shifting the curve to the right or the left. For example, in graphing $y = \sin \theta$, the origin is considered the starting point for

TABLE 8.9 $y = \sin\left(\theta + \dfrac{\pi}{4}\right)$

θ	$\theta + \dfrac{\pi}{4}$	$\sin\left(\theta + \dfrac{\pi}{4}\right)$
$-\dfrac{\pi}{4}$	0	0
$\dfrac{\pi}{4}$	$\dfrac{\pi}{2}$	1
$\dfrac{3\pi}{4}$	π	0
$\dfrac{5\pi}{4}$	$\dfrac{3\pi}{2}$	-1
$\dfrac{7\pi}{4}$	2π	0

TABLE 8.10 $y = \sin\left(\theta - \dfrac{\pi}{4}\right)$

θ	$\theta - \dfrac{\pi}{4}$	$\sin\left(\theta - \dfrac{\pi}{4}\right)$
$\dfrac{\pi}{4}$	0	0
$\dfrac{3\pi}{4}$	$\dfrac{\pi}{2}$	1
$\dfrac{5\pi}{4}$	π	0
$\dfrac{7\pi}{4}$	$\dfrac{3\pi}{2}$	-1
$\dfrac{9\pi}{4}$	2π	0

sketching the curve (Figure 8.27). Now sketch $y = \sin\left(\theta + \dfrac{\pi}{4}\right)$ and $y = \sin\left(\theta - \dfrac{\pi}{4}\right)$ on the same set of axes with $y = \sin\theta$. Use a graphics calculator

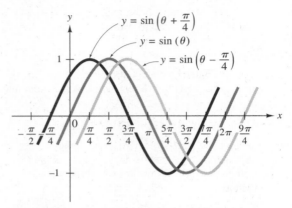

FIGURE 8.27

to sketch the curves. Tables 8.9 and 8.10 provide key points for graphing the functions.

From the graph and the tables, we can see that the starting point is shifted, or horizontally displaced, $\dfrac{\pi}{4}$ units to the left or right of the origin, depending on whether $\dfrac{\pi}{4}$ is positive or negative. (Note that the curves are "parallel.") This horizontal displacement is called the **phase shift** of the function.

The phase shift for $y = \sin(B\theta + C)$ or $y = \cos(B\theta + C)$ is found by setting $B\theta + C = 0$ and solving for θ. Therefore, $\theta = \dfrac{-C}{B}$; that is, phase shift $= \dfrac{-C}{B}$. Note that if $C > 0$, the shift is to the left, and if $C < 0$, the shift is to the right.

Because the graph of a sine or cosine function with a phase shift does not have a period that starts at the origin, when sketching the graph of a function, we need to know the left and right end points of the period. We can find these by solving the equations $B\theta + C = 0$ and $B\theta + C = 2\pi$.

$$B\theta + C = 0 \qquad\qquad B\theta + C = 2\pi$$

$$\theta = -\dfrac{C}{B} \qquad\qquad \theta = -\dfrac{C}{B} + \dfrac{2\pi}{B}$$

Therefore, the left end of the period is the phase shift, and the right end of the period is the phase shift plus the period of the function.

To sketch a sinusoid with a phase shift, one more step is added to the guidelines from Section 8.3. (What was Step 3 there becomes Step 4 here.)

GUIDELINES FOR SKETCHING $y = A \sin (B\theta + C)$ AND $y = A \cos (B\theta + C)$

Step 1. Determine the amplitude. (Amplitude $= |A|$)

Step 2. Determine the period. $\left(\text{Period} = \dfrac{360°}{|B|} \text{ or } \dfrac{2\pi}{|B|}\right)$

Step 3. Determine the phase shift. $\left(\text{Phase shift} = \dfrac{-C}{B}\right)$

Step 4. Determine the key points for the period. Determine the left and right end points of the period. Each period is divided into four equal intervals corresponding to the four quadrants. The end points of these intervals are the five key values, which are found by dividing the period by 4, multiplying the result by 0, 1, 2, 3, and 4, and adding each of the products to the left end point of the interval. Substitute the key values into the function to determine the functional values. With these results we can write the ordered pairs that we call the key points.

These guidelines are just as helpful to the individual with a graphics calculator as the person who sketches the curve by hand. Knowing the amplitude, period, phase shift, and key points helps us set the range on the calculator. These facts are useful in checking the graph on the calculator. Using the trace or zoom function provides a visual check on whether the graph is correct.

EXAMPLE 1 **Graphing Using Amplitude, Period, and Phase Shift**

Sketch one cycle of the function $y = 3 \cos\left(\theta - \dfrac{\pi}{3}\right)$.

Solution Use the guidelines to help sketch the function $y = 3 \cos\left(\theta - \dfrac{\pi}{3}\right)$.

a. Amplitude $= |3| = 3$.   **Step 1**

b. Period $= \dfrac{2\pi}{|1|} = 2\pi$.   **Step 2**

c. Phase shift $= -\dfrac{\left(-\dfrac{\pi}{3}\right)}{1}$ **Step 3**

d. 1. The left end value of the period is $\dfrac{\pi}{3}$, and the right end value is $\dfrac{\pi}{3} + 2\pi = \dfrac{7\pi}{3}$.   **Step 4**

   2. Divide the period by 4: $\dfrac{2\pi}{4} = \dfrac{\pi}{2}$.

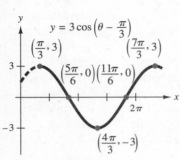

FIGURE 8.28

3. Multiply $\frac{\pi}{2}$ by 0, 1, 2, 3, and 4. The results are 0, $\frac{\pi}{2}$, π, $\frac{3\pi}{2}$, and 2π. Now add the left end of the period, $\frac{\pi}{3}$, to each of these products to obtain the five key values $\frac{\pi}{3}$, $\frac{5\pi}{6}$, $\frac{4\pi}{3}$, $\frac{11\pi}{6}$, and $\frac{7\pi}{3}$.

$$\left(\frac{\pi}{3} + \frac{\pi}{2} = \frac{5\pi}{6}, \frac{5\pi}{6} + \frac{\pi}{2} = \frac{4\pi}{3}\right).$$

4. Calculate the corresponding values for y by substituting the key values for θ in the function $y = 3\cos\left(\theta - \frac{\pi}{3}\right)$. The key points are $\left(\frac{\pi}{3}, 3\right)$, $\left(\frac{5\pi}{6}, 0\right)$, $\left(\frac{4\pi}{3}, -3\right)$, $\left(\frac{11\pi}{6}, 0\right)$, and $\left(\frac{7\pi}{3}, 3\right)$.

Now use these ordered pairs to sketch one period of $y = 3\cos\left(\theta - \frac{\pi}{3}\right)$. Figure 8.28 illustrates the phase shift of the curve, $\frac{\pi}{3}$ units to the right. ◆

CAUTION The phase shift indicates a point where a cycle of the curve begins, while the phase shift plus the period of the function indicates where a cycle ends or where the curve begins a new cycle. In Example 1, the phase shift is $\frac{\pi}{3}$, and the point $\left(\frac{\pi}{3}, 3\right)$ is a starting point for a cycle of the curve. The phase shift plus the period, $\frac{\pi}{3} + 2\pi$, is $\frac{7\pi}{3}$, and the point $\left(\frac{7\pi}{3}, 3\right)$ is the end of the cycle or the beginning of a new cycle. Therefore, the curve recycles at the point $\left(\frac{7\pi}{3}, 3\right)$. ■

EXAMPLE 2 **Graphing Using Amplitude, Period, and Phase Shift**

Sketch one cycle of the function $y = \cos\left(3\theta + \frac{3\pi}{2}\right)$.

Solution

a. Amplitude $= |1| = 1$. Step 1

b. Period $= \dfrac{2\pi}{|3|}$. Step 2

c. Phase shift $= \dfrac{-\left(\dfrac{3\pi}{2}\right)}{3} = \dfrac{-\pi}{2}$ Step 3

d. 1. The left end value of a period is $\dfrac{-\pi}{2}$, and the right end value is

$$\dfrac{-\pi}{2} + \dfrac{2\pi}{3} = \dfrac{\pi}{6}.$$ **Step 4**

2. Divide the period by 4: $\dfrac{2\pi}{3} \div 4 = \dfrac{\pi}{6}$.

3. Multiply $\dfrac{\pi}{6}$ by 0, 1, 2, 3, and 4. Add each of the products to the phase

shift $\dfrac{-\pi}{2}$ to obtain the key values $\dfrac{-\pi}{2}$, $\dfrac{-2\pi}{6}$, $\dfrac{-\pi}{6}$, 0 , and $\dfrac{\pi}{6}$.

4. The key points obtained by substituting the key values for θ in the

function are $\left(\dfrac{-\pi}{2}, 1\right)$, $\left(\dfrac{\pi}{3}, 0\right)$, $\left(\dfrac{-\pi}{6}, -1\right)$, $(0, 0)$, and $\left(\dfrac{\pi}{6}, 1\right)$.

5. Use the ordered pairs to sketch one period of the function $y = \cos\left(3x + \dfrac{3\pi}{2}\right)$. Figure 8.29 illustrates the phase shift of the curve,

which is $\dfrac{\pi}{2}$ units to the left.   ◆

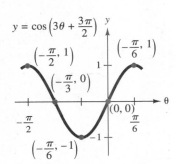

$y = \cos\left(3\theta + \dfrac{3\pi}{2}\right)$

FIGURE 8.29

EXAMPLE 3   **Graphing Using Amplitude, Period, and Phase Shift**

Sketch one cycle of the function $y = 3 \sin (2\theta + 60°)$.

Solution

a. Amplitude $= |3| = 3$.   **Step 1**

b. Period $= \dfrac{360°}{2} = 180°.$ **Step 2**

c. Phase shift $= \dfrac{-60°}{2} = -30°.$ **Step 3**

d. 1. The left end value of a period is $-30°$, and the right end value is
$-30° + 180° = 150°.$ **Step 4**

2. Divide the period by 4: $\dfrac{180°}{4} = 45°$.

3. Multiply $45°$ by 0, 1, 2, 3, and 4. Add each of the products to the phase
shift $-30°$ to obtain the key values $-30°$, $15°$, $60°$, $105°$, and $150°.$

4. The key points obtained by substituting the key values for θ in the
function are $(-30°, 0)$, $(15°, 3)$, $(60°, 0)$, $(105°, -3)$, and $(150°, 0).$

Use the key points to sketch one period of the function $y = 3 \sin (2\theta + 60°)$.
Figure 8.30 illustrates that the curve is shifted $30°$ to the left of the origin.   ◆

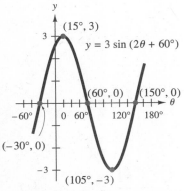

$y = 3 \sin (2\theta + 60°)$

FIGURE 8.30

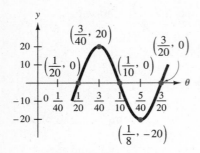

FIGURE 8.31

EXAMPLE 4   Determine the Function Given the Graph

Determine the amplitude, period, and phase shift for the sine function whose graph is shown in Figure 8.31. Write an equation for the graph.

Solution   Recall that the general form of the equation of the sine function is $y = A \sin (B\theta + C)$. So to write the equation of the graph, we must determine A, B, and C.

The highest and lowest points of the graph are $\left(\frac{3}{40}, 20\right)$ and $\left(\frac{1}{8}, -20\right)$, respectively. Therefore, the amplitude is $A = 20$.

The curve goes through one complete cycle between $\left(\frac{1}{20}, 0\right)$ and $\left(\frac{3}{20}, 0\right)$.

The period is $\frac{3}{20} - \frac{1}{20} = \frac{2}{20}$ or $\frac{1}{10}$. Recall that the period $= \frac{2\pi}{|B|}$. Solving for B, we have $|B| = \frac{2\pi}{\text{period}}$. Substituting $|B| = \frac{2\pi}{\frac{1}{10}}$ or $B = 20\pi$.

Read the phase shift from the graph. The curve is shifted to the right and so the phase shift is $+\frac{1}{20}$. Recall that the phase shift $= \frac{-C}{B}$. Solving for C, we have $C = (-\text{phase shift})(B)$. Substitute

$$C = -\left(\frac{+1}{20}\right)(20\pi)$$

$$= -\pi.$$

Substituting the values for A, B, and C, the equation is $y = 20 \sin (20\pi \, \theta - \pi)$. ◆

EXAMPLE 5   Application: Amplitude, Period, and Phase Shift

A mass-spring system is oscillating vertically with displacement given by $y = 24 \sin \left(3\pi t - \frac{\pi}{6}\right)$. Sketching one cycle of the function.

Solution   For the function $y = 24 \sin \left(3\pi t - \frac{\pi}{6}\right)$, the amplitude $= |24|$ or 24, the period $= \frac{2\pi}{|3\pi|}$ or $\frac{2}{3}$, and the phase shift $= \frac{-\left(-\frac{\pi}{6}\right)}{3\pi}$ or $\frac{1}{18}$. To simplify the arithmetic, perform the calculations with a calculator.

One cycle of the curve starts at $(0.056, 0)$, and the end point of that cycle is $(0.722, 0)$.

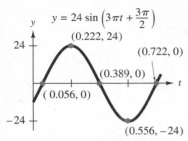

$$y = 24 \sin \left(3\pi t + \frac{3\pi}{2}\right)$$

FIGURE 8.32

The key values are 0.056, 0.056 + 0.166, 0.056 + 2(0.166), 0.056 + 3(0.166) + 0.056 + 4(0.166) or 0.056, 0.222, 0.389, 0.556, 0.722. To do these calculations, divide the period by 4 and store this value. Then change $\frac{1}{18}$ to the decimal value, which is 0.056, and add the stored value to the display (0.056). This is the second key value. Keep adding the stored value to the display to obtain the other key values. The key points are (0.056, 0), (0.222, 24), (0.389, 0), (0.556, −24), and (0.722, 0). The curve is sketched in Figure 8.32. ◆

For the spring-mass system in Example 5, the calculations and the graph in Figure 8.32 show that the curve is shifted to the right. The phase shift is $\frac{1}{18}$ of a unit of time. This means that for time $t = 0$, the spring-mass system is not at zero; it has a negative value. At time $t = 0$, the curve is increasing, and the spring-mass system is moving from a negative value to zero and then to a positive value.

Electronics

In technical application, T is often used to represent the period of the function $[y = A \sin (\omega\theta + C)]$. The period T of the function is the time required for the function to complete one cycle. Therefore, $T = \frac{2\pi}{\omega}$. Solving for ω, the angular frequency, we have $\omega = \frac{2\pi}{T}$. The relationship between the frequency (or linear frequency) f of the function and the period T can be expressed as $f = \frac{1}{T}$. Solving for T, we have $T = \frac{1}{f}$. Now substituting for T in the equation for angular frequency, we can express the angular frequency in terms of the linear frequency f, that is, $\omega = 2\pi f$. The phase shift of the function is given by $\frac{-C}{\omega}$. (If the units require, we can use 360° instead of 2π.)

If two sine functions or cosine functions have the same angular frequency, it is said that the second leads or lags the first by the amount $C_2 - C_1$, depending on whether $C_2 - C_1$ is positive or negative.

For example, for the functions $y_1 = 3 \sin (\omega t + 45°)$ and $y_2 = 3 \sin (\omega t + 60°)$, 60° − 45° = 15°. Therefore, the second function leads the first by 15°.

An impressed voltage (one that comes from a source such as a generator) that varies sinusoidally is called an alternating voltage. An alternating voltage gives rise to an alternating current (AC). If there is just resistance in a wire, then the variable current i is a function of time: $i = I_{max} \sin (2\pi ft)$ where I_{max} = peak current, and f = frequency.

However, if the wire is coiled in any way, then the current is shifted from the standard equation to $i = I_{max} \sin (2\pi ft + \theta)$ where θ = phase angle. As the amount of coiling increases so does the size of the phase angle. In the same manner, the voltage is given by the equation $v = V_{max} \sin (2\pi ft + \theta)$ where v_{max} = peak voltage, f = frequency, and θ = phase angle.

EXAMPLE 6 Application: Equations for Voltage and Current

In a particular circuit, the maximum values of the voltage and current are 175 V and 113 A, respectively. The frequency is 50 Hz, and the current leads the voltage by 45°.

a. Write the equation for the voltage at any time t, when $C = 0$.
b. Write the equation for the current at any time t.
c. Plot one cycle of both curves starting with $t = 0$, using the equations from parts a and b.

Solution We are given the following information:

$$\text{maximum voltage} = V_{max} = \boxed{175} \text{ V}$$
$$\text{maximum current} = I_{max} = \boxed{113} \text{ A}$$
$$\text{frequency} = f = \boxed{50} \text{ Hz}$$
$$\text{phase angle} = \theta = \boxed{45} \text{ or } \boxed{\frac{\pi}{4}}$$
$$\text{angular frequency} = \omega = \boxed{2\pi f}.$$

a. Substitute the values in the equation:

$$v = V_{max} \sin \omega t$$
$$= \boxed{175} \sin \boxed{(2\pi)} \boxed{(50)} t$$
$$= 175 \sin 100\pi t.$$

b. Substitute the values in the equation:

$$i = I_{max} \sin (\omega t + \theta)$$
$$= \boxed{113} \sin \left[\boxed{(2\pi)} \boxed{(50)} t + \boxed{\frac{\pi}{4}} \right]$$
$$= 113 \sin \left(100\pi t + \frac{\pi}{4} \right).$$

c. Figure 8.33 shows one cycle for each curve.

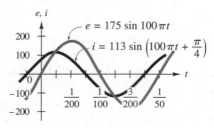

FIGURE 8.33

8.4 Exercises

In Exercises 1–20, **a.** determine the period, **b.** determine the phase shift, **c.** determine the amplitude, **d.** determine where the curve will recycle, and **e.** sketch one cycle of the function.

1. $y = \sin \left(\theta - \dfrac{\pi}{2} \right)$

2. $y = \cos (\theta - \pi)$

3. $y = \sin (\theta - 30°)$

4. $y = \cos (\theta - 90°)$

5. $y = \cos \left(3\theta - \dfrac{\pi}{3} \right)$

6. $y = \sin \left(2\theta - \dfrac{\pi}{4} \right)$

7. $y = \cos (5\theta + 15°)$

8. $y = \sin (4\theta + 12°)$

9. $y = 3 \sin \left(\theta - \dfrac{\pi}{2} \right)$

10. $y = \dfrac{1}{2} \cos \left(3\theta - \dfrac{\pi}{2} \right)$

11. $y = 10 \sin (2\pi\theta - 2\pi)$

12. $y = 0.01 \cos (0.1\theta - 20°)$

13. $y = -30 \cos (15\pi\theta + \pi)$

14. $y = -60 \sin \left(30\pi\theta + \dfrac{\pi}{2} \right)$

15. $y = 25 \sin (3\theta - 120°)$

16. $y = -25 \sin (3\theta - 120°)$

17. $y = 15 \cos \left(\dfrac{\theta}{2} + 360° \right)$

18. $y = -\cos \left(\dfrac{\theta}{3} - 180° \right)$

19. $y = 35 \cos (\theta - 720°)$

20. $y = 100 \sin (\theta + 90°)$

In Exercises 21–24, write an equation for each specified trigonometric function with given amplitude, period, and phase shift, respectively.

21. $\sin$ e, 5, 3π, $\dfrac{\pi}{4}$ to the right.

22. $\sin$ e, 3, $\dfrac{2\pi}{3}$, $\dfrac{\pi}{3}$ to the left.

23. $\cos$ ine, 2, $\dfrac{\pi}{2}$, $\dfrac{\pi}{6}$ to the left.

24. $\cos$ ine, 4, π, $\dfrac{\pi}{2}$ to the right.

In Exercises 25–26, write an equation for the graph of the sine curve.

25.

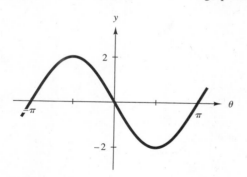

26.

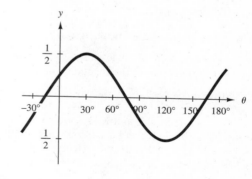

In Exercises 27–28, write an equation for the graph of the cosine curve.

27.

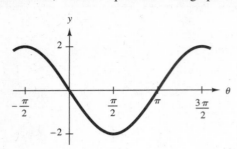

28.

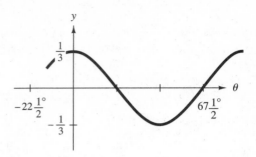

29. At a time t (in seconds), the distance y (in meters) of a point P above or below the edge of the supporting beam of a water wheel of radius 8 is given by $y = 8 \cos \left(\dfrac{\pi t}{6} - \dfrac{\pi}{3} \right)$. (See the sketch.)

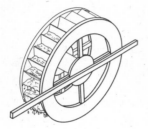

a. Determine the amplitude, period, and phase shift.
b. Graph one cycle, starting with $t = 0$.
c. Determine the position of P when $t = 4$ s and when $t = 7$ s.

30. The pressure P in a particular traveling sound wave is given by $P = -1.8 \sin (2\pi t - \pi)$.
a. Determine the amplitude, period, and phase shift.
b. Sketch one cycle, starting with $t = 0$.

31. The displacement of a particle is given by $y = 8 \cos \left(30\pi t + \dfrac{\pi}{4} \right)$.
a. Determine the amplitude, period, and phase shift.
b. Sketch one cycle of the function.
c. What is the position of the particle after $\dfrac{1}{130}$ s?

32. Radio waves, red light, and X rays are identical forms of radiation except for their frequencies. Radio waves have the smallest frequency, and X rays have the largest. Assume that the intensity of each wave is the same. Sketch the three types of radiations on the same graph to illustrate the effect of the difference in frequencies.

33. A capacitor voltage is given by $v = 1.5 \cos (3t + 2)$. Sketch one cycle of the function.

34. A wire vibrates according to the function $y = 0.2 \sin (4t - 1)$. Sketch one cycle of the function.

35. A 30-Hz alternator generates 7.2 kV at 750 A. The current leads the voltage by an angle of 25°. (*Hint:* See Example 6.)
  a. Write the equation for the current at any time t.
  b. How much of the voltage cycle will have been completed the first time that the instantaneous current rises to 465 A?
  c. Sketch the graph of the equation in **a.** for one cycle starting with $t = 0$.

36. A 45-Hz alternator generates 3.2 kV with a current of 190 A. The voltage is lagging the current by 20°. (*Hint:* See Example 6.)
  a. Write the equation for the current at any time t.
  b. What is the instantaneous value of the current when the voltage has completed 182° of its cycle?

37. The maximum values of a voltage and a current are 150 V and 12 A, respectively. The frequency is 700 Hz, and the current lags the voltage by 30° at any time t.
  a. Write the equation for the voltage at any time t.
  b. Write the equation for the current at any time t.
  c. Plot one cycle of both curves starting with $t = 0$, using the equations from **a.** and **b.**

38. The hum you hear on a radio when it is not tuned to a station is a sound wave of 60 cycles per second.
  a. Is the 60 cycles per second the period or the frequency?
  b. If it is the period, find the frequency. If it is the frequency, find the period.
  c. The wavelength of a sound wave is defined to be the distance the wave travels in one period. If sound travels at 1100 ft/s, find the wavelength of a 60-cycle-per-second sound wave.

39. The lowest musical note the human ear can hear is about 16 cycles per second. In order for the pipe on a church organ to generate this note, the pipe must be exactly half as long as the wavelength. What length organ pipe would be required to generate a 16-cycle-per-second note?

8.5

Graphs of Tangent, Cosecant, Secant, and Cotangent

The graphs and functions of $y = \tan \theta$, $y = \csc \theta$, $y = \sec \theta$, and $y = \cot \theta$ were introduced in Section 8.1. This section discusses the techniques for graphing $y = A \tan (B\theta + C)$, $y = A \csc (B\theta + C)$, $y = A \sec (B\theta + C)$, and $y = A \cot (B\theta + C)$.

The graphs of these functions show some characteristics (period and phase shift) similar to the graphs of sine and cosine.

One of the differences between these and the earlier graphs is that not one of the graphs in this section is a smooth, continuous curve as are the sine and cosine curves. Each is discontinuous. In other words, each has points where the function is undefined. At the points where the curves are discontinuous, vertical asymptotes occur. Also, due to the fact that the curves are discontinuous, the cycles do not have end points.

Using the graphics calculator, graph $y = \tan \theta$, with the calculator in the degree mode and range settings of $(-270, 270, 45, -10, 10, 5)$. Study the graph of the function shown in Figure 8.34, to observe several things. For these range settings, the curve goes through three cycles; each cycle is 180°. The function is undefined when θ is $-270°$, $-90°$, $90°$, and $270°$. The curve has no amplitude since there is no maximum value for the curve.

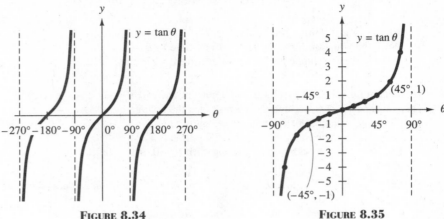

FIGURE 8.34   **FIGURE 8.35**

Now let's focus on one cycle of the tangent curve in the open interval $(-90°, 90°)$, as shown in Figure 8.35. The period of the function is 180°. At the midpoint of the interval, the curve crosses the horizontal axis at the point $(0°, 0)$. There are no end points to the cycle, the curve is asymptotic to the lines $\theta = -90°$ and $\theta = 90°$. In the two subintervals to the left and right of zero, $(-90°, 0°)$ and $(0°, 90°)$, the midpoints are $(-45°, -1)$ and $(45°, 1)$, respectively. These two midpoints are helpful; in sketching, they are key points. In fact, for the curve $y = A \tan \theta$, the points would be $(-45°, -A)$, $(0, 0)$ and $(45°, A)$.

The graph of the function $y = A \tan (B\theta + C)$ is effected by the values of A, B, and C. The value of A causes the curve either to rise very rapidly or to rise more gently. The smaller the value of A, the flatter the curve is in the region where it crosses the horizontal axis. For graphs of $y = A \tan \theta$ where $A = \dfrac{1}{2}$, 1, and 3, see Figure 8.36.

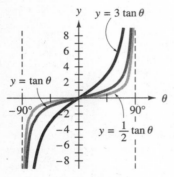

FIGURE 8.36

Determining key values as we did for sine and cosine, makes it easier to sketch the graph of a general tangent function $y = A \tan (B\theta + C)$. The technique for sketching a graph of the general tangent function is illustrated by graphing one cycle of $y = 2 \tan (3\theta + 45°)$.

Guidelines for Sketching Tangent Function
$y = A \tan (B\theta + C)$

Step 1. First determine the interval of one cycle of the tangent function that contains the origin. For the function $y = \tan \theta$, the interval about the origin is $(-90°, 90°)$. Therefore, to determine the end points of the open interval for one cycle of the general tangent function, we must solve the following equations.

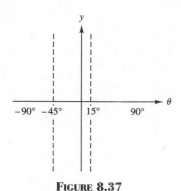

FIGURE 8.37

$$B\theta_1 + C = -90° \qquad \text{and} \quad B\theta_2 + C = 90°$$

$$\theta_1 = \frac{1}{B}(-C - 90°) \qquad\qquad \theta_2 = \frac{1}{B}(-C + 90°)$$

Substituting in the equation results in:

$$\theta_1 = \frac{1}{3}(\boxed{-45°} - 90°) \qquad \theta_2 = \frac{1}{3}(\boxed{-45°} + 90°)$$

$$= \boxed{-45°} \qquad\qquad = \boxed{15°.}$$

Not only does one cycle of the curve lie between $-45°$ and $15°$, but also the lines $\theta = -45°$ and $\theta = 15°$ are asymptotes for the curve (see Figure 8.37). Knowing the interval, we can determine that the period is $15° - (-45°)$ or $60°$.

Step 2. Determine the midpoint of the interval found in Step 1, using the equation:

$$\text{Midpoint} = \frac{\theta_1 + \theta_2}{2} = \theta_3.$$

Substituting results in:

$$\text{Midpoint} = \frac{\boxed{-45°} + \boxed{15°}}{2}$$

$$= \boxed{-15°}.$$

The midpoint of the interval is where the curve crosses the horizontal axis. The curve $y = 2 \tan (3\theta + 45°)$ crosses the horizontal axis at $\boxed{(-15°, 0).}$ The midpoint also divides the interval into two subintervals, $(-45°, -15°)$ and $(-15°, 15°)$.

Step 3. Now determine the midpoints of the the two subintervals created in Step 2; that is, determine the midpoints of (θ_1, θ_3) and (θ_3, θ_2).

$$\text{Midpoint} = \frac{\theta_1 + \theta_3}{2} \qquad \text{Midpoint} = \frac{\theta_3 + \theta_2}{2}$$

Substituting results in:

$$\text{Midpoint} = \frac{\boxed{-45°} + \boxed{(-15°)}}{2} \qquad \text{Midpoint} = \frac{\boxed{-15°} + \boxed{15°}}{2}$$

$$= \boxed{-30°} \qquad\qquad = 0°.$$

$y = 2 \tan (3\theta + 45°)$

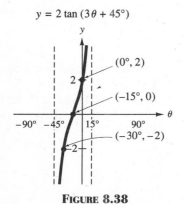

FIGURE 8.38

Substituting these midpoints into the function, we determine the points on the graph: $\boxed{(-30°, -2)}$ and $\boxed{(0°, 2).}$

Step 4. Using the information from Steps 1–3, we can sketch a graph, like the one in Figure 8.38.

Another way of determining the period of $y = A \tan (B\theta + C)$ is by dividing the period (180° or π) by the absolute value of B.

$$\text{Period} = \frac{180°}{|B|}$$

Calculating the period of the function $y = 2 \tan (3\theta + 45°)$ using the formula gives $\dfrac{180°}{3} = 60°$. This result, 60°, checks with the earlier calculations.

Recall that the phase shift is determined by dividing $-C$ by B.

$$\text{Phase shift} = \frac{-C}{B}.$$

Calculating the phase shift using the formula, the result is $\dfrac{-45°}{3} = -15°.$

This result tells us that each cycle of the function $y = 2 \tan (3\theta + 45°)$ is shifted 15° to the left. Recall that each cycle of the function has a period of 60°. With the cycle centered about the origin, if it is shifted 15° to the left, the left end of the open interval would be $(-30° - 15°)$ or $-45°$. The right end of the open interval would be $(30° - 15°)$ or 15°. Look at Figure 8.38. This checks with our previously obtained results.

To sketch one cycle of $y = A \cot (B\theta + C)$, use the same general procedure used for sketching the tangent function. The function $y = \cot \theta$ has a period of 180° (or π) and goes through one complete cycle in the interval $(0°, 180°)$. Generally, when one cycle of the curve is sketched, the interval $(0°, 180°)$ is selected. The cotangent function is undefined at 0° and 180°; $\theta = 0°$ and $\theta = 180°$ are asymptotes.

EXAMPLE 1   Sketching the Graph of a Cotangent Function

Sketch one cycle of $y = \cot (2\theta - 30°)$.

Solution   To sketch one cycle of the graph, determine an interval that contains one cycle by solving the following equations for θ. The reason for selecting 0° and 180° is that one cycle of the contangent function lies in the interval $0° < \theta < 180°$.

$$B\theta + C = 0° \quad \text{and} \quad B\theta + C = 180°$$
$$2\theta - 30° = 0° \quad \text{and} \quad 2\theta - 30° = 180°$$
$$\theta = 15° \qquad\qquad \theta = 105°$$

The function $y = \cot (2\theta - 30°)$ goes through one cycle in the interval $(15°, 105°)$; $\theta = 15°$ and $\theta = 105°$ are asymptotes to the curve.

Knowing the interval $15° < \theta < 105°$, we can set a range on the graphics

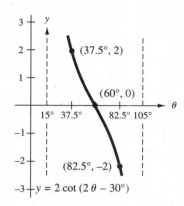

FIGURE 8.39

calculator and graph the function. Use range settings of $(15, 105, 30, -20, 20, 5)$. When entering the function in the calculator we use the fact that $\cot \theta = \dfrac{1}{\tan \theta}$. The graph of the function is given in Figure 8.39. ◆

You can sketch the graphs of $y = A \sec (B\theta + C)$ and $y = A \csc (B\theta + C)$ by applying the guidelines for sketching the sinusoids with a few exceptions. First recall that $\csc \theta = \dfrac{1}{\sin \theta}$ and $\sec \theta = \dfrac{1}{\cos \theta}$. For the values when $\sin \theta$ and $\cos \theta$ are zero, the cosecant and secant functions are undefined and have asymptotes. Also, it is impossible to determine the amplitude for cosecant and secant since there are no maximum points on the curves. The absolute value of A determines the distances above and below the horizontal axis where the curve does not exist.

Determine the period and the phase shift by using the formulas:

$$\text{Period} = \frac{2\pi}{|B|} \left(\text{or } \frac{360°}{|B|} \right) \qquad \text{Phase shift} = \frac{-C}{B}$$

EXAMPLE 2   Sketching the Graph of a Secant Function

Sketch one cycle of the graph of $y = \sec \pi\theta$. Determine the period and indicate the key values on the horizontal axis.

Solution   To help sketch $y = \sec \pi\theta$, first sketch $y = \cos \pi\theta$. The period and phase shift is the same for both functions.

$$\text{Period} = \frac{2\pi}{|\pi|} = 2 \qquad \text{Phase shift} = \frac{-0}{\pi} = 0$$

Sketch the cosine curve using the guidelines for sketching sine and cosine. Then, amplitude is 1; period is 2; and phase shift is 0. The key points are $(0, 1)$, $\left(\dfrac{1}{2}, 0\right)$, $(1, -1)$, $\left(\dfrac{3}{2}, 0\right)$, and $(2, 1)$. Remember that $\sec \theta = \dfrac{1}{\cos \theta}$. The function $y = \sec \pi\theta$ is undefined when $\cos \theta = 0$. Therefore, the lines $\theta = \dfrac{1}{2}$ and $\theta = \dfrac{3}{2}$ are asymptotes. The amplitude of cosine function is 1. The lowest point on the curve of the secant function that is above the θ-axis is 1 unit above the axis. Also, the highest point below the θ-axis is 1 unit below the axis. The sketch of $y = \sec \pi\theta$ is given in Figure 8.40. ◆

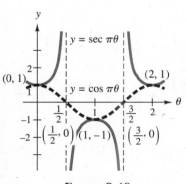

FIGURE 8.40

EXAMPLE 3 **Sketching the Graph of a Cosecant Function**

Sketch one cycle of the graph of $y = 2 \csc \left(\dfrac{\theta}{3} + \dfrac{\pi}{3} \right)$.

Solution To help sketch $y = 2 \csc \left(\dfrac{\theta}{3} + \dfrac{\pi}{3} \right)$, first sketch $y = 2 \sin \left(\dfrac{\theta}{3} + \dfrac{\pi}{3} \right)$.
To sketch the sine function, use the guidelines for sine and cosine. The amplitude

is 2; period $= \dfrac{2\pi}{\frac{1}{3}}$ or 6π; and the phase shift $= \dfrac{-\dfrac{\pi}{3}}{\dfrac{1}{3}}$ or $-\pi$. The key points are

$(-\pi, 0)$, $\left(\dfrac{\pi}{2}, 1 \right)$, $(2\pi, 0)$, $\left(\dfrac{7\pi}{2}, -1 \right)$, and $(5\pi, 0)$. Remember that csc

$\theta = \dfrac{1}{\sin \theta}$. The function $y = 2 \csc \left(\dfrac{\theta}{3} + \dfrac{\pi}{3} \right)$ is undefined when $\sin \theta = 0$.
Therefore, the lines $\theta = -\pi$, $\theta = 2\pi$, and $\theta = 5\pi$ are asymptotes. The
coefficient of 2 has the effect of raising the cosecant curve that lies above the
θ-axis to 2 units above the axis. It also has the effect of lowering the highest point
of the curve below the θ-axis to 2 units below the axis. The sketch of
$y = 2 \csc \left(\dfrac{\theta}{3} + \dfrac{\pi}{3} \right)$ is given in Figure 8.41.

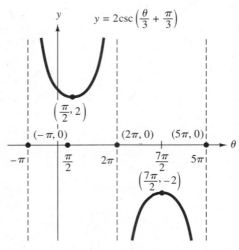

FIGURE 8.41

8.5 Exercises

In Exercises 1–10, match the function with the graph of the function.

1. $y = \tan 2\theta$

2. $y = 2 \cot 3\theta$

3. $y = \sec 2\theta$

4. $y = \csc 3\theta$

5. $y = \tan (\theta + 30°)$

6. $y = \cot (\theta - 30°)$

a.
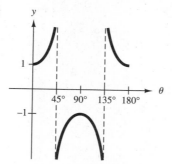

b.
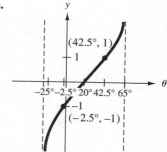

c.
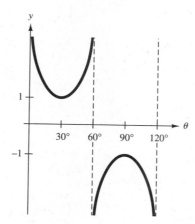

d.
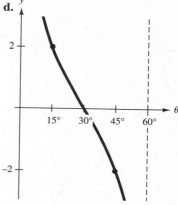

e.
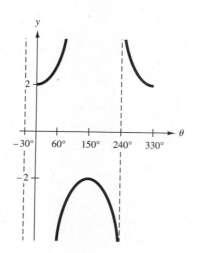

f.
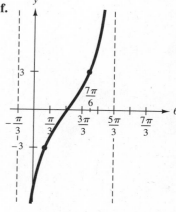

7. $y = 2 \sec (\theta + 30°)$

g.

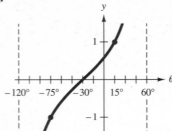

8. $y = 3 \csc \left(\theta - \dfrac{\pi}{4} \right)$

h.

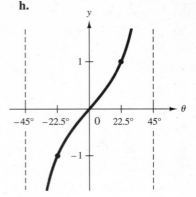

9. $y = \tan (2\theta - 40°)$

i.

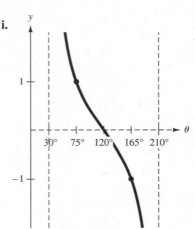

10. $y = 3 \tan \left(\dfrac{\theta}{2} - \dfrac{\pi}{3} \right)$

j.

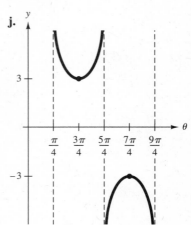

In Exercises 11–34, sketch one cycle of the function.

11. $y = 2 \tan \theta$

12. $y = -\dfrac{1}{4} \tan \theta$

13. $y = -\dfrac{1}{3} \cot \theta$

14. $y = 3 \cot \theta$

15. $y = 3 \csc \theta$

16. $y = \dfrac{1}{2} \csc \theta$

17. $y = 4 \sec \theta$

18. $y = -\dfrac{1}{3} \sec \theta$

19. $y = \dfrac{3}{2} \tan \theta$

20. $y = \dfrac{3}{2} \cot \theta$

21. $y = \dfrac{4}{5} \csc \theta$

22. $y = \dfrac{2}{3} \sec \theta$

23. $y = \tan \left(\theta + \dfrac{\pi}{4} \right)$

24. $y = \dfrac{1}{2} \tan \left(\dfrac{\theta}{2} + \dfrac{\pi}{12} \right)$

25. $y = \dfrac{3}{2} \cot \left(2\theta - \dfrac{2\pi}{3} \right)$

26. $y = \cot \left(2x + \dfrac{2\pi}{3} \right)$

27. $y = -\csc \left(2\theta + \dfrac{\pi}{2} \right)$

28. $y = \dfrac{1}{2} \csc \left(3x + \dfrac{3\pi}{4} \right)$

29. $y = \sec \left(\theta - \dfrac{\pi}{6} \right)$

30. $y = -3 \sec (2\theta + 2\pi)$

31. $y = 2 \tan \left(\theta - \dfrac{\pi}{4} \right)$

32. $y = 3 \cot \left(\dfrac{\theta}{2} - \dfrac{\pi}{3} \right)$

33. $y = 3 \csc \left(2\theta - \dfrac{\pi}{2} \right)$

34. $y = 4 \sec \left(\theta + \dfrac{\pi}{6} \right)$

In Exercises 35–42, sketch the graph of the function over the given interval.

35. $y = 2 \tan \left(\dfrac{\theta}{2} + \dfrac{\pi}{6} \right)$, $-4\pi \le \theta \le 4\pi$

36. $y = \dfrac{1}{2} \tan \left(\dfrac{\theta}{3} - \dfrac{\pi}{6} \right)$, $-2\pi \le \theta \le 2\pi$

37. $y = 3 \cot (2\theta - 135°)$, $-180° \le \theta \le 180°$

38. $y = 2 \cot \left(\dfrac{\theta}{3} + 30° \right)$, $-540° \le \theta \le 540°$

39. $y = -3 \csc \left(3\theta - \dfrac{3\pi}{2} \right)$, $-2\pi \le \theta \le 2\pi$

40. $y = 2 \csc \left(2\theta + \dfrac{\pi}{3} \right)$, $-\pi \le \theta \le \pi$

41. $y = \dfrac{1}{3} \sec (2\theta - 180°)$, $-180° \le \theta \le 180°$

42. $y = 2 \sec \left(\dfrac{1}{3} x - 60° \right)$, $0° \le \theta \le 540°$

In Exercises 43–50, write an equation for each of the trigonometric functions with the given coefficient A, period, and phase shift, respectively.

43. tangent, $\dfrac{1}{3}$, $\dfrac{\pi}{2}$, $\dfrac{\pi}{12}$ to the left

44. cotangent, 2, 2π, $\dfrac{\pi}{3}$ to the right

45. secant, -3, $720°$, $30°$ to the left

46. cosecant, $\dfrac{2}{3}$, $\dfrac{\pi}{4}$, $\dfrac{\pi}{5}$ to the left

47. cosecant, 3, π, $\dfrac{\pi}{2}$ to the right

48. secant, $\dfrac{1}{2}$, $90°$, $36°$ to the right

49. cotangent, $-\dfrac{3}{4}$, $\dfrac{\pi}{3}$, $\dfrac{\pi}{4}$ to the left

50. tangent, $\dfrac{3}{2}$, 3π, π to the right

8.6

Combinations of Trigonometric Functions

Many applications involve combinations of trigonometric functions or combinations of other functions with trigonometric functions. For example, musical notes are composed of a sum of many sinusoids. Therefore, those who design the amplification systems that carry the musical notes deal with functions that are sums and differences of sine and cosine functions.

The graph of such combinations of trigonometric functions may be sketched by a technique known as **addition of ordinates.** This technique will be illustrated in Examples 1 and 2.

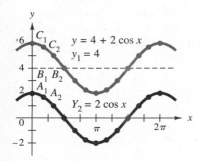

FIGURE 8.42

EXAMPLE 1 **Sketching the Sum of a Constant Function and a Cosine Function**

Sketch the graph of $y = 4 + 2 \cos \theta$ for the interval $0 \le \theta \le 2\pi$.

Solution The equation $y = 4 + 2 \cos \theta$ can be thought of as the sum of two functions: $y = y_1 + y_2$, where $y_1 = 4$ and $y_2 = 2 \cos \theta$.

To illustrate the concept of addition of ordinates, begin by sketching graphs of $y_1 = 4$ and $y_2 = 2 \cos \theta$ on the same set of axes as shown in Figure 8.42.

TABLE 8.11 $y = 4 + 2 \cos \theta$

θ	y_1	y_2	$y = y_1 + y_2$
0	4	2.0	6.0
$\dfrac{\pi}{4}$	4	1.4	5.4
$\dfrac{\pi}{2}$	4	0	4.0
$\dfrac{3\pi}{4}$	4	-1.4	2.6
π	4	-2.0	2.0
$\dfrac{5\pi}{4}$	4	-1.4	2.6
$\dfrac{3\pi}{2}$	4	0	4.0
$\dfrac{7\pi}{4}$	4	1.4	5.4
2π	4	2.0	6.0

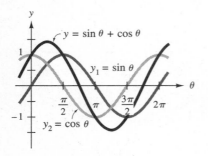

FIGURE 8.43

To help us better understand the concept, Table 8.11 contains the values y_1, y_2, and $y = y_1 + y_2$. On the graph, the points for y_1, y_2, and y when $x = 0$ are indicated by A_1, B_1, and C_1, respectively. If we measure the distance from the θ-axis to A_1 and add this distance to B_1, we would be at C_1. ◆

In Example 1, the graph of the combinations of the functions is the graph of the cosine function raised 4 units. The constant function has the effect of raising the graph of the combination function. If it had been a negative constant, it would have lowered the graph that number of units. Also, since $y_2 = 2 \cos \theta$ is a periodic function, the combination function is periodic.

EXAMPLE 2 **Sketching the Sum of Two Periodic Functions**

Sketch the graph of $y = \sin \theta + \cos \theta$ for the interval $0° \le \theta \le 360°$.

Solution The function $y = \sin \theta + \cos \theta$ can be separated into two functions: $y_1 = \sin \theta$, and $y_2 = \cos \theta$. To help sketch the function, look at Table 8.12, which provides values for y_1, y_2, and $y = y_1 + y_2$. The graph of the combination of the functions is given in Figure 8.43.

TABLE 8.12 $y = \sin \theta + \cos \theta$

θ	0°	45°	90°	135°	180°	225°	270°	315°	360°
y_1	0	0.7	1	0.7	0	-0.7	-1	-0.7	0
y_2	1	0.7	0	-0.7	-1	-0.7	0	0.7	1
$y = y_1 + y_2$	1	1.4	1	0	-1	-1.4	-1	0	1

◆

The function $y = \sin \theta + \cos \theta$ is a periodic function. In fact any function of the form $y = A \sin B\theta + A \sin B\theta$ is a periodic function.

EXAMPLE 3 **Sketching Two Periodic Functions with Different Periods**

Sketch the graph of $y = \sin 2\theta + 2 \sin \theta$ for the interval $0° \le \theta \le 720°$.

Solution The function $y = \sin 2\theta + 2 \sin \theta$ can be separated into two functions: $y_1 = \sin 2\theta$, and $y_2 = 2 \sin \theta$. To sketch the curve, use Table 8.13, which provides values for y_1, y_2, and $y = y_1 + y_2$. The graph of the composite function is given in Figure 8.44.

TABLE 8.13 $y = \sin 2\theta + 2 \sin \theta$

θ	0°	45°	90°	135°	180°	225°	270°	315°	360°
y_1	0	1.0	0	-1.0	0	1.0	0	-1.0	0
y_2	0	1.4	2.0	1.4	0	-1.4	-2.0	-1.4	0
$y = y_1 + y_2$	0	2.4	2.0	0.4	0	-0.4	-2.0	-2.4	0

θ	405°	450°	495°	540°	585°	630°	675°	720°
y_1	1.0	0	-1.0	0	1.0	0	-1.0	0
y_2	1.4	2.0	1.4	0	-1.4	-2.0	-1.4	0
$y = y_1 + y_2$	2.4	2.0	0.4	0	-0.4	-2.0	-2.4	0

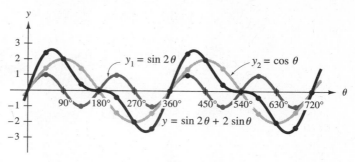

FIGURE 8.44 ◆

If the graph in Figure 8.44 were extended in either direction, we could see easily that the composite function is periodic.

CAUTION Are all combinations of trigonometric functions periodic? No. If any of the functions forming the sum has a variable factor (for example, if $y_1 = x \sin \theta$ or $y_2 = x^2 \cos \theta$), then the sum of the functions is not periodic. Also, combination functions of the type $y = \sin \pi t + \cos 2t$ are not periodic. ∎

EXAMPLE 4 **Determining the Period of Sums of Functions**

Determine the period of each of the combination of the functions.

a. $y = 2 \sin 3\theta + \cos \theta$ **b.** $y = \cos \dfrac{\pi}{2}\theta + \sin \dfrac{\pi}{3}\theta$

Solution

a. For the function $y = 2 \sin 3\theta + \cos \theta$, the periods of the component functions are 120° and 360°, respectively. The lcm (the smallest number that is divisible by x and y) of 120° and 360° is 360°. Thus, the period of the combination function is 360°. With a period of 360° for the combination function, the function $y_1 = 2 \sin 3\theta$ goes through three cycles, and $y_2 = \cos \theta$ goes through one cycle. The functions are graphed in Figure 8.45.

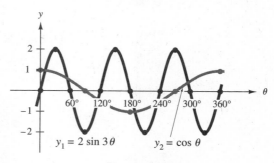

FIGURE 8.45

b. For the function $y = \cos \frac{\pi}{2}\theta + \sin \frac{\pi}{3}\theta$, the periods of the component functions are 4 and 6, respectively. The lcm of 4 and 6 is 12. Thus, the period of the combination function is 12. With a period of 12 for the combination function, the function $y_1 = \cos \frac{\pi}{2}\theta$ goes through three cycles, and $y_2 = \sin \frac{\pi}{3}\theta$ goes through two cycles. The functions are graphed in Figure 8.46.

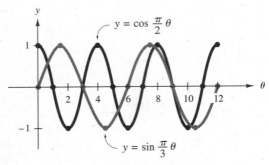

FIGURE 8.46

Use the addition method or a graphics calculator to sketch two periods of the combination functions $y = 2 \sin 3\theta$ and $y = \cos \frac{\pi}{2}\theta + \sin \frac{\pi}{3}\theta$. Doing so will show you that the combination function is periodic.

EXAMPLE 5   **Application: Spring Problem**

If a weight hanging on a spring is given an initial displacement of y_0 and an initial velocity of v_0, then its position at any time t is given by $y = y_0 \cos \omega t + \frac{v_0}{\omega} \sin \omega t$, where the angular frequency ω depends on the weight and the elasticity of the spring. Use a graph to illustrate the spring action for one cycle of the combination function when $y_0 = 0.75$ m, $v_0 = 2.4$ m/s, and $\omega = 3.0$ rad/s.

Solution   Substituting the given values into the function:

$$y = y_0 \cos \omega t + \frac{v_0}{\omega} \sin \omega t,$$

we have:

$$y = \boxed{0.75 \text{ m}} \cos \boxed{(3.0 \text{ rad/s})}\, t + \frac{\boxed{2.4 \text{ m/s}}}{\boxed{3.0 \text{ rad/s}}} \sin \boxed{(3.0 \text{ rad/s})}\, t,$$

or,

$$y = 0.75 \cos 3.0t + 0.8 \sin 3.0t.$$

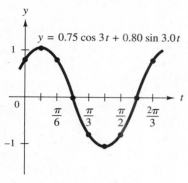

FIGURE 8.47

$y = 0.75 \cos 3t + 0.80 \sin 3.0t$

This equation shows that the periods of the individual functions are the same; that is, $\dfrac{2\pi}{3}$. Thus, the period of the combination function is $\dfrac{2\pi}{3}$. Functional values calculated for intervals of $\dfrac{\pi}{12}$ are recorded in Table 8.14. The curve is sketched in Figure 8.47. ◆

TABLE 8.14 $y = 0.75 \cos 3.0t + 0.8 \sin 3.0t$

t	0	$\dfrac{\pi}{12}$	$\dfrac{\pi}{6}$	$\dfrac{\pi}{4}$	$\dfrac{\pi}{3}$	$\dfrac{5\pi}{12}$	$\dfrac{\pi}{2}$	$\dfrac{7\pi}{12}$	$\dfrac{2\pi}{3}$
y	0.75	1.10	0.80	0.04	-0.75	-1.10	-0.80	-0.04	0.75

8.6 Exercises

In Exercises 1–20, for the combination function perform the following.
a. Determine the period of each component function.
b. Determine the period of the combination function.
c. Determine the number of cycles each component function will go through for one period of the combination function.
d. Sketch the curve for one period of the combination function.

1. $y = 3 + 2 \sin x$

2. $y = 5 - \cos 2x$

3. $y = -0.5 + 0.4 \cos \dfrac{\theta}{2}$

4. $y = -0.8 + 0.9 \sin \dfrac{\theta}{3}$

5. $y = x + 2 \sin x$

6. $y = x - 4 \cos x$

7. $y = 2x + 3 \cos x$

8. $y = 5x - 3 \sin x$

9. $y = \sin x - \cos x$

10. $y = \sin x + 2 \cos x$

11. $y = 2 \cos x + 2 \sin x$

12. $y = 3 \cos x - \sin x$

13. $y = 4 \cos x + 3 \sin x$

14. $y = 5 \sin x - 6 \cos x$

15. $y = \cos 2x + \sin 2x$

16. $y = \sin \dfrac{x}{2} + \cos \dfrac{x}{2}$

17. $y = \cos 3x + 2 \cos (x + 45°)$

18. $y = \sin 2x + 3 \sin (x + 30°)$

19. $y = \sin 3x + \sin \left(x + \dfrac{\pi}{4}\right)$

20. $y = 3 \cos x + \cos \left(x - \dfrac{\pi}{3}\right)$

In Exercises 21–24, use point-by-point plotting to sketch the graph of each function over $0 \le x \le 4.5$ for values of x for each quarter unit $(0, 0.25, 0.50, \ldots)$.

21. $y = \sin \pi x - 2 \cos x$

22. $y = 0.5 \cos x + 2 \sin \pi x$

23. $y = 2 \sin \dfrac{\pi}{2} x + \cos 3x$

24. $y = 4 \sin 2x - 2 \cos \dfrac{\pi}{4} x$

In Exercises 25–28, graph each function over the specified interval.

25. $y = \dfrac{3}{x+3} \cos \pi x,\ 0 \le x \le 4$

26. $y = \dfrac{4}{x+3} \sin 2x,\ 0 \le x \le 3\pi$

27. $y = \dfrac{6+x}{5} \sin \dfrac{\pi}{2}x,\ 0 \le x \le 8$

28. $y = \dfrac{3.4-x}{2.1} \cos \dfrac{\pi}{3}x,\ 0 \le x \le 8$

29. In a graph, illustrate the spring action for one cycle of the combination function when $y_0 = 2.3$ m, $v_0 = 3.4$ m/s, and $\omega = 1.7$ rad/s. Use the function given in Example 5.

30. The instantaneous power in an AC circuit containing an inductance is given by $P = VI$, where $V = 150 \cos \omega t$ and $I = 15 \sin \omega t$. (Inductance is a varying current in a circuit that induces voltages in the same or a nearby circuit.) **a.** Sketch P versus the angle ωt for one cycle by finding values for E and I and computing the product VI. **b.** Show that the curve in part a is equivalent to the one described by $P = 1125 \sin 2\,\omega t$ by sketching this curve and comparing it with the curve in part a.

31. The motion of a vibrating spring is given by $d = 20 \cos (t + \pi) + 10 \sin \left(t - \dfrac{\pi}{2}\right)$. Sketch one cycle. When does it reach its maximum displacement? (See the illustration.)

32. The horizontal position of a point on a rolling wheel is given by $x = -r\phi + r\cos \phi$, where r is the radius and ϕ is the angle of rotation in radians. Sketch x vs. ϕ for one cycle when $r = 0.25$ m. (See the drawing.)

33. Two waves in a medium (a substance in which bodies exist) will reinforce or weaken each other depending on their phase difference. Show by graphing $y_1 + y_2$ that when one water wave described by $y_1 = 2.5 \cos x$ meets another wave, $y_2 = 3 \sin \left(\dfrac{\pi}{2} + x\right)$, the result is a stronger wave.

34. The acceleration of a piston is expressed as $a = 10\omega^2 \left(\cos \omega t + \dfrac{1}{2} \cos 2\omega t\right)$, where $\omega = 10^2$ rad/s. Sketch one cycle of the function.

35. The power dissipated by an electric resistor is $P = vi$, where P is the power in watts, v is the instantaneous voltage in volts, and i is the instantaneous current in amps. If $v = 120 \sin 120\pi t$ and $i = 5 \sin 120\pi t$, graph the power loss for one voltage cycle.

36. A square wave established on an oscilloscope is often used in electronics to check circuitry. While it is impossible to form a pure square wave from the sinusoidal transmissions of an oscilloscope, a nearly perfect square wave may be formed by adding all the sinusoidal terms in the series: $y = A \sin (kx) + \dfrac{A}{3} \sin (3kx) +$

$\dfrac{A}{5} \sin (5kx) + \cdots$. (k is a constant. Let $k = 1$.) Sketch the graph of the function using the first three terms. Discuss how closely the resultant wave resembles a square wave. If you increase or decrease the value of k, what is the affect on the sum of the terms of the series?

37. A trigonometric function of the form $y = M \sin Bt + N \cos Bt$ may be used to define a vibrating system, such as an automobile suspension system. (See the illustration.) Analysis of a vibrating system is often easier if the function is written in the equivalent form $y = A \sin (Bt + C)$.

A function that defines a vibrating system is $y = 15 \sin 7t + 8 \cos 7t$. An equivalent function for the same system is $y = 17 \sin (7t + 0.490)$.
a. Graph the two functions for the interval $0 \le t \le 1$.
b. Are the graphs equivalent?
c. When the vibrating system is slowed down by an outside force, we say that there is a damping effect. Does the graph indicate a damping effect for this system?

Review Exercises

In Exercises 1–10, graph each of the functions for the indicated interval.

1. $y = 3 \sin \theta$ for values of θ at intervals of $15°$, $-270° \le \theta \le 270°$

2. $y = \dfrac{1}{2} \cos \theta$ for values of θ at intervals of $\dfrac{\pi}{6}$, $-2\pi \le \theta \le 2\pi$

3. $y = 2 \tan \theta$ for values of θ at intervals of $\dfrac{\pi}{4}$, $-\pi \le \theta \le 3\pi$

4. $y = 0.5 \sin \dfrac{\pi}{4} \theta$, $-\dfrac{\pi}{4} \le \theta \le 3\pi$

5. $y = -3 \cos 3\theta$, $-\dfrac{\pi}{2} \le \dfrac{5}{2}\pi$

6. $y = 5 \tan \dfrac{\theta}{2}$, $0 < \theta \le 4\pi$

7. $y = 2 \csc 0.5\theta$, $-360° \le \theta \le 360°$

8. $y = 3 \sec 0.75\theta$, $-45° \le \theta \le 360°$

9. $y = \dfrac{1}{2} \cot 3\theta$, $-\dfrac{3\pi}{2} \le \theta \le \dfrac{3\pi}{2}$

10. $y = -2 \tan \dfrac{\theta}{2}$, $0 \le \theta \le 180°$

In Exercises 11–16, determine **a.** the period, **b.** the phase shift, **c.** the amplitude, **d.** where the curve will recycle, and **e.** sketch the graph.

11. $y = 2 \sin \left(3\theta + \dfrac{\pi}{4} \right)$

12. $y = \dfrac{1}{2} \sin \left(\pi\theta - \dfrac{\pi}{2} \right)$

13. $y = 4 \cos \left(\dfrac{1}{3}\theta + 30° \right)$

14. $y = 2.4 \cos \left(\dfrac{\pi}{2}\theta - \dfrac{\pi}{4} \right)$

15. $y = 16 \sin (0.01\theta + 15°)$

16. $y = 13 \cos (0.03\theta - 18°)$

In Exercises 17–21, sketch the graph of each function over the given interval.

17. $y = \dfrac{1}{2} \tan \left(3\theta + \dfrac{\pi}{4} \right),\ -\pi \le \theta \le \pi$

18. $y = 2 \tan \left(\dfrac{1}{2}\theta - \dfrac{\pi}{3} \right),\ 0 \le \theta \le 2\pi$

19. $y = \dfrac{1}{3} \cot (2\theta - 60°),\ -90° \le \theta \le 270°$

20. $y = -4 \csc (2\theta - 60°),\ -90° \le \theta \le 270°$

21. $y = 0.4 \sec \left(0.4\theta - \dfrac{\pi}{12} \right),\ -4\pi \le \theta \le 4\pi$

In Exercise 22–27, for the combination function perform the following.
a. Determine the period of each component function.
b. Determine the period of the combination function.
c. Determine the number of cycles each component function will go through for one period of the combination function.
d. Sketch the curve for two periods of the combination function.

22. $y = -2 + 0.5 \sin 2x$

23. $y = 3 - 2 \cos \dfrac{x}{2}$

24. $y = \dfrac{1}{2} \sin (2\theta + 22°) + 2 \cos (\theta - 15°)$

25. $y = 2 \sin \left(\dfrac{\pi}{3}\theta + \dfrac{\pi}{2} \right) - \dfrac{1}{2} \cos \left(\dfrac{\pi}{4}\theta \right)$

26. $y = 2 \cos 2\theta - 3 \cos \left(\dfrac{\theta}{2} + 30° \right)$

27. $y = 3 \sin \left(\dfrac{\pi}{6}\theta - \dfrac{\pi}{3} \right) + 2 \sin \left(\dfrac{\pi}{3}\theta - \dfrac{\pi}{6} \right)$

28. The motion of a vibrating spring is given by $d = 15 \sin (\theta + \pi) + 5 \cos \left(\theta - \dfrac{\pi}{2} \right)$.
Sketch one cycle. When does the spring reach its maximum displacement?

29. The motion of a piston is given by $s = \sin 5t + \cos (5t - \pi)$. Sketch one period. When is the maximum displacement reached?

30. A current in an AC circuit is given by $i = -30 \sin (30\pi t + \pi)$, where i is in amperes and 30π is in rad/s. What are the amplitude, period, phase shift, and frequency of this waveform?

31. A voltage in an AC circuit is given by $v = 5 \sin 60\pi t$. Find the amplitude and period, and sketch one cycle.

32. The range r of a cannon on a tank with muzzle velocity v, fired at an angle of elevation ϕ is given by $r = (v^2 \sin 2\phi)/g$ where g is a constant. Sketch a graph of r vs. ϕ. What angle determines the maximum value for r?

33. The waves on the surface of a lake when viewed horizontally appear to be approximately sinusoidal curves. Assuming that the x-axis coincides with the surface of the lake during a perfect calm, write the equation when the waves are 5 ft from crest to trough and 20 ft in wavelength. Draw a graph of the sinusoidal curve.

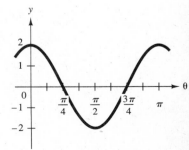

In Exercises 34–38, there may be more than one answer for the equation of each of the graphs.

34. **a.** What is the amplitude of the graph in the figure?
     b. What is the period?
     c. Write an equation for the graph.

35. **a.** What is the amplitude of the graph in the illustration?
  b. What is the period?
  c. Write an equation for the graph.

36. Write an equation that describes the function graph in the illustration.

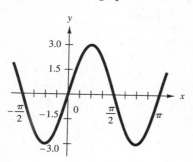

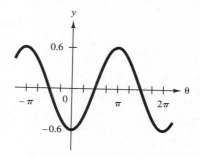

37. Write an equation that describes the function graphed in the figure.

38. Write an equation that describes the function graphed in the figure.

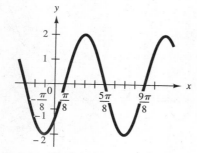

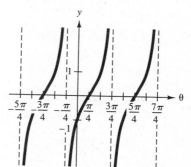

39. Two AC generators produce currents that are given in terms of time t by the equations:

$$I_1 = \sqrt{3}\ \sin 120\pi t \quad \text{and} \quad I_2 = -\cos 120\pi t.$$

If the output of the second is added to the output of the first, determine **a.** the maximum current output, **b.** when it occurs.

40. During the respiratory cycle (inhale and then exhale) of a person at rest, the rate of flow y (in liters per second) of air in and out of a person's lungs at time t (in seconds) is:

$$y = \frac{1}{2} \sin \frac{2\pi}{5} t.$$

  a. How long does it takes to complete one respiratory cycle?
  b. How many respiratory cycles are completed in one-half minute?
  c. Graph two complete cycles starting with $t = 0$.
  d. Explain the positive and negative values of y.
  e. Determine y when $t = 2.5$ s, to the nearest hundredth.

41. The motion of each piston in a reciprocating engine closely approximates simple harmonic motion. The speed of each piston is $v = \omega y_0 \sin\left(\omega t + \dfrac{\pi}{2}\right)$, where ω is the angular speed of the crank shaft in radians per unit time and y_0 is the maximum displacement of the piston from the midpoint of its path. Graph the speed vs. time in seconds of the engine at 8500 rpm. The stroke (distance between extreme positions) of each piston is 2.3 s in.

42. A pebble becomes wedged between the treads of your tire. As you drive away, the distance of the pebble from the road surface varies sinusoidally with the distance you have traveled. The period is the time needed for the wheel to make one complete revolution. Assume that the diameter of the wheel is 0.600 m and it makes one revolution in 0.305 s. (Use cosine function).
    a. Write an equation of the function.
    b. Sketch a graph of the function.
    c. Predict the distance of the pebble from the surface of the road when you have traveled for 3 s.

43. The distance from the ground to the tips of the blades of a windmill varies sinusoidally with time. When the windmill starts, P is in the position shown in the illustration. Let $t = $ the number of seconds that have elapsed since the windmill started. You find that it takes point $P \dfrac{1}{25}$ second to reach the top, 60 ft above the ground, and that the tip of the blade makes a revolution once every one-tenth of a second. The diameter of the circle described by the blades of the windmill is 50 ft. (Use sine function.)
    a. What is the lowest that point P will go as the blades of the windmill rotate? Why is the number greater than zero?
    b. Write an equation of this sinusoid.
    c. Sketch a graph of this sinusoid.
    d. Determine the height of point P above the ground when (1) $t = \dfrac{1}{5}$s, (2) $t = \dfrac{1}{4}$s.
    e. What is the value of t the second time point P is 30 ft above the ground?

44. A rotating coil of wire in a magnetic field develops an elecromotive force (emf) that is expressed by emf $= nBA \cos\left(\omega t - \dfrac{\pi}{2}\right)$ where $n = $ number of turns of wire in the coil, $B = $ magnetic density, and $A = $ area of the coil. If $n = 100$, $B = 3 \times 10^{-3} t$, $A = 2$ m^2, and $\omega = 60$ rad/s:
    a. sketch the graph of emf vs. t for one cycle, and
    b. express emf in terms of the sine function.

45. The angular acceleration of a simple pendulum is given by $\alpha = 2 \sin\left(2t + \dfrac{\pi}{2}\right)$ where $t = $ time in seconds.
    a. Determine the period of the pendulum's acceleration curve and sketch α vs. t for one cycle.
    b. Express α in terms of the cosine function.

46. On a windy day the waves on the surface of Lake Ontario appear to be approximately a sine or cosine curve if viewed horizontally. Determine the equation describing the surface waves that are 4 ft from crest to trough and 30 ft from one crest to the next. See the diagram.

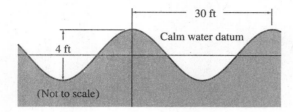

47. Show that $\cos(90° - \theta) = \sin \theta$ by sketching graphs of both functions.

48. In the illustration, suppose that $r = 0.1$ m, $\theta = 0$ when $t = 0$, and point P completes 1 revolution in 3 s with uniform speed.

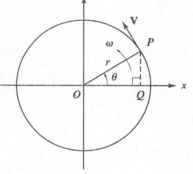

   a. What is the angular speed of the radius to point P in rad/s?

   b. What is the equation for the position of Q (the point directly under P on the x-axis) as a function of time?

   c. What is the frequency of the motion of Q?

   d. What is the position of Q when $t = 0.60$ s? When $t = 1.50$ s? When $t = 1.75$ s?

   e. Graph two cycles of the position of Q.

49. A saw-tooth waveform can be closely approximated by the following sum of sine functions:

$$\sin x - \frac{\sin 3x}{3^2} + \frac{\sin 5x}{5^2} - \cdots$$

To see the saw tooth forming, graph the first term, then the first two terms, and finally the first three terms on one set of axes. Write a program for your calculator or computer to facilitate the graphing.

50. The equation $x = 3^{-1} \cos \pi t$ is an algebraic model for a certain damped harmonic oscillator where x is displacement and t is time. Graph the equation and comment on the nature of the motion ($t \geq 0$). Write a program for the calculator or computer to determine coordinates and to make the graphing easier.

✎ Writing About Mathematics

1. After you submit a report, your supervisor says that you could have simplified it by using only the sine function. He makes this claim because of the relationship between $\sin(\theta + 90°)$ and $\cos \theta$. Can you write all sinusoidal functions in terms of the sine function? Write a paper defending your position.

2. Write a letter to a friend explaining amplitude, period, and phase shift. Tell why each of them is important in sketching the graph of a sinusoid.

Chapter Test

For each of the functions determine amplitude, period, and phase shift where applicable.

1. $y = 3 \sin \theta$ **2.** $y = 5 \cos 3\theta$ **3.** $y = \cos (2\theta - 45°)$ **4.** $y = 3 \tan (4\theta + 60°)$

For each function in Exercises 5–8, determine **a.** the amplitude, **b.** the period, **c.** the phase shift, **d.** where the curve will recycle, and **e.** sketch one cycle.

5. $y = 2 \sin (3\theta - 60°)$ **6.** $y = \dfrac{1}{2} \cos \left(\theta + \dfrac{\pi}{6}\right)$ **7.** $y = \tan (\theta - 45°)$ **8.** $y = 2 \csc (\theta + 30°)$

For the combination function in Exercises 9 and 10, perform the following.
a. Determine the period of each component of the combination function.
b. Determine the period of the combination function.
c. Find the number of cycles each component function will go through for one period of the combination function.
d. Sketch the curve for one period of the combination function.

9. $y = 3 + 2 \sin x$ **10.** $y = 4 \cos x + 3 \sin x$

11. The waves on the surface of a lake when viewed horizontally appear to be approximately sinusoidal curves. Assuming that the x-axis coincides with the surface of the lake during a perfect calm, write the equation when the waves are 4 ft from crest to trough and 15 ft in wavelength.
   a. Write an equation that describes the sinusoidal curve.
   b. Draw a graph of the sinusoidal curve.

12. During the respiratory cycle (inhale and then exhale) of a person at rest, the rate of flow y (in liters per second) of air in and out of a person's lungs at time t (in seconds) is $y = \dfrac{2}{5} \sin \dfrac{3\pi}{5} t$.
   a. How long does it take to complete one respiratory cycle?
   b. How many respiratory cycles are completed in one-half minute?
   c. Graph two complete cycles starting with $t = 0$.
   d. Explain what could be happening during the positive and negative values of y.
   e. Determine y when $t = 2.3$ seconds, to the nearest hundredth.

Imaginary and Complex Numbers

9

I f you open the case of almost any electronic device, you see boards with objects attached and what appear to be paths going from one object to the other. These boards are called circuit boards, the objects on the boards may be resistors and transistors, and the paths on the boards are electric circuits. When the boards are designed, components may be described with mathematical equations. In fact, technicians usually combine several components into a single mathematical equivalent so as to work with simpler equations. (See Section 9.2, Example 6, and Section 9.5, Exercise 28.)

9.1

Introduction

In Chapter 1, as part of the discussion of number systems, we were introduced to the number $\sqrt{-9}$, which was defined as an **imaginary number.** We also learned that an imaginary number, such as $\sqrt{-9}$, could be expressed as:

$$\sqrt{-9} = \sqrt{(-1)9}$$
$$= \sqrt{-1}\,\sqrt{9}$$
$$= (\sqrt{-1})(3)$$
$$= 3j.$$

DEFINITION 9.1

$$j = \sqrt{-1}.$$

In many mathematics textbooks $\sqrt{-1} = i$. This text uses the letter j to avoid confusing the i for imaginary numbers with the i used for current in electrical-circuit theory, which is standard practice in electrical textbooks. The reason for this becomes evident in the last section of this chapter when applications dealing with alternating currents are discussed.

DEFINITION 9.2

If n is a positive real number,

$$\sqrt{-n} = \sqrt{-1}\,\sqrt{n} = j\sqrt{n},$$

and $j\sqrt{n}$ is called an **imaginary** number.

EXAMPLE 1   Writing Imaginary Numbers with j-Notation

Write the following as a product of a real number and the imaginary unit j.

a. $\sqrt{-25}$ **b.** $\sqrt{-7}$ **c.** $\sqrt{-a}, a > 0$

Solution

a. $\sqrt{-25} = \sqrt{-1}\,\sqrt{25}$ Definition 9.2
       $= j5$ or $5j$ Definition 9.1

b. $\sqrt{-7} = \sqrt{-1}\,\sqrt{7}$
       $= j\sqrt{7}$

c. Since $a > 0$,

$$\sqrt{-a} = \sqrt{-1}\,\sqrt{a}$$
$$= j\sqrt{a} \quad \blacklozenge$$

We will now define j^2.

> **DEFINITION 9.3**
>
> $$j^2 = -1$$

EXAMPLE 2 **Evaluating j^2**

Determine the products of the following. **a.** $(3j)^2$ **b.** $\sqrt{-9}\sqrt{-25}$

Solution

a. $(3j)^2 = (3^2)(j^2)$ **Rules of exponents**
$\qquad\quad = (9)(-1)$ **Definition 9.3**
$\qquad\quad = -9$

b. Before we perform any operations on imaginary numbers, we must write them in j-notation.

$$\sqrt{-9}\sqrt{-25} = \sqrt{-1}\sqrt{9}\sqrt{-1}\sqrt{25}$$
$$= j\sqrt{9}\,j\sqrt{25} \qquad\qquad \textbf{Definition 9.2}$$
$$= (3j)(5j)$$
$$= 15j^2 \qquad\qquad \textbf{Commutative and associative laws}$$
$$= 15(-1) \qquad\qquad \textbf{Definition 9.3}$$
$$= -15 \;\blacklozenge$$

CAUTION   Recall from Chapter 2 that the following multiplication could be performed before we take the square root.

$$\sqrt{9}\,\sqrt{25} = \sqrt{9(25)}$$
$$= \sqrt{225}$$
$$= 15.$$

This procedure does not work when both numbers under the radicals are negative. For example:

$$\sqrt{-9}\,\sqrt{-25} \neq \sqrt{(-9)(-25)}$$
$$\neq \sqrt{225}$$
$$\neq 15. \;\blacksquare$$

The rules for order of operations tell why we must convert the numbers to j-notation before we multiply them. The rules state that, in working from left to right, we perform the operations in the following order:

1. Take the root or raise the number to a power.
2. Perform all multiplications and divisions.
3. Perform all additions and subtractions.

If grouping symbols (parentheses, braces, brackets, etc.) are part of the expression, always perform the operations inside the grouping symbols first.

The radical ($\sqrt{}$) is a grouping symbol. So by performing the operations first and then taking the root of the quantity, we avoid problems. For example,

$$\sqrt{(-4)(-9)} = \sqrt{36} = 6.$$

The product of two negative integers is a positive integer, and the root of a positive integer is a real number. Therefore, $\sqrt{(-4)(-9)}$ cannot be written as:

$$\sqrt{-4}\,\sqrt{-9} \neq (2j)(3j)$$
$$\neq -6.$$

By writing $\sqrt{(-4)(-9)} = \sqrt{-4}\sqrt{-9}$, we are changing a real number to an imaginary number, which we cannot do.

Definition 9.3 says that $j^2 = -1$. Let us now examine what happens when j is raised to a higher power. To evaluate the higher powers of j, we will use Definitions 9.1, 9.2, and 9.3 and the laws of exponents.

For example, can we write j^{-1} without a negative exponent?

$$j^{-1} = \frac{1}{j} \qquad \text{Definition 2.8: } a^{-n} = \frac{1}{a^n}$$

$$= \frac{1}{\sqrt{-1}} \qquad \text{Definition 9.1}$$

Since mathematicians prefer to have no radicals in the denominator, we rationalize the denominator.

$$j^{-1} = \frac{1}{\sqrt{-1}} \cdot \frac{\sqrt{-1}}{\sqrt{-1}}$$

$$= \frac{\sqrt{-1}}{-1} \text{ or } -j \qquad \text{Definition 9.1}$$

Table 9.1 contains 20 powers of j.

TABLE 9.1 Powers of j

$j^{-3} = j$	$j^1 = j$	$j^5 = j$	$j^9 = j$	$j^{13} = j$
$j^{-2} = -1$	$j^2 = -1$	$j^6 = -1$	$j^{10} = -1$	$j^{14} = -1$
$j^{-1} = -j$	$j^3 = -j$	$j^7 = -j$	$j^{11} = -j$	$j^{15} = -j$
$j^0 = 1$	$j^4 = 1$	$j^8 = 1$	$j^{12} = 1$	$j^{16} = 1$

Table 9.1 shows us that for any power of j there is only one of four possible solutions: j, -1, $-j$, and 1. The exponents of j in the second column are 1, 2, 3, and 4. If we divide those exponents by 4, the remainders are 1, 2, 3, and 0. The exponents in the third column are 5, 6, 7, and 8. If we divide them by 4 the remainders are 1, 2, 3, and 0. Similarly if we divide the exponents in the fourth column (9, 10, 11, 12) and the exponents in the fifth column (13, 14, 15, 16) by 4, the remainders are 1, 2, 3, and 0. The relationship among the exponent, remainder, and value of j follows.

RELATIONSHIP AMONG EXPONENT, REMAINDER, AND VALUE OF j

For j^n, n is a natural number.

If $n \div 4$ has a remainder of 1, then $j^n = j^1 = j$.
If $n \div 4$ has a remainder of 2, then $j^n = j^2 = -1$.
If $n \div 4$ has a remainder of 3, then $j^n = j^3 = -j$.
If $n \div 4$ has a remainder of 0, then $j^n = j^0 = 1$.

EXAMPLE 3   **Writing an Equivalent Symbol for j^n**

Simplify the following.

a. j^{17} **b.** j^{23} **c.** $(-2j^{18})(5j^6)$

Solution

a. To simplify j^{17}, divide 17 by 4 and find that the remainder is 1. Thus, $j^{17} = j^1 = j$.

b. To simplify j^{23}, divide 23 by 4 and find that the remainder is 3. Thus, $j^{23} = j^3 = -j$.

c. To simplify $(-2j^{18})(5j^{24})$, divide 18 and 24 by 4 and find that the remainders are 2 and 0, respectively. With these remainders, $j^{18} = j^2 = -1$ and $j^{24} = j^0 = 1$. Therefore, substitute -1 for j^{18} and 1 for j^{24}. The result is $(-2j^{18})(5j^{24}) = [-2(-1)][5(1)] = 10$.   ◆

Division of imaginary numbers is slightly different from conventional long division. The objective is to have no imaginary numbers in the denominator of a fraction. This is illustrated in Example 4.

PROCEDURE FOR DIVIDING IMAGINARY NUMBERS

To divide a number of the form $\dfrac{k}{\sqrt{-a}}$ where a and k are constants, proceed as follows.

1. Write $\sqrt{-a}$ in the form $j\sqrt{a}$.
2. Multiply the numerator and the denominator by j.
3. The product of Step 2 will have a factor of j^2 in the denominator. Since $j^2 = -1$, replace j^2 with -1 and simplify the fraction.

EXAMPLE 4   **Dividing a Real Number by an Imaginary Number**

Divide the following.

a. 28 by $\sqrt{-7}$ **b.** 17 by $\sqrt{-5}$

Solution Because it is more convenient to work with numbers that do not contain imaginary numbers in the divisor (denominator), our objective is to remove the imaginary number from the denominator. To accomplish this, use the definition of j^2.

a. $28 \div \sqrt{-7} = \dfrac{28}{\sqrt{-7}}$. Since the denominator contains an imaginary number, can we change it to a real number? Yes, since $\sqrt{-7} = j\sqrt{7}$, multiply j by j; the result is -1. Thus,

$$\dfrac{28}{\sqrt{-7}} = \dfrac{28}{j\sqrt{7}} \qquad \textbf{Definition 9.2}$$

$$= \dfrac{28}{j\sqrt{7}} \cdot \dfrac{j}{j} \qquad \textbf{Multiplying by 1}$$

$$= \dfrac{28j}{j^2\sqrt{7}}$$

$$= \dfrac{28j}{-1\sqrt{7}} \quad \text{or} \quad j^2 = -1.$$

At this stage we have accomplished our goal. However, we can also eliminate the radical in the denominator by multiplying numerator and denominator by $\sqrt{7}$.

$$\dfrac{28}{\sqrt{-7}} = \dfrac{28j}{-\sqrt{7}} \cdot \dfrac{\sqrt{7}}{\sqrt{7}} \qquad \textbf{Rationalizing the denominator}$$

$$= \dfrac{28j\sqrt{7}}{-7} \qquad (\sqrt{a})(\sqrt{a}) = a$$

$$= -4j\sqrt{7} \qquad \textbf{Reducing the fraction}$$

Therefore, $28 \div \sqrt{-7} = -4j\sqrt{7}$.

b. Following the procedure developed in part a we have:

$$\dfrac{17}{\sqrt{-5}} = \dfrac{17}{j\sqrt{5}} \qquad \textbf{Definition 9.2}$$

$$= \dfrac{17}{j\sqrt{5}} \cdot \dfrac{j\sqrt{5}}{j\sqrt{5}} \qquad \textbf{Rationalizing the denominator}$$

$$= \dfrac{17j\sqrt{5}}{-5} \qquad \textbf{Multiplication of fractions and } j^2 = -1$$

$$= \dfrac{-17j\sqrt{5}}{5}. \qquad \begin{array}{l}\textbf{Multiplying numerator and} \\ \textbf{denominator by } -1 \quad \blacklozenge\end{array}$$

We can add and subtract imaginary numbers if the terms are like terms. That is, $7j\sqrt{3}$ and $4j\sqrt{3}$ are like terms; they both have factors of $j\sqrt{3}$. Thus, the sum of $7j\sqrt{3}$ and $4j\sqrt{3}$ is $(7 + 4)(j\sqrt{3})$ or $11j\sqrt{3}$.

Before we can perform the addition or subtraction, we may need to write the imaginary number with j-notation and simplify the radicals to see if there are like terms. This is illustrated in Example 5.

EXAMPLE 5   **Adding Imaginary Numbers**

Determine the sum of the following numbers: $\sqrt{-27}$, $\sqrt{-8}$, and $\sqrt{-12}$.

Solution

$$\sqrt{-27} + \sqrt{-8} + \sqrt{-12}$$

$$= j\sqrt{27} + j\sqrt{8} + j\sqrt{12} \qquad \text{Definition 9.2}$$

$$= j \cdot 3\sqrt{3} + j \cdot 2\sqrt{2} + j \cdot 2\sqrt{3} \qquad \text{Simplifying radicals}$$

$$= j(3\sqrt{3} + 2\sqrt{2} + 2\sqrt{3}) \qquad j \text{ is a common-term factor}$$

$$= j(5\sqrt{3} + 2\sqrt{2}). \qquad \text{Adding "like" terms}$$

$$= 5j\sqrt{3} + 2j\sqrt{2} \quad \blacklozenge$$

9.1 Exercises

In Exercises 1–8, write each of the following as a product of a real number and the imaginary unit j. Leave the answers in simplest radical form.

1. $\sqrt{-64}$ **2.** $\sqrt{-81}$ **3.** $\sqrt{-32}$ **4.** $\sqrt{-125}$

5. $\sqrt{-a^4}$ **6.** $\sqrt{-x^8}$ **7.** $\sqrt{-p^6}$ **8.** $\sqrt{-q^{12}}$

In Exercises 9–20, simplify the expression. (The highest power of j in the answer should be 1.)

9. j^{13} **10.** j^{14} **11.** j^{31} **12.** j^{56}

13. j^{63} **14.** j^{175} **15.** j^{302} **16.** j^{1001}

17. j^0 **18.** j^{-2} **19.** j^{-29} **20.** j^{-58}

In Exercises 21–46, perform the indicated operations and simplify. Leave the answers in radical form.

21. $(\sqrt{-121})(\sqrt{-144})$ **22.** $(\sqrt{-72})(\sqrt{-50})$ **23.** $(\sqrt{-96})(\sqrt{-24})$ **24.** $(\sqrt{-15})(\sqrt{-12})$

25. $(\sqrt{-8})(\sqrt{-24})$ **26.** $(\sqrt{-51})(\sqrt{17})$ **27.** $(\sqrt{-4})(\sqrt{-9})$ **28.** $(\sqrt{-12})(\sqrt{-48})$

29. $(\sqrt{-8})(\sqrt{-15})(\sqrt{-24})$ **30.** $(\sqrt{-14})(\sqrt{+21})(\sqrt{-18})$ **31.** $(\sqrt{-11})(\sqrt{-44})(\sqrt{-20})$

32. $(\sqrt{-45})(\sqrt{-35})(\sqrt{-7})$ **33.** $\dfrac{\sqrt{5}}{\sqrt{-14}}$ **34.** $\dfrac{\sqrt{-7}}{\sqrt{28}}$

35. $\dfrac{\sqrt{35}}{\sqrt{5}}$ **36.** $\dfrac{\sqrt{1001}}{\sqrt{-11}}$ **37.** $\dfrac{\sqrt{-8}}{\sqrt{-32}}$

38. $\dfrac{\sqrt{-28}}{\sqrt{-14}}$ **39.** $\sqrt{-18} + \sqrt{-50} + \sqrt{-32}$ **40.** $\sqrt{-24} + \sqrt{-150} + \sqrt{-96}$

41. $3\sqrt{-80} - \sqrt{-45}$

42. $4\sqrt{-4} + 4\sqrt{-9} - 5\sqrt{16}$

43. $\sqrt{-25} + \sqrt{-144} - \sqrt{169}$

44. $\sqrt{-289} + \sqrt{-361} - \sqrt{-400}$

45. $j + j^2 + j^3 + j^4$

46. $\dfrac{1}{j} + \dfrac{1}{j^2} + \dfrac{1}{j^3} + \dfrac{1}{j^4}$

In Exercises 47–54, use the following technique to simplify the expression. Another way of writing j^n in a simpler form is to use the rules of exponents. In general for any positive integer n, j^n can be expressed as $(j^2)^{n/2}$ if n is an even positive integer or as $(j^2)^{(n-1)/2}(j)$ if n is an odd positive integer. The object is never to leave j to a higher power than one. For example,

$$j^{23} = (j^2)^{(23-1)/2}j \quad \text{n is an odd integer}$$
$$= (j^2)^{11}(j)$$
$$= (-1)^{11}(j)$$
$$= -j.$$

47. j^{13}

48. j^{24}

49. j^{103}

50. j^{126}

51. j^{73}

52. j^{84}

53. j^{90}

54. j^{35}

9.2

Basic Operations with Complex Numbers

When an imaginary number and a real number are combined, the result is a number called a complex number.

DEFINITION 9.4

A **complex number** is a number of the form $a + bj$, where a and b are real numbers and $j = \sqrt{-1}$.

Let $Z = a + bj$; a is called the **real part** of Z, and b is called the **imaginary part** of Z. When $a = 0$ and $b \neq 0$, the complex number $(a + bj)$ may be expressed in the form $0 + bj = bj$ and is called a **pure imaginary number.** When $b = 0$ and $a \neq 0$, the complex number $(a + bj)$ represents the real number $a + 0j = a$.

Complex numbers that differ only in the signs of the imaginary parts are called **conjugate complex numbers** (or complex conjugates), and either is called the conjugate of the other. Thus, if $Z_1 = a + bj$ and $Z_2 = a - bj$ then Z_1 is the complex conjugate of Z_2, and Z_2 is the complex conjugate of Z_1. (Also, if $Z_1 = bj$ and $Z_2 = -bj$, then Z_1 and Z_2 are conjugates of each other.) If there is no imaginary part, then the remaining number is real, and that number is its own conjugate.

DEFINITION 9.5

Two complex numbers are equal if, and only if, their real parts are equal and their imaginary parts are equal. That is $a + bj = c + dj$ if, and only if, $a = c$ and $b = d$.

From Definition 9.5 we can conclude that if the complex number $a + bj = 0$, then $a = 0$ and $b = 0$.

EXAMPLE 1 Finding Equal Complex Numbers

For what real values of x and y will the complex numbers Z_1 and Z_2 be equal?

a. $Z_1 = x + 3j$, $Z_2 = 2 - 3yj$ **b.** $Z_1 = 2x + 7yj$, $Z_2 = 6 - 49j + xj$

Solution

a. For the complex numbers $Z_1 = x + 3j$ and $Z_2 = 2 - 3yj$ to be equal, Definition 9.5 states that the corresponding parts must be equal. Thus,

$$x = 2 \quad \text{and} \quad 3 = -3y \qquad \text{Definition 9.5}$$
$$y = -1.$$

Substituting these values for x and y we have:

$$Z_1 = 2 + 3j \quad \text{and} \quad Z_2 = 2 - 3(-1)j$$
$$= 2 + 3j.$$

b. For $Z_1 = 2x + 7yj$ and $Z_2 = 6 - 49j + xj$ to be equal, the corresponding parts must be equal; that is, real parts are equal and the imaginary parts are equal. Thus:

$$2x = 6 \quad \text{and} \quad 7y = -49 + x \qquad \text{Definition 9.5}$$
$$x = 3 \quad \text{and} \quad 7y = -49 + 3$$
$$7y = -46$$
$$y = \frac{-46}{7}$$

Therefore, with $x = 3$ and $y = \dfrac{-46}{7}$,

$$Z_1 = 2(3) + 7\left(\frac{-46}{7}\right)j \quad \text{and} \quad Z_2 = 6 - 49j + 3j$$
$$Z_1 = 6 - 46j \qquad \qquad \text{and} \quad Z_2 = 6 - 46j. \quad \blacklozenge$$

It can be demonstrated that the definition of equality of complex numbers is the only possible definition that is consistent with the definition of real numbers. For example, if:

$$a + bj = c + dj$$

then:

$$a - c = -bj + dj \qquad \textbf{Subtracting } c \textbf{ and } bj \textbf{ from both}$$
$$a - c = -(b - d)j \qquad \textbf{sides of the equation}$$
$$(a - c)^2 = [-(b - d)j]^2$$
$$(a - c)^2 = (b - d)^2.$$

This is only possible when $(a - c) = 0$ and $(d - b) = 0$. Thus $a = c$ and $b = d$ when two complex numbers are equal.

Since the complex numbers are a new set of numbers, we must define how to perform the fundamental operations of addition, subtraction, multiplication, and division with complex numbers.

Addition and Subtraction   In order to add (or subtract) two complex numbers, such as $a + bj$ and $c + dj$, add (or subtract) the real parts and the imaginary parts separately.

> **DEFINITION 9.6**
> $$(a + bj) + (c + dj) = (a + c) + (b + d)j.$$
> $$(a + bj) - (c + dj) = (a - c) + (b - d)j.$$

EXAMPLE 2   Adding and Subtracting Complex Numbers

Perform the indicated operations.

a. $(6 + 3j) + (8 - 5j)$ **b.** $(5 + 4j) - (6 - 7j)$

Solution

a. $(6 + 3j) + (8 - 5j) = (6 + 8) + [3 + (-5)]j$ Definition of addition

$= 14 - 2j$ Combining like terms

b. $(5 + 4j) - (6 - 7j) = (5 - 6) + [4 - (-7)]j$ Definition of subtraction

$= -1 + 11j$ Combining like terms

Multiplication   To multiply two complex numbers, such as $a + bj$ and $c + dj$, find the product as you would for any product of two binomials. Then evaluate the powers of j according to the rules in Section 9.1 and combine the real and imaginary parts.

$$(a + bj)(c + dj) = ac + adj + bcj + bdj^2$$
$$= ac + (ad + bc)j + (-1)bd$$
$$= (ac - bd) + (ad + bc)j$$

> **DEFINITION 9.7**
> $$(a + bj)(c + dj) = (ac - bd) + (ad + bc)j$$

EXAMPLE 3   Multiplying Complex Numbers

Determine the product of $(3 + 7j)$ and $(4 - 5j)$.

Solution

$$(3 + 7j)(4 - 5j) = 12 - 15j + 28j - 35j^2 \qquad \text{Definition of multiplication}$$

$$= 12 + 13j - 35(-1) \qquad \text{Definition of } j^2$$
$$= 12 + 35 + 13j \qquad \text{Commutative law}$$
$$= 47 + 13j \qquad \text{Combining like terms} \quad \blacklozenge$$

The product of the complex number $3 + 4j$ and its conjugate $3 - 4j$ is a real number. For example,

$$(3 + 4j)(3 - 4j) = (3)(3) + (3)(-4j) + (4j)(3) + (4j)(-4j)$$
$$= 9 - 12j + 12j - 16j^2$$
$$= 9 + 16$$
$$= 25.$$

In general, $(a + bj)(a - bj) = a^2 + b^2$. This product is used in defining division of complex numbers.

Division

To express the quotient of $a + bj$ and $c + dj$ where $c + dj \neq 0$ as a single complex number, multiply both numerator and denominator of the indicated quotient $\dfrac{a + bj}{c + dj}$ by $c - dj$, the conjugate of the denominator.

$$\frac{a + bj}{c + dj} = \frac{a + bj}{c + dj} \cdot \frac{c - dj}{c - dj}$$

$$= \frac{ac - adj + bcj - bdj^2}{c^2 - cdj + cdj - d^2 j^2} \qquad \begin{array}{l}\textbf{Multiply and recall that}\\ \boldsymbol{(c + dj)(c - dj) = c^2 + d^2}\end{array}$$

$$= \frac{(ac + bd) + (bc - ad)j}{c^2 + d^2}$$

$$= \frac{ac + bd}{c^2 + d^2} + \frac{bc - ad}{c^2 + d^2}j. \qquad \begin{array}{l}\textbf{Writing in complex number}\\ \textbf{form } \boldsymbol{a + bj}\end{array}$$

DEFINITION 9.8

$$\frac{a + bj}{c + dj} = \frac{ac + bd}{c^2 + d^2} + \frac{bc - ad}{c^2 + d^2}j \quad \text{if } c + dj \neq 0.$$

EXAMPLE 4 Dividing Complex Numbers

Determine the quotient of $6j$ and $5 - 7j$.

Solution

$$\frac{6j}{5-7j} = \frac{6j}{5-7j} \cdot \frac{5+7j}{5+7j}$$ Multiply numerator and denominator by the conjugate of $5 - 7j$

$$= \frac{30j + 42j^2}{25 + 49}$$ Multiply and recall that $(c + dj)(c - dj) = c^2 + d^2$

$$= \frac{-42 + 30j}{74}$$

$$= \frac{-21}{37} + \frac{15}{37}j. \quad \blacklozenge$$

EXAMPLE 5 Dividing Complex Numbers

Determine the quotient of $4 - 5j$ and $2 + 3j$.

Solution

$$\frac{4-5j}{2+3j} = \frac{4-5j}{2+3j} \cdot \frac{2-3j}{2-3j}$$ Multiply numerator and denominator by the conjugate of $2 + 3j$

$$= \frac{8 - 12j - 10j + 15j^2}{4 + 9}$$ Multiply and recall that $(c + dj)(c - dj) = c^2 + d^2$

$$= \frac{(8 - 15) + (-12j - 10j)}{4 + 9}$$ Collecting like terms, $j^2 = -1$

$$= \frac{-7 - 22j}{13}$$

$$= \frac{-7}{13} - \frac{22}{13}j. \quad \blacklozenge$$

EXAMPLE 6 Application: Parallel Circuit

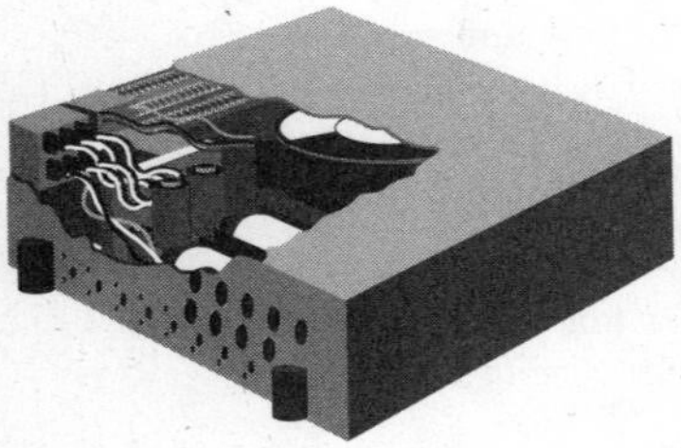

A commonly used technique when analyzing a VCR circuit (see Figure 9.1) is to replace a portion of a circuit with its simplified electrical equivalent. The equivalent electrical impedance Z of two impedances Z_1 and Z_2 connected in parallel is given by:

$$Z = \frac{Z_1 Z_2}{Z_1 + Z_2}.$$

If $Z_1 = 3.0 + 4.0j$ and $Z_2 = 5.0 + 12.0j$, calculate Z.

FIGURE 9.1

Solution For $Z_1 = 3.0 + 4.0j$ and $Z_2 = 5.0 + 12.0j$:

$$z = \frac{(3.0 + 4.0j)(5.0 + 12.0j)}{(3.0 + 4.0j) + (5.0 + 12.0j)} \qquad \text{Substituting}$$

$$= \frac{-33.0 + 56.0j}{8.0 + 16.0j} \qquad \begin{array}{l}\text{Multiplying numerator;} \\ \text{adding denominator}\end{array}$$

To divide two complex numbers, multiply both the numerator and the denominator by the complex conjugate of the denominator.

$$Z = \left(\frac{-33.0 + 56.0j}{8.0 + 16.0j}\right)\left(\frac{8.0 - 16.0j}{8.0 - 16.0j}\right) \qquad \text{Multiply by complex conjugate}$$

$$= \frac{632 + 976j}{320}$$

$$= \frac{79}{40} + \frac{61}{20}j$$

The equivalent impedance is $2.0 + 3.1j$. ◆

A BASIC computer program which can be used to find the equivalent of two impedances connected in parallel is found on the tutorial disk.

9.2 Exercises

In Exercises 1–6, determine the real values that may be substituted for x and y so that $Z_1 = Z_2$.

1. $Z_1 = 2x + yj, Z_2 = 3 + 5j$

2. $Z_1 = 4x + 6yj, Z_2 = 12 - 42j + xj$

3. $Z_1 = (x - 4) + (3y - 5)j, Z_2 = -3y + 7j$

4. $Z_1 = (3x - 5) + (7y - 1)j, Z_2 = (16x + 4) + (11y - 5x)j$

5. $Z_1 = (2x - 3) + (5x - y)j, (Z_2 = (x - 4) + 4yj$

6. $Z_1 = (5x - 3) + (7y - 11)j, Z_2 = (2x + 4y) + 10j$

In Exercises 7–28, perform the indicated operations and write the answer in the form $a + bj$.

7. $(5 - 6j) + (6 + 5j)$

8. $(4 - j) + (-9 - 3j)$

9. $(4 + 11j) + (8 + 6j)$

10. $(8 + j) - (4 + j)$

11. $(8 - \sqrt{-16}) - (3 + \sqrt{-36})$

12. $(9 + \sqrt{-18}) - (\sqrt{-72} + 1)$

13. $(2 + 5j)(4 + 2j)$

14. $(8 - 7j)(7 + 8j)$

15. $(3 + 6j)^2$

16. $(-5 + 4j)^2$

17. $(2 - \sqrt{-3})(3 + \sqrt{-12})$

18. $(8 - 6\sqrt{-5})(-2 + \sqrt{-20})$

19. $(4j - 2j^2 + 3j^3)(2j^4 - j^5)$

20. $(2j^5 - 5j^4 + 2j^3 - 5j^2)^2$

21. $(3 + 2j) \div (1 - 2j)$

22. $(4 - 5j) \div (4 + 6j)$

23. $(3 - \sqrt{-5}) \div (5 + 2\sqrt{-5})$

24. $(9 + \sqrt{-32}) \div (\sqrt{-2} + 3)$

25. $(8 - j^3) \div (3 - j)$

26. $(64 + j^3) \div (4 + j)$

27. $1 \div (4 + 3j)$

28. $1 \div (6 + 3\sqrt{-50})$

29. In the theory of optics it is convenient to locate a point in the complex plane. The mirror image of this point is its conjugate. Determine the mirror image of the point whose complex number representation is given by $7 - 5j$.

30. Two AC voltages are given by the expressions $\frac{3}{2} + (y + 1)j$ and $(7 + x) - 3j$. If the voltages are equal, what are the values of x and y?

31. What can be said about a complex number that is equal to its conjugate?

32. The impedance Z in an AC circuit is given by $Z = \frac{V}{I}$, where $V =$ voltage and $I =$ current. Find Z when $V = 2.3 - 0.4j$ and $I = -1.2j$.

33. Determine V based on the formula in Exercise 32 if $Z = +0.40 + 1.30j$ and $I = 2.20j$.

34. What type of a number is
    a. the sum of a complex number and its conjugate?
    b. The difference of a complex number and its conjugate?

35. The total impedance Z_T of an AC circuit containing two impedances Z_1 and Z_2 in parallel is given by:

$$Z_T = \frac{Z_1 Z_2}{Z_1 + Z_2}.$$

Determine Z_T when $Z_1 = 2 - j$ and $Z_2 = 3 + j$.

36. Determine Z_T using the formula in Exercise 35 if $Z_1 = 1.73 - j$ and $Z_2 = 0.5 + 0.9j$.

37. If three resistors in an AC circuit are connected in parallel, the total impedance Z_T may be found by using the formula:

$$\frac{1}{Z_T} = \frac{1}{Z_1} + \frac{1}{Z_2} + \frac{1}{Z_3},$$

or by the formula:

$$Z_T = \frac{Z_1 Z_2 Z_3}{Z_1 Z_2 + Z_1 Z_3 + Z_2 Z_3}.$$

If $Z_1 = 2 - j$, $Z_2 = 2 + j$, and $Z_3 = 4j$, determine Z_T using each formula.

9.3

Polar and Exponential Forms of a Complex Number

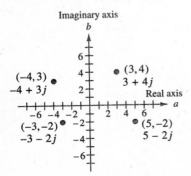

FIGURE 9.2

Before discussing the polar form of complex numbers, let's look at the technique for graphing complex numbers.

A complex number is graphed on a complex plane. A **complex plane** is divided into four quadrants with a horizontal and a vertical axis similar to those in a real plane. By convention, the horizontal axis is called the **real axis**, and the vertical axis is called the **imaginary axis**. Figure 9.2 shows these axes and the graph for the complex numbers $3 + 4j$, $5 - 2j$, $-4 + 3j$, and $-3 - 2j$.

A complex number is not a vector because no direction is needed for a complex number. However, the vector form is sometimes used since a complex number appears to be of a vector form and can be illustrated easily as a vector on a graph.

For any complex number $(a + bj)$, let P be the point on the complex plane that represents the number and O be the origin. Draw the line segment OP

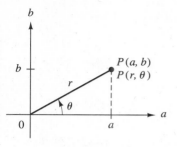

FIGURE 9.3

(Figure 9.3). From point P, drop a perpendicular to the real axis and label the point a. Now using the Pythagorean relation, we can determine the length r or $\overline{OP}$. The length r is determined using the formula:

$$r = \sqrt{a^2 + b^2}.$$

The length r is called the **magnitude, asbolute value,** or modulus of the complex number $(a + bj)$. The angle θ, with the a-axis as the initial side, is called the **argument** or the **amplitude** of the complex number.

The absolute value of any complex number $(a + bj)$ is denoted by $|a + bj|$. Since the absolute value is another name for the magnitude,

$$|a + bj| = \sqrt{a^2 + b^2}.$$

Thus, the absolute value of any complex number is a nonnegative real number.

The argument θ may be replaced by any angle that is coterminal with it: $\theta + k \cdot 360°$ where k is an integer. For convenience, select the smallest nonnegative value, usually between $0°$ and $360°$, for the argument of the complex number.

Polar Form

Any point P on the complex plane can be identified with its rectangular coordinates (a, b) or by its **polar coordinates (r, θ)** (see Figure 9.3). We will see that it is easier to perform certain operations if the complex numbers are written in polar form.

Using Figure 9.3 and our knowledge of right-triangle trigonometry, let us develop formulas that relate the polar and rectangular coordinates for complex numbers.

$$\cos \theta = \frac{a}{r} \qquad \text{and} \quad \sin \theta = \frac{b}{r}$$

$$a = r \cos \theta \quad \text{and} \qquad b = r \sin \theta \quad \textbf{Solving for } \textit{a} \textbf{ and } \textit{b} \qquad (9.1)$$

Now substitute for a and b in the complex number $a + bj$.

$$a + jb = r \cos \theta + jr \sin \theta,$$

or

$$a + jb = r(\cos \theta + j \sin \theta). \qquad (9.2)$$

The right side of Equation 9.2 is known as the **trigonometric form** or the **polar form** of the complex number, and the left side is the **rectangular form** of the complex number. The polar form may be abbreviated as:

$$r \text{ cis } \theta \quad \text{or} \quad r\underline{/\theta}.$$

That is, cis θ may be used in place of $\cos \theta + j \sin \theta$ to shorten the expression.

To convert from polar to rectangular form or from rectangular to polar form, use the following formulas.

> **CONVERSION FORMULAS**
>
> | Polar to Rectangular | Rectangular to Polar |
> |---|---|
> | $a = r \cos \theta$ | $r = \sqrt{a^2 + b^2}$ |
> | $b = r \sin \theta$ | $\theta = \tan^{-1} \dfrac{b}{a}$ |
>
> (Recall that a is the real part and b is the coefficient of j of the complex number.) See Figure 9.3.

EXAMPLE 1 **Graphing Complex Numbers, Convert from Polar Form to Rectangular Form**

Graph $3.0(\cos 210° + j \sin 210°)$ and write the complex number in rectangular form.

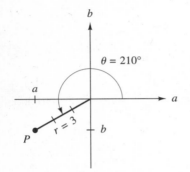

$\theta = 210°$

$r = 3$

FIGURE 9.4

Solution To graph $3.0\underline{/210°}$, first construct the angle $\theta = 210°$, as shown in Figure 9.4. On the terminal side of the angle, mark off 3 units from the vertex and locate point P. The point P represents the complex number. From the information on the figure and using Equation 9.1:

$$a = 3.0 \cos 210° \quad \text{and} \quad b = 3.0 \sin 210°$$
$$= -2.60 \qquad\qquad = -1.50.$$

Therefore, the rectangular form of $3.0\underline{/210°}$ is $-2.6 - 1.5\,j$. ◆

CAUTION Be careful in converting from rectangular to polar form; the signs of a and b are determined by the quadrant in which the terminal side of θ lies. If a and b are both negative, the tangent ratio is positive, and the calculator result is a first-quadrant angle, not a third-quadrant angle. If a and b have opposite signs, the tangent ratio is negative, and the calculator result is a second-quadrant angle, when the result could be either a second- or fourth-quadrant angle. Sketching a graph of the complex number is helpful in determining the correct angle. ∎

EXAMPLE 2 **Graph the Complex Number, Convert from Rectangular Form to Polar Form**

Graph the complex numbers and write the polar form of each using the smallest non-negative argument.

a. $1 + j\sqrt{3}$ **b.** $2 - 2j\sqrt{3}$

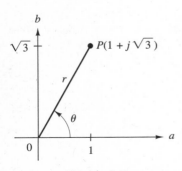

$P(1 + j\sqrt{3})$

FIGURE 9.5

Solution

a. In Figure 9.5, point p represents $(1 + j\sqrt{3})$. The point is in the first quadrant. Knowing that $a = 1$ and $b = \sqrt{3}$ and using the conversion for-

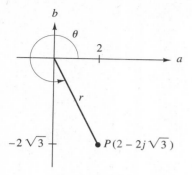

FIGURE 9.6

mulas from rectangular to polar we have:

$$r = \sqrt{a^2 + b^2} \qquad \text{and} \qquad \theta = tan^{-1}\frac{b}{a}$$

$$= \sqrt{1^2 + (\sqrt{3})^2} \qquad\qquad = tan^{-1}\frac{\sqrt{3}}{1}$$

$$= 2 \qquad\qquad = 60°.$$

Since the complex number is in the first quadrant $\theta = \alpha$. Therefore, $1 + j\sqrt{3}$ in polar form is $2(\cos 60° + j \sin 60°)$ or $2/60°$.

b. In Figure 9.6, the point P represents $(2 - 2j\sqrt{3})$. The point is in the fourth quadrant. Knowing that $a = 2$ and $b = -2\sqrt{3}$ and using the conversion formulas from rectangular to polar we have:

$$r = \sqrt{a^2 + b^2} \qquad \text{and} \qquad \theta = tan^{-1}\frac{b}{a}$$

$$= \sqrt{2^2 + (-2\sqrt{3})^2} \qquad\qquad = tan^{-1}\frac{-2\sqrt{3}}{2}$$

$$= 4 \qquad\qquad = -60° \quad \text{or} \quad \alpha = 60°.$$

Since the complex number is in the fourth quadrant, $\theta = 360° - \alpha$ or $\theta = 360° - 60°$ or $300°$. Therefore, $2 - 2j\sqrt{3}$ in polar form is $4(\cos 300° + j \sin 300°)$ or $4/300°$. ◆

Changing complex numbers from polar to rectangular coordinates and from rectangular to polar coordinates can be done using a calculator. Directions for Example 2a with the fx–7700G and for Example 2b with the TI–82 are given. Check the instruction book for your calculator for specific instructions.

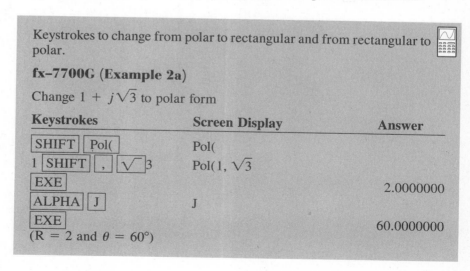

Keystrokes to change from polar to rectangular and from rectangular to polar.

fx–7700G (Example 2a)

Change $1 + j\sqrt{3}$ to polar form

Keystrokes	Screen Display	Answer
SHIFT Pol(	Pol(	
1 SHIFT , √3	Pol(1, √3	
EXE		2.0000000
ALPHA J	J	
EXE		60.0000000
(R = 2 and θ = 60°)		

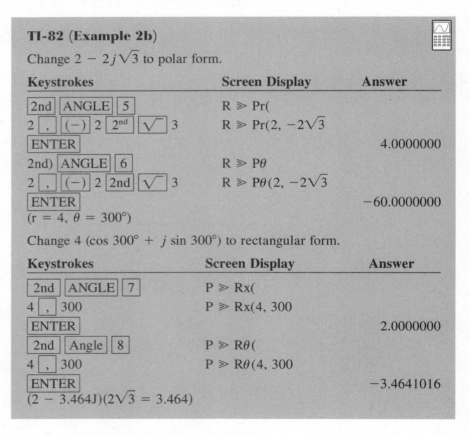

Change $2 (\cos 60° + j \sin 60°)$ to rectangular form.

Keystrokes	Screen Display	Answer
SHIFT Rec(	Rec(	
2 SHIFT , 60	Rec(2, 60	
EXE		1.0000000
ALPHA J	J	
EXE		1.7320508
(1 + 1.732J)		

TI-82 (Example 2b)

Change $2 - 2j\sqrt{3}$ to polar form.

Keystrokes	Screen Display	Answer
2nd ANGLE 5	R ≫ Pr(	
2 , (−) 2 2nd √ 3	R ≫ Pr(2, −2√3	
ENTER		4.0000000
2nd) ANGLE 6	R ≫ Pθ	
2 , (−) 2 2nd √ 3	R ≫ Pθ(2, −2√3	
ENTER		−60.0000000
(r = 4, θ = 300°)		

Change $4 (\cos 300° + j \sin 300°)$ to rectangular form.

Keystrokes	Screen Display	Answer
2nd ANGLE 7	P ≫ Rx(	
4 , 300	P ≫ Rx(4, 300	
ENTER		2.0000000
2nd Angle 8	P ≫ Rθ(	
4 , 300	P ≫ Rθ(4, 300	
ENTER		−3.4641016
(2 − 3.464J)(2√3 = 3.464)		

Exponential Form

A third way of writing a complex number frequently used in electronics and physics, is using exponential form. The **exponential form** of a complex number is defined as:

$$re^{j\theta} = r(\cos \theta + j \sin \theta) = r\underline{/\theta}.$$

In exponential form, $re^{j\theta}$, r is the **modulus** and θ is the **argument in radians.** The base e is an irrational number that has an approximate value of 2.71828 . . . . When this form of a complex number is used, the angle θ must be expressed in radians so that the exponent $j\theta$ is a dimensionless quantity and obeys all the rules of exponents. The exponential form of a complex number is also a convenient form for multiplication and division.

To convert a complex number from polar to exponential form, you must change the argument from degrees to radians, recognize the modulus, and then substitute for r and θ.

EXAMPLE 3   **Converting from Polar to Exponential Form**

Change the complex number $34.7(\cos 38.4° + j \sin 38.4°)$ to exponential form.

Solution   Recall from Chapter 6 that to convert from degrees to radians we use the conversion factor $\dfrac{\pi}{180°}$. Thus,

$$38.4° = 38.4°\left(\frac{\pi}{180°}\right) \text{ rad}$$

$$= 0.670.$$

From the polar form, we recognize the modulus $r = 34.7$. Thus,

$$re^{j\theta} = \boxed{34.7}\, e^{j\,\boxed{0.670}}. \quad \textbf{Substituting for } r \textbf{ and } \theta$$

The complex number $34.7(\cos 38.4° + j \sin 38.4°)$ in exponential form is $34.7e^{j0.670}$.   ◆

To change a complex number from rectangular to exponential form, we must determine r and θ. To do this, use the formulas $r = \sqrt{a^2 + b^2}$ and $\theta = \tan^{-1}\dfrac{b}{a}$, which are given at the beginning of this section.

EXAMPLE 4   **Converting from Rectangular to Exponential Form**

Change the complex number $5 - 4j$ to exponential form.

Solution   To change $5 - 4j$ to exponential form, we must determine r and θ.

$$r = \sqrt{a^2 + b^2} \qquad \theta = \tan^{-1}\frac{b}{a}$$

Substituting for a and b in the formulas we have:

$$= \sqrt{\boxed{5}\,^2 + (\,\boxed{-4}\,)^2} \qquad = \tan^{-1}\frac{\boxed{-4}}{5}$$

$$= 6.40 \qquad\qquad\qquad = -0.675 \quad \text{or} \quad \alpha = 0.675.$$

Since $5 - 4j$ is in the fourth quadrant, $\theta = 2\pi - \alpha$. Thus, $\theta = 2\pi - 0.675$ or $\theta = 5.608$. Substituting for r and θ, $5 - 4j = \boxed{6.40}\, e^{j\,\boxed{5.608}}$ in exponential form.   ◆

EXAMPLE 5 **Application: Converting Rectangular Form to Exponential Form**

The current i of an AC circuit may be represented by a complex number. For a specific circuit the current is $i = 0.314 + 0.275j$ A. Write the current in exponential form, $i = i_0 e^{j\theta}$ and determine the magnitude i_0 of the current in the circuit.

Solution   Since i_0 is the modulus and θ is in the first quadrant we have:

$$i_0 = \sqrt{(0.314)^2 + (0.275)^2} \qquad \theta = \tan^{-1}\left(\frac{0.275}{0.314}\right)$$

$$= 0.417 \text{ A} \qquad\qquad\qquad = 0.719 \text{ rad}$$

Therefore, $i = \boxed{0.417}\, e^{j\,\boxed{0.719}}$. The magnitude of the current is approximately 0.417 A. ◆

CAUTION   In technical applications, the argument θ of a complex number in polar form is usually given in degrees. Therefore, to convert a number from exponential form to polar form, we first convert the angle to degree measure and then write the number in polar form. ■

EXAMPLE 6   **Converting from Exponential to Polar Form**

Change the complex number $1.45e^{j3.42}$ to polar form.

Solution   To change 3.42 radians to degrees, use the conversion factor $\dfrac{180°}{\pi}$.

Thus $3.42 = 3.42\left(\dfrac{180°}{\pi}\right)$ or 196.0°. The modulus is 1.45. Therefore, $1.45e^{j3.42} = \boxed{1.45}\,(\cos \boxed{196.0°} + j \sin \boxed{196.0°})$ in polar form. ◆

To convert a complex number from exponential to rectangular form, we must determine values for a and b. Since in exponential form we are given r and θ, we can use the formulas $a = r \cos \theta$ and $b = r \sin \theta$.

EXAMPLE 7   **Converting from Exponential to Rectangular Form**

Convert $6.21e^{j2.45}$ to rectangular form.

Solution   There is no need to convert the angle measure to degrees.

$$\begin{aligned}
a &= r \cos \theta & b &= r \sin \theta \\
&= \boxed{6.21}\, \cos \boxed{2.45} & &= \boxed{6.21}\, \sin \boxed{2.45} \quad \textbf{Substituting} \\
&= -4.78 & &= 3.96
\end{aligned}$$

Therefore, $6.21e^{j2.45} = \boxed{-4.78} + \boxed{3.96}\,j$ in rectangular form. ◆

THREE FORMS OF A COMPLEX NUMBER

Rectangular		Polar		Exponential
$a + bj$	$=$	$r(\cos\theta + j\sin\theta)$	$=$	$re^{j\theta}$

In polar form θ is expressed in degrees. In exponential form θ is expressed in radians. To convert from one form to the other, use the following formulas.

CONVERSION FORMULAS

Polar or Exponential to Rectangular	Rectangular to Polar or Exponential
$a = r\cos\theta$	$r = \sqrt{a^2 + b^2}$
$b = r\sin\theta$	$\theta = \tan^{-1}\dfrac{b}{a}$

9.3 Exercises

In Exercises 1–12, graph each complex number in the complex plane.

1. $4(\cos 85° + j\sin 85°)$ **2.** $3(\cos 300° + j\sin 300°)$ **3.** $2 + 3j$ **4.** $-3 + 5j$

5. $-6 - 4j$ **6.** $3 - 5j$ **7.** $-3j$ **8.** 4.7

9. -6.3 **10.** $2.3j$ **11.** $-3 - 3j\sqrt{3}$ **12.** $3 + 2j - 4j^2$

In Exercises 13–24, express each complex number in polar form.

13. $2 + 2j$ **14.** $-4 + 3j$ **15.** $\sqrt{3} - j$ **16.** $4 + 3j$

17. $-11 - 6j$ **18.** $-4j$ **19.** $3.4e^{j1.05}$ **20.** $5.14e^{j2.27}$

21. $7.34e^{j3.93}$ **22.** $4.00e^{j5.76}$ **23.** $3.00e^{j5.50}$ **24.** $1.24e^{j0.654}$

In Exercises 25–36, express each complex number in rectangular form.

25. $3\underline{/45°}$ **26.** $5\underline{/90°}$ **27.** $2.34\underline{/193.4°}$ **28.** $2.5\underline{/73°}$

29. $3\underline{/135°}$ **30.** $1.3\underline{/118.7°}$ **31.** $2e^{j5.24}$ **32.** $\sqrt{5}e^{j3.14}$

33. $8.7e^{j1.65}$ **34.** $e^{j2.62}$ **35.** $5.0e^{j0.044}$ **36.** $2e^{j0.785}$

In Exercises 37–48, express each complex number in exponential form.

37. $5\underline{/60°}$ **38.** $6\underline{/120°}$ **39.** $2\underline{/180°}$ **40.** $5.0\underline{/2.5°}$

41. $8.7\underline{/94.7°}$ **42.** $\sqrt{3.20}\underline{/275.6°}$ **43.** $3 - 3j$ **44.** $-4 - 4j$

45. $5 + 7j$ **46.** $3 + 11j$ **47.** $4j$ **48.** $-3 - 5j$

49. In an AC circuit the current is represented by the complex number $i = 0.354 - 0.542j$ A. Write the complex number in exponential form and determine the magnitude of the current.

50. For a current or voltage represented by a complex number in rectangular form, the real part of the number is the in-phase component. Find the in-phase component of voltage represented by $V = 240e^{j2.63}$ V.

51. A force is represented by the complex number $20e^{j2.54}$. What is the magnitude of the force? What angle in degrees does the force make with the positive x-axis? What are the horizontal and vertical components of the force?

52. The expression $a = \dfrac{5}{1 + 0.8j}$ occurs in the study of transistors. Express a in rectangular and exponential forms.

9.4

Products, Quotients, Powers, and Roots of Complex Numbers

When complex numbers are used in the field of electronics they may need to be multiplied, divided, or raised to a power. For example, several operations may have to be combined. When two impedances Z_1 and Z_2 are connected in parallel, the equivalent impedance is:

$$Z_p = \frac{Z_1 Z_2}{Z_1 + Z_2}.$$

To find Z_p, we must be able to add, multiply, and divide complex numbers. These operations were discussed in 9.2. However, the operations of multiplication and division can now be simplified by using exponential or polar forms.

Multiplication

To multiply two complex numbers in exponential form, apply the rules for exponents. For example, to multiply any two complex numbers $r_1 e^{j\theta_1}$ and $r_2 e^{j\theta_2}$ we do the following.

$$
\begin{aligned}
r_1 e^{j\theta_1} \cdot r_2 e^{j\theta_2} &= r_1 \cdot r_2 \cdot e^{j\theta_1} \cdot e^{j\theta_2} && \textbf{Commutative property} \\
&= r_1 \cdot r_2 e^{(j\theta_1 + j\theta_2)} && \textbf{Product rule for exponents} \\
&= r_1 \cdot r_2 e^{(\theta_1 + \theta_2)j} && \textbf{Common term factoring}
\end{aligned}
$$

> **DEFINITION 9.9 Multiplication in Exponential Form**
> If $Z_1 = r_1 e^{j\theta_1}$ and $Z_2 = r_2 e^{j\theta_2}$, then
> $$Z_1 \cdot Z_2 = r_1 \cdot r_2 e^{j(\theta_1 + \theta_2)}.$$

EXAMPLE 1 **Products: Complex Numbers in Exponential Form**

Determine the product of $5e^{3j}$ and $6e^{4j}$.

Solution

$$5e^{3j} \cdot 6e^{4j} = \boxed{5} \cdot \boxed{6} \cdot e^{(3+4)j} \quad \text{Definition 9.9}$$
$$= 30e^{7j}. \quad \blacklozenge$$

The procedure for determining the product of two complex numbers in polar form is similar to the procedure for determining the product in exponential form. To find the product in polar form, determine the product of the moduli ($r_1 \cdot r_2$) and determine the sum of the arguments ($\theta_1 + \theta_2$).

DEFINITION 9.10   Multiplication in Polar Form

If $Z_1 = r_1(\cos \theta_1 + j \sin \theta_1)$ and $Z_2 = r_2(\cos \theta_2 + j \sin \theta_2)$, then
$Z_1 \cdot Z_2 = r_1 \cdot r_2[\cos(\theta_1 + \theta_2) + j \sin(\theta_1 + \theta_2)]$ or

$$r_1 \underline{/\theta_1} \cdot r_2 \underline{/\theta_2} = r_1 \cdot r_2 \underline{/\theta_1 + \theta_2}.$$

EXAMPLE 2   **Application: Multiplying Complex Numbers in Polar Form**

For a particular AC circuit, the current $I = 3$ A $\underline{/25°}$, and the impedance $Z = 4\ \Omega \underline{/42°}$. Determine the voltage E where $E = IZ$.

Solution

$$E = IZ$$
$$= (3\text{ A }\underline{/25°})(4\ \Omega\ \underline{/42°}) \quad \text{Substituting}$$
$$= \boxed{3} \cdot \boxed{4}\ \underline{/25° + 42°} \quad \text{Definition 9.10}$$
$$= 12\underline{/67°}\text{ V} \quad \blacklozenge$$

In the situation where θ_1 and θ_2 is greater than or equal to 360°, write the argument as a nonnegative coterminal angle less than 360°.

Thus, in general:

$$r(\cos \theta + j \sin \theta) = r[\cos (\theta + 360k°) + j \sin(\theta + 360k°)].$$

Division

As in multiplication, to divide two complex numbers in exponential form, apply the rules of exponents. For example, to divide any two complex numbers $r_1e^{j\theta_1}$ and $r_2e^{j\theta_2}$ we do the following.

$$\frac{r_1e^{j\theta_1}}{r_2e^{j\theta_2}} = \frac{r_1}{r_2}e^{(j\theta_1 - j\theta_2)} \quad \text{Quotient rule for exponents}$$

$$= \frac{r_1}{r_2}e^{(\theta_1 - \theta_2)j} \quad \text{Common-term factoring}$$

DEFINITION 9.11 Division in Exponential Form

If $Z_1 = r_1 e^{j\theta_1}$ and $Z_2 = r_2 e^{j\theta_2}$, then $\dfrac{Z_1}{Z_2} = \dfrac{r_1}{r_2} e^{j(\theta_1 - \theta_2)}$.

EXAMPLE 3   Dividing Complex Numbers in Exponential Form

Determine the quotient of $14e^{15j}$ and $2e^{9j}$.

Solution

$$\frac{14e^{15j}}{2e^{9j}} = \frac{14}{2} e(\boxed{15} - \boxed{9})j \qquad \text{Definition 9.11}$$

$$= 7e^{6j} \quad \blacklozenge$$

The procedure for determining the quotient of two complex numbers in polar form is similar to the method of determining the quotient in exponential form. To find the quotient in polar form, determine the quotient of the moduli ($r_1 \div r_2$) and the difference of the arguments ($\theta_1 - \theta_2$).

DEFINITION 9.12 Division in Polar Form

If $Z_1 = r_1(\cos\theta_1 + j\sin\theta_1)$ and $Z_2 = r_2(\cos\theta_2 + j\sin\theta_2)$, then $\dfrac{Z_1}{Z_2} = \dfrac{r_1}{r_2}[\cos(\theta_1 - \theta_2) + j\sin(\theta_1 - \theta_2)]$, or $\dfrac{Z_1}{Z_2} = \dfrac{r_1}{r_2} \underline{/(\theta_1 - \theta_2)}$.

EXAMPLE 4   Dividing Complex Numbers in Polar Form

Find the quotient of Z_1 divided by Z_2.

a. $Z_1 = 18\underline{/75°}$, $Z_2 = 5.0\underline{/41°}$ **b.** $Z_1 = 84.1\underline{/18°}$, $Z_2 = 15.0\underline{/-34°}$.

Solution

a. $\dfrac{Z_1}{Z_2} = \dfrac{18\underline{/7.5°}}{5.0\underline{/41°}}$

$= 3.6\underline{/75° - 41°}$ Definition 9.12

$= 3.6\underline{/34°}$

b. $\dfrac{Z_1}{Z_2} = \dfrac{84.1\underline{/18°}}{15.0\underline{/-34°}}$

$= 5.61\underline{/18° - (-34°)}$ Definition 9.12

$= 5.61\underline{/52°} \quad \blacklozenge$

CAUTION   We are now ready to determine the impedance Z_p for a parallel circuit; that is, $Z_p = \dfrac{Z_1 Z_2}{Z_1 + Z_2}$. In this problem we perform three operations (addition, multiplication, and division) on the complex numbers. We have learned that it is easier to perform the multiplication and division when the numbers are expressed in exponential or polar form. However, the operations of addition and subtraction of complex numbers can be performed only when the numbers are expressed in rectangular form. ■

EXAMPLE 5   Applications: Operations on Complex Numbers

Impedances $Z_1 = 2 - 3j$ and $Z_2 = 4 + 5j$ are connected in parallel. Determine the equivalent impedance $Z_p = \dfrac{Z_1 Z_2}{Z_1 + Z_2}$. Express the answer to two significant digits.

Solution   First find the sum of Z_1 and Z_2.

$$Z_1 + Z_2 = (2 - 3j) + (4 + 5j) = 6 + 2j.$$

Then convert Z_1, Z_2, and the sum to exponential form. Begin with $Z_1 = 2 - 3j$:

$$r = \sqrt{2^2 + (-3)^2} \qquad \theta = \tan^{-1} \frac{-3}{2}$$

$$= 3.606 \qquad\qquad = -0.983 \quad \text{or} \quad \alpha = 0.983$$

Since Z_1 is in the fourth quadrant, $\theta = 2\pi - 0.983$ or 5.300. Thus, $Z_1 = 3.606 e^{5.300 j}$ in exponential form.

Now convert $Z_2 = 4 + 5j$.

$$r = \sqrt{4^2 + 5^2} \qquad \theta = \tan^{-1} \frac{5}{4}$$

$$= 6.403 \qquad \theta = 0.8961$$

Thus, $Z_2 = 6.403 e^{0.8961 j}$ in exponential form.

Finally, convert $Z_1 + Z_2 = 6 + 2j$.

$$r = \sqrt{6^2 + 2^2} \qquad \theta = \tan^{-1} \frac{2}{6}$$

$$= 6.325 \qquad\qquad = 0.3218$$

Thus, $Z_1 + Z_2 = 6.325 e^{0.3218 j}$ in exponential form. Now substitute in the formula for Z_p and find the product and the quotient.

$$Z_p = \frac{(3.606 \, e^{5.300 \, j})(6.403 \, e^{0.8961 \, j})}{6.325 \, e^{0.3218 \, j}}$$

$$= \frac{23.09 e^{6.1961 j}}{6.325 e^{0.3218 j}} \qquad\qquad \textbf{Definition 9.9}$$

$$= 3.65 e^{5.874 j} \qquad\qquad \textbf{Definition 9.11}$$

Thus, the impedance for the parallel circuit is $3.7 e^{5.9 j}$ ◆

Powers and Roots

Using the rule developed for multiplication of complex numbers in polar form we can find Z^2, Z^3, Z^4, and Z^5. For example, if $Z = r\underline{/\theta}$, then:

$$Z^2 = (r\underline{/\theta})(r\underline{/\theta}) \qquad\qquad Z^3 = (r^2\underline{/2\theta})(r\underline{/\theta})$$
$$= r^2\underline{/\theta + \theta} \qquad\qquad\quad = r^3\underline{/2\theta + \theta}$$
$$= r^2\underline{/2\theta} \qquad\qquad\qquad = r^3\underline{/3\theta}$$
$$Z^4 = (r^3\underline{/3\theta})(r\underline{/\theta}) \qquad\quad Z^5 = (r^4\underline{/4\theta})(r\underline{/\theta})$$
$$= r^4\underline{/3\theta + \theta} \qquad\qquad\quad = r^5\underline{/4\theta + \theta}$$
$$= r^4\underline{/4\theta} \qquad\qquad\qquad = r^5\underline{/5\theta}.$$

From these few examples we may conclude that $Z^n = r^n\underline{/n\theta}$. This result is known as De Moivre's theorem.

The theorem is named for Abraham De Moivre (1667–1754), who was born in France and spent most of his life in London. He was a friend of Newton and a fellow of the Royal Society.

THEOREM 9.1 De Moivre's Theorem

Polar form: $Z^n = r^n(\cos n\theta + j \sin n\theta)$ or $Z^n = r^n\underline{/n\theta}$

Exponential form: $Z^n = r^n e^{jn\theta}$

De Moivre's theorem can be extended to determine the roots of a number by accepting the fact that it will work for fractional exponents. That is,

$$\sqrt[n]{Z} = Z^{1/n} = r^{1/n}\underline{/\dfrac{\theta}{n}}.$$

To determine whether or not this is correct, raise both sides to the nth power.

$$(\sqrt[n]{Z})^n = (r^{1/n})^n\underline{/n \cdot \dfrac{\theta}{n}}.$$

The result is $Z = r\underline{/\theta}$, which is what we would expect.

DEFINITION 9.13 nth Root of a Complex Number

Polar form: $\sqrt[n]{Z} = r^{1/n}\left(\cos\dfrac{\theta}{n} + j\sin\dfrac{\theta}{n}\right)$ or $\sqrt[n]{Z} = r^{1/n}\underline{/\dfrac{\theta}{n}}$

Exponential form: $\sqrt[n]{Z} = r^{1/n}e^{j(\theta/n)}$

The only problem is that $r^{1/n}\underline{/\dfrac{\theta}{n}}$ is not the only nth root of Z. Since there are many angles that are coterminal with θ, other roots can be found. Coterminal angles for θ may be $\theta + 360°$, $\theta + 720°$, or in general, $\theta + k \cdot 360°$ where k is an integer.

EXAMPLE 6   Determining Roots of Complex Numbers

Determine all the cube roots of $Z = 27\underline{/30°}$ and plot the roots on a complex plane.

Solution   To determine the first cube root of $Z = 27\underline{/30°}$, use Definition 9.13.

$$\sqrt[3]{Z} = \sqrt[3]{27}\Big/\frac{30°}{3}$$

$$= 3\underline{/10°}$$

Thus, $3\underline{/10°}$ is one cube root of $27\underline{/30°}$.

Since $30° + 1(360°) = 390°$ is coterminal with $30°$, another value for Z is $Z = 27\underline{/390°}$. A second possible cube root of Z is:

$$\sqrt[3]{Z} = \sqrt[3]{27}\Big/\frac{390°}{3}$$

$$= 3\underline{/130°}$$

This value is different from the first value obtained and is also a cube root of $27\underline{/30°}$.

Another coterminal angle is $30° + 2(360°) = 750°$, and another value for Z is $Z = 27\underline{/750°}$. Again taking the cube root, we have

$$\sqrt[3]{Z} = \sqrt[3]{27}\Big/\frac{750°}{3}$$

$$= 3\underline{/250°}.$$

Thus, $3\underline{/250°}$ is a third cube root of $27\underline{/30°}$.

The next coterminal angle would be $30° + 3(360°) = 1110°$, which may produce a fourth cube root of Z.

$$\sqrt[3]{Z} = \sqrt[3]{27}\Big/\frac{1110°}{3}$$

$$= 3\underline{/370°}.$$

However, $370°$ and $10°$ are coterminal angles. Therefore, $3\underline{/370°}$ is not a unique root. The three unique cube roots of $27\underline{/30°}$ are $3\underline{/10°}$, $3\underline{/130°}$, and $3\underline{/250°}$. In rectangular form the roots are $2.95 + 0.52j$, $-1.93 + 2.30j$, and $-1.03 - 2.82j$, respectively. The roots are plotted in Figure 9.7.   ◆

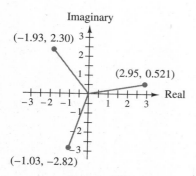

FIGURE 9.7

In general, there are n distinct nth roots of a complex number. If $Z = r\underline{/\theta}$, then $\sqrt[n]{Z} = \sqrt[n]{r}\Big/\dfrac{\theta + k \cdot 360°}{n}$.

EXAMPLE 7   Determining Roots of Complex Numbers

Determine the five roots of $Z = 32\underline{/-30°}$ and plot the points on the complex plane.

Solution   Determine the five roots of $Z = 32\underline{/-30°}$ using the formula

$$\sqrt[n]{Z} = \sqrt[n]{r}\;\underline{\bigg/\dfrac{\theta + k \cdot 360°}{n}}.$$

For $k = 0$:

$$\sqrt[5]{Z} = \sqrt[5]{32}\;\underline{\bigg/\dfrac{-30°}{5}} = 2\underline{/-6°} \quad \text{or} \quad 2\underline{/354°}.$$

For $k = 1$:

$$\sqrt[5]{Z} = \sqrt[5]{32}\;\underline{\bigg/\dfrac{-30° + 360°}{5}} = 2\underline{/66°}.$$

For $k = 2$:

$$\sqrt[5]{Z} = \sqrt[5]{32}\;\underline{\bigg/\dfrac{-30° + 2(360°)}{5}} = 2\underline{/138°}.$$

For $k = 3$:

$$\sqrt[5]{Z} = \sqrt[5]{32}\;\underline{\bigg/\dfrac{-30° + 3(360°)}{5}} = 2\underline{/210°}.$$

For $k = 4$:

$$\sqrt[5]{Z} = \sqrt[5]{32}\;\underline{\bigg/\dfrac{-30° + 4(360°)}{5}} = 2\underline{/282°}.$$

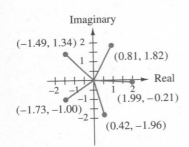

FIGURE 9.8

Thus, the five roots of $32\underline{/-30°}$ are $2\underline{/-6°}$, $2\underline{/66°}$, $2\underline{/138°}$, $2\underline{/210°}$, and $2\underline{/282°}$. In rectangular form the roots are $1.99 - 0.21j$, $0.81 + 1.83j$, $-1.49 + 1.34j$, $-1.73 - j$, $0.42 - 1.96j$, respectively. The roots are plotted in Figure 9.8.   ◆

In Example 7, the modulus is the same for all five roots; however, the arguments differ by $\dfrac{360°}{5}$ or $72°$. Therefore, we could find the first root and then add $72°$ to the argument to obtain the second root. For example, the first argument, $-6°$, plus $72°$ is $66°$, the second argument. The third argument is $72°$ plus $66°$ or $138°$, and so on. Thus, in general, we can find the nth root of the modulus and the first argument with the formula used in Example 7. Then divide $360°$ by n and add this result to the first argument to obtain the second argument. We keep adding $(360°/n)$ to each new argument until we have n arguments and can write n distinct roots.

9.4 Exercises

In Exercises 1–12, determine $Z_1 \cdot Z_2$.

1. $Z_1 = 4\underline{/35°}$, $Z_2 = 5\underline{/42°}$

3. $Z_1 = 2e^{2.531j}$, $Z_2 = e^{2.007j}$

5. $Z_1 = 4.7\underline{/216°}$, $Z_2 = 5.6\underline{/285°}$

7. $Z_1 = 1 + 2j$, $Z_2 = 3 - j$

9. $Z_1 = 8.5(\cos 85.6° + j \sin 85.6°)$,
$Z_2 = 9.4(\cos 105.4° + j \sin 105.4°)$

11. $Z_1 = 24.5e^{1.944j}$, $Z_2 = 27.6e^{2.64j}$

2. $Z_1 = 8\underline{/127°}$, $Z_2 = 5\underline{/75°}$

4. $Z_1 = 6e^{2.618j}$, $Z_2 = e^{3.665j}$

6. $Z_1 = 6.4\underline{/290.°}$, $Z_2 = 7.3\underline{/320.°}$

8. $Z_1 = 5 + 7j$, $Z_2 = -2 - 3j$

10. $Z_1 = 7.4\,[\cos(-48.3°) + j \sin(-48.3°)]$,
$Z_2 = 9.5\,[\cos(-107.3°) + j \sin(-107.3°)]$

12. $Z_1 = 2.017e^{1.995j}$, $Z_2 = 3.165e^{1.873j}$

In Exercises 13–24, determine $Z_1 \div Z_2$.

13. $Z_1 = 8\underline{/47°}$, $Z_2 = 2\underline{/38°}$

15. $Z_1 = 3.8(\cos 138° + j \sin 138°)$,
$Z_2 = 1.9(\cos 87° + j \sin 87°)$

17. $Z_1 = 14.71e^{2.040j}$, $Z_2 = 17.85e^{2.230j}$

19. $Z_1 = 1 + 2j$, $Z_2 = -2 - 3j$

21. $Z_1 = 0.014\underline{/108.7°}$, $Z_2 = 0.041\underline{/-164.2°}$

23. $Z_1 = 0.214(\cos 782° + j \sin 782°)$,
$Z_2 = 0.317(\cos 325° + j \sin 325°)$

14. $Z_1 = 15\underline{/85°}$, $Z_2 = 3\underline{/72°}$

16. $Z_1 = 9.6(\cos 182° + j \sin 182°)$,
$Z_2 = 3.2(\cos 108° + j \sin 108°)$

18. $Z_1 = 9.08e^{3.360j}$, $Z_2 = 8.40e^{3.688j}$

20. $Z_1 = 5 + 7j$, $Z_2 = 3 - j$

22. $Z_1 = 0.094\underline{/137.2°}$, $Z_2 = 0.347\underline{/113.2°}$

24. $Z_1 = 98.72[\cos(-142.7°) + j \sin(-142.7°)]$,
$Z_2 = 46.89[\cos(-138.6°) + j \sin(-138.6°)]$

In Exercises 25–36, determine Z_1^2 for Exercises 1–12.
In Exercises 37–48, determine Z_2^3 for Exercises 1–12.
In Exercises 49–56, determine the indicated roots of the complex number. Graph the results.

49. Cube roots of $64\underline{/48°}$

51. Fourth roots of $16\underline{/96°}$

53. Square roots of j

55. Cube roots of 8

50. Cube roots of $125\underline{/63°}$

52. Fourth roots of $81\underline{/48°}$

54. Square roots of $-j$

56. Cube roots of -27

57. In an AC circuit the current $I = 8$ A$\underline{/32°}$ and the impedance $Z = 4\ \Omega\underline{/72°}$. Determine the voltage E where $E = IZ$.

58. In an AC circuit the current $I = 3.0$ A$\underline{/17.2°}$ and the voltage $E = 13$ V $\underline{/18.4°}$.
Determine the impedance Z where $Z = \dfrac{E}{I}$.

59. Impedances $Z_1 = 4 + 5j$ and $Z_2 = 1 + 3j$ are connected in parallel. Determine the equivalent impedance $Z = \dfrac{Z_1 Z_2}{Z_1 + Z_2}$.

60. Impedances $Z_1 = 7 + 5j$ and $Z_2 = 11 - 3j$ are connected in parallel. Determine the equivalent impedance $Z = \dfrac{Z_1 Z_2}{Z_1 + Z_2}$.

9.5

An Application to Alternating Current (AC Circuits)

As we have learned, currents and voltages in most AC circuits are sinusoidal quantities. To analyze the behavior of these circuits, we must add, subtract, multiply, and divide quantities of the form $y = A \sin (Bt + C)$. Recall from Chapter 8 that $|A|$ is the amplitude, $\frac{2\pi}{|B|}$ is the period, and $\frac{-C}{B}$ is the phase shift.

It has been found that as long as the frequencies (the B terms) of the sinusoids to be manipulated are the same, then the sinusoids can be represented as complex numbers called *phasors*. Writing the sinusoids representing circuits as phasors makes the mathematical manipulations simpler.

A sinusoid $y = A \sin (Bt + C)$ can be written as a **phasor** (complex number) $r\underline{/\theta}$ by replacing the modulus with the amplitude and the argument with C (usually in degrees even if Bt is in radians.) Thus, $y = A \sin (Bt + C)$ becomes $A\underline{/C}$ as a phasor. For $y = A \sin Bt$, the phase shift is zero. It follows that the argument is $0°$, and the phasor is $A\underline{/0°}$. Here the common term in the phasor is understood to be $\sin Bt$.

CAUTION   Remember that when using phasors to manipulate sinusoids, the frequencies must be the same. Also, the sinusoids must be of the same function (all sines or all cosines). ∎

EXAMPLE 1   Application: Sum of Two Phasors

Two voltage sources are connected in series. The resulting output is the sum of the two voltages: $v_1 = 117 \sin 377t$ volts and $v_2 = 117 \sin (337t + 60°)$. What is the output?

Solution   Since the periods are the same, we can write the sinusoids as phasors and add. Therefore, the phasor representation of $v_1 = 117 \sin 377t$ is $V_1 = 117\underline{/0°}$, and the phasor representation of $v_2 = 117 \sin(377t + 60.0°)$ is $V_2 = 117\underline{/60.0°}$. Recall that to add two complex numbers they must be in rectangular form. Therefore, convert these numbers.

$$
\begin{aligned}
V_1 &= 117\underline{/0°} &= 117 + 0.j \\
V_2 &= 117\underline{/60.0°} &= 58.50 + 101.3j \\
& & V_T = 175.5 + 101.3j.
\end{aligned}
$$

Converting V_T to polar form, the result is $V_T = 203\underline{/30.0°}$. Thus, the sum of $v_1 = 117 \sin 377t$ and $v_2 = 117 \sin(377t + 60.0°)$ is $v_T = 203 \sin (377t + 30.0°)$. ◆

In addition to treating currents and voltages as phasors, the passive circuit elements (resistor, inductor, and capacitor) may be represended by complex numbers. A resistor converts electrical energy to heat. Examples are filaments of light bulbs and heater coils. An inductor is a continuous loop or coil of wire that opposes any change in the current of the circuit. A transformer is essentially two inductors wound together. A capacitor is a device designed to store a charge and oppose the change in voltage. A simple capacitor consists of two conducting plates placed very close together. One plate holds a positive charge, while the other holds an equal amount of negative charge.

When these common circuit components are expressed as complex numbers they can be treated in the same way as pure resistances are treated in Ohm's law.

In such a case, Ohm's law is stated is $i = \dfrac{v}{Z}$ where i is the effective current, v is the applied voltage, and Z is the impedance or the resultant "resistance" of all circuit elements. Such circuit elements include normal resistance of a resistor, the "resistance" of an inductor as it opposes the change in current, and the "resistance" of a capacitor as it opposes the change in voltage. The common circuit elements and their impedances are shown in Table 9.2.

TABLE 9.2

Circuit Element	Symbol and Units	Resistance or Reactance	Impedance in Rectangular Form	Impedance in Polar Form
Resistor	R (ohm)	Resistance: $R = R$	$Z_R = R = R + 0j$	$Z_R = R\underline{/0°}$
Inductor	L (henry)	Inductive reactance: $XL = \omega L$	$Z_L = jX_L = 0 + jX_L$	$Z_L = X_L\underline{/90°}$
Capacitor	C (farad)	Capacitive reactance: $X_C = \dfrac{1}{\omega C}$	$Z_C = -jX_C = 0 - jX_C$	$Z_C = X_C\underline{/-90°}$

The SI unit for frequency, the hertz (Hz), is defined as 1 Hz = 1 cycle/s. Thus a 60 cycle/s alternating current has a frequency f of 60 Hz. The angular frequency, $\omega = 2\pi$ (60 Hz) or 377. rad/s.

The impedances of combined circuit elements are treated the same as combined resistors in DC circuits. If the elements are hooked in series (one after the other) the combined impedance is the sum of the individual impedances: $Z_s = Z_1 + Z_2 + Z_3 + \cdots$. If the elements are hooked in parallel (one next to the other) the combined impedance is given by $\dfrac{1}{Z_p} = \dfrac{1}{Z_1} + \dfrac{1}{Z_2} + \dfrac{1}{Z_3} + \cdots$. In the special case of just two impedances hooked in parallel, the preceding equation is rearranged to give

$$Z_p = \frac{Z_1 Z_2}{Z_1 + Z_2}.$$

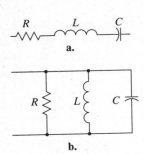

a.

b.

FIGURE 9.9

EXAMPLE 2 **Application: Determining Impedance for a Series and a Parallel Circuit**

Given a resistor $R = 100\ \Omega$, an inductor $L = 7.00$ millihenry $(7.00 \times 10^{-3}\ H)$, and a capacitor $C = 47.0$ microfarad $(47.0 \times 10^{-6}\ F)$ in a 60-Hz circuit. What is the combined impedance under the following conditions?

a. They are connected in series as shown in Figure 9.9a.
b. They are connected in parallel as shown in Figure 9.9b. (Assume three significant digits.)

Solution Find the impedances of the individual elements in Table 9.2 and substitute.

$$Z_R = R = \boxed{100} + \boxed{0}j\ \Omega$$
$$Z_L = j\omega L = j[\ \boxed{2\pi(60)}\ ][\ \boxed{7 \times 10^{-3}}\ ] = 0 + 2.639j\ \Omega.$$
$$Z_C = -j\frac{1}{\omega C} = -j\frac{1}{[\ \boxed{2\pi(60)}\ ][\ \boxed{47 \times 10^{-6}}\ ]} = 0 - 56.44j\ \Omega.$$

a. The combined impedance for the series circuit is

$$Z_S = Z_R + Z_L + Z_C$$
$$= (100 + 0j) + (0 + 2.639j) + (0 - 56.44j)$$
$$= 100 - 53.80j\ \Omega$$
$$= 114\underline{/-28.3°}\ \Omega.$$

b. The combined impedance for the parallel circuit is:

$$\frac{1}{Z_P} = \frac{1}{Z_R} + \frac{1}{Z_L} + \frac{1}{Z_C}$$

$$= \frac{1}{\boxed{100}} + \frac{1}{\boxed{2.639}\ j} + \frac{1}{\boxed{-56.44}\ j} \qquad \text{Substituting}$$

$$= \frac{1\underline{/0°}}{100\underline{/0°}} + \frac{1\underline{/0°}}{2.639\underline{/90°}} + \frac{1\underline{/0°}}{56.44\underline{/-90°}} \qquad \begin{array}{l}\textbf{Changing numerator and}\\\textbf{denominator to polar form}\end{array}$$

$$= 0.0100\underline{/0°} + \frac{1}{2.639}\underline{/-90°} + \frac{1}{56.44}\underline{/90°} \qquad \text{Dividing}$$

$$= 0.0100 - 0.3789j + 0.01772j \qquad \begin{array}{l}\textbf{Changing back to}\\\textbf{rectangular form}\end{array}$$

$$= 0.0100 - 0.3612j \qquad \textbf{Combining like terms}$$

$$Z_P = \frac{1}{0.0100 - 0.3612j} \qquad \begin{array}{l}\textbf{Taking the reciprocal}\\\textbf{of both sides}\end{array}$$

$$= \frac{1\underline{/0°}}{0.3613\underline{/-88.4°}} \qquad \textbf{Changing to polar form}$$

$$= 2.768\underline{/88.4°}\ \Omega \qquad \textbf{Dividing}$$

$$= 0.0773 + 2.77j\ \Omega$$

Therefore, the combined impedance for the series circuit is:

$$Z_c = 114\underline{/-28.3°} \ \Omega$$

and for the parallel circuit is:

$$Z_p = 0.0773 + 2.77j \ \Omega. \quad \blacklozenge$$

EXAMPLE 3   Application: Impedance Parallel Circuit

Figure 9.10 shows a parallel circuit. In determining the resultant wave equation defined by this circuit, a technician usually changes the sinusoidal terms into imaginary expressions (complex numbers). By using complex numbers the solution is a relatively easy one to compute. Note that all the data are given with two significant digits.

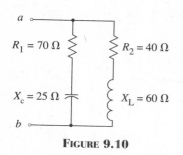

$R_1 = 70 \ \Omega$ $R_2 = 40 \ \Omega$

$X_c = 25 \ \Omega$ $X_L = 60 \ \Omega$

FIGURE 9.10

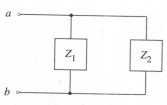

Z_1 Z_2

FIGURE 9.11

Solution   Because R_1 and X_c are hooked in series, they can be replaced by $Z_1 = R_1 - jX_c = \boxed{70} - \boxed{25} \, j$. R_2 and X_L are also in series so they can be replaced by $Z_2 = R_2 + jX_L = \boxed{40} + \boxed{60} \, j$. The circuit can then be represented as shown in Figure 9.11. A single impedance Z that represents parallel impedances Z_1 and Z_2 is given by the formula:

$$Z = \frac{Z_1 Z_2}{Z_1 + Z_2}$$

$$= \frac{(70 - 25j)(40 + 60j)}{(70 - 25j) + (40 + 60j)} \qquad \text{Substituting}$$

$$= \frac{(74.3\underline{/-19.7°})(72.1\underline{/56.3°})}{110 + 35.0j} \qquad \begin{array}{l}\text{Changing numerator to polar form and} \\ \text{finding the sum in the denominator}\end{array}$$

$$= \frac{5357\underline{/36.6°}}{115\underline{/17.7°}} \qquad \begin{array}{l}\text{Finding product in the numerator and} \\ \text{changing denominator to polar form}\end{array}$$

$$= 46.6\underline{/18.9°}. \qquad \text{Dividing}$$

The combined impedance $Z = 47 \ \Omega\underline{/19°}$. $\quad \blacklozenge$

Since current and voltage can be represented as sine or cosine functions (Section 8.4), their graphs have the same form as a sine or cosine curve. Thus, the voltage and current continually pass through high and low points. If the voltage and the current hit the high and low points at the same time, they are said to be **in phase.** If the voltage reaches the high point before the current, then the voltage leads the current. If the voltage reaches the high point after the current, it lags the current.

EXAMPLE 4   Application: Determining Amount of Current

If an AC voltage source $v = 120$ volt is connected to the circuit of Example 3 at point a, what is the current supplied to the circuit? (Assume two significant digits.)

Solution   From Ohm's law, we know $i = \dfrac{v}{Z}$. The voltage $v = 120$ volt can be written in phasor form as $V = 120\underline{/0°}$. From Example 3, $Z = 46.6\underline{/18.9°}$.

$$i = \frac{120\underline{/0°}}{46.6\underline{/18.9°}}$$
$$= 2.6\underline{/-19.°}$$

The current supplied will be 2.6 A at the same frequency as the voltage and lagging the voltage by 19°.   ◆

9.5 Exercises

In Exercises 1–4, consider two voltage sources that are connected in series. Determine the sum of the two voltages.

1. $v_1 = 193 \sin 337t$, $v_2 = 123 \sin(377t + 30°)$

2. $v_1 = 124 \sin 314t$, $v_2 = 175 \sin(314t + 30°)$

3. $v_1 = 245 \cos(251t + 45°)$, $v_2 = 165 \cos(251t + 30°)$

4. $v_1 = 48 \; \cos(440t + 45°)$, $v_2 = 96 \; \cos(440t + 75°)$

In Exercises 5–12, determine the following for each circuit pictured.
a. The equivalent or combined impedance of the circuit
b. The current supplied
c. The phase angle between the voltage and the current

5.

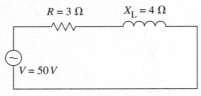

$R = 3\,\Omega$ $X_L = 4\,\Omega$
$V = 50\,V$

6.

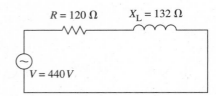

$R = 120\,\Omega$ $X_L = 132\,\Omega$
$V = 440\,V$

7.

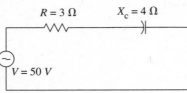

$R = 3\,\Omega$ $X_c = 4\,\Omega$
$V = 50\,V$

8.

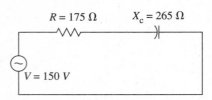

$R = 175\,\Omega$ $X_c = 265\,\Omega$
$V = 150\,V$

9.

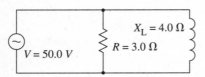

$X_L = 4.0\,\Omega$
$R = 3.0\,\Omega$
$V = 50.0\,V$

10.

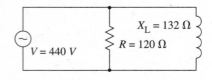

$X_L = 132\,\Omega$
$R = 120\,\Omega$
$V = 440\,V$

11.

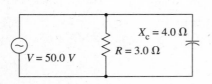

$X_c = 4.0\,\Omega$
$R = 3.0\,\Omega$
$V = 50.0\,V$

12.

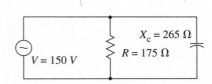

$X_c = 265\,\Omega$
$R = 175\,\Omega$
$V = 150\,V$

In Exercises 13–20, determine the impedance Z_p for the impedances Z_1 and Z_2 connected in parallel.

13. $Z_1 = 161\ \Omega\ \underline{/9.3°}$, $Z_2 = 60.0\ \Omega\ \underline{/58.3°}$

14. $Z_1 = 158.6\ \Omega\ \underline{/47.2°}$, $Z_2 = 155.0\ \Omega\ \underline{/-17.7°}$

15. $Z_1 = 83.9 - 44.4j\ \Omega$, $Z_2 = 40.0 + 50.0j\ \Omega$

16. $Z_1 = 268 - 190j\ \Omega$, $Z_2 = 107 - 28.6j\ \Omega$

17. $Z_1 = 50.5\ \Omega\ \underline{/15°}$, $Z_2 = 110 + 0j\ \Omega$

18. $Z_1 = 345\ \Omega\ \underline{/18°}$, $Z_2 = 0 - 90j\ \Omega$

19. $Z_1 = 241\ \Omega\ \underline{/-8°}$, $Z_2 = 0 + 80j\ \Omega$

20. $Z_1 = 40.3\ \Omega\ \underline{/-10°}$, $Z_2 = 28\ \Omega\ \underline{/0°}$

In Exercises 21–24, determine the equivalent impedance of the circuit shown.

21.

$R = 52\ \Omega$ $R = 46\ \Omega$

$X_L = 28\ \Omega$ $X_c = 13\ \Omega$

22.

$R = 45\ \Omega$ $R = 70\ \Omega$

$X_L = 36\ \Omega$ $X_c = 42\ \Omega$

23.

$X_c = 75\ \Omega$ $R = 98\ \Omega$

$X_L = 104\ \Omega$

24.

$R = 115\ \Omega$ $X_c = 94\ \Omega$

$X_L = 132\ \Omega$

25. The joint impedance of two parallel impedances is $58.3\ \Omega\ \underline{/-38.2°}$. One of the impedances is $138.\ \Omega\ \underline{/32°}$. What is the other impedance?

26. What impedance must be connected in parallel with $74.3 + j54.5\ \Omega$ to produce $33.7 + j145.4\ \Omega$?

27. Light intensity varies as a sinusoidal expression. When light penetrates through a thin layer of oil on water, many reflections reflect back to the eye of the observer. Such multiple reflections interfere with one another causing some reinforcement of the light in one direction and partial cancellation of the light in another direction.

When the sine function is replaced with an exponential form, it is easier to determine the sum of the various reflections. Consider the two waves that are reflected to the eye in the diagram: $Y_1 = 4 \sin(\omega t + 0°)$ and $Y_2 = 1 \sin(\omega t + 90°)$. Change the sinusoids into phasors and determine the sum of the two intensities.

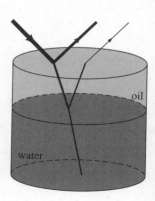

28. If you open the case of any electronic device, you will see a bewildering array of interconnected components. Each component may be defined mathematically by a sinusoidal equation. The symbols for the three common components are given in Table 9.2. An equation that may represent a resistor is $y = 7 \sin (\omega t + 90°)$, an inductor is $y = 8 \sin (\omega t + 0°)$ and a capacitor is $y = 3 \sin (\omega t + 90°)$. Of course most circuits have many resistors, many inductors, and many capacitors, each with slightly different equations. To determine how the combinations of such elements will work together, technicians have to add the sinusoidal equations. Determine the sum of the three sinusoids given for the resistor, inductor, and capacitor.

Review Exercises

In Exercises 1–12, carry out the indicated operations and simplify. Leave the answers in radical form.

1. j^{15}

2. j^{-14}

3. $(\sqrt{-324})(\sqrt{-361})$

4. $(\sqrt{-1058})(\sqrt{-192})$

5. $\sqrt{-484}\,\sqrt{507}$

6. $\sqrt{(-6)(-24)}$

7. $\sqrt{(-8)(-98)}$

8. $(\sqrt{-48})(\sqrt{-75})(\sqrt{-90})$

9. $\dfrac{\sqrt{343}}{\sqrt{-49}}$

10. $\dfrac{\sqrt{-135}}{\sqrt{-14}}$

11. $\sqrt{-288} + \sqrt{-363} + \sqrt{-162}$

12. $\sqrt{-578} - \sqrt{-676} + \sqrt{-729}$

In Exercises 13–18, perform the indicated operations and write the answer in the form $a + bj$.

13. $(3 - 5j) + (8 + 7j)$

14. $(-19 + 31j)^2$

15. $(8 + 5\sqrt{-6})(17 - 7\sqrt{-216})$

16. $(3j^7 - 8j^{13} + 7j^{17} - 5j^{15})^2$

17. $(5 - j^{14}) \div (18 - j^{25})$

18. $1 \div (13 + 4\sqrt{-72})$

In Exercises 19–24, graph each complex number and change to polar form. (Answer to the nearest tenth.)

19. $-8 + 11j$

20. $4 - 8j$

21. $5 + 14j$

22. $5\sqrt{3} - 4j\sqrt{3}$

23. $11j$

24. -3

In Exercises 25–30, change each complex number in polar form to rectangular form. Determine exact answers if the reference angle is 30°, 45°, 60°, or quadrantal; otherwise give answers to three significant digits.

25. $3.4/240°$

26. $3/138°$

27. $1.40/120.°$

28. $5.0/258.4°$

29. $5.0/3.4°$

30. $9.3/113.8°$

In Exercises 31–36, perform the indicated operation on the complex numbers.

31. $(5/28°) \cdot (15/87°)$

32. $(133/173°) \div (15/29°)$

33. $(7.5/83.4°)^2$

34. $(5.6/148.3°)^3$

35. $(14.7/-32.1°)\,(19.8/51.7°)$

36. $(0.040/122.3°)(0.016/-37.4°)$

37. Determine the cube roots of $216/114°$.

38. Determine the cube root of -512.

39. Use the illustration of the circuit to determine the following.
   a. The phase angle between voltage and current
   b. The impedance of the circuit
   c. The current through the circuit.

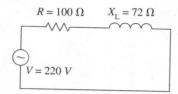

$R = 100 \ \Omega$ $X_L = 72 \ \Omega$

$V = 220 \ V$

40. The equivalent impedance of a circuit may be found using the formula:

$$Z_T = Z_S + \frac{Z_1 Z_2}{Z_1 + Z_2}.$$

Determine Z_T when $Z_S = 15.4 + 20.6j \ \Omega$, $Z_1 = 50 + 11.8j \ \Omega$, and $Z_2 = 40.0 - 80.0j \ \Omega$. Sketch the circuit.

41. The characteristic impedance of a T-section filter is:

$$Z_{OT} = \sqrt{Z_1 Z_2 + \frac{Z_1^{\,2}}{4}},$$

where Z_1 is the full series arm impedance and Z_2 is the shunt impedance of the filter section. If $Z_1 = 30.0\underline{/86.0°}$ and $Z_2 = 10.0\underline{/-90°} \ \Omega$, determine Z_{OT}.

42. Under what circumstances does the conjugate of a complex number equal its reciprocal?

43. Although $\dfrac{2}{6} = \dfrac{1}{3}$, $(-27)^{\frac{2}{6}} \neq (-27)^{\frac{1}{3}}$. This awkward situation arises because $(-27)^{\frac{1}{3}} = -3$, but $(-27)^{\frac{2}{6}} = [(-27)^2]^{\frac{1}{6}} = (729)^{\frac{1}{6}} = 3$. Use De Moivre's theorem to unravel the mystery of this problem. (*Hint:* Determine the three distinct roots of -27 and the six distinct roots of 729.)

✍ Writing About Mathematics

1. In elementary algebra you were told that the expression $a^2 + b^2$ is prime. Write a paper explaining that $a^2 + b^2$ is not prime for the set of complex numbers.

2. Write a paper explaining how to determine the three cube roots of 125. Include in your discussion an explanation of why there are three cube roots.

Chapter Test

In Exercises 1–4, perform the indicated operations and simplify.

1. j^{29} **2.** j^{138} **3.** $(\sqrt{-49})(\sqrt{-49})$ **4.** $\sqrt{(-25)(-16)}$

In Exercises 5–7, perform the indicated operations and write the answer in the form
$a + bj$.

5. $(3 + 7j) + (-8 + 5j)$

6. $(3 + 7j)(3 - 7j)$

7. $(5 - 2\sqrt{-9}) - (3 + \sqrt{-16})$

In Exercises 8 and 9, graph the complex number.

8. $-3 + 7j$

9. $2(\cos 45° + j \sin 45°)$

10. Convert $-4 + 5j$ to polar and exponential forms.

11. Convert $5(\cos 315° + j \sin 315°)$ to rectangular and exponential forms.

In Exercises 12–15, perform the indicated operations given $Z_1 = 22(\cos 85° + j \sin 85°)$, $Z_2 = 4(\cos 140° + j \sin 140°)$, $Z_3 = 4e^{2j}$, $Z_4 = 5e^{1.5j}$, $Z_5 = 2 + 7j$, $Z_6 = 3 - 5j$.

12. $Z_1 \div Z_2$

13. $Z_3 \cdot Z_4$

14. $(Z_4)^4$

15. $\dfrac{Z_5 Z_6}{Z_5 + Z_6}$

16. Determine the four fourth roots of 81.

17. The impedance Z in an AC circuit is given by $Z = \dfrac{v}{I}$ where v is the voltage and I is the current. Determine Z when $v = 3.2 - 0.42j$ and $I = -1.4j$.

18. For the illustrated circuit, determine the following.
   a. The equivalent or combined impedance of the circuit
   b. The current supplied
   c. The phase angle between the voltage and the current

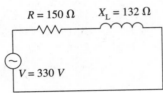

$R = 150\ \Omega$ $X_L = 132\ \Omega$

$V = 330\ V$

Systems of Linear Equations and Determinants

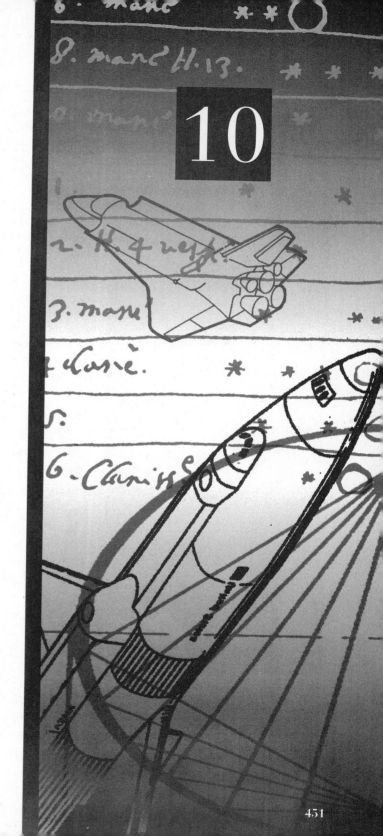

I f you look out an airplane window during a storm, you might see the wing tips flapping up and down through an arc of several feet. The wings may seem about to break under the strain; but don't worry. In testing the plane's safety, the designers used strain gauges to test the ability of the wings to withstand both horizontal and vertical stress. To determine the maximum stress, designers solved equations involving independent relationships. Methods for solving systems of equations are discussed in this chapter. (See Section 10.3, Example 6.)

10.1

Introduction to Linear Equations

In Chapter 4, we learned how to solve linear equations with one variable, and in Chapter 5, we graphed linear equations in two variables. In technical fields, people commonly work with linear equations with two variables. The general form of the linear equation in two variables is $ax + by + c = 0$. A specific example is:

$$3x + 4y - 12 = 0, \quad \text{where } a = 3, b = 4 \quad \text{and} \quad c = -12.$$

We can tell that the equation is linear since the exponents of both the variables x and y are 1. For example, $3x^2 - 5y + 2 = 0$ is not a linear equation because the exponent of the variable x is 2. Another way of recognizing a linear equation is by its graph, which is always a straight line (as illustrated in Chapter 5).

A linear equation may take the form $y = mx + b$, where m and b are constants and x and y are the independent and dependent variables, respectively. This form of the equation of a line is called the **slope-intercept form** because m is the slope of the line, and b is the ordinate of the y-intercept. The **slope** of the line is a ratio of two numbers that indicates the slant of the line with respect to the positive x-axis. The slope is also defined as the angle the line forms with respect to the positive x-axis. The **y-intercept** is the point where the line crosses the y-axis.

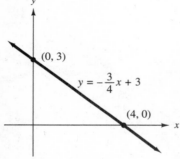

FIGURE 10.1

> **DEFINITION 10.1**
>
> The **slope-intercept form** of an equation of a line is:
>
> $$y = mx + b.$$

The equation $3x + 4y - 12 = 0$ can be written in the form $y = mx + b$.

$$4y = -3x + 12 \quad \text{Subtracting } 3x \text{ and adding 12 to both sides of the equation}$$

$$y = \frac{-3}{4}x + 3 \quad \text{Dividing both sides of the equation by 4.}$$

By rewriting the equation in this form, we see that $m = \dfrac{-3}{4}$ and $b = 3$. The graph is sketched in Figure 10.1. The line crosses the y-axis at the point $(0, 3)$, the y-intercept.

Now we will look at the slope. As shown in Figure 10.2, draw a line parallel to the y-axis through the point $A(-4, 6)$ and another line parallel to the x-axis through the point $B(8, -3)$. These lines intersect at the point $C(-4, -3)$. The resulting figure is right triangle ABC, with a right angle at $C(-4, -3)$. The length of the line segment BC is:

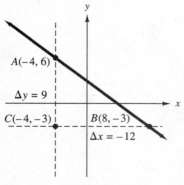

FIGURE 10.2

$$\overline{BC} = -4 - 8 = \boxed{-12,}$$

which is called the **change in x, or Δx** ($\Delta x = -12$). The length of the segment CA is:

$$\overline{CA} = 6 - (-3) = \boxed{-9,}$$

which is called the **change in y, or Δy.** The ratio $\dfrac{\Delta y}{\Delta x} = \dfrac{\boxed{9}}{\boxed{-12}}$, or $\dfrac{-3}{4}$, which is

the slope of the equation $y = \dfrac{-3}{4}x + 3$. The slope may be expressed as $\boldsymbol{m = }$

$\dfrac{\boldsymbol{\Delta y}}{\boldsymbol{\Delta x}}$. For any points $P_1(x_1, y_1)$ and $P_2(x_2, y_2)$, $\Delta x = (x_2 - x_1)$, and $\Delta y = (y_2 - y_1)$. Δy and Δx frequently are referred to as the **rise** and the **run,** or $\dfrac{\boldsymbol{\Delta y}}{\boldsymbol{\Delta x}} = \dfrac{\textbf{rise}}{\textbf{run}}.$

DEFINITION 10.2

The **slope** of a straight line containing points $P_1(x_1, y_1)$ and $P_2(x_2, y_2)$ is defined by the equation:

$$m = \frac{y_2 - y_1}{x_2 - x_1},$$

where $x_2 - x_1 \neq 0$.

EXAMPLE 1   Determining the Slope Given Two Points

Determine the slope of the line passing through the points $(-2, 3)$ and $(5, 4)$.

Solution   The equation used to determine the slope of the line given two points is:

$$m = \frac{y_2 - y_1}{x_2 - x_1}$$

$$= \frac{\boxed{4} - \boxed{3}}{\boxed{5} - (\boxed{-2})}. \quad \textbf{Substituting}$$

$$= \frac{1}{7}$$

The slope of the line is $\dfrac{1}{7}$, which tells us that y increases one unit for every seven units of increase for x. Another way of saying this is that the line rises one unit for every seven units of run.   ◆

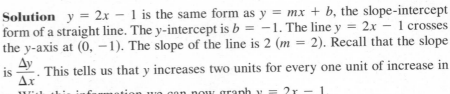

EXAMPLE 2 **Sketch the Graph Given Slope and y-Intercept**

Sketch the graph of $y = 2x - 1$ using the slope and the y-intercept.

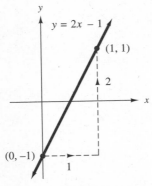

FIGURE 10.3

Solution $y = 2x - 1$ is the same form as $y = mx + b$, the slope-intercept form of a straight line. The y-intercept is $b = -1$. The line $y = 2x - 1$ crosses the y-axis at $(0, -1)$. The slope of the line is 2 ($m = 2$). Recall that the slope is $\dfrac{\Delta y}{\Delta x}$. This tells us that y increases two units for every one unit of increase in x. With this information we can now graph $y = 2x - 1$.

As shown in Figure 10.3, first plot the y-intercept $(0, -1)$. Since the slope is positive, starting with the point $(0, -1)$ move one unit to the right and then two units up. As a result of moving in both directions, we end up at the point with coordinates $(1, 1)$. The point $(1, 1)$ is also a point on the line $y = 2x - 1$. Check this by substituting $(1, 1)$ in the equation for x and y, respectively. Now we have two points and can draw the line. (Note that if we moved two units up and then one unit to the right, we will also end up at the point $(1, 1)$.) ◆

Using the fact that $\dfrac{2}{1} = \dfrac{-2}{-1}$, we can find a third point on the line. Starting at the point $(0, -1)$, move one unit to the left and then two units down (both negative directions), we end up at the point $(-1, -3)$. Check to see if the order pair $(-1, -3)$ is a point on the line.

EXAMPLE 3 **Sketch the Graph Given the Slope and y-Intercept**

Sketch the graph of $y = -2x - 1$ using the slope and the y-intercept.

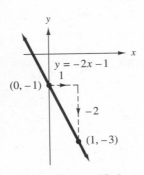

FIGURE 10.4

Solution The graph of the linear equation $y = -2x - 1$ is a straight line. With $b = -1$, the y-intercept of the graph is the point $(0, -1)$. This equation differs from the equation of Example 1 in that the slope is a negative number. Equivalent ways of writing the slope are $m = -2$, $m = \dfrac{-2}{1}$, and $m = \dfrac{2}{-1}$.

The slope $m = \dfrac{-2}{1}$ tells us that y decreases by two units ($\Delta y = -2$) for each unit x increases ($\Delta x = 1$). With this information in mind and starting at the point $(0, -1)$, move one unit to the right and two units down, we end up at the point $(1, -3)$. Since two points determine a straight line, we can draw the graph of the equation as in Figure 10.4. ◆

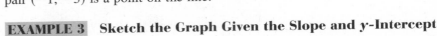

CAUTION For the linear equation $3y - 4x + 5 = 0$, the slope and y-intercept are not -4 and 5. We always must solve the equation for y before determining the slope and y-intercept. Solving the equation for y, $y = \dfrac{4}{3}x - \dfrac{5}{3}$, and we see that $m = \dfrac{4}{3}$ and $b = \dfrac{-5}{3}$. ■

In Example 3, the slope in the form $m = \dfrac{2}{-1}$, tells us that y increases two units ($\Delta y = 2$) for each unit x decreases ($\Delta x = -1$). Using this information and starting at the point $(0, -1)$, move one unit to the left and two units up, to end up at the point $(-1, 1)$. The point $(-1, 1)$ is a third point on the line. This illustrates that we can use either form of the slope to determine points on the graph.

EXAMPLE 4   Sketching Parallel Lines

Sketch the graphs of $2x + y + 1 = 0$ and $-6x - 3y + 2 = 0$ using the slope and the y-intercept.

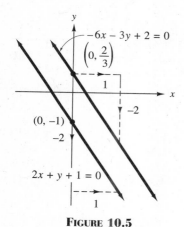

FIGURE 10.5

Solution   Both equations are linear and can be written in the form $y = mx + b$.

$$2x + y + 1 = 0 \qquad -6x - 3y + 2 = 0$$

Solve both equations for y.

$$y = \boxed{-2}\, x - 1 \qquad -3y = 6x - 2$$

$$y = \boxed{-2}\, x + \frac{2}{3} \quad \textbf{Dividing by } -3$$

Both lines have the same slope, $m = \boxed{-2}$, but their y-intercepts are $(0, -1)$ and $\left(0, \dfrac{2}{3}\right)$, respectively. As you can see from the graph of the lines in Figure 10.5, they are parallel. ◆

DEFINITION 10.3

Two lines are **parallel** if, and only if, their slopes are equal. That is, $L_1 \parallel L_2$, if and only if, $m_1 = m_2$.

EXAMPLE 5   Sketching Perpendicular Lines

Sketch the graphs of $y = -2x - 1$ and $y = \dfrac{1}{2}x + 3$.

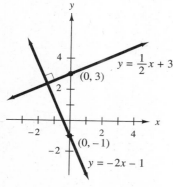

FIGURE 10.6

Solution   Both lines are linear. $y = -2x - 1$ has a slope of -2 and y-intercept $(0, -1)$. $y = \dfrac{1}{2}x + 3$ has a slope of $\dfrac{1}{2}$ and y-intercept $(0, 3)$. The slopes of the lines $\left(-2 \text{ and } \dfrac{1}{2}\right)$ are negative reciprocals. As you can see in Figure 10.6, the lines are perpendicular. ◆

TABLE 10.1

x	y
-3	-2
0	-2
3	-2

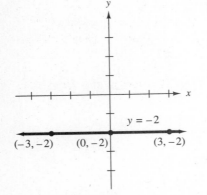

FIGURE 10.7a

DEFINITION 10.4

Two lines are **perpendicular** if, and only if, their slopes are negative reciprocals. That is, $L_1 \perp L_2$ if, and only if, $m_1 = -\dfrac{1}{m_2}$.

EXAMPLE 6   **Sketching a Line Parallel to the x-Axis and the y-Axis**

a. Sketch the graph of $y = -2$.   **b.** Sketch the graph of $x = 3$.

Solution

a. We can write the linear equation $y = -2$ in the slope-intercept form: $y = 0x - 2$. This form makes it easy to recognize that $m = 0$ and $b = -2$. With $m = 0$, there is no slope, and the line is parallel to the x-axis. To sketch the graph, determine sets of ordered pairs as shown in Table 10.1. The sketch of the line is given in Figure 10.7a.

b. We can write the linear equation $x = 3$ in the slope-intercept form: $0y = -x + 3$, making it easy to see that for any value of y, $x = 3$. If we solve the equation, $0y = -x + 3$ for y, we would be dividing each term by 0, which we cannot do. Therefore, the slope and y-intercept are undefined. To sketch the graph, determine sets of ordered pairs as shown in Table 10.2. The graph is a straight line parallel to the y-axis, as shown in Figure 10.7b.   ◆

The following two statements generalize the results from Example 6.

TABLE 10.2

x	y
3	-2
3	0
3	2

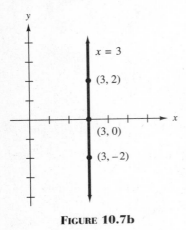

FIGURE 10.7b

GRAPH OF $y = a$ AND $x = a$

1. The graph of any equation of the form $y = a$ (where a is a real number) is parallel to the x-axis and passes through the point $(0, a)$.
2. The graph of any equation of the form $x = a$ (where a is any real number) is parallel to the y-axis and passes through the point $(a, 0)$.

EXAMPLE 7   **Write the Equation of a Line Given the Graph**

Write the equation of the line in Figure 10.8 in slope-intercept form.

Solution   The line crosses the y-axis at $(0, 2)$, the y-intercept. Therefore, $b = 2$. A second point on the line is $(3, 4)$. With two points, we can use Definition 10.2 to determine m.

$$m = \frac{y_2 - y_1}{x_2 - x_1} = \frac{4 - 2}{3 - 0} \quad \text{or} \quad m = \frac{2}{3}$$

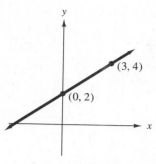

FIGURE 10.8

Substituting the values $\frac{2}{3}$ and 2 for m and b, respectively, in the equation $y = mx + b$, we determine that the equation of the line is $y = \boxed{\frac{2}{3}}x + \boxed{2.}$ ◆

EXAMPLE 8   Application: Equation of a Falling Object

A hammer falls from a scaffold. Its speed by the end of the first second of fall is −9.8 meters per second. Its speed at $t = 2.0$ seconds is −19.6 meters per second. The minus sign indicates that the hammer is moving in the downward direction.

a. Plot these points on velocity-versus-time graph.
b. The slope of a velocity-versus-time graph is **acceleration** (in this case, acceleration due to gravity). What is the acceleration of the hammer due to gravity?
c. Write the linear equation that describes the motion of the hammer.
d. What is the speed of the hammer at 1.75 seconds?

Solution

a. A velocity-versus-time graph uses the dependent (or vertical) axis as the velocity axis and the independent (or horizontal) axis as the time axis. We can determine three points from the problem: $(1.0, -9.8)$, $(2.0, -19.6)$, and $(0, 0)$. When the hammer falls from the scaffold, time is zero, and the velocity is zero, giving the point $(0, 0)$. Plot the points and draw the line as shown in Figure 10.9.

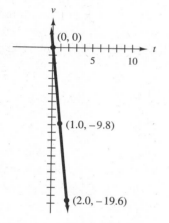

FIGURE 10.9

b. Since the slope is the acceleration in this situation, we can determine the acceleration using the formula for determining the slope.

$$m = \frac{y_2 - y_1}{x_2 - x_1}$$

$$= \frac{-19.6 - (-9.8)}{2 - 1}$$

$$= -9.8$$

Thus, the acceleration is −9.8 m/s².

c. From part a we know that the y-intercept is $(0, 0)$ and that $b = 0$. In part b we found the slope, $m = -9.8$. Substituting in the equation $v = mt + b$ (substituting v for y and t for x), we have the equation of the line.

$$v = -9.8t$$

d. We can determine the velocity of the hammer at 1.75 seconds by substituting 1.75 for t in the equation, $v = -9.8t$.

$$v = -9.8(1.75)$$

$$= -17.15$$

The velocity when $t = 1.75$ seconds is −17 ft/s. ◆

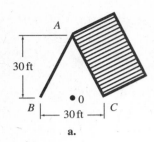

30 ft

B | 30 ft | C

a.

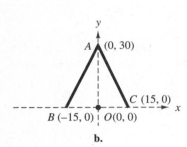

b.

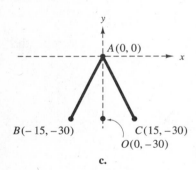

c.

FIGURE 10.10

EXAMPLE 9   **Application: Writing the Equation of a Line Given the Graph**

An A-frame house has a roof composed of lines *AB* and *AC* as shown in Figure 10.10a. Write the equation that represents the line *AB*, if the origin of the axis system is at the following points. (0 is the midpoint between points B and C.)

a. 0   **b.** *A*

Solution

a.   With the origin of the axis system at point 0, Figure 10.10b shows an *x*, *y*-axis system with the dotted lines. Using the measurements from Figure 10.10a we can determine the coordinates for the points *O*, *A*, *B*, and *C*. Point *O*, the origin, has coordinates (0, 0). Points *B* and *C* are 15 units to the left and the right of the origin, respectively; their coordinates are $(-15, 0)$ and $(15, 0)$. Point *A* is 30 units above the origin; therefore, its coordinates are $(0, 30)$. Knowing the coordinates of the points, we can determine the slope and the *y*-intercept for the line *AB*.

   Starting at point *B* and moving to point *A*, *x* increases 15 units $(\Delta x = 15)$ and *y* increases 30 units $(\Delta y = 30)$. The slope is:

$$m = \frac{\Delta y}{\Delta x} = \frac{30}{15} = 2.$$

Since point *A* is on the *y*-axis, it is the *y*-intercept, and $b = 30$. Substituting for *m* and *b* in the equation $y = mx + b$, the equation of the line through points *A* and *B* with the origin of axis system located at point 0 is:

$$y = \boxed{2}\, x + \boxed{30.}$$

b.   Now sketch an axis system on the figure with the origin at point *A*, as shown in Figure 10.10c. The coordinates of point *A* are (0, 0). With point *O* on the vertical axis and 30 units below point *A*, its coordinates are $(0, -30)$. Since *B* and *C* are 15 units to the left and right of point *O*, their coordinates are $(-15, -30)$ and $(15, -30)$, respectively. Starting at point *B* and moving to point *A*, *x* increases 15 units $(\Delta x = 15)$, and *y* increases 30 units $(\Delta y = 30)$. The slope is:

$$m = \frac{\Delta y}{\Delta x} = \frac{30}{15} = 2.$$

With point *A* at the origin, the *y*-intercept is (0, 0) and $b = 0$. Substituting for *m* and *b* in the equation $y = mx + b$, the equation of the line *AB*, with point *A* as the origin, is:

$$y = \boxed{2}\, x. \quad \blacklozenge$$

When you are asked to write the equation of a line by looking at a graph, remember that it is possible to tell whether the slope is positive or negative from the graph. Look at Figure 10.8 again. Here the line rises as we move from left to right so the slope is positive. Look at Figure 10.4. Here the line falls as we move from left to right so the slope is negative.

Other Forms of the Equation of the Line

Since the slope of a line can be found using any two points on the line, we may assume an unknown point (x, y), lying on the line and write $m = \dfrac{y - y_1}{x - x_1}$. When we multiply both sides of the equation by $(x - x_1)$, we have $y - y_1 = m(x - x_1)$, which is called the *point-slope* form of the equation of a straight line.

> **DEFINITION 10.5**
> The **point-slope form** of an equation of a line is:
> $$y - y_1 = m(x - x_1).$$

If the slope m in the point-slope form of the equation is replaced with $\dfrac{y_2 - y_1}{x_2 - x_1}$, then the equation becomes $y - y_1 = \dfrac{y_2 - y_1}{x_2 - x_1}(x - x_1)$. This equation is called the **two-point form** of the equation of a line.

EXAMPLE 10   Application: Given Two Points Write the Equation of the Line

The floor of a circular wading pool with a diameter of 50 ft is sloped to the drain, which is in the center. The surface of the floor is a straight line from the center to the outer edge of the pool. If the slope is 1 ft of rise for each 25 ft of run, write an equation that describes the surface of the floor to the right of the drain. Assume that the origin is at the drain, as shown in Figure 10.11. (Assume two significant digits.)

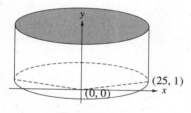

FIGURE 10.11

Solution   With the origin at the drain, the coordinates of this point are $(0, 0)$. With a rise of 1 ft for each 25 ft of run, the point $(25, 1)$ would be on the bottom outer edge of the pool (see Figure 10.11). Substitute the two points in the point-slope form of the equation of a line.

$$y - y_1 = \frac{y_2 - y_1}{x_2 - x_1}(x - x_1)$$

$$y - 0 = \frac{1 - 0}{25 - 0}(x - 0)$$

$$y = \frac{1}{25}x$$

The equation that describes the surface of the floor to the right of the drain is $y = 0.040x$. ◆

Changing Axes

In all of the examples to this point, the independent variable (x) is on the horizontal axis and the dependent variable is on the vertical axis. However, this convention is broken occasionally as when economists work with demand and supply functions. They agree that quantity (demanded or supplied), which is the dependent variable, is depicted on the x-axis and that price, the independent variable, is depicted on the y-axis. In writing the linear equation $y = mx + b$, representing the line, the slope $m = \dfrac{\Delta y}{\Delta x}$ is in this case the change in price divided by the change in quantity.

The **law of demand** states that there is an inverse relationship between price (y) and quantity demanded (x). That is, as the price increases quantity decreases. With the inverse relation, the slope of the line representing a demand curve is always negative. An example of a demand curve is $y = 25 - 5x$. From the function we can see that the slope is negative. This is also illustrated in Figure 10.12.

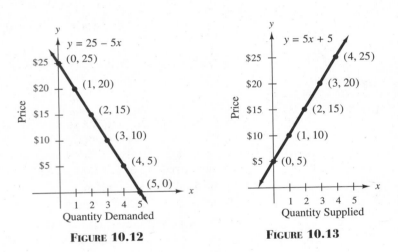

FIGURE 10.12   FIGURE 10.13

On the other hand, the law of supply represents a direct relationship. The **law of supply** states that there is a direct relationship between price (y) and quantity supplied (x). That is, as price increases quantity supplied increases, and as price decreases quantity supplied decreases. Since the law of supply is a direct relationship, the slope is never negative. An example of a supply curve is $y = 5x + 5$. From the function, we see that the slope is positive. This is illustrated in Figure 10.13.

10.1 Exercises

In Exercises 1–8, match the equation with its graph.

1. $y = \dfrac{1}{2}x + 1$

a.

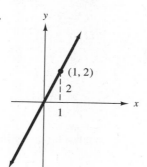

b.

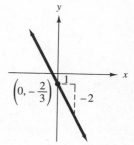

2. $y = 1$

3. $x = -1$

c.

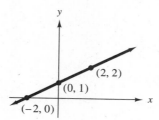

d.

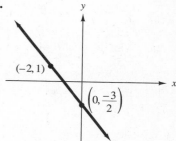

4. $6x + 3y + 2 = 0$

5. $\dfrac{1}{2}y = \dfrac{-3}{4}x + 2$

e.

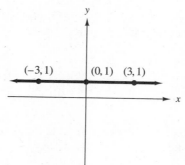

f.

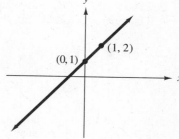

6. $y = x + 1$

7. $2y + \dfrac{5}{2}x + 3 = 0$

g.

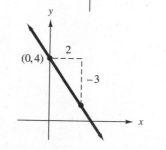

h.

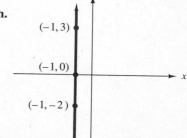

8. $2x - y = 0$

In Exercises 9–22, identify the slope and the y-intercept of the lines. Sketch the graph of the lines.

9. $y = 3x + 2$ **10.** $y = x - 1$ **11.** $y = \dfrac{1}{2}x - 1$ **12.** $y = \dfrac{3}{4}x + 2$

13. $y = 3x - 2$ **14.** $y = x + 1$ **15.** $y = 4$ **16.** $y = \dfrac{3}{4}x - 2$

17. $2x + 4y + 1 = 0$ **18.** $x = \dfrac{1}{2}$ **19.** $-2x + 4y + 1 = 0$ **20.** $-6x + 3y + 2 = 0$

21. $y = \dfrac{1}{3}x$ **22.** $y = 3x$

In Exercises 23–28, determine whether the lines are parallel, perpendicular, or neither. Do this without graphing the lines.

23. $\quad x + y + 1 = 0$
$\quad\ \ 2x + 2y - 2 = 0$

24. $\quad y = \dfrac{1}{2}x - 3$
$\quad\ \ 2y = x - 6$

25. $\quad 2x + 3y = -4$
$\quad\ \ -3x + 2y = -1$

26. $\quad 2x + 3y - 6 = 0$
$\quad\ \ -3x + 2y - 4 = 0$

27. $\quad \dfrac{1}{2}y = 2x + 3$
$\quad\ \ 2y = 8x + 12$

28. $\quad y - 2x = 3$
$\quad\ \ 4y - 2x = 5$

In Exercises 29–36, write the equation of the line given the slope and the y-intercept.

29. $m = 2, b = 2$ **30.** $m = 3, b = -1$ **31.** $m = -6, b = 3$

32. $m = -2, b = 8$ **33.** $m = 1, b = 0$ **34.** $m = -1, b = 0$

35. $m = 0, b = 4$ **36.** m is undefined, $x = -4$

In Exercises 37–42, given a y-intercept and the equation of a line, write the equation of a line that is parallel to the given line and passes through the given y-intercept.

37. $b = 2; y = \dfrac{1}{2}x - 3$ **38.** $b = 2; y = 2x + 3$ **39.** $b = \dfrac{3}{2}; 3x + 2y + 1 = 0$

40. $b = 0; 2x + 3y + 4 = 0$ **41.** $b = 4; y = -1$ **42.** $b = \dfrac{2}{3}; y = -\dfrac{2}{3}$

In Exercises 43–48, given a y-intercept and the equation of a line, write the equation of a line that is perpendicular to the given line and passes through the given y-intercept.

43. $b = 1; y = \dfrac{1}{3}x - 2$ **44.** $b = -2; y = 2x + 4$ **45.** $b = \dfrac{1}{2}; 2x - y - 5 = 0$

46. $b = 0; x + y + 7 = 0$ **47.** $b = 3; x = -2$ **48.** $b = \dfrac{4}{3}; x = 5$

49. Write the equation of a line that passes through the points $(5, 2)$ and $(-1, 8)$.

50. Write the equation of a line that is perpendicular to $3x + 4y - 7 = 0$ and that passes through the point $(-4, 3)$.

51. Write the equation of a line that is perpendicular to $2x - 5y - 8 = 0$ and that passes through the point $(2, 3)$.

52. Write the equation of the line that is the perpendicular bisector of the line segment defined by $(-4, 3)$ and $(1, -7)$.

In Exercises 53–62, write the equation of the straight line illustrated in each sketch.

53.

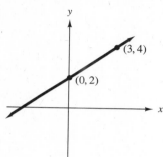

54.

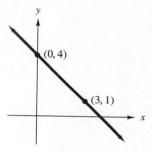

55.

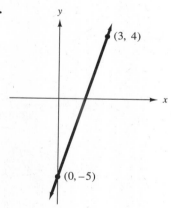

56.

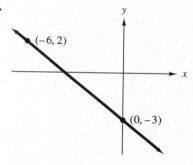

57.

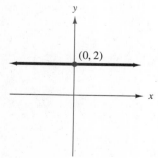

58.

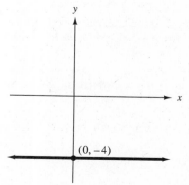

59.

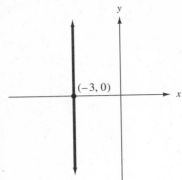

60.

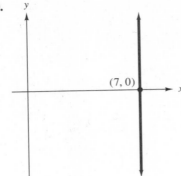

61.

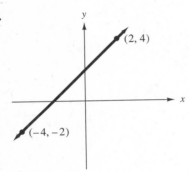

62.

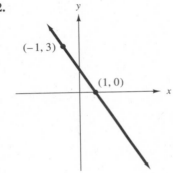

63. If the resistance R is a constant, is Ohm's law, $i = \dfrac{v}{R}$, a straight line? Explain.

64. Standing at the end of a parking lot, the engineer explains that the left and right sides are built higher so that water drains to a center drain line. The engineer states that the parking lot has a drainage slope of 1 ft in 50 ft. Write an equation that describes the surface of the parking lot to the right of the drain. Assume that the axis system is in the center of the lot.

65. Use the information given in Exercise 64 to write an equation that describes the surface of the lot to the left of the drain.

66. For the A-frame house in Example 9, write the equation that represents the line AC if the origin of the axis system is at each point given.
   a. A **b.** C **c.** O.

67. A construction worker throws a tape measure to a friend who is high on a scaffold. The tape measure is thrown straight up with an initial velocity of 30.0 m/s. In other words, at $t = 0.0$ s, the velocity of the tape measure is $+30.0$ m/s. After 3.05 s, it reaches the top of its path and *momentarily stops;* hence, the velocity is zero. The tape measure then begins to fall, changing its direction. The velocity changes at a constant rate.

a. Using the two points, sketch a velocity-versus-time graph. Unlike the graph in Example 8, this one starts with positive velocities.
b. Write a linear equation for the motion of the tape measure.
c. What is the acceleration for this motion? What is the significance of the negative sign in your answer?

68. Using the information in Exercise 67, suppose the worker on the scaffold cannot catch the tape measure and that it falls back down to the ground.
a. Draw the graph using the points given in Exercise 67. Extend the line beyond $t = 3.05$ s to 6.10 s.
b. What is the velocity at 4.0 s? What does the negative sign in your answer suggest about the motion of the tape measure?
c. What is the acceleration of the tape measure for values of t greater than 3.05 s? How does this acceleration compare with the acceleration in Exercise 67?
d. What is the acceleration of the tape measure when it momentarily stops at the top? (*Hint:* The answer is *not* zero. Rely on your graph.)

69. As the temperature of a solid increases, its length increases proportionately. The equation of expansion is $\Delta L = \alpha L_0 \Delta T$, where ΔT is the change in temperature; ΔL is the metal's increase in length when subjected to a ΔT; L_0 is the initial length of the metal; and α is the coefficient of linear expansion that determines the expansion per degree change in temperature. The magnitude α for steel is 12×10^{-6} $(C°)^{-1}$. Suppose the initial length of a steel girder is 200 m. Graph the increase in length versus the change in temperature.

70. An engineer in Florida designs a bridge with steel girders that are 200 m long. The bridge is to be constructed in winter when the coldest temperature could be $-5°C$. In the summer temperatures are no higher than $30°C$. How much room for expansion should be built into each joint? Long steel girders often have bolts placed in slotted holes at the ends. Why?

71. Concrete has a coefficient of expansion which is about the same as steel. Why are there gaps between sidewalk slabs? If the length of the slabs are 10 ft, determine how big the gap between slabs should be. (*Hint:* See Exercise 70.)

72. The river was rising at a constant rate of 3 in./hr when the weather bureau issued a flood watch. At midnight, the river level gauge indicated 8 in. above the normal level. The table shows the height of the river above the normal level for next 12 hours after midnight. (Midnight is taken as 0 hours). Plot the points and write an equation that describes the series of points.

Time (hr)	Level (in.)
0	8
1	11
2	14
3	17
4	20
5	23
6	26
7	29
8	32
9	35
10	38
11	41
12	44

73. A carpenter purchases a tool for $375. After five years, the tool will be worn out and have no value. Write a linear equation that gives the value of the tool during its five years of use.

74. Brand X wood stove sells for $1150, and a face cord of wood sells for $50.
a. Write an equation giving the total cost C of operating the stove in terms of the number x of face cords of wood purchased.
b. Determine the total cost of burning 5 face cords of wood in the stove.

In Exercises 75 and 76 the table provides the quantity demanded of a certain item at various prices. **a.** Plot the points. **b.** Determine the slope of the curve. **c.** Write the equation in the form $y = mx + b$; x represents demand for Exercise 75 and supply for Exercise 76.

75.	Price (y)	Quantity Demanded (x)
	$15	0
	$18	1
	$21	2
	$24	3
	$27	4
	$30	5

76.	Price (y)	Quantity Supplied (x)
	$15	0
	$18	1
	$21	2
	$24	3
	$27	4
	$30	5

10.2

Solving Two Linear Equations in Two Unknowns Algebraically

In Section 10.1, we illustrated the ordered pairs that are solutions to a linear equation by drawing a graph. In this section, we look at techniques of determining the solution to a system of linear equations:

$$0.8A - 0.6B = 4,$$
$$0.6A + 0.8B = 0.$$

This pair of equations forms a system of two linear equations in two unknowns (or two variables; in this case, A and B). The **solution** to a system of two linear equations in two unknowns is the ordered pair (or pairs) that makes the equations true simultaneously. The solution of a system of equations in two unknowns may be found either by graphing or by algebraic methods. Graphical solutions to systems of linear equations are useful because they provide a visual insight into the type of system of equations. However, graphs are sometimes hard to read. Algebraic solutions let you obtain more accurate solutions. The two algebraic methods of solving equations discussed in this section are substitution and elimination.

Substitution Method

When we use the substitution method, we first solve one of the equations for one of the variables, say x. Generally, we will find it easier to begin with the equation that contains a variable whose coefficient is 1 or -1, if this case exists. Having solved for x, substitute this value or expression for x in the other equation. Then solve the equation for y. Having found a value for y, we can go back and determine a value for x. This method is illustrated in Example 1.

EXAMPLE 1 **Solving a System of Equations by the Substitution Method**

Solve the following system of equations using the substitution method.

$$2x + y = 4 \tag{1}$$
$$x - 2y = -3 \tag{2}$$

Solution   To solve by the substitution method, first solve either one of the equations for one of the variables. In the second equation the coefficient of x is 1. Remembering that makes things easier so solve the second equation for x first.

$$x = 2y - 3 \quad \text{\small Solving second equation for } x$$

Now substitute the value of x (in this case, the expression $2y - 3$) for x into the first equation.

$$2(\,2y - 3\,) + y = 4 \quad \text{\small Substituting for } x \text{ in (1)}$$
$$4y - 6 + y = 4$$
$$5y - 6 = 4$$
$$5y = 10$$
$$y = 2$$

Substitute the value for y (2) into the equation we started with.

$$x = 2y - 3$$
$$x = 2(\,2\,) - 3$$
$$x = 1$$

The solution for the system $2x + y = 4$ and $x - 2y = -3$ is the ordered pair $(1, 2)$. To check the solution, substitute $(1, 2)$ into both equations.
Check:

$$2x + y = 4$$
$$2(\,1\,) + 2 = 4$$
$$4 = 4. \checkmark$$
$$x - 2y = -3$$
$$1 - 2(\,2\,) = -3$$
$$-3 = -3. \checkmark$$

The results of the check show that the ordered pair is a solution for the system of equations; and we say that the solution checks. For practice, try solving for y first instead of x, or try starting with the first equation instead of the second equation. Either approach should produce the same solution. ◆

The graph of the equations in Example 1 is shown in Figure 10.14. The lines intersect in a unique point, $(1, 2)$. The unique point is the solution set of the system of equations. The graph is a visual check. The graphics calculator can be used to estimate the point of intersection, if one exists.

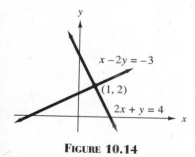

FIGURE 10.14

PROCEDURE FOR USING SUBSTITUTION METHOD

1. Solve one of the equations for one of the variables. To simplify the procedure, assume that variable is x.
2. Substitute the value of x (found in Step 1) into the other equation.
3. Now the only variable in the equation from Step 2 is y. Solve for y.
4. Substitute the value we found in Step 3 for y, into the equation we started with. Solve for x.
5. Check and record the solution.

EXAMPLE 2   **Application: Solving a System of Equations by the Substitution Method**

Two forces A and B in a roof truss are related in the system of equations:

$$0.800A - 0.600B = 4000,$$
$$0.600A + 0.800B = 0.$$

Solve the system of equations using the substitution method.

Solution   The rules for significant digits tell us that the solution of the equations should contain three significant digits. Keep this in mind even though, for convenience, we will not carry all the trailing zeros. The system of equations is:

$$0.8A - 0.6B = 4000, \tag{1}$$
$$0.6A + 0.8B = 0. \tag{2}$$

Since Equation (2) has no constant term, it appears to be easier to solve for one of the variables in terms of the other. Solve for A.

$$0.6A + 0.8B = 0$$
$$A = \frac{-0.8B}{0.6}.$$

Now substitute into Equation (1).

$$0.8\left(\frac{-0.8B}{0.6}\right) - 0.6B = 4000$$
$$-1.067B - 0.6B = 4000$$
$$-1.667B = 4000$$
$$B = -2400$$
$$A = \frac{-0.8}{0.6}(-2400)$$
$$A = 3200.$$

Thus, the force in member A is 3200 lb, and the force in member B is -2400 lb. Here the minus sign has a physical significance; it indicates that force B is in a direction opposite of force A.   ◆

EXAMPLE 3   **Solving a System of Equations by the Substitution Method**

Solve the system of equations using the substitution method.

$$2x + y = 2$$
$$4x + 2y = -1$$

Solution   Since the y term in the first equation has a coefficient of 1, solve for y in that equation.

$$2x + y = 2$$
$$y = 2 - 2x.$$

Substitute $2 - 2x$ for y in the other equation.

$$4x + 2(\boxed{2 - 2x}) = -1$$
$$4x + 4 - 4x = -1$$
$$4 = -1.$$

This result is a contradiction since 4 cannot equal -1. Therefore, there is no solution to the system of equations. Indicate this by writing the words "no solution" or using the symbol Ø, which means the empty set in set notation.   ◆

The graphs of the equations in Example 3, are shown in Figure 10.15. The lines are parallel, which means they have no points in common. This confirms the result (no solution) arrived at in Example 3.

A system of two linear equations in two unknowns in which the lines have no points in common is called an *inconsistent system*. An **inconsistent system** has no solution (no ordered pairs of numbers that will satisfy both equations.) The graphs of all inconsistent systems are parallel lines.

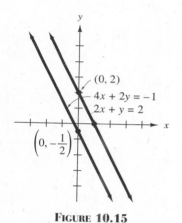

FIGURE 10.15

Elimination Method

The elimination method is a procedure that uses the operations of addition and subtraction to eliminate one of the variables in a system of equations. For example, look at the system:

$$x + 3y = 5$$
$$-x + y = 3.$$

If we add the two equations, the result is:

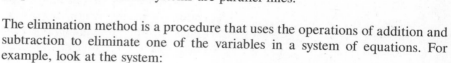

$$x + 3y = 5$$
$$\underline{-x + y = 3}$$
$$4y = 8.$$

This is a linear equation in terms of one variable, y in this case. Now we can solve for y ($y = 2$) and substitute the value 2 for y in either of the two original equations to obtain the value for x ($x = -1$). Thus, the solution is $(-1, 2)$.

In some cases, we may need to multiply one or both equations by a constant to make the coefficients of one of the variables the same. A set of procedures follow, and they are applied in Example 4.

PROCEDURE FOR USING THE ELIMINATION METHOD

1. If necessary, rewrite each equation so that the terms containing variables appear on the left side of the equal sign and any constants are on the right side of the equal sign.
2. If necessary, multiply one or both equations by a constant to make the coefficients of one of the variables the same for both equations. (Their signs may differ.)
3. Add or subtract the equations to obtain a single equation containing only one variable.
4. Solve for the variable in the equation from Step 3.
5. Substitute the value found in Step 4 into either of the original equations. Solve that equation to determine the value of the other variable.
6. Check and record the solution.

EXAMPLE 4 **Solving a System of Equations by the Elimination Method**

Determine the solution of the system of equations using the elimination method.

$$6x + 5y = 12$$
$$3x - 2y = -3$$

Solution Observe that if the x term in the second equation has a coefficient of -6, we can add the equations and eliminate the variable x. The result is a linear equation with a single variable y. To do this, multiply the second equation by -2. (Remember that the equality is preserved when both sides of an equation are multiplied by the same nonzero factor.)

$$6x + 5y = 12$$
$$-2\,(3x - 2y) = -2\,(-3) \qquad \text{Multiplying both sides of the equation by } -2$$
$$-6x + 4y = 6$$
$$\underline{6x + 5y = 12} \qquad\qquad \text{Adding}$$
$$9y = 18$$
$$y = 2$$

Now we can substitute the value obtained for y into either of the original equations to solve for x.

$$6x + 5(\boxed{2}) = 12 \quad \text{Substituting 2 for y in the first equation}$$
$$6x = 2$$
$$x = \frac{1}{3}$$

To verify the solution $\left(\frac{1}{3}, 2\right)$, substitute the value for x and y into the second equation.

$$3\left(\boxed{\frac{1}{3}}\right) - 2(\boxed{2}) = -3 \; \checkmark \quad \blacklozenge$$

For practice, graph the system in Example 4 and show that the lines intersect at a unique point.

EXAMPLE 5   **Solving a System of Equations by the Elimination Method**

Determine the solution of the given system of equations using the elimination method.

$$x - 2y = -1$$
$$-2x + 4y = 2$$

Solution   To have a coefficient of 2 for the x term in the first equation, multiply both sides of the equation by 2.

$$\boxed{2}\,(x - 2y) = \boxed{2}\,(-1) \quad \text{Multiplying both sides of the equation by 2}$$

so:

$$\begin{array}{r} 2x - 4y = -2 \\ -2x + 4y = 2 \\ \hline 0 = 0 \end{array} \quad \text{Adding}$$

Although this result is true it does not provide a unique solution. Instead, it tells us that any ordered pair (x, y) that satisfies one equation also satisfies the other equation. In other words, there are infinitely many ordered pairs that satisfy both equations.   ◆

When the equations in Example 5 are graphed (shown in Figure 10.16), we see that the lines coincide, forming a single line.

A system of two linear equations in two unknowns in which the graphical solution is a single line is a dependent system. A **dependent system** of equations has an infinite number of ordered pairs that satisfy both equations.

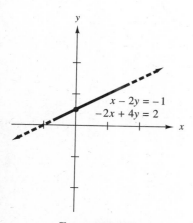

FIGURE 10.16

EXAMPLE 6   Application: Writing and Solving a System of Linear Equations

Twice the current in electric circuit A is 3 amperes greater than three times the current in the adjoining circuit B. When current B is subtracted form current A and the difference of the two currents is multiplied by a resistance of 5 Ω, the result is a voltage drop of 7 V. What are the currents?

Solution   Let:

$$i_A = \text{current in circuit } A,$$
$$i_B = \text{current in circuit } B,$$
$$2i_A = \text{twice the current in circuit } A, \text{ and}$$
$$3i_B = \text{three times the current in circuit } B.$$

Knowing that twice the current in circuit A is 3 amperes greater than three times the current in circuit B, we can write:

$$2i_A = 3i_B + 3.$$

Knowing $5(i_A - i_B) = $ voltage drop of 7 V, we can write

$$5i_A - 5i_B = 7.$$

Now we can solve the system of two equations in two unknowns using the elimination method.

$$2i_A - 3i_B = 3 \qquad\qquad (1)$$
$$5i_A - 5i_B = 7 \qquad\qquad (2)$$

To eliminate i_A, multiply Equation (1) by 5 and Equation (2) by -2. Then add.

$$10i_A - 15i_B = 15 \qquad \mathbf{5 \times (1)} \qquad (3)$$
$$-10\, i_A + 10\, i_B = -14 \qquad \mathbf{-2 \times (2)} \qquad (4)$$
$$-5i_B = 1 \qquad \mathbf{(3) + (4)}$$

$$i_B = \frac{-1}{5}$$

Substituting for i_B in Equation (1), we have:

$$2i_A - 3\left(-\frac{1}{5}\right) = 3$$

$$2i_A + \frac{3}{5} = 3$$

$$2i_A = \frac{12}{5}$$

$$i_A = \frac{6}{5}.$$

The currents are $i_A = \dfrac{6}{5}$ amperes and $i_B = -\dfrac{1}{5}$ amperes.   ◆

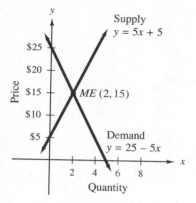

FIGURE 10.17

In Section 10.1 we introduced the ideas of supply and demand. When the lines that represent supply and demand are sketched on the same set of axes, the two lines intersect in a unique point called **market equilibrium** (ME). This point indicates a special price that results in the same value of total quantity (number of items) supplied and the total quantity (number of items) demanded. If the price increases, then quantity supplied would exceed quantity demanded. If the price decreases, then quantity demanded would exceed quantity supplied. See Figure 10.17.

EXAMPLE 7   **Application: Determining the Market Equilibrium**

a. Determine the point of market equilibrium for the demand function $y = 25 - 5x$ and the supply function $y = 5x + 5$.
b. Graph the functions.

Solution

a. We can determine the point of market equilibrium by solving the system of equations using the substitution method. Since $y = 5x + 5$, if we substitute $5x + 5$ for y in the function $y = 25 - 5x$, we have:

$$5x + 5 = 25 - 5x$$
$$x = 2 \qquad \text{Solving for } x$$

Now substitute $x = 2$ in the equation $y = 25 - 5x$.

$$y = 25 - 5(2)$$
$$= 15$$

The point of market equilibrium is (2, 15).
b. The graph is sketched in Figure 10.17.   ◆

10.2 Exercises

 In Exercises 1–10, use the substitution method to solve the system of equations. As a visual check, use a graphics calculator to graph the system of equations and use its trace function to estimate the point of intersection, if one exists.

1. $2x + y = -10$
$\quad x - 4y = -14$

2. $x + y = 21$
$\quad 3x - 2y = 8$

3. $-5x + y = 1$
$\quad 7x - y = 5$

4. $x - y = 4$
$\quad 3x - 2y = 1$

5. $7x - y = 1$
$\quad -2x + y = 19$

6. $x + 2y = -3$
$\quad -x + y = -30$

7. $-5x + 2y = -11$
$\quad x - 3y = -3$

8. $3x - y = -4$
$\quad -5x + 4y = 7$

9. $0.4x + 0.5y = 3.7$
$\quad x - 0.7y = -0.5$

10. $0.07x + 0.09y = 0.011$
$\quad -0.03x + y = 0.203$

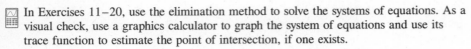 In Exercises 11–20, use the elimination method to solve the systems of equations. As a visual check, use a graphics calculator to graph the system of equations and use its trace function to estimate the point of intersection, if one exists.

11. $-5x + 2y = -11$
$2x - 3y = 22$

12. $3x - 2y = -4$
$-5x + 4y = 0$

13. $2x - 3y = -3$
$4x + 3y = 39$

14. $5x + 2y = 6$
$9x + 3y = 3$

15. $2x + 5y = 12$
$3x - 2y = 18$

16. $4x + 7y = -16$
$-5x + 3y = 20$

17. $8x - 5y = 15$
$-3x + 7y = -21$

18. $-3x + 8y = 56$
$5x + 2y = 14$

19. $2x + 5y = 13$
$4x - 7y = -18$

20. $4x - 7y = 7$
$2x - 14y = -7$

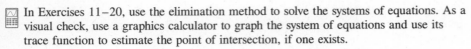 In Exercises 21–30, use algebraic methods to solve the systems of linear equations. Are any of the systems inconsistent or dependent? As a visual check, use a graphics calculator to graph each system of equations and use its trace function to estimate the point of intersection, if one exists.

21. $2x + 3y = 6$
$-4x - 6y = -8$

22. $x - y = 3$
$-x + y = 0$

23. $2x + 5y = 2$
$4x + 10y = 4$

24. $3x - 2y = -1$
$-3x + 2y = 1$

25. $3x + 2y = -4$
$-3x + 2y = -4$

26. $6x + 5y = 18$
$6x - 5y = 18$

27. $3.6x - 1.4y = 0.5$
$-1.8x + 0.7y = 0.2$

28. $1.8x + 2.0y = 2.3$
$3.6x + 4.0y = -4.6$

29. $-1.8x + 1.1y = 2.1$
$5.4x - 3.3y = -6.2$

30. $5.1x + 7.2y = -0.6$
$1.7x + 2.4y = -0.2$

Exercises 31–32 are the equilibrium equations for the joints in the truss shown in the illustration. Solve for the unknown forces in each system of equations.

31. $0.866A + B = 0$
$7500 + 0.500A = 0$

32. $-0.866C + 0.866H = 0$
$-0.500C - 5000 - 0.500H - 5000 = 0$

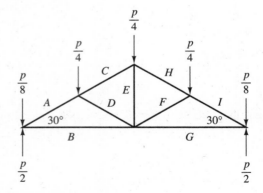

In Exercises 33–40, examine the graphs of the systems of equations and use the information given to write the equations that represent the lines in the graphs. (*Hint:* See Section 10.1.)

33.

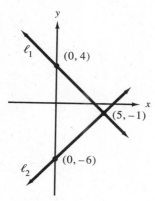

34.

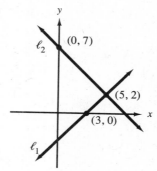

35.

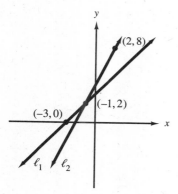

36.

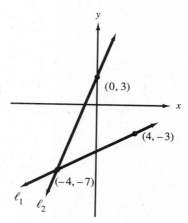

37.

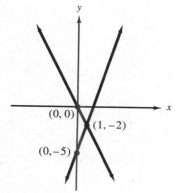

38.

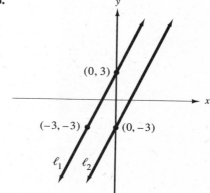

39.

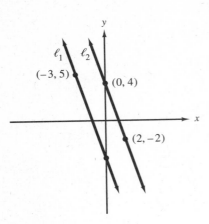

40.

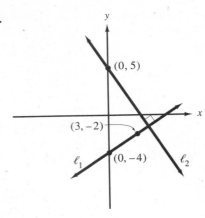

41. A car, traveling along the highway at a speed of 70 mph, passes a police car traveling at 50 mph. The driver slows, while the police officer speeds up in pursuit. The acceleration of the car is -2.3 mph per second squared. The acceleration of the police car is $+3.4$ mph per second squared. The general equation for uniformly accelerated motion is:

$$v = v_0 + at$$

where v_0 is the initial speed of the vehicle; a is the acceleration (change in speed for a given time interval); t is the time elapsed; and v is the instantaneous velocity of the car at time t. The equation showing the speed of the car is: $v = 70 + (-2.3)t$.
   a. Write an equation showing the speed of the police car.
   b. How long will it take the two vehicles to reach the same speed?

42. In the circuit shown in the sketch, all of the resistances are the same 10 Ω. The current through the first resistor out of the 12 volt battery breaks up into two smaller "tributaries": I_1 and I_2. Hence, $I = I_1 + I_2$. But, since the resistances are the same, $I_1 = I_2$. Therefore, $I = 2I_1$. Another equation to describe the given circuit is $12 = 10I_1 + 10I_2$. This says that the voltage output of the battery is the voltage used up by the first resistor and the branch. Determine I, the current that flows out of the battery.

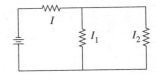

43. Scientists can determine the speed of a subatomic particle by the amount of bending it experiences in a magnetic field. To determine the mass of the particle, they often use momentum. Momentum can be separated into a y-equation and an x-equation as shown.

$$x\text{-axis:}\quad m_1v_0 = m_1v_1' \cos \theta_1 + m_2v_2' \cos \theta_2$$
$$y\text{-axis:}\quad 0 = m_1v_1' \sin \theta_1 + m_2v_2' \sin \theta_2$$

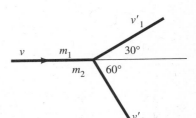

In this case the prime, $'$, indicates speed after the collision. The above equation assumes that the unknown particle m_1 hits a different unknown particle m_2 that is initially at rest (see the drawing). The slight bending of the tracks after collision indicate post collision speeds of 5.2×10^6 m/s and 3.0×10^6 m/s for m_1 and m_2, respectively. You can measure the angles with a protractor. If the initial speed of the incident particle is 6.0×10^6 m/s, how much larger is m_1 than m_2?

44. A simple camera uses just one lens. When you focus this camera, it adjusts the position of the lens so that a sharp image forms on the film. The equations that describe the behavior of the image are:

$$\frac{h_i}{h_o} = \frac{d_i}{d_o} \qquad \frac{1}{d_o} + \frac{1}{d_i} = \frac{1}{f},$$

where h_1 is the height of the image; h_0 is the height of the object; d_i is the distance from the image to the lens; d_0 is the distance from the object to the lens; and f is the focal length of the lens. The focal length for any given lens is a constant that depends upon the shape of the lens and the material with which it is made.

Suppose you took a picture of a friend with your 35 mm camera (i.e, a camera with a focal length of 35 mm). Someone asks how far away your friend was from the camera lens when you took the picture. Since you know your friend is 1.5 meters tall and that the height of the image on the photograph is 2.2 cm, you can answer the question without returning to the original scene. Determine how far away your friend was from the lens of the camera. (*Hint:* Don't forget to convert to common units.)

45. A chemist wants to mix a solution containing 10% sulfuric acid with a solution containing 25% sulfuric acid. She needs 30 cc in all, and she wants the final mixture to be 18% sulfuric acid. How much of each concentration should she use?

46. A cardiac patient covered a distance of 3 miles by alternating between jogging and walking. He walked a total of 10 min and jogged a total of 30 min. The next day, he walked 20 min and jogged 20 min, covering a total distance of $4\frac{1}{3}$ mi. Assuming his rates of walking and jogging are the same on both days, find these rates in miles/hr.

47. A restaurant serves a salad that contains lettuce and bean sprouts in a ratio of 4 : 1 by weight. If the chef wants to make a 100-lb batch of salad, how much produce will he order?

48. Twice the water flow in the hot-water pipe is the same as three times the flow in the cold-water pipe. The combined flow is 1600 liters/hr. What is the flow in each pipe?

49. The total cost of producing a metal toy truck consists of a fixed charge for the mold and an additional cost for the metal used in each truck. If the total cost to produce 1000 toy trucks is $550.00 and the cost to produce 2000 trucks is $800.00, determine the mold cost and the cost of metal for each truck.

50. The perimeter of a rectangle is 24 units. If the width is doubled and the length is tripled, the perimeter is 64 units. Determine the original length and width.

51. Determine the values of a and b if the line $ax + by = 12$ passes through the points $(8, 3)$ and $(-4, -6)$.

52. An apartment buildings contains 20 rental units with one and two bedrooms. The rental price for a one-bedroom is $350 per month and for a two-bedroom $425 per month. If the total rental income from all units is $7600 per month, determine the number of each type of rental unit in the building.

In Exercises 53–56, **a.** determine the market equilibrium (ME) for the supply and demand functions, and **b.** plot the graphs. In these exercises, y represents price, and x represents quantity.

53. $y = 30 - 3x$ (demand)
$y = 4x - 5$ (supply)

55. $4y + 5x = 2$ (demand)
$2y - 3x = -5$ (supply)

54. $y = 15x - 3$ (demand)
$y = 12$ (supply)

56. $x = -4y + 12$ (demand)
$x = 3y - 2$ (supply)

10.3

Solving Two Linear Equations in Two Unknowns by Determinants

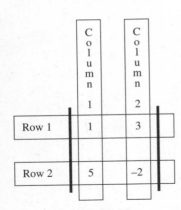

FIGURE 10.18

Another method for solving systems of linear equations is using determinants. A **determinant** is a unique value that is associated with a square array of numbers. A square array is an arrangement of the same number of numbers in each row and each column. A 2×2 (read 2-by-2) array of four numbers arranged in two rows and two columns and bordered by two vertical lines, as shown in Figure 10.18, is called a 2×2 determinant. Each number in the determinant is an **element** of the determinant and is designated by its position at the intersection of a row and a column. For example, 3 appears in the first row and the second column. To indicate the position of 3 in the determinant, we would write 3_{12}. The subscript 12 indicates first the row and then the column of the element 3. If we use i to designate a row in the determinant ($i = 1, 2, \ldots$, indicates $1^{st}, 2^{nd}, \ldots$ row) and use j to designate a column in the determinant ($j = 1, 2, \ldots$, indicates $1^{st}, 2^{nd}, \ldots$, column), we can specify the location of any element in the determinant by using subscripts. For example, the element a_{ij} is located in the ith row and j^{th} column.

EXAMPLE 1 **Designating Elements of a Determinant**

Use the a_{ij} notation to designate the elements of the determinant.

$$\begin{vmatrix} 1 & 3 \\ 5 & -2 \end{vmatrix}$$

Solution a_{11} indicates the element at the intersection of the first row and first column of the determinant. Therefore, $a_{11} = 1$. a_{12} indicates the element at the intersection of the first row and second column of the determinant, so $a_{12} = 3$. In similar fashion, $a_{21} = 5$ and $a_{22} = -2$. ◆

Using the subscript notation, a 2×2 determinant can be written as:

$$\begin{vmatrix} a_{11} & a_{12} \\ a_{21} & a_{22} \end{vmatrix}.$$

DEFINITION 10.6

A 2×2 determinant A has a unique value given by:

Principal Diagonal

$$A = \begin{vmatrix} a_{11} & a_{12} \\ a_{21} & a_{22} \end{vmatrix} = a_{11}a_{22} - a_{21}a_{12}.$$

Secondary Diagonal

($a_{11}a_{22}$ is the product of the elements of the principal diagonal and $a_{21}a_{12}$ is the product of the elements of the secondary diagonal.)

EXAMPLE 2 **Determining the Value of a 2×2 Determinant**

Evaluate the 2×2 determinants.

a. $\begin{vmatrix} 1 & 3 \\ 5 & -2 \end{vmatrix}$ **b.** $\begin{vmatrix} 1 & -2 \\ 3 & 5 \end{vmatrix}$

Solution Use Definition 10.6.

a. $\begin{vmatrix} 1 & 3 \\ 5 & -2 \end{vmatrix} = (1)(-2) - (5)(3)$ **b.** $\begin{vmatrix} 1 & -2 \\ 3 & 5 \end{vmatrix} = (1)(5) - (3)(-2)$

$\qquad\qquad\qquad = -17 \qquad\qquad\qquad\qquad\qquad = 11.$

Notice that even though the numbers in the determinants of parts a and b are the same, the values of the determinants are different because the numbers are located in different positions in the determinants. ◆

We can use determinants to solve systems of equations with a technique called *Cramer's rule*. To show this, let us consider a system of two linear equations in two unknowns.

$$ax + by = c, \tag{1}$$
$$dx + ey = f, \tag{2}$$

where a, b, c, d, e, and f are constants. To solve such a system of equations by the elimination method, we could eliminate the y-term by multiplying Equation (1) by e, multiplying Equation (2) by $-b$, and adding the results:

$$
\begin{array}{ll}
aex + bey = ce & e \times (1) \tag{3} \\
\underline{-bdx - bey = -bf} & -b \times (2) \tag{4} \\
aex - bdx = ce - bf & \textbf{(3)} + \textbf{(4)} \\
(ae - bd)x = ce - bf & \textbf{Factoring} \\
\end{array}
$$

$$x = \frac{ce - bf}{ae - bd}.$$

We can find the value for y using the same techniques we used to solve for x:

$$y = \frac{af - cd}{ae - bd}.$$

When we examine the solutions for x and y, we see the similarity between the expressions in the numerators and denominators of the fractions and the value of a 2×2 determinant. The denominators for x and y are the same, so if we call them D and call the numerators N_x and N_y, respectively, we have:

$$x = \frac{N_x}{D},$$

and

$$y = \frac{N_y}{D},$$

where:

$$D = ae - bd, \text{ which is the value of the determinant } \begin{vmatrix} a & b \\ d & e \end{vmatrix};$$

$$N_x = ce - fb, \text{ which is the value of the determinant } \begin{vmatrix} c & b \\ f & e \end{vmatrix}; \text{ and}$$

$$N_y = af - dc, \text{ which is the value of the determinant } \begin{vmatrix} a & c \\ d & f \end{vmatrix}.$$

This result is **Cramer's rule.**

DEFINITION 10.7 Cramer's Rule (Two Equations in Two Unknowns)

The solution to a system of two linear equations in two unknowns in standard form:

$$ax + by = c,$$
$$dx + ey = f,$$

is given by:

$$x = \frac{N_x}{D} \qquad y = \frac{N_y}{D},$$

where $D = \begin{vmatrix} a & b \\ d & e \end{vmatrix}$, $N_x = \begin{vmatrix} c & b \\ f & e \end{vmatrix}$, $N_y = \begin{vmatrix} a & c \\ d & f \end{vmatrix}$, and $D \neq 0$.

CAUTION   Remember to write the equations in standard form. With the equations in standard form, elements of D are the coefficients of the unknowns x and y. The elements of N_x are the elements of D with the x-coefficients replaced by the constants. The elements of N_y are the elements of D with the y-coefficients replaced by the constants. ■

EXAMPLE 3   **Solving a System of Equations Using Determinants**

Use Cramer's rule to solve the system of equations.

$$2x + y = 4$$
$$x - 2y = -3$$

Solution   To use Cramer's rule, we must write the equations of the system with the constant terms on the right side of the equal sign. That is, we write the equations in standard form.

$$2x + 1y = 4$$
$$1x - 2y = -3$$

Now substitute in the determinants.

$$D = \begin{vmatrix} 2 & 1 \\ 1 & -2 \end{vmatrix} \qquad \text{Determinant of coefficients of variables}$$

$$= (2)(-2) - (1)(1)$$

$$= -5$$

$$N_x = \begin{vmatrix} 4 & 1 \\ -3 & -2 \end{vmatrix} \qquad \text{x-column of D replaced by constant column}$$

$$= (4)(-2) - (-3)(1)$$

$$= -5$$

$$N_y = \begin{vmatrix} 2 & 4 \\ 1 & -3 \end{vmatrix} \qquad \text{y-column of D replaced by constant column}$$

$$= (2)(-3) - (1)(4)$$

$$= -10$$

$$x = \frac{N_x}{D} \qquad y = \frac{N_y}{D}$$

$$= \frac{-5}{-5} \qquad y = \frac{-10}{-5}$$

$$x = 1 \qquad y = 2$$

The solution to the system of equations is the ordered pair $(1, 2)$. To check the solution, substitute $(1, 2)$ into both of the original equations. ◆

EXAMPLE 4   **Solving a System of Equations Using Determinants**

Use Cramer's rule to solve the system of equations.

$$2x + y - 2 = 0$$
$$4x + 2y + 1 = 0$$

Solution   To use Cramer's rule, we must write the equations of the system with the constant terms on the right side of the equal sign. That is, we write the equations in standard form.

$$2x + 1y = 2$$
$$4x + 2y = -1$$

Now substitute in the determinants.

$$D = \begin{vmatrix} 2 & 1 \\ 4 & 2 \end{vmatrix}$$
$$= (2)(2) - (4)(1)$$
$$= 0$$

Because the result is 0, we cannot solve this system of equations using Cramer's rule. If we attempt to continue the solution procedure, we find:

$$N_x = \begin{vmatrix} 2 & 1 \\ -1 & 2 \end{vmatrix}$$
$$= (2)(2) - (-1)(1)$$
$$= 5.$$
$$N_y = \begin{vmatrix} 2 & 2 \\ 4 & -1 \end{vmatrix}$$
$$= (2)(-1) - (4)(2)$$
$$= -10.$$

$$x = \frac{N_x}{D} \qquad\qquad y = \frac{N_y}{D}$$

$$= \frac{5}{0} \qquad\qquad = \frac{-10}{0}$$

$$x = \text{undefined.} \qquad y = \text{undefined.}$$

With x and y both undefined, the system of equations has no solution. It is an inconsistent system of equations.   ◆

CAUTION   When using Cramer's rule, a solution of the form $\left(\dfrac{a}{0}, \dfrac{b}{0}\right)$, where a and b are nonzero constants, indicates an inconsistent system of equations. The graphs of the linear equations are parallel lines. (Remember that parallel lines have the same slope, but different y-intercepts.) If the solution is of the form $\left(\dfrac{0}{0}, \dfrac{0}{0}\right)$, it indicates a dependent system of equations. That is, the graphs of the two linear equations are one line. (The equations have the same slope and the same y-intercept.) ■

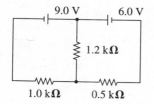

9.0 V   6.0 V

1.2 kΩ

1.0 kΩ   0.5 kΩ

FIGURE 10.19

EXAMPLE 5   Application: Electric Circuit

Use Cramer's rule to determine the loop currents in the system of equations representing the circuit shown in Figure 10.19. Recall that current is measured in amperes.

$$2.2i_1 - 1.2i_2 = 9.0$$
$$-1.2i_1 + 1.7i_2 = -6.0$$

Solution

$$D = \begin{vmatrix} 2.2 & -1.2 \\ -1.2 & 1.7 \end{vmatrix}$$
$$= (2.2 \times 1.7) - [-1.2 \times (-1.2)]$$
$$= 2.3$$

$$N_{i_1} = \begin{vmatrix} 9.0 & -1.2 \\ -6.0 & 1.7 \end{vmatrix}$$
$$= (9.0 \times 1.7) - [-6.0 \times (-1.2)]$$
$$= 8.1$$

$$N_{i_2} = \begin{vmatrix} 2.2 & 9.0 \\ -1.2 & -6.0 \end{vmatrix}$$
$$= [2.2 \times (-6.0)] - (-1.2 \times 9.0)$$
$$= -2.4$$

$$i_1 = \frac{N_{i_1}}{D} \qquad i_2 = \frac{N_{i_2}}{D}$$

$$= \frac{8.1}{2.3} \qquad\qquad = \frac{-2.4}{2.3}$$

$$= 3.522 \qquad\qquad = -1.043$$

Therefore, $i_1 = 3.5$ A and $i_2 = -1.0$ A. ◆

EXAMPLE 6   Application: Strain Gauges

A strain gauge measures the stress imposed on a body by some external load. Measurements on a strain gauge are proportional to the stress.

    A pair of crossed strain gauges on a steel member yielded data for the following equations relating the stresses s_x and s_y in the member.

$$s_x - 0.30s_y = 68,100 \qquad -0.30s_x + s_y = 15.90$$

Use Cramer's rule to solve for s_x and s_y.

Solution Using Cramer's rule:

$$s_x = \frac{\begin{vmatrix} 68100 & -0.30 \\ 15.90 & 1 \end{vmatrix}}{\begin{vmatrix} 1 & -0.30 \\ -0.30 & 1 \end{vmatrix}} \qquad s_y = \frac{\begin{vmatrix} 1 & 68100 \\ -0.30 & 15.90 \end{vmatrix}}{\begin{vmatrix} 1 & -0.30 \\ -0.30 & 1 \end{vmatrix}}$$

$$s_x = 75000. \text{ psi} \qquad s_y = 22000. \text{ psi} \qquad \textbf{2 significant digits} \blacklozenge$$

A computer program that can be used to calculate the stress is found on the tutorial disk.

Evaluating Determinants with a Calculator

We defined a determinant as a unique value that is associated with a square array of numbers. A determinant also is the value of a special subset of the rectangular array of numbers called *matrices* (singular matrix). Matrices are discussed in detail in Sections 10.6 and 10.7. However, we need to know the term now because the mode and/or keys that you use to evaluate determinants are labeled MATRIX on your graphics calculator.

fx–7700G

To evaluate a determinant, we must place the calculator in the matrix mode by pressing MODE 0 . The display appears as shown in Figure 10.20. The notation A (2 × 2) indicates that the calculator is prepared to accept a square matrix that has two rows and two columns. Now press F₁ (A), and the display is like that as shown in Figure 10.21. Determinant A appears with the cursor over the element in the first row and first column, ready for us to enter the first element into the determinant. However, before we enter values, there are two instructions on the bottom of the screen that are important at this time: the F3 key, (|A|), and the F6 key, (○). The F3 key gives the value of determinant A. The F6 key provides a new list of instructions at the bottom of the screen. The one that concerns us is F1 (DIM), which makes it possible to change the dimensions of the determinant.

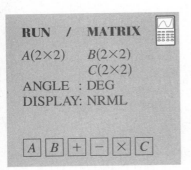

FIGURE 10.20

EXAMPLE 7 **Evaluate a 2 × 2 Determinant with a Calculator**

Evaluate the determinant $A = \begin{vmatrix} 6 & 3 \\ 4 & 5 \end{vmatrix}$.

Solution Place the calculator in the matrix mode by pressing MODE 0 , and then press F1 (A). If the dimensions of A are not 2 × 2, press F6 (○). Then press F1 (DIM) and change the number of rows to 2 and the number of columns to 2. Now enter the elements for the determinant with the following keystrokes: 6 EXE 3 EXE 4 EXE 5 EXE . To determine the value of the determinant, press F3 (|A|), and the calculator will display:

$$\det A =$$

$$18.$$

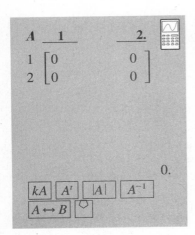

FIGURE 10.21

The value of the determinant is 18. $\blacklozenge$

We now can evaluate other determinants by pressing $\boxed{\text{F1}}$ (*A*) and entering new values. If you press a wrong key and want to go back to a previous screen, press $\boxed{\text{PRE}}$.

TI–82

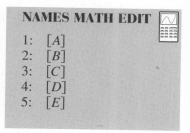

NAMES MATH EDIT

1: [*A*]
2: [*B*]
3: [*C*]
4: [*D*]
5: [*E*]

FIGURE 10.22

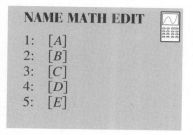

NAME MATH EDIT

1: [*A*]
2: [*B*]
3: [*C*]
4: [*D*]
5: [*E*]

FIGURE 10.23

MATRIX [*A*]   2 × 2

[6   3   ]
[4   5   ]

2,2 = 5

FIGURE 10.24

To evaluate a determinant use the matrix menu. When we press $\boxed{\text{MATRIX}}$, the menu shown in Figure 10.22 appears. To enter elements in a determinant, we must be in the edit mode. To enter the edit mode, press the right arrow key twice, and $\boxed{1:}$ and EDIT are highlighted. Now press $\boxed{1:}$ to select matrix *A*. Figure 10.23. To enter elements in determinant *A*, press $\boxed{\text{ENTER}}$, and a new menu appears as shown in Figure 10.24. The numbers to the right of *A* indicate the dimension of the determinant (in this case, 2 rows and 2 columns). To change the number of rows, place the cursor on the first 2, enter the desired number of rows, and press $\boxed{\text{ENTER}}$. Then enter the desired number of columns and press $\boxed{\text{ENTER}}$. The element in the a_{11} position is highlighted, and at the bottom of the screen, 1, 1 = is displayed. When we enter a 6, that number appears at the bottom of the screen (1,1 = 6). To enter 6 in the a_{11} position of the matrix, press $\boxed{\text{ENTER}}$. Now 6 appears in the a_{11} position. The a_{12} position then is highlighted, and 1,2 = appears at the bottom of the screen. To enter elements in the other three positions, press the following keys: 3 $\boxed{\text{ENTER}}$ 4 $\boxed{\text{ENTER}}$ 5 $\boxed{\text{ENTER}}$. Figure 10.24 shows how the final display appears.

To evaluate the determinant, we must return to the Home screen by pressing $\boxed{\text{2nd}}$ $\boxed{\text{QUIT}}$. Now to determine the value, press $\boxed{\text{MATRIX}}$ $\boxed{\blacktriangleright}$ (right arrow) $\boxed{1}$ (for det) $\boxed{\text{MATRIX}}$ $\boxed{1}$ $\boxed{\text{ENTER}}$, and the result is 18. Thus, the value of the determinant is 18.

EXAMPLE 8   **Cramer's Rule Using a Calculator**

Solve the system of equations applying Cramer's rule and using a calculator.

$$2x + 3y = 7$$
$$x - 3y = 8$$

Solution   The keystrokes that are used for solving the system of equations on an fx–7700G and a TI-82 will be demonstrated. The system has three determinants

$$D = \begin{vmatrix} 2 & 3 \\ 1 & -3 \end{vmatrix} \qquad N_x = \begin{vmatrix} 7 & 3 \\ 8 & -3 \end{vmatrix} \qquad N_y = \begin{vmatrix} 2 & 7 \\ 1 & 8 \end{vmatrix}$$

The keystrokes for each type of calculator are given. Because the previous discussion showed how to enter values into the matrices for each calculator, these steps are not included in this solution. The fx–7700G does not have three independent matrices, so we must leave the matrix MODE to perform division. Therefore, we will store the values of the determinants N_x, N_y, and D in storage *A*, *B*, and *C* respectively.

fx–7700G

Keystrokes	Screen Display	Answer
MODE 0 A	$A \begin{matrix} 1 & 2 \end{matrix}$ $\begin{matrix} 1 \\ 2 \end{matrix} \begin{bmatrix} 7 & 3 \\ 8 & -3 \end{bmatrix}$	
|A|	det A =	−45.00
→ ALPHA A EXE	−45. → A	−45.00
B	$B \begin{matrix} 1 & 2 \end{matrix}$ $\begin{matrix} 1 \\ 2 \end{matrix} \begin{bmatrix} 2 & 7 \\ 1 & 8 \end{bmatrix}$	
|B|	det B	9.00
→ ALPHA B EXE	9. → B	9.00
C C→B	$B \begin{matrix} 1 & 2 \end{matrix}$ $\begin{matrix} 1 \\ 2 \end{matrix} \begin{bmatrix} 2 & 3 \\ 1 & -3 \end{bmatrix}$	
|B|	det B =	−9.00
→ ALPHA C EXE	−9. → C	−9.00
MODE + ALPHA A ÷ ALPHA C		
EXE	A ÷ C	5.00
ALPHA B ÷ ALPHA C EXE		−1.00

Thus, the solution to the system of equations is the ordered pair $(5, -1)$.

TI–82

The TI-82 has three independent matrices. Therefore, we can have $N_x = A$, $N_y = B$ and $D = C$.

Keystrokes	Screen Display	Answer
MATRIX ► ► (right arrow) 1 (A)	MATRIX [A] 2×2 [7 3] [8 −3]	
MATRIX ► ► 2 (B)	MATRIX [B] 2×2 [2 7] [1 8]	
MATRIX ► ► 3 (C)	MATRIX [C] 2×2 [2 3] [1 −3]	
2ⁿᵈ QUIT		
MATRIX ► 1 MATRIX 1 ÷		
MATRIX ► 1 MATRIX 3 ENTER	det [A]/det[B]	5.00
MATRIX ► 1 MATRIX 2 ÷		
MATRIX ► 1 MATRIX 3 ENTER	det[B]/det[C]	−1.00

The solution set of the system of equations is $(5, -1)$. ◆

10.3 Exercises

In Exercises 1–20, use Cramer's rule to solve the systems of linear equations using a graphics calculator.

1. $3x + y = 6$
$-2x + y = 1$

2. $-x + 3y = 3$
$x - 2y = 0$

3. $5x - 2y = -9$
$3x + 7y = 11$

4. $3x + 5y = 16$
$-3x + 2y = 19$

5. $5x + y = -17$
$2x - 7y = 8$

6. $x - 5y = 0$
$x + 5y = -10$

7. $2x + y = 40$
$-2x + y = 0$

8. $2x + 7y = 1$
$2x + 9y = -5$

9. $y = -4x - 2$
$y = -7x + 1$

10. $y = -3x - 3$
$y = -2x - 11$

11. $y = 2x + 1$
$y = \frac{1}{2}x - 4$

12. $y = -\frac{1}{2}x - \frac{1}{2}$
$y = -\frac{1}{4}x - \frac{9}{4}$

13. $3x - y = 2.0$
$2x + y = 5.5$

14. $2x + y = 0.0$
$4x - y = 1.0$

15. $x + 4y = 23.5$
$2x - 12y = 17.0$

16. $2x - 10y = -15.0$
$-x + 4y = 9.0$

17. $1.6x + 2.4y = 3.0$
$8.2x - 1.3y = 5.5$

18. $3.1x + 5.7y = 0.0$
$-6.5x + 4.7y = 3.7$

19. $4.4x - 6.1y = 1.6$
$3.2x + 4.0y = -2.7$

20. $6.2x + 4.8y = -7.7$
$8.3x + 2.6y = -5.1$

In Exercises 21–24, use Cramer's rule to solve the systems of linear equations. From the results, state whether the systems are inconsistent, dependent, or neither.

21. $1.6x + 2.1y = 7.3$
$4.8x + 6.3y = 3.7$

22. $2.0x - 5.0y = 4.0$
$3.0x - 7.5y = 6.0$

23. $2.7x + 3.9y = -1.5$
$0.9x + 1.3y = -0.5$

24. $2.7x - 3.1y = 1.6$
$8.1x - 9.3y = 4.8$

25.   Stresses in an aluminum bar connected to a prestressed steel sleeve are related by:

$$5S_S + 2\frac{1}{2}S_A = 0,$$

$$S_S - 2S_A = 80{,}000 \text{ MPa}.$$

Determine the stresses S_S and S_A in MPa.

26.   Force transfer due to tightening a bolt in a sleeve produces stresses S_B in the bolt and S_S in the sleeve related by:

$$\frac{3\pi}{16}S_S + \frac{\pi}{16}S_B = 0,$$

$$S_B = 3.0S_S + 6600.$$

Determine the stresses S_B and S_S. The stresses are measured in pounds per square inch.

27.   Forces on a beam produce the following equilibrium equations.

$$R_1 + R_2 = 5500$$
$$-R_1 + R_2 = 2310$$

What are the reaction forces R_1 and R_2 (in newtons)?

28. Forces on a beam produce the following equilibrium equations.

$$R_1 + R_2 = 1000$$
$$-5.00R_1 + 3.00R_2 = -2990$$

What are the reaction forces R_1 and R_2 (in newtons)?

29. The loop currents in a circuit are given by:

$$11i_1 - 6.0i_2 = 4.0,$$
$$-6.0i_1 + 8.5i_2 = -3.0.$$

What are the currents i_1 and i_2 (in amperes)?

30. The loop equations for a circuit are:

$$15i_1 - 7.0i_2 = 6.0,$$
$$-7.0i_1 + 17i_2 = -16.$$

What are the currents i_1 and i_2 (in amperes)?

31. Four bolts with one nut on each weigh a total of 3 oz. Six bolts with two nuts on each weigh a total of 5 oz. What are the individual weights of the nuts and bolts?

32. A dozen bolts with single nuts weigh a total of 12 oz. A gross of the same bolts with double nuts weighs 160 oz. What are the individual weights of the nuts and bolts?

33. An asymmetrical parking lot is 100 m wide and has both sides at the same height. The left side slopes down with a drop of 1 m in 50 m. The right side slopes down with a drop of 1 m in 75 m. With respect to the left side, where is the low point for the location of the drain?

34. If the right side of the parking lot in Exercise 33 is $\frac{1}{2}$ m higher than the left side, where should the drain be located?

35. Adam invested $35,000 in certificates of deposit, part at 9% and part at 11%. The total income for the year from the investments is $3450. How much did Adam invest at each rate?

36. A plane traveled 1500 miles with the wind in 4 hours and made the return trip against the wind in 5 hours. Determine the speed of the plane and the speed of the wind.

37. Emily needs to purchase a new air conditioner unit for the office. Model A costs $950 to purchase and $32 per month to operate, while model B, a more efficient unit, costs $1275 to purchase and $22 per month to operate.
   a. Find the number of months it will take for the total cost of both units to be equal.
   b. Determine which model, A or B, will be the most cost effective if the life of both units is guaranteed for 10 years.

38. A wholesaler sells two types of mechanical pencils to stationery stores. It sells type A for $1.50 each and type B for $2.00 each. The wholesaler receives an order for 200 pencils, together with a check for $337.50. In filling out the order, the stationery store clerk failed to specify how many of each type. Is it possible for the wholesaler to fill the order with the given information. If so, what is the order?

39. A car traveling along the highway at a speed of 70 mph, passes a police car traveling at 50 mph. The driver tries to get away, while the police officer speeds up in pursuit. The acceleration of the car is +2.3 miles per hour per second squared. The acceleration of the police car is +3.4 miles per hour per second squared. The general equation for this kind of motion is:

$$v = v_0 + at.$$

The equation representing the motion of the car is:

$$v = 70 + 2.3t.$$

a. Write the equation for the police car.

b. After some time has elapsed, the police car's velocity will be momentarily the same as the speeder's. How fast will the two cars be traveling when they have this common velocity?

10.4

Solving Three Linear Equations in Three Unknowns Algebraically

A group of equations such as the following is a system of three linear equations in three unknowns.

$$x + y + z = -2$$
$$2x - y - z = 5$$
$$-x + 3y + 4z = -11$$

Graphical methods are not practical for the solution of equations in three unknowns because of the complexity of drawing figures in three-dimensional space. However, the algebraic techniques used for solving systems of two equations also can be used to solve systems of three equations in three unknowns. With three equations and three unknowns the answer is not an ordered pair (x, y), but rather an ordered triple (x, y, z). An ordered triple means that there are three answers and that the order in which they are expressed is important.

Substitution Method

EXAMPLE 1 **Substitution Method**

Use the substitution method to solve the following system of three linear equations in three unknowns

$$x + y + z = -2 \tag{1}$$
$$2x - y - z = 5 \tag{2}$$
$$-x + 3y + 4z = -11 \tag{3}$$

Solution First solve any one of the equations for any variable that is convenient. Remember that generally the most convenient variable to solve for is a variable with a coefficient of 1. Choose to solve Equation (3) for x.

$$x = 3y + 4z + 11 \tag{4}$$

Substitute this value of x into Equations (1) and (2).

$$3y + 4z + 11 + y + z = -2 \qquad \text{Substituting for } x \text{ in (1)}$$
$$4y + 5z = -13 \qquad \text{Rearranging} \qquad (5)$$
$$2(3y + 4z + 11) - y - z = 5 \qquad \text{Substituting for } x \text{ in (2)}$$
$$5y + 7z = -17 \qquad \text{Rearranging} \qquad (6)$$

Equations (5) and (6) form a system of two linear equations in two unknowns. Solve this system using the substitution method. If you prefer, of course, you could use the elimination method.

$$y = \frac{-5z - 13}{4} \qquad \text{Solving (5) for } y \qquad (7)$$

$$5\left(\frac{-5z - 13}{4}\right) + 7z = -17 \qquad \text{Substituting for } y \text{ in (6)}$$

$$\frac{-25z - 65}{4} + 7z = -17 \qquad \text{Removing parentheses}$$

$$-25z - 65 + 28z = -68 \qquad \text{Multiplying each term by 4}$$

$$3z = -3$$

$$z = -1$$

$$y = \frac{-5(-1) - 13}{4} \qquad \text{Substituting for } z \text{ in (6)}$$

$$y = -2$$

$$x = 3(-2) + 4(-1) + 11 \qquad \text{Substituting for } y \text{ and } z \text{ in (4)}$$

$$x = 1$$

The solution to the system of three linear equations is the ordered triple $(1, -2, -1)$. We can check the solution by substituting the values for x, y, and z into the three original equations to show that the ordered triple satisfies all three equations. ◆

Elimination Method

EXAMPLE 2 Elimination Method

Use the elimination method to solve the following system of three linear equations in three unknowns.

$$x + y + z = -2 \qquad (1)$$
$$2x - y - z = 5 \qquad (2)$$
$$-x + 3y + 4z = -11. \qquad (3)$$

Solution If we can eliminate one of the variables, then we will have a system of two equations in two unknowns.

Eliminate the x-terms first. (we could just as well begin with either the y- or z-terms.) Since the coefficient of the x-term in Equation (1) is 1 and the coefficient of the x-term in Equation (3) is -1, we can add the two equations to eliminate the x from the resulting equation.

$$
\begin{array}{rl}
x + y + z = -2 & \textbf{(1)} \\
-x + 3y + 4z = -11 & \textbf{(3)} \\
\hline
4y + 5z = -13 & \textbf{Adding} \qquad \textbf{(4)}
\end{array}
$$

Now produce another equation with the x-term eliminated by working with either of the unused combinations, Equations (1) and (2) or Equations (2) and (3). Let us multiply Equation (3) by 2 and add the result to Equation (2).

$$
\begin{array}{rl}
2x - y - z = 5 & \textbf{(2)} \\
-2x + 6y + 8z = -22 & \textbf{2} \times \textbf{(3)} \\
\hline
5y + 7z = -17 & \textbf{Adding} \qquad \textbf{(5)}
\end{array}
$$

Equations (4) and (5) now form a system of two linear equations in two unknowns.

$$
\begin{array}{rl}
4y + 5z = -13 & \textbf{(4)} \\
5y + 7z = -17 & \textbf{(5)} \\
20y + 25z = -65 & \textbf{5} \times \textbf{(4)} \\
20y + 28z = -68 & \textbf{4} \times \textbf{(5)} \\
\hline
-3z = 3 & \textbf{Subtracting} \\
z = -1 & \\
5y + 7(\boxed{-1}) = -17 & \textbf{Substituting for } z \textbf{ in (5)} \\
5y = -10 & \\
y = -2 & \\
-x + 3(\boxed{-2}) + 4(\boxed{-1}) = -11 & \textbf{Substituting for } y \textbf{ and } z \textbf{ in (3)} \\
-x = -1 & \\
x = 1 &
\end{array}
$$

The solution to the system of three linear equations in three unknowns is $x = 1$, $y = -2$, $z = -1$, or the ordered triple $(1, -2, -1)$. The check is left to the reader. ◆

EXAMPLE 3 Application: Chemical Problem

A biochemist makes a 50 liter solution containing 45% acid by using three solutions containing 30%, 40%, and 80% acid. The amount of 30% solution must be twice that of the 80% solution. How many liters of each solution should she use? (Assume three significant digits.)

Solution Let:

$$x = \text{number of liters of 30\% solution to be added;}$$
$$y = \text{number of liters of 40\% solution to be added;}$$
$$z = \text{number of liters of 80\% solution to be addded.}$$

$$x + y + z = 50 \text{ liters} \qquad \textbf{Total amount of solution} \qquad (1)$$

$0.30x$ = number of liters of acid in 30% solution

$0.40y$ = number of liters of acid in 40% solution

$0.80z$ = number of liters of acid in 80% solution

$0.45(50)$ = number of liters of acid in the mixture

$$0.30x + 0.40y + 0.80z = 0.45(50) \qquad \begin{array}{l} \textbf{Sum of the liters of acid in the} \\ \textbf{individual solutions must equal the} \\ \textbf{number of liters of acid in the mixture} \end{array} \qquad (2)$$

$$x = 2z \qquad \begin{array}{l} \textbf{Amount of 30\% solution is twice} \\ \textbf{amount of 80\% solution} \end{array} \qquad (3)$$

There are three equations in three unknowns.

$$x + y + z = 50 \qquad (1)$$

$$0.30x + 0.40y + 0.80z = 22.5 \qquad (2)$$

$$x + 0y - 2z = 0 \qquad (3)$$

Multiply Equation (1) by $-.40$ and add to Equation (2).

$$
\begin{array}{ll}
-0.40x - 0.40y - 0.40z = -20.0 & \textbf{(}\mathbf{-0.40)(1)} \\
\underline{0.30x + 0.40y + 0.80z = 22.5} & \qquad\qquad (2) \\
-0.10x + 0.40z = 2.5 & \qquad\qquad (4)
\end{array}
$$

Multiply Equation (3) by (0.10) and add to equation (4).

$$
\begin{array}{ll}
-0.10x + 0.40z = 2.5 & \qquad\qquad (4) \\
\underline{0.10x - 0.20z = 0} & \textbf{(0.10)(3)} \\
0.20z = 2.5 & \\
z = 12.5 &
\end{array}
$$

Substitute in Equation (3) for z.

$$x = 2(\boxed{12.5})$$

$$= 25.0$$

Substitute in Equation (1) for x and z.

$$\boxed{25} + y + \boxed{12.5} = 50$$

$$y = 12.5$$

The number of liters of 30%, 40%, and 80% acid used to create the 50 liters of 45% acid solution are 25.0, 12.5, and 12.5, respectively. ◆

For practice, try the substitution method to solve the system of equations in Example 3.

CAUTION Writing equations in an orderly fashion helps reduce the number of erros. Place like terms in similar positions and insert zero coefficients for missing terms. In Example 3, examine the manner in which Equations (1), (2), and (3) are displayed as a system. ∎

10.4 Exercises

In Exercises 1–20, use algebraic methods to solve the systems of three linear equations in three unknowns.

1.
$$6x - 7y = 6$$
$$8x + 3y + 5z = 3$$
$$-5x - 3y + 2z = -7$$

2.
$$7x - z = -5$$
$$4x + 2y - 5z = 2$$
$$-2x + y = 0$$

3.
$$7x + 4y = 8$$
$$x + 3z = -5$$
$$2x + y = 0$$

4.
$$-2y - 3z = 7$$
$$8x - 3y - 9z = 7$$
$$-5x + 2y - 8z = 9$$

5.
$$-2x + 5y + 8z = -2$$
$$x + 4z = 4$$
$$-x - y - z = 1$$

6.
$$6x - 6y + 7z = 7$$
$$-5y - z = 4$$
$$-7x + 7y + 8z = 8$$

7.
$$-8x - y - 8z = -1$$
$$-4x - 8y + 3z = 3$$
$$-2x - 3y + z = 8$$

8.
$$-3x - 2y + 6z = 1$$
$$3x + y + 4z = 8$$
$$6x + y - 5z = 2$$

9.
$$-8x + 4y + 5z = -6$$
$$-3x + 9y + z = -8$$
$$-3x - 4y + 7z = -7$$

10.
$$-5x + 4y + 3z = -4$$
$$5x + 3y + 3z = 1$$
$$-3x - 2y - 2z = 4$$

11.
$$2x + y - z = -1$$
$$x - 3y - 3z = 2$$
$$7x - 9y - 4z = -3$$

12.
$$5x + 4y + 3z = 9$$
$$-2x + y + 5z = -2$$
$$2x - y + 6z = -9$$

13.
$$3x + 7y + 8z = -6$$
$$-2x - y - 9z = -7$$
$$-3x + 5y + 6z = -4$$

14.
$$-2x + y + 4x = 2$$
$$5x - y + 8z = -2$$
$$4x - y + 3z = -2$$

15.
$$-2x - 3y - 4z = 3$$
$$5x - 3y + 3z = 3$$
$$5x - 4y - 9z = 4$$

16.
$$2x - 3y - 4z = 3$$
$$4x + y + 2z = -5$$
$$3x - 2y - z = 2$$

17.
$$2x - 2y + z = -2$$
$$-x - 6y - 8z = 1$$
$$6x - 6y + 7z = -6$$

18.
$$-4x - 8y + 9z = -1$$
$$7x + 4y - 8z = 4$$
$$-4x - 3y + 4z = -1$$

19.
$$-8x - 4y + 5z = -6$$
$$-3x + 9y + z = -8$$
$$-3x - 4y + 7z = -7$$

20.
$$7x - 7y - z = -9$$
$$-4x - 9y + z = -7$$
$$-x + 4y - 4z = -4$$

21. The loop equations for a circuit follow.

$$220i_1 - 100i_2 = 300$$
$$-100i_1 + 250i_2 - 150i_3 = 0$$
$$-50i_2 + 70i3 = 0$$

What are the loop currents i_1, i_2, and i_3?

22. The loop equations for a circuit follow.

$$325i_1 - 300i_2 - 200 = 0$$
$$-300i_1 + 410i_2 - 30i_3 = 0$$
$$-30i_2 + 60i_3 = 0$$

What are the loop currents i_1, i_2 and i_3? (Assume two significant digits.)

23. The mode equations for a circuit follow.

$$32.5v_1 - 7.5v_2 - 2500 = 0$$
$$-29v_1 + 29v_2 = 0$$
$$-5.0v_2 + 8.0v_3 = 0$$

What are the node voltages v_1, v_2, and v_3?

24. The mode equations for a circuit follow.

$$v_1 - 20v_2 - 15v_3 = 400$$
$$-7v_1 + 10v_2 = 0$$
$$-3v_1 + 5v_3 = 0$$

What are the voltages v_1, v_2, and v_3?

25. The section forces in a truss are related by the following equations.

$$F_1 + F_2 \cos 30° + F_3 \cos 30° = -500$$
$$F_3 \sin 30° - F_2 \sin 30° + 1000 = 0$$
$$(\tan 30°)(F_2 \cos 30°) + (\tan 30°)(F_3 \cos 30°) - 1000 = 0$$

What are the forces F_1, F_2, and F_3?

26. The section forces in a truss are related by the following equations.

$$500 + F_1 \cos 30° + F_3 = 0$$
$$F_1 \sin 30° + F_2 + 500 = 0$$
$$1500 + (\tan 30°)(F_1 \cos 30°) + F_2 = 0$$

What are the forces F_1, F_2, and F_3?

27.   The table shows the carbohydrates, fat, and protein content of three different foods.

Food Type	Grams per 100 grams		
	Carbohydrate	Fat	Protein
A	5	16	9
B	9	3	1
C	10	2	1

A nutritionist is preparing a diet that has three different foods. The diet must contain 34 g of carbohydrates, 23 g of fat, and 12 g of protein. How many grams of each food type will she include in the diet?

28.   A compound is made up of parts x, y, and z. The proportion of part x to part y is 5 to 1, and the proportion of part y to part z is 3 to 1. What percentage of the total is furnished by each part?

29.   A nut, bolt, and washer assembly has a mass of 0.044 kg. Five nuts and bolts have a mass of 0.200, and ten assemblies of a bolt with double washers and double nuts has a mass of 0.560 kg. What are the masses of the individual nuts, bolts, and washers?

30.   Design the height dimensions for three supports such that S_3 is twice as tall as S_1, the height of S_2 is 3 m more than S_1, and the height of S_3 is 5 m more than S_2.

31.   A chemist has three acid solutions with concentrations of 20%, 32%, and 50%. The chemist needs 100 liters of a 30% solution to create product X. Since there is a greater supply of the 20% solution, the chemist has decided to use three times as much of the 20% solution as the 32% solution. How many liters of each solution will the chemist use?

32.   In the circuit shown in the figure, the first resistor has a current I traversing through it; its resistance is 10 Ω. The other two resistors share the current I; their resistances are 15 Ω and 25 Ω. More current goes through the smaller resistor; therefore, $I_1 > I_2$. Recall that the sum of the smaller currents is the same as I; that is, $I = I_1 + I_2$. With three unknowns, we need two more independent equations. Voltage considerations supply these additional equations.

$$12 = 10I + 15I_1$$
$$0 = 25I_2 + 15I_1$$

Determine I.

33.   A company has budgeted $300,000 for developing a new product. The money is to be divided among salaries, equipment, and overhead. The equipment amount is to be twice the total amount for salaries and overhead. At the same time, the cost of salaries is to be three times the cost of overhead. How much of the budget is allocated for each category amount?

34.   A vice-president in charge of production decides on the following production quotas for the next six weeks: three times as many of model A as of model B; 500 more of model C than of model B. The total output of all three models is to be 960 for the period. How many of each model is to be produced in the next six weeks?

35.   The computer club bought 15 hot dogs, 20 cheeseburgers, and 30 sandwiches for a party. The total bill was $98.25. A cheeseburger costs $0.40 more than a hot dog, and a sandwich costs $0.30 more than a cheeseburger. Determine the price of each.

10.5

Solving Three Linear Equations in Three Unknowns by Determinants

We can use Cramer's rule to solve systems of three linear equations in three unknowns. When the unknowns are arranged in order on the left of the equal sign and the constants on the right, the determinants for the denominator and numerators of the solution are formed in a pattern similar to that for systems of two linear equations (Section 10.3). However, before we can solve a system of three equations in three unknowns, we need to know how to evaluate a 3×3 determinant.

The technique for evaluating a 2×2 determinant cannot be used to evaluate a 3×3 determinant. Instead, we need to use a procedure called **expansion by cofactors,** which reduces the 3×3 determinant to a series of 2×2 determinants. Then we can evaluate the 2×2 determinants by Definition 10.6.

Consider the 3 following 3×3 determinant.

$$A = \begin{vmatrix} a_{11} & a_{12} & a_{13} \\ a_{21} & a_{22} & a_{23} \\ a_{31} & a_{32} & a_{33} \end{vmatrix}$$

To evaluate this determinant, we must be able to determine the minors of the determinant. A **minor** of any determinant is a determinant that has one less row and one less column than the original determinant. Thus, for determinant A, one minor is:

$$M_{11} = \begin{vmatrix} a_{22} & a_{23} \\ a_{32} & a_{33} \end{vmatrix}.$$

A minor is created in relationship to a particular element. The minor M_{11} of the element a_{11} is the determinant formed by the remaining elements when the first row and the first column of determinant are deleted. That is, we eliminate the row and the column in which the element appears.

$$A = \begin{vmatrix} a_{11} & a_{12} & a_{13} \\ a_{21} & a_{22} & a_{23} \\ a_{31} & a_{32} & a_{33} \end{vmatrix} \qquad M_{11} = \begin{vmatrix} a_{22} & a_{23} \\ a_{32} & a_{33} \end{vmatrix}$$

The **cofactor** A_{11} of element a_{11} is the product $(-1)^{1+1} \times M_{11}$.

In general, the minor M_{ij} of element a_{ij} is the determinant formed by the remaining elements when the ith row and the jth column are deleted. The cofactor A_{ij} of element a_{ij} is the product $(-1)^{i+j} \times M_{ij}$.

EXAMPLE 1 **Determining Cofactors and Minors of a Determinant**

Given the 3×3 determinant, determine the following.

$$\begin{vmatrix} 2 & -6 & -2 \\ -5 & -3 & 1 \\ -1 & 4 & 3 \end{vmatrix}$$

a. minor M_{13} **b.** cofactor A_{13} **c.** minor M_{32} **d.** cofactor A_{32}

2	−6	−2
−5	−3	1
−1	4	3

FIGURE 10.25

Solution

a. To determine minor M_{13}, delete the first row and the third column as shown in Figure 10.25. The minor is the determinant of the remaining elements.

$$M_{13} = \begin{vmatrix} -5 & -3 \\ -1 & 4 \end{vmatrix}$$
$$= (-5)(4) - (-1)(-3)$$
$$= \boxed{-23}$$

b. The cofactor, then, is:

$$A_{ij} = (-1)^{i+j} \times M_{ij}$$
$$A_{13} = (-1)^{1+3} \times M_{13}$$
$$= (-1)^4(\boxed{-23})$$
$$= -23$$

c. The minor M_{32} is the determinant of the remaining elements after we delete the third row and the second column as shown in Figure 10.26.

2	−6	−2
−5	−3	1
−	4	3

FIGURE 10.26

$$M_{32} = \begin{vmatrix} 2 & -2 \\ -5 & 1 \end{vmatrix}$$
$$= (2)(1) - (-5)(-2)$$
$$= \boxed{-8}$$

d. The cofactor is:

$$A_{32} = (-1)^{3+2} \times M_{32}$$
$$= (-1)^5(\boxed{-8})$$
$$= 8. \quad \blacklozenge$$

> **DEFINITION 10.8 Minor of a_{ij}**
>
> If A is any $n \times n$ determinant, and M_{ij} is the $(n-1) \times (n-1)$ determinant formed from A by deleting the ith row and the jth column of A, then M_{ij} is called the minor of the element a_{ij} of A.

> **DEFINITION 10.9 Cofactor of a_{ij}**
>
> If A is any $n \times n$ determinant and M_{ij} is the minor of the element a_{ij} of A, then the cofactor of the element a_{ij} of A is
>
> $$A_{ij} = (-1)^{i+j} \times M_{ij}.$$

You can evaluate any determinant by using expansion by cofactors. Simply select any row or column and then multiply each element in that row or column by its own cofactor. The sum of the products is the value of the determinant.

DEFINITION 10.10

To evaluate an $n \times n$ determinant A by expansion by cofactors using the ith row, use the following equation

$$A = a_{i1}A_{i1} + a_{i2}A_{i2} + \cdots + a_{in}A_{in}$$

To evaluate an $n \times n$ determinant A by expansion by cofactors using the ith column, use the following equation

$$A = a_{1j}A_{1j} + a_{2j}A_{2j} + \cdots + a_{nj}A_{nj}$$

EXAMPLE 2   **Evaluating a 3 × 3 Determinant with Cofactors**

Evaluate the following determinant using expansion by cofactors.

$$\begin{vmatrix} 2 & -6 & -2 \\ -5 & -3 & 1 \\ -1 & 4 & 3 \end{vmatrix}$$

a. Expand about the first row.
b. Expand about the second column.

Solution

a. Evaluating the 3×3 determinant $\begin{vmatrix} 2 & -6 & -2 \\ -5 & -3 & 1 \\ -1 & 4 & 3 \end{vmatrix}$ by expanding about the first row, from Definition 10.10, we have:

$$
\begin{aligned}
A &= a_{i1}A_{i1} + a_{i2}A_{i2} + \cdots + a_{in}A_{in},\ (n = 3,\ i = 1) \\
&= a_{11} \cdot A_{11} + a_{12} \cdot A_{12} + a_{13} \cdot A_{13} \\
&= (2)\left[(-1)^{1+1}\begin{vmatrix} -3 & 1 \\ 4 & 3 \end{vmatrix}\right] + (-6)\left[(-1)^{1+2}\begin{vmatrix} -5 & 1 \\ -1 & 3 \end{vmatrix}\right] \\
&\quad + (-2)\left[(-1)^{1+3}\begin{vmatrix} -5 & -3 \\ -1 & 4 \end{vmatrix}\right] \\
&= (2)[(-1)^2(-9 - 4)] + (-6)[(-1)^3(-15 - (-1))] \\
&\quad + (-2)[(-1)^4(-20 - 3)] \\
&= (2)[(1)(-13)] + (-6)[(-1)(-14)] + (-2)[(1)(-23)] \\
&= (2)[-13] + (-6)[14] + (-2)[-23] \\
&= -26 - 84 + 46 \\
&= -64.
\end{aligned}
$$

Therefore, $\begin{vmatrix} 2 & -6 & -2 \\ -5 & -3 & 1 \\ -1 & 4 & 3 \end{vmatrix} = -64.$

b. Expanding about the second column, we have

$$A = a_{ij}A_{ij} + a_{2j}A_{2j} + \cdots + a_{nj}A_{nj}, \ (n = 3, j = 2)$$

$$= a_{12} \cdot A_{12} + a_{22} \cdot A_{22} + a_{32} \cdot A_{32}$$

$$= (-6)\left[(-1)^{1+2}\begin{vmatrix} -5 & 1 \\ -1 & 3 \end{vmatrix}\right] + (-3)\left[(-1)^{2+2}\begin{vmatrix} 2 & -2 \\ -1 & 3 \end{vmatrix}\right]$$

$$+ (4)\left[(-1)^{3+2}\begin{vmatrix} 2 & -2 \\ -5 & 1 \end{vmatrix}\right]$$

$$= (-6)[(-1)(-14)] + (-3)[(1)(4)] + (4)[(-1)(-8)]$$

$$= (-6)[14] + (-3)[4] + (4)[8]$$

$$= -84 - 12 + 32$$

$$= -64. \ \blacklozenge$$

Notice that the value of a given determinant is the same no matter what row or what column we use for the expansion.

As mentioned at the beginning of this section, Cramer's rule can also be used to solve systems of three linear equations in three unknowns.

DEFINITION 10.11 Cramer's Rule

The solution to a system of three linear equations in three unknowns with the equations in standard form

$$a_{11}x + a_{12}y + a_{13}z = c_1$$
$$a_{21}x + a_{22}y + a_{23}z = c_2$$
$$a_{31}x + a_{32}y + a_{33}z = c_3$$

is given by

$$x = \frac{N_x}{D}, \qquad y = \frac{N_y}{D}, \qquad z = \frac{N_z}{D},$$

where

$$D = \begin{vmatrix} a_{11} & a_{12} & a_{13} \\ a_{21} & a_{22} & a_{23} \\ a_{31} & a_{32} & a_{33} \end{vmatrix}, \qquad N_x = \begin{vmatrix} c_1 & a_{12} & a_{13} \\ c_2 & a_{22} & a_{23} \\ c_3 & a_{32} & a_{33} \end{vmatrix},$$

$$N_y = \begin{vmatrix} a_{11} & c_1 & a_{13} \\ a_{21} & c_2 & a_{23} \\ a_{31} & c_3 & a_{33} \end{vmatrix}, \qquad N_z = \begin{vmatrix} a_{11} & a_{12} & c_1 \\ a_{21} & a_{22} & c_2 \\ a_{31} & a_{32} & c_3 \end{vmatrix},$$

and $D \neq 0$.

EXAMPLE 3 **Solving Systems of Equations Using Determinants**

Solve the following system of three linear equations in three unknowns using determinants (Cramer's rule).

$$x + y + z = -2$$
$$2x - y - z = 5$$
$$-x + 3y + 4z = -11$$

Solution We already know the solution, since we solved the system of equations using algebraic methods. The important thing here is to develop an understanding of Cramer's rule. Applying Cramer's rule to the system of three equations in three unknowns, we have $x = \dfrac{N_x}{D}, y = \dfrac{N_y}{D}$, and $z = \dfrac{N_z}{D}$. Begin by setting up the matrix for D and evaluating it.

$$D = \begin{vmatrix} 1 & 1 & 1 \\ 2 & -1 & -1 \\ -1 & 3 & 4 \end{vmatrix} \qquad \textbf{Determinant of the coefficients}$$

Expanding about the first row:

$$D = (1)\left[(-1)^{1+1}\begin{vmatrix} -1 & -1 \\ 3 & 4 \end{vmatrix}\right] + (1)\left[(-1)^{1+2}\begin{vmatrix} 2 & -1 \\ -1 & 4 \end{vmatrix}\right]$$
$$+ (1)\left[(-1)^{1+3}\begin{vmatrix} 2 & -1 \\ -1 & 3 \end{vmatrix}\right]$$
$$= (1)[(1)(-1)] + (1)[(-1)(7)] + (1)[(1)(5)]$$
$$= -1 - 7 + 5$$
$$= -3.$$

$$N_x = \begin{vmatrix} -2 & 1 & 1 \\ 5 & -1 & -1 \\ -11 & 3 & 4 \end{vmatrix} \qquad \textbf{\textit{D}, with the \textit{x}-column replaced by the column of constants}$$

Expanding about the first row:

$$N_x = (-2)\left[(-1)^{1+1}\begin{vmatrix} -1 & -1 \\ 3 & 4 \end{vmatrix}\right] + (1)\left[(-1)^{1+2}\begin{vmatrix} 5 & -1 \\ -11 & 4 \end{vmatrix}\right]$$
$$+ (1)\left[(-1)^{1+3}\begin{vmatrix} 5 & -1 \\ -11 & 3 \end{vmatrix}\right]$$
$$= (-2)[(1)(-1)] + (1)[(-1)(9)] + (1)[(1)(4)]$$
$$= 2 - 9 + 4$$
$$= -3.$$

$$N_y = \begin{vmatrix} 1 & -2 & 1 \\ 2 & 5 & -1 \\ -1 & -11 & 4 \end{vmatrix}$$ **D, with the y-column replaced by the column of constants**

Expanding about the first row:

$$N_y = (1)\left[(-1)^{1+1}\begin{vmatrix} 5 & -1 \\ -11 & 4 \end{vmatrix}\right] + (-2)\left[(-1)^{1+2}\begin{vmatrix} 2 & -1 \\ -1 & 4 \end{vmatrix}\right]$$

$$+ (1)\left[(-1)^{1+3}\begin{vmatrix} 2 & 5 \\ -1 & -11 \end{vmatrix}\right]$$

$$= (1)[(1)(9)] + (-2)[(-1)(7)] + (1)[(1)(-17)]$$

$$= 9 + 14 - 17$$

$$= 6.$$

$$N_z = \begin{vmatrix} 1 & 1 & -2 \\ 2 & -1 & 5 \\ -1 & 3 & -11 \end{vmatrix}$$ **D, with the z-column replaced by the column of constants**

Expanding about the first row:

$$N_z = (1)\left[(-1)^{1+1}\begin{vmatrix} -1 & 5 \\ 3 & -11 \end{vmatrix}\right] + (1)\left[(-1)^{1+2}\begin{vmatrix} 2 & 5 \\ -1 & -11 \end{vmatrix}\right]$$

$$+ (-2)\left[(-1)^{1+3}\begin{vmatrix} 2 & -1 \\ -1 & 3 \end{vmatrix}\right]$$

$$= (1)[(1)(-4)] + (1)[(-1)(-17)] + (-2)[(1)(5)]$$

$$= -4 + 17 - 10$$

$$= 3.$$

Now, substitute the values into the original equations.

$$x = \frac{N_x}{D} \qquad y = \frac{N_y}{D} \qquad z = \frac{N_z}{D}$$

$$= \frac{-3}{-3} \qquad = \frac{6}{-3} \qquad = \frac{3}{-3}$$

$$= 1. \qquad = -2. \qquad = -1.$$

The solution to the system of three linear equations in three unknowns is the ordered triple $(1, -2, -1)$. The check is left to you. ◆

The technique of solving a system of two equations with two unknowns using the fx–7700G and the T1–82 calculators was demonstrated in Section 10.4. We can use these calculators to solve a system of equations with three equations and three unknowns. The technique is the same with the exception that before entering values into the matrix, we must change the dimensions from 2 × 2 to 3 × 3. Use your calculator to solve for x, y, and z in Example 3.

10.5 Exercises

In Exercises 1–12, evaluate the 3×3 determinant using expansion by cofactors.

1. $\begin{vmatrix} -8 & -5 & 5 \\ -6 & -3 & 5 \\ -9 & 5 & 4 \end{vmatrix}$

2. $\begin{vmatrix} 4 & 4 & -3 \\ 0 & -5 & -2 \\ 4 & -2 & -3 \end{vmatrix}$

3. $\begin{vmatrix} -1 & -7 & -3 \\ -1 & 4 & -8 \\ 2 & 1 & -2 \end{vmatrix}$

4. $\begin{vmatrix} -2 & 5 & 6 \\ 1 & 2 & 1 \\ 0 & -2 & -1 \end{vmatrix}$

5. $\begin{vmatrix} 0 & 0 & -1 \\ 2 & 4 & 1 \\ -1 & 6 & 2 \end{vmatrix}$

6. $\begin{vmatrix} 2 & -3 & 0 \\ -1 & -2 & 0 \\ -3 & 4 & 5 \end{vmatrix}$

7. $\begin{vmatrix} 5 & 3 & -2 \\ 9 & -6 & -7 \\ 7 & -9 & 8 \end{vmatrix}$

8. $\begin{vmatrix} 9 & 7 & 0 \\ 5 & -2 & -7 \\ 2 & 6 & -3 \end{vmatrix}$

9. $\begin{vmatrix} -8 & -7 & -2 \\ -5 & -2 & 0 \\ -8 & -3 & 4 \end{vmatrix}$

10. $\begin{vmatrix} -4 & -3 & -1 \\ -7 & 3 & 4 \\ 0 & 8 & -3 \end{vmatrix}$

11. $\begin{vmatrix} -9 & -1 & 1 \\ 2 & 1 & 2 \\ -3 & 8 & 0 \end{vmatrix}$

12. $\begin{vmatrix} 2 & -3 & -2 \\ -8 & 0 & -5 \\ -1 & -7 & 6 \end{vmatrix}$

In Exercises 13–33, use Cramer's rule and your calculator to solve the systems of equations.

13. $7x + 4y = 8$
$x + 3z = -5$
$2x + y = 0$

14. $-2y - 3x = 7$
$8x - 3y - 9z = 7$
$-5x + 2y - 8z = 9$

15. $6x - 7y = 6$
$8x + 3y + 5z = 3$
$-5x - 3y + 2z = -7$

16. $7x - z = -5$
$4x + 2y - 5z = 2$
$-2x + y = 0$

17. $-2x + 5y + 8z = -2$
$x + 4z = 4$
$-x - y - z = 1$

18. $6x - 6y + 7z = 7$
$-5y - z = 4$
$-7x + 7y + 8z = 8$

19. $-8x - y - 8z = -1$
$-4x - 8y + 3z = 3$
$-2x - 3y + z = 8$

20. $-3x - 2y + 6z = 1$
$3x + y + 4z = 8$
$6x + y - 5z = 2$

21. $-8x + 4y + 5z = -6$
$-3x + 9y + z = -8$
$-3x - 4y + 7z = -7$

22. $-5x + 4y + 3z = -4$
$5x + 3y + 3z = 1$
$-3x - 2y - 2z = 4$

23. $2x + y - z = -1$
$x - 3y - 3z = 2$
$7x - 9y - 4z = -3$

24. $5x + 4y + 3z = 9$
$-2x + y + 5z = -2$
$2x - y + 6z = -9$

25. $3x + 7y + 8z = -6$
$-2x - y - 9z = -7$
$-3x + 5y + 6z = -4$

26. $-2x + y + 4z = 2$
$5x - y + 8z = -2$
$4x - y + 3z = -2$

27. A network of guy wires as shown in the figure, is used to support the 50 kg mass unitl it can finally be anchored in position. The forces in the wires are described by the system of equations.

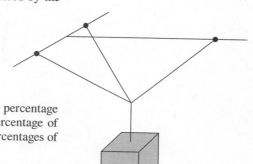

$$0.392T_1 + 0.314T_2 + 0.500T_3 - 491. = 0$$
$$0.235T_1 - 0.628T_2 = 0$$
$$-0.889T_1 - 0.712T_2 + 0.866T_3 = 0$$

Determine the forces in the wires.

28. A chemical compound is composed of three elements, A, B, and C. The percentage of element A present is triple the percentage of element C, and the percentage of element B present is twice the percentage of element A. What are the percentages of the individual elements?

29. A mixture is made up of parts x, y, and z. The proportion of part x to part y is 4 to 1, and the proportion of part y to part z is 5 to 1. What percentage of the total is furnished by each part?

30. An assembly using three different sizes of bolts, B_1, B_2, and B_3 is capable of supporting 120,000 N. The assembly may be made up of combinations of $2B_1$, $3B_2$, and $3B_3$ or of $3B_1$, $2B_2$, and $2B_3$ or of $2B_1$, $2B_2$, and $5B_3$. What are the load ratings of the individual bolts?

31. A medical treatment requires a total dosage of 16 ml of medicine. It can be given as one massive dose followed by ten small doses or as one massive dose followed by three intermediate doses and four small doses or as five intermediate doses followed by six small doses. What size are the doses?

32. A 15-m length of 0.1-m-×-0.1-m steel angle is to be cut into three lengths. The first length is 1 m longer than the second length and is 2 m shorter than twice the third length. What are the individual lengths?

33. A car with an adult passenger and a full fuel tank weighs 2703 lb. The same car with the same adult passenger and a half tank of fuel weighs 2668 lb. The car with the adult, one child, and a fifth of a tank of fuel weighs 2690 lb. The child weighs one-quarter as much as the adult.
   a. What is the weight of the car?
   b. What is the weight of the gasoline when the tank is full?
   c. What is the weight of the adult?
   d. What is the weight of the child?

10.6

Matrix Operations

When we work with fluid-flow problems, electrical-network problems, or structural-truss problems, we often need to solve systems of simultaneous equations that are larger than 3 × 3. In the preceding sections, when we increase the number of simultaneous equations from two to three, the number of steps required for the solution increases dramatically. A more efficient method for solving larger systems of equations is using matrices. This section contains an introduction to some elementary matrix operations. The next section introduces the technique of solving a system of equations using the inverse matrix.

A **matrix** (plural: matrices) is a rectangular array of numbers (called elements) arranged in rows and columns. It may consist of a single element, a single row, a single column, equal numbers of rows and columns, or unequal numbers of rows and columns. An $m \times n$ (**or a matrix of order** $m \times n$) **has m rows and n columns.** Table 10.3 contains examples of matrices.

As in determinants, an element of a matrix may be denoted using subscript notation a_{ij}, where i represents the row in which the element is located and j represents the column. The element a_{ij} lies at the intersection of the ith row and the jth column. A square matrix and the determinant consisting of the same array of elements are not the same. **The determinant assigns a single value to the array, whereas the matrix is the array.**

TABLE 10.3

Matrix	Dimension	Special Name
(27)	1×1	Single element
$(x \quad y \quad z)$	1×3	Row matrix (or row vector)
$\begin{pmatrix} 5 \\ -3 \\ 0 \end{pmatrix}$	3×1	Column matrix (or column vector)
$\begin{pmatrix} 1 & -6 & 3 \\ -2 & 1 & 4 \\ 7 & 5 & 8 \end{pmatrix}$	3×3	Square matrix
$\begin{pmatrix} 1 & a & -7 \\ b & d & 5 \end{pmatrix}$	2×3	

For two matrices to be equal they must be of the same dimension and each element of one must equal the corresponding element of the other.

EXAMPLE 1 **Equal Matrices**

Are the following matrices equal?

a. $A = (7 \quad 9), B = \begin{pmatrix} 7 \\ 9 \end{pmatrix}$

b. $A = \begin{pmatrix} 4 & -6 & 3 \\ 2 & 1 & -5 \end{pmatrix}, B = \begin{pmatrix} 4 & -6 & 3 \\ 2 & -1 & -5 \end{pmatrix}$

c. $A = \begin{pmatrix} 1 & 2 \\ a & 3 \\ 4 & b \end{pmatrix}, B = \begin{pmatrix} c & 2 \\ 5 & 3 \\ d & 6 \end{pmatrix}$

Solution

a. $(7 \quad 9) \neq \begin{pmatrix} 7 \\ 9 \end{pmatrix}$

The matrices are not of the same dimension. A is a 1×2 matrix, while B is a 2×1 matrix.

b. $\begin{pmatrix} 4 & -6 & 3 \\ 2 & 1 & -5 \end{pmatrix} \neq \begin{pmatrix} 4 & -6 & 3 \\ 2 & -1 & -5 \end{pmatrix}$

The matrices are both of dimension 2×3, but the elements of A are not equal to the corresponding elements of B ($a_{22} \neq b_{22}$).

c. $\begin{pmatrix} 1 & 2 \\ a & 3 \\ 4 & b \end{pmatrix} \overset{?}{=} \begin{pmatrix} c & 2 \\ 5 & 3 \\ d & 6 \end{pmatrix}$

Matrix A will be equal to matrix B if $a = 5$ and $b = 6$ and $c = 1$ and $d = 4$. If all of these conditions are not met, the matrices are not equal. ◆

To add two matrices A and B, they must be of the same dimension. If they are, then the sum of matrix C is also of that dimension, and each element of C is the sum of the corresponding elements of A and B. In other words, $c_{ij} = a_{ij} + b_{ij}$ for all possible combinations of $1 \leq i \leq m$ and $1 \leq j \leq n$.

EXAMPLE 2   The Sum of Matrices

If possible, add the following matrices.

a. $\begin{pmatrix} 1 & 4 \\ 5 & -3 \end{pmatrix} + \begin{pmatrix} 2 \\ -1 \end{pmatrix}$ **b.** $\begin{pmatrix} 1 & -5 \\ 2 & 3 \\ -1 & 6 \end{pmatrix} + \begin{pmatrix} 2 & 1 \\ -5 & 0 \\ 7 & 3 \end{pmatrix}$

Solution

a. $\begin{pmatrix} 1 & 4 \\ 5 & -3 \end{pmatrix} + \begin{pmatrix} 2 \\ 1 \end{pmatrix}$

These cannot be added. The matrices are not of the same dimension.

b. $\begin{pmatrix} 1 & -5 \\ 2 & 3 \\ -1 & 6 \end{pmatrix} + \begin{pmatrix} 2 & 1 \\ -5 & 0 \\ 7 & 3 \end{pmatrix} = \begin{pmatrix} 1+2 & -5+1 \\ 2+(-5) & 3+0 \\ -1+7 & 6+3 \end{pmatrix}$

$$= \begin{pmatrix} 3 & -4 \\ -3 & 3 \\ 6 & 9 \end{pmatrix}. \quad \blacklozenge$$

The addition of matrices can be performed with a calculator as illustrated in the box in the margin.

The negative of a matrix is formed by multiplying each element of the matrix by the scalar -1. A **scalar** is a real number. Matrices are subtracted by adding the negative matrix. A matrix may be multiplied by a scalar by multiplying every element of the matrix by the scalar.

EXAMPLE 3   A Scalar Times a Matrix and Difference of Matrices

Let $a = \begin{pmatrix} 1 & 0 \\ 4 & 2 \end{pmatrix}$ and $B = \begin{pmatrix} -1 & 3 \\ 5 & 6 \end{pmatrix}$

Determine the following.

a. $2A$ **b.** $3A - 4B$

Solution

a. To determine $2A$, multiply every element of A by 2.

$$2A = \begin{pmatrix} 2(1) & 2(0) \\ 2(4) & 2(2) \end{pmatrix} \quad \text{or} \quad \begin{pmatrix} 2 & 0 \\ 8 & 4 \end{pmatrix}$$

b. To determine $3A - 4B$, use the definition of subtraction and write $3A - 4B$ as $3A + (-4B)$.

$$3A + (-4B) = \begin{pmatrix} 3(1) & 3(0) \\ 3(4) & 3(2) \end{pmatrix} + \begin{pmatrix} -4(-1) & -4(3) \\ -4(5) & -4(6) \end{pmatrix}$$

Keystrokes for
Example 3b with
TI–82

Place calculator in matrix Edit mode.
| MATRIX | ▷ | ▷ | 1 |

Enter DIM of matrix A.
2 | ENTER | 2 | ENTER |

Enter elements of matrix A.
1 | ENTER | 0 | ENTER |
4 | ENTER | 2 | ENTER |

Enter DIM of Matrix B.
| MATRIX | ▷ | ▷ | 2 | 2
| ENTER | 2 | ENTER |

Enter elements of matrix B.
| (−) | 1 | ENTER | 3
| ENTER | 5 | ENTER | 6
| ENTER |

Determining $3A − 4B$.
| 2ⁿᵈ | QUIT | 3 | × |
| MATRIX | 1 | (A) | − | 4
| × | MATRIX | 2 | (B)
| ENTER |

$$[[7 \qquad -12]$$
$$[-8 \qquad -18]]$$

$$= \begin{pmatrix} 3 & 0 \\ 12 & 6 \end{pmatrix} + \begin{pmatrix} 4 & -12 \\ -20 & -24 \end{pmatrix}$$

$$= \begin{pmatrix} 3+4 & 0-12 \\ 12-20 & 6-24 \end{pmatrix}$$

$$= \begin{pmatrix} 7 & -12 \\ -8 & -18 \end{pmatrix}. \quad \blacklozenge$$

Scalar multiplication and addition or subtraction of matrices can be performed with a calculator as illustrated in the box in the margin.

In order to multiply two matrices, the number of columns of the first must be the same as the number of rows of the second. The result of multiplying two matrices is a matrix that has the number of rows of the first and the number of columns of the second. For example, if A is a 3×4 matrix and B is a 4×2 matrix, we can write symbolically:

$$3 \times 4 \quad 4 \times 2.$$

The upper bracket indicates that A has four columns and B has four rows, so it is possible to multiply A times B. The lower bracket indicates that the product AB will be a 3×2 matrix.

In general, we can multiply two matrices of dimension $m \times n$ and $n \times p$, respectively. (The number of columns of the first, n, is equal to the number of rows of the second.) The product matrix will be of dimension $m \times p$.

EXAMPLE 4   The Product of Matrices

Given the row vector (matrix) $A = (1 \quad 2 \quad 3)$ and the column vector (matrix) $B = \begin{pmatrix} 4 \\ 5 \\ 6 \end{pmatrix}$ determine the following.

a. AB **b.** BA.

Solution

a. $AB = (1 \quad 2 \quad 3) \begin{pmatrix} 4 \\ 5 \\ 6 \end{pmatrix}$,

$$1 \times 3 \quad 3 \times 1 \qquad \text{Dimension of matrices}$$

The dimensions of the matrices tell us that we can multiply them and the product will be a 1×1 matrix.

$$(1 \quad 2 \quad 3)\begin{pmatrix} 4 \\ 5 \\ 6 \end{pmatrix} = (1(4) + 2(5) + 3(6))$$

$$= (4 + 10 + 18)$$

$$= (32).$$

Therefore, $AB = (32)$.

b. $BA = \begin{pmatrix} 4 \\ 5 \\ 6 \end{pmatrix}(1 \quad 2 \quad 3)$

$$\overbrace{3 \times 1 \quad 1 \times 3}^{} \quad \textbf{Dimension of matrices}$$

The dimensions of the matrices tell that we can multiply them and that the product will be a 3×3 matrix.

$$\begin{pmatrix} 4 \\ 5 \\ 6 \end{pmatrix}(1 \quad 2 \quad 3) = \begin{pmatrix} 4(1) & 4(2) & 4(3) \\ 5(1) & 5(2) & 5(3) \\ 6(1) & 6(2) & 6(3) \end{pmatrix}$$

$$= \begin{pmatrix} 4 & 8 & 12 \\ 5 & 10 & 15 \\ 6 & 12 & 18 \end{pmatrix}.$$

Therefore, $BA = \begin{pmatrix} 4 & 8 & 12 \\ 5 & 10 & 15 \\ 6 & 12 & 18 \end{pmatrix}.$ ◆

CAUTION In general, matrix multiplication is not commutative; that is, $AB \neq BA$. Example 4 illustrates this. ■

EXAMPLE 5 **Multiplying Two Matrices**

Determine the product of the following. $\begin{pmatrix} 1 & 3 \\ 2 & -1 \\ 0 & 9 \end{pmatrix}\begin{pmatrix} 4 & 7 & -3 & -2 \\ 5 & -1 & 0 & 8 \end{pmatrix}$

Solution Since the matrices are of dimension 3×2 and 2×4, we can multiply them. Their product will be a 3×4 matrix.

$$\begin{pmatrix} 1 & 3 \\ 2 & -1 \\ 0 & 9 \end{pmatrix}\begin{pmatrix} 4 & 7 & -3 & -2 \\ 5 & -1 & 0 & 8 \end{pmatrix}$$

$$= \begin{pmatrix} 1(4) + 3(5) & 1(7) + 3(-1) & 1(-3) + 3(0) & 1(-2) + 3(8) \\ 2(4) + (-1)(5) & 2(7) + -1(-1) & 2(-3) + (-1)(0) & 2(-2) + -1(8) \\ 0(4) + 9(5) & 0(7) + 9(-1) & 0(-3) + 9(0) & 0(-2) + 9(8) \end{pmatrix}$$

$$= \begin{pmatrix} 19 & 4 & -3 & 22 \\ 3 & 15 & -6 & -12 \\ 45 & -9 & 0 & 72 \end{pmatrix}.$$

Note that the element in the first-row, first-column position of the result is formed by multiplying the first row of the first matrix term by term times the first column of the second matrix and adding the results. The second-row, third-column element is formed by multiplying the second row of the first matrix times the third column of the second matrix. Each element of the product is formed in this way. Therefore,

$$\begin{pmatrix} 1 & 3 \\ 2 & -1 \\ 0 & 9 \end{pmatrix} \begin{pmatrix} 4 & 7 & -3 & -2 \\ 5 & -1 & 0 & 8 \end{pmatrix} = \begin{pmatrix} 19 & 4 & -3 & 22 \\ 3 & 15 & -6 & -12 \\ 45 & -9 & 0 & 72 \end{pmatrix}.$$

Note that it is impossible to multiply these matrices in the reverse order, since the number of columns in B is not equal to the number of rows in A. ($B(2 \times 4)$ and $A(3 \times 2)$, $4 \neq 3$). ◆

CAUTION In performing operations with matrices, the number of rows and number of columns is important. To be able to add or subtract two matrices, both matrices must have the same number of rows and the same number of columns. The sum (difference) matrix has the same number of rows and columns as the matrices being added (subtracted). That is, $(2 \times 3) + (2 \times 3) = (2 \times 3)$.

To be able to multiply two matrices, the matrix on the left must have the same number of columns as the matrix on the right has rows. The product matrix has the same number of rows as the matrix on the left and the same number of columns as the matrix on the right. That is, $(3 \times 2)(2 \times 3) = (3 \times 3)$. ■

EXAMPLE 6 **Application: Product of Matrices—Pricing Grass Mixtures**

A seed company has distribution centers in the New England and Mid-Atlantic states. Matrix A represents the number of ounces of each type of grass seed that are mixed to be sold under brand names J, R, and P. Matrix B represents the price per ounce of each type of seed at each distribution center. Determine the price of each mixture of seed at each distribution center using your calculator.

	Rye Grass	Fescue	Blue Grass
Brand J	3	3	10
$A =$ Brand R	6	3	7
Brand P	0	7	9

		New England	Mid-Atlantic
	Rye Grass	.07	.06
$B =$	Fescue	.18	.19
	Blue Grass	.26	.27

Solution   To determine the price of each brand of grass seed in each region of the country, determine the product of matrix A and B. To multiply the matrices, use the following keystrokes on the calculators.

fx–7700G	TI–82
Place calculator into MATRIX mode.	Place calculator into MATRIX mode.
MODE 0	MATRIX ▷ ▷ 1
Enter DIM of matrix A.	Enter DIM of matrix A.
F1 (A) F6 (◌) F1 (DIM) 3	3 ENTER 3 ENTER
EXE 3 EXE	
Enter elements of matrix A.	Enter elements of matrix A.
3 EXE 3 EXE 10 EXE	3 ENTER 3 ENTER 10 ENTER
6 EXE 3 EXE 7 EXE	6 ENTER 3 ENTER 7 ENTER 0
0 EXE 7 EXE 9 EXE	ENTER 7 ENTER 9 ENTER
Enter DIM of matrix B.	Enter DIM of matrix B.
PRE F2 (B) F6 (◌) F1	MATRIX ▷ ▷ 2 3 ENTER 2
(DIM) 3 EXE 2 EXE	ENTER
Enter elements of matrix B.	Enter elements of matrix B.
.07 EXE .06 EXE .18 EXE	.07 ENTER .06 ENTER .18
ENTER	
.19 EXE .26 EXE .27 EXE	.19 ENTER .26 ENTER
.27 ENTER	
Determining $A \times B$, $A \times B = C$	Determining $A \times B$.
PRE F5 (X)	2nd QUIT MATRIX 1
C 1. 2.	(A) × MATRIX 2 (B) ENTER
1 ⎛ 3.35 3.45 ⎞	$[A] * [B]$
2 ⎜ 2.78 2.82 ⎟	[[3.35 3.45]
3 ⎝ 3.60 3.76 ⎠	[2.78 2.82]
	[3.60 3.76]]

The product matrix is:

	New England	Mid-Atlantic
Brand J	$3.35	$3.45
$C =$ Brand R	$2.78	$2.82
Brand P	$3.60	$3.76

The product matrix tells us the price of 16 ounces of a particular brand of grass seed at a distribution center. For example, the price of Brand J in the New England states distribution center is $3.35 for 16 ounces. ◆

10.6 Exercises

In Exercises 1–4, determine the values of the variables for which the matrices will be equal.

1. $\begin{pmatrix} 1 & 2 \\ 0 & -1 \end{pmatrix} = \begin{pmatrix} a & b \\ d & c \end{pmatrix}$

2. $\begin{pmatrix} 1 & -6 & 3 \\ 2 & 0 & -2 \end{pmatrix} = \begin{pmatrix} a & d & f \\ c & b & e \end{pmatrix}$

3. $\begin{pmatrix} 1 \\ a \\ 4 \end{pmatrix} = \begin{pmatrix} a \\ x \\ 4 \end{pmatrix}$

4. $(5 \quad 1 \quad b) = (b \quad c \quad b)$

In Exercises 5–12, state the dimension of the indicated matrix.

$$A = (1 \quad 0 \quad -3), B = (4 \quad -1 \quad 2), C = \begin{pmatrix} 1 & 2 \\ 4 & 3 \end{pmatrix},$$

$$D = \begin{pmatrix} -3 & 1 \\ 6 & 5 \end{pmatrix}, E = \begin{pmatrix} -2 & 6 & 3 \\ 1 & 0 & -1 \\ 5 & 8 & 4 \end{pmatrix}.$$

$$F = \begin{pmatrix} 9 & 0 & -7 \\ 2 & -2 & 0 \\ 1 & 6 & 5 \end{pmatrix}, G = \begin{pmatrix} 2 & 6 \\ 1 & 3 \\ -1 & 5 \end{pmatrix},$$

$$H = \begin{pmatrix} 0 & -6 \\ -1 & 0 \\ 3 & 8 \end{pmatrix}$$

5. A

6. B

7. D

8. C

9. E

10. F

11. H

12. G

In Exercises 13–20, perform the operations using the matrices given in Exercises 5–12.

13. $E + F$

14. $G + H$

15. $H - G$

16. $F - E$

17. $5A + 2B$

18. $3A - 2B$

19. $2D - 3C$

20. $4C + 2D$

In Exercises 21–30, **a.** state the dimension of each matrix, **b.** multiply the matrices in the order given (if possible), and **c.** multiply the matrices in the opposite order (if possible).

21. $\begin{pmatrix} 2 & 4 \\ 1 & 3 \end{pmatrix} \begin{pmatrix} 5 & 6 \\ 0 & 9 \end{pmatrix}$
   22. $\begin{pmatrix} -4 & 6 \\ 2 & -1 \end{pmatrix} \begin{pmatrix} 0 & 3 \\ 5 & 7 \end{pmatrix}$
   23. $(1 \quad 0 \quad 3) \begin{pmatrix} 3 & 2 \\ 1 & 4 \\ 5 & 7 \end{pmatrix}$
   24. $\begin{pmatrix} 1 & 0 & 3 \\ 3 & 1 & 5 \end{pmatrix} \begin{pmatrix} 2 \\ 4 \\ 7 \end{pmatrix}$

25. $\begin{pmatrix} 2 & 8 & 7 \\ 6 & 2 & 1 \end{pmatrix} \begin{pmatrix} -1 & 3 \\ 5 & 4 \\ 9 & 0 \end{pmatrix}$
   26. $\begin{pmatrix} 2 & -3 \\ 6 & 1 \\ -4 & 5 \end{pmatrix} \begin{pmatrix} 8 & 0 & 3 \\ -1 & 5 & 9 \end{pmatrix}$
   27. $(3 \quad 0 \quad 8) \begin{pmatrix} 1 & 2 & 3 \\ -5 & 0 & 4 \\ 2 & 6 & -2 \end{pmatrix}$

28. $\begin{pmatrix} 2 & -4 & 5 \\ 3 & 6 & -2 \\ -5 & 1 & 1 \end{pmatrix} \begin{pmatrix} 9 \\ 2 \\ -1 \end{pmatrix}$
   29. $\begin{pmatrix} 1 & 2 & 3 \\ -2 & 1 & 0 \\ 5 & -4 & 6 \end{pmatrix} \begin{pmatrix} x \\ y \\ z \end{pmatrix}$
   30. $\begin{pmatrix} 3 & 4 & -2 \\ 1 & -3 & 5 \\ 2 & -1 & 6 \end{pmatrix} \begin{pmatrix} a \\ b \\ c \end{pmatrix}$

31. The food and drink orders for three student groups on a college campus are summarized in matrix A.

	Pizza	Soda	Salad
Drama Club	3	8	4
A = Dance Band	5	12	15
Rugby Team	7	28	5

The price (in dollars) of pizza, soda, and salad at three local pizza parlors are summarized in matrix B.

	Tony's	Elaine's	Ruth's
Pizza	7.95	8.35	8.15
B = Soda	.60	.65	.75
Salad	1.35	1.25	1.15

   a. Multiply the two matrices $(A \times B)$ to form a 3×3 matrix that shows the amount each student group would be charged for its order at each pizza parlor.
   b. Decide at which pizza parlor each student group will get the best price for their order.

32. A garment manufacturer makes a particular shirt in three sizes: small (S), medium (M), and large (L). The shirts require fabric (F), buttons (B), trimming (T), and labor (L). They are sold in twelve-shirt lots. The number of units of each item required to make twelve shirts is given in the matrix A. The cost in dollars of a unit is given in matrix B.

	F	B	T	L
S	7	36	6	10
A = M	10	48	8	11
L	20	60	10	15

	Cost
F	4.00
B	0.11
B = T	0.35
L	5.00

   a. Calculate $A \times B$.
   b. What information is displayed in the product matrix $A \times B$?
   c. What is the production cost of the following?
      i. Twelve small shirts
      ii. Three dozen large shirts

10.7

The Inverse Matrix: Simultaneous Equations

In this section we use matrix methods to solve systems of linear equations. Begin by writing this system of equations in matrix form.

$$2x - y - 2z = -6$$
$$4x + 2y + z = 1$$
$$2x + y - z = -1$$

The system of equations can be written as a matrix equation $AX = C$, where:

$$A = \begin{pmatrix} 2 & -1 & -2 \\ 4 & 2 & 1 \\ 2 & 1 & -1 \end{pmatrix}, \quad X = \begin{pmatrix} x \\ y \\ z \end{pmatrix}, \quad \text{and} \quad C = \begin{pmatrix} -6 \\ 1 \\ -1 \end{pmatrix}.$$

Here matrix A represents the coefficients of the variables: X is a matrix of just the variables; the matrix C represents the constants. Substituting, we have:

$$\begin{pmatrix} 2 & -1 & -2 \\ 4 & 2 & 1 \\ 2 & 1 & -1 \end{pmatrix} \begin{pmatrix} x \\ y \\ z \end{pmatrix} = \begin{pmatrix} -6 \\ 1 \\ -1 \end{pmatrix}.$$

(At this time multiply A and X to verify that $AX = C$.)

If it were possible to divide both sides of the equation by A, we would obtain the matrix of unknowns X (equal to a matrix of constants). Unfortunately, matrix division is not defined. Instead, we have to use an alternate approach. *First we define the* **multiplicative identity matrix I** *as a square matrix having ones as the elements on the principal diagonal (from upper left to lower right) and zeros as all the remaining elements.* Thus, for the set of 3×3 matrices the identity matrix I is:

$$I = \begin{pmatrix} 1 & 0 & 0 \\ 0 & 1 & 0 \\ 0 & 0 & 1 \end{pmatrix}$$

We can demonstrate that for any 3×3 matrix A, $AI = A$ and $IA = A$. If a system of simultaneous equations has a solution, we can find an inverse matrix A^{-1}, which has the property $A^{-1}A = I$ and $AA^{-1} = I$. For this to happen A and A^{-1} must be square matrices of the same order. We can use the inverse matrix A^{-1} to solve the matrix equation that represents the system of simultaneous equations.

$$AX = C$$
$$A^{-1}AX = A^{-1}C \qquad \textbf{Premultiply by the inverse of } A$$
$$IX = A^{-1}C \qquad \textbf{Recall } A^{-1}A = I$$
$$X = A^{-1}C \qquad \textbf{Recall } IX = X$$

For a system of n simultaneous equations, A is of order $n \times n$; X is of order $n \times 1$; C is of order $n \times 1$. The inverse of A, A^{-1} must also be of order $n \times n$. Therefore, the product $A^{-1}C$ must exist, and it is an $n \times 1$ matrix of constants. This, then, must correspond element by element with the $n \times 1$ matrix of unknowns X, and the system of equations is solved.

There are several methods for finding the inverse matrix. This text adopts the Gauss elimination method because it requires the fewest number of operations and because it is readily adaptable to repetitive operations.

To perform the Gauss elimination method of matrix inversion, write the matrix to be inverted followed by an identity matrix of the some order. For the matrix

$$A = \begin{pmatrix} 2 & -1 & -2 \\ 4 & 2 & 1 \\ 2 & -1 & -1 \end{pmatrix}, \quad \text{write} \quad \left(\begin{array}{ccc|ccc} 2 & -1 & -2 & 1 & 0 & 0 \\ 4 & 2 & 1 & 0 & 1 & 0 \\ 2 & 1 & -1 & 0 & 0 & 1 \end{array} \right).$$

The objective is to obtain the inverse of matrix A. If A has an inverse, we can determine the inverse by performing the same operations on the rows of both matrices A and I. As a result of the operations, matrix A is changed to the identity matrix, and the identity matrix is changed to A^{-1}.

OPERATIONS THAT CAN BE PERFORMED ON ROWS OF MATRICES

1. Interchange any two rows.
2. Multiply any row by a non-zero constant.
3. Add a multiple of one row to any other row.

To change A to the identity matrix, we must obtain a 1 in the a_{11} position and then get 0s in all the other rows of the first column. Next, we want to obtain a 1 in the a_{22} position and get 0s in all the other rows of the second column. We continue this process until there are 1s on the principal diagonal and 0s in every other position.

To find the inverse of the example, first multiply the first row by $\frac{1}{2}$, which is the multiplicative inverse of the a_{11} element, 2. Doing so produces a 1 in the a_{11} position.

$$\begin{pmatrix} 2 & -1 & -2 & \bigm| & 1 & 0 & 0 \\ 4 & 2 & 1 & \bigm| & 0 & 1 & 0 \\ 2 & 1 & -1 & \bigm| & 0 & 0 & 1 \end{pmatrix} \quad \textbf{Given}$$

$$\begin{pmatrix} \frac{1}{2}(2) & \frac{1}{2}(-1) & \frac{1}{2}(-2) & \bigm| & \frac{1}{2}(1) & \frac{1}{2}(0) & \frac{1}{2}(0) \\ 4 & 2 & 1 & \bigm| & 0 & 1 & 0 \\ 2 & 1 & -1 & \bigm| & 0 & 0 & 1 \end{pmatrix}$$

$$\begin{pmatrix} 1 & -\frac{1}{2} & -1 & \bigm| & \frac{1}{2} & 0 & 0 \\ 4 & 2 & 1 & \bigm| & 0 & 1 & 0 \\ 2 & 1 & -1 & \bigm| & 0 & 0 & 1 \end{pmatrix}$$

To obtain a 0 in the a_{21} position, replace row 2 with the sum of itself and -4 times row 1. (-4 is the additive inverse of the element in the second row of column 1. Leave row 1 unchanged while doing this.)

$$\begin{pmatrix} 1 & -\frac{1}{2} & -1 & \bigm| & \frac{1}{2} & 0 & 0 \\ 4+(-4)(1) & 2+(-4)\left(-\frac{1}{2}\right) & 1+(-4)(-1) & \bigm| & 0+(-4)\left(\frac{1}{2}\right) & 1+(-4)(0) & 0+(-4)(0) \\ 2 & 1 & -1 & \bigm| & 0 & 0 & 1 \end{pmatrix}$$

$$\begin{pmatrix} 1 & -\frac{1}{2} & -1 & \bigm| & \frac{1}{2} & 0 & 0 \\ 0 & 4 & 5 & \bigm| & -2 & 1 & 0 \\ 2 & 1 & -1 & \bigm| & 0 & 0 & 1 \end{pmatrix}$$

Next, to obtain a 0 in the a_{31} position, replace row 3 with the sum of itself and -2 times row 1. (-2 is the additive inverse of the element in the third row of column 1. Leave row 1 unchanged while doing this.)

$$\begin{pmatrix} 1 & -\frac{1}{2} & -1 & \bigm| & \frac{1}{2} & 0 & 0 \\ 0 & 4 & 5 & \bigm| & -2 & 1 & 0 \\ 2+(-2)(1) & 1+(-2)\left(-\frac{1}{2}\right) & -1+(-2)(-1) & \bigm| & 0+(-2)\left(\frac{1}{2}\right) & 0+(-2)(0) & 1+(-2)(0) \end{pmatrix}$$

$$\begin{pmatrix} 1 & -\frac{1}{2} & -1 & \bigm| & \frac{1}{2} & 0 & 0 \\ 0 & 4 & 5 & \bigm| & -2 & 1 & 0 \\ 0 & 2 & 1 & \bigm| & -1 & 0 & 1 \end{pmatrix}$$

We see that the result of these operations is a 1 on the principal diagonal and zeros in all the other positions in the first column. Now repeat the same sequence of operations on the second column. First, multiply the second row by $\frac{1}{4}$, which is the multiplicative inverse of the a_{22} element, 4. This gives us a 1 on the principal diagonal in the second column.

$$\left(\begin{array}{ccc|ccc} 1 & -\dfrac{1}{2} & -1 & \dfrac{1}{2} & 0 & 0 \\ 0 & 1 & \dfrac{5}{4} & -\dfrac{1}{2} & \dfrac{1}{4} & 0 \\ 0 & 2 & 1 & -1 & 0 & 1 \end{array}\right)$$

Now replace row 3 with the sum of itself and -2 times row 2. (Row 2 is the principal diagonal row for the second column; -2 is the additive inverse of the element in the third row of column 2.)

$$\left(\begin{array}{ccc|ccc} 1 & -\dfrac{1}{2} & -1 & \dfrac{1}{2} & 0 & 0 \\ 0 & 1 & \dfrac{5}{4} & -\dfrac{1}{2} & \dfrac{1}{4} & 0 \\ 0 & 0 & -\dfrac{3}{2} & 0 & -\dfrac{1}{2} & 1 \end{array}\right)$$

Now replace row 1 with the sum of itself and $\frac{1}{2}$ times row 2. $\left(\frac{1}{2}\right.$ is the additive inverse of the element in the first row of column 2.$\big)$

$$\left(\begin{array}{ccc|ccc} 1 & 0 & -\dfrac{3}{8} & \dfrac{1}{4} & \dfrac{1}{8} & 0 \\ 0 & 1 & \dfrac{5}{4} & -\dfrac{1}{2} & \dfrac{1}{4} & 0 \\ 0 & 0 & -\dfrac{3}{2} & 0 & -\dfrac{1}{2} & 1 \end{array}\right)$$

We have now obtained a 1 on the principal diagonal and zeros in all the other positions of column 2. Now perform a similar series of operations on the third column. First multiply row 3 by $-\frac{2}{3}$.

$$\left(\begin{array}{ccc|ccc} 1 & 0 & -\dfrac{3}{8} & \dfrac{1}{4} & \dfrac{1}{8} & 0 \\ 0 & 1 & \dfrac{5}{4} & -\dfrac{1}{2} & \dfrac{1}{4} & 0 \\ 0 & 0 & 1 & 0 & \dfrac{1}{3} & -\dfrac{2}{3} \end{array}\right)$$

Next replace row 1 with the sum of itself and $\dfrac{3}{8}$ times row 3.

$$\left(\begin{array}{ccc|ccc} 1 & 0 & 0 & \dfrac{1}{4} & \dfrac{1}{4} & -\dfrac{1}{4} \\[2mm] 0 & 1 & \dfrac{5}{4} & -\dfrac{1}{2} & \dfrac{1}{4} & 0 \\[2mm] 0 & 0 & 1 & 0 & \dfrac{1}{3} & -\dfrac{2}{3} \end{array}\right)$$

Finally, replace row 2 with the sum of itself and $-\dfrac{5}{4}$ times row 3.

$$\left(\begin{array}{ccc|ccc} 1 & 0 & 0 & \dfrac{1}{4} & \dfrac{1}{4} & -\dfrac{1}{4} \\[2mm] 0 & 1 & 0 & -\dfrac{1}{2} & -\dfrac{1}{6} & \dfrac{5}{6} \\[2mm] 0 & 0 & 1 & 0 & \dfrac{1}{3} & -\dfrac{2}{3} \end{array}\right)$$

The operations have transformed the matrix A into an identity matrix and the identity matrix into the inverse matrix A^{-1}.

$$A^{-1} = \left(\begin{array}{ccc} \dfrac{1}{4} & \dfrac{1}{4} & -\dfrac{1}{4} \\[2mm] -\dfrac{1}{2} & -\dfrac{1}{6} & \dfrac{5}{6} \\[2mm] 0 & \dfrac{1}{3} & -\dfrac{2}{3} \end{array}\right)$$

Perform the multiplications to verify that $A^{-1}A = I$ and $AA^{-1} = I$.

To solve the set of simultaneous equations $AX = C$, we now need only to solve $X = A^{-1}C$. Recalling X and C, we have:

$$\begin{pmatrix} x \\ y \\ z \end{pmatrix} = \left(\begin{array}{ccc} \dfrac{1}{4} & \dfrac{1}{4} & -\dfrac{1}{4} \\[2mm] -\dfrac{1}{2} & -\dfrac{1}{6} & \dfrac{5}{6} \\[2mm] 0 & \dfrac{1}{3} & -\dfrac{2}{3} \end{array}\right) \begin{pmatrix} -6 \\ 1 \\ -1 \end{pmatrix}$$

$$= \left(\begin{array}{c} \left(\dfrac{1}{4}\right)(-6) + \left(\dfrac{1}{4}\right)(1) + \left(-\dfrac{1}{4}\right)(-1) \\[2mm] \left(-\dfrac{1}{2}\right)(-6) + \left(-\dfrac{1}{6}\right)(1) + \left(\dfrac{5}{6}\right)(-1) \\[2mm] (0)(-6) + \left(\dfrac{1}{3}\right)(1) + \left(-\dfrac{2}{3}\right)(-1) \end{array}\right) = \begin{pmatrix} -1 \\ 2 \\ 1 \end{pmatrix}$$

Therefore, $x = -1$, $y = 2$, and $z = 1$ is the solution set of the system of simultaneous equations:

$$2x - y - 2z = -6$$
$$4x + 2y + z = 1.$$
$$2x + y - z = -1$$

STEPS FOR FINDING THE INVERSE OF A SQUARE MATRIX

1. Write down the matrix to be inverted and write an identity matrix of the same order next to it.
2. Start with the left column of the matrix A to be inverted and complete Steps 3 through 5 before proceeding to the next column.
3. Identify the column to be operated on. Call it the jth column.
4. Multiply each element of the principle diagonal row (the jth row) by the multiplicative inverse of the principle diagonal element (the a_{jj} element). Replace the row by the results of the multiplication.
5. For each remaining row in the column (call it the ith row), replace each element of the row by the sum of itself and the product of the additive inverse of a_{ij} times the corresponding element of the principle diagonal row (the one in the same column as the element being replaced). Repeat Step 5 for each row in the matrix except the jth row. Then return to Step 3 and select the next column in the matrix. Repeat until all columns are completed.
6. After we perform Steps 3 through 5 on each column of the original matrix A, the matrix is transformed into an identity matrix, and the original identity matrix is transformed into the inverse matix A^{-1}.

A square matrix does not have an inverse when a zero on the main diagonal cannot be removed. If the matrix does not have an inverse, it is called a **singular matrix.** One method of determining whether or not a zero on the main diagonal is removable is to start the inversion process over after rearranging the order of the rows of the matrix. If, after all possible combinations of rearrangement of the rows, a zero still exists on the main diagonal, then the matrix is singular. The matrix of coefficients of a solvable system of simultaneous equations always has an inverse. Remember that if the rows of the matrix of coefficients must be rearranged, the matrix of constants must be rearranged accordingly. (The matrix of unknowns remains the same.) It is also possible to determine whether or not a matrix is singular by evaluating the determinant of the matrix. If the determinant of the matrix is not equal to zero, then the matrix is **nonsingular.**

EXAMPLE 1 **Solving Systems of Equations with Matrices**

Solve the following system of simultaneous linear equations using the inverse matrix.

$$\begin{aligned} x + y + z &= -2 \\ 2x - y - z &= 5 \\ -x + 3y + 4z &= -11 \end{aligned}$$

Solution The system of equations can be written in matrix form as:

$$\begin{pmatrix} 1 & 1 & 1 \\ 2 & -1 & -1 \\ -1 & 3 & 4 \end{pmatrix} \begin{pmatrix} x \\ y \\ z \end{pmatrix} = \begin{pmatrix} -2 \\ 5 \\ -11 \end{pmatrix},$$

or, $AX = C$.

Apply the steps for matrix inversion to A.

$$\left(\begin{array}{ccc|ccc} 1 & 1 & 1 & 1 & 0 & 0 \\ 2 & -1 & -1 & 0 & 1 & 0 \\ -1 & 3 & 4 & 0 & 0 & 1 \end{array} \right) \quad \textbf{Step 1}$$

For Step 3, start with column 1. Since there is a 1 in position a_{11}, we can skip the multiplication called for in Step 4. For the second row of the first column, the additive inverse of a_{21} is -2, and for the third row of the first column, the additive inverse of a_{31} is 1.

$$\left(\begin{array}{ccc|ccc} 1 & 1 & 1 & 1 & 0 & 0 \\ 2 + (-2)(1) & -1 + (-2)(1) & -1 + (-2)(1) & 0 + (-2)(1) & 1 + (-2)(0) & 0 + (-2)(0) \\ -1 + (1)(1) & 3 + (1)(1) & 4 + (1)(1) & 0 + (1)(1) & 0 + (1)(0) & 1 + (1)(0) \end{array} \right) \quad \textbf{Step 5}$$

$$\left(\begin{array}{ccc|ccc} 1 & 1 & 1 & 1 & 0 & 0 \\ 0 & -3 & -3 & -2 & 1 & 0 \\ 0 & 4 & 5 & 1 & 0 & 1 \end{array} \right) \quad \textbf{Combining terms}$$

Going to the second column, the multiplicative inverse of the a_{22} element, -3, is $\dfrac{1}{-3}$. Replace row 2 with the product of $\dfrac{-1}{3}$ and row 2.

$$\left(\begin{array}{ccc|ccc} 1 & 1 & 1 & 1 & 0 & 0 \\ 0 & 1 & 1 & \dfrac{2}{3} & -\dfrac{1}{3} & 0 \\ 0 & 4 & 5 & 1 & 0 & 1 \end{array} \right) \quad \textbf{Step 4}$$

For the first row of the second column, the additive inverse of a_{12} is -1, and for the third row of the second column, the additive inverse of a_{32} is -4.

$$\left(\begin{array}{ccc|ccc} 1+(-1)(0) & 1+(-1)(1) & 1+(-1)(1) & 1+(-1)\left(\frac{2}{3}\right) & 0+(-1)\left(-\frac{1}{3}\right) & 0+(-1)(0) \\ 0 & 1 & 1 & \frac{2}{3} & -\frac{1}{3} & 0 \\ 0+(-4)(0) & 4+(-4)(1) & 5+(-4)(1) & 1+(-4)\left(\frac{2}{3}\right) & 0+(-4)\left(-\frac{1}{3}\right) & 1+(-4)(0) \end{array}\right) \quad \text{Step 5}$$

$$\left(\begin{array}{ccc|ccc} 1 & 0 & 0 & \frac{1}{3} & \frac{1}{3} & 0 \\ 0 & 1 & 1 & \frac{2}{3} & -\frac{1}{3} & 0 \\ 0 & 0 & 1 & -\frac{5}{3} & \frac{4}{3} & 1 \end{array}\right) \quad \text{Combining terms}$$

Since the element in the a_{33} position is a 1, all that needs to be done is to get a zero in the a_{23} position. The additive inverse of a_{23} is -1.

$$\left(\begin{array}{ccc|ccc} 1 & 0 & 0 & \frac{1}{3} & \frac{1}{3} & 0 \\ 0+(-1)(0) & 1+(-1)(0) & 1+(-1)(1) & \frac{2}{3}+(-1)\left(-\frac{5}{3}\right) & -\frac{1}{3}+(-1)\left(\frac{4}{3}\right) & 0+(-1)(1) \\ 0 & 0 & 1 & -\frac{5}{3} & \frac{4}{3} & 1 \end{array}\right)$$

$$\left(\begin{array}{ccc|ccc} 1 & 0 & 0 & \frac{1}{3} & \frac{1}{3} & 0 \\ 0 & 1 & 0 & \frac{7}{3} & -\frac{5}{3} & -1 \\ 0 & 0 & 1 & -\frac{5}{3} & \frac{4}{3} & 1 \end{array}\right)$$

Now matrix A is transformed completely to an identity matrix, so:

$$A^{-1} = \begin{pmatrix} \frac{1}{3} & \frac{1}{3} & 0 \\ \frac{7}{3} & -\frac{5}{3} & -1 \\ -\frac{5}{3} & \frac{4}{3} & 1 \end{pmatrix}$$

Recall that:

$$C = \begin{pmatrix} -2 \\ 5 \\ -11 \end{pmatrix}.$$

To solve the equations, use $X = A^{-1}C$.

$$\begin{pmatrix} x \\ y \\ z \end{pmatrix} = \begin{pmatrix} \frac{1}{3} & \frac{1}{3} & 0 \\ \frac{7}{3} & -\frac{5}{3} & -1 \\ -\frac{5}{3} & \frac{4}{3} & 1 \end{pmatrix} \begin{pmatrix} -2 \\ 5 \\ -11 \end{pmatrix}$$

$$\begin{pmatrix} x \\ y \\ z \end{pmatrix} = \begin{pmatrix} \left(\frac{1}{3}\right)(-2) + \left(\frac{1}{3}\right)(5) + (0)(-11) \\ \left(\frac{7}{3}\right)(-2) + \left(-\frac{5}{3}\right)(5) + (-1)(-11) \\ \left(-\frac{5}{3}\right)(-2) + \left(\frac{4}{3}\right)(5) + (1)(-11) \end{pmatrix}$$

$$= \begin{pmatrix} 1 \\ -2 \\ -1 \end{pmatrix}.$$

Therefore, $x = 1$, $y = -2$, and $z = -1$ is the solution set for the given system of simultaneous equations. ◆

EXAMPLE 2 Solving Systems of Equations with Matrices

Solve the following system of simultaneous linear equations using the inverse matrix.

$$2x + y - z = -1$$
$$4x + 2y + z = 1$$
$$2x - y - 2z = -6.$$

Solution

$$\begin{pmatrix} 2 & 1 & -1 & | & 1 & 0 & 0 \\ 4 & 2 & 1 & | & 0 & 1 & 0 \\ 2 & -1 & -2 & | & 0 & 0 & 1 \end{pmatrix} \quad \textbf{Step 1}$$

We need a 1 in the a_{11} position, so multiply row 1 by $\frac{1}{2}$.

$$\begin{pmatrix} 1 & \frac{1}{2} & -\frac{1}{2} & | & \frac{1}{2} & 0 & 0 \\ 4 & 2 & 1 & | & 0 & 1 & 0 \\ 2 & -1 & -2 & | & 0 & 0 & 1 \end{pmatrix}$$

For the first column $a_{21} = 4$ and $a_{31} = 2$, so the appropriate additive inverses are -4 and -2.

$$
\begin{pmatrix}
1 & \frac{1}{2} & -\frac{1}{2} & \bigg| & \frac{1}{2} & 0 & 0 \\[2mm]
4 + (-4)(1) & 2 + (-4)\left(\frac{1}{2}\right) & 1 + (-4)\left(-\frac{1}{2}\right) & \bigg| & 0 + (-4)\left(\frac{1}{2}\right) & 1 + (-4)(0) & 0 + (-4)(0) \\[2mm]
2 + (-2)(1) & -1 + (-2)\left(\frac{1}{2}\right) & -2 + (-2)\left(-\frac{1}{2}\right) & \bigg| & 0 + (-2)\left(\frac{1}{2}\right) & 0 + (-2)(0) & 1 + (-2)(0)
\end{pmatrix}
$$

$$
\begin{pmatrix}
1 & \frac{1}{2} & -\frac{1}{2} \\[1mm]
0 & 0 & 3 \\[1mm]
0 & -2 & -1
\end{pmatrix}
\begin{pmatrix}
\frac{1}{2} & 0 & 0 \\[1mm]
-2 & 1 & 0 \\[1mm]
-1 & 0 & 1
\end{pmatrix}
$$

For the second column $a_{22} = 0$, so we cannot proceed because division by zero is undefined. However if we change the order of the rows by interchanging the first and third rows of the original matrix and change the matrix of constants correspondingly, we obtain the system of equations used to illustrate the first part of this section. As was shown, this system can be solved by matrix inversion giving the results $x = -1$, $y = 2$, and $z = 1$. ◆

EXAMPLE 3 **Application: Solving a Dietary Problem**

A dietician is planning a meal consisting of three foods. A 30-gram serving of the first food contains 5 units of protein, 2 units of carbohydrates, and 3 units of iron. A 30-gram serving of the second food contains 10 units of protein, 3 units of carbohydrates, and 6 units of iron. And a 30-gram serving of the third food contains 15 units of protein, 2 units of carbohydrates, and 1 unit of iron. How many grams of each food would be used to create a meal containing 55 units of protein, 13 units of carbohydrates, and 17 units of iron?

Solution Start by assigning variables for the unknown quantities and then writing a system of equations. The unknown quantities are the number of 30-gram units of food 1, food 2, and food 3. Let:

$$F_1 = \text{the number of 30-gram units of food 1;}$$
$$F_2 = \text{the number of 30-gram units of food 2;}$$
$$F_3 = \text{the number of 30-gram units of food 3.}$$

Now write an equation to determine the number of 30-gram units of each type of food needed to create a meal that has the required number of units of protein, carbohydrates, and iron.

$$
\begin{aligned}
5F_1 + 10F_2 + 15F_3 &= 55 \quad &&\textbf{Protein} \\
2F_1 + 3F_2 + 2F_3 &= 13 \quad &&\textbf{Carbohydrates} \\
3f_1 + 6F_2 + 1F_3 &= 17 \quad &&\textbf{Iron}
\end{aligned}
$$

To solve the system of equations, use the technique of inverse matrices and a calculator to determine the inverse matrices and the product of the matrices. The matrices for the system of equations follow.

$$A = \begin{pmatrix} 5 & 10 & 15 \\ 2 & 3 & 2 \\ 3 & 6 & 1 \end{pmatrix} \quad X = \begin{pmatrix} F_1 \\ F_2 \\ F_3 \end{pmatrix} \quad C = \begin{pmatrix} 55 \\ 13 \\ 17 \end{pmatrix}$$

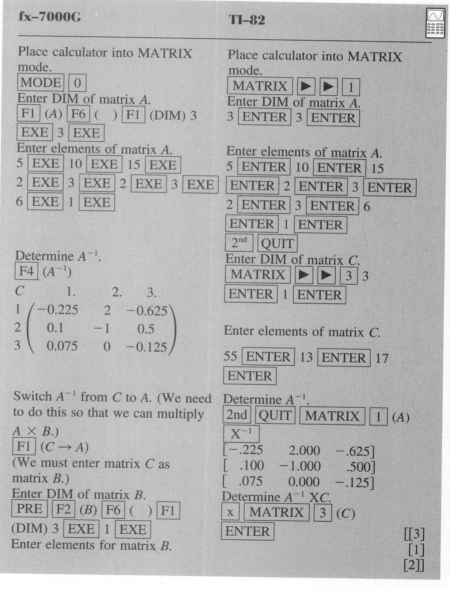

fx–7000G

Place calculator into MATRIX mode.
[MODE] [0]
Enter DIM of matrix A.
[F1] (A) [F6] ( ) [F1] (DIM) 3
[EXE] 3 [EXE]
Enter elements of matrix A.
5 [EXE] 10 [EXE] 15 [EXE]
2 [EXE] 3 [EXE] 2 [EXE] 3 [EXE]
6 [EXE] 1 [EXE]

Determine A^{-1}.
[F4] (A^{-1})

$$\begin{matrix} C & 1. & 2. & 3. \\ 1 & \begin{pmatrix} -0.225 & 2 & -0.625 \\ 2 & 0.1 & -1 & 0.5 \\ 3 & 0.075 & 0 & -0.125 \end{pmatrix} \end{matrix}$$

Switch A^{-1} from C to A. (We need to do this so that we can multiply $A \times B$.)
[F1] $(C \rightarrow A)$
(We must enter matrix C as matrix B.)
Enter DIM of matrix B.
[PRE] [F2] (B) [F6] ( ) [F1]
(DIM) 3 [EXE] 1 [EXE]
Enter elements for matrix B.

TI–82

Place calculator into MATRIX mode.
[MATRIX] [▶] [▶] [1]
Enter DIM of matrix A.
3 [ENTER] 3 [ENTER]
Enter elements of matrix A.
5 [ENTER] 10 [ENTER] 15
[ENTER] 2 [ENTER] 3 [ENTER]
2 [ENTER] 3 [ENTER] 6
[ENTER] 1 [ENTER]
[2nd] [QUIT]
Enter DIM of matrix C.
[MATRIX] [▶] [▶] [3] 3
[ENTER] 1 [ENTER]

Enter elements of matrix C.

55 [ENTER] 13 [ENTER] 17
[ENTER]
Determine A^{-1}.
[2nd] [QUIT] [MATRIX] [1] (A)
[X^{-1}]
[−.225   2.000   −.625]
[.100   −1.000   .500]
[.075   0.000   −.125]
Determine $A^{-1} XC$.
[x] [MATRIX] [3] (C)
[ENTER]   [[3]
                                [1]
                                [2]]

Continued on page 522.

55 [EXE] 13 [EXE] 17 [EXE]
Determining $A^{-1} XB$.
[PRE] [F5] (X)
C *1.*
1 $\begin{pmatrix} 3 \\ 1 \\ 2 \end{pmatrix}$
2
3

The product matrix shows that $F_1 = 3$, $F_2 = 1$, and $F_3 = 2$. To have a meal with the desired units of protein, carbohydrates, and iron we must have:

$$3 \times 30 = 90 \text{ grams of food 1;}$$
$$1 \times 30 = 30 \text{ grams of food 2;}$$
$$2 \times 30 = 60 \text{ grams of food 3.} \quad \blacklozenge$$

A BASIC program that uses matrix methods to solve systems of equations is found on the tutorial disk.

10.7 Exercises

In Exercises 1–4, **a.** write the systems of simultaneous equations in matrix form; **b.** determine the inverse of the matrices of the coefficients; and **c.** determine the solution sets of the systems of equations using matrix methods.

1.
$$x + y + z = -2$$
$$2x + y + 2z = 0$$
$$-x - 2y - 3z = 5$$

2.
$$2x + 3y + z = -1$$
$$x - 2y - z = -8$$
$$3x - 3y + 2z = 1$$

3.
$$x + y + z = 7$$
$$y - z = 7$$
$$2x - 2y - z = 0$$

4.
$$2x + y - z = 2$$
$$4x + y - 2z = -3$$
$$x + z = 2$$

In Exercises 5–27, determine the solution sets for the system of equations using matrix methods.

5.
$$-x + 4y - z = 7$$
$$x - y + 4z = -1$$
$$3x + 9y - 2z = -5$$

6.
$$x + 2y + 3z = 5$$
$$3x + 3y - z = 0$$
$$x + 3y - 2z = -3$$

7.
$$4x - 5y + 5z = 2$$
$$x + y + 3z = 1$$
$$9x - 3y + 2z = -1$$

8.
$$x + 5z = -6$$
$$2x + 3y + 4z = -9$$
$$3x + 4y = -7$$

9.
$$2x - y + 2z = 2$$
$$5x + 4y + 3z = -7$$
$$x + y + z = -1$$

10.
$$4x - y + z = 9$$
$$5x + y + 9z = 9$$
$$6x + 5y - z = 7$$

11.
$$x + 5z = -4$$
$$8x + 7y - z = 9$$
$$9x - 4y + z = 8$$

12.
$$4x + 4y + 5z = -7$$
$$4x + 3y + 6z = 8$$
$$3x + y + 9z = 4$$

13.
$$8x - 3y - 9z = 7$$
$$-5x + 2y - 8z = 9$$
$$-2y - 3x = 7$$

14.
$$4x + 2y - 5z = 2$$
$$7x - z = -5$$
$$-2x + y = 0$$

15.
$$-7x + 7y + 8z = 8$$
$$-5y - z = 4$$
$$6x - 6y + 7z = 7$$

16.
$$3x + y + 4z = 8$$
$$-3x - 2y + 6z = 1$$
$$6x + y - 5y = 2$$

17.
$$5x + 3y + 3z = 1$$
$$-5x + 4y + 3z = -4$$
$$-3x - 2y - 2z = 4$$

18.
$$2x - y + 6z = -9$$
$$5x + 4y + 3z = 9$$
$$-2x + y + 5z = -2$$

19.
$$x + y + z = -1$$
$$2x - 5y - 8z = 2$$
$$x + 4z = 4$$

20.
$$-4x + y - 3z = 2$$
$$-2x + y + 4z = 2$$
$$5x - y + 8z = -2$$

21. The loop equations for a circuit are:

$$30i_1 + 41i_2 - 3i_3 = 0$$
$$3i_3 - i_2 = 0.$$
$$20i_1 - 9i_2 - 3i_3 = 0$$

What are the currents i_1, i_2, and i_3?

22. The node equations for a circuit are:

$$-7v_1 + 10v_2 = 0$$
$$39v_1 - 20v_2 - 15v_3 = 400$$
$$-3v_1 + 5v_3 = 0$$

What are the node voltages v_1, v_2, and v_3?

23. The section forces in a truss are related by:

$$F_1 \sin 30° + F_2 + 500 = 0$$
$$(\tan 30°)(F_1 \cos 30°) + F_3 = 1500.$$
$$F_2 \sin 30° + F_3 = 500$$

What are the forces F_1, F_2, and F_3?

24. Design the height dimensions for three supports such that S_3 is three times as tall as S_1; the height of S_2 is 4 more than S_1; the height of S_3 is 3 more than S_2.

25. The college bookstore buys sweatshirts from three sources A, B, and C. It buys as many sweatshirts from source A as from the other two sources combined. The bookstore must pay source A at the rate of $4 per unit, source B at $5 per unit, and C at the rate of $6 per unit. If the bookstore needs 100 units per month to satisfy student demand and if the cost of one month's supply is $470, how many sweatshirts does it buy from each source?

26. The sum of the angles of a triangle is 180°. The largest angle is 30° larger than the sum of the other two and is 25° larger than 3 times the smallest angle. What is the measure of each angle?

27. You dilute a solution containing 40% alcohol by adding pure water, resulting in a 25% alcohol solution. Then you add 20 more gallons of pure water, diluting the mixture to a 20% alcohol solution. Determine how much solution there was originally and how many gallons of water you added the first time.

Review Exercises

In Exercises 1–8, identify the slope and y-intercept of the lines and sketch the graph of the lines.

1. $y = -x + 4$

2. $y = x - 3$

3. $y = \dfrac{1}{2}x + 2$

4. $y = -\dfrac{1}{4}x - 2$

5. $y = 3$

6. $x = -4$

7. $3x + 6y - 2 = 0$

8. $2x + 4y + 1 = 0$

In Exercises 9–16, write the equation of the line.

9. $m = -4, b = 0$

10. $m = 5, b = 0$

11. $m = \dfrac{1}{2}, b = 3$

12. $m = \dfrac{1}{6}, b = 1$

13. $m = 0, b = -2$

14. $m = 0, b = 4$

15. $m = 1.32, b = 4.17$

16. $m = -2.61, b = 4.88$

In Exercises 17–24, solve the system of two linear equations in two unknowns using algebraic methods.

17. $x - y = 0$
  $2x - y = 6$

18. $-x - 3y = 3$
  $2x - 3y = 3$

19. $5x - y = -1$
  $x + y = 1$

20. $4x + y = 1$
  $9x + 5y = -6$

21. $-x - y + 2 = 0$
  $x + 2y = 0$

22. $3x - y + 4 = 0$
  $x + y = 0$

23. $2x - y - 8 = 0$
  $-3x - 2y + 5 = 0$

24. $x - y - 4 = 0$
  $3x - y - 2 = 0$

In Exercises 25–32, use algebraic methods to solve the system of linear equations. From the results determine whether any of the systems are inconsistent or dependent. Graph the equations with a graphics calculator as a visual check.

25. $8x - 5y = 6$
  $3x - 2y = 8$

26. $7x + 8y = -3$
  $6x + 7y = 8$

27. $x + 9y + 1 = 0$
  $x + 9y = 1$

28. $-7x - y + 4 = 0$
  $14x + 2y - 16 = 0$

29. $x + 2y = -1$
  $4x + 3y = 4$

30. $x + 3y = 0$
  $x + 2y = -6$

31. $x - y + 3 = 0$
  $-3x + 3y - 9 = 0$

32. $x + 3y - 1 = 0$
  $-x - 3y = -1$

In Exercises 33–38, evaluate the 2×2 determinants.

33. $\begin{vmatrix} 3 & -5 \\ 9 & 2 \end{vmatrix}$

34. $\begin{vmatrix} -9 & 1 \\ -5 & 0 \end{vmatrix}$

35. $\begin{vmatrix} -1 & -3 \\ 0 & 8 \end{vmatrix}$

36. $\begin{vmatrix} -1 & 1 \\ -7 & 5 \end{vmatrix}$

37. $\begin{vmatrix} 1 & 0 \\ -4 & -1 \end{vmatrix}$

38. $\begin{vmatrix} 9 & 0 \\ -5 & -1 \end{vmatrix}$

In Exercises 39–50, solve the systems of equations using Cramer's rule.

39. $8x - 3y = -3$
  $7x - 2y = 3$

40. $x + 3y = 1$
  $x + 2y = 4$

41. $2x - 3y + 2 = 0$
  $3x - 4y - 8 = 0$

42. $-3x + 5y - 2 = 0$
  $3x - 4y + 4 = 0$

43. $7x - y - 8 = 0$
  $-6x + y + 4 = 0$

44. $x - y + 8 = 0$
  $-7x + 8y - 4 = 0$

45. $6x - 7y = 4$
  $-2x + y = -4$

46. $-2x - 5y = 8$
  $-7x + y = -9$

47. $-3x - 4y = -4$
  $2x + 3y = 2$

48. $-2x - 9y = -7$
  $5x - y = -6$

49. $x + 2y = -2$
  $-x + 5y = 9$

50. $-4x - 5y = 7$
  $-9x - 7y = 3$

In Exercises 51–58, solve the systems of three equations in three unknowns using algebraic methods.

51. $-7x - 6y + 3z = 3$
  $-x - 9y + 6z = 3$
  $-4x - 9y = 9$

52. $-x - 5z = 4$
  $8x + 7y - z = 9$
  $9x - 4y + z = 8$

53. $5x + y + 9z = 9$
  $6x + 5y - z = 7$
  $4x - y + z = 9$

54. $-8x + 4y + 5z = -6$
  $-3x + 9y + z = -8$
  $-3x - 4y + 7z = -7$

55. $8x + y - 8z = 1$
  $-4x - 8y + 3z = 3$
  $-2x - 3y + z = 8$

56. $x - y + 4z = -1$
  $3x + 9y - 2z = -5$
  $-x + 4y - z = 7$

57. $-5x + 4y + 3z = 6$
  $6x + 3y + 7z = -5$
  $3x + 2y + 2x = 2$

58. $5x + y + 6z = 9$
  $x + 6y - 8z = -5$
  $-7x + 9y - 8z = 9$

In Exercises 59–66, evaluate the 3×3 determinant using expansion by cofactors.

59. $\begin{vmatrix} -1 & -9 & 0 \\ -4 & -3 & 7 \\ 3 & 0 & -5 \end{vmatrix}$ **60.** $\begin{vmatrix} -1 & -1 & 3 \\ 0 & -1 & -1 \\ 7 & -6 & 0 \end{vmatrix}$ **61.** $\begin{vmatrix} -2 & -9 & 5 \\ 3 & 5 & 3 \\ 1 & 3 & 3 \end{vmatrix}$ **62.** $\begin{vmatrix} 7 & -8 & 1 \\ -3 & -2 & -4 \\ 4 & 3 & -3 \end{vmatrix}$

63. $\begin{vmatrix} 4 & -5 & 0 \\ -5 & 6 & 0 \\ 5 & -6 & 5 \end{vmatrix}$ **64.** $\begin{vmatrix} 2 & 7 & 2 \\ 9 & 8 & 2 \\ 0 & 0 & -3 \end{vmatrix}$ **65.** $\begin{vmatrix} 3 & 4 & -8 \\ 3 & -9 & 4 \\ 4 & -1 & -3 \end{vmatrix}$ **66.** $\begin{vmatrix} 2 & -3 & -1 \\ -7 & 5 & 3 \\ -1 & 3 & 4 \end{vmatrix}$

In Exercises 67–74, solve the systems of three linear equations in three unknowns using Cramer's rule.

67.
$$2r - t = 0$$
$$r + 2s + t = 3$$
$$-2s - 3t = 0$$

68.
$$3r + 4s = -7$$
$$r + 5t = -6$$
$$2r + 3s + 4t = -9$$

69.
$$2x - 3y + 4z = 1$$
$$4x - 5y + 2z = -1$$
$$x - y + 5z = 2$$

70.
$$6x - y - 3z = -8$$
$$-7x + 9y - 7z = -9$$
$$-x + 4y + 2z = -1$$

71.
$$9x + 5y - z = 5$$
$$-2x + y - 7z = 4$$
$$x - 3y + 3z = 1$$

72.
$$-9x + 3y - 2z = 1$$
$$x + y + 3z = 1$$
$$4x - 5y + 5z = 2$$

73.
$$4x + 4y + 5z = -7$$
$$4x - 3y + 6z = 8$$
$$3x + y + 9z = 4$$

74.
$$5x + 3y - 2z = 1$$
$$2x + 2y + z = 4$$
$$x - 2y + 4z = 9$$

Exercises 75–80 are systems of joint equations used for finding the forces in members of trusses. Solve for the unknown forces.

75. $-F_1 \cos 56° + F_2 \cos 56° + 4000 = 0$
$\quad\quad F_1 \sin 56° + F_2 \sin 56° - 5000 = 0$

76. $-F_5 \cos 45° - F_6 \cos 45° + 10{,}000 = 0$
$\quad\quad F_5 \sin 45° - F_6 \sin 45° - 10{,}000 = 0$

77. $-F_A \cos 58° + 5000 + 4720 \cos 58° - F_B \cos 58° = 0$
$\quad\quad F_A \sin 58° - 4720 \sin 58° - F_B \sin 58° = 0$

78. $-5000 - F_2 \cos 58° + F_3 \cos 58° + 5000 = 0$
$\quad\quad -F_1 - F_2 \sin 58° - F_3 \sin 58° = 0$
$\quad\quad 8(5000) - 5F_2 \sin 58° - 5F_3 \sin 58° = 0$

Exercises 79–83 are systems of loop or node equations used to find the currents or voltages in electric circuits. Solve for the unknown currents or voltages.

79. $-100 + 4i_1 + 40(i_1 - i_2) + 6i_1 + 10i_1 = 0$
$\quad\quad 5i_2 + 15i_2 + 40(i_2 - i_1) = 0$

80. $-100i_1 + 4i_1 + 40(i_1 - i_2) + 6i_1 + 40 + 10i_1 = 0$
$\quad\quad 5i_2 + 15i_2 + 40(i_2 - i_1) = 0$

81. $\dfrac{1}{10}(v_1 - 100) + \dfrac{1}{20}v_1 + \dfrac{1}{5}(v_1 - v_2) + \dfrac{1}{20}(v_1 - v_2) = 0$

$\quad\quad \dfrac{1}{5}(v_2 - v_1) + \dfrac{1}{20}(v_2 - v_1) + \dfrac{1}{16}v_2 = 0$

82. $\quad\quad -60 + 60i_1 + 30(i_1 - i_2) = 0$
$30(i_2 - i_1) - 50 + 30i_2 + 60(i_2 - i_3) = 0$
$\quad\quad 60(i_3 - i_2) + 40i_3 + 100 = 0$

83. $\quad\quad -120 + 30i_1 + 60(i_1 - i_2) = 0$
$60(i_2 - i_1) - 40 + 10i_2 + 30(i_2 - i_3) = 0$
$\quad\quad 30(i_3 - i_2) + 60i_3 - 10 = 0$

84. An architect is laying out a parking lot. She finds she can put five parking spaces for compact cars and six spaces for standard cars into a 100-ft row. She alternatively can put 10 compact and 2 standard spaces in the same row. What widths is she using as the basis for her layout?

85. "Grab bag" assortments of fishing lures all contain $6 worth of lures. Selection A contains three plugs and one jig. Selection B has a plug, two spoons, and a jig. Selection C has five jigs and two spoons. What are the types of lures worth?

86. Grandpa Joe's walking stick is 2 cm longer than Uncle Paul's, whose stick, in turn, is 5 cm shorter than Brother Will's. The sticks average 1 m in length. How long are the individual sticks?

In Exercises 87–92, perform the indicated operations on the matrices, if possible.

$$A = \begin{pmatrix} 4 & 2 \\ -1 & 0 \\ 3 & 6 \end{pmatrix} \qquad B = \begin{pmatrix} -1 & 0 \\ 2 & 4 \\ 5 & 3 \end{pmatrix}$$

$$C = \begin{pmatrix} 2 & 61 \\ 4 & -13 \end{pmatrix} \qquad D = \begin{pmatrix} 2 & 6 & -1 \\ 4 & -1 & 5 \end{pmatrix}$$

$$E = \begin{pmatrix} 1 & 4 & 2 \\ 6 & 3 & 1 \\ 2 & -5 & 3 \end{pmatrix} \qquad F = \begin{pmatrix} -3 & 0 & 2 \\ 1 & 1 & 0 \\ 5 & 4 & 3 \end{pmatrix}$$

87. $A + B$ **88.** $2C - 3D$ **89.** $A \times D$ **90.** $D \times E$

91. $2E - F$ **92.** $B \times F$

In Exercises 93–96, determine the solution set for the systems of equations using matrix methods.

93. $x + 6y - 8z = -5$
      $7z - 9y + 8z = -9$
      $5x + y + 6z = 9$

94. $4x - 5y + 2z = -1$
      $x - y + 5z = 2$
      $2x - 3y + 4z = 1$

95. $x - 4y + z = -7$
      $-x + y - 4z = 1$
      $3x + 9y - 2z = -5$

96. $6x - y - 9z = 7$
      $x + 2y = -1$
      $x - 4y - z = 5$

97. A line that goes up and to the right has a positive slope. Use the definition of slope to show that a line going down and to the left also has a positive slope.

98. Write the equation of the line that has a slope $m = \dfrac{3}{5}$ and passes through point $(2, -1)$. Write the general equation of a line with slope m and passing through the point (x_1, y_1).

99. Write the equation of the line that passes through points $(3, 2)$ and $(5, -1)$. Write the general equation of a line that passes through the points (x_1, y_1) and (x_2, y_2).

100. Show that the order in which a system of equations is written does not affect its solution by Cramer's rule. In other words, show that the system

$$ax + by = c$$
$$dx + ey = f$$

has the same solution for x and y as the system:

$$dx + ey = f$$
$$ax + by = c.$$

101. A determinant containing just one element is equal to that element. Using this information show that a 2×2 determinant can be evaluated using expansion by cofactors.

102. Interchaning any two rows or any two columns of a determinant changes the sign of its value but not its magnitude. Using this information, show that the order of a system of three equations in three unknowns makes no difference in its solution by Cramer's rule.

✐ Writing About Mathematics

1. You are in charge of mixing the ingredients at a cement plant. For one product line the three basic ingredients are cement, sand, and stone; however, amounts of each can vary. To develop a chart of ingredients for certain strengths of concrete, you need to solve systems of equations with three unknowns. Write a report to the plant manager explaining how a system of three equations with three unknowns can be solved.

2. A classmate has difficulty evaluating determinants. Write a paper explaining how to evaluate a determinant using cofactors.

3. In your position as assistant group leader in a plant, you are required to purchase certain supplies. To set up a chart of various combinations, you must solve systems of three equations in three unknowns. Write instructions that show how a system of equations can be solved using Cramer's rule.

Chapter Test

In Exercises 1 and 2, identify the slope and the y-intercept of the lines and sketch the graph of the lines.

1. $y = 2x - 3$

2. $2y + x + 1 = 0$

In Exercises 3 and 4, write the equation of the line.

3. $m = \dfrac{2}{3}$ and $b = 2$

4. The line passes through $(-1, 3)$ and $(-4, 1)$.

Using the substitution method to solve the following system of equations.

5. $2x + y = -1$
    $3x + 2y = 1$

Use the elimination method to solve the following system of equations.

6. $2x + y - 3z = 9$
    $x + y + z = 3$
    $3x - 2y + 3z = -1$

In Exercises 7 and 8, use Cramer's rule to solve the following system of equations.

7. $3x + 2y = -13$
    $x + y = -5$

8. $8x + 3y + 5z = 3$
    $5x + 3y - 2z = 7$
    $6x - 7y = 6$

9. Multiply the following matrices.

$$\begin{pmatrix} 2 & 5 & -1 \\ 3 & 1 & 4 \end{pmatrix} \begin{pmatrix} -1 & 7 \\ 6 & 0 \\ 9 & 3 \end{pmatrix}$$

10. Use matrix methods to solve the following system of equations.

$$2x - y + 6z = -9$$
$$2x - y - 5z = 2$$
$$5x + 4y + 3z = 9$$

11. Write the equation of the line that passes through $(-3, 4)$ and is parallel to the line $2x - 5y + 4 = 0$.

12. Write the equation of the line that passes through $(2, 5)$ and is perpendicular to the line $3x + 2y - 6 = 0$.

13. A silversmith uses two silver alloys. The first alloy is 40% silver; the second alloy is 60% silver. How many grams of each should be mixed together to produce 20 grams of alloy that is 52% silver?

14. The sum of the angles of a triangle is 180°. The largest angle is 40° larger than the sum of the other two, and it is 10° less than 4 times the smallest. What is the measure of each angle?

Quadratic Equations

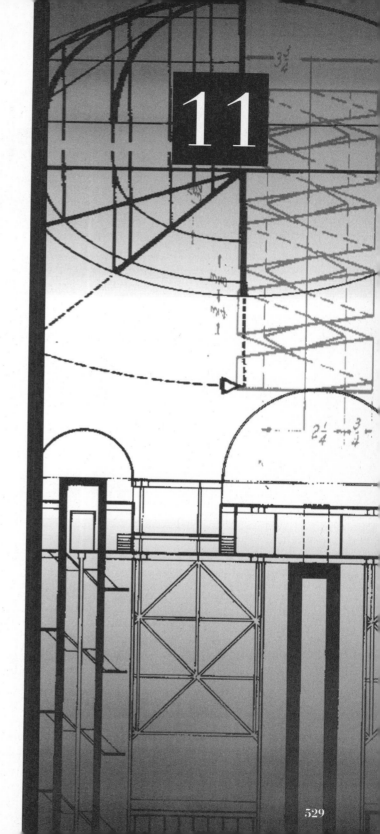

among the initial design requirements for a doctor's examining table is that the table be 7.00 ft in length. The designer assumes that the load on the table due to the patient's weight is evenly distributed along the table top at 70.0 lb/ft of length. (A safety factor is included.) The designer plans initial positions for the legs at 1.00 ft from the ends of the table. The table must also bend in the center so that the upper part of the patient's body may be raised. The bending motion could cause the table to collapse in the middle; therefore, the designer must determine the amount of bend possible before collapse.

The bending moment (a force) induced in the center section of the table top by the patient's weight is given by the following quadratic equation.

$$M = -\frac{W}{2L}x^2 + \frac{W}{2x} - \frac{W}{2}d \qquad (d \leq x \leq (L - d))$$

When excess weight is applied at some point, the table tends to bend down, while at other places, the table is forced up. The designer must determine where this occurs. To determine these points in the table top where the bending changes from down to up or from up to down (i.e., where the bending moment is zero), the designer solves the quadratic equation. (Example 6 in Section 11.4 solves this problem.)

11.1

**The Quadratic
Equation**

> **DEFINITION 11.1**
>
> An equation of the form
> $$ax^2 + bx + c = 0,$$
> where a, b, and c are real numbers, $a \neq 0$, is called a **general quadratic equation** in the variable x.

Any quadratic equation of the form of $ax^2 + bx + c = 0$ is said to be of **standard form.** For example, the following are quadratic equations in standard form.

$$6x^2 + 7x + 13 = 0 \qquad (a = 6, b = 7, \text{ and } c = 13)$$

$$\frac{3}{2}x^2 + 7 = 0 \qquad (a = \frac{3}{2}, b = 0, \text{ and } c = 7)$$

$$4x^2 - 3x = 0 \qquad (a = 4, b = -3, \text{ and } c = 0)$$

When we work with quadratic equations, we must write the equation in standard form. The quadratic equation $\frac{1}{3}x^2 = \frac{1}{4}x + 13$ can be expressed in standard form.

$$\frac{1}{3}x^2 - \frac{1}{4}x - 13 = 0$$

The equation

$$(x + 3)(x - 5) = 3x$$

is a quadratic and can be expressed in standard form.

$$(x + 3)(x - 5) = 3x \qquad \textbf{Multiplying terms}$$
$$x^2 - 5x - 15 = 0 \qquad \textbf{Combining like terms}$$

Such equations also are called second-degree equations in the variable x, since the polynomial $ax^2 + bx + c$ is of degree 2. The name *quadratic* comes from the Latin word *quadrare* for square, as in x^2.

On the other hand,

$$3x^2 + 5x = \frac{3}{x}$$

is not a quadratic equation because of the x in the denominator. When we multiply each term by x, $x \neq 0$, we arrive at the equation:

$$3x^3 + 5x^2 = 3.$$

Also,

$$(x^2 + 5)(x - 3) = 3x$$

is not a quadratic equation. When we multiply the left side, we have:

$$x^3 - 3x^2 + 5x - 15 = 3x,$$

which is an equation of the third degree.

This chapter discusses the quadratic equation and the various ways of finding the solutions (roots) of such equations. The individual elements of the solution of a quadratic equation are called the roots of the equation. Then we use quadratic equations to solve applied problems.

Our ability to work the problems in Section 11.2 depends on being able to factor quadratic expressions. Factoring techniques are presented in Section 2.7, so you may want to review that section before continuing.

11.2

Solving Quadratic Equations by Factoring

The solution of a quadratic equation by factoring depends upon the following theorem of algebra.

> **THEOREM 11.1**
>
> If the product of two numbers is zero, then one (or both) of the numbers is zero; that is, if $a \cdot b = 0$, then $a = 0$ or $b = 0$ (or both $a = 0$ and $b = 0$).

This theorem says that if the product of two expressions is zero, at least one of the two expressions must be zero. For example, if

$$(x - 5)(x + 7) = 0,$$

then

$$x - 5 = 0 \quad \text{or} \quad x + 7 = 0.$$

Now we have two linear expressions that can be solved. Thus, the values $x = 5$ and $x = -7$ are the solution of the original equation.

CAUTION A common mistake is to take an equation such as $(x + 5)(x - 7) = 3$ and set it up as follows.

$$x + 5 = 3 \quad \text{or} \quad x - 7 = 3$$

If we solve both equalities for x, the results are $x = -2$ and $x = 10$. When we substitute these values back into the original equation, we have:

$$(-2 + 5)(-2 - 7) \neq 3 \quad \text{or} \quad (10 + 5)(10 - 7) \neq 3$$
$$-27 \neq 3 \qquad\qquad\qquad\qquad 45 \neq 3. \quad \blacksquare$$

Remember that the first step in solving the equation is to express the quadratic equation in standard form.

The following set of rules is a guide for solving a quadratic equation by factoring.

SOLVING QUADRATIC EQUATIONS BY FACTORING

1. Write the quadratic equation in standard form.

$$ax^2 + bx + c = 0 \text{ (must be equal to zero)}$$

2. Factor $ax^2 + bx + c$.
3. Set each factor equal to 0 and solve the resulting linear equations.
4. Check the results by substituting the solution (the roots) back into the original equation.

EXAMPLE 1 **Determine the Solution of the Quadratic Equation by Factoring**

Solve $x^2 - 3x - 10 = 0$.

Solution The equation is in standard form, so we must factor $x^2 - 3x - 10 = 0$.

$$x^2 - 3x - 10 = 0$$
$$(x - 5)(x + 2) = 0 \qquad \text{Factoring}$$
$$x - 5 = 0 \quad \text{or} \quad x + 2 = 0 \qquad \text{Setting each factor equal to 0}$$
$$x = 5 \quad \text{or} \qquad x = -2 \quad \text{Solving for } x$$

The two roots, $x = 5$ and $x = -2$, are the solution of the quadratic equation $x^2 - 3x - 10 = 0$. To check the solution, substitute each root back into the original equation.

$$x = 5: \qquad (\boxed{5})^2 - 3(\boxed{5}) - 10 = 0$$
$$25 - 15 - 10 = 0$$
$$0 = 0. \checkmark$$

$$x = -2: \quad (\boxed{-2})^2 - 3(\boxed{-2}) - 10 = 0$$
$$4 + 6 - 10 = 0$$
$$0 = 0. \checkmark$$

Both roots check; therefore, $x = 5$ and $x = -2$ are the solution to the quadratic equation. ◆

We can make a visual check of Example 1 with the graphics calculator. Figure 11.1 shows the graph crossing the x-axis at -2 and at 5. Recall from Chapter 5 that the curve in Figure 11.1 is called a parabola. In fact, the solutions

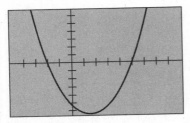

FIGURE 11.1

of all quadratic equations that have real number solutions can be checked visually with a graphics calculator.

EXAMPLE 2 **Solving a Quadratic Equation That Has No Constant Term**

Solve $I^2 + 5I = 0$

Solution The equation is in standard form and each term has a common factor of I.

$$I(I + 5) = 0 \qquad \text{Factoring common factor } I$$
$$I = 0 \quad \text{or} \quad I + 5 = 0 \qquad \text{Setting factors equal to zero}$$
$$I = 0 \quad \text{or} \qquad I = -5 \qquad \text{Solving for } I$$

The roots of the equation are 0 and -5. To check, substitute each root back into the original equation.

$$I = 0: \quad (\boxed{0})^2 + 5(\boxed{0}) = 0 \qquad I = -5: \quad (\boxed{-5})^2 + 5(\boxed{-5}) = 0$$
$$0 = 0. \checkmark \qquad\qquad\qquad 25 - 25 = 0$$
$$0 = 0. \checkmark$$

Both values check; therefore, $I = 0$ and $I = -5$ are roots of the quadratic equation. ◆

EXAMPLE 3 **Determining the Solution for the Difference of Two Squares**

Solve $R^2 - 9 = 0$.

Solution The equation $R^2 - 9 = 0$ is in standard form and it is another special case of the quadratic equation. The coefficient of the linear term is zero; that is, $b = 0$.

$$R^2 - 9 = 0$$
$$(R - 3)(R + 3) = 0 \qquad \text{Factoring difference of two squares}$$
$$R - 3 = 0 \quad \text{or} \quad R + 3 = -0 \qquad \text{Setting the factors equal to zero}$$
$$R = 3 \quad \text{or} \qquad R = -3 \qquad \text{Solving for } R$$

The solution of the equation is $R = 3$ and $R = -3$. ◆

A second method of solving for R in Example 3 is to solve for R^2 and take the square root of both sides. Try this method; the roots are the same.

EXAMPLE 4 **Determining the Solution for the Sum of Two Squares**

Solve $x^2 + 81 = 0$.

Solution   The equation $x^2 + 81 = 0$ is in standard form. The sum of two squares cannot be factored for the set of real numbers. That is, the roots of this equation are imaginary numbers.

$$x^2 + 81 = 0$$
$$x^2 + 81 - 81 = 0 - 81 \qquad \text{Solving for } x^2$$
$$x^2 = -81$$
$$x = \pm\sqrt{-81} \qquad \text{Taking the square root of both sides}$$

The solution is two imaginary numbers. Using rules from Chapter 9, we can write $x = \pm 9j$. The roots of the equation are $x = 9j$ and $x = -9j$. ◆

EXAMPLE 5   **Determining the Solution of a Quadratic Equation Not in Standard Form**

Solve $9x^2 + 12x = -4$.

Solution   First write the equation in standard form.

$$9x^2 + 12x = -4$$
$$9x^2 + 12x + 4 = 0 \qquad \text{Adding 4 to both sides of the equation}$$
$$(3x + 2)(3x + 2) = 0 \qquad \text{Factoring (perfect square trinomial)}$$
$$3x + 2 = 0 \quad \text{or} \quad 3x + 2 = 0 \qquad \text{Setting each factor equal to zero}$$
$$x = -\frac{2}{3} \quad \text{or} \quad x = -\frac{2}{3} \qquad \text{Solving for } x$$

For this problem, the solution to the equation is one unique value, $x = -\frac{2}{3}$. This is always the case when the factors are identical. We say that the root, $-\frac{2}{3}$, is of multiplicity two (i.e., the root appears twice). ◆

CAUTION   We may divide both sides of an equation with the same quantity only if we are positive that the quantity is not zero. In practice, this means that we can only divide both sides with a constant factor and not with an expression containing a variable. If we divide both sides with a variable, we may not arrive at the complete solution. This situation is illustrated in Example 6. ■

EXAMPLE 6   **Solving an Equation That Has a Common Factor on Both Sides of the Equal Sign**

Solve $k(3k + 2) = 5(3k + 2)$.

Solution   When we see the factor $3k + 2$ on both sides of the equal sign, we may be tempted to divide both sides of the equation by $3k + 2$. If we solve $3k + 2 = 0$ for k, $k = -\frac{2}{3}$. When $k = -\frac{2}{3}$, the expression $3k + 2$ is equal to zero.

Thus, when we divide both sides of the equation by $3k + 2$, we are dividing by zero, which is not permitted. To solve the equation, we need to have the equation in standard form.

$$k(3k + 2) - 5(3k + 2) = 0 \qquad \text{Subtracting } 5(3x + 2) \text{ from both sides}$$
$$(3k + 2)(k - 5) = 0 \qquad \text{Common factor}$$
$$3k + 2 = 0 \quad \text{or} \quad k - 5 = 0 \qquad \text{Setting both factors equal to zero}$$
$$k = -\frac{2}{3} \quad \text{or} \qquad k = 5 \qquad \text{Solving for } k$$

Thus, the solution of the equation is $k = -\frac{2}{3}$ and $k = 5$. We can obtain the same results by removing the parentheses, expressing the result in the form $ax^2 + bx + c = 0$, and then factoring. ◆

In Example 6, if we disregarded the rule about dividing by zero and performed the division, the result would have been $k = 5$. Had we done this, we may have concluded that there was only one root and never have discovered the second root.

From the previous examples, we may have concluded that the solution of a quadratic equation may have zero , one , or two distinct real roots. Section 11.4 discusses how to determine whether there will be zero , one , or two distinct real roots.

EXAMPLE 7 **Given the Roots, Determine the Quadratic Equation**

Determine a quadratic equation that has roots 3 and $-\frac{2}{3}$.

Solution To determine a quadratic equation whose roots are 3 and $-\frac{2}{3}$ think

of the procedure used to determine the roots in reverse. Let $x = 3$ or $x = -\frac{2}{3}$.

$$x = 3 \quad \text{or} \qquad 3x = -2 \qquad \text{Multiplying both sides of } x = -\frac{2}{3} \text{ by}$$
$$\text{3 to eliminate the fraction}$$
$$x - 3 = 0 \quad \text{or} \quad 3x + 2 = -0 \qquad \text{Subtracting 3 and adding 2 to both sides of}$$
$$\text{the respective expressions}$$
$$(x - 3)(3x + 2) = 0$$
$$3x^2 - 9x + 2x - 6 = 0 \qquad \text{Since both expressions equal zero, their}$$
$$\text{product equals zero}$$
$$3x^2 - 7x - 6 = 0 \qquad \text{Product of two factors}$$

A quadratic equation whose roots are 3 and $-\frac{2}{3}$ is $3x^2 - 7x - 6 = 0$. ◆

A quadratic function is a function of the form:

$$f(x) = ax^2 + bx + c \quad \text{where} \quad a \neq 0.$$

Thus, $f(x) = 3x^2 - 7x - 6$ is a quadratic function that takes on the value 0 when $x = 3$ and when $x = -\frac{2}{3}$ (see Example 8).

$$f(\boxed{3}) = 3(\boxed{3})^2 - 7(\boxed{3}) - 6 = 0 \checkmark$$

$$f\left(-\frac{2}{3}\right) = 3\left(\boxed{-\frac{2}{3}}\right)^2 - 7\left(\boxed{-\frac{2}{3}}\right) - 6 = 0 \checkmark$$

The roots of a quadratic function are also called **zeros** of the function. When the roots are substituted for the variable x, the functional value is zero.

EXAMPLE 8   Determine the Zeros of a Function

Determine the value(s) of x so that $f(x) = 0$.

$$f(x) = (2x - 3)(x - 7)$$

Solution   To determine the values of x so that $f(x) = 0$, set the factors $2x - 3$ and $x - 7$ equal to zero.

$$2x - 3 = 0 \quad \text{or} \quad x - 7 = 0$$

$$x = \frac{3}{2} \quad \text{or} \quad x = 7$$

Now substitute these values in the function, and we have:

$$f\left(\frac{3}{2}\right) = \left(2\left(\boxed{\frac{3}{2}}\right) - 3\right)\left(\boxed{\frac{3}{2}} - 7\right)$$

$$= (3 - 3)\left(\frac{3}{2} - 7\right)$$

$$= 0$$

and

$$f(\boxed{7}) = (2(\boxed{7}) - 3)(\boxed{7} - 7)$$

$$= 0.$$

Therefore, for $x = \frac{3}{2}$ and $x = 7$, $f(x) = 0$ or $\frac{3}{2}$ and 7 are zeros of $f(x) = (2x - 3)(x - 7)$. ◆

EXAMPLE 9   Application: Effect of Ticket Price on Game Attendance

The average attendance at a high school basketball game is 800 with a current ticket price of $4.00. A survey indicates that for each $0.20 decrease in the ticket price, the average attendance will increase by 80 persons. What price should be charged to increase gross income by $400? What will be the new average attendance?

Solution With the given information, it is better to use as the unknown something related to the change in ticket price rather than the price of the ticket. Let:

$$x = \text{the number of \$0.20 reductions that will give the desired increase in attendance,}$$

$$4.00 - 0.20x = \text{new ticket price in dollars,}$$

$$80x = \text{increase in average attendance;}$$

$$800 + 80x = \text{new average attendance;}$$

$$\text{Income in dollars} = \text{price of each ticket in dollars times average attendance;}$$

$$\text{Present income in dollars} = (4.00)(800) = 3200;$$

$$\text{Desired income in dollars} = 3200 + 400 = 3600.$$

Substitute in the formula:

$$\text{Desired income in dollars} = \text{New ticket price in dollars times average attendance}$$

$$3600 = (4.00 - 0.20x)(800 + 80x)$$

$$3600 = 3200 + 160x - 16x^2$$

$$16x^2 - 160x + 400 = 0$$

$$x^2 - 10x + 25 = 0$$

$$(x - 5)(x - 5) = 0$$

$$x = 5.$$

Since x is the number of reductions at $0.20 each, the ticket price should be:

$$\$4.00 - 0.20(5) = \$3.00.$$

With this reduction in ticket price, the new average attendance per game should be:

$$800 + 80(5) = 1200. \quad \blacklozenge$$

The technique used to solve the problem in Example 9 also can be used in solving similar problems in sales and manufacturing.

EXAMPLE 10 **Application: Constructing a Square Corner**

A contractor is preparing a corner for the footing (the base on which the walls are built) of a new building. He has a board 12 ft $\frac{1}{4}$ in. long. He plans to use the board to construct a right triangle as a guide to insure a square corner. His plan calls for making the hypotenuse 5 ft long. After he cuts a piece 5 ft long from the board for the hypotenuse, how should be cut the remaining piece of the board to insure that he ends up with a right triangle? (Assume that the width of each saw cut is $\frac{1}{8}$ in.; thus, the two saw cuts take up $\frac{1}{4}$ in.)

Solution   The problem provides the following facts (see Figure 11.2).

$\quad$ 12 ft = usable length of the board after saw cuts;

$\quad$ 5 ft = length of the hypotenuse of the right triangle;

$\quad$ 7 ft = sum of the length of the other two sides of the triangle.

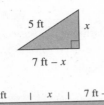

FIGURE 11.2

Let:

$\quad\quad x$ = the length of one of the sides of the triangle.

Then $7 - x$ = length of the third side. Since we are working with a right triangle, use the Pythagorean relation $c^2 = a^2 + b^2$.

$$5^2 = x^2 + (7 - x)^2 \quad\quad \text{Substituting}$$
$$25 = x^2 + 49 - 14x + x^2 \quad\quad \text{Squaring}$$
$$2x^2 - 14x + 24 = 0 \quad\quad \text{Collecting like terms}$$
$$x^2 - 7x + 12 = 0 \quad\quad \text{Dividing each term by 2}$$
$$(x - 4)(x - 3) = 0 \quad\quad \text{Factoring}$$
$$x - 4 = 0 \quad \text{or} \quad x - 3 = 0 \quad\quad \text{Setting each factor equal to zero}$$
$$x = 4 \quad \text{or} \quad x = 3 \quad\quad \text{Solving for } x$$

With the roots of the quadratic equation being 3 and 4, the carpenter cuts the remaining piece into lengths of 3 ft and 4 ft. With lengths of 3 ft, 4 ft, and 5 ft, the figure is a right triangle. ◆

11.2 Exercises

In Exercises 1–30, solve the quadratic equations and check the solutions algebraically and visually with a graphics calculator.

1. $x^2 - 5x + 6 = 0$

2. $x^2 - 11x + 18 = 0$

3. $x^2 - 2x = 15$

4. $x^2 + 2x = 15$

5. $2x^2 - 7x - 15 = 0$

6. $3x^2 - 4x - 15 = 0$

7. $5x^2 - 18x = 8$

8. $6x^2 - 19x = 20$

9. $9R^2 + 2 = -11R$

10. $7R^2 - 2 = -5R$

11. $R^2 - 64 = 0$

12. $R^2 - 121 = 0$

13. $I^2 - 125 = 0$

14. $I^2 - 13 = 0$

15. $Y^2 + 18Y + 81 = 0$

16. $Y^2 + 12Y + 36 = 0$

17. $16Y^2 + 24Y + 9 = 0$

18. $9Y^2 + 30Y + 25 = 0$

19. $Z^2 + 169 = 0$

20. $Z^2 + 289 = 0$

21. $n^2 + 13n = 0$

22. $n^2 + 11 = 0$

23. $5t^2 + 15t = 0$

24. $11t^2 + 1001t = 0$

25. $3c^2 = 7c + 6$

26. $c(c - 2) = 3c - 2(c + 1)$

27. $c(c + 4) - 4 = c - 3(c - 1)$

28. $3(x - 4) = x(x - 4)$

29. $\dfrac{x^2}{8} + \dfrac{3x}{4} = -1$

30. $\dfrac{x^2}{4} + x = 3$

In Exercises 31–40, determine the quadratic equation that has the indicated roots.

31. 3 and -5

32. 5 and -3

33. $\dfrac{2}{5}$ and 3

34. $\dfrac{2}{3}$ and 4

35. only 5

36. only $\dfrac{3}{4}$

37. $\dfrac{1}{3}$ and $\dfrac{1}{2}$

38. -0.2 and -0.4

39. $-1 - 2j$ and $-1 + 2j$

40. only $-\dfrac{3}{2}$

In Exercises 41–50, determine each value(s) of x for which $f(x) = 0$.

41. $f(x) = (x - 5)(x + 3)$

42. $f(x) = (x + 11)(x - 7)$

43. $f(x) = (2x + 11)(3x - 5)$

44. $f(x) = (3x + 8)(2x - 9)$

45. $f(x) = x^2 - 7x + 12$

46. $f(x) = x^2 - x - 12$

47. $f(x) = x^2 + 22x + 121$

48. $f(x) = x^2 - \dfrac{x}{4} - \dfrac{1}{8}$

49. $f(x) = \dfrac{x^2}{9} - \dfrac{2x}{3} + 1$

50. $f(x) = 13x^2 + 169x$

Perform each of the following steps in Exercises 51–60.
  a. If the quadratic equation is not given, write the equation.
  b. Solve the equation.
  c. Check the solution.
  d. Answer the question.

51. A rectangular heating duct is to be formed by bending an 80-cm-wide piece of sheet metal into a rectangular-shaped opening. The cross-sectional area of the duct opening must be 300 cm². What are the height and width of the duct? Use the drawing.

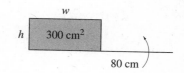

52. The Pepper family uses about 10 tons of coal each winter to heat their home. They want to build a bin 10 ft high and with the length 3 ft more than the width. Allowing 40 ft³ to the ton, what dimensions should they use for the bin so that it will hold 10 tons of coal?

53. Determine the length of a side of a right triangle that is 4.00 m shorter than the other side and 8.00 m shorter than the hypotenuse.

54. The net power output $P(I)$ from an electric generator is the gross power produced VI less the heat loss generated in the wires (I^2R). A quadratic with respect to the current can be expressed as $P(I) = VI - I^2R$.
  a. What is the current I in the generator that gives a net output of 2400 watts if the voltage V is set at the standard 120 volts and the resistance R is 1.5 Ω?
  b. If the current gets too high, more and more power is lost to heat. What is the value for the current when $P(I) = 0$, $V = 120$ volts, and $R = 1.5$ Ω?

55. The number of calories burned by running at a speed of x mph is determined by the function $C(x) = 5x^2 + 20x + 25$. How fast should someone run to burn 505 cal in one hour?

56. A retailer sells, on average, 1200 gadgets a month. The current selling price of the gadget is $6.00. It is estimated that for each $0.30 decrease in the price, the average monthly sales will increase by 120. What price should be charged to increase gross income by $900.00? What will be the new average monthly sales of gadgets?

57. The rotational motion of a body is given by $\phi = \alpha t^2 + \beta t$. Find the value of t (in seconds) for $\alpha = 4$, $\beta = 12$, and $\phi = 16$.

58. Determine the dimensions of a rug that covers 60% of the floor of a room and is equidistant from the walls. The dimensions of the floor of the room are 12 ft by 15ft.

59. When an object is propelled in one direction and it has a constant acceleration in another direction, the path is a parabola. For example, a cannon ball fired at an angle of 30° very nearly follows the path of a parabola. If air resistance is neglected the equation that describes this motion is $y = 0.58x - 6.5x^2$. Let $y = 0$ and solve for x. The roots you obtain are the starting point of the projectile on the x-axis and where it finally hits along the x-axis.

60. Suppose a projectile is aimed straight up. Although the path of the projectile is a linear one, straight up and straight down, a quadratic equation still plays a crucial role. Suppose a ball is thrown up with an initial speed of 64 ft/s. In this case, the relationship between the distance and the time is given by the equation $y = 64t - 16t^2$.
   a. How long does the ball stay in the air? (*Hint:* Let $y = 0$.)
   b. Assuming that time up to the highest point is the same as the time down, what is the maximum height the ball reaches?

11.3

Completing the Square

In Chapter 2 we learned how to solve $x^2 = a$ by taking the square root of both sides:

$$\sqrt{x^2} = \sqrt{a}$$
$$x = \pm\sqrt{a},$$

where $x = \pm\sqrt{a}$ means $x = +\sqrt{a}$ or $x = -\sqrt{a}$. What this means is that we must take both the positive and negative square root to get the complete solution to the equation.

The solution of the equation $(x + 5)^2 = 64$ can be found by taking the square root of both sides of the equation:

$x + 5 = \pm 8$	**Taking square root of both sides**
$x = -5 \pm 8$	**Solving for x**
$x = -5 + 8$ or $x = -5 - 8$	
$x = 3$ or $x = -13$.	

The roots of the equation $(x + 5)^2 = 64$ are 3 and -13.

The right side of the equation $(x + 5)^2 = 64$ is a perfect square. Thus, if we can take any quadratic equation and write it in the form

$$(x + h)^2 = k^2,$$

where h and k are constants, we can solve for x by taking the square root of both sides. Quadratic equations that are not in this form can be put into this form by a technique called **completing the square.** That is, we make the left side of the equation a perfect square. This is illustrated in Example 1.

EXAMPLE 1   Solving a Quadratic by Completing the Square

Solve the quadratic equation $x^2 + 8x - 9 = 0$ by the method of completing the square.

Solution   To solve the equation $x^2 + 8x - 9 = 0$ by completing the square, we do the following:

$$x^2 + 8x - 9 + 9 = 0 + 9 \qquad \text{Adding 9 to both sides}$$
$$x^2 + 8x = 9. \qquad \text{Simplifying}$$

The left side becomes a perfect square when we add the correct constant. We can find this constant by dividing the coefficient of the linear term by 2, which leads us to the coefficient $\dfrac{8}{2}$ or 4. Then, when we square this quotient $(4)^2$, the product is 16. The square of the quotient is then added to both sides of the equation. On the left side, we have a perfect square trinomial. (Remember that we add the number to the right side so that the equality is not changed.)

$$x^2 + 8x + (4)^2 = 9 + (4)^2$$
$$x^2 + 8x + (4)^2 = 9 + 16$$
$$(x + 4)^2 = 25 \qquad \text{Writing the left side as a perfect square}$$
$$\sqrt{(x + 4)^2} = \pm\sqrt{25} \qquad \text{Taking the square root of both sides}$$
$$x + 4 = \pm 5$$

$$x + 4 = 5 \qquad \text{or} \quad x + 4 = -5$$
$$x = -4 + 5 \quad \text{or} \qquad x = -4 - 5 \qquad \text{Solving for } x$$
$$x = 1 \qquad \text{or} \qquad x = -9$$

The roots of the equation $x^2 + 8x - 9 = 0$ are 1 and -9. We could arrive at the same result by factoring. ◆

We can check the solution for Example 1 by substituting 1 and -9 into the original equation, or we may check it visually with a graphics calculator. Show that the solution is correct by checking now.

To use the method of completing the square, the coefficient of the second-degree term must be 1. This is illustrated in Example 2.

EXAMPLE 2   Solve the Equation by Completing the Square

Solve the quadratic equation $3x^2 + 12x - 15 = 0$ by the method of completing the square.

Solution In Example 1 the coefficient of the second-degree term was 1. For the equation $3x^2 + 12x - 15 = 0$ the coefficient is 3. To use the method of completing the square, the coefficient of the second-degree term must be 1.

$$\frac{3x^2}{3} + \frac{12x}{3} - \frac{15}{3} = \frac{0}{3}$$ **Dividing each term by 3**

$$x^2 + 4x = 5$$ **Adding 5 to both sides**

$$x^2 + 4x + \left(\frac{4}{2}\right)^2 = 5 + \left(\frac{4}{2}\right)^2$$ **Dividing 4 by 2, squaring, and adding to both sides**

$$x^2 + 4x + (2)^2 = 5 + 4$$

$$(x + 2)^2 = 9$$ **Factoring**

$$x + 2 = \pm 3$$ **Taking square root of both sides**

$$x = -2 \pm 3$$ **Solving for x**

$$x = 1 \quad \text{and} \quad x = -5$$

The roots of the quadratic equation $3x^2 + 12x - 15 = 0$ are 1 and -5. ◆

SOLVING QUADRATIC EQUATIONS BY COMPLETING THE SQUARE

1. Divide both sides of the equation by the coefficient of x^2 (if it is not already one).
2. Move the constant term to the right side of the equal sign and all terms containing xs to the left.
3. Complete the square by adding the square of one-half the coefficient of the linear term (the first-degree term) to both sides.
4. Factor the left side and combine the terms on the right side.
5. Take the square root of both sides and solve for x.

EXAMPLE 3 **Application: Dimensions of a Storage Area**

A company decides on a rectangular storage area for its chemicals. The best design for the material to be stored is that the length of the area be twice the width. The storage area will be surrounded by a concrete wall that is 4 m wide. If the area of the concrete wall is exactly twice the area of the storage area, determine the dimensions of the storage area.

Solution Figure 11.3 will provide a better understanding of the problem. Let:

$$x = \text{width of the storage area in meters,}$$

then

$$2x = \text{length of the storage area in meters;}$$

Area of storage $= (\text{length}) (\text{width}) = (2x)(x) = 2x^2;$

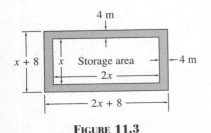

FIGURE 11.3

Area of wall = area of two end rectangles + area of two side rectangles;

$$= 2(\text{width of wall})(\text{length of wall})$$
$$+ 2(\text{width of wall})(\text{length of wall})$$
$$= 2(\boxed{4})(\boxed{x+8}) + 2(\boxed{4})(\boxed{2x}) \qquad \textbf{Area of four rectangles}$$
$$\textbf{forming the wall}$$
$$= 8x + 64 + 16x$$
$$= 24x + 64.$$

Therefore, since the area of the concrete wall is twice the storage area:

$$\text{Area of concrete wall} = 2(\text{area of storage area})$$

$$24x + 64 = 2(2x^2)$$

$$4x^2 - 24x - 64 = 0 \qquad\qquad \textbf{Writing in standard form}$$

$$x^2 - 6x - 16 = 0 \qquad\qquad \textbf{Dividing each term by 4}$$

$$x^2 - 6x - 16 + 16 = 0 + 16$$

$$x^2 - 6x + (-3)^2 = 16 + 9$$

$$(x - 3)^2 = 25$$

$$x - 3 = \pm 5 \qquad\qquad \textbf{Square root of both sides}$$

$$x = -2 \quad \text{or} \quad x = 8. \qquad \textbf{Solving for } x$$

The width of the storage area cannot be negative, so the width must be 8 m, and the length is $2(8) = 16$ m. ◆

EXAMPLE 4 Complex Roots

Solve the quadratic equation $3x^2 - 5x + 7 = 0$ by completing the square.

Solution   Solve the equation $3x^2 - 5x + 7 = 0$ by following the suggested steps.

$$x^2 - \frac{5}{3}x + \frac{7}{3} = \frac{0}{3} \qquad\qquad \textbf{Step 1, divide each term by 3}$$

$$x^2 - \frac{5}{3}x = -\frac{7}{3} \qquad\qquad \textbf{Step 2, move constant term to the right side}$$

$$x^2 - \frac{5}{3}x + \left[\frac{1}{2}\left(\frac{-5}{3}\right)\right]^2 = -\frac{7}{3} + \left[\frac{1}{2}\left(\frac{-5}{3}\right)\right]^2 \qquad \textbf{Step 3, complete the square}$$

$$x^2 - \frac{5}{3}x + \left(\frac{-5}{6}\right)^2 = -\frac{7}{3} + \frac{25}{36}$$

$$\left(x - \frac{5}{6}\right)^2 = -\frac{59}{36} \qquad\qquad \textbf{Step 4, factor left side and combine like terms on the right}$$

$$x - \frac{5}{6} = \pm \frac{\sqrt{-59}}{6} \qquad\qquad \textbf{Step 5, take the square root of both sides and solve for } x$$

$$x = \frac{5}{6} \pm j\frac{\sqrt{59}}{6}$$

The roots of the equation $3x^2 - 5x + 7 = 0$ are

$$\frac{5}{6} + j\frac{\sqrt{59}}{6} \quad \text{and} \quad \frac{5}{6} - j\frac{\sqrt{59}}{6}. \quad \blacklozenge$$

11.3 Exercises

In Exercises 1–10, determine the roots of the quadratic equation by taking the square root of both sides.

1. $(x - 8)^2 = 36$ **2.** $(x - 4)^2 = 64$ **3.** $2x^2 = 18$ **4.** $2x^2 = 50$

5. $(x + 5)^2 = 15$ **6.** $(x + 7)^2 = 13$ **7.** $(3x + 5)^2 = 49$ **8.** $(4x + 1)^2 = 121$

9. $(y - 3)^2 = -16$ **10.** $(y + 5)^2 = -81$

In Exercises 11–30, solve each quadratic equation by completing the square. Check the results algebraically and visually with a graphics calculator.

11. $x^2 + 3x - 4 = 0$ **12.** $x^2 + 2x - 8 = 0$ **13.** $x^2 + 7x - 3 = 0$ **14.** $x^2 + 10x - 4 = 0$

15. $2R^2 + 3R = 9$ **16.** $2R^2 - 4R - 6 = 0$ **17.** $5A^2 = 1 - A$ **18.** $\frac{1}{3} - 2A - 3A^2 = 0$

19. $3x^2 - 9x = 4$ **20.** $3x^2 + 15x = 2$ **21.** $x(x - 5) = 3(1 - x)$ **22.** $(x - 3)(x + 4) = x$

23. $y^2 - 6y + 9 = 0$ **24.** $3y^2 - 30y + 75 = 0$ **25.** $Z^2 + 5Z + 7 = 0$ **26.** $3Z^2 + 2Z + 1 = 0$

27. $(x - 4)(x + 5) = x$ **28.** $x(x + 7) = 5(1 - x)$ **29.** $ax^2 + bx + c = 0$ **30.** $mx^2 + nx + c = 0$

31. The cost c (in dollars) of building a counter is given by:

$$c = \frac{\ell^2}{10} - 3\ell,$$

where ℓ is the length of the counter (in meters). How long a counter can be made if the cost is to be kept to $280?

32. The Christophers are planning a garden on a rectangular piece of land that is 50 ft by 60 ft. The plans include a border of uniform width planted with flowers around the outside of the garden. How wide should the border of flowers be if the Christophers plan to use two-thirds of the available space for produce?

33. In business, the cost of producing a product may be represented by an equation that is quadratic in nature. These equations are called cost equations. For example, a cost equation for manufacturing a gadget may be $C = x^2 + 16x + 114$, where C is the cost of manufacturing x units per week. Determine the number of units that can be manufactured with a weekly cost of $6450.

34. A farmer has a rectangular field that is 100 m by 200 m. He wants to plant exactly half the field in alfalfa and the other half in corn. In his planning for future crop rotation and conservation, he decides to plant the corn in the center of the field and the alfalfa around the outside. The band of alfalfa around the corn will be the same on each side of the field. How wide will the strip of alfalfa be on the four sides of the field? What are the dimensions of the corn field?

11.4

The Quadratic Formula

The method of solving a quadratic equation by completing the square provides us with a technique for solving any quadratic equation. In this section, we use the method of completing the square to solve the general quadratic equation:

$$ax^2 + bx + c = 0 \quad (a, b, \text{ and } c \text{ are any real numbers and } a \neq 0).$$

The final result is the powerful **quadratic formula.** The quadratic formula is not always the simplest tool to use to obtain a solution; however, it always works. Because of its importance in solving quadratic equations, it is a useful formula to memorize. The derivation of this formula follows.

$$ax^2 + bx + c = 0$$

$$x^2 + \frac{b}{a}x + \frac{c}{a} = \frac{0}{a}$$ Step 1, divide each term by a

$$x^2 + \frac{b}{a}x = -\frac{c}{a}$$ Step 2, subtract constant term from both sides of equation

$$x^2 + \frac{b}{a}x + \left[\frac{1}{2}\left(\frac{b}{a}\right)\right]^2 = -\frac{c}{a} + \left[\frac{1}{2}\left(\frac{b}{a}\right)\right]^2$$ Step 3, complete the square

$$x^2 + \frac{b}{a}x + \left(\frac{b}{2a}\right)^2 = -\frac{c}{a} + \frac{b^2}{4a^2}$$

$$\left(x + \frac{b}{2a}\right)^2 = \frac{-4ac + b^2}{4a^2}$$ Step 4, factor left side, and combine like terms on the right

$$x + \frac{b}{2a} = \pm\frac{\sqrt{b^2 - 4ac}}{2a}$$ Step 5, take square root of both sides

Therefore,

$$x = -\frac{b}{2a} + \frac{\sqrt{b^2 - 4ac}}{2a} \quad \text{or} \quad x = \frac{-b}{2a} - \frac{\sqrt{b^2 - 4ac}}{2a}$$

$$x = \frac{-b + \sqrt{b^2 - 4ac}}{2a} \quad \text{or} \quad x = \frac{-b - \sqrt{b^2 - 4ac}}{2a}$$

The solutions to $ax^2 + bx + c = 0$ are

$$x = \frac{-b + \sqrt{b^2 - 4ac}}{2a} \quad \text{or} \quad x = \frac{-b - \sqrt{b^2 - 4ac}}{2a}.$$

DEFINITION 11.2

The **quadratic formula** for solving $ax^2 + bx + c = 0$, where $a \neq 0$, is:

$$x = \frac{-b \pm \sqrt{b^2 - 4ac}}{2a}.$$

The formula can be used to solve any quadratic equation in standard form without going through the steps of completing the square. All we need do is to substitute the appropriate values for a, b, and c in the formula.

EXAMPLE 1 Solving a Quadratic Equation with the Quadratic Formula

Solve $2x^2 + 3x - 1 = 0$ using the quadratic formula.

Solution For the equation $2x^2 + 3x - 1 = 0$, $a = 2$, $b = 3$, and $c = -1$. Substituting in the quadratic formula.

$$x = \frac{-b \pm \sqrt{b^2 - 4ac}}{2a},$$

we have:

$$x = \frac{-3 \pm \sqrt{(3)^2 - 4(2)(-1)}}{2(2)}$$

$$= \frac{-3 \pm \sqrt{9 + 8}}{4} \quad \text{or} \quad \frac{-3 \pm \sqrt{17}}{4}.$$

Therefore,

$$x = \frac{-3 + \sqrt{17}}{4} \quad \text{and} \quad x = \frac{-3 - \sqrt{17}}{4}$$

is the solution of the quadratic equation. We can check the solution by substituting the roots back into the original equation.

Check: $2x^2 + 3x - 1 = 0$ if $x = \dfrac{-3 + \sqrt{17}}{4}$

$$2\left(\frac{-3 + \sqrt{17}}{4}\right)^2 + 3\left(\frac{-3 + \sqrt{17}}{4}\right) - 1 = 0$$

$$\overset{}{\cancel{2}}\left(\frac{9 - 6\sqrt{17} + 17}{\underset{8}{\cancel{16}}}\right) + \frac{-9 + 3\sqrt{17}}{4} - 1 = 0$$

$$\frac{9 - 6\sqrt{17} + 17 - 18 + 6\sqrt{17} - 8}{8} = 0 \quad 0 = 0. \checkmark$$

Now check the other root. ◆

EXAMPLE 2 Determining the Roots of a Quadratic Equation Using the Quadratic Formula

Solve $5x^2 - 6x - 2 = 0$ using the quadratic formula.

Solution   For the equation $5x^2 - 6x - 2 = 0$, $a = 5$, $b = -6$, and $c = -2$. Substituting in the equation

$$x = \frac{-b \pm \sqrt{b^2 - 4ac}}{2a}.$$

we have:

$$x = \frac{-(-6) \pm \sqrt{(-6)^2 - 4(5)(-2)}}{2(5)}$$

$$= \frac{6 \pm \sqrt{36 + 40}}{10}$$

$$= \frac{6 \pm \sqrt{76}}{10}$$

$$= \frac{6 \pm \sqrt{(4)(19)}}{10} \qquad \text{Factoring}$$

$$= \frac{6 \pm 2\sqrt{19}}{10} \qquad \text{Taking square root of 4}$$

$$= \frac{2(3 \pm \sqrt{19})}{10} \qquad \text{Common-term factoring}$$

$$= \frac{3 \pm \sqrt{19}}{5} \qquad \text{Dividing numerator and denominator by 2}$$

Therefore,

$$x = \frac{3 + \sqrt{19}}{5}, \text{ or } 1.472 \quad \text{and} \quad x = \frac{3 - \sqrt{19}}{5}, \text{ or } -0.272$$

are the roots of the quadratic equation. Assuming that 2, 5, and 6 are exact numbers, we can express the answer to the thousandths place. ◆

The solutions of some quadratic equations are complex roots, as illustrated in Example 3.

EXAMPLE 3   **Complex Roots with the Quadratic Formula**

Solve $x^2 + x + 1$ using the quadratic formula.

Solution   In the equation $x^2 + x + 1 = 0$, $a = 1$, $b = 1$, and $c = 1$. Substituting in the equation we have:

$$x = \frac{-1 \pm \sqrt{(1)^2 - 4(1)(1)}}{2(1)}$$

$$= \frac{-1 \pm \sqrt{1 - 4}}{2}$$

$$= \frac{-1 \pm \sqrt{-3}}{2}.$$

Recall that $\sqrt{-1} = j$. Therefore,

$$x = \frac{-1}{2} + \frac{j\sqrt{3}}{2} \quad \text{and} \quad x = \frac{-1}{2} - \frac{j\sqrt{3}}{2}$$

are the roots of the equations $x^2 + x + 1 = 0$. ◆

EXAMPLE 4   **Application: Determining Dimensions of a Pool and Pool Area**

An architect is designing a solar house with a glass enclosed area on the south side. The client decides that she wants a swimming pool (for recreation and a heat-storage mass) in this enclosed area.

The client wants the length of the pool area to be $2\frac{1}{2}$ times the width. She also wants a walk around the pool that is 1 m wide on one side and one end and $1\frac{1}{2}$ m wide on the other side and end. With these restrictions, the architect has an area of $43\frac{3}{4}$ m² for the pool. What are the dimensions of the pool and the pool area (pool and walk)?

Solution   First, sketch the pool and the surrounding walk as in Figure 11.4. Let x = width of the pool area,

$$2\frac{1}{2}x = \text{length of the pool area;}$$

$$x - 2\frac{1}{2} = \text{width of pool;}$$

$$2\frac{1}{2}x - 2\frac{1}{2} = \text{length of pool;}$$

$$\left(x - 2\frac{1}{2}\right)\left(2\frac{1}{2}x - 2\frac{1}{2}\right) = \text{area of the pool (width × length).}$$

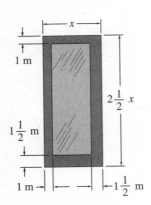

FIGURE 11.4

Since the area to be covered by the pool is $43\frac{3}{4}$ m², we have:

$$\left(x - \frac{5}{2}\right)\left(\frac{5}{2}x - \frac{5}{2}\right) = 43\frac{3}{4}$$

$$\frac{5}{2}x^2 - \frac{35}{4}x + \frac{25}{4} = \frac{175}{4} \qquad \textbf{Multiplying}$$

$$2x^2 - 7x + 5 = 35 \qquad \textbf{Multiplying each term by } \frac{4}{5}$$

$$2x^2 - 7x - 30 = 0.$$

Solve for x using the quadratic formula with $a = 2$, $b = -7$, and $c = -30$:

$$x = \frac{-(\boxed{-7}) \pm \sqrt{(\boxed{-7})^2 - 4(\boxed{2})(\boxed{-30})}}{2(\boxed{2})}$$ **Substituting**

$$= \frac{7 \pm \sqrt{289}}{4}$$

$$= \frac{7 + 17}{4} \quad \text{or} \quad x = \frac{7 - 17}{4}$$

$$x = 6 \quad \text{or} \quad = -\frac{10}{4}.$$

Reject $-\frac{10}{4}$ since x is a positive measurement. Therefore, the width of the pool area is 6 m and the length is 15 m. The width of the pool is $3\frac{1}{2}$ m and the length is $12\frac{1}{2}$ m.

Check: Width of the pool $\times$ length of pool $= 43\frac{3}{4}$.

$$3\frac{1}{2} \times 12\frac{1}{2} = 43\frac{3}{4} \checkmark$$

Also, the length of the pool area is $2\frac{1}{2}$ times the width. ◆

EXAMPLE 5 **Application: Determining the Speed of Two Boats**

Two motorboats travel at right angles to each other after leaving the same dock at the same time. One hour after starting, the boats are 15 km apart. If one travels 9.0 km/hr faster than the other, what is the speed of each boat?

Solution Let:

$$x\frac{\text{km}}{\text{hr}} = \text{speed of one boat and } (x + 9)\frac{\text{km}}{\text{hr}} = \text{speed of other boat.}$$

The first boat travels x km in one hour, and the second boat travels $(x + 9)$ km in one hour. Since the boats are traveling at right angles to each other, we can illustrate their relative positions with a right triangle as in Figure 11.5. The hypotenuse of the right triangle is the distance the two boats are apart at the end of one hour. Because we assigned values to the two legs and the hypotenuse of the right triangle, we can solve for x using the Pythagorean theorem.

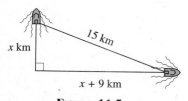

FIGURE 11.5

$$a^2 + b^2 = c^2$$
$$(\boxed{x})^2 + (\boxed{x + 9})^2 = (\boxed{15})^2 \quad \textbf{Substituting}$$
$$x^2 + x^2 + 18x + 81 = 225 \quad \textbf{Squaring}$$
$$2x^2 + 18x - 144 = 0 \quad \textbf{Collecting like terms}$$
$$x^2 + 9x - 72 = 0 \quad \textbf{Dividing each term by 2}$$

To solve, use the quadratic formula.

$$x = \frac{-\boxed{9} \pm \sqrt{(\boxed{9})^2 - 4(\boxed{1})(-\boxed{72})}}{2(\boxed{1})} \quad \text{or} \quad \frac{-9 \pm \sqrt{369}}{2}$$

$$x = 5.10 \quad \text{or} \quad x = -14.1.$$

Since rate of speed is positive, reject -14.1. Therefore, one boat traveled at the rate of 5.1 km/hr, and the other boat traveled at the rate of 14 km/hr. Check the answer. (It may not check exactly due to rounding.) ◆

EXAMPLE 6   Application: Zero Bending Moments for an Examining Table

A doctor's examining table has a length L of 7.00 ft. The load W on the table top due to the patient's weight is assumed to be distributed evenly along the table top at 70.0 lb/ft of length. The positions of the legs d are planned at 1.00 ft from the ends of the table. The bending moment M induced in the center section of the table top by the patient's weight is given by:

$$M = -\frac{W}{2L}x^2 + \frac{W}{2}x - \frac{W}{2}d \quad [d \le x \le (L - d)].$$

The origin is at the foot end of the table. Determine the location(s) in the table where the bending moment is zero.

Solution   To determine the locations where the bending motion is zero, let $M = 0$ in the equation for the bending moment and solve the quadratic equation for x.

$$0 = -\frac{W}{2L}x^2 + \frac{W}{2}x - \frac{W}{2}d$$

From the information given, we know that $L = 7.00$, $W = 70.0$, and $d = 1.00$. Substituting these values into the equation, we have:

$$0 = -\frac{70.0}{2(7.00)}x^2 + \frac{70.0}{2}x - \frac{70.0}{2}(1.00)$$

$$0 = -5.00x^2 + 35.0x - 35.0$$

$$1.00x^2 - 7.00x + 7.00 = 0 \quad \text{Dividing each term by } -5$$

Since the quadratic is not factorable, use the quadratic formula to solve for x.

$$x = \frac{-b \pm \sqrt{b^2 - 4ac}}{2a}$$

$$= \frac{-(-7.00) \pm \sqrt{(-7.00)^2 - 4(1.00)(7.00)}}{2(1.00)}$$

$$= 1.21 \quad \text{or} \quad 5.79$$

The points where the table top has zero bending moments are 1.21 ft and 5.79 ft from the foot of the table. ◆

By writing a program for calculator or computer, we can save a great deal of time finding the solution of a quadratic equation. The following programs for the fx–7700G and the TI–82 give real number solutions.

Program on the fx–7700G for the Quadratic Formula

Keystrokes	Screen Display

Place calculator in WRITE mode:

MODE 2

SHIFT PRGM

F1 (JMP)   JMP REL Prg ? ◢ :
                                  ⇒ Goto Lbl Dsz Isz

F3 (Lb1) 1 PRE F6 (:)*   Lbl 1:

F4 (?) → ALPHA A F6 (:)   ? → A:

F4 (?) → ALPHA B F6 (:)   ? → B:

F4 (?) → ALPHA C F6 (:)   ? → C:

(ALPHA B SHIFT x^2 − 4
× ALPHA A × ALPHA C)) $(B^2 − 4 \times A \times C)$

F2 (REL) F4 (<) 0 PRE F1
(JMP) F1 (⇒) F2 (Goto) 2

PRE F6 (:) $0 < ⇒ \text{Goto } 2:$

(SHIFT (−) ALPHA B +
√ (ALPHA B SHIFT X^2
− 4 × ALPHA A × ALPHA
C)) ÷ (2 × ALPHA
A) F5 (◢) $(−B + \sqrt{\ }(B^2 − 4 \times A \times C))$
                                  $÷ (2 \times A)◢$

(SHIFT (−) ALPHA B −
√ (ALPHA B SHIFT X^2
− 4 × ALPHA A × ALPHA
C)) ÷ (2 × ALPHA
A) F5 (◢) $(−B − \sqrt{\ }(B^2 − 4 \times A \times C)$
                                  $÷ (2 \times A)$ ◢

F2 (Goto) 1 F5 (◢)   Goto 1◢

F3 (Lbl) 2 PRE F6 (:) ALPHA
F2 (″) SHIFT ALPHA C O

M P L E × SPACE R O O T
S F2 (″) PRE F5 (◢)   Lbl 2: "COMPLEX ROOTS"◢

(Take calculator out of WRITE mode):

MODE 1

*Do not press EXE ; the program will wrap around.

To see how the program runs, use the quadratic equation from Example 6. Recall that $a = 1$, $b = -7$, and $c = 7$.

Keystrokes	Screen Display
SHIFT PRGM F3 (Prg) 0	Prog 0 (0 is the place where program is stored.)
EXE	?
1	1
EXE	?
(−) 7	−7
EXE	?.
7	7
EXE	5.791288 –Disp–
EXE	1.208712 –Disp–

In executing the program, three question marks that represent the coefficients of the quadratic equation are displayed. The calculator is asking you to enter the values for a, b, and c in that order. The word "Disp" indicates that the answer is displayed. To solve additional quadratic equations, all we need do is press EXE . A question mark appears requesting a value for a. Enter the value and follow the procedure that was just illustrated.

Program on the TI-82 for the Quadratic Formula

Keystrokes	Screen Display
PRGM	
▶ ▶	EXEC EDIT NEW
ENTER	1: Create New
2nd ALPHA Q U A D	PROGRAM Name = QUAD
ENTER	PROGRAM: QUAD
PRGM ▶ 3 (Display) 2ND	
A-LOCK " A " ENTER	: Disp "A"
PRGM ▶ 1 (Input)	
ALPHA A ENTER	: Input A
PRGM ▶ 3 (Display) 2ND	
A-LOCK " B " ENTER	: Disp "B"

PRGM ▶ 1 (Input) ALPHA	
B ENTER	: Input B
PRGM ▶ 3 (Display) 2nd	
A-LOCK " C " ENTER	: Display "C"
PRGM ▶ 1 (Input) ALPHA	
C ENTER	: Input C
PRGM 1 (IF) (ALPHA	
B x^2 − 4 × ALPHA A	
× ALPHA C) 2ND TEST	
5 (<) 0 ENTER	: If ($B^2 − 4*A*C$) < 0
PRGM 0 (Goto) 2 ENTER	: Goto 2
2nd Y-VARS 1	
(Function) 1 (Y_1) ENTER	: Y_1
PRGM ▶ 3 (Disp) ALPHA	
" 2nd Y-VARS 1 1	
(Y_1) 2nd TEST 1 (=)	
ALPHA " ENTER	: Disp "Y_1 = "
PRGM ▶ 3 (Disp) 2nd	
Y-VARS 1 1 (Y_1)	
ENTER	: Disp Y_1
2nd Y-VARS 1 2 (Y_2)	
ENTER	: Y_2
PRGM ▶ 3 (Disp) ALPHA	
" 2nd Y-VARS 1 2	
(Y_2) 2nd TEST 1	
(=) ALPHA " ENTER	: Disp "Y_2="
PRGM ▶ 3 (Disp) 2nd	
Y-VARS 1 2 (Y_2)	
ENTER	: Disp Y_2
PRGM 0 (Goto) 3 ENTER	: Goto 3
PRGM 9 (Lbl) 2 ENTER	: Lb1 2
PRGM ▶ 3 (Disp) 2nd	
ALPHA " C O M P	
L E X ⌴ R O O T	
S " ENTER	: Disp "COMPLEX ROOTS"
PRGM 9 (Lbl) 3 ENTER	: Lbl 3
2nd QUIT	: (Return to home screen)

The Y_1 and Y_2 in the program for the TI–82 represent equations. The equations in this case are:

$$Y_1 = \frac{-b + \sqrt{b^2 - 4ac}}{2a} \qquad Y_2 = \frac{-b - \sqrt{b^2 - 4ac}}{2a}.$$

We press $\boxed{Y=}$ and then enter the equations to the right of $Y_1 =$ and $Y_2 =$.

To demonstrate how the program for solving a quadratic equation on the TI–82 runs, solve the following equation.

$$3x^2 + 7x - 5 = 0$$

For this equation $a = 3$, $b = 7$, and $c = -5$.

Keystrokes	Screen Display
$\boxed{\text{PRGM}}$ $\boxed{\blacktriangledown}$ (2) $\boxed{\text{ENTER}}$	Prgm QUAD
	A
$\boxed{\text{ENTER}}$ 3	?3
	B
$\boxed{\text{ENTER}}$ 7	?7
	C
$\boxed{\text{ENTER}}$ $\boxed{(-)}$ 5	?−5
$\boxed{\text{ENTER}}$	$Y_1 = .573$
	$Y_2 = -2.907$
	Done

To solve additional equations, press $\boxed{\text{ENTER}}$. A appears followed by a question mark. Follow the procedure that was just demonstrated.

11.4 Exercises

In Exercises 1–34, solve the quadratic equations using the quadratic formula. Use the calculator to check the solution.

1. $x^2 - 16x - 36 = 0$

2. $x^2 + 16x - 36 = 0$

3. $x^2 + 14x + 24 = 0$

4. $x^2 - 14x + 24 = 0$

5. $6x^2 - 12x - 5 = 0$

6. $6x^2 - 7x - 5 = 0$

7. $2y^2 + 4y = -3$

8. $2y^2 + 7y = -3$

9. $6y^2 = 19y - 3$

10. $6y^2 = 9y - 3$

11. $8x^2 + 15 = 22x$

12. $8x^2 + 15 = 26x$

13. $-4t + 1 = -t^2$

14. $-7t + 2 = -t^2$

15. $2t^2 - 10t + 11 = 0$

16. $3t^2 - 6t + 1 = 0$

17. $R^2 + 2R - 4 = 0$

18. $2R^2 + 3R - 7 = 0$

19. $Z^2 - 2Z + 5 = 0$

20. $Z^2 - 6Z + 34 = 0$

21. $Z^2 + 4Z + 13 = 0$

22. $Z^2 + Z + 2 = 0$

23. $A^2 - 5A + 1 = 0$

24. $2A^2 - 5A - 3 = 0$

25. $A(3A - 11) = 20$

26. $A(6A + 7) = 3$

27. $(B + 1)(B + 5) = 13$

28. $(3B + 1)(B - 3) = 7$

29. $2B^2 - \sqrt{3}\,B + 1 = 0$

30. $\sqrt{2}\,B^2 - 3B + \sqrt{18} = 0$

31. $9.8x^2 + 14.0x - 8.0 = 0$

32. $-4x^2 + 13x - 10 = 0$

33. $980x^2 + 100x - 50 = 0$

34. $1.2x^2 + 2.5x - 1.0 = 0$

35. The rotational motion of a body is given by $\theta = \dfrac{1}{2}\alpha t^2 + \omega_0 t$. Determine the values of t (in seconds) for $\alpha = 3$, $\omega_0 = 5$, and $\theta = 4$.

36. The reactance X of an AC circuit is given by $X = \omega L - \dfrac{1}{\omega C}$. For $X = 5$, $L = 3$, and $C = 0.02$, determine the values of ω in rad/s.

37. The cost (in dollars) of building a shelf is given by $C = \dfrac{\ell^2}{10} - 3\ell$, where ℓ is the length of the shelf (in inches). How long can the shelf be made if the cost is to be kept under \$125?

38. An object is thrown downward with an initial velocity of 5 ft/s. The relation between the distance s it travels and time t is given by $s = 5t + 16t^2$. How long does it take the object to fall 74 ft?

39. A filter has a transfer function given as:

$$\frac{E_0}{E_1} = S^2 - S - 1.293.$$

What positive value of S will cause the value of $\dfrac{E_0}{E_1}$ to be 0.707?

40. The ionization constant of acids is given by:

$$K = \frac{\alpha^2}{1 - \alpha}$$

Solve for α.

41. If the weekly profit p (in thousands of dollars) obtained from manufacturing computers is given by $p = -x^2 + 16x - 24$, where x is the number of computers produced, how many computers must be produced to derive a weekly profit of \$40,000?

42. The radius r of a circular arch of height h and span b is given by the formula: $r = \dfrac{(b^2 + 4h^2)}{8h}$. Determine h if $b = 20$ and $r = 26$.

43. A rectangular swimming pool is surrounded by a 2-m-wide walk on two ends and one side. The walk is 1 m wide on the other side. The length of the pool area (pool and walk) is 3 m more than twice the width, and the area of the pool is 52 m².
    a. Determine the length and the width of the pool.
    b. Determine the total area covered by the walk and the pool.

44. A page being designed for a textbook is to have 20-mm margins on the top and bottom and right side. The left side of the page is to have a margin of 40 mm. The height of the page is to be 80 mm more than the width. The area of the printed part of the page is to be 8800 mm². Determine the width and height of the entire page.

45. A rectangular piece of metal is 60 mm longer than it is wide. An open box containing 405,000 mm³ is made by cutting a 30-mm square from each corner and turning up the sides and ends (see the drawing). Determine the dimensions of the box.

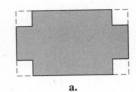

a.

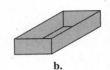

b.

46. An open box is made from a square piece of tin by cutting a 4-in. square from each corner and turning up the sides. Determine the area of the original square if the box is to contain
   a. 16 m³,
   b. 400 in³.

47. On the bottom floor, an elevator must have its fall slowed by a spring. The equation for the gradual descent of the elevator is governed by the following equation:

$$d = 36,000 - 5000x^2 - 4000x,$$

where x is the distance above the base of the spring. Knowing the approximate equation and the particular constants associated with this elevator, determine how high above the base of the spring the ground floor should be built so that elevator will come to rest level with the floor.

48. A population of birds grows at a rate of $0.003x^2$ per year. In other words, if the present population of birds is 1000, there will be 3000 by the end of the year. These prolific birds suffer from various diseases and are hunted by natural predators such that their death rate can be represented as $0.84x$. Determine the number of birds that make a stable population. (*Hint:* Let the function be expressed as $P(x)$ and set it equal to zero.)

49. A rancher wants to plant some hay in an open pasture. To protect the hay, he must fence in the hay field. The area of the hay field must be large enough to pay for the expenses of the fence. The fence costs $3.35 per linear foot, while the hay crop nets about $0.06 per square foot. The labor to install the fence is going to cost $265.00 If the shape of the field is to be a square, what should be the size of the field so that the income from the hay equals the cost of the fence?

50. Write a program for the calculator that gives both real and complex roots for a quadratic equation.

11.5

The Roots of a Quadratic Equation

In problem solving it may be helpful to have information about the roots of a quadratic equation. That is, are the roots real or imaginary, equal or unequal, rational or irrational?

Let r_1 and r_2 denote the roots of the general quadratic equation $ax^2 + bx + c = 0$. Using the quadratic formula, we have:

$$r_1 = \frac{-b + \sqrt{b^2 - 4ac}}{2a} \qquad r_2 = \frac{-b - \sqrt{b^2 - 4ac}}{2a}.$$

We can see by examining these two roots that the expression under the radical, $b^2 - 4ac$, affects the roots. The expression $b^2 - 4ac$ is called the **discriminant** of the quadratic formula quite simply because it discriminates between the kinds of roots we can expect.

EXAMPLE 1 **Rational, Real, and Unequal Roots**

Solve the quadratic equation, $x^2 + 5x + 6 = 0$ using the quadratic formula. When we find the roots, indicate whether they are real, rational, irrational, or complex numbers. Then note whether the value of the discriminant is a negative number, zero, or a positive number.

Solution Substitute the values in the quadratic equation.

$$r_1 = \frac{-(5) + \sqrt{5^2 - 4(1)(6)}}{2(1)} \qquad r_2 = \frac{-(5) - \sqrt{5^2 - 4(1)(6)}}{2(1)}$$

$$= \frac{-5 + \sqrt{1}}{2} \qquad\qquad\qquad = \frac{-5 - \sqrt{1}}{2}$$

$$= -2 \qquad\qquad\qquad\qquad = -3$$

The roots of the equation are real, unequal, rational numbers. This result occurs since the coefficients a, b, and c of the equation are rational numbers and the discriminate $b^2 - 4ac$ is positive. It is also a perfect square: $b^2 - 4ac = 1$. Since $b^2 - 4ac$ is a perfect square, the square root of the discriminate is a rational number, and thus, the roots are rational numbers. ◆

EXAMPLE 2 **Rational, Real, and Equal Roots**

Solve the quadratic equation, $x^2 + 6x + 9 = 0$ using the quadratic formula. When we find the roots, indicate whether they are real, rational, irrational, or complex numbers. Then note whether the value of the discriminant is a negative number, zero, or a positive number.

Solution Solving the equation using the quadratic formula, the result is:

$$r_1 = \frac{-6 + \sqrt{36 - 36}}{2} \qquad r_2 = \frac{-6 - \sqrt{36 - 36}}{2}$$

$$= -3. \qquad\qquad\qquad = -3.$$

The roots of the equation are real, equal, rational numbers. This result occurs since the coefficients a, b, and c of the equation are rational numbers and the discriminant $b^2 - 4ac$ equals 0. ◆

EXAMPLE 3 **Irrational, Real, and Unequal Roots**

Solve the quadratic equation, $x^2 + 5x + 3 = 0$ using the quadratic formula. When we find the roots, indicate whether they are real, rational, irrational, or complex numbers. Then note whether the value of the discriminant is a negative number, zero, or a positive number.

Solution   Solving the equation using the quadratic equation, the result is:

$$r_1 = \frac{-5 + \sqrt{13}}{2} \qquad r_2 = \frac{-5 - \sqrt{13}}{2}$$

$$= -.697224. \qquad\qquad = -4.302776.$$

The roots of the equation are real, unequal, irrational numbers. This result occurs since the coefficients a, b, and c of the equation are rational numbers and the discriminant $b^2 - 4ac$ equals 13 is a positive number. The number 13 is not a perfect square. The square root of the discriminant is an irrational number; thus, the roots are irrational numbers.   ◆

EXAMPLE 4   Imaginary and Unequal Roots

Solve the quadratic equation, $x^2 + 3x + 3 = 0$ using the quadratic formula. When we find the roots, indicate whether they are real, rational, irrational, or complex numbers. Then note whether the value of the discriminate is a negative number, zero, or a positive number.

Solution   Solving the equation using the quadratic formula the result is:

$$r_1 = \frac{-3 + \sqrt{-3}}{2} \qquad r_2 = \frac{-3 - \sqrt{-3}}{2}$$

$$= -\frac{3}{2} + \frac{j\sqrt{3}}{2}. \qquad = -\frac{3}{2} - \frac{j\sqrt{3}}{2}.$$

The roots of the equation are complex and unequal. This result occurs since the coefficients a, b, and c of the equation are rational numbers and the discriminant $b^2 - 4ac = -3$ is negative.   ◆

These results are summarized in Tables 11.1 and 11.2.

TABLE 11.1   Quadratic Equations with Real Coefficients

If $b^2 - 4ac$ is	Positive	Zero	Negative
The roots of $ax^2 + bx + c = 0$ are	real and unequal	real and equal	complex and unequal

TABLE 11.2   Quadratic Equations with Rational Coefficients

If $b^2 - 4ac$ is	Positive and a Perfect Square	Positive but not a Perfect Square	Zero	Negative
The roots of $ax^2 + bx + c = 0$ are	real, rational, and unequal	real, irrational, and unequal	real, rational, and equal	complex and unequal

EXAMPLE 5   **Determining Coefficients of a Quadratic Equation**

Determine the values of the constant m for which the equation $mx^2 - 3x = m - 3$ will have equal roots.

Solution   Write the given equation in standard form.

$$mx^2 - 3x = m - 3$$
$$mx^2 - 3x - m + 3 = 0$$

If the roots are to be equal, the discriminant $b^2 - 4ac$ must be equal to zero. In the equation, $a = m$, $b = -3$, and $c = -m + 3$.

$$(-3)^2 - 4(m)(-m + 3) = 0 \quad \text{Substituting in } b^2 - 4ac = 0$$
$$9 + 4m^2 - 12m = 0 \quad \text{Multiplying}$$
$$4m^2 - 12m + 9 = 0$$
$$(2m - 3)(2m - 3) = 0 \quad \text{Factoring}$$
$$m = \frac{3}{2}.$$

Therefore, if $m = \dfrac{3}{2}$, the equation $mx^2 - 3x = m - 3$ has equal roots.   ◆

11.5 Exercises

In Exercises 1–20, state whether the discriminate is positive, negative, or zero and describe the roots of each equation (i.e., real, rational, and unequal).

1. $3x^2 + 5x - 2 = 0$

2. $9x^2 + 12x + 4 = 0$

3. $0.3y^2 - 0.1y + 1 = 0$

4. $2y^2 - 0.2y + 0.6 = 0$

5. $\dfrac{1}{3}t^2 + \dfrac{1}{4}t - \dfrac{1}{7} = 0$

6. $\dfrac{3}{5}t^2 + \dfrac{2}{7}t - \dfrac{1}{3} = 0$

7. $4R^2 - 5 = R + 3$

8. $6R^2 + 13 = 13R$

9. $9x^2 + 12x + 4 = 0$

10. $x^2 - 8x + 16 = 0$

11. $y^2 - 7y = 0$

12. $3y^2 + 11y = 0$

13. $2x^2 - \sqrt{3}\,x + 1 = 0$

14. $\sqrt{2}\,x^2 + \sqrt{3}\,x - \sqrt{8} = 0$

15. $x^2 + \sqrt{5}\,x - 5 = 0$

16. $\sqrt{2}\,x^2 - x + \sqrt{2} = 0$

17. $x^2 + 2jx - 5 = 0$

18. $2jx^2 + 5x + j = 0$

19. $x^2 + 4jx - 4 = 0$

20. $9x^2 - 6jx - 1 = 0$

In Exercises 21–26, determine the value(s) of k for which the equations will have equal roots.

21. $kx^2 - 12x + 4 = 0$

22. $2kx^2 - 20x + 25 = 0$

23. $9x^2 - 30x + k = 0$

24. $kx^2 - 8x + 16k = 0$

25. $k(y^2 + y + 1) = y + 1$

26. $y^2 + 2(k + 4)y + 16k = 0$

27. In solving for the area of a rectangular region, John arrived at the equation $x^2 + 5x + 7 = 0$. His instructor informed him that this could not be the correct equation. Why not?

In Exercises 28 and 29, determine B so that the given equation has equal roots.

28. $x^2 + Bx + 16 = 0$

29. $3x^2 + Bx + 7 = 0$

30. Show that the equation $x^2 + Bx - 2 = 0$ always has real roots (regardless of the coefficient B).

11.6

Graphical Solutions of Quadratic Functions

In Chapter 5 the concept of function was introduced. A polynomial of degree 2, such as $ax^2 + bx + c$, defines a class of functions called quadratic functions.

> **DEFINITION 11.3**
> A function f defined by $f(x) = ax^2 + bx + c$, where a, b, and c are real numbers and $a \neq 0$, is called a **quadratic function.**

EXAMPLE 1   Quadratic Functions

Determine whether the function is a quadratic function. If it is a quadratic function, determine the values for a, b, and c.

a. $g(x) = x^2$

b. $f(x) = x^2 + x - 20$

c. $h(x) = \dfrac{1}{3x^2 + 7x}$

d. $k(x) = -2x^2 + 3x + 20$

e. $f(x) = 3x + 20$

Solution

a. Yes, $g(x) = x^2$ is a quadratic function. It is a polynomial of degree 2 with $a = 1$, $b = 0$, and $c = 0$.

b. Yes, $f(x)$ is a quadratic function. It is a polynomial of degree 2 with $a = 1$, $b = 1$, and $c = -20$.

c. No, $h(x)$ is not a quadratic function, since $\dfrac{1}{3x^2 + 7x}$ cannot be expressed as a second degree polynomial.

d. Yes, $k(x)$ is a quadratic function. It is a polynomial of degree 2 with $a = -2$, $b = 3$, and $c = 20$.

e. No, since the coefficient of a is zero, it is a linear function.   ◆

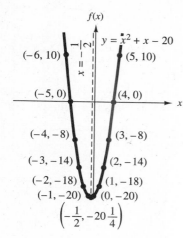

$f(x)$

$(-6, 10)$ $y = x^2 + x - 20$ $(5, 10)$

$(-5, 0)$ $(4, 0)$

x

$(-4, -8)$ $(3, -8)$

$(-3, -14)$ $(2, -14)$

$(-2, -18)$ $(1, -18)$

$(-1, -20)$ $(0, -20)$

$\left(-\dfrac{1}{2}, -20\dfrac{1}{4}\right)$

FIGURE 11.6

The graph of every quadratic function is a special curve called a **parabola.** In Section 5.3, we learned the technique of graphing parabolas. Let us review here by graphing a parabola to illustrate important properties and key points.

To graph the parabola defined by the function $f(x) = x^2 + x - 20$, determine the set of ordered pairs listed in Table 11.3. Plot ordered pairs and then sketch the curve as shown in Figure 11.6.

TABLE 11.3

x	$y = f(x) = x^2 + x - 20$	$y = f(x)$	Ordered Pairs
-6	$f(-6) = (-6)^2 + (-6) - 20$	10	$(-6, 10)$
-5	$f(-5) = (-5)^2 + (-5) - 20$	0	$(-5, 0)$
-4	$f(-4) = (-4)^2 + (-4) - 20$	-8	$(-4, -8)$
-3	$f(-3) = (-3)^2 + (-3) - 20$	-14	$(-3, -14)$
-2	$f(-2) = (-2)^2 + (-2) - 20$	-18	$(-2, -18)$
-1	$f(-1) = (-1)^2 + (-1) - 20$	-20	$(-1, -20)$
$-\dfrac{1}{2}$	$f\left(-\dfrac{1}{2}\right) = \left(-\dfrac{1}{2}\right)^2 + \left(-\dfrac{1}{2}\right) - 20$	$-20\dfrac{1}{4}$	$\left(-\dfrac{1}{2}, -20\dfrac{1}{4}\right)$
0	$f(0) = (0)^2 + (0) - 20$	-20	$(0, -20)$
1	$f(1) = (1)^2 + (1) - 20$	-18	$(1, -18)$
2	$f(2) = (2)^2 + (2) - 20$	-14	$(2, -14)$
3	$f(3) = (3)^2 + (3) - 20$	-8	$(3, -8)$
4	$f(4) = (4)^2 + (4) - 20$	0	$(4, 0)$
5	$f(5) = (5)^2 + (5) - 20$	10	$(5, 10)$

As we examine the graph of $y = x^2 + x - 20$, look for special properties that are also true for all quadratic functions. Note the sets of points $(-6, 10)$, $(5, 10)$; $(-5, 0)$, $(4, 0)$; $(-4, -8)$, $(3, -8)$; $(2, -18)$, $(1, -18)$; and $(-1, -20)$, $(0, -20)$. The first point of each ordered pair is on the left branch of the curve, and the second point is directly opposite it on the right branch of the curve. We can almost imagine a line dividing the curve into two identical halves. In fact, $x = -\dfrac{1}{2}$ is the **line of symmetry** that cuts through the center of the parabola at the point $\left(-\dfrac{1}{2}, -20\dfrac{1}{4}\right)$. This low point is called the **vertex** of the parabola, and it is on the line of symmetry. The vertex is either the maximum point (highest point) or the minimum point (lowest point) on the curve. For $f(x) = x^2 + x - 20$, the vertex is the minimum point.

DETERMINING THE VERTEX OF THE PARABOLA

The abscissa (*x*-value) of the vertex of the parabola:

$$f(x) = ax^2 + bx + c$$

can be determined using the formula $x = -\dfrac{b}{2a}$. The ordinate (*y*-value) of the vertex is found by substituting the value for *x* into the function. That is, $y = f\left(\dfrac{-b}{2a}\right)$.

Look back at Figure 11.6 to see that the parabola crosses the *x*-axis at two points, (4, 0) and (−5, 0). Note that the *y*-value of these points is zero. To determine these points, solve the quadratic equation $x^2 + x - 20 = 0$. The roots, or zeros, of the quadratic equation $x^2 + x - 20 = 0$ are the abscissas of the points on the graph where the parabola crosses the *x*-axis. These points are called the **x-intercepts.** The abscissas of the *x*-intercepts are those values of *x* for which $f(x) = 0$. For this function, 4 and −5 are the roots (or zeros) since $f(4) = 0$ and $f(-5) = 0$.

Another key point on the graph is (0, −20), which is the **y-intercept,** the point where the curve crosses the *y*-axis. For the general function $f(x) = ax^2 + bx + c$, the *y*-intercept is found by letting $x = 0$. Then $f(0) = c$, giving the ordered pair (0, *c*).

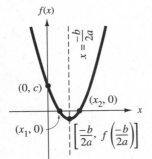

FIGURE 11.7

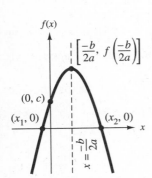

FIGURE 11.8

KEY INFORMATION FOR GRAPHING A QUADRATIC FUNCTION

1. The vertex of the parabola is $\left[\dfrac{-b}{2a}, f\left(\dfrac{-b}{2a}\right)\right]$.
2. If *a* (the coefficient of the second-degree term) is positive ($a > 0$), the vertex is a minimum so that the parabola opens up (see Figure 11.7). If *a* is negative ($a < 0$), then the vertex is a maximum and the parabola opens down (see Figure 11.8).
3. Determine the *x*-intercepts by solving $ax^2 + bx + c = 0$ for *x*. (As illustrated in the following examples, there may be 2, 1, or 0 intercepts.)
4. The *y*-intercept is (0, *c*).
5. The line of symmetry is $x = \dfrac{-b}{2a}$.
6. To graph additional points, select *x*-values on one side of the line of symmetry. Then use the definition of symmetry to graph points on the other side of the line.

EXAMPLE 2   Vertex of Quadratic Function Is a Maximum

Graph the quadratic function $f(x) = -2x^2 + 3x + 20$ and determine the roots of the quadratic function.

Solution   First determine the key information as suggested.

a.   *Step 1,* the vertex is (0.75, 21.125).

$$\text{Since } \frac{-b}{2a} = \frac{-(\boxed{3})}{2(\boxed{-2})}$$
$$= 0.75,$$
$$f(0.75) = -2(\boxed{0.75})^2 + 3(\boxed{0.75}) + 20$$
$$= 21.125.$$

b.   *Step 2,* the parabola opens down since $a = -2 < 0$. Therefore, the vertex (0.75, 21.125) is the maximum point on the curve.

c.   *Step 3,* the x-intercepts are (−2.5, 0) and (4, 0).

$$-2x^2 + 3x + 20 = 0$$
$$(-x + 4)(2x + 5) = 0 \qquad \text{**Factoring**}$$
$$x = 4, \ x = -\frac{5}{2} \quad \text{or} \quad -2.5 \quad \text{**Solving for x**}$$

d.   *Step 4,* the y-intercept is ($\boxed{0, 20}$).

e.   *Step 5,* the line of symmetry is $\boxed{x = 0.75}$, since $x = \frac{-b}{2a}$.

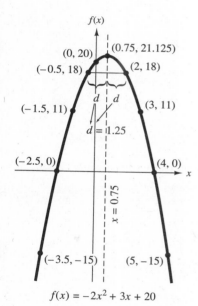

$f(x) = -2x^2 + 3x + 20$

FIGURE 11.9

With this information, we can sketch the curve. To develop a smoother curve, determine the points on the curve for $x = 2$, 3, and 5. The corresponding ordered pairs are (2, 18), (3, 11), and (5, −15). Now that we have these points and know that the line of symmetry is $x = 0.75$, we can determine the respective symmetric points (−0.5, 18), (−1.5, 11), and (−3.5, −15).

Note that the ordinates of (2, 18) and (−0.5, 18) are both the same, 18, and the abscissas 2 and −0.5 are both the same distance, 1.25 units, from the line of symmetry.

The roots of the quadratic equation are (−2.5, 0) and (4, 0). The curve is sketched in Figure 11.9.   ◆

EXAMPLE 3   Quadratic Function with No x-Intercepts

Graph the quadratic function $f(x) = 2x^2 + 4x + 3$.

Solution   First determine the key information as suggested.

a.   *Step 1,* the vertex is (−1, 1).

b.   *Step 2,* the parabola opens up since $a = 2 > 0$. Thus, the vertex (−1, 1) is a minimum point on the curve.

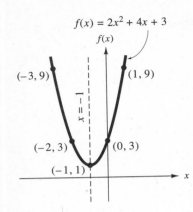

$$f(x) = 2x^2 + 4x + 3$$

FIGURE 11.10

c. *Step 3*, since $2x^2 + 4x + 3$ is not factorable, use the quadratic formula.

$$x = \frac{-4 \pm \sqrt{16 - 24}}{4} \quad \text{or} \quad x = \frac{-4 \pm \sqrt{-8}}{4}$$

since $b^2 - 4ac < 0$, the roots are not real numbers and there are no x-intercepts.
d. *Step 4*, the y-intercept is $(0, 3)$.
e. *Step 5*, the line of symmetry is $x = -1$.
f. *Step 6*, calculate points for $x = 0, 1$, and their respective symmetric points. The result is $(0, 3), (-2, 3), (1, 9)$, and $(-3, 9)$.

The curve is sketched in Figure 11.10. The quadratic equation $2x^2 + 4x + 3 = 0$ does not have any real roots since the curve does not cross the x-axis. ◆

In Examples 2 and 3, the quadratic functions have 2 and 0 real distinct roots, respectively. The curve crosses the horizontal axis at two points in Example 2, while in Example 3, it does not cross the horizontal axis. If we had calculated the discriminant, the results would have been $b^2 - 4ac > 0$ and $b^2 - 4ac < 0$, respectively. When the discriminant is $b^2 - 4ac = 0$, the graph touches only one point on the horizontal axis.

EXAMPLE 4 **Application: Determining Maximum Area**

A horse breeder plans to build a rectangular corral with 300 ft of fencing. What should be the dimensions of the corral if the area is to be a maximum?

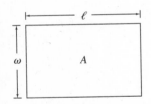

FIGURE 11.11

Solution   Let the length and the width of the corral be denoted by ℓ and w, respectively as shown in Figure 11.11. The area of the rectangle is $A = \ell w$. Since we know that the perimeter is 300 ft, we can express it in terms of the width and the length.

$$2\ell + 2w = 300$$
$$2\ell = 300 - 2w \quad \text{Solving for } \ell$$
$$\ell = 150 - w$$

Substitute the value for ℓ into the equation for area.

$$A = (150 - w)w$$
$$A = 150w - w^2$$

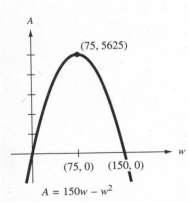

$$A = 150w - w^2$$

FIGURE 11.12

The graph of the function in Figure 11.12 shows that the parabola opens down, which is also indicated by the fact that $a < 0$. Therefore, the function has a maximum value that occurs at the vertex. The vertex is the point (w, A). The value of w is found using the formula $w = \dfrac{-b}{2a}$.

$$w = \frac{-150}{2(-1)} = 75$$

Knowing the value of w, we can determine the value of ℓ using the fact that $\ell = 150 - w$.

$$\ell = 150 - \boxed{75} = 75$$

The desired dimensions of the corral are 75 ft by 75 ft.   ◆

EXAMPLE 5   **Application: Flight of Toy Rocket**

A toy rocket is launched from a deck. The path of the rocket is defined by the quadratic function $s(t) = -16t^2 + 96t + 28$, where s is the distance traveled and t is the time in seconds. Determine the following.

a.   The maximum height attained by the rocket.
b.   The time at which it strikes the ground.
c.   The height of the deck from which the rocket is launched.

Solution   For the quadratic function $s(t) = -16t^2 + 96t + 28$, $a = -16$, $b = 96$, and $c = 28$. Since a is negative, the graph of the parabola opens downward, and the vertex is the maximum point on the curve.

a.   The maximum height is the vertex, and when

$$t = \frac{-b}{2a}$$

$$= \frac{\boxed{-96}}{\boxed{-32}}$$

$$= 3,$$

then:

$$s(\boxed{3}) = -16(\boxed{3})^2 + 96(\boxed{3}) + 28$$
$$= 172 \text{ ft.}$$

The maximum height of 172 ft is attained after 3 seconds.

b.   The toy rocket strikes the ground when $s(t) = 0$.

$$-16t^2 + 96t + 28 = 0$$
$$-4t^2 + 24t + 7 = 0 \quad \textbf{Dividing each term by 4}$$
$$t = \frac{-24 \pm \sqrt{(24)^2 - 4(-4)(7)}}{-2(-4)}$$
$$t = 6.3 \text{ s} \quad \text{or} \quad t = -0.28 \text{ s}$$

Since the time t represents the number of seconds after the rocket is launched, the solution $t = -0.28$ is not applicable. Therefore, the rocket strikes the ground 6.3 s after being launched.

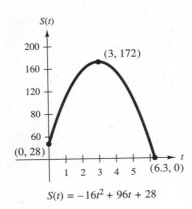

$$S(t) = -16t^2 + 96t + 28$$

FIGURE 11.13

c. The height of the deck is given when $t = 0$, the initial position.

$$s(\boxed{0}) = 28 \text{ ft.}$$

The path of the rocket is sketched in Figure 11.13. Since s represents the distance above the ground, the domain is restricted to $0 \le t < 6.3$. The ordinate value of the vertex is the maximum height of the rocket. The time the flight ends (when the rocket strikes the ground) is the t-intercept $(6.3, 0)$. The s-intercept $(0, 28)$ is the height of the deck—the initial height. ◆

We can sketch each of the parabolas in Examples 1 through 5 using a graphics calculator. The coordinates of the vertex and the roots of the parabola can be approximated with the trace and/or zoom function.

11.6 Exercises

In Exercises 1–20, graph the quadratic function and determine the roots of the quadratic equation to the nearest tenth. Check the answers using a graphics calculator.

1. $f(x) = x^2 + 6x + 5$ **2.** $f(x) = x^2 - 6x + 5$ **3.** $f(x) = x^2 + 5x + 4$

4. $f(x) = 4x^2 + 12x + 9$ **5.** $f(x) = -3x^2 + 2x + 5$ **6.** $f(x) = x^2 + x + 2$

7. $f(x) = -2x^2 + 3x + 5$ **8.** $f(x) = -4x^2 - 4x + 3$ **9.** $f(x) = 2x^2 + 3x + 2$

10. $f(x) = -2x^2 + 5x - 6$ **11.** $f(x) = x^2 - 4x$ **12.** $f(x) = 3x - x^2$

13. $f(x) = 4x^2 - 20x + 25$ **14.** $f(x) = -9x^2 + 42x - 49$ **15.** $f(x) = 3(x + x^2) + x - 5$

16. $f(x) = 4(1 - x^2) + 2x - 3$ **17.** $f(x) = -9x^2 + 12x - 4$ **18.** $f(x) = 4x^2 - 9$

19. $f(x) = 2x^2 + 5x + 2$ **20.** $f(x) = \frac{1}{3}x^2 + 3$

In Exercises 21–30, without plotting the graph, decide by means of the discriminant whether the graph will cross the x-axis at two distinct points, will touch the x-axis at one distinct point, or will not touch the x-axis at all.

21. $f(x) = 4x^2 + 12x + 9$ **22.** $f(x) = -x^2 + 7x - 12$ **23.** $f(x) = x^2 - 2x - 15$

24. $f(x) = 3x^2 + 3x + 4$ **25.** $f(x) = 4x^2 + 3x + 2$ **26.** $f(x) = -x^2 + 2x + 5$

27. $f(x) = 2x^2 + 3x - 4$ **28.** $f(x) = 16x^2 + 25$ **29.** $f(x) = 13x^2 + 11x - 5$

30. $f(x) = 8x^2 + 11x - 8$

In Exercises 31–34, determine the value of m for which the graph of each function will **a.** touch the x-axis at one point, **b.** cross the x-axis in two distinct points, or **c.** will not touch the x-axis at all.

31. $f(x) = mx^2 - 2x + 1$ **32.** $f(x) = 2x^2 - 4x + m$ **33.** $f(x) = 4x^2 - 3mx + 9$

34. $f(x) = 4x^2 + 4mx + 1$

35. The height of a ball after it is thrown vertically upward is given in feet by
$$h = 4 + 64t - 16t^2,$$
where h is the height above the ground, 4 represents the release point of the ball, and -16 is one-half of the acceleration due to gravity, which is -32 ft/s^2.
  a. Sketch the parabola for this curve.
  b. From the sketch, determine at what time the ball is at its peak.
  c. Determine the maximum height of the ball.

36. The formula $s = v_0 t + \dfrac{1}{2}gt^2$ gives the distance s in feet a body has fallen in t seconds in a vacuum with an initial velocity downward of v_0 feet per second. With $g = 32$ ft/s^2 and $v_0 = 80$ ft/s, how long will it take a body to fall the following distances?
  a. 384 ft
  b. 1200 ft

37. Matthew is enclosing a rectangular area for his dog next to the garage. The garage wall will be on one side; the other three sides will be fencing. He has 60 ft of fencing. Determine the maximum area he can enclose without buying more fence. What are the dimensions of the fenced-in area?

38. A builder has a rectangular piece of sheet metal that is 6 ft long and 12 in. wide. He is constructing a chute for pouring concrete by bending up the sides perpendicular to the bottom as shown in the sketch. How many inches should be bent up in order to make the capacity of the chute a maximum?

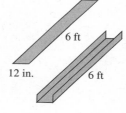

39. A Norman window consists of a rectangle surmounted by a semicircle as shown in the drawing. Determine the window with the largest area if the perimeter is 15 ft.

40. A homemade rocket is launched vertically from the top of a hill 800 ft above the floor of a valley with an initial upward velocity of 100 ft per second. This problem is related to Exercise 35.
  a. What is the equation for this motion?
  b. Sketch the path of the rocket from launch until it lands.
  c. What is the maximum height of the rocket?
  d. When will the rocket land in the valley?

41. Suppose a cannonball is fired at an angle of 30°. If air resistance is neglected, the equation that describes this motion is $y = 0.58x - 6.5x^2$.
  a. Sketch the graph described by this equation from $x = 0$ to $x = 0.1$ in increments of 0.01.
  b. What is the maximum height of the projectile?
  c. From your graph, estimate how far out the cannonball will fall along the x-axis.
  d. To obtain a more exact distance the cannonball will fall along the x-axis, let $y = 0$ and solve for x.

11.7

Equations That Can Lead to Quadratic Equations

Many equations are not quadratic. However, with algebraic techniques, we can change some of these equations to quadratic form and solve them. In general, these equations resemble quadratic equations, but they are not to the second degree. By using substitution techniques, we can modify such equations to be of quadratic form.

In Section 11.1, we learned that an algebraic equation of the form:

$$ax^2 + bx + c = 0, \quad a \neq 0$$

is a quadratic equation in standard form. Note that the quadratic term ax^2 has an exponent of 2, which is twice the exponent of the linear term bx^1. This is the key to identifying equations that can be transformed to quadratic form. For example, we can transform the equation $x^4 + 7x^2 - 13 = 0$ to quadratic form since the degree of the first term, 4, is twice the degree of the second term. Use rules of exponents to write the equation as:

$$(x^2)^2 + 7x^2 - 13 = 0.$$

Now let $q = x^2$ and substitute so that

$$q^2 + 7q - 13 = 0,$$

which is a quadratic equation.

EXAMPLE 1   Writing Nonquadratic Equations in Quadratic Form

Transform each equation into an equation in quadratic form.

a. $2x^{\frac{1}{3}} + 5x^{\frac{1}{6}} + 13 = 0.$ b. $2x + 3\sqrt{x} + 2 = 0$

c. $\dfrac{1}{x^4} - \dfrac{5}{x^2} - 8 = 0$ d. $(x + 4)^2 - 5(x + 4) + 7 = 0$

Solution

a.   Since $\dfrac{1}{3}$ is twice $\dfrac{1}{6}$, we can write the equation as:

$$2(\,x^{\frac{1}{6}}\,)^2 + 5x^{\frac{1}{6}} + 13 = 0.$$

Let $q = x^{\frac{1}{6}}$ and substitute, and the equation in quadratic form is:

$$2q^2 + 5q + 13 = 0.$$

b.   If we rewrite the equation with exponents, we have $2x + 3x^{\frac{1}{2}} + 2 = 0$. Since 1 is twice $\dfrac{1}{2}$, we can write the equation as:

$$2(\,x^{\frac{1}{2}}\,)^2 + 3x^{\frac{1}{2}} + 2 = 0.$$

Let $q = x^{\frac{1}{2}}$ and substitute, and the equation in quadratic form is:

$$2q^2 + 3q + 2 = 0.$$

c.   Rewriting the equation, we have $\left(\dfrac{1}{x}\right)^4 - 5\left(\dfrac{1}{x}\right)^2 - 8 = 0$. Since 4 is twice 2, we can write the equation as:

$$\left[\left(\dfrac{1}{x}\right)^2\right]^2 - 5\left(\dfrac{1}{x}\right)^2 - 8 = 0.$$

Let $q = \left(\dfrac{1}{x}\right)^2$ and substitute, and the equation is:

$$q^2 - 5q - 8 = 0.$$

d. The exponent of the first quantity is twice that of the second. Let $q = x + 4$ and substitute. Then the equation is:

$$q^2 - 5q + 7 = 0. \quad \blacklozenge$$

Example 1 illustrates how to change nonquadratic equations to quadratic form. With equations in quadratic form, we can use the rules for solving quadratic equations to determine the solution of the nonquadratic equation. The technique of solving quadraticlike equations is illustrated in Examples 2–6.

EXAMPLE 2   Solving a Fourth-Degree Equation Using Quadratic Techniques

Solve the equation $x^4 - 5x^2 + 6 = 0$.

Solution   Because the equation is of degree four, a possibility of four distinct roots exists. In fact, the solution could have 4, 3, 2, 1, or 0 real distinct roots. Since the equation:

$$x^4 - 5x^2 + 6 = 0$$

can be rewritten as:

$$(x^2)^2 - 5x^2 + 6 = 0,$$

we can change the variable by letting $q = x^2$. Substituting, we have:

$$q^2 - 5q + 6 = 0.$$

The new equation is in quadratic form in terms of the variable q. Now we can solve for q by factoring the equation:

$$(q - 3)(q - 2) = 0$$
$$q = 3 \quad \text{or} \quad q = 2.$$

Since $q = x^2$, we can substitute to determine the value of x.

$$x^2 = 3 \quad \text{or} \quad x^2 = 2.$$

Thus, $x = \pm\sqrt{3}$ or $x = \pm\sqrt{2}$. We started with a fourth-degree equation, and in this case, there are four distinct roots. The roots are $\sqrt{3}$, $-\sqrt{3}$, $\sqrt{2}$, and $-\sqrt{2}$. $\quad \blacklozenge$

A visual check is possible by sketching the graph with a graphics calculator, as in Figure 11.14, and zooming to see that the curve crosses the axis at $-\sqrt{3}$, $-\sqrt{2}$, $\sqrt{2}$, and $\sqrt{3}$.

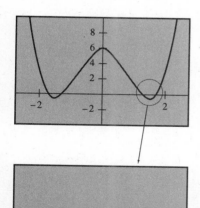

FIGURE 11.14

EXAMPLE 3   **Solving an Equation with Fractional Exponents Using Quadratic Techniques**

Solve the equation $x^{\frac{1}{3}} - 3x^{\frac{1}{6}} + 2 = 0$.

Solution   The equation can be rewritten as:

$$(x^{\frac{1}{6}})^2 - 3x^{\frac{1}{6}} + 2 = 0.$$

We can write the equation in quadratic form by letting $q = x^{\frac{1}{6}}$. Substituting, we have:

$$q^2 - 3q + 2 = 0.$$

Solving the equation by factoring, we have:

$$(q - 2)(q - 1) = 0$$
$$q = 2 \quad \text{or} \quad q = 1.$$

Since $q = x^{\frac{1}{6}}$, then $x^{\frac{1}{6}} = 2$ or $x^{\frac{1}{6}} = 1$. Now we can find x by raising both sides of the equation to the sixth power.

$$(x^{\frac{1}{6}})^6 = (2)^6 \quad \text{or} \quad (x^{\frac{1}{6}})^6 = (1)^6$$
$$x = 64 \quad \text{or} \quad x = 1$$

Check:

$x = 64$	$x = 1$
$(64)^{\frac{1}{3}} - 3(64)^{\frac{1}{6}} + 2 = 0.$	$(1)^{\frac{1}{3}} - 3(1)^{\frac{1}{6}} + 2 = 0.$
$4 - 3(2) + 2 = 0$	$1 - 3(1) + 2 = 0$
$0 = 0.$ ✓	$0 = 0.$ ✓

Thus, the roots of the equation $x^{\frac{1}{3}} - 3x^{\frac{1}{6}} + 2 = 0$ are 1 and 64.   ◆

EXAMPLE 4   **Writing an Equation In Quadratic Form and Solving**

Solve the equation $x + 3x^{\frac{1}{2}} = 4$.

Solution   Rewriting the equation, we have:

$$(x^{\frac{1}{2}})^2 + 3x^{\frac{1}{2}} - 4 = 0.$$

Let $q = x^{\frac{1}{2}}$, then $q^2 = x$; substituting, we have:

$$q^2 + 3q - 4 = 0.$$

The equation is quadratic in q, and we can solve it by factoring.

$$(q + 4)(q - 1) = 0$$
$$q = -4 \quad \text{or} \quad q = 1$$

Now, substituting for q, we have:

$$x^{\frac{1}{2}} = -4 \quad \text{or} \quad x^{\frac{1}{2}} = 1$$
$$x = 16 \quad \text{or} \quad x = 1. \quad \textbf{Squaring both sides}$$

Check:

$$x = 16 \qquad\qquad\qquad x = 1$$
$$\boxed{16} + 3(\boxed{16})^{\frac{1}{2}} = 4 \qquad \boxed{1} + 3(\boxed{1})^{\frac{1}{2}} = 4$$
$$16 + 12 \neq 4. \; \textbf{x} \qquad\qquad 1 + 3 = 4. \; \checkmark$$

Thus, the only root of the equation $x + 3x^{\frac{1}{2}} = 4$ is 1. ◆

CAUTION   In solving equations of the type given in Example 4, we change the form of the equation to obtain a solution. However, as we saw, the solution obtained may not satisfy the original equation. When this happens, such a solution is called an **extraneous root.** We can see that it is essential that we check all solutions in the original equation. In Example 4, 16 is an extraneous root. ■

EXAMPLE 5   **Solving an Equation Whose Roots Are Complex Numbers**

Solve the equation $\dfrac{1}{x^4} - \dfrac{2}{x^2} - 8 = 0$.

Solution   Rewriting the equation, we have:

$$\left(\frac{1}{x^2}\right)^2 - 2\left(\frac{1}{x^2}\right) - 8 = 0.$$

The equation is in quadratic form. Let $q = \dfrac{1}{x^2}$; substituting, we have:

$$\boxed{q}^2 - 2\boxed{q} - 8 = 0.$$

Now we can solve the equation by factoring.

$$(q - 4)(q + 2) = 0$$
$$q = 4 \quad \text{or} \quad q = -2$$

Since $q = \dfrac{1}{x^2}$, we have:

$$\frac{1}{x^2} = 4 \qquad \text{or} \qquad \frac{1}{x^2} = -2$$

$$x^2 = \frac{1}{4} \qquad \text{or} \quad x^2 = -\frac{1}{2} \qquad\qquad \textbf{Equivalently}$$

$$x = \pm\frac{1}{2} \qquad \text{or} \quad x = \pm j\sqrt{\frac{1}{2}}. \qquad \textbf{Taking square root of both sides}$$

Check: $x = \pm\dfrac{1}{2}$

$$\frac{1}{\left(\pm\dfrac{1}{2}\right)^4} - \frac{2}{\left(\pm\dfrac{1}{2}\right)^2} - 8 = 0.$$

$$16 - 2(4) - 8 = 0$$

$$0 = 0. \checkmark$$

$$x = \pm j\sqrt{\frac{1}{2}}$$

$$\frac{1}{\left(\pm j\sqrt{\dfrac{1}{2}}\right)^4} - \frac{2}{\left(\pm j\sqrt{\dfrac{1}{2}}\right)^2} - 8 = 0.$$

$$4 - 2(-2) - 8 = 0$$

$$0 = 0. \checkmark$$

The roots of the equation $\dfrac{1}{x^4} - \dfrac{2}{x^2} - 8 = 0$ are $\pm\dfrac{1}{2}$ and $\pm j\sqrt{\dfrac{1}{2}}$. ◆

We also can solve fractional equations like the one in Example 5 by multiplying each term by the lowest common denominator and then use the techniques discussed in examples 2, 3, and 4.

EXAMPLE 6   **Writing an Equation in Quadratic Form and Solving**

Solve the equation $3(y - 2)^2 - 5(y - 2) - 2 = 0$.

Solution   Let $u = y - 2$. Substituting into the equation, we have:

$$3u^2 - 5u - 2 = 0.$$

The new equation is quadratic in u, and we can solve it by factoring.

$$(3u + 1)(u - 2) = 0$$

$$u = -\frac{1}{3} \quad \text{or} \quad u = 2$$

Now, substituting for u, we can solve for y.

$$-\frac{1}{3} = y - 2 \quad \text{or} \quad 2 = y - 2$$

$$y = 1\frac{2}{3} \qquad \text{or} \quad y = 4$$

Check:

$$y = 1\frac{2}{3}$$

$$3\left(1\frac{2}{3} - 2\right)^2 - 5\left(1\frac{2}{3} - 2\right) - 2 = 0$$

$$\frac{1}{3} + \frac{5}{3} - 2 = 0. \checkmark$$

$$y = 4$$

$$3(4 - 2)^2 - 5(4 - 2) - 2 = 0.$$

$$12 - 10 - 2 = 0. \checkmark$$

The roots of the equation $3(y - 2)^2 - 5(y - 2) - 2 = 0$ are $1\frac{2}{3}$ and 4. ◆

Equations Involving Radical Expressions

Solving equations with radicals is easier when we remember the definitions of exponents from Section 2.2 along with the definitions of radicals from Section 2.4. A method of solving equations with radicals is to raise both sides of the equation to the same power to eliminate the radicals. In some cases, we may be able to solve an equation with radicals by using substitution.

EXAMPLE 7   **Solving an Equation with Radicals Using Substitution**

Solve the equation $\sqrt{x + 3} + 2x = 0$.

Solution   Let $y = x + 3$. Then, solving for x, we have $x = y - 3$. Rewriting the original equation in terms of y, we have:

$$\sqrt{y} + 2(y - 3) = 0 \quad \text{or} \quad y^{\frac{1}{2}} + 2y - 6 = 0$$

We already know how to work this kind of equation. (Refer back to Example 4.) If we let $q = y^{\frac{1}{2}}$ and substitute, then we have the following quadratic expression.

$$q + 2q^2 - 6 = 0 \quad \text{or} \quad 2q^2 + q - 6 = 0$$

The factors of the equation are $(2q - 3)$ and $(q + 2)$. Thus, $q = \frac{3}{2}$ and $q = -2$.

Since $q = y^{\frac{1}{2}}$, square both sides of the equation and determine that $q^2 = y$. Substituting for q, we have:

$$y = \left(\frac{3}{2}\right)^2 = \frac{9}{4} \quad \text{or} \quad y = (-2)^2 = 4.$$

To determine the values of x, substitute the values of y in the equation $x = y - 3$.

$$x = \frac{9}{4} - 3 = -\frac{3}{4} \quad \text{or} \quad x = 4 - 3 = 1.$$

Check:

$$x = -\frac{3}{4} \qquad\qquad\qquad x = 1$$

$$\sqrt{-\frac{3}{4} + 3} + 2\left(-\frac{3}{4}\right) = 0 \qquad \sqrt{1 + 3} + 2(1) = 0$$

$$\frac{3}{2} - \frac{3}{2} = 0. \checkmark \qquad\qquad 2 + 2 \neq 0. \text{ x}$$

Thus, the only root of the equation $\sqrt{x + 3} + 2x = 0$ is $-\frac{3}{4}$. The value $x = 1$ is an extraneous root. ◆

EXAMPLE 8 **Solving an Equation with Radicals Using the Squaring Technique**

Solve the equation $\sqrt{u - 2} = 2 + \sqrt{2u + 3}$.

Solution Since there are two radicals in the equation, we cannot isolate the radical. The best approach in this situation is to have one radical on each side of the equal sign. By doing this, we have a simpler expression with which to work. So squaring both sides of $\sqrt{u - 2} = 2 + \sqrt{2u + 3}$ we have:

$$(\sqrt{u - 2})^2 = (2 + \sqrt{2u + 3})^2$$

$$u - 2 = 4 + 4\sqrt{2u + 3} + 2u + 3$$

$$-u - 9 = 4\sqrt{2u + 3} \qquad \text{Collecting like terms and isolating the radical}$$

$$u^2 + 18u + 81 = 16(2u + 3) \qquad \text{Squaring both sides}$$

$$u^2 - 14u + 33 = 0 \qquad \text{Collecting like terms}$$

$$(u - 11)(u - 3) = 0 \qquad \text{Factoring}$$

$$u = 11 \quad \text{or} \quad u = 3.$$

Check:

$$u = 11 \qquad\qquad\qquad u = 3$$

$$\sqrt{11 - 2} = 2 + \sqrt{2(11) + 3} \qquad \sqrt{3 - 2} = 2 + \sqrt{2(3) + 3}$$

$$3 \neq 2 + 5. \text{ x} \qquad\qquad 1 \neq 2 + 3. \text{ x}$$

Since neither value checks, the equation $\sqrt{u - 2} = 2 + \sqrt{2u + 3}$ has no solution. ◆

In Examples 2–7, we were able to use substitution to rewrite the equations in quadratic form. In Example 7, however, we used the substitution technique to solve the **original** equation.

GENERAL PROCEDURES FOR SOLVING QUADRATICLIKE EQUATIONS

1. Identify the type of problem.
2. Identify the quantity for which you can substitute to change the equation to a quadratic equation.
3. Write the quadratic equation.
4. Solve the quadratic equation.
5. Determine the possible roots of the original equation.
6. Check the solution. Are there any extraneous roots?

11.7 Exercises

In Exercises 1–20, solve the equations and check algebraically or visually with the graphics calculator.

1. $x^4 - 11x^2 + 10 = 0$

2. $x^4 - 7x^2 + 10 = 0$

3. $x^4 - 9x^2 + 14 = 0$

4. $x^4 - 14x^2 + 33 = 0$

5. $x^5 - 8x^3 + 15x = 0$

6. $x^6 - 12x^4 + 11x^2 = 0$

7. $4x^4 - 13x^2 + 3 = 0$

8. $5x^4 - 16x^2 + 11 = 0$

9. $2x + 8\sqrt{x} - 10 = 0$

10. $x + 7\sqrt{x} - 8 = 0$

11. $\dfrac{1}{x^2} + \dfrac{9}{x} + 8 = 0$

12. $\dfrac{1}{x^2} - \dfrac{8}{x} + 7 = 0$

13. $\dfrac{1}{x^4} - \dfrac{13}{x^2} + 12 = 0$

14. $\dfrac{1}{x^4} - \dfrac{2}{x^2} - 15 = 0$

15. $x^{\frac{1}{3}} - 7x^{\frac{1}{6}} + 6 = 0$

16. $x^{\frac{2}{3}} - 6x^{\frac{1}{3}} - 7 = 0$

17. $2x - 5x^{\frac{1}{2}} + 2 = 0$

18. $x^{\frac{5}{3}} - 16x^{\frac{4}{3}} + 15x = 0$

19. $(x - 5)^2 + 10(x - 5) + 16 = 0$

20. $(x + 3)^2 - 5(x + 3) + 6 = 0$

In Exercises 21–34, solve the fractional equations and check algebraically or visually with the graphics calculator.

21. $\dfrac{x}{4} + \dfrac{2}{x} = 1$

22. $x = \dfrac{4}{x - 3}$

23. $\dfrac{x}{5 - x} = \dfrac{2}{x - 5}$

24. $\dfrac{x}{3 - x} = \dfrac{3}{x - 3}$

25. $\dfrac{3}{x - 2} + \dfrac{x - 4}{x} = 5$

26. $\dfrac{x - 2}{x + 1} = \dfrac{4x - 3}{2x}$

27. $\dfrac{2}{3x + 4} - \dfrac{1}{3x - 4} = \dfrac{2x}{9x^2 - 16}$

28. $\dfrac{1}{2x + 5} - \dfrac{1}{2x - 5} = \dfrac{2}{4x^2 - 25}$

29. $\dfrac{4}{x + 5} + \dfrac{9}{x + 1} = 1$

30. $\dfrac{3}{3x + 7} + \dfrac{2}{5x + 5} = \dfrac{1}{3x - 1}$

31. $\dfrac{1}{2x + 3} + \dfrac{2}{x - 4} = \dfrac{3}{2x^2 - 5x - 12}$

32. $\dfrac{1}{3x + 4} + \dfrac{2}{x - 5} = \dfrac{3}{3x^2 - 11x - 20}$

33. $\dfrac{x+2}{x} + \dfrac{1}{x+4} = \dfrac{-2x-6}{x(x+4)}$

34. $\dfrac{2}{x^2-9} - \dfrac{1}{x(x-3)} = \dfrac{2}{x^2}$

In Exercises 35–46, solve the radical equation and check algebraically or visually with the graphics calculator.

35. $\sqrt{x+3} = 4$

36. $4 - \sqrt{3x} = 1$

37. $\sqrt{x} = 2 - x$

38. $\sqrt{x} = x - 2$

39. $x + 2 = 2\sqrt{2x-7}$

40. $\sqrt{x+4} - \sqrt{2x} = 1$

41. $\sqrt{2x+4} = 1 + \sqrt{2x}$

42. $\sqrt{3x-1} + \sqrt{3x+6} = 7$

43. $\sqrt{2x+1} + \sqrt{2x-1} = 4$

44. $\sqrt{3-2x} - 3 = \sqrt{2+2x}$

45. $3\sqrt{2x+1} = 3$

46. $3 + 4\sqrt{2x-3} = 3 + 4\sqrt{17}$

47. An object is 60 cm from a screen. In order to focus the image of the object on the screen, a converging lens with a focal length of 12 cm is to be used. The lens is placed between the object and the screen at a distance of p cm from the object, where:

$$\frac{1}{p} + \frac{1}{60-p} = \frac{1}{12}.$$

Determine p.

48. If a resistance R and reactance X are connected in parallel, the impedance Z of the circuit is given by:

$$Z = \frac{RX}{\sqrt{R^2 + X^2}}.$$

Solve the equation for R.

49. A trucking company offers daily shipping service between Rochester and Syracuse. The total monthly cost for the shipping service is given by:

$$C = \sqrt{0.5x + 2},$$

where C is measured in thousands of dollars and x is measured in hundreds of parcels. Determine the number of parcels if the monthly cost is 25.5 thousand dollars.

50. The demand equation for a certain product is given by:

$$Q = \sqrt{35 + P} + \sqrt{P - 1},$$

where Q is the quantity (number of units) and P is the price per unit. Determine the price per unit if there is a demand for 150 units.

51. The illumination from any light at a stated distance may be expressed as a constant multiple k of the intensity of the light divided by the square of the distance. Two lamps of intensity 4 and 9 candles are placed 20 in. apart. Where should a screen be placed between them in order to be equally illuminated on both sides?

52. When an object with mass travels near the speed of light, its mass (often referred to as the relativistic mass) from the point of view of a stationary observer increases by the following equation:

$$m = \frac{m_0}{\left[1 - \left(\dfrac{v}{c}\right)^2\right]^{\frac{1}{2}}},$$

where m_0 is the mass when the object was at rest in the laboratory frames, v is the speed of the object, and c is the speed of light —3×10^8 m/s.
   a. Solve for v.
   b. At what speed does the relativistic mass m double the rest mass m_0? Triple? Increase by a factor of 100?

53. A thin nonconducting rod of finite length L carries a total charge of Q spread uniformly along it. The electric influence or field set up by this rod decreases along the perpendicular bisector of the rod by the following equation.

$$E = \frac{Q}{ky(L^2 + 4y^2)^{\frac{1}{2}}}$$

Suppose that an electric field is measured along the perpendicular bisector, and thus, is a known quantity. Solve for y if $E = 0.56$, $k = 0.0045$, $Q = 0.000034$, and $L = 1.2$.

Review Exercises

In Exercises 1–10, solve the quadratic equations and check the solutions algebraically or visually with a graphics calculator.

1. $x^2 - 7x - 44 = 0$ **2.** $x^2 + x - 72 = 0$ **3.** $6R^2 - R - 35 = 0$ **4.** $5R^2 - 13R - 6 = 0$

5. $3t^2 + 7t - 5 = 0$ **6.** $2t^2 + 5t - 4 = 0$ **7.** $\frac{1}{9}R^2 - \frac{5}{3}R + 6 = 0$ **8.** $\frac{1}{4}R^2 + 4R + 15 = 0$

9. $3x^2 + 7x + 5 = 0$ **10.** $2x^2 + 4x + 2 = 0$

In Exercises 11–16, determine the quadratic equation that has the indicated roots.

11. 4 and 3   **12.** $\frac{2}{3}$ and -3 **13.** 4 and 4   **14.** $\frac{1}{5}$ and $\frac{1}{7}$

15. $-3 + 4j$ and $-3 - 4j$ **16.** $3j$ and $-3j$

In Exercises 17–20, solve each quadratic equation by completing the square.

17. $x^2 + 6x - 55 = 0$ **18.** $2x^2 - x - 21 = 0$ **19.** $10x^2 + 7x - 12 = 0$ **20.** $\alpha x^2 + \beta x + \gamma = 0$

In Exercises 21–26, state whether the discriminant is positive, negative, or zero and describe the roots of each equation.

21. $2x^2 + 5x - 7 = 0$ **22.** $\frac{1}{9}x^2 + \frac{1}{6}x + \frac{1}{16} = 0$ **23.** $0.1R^2 + 0.01R + 0.02 = 0$

24. $x^2 + 2x - 35 = 0$ **25.** $6x^2 + 13x - 28 = 0$ **26.** $3x^2 + 4x + 4 = 0$

In Exercises 27–36, graph the quadratic function and determine the roots of the quadratic equation to the nearest tenth.

27. $f(x) = x^2 - 11x + 18$ **28.** $f(x) = x^2 + 4x - 21$ **29.** $f(t) = t^2 + 7t + 5$ **30.** $f(t) = t^2 + 5t + 7$
31. $f(R) = 8R^2 + 2R - 15$ **32.** $f(R) = 15R^2 + R - 2$ **33.** $f(t) = 2t^2 + 2t + 3$ **34.** $f(t) = 2t^2 + 3t - 2$
35. $f(x) = 9x^2 - 16$ **36.** $f(x) = \frac{1}{4}x^2 - 4$

In Exercises 37–50, solve the equations and check the solutions.

37. $x^4 - 13x^2 + 36 = 0$ **38.** $9x^4 - 37x^2 + 4 = 0$ **39.** $x^{\frac{1}{3}} - 10x^{\frac{1}{6}} + 9 = 0$ **40.** $\dfrac{1}{x^2} + \dfrac{8}{x} + 7 = 0$

41. $\dfrac{x}{3} - \dfrac{1}{x} = 1$ **42.** $\dfrac{2}{x-1} + \dfrac{x-3}{x} = 6$ **43.** $\dfrac{2}{x^2-4} - \dfrac{1}{x(x-2)} = \dfrac{1}{x^2}$ **44.** $\sqrt{x-5} = 4x$

45. $\sqrt{2x+7} = 5x$ **46.** $\sqrt{3x-4} = 11x$ **47.** $\sqrt{2x+1} = 1 + 3x$

48. $\sqrt{2x+3} + \sqrt{x-5} = 4$ **49.** $\sqrt{3x-4} - \sqrt{x-5} = 1$ **50.** $\sqrt{x-1} + \sqrt{2x+1} = 2$

51. If two electrical resistors are connected in series, the equivalent resistance is $R_s = R_1 + R_2$. If they are connected in parallel, the equivalent resistance is $R_p = \dfrac{R_1 R_2}{R_1 + R_2}$. What two resistors should be used so that the series resistance is 2 Ω and the parallel resistance $= \dfrac{1}{2}\,\Omega$?

52. If two springs are connected in parallel as shown in the illustration, the equivalent spring coefficient is $K_p = K_1 + K_2$. If they are connected in series as shown, the equivalent spring coefficient is $K_g = \dfrac{K_1 K_2}{K_1 + K_2}$. What two springs should be used if the parallel spring factor is 3 lb/in. and the series spring factor is $\dfrac{1}{2}$ lb/in.?

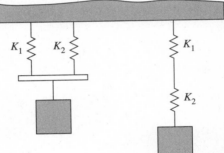

53. A wire stretches from the top of an 80 ft high pole to a stake in the ground 120 ft from the foot of the pole. Determine the length of the wire.

54. A 10-m ladder is placed against a wall with the foot of the ladder 2 m from the base of the wall. (See the drawing.) How high above the ground will the ladder reach?

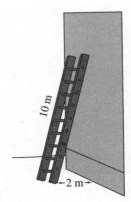

55. A doorway measures 3 ft by 7 ft. What is the diameter of the largest thin metal circular disk that will pass through the doorway? (See the sketch.)

56. An electrician's tool box measures 16 in. in length, 12 in. in width and 10 in. in height. What is the length of the longest extension bit holder that may be placed in the tool box?

57. A rectangular heating duct is to be formed by bending a 68-cm-wide piece of sheet metal. The area inside the rectangle is to be 228 cm². (See the drawing.) Determine the width and height of the duct.

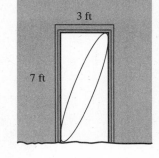

228 cm²

68 cm

58. A page being designed for a textbook is to have 30-mm margins on the top, bottom, and right side. The left side of the page is to have a margin of 50 mm. The height of the page is to be 30 mm more than the width. The area of the printed part of the page is to be 23,400 mm². Determine the width and height of the entire page.

59. A computer programmer and her assistant, working together, complete a special program in four days. The programmer working alone can finish the job in six days less than the assistant working alone. How long does it take each person to do the job alone?

60. A free-lance writer works a certain number of days for a newspaper to earn $480. If she were paid $8 less per day, she would earn the same amount in two more days. What is her daily rate of pay?

61. The equation $\dfrac{1}{d} = \dfrac{pq}{p + q}$ is used in optometry. Solve the equation for q.

62. The time T it takes a pendulum of length ℓ to swing through one complete oscillation is given in the algebraic model:

$$T = 2\pi\sqrt{\frac{\ell}{9.8}}.$$

a. Solve the equation for ℓ.
b. If T is measured in seconds and ℓ in meters, how long would a pendulum have to be to have an oscillation time of 1 s?

63. For a car traveling at a speed of v mph the least number of feet d necessary to stop the car is given by the formula $d = 0.044v^2 + 1.1v$. Estimate the speed of a car requiring 165 ft to stop after the driver becomes aware of danger.

64. In calculating the current in a simple-series electric circuit with inductance L henrys, resistance R ohms, and capacitance C farads, it becomes necessary to solve the equation $Lx^2 + Rx + \dfrac{1}{C} = 0$. Determine x in terms of L, R, and C.

65. From a proportion in geometry, a relationship can be developed between the span of the arch s, the height of the span h, and the radius of the arch r as shown in the sketch. This relation is:

$$h(2r - h) = \frac{s^2}{4}.$$

Determine h if $r = 16.00$ and $s = 25.00$.

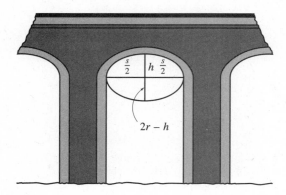

66. To get to the cottage, Barbara rides her moped for 5 km and then jogs the remaining 3 km through the woods. She rides 14 km/hr faster than she jogs (both at constant rates). If it takes 45 min to make the entire trip, how fast does she jog?

67.* Two pipes can fill a tank in seven hours when used together. Alone, one can fill the tank three hours faster than the other. How long will it take each pipe to fill the tank alone?

68.* A new printing press can do a job in one hour less than an older press. Together they can do the same job in 1.2 hours. How long will it take each to do the job alone?

69.* The formula $S = \dfrac{n(n + 1)}{2}$ gives the sum of the first n natural numbers $1, 2, 3, \ldots$. How many consecutive numbers must be added to obtain a sum of 325?

70.* The sum of the roots of a quadratic equation is $r_1 + r_2 = -\dfrac{b}{a}$. The product of the roots of a quadratic equation is $r_1 r_2 = \dfrac{c}{a}$. Use the quadratic formula to develop these two formulas.

71.* Show that if $y = ax^2 + bx + c \ (a \neq 0)$ has real coefficients, then the maximum or minimum point of its graph has the coordinates $x = -\dfrac{b}{2a}, y = \dfrac{4ac - b^2}{4a}$, and that the point is a maximum when a is negative or is a minimum when a is positive.

72.* If an isosceles triangle has an altitude h and a perimeter p, determine an expression in terms of h and p for the base b.

73.* A monument consists of two cubical blocks of granite, the smaller resting on the larger. The total height of the monument is 10 m. The total volume of the monument is 270 m^3. Determine the exposed surface area of the monument.

✎ Writing About Mathematics

1. You are working on a special project at work that involves solving a quadratic equation. The technical information clearly indicates that there are two distinct roots. You calculate the discriminate and tell your co-worker that this cannot be the correct equation. Write a memo explaining how and why you arrived at that conclusion.

2. Your friend has a difficult time understanding why a quadratic equation can have 0, 1, or 2 distinct real roots. Write a paper, including examples, explaining how and why it is possible to have 0, 1, or 2 distinct real roots.

3. In Example 6 of Section 11.4, we determined that the zero bending moments were 1.21 ft and 5.79 ft from the foot of the table. If the bending motion is zero for these values, what happens in between and on both ends? Is the bending motion positive or negative? Write a short paper explaining your answer and how you arrived at your conclusion.

Chapter Test

1. Solve $x^2 - 7x - 18 = 0$ by factoring.

In Exercises 2 and 3, determine the quadratic equation given the roots.

2. 3 and -4 3. 4 and 4

In Exercises 4 and 5, calculate the discriminate of the quadratic equation and use the result to describe the roots of the equation.

4. $x^2 + 3x - 8 = 0$ 5. $2x^2 + 4x + 5 = 0$

6. Sketch the graph of $f(x) = x^2 + 2x - 15$. On the graph, label the y-intercept, x-intercepts, line of symmetry, and the vertex.

7. Solve $x^{\frac{1}{2}} + 6x^{\frac{1}{4}} + 8 = 0$ and check.

8. Solve the quadratic equation $ax^2 + bx + c = 0$ by completing the square.

9. The length of a rectangle exceeds its width by 5 ft. The area of the rectangle is 50 square feet. What are the dimensions of the rectangle?

10. A ball is propelled from a deck. The path of the ball is defined by the quadratic function $s(t) = -16t^2 + 128t + 32$, where s is the distance traveled and t is the time in seconds. Determine the following
    a. The maximum height of the ball.
    b. The time the ball strikes the ground.
    c. The height of the deck from which the ball was propelled.

Equations of Higher Degree

In 1825 Michael Faraday, a research chemist, discovered a new hydrocarbon—a molecule consisting of only hydrogen and carbon atoms—and called it bicarburet of hydrogen. The substance now is called benzene and is an example of a class of molecules known as aromatic compounds. Such compounds exhibit a very important property; they are very stable. Aromatic compounds are primary components in gasoline, paint thinner, mothballs, and they are essential in producing synthetic fibers, resins, and dyes. Chemists have had difficulty determining why these compounds are so stable. However, a breakthrough has occurred. One step in determining the stability of one of these molecules is to solve the following polynomial.

$$P(x) = x^{10} - 11x^8 + 41x^6 - 65x^4 + 43x^2 - 9*$$

In this chapter we learn how to solve polynomials of degree higher than 2. In Example 3 of Section 12.4, we will determine the roots of the polynomial above.

*Jun-ichi Aihara. 1992. "Why Aromatic Compounds Are Stable." *Scientific American* 266 (March): 62.

12.1

Polynomial Functions and Synthetic Division

In Chapters 4 and 10 you learned methods for solving linear equations. In Chapter 11 you learned the various techniques for determining the solutions of quadratic equations. Linear and quadratic equations (first-degree and second-degree equations, respectively) belong to a class of equations called **polynomial equations.** In this chapter you will learn techniques for determining the roots of a polynomial equation of degree greater than 2.

DEFINITION 12.1

Any function of the form:

$$P(x) = a_n x^n + a_{n-1} x^{n-1} + \cdots + a_1 x + a_0, \qquad (1)$$

where n is a nonnegative integer and the coefficients $a_n, a_{n-1}, \ldots, a_2, a_1, a_0$ are real or complex numbers, $a_n \neq 0$, is called a **polynomial function.**

If the polynomial function (1) is set equal to zero, then it is a polynomial equation of **degree** n.

$$a_n x^n + a_{n-1} x^{n-1} + a_{n-2} x^{n-2} + \cdots + a_1 x + a_0 = 0. \qquad (2)$$

EXAMPLE 1 **Degree of Polynomials**

Determine the degree of the following polynomials.

a. $3x + 1$ **b.** $4x^2 - 7x + 5$ **c.** $6x^3 + 7x^2 + 5x + 6$ **d.** 5

Solution

a. $3x^{1} + 1$ is of degree 1.
b. $4x^{2} - 7x + 5$ is of degree 2.
c. $6x^{3} + 7x^2 + 5x + 6$ is of degree 3.
d. 5^{0} is of degree 0.

The exponents 1, 2, 3, and 0 indicate the highest power of the variable, and thus, the degree of the polynomial. ◆

Recall that the roots of the polynomial function are those values of x for which the function is zero. For example, if r is a root of $f(x)$, then $f(r) = 0$. (Recall that roots were defined in Chapter 11.) Thus, the roots of an equation are also called zeros of the polynomial function.

EXAMPLE 2 **Roots of Polynomial Equation Are Zeros of the Polynomial Function**

Show that the roots of the polynomial equation $x^2 + 5x + 6 = 0$ are also the zeros of the polynomial function $P(x) = x^2 + 5x + 6$.

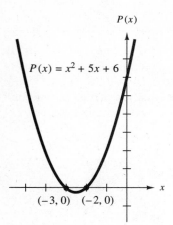

$P(x)$

$P(x) = x^2 + 5x + 6$

x

$(-3, 0)$ $(-2, 0)$

FIGURE 12.1

Solution   A sketch of the polynomial function is given in Figure 12.1. From the figure, we can see that the curve crosses the x-axis at the points $(-3, 0)$ and $(-2, 0)$. The roots of the polynomial equation, then, are -3, and -2. You also may determine the roots by factoring the polynomial equation $x^2 + 5x + 6 = 0$: $(x + 2)(x + 3) = 0$.

Substituting the roots into the polynomial function, we have:

$$P(-2) = (\,-2\,)^2 + 5(\,-2\,) + 6$$
$$= 0$$

and

$$P(-3) = (\,-3\,)^2 + 5(\,-3\,) + 6$$
$$= 0$$

Thus, the roots -2 and -3 of the polynomial equation $x^2 + 5x + 6 = 0$ are zeros of the polynomial function. ◆

QUESTIONS TO ASK ABOUT A POLYNOMIAL

We may need to ask one or more of the following questions to find the root (s) of an nth-degree polynomial.

1. How many roots does a polynomial equation have within the set of complex numbers?
2. How many roots of the polynomial equation are real numbers?
3. Are the roots of the equation rational or irrational numbers?
4. If the coefficients of the polynomials are integers, how many roots are rational numbers?
5. What relation (if any) exists between the roots and the factors of the polynomial equation?

The next sections provide methods to help you answer these questions.

In Example 2, techniques for determining a solution to a particular polynomial function were discussed. Now let us consider the general case. Suppose that $x - r$ is a factor of a polynomial of degree n. By the process of division, we can express the polynomial as the product of the factor $x - r$ and a new polynomial. The new polynomial will be of degree $n - 1$. This is true since the factor $x - r$ is of degree 1. For example, if $x - 1$ is a factor of the polynomial equation:

$$x^3 + 4x^2 + x - 6 = 0,$$

then this equation can be expressed as:

$$(x - 1)(x^2 + 5x + 6) = 0.$$

From Example 2, we also know that the roots to this polynomial equation are $x = -1$, $x = -3$, and $x = -2$. This new technique is useful when dealing with polynomials with $n \geq 3$.

To simplify the following discussion, this text uses the following notation.

$$P(x) = \text{the dividend}$$
$$D(x) = \text{the divisor}$$
$$Q(x) = \text{the quotient}$$
$$R(x) = \text{the remainder}$$

The result of dividing two polynomials can be stated as:

$$\frac{P(x)}{D(x)} = Q(x) + \frac{R(x)}{D(x)}, \quad \text{if } D(x) \neq 0$$

or

$$P(x) = Q(x)D(x) + R(x). \tag{3}$$

Equation (3) is the definition of division, which states that the dividend is equal to the quotient times the divisor plus the remainder. We can use this equation to check the results of the division. If we divide the polynomial $P(x) = x^3 + 4x^2 + x - 6$ by $D(x) = x - 1$, the result is $Q(x) = x^2 + 5x + 6$ and $R(x) = 0$. Checking the results by substitution into Equation (1), we have:

$$P(x) = Q(x)D(x) + R(x)$$
$$x^3 + 4x^2 + x - 6 = (x^2 + 5x + 6)(x - 1) + 0$$
$$= x^3 + 4x^2 + x - 6. \checkmark \qquad \textbf{Multiplying and collecting like terms}$$

Synthetic Division

If the divisor is of the form $x - r$, the process of division of polynomials may be simplified by a method called **synthetic division.** The method is illustrated by dividing $P(x) = 2x^3 + 9x^2 + 7x - 6$ by $D(x) = x + 3$. Begin by arranging the coefficients of $P(x)$ in order of descending power on a line. Insert a zero for the coefficient of any missing power of x:

$$\boxed{2} \quad \boxed{9} \quad \boxed{7} \quad \boxed{-6.}$$

Now set the divisor equal to zero and solve for x.

$$x + 3 = 0 \quad \text{or} \quad x = \boxed{-3}$$

Use the number -3 as the divisor for this method. Place the divisor on the original line to the right of the -6:

$$2 \quad 9 \quad 7 \quad -6 \qquad \underline{|\,\boxed{-3}} \text{ (divisor).}$$

Skip a line and draw a line under the dividend. Then bring 2 down to line 3.

Line 1   2   9   7   −6 |−3

Line 2   ——————

Line 3 $\boxed{2}$

Multiply the 2 on line 3 times the divisor, -3, and place the product, -6, under the 9 on line 2.

	Line 1	2	9	7	-6	$\lfloor -3$
	Line 2		-6			
	Line 3	2				

Find the sum of 9 and -6; place the sum on line 3 under the -6.

	Line 1	2	9	7	-6	$\lfloor -3$
	Line 2		-6			
	Line 3	2	3			

Now multiply 3 by the divisor, -3, and place the product under the 7 on line 2.

	Line 1	2	9	7	-6	$\lfloor -3$
	Line 2		-6	-9		
	Line 3	2	3			

Find the sum of 7 and -9; place the sum on line 3 under the -9.

	Line 1	2	9	7	-6	$\lfloor -3$
	Line 2		-6	-9		
	Line 3	2	3	-2		

Now multiply the -2 on line 3 by the divisor, -3, and place the product on line 2 under the -6.

	Line 1	2	9	7	-6	$\lfloor -3$
	Line 2		-6	-9	6	
	Line 3	2	3	-2		

Find the sum of -6 and 6 and place under the 6 on line 3.

	Line 1	2	9	7	-6	$\lfloor -3$
	Line 2		-6	-9	6	
	Line 3	2	3	-2	0	

At this point, we have completed the division. How do we read the result? The dividend, $P(x) = 2x^3 + 9x^2 + 7x - 6$, is a polynomial of degree 3, and the divisor, $D(x) = x + 3$, is of degree 1; therefore, the quotient is of degree 2. The first three numbers in line 3 are the coefficients of $Q(x)$. Therefore, the quotient $Q(x) = 2x^2 + 3x - 2$. The fourth, or last, number in line 3 is the remainder; therefore, $R(x) = 0$.

If the dividend is a fourth-degree polynomial, line 1 is five numbers across, and so is line 3. In this case, the first four numbers of line 3 are the coefficients of the third-degree polynomial, which is the quotient. The last, or fifth, number is the remainder.

Writing the results of the problem using the definition of division, we have:

$$P(x) = Q(x)D(x) + R(x)$$
$$2x^3 + 9x^2 + 7x - 6 = (2x^2 + 3x - 2)(x + 3) + 0$$
$$= 2x^3 + 9x^2 + 7x - 6. \checkmark$$

In this particular case the divisor of the synthetic division, -3, is also a zero of the polynomial. This is always true when the remainder is zero.

CAUTION   If the dividend has missing terms, remember to fill in the missing coefficients with zeros when we set up for synthetic division. For example, if $P(x) = 3x^4 + 2x - 1$ and $D(x) = x - 1$, we set up line 1 as follows.

$$3 \quad 0 \quad 0 \quad 2 \quad -1 \qquad\qquad (1) \quad \blacksquare$$

EXAMPLE 3   Synthetic Division: Divisor of Form $x - r$

Divide $x^3 + 4x - 7$ by $x - 3$.

Solution   Arrange the coefficients of $P(x)$, $P(x) = x^3 + 4x - 7$ in order of descending power of the variable, inserting a zero for the coefficient of any missing power of x. The polynomial function, then, is $P(x) = x^3 + 0x^2 + 4x - 7$. We determine that the divisor is 3 by letting $x - 3 = 0$ and solving for x. Now write on line 1, the dividend $P(x)$ and the divisor, leave space for line 2, draw a line, and place the first coefficient of $P(x)$ on line 3.

Line 1   1   0   4   −7 |3
Line 2
Line 3   1

Next multiply 1 by 3 and write the product under the 0 on line 2.

Line 1   1   0   4   −7 |3
Line 2   3
Line 3   1

Now add 0 and 3, placing the sum on line 3. Multiply the sum, 3, by 3 and place the product under 4 on line 2.

Line 1   1   0   4   −7 |3
Line 2   3   9
Line 3   1   3

Add 4 and 9, placing the sum, 13, on line 3. Multiply 13 by 3 and place the product under −7 on line 2. Do the final addition to complete the division.

Line 1   1   0   4   −7 |3
Line 2   3   9   39
Line 3   1   3   13   32

The result of the synthetic division is a quotient of $x^2 + 3x + 13$ and a remainder of 32.

Check:

$$x^3 + 4x - 7 = (x^2 + 3x + 13)(x - 3) + 32$$
$$= x^3 + 4x - 7. \ \checkmark \ \blacklozenge$$

The opening sentence of this section and the examples up to this point show divisors of the form $x - r$, where the coefficient of x is 1. For synthetic division we have no problems performing the division when the divisor is in the form $x - r$. Example 4 will address the problem when the divisor is of the form $ax - r$. The process is the same, but the divisor is no longer an integer.

EXAMPLE 4   **Synthetic Division: Divisor of Form $ax - r$**

Divide $2x^4 + 11x^3 + 2x^2 - 55x - 60$ by $2x + 3$.

Solution   The divisor, $2x + 3$, is not of the form $x - r$ since the coefficient of the x-term is 2. Whenever this happens, we can set the divisor equal to zero and solve for x. Thus, $2x + 3 = 0$, or $x = -\dfrac{3}{2}$. Now we are ready to divide using synthetic division.

| Line 1 | 2 | 11 | 2 | −55 | −60 | $\boxed{-\frac{3}{2}}$ |

Line 2: −3 −12 15 60
Line 3: 2 8 −10 −40 0

$(2)\left(-\frac{3}{2}\right)$ $(8)\left(-\frac{3}{2}\right)$ $(-10)\left(-\frac{3}{2}\right)$ $(-40)\left(-\frac{3}{2}\right)$

The result of the synthetic division is the quotient $2x^3 + 8x^2 - 10x - 40$ and the remainder is 0.

Check:

$$2x^4 + 11x^3 + 2x^2 - 55x - 60 = (2x^3 + 8x^2 - 10x - 40)\left(x + \frac{3}{2}\right) + 0$$
$$= 2x^4 + 11x^3 + 2x^2 - 55x - 60. \ \checkmark \ \blacklozenge$$

We may encounter problems that have divisors that are not of the form $x - r$. They may be of the form $(x - r)^2$ or $(x - r)(x + s)$ or $(x + r)(x - s)$. Example 5 illustrates how we can do the division with synthetic division.

EXAMPLE 5   **Synthetic Division: Divisor of Form $(x - r)^2$**

Divide $2x^4 + 3x^3 - 18x^2 - 44x - 24$ by $(x + 2)^2$.

Solution   The divisor $(x + 2)^2$ can be expressed as $(x + 2)(x + 2)$. From arithmetic we know that we can find the quotient for $12 \div [(2)(3)]$ by dividing

12 by 2 and that quotient by 3 (i.e., 12 ÷ 2 = 6 and 6 ÷ 3 = 2; thus, 12 ÷ [(2)(3)] = 2). Using this idea, set up the problem in the usual manner for synthetic division and proceed in the usual way.

Line 1	2	3	−18	−44	−24	$\lfloor$−2
Line 2		−4	2	32	24	
Line 3	2	−1	−16	−12	0	

The quotient from the first division becomes the dividend for the second division.

Line 1	2	−1	−16	−12	$\lfloor$−2
Line 2		−4	10	12	
Line 3	2	−5	−6	0	

Instead of beginning a new setup, we could have continued with the original problem; it would look like this.

$$
\begin{array}{rrrrr}
2 & 3 & -18 & -44 & -24 \quad \lfloor-2 \\
& -4 & 2 & 32 & 24 \\
\hline
2 & -1 & -16 & -12 & 0 \quad \lfloor-2 \\
& -4 & 10 & 12 \\
\hline
2 & -5 & -6 & 0
\end{array}
$$

The result of the division is:

$$2x^4 + 3x^3 - 18x^2 - 44x - 24 = (2x^2 - 5x - 6)(x + 2)(x + 2) + 0$$
$$= 2x^4 + 3x^3 - 18x^2 - 44x - 24. \quad \checkmark \quad \blacklozenge$$

CAUTION   We can use the technique shown in Example 5 when the divisors for synthetic division are not zeros of the polynomial. However, be careful when reading the answer to read the remainder correctly.   ■

12.1 Exercises

In Exercises 1–25, use synthetic division to determine the quotient and remainder. Check each of the results using the equation $P(x) = Q(x)D(x) + R(x)$.

1. $(6x^2 - 2x - 20) \div (x - 4)$

2. $(2x^2 - 7x - 1) \div (x - 4)$

3. $(x^3 + 2x^2 - 3x) \div (x + 1)$

4. $(z^3 - 6z^2 + 13z + 2) \div (z - 3)$

5. $(-x^3 + 2x + 2) \div (x + 2)$

6. $(-y^3 + 3y + 1) \div (y + 2)$

7. $(2x^3 - 6x^2 + 3x + 2) \div (x - 2)$

8. $(3x^3 - x^2 + 4) \div \left(x - \frac{1}{2}\right)$

9. $(x^4 + 3x^3 + x^2 + 4) \div (x + 2)$

10. $(x^4 + x^3 + 4) \div (x + 2)$

11. $(2z^4 + 4z^3 + 3z^2 + 11z - 7) \div (z + 4)$

12. $(3x^5 - 5x^3 + 1) \div (x - 5)$

13. $(3x^5 - 9x^2 + 6x) \div (x - 1)$

14. $(-y^5 + 20y^2 + 8) \div (y - 3)$

15. $(9R^4 - 15R^3 + 13R^2 + 3R + 1) \div \left(R - \frac{1}{3}\right)$

16. $(2R^3 + 0.3R^2 - 0.45R - 0.167) \div (R - 0.2)$

17. $(2x^4 + 6x^3 - 8x + 10) \div \left(x + \dfrac{1}{2}\right)$

18. $(6x^4 - 10x^3 - 4x^2 + 3x + 3) \div \left(x + \dfrac{1}{3}\right)$

19. $(y^5 - 32) \div (y - 2)$

20. $(y^6 + a^6) \div (y + a)$

21. $(-Z^7 + 4.3Z^4 - 7.2Z^2) \div (Z + 2)$

22. $(4Z - 3Z^3 - 0.384) \div (Z + 1.2)$

23. $(-3c^3z + z^4 - c^2z^2 - 81z^4) \div (z + 3c)$

24. $(2x^4 - 12x^3 + 27x^2 - 54x + 81) \div (x - 3)^2$

25. $(2x^5 + 3x^4 - 11x^3 - 5x^2 + 2x + 6) \div [(x - 2)(x + 3)]$

12.2

The Remainder Theorem and the Factor Theorem

Synthetic division is a mechanical method of determining whether or not a number is a root of a polynomial equation. It unavoidably involves some guessing. However, the theorems introduced in this section help keep guessing to a minimum. These theorems not only help us solve polynomial equations, but they also help us graph polynomial functions.

We begin by looking at a division problem that helps us understand the remainder theorem. When we divide $(x^3 - 2x^2 + 2x - 3)$ by $(x - 2)$, the quotient is $(x^2 + 2)$ with a remainder of $+1$.

$$
\begin{array}{rrrr|l}
1 & -2 & 2 & -3 & \underline{2} \\
  & 2 & 0 & 4 & \\
\hline
1 & 0 & 2 & 1 &
\end{array}
$$

From the definition of division, we can write the following statement of equality.

$$(x^3 - 2x^2 + 2x - 3) = (x^2 + 2)(x - 2) + 1$$

This statement is an identity that is true for all values of x. If, in the same way, any division of the polynomial $P(x)$ by $(x - r)$, where r is a constant, results in a quotient $Q(x)$ and a constant remainder R, then we can write:

$$P(x) = Q(x)(x - r) + R.$$

This statement is true for all values of x, and in particular, it is true when $x = r$.

Substituting r in the equation, we obtain:

$$P(r) = (r - r)Q(r) + R$$
$$P(r) = 0 \cdot Q(r) + R$$

or

$$P(r) = R.$$

The general discussion of division of a polynomial by $(x - r)$ provides a proof of an important theorem called the remainder theorem.

THEOREM 12.1 The Remainder Theorem

If a polynomial $P(x)$ is divided by $(x - r)$, then the remainder is $R = P(r)$.

EXAMPLE 1   Applying the Remainder Theorem

Let $P(x) = x^3 + 4x - 7$ and $r = 3$. Determine the remainder R using the remainder theorem.

Solution   Since $r = 3$, a constant, a condition of the hypothesis of the remainder theorem is satisfied. Now, substituting $x = 3$ into the polynomial, we have:

$$P(x) = x^3 + 4x - 7$$
$$P(3) = (3)^3 + 4(3) - 7$$
$$= 32$$

Thus the remainder of the polynomial is 32.   ◆

In Example 3 of Section 12.1 we found, using synthetic division, that the remainder for the polynomial $P(x) = x^3 + 4x - 7$ is 32. Thus the remainder of 32 obtained in Example 1 using the remainder theorem checks with the result obtained by synthetic division.

EXAMPLE 2   Remainder Theorem as a Check on Synthetic Division

Let $P(x) = 2x^3 + 11x^2 - 7x - 7$ and $D(x) = 2x + 1$.

a.   Determine the remainder by synthetic division.
b.   Check the result of part a using the remainder theorem.

Solution

a.   To determine the remainder by synthetic division, arrange the coefficients of the dividend in proper order. Let $2x + 1 = 0$ and solve for x, $x = -\dfrac{1}{2}$.

Thus, $-\dfrac{1}{2}$ is the divisor.

$$
\begin{array}{rrrr|}
2 & 11 & -7 & -7 \quad \underline{-\frac{1}{2}} \\
 & -1 & -5 & 6 \\
\hline
2 & 10 & -12 & -1
\end{array}
$$

From the synthetic division, we see that the remainder is -1.
b.   Using the remainder theorem we obtain the following result.

$$P(x) = 2x^3 + 11x^2 - 7x - 7$$
$$P\left(-\frac{1}{2}\right) = 2\left(-\frac{1}{2}\right)^3 + 11\left(-\frac{1}{2}\right)^2 - 7\left(-\frac{1}{2}\right) - 7$$
$$= -1.$$

Since the remainder for parts a and b are both -1, we can assume that the synthetic division is correct.   ◆

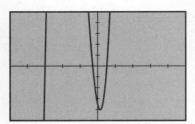

FIGURE 12.2

CAUTION   In determining the remainder using the remainder theorem in Example 2, a common error is to calculate $P(-1) = 9$, rather than $P\left(-\dfrac{1}{2}\right) = -1$. Remember that the hypothesis of the remainder theorem says that the divisor must be of the form $x - r$. In Example 2 the divisor is not of this form—the x-term has a coefficient of 2. The best way to avoid the problem, whether the divisor is $x - 4$ or $3x - 5$, is to solve for x and substitute this value for x into $P(x)$. ■

We can check the result of the division in Example 2 by graphing the function. Figure 12.2 shows the graph of $P(x) = 2x^3 + 11x^2 - 7x - 7$ and illustrates that the curve does not cross the horizontal axis at $x = -\dfrac{1}{2}$. In fact, by using the trace function, we can estimate the when $x = -\dfrac{1}{2}$, $f\left(\boxed{-\dfrac{1}{2}}\right) = \boxed{-1.}$ For the cases where $R = 0$, the graph crosses the horizontal axis.

The special case of the remainder theorem when $R = 0$ is called the **factor theorem.** For example, for the polynomial function $P(x) = x^2 + 5x + 6$, we know from the study of quadratic equations that $x + 2$ is a factor of $x^2 + 5x + 6$. Using the remainder theorem, we see that:

$$P(-2) = (-2)^2 + 5(-2) + 6$$
$$= 0.$$

This leads to a formal statement of the factor theorem.

> **THEOREM 12.2 The Factor Theorem**
> A polynomial $P(x)$ has a factor $(x - r)$ if, and only if, $P(r) = 0$.

The example above demonstrates the factor theorem for one case. We will write a proof that states it is true in all cases.

Proof:
By hypothesis, r is a zero of $P(x)$; that is, $P(r) = 0$ and $R = 0$. Thus,

$$P(x) = (x - r)Q(x) + 0.$$

Therefore $x - r$ is a factor of $P(x)$. ◆

EXAMPLE 3   **Applying the Factor Theorem**

Determine whether $x - 5$ is a factor of $x^2 - 6x + 5$.

Solution   To use the factor theorem, solve $x - 5 = 0$ for x. Substituting $x = 5$ in $P(x)$:

$$P(x) = x^2 - 6x + 5$$
$$P(\boxed{5}) = (\boxed{5})^2 - 6(\boxed{5}) + 5)$$
$$= 0.$$

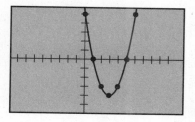

FIGURE 12.3

Since $R = 0$, there is no remainder when we divide $x^2 - 6x + 5$ by $x - 5$. Thus, the factor theorem tells us that $x - 5$ is a factor of $x^2 - 6x + 5$. ◆

If we sketch the function with a graphics calculator as in Figure 12.3, we see that the curve crosses the horizontal axis at $x = 5$. Doing this provides a visual check.

EXAMPLE 4 **The Factor Theorem Shows $(x + 1)$ Is Not a Factor**

Determine whether $x + 1$ is a factor of $x^5 - 1$.

Solution Solving $x + 1 = 0$ for x, we obtain $x = -1$. Substituting for x in $P(x)$, we have:

$$P(x) = x^5 - 1$$
$$P(\boxed{-1}) = (\boxed{-1})^5 - 1$$
$$= -2.$$

Because the remainder is not equal to zero, $x + 1$ is not a factor of $x^5 - 1$. ◆

EXAMPLE 5 **Is $(x + a)$ a Factor?**

Determine under what conditions $x + a$ is a factor of $x^n + a^n$.

Solution Solving $x + a$ for x, we have $x = -a$. By the factor theorem, for $x + a$ to be a factor of $x^n + a^n$, $(-a)^n + a^n = 0$. For this sum to equal zero, $(-a)^n$ must equal $-(a)^n$, which is true only when n is odd. For example, $(-2)^3 = -(2)^3$. Therefore, the condition for $x + a$ to be a factor of $x^n + a^n$ is that n be odd. ◆

EXAMPLE 6 **Determining Roots of $P(x)$**

a. If $P(x) = x^3 + 3x^2 - x - 3$ and $P(x) = (x + 1)Q(x)$, determine $Q(x)$.
b. Determine all roots of $P(x)$.

Solution

a. If $(x + 1)$ is a factor of $P(x)$, then by the factor theorem, $P(-1) = 0$.

$$P(-1) = (-1)^3 + 3(-1)^2 - (-1) - 3 = 0$$

We can determine $Q(x)$ using synthetic division.

$$
\begin{array}{rrrr|r}
1 & 3 & -1 & -3 & \underline{-1} \\
  & -1 & -2 & +3 & \\
\hline
1 & 2 & -3 & 0 &
\end{array}
$$

Thus, $Q(x) = x^2 + 2x - 3$ and $P(x) = (x + 1)(x^2 + 2x - 3)$.

b. To determine the roots of $P(x)$, we need to factor $x^2 + 2x - 3$. Then set the factors of $P(x)$ equal to zero and solve for x.

$$P(x) = (x + 1)(x + 3)(x - 1)$$

$$x + 1 = 0 \qquad x + 3 = 0 \qquad x - 1 = 0$$
$$x = -1 \qquad x = -3 \qquad x = 1$$

The roots of $P(x)$ are -3, -1, and 1. ◆

EXAMPLE 7 **Is $(x - 2j)$ a Factor?**

Determine whether $x - 2j$ is a factor of $x^2 + 4$.

Solution When we solve $x - 2j = 0$ for x, the result is $x = \boxed{2j}$. Substituting for x in $P(x)$, we have:

$$P(x) = x^2 + 4$$
$$P(2j) = (\,\boxed{2j}\,)^2 + 4$$
$$= 4j^2 + 4$$
$$= 0.$$

The remainder is zero; therefore, $x - 2j$ is a factor of $x^2 + 4$. ◆

12.2 Exercises

In Exercises 1–10, perform the following.
a. Determine the remainder by synthetic division.
b. Check the result of part a using the remainder theorem and/or the graphics calculator.

1. $P(x) = 3x^2 - 2x - 4$, $D(x) = x - 2$

2. $P(x) = x^3 - 2x^2 + 4x - 7$, $D(x) = x - 3$

3. $P(x) = x^3 - 3x^2 + 9$, $D(x) = x + 2$

4. $P(x) = 3x^3 - 5x^2 + x + 2$, $D(x) = x - 2$

5. $P(x) = 2x^4 - x^3 + 2x^2 + x - 8$, $D(x) = x + 1$

6. $P(x) = x^4 - 3x^3 - 3x^2 + x - 4$, $D(x) = x - 3$

7. $P(x) = x^4 - 3x^3 - 3x^2 + x + 4$, $D(x) = x - 1$

8. $P(x) = 4x^5 - 5x^4 - x + 8$, $D(x) = x - \dfrac{1}{2}$

9. $P(x) = 7x^3 - x^2 + 7x - 1$, $D(x) = x - j$

10. $P(x) = x^3 + 4x^2 - 2x - 5$, $D(x) = x + 3j$

In Exercises 11–20, use the factor theorem to determine whether the first quantity is a factor of the second quantity.

11. $x - 2$, $x^3 + 2x^2 - 5x - 6$

12. $x + 3$, $x^3 - 4x^2 + 9$

13. $x + 1$, $x^3 - 4x^2 - 2x - 2$

14. $x - 2$, $x^3 - x^2 - 5x + 6$

15. $x + 2$, $x^4 - 3x - 5$

16. $x + 3$, $x^3 + 27$

17. $x + 2$, $x^4 + 16$

18. $x - 3$, $x^3 + 27$

19. $x + 2j$, $x^3 - x^2 + 4x - 4$

20. $x - j$, $x^4 + 4x^3 + 2x^2 + 4x + 1$

21. **a.** If $P(x) = 2x^3 + 10x^2 - 13x - 30$ and $P(x) = (x - 2)Q(x)$, determine $Q(x)$.
  b. Determine all the roots of $P(x)$.

22. **a.** If $P(x) = 3x^3 + 7x^2 - 3x - 7$ and $P(x) = (x + 1)Q(x)$, determine $Q(x)$.
  b. Determine all the roots of $P(x)$.

23. **a.** If $P(x) = 2x^3 - 21x^2 + 64x - 105$ and $P(x) = (x - 7)\,Q(x)$, determine $Q(x)$.
  b. Determine all the roots of $P(x)$.

24. **a.** If $P(x) = x^3 + 12x^2 + 39x + 18$ and $P(x) = (x + 6)Q(x)$, determine $Q(x)$.
  b. Determine all the roots of $P(x)$.

25. Determine the values of r for which division of $x^2 - 3x - 1$ by $x - r$ has a remainder of 3.

26. Determine the values of r for which division of $3x^2 + 5x - 12$ by $x - r$ has a remainder of -4.

In Exercises 27–32, use the factor theorem to determine whether or not each of the following statements is true.

27. $x - y$ is a factor of $x^n - y^n$ if n is any natural number.

28. $x - y$ is not a factor of $x^n + y^n$ if n is any natural number.

29. $x + y$ is a factor of $x^n - y^n$ if n is even.

30. $x + y$ is a factor of $x^n + y^n$ if n is odd.

31. $x + y$ is not a factor of $x^n - y^n$ if n is odd.

32. $x + y$ is not a factor of $x^n + y^n$ if n is even.

33. Verify that $3x - 1$ is a factor of $P(x) = 3x^3 + 14x^2 + 7x - 4$. May we conclude that $P\left(\dfrac{1}{3}\right) = 0$?

34. Verify that $x^2 + 1$ is a factor of $P(x) = 3x^3 - 5x^2 + 3x - 5$. May we conclude that $P(-1) = 0$?

35. Determine the value of k for which $x - 3$ is a factor of $kx^3 - 6x^2 + 2kx - 12$.

36. Determine the value of k for which $x - 9$ is a factor of $3x^4 - 40x^3 + 130x^2 + kx + 27$.

12.3

Factors and Roots of Polynomial Equations

This section deals with theorems that help determine the factors and roots of polynomial functions.

Does every polynomial have a root? This question was answered by Karl Friedrich Gauss in 1799. At the age of 22, Gauss proved the following theorem.

> **THEOREM 12.3 The Fundamental Theorem of Algebra**
> Every polynomial function of the form below of degree n has at least one complex root.
> $$P(x) = a_n x^n + a_{n-1} x^{n-1} + a_{n-2} x^{n-2} + \cdots + a_1 x + a_0$$

The proof of the theorem is beyond this text; however, the theorem is important in helping us to understand the following theorem.

> **THEOREM 12.4**
> Every polynomial function of the form below of degree n has exactly n roots.
>
> $$P(x) = a_n x^n + a_{n-1} x^{n-1} + a_{n-2} x^{n-2} + \cdots + a_1 x + a_0$$

EXAMPLE 1   Number of Roots

Determine the number of roots of $P(x) = (x - 5)^4 (x - 3)(x + 7)^3$.

Solution $P(x)$ is a polynomial function of *eighth degree* because it contains *eight roots:* $5, 5, 5, 5, 3, -7, -7,$ and -7. Note than any root that occurs m times counts as m roots. We say that 5 is a root of multiplicity 4 and that -7 is a root of multiplicity 3.   ◆

EXAMPLE 2   Given Three Roots, Write an Equation of Degree 3

Given the roots $\frac{1}{2}$, $-\frac{2}{3}$, and 4, write a polynomial equation that has integral coefficients.

Solution   An equation having the desired roots is:

$$\left(x - \frac{1}{2}\right)\left(x + \frac{2}{3}\right)(x - 4) = 0.$$

The equation we obtain by multiplying these factors together has fractional coefficients. We may clear away the fractional coefficients by multiplying the equation by the lcd. However, there is less work involved if we introduce the lcd immediately. The lcd of 2 and 3 is 6. The lcd turns out to be the same value as a_n. We now can write:

$$6\left(x - \frac{1}{2}\right)\left(x + \frac{2}{3}\right)(x - 4) = 0,$$

or

$$(2)(3)\left(x - \frac{1}{2}\right)\left(x + \frac{2}{3}\right)(x - 4) = 0,$$

or

$$2\left(x - \frac{1}{2}\right)3\left(x + \frac{2}{3}\right)(x - 4) = 0,$$

or

$$(2x - 1)(3x + 2)(x - 4) = 0.$$

When we multiply these factors, the product is the desired equation. Therefore, a polynomial equation with integral coefficients with roots of $\frac{1}{2}$, $-\frac{2}{3}$, and 4 is $6x^3 - 23x^2 - 6x + 8 = 0$. The result, a polynomial of degree 3, is exactly what we would expect from Theorem 12.4. ◆

The roots of a polynomial equation may be real numbers or complex numbers. The roots of the quadratic equation $x^2 - 4x + 5 = 0$ are $x = 2 + j$ and $x = 2 - j$. We can see from this polynomial equation with real coefficients that the solution is made up of a complex number and its conjugate. (The roots of the polynomial equation were determined using the quadratic formula.)

This brings us to the next theorem.

THEOREM 12.5 Conjugate-Roots Theorem

If $P(x) = a_nx^n + a_{n-1}x^{n-1} + a_{n-2}x^{n-2} + \cdots + a_1x + a_0$ is a polynomial of degree $n \geq 1$ with real coefficients and if $a + bj$, $b \neq 0$ is a root of $P(x)$, then the complex conjugate $a - bj$ is also a root of $P(x)$.

Theorems 12.4 and 12.5 help us determine the number of roots of a polynomial function. The question now is, how can we determine the roots given the polynomial function? Although the following theorems (Theorems 12.6, 12.7 and 12.8), do not provide the roots, they help us to narrow the choices. The goal here is to cut down on the amount of guessing in determining the roots.

THEOREM 12.6 Descartes' Rule of Signs

Let $P(x) = a_nx^n + a_{n-1}x^{n-1} + a_{n-2}x^{n-2} + \cdots + a_1x + a_0$ be a polynomial with real cofficients and $a_n \neq 0$.

1. The number of positive real roots of $P(x)$ is either equal to the number of variations in sign in $P(x)$ or less than that number by a positive even integer.
2. The number of negative real roots is either equal to the number of variations in sign in $P(-x)$ or less than that number by a positive even integer.

EXAMPLE 3   Applying Descartes' Rule of Signs

Use Descartes' rule of signs to determine the possible number of positive and negative roots of the polynomial equations.

a. $P(x) = x^3 + 6x^2 - 13x - 42$
b. $P(x) = 2x^3 + 9x^2 - 17x + 6$
c. $P(x) = 4x^4 + 16x^3 - 31x^2 - 49x + 30$
d. $P(x) = x^4 + 3x^2 + 7x + 6$

TABLE 12.1

Possibility	Positive Real	Negative Real	Complex
1	1	2	0
2	1	0	2

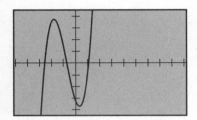

FIGURE 12.4

TABLE 12.2

Possibility	Positive Real	Negative Real	Complex
1	2	1	0
2	0	1	2

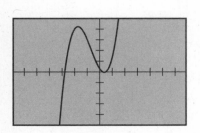

FIGURE 12.5

Solution

a. To determine the number of possible positive real roots of $P(x) = x^3 + 6x^2 - 13x - 42$, determine the number of changes of sign of $P(x)$.

$$P(x) = \quad + \quad + \quad - \quad - \qquad \textbf{Looking only at signs}$$
$$\underset{\text{change}}{\vee}$$

We see that there is only one change of sign. Therefore, according to Theorem 12.6, there is one positive real root.

To determine the possible number of negative real roots, replace x with $-x$ and determine $P(-x)$.

$$P(-x) = (-x)^3 + 6(-x)^2 - 13(-x) - 42$$
$$= -x^3 + 6x^2 + 13x - 42$$

Now determine the the number of changes of sign of $P(-x)$.

$$P(-x) = \quad - \quad + \quad + \quad - \qquad \textbf{Looking only at signs}$$
$$\underset{\text{change}}{\vee} \quad \underset{\text{change}}{\vee}$$

There are two changes of sign. Thus, according to Theorem 12.6, there may be two real negative roots or no negative real roots.

Since $P(x)$ has three roots (not more than one positive and not more than two negative), the following possibilities exist: one positive real root and two negative real roots, or one positive real root and two complex roots (see Table 12.1). If we sketch the graph of the function as shown in Figure 12.4, we see that the curve crosses the x-axis twice to the left of the origin and once to the right of the origin. This illustrates that there are two negative real roots and one positive real root—thereby verifying one of the conclusions we reached using Descartes' rule.

b. For the function $P(x) = 2x^3 + 9x^2 - 17x + 6$, we have:

$$P(x) = + \quad + \quad - \quad +.$$
$$\underset{\text{change}}{\vee} \ \underset{\text{change}}{\vee}$$

The two changes of sign mean we could have two positive real roots or no positive real roots. When we replace x with $-x$, we have:

$$P(-x) = 2(-x)^3 + 9(-x)^2 - 17(-x) + 6$$
$$= -2x^3 + 9x^2 + 17x + 6$$
$$P(-x) = - \quad + \quad + \quad +.$$
$$\underset{\text{change}}{\vee}$$

There is one change of sign; therefore, there is one negative real root.

Since $P(x) = 0$ has three roots, the following possibilities exist: two positive real roots and one negative real root, or two complex roots and one negative real root (see Table 12.2). Sketching the graph of the function as shown in Figure 12.5, we can see that the function has one negative real and two positive real roots—one of the possibilities from our analysis.

TABLE 12.3

Possibility	Positive Real	Negative Real	Complex
1	2	2	0
2	2	0	2
3	0	2	2
4	0	0	4

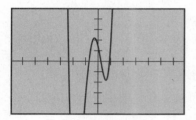

FIGURE 12.6

TABLE 12.4

Possibility	Positive Real	Negative Real	Complex
1	0	2	2
2	0	0	4

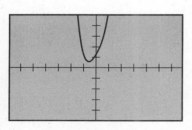

FIGURE 12.7

c. For the function $P(x) = 4x^4 + 16x^3 - 31x^2 - 49x + 30$, we have:

$$P(x) = +\quad +\quad -\quad -\quad +.$$
$$\underset{\text{change}}{\vee}\ \underset{\text{change}}{\vee}$$

With two changes of sign, we could have two positive real roots or no positive real roots. When we replace x with $-x$, we have:

$$P(-x) = 4(-x)^4 + 16(-x)^3 - 31(-x)^2 - 49(-x) + 30$$
$$= x^4 - 16x^3 - 31x^2 + 49x + 30$$
$$P(-x) = +\quad -\quad -\quad +\quad +.$$
$$\underset{\text{change}}{\vee}\ \underset{\text{change}}{\vee}$$

With two changes of sign, we could have two negative real roots or no negative real roots.

Since $P(x) = 0$ has four roots, the following possibilities exist: two positive and two negative real roots, or two positive real and two complex roots, or two complex and two negative real roots, or four complex roots (see Table 12.3). Sketching the graph of the function as shown in Figure 12.6, we can see that the function has two negative and two positive real roots, one of the possibilities from the analysis.

d. For the function $P(x) = x^4 + 3x^2 + 7x + 6$, we have:

$$P(x) = +\quad +\quad +\quad +.$$

There are no changes in sign; therefore, there are no positive real roots. When we replace x with $-x$, we have:

$$P(-x) = (-x)^4 + 3(-x)^2 + 7(-x) + 6$$
$$= x^4 + 3x^2 - 7x + 6$$
$$P(-x) = +\quad +\quad -\quad +.$$
$$\underset{\text{change}}{\vee}\ \underset{\text{change}}{\vee}$$

There are two changes in sign. Thus, there may be two negative real roots and two complex roots, or no negative real roots and four complex roots (see Table 12.4). Sketching the graph of the function as in Figure 12.7, we can see that the function has no negative and no positive roots, which is one of the possibilities from the analysis. Thus, there are four complex roots for this function. ◆

Now that we can determine the possible number of real roots of a polynomial function, the following theorem helps us determine which rational numbers are possible roots.

> ### THEOREM 12.7 Rational-Root Theorem
>
> If the rational number $\dfrac{p}{q}$, a fraction in lowest terms, is a root of the equation,
>
> $$P(x) = a_n x^n + a_{n-1}x^{n-1} + a_{n-2}x^{n-2} + \cdots + a_1 x + a_0,$$
>
> where a_i $(i = 0, 1, 2, \ldots, n)$ are integral coefficients, then p is a factor of a_0, and q is a factor of a_n.

Theorem 12.7 says that for $\dfrac{-3}{2}$ to be a possible root of the polynomial equation $2x^3 + 5x + 3 = 0$, -3 must be a factor of 3, the constant term, and 2 must be a factor of the coefficient of x^3, 2. In this case, they are factors, and $\dfrac{-3}{2}$ is a possible root. If $a_n = 1$ for the polynomial equation:

$$a_n x^n + a_{n-1}x^{n-1} + a_{n-2}x^{n-2} + \cdots + a_1 x + a_0 = 0,$$

then any rational roots of the equation are integers, and they are factors of a_0. This idea is illustrated in Example 4.

EXAMPLE 4   Applying the Rational Root Theorem When $a_n = 1$

Determine the roots of the polynomial function $P(x) = x^3 + 6x^2 - 13x - 42$.

Solution   The function $P(x) = x^3 + 6x^2 - 13x - 42$ is of degree 3. By Theorem 12.4, there must be exactly three roots. By applying Descartes' rule of signs in Example 3a, we learned that there is one positive real root and possibly two negative real roots. Furthermore, by graphing the function, Figure 12.4, we can see that there are two negative and one positive real roots. From the rational-root theorem, we know that any rational root must be of the form p/q. For the polynomial function $P(x)$, the factors of p are ± 1, ± 2, ± 3, ± 6, ± 7, ± 14, ± 42, and the factors of q are ± 1. Thus, the possible rational roots, p/q are ± 1, ± 2, ± 3, ± 6, ± 7, ± 14, ± 42. The theorems have helped us narrow the search, but have not told us the roots of the polynomial.

To determine the roots, use synthetic division. From the sketch, Figure 12.8, we see that the curve crosses the horizontal axis close to what appears to be 1. Knowing that there is one positive real root and 1 is one of the possibilities, select 1 and begin the synthetic division.

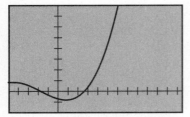

FIGURE 12.8

$$
\begin{array}{rrrr|l}
1 & 6 & -13 & -42 & \underline{1} \\
  & 1 &   7 & -6 & \\
\hline
1 & 7 & -6 & -48 &
\end{array}
$$

Since there is a remainder of -48, 1 cannot be a root. Now let us try 6.

$$
\begin{array}{rrrr|r}
1 & 6 & -13 & -42 & \underline{6} \\
  & 6 & 72 & 354 & \\
\hline
1 & 12 & 59 & 312 &
\end{array}
$$

In this case, the remainder is 312. The results of the synthetic division give us two ordered pairs $(1, -48)$ and $(6, 312)$ and each is a point on the curve defined by the function. Graphing these points shows that for a line to join the two points, it must cut across the x-axis (see Figure 12.8). Therefore, there must be a root between 1 and 6. The possible rational roots between 1 and 6 are 2 and 3. Since the line on the graph crosses the x-axis closer to 3, try 3.

$$
\begin{array}{rrrr|r}
1 & 6 & -13 & -42 & \underline{3} \\
  & 3 & 27 & 42 & \\
\hline
1 & 9 & 14 & 0 &
\end{array}
$$

There is no remainder; therefore, 3 is a root and $x - 3$ is a factor of the polynomial. We can now write the function as the product of two factors.

$$P(x) = (x^2 + 9x + 14)(x - 3)$$

We could find the remaining roots by synthetic division. However, since the quotient is a quadratic expression, we can also find the roots by factoring (or by using the quadratic formula).

$$P(x) = (x + 7)(x + 2)(x - 3).$$

Thus, the roots of $P(x) = x^3 + 6x^2 - 13x - 42$ are -7, -2, and 3. The solution consists of two negative roots and a positive root, one of the possibilities determined by Descartes' theorem. ◆

In Example 4, when we performed the synthetic division with 6 as the divisor, the results in the third line were all positive numbers (1, 12, 59, and 312). This fact can help us reduce the number of choices we must make as indicated in Theorem 12. 8.

THEOREM 12.8 Lower-Bound and Upper-Bound Rule

In the synthetic division of:

$$P(x) = a_n x^n + a_{n-1} x^{n-1} + a_{n-2} x^{n-2} + \cdots + a_1 x + a_0$$

by $(x - r)$, where $a_n > 0$,

1. if $r > 0$ and all the numbers in the third row are nonnegative, then r is an upper bound for the positive zeros (roots) of $P(x)$;
2. if $r < 0$ and the signs of the numbers in the third row alternate in sign (zero entries count as positive or negative), then r is a lower bound for the negative zeros (roots) of $P(x)$.

If the conditions of the hypotheses of the theorem are met, then we can conclude the following when performing synthetic division:

1. If all the numbers in the third row are positive, then there is no positive real number larger than r that is a zero (root) of the equation.
2. If the numbers in the third row alternate signs, then there is no negative real number smaller than r that is a zero (root) of the equation.

In Example 4, with 6 as a divisor, the signs of the numbers in the third row of the synthetic division are all positive; thus, 6 is an upper bound. If we try -14 as a divisor, we get the following result.

$$
\begin{array}{rrrr|r}
1 & 6 & -13 & -42 & \underline{-14} \\
  & -14 & 112 & -1386 & \\
\hline
1 & -8 & 99 & -1428 &
\end{array}
$$

The signs in row three alternate, $+\ -\ +\ -$. Thus, -14 is a lower bound. Now we know that the roots of $P(x)$ are between -14 and 6. This is true since we found the roots of $P(x)$ in Example 4 to be -7, -2, and 3.

TECHNIQUES FOR DETERMINING RATIONAL ROOTS OF POLYNOMIAL FUNCTIONS

The question remains. In what order should a trial root be selected from the list of possible roots? The only way is to guess; however, the search can be narrowed if we consider the following.

1. Use the degree of the polynomial to determine the number of roots.
2. Use Descartes' rule of signs to determine the possible number of positive, negative, and complex roots.
3. Use the rational-root theorem to prepare a list of the possible rational roots of the function.
4. Use a graphics calculator to sketch the graph of the function to help reduce the list of choices from Step 3.
5. Select a number from the possible subset of rational roots for the starting guess. If Step 2 indicates an odd number of positive or negative roots, our first guess should be from the subset of odd roots.
6. Use synthetic division.
   a. If the guess is not a root, check whether it is a lower bound or an upper bound.
   b. If the remainders of two possible roots have opposite signs, we know that there must be a root between these two numbers.
7. If we find a root and the possibility of a multiple root exists, always check for the multiple root before trying another number.
8. Try numbers that may be possible roots in order. If the remainder keeps getting farther away from zero, we know that this will not lead to a root of the polynomial.
9. Make a list of divisors and remainders. These ordered pairs satisfy the polynomial function, and we can use them in graphing the function.
10. Inspect the results. And, as in life, use common sense.

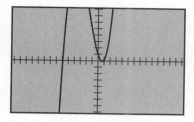

FIGURE 12.9

In Example 5, apply the techniques for determining the roots of a polynomial function.

EXAMPLE 5   Determining Roots of a Polynomial Function

Determine the roots of $P(x) = 2x^3 + 9x^2 - 17x + 6$.

Solution

a. *Step 1*, the degree of the polynomial function is 3; therefore, there are three roots.

b. *Step 2*, in Example 3b, we found that there could be one negative and two positive roots for $P(x)$.

c. *Step 3*, for the polynomial function $P(x)$, the factors of p are ± 1, ± 2, ± 3, ± 6, and the factors of q are ± 1 and ± 2. Thus, the possible rational roots $\frac{p}{q}$, are ± 1, ± 2, ± 3, ± 6, $\pm \frac{1}{2}$, $\pm \frac{3}{2}$.

d. *Step 4*, sketch the function as in Figure 12.9. We can see that the curve crosses the horizontal axis close to -6 and 1.

e. *Step 5*, since there is only one negative root, select -6 as a possible root.

f. *Step 6*, now use synthetic division.

$$
\begin{array}{rrrr|l}
2 & 9 & -17 & 6 & \underline{-6} \\
  & -12 & 18 & -6 & \\
\hline
2 & -3 & 1 & 0 &
\end{array}
$$

A remainder of 0 indicates that -6 is a root. Examinining the reduced polynomial, $2x^2 - 3x + 1$, we can see that the possible rational roots are ± 1 and $\pm \frac{1}{2}$. Since the remaining roots must be positive, we have two choices as possible positive rational roots of the reduced polynomial: 1 and $\frac{1}{2}$.

$$
\begin{array}{rrr|l}
2 & -3 & 1 & \dfrac{1}{2} \\
  & 1 & -1 & \\
\hline
2 & -2 & 0 & \underline{1} \\
  & 2 & & \\
\hline
2 & 0 & &
\end{array}
$$

From the results of the division, we can see that the roots are $\frac{1}{2}$, 1, and -6. ◆

The application of theorems up to this point has been for the purpose of finding the roots of a polynomial. These roots are also the zeros of the polynomial functions and are extremely helpful in graphing the function.

EXAMPLE 6   Graphing Polynomial Function

Sketch the graph of $P(x) = 2x^3 + 9x^2 - 17x + 6$ for the interval $-7 \leq x \leq 3$.

Solution   The x-intercepts were found in Example 5. They are $\left(\frac{1}{2}, 0\right)$, $(1, 0)$, and $(-6, 0)$. The y-intercept, $(0, 6)$, is found by letting $x = 0$. In addition to the

TABLE 12.5 $P(x) = 2x^3 + 9x^2 - 17x + 6$

x	$P(x)$
-7	-120
-6	0
-5	66
-4	90
-3.75	90.8
-3	84
-2	60
-1	30
0	6
$\frac{1}{2}$	0
0.75	-0.84
1	0
2	24
3	90

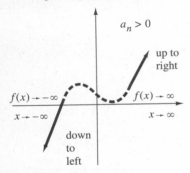

Degree of polynomial is odd

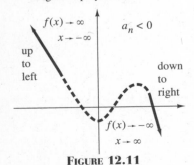

FIGURE 12.11

Behavior of Large Values for x

intercept there are additional ordered pairs shown in Table 12.5 to help us graph the function. The function is graphed in Fig. 12.10. ◆

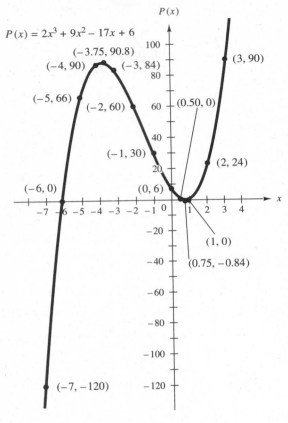

FIGURE 12.10

Now that we have the graph, what additional information can we obtain from it? The curve has a maximum point (high point) close to $(-3.75, 90.8)$ and a minimum point (low point) close to $(0.75, -0.84)$. As we move from left to right in examining the curve, we find the following. The curve is increasing (going up) to the left of -3.75. The curve is decreasing (going down) between -3.75 and 0.75. The curve is **increasing** to the right of 0.75. We also can see that where the curve lies above the horizontal axis, the functional value are positive. Where the curve lies below the horizontal axis, the functional values are negative.

The degree of a polynomial and the coefficient of the nth-degree term provides information about the behavior of the polynomial function for extremely large positive and negative values for x. In Example 4, $a_n = 2 > 0$, the degree of the polynomial is odd (degree 3). For large negative values of x, $f(x)$ decreases [$f(x) \to -\infty$ as $x \to -\infty$]. For large positive values of x, $f(x)$ increases

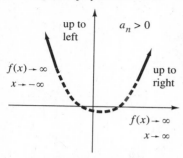

Degree of polynomial is even

$f(x) \to \infty$
$x \to -\infty$

up to
left

$a_n > 0$

up to
right

$f(x) \to \infty$
$x \to \infty$

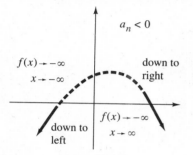

Degree of polynomial is even

$a_n < 0$

$f(x) \to -\infty$
$x \to -\infty$

down to
right

down to
left

$f(x) \to -\infty$
$x \to \infty$

FIGURE 12.12

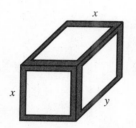

FIGURE 12.13

$[f(x) \to \infty$ as $x \to \infty]$. The symbols $f(x) \to \infty$ as $x \to \infty$ mean that $f(x)$ increases without bound as x increases without bound. The diagrams in Figure 12.11 and Figure 12.12 illustrate the behavior of polynomial functions for extremely large positive and negative value for x.

EXAMPLE 7 **Application: Determine the Volume of a Shipping Crate**

The frame for a shipping crate for wood pellets is constructed from 48 ft of 2×2 lumber (see Figure 12.13). The volume of the crate must be 54 ft³. For shipping purposes, the ends of the crate must be a square, and each side of the square must be between 2.5 ft and 3.5 ft. Determine the dimensions of the crate.

Solution Figure 12.13 shows the length, width, and height of the crate as y ft, x ft, and x ft. The volume for a crate is length times width times height.

$$V = x^2y \quad \text{or} \quad 54 = x^2y$$

We now must eliminate one of the variables. We can accomplish this by determining the amount of linear feet of lumber in the frame. The frame of the crate consists of 4 pieces of 2×2 for each end (x ft by x ft) and 4 pieces of length y feet for the sides. The number of linear feet of 2×2 for the frame is:

$$2(4x) + 4y = 48$$

Now that we know the number of linear feet, to solve for y, substitute for y in the volume equation.

$$y = 12 - 2x \qquad \text{Solving for } y$$
$$54 = x^2(12 - 2x). \qquad \text{Substituting for } y \text{ in the volume equation}$$

$$2x^3 - 12x^2 + 54 = 0$$
$$x^3 - 6x^2 + 27 = 0. \qquad \text{Dividing each term by 2}$$

Now use the rules for solving polynomial equations to determine the values of x. Using Descartes' rule of signs, we determine that there are possibly two positive roots and one negative root. The possible rational roots are ± 1, ± 3, ± 9, and ± 27. Since x is a length, the roots must be positive, so we can eliminate the negative values. x cannot be 9 ft or 27 ft since 8(9) and 8(72) are greater than 48 ft. Try 3.

$$
\begin{array}{r|rrrr}
3 & 1 & -6 & 0 & 27 \\
  &   & 3 & -9 & -27 \\
\hline
  & 1 & -3 & -9 & 0
\end{array}
$$

The factors are $(x^2 - 3x - 9)(x - 3)$. By using the quadratic formula, we can determine that the other two roots are -1.9 and 4.9. However, these two values do not meet the criteria of being between 2.5 ft and 3.5 ft. Therefore, the height and the width of the crate must be 3 ft. Substituting this value into the equation $54 = x^2y$, we have:

$$54 = (3)^2y$$
$$54 = 9y$$
$$6 = y.$$

Thus, the dimensions of the crate that meet the given criteria are 6 ft by 3 ft by 3 ft. ◆

12.3 Exercises

In Exercises 1–6, perform the following.
a. Determine by inspection the zeros of each of the functions.
b. Give the multiplicity of each root.

1. $P(x) = (x + 3)(x - 2)^3(x + 5)^2$

2. $P(x) = (x + 1)^4(x - 4)^5$

3. $P(x) = (2x + 7)^2(3x - 5)^3(5x - 3)^4$

4. $P(x) = (2x + 7)^2(3x - 5)^3(5x - 3)^4$

5. $P(x) = (x^2 + 10x + 25)(x^2 - 2x - 35)$

6. $P(x) = (x \pm j)(2x^2 - x - 15)$

In Exercises 7–16, write polynomial equations with integral coefficients for the given roots. Write the answer in the form $a_nx^n + a_{n-1}x^{n-1} + \cdots + a_1x + a_0 = 0$.

7. 3, 13,

8. $-2, 1, 3$

9. 1, 2, 3,

10. $-3, 0, 1, 1$

11. $-1, -1, 2, 2$

12. $\dfrac{1}{2}, \dfrac{1}{3}, -1$

13. $-\dfrac{3}{2}, -\dfrac{3}{2}, \dfrac{1}{3}$

14. $2, 1, \pm\sqrt{3}$

15. 2 is a double root, and -1 is a triple root.

16. 1 is a triple root and $\pm\ j$ are roots.

17. The equation $R^4 + 6R^3 + 14R^2 + 14R + 5 = 0$ has -1 as a double root. Determine the other roots.

18. The equation $I^4 - 8I^3 + 30I^2 - 76I + 80 = 0$ has 2 and 4 as roots. Determine the other roots.

In Exercises 19–38, perform the following for each of the polynomial functions.
a. Determine the possible number of positive roots.
b. Determine the possible number of negative roots.
c. List all of the possible rational roots.
d. Determine the rational roots.
e. Sketch the function.

19. $P(x) = x^3 - x^2 - 5x - 3$

20. $P(x) = x^3 + x^2 - 5x + 3$

21. $P(x) = x^3 + 2x^2 - x - 2$

22. $P(x) = x^3 - 2x^2 - 5x + 6$

23. $P(x) = 2x^3 + 9x^2 + 7x - 6$

24. $P(x) = 6x^3 + 19x^2 + x - 6$

25. $P(z) = z^3 - 2z^2 - 11z - 12$

26. $P(z) = z^3 - 5z^2 - 12z + 36$

27. $P(v) = v^3 - 8v^2 + 14v - 4$

28. $P(v) = 2v^3 + 11v^2 + 44v - 25$

29. $P(t) = 3t^3 - 10t^2 + 25t + 22$

30. $P(t) = 5t^3 - 7t^2 + 19t + 15$

31. $P(x) = x^4 - 16x^3 + 86x^2 - 176x + 105$

32. $P(x) = x^4 - x^3 - 19x^2 + 49x - 30$

33. $P(x) = 4x^4 + 8x^3 - 7x^2 - 21x - 9$

34. $P(x) = x^4 - 4x^3 + 6x^2 - 4x + 1$

35. $P(x) = 8x^5 - 12x^4 + 14x^3 - 12x^2 + 6x - 1$

36. $P(x) = x^4 - 2x^3 - 8x^2 + 10x + 15$

37. $P(x) = 6x^4 - 31x^3 + 19x^2 + 19x + 3$

38. $P(x) = x^4 + x^3 - x^2 + x - 2$

In Exercises 39–44, perform the following for each of the polynomial functions.
a. Graph each using a graphics calculator.
b. Label all the intercepts on the graph.
c. Discuss where the curve is increasing and decreasing.
d. Discuss what happens for large positive and negative values of x.

39. $P(x) = x^3 + 4x^2 - x - 4$

40. $P(x) = x^3 + x^2 - 4x - 4$

41. $P(x) = x^3 + 2x^2 - 9x - 18$

42. $P(x) = x^3 + 5x^2 - x - 5$

43. $P(x) = x^4 - 5x^2 + 4$

44. $P(x) = -x^4 + 5x^2 - 4$

45. How do we know that every polynomial of odd degree with real coefficients has at least one real root?

46. The bending $V(x)$ of a beam in the floor of a house occurs along a line in the beam where the center of gravity is located. The equation describing this bending is $V(x) = x^4 - 108x + 243$, where x is the position on the beam. Determine the position x where the deflection V is 0.

47. The angle θ (in radians) through which the flywheel of a generator in a locomotive turns is related to time t (in seconds) by $\theta = 4t - 4t^2 + 4t^3 - t^4$. Determine t when $\theta = 3$ rad.

48. A contractor is designing a silo that is in the shape of a right circular cylinder with a hemisphere attached to the top as shown in the sketch. For a silo of height 25 ft, the contractor needs to determine the radius of the cylinder so that the volume is $828\,\pi$ ft³.

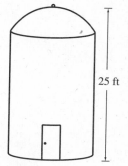

25 ft

49. The frame for a crate used to ship a wood-burning fireplace is constructed from 24 ft of 2×2 lumber. The crate has square ends of x feet. Determine the value(s) of x that result in a volume of 8 ft³.

50. The cost equation $c = x^3 - 18x^2 - 165x$ is developed to estimate the cost c of manufacturing x gadgets. If the marketing cost is \$250, how many gadgets must be produced to pay for the marketing?

51. A surveyor lays out a triangular plot such that one side is 2 m longer than the shortest side and the remaining side is 1 m longer than the shortest side. She determines that the product of the lengths of the three sides equal five times their sum. Find the lengths of the sides of the triangular plot.

52. A school psychologist finds that the response to a certain set of questions varies with the mental age of the child according to the equation $R = x^3 - 3x^2 - 25x - 10$, where R is the response in seconds and x is the mental age of the child in years. For what mental age is the response time equal to 11 s?

53. The total impedance Z of three impedances in series is given by $Z = Z_1 + Z_2 + Z_3$. If $Z = 3.0\ \Omega$, $Z_2 = Z_1^3$, and $Z_3 = 2 - Z_1^3$, what are the possible values for Z_1?

54. The radius of a cylindrical tank is 3 m less than its height. If the volume of the tank is 200π m³, find the height of the tank.

55. A variable electric current is given by the formula $i = 2t^3 - 13t^2 + 27t - 15$. If t is in seconds, at what time is the current equal to 3 A?

56. A variable voltage is given by the formula $e = 2t^3 - 12t^2 + 22t - 3$. At what time t (in seconds) is the voltage equal to 21 V?

12.4

Irrational Roots

In the previous sections you learned techniques for determining the rational roots of polynomial equations. Now let us look at how to determine irrational roots.

The roots of the quadratic equation $x^2 + x - 3 = 0$ are:

$$x = \frac{-1 + \sqrt{13}}{2} \qquad x = \frac{-1 - \sqrt{13}}{2}$$

$$= 1.3027756 \qquad = -2.3027756.$$

These roots are irrational numbers since they are not terminating or repeating decimals.

One procedure for finding irrational roots is by linear approximation. This process is illustrated in Example 1. But first we need to examine a theorem whose conclusion is used frequently in finding irrational roots.

> **THEOREM 12.9**
> If $P(x)$ is a polynomial with real coefficients, and if for the real numbers a and b, $P(a)$ and $P(b)$ have opposite signs, the equation $P(x) = 0$ has at least one number r between a and b such that $P(r) = 0$.

The theorem asserts that if at $x = a$ the curve is on one side of the x-axis and at $x = b$ the curve is on the other side, the curve must cross the x-axis at least one time between a and b. This says that there is at least one real root between a and b.

EXAMPLE 1 **Determining Roots by Linear Approximation**

Determine the real root(s) of the polynomial function $P(x) = x^3 + 3x^2 + 7x + 6$ to the nearest hundredth.

Solution Using the theorems from the previous sections we can determine several things about the function.

a. Using Descartes' rule of signs we determine that there are no positive roots and that there may be three negative roots or one negative root and two complex roots.

b. Using the rational-root theorem we find that if there are any rational roots, they must be $-1, -2, -3,$ and -6. (Remember there are no positive roots.)

Now use synthetic division to determine whether any of these are roots of $P(x)$.

```
1    3    7    6   |-1      1    3    7    6   |-2
    -1   -2   -5              -2   -2  -10
1    2    5    1          1    1    5   -4
1    3    7    6   |-3      1    3    7    6   |-6
    -3    0  -21              -6   18 -150
1    0    7  -15          1   -3   25 -144
```

None of the divisions has the remainder zero. Therefore, there is no rational root. The results of the synthetic division indicate that -3 is a lower bound since the signs in the third row alternate. Thus, there are no real roots smaller than -3. Also note that the remainder for the divisor -1 is positive and the remainder for the divisor -2 is negative. Theorem 12.9 tells us that there exists a real number between -1 and -2 that is a root of the equation. To illustrate this, the function is sketched in Figure 12.14.

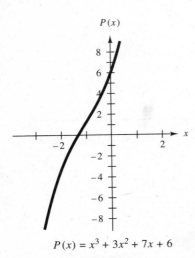

$P(x) = x^3 + 3x^2 + 7x + 6$

FIGURE 12.14

We know that the root between -1 and -2 is an irrational number since the synthetic division demonstrated that it cannot be rational. Now we use the method of linear approximation to determine the root.

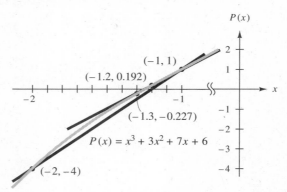

FIGURE 12.15

In Figure 12.15 the region around the points $(-1, 1)$ and $(-2, -4)$ is enlarged and the two points joined with a straight line. The point where the straight line crosses the x-axis is a first linear approximation of the root. The line crosses between -1.2 and -1.3. So test -1.2 and -1.3 by synthetic division.

$$
\begin{array}{rrrr|r}
1 & 3 & 7 & 6 & \underline{-1.2} \\
 & -1.2 & -2.16 & -5.808 & \\
\hline
1 & 1.8 & 4.84 & \boxed{0.192} &
\end{array}
$$

$$
\begin{array}{rrrr|r}
1 & 3 & 7 & 6 & \underline{-1.3} \\
 & -1.3 & -2.21 & -6.227 & \\
\hline
1 & 1.7 & 4.79 & \boxed{-0.227} &
\end{array}
$$

The positive remainder of 0.192 indicates that $(-1.2, 0.192)$ is a point on the curve above the x-axis. The negative remainder of -0.227 indicates that $(-1.3, -0.227)$ is a point on the curve below the x-axis. To the nearest tenth, the root is -1.2 since its remainder is closer to zero. To obtain a root to the nearest hundredth, we continue the process. Plot $(-1.2, 0.192)$ and $(-1.3, -0.227)$ and join the two points with a straight line as shown in Figure 12.15. The line appears to cut the x-axis at -1.25. Test this value with synthetic division.

$$
\begin{array}{rrrr|r}
1 & 3 & 7 & 6 & \underline{-1.25} \\
 & -1.25 & -2.1875 & -6.015625 & \\
\hline
1 & 1.75 & 4.8125 & \boxed{-0.015625} &
\end{array}
$$

The remainder is negative, which indicates that $(-1.25, -0.0156)$ is below the x-axis. This remainder also has the opposite sign of the remainder for -1.2.

Thus to find a root to the nearest hundredth, we must select larger values (values to the right of −1.25) until the sign of the remainder changes. Select −1.24.

$$
\begin{array}{cccc|c}
1 & 3 & 7 & 6 & \underline{-1.24} \\
 & -1.24 & -2.1824 & -5.973824 & \\
\hline
1 & 1.76 & 4.8176 & \boxed{0.026176} &
\end{array}
$$

Since the sign of the remainder changed, the root must be between −1.25 and −1.24. The remainder for −1.25 is closer to zero than the remainder for −1.24. Therefore, $x = -1.25$ is the root to the nearest hundredth. ◆

▨ EXAMPLE 2 Determining Roots with a Graphics Calculator

Determine the real roots for the polynomial function $P(x) = x^3 - 3x - 1$.

Solution By examining the function, we can see that for values of $x > 3$, the function is positive and increases in value. Similarly for values of $x < -3$, the function is negative and decreases in value. Therefore, if the function has any roots, they must be values of x between −3 and 3. With this information, set the range of your calculator as $[-3, 3, 0.5, -10, 10, 1]$ and graph the function. The graph in Figure 12.16 shows that there are three roots, two negative and one positive. To determine the roots to the nearest hundredths we use the ZOOM and the BOX keys. When we press ZOOM BOX, the cursor blinks at the origin. Use the right arrow key to move the cursor to a position slightly above and to the left of where the curve crosses the positive x-axis. Then press either ENTER or EXE and move the cursor down and to the right. A small rectangle forms around the point as shown in Figure 12.16. Now press either ENTER or EXE and that part of the graph inside the rectangle fills the screen as shown in Figure 12.17. To determine the coordinates of the point, press TRACE. Two things happen: values for x and y appear at the bottom of the screen, and the cursor blinks. Use the left and right arrow keys to move the cursor to the point where the line intersects the x-axis as shown in Figure 12.17. (Your picture may not be exactly the same as that in Figure 12.17, since your rectangle about the point may have been larger or smaller.) The values shown for the coordinates, $x = 1.8795835$ and $y = 1.5044\text{E-}3$, are approximate. To obtain a more precise value, repeat the process: ZOOM BOX and TRACE. The new results are shown in Figure 12.18. Now compare the results shown in Figures 12.17 and 12.18 to see that the x-coordinate is the same is both cases up to the thousandths place, $x = 1.879$. Thus, the root to the nearest hundredth is 1.88. Now return to the original graph and use the same technique to obtain values for the two negative roots. The three roots of the function are 1.88, −0.35, and −1.53. ◆

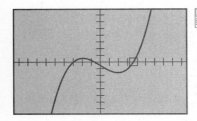

FIGURE 12.16

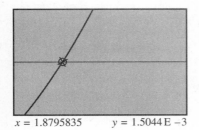

$x = 1.8795835$ $y = 1.5044\text{E}-3$

FIGURE 12.17

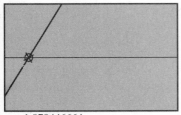

$x = 1.879446664$
$y = 1.6783592\text{E}-4$

FIGURE 12.18

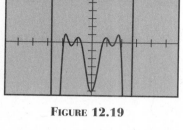

FIGURE 12.19

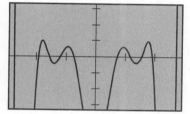

FIGURE 12.20

EXAMPLE 3 Application: Chemistry Problem

As you learned at the beginning of this chapter, one step chemists use in determining the stability of a molecule such as benzene is to determine the roots of the polynomial function:

$$P(x) = x^{10} - 11x^8 + 41x^6 - 65x^4 + 43x^2 - 9.$$

Determine the roots to the nearest hundredth.

Solution   The polynomial is of degree 10; therefore, we know that there are 10 roots. To get an idea of how many distinct real roots, graph the function on a graphics calculator. Use the range $[-10, 10, 1, -10, 10, 1]$. As we see from the graph in Figure 12.19, the curve is increasing and never decreasing for values of $x > 3$ and $x < -3$. To get a clearer picture of the number of negative and positive roots, change the range values to $[-3, 3, 0.5, -3, 3, 0.5]$. Figure 12.20 shows the graph. We can see that there are five negative and five positive real roots. Use the $\boxed{\text{ZOOM}}$ and $\boxed{\text{BOX}}$ keys as we did in Example 2 to determine that the roots are $-2.30, -1.62, -1.30, -1.00, -0.62, 0.62, 1.00, 1.30, 1.62, 2.30$. These values are used to determine other values that determine the stability of the molecule. ◆

By repeating the process of zooming, we can obtain roots to the nearest thousandths or ten-thousandths of a unit.

CAUTION   Theorem 12.9 states that there is at least one solution between a and b if $P(a)$ and $P(b)$ are of opposite sign. However, it does not say that a root cannot exist if $P(a)$ and $P(b)$ have the same sign (see Figure 12.21). Also, there may be more than one root between a and b if $P(a)$ and $P(b)$ have opposite signs (see Figure 12.22). ■

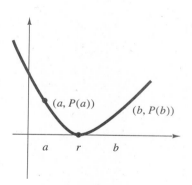

FIGURE 12.21

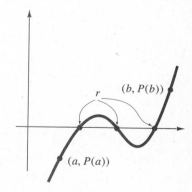

FIGURE 12.22

12.4 Exercises

In Exercises 1–10, use Theorem 12.9 to show that the polynomial equations have at least one root between the indicated values.

1. $P(x) = 2x^3 + x^2 - x - 1$ $[0, 1]$
2. $P(x) = x^4 - 6x^3 + 7x^2 + 5x - 1$ $[-2, 0]$
3. $P(x) = x^4 - 6x^3 + 8x^2 + 2x - 2$ $[-0.6, -0.5]$
4. $P(x) = x^3 - 7x + 1$ $[-2.72, -2.71]$
5. $P(x) = x^3 - 7x + 1$ $[0.14, 0.15]$
6. $P(x) = x^3 - 7x + 1$ $[2.57, 2.58]$
7. $P(x) = x^3 - 5x + 1$ $[2.12, 2.13]$
8. $P(x) = x^3 - 2x + 5$ $[-2.09, -2.10]$
9. $P(x) = x^3 + 3x^2 - 4x - 11$ $[1.94, 1.95]$
10. $P(x) = x^3 + 4x^2 - 2$ $[-0.79, -0.78]$

In Exercises 11–20, determine a root of $P(x)$ to the nearest tenth in the indicated intervals.

11. $P(x) = x^3 + 3x^2 + 4x - 1$ $[0, 1]$
12. $P(x) = x^3 + x^2 - 7x + 1$ $[2, 3]$
13. $P(x) = x^3 - 3x^2 - 5x + 9$ $[3, 4]$
14. $P(x) = x^3 + 3x^2 - 6x - 2$ $[1, 2]$
15. $P(x) = x^3 + 6x^2 + 5x - 1$ $[-1, -2]$
16. $P(x) = 2x^3 - 6x + 1$ $[-1, -2]$
17. $P(x) = x^3 - 3x^2 - 3x + 3$ $[-1, 1]$
18. $P(x) = x^4 - 11x^2 + 15x - 2$ $[0, 1]$
19. $P(x) = x^4 + x^3 - x^2 - 3x - 6$ $[1, 2]$
20. $P(x) = x^4 + 5x^3 + 8x^2 + 2x - 8$ $[0, 1]$

In Exercises 21–30, sketch the graphs of the function and determine all the real roots of the function to the nearest hundredth.

21. $P(x) = x^3 + 4x - 3$
22. $P(x) = x^3 + 5x - 2$
23. $P(x) = x^3 + 7x - 1$
24. $P(x) = x^3 + 3x^2 + 4$
25. $P(x) = x^3 + x^2 + 1$
26. $P(x) = x^4 + 3x^2 + 5x + 3$
27. $P(x) = x^4 + 5x^3 + 6x^2 + 4x + 12$
28. $P(x) = x^4 - 2x^3 + 3x^2 - 8x + 4$
29. $P(x) = x^4 - 2x^3 + x^2 - 1$
30. $P(x) = x^4 - 10x^2 - 5x + 7$

31. The depth x to which a buoy in the shape of a sphere of radius r and density d sinks in water is a positive root of the equation $x^3 - 3rx^2 + 4dr^3 = 0$. Determine the depth, correct to the nearest hundredth, to which a sphere of 1 m in diameter will sink if it is made of cork whose density is 0.23.

32. If its breaking strength is not exceeded, a 2-in. × 4-in. oak beam L ft long supported at its ends has a maximum sag of y in. if two concentrated loads of P lb each are placed on it, one load d ft from one end and the other load d ft from the other end, where:

$$y = \frac{12^3 Pd(3L^2 - 4d^2)}{3.84 \times 10^8}.$$

To the nearest tenth, determine how many feet from the ends of a 6-ft beam loads of 500 lb each can be put without causing sag of more than $\frac{1}{4}$ in.

33. The total capacitance C of three capacitors connected in series is given by:

$$\frac{1}{C} = \frac{1}{C_1} + \frac{1}{C_2} + \frac{1}{C_3}.$$

If $C = 0.25$ μF, $C_2 = C_1 - 1.3$ μF, and $C_3 = C_1 + 1.2$ μF, what are the two possible values for C_1?

34. A rectangular-shaped swimming pool has the dimensions 6 m × 4 m × 1 m. If the length, width, and height are increased by the same amount, the volume will double. To the nearest tenth, determine the amount by which the dimensions are to be increased.

35. A cube has a volume equal to that of a sphere whose radius is 10 mm shorter than an edge of the cube. What is the radius of the sphere?

36. The unknown h in the equation $h^4 - 12h^3 + 300h^2 - 3600h + 10{,}800 = 0$ defines a height in a construction problem. If the height must be in the interval $6 \le h \le 10$, determine the height to the nearest hundredth.

37. According to the historian Carl Boyer, no published cubic equations existed in 1544 because there were no known methods of finding a solution. However, in 1545 Geronimo Cardano (1501–1576) published a book that contained a solution for the cubic and the quartic equation. The methods for solving these equations were not his but the work of Niccolo Tartaglia and Lodovico Ferrari. The cubic equation $x^3 + px = q$, with positive coefficients, was one of the first forms for which a solution was found. Determine the number of real solutions for the equation for the following values.
a. $p = 4, q = -5$
b. $p = -4, q = -5$
c. $p = -4, q = 5$
d. $p = -8, q = 5$
e. $p = 8, q = 5$
f. $p = 8, q = -5$
g. Can you predict the number of solutions by knowing the values for p and q? Explain how you arrived at your method of predicting the number of solutions.

Review Exercises

In Exercises 1–10, use synthetic division to find the quotient and remainder. Check each of the results using the equation $P(x) = Q(x)D(x) + R(x)$.

1. $(2x^3 - 21x^2 + 43x + 60) \div (x - 5)$

2. $(2x^4 - 11x^3 - 111x^2 - 108) \div (x + 4)$

3. $(2x^3 + 25x^2 + 9x - 38) \div (x + 12)$

4. $(x^3 + 14x^2 + 25x + 22) \div (x + 2)$

5. $(2x^4 + 27x^3 + 34x^2 - 27x - 38) \div (x + 1)$

6. $(6x^4 + 5x^3 - 6x^2 - 45x - 27) \div (x + 3)$

7. $(6x^4 + 5x^3 - 6x^2 - 45x - 27) \div (3x + 1)$

8. $(x^5 - 5x^4 - 13x^3 + 65x^2 + 36x - 180) \div (x - 5)$

9. $(x^5 - 5x^4 - 13x^3 + 65x^2 + 36x - 183) \div (x - 3)$

10. $(x^8 + a^8) \div (x + a)$

In Exercises 11–16, form polynomial equations with integral coefficients with the given roots.

11. $3, -4, 5$

12. $-4, 0, 2, 2$

13. $\dfrac{2}{3}, \dfrac{1}{5}, -4$

14. $3 \pm \sqrt{3}, -4$

15. $2 \pm \sqrt{3}, \pm 5j$

16. $\dfrac{1}{3}, 1 \pm j\sqrt{7}, \pm\sqrt{3}$

In Exercises 17–26, perform the following for each of the polynomial functions.
a. Determine the possible number of positive roots.
b. Determine the possible number of negative roots.
c. List all of the possible rational roots.
d. Determine the rational roots.
e. Sketch the function.

17. $P(x) = x^3 - 4x^2 - 7x + 10$

18. $P(z) = z^3 + 2z^2 - 11z - 12$

19. $P(v) = 2v^3 + 11v^2 + 44v - 25$

20. $P(y) = 5y^3 - 7y^2 + 19y + 15$

21. $P(v) = v^4 - 10v^3 + 32v^2 - 30v - 9$

22. $P(y) = 2y^4 + 13y^3 + 19y^2 - 18y - 36$

23. $P(y) = 15y^4 + 4y^3 + 56y^2 + 16y - 16$

24. $P(x) = 36x^5 - 37x^3 + 12x^2 - 73x + 12$

25. $P(x) = x^4 - 2x^3 + 29x^2 - 64x - 96$

26. $P(x) = x^6 + x^5 + 20x^4 + 4x^3 + 59x^2 - 5x - 80$

In Exercises 27–36, sketch the graph of each of the functions for the indicated intervals and find all the real roots of the function to the nearest hundredth.

27. $P(x) = x^3 + 2x^2 - 7, \quad -3 < x < 3$

28. $P(x) = x^3 + 2x^2 + 7, \quad -4 < x < 2$

29. $P(x) = x^4 + 6x^3 + 8x^2 - 6x - 24, \quad -5 < x < 3$

30. $P(x) = 2x^4 + 5x^3 + 2x^2 + 14x + 7, \quad -4 < x < 1$

31. $P(R) = 2R^4 - 3R^3 + 6R^2 - 19R + 15, \quad -1 < R < 2$

32. $P(t) = t^4 + t^3 - 7t^2 - 5t + 2, \quad -4 < t < 4$

33. $P(Z) = 2Z^3 + 7Z^2 - Z - 15, \quad -3 < Z < 2$

34. $P(Z) = Z^4 + 4Z^3 - 37, \quad -6 < Z < 4$

35. $P(t) = t^3 - 5t^2 + 12t + 26, \quad -3 < t < 3$

36. $P(x) = x^4 - 6x^3 + 8x^2 + 2x - 1, \quad -1 < x < 4$

37. After a 10-mm-thick slice is cut from one side of a cube, the new volume is 48 cm^3. Determine the length of the side of the original cube.

38. An open box is to be made from a square sheet of cardboard 180 mm on a side by cutting equal squares from the corners and folding up the sides. How long should the edge of the cut-out square be if the volume of the box is to be 432 cm^3?

39. Determine the value of k for which $x - 3$ is a factor of $kx^3 - 6x^2 + 2kx - 12$.

40. Determine the value of k for which $x + 2$ is a factor of $3x^3 + 2kx^2 - 4x - 8$.

41. Tanya wants to triple the volume of a rectangular wood bin by increasing each dimension by the same amount. If the wood bin is initially 2 by 3 by 5 m, what must be the amount of the increase in each dimension?

42. The motion of a particle is such that the distance (or distances) traveled from a fixed reference point is given by:

$$s = 2t^4 - 18t^3 - 18t^2 + 36t + 20, \quad t \geq 0.$$

For what values of t will the particle return to the reference point?

43. Deano's Ice Cream Company determines that the cost $C(x)$ of manufacturing x units of chocolate peanut butter fudge is given by:

$$C(x) = 4x^3 - 100x^2 + 344x - 420.$$

If the company allocates \$700 daily to the production of this kind of ice cream how many units per day can be produced?

44. Use the factor theorem to show that $x - y$ is a factor of $x^n - y^n$, where n is a natural number.

45. Write a polynomial $P(x)$ with complex coefficients that has the root $a + bj, b \neq 0$ and that does not have $a - bj$ as a root.

46. Prove that a polynomial of odd degree with real coefficients has at least one real root.

47. Prove that if $P(x)$ involves only even powers of x and if all of its coefficients including the constant term are positive, then $P(x) = 0$ has no real roots.

48. Show that $\sqrt{3}$ is irrational.

49. Determine a polynomial function $y = P(x)$ with all of the following properties
   a. $P(x)$ is of degree 4.
   b. $P(0) = 1$.
   c. The graph is symmetric with respect to the y-axis.
   d. As x becomes large, then y becomes large (as $x \to \infty$, then $y \to \infty$).

50. In studying the lateral stability of a certain airplane, it is necessary to solve the equation:

$$x^4 + 45x^3 + 214x^2 + 1492x + 595 = 0.$$

Determine the real roots to the nearest tenth.

51. Solve the following equation for all positive values of θ for the interval $0 \leq \theta < 2\pi$.

$$4 \sin^2 \theta - 12 \sin \theta \cos^2 \theta - 7 \cos^2 \theta + 9 \sin \theta + 5 = 0.$$

[*Hint:* Use the identity $\cos^2 \theta = 1 - \sin^2 \theta$.]

52. In a hollow rectangular tube with outside dimensions a in. and b in. and wall thickness t in., subject to a twisting moment of T lb-in., the shearing stress in the walls is S lb/in², where $S = \dfrac{T}{[2t(a - t)(b - t)]}$. In a tube with $a = 2$ in. and $b = 1$ in., for what wall thickness would a twisting moment of 5760 lb-in. produce $S = 10,000$ psi?

53. A ball is a spherical shell of uniform thickness. Determine the thickness of the spherical shell if the outer diameter of the ball is 86 mm and the volume of the material from which the ball is made is 324 cm³.

✐ Writing About Mathematics

1. Write a sentence or two defining a polynomial function.

2. In a paragraph or two interpret the meaning of the results when a polynomial function is divided using synthetic division.

3. Given a polynomial function, explain in your own words how you go about determining the roots.

4. Write a paragraph or two describing the use of Descartes' rule of signs to help solve a polynomial function.

Chapter Test

1. Use synthetic division to divide $3x^4 + x^3 - 2x^2 - 5x + 3$ by $x + 1$.

In Exercises 2 and 3, use the remainder and/or factor theorem to determine whether the given number is a root of the function.

2. $f(x) = 4x^2 - 7x + 5$; $x = 2$

3. $f(x) = 2x^3 - 4x^2 + 5x - 10$; $x = 2$

4. Is $x = 3$ an upper bound, a lower bound, or neither for the function $f(x) = x^4 - 2x^3 + x^2 - 3x$?

Use the function $f(x) = 2x^3 - 3x^2 - 11x + 6$ to answer Exercises 5–8.

5. Determine the possible number of positive roots.

6. Determine the possible number of negative roots.

7. List all of the possible rational roots of the functions.

8. Determine the rational roots of the function.

9. If $x = -3j$ is a root of $f(x) = x^3 + 2x^2 + 9x + 18$, what are the other roots?

10. Does the function $f(x) = 3x^3 - 5x^2 + 2x + 1$ have a root between $x = -0.8$ and $x = 0.8$? Explain why or why not.

11. Determine the positive roots of the function $f(x) = x^3 - 6x + 5$ to the nearest tenth.

12. The equation $c = x^3 + 2x^2 - 400x - 2400$ is used to determine the cost of manufacturing x units of product Z. Determine the number of units of product Z that can be manufactured for $1600.

13. A rectangular box is made from a piece of metal 10 cm by 15 cm by cutting a square piece of metal from each corner and bending up the sides and welding the seams. If the volume of the box is 210 cm^3, what is the size of the square that is cut from each corner? The length of the side of the square is a rational number.

14. A cost equation of the form $C = x^3 + 2x^2 - 23x + 80$ is developed to estimate the cost, C, of a new product, where x is the number of products to be manufactured. If the cost for marketing is $140, how many products should be produced to pay for the marketing.

Exponential and Logarithmic Functions

13

In the 1960s a group of medical doctors became aware of a tribe of Indians in the Peruvian rain forest that had an extremely high infant mortality rate. To compensate for the high death rate, couples had many children; in fact, the norm was six children per family. With medical help, the infant mortality rate has declined in the past 30 years and the population has grown from 15,000 to over 600,000.

This rate of growth is known as exponential growth, which can be determined by an exponential function. Different forms of bacteria also grow at a exponential rate. The exponential function is discussed in this chapter and is used in Example 6 in Section 13.1 to determine the rate of growth of bacteria.

13.1

The Exponential Function

In this chapter, we learn about both exponential functions and the logarithmic functions. Just like the human growth rate, the rate of growth of money may be determined with an exponential function. In geology, logarithms are used to measure the intensity of earthquakes on the Richter scale. In chemistry, a base and an acid are determined by the PH factor, which is logarithmic in nature. In engineering, many mechanical systems can be best understood using these functions. The intensity of a sound is measured with a logarithmic function. In electronics, circuit current charges up and later decays exponentially. Also, the rate of change in the value of a new car a few years after purchase is an example of exponential decay.

A general form of the equation that represents exponential growth is the **exponential function** $y = b^x$. The base of the exponential is the constant b, and the exponent is the variable x.

DEFINITION 13.1

The **exponential function** with base b is denoted by:
$$f(x) = b^x,$$
where $b > 0$, $b \neq 1$, and x is any real number.

If we let $b = 1$, then $y = 1^x = 1$ for all real numbers x. Why must b be greater than zero?

With the restriction that $b > 0$, regardless of what values we substitute for x, y is always greater then zero. Thus, the domain of the exponential function is the set of all real numbers, and the range is the set of all positive real numbers. The graphs in Example 1 illustrate this.

Recall from the definition of exponents in Chapter 2 that $b^0 = 1$ and also that $b^{-x} = \dfrac{1}{b^x}$.

 EXAMPLE 1   Graphs of Exponential Functions

Graph the exponential functions of $y = 2^x$, $-2 \leq x \leq 3$, and of $y = 3^x$, $-2 \leq x \leq 3$

Solution   With bases of 2 and 3, we will find it easy to calculate the value of y when we substitute a positive integer for x. However, when we substitute a negative value for x, remember to use the rule $b^{-x} = \dfrac{1}{b^x}$. Substitute -2 for x in the function $y = 3^x$.

$$3^{-2} = \frac{1}{3^2}$$
$$= \frac{1}{9}$$
$$= 0.111$$

With your calculator, use either $\boxed{x^y}$ or $\boxed{\wedge}$ to determine values for the function $y = 3^x$. For example, to evaluate 3^5 and 3^{-6}, use the following keystrokes.

$$3^5\text{: } 3 \boxed{x^y} 5 \boxed{\text{EXE}} \text{ } 243 \quad \text{or} \quad 3 \boxed{\wedge} 5 \boxed{\text{ENTER}} \text{ } 243$$
$$3^{-6}\text{: } 3 \boxed{x^y} \boxed{\text{SHIFT}} \boxed{(-)} 6 \boxed{\text{EXE}} \text{ } 0.001371742 \quad \text{or}$$
$$3 \boxed{\wedge} \boxed{(-)} 6 \boxed{\text{ENTER}} \text{ } .001371742$$

The functional values for both functions are shown in the tables next to the graphs in Figures 13.1 and 13.2. Note that the scales on the x-axes and the y-axes are different. This is necessary because of the large differences between the x-values and the y-values. Observe that $y = 3^x$ increases more rapidly for each larger value of x than does $y = 2^x$.   ◆

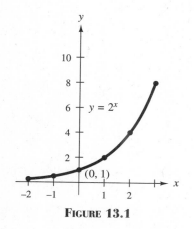

FIGURE 13.1

$y = 2^x$	
x	2^x
-2	0.25
-1	0.5
0	1.0
1	2.0
2	4.0
3	8.0

FIGURE 13.2

$y = 3^x$	
x	3^x
-2	0.11
-1	0.33
0	1.0
1	3.0
2	9.0
3	27.0

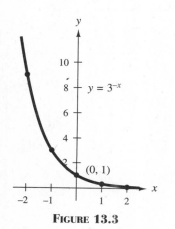

FIGURE 13.3

What does the graph look like when the exponent is negative (when the function is of the form $y = b^{-x}$)? Sketch the graph of $y = 3^{-x}$ for $-3 \le x \le 2$, shown in Figure 13.3, and compare the results with those in Figure 13.2. In Figure 13.2 we see that the curve increases as we move from left to right. This graph represents **exponential growth.** In Figure 13.3 we see that the curve decreases as we move from left to right. This graph represents **exponential decay.**

Now what happens when the base is a fractional value $\left(\text{when } b = \dfrac{1}{2} \text{ or } \dfrac{1}{3}\right)$? This is illustrated in Example 2.

EXAMPLE 2   Exponential Function Where Base Is a Fraction

Graph the exponential functions of $y = \left(\dfrac{1}{2}\right)^x$, $-3 \le x \le 2$, and of $y = \left(\dfrac{1}{3}\right)^x$, $-3 \le x \le 2$.

Solution   Use the calculator to determine functional values for the tables of values and then graph the functions.

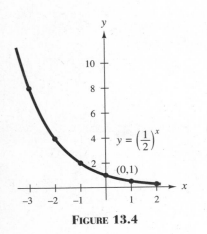

$y = \left(\dfrac{1}{2}\right)^x$

x	$\left(\dfrac{1}{2}\right)^x$
-3	8.0
-2	4.0
-1	2.0
0	1.0
1	0.5
2	0.3

FIGURE 13.4

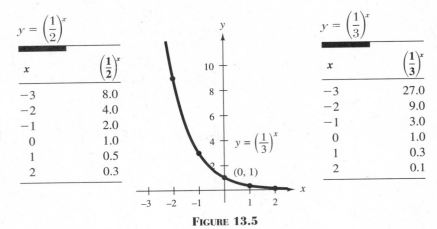

$y = \left(\dfrac{1}{3}\right)^x$

x	$\left(\dfrac{1}{3}\right)^x$
-3	27.0
-2	9.0
-1	3.0
0	1.0
1	0.3
2	0.1

FIGURE 13.5

The graphs in Figures 13.4 and 13.5 show that the exponential function with a fractional base is a decreasing function.   ◆

Examine Figures 13.1, 13.2, 13.3, 13.4 and 13.5 to gain a better understanding of the properties of the exponential functions listed in the box.

PROPERTIES OF EXPONENTIAL FUNCTIONS, $(y = b^x)$

1. The functions are asymptotic to the x-axis.
2. The functions are not asymptotic to any line $x = c$.
3. Since $b^0 = 1$, the graph of $f(x) = b^x$ passes through the point $(0, 1)$ for any value of b. It also passes through $(1, b)$.
4. The domain of $f(x) = b^x$ consists of the set of all real numbers; the range is the set of all positive real numbers.
5. If $b > 1$, b^x is an increasing function; if $b < 1$, b^x is a decreasing function.
6. If $a < b$, then $a^x < b^x$ for all $x > 0$, and $a^x > b^x$ for all $x < 0$.
7. Since b^x is either increasing or decreasing, it never assumes the same value twice ($b \ne 1$). The leads to the conclusion: if $b^x = b^y$, then $x = y$.
8. We can also state that: if $a^x = b^x$ for all $x \ne 0$, then $a = b$.

A special case of the exponential function that is often found in technical applications is the function $y = e^x$ or $y = e^{-x}$, $x \geq 0$, where $e = \mathbf{2.71828} \ldots$ is the base of the natural logarithms. The natural (or Naperian) logarithms are discussed in Section 13.4. (In electrical applications, the letter e is often used to represent electrical potential or voltage. To prevent confusion in those circumstances, the Greek letter epsilon (ϵ) is sometimes used to represent the base of the natural logarithms.)

EXAMPLE 3 Graphs of the Exponential Function

Graph $y = e^x$ and $y = e^{-x}$, $0 \leq x \leq 5$

Solution To calculate values for the function $y = e^x$, the keystrokes we use are similar to the ones in Example 1. For example, the keystrokes used to calculate a value for $y = e^x$ when $x = 3$ are:

$$\boxed{\text{SHIFT}} \;\; \boxed{e^x} \;\; 3 \;\; \boxed{\text{EXE}} \;\; 20.08553692$$

or

$$\boxed{\text{2nd}} \;\; \boxed{e^x} \;\; 3 \;\; \boxed{\text{ENTER}} \;\; 20.08553692.$$

We find e^x above the ln x key. (The relation between e^x and ln x is that ln x is the inverse of e^x. This relationship is discussed in Section 13.4.) Calculating $y = e^x$ to the nearest unit and $y = e^{-x}$ to the nearest hundredth produces the tables next to the graphs in Figures 13.6 and 13.7. ◆

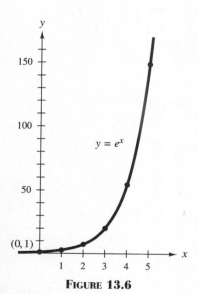

FIGURE 13.6

$y = e^x$	
x	e^x
0	1.
1	3.
2	7.
3	20.
4	55.
5	148.

FIGURE 13.7

$y = e^{-x}$	
x	e^{-x}
0	1.00
1	0.37
2	0.14
3	0.05
4	0.02
5	0.01

From Figure 13.7 we see that the function $y = e^{-x}$ has a rapid rate of decrease. It is interesting to observe that by the time $x = 5$, e^{-x} has decreased from 1.00 to 0.006738, or to less than 1% of its starting value. For practical purposes, functions of the form $y = Ae^{-x}$ often are considered to be negligible, or shut off, for $x \geq 5$.

EXAMPLE 4   Decay Time Less than 1% of Starting Value

Determine how long it takes for the function $v = 7e^{-\frac{t}{8}}$ to decay to less than 1% of its starting value. Time t is in seconds.

Solution   A function of the form $y = Ae^{-x}$ decays to less than 1% of its starting value in 5 s. To determine the value of t for $v = 7e^{-\frac{t}{8}}$, set the exponent equal to 5.

$$\boxed{\frac{t}{8}} = \boxed{5} \quad \text{or} \quad t = 40 \text{ s}$$

Thus, the function $v = 7e^{-\frac{t}{8}}$ decays to less than 1% of its starting value in 40 s. ◆

Often the exponent of the exponential function is not expressed just as x, but rather as a constant times x or as x divided by a constant. Graph the functions $y = 3^{kx}$ and $3^{\frac{x}{k}}$ for different values of k (i.e., $k = 1, 2, 3, 4, 5, \ldots$). What effects do the different values of k have on the shape of the curve?

EXAMPLE 5   Application: Chemical Compound

A chemical compound decays according to the relation $S = S_0 e^{-\frac{t}{4}}$, where S_0 is the initial amount of the compound present in kilograms and t is the time in hours. Say that S_0 is 6 kg.

a.   What is the domain of the function with the indicated restrictions?
b.   What is the range of the function with the indicated restrictions?
c.   Graph the function.

Solution

a.   In problems for which time is the independent variable, the initial, or starting, conditions usually are considered to occur at $t = 0$. Since time does not run backward, we must have $t \geq 0$. Because functions of the form $y = Ae^{-x}$ are considered shut off for $x \geq 5$, the minimum shutoff value for $6e^{-\frac{t}{4}}$ is when $\frac{t}{4} = 5$, or when $t = 20$. The required domain for this function, then, is $0 \leq t \leq 20$.

b.   The function $y = e^{-x}$, $x \geq 0$, was shown to have a maximum of 1 and a minimum approaching 0. Since $6e^{-\frac{t}{4}}$ is the same form multiplied by 6, the range of $6e^{-\frac{t}{4}}$ is $6 \cdot 1 = 6$ to $6 \cdot 0 = 0$, so the range of the function is $0 < S \leq 6$.

c.   The values of $S = 6e^{-\frac{t}{4}}$ for $0 \leq t \leq 20$ are shown in the table and are graphed in Figure 13.8. ◆

$S = 6e^{-\frac{t}{4}}$

t	$\dfrac{t}{4}$	$e^{-\frac{t}{4}}$	$6e^{-\frac{t}{4}}$
0	0	1.00	6.0
4	1	0.37	2.2
8	2	0.14	0.8
12	3	0.05	0.3
16	4	0.02	0.1
20	5	0.01	0.0*

*Rounded

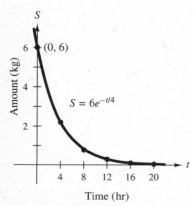

Time (hr)

FIGURE 13.8

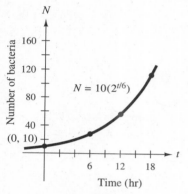

$$N = N_0 r^{\frac{t}{T}}$$

t	$\dfrac{t}{6}$	$10(2^{\frac{t}{6}})$
0	0	10
6	1	20
12	2	40
18	3	80
24	4	160

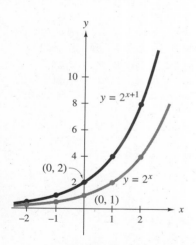

FIGURE 13.9

EXAMPLE 6   Application: Growth Rate of Bacteria

A bacteria culture starts with 10 bacteria and doubles every 6 hours. The equation describing the number of bacteria in the colony is of the form $N = N_0 r^{\frac{t}{T}}$, where N_0 is the initial number of bacteria, r is the growth factor per unit of time T, and t is the time of growth. Graph the number of bacteria present at any time during the course of one day.

Solution   Since the culture starts with 10 bacteria, $N_0 = 10$. The colony doubles in 6 hours so $r = 2$ and $T = 6$. Therefore, the equation for the average number of bacteria present during a 24-hour day is $N = 10\,(2^{\frac{t}{6}})$, where t is in hours because T is in hours. The culture size is graphed in Figure 13.9.   ◆

What effect do changes have on the graph of a function? What happens if we add a constant to the variable, subtract a constant from the variable, or change the sign of the function? To see the results, graph each pair of the functions $y = 2^x$ and $y = 2^{x+1}$, $y = 2^x$ and $y = 2^{x-1}$, $y = 2^x$ and $y = -2^x$ on the same set of axes. Figure 13.10 shows that the effect of adding 1 to the variable shifts the curve to the left, and it crosses the y-axis at $(0, 2)$. Try to predict the path of the curve $y = 2^{x+2}$. Figure 13.11 shows that the effect of subtracting 1 from the variable shifts the curve to the right. Try to visualize the graph of $y = 2^{x-2}$.

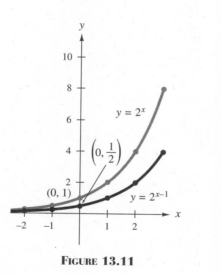

FIGURE 13.11

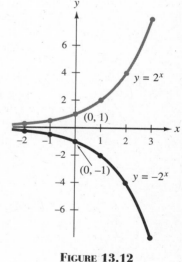

FIGURE 13.12

FIGURE 13.10

As Figure 13.12 shows, taking the negative of the function results in the mirror image of the function with respect to the x-axis. Try to visualize the graphs of $y = 3^x$ and $y = -3^x$. Figures 13.10, 13.11, and 13.12 should help us predict the graphs of exponential functions.

Compound Amounts

The newspapers continually warn us about the rate of inflation. How does inflation affect the average person? The **rate of inflation** is a percent of increase of the price of items we purchase over a certain period of time. Although the rate of inflation may be measured for any period of time, the usual time periods are monthly, quarterly, or annually.

To illustrate how we can predict the effect of inflation, let us assume that the price of a pizza is $3.50 today. If the rate r of inflation is 3% and remains the same for two years, what will be the price of the same pizza two years from now?

Let:

$$P = \$3.50 \quad \text{The price of the pizza today}$$

$$r = 0.03 \quad \text{the rate of inflation}$$

To determine the price at the end of one year P_1, determine the increase in the price (rP, or 3% of P) and add this amount to the original price.

$$P_1 = P + rP$$
$$= \$\,3.50\, + (\,0.03\,)(\,3.50\,)$$
$$= \$3.61$$

To determine the price at the end of two years P_2, determine the increase in price (3% of P_1) and add this amount to P_1, the price at the end of year 1.

$$P_2 = P_1 + rP_1$$
$$= \$\,3.61\, + (\,0.03\,)(\$\,3.61\,)$$
$$= \$3.72$$

Thus, the pizza that costs $3.50 today will cost $3.72 two years from now if the rate of inflation is constant at 3%.

What if we want to know the price of the pizza in ten years? We could go through the same procedure ten times or we could develop a general formula. To develop the general formula, use the following notation: r is the rate of inflation, n is the number of years, and P is the principal (original cost of the pizza). Thus, P_n represents the principal after n years. The principal after one year is:

$$P_1 = P + rP$$
$$= P(1 + r). \quad \text{Common-term factoring}$$

After two years, it is:

$$P_2 = P_1 + rP_1$$
$$= P_1(1 + r) \qquad \text{Common-term factoring}$$
$$= P(1 + r)(1 + r) \quad \text{Substituting for } P_1$$
$$= P(1 + r)^2.$$

After three years, it is:

$$P_3 = P(1 + r)^3.$$

And, after n years, it is:

$$P_n = P(1 + r)^n$$

> The value of an item in n years with a constant annual rate of inflation is:
> $$P_n = P(1 + r)^n.$$

To calculate the price of the pizza in ten years, substitute $n = 10$, $P = \$3.50$, and $r = .03$ into the equation.

$$P_{10} = \$ \; 3.50 \; (1 + \; 0.03 \;)^{10}$$
$$= \$4.70$$

Thus, the price of the pizza in ten years will be $4.70 if the annual rate of inflation remains constant at 3%.

If the value of an item decreases at a constant rate over a period of years, we can use the formula with one slight modification, subtract r from 1.

> The value of an item that decreases at a constant annual rate is:
> $$P_n = P(1 - r)^n.$$

To calculate the effects of inflation for any period of time (monthly, quarterly, etc.), a more general form of this equation is used. To do this, some terms must be redefined and some new symbols introduced: n is the number of compounding periods per year, t is the number of years, and A is the amount after t years. Since the compounding periods may be monthly or quarterly, the rate must be for the same period, so the rate is expressed as $\dfrac{r}{n}$. The formula for compounding for n periods per year is shown below.

> **The compound interest formula** for n compounding periods per year is:
> $$A = P\left(1 + \frac{r}{n}\right)^{nt}.$$

Example 7 uses this formula to calculate compound interest.

EXAMPLE 7 **Application: Compound Interest**

Maria has $3000 in a savings account that is paying interest at the rate of 3.5% compounded quarterly. Assuming that she makes no deposits or withdrawals, what will be the amount in the account at the end of six years?

Solution Because the interest is compounded quarterly, there are four periods in a year, and $n = 4$. The other values we need to substitute into the formula are principal $P = \$3000$, rate $r = 0.035$, and the time $t = 6$ years.

$$A = P\left(1 + \frac{r}{n}\right)^{nt}$$

$$= \$3000\left(1 + \frac{0.035}{4}\right)^{(4)(6)}$$

$$= \$3697.66$$

The amount in Maria's savings account after six years of compounding quarterly would be $3697.66. ◆

13.1 Exercises

In Exercises 1–10, use a calculator to evaluate each of the expressions to the nearest hundredth.

1. 2.75^9
2. 4.8^{11}
3. $3.7^{1.04}$
4. $7.5\sqrt{5}$
5. $3e^{1.6}$
6. $5.7e^7$
7. $45(2.65)^{3.4}$
8. $(\sqrt{34.5})^5$
9. $e^{4.7}$
10. $4.8e^{-1.2}$

In Exercises 11–18, match the function with the graph. The graph is identified by a letter.

11. $y = 4^{x-2}$

12. $y = 4^x + 3$

13. $y = -4^{-x}$

14. $y = 2^{x-1}$

a.

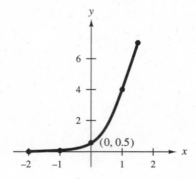

b.

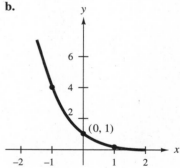

c.

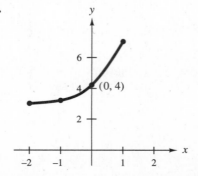

d.

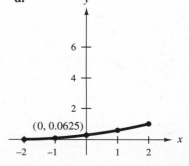

15. $y = -4^x$ **e.**

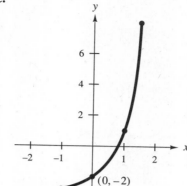

16. $y = 4^{-x}$

f.

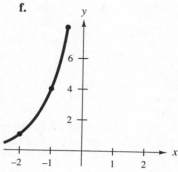

17. $y = 4^{x+2}$ **g.**

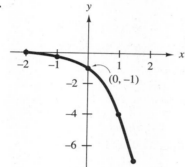

18. $y = 4^x - 3$

h.

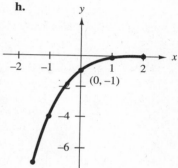

In Exercises 19–34, graph the exponentional functions for the given interval.

19. $y = 4^x,\ -2 \le x \le 2$

20. $y = 5^x,\ -2 \le x < 2$

21. $y = 4^{x+1},\ -2 \le x \le 2$

22. $y = 5^{x+1},\ -2 \le x < 2$

23. $y = \left(\dfrac{3}{2}\right)^x,\ -3 \le x \le 5$

24. $y = \left(\dfrac{5}{2}\right)^x,\ -3 \le x \le 5$

25. $y = \left(\dfrac{1}{4}\right)^x,\ -3 \le x \le 2$

26. $y = \left(\dfrac{1}{5}\right)^x,\ -3 \le x \le 2$

27. $y = 3^{x+2},\ -3 \le x \le 2$

28. $y = 3^{x-2},\ -1 \le x \le 4$

29. $y = 3^x - 2,\ -3 \le x \le 2$

30. $y = 3^x + 2,\ -3 \le x \le 2$

31. $p = 0.625^t,\ -6 \le t \le 3$

32. $p = 0.75^t,\ -6 \le t \le 3$

33. $p = 2(3^{-t}) -3 \le t \le 3$

34. $p = 3(2^{-t}),\ -3 \le t \le 3$

In Exercises 35–40, if t is in seconds, how long does it take for the function to decay to less than 1% of its starting value? (*Hint:* See Example 4.)

35. $v = e^{-\frac{t}{6}}$

36. $i = e^{\frac{-t}{5}}$

37. $p = 17e^{-2t}$

38. $p = -21e^{-4t}$

39. $y = 4 \times 10^{-19}e^{-7.2t}$

40. $s = 0.327e^{-1.6t}$

41. As you read at the beginning of this chapter, a tribe of Peruvian Indians has grown from 15,000 to over 600,000 in the past 30 years. A function that can be used to determine exponentional growth is $N = N_0 2^{\frac{t}{5}}$, where N_0 is the original population, t is the time in years, and N is the new population after t years.
    a. Show that the population grew from 15,000 to over 600,000 using the exponentional function. Assume that the time period t for the actual growth is 27 years.
    b. If the population growth continues at the same rate, determine the population in 20 years. Use the results in part a as the original population.

42. An organic compound decays according to the equation $M = M_0\, 6.1^{-\frac{t}{56}}$, where t is in hours and M_0 is the original amount in kilograms. How much of a 1-kg batch of the compound remains after one week?

43. A bacteria culture triples every three hours. If it starts with two bacteria, how many are there after one day? (See Example 6.)

44. The population of a city is decreasing according to the relation $p = p_0(0.9)^t$, where t is in years and p_0 is the original population. What will be the population of the city of 100,000 people after 10 years?

45. The amount of a radioactive substance remaining after a period of time is given by $A = A_0\left(\dfrac{1}{2}\right)^{\frac{t}{L}}$, where A_0 is the initial amount of substance present and L is the half-life of the material. How much cobalt-60 remains from a 1-kg sample after six months? (Cobalt-60 has a half-life of 5.25 yr.)

46. How much of a 10-kg sample of uranium 238 is present after one million years if it has a half-life of 4.5×10^9 yr? (See Exercise 45.)

47. If the cost of a hamburger today is $4.00 and the annual rate of inflation is 5%, determine the cost of the hamburger in six years. Assume that the inflation rate remains constant for the six years.

48. If groceries cost $100 at the beginning of the year and the rate of inflation is $\dfrac{1}{2}$% per year, what would be the cost of groceries at the end of the year?

49. If the ABC computer costs $1580 today and the price is decreasing at an annual rate of 15%, what will be the price of the computer in five years?

50. If a new car costs $18,500 and the value of the car decreases 20% per year, determine the value of the car in five years.

51. The price of firewood four years ago was $40 per cord. Today a cord costs $65. To the nearest percent, what has been the annual rate of increase of cost of a cord of firewood?

52. The price of a house in 1985 was $80,000; the price of the same house in 1995 is $130,000. What has been the annual rate of increase of the house?

In Exercises 53–56, $2000 is placed in a savings account at an annual rate of interest of 8%. Find the amount accumulated after four years when the interest is compounded for the period specified.

53. Annually

54. Semiannually

55. Quarterly

56. Daily (365 days)

57. The formula for continuous compounding of interest is $A = Pe^{rt}$. If $2000 are placed in a savings account with an annual interest rate of 8%, determine the amount accumulated after four years if the interest is compounded continuously. Compare the amount for continuous compounding with the amount for daily compounding in Exercise 56.

58. Assume that you invest \$2000 in a tax-free investment, such as an IRA, that pays a fixed annual rate of interest. Also assume that you do not make any withdrawals for 20 years. What will be the value of your investment after 20 years under the following conditions?
   a. The interest rate is 6%.
   b. The interest rate is 9%.
   c. The interest rate is 12%.
   d. When the interest rate doubles, by what factor does the amount increase?

59. In radiotherapy, the treatment of tumors by radiation, many factors, such as energy, the nature of the radiation, and the depth, size, and characteristics of the tumor need to be considered. When radiation destroys a tumor cell, it is called a "hit." If k denotes the average target size of a tumor cell and if x is the size of the dose of radiation, the surviving fraction of the tumor is defined by $f(x) = e^{-kx}$. This formula determines the surviving fraction of a tumor cell when there is only one place that will destroy a cell. When $x = 0$, then $f(0) = 1$; that is, there is no radiation, and all cells survive. Sketch the graph of the function for $k = 1$ and discuss the possible effects of larger doses of radiation.

60. If a cell has many points where the radiation can score a hit, the survival rate is defined by the function $f(x) = 1 - (1 - e^{-kx})^n$, $x \geq 0$. Sketch the graph of the function for $k = 1$ and $n = 2$ and discuss the possible effects of larger doses of radiation. (See Exercise 59.)

61. The approximate number of fish from one annual reproduction still alive after t years is given by an exponential function. For Pacific halibut, the function is $N(t) = N_0 e^{-0.2t}$, where N_0 is the initial number of fish born in one annual reproduction. What percentage of the original number is still alive after 10 years?

62. The length (in cm) of many common commercial fish t years old is closely approximated by a von Bertalanffy growth function $f(t) = a(1 - be^{-kt})$, where a, b, and k are constants. If for Pacific halibut, $a = 200$, $b = 0.956$, and $k = 0.18$, estimate the length of a typical 8-year-old halibut.

13.2

The Logarithmic Function

Though parents and children may debate whether a radio is too loud, physicists are able to give a precise number that measures the intensity of sound. The unit for **sound intensity** is the **decibel (dB),** the power passing through a unit area.

Table 13.1 lists some sound levels. The intensity of sound from a quiet radio is 40 dB, while the sound of an office is 50 dB. Though the sound intensity has increased by only 10 dB, the power transferred to the ear has increased by a factor of 10. The intensity of a sound wave in decibels is given by:

$$dB = 10 \log_{10} \frac{I}{10^{-12}},$$

where I is the intensity of the wave in Watts per square meter, written W/m². (1×10^{-12} W/m² is the faintest audible sound. It has been adopted by acoustics experts as the zero of intensity. This intensity is referred to as the hearing threshold.) The function that measures decibels is of the form $y = \log_b x$, which is a **logarithmic function.**

TABLE 13.1   Sound Levels

(dB)	Unit
10	Soundproof room
20	Rustle of leaves
30	City street, no traffic
40	Quiet radio
50	Office
60	Normal conversation
70	Busy street
80	Police whistle
90	Pneumatic drill
100	Boiler factory
110	Loud stereo
120	Threshold of pain
130	Jet engine

DEFINITION 13.2 Logarithmic Function

If $b > 0$, $b \neq 1$ and $x > 0$, then the function f defined by:

$$f(x) = \log_b x$$

is the logarithmic function with base b.

Understanding the logarithmic function is easier if you realize that the logarithm is just another word for exponent. The logarithmic function $y = \log_b x$ is the equivalent of the exponential function $x = b^y$. In other words, if we have the exponential function $2^3 = 8$ (the base 2 raised to the power of 3 equals 8), then the equivalent is the logarithmic function $\log_2 8 = 3$ (the logarithm to the base 2 of 8 equals 3).

DEFINITION 13.3

$$y = \log_b x \text{ if, and only if, } x = b^y$$

The definition of the logarithmic function states that $x > 0$. Why must x be greater than 0?

EXAMPLE 1   Writing an Equivalent Form of the Function

Write the equivalent logarithmic function of the following.

a. $9 = 3^2$ **b.** $y = 4^3$

Write the equivalent exponential function of the following.

c. $5 = \log_2 32$ **d.** $y = \log_3 81$

Solution   Use Definition 13.3 to write each equivalent function.

a. Given the exponential function $9 = 3^2$, the equivalent logarithmic function is:

$$2 = \log_3 9.$$

b. Given the exponential function $y = 4^3$, the equivalent logarithmic function is:

$$3 = \log_4 y.$$

c. Given the logarithmic function $5 = \log_2 32$, the equivalent exponential function is:

$$32 = 2^5.$$

d. Given the logarithmic function $y = \log_3 81$, the equivalent exponential function is:

$$81 = 3^y. \quad \blacklozenge$$

For the logarithmic function $y = \log_b x$, determining the values for y, b, or x may be simplified by writing the logarithmic function in exponential form $x = b^y$. This is illustrated in Examples 2 and 3.

EXAMPLE 2   Solving for *y*, *b* or *x*

For each of the logarithmic functions determine the unknown b, x, or y.

a. $\log_3 x = -2$ **b.** $\log_b 81 = 4$ **c.** $y = \log_7 \dfrac{1}{49}$

Solution   With the functions in logarithmic form, we may find it difficult to determine a solution. However, if we write the equivalent exponential form, determining the solution is easier.

a.   If $\log_3 x = -2$, then:

$$x = 3^{-2} \quad \textbf{Definition 13.3}$$

$$= \frac{1}{3^2} \quad \textbf{Properties of exponents}$$

$$= \frac{1}{9}.$$

Therefore, we see that $\log_3 \dfrac{1}{9} = -2$.

b.   If $\log_b 81 = 4$, then:

$$b^4 = 81 \quad \textbf{Definition 13.3}$$

$$= 3^4 \quad \textbf{Recalling } 3^4 = 81$$

$$b = 3 \quad \textbf{If } a^x = b^x\textbf{, then } a = b$$

Therefore, we see that $\log_3 81 = 4$.

c.   If $y = \log_7 \dfrac{1}{49}$, then:

$$7^y = \frac{1}{49} \quad \textbf{Definition 13.3}$$

$$= \frac{1}{7^2} \quad \textbf{Recalling } 7^2 = 49$$

$$7^y = 7^{-2} \quad \textbf{Properties of exponents}$$

$$y = -2. \quad \textbf{If } b^x = b^y\textbf{, then } x = y$$

Therefore, we see that $\log_7 \dfrac{1}{49} = -2$.   ◆

EXAMPLE 3   Solving for y

For each function, determine the value of y.

a. $y = \log_b b$ **b.** $y = \log_b 1$, $(b > 0)$

Solution

a. If $y = \log_b b$, then:

$$b^y = b \qquad \text{Definition 13.3}$$
$$b^y = b^1 \qquad \text{Properties of exponents}$$
$$y = 1. \qquad \text{If } b^x = b^y, \text{ then } x = y$$

Therefore, we see that $\log_b b = 1$.

b. If $y = \log_b 1$, then:

$$b^y = 1 \qquad \text{Definition 13.3}$$
$$b^y = b^0 \qquad \text{Properties of exponents}$$
$$y = 0. \qquad \text{If } b^x = b^y, \text{ then } x = y.$$

Therefore, we see that $\log_b 1 = 0$. ◆

The solutions to Example 3 are two important results that we will use frequently in our work with logarithms.

$$\log_b b = 1 \quad \text{and} \quad \log_b 1 = 0$$

To help us to better understand the properties of the logarithmic function, we sketch the graphs of logarithmic functions in Examples 4 and 5.

EXAMPLE 4   Graph of $y = \log_b x$

Graph the logarithmic functions $y = \log_2 x$ for the interval $0 < x \le 16$ and of $y = \log_3 x$ for the interval $0 < x \le 9$.

Solution   We do not know how to calculate values of $y = \log_2 x$, but by transforming the log function into the exponential format, we can obtain values for x and y. Use the fact that $x = 2^y$ and $y = \log_2 x$ are equivalent functions to provide values for $y = \log_2 x$ and to construct a table of values like the one near Figure 13.3. Substituting some of the values for y into the function $x = 2^y$, we have:

$$
\begin{array}{ccc}
y = -4 & y = -3 & y = 3 \\
x = 2^{-4} & x = 2^{-3} & x = 2^{3} \\
= \dfrac{1}{16} & = \dfrac{1}{8} & = 8.
\end{array}
$$

Having found values for x and y, we can substitute these values into the equation $y = \log_2 x$. For example, when $x = 8$ and $y = 3$, the equation is $3 = \log_2 8$. Using the technique demonstrated above and the results of Example 3, we can construct the tables and the graphs as shown in Figures 13.13 and 13.14.

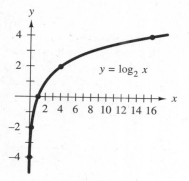

FIGURE 13.13

$y = \log_2 x$	
x	$\log_2 x$
$\dfrac{1}{16}$	-4
$\dfrac{1}{4}$	-2
1	0
4	2
16	4

FIGURE 13.14

$y = \log_3 x$	
x	$\log_3 x$
$\dfrac{1}{9}$	-2
$\dfrac{1}{3}$	-1
1	0
3	1
9	2

EXAMPLE 5   **Graph of $y = \log_{a/b} x$**

Graph the logarithmic functions $y = \log_{1/2} x$ for the interval $0 < x \le 16$ and $y = \log_{1/3} x$ for the interval $0 < x \le 9$.

Solution   Using Definition 13.3, we can write the equivalent exponentional function for $y = \log_{1/2} x$ as $x = \left(\dfrac{1}{2}\right)^y$. Use this exponentional to determine values for $y = \log_{1/2} x$. The values appear in the table next to Figure 13.15, where the curve is sketched. Using the equivalent exponential function $x = \left(\dfrac{1}{3}\right)^y$, we can determine values for $y = \log_{1/3} x$ that are given in the table near Figure 13.16, where the curve is sketched.

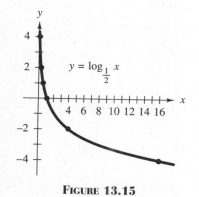

FIGURE 13.15

$y = \log_{1/2} x$	
x	$\log_{1/2} x$
$\dfrac{1}{16}$	4
$\dfrac{1}{4}$	2
$\dfrac{1}{2}$	1
4	-2
16	-4

FIGURE 13.16

$y = \log_{1/3} x$	
x	$\log_{1/3} x$
$\dfrac{1}{9}$	2
$\dfrac{1}{3}$	1
1	0
3	-1
9	-2

Table 13.2

$2^3 = 8$

$2^2 = 4$

$2^1 = 2$

$2^0 = 1$

$2^{-1} = \dfrac{1}{2}$

$2_2 = \dfrac{1}{2}$

Notice that all of the graphs in Examples 4 and 5 have both positive and negative values for y. However, there are no negative x values. Suppose we try to find the log of a negative number. That is, what is the value of:

$$y = \log_2 (-4)?$$

Writing the equivalent exponential equation, we have:

$$2^y = -4.$$

Substituting positive values for y into the exponentional equation gives results greater than one, as shown in Table 13.2. When we substitute negative values for y, the results are smaller than one. Recall that a zero exponent to any base gives the identity one. However, there is no exponent y that can give a negative number. *Hence, we simply cannot determine the log of a negative number.*

The graphs of Figures 13.13, 13.14, 13.15, and 13.16 lead to some properties of logarithmic functions.

PROPERTIES OF LOGARITHMIC FUNCTIONS ($y = \log_b x$)

1. The functions are asymptotic to the y-axis.
2. The functions are not asymptotic to any line $y = c$.
3. Since $\log_b 1 = 0$, the graph of $f(x) = \log_b x$ passes through the point $(1, 0)$ for any value for b. It also passes through $(b, 1)$.
4. The domain of $f(x) = \log_b x$ consists of the set of all positive real numbers; the range is the set of all real numbers.
5. If $b > 1$, $\log_b x$ is an increasing function; if $b < 1$, $\log_b x$ is a decreasing function.
6. If $a < b$, then $\log_a x > \log_b x$ for all $x > 1$ and $\log_a x < \log_b x$ for $0 < x < 1$.
7. Since $\log_b x$ is either increasing or decreasing, it never assumes the same value twice ($b \neq 1$). This leads to the conclusion that if $\log_b x = \log_b y$, then $x = y$.
8. Since the graphs of $\log_a x$ and $\log_b x$ intersect only at $x = 1$, if $\log_a x = \log_b x$ and $x \neq 1$, then $a = b$.

Figure 13.13 through 13.16 illustrate the first five properties listed in the box. To help us understand Property 6, consider the special cases $y = \log_2 x$ and $y = \log_{16} x$, where $a = 2$ and $b = 16$. The equivalent exponential functions are $2^y = x$ and $16^y = x$, respectively. When $x = 16$, then $2^y = 16$ or $y = 4$ and $16^y = 16$ or $y = 1$. This illustrates that when:

$$a < b \ (2 < 16), \text{ then } \log_a x > \log_b x \ (\log_2 x > \log_{16} x).$$

This works only for $x > 1$. In considering the situation when $0 < x < 1$, let $x = \dfrac{1}{2}$, then $2^y = \dfrac{1}{2}$ and requires $y = -1$. In the second equation $16^y = \dfrac{1}{2}$, $y = $ must equal $-\dfrac{1}{4}$. Therefore, $\log_2 x < \log_{16} x$, which confirms the second part of Property 6.

Inverse Function

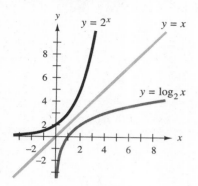

FIGURE 13.17

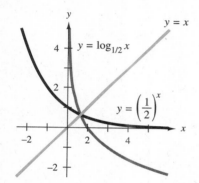

FIGURE 13.18

We can find the inverse of the exponential function $y = b^x$ by interchanging the independent and dependent variables and solving for y.

$$y = b^x \quad \textbf{Exponential function}$$
$$x = b^y \quad \textbf{Interchanging variables}$$

Since there is no algebraic technique to solve $x = b^y$, use Definition 13.3 to write the inverse function as $y = \log_b x$. Recall that when any function and its inverse are graphed on the same axes, the two function are reflections, or mirror images, of each other about the line $y = x$. So redraw the graphs of $y = 2^x$ (Figure 13.1) and $y = \log_2 x$ (Figure 13.13) on the same axes as shown in Figure 13.17. Also redraw the graphs of $y = \left(\dfrac{1}{2}\right)^x$ (Figure 13.4) and $y = \log_{1/2} x$ (Figure 13.15) on the same axes as shown in Figure 13.18. We can observe several things from these combined graphs.

> 1. The domain of the exponential function is all real numbers, and its range is the real numbers greater than 0. The domain of the logarithmic function is the real numbers greater than 0, and its range is all real numbers.
> 2. The exponential function is asymptotic to the x-axis; the logarithmic function is asymptotic to the y-axis.
> 3. The exponential function has y-intercept $(0, 1)$; the logarithmic function has x-intercept $(1, 0)$.

EXAMPLE 6   Application: Rate of Growth of Bacteria

A bacteria culture starts with 10 bacteria and doubles every 6 hours. The equation describing the number of bacteria in the colony is $N = 10(2^{\frac{t}{6}})$ (see Example 6 in Section 13.1). How long will it take the colony to reach 80 bacteria?

Solution   When $N = 80$, the population equation becomes:

$$80 = 10(2^{\frac{t}{6}})$$
$$\frac{80}{10} = \frac{10(2^{\frac{t}{6}})}{10}$$
$$8 = 2^{\frac{t}{6}},$$

which is an exponential function. When we write the equivalent log form, we have:

$$\frac{t}{6} = \log_2 8.$$

Now we can rewrite the right side of the equation as $\log_2 8$ is equal to some number x. Thus, $\log_2 8 = x$. Rewrite this in exponential form. We have:

$$2^x = 8 \quad \text{or} \quad 2^x = 2^3 \quad \text{and} \quad x = 3 \quad \text{and} \quad \log_2 8 = 3.$$

Substituting we have:

$$\frac{t}{6} = 3 \quad \text{and} \quad t = 18.$$

Therefore, it takes 18 hours for the colony to increase to 80 bacteria. ◆

13.2 Exercises

In Exercises 1–20, write the equivalent function in exponential or logarithmic form.

1. $25 = 5^2$
2. $36 = 6^2$
3. $64 = 4^3$
4. $125 = 5^3$

5. $y = 5^2$
6. $y = 7^6$
7. $p = 10^t$
8. $g = 8^{2x}$

9. $m = n^x$
10. $m = n^{-x}$
11. $2 = \log_7 49$
12. $3 = \log_6 216$

13. $4 = \log_5 625$
14. $3 = \log_{12} 1728$
15. $y = \log_{12} 144$
16. $y = \log_{10} 10{,}000$

17. $v = \log_{6.3} 123$
18. $s = \log_{1.7} 213.$
19. $r = \log_m t$
20. $q = \log_y x$

In Exercises 21–52, rewrite the logarithmic function in exponential form and solve for the unknown.

21. $\log_4 x = 2$
22. $\log_4 x = 3$
23. $\log_5 x = -1$
24. $\log_6 x = -2$

25. $\log_7 x = 2$
26. $\log_5 x = -3$
27. $\log_x 125 = 3$
28. $\log_x 256 = 4$

29. $\log_x 216 = 3$
30. $\log_x 625 = 4$
31. $\log_x 128 = 7$
32. $\log_x 243 = 5$

33. $\log_8 64 = x$
34. $\log_7 343 = x$
35. $x = \log_6 36$
36. $x = \log_{10} 1000$

37. $x = \log_{10} 100{,}000$
38. $x = \log_5 625$
39. $x = \log_{10} \frac{1}{100}$
40. $x = \log_{10} \frac{1}{1000}$

41. $x = \log_4 \frac{1}{16}$
42. $x = \log_5 \frac{1}{25}$
43. $x = \log_2 0.25$
44. $x = \log_2 0.50$

45. $x = \log_7 7$
46. $x = \log_8 8$
47. $x = \log_{6.3} 6.3$
48. $x = \log_{2.7} 2.7$

49. $x = \log_8 1$
50. $x = \log_7 1$
51. $x = \log_{5.2} 1$
52. $x = \log_{18.1} 1$

In Exercises 53–55, graph the following pairs of functions on the same set of axes. Discuss the similarities and differences.

53. $y = \log_4 x, \ 0 < x \le 2$
  $y = \log_2 x, \ 0 < x \le 2$

54. $y = \log_4 x, \ 0 < x \le 1$
  $y = \log_{1/2} x, \ 0 < x \le 1$

55. $y = \log_4 x, \ 0 < x \le 1$
  $y = \log_{1/4} x, \ 0 < x \le 1$

In Exercises 56–60, graph the logarithmic function.

56. $y = \log_5 x, \ 1 \le x \le 625$
57. $y = \log_6 x, \ 0 < x \le 1$
58. $y = \log_7 x, \ 0 < x \le 1$

59. $y = \log_6 x, \ 1 \le x \le 216$
60. $y = \log_7 x, \ 1 \le x \le 343$

61. How long does it take a bacteria colony that doubles every 10 hours to reach 16 bacteria if it starts with 4? (*Hint:* See Example 6.)

62. How long does it take 100 bacteria to become 1600 bacteria if they double every 3 hours? (*Hint:* See Example 6.)

13.3

Laws of Logarithms

Now that we know that logarithms are just another way of expressing exponents, the law of exponents can be used to develop the laws of logarithms. Recall from Chapter 2 the exponential laws, which are related below.

$$b^k \cdot b^\ell = b^{k+\ell} \tag{1}$$

$$\frac{b^\ell}{b^k} = b^{\ell-k} \tag{2}$$

$$(b^k)^n = b^{nk} \tag{3}$$

If we let $R = b^k$ and $S = b^\ell$, we can use Definition 13.3 to write the equivalent functions $k = \log_b R$ and $\ell = \log_b S$. Forming the product of $(R = b^k)$ and $(S = b^\ell)$, we have:

$$R \cdot S = b^k \cdot b^\ell$$
$$R \cdot S = b^{k+\ell} \qquad \text{Applying Equation [1]}$$
$$k + \ell = \log_b(R \cdot S) \qquad \text{Using Definition 13.3}$$
$$\log_b R + \log_b S = \log_b(R \cdot S). \qquad \text{Substituting for } k \text{ and } \ell$$

Reversing the expression, we have:

LAW 13.1

$$\log_b(R \cdot S) = \log_b R + \log_b S.$$

We can apply Law 13.1 in two different ways.

1. The logarithm of the product of two numbers is equal to the sum of the logarithm of the factors.

$$\log_3 5x = \log_3 5 + \log_3 x$$

2. If the bases are alike, the sum of the logarithms of the numbers is equal to the logarithm of the product of the numbers.

$$\log_2 7 + \log_2 5x = \log_2 7(5x) = \log_2 35x$$

EXAMPLE 1 **Rewriting a Logarithmic Expression**

Use Law 13.1 to write an equivalent expression for the following.

a. $\log_5 3 + \log_5 8$ **b.** $\log_2(2x)$

Solution

a. $\log_b R + \log_b S = \log_b(R \cdot S)$ **Law 13.1**

  $\log_5 \boxed{3} + \log_5 \boxed{8} = \log_5 \boxed{3} \cdot \boxed{8}$

  $= \log_5 24$

b. $\log_b(R \cdot S) = \log_b R + \log_b S$ **Law 13.1**

  $\log_2(\boxed{2x}) = \log_2 \boxed{2} + \log_2 \boxed{x}$

We can further simplify the function by recalling that $\log_b b = 1$. Therefore, $\log_2 2 = 1$, and the function becomes $\log_2 2x = 1 + \log_2 x$. ◆

The quotient rule for exponents can be used to develop a law for the logarithm of a quotient. Again using the equalities that were used to develop Law 13.1 ($R = b^k$, $S = b^\ell$, $k = \log_b R$, and $\ell = \log_b S$), start by writing the quotient of $S = b^\ell$ and $R = b^k$.

$$\frac{S}{R} = \frac{b^\ell}{b^k}$$

$$\frac{S}{R} = b^{\ell-k}$$ **Law of exponents, Equation (2)**

$$\ell - k = \log_b \frac{S}{R}$$ **Using Definition 13.3**

$$\log_b S - \log_b R = \log_b \frac{S}{R}$$ **Substituting for ℓ and k**

Reversing the expression, we have:

LAW 13.2

$$\log_b \frac{S}{R} = \log_b S - \log_b R.$$

We can look at Law 13.2 in two different ways.

1. The logarithm of the quotient of two numbers is equal to the difference of the logarithm of the numerator minus the logarithm of the denominator.

$$\log_2 \left(\frac{4x}{5} \right) = \log_2 \boxed{4x} - \log_2 \boxed{5}$$

2. If the bases are alike, then the difference of the logarithms of the numbers is equal to the logarithm of the quotient of the numbers.

$$\log_3 \boxed{7y} - \log_3 \boxed{5x} = \log_3 \left(\frac{7y}{5x} \right)$$

EXAMPLE 2   **Rewriting a Logarithmic Expression**

Use Law 13.2 to write an equivalent expression for the following.

a. $\log_3\left(\dfrac{1}{9}\right)$ **b.** $\log_2 x - \log_2 5$

Solution

a. $\log_b \dfrac{S}{R} = \log_b S - \log_b R$ **Law 13.2**

$\log_3 \dfrac{1}{9} = \log_3 1 - \log_3 9$

If we let $x = \log_3 1$ and $y = \log_3 9$, the equivalent statements are $3^x = 1$ and $3^y = 9$ from which we can see that $x = 0$ and $y = 2$. Therefore,

$$\log_3 \frac{1}{9} = x - y$$
$$= 0 - 2.$$

Thus,

$$\log_3 \frac{1}{9} = -2.$$

b. $\log_b S - \log_b R = \log_b \dfrac{S}{R}$ **Law 13.2**

$\log_2 x - \log_2 5 = \log_2 \dfrac{x}{5}$ ◆

The power rule for exponents can be used to develop a corresponding law for logarithms. If the exponential expression $R = b^k$ is raised to the nth power, we have:

$$R^n = (b^k)^n$$
$$R^n = b^{nk} \qquad \text{\textbf{Law of exponents, Equation (3)}}$$
$$nk = \log_b R^n \qquad \text{\textbf{Definition 13.3}}$$
$$n \log_b R = \log_b R^n \qquad \text{\textbf{Substituting for } } k \ (k = \log_b R)$$

Reversing the expression, we have:

LAW 13.3

$$\log_b R^n = n \log_b R.$$

Again, we can look at Law 13.3 in two different ways.

1. The logarithm of a quantity to a power is equal to the power times the logarithm of the quantity.

$$\log_3 x^4 = 4 \log_3 x$$

2. A number times the logarithm of a quantity is equal to the logarithm of the quantity raised to the power of that number.

$$7 \log_2 3x = \log_2 (3x)^7$$

EXAMPLE 3   Rewriting a Logarithmic Expression

Use Law 13.3 to write an equivalent expression for the following.

a. $\log_{10} 10^6$ **b.** $-\log_3 81$

Solution

a. $\log_b R^n = n \log_b R$ **Law 13.3**
   $\log_{10} 10^6 = 6 \log_{10} 10$
   $\qquad\qquad = 6(1)$ **$\text{Log}_b\ b = 1$**
   $\qquad\qquad = 6.$

b. $-\log_3 81 = -\log_3 3^4$ **Recalling $3^4 = 81$**
   $\qquad\qquad = -4 \log_3 3$ **Law 13.3**
   $\qquad\qquad = -4(1)$ **Recalling $\log_b b = 1$**
   $\qquad\qquad = -4.$ ◆

Using Law 13.3 and letting $R = b$ we have:

$$\log_b R^n = n \log_b R$$
$$\log_b b^n = n \log_b b,$$

but since:

$$\log_b b = 1$$
$$\log_b b^n = n(1).$$

LAW 13.4

$$\log_b b^n = n.$$

Law 13.4 is illustrated in Example 4.

EXAMPLE 4   Writing an Equivalent Expression

Use Law 13.4 to write an equivalent expression for the following.

a. $\log_{2.5} 2.5^{7.3}$ **b.** 5

Solution

a.
$$\log_b b^n = n \qquad \text{Law 13.4}$$
$$\log_{2.5} 2.5^{7.3} = 7.3.$$

b. Using Law 13.4, all of the following expressions are equivalent to 5.

$$\log_2 2^5 = 5$$

or

$$\log_8 8^5 = 5$$

or

$$\log_{6.2} 6.2^5 = 5$$

$$\vdots$$

The laws of logarithms are useful when we want to rewrite algebraic expressions so that the expressions are easier to work with. We can write complex expressions in simpler sums, diferences, and products.

EXAMPLE 5 **Rewriting a Logarithm as a Sum or Difference**

Using the laws of logarithms, write each expression as the sum or difference of logarithms.

a. $\log_3 \sqrt{3x^2y}$ **b.** $\log_4\left(\dfrac{x^2}{16y}\right)$

Solution

a.
$$\log_3 \sqrt{3x^2y} = \log_3(3x^2y)^{\frac{1}{2}}$$
$$= \frac{1}{2}\log_3 3x^2y \qquad \text{Law 13.3}$$
$$= \frac{1}{2}(\log_3 3 + \log_3 x^2 + \log_3 y) \qquad \text{Law 13.1}$$
$$= \frac{1}{2}(1 + 2\log_3 x + \log_3 y) \qquad \text{Law 13.3; } \log_b b = 1$$
$$= \frac{1}{2} + \log_3 x + \frac{1}{2}\log_3 y$$

b.
$$\log_4\left(\frac{x^2}{16y}\right) = \log_4 x^2 - \log_4 16y \qquad \text{Law 13.2}$$
$$= 2\log_4 x - (\log_4 16 + \log_4 y) \qquad \text{Laws 13.3, 13.1}$$
$$= 2\log_4 x - (\log_4 4^2 + \log_4 y)$$
$$= 2\log_4 x - (2 + \log_4 y) \qquad \text{Law 13.4}$$
$$= 2\log_4 x - \log_4 y - 2 \quad \blacklozenge$$

EXAMPLE 6   **Write an Expression as a Single Logarithm**

Use the laws of logarithms to write $2 \log_b x - 4 \log_b(2x + 3) \log_b \sqrt{2x - 7}$ as a single logarithm.

Solution

$2 \log_b x - 4 \log_b(2x + 3) + \log_b \sqrt{2x - 7}$

$= \log_b x^2 - \log_b(2x + 3)^4 + \log_b \sqrt{2x - 7}$

$= (\log_b x^2 + \log_b \sqrt{2x - 7}) - \log_b(2x + 3)^4$ **Law 13.3**

$= \log_b \dfrac{x^2 \sqrt{2x - 7}}{(2x + 3)^4}.$ **Laws 13.1, 13.2** ◆

EXAMPLE 7   **Write an Expression as a Single Logarithm**

Use the laws of logarithms to write $n \log_3(a + b) - \log_3 c$ as a single logarithm.

Solution   To write $n \log_3(a + b) - \log_3 c$ as a single logarithm, first recognize that:

$$n \log_3(a + b) = \log_3(a + b)^n.$$ **Law 13.3**

Then:

$$n \log_3(a + b) - \log_3 c = \log_3(a + b)^n - \log_3 c$$ **Substituting**

$$= \log_3 \frac{(a + b)^n}{c}.$$ **Law 13.2** ◆

CAUTION   Table 13.3 shows errors that are commonly made. ■

TABLE 13.3   **Common Errors When Using the Laws of Logarithms**

Law	Correct Example	Incorrect Variation
13.1	$\log_2(3 \cdot 4) = \log_2 3 + \log_2 4$	$\log_2(3 \cdot 4) \neq \log_2(3 + 4)$
		$(\log_2 3)(\log_2 4) \neq \log_2(3 + 4)$
		$\log_2 3 + \log_3 4 \neq \log_2(3 \cdot 4)$
		$\log_2(3 + 4) \neq \log_2 3 + \log_2 4$
13.2	$\log_2 \frac{3}{4} = \log_2 3 - \log_2 4$	$\log_2 \frac{3}{4} \neq \log_2(3 - 4)$
		$\frac{\log_2 3}{\log_2 4} \neq \log_2(3 - 4)$
		$\log_2 3 - \log_3 4 \neq \log_2 \frac{3}{4}$
		$\log_2(3 - 4) \neq \log_2 3 - \log_2 4$

EXAMPLE 8 **Evaluating Logarithms**

If $\log_b 1.5 = 0.37$, $\log_b 2.0 = 0.63$, and $\log_b 5.0 = 1.46$, find the following.

a. $\log_b 7.5$ **b.** $\log_b\left[1.5^4 \sqrt[3]{\dfrac{2}{5}}\right]$

Solution

a. $\log_b 7.5 = \log_b[(5)(1.5)]$

$= \log_b 5 + \log_b 1.5$ Law 13.1

$= 1.46 + 0.37$ Substituting

$= 1.83.$

b.

$$\log_b\left[1.5^4 \sqrt[3]{\frac{2}{5}}\right] = \log_b\left[(1.5)^4\left(\frac{2}{5}\right)^{\frac{1}{3}}\right]$$

$$= \log_b (1.5)^4 + \log_b \left(\frac{2}{5}\right)^{\frac{1}{3}} \quad \text{Law 13.1}$$

$$= 4\log_b 1.5 + \frac{1}{3}\log_b \frac{2}{5} \quad \text{Law 13.3}$$

$$= 4\log_b 1.5 + \frac{1}{3}(\log_b 2 - \log_b 5) \quad \text{Law 13.2}$$

$$= 4(0.37) + \frac{1}{3}(0.63 - 1.46) \quad \text{Substituting}$$

$$= 1.20. \quad \blacklozenge$$

13.3 Exercises

In Exercises 1–20, use the laws of logarithms to write the expression as a sum or a difference of logarithms. Your answer should not contain any exponents or radicals. Simplify where possible.

1. $\log_b 7x$

2. $\log_b x^7$

3. $\log_b \dfrac{x}{7}$

4. $\log_b(3 + x)^7$

5. $\log_5 5xy$

6. $\log_5 x^5 y$

7. $\log_5\left(\dfrac{x}{5}\right)y$

8. $\log_5 x\left(\dfrac{y}{5}\right)$

9. $\log_3 \sqrt{4xy}$

10. $\log_3(4xy)^2$

11. $\log_3 \dfrac{x-4}{y+5}$

12. $\log_2 \dfrac{1}{8}$

13. $\log_5 \sqrt[3]{125}$

14. $\log_a \dfrac{3x^2}{y}$

15. $\log_a \dfrac{x^2 y^2}{4}$

16. $\log_3 \dfrac{9}{x^2}$

17. $\log_b \dfrac{b^3}{6}$

18. $\log_{10} \dfrac{x^2 y^3}{1000}$

19. $\log_{10} \dfrac{100}{x+1}$

20. $\log_b \dfrac{x^3 y^4}{w\sqrt[3]{z^2}}$

In Exercises 21–38, use the laws of logarithms to write each expression as a single logarithm.

21. $3 \log_2 x + \log_2 y$

22. $2 \log_2(x + 1) - \log_2 y$

23. $\log_3 x + 1$

24. $4 \log_3 x - 1$

25. $\log_2 x + \log_2 y - \log_2 z$

26. $\log_b(x + y) - \log_b z$

27. $\log_b(2 - x) - \log_b(x + 2)$

28. $2 \log_b x + \frac{1}{2} \log_b y$

29. $5 \log_3(2x - 1) + 5 \log_3(x - 2)$

30. $n \log_a p + (n - 1) \log_a V$

31. $\frac{1}{2} \log_2 x^4 + \frac{1}{4} \log_2 x^4 + \frac{1}{8} \log_2 x^8$

32. $\log_4 x^4 - 4 \log_4 x^4 + \frac{1}{2} \log_4 x$

33. $2.1 \log_{6.3} y + 3.5 \log_{6.3} x$

34. $-1.8 \log_{2.5} x + 2.7 \log_{2.5} y - 1.6 \log_{2.5} z$

35. $4 \log_b x - 5 \log_b y + 13 \log_b(x + 1)$

36. $2 \log_b(x - 3) + 5 \log_b(x - 2) - 7 \log_b(x + 7)$

37. $5 \log_b(y + 3) - \frac{1}{4} \log_b(y - 5) + \frac{1}{3} \log_b(y + 2)$

38. $6 \log_b x + 3 \log_b z - \frac{1}{3} \log_b w$

In Exercises 39–46, let $\log_b 1.5 = 0.37$, $\log_b 2.0 = 0.63$, and $\log_b 5.0 = 1.46$. Find the value of the logarithms of the expressions.

39. $\log_b 3.0$

40. $\log_b 10.0$

41. $\log_b 6.0$

42. $\log_b 40.$

43. $\log_b(1.5)^3 \sqrt{5}$

44. $\log_b \frac{125}{4}$

45. $\log_b \sqrt[5]{\frac{1.5}{2.0}}$

46. $\log_b \sqrt[5]{\frac{2.0}{1.5}}$

47. The intensity level in decibels of any sound of intensity can be found from the general relation $\beta = 10(\log_{10} I - \log_{10} I_0)$, where I_0 is the hearing threshold. Write the logarithm on the right side of the equation as a single expression.

13.4

Common Logarithms and Natural Logarithms

Common Logarithm

The previous sections discussed logarithms in general, used graphs to illustrate properties of logarithms and then discussed the laws of logarithms. In going through the material the logarithm function was converted to the equivalent exponential form to obtain values for the logarithm function. You may wonder if there is a method of obtaining values directly from the logarithm function. The answer is yes.

On most scientific and graphing calculators, there are two keys that relate to logarithms. One is the ☐ Log ☐ key that indicates the logarithm to the base 10, which is called the **common logarithm.** When a logarithm to the base 10 is written, the base is generally omitted; i.e.,

$$\log_{10} x = \log x.$$

Example 1 illustrates how to determine values of common logarithms using a calculator.

EXAMPLE 1 **Determining log x**

Use a calculator to determine a value for the common logarithms.

a. log 234 **b.** log 23.4 **c.** log 2.34 **d.** log 0.234

Solution The keystrokes for both the fx–7700G and the TI–82 are shown along with answers expressed to five decimal places.

fx–7700G		TI–82	
a. ⬚log⬚ 234 ⬚EXE⬚ 2.36922		⬚LOG⬚ 234 ⬚ENTER⬚ 2.36922	
b. ⬚log⬚ 23.4 ⬚EXE⬚ 1.36922		⬚LOG⬚ 23.4 ⬚ENTER⬚ 1.36922	
c. ⬚log⬚ 2.34 ⬚EXE⬚ 0.36922		⬚LOG⬚ 2.34 ⬚ENTER⬚ .36922	
d. ⬚log⬚ .234 ⬚EXE⬚ −0.63078		⬚LOG⬚ .234 ⬚ENTER⬚ −.63078 ◆	

In the answers for Example 1, the numbers to the right of the decimal point in a–c are the same: 0.36922. The portion of the logarithm to the right of the decimal point is called the **mantissa.** When the original numbers are written in scientific notation, they are: $234 = 2.34 \times 10^2$, $23.4 = 2.34 \times 10^1$, and 2.34×10^0 . Again, look at the answers in Example 1. Observe that the digit(s) of the logarithm to the left of the decimal point is the same as the power of ten when the original number is written in scientific notation. For 23.4, the power of ten in scientific notation is 1 and the number to the left of the decimal point of the log 2.34 is a 1, for example. The number to the left of the decimal, 1 in this case, is called the **characteristic** of the logarithm.

Some technical equations require the use of logarithms to determine their solution. The results of the calculations may be a logarithm; that is, $\log N = Q$. Then we must find the **antilog** of Q, which is N. For example, if the log $N = 1.63485$, what is N, the antilog of 1.63485? We can determine N (the antilog of 1.63485) by using the following keystrokes.

$$\boxed{\text{SHIFT}} \ \boxed{10^x} \ 1.63485 \ \boxed{\text{EXE}} \ 43.13700612$$

or

$$\boxed{\text{2nd}} \ \boxed{10^x} \ 1.63485 \ \boxed{\text{ENTER}} \ 43.13700612$$

To the nearest tenth, $N = 43.1$.

EXAMPLE 2 **Performing a Calculation Using Logarithms**

Determine $\sqrt[3]{17.3}$ using logarithms.

Solution   To determine $\sqrt[3]{17.3}$ using logarithms, first take the logarithm of the function.

$$\log (\sqrt[3]{17.3}) = \log(17.3)^{\frac{1}{3}}$$

$$= \frac{1}{3} \log 17.3 \qquad \textbf{Properties of logarithms}$$

$$= \frac{1}{3} (\boxed{1.238})$$

$$= 0.4127.$$

Then determine the antilog.

$$\text{antilog}(0.4127) = \boxed{2.586} \quad \textbf{Taking antilog}$$
$$\sqrt[3]{17.3} = 2.59. \quad \blacklozenge$$

From the properties of logarithms discussed in Section 13.2, we would expect that the graph of $y = \log x$ would be asymptotic to the y-axis, cross the x-axis at $(1, 0)$, and be an increasing function. The graph of the function is given in Figure 13.19. In Example 3 we examine some variations of the logarithm function and the effects on the graph.

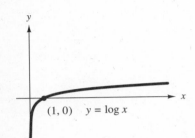

FIGURE 13.19

EXAMPLE 3   **Graphs of log x**

On the same set of axes sketch the graph of the following and compare each graph with the graph of $y = \log x$ in Figure 13.19.

a. $y = \log(x + 1)$ **b.** $y = \log(x - 1)$ **c.** $y = \log 2x$

Solution   The graphs of the functions are given in Figure 13.20.

a. Notice that the graph of the function $y = \log(x + 1)$ is shifted to the left and that the curve intersects the x-axis at the point $(0, 0)$ rather than at $(1, 0)$ when compared to the graph of $y = \log x$. Also the curve $y = \log(x + 1)$ is asymptotic to the line $x = -1$, which is one unit to the left of the asymptote for $y = \log x$. When $x = -1$,

$$y = \log(\boxed{-1} + 1) = \log 0.$$

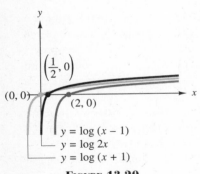

FIGURE 13.20

Recall that the log 0 is undefined. It is possible to see graphically that the curve is asymptotic to the line, by drawing the line $x = -1$ and using the zoom feature on the graphics calculator.

b. The function $y = (x - 1)$ is shifted one unit to the right and crosses the x-axis at the point $(2, 0)$. This curve is asymptotic to the line $x = 1$.

c. The factor of 2 in the function $y = \log 2x$ causes the curve to rise at a more rapid rate; thus, it crosses the x-axis at the point $\left(\frac{1}{2}, 0\right)$. The curve is asymptotic to the y-axis.

All three functions are strictly increasing without bound. $\blacklozenge$

Natural Logarithm

The second key on the calculator that relates to logarithms is the natural logarithm key, generally indicated $\boxed{\ln}$. The **natural (Naperian) logarithm** has a base of e. Recall that $e = 2.718281828$.

The natural logarithm may be written as $y = \log_e x$. Because it has a special base and is commonly used, it has a special name, $y = \ln x$.

In Example 4 we illustrate how to determine values of the natural logarithm using a calculator.

EXAMPLE 4 **Determining Values of ln x**

Use a calculator to determine a value for the natural logarithms.

a. $\ln 234$ **b.** $\ln 23.4$ **c.** $\ln 2.34$ **d.** $\ln 0.234$

Solution The keystrokes for both the fx–7700G and the TI–82 are shown. The answers are expressed to five decimal places.

	fx–7700G				TI–82		
a.	$\boxed{\ln}$ 234 $\boxed{\text{EXE}}$	5.45532		$\boxed{\text{LN}}$ 234 $\boxed{\text{ENTER}}$	5.45532
b.	$\boxed{\ln}$ 23.4 $\boxed{\text{EXE}}$	3.15274		$\boxed{\text{LN}}$ 23.4 $\boxed{\text{ENTER}}$	3.15274
c.	$\boxed{\ln}$ 2.34 $\boxed{\text{EXE}}$	0.85015		$\boxed{\text{LN}}$ 2.34 $\boxed{\text{ENTER}}$	.85015
d.	$\boxed{\ln}$ .234 $\boxed{\text{EXE}}$	-1.45243		$\boxed{\text{LN}}$ .234 $\boxed{\text{ENTER}}$	-1.45243 ◆

When we perform calculations with natural logarithms, we may need to determine the antilogarithm of a number. For example, if $\ln N = 2.48490$, what is N (the antilogarithm of 2.48490)? Determine N using the following keystrokes.

$$\boxed{\text{SHIFT}}\ \boxed{e^x}\ 2.48490\ \boxed{\text{EXE}}\ 11.99992$$

or

$$\boxed{\text{2nd}}\ \boxed{e^x}\ 2.48490\ \boxed{\text{ENTER}}\ 11.99992$$

To the nearest tenth, $N = 12.0$.

The graph of $y = \ln x$, shown in Figure 13.21, is similar to the graph of the common logarithm in that it crosses the x-axis at $(1, 0)$, is asymptotic to the y-axis, and is an increasing function.

Since $y = e^x$ and $y = \ln x$ behave in the same fashion as other exponential and logarithmic functions, they obey the same laws. Reviewing the laws of logarithms, specifically for the case of the logarithms, we have:

$$\ln(R \cdot S) = \ln R + \ln S \quad \textbf{Law 13.1}$$

$$\ln \frac{S}{R} = \ln S - \ln R \quad \textbf{Law 13.2}$$

$$\ln R^n = n \ln R \quad \textbf{Law 13.3}$$

$$\ln e^n = n. \quad \textbf{Law 13.4}$$

The foundation and early development of natural logarithms was the work of John Napier (1550–1617), a Scottish Laird, the Baron of Murchiston, who considered his interest in mathematics a recreational activity. Therefore, his name is associated with the logarithm.

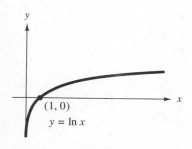

$(1, 0)$

$y = \ln x$

FIGURE 13.21

Change-of-Base Formula

Now we can determine values for log x and ln x and graph these functions with a graphics calculator. However, there are times when we may need to determine values for logarithms with bases other than 10 or e or need to graph these functions. We can accomplish this with the change-of-base formula.

CHANGE-OF-BASE FORMULA

$$\log_a x = \frac{\log_b x}{\log_b a}$$

Examples 5 and 6 illustrate how the formula is used with bases 10 and e.

EXAMPLE 5 **Evaluating Logarithms Using the Change-of-Base Formula**

Determine the values of the logarithms using common logarithms.

a. $\log_3 5.4$ **b.** $\log_7 42.7$

Solution

a. $\log_3 5.4 = \dfrac{\log 5.4}{\log 3} \approx \dfrac{0.73239}{0.47712} \approx 1.53503$

b. $\log_7 42.7 = \dfrac{\log 42.7}{\log 7} \approx \dfrac{1.63043}{0.84510} \approx 1.92928$ ◆

EXAMPLE 6 **Evaluating Logarithms Using the Change-of-Base Formula**

Determine the values of the logarithms using natural logarithms.

a. $\log_3 5.4$ **b.** $\log_7 42.7$

Solution

a. $\log_3 5.4 = \dfrac{\ln 5.4}{\ln 3} \approx \dfrac{1.68640}{1.09861} \approx 1.53503$

b. $\log_7 42.7 = \dfrac{\ln 42.7}{\ln 7} \approx \dfrac{3.75420}{1.94591} \approx 1.92928$ ◆

We can use the change-of-base formula to graph a logarithm function to any base. For example, in Section 13.2, Example 4, we graphed $y = \log_3 x$ by calculating ordered pairs using the equivalent exponential function. Using the change-of-base formula, we have:

$$\log_3 x = \frac{\log x}{\log 3}.$$

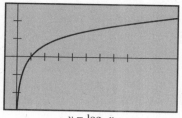

$y = \log_3 x$

FIGURE 13.22

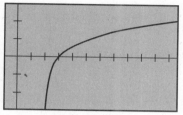

$y = \log_3 (x - 2)$

FIGURE 13.23

To graph the function, set the range values at $[-1, 16, 1, -3, 4, 1]$, enter the function $y = \dfrac{\log x}{\log 3}$. The graph is given in Figure 13.22. When we compare this graph with the one in Figure 13.14, we can see that the results are the same.

Now graph $y = \log_3 x$ using the change-of-base formula to change the base of the logarithm to base e. The result should be the same as in Figure 13.22.

If we replace the x in the function $y = \log_3 x$ with $(x - 2)$, what changes take place in the graph? Are the graphs of the functions the same? Different? This is illustrated in Example 7.

EXAMPLE 7 **Graph of $y = \log_b (x - c)$ versus Graph of $y = \log_b x$**

Graph $y = \log_3(x - 2)$ and compare the result of the graph with $y = \log_3 x$ (Figure 13.22).

Solution To graph the function, use the change-of-base formula to change the base of the function to e.

$$y = \log_3(x - 2) = \frac{\ln(x - 2)}{\ln 3}$$

Use range values of $[-1, 16, 1, -3, 4, 1]$ to graph the function. The result is shown in Figure 13.23. Comparing the graphs, we see that when two is subtracted from the x, the curve is shifted two units to the right; it crosses the x-axis at the point $(3, 0)$. Also it is asymptotic to the line $x = 2$. The general shape of both curves is the same, and both curves are strictly increasing functions. ◆

EXAMPLE 8 **Application: Effect of Temperature Change on Metal**

The time in hours needed for the gain size of a metal to double is related to the temperature T in degrees Kelvin by the equation $\ln t = -40 + \dfrac{40,000}{T}$.

a. If $t = 18$ hr, calculate T.
b. If $T = 950°$K, calculate t.

Solution

a. $\ln t = -40 + \dfrac{40,000}{T}$

   $\ln 18 = -40 + \dfrac{40,000}{T}$ **Substituting $t = 18$**

   $2.890 = -40 + \dfrac{40,000}{T}$

   $42.890 = \dfrac{40,000}{T}$

   $T = \dfrac{40,000}{42.890}$ or 932.6.   **Solving for T**

Therefore, $T = 930°$K.

b. $\ln t = -40 + \dfrac{40,000}{T}$

$\qquad\quad = -40 + \dfrac{40,000}{\boxed{950}}$ **Substituting $T = 950$**

$\qquad\quad = -40 + 42.11$

$\qquad\quad = 2.11$

$\qquad t = \text{antiln } 2.11$

$\qquad\quad = 8.248.$

Therefore, $t = 8.2$ hours. ◆

13.4 Exercises

In Exercises 1–4, determine the logarithm, in base 10, for each number to five decimal places. In each grouping, does the characteristic and mantissa change? If there is a change, why does it change?

1.	**a.** 1.63	2.	**a.** 7.62	3.	**a.** 5.4×10^6	4.	**a.** 6.02×10^{-21}
	b. 16.3		**b.** 0.762		**b.** 5.4×10^7		**b.** 6.02×10^{-22}
	c. 163		**c.** 0.0762		**c.** 5.4×10^8		**c.** 6.02×10^{-23}

In Exercises 5–10, determine the logarithm, in base 10, of the given number to five decimal places.

5. 2.51	6. 286	7. 5280	8. 0.683
9. 0.000721	10. 1.71×10^{-5}		

In Exercises 11–18, determine the antilog, in base 10, of the given number to four significant digits.

11. 0.67300	12. 0.69205	13. 1.76466	14. 2.35402
15. −2.08888	16. −1.03571	17. 19.82805	18. −18.10002

In Exercises 19–28, determine the natural logarithm of the number to five decimal places.

19. 1.63	20. 5.84	21. 58.2	22. 41.9
23. 286	24. 5280	25. 0.683	26. 0.174
27. 1.17×10^{-5}	28. 6.02×10^{-21}		

In Exercises 29–36, determine the antilog, base e, of the given number to five significant digits.

29. 0.67300	30. 0.69242	31. 1.35746	32. 2.98767
33. −2.08888	34. −1.03576	35. 19.82805	36. −18.10023

In Exercises 37–46, evaluate the expression using logarithms.

37. $14.2 \sqrt[3]{21.6}$

38. $53.8 \sqrt[5]{1.33}$

39. $\dfrac{19.3^5}{178.}$

40. $\dfrac{(81.0)(27.6)}{(18.2)^3}$

41. $[(2.6 \times 10^{-3})(1.8 \times 10^4)]^2$

42. $\sqrt{\sqrt[3]{32} \sqrt[5]{81}}$

43. $\dfrac{\sqrt[3]{14.8}}{\sqrt{31.5}}$

44. $(4.63 \sin 42.1°)^{\frac{3}{2}}$

45. $\left(\dfrac{\tan^2 61.7°}{4.87}\right)^{\frac{1}{3}}$

46. $(87.2)^{1.73}$

In Exercises 47–52, determine the value of the logarithm to four decimal places using common logarithms.

47. $\log_3 5$

48. $\log_7 11$

49. $\log_4 46.1$

50. $\log_6 51.8$

51. $\log_5 32$

52. $\log_4 81$

In Exercises 53–58, determine the value of the logarithm using natural logarithms.

53. $\log_7 100$

54. $\log_5 100$

55. $\log_6 72$

56. $\log_3 64$

57. $\log_{1/2} 9$

58. $\log_{1/3} 8$

59. Sound intensity in decibels is given by:

$$dB = 10 \log \frac{I}{I_0},$$

where $I_0 = 10^{-16}$ W/cm^2. The maximum sound intensity the ear can stand is about 10^{-4} W/cm^2. What is this intensity in decibels?

60. The sound intensity in a luxury automobile is approximately 40 dB. What is the sound intensity in W/cm^2?

61. Sound pressure level in decibels is given by:

$$S_p = 20.0 \log \frac{p_r}{0.000200}.$$

If the sound pressure p_r is 0.00360 dyn/cm^2, what is the sound pressure level?

62. The frequency response of a mid-priced stereo speaker is ± 3.0 dB over its rated frequency range. What is the sound pressure level in dyn/cm^2?

63. The viscosity of a glass N in Pa · s as a function of temperature in °K is given by:

$$\log N = -7.0 + \frac{14,000}{T}.$$

If $N = 10^{11}$ Pa · s, what is the glass temperature?

64. The temperature of a glass as described in Exercise 63 is 600°K. What is its viscosity at that temperature?

65. The pH of a solution is defined as:

$$pH = -\log(M_H^+),$$

where M_H^+ is the hydrogen-ion concentration in moles/liter. The hydrogen ion concentration of a shampoo is 3.16×10^{-6} mol/l. What is its pH?

66. An aquarium-water test kit shows the pH of a guppy tank to be 6.3. What is the hydrogen-ion concentration of the water? (See Exercise 65.)

67. As light passes through a transparent material, it is gradually absorbed. The equation relating distance and intensity is:

$$x = k(\ln I_0 - \ln I),$$

where k is a constant related to the material, I_0 is the intensity of the light at the edge of the material, and I is the intensity at some depth x. A photometer uses this equation to determine the depth of water. Let $k = 10.0$. Calculate the depth of the water if the intensity at the surface is twice the intensity at the new position.

68. As a boat coasts through water, the distance traveled is related by the following equation.

$$x = k \ln\left(\frac{v_0 t}{k} + 1\right),$$

where v_0 is the speed of the boat just as it starts to coast. If k is 0.0010 and v_0 is 7.5 m/s, how far will the boat coast in 15 seconds?

69. The electrode potential E volts in a galvanic cell with copper concentration c is given by:

Determine E if $c = 0.025$. $\qquad E = 0.34 + \dfrac{0.0257}{2} \ln c$ volts.

70. In Exercise 69, if $E = 0.275$ V, determine c.

71. Wind speed u_h in mph at height h in ft above the ocean surface is given by $\dfrac{u_h}{u_{43}} = 0.266 \ln h$, where u_{43} is the wind speed at 43 ft. If $u_{43} = 30.0$ mph, what is the wind speed at 5 ft above the surface?

72. In Exercise 71, at what height above the ocean is the wind speed 15 mph?

73. The work done in expanding a gas isothermally (at constant temperature) is given by $W = k \ln \dfrac{V_f}{V_o}$, where V_o is the initial volume, V_f is the final volume, and k is a constant. If k is 15.2 J, what is the work done by a gas expanding from a volume of 81.3 m³ to a volume of 98.6 m³?

74. The time in seconds required for the voltage drop across a capacitor in a series RC circuit to reach a value v_c is given by:

$$t = RC \ln\left(\frac{E}{E - v_c}\right),$$

where E is the supply voltage, R is the resistance, and C is the capacitance. If E is 110 V, R is 10^6 Ω, and C is 10^{-6} F, how long does it take for v_c to become 55 V?

75. If an object starts from rest and falls such that its air resistance is proportional to the square of its velocity, the time in seconds necessary for it to reach a velocity v is given by:

$$t = \frac{v_L}{2g} \ln \frac{v_L + v}{v_L - v},$$

where v_L is the object's terminal velocity and g is the gravitational acceleration. If $v_L = 70.0$ m/s and $g = 9.81$ m/s², how long does it take for the object to attain a speed of 35.0 m/s?

13.5

Exponential and Logarithmic Equations

In Section 13.2, we solved *exponential equations* of the form, $3^x = 9$. Using the rules of exponents—that is, $3^x = 3^2$—we found the solution to be 2. To determine the solution of a *logarithmic equation* of the form $\log_3 x = 2$, we wrote the equivalent exponential equation; $3^2 = x$, or $x = 9$. But how would we solve $3^x = 1.256$? This section shows how we can solve more complex exponential and logarithm equations using rules of exponents and logarithms.

Exponential Equations

A general approach for solving exponential equations follows.

---
1. Take the logarithm of both sides of the equation.
2. Use the law of logarithms ($\log_b R^n = n \log_b R$) to change the exponent to a multiple.
3. Solve for the variable.

Example 1 illustrates this technique.

EXAMPLE 1 **Solving an Exponential Equation**

If $3^{2x-1} = 17$, determine x.

Solution

$$3^{2x-1} = 17$$
$$\log 3^{2x-1} = \log 17 \qquad \text{Taking common logarithms of both sides, Step 1}$$
$$(2x - 1) \log 3 = \log 17 \qquad \log_b R^n = n \log_b R, \text{ Step 2}$$
$$(2x - 1) = \frac{\log 17}{\log 3}$$
$$= 2.5789$$
$$x = \frac{2.5789 + 1}{2} \qquad \text{Solving for } x, \text{ Step 3}$$
$$= 1.79.$$

Determine for yourself that it makes no difference whether logarithms to base 10 or to base e are used to calculate the solution. ◆

In Example 2 we demonstrate how to solve an exponentional equation that has an exponentional on both sides of the equal sign.

EXAMPLE 2   **Solving an Exponentional Equation**

If $2^{(2x+1)} = 3^{(2x-3)}$, determine x.

Solution

$$2^{(2x+1)} = 3^{(2x-3)}$$

$$\boxed{\ln}\ 2^{(2x+1)} = \boxed{\ln}\ 3^{(2x-3)} \qquad \text{Taking natural logarithms of both sides}$$

$$(\ \boxed{2x + 1}\ )\ln 2 = (\ \boxed{2x - 3}\ )\ln 3 \qquad \log_b R^n = n \log_b R$$

$$(2x + 1)0.6931 = (2x - 3)1.099$$

$$0.6931x + 0.6931 = 2.198x - 3.297 \qquad \text{Distributive property}$$

$$0.6931x - 2.198x = -3.297 - 0.6931 \qquad \text{Collecting like terms}$$

$$-1.505x = -3.990$$

$$x = 2.65. \quad \blacklozenge$$

EXAMPLE 3   **Application: Electric Circuit Problem**

The voltage in a circuit is given by $v = 110e^{-\frac{t}{3}}$ with t in seconds. Determine the value of t when $v = 55$ V.

Solution   When $v = 55$ V, $110e^{-\frac{t}{3}} = 55$. Take logarithms of both sides of the equation using logarithms of any base. Since the equation contains e, the base of the natural logarithms, we determine the solution faster and easier by using natural logarithms.

$$110e^{-\frac{t}{3}} = 55$$

$$e^{-\frac{t}{3}} = \frac{55}{110} \qquad \text{Dividing both sides by 110}$$

$$e^{-\frac{t}{3}} = 0.5$$

$$\boxed{\ln}\ e^{-\frac{t}{3}} = \boxed{\ln}\ 0.5 \qquad \text{Taking natural logarithms of both sides}$$

$$-\frac{t}{3}\ln e = \ln 0.5$$

$$-\frac{t}{3}(1) = -0.6931$$

$$t = 2.08 \text{ s.} \quad \blacklozenge$$

Logarithmic Equations

There are two general types of logarithmic equations. Type I is an equation that involves only logarithms. For this type of equation, we can combine terms so that we have an equation of the form:

$$\text{If}\quad \log R = \log S,$$
$$\text{then}\quad R = S$$

This type of logarithmic equation is illustrated in Example 4.

EXAMPLE 4   **Solving a Logarithmic Equation**

If $\log(x + 3) = \log x + \log 7$, determine the value for x.

Solution

$$\log(x + 3) = \log x + \log 7$$
$$\log(x + 3) = \log 7x \qquad\qquad \mathbf{\log(R \cdot S) = \log R + \log S}$$
$$x + 3 = 7x \qquad\qquad \mathbf{If \log R = \log S, then R = S}$$
$$x = \frac{1}{2}$$

Check: When $x = \frac{1}{2}$,

$$\log\left(\frac{1}{2} + 3\right) = \log \frac{1}{2} + \log 7$$

$$\log \frac{7}{2} = \log \frac{7}{2}. \quad\blacklozenge$$

The Type II equation involves the combination of logarithms and constants. For this type of equation, we can combine terms so that we have an equation of the form $\log A = $ constant. This type of equation is illustrated in Example 5.

EXAMPLE 5   **Solving a Logarithmic Equation**

If $\log(2x + 8) = 1 + \log(x - 4)$, determine x.

Solution   In order to solve $\log(2x + 8) = 1 + \log(x - 4)$ for x, we need to regroup terms so that all terms containing x can be combined. (Remember $\log(2x + 8) \neq \log 2x + \log 8$.)

$$\log(2x + 8) = 1 + \log(x - 4)$$
$$\log(2x + 8) - \log(x - 4) = 1 \qquad\qquad \mathbf{Rearranging}$$
$$\log\left(\frac{2x + 8}{x - 4}\right) = 1 \qquad\qquad \mathbf{\log \frac{S}{R} = \log S - \log R}$$
$$\frac{2x + 8}{x - 4} = 10^1 \qquad\qquad \mathbf{y = \log_b x\ if\ x = b^y}$$
$$2x + 8 = 10(x - 4) \qquad\qquad \mathbf{Multiplying\ by\ (x - 4)}$$
$$2x + 8 = 10x - 40$$
$$-8x = -48 \quad \text{or} \quad x = 6$$

Check the answers. $\blacklozenge$

CAUTION   If $\log(x + 1) + \log x = \log x^2$, then:
$$(x + 1) + x \neq x^2. \quad\blacksquare$$

Example 6 illustrates the Caution.

EXAMPLE 6   **Solving a Logarithmic Equation**

If $\log_{\frac{1}{2}}(x + 1) + \log_{\frac{1}{2}} x^2 = \log_{\frac{1}{2}} x$, determine x.

Solution

$$\log_{\frac{1}{2}}(x + 1) + \log_{\frac{1}{2}} x^2 = \log_{\frac{1}{2}} x$$

$$\log_{\frac{1}{2}}(x + 1)x^2 = \log_{\frac{1}{2}} x \qquad \log_b R \cdot S = \log_b R + \log_b S$$

$$(x + 1)x^2 = x \qquad \text{If } \log_b x = \log_b y, \text{ then } x = y$$

$$x^3 + x^2 - x = 0$$

$$x(x^2 + x - 1) = 0$$

$$x = 0 \quad \text{or} \quad x^2 + x - 1 = 0$$

Use the quadratic formula to solve $x^2 + x - 1 = 0$, to determine that:

$$x = \frac{-1 + \sqrt{5}}{2} = 0.618 \quad \text{and} \quad x = \frac{-1 - \sqrt{5}}{2} = -1.618.$$

Thus 0, 0.168, and -1.618 are possible solutions of the equation $\log_{\frac{1}{2}}(x + 1) + \log_{\frac{1}{2}} x^2 = \log_{\frac{1}{2}} x$. Because the definition of logarithm, $y = \log_b x, x > 0$, there are no solutions to the equation if $x \leq 0$. Therefore, $x = 0$ and $x = -1.618$ cannot be solutions of the equation. The only acceptable solution is $x = 0.618$.

Check the solution. ◆

EXAMPLE 7   **Application: Time to Double Investment**

An individual deposits $2000 in a tax-free savings account compounded continuously. How long will it take to double the principal for the following interest rates? (Recall that the formula for continuous compounding is $A = Pe^{rt}$.) Time is in terms of years.

a. 6%   **b.** 9%

Solution   In the formula $A = Pe^{rt}$,

$$P = \text{principal} = \$2000 \text{ (deposit)}$$

$$A = \text{amount} = \$4000 \text{ (deposit} + \text{interest)}$$

$$r = \text{rate of interest} = 6\% \text{ or } 9\%$$

$$t = \text{time (number of years)}$$

a.   Substituting in the equation, $A = \$4000$, $P = \$2000$, and $r = 6\%$:

$$\$4000 = \$2000 \, e^{0.06 \, t}$$

$$2 = e^{0.06t}$$

$$\ln 2 = \ln e^{0.06t} \qquad \text{Taking natural logarithms of both sides}$$

$$\ln 2 = 0.06t \ln e \qquad \ln R^n = n \ln R$$
$$\ln 2 = 0.06t \qquad \ln e = 1$$
$$\frac{\ln 2}{0.06} = t$$
$$t = 11.6 \text{ yr}$$

b. Substituting in the equation, $A = \$4000$, $P = \$2000$, and $r = 9\%$:

$$\$\boxed{4000} = \$\boxed{2000}\, e^{0.09\, t}$$

Using the same procedure as part a, we can reduce the equation to:

$$0.09t = \ln 2$$
$$t = \frac{\ln 2}{0.09}$$
$$= 7.7 \text{ yr} \quad \blacklozenge$$

A quick way to approximate the answers to Example 7 is to use the **Rule of 72,** which states that if you divide the number 72 by the interest rate, then the quotient is the number of years needed for doubling. For example, $72 \div 6 = 12$. Our answer in 7a was 11.6 yr. Determine the number of years for Example 7 when the interest rate is 12%. Then estimate the time using the Rule of 72.

13.5 Exercises

In Exercises 1–30, determine x.

1. $2^{3x} = 5$

2. $5^{2x} = 3$

3. $4^{x+1} = 20$

4. $2^{2x+1} = 16$

5. $3^{2x-3} = 14 + \log 5$

6. $5^{x+5} = 125 + \log 125$

7. $\left(\dfrac{1}{2}\right)^{x^2} = 16$

8. $\left(\dfrac{1}{3}\right)^{x^2} = 81$

9. $4^{x+1} = 5^{x+1}$

10. $5 \cdot 4^{x-3} = 2^x$

11. $\dfrac{1}{2} \cdot 3^{2x} = \left(\dfrac{1}{2}\right)^{3x}$

12. $4^{3-x} = 7^{6-2x}$

13. $6^{2x} = 10^{x-3}$

14. $10^{x+1} = 2^{2x}$

15. $\log(2x + 3) = \log(2x - 1)$

16. $\ln(2x + 3) = \ln(2x - 1)$

17. $\ln x + \ln(x - 1) = 5$

18. $\log x + \log(x - 1) = 5$

19. $\log_2(2x + 3) = \log_2(2x - 1) + 3$

20. $\log_2 x + \log_2(x - 1) = 5$

21. $\ln(x + 2) + \ln x^2 = \ln x$

22. $\ln(x + 5) + \ln x^2 = \ln x$

23. $\log(x + 2) + \log x^2 = \log x$

24. $\log(x + 5) + \log x^2 = \log x$

25. $\log 4x = \log(x + 1) + e$

26. $\log(x - 2) = \log(x + 1) - e$

27. $\ln 2x = \ln(x - 1) + 10$

28. $\ln(3 - x) = \ln(x + 2) - 10$

29. $\log_7 x^3 + \log_7(x + 2) = \log_7 x^2$

30. $\log_5(x + 3)^2 = \log_5(x + 3) - \log_5 x$

31. The half-life of uranium-283 is 4.5×10^9 yr. How long does it take a 1-kg sample of uranium-238 to decay to $\dfrac{1}{5}$ its original mass? Use:

$$A = A_0 \left(\frac{1}{2}\right)^{\frac{t}{L}},$$

where A_0 is the initial amount, L is the half-life, and t is the time.

32. A bacteria colony doubles in 6 hr. How long does it take the colony to triple? Use $N = N_0 \, 2^{\frac{t}{T}}$, where N_0 is the initial number of bacteria and T is the time in hours it takes the colony to double.

33. The siren for a home security system produces a noise level of 108 dB at a distance of 3 m. What is the sound intensity in W/cm^2 at that distance? Use:

$$\left(dB = 10 \log \frac{I}{10^{-16}} \right)$$

34. The stopping tension T_1 applied to a rope holding a load T_2 when wrapped around a capstan is given by $\dfrac{T_2}{T_1} = e^{\mu\beta}$, where μ is the coefficient of friction and β is the total angle of wrap in radians. What must the coefficient of friction be for a stopping tension of 20.0 N to hold a load of 1000 N if there are three complete wraps around the capstan?

35. True strain e_t is related to engineering strain e by $e_t = \ln(1 + e)$. If $e_t = 0.020$, what is e?

36. In hardening the surface of steel by diffusing carbon into its surface at elevated temperatures, the diffusivity D is given by $\ln D = \ln A - \dfrac{Q}{RT}$, where $A = 0.250$ cm^2/s, $Q = 34{,}500$ cal/mole, and R is the gas constant 1.987 cal/mole °K. For a diffusivity of 0.300×10^{-6} cm^2/s, what must the temperature be?

37. An equation relating the required height of a cooling tower to the humidity of the entering and exiting air is $Z = K \ln\left(\dfrac{y - y_1}{y - y_2}\right)$. Solve for y.

38. When a valve is released in a tank of compressed gas, the gas escapes in proportion to the amount of gas remaining in the tank. During each second the value is released, 12% of the gas escapes. This means that 88% of the gas remains after each second.
  a. Which of the following expressions best expresses the amount of gas remaining as a function of time?

$$N = N_0(0.12)^t$$
$$N = N_0(0.88)^t$$
$$N = N_0(t)^{0.12}$$
$$N = N_0(t)^{0.88}$$

  b. An engineer wants to release 40% of the gas in the tank. How much time is required to release 40% of the gas?

39. The relation between pressure and volume during the compression stroke of a gasoline or diesel engine is given by $\ln p + \gamma \ln V = C$. Determine V.

13.6

Data Analysis Graphs on Logarithmic and Semilogarithmic Paper

In Chapter 10 we learned that a graph of a straight line can be represented by the equation $y = mx + b$, where m is the slope of the line and b is the y-intercept. (Remember that the y-intercept is the point where the line crosses the y-axis.) We also learned that given the graph of a straight line, we can relatively easily write an equation that represents the straight line. Unfortunately, in many situations, when experimental data are graphed they yield curves instead of straight lines.

For example, the graph of data collected in an experiment may be as shown in Figure 13.24. When scientists observe the graph of the data, they try to find the **curve of best fit.** A function that best approximates a set of data is referred to as the curve of best fit for that data. In many situations they use a logarithmic function to change the data from a nonlinear relationship to a linear relationship.

Assume that the data in Figure 13.24 is described by the function $y = bx^m$, where b is a positive constant and x is a positive independent variable. This type of function is called a **power function.** Now use the laws of logarithms to write equivalent equations.

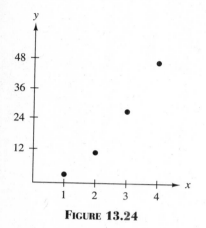

FIGURE 13.24

$$y = bx^m$$
$$\log y = \log bx^m$$
$$\log y = \log b + \log x^m$$
$$\log y = \log b + m \log x$$

The result is a linear function: that is, $\log y$ is a linear function of $\log x$ with slope m and y-intercept $\log b$. In the logarithm function, if we let $Y = \log y$, $X = \log x$, and $B = \log b$ and make the appropriate substitutions, the result is the linear function

$$Y = mX + B.$$

Example 1 illustrates a technique for writing a power function that describes a set of data.

EXAMPLE 1 **Writing a Function That Describes a Set of Data**

Given the set of data, determine a function that approximates the data.

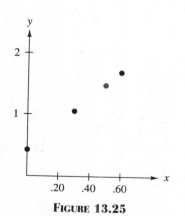

FIGURE 13.25

x	1	2	3	4
y	3	12	27	48

Solution These data may be approximated with a power function. To do this, we need to calculate and graph the data points $(\log x, \log y)$. See Figure 13.25 and the data table.

$\log x$	0	0.30	0.48	0.60
$\log y$	0.48	1.08	1.43	1.68

The points in the graph of Figure 13.25 appear to be in a straight line. Now select any of the two points and use them to find an equation of the line. Use the points $(1, 3)$ and $(3, 27)$ to determine m and b in the equation $\log y = m \log x + \log b$. First solve for b.

$$\log 3 = m \log 1 + \log b \qquad \text{Substituting (1, 3) for } (x, y)$$
$$\log 3 = m(0) + \log b$$
$$\log 3 = \log b$$
$$3 = b$$

Now solve for m.

$$\log 27 = m \log 3 + \log 3 \qquad \text{Substituting (3, 27) for } (x, y) \text{ and } b = 3$$
$$\log 27 - \log 3 = m \log 3$$
$$\log \frac{27}{3} = m \log 3$$
$$\log 9 = m \log 3$$
$$\frac{\log 3^2}{\log 3} = m$$
$$\frac{2 \log 3}{\log 3} = m$$
$$2 = m$$

Now use the values of m and b to write y as a power function of x.

$$\log y = m \log x + \log b$$
$$\log y = 2 \log x + \log 3$$
$$\log y = \log x^2 + \log 3$$
$$\log y = \log 3x^2$$
$$y = 3x^2$$

Thus, the power function that represents the data is $y = 3x^2$. ◆

Before the advent of calculators, the data would have been plotted on **log-log graph paper** as shown in Figure 13.26. Both the vertical and the horizontal scales of the log-log paper are logarithmic scales. Table 13.4 shows the common logarithms of some of the numbers from 1 to 100.

In Table 13.4, we see that the difference between $\log 3$ and $\log 4$ is the same as the difference between $\log 30$ and $\log 40$ ($0.60 - 0.48 = 1.60 - 1.48$). We also note that all the logs of the tens digits are the same amount greater than the

TABLE 13.4

x	$\log x$
1	0.00
2	0.30
3	0.48
4	0.60
5	0.70
6	0.78
7	0.85
8	0.90
9	0.95
10	1.00
20	1.30
30	1.48
40	1.60
50	1.70
60	1.78
70	1.85
80	1.90
90	1.95
100	2.00

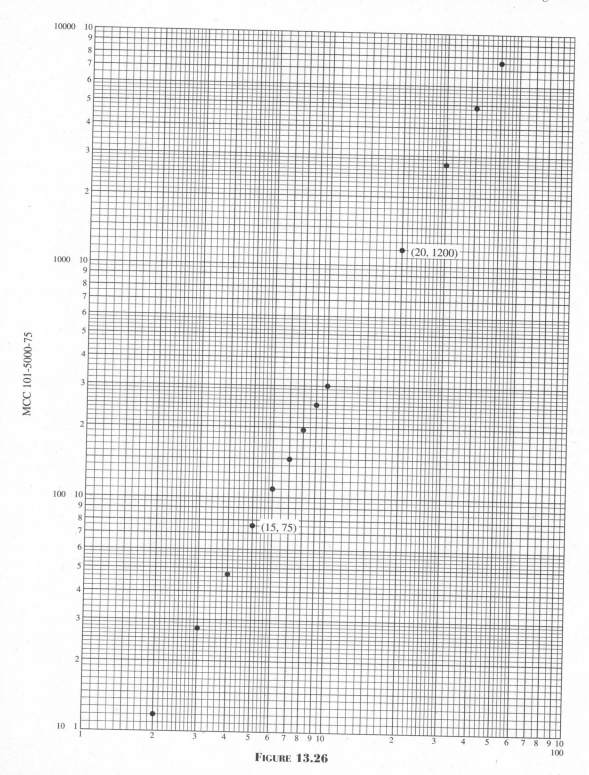

FIGURE 13.26

logs of the units digits. The results of Table 13.4 can be graphed on a number line as in Figure 13.27. Notice that the logarithms of the numbers are represented by their spacing along the line.

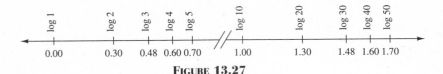

FIGURE 13.27

When we use a number line with the proportional spacing already marked on it and numbers at the appropriate place on the scale, they are automatically being spaced on the line as the log of the numbers. This principle is used to form log-log graph paper, which is printed with two perpendicular logarithmic scales (see Figure 13.26) and to form semilog graph paper, which is printed with an evenly divided scale perpendicular to a logarithmic scale (such a scale is used for Figure 13.31). Because logarithmic scales display several "cycles" or powers of ten in a relatively short length, log-log graph paper is also used to graph functions that have large domains and/or ranges.

If the graph on the log-log graph paper is a straight line, then the data may be represented by the linear equation $y = bx^m$.

DEFINITION 13.4

A straight line on log-log graph paper represents an equation of the form $y = bx^m$. For any two points on the graph (x_1, y_1) and $(x_2, y_2,)$ values for m and b can be determined with the following formulas.

$$m = \frac{\log y_2 - \log y_1}{\log x_2 - \log x_1} \quad \text{and} \quad \log b = \log y_1 - m \log x_1$$

For example, if we select two points $(5, 75)$ and $(20, 1200)$ from Figure 13.26 and substitute the values in the equations the results are:

$$m = \frac{\log 1200 - \log 75}{\log 20 - \log 5} \quad \text{and} \quad \log b = \log 75 - 2 \log 5$$

$$= 2 \qquad\qquad\qquad = 0.47712$$

$$b = 3.$$

When we substitute the values for m and b into the equation $y = bx^m$, the result is the equation $y = 3x^2$. The result is the same as in Example 1, which we would expect since the data are the same.

EXAMPLE 2 **Writing an Equation for Experimental Data**

An experiment produced the data given in the following table.

x	1	2	5	10	20	50	100
y	2.00	1.00	0.40	0.20	0.10	0.04	0.02

a. Graph the data on Cartesian graph paper.
b. Graph the data on log-log graph paper.
c. Determine the equation that represents the data.

Solution

a. The data are graphed on Cartesian graph paper in Figure 13.28.
b. The data are graphed on log-log graph paper in Figure 13.29.

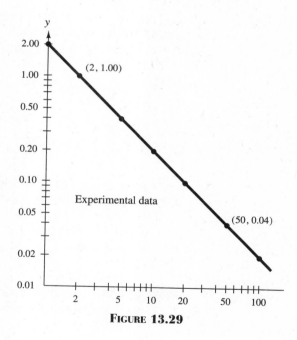

FIGURE 13.29

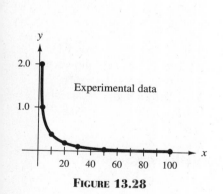

FIGURE 13.28

c. Since the data graph as a straight line on log-log graph paper, the equation will be of the form $y = bx^m$. Therefore, we must determine m and b. Picking two points on the line, say (2, 1.00) and (50, 0.04) and using the formulas, we have:

$$m = \frac{\log y_2 - \log y_1}{\log x_2 - \log x_1}$$

$$= \frac{\log 0.04 - \log 1.00}{\log 50 - \log 2}$$

$$= \frac{-1.398 - 0.000}{1.699 - 0.3010} \quad \text{or} \quad -1.000.$$

$$\log b = \log y_1 - m \log x_1$$

$$= \log 1.00 - (-1.000) \log 2$$

$$= 0.0000 - (-1.000)(0.3010)$$

$$= 0.3010$$

$$b = 1.999.$$

Therefore, the equation that best fits the data is $y = 2x^{-1}$. ◆

There are times when a function of the form $y = bx^m$ does not provide the best fit for the data. The best fit for the data may be provided with an exponential function, $y = ba^{mx}$, where m, b, and a are constants. If we take logarithms of both sides of the equation $y = ba^{mx}$, we have the following equivalent equations.

$$\log y = \log ba^{mx}$$

$$= \log b + \log a^{mx} \qquad \log_b R \cdot S = \log_b R + \log_b S$$

$$= mx \log a + \log b \qquad \log_b R^n = n \log_b R$$

$$= (m \log a)x + \log b.$$

Now let $Y = \log y$, $M = m \log a$ and $B = \log b$ and substitute to obtain $Y = Mx + B$. This equation is now the form of a straight line. In this case, however, we do not have $\log x$ in the linear form of the equation, only $\log y$. Therefore for this equation to plot as a straight line, we need to plot x using a regular Cartesian scale division and to plot y on a logarithmic scale. That is, we must use semilog graph paper.

EXAMPLE 3 **Graphing Exponentional Function on Semilog Graph Paper**

Graph $y = \frac{1}{4}(2)^{3x}$, $0 \le x \le 5$ on the following.

a. Cartesian graph paper
b. Semilog graph paper.

Solution   The table accompanying Figure 13.30 shows values of $y = \frac{1}{4}(2)^{3x}$, $0 \le x \le 5$. The function is graphed on Cartesian graph paper in Figure 13.30 and on semilog graph paper in Figure 13.31.   ◆

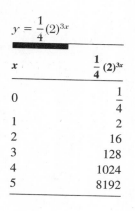

$$y = \frac{1}{4}(2)^{3x}$$

x	$\frac{1}{4}(2)^{3x}$
0	$\frac{1}{4}$
1	2
2	16
3	128
4	1024
5	8192

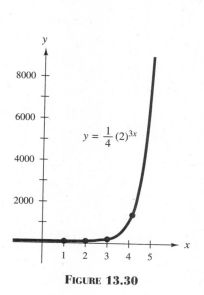

FIGURE 13.30

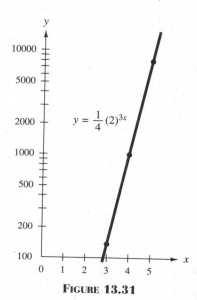

FIGURE 13.31

DEFINITION 13.5

A straight line on semilog graph paper represents the equation $y = be^{mx}$, where given the points (x_1, y_1) and (x_2, y_2):

$$m = \frac{\log y_2 - \log y_1}{(x_2 - x_1) \log e} \quad \text{and} \quad \log b = \log y_1 - (m \log e)x_1.$$

EXAMPLE 4   Application: Writing an Equation For Experimental Data

The distance in feet that a car travels in t seconds, starting from rest, is given in the table.

t	1	2	3	4	5	6
d	1.00	2.03	4.08	8.22	16.6	33.3

a.   Graph the data on Cartesian graph paper.
b.   Graph the data on semilog graph paper.
c.   Determine the equation that represents the data.

Solution

a. The data are graphed on Cartesian graph paper in Figure 13.32.
b. The data are graphed on semilog graph paper in Figure 13.33.

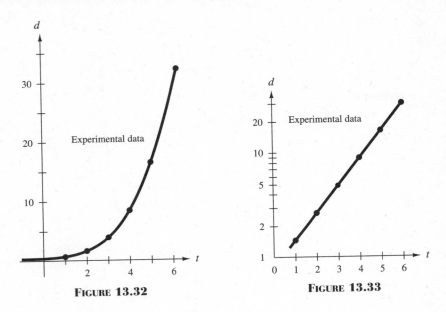

FIGURE 13.32   **FIGURE 13.33**

c. Picking two points on the line from tha table, say $(1, 1.00)$ and $(4, 8.22)$, and using Definition 13.5 we have:

$$m = \frac{\log y_2 - \log y_1}{(x_2 - x_1) \log e}$$

$$= \frac{\log 8.22 - \log 1.00}{(4 - 1) \log e} \qquad \textbf{Substituting}$$

$$= \frac{0.9149 - 0.0000}{3(0.4343)} \qquad \textbf{Taking logarithms}$$

$$= 0.7022.$$

$$\log b = \log y_1 - (m \log e)x_1$$

$$= \log 1.00 - (0.7022)(\log e)(1) \qquad \textbf{Substituting}$$

$$= 0.0000 - 0.7022(0.4343)(1) \qquad \textbf{Taking logarithms}$$

$$= -0.3050$$

$$b = 0.4955. \qquad\qquad y = \log_b x, \text{ if } x = b^y$$

Therefore, the equation that represents the data is $d = 0.5e^{0.7t}$ ◆

13.6 Exercises

In Exercises 1–6, write a function that describes the data. (See Example 1.)

1.

x	1.0	2.0	3.0	4.0
y	2.0	16.0	54.0	128.0

2.

x	1.0	2.0	3.0	4.0
y	0.50	4.00	13.50	32.00

3.

x	2.0	3.0	4.0	5.0
y	16.0	36.0	64.0	100.0

4.

x	2.00	3.00	4.00	5.00
y	1.00	2.25	4.00	6.25

5.

x	1.00	2.00	3.00	4.00
y	0.200	0.317	0.416	0.504

6.

x	1.00	2.00	3.00	4.00
y	0.167	0.280	0.380	0.471

In Exercises 7–12, plot the function for $0 < x \le 5$ on log-log graph paper.

7. $y = x^3$

8. $y = x^2$

9. $y = x^{-1}$

10. $y = x^{-0.5}$

11. $y = 2x^{0.5}$

12. $y = 4x^{-\frac{1}{3}}$

In Exercises 13–18, plot the equation for $0 < x \le 5$ on semilog graph paper.

13. $y = 2^x$

14. $y = 3 \cdot 2^{-x}$

15. $y = 2 \cdot 3^{-x}$

16. $y = 3e^{2x}$

17. $y = 2e^{3x}$

18. $y = 2e^{-3x}$

In Exercises 19–24, plot the experimental data on the following.
a. Cartesian graph paper
b. Log-log graph paper
c. Semilog paper
d. Select the graph that represents a straight line and determine the equation that best approximates the data.

19.

x (m)	1	2	3	4	5	6
y (m)	2.5	7.1	13.0	20.0	28.0	36.7

20.

H (m)	3	6	9	12	15	24	30
Q (m³/s)	0.11	0.18	0.23	0.28	0.33	0.46	0.54

21.

x (m)	1	2	3	4	5	6	7
y (m)	1.01	2.03	4.08	8.22	16.56	33.34	67.14

22.

t (s)	0	3	6	9	12	15	18
v (m/s)	1.40	1.75	2.20	2.75	3.44	4.31	5.40

23.

H (ft)	0	2	4	6	8	10
P (psi)	2000	1850	1700	1570	1450	1340

24.

r (cm)	1	2	3	4	5	6
T (°C)	491	402	329	270	221	181

In Exercises 25–30, given the type of graph paper and two points on the graph, write the equation of the straight line passing through the given points.

25. log-log; $(100, 0.02)$, $(900, 10^2)$

26. log-log; $(800, 0.02)$, $(9000, 10^2)$

27. log-log; $(10^{-3}, 50)$, $(3, 100)$

28. semilog; $(0.5, 0.6)$, $(1.1, 0.3)$

29. semilog; $(0.4, 0.6)$, $(1.0, 0.3)$

30. semilog; $(2.0, 8 \times 10^2)$, $(3.6, 1)$

31. Martha recorded her pulse rate while exercising on her stationary bicycle. The time t is measured in minutes, and the rate r is expressed in heartbeats per minute. Determine an equation that best represents the data.

t	2.0	4.0	6.0	8.0
r	95	107	120	135

32. A laboratory worker records the number N of insects that hatched at time intervals of 1, 4, 9, and 16 minutes. Determine an equation that best represents the data.

t	1.0	4.0	9.0	16.0
N	2.00	16.00	54.00	128.00

Review Exercises

 In Exercises 1–12, graph the exponential functions.

1. $y = 3.5^x, -1 \le x \le 4$

2. $y = 2.5^x, -1 \le x \le 4$

3. $y = 0.6^x, -5 \le x \le 1$

4. $y = 0.8^x, -5 \le x \le 1$

5. $v = 2.5 \cdot 0.7^x, -4 \le x \le 1$

6. $p = 0.4 \cdot 2^x, -1 \le x \le 5$

7. $a = 2.5^{-x}, 0 \le x \le 5$

8. $S = 0.8^{-t}, 0 \le t \le 5$

9. $V = 10e^{-2t}, 0 \le t \le 3$

10. $Q = 5e^{-1.5t}, 0 \le t \le 4$

11. $i = -9e^{-t}, 0 \le t \le 5$

12. $i = -6e^{-\frac{1}{2}t}, 0 \le t \le 10$

In Exercises 13–16, how long does it take the function to decay to less than 1% of its starting value if t is in seconds?

13. $i = -9e^{-3t}$

14. $v = 10e^{-\frac{t}{4}}$

15. $v = 12e^{-\frac{t}{6}}$

16. $i = -110e^{-4t}$

17. An amount of money P compounded annually at an interest rate r yields an amount A after n years. What would \$100 placed at 12% yield after 15 years? $\left(A = P\left[1 + \dfrac{r}{100}\right]^n\right)$

18. The decay of a radioactive substance is governed by $A = A_0 \left(\dfrac{1}{2}\right)^{\frac{t}{L}}$. Radium has a half-life of 1620 yr. How much is left of a 1-kg sample after 1000 years?

19. A bacteria culture doubles every 12 days. If it starts with 2 bacteria, how many bacteria will be present after 30 days?

20. A compound decays according to $S = S_0 4.7^{-\frac{t}{4}}$. What percentage of the compound will be left when $t = 18$?

In Exercises 21–28, determine the inverse of the function.

21. $49 = 7^2$

22. $1728 = 12^3$

23. $P = 5^t$

24. $Q = 6^{2h}$

25. $\log_5 125 = 3$

26. $\log_8 4096 = 4$

27. $s = \log_{1.3} 416$

28. $m = \log_{2.7} 114$

In Exercises 29–38, solve for x.

29. $\log_6 x = 2$

30. $\log_{1.3} x = 3$

31. $\log_2 x = -5$

32. $\log_3 x = -2$

33. $x = \log_{1.7} 1.7$

34. $x = \log_{13} 13$

35. $x = \log_4 \dfrac{1}{64}$

36. $x = \log_{10} 10^5$

37. $x = \log_{\frac{1}{3}} 1$

38. $x = \log_{\frac{1}{4}} 1$

In Exercises 39–42, graph the logarithmic functions.

39. $y = \log_{\frac{3}{2}} x, 0 < x \le 1$

40. $y = \log_{\frac{2}{3}} x, 0 < x \le 1$

41. $y = \log_{\frac{3}{2}} x, 1 \le x \le \dfrac{81}{8}$

42. $y = \log_{\frac{2}{3}} x, \dfrac{8}{81} \le x \le 1$

43. How long does it take a bacteria colony starting with 2 bacteria to reach 100 bacteria if it doubles every hour?

44. How long does it take 0.5 kg of radium (half-life 1620 yr) to decay to 0.2 kg?

In Exercises 45–50, write each expression as the sum or difference of logarithms. The answer should not contain any exponents or radicals. Simplify where possible.

45. $\log_7(5x^3)^2$
46. $\log_b \dfrac{12}{b^4}$
47. $\log_2 \dfrac{x - 2}{x + 2}$
48. $\log_3 \dfrac{27}{1 - x}$

49. $\log_{10} \dfrac{x^2 + y^2}{1000}$
50. $\log_{10} \dfrac{2x^2 - 3y}{10}$

In Exercises 51–56, use the properties of logarithms to write the expressions as a single logarithm.

51. $\log_{10} 1 + \log_{10} x$
52. $\log_2(x + y) + \log_2 y$
53. $\log_2(x + 3) - \log_2(x - 3)$

54. $2 \log_a x - \dfrac{1}{2} \log_a y$
55. $\log_e e + \log_e 1$
56. $2 \log_e 1 - \log_e e^2$

In Exercises 57–62, let $\log_a 1.5 = 0.37$, $\log_a 2.0 = 0.63$, and $\log_a 5.0 = 1.46$. Determine the logarithms of the expressions.

57. $\log_a 125$
58. $\log_a 128$
59. $\log_a \dfrac{1}{25}$

60. $\log_a \dfrac{1}{16}$
61. $\log_a 50$
62. $\log_a 9$

63. If $A = P(1 + r)^n$, use logarithms to solve for n.

64. If $\dfrac{P_1{}^{n-1}}{V_1} = \dfrac{P_2{}^{n-1}}{V_2}$, use logarithms to solve for n.

In Exercises 65–72, determine the logarithm to the base 10 of the given number to four decimal places. Identify the characteristic and the mantissa of the logarithm.

65. 14.2
66. 17.6
67. 0.0113
68. 0.0276

69. 1.5×10^6
70. 2.2×10^4
71. 8.1×10^{-2}
72. 7.6×10^{-3}

In Exercises 73–78, determine the antilog, base 10, of the given number to three significant digits.

73. 2.5110
74. 3.6271
75. -2.8173
76. -1.2845

77. 1.6374
78. -11.1111

In Exercises 79–82, evaluate the expressions using logarithms.

79. $3.61(18.2)^{1.7}$
80. $4.25(8.63)^{2.3}$
81. $\sqrt[5]{(16.3)^{\frac{3}{2}}}$
82. $\sqrt[7]{(8.96)^{1.5}}$

83. The Richter scale for measuring earthquakes is $R = \log \dfrac{I}{I_0}$, where I_0 is a constant. How many times greater is the intensity of an earthquake for which $R = 4.2$ than one for which $R = 3.7$?

84. How many times greater is a sound intensity of 90 dB (the threshold of permanent hearing damage) than an intensity of 65 dB (ordinary conversation)? $\left(dB = 10 \log \dfrac{I}{10^{-16}} \right)$

In Exercises 85–92, transform the logarithm to the base indicated.

85. $\log_5 125$; 10

86. $\log_3 0.370$; 10

87. $\log_4 0.0156$; e

88. $\log_5 625$; e

89. $\log 14.3$; e

90. $\log 0.462$; e

91. $\ln 4.17$; 10

92. $\ln 0.285$; 10

In Exercises 93–96, calculate the natural logarithm of the expression.

93. $\dfrac{8}{e^2}$

94. $\dfrac{e^{\frac{1}{2}}}{4}$

95. $\dfrac{e^{-\frac{1}{6}}}{\dfrac{1}{12}}$

96. $\dfrac{e^{-3}}{27}$

97. The recrystallization of aluminum is governed by the expression $\ln 0.14 = C + \dfrac{B}{T}$. If $C = -52$ and $B = 30{,}000°$K, what is T in °K?

98. The relaxation time t in days of a rubber band is given by $\ln \dfrac{5.5}{11} = -\dfrac{42}{t}$. What is t?

In Exercises 99–106, determine x.

99. $\left(\dfrac{1}{2}\right)^{2x} = 10$

100. $3^{x-1} = 12$

101. $5^{x+2} = 4^{x-1}$

102. $2^x = 64 + \log 64$

103. $\log(x + 1) = \log(x - 1)$

104. $\log x - \log(x + 2) - 4 = 0$

105. $\ln(x - 2) - \ln x = 8$

106. $\ln(2x - 1) = \ln(x + 2)$

107. If the half-life of radium is 1620 yr, how long does it take a sample to decay to $\dfrac{3}{4}$ of its original mass?

108. If the pH of tomatoes is 4.2, what is the hydrogen-ion concentration, H^+? Use pH $= -\log(H^+)$.

109. Belt tensions on either side of a flat pulley are related by $\dfrac{T_2}{T_1} = e^{\mu\beta}$. If $T_1 = 100$ N, $T_2 = 250$ N, and the coefficient of friction μ is 0.2, what is the angle of the wrap β in radians?

110. V-belt tensions driving a pulley are related by $\dfrac{T_2}{T_1} = e^{\left(\frac{\mu\beta}{\sin\alpha/2}\right)}$. For $T_2 = 12T_1$, $\mu = 0.3$, $\beta = \pi$, what is the belt angle α in degrees?

In Exercises 111 and 112, plot the equation for $0 < x \le 5$ on **a.** Cartesian graph paper, **b.** log-log graph paper, **c.** semilog graph paper.

111. $y = 5x^{\frac{1}{4}}$

112. $y = 12x^{-\frac{1}{5}}$

In Exercises 113–116, plot the equation for $0 < x \le 5$ on **a.** Cartesian graph paper, **b.** log-log graph paper, **c.** semilog graph paper.

113. $y = 5\left(\dfrac{1}{4}\right)^x$

114. $y = 12\left(\dfrac{1}{5}\right)^{-x}$

115. $y = \dfrac{1}{4}e^{-2x}$

116. $y = 4e^{-x}$

In Exercises 117–120, plot the data on log-log graph paper or semilog graph paper, as appropriate and determine the equation that describes the graph.

117.

Boat dis-placement d (ft)	4.0	5.0	6.0	7.0	8.0	9.0
Plank thickness t (in.)	0.53	0.69	0.86	1.03	1.21	1.40

118.

scan time t	0.03	0.04	0.05	0.06	0.07
screen intensity I	0.9	1.5	2.4	4.0	6.6

119.

time t (s)	10	12	14	16	18	20
current i (A)	7.9	6.9	6.0	5.2	4.5	3.9

120.

pressure p (MPa)	10	20	30	40	50	60	70	80	90	100
volume V (m³)	0.100	0.081	0.072	0.066	0.062	0.058	0.056	0.054	0.052	0.050

121. Show that $\log_b RST = \log_b R + \log_b S + \log_b T$.

122. Assume that your calculator can calculate only $\log n$ for $1 \leq n < 10$ and antilog N for $0 \leq N < 1$. Show how to find the antilog of any number N by using its characteristic and mantissa.

123. Show why a $\boxed{10^x}$ key can be used to find antilogs if your calculator has no $\boxed{\text{INV}}$ $\boxed{\text{LOG}}$ function.

124. Derive the properties of logarithms for the natural logarithms.

125. Show that if $y = b^{\log_b n}$, then $y = n$.

✎ Writing About Mathematics

1. As an employee of an investment company, you frequently are asked to determine the number of years it will take for a sum of money invested at a fixed rate to double.

Using the formula $A = P\left(1 + \dfrac{r}{n}\right)^{nt}$, write a formula to handle that computation. Explain how to use the formula.

2. Research the development of logarithms by John Napier and write a brief paper explaining his work.

3. The slide rule was an early computing machine. Do research and write a brief paper explaining how logarithmic scales were used on the slide rule.

Chapter Test

In Exercises 1 and 2, use a calculator to evaluate each of the expressions to the nearest hundredths.

1. 3.89^6

2. $5e^{1.85}$

3. Sketch the graph of the exponential function $y = 3(2^x)$ on Cartesian graph paper for the interval $-2 \leq x \leq 2$.

In Exercises 4 and 5, write the equivalent function in logarithmic or exponential form.

4. $121 = 11^2$

5. $z = \log_{1.3} 234$

In Exercises 6–8, solve for x.

6. $\log_4 x = 3$

7. $\log(x + 1) + \log(x - 1) = 1$

8. $4^{(x+1)} = 12$

In Exercises 9 and 10, use the fact that $\log_b 1.5 = 3.70$, $\log_b 2.0 = 2.63$, and $\log_b 5.0 = 1.46$ to determine the logarithms of the expressions.

9. $\log_b 3.0$

10. $\log_b 7.5$

In Exercises 11 and 12, determine the logarithm, in base 10, for each number to five decimal places.

11. $\log 34.7$

12. $\log 8.4 \times 10^{-7}$

In Exercises 13 and 14, determine the antilog, in base 10, for each number to three significant digits.

13. 1.4314

14. -2.4685

In Exercises 15 and 16, evaluate the expression using logarithms.

15. $37.2 \sqrt[3]{2.47}$

16. $\dfrac{(73.4)(56.8)}{(11.2)^2}$

In Exercises 17 and 18, determine the value of the logarithms using the change-of-base formula.

17. $\log_7 63.9$

18. $\log_4 79.5$

In Exercises 19 and 20, graph the logarithmic functions.

19. $y = \log(x - 1)$

20. $y = \log_5 2x$

21. The diastolic pressure in an artery during a heartbeat is given by $P = P_o e^{-Rt}$. Solve for t.

22. The population of ladybugs rapidly multiplies so that the population t days from now is given by $A(t) = 3000e^{0.01t}$.

    a. How many ladybugs are present now?

    b. How many will there be in seven days?

23. Given the set of data, determine a power function that approximates the data.

x	1	2	3	4
y	5	40	135	320

Inequalities and Absolute Values

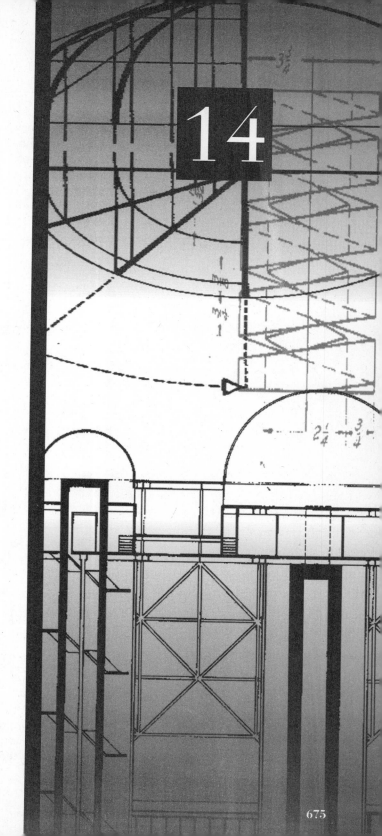

14

66 The average number of miles per gallon of gasoline for all the cars manufactured by American auto manufacturers must be greater than 27.5 mpg by 1993." "The load on a helicopter must be less than x pounds for the helicopter to be an efficient machine." "The post office limits the size of rectangular boxes sent through the mail by stating that length plus girth cannot exceed 100 inches." (Recall that girth is the perimeter of a cross-section.) "A manufacturer makes a profit when the revenue is greater than the cost."

Each of these statements indicates a relationship between two things that may not be equal. Such a statement is called an *inequality*. Example 6 in Section 4.1 illustrates that when revenue is greater than cost, the manufacturer earns a profit.

14.1

Linear Inequalities

The basic symbols of inequality were introduced in Section 1.2. To this point in the text, the symbols of inequality have been used to indicate relationships between numbers and to indicate intervals. For future topics, we must learn how to solve statements of inequality. However, we already know something about the process since many of the procedures we use to solve equalities can be used to solve statements of inequality. This section is devoted to the definitions of inequalities and to solving linear inequalities.

Table 14.1 contains a list of the basic inequality symbols, the meanings of the symbols, and their geometrical interpretations.

TABLE 14.1   Inequality Notation

Symbol*	Meaning	Graph
$x < a$	x is less than a	
$x \leq a$	x is less than or equal to a	
$x > a$	x is greater than a	
$x \geq a$	x is greater than or equal to a	
$a < x < b$	x is greater than a and x is less than b	
$a \leq x \leq b$	x is greater than or equal to a and x is less than or equal to b	
$a < x \leq b$	x is greater than a and x is less than or equal to b	
$a \leq x < b$	x is greater than or equal to a and less than b	

*a and b represent real numbers, and x is a variable.

As we study the graphs in Table 14.1, recall that a *solid dot* indicates that the solution includes that point. An *open circle,* on the other hand, means that the solution approaches the point, but that the point itself is not part of the solution set.

EXAMPLE 1 Graphing Inequalities

Graph the inequality statements on a real number line.

a. $x < 5$ **b.** $-2 \le x < 4$

Solution

a.

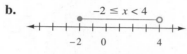

FIGURE 14.1

In Figure 14.1, the open circle indicates that the 5 is not part of the solution, but that all the points to the left of 5 are part of the solution.

b.

FIGURE 14.2

In Figure 14.2, the solid dot indicates that -2 is part of the solution. The open circle indicates that 4 is not part of the solution. The line indicates that all the points between -2 and 4 are part of the solution. ◆

Inequality notation can be used to define regions of the coordinate plane. Example 2 illustrates this concept.

EXAMPLE 2 Writing an Inequality Statement

Write a statement or statements of inequality involving x or y or both that provide the location of all the points in the coordinate plane that satisfy the following statements.

a. The region below the x-axis.
b. The region to the left of the line $x = 4$ and the region above the line $y = 2$.

Solution Sketching the region described helps us to write the required statement(s) of inequality.

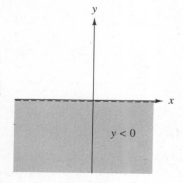

FIGURE 14.3

a. The shaded region of Figure 14.3 indicates the designated region of the xy-plane. We see from the shaded figure that y is less than zero. Therefore, the inequality $y < 0$ describes the location of all the points $P(x, y)$ that are below the x-axis.

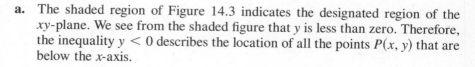

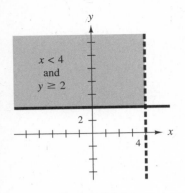

FIGURE 14.4

b. The shaded region of Figure 14.4 indicates the designated region of the xy-plane. We see from the figure that two statements of inequality are necessary to describe the shaded region. Therefore, the inequalities $x < 4$ and $y \geq 2$ describe the location of all the points $P(x, y)$ that are to the left of the line $x = 4$ and above and on the line $y = 2$. ◆

Similar to the convention for points, a solid line indicates that the border line is part of the solution. A dotted line is not part of the solution. This convention also is illustrated in Figure 14.4.

We know that if $6 > 4$, then $6 - 4$ is a real positive number and that real positive number is 2. Also, if $-7 > -16$, then $-7 - (-16)$ is a real number and that real number is 9. This property is stated formally as Definition 14.1.

DEFINITION 14.1

For all real numbers a and b, if $a > b$, then there exists a real positive number c such that $a - b = c$.

The sense of an inequality compares the direction of the inequality symbol in one statement to the direction at the inequality symbol in another statement.

DEFINITION 14.2 Sense of Inequality

If $a < b$ and $c < d$, then the two inequalities are unequal in the same sense. If $a < b$ and $c > d$, then the two inequalities are unequal in the opposite sense.

For example, if $3 < 5$ and $6 < 7$ then the inequalities are unequal in the same sense; that is, both are less than. If $5 < 8$ and $9 > 4$ then the two inequalities are unequal in the opposite sense; one is less than and the other is greater than.

An inequality of the form $x^2 + 2 > 0$ is an **absolute inequality** since any real number substituted for x makes the statement true. An inequality of the form $x > 1$ is a **conditional inequality.** That is, the statement is not true for all real numbers substituted for x.

When adding, subtracting, multiplying, dividing, or taking the reciprocal of inequality statements we apply certain rules. These rules are stated in the following box.

PROPERTIES OF INEQUALITIES

Let a, b, and c be any real numbers.

Addition property

    1. If $a < b$, then $a + c < b + c$.

    2. If $a > b$, then $a + c > b + c$.

Subtraction property

    1. If $a < b$, then $a - c < b - c$.

    2. If $a > b$, then $a - c > b - c$.

Multiplication property

    1. If $a < b$, and $c > 0$, then $ac < bc$.

    2. If $a < b$, and $c < 0$, then $ac > bc$.

Division property

    1. If $a < b$, and $c > 0$, then $a \div c < b \div c$.

    2. If $a < b$, and $c < 0$, then $a \div c > b \div c$.

Reciprocal property

If $ab > 0$ and $a < b$, then $\dfrac{1}{a} > \dfrac{1}{b}$.

CAUTION   The sense of an inequality is reversed when multiplying or dividing both sides of the inequality by a negative quantity.

$$\text{If } 3 < 4, \text{ then } 3(-5) > 4(-5)$$
$$\text{If } 24 > 15, \text{ then } 24 \div (-3) < 15 \div (-3).$$

Note that in each case the inequality signs are reversed.   ∎

An inequality is a **linear inequality** if the exponents of all the variables are 1 (or the variables are of degree 1). Examples of linear inequalities are:

$$2x + 1 \leq 6, \quad x \geq 3, \quad x + 7y \leq 5, \quad 3x - 7 < 6y.$$

Examples of inequalities that are not linear are:

$$x^2 + 2 > 4, \quad x^2 + 5x \leq 11, \quad 2xy + 7x^2 < 3, \quad 2xy \leq 3, \quad 2\sqrt{x} \leq 5.$$

The solution of a linear inequality is the set of all numbers that make the statement true. For example, a solution for $3x + 1 > 4$ is $x = 2$; substituting 2 for x we see that $7 > 4$. In fact 3, 10, and 1.1 are also solutions for the inequality. All real numbers greater than 1 make the statement true. Thus, $x > 1$ is the solution to $3x + 1 > 4$.

EXAMPLE 3 **Algebraic and Graphic Solutions to a Linear Inequality**

Determine the solution of the linear inequality $2x - 6 \geq -3 + x$ and graph the solution set.

Solution The general procedure here is the same as in solving an equality: we want to isolate the variable on one side of the inequality symbol and the constants on the other.

$$2x - 6 \geq -3 + x$$
$$2x - 6 \boxed{+ 6} \geq -3 \boxed{+ 6} + x \qquad \textbf{Addition property}$$
$$2x \geq 3 + x$$
$$2x \boxed{- x} \geq 3 + x \boxed{- x} \qquad \textbf{Subtraction property}$$
$$x \geq 3.$$

Thus, the solution set of $2x - 6 \geq -3 + x$ is all real numbers greater than or equal to 3. The graph of the solution set is given in Figure 14.5.

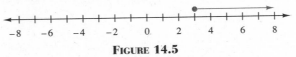

FIGURE 14.5

◆

EXAMPLE 4 **Algebraic and Graphic Solutions to a Linear Inequality**

Determine the solution of the linear inequality $3(x + 4) - 2 < 4x + 12$ and graph the solution set.

Solution First remove the parentheses and combine like terms.

$$3(x + 4) - 2 < 4x + 12$$
$$3\boxed{x} + 12 - 2 < 4x + 12 \qquad \textbf{Distributive property}$$
$$3x + 10 < 4x + 12$$
$$3x + 10 \boxed{-10} < 4x + 12 \boxed{-10} \qquad \textbf{Subtraction property}$$
$$3x < 4x + 2$$
$$3x \boxed{- 4x} < 4x \boxed{- 4x} + 2 \qquad \textbf{Subtraction property}$$
$$-x < 2$$
$$(\boxed{-1})(-x) > (\boxed{-1})(2) \qquad \textbf{Multiplication property (note change in inequality sign because of multiplication by a negative number)}$$
$$x > 2.$$

Thus, the solution set of $3(x + 4) - 2 < 4x + 12$ is all real numbers greater than -2 Figure 14.6 shows the graph of the solution set.

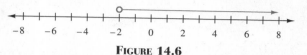

FIGURE 14.6 ◆

CAUTION People sometimes misread results when the variable is on the right side of the inequality symbol $(3 < x)$. To prevent such a misunderstanding, place the variable on the left side of the inequality symbol $(x > 3)$ when solving, or remember to read the result backward (right to left). ■

An inequality of the form $a < x < b$ where there are two inequality symbols in an expression is called a **double inequality.** A technique for determining the solution of a double inequality is illustrated in Example 5.

EXAMPLE 5 **Algebraic and Graphic Solutions to a Double Inequality**

Determine the solution of the double inequality $-1 < \dfrac{x + 3}{2} \le 5$ and graph the solution.

Solution To solve a double inequality we must isolate x as the middle term. To do this, use the same properties used to solve single inequalities.

$$-1 < \frac{x + 3}{2} \le 5$$

$$2\,(-1) < 2\left(\frac{x + 3}{2}\right) \le 2\,(5) \qquad \text{Multiply the three terms by 2}$$

$$-2 < x + 3 \le 10$$

$$-2 - 3 < x + 3 - 3 \le 10 - 3 \qquad \text{Subtract 3 from all three terms}$$

$$-5 < x \le 7$$

Thus, the solution set of $-1 < \dfrac{x + 3}{2} \le 5$ is all the real numbers greater than -5 and less than or equal to 7. The graph of the solution set is given in Figure 14.7.

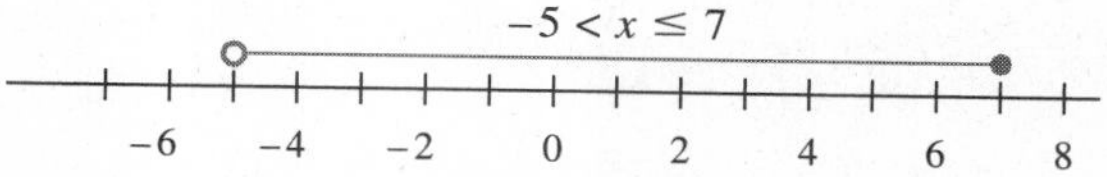

$$-5 < x \le 7$$

FIGURE 14.7 ◆

CAUTION   We cannot combine two inequalities like $x \leq -4$ and $x > 5$ into the single statement $-4 \geq x > 5$. This is a false statement since there is no real number that is less than or equal to -4 and at the same time greater than 5. ∎

EXAMPLE 6   **Application: Revenue Greater Than Cost**

The revenue R a manufacturer receives for selling x pieces of a specific type of hardware is $R = 239.79x$. The cost C of manufacturing x units is $C = 142\ x + 675$. As mentioned in the beginning of the chapter, a manufacturer makes a profit when R is greater than C. How many pieces must the manufacturer sell to make a profit?

Solution   To make a profit $R > C$.

$$239.79x > 142x + 675 \qquad \text{Substituting for } R \text{ and } C$$
$$239.79x\ -142x > 142x\ -\ 142x + 675 \qquad \text{Subtraction property}$$
$$97.79x > 675$$
$$x > 6.90$$

Since x must be greater than 6.90, the manufacturer must sell 7 pieces of hardware to make a profit. ◆

EXAMPLE 7   **Application: Drying Temperature**

The most effective drying temperatures for a special paint are between 60°F and 75°F. Determine the interval, to the nearest whole degree, in degrees Celsius that are the most effective drying temperatures. Use the formula $F = \dfrac{9}{5}C + 32$ to convert temperatures from one measure to the other.

Solution   The desired interval in degrees Fahrenheit is:

$$60 < F < 75$$
$$60 < \frac{9}{5}C + 32 < 75 \qquad \text{Substituting for F}$$
$$60\ -\ 32 < \frac{9}{5}C + 32\ -\ 32 < 75\ -\ 32 \qquad \text{Subtraction property}$$
$$28 < \frac{9}{5}C < 43$$
$$\frac{5}{9}(28) < \frac{5}{9} \cdot \frac{9}{5}C < \frac{5}{9}(43) \qquad \text{Multiplication property}$$
$$16 < C < 24.$$

The best drying temperatures for the paint are between 16°C and 24°C, to the nearest degree. ◆

We can use inequalities to determine the numerical value we need on the final exam to achieve the desired letter grade. This is illustrated in Example 8.

EXAMPLE 8   **Application: Grade Needed for a B**

A student must have an average grade greater than or equal to 80% but less than 90% on five tests to receive a final grade of B. The student's grades on the first four tests were 98, 76, 86, and 92%. What range of grades on the fifth test would give the student a B in the course?

Solution   First construct an inequality that can be used to determine the range of grades on the fifth exam. The average (mean) is found by adding the grades and dividing the sum by the number of exams. Let x = the fifth grade:

$$\text{average} = \frac{98 + 76 + 86 + 92 + x}{5}.$$

For the student to obtain a B, the average must be greater than or equal to 80 but less than 90%.

$$80 \leq \frac{98 + 76 + 86 + 92 + x}{5} < 90$$

$$80 \leq \frac{352 + x}{5} < 90$$

$$5\,(80) \leq 5\left(\frac{352 + x}{5}\right) < 5\,(90) \qquad \begin{array}{l}\textbf{Multiply the}\\\textbf{three terms by 5}\end{array}$$

$$400 \leq 352 + x < 450$$

$$400 - 352 \leq 352 - 352 + x < 450 - 352 \qquad \begin{array}{l}\textbf{Subtract 352 from}\\\textbf{all three terms}\end{array}$$

$$48 \leq x < 98$$

Thus, a grade of 48% up to but not including a grade of 98% will result in a B.

$\blacklozenge$

14.1 Exercises

In Exercises 1–10, graph the solution set on the real number line.

1. $x > 7$ **2.** $x < -3$ **3.** $-3 < x < 4$ **4.** $-5 \leq x < 4$

5. $-13 < x \leq -2$ **6.** $-4 \leq x \leq 7$ **7.** $-x > 3$ **8.** $4 < -x$

9. $3 > x > 1$ **10.** $\dfrac{1}{x} < \dfrac{1}{3}$

In Exercises 11–20, express each statement using the appropriate inequality symbol(s).

11. -5 is greater than or equal to -16.   **12.** 11 is less than or equal to 24.

13. $R + 6$ is greater than 11.   **14.** t is between 0 and 6.

15. Z is greater than 2 and less than or equal to 11.

16. $2 + t$ is positive.

17. $6 - x$ is negative.

18. $x + 3$ is nonnegative.

19. $R - 2$ is nonpositive.

20. $x + 5$ is greater than 4 and less than 10.

In Exercises 21–30, insert the proper inequality symbol in the blank between the two members.

21. If $5 < y$, then $5 + 4$ _____ $y + 4$.

22. If $a > b$, then $a - x$ _____ $b - x$.

23. If $-3 < 11$, then $-3 + z$ _____ $11 + z$.

24. If $3 < 11$, then $3(-2)$ _____ $11(-2)$.

25. If $x + 3 < -2$, then $(x + 3) \div (-4)$ _____ $(-2) \div (-4)$.

26. If $t + 3 > t - 5$, then $(t + 3)(-5)$ _____ $(t - 5)(-5)$.

27. If $x + y > x - z$ and $w < 0$, then $(x + y)w$ _____ $(x - z)w$.

28. If $a + b + c < 2a + b - c$ and $d < 0$, then $(a + b + c) \div d$ _____ $(2a + b - c) \div d$.

29. If $7 < 13$, then $\dfrac{1}{7}$ _____ $\dfrac{1}{13}$.

30. If $(x + y)(x - y) > 0$ and $x + y > x - y$, then $\dfrac{1}{x + y}$ _____ $\dfrac{1}{x - y}$.

In Exercises 31–36, for the point $P(x, y)$ in the coordinate plane, write a statement (or statements) of inequality involving x or y or both that provides the location of the point P according to the following statements.

31. above the x-axis

32. to the left of the y-axis

33. in the second quadrant

34. below the line $y = -3$

35. to the right of the line $x = 4$

36. between the lines $x = -3$ and $x = 4$

In Exercises 37–54, solve the inequality and graph the solution set on a number line.

37. $x + 3 < 6$

38. $x + 7 \le 3$

39. $r + 3 \le 2r - 4$

40. $\dfrac{Z}{5} \ge -\dfrac{3}{5}$

41. $\dfrac{R}{2} \ge 0$

42. $\dfrac{R}{2} + 3 \ge 2R + 4$

43. $\dfrac{R}{4} + \dfrac{1}{2} \le \dfrac{3}{8}$

44. $2(t - 1) + 3(t + 2) \ge t - 5$

45. $2(3t - 2) \le 4(1 - t)$

46. $4(t + 1) - 3 \le 2(1 - 2t) - 1$

47. $\dfrac{3z + 2}{5} \le -\dfrac{2}{3}$

48. $\dfrac{5y - 1}{2} + \dfrac{1}{5} > \dfrac{y + 1}{10}$

49. $3 < y + 4 < 9$

50. $0 < y - 5 < 7$

51. $-4 \le \dfrac{1 + x}{2} \le 3$

52. $-3 \le \dfrac{4x - 5}{3} \le 5$

53. $0.05 < 0.75 - 0.55x < 1$

54. $0.05 \le 0.75 - 2.00x \le 1.25$

In Exercises 55–71, write a statement of inequality and solve it.

55. You can rent a compact car from firm A for \$190 per week with no charge for mileage or from firm B for \$110 per week plus 20 cents for each mile driven. At what mileage does it cost less to rent from firm A?

56. A machine that packages 200 screws per box can make an error of 2 screws per box. If x is the number of screws in a box, write an inequality that indicates a maximum error of 2 screws per box.

57. Lead for a pencil must have a diameter of 0.5 mm with a tolerance of ±0.005 mm. Find the acceptable range of diameters of the lead.

58. The altitudes for a certain commercial airline route are between 25,370 ft and 32,450 ft. Express the interval in miles.

59. The relationship between Fahrenheit temperature F and Celsius temperature C is given by the formula $C = \frac{5}{9}(F - 32)$. Find the range of values of F when C is between 35 and 39.

60. The velocity v (in ft/s) of an object fired directly upward is given as a function of time t (seconds) by the formula $v = 92 - 32t$. For what positive time interval will the velocity be positive?

61. In Exercise 60, when will the velocity be between 19 and 32 ft/s?

62. Seven washers, all of the same thickness, must have a combined thickness of 15.05 mm with a tolerance of ±0.02 mm. Find the acceptable range of values for the thickness of a single washer.

63. Two resistors in parallel have a total resistance R given by $\frac{1}{R} = \frac{1}{R_1} + \frac{1}{R_2}$. If $R_1 = 15\ \Omega$, find the positive values for R_2 so that R is between 5 and 8 Ω.

64. In Example 8, what grade on the fifth test would result in the student receiving the following.:
  a. A grade of B if her individual grades on the first four tests were 95, 72, 85, and 89%?
  b. A grade of C with the same test grades? To obtain a C, she must have an average greater than or equal to 70 but less than 80%.
  c. Is it at all possible for the student to obtain an A (average $\geq$ 90)?

65. The minimum speed for vehicles on a highway is 40 mph, the maximum speed is 55 mph. If Tim has been driving his truck along the highway for four hours, what is the range in miles that he legally could have traveled?

66. Idelisa's Foundry manufactures wood-burning stoves. The revenue equation is $R = 2x$, where x is the number of wood-burning stoves of type A sold per year. The cost equation is $C = 140 + 1.2x$. How many type A wood-burning stoves must be sold each year to make a profit? Write a statement of inequality and solve it.

67. A fan motor at high speed must rotate at least 24 radians per second. However, the motor seriously overheats at 32 rad/s. Use an inequality statement to express this acceptance range for high speed.

68. Satellites revolve around central bodies in elliptical paths (not necessarily oval paths). For example, the radius of the earth's orbit about the sun varies from 1.471×10^{11} m (closest approach called *perihelion*) to 1.521×10^{11} m (farthest approach called *aphelion*). The speed at perihelion is 3.027×10^4 m/s, while the speed at aphelion is 2.954×10^4 m/s. State both the radius and the speed as inequalities.

69. The dinosaurs and 70% of the plankton (and many other families of flora and fauna) died out in a relatively short period of time (geologically speaking)—about 100,000 years. The rough date given for the dinosaur extinction is 65 million years ago. Express this value as an inequality taking into account the information that the extinction took place in less than 100,000 years.

70. A surveyor measures a parcel of land and reports the length to be 82.34 m and the width to be 24.32 m. As is true with any measurement, a degree of uncertainty is always present and should always be expressed. In the case of the surveyor, the probable error for both measurements is ±0.03 m.
  a. Express the length and width as an inequality.
  b. Compute the maximum area of the parcel of land.
  c. Compute the minimum area of the plot of land.
  d. Express the area as an inequality.

14.2

Inequalities in One Variable

Quadratic Inequalities

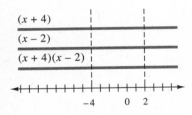

FIGURE 14.8

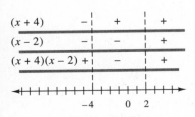

FIGURE 14.9

Our ability to solve quadratic inequalities is dependent upon our ability to solve quadratic equations. We learned the techniques for solving quadratic equations in Chapter 11. (Quickly look through that chapter if you need to brush up.)

For example, if we express the inequality $(x + 4)(x - 2) > 0$ as an equality, we have $(x + 4)(x - 2) = 0$. Solving for x, we have $x = -4$ or $x = 2$. *The values of x that make the statement of equality zero are called **critical values** of the inequality. Thus,* $x = 2$ *and* $x = -4$ *are critical values of the inequality* $(x + 4)(x - 2) > 0$.

In order for the statement of inequality in this example to be true, the product of the two quantities $(x + 4)$ and $(x - 2)$ must be positive. To express this in a simpler form, let $a = x + 4$ and $b = x - 2$, then, substituting, we have $a \cdot b > 0$. For this statement to be true, if $a \cdot b > 0$, then either $a > 0$ and $b > 0$ or $a < 0$ and $b < 0$. Therefore, we are looking for values of x that either will make both $x - 2$ and $x + 4$ positive quantities or both $x - 2$ and $x + 4$ negative quantities. To determine these values of x let us use a graphical method.

To determine the values of x by the graphical method, first draw a number line as in Figure 14.8. On the number line indicate the critical values:

$$x = \boxed{-4} \quad \text{and} \quad x = \boxed{2.}$$

Draw dashed vertical lines through $x = 2$ and $x = -4$. Then draw three lines parallel to the number line and label these lines with the factors $(x + 4)$, $(x - 2)$, and $(x + 4)(x - 2)$.

When $x = -4$, the quantity $x + 4 = 0$. For this quantity, the value -4 divides the number line into three parts: the points to the left of $x = -4$, which will make the expression $(x + 4)$ negative; $x = -4$, for which the expression $(x + 4)$ is zero; and the points to the right of $x = -4$ for which the expression $(x + 4)$ is positive. As shown in Figure 14.9, indicate this on the line labeled

$(x + 4)$ by placing $+$ and $-$ signs on the line. The factor $(x - 2)$ is zero when $x = 2$; $x - 2$ is negative to the left of $x = 2$ and positive to the right of $x = 2$. Again, place the appropriate $+$ and $-$ signs on the line labeled $(x - 2)$ as shown in Figure 14.9. Since we want to determine the values of x for which the product $(x + 4)(x - 2)$ is positive, assume that the $+$ and $-$ signs are $+1$ and -1 and multiply them using the rules of algebra. Thus, to the left of -4 we have $(-1)(-1) = +1$, so place a $+$ sign on the third line opposite the product $(x + 4)(x - 2)$. Between -4 and 2, we have $(+1)(-1) = -1$, so record the $-$ sign on the line labeled $(x + 4)(x - 2)$. To the right of 2, we have $(+1)(+1) = +1$; thus in this space, record $+$. Now looking at Figure 14.9, we see that $(x + 4)(x - 2)$ is positive (greater than zero) for values of $x < -4$ or $x > 2$. The solution set is graphed in Figure 14.10.

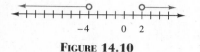

FIGURE 14.10

Solutions for inequalities can be written in several different forms. From our previous discussion we saw that the solution could be represented two ways, by graphing and by statements of inequality. *The solution could also be given in* **interval notation,** *which uses parentheses and/or square brackets. The parentheses correspond to the symbols for less than and greater than, while the square brackets correspond to the symbols for greater than and equal to and less than and equal to.* Table 14.2 illustrates these symbols and their use.

TABLE 14.2

Inequality Notation	Graph	Interval Notation*	Meaning
$a < x$		(a, ∞)	Open interval
$x < a$		$(-\infty, a)$	Open interval
$a \leq x$		$[a, \infty)$	Half-closed, half-open interval
$x \leq a$		$(-\infty, a]$	Half-open, half-closed interval
$a < x < b$		(a, b)	Open interval
$a \leq x \leq b$		$[a, b]$	Closed interval

*a and b are real numbers; ∞ is the symbol for infinity.

The **open interval** $a < x < b$ or (a, b) contains all the points or real numbers between a and b but does not include the end points a and b. The **closed interval** $a \leq x \leq b$ or $[a, b]$ contains all the points between a and b, including the end points a and b.

An inequality is in **standard form** if all the terms are on one side of the inequality symbol and there is a zero on the other side. For example, $x^2 + 3x - 5 \geq 0$ is in standard form, and $x^2 + 3x \geq 5$ is not in standard form.

EXAMPLE 1   Quadratic Inequality ≤ 0

Determine the solution of the quadratic inequality $2x^2 + 6x \leq 0$.

Solution

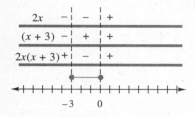

FIGURE 14.11

$$2x^2 + 6x \leq 0$$
$$2x(x + 3) \leq 0 \quad \text{Common factor}$$

Solving for x, we see that $x = 0$ and $x = -3$ are the critical points. Indicate the critical points on the number line as shown in Figure 14.11. Now determine where the individual factors $2x$ and $x + 3$ and the product $2x(x + 3)$ are positive or negative. The graph indicates that $2x(x + 3)$ is negative (less than zero) between -3 and 0. Therefore, all the points between -3 and 0 are part of the solution set. Since the inequality includes "equals," the critical values are also included. Therefore, the solution set for $2x^2 + 6x \leq 0$ is $-3 \leq x \leq 0$, or in interval notation, $[-3, 0]$. ◆

EXAMPLE 2   Quadratic Inequality > 0

Determine the solution of the quadratic inequality $x^2 + 6x > 7$.

Solution   Recall that when we solve quadratic equations, we must begin with the equation in standard form. The same thing holds true for quadratic inequalities. So begin by transforming the inequality into standard form.

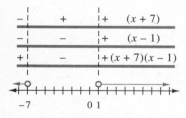

FIGURE 14.12

$$x^2 + 6x > 7$$
$$x^2 + 6x - 7 > 0 \quad \text{Standard form}$$
$$(x + 7)(x - 1) > 0. \quad \text{Factoring}$$

Setting the factors equal to zero and solving we have the following critical values.

$$x = \boxed{-7} \quad \text{and} \quad x = \boxed{1}$$

Indicate the critical values on the number line as shown in Figure 14.12. The critical points divide the graph into three regions. The graph indicates that the product $(x + 7)(x - 1)$ is positive to the left of -7 and to the right of 1. Therefore, the solution set for $x^2 + 6x > 7$ is $x < -7$ or $x > 1$, or in interval notation, $(-\infty, -7)$ or $(1, \infty)$. ◆

We can determine the solutions to Examples 1 and 2 using a graphics calculator. Because the techniques for the fx–7700G and the TI–82 differ, each is discussed individually.

fx–7700G

The first step in graphing the inequality is to change the graphing mode of the calculator. The following steps change the mode. Press $\boxed{\text{MODE}}$ and $\boxed{\text{SHIFT}}$. A menu appears with a box labeled "Graph type." The last entry in the box is "÷ INEQ." Press $\boxed{÷}$. The graph type in the mode display changes to:

$$\text{G-type: INQ/CON.}$$

This mode provides the instructions for the calculator to graph an inequality. Now press $\boxed{\text{GRAPH}}$, and a menu appears across the bottom of the screen.

$$\boxed{Y >} \quad \boxed{Y <} \quad \boxed{Y \geq} \quad \boxed{Y \leq}$$

When you press $\boxed{\text{F1}}$, "Graph Y >" appears at the top of the screen. The calculator is ready for us to enter the rest of the inequality. However, the inequality must be written in the correct form: $y > ax^2 + bx + c$, $y < ax^2 + bx + c$, $y \geq ax^2 + bx + c$, or $y \leq ax^2 + bx + c$. To write the inequalities for Examples 1 and 2 in an equivalent form, first rewrite the inequality with a zero to the left of the inequality symbol.

$$0 \geq 2x^2 + 6x \quad \text{and} \quad 0 < x^2 + 6x - 7$$

Replace the zero with a y, and write the inequalities as:

$$y \geq 2x^2 + 6x \quad \text{and} \quad y < x^2 + 6x - 7.$$

Now graph $y \geq 2x^2 + 6x$ using the range settings $[-10, 10, 1, -10, 10, 1]$. The keystrokes are: $\boxed{Y \geq}$ (F 3) 2 $\boxed{x,\Theta,T}$ $\boxed{\text{SHIFT}}$ $\boxed{x^2}$ $\boxed{+}$ 6 $\boxed{x,\Theta,T}$ $\boxed{\text{EXE}}$. The display we see before we press $\boxed{\text{EXE}}$ is: "Graph $Y \geq 2x^2 + 6x$. The graph, shown in Figure 14.13, is a parabola with the interior of the curve shaded. The interval on the x-axis that is shaded is the desired solution to the inequality, $2x^2 + 6x \geq 0$. By using the trace and/or zoom functions, we can determine that a good approximation to the solution is the closed interval $[-3, 0]$, as we found in Example 1.

Now graph $y < x^2 + 6x - 7$ and determine the solution, which is shown in Figure 14.14. In this case, the exterior of the parabola is shaded. The shading indicates that the solution is that portion of the x-axis that lies to the left and to the right of the parabola. Again, using the trace and/or the zoom functions, we can determine that the solution is $(-\infty, -7)$ or $(1, \infty)$.

CAUTION   If we do not clear the screen before graphing the second inequality, the screen shows two parabolas with no shading. ■

TI–82

The TI–82 does not have a special graphing mode nor does it require that we rewrite the inequalities. To illustrate the procedure, let us use the inequality $x^2 + 6x > 7$. First set the range as $[-10, 10, 1, -10, 10, 1]$. Then press $\boxed{Y =}$.

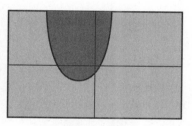

FIGURE 14.13

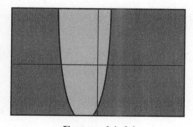

FIGURE 14.14

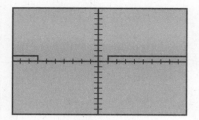

FIGURE 14.15

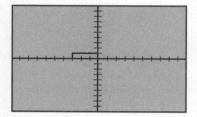

FIGURE 14.16

Enter the expression to the left of the inequality sign ($x^2 + 6x$) as the first equation, $Y1$. Enter the expression to the right of the inequality sign (7) as the second equation, $Y2$. Since $x^2 + 6x > 7$, we want $Y1 > Y2$. To accomplish this, for the third equation, $Y3$, enter $Y1 > Y2$, using the following keystrokes.

$$\boxed{2\text{nd}}\ \boxed{Y\text{-vars}}\ \boxed{1}\ \boxed{1}\ \boxed{2\text{nd}}\ \boxed{\text{TEST}}\ 3\ \boxed{2\text{nd}}\ \boxed{Y\text{-vars}}\ \boxed{1}\ \boxed{2}\ .$$

We only want to graph $Y1 > Y2$, so the only equal sign that is shaded is opposite $Y3$. Now press $\boxed{\text{GRAPH}}$. The result is shown in Figure 14.15. The lines on the graph show the desired solution: $(-\infty, -7)$ or $(1, \infty)$ as we found in Example 2.

Now try solving the inequality for Example 1. Note, since the inequality is $2x^2 + 6x \leq 0$, for the second equation $Y2$, enter 0. The solution is graphed in Figure 14.16.

<table>
<tr><td>**EXAMPLE 3**</td><td>**Application: Time Interval That a Projectile Is Above a Specific Height**</td></tr>
</table>

A projectile is fired from ground level. After t seconds its height, $h(t) = 320t - 16t^2$, is $h(t)$ feet above the ground. During what period of time is the projectile higher than 1536 ft?

Solution   The height of the projectile is higher than 1536 ft when $h(t) > 1536$. Thus,

$$320t - 16t^2 > 1536 \qquad \text{Substituting}$$
$$-16t^2 + 320t - 1536 > 1536 > 0 \qquad \text{Standard form}$$
$$t^2 - 20t + 96 < 0 \qquad \text{Dividing each term by } -16$$
$$(t - 12)(t - 8) < 0. \qquad \text{Factoring}$$

Setting the factors $t - 8 = 0$ and $t - 12 = 0$, we can determine the critical values:

$$t = 8 \quad \text{and} \quad t = 12.$$

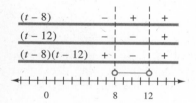

FIGURE 14.17

After constructing a number line, as shown in Figure 14.17, we can see that the product $(t - 8)(t - 12) < 0$ for values of t occur in the open interval between 8 and 12. Therefore, the projectile is higher than 1536 ft for the time interval between 8 and 12 s. To check the results, select a few values in the interval as follows.

$$t = 9: \qquad 320(9) - 16(9)^2 > 1536$$
$$2880 - 1296 > 1536$$
$$1584 > 1536. \checkmark$$
$$t = 11: \quad 320(11) - 16(11)^2 > 1536$$
$$3520 - 1936 > 1536$$
$$1584 > 1536 \checkmark$$

These two values check, and we can determine that any value selected from the interval also checks.   ◆

EXAMPLE 4   **Solution Set Is the Empty Set**

Determine the solution set of the quadratic inequality $x^2 + 3x + 5 < 0$

Solution   The expression $x^2 + 3x + 5$ is not factorable. Therefore, use the quadratic formula to determine the critical values. With $a = 1$, $b = 3$, and $c = 5$, substituting yields:

$$x = \frac{-3 \pm \sqrt{(3)^2 - 4(1)(5)}}{2(1)}$$

$$= \frac{-3 \pm \sqrt{-11}}{2}.$$

Since the roots of the equation are not real numbers, there are no critical values. This means that we cannot use the graph method. Examining the inequality, we see that no positive real numbers can possibly make the statement $x^2 + 3x + 5 < 0$ true. Therefore, try negative values. Let $x = -5, -3$, and -1. Substituting in $x^2 + 3x + 5 < 0$, we have:

$$(-5)^2 + 3(-5) + 5 = 25 - 15 + 5 = 15$$
$$(-3)^2 + 3(-3) + 5 = 9 - 9 + 5 = 5$$
$$(-1)^2 + 3(-1) + 5 = 1 - 3 + 5 = 3.$$

Try a few more negative values to see that none of them make $x^2 + 3x + 5 < 0$ a true statement of inequality. In fact, no negative value does. Therefore, the solution set of $x^2 + 3x + 5 < 0$ is the empty set.   ◆

Graph the inequality in Example 4. What does the graph look like?

In Example 4, when we solved the equality to determine the critical values for x, we found that the discriminant $b^2 - 4ac < 0$ indicates that there are no real roots. However, in problems where $b^2 - 4ac < 0$, we can use the following theorem to determine the solution set of the inequality.

THEOREM 14.1

If $ax^2 + bx + c = 0$ $(a > 0)$ has no real roots, then the solution set for:

$$ax^2 + bx + c > 0 \text{ is all real values of } x;$$
$$ax^2 + bx + c < 0 \text{ is no real values of } x.$$

EXAMPLE 5   Solution Set Is the Set of All Real Numbers

Determine the solution set of the quadratic inequality $x^2 + 5x + 7 > 0$.

Solution   To determine the critical values of $x^2 + 5x + 7 = 0$, use the quadratic formula:

$$x = \frac{-5 \pm \sqrt{(5)^2 - 4(1)(7)}}{2(1)}$$

$$= \frac{-5 \pm \sqrt{-3}}{2}.$$

The discriminant, -3, is less than zero; therefore, there are no real roots. The hypothesis of Theorem 14.1 is satisfied; the coefficient of the x^2-term is positive and the equation $x^2 + 5x + 7 = 0$ has no real roots. The conclusion of the theorem states that if the inequality $x^2 + 5x + 7 > 0$, then the solution set is all real values of x, or in interval notation, $(-\infty, \infty)$. ◆

Graph the inequality in Example 5. What does the graph look like?

Inequalities of Higher Degree and Rational Inequalities

The method for solving quadratic inequalities can be extended to solve inequalities of higher degrees and rational inequalities.

EXAMPLE 6   Inequality of Degree 3

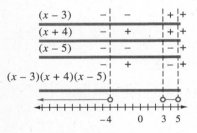

FIGURE 14.18

Determine the solution set of the inequality $(x - 3)(x + 4)(x - 5) < 0$.

Solution   To solve the inequality $(x - 3)(x + 4)(x - 5) < 0$, first determine the critical values. Setting the factors equal to zero, we have $x - 3 = 0$, $x + 4 = 0$, and $x - 5 = 0$. Solving these equalities for x, we find the critical values 3, -4, and 5. Now we can set up the graph as we did for solving quadratic inequalities (see Figure 14.18). Note that, with three critical values, the graph is divided into four regions: values of x less than -4, values of x between -4 and 3, values of x between 3 and 5, and values of x greater than 5. Examine each factor in each of these regions and determine whether the factor is positive or negative in each region. For the factor $x - 3$ we see that it is $-$, $-$, $+$, $+$ as we move from left to right. Having found the value ($+$ or $-$) for each factor in an interval, we then find the respective products. For $x < -4$, $(-1)(-1)(-1) = -1$. The signs indicate that the product is less than zero for $x < -4$ or $3 < x < 5$. This solution is indicated on the graph. In interval notation the solution for $(x - 3)(x + 4)(x - 5) < 0$ is stated $(-\infty, -4)$ or $(3, 5)$. ◆

Check the solution for Example 6 with your graphics calculator. Can you predict the graph that will appear on the screen?

In solving an inequality, we must determine whether the factors are positive or negative. In reexamining the graph in Figure 14.18, we see that an odd number of negative factors produces a negative product or a product that is less than zero. In cases where there are even numbers of negative factors, the products are positive or greater than zero.

EXAMPLE 7 Inequality of Degree 3

Determine the solution set of the inequality $x^3 + x^2 - 12x \geq 0$.

Solution To determine the solution set of $x^3 + x^2 - 12x \geq 0$, write the polynomial as a product of its factors:

$$x^3 + x^2 - 12x \geq 0$$
$$x(x^2 + x - 12) \geq 0 \qquad \text{Common factor}$$
$$x(x + 4)(x - 3) \geq 0. \qquad \text{Factoring}$$

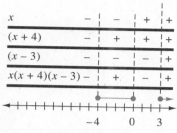

FIGURE 14.19

In this form we can set each of the factors equal to zero and determine the critical values. The critical values are 0, −4, and 3. Place these values on the graph as in Figure 14.19. Filling in the graph, we can see that the product of the quantities is positive for $-4 \leq x \leq 0$ or for $x \geq 3$. Thus, the solution set for the inequality $x^3 + x^2 - 12x \geq 0$ is $[-4, 0]$ or $[3, \infty)$. ◆

EXAMPLE 8 Rational Inequality

Determine the solution set of the inequality $\dfrac{(x + 2)(x - 3)}{x + 5} < 2x - 7$.

Solution To solve the inequality, use the method of collecting all the terms on one side of the inequality sign.

$$\frac{(x + 2)(x - 3)}{x + 5} < 2x - 7$$

$$\frac{(x + 2)(x - 3)}{x + 5} - (2x - 7) < 0$$

$$\frac{x^2 - x - 6}{x + 5} - \frac{2x^2 + 3x - 35}{x + 5} < 0 \qquad \text{Common denominator}$$

$$\frac{-x^2 - 4x + 29}{x + 5} < 0 \qquad \text{Collecting like terms}$$

$$\frac{x^2 + 4x - 29}{x + 5} > 0. \qquad \begin{array}{l}\text{Multiplying both sides}\\ \text{by } -1\end{array}$$

Since $x^2 + 4x - 29$ is not factorable, find the roots using the quadratic formula. Thus, the factors to the nearest hundredth are:

$$\frac{(x - 3.74)(x + 7.74)}{x + 5} > 0.$$

The critical values for the inequality are -7.44, -5, and 3.74. The denominator equals zero when $x = -5$; thus, -5 is a critical value. Figure 14.20

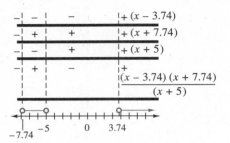

FIGURE 14.20

contains the graph of the critical points and the intervals where the factors are positive or negative. Reading the results from the figure, we see that the solution set for $\dfrac{(x + 2)(x - 3)}{x + 5} < 2x - 7$ is $(-7,74, -5)$ or $(3.74, \infty)$. ◆

In Example 8, if the $<$ symbol is replaced by the $\leq$ symbol, the problem becomes:

$$\frac{(x + 2)(x - 3)}{x + 5} \leq 2x - 7.$$

How does the replacement change the solution? It means that the solution includes -7.74 and 3.74 but does not include -5. Thus, the solution is $[-7.74, -5)$ or $[3.74, \infty)$. Why is -5 not part of the solution?

In business, the terms *profit*, *revenue*, and *cost* are very important. These terms can be expressed in a formula. **Profit** (or net income) is the amount of money remaining after all costs are deducted from the income. **Revenue** is the amount of income generated from selling a product or service, **Cost** is the amount of money spent in producing or manufacturing a commodity. Therefore,

$$\text{Profit} = \text{Revenue} - \text{Cost}$$
$$P = R - C.$$

Example 9 illustrates how to determine the number of items that must be sold at various prices to insure a desired profit.

EXAMPLE 9   Application: Inequality in Business

The sales department of a roller blade manufacturing company determines that the demand for a new model of roller blades is given by:

$$p = \$200 - 0.0005x \qquad 0 \leq x < 400,000. \qquad \textbf{Demand equation}$$

The price p of a pair of roller blades is given in dollars. The inequality $0 \leq x \leq 400,000$ tells them that no one will pay $200 for a pair of roller blades and that the company must make a profit to stay in business (since $400,000 \times 0.0005 = \$200$). The revenue is the product of the number units sold times the demand or:

$$R = xp = x(\$200 - 0.0005x). \quad \text{Revenue equation}$$
$$= \$200x - 0.0005x^2$$

The cost of making x pairs of roller blades is $25 per pair plus an initial cost of $800,000. The total cost is given by:

$$C = 25x + \$800,000 \quad \text{Cost equation}$$

What price per pair should the company charge to have a profit of at least $3,262,500?

Solution We first must determine the profit. Use the formula Profit = Revenue − Cost.

$$P = R - C$$
$$P = (\$200x - 0.0005x^2) - (\$25x + \$800,000)$$
$$= -0.0005x^2 + 175x - 800,000$$

To determine the number of pairs of roller blades for the desired profit, we must solve the inequality:

$$-0.0005x^2 + 175x - 800,000 \geq 3,262,500$$
$$-0.0005x^2 + 175x - 4,062,500 \geq 0 \quad \text{Standard form}$$

Now, using the quadratic formula, we can determine that the critical values are 25,000 and 325,000. A quick check shows that the number they must produce is in the closed interval:

$$25,000 \leq x \leq 325,000.$$

Using the demand equation, we can determine the corresponding range of prices that must be charged for each pair of roller blades to insure the desired profit. For $x = 25,000$:

$$p = \$200 - 0.0005(25,000) = \$187.50.$$

For $x = 325,000$:

$$p = \$200 - 0.0005(325,000) = \$37.50.$$

Thus, the corresponding range of prices per pair is represented by the closed interval $\$37.50 \leq p \leq \$187.50$. ◆

14.2 Exercises

In Exercises 1–20, perform the following.
a.   Determine the critical values for each inequality.
b.   Use the graphical method to determine the solution.
c.   Write the solution set in interval notation.
d.   Check the solution with the graphics calculator.

1. $(x + 2)(x - 3) > 0$ **2.** $(x - 2)(x + 3) \geq 0$ **3.** $(x + 3)(x - 5) \leq 0$ **4.** $(x - 3)(x + 5) < 0$

5. $2x^2 - 6x \geq 0$ **6.** $3x^2 - 15x \leq 0$ **7.** $3x^2 + 15x > 0$ **8.** $2x^2 + 6x > 0$

9. $t^2 - 4t - 5 < 0$ **10.** $t^2 - 7t - 8 < 0$ **11.** $2Z^2 + 5Z + 2 \geq 0$ **12.** $\frac{1}{2}Z^2 - 8Z + \frac{15}{2} > 0$

13. $2Z^2 + 7Z + 2 \leq 0$ **14.** $3Z^2 + 5Z + 1 < 0$ **15.** $x^2 + 3x + 4 < 0$ **16.** $3x^2 + 3x + 1 \geq 0$

17. $R^2 + R + 3 > 0$ **18.** $2R^2 + R + 1 \leq 0$ **19.** $3x^2 - 8x + 2 < 0$ **20.** $3x^2 + 5x - 3 > 0$

21.   A manufacturer of wood-burning stoves finds that when x units, $x \geq 10$, are made and sold per month, its profit (in thousands of dollars) is given by $P(x) = x^2 - 22x + 105$.
   a.   What is the least number of units that the company can manufacture each month and make a profit?
   b.   For what values of x is the company losing money?

22.   Repeat Exercise 21 with $P(x) = x^2 - 40x + 375$ ($x \geq 18$).

23.   If a slingshot shoots a stone straight up from the ground with an initial velocity of 160 m/s, its distance d in meters above the ground at the end of t seconds (neglecting air resistance) is given by $d = 160t - 16t^2$. Determine the duration of time for which $d \geq 256$.

24.   A projectile is fired from ground level. After t seconds, it is $h(t) = 320t - 16t^2$ meters above the ground. During what period is the projectile higher than 1400 m?

25.   A designer is laying out a page in a textbook with a page size of 170 mm × 190 mm. She wants a uniform margin around the printed material. At most how wide can the margin be if the printed material must cover at least 224 cm²?

26.   A carpenter is designing a box with a lid. The box will have a square base and a height of 200 mm. If the total surface area must be less than 3072 cm², what is the maximum length of the base?

27.   Pat is purchasing a rug for a living room that is 4 m × 5 m. She wants a uniform border around the rug. At most how wide can the border be if the rug must cover 12 m²?

28.   The deflection of a beam is given by $d = x^2 - 0.5x + 0.06$, where x is the distance from one end. For what values of x is $d > 0.02$?

29.   The formula for the total surface area of a right circular cylinder is $S = 2\pi r^2 + 2\pi rh$. If the height h of a certain cylinder is 200 mm, what values of the radius will yield a surface area less than 250π cm²?

30.   A carpenter is asked to construct two picture frames. One frame will have a rectangular shape with a length 40 mm more than its width. The second frame will have a square shape, each of whose sides will be twice as large as the width of the rectangle. If the total area enclosed by the outer edges of the two frames is at least 2508 cm², determine the least amount of picture frame molding needed for the job. (Do not allow for cutting or waste.)

In Exercises 31–50, determine the solution set for the inequality. Check the solution with the graphics calculator.

31. $(x + 2)(x - 1)(x - 4) < 0$

32. $(x + 3)(x + 4)(x - 5) > 0$

33. $(x - 3)(x - 4)(x + 7) > 0$

34. $(x + 5)(x - 1)(x + 6) < 0$

35. $(x + 5)^2(x - 4)^2 \geq 0$

36. $(x - 9)^2(x + 3)^2 \leq 0$

37. $4t^3 + 17t^2 - 15t > 0$

38. $3t^3 + 20t^2 - 7t < 0$

39. $2R^4 - 11R^3 + 30R^2 > 0$

40. $2t^4 - 13t^3 - 7t^2 > 0$

41. $\dfrac{3}{z} < 1$

42. $1 - \dfrac{5}{Z} \geq \dfrac{1}{2}$

43. $\dfrac{1}{z + 1} < 1$

44. $\dfrac{z}{z + 2} < 1$

45. $\dfrac{y + 2}{y + 3} < 1$

46. $\dfrac{y + 2}{y + 7} \geq 1$

47. $\dfrac{4x}{x^2 + 4} \leq 1$

48. $\dfrac{x^2 - 25}{x^2 - 16} < 0$

49. $\dfrac{x}{x + 3} \leq \dfrac{2x}{x - 2}$

50. $\dfrac{x + 3}{x - 5} \geq \dfrac{2x}{x - 3}$

51. For what values of x will $\sqrt{x^2 - 8x + 15}$ be a real number?

52. For what values of x will $\sqrt{\dfrac{x - 4}{x + 3}}$ be a real number?

53. A garden plot must have an area less than 50 ft^2. The width of the plot must be between 2 ft and 5 ft. What will be the acceptable length of the plot?

54. A manufacturer of stockings determines that the demand for light-colored stockings is given by:

$$p = \$10 - 0.00008x, \qquad 0 \leq x \leq 125,000.$$

The cost of making x pairs of stockings is $1 per pair plus an initial cost of $100,000. What price per pair should the company charge to have a profit of at least $100,000? (See Example 9.)

14.3

Equalities and Inequalities Involving Absolute Values

The concept of absolute value was introduced in Section 1.2. This section more formally defines the concept and discusses techniques for solving equations and inequalities containing absolute values. Let us begin with two definitions.

> **DEFINITION 14.3**
>
> The absolute value of the number n, written $|n|$, is the distance that the point representing n is from the point representing zero on a number line (Figure 14.21).
>
>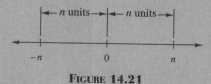
>
> **FIGURE 14.21**

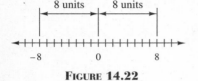

FIGURE 14.22

> **DEFINITION 14.4**
>
> For every number n, $|n| = |-n|$.

Definition 14.4 is true since n and its additive inverse $-n$ are the same distance from the origin. The absolute value measures distance from the origin, not taking direction into account. For example, $|8| = 8$ since 8 is eight units from the origin; $|-8| = 8$ since -8 is also eight units from the origin (Figure 14.22).

Having briefly considered the geometric approach to absolute value, let us revisit the definition from Section 1.2.

> **DEFINITION 1.5**
>
> $$|x| = \begin{cases} x, & \text{provided } x > 0. \\ 0, & \text{provided } x = 0. \\ -x, & \text{provided } x < 0. \end{cases}$$

When we apply the definition to solving the equation $|x - 4| = 7$, we have two cases as shown in Table 14.3.

TABLE 14.3

Case 1: $x - 4 > 0$	Case 2: $x - 4 < 0$
Applying the definition	Applying the definition
$|x - 4| = x - 4,$	$|x - 4| = -(x - 4),$
then	then
$x - 4 = 7$	$-(x - 4) = 7$
$x = 11.$	$-x + 4 = 7$
	$x = -3.$

We can check these results by substituting into the original statement

$$x = 11: \quad |11 - 4| = 7 \qquad x = -3: \quad |-3 - 4| = 7$$
$$|7| = 7 \qquad\qquad\qquad |-7| = 7$$
$$7 = 7 \ \checkmark \qquad\qquad\qquad 7 = 7 \ \checkmark$$

◆

EXAMPLE 1   **Solution of an Equality Containing an Absolute Value**

Determine the solution of $|3x - 7| = 14$.

Solution   Using the definition of absolute value, we must consider the two cases: $(3x - 7) > 0$ and $(3x - 7) < 0$. Using these two cases we can write and solve the two equations.

$$3x - 7 = 14 \qquad \text{and} \qquad -(3x - 7) = 14$$
$$3x = 21 \qquad\qquad\qquad -3x + 7 = 14$$
$$x = 7 \qquad\qquad\qquad\qquad -3x = 7$$
$$x = -\frac{7}{3}$$

Thus, the solution of $|3x - 7| = 14$ is $x = 7$ and $x = -\frac{7}{3}$. Check the solution.

◆

We also can determine the solution for Example 1 graphically with a calculator. First write the equality in the form $y = |3x - 7| - 14$. Now use the following keystrokes to graph the equality.

fx–7700G: GRAPH SHIFT GRAPH NUM (F 3) Abs (F 1) (3
x,Θ,T − 7) − 14 EXE

TI–82: Y= 2nd ABS (3 x,T,Θ − 7) − 14 GRAPH

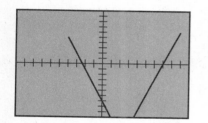

FIGURE 14.23

The resulting graph (see Figure. 14.23) shows that the curve crosses the x-axis in two places. (If you cannot see this, use the zoom function.) The curve crosses the x-axis at approximately $-\frac{7}{3}$ and 7. We obtained the same result in Example 1.

For certain problems, we may find it more convenient to write the absolute value of a product or quotient as the product or quotient of the absolute values. Definitions 14.5 and 14.6 provide us with the information to do this.

DEFINITION 14.5

For any real numbers a and b,

$$|ab| = |a||b|$$

$$\text{and} \quad \left|\frac{a}{b}\right| = \frac{|a|}{|b|}, \quad \text{if } b \neq 0,$$

EXAMPLE 2   **Illustrating Definition 14.5**

a.   Let $a = 4$ and $b = -5$.

$$|ab| = |(4)(-5)| = |-20| = 20$$
$$|a||b| = |4||-5| = 4 \cdot 5 = 20$$

b. Let $a = -18$ and $b = 5$.

$$\left|\frac{a}{b}\right| = \left|\frac{-18}{5}\right| = \frac{18}{5}.$$

$$\frac{|a|}{|b|} = \frac{|-18|}{|5|} = \frac{18}{5}. \quad \blacklozenge$$

Definition 14.6 provides us with a tool for solving an equality of the form $|12x + 3| = |4x|$.

DEFINITION 14.6

If $|a| = |b|$, then $a = \pm b$.

EXAMPLE 3   Absolute Value of a Quotient

Determine the solution of $\left|\dfrac{3x + 5}{5x}\right| = 3$.

Solution   Use Definitions 14.5 and 14.6 to solve for x.

$$\left|\frac{3x + 5}{5x}\right| = 3$$

$$\frac{|3x + 5|}{|5x|} = 3 \qquad \textbf{By Definition 14.5}$$

$$|3x + 5| = 3|5x| \qquad \textbf{Multiplying both sides by } |5x|$$

$$3x + 5 = 3(\pm 5x) \qquad \textbf{Definition 14.6}$$

$$3x + 5 = \pm 15x$$

Now solve for x.

$$3x + 5 = 15x \qquad\qquad 3x + 5 = -15x$$
$$5 = 12x \qquad\qquad\quad 5 = -18x$$
$$\frac{5}{12} = x. \qquad\qquad\quad \frac{-5}{18} = x.$$

The roots of the equation $\left|\dfrac{3x + 5}{5x}\right| = 3$ are $\dfrac{5}{12}$ and $-\dfrac{5}{18}$. $\blacklozenge$

We can obtain a visual check for Example 3 by graphing the function $y = \left|\dfrac{3x + 5}{5x}\right| - 3$. Since the values of x are small, use the following range: $[-2, 2, 0.5, -10, 10, 1]$. What happens when $x = 0$? Is the y-axis an asymptote?

Absolute Values and Inequalities

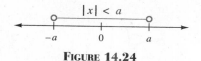

FIGURE 14.24

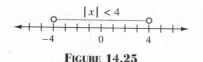

FIGURE 14.25

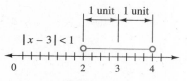

FIGURE 14.26

To solve the inequality $|x| < a$ means geometrically that we are looking for all the points on the number line that are less than a units from the origin. Thus, on a number line, the solution is as indicated in Figure 14.24.

For example, for the statement $|x| < 4$, we are looking for all the values of x that are within four units of the origin (Figure 14.25). Any value of x in the open interval $(-4, 4)$ we select satisfies the inequality. For example, if $x = -3$, $|-3| = 3$, which is less than 4.

DEFINITION 14.7

For any positive real number a, $|x| < a$ is equivalent to $-a < x < a$.

EXAMPLE 4   Solving $|x| < a$

Determine the solution of $|x| < 7$.

Solution $|x| < 7$ means that $-7 < x < 7$. The solution is the open interval $(-7, 7)$. ◆

Now let us consider the statement $|x - 3| < 1$. From the definition of absolute value $|x - 3|$ is the distance between x and 3. Therefore, $|x - 3| < 1$ means that x lies within one unit of 3 (Figure 14.26). Thus, the numbers 2.1, 2.7, 2.9, 3.1, 3.4, and 3.9, are all within one unit of 3 and are solutions to the problem. To determine all the numbers that satisfy $|x - 3| < 1$, apply the definition of absolute value.

$$x - 3 < 0 \qquad \text{and} \qquad x - 3 > 0$$
$$|x - 3| = -(x - 3) \qquad\qquad |x - 3| = x - 3$$

then   then

$$-(x - 3) < 1 \qquad\qquad\qquad x - 3 < 1$$

multiplying both sides by -1

$$x - 3 > -1$$

or

$$-1 < x - 3 \quad \text{and} \quad x - 3 < 1$$

Since $x - 3$ is between -1 and 1, we can write a single statement $-1 < x - 3 < 1$. Thus, $|x - 3| < 1$ is equivalent to $-1 < x - 3 < 1$. Now solve for x by adding 3 to all three terms.

$$-1 + 3 < x - 3 + 3 < 1 + 3 \quad \text{or} \quad 2 < x < 4.$$

> **DEFINITION 14.8**
>
> For real numbers a and b with $b > 0$, $|x - a| < b$ is equivalent to $-b < x - a < b$.

EXAMPLE 5 Solving $|kx - a| < b$

Determine the solution set of $|2x - 7| < 11$ and graph the solution set.

Solution Using Definition 14.8, we can write $|2x - 7| < 11$ as the double inequality:

$$-\boxed{11} < 2x - 7 < \boxed{11}$$
$$-11 \boxed{+\,7} < 2x - 7 \boxed{+\,7} < 11 \boxed{+\,7} \qquad \text{Adding 7 to all three terms}$$
$$-4 < 2x < 18$$
$$\frac{-4}{\boxed{2}} < \frac{2x}{\boxed{2}} < \frac{18}{\boxed{2}} \qquad \text{Dividing each term by 2}$$
$$-2 < x < 9.$$

The graph of the solution set is given in Figure 14.27.
The solution set of $|2x - 7| < 11$ is the open interval $(-2, 9)$. ◆

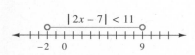

FIGURE 14.27

To make a visual check for Example 5 using the fx–7700G, graph $y < -|2x - 7| + 11$. The solution is the open interval between the points, where the graph crosses the x-axis. If we use the TI–82, we can determine the interval by graphing $y_1 = |2x - 7|$, $y_2 = 11$, and $y_3 = y_1 < y_2$.

To solve the inequality $|x| > a$ means geometrically that we are looking for all the points on the number line that are more than a units from the origin. Thus, on a number line, the solution is as shown in Figure 14.28.

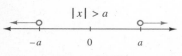

FIGURE 14.28

For example, for the statement $|x| > 5$, we are looking for all the values of x that are more than five units from the origin (Figure 14.29.) If we select any value of x in the open interval $(-\infty, -5)$ or in the open interval $(5, \infty)$, the value of x satisfies the inequality.

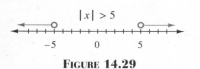

FIGURE 14.29

> **DEFINITION 14.9**
>
> For any positive real number a, $|x| > a$ is equivalent to $x > a$ or $x < -a$.

EXAMPLE 6 Solving $|x| > a$

Determine the solution set of $|x| > 6$.

Solution $|x| > 6$ means $x > 6$ or $x < -6$. The solution is the open intervals $(-\infty, -6)$ or $(6, \infty)$. ◆

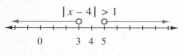

FIGURE 14.30

Now let us consider the statement $|x - 4| > 1$. From the definition of absolute value $|x - 4|$ is the distance between x and 4. Therefore, $|x - 4| > 1$ means that x is more than one unit from 4 (Figure 14.30). Thus, the numbers

$-8.2, -6, -4.1, -3.2, 5.3, 6.7, 11$ are all more than one unit from 4 and are solutions to the problem. To determine all the numbers that satisfy $|x - 4| > 1$, apply the definition of absolute value.

$$x - 4 < 0 \qquad \text{and} \qquad x - 4 > 0$$
$$|x - 4| = -(x - 4) \quad \text{and} \quad |x - 4| = x - 4$$

then

$$-(x - 4) > 1$$
$$x - 4 < -1 \qquad \text{or} \qquad x - 4 > 1$$

Since $x - 4$ is less than -1 or $x - 4$ is greater than 1, we cannot write the solution as a single statement. Thus, $|x - 4| > 1$ is equivalent to $x - 4 < -1$ or $x - 4 > 1$. Now solving for x by adding 4 to both sides of both inequalities, we have:

$$x - 4 + 4 < -1 + 4 \quad \text{or} \quad x - 4 + 4 > 1 + 4 \qquad \textbf{Adding 4 to}$$
$$x < 3 \qquad\qquad\qquad x > 5. \qquad \textbf{both sides} \qquad \blacklozenge$$

CAUTION By the definition of absolute value, $|x - 4| > 1$ is written as $x - 4 < -1$ or $x - 4 > 1$. Often students write $1 < x - 4 < -1$. This is an incorrect statement; 1 cannot be less than -1. So remember, when rewriting a statement of absolute value associated with a "greater than" sign, always express the result as two inequalities and **never** express it as a single inequality. ■

> **DEFINITION 14.10**
> For real numbers a and b with $b > 0$, $|x - a| > b$ is equivalent to $x - a < -b$ or $x - a > b$.

EXAMPLE 7   **Solving $|kx - a| > b$**

Determine the solution set of $|3x + 7| > 5$ and graph the solution set.

Solution   Using Definition 14.10, we can write $|3x + 7| > 5$ equivalently as follows.

$$3x + 7 < -5 \qquad \text{or} \qquad 3x + 7 > 5$$
$$3x + 7 - 7 < -5 - 7 \qquad 3x + 7 - 7 > 5 - 7 \qquad \textbf{Subtracting 7 from both sides}$$
$$3x < -12 \qquad\qquad\qquad 3x > -2$$
$$x < -4 \qquad\qquad\qquad x > -\frac{2}{3} \qquad \textbf{Dividing both sides by 3}$$

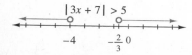

FIGURE 14.31

The solution set of $|3x + 7| > 5$ is $(-\infty, -4)$ or $\left(-\frac{2}{3}, \infty\right)$. Its graph is given in Figure 14.31. ◆

EXAMPLE 8   **Solving** $|a - kx| \geq b$

Determine the solution set of $|4 - 7x| \geq -3$.

Solution   The absolute value of a number is always non-negative. Therefore, $|4 - 7x| \geq -3$ is always true. The solution set is the set of all real numbers, or in interval notation, $(-\infty, \infty)$. ◆

Example 9 illustrates how to translate verbal statements that indicate an absolute value and/or an inequality relation into symbolic form.

EXAMPLE 9   **Rewriting a Statement Using the Absolute Value and Inequality Symbols**

Write the following statements using the absolute value symbol.

a. R is less than nine units from 25.
b. Z differs from 7 by at least 5.
c. y is within four units of -7.
d. $x < 5$ and $x > -5$

Solution

a. R is less than nine units from 25 can be written;

$$|R - 25| < 9.$$

b. Z differs from 7 by at least 5 can be written:

$$|Z - 7| \geq 5.$$

c. y is within four units of -7 can be written:

$$|y - (-7)| \leq 4$$
$$|y + 7| \leq 4.$$

d. $x < 5$ and $x > -5$ can be written:

$$-5 < x < 5 \quad \text{or} \quad |x| < 5. \quad ◆$$

Knowing how to translate statements into symbolic forms is helpful in solving applied problems as illustrated in Examples 10 and 11.

EXAMPLE 10 **Application: Diameter of a Shaft**

In manufacturing parts for machines, the parts must be produced to meet specific standards of precision. For example, in manufacturing shafts and bearings, the standards ensure that the bearings will fit properly on the shafts. Assume that a shaft must have a diameter of 12.57 mm with a tolerance of 0.02 mm (or ±0.02 mm as it is sometimes written). Determine the interval for the actual diameter of the shaft.

Solution If we let x represent the actual diameter of the shaft, we can express the required dimensions in the following form:

$$|x - 12.57| \leq 0.02.$$

To find the acceptable limits on the diameter of the shaft, solve for x.

$$-0.02 \leq x - 12.57 \leq 0.02 \quad \text{Definition 14.8}$$
$$-0.02 + 12.57 \leq x - 12.57 + 12.57 \leq 0.02 + 12.57$$
$$12.55 \leq x \leq 12.59.$$

The resulting inequality states that the actual diameter of the shaft must be between 12.55 mm and 12.59 mm. ◆

EXAMPLE 11 **Application: Cost for Variation in Measurement**

Assume that a 12 oz can of soda costs $0.75. Furthermore, assume that the machine that fills the cans is set to be accurate within $\frac{1}{4}$ oz. How much might the customer be overcharged and the company underpaid due to an error in the filling machine?

Solution Begin by letting x equal the exact number of ounces of soda in a can. Since the bottling company claims that the machine is accurate within $\frac{1}{4}$ oz, we can say that the difference between the exact number of ounces (x) and the number of ounces dispensed by the machine (12 oz) is less than or equal to $\frac{1}{4}$ oz with the equation:

$$|x - 12| \leq \frac{1}{4}.$$

To determine the variation, we must solve the following inequality.

$$-\frac{1}{4} \le x - 12 \le \frac{1}{4} \qquad \text{Definition 14.8}$$

$$12 - \frac{1}{4} \le x \le 12 + \frac{1}{4}$$

$$11\frac{3}{4} \le x \le 12\frac{1}{4}$$

This statement tells us that there could be as little as $11\frac{3}{4}$ oz or as much as $12\frac{1}{4}$ oz in a can. With a cost of $0.75 for 12 oz, the cost of 1 oz is $0.0625. Thus, for $11\frac{3}{4}$ oz, the cost is $0.73 and for $12\frac{1}{4}$ oz, the cost is $0.77. If a container contains less than 12 oz of soda, the customer may have been overcharged by as much as $0.02 per can. On the other hand, if the can contains more than 12 oz, the company may have lost as much as $0.02 per can. ◆

14.3 Exercises

In Exercises 1–15, determine the solution set of each statement of equality and check it.

1. $|x + 3| = 5$

2. $|x - 5| = 7$

3. $|Z - 2| = \frac{1}{3}$

4. $|3Z + 11| = \frac{1}{4}$

5. $|5R + 3| = 1$

6. $|-4R + 2| = 7$

7. $|0.03t - 0.05| = 0.17$

8. $|0.02t - 0.01| = 0.15$

9. $|3t - 4| - 5 = 6$

10. $\left|\frac{1}{2}y + \frac{1}{4}\right| + \frac{1}{3} = \frac{2}{3}$

11. $\left|\frac{2}{3}y - \frac{1}{5}\right| - \frac{1}{7} = 0$

12. $\left|-\frac{1}{4}y - \frac{1}{6}\right| - \frac{1}{8} = 0$

13. $\left|\frac{2x + 5}{3x}\right| = 7$

14. $\left|\frac{2x - 4}{5x}\right| = -2$

15. $\left|\frac{5x - 6}{3x}\right| = 5$

In Exercises 16–42, determine the solution set and graph the inequality.

16. $|x| < 3$

17. $|x| \le 8$

18. $|x| \ge 11$

19. $|x| > 13$

20. $|t| > -5$

21. $|t| > 0$

22. $|t + 2| < 5$

23. $|t - 4| \le 7$

24. $|R + 5| \ge 3$

25. $|R - 7| \ge 4$

26. $|2R - 7| \le 1$

27. $|3Z + 5| < 2$

28. $|5Z + 6| > 3$

29. $|3Z - 4| \ge 5$

30. $|1 - 3Z| \le 5$

31. $|2 - 5x| < -7$

32. $|3 - 4x| \ge 2$

33. $|5 - 7x| > 4$

34. $|1 + 11x| > 0$

35. $\left|\frac{x - 3}{2}\right| < 2$

36. $\left|\frac{2x + 1}{3}\right| \le 4$

37. $\frac{|7x + 3|}{2} < 0$

38. $\frac{|2x - 14|}{4} < 2$

39. $\left|\frac{1}{3} + x\right| < \frac{1}{4}$

40. $\left|\frac{2x + 1}{3}\right| < 0.03$

41. $\left|\frac{1}{5} + x\right| < 0.05$

42. $\left|\frac{x - 4}{2}\right| > 0.02$

In Exercises 43–48, use absolute value notation to define each interval, or intervals, on the real number line.

43.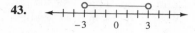

44.

45.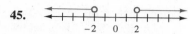

46.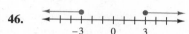

47.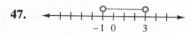

48.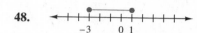

49. A bolt must have a diameter of 13.45 mm with a tolerance of ±0.01 mm. Determine the acceptable range of diameters of the bolt.

50. A machine that packages 250 aspirin tablets per container can make an error of three pills per container. If x is the number of tablets in a bottle, write an inequality using absolute value that indicates a maximum error of three tablets per container. Solve the inequality to determine the possible number of tablets in the container.

51. A plastic disk must have a thickness of 4 mm with a tolerance of ±0.03 mm. Determine the acceptable range of thickness of the plastic disk.

52. Ten boxes of table salt have a combined "weight" of 7370 g with the measurement accurate to ±2 g. Determine the range of probable "weights" of a single box of salt.

In Exercise 53–56, perform the following.
a.   Write an inequality with absolute values that describes the statement.
b.   Determine the solution set of the statement.

53. y is less than nine units from $\dfrac{4}{3}$.

54. $5x$ is more than 7/3 units from $\dfrac{11}{2}$.

55. R is within five units of -13.

56. $Z \le 14$ and $Z \ge -14$.

57. If the length of a rectangle is 80 in. and the perimeter is between 240 in. and 280 in., what is the range of values for its width?

58. Heat loss from a building depends upon the exterior surface area of the structure. Assuming that the structure is a cube, the heat loss is given by the equation:

$$HL = 5s^2,$$

where s is the length of the side of the cube. In order for the temperature of the structure to remain the same, the heat loss must be balanced by the fuel consumption. If the fuel consumption for a furnace lies between 500.0 and 750.0 units, what is the range for the length of the structure?

59. Assume that the thermostat in your house is set at 68°F and that, with that setting, the cost of heating your house is $3.06 per day on a 30°F day. If the thermostat is accurate to $\dfrac{1}{4}$ degree, how much more or less might you pay for heating the house for a 30 day period of 30°F temperatures?

60. The fish you purchase is marked at $4.75 per pound, and the weight marked on the package is 7.45 lb. If the scale that weighted the fish is accurate to $\dfrac{1}{4}$ ounce, how much might you have been overcharged or the store underpaid?

14.4

Systems of Inequalities

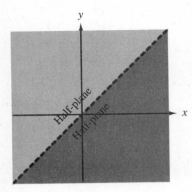

FIGURE 14.32

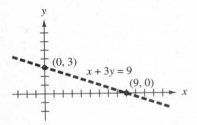

FIGURE 14.33

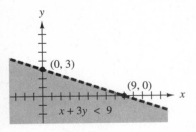

FIGURE 14.34

In Section 14.2 we learned how to solve inequalities in one variable. This section discusses inequalities in two variables. Examples of inequalities in two variables are $2x + 3y < 6$, $x - y \leq 3$, and $x^2 + 7y^2 > 4$.

We were able to indicate the solution set of an inequality in one variable on a number line. However, the solution set of an inequality in two variables requires us to use the coordinate plane.

A linear inequality that is strictly $<$ or $>$ has as its solution a half-plane. A **half-plane** is represented by all the points on one side of a line (Figure 14.32). An inequality that is $\leq$ or $\geq$ has as its solution set the set of points that consists of a half-plane and a line. To indicate that the line is part of the solution set, draw a solid line. To indicate that the line is not part of the solution set, draw a dashed line. (This convention is used in the text figures.)

EXAMPLE 1   Graph of a Linear Inequality

Graph the solution set of $x + 3y < 9$.

Solution   To obtain the solution set, graph $x + 3y = 9$. Since the original statement is strictly "less than," draw a dashed line (Figure 14.33). (Remember that the dashed line indicates that the line is not a part of the solution set.)

The line $x + 3y = 9$ divides the plane into two half-planes. The points in one half-plane satisfy the inequality $x + 3y < 9$. The points in the other half-plane satisfy the inequality $x + 3y > 9$.

To determine the solution set of the inequality $x + 3y < 9$, pick any point on the plane that is not on the line. The simplest point to work with is the origin, $(0, 0)$. Substituting $x = 0$ and $y = 0$ into $x + 3y < 9$ gives $0 + 3(0) < 9$, or $0 < 9$. Since 0 is less than 9, the point $(0, 0)$ is part of the solution set. All the points on the same side of the line $x + 3y = 9$ as the point $(0, 0)$ are members of the solution set. We indicate this by shading the half-plane that contains $(0, 0)$ as illustrated in Figure 14.34.   ◆

In Chapter 10 we learned how to find the solution of two linear equations with two variables. Here we will explore the techniques for finding the solutions of two linear inequalities with two variables.

EXAMPLE 2   Solution of a System of Linear Inequalities

Graph the solution set of the following system of linear inequalities.

$$x + y < 3$$
$$x - y < 7$$

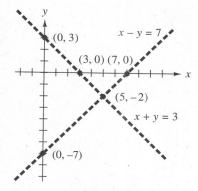

FIGURE 14.35

Solution The solution set is the set of all points that make both statements true. Graph both statements on the same axes. First, draw a graph of the lines $x + y = 3$ and $x - y = 7$. Both lines are dashed since each boundary is not part of the solution set. As we see in Figure 14.35, the two lines intersect at the point $(5, -2)$.

Now determine the set of points that make $x + y < 3$ a true statement of inequality. Substitute the ordered pair $(0, 0)$ for x and y. The result is $0 + 0 < 3$, a true statement of inequality. Thus, the point $(0, 0)$ and all the points in the same half-plane as $(0, 0)$ are members of the solution set. The solution set is shaded in Figure 14.36. Next find the half-plane determined by the inequality $x - y < 7$. Substituting $(0, 0)$ for x and y in the inequality results in $0 - 0 < 7$, a true statement of inequality. The point $(0, 0)$ and all the points in the same half-plane as $(0, 0)$ are members of the solution set. The solution set is shaded in Figure 14.37. The solution set of the system of linear inequalities consists of all the points common to the two half-planes shaded in Figures 14.36 and 14.37. The solution set of the system is the more darkly shaded region in Figure 14.38. ◆

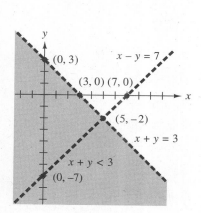

FIGURE 14.36

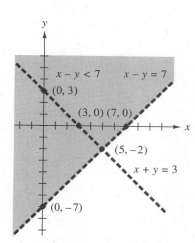

FIGURE 14.37

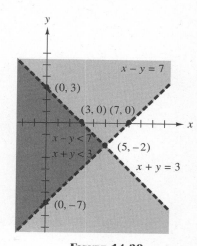

FIGURE 14.38

Systems of inequalities need not be restricted to systems of linear inequalities. They may also include absolute values and nonlinear systems, which are illustrated in Examples 3 and 4.

EXAMPLE 3 **Graph of an Absolute Value Inequality**

Graph the solution set of $|y + 2x| \geq 14$

Solution To graph the solution set of $|y + 2x| \geq 14$, first write the inequality without the absolute value sign. Do this by using Definition 14.10.

$$|y + 2x| \geq 14$$
$$y + 2x \geq 14 \quad \text{or} \quad y + 2x \leq -14.$$

The following two statements of equality are graphed in Figure 14.39.

$$y + 2x = 14 \quad \text{and} \quad y + 2x = -14$$

Solid lines are drawn since the inequality symbols $\leq$ and $\geq$ tell us that the line is part of the solution set. The two equations have the same slope, therefore the lines are parallel. Testing the point $(0, 0)$ we see that this point does not satisfy $y + 2x \geq 14$ or $y + 2x \leq -14$. Therefore, the solution set is the set of all the points outside the lines and on the lines. The solution set is indicated by shading and the solid lines in Figure 14.40.

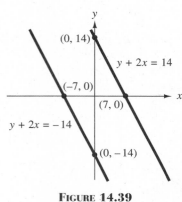

FIGURE 14.39

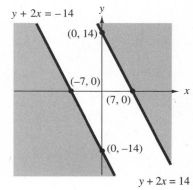

FIGURE 14.40

EXAMPLE 4　Solution of a System of Inequalities

Graph the solution set of the following system of inequalities.

$$y > x^2 - 9$$
$$y - x \leq 3$$

Solution　To find the solution of the system, we must graph the equality $y = x^2 - 9$. Using the techniques we learned in Section 11.6, we quickly can determine the following key facts: the parabola opens up, $(-3, 0)$ and $(3, 0)$ are x-intercepts, $(0, -9)$ is the y-intercept, the y-axis is the line of symmetry, and $(0, -9)$ is the vertex of the parabola. The table with Figure 14.41 contains sets of ordered pairs that were used to sketch the graph.

Figure 14.41 also shows the straight line representing the equation $y - x = 3$. We see that the curves intersect at $(-3, 0)$ and $(4, 7)$. The points of intersection can be determined either from the graph or by algebraic methods.

The solution set for the inequality $y > x^2 - 9$ is either all the points inside or outside the curve. Select the point $(0, 0)$; substituting $(0, 0)$ into $y > x^2 - 9$ yields $0 > -9$, a true statement of inequality. Since $(0, 0)$ is inside the curve, all the points inside the curve satisfy the inequality. Also, testing $y - x \leq 3$ with the point $(0, 0)$ yields $0 \leq 3$, a true statement of inequality. We see that the solution set is the half-plane containing the point $(0, 0)$ and the line $y - x = 3$. Therefore the solution set for the system is the set of all the points inside

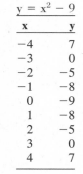

$y = x^2 - 9$	
x	**y**
-4	7
-3	0
-2	-5
-1	-8
0	-9
1	-8
2	-5
3	0
4	7

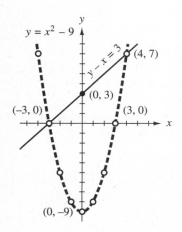

FIGURE 14.41

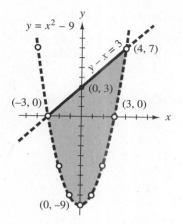

$y = x^2 - 9$

$y - x = 3$

$(4, 7)$

$(0, 3)$

$(-3, 0)$

$(3, 0)$

$(0, -9)$

FIGURE 14.42

the parabola, on that part of the line inside the parabola, and below the line. The solution set does not include the points of intersection of the line and the parabola, since the boundary of the parabola is not part of the solution set. The open circles at $(-3, 0)$ and $(4, 7)$ in Figure 14.42 indicate that the points of intersection are not part of the solution set. The solution set of the system is indicated by the solid line and the shaded region. ◆

The following box contains a summary of the steps we follow in determining the solution of a system of inequalities.

To find the solution set for a system of inequalities, do the following.

1. Write the statements of inequality as statements of equality.
2. Graph the statements of equality.
3. Determine the points of intersection of the curves.
4. Determine the points on the plane that satisfy each statement of inequality.
5. Indicate the solution set on a graph by shading and by solid or dashed lines.

It is possible to display the solution set for a system of inequalities with a graphics calculator. To illustrate the keystrokes, use Example 4.

 fx–7700G

The first step is to change to the inequality mode. (Look back at Section 14.2, Example 2 if you need to review those keystrokes.) Then press

| GRAPH | Y> | x,Θ,T | SHIFT | x^2 | − | 9 | SHIFT | PRGM | : |

| GRAPH | Y≤ | x,Θ,T | + | 3 | EXE |

The display on the screen is the same as in Figure 14.42.

 TI–82

To graph the solution set of the inequality, begin by graphing the two equalities and determine the points of intersection. Then determine which curve is the upper boundary of the solution set and which curve is the lower boundary. For Example 4, the points of intersection are $(-3, 0)$ and $(4, 7)$. The upper boundary is $y = x + 3$, and the lower boundary is $y = x^2 - 9$. To graph the solution set, all $y=$ must be turned off.

| 2nd | DRAW | 7 | (Shade) | x,T,Θ | x^2 | − | 9 | , | x,T,Θ | + | 3 | , |

| 2 | , | − | 3 | , | 4 |) | ENTER |

The entries after (shade) are (lower function, upper function, degree of shading, left boundary, right boundary). The display is the same as Figure 14.42.

EXAMPLE 5   Application: Determining Safe Stress Levels

A special titanium alloy is being considered for use as the turbine impeller shaft that drives the fuel-circulating pumps for a space vehicle's launch rocket engine. The alloy has been extensively tested, and the results show that, for an expected life of 20×10^6 cycles, the safe combinations of the mean stress m and the alternating stress a are given by the solution set of the system of inequalities:

$$m \geq 0$$
$$a \geq 0$$
$$a \leq -0.50m + 45,$$

where the stresses are in ksi (thousand pounds per square inch). Graph the solution set to determine what combinations of mean stress and alternating stress are acceptable for the shaft.

Solution   The system of inequalities to be graphed is:

$$m \geq 0$$
$$a \geq 0$$
$$a \leq -0.50 \, m + 45$$

The last equation is a straight line with slope $= -0.50$ and a-intercept $= 45$. The point $(1, 1)$ satisfies the equation and is, therefore, part of the solution set. The system of inequalities is graphed in Figure 14.43. The solution set for the system is shaded. The shaded region tells all the combinations of mean stress and alternating stress that are acceptable for the shaft. ◆

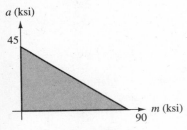

a (ksi)

45

90   m (ksi)

FIGURE 14.43

14.4 Exercises

In Exercises 1–10, graph the solution set of the inequality.

1. $y \geq 3$

2. $x \leq 4$

3. $y \leq x$

4. $x + y > 0$

5. $x + 3y \leq 6$

6. $2x + 3y \geq 12$

7. $3x + 4y < 12$

8. $y < 9 - x^2$

9. $y \leq 4x - x^2$

10. $x^2 + y^2 < 9$

In Exercises 11–40, graph the solution set of the system of inequalities.

11. $x + y < 5$
$x - y < 7$

12. $3x + 2y < 6$
$x + 2y < 4$

13. $x + y \leq 3$
$2x + 3y < 6$

14. $2x - y \leq 3$
$x + y < 3$

15. $x - 3y \leq 3$
$x - 2y \geq -4$

16. $x + 2y \leq 4$
$3x - y \geq 6$

17. $y \leq 3x$
$x \geq 3y$

18. $x \geq 4$
$x + y < 2$

19. $x \geq 2$
$y \leq 2$

20. $x \leq 0$
$y \leq 0$

21. $x \geq 0$
$y \geq 0$

22. $x \leq 4$
$x + 2y < -1$

23. $|y + x| \leq 10$

24. $|y - 3x| < 8$

25. $|y + 2x| \geq 4$

26. $|2y - 4x| > 10$

27. $|3y + 6x| < 12$

28. $\left| y - \dfrac{1}{2}x \right| \leq 4$

29. $\left| y - \dfrac{1}{3}x \right| > 4$

30. $|y + .5x| \geq 9$

31. $x + y < 2$
$y < x^2 - 4$

32. $x + y > 2$
$y < x^2 - 4$

33. $2x + y \leq 3$
$y < x^2 - 5$

34. $y - 2x \leq -7$
$y \geq x^2 + 2x - 8$

35. $y \geq x + 1$
$y \geq x^2 - 4x - 5$

36. $y < 9x - 45$
$y > x^2 + 2x - 15$

37. $x - y < 5$
$x > 0$
$y < 0$

38. $4x + 5y \leq 20$
$x \geq 0$
$y \geq 0$

39. $4x + 5y \leq 20$
$y \leq x + 5$
$y \geq 0$

40. $3y < 2x + 12$
$y < x + 4$
$y > 0$

41. A plumbing store manager pays \$1.00 for a small can of plastic pipe cement and \$1.50 for a larger one. The manager wants to stock at least 200 cans and can spend up to \$300 on the cement. Determine a system of inequalities that describes all possible combinations of can sizes that could be stocked. Determine the solution set.

42. Tim's truck is to be loaded with two types of cargo. Each unit of type A weighs 100 kg, is 3 m³ in volume, and earns \$8 for the driver. Each unit of type B weighs 80 kg, is 4 m³ in volume, and earns \$11.00 for the driver. The truck can carry no more than 6000 kg and no more than 300 m³ of cargo. Determine a system of inequalities that describes all possibilities and sketch the graph.

43. A racketball racket manufacturer makes two types of rackets, Quick Serve and Hard Return. Because of demand he wants to make at least twice as many Quick Serves as Hard Returns. It takes three hours to make a Quick Serve and four hours to make a Hard Return. Altogether he wants to make at least 10 rackets and work no more than 80 hours. Determine a system of inequalities that describes all possibilities and sketch the graph.

44. A fast food restaurant is known for two sandwiches. Because of demand at peak times, sandwiches must be prepared in advance. At least four times as many type A sandwiches as type B are made. The cost to the store is \$0.72 and \$0.83, respectively. The manager does not want more than \$100 in sandwiches prepared ahead at any time, but she wants at least 40 type A sandwiches and 10 type B sandwiches ready for peak times. Devise a system of inequalities that describes all possible inventory levels and sketch the graph of the system.

Review Exercises

In Exercises 1–49, determine the solution set of the inequality. Indicate the solution
a. on a number line, and **b.** in interval notation.

1. $x + 1 < 4$

2. $x - 1 < 3$

3. $x - 1 > 1$

4. $-2x < -6$

5. $-3x > 12$

6. $x + 1 \leq 3x - 1$

7. $2Z - 1 \geq 3Z - 1$

8. $2 - Z \geq 2Z + 5$

9. $1 - 3Z \leq Z + 5$

10. $3(y + 1) < 5y + 3$

11. $2(3y + 2) > 3(2y + 1)$

12. $3(4y + 1) < 12y$

13. $\dfrac{R}{3} - R \leq 1$

14. $\dfrac{2R - 1}{-2} < R$

15. $\dfrac{R + 3}{-4} \geq \dfrac{R + 2}{-3}$

16. $-1 < R + 3 < 2$

17. $0 \leq R + 1 \leq 5$

18. $-1 \leq 2R - 1 < 3$

19. $-5 < 2x + 1 \leq 3$

20. $|x| \leq 2$

21. $|x| < 4$

22. $|x| \geq 3$

23. $|x| > 5$

24. $|x + 1| < 3$

25. $|x - 1| \geq 2$

26. $|x - 1| \leq 1$

27. $|2x + 3| \leq 3$

28. $|3Z - 1| > 11$ **29.** $Z^2 + 3Z < 0$ **30.** $Z^2 - 4 > 0$

31. $4t - t^2 < 0$ **32.** $t(1 - t) > t - 1$ **33.** $t^2 + 15 \le 8t$

34. $2t^2 - t \ge 16$ **35.** $t^2 - t \ge 6$ **36.** $2t(t + 1) < 3t$

37. $(2y)^2 < 1 - 5y^2$ **38.** $2y^2 + 5y > 3$ **39.** $7y > 6y^2 + 2$

40. $(x - 2)(x + 5)(x - 11) < 0$ **41.** $(x - 5)(x + 3)(x - 7) > 0$,   **42.** $(x - 5)^2(x + 3)^2 \le 0$

43. $(x - 8)^2(x - 3)^2 \ge 0$ **44.** $4t^3 - 7t^2 - 2t \ge 0$ **45.** $6t^3 + 11t^2 - 2t \le 0$

46. $\dfrac{x}{(x + 3)(x - 4)} < 0$ **47.** $\dfrac{(x - 1)(x + 3)}{(x - 4)(x + 5)} \le 0$ **48.** $\dfrac{2}{x^2 + 5x + 4} - \dfrac{3}{x^2 + x - 12} \ge 0$

49. $\sqrt{x^2 + 7} \le 4$

In Exercises 50–59, graph the solution set of the system of inequalities.

50. $x + y < 9$
$x - y < 11$

51. $2x + 5y \le 10$
$x + 2y \le 4$

52. $2x + 3y \ge 5$
$x - 3y \ge 7$

53. $3x + 2y > 11$
$x - 2y > 5$

54. $y \le 4x$
$x \ge 4y$

55. $x \ge 7$
$x + y < 11$

56. $|y + x| \le 14$

57. $|y - 3x| < 15$

58. $\left| y - \dfrac{2}{3}x \right| > 4$

59. $\left| y + \dfrac{1}{5}x \right| \le 1$

60. After t seconds the velocity of a ball that is thrown upward with an initial velocity of 80 ft/s will be v ft/s, with $v = 80 - 32t$. What is the time interval during which the velocity is between 16 and 64 ft/s?

61. If F degrees is the temperature using the Fahrenheit scale and C degrees is the temperature using the Celsius scale, then $C = \dfrac{5}{9}(F - 32)$. If the temperature is between 20°C and 30°C, what is the corresponding interval on the Fahrenheit scale?

62. A machine that packages 100 bolts per box can make an error of two bolts per box. If b is the number of bolts in a box, write an inequality that indicates a maximum error of two bolts per box. Solve the inequality.

63. Two resistors in parallel have total resistance R given by $\dfrac{1}{R} = \dfrac{1}{R_1} + \dfrac{1}{R_2}$. If $R_1 = 21 \ \Omega$, determine the possible values for R_2 so that R is between 4 Ω and 11 Ω.

64. The length of a rectangular field is 1 m more than three times its width. If its area is to be at least 10 m², what is the smallest amount of fencing needed to enclose the field? Set up an inequality and solve.

65. The minimum speed for a truck on a highway is 64 km/hr; the maximum speed is 88 km/hr. If Tod has been driving his mini-motorhome along the highway for five hours, what is the range in kilometers that he could legally have traveled? Set up an inequality and solve.

66. The deflection of a beam is given by:

$$d = x^2 - 0.07x - 0.11,$$

where x is the distance from the end. For what values of x is $d > 0.03$?

67. A floppy disk for a computer must have a thickness of 2 mm with a tolerance of ±0.04 mm. Determine the acceptable range of thickness of the disk. Set up a statement of inequality with absolute value and solve.

68. Determine the values of x for which $\sqrt{3x^2 - 14x + 8}$ has real values.

69.* It can be proved using calculus that the curve defined by the equation $y = ax^3 + bx^2 + cx + d$ is concave up (opens upward) for those values of x for which $3ax + b > 0$ and is concave down when $3ax + b < 0$. Use this to find the values of x for which the graph of
   a. $y = 2x^3 + 18x^2 - 13x + 4$ and
   b. $y = 7 + 11x + 9x^2 - 2x^3$ is concave up and the values for which it is concave down.

70.* Prove that $\dfrac{x^2 + y^2}{2} \geq \left(\dfrac{x + y}{2}\right)^2$ for all x and y.

71.* The brake horsepower of a pump can be defined by the equation $BHP = \dfrac{QP}{1715NT}$, where Q is the flow in gallons per minute, P is the pressure at discharge in psi, and NT is the total system efficiency. Given that $Q = 10$ gpm and $P = 40$ psi, for what value of NT will the BHP be greater than 0.31?

72.* Find the values of θ for which the following inequality holds. Limit the answer to values of θ such that $0 \leq \theta < 2\pi$.
$$\sqrt{3}\sin^2 \theta + 2\sqrt{3} > 5 \sin \theta$$

73.* Two positive real numbers differ by 2. If the sum of their reciprocals is greater than 1, show that the smaller of the two numbers must be less than $\sqrt{2}$.

74.* **a.** Write the compound inequality $-5 < x < 14$ as a single inequality involving absolute value. (*Hint:* Look for a number to add to all three parts of the compound inequality that will make the two outside numbers opposite.)
   b. Write a general rule for writing a compound inequality as a single inequality involving absolute value.

75.* Graph the solution set of $|x^2 - 9| \geq 7$ over the range $-6 < x < 6$.

✎ Writing About Mathematics

1. In your position as assistant manager of a candy store, one of your jobs is to check the accuracy of the scales. You establish that there is a maximum variance in the scales of $\frac{1}{8}$ 1b. Using this information and your knowledge of inequalities, write a report to your supervisor explaining how this variance may affect the consumer and the profits of the store.

2. **a.** For the statement $|x - 4| \leq 5$, use the definition of absolute value to explain how the statement may be expressed without the absolute value symbol.
   b. Describe two specific examples that illustrate the use of a statement of the form $|x - 4| \leq 5$.

3. **a.** For the statement $|x - 4| \geq 5$, use the definition of absolute value to explain how the statement may be expressed without the absolute value symbol.
   b. Describe two specific examples that illustrate the use of a statement of the form $|x - 4| \geq 5$.

Chapter Test

In Exercises 1 and 2, graph the solution set on the number line.

1. $-5 \le x \le 1$ **2.** $\dfrac{1}{x} < \dfrac{1}{2}$

In Exercises 3 and 4, express the statement using the appropriate inequality symbol.

3. t is less than 7 and greater than or equal to 0.   **4.** $t - 5$ is non-negative.

In Exercises 5 and 6, solve the inequality and graph the solution set.

5. $x - 3 \le 2$ **6.** $\dfrac{1 - x}{3} < \dfrac{2x}{5}$

In Exercises 7 and 8, determine the solution set for the inequality statement.

7. $(x + 2)(x)(x - 3) < 0$ **8.** $\dfrac{x + 2}{x - 3} > 5$

In Exercises 9 and 10, determine the solution set.

9. $|3x + 2| = \dfrac{1}{2}$ **10.** $\left| \dfrac{2x - 1}{3x} \right| \le -4$

In Exercises 11 and 12, determine the solution set and graph the inequality.

11. $|t - 2| \ge 0$ **12.** $\left| \dfrac{x + 2}{3} \right| < -4$

In Exercises 13 and 14, graph the solution set of the system of inequalities.

13. $x + 2y \le 0$ **14.** $|x - 3y| > 2$
      $x - y \ge 3$

15. A garden store manager pays $2.00 for a small bag of potting soil and $3.00 for a larger one. The manager wants to stock at least 200 bags and can spend up to $600 on the potting soil. Find a system of inequalities that describes all possible combinations of bag sizes that could be stocked. Determine the solution set.

Additional Topics in Trigonometry

T he electrical energy coming into the antenna of a radiotelephone and the sound energy going out of the speaker can be represented by a sinusoidal (sine or cosine) function. For example, when the weak signal from the air waves hits the antenna it is amplified and turned into another signal. Thus, the resulting signal may be the sum of two sine waves with the same frequency. The two waves may be $A \cos \omega t$ and $B \sin (\omega t + R)$ whose sum could be represented by $C \sin (\omega t + S)$. Example 7 in Section 15.4 examines a problem concerning such an energy system.

15.1

Basic Trigonometric Identities

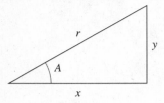

FIGURE 15.1

TABLE 15.1   Trigonometric Functions

$\sin A = \dfrac{y}{r}$	$\cos A = \dfrac{x}{r}$
$\tan A = \dfrac{y}{x}$	$\cot A = \dfrac{x}{y}$
$\csc A = \dfrac{r}{y}$	$\sec A = \dfrac{r}{x}$

In the discussion of equations in Section 4.1 we learned that the solution to a special equation called an **identity** is the set of all real numbers. In this section we are introduced to trigonometric identities that, later in the chapter, are used to solve systems of trigonometric equations. We also learn to use them to simplify equations containing several trigonometric terms. Historically, before calculators and complete tables became available, such identities were used to calculate the trigonometric functions of the less common angles.

To demonstrate that a statement of equality is an identity, we must show that the statement is true for all values of the variable. To do this, we must use "trig" tools. The basic tools are the definitions of the six trigonometric functions that we learned in Section 6.1. The definitions are based on the ratios of the sides of a right triangle (see Figure 15.1). These six functions are repeated in Table 15.1. Using the six definitions the following identities can be developed.

$$\sin A = \frac{1}{\csc A} \qquad \csc A = \frac{1}{\sin A} \qquad \sin A \csc A = 1$$

$$\cos A = \frac{1}{\sec A} \qquad \sec A = \frac{1}{\cos A} \qquad \cos A \sec A = 1$$

$$\tan A = \frac{1}{\cot A} \qquad \cot A = \frac{1}{\tan A} \qquad \tan A \cot A = 1$$

Example 1 demonstrates how to verify an identity.

EXAMPLE 1   **Verifying an Identity**

Show that $\sin A = \dfrac{1}{\csc A}$.

Solution   Using the definition of csc A, we have:

$$\csc A = \frac{r}{y}$$

$$\frac{1}{\csc A} = \frac{1}{\dfrac{r}{y}} \qquad \textbf{Taking the reciprocal of both sides}$$

$$= \frac{y}{r} \qquad \textbf{Division of fractions}$$

$$= \sin A. \qquad \textbf{Definition of sine}$$

Therefore, $\sin A = \dfrac{1}{\csc A}$. ◆

Two other identities that can be verified by using the six basic definitions are:

$$\tan A = \frac{\sin A}{\cos A} \quad \text{and} \quad \cot A = \frac{\cos A}{\sin A}.$$

EXAMPLE 2   **Verifying An Identity**

Show that $\tan A = \dfrac{\sin A}{\cos A}$, $A \neq \dfrac{\pi}{2} + n\pi$, n is an integer.

Solution   Using the definitions of $\sin A$ and $\cos A$, we have:

$$\frac{\sin A}{\cos A} = \frac{\dfrac{y}{r}}{\dfrac{x}{r}}$$

$$= \left(\frac{y}{r}\right)\left(\frac{r}{x}\right) \quad \textbf{Division of fractions}$$

$$= \frac{y}{x}$$

$$= \tan A. \quad \textbf{Definition of tangent}$$

Therefore, $\tan A = \dfrac{\sin A}{\cos A}$.   ◆

There are three identities that can be developed using the six basic definitions and the Pythagorean theorem. They are:

$$\sin^2 A + \cos^2 A = 1 \qquad 1 + \tan^2 A = \sec^2 A \qquad 1 + \cot^2 A = \csc^2 A.$$

These identities are called the Pythagorean identities.

To develop the identity $\sin^2 A + \cos^2 A = 1$, refer to the right triangle in Figure 15.2. From the definitions of the trigonometric functions and the Pythagorean relation, we have the following:

$$\sin A = \frac{y}{r} \quad \text{or} \quad y = r \sin A$$

and

$$\cos A = \frac{x}{r} \quad \text{or} \quad x = r \cos A$$

and

$$x^2 + y^2 = r^2.$$

$$(\ r \cos A\ )^2 + (\ r \sin A\ )^2 = r^2 \quad \begin{array}{l}\textbf{Substituting for } x \textbf{ and } y \textbf{ in}\\ \textbf{the Pythagorean relation}\end{array}$$

$$r^2 \cos^2 A + r^2 \sin^2 A = r^2 \quad \textbf{Remember that } \cos^2 A = (\cos A)^2$$

$$r^2 (\cos^2 A + \sin^2 A) = r^2 \quad \textbf{Common factor}$$

$$\cos^2 A + \sin^2 A = 1 \quad \textbf{Dividing by } r^2$$

The other two Pythagorean identities can be developed in a similar manner.

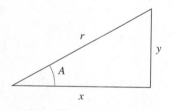

FIGURE 15.2

The basic trigonometric identities, which we will use most frequently and which should be memorized, are listed in Table 15.2.

TABLE 15.2 Basic Trigonometric Identities

Fundamental Identities	Quotient Identities	Pythagorean Identities
$\sin A = \dfrac{1}{\csc A}$	$\tan A = \dfrac{\sin A}{\cos A}$	$\sin^2 A + \cos^2 A = 1$
$\cos A = \dfrac{1}{\sec A}$	$\cot A = \dfrac{\cos A}{\sin A}$	$1 + \tan^2 A = \sec^2 A$
$\tan A = \dfrac{1}{\cot A}$		$1 + \cot^2 A = \csc^2 A$

Why memorize the identities in Table 15.2? The answer is that we save time by remembering the identities. We use these basic trigonometric identities, as well as other identities discussed in this chapter, to simplify equations containing trigonometric terms. Or we may use them to simplify trigonometric statements so that we can find a solution to a calculus problem. Also, we can use these identities to prove other identities.

How do we prove identities? Although we must treat every identity as an individual case, since each is different, proving an identity is similar to solving a word problem. Here are some guidelines.

SUGGESTED GUIDELINES FOR PROVING IDENTITIES

1. Work on one side of the equal sign. Start working on the side with the most complex statement, trying to reduce it to a simpler statement.
2. Whenever possible, use the basic identities to simplify the statement.
3. When it appears that nothing else will work, change all statements to sine and/or cosine functions.
4. Perform an algebraic operation—combine like terms, factor, find a common denominator, or multiply a numerator or denominator by its conjugate to get a special expression in the numerator or denominator.
5. Look for familiar trigonometric expressions such as:

$$1 - \sin^2 x \quad \text{or} \quad \sin x \csc x \quad \text{or} \quad \frac{\sin x}{\cos x}.$$

6. Always keep the other side, the answer, in mind. Doing so may provide ideas to help you achieve the answer.

EXAMPLE 3   Proving an Identity

Use identities to show that $\csc A = \sin A + \cos A \cot A$ is true for all values of A.

Solution   Following the guidelines, try to reduce the right side of the equality to $\csc A$.

$$\csc A = \sin A + \cos A \cot A$$

$$= \sin A + \cos A \,\boxed{\frac{\cos A}{\sin A}} \qquad \textbf{Identity: } \cot A = \frac{\cos A}{\sin A}$$

$$= \frac{\sin^2 A + \cos^2 A}{\sin A} \qquad \textbf{Step 4, common denominator}$$

$$= \frac{1}{\sin A} \qquad \textbf{Identity: } \sin^2 A + \cos^2 A = 1$$

$$= \csc A \qquad \textbf{Identity: } \csc A = \frac{1}{\sin A}$$

We have proved the equality is an identity by changing the right side to $\csc A$ using basic identities and algebra.   ◆

In Chapter 8, we discussed the graphs of the trigonometric functions. Can you recall the graph of $y = \csc x$? Set the range on your graphics calculator to $[-10, 10, 1, -10, 10, 1]$ and switch the angle measure to radian mode. Now graph the function $y = \sin A + \cos A \cot A$, the right side of the equality in Example 3.

When entering the function in the calculator, recall that $\cot A = \dfrac{1}{\tan A}$. The graph is shown in Figure 15.3. Does it look familiar? Yes, it is the graph of $y = \csc A$. This illustrates that graphing is another technique of showing that the equality in Example 3 is an identity.

To prove that some statements of equality are identities, we may have to use some different forms of the basic identities.

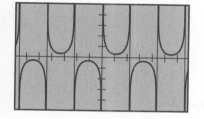

FIGURE 15.3

EXAMPLE 4   Proving an Identity

Use identities to show that $\sec B = \dfrac{\csc B}{\cot B}$.

Solution   Following the guidelines, start working with the most complicated expression, $\dfrac{\csc B}{\cot B}$.

$$\sec B = \frac{\csc B}{\cot B}$$

$$= \frac{\dfrac{1}{\sin B}}{\dfrac{\cos B}{\sin B}} \qquad \textbf{Identities: } \csc B = \frac{1}{\sin B},$$

$$\cot B = \frac{\cos B}{\sin B}$$

$$= \left(\frac{1}{\sin B}\right)\left(\frac{\sin B}{\cos B}\right) \qquad \textbf{Dividing}$$

$$= \frac{1}{\cos B}$$

$$= \sec B \qquad \textbf{Identity: } \sec B = \frac{1}{\cos B}$$

We have proved the equality is an identity by changing the right side to $\sec B$ using basic identities and algebra. ◆

It is possible to prove or develop an identity from a known identity. This is illustrated in Example 5.

EXAMPLE 5 Developing an Identity

Derive the identity $1 + \tan^2 A = \sec^2 A$ using $\sin^2 A + \cos^2 A = 1$.

Solution

$$\sin^2 A + \cos^2 A = 1 \qquad \textbf{Given identity}$$

$$\frac{\sin^2 A}{\cos^2 A} + \frac{\cos^2 A}{\cos^2 A} = \frac{1}{\cos^2 A} \qquad \textbf{Dividing each term by } \cos^2 A$$

$$\left(\frac{\sin A}{\cos A}\right)^2 = \left(\frac{1}{\cos A}\right)^2 \qquad \textbf{Rules of exponents}$$

$$(\tan A)^2 + 1 = (\sec A)^2 \qquad \textbf{Identities: } \frac{\sin A}{\cos A} = \tan A,$$

$$\frac{1}{\cos A} = \sec A$$

We have derived the identity $1 + \tan^2 A = \sec^2 A$ from another identity by using basic identities and rules of algebra. ◆

Sometimes we need to rewrite an expression as a single trigonometric function. To accomplish this, we use the same techniques that are used to prove an identity. This is illustrated in Example 6.

EXAMPLE 6 Rewriting Expressions

Rewrite the expression as a single trigonometric function. (The answer may be multiplied by a constant and/or raised to a power.)

$$\frac{1 + \csc \theta}{\cos \theta + \cot \theta}$$

Solution

$$\frac{1 + \csc \theta}{\cos \theta + \cot \theta} = \frac{1 + \dfrac{1}{\sin \theta}}{\cos \theta + \dfrac{\cos \theta}{\sin \theta}}$$

Step 3, changing all functions to sine or cosine

$$= \frac{\dfrac{\sin \theta + 1}{\sin \theta}}{\dfrac{\sin \theta \cos \theta + \cos \theta}{\sin \theta}}$$

Common denominator for numerator and denominator

$$= \frac{\dfrac{\sin \theta + 1}{\sin \theta}}{\dfrac{\cos \theta(\sin \theta + 1)}{\sin \theta}}$$

Common factor

$$= \frac{\sin \theta + 1}{\sin \theta} \cdot \frac{\sin \theta}{\cos \theta (\sin \theta + 1)}$$

Division of fractions

$$= \frac{1}{\cos \theta}$$

$$= \sec \theta$$

Identity: $\sec \theta = \dfrac{1}{\cos \theta}$

With our knowledge of algebra and a few identities we have shown that $\dfrac{1 + \csc \theta}{\cos \theta + \cot \theta} = \sec \theta$. ◆

In Example 6, the statement is not true for all values of θ. When $\theta = 90° \pm 180°n$ (where $n = 0, 1, 2, \ldots$), both sides of the equality are undefined. For all other values of θ the statement is true. Recall from Chapter 8 that $y = \sec \theta$ is undefined at $\ldots, -90°, 90°, 270°, \ldots$. Graph the function $y = \dfrac{1 + \csc \theta}{\cos \theta + \cot \theta}$. This graph is shown in Figure 15.4. Is this the graph of $y = \sec \theta$?

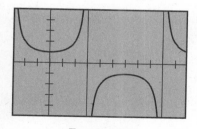

FIGURE 15.4

15.1 Exercises

In Exercises 1 and 2 use the definitions of the trigonometric functions in Table 15.1 to show that the right side is equal to the left side.

1. $\dfrac{1}{\cos A} = \sec A$

2. $\dfrac{1}{\tan A} = \cot A$

In Exercises 3–12, use the identities in Table 15.2 to prove the identity.

3. $\sin A = \tan A \cos A$ **4.** $\cos A = \cot A \sin A$ **5.** $\csc B \tan B = \sec B$ **6.** $\sec B \cot B = \csc B$

7. $\sin A = \dfrac{\tan A}{\sec A}$ **8.** $\cos A = \dfrac{\cot A}{\csc A}$ **9.** $\dfrac{\sec A}{\tan A} = \csc A$ **10.** $\dfrac{\cos B}{\sin B} = \cot B$

11. $\cos A \sec A + \cot A \tan A = 2$ **12.** $\sin A \csc A + \cos A \sec A = 2$

13. Prove that $1 + \tan^2 A = \sec^2 A$ using $r^2 = x^2 + y^2$. (*Hint:* Examine the development of $\sin^2 A + \cos^2 A = 1$.)

14. Prove that $1 + \cot^2 A = \csc^2 A$ using $\sin^2 A + \cos^2 A = 1$.

15. Prove that $1 + \cot^2 A = \csc^2 A$ using $r^2 = x^2 + y^2$.

In Exercises 16–31, rewrite each of the expressions in terms of a single trigonometric function. (The answer may be multiplied by a constant and/or raised to a power.) Check the solution by comparing the graphs of the original statement with the solution.

16. $\sin A + \cos A \cot A$ **17.** $(\sec A + \tan A)(1 - \sin A)$ **18.** $\dfrac{\csc^2 A}{1 + \tan^2 A}$

19. $\cot A \sec^2 A - \cot A$ **20.** $\cos A (\tan A + \cot A)$ **21.** $\dfrac{\sec^2 A - 1}{\sec^2 A}$

22. $\dfrac{\cot \theta + 1}{\cot \theta} - 1$ **23.** $\sec \theta (1 - \sin^2 \theta)$ **24.** $\dfrac{1 + \sec \theta}{\sin \theta + \tan \theta}$

25. $\sin \theta \cos^2 \theta - \sin \theta$ **26.** $\dfrac{1 - \cot \theta}{\tan \theta - 1}$ **27.** $\dfrac{\sin \theta}{1 - \cos \theta} - \csc \theta$

28. $\dfrac{1 + \tan A}{\sin A} - \csc A$ **29.** $\csc A (\csc A - \sin A)$ **30.** $\dfrac{\tan B + \cot B}{\sec B}$

31. $\dfrac{\tan B + \cot B}{\tan B}$

In Exercises 32–37, show that each expression is equal to a constant.

32. $\csc^2 \theta (1 - \cos^2 \theta)$ **33.** $2 \cos^2 \theta - 1 + 2 \sin^2 \theta$ **34.** $\cot^2 \theta \sin^2 \theta + \tan^2 \theta \cos^2 \theta$

35. $\cos^2 \theta + \sin \theta \cos \theta \tan \theta$ **36.** $(1 - \sin^2 \theta)(1 + \tan^2 \theta)$ **37.** $(1 + \tan^2 \theta) \cos^2 \theta$

In Exercises 38–48, prove the identity.

38. $\tan^2 \omega t \sin^2 \omega t = \tan^2 \omega t - \sin^2 \omega t$ **39.** $\tan^2 \omega t + \tan^4 \omega t = \tan^2 \omega t \sec^2 \omega t$

40. $\dfrac{\sin A}{1 - \cos A} = \csc A + \cot A$ **41.** $\dfrac{\cos^3 A - \sin^3 A}{1 + \sin A \cos A} = \cos A - \sin A$

42. $\dfrac{\sin A + 1}{\tan A + \cot A + \sec A} = \sin A \cos A$ **43.** $\dfrac{\sin A - \cos A}{\cos^2 A} = \dfrac{\tan^2 A - 1}{\sin A + \cos A}$

44. $\dfrac{\tan^2 \theta + \sec^2 \theta}{\sec^4 \theta} = 1 - \sin^4 \theta$ **45.** $\dfrac{\cot(-t) \cos(-t)}{\cot(-t) + \cos(-t)} = \dfrac{\cot(-t) - \cos(-t)}{\cot(-t) \cos(-t)}$

46. $\csc^2 2\theta + \sec^2 2\theta = \csc^2 2\theta \sec^2 2\theta$ **47.** $\sin^4 \pi t + \cos^2 \pi t = \sin^2 \pi t + \cos^4 \pi t$

48. $\dfrac{\sec \theta + \csc \theta}{\sin \theta + \cos \theta} = \tan \theta + \cot \theta$

49. In calculus, the derivative of $\tan x$ with respect to x is found to be $\dfrac{\cos x \cos x - \sin x\,(-\sin x)}{\cos^2 x}$. Show that this expression is equivalent to $\sec^2 x$.

50. In calculus, the derivative of $\cot x$ with respect to x is found to be $\dfrac{\sin x\,(-\sin x) - \cos x \cos x}{\sin^2 x}$. Show that this expression is equivalent to $-\csc^2 x$.

51. Show that the system of equations

$$x = a \cos A$$
$$y = b \sin A$$

is equivalent to the equation of an ellipse

$$\frac{x^2}{a^2} + \frac{y^2}{b^2} = 1.$$

52. The solution to a problem in the back of the book is $(\csc x - \cot x)^{-1}$. Your solution to the problem is $\csc x + \cot x$. Is your answer the same as the answer in the back of the book? Demonstrate why or why not.

53. The efficiency of an inclined plane with a coefficient of friction of 0.40 may be described by the equation:

$$\text{Eff} = \frac{\sin \theta}{\sin \theta + 0.40 \cos \theta}.$$

θ is the angle between the gravitational force and the force of the block against the plane. Show that the equation can be expressed as:

$$\text{Eff} = \frac{1}{1 + 0.40 \cot \theta}.$$

54. When an airplane wing with a flat bottom is inclined at an angle θ with the horizontal, the lifting power LP of the wing is given by:

$$LP = k \sin^2 \theta \cos \theta,$$

where k is a constant. Show that LP is also given by:

$$LP = k(\cos \theta - \cos^3 \theta).$$

55. When light travels from air to water, it bends. The path of light is so constructed that it traverses a set distance in the least amount of time. In proving this, the following expression was derived:

$$\left(\sqrt{\frac{1 + \cos \theta}{1 - \cos \theta}} \right) \sin \theta.$$

Use algebra and identities to write the expression as an expression in terms of cosine.

56. The potential energy expression for a spring in simple harmonic motion is:

$$U = \frac{1}{2} \frac{k}{m} A^2 \cos^2 \omega t.$$

The kinetic energy expression for a spring in simple harmonic motion is:

$$K = \frac{1}{2} \omega^2 A^2 \sin^2 \omega t.$$

If $\omega^2 = \dfrac{k}{m}$, derive an expression for the total energy $(U + k)$ that is independent of time.

15.2

Sum and Difference of Two Angles

In this section we will develop formulas for the sum and difference of two angles for the sine and cosine functions. These formulas will help us solve problems with springs or light refraction that may contain expressions such as sin $(A + B)$ or cos $(A + B)$. They also are essential in the development of the double- and half-angle formulas in Section 15.3.

To develop an expression for the cosine of the sum of two angles A and B, cos $(A + B)$, consider Figure 15.5. It shows the three angles A, B, and $A + B$. Choose any point U on the terminal side of angle $A + B$, and draw the line UV so that it is perpendicular to $\overline{OV}$, as shown in Figure 15.6. Now, by the definition of cosine,

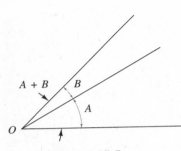

FIGURE 15.5

$$\cos (A + B) = \frac{\overline{OV}}{\overline{OU}},$$

to complete the derivation, we need several more lines. Draw line $\overline{UW}$ so that it is pependicular to the initial side of angle B, which is also the terminal side of angle A (see Figure 15.7). Then draw the line $\overline{WX}$ perpendicular to the extension of line $\overline{OV}$. Now draw line $\overline{WY}$ perpendicular to line $\overline{UV}$. With these lines in place, we can proceed.

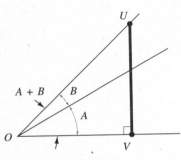

FIGURE 15.6

$$\cos (A + B) = \frac{\overline{OV}}{\overline{OU}} \qquad \text{Definition of cosine} \tag{1}$$

$$= \frac{\overline{OX} - \overline{VX}}{\overline{OU}} \qquad \text{See Figure 15.7} \tag{2}$$

$$\overline{OX} = \overline{OW} \cos A \qquad \text{From right triangle } OXW \tag{3}$$

$$\overline{OW} = \overline{OU} \cos B \qquad \text{From right triangle } OWU \tag{4}$$

$$\overline{OX} = \overline{OU} \cos A \cos B \qquad \begin{array}{l}\text{Substituting Equation (4) for}\\ \overline{OW} \text{ in Equation (3).}\end{array} \tag{5}$$

$$\overline{VX} = \overline{YW}. \qquad \begin{array}{l}\text{Opposite sides of}\\ \text{rectangle } VXWY\end{array} \tag{6}$$

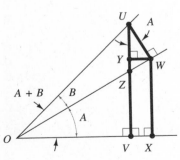

FIGURE 15.7

From Figure 15.7, we can see that:

$$\angle YWZ + \angle YWU = 90°$$

and

$$\angle YUW + \angle YWU = 90°.$$

Substituting and solving, we find that:

$$\angle YWZ = \angle YUW.$$

Also:

$$\angle A = \angle YWZ$$

because they are alternate interior angles formed by parallel lines. Thus, $\angle YUW = \angle A$.

$\overline{YW} = \overline{UW} \sin A$	**From right triangle** UYW	(7)
$\overline{UW} = \overline{OU} \sin B$	**From right triangle** OWU	(8)
$\overline{YW} = \overline{OU} \sin A \sin B$	**Substituting Equation (8) for** $\overline{UW}$ **in Equation (7)**	(9)
$\overline{VX} = \overline{OU} \sin A \sin B$	**Substituting Equation (9) for** $\overline{YW}$ **in Equation (6)**	(10)

$$\cos (A + B) = \frac{\overline{OU} \cos A \cos B - \overline{OU} \sin A \sin B}{\overline{OU}}$$

Substituting Equation (5) for $\overline{OX}$ **and Equation (10) for** $\overline{VX}$ **in Equation (2)**

$$= \frac{\overline{OU}(\cos A \cos B - \sin A \sin B)}{\overline{OU}}.$$

Common factor

Therefore, $\cos (A + B) = \cos A \cos B - \sin B \sin B$.

We can develop the identity for $\sin (A + B)$ using the same figure and similar procedures. Using the identities for cosine and sine of the sum of two angles, we can develop the identities for cosine and sine of difference of two angles. Using the sum and difference of two angles for sine and cosine, we can develop the identities for the sum and difference of two angles for the tangent function.

TABLE 15.3 Sum and Difference Identities

$\sin (A + B) = \sin A \cos B + \cos A \sin B$

$\sin (A - B) = \sin A \cos B - \cos A \sin B$

$\cos (A + B) = \cos A \cos B - \sin A \sin B$

$\cos (A - B) = \cos A \cos B + \sin A \sin B$

$\tan (A + B) = \dfrac{\tan A + \tan B}{1 - \tan A \tan B}$

$\tan (A - B) = \dfrac{\tan A - \tan B}{1 + \tan A \tan B}$

EXAMPLE 1   **Sum of Two Angles: Sine and Cosine**

Use the identities for the cosine of the sum of two angles and the sine of the sum of two angles to obtain exact values for **a.** cos 75° and **b.** sin 75°. (Exact values means that all fractions and radicals are reduced to their lowest possible terms without producing a decimal.)

Solution   To find exact values for cos 75° and sin 75° we need to express 75° in terms of 0°, 30°, 45°, 60°, or 90° since we know exact values for the trigonometric functions of these angles and their multiples. These values were discussed in Section 6.2 (see Table 6.3). Therefore, express 75° as 45° + 30°.

a. $\cos (A + B) = \cos A \cos B - \sin A \sin B$

$\cos 75° = \cos (45° + 30°)$

$= \cos 45° \cos 30° - \sin 45° \sin 30°$ **Substituting**

$= \left(\dfrac{1}{\sqrt{2}}\right)\left(\dfrac{\sqrt{3}}{2}\right) - \left(\dfrac{1}{\sqrt{2}}\right)\left(\dfrac{1}{2}\right)$ **Substituting (Table 6.3)**

$= \dfrac{\sqrt{3}}{2\sqrt{2}} - \dfrac{1}{2\sqrt{2}}$

$= \left(\dfrac{\sqrt{3} - 1}{2\sqrt{2}}\right)\left(\dfrac{\sqrt{2}}{\sqrt{2}}\right)$ **Combining terms and rationalizing denominator**

$\cos 75° = \dfrac{\sqrt{2}(\sqrt{3} - 1)}{4}$ or $\dfrac{\sqrt{6} - \sqrt{2}}{4}$.

b. $\sin (A + B) = \sin A \cos B + \cos A \sin B$

$\sin 75° = \sin (45° + 30°)$

$= \sin 45° \cos 30° + \cos 45° \sin 30°$ **Substituting**

$= \left(\dfrac{1}{\sqrt{2}}\right)\left(\dfrac{\sqrt{3}}{2}\right) + \left(\dfrac{1}{\sqrt{2}}\right)\left(\dfrac{1}{2}\right)$ **Substituting (Table 6.3)**

$= \left(\dfrac{\sqrt{3} + 1}{2\sqrt{2}}\right)\left(\dfrac{\sqrt{2}}{\sqrt{2}}\right)$ **Combining terms and rationalizing denominator**

$\sin 75° = \dfrac{\sqrt{2}(\sqrt{3} + 1)}{4}$ or $\dfrac{\sqrt{6} + \sqrt{2}}{4}$.   ◆

CAUTION   The cos (45° + 30°) ≠ cos 45° + cos 30°. From Example 1 we have:

$$\cos (45° + 30°) = \dfrac{\sqrt{2}(\sqrt{3} - 1)}{4} \approx 0.25882$$

$$\cos 45° + \cos 30° = \dfrac{1}{\sqrt{2}} + \dfrac{\sqrt{3}}{2} \approx 1.57313.$$

Select values for A and B and show that

$$\sin (A + B) \neq \sin A + \sin B. \quad \blacksquare$$

It is possible to determine the exact value of angles using the difference of angles as Example 2 illustrates.

EXAMPLE 2 Difference of Two Angles: Cosine

Use the identity for the cosine of the difference of two angles to find an exact value for the cosine of 15°.

Solution   There are several possible ways to find cos 15° using the difference of two angles with known exact values for their trigonometric functions; for this solution use $15° = 45° - 30°$.

$$\cos (A - B) = \cos A \cos B + \sin A \sin B$$

Let $A = 45°$ and $B = 30°$.

$$\cos(\ 45° \ - \ 30° \ ) = \cos \ 45° \ \cos \ 30° \ + \sin \ 45° \ \sin \ 30° \qquad \textbf{Substituting}$$

$$\cos 15° = \left(\frac{1}{\sqrt{2}}\right)\left(\frac{\sqrt{3}}{2}\right) + \left(\frac{1}{\sqrt{2}}\right)\left(\frac{1}{2}\right) \qquad \begin{matrix}\textbf{Substituting}\\ \textbf{(Table 6.3)}\end{matrix}$$

$$= \frac{\sqrt{3} + 1}{2\sqrt{2}}$$

$$= \frac{\sqrt{2}(\sqrt{3} + 1)}{4} \quad \text{or} \quad \frac{\sqrt{6} + \sqrt{2}}{4}. \quad \blacklozenge$$

The formula for the sum of two angles can be used to simplify expressions.

EXAMPLE 3 Simplifying an Expression

Simplify the expression $\sin (\theta + \pi)$ using an appropriate trigonometric identity.

Solution   To simplify the expression, use the sine function for the sum of the two angles.

$$\sin (A + B) = \sin A \cos B + \cos A \sin B$$
$$\sin (\ \theta \ + \ \pi \ ) = \sin \ \theta \ \cos \ \pi \ + \cos \ \theta \ \sin \ \pi$$
$$= \sin \theta(\ -1 \ ) + \cos \theta(\ 0 \ )$$
$$= -\sin \theta \quad \blacklozenge$$

EXAMPLE 4 Application: Out-of-Phase Sinusoids

A mathematical representation for 120-V, 60-Hz house voltage is $v_1 = 170 \sin 377t$. A representation for 240-V, 60-Hz voltage, which is out of phase with v_1, is $v_2 = 340 \cos (377t + 60°)$ (Figure 15.8). Show that the sum of v_1

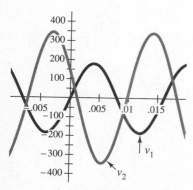

FIGURE 15.8

and v_2 can be written as the sum of two sinusoids with no phase shift. (Even though $377t + 60°$ is the sum of an angle in radians and an angle in degrees, it is a form often used in practice. As we will see, the part of the expression in radians does not enter directly into the solution, and the phase shift usually is easier to handle when it is expressed in degrees.)

Solution   To determine the sum of the two functions v_1 and v_2, use the identity $\cos (A + B) = \cos A \cos B - \sin A$ and B.

$$v_1 + v_2 = 170 \sin 377t + 340 \cos(377t + 60°)$$

$$= 170 \sin 377t + 340(\cos 377t \; \cos 60° - \sin 377t \; \sin 60°)$$

Substituting

$$= 170 \sin 377t + 340 \cos 377t \cos 60° - 340 \sin 377t \sin 60°$$

Distributive property

$$= 170 \sin 377t + 340\left(\frac{1}{2}\right) \cos 377t - 340\left(\frac{\sqrt{3}}{2}\right) \sin 377t$$

Substituting values for
$\cos 60°$ and $\sin 60°$

$$= 170 \sin 377t + \overset{170}{\cancel{340}}\left(\frac{1}{\cancel{2}}\right) \cos 377t - \overset{170}{\cancel{340}}\left(\frac{\sqrt{3}}{\cancel{2}}\right) \sin 377t$$

$$= 170 \sin 377t + 170 \cos 377t - 170 \sqrt{3} \sin 377t$$

$$= (170 - 170 \sqrt{3}) \sin 377t + 170 \cos 377t$$

Common factor

$$= -124 \sin 377t + 170 \cos 377t.$$

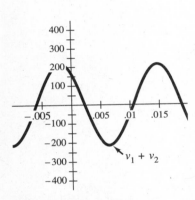

FIGURE 15.9

Figure 15.9 shows the sum of the two sinusoids with no phase shift.   ◆

Example 4 demonstrates how two out-of-phase sinusoids (with a phase shift) can be written as the sum of two in-phase sinusoids (with no phase shift).

15.2 Exercises

In Exercises 1–22, use the sum and difference identities to determine the exact values without a table or a calculator.

1. $\cos 105°$ $(60° + 45°)$

2. $\cos 165°$ $(120° + 45°)$

3. $\cos 255°$ $(210° + 45°)$

4. $\cos 210°$ $(180° + 30°)$

5. $\sin 105°$

6. $\sin 165°$

7. $\sin 255°$

8. $\sin 195°$

9. $\cos (-15°)$

10. $\sin (-15°)$

11. $\tan 105°$

12. $\tan 165°$

13. $\sin 135°$

14. $\cos 225°$

15. $\cos 315°$

16. $\sin 225°$

17. $\sin 42° \cos 18° + \cos 42° \sin 18°$

18. $\cos 23° \cos 22° - \sin 23° \sin 22°$

19. $\cos 73° \cos 13° + \sin 73° \sin 13°$

20. $\sin 63° \cos 18° - \cos 63° \sin 18°$

21. $\sin \dfrac{7\pi}{12} \cos \dfrac{\pi}{12} - \cos \dfrac{7\pi}{12} \sin \dfrac{\pi}{12}$

22. $\cos \dfrac{3\pi}{5} \cos \dfrac{7\pi}{20} + \sin \dfrac{3\pi}{5} \sin \dfrac{7\pi}{20}$

In Exercises 23–30, simplify the expression using the appropriate idenity for the sum and difference of two angles.

23. $\cos \left(\theta + \dfrac{\pi}{2} \right)$

24. $\sin \left(\theta + \dfrac{\pi}{2} \right)$

25. $\sin \left(\theta - \dfrac{\pi}{2} \right)$

26. $\cos \left(\theta - \dfrac{\pi}{2} \right)$

27. $\cos \left(\dfrac{\pi}{3} - \theta \right)$

28. $\sin \left(\dfrac{\pi}{3} - \theta \right)$

29. $\sin \left(\theta + \dfrac{\pi}{6} \right)$

30. $\cos \left(\theta + \dfrac{\pi}{6} \right)$

In Exercises 31–34, use specific values to show that the statements are not equal.

31. $\cos (\theta + \phi) \neq \cos \theta + \cos \phi$.

32. $\sin (\theta + \phi) \neq \sin \theta + \sin \phi$.

33. $\sin (\theta - \phi) \neq \sin \theta - \sin \phi$.

34. $\cos (\theta - \phi) \neq \cos \theta + \cos \phi$.

In Exercises 35–42, write the sum of the two functions as the sum of two sinusoids with no phase shift.

35. $10 \sin 20t + 10 \cos (20t + 30°)$

36. $9 \sin 20t + 12 \cos (20t + 45°)$

37. $12 \cos (t + 120°) + 9 \cos t$

38. $24 \cos (6t + 30°) + 15 \sin 6t$

39. $\sin (377t - 45°) + 2 \cos (377t - 20°)$

40. $\cos (377t + 30°) + \cos (377t - 60°)$

41. $170 \cos (4t - 22°) + 170 \sin (4t + 38°)$

42. $340 \cos (377t - 120°) + 340 \sin (377t - 60°)$

43. A wave action on the surface of a body of water is sinusoidal and may be described by the equation $y = 20 \sin \left(\dfrac{\pi t}{4} - 3 \right)$. Write an equivalent expression by expanding the right member and then evaluating the result where possible.

In Exercises 44 and 45, use $\tan A = \dfrac{\sin A}{\cos A}$ and $\cot A = \dfrac{\cos A}{\sin A}$ and the identities for $\cos (A + B)$ and $\sin (A + B)$ to show that the equalities are valid.

44. $\tan (A + B) = \dfrac{\tan A + \tan B}{1 - \tan A \tan B}$

45. $\cot (A + B) = \dfrac{\cot A \cot B - 1}{\cot A + \cot B}$

In Exercises 46 and 47, use the equations of Exercises 44 and 45 to find expressions for them.

46. $\tan (A - B)$

47. $\cot (A - B)$

48.* Prove the theorem: If C is an angle such that $\sin C = \dfrac{b}{\sqrt{a^2 + b^2}}$ and $\cos C = \dfrac{a}{\sqrt{a^2 + b^2}}$, then $a \sin Bt + b \cos Bt = \sqrt{a^2 + b^2} \sin (Bt + C)$. Use the figure.

In Exercises 49 and 50, use Exercise 48 to transform the function into the form $y = \sqrt{a^2 + b^2} \sin (Bx + C)$.

49. $y = \sqrt{3} \sin x + \cos x$

50. $y = \sin 2x - \sqrt{3} \cos 2x$

In Exercises 51 and 52, use a geometric approach similar to that used to derive the identity for cos $(A + B)$ to derive identities.

51. sin $(A + B)$ **52.** sin $(A - B)$

53. A sign of weight W is held motionless by two connecting cables, one making an angle α with the vertical and the other making an angle β. (See the illustration.) In order for the weight to be held motionless (equilibrium), the following equations must be satisfied.

$$T_1 \sin \alpha = T_2 \sin \beta$$
$$W = T_1 \cos \alpha + T_2 \cos \beta$$

Express W in terms of sin α and sin $(\alpha + \beta)$.

Ann's

54. Consider these three sinusoidal waves of constant frequency but different amplitudes and phase shifts.

$$y_1 = A \sin \theta$$
$$y_2 = \frac{A}{2} \sin \left(\theta + \frac{\pi}{2} \right)$$
$$y_3 = \frac{A}{2} \sin (\theta + \pi)$$

Using a trigonometric identity, change y_2 and y_3 into a simpler format.

15.3

Double-Angle and Half-Angle Formulas

We can use the formulas for the sum of two angles to derive formulas for angles that are the integer multiples of an angle. The derived formulas are important tools that help us to simplify more complex equalities and/or to prove identities.

EXAMPLE 1   Identity cos 2A

Show that cos $2A = \cos^2 A - \sin^2 A$.

Solution   We can obtain an expression for cos $2A$ by using the formula for the cosine of the sum of two angles.

$$\cos (A + B) = \cos A \cos B - \sin A \sin B$$

Let $B = A$.

$$\cos (A + A) = \cos A \cos A - \sin A \sin A.$$
$$\cos (2A) = \cos^2 A - \sin^2 A \quad \blacklozenge$$

Using the result of Example 1 and the identity $\sin^2 A + \cos^2 A = 1$, two additional identities for cos $2A$ can be developed. The double-angle identity for sine can be developed using the identity for the sum of two angles. Using double-angle formulas for sine and cosine, we can develop a double-angle formula for tangent. These double-angle formulas are listed in Table 15.4.

TABLE 15.4 Double-Angle Identities

$$\sin 2A = 2 \sin A \cos A$$
$$\cos 2A = 2 \cos^2 A - 1$$
$$\cos 2A = 1 - 2 \sin^2 A$$
$$\cos 2A = \cos^2 A - \sin^2 A$$
$$\tan 2A = \frac{2 \tan A}{1 - \tan^2 A}$$

CAUTION $\sin 2A \neq 2 \sin A$, $\cos 2A \neq 2 \cos A$, and $\tan 2A \neq 2 \tan A$. To show that these expressions are not equal, substitute values for A and check with your calculator. ∎

EXAMPLE 2 Using Sine Double-Angle Identity

Find the exact value of $\sin 120°$ using the double-angle identity.

Solution The double-angle identity for sine is:

$$\sin 2A = 2 \sin A \cos A.$$

$$\sin 120° = \sin 2\,(60°) = 2 \sin 60° \cos 60° \quad \text{Substituting}$$

$$= 2\left(\frac{\sqrt{3}}{2}\right)\left(\frac{1}{2}\right) \quad \textbf{Recall: } \sin 60° = \frac{\sqrt{3}}{2}$$

$$= \frac{\sqrt{3}}{2} \quad \textbf{and } \cos 60° = \frac{1}{2}$$

We can check the result using the reference angle. The reference angle for a $120°$ angle is $60°$; the $\sin 60°$ is $\dfrac{\sqrt{3}}{2}$. Thus, the result $\sin 120° = \dfrac{\sqrt{3}}{2}$ is correct. ◆

We also can check the result in Example 2 using a calculator. Such a check shows that $\sin 120° = 0.866025$ and $\dfrac{\sqrt{3}}{2} = 0.866025$.

Half-angle identities can be developed using the double-angle identities. This is illustrated in Example 3.

TABLE 15.5 Half-Angle Identities

$$\sin \frac{B}{2} = \pm \sqrt{\frac{1 - \cos B}{2}}$$

$$\cos \frac{B}{2} = \pm \sqrt{\frac{1 + \cos B}{2}}$$

$$\tan \frac{B}{2} = \frac{\sin B}{1 + \cos B}$$

$$\tan \frac{B}{2} = \frac{1 - \cos B}{\sin B}$$

EXAMPLE 3 Cosine Half-Angle Identity

Determine an expression for $\cos \dfrac{B}{2}$ in terms of the angle B.

Solution Recall that:

$$\cos 2A = 2 \cos^2 A - 1.$$

Let:

$$A = \frac{B}{2}.$$

$$\cos 2\left(\frac{B}{2}\right) = 2 \cos^2 \frac{B}{2} - 1$$

$$\cos B = 2 \cos^2 \frac{B}{2} - 1$$

$$2 \cos^2 \frac{B}{2} = 1 + \cos B \quad \textbf{Rearranging}$$

$$\cos^2 \frac{B}{2} = \frac{1 + \cos B}{2}$$

Taking the square root (remember to take both roots), we get:

$$\cos \frac{B}{2} = \pm \sqrt{\frac{1 + \cos B}{2}}.$$

◆

The use of the $\pm$ in trigonometric identities is different from the use of $\pm$ in algebra. When we solve the quadratic formula in algebra, the $\pm$ indicates two possible correct roots. However, trigonometric identities have only one correct answer which is determined by the terminal side of the angle $\frac{B}{2}$ for the half-angle formula.

The quadrant in which the angle $\frac{B}{2}$ appears determines the sign (plus or minus) for the sine and cosine identities. For the cosine identity, if the angle is in the first or fourth quadrant, the sign is positive. If the angle is in the second or third quadrants, the angle is negative and the sign is negative. For the sine identity, if the angle is in the first or second quadrant, the sign is positive. If the angle is in the third or fourth quadrants, the angle is negative and the sign is negative.

EXAMPLE 4   Sine Half-Angle Identity

Determine the exact value for sine $75°$.

Solution   Recognizing that half of $150°$ is $75°$, use the half-angle formula for sine.

$$\sin \frac{B}{2} = \pm \sqrt{\frac{1 - \cos B}{2}}$$

$$\sin 75° = \pm \sqrt{\frac{1 - \cos 150°}{2}} \qquad \textbf{Substituting}$$

$$= \pm \sqrt{\frac{1 - (-\cos 30°)}{2}} \qquad \begin{array}{l}\textbf{Cosine in second quad so negative 30°}\\\textbf{is reference angle}\end{array}$$

$$= \pm \sqrt{\frac{1 + \dfrac{\sqrt{3}}{2}}{2}} \qquad \textbf{Recall: } \cos 30° = \frac{\sqrt{3}}{2}$$

$$= \pm \frac{\sqrt{2 + \sqrt{3}}}{2}$$

Since the angle $75°$ is in the first quadrant we select the positive value. Thus,

$$\sin 75° = \frac{\sqrt{2 + \sqrt{3}}}{2} \approx 0.96593. \quad ◆$$

We can develop an identity for $\sin A$ in terms of the double angle $2A$. This is illustrated in Example 5.

EXAMPLE 5 **Sin A in Terms of a Double Angle**

Determine an expression for $\sin A$ in terms of a function of the double angle $2A$ using the identity $\cos 2A = 1 - 2 \sin^2 A$.

Solution

$$\cos 2A = 1 - 2 \sin^2 A$$
$$2 \sin^2 A = 1 - \cos 2A$$
$$\sin^2 A = \frac{1 - \cos 2A}{2}$$

TABLE 15.6

$$\sin A = \pm \sqrt{\frac{1 - \cos 2A}{2}}$$

$$\cos A = \pm \sqrt{\frac{1 + \cos 2A}{2}}$$

Taking the square root of both sides,

$$\sin A = \pm \sqrt{\frac{1 - \cos 2A}{2}} \quad \blacklozenge$$

Using a similar technique, a formula for $\cos A$ as a function of the double angle $2A$ can be developed. Table 15.6 provides both expressions.

EXAMPLE 6 **Sin A Using Double-Angle Identity**

Determine the value of $\sin 120°$ using the identity:

$$\sin A = \pm \sqrt{\frac{1 - \cos 2A}{2}}.$$

Solution Substitute $120°$ for A in the identity.

$$\sin \boxed{120°} = \sqrt{\frac{1 - \cos 2(\boxed{120°})}{2}} \qquad \text{Second quadrant angle, so positive root } 120°$$

$$= \sqrt{\frac{1 + \frac{1}{2}}{2}} \qquad \cos 240° = -\frac{1}{2}$$

$$= \frac{\sqrt{3}}{2} \quad \blacklozenge$$

EXAMPLE 7 **Application: Second Moment of An Area**

When the second moment of an area ("moment of inertia" of the area) is determined, an expression that must be simplified is:

$$I_A = I_b \sin^2 \theta - 2I_c \sin \theta \cos \theta + I_a \cos^2 \theta.$$

Write the expression for I_A as a function of the double angle 2θ.

Solution   To express I_A as a function of a double angle, we must recall from the list of identities those identities that help change $\sin^2 A$, $\sin A \cos A$, and $\cos^2 A$ into double-angle expressions.

$$I_A = I_b \sin^2 \theta - 2I_c \sin \theta \cos \theta + I_a \cos^2 \theta \tag{1}$$

$$\sin \theta = \pm \sqrt{\frac{1 - \cos 2\theta}{2}} \quad \textbf{Squaring} \quad \sin^2 \theta = \frac{1 - \cos 2\theta}{2} \tag{2}$$

$$\cos \theta = \pm \sqrt{\frac{1 + \cos 2\theta}{2}} \quad \textbf{Squaring} \quad \cos^2 \theta = \frac{1 + \cos 2\theta}{2} \tag{3}$$

$$\sin 2\theta = 2 \sin \theta \cos \theta \tag{4}$$

Using Equations (2), (3), and (4) to substitute for $\sin^2 \theta$, $\cos^2 \theta$, and $\sin \theta \cos \theta$, respectively, in Equation (1), we obtain:

$$I_A = I_b \left(\frac{1 - \cos 2\theta}{2} \right) - I_c (\sin 2\theta) + I_a \left(\frac{1 + \cos 2\theta}{2} \right)$$

$$= \frac{I_b}{2} - \frac{I_b \cos 2\theta}{2} - I_c \sin 2\theta + \frac{I_a}{2} + \frac{I_a \cos 2\theta}{2}.$$

Grouping and the common factor gives:

$$I_A = \frac{I_a + I_b}{2} + \frac{I_a - I_b}{2} \cos 2\theta - I_c \sin 2\theta.$$

Thus, I_A is expressed as a function of a double angle. The expression in this form is used to find the maximum second moment.   ◆

15.3 Exercises

In Exercises 1–6, use the double-angle identities listed in Table 15.4 to determine the value of the trigonometric function.

1. $\sin 60°$ **2.** $\cos 90°$ **3.** $\tan 120°$ **4.** $\sin 300°$

5. $\cos 120°$ **6.** $\tan 270°$

In Exercises 7–12, use the half-angle identities listed in Table 15.5 to determine the value of the trigonometric function.

7. $\sin 15°$ **8.** $\cos 22.5°$ **9.** $\tan 75°$ **10.** $\sin 22.5°$

11. $\cos 15°$ **12.** $\tan 22.5°$

In Exercises 13–16, use the identities listed in Table 15.6 to determine the value of the trigonometric function.

13. $\sin 165°$ **14.** $\cos 120°$ **15.** $\cos 75°$ **16.** $\sin 75°$

17. Find $\tan 2A$ as a function of $\tan A$.   **18.** Find $\cot 2A$ as a function of $\cot A$.

19. Show that $\sin 2A = \dfrac{2 \tan A}{1 + \tan^2 A}$.   **20.** Show that $\cos 2A = \dfrac{1 - \tan^2 A}{1 + \tan^2 A}$.

21.   Show that $\tan \dfrac{A}{2} = \dfrac{1 - \cos A}{\sin A}$.

22.   Show that $\tan \dfrac{A}{2} = \dfrac{\sin}{1 + \cos A}$.

23.   Show that $(\sin A + \cos A)^2 = 1 + \sin 2A$.

24.   Show that $\cos^4 A - \sin^4 A = \cos 2A$.

25.   Normal stress at a point can be written as:

$$N = s_1 \cos^2 \theta - 2s_2 \sin \theta \cos \theta + s_3 \sin^2 \theta.$$

Write the normal stress as a function of 2θ.

26.   Shear stress at a point can be written as:

$$S = -s_1 \sin \theta \cos \theta + s_2(\cos^2 \theta - \sin^2 \theta) + s_3 \sin \theta \cos \theta.$$

Write the shear stress as a function of 2θ.

27.   As a projectile takes off, the position along the plane is given by:

$$x = (V_0 \cos \alpha)t,$$

where V_0 is the initial speed of the projectile at take-off and α is the angle at which it was launched. The term t represents the time at any given moment, up to the time the projectile impacts the ground. When the projectile does impact the ground, the time in flight can be represented by $t = \dfrac{2V_0 \sin \alpha}{g}$. Using this equation, substitute for t in the position equation and solve for x. (This equation is referred to as the range equation.) Make sure that the final result is simplifed.

28.   The instantaneous power of an inductor has the equation:

$$P = VI \sin \left(\omega t - \frac{\pi}{2} \right).$$

The $\dfrac{\pi}{2}$ is the characteristic phase shift between the current and the voltage for an inductor. Rewrite this expression as a function of $\sin 2\omega t$.

29.   **a.**   A successful punt in football depends upon two factors: the hang time and the range. To give defenders enough time to surround the receiver, A minimum hang time of 3.0 is needed. In an average professional punt, the ball leaves the kicker's foot at a speed of 26 yd/s. The hang time is given by the equation:

$$t = \frac{2V_0 \sin \theta}{9.8 \text{ m/s}^2},$$

where V_0 is the initial speed of the punt and θ is the angle of the punt. Express the range of angles that gives an acceptable hang time.

b.   A successful punt should travel at least 40 yd. The range of a punt is given by:

$$R = \frac{V_0^2 \sin 2\theta}{9.8 \text{ m/s}^2}.$$

Express the range of angles that gives an acceptable range.

c.   Express the range of angles that gives both an acceptable hang time and an acceptable range.

15.4

Trigonometric Equations

In Chapters 10 and 11, we learned methods for solving equations. In solving trigonometric equations, we can substitute a variable for the trigonometric function (for instance, we can replace $\sin \theta$ with x). We then solve the equation in terms of the new variable. Finally, we use that answer to determine the value of the function in the original equation.

EXAMPLE 1   **Solving a Trigonometric Equation: Sine**

Solve $4 \sin^2 \theta - 3 = 0$ for θ, $0° \leq \theta < 360°$.

Solution

$$4 \sin^2 \theta - 3 = 0$$

Let $x = \sin \theta$. Substituting yields:

$$4x^2 - 3 = 0$$

$$x = \pm \sqrt{\frac{3}{4}} \qquad \text{Solving for } x$$

$$= \pm 0.8660$$

$$\sin \theta = \pm 0.8660 \qquad \text{Substituting } \sin \theta \text{ for } x$$

$$\theta = 60°, 120°, 240°, \text{ or } 300°.$$

Check to see that all four values of θ satisfy the original equation.   ◆

Now, however, we also can solve problems of this type directly, without substitution. This is illustrated in Examples 2 and 3.

EXAMPLE 2   **Solving a Trigonometric Equation: Cosine**

Solve $2 \cos^2 x - \cos x = 0$ for x, $0 \leq x < 2\pi$.

Solution

$$2 \cos^2 x - \cos x = 0$$

$$\cos x (2 \cos x - 1) = 0 \qquad \text{Common factor}$$

Setting the factors equal to zero, we have:

$$\cos x = 0 \qquad\qquad 2 \cos x - 1 = 0$$

$$x = \frac{\pi}{2}, \frac{3\pi}{2} \qquad\qquad \cos x = \frac{1}{2}$$

$$x = \frac{\pi}{3}, \frac{5\pi}{3}$$

Therefore, the solutions for $2 \cos^2 x - \cos x = 0$ are $x = \frac{\pi}{3}, \frac{\pi}{2}, \frac{3\pi}{2},$ and $\frac{5\pi}{3}$.

◆

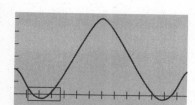

FIGURE 15.10

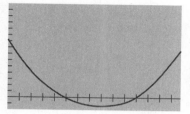

FIGURE 15.11

We also may determine the solution to the equation in Example 2 by graphing the function $y = 2 \cos^2 x - \cos$. Figure 15.10 shows the calculator screen. We see that the curve crosses the x-axis at four points, which indicates four solutions for the indicated interval. Use the zoom and box features of the calculator to enlarge that part of the function shown in the rectangle in Figure 15.10. The result, Figure 15.11, shows the enlarged curve cutting the x-axis at two points. Use the trace function to see that the curve crosses the x-axis at approximately 1.047 and 1.573. These values are good approximations to $\dfrac{\pi}{3}$ and $\dfrac{\pi}{2}$ to three significant digits. Graph the equation and check the other two solutions.

In some situations, methods like those in Examples 1 and 2 will not work. However, in such cases, we may be able to write the equation in a simpler form by using trigonometric identities and then be able to solve the equation.

EXAMPLE 3 **Application: Shear Stress**

To determine the maximum shear stress in an electric-motor shaft, it is necessary to solve the equation $-30 \cos 2\theta + 8 \sin 2\theta = 0$. Find θ, $0° \leq \theta < 360°$.

Solution There is no standard method for solving a problem of this type. Usually the only approach is to try several methods until you find one that works. (However, with practice and familiarity, you will become faster at selecting the correct method.)

In this instance, we might let $\cos 2\theta = x$ and $\sin 2\theta = y$ and write the equation as $-30x + 8y = 0$. We can see this is not the best procedure since we get an equation in two unknowns, x and y, where before we only had a single unknown, θ. Since we cannot solve a single equation with two unknowns, this is a blind alley.

Taking another look at the equation $-30 \cos 2\theta + 8 \sin 2\theta = 0$, we see that the terms are either multiplied by $\cos 2\theta$ or $\sin 2\theta$, and we recall that $\tan 2\theta = \dfrac{\sin 2\theta}{\cos 2\theta}$. Therefore, we might try dividing the equation by $\cos 2\theta$ ($\cos 2\theta \neq 0$). The division by $\cos 2\theta$ excludes $45°$, $135°$, $225°$, and $315°$ as possible values for θ. At this point we must check these values in the original equation to make sure they are not solutions. Having confirmed that these angles are not solutions, divide by 2θ.

$$\frac{-30 \cos 2\theta + 8 \sin 2\theta}{\cos 2\theta} = \frac{0}{\cos 2\theta} \qquad \textbf{Dividing both sides of the equation by } 2\theta$$

$$\frac{-30 \cos 2\theta}{\cos 2\theta} + \frac{8 \sin 2\theta}{\cos 2\theta} = 0$$

$$-30 + 8\ \boxed{\tan 2\theta} = 0 \qquad \textbf{Substituting}$$

$$\tan 2\theta = \frac{30}{8}$$

$$\tan 2\theta = 3.750$$

$$2\theta = 75.1°$$

Since the tangent is positive in the first and third quadrants, $2\theta = 75.1°$ or $255.1°$. Therefore, the shear stress in the motor may occur when $\theta = 37.5°$ or $127.5°$. ◆

We can check both values either by substituting back into the original equation or by graphing the function and using the zoom or tracing functions of the calculator.

EXAMPLE 4   **Using Identities to Solve Equations**

Given $\sin 2t + \cos t = 0$, solve for t, $0 \le t < 2\pi$.

Solution   We cannot use the approach we used in Example 3 to solve $\sin 2t + \cos t = 0$ because the arguments for the sine and cosine functions, $2t$ and t, respectively, are different. In this instance, we can obtain an equation with the same arguments for all the trigonometric functions by using the formula for $\sin 2t$.

$\sin 2t + \cos t = 0$	**Given**
$\sin 2t = 2 \sin t \cos t$	**Identity**
$2 \sin t \cos t + \cos t = 0$	**Substituting for** $\sin 2t$
$\cos t (2 \sin t + 1) = 0$	**Common factor**
$\cos t = 0 \qquad\qquad 2 \sin t + 1 = 0$	**Setting factors equal to 0**

$$t = \frac{\pi}{2}, \frac{3\pi}{2} \qquad\qquad \sin t = -\frac{1}{2}$$

$$t = \frac{7\pi}{6}, \frac{11\pi}{6}$$

Verify by substitution that $t = \dfrac{\pi}{2}, \dfrac{3\pi}{2}, \dfrac{7\pi}{6}$, and $\dfrac{11\pi}{6}$ are all solutions for $\sin 2t + \cos t = 0$. ◆

EXAMPLE 5   **Using the Pythagorean Identity to Solve Equations**

Solve $2 \sin^2 \theta + 3 \cos \theta + 5 = 0$ for $0 \le \theta < 2\pi$.

Solution   To solve the equation $2 \sin^2 \theta + 3 \cos \theta + 5 = 0$, we must write it in terms of $\sin \theta$ or $\cos \theta$. It is easiest to change $\sin^2 \theta$ to $\cos^2 \theta$ using the identity $\sin^2 \theta + \cos^2 \theta = 1$. The identity can be rewritten as $\sin^2 \theta = 1 - \cos^2 \theta$, and we can write the equation in terms of the same function, $\cos \theta$.

$2 \sin^2 \theta + 3 \cos \theta + 5 = 0$	**Given**
$2 (1 - \cos^2 \theta) + 3 \cos \theta + 5 = 0$	**Substituting for** $\sin^2 \theta$
$2 - 2 \cos^2 \theta + 3 \cos \theta + 5 = 0$	
$-2 \cos^2 \theta + 3 \cos \theta + 7 = 0.$	

Because this equation is quadratic in cos θ, we can use the quadratic formula.

$$\cos\theta = \frac{-3 \pm \sqrt{(3)^2 - 4(-2)(7)}}{2(-2)}$$

$$= \frac{-3 \pm \sqrt{65}}{-4}$$

$$= \frac{-3 \pm 8.062}{-4}$$

$$= -1.27, 2.77.$$

Since -1.27 and 2.77 are outside the range of the cosine function, there is no angle θ that is a solution to the given equation. So although the method of obtaining the solution is valid, no solution exists for the equation. ◆

EXAMPLE 6   Application: Designing a Truss

To design a simple truss we must solve the equation $4000 \sin\phi + 3000 \cos\phi = 5000$. Determine ϕ, $0° \leq \phi < 360°$.

Solution   Given the equation:

$$4000 \sin\phi + 3000 \cos\phi = 5000,$$

we first make the numbers easier to work with by dividing the equation by 1000, giving:

$$4 \sin\phi + 3 \cos\phi = 5.$$

We cannot solve the equation in this form, we must change it to a form that contains only one of the trigonometric functions.

$4 \sin\phi + 3 \cos\phi = 5$ **Given**

$\sin^2\phi + \cos^2\phi = 1$ **Identity**

$\sin^2\phi = 1 - \cos^2\phi$

$\sin\phi = \pm\sqrt{1 - \cos^2\phi}$

$\pm 4\sqrt{1 - \cos^2\phi} + 3\cos\phi = 5$ **Substituting for sin ϕ in the given equation**

$\pm 4\sqrt{1 - \cos^2\phi} = 5 - 3\cos\phi$ **Rearranging before squaring to remove the radical**

$16(1 - \cos^2\phi) = 25 - 30\cos\phi + 9\cos^2\phi$

     Squaring the expressions on both sides of the equal sign

$$16 - 16 \cos^2 \phi = 25 - 30 \cos \phi + 9 \cos^2 \phi$$

$$0 = 25 - 16 - 30 \cos \phi$$
$$+ 9 \cos^2 \phi + 16 \cos^2 \phi$$

$$0 = 9 - 30 \cos \phi + 25 \cos^2 \phi$$

$$0 = (3 - 5 \cos \phi)(3 - 5 \cos \phi) \quad \textbf{Factoring}$$

$$3 - 5 \cos \phi = 0 \qquad \textbf{Setting factor equal}$$
$$\textbf{to zero}$$

$$\cos \phi = \frac{3}{5}$$

$$= 0.6$$

$$\phi = 53.13°, 306.9°.$$

Checking the solutions, we have the following.

For $\phi = 53.13°$ For $\phi = 306.9°$

$4 \sin 53.13° + 3 \cos 53.13° = 5$ $4 \sin 306.9° + 3 \cos 306.9° = 5$

$\quad 4(0.8000) + 3(0.6000) \quad = 5$ $4(-0.8000) + 3(0.6000) \quad = 5$

$\qquad 3.200 + 1.800 \qquad = 5$ $-3.200 + 1.800 \qquad = 5$

$\qquad\qquad\qquad\qquad 5 = 5 \checkmark$ $-1.4 = 5 \textbf{ x.}$

Only one of the solutions checks in the original equation, so the only solution for $4 \sin \phi + 3 \cos \phi = 5$ in the given interval is $\phi = 53.1°$. ◆

EXAMPLE 7   Application: System of Equations

As a part of a design proposal for a radiotelephone the following system of equations must be solved for R and S.

$$2 + 4 \sin R = 3 \sin S \qquad\qquad\qquad (1)$$
$$4 \cos R = \cos S \qquad\qquad\qquad\qquad (2)$$

Determine values for R and S to the nearest tenth of a degree.

Solution   Given $2 + 4 \sin R = 3 \sin S$ and $4 \cos R = \cos S$, first rewrite the first equation.

$$\frac{2 + 4 \sin R}{3} = \sin S$$

$$\frac{4 + 16 \sin R + 16 \sin^2 R}{9} = \sin^2 S \qquad \textbf{Squaring}$$
$$\textbf{equation (1)}$$

$$16 \cos^2 R = \cos^2 S \qquad \textbf{Squaring}$$
$$\textbf{equation (2)}$$

$$\frac{4 + 16 \sin R + 16 \sin^2 R}{9} + 16 \cos^2 R = \sin^2 S + \cos^2 S \qquad \textbf{Sum}$$

$$\frac{4 + 16 \sin R + 16 \sin^2 R}{9} + 16 \cos^2 R = 1 \qquad \textbf{Recall sin}^2\ S + \cos^2 S = 1$$

$$4 + 16 \sin R + 16 \sin^2 R + 144 \cos^2 R = 9$$

$$4 + 16 \sin R + 16 \sin^2 R + 144\,(\,1 - \sin^2 R\,) = 9 \qquad \textbf{Substituting}$$

$$-128 \sin^2 R + 16 \sin R + 139 = 0$$

Using the quadratic formula (and carrying several extra significant digits) to solve for sin R, the results are:

$$\sin R = 1.10646 > 1 \text{ cannot be a solution}$$

or

$$\sin R = -0.98146.$$

The negative value gives a third- and fourth-quadrant answer. The angles are $R = 258.95°$ or $R = 281.05°$. Substituting in the second equation, we have:

$$4 \cos 258.95° = \cos S$$

$$S = 140.06° \quad \text{or} \quad 219.94°;$$

$$4 \cos 281.05° = \cos S$$

$$S = 39.94° \quad \text{or} \quad 320.06°$$

Checking the possible solutions in the original equations shows that we have two extraneous solutions and two valid solutions. The valid solutions, rounded to the nearest tenth of a degree, are: $R = 258.9°$ and $S = 219.9°$ or $R = 281.1°$ and $S = 320.1°$. ◆

15.4 Exercises

In Exercises 1–36, determine θ, $0° \leq \theta < 360°$.

1. $\tan^2 \theta - \tan \theta = 0$

2. $\cot^2 \theta + \cot \theta = 0$

3. $\sin^2 \theta - \dfrac{\sqrt{2}}{2} \sin \theta = 0$

4. $\cos^2 \theta + \dfrac{\sqrt{2}}{2} \cos \theta = 0$

5. $\sin^2 \theta - \sin \theta - 2 = 0$

6. $\cos^2 \theta + \cos \theta - 2 = 0$

7. $\sin \theta = \cos \theta$

8. $\cos \theta - \sin \theta = 0$

9. $\cos 2\theta = \sin 2\theta$

10. $\csc 2\theta = \sec 2\theta$

11. $\cos 2\theta = \sin \theta$

12. $\sin 2\theta = \cos \theta$

13. $\cos \theta - \sin \dfrac{1}{2} \theta = 0$

14. $\sin \dfrac{1}{2} \theta - \cos \dfrac{1}{4} \theta = 0$

15. $\cos 2\theta + \cos \theta = 0$

16. $\sin 2\theta + \cos \theta = 0$

17. $\cos \theta + \sin \theta = 1$

18. $\sin \theta = \cos \theta - \sqrt{2}$

19. $\cos 2\theta + \cos \theta + 1 = 0$

20. $\cos \theta + \sqrt{3} \sin \theta = 1$

21. $\cos 2\theta + \sin 2\theta = 1$

22. $\cos \theta - \sin \theta = 1$

23. $\tan \theta + \sec \theta - 1 = 0$

24. $2 \sin \theta + \cot \theta = \csc \theta$

25. $\tan^2 \theta + 3 = 3 \sec \theta$

26. $\tan 2\theta + \cot 2\theta - 2 = 0$

27. $3 \sin^2 \theta - \cos^2 \theta = 0$

28. $2 \tan \theta + 3 = 2 \sec^2 \theta$

29. $\sin 2\theta \cos \theta + \cos 2\theta \sin \theta = 1$

30. $\cos \theta \cos 2\theta + \sin \theta \sin 2\theta = -1$

31. $\sin^2 \theta + 2 \cos \theta = 2$

32. $2 \cos^2 \theta - \sin^2 \theta = 0$

33. $4 \sin^2 \theta = 2 + \cos \theta$

34. $3 - 2 \cos \dfrac{1}{2} \theta = 4 \sin^2 \dfrac{1}{2} \theta$

35. $\cos (\theta - 45°) = \sin (\theta - 45°)$

36. $5 \sin \theta - 12 \cos \theta = 13$

37. The equilibrium equations for an object resting on an inclined plane are:

$$f - w \sin \theta = 0$$
$$N - w \cos \theta = 0.$$

If the object is just about to slip on the plane, the coefficient of friction between the object and the plane is $\mu = \dfrac{f}{N}$. What is the coefficient of friction if the object just starts to slip when the plane is at 20° to the horizontal?

38. Solve Exercise 37 if an object just starts to slip when the plane is at 27° to the horizontal.

In Exercises 39–44, an equation from an electrical phasor diagram can be:

$$I = I_1 \cos \theta + I_2 \sin \theta \quad \text{or} \quad V = V_1 \cos \theta + V_2 \sin \theta.$$

Solve for θ, $0 \le \theta < 360°$ for the values given. (Similar equations are found in mechanics.)

39. $I_1 = 3.00,\ I_2 = -8.00,\ I = -2.00$

40. $I_1 = 0.261,\ I_2 = -0.115,\ I = 0.169$

41. $I_1 = -1.36,\ I_2 = 2.81,\ I = 2.18$

42. $I_1 = -1 \times 10^{-3},\ I_2 = 3 \times 10^{-3},\ I = 1.66 \times 10^{-3}$

43. $V_1 = 9,\ V_2 = 9,\ V = 1.11$

44. $V_1 = 12,\ V_2 = 13,\ V = -6.83$

15.5

Inverse Trigonometric Functions

The procedure in Section 5.4 tells how the inverse of a function can be found by interchanging the role of the independent variable and the dependent variable and then solving for the dependent variable. The result is the inverse function. For example, to find the inverse function of $y = f(x) = 3x - 5$, we interchanged x and y in the equation $(x = 3y - 5)$ and then solved for y to obtain $y = \dfrac{x + 5}{3}$. Therefore, the inverse function of $f(x) = 3x - 5$ is the function $f^{-1}(x) = \dfrac{x + 5}{3}$.

This method for finding an inverse algebraic function cannot be used for finding inverse trigonometric functions. For example, let us look at $y = \sin x$. If we interchange variables $(x = \sin y)$, there is no direct way we can solve for y using algebraic procedures. So how do we determine inverse trigonometric functions? Let us begin by examining what is meant by inverse trigonometric functions.

In Chapter 8 and again in this chapter, we have solved equations of the form

$\cos \theta = \dfrac{1}{2}$. The solution to this equation is "the angle whose cosine is $\dfrac{1}{2}$," or symbolically,

$$\theta = \cos^{-1}\left(\frac{1}{2}\right).$$

To solve this equation using a graphics calculator, press the keys $\boxed{\cos^{-1}}$ $\boxed{(}$ 1 $\boxed{\div}$ 2 $\boxed{)}$, and the screen display is $\cos^{-1}(1 \div 2)$ or $\cos^{-1}\left(\dfrac{1}{2}\right)$. When we looked at this equation before, we found that there is no unique solution. The solutions are the points of intersection of the curve and the dashed line as shown in Figure 15.12.

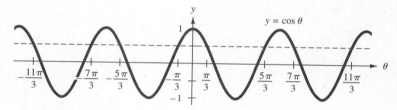

FIGURE 15.12

It is helpful to be able to express "the angle whose cosine is $\dfrac{1}{2}$" in a more compact fashion. One way is to use the symbols $\theta = \cos^{-1}\dfrac{1}{2}$, as we have just seen. An equivalent expression is $\theta = \arccos\dfrac{1}{2}$. If $x = \cos \theta$, then the equivalent relation is written $\theta = \arccos x$.

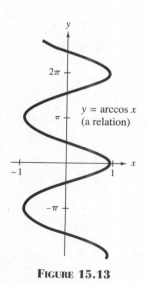

$y = \arccos x$
(a relation)

FIGURE 15.13

CAUTION $\theta = \cos^{-1} x$ does not mean $\theta = \dfrac{1}{\cos x}$ (the reciprocal). Rather the -1 exponent just reminds us that we are truly dealing with an inverse function. It is a notational convention that is used only as an alternate way of writing $\theta = \arccos x$. ■

If the reciprocal of $\cos x$ is meant, this text uses $(\cos x)^{-1}$. Also, to avoid confusion, the text uses the "arc" form to denote the relation.

The equation $y = \arccos x$ (Figure 15.13) is a relation because there are many values of y that satisfy the equation for any given value of x. In order to have a function, the range of the relation is restricted so that there is only one possible value of y for any given value of x. The restriction defines the inverse trigonometric function.

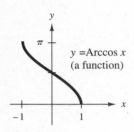

FIGURE 15.14

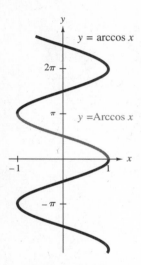

FIGURE 15.15

$$y = \text{Arccos } x$$

if, and only if,

$$x = \cos y,$$

where $\qquad -1 \le x \le 1, \qquad 0 \le y \le \pi.$

(The capital A in "arc" designates a function. A lowercase a designates a relation.)

$y = $ Arccos x (Figure 15.14) is a function because it represents a one-to-one relationship. There is one value of y for each value of x, whereas $y = $ arccos x is a relation because there are many values of y for each value of x.

EXAMPLE 1   Graph of Inverse Cosine

Graph $y = $ arccos x and $y = $ Arccos x

Solution   The relation $y = $ arccos x is equivalent to writing $x = \cos y$. Since we are already familiar with the graph of $y = \cos x$, all we have to do to obtain the graph of $x = \cos y$ is to interchange the axes (giving the same effect as interchanging the variables). This has been done to produce Figure 15.15. The darker curve is the graph of $y = $ arccos x. The lighter portion of the curve represents the inverse trigonometric function $y = $ Arccos x. The graph emphasizes that the range of $y = $ Arccos x must be restricted to $0 \le y \le \pi$ in order for $y = $ Arccos x to be a function.   ◆

The inverse trigonometric functions are defined for all of the trigonometric functions. The most commonly used inverse trigonometric functions (in addition to Arccos x) are $y = $ Arcsin x, $y = $ Arctan x, and $y = $ Arccot x. These functions along with their corresponding relations are listed in Table 15.7 and graphed in Figure 15.16.

TABLE 15.7   The Inverse Trigonometric Functions

Function	Domain	Range
$y = $ Arccos x	$-1 \le x \le 1$	$0 \le y \le \pi$
$y = $ Arcsin x	$-1 \le x \le 1$	$-\dfrac{\pi}{2} \le y \le \dfrac{\pi}{2}$
$y = $ Arctan x	$-\infty < x < \infty$	$-\dfrac{\pi}{2} < y < \dfrac{\pi}{2}$
$y = $ Arccot x	$-\infty < x < \infty$	$0 < y < \pi$
$y = $ Arccsc x	$x \le -1$ or $x \ge 1$	$-\dfrac{\pi}{2} \le y \le \dfrac{\pi}{2}, y \ne 0$
$y = $ Arcsec x	$x \le -1$ or $x \ge 1$	$0 \le y \le \pi, y \ne \dfrac{\pi}{2}$

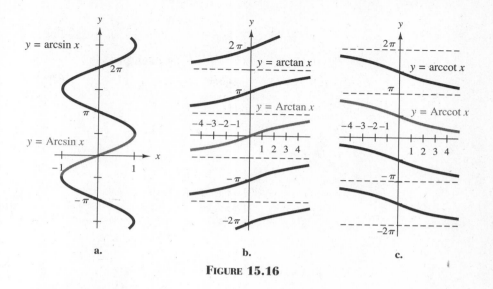

FIGURE 15.16

EXAMPLE 2   Evaluating Inverse Sine Functions

Evaluate the following.

a. $y = \arcsin \dfrac{1}{2}$ **b.** $y = \text{Arcsin } \dfrac{1}{2}$

Solution

a. $y = \arcsin \dfrac{1}{2}$ is equivalent to writing $\sin y = \dfrac{1}{2}$. Therefore, $y = \ldots -210°$, $30°$, $150°$, $390°$, ...

b. $y = \text{Arcsin } \dfrac{1}{2}$ restricts y to the range $-90° \le y \le 90°$. Therefore, $y = 30°$.

◆

Because graphics calculators graph only functions, graphing with the calculator is restricted to graphing the "Arc" functions.

EXAMPLE 3   Evaluating Inverse Functions

Evaluate the following with a calculator.

a. $y = \text{Arctan } 4$ **b.** $y = \text{Arccot } (-6.21)$ **c.** $y = \text{Arccos } 2$

Solution

a. $y = \text{Arctan } 4$ is equivalent to $\tan y = 4$, $-\dfrac{\pi}{2} < y < \dfrac{\pi}{2}$. Therefore, $y = 1.33$ rad.

b. $y = \text{Arccot}\ (-6.21)$ is equivalent to $\cot y = -6.21$, $0 < y < \pi$. Most calculators do not have keys for finding cotangent; but if we remember that $\cot y = \dfrac{1}{\tan y}$, we can proceed.

$$\cot y = -6.21$$
$$\frac{1}{\cot y} = \frac{1}{-6.21}$$
$$\tan y = -0.1610$$
$$y = -0.160 \text{ rad.}$$

This value of y cannot be a solution to $y = \text{Arccot}\ (-6.21)$ because it is outside the required range $0 < y < \pi$. Recalling the behavior of the tangent function, we remember that $y = -0.160 + \pi$ is also a solution of $\tan y = -0.1610$. This value of y falls within the required range so $y = 2.98$ rad.

c. $y = \arccos 2$ is equivalent to $\cos y = 2$, but we know that $-1 \le \cos y \le 1$, so $y = \text{Arccos}\ 2$ is undefined. ◆

There are times when we must evaluate an expression like $\sin\ (\text{Arccos}\ 6.5)$. This is illustrated in Example 4.

EXAMPLE 4   Evaluating Inverse Functions

Evaluate $y = \sin\left(\text{Arccos}\ \dfrac{\sqrt{3}}{2}\right)$.

Solution   Recalling that $\text{Arccos}\ \dfrac{\sqrt{3}}{2}$ is the angle whose cosine is $\dfrac{\sqrt{3}}{2}$, let $\theta = \text{Arccos}\ \dfrac{\sqrt{3}}{2}$. Then the given expression becomes $y = \sin \theta$.

$$y = \sin\left(\text{Arccos}\ \frac{\sqrt{3}}{2}\right)$$
$$\theta = \boxed{\text{Arccos}\ \frac{\sqrt{3}}{2}}$$
$$= \boxed{30°} \qquad\qquad \textbf{Evaluate Arccos}$$
$$y = \sin \theta$$
$$= \sin\ \boxed{30°}$$
$$= \frac{1}{2} \qquad\qquad \textbf{Evaluate sine} \quad ◆$$

EXAMPLE 5 Application: Calculating Strut Lengths

When calculating the strut lengths to construct a geodesic dome, it is necessary to evaluate:

$$y = \cos\left[2\,\text{Arcsin}\left(\frac{1}{2\sin 60°}\right)\right].$$

Determine y.

Solution

$$y = \cos\left[2\,\text{Arcsin}\left(\frac{1}{2\sin 60°}\right)\right]$$

$$= \cos\left[2\,\text{Arcsin}\left(\frac{1}{2\,(\,0.866\,)}\right)\right] \quad \textbf{Evaluate sine}$$

$$= \cos\left[2\,\text{Arcsin}\ 0.5774\ \right]$$

$$= \cos\left[2(\,35.26°\,)\right] \qquad\qquad \textbf{Evaluate Arcsin}$$

$$= \cos\left[\,70.53°\,\right]$$

$$= 0.333. \qquad\qquad\qquad \textbf{Evaluate cosine} \quad\blacklozenge$$

EXAMPLE 6 Application: Maximum Stress of Landing Gear

In developing the equations for maximum stress at a point in the landing gear of a space shuttle, it is neccessary to evaluate $q = \cos\left(\text{Arctan}\,\dfrac{a}{b}\right)$. What is q?

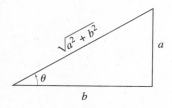

FIGURE 15.17

Solution To solve $q = \cos\left(\text{Arctan}\,\dfrac{a}{b}\right)$, we first must evaluate $\text{Arctan}\,\dfrac{a}{b}$.

$\text{Arctan}\,\dfrac{a}{b} = \theta$ or, equivalently, $\tan\theta = \dfrac{a}{b}$. Construct a right triangle and label it as shown in Figure 15.17. From the Pythagorean theorem the length of the hypotenuse of the triangle must be $\sqrt{a^2 + b^2}$. Therefore, we have:

$$q = \cos\left(\text{Arctan}\,\frac{a}{b}\right)$$

$$\theta = \text{Arctan}\,\frac{a}{b}$$

$$= \cos\theta \qquad\qquad \textbf{Where } \cos\theta = \frac{b}{\sqrt{a^2+b^2}} \textbf{ from Figure 15.17}$$

$$= \frac{b}{\sqrt{a^2+b^2}} \qquad\qquad \textbf{Substituting}$$

The solution to $q = \cos\left(\text{Arctan}\,\dfrac{b}{a}\right)$ is $q = \dfrac{b}{\sqrt{a^2+b^2}}$ $\quad\blacklozenge$

15.5 Exercises

In Exercises 1–6, **a.** graph the relation and the function specified, **b.** state the domain of each, and **c.** state the range of each.

1. $y = \arccos x$, $y = \text{Arccos } x$
2. $y = \arcsin x$, $y = \text{Arcsin } x$
3. $y = \arctan x$, $y = \text{Arctan } x$
4. $y = \text{arccot } x$, $y = \text{Arccot } x$
5. $y = \text{arccsc } x$, $y = \text{Arccsc } x$
6. $y = \text{arcsec } x$, $y = \text{Arcsec } x$

In Exercises 7–44, evaluate the expression.

7. $\text{Arcsin } 0$
8. $\text{Arccos } 0$
9. $\text{Arccos } \dfrac{1}{\sqrt{2}}$
10. $\text{Arcsin } \dfrac{1}{\sqrt{2}}$

11. $\text{Arctan } 1$
12. $\text{Arctan } (-1)$
13. $\text{Arctan } (-4)$
14. $\text{Arctan } 4$

15. $\text{Arcsin } \left(-\dfrac{1}{2}\right)$
16. $\text{Arccos } \left(-\dfrac{1}{2}\right)$
17. $\text{Arccos } \left(-\dfrac{\sqrt{3}}{2}\right)$
18. $\text{Arcsin } \left(-\dfrac{\sqrt{3}}{2}\right)$

19. $\text{Arccot } \dfrac{1}{\sqrt{3}}$
20. $\text{Arccot } \sqrt{3}$
21. $\text{Arccot } \left(-\dfrac{4}{3}\right)$
22. $\text{Arccot } \left(-\dfrac{3}{4}\right)$

23. $\text{Arcsin } 0.831$
24. $\text{Arccos } (-0.463)$
25. $\text{Arctan } (-21.6)$
26. $\text{Arccot } (16.3)$

27. $\cos \left(\text{Arcsin } \dfrac{1}{2}\right)$
28. $\sin \left(\text{Arccos } \dfrac{1}{2}\right)$
29. $\sin (\text{Arctan } 1)$
30. $\cos [\text{Arctan } (-1)]$

31. $\tan (\text{Arccos } 0.6)$
32. $\tan (\text{Arcsin } 0.8)$
33. $\cot (\text{Arccos } 0.6)$
34. $\cot (\text{Arccos } 0.8)$

35. $\cos (\text{Arccos } 0.621)$
36. $\sin [\text{Arcsin } (-0.527)]$
37. $\cos \left(2 \text{ Arcsin } \dfrac{4}{5}\right)$
38. $\cos \left(\dfrac{1}{2} \text{ Arccos } \dfrac{1}{2}\right)$

39. $\tan \left(3 \text{ Arctan } \dfrac{1}{\sqrt{3}}\right)$
40. $\cot \left(\dfrac{1}{2} \text{ Arctan } \sqrt{3}\right)$

41. $\tan (\text{Arccos } 0.112 + \text{Arcsin } 0.736)$
42. $\cos (\text{Arctan } 3.00 + \text{Arcsin } 0.500)$
43. $\cos (\text{Arccos } 0.416 + \text{Arcsin } 0.416)$
44. $\sin (\text{Arcsin } 0.643 + \text{Arccos } 0.643)$

In Exercises 45 and 46, find y as a function of a and b, $|a| \le |b|$, $b \ne 0$.

45. $y = \cos \left(\text{Arcsin } \dfrac{a}{b}\right)$
46. $y = \sin \left(\text{Arccos } \dfrac{a}{b}\right)$

In Exercises 47 and 48, find y as a function of x, $|x| \le 1$.

47. $y = \tan (\text{Arccos } x)$
48. $y = \cos (\text{Arctan } x)$

When constructing geodesic domes, it is necessary to evaluate functions such as those in Exercises 49–52. In each exercise, determine y.

49. $y = \text{Arctan } \left[\dfrac{\tan 10.57°}{\cos 36°}\right]$
50. $y = \text{Arctan } \left[\dfrac{\tan 21.15°}{\cos 36°}\right]$

51. $y = \text{Arctan } \left(\dfrac{1}{3} \sin 45°\right)$
52. $y = \cos \left\{2 \text{ Arcsin } \left[\dfrac{1}{(1 + \sqrt{5}) \sin 36°}\right]\right\}$

53. Why is the range of $y = \text{Arctan } x$ such that $-\dfrac{\pi}{2} < y < \dfrac{\pi}{2}$ instead of

$$-\dfrac{\pi}{2} \le y \le \dfrac{\pi}{2}?$$

54. Sharp curves are often banked at some angle β. The maximum speed V that a car can have without slipping is given by:

$$N \sin \beta = \frac{mV^2}{R},$$

where R is the radius of the turn. Another equation guarantees that the car doesn't sink into the surface is:

$$N \cos \beta = mg.$$

As an engineer you must solve for β and determine the angle of banking such that a car can turn a 230.0 m radius at a speed of 28.0 m/s (63.0 mph).

55. The range equation for a football may be given as:

$$R = \frac{V_0 \cos \alpha \, (2V_0 \sin \alpha)}{g},$$

where V_0 is the initial speed of the ball and α is the angle at which it is thrown. If $g = 9.8$ m/s², at what angle should a quarterback throw the ball so that it can be caught at a distance of 30 m down the field. The initial speed of the ball is 20 m/s.

56. It is often stated that a person on the surface of the earth has a speed of 1000 mph. This is only true if the person happens to be on the equator (latitude of 90°). The speed depends upon the distance between the axis of the earth and the surface given by:

$$V = 1041 \sin \theta,$$

where θ is the latitude of the person. (See the illustration.) At what latitude would the speed be 500 mph?

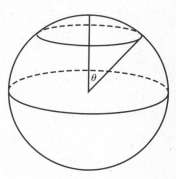

Review Exercises

In Exercises 1–4, prove the identities.

1. $\cot \theta = \cos \theta \csc \theta$

2. $\cos \theta \sec \theta = 1$

3. $\dfrac{\sin \theta}{\csc \theta} + \dfrac{\cos \theta}{\sec \theta} = 1$

4. $\sin \theta \csc \theta + \tan \theta \cot \theta = 2$

In Exercises 5–16, write the expression in terms of a single trigonometric function. (The function may be multiplied by a constant and/or raised to a power). Check the solution by comparing the graphs of the original statement with the solution.

5. $\sin \theta \tan \theta + \cos \theta$

6. $\csc \theta \sec \theta - \tan \theta$

7. $\sin \theta \, (\csc \theta - \sin \theta)$

8. $\tan \theta \, (\cot \theta + \tan \theta)$

9. $\dfrac{1 + \cos^2 A}{\sin^2 A} + 1$

10. $\dfrac{1 + \tan^2 A}{\tan^2 A}$

11. $\sin B \, (\csc B - \sin B)$

12. $\sin^2 B - \cos^2 B + 1$

13. $\sec A \csc A \tan A - 1$

14. $1 - \tan A \sin A \cos A$

15. $(1 + \sin \theta)(1 - \sin \theta)$

16. $(1 - \cos \theta)(1 + \cos \theta)$

In Exercises 17–20, show that each expression is equal to a constant.

17. $(\cos^2 \omega t - 1)(\tan^2 \omega t + 1) + \sec^2 \omega t$

18. $(\sin^2 \pi t + \cos^2 \pi t)^{\frac{1}{2}}$

19. $(1 + \cos \pi t)(\csc^2 \pi t - \csc^2 \pi t \cos \pi t)$

20. $(\csc \omega t + \cot \omega t)^2 (\csc \omega t - \cot \omega t)^2$

In Exercises 21–28, prove the identity.

21. $\dfrac{\sin \phi}{1 - \cos \phi} = \dfrac{1 + \cos \phi}{\sin \phi}$

22. $\dfrac{1}{\csc \phi - \cot \phi} = \csc \phi + \cot \phi$

23. $\cot^2 \omega t + \csc^2 \omega t = \csc^4 \omega t - \cot^4 \omega t$

24. $\dfrac{\cot \omega t}{1 + \csc \omega t} = \sec \omega t - \tan \omega t$

25. $\dfrac{1 + \tan \theta}{1 - \tan \theta} = \dfrac{\cot \theta + 1}{\cot \theta - 1}$

26. $\dfrac{\cos \omega t}{1 - \cos \omega t} = \dfrac{1 + \sec \omega t}{\tan^2 \omega t}$

27. $\dfrac{\cos 6t}{1 - \sin 6t} = \tan 6t + \sec 6t$

28. $(\tan \phi + 1) \csc \phi = \sec \phi + \csc \phi$

In Exercises 29–38, use the trigonometric identities to find exact values.

29. $\cos 285°$ **30.** $\cos 315°$ **31.** $\sin 315°$ **32.** $\sin 285°$

33. $\tan 285°$ **34.** $\tan 315°$ **35.** $\cot 315°$ **36.** $\cot 285°$

37. $\tan (-15°)$ **38.** $\cot (-15°)$

39. Show that $\cos (\theta - \phi) \neq \cos \theta - \cos \phi$.

40. Show that $\sin (\theta - \phi) \neq \sin \theta - \sin \phi$.

41. Use the identity for $\tan (\theta + \phi)$ to find an identity for $\tan 2\theta$.

42. Use the identity for $\cot (\theta + \phi)$ to find an identity for $\cot 2\theta$.

43. Find $\sin 2A$ as a function of $\tan A$.

44. Find $\cos 2A$ as a function of $\tan A$.

45. Show that $\cot \dfrac{\theta}{2} = \dfrac{1 + \cos \theta}{\sin \theta}$.

46. Show that $\cot \dfrac{\theta}{2} = \dfrac{\sin \theta}{1 - \cos \theta}$.

In Exercises 47–64, find θ, $0° \leq \theta < 360°$.

47. $\sec \theta = \csc \theta$

48. $\sin \theta - \cos \theta = 0$

49. $\sin \dfrac{1}{2}\theta - \cos \dfrac{1}{2}\theta = 0$

50. $\csc 2\theta + \sec 2\theta = 0$

51. $\cos \theta - \sin 2\theta = 0$

52. $\cos 4\theta = \sin 2\theta$

53. $\sin \dfrac{1}{2}\theta + \cos \theta - 1 = 0$

54. $4 \sin^2 \theta + \cos 2\theta - 2 = 0$

55. $\cot^2 \theta = 7 - 3 \sec^2 \theta$

56. $2 \tan \theta = \sec^2 \theta$

57. $2 \sin^2 \theta + \cos \theta - 1 = 0$

58. $\cos 2\theta + 3 \cos \theta + 2 = 0$

59. $3 \sec^2 \theta + \tan \theta = 5$

60. $2 \sin 2\theta \cos 2\theta = -1$

61. $\cos (\theta + 60°) = \sin (\theta + 60°)$

62. $\tan (\theta + 45°) - \tan (\theta - 45°) = 2$

63. $\cos \theta = 2 + \sqrt{3} \sin \theta$

64. $3 \cos \theta + 4 \sin \theta = 2$

The equations in Exercises 65–68 are taken from mechanical structures. Solve for θ, $0° \leq \theta \leq 360°$.

65. $4000 \sin \theta - 5000 \cos \theta = 1510$.

66. $-1580 \sin \theta + 2760 \cos \theta = 289$.

67. $-2000 \cos \theta - 4000 \sin \theta = -2460$.

68. $1800 \cos \theta + 1500 \sin \theta = 1180$.

In Exercises 69–90, evaluate the expression.

69. $\text{Arccos } \dfrac{\sqrt{3}}{2}$

70. $\text{Arccos } \dfrac{3}{5}$

71. $\text{Arcsin } \dfrac{3}{5}$

72. $\text{Arcsin } \left(-\dfrac{4}{5}\right)$

73. $\text{Arctan } \sqrt{3}$

74. $\text{Arctan } \left(-\dfrac{1}{3}\right)$

75. $\text{Arccot } \sqrt{3}$

76. $\text{Arccot } \left(-\dfrac{1}{\sqrt{3}}\right)$

77. $\sin\left(\text{Arccos } \dfrac{12}{13}\right)$

78. $\cos\left(\text{Arcsin } \dfrac{5}{13}\right)$

79. $\cos\left(\text{Arctan } \dfrac{15}{8}\right)$

80. $\sin\left(\text{Arctan } \dfrac{8}{15}\right)$

81. $\tan\left(\text{Arccos } \dfrac{24}{25}\right)$

82. $\tan\left(\text{Arcsin } \dfrac{7}{25}\right)$

83. $\tan\left(\text{Arctan } 0.114\right)$

84. $\cot\left[\text{Arccot } (-14.3)\right]$

85. $\sin\left(2\,\text{Arccos } \dfrac{4}{5}\right)$

86. $\sin\left(\dfrac{1}{2}\,\text{Arcsin } \dfrac{4}{5}\right)$

87. $\cos\left(\text{Arcsin } 0.715 + \text{Arccos } 0.883\right)$

88. $\sin\left(\text{Arctan } 4.12 - \text{Arccos } 0.347\right)$

89. $\cot\left(\text{Arccos } x\right)$

90. $\tan\left(\text{Arcsin } x\right)$

91. In deriving $\cos^2\theta + \sin^2\theta = 1$ (Section 15.1), why don't we have to worry about $r = 0$ when dividing by r^2?

92. When deriving $1 + \tan^2\theta = \sec^2\theta$ (Section 15.1), why don't we have to worry about $\theta = \dfrac{\pi}{2} + k\pi$ when dividing by $\cos^2\theta$?

93. Show that the sum of two sinusoids is itself a sinusoid, given

$$A_1 \cos(\omega t + \phi_1) + A_2 \cos(\omega t + \phi_2)$$

94. In geodesic dome design, it is necessary to solve the trigonometric equation:

$$\sin\left(\dfrac{360°}{n} + a\right) = \sin a,$$

(n is a constant) for a. Find a.

95. What is the difference between $y = \sin(\arcsin x)$ and $y = \sin(\text{Arcsin } x)$?

96. In calculating maximum normal stress, it is necessary to find $\cos 2\theta$ when it is known that:

$$\tan 2\theta = \dfrac{2b}{a - c}.$$

Find 2θ in terms of a, b, and c.

97. When light is cast symmetrically through a prism with vertex angle A, a constant angle of deviation ϕ is obtained. (See the drawing.) This angle of deviation depends only upon the index of refraction n and the vertex angle A.

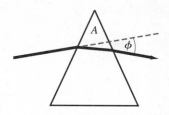

$$n = \sqrt{\frac{1 - \cos A \cos \phi + \sin A \sin \phi}{1 - \cos A}}$$

Rewrite the equation for n in the more commonly accepted form:

$$n = \frac{\sin \dfrac{(A + \phi)}{2}}{\sin \dfrac{A}{2}}.$$

98. The instantaneous current in a circuit may be expressed as:

$$i = I \cos \omega t.$$

To express only positive values, this equation can be modified as:

$$i^2 = I^2 \cos^2 \omega t.$$

Show how the second expression can be further modified to:

$$i^2 = \frac{1}{2}I^2 + \frac{1}{2} \cos \omega t.$$

99.* When x is in radians, $\sin x$ can be approximated by:

$$\sin x = x - \frac{x^3}{6} + \frac{x^5}{120} - \frac{x^7}{5040}.$$

Write a computer program to calculate approximate values for the sine of x when $0° \le x \le 360°$.

✐ Writing About Mathematics

1. In Section 15.5, two answers could be obtained for Exercises 55. Explain why one answer is more acceptable than the other.

2. A skier rides up a slope on a training rope as shown in the margin. The equation that best expresses the power required is:

$$P = k \frac{\sin \theta + \mu \cos \theta}{1 + \mu \tan \alpha}.$$

From any patterns you have noticed about trigonometric identities, explain why this expression cannot be further simplified.

3. Write a paragraph explaining why the solution to $\theta = \arccos 0.5$ is not the same as the solution to $\theta = \text{Arccos } 0.5$.

4. We have learned that $\sin^2 \theta + \cos^2 \theta = 1$ is an identity. However, $\sin \theta = \sqrt{1 - \cos^2 \theta}$ is not an identity. Explain why this is true.

Chapter Test

1. Rewrite the expression $\dfrac{1 + \cot x}{\cos x} - \sec x$ as a single trigonometric function.

In Exercises 2 and 3, prove the identities.

2. $\dfrac{\cos x}{1 - \sin x} = \tan x + \sec x$

3. $\dfrac{\cos x}{1 - \cos x} = \dfrac{1 + \sec x}{\tan^2 x}$

4. Use an identity for the sum of two angles to find the exact value for $\sin 225°$.

5. Use a double-angle identity to find the exact value of $\cos 60°$.

In Exercises 6–8, solve the trigonometric equations for $0° \le x \le 360°$.

6. $\cos^2 x - 3 \sin^2 x + 1 = 0$

7. $\tan^2 x - 4 \tan x = 0$

8. $\tan^2 x = 2 \sec x$

In Exercises 9–11, evaluate the expression.

9. $\text{Arcsin } \dfrac{\sqrt{2}}{2}$

10. $\tan \left(\text{Arccos } \dfrac{\sqrt{3}}{2} \right)$

11. $\text{Arctan } (-\sqrt{3})$

12. Use the identity for $\sin (A + B)$ to find an identity for $\sin 2x$.

13. Prove the identity $\sin (\pi - \theta) = \sin \theta$.

14. Use your graphing calculator to determine whether the following equation is an identity. If not, determine a value of x to show why not.

$$\sec x - \sin x \tan x = \csc x$$

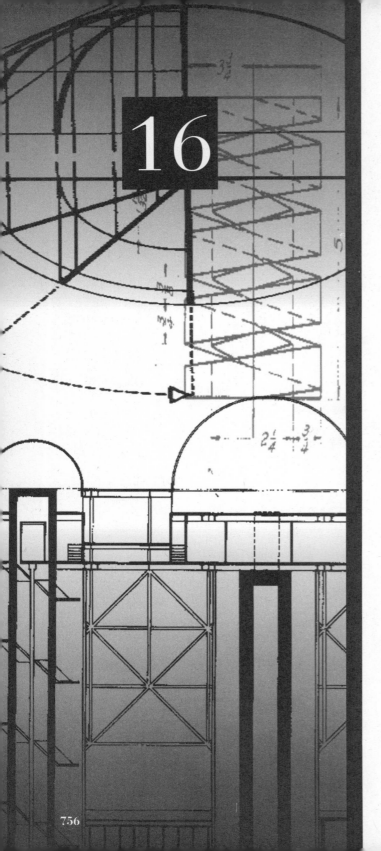

16

Sequences and Series

When purchasing screws, we probably are looking for screws with specific diameters. The diameters of screws increase in discrete amounts, starting with size number 1 (which has a diameter of 0.073 in.). The sequence of sizes is interrupted because the two sizes closest to $\frac{1}{4}$ in. and $\frac{5}{16}$ in. are omitted so that these more common fractional sizes may be included instead.

When we analyze the sequence of screw sizes, as we are asked to do in Example 7 of Section 16.3, we can find the common difference in sizes and write the remaining terms of the sequence. This enables us to identify the two number sizes that are omitted.

16.1

Sequences

Sequences are used frequently in many branches of mathematics and science. A sequence can be thought of as a finite or infinite list of numbers. Examples of sequences include the following.

$$1, 3, 5, 7, 9, 11, 13 \qquad \textbf{Finite}$$

$$-1, 1, 3, 5, 7, 9, 11, 13, \ldots \quad \textbf{Infinite}$$

$$0, \frac{1}{2}, 1, \frac{3}{2}, 2, \frac{5}{2}, \ldots \qquad \textbf{Infinite}$$

$$-5, 0, 5, 10, 15 \qquad \textbf{Finite}$$

Many real-life situations are described in terms of an arithmetic or geometric sequence. The swing of a pendulum that is decreasing at a set rate is an example of an arithmetic sequence. The yearly amount of money accumulated in a savings account that compounds interest at regular intervals is an example of a geometric sequence.

A more precise definition of a sequence is based on the understanding of the concept of function. Recall from Chapter 5 that a function is a rule that determines a unique element of the range for each element of the domain.

DEFINITION 16.1

A **sequence** is a function whose domain is a subset of the set of positive integers (1, 2, 3, . . .).

The elements of the range of any sequence are called its **terms.** To represent a sequence with functional notation we write $f(1) = a_1, f(2) = a_2, f(3) = a_3, f(4) = a_4$, and in general, $f(i) = a_i$. When these terms are written in their natural order, the following sequences occur.

$a_1, a_2, a_3, \ldots, a_i, \ldots, a_n$ **Represents a finite sequence ending at a_n**

$a_1, a_2, a_3, \ldots, a_i, \ldots$ **Represents an infinite sequence**

It is important to be able to write the **ith term** or the **general term of a sequence.** The ith term can be used to generate other terms of that sequence. In Example 1 we see that a_i is defined as an algebraic expression.

EXAMPLE 1 Writing Terms of a Sequence

Write terms for each of the sequences for the indicated domain.

a. For $a_i = i^3 + 1$, $D = 1, 2, 3, 4$,

b. For $a_i = \dfrac{1}{i^2}$, $D =$ set of positive integers.

Solution

a. For $a_i = i^3 + 1$, with the domain limited to the first four counting numbers, the result is a finite sequence. Substituting we find the following terms of the sequence.

$$a_1 = 1^3 + 1 \quad a_2 = 2^3 + 1 \quad a_3 = 3^3 + 1 \quad a_4 = 4^3 + 1$$
$$= 2 \qquad\qquad = 9 \qquad\qquad = 28 \qquad\qquad = 65$$

Thus, the finite sequence whose ith term is $a_i = i^3 + 1$ is 2, 9, 28, 65.

b. For $a_i = \dfrac{1}{i^2}$, with the domain the set of positive integers, the result is an infinite sequence. Substituting we find the following terms of the sequence.

$$a_1 = \frac{1}{1^2} \quad a_2 = \frac{1}{2^2} \quad a_3 = \frac{1}{3^2}$$
$$= 1 \qquad\quad = \frac{1}{4} \qquad\quad = \frac{1}{9}$$

$$a_4 = \frac{1}{4^2} \quad a_5 = \frac{1}{5^2} \quad a_6 = \frac{1}{6^2}$$
$$= \frac{1}{16} \qquad = \frac{1}{25} \qquad = \frac{1}{36}$$

Thus, the infinite sequence whose ith term is $a_i = \dfrac{1}{i^2}$ is

$$1, \frac{1}{4}, \frac{1}{9}, \frac{1}{16}, \frac{1}{25}, \frac{1}{36}, \ldots, \frac{1}{i^2}, \ldots \quad \blacklozenge$$

Since a sequence is a function, it can be graphed. The graph consists of the points $(1, a_1)$, $(2, a_2)$, $(3, a_3)$, . . . , determined from the general term and the domain. It is impossible to show all the points of an infinite sequence. However, a finite set of points usually can provide a picture of the sequence.

EXAMPLE 2   Writing Terms of and Graphing a Sequence

a. Determine the first six terms of the sequence given by $a_i = 3i - 12$.
b. Graph the first six terms of the sequence in part a.

Solution

a. Replacing i with 1, 2, 3, 4, 5, and 6, we have the first six terms.

$$a_1 = 3(1) - 12 \quad a_2 = 3(2) - 12 \quad a_3 = 3(3) - 12$$
$$= -9 \qquad\qquad = -6 \qquad\qquad = -3$$
$$a_4 = 3(4) - 12 \quad a_5 = 3(5) - 12 \quad a_6 = 3(6) - 12$$
$$= 0 \qquad\qquad = 3 \qquad\qquad = 6$$

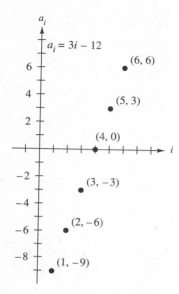

$a_i = 3i - 12$

FIGURE 16.1

b. The six points of the sequence to be graphed are $(1, -9)$, $(2, -6)$, $(3, -3)$, $(4, 0)$, $(5, 3)$, and $(6, 6)$. The sequence is graphed in Figure 16.1. ◆

EXAMPLE 3 **Writing Terms of and Graphing a Sequence**

a. Determine the first six terms of the sequence given by $a_i = 3 - \dfrac{i}{4}$.

b. Graph the first six terms of the sequence in part a.

Solution

a. Replacing i with $1, 2, 3, \ldots, 6$, we have the first six terms.

$$a_1 = 3 - \frac{1}{4} \qquad a_2 = 3 - \frac{2}{4} \qquad a_3 = 3 - \frac{3}{4}$$

$$= 2\frac{3}{4} \qquad\qquad = 2\frac{1}{2} \qquad\qquad = 2\frac{1}{4}$$

$$a_4 = 3 - \frac{4}{4} \qquad a_5 = 3 - \frac{5}{4} \qquad a_6 = 3 - \frac{6}{4}$$

$$= 2 \qquad\qquad = 1\frac{3}{4} \qquad\qquad = 1\frac{1}{2}$$

b. The six terms of the sequence to be graphed are $\left(1, 2\frac{3}{4}\right)$, $\left(2, 2\frac{1}{2}\right)$, $\left(3, 2\frac{1}{4}\right)$, $(4, 2)$, $\left(5, 1\frac{3}{4}\right)$, and $\left(6, 1\frac{1}{2}\right)$. The sequence is graphed in Figure 16.2. ◆

We can see from the terms of the sequence in Example 2 and from Figure 16.1 that each succeeding term is larger than the preceding term. When this happens, the sequence is said to be an **increasing sequence.** In Example 3 and Figure 16.2 each succeeding term is smaller than the preceding term. This type of sequence is a **decreasing sequence.**

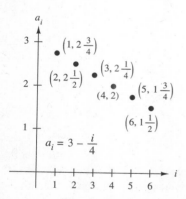

$a_i = 3 - \dfrac{i}{4}$

FIGURE 16.2

DEFINITION 16.2

A sequence $a_1, a_2, a_3, \ldots, a_i, \ldots$, is said to be **increasing** if $a_i < a_{i+1}$ for all $i = 1, 2, 3, \ldots$, and **decreasing** if $a_i > a_{i+1}$ for all $i = 1, 2, 3, \ldots$.

EXAMPLE 4 **Increasing or Decreasing Sequence?**

Determine whether the sequence is increasing or decreasing.

a. $a_i = -2i$ **b.** $a_i = 2i + 3$

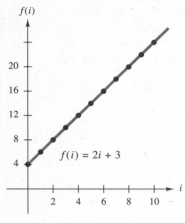

FIGURE 16.3

Solution

a. The first four terms of the sequence $a_i = -2i$ are:

$$a_1 = -2,\ a_2 = -4,\ a_3 = -6,\ \text{and}\ a_4 = -8.$$

We see that $a_1 > a_2$, $a_2 > a_3$, $a_3 > a_4$, and in general, $a_i > a_{i+1}$ for all i. Therefore, the sequence $a_i = -2i$ is a *decreasing sequence*.

b. The first four terms of the sequence $2i + 3$ are:

$$a_1 = 5,\ a_2 = 7,\ a_3 = 9,\ \text{and}\ a_4 = 11.$$

We see that $a_1 < a_2$, $a_2 < a_3$, $a_3 < a_4$, and in general, $a_i > a_{i+1}$ for all i. Therefore, the sequence $a_i = 2i + 3$ is an *increasing sequence*. ◆

Another way of determining whether a sequence is increasing or decreasing is to graph the function. For example, if we graph the function $f(i) = 2i + 3$, as shown in Figure 16.3, we can see that the function is increasing. As we move from left to right, a line drawn through the points is moving upward, and the line has a positive slope. If a line drawn through the set of points is moving downward, the slope of the line is negative, and the sequence is decreasing.

There are sequences that do not increase or decrease; one is illustrated in Example 5.

EXAMPLE 5   Nonincreasing Nondecreasing Sequence

Determine the first six terms of the sequence $a_i = 1 + (-1)^i$.

Solution   The first six terms of the sequence are:

$$a_1 = 1 + (-1)^{1} \qquad a_3 = 1 + (-1)^{3} \qquad a_5 = 1 + (-1)^{5}$$
$$= 0 \qquad\qquad\quad = 0 \qquad\qquad\quad = 0$$
$$a_2 = 1 + (-1)^{2} \qquad a_4 = 1 + (-1)^{4} \qquad a_6 = 1 + (-1)^{6}$$
$$= 2 \qquad\qquad\quad = 2 \qquad\qquad\quad = 2.$$

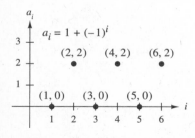

FIGURE 16.4

We see from the results that $a_1 < a_2$, but that $a_2 > a_3$. If i is odd, then a_i is 0, and if i is even, then a_i is 2. Thus, the sequence does not increase nor does it decrease. The sequence is 0, 2, 0, 2, 0, ... The graph of the first six terms of the sequence is given in Figure 16.4. ◆

Finding the terms of a sequence when given the general term (ith term) is a relatively simple task. Can we reverse the process? Can we determine the general term (or a formula for the general term) when we know only a few terms of the sequence? Example 6 illustrates a method of determining the ith term.

EXAMPLE 6   Determining the ith Term

Determine a formula for the ith term of the sequence 4, 7, 10, 13, 16, 19, ...

Solution   The technique for finding the ith term may involve some guessing. When we examine the terms of the sequence, we see that each term can be written as the previous number plus 3. We also see that the first term, 4, can be written as $(1 + 3)$. The second term, 7, can be written as $4 + 3 = 1 + 3 + 3$. Each of the succeeding terms can be written as:

$$1 + 3 + 3 + \cdots + 3.$$

$$4 = 1 + 3$$
$$7 = 4 + 3 \ = 1 + 3 + 3$$
$$10 = 7 + 3 \ = 1 + 3 + 3 + 3$$
$$13 = 10 + 3 = 1 + 3 + 3 + 3 + 3$$
$$16 = 13 + 3 = 1 + 3 + 3 + 3 + 3 + 3$$
$$19 = 16 + 3 = 1 + 3 + 3 + 3 + 3 + 3 + 3.$$

Having written a few terms of the sequence, we see that a pattern is present. Each term can be written as 1 plus 3 times a counting number. The first term would be $1 + 3(1)$ and the second term would be $1 + 3(2)$. The counting number corresponds to the number of the term in the sequence:

$$a_1 = 1 + 3(1) = 4$$
$$a_2 = 1 + 3(2) = 7$$
$$a_3 = 1 + 3(3) = 10$$
$$a_4 = 1 + 3(4) = 13$$
$$a_5 = 1 + 3(5) = 16$$
$$a_6 = 1 + 3(6) = 19$$
$$\vdots$$
$$a_i = 1 + 3(i).$$

The formula for the ith term of the sequence 4, 7, 10, 13, 16, 19, . . . , is $a_i = 1 + 3i$. ◆

EXAMPLE 7 **Determining the ith Term**

Determine a formula for the ith term of the following sequences.

a. $-1, 1, -1, 1, -1, \ldots$
b. $-2, 4, -8, 16, -32, 64, \ldots$

Solution

a. For the sequence $-1, 1, -1, 1, -1, \ldots$, we see that the terms are all the same except for the sign. The negative signs appear in every other term, which corresponds to raising (-1) to an odd power. Likewise, raising (-1) to even power will give us the positive terms. Therefore,

$$a_1 = (-1)^1 = -1$$
$$a_2 = (-1)^2 = 1$$
$$a_3 = (-1)^3 = -1$$
$$\vdots$$
$$a_i = (-1)^i.$$

The formula for the ith term of the sequence $-1, 1, -1, 1, -1, \ldots$ is $a_i = (-1)^i$.

b. Each of the terms of the sequence $-2, 4, -8, 16, -32, 64, \ldots$, can be written as follows:

$$-2 = -1 \cdot 2, \qquad 4 = 1 \cdot 4, \qquad -8 = -1 \cdot 8,$$
$$6 = 1 \cdot 16, \qquad -32 = -1 \cdot 32, \qquad 64 = 1 \cdot 64, \ldots.$$

Note that the terms alternate in sign as does the series in part a. Therefore, we can write the first factor as $(-1)^1, (-1)^2, (-1)^3 \ldots$. In each case, the second factor is the second factor of the previous term multiplied by 2. In fact, each of these factors can be expressed as 2 to a power. The power of each factor is the same as the number of that term (i.e., 2^3 is the factor of the third term).

$$2 \qquad\qquad 2 \qquad\qquad\qquad = 2^1$$
$$4 = 2 \cdot 2 \quad = 2 \cdot 2 \qquad\qquad\quad = 2^2$$
$$8 = 4 \cdot 2 \quad = 2 \cdot 2 \cdot 2 \qquad\qquad = 2^3$$
$$16 = 8 \cdot 2 \quad = 2 \cdot 2 \cdot 2 \cdot 2 \qquad\quad = 2^4$$
$$32 = 16 \cdot 2 = 2 \cdot 2 \cdot 2 \cdot 2 \cdot 2 \qquad = 2^5$$
$$64 = 32 \cdot 2 = 2 \cdot 2 \cdot 2 \cdot 2 \cdot 2 \cdot 2 = 2^6$$
$$\vdots$$

Therefore, we can write the terms of the sequence:

$$a_1 = \quad -2 = (-1)^1(2)^1 = (-2)^1$$
$$a_2 = \quad\;\; 4 = (-1)^2(2)^2 = (-2)^2$$
$$a_3 = \quad -8 = (-1)^3(2)^3 = (-2)^3$$
$$a_4 = \quad\;\; 16 = (-1)^4(2)^4 = (-2)^4$$
$$a_5 = -32 = (-1)^5(2)^5 = (-2)^5$$
$$a_6 = \quad\;\; 64 = (-1)^6(2)^6 = (-2)^6$$
$$\vdots$$
$$a_i = \qquad\quad (-1)^i(2)^i = (-2)^i.$$

The formula for the ith term of the sequence $-2, 4, -8, 16, -32, 64, \ldots$, is $a_i = (-2)^i$. ◆

Because there may be more than one way of expressing the formula for the general term, after we write the term, test the formula and determine whether or not it is generating the terms of the sequence.

16.1 Exercises

In Exercises 1–20, **a.** write the first six terms of the sequence; **b.** indicate whether the sequence is increasing, decreasing, or neither; and **c.** graph the sequence.

1. $a_i = 2i$

2. $a_i = -3i$

3. $a_i = i + 3$

4. $a_i = 4 - i$

5. $a_i = 3i - 5$

6. $a_i = 6 - 4i$

7. $a_i = -\dfrac{3i}{4}$

8. $a_i = \dfrac{4i}{3}$

9. $a_i = -1 + (-1)^i$

10. $a_i = 2 + (-1)^i$

11. $a_i = -1 + 2(-1)^i$

12. $a_i = 2 + 2(-1)^i$

13. $a_i = i^2 - 1$

14. $a_i = i^2 - 6$

15. $a_i = \dfrac{i + 2}{i + 3}$

16. $a_i = \dfrac{i - 2}{i - 3}$

17. $a_i = \dfrac{i^2 - 1}{i^2 + 1}$

18. $a_i = \dfrac{i^2 + 1}{i^2 - 1}$

19. $a_i = \dfrac{(-1)^{i+1}}{i}$

20. $a_i = \dfrac{(-1)^{i-1}}{i}$

In Exercises 21–40, determine a formula for the ith term of each of the sequences.

21. $0, 1, 2, 3, 4, 5, \ldots$

22. $3, 4, 5, 6, 7, 8, \ldots$

23. $2, 4, 6, 8, 10, 12, \ldots$

24. $3, 6, 9, 12, 15, 18, \ldots$

25. $1, 3, 5, 7, 9, 11, \ldots$

26. $3, 5, 7, 9, 11, 13, \ldots$

27. $3, 7, 11, 15, 19, 23, \ldots$

28. $2, 5, 8, 11, 14, 17, \ldots$

29. $-2, -4, -6, -8, -10, -12, \ldots$

30. $-3, -6, -9, -12, -15, -18, \ldots$

31. $4, 5, 6, 7, 8, 9, \ldots$

32. $-2, -1, 0, 1, 2, 3, \ldots$

33. $-4, 0, 4, 8, 12, 16, \ldots$

34. $-3, 9, -27, 81, -243, 729, \ldots$

35. $1, \dfrac{1}{2}, \dfrac{1}{3}, \dfrac{1}{4}, \dfrac{1}{5}, \dfrac{1}{6}, \ldots$

36. $\dfrac{1}{2}, \dfrac{1}{4}, \dfrac{1}{6}, \dfrac{1}{8}, \dfrac{1}{10}, \dfrac{1}{12}, \ldots$

37. $1, \dfrac{1}{4}, \dfrac{1}{9}, \dfrac{1}{16}, \dfrac{1}{25}, \dfrac{1}{36}, \ldots$

38. $1, \dfrac{1}{8}, \dfrac{1}{27}, \dfrac{1}{64}, \dfrac{1}{125}, \dfrac{1}{216}, \ldots$

39. $2, \dfrac{3}{2}, \dfrac{4}{3}, \dfrac{5}{4}, \dfrac{6}{5}, \dfrac{7}{6}, \ldots$

40. $0, \dfrac{-1}{2}, \dfrac{-2}{3}, \dfrac{-3}{4}, \dfrac{-4}{5}, \dfrac{-5}{6}, \ldots$

In Exercises 41–46, determine the 6th term of the sequence and write a formula for the ith term.

41. $x, 2x, 3x, 4x, 5x$

42. $x + 2, x + 3, x + 4, x + 5, x + 6$

43. $x + 3y, x + 6y, x + 9y, x + 12y, x + 15y$

44. $2z - 4y, 4z - 4y, 6z - 4y, 8z - 4y, 10z - 4y$

45. $x, x + 5y, x + 10y, x + 15y, x + 20y$

46. $2x + 3y, 4x + 6y, 6x + 9y, 8x + 12y, 10x + 15y$

47. A newspaper carrier earning $15.50 per week is depositing $3.50 per week in savings. The carrier decides to increase the amount deposited by 10% per week. Write a sequence that gives the amount being deposited the 1st week, 2nd week, 3rd week, 4th week, 5th week, and 6th week.

48. Assume that the carrier in Exercise 47 is making $22.00 per week and is spending $21.00 per week. The carrier decides to decrease the amount spent each week by 5%. Write a sequence that represents the amount that is being spent each week for the first six weeks.

16.2

Series and Summation

The sum of the terms of a sequence is called a series. For the sequence 1, 3, 5, 7, 9, the series associated with it is:

$$S_5 = 1 + 3 + 5 + 7 + 9$$
$$= 25.$$

The capital letter S indicates sum, and the subscript 5 indicates that we are finding the sum of the first five terms.

> **DEFINITION 16.3**
>
> The sum S_n of an indicated number n of terms of a sequence is called a **series**.

If the sequence has a finite number of terms, then the corresponding series is called a **finite series**:

$$S_n = a_1 + a_2 + a_3 + \cdots + a_n.$$

If the sequence has an infinite number of terms, then the corresponding series is called an **infinite series**:

$$a_1 + a_2 + a_3 + a_4 + \cdots + a_i + \cdots.$$

EXAMPLE 1 **Determining the Series of a Sequence**

Determine the series associated with the sequence 3, 9, 27, 81, 243, . . . , 3^n for the following.

a. $n = 3$ **b.** $n = 6$

Solution

a.

$$S_3 = (3)^1 + (3)^2 + (3)^3$$
$$= 3 + 9 + 27$$
$$= 39.$$

b.

$$S_6 = (3)^1 + (3)^2 + (3)^3 + (3)^4 + (3)^5 + (3)^6$$
$$= 3 + 9 + 27 + 81 + 243 + 729$$
$$= 1092. \quad \blacklozenge$$

A series for which the general term is known can be written in a compact form using **summation notation** or **sigma notation.** (Sigma is the Greek letter for capital S.) For example, the sum S_n of the sequence with general term a_i can be written using summation notation as follows:

$$S_n = \sum_{i=1}^{n} a_i = a_1 + a_2 + a_3 + a_4 + \cdots + a_n.$$

The capital Greek letter sigma, Σ, stands for the "sum of." The letter or number under the Σ is called the *index of summation*. The **index of summation** indicates the starting term of the sequence. The letter or number on top of the Σ indicates the ending term of the sequence. The letter i under the Σ is called a *dummy variable* and could be replaced by any letter. Thus, we could write:

$$\sum_{i=1}^{n} a_i \quad \text{or} \quad \sum_{j=1}^{n} a_j \quad \text{or} \quad \sum_{k=1}^{n} a_k.$$

The important parts of the summation notation are 1, n, and the expression for the general term. Thus, the series on the right of the equal sign is the sum of the terms a_1 through a_n.

Knowing the general term of the sequence

$$\frac{2}{3}, \frac{3}{4}, \frac{4}{5}, \frac{5}{6}, \frac{6}{7}, \frac{7}{8}, \ldots, \frac{1+n}{2+n},$$

We can write the corresponding series in summation notation as:

$$S_n = \sum_{i=1}^{n} \frac{1+i}{2+i},$$

or in expanded form as:

$$S_n = \frac{2}{3} + \frac{3}{4} + \frac{4}{5} + \frac{5}{6} + \cdots + \frac{1+n}{2+n}.$$

This will be illustrated in Examples 2 and 3.

EXAMPLE 2 **Series in Expanded Form**

Write the terms of the series $\sum_{i=1}^{6} 3i$ in expanded form.

Solution

$$\sum_{i=1}^{6} 3i = 3(\boxed{1}) + 3(\boxed{2}) + 3(\boxed{3}) + 3(\boxed{4}) + 3(\boxed{5}) + 3(\boxed{6})$$

$$= 3 + 6 + 9 + 12 + 15 + 18 \quad \blacklozenge$$

EXAMPLE 3   Series in Expanded Form

Write the series $\sum_{i=1}^{5} (-1)^i \left(\dfrac{3}{i}\right)$ in expanded form.

Solution

$$\sum_{i=1}^{5} (-1)^i \left(\frac{3}{i}\right) = (-1)^1 \left(\frac{3}{1}\right) + (-1)^2 \left(\frac{3}{2}\right) + (-1)^3 \left(\frac{3}{3}\right)$$
$$+ (-1)^4 \left(\frac{3}{4}\right) + (-1)^5 \left(\frac{3}{5}\right)$$
$$= (-1)(3) + (1)\left(\frac{3}{2}\right) + (-1)(1) + (1)\left(\frac{3}{4}\right) + (-1)\left(\frac{3}{5}\right)$$
$$= -3 + \frac{3}{2} - 1 + \frac{3}{4} - \frac{3}{5}. \quad \blacklozenge$$

Examples 4 and 5 demonstrate how to write a series in summation notation.

EXAMPLE 4   Writing the Series in Summation Notation

Write the series $2 + 4 + 8 + 16 + 32 + 64 + 128$ using summation notation.

Solution   Each term of the series is 2 raised to a power:

$$2 = 2^1, 4 = 2^2, 8 = 2^3, \ldots, 128 = 2^7.$$

With $i = 1$ and $n = 7$, we can write the series in summation notation as $\sum_{i=1}^{7} 2^i$. $\blacklozenge$

EXAMPLE 5   Writing the Series in Summation Notation

Write the series $3 + 5 + 7 + 9 + \cdots + (2n + 1)$ using summation notation.

Solution   Since we are given the general term, all we need do is determine the index and the value for n. If we substitute 1, 2, and 3 for n in the general term, the results are 3, 5, and 7, respectively. The index on the summation symbol is 1, and the letter indicating the ending term is n. The series written in summation notation is $\sum_{k=1}^{n} (2k + 1)$. $\blacklozenge$

EXAMPLE 6 Writing an Alternating Series in Summation Notation

Write the following series using summation notation.

$$1 - \frac{3}{2} + \frac{9}{5} - 2 + \frac{15}{7} - \frac{9}{4} + \cdots + \frac{(-1)^{n+1}3n}{n+2}$$

Solution Since we are given the general term, as in Example 5, we need to determine the index. The factor $(-1)^{n+1}$ in the general term causes the signs of the terms to alternate from positive to negative. For example when $n = 1, 2$ and 3, the respective terms of the series are $1, -\frac{3}{2}$, and $\frac{9}{5}$. The index is 1, and the letter indicating the ending term is n. The series written in summation notation is $\sum\limits_{k=1}^{n} \frac{(-1)^{k+1}3k}{k+2}$ ◆

EXAMPLE 7 Series in Expanded Form

Write the series $\sum\limits_{i=4}^{7} (2i^2 + 3)$ in expanded form and determine the sum.

Solution

$$\sum_{i=4}^{7} (2i^2 + 3) = [2(4)^2 + 3] + [2(5)^2 + 3] + [2(6)^2 + 3]$$
$$+ [2(7)^2 + 3]$$
$$= 35 + 53 + 75 + 101$$
$$= 264 \quad ◆$$

EXAMPLE 8 Writing a Series in Summation Notation

Write the series $(y^2 + 5) + (y^2 + 6) + (y^2 + 7) + (y^2 + 8)$ using summation notation.

Solution Each term of the series is made up of the constant y^2 and a counting number that increases by one with each term. Since the first term starts with 5, the index is 5 and the top number is 8. Thus, writing the series in summation notation, we have $\sum\limits_{i=5}^{8} (y^2 + i)$. ◆

16.2 Exercises

In Exercises 1–10, write the series associated with each of the sequences for **a.** S_3 and
b. S_8.

1. $2, 4, 8, 16, \ldots, 2^n$

2. $-2, 4, -8, 16, \ldots, (-1)^n 2^n$

3. $\dfrac{1}{3}, \dfrac{1}{9}, \dfrac{1}{27}, \dfrac{1}{81}, \ldots, \left(\dfrac{1}{3}\right)^n$

4. $1, -\dfrac{1}{3}, \dfrac{1}{9}, -\dfrac{1}{27}, \ldots, \left(\dfrac{-1}{3}\right)^{n-1}$

5. $5, 7, 9, 11, \ldots, (2n + 3)$

6. $\dfrac{1}{4}, 0, -\dfrac{1}{6}, -\dfrac{2}{7}, \ldots, \dfrac{2 - n}{3 + n}$

7. $-\dfrac{1}{4}, \dfrac{0}{5}, \dfrac{1}{6}, \dfrac{2}{7}, \ldots, \dfrac{n - 2}{n + 3}$

8. $-2, 16, -216, 4096, \ldots, (-2n)^n$

9. $1, -2, 3, -4, \ldots, (-1)^{n+1} n$

10. $1, -4, 9, -16, \ldots, (-1)^{n+1} n^2$

In Exercises 11–20, write each of the series in expanded form and determine the sum.

11. $S_5 = \displaystyle\sum_{i=1}^{5} \dfrac{(-2)^{i+1}}{i}$

12. $S_4 = \displaystyle\sum_{i=1}^{4} (-1)^i (2i - 1)^2$

13. $S_4 = \displaystyle\sum_{i=1}^{4} \dfrac{1}{i} 3^{i+1}$

14. $S_6 = \displaystyle\sum_{i=1}^{6} e^{i-2}$

15. $S_6 = \displaystyle\sum_{i=1}^{6} \dfrac{(-1)^i 2^i}{i}$

16. $S_6 = \displaystyle\sum_{i=2}^{7} \dfrac{(-1)^i 2^{i+1}}{2i + 1}$

17. $S_5 = \displaystyle\sum_{i=1}^{5} \log(2 + i)$

18. $S_7 = \displaystyle\sum_{i=1}^{7} (|i - 2| - |3 - i|)$

19. $\displaystyle\sum_{i=2}^{8} \dfrac{i + 1}{i + 5}$

20. $\displaystyle\sum_{i=1}^{5} (2i - 1)(i + 2)$

In Exercises 21–36, write each series using summation notation.

21. $3 + 6 + 9 + 12 + 15 + \cdots + 3n$

22. $\dfrac{2}{2} + \dfrac{4}{4} + \dfrac{6}{8} + \dfrac{8}{16} + \dfrac{10}{32} + \cdots + \dfrac{2n}{2^n}$

23. $\dfrac{-1}{1} + \dfrac{2}{2} - \dfrac{3}{4} + \dfrac{4}{8} + \dfrac{5}{16} + \cdots + \dfrac{(-1)^n n}{2^{n-1}}$

24. $\dfrac{1}{1} + \dfrac{2}{3} + \dfrac{3}{9} + \dfrac{4}{27} + \dfrac{5}{81} + \cdots + \dfrac{n}{3^{n-1}}$

25. $\dfrac{1}{1^2} + \dfrac{1}{2^2} + \dfrac{1}{3^2} + \dfrac{1}{4^2} + \dfrac{1}{5^2}$

26. $\dfrac{1}{1^3} + \dfrac{1}{2^3} + \dfrac{1}{3^3} + \dfrac{1}{4^3} + \dfrac{1}{5^3}$

27. $\dfrac{1}{1^2} - \dfrac{1}{2^2} + \dfrac{1}{3^2} - \dfrac{1}{4^2} + \dfrac{1}{5^2}$

28. $\dfrac{1}{1^3} - \dfrac{1}{2^3} + \dfrac{1}{3^3} - \dfrac{1}{4^3} + \dfrac{1}{5^3}$

29. $\dfrac{1 \cdot 2}{2} + \dfrac{2 \cdot 3}{2} + \dfrac{3 \cdot 4}{2} + \dfrac{4 \cdot 5}{2} + \dfrac{5 \cdot 6}{2} + \cdots + \dfrac{n(n + 1)}{2}$

30. $\dfrac{3}{1 \cdot 1} + \dfrac{3}{2 \cdot 3} + \dfrac{3}{3 \cdot 5} + \dfrac{3}{4 \cdot 7} + \dfrac{3}{5 \cdot 9} + \cdots + \dfrac{3}{n(2n - 1)}$

31. $(z + 4) + (z + 5) + (z + 6) + (z + 7) + (z + 8)$

32. $z^2 + z^3 + z^4 + z^5 + z^6$

33. $\dfrac{z - 3}{z + 3} + \dfrac{z - 4}{z + 4} + \dfrac{z - 5}{z + 5} + \dfrac{z - 6}{z + 6} + \dfrac{z - 7}{z + 7}$

34. $\dfrac{z^2}{(z + 2)^2} + \dfrac{z^3}{(z + 3)^2} + \dfrac{z^4}{(z + 4)^2} + \dfrac{z^5}{(z + 5)^2} + \dfrac{z^6}{(z + 6)^2}$

35. $z^3(z + 3) + z^4(z + 4) + z^5(z + 5) + z^6(z + 6) + z^7(z + 7)$

36. $(-1)^5(z^5 - 5) + (-1)^6(z^6 - 6) + (-1)^7(z^7 - 7) + (-1)^8(z^8 - 8)$

In Exercises 37–46, write each of the series in expanded form.

37. $\displaystyle\sum_{k=1}^{5} \dfrac{k^2}{2k + 3}$

38. $\displaystyle\sum_{k=1}^{n} \dfrac{y + k}{k^2}$

39. $\displaystyle\sum_{k=1}^{4} k \ln (k)$

40. $\displaystyle\sum_{k=1}^{5} \dfrac{k - 1}{\sqrt{k + 1}}$

41. $\displaystyle\sum_{k=1}^{n} k(k + 1)(k - 2)$

42. $\displaystyle\sum_{k=1}^{n} \dfrac{k(k + 1)(2k + 1)}{6}$

43. $\displaystyle\sum_{k=1}^{n} \left(\dfrac{2k^2 - k + 1}{2}\right)^2$

44. $\displaystyle\sum_{k=1}^{6} \dfrac{(-1)k}{\sqrt[3]{k}}$

45. $\displaystyle\sum_{k=2}^{5} (y^2 + 2ky + k)$

46. $\displaystyle\sum_{k=3}^{7} \dfrac{y + 2}{\sqrt{k}}$

47. A truck starting from rest travels at the rate of 2 ft the first second, 3 ft the second second, 4 ft the third second, 5 ft the fourth second, and 6 ft the fifth second. If the truck continues to accelerate at the same rate,
a. how many feet will the truck cover in the ninth second, and
b. what is the total distance the truck moves in nine seconds?

48. The temperature at 9:00 A.M. was 3°C and increased uniformly at the rate of 1.5° per hour.
a. Write the first five terms of the sequence.
b. Write a general term and express in summation notation.
c. Using the results from part b determine the sum of the hourly temperatures for the first five hours.

49. Dale repays a $10,000 loan by paying $1000 at the end of each year plus a yearly interest charge of 10% of the unpaid balance. Hence, her payment at the end of the first year is:

$$\$1000 + 10\% \text{ of } \$10,000 = \$1000 + \$1000 \text{ or } \$2000.$$

a. What is her payment at the end of the second year?
b. What are the interest payments for each of the 10 years it takes for Dale to pay back the loan?
c. These interest payments define an arithmetic series. Write the terms of the series and determine the amount of interest paid.
d. Write the series using summation notation.

16.3

Arithmetic Sequences and Series

In examining the terms of the sequence 1, 4, 7, 10, 13, 16, . . . , we see that any two successive terms have a difference of three. That is, $4 - 1 = 3, 7 - 4 = 3$, $10 - 7 = 3$, etc. All the successive terms of the sequence 1, 5, 9, 13, 17, 21, . . . , differ by four. That is, $5 - 1 = 4, 9 - 5 = 4, 13 - 9 = 4$, etc. Any sequence where the terms have a **common difference** is called an **arithmetic sequence** or **arithmetic progression.**

EXAMPLE 1   **Application: Distance Traveled**

A child is playing inside a car and accidentally releases the hand brake. The car is on an incline and starts rolling down the hill. At the end of the first second it has rolled 2 m, at the end of the second second it has rolled 4 m, and at the end of the third second it has rolled 6 m. (These figures include friction.)

a.   How many meters did the car travel each second?
b.   How far had the car traveled at the end of the six seconds?

Solution

a.   We can determine how far the car goes each second by determining the common difference d. That is, $d = 4 - 2$ or $d = 2$ m.
b.   To determine how far the car has traveled in six seconds, all we need do is multiply the common difference times the number of seconds. That is,

$$\text{distance after six seconds} = (d)(t)$$
$$= (\;2\;)(\;6\;)\quad\textbf{Substituting}$$
$$= 12 \text{ m.}\quad\blacklozenge$$

DEFINITION 16.4
An arithmetic sequence is a sequence of numbers called terms, such that any two successive terms differ by the same number, called the **common difference** d.

EXAMPLE 2   **Common Difference and the Fifth Term**

For the arithmetic sequence 15, 12, 9, 6, . . . , determine the following.

a.   The common difference
b.   The fifth term

Solution

a.   The common difference is found by subtracting the preceding term from a succeeding term. Thus,

$$d = \boxed{12} - \boxed{15} \quad\text{or}\quad d = \boxed{-3.}$$

b. To determine the fifth term we add the common difference to the fourth term:

$$6 + (\boxed{-3}) = 3.$$

The fifth term of the sequence is 3.   ◆

In the arithmetic sequence 4, 11, 18, 25, 32, the common difference is seven. If the order of the terms is reversed, a new arithmetic sequence is formed, 32, 25, 18, 11, 4. The common difference of the new arithmetic sequence is -7.

To develop a rule to determine any term of an arithmetic sequence, let us use the arithmetic sequence 32, 25, 18, 11, 4. From the previous discussion we know that the sequence has a first term of 32 and a common difference of -7.

$$a_2 = 32 + (-7) = 25$$
$$a_3 = 32 + 2(-7) = 18$$
$$a_4 = 32 + 3(-7) = 11$$
$$a_5 = 32 + 4(-7) = 4$$

Note that to find any term, we add to the first term the product of the common difference and one less than the number of the term we are finding. For example, to get the fourth term we do the following.

$$a_4 = 32 + (4 - 1)(-7).$$

Thus, if a_1 denotes the first term of an arithmetic sequence and d denotes the common difference, the terms of the sequence are:

$$a_2 = a_1 + (2 - 1)d$$
$$a_3 = a_1 + (3 - 1)d$$
$$a_4 = a_1 + (4 - 1)d$$
$$\vdots$$
$$a_n = a_1 + (n - 1)d$$

DEFINITION 16.5

The nth term of an arithmetic sequence is given by the formula

$$a_n = a_1 + (n - 1)d,$$

where a_1 is the first term and d is the common difference.

EXAMPLE 3   **Given First Term and Common Difference, Determine Term *x***

Determine the 13th term of an arithmetic sequence whose first term is 15 and whose common difference is 3.

Solution We are given that $a_1 = 15$, $d = 3$, and $n = 13$, and we want to find a_{13}. Substituting into the formula:

$$a_n = a_1 + (n - 1)d, \quad \text{Definition 16.5}$$

we have:

$$a_{13} = 15 + (13 - 1)3$$
$$= 51.$$

Thus, the 13th term of the sequence is 51. ◆

EXAMPLE 4 Given First Term and Common Difference, Determine Terms of the Sequence

Determine the first four terms of the arithmetic sequence in which the first term is 17 and the common difference is 4.

Solution With $a_1 = 17$ and $d = 4$, we can establish the following formula for the sequence:

$$a_n = 17 + (n - 1)4.$$

Now let $n = 1, 2, 3$, and 4, respectively, and we obtain the first four terms of the sequence.

$$a_1 = 17 + (1 - 1)4 = 17 \quad a_3 = 17 + (3 - 1)4 = 25$$
$$a_2 = 17 + (2 - 1)4 = 21 \quad a_4 = 17 + (4 - 1)4 = 29$$

The first four terms of the arithmetic sequence with $a_1 = 17$ and $d = 4$ are 17, 21, 25 and 29. ◆

In Example 4, the nth term of the sequence is $[17 + (n + 1)4]$.

EXAMPLE 5 Determine the First Term

Determine the first term of a seven-term arithmetic sequence in which the common difference is 5 and the seventh term is 13.

Solution We know that $n = 7$, $a_7 = 13$, and $d = 5$, and we want to find a_1. Thus,

$$a_n = a_1 + (n - 1)d \quad \text{Definition 16.5}$$
$$13 = a_1 + (7 - 1)5$$
$$13 = a_1 + 30$$
$$-17 = a_1.$$

Thus, the first term of the sequence is -17. ◆

EXAMPLE 6   **Determining the Number of Terms**

How many terms are there in an arithmetic sequence in which the first term is $2\frac{1}{2}$, the last term is $5\frac{3}{4}$, and the common difference is $\frac{1}{4}$?

Solution   We know that $a_1 = \frac{5}{2}$, $a_n = \frac{23}{4}$, and $d = \frac{1}{4}$, and we want to find n.

$$a_n = a_1 + (n-1)d \qquad \text{Definition 16.5}$$

$$\boxed{\frac{23}{4}} = \boxed{\frac{5}{2}} + (n-1)\boxed{\frac{1}{4}}.$$

Multiplying each term by 4, we have:

$$23 = 10 + n - 1$$
$$14 = n. \qquad \text{Solving for } n$$

Thus, there are 14 terms in the sequence.   ◆

EXAMPLE 7   **Application: Screw Sizes**

In the chapter opener, we read that screw sizes are indicated by a number. The smallest size screw is a number 1, which has a diameter of 0.073 in. A number 24 size screw is the 24th in size and has a diameter of 0.372 in. Determine the common difference d in screw sizes and identify the two omitted that are closest to $\frac{1}{4}$ and $\frac{5}{16}$.

Solution   We know that $a_1 = 0.073$ in., $n = 24$, and $a_{24} = 0.372$ in., and we want to determine the common difference d.

$$a_n = a_1 + (n-1)d \qquad \text{Definition 16.5}$$

$$\boxed{0.372} = \boxed{0.073} + (\boxed{24} - 1)d$$

$$d = \frac{0.372 - 0.073}{23}$$

$$= 0.013 \text{ in.}$$

We know that $\frac{1}{4} = 0.25$ and $\frac{5}{16} = 0.3125$. Knowing the common difference and the first term of the sequence, we can calculate various terms of the sequence. Calculating the values for the 14th through the 20th terms, we have:

$$a_{14} = 0.242 \text{ in.} \qquad a_{18} = 0.294 \text{ in.}$$
$$a_{15} = \boxed{0.255} \text{ in.} \qquad a_{19} = \boxed{0.307} \text{ in.}$$
$$a_{16} = 0.268 \text{ in.} \qquad a_{20} = 0.320 \text{ in.}$$
$$a_{17} = 0.281 \text{ in.}$$

From the values shown, we can see that a_{15} is closest to $\frac{1}{4}$ in. and a_{19} is closest to $\frac{5}{16}$ in. Therefore, screws in sizes 15 and 19 are not produced.   ◆

As we learned in the previous section, associated with every sequence is a series. The series associated with the arithmetic sequence

$$a_1, (a_1 + d), (a_1 + 2d), (a_1 + 3d), \ldots, [a_1 + (n-1)d]$$

is

$$S_n = a_1 + (a_1 + d) + (a_1 + 2d) + (a_1 + 3d) \\ + \cdots + [a_1 + (n-1)d] \qquad (1)$$

or

$$S_n = \sum_{i=1}^{n} [a_1 + (i-1)d].$$

The sum could just as well be written in reverse order, or as

$$S_n = a_n + (a - d) + (a_n - 2d) + (a_n - 3d) \\ + \cdots + [a_n - (n-1)d]. \qquad (2)$$

A formula to determine the sum of the terms of a sequence is developed by finding the sum of Equations (1) and (2).

$$2S_n = (a_1 + a_n) + (a_1 + d + a_n - d) + (a_1 + 2d + a_n - 2d) \\ + \cdots + [a_1 + (n-1)d + a_n - (n-1)d] \\ = \underbrace{(a_1 + a_n) + (a_1 + a_n) + (a_1 + a_n) + \cdots + (a_1 + a_n)}_{n \text{ terms}} \\ = n(a_1 + a_n) \\ S_n = \frac{n}{2}(a_1 + a_n).$$

This demonstration is the proof of the following theorem, which gives the formula for finding S_n, called the **nth partial sum.**

THEOREM 16.1

The sum of the first n terms of an arithmetic sequence whose first term is a_1 and whose nth term is a_n is given by

$$S_n = \frac{n}{2}(a_1 + a_n).$$

In Example 8, we use the formula given in Theorem 16.1 to demonstrate how to determine the sum of the terms of an arithmetic sequence.

EXAMPLE 8 Sum of the Terms of an Arithmetic Sequence

Determine the sum of the terms of the arithmetic sequence with eight terms, whose first term is 3 and whose last term is 19.

Solution In this problem, $n = 8$, $a_1 = 3$, and $a_n = 19$. Substituting in the equation, we have:

$$S_n = \frac{n}{2}(a_1 + a_n) \qquad \text{Theorem 16.1}$$

$$S_8 = \frac{8}{2}(\boxed{3} + \boxed{19}) \quad \text{or} \quad S_8 = 88.$$

Thus, the sum of the first eight terms of the sequence is 88. ◆

EXAMPLE 9 Sum of the Terms of an Arithmetic Sequence

Determine the sum of the first 12 terms of the sequence 3, 9, 15, 21, ...

Solution To determine the sum of the first 12 terms, we must know n, which is 12, a_1, which is 3, and a_{12}, which we do not know. To determine a_{12}, use the formula for the nth term:

$$a_n = a_1 + (n - 1)d$$
$$a_{12} = \boxed{3} + (\boxed{12} - 1)\boxed{6}$$
$$= 69.$$

Now substituting $n = 12$, $a_1 = 3$, and $a_n = 69$ into the formula:

$$S_n = \frac{n}{2}(a_1 + a_n), \quad \text{Theorem 16.1}$$

we have:

$$S_{12} = \frac{12}{2}(\boxed{3} + \boxed{69}) \quad \text{or} \quad S_{12} = 432.$$

Thus, the sum of the first 12 terms is 432. ◆

The sum of the first 12 terms in Example 9 could have been found without knowing a_n. Since

$$a_n = a_1 + (n - 1)d,$$

if we substitute for a_n in the formula

$$S_n = \frac{n}{2}(a_1 + a_n),$$

the result is

$$S_n = \frac{n}{2}[a_1 + a_1 + (n - 1)d]$$

or

$$S_n = \frac{n}{2}[2a_1 + (n - 1)d],$$

a formula for the sum that does not contain a_n. This is stated as Theorem 16.2.

> ### THEOREM 16.2
>
> The sum of the first n terms of an arithmetic sequence whose first term is a_1 and whose common difference is d is given by:
>
> $$S_n = \frac{n}{2}[2a_1 + (n - 1)d].$$

In Example 9 we were given the values for $a_1 = 3$, $d = 6$, and $n = 12$. Substituting in the formula:

$$S_n = \frac{n}{2}[2a_1 + (n - 1)d],$$

we have:

$$S_{12} = \frac{12}{2}[2(3) + (12 - 1)6] \quad \text{or} \quad S_{12} = 432.$$

Using Theorem 16.2, the sum of the first 12 terms is 432.

EXAMPLE 10   Application: Yearly Increase in Population

The number of citizens in a town increases from 32,000 to 60,000 in six years. Determine the yearly increase if this rate is assumed to be constant.

Solution   Since the yearly increase is constant, it is an arithmetic sequence with the constant difference d being the yearly increase. We know the initial term $a_1 = 32,000$, the number of years $n = 6$, and the nth term $a_n = 60,000$. Substituting in the formula:

$$a_n = a_1 + (n - 1)d$$
$$60,000 = 32,000 + (6 - 1)d$$
$$28,000 = 5d$$
$$d = 5600.$$

Thus, the population increase is 5600 people per year.   ◆

The formulas for the nth term and the sum of terms for an arithmetic sequence are listed in the summary box.

FORMULAS FOR ARITHMETIC SEQUENCES

$$a_n = a_1 + (n - 1)d$$

$$S_n = \frac{n}{2}(a_1 + a_n)$$

$$S_n = \frac{n}{2}[2a_1 + (n - 1)d]$$

16.3 Exercises

In Exercises 1–10, determine whether or not the sequence is an arithmetic sequence. If it is an arithmetic sequence, determine the common difference.

1. 2, 4, 6, 8, 10, . . .

2. 1, 3, 5, 7, 9, . . .

3. 1, 4, 6, 9, 11, . . .

4. 5, 10, 15, 25, 30, . . .

5. $\frac{1}{2}, 1, \frac{3}{2}, 2, \frac{5}{2}, \ldots$

6. $\frac{1}{5}, \frac{2}{5}, \frac{3}{5}, \frac{4}{5}, 1, \ldots$

7. $\frac{1}{2}, \frac{1}{4}, -\frac{1}{4}, -\frac{1}{2}, -1, \ldots$

8. 0.1, 1.0, 1.9, 2.8, 3.7, . . .

9. 1, 5, 10, 16, 23, . . .

10. 6, 11, 16, 21, 26, . . .

In Exercises 11–20, perform the following.
a. Determine the first five terms of the arithmetic sequence.
b. Determine the sum of the first five terms.

11. $a_1 = 13$ and $d = 5$

12. $a_1 = -11$ and $d = 7$

13. $a_1 = 15$ and $d = -3$

14. $a_1 = \frac{1}{3}$ and $d = \frac{2}{3}$

15. $a_3 = 9$ and $d = -1$

16. $a_2 = 8$ and $d = 2$

17. $a_4 = -4$ and $d = \frac{1}{2}$

18. $a_3 = 6$ and $d = -\frac{1}{2}$

19. $a_2 = \frac{3}{5}$ and $d = \frac{3}{5}$

20. $a_2 = \frac{1}{4}$ and $d = -1$

In Exercises 21–30, some of the quantities a_1, d, a_n, n, S_n, are given; determine the quantities that are not given.

21. $a_1 = 1, d = 2, n = 8$

22. $a_1 = 3, d = 5, n = 7$

23. $a_1 = 3, a_n = 21, d = 3$

24. $a_1 = 26, a_n = 2, d = -4$

25. $n = 8, a_1 = -3, a_n = 18$

26. $n = 6, a_1 = 17, a_n = 2$

27. $S_6 = 6.9$, $n = 6$, $a_1 = 0.4$

28. $S_8 = 11$, $n = 8$, $a_1 = \dfrac{1}{2}$

29. $S_6 = 78$, $n = 6$, $d = -4$

30. $S_n = 108$, $a_1 = 24$, $d = -3$

31. **a.** How many even numbers lie between 13 and 71?
    b. Determine their sum.

32. **a.** How many numbers that are divisible by 4 lie between 30 and 182?
    b. Determine their sum.

33. **a.** How many numbers that are divisible by 3 lie between 32 and 184?
    b. Determine their sum.

34. **a.** How many numbers that are divisible by 7 lie between 19 and 143?
    b. Determine their sum.

35. The first year Karen worked she saved $1,000. In each succeeding year she saved $100 more than the preceding year. How much has she accumulated at the end of 12 years? (Ignore interest.)

36. An employee receives a salary of $16,000 per year with the understanding that the salary will increase by $1,000 per year up to and including the 12th year.
    a. What will the employee's salary be in the 12th year?
    b. What are the total earnings for the 12 years?

37. A body falls 16 ft during the first second, 48 ft during the second second, 80 ft during the third second, and so on, in arithmetic sequence.
    a. How far does it fall during the eighth second?
    b. How far does it fall in the first eight seconds?

38. Three workers X, Y, and Z are hired by a company at the same time, each at a starting salary of $16,000. X agrees to accept an annual raise of $2,000. Y agrees to accept a semiannual increase of $400. Z agrees to accept a quarterly raise of $100. After four years have elapsed, who has earned more money, and how much more has that worker earned?

39. An entry in a soap-box derby coasts downhill covering 5 ft the first second, 12 ft the second second, and an average of 7 ft more each second than the previous second. If the racer reaches the bottom of the hill in 10 seconds, how far did he coast? How far did he travel in the eighth second?

40. A taxi company charges $2.00 for the first $\dfrac{1}{4}$ mi and 35 cents for each $\dfrac{1}{4}$ mi thereafter. How much is the fare for a 12-mi ride?

41. The first and last numbers of an arithmetic sequence are 6 and 8, respectively. Then five numbers are placed between 6 and 8 so that the seven numbers form an arithmetic sequence. Determine the five numbers.

42. The first and last numbers of an arithmetic sequence are 4 and 25, respectively. If six numbers are placed between 4 and 25 so that the eight numbers form an arithmetic sequence, determine the six numbers.

43. The number of bacteria in a culture increases from 32,000 to 70,000 in six days. Find the daily rate of increase if this rate is assumed to be constant.

44. a. Leonardo da Vinci, the great Renaissance artist, thought that falling bodies fell 4.9 m in the first second, 9.8 m in the second second, 14.7 m in the third second, and so forth. Write these numbers in the form of an arithmetic series. Using da Vinci's theory, how far will a body fall in the sixth second? How far will a body fall in a total of six seconds?

b. Galileo Galilei, the great Renaissance scientist, thought that falling bodies fell 4.9 m in the first second, 14.67 m in the second second, and 24.5 m in the third second, and so forth. Write these numbers in the form of an arithmetic series. Using Galileo's theory, how far will a body fall during the sixth second? How far will a body fall in a total of six seconds?

c. Falling bodies that start from rest fall a distance d after t seconds. Given the time t the distance d can be determined using the formula $d = 4.9t^2$. Using this formula, determine the distance a body will fall in six seconds. Which of the theories in parts a or b is correct?

45. A car coasts down a hill. Its initial speed is 3.0 m/s. Every second, the car gains 0.8 m/s of speed. In other words, after two seconds, its speed is 4.6 m/s. After how many seconds will its speed be 11.8 m/s?

46. In building tall buildings, costs rise as the floors rise. Suppose the first floor of a building costs $300,000 and that each additional floor costs an additional $30,000. Determine the total cost of a ten-story building.

16.4

Geometric Sequences and Series

In the study of population increases, the change from year to year is not usually a common difference. Also, in the study of atomic decay of an element, the changes over set periods of time are not the same for each period. Sequences, called *geometric sequences,* show such changes. For example,

$$2, 6, 18, 54, 162, \ldots ,$$

where each succeeding term is obtained by multiplying each preceding term by a common factor is a geometric sequence. In the sequence 2, 6, 18, 54, 162, . . . , the common factor is 3. That is, $2 \cdot 3 = 6, 6 \cdot 3 = 18, 18 \cdot 3 = 54$, and so on. The 3 in this case is called the *common ratio.*

The common ratio can be determined in the following way. The second term 6 divided by the first term 2 is 3:

$$\left(\frac{6}{2} = 3 \right).$$

The third term, 18, divided by the second term 6 is 3:

$$\left(\frac{18}{6} = 3 \right).$$

We can continue on, always taking one term and dividing by the preceding term. This process gives us the common ratio.

> **DEFINITION 16.6**
>
> A **geometric sequence** $a_1, a_2, a_3, a_4, \ldots, a_{i-1}, a_i \ldots$, is a sequence of numbers, called terms, such that the ratio of any term to the preceding term is a fixed number called the **common ratio** r:
>
> $$\left(r = \frac{a_i}{a_{i-1}} \right)$$

EXAMPLE 1   Common Ratio and Terms of a Geometric Sequence

Determine the following for the geometric sequence 4, 12, 36, 108, ...

a.   The common ratio
b.   The fifth and sixth terms

Solution

a.   The common ratio is found by setting up the ratio of any succeeding term to the preceding term. For example,

$$r = \frac{12}{4} \quad \text{or} \quad r = \frac{36}{12} \quad \text{or} \quad r = \frac{108}{36}.$$

In each case the quotient is 3. Thus, the common ratio r for the geometric sequence 4, 12, 36, 108, . . . , is 3.

b.   To determine the fifth term of the sequence, multiply the fourth term 108 by the common ratio 3. Thus, the fifth term is

$$108 \times 3 = 324.$$

The sixth term is $324 \times 3 = 972$.   ◆

In the geometric sequence $1, \frac{1}{3}, \frac{1}{9}, \frac{1}{27}, \frac{1}{81}, \frac{1}{243}$, the common ratio is $\frac{1}{3}$. If the order of the terms of the sequence is reversed, a new geometric sequence is formed $\frac{1}{243}, \frac{1}{81}, \frac{1}{27}, \frac{1}{9}, \frac{1}{3}, 1$. The common ratio of the new geometric sequence is 3.

To develop a rule to determine any term of a geometric sequence let us use the geometric sequence $\frac{1}{243}, \frac{1}{81}, \frac{1}{27}, \frac{1}{9}, \frac{1}{3}, 1$. From the previous discussion we know that the sequence has a first term of $\frac{1}{243}$ and a common ratio of 3.

$$a_2 = \frac{1}{243} \cdot 3 \qquad\qquad = \frac{1}{243} \cdot 3^1 = \frac{1}{81}$$

$$a_3 = \frac{1}{81} \cdot 3 \;\; = \left(\frac{1}{243} \cdot 3\right) \cdot 3 \;\; = \frac{1}{243} \cdot 3^2 = \frac{1}{27}$$

$$a_4 = \frac{1}{27} \cdot 3 \;\; = \left(\frac{1}{243} \cdot 3^2\right) \cdot 3 = \frac{1}{243} \cdot 3^3 = \frac{1}{9}$$

$$a_5 = \frac{1}{9} \cdot 3 \;\; = \left(\frac{1}{243} \cdot 3^3\right) \cdot 3 = \frac{1}{243} \cdot 3^4 = \frac{1}{3}$$

$$a_6 = \frac{1}{3} \cdot 3 \;\; = \left(\frac{1}{243} \cdot 3^4\right) \cdot 3 = \frac{1}{243} \cdot 3^5 = 1$$

To find the ith term, multiply the first term by the common ratio raised to the $i - 1$ power. For example, to find the fourth term, do the following.

$$a_4 = \frac{1}{243} \cdot 3^{4-1}$$

Thus if a_1 denotes the first term of a geometric sequence and r the common ratio, then the terms of the sequence are:

$$a_2 = a_1 r^{2-1},$$
$$a_3 = a_1 r^{3-1},$$
$$a_4 = a_1 r^{4-1},$$
$$\vdots$$
$$a_n = a_1 r^{n-1}.$$

DEFINITION 16.7

The nth term of a geometric sequence is given by the formula:

$$a_n = a_1 r^{n-1},$$

where a_1 is the first term of the sequence and r is the common ratio.

EXAMPLE 2   **Terms of a Geometric Sequence**

Determine the first four terms of the geometric sequence in which the first term is 5 and the common ratio is 2.

Solution   With $a_1 = 5$ and $r = 2$, we can establish the following general formula for the sequence.

$$a_n = \boxed{5} \cdot \boxed{2}^{\,n-1} \qquad \text{Definition 16.7}$$

Now let $n = 1, 2, 3$, and 4, respectively, and we obtain the first four terms of the sequence.

$$a_1 = 5 \cdot 2^{1-1} \qquad a_3 = 5 \cdot 2^{3-1}$$
$$= 5 \cdot 2^0 \qquad\quad = 5 \cdot 2^2$$
$$= 5 \qquad\qquad = 20$$
$$a_2 = 5 \cdot 2^{2-1} \qquad a_4 = 5 \cdot 2^{4-1}$$
$$= 5 \cdot 2^1 \qquad\quad = 5 \cdot 2^3$$
$$= 10 \qquad\qquad = 40$$

Thus, the first four terms of the geometric sequence with first term 5 and common ratio 2 are 5, 10, 20, and 40. ◆

EXAMPLE 3 The Eighth Term of a Geometric Sequence

Determine the eighth term of the geometric sequence $1, \dfrac{1}{2}, \dfrac{1}{4}, \dfrac{1}{8}, \ldots$.

Solution We know that $a_1 = 1$, $n = 8$, and $r = \dfrac{1}{2}$ $\left(\text{since } \dfrac{\frac{1}{2}}{1} = \dfrac{1}{2}\right)$. Substituting into the formula:

$$a_n = a_1 r^{n-1}, \quad \textbf{Definition 16.7}$$

we have:

$$a_8 = \boxed{1} \cdot \left(\boxed{\frac{1}{2}}\right)^{8-1} = 1 \cdot \left(\frac{1}{2}\right)^7 = \frac{1}{128}.$$

Thus, the eighth term of the sequence is $\dfrac{1}{128}$. ◆

EXAMPLE 4 The First Term of a Geometric Sequence

Determine the first term of a geometric sequence in which the common ratio is 3 and the seventh term is 162.

Solution Given $r = 3$ and the seventh term $a_n = a_7 = 162$, we can determine a_1 using the formula:

$$a_n = a_1 r^{n-1}. \quad \textbf{Definition 16.7}$$

Substituting, we have:

$$162 = a_1 \cdot 3^{7-1}$$

$$162 = a_1(729)$$

$$\frac{162}{729} = a_1$$

$$\frac{2}{9} = a_1.$$

Thus, the first term of the geometric sequence is $\frac{2}{9}$. ◆

As stated earlier, associated with every sequence is a series. The series associated with the geometric sequence:

$$a_1, a_1r, a_1r^2, a_1r^3, a_1r^4, \ldots, a_1r^{n-1},$$

is:

$$S_n = a_1 + a_1r + a_1r^2 + a_1r^3 + a_1r^4 + \cdots + a_1r^{n-1}, \qquad (3)$$

or:

$$S_n = \sum_{i=1}^{n} a_1 r^{i-1}.$$

We want a formula to determine the sum S_n in terms of a_1, n, and r. If we multiply Equation (3) by r, we have:

$$rS_n = a_1r + a_1r^2 + a_1r^3 + a_1r^4 + \cdots + a_1r^n. \qquad (4)$$

Now subtract Equation (4) from Equation (3).

$$S_n = a_1 + a_1r + a_1r^2 + a_1r^3 + a_1r^4 + \cdots + a_1r^{n-1} \qquad (3)$$

$$rS_n = a_1r + a_1r^2 + a_1r^3 + a_1r^4 + a_1r^5 + \cdots + a_1r^n \qquad (4)$$

$$S_n - rS_n = a_1 - a_1r^n \qquad \textbf{Equation (3)–Equation (4)}$$

$$S_n(1 - r) = a_1(1 - r^n) \qquad \textbf{Common factor}$$

$$S_n = \frac{a_1(1 - r^n)}{1 - r},$$

provided $r \neq 1$.

This demonstration is the proof of the following theorem, which gives the formula for finding S_n, called the **nth partial sum** of the geometric sequence.

> **THEOREM 16.3**
>
> The sum of the first n terms of a geometric sequence whose first term is a_1 and whose common ratio is r is given by:
>
> $$S_n = \frac{a_1(1 - r^n)}{1 - r},$$
>
> provided $r \neq 1$.

EXAMPLE 5 **The Sum of a Finite Geometric Sequence**

Determine the sum of the geometric sequence with six terms, whose first term is 1 and whose common ratio is 5.

Solution In this problem $n = 6$, $a_1 = 1$, and $r = 5$. Substituting in the equation:

$$S_n = \frac{a_1(1 - r^n)}{1 - r} \qquad \text{Theorem 16.2}$$

we have:

$$S_n = \frac{1(1 - 5^6)}{1 - 5}$$

$$= \frac{-15624}{-4}$$

$$= 3906.$$

Thus, the sum of the first six terms of the sequence is 3906. ◆

EXAMPLE 6 **Application: Distance Traveled by a Pendulum in n Swings**

The length of the first swing of a pendulum is 15<u>0</u> mm. If on each succeeding swing the pendulum travels $\frac{4}{5}$ as far as on the preceding swing, how far will the pendulum have traveled after six swings?

Solution The length of the first swing is $a_1 = 150$, the constant decrease in length of each swing is $r = \frac{4}{5}$, and the number of swings is $n = 6$. Substituting into the formula:

$$S_n = \frac{a_1(1 - r^n)}{1 - r}$$ **Theorem 16.2**

$$= \frac{150\left(1 - \left(\frac{4}{5}\right)^6\right)}{1 - \frac{4}{5}}$$

$$= 553 \text{ mm}.$$

The distance the pendulum has traveled in six swings is 553 mm. ◆

The formulas discussed for geometric sequences are listed in the summary box.

FORMULAS FOR GEOMETRIC SEQUENCES

$$r = \frac{a_i}{a_i - 1}$$

$$a_n = a_1 r^{n-1}$$

$$S_n = \frac{a_1(1 - r^n)}{1 - r}, \quad r \neq 1$$

16.4 Exercises

In Exercises 1–10, determine whether the sequence is a geometric sequence. If it is, determine the common ratio and find the seventh term.

1. 1, 2, 4, 8, 16, . . .

2. 1, 3, 9, 27, 81, . . .

3. 1, 4, 8, 12, 16, . . .

4. 1, 6, 12, 18, 24, . . .

5. $1, \dfrac{1}{4}, \dfrac{1}{16}, \dfrac{1}{64}, \dfrac{1}{256}, \ldots$

6. $1, \dfrac{1}{5}, \dfrac{1}{25}, \dfrac{1}{125}, \dfrac{1}{625}, \ldots$

7. 16,807, 2401, 343, 49, 7, . . .

8. $\dfrac{2}{81}, \dfrac{2}{27}, \dfrac{2}{9}, \dfrac{2}{3}, 2, \ldots$

9. $\dfrac{1}{3}, \dfrac{1}{6}, \dfrac{1}{12}, \dfrac{1}{24}, \dfrac{1}{48}, \ldots$

10. $5, \dfrac{5}{5}, \dfrac{10}{5}, \dfrac{15}{5}, \dfrac{20}{5}, \ldots$

In Exercises 11–20, perform the following:
a. Determine the first five terms of the geometric sequence.
b. Determine the sum of the first five terms.

11. $a_1 = 3, r = 2$

12. $a_1 = -3, r = 2$

13. $a_1 = 2, r = \dfrac{1}{3}$

14. $a_1 = 5, r = \dfrac{1}{2}$

15. $a_1 = \dfrac{2}{5}, r = \dfrac{1}{4}$

16. $a_1 = \dfrac{3}{7}, r = \dfrac{1}{5}$

17. $a_1 = -4, r = 0.1$

18. $a_1 = -5, r = 0.2$

19. $a_1 = 4, r = -2$

20. $a_1 = \dfrac{1}{5}, r = -3$

In Exercises 21–30, some of the quantities a_1, r, a_n, n, S_n are given; determine the quantities that are not given.

21. $a_1 = \dfrac{1}{2}$, $r = \dfrac{1}{3}$, $n = 4$

22. $a_1 = \dfrac{1}{3}$, $r = \dfrac{1}{2}$, $n = 4$

23. $a_6 = 96$, $a_1 = 3$, $n = 6$

24. $a_5 = 162$, $a_1 = 2$, $n = 5$

25. $S_6 = 189$, $n = 6$, $r = 2$

26. $S_5 = 284$, $n = 5$, $r = 3$

27. $a_1 = \dfrac{1}{3}$, $r = -3$, $n = 4$

28. $a_1 = \dfrac{5}{3}$, $r = -\dfrac{2}{3}$, $n = 5$

29. $S_4 = \dfrac{117}{125}$, $r = \dfrac{1}{5}$, $n = 4$

30. $S_5 = \dfrac{242}{81}$, $n = 5$, $r = \dfrac{1}{3}$

31. Tom decides that he wants to become a weight-lifting champion. He starts his weight-lifting program by lifting 500 N. His plan is to increase the weight by 10% per week. How much weight will he be lifting at the end of 5 weeks? What is the total amount of weight he has lifted over the 5 weeks?

32. Jane decides that she wants to save money by making weekly deposits in her savings account and doubling the amount of her deposit from the week before. If she begins by depositing $1 the first week in January, how much will she have saved at the end of 10 weeks?

33. The number of bacteria in a culture increases by 20% each hour. If the culture contains 40,000 bacteria at 11:00 A.M., how many bacteria are present at 2:00 P.M.?

34. If the sum of the first three terms of a geometric sequence is 13 times as large as the first term, determine the common ratio. (Two possible answers.)

35. If the sum of the first three terms of a geometric sequence is 21 times as large as the first term, determine the common ratio. (Two possible answers.)

36. A sheet-metal worker took a job with a starting salary of $15,000 and a guaranteed increase of 8% per year, based on the previous year's salary.
  a. Determine the salary for the seventh year.
  b. Determine the amount of money earned for the first seven years.

37. A carpenter was offered a job for 20 days with a choice between two methods of being paid. One choice is $100 per day for the 20-day period; the other is to be paid 1 cent the first day, 2 cents the second day, 4 cents the third day, and so on doubling the amount paid from each preceding day for the 20-day period.
  a. Determine the total amont the worker would be paid for 20 days at $100 per day.
  b. The second method of payment is a geometric sequence. Write the first five terms of the sequence.
  c. Determine the total amount the worker would be paid using the second method of payment.
  d. Which method of payment would you choose and why?

38. An assembly-line worker is switched from hourly rate to piecework. The first day of piecework the worker produces 300 items. In order to make the same rate of pay, the worker must produce 500 items per day. At what rate must the worker increase production over the previous day in order to achieve the quota of 500 items per day in 20 days?

39. In a chemical plant, a tank holds 60 gal of a liquid soap, which mixes with water. After 12 gal of the soap are drawn out, the tank is filled by replacing the soap with water. Then 12 gal of the resulting mixture are drawn out, and the tank is again filled with water. If this operation is performed until eight batches have been drawn from the tank, how much of the original liquid remains?

40. Cooling seems to obey a geometric sequence. A hot cup of coffee has an initial temperature of 92°C. If we gently stir the coffee, the temperature drops 7% of its preceding temperature every minute. What will be its temperature after 10 minutes?

41. When light is cast through a plate of glass, some of the light is reflected back. Suppose that a plate of glass reflects back 8% of the light that is normally incident upon it. This means that only 92% of the light gets through. How much light gets through if 10 plates of glass are lined up? (Although glass is considered transparent, like water, a stack of glass can reduce incoming light quite a bit.)

42. A pendulum reduces its swing in a way that can be explained by a geometric sequence. The arc length of the swing is initially 1.00 m. However, the second swing has an arc of only 0.95 m. What will be the arc length of the sixth swing of the pendulum?

43. Your genes are made up from your father and your mother, making you the product of two people. Additionally, your father and mother are each the product of two people, your grandparents. Hence, your genetic makeup is tied to four people. If your two parents represent the first generation and your four grandparents represent the second generation, how many people represent the sixth generation?

44. Warmer temperatures tend to promote chemical activity. Chemists often observe that a 10°C temperature rise approximately doubles the reaction rate. Suppose a chemist needs the reaction to proceed 4.5 times faster than it normally does at room temperature. By how much should the temperature of the solution be raised?

45. a. Suppose the world population is increasing at the rate of 3% per year. If the population is now 5 billion people, what will it be in 20 years?

b. Suppose that the world population is increasing at the rate of 6% per year. If the population is now 5 billion people, what will it be in 20 years?

16.5

Infinite Geometric Series

In Section 16.4, we learned how to determine the sum of a finite geometric series. In this section we learn how to find the sum of an infinite geometric series. But how can an infinite geometric series lead to a finite number?

Consider the sequence $\frac{1}{3}, \frac{1}{9}, \frac{1}{27}, \frac{1}{81}, \frac{1}{243}, \ldots, \frac{1}{3^n}, \ldots$, which has a common ratio r of $\frac{1}{3}$. Use the formula for finding the sum of a geometric sequence to write the sum for $n = 1$ to 10 inclusive and see if any pattern emerges. Substituting in the formula $S_n = \dfrac{a_1(1 - r^n)}{1 - r}$, $r \neq 1$, we have the following sums.

$$S_1 = \frac{1}{3} = 0.33333\ldots \qquad S_6 = \frac{364}{729} = 0.49931\ldots$$

$$S_2 = \frac{4}{9} = 0.44444\ldots \qquad S_7 = \frac{1093}{2187} = 0.49977\ldots$$

$$S_3 = \frac{13}{27} = 0.48148\ldots \qquad S_8 = \frac{3280}{6561} = 0.49992\ldots$$

$$S_4 = \frac{40}{81} = 0.49382\ldots \qquad S_9 = \frac{9841}{19683} = 0.49997\ldots$$

$$S_5 = \frac{121}{243} = 0.49794\ldots \qquad S_{10} = \frac{29524}{59049} = 0.49999\ldots$$

With the calculations for the first ten sums we can predict that the final sum is 0.5000 since n increases without bound.

Now let us examine the sequence 3, 9, 27, 81, 243, . . . , where $r = 3$. The first six sums are:

$$S_1 = 3, \qquad S_3 = 39, \qquad S_5 = 363,$$
$$S_2 = 12, \qquad S_4 = 120, \qquad S_6 = 1092.$$

From the pattern developing we can see that as n increases, the sum increases without bound.

Having examined two cases, one where $|r| < 1$ and the other where $|r| > 1$, let us look at the general formula for finding the sum of a finite geometric series.

$$S_n = \frac{a_1(1 - r^n)}{1 - r}, \qquad r \neq 1.$$

We can write this equation as:

$$S_n = \frac{a_1 - a_1 \cdot r^n}{1 - r}, \qquad r \neq 1$$

or:

$$S_n = \frac{a_1}{1 - r} - \left(\frac{a_1}{1 - r}\right)r^n, \qquad r \neq 1.$$

This form of the equation makes it is easier for us to understand how certain series can approach a finite value.

If the values of r are restricted to the open interval $(-1, 1)$ or $|r| < 1$, then as n takes on larger and larger values, r^n gets smaller and smaller and eventually approaches zero. For example, let $r = \dfrac{1}{4}$. Then r^n forms the sequence:

$$\left(\frac{1}{4}\right)^1, \left(\frac{1}{4}\right)^2, \left(\frac{1}{4}\right)^3, \left(\frac{1}{4}\right)^4, \left(\frac{1}{4}\right)^5, \ldots, \left(\frac{1}{r}\right)^n, \ldots$$

or:

$$\frac{1}{4}, \frac{1}{16}, \frac{1}{64}, \frac{1}{256}, \frac{1}{1024}, \ldots, \left(\frac{1}{r}\right)^n, \ldots$$

We can see that as n becomes larger, the terms of the sequence become smaller and smaller. For all practical purposes, they become so small that the terms converge or approach zero. In a more formal manner, using limits (which you may study later in calculus), it can be shown that if $-1 < r < 1$, then as n grows without bound, r^n becomes arbitrarily small. Thus, as n grows without bound:

$$\left(\frac{a_1}{1 - r}\right)r^n \to 0^* \quad \text{and} \quad S_n \to \frac{a}{1 - r}.$$

Thus, we can state the sum of the infinite geometric series is:

$$S = \frac{a_1}{1 - r} \text{ for } -1 < r < 1 \quad (\text{or } |r| < 1).$$

To check the assumption (made at the beginning of this section) that the sum of the series $\dfrac{1}{3}, \dfrac{1}{9}, \dfrac{1}{27}, \dfrac{1}{81}, \ldots, \dfrac{1}{3^n}, \ldots$ is 0.5000, substitute $a_1 = \dfrac{1}{3}$ and $r = \dfrac{1}{3}$ in the formula for sum of an infinite series.

$$S = \frac{a_1}{1 - r}$$

$$= \frac{\dfrac{1}{3}}{1 - \dfrac{1}{3}}$$

$$= \frac{\dfrac{1}{3}}{\dfrac{2}{3}}$$

$$= \frac{1}{2} \quad \text{or} \quad 0.5000.$$

*$\to$ is read "approaches."

> **THEOREM 16.4**
>
> The sum of the terms of an infinite geometric series whose first term is a_1 and whose common ratio is r is given by:
>
> $$S = \frac{a_1}{1 - r},$$
>
> provided $-1 < r < 1$.

EXAMPLE 1   Sum of an Infinite Series

Determine the sum of the following series.

$$1 - \frac{3}{5} + \frac{9}{25} - \frac{27}{125} + \frac{81}{625} - \cdots$$

Solution   The series $1 - \frac{3}{5} + \frac{9}{25} - \frac{27}{125} + \frac{81}{625} - \cdots$ has a common ratio r of $-\frac{3}{5}$ and $a_1 = 1$. Substituting into the formula:

$$S = \frac{a_1}{1 - r}, \quad \textbf{Theorem 16.4}$$

we have:

$$S = \frac{1}{1 - \left(-\frac{3}{5}\right)} \quad \text{or} \quad S = \frac{5}{8}.$$

The sum of the infinite geometric series is $\frac{5}{8}$. ◆

Note that we are not adding an infinite number of terms. Doing so is impossible even with a calculator. Instead, we are relying on the notion that r^n is essentially zero, if $|r| < 1$ and n is extremely large. The sum S is a value that is attained with large values of n. In calculus we expand on this idea and call S a limit.

The series in Example 1 can be written in summation notation.

$$\sum_{k=1}^{\infty} 1\left(-\frac{3}{5}\right)^{k-1} = \sum_{k=0}^{\infty} 1\left(-\frac{3}{5}\right)^{k} = 1 - \frac{3}{5} + \frac{9}{25} - \frac{27}{125}$$
$$+ \cdots + 1\left(\frac{-3}{5}\right)^{n-1} + \cdots.$$

EXAMPLE 2   **Sum of a Series in Summation Notation**

Determine the sum of the given series.

a. $\displaystyle\sum_{k=1}^{\infty} 2\left(\frac{3}{7}\right)^{k-1}$ **b.** $\displaystyle\sum_{k=0}^{\infty} -4\left(\frac{2}{5}\right)^{k}$

Solution

a.   For the series

$$\sum_{k=1}^{\infty} 2\left(\frac{3}{7}\right)^{k-1},$$

$a_1 = 2$ and $r = \dfrac{3}{7}$.

Substituting in the formula:

$$S = \frac{a_1}{1 - r},$$

we have:

$$S = \frac{2}{1 - \left(\dfrac{3}{7}\right)} = \frac{14}{4} = \frac{7}{2}.$$

Thus, the sum of the infinite geometric series is $\dfrac{7}{2}$.

b.   For the series

$$\sum_{k=0}^{\infty} -4\left(\frac{2}{5}\right)^{k},$$

$a_1 = -4$ and $r = \dfrac{2}{5}$.

Substituting in the formula:

$$S = \frac{a_1}{1 - r},$$

we have:

$$S = \frac{-4}{1 - \left(\dfrac{2}{5}\right)} \quad \text{or} \quad S = -\frac{20}{3}.$$

Thus, the sum of the infinite geometric series is $-\dfrac{20}{3}$.   ◆

We often may find it useful and sometimes necessary to express repeating decimals like 0.333 . . . as fractions. From past experience we know that

$$\frac{1}{3} = 0.333 \ldots .$$

A definition of rational numbers in Chapter 1 stated that a rational number can be expressed as a repeating decimal. Another definition also stated that a rational number can be expressed in the form $\frac{a}{b}$, where $b \neq 0$. We now can use the method for finding the sum of an infinite series to determine the fraction that is equivalent to the repeating decimal. This is illustrated in Example 3.

EXAMPLE 3   **Converting a Repeating Decimal to Fractional Form**

Determine the fraction that is equivalent to the repeating decimal.

a. 0.376376376 . . .   **b.** 0.29878787 . . .

Solution

a. The first thing we do is to write the repeating decimal $0.\overline{376}$ as an infinite geometric series.

$$0.\overline{376} = 0.376 + 0.000376 + 0.000000376 + \cdots$$
$$= 0.376 + 0.376(0.001) + 0.376(0.001)^2 + \cdots .$$

We can see that $a_1 = \boxed{0.376}$ and $r = \boxed{0.001}$. We can find the sum S since $|r| < 1$.

$$S = \frac{a_1}{1 - r}$$

$$= \frac{\boxed{0.376}}{1 - \boxed{0.001}} \qquad \textbf{Substituting}$$

$$= \frac{0.376}{0.999} \quad \text{or} \quad S = \frac{376}{999}.$$

Thus, an equivalent expression for the rational number $0.\overline{376}$ is $\frac{376}{999}$.

b. In the decimal expression 0.29878787 . . . , the digits 2 and 9 are not part of the digits that repeat. Thus, we must write the repeating decimal as an initial term plus an infinite geometric series:

$$0.29\overline{87}878787 \cdots = 0.29 + (0.0087 + 0.000087 + 0.00000087 + \cdots)$$
$$= 0.29 + (0.0087 + 0.0087(0.01)$$
$$+ 0.0087(0.01)^2 + \cdots)$$
$$= 0.29 + S,$$

where S is the sum of the series. For the infinite series, $a_1 = 0.0087$ and $r = 0.01$. We can determine the sum S since $|r| < 1$.

$$S = \frac{a_1}{1 - r}$$

$$= \frac{\boxed{0.0087}}{1 - \boxed{0.01}} \qquad \textbf{Substituting}$$

$$= \frac{0.0087}{0.99} \quad \text{or} \quad \frac{87}{9900}.$$

Since we can write 0.29 as $\frac{29}{100}$, we can write:

$$0.29878787\ldots = \frac{29}{100} + \frac{87}{9900}$$

$$= \frac{2871}{9900} + \frac{87}{9900}$$

$$= \frac{2958}{9900}.$$

Thus, an equivalent expression for the rational number $0.29\overline{87}$ is $\dfrac{2958}{9900}$. ◆

EXAMPLE 4 **Application: Vertical Distance Traveled by a Bouncing Ball**

A ball is dropped from a height of 2 m, and each time it hits the ground, it bounces back $\frac{5}{6}$ of the previous height. What is the total vertical motion of the ball? (Do not take into consideration the horizontal component of the motion.)

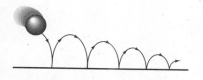

FIGURE 16.5

Solution An illustration of the path of the ball is given in Figure 16.5. The path of the ball can be represented by the following series:

$$2 + 2\left[2\left(\frac{5}{6}\right)\right] + 2\left[2\left(\frac{5}{6}\right)^2\right] + 2\left[2\left(\frac{5}{6}\right)^3\right] + \cdots. \qquad (5)$$

This is not a geometric series, but we can modify the form of this series so that it contains an infinite geometric series. Rewrite Equation (5) as

$$2 + 4\left(\frac{5}{6}\right) + 4\left(\frac{5}{6}\right)^2 + 4\left(\frac{5}{6}\right)^3 + \cdots. \qquad (6)$$

Starting with the second term we have an infinite geometric series. Thus, the sum of the series is:

$$S = 2 + \frac{a_1}{1 - r}.$$

Substituting $a_1 = 4\left(\dfrac{5}{6}\right)$ and $r = \dfrac{5}{6}$ we have:

$$S = 2 + \dfrac{4\left(\dfrac{5}{6}\right)}{1 - \dfrac{5}{6}}$$

$$= 2 + 20 \quad \text{or} \quad S = 22.$$

Thus, the total distance traveled by the ball is 22 m.   ◆

16.5 Exercises

In Exercises 1–20, if a sum exists, determine the sum of the infinite geometric series.

1. $\dfrac{1}{2} + \dfrac{1}{4} + \dfrac{1}{8} + \dfrac{1}{16} + \cdots$

2. $\dfrac{1}{2} - \dfrac{1}{4} + \dfrac{1}{8} - \dfrac{1}{16} + \cdots$

3. $\dfrac{2}{3} - \dfrac{4}{9} + \dfrac{8}{27} - \dfrac{24}{81} + \cdots$

4. $\dfrac{2}{3} + \dfrac{4}{9} + \dfrac{8}{27} + \dfrac{24}{81} + \cdots$

5. $1 + 0.003 + 0.0009 + 0.000027 + \cdots$

6. $2 + 0.04 + 0.0008 + 0.000016 + \cdots$

7. $3 - 3 + 3 - 3 + \cdots$

8. $5 + 5 + 5 + 5 + \cdots$

9. $27 + 3 + \dfrac{1}{3} + \dfrac{1}{9} + \cdots$

10. $27 - 3 + \dfrac{1}{3} - \dfrac{1}{9} + \cdots$

11. $\displaystyle\sum_{k=1}^{\infty} 16\left(-\dfrac{1}{2}\right)^{k-1}$

12. $\displaystyle\sum_{k=1}^{\infty} 16\left(\dfrac{1}{4}\right)^{k-1}$

13. $\displaystyle\sum_{k=1}^{\infty} \dfrac{1}{3}(-3)^{k-1}$

14. $\displaystyle\sum_{k=1}^{\infty} \left(-\dfrac{1}{3}\right)(3)^{k-1}$

15. $\displaystyle\sum_{k=0}^{\infty} \left(\dfrac{3}{5}\right)^{k}$

16. $\displaystyle\sum_{k=0}^{\infty} \left(-\dfrac{5}{3}\right)^{k}$

17. $\displaystyle\sum_{k=0}^{\infty} 4(0.3)^{k}$

18. $\displaystyle\sum_{k=1}^{\infty} 150\left(\dfrac{1}{10}\right)^{k}$

19. $\displaystyle\sum_{k=1}^{\infty} \dfrac{3^k + 4^k}{7^k}$

20. $\displaystyle\sum_{k=1}^{\infty} \dfrac{7^k}{3^k + 4^k}$

In Exercises 21–32, determine the fraction that is equivalent to the repeating decimal.

21. $0.3333\ldots$

22. $0.6666\ldots$

23. $0.9999\ldots$

24. $0.5555\ldots$

25. $0.45454545\ldots$

26. $0.73737373\ldots$

27. $0.137137137\ldots$

28. $0.219219219\ldots$

29. $0.48764876\ldots$

30. $0.5962159621\ldots$

31. $1.\overline{431}$

32. $5.\overline{762}$

In Exercises 33–38, two of the three quantities a_1, r, and S are given. Determine the third quantity using the formula for finding the sum of a geometric series.

33. $S = 1, r = \dfrac{1}{2}$

34. $S = \dfrac{1}{3}, a_1 = \dfrac{1}{2}$

35. $a_1 = \dfrac{2}{3}, r = -\dfrac{2}{3}$

36. $a_1 = \dfrac{2}{3}, r = \dfrac{2}{3}$

37. $S = \dfrac{100}{7}, a_1 = 1$

38. $S = \dfrac{3}{8}, r = -\dfrac{5}{3}$

39. A new superball is manufactured that, when dropped from a height of 5 m, rebounds 0.95 of the height from which it is dropped.
   a. Determine the height the ball reaches on the tenth bounce.
   b. Determine the total vertical distance the ball travels before it comes to rest.

40. A pendulum swings through an arc of 750 mm. On each successive swing, the pendulum covers an arc $\dfrac{14}{15}$ of the preceding swing.
   a. How far does the pendulum travel on the fifth swing?
   b. How far does the pendulum travel before it comes to rest?

41. In the race of the tortoise and the hare, the hare in the first minute after he wakes up reduces the distance between them to half what it was, in the next half minute reduces the remaining distance by half of itself, in the next quarter minute reduces the still-remaining distance by half, and so on. How long does it take the hare to overtake the tortoise?

42. On a steep mountain road, the highway department plans to build an artificial incline for cars and trucks to use in the event of brake failure. The artificial incline is designed so that if a truck were traveling at the rate of 70 mph, its speed would be reduced by 10 mph for each 100 ft traveled. What should be the minimum length of the artificial incline in order for a truck traveling at 70 mph to stop safely?

43. Consider an equilateral triangle of area A. Suppose another equilateral triangle is inscribed in the first. Further suppose that other equilateral triangles are inscribed, one inside the other as shown in the illustration.
   a. What is the area of the second largest triangle?
   b. What is the sum of the areas of all the equilateral triangles?

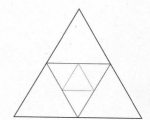

44. Zeno's paradox confused the ancient Greeks for many centuries. Zeno claimed that Achilles would never finish the 100 yard dash! Zeno's argument went something like this. First Achilles would have to run to the 50 yard mark, which was halfway to the end. Then he would have to run 25 more yards, which was half of the remaining distance. During each segment of the run, he could only cover half of the remaining distance. Since there are an infinite number of these half segments, it would take an infinite amount of time to finish the race. Hence, Achilles could never finish!
   a. To resolve this paradox, express Zeno's paradox as an infinite series.
   b. Now, determine the sum of the series in part a.

16.6

The Binomial Theorem

In working with algebraic expressions, it is frequently necessary to expand expressions of the form $(a + b)^n$, where n is a positive integer. For example,

$$(a + b)^1 = \qquad 1a^1b^0 + 1a^0b^1 \qquad\qquad = \qquad\qquad a + b$$

$$(a + b)^2 = \qquad 1a^2b^0 + 2a^1b^1 + 1a^0b^2 \qquad = \qquad a^2 + 2ab + b^2$$

$$(a + b)^3 = \qquad 1a^3b^0 + 3a^2b^1 + 3a^1b^2 + 1a^0b^3 \qquad = \qquad a^3 + 3a^2b + 3ab^2 + b^3$$

$$(a + b)^4 = 1a^4b^0 + 4a^3b^1 + 6a^2b^2 + 4a^1b^3 + 1a^0b^4 = a^4 + 4a^3b + 6a^2b^2 + 4ab^3 + b^4.$$

A discussion of a general method of expanding a binomial follows.

FACTS ABOUT THE EXPANSION OF A BINOMIAL

1. For any expansion as we move from left to right, we write:

$$a^nb^0, \ a^{n-1}b^1, \ a^{n-2}b^2, \ \ldots, \ a^1b^{n-1}, \ a^0b^n.$$

   Since $a^0 = 1$ and $b^0 = 1$, we normally do not write a^0 and b^0. This was done for illustration purposes.
2. The sum of the exponents for each term must equal the degree of the expansion. For example, in the third-row terms (a^3b^0, a^2b^1, a^1b^2, a^0b^3), the sum of the exponents in each term is 3. For an nth degree expansion, the sum of the exponents of each term is n (i.e., $a^{n-2}b^2$, $n - 2 + 2 = n$).
3. The coefficient of the first and last term is always 1.
4. There are $n + 1$ terms in the expression of $(a + b)^n$.

If we had a method of writing coefficients for each term, with the previous information, we could write the expansion of any binomial.

If we write only the coefficients of the expansions previously shown, we have the following array.

$$
\begin{array}{ccccccccc}
 & & & 1 & & 1 & & & \textbf{Row 1} \\
 & & 1 & & 2 & & 1 & & \textbf{Row 2} \\
 & 1 & & 3 & & 3 & & 1 & \textbf{Row 3} \\
1 & & 4 & & 6 & & 4 & & 1 \quad \textbf{Row 4}
\end{array}
$$

This array of numbers is called **Pascal's triangle.** Examining the pattern of Pascal's triangle, we can see that each succeeding row is obtained from each preceding row. For example, to create the fifth row from the fourth, do the following. Write a 1 at each end of the row. To obtain the other elements in the row, moving from left to right, to get the second element in row 5, add the first and second elements in row 4:

$$(\boxed{1} + \boxed{4} = \boxed{5}).$$

The third element in row 5 is found by adding the second and third elements in row 4:

$$(\boxed{4} + \boxed{6} = \boxed{10}).$$

The fourth and fifth elements in row 5 are the sums of the third and fourth and the fourth and fifth elements, respectively:

$$(\boxed{6} + \boxed{4} = \boxed{10} \quad \text{and} \quad \boxed{4} + \boxed{1} = \boxed{5}).$$

The following diagram illustrates the techniques.

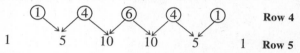

| | 1 | | 5 | | 10 | | 10 | | 5 | | 1 | | **Row 5** |

We can create any succeeding rows of the triangle in the same manner. Here is the triangle with the next two lines added.

$(a + b)^1$					1	1						**Row 1**
$(a + b)^2$					1	2	1					**Row 2**
$(a + b)^3$				1	3	3	1					**Row 3**
$(a + b)^4$			1	4	6	4	1					**Row 4**
$(a + b)^5$		1	5	10	10	5	1					**Row 5**
$(a + b)^6$	1	6	15	20	15	6	1					**Row 6**
$(a + b)^7$	1	7	21	35	35	21	7	1				**Row 7**

Using Pascal's triangle, we can write the expansion of $(x + y)^6$:

$$(a + b)^6 = a^6 + 6a^5b + 15a^4b^2 + 20a^3b^3 + 15a^2b^4 + 6ab^5 + b^6$$

We can use Pascal's triangle to expand a binomial when the coefficients of the variables are real numbers other than 1. This is illustrated in Examples 1 and 2.

EXAMPLE 1   **Binomial Expansion Using Pascal's Triangle**

Expand $(2x + 3y)^4$ using Pascal's triangle.

Solution   Writing the binomial $(2x + 3y)^4$ in the form $[(2x) + (3y)]^4$ and treating $(2x)$ and $(3y)$ as x and y, respectively, we can write:

$$[(2x) + (3y)]^4 = 1(2x)^4 + 4(2x)^3(3y) + 6(2x)^2(3y)^2 + 4(2x)(3y)^3 + 1(3y)^4$$

$$= 16x^4 + 96x^3y + 216x^2y^2 + 216xy^3 + 81y^4. \quad \blacklozenge$$

EXAMPLE 2   Binomial Expansion Using Pascal's Triangle

Expand $(0.2x - 0.4y)^5$ using Pascal's triangle.

Solution   Writing the binomial $(0.2x - 0.4y)^5$ in the form $[(0.2x) + (-0.4y)]^5$ and treating $(0.2x)$ and $(-0.4y)$ as x and y, respectively, we write:

$$[(0.2x) + (-0.4y)]^5 = 1(0.2x)^5 + 5(0.2x)^4(-0.4y)^1$$
$$+ 10(0.2x)^3(-0.4y)^2 + 10(0.2x)^2(-0.4y)^3$$
$$+ 5(0.2x)(-0.4y)^4 + 1(-0.4y)^5$$

$$= 0.00032x^5 - 0.0032x^4y + 0.0128x^3y^2$$
$$- 0.0256x^2y^3 + 0.0256xy^4 - 0.01024y^5. \quad \blacklozenge$$

These two examples show how Pascal's triangle can be used to determine the coefficients when expanding a binomial expression. The problem with the technique is in finding the coefficients for large values of n or in being able to write a specific term of an expansion.

Each element in Pascal's triangle can be replaced with the symbol $\binom{n}{r}$. The letter n is the degree of the binomial expansion, and it identifies the row of Pascal's triangle. The letter r $(r \le n)$ is the position of the element in a particular row, or the rth term of a binomial expansion. The possible values for r are 0, 1, 2, 3, 4, . . . , n. For the first element in a row $r = 0$, for the second element $r = 1$, etc. Thus, $\binom{7}{3}$ indicates a binomial of degree 7, which is the fourth element in the row. The following display illustrates the first four rows of Pascal's triangle.

$$
\begin{array}{ccccccccc}
 & & & 1 & \quad & 1 & & & \binom{1}{0} \quad \binom{1}{1} \\
 & & 1 & \quad & 2 & \quad & 1 & & \binom{2}{0} \quad \binom{2}{1} \quad \binom{2}{2} \\
 & 1 & \quad & 3 & \quad & 3 & \quad & 1 & \binom{3}{0} \quad \binom{3}{1} \quad \binom{3}{2} \quad \binom{3}{3} \\
1 & \quad & 4 & \quad & 6 & \quad & 4 & \quad & 1 \quad \binom{4}{0} \quad \binom{4}{1} \quad \binom{4}{2} \quad \binom{4}{3} \quad \binom{4}{4}
\end{array}
$$

To be able to write any term of Pascal's triangle, we must be able to evaluate $\binom{n}{r}$. However, to understand the evaluation of $\binom{n}{r}$, we must have the definition of $n!$ (read n factorial).

DEFINITION 16.8
The expression $n!$ is read n **factorial** and means the product of the first n counting numbers.

For example,

$$1! = 1$$
$$2! = 2 \cdot 1 = 2$$
$$3! = 3 \cdot 2 \cdot 1 = 6$$
$$4! = 4 \cdot 3 \cdot 2 \cdot 1 = 24$$
$$5! = 5 \cdot 4 \cdot 3 \cdot 2 \cdot 1 = 120$$
$$6! = 6 \cdot 5 \cdot 4 \cdot 3 \cdot 2 \cdot 1 = 720$$
$$\vdots \qquad \vdots \qquad \vdots$$
$$n! = n(n - 1)(n - 2)(n - 3) \cdots 3 \cdot 2 \cdot 1.$$

DEFINITION 16.9

$$0! = 1$$

DEFINITION 16.10

$$\binom{n}{r} = \frac{n!}{r!(n - r)!}$$

EXAMPLE 3   Evaluating $\binom{n}{r}$

Evaluate the following.

a. $\binom{4}{3}$ **b.** $\binom{9}{3}$ **c.** $\binom{11}{0}$

Solution

a. $\binom{4}{3} = \dfrac{4!}{3!(4 - 3)!}$ **Definition 16.10**

$$= \frac{4!}{3! \, 1!}$$

$$= \frac{4 \cdot \overset{1}{\cancel{3}} \cdot \overset{1}{\cancel{2}} \cdot 1}{\underset{1}{\cancel{3}} \cdot \underset{1}{\cancel{2}} \cdot 1 \cdot 1} \qquad \textbf{Definition 16.8}$$

$$= 4$$

Check this result with Pascal's triangle shown on p. 797.

b. $\dbinom{9}{3} = \dfrac{9!}{3!\,(\,9\,-\,3\,)!}$ **Definition 16.10**

$= \dfrac{9!}{3!\,6!}$

$= \dfrac{\overset{3}{\cancel{9}} \cdot \overset{4}{\cancel{8}} \cdot 7 \cdot \overset{1}{\cancel{6}} \cdot \overset{1}{\cancel{5}} \cdot \overset{1}{\cancel{4}} \cdot \overset{1}{\cancel{3}} \cdot \overset{1}{\cancel{2}} \cdot 1}{(\underset{1}{\cancel{3}} \cdot \underset{1}{\cancel{2}} \cdot 1)(\underset{1}{\cancel{6}} \cdot \underset{1}{\cancel{5}} \cdot \underset{1}{\cancel{4}} \cdot \underset{1}{\cancel{3}} \cdot \underset{1}{\cancel{2}} \cdot 1)}$ **Definition 16.8**

$= 84$

c. $\dbinom{11}{0} = \dfrac{11!}{0!\,(\,11\,-\,0\,)!}$ **Definition 16.10**

$= \dfrac{11!}{0!\,11!}$

$= \dfrac{\overset{1}{\cancel{11!}}}{1\,(\underset{1}{\cancel{11!}})}$

$= 1$ ◆

HINTS FOR EVALUATING FACTORIALS

1. In the second step of each problem, the sum of the numbers in the denominator must always equal the number in the numerator. In $\dfrac{9!}{3!\,6!}$, we see that $3 + 6 = 9$.

2. When calculating $9! = 9 \cdot 8 \cdot 7 \cdot 6 \cdot 5 \cdot 4 \cdot 3 \cdot 2 \cdot 1$, we see that it also can be written as $9! = 9 \cdot 8 \cdot 7 \cdot 6!$. Therefore, in evaluating $\dfrac{9!}{3!\,6!}$, we can write:

$$\frac{9!}{3!\,6!} = \frac{\overset{3}{\cancel{9}} \cdot \overset{4}{\cancel{8}} \cdot 7 \cdot \cancel{6!}}{\underset{1}{\cancel{3}} \cdot \underset{1}{\cancel{2}} \cdot 1 \cdot \underset{1}{\cancel{6!}}} = 84.$$

This is extremely helpful when large numbers are involved.

This technique for determining the coefficients leads to the binomial theorem.

THEOREM 16.5 Binomial Theorem

$$(x + y)^n = \binom{n}{0}x^n + \binom{n}{1}x^{n-1}y + \binom{n}{2}x^{n-2}y^2 + \binom{n}{3}x^{n-3}y^3$$

$$+ \cdots + \binom{n}{n-1}xy^{n-1} + \binom{n}{n}y^n,$$

where n is a positive integer.

With the coefficients of the binomial in this form, we can expand a binomial to any power and we can write a single term of the expansion.

EXAMPLE 4   **Expansion of a Binomial Using the Binomial Theorem**

Expand $(x - 3)^4$ using the binomial theorem.

Solution   We can write $(x - 3)^4$ as $[x + (-3)]^4$ and apply the binomial theorem.

$$[x + (-3)]^4 = \binom{4}{0}x^4 + \binom{4}{1}x^3(-3) + \binom{4}{2}x^2(-3)^2$$

$$+ \binom{4}{3}x(-3)^3 + \binom{4}{4}(-3)^4$$

$$= \frac{4!}{0!\,4!}x^4 + \frac{4!}{1!\,3!}x^3(-3) + \frac{4!}{2!\,2!}x^2(-3)^2$$

$$+ \frac{4!}{3!\,1!}x(-3)^3 + \frac{4!}{4!\,0!}(-3)^4$$

$$= 1x^4 + 4(-3)x^3 + (6)(9)x^2 + 4(-27)x + (1)(81)$$

$$= x^4 - 12x^3 + 54x^2 - 108x + 81. \quad \blacklozenge$$

EXAMPLE 5   **Writing Specific Terms of a Binomial**

Write the first three terms of the expansion of $(2a + 5b)^{10}$.

Solution   The coefficients determined by Pascal's triangle of the first three terms of the expansion of $(2a + 5b)^{10}$ are $\binom{10}{0}$, $\binom{10}{1}$, and $\binom{10}{2}$.

$$\binom{10}{0} = \frac{10!}{0!\,10!} \qquad \binom{10}{1} = \frac{10!}{1!\,9!} \qquad \binom{10}{2} = \frac{10!}{2!\,8!}$$

$$= 1 \qquad\qquad = \frac{10 \cdot 9!}{1!\,9!} \qquad = \frac{\overset{5}{\cancel{10}} \cdot 9 \cdot \overset{1}{\cancel{8!}}}{\underset{1}{\cancel{2}} \cdot \underset{1}{\cancel{1}} \cdot \cancel{8!}}$$

$$= 10 \qquad\qquad = 45$$

Using the binomial theorem, we have:

$$(2a + 5b)^{10} = 1(2a)^{10} + 10(2a)^9(5b) + 45(2a)^8(5b)^2 + \cdots$$

$$= (1)(1024a^{10}) + 10(512a^9)(5b) + 45(256a^8)(25b^2) + \cdots$$

$$= 1024a^{10} + 25{,}600a^9b + 288{,}000a^8b^2 + \cdots \quad \blacklozenge$$

EXAMPLE 6 Determining the Value of a Decimal

Determine the value of $(0.99)^7$ to five significant digits by expanding a binomial whose terms have the sum 0.99.

Solution We can write 0.99 as $1 - 0.01$. Thus,

$$(0.99)^7 = (1 - 0.01)^7$$

$$= \binom{7}{0}1^7 + \binom{7}{1}1^6(-0.01) + \binom{7}{2}1^5(-0.01)^2$$

$$+ \binom{7}{3}1^4(-0.01)^3 + \binom{7}{4}1^3(-0.01)^4$$

$$+ \binom{7}{5}1^2(-0.01)^5 + \binom{7}{6}1(-0.01)^6$$

$$+ \binom{7}{7}(-0.01)^7$$

$$= 1(1) + 7(1)(-0.01) + 21(1)(0.0001)$$
$$+ 35(1)(-0.000001) + 35(1)(0.00000001)$$
$$+ 21(1)(-0.0000000001)$$
$$+ 7(1)(0.000000000001)$$
$$+ 1(-0.00000000000001)$$
$$= 1 - 0.07 + 0.0021 - 0.000035 + 0.00000035 + \cdots *$$
$$= 0.932065.$$

Thus, $(0.99)^7$ to five significant digits is 0.93207. ◆

It is possible to determine a specific term of a binomial expansion without writing every term of the expansion. For example, the term containing x^a in the expansion of the binomial $(x + y)^n$ is $\binom{n}{n-a} x^a y^{n-a}$. This is illustrated in Example 7.

EXAMPLE 7 Determining a Single Term

Find the term containing x^8 in the expansion of $(2x + 3y)^{12}$.

Solution To find a specific term of an expression we must remember that the sum of the exponents must equal the degree of the expansion. The term containing x^8 is:

$$\binom{12}{4}(2x)^8(3y)^4 = \frac{\overset{1}{\cancel{12}} \times 11 \times \overset{5}{\cancel{10}} \times 9 \times \overset{1}{\cancel{8!}}}{\underset{1}{\cancel{4}} \times \underset{1}{\cancel{3}} \times \underset{1}{\cancel{2}} \times 1 \times \underset{1}{\cancel{8!}}}(256x^8)(81y^4)$$

$$= 10{,}264{,}320\, x^8y^4. ◆$$

* We can see from the pattern that is developing that the addition of more terms would not make any difference in the fifth significant digit.

16.6 Exercises

In Exercises 1–20, evaluate the following.

1. $8!$

2. $6!$

3. $\dfrac{7!}{5!}$

4. $\dfrac{11!}{9!}$

5. $\dfrac{7!}{4!\,3!}$

6. $\dfrac{8!}{6!\,2!}$

7. $\dfrac{9!}{2!\,7!}$

8. $\dfrac{12!}{10!\,2!}$

9. $\dfrac{0!}{0!}$

10. $\dfrac{14!}{0!\,14!}$

11. $\dfrac{15!}{3!\,12!}$

12. $\dbinom{4}{3}$

13. $\dbinom{6}{2}$

14. $\dbinom{9}{0}$

15. $\dbinom{13}{2}$

16. $\dbinom{15}{4}$

17. $\dbinom{100}{99}$

18. $\dbinom{155}{154}$

19. $\dbinom{n}{n-1}$

20. $\dbinom{n}{n-2}$

In Exercises 21–35, expand the following expressions using the binomial theorem.

21. $(x + 1)^4$

22. $(Z + 3)^4$

23. $(a + b)^4$

24. $(x^2 - 2y^2)^5$

25. $(3a^2 + 2b^2)^5$

26. $\left(\dfrac{x}{2} + \dfrac{y}{3}\right)^5$

27. $(2x^2 - 3t^2)^4$

28. $(x^{-1} + y^{-1})^5$

29. $\left(x\sqrt{2} - \dfrac{y}{3}\right)^4$

30. $(y^2 - 2)^6$

31. $(a^3 - 2b)^6$

32. $(1 - y^{-1})^8$

33. $(a^{\frac{1}{2}} + b^{\frac{1}{2}})^7$

34. $(0.3x - 0.4y)^4$

35. $(x^{\frac{2}{3}} - 2y^{\frac{2}{3}})^7$

In Exercises 36–44, write the first four terms of the expansion.

36. $(x + y)^{16}$

37. $(1 - x^2)^{17}$

38. $(R - 2I)^{14}$

39. $\left(u^2 + \dfrac{1}{2}v^3\right)^{18}$

40. $(x^{\frac{1}{3}} - y^{\frac{1}{3}})^{20}$

41. $(2x - 3y)^{22}$

42. $(x^{\frac{1}{2}} - y^{\frac{1}{2}})^{15}$

43. $(1 + y)^{96}$

44. $(Z - 3)^{110}$

In Exercises 45–50, use the binomial theorem to determine the value of each of the following expressions correct to six significant digits.

45. $(1.01)^7$

46. $(1.03)^8$

47. $(1.3)^6$

48. $(0.98)^8$

49. $(0.99)^9$

50. $(0.97)^6$

51. Find the term containing x^6 in the expansion of $(a + x)^{12}$.

52. Find the term containing x^8 in the expansion of $(x - 2)^{12}$.

53. Find the term containing u^{12} in the expansion of $(u^2 + v^2)^{18}$.

54. Find the term containing y^3 in the expansion of $(y^{\frac{1}{2}} - a^{\frac{1}{2}})^{16}$.

55. Find the term independent of x in the expansion of $(3x + x^{-1})^{12}$.

56. Find the value of n if the coefficients of the fifth and the thirteenth terms in the expansion of $(Z + 1)^n$ are equal.

57. Find the value of Z such that the eighth term of $(3 + Z)^{18}$ equals the seventh term in the expansion of $(3 + Z)^{17}$.

58. Show that $\binom{n}{r} = \binom{n}{n-r}$.

Review Exercises

In Exercises 1–6, determine a formula for the ith term of each of the sequences.

1. $1, 4, 7, 10, \ldots$

2. $1, 3, 5, 7, 9, \ldots$

3. $1, -1, -3, -5, \ldots$

4. $\dfrac{2}{3}, \dfrac{4}{3}, \dfrac{8}{3}, \dfrac{12}{3}, \dfrac{16}{3}, \ldots$

5. $2, 4, 2, 4, \ldots$

6. $0, 3, 8, 15, \ldots$

In Exercises 7–16, perform the following.
a. Write the first six terms of the sequence.
b. Indicate whether the sequence is increasing, decreasing or neither.
c. Graph the sequence.

7. $a_i = 4i$

8. $a_i = -2i$

9. $a_i = i - 5$

10. $a_i = 3i + 2$

11. $a_i = -\dfrac{2}{3}i$

12. $a_i = \dfrac{2}{3}i$

13. $a_i = 3 + (-1)^i$

14. $a_i = 3 + (-2)^i$

15. $a_i = \dfrac{i-3}{i+4}$

16. $a_i = \dfrac{i^2 - 2}{i^2 + 2}$

In Exercises 17–24, write the series associated with each of the sequences for the following. **a.** S_4 **b.** S_6

17. $4, 16, 64, 256, \ldots, 4^n$

18. $-4, 16, -64, 256, \ldots, (-1)^n 4^n$

19. $1, 3, 5, 7, \ldots, (2n - 1)$

20. $-1, 1, 3, 5, \ldots, 2n - 3$

21. $0, \dfrac{1}{4}, \dfrac{2}{5}, \dfrac{3}{6}, \ldots, \dfrac{n-1}{n+2}$

22. $\dfrac{2}{3}, \dfrac{3}{4}, \dfrac{4}{5}, \dfrac{5}{6}, \ldots, \dfrac{n+1}{n+2}$

23. $2, -3, 4, -5, \ldots, (-1)^{n+1}(n + 1)$

24. $1, -8, 27, -64, \ldots, (-1)^{n-1} n^3$

In Exercises 25–30, write each of the series in expanded form and determine the sum.

25. $S_5 = \displaystyle\sum_{i=1}^{5} \dfrac{(-2)^{i+1}}{i+1}$

26. $S_5 = \displaystyle\sum_{i=1}^{5} (-1)^i (2i - 1)^2$

27. $S_5 = \displaystyle\sum_{i=1}^{5} \dfrac{1}{i} 2^{i+1}$

28. $S_7 = \displaystyle\sum_{i=1}^{7} \sin(2i)$

29. $S_6 = \displaystyle\sum_{i=1}^{6} \dfrac{i-5}{i+1}$

30. $S_7 = \displaystyle\sum_{i=1}^{7} (3i - 1)(i + 3)$

In Exercises 31–36, write each series using summation notation.

31. $5 + 9 + 13 + 17 + \cdots + (4 + n)$

32. $2 + 1 + 0 + (-1) + \cdots + (3 - n)$

33. $-2 + \dfrac{3}{2} - \dfrac{4}{4} + \dfrac{5}{8} - \cdots + \dfrac{(-1)^n(n + 1)}{2^{n-1}}$

34. $\dfrac{1}{2^1} + \dfrac{2}{2^3} + \dfrac{3}{2^5} + \dfrac{4}{2^7} + \dfrac{5}{2^9}$

35. $\dfrac{1}{3} + \dfrac{2}{3^3} + \dfrac{3}{3^5} + \dfrac{4}{3^7} + \dfrac{5}{3^9}$

36. $\dfrac{z-5}{z+5} + \dfrac{z-6}{z+6} + \dfrac{z-7}{z+7} + \dfrac{z-8}{z+8}$

In Exercises 37–42, write each of the series in expanded form.

37. $\displaystyle\sum_{k=0}^{6} \frac{k}{2k-1}$

38. $\displaystyle\sum_{k=1}^{n} \frac{2k-1}{k}$

39. $\displaystyle\sum_{k=1}^{n} \frac{2k+3}{k^2}$

40. $\displaystyle\sum_{k=1}^{6} ke^k$

41. $\displaystyle\sum_{k=1}^{4} (z^2 + 3kz - k)$

42. $\displaystyle\sum_{k=1}^{6} \frac{(-1)^k}{\sqrt[5]{k}}$

In Exercises 43–48, write each of the infinite series in expanded form using the indicated term and the common difference.

43. $a_1 = 11, d = -5$

44. $a_1 = -7, d = 11$

45. $a_3 = 7, d = -1$

46. $a_4 = 8, d = 3$

47. $a_6 = -4, d = \dfrac{1}{2}$

48. $a_2 = \dfrac{4}{5}, d = \dfrac{4}{5}$

In Exercises 49–54, some of the quantities of an arithmetic sequence are given. Determine the quantities that are not given (a_1, d, a_n, n, S_n).

49. $n = 11, a_1 = 3, a_{11} = 13$

50. $n = 10, a_1 = -1, a_{10} = 19$

51. $S_6 = 48, a_1 = -3$

52. $S_7 = 56, a_7 = 22$

53. $a_2 = \dfrac{4}{5}, d = \dfrac{4}{5}$

54. $S_8 = -2, d = \dfrac{1}{2}$

In Exercises 55–60, perform the following.
a. Construct the first six terms of the geometric sequence.
b. Determine the sum of the first six terms given the first term and the ratio.

55. $a_1 = \dfrac{1}{3}, r = 2$

56. $a_1 = -\dfrac{1}{3}, r = 2$

57. $a_1 = 2, r = \dfrac{1}{4}$

58. $a_1 = \dfrac{3}{4}, r = \dfrac{1}{5}$

59. $a_1 = -2, r = 0.1$

60. $a_1 = \dfrac{1}{4}, r = -2$

In Exercises 61–66, some of the quantities of a geometric sequence are given. Determine the quantities that are not given (a_1, r, a_n, n, S_n).

61. $a_1 = \dfrac{2}{3}, r = \dfrac{1}{2}, n = 4$

62. $a_1 = \dfrac{1}{2}, r = \dfrac{2}{3}, n = 4$

63. $S_5 = -3, n = 5, r = 2$

64. $S_5 = 3, n = 5, r = 2$

65. $a_7 = 243, a_1 = \dfrac{1}{3}, n = 7$

66. $a_1 = \dfrac{3}{5}, r = \dfrac{-2}{3}, n = 5$

In Exercises 67–74, if a sum exists, determine the sum of the infinite geometric series.

67. $\dfrac{1}{5} + \dfrac{1}{25} + \dfrac{1}{125} + \dfrac{1}{625} + \cdots$

68. $1 + \dfrac{1}{6} + \dfrac{1}{36} + \dfrac{1}{216} + \cdots$

69. $1 + 0.02 + 0.0004 + 0.000008 + \cdots$

70. $1 - 0.03 + 0.0009 - 0.000027 + \cdots$

71. $7 - 7 + 7 - 7 + \cdots$

72. $81 - 27 + 9 - 3 + \cdots$

73. $\displaystyle\sum_{k=0}^{\infty} \left(\frac{4}{7}\right)^k$

74. $\displaystyle\sum_{k=1}^{\infty} \frac{5^k}{3^k + 6^k}$

In Exercises 75–78, determine the fraction that is equivalent to the repeating decimal.

75. $0.8888\ldots$

76. $0.232323\ldots$

77. $1.237237\ldots$

78. $3.471471\ldots$

In Exercises 79–84, expand the binomial expressions.

79. $(Z + 5)^4$ **80.** $(x^2 - 3y^2)^4$ **81.** $(3u^2 - 2w^2)^4$

82. $(2a^3 + 3b)^6$ **83.** $(a^{\frac{1}{3}} + b^{\frac{1}{3}})^6$ **84.** $(1 - z^{-1})^7$

In Exercises 85–88, write the first four terms of the expansion.

85. $(x + y)^{18}$ **86.** $(x^2 - 1)^{20}$ **87.** $(a^{\frac{1}{3}} + b^{\frac{1}{3}})^{16}$ **88.** $(Z + 1)^{92}$

89. Find the term containing x^{11} in the expansion of $(x - 3)^{15}$.

90. Find the term containing y^5 in the expansion of $(y^{\frac{1}{3}} - a^{\frac{1}{3}})^{24}$.

91. The first swing of a pendulum is 250 mm, and each succeeding swing is reduced by $\frac{1}{5}$ mm. What is the length of the tenth pendulum swing?

92. A grandfather's clock strikes once at 1 o'clock, twice at 2 o'clock, and so on. What is the total number of times the clock strikes in 12 hours?

93. If John saves $\frac{1}{2}$¢ the first day, 1¢ the next day, 2¢ the next, 4¢ the next, and so on, how much money will he have saved after 30 days?

94. How many ancestors are listed in your family tree in the last ten generations, counting your two parents, four grandparents, eight great-grandparents, and so on?

95. A florist started eight cannas bulbs. At the end of the year they had grown to 16 bulbs. If they continue dividing in this manner, how many bulbs will he have in five years?

96. If $1,000 is placed in an account that earns 8% simple interest each year, it will amount to $1,080 in one year, $1,160 in two years, $1,240 in three years, and so on. How much will the principal and the interest amount to in 25 years?

97. The circumference of a circular container around a clock spring is 2.82 in. The spring is in the form of a spiral having 12 complete turns. Find the length of the spring assuming that each spiral is 0.15 in. shorter than the next larger spiral.

98. Show that 998^2 equals the binomial $(1000 - 2)^2$. Calculate the square of 998 using the binomial expansion.

99. A $275 bicycle depreciates 20% of its value each year. What is the value of the bicycle at the end of the sixth year?

100. A chemical substance decomposes in such a way that, at the end of each hour, it is only 80% as heavy as it was at the beginning of the hour.
 a. How much would 10 lb of this substance weigh after 5 hr?
 b. How long would it take for 10 lb to decompose completely?

101.* Show that the reciprocals of the terms of a geometric sequence also form a geometric sequence.

102.* A harmonic sequence is a type of sequence not mentioned in this chapter.
 a. Do research and write a definition for a harmonic sequence.
 b. Write two examples of harmonic sequences.

103.* The Fibonacci sequence was discovered in the Middle Ages by the mathematician Fibonacci. The first eleven terms of the sequence are 1, 1, 2, 3, 5, 8, 13, 21, 34, 55, and 89.

    a. Is this an arithmetic or geometric sequence?

    b. Determine how each succeeding term is determined and find the twelfth and thirteenth terms of the sequence.

104.* If the midpoints of the sides of a square are connected, the resulting figure is a square. If this process is continued indefinitely, the result is a set of squares as illustrated. We can use the Pythagorean theorem to compute the length of the side of any square in the set if we know the length of the side of the square that immediately contains it. Consider a set of squares where the outermost square has a side length of 5 mm.

    a. Determine the sum of the areas of all the squares in the nest.

    b. Determine the sum of the perimeters of all the squares in the nest.

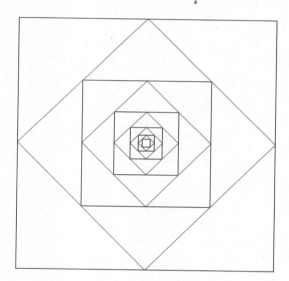

105.* A geometric sequence has a first term of 3 and a common ratio of r, while an arithmetic sequence has a first term of 3 and a common difference of d. If S_4 is the same for both sequences, express d in terms of r.

106.* Show that a geometric sequence with ratio $r = 1$ is the same as an arithmetic sequence with difference $d = 0$.

107.* Write a computer program to determine the sum of the terms of an arithmetic sequence.

108.* Write a computer program to determine the sum of the terms of a geometric series.

109.* Write a computer program to determine the sum of an infinite geometric series.

✎ Writing About Mathematics

1. Write a paragraph or two discussing the differences between an arithmetic sequence and a geometric sequence.

2. Exercise 40 in Section 16.4 concerns the rate of cooling for a cup of coffee. Using the numbers given in the exercise, determine the temperature of the cup of coffee in 20 minutes. What is the temperature of the cup of coffee in one hour? Does it make sense that the coffee would be colder? Is the math wrong here? If not, then what is wrong?

3. Does a bouncing superball come to rest? Using the values given in Exercise 39 of Section 16.5, calculate the length of time it will take the ball to come to rest. Discuss your conclusion.

4. Exercise 44 in Section 16.5 asked us to express Zeno's paradox as an infinite series and to find the sum of the series. Write a paragraph or two answering the following questions. Does finding the sum disprove Zeno's claim? If so, how?

Chapter Test

In Exercises 1 and 2, determine a formula for the ith term of the sequences.

1. $2, -2, -6, -10, \ldots$

2. $\dfrac{1}{5}, \dfrac{2}{5}, \dfrac{4}{5}, \dfrac{8}{5}, \ldots$

In Exercises 3 and 4, write the first five terms of the sequence and indicate whether it is increasing, decreasing, or neither. Also graph the sequence.

3. $a_i = 2i + 3$

4. $a_i = -\dfrac{1}{2}i$

In Exercises 5 and 6, write the series associated with each of the sequences for S_5.

5. $1, 4, 7, 10, \ldots, 3n - 2$

6. $-\dfrac{1}{2}, 0, \dfrac{1}{4}, \dfrac{2}{5}, \ldots, \dfrac{n-2}{n+1}$

7. Write the given series in expanded form and determine the sum.

$$\sum_{i=1}^{5} \frac{i-2}{2+i}$$

8. Write the given series using summation notation.

$$-1 + 1 - \frac{3}{4} + \frac{1}{2} - \cdots + \frac{(-1)^n n}{2^{n-1}}$$

9. Write the given series in expanded form.

$$\sum_{k=1}^{n} \frac{1-k}{1+k}$$

10. Write the infinite series in expanded form using the indicated term $a_5 = 2$ and the common difference $d = -2$.

11. If $d = -2$ and $S_5 = 6$, determine $a_1 a_n$, and n.

12. **a.** Given $a_1 = 27$ and $r = \dfrac{1}{3}$, write the first five terms of the geometric sequence.

    b. Determine the sum of the first five terms determined in part a.

13. If $r = 2$, $n = 4$, and $S_n = 6$, determine a_1 and a_n.

14. Determine the sum of the infinite geometric series if a sum exists

$$1 + \frac{1}{2} + \frac{1}{4} + \frac{1}{8} + \cdots .$$

15. Determine the fraction that is equivalent to the repeating decimal $2.565656\ldots$.

16. Expand the binomial expression $(2a^3 - b^2)^4$.

17. Write the first three terms of the expansion of $(a^2 - 3)^7$.

18. Determine the term containing a^5 in the expansion of $(a + 3b^2)^9$.

19. The first year Yolanda worked she saved \$2000. In each succeeding year she saved \$100 more than the preceding year. How much has she accumulated at the end of ten years? (Ignore interest.)

20. Suppose that the population of Small Town USA is increasing at the rate of 5% per year. If the population today is 2500 people, what will the population be in 20 years?

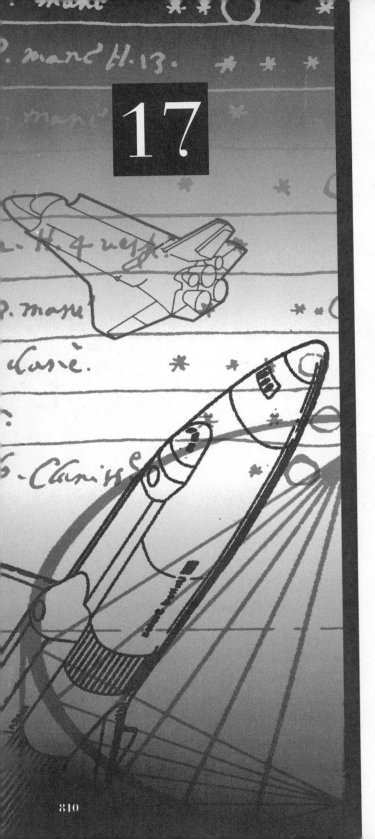

17

Introduction to Statistics

Statistics has become a very important ingredient in the manufacturing process in the United States. By developing a proper quality control program that reduced defects, Motorola saved $500 million in one year. Having been out-performed by foreign manufacturers in the 60s and 70s, many U.S. manufacturers are following Motorola's lead, setting a goal of six sigma defects. Six sigma is a statistical term attached to the concept of having approximately 3.4 defects per million items produced. See Section 17.6 for more information on six sigma.

17.1

Frequency Distribution

The results of an experiment conducted by a statistician or a quality-control person may consist of many pieces of data or just a few. **A piece of datum** is a single response to an experiment. The individual conducting the experiment must know how to organize and analyze the data, and be able to draw conclusions. When the amount of data is large, it is usually advantageous to construct a frequency distribution. A **frequency distribution** is a listing of the observed values and the corresponding frequency of each value.

EXAMPLE 1   Tally Sheet and Frequency Distribution

Table 17.1 shows the number of gadgets manufactured by ABC Corporation on each of 100 days. Construct a tally sheet and a frequency distribution of the data.

TABLE 17.1

23	13	20	24	22	19	24	22	24	21
32	20	16	31	24	27	16	27	25	32
29	18	25	22	12	24	23	28	19	24
17	27	22	33	25	22	20	22	32	25
20	25	19	26	17	19	26	29	13	28
19	32	26	29	24	28	21	24	31	24
28	20	18	22	18	21	28	17	22	30
23	18	26	29	28	19	23	27	15	22
27	23	21	19	24	27	20	27	21	33
20	19	23	24	30	22	31	18	32	26

Solution   A tally sheet is made and a frequency distribution is formed as shown in Table 17.2 ◆

TABLE 17.2

Number of Gadgets (Observation)	Tally	Number of Days (Frequency)	Number of Gadgets (Observation)	Tally	Number of Days (Frequency)
12	I	1	24	ⅢⅢ I	11
13	II	2	25	Ⅲ	5
14		0	26	Ⅲ	5
15	I	1	27	Ⅲ II	7
16	II	2	28	Ⅲ I	6
17	III	3	29	Ⅲ	4
18	Ⅲ	5	30	II	2
19	Ⅲ III	8	31	III	3
20	Ⅲ II	7	32	Ⅲ	5
21	Ⅲ	5	33	II	2
22	Ⅲ Ⅲ	10		Total	100
23	Ⅲ I	6			

Although the distribution may be studied from Table 17.2, if the frequencies are grouped into **classes,** the distribution may be more meaningful and easier to study.

When data are grouped in classes, certain rules must be followed.

> 1. All classes must be of the same "width."
> 2. No classes may overlap.
> 3. Each piece of datum must belong to one and only one class.

In addition, it is often suggested that a frequency distribution should be constructed with 5 to 12 classes. If there are too few or too many classes, the distribution may be difficult to interpret.

Following the suggested rules and guidelines, Table 17.3 illustrates one way of grouping the data from Table 17.2. It is obvious that the original distribution can be grouped in other ways. For instance, the classes might be 10 to 14, 15 to 19, 20 to 24, etc., or 12 to 13, 14 to 15, 16 to 17, etc.

The **class limits** are shown in Table 17.3. The lower limit of the second class is 14. The upper class limit of this class is 16. **Class boundaries** are found by computing values midway between the upper and lower class limits of consecutive classes. Thus, the class boundaries of the first three classes of Table 17.3 are 10.5 to 13.5, 13.5 to 16.5, and 16.5 to 19.5 (see Table 17.4). The values halfway between successive class boundaries are called **class marks** or **midpoints.** The class marks for the first three classes of the table are 12, 15, and 18 (Table 17.4). The **class interval** or **class width** is the distance between two successive class boundaries or class marks. The class width for Table 17.3 is 3.

TABLE 17.3

Class Limits	Frequency
11–13	3
14–16	3
17–19	16
20–22	22
23–25	22
26–28	18
29–31	9
32–34	7
Total	100

TABLE 17.4

Class Boundaries	Class Limits	Class Marks	Frequencies
10.5–13.5	11–13	12	3
13.5–16.5	14–16	15	3
16.5–19.5	17–19	18	16
19.5–22.5	20–22	21	22
22.5–25.5	23–25	24	22
25.5–28.5	26–28	27	18
28.5–31.5	29–31	30	9
31.5–34.5	32–34	33	7

Histograms and frequency polygons are statistical graphs used to illustrate a frequency distribution. A **histogram** is a bar graph with observed values on its horizontal scale and frequencies on its vertical scale. The bar is constructed so that the class mark is in the center of the bar on the horizontal scale. The height of the bar is determined by the frequency of that class.

Figure 17.1 illustrates a histogram for the set of data in Example 1. There are several things to observe from Figure 17.1. The horizontal and vertical scales do not have the same units of measurement. There is a label for each scale, as well as a general label for the complete graph. The labels are an essential part of the graph. How can any meaning be attached to the graph if it is not labeled?

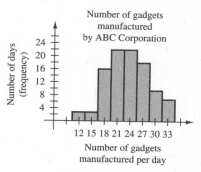

FIGURE 17.1

Frequency polygons are line graphs with their scales the same as those of the histogram; that is, the horizontal scale indicates observed values, and the vertical scale indicates frequency. To construct a frequency polygon, place a dot above the class mark at the corresponding frequency. Then connect the dots with straight line segments. In constructing frequency polygons always put in two additional class marks, one at the left end and one at the right end on the horizontal scale. Since the frequency at these added class marks is 0, the end points of the frequency polygon are always on the horizontal scale.

EXAMPLE 2   Frequency Polygon

Construct a frequency polygon for the frequency distribution in Table 17.4.

Solution   Figure 17.2 illustrates the frequency polygon. Notice that the scales are the same as those in Figure 17.1, as is the labeling.

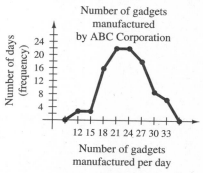

FIGURE 17.2

The class boundaries, the class limits, or the class marks each can be used to indicate the horizontal scale for the histogram or frequency polygon. In fact, since the class mark was used to construct the frequency polygon, it may have been more convenient to use the class mark on the horizontal scale.

A frequency polygon can be constructed over a histogram by connecting the midpoints of the the tops of the rectangles in the histogram.

17.1 Exercises

In Exercises 1–7, use the information in the table on algebra test results.

Algebra Test Results (Maximum Possible Score 50 Points)

Class Limits (Scores)	Frequency (No. of Students)
15–19	6
20–24	0
25–29	8
30–34	14
35–39	36
40–44	20
45–49	16

1. Determine the total number of observations.
2. Determine the width of each class.
3. Determine the class boundaries for each class.
4. Determine the midpoint of each class.
5. Determine the class limits of the next class if an additional class has to be added.
6. Construct and properly label a histogram.
7. Construct and properly label a frequency polygon.

In Exercises 8–14, use the information in the table on ABC Corporation.

Age of Employees at ABC Corporation

Class Limits (Ages)	Frequency (No. of Employees)
20–29	27
30–39	26
40–49	48
50–59	58
60–69	36
70–79	5

8. Determine the total number of observations.
9. Determine the width of each class.
10. Determine the class boundaries for each class.
11. Determine the midpoint of each class.
12. Determine the class limits of the next class if an additional class has to be added.
13. Construct and properly label a histogram.
14. Construct and properly label a frequency polygon.

In Exercises 15–19, use the information in the table on quiz scores.

Quiz Scores of a Mathematics Class

1	7	10	9	8	7	10	10
9	7	8	9	8	7	10	10
8	8	8	8	6	7	10	8
9	8	7	8	5	7	9	8
10	6	7	7	4	6	1	9
7	5	6	7	2	6	3	9
6	3	2	3	1	6	9	7
10	10	5	10	1	5	9	7
9	9	4	2	2	5	9	5
8	9	10	10	9	5	9	4

15. Construct a tally chart and a frequency distribution with a class width of 2 (first class, 1 to 2).

16. Determine the class boundaries for each class.

17. Determine the midpoint of each class.

18. Construct and properly label a histogram.

19. Construct and properly label a frequency polygon.

In Exercises 20–23, use the information in the table on XYZ Corporation.

Number of Hours Worked in One Week by 25 Secretaries of XYZ Corporation

30	33	34	36	37
32	34	35	36	37
32	34	35	37	38
33	34	35	37	38
33	34	35	37	38

20. Construct a frequency distribution with a class width of 1.

21. Determine the class boundaries for each class.

22. Construct and properly label a histogram.

23. Construct and properly label a frequency polygon.

In Exercises 24–29, use the information in the table on sixth graders.

IQ of 50 Sixth Graders

80	89	92	95	97	100	102	106	110	120
81	89	93	95	98	100	103	108	113	120
87	90	94	97	99	100	103	108	114	122
88	91	94	97	100	100	103	108	114	128
89	92	94	97	100	101	104	109	119	135

24. Construct a frequency distribution with a first class of 80 to 84.

25. Construct a histogram for the frequency distribution in Exercise 24.

26. Construct a frequency distribution with a first class of 80 to 88.

27. Construct a histogram for the frequency distribution in Exercise 26.

28. Construct a frequency distribution with a first class of 80 to 90.

29. Construct a histogram for the frequency distribution in Exercise 28.

In Exercises 30–34, use the information in the table on U.S. presidents.

Ages of U.S. Presidents at Their First Inauguration

57	57	49	52	50	51	51	55	64
61	61	64	56	47	56	60	61	46
57	54	50	46	55	55	62	52	
57	68	48	54	54	51	43	70	
58	51	65	49	42	54	55	73	

30. Construct a frequency distribution with a first class of 40 to 45.

31. Construct a frequency polygon for the frequency distribution in Exercise 30.

32. Construct a frequency distribution with a first class of 42 to 47.

33. Construct a frequency polygon for the frequency distribution in Exercise 32.

34. Construct a frequency distribution with a first class of 42 to 46.

35. Construct a frequency polygon for the frequency distribution in Exercise 34.

In Exercises 36–38, use the information in the table on starting salaries.

Starting Salaries for Electronic Technologists for 25 Different Companies (in thousands)

20	21	22	24	25
21	21	22	24	25
21	21	23	25	25
21	22	23	25	26
21	22	23	25	26

36. Construct a frequency distribution.

37. Construct a histogram.

38. Construct a frequency polygon.

In Exercises 39–43, use the information in the histogram about garages.

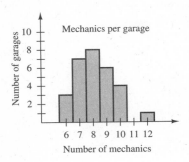

39. How many garages were surveyed?

40. How many garages employed nine mechanics?

41. How many mechanics were observed?

42. Construct a frequency distribution from the histogram.

43. Determine the class boundaries of the classes.

In Exercises 44–48, use the information in the histogram on students.

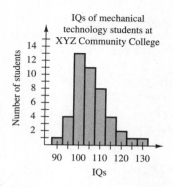

IQs of mechanical
technology students at
XYZ Community College

44. How many students were surveyed?

45. How many students had IQs in the class with a class mark of 100?

46. Determine the class limits for each of the classes.

47. Determine the class boundaries for each of the classes.

48. Construct a frequency distribution.

In Exercises 49–51, use the information in the frequency polygon about pens.

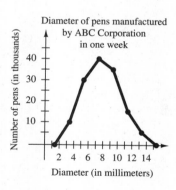

Diameter of pens manufactured
by ABC Corporation
in one week

49. How many pens were manufactured?

50. How many pens were manufactured with a diameter of 9 mm?

51. Construct a frequency distribution.

17.2

Measures of Central Tendency

The term "average" is used daily in many ways: The truck averages a mileage of 4.2 km/liter. The machine fills an average of 300 bottles/hr. The machine dispenses an average of 7.4 oz of coffee per cup. On the average one out of four students will pass a test.

An **average** is a number that is representative of a group of data. Averages discussed in this section are the *mean, median,* and *mode.* Each is calculated differently and may yield different results for the same set of data. Each usually results in a number near the center of the data, and for this reason, averages are commonly referred to as **measures of central tendency.**

"Mean"

The arithmetic mean, or simply the mean, is symbolized either by $\bar{X}$ (read X bar) or by the Greek letter μ (mu). The symbol $\bar{X}$ is used when the mean of a sample (subset) of the population is calculated. The symbol μ is used when the mean of the entire population is calculated.

> **DEFINITION 17.1**
>
> The **mean** of a set of data is found by summing all the pieces of data and dividing the sum by the number of pieces of data. That is,
>
> $$\text{mean} = \frac{\text{sum of the data}}{\text{number of pieces of data}}$$

In notation the equation is:

$$\bar{X} = \frac{\sum\limits_{i=1}^{n} x_i}{n}.$$

As you recall, the symbol Σ is the Greek letter sigma—the letter used to indicate summation. This notation is used to indicate the sum of the data for $i = 1$ to n. That is,

$$\sum_{i=1}^{n} x_i = x_1 + x_2 + x_3 + \cdots + x_n,$$

where i is the index that indicates the number we start with (in this case, 1), and n is the number of pieces of data.

EXAMPLE 1 Calculating the Mean

The lengths of five bolts in a sample are 24.0 mm, 25.0 mm, 26.0 mm, 25.0 mm and 27.0 mm. Determine the mean length of the bolts.

Solution

$$\overline{X} = \frac{\sum x_i}{n}$$

$$= \frac{24.0 \; + \; 25.0 \; + \; 26.0 \; + \; 25.0 \; + \; 27.0}{5}$$

$$= \frac{127.0}{5}$$

$$= 25.4 \text{ mm}$$

Therefore, the mean length of the bolts $\overline{X}$ is 25.4 mm.   ◆

The mean represents "the balancing point" of a set of scores. Consider a seesaw (teeter-totter). If a seesaw is pivoted at the mean and uniform weights are placed at points corresponding to the lengths of the bolts, as illustrated in Figure 17.3, the seesaw balances.

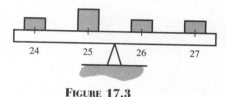

24   25   26   27

FIGURE 17.3

The mean is the most common measure of central tendency. In order for the mean to represent a whole set of data, it must have some meaningful properties.

PROPERTIES OF THE MEAN

1.   The mean always exists; you can calculate it with any set of data.
2.   The mean is unique; that is, a set of data has one, and only one, mean.
3.   Each piece of datum is used in calculating the mean.
4.   The mean is used in other statistical calculations. For example, we can find the mean of a set of means, and as we shall see in the next section, it is used in finding the standard deviation.
5.   The mean is more reliable than other measures of central tendency in that it is not as strongly affected by chance.

To determine the mean of a frequency distribution, we multiply the frequency of each class by the class mark of the class and divide that sum by the sum of the frequencies. For example, if we let $x_1, x_2, \ldots , x_n$ be the class marks and f_1, $f_2, \ldots , f_n$ be the frequency of each class, then the mean is:

$$\overline{X} = \frac{x_1 f_1 \; + \; x_2 f_2 \; + \; x_3 f_3 \; + \; \cdots \; + \; x_n f_n}{n},$$

where $n = f_1 + f_2 + f_3 + \cdots + f_n$ which is the sum of the frequencies.

EXAMPLE 2   Mean of a Frequency Distribution

Use the frequency distribution from Example 1 in Section 17.1 to determine the mean number of gadgets manufactured daily by ABC Corporation.

Solution   First construct a table like that shown in Table 17.5. Looking at the table (moving from left to right), we see that in the first column are the class limits, in the second column are the class marks, in the third column are the frequencies for each class, and in the fourth column are the products of the class marks and the frequencies $(x \cdot f)$.

TABLE 17.5

Number of Gadgets (Class Limits)	Class Mark (x)	Frequencies (f)	$x \cdot f$
11–13	12	3	36
14–16	15	3	45
17–19	18	16	288
20–22	21	22	462
23–25	24	22	528
26–28	27	18	486
29–31	30	9	270
32–34	33	7	231
		$\sum f = 100$	$\sum xf = 2346$

We determine the mean of the distribution by substituting in the formula the sum of the third column and the sum of the fourth column.

$$\overline{X} = \frac{\sum xf}{\sum f} = \frac{2346}{100} = 23.46$$

The mean, to the nearest tenth, is 23.5.   ◆

The mean of the ungrouped data for Example 2 is 23.52. We can see that the error in calculating a mean with grouped data is relatively small. We can calculate the mean with a calculator; see Example 5 in Section 17.3 and Appendix B.

Weighted Mean

There are situations where we must take into account the relative importance of each piece of data in relation to the entire set of data. For example, say a student's final grades in English, biology, mathematics, and physical education are, respectively, 85, 91, 88, and 75. If the respective credits received for these courses are 3, 4, 3, and 1, determine an average grade. To do this, we cannot just find the sum of 85, 91, 88, and 75 and divide by 4. If we do this, we are not taking into account that English is a three-credit course, biology is a four-credit course, and so on. We must find a special mean called a **weighted arithmetic mean.**

$$\text{Weighted arithmetic mean} = \frac{w_1x_1 + w_2x_2 + w_3x_3 + \cdots + w_kx_k}{w_1 + w_2 + w_3 + \cdots + w_k}$$

$$\overline{X} = \frac{\sum w_ix_i}{\sum w}$$

where w is the weight and x is the score or piece of data. Thus, the student's average grade is:

$$\overline{X} = \frac{3 \times 85 + 4 \times 91 + 3 \times 88 + 1 \times 75}{3 + 4 + 3 + 1}$$

$$= 87.$$

Thus, the weighted average grade or weighted mean grade is 87.

Do you recognize that the mean of a set of grouped data, such as in Example 2, is a weighted arithmetic mean?

"Median"

> **DEFINITION 17.2**
> The **median** is the value in the middle of a set of ranked data.

Ranked Data are data that are listed from smallest values to largest values, or vice versa. To find the median of a set of data, rank the data and determine the value in the middle; this value is the median. In this text, the median is indicated by the symbol **md.**

EXAMPLE 3   **Median: Odd Number of Pieces of Data**

Determine the median length of the five bolts in Example 1 (that is, 24.0 mm, 25.0 mm, 26.0 mm, 25.0 mm, and 27.0 mm).

Solution   Ranking the data yields:

$$\underbrace{24.0, \quad 25.0,}_{2 \text{ pieces}} \quad \boxed{25.0,} \quad \underbrace{26.0, \quad 27.0.}_{2 \text{ pieces}}$$

Since 25.0 mm is the middle length of this set of ranked data (there are two pieces above 25.0 mm, and two pieces below 25.0 mm), 25.0 mm is the median. ◆

If there is an even number of pieces of data, the median is halfway between the two middle pieces. In this case, we find the median by adding the two middle pieces and dividing by 2.

EXAMPLE 4   Median: Even Number of Pieces of Data

Determine the median salary of six workers whose salaries are $28,000, $18,000, $23,000, $15,000, $12,000, and $15,000.

Solution   Ranking the data, we have:

$12,000, $15,000, $15,000, $18,000, $23,000, $28,000.

Since there are six pieces of data, the median lies halfway between the two middle pieces, $15,000 and $18,000. The median is:

$$md = \frac{15,000 + 18,000}{2} = \$16,500.$$

Thus, the median of the salaries is $16,500.   ◆

The median of a grouped set of data is a little more difficult to determine and is an approximate value. The median of a frequency distribution is a point on the horizontal scale that divides the area of the histogram so that half the area falls to the right of the point and the other half to the left of the point. To determine this point, we must calculate $n/2$ of the frequency distribution. To illustrate how this is done, let us use the frequency distribution on gadgets manufactured by ABC Corporation (Table 17.6). Since $n = 100$, divide by 2: $100/2 = 50$. We want to find the 50th day so that we can start counting from either end. If we begin with the smallest value of the distribution, we find 3 values in the first class, a total of 6 values in the first two classes, a total of 22 values in the first three classes, and a total of 44 values in the first four classes. The median falls in the fifth class since we need six more values to make 50 and the next class contains 22, which would make a total of 66. We just need 6 of the 22 in the fifth class to make a total of 50. To find the median, we must know the class boundary

TABLE 17.6

Number of Gadgets (Class Limits)	Frequency	Total of Successive Classes	
		Adding Down	**Adding Up**
11–13	3	3	
14–16	3	6 $j = 50 - 44 = ⑥$	
17–19	16	22	
20–22	22	44	
23–25	22	⑥ needed	50
		50	⑯ needed
26–28	18		34 $k = 50 - 34 = ⑯$
29–31	9		16
32–34	7		7
	$\sum f = 100$		

between the fourth and fifth classes and the class width. The boundary is 22.5 and the width is 3. We can now find the median by adding 6/22 of the class width, 3, to the lower class boundary or:

$$md = \boxed{22.5} + \frac{6}{22}(\boxed{3})$$

$$= 23.3.$$

In general, if l is the lower boundary of the class into which the median must fall, f is the frequency of the class, c is the class width, and j is the number of pieces of data still needed in that class to reach $n/2$, then the **median of the distribution** is given by the formula:

$$md = l + \left(\frac{j}{f}\right)(c).$$

The median of a frequency distribution can also be found by starting with the largest values of the distribution and subtracting an appropriate fraction of the class width from the upper-class boundary of the class into which the median must fall. For the distribution of gadgets manufactured by ABC Corporation, we have:

$$md = \boxed{25.5} - \frac{16}{22}(\boxed{3})$$

$$= 23.3,$$

which agrees with our previous answer. A general formula for the median of a distribution when we start with the largest values is given by:

$$md = u - \left(\frac{k}{f}\right)(c),$$

where k is the pieces of data needed in the class in which the median falls and u is the upper boundary of that class.

"Mode"

> **DEFINITION 17.3**
> The **mode** is the piece of data that occurs most frequently.

EXAMPLE 5 **Determining the Mode**

Determine the mode of the lengths of the five bolts in Example 1 (that is, 24.0 mm, 25.0 mm, 26.0 mm, 25.0 mm, and 27.0 mm).

Solution The mode of the set of data is 25.00 mm since it occurs twice and all the other lengths occur only once. ◆

If there is no piece of datum that occurs more frequently than any other piece of datum, there is no mode. This is a definite disadvantage of the mode.

EXAMPLE 6   No Mode

Determine the mode for the hourly rates of six mechanics who earn $8, $10, $7, $14, $12, and $16, respectively.

Solution   Each hourly rate is different; therefore there is no mode.   ◆

For a frequency distribution, the **modal class** is the class that has the largest number of frequencies. For the frequency distribution in the Table for Exercises 8–14 of Section 17.1, the modal class has class limits of 50 to 59.

17.2 Exercises

In Exercises 1–14, determine the mean, median, and mode for the set of data.

1. 3, 5, 5, 6, 7, 9, 10, 10, 10

2. 2, 3, 8, 10, 8, 7, 6, 370, 38, 35

3. 58, 70, 78, 82, 84, 43, 94

4. 3, 1, 4, 4, 4, 7, 9

5. 1, 3, 5, 7, 9, 11, 13, 15

6. 3, 9, 13, 29, 38, 16, 14, 11, 3

7. 40, 50, 30, 60, 90, 100, 140

8. 2, 2, 2, 2, 5, 5, 5, 5, 7, 9, 12, 14, 17, 23

9. 7, 9, 13, 14, 12, 11, 14, 16, 18

10. 2, 2, 2, 2, 10, 10, 10, 10

11. 230, 456, 805, 480, 380

12. 1001, 2430, 16201, 960, 70

13. 87, 70, 62, 93, 82, 69

14. 70, 80, 85, 85, 90, 100

15. The mean is the most sensitive average because it is most affected by a change in data.
  a. Determine the mean, median, and mode for 1, 2, 4, 4, 4, 8, 12.
  b. Change the 8 in part a to an 11. Then find the mean, median, and mode.
  c. Which averages were affected by changing the 8 to an 11 in part b?
  d. Will any averages be affected if we change the 12 in part a to an 8?

16. In order to get a grade of A, a student must have a mean average of 90. Jane has a mean average of 89 for ten quizzes. She approaches her teacher and asks for an A, explaining that she missed an A by only one point. What is wrong with Jane's reasoning?

17. The Riveral's monthly gas and electric bills for the year are listed in the following table. Determine the mean, median, and mode.

$98.64	$99.72	$87.18	$97.74	$95.05	$91.96
$90.96	$72.95	$64.41	$96.08	$95.06	$96.72

18. The Hampton's telephone bills for the year are listed in the following table. Determine the mean, median and mode.

$147.91	$275.80	$349.82	$652.31
$329.35	$292.81	$652.31	$285.31
$351.29	$198.41	$247.85	$329.72

19. The mean length of six pieces of pipe is 0.78 m. Determine the total length of the six pieces.

20. The mean length of eight pieces of 2-by-4 lumber is 23 in. Determine the total length of the eight pieces.

21. Construct a set of five pieces of data with a mean of 70, which has no two pieces of data the same.

22. A mean score of 80 for five exams is needed for a final grade of B. John's first four exam grades are 72, 74, 85, and 81. What grade does John need on the fifth exam to get a B in the course?

23. A mean grade of 60 on seven exams is needed to pass a course. On her first six exams Carol received grades of 49, 72, 80, 60, 57, and 69.
  a. What grade must she receive on her last exam to pass the course?
  b. An average of 70 is needed to get a C in the course. Is it possible for Carol to get a C? If so, what grade must she receive on the seventh exam?
  c. If her lowest grade is to be dropped and only her six best exams are counted, what grade must she receive on her last exam to pass the course?
  d. If her lowest grade is to be dropped and only her six best exams are counted, what grade must she receive on her last exam to get a C in the course?

24. A student's final grades in history, English, chemistry, calculus, and health are 95, 80, 75, 78, and 86. If the respective credits received for the five courses are 3, 3, 4, 4, and 1, determine the weighted average grade.

25. Company ABC has 75 employees; 30 earn $15.00 per hour, and 45 earn $9.50 per hour. Determine the mean earnings per hour.

26. Five groups of students, consisting of 10, 8, 11, 13, and 9 individuals, reported mean computer times of 18, 33, 11, 42, and 53 hours, respectively. Determine the mean computer time of all the students.

27. The scores of 120 technical mathematics students on a 50-point test are given in the following table.

Class Limits (Score)	Frequency (No. of Students)
46–50	3
41–45	12
36–40	36
31–35	40
26–30	23
21–25	6
	$\Sigma f = 120$

  a. Determine the mean score of the students.
  b. Determine the median score of the students.
  c. Determine the modal class.

28. The length of time in minutes it takes 120 different students to complete a test is given in the following table.

Class Limits (Time)	Frequency (No. of Students)
47–48	1
45–46	4
43–44	3
41–42	7
39–40	19
37–38	12
35–36	11
33–34	17
31–32	17
29–30	13
27–28	5
25–26	8
23–24	2
21–22	1
	$\Sigma f = 120$

a. Determine the mean length of time it took the students to complete the test.
b. Determine the median length of time it took the students to complete the test.
c. Determine the modal class.

29. The length of time in hours spent in training to learn a skill is given in the following table.

Length of Time	No. of Individuals
41–43	1
38–40	3
35–37	4
32–34	4
29–31	5
26–28	8
23–25	10
20–22	18
17–19	23
14–16	17
11–13	10
	$\Sigma f = 103$

a. Determine the mean length of time it took the individuals to learn the skill.
b. Determine the median number of hours it took the individuals to learn the skill.
c. Determine the modal class.

30. For the frequency distribution in Exercise 15 in Section 17.1, determine the following.
  a. The mean quiz grade
  b. The median quiz grade
  c. The modal class

31. For the frequency distribution in the table for Exercises 8-14 of Section 17.1, determine the following.
  a. The mean age of the employees
  b. Median age of the employees

17.3

Measures of Dispersion

Sufficient information to analyze a set of data is not always obtained from the measures of central tendency. As an example, say a hospital is seeking bids for a generator to provide emergency power. There are two manufacturers who have the product. Manufacturer A's generator has an average (mean) life of 500 hours of generating power before it must be rebuilt. Manufacturer B's generator has an average life of 350 hours before it must be rebuilt. If you assume that both generators cost the same, which generator should the hospital purchase? There are many things that go into this decision. The average generator life may not be the most important factor. Manufacturer A's generators have an average life of 500 hours. This could mean that half will last 250 hours and the other half will last about 750 hours. If, in fact, all of Manufacturer B's generators have a life span of between 325 and 375 hours, then B's generators are most consistent and reliable. If Manufacturer A's generator were purchased, it would have to be rebuilt every 200 hours or so because it is impossible to tell which ones would fail first. If B's generator were purchased, it could go longer before being rebuilt. This example is an exaggeration used to illustrate the importance of the spread on the variability of data.

The measures of dispersion are used to give indications of the spread of data.

Range

The **range** is the difference between the highest and lowest values of a set of data; it indicates the total spread of the data.

$$\text{Range} = \text{highest value} - \text{lowest value.}$$

EXAMPLE 1 Range

To determine the quality of stainless steel, it is tested with an acid solution. The times required to test five pieces of steel are 3.0, 3.0, 4.0, 4.0, and 6.0 min. Determine the range of the testing times.

Solution

$$\text{range} = \text{highest} - \text{lowest}$$
$$= 6.0 - 3.0$$
$$= 3.0$$

The range of the test times is 3 min. ◆

Standard Deviation

The standard deviation is symbolized either by the letter s or the Greek letter sigma, σ. The s is used when the standard deviation of a sample is calculated. The σ is used when the standard deviation of an entire population is calculated. The **standard deviation** is a measure of how much the data differs from the mean. The farther the data is spread from the mean, the larger the standard deviation is. The formula for finding the standard deviation of a sample is:

Standard Deviation of a Sample

$$s = \sqrt{\frac{\sum (x_i - \overline{X})^2}{n - 1}}.$$

Example 2 illustrates how to calculate the standard deviation of a sample using the formula.

EXAMPLE 2   Standard Deviation of a Sample

Determine the standard deviation of the set of data: 2, 3, 4, 8, 13.

Solution   The first thing we do is calculate the mean.

$$\overline{X} = \frac{2 + 3 + 4 + 8 + 13}{5}$$
$$= 6.$$

Then construct a table with three columns as illustrated in Table 17.7. List the data in the first column in ascending or descending order. This helps us check our calculations. Complete the second column by subtracting the mean, 6, from each piece of data in the first column as shown in Table 17.8.

TABLE 17.7

Data x_i	Data − Mean $x_i - \overline{X}$	(Data − Mean)^2 $(x_i - \overline{X})^2$
2		
3		
4		
8		
13		

TABLE 17.8

x_i	$x_i - \overline{X}$	$(x_i - \overline{X})^2$
2	$2 - 6 = -4$	
3	$3 - 6 = -3$	
4	$4 - 6 = -2$	
8	$8 - 6 = 2$	
13	$13 - 6 = 7$	
	$\sum (x_i - \overline{X}) = 0$	

The sum of the values in the $x_i - \bar{X}$ column must always be zero; if it is not, you have made an error. (If the mean is a decimal, there may be a slight round-off error, and the sum may not be exactly zero.) Next square the value in column 2 and place the result in column 3 (Table 17.9).

TABLE 17.9

x_i	$x_i - \bar{X}$	$(x_i - \bar{X})^2$
2	-4	$(-4)^2 = 16$
3	-3	$(-3)^2 = 9$
4	-2	$(-2)^2 = 4$
8	2	$(2)^2 = 4$
13	7	$(7)^2 = 49$
	$\sum (x_i - \bar{X}) = 0$	$\sum (x_i - \bar{X})^2 = 82$

The sum of the numbers in column 3 is 82. Substituting in the formula for the standard deviation, we have:

$$s = \sqrt{\frac{\sum (x_i - \bar{X})^2}{n-1}}$$
$$= \sqrt{\frac{82}{5-1}}$$
$$= \sqrt{20.5}$$
$$= 4.53.$$

Thus, the standard deviation of the set of data to the nearest tenth is 4.5. ◆

In Example 2, $s = 4.5$. This value indicates how the data differ from the mean. For example, if we replace the 13 with a 20, then $s = 7.4$. The larger value for s indicates a greater difference between the data and the mean. What happens if the 13 in the original set of data is replaced with a 9? Calculate s and see if your conclusion is correct.

EXAMPLE 3   **Standard Deviation of a Sample**

Determine the standard deviation of the set of data in Example 1. The times were 3.0, 3.0, 4.0, 4.0 and 6.0 min.

Solution   To determine the standard deviation, first calculate the mean of the times.

$$\bar{X} = \frac{3.0 + 3.0 + 4.0 + 4.0 + 6.0}{5} = 4.0$$

Having calculated the mean to be 4.0 min., now set up a table like that in Table 17.10.

TABLE 17.10

x_i	$x_i - \overline{X}$	$(x_i - \overline{X})^2$
3.0	-1	1
3.0	-1	1
4.0	0	0
4.0	0	0
6.0	2	4
	$\sum (x_i - \overline{X}) = 0$	$\sum (x_i - \overline{X})^2 = 6.0$

Substituting number of test times, 5, for n and 6.0 for the $\Sigma(x_i - \overline{X})^2$ in the formula, we have:

$$s = \sqrt{\frac{\Sigma (x_i - \overline{X})^2}{n - 1}}$$

$$= \sqrt{\frac{6.0}{5 - 1}}$$

$$= 1.2.$$

Therefore, the standard deviation of the test times is 1.2 min. ◆

With larger sets of data, we will find it is more convenient to work with a frequency distribution when calculating the standard deviation. The formula to use when calculating the standard deviation of grouped data (frequency distribution) is given in the box.

Standard Deviation of Grouped Data

$$s = \sqrt{\frac{n(\Sigma x_i^2 f_i) - (\Sigma x_i f_i)^2}{n(n - 1)}},$$

where $n = \Sigma f_i$.

This formula is illustrated in Example 4.

EXAMPLE 4 **Standard Deviation of Grouped Data**

Determine the standard deviation of the number of gadgets manufactured daily by ABC Corporation to the nearest tenth. Table 17.11 summarizes the data from previous material.

TABLE 17.11

Number of Gadgets (Class Limits)	Class Mark x_i	Frequencies f_i	$x_i f_i$	$x_i^2 f_i$
11–13	12	3	36	432
14–16	15	3	45	675
17–19	18	16	288	5,184
20–22	21	22	462	9,702
23–25	24	22	528	12,672
26–28	27	18	486	13,122
29–31	30	9	270	8,100
32–34	33	7	231	7,623
		$\sum f_i = 100$	$\sum x_i f_i = 2346$	$\sum x_i^2 f_i = 57{,}510$

Solution The formula

$$s = \sqrt{\frac{n(\sum x_i^2 f_i) - (\sum x_i f_i)^2}{n(n-1)}}$$

tells us that we need two new quantities, $\sum x_i^2 f_i$ and $(\sum f_i x_i)^2$, which we did not have when we found the mean of a frequency distribution. We find them by forming the columns headed $x_i f_i$ and $x_i^2 f_i$ as shown in Table 17.11. We determine the entries in the $x_i f_i$ column by finding the product of the class marks x_i, and the frequencies f_i. We determine the entries in the $x_i^2 f_i$ column by multiplying each entry in the $x_i f_i$ column by the respective class mark x_i. Then sum the $x_i f_i$ and the $x_i^2 f_i$ columns.

Substituting these values in the formula, we have:

$$s = \sqrt{\frac{100\,(57{,}510) - (2346)^2}{100\,(100 - 1)}}$$

$$= \sqrt{\frac{5{,}751{,}000 - 5{,}503{,}716}{9900}}$$

$$= 5.0.$$

Thus, the standard deviation of the distribution is 5.0. ◆

Using a calculator simplifies the calculations for determining the arithmetic mean and the standard deviation. The specific keys for determining the mean and standard deviation using the fx–7700G and the TI–82 are demonstrated in Example 5.

EXAMPLE 5 **Calculator Techniques for Determining the Mean and the Standard Deviation**

The hourly wages of fifteen cooks are shown in Table 17.12. Use a calculator to determine the mean salary and the standard deviation of the salaries.

TABLE 17.12

No. of Cooks	Hourly Wage (in Dollars)
2	21
5	13
3	17
4	15
1	24

Solution Keystrokes are given for the fx–7700G and the TI–82.

fx-7700G

To determine the mean and standard deviation, it is necessary to place the calculator in the correct mode. To change to statistical mode we press the following keys:

Keystrokes	Screen Display
SHIFT MODE 1 SHIFT MODE	RUN / SD
4 MODE × SHIFT DISP	S-data : STO
Nrm (F3) EXE PRE M Disp	S-graph : NON-
	G-type : REC/CON
	angle : Rad
	display : Nrm2
	DT EDIT ; DEV Σ PQR

To clear any values that may be stored in the statistical memory we use the following key strokes.

Keystrokes	Screen Display
EDIT (F2)	DEL INS ERS
ERS (F1) YES	DT EDIT ; DEV Σ PQR

Now we are ready to enter values. We can enter each piece of data individually, or we may enter multiple scores by using the ; (F3) keys.

Keystrokes	Screen Display	Answer
21 ; 2 DT	21; 2	21.
13 ; 5 DT	13; 5	13.
17 ; 3 DT	17; 3	17.
15 ; 4 DT	15; 4	15.
24 DT	24	24

Having entered all the data, we can check whether the data was entered correctly by pressing EDIT. The screen display shows the order in which each piece of data was entered, data entered and the frequency of each piece of data (Figure 17.4).

	x	f
1	21	2
2	13	5
3	17	3
4	15	4
5	24	1
		21.

FIGURE 17.4

If there are any incorrect entries, use the arrow keys to move the cursor to the position, enter the correct number, and press EXE . Having made all corrections, we can determine the mean, the standard deviation, median, maximum value, and minimum value by pressing the following keys.

Keystrokes	Screen Display	Answer
DEV (F4)	$\overline{x}$ $x\sigma n$ $x\sigma n-1$ ◇	
$\overline{x}$ (F1) EXE	$\overline{x}$	16.13333333
$x\sigma n^{-1}$ (F3) EXE	$x\sigma n - 1$	3.440653316
◇ (F4)	Mod Med Max Min	
Med (F2) EXE	Med	15
Max (F3) EXE	Max	24
Min (F4) EXE	Min	13

The results are: mean is 16.13333333, standard deviation is 3.440653316, median is 15, maximum value is 24, and minimum value is 13.

TI-82

Before data can be entered, it is necessary to clear any data on the lists (storage locations for statistical data).

Keystrokes	Screen Display	Answer
STAT 4 2nd L1 (1) , 2nd L2 (2) ENTER	ClrList L1, L2	Done

The next step is to enter the data, the hourly wages in list 1 and the number of cooks (the frequency) in list 2.

Keystrokes			
STAT 1 (Edit) 21 ENTER 13 ENTER 17 ENTER 15 ENTER 24 ENTER ▷ 2 ENTER 5 ENTER 3 ENTER 4 ENTER 1 ENTER			

L1	L2	L3
21	2	- -
13	5	- -
17	3	- -
15	4	- -
24	1	- -

L1(1) = 21

Chapter 17   Introduction to Statistics

The next step is to tell the calculator which list contains data and which list contains the frequencies. Remember that we are dealing with one-variable statistics, so under this heading we shade L1 to the right of *x*list and L2 to the right of Freq.

Keystrokes	Screen Display
STAT ▷ 3 (SetUp)	SET UP CALCS 1-Var Stats Xlist: L1 L2 L3 L4 Freq: L1 L2 L3 L4. 2-Var Stats Xlist: L1 L2 L3 L4 Ylist: L1 L2 L3 L4 Freq: L1 L2 L3 L4

The next step is to have the calculator display the mean and standard deviation.

Keystrokes	Screen Display
STAT ▷ (CALC) 1 (1-Var Stats) ENTER	1-Var $\overline{X} = 16.13333333$ $\Sigma x = 242$ $\Sigma x^2 = 4070$ $Sx = 3.440653316$ $\sigma x = 3.323986897$ $n = 15.$ $minX = 13$ $Q_1 = 13$ $Med = 15$ $Q_3 = 17$ $maxX = 24$

The calculator not only gives the mean and standard deviation, it also gives number of pieces of data (n), minimum value ($minX$), first quartile (Q_1), median (Med), third quartile (Q_3), and maximum value ($maxX$).

From the calculator results we see that the mean of the hourly wages of the cooks is $16.13, and the standard deviation is $3.44. ◆

17.3 Exercises

1. Without actually doing the calculations, decide which of the following two sets of data has the greater standard deviation: 3, 6, 7, 8, 10, 14; or 10, 11, 11, 12, 12, 13. Explain why.

2. Of the two sets of data, 1, 3, 5, 7, 9, and 101, 103, 105, 107, 109, which would you expect to have a larger standard deviation? Explain.

3. What does it mean when the standard deviation of a set of data is zero?

In Exercises 4–14, determine the range and standard deviation of the sets of data.

4. 8, 8, 12, 14, 5, 7

5. 111, 113, 113, 114, 117, 121, 123

6. 4, 0, 2, 5, 7, 9, 1, 3, 5

7. 5, 8, 13

8. 4, 4, 4, 4, 4, 4, 4

9. 15, 18, 21, 13, 19

10. 4, 5, 3, 3, 4, 5

11. 31, 29, 33, 38, 19, 22, 43, 41

12. 6, 3, 6, 3, 6, 3, 7

13. 1, 3, 8, 16, 13, 4, 9, 6, 6, 4

14. 70, 82, 94, 98, 86

15. In shopping for a telephone Ann found that, for the same style phone, the prices in six stores were $19.95, $21.95, $23.95, $25.95, $29.95, and $32.95. Determine the range and standard deviation of the prices.

16. Machine A is making a special screw that is to measure 51 mm in length. The inspector randomly selects 10 screws from a three-hour output and records the measurements as: 48.2 mm, 49.0 mm, 49.7 mm, 48.7 mm, 52.1 mm, 51.1 mm, 50.9 mm, 51.0 mm, 51.1 mm, and 50.8 mm. Determine the range and standard deviation of the screw lengths.

17. In a sample of six felt-tipped pens, the inspector finds that the writing time is 3.4 hr, 3.9 hr, 4.1 hr, 4.0 hr, 3.8 hr, and 4.2 hr. Determine the mean and standard deviation of the writing time for the pens.

18. Maggie, a wirehaired terrier, has had six litters of pups. The number of pups in each litter was two, two, six, five, three, and four. Determine the mean and standard deviation of the number of pups.

19. A manufacturer is trying to keep the tensile strength of its plastic at a minimum of 340 psi. In a test group of 10 samples the tensile strength was 360 psi, 341 psi, 338 psi, 340 psi, 343 psi, 339 psi, 341 psi, 355 psi, 342 psi, and 339 psi. Determine the mean and the standard deviation of tensile strengths of the 10 sample pieces.

20. A company is testing the length of time a toy will run using a Brand X D-cell battery. For the 10 batteries tested the toy ran for 8 min, 8 min, 14 min, 11 min, 9 min, 15 min, 12 min, 14 min, 15 min, and 17 min. Determine the mean and standard deviation of the times for the 10 batteries.

21. The frequency distribution shown in the following table represents a sample of the ages of 100 randomly selected spectators at a college football game.

Class Limits (Ages)	Frequency (Number)
0–8	3
9–17	12
18–26	22
27–35	12
36–44	15
45–53	16
54–62	10
63–71	6
72–80	4

a. Construct a histogram of the distribution.
b. Calculate the mean age of the spectators.
c. Calculate the standard deviation of the ages.

22. The results of testing 220 incandescent 75-W bulbs provided the results given in the frequency distribution in the table at right.
   a. Construct a histogram of the distribution.
   b. Calculate the mean life of a 75-W bulb.
   c. Calculate the standard deviation of the life of a 75-W bulb.

Class Limit (Hours)	Frequency
600–699	4
700–799	6
800–899	28
900–999	69
1000–1099	65
1100–1199	25
1200–1299	14
1300–1399	6
1400–1499	3

23. To test the quality of steel used for ball bearings and tools a mixture of muriatic acid and water is placed on the steel. The reaction time in minutes of 100 samples is given in the following table.

1	5	3	4	2	3	4	6	5	8
2	4	5	1	4	8	7	1	2	4
3	3	7	6	3	8	2	5	3	1
4	2	8	3	4	6	7	9	5	4
8	4	6	5	4	5	4	2	1	9
6	3	4	6	4	5	4	3	3	8
4	7	5	7	3	6	4	5	3	7
4	9	4	8	3	6	3	6	3	6
2	4	3	9	2	3	3	5	4	3
3	3	3	4	1	4	3	4	4	2

   a. Construct a grouped frequency distribution that has a class width of 2. Start with 0 to 1.
   b. Construct a histogram of the frequency distribution.
   c. Calculate the mean of the frequency distribution.
   d. Calculate the standard deviation of the distribution.

24. The heart rates at the end of a physical fitness class for cardiac patients are listed in the following table.

112	96	108	136	132	128	112	100	118
100	100	112	112	108	116	108	96	108
116	120	100	108	104	112	112	128	112
120	124	92	92	112	108	104	132	116
92	128	128	116	108	104	100	124	104
108	116	132	104	116	108	132	124	108
104	112	108	104	100	96	96	112	112
108	108	98	92	104	92	96	108	116
108	100	92	112	120	124	112	120	104

   a. Construct a grouped frequency distribution that has a class width of 6. Start with 92–97.
   b. Construct a frequency polygon of the frequency distribution.
   c. Calculate the mean of the frequency distribution.
   d. Calculate the standard deviation of the distribution.

17.4

Normal Curve

Many different populations of data have one thing in common—the data can be represented by a bell-shaped curve. This bell-shaped curve is called the **normal curve,** shown in Figure 17.5. A line through the mean $\overline{X}$, as shown in Figure 17.5, is a line of symmetry. That is, half the area under the curve is to the left of this line, and the other half of the area is to the right of this line. Also, for a normal distribution, the mean, median, and mode are equal.

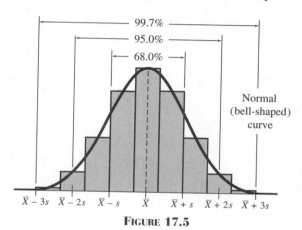

FIGURE 17.5

Although all normal curves have the same basic bell shape, some curves are thinner, while others are fatter. This change is due to a change in the standard deviation. To better understand the effect of the standard deviation on a normal curve, consider the following rule.

The Empirical Rule

Within one standard deviation of the mean there will be approximately 68.0% of the data. Within two standard deviations of the mean there will be approximately 95.0% of the data. Within three standard deviations there will be approximately 99.7% of the data.

This is also illustrated in Figure 17.5. Remember that this rule applies only to a normal distribution.

To illustrate the empirical rule, let us consider again the number of gadgets manufactured by ABC Corporation. We calculated the mean, $\overline{X} = 23.5$, and the standard deviation, $s = 5.0$. To determine the values one, two, and three standard deviations from the mean, perform the following calculations.

$$\overline{X} - s = 23.5 - 5.0 = 18.5 \qquad \overline{X} + s = 23.5 + 5.0 = 28.5$$
$$\overline{X} - 2s = 23.5 - 10.0 = 13.5 \qquad \overline{X} + 2s = 23.5 + 10.0 = 33.5$$
$$\overline{X} - 3s = 23.5 - 15.0 = 8.5 \qquad \overline{X} + 3s = 23.5 + 15.0 = 38.5$$

Each of these values, as well as the number of gadgets for each interval, is shown in Figure 17.6. The figure shows that within one standard deviation above the mean there are 34 gadgets. Of the 100 gadgets manufactured, 70, or 70%, fall within one standard deviation of the mean, 97% fall within two standard deviations of the mean, and 100% fall within three standard deviations of the mean. Thus, the number of gadgets manufactured by ABC Corporation over a 100-day period is normally distributed.

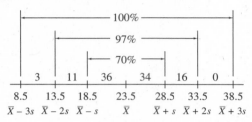

FIGURE 17.6

EXAMPLE 1   Application: Diameter of Garden Hose

A company manufactures garden hose. It determines from a large sample that the mean diameter of the hose is 20 mm and the standard deviation of the diameter is $\frac{1}{2}$ mm. If they assume a normal distribution, what is the range of measurements for 95% of the hose in the sample?

Solution   From Figure 17.5 we know that, for a normal distribution, 95% of the scores lie within 2s of the mean. That is,

$$\overline{X} + 2s = 20 + 2\left(\frac{1}{2}\right) \quad \text{or} \quad 21,$$

and

$$\overline{X} - 2s = 20 - 2\left(\frac{1}{2}\right) \quad \text{or} \quad 19.$$

Therefore, the diameters of 95% of the hoses in the sample are between 19 mm and 21 mm.   ◆

In statistical work it is important to be able to determine the area under a normal curve. To develop the technique, let us first consider a special curve called the **standard normal curve.** The scale for this curve is a Z-scale with mean $\mu = 0$ and standard deviation $\sigma = 1$, as shown in Figure 17.7. The curve is also known as the z-curve.

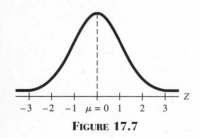

FIGURE 17.7

> **BASIC PROPERTIES OF THE STANDARD NORMAL CURVE**
>
> 1. The area under the standard normal curve is 1.
> 2. The standard normal curve extends indefinitely in both directions and is **asymptotic** to the horizontal axis.
> 3. The standard normal curve is **symmetric** about the mean 0. That is, the part of the curve to the left of the mean is the mirror image of the part of the curve to the right of the mean.
> 4. About 99.74% of the area under the standard normal curve lies between −3 and 3.

The equation for the standard normal curve is $y = \dfrac{1}{\sqrt{2\pi}} e^{-x^2/2}$. To determine areas under the standard normal curve using this formula, we need a working knowledge of calculus. Since this is beyond the level of this text, we will use the table that comprises Appendix C.

Because the normal curve is symmetric about the mean, it is important to recognize that 50% (or half the area) under the curve is to the right of the mean and 50% is to the left of the mean. This information is needed when solving problems dealing with a normal distribution.

EXAMPLE 2   **Area Between 0 and a Positive Value of Z**

Determine the area under the standard normal curve between $Z = 0$ and $Z = 1.96$.

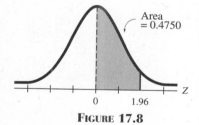

Area = 0.4750

FIGURE 17.8

Solution   Sketch the normal curve and shade the desired region, as shown in Figure 17.8. Now turn to Appendix C. Read down the left column, labeled Z, to 1.9. Read across that row until you are under the column labeled 0.06. The number in the table where the row and column intersect is 0.4750.

Thus, the area under the standard normal curve between $Z = 0$ and $Z = 1.96$ is 0.4750. See Figure 17.8.   ◆

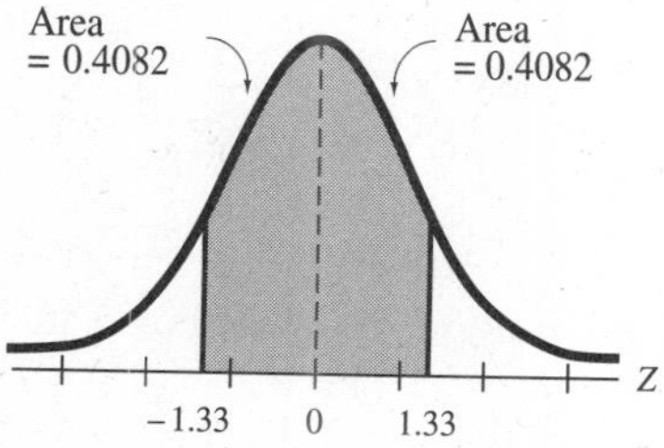

Area = 0.4082   Area = 0.4082

FIGURE 17.9

To determine the area under the standard normal curve between 0 and a negative value of Z, use the fact that the normal curve is symmetric about 0. For example, as shown in Figure 17.9 the area between $Z = 0$ and $Z = 1.33$ is the same as the area between $Z = 0$ and $Z = -1.33$. That is so since both points are the same distance from 0. In Appendix C, the value for $Z = 1.33$ is 0.4082. Thus, the area for $Z = 0$ to $Z = 1.33$ is 0.4082, and the area for $Z = -1.33$ to $Z = 0$ is 0.4082.

EXAMPLE 3 Area to the Right of a Positive Z-Value

Determine the area to the right of $Z = 1.25$.

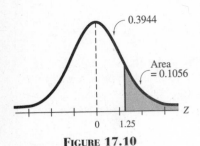

FIGURE 17.10

Solution Sketch the curve and shade the desired area, as shown in Figure 17.10. The value for $Z = 1.25$ in Appendix C is 0.3944. Therefore, the area between $Z = 0$ and $Z = 1.25$ is 0.3944. The total area under the curve to the right of $Z = 0$ is 0.5000. To determine the area to the right of $Z = 1.25$, subtract 0.3944 from 0.5000.

$$0.5000 - 0.3944 = 0.1056$$

The desired area is 0.1056. ◆

By going one more step with the information in Example 3, we can determine the area to the left of $Z = -1.25$. Applying the symmetry property, that area is 0.1056, as shown in Figure 17.11.

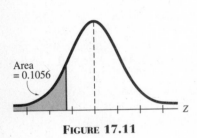

FIGURE 17.11

Since the total area under the standard normal curve is 1 and the sum of the probabilities of the outcomes of an experiment is 1, we may interpret the values from the Z-table as probabilities. For instance, we may interpret the result in Example 3 as a probability of 0.1056 that a piece of data will be greater than $Z = 1.25$.

Is it possible for us to determine the area under any normal curve? Yes, all we need to know is the mean μ and the standard deviation σ of the set of data. Knowing these two values enables us to change any x-value to a Z-score.

Formula to Change an x-Value to a Z-Score

$$Z = \frac{x - \mu}{\sigma}$$

To determine the desired area under the normal curve, first convert the x-value to a Z-score and then use the same procedures that we used in Examples 1–3.

EXAMPLE 4 Area on Both Sides of the Mean

Determine the total area under the normal curve with mean $\mu = 4$ and $\sigma = 3$ that lies between $x = -3$ and $x = 9$.

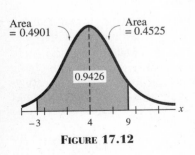

FIGURE 17.12

Solution First sketch the normal curve with $\mu = 4$ and $\sigma = 3$, as shown in Figure 17.12. In this figure the tick marks are 3 units apart, since $\sigma = 3$. Now locate $x = -3$ and $x = 9$ on the x-scale and shade the desired area. Now convert the x-values, $x = -3$ and $x = 9$ to Z-scores using the formula $Z = \frac{x - \mu}{\sigma}$ as shown in the table.

x-Value	Z-Value	Area Between 0 and Z
x = −3	$Z = \dfrac{-3 - 4}{3} = -2.33$	0.4901
x = 9	$Z = \dfrac{9 - 4}{3} = 1.67$	0.4525

The total area is the sum of the two areas.

$$0.4901 + 0.4525 = 0.9426. \quad \blacklozenge$$

EXAMPLE 5   Application: Serum Cholesterol Levels

The serum cholesterol levels in men aged 18 to 74 are normally distributed with a mean of 175.2 and a standard deviation of 41.2. All units are in mg/100 ml and the data are based on a national survey. If a man is randomly selected, find the probability that his serum cholesterol level is each of the following.

a.   Below 200   **b.**   Above 200

Solution

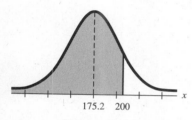

FIGURE 17.13

a.   A sketch showing the area of the curve that gives the desired probability is shown in Figure 17.13. Since we want the probability on both sides of 0, we must find the sum as we did in Example 4. For x = 200, the corresponding Z-score is:

$$Z = \frac{200 - 175.2}{41.2} = 0.60.$$

From Appendix C we see that the probability is 0.2257. Knowing that the probability to the left of 0 is 0.5000, we determine that the probability that a man has a cholesterol level below 200 is 0.5000 + 0.2257, or 0.7257.

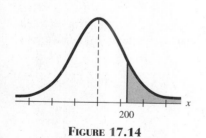

FIGURE 17.14

Six Sigma

b.   A sketch showing the area of the curve that gives the desired probability is shown in Figure 17.14. We found the probability for x = 200 in part a. To determine the probability that the individual selected is in the area to the right of 200, subtract 0.2257 from 0.5000. Thus, the probability that a person has a cholesterol serum level greater than 200 is 0.5000 − 0.2257 or 0.2743. $\blacklozenge$

The success of the industrial machine in Japan after the Second World War is due partially to the influence of W. Edwards Deming. Deming believed that many different elements contribute to a quality product. In fact, many people credit Deming for the process known as **Total Quality Management** or **TQM**. Part of the **TQM** program is to reduce the number of defects in the production process.

As mentioned at the beginning of this chapter, Motorola has set a goal of six sigma defects. Now that we know the normal curve and some of its properties, we can understand the six sigma level more fully. Remember that, under a standard normal curve, 99.7% of the scores are within three standard deviations of the mean. In terms of defects, this means that there are 23 defects per thousand. Six sigma, then, means that only products within six standard deviations of the mean are acceptable. That results in only 3.4 defects per million. While this is the goal of some companies, none have reached it yet. To determine this degree of accuracy requires Z-tables with a greater degree of accuracy than the 4-decimal-place table in Appendix C.

17.4 Exercises

In Exercises 1–22, assume a standard normal curve, which means that $\mu = 0$ and $\sigma = 1$. For each exercise, sketch the curve indicating the desired area and determine the area.

1. Between $Z = 0$ and $Z = 1$

2. Between $Z = 0$ and $Z = 2$

3. Between $Z = 0$ and $Z = 3$

4. Between $Z = 0$ and $Z = 1.64$

5. Between $Z = 0$ and $Z = 1.96$

6. Between $Z = 0$ and $Z = 2.33$

7. Between $Z = 0$ and $Z = 2.58$

8. Between $Z = 0$ and $Z = -1$

9. Between $Z = 0$ and $Z = -2$

10. Between $Z = 0$ and $Z = -3$

11. Between $Z = 0$ and $Z = -2.33$

12. To the right of $Z = 1.5$

13. To the right of $Z = 2.5$

14. To the right of $Z = 3.5$

15. To the left of $Z = -1.5$

16. To the left of $Z = -2.5$

17. To the left of $Z = -3.5$

18. Between $Z = -2$ and $Z = 1.5$

19. Between $Z = -1.75$ and $Z = 2.33$

20. Between $Z = -1.35$ and $Z = 1.76$

21. Either to the left of $Z = -1.33$ or to the right of $Z = 2.65$

22. Either to the left of $Z = -2.75$ or to the right of $Z = 0.97$

23. Which normal curve has a wider spread, the one with $\mu = 2$ and $\sigma = 3$ or the one with $\mu = 3$ and $\sigma = 2$? Explain your answer.

24. Do the normal curves with $\mu = 7$ and $\sigma = 4$ and $\mu = 13$ and $\sigma = 4$ have the same shape? Explain your answer.

In Exercises 25–34, sketch the normal curve and indicate the desired area on the curve.

25. Given $\mu = 2$ and $\sigma = 1$, determine the area under the curve for each of the following.
    a. Between 2.0 and 3.5
    b. To the right of 3.2
    c. Between -3.2 and 3.5

26. Given $\mu = 7$ and $\sigma = 2$, determine the area under the curve for each of the following.
    a. Between 5 and 9
    b. To the right of 8
    c. To the left of 6

27. Given $\mu = -4$ and $\sigma = 1$, determine the area under the curve for each of the following.
  a. Between -4 and -2
  b. Between -5.0 and -1.5
  c. To the right of -1

28. Given $\mu = 15$ and $\sigma = 3$, determine the area under the curve for each of the following.
  a. To the left of 12.5
  b. To the right of 15.5
  c. Between 14 and 16

29. Given $\mu = 3$ and $\sigma = \dfrac{1}{2}$, determine the area under the curve for each of the following.
  a. To the right of 2.75
  b. Between $1\dfrac{3}{4}$ and $3\dfrac{1}{4}$
  c. Between $1\dfrac{1}{2}$ and 3

30. Given $\mu = 0.45$ and $\sigma = 0.15$, determine the area under the curve for each of the following.
  a. Between 0.45 and 0.65
  b. Between 0.35 and 0.75
  c. To the left of 0.15

31. Given $\mu = 75$ and $\sigma = 11$, determine the area under the curve for each of the following.
  a. Between 70 and 80
  b. To the right of 90
  c. To the left of 60

32. Given $\mu = 75$ and $\sigma = 7$, determine the area under the curve for each of the following.
  a. Between 70 and 80
  b. To the right of 90
  c. To the left of 60

33. Given $\mu = 84$ and $\sigma = 7$, determine the area under the curve for each of the following.
  a. Between 70 and 80
  b. To the right of 90
  c. To the left of 60

34. Given $\mu = 65$ and $\sigma = 7$, determine the area under the curve for each of the following.
  a. Between 70 and 80
  b. To the right of 90
  c. To the left of 60

35. Suppose that the monthly rents for two-bedroom apartments in a region is approximately normally distributed with mean μ = $450 and standard deviation σ = $50.
   a. What is the probability that an apartment rents for more than $525?
   b. What is the probability that an apartment rents for less than $320?
   c. What is the probability that an apartment rents for less than $475?

36. Bond ratings and bond interest rates depend on the current level of the cost of borrowing and on the credit ratings of the respective community. Interest rate on a class of municipal tax-free bonds currently is normally distributed with a mean of 7.5% and a standard deviation of 0.4%. A community's bonds are rated within this class. Determine the probability that the interest rate the community has to pay is each of the following.
   a. More than 8%
   b. More than 8.5%
   c. Less than 7%

37. A regional weather station claims that the annual average number of inches of snowfall is 90.4 in. and that the standard deviation is 5.5 in. Determine the probability that annual snowfall is each of the following.
   a. Greater than 94 in.
   b. Less than 75 in.
   c. Between 80 and 100 in.

38. An instructor determines that, over 20 years of teaching, grades for a mathematics class are normally distributed with μ = 74 and σ = 6. Determine the probability that a student's grade will be each of the following.
   a. Greater than 80
   b. Less than 70
   c. Greater than 70

39. The lengths of panfish caught over a summer by a group of young people has a mean length of 6.3 in. and a standard deviation of 0.5 in. It the distribution of these lengths has roughly the shape of a normal distribution, what percentage of all the panfish are each of the following?
   a. Longer than 7.0 in.
   b. Shorter than 6.0 in.
   c. Between 5.0 and 8.0 in.

40. A random variable has a normal distribution with the standard deviation σ = 6.2. Determine its mean if the probability is 0.1539 that the random variable is greater than 55.0.

Review Exercises

In Exercises 1–6, determine the mean, median, mode, range, and standard deviation for each set of data.

1. 1, 3, 4, 7, 11

2. 3, 11, 14, 17, 17, 21, 23

3. 5, 7, 9, 9, 7, 6, 4, 2

4. 1, 3, 5, 5, 6, 6, 6, 7, 9, 9

5. 23, 24, 25, 25, 27, 27, 27, 29

6. 19, 23, 31, 32, 41, 41, 43

In Exercises 7–12, refer to the table which is the frequency distribution of test results for 150 students.

Grade	Number of Students
90–99	10
80–89	36
70–79	54
60–69	31
50–59	13
40–49	4
30–39	2

7. Determine the mean.

8. Determine the median.

9. Determine the modal class.

10. Determine the standard deviation.

11. Construct a histogram.

12. Construct a frequency polygon.

13. Three mathematics teachers reported mean examination grades of 83, 72, and 85 in their classes, which consisted of 35, 41, and 28 students, respectively. Determine the mean grade for all the classes.

In Exercises 14–19, refer to the table, which is the frequency distribution of the weight in pounds of bags of groceries.

Weight	Frequency
0–3	8
4–7	14
8–11	18
12–15	17
16–19	13
20–23	5

14. Determine the mean.

15. Determine the median.

16. Determine the modal class.

17. Determine the standard deviation.

18. Construct a histogram.

19. Construct a frequency polygon.

In Exercises 20–26, use the set of data in the table.

40	50	62	68	73	82	88	94
42	54	63	70	77	82	88	96
46	56	64	70	78	84	88	98
48	58	65	70	80	86	90	102
49	60	67	72	82	88	92	108

20. Construct a frequency distribution with the first class 40 to 48.

21. Construct a histogram.

22. Construct a frequency polygon.

23. Determine the mean.

24. Determine the median.

25. Determine the standard deviation.

26. Determine the modal class.

In Exercises 27–30, use the following information. The diameters of red blood cells are normally distributed with a mean diameter of 0.008 mm and a standard deviation of 0.002 mm. Determine the probability of selecting red blood cells for each of the diameters.

27. Greater than 0.009 mm

28. Between 0.004 mm and 0.013 mm

29. Greater than 0.003 mm

30. Less than 0.005 mm

In Exercises 31–34, use the following information. A vending machine is designed to dispense a mean of 6.7 oz of coffee into a 7-oz cup. If the standard deviation of the amount of coffee dispensed is 0.4 oz and the amount of coffee dispensed is normally distributed, determine each of the following.

31. The percent of times it will dispense less than 6.0 oz.

32. The percent of times it will dispense less than 6.8 oz.

33. The percent of times the 7-oz cup will overflow.

34. The percent of times the 7-oz cup will be three-fourths full.

*35.   Write a computer program to determine the mean of a set of data.

*36.   Write a computer program to determine the standard deviation of a set of data.

*37.   The following formulas can both be used to calculate the standard deviation of ungrouped data. Show that these equations are equal.

$$s = \sqrt{\frac{\sum (x_i - \overline{X}_i)^2}{n - 1}} \qquad s = \sqrt{\frac{n(\sum x_i^2) - (\sum x_i)^2}{n(n - 1)}}$$

✏ Writing About Mathematics

1.   Your friend does not understand how your school determines grade point averages. You recognize that the grade point average is the same as a weighted mean. Write a few paragraphs, including examples, to explain how to calculate a grade point average so that your friend does understand.

2.   In labor negotiations the union states that the average pay of company employees is $35,000 per year. On the other hand, management claims that the average pay of employees is $42,000 per year. Both claims are correct because one group is using the median and the other is using the mean. Discuss why the difference exists and why each used a different average to make their point.

3.   A psychology instructor assumes that her test grades are normally distributed. That is, she assigns letter grades by the method shown in the diagram. On a particular exam a student scores 116 out of a possible 120. The letter grade the teacher assigns is a C. Explain how that grade is possible using her technique.

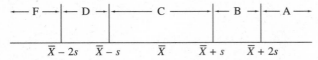

4.   Using the grading technique shown in Exercise 3, discuss the following questions. Is it possible that no student will receive an A? Is is possible that no student will fail?

Chapter Test

In Exercises 1–5, use the data set 2, 2, 3, 4, 4, 4, 5, 6, 8, 8, 9.

1.   Determine the mean.

2.   Determine the median.

3.   Determine the mode.

4.   Determine the range.

5.   Determine the standard deviation.

In Exercises 6–12, use the data in the table.

0	4	7	11	13	17	21	23	27
1	4	8	12	13	18	21	24	27
2	5	9	12	14	19	21	25	28
2	6	9	12	15	20	22	26	28
3	6	10	13	16	20	22	26	29

6. Construct a frequency distribution with the first class 0–5.   7. Construct a histogram.

8. Construct a frequency polygon.

9. Determine the mean.

10. Determine the median.

11. Determine the standard deviation.

12. Determine the modal class.

In Exercises 13–15, sketch the normal curve and indicate the desired region(s) under the curve using the following information. The average life of a certain steel-belted radial tire is advertised as 60,000 mi. Assume that the life of the tire is normally distributed with a standard deviation of 2,500 mi. (The "life" of a tire is defined as the number of miles the tire is driven before blowing out.)

13. Determine the probability that a randomly selected steel-belted radial tire will have a life of at least 61,800 mi.

14. Determine the probability that a randomly selected steel-belted radial tire will have a life less than 58,000 mi.

15. Determine the probability that a randomly selected steel-belted radial tire will have a life between 55,000 and 65,000 mi.

18

Plane Analytic Geometry

This chapter explores a special set of curves called *conic sections*. There are many applications of these curves—from designing automobile headlights to solving sound problems.

Two people are fishing offshore 2.0 miles apart. As they talk to each other on a radiotelephone, they hear thunder. One person hears the thunder 3.0 seconds after the other hears it. Example 5 of Section 18.4 discusses how to graph the possible locations of the lightning bolt that produced the thunder.

18.1

The Circle

Introduction

Analytic geometry is the bridge between geometry and algebra. When we study geometric figures, we learn to write algebraic statements that define the figures. When we study an algebraic statement, we learn to identify properties that help us construct a geometric figure that illustrates the algebraic statement.

In the Chapter 5 discussion of graphing functions, we used analytic geometry for the first time. There we discovered that the algebraic equation $y = mx + b$ describes the straight line, a geometric figure. Then, in Chapter 10, we examined the equation of the straight line in great detail. You may want to review the material on the line in Section 10.1.

In more general terms, the equation of every straight line can be written in the form

$$Dx + Ey + F = 0,$$

where neither D nor E are zero. In this form, the exponents of x and y are understood to be one. Therefore, recall that this is a first degree equation.

The second degree equation:

$$Ax^2 + Bxy + Cy^2 + Dx + Ey + F = 0$$

is the equation of a circle, a parabola, an ellipse, or a hyperbola depending on the values of A, B, and C. Such figures are called **conic sections** since they are formed by the intersection of a plane with a right circular cone, as shown in Figure 18.1. This chapter discusses the circle, parabola, ellipse, and hyperbola.

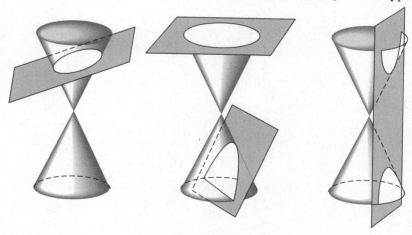

FIGURE 18.1

Circle

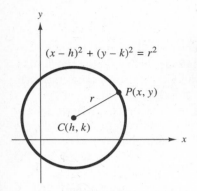

$$(x - h)^2 + (y - k)^2 = r^2$$

$P(x, y)$

r

$C(h, k)$

FIGURE 18.2

A circle is a shape that has a continuous and constant curve. Though it is always curving, it too has an algebraic expression that describes its nature.

> **DEFINITION 18.1**
>
> A **circle** is the collection of all the points in a plane that are a given distance from a fixed point. The fixed point is called the **center.** The given distance from the points to the center is called the **radius.**

This definition helps develop a standard form of the equation of a circle.

Let $C(h, k)$ be the fixed point at the center of the circle as shown in Figure 18.2.
Let r be the given distance (or the radius of the circle).
Let $P(x, y)$ be any one of the collection of points that is a given distance from the fixed point. This set of points is on the **circumference** of the circle.

The given distance is the distance from point C to point P, or $d(C, P)$. Read the notation $d(C, P)$ as "the distance from point C to point P." Because this distance is the radius of the circle, we have the equality:

$$d(C, P) = r.$$

We can determine $d(C, P)$ using the distance formula.

$$d(C, P) = \sqrt{(x - h)^2 + (y - k)^2}$$
$$r = \sqrt{(x - h)^2 + (y - k)^2} \quad \text{Substituting}$$
$$r^2 = (x - h)^2 + (y - k)^2 \quad \text{Squaring both sides}$$

> **DEFINITION 18.2**
>
> The **standard form of the equation of a circle** with radius r and center (h, k) is:
>
> $$(x - h)^2 + (y - k)^2 = r^2.$$

If the center of the circle is at the origin, then $h = 0$ and $k = 0$. In this case, the equation is $x^2 + y^2 = r^2$.

EXAMPLE 1   **Equation of a Circle Given Center and Radius**

Determine the equation of the circle with center at $(-2, 1)$ and radius 3.

Solution   The equation for a circle with center at (h, k) and radius r is:

$$(x - h)^2 + (y - k)^2 = r^2$$
$$[x - (\boxed{-2})]^2 + (y - \boxed{1})^2 = \boxed{3}^2 \quad \text{Substituting } h = -2,$$
$$k = 1, \text{ and } r = 3$$
$$(x + 2)^2 + (y - 1)^2 = 9. \quad \blacklozenge$$

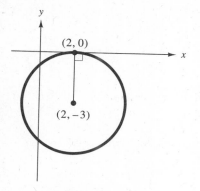

(2, 0)

(2, −3)

FIGURE 18.3

EXAMPLE 2   **Equation of a Circle Given Center and Tangent Line**

Determine the equation of the circle with center at $C(2, -3)$ that is tangent to the x-axis.

Solution   For the circle to be tangent to the x-axis, we know from geometry that the radius must be perpendicular to the x-axis. Therefore, the radius is a vertical line and intersects the x-axis at the point $P(2, 0)$, as shown in Figure 18.3. From the figure, we can see that $d(C, P)$ is 3 units. Thus, $r = 3$. Substituting $h = 2$, $k = -3$ and $r = 3$ in the standard form of the equation, we have:

$$(x - \boxed{2})^2 + [y - (\boxed{-3})]^2 = (\boxed{3})^2$$
$$(x - 2)^2 + (y + 3)^2 = 9. \quad \blacklozenge$$

The circle can be represented with the general expression

$$Ax^2 + Cy^2 + Dx + Ey + F = 0,$$

where $A = C, A \neq 0$, and $C \neq 0$. This form is illustrated in Example 3.

EXAMPLE 3   **Center and Radius Using General Form of the Equation**

a.   Determine the center and radius of the circle $x^2 + y^2 + 6x - 4y = 12$.
b.   Sketch the graph of the circle.

Solution

a.   To determine the center and radius of the circle, we first must transform the equation to the standard form $(x - h)^2 + (y - k)^2 = r^2$. To accomplish this, complete the square on the x-terms and the y-terms. Start the process by grouping the x- and y-terms.

$$(x^2 + 6x \qquad) + (y^2 - 4y \qquad) = 12$$

Now multiply the coefficient of x by $\dfrac{1}{2}$.

$$\left(\frac{1}{2} \times \boxed{6} = 3\right)$$

Square the result ($3^2 = 9$) and add 9 to both sides of the equation. Next multiply the coefficient of y by $\dfrac{1}{2}$.

$$\left(\frac{1}{2} \times (\boxed{-4}) = -2\right)$$

Square the result $[(-2)^2 = 4]$ and add 4 to both sides of the equation. The result is:

$$(x^2 + 6x + \boxed{9}) + (y^2 - 4y + \boxed{4}) = 12 + \boxed{9} + \boxed{4}$$
$$(x + 3)^2 + (y - 2)^2 = 25.$$

Comparing this equation to the standard form, we see that $h = -3, k = 2$, and $r = \sqrt{25} = 5$. Therefore, the center of the circle is $(-3, 2)$, and the radius is 5.

b. Graph the circle with a graphics calculator. For the equation $(x + 3)^2 + (y - 2)^2 = 25$, solve the equation for y.

$$y = 2 \pm \sqrt{25 - (x + 3)^2}$$

Set the range $[-12, 12, 1, -8, 8, 1]$ and graph the two functions

$$y = 2 + \sqrt{25 - (x + 3)^2} \quad \text{and} \quad y = 2 - \sqrt{25 - (x + 3)^2}$$

The results on the screen should be similar to Figure 18.4. ◆

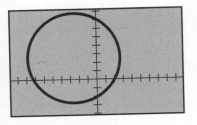

FIGURE 18.4

CAUTION   The expression $Ax^2 + Ay^2 + Dx + Ey + F = 0$ may not be a circle in every case. When we transform the equation to standard form, the result may be $(x - 3)^2 + (y - 5)^2 = 0$ or $(x - 3)^2 + (y - 5)^2 = -16$. In the first case, $r = 0$, and there is no circle. In the second case, $r^2 = -16$ and r is not a real number; again there is no circle. ■

EXAMPLE 4   Application: Diameter of a Drum

The design of a portable room humidifier specifies that the rotating drum of the humidifier be supported on two rollers (idlers) that help keep the drum in a

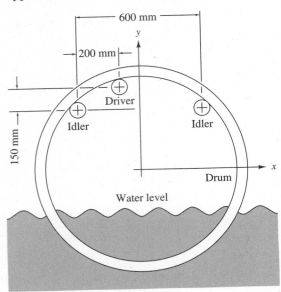

FIGURE 18.5

particular position. The drum is driven by frictional contact with a third roller (driver). If each roller is 50 mm in diameter and is located as shown in Figure 18.5, what is the required inside diameter of the drum?

Solution   The rollers are all the same size, so the circle passing through the center of the rollers is concentric with the inner circle of the drum. Therefore, first determine the equation of the circle that passes through the centers of the rollers. Choose an axis that passes through the two idler rollers, with its origin midway between them. (The location of the axis system is arbitrary. Any location can be chosen, but this one simplifies the calculations somewhat.) To further simplify calculations, convert all dimensions to centimeters. In centimeters, the coordinates of the centers of the three rollers are $(-30, 0)$, $(-10, 15)$, and $(30, 0)$.

Now begin with the standard equation of a circle for each of the three rollers.

$$(-30 - h)^2 + (0 - k)^2 = r^2 \quad \text{Substituting for } x, y \text{ (1)}$$
$$(-10 - h)^2 + (15 - k)^2 = r^2 \quad \text{Substituting for } x, y \text{ (2)}$$
$$(30 - h)^2 + (0 - k)^2 = r^2 \quad \text{Substituting for } x, y \text{ (3)}$$
$$h^2 + 60h + 900 + k^2 = r^2 \quad \text{Squaring (1)}$$
$$h^2 + 20h + 100 + k^2 - 30k + 225 = r^2 \quad \text{Squaring (2)}$$
$$h^2 - 60h + 900 + k^2 = r^2 \quad \text{Squaring (3)}$$

Use the elimination method to solve the system of three equations.

$$120h = 0 \quad \text{(1)} - \text{(3)}$$
$$h = 0$$
$$80h - 30k - 575 = 0 \quad \text{(2)} - \text{(3)}$$
$$80(0) - 30k - 575 = 0 \quad \text{Substituting } h = 0$$
$$k = -19.17 \quad \text{Solving for } k$$
$$(0)^2 + 60(0) + 900 + (-19.17)^2 = r^2 \quad \text{Substituting in Equation (1)}$$
$$r = 35.60 \text{ cm.}$$

Adding the radius of the circle passing through the center of the 5-cm-diameter rollers to the radius of the rollers we have:

$$r_{drum} = 35.60 + 2.50$$
$$= 38.1 \text{ cm.}$$

Therefore, the required inside diameter of the drum is 762 mm. ◆

18.1 Exercises

In Exercises 1–12, determine the equation of a circle using the given information.

1. $C(0, 0), r = 3$

2. $C(0, 4), r = \dfrac{1}{2}$

3. $C(3, 0), r = 3$

4. $C(-3, 2), r = 7$

5. $C(0, 0)$, tangent to the line $x = -5$

6. $C(0, 0)$, tangent to the line $y = 6$

7. $C(6, -6)$, circumference passes through the origin

8. $C(0, 5)$, circumference passes through the point $(5, 0)$

9. $C(-9, -6)$, circumference passes through $(-20, 8)$

10. $C(2, -8)$, circumference passes through $(-10, -6)$

11. $C(-5, 4)$, tangent to the x-axis

12. $C(5, 3)$, tangent to the y-axis

In Exercises 13–20, determine the equation of the circle shown.

13.

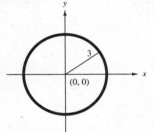

14.

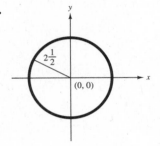

15.

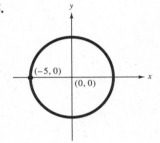

16.

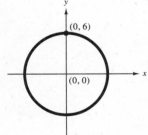

17.

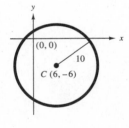

18.

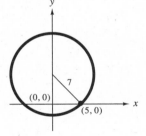

19.

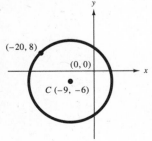

20.

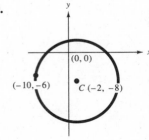

In Exercises 21–26, find the center and the radius, then sketch the graph of the circle.

21. $x^2 + y^2 = 9$

22. $x^2 + y^2 = 16$

23. $(x - 2)^2 + (y - 4)^2 = 16$

24. $(x + 1)^2 + (y - 2)^2 = 4$

25. $(x - 1)^2 + (y + 2)^2 = 9$

26. $(x + 3)^2 + (y + 1)^2 = 9$

27. Write the equation of a circle that passes through the points with coordinates $(3, -2)$, $(0, 1)$, and $(4, 2)$.

28. Write the equation of a circle that passes through the points with coordinates $(-4, -3)$, $(2, -1)$, and $(6, 5)$.

In Exercises 29–38, determine whether the given equation represents a circle. If not, state why not. If it does, state the coordinates of the center and the radius.

29. $x^2 + y^2 - 8x - 4y + 16 = 0$

30. $x^2 + y^2 + 8y + 6x = 0$

31. $x^2 + y^2 - y - 2 = 0$

32. $x^2 + y^2 + 5x = 0$

33. $3x^2 + 3y^2 + 6x - 6y = 0$

34. $2x^2 + 2y^2 - 8x + 12y + 8 = 0$

35. $x^2 + 2y^2 - 2x - 2y = 0$

36. $3x^2 + 2y^2 + 3x + 2y = 0$

37. $x^2 + y^2 + 25 = 0$

38. $x^2 + y^2 + 16 = 0$

39. Write the equation of the line that is tangent to the circle $(x - 1)^2 + (y + 2)^2 = 4$ at the point $(3, -2)$.

40. Write the equation of the line that is tangent to the circle $(x + 2)^2 + (y - 3)^2 = 25$ at the point $(1, 7)$.

41. The magnetic field surrounding a current-carrying conductor can be depicted as concentric circles that diminish in strength with an increase in radius. If a wire with a current flowing in it passes through a circuit board 3.00 in. to the right and 2.00 in. above the lower left corner, will the field strength at the corner be the same as the strength at a point 4.00 in. to the right and 6.00 in. above the corner?

42. In Exercise 41, where on the bottom edge and on the left edge of a circuit board will the field strength be the same as it is at the corner?

43. A 1.00 m diameter steel conduit is to be supported by an A-frame structure so that its lower edge is 0.500 m above the ground. If the outer legs of the support make angles of 60° with the ground, how long must they be? (See the diagram.)

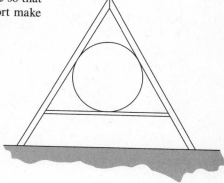

44. Two pipelines are to be joined by a smooth circular arc with a 2.00 ft radius. (A smooth arc is tangent to both of the lines (see the drawing.) If one line is described by $x - y + 3 = 0$ and the other is described by $x + 2y = 5$, what is the equation of the circle whose arc joins the two?

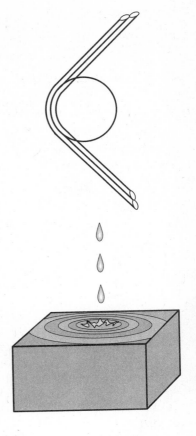

45. The surface waves from a drop falling on the surface of a square tank of smooth water (see the figure) are described at time t by the equation:

$$x^2 + y^2 - 14.40x + 6.20y + 55.20 = 0,$$

which uses one corner of the tank as the origin of the axis system. Determine the center and the radius of the circle by completing the squares.

46. A conical pendulum traces out a path over the x, y-plane described by:

$$2x^2 + 2y^2 + 21.20x + 18.80y + 13.24 = 0.$$

Locate the center and the radius of the circle by completing the squares.

47. **a.** Use your graphics calculator to draw the circle $(x + 3)^2 + (y - 2)^2 = 23$. (*Hint:* Since the calculator only graphs functions, solve for y.

$$y = 2 \pm \sqrt{23 - (x + 3)^2}$$

Use range settings $[-13, 5, 1, -4, 8, 1]$.)
   b. Place the center in the circle.
   c. Draw a radius in the circle.
   d. Draw the circle again using the range settings $[-10, 10, 1, -10, 10, 1]$. Which figure (this one or the one in part a) looks like a circle? Explain why one looks more like a circle than the other.

48. Architects now can use CAD (Computer Assisted Design) programs to accurately incorporate arches into their drawings. To locate circular arches in a CAD diagram, an equation is required.
   a. Assuming that the origin on an auxiliary axis is as shown in the diagram, write an equation for each of the two arches.
   b. Note that the two arches intersect at point P. Determine this point of intersection and describe it in terms of an (x, y) statement.
   c. Determine the height of the column needed to support arches at point P.

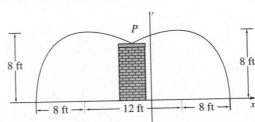

49. Consider the two points $R(3, -4)$ and $S(2, 1)$.
   a. How many circles can be formed with the points R and S on the circumference?
   b. Write the equation of the line that contains the centers of all the circles determined in part a.

18.2

The Parabola

We often see and use lines and circles to form pipes, coffee cups, and wheels. Less obvious to our eye, but just as important, is the parabolic shape. We can observe the parabola in the path of a home run or in the gradual curve of a telescopic mirror.

> **DEFINITION 18.3**
>
> The **parabola** is the set of all the points in a plane equidistant from a fixed point, called the **focus,** and a fixed line, called the **directrix.**

This definition can be used to derive a general equation.

Assume a focus F with coordinates $(p, 0)$ and a directrix parallel to the y-axis with equation $x = -p$ (see Figure 18.6). Let $P(x, y)$ be any point on the curve

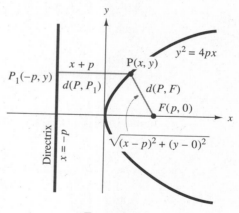

FIGURE 18.6

and $P_1(-p, y)$ a point on the directrix such that $d(P, P_1) = d(P, F)$. The points P and P_1 have the same y-coordinate. Thus,

$$d(P, P_1) = |x - (-p)| = |x + p|.$$

The distance from P to F is given by:

$$d(P, F) = \sqrt{(x - p)^2 + (y - 0)^2}.$$

The definition of the parabola states that $d(P, P_1) = d(P, F)$. Therefore,

$$|x + p| = \sqrt{(x - p)^2 + (y - 0)^2}$$ **Substituting**

$$(x + p)^2 = (x - p)^2 + y^2$$ **Squaring both sides**

$$x^2 + 2px + p^2 = x^2 - 2px + p^2 + y^2$$

$$4px = y^2$$ **Combining like terms**

The result, $y^2 = 4px$ is the general form of the equation of a parabola with vertex at $(0, 0)$, focus at $(p, 0)$, and directrix the line $x = -p$. The parabola is symmetric with respect to the positive x-axis if $p > 0$ and symmetric with respect to the negative x-axis if $p < 0$. The vertex $V(0, 0)$ of the parabola is on the line of symmetry midway between the focus and the directrix.

DEFINITION 18.4

The **standard form of the equation of a parabola** that is symmetric with respect to the x-axis, with vertex $(0, 0)$, focus $(p, 0)$, and directrix the line $x = -p$ is:

$$y^2 = 4px.$$

EXAMPLE 1 **Graphing a Parabola $y^2 = 4px$**

Sketch the graph of $y^2 = 8x$, indicating the focus, vertex, and directrix.

Solution The parabola $y^2 = 8x$ is of the form $y^2 = 4px$, with vertex at the origin and the focus on the positive x-axis. Thus,

$$4p = 8 \quad \text{and} \quad p = 2.$$

Since $p = 2$, we know that the focus is at the point $(2, 0)$, and the directrix is the line $x = -2$. The parabola is symmetric with respect to the positive x-axis. The vertex is $V(0, 0)$ and focus is $F(2, 0)$; both lie on the line of symmetry, the x-axis. With this information, we need calculate only a few key points, such as $(2, 4)$, $(2, -4)$, $(8, 8)$, and $(8, -8)$ in order to sketch the curve, which is shown in Figure 18.7. ◆

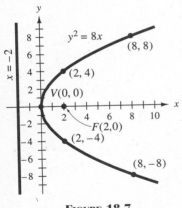

FIGURE 18.7

EXAMPLE 2 **Graphing a Parabola $y^2 = 4px$**

Sketch the graph of $y^2 = -2x$. Show the focus, vertex, and directrix.

Solution To determine the focus of the parabola $y^2 = -2x$, let $4p = -2$. Therefore, $p = -\dfrac{1}{2}$. Since $p < 0$, the parabola opens to the left and the focus is on the negative x-axis. The focus is $F\left(-\dfrac{1}{2}, 0\right)$, the vertex is $V(0, 0)$ at the origin, the directrix is $x = \dfrac{1}{2}$, and the curve is symmetric with respect to the

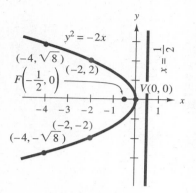

$y^2 = -2x$

$(-4, \sqrt{8})$

$(-2, 2)$

$F\left(-\frac{1}{2}, 0\right)$

$V(0, 0)$

$(-2, -2)$

$(-4, -\sqrt{8})$

FIGURE 18.8

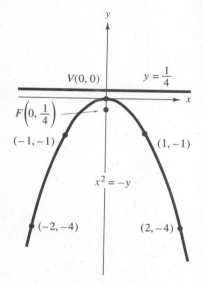

$V(0, 0)$ $y = \frac{1}{4}$

$F\left(0, \frac{1}{4}\right)$

$(-1, -1)$ $(1, -1)$

$x^2 = -y$

$(-2, -4)$ $(2, -4)$

FIGURE 18.9

negative x-axis. With this information, we need calculate only a few key points such as $(-2, 2)$, $(-2, -2)$, $(-4, \sqrt{8})$, and $(-4, -\sqrt{8})$ in order to sketch the curve as shown in Figure 18.8. ◆

Graph the parabola $y^2 = -2x$, using your graphing calculator. Recall that we can only graph functions. Therefore, solve for y and graph $y = \sqrt{-2x}$ and $y = -\sqrt{-2x}$. Is the graph the same as Figure 18.8?

If we place the focus of a parabola on the y-axis a distance p from the origin and make its directrix perpendicular to the y-axis (parallel to the x-axis) the same distance p from the other side of the origin, the equation we can derive by using the definition of a parabola is $x^2 = 4py$.

DEFINITION 18.5

The **standard form of the equation of a parabola** that is symmetric with respect to the y-axis and with vertex $(0, 0)$, focus $(0, p)$, and directrix the line $y = -p$ is:

$$x^2 = 4py.$$

EXAMPLE 3   **Graphing a Parabola $x^2 = 4py$**

Sketch the graph of the parabola $x^2 = -y$. Show the vertex, focus, and directrix.

Solution   The parabola $x^2 = -y$ is of the form $x^2 = 4py$, which tells us that $4p = -1$, or $p = -\frac{1}{4}$. From this, we know that the focus is at the point $\left(0, -\frac{1}{4}\right)$, the vertex is $V(0, 0)$, and the directrix is the line $y = \frac{1}{4}$. The line of symmetry is the negative y-axis. With this information, we need calculate only a few key points, such as $(-1, -1)$, $(1, -1)$, $(-2, -4)$, and $(2, -4)$ in order to sketch the curve as shown in Figure 18.9. ◆

CAUTION   The equations $y = 3x^2$ and $x = 4y^2$ are equations of parabolas. Note for each equation that there is one variable which has an exponent of 2 (second degree) and one variable has an exponent of 1 (first degree). If the variable x is of second degree, the graph of the parabola opens up or down. The graph of $y = 3x^2$ opens up. If the variable y is of second degree, the graph of the parabola opens to the left or the right. The graph of $x = 4y^2$ opens to the right. ■

If we select any point on the plane, except the origin, as the vertex of the parabola and select a directrix parallel to the x- or y-axis, then with the geometrical definition a new set of equations for the parabola can be derived.

DEFINITION 18.6

The **standard form of the equation of a parabola** that is symmetric with respect to the line $y = k$ and with vertex (h, k), focus $(h + p, k)$, and directrix the line $x = h - p$ is:

$$(y - k)^2 = 4p(x - h).$$

See Figure 18.10.

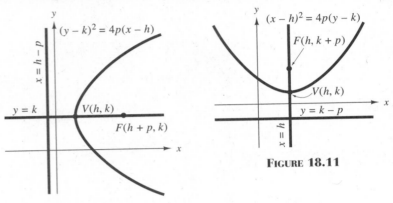

FIGURE 18.10

FIGURE 18.11

DEFINITION 18.7

The **standard form of the equation of a parabola** that is symmetric with respect to the line $x = h$ and with vertex (h, k), focus $(h, k + p)$, and directrix the line $y = k - p$ is

$$(x - h)^2 = 4p(y - k).$$

See Figure 18.11.

The parabolas in Definitions 18.6 and 18.7 open to the right or up if p is positive. They open to the left or down if p is negative.

An aid used to sketch the curve of a parabola is the **latus rectum**, which is a chord of the parabola that passes through the focus and is perpendicular to the axis of symmetry. The end points of the latus rectum are a distance of $2p$ from the focus. The latus rectum is discussed in Example 4.

EXAMPLE 4   Graphing a Parabola, Vertex at (h, k)

Sketch the graph of the parabola whose equation is $(y + 1)^2 = 6(x - 2)$.

Solution   The equation $(y + 1)^2 = 6(x - 2)$ is similar to the standard form $(y - k)^2 = 4p(x - h)$. Therefore, $h = 2$, $k = -1$, and $p = \dfrac{3}{2}$. Since $p > 0$, the parabola opens to the right. With the parabola opening to the right, the line of symmetry is parallel to the x-axis, and it passes through the vertex $(2, -1)$ and the focus. The line of symmetry is $y = -1$. Knowing that the parabola opens to the right, we also know that the focus is to the right of the vertex and is found using the formula

$$(h + p, k) = \left(2 + \frac{3}{2}, \ -1 \right) \quad \textbf{Substituting}$$

$$= \left(\frac{7}{2}, -1 \right)$$

The latus rectum is on the line through the focus, $x = \frac{7}{2}$. The end points of the latus rectum are

$$2p = 2\left(\frac{3}{2}\right) = 3 \text{ units.}$$

That is, they are 3 units above and below the focus. The coordinates of the end points of the latus rectum are $\left(\frac{7}{2}, -1 + 3 \right)$ or $\left(\frac{7}{2}, 2 \right)$ and $\left(\frac{7}{2}, -1 - 3 \right)$ or $\left(\frac{7}{2}, -4 \right)$. With these key points, we can sketch the parabola as shown in Figure 18.12. ◆

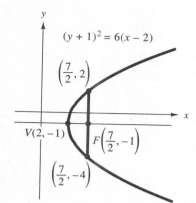

FIGURE 18.12

Is it possible to graph the parabola in Example 4 on the graphics calculator? (*Hint:* For the relation $(y + 1)^2 = 6(x - 2)$, take the square root of both sides.)

A general quadratic expression of the following form may represent a parabola.

$$Ax^2 + Dx + Ey + F = 0 \quad \text{or} \quad Cy^2 + Dx + Ey + F = 0$$

Example 5 illustrates how an equation of this form can be changed to the standard form.

EXAMPLE 5 **Changing to Standard Form and Graphing**

For the parabola $3x^2 - 24x + 12y + 36 = 0$, determine the coordinates of the vertex, the focus, the end points of the latus rectum, the directrix, and the line of symmetry. Sketch the curve.

Solution Since the equation is quadratic in x, we want to match it to the standard form $(x - h)^2 = 4p(y - k)$ from Definition 18.7.

$$3x^2 - 24x + 12y + 36 = 0$$
$$3x^2 - 24x = -12y - 36$$
$$x^2 - 8x = -4y - 12$$

Dividing each term by 3 to obtain coefficient of 1 for x^2

$$x^2 - 8x + 16 = -4y - 12 + 16$$

Completing the square on the left and adding 16 to both sides

$$(x - 4)^2 = -4y + 4$$

Factoring on left, combining terms on right

$$(x - 4)^2 = -4(y - 1)$$

Factor common factor of -4 on right to obtain a coefficient of 1 for y

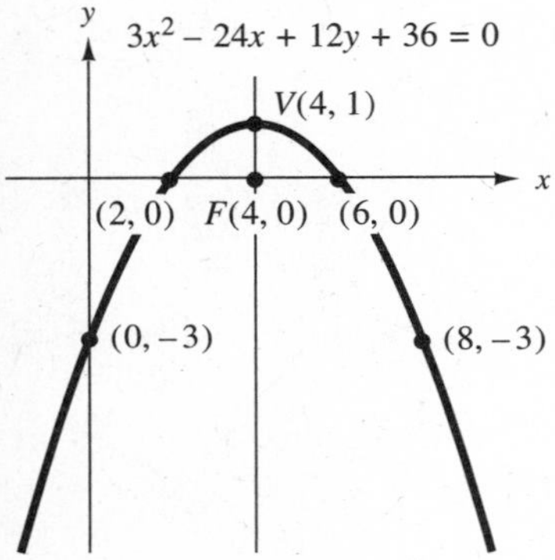

Comparing the result with the standard form, we see that $h = 4$, $k = 1$, and $p = -1$. The parabola is symmetric with respect to the line $x = 4$ and opens down since $p = -1$.
Therefore, we arrive at the following values.

$$\text{Vertex} = V(h, k) = V(\boxed{4, 1})$$
$$\text{Focus} = F(h, k + p) = F(4, 1 - 1) = F(4, 0)$$
$$\text{End points of latus rectum} =$$
$$(h + 2p, k + p) = [\,\boxed{4} + 2\boxed{(-1)}, \boxed{1}\ \boxed{-1}\,] = (2, 0)$$
$$(h - 2p, k + p) = [\,\boxed{4} - 2(\boxed{-1}), \boxed{1}\ \boxed{-1}\,] = (6, 0)$$

Using these facts and the additional points $(0, -3)$ and $(8, -3)$, we can sketch the parabola as shown in Figure 18.13. ◆

Graph the parabola in Example 5 on the graphics calculator. Is the result the same as Figure 18.13? Check your range settings!

The parabola is more than just a geometric concept. It has many uses in the physical world.

1. Projectiles in the air, such as a ball, or a missile, or water sprayed from a hose, describe a parabolic path when acted on only by gravity.
2. Many arches of bridges or buildings are parabolic in shape. With this shape, the arch can support the structure above it.
3. Rotating a parabola about its line of symmetry creates a bowl type surface called a paraboloid of revolution. A paraboloid has an important reflection property. Any ray or wave that originates at the focus and strikes the surface of the paraboloid is reflected parallel to the line of symmetry. (See Figure 18.14.) This forms the basic design of the reflectors for automobile headlights, flashlights, searchlights, telescopes, etc. This also is an excellent collecting device and is the basic design of TV, radar, and radio antennas.

EXAMPLE 6   Application: Cables on a Bridge

The cables of a bridge form a parabolic arc. The low point of the cable is 10 ft above the roadway midway between two towers. The distance between the

FIGURE 18.13

FIGURE 18.14

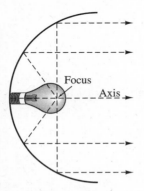

In the upper-left figure: $3x^2 - 24x + 12y + 36 = 0$, $V(4, 1)$, $(2, 0)$, $F(4, 0)$, $(6, 0)$, $(0, -3)$, $(8, -3)$

In the lower-left figure (Figure 18.14): Focus, Axis

towers is 400 ft. The cable is attached to the towers 50 ft above the roadway. Determine the equation of the parabola that describes the path of the cable. (See Figure 18.15.)

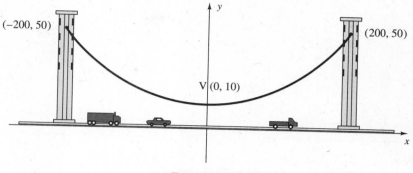

FIGURE 18.15

Solution   The parabola is formed by the cable between the two towers. The low point on the cable is midway between the towers, and 10 ft above the roadway. In order to write an equation, we must locate the x-axis and the y-axis. Select the roadway as the x-axis and the line perpendicular to the roadway through the lowest point of the curve as the y-axis. The parabola opens up with the vertex at the point $(0, 10)$. Two other points on the parabola are $(200, 50)$ and $(-200, 50)$. The standard form for this equation is $(x - h)^2 = 4p(y - k)$. Knowing the vertex $V(0, 10)$ and a point on the curve $(200, 50)$, we can substitute for h, k, x, and y in the equation to obtain a value for p.

$$(\boxed{200} - \boxed{0})^2 = 4p(\boxed{50} - \boxed{10})$$
$$40{,}000 = 160p$$
$$250 = p$$

Substituting for h, k, and p, we have the equation:

$$(x - \boxed{0})^2 = 4(\boxed{250})(y - \boxed{10}).$$

Thus, the equation that describes the path of the cable is $x^2 - 1000y + 10000 = 0$. ◆

The key information about the parabola with vertices at $(0, 0)$ and (h, k) is summarized in the two boxes.

STANDARD FORM OF EQUATIONS FOR PARABOLA WITH VERTEX AT ORIGIN

Equation		Curve	Directrix	Vertex	Focus
$x^2 = 4py$	if $p > 0$	Opens up	$y = -p$	$(0, 0)$	$(0, p)$
	if $p < 0$	Opens down	$y = -p$	$(0, 0)$	$(0, p)$
$y^2 = 4px$	if $p > 0$	Opens right	$x = -p$	$(0, 0)$	$(p, 0)$
	if $p < 0$	Opens left	$x = -p$	$(0, 0)$	$(p, 0)$

STANDARD FORM OF EQUATIONS FOR PARABOLA WITH VERTEX NOT AT ORIGIN					
Equation		Curve	Directrix	Vertex	Focus
$(x - h)^2 = 4p(y - k)$	if $p > 0$	Opens up	$y = k - p$	(h, k)	$(h, k + p)$
	if $p < 0$	Opens down	$y = k - p$	(h, k)	$(h, k + p)$
$(y - k)^2 = 4p(x - h)$	if $p > 0$	Opens right	$x = h - p$	(h, k)	$(h + p, k)$
	if $p < 0$	Opens left	$x = h - p$	(h, k)	$(h + p, k)$

18.2 Exercises

In Exercises 1–3, sketch the three curves on the same axis system. Examine the curves and discuss the differences among the curves and what changes in the equations caused the differences.

1. $y = x^2 - 1$ **2.** $y = x^2 + 1$ **3.** $y = x^2 + 2$

In Exercises 4–6, sketch the three curves on the same axis system. Examine the curves and discuss the differences among the curves and what changes in the equations caused the differences.

4. $y = \dfrac{1}{2} x^2$ **5.** $y = 2x^2$ **6.** $y = 4x^2$

In Exercises 7–9, sketch the three curves on the same axis system. Examine the curves and discuss the differences among the curves and what changes in the equations caused the differences.

7. $2y = x^2$ **8.** $4y = x^2$ **9.** $8y = x^2$

In Exercises 10–12, sketch the three curves on the same axis system. Examine the curves and discuss the differences among the curves and what changes in the equations caused the differences.

10. $(x - 2)^2 = y$ **11.** $(x - 0)^2 = y$ **12.** $(x + 2)^2 = y$

In Exercises 13–24, sketch the parabola represented by the equation. Indicate the vertex, the focus, the end points of the latus rectum, and the axis of symmetry.

13. $x^2 = 2y$ **14.** $y^2 = 16x$ **15.** $y^2 = -3(x + 1)$

16. $x^2 = -2(y - 3)$ **17.** $(y - 3)^2 = x$ **18.** $(x + 3)^2 = -y$

19. $(x - 2)^2 = 8(y + 1)$ **20.** $(y + 1)^2 = 8(x - 2)$ **21.** $(y - 2)^2 = -\dfrac{1}{2}(x - 2)$

22. $2x^2 + 3x + 4y - 4 = 0$ **23.** $x^2 + 2x - 3y + 1 = 0$ **24.** $0.5y^2 + 1.5x + 2.0y + 1.0 = 0$

In Exercises 25–30, write the equation of the parabola shown.

25.

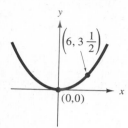

26.

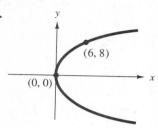

27.

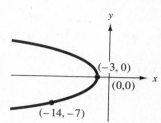

28.

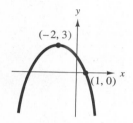

29.

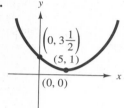

30.

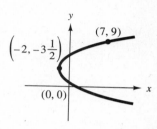

In Exercises 31–36, write the equation of the parabola using the given information.

31. Focus at $(0, 3)$, directrix $y = -3$

32. Focus at $(4, 0)$, directrix $x = -4$

33. Vertex at $(0, 0)$, x-axis is the line of symmetry, passes through $(3, 6)$.

34. Vertex at $(0, 0)$, axis of symmetry is the y-axis, passes through $(-12, -3)$.

35. Line of symmetry vertical, passes through $(-3, 4)$, vertex $(5, 1)$.

36. Line of symmetry horizontal, passes through $(7, 9)$, vertex at $(3, -7)$.

37. The side-to-side curve of a ship's deck is in the shape of a parabola that opens downward. What is the equation of the deck section where the ship is 20 ft wide and the center of the deck is 6 in. above the sides? (See the illustration.) Take the origin of the axis system at the center of the deck.

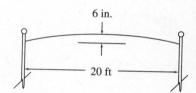

38. The cables of suspension bridges form parabolic curves. What is the equation of the cable for a bridge with supports 400 ft apart and a maximum center sag of 50 ft? (See the figure.) Take the origin of the axis system at the low point of the cable.

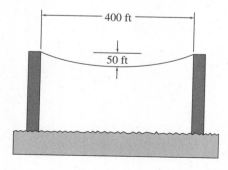

39. What is the equation of the parabolic cable in Exercise 38 if the origin of the axis system is taken at the point of attachment of the cable on the left support?

40. A cannon on top of a tower is aimed parallel to the ground and is fired. The distance the shell travels along the horizontal axis is called the range. The range is given by $x = v_0 t$, where v_0 is the initial speed of the shell and t is the time the shell spends in the air. While moving out along the x-axis, the shell is also falling. The falling motion is described by $y = \frac{1}{2} at^2$, where y is the shell's falling position as a function of time t and a is the acceleration due to gravity. It is suspected that this motion yields a parabolic path. By solving for the time in the first equation and plugging that expression into the second, prove that the path along the x, y-plane is indeed parabolic.

41. The collecting microphone of a radio telescope is placed in the same plane as the outer rim of the dish. The dish has a diameter of 120 ft. If the microphone is 45 ft off the ground when the telescope is pointed straight up, what is the equation of the parabolic cross-section with respect to an axis directly under the telescope on the ground? The microphone is at the focus of the parabola. (See the illustration.)

42. If a light source is placed at the focus of a parabolic reflector, all the light rays that strike the surface of the reflector are reflected parallel to the axis of symmetry of the parabola. A headlight has a filament 20 mm from the back of the reflector, and the reflector is 80 mm wide in the plane of the filament. What is the equation of the parabolic reflector if the origin of the axis system is placed at the filament? (See the sketch.)

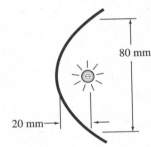

43. The water from the faucet of a watercooler is a parabolic curve. In designing water-cooler A, the designer specifies the maximum height of the water above the top edge of the cooler is to be 6 in. For the bowl to collect the water, what is the minimum width of the bowl? Assume that the horizontal axis goes through the center of the faucet and the vertical axis goes through the vertex to the curve. (See the diagram.) Also assume that the focus is at the point $(0, -3)$.

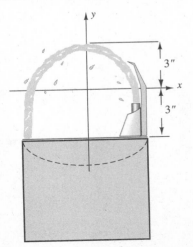

18.3

The Ellipse

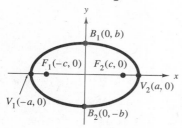

FIGURE 18.16

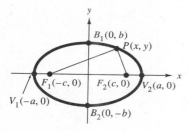

FIGURE 18.17

In shape and in format, the ellipse is different from the parabola. Although the parabola is open at one end, the ellipse is entirely closed. The parabola has one focus and one vertex, while the ellipse has two foci and two vertices (Figure 18.16).

> **DEFINITION 18.8**
>
> An **ellipse** is the set of all points in a plane such that the sum of the distances from two fixed points to a point on the ellipse is a constant. The fixed points are called the **foci** (singular focus).

Let us choose two points for the foci of an ellipse such that they lie on the x-axis a distance c on each side of the origin, as shown in Figure 18.16. Let us also assume that the ellipse crosses the x-axis a distance a on either side of the origin and that it crosses the y-axis a distance b above and below the origin.

Applying the definition of an ellipse to the information given with the curve of Figure 18.17, we see that the sum of the distances from the two foci to the point on the right side where the ellipse crosses the x-axis is:

$$d(F_1, V_2) + d(F_2, V_2) = [a - (-c)] + (a - c)$$
$$= (a + c) + (a - c)$$
$$= 2a.$$

Therefore, the constant specified in the definition must equal $2a$. The sum of the distances from the foci to the point where the ellipse crosses the y-axis is:

$$d(F_1, B_1) + d(F_2, B_1) = \sqrt{[0 - (-c)]^2 + (b - 0)^2}$$
$$+ \sqrt{(0 - c)^2 + (b - 0)^2}$$
$$= \sqrt{c^2 + b^2} + \sqrt{c^2 + b^2}$$
$$= 2a. \quad \text{Definition of ellipse}$$

Therefore, $2\sqrt{c^2 + b^2} = 2a$, or by squaring and solving for c^2 we have:

$$c^2 = a^2 - b^2.$$

Let $P(x, y)$ be any point on the ellipse, then the distances from the two foci to the point are:

$$d(F_1, P) = \sqrt{[x - (-c)]^2 + (y - 0)^2} \quad \text{and}$$
$$d(F_2, P) = \sqrt{(x - c)^2 + (y - 0)^2}.$$

Using the definition of an ellipse, we can write:

$$d(F_1, P) + d(F_2, P) = 2a$$

$$\sqrt{(x + c)^2 + y^2} = \sqrt{(x + c)^2 + y^2} = 2a \quad \text{Substituting}$$

$$\sqrt{(x + c)^2 + y^2} = 2a - \sqrt{(x - c^2) + y^2}$$

$$(x + c)^2 + y^2 = 4a^2 - 4a\sqrt{(x - c)^2 + y^2} + (x - c)^2 + y^2$$

Squaring both sides

$$x^2 + 2cx + c^2 + y^2 = 4a^2 - 4a\sqrt{(x - c)^2 + y^2} + x^2 - 2cx + c^2 + y^2$$

Binomial expansion

$$4a\sqrt{(x - c)^2 + y^2} = 4a^2 - 4cx \quad \text{Combining like terms}$$

$$a\sqrt{(x - c)^2 + y^2} = a^2 - cx \quad \text{Dividing by 4}$$

$$a^2[(x - c)^2 + y^2] = a^4 - 2a^2cx + c^2x^2 \quad \text{Squaring both sides}$$

$$a^2(x^2 - 2cx + c^2 + y^2) = a^4 - 2a^2cx + c^2x^2 \quad \text{Binomial expansion}$$

$$a^2x^2 - 2a^2cx + a^2c^2 + a^2y^2 = a^4 - 2a^2cx + c^2x^2$$

$$a^2x^2 - c^2x^2 + a^2y^2 = a^4 - a^2c^2$$

$$(a^2 - c^2)x^2 + a^2y^2 = a^2(a^2 - c^2) \quad \text{Factoring}$$

$$b^2x^2 + a^2y^2 = a^2b^2$$

Recalling $c^2 = a^2 - b^2$ or $a^2 - c^2 = b^2$ and substituting b^2 for $a^2 - c^2$

$$\frac{x^2}{a^2} + \frac{y^2}{b^2} = 1. \quad \text{Dividing by } a^2b^2$$

The portion of the x-axis bounded by the ellipse is its **major axis,** so the distance a is called the length of the **semimajor axis.** The end points of the major axis are called the vertices of the ellipse. The portion of the y-axis bounded by the ellipse is its **minor axis,** so the distance b is called the length of the **semiminor axis.**

DEFINITION 18.9

The **standard form of the equation of an ellipse** with center at the origin, length of the semimajor axis a and semiminor axis b, and major axis along the x-axis is:

$$\frac{x^2}{a^2} + \frac{y^2}{b^2} = 1.$$

See Figure 18.18a.

If we place the foci on the y-axis, center at the origin, and pick any point $P(x, y)$ on the plane, we can develop the equation of the ellipse given in Definition 18.10.

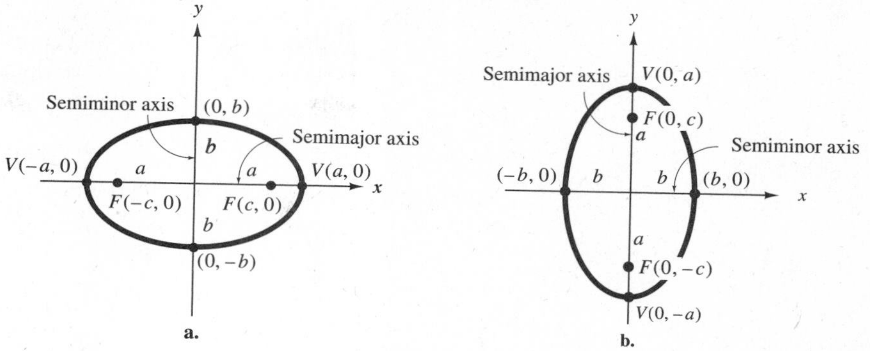

FIGURE 18.18

DEFINITION 18.10

The **standard form of the equation of an ellipse** with center at the origin, length of the semimajor axis a and semiminor axis b, and major axis along the y-axis is:

$$\frac{x^2}{b^2} + \frac{y^2}{a^2} = 1.$$

See Figure 18.18b.

CAUTION   With the equation of the ellipse in standard form, the values of a^2 and b^2 indicate whether the major axis is vertical or horizontal. Since a^2 is always greater than b^2, the variable with the largest denominator is the major axis of the ellipse. For example, for the ellipse $\dfrac{x^2}{5} + \dfrac{y^2}{7} = 1$, 7 is greater than 5; therefore, since the numerator of the fraction with 7 as the denominator is y^2, the major axis is vertical (coincides with the y-axis), and the minor axis is horizontal (coincides with the x-axis). For the ellipse $\dfrac{x^2}{25} + \dfrac{y^2}{4} = 1$, 25 is greater then 4; therefore, the major axis is horizontal (coincides with the x-axis), and the minor axis is vertical (coincides with the y-axis). ■

EXAMPLE 1   Graph of Ellipse with Center at (0, 0)

For the ellipse $9x^2 + 4y^2 = 36$, determine the vertices, end points of the minor axis, and the coordinates of the foci, and sketch the curve.

Solution   To determine the vertices and the end points of the minor axis, first find a and b. Begin by placing the equation $9x^2 + 4y^2 = 36$ in standard form.

$$\frac{9x^2}{36} + \frac{4y^2}{36} = \frac{36}{36} \qquad \textbf{Dividing both sides by 36}$$

$$\frac{x^2}{4} + \frac{y^2}{9} = 1$$

With the equation in this form, the center of the ellipse is at the origin. We know that a must be greater than b, and since the larger numerical value is under y^2, the vertices of the major axis are on the y-axis. Thus, $a^2 = 9$ or $a = 3$, and $b^2 = 4$ or $b = 2$. We also know that $c^2 = a^2 - b^2$; substituting we have:

$$c^2 = 9 - 4 \quad \text{or} \quad c = \sqrt{5}.$$

Knowing a, b, and c, we can write the coordinates:

$C(\ 0, 0\ )$	**Center**
$V_1(\ 0, 3\ )$, $V(\ 0, -3\ )$	**End points of the major axis**
$(\ 2, 0\ )$, $(\ -2, 0\ )$	**End points of the minor axis**
$F_1(\ 0, \sqrt{5}\ )$, $F_2(\ 0, -\sqrt{5}\ )$.	**Foci**

With this information and a few key points, we can sketch the ellipse. When $y = 1$,

$$\frac{x^2}{4} + \frac{(1)^2}{9} = 1 \quad \text{and} \quad x = \pm 4\frac{\sqrt{2}}{3}.$$

When $y = 2$,

$$\frac{x^2}{4} + \frac{(2)^2}{9} = 1 \quad \text{and} \quad x = \pm 2\frac{\sqrt{5}}{3}.$$

Since the ellipse is symmetric with respect to the major axis, minor axis, and its center, and knowing the points $\left(\frac{4\sqrt{2}}{3}, 1\right)$ and $\left(\frac{2\sqrt{5}}{3}, 2\right)$ are on the ellipse, we can write the following points: $\left(\frac{4\sqrt{2}}{3}, -1\right) \left(\frac{-4\sqrt{2}}{3}, 1\right) \left(\frac{-4\sqrt{2}}{3}, -1\right)$, $\left(\frac{2\sqrt{5}}{3}, -2\right)$, $\left(\frac{-2\sqrt{5}}{3}, 2\right)$ and $\left(\frac{-2\sqrt{5}}{3}, -2\right)$. The curve is sketched and labeled in Figure 18.19. ◆

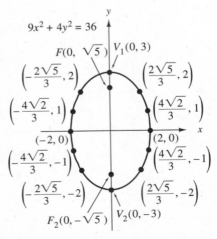

FIGURE 18.19

Can you graph the ellipse in Example 1 on your graphics calculator? Can you determine the vertices?

EXAMPLE 2   **Application: Race Track**

Molly is designing an elliptical track for go-cart races. The maximum lengths for the major and minor axes are 40 and 32 m, respectively. See Figure 18.20. Write the equation of the ellipse that describes the path of the track.

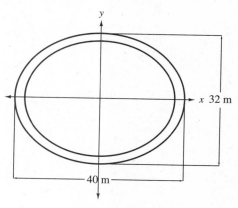

FIGURE 18.20

Solution   Select the center of the track as the origin of the coordinate system. Also select the major axis as the horizontal axis so that it coincides with the x-axis. If the track is a maximum length of 40 m, then the length of the semimajor axis is 20 m; and a is 20 m. With a maximum width of 32 m, the semiminor axis is 16 m, and b is 16 m. We can use the standard form of the ellipse $\dfrac{x^2}{a^2} + \dfrac{y^2}{b^2} = 1$. Substituting for a and b, the equation is:

$$\frac{x^2}{(\boxed{20})^2} + \frac{y^2}{(\boxed{16})^2} = 1.$$

The equation that describes the path of the track is $\dfrac{x^2}{400} + \dfrac{y^2}{256} = 1$.   ◆

We can identify the major and minor axis in the same manner that we have been even when the center of the ellipse is not at the origin.

If any point on the plane other than the origin is selected as the center of the ellipse and if a major axis parallel to the x- or y-axis is selected, then using the geometrical definition, a new set of equations for the ellipse can be derived.

DEFINITION 18.11

The **standard form of the equation of an ellipse** with center at $C(h, k)$, length of the semimajor axis a, and semiminor axis b, and major axis parallel to the x-axis is:

$$\frac{(x - h)^2}{a^2} + \frac{(y - k)^2}{b^2} = 1.$$

See Figure 18.21a.

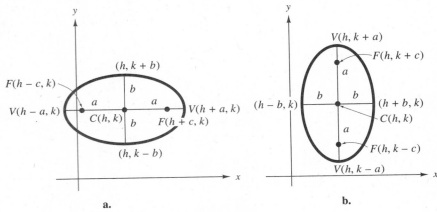

FIGURE 18.21

DEFINITION 18.12

The **standard form of the equation of an ellipse** with center at $C(h, k)$, length of the semimajor axis a, and semiminor axis b, and major axis parallel to the y-axis is:

$$\frac{(x - h)^2}{b^2} + \frac{(y - k)^2}{a^2} = 1.$$

See Figure 18.21b.

EXAMPLE 3 **Equation of Ellipse with Center at (h, k)**

Determine the equation of the ellipse with vertices at $(-1, 2)$ and $(7, 2)$ and with 2 as the length of the semiminor axis.

Solution With vertices of the ellipse at $(-1, 2)$ and $(7, 2)$, we know that the center is at the midpoint of the line segment joining these points. The midpoint of the line segment is the point $(3, 2)$, Figure 18.22.

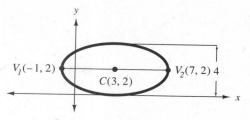

FIGURE 18.22

The distance from the midpoint to either vertex is 4 units, and a = 4. Given that the semiminor axis has length of 2, we know that b = 2. From Figure 18.22, we see that the major axis is parallel to the x-axis. Substituting into the standard form of the equation, we have:

$$\frac{(x-\boxed{3})^2}{\boxed{4}^2} + \frac{(y-\boxed{2})^2}{\boxed{2}^2} = 1 \quad \text{or} \quad \frac{(x-3)^2}{16} + \frac{(y-2)^2}{4} = 1. \quad \blacklozenge$$

EXAMPLE 4   Graph of Ellipse Center (h, k)

Sketch the ellipse whose equation is $4x^2 + 25y^2 - 8x + 100y + 4 = 0$ and label the center, foci, and vertices.

Solution   First change the equation into the standard form by completing the square.

$$\boxed{4x^2 - 8x} + \boxed{25y^2 + 100y} = -4 \qquad \text{Group } x \text{ and } y \text{ terms}$$
$$\boxed{4}(x^2 - 2x) + \boxed{25}(y^2 + 4y) = -4 \qquad \text{Common factors}$$
$$4(x^2 - 2x + \boxed{1}) + 25(y^2 + 4y + \boxed{4}) = -4 + \boxed{4} + 100$$

Completing the square

$$4(\boxed{x-1})^2 + 25(\boxed{y+2})^2 = 100 \qquad \text{Factoring trinomial}$$
$$\frac{4(x-1)^2}{\boxed{100}} + \frac{25(y+2)^2}{\boxed{100}} = \frac{100}{\boxed{100}} \qquad \text{Divide both sides by 100}$$
$$\frac{(x-1)^2}{25} + \frac{(y+2)^2}{4} = 1 \quad \text{or} \quad \frac{(x-1)^2}{5^2} + \frac{(y+2)^2}{2^2} = 1$$

Since the larger numerical value is in the denominator of the fraction with x^2 in the numerator, the major axis is horizontal, $a = 5$, $b = 2$, $c = \sqrt{21}$, $h = 1$, and $k = -2$. The ellipse has:

$$(h, k) = \boxed{(1, -2)} \qquad \text{Center}$$
$$(h - c, k) = \boxed{(1 - \sqrt{21}, -2)} \qquad F_1$$
$$(h + c, k) = \boxed{(1 + \sqrt{21}, -2)} \qquad F_2$$
$$(h - a, k) = (1 - 5, -2) \quad \text{or} \quad \boxed{(-4, -2)} \qquad V_1$$
$$(h + a, k) = (1 + 5, -2) \quad \text{or} \quad \boxed{(6, -2)} \qquad V_2$$

In decimal form the foci are at $(-3.58, -2)$ and $(5.58, -2)$. Figure 18.23 is a sketch of the ellipse properly labeled. ◆

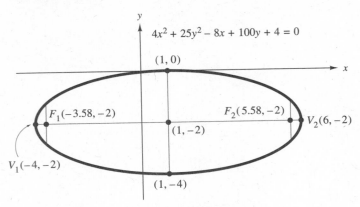

$$4x^2 + 25y^2 - 8x + 100y + 4 = 0$$

FIGURE 18.23

The relative shape of an ellipse can be determined by its **eccentricity, *e*.** The distance from the center of the ellipse to a foci is c, and the distance from the center to a vertex is a. The eccentricity is given by the equation:

$$e = \frac{c}{a}.$$

The eccentricity of all ellipses are in a range between 0 and 1, $0 < e < 1$, as shown in Figure 18.24. An ellipse with an eccentricity close to 1 is long and thin, and the foci are relatively far apart. If the eccentricity is small, close to 0, then the ellipse resembles a circle. It can be shown that the circle is a special case of the ellipse when $e = 0$.

Like other conic sections, the ellipse plays a significant part in our universe.

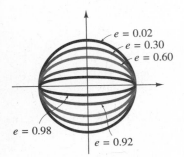

FIGURE 18.24

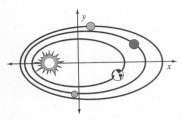

FIGURE 18.25

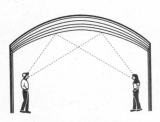

FIGURE 18.26

1. The planets in our solar system revolve in an elliptical oribit about the sun, which is located at one focus, (see Figure 18.25). The closer planets have a low eccentricity and, thus, appear to have a circular path. It turns out that all orbiting bodies—from the moon and our satellites to the electrons moving about the nucleus of the atom—assume elliptical paths.

2. Within an elliptical shell, a ray or beam originating at one focus reflects off the shell and passes through the other focus. In a room with an elliptical shape, a slight sound at one focus can be heard at the other focus, but cannot be heard between the foci (see Figure 18.26). Rooms with this feature, known as a whispering gallery, are in the Capitol in Washington, D.C., at Sonnenberg Gardens in Canandaiqua, New York, and also at the Museum of Science and Industry in Chicago.

3. Dental and surgical lights use the focal properties of an ellipse. With one focus being the light source, the light becomes concentrated at the other focus of the ellipse. By adjusting the position of the light so that the point of concentration is at the opening of the patient's mouth, the patient is not

bothered by the glare from the reflector, a small area inside the patient's mouth is illuminated brightly, and the dentist's hands do not interfere with the light.

4. Removing kidney stones with shock waves is another application of ellipsoidal properties. A lithotripter—a special device, which is in the shape of a three-dimensional ellipse called a ellipsoid—emits ultra-high-frequency (UHF) shock waves that move through water to break up kidney stones. The ellipsoid at one focus emits the UHF, and the patient's kidney is located at the other focus. The shock waves reflect off the inner surface of the ellipsoid.

The key information about the ellipse with vertices at $(0, 0)$ and (h, k) is summarized in the two boxes.

STANDARD FORM OF EQUATIONS FOR ELLIPSE WITH CENTER AT ORIGIN (WHERE $a^2 = b^2 + c^2$)

Major Axis	Center	Focus	Intercepts — Semimajor Axis	Intercepts — Semiminor Axis	Equation
Horizontal	$(0, 0)$	$(c, 0)$	$(a, 0)$	$(0, b)$	$\dfrac{x^2}{a^2} + \dfrac{y^2}{b^2} = 1$
		$(-c, 0)$	$(-a, 0)$	$(0, -b)$	
Vertical	$(0, 0)$	$(0, c)$	$(0, a)$	$(b, 0)$	$\dfrac{x^2}{b^2} + \dfrac{y^2}{a^2} = 1$
		$(0, -c)$	$(0, -a)$	$(-b, 0)$	

STANDARD FORM OF EQUATIONS FOR ELLIPSE WITH CENTER NOT AT ORIGIN

Major Axis	Center	Focus	Intercepts — Semimajor Axis	Intercepts — Semiminor Axis	Equation
Horizontal	(h, k)	$(h + c, k)$	$(h + a, k)$	$(h, k + b)$	$\dfrac{(x - h)^2}{a^2} + \dfrac{(y - k)^2}{b^2} = 1$
		$(h - c, k)$	$(h - a, k)$	$(h, k - b)$	
Vertical	(h, k)	$(h, k + c)$	$(h, k + a)$	$(h + b, k)$	$\dfrac{(x - h)^2}{b^2} + \dfrac{(y - k)^2}{a^2} = 1$
		$(h, k - c)$	$(h, k - a)$	$(h - b, k)$	

18.3 Exercises

In Exercises 1–2, sketch the two curves on the same axis system. Examine the curves and discuss the difference between the curves and what changes in the equations caused the differences.

1. $\dfrac{x^2}{9} + \dfrac{y^2}{4} = 1$

2. $\dfrac{x^2}{4} + \dfrac{y^2}{9} = 1$

In Exercises 3–4, sketch the two curves on the same axis system. Examine the curves and discuss the difference between the curves and what changes in the equations caused the differences.

3. $\dfrac{x^2}{16} + \dfrac{y^2}{4} = 1$

4. $\dfrac{x^2}{16} + \dfrac{y^2}{15} = 1$

In Exercises 5–14, for each ellipse, determine the coordinates of **a.** center, **b.** foci, **c.** end points of major axis, and **d.** end points of minor axis. Then sketch.

5. $\dfrac{x^2}{9} + \dfrac{y^2}{4} = 1$

6. $\dfrac{x^2}{16} + \dfrac{y^2}{9} = 1$

7. $\dfrac{x^2}{16} + \dfrac{y^2}{25} = 1$

8. $\dfrac{x^2}{4} + \dfrac{y^2}{9} = 1$

9. $\dfrac{(x-1)^2}{9} + \dfrac{(y-2)^2}{4} = 1$

10. $\dfrac{(x-2)^2}{16} + \dfrac{(y-1)^2}{9} = 1$

11. $\dfrac{(x+1)^2}{16} + \dfrac{(y-2)^2}{9} = 1$

12. $\dfrac{(x-3)^2}{25} + \dfrac{(y+1)^2}{4} = 1$

13. $\dfrac{(x+5)^2}{4} + \dfrac{y^2}{\sqrt{5}} = 1$

14. $\dfrac{(x+2)^2}{7} + \dfrac{y^2}{4} = 1$

In Exercises 15–26, write the standard form of the equation of the ellipse.

15.

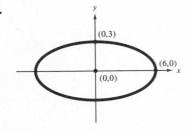

16.

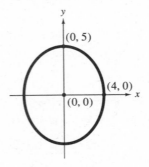

17.

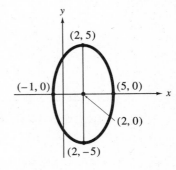

18.

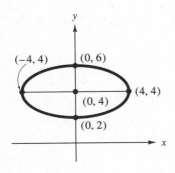

19.

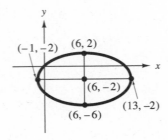

20.

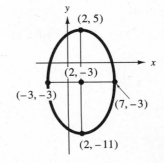

21. Center $(-3, 2)$ $a = 2$, $b = 1$, major axis horizontal

22. Center $(-2, -1)$, $a = 4$, $b = 2$, major axis vertical

23. Vertices $(4, 2)$ and $(12, 2)$, $b = 2$

24. Foci $(-2, 3)$ and $(6, 3)$, length of minor axis $= 6$

25. A focus at $(-2, 3)$, a vertex at $(6, 3)$, length of minor axis $= 6$

26. A vertex at $(-8, 0)$, one end of minor axis at $(-4, 3)$, major axis horizontal

27. What are the coordinates of the foci of the ellipse in Exercises 15, 17, and 19?

28. What are the coordinates of the foci of the ellipse in Exercises 16, 18, and 20?

29. Determine the equation of the set of points that moves so that the sum of its distances from the points $(1, 3)$ and $(9, 3)$ is 10.

30. The earth travels in an elliptical orbit around the sun, which is located at one of the foci. The largest distance between the earth and the sun is 94,500,000 mi and the shortest distance is 91,500,000 mi.
    a. Determine the length of the major axis of the ellipse.
    b. Determine the distance between the foci of the orbit.

31. Viewed from the top, the wingtip of a crop-duster aircraft is shaped like half an ellipse. As shown in the diagram, the chord of the wing (distance from front to back) is 1.20 m, and the tip extends 0.20 m beyond the parallel leading and trailing edges (front and back edges) of the wing. Using the center line of the wing as an axis, what is the equation of the ellipse that forms the wingtip?

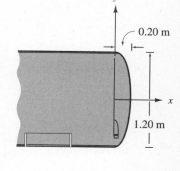

32. A furniture manufacturer is asked to build a table with an elliptical top (see drawing). To draw the curve on the piece of wood, the designer uses a piece of string and two nails. The length and width of the top are to be 8 ft and 4 ft, respectively. **a.** How far apart should the two nails (foci) be? **b.** What length of string does the designer need to lay out the elliptical top?

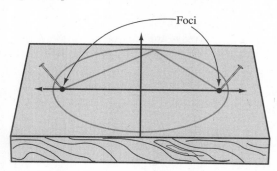

33. The circumference of an ellipse can be approximated with the formula :
$$C = 2\pi\sqrt{\frac{a^2 + b^2}{2}},$$
where a is length of the semimajor axis and b is the length of the semiminor axis.
    a. Use the information given in Example 2 to determine the distance a car travels in one loop around the track.
    b. Determine the number of laps that a person must drive to travel 1 km.

34. Molly would like to have a better surface for her go-cart race track in Example 2. She is told that she must excavate to a depth of 25 cm and then replace the dirt with gravel and blacktop. If the track is 2 m wide, how many cubic meters of dirt must she remove? The area of an ellipse is estimated with the formula $A = \pi ab$. The lengths of the semimajor and semiminor axes are a and b, respectively.

35. The shape of an ellipse depends on the eccentricity of the ellipse; that is, $e = \dfrac{c}{a}$.

Determine the equation of the ellipse with vertices at $(\pm 5, 0)$ and $e = \dfrac{3}{5}$.

36. The width of the ellipse through a focus is called the focal width of the ellipse or the latus rectum (see the sketch). Knowing the length of the latus rectum is helpful in sketching an ellipse, since it gives additional points on the curve. Show that the length of each latus rectum is $\dfrac{2b^2}{a}$.

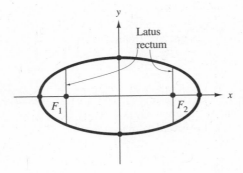

In Exercises 37 and 38, sketch the graph of each ellipse by finding the coordinates of center, foci, vertices, and the end points of the latus rectum.

37. $\dfrac{(x + 3)^2}{25} + \dfrac{(y + 1)^2}{9} = 1$

38. $\dfrac{(x - 12)^2}{144} + \dfrac{(y + 16)^2}{169} = 1$

39. A flat elliptical shaped piece of metal from which the blades of outboard motor propellers are shaped is called a blank (see the drawing). The blank is cut along a latus rectum and this is where the finished blade is fastened to the propeller hub. If the width of a blank is to be 3 in. and the length from hub to the tip of the blade is to be 5 in., what is the equation of the ellipse used in cutting the blank? For convenience, assume that the origin of the axis system is at the center of the required ellipse.

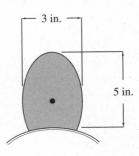

40. A bridge support, which bears a large load, forms the upper half of an ellipse with its center at $(40, -3)$ and passing through points $(20, 1)$ and $(50, 2)$. (See the drawing.) What is the equation of the ellipse?

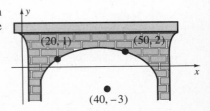

41. When the Vanguard satellite was launched it went into a slightly elliptical orbit around the earth. Selecting an axis system with its origin at the center of the ellipse, the satellite passed through points (1700, 4740) and (4500, 2334). The distances are in miles. The center of the earth was located at one of the foci of the elliptical orbit. If the radius of the earth is 3960 mi, how close did the satellite come to the surface?

42. A magician seeks to create the illusion of light in an elliptical room. He places a light source (disguised with a baffle) at one focal point, and the reflected light merges at the other focal point. Suppose the distance between the focal points is designed to be 7.00 m, while the length of the elliptical chamber is 12.0 m. (See the sketch.) What is the required width of the chamber?

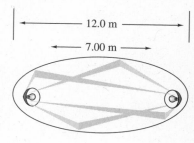

43. The center of the ellipse that is rotated to form the elliptical shaped lithotripter is the origin (0, 0). The length of the major axis of the ellipse is 12.0 ft, and the length of the minor axis is 5.00 ft. Determine the points (foci) where the shock waves originate and the location of the kidney stones.

44. As we know, the equation for the ellipse with its center at the origin is given as $\dfrac{x^2}{a^2} + \dfrac{y^2}{b^2} = 1$. This closely resembles the equation for a circle with its center at the origin: $x^2 + y^2 = r^2$. In fact, the circle is an ellipse with an eccentricity of $e = 0$.
  a. If the eccentricity is zero, then what must be the relationship between a and b?
  b. Show how the equation for a circle can be derived from the equation for an ellipse.

45. The planets revolve around the sun in elliptical orbits. The closest that the planet Pluto approaches the sun is 2.8 billion mi. It gets as far away as 4.6 billion mi. With this orbiting path, what is its eccentricity of the elliptical path?

18.4

The Hyperbola

In the previous section the ellipse was expressed in terms of the sum of two distances being a constant. The hyperbola is expressed in terms of the difference of two distances being a constant.

> **DEFINITION 18.13**
> A **hyperbola** is the set of all points such that the absolute value of the difference of the distances from two fixed points (the foci) to a point on the hyperbola is a constant.

The hyperbola is the basis for military sound-ranging systems and for the Loran-C navigation system.

Choose two points for the foci of a hyperbola such that they lie on the x-axis a distance c on each side of the origin, as shown in Figure 18.27. Also assume that the hyperbola crosses the x-axis a distance a on either side of the origin. These points of intersection are the **vertices** of the hyperbola. The **transverse axis** (or major axis) of the hyperbola coincides with the x-axis. The **conjugate axis** (or minor axis) of the hyperbola coincides with the y-axis.

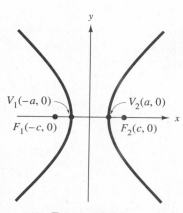

FIGURE 18.27

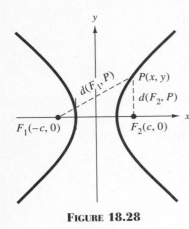

FIGURE 18.28

Applying the definition of a hyperbola to the information given with the curve in Figure 18.27, we see that the absolute value of the differences of the distances from the two foci to the right vertex is:

$$|d(F_1, V_2) - d(F_2, V_2)| = |[a - (-c)] - (c - a)|$$
$$= |2a|$$

Since a is a measured distance and is always positive, the constant specified in the definition of a hyperbola equals $2a$. The distances from the two foci to any point $P(x, y)$ on the hyperbola (Figure 18.28) are given by the following formulas.

$$d(F_1, P) = \sqrt{[x - (-c)]^2 + (y - 0)^2}$$
$$d(F_2, P) = \sqrt{(x - c)^2 + (y - 0)^2}$$

According to the definition of a hyperbola,

$$|d(F_1, P) - d(F_2, P)| = 2a$$
$$\left|\sqrt{[x - (\boxed{-c})]^2 + (y - \boxed{0})^2} - \sqrt{(x - \boxed{c})^2 + (y - \boxed{0})^2}\right| = 2a.$$

          Substituting

Since we are going to square both sides of the equation, we may drop the absolute value sign.

$$\sqrt{(x + c)^2 + y^2} = 2a + \sqrt{(x - c)^2 + y^2}$$ **Simplifying**

$$(x + c)^2 + y^2 = 4a^2 + 4a\sqrt{(x - c)^2 + y^2} + (x - c)^2 + y^2$$

          Squaring both sides

$$x^2 + 2cx + c^2 + y^2 = 4a^2 + 4a\sqrt{(x - c)^2 + y^2} + x^2 - 2cx + c^2 + y^2$$

          Squaring

$$4cx - 4a^2 = 4a\sqrt{(x - c)^2 + y^2}$$ **Combining terms**

$$cx - a^2 = a\sqrt{(x - c)^2 + y^2}$$ **Dividing by 4**

$$c^2x^2 - 2a^2cx + a^4 = a^2[(x - c)^2 + y^2]$$ **Squaring both sides**

$$c^2x^2 - 2a^2cx + a^4 = a^2(x^2 - 2cx + c^2 + y^2)$$

$$c^2x^2 - 2a^2cx + a^4 = a^2x^2 - 2a^2cx + a^2c^2 + a^2y^2$$

$$(c^2 - a^2)x^2 - a^2y^2 = a^2(c^2 - a^2)$$ **Factoring**

Now let $b^2 = c^2 - a^2$; note that $b^2 = c^2 - a^2$ requires that $c > a$.

$$b^2x^2 - a^2y^2 = a^2b^2$$ **Substituting for c^2-a^2**

$$\frac{x^2}{a^2} - \frac{y^2}{b^2} = 1$$ **Dividing by a^2b^2**

DEFINITION 18.14

The **standard form of the equation of a hyperbola** with center at the origin and the *x*-axis as the transverse axis is:

$$\frac{x^2}{a^2} - \frac{y^2}{b^2} = 1.$$

See Figure 18.29a.

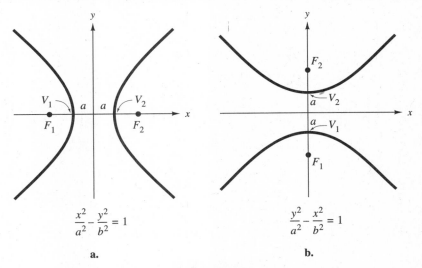

$$\frac{x^2}{a^2} - \frac{y^2}{b^2} = 1$$

a.

$$\frac{y^2}{a^2} - \frac{x^2}{b^2} = 1$$

b.

FIGURE 18.29

If we place the foci on the *y*-axis, center at the origin, and pick any point $P(x, y)$ on the plane, we can develop the equation of the hyperbola given in Definition 18.15.

DEFINITION 18.15

The **standard form of the equation of a hyperbola** with center at the origin and the *y*-axis as the transverse axis is:

$$\frac{y^2}{a^2} - \frac{x^2}{b^2} = 1.$$

See Figure 18.29b.

In the development of the standard form of the hyperbola as given in Definition 18.14, the terms *transverse axis* and *conjugate axis* were introduced. The end points of the transverse axis are the vertices of the hyperbola at $(a, 0)$ and $(-a, 0)$. The axis has a length of $2a$. The conjugate axis coincides with the *y*-axis and has its end points at $(0, b)$ and $(0, -b)$. If we set $x = 0$, we see that there are no y-intercepts, so why should we be concerned about the conjugate axis or the length b? The significance of b is determined by solving the equation of the hyperbola for y.

$$\frac{x^2}{a^2} - \frac{y^2}{b^2} = 1$$

$$\frac{y^2}{b^2} = \frac{x^2}{a^2} - 1$$

$$\frac{y^2}{b^2} = \frac{x^2 - a^2}{a^2} \qquad \textbf{LCD}$$

$$y^2 = \boxed{b^2}\left(\frac{x^2 - a^2}{a^2}\right) \qquad \textbf{Multiplying by } b^2$$

$$y^2 = \frac{b^2}{\boxed{a^2}}(x^2 - a^2) \qquad \textbf{Factoring}$$

$$y^2 = \frac{b^2}{a^2}\left(x^2 - \frac{a^2\boxed{x^2}}{\boxed{x^2}}\right) \qquad \textbf{Multiplying } a^2 \textbf{ by } \frac{x^2}{x^2}$$

$$y^2 = \frac{b^2\boxed{x^2}}{a^2}\left(1 - \frac{a^2}{x^2}\right) \qquad \textbf{Common factor of } x^2$$

$$y = \pm\frac{bx}{a}\sqrt{1 - \frac{a^2}{x^2}} \qquad \textbf{Square root of both sides}$$

Now let us examine the fraction $\frac{a^2}{x^2}$. We know a is a constant. If we substitute larger and larger values for x, then $\frac{a^2}{x^2}$ becomes smaller and smaller. In fact, the fraction eventually gets very close to zero. Thus, for large values of x, $1 - \frac{a^2}{x^2}$ approaches 1. Therefore, for large values of x, y approaches the value $\pm\frac{b}{a}x$, and the value of the hyperbola gets closer and closer to the lines:

$$y = \pm\frac{b}{a}x.$$

These lines are called the *asymptotes* of the hyperbola. Recall that asymptotes were discussed in Chapter 5. As x takes on values that are greater distances from the center of the hyperbola, the values of y (of the hyperbola) become closer and closer to the asymptotes even though they never actually reach the corresponding y-values of the asymptotes. Since these lines are easy to graph, the asymptotes are valuable aids in sketching the hyperbola. For a hyperbola of the form $\frac{x^2}{a^2} - \frac{y^2}{b^2} = 1$, the asymptotes are the lines:

$$y = \pm\frac{b}{a}x.$$

EXAMPLE 1   **Sketch Hyperbola Center (0, 0) with Transverse Axis Horizontal**

Sketch the hyperbola $\dfrac{x^2}{9} - \dfrac{y^2}{16} = 1$.

Solution   The equation $\dfrac{x^2}{9} - \dfrac{y^2}{16} = 1$ is in the standard form of the equation of a hyperbola with transverse axis along the x-axis. This tells us that $a^2 = 9$ or $a = 3$ and $b^2 = 16$ or b = 4. The vertices of the hyperbola are $V_1(-3, 0)$ and $V_2(3, 0)$. We can determine the foci by using the fact that $c^2 = a^2 + b^2$ and $c = 5$.

Therefore, the foci are $F_1(-5, 0)$ and $F_2(5, 0)$.

The asymptotes are the lines $y = \pm \dfrac{b}{a} x$ or:

$$y = \frac{4}{3}x \quad \text{and} \quad y = -\frac{4}{3}x.$$

We can simplify sketching the asymptotes by plotting the end points of the conjugate axis $(0, -4)$ and $(0, 4)$. Then draw the lines through the points $(0, -4)$ and $(0, 4)$ parallel to the x-axis and draw the lines through the end points of the transverse axis $(-3, 0)$ and $(3, 0)$ parallel to the y-axis. The result is a rectangle and the extended diagonals of the rectangle are the asymptotes of the hyperbola. Having completed the asymptote rectangle, we easily can sketch in the hyperbola, as shown in Figure 18.30. ◆

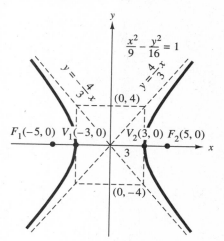

FIGURE 18.30

If we solve the equation $\dfrac{y^2}{a^2} - \dfrac{x^2}{b^2} = 1$ for y, we can show that the asymptotes are $y = \pm \dfrac{a}{b} x$. Example 2 uses these equations.

EXAMPLE 2   **Sketch Hyperbola Center $(0, 0)$ with Transverse Axis Vertical**

Sketch the hyperbola $16y^2 - 9x^2 = 144$.

Solution   The first thing to do is to write the equation in standard form by dividing each term by 144.

$$\frac{16y^2}{144} - \frac{9x^2}{144} = 1 \quad \text{or} \quad \frac{y^2}{9} - \frac{x^2}{16} = 1$$

With a positive y^2 term, the transverse axis is along the y-axis, and $a = 3$, $b = 4$. The vertices are $V_1(0, -3)$ and $V_2(0, 3)$. The end points of the conjugate axis are $(-4, 0)$ and $(4, 0)$. To determine the coordinates of the foci, first find c by using the equation $c^2 = a^2 + b^2$.

$$c^2 = 9 + 16$$
$$c^2 = 25 \quad \text{or} \quad c = 5.$$

Thus, the foci are $F_1(0, -5)$ and $F_2(0, 5)$. Sketch the rectangle formed by the points $(0, \pm 3)$ and $(\pm 4, 0)$. Then sketch the asymptotes using the diagonals of the rectangle as a guide. With the transverse axis vertical, the equations of the asymptotes are of the form $y = \pm\dfrac{a}{b}x$. Substituting for a and b, the equations of the asymptotes are:

$$y = \pm\frac{3}{4}x.$$

With the asymptotes and vertices in place, it is easy to sketch the hyperbola, as shown in Figure 18.31.

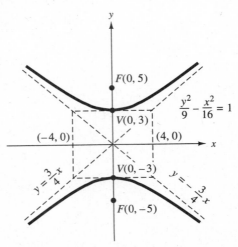

FIGURE 18.31

Use your graphics calculator to sketch the hyperbola in Example 2. Does your curve have the same shape? If not, why not? Can you include the asymptotes in your sketch? Can you determine where the curve crosses the x-axis?

If any point (h, k) on the plane is selected as the center of the hyperbola and a major axis parallel to the x- or y-axis is selected, then with the geometrical definition, a new set of equations for the hyperbola can be derived.

DEFINITION 18.16

The **standard form of the equation of a hyperbola** with a horizontal transverse axis, center at (h, k), vertices at $(h + a, k)$ and $(h - a, k)$, and foci at $(h + c, k)$ and $(h - c, k)$ is:

$$\frac{(x - h)^2}{a^2} - \frac{(y - k)^2}{b^2} = 1.$$

See Figure 18.32a.

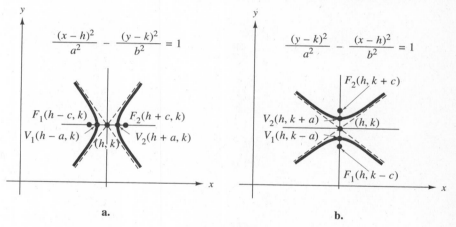

a.

b.

FIGURE 18.32

DEFINITION 18.17

The **standard form of the equation of a hyperbola** with a vertical transverse axis, center at (h, k), vertices at $(h, k + a)$ and $(h, k - a)$, and foci at $(h, k + c)$ and $(h, k - c)$ is:

$$\frac{(y - k)^2}{a^2} - \frac{(x - h)^2}{b^2} = 1.$$

See Figure 18.32b.

The hyperbola with a horizontal transverse axis has asymptotes with slope $\pm\dfrac{b}{a}$, and the asymptotes pass through (h, k), Figure 18.32a. The hyperbola with a vertical transverse axis has asymptotes with slope $\pm\dfrac{a}{b}$, and the asymptotes pass through (h, k), Figure 18.32b.

EXAMPLE 3 Sketch Hyperbola Center (h, k) with Transverse Horizontal

Sketch the hyperbola $\dfrac{(x - 1)^2}{144} - \dfrac{(y + 2)^2}{25} = 1$.

Solution By comparing the given equation with the form of Definition 18.21, we see that the hyperbola has a horizontal transverse axis and that $a^2 = 144$. Therefore, $a = 12$ and $b^2 = 25$, so $b = 5$. From the equation, we see that $h = 1$ and $k = -2$; therefore, the center of the hyperbola has coordinates $(1, -2)$. Since the transverse axis is horizontal, the vertices are a units to the left and right of center, $(h - a, k)$ and $(h + a, k)$. Substituting, we have $(1 - 12, -2)$ and $(1 + 12, -2)$ or $(-11, -2)$ and $(13, -2)$. The foci are c units to either side of center. Using the formula

$$c = \sqrt{a^2 + b^2} = \sqrt{144 + 25} \quad \text{or} \quad c = 13.$$

The foci are at $(h - c, k)$ and $(h + c, k)$; substituting, we have $(1 - 13, -2)$ and $(1 + 13, -2)$ or $(-12, -2)$ and $(14, -2)$. Locate the center, vertices, and foci on the graph. The transverse axis is on the line $y = -2$, and the conjugate axis is on the line $x = 1$. The asymptotes are lines that cross at the center of the hyperbola and have slopes of $\dfrac{b}{a} = \dfrac{5}{12}$ and $-\dfrac{b}{a} = -\dfrac{5}{12}$. Knowing the slopes, center, and vertices, we can sketch and label the asymptotes. The equations of the asymptotes are $12y + 5x + 19 = 0$ and $12y - 5x + 29 = 0$. Having sketched the asymptotes and the vertices, we can sketch the hyperbola, as shown in Figure 18.33.

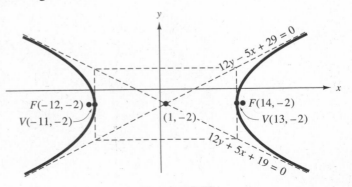

FIGURE 18.33

EXAMPLE 4 **Sketch Hyperbola Center (h, k) with Transverse Axis Vertical**

Sketch the hyperbola $\dfrac{(y - 2)^2}{9} - \dfrac{x^2}{4} = 1$.

Solution The given hyperbola is in the form of Definition 18.17, so the transverse axis is parallel to the y-axis. By comparing this equation with Definition 18.17, $h = 0$; $k = 2$; $a^2 = 9$, so $a = 3$; and $b^2 = 4$, so $b = 2$. Using $c^2 = a^2 + b^2$, we have $c = \sqrt{9 + 4} = 3.61$. The vertices are the points $(h, k + a)$ and $(h, k - a)$, which are, respectively, $(0, 5)$ and $(0, -1)$. The foci are the points $(h, k + c)$ and $(h, k - c)$, which are, respectively, $(0, 5.61)$ and $(0, -1.61)$. The asymptotes have slopes $\pm \dfrac{a}{b}$ or $\dfrac{3}{2}$ and $-\dfrac{3}{2}$. The equations of the asymptotes are $y = \dfrac{3}{2}x + 2$ and $y = -\dfrac{3}{2}x + 2$. Sketch the foci, vertices, and asymptotes. Then sketch the hyperbola, as shown in Figure 18.34.

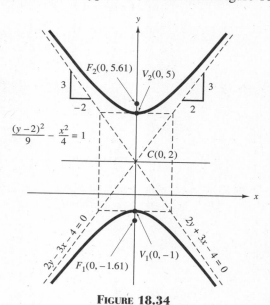

FIGURE 18.34 ◆

EXAMPLE 5 **Application: Location of Lightning Bolt**

Two people are fishing offshore 2.0 miles apart. As they talk to each other on a radiotelephone, they hear thunder. One person hears the thunder 3.0 seconds after the other hears it as discussed on the first page of this chapter. Sketch a graph of the possible locations of the lightning bolt that produced the thunder.

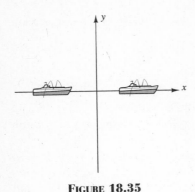

FIGURE 18.35

Solution   A hyperbola is the set of points whose difference in position from two fixed points is a constant. Sound travels in air approximately 1100 ft/s, so, wherever the thunder originated, the location must be farther from one person than the other. Since one person hears the sound three seconds later than the other, there is a known distance—the distance between the two people. This distance is (3s) (1100 ft/s) or 3300 ft. We have two fixed points: the location of the two people and a constant distance of 3300 ft. The lighting bolt must be located on a branch of a hyperbola.

Let the transverse axis of the hyperbola pass through the positions of the two people and the center of the hyperbola lie midway between them, as shown in Figure 18.35. By taking the origin of an axis system at the center of the hyperbola, we can use the standard form of the hyperbolic equation with $(h, k) = (0, 0)$. Let the people, who are 2.0 mi apart, be at the foci of the hyperbola. With the people at the foci, the distance between them is $2c$. Therefore,

$$2 \, c = 2 \text{ mi} \left(\frac{5280}{\text{mi}} \right) = 10{,}560 \text{ ft} \quad \text{or} \quad c = 5280 \text{ ft.}$$

From the definition of the hyperbola, we know that the constant difference is $2a$; therefore,

$$2a = 3300 \text{ ft} \quad \text{or} \quad a = 1650 \text{ ft.}$$

For a hyperbola, $c^2 = a^2 + b^2$ and:

$$b^2 = (5280)^2 - (1650)^2 = (5020)^2.$$

Therefore, the possible locations of the lightning lie on the branch of the hyperbola:

$$\frac{x^2}{(1650)^2} - \frac{y^2}{(5020)^2} = 1,$$

which is closest to the fisherman who heard the sound first. The graph of the hyperbola is shown in Figure 18.36.

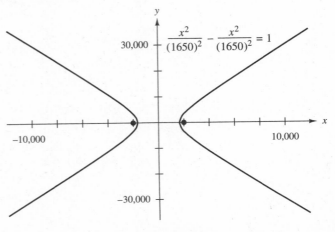

FIGURE 18.36

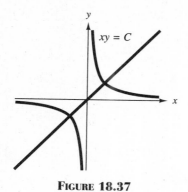

FIGURE 18.37

A function of the form $y = \dfrac{C}{x}$ is an inverse function. The graph of the function is shown in Figure 18.37. The shape of the graph is hyperbolic, with the transverse axis along the line $y = x$. The curve is asymptotic to both the x- and the y-axes. The equation of the hyperbola is generally expressed in the form:

$$xy = C.$$

This particular format is used extensively in all parts of physics, chemistry, and engineering.

The hyperbola $xy = -C$ has its transverse axis along the line $y = -x$.

EXAMPLE 6 Coordinates of Vertices

Determine the coordinates of the vertices of the hyperbola $xy = C$.

Solution   Since the vertices of a hyperbola lie on its transverse axis and the transverse axis of $xy = C$ is the line $y = x$, solving the two equations simultaneously produces the coordinates of the vertices.

$xy = C$	**Given**	(4)
$y = x$	**Given**	(5)
$xx = C$	**Substituting for y in Equation (4)**	
$x^2 = C$		
$x = \pm\sqrt{C}$		
$y = \pm\sqrt{C}$	**Substituting for x in Equation (5)**	

Thus, the coordinates of the vertices of the hyperbola $xy = C$ are $(\sqrt{C}, \sqrt{C})$ and $(-\sqrt{C}, -\sqrt{C})$. ◆

DEFINITION 18.18

The equation $xy = C$ represents a **hyperbola** with its transverse axis on the line $y = x$, vertices $(\sqrt{C}, \sqrt{C})$ and $(-\sqrt{C}, -\sqrt{C})$, and the x- and y-axes are the asymptotes.

EXAMPLE 7 Application: Current Versus Resistance Curve

The current I through a resistance R with a voltage drop of 9 V is given by $IR = 9$. Sketch the current versus resistance curve.

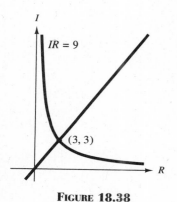

FIGURE 18.38

Solution $IR = 9$ is a hyperbolic equation of the form $xy = C$. Its transverse axis lies on the line $I = R$, and it has the R-axis and the I-axis as asymptotes. Its vertex is at $(3, 3)$. There is no negative portion of the curve because a resistance cannot be negative. The curve is sketched in Figure 18.38. ◆

The key information about the hyperbola with centers at $(0, 0)$ and (h, k) is given in the two boxes.

STANDARD FORM OF EQUATIONS FOR HYPERBOLA WITH CENTER AT ORIGIN (WHERE $c^2 = a^2 + b^2$)

Transverse Axis	Center	Vertices	Foci	Equation
Horizontal	$(0, 0)$	$\begin{cases} (-a, 0) \\ (a, 0) \end{cases}$	$\begin{cases} (-c, 0) \\ (c, 0) \end{cases}$	$\dfrac{x^2}{a^2} - \dfrac{y^2}{b^2} = 1$
Vertical	$(0, 0)$	$\begin{cases} (0, -a) \\ (0, a) \end{cases}$	$\begin{cases} (0, -c) \\ (0, c) \end{cases}$	$\dfrac{y^2}{a^2} - \dfrac{x^2}{b^2} = 1$

STANDARD FORM OF EQUATIONS FOR HYPERBOLA WITH CENTER AT (h, k) (WHERE $c^2 = a^2 + b^2$)

Transverse Axis	Center	Vertices	Foci	Equation
Horizontal	(h, k)	$\begin{cases} (h - a, k) \\ (h + a, k) \end{cases}$	$\begin{cases} (h - c, k) \\ (h + c, k) \end{cases}$	$\dfrac{(x - h)^2}{a^2} - \dfrac{(y - k)^2}{b^2} = 1$
Vertical	(h, k)	$\begin{cases} (h, k - a) \\ (h, k + a) \end{cases}$	$\begin{cases} (h, k - c) \\ (h, k + c) \end{cases}$	$\dfrac{(y - k)^2}{a^2} - \dfrac{(x - h)^2}{b^2} = 1$

18.4 Exercises

In Exercises 1–2, sketch the two curves on the same axis system. Examine the curves and discuss the difference between the curves and what changes in the equations caused the differences.

1. $\dfrac{x^2}{9} - \dfrac{y^2}{4} = 1$

2. $\dfrac{x^2}{4} - \dfrac{y^2}{9} = 1$

In Exercises 3–4, sketch the two curves on the same axis system. Examine the curves and discuss the difference between the curves and what changes in the equations caused the differences.

3. $\dfrac{y^2}{25} - \dfrac{x^2}{4} = 1$

4. $\dfrac{y^2}{25} - \dfrac{x^2}{24} = 1$

In Exercises 5–16, determine the coordinates of the centers, vertices, foci, and the equations of the asymptotes. Then sketch the hyperbola.

5. $\dfrac{x^2}{4} - \dfrac{y^2}{9} = 1$

6. $\dfrac{x^2}{16} - \dfrac{y^2}{9} = 1$

7. $\dfrac{y^2}{4} - \dfrac{x^2}{9} = 1$

8. $\dfrac{y^2}{25} - \dfrac{x^2}{36} = 1$

9. $4x^2 - y^2 = 16$

10. $x^2 - 9y^2 = 9$

11. $4y^2 - 25x^2 = 100$

12. $3y^2 - 2x^2 = 24$

13. $\dfrac{(x - 2)^2}{9} - \dfrac{(y - 3)^2}{16} = 1$

14. $\dfrac{(x + 1)}{36} - \dfrac{(y - 2)^2}{4} = 1$

15. $\dfrac{(y + 1)^2}{16} - \dfrac{(x + 3)^2}{25} = 1$

16. $\dfrac{(y - 2)^2}{4} - (x - 1)^2 = 1$

In Exercises 17–24, write the standard equation for the hyperbola shown.

17.

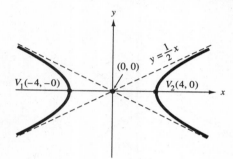

18.

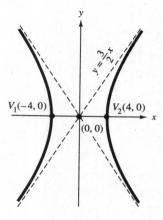

19.

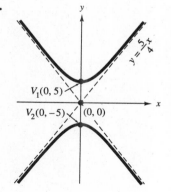

20.

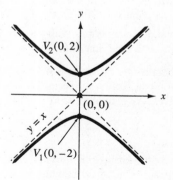

21.

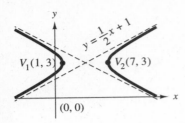

22.

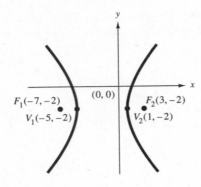

23.

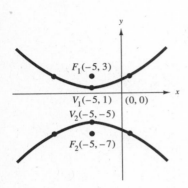

24.

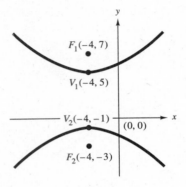

25. Determine the path of a point that moves so that the difference of its distances from the points $(-5, 0)$ and $(5, 0)$ is 8.

26. Determine the path of a point that moves so that the difference of its distances from the points $(0, -13)$ and $(0, 13)$ is 10.

27. Write the equation of the hyperbola with vertices at $(2, -2)$, $(-4, -2)$ and that passes through the point with coordinates $(5, 1)$

28. Write the equation of the hyperbola with vertices at $(-3, 1)$, $(-3, 3)$ and that passes through the point with coordinates $(0, 4)$.

A rectangular hyperbola is one for which $a = b$ and the asymptotes, therefore, are perpendicular to each other. Sketch the rectangular hyperbolas in Exercises 29 and 30 and identify the vertices, the foci, and the asymptotes.

29. $(x + 1)^2 - (y - 2)^2 = 1$

30. $\dfrac{(x - 3)^2}{4} - \dfrac{(y + 1)^2}{4} = 1$

31. Two seismographs 400 km apart record the tremors from an earthquake at 14-50-48 and 14-51-18 (read 14 hours 50 minutes 48 seconds and 14 hours 51 minutes 18 seconds). The tremors travel through the earth at an average speed of approximately 5000 m/s. What is the locus of points where the quake might have occurred? (*Hint:* See Example 5.)

32. The velocity of sound in seawater is approximately 1530. m/s. Two sonar stations on either side of a harbor are 9.40 km apart and pick up the sound of a submarine 4.73 seconds apart.
    a. Where are the possible locations of the submarine with respect to the stations?
    b. Indicate how the location of the submarine can be pinpointed if a third station also picks up the sound of the submarine.

33. The roof of a beach cabana at a resort hotel is to be shaped like part of a hyperbola. The joist (a beam in a ceiling) of the structure is to be at the latus rectum (a line segment through the focus and perpendicular to the transverse axis) and is to be 8 ft above the beach and 20 ft long. The height of the roof above the joist is to be 3 ft to allow room for installation of ceiling fans. (See the drawing.) Determine the equation of the hyperbola so that the curved rafters can be prefabricated.

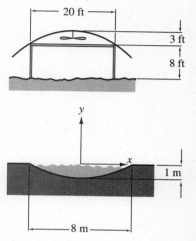

34. A roof is to be shaped like the one in Exercise 33 with a joist length of 10 m and a distance from joist to peak of 2 m. If the origin of the axis system used in this case is to be located at the center of the joist, what is the equation of the hyperbola that describes the rafter shape?

35. The cross-section of a reflecting pool is to be dug in the shape of a hyperbola in an experiment to try to minimize ripple formation. The pool will be 8 m across by 1 m deep with its focus at ground level in its center. Locate the origin of an axis system at ground level at the center of the pool and determine the necessary depths at 1-m intervals across the pool. (See the diagram.)

36. Two Loran-C stations broadcast a signal at exactly the same time. Show that the locus of positions that receive the signals t seconds apart is a hyperbola. (The stations can be represented by two points that are a fixed distance apart.)

18.5

Translation of Axes

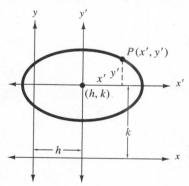

FIGURE 18.39

In this chapter the equations for the parabola, ellipse, and hyperbola were developed with the center or the vertex at the origin. Then, in each case, a standard form of the equation with center or vertex at the point (h, k), where (h, k) could be any point on the plane were defined. This section examines how this second set of equations can be simplified. Take, for example, the equation $\dfrac{(x-3)^2}{16} + \dfrac{(y-4)^2}{9} = 1$. This is an equation of an ellipse with center $(3, 4)$, $a = 4$ and $b = 3$, as shown in Figure 18.39. This graph shows lines through the center of the ellipse and parallel to the x- and y-axes, called the x'- and y'-axes, respectively. With this new axis system, the values for a and b do not change. The only change is that (h, k) becomes $(0, 0)$ with the x'-, y'-axis system. The equation for the x'-, y'-axis system is:

$$\frac{(x')^2}{16} + \frac{(y')^2}{9} = 1.$$

In relation to the new axis system (x', y') we can locate any point $P'(x', y')$ on the ellipse. If we express the point in terms of xy-axis system, we have:

$$x = h + x' \quad \text{and} \quad y = k + y'.$$

Solving the equations for x' and y', we have:

$$x' = x - h \quad \text{and} \quad y' = y - k.$$

With these formulas we can change any equation to a form where the center or vertex is at the origin, or we can relocate the center or the vertex to any point on the plane. This process is known as translation of axes.

RULE 18.1 Translation of Axes

$$x = x' + h \quad \text{or} \quad x' = x - h$$
$$y = y' + k \quad \text{or} \quad y' = y - k$$

CAUTION   When sketching the curve, the center, or the vertex, (h, k) is the origin of the x', y'-system. For example, constructing the figure for the parabola $(y + 1)^2 = 6(x - 2)$ with a translation of axes, the point $(2, -1)$ becomes the origin of the x', y'-system. ∎

EXAMPLE 1   **Parabola Translation of Axes**

In Section 18.2 Example 4, we sketched the parabola $(y + 1)^2 = 6(x - 2)$. Use axis translation to simplify the form of the equation of the parabola.

Solution   Figure 18.40 shows the parabola $(y + 1)^2 = 6(x - 2)$ with its vertex and focus shown in parentheses as it was sketched in Figure 18.12. To

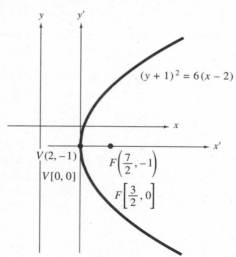

FIGURE 18.40

simplify the equation, choose an x', y'-axis system such that its origin falls at the vertex of the parabola, point (h, k). The translation equations are:

$$x = x' + 2 \quad \text{and} \quad y = y' + (-1).$$

By substitution, the equation of the parabola becomes:

$$(y' - 1 + 1)^2 = 6(x' + 2 - 2)$$
$$y'^2 = 6x',$$

which is the equation of the parabola in the x', y'-system. We can find the vertex and focus using the methods of Section 18.2. Or, since in this case they are already given in the x, y-system, we can calculate them directly using the translation equations. (Brackets are not normally used to indicate ordered pairs. However, for the sake of clarity, the points in the x', y'-system are shown in brackets: $[x', y']$. Using the translation equations, the point $(2, -1)$ becomes $[0, 0]$ as shown by the substitution.

$$2 = x' + 2 \quad \text{and} \quad -1 = y' - 1$$
$$x' = 0 \qquad\qquad y' = 0$$

For the focus, $\dfrac{7}{2} = x' + 2$ and $-1 = y' - 1$, so $\left(\dfrac{7}{2}, -1\right)$ translates to $\left[\dfrac{3}{2}, 0\right]$. The x', y'-axes are shown in Figure 18.40, and the appropriate points are indicated. ◆

EXAMPLE 2   Ellipse Axis Translation

Use axis translation to justify the standard form of the equation of the ellipse as $\dfrac{(x - h)^2}{a^2} + \dfrac{(y - k)^2}{b^2} = 1$ after the equation was derived from the ellipse with center at the origin as $\dfrac{x^2}{a^2} + \dfrac{y^2}{b^2} = 1$.

Solution   The equation of the ellipse shown in Figure 18.41 can be written as $\dfrac{x'^2}{a^2} + \dfrac{y'^2}{b^2} = 1$. From Rule 18.1 we can write $x = x' + h$ and $y = y' + k$, which we can rearrange to $x' = x - h$ and $y' = y - k$.

$$\frac{x'^2}{a^2} + \frac{y'^2}{b^2} = 1 \qquad \text{Equation of ellipse in } x', y'\text{-system}$$

$$\frac{(x - h)^2}{a^2} + \frac{(y - k)^2}{b^2} = 1 \qquad \begin{array}{l}\text{Substituting translation}\\\text{relations for } x' \text{ and } y'\end{array}$$

Therefore, we can justify going from:

$$\frac{x'^2}{a^2} + \frac{y'^2}{b^2} = 1 \quad \text{to} \quad \frac{(x - h)^2}{a^2} + \frac{(y - k)^2}{b^2} = 1$$

when the center of the ellipse is shifted from $(0, 0)$ to (h, k) by using axis translation. ◆

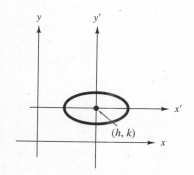

FIGURE 18.41

EXAMPLE 3   General Form to Standard Form

Use a translation of coordinates to transform $9x^2 - 4y^2 + 36x + 8y - 4 = 0$ into one of the standard equations of a conic. Identify the curve and graph the conic.

Solution   First write the conic in a standard form using the method of completing the square.

$$9x^2 + 36x - 4y^2 + 8y = 4$$
$$9(x^2 + 4x + 4) - 4(y^2 - 2y + 1) = 4 + 36 - 4 \quad \text{Completing the square}$$
$$9(x + 2)^2 - 4(y - 1)^2 = 36 \quad \text{Factoring}$$
$$\frac{(x + 2)^2}{4} - \frac{(y - 1)^2}{9} = 1 \quad \text{Dividing each term by 36}$$

Using Rule 18.1 the following substitutions can be made:

$$x' = x + 2 \quad \text{and} \quad y' = y + 1.$$

These equations tell us that the origin of the x', y'-axis system is located at the point $(-2, 1)$. The equation in translated form is:

$$\frac{x'^2}{4} - \frac{y'^2}{9} = 1.$$

This is the equation of a hyperbola with transverse axis coinciding with the x'-axis. Knowing that $a = 2$ and $b = 3$, we can sketch the asymptotes using the asymptote rectangle and then sketch the hyperbola, as shown in Figure 18.42. ◆

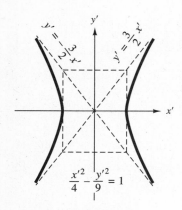

FIGURE 18.42

EXAMPLE 4   Application: AC Voltage

The voltage waveform produced by an AC voltage superimposed on a DC voltage is given by the function $V = 9 + 4 \sin (5 \times 10^4 t)$. Translate the axis system.

Solution   The function $V = 9 + 4 \sin (5 \times 10^4 t)$ is sketched on the tV-axis system in Figure 18.43 (shown on the next page). If we place the $t'V'$-system as shown with its origin at $(0, 9)$ of the tV-system, we see that the function becomes a pure sinusoid with respect to the $t'V'$-system. To solve the equation in terms of V' and t', use Rule 18.1. Then $t = t' + 0$ and $V = V' + 9$. Substituting yields:

$$V' + 9 = 9 + 4 \sin [5 \times 10^4 (t' + 0)]$$
$$V' = 4 \sin (5 \times 10^4 t'),$$

which is the equation of the function $V = 9 + 4 \sin (5 \times 10^4 t)$ when translated from the tV-axis to the $t'V'$-axis. ◆

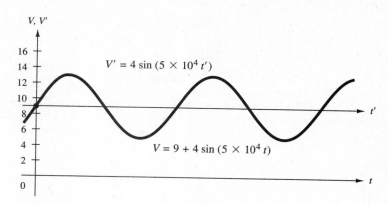

FIGURE 18.43

18.5 Exercises

In Exercises 1–4, use axis translation to rewrite the equation of the lines so that the lines pass through the origin of the translated system. (A sketch of each line may help simplify finding the solution.)

1. $y = x + 2$

2. $y = \frac{1}{2}x - 3$

3. $-y + 3t - 1 = 0$

4. $0.3x + 0.4y - 1.2 = 0$

In Exercises 5–8, use axis translation to rewrite the equations of the lines in terms of a parallel system with the origin of the $x'y'$-system at the indicated point.

5. $y = -2x$; $(3, 1)$

6. $x = 4y$; $(3, -1)$

7. $y = -\frac{1}{2}x - 2$; $(-1, 3)$

8. $2y - x = 4$; $(-3, -1)$

In Exercises 9–20, rewrite the equation of the given figure for a parallel x', y'-axis system with its origin at the center of the figure (vertex in the case of a parabola). Give the coordinates of foci and vertices where appropriate.

9. $(x - 2)^2 + (y - 4)^2 = 16$

10. $(x + 1)^2 + (y - 2)^2 = 4$

11. $x^2 = -2(y - 3)$

12. $(x + 1)^2 - (y - 2)^2 = 1$

13. $\dfrac{(x + 3)^2}{25} + \dfrac{(y + 1)^2}{9} = 1$

14. $\dfrac{(x - 2.41)^2}{14.2} + \dfrac{(y + 3.51)^2}{27.2} = 1$

15. $(y - 2)^2 = -\dfrac{1}{2}(x - 2)$

16. $\dfrac{(x - 3)^2}{4} - \dfrac{(y - 1)^2}{4} = 1$

17. $x^2 + 8x + 8y = 0$

18. $x^2 + y^2 - 8x - 6y = 0$

19. $4x^2 + 9y^2 - 16x - 36y + 16 = 0$

20. $-9x^2 + 16y^2 - 72x - 96y - 144 = 0$

In Exercises 21–28, rewrite the equation of the figure for a translated x', y'-axis system with its origin at the indicated point. Give the coordinates in the x', y'-system for the center of each figure (vertex in the case of a parabola).

21. $(x - 2)^2 + (y - 4)^2 = 16$; $(1, 1)$

22. $(x + 1)^2 + (y - 2)^2 = 4$; $(3, 1)$

23. $x^2 = -2(y - 3)$; $(0, 1)$

24. $(y - 2)^2 = -\dfrac{1}{2}(x - 2)$; $(-2, 3)$

25. $\dfrac{(x + 3)^2}{25} + \dfrac{(y + 1)^2}{9} = 1$; $(-1, -2)$

26. $\dfrac{(x - 2.41)^2}{14.2} + \dfrac{(y + 3.51)^2}{27.2} = 1$; $(0.50, -1.00)$

27. $(x + 1)^2 - (y - 2)^2 = 1$; $(4, -1)$

28. $\dfrac{(x - 3)^2}{4} - \dfrac{(y + 1)^2}{4} = 1$; $(3, 0)$

29. A suspension bridge with a parabolic cable has supports 400 ft apart and a maximum center sag of 50 ft. Write the equation for the cable with respect to an axis system at the right-hand support by first writing the equation of the curve for an axis system with its origin at the point of maximum sag and then using axis translation.

30. The cross-section of a ship's deck is a parabola. The deck is 20 ft wide, and the center of the deck is 6 in. above the sides. (See the drawing.) Write the equation of the parabolic section with respect to an axis at the port (left) side of the deck by writing the equation for an origin at the center of the deck and then using axis translation.

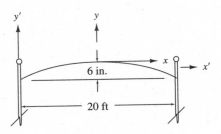

31. Refer to Example 5 in Section 18.4. Assuming that the center of the axis system is located at one of the boats, use axis translation to write the equation of possible lightning locations.

Review Exercises

In Exercises 1–4, use the given information to write the equation of the circle.

1. Center $(-3, 4)$; $r = 8$ cm

2. Center $(-4, -5)$; $r = 6$ cm

3. Center $(4, -7)$; $r = 5$ cm

4. Center $(5, 6)$; $r = 7$ cm

In Exercises 5–12, write the standard form of the equation that represents the geometric figure.

5.

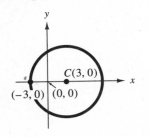

6.

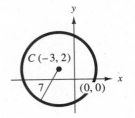

7.

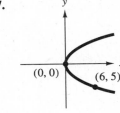

8.

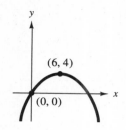

9.

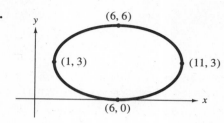

10.

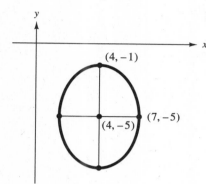

11.

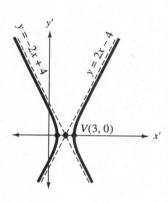

12.

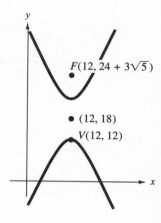

In Exercises 13–16, sketch the lines.

13. $x = 4$

14. $y = -2$

15. $y = -2x + 3$

16. $4x + 3y + 2 = 0$

17. Determine the equation of the circle that has the endpoints of its diameter at $A(4, -3)$ and $B(-2, 7)$.

18. Determine the equation of the circle that is tangent to both axes, has its center in the first quadrant, and has a radius of 3.

19. The blade of a saw that will cut an 80 mm diameter hole is centered 20 mm to the right and 30 mm above the lower left corner of a rectangular plate (see the diagram). Where will the saw cut across the edge of the plate?

20. Where is it necessary to center the saw that will cut an 80 mm hole so that it will cut the edges of a rectangular plate 10 mm above and 20 mm to the right of the lower left corner of the plate?

21. Three observers stationed at $(1, 4)$, $(3, -2)$ and $(-2, -4)$ all see the flash of an explosion at the same time. How far away is the explosion, and where is it located?

22. What is the equation of a $\dfrac{1}{2}$ m diameter circle that is tangent to the line $y - 2x = 0$ at the origin?

In Exercises 23–26, sketch the circles showing their centers and radii.

23. $x^2 + y^2 = 2.25$

24. $(x - 1)^2 + (y - 3)^2 = 16$

25. $x^2 + y^2 - 4x - 8y + 4 = 0$

26. $x^2 + y^2 + 2x - 4y + 1 = 0$

In Exercises 27–30, sketch the parabolas and locate the vertex and focus for each.

27. $y^2 = 12x$

28. $x^2 = -3(y + 2)$

29. $-3x + y^2 + 2y + 1 = 0$

30. $0.5x^2 + 2.0x + 1.5y + 1.0 = 0$

31. What is the equation of a parabolic suspension bridge cable with 500 m between supports and with a sag of 75 m? Locate the axis system at the low point of the cable.

32. Write the equation for the cable of Exercise 31 using the point of attachment of the left support for the origin of the axis system.

33. A parabolic reflector is to be 80 mm across at the focus, and the vertex is to be 20 mm to the left of the focus. What is the equation of the shape of the reflector if the origin of the axis system is located at the focus?

34. A constant-stress cantilever beam has its vertical longitudinal section shaped like a parabola (see the sketch). What is the equation of the beam?

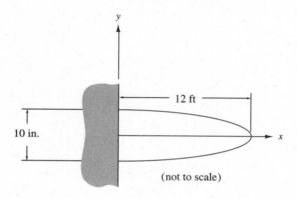

(not to scale)

35. Determine the equation of the ellipse with $F_1(-3, -1)$, $F_2(-3, 1)$ and length of major axis 10 units.

36. Determine the equation of the ellipse with $V_1(-2, 0)$, $V_2(-2, 26)$ and length of minor axis 10 units.

37. Determine the equation of the ellipse with a focus at $(-3, -1)$, its major axis vertical, and one end of its minor axis at $(0, 3)$.

38. Determine the equation of the path of a point that moves so that the sum of its distances from the points $(-4, -1)$ and $(-4, 5)$ is 10.

In Exercises 39–42, sketch the ellipses and locate the center, the foci, and the vertices for each.

39. $\dfrac{(x + 1)^2}{4} + \dfrac{y^2}{16} = 1$

40. $\dfrac{(x - 2)^2}{25} + \dfrac{(y + 2)^2}{9} = 1$

41. $x^2 + 4y^2 + 6x + 8y + 9 = 0$

42. $9x^2 + y^2 - 36x + 8y + 43 = 0$

43. The underside of a masonry bridge is shaped like an ellipse to provide a constant stress distribution. Write the equation for the ellipse shown in the illustration.

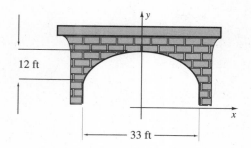

12 ft

33 ft

44. The elliptical orbit of a satellite is 6000 km wide at its widest and 2000 km wide at its narrowest part. What is an equation for the orbit if the origin is located at the center of the ellipse?

45. What is the equation for the orbit of Exercise 44 if the origin of the axis system is located at one of the foci of the ellipse?

46. A vent pipe, 4 in. in diameter, is extended straight up through the roof. The roof has a slope of 45°. What are the dimensions of the elliptical hole that must be cut in the roof for the pipe? What is the equation that defines the elliptical hole in the roof?

In Exercises 47–50, determine the coordinates of the center, the foci, the vertices, and the equations of the asymptotes for each hyperbola. Sketch the hyperbola.

47. $\dfrac{(x-3)^2}{9} - \dfrac{(y-2)^2}{16} = 1$

48. $\dfrac{y^2}{16} - \dfrac{(x-2)^2}{25} = 1$

49. $-4x^2 + y^2 + 8x - 4y - 4 = 0$

50. $x^2 - 9y^2 + 2x + 18y - 44 = 0$

51. Write the equation of the hyperbola whose foci are at $(0, 0)$ and $(0, 10)$ and whose vertices are at $(0, 2)$ and $(0, 8)$.

52. Write the equation of the hyperbola whose foci are at $(2, -1)$ and $(12, -1)$ and whose vertices are at $(5, -1)$ and $(9, -1)$.

In Exercises 53–56, rewrite the equation of the given figure for a parallel x', y'-axis system with its origin at the center of the figure (vertex in the case of a parabola). Give the coordinates of the foci and vertices where appropriate.

53. $(y-3)^2 = 12(x+1)$

54. $(x-1)^2 + (y-3)^2 = 16$

55. $x^2 + y^2 + 2x - 4y + 1 = 0$

56. $4x^2 - y^2 - 8x + 4y + 4 = 0$

57. Boyle's law states that for ideal gases at a constant temperature, the product of the pressure and the volume is a constant. Identify and sketch the graph of $pV = 2.76 \times 10^7$.

*58. The asymptotes of the hyperbola $\dfrac{x^2}{a^2} - \dfrac{y^2}{b^2} = 1$ are the lines $y = \pm \dfrac{b}{a}x$. Write a computer program to show that as x becomes larger, the difference between asymptote and hyperbola becomes smaller.

*59. Does $x^2 + y^2 + 1x + my + n = 0$ always represent a circle? Justify your answer.

*60. An equation of the form $Ax^2 + Bxy + Cy^2 + Dx + Ey + F = 0$ can be transformed to a form without the x, y-term by rotating the axis system counterclockwise by an angle θ. The required angle of rotation is found by using $\theta = \dfrac{1}{2}\,\text{Arctan}\,\dfrac{B}{A-C}$ and the transformation equations $x = x'\cos\theta - y'\sin\theta$ and $y = x'\sin\theta + y'\cos\theta$. Use this information to show that the equation $xy = C$ represents a hyperbola.

*61. Use the equations from Exercise 60 to write the equation $5x^2 - 6xy + 5y^2 = 8$ without an xy-term. Identify the conic section and sketch it.

✎ Writing About Mathematics

1. Compare the equations and figures of all the conic sections in this chapter. Discuss what elements are common to the conic sections and identify those elements that cause differences in the conic sections.

2. Your assignment in an astronomy class is to discuss the elliptical paths of the planets that orbit about the sun. Explain how the shape of the path can be determined if one knows eccentricity of the elliptical path. Show how the eccentricity can be determined for several planets.

3. Because of their reflective power hyperbolas are important in constructing lenses. In fact, more complex lenses are constructed using both parabolic- and hyperbolic-shaped lenses. Do research and write a short paper on how the parabola and the hyperbola are used in lense construction.

4. Do research and write a short paper on the development of the lithotripter.

5. You are visiting the Capitol in Washington, D.C., with your younger brother or sister who cannot understand how the whispering gallery works. Write a short paper with diagrams explaining how the elliptical shape of the room produces this effect.

Chapter Test

1. Write the equation of the circle with center $(-4, 3)$ and radius of 7 in.

2. For the parabola $x^2 + 4x + 4y - 4 = 0$, perform the following.
   a. Determine coordinates of the vertex.
   b. Determine coordinates of the focus.
   c. Determine end points of the latus rectum.
   d. Determine the axis of symmetry.
   e. Sketch the curve and label it.

3. For the ellipse $\dfrac{(x-3)^2}{64} + \dfrac{(y-1)^2}{36} = 1$, perform the following.
   a. Determine the coordinates of the center.
   b. Determine the coordinates of the foci.
   c. Determine the coordinates of the vertices.
   d. Determine the coordinates of the end points of the semiminor axis.
   e. Sketch the ellipse and label it.

4. For the hyperbola $\dfrac{y^2}{16} - \dfrac{x^2}{9} = 1$, perform the following.

    a. Determine the coordinates of the center.
    b. Determine the coordinates of the foci.
    c. Determine the coordinates of the vertices.
    d. Determine the equations of the asymptotes.
    e. Sketch the hyperbola and label it.

In Exercises 5 and 6, rewrite the equation of the given figure for a parallel x', y'-axis system with its origin at the center of the figure (the vertex in case of a parabola). Give the coordinates of the foci and vertices where appropriate.

5. $(x - 2)^2 - (y - 5)^2 = 25$ **6.** $x^2 - 8x - 4y + 28 = 0$

7. A fireplace is designed with a semielliptical arch. If the arch has a span of 3.0 ft and a height of 1.0 ft, where should the foci be placed in order to sketch the curve? (See the diagram.)

8. Determine the path of a point that moves so that the difference of its distances from the points $(-2, 0)$ and $(2, 0)$ is 3.

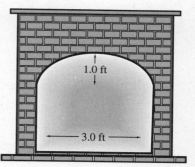

In Exercises 9 and 10, use your graphics calculator to sketch the two curves on the same axis system. Examine the curves and discuss the differences between the two curves and what changes in the equations caused the differences.

9. $\dfrac{x^2}{9} + \dfrac{y^2}{4} = 1$ **10.** $\dfrac{x^2}{16} + \dfrac{y^2}{4} = 1$

In Exercises 11 and 12, use your graphics calculator to sketch the two curves on the same axis system. Examine the curves and discuss the differences between the two curves and what changes in the equations caused the differences.

11. $\dfrac{x^2}{9} - \dfrac{y^2}{4} = 1$ **12.** $\dfrac{x^2}{16} - \dfrac{y^2}{4} = 1$

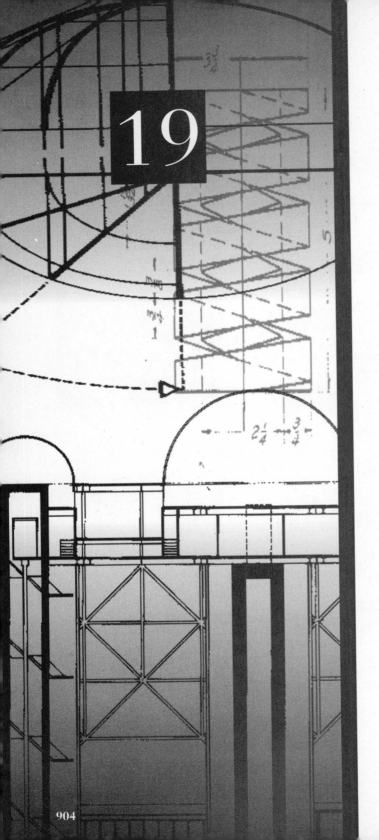

The Derivative

O n her way to work a carpenter notices a small crack in the corner of her windshield. To temporarily save the windshield she uses a battery-powered drill to make a small hole at the end of the crack.

The technique works because the small hole she drills at the end of the crack relieves some of the stress on the glass. Example 9 of Section 19.6 shows that, as the width of the crack is changed, there is a change in the stress concentration.

19.1

Limits and Continuity

The branch of mathematics called calculus is divided into two general areas, differential calculus and integral calculus. We use differential calculus to find the speed of a moving object at a particular instant in time. We use integral calculus to find, among other things, the area of an irregularly shaped region, the volume of a solid or the length of a curve. The processes of differentiation and integration are related in a manner similar to those of multiplication and division. The one is the inverse process of the other.

But before launching into a study of calculus, its a good idea to review the basics of functions. For a review of functions refer back to Chapter 5.

Limit

The concept of limit is the cornerstone of calculus. Both differential and integral calculus are defined in terms of limits.

To develop an intuitive idea of limit, let us examine the values of the function $f(x) = 2x - 5$ as x approaches 3. Table 19.1 shows the values of $x < 3$, and Table 19.2 shows the values of $x > 3$.

TABLE 19.1 $f(x) = 2x - 5$, $x < 3$

x	$f(x)$
2	-1
2.5	0
2.9	0.8
2.99	0.98
2.999	0.998
2.9999	0.9998
2.99999	0.99998
2.999999	0.999998

TABLE 19.2 $f(x) = 2x - 5$, $x > 3$

x	$f(x)$
4	3
3.5	2
3.1	1.2
3.01	1.02
3.001	1.002
3.0001	1.0002
3.00001	1.00002
3.000001	1.000002

FIGURE 19.1

In Table 19.1 we select an initial value of x ($x = 2$) that is smaller than 3 (to the left of 3 on the x-axis). We select additional values of x that are closer and closer to 3. This shows us that as x approaches 3 from the left, the value of $f(x)$ approaches 1. In a similar manner we select an initial value to the right of 3 ($x = 4$). Table 19.2 shows that as x approaches 3 from the right, the value of $f(x)$ approaches 1. The graph in Figure 19.1 illustrates that as x approaches 3 from either side, $f(x)$ approaches 1. That is, the value of the function, as x gets closer and closer to 3 from either side, is 1. In Figure 19.1 the open circle at (3, 1) indicates that we are examining values of the function when x is close to 3, but not for $x = 3$. Another way of saying this is *"the limit of the function, as x approaches 3, is 1."* Symbolically, we write:

$$\lim_{x \to 3} (2x - 5) = 1.$$

TABLE 19.3 $f(x) = \dfrac{x^2 - 4}{x - 2}$, $x < 2$

x	$f(x)$
1	3
1.5	3.5
1.9	3.9
1.99	3.99
1.999	3.999
1.9999	3.9999
1.99999	3.99999
1.999999	3.999999

TABLE 19.4 $f(x) = \dfrac{x^2 - 4}{x - 2}$, $x > 2$

x	$f(x)$
3	5
2.5	4.5
2.1	4.1
2.01	4.01
2.001	4.001
2.0001	4.0001
2.00001	4.00001
2.000001	4.000001
2.0000001	4.0000001

Why all this fuss since we can see that $f(3) = 1$? Well, the challenge is to determine the value of the function when x is close to 3, not when $x = 3$. For many functions, the function is not defined for a specific value of x, but it is defined for values close to the specific value. This is illustrated in Example 1.

EXAMPLE 1 Limit of a Rational Function

Show how $f(x) = \dfrac{x^2 - 4}{x - 2}$ behaves as x approaches 2.

Solution Now at first glance, it appears that we have a serious mathematical infraction. What happens to the denominator of the polynomial when x approaches 2? As the value of x gets closer and closer to 2, the denominator approaches zero, and division by zero is undefined. When $x = 2$, $f(2) = \dfrac{0}{0}$, which is indeterminate. We are asked to show how the function behaves as x gets closer and closer to 2. To do this, use Table 19.3 for $x < 2$ and Table 19.4 for $x > 2$.

From the tables we can see that as x approaches 2 from the left and the right, $f(x)$ approaches 4. The graph in Figure 19.2 also shows that as x gets closer and closer to 2, $f(x)$ approaches 4. Thus, we say that the limit of the function, as x approaches 2, is 4. This is expressed symbolically as:

$$\lim_{x \to 2} \frac{x^2 - 4}{x - 2} = 4. \; \blacklozenge$$

From these examples we can see that there are times direct substitution works and other times direct substitution will not work. The following guidelines and examples will help determine the proper technique to use to find the limit.

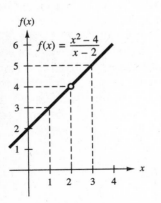

FIGURE 19.2

GUIDELINES FOR FINDING LIMITS

1. The limit as $x \to a$, for all polynomial functions of the following form is $f(a)$.
$$f(x) = a_n x^n + a_{n-1} x^{n-1} + \cdots + a_1 x + a_0$$

2. The limit as $x \to a$ of a rational function:
$$f(x) = \frac{p(x)}{q(x)}$$
is $f(a) = \dfrac{p(a)}{q(a)}$ if $q(a) \neq 0$.

3. If the limit cannot be found by direct substitution, use algebraic techniques to write a function g such that $g(x) = f(x)$ for all x except when $x = a$. This selection of g allows us to evaluate the limit of $g(x)$ by direct substitution.

We determined earlier that the $\lim_{x \to 3} (2x - 5) = 1$. The functional value of $f(x) = 2x - 5$ when $x = 3$ is $f(3) = 1$, which is the limit of the function. (Here we are using Guideline 1.)

The denominator of the rational function $f(x) = \dfrac{x^2 - 4}{x - 2}$ is not equal to zero when $x = 3$. Thus, using Guideline 2, the $\lim_{x \to 3} \dfrac{x^2 - 4}{x - 2}$ is:

$$f(3) = \frac{3^2 - 4}{3 - 2} = 5.$$

The rational function $f(x) = \dfrac{x^2 - 4}{x - 2}$ is not defined when $x = 2$; but, as demonstrated by Table 19.3 and Table 19.4, the limit does exist as x approaches 2. We can determine the limit using algebraic techniques (Guideline 3).

$$\lim_{x \to 2} \frac{x^2 - 4}{x - 2} = \lim_{x \to 2} \frac{(x - 2)(x + 2)}{x - 2} \qquad \textbf{Factoring}$$

$$= \lim_{x \to 2} \frac{\overset{1}{(\cancel{x - 2})}(x + 2)}{\underset{1}{\cancel{x - 2}}} \qquad x \neq 2$$

$$= \lim_{x \to 2} (x + 2)$$

$$= 2 + 2 \qquad \textbf{Substitution}$$

$$= 4$$

We can find the $\lim_{x \to 2} \dfrac{x^2 - 4}{x - 2} = 4$ in this manner as long as we retain the restriction $x \neq 2$.

> **DEFINITION 19.1**
>
> If a function has **limit L as x approaches a,** it can be written symbolically as:
>
> $$\lim_{x \to a} f(x) = L.$$

The illustrations and the definition indicate that a limit exists if $f(x)$ gets closer to a number L as x approaches a from either side. Two points need to be stressed here.

1. The limit is independent of the way in which $x \to a$. In other words, the limit must be the same whether x approaches a from the left or the right (for values of $x < a$ and $x > a$).
2. In finding the $\lim_{x \to a}$, $x \neq a$, x can be as close to a as we like, but x is never equal to a.

For practical purposes, when the function is defined at $x = a$, the difference between $\lim\limits_{x \to a} f(x)$ and $f(a)$ is *so small* that we can find the limit by direct substitution (Guidelines 1 and 2). We can also use the calculator to determine the limit if it exists. This is illustrated in Example 2.

EXAMPLE 2   Limit of a Polynomial Function

Determine $\lim\limits_{x \to 2} (x^2 - 3)$ from the graph of the function.

Solution   The function is graphed in Figure 19.3. The graph of the function suggests that as x approaches 2, $f(x) = (x^2 - 3)$ approaches 1.   ◆

In Example 2, since $f(x)$ is a polynomial function, we can use direct substitution and find:

$$\lim_{x \to 2} (x^2 - 3) = 2^2 - 3$$
$$= 1.$$

This is the result we estimated using the graphing method.

There are times when we cannot determine the limit of the function as x approaches from both sides. For example, consider:

$$\lim_{x \to 3} \sqrt{x - 3}.$$

For values of $x < 3$, $\sqrt{x - 3}$ is not a real number, and the limit does not exist.

$$\lim_{x \to 3^-} \sqrt{x - 3} \qquad \text{does not exist}$$

On the other hand, for values of $x > 3$, the $\sqrt{x - 3}$ is a real number, and:

$$\lim_{x \to 3^+} \sqrt{x - 3} = 0.$$

The small plus and minus superscripts in the statements:

$$\lim_{x \to a^+} \quad \text{and} \quad \lim_{x \to a^-}$$

indicate that the limit is taken from the right and the left sides, respectively. For values of $x > 3$, since the limit exists as x is approached from the right, we say the function has a right-hand limit.

DEFINITION 19.2

A function $f(x)$ has a right-hand limit if $f(x)$ can be made as close to the number L as we please for all values of $x > a$:

$$\lim_{x \to a^+} f(x) = L.$$

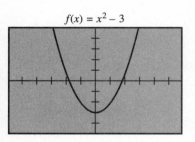

$f(x) = x^2 - 3$

FIGURE 19.3

> **DEFINITION 19.3**
>
> A function $f(x)$ has a left-hand limit if $f(x)$ can be made as close to the number L as we please for all values of $x < a$:
>
> $$\lim_{x \to a^-} f(x) = L.$$

RELATIONSHIP BETWEEN ONE-SIDED AND TWO-SIDED LIMITS

The function has a limit as x approaches a if, and only if, both the left-hand limit and the right-hand limit at a exist and are equal. That is:

$$\lim_{x \to a} f(x) = L \text{ if, and only if, } \lim_{x \to a^-} = L \text{ and } \lim_{x \to a^+} = L.$$

TABLE 19.5 $f(x) = \dfrac{1}{x-1}$, $x < 1$

x	$f(x)$
0	-1
0.5	-2
0.9	-10
0.99	-100
0.999	-1000
0.9999	$-10,000$

TABLE 19.6 $f(x) = \dfrac{1}{x-1}$, $x > 1$

x	$f(x)$
2	1
1.5	2
1.1	10
1.01	100
1.001	1000
1.0001	10,000

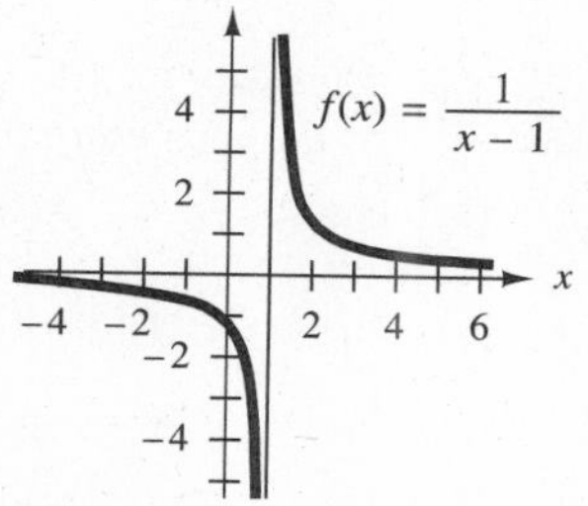

$f(x) = \dfrac{1}{x-1}$

FIGURE 19.4

EXAMPLE 3 **Limit Does not Exist**

Determine the $\lim\limits_{x \to 1} \dfrac{1}{x-1}$.

Solution The function $f(x) = \dfrac{1}{x-1}$ is not defined for $x = 1$. We cannot write a function g, such that $g(x) = f(x)$ when $x \neq 1$. One approach to arrive at a solution is to use two tables. Tables 19.5 and 19.6 are for values of $x < 1$ and $x > 1$, respectively. This tells us what happens to $f(x)$ as x approaches 1.

From Table 19.5 we see that as x approaches 1 from the left, the absolute value of $f(x)$ increases. In fact $f(x)$ decreases without bound, and we write:

$$\lim_{x \to 1^-} \frac{1}{x-1} = -\infty.$$

The minus sign in front of the symbol indicates that the infinitely large value is negative.

From Table 19.6 we see that as x approaches 1 from the right, $f(x)$ becomes a larger and larger positive value. Therefore,

$$\lim_{x \to 1^+} \frac{1}{x-1} = \infty.$$

Since the limit is not the same for $x < 1$ and $x > 1$, the limit does not exist. Symbolically,

$$\lim_{x \to 1} \frac{1}{x-1} \qquad \text{does not exist.}$$

In Figure 19.4 we see that as the curve approaches 1, it does not approach the same functional value on both sides of $x = 1$; therefore, the limit does not exist. ◆

Is it possible to determine a finite value for the function $f(x)$ when the limit is "x approaches infinity"? In Figure 19.4, what values does the function approach as x takes on larger and larger values? Definition 19.5 helps answer this question.

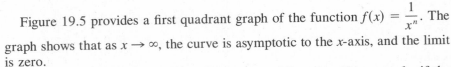

> **DEFINITION 19.4**
>
> $\lim\limits_{x\to\infty} \dfrac{1}{x^n} = 0$, when n is a positive integer.

Figure 19.5 provides a first quadrant graph of the function $f(x) = \dfrac{1}{x^n}$. The graph shows that as $x \to \infty$, the curve is asymptotic to the x-axis, and the limit is zero.

There are properties of limits that make our job easier. For example, if the limits of two functions are known as $x \to a$, then we automatically know the limit of the sum; it is the sum of the limits. Likewise, the limit of the product is the product of the limits. These properties are described in Theorem 19.1.

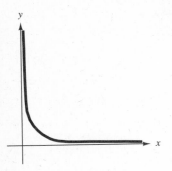

FIGURE 19.5

> **THEOREM 19.1   Properties of Limits**
>
> 1. If f is the **identity function** $f(x) = x$, then for any value a:
> $$\lim_{x\to a} f(x) = \lim_{x\to a} (x) = a.$$
>
> 2. If f is the **constant function** $f(x) = c$ (the function whose functional values have the constant value c), then for any value a:
> $$\lim_{x\to a} f(x) = \lim_{x\to a} (c) = c.$$
>
> 3. If $\lim\limits_{x\to a} f(x) = L$ and $\lim\limits_{x\to a} g(x) = M$, then:
>   a. **Sum rule:** $\lim\limits_{x\to a} [f(x) + g(x)] = L + M$
>   b. **Difference rule:** $\lim\limits_{x\to a} [f(x) - g(x)] = L - M$
>   c. **Product rule:** $\lim\limits_{x\to a} [f(x) \cdot g(x)] = L \cdot M$
>   d. **Constant multiple rule** (for any number c): $\lim\limits_{x\to a} c \cdot f(x) = c \cdot L$
>   e. **Quotient rule** (L and M are real numbers): $\lim\limits_{x\to a} \dfrac{f(x)}{g(x)} = \dfrac{L}{M}$, if
> $M \neq 0$.

Examples 4 through 7 use the definitions and properties of limits to determine the limits of functions.

EXAMPLE 4 **Limit of a Sum**

Determine the $\lim\limits_{x \to 2} (3x^4 + 7)$.

Solution The property states that the limit of the sum is the sum of the limits. Using this property we determine the limit of each piece.

$$\lim\limits_{x \to 2} 3x^4 = 3\lim\limits_{x \to 2} x^4 \qquad \text{Constant multiple rule}$$

$$= 3(\;2\;)^4 \qquad \text{Substitution}$$

$$= 48$$

$$\lim\limits_{x \to 2} 7 = 7 \qquad \text{Constant function}$$

$$\lim\limits_{x \to 2} (3x^4 + 7) = 48 + 7 \qquad \text{Sum rule}$$

$$= 55. \quad \blacklozenge$$

EXAMPLE 5 **Limit of a Quotient, $x \to a$**

Determine the $\lim\limits_{x \to 3} \left(\dfrac{x^2 + 5}{x + 4} \right)$.

Solution

$$\lim\limits_{x \to 3} (x^2 + 5) = 14$$

$$\lim\limits_{x \to 3} (x + 4) = 7$$

Thus,

$$\lim\limits_{x \to 3} \left(\frac{x^2 + 5}{x + 4} \right) = \frac{14}{7}$$

$$= 2. \quad \text{Quotient rule} \quad \blacklozenge$$

There are times when we cannot apply the quotient rule directly. This is illustrated in Examples 6 and 7.

EXAMPLE 6 **Limit of a Quotient, $x \to \infty$**

Determine the $\lim\limits_{x \to \infty} \left(\dfrac{5x}{x + 3} \right)$.

Solution If we try the quotient rule the result is:

$$\frac{\lim\limits_{x \to \infty} 5x}{\lim\limits_{x \to \infty} (x + 3)} = \frac{\infty}{\infty}.$$

This is meaningless.

Let us try another approach—divide the numerator and the denominator by x.

$$\frac{5x}{x+3} = \frac{\dfrac{5x}{x}}{\dfrac{x}{x} + \dfrac{3}{x}}$$

$$= \frac{5}{1 + \dfrac{3}{x}}$$

Now taking the limit of the function in this form, we have:

$$\lim_{x\to\infty} \frac{5}{1 + \dfrac{3}{x}} = \frac{\displaystyle\lim_{x\to\infty} 5}{\displaystyle\lim_{x\to\infty} 1 + \lim_{x\to\infty} \frac{3}{x}} \qquad \textbf{Quotient rule}$$

$$= \frac{5}{1 + 0} \qquad \begin{array}{l}\textbf{Constant multiple rule}\\ \textbf{and Definition 19.4}\end{array}$$

$$= 5.$$

Therefore,

$$\lim_{x\to\infty} \frac{5x}{x+3} = 5. \quad \blacklozenge$$

GUIDELINES FOR DETERMINING LIMITS OF RATIONAL FUNCTIONS

To determine the $\displaystyle\lim_{x\to\infty} \frac{f(x)}{g(x)}$ when it is impossible to simplify the rational function $\dfrac{f(x)}{g(x)}$, perform the following.

1. Divide both numerator and denominator by the variable to the highest degree.
2. Then take the limit.

EXAMPLE 7   Quotient Rule, $x \to \infty$

Determine the $\displaystyle\lim_{x\to\infty} \frac{3x^2 + 2}{x^2 + 1}$.

Solution Using the guidelines, divide both numerator and denominator by x^2 (the variable to the highest degree).

$$\lim_{x \to \infty} \frac{3x^2 + 2}{x^2 + 1} = \lim_{x \to \infty} \frac{\dfrac{3x^2}{x^2} + \dfrac{2}{x^2}}{\dfrac{x^2}{x^2} + \dfrac{1}{x^2}}$$

$$= \lim_{x \to \infty} \frac{3 + \dfrac{2}{x^2}}{1 + \dfrac{1}{x^2}}$$

$$= \frac{\displaystyle\lim_{x \to \infty} 3 + \lim_{x \to \infty} \dfrac{2}{x^2}}{\displaystyle\lim_{x \to \infty} 1 + \lim_{x \to \infty} \dfrac{1}{x^2}} \qquad \textbf{Quotient rule}$$

$$= \frac{3 + 0}{1 + 0} \qquad \textbf{Definition 19.4}$$

$$= 3$$

Therefore,

$$\lim_{x \to \infty} \frac{3x^2 + 2}{x^2 + 1} = 3.$$

(See Figure 19.6) ◆

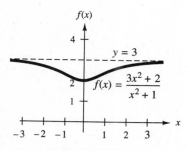

$f(x)$

$y = 3$

$f(x) = \dfrac{3x^2 + 2}{x^2 + 1}$

FIGURE 19.6

The function $f(x) = \dfrac{3x^2 + 2}{x^2 + 1}$ is graphed in Figure 19.6 where we can see that the line $y = 3$ is an asymptote to the curve. We found that the $\lim\limits_{x \to \infty} \dfrac{3x^2 + 2}{x^2 + 1} = 3$.

All **horizontal asymptotes** can be found by finding $\lim\limits_{x \to \infty} f(x)$. If the limit exists, then $y = L$ is the horizontal asymptote.

CAUTION For the rational function $\dfrac{f(x)}{g(x)}$, if the degree of $f(x)$ is greater then the degree of $g(x)$ and the functions cannot be factored, then the technique used in Examples 7 and 8 may not work, and the limit does not exist. For example, show that

$$\lim_{x \to \infty} \frac{2x^2 + x}{3x + 1}$$

does not exist.

$$\lim_{x \to \infty} \frac{\dfrac{2x^2}{x} + \dfrac{x}{x}}{\dfrac{3x}{x} + \dfrac{1}{x}} = \lim_{x \to \infty} \frac{2x + 1}{3 + \dfrac{1}{x}} = \frac{\infty}{3} \qquad \text{does not exist.} \quad \blacksquare$$

Continuity

A computer or graphics calculator draws an unbroken curve for the function $f(x) = -3x^2 + 8x + 4$ in the interval from $[0, 3]$. The path of the curve may be compared to the flight of a baseball, shown in Figure 19.7. This path is smooth with no breaks; thus, we say the curve is *continuous*.

FIGURE 19.7

In the world around us equations normally define numerical data that, if graphed, provide continuous curves. These curves may be the flight path of a space shuttle or the results of a chemical reaction over time. In fact, up until the early 1900s, it seemed that all physical processes behave in a continuous manner. In the 1920s, however, physicists discovered that vibrating atoms in a hydrogen atom can oscillate at discrete energy levels. **Discrete** means that the energy levels are not connected; they are discontinuous.

With these discoveries and the use of discrete functions in computer science and statistics, the concept of continuity has become extremely important. It is important to be able to test for continuity.

DEFINITION 19.5

A function f is said to be **continuous** for $x = a$ if all three of the following conditions are satisfied.

1. The function is defined at $x = a$; that is, $f(a)$ exists.
2. The function approaches a definite limit as x approaches a; that is, $\lim\limits_{x \to a} f(x)$ exists.
3. The limit of the function is equal to the value of the function when $x = a$; that is, $\lim\limits_{x \to a} f(x) = f(a)$.

A function is said to be **discontinuous** at $x = a$ if any one of the three conditions is not satisfied.

EXAMPLE 8 **Continuity of a Linear Function**

Determine whether the function $f(x) = 2x + 4$ is continuous at $x = 2$.

Solution To determine whether the function $f(x) = 2x + 4$ is continuous at $x = 2$ we apply Definition 19.5 and determine whether or not all three properties are satisfied.

a.

$$f(\boxed{2}) = 2(\boxed{2}) + 4 \qquad \textbf{Property 1}$$
$$= 8$$

b.

$$\lim_{x \to 2} f(x) = 2(\boxed{2}) + 4 \qquad \textbf{Property 2}$$
$$= 8$$

c.

$$\lim_{x \to 2} f(x) = f(2) \qquad \textbf{Property 3}$$
$$8 = 8$$

All three properties are satisfied, so the function $f(x) = 2x + 4$ is continuous at $x = 2$, as illustrated in Figure 19.8. ◆

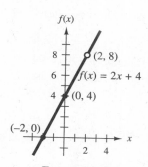

FIGURE 19.8

Continuity is a point-wise property. The ○ on Figure 19.8 indicates the point at which we are determining the continuity of the function.

EXAMPLE 9 Continuity of a Rational Function

Determine whether the function $f(x) = \dfrac{x^2 - 4}{x - 2}$ is continuous at $x = 2$.

Solution The function $f(x) = \dfrac{x^2 - 4}{x - 2}$ is not continuous at $x = 2$ because $f(x)$ is not defined for $x = 2$ (Figure 19.9). However, if we change the original function to a compound function, we have:

$$f(x) = \begin{cases} \dfrac{x^2 - 4}{x - 2} & \text{for } x \neq 2 \\ 4 & \text{for } x = 2. \end{cases}$$

$f(x)$ is now defined at 2. $f(2) = 4$. Also,

$$\lim_{x \to 2} \frac{x^2 - 4}{x - 2} = \lim_{x \to 2} \frac{\overset{1}{\cancel{(x - 2)}}(x + 2)}{\underset{1}{\cancel{(x - 2)}}}$$
$$= \lim_{x \to 2} (x + 2)$$
$$= 4.$$

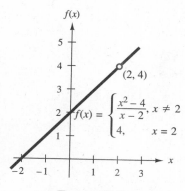

FIGURE 19.9

The compound function

$$f(x) = \begin{cases} \dfrac{x^2 - 4}{x - 2}, & x \neq 2 \\ 4, & x = 2 \end{cases}$$

is continuous at $x = 2$ since all three conditions are met. Note, however, that the original function is not continuous at $x = 2$. ◆

Example 9 showed that if a function is discontinuous, we may be able to change the restrictions and make the function continuous. When this is possible we say that the *discontinuity can be removed.*

EXAMPLE 10   Continuity of a Compound Function

Determine whether the following function is continuous at $x = -5$.

$$f(x) = \begin{cases} \dfrac{x^2 - 25}{x + 5} & \text{for } x \neq -5 \\ 18 & \text{for } x = -5 \end{cases}$$

Solution   Apply Definition 19.5.

a. $f(-5) = 18$ **Property 1**

b. $\displaystyle\lim_{x \to -5} f(x) = \lim_{x \to -5} \dfrac{x^2 - 25}{x + 5}$ **Property 2**

$$= \lim_{x \to -5} \frac{(x + 5)\overset{1}{(x + 5)}}{\underset{1}{x + 5}}$$

$$= -10$$

This function is not continuous at $x = -5$ since

$$f(-5) \neq \lim_{x \to -5} f(x). \quad ◆$$

The discontinuity in Example 11 can be removed if the compound function is defined as follows.

$$f(x) = \begin{cases} \dfrac{x^2 - 25}{x + 5} & \text{for } x \neq -5 \\ -10 & \text{for } x = -5 \end{cases}$$

Test the redefined compound function to determine if it meets the criteria of Definition 19.5 for continuity.

19.1 Exercises

In Exercises 1–10, determine the limits using Example 1 as a guide. Use a calculator to develop the tables of values. Sketch the graph of the function.

1. $\lim\limits_{x \to 4} (3x - 11)$

2. $\lim\limits_{x \to 3} (5x + 11)$

3. $\lim\limits_{x \to 4} \dfrac{x^2 - 16}{x - 4}$

4. $\lim\limits_{x \to -4} \dfrac{x^2 - 16}{x - 4}$

5. $\lim\limits_{x \to -3} \dfrac{x^2 + 5x + 6}{x + 3}$

6. $\lim\limits_{x \to 1} \dfrac{x^2 - 9x + 8}{x - 1}$

7. $\lim\limits_{x \to 4} \dfrac{x - 4}{5(\sqrt{x} - 2)}$

8. $\lim\limits_{x \to -1} \dfrac{4(x^2 - 1)}{x + 1}$

9. $\lim\limits_{x \to 0} \dfrac{\sin x}{x}$

10. $\lim\limits_{x \to 0} \dfrac{\tan x}{x}$

In Exercises 11–30, determine the indicated limits.

11. $\lim\limits_{x \to 4} 5x$

12. $\lim\limits_{x \to 0} (7x - 4)$

13. $\lim\limits_{x \to 3} (2x^2 - 3x + 1)$

14. $\lim\limits_{x \to 2} (4x^2 - 5x + 11)$

15. $\lim\limits_{x \to 3} \dfrac{x}{x + 4}$

16. $\lim\limits_{x \to -3} \dfrac{x + 4}{3}$

17. $\lim\limits_{y \to 1} \dfrac{(y - 1)^2}{y - 1}$

18. $\lim\limits_{x \to -1} \dfrac{y^2 + y}{y + 1}$

19. $\lim\limits_{t \to 0} \dfrac{t^4 + 2t^2 + 3}{t^3 + 1}$

20. $\lim\limits_{t \to 0} \dfrac{t^3 + 3t^2}{t^3 + t}$

21. $\lim\limits_{x \to 2} \dfrac{x^2 - 4}{x^2 - x - 2}$

22. $\lim\limits_{x \to 3} \dfrac{x^2 - 9}{x^2 - x - 6}$

23. $\lim\limits_{x \to 4^+} \sqrt{x - 4}$

24. $\lim\limits_{x \to 5^+} \sqrt{x - 5}$

25. $\lim\limits_{x \to 4^-} 6^{x-4}$

26. $\lim\limits_{x \to 6^-} 4^{x-6}$

27. $\lim\limits_{x \to \infty} \dfrac{3x^2 - 5}{2x^2 - 7x + 6}$

28. $\lim\limits_{x \to \infty} \dfrac{3x^3 + 7x - 1}{3x^3 + 2x^2 - 7}$

29. $\lim\limits_{x \to -\infty} \dfrac{\sqrt{x^2 + 3}}{x + 2}$

30. $\lim\limits_{x \to \infty} \dfrac{\sqrt{x + 3}}{x + 4}$

In Exercises 31–42, determine whether the statements are true or false for the function $y = f(x)$ that is shown in the graph.

31. $\lim\limits_{x \to 2^-} f(x) = 4$

32. $\lim\limits_{x \to 2^+} f(x) = 4$

33. $\lim\limits_{x \to -1^+} f(x) = 1$

34. $\lim\limits_{x \to -1^-} f(x) = 1$

35. $\lim\limits_{x \to 0^-} f(x) = 0$

36. $\lim\limits_{x \to 0^+} f(x) = 0$

37. $\lim\limits_{x \to 0} f(x) = 0$

38. $\lim\limits_{x \to 0} f(x) = f(0)$

39. $\lim\limits_{x \to 1^-} f(x) = 1$

40. $\lim\limits_{x \to 1^+} f(x) = 1$

41. $\lim\limits_{x \to 1} f(x) = 1$

42. $\lim\limits_{x \to 1} f(x) = f(1)$

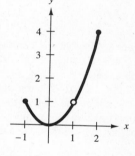

43. Use the graph to determine the limits as $x \to a$, when a is equal to
  a. -1 **b.** 0 **c.** 1 **d.** 2 **e.** 3 **f.** 4

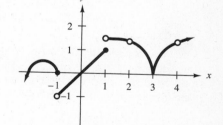

44. Use the graph to determine the limits as $x \to a$, when a is equal to
  a. -2 **b.** 0 **c.** 1 **d.** 2

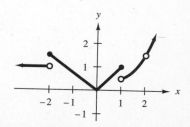

In Exercises 45–54, determine whether the function is continuous at the given value of *x*. If the function is not continuous, and if it is possible, remove the discontinuity.

45. $f(x) = 2x + 4, x = 2$

46. $f(x) = x^2 - 3x, x = -1$

47. $f(x) = \dfrac{x - 5}{x + 6}, x = 2$

48. $f(x) = \dfrac{x + 7}{x - 3}, x = 4$

49. $f(x) = \dfrac{x^2 - 36}{x - 6}, x = 6$

50. $f(x) = \dfrac{x^2 - 7}{x + \sqrt{7}}, x = -\sqrt{7}$

51. $f(x) = \begin{cases} \dfrac{x^2 - 9}{x - 3} & \text{for } x \neq 3 \\ 7 & \text{for } x = 3 \end{cases} x = 3$

52. $f(x) = \begin{cases} \dfrac{x^2 - 5x + 4}{x - 4} & \text{for } x \neq 4 \\ 13 & \text{for } x = 4 \end{cases} x = 4$

53. $f(x) = \begin{cases} x^2 + 3 & \text{for } x > 3 \\ 2x - 1 & \text{for } x \leq 3 \end{cases} x = 3$

54. $f(x) = \begin{cases} x^2 - 5 & \text{for } x > 2 \\ x - 3 & \text{for } x \leq 2 \end{cases} x = 2$

55. The population of an ant colony *n* months from now is predicted to be:

$$N = 30,000 + \frac{15,000}{(n + 2)^2}.$$

Determine the long-term population trend; that is, determine $\lim_{n \to \infty} N$.

56. The wave number of visible waves emitted by hydrogen is given by:

$$k = R\left(\frac{1}{4} - \frac{1}{n^2}\right)$$

where *R* (the Rydberg constant) is 109.678 cm² and *n* represents the number of the orbit or shell which the electron initially occupies. In this case *n* is an integer greater than 2, hence *n* = 3, 4, 5, ... The wave number is proportional to the energy of the light emitted when the electron goes from a higher energy level to the *n* = 2 level or shell. Find the limit of the wave number as *n* approaches infinity. This number is proportional to the maximum energy which light can have as the electron drops from a higher orbit to the *n* = 2 orbit.

57. The current in a *RL* circuit is given by:

$$i = \frac{v}{r}(1 - e^{-Rt/L}),$$

where *v* is the applied voltage, *R* is the total resistance of the circuit, *L* is the inductance, and *t* is the time. This equation implies that, as time passes, the current climbs in such a circuit. But does it climb to infinity? Determine the current in the circuit as *t* approaches infinity (lim). This is called the steady state current.

19.2

Average Rate of Change

The idea of **average rate of change** is something we encounter every day. For example, if a car accelerates from 0 to 96 km/h in 8.0 s, we say that it accelerates at an average rate of 12 km/h/s. If a spaceship climbs from 0 to 10,000 m in 2.5 s, we say that the ship climbs at an average velocity of 4000 m/s. If corn grows a total of 28 in. in 2 wk, it grows an average of 2 in./day.

For each of these examples, the average rate of change is found by dividing the change in the dependent variable by the change in the independent variable. For a function expressed in the form $y = f(x)$, the average rate of change of y per unit change in x is expressed by the ratio $\dfrac{\Delta y}{\Delta x}$, which is read "change in y divided by change in x." Therefore, for any two points (x_1, y_1) and (x_2, y_2),

$$\frac{\Delta y}{\Delta x} = \frac{y_2 - y_1}{x_2 - x_1}.$$

Recall that this is the same expression we used for finding the slope of a line.

EXAMPLE 1   **Average Rate of Change of a Function**

Determine the average rate of change of the function $y = x^2 - 3$ as x increases from $x = 1$ to $x = 3$.

Solution   To find the change in y we must first determine y when $x = 1$ and $x = 3$.
When $x = 1$,

$$y = (\boxed{1})^2 - 3 = -2.$$

When $x = 3$,

$$y = (\boxed{3})^2 - 3 = 6.$$

Thus,

$$\Delta y = \boxed{6} - (\boxed{-2}) = 8,$$

and

$$\Delta x = \boxed{3} - \boxed{1} = 2.$$

Therefore, the average rate of change of y is

$$\frac{\Delta y}{\Delta x} = \frac{\boxed{8}}{\boxed{2}}$$

$$= 4 \text{ units per unit change in } x. \quad \blacklozenge$$

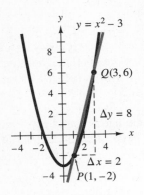

$y = x^2 - 3$

$\Delta y = 8$

$Q(3, 6)$

$\Delta x = 2$

$P(1, -2)$

FIGURE 19.10

The sketch of the function $y = x^2 - 3$ in Figure 19.10 illustrates the idea of average rate of change. The dashed line through the point $P(1, -2)$ parallel to the x-axis and the dashed line through point $Q(3, 6)$ parallel to the y-axis intersect at point $R(3, -2)$ to form the triangle PQR.

Recalling the discussion of the slope of a line, we can see that $\Delta x = 2$ and $\Delta y = 8$. Thus, the slope of the line PQ is $\dfrac{\Delta y}{\Delta x} = \dfrac{8}{2}$, or 4. The average rate of change of y per unit change in x (see Example 1) is the same as the slope of the secant line PQ. Recall that a secant line is a line that joins two points on the curve, in this case P and Q.

The slope of a secant line is one example of an average rate of change. Another example is *average velocity,* which is illustrated in Example 2. **Average velocity** is the average rate of change of the distance s per unit of time t.

EXAMPLE 2   **Application: Average Velocity**

A ball is thrown straight up, and its distance s (in feet) above the ground after t seconds is given by the formula $s = 64t - 16t^2$. Determine the average velocity of the ball $\dfrac{\Delta s}{\Delta t}$ for values given.

a.   From $t = 2$ to $t = 2.1$ **b.**   From $t = 2$ to $t = 2.01$

Solution

a.   When $t = 2$,

$$s = 64(2) - 16(2)^2$$
$$= 64;$$

when $t = 2.1$,

$$s = 64(2.1) - 16(2.1)^2$$
$$= 63.84.$$

Therefore, $\Delta t = 2.1 - 2 = 0.1$ and $\Delta s = 63.84 - 64 = -0.16$. The average rate of change of s per unit of change in t is:

$$\frac{\Delta s}{\Delta t} = \frac{-0.16}{0.1} \qquad \textbf{Definition of average velocity}$$

$$= -1.6 \text{ ft/s.}$$

b.   When $t = 2.01$, $s = 63.9984$. Therefore $\Delta t = 2.01 - 2 = 0.01$ and $\Delta s = 63.9984 - 64 = -0.0016$. The average rate of change of s per unit change in t is:

$$\frac{\Delta s}{\Delta t} = \frac{-0.0016}{0.01}$$

$$= -0.16 \text{ ft/s.}$$

The minus sign in the answer indicates that the average velocity is downward.   ◆

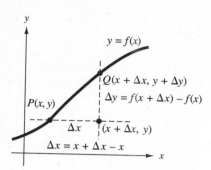

FIGURE 19.11

Now let us examine the process of finding the average rate of change of any function $y = f(x)$. If we select any value of x and increase it by the amount Δx, then a second value of the independent variable is $x + \Delta x$. As x changes from x to $x + \Delta x$, y will change a corresponding amount Δy. The corresponding new value of the dependent variable is $y + \Delta y$. Thus, we have the ordered pairs (x, y) and $(x + \Delta x, y + \Delta y)$, which must satisfy the function (i.e., they are points on the graph of the function, as shown in Figure 19.11). Therefore, if

$$y = f(x), \tag{1}$$

then

$$y + \Delta y = f(x + \Delta x). \tag{2}$$

Now, if we subtract Equation (1) from Equation (2), we can determine the change in y.

$$y + \Delta y - y = f(x + \Delta x) - f(x)$$
$$\Delta y = f(x + \Delta x) - f(x).$$

Then, to find the average rate of the change in y per unit change in x, we divide both sides by the change in x, Δx.

$$\frac{\Delta y}{\Delta x} = \frac{f(x + \Delta x) - f(x)}{\Delta x}.$$

DEFINITION 19.6

The **average rate of change** of y per unit change in x is given by:

$$\frac{\Delta y}{\Delta x} = \frac{f(x + \Delta x) - f(x)}{\Delta x}.$$

EXAMPLE 3   **Average Rate of Change**

a. Determine the average rate of change of y per unit change in x for $y = x^2 - 6x + 5$.
b. Determine the average rate of change of the function in part a when $x = 4$ and $\Delta x = 3$.

Solution

a. We determine the average rate of change using Definition 19.6

$$\frac{\Delta y}{\Delta x} = \frac{f(x + \Delta x) - f(x)}{\Delta x}$$ **Definition 19.6**

$$= \frac{[(\boxed{x + \Delta x})^2 - 6(\boxed{x + \Delta x}) + 5] - (x^2 - 6x + 5)}{\Delta x}$$

Substituting

$$= \frac{(x^2 + 2x\Delta x + \overline{\Delta x}^2* - 6x - 6\Delta x + 5) - (x^2 - 6x + 5)}{\Delta x}$$

$$= \frac{2x\Delta x + \overline{\Delta x}^2 - 6\Delta x}{\Delta x}$$

$$= \frac{(2x - 6 + \Delta x)\Delta x}{\Delta x}.$$

Therefore, the average rate of change for any value of x is:

$$\frac{\Delta y}{\Delta x} = 2x - 6 + \Delta x.$$

b. To find the average rate of change when $x = 4$ and $\Delta x = 3$ we substitute these values into the result obtained in part a.

$$\frac{\Delta y}{\Delta x} = 2x - 6 + \Delta x$$

$$= 2(\boxed{4}) - 6 + \boxed{3}$$ **Substituting**

$$= 5.$$

Therefore, the average rate of change is:

$$\frac{\Delta y}{\Delta x} = \frac{5 \text{ units of } y}{1 \text{ unit of } x}. \quad \blacklozenge$$

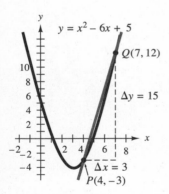

FIGURE 19.12

The graph of $y = x^2 - 6x + 5$ is sketched in Figure 19.12. The slope of the secant line PQ is given by $\dfrac{\Delta y}{\Delta x} = \dfrac{15}{3} = 5$. We see again that the average rate of change of the function is illustrated geometrically by the secant line.

---

* The bar above the Δx indicates that we mean $(\Delta x)^2$ and not just x^2.

EXAMPLE 4 **Application: Average Growth Rate**

The height h of a certain brand of corn t days ($t \geq 1$) after the seed germinates is $h(t) = \sqrt{t} - 1$.

a. Determine the average growth rate $\dfrac{\Delta h}{\Delta t}$.

b. Determine the average growth rate between days 4 and 9.

Solution

a. To find the change at any time t we use Definition 19.6.

$$\frac{\Delta h}{\Delta t} = \frac{f(t + \Delta t) - f(t)}{\Delta t} \qquad \text{Definition 19.6}$$

$$= \frac{(\sqrt{t + \Delta t} - 1) - (\sqrt{t} - 1)}{\Delta t} \qquad \text{Substituting}$$

$$= \frac{\sqrt{t + \Delta t} - \sqrt{t}}{\Delta t}.$$

b. For $t = 4$ to $t = 9$, $\Delta t = 5$. Substituting $t = 4$ and $\Delta t = 5$ in.,

$$\frac{\Delta h}{\Delta t} = \frac{\sqrt{t + \Delta t} - \sqrt{t}}{\Delta t},$$

we have:

$$\frac{\Delta h}{\Delta t} = \frac{\sqrt{4 + 5} - \sqrt{4}}{5}$$

$$= \frac{1}{5}.$$

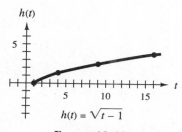

$h(t) = \sqrt{t - 1}$

FIGURE 19.13

Therefore, the average rate of change of the height of the corn with respect to time (between days 4 and 9) is $\dfrac{\Delta h}{\Delta t} = \dfrac{1}{5}$ (or 1 unit change in height for each 5 units change in time). The graph is shown in Figure 19.13. ◆

Why did we discuss this technique of finding the average rate of change? The technique of determining the average rate of change is helpful in understanding the instantaneous *rate of change*, the derivative. The derivative is discussed in the next section.

19.2 Exercises

In Exercises 1–10, find the average rate of change of the function over the indicated interval.

1. $y = x^2 + 4$ from $x = 2$ to $x = 3$

2. $y = x^2 - 7$ from $x = 1$ to $x = 3$

3. $y = x^2 + \dfrac{1}{3}x$ from $x = -3$ to $x = 3$

4. $y = x^2 - \dfrac{1}{2}x$ from $x = -2$ to $x = 4$

5. $s = 2t^3 - 5t + 7$ from $t = 1$ to $t = 3$

6. $s = t^3 - 2t^2 + 5$ from $t = 2.0$ to $t = 2.6$

7. $s = \dfrac{5}{t}$ from $t = 1.0$ to $t = 1.4$

8. $s = t + \dfrac{2}{t}$ from $t = 3.0$ to $t = 3.1$

9. $h = \sqrt{2t} - 7$ from $t = 8.0$ to $t = 8.5$

10. $h = \sqrt{3t - 1} + 2$ from $t = 4.0$ to $t = 4.2$

In Exercises 11–16, draw a graph of the function and indicate the average rate of change by drawing the secant line. Also, on the graph indicate the change of the independent variable and the dependent variable, using Figure 19.12 as a model.

11. $y = 2x^2 - 5$, $x = 1$ to $x = 3$

12. $y = x^2 - 6x - 7$, $x = 2$ to $x = 5$

13. $s = t^3 - 3t - 1$, $t = 1$ to $= 3$

14. $s = 2t^3 - t^2 - 2$, $t = 0$ to $t = 2$

15. $h = \sqrt{2t + 1}$, $t = 4$ to $t = 12$

16. $h = \sqrt{2t - 7}$, $t = 4$ to $t = 8$.

In Exercises 17–26, perform the following.
a. Find the average rate of change for any interval.
b. Find the average rate of change for the specified interval.

17. $s = 2t - 3$ from $t = 2$ to $t = 5$

18. $s = 2t^2 + 2$ from $t = 2$ to $t = 4$

19. $y = x^2 - 6x + 8$ from $x = 3.0$ to $x = 3.1$

20. $y = 2x + \dfrac{2}{x}$ from $x = 2.0$ to $x = 2.1$

21. $y = 2x^2 - 5x + 4$ from $x = 2.0$ to $x = 2.1$

22. $y = 4x - x^2$ from $x = 3.0$ to $x = 3.5$

23. $A = \pi r^2$ from $r = 2.0$ to $r = 2.1$

24. $V = \dfrac{4}{3}\pi r^3$ from $r = 3.0$ to $r = 3.1$

25. $h = \sqrt{t} - 9$ from $t = 9$ to $t = 16$

26. $h = \sqrt{t + 1}$ from $t = 3$ to $t = 8$

27. A ball is thrown straight up. Its height after t seconds is given by the formula $h = -16t^2 + 80t$. Determine its average velocity for the specified intervals.
a. from $t = 2$ to $t = 2.1$
b. from $t = 2.00$ to $t = 2.01$

28. A ball rolls down an inclined plane in such a way that its distance in millimeters from the top of the plane after t seconds is given by the formula $s = 2 + 10t + 5t^2$. Determine the average velocity for the specified intervals.
a. from $t = 3$ to $t = 4$
b. from $t = 3.0$ to $t = 3.1$

29. Show that the average rate of change for the function $y = x^2 + 5x - 14$ is greater from $x = 3.0$ to $x = 3.2$ than it is from $x = 2.0$ to $x = 2.2$. What is the graphical meaning of this fact?

30. The intensity of light on an object at a distance x units from the source is given by the formula $I = \dfrac{25}{x^2}$. Determine the average rate of change of I as the distance changes from $x = 2.5$ to $x = 2.6$ m.

31. If a farmer plants x acres of sugar beets, her profit is $f(x)$ dollars, where $f(x) = 1800x - 9x^2$. Determine the average rate of change of the profit between $x = 20$ acres and $x = 50$ acres. Explain what the rate of change means.

32. Recall that the rate of change of velocity is called acceleration. If velocity v in meters per second of a particle t seconds after it begins moving is given by the formula $v = 3t^2 + t$, determine the average acceleration between $t = 2$ s and $t = 5$ s.

33. The rate of change of price is called inflation. If the price p in dollars after t years is $p = 3t^2 + t + 1$, determine the average change of inflation from $t = 3$ to $t = 5$ years. Explain what the rate of change means.

34. An antibiotic is introduced into a petri dish. The number of bacteria in the dish t hours later can be determined by the function $B(t) = 15{,}200 + 1{,}200t - 120t^2$ for values of t with $0 \le t \le 45$. Determine the average change in the number of bacteria from $t = 3$ to $t = 5$ hr. Explain what the rate of change means.

35. Suppose that t seconds after lift-off a rocket has traveled s m. If s is determined by the function $s = 0.90t^2 + 1.8t$, determine the average velocity from $t = 2$ to $t = 5$ s. Explain what the rate of change means.

36. A coal-burning electrical generating plant emits sulfur dioxide into the surrounding air. The concentration $c(x)$ in parts per million is given by the formula $c(x) = \dfrac{0.1}{x^2}$, where x is the distance from the plant in kilometers. Find the average rate of change of concentration from $x = 2$ to $x = 3$ km. Explain what the rate of change means.

37. The average velocity of an object can be found by taking the distance traveled over the elapsed time $\dfrac{\Delta d}{\Delta t}$. The speed at an instant can be approximated as $\dfrac{\Delta d}{\Delta t}$ as long as the Δt is made very, very small. If the instantaneous speed is equal to $\dfrac{\Delta d}{\Delta t}$, then Δt approaches zero. In the table below the distance covered for progressively smaller intervals of time is given. The ratio of the distance traveled over the elapsed time is an approximation of the instantaneous speed. Evaluate $\dfrac{\Delta d}{\Delta d}$ for each of these time intervals and determine the limit of the velocity. Note that even though both the change in time and the change in distance approach zero, their quotient does not approach zero.

Δd (cm)	0.480000	0.280000	0.298000	0.002998	0.00029998
Δt (s)	0.200000	0.100000	0.010000	0.001000	0.00010000

19.3

The Derivative

In the previous section, we discussed the average rate of change, and learned that the average rate of change is illustrated geometrically by determining the slope of the secant line joining two points on the curve. More commonly, we are asked to determine the exact, or instantaneous, rate of change at a particular time. For example, what is the velocity of a spaceship exactly three seconds after lift-off? For an airplane, what is the instantaneous rate of change of the distance that occurs at a specific time? What is the slope of a line tangent to a curve at a specific point?

To illustrate this idea let us examine the graph of the function $y = x^2 - 6x + 5$ (from Example 3 in Section 19.3) with different values for Q. The

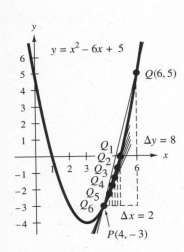

FIGURE 19.14

TABLE 19.7

P	Q	Δx	Δy	$\dfrac{\Delta y}{\Delta x}$
$(4, -3)$	$(6, 5)$	2	8	$\dfrac{8}{2} = 4$
$(4, -3)$	$(5, 0)$	1	3	$\dfrac{3}{1} = 3$
$(4, -3)$	$(4.8, -0.76)$	0.8	2.24	$\dfrac{2.24}{0.8} = 2.8$
$(4, -3)$	$(4.6, -1.44)$	0.6	1.56	$\dfrac{1.56}{0.6} = 2.6$
$(4, -3)$	$(4.4, -2.04)$	0.4	0.96	$\dfrac{0.96}{0.4} = 2.4$
$(4, -3)$	$(4.2, -2.56)$	0.2	0.44	$\dfrac{0.44}{0.2} = 2.2$
$(4, -3)$	$(4.1, -2.79)$	0.1	0.21	$\dfrac{0.21}{0.1} = 2.1$
$(4, -3)$	$(4.01, -2.9799)$	0.01	0.0201	$\dfrac{0.0201}{0.01} = 2.01$
$(4, -3)$	$(4.001, -2.997999)$	0.001	0.002001	$\dfrac{0.002001}{0.001} = 2.001$

function is sketched in Figure 19.14. The slope of the secant line $PQ = 4$ $\left(\dfrac{\Delta y}{\Delta x} = \dfrac{8}{2}\right)$. We see that as we take values of Q closer to P $(Q_1, Q_2, Q_3, \ldots)$ and construct new triangles, Δx gets smaller and smaller. With these changes, the slope of the secant line becomes a better approximation for the slope of the tangent line to the curve at point P. The exact (or instantaneous) change can be illustrated geometrically as the slope of a line tangent to a curve at a specific point. Table 19.7 contains coordinates for P and Q, Δx, Δy, and the slope of the secant lines $\dfrac{\Delta y}{\Delta x}$. The table clearly illustrates that, as Q approaches P, Δx approaches 0, and the slope of the secant line approaches 2.

Look again at Figure 19.14. We see that the secant line gets closer and closer to one particular line. This line is called a **tangent line.** Thus, we define the slope of the line tangent to a curve at a given point to be the number (if it exists) that the slopes of the secant lines approach. From our discussions on limit, it follows *that the slope of the tangent line to a curve at a point corresponds to the **instantaneous rate of change** at that point.* That is,

$$\frac{\Delta y}{\Delta x} = \text{slope of the secant line } PQ$$

$$\lim_{\Delta x \to 0} \left(\frac{\Delta y}{\Delta x}\right) = \text{slope of the tangent line at } P.$$

The statement $\displaystyle\lim_{\Delta x \to 0} \frac{\Delta y}{\Delta x}$ is read "the limit as delta x approaches zero of delta y divided by delta x." If the limit exists, then the result is the slope of the tangent line or the instantaneous rate of change of y with respect to x, which we call the *derivative of the function.* The derivative of a function may be indicated with the symbol $\dfrac{dy}{dx}$ or $f'(x)$.

DEFINITION 19.7 Definition of the Derivative
The derivative of f at x, denoted by $f'(x)$, is defined as:

$$f'(x) = \lim_{\Delta x \to 0} \frac{f(x + \Delta x) - f(x)}{\Delta x},$$

if the limit exists.

EXAMPLE 1　Derivative of a Function Using Definition 19.8

Determine the derivative of $f(x) = x^2 - 6x + 5$.

Solution　We find the derivative using the definition of the derivative.

$$f'(x) = \lim_{\Delta x \to 0} \frac{f(x + \Delta x) - f(x)}{\Delta x} \qquad \text{Definition 19.7}$$

$$= \lim_{\Delta x \to 0} \frac{[(x + \Delta x)^2 - 6(x + \Delta x) + 5] - (x^2 - 6x + 5)}{\Delta x}$$

Substituting

$$= \lim_{\Delta x \to 0} \frac{(x^2 + 2x\Delta x + \overline{\Delta x}^2 - 6x - 6\Delta x + 5) - (x^2 - 6x + 5)}{\Delta x}$$

$$= \lim_{\Delta x \to 0} \frac{2x\Delta x + \overline{\Delta x}^2 - 6\Delta x}{\Delta x} \qquad \text{Simplifying}$$

$$= \lim_{\Delta x \to 0} \frac{(2x + \Delta x - 6)\overset{1}{\cancel{\Delta x}}}{\underset{1}{\cancel{\Delta x}}} \qquad \text{Common factor}$$

$$= \lim_{\Delta x \to 0} (2x + \Delta x - 6) \qquad \text{Dividing by } \Delta x$$

$$= \lim_{\Delta x \to 0} 2x + \lim_{\Delta x \to 0} \Delta x - \lim_{\Delta x \to 0} 6.$$

Theorem 19.1
Limit of sum or
difference is equal to the
sum or difference of the
limits.

But $2x$ and 6 do not change as Δx approaches zero, so the

$$\lim_{\Delta x \to 0} 2x = 2x \quad \text{and} \quad \lim_{\Delta x \to 0} 6 = 6,$$

while the

$$\lim_{\Delta x \to 0} \Delta x = 0.$$

Therefore, $f'(x) = 2x - 6$. ◆

The result $f'(x) = 2x - 6$ is the slope of the tangent line to the curve at any point on the curve. To find the slope at a particular point, say $(4, -3)$, we substitute for x in the derivative. Thus,

$$f'(4) = 2(4) - 6$$
$$= 2.$$

The slope of the line tangent to the curve $y = x^2 - 6x + 5$ at the point $(4, -3)$ is 2, the value implied by the illustration at the beginning of this section.

EXAMPLE 2　Derivative of a Function Using Definition 19.8

Determine the derivative of $f(x) = 3x^3 + 2x^2 + 2$.

Solution

$$f'(x) = \lim_{\Delta x \to 0} \frac{[3(x + \Delta x)^3 + 2(x + \Delta x)^2 + 2] - (3x^3 + 2x^2 + 2)}{\Delta x} \qquad \text{Definition 19.8}$$

$$= \lim_{\Delta x \to 0} \frac{(3x^3 + 9x^2\Delta x + 9x\overline{\Delta x}^2 + 3\overline{\Delta x}^3 + 2x^2 + 4x\Delta x + 2\overline{\Delta x}^2 + 2) - (3x^3 + 2x^2 + 2)}{\Delta x}$$

$$= \lim_{\Delta x \to 0} \frac{(9x^2\Delta x + 9x\overline{\Delta x}^2 + 3\overline{\Delta x}^3 + 4x\Delta x + 2\overline{\Delta x}^2}{\Delta x} \qquad \text{Combining like terms}$$

$$= \lim_{\Delta x \to 0} \frac{(9x^2 + 9x\Delta x + 3\overline{\Delta x}^2 + 4x + 2\Delta x)\overset{1}{\cancel{\Delta x}}}{\underset{1}{\cancel{\Delta x}}} \qquad \text{Common factor}$$

$$= \lim_{\Delta x \to 0} (9x^2 + 9x\Delta x + 3\overline{\Delta x}^2 + 4x + 2\Delta x) \qquad \text{Dividing by } \Delta x$$

$$= \lim_{\Delta x \to 0} 9x^2 + \lim_{\Delta x \to 0} 9x\Delta x + \lim_{\Delta x \to 0} 3\overline{\Delta x}^2 + \lim_{\Delta x \to 0} 4x + \lim_{\Delta x \to 0} 2\Delta x \qquad \text{Theorem 19.1}$$

$$= \qquad 9x^2 + \qquad 0 \qquad + \qquad 0 \qquad + \qquad 4x + \qquad 0 \qquad \text{Limit of sum equals sum of limits.}$$

$$= 9x^2 + 4x. \quad \blacklozenge$$

EXAMPLE 3   **Slope of a Tangent Line at a Given Point**

a.   Determine the slope of the tangent line at any point on the curve $f(x) = \dfrac{1}{x}$.

b.   Determine the slope of the tangent line when $x = 2$.

Solution

a.   Using the definition of the derivative, we have:

$$f'(x) = \lim_{\Delta x \to 0} \frac{\dfrac{1}{x + \Delta x} - \dfrac{1}{x}}{\Delta x} \qquad \text{Definition 19.8}$$

$$= \lim_{\Delta x \to 0} \frac{\dfrac{-\Delta x}{x(x + \Delta x)}}{\Delta x} \qquad \begin{array}{l}\text{Subtracting, placing numerator}\\ \text{over common denominator}\end{array}$$

$$= \lim_{\Delta x \to 0} \frac{-\overset{1}{\cancel{\Delta x}}}{x(x + \Delta x)} \cdot \frac{1}{\underset{1}{\cancel{\Delta x}}} \qquad \text{Dividing by } \Delta x$$

$$= \lim_{\Delta x \to 0} \frac{-1}{x(x + \Delta x)}$$

$$= \left(\lim_{\Delta x \to 0} \frac{1}{x}\right)\left(\lim_{\Delta x \to 0} \frac{-1}{x + \Delta x}\right) \qquad \begin{array}{l}\text{Theorem 19.1 Limit of product}\\ \text{equals product of limits}\end{array}$$

$$= \frac{1}{x}\left(\frac{-1}{x}\right)$$

$$= \frac{-1}{x^2}.$$

Therefore, the slope of the tangent line at any point on the curve is

$$f'(x) = \frac{-1}{x^2}.$$

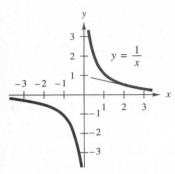

FIGURE 19.15

b. To determine the slope when $x = 2$, we substitute into the result obtained in a.; thus,

$$f'(\boxed{2}) = \frac{-1}{(\boxed{2})^2}$$

$$= \frac{-1}{4}.$$

The slope of the tangent line to the curve $y = \frac{1}{x}$ when $x = 2$ is $-\frac{1}{4}$ (See Figure 19.15).   ◆

If we can determine the derivative of a function then we say the function is **differentiable.** The method of finding the derivative is called **differentiation.**
To this point, we have used the symbol $f'(x)$ to indicate the derivative of $f(x)$ with respect to x. Sometimes other symbols are used to indicate the derivative. Each of the symbols in the following box indicates the derivative of the dependent variable with respect to the independent variable.

SYMBOLS INDICATING THE DERIVATIVE

1. $f'(x)$: read "f prime of x" (derivative of $f(x)$ with respect to x).
2. $\frac{dy}{dx}$: read "dee y, dee x" (the derivative of y with respect to x).
3. f': read "f prime" (derivative of the function with respect to the independent variable).
4. $D_x y$: read "D sub x, y" (the derivative of y with respect to x).
5. y': read "y prime" (derivative of y with respect to x).

Supply and Demand

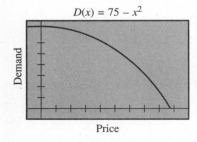

$D(x) = 75 - x^2$

Demand

Price

FIGURE 19.16

The concept of instantaneous change can be applied to supply and demand curves in economics. Recall that the demand $D(x)$ for a product is the number of articles the consumer buys at a given price per unit. For example, assume that the demand for a new video game called "Slow Draw" is $D(x) = 75 - x^2$, where x is in dollars per game and $D(x)$ is in hundreds of games per day. Figure 19.16 shows that, as the price increases, the demand decreases. For example, if the price of each game is raised from $5 to $8 the average rate of change of the demand is:

$$\frac{D(8) - D(5)}{8 - 5} = \frac{11 - 50}{3} = -13.$$

The result of -13 indicates that the demand will decrease by 13 (or 1300 games per day) if the price of "Slow Draw" is raised from $5 to $8. The instantaneous rate of change of the demand at any price is called the **marginal demand,** $D'(x)$. In Example 4, we determine the marginal demand for this function.

EXAMPLE 4 **Application: Marginal Demand**

Determine the marginal demand for the demand function $D(x) = 75 - x^2$.

Solution We can find the marginal demand by applying Definition 19.7.

$$D'(x) = \lim_{\Delta x \to 0} \frac{[75 - (x + \Delta x)^2] - (75 - x^2)}{\Delta x} \qquad \text{Definition 19.8}$$

$$= \lim_{\Delta x \to 0} \frac{(75 - x^2 - 2x\Delta x - \overline{\Delta x}^2) - (75 - x^2)}{\Delta x}$$

$$= \lim_{\Delta x \to 0} \frac{-2x\Delta x - \overline{\Delta x}^2}{\Delta x} \qquad \text{Combining like terms}$$

$$= \lim_{\Delta x \to 0} \frac{(-2x - \Delta x)\overset{1}{\Delta x}}{\underset{1}{\Delta x}} \qquad \text{Common factor}$$

$$= \lim_{\Delta x \to 0} (-2x - \Delta x) \qquad \text{Dividing by } \Delta x$$

$$= -2x$$

The result $-2x$ indicates that the demand for "Slow Draw" decreases at a rate that is double the price of each game. ◆

Numerical Derivatives

We can approximate accurately the derivative of a differentiable function at a particular value with a graphics calculator.

To determine the numerical approximation on the TI–82 we either select a tolerance level (i.e., a value for Δx) or accept the default value in the calculator. The default value is 0.001. In this text we use the default value of 0.001. To determine the numerical approximation of $f(x) = x^3$ at $x = 2$, with $\Delta x = 0.001$, use the keystrokes shown in the box,

Keystrokes	Screen Display
MATH 8 x,T,Θ ^ 3 ,	nDeriv($X^3,X,2$)
x,T,Θ , 2) ENTER	12.000001

The result is 12.000001. Using Definition 19.8, the result is 12. The calculator is accurate to five decimal places. If we had used a smaller tolerance, say $\Delta x = 0.0001$, the result would have been 12.00000001, which is accurate to seven decimal places. Thus, the smaller the tolerance, the more accurate the answer. In general, the calculator display is in the form of nDeriv (function, variable, value, tolerance).

EXAMPLE 5   **Numerical Approximation of f'(a)**

Use a calculator to calculate $f'(2)$ for the function $f(x) = 3x^3 - 5x^2 - 8x + 7$.

Solution

TI-82:

Keystrokes	Screen Display
MATH 8 3 $\boxed{x,T,\Theta}$ $\boxed{^{\wedge}}$ 3 $\boxed{-}$ 5 $\boxed{x,T,\Theta}$ $\boxed{x^2}$	nDeriv $(3x\wedge3-5x^2-$
$\boxed{-}$ 8 $\boxed{x,T,\Theta}$ $\boxed{+}$ 7 $\boxed{,}$ $\boxed{x,T,\Theta}$ $\boxed{,}$	$8x + 7, x, 2)$
2 $\boxed{)}$ ENTER	8.000003

Using the definition of the derivative, $f'(2) = 8$. ◆

19.3   Exercises

In Exercises 1–14, use Definition 19.8 to determine the derivative of the function.

1. $f(x) = 3x$

2. $f(x) = 4x$

3. $f(x) = 5x + 6$

4. $f(x) = 6x - 7$

5. $f(x) = x^2 + 1$

6. $f(x) = 2x^2 - 7$

7. $f(x) = 12 - x^2$

8. $f(x) = 7x + 2x^2$

9. $f(x) = 16x^2 - 7x$

10. $f(x) = 3x^2 + 4x - 9$

11. $f(x) = \dfrac{7}{x}$

12. $f(x) = \dfrac{3}{x}$

13. $f(x) = \dfrac{3}{x + 3}$

14. $f(x) = \dfrac{5}{2x - 4}$

15. To answer parts a–d use the graph of the function, $f(x) = x^2 - 7x + 6$. For part e use the function.

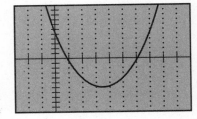

   a. Use a straight edge to draw a line tangent to the curve at $(1, 0)$.
   b. From the sketch in part a estimate the slope of the tangent line at $(1, 0)$.
   c. Use a straight edge to draw a line tangent to the curve at $(5, -4)$.
   d. From the sketch in part c estimate the slope of the tangent line at $(5, -4)$.
   e. The estimates of the slopes in parts b and c are also estimates of the numerical value of the derivative at those points. Are your estimates of the derivative close to the actual value?

16. To answer parts a–d use the graph of the function, $f(x) = x^3 - 5x + 4$. For part e use the function.

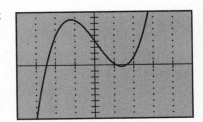

   a. Use a straight edge to draw a line tangent to the curve at $(-2, 6)$.
   b. From the sketch in part a estimate the slope of the tangent line at $(-2, 6)$.
   c. Use a straight edge to draw a line tangent to the curve at $(0, 4)$.
   d. From the sketch in part c estimate the slope of the tangent line at $(0, 4)$.
   e. The estimates of the slopes in parts b and c are also estimates of the numerical value of the derivative at those points. Are your estimates of the derivative close to the actual value?

In Exercises 17–24, determine the derivative of the function at the stated value of x. Use the calculator to check answers.

17. $f(x) = 7x - 5, x = 3$

18. $f(x) = -9x + 11, x = 2$

19. $f(x) = 3x^2 - 4, x = 7$

20. $f(x) = 5x^2 + 13, x = -2$

21. $f(x) = 5 - x^2, x = 4$

22. $f(x) = 7 - 2x^2, x = 5$

23. $f(x) = x^2 + 5x - 4, x = 3$

24. $f(x) = 3x^2 - 11x + 15, x = -3$

In Exercises 25–28, perform the following.
a. Determine the slope of the tangent line to the curve at any point.
b. Determine the slope of the tangent line for the given value of x.
c. Sketch the graph of the curve and the tangent line at the point on the curve for the given value of x.

25. $y = -x^2 + 7x, x = 3$

26. $y = 6x^2 - 11x - 10, x = 1$

27. $y = 3x^2 - 6x - 10, x = 0$

28. $y = 2x^2 + 3x - 4, x = 1$

In Exercises 29–32, the path of a ball thrown in the air is given by $s(t)$. Determine the slope of the line tangent to the graph of the function $s(t)$ at any point on the curve.

28. $s(t) = -16t^2 + 120t$

30. $s(t) = 84t - 16t^2$

31. $s(t) = -16t^2 + 32t + 128$

32. $s(t) = 64 + 128t - 16t^2$

In Exercises 33–36, for the demand function $D(x)$, determine the marginal demand for any arbitrary price x.

33. $D(x) = 8 + 5x - 2x^2$

34. $D(x) = 55 - 14x - 3x^2$

35. $D(x) = -7x^2 + 21x + 28$

36. $D(x) = 72 + 16x - 5x^2$

37. Determine $f'(x)$ for each of the following.
a. $f(x) = x$
b. $f(x) = x^2$
c. $f(x) = x^3$
d. $f(x) = x^4$

19.4

Derivatives of Polynomials

In Section 19.3 we defined the derivative of a function $f(x)$ to be

$$f'(x) = \lim_{\Delta x \to 0} \frac{f(x + \Delta x) - f(x)}{\Delta x}.$$

We learned that the derivative is found by applying the definition. Now, after doing the exercises for Section 19.3, you may be wondering whether there is a shorter way of finding the derivative. In this and the next several sections we will discuss the theorems that provide us with easier ways of finding derivatives.

Derivative of a Constant

If $f(x) = c$, where c is a constant, determine $f'(x)$.

$$f'(x) = \lim_{\Delta x \to 0} \frac{c - c}{\Delta x}$$

$$= 0$$

Therefore, the derivative of a constant is zero.

> **Constant Rule**
> Where c is any constant if $f(x) = c$, then:
> $$f'(x) = 0.$$

EXAMPLE 1   Derivative of a Constant

Determine the derivative of the following functions.

a. $f(x) = 3$ **b.** $f(x) = 13$ **c.** $f(x) = 7\pi$ **d.** $f(x) = \sqrt{149}$

Solution   Since each is a constant function, the graphs of the functions are horizontal lines, and the derivative in each case is zero.   ◆

Derivative of x^n

Exercise 37 in Section 19.3 asked you to determine the derivative of a series of functions. Those functions and their derivatives are listed below.

$$f(x) = x \qquad f'(x) = 1$$
$$f(x) = x^2 \qquad f'(x) = 2x$$
$$f(x) = x^3 \qquad f'(x) = 3x^2$$
$$f(x) = x^4 \qquad f'(x) = 4x^3$$

A pattern is evident in the exponents. The derivative's exponent is always one less than the original function. The full pattern yields a simple relationship:

$$\text{when } f(x) = x^n, \quad \text{then} \quad f'(x) = nx^{n-1}.$$

In the functions we just examined, the pattern of exponents runs from 1 ($x = x^1$) to 4. But, what about zero as an exponent?

$$f(x) = x^0 = 1 \quad \text{and} \quad f'(x) = 0x^{-1} = 0,$$

which is another way of showing that the derivative of a constant is zero.

We will show, using the definition of the derivative, that if $f(x) = x^n$, then $f'(x) = nx^{n-1}$.

$$f'(x) = \lim_{\Delta x \to 0} \frac{(x + \Delta x)^n - x^n}{\Delta x} \qquad \text{Definition 19.7}$$

$$= \lim_{\Delta x \to 0} \frac{x^n + nx^{n-1}\Delta x + \dfrac{n(n-1)}{2}x^{n-2}\overline{\Delta x}^2 + \cdots + \overline{\Delta x}^{n*} - x^n}{\Delta x}$$

$$= \lim_{\Delta x \to 0} \frac{1}{\Delta x}\left[nx^{n-1}\Delta x + \frac{n(n-1)}{2}x^{n-2}\overline{\Delta x}^2 + \cdots + \overline{\Delta x}^n \right]$$

*Recall expansion of the binomial $(a + b)^n$; see Section 16.6.

$$= \lim_{\Delta x \to 0} \frac{\Delta x}{\Delta x} \left[nx^{n-1} + \frac{n(n-1)}{2} x^{n-2} \Delta x + \cdots + \overline{\Delta x}^{n-1} \right]$$

Common factor

$$= \lim_{\Delta x \to 0} [nx^{n-1}] + \lim_{\Delta x \to 0} \left[\frac{n(n-1)}{2} x^{n-2} \Delta x + \cdots + \overline{\Delta x}^{n-1} \right]$$

$$= nx^{n-1} \qquad + \qquad 0 \qquad + \cdots + 0$$

$$= nx^{n-1}$$

The result is the power rule.

> **Power Rule**
>
> For any real number n if $f(x) = x^n$, then:
> $$f'(x) = nx^{n-1}.$$

We proved the power rule for the case where n is a positive integer. Later in this chapter we provide a proof for the case where n is a real number.

EXAMPLE 2 **Power Rule**

Determine the derivative of each function.

a. $f(x) = x^5$ **b.** $f(x) = x^{12}$

Solution

a. If $f(x) = x^5$, then:

$$f'(x) = \boxed{5} x^{\boxed{5-1}} \quad \text{Power rule}$$
$$= 5x^4.$$

b. If $f(x) = x^{12}$, then:

$$f'(x) = \boxed{12} x^{\boxed{12-1}} \quad \text{Power rule}$$
$$= 12x^{11} \quad \blacklozenge$$

Derivative of cx^n

If $f(x) = cx^n$, where c is a constant and n a positive integer, find $f'(x)$.

$$f(x) = \lim_{\Delta x \to 0} \frac{c(x + \Delta x)^n - cx^n}{\Delta x} \qquad \text{Definition 19.8}$$

$$= \lim_{\Delta x \to 0} c \left[\frac{(x + \Delta x)^n + x^n}{\Delta x} \right] \qquad \text{Common factor}$$

$$= \lim_{\Delta x \to 0} c \lim_{\Delta x \to 0} \frac{(x + \Delta x)^n + x^n}{\Delta x} \qquad \begin{array}{l} \text{Theorem 19.1 Limit of product is equal} \\ \text{to product of limits} \end{array}$$

$$\underbrace{= \quad c}_{} \qquad \underbrace{nx^{n-1}}_{}$$

$$= cnx^{n-1}$$

Thus, if $g(x) = x^n$ and $f(x) = cg(x)$, then $f'(x) = cg'(x)$.

> **Constant Times a Function Rule**
>
> For $g(x) = x^n$ and any real number c if $f(x) = cg(x)$, then:
> $$f'(x) = cg'(x) = cnx^{n-1}.$$

The rule states that the derivative of a constant times a function is the constant times the derivative of the function. The statement can also be proved for the case where n is a real number.

EXAMPLE 3   **Derivative of a Constant Times a Function**

Determine the derivative of the following.

a. $f(x) = 4x^3$ **b.** $f(x) = -\dfrac{2}{3}x^6$

Solution

a. If $f(x) = 4x^3$, then:
$$f'(x) = 4 \cdot 3\, x^{\,3-1} \qquad \text{Constant times a function rule}$$
$$= 12x^2.$$

b. If $f(x) = -\dfrac{2}{3}x^6$, then:
$$f'(x) = -\frac{2}{3} \cdot 6\, x^{\,6-1}$$
$$= -4x^5. \quad \blacklozenge$$

Derivative of a Sum

To be able to determine the derivative of a polynomial, we must be able to determine the derivative of the sum or difference of two or more functions. For example, at this time we can determine the derivative of $f(x) = 3x^4$, but are not prepared to determine the derivative of $f(x) = 3x^5 + 2x^2 + 3$. If $h(x) = f(x) + g(x)$, our task is to determine $h'(x)$.

$$h'(x) = \lim_{\Delta x \to 0} \frac{[f(x + \Delta x) + g(x + \Delta x)] - [f(x) + g(x)]}{\Delta x}$$

Definition 19.8

$$= \lim_{\Delta x \to 0} \underbrace{\frac{f(x + \Delta x) - f(x)}{\Delta x}}_{f'(x)} + \lim_{\Delta x \to 0} \underbrace{\frac{g(x + \Delta x) - g(x)}{\Delta x}}_{g'(x)}$$

$$= \qquad\qquad f'(x) \qquad + \qquad g'(x)$$

$$= f'(x) + g'(x)$$

> **Sum Rule**
>
> If $h(x) = f(x) + g(x)$, then:
>
> $$h'(x) = f'(x) + g'(x).$$

The sum rule states that the derivative of a sum is the sum of the derivatives. Since the difference of two functions $f(x) - g(x)$ can be written as the sum $f(x) + [-g(x)]$, the derivative of the difference of two functions is the difference of their derivatives.

The sum rule can be restated using the symbols $\dfrac{du}{dx}$ and $\dfrac{dv}{dx}$, if we let $u = f(x)$ and $v = g(x)$.

> $$\frac{d}{dx}(u + v) = \frac{du}{dx} + \frac{dv}{dx}$$

Example 4 demonstrates this form of the sum rule.

EXAMPLE 4　**Sum Rule**

a.　Determine $f'(x)$, if $f(x) = 3x^2 + 7x$.

b.　Determine $\dfrac{dy}{dx}$ if $y = 4x^3 - 2x^2 + 5x + 3$

Solution　Using the sum rule we can determine the derivative of each term separately.

a.　If $f(x) = 3x^2 + 7x$, then:

$$f'(x) = 3 \cdot 2x^{2-1} + 7 \cdot 1x^{1-1} \quad \text{Sum rule; Constant times a function rule; Power rule}$$

$$= 6x + 7.$$

b.　If $y = 4x^3 - 2x^2 + 5x + 3$, then:

$$\frac{dy}{dx} = \frac{d(4x)^3}{dx} - \frac{d(2x^2)}{dx} + \frac{d(5x)}{dx} + \frac{d(3)}{dx} \quad \text{Sum rule}$$

$$= 4\frac{d(x^3)}{dx} - 2\frac{d(x^2)}{dx} + 5\frac{d(x)}{dx} + \frac{d(3)}{dx} \quad \text{Constant times a function rule}$$

$$= 4(3x^2) - 2(2x) + 5(1) + 0 \quad \text{Power rule}$$

$$= 12x^2 - 4x + 5. \quad \blacklozenge$$

The derivative of a function at a point is the slope of the tangent line to the curve at that point on the curve. Example 5 illustrates this.

EXAMPLE 5   Application: Slope of a Tangent Line

Determine the slope of the tangent line to the curve $f(x) = x^3 + 3x^2 - 2$ at the point $(-1, 0)$.

Solution   To determine the slope of the tangent line at the point $(-1, 0)$ we must determine $f'(-1)$.

$$f'(x) = 3x^2 + 6x$$
$$f'(-1) = 3(\boxed{-1})^2 + 6(\boxed{-1})$$
$$= -3.$$

Thus, the slope of the tangent line to the curve $f(x) = x^3 + 3x^2 - 2$ at the point $(-1, 0)$ is -3; see Figure 19.17. ◆

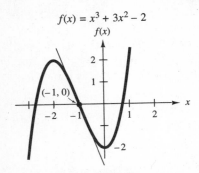

$f(x) = x^3 + 3x^2 - 2$

FIGURE 19.17

A tangent line that is parallel to the x-axis (horizontal tangent line) has a slope of zero. The points on the curve where the tangent line is parallel to the x-axis can be determined by finding the derivative, setting the derivative equal to zero, and solving the equation. With these solutions we can determine point(s) on the curve where the tangent line is horizontal.

EXAMPLE 6   Points of Horizontal Tangent Lines

For the function $f(x) = x^3 + 3x^2 - 2$, determine the points on the graph where the tangent lines are horizontal.

Solution   A horizontal tangent line exists when the slope is equal to zero. To determine these points, first determine the derivative of the function. Then set the derivative equal to zero, and solve the equation. From Example 5, the derivative of the function is $f'(x) = 3x^2 + 6x$, so:

$$3x^2 + 6x = 0$$
$$3x(x + 2) = 0$$
$$x = 0 \quad \text{or} \quad x = -2.$$

Substituting these values into the original function:

$$f(\boxed{0}) = -2 \quad \text{and} \quad f(\boxed{-2}) = (\boxed{-2})^3 + 3(\boxed{-2})^2 - 2$$
$$= 2.$$

Therefore, the points on the curve where the tangent lines to the curve are horizontal are $(0, -2)$ and $(-2, 2)$. The tangent lines are drawn in Figure 19.18. ◆

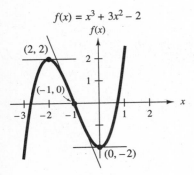

$f(x) = x^3 + 3x^2 - 2$

FIGURE 19.18

19.4 Exercises

In Exercises 1–20, determine the derivative of each of the functions.

1. $f(x) = x + 4$

2. $f(x) = 3x + 7$

3. $f(x) = x^2 + 3x$

4. $f(x) = 2x^2 + 11$

5. $f(x) = 2x^2 + 3x$

6. $f(x) = 4x^2 - 7x$

7. $y = x^3 - 5x + 11$

8. $y = x^4 - 11x^3 + 7$

9. $y = x^7 + 5x^4 - 9x$

10. $y = 3x^9 - 8x^5 - 7x^4$

11. $y = x^5 + 6x^3 + 5x$

12. $y = x^9 - 3x^8 + 7x^2 + 4$

13. $y = \dfrac{1}{4}x^4 - \dfrac{2}{3}x^3 + 3$

14. $y = \dfrac{3}{7}x^7 - \dfrac{3}{5}x^2 - 7x$

15. $y = \sqrt{3}\,x^5 + \dfrac{1}{3}x^2 - 7x$

16. $y = \dfrac{1}{\sqrt{3}}x^5 - \dfrac{1}{\sqrt{2}}x^4 + \pi x^3$

17. $f(r) = 2\pi r$

18. $f(r) = \pi r^2$

19. $f(r) = \dfrac{4}{3}\pi r^3$

20. $f(r) = 4\pi r^2$

21. For the function $f(x) = 2x^2 + 5x - 1$, perform the following.
   a. Determine $f'(x)$.
   b. Determine $f'(1)$.
   c. Sketch the graph and on the graph, sketch the tangent line when $x = 1$.

22. For the function $f(x) = x^3 - 4x - 3$, perform the following.
   a. Determine $f'(x)$.
   b. Determine $f'(-1)$.
   c. Sketch the graph and on the graph, sketch the tangent line when $x = -1$.

23. For the function $f(x) = 3x^2 - 12x + 5$, determine the value(s) of x for which $f'(x) = 0$.

24. For the function $f(x) = -\dfrac{x^3}{3} - x^2 - 15x + 11$, determine the value(s) of x for which $f'(x) = 0$.

25. For the function $f(x) = 2x^3 - 3x^2 - 36x + 30$, determine the points on the curve at which the tangent line is parallel to the x-axis.

26. Determine the points on the curve $y = 2x^3 - 3x^2 - 12x + 3$ where the tangent lines are horizontal.

27. For what values of x does the graph of $f(x) = \dfrac{x^4}{4} + \dfrac{2x^3}{3} - \dfrac{x^2}{2} + 5$ have a horizontal tangent?

19.5

Instantaneous Rate of Change

The two most common physical interpretations of the derivative are the slope of a line tangent to a curve and the instantaneous rate of change of one thing with respect to another, as we have learned. This section explores the powerful uses of the concept of instantaneous rate of change in the real world.

When traveling from Rochester to Philadelphia we could determine our average speed by dividing the distance traveled (change in distance Δs) by the length of time the trip takes (change of time Δt). That is, the average speed for the trip is $\dfrac{\Delta s}{\Delta t}$. What we may want to know is what our velocity was at the exact moment we passed the police officer with the radar gun. We need to know the instantaneous rate of change of distance with respect to time or **velocity.** That is, if $s = f(t)$, then $\dfrac{ds}{dt} = f'(t)$, or we say, $v = \dfrac{ds}{dt}$.

> **DEFINITION 19.8 Velocity**
>
> $$v = \frac{ds}{dt}$$

EXAMPLE 1   Velocity

If a distance s in meters is given by $s = 4t^2 - 3t$, determine the following.

a.   The velocity for any time t
b.   The velocity when $t = 2$ seconds

Solution

a.   To determine the velocity for any time t, we determine the derivative of s with respect to t, $\dfrac{ds}{dt}$.

$$s = 4t^2 - 3t$$
$$\frac{ds}{dt} = \frac{d(4t^2)}{dt} - \frac{d(3t)}{dt}$$
$$= 8t - 3.$$

The velocity at any time t is $8t - 3$.

b. To determine the velocity when $t = 2$, we substitute 2 for t in the equation.

$$v = 8t - 3$$
$$v = 8(\boxed{2}) - 3 \quad \textbf{Substituting for } t$$
$$v = 13$$

Therefore, the instantaneous rate of change of distance with respect to time (velocity) when $t = 2$ is $v = 13$ m/s. ◆

We defined velocity to be the instantaneous change of distance with respect to time. What happens when we are driving down the street at a constant velocity and we very quickly press down on the accelerator to pass another car? This action causes an instantaneous change in the velocity of the car with respect to time, and is called **acceleration.** Acceleration is represented with the letter a.

DEFINITION 19.9 Acceleration

$$a = \frac{dv}{dt}$$

EXAMPLE 2 Application: Acceleration of a Rocket

After blast-off, a rocket has a velocity of $v = t^3 - 4t^2 + 4t$, where t is in seconds and v is in kilometers per second. Determine the following.

a. The acceleration for any time t
b. The acceleration 4 seconds ($t = 4$) after blast-off.

Solution

a. From Definition 19.10 we know that $a = \dfrac{dv}{dt}$. Thus, the acceleration at any time t for $v = t^3 - 4t^2 + 4t$ is:

$$\frac{dv}{dt} = 3t^2 - 8t + 4.$$

Therefore for any time t, $a = 3t^2 - 8t + 4$ km/s².
b. We can find the acceleration when $t = 4$ by substituting $t = 4$ in the equation:

$$a = 3t^2 - 8t + 4$$
$$= 3(\boxed{4})^2 - 8(\boxed{4}) + 4 \quad \textbf{Substituting for } t$$
$$= 20.$$

The acceleration of the rocket 4 seconds after blast-off is 20 km/s². ◆

Instantaneous rate of change occurs in electricity. A frequently used equation in electricity is *Ohm's Law,* which is stated as $i = \dfrac{V}{R}$. The current i is in amperes; it is created by the voltage v in volts and regulated by the resistance R, which is measured in ohms. A *capacitor* is a device that stores charge on parallel plates and thereby inhibits current. The *Law of capacitance* specifies that the current is limited by the capacitance and the rate of change of voltage with respect to time. The unit of capacitance is the farad (F), and one farad is a large capacitance. Commonly used capacitors usually are rated in terms of microfarads ($1 \ \mu F = 10^{-6}$ F).

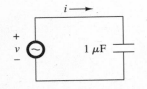

FIGURE 19.19

> **DEFINITION 19.10 Law of Capacitance**
>
> $$i = C\frac{dv}{dt}$$

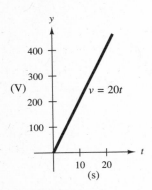

FIGURE 19.20

EXAMPLE 3   Application: Current

The voltage source in Figure 19.19 generates a voltage of $v = 20t$ for $0 < t < 20$, where v is in volts and t is in seconds. Determine the current and sketch the current versus time.

Solution   Note that $v = 20t$ applies during the first 20 seconds. The voltage during this time increases linearly with respect to time. This means that the voltage is increasing at a fixed rate for each second that elapses. For $t = 0$, $v = 0$; for $t = 1$, $v = 20$; and so on.

Figure 19.20 shows the graph of capacitor versus time. The slope of the line is $20 \ \dfrac{V}{s}$. Therefore, the derivative $\dfrac{dv}{dt} = 20$. To find the current we must use Definition 19.10, which says that $i = C\dfrac{dv}{dt}$.

From Figure 19.19 we know that $C = 1 \ \mu F$. By substituting, we obtain:

$$i = (10^{-6})(20)$$
$$= 20 \ \mu A.$$

With this result we can sketch the current versus time as illustrated in Figure 19.21. ◆

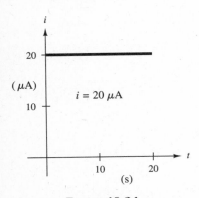

FIGURE 19.21

EXAMPLE 4   **Application: Current in a Capacitor**

Determine the equation for the current in a capacitor if $C = 10^{-6}$ F and voltage is given by $v = t^2 - 8t + 11$ V.

Solution   From the definition we know that $i = C\dfrac{dv}{dt}$; thus:

$$i = 10^{-6}\frac{d(t^2 - 8t + 11)}{dt}$$
$$= 10^{-6}(2t - 8).$$

Therefore, the current in the capacitor of 1 μF with $v = t^2 - 8t + 11$ V is $i = 10^{-6}(2t - 8)$ A.   ◆

In an electric circuit containing a coil of wire, voltage and thus current are inhibited by a changing magnetic field. In this case, the changing current limits the voltage across the coil. The instantaneous change of current i with respect to time and voltage v of the coil are related by the Law of inductance.

Inductance L is defined as a measure of a coil's opposition to change in current flow. The unit of inductance, the *henry* (H) is one volt/(ampere/second).

DEFINITION **19.11 Law of Inductance**

$$v = L\frac{di}{dt}$$

EXAMPLE 5   **Application: Voltage Across the Inductor**

An 8-mH inductor has a current of $i = 30t - 48$ flowing through it. What is the induced voltage across the inductor?

Solution   We can find the voltage across the inductor using the formula

$$v = L\frac{di}{dt}.$$

Since

$$i = 30t - 48$$
$$\frac{di}{dt} = 30.$$

Substituting, we have:

$$v = (\,0.008\,)(\,30\,) \quad \text{(8-mH = 0.008)}$$
$$= 0.240.$$

Therefore, the induced voltage across the inductor, $v = 0.240$ V.   ◆

19.5   Exercises

In Exercises 1–6, determine the velocity of each function for any time t.

1. $s = 2t + 5$

2. $s = t^2 - 7t + 6$

3. $s = t^2 + 7t + 10$

4. $s = -2t^3 + 2t^2 + 16t - 1$

5. $s = t^7 + 5t^4 - 5t^2$

6. $s = 3t^8 + 8t^5 - 11t^3 + 19$

In Exercises 7–12, determine the acceleration of each function for any time t.

7. $v = 3t^2 - 7t + 5$

8. $v = 4t^2 + 11t - 11$

9. $v = 2t^3 - 5t^2 + 11t + 1$

10. $v = 4t^3 - 4t^2 + 5t - 3$

11. $v = 3t^7 + 5t^5 - 2t^2$

12. $v = 13t^4 + 9t^3 - 6t^2 + 11t$

13. After t hr, a moving object has gone $s = 4t^3 + 3t^2 + t$ mi, for $t \geq 0$. Determine the following.
  a. The velocity of the object for any time t
  b. The velocity for $t = 3.5$ hr

14. If the distance traveled by an object is $s = t^3 + 2t$ m, determine the following.
  a. The velocity of the body for any time t
  b. The velocity of the body for $t = 5.3$ min

15. The velocity of an object is $v = 15t^2 + 7t + 3$ m/s. Determine the following.
  a. The acceleration of the body for any time t
  b. The acceleration of the body for $t = 3.4$ s

16. The velocity of an object is given by the function $v = t^3 - 3t^2 + 7t + 13$. Determine the following.
  a. The acceleration of the object for any time t
  b. The acceleration for $t = 8.5$

17. When a ball rolls down an incline its distance from the top after t seconds is given by the formula $s = 2 + 10t + 5t^2$ m. Perform the following.
  a. Determine the velocity for any time t.
  b. Determine the velocity for $t = 3$ and $t = 4$ s.
  c. What is the average velocity for $t = 3$ to $t = 4$ s?
  d. Would you expect the average velocity found in part c to be more or less than the instantaneous velocity at $t = 3$? Why?

18. Determine the acceleration of an object when $v = 3t^3 + 5t^2 - 2t + 5$ and $t = 3$ s.

19. A 15-mH inductor has a current of $i = 15t^2 - 14t + 5$ mA flowing through it. What is the induced voltage across the inductor?

20. Given the function $i = 3t^2 + 12t$ mA.
  a. Determine the induced voltage v for any value of t in an inductor L.
  b. If $L = 10^{-3}$ H, what would be the value of v when $t = 3$ s?

21. The current in a 500-mH inductor is given by $i = 0.03t$ mA. What is the induced voltage across the inductor?

22. The voltage across a capacitor is given by $v = 45t + 30$. The capacitor has a value of 50 μF. How much current flows into the capacitor?

23. The voltage across a 470-pF capacitor is given by $v = 3t^2 + 2t$, where t is in seconds.

    a. Determine the value of the charging current in the capacitor when $t = 1$ s.
    b. Determine the value when $t = 2$ s.

24. A triangular wave form drives the 8-μF capacitor shown in the illustration. Sketch current versus time.

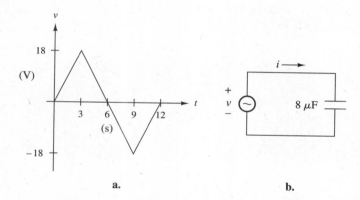

     a.   **b.**

25. A coil of copper wire has an inductance $L = 18$ H. What is the expression for the voltage in the coil if the current varies as $i = 9t^6 + 30t^3 + 25$ A?

26. For an object freely falling near the surface of the earth, the equation that relates speed and time is given by $v = v_0 - 9.8t$, where v_0 is the initial speed of the object. Assume that the initial speed is 10 m/s.

    a. Determine the derivative of speed with respect to time. The result is acceleration.
    b. Determine the acceleration when $t = 0.2$ s and when $t = 2.0$ s.

27. The electric current is the rate of charge flow and is given by $i = \dfrac{dq}{dt}$.

    a. What is the current if the charge varies in the following fashion: $q = 1.2t$? Determine the current after 3.0 seconds have elapsed.
    b. In a certain complex circuit, the charge varies as $q = 2.2t + 0.22t^2$. What is the current after 3.0 seconds have elapsed?

28. It is predicted that the number N of bacteria in an experimental sample will change with respect to time (in minutes).

    a. Determine the rate of change of N with respect to t when $N = 0.32t^2 - 0.12t + 1000$.
    b. What rate of increase would be expected after 2 min? After 2 hr? After 2 days?

19.6

Derivatives of Products and Quotients of Functions

The rules discussed so far on finding derivatives do not provide us with a convenient way of finding the derivatives of such functions as $y = (3x^2 + 7x)(x^3 - 6x^5)$ and $y = \dfrac{x^2 + 5x}{x^3 + 3x - 2}$. In this section the rules that deal with the derivative of a product and the derivative of a quotient are discussed. These rules provide the tools to handle more complicated functions.

Derivative of a Product

If $h(x) = f(x)g(x)$ and $f(x)$ and $g(x)$ are differentiable functions of x (that is, we can determine the derivatives of each with respect to x), then the derivative of the product $h(x)$ is found by using the *product rule*.

> **Product Rule**
>
> If $f(x)$ and $g(x)$ are differentiable functions of x and $h(x) = f(x)g(x)$, then:
>
> $$h'(x) = f(x)g'(x) + g(x)f'(x).$$

The rule states that the derivative of a product of two differentiable functions is the first times the derivative of the second plus the second times the derivative of the first. We can restate the product rule using the symbols $\dfrac{dy}{dx}$. If we let $u = f(x)$, $v = g(x)$, and $y = u \cdot v$, then:

$$\frac{dy}{dx} = u \cdot \frac{dv}{dx} + v \cdot \frac{du}{dx}.$$

Example 1 demonstrates this form of the product rule.

EXAMPLE 1 **Product Rule**

Determine $\dfrac{dy}{dx}$ for the following functions.

a. $y = (5x + 6)(9x - 7)$ **b.** $y = (x^2 - 2x)(x^3 - 3)$

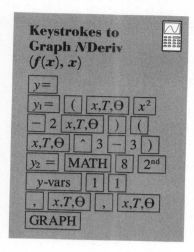

Keystrokes to Graph NDeriv ($f(x)$, x)

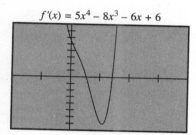

$f'(x) = 5x^4 - 8x^3 - 6x + 6$

FIGURE 19.22

Solution

a.   Let $u = 5x + 6$ and $v = 9x - 7$, then $\dfrac{dv}{dx} = 9$ and $\dfrac{du}{dx} = 5$.

$$\frac{dy}{dx} = u \cdot \frac{dv}{dx} + v \cdot \frac{du}{dx} \qquad \text{Product rule}$$

$$= (5x + 6)(9) + (9x - 7)(5) \qquad \text{Substituting}$$

$$= 45x + 54 + 45x - 35$$

$$= 90x + 19$$

b.   Let $u = x^2 - 2x$ and $v = x^3 - 3$, then $\dfrac{dv}{dx} = 3x^2$ and

$$\frac{du}{dx} = 2x - 2.$$

$$\frac{dy}{dx} = u \cdot \frac{dv}{dx} + v \cdot \frac{du}{dx} \qquad \text{Product rule}$$

$$= (x^2 - 2x)(3x^2) + (x^3 - 3)(2x - 2) \qquad \text{Substituting}$$

$$= 3x^4 - 6x^3 + 2x^4 - 2x^3 - 6x + 6$$

$$= 5x^4 - 8x^3 - 6x + 6 \quad \blacklozenge$$

The graph of the derivative of the function can be drawn directly from the function with the TI-82. The specific keystrokes used in creating the graph for Example 1b, Figure 19.22, are given in the box in the margin.

EXAMPLE 2   **Product Rule. Determine $h'(a)$**

If $h(x) = (2x^3 - 7)(x^4 + 3)$, determine $h'(2)$.

Solution   To determine $h'(2)$, we first must determine $h'(x)$ and then substitute 2 for x in $h'(x)$. Let $f(x) = 2x^3 - 7$ and $g(x) = x^4 + 3$, then $f'(x) = 6x^2$ and $g'(x) = 4x^3$.

$$h'(x) = f(x)g'(x) + g(x)f'(x) \qquad \text{Product rule}$$

$$= (2x^3 - 7)(4x^3) + (x^4 + 3)(6x^2) \qquad \text{Substituting}$$

$$= 8x^6 - 28x^3 + 6x^6 + 18x^2$$

$$= 14x^6 - 28x^3 + 18x^2$$

Now, substituting 2 for x:

$$h'(2) = 14(2)^6 - 28(2)^3 + 18(2)^2$$

$$= 744. \quad \blacklozenge$$

CAUTION   If $h(x) = f(x)g(x)$, then $h'(x) \neq f'(x)g'(x)$. For example, if $h(x) = (3x^2)(4x)$, then $h'(x) \neq (6x)(4)$. Show that $h'(x) = 36x^2$.   ∎

EXAMPLE 3   Proof of the Product Rule

If $h(x) = f(x)g(x)$ and $f(x)$ and $g(x)$ are differentiable functions of x, then the derivative of the product $h(x)$ can be found by using the definition of the derivative.

Solution

$$h'(x) = \lim_{\Delta x \to 0} \frac{h(x + \Delta x) - h(x)}{\Delta x} \qquad \text{Definition 19.7}$$

$$= \lim_{\Delta x \to 0} \frac{f(x + \Delta x)g(x + \Delta x) - f(x)g(x)}{\Delta x} \qquad \text{Substituting}$$

To be able to take the limit of the expression on the right side of the equal sign, we rewrite the expression in a different form, subtracting and adding $f(x + \Delta x)g(x)$. This process changes the form of the expression, not the value.

$$h'(x) = \lim_{\Delta x \to 0} \frac{f(x + \Delta x)g(x + \Delta x) - f(x + \Delta x)g(x) + f(x + \Delta x)g(x) - f(x)g(x)}{\Delta x}$$

$$= \lim_{\Delta x \to 0} \frac{f(x + \Delta x)[g(x + \Delta x) - g(x)] + g(x)[f(x + \Delta x) - f(x)]}{\Delta x}$$

$$\text{Common factor}$$

$$= \lim_{\Delta x \to 0} \left\{ f(x + \Delta x)\left[\frac{g(x + \Delta x) - g(x)}{\Delta x}\right] + g(x)\left[\frac{f(x + \Delta x) - f(x)}{\Delta x}\right]\right\}$$

$$= \underbrace{\lim_{\Delta x \to 0} f(x + \Delta x)}_{f(x)} \underbrace{\lim_{\Delta x \to 0} \frac{g(x + \Delta x) - g(x)}{\Delta x}}_{g'(x)} + \underbrace{\lim_{\Delta x \to 0} g(x)}_{g(x)} \underbrace{\lim_{\Delta x \to 0} \frac{(x + \Delta x) - f(x)}{\Delta x}}_{f(x)}$$

$$\text{Use Definitions 19.1 and 19.7 to find limits}$$

The $\lim_{\Delta x \to 0} f(x + \Delta x) = f(x)$ as Δx approaches 0. Also, $\lim_{\Delta x \to 0} g(x) = g(x)$. Therefore, $h'(x) = f(x)g'(x) + g(x)f'(x)$. ◆

Derivative of a Quotient

If $h(x) = \dfrac{f(x)}{g(x)}$, where $g(x) \neq 0$ and $f(x)$ and $g(x)$ are differentiable functions of x, then the derivative of the quotient $h(x)$ can be found by using the *Quotient rule*.

Quotient Rule

If $f(x)$ and $g(x)$ are differentiable functions of x, where $g(x) \neq 0$ and $h(x) = \dfrac{f(x)}{g(x)}$, then:

$$h'(x) = \frac{g(x)f'(x) - f(x)g'(x)}{[g(x)]^2}.$$

As we did for the product rule, we can write the quotient rule in symbols using $\dfrac{dy}{dx}$. If we let $u = f(x)$ and $v = g(x)$ and $y = \dfrac{u}{v}$ ($v \neq 0$), then:

$$\frac{dy}{dx} = \frac{v \cdot \dfrac{du}{dx} - u \cdot \dfrac{dv}{dx}}{v^2}.$$

Example 4 demonstrates this form of the quotient rule.

EXAMPLE 4 **Quotient Rule**

Determine $\dfrac{dy}{dx}$ for the following functions.

a. $y = \dfrac{3x}{x + 2}$ **b.** $y = \dfrac{x - 3}{x^2}$

Solution

a. Given $y = \dfrac{3x}{x + 2}$, let $u = 3x$ and $v = x + 2$, then $\dfrac{du}{dx} = 3$ and $\dfrac{dv}{dx} = 1$.

$$\frac{dy}{dx} = \frac{v \cdot \dfrac{du}{dx} - u \cdot \dfrac{dv}{dx}}{v^2} \qquad \text{Quotient rule}$$

$$= \frac{(x + 2)(\boxed{3}) - 3x(\boxed{1})}{(x + 2)^2} \qquad \text{Substituting}$$

$$= \frac{3x + 6 - 3x}{x^2 + 4x + 4}$$

$$= \frac{6}{x^2 + 4x + 4}$$

b. Given $y = \dfrac{x - 3}{x^2}$, let $u = x - 3$ and $v = x^2$, then $\dfrac{du}{dx} = 1$ and $\dfrac{dv}{dx} = 2x$.

$$\frac{dy}{dx} = \frac{v \dfrac{du}{dx} - u \dfrac{dv}{dx}}{v^2} \qquad \text{Quotient rule}$$

$$= \frac{x^2(\boxed{1}) - (x - 3)(\boxed{2x})}{(x^2)^2} \qquad \text{Substituting}$$

$$= \frac{x^2 - 2x^2 + 6x}{x^4}$$

$$= \frac{x(-x + 6)}{x^4}$$

$$= \frac{-x + 6}{x^3} \qquad \blacklozenge$$

EXAMPLE 5   **Quotient Rule: Determine $h'(a)$**

If $h(x) = \dfrac{x^2 + 2x}{2x^3 - 1}$, determine $h'(1)$.

Solution   To determine $h'(1)$, we first determine $h'(x)$ and then substitute 1 for x in $h'(x)$. Let $f(x) = x^2 + 2x$ and $g(x) = 2x^3 - 1$, then $f'(x) = 2x + 2$ and $g'(x) = 6x^2$.

$$
\begin{aligned}
h'(x) &= \frac{g(x)f'(x) - f(x)g'(x)}{[g(x)]^2} && \text{Quotient rule} \\[2mm]
&= \frac{(2x^3 - 1)(\,2x + 2\,) - (x^2 + 2x)(\,6x^2\,)}{(2x^3 - 1)^2} && \text{Substituting} \\[2mm]
&= \frac{4x^4 + 4x^3 - 2x - 2 - 6x^4 - 12x^3}{(2x^3 - 1)^2} \\[2mm]
&= \frac{-2x^4 - 8x^3 - 2x - 2}{(2x^3 - 1)^2}
\end{aligned}
$$

Substituting 1 for x, we have:

$$
\begin{aligned}
h'(1) &= \frac{-2(\,1\,)^4 - 8(\,1\,)^3 - 2(\,1\,) - 2}{[2(\,1\,)^3 - 1]^2} \\[2mm]
&= -14. \quad \blacklozenge
\end{aligned}
$$

CAUTION   If $h(x) = \dfrac{f(x)}{g(x)}$, then $h'(x) \neq \dfrac{f'(x)}{g'(x)}$. For example, if $h(x) = \dfrac{3x + 2}{x^2}$, then $h'(x) \neq \dfrac{3}{2x}$. Show that $h'(x) = \dfrac{-3x - 4}{x^3}$. ■

EXAMPLE 6   **Proof of the Quotient Rule**

If $h(x) = \dfrac{f(x)}{g(x)}$, where $g(x) \neq 0$ and $f(x)$ and $g(x)$ are differentiable functions of x, then the derivative of the quotient $h(x)$ can be found by using the definition of the derivative.

Solution

$$
\begin{aligned}
h'(x) &= \lim_{\Delta x \to 0} \frac{1}{\Delta x}\left\{\frac{f(x + \Delta x)}{g(x + \Delta x)} - \frac{f(x)}{g(x)}\right\} && \text{Definition 19.7} \\[2mm]
&= \lim_{\Delta x \to 0} \frac{1}{\Delta x}\left\{\frac{f(x + \Delta x)g(x) - f(x)g(x + \Delta x)}{g(x + \Delta x)g(x)}\right\} && \text{Common denominator}
\end{aligned}
$$

To be able to take the limit of the expression on the right-hand side of the equal sign, we use a similar approach to that used in finding the derivative of a product. When we add and subtract $f(x)g(x)$ in the numerator, the form of the expression changes, but the value does not.

$$h'(x) = \lim_{\Delta x \to 0} \frac{1}{\Delta x} \left\{ \frac{f(x + \Delta x)g(x) - f(x)g(x) - f(x)g(x + \Delta x) + f(x)g(x)}{g(x + \Delta x)g(x)} \right\}$$

$$= \lim_{\Delta x \to 0} \frac{1}{\Delta x} \left\{ \frac{[f(x + \Delta x) - f(x)]g(x) - [g(x + \Delta x) - g(x)]f(x)}{g(x + \Delta x)g(x)} \right\} \quad \textbf{Common factor}$$

$$= \lim_{\Delta x \to 0} \left\{ \frac{\left[\dfrac{f(x + \Delta x) - f(x)}{\Delta x} \right]g(x) - \left[\dfrac{g(x + \Delta x) - g(x)}{\Delta x} \right]f(x)}{g(x + \Delta x)g(x)} \right\}$$

$$= \frac{\displaystyle\lim_{\Delta x \to 0} \left[\frac{f(x + \Delta x) - f(x)}{\Delta x} \right]g(x) - \lim_{\Delta x \to 0} \left[\frac{g(x + \Delta x) - g(x)}{\Delta x} \right]f(x)}{\displaystyle\lim_{\Delta x \to 0} [g(x + \Delta x)g(x)]} \quad \textbf{Definition 19.2}$$

Let us look at each piece of the fourth expression separately. (They are numbered for easy reference.) ① $\lim_{\Delta x \to 0} \left[\dfrac{f(x + \Delta x) - f(x)}{\Delta x} \right]g(x)$, by definition is $f'(x)g(x)$. Similarly, ② $\lim_{\Delta x \to 0} \left[\dfrac{g(x + \Delta x) - g(x)}{\Delta x} \right]f(x)$ is $g'(x)f(x)$, ③ $\lim_{\Delta x \to 0} [g(x + \Delta x)g(x)] = \lim_{\Delta x \to 0} g(x + \Delta x) \lim_{\Delta x \to 0} g(x) = g(x)g(x)$ or $[g(x)]^2$. Therefore,

$$h'(x) = \frac{f'(x)g(x) - f(x)g'(x)}{[g(x)]^2}. \quad \blacklozenge$$

In Section 19.5 we learned that if $f(x) = x^n$, the power rule holds when n is a positive integer. That is,

$$f'(x) = nx^{n-1}. \quad \textbf{Power rule}$$

We can show that the power rule also holds when n is a negative integer using it and the quotient rule. If n is a negative integer, then $-n$ is positive.

$$\frac{d(x^n)}{dx} = \frac{d}{dx}\left(\frac{1}{x^{-n}}\right)$$

$$= \frac{(x^{-n})\dfrac{d(1)}{dx} - 1\dfrac{d(x^{-n})}{dx}}{(x^{-n})^2} \quad \textbf{Quotient rule}$$

$$= \frac{(x^{-n})(0) - 1(-nx^{-n-1})}{x^{-2n}} \quad \begin{array}{l}\textbf{Using power rule}\\ \textbf{to differentiate } x^{-n}\end{array}$$

$$= \frac{nx^{-n-1}}{x^{-2n}}$$

$$= nx^{-n-1+2n}$$

$$= nx^{n-1}$$

Therefore, if $f(x) = x^n, f'(x) = nx^{n-1}$ holds for any integer n.

EXAMPLE 7 **Derivative of x^n When n Is Negative**

If $f(x) = \dfrac{4}{x^3}$, determine $f'(x)$.

Solution We can write $f(x) = 4x^{-3}$ and differentiate.

$$f'(x) = 4(-3)x^{-3-1}$$
$$= -12x^{-4}$$
$$= \frac{-12}{x^4}$$

This approach is often easier than using the quotient rule. ◆

In general, if $f(x) = x^n$, then $f'(x) = nx^{n-1}$ for any real number n.

EXAMPLE 8 **Derivative of x^n Where n Is a Rational Number**

If $f(x) = x^{\frac{3}{2}}$, determine $f'(x)$.

Solution

$$f'(x) = \frac{3}{2} x^{(\frac{3}{2}-1)}$$
$$= \frac{3}{2} x^{\frac{1}{2}} ◆$$

To determine the derivative of a function involving a radical, we must change the radical to exponential form. For example, to find the derivative of $f(x) = \sqrt[3]{x^4}$, rewrite the function as $f(x) = x^{\frac{4}{3}}$. The derivative, then, is $f'(x) = \dfrac{4}{3} x^{\frac{1}{3}}$.

EXAMPLE 9 **Application: Saving the Windshield**

For a cracked windshield, the stress concentration k at the end of the crack is given by $k = \sqrt{\dfrac{8c}{w}}$, where c is the length of the crack and w is the width of the crack. Determine the change of k with respect to w. Use this result to explain why the carpenter reached for her drill at the beginning of the chapter to make the crack wider.

Solution   Given $k = \sqrt{\dfrac{8c}{w}}$, we take the derivative $\dfrac{dk}{dw}$ to find the change of k with respect to w. Since we can treat c as a constant, we can write the expression as:

$$k = \sqrt{8c}\; w^{-\frac{1}{2}}$$

$$\frac{dk}{dw} = \sqrt{8c}\left(-\frac{1}{2}\right)w^{-\frac{3}{2}}$$

$$= -\frac{1}{2}\sqrt{\frac{8c}{w^3}}$$

$$= -\sqrt{\frac{2c}{w^3}}.$$

Since $\dfrac{dk}{dw}$ is negative, positive values of w (which are the only physically possible values) cause k to decrease. (Notice also that the rate of decrease of k becomes less as w increases.) ◆

19.6 Exercises

In Exercises 1–12, determine the derivative $\dfrac{dy}{dx}$ using either the product or quotient rule.

1. $y = (x + 1)(x + 3)$

2. $y = (x - 3)(x + 11)$

3. $y = (3x + 7)(5x + 9)$

4. $y = (13x - 4)(9x - 11)$

5. $y = (x^2 + 7x + 5)(5x - 9)$

6. $y = (4x^2 - 7x + 9)(2x - 13)$

7. $y = \dfrac{x + 4}{x - 3}$

8. $y = \dfrac{x + 11}{x + 9}$

9. $y = \dfrac{2x - 9}{3x - 11}$

10. $y = \dfrac{5x - 17}{11x + 13}$

11. $y = \dfrac{x^2 - 5x + 7}{x - 3}$

12. $y = \dfrac{x - 5}{x^2 + 11x - 6}$

In Exercises 13–24, determine the derivative $f'(x)$ using either the product or quotient rule.

13. $f(x) = (x^2 + 7x + 3)(x^2 + 5x + 6)$

14. $f(x) = (3x^2 - 2x + 1)(5x^2 + 7x + 4)$

15. $f(x) = (x^6 - 13)(2x^4 + 9)$

16. $f(x) = (x^7 - 9)(3x^{11} + 13)$

17. $f(x) = (x^3 + 7x + 1)(x^{-3} + x^2 + x)$

18. $f(x) = (x^4 - 9)\left(x^5 + \dfrac{3}{x^4}\right)$

19. $f(x) = \dfrac{x^2 + 13x + 9}{x^2 + 11x + 3}$

20. $f(x) = \dfrac{x^2 + 9x + 11}{x^2 - 13x - 5}$

21. $f(x) = \dfrac{x^3 - 3x^2}{x^7 + 4}$

22. $f(x) = \dfrac{x^6}{x^5 - 7}$

23. $f(x) = \dfrac{(x + 2)(x + 3)}{x^2 + 5x + 11}$

24. $f(x) = \dfrac{x^3 - 7}{(x + 3)(2x + 1)}$

In Exercises 25–34, determine the derivative.

25. $y = x^{\frac{2}{3}}$

26. $y = 3x^{\frac{5}{2}}$

27. $y = x^{-\frac{2}{3}}$

28. $y = \dfrac{4}{x^5}$

29. $y = \dfrac{-2}{x^3}$

30. $y = \dfrac{3}{x^{-4}}$

31. $y = \sqrt{3x^5}$

32. $y = \sqrt[3]{x^2}$

33. $y = \dfrac{x^3}{\sqrt{x^7}}$

34. $y = \dfrac{\sqrt[5]{x^3}}{x^4}$

In Exercises 35–40, determine the values of x for which the tangent line to the curve has a slope of zero.

35. $y = (x^2 - 2x - 5)(x + 1)$

36. $y = (x^2 - 2x + 1)(2x - 5)$

37. $y = (x^2 - 4x + 3)(x - 2)$

38. $y = \dfrac{3x^2}{x - 1}$

39. $y = \dfrac{x - 1}{3x^2}$

40. $y = \dfrac{x^2 + 2x + 1}{x^2 - 3x + 6}$

41. Determine the equation of the line tangent to the curve $y = (x^3 - 1)(x^2 + 3x + 2)$ at the point $(1, 0)$.

42. Let $C = \dfrac{q}{30 - q^2}$ be the total cost to produce q units of a commodity. Determine the marginal cost when the production is five units. (That is, determine $\dfrac{dC}{dq}$ when $q = 5$.)

43. One model of automobile depreciates according to the formula $V = \dfrac{9600}{1 + 0.4t + 0.1t^2}$, where t represents the time since purchase in years. Determine the instantaneous rate at which the car is depreciating for each of the following.
   a. $t = 1$ yr
   b. $t = 2$ yr

44. As blood moves from the heart through the major arteries out to the capillaries and back through the veins, the systolic pressure continuously drops. Suppose that this pressure is given by $P = \dfrac{25t^2 + 125}{t^2 + 1}$, $0 < t \leq 10$, where P is measured in millimeters of mercury and t is measured in seconds. At what rate is the pressure dropping 5 s after leaving the heart?

45. The bending moment M of a simply supported beam is a function of the distance x measured from the left end of the beam. (See the drawing.) The change in bending moment M with respect to the distance x is called the shear force and is denoted by $V = \dfrac{-dM}{dx}$. The bending moment in a certain I-beam is given by $M = \dfrac{4x^3}{2(x^2 - 7)}$. Determine the equation for the shear force.

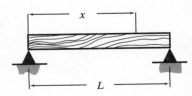

46. Suppose that the temperature T of food placed in a freezer drops according to the equation $T = \dfrac{700}{t^2 + 4t + 10}$, where t is the time in hours. Determine the rate of change of T with respect to t for the following.
   a. $t = 1$ hr
   b. $t = 2$ hr

47. For a thin lens of constant focal length P, the object distance x and the image distance y are related by the formula $\dfrac{1}{x} + \dfrac{1}{y} = \dfrac{1}{p}$.
   a. Solve for y in terms of x and P.
   b. Determine the rate of change of y with respect to x.

48. An object moves along a straight line in such a way that at the end of t seconds its distance s in feet from the starting point is given by $s = 8t + \dfrac{t^2}{2}$, with $t > 0$. Determine the velocity of the object at the instant when $t = 2$ s.

49. Ohm's law is expressed as $v = iR$. In a certain circuit, the current varies as $i = 2.00 + 0.0032t^3$. Due to temperature changes, the resistor varies in a complicated fashion too: $R = 22 + 0.42t^2$. All of this ultimately is the result caused by a changing voltage. Determine the rate of change of voltage at $t = 2.0$ s and $t = \underline{20}$ s.

19.7

Chain Rule—Power Function

Can we find the derivative of $f(x) = (4x - 3)^2$? The answer is yes, if we expand the binomial and write:

$$f(x) = 16x^2 - 24x + 9,$$

then

$$f'(x) = 32x - 24. \quad \textbf{Derivative of a sum}$$

Is it possible to take the derivative of $f(x) = (4x - 3)^3$? The answer is yes, if we expand the binomial expression. Applying the binomial expansion, we have:

$$f(x) = 64x^3 - 144x^2 + 108x - 27$$
$$f'(x) = 192x^2 - 288x + 108. \quad \textbf{Derivative of a sum}$$

However, we encounter a real problem when we try to find the derivative of $f(x) = (4x - 3)^6$. The process of expanding the binomial in this case is a terrible, time-consuming task. So let us see if there is an easier way. First we need to re-examine the result of the two derivatives we found in the first two cases.

$$f(x) = (4x - 3)^2 \qquad \textbf{Original function}$$
$$f'(x) = 32x - 24 \qquad \textbf{Derivative of the function}$$
$$= 2(16x - 12) \qquad \textbf{Factor a common factor of 2}$$
$$= 2(4x - 3) \cdot 4 \qquad \textbf{Factor a common factor of 4}$$

Why didn't we factor out a common term of 8? We wanted to have the result expressed in a special way. For $f(x) = (4x - 3)^2$, if we let $g(x) = 4x - 3$, then:

$$f(x) = [g(x)]^2 \quad \text{Substituting}$$

and

$$f'(x) = \boxed{2} [g(x)] \cdot \boxed{4}.$$

Also note that since $g(x) = 4x - 3$, then $g'(x) = 4$. Thus,

$$f'(x) = \boxed{2} [g(x)] \cdot \boxed{g'(x)}. \quad \text{Substituting}$$

Now let us examine $f(x) = (4x - 3)^3$ in the same manner.

$f(x) = (4x - 3)^3$	**Original function**
$f'(x) = 192x^2 - 288x + 108$	**Derivative of the function**
$= 3(64x^2 - 96x + 36)$	**Factor a common factor of 3**
$= 3(16x^2 - 24x + 9)4$	**Factor a common factor of 4**
$= \boxed{3} (4x - 3)^2 \boxed{4}$	**Factor $16x^2 - 24x + 9$**

If we again let $g(x) = 4x - 3$, then:

$$f(x) = [g(x)]^3 \quad \text{Substituting}$$

and

$$f'(x) = \boxed{3} [g(x)]^2 \cdot \boxed{4}.$$

Thus,

$$f'(x) = 3[g(x)]^2 \cdot \boxed{g'(x)}. \quad \text{Substituting, since } g(x) = 4x - 3,\ g'(x) = 4$$

The result in both cases takes the following form: If $f(x) = [g(x)]^n$, then $f'(x) = n[g(x)]^{n-1} \cdot g'(x)$. In words, if $f(x)$ is equal to an expression in x raised to a power of n, then $f'(x)$ is equal to the product of n times the expression to the $n - 1$ power times the derivative of the expression with respect to the variable x. The statement is known as the *General power rule*.

> **General Power Rule**
>
> Let $g(x)$ be a differentiable function and let $f(x) = [g(x)]^n$; then:
>
> $$f'(x) = n[g(x)]^{n-1} \cdot g'(x),$$
>
> where n is a real number.

The rule can also be written as "Let $f(x)$ be a differentiable function and let $y = [f(x)]^n$; then $\dfrac{dy}{dx} = n[f(x)]^{n-1} \cdot \dfrac{d[f(x)]}{dx}$, where n is a real number."

EXAMPLE 1   **General Power Rule**

Given $f(x) = (11x^2 - 7)^8$, determine $f'(x)$.

Solution   To determine $f'(x)$, we apply the general power rule.

$$f(x) = (11x^2 - 7)^8 \qquad \text{Given}$$
$$f'(x) = n[g(x)]^{n-1} \cdot g'(x) \qquad \text{General power rule}$$

Let $g(x) = 11x^2 - 7$ and $g'(x) = 22x$.

$$f'(x) = 8 (11x^2 - 7)^{8-1} \cdot (22x) \qquad \text{Substituting}$$
$$= 176x(11x^2 - 7)^7 \quad \blacklozenge$$

Determining the derivative using the general power rule may be simpler if we think of it in the following manner.

$$f(x) = \underbrace{(11x^2 - 7)}_{\text{Inside}} \underbrace{{}^8}_{\text{Outside}}$$

$$f'(x) = \underbrace{8(11x^2 - 7)^{8-1}}_{\substack{\downarrow \\ \text{Derivative of outside}}} \cdot \underbrace{(2 \cdot 11x^{2-1})}_{\text{Derivative of inside}}$$

EXAMPLE 2   **General Power Rule**

Given $f(x) = 7(5x^2 - 3x + 1)^4$, determine $f'(x)$.

Solution   The constant 7 does not change our approach to the problem. We use the power rule to find the derivative. Recall that in Section 19.4 we learned that if $f(x) = cx^n$, where c is a constant, then $f'(x) = c(nx^{n-1})$.

$$f(x) = 7(5x^2 - 3x + 1)^4 \qquad \text{Given}$$

Letting $g(x) = 5x^2 - 3x + 1$, then $g'(x) = 10x - 3$.

$$f'(x) = c\{n[g(x)]^{n-1} \cdot g'(x)\} \qquad \text{Power rule}$$
$$= 7[\, 4 \,(5x^2 - 3x + 1)^{4-1} (10x - 3)\,] \qquad \text{Substituting}$$
$$= 28(10x - 3)(5x^2 - 3x + 1)^3 \qquad \text{Rearranging} \quad \blacklozenge$$

The power function is a special case of a more general rule called the *chain rule*. The chain rule is a technique for finding the derivative of a composite function.

Chain Rule

Let $y = f(u)$ and $u = g(x)$ be differentiable functions. Then:

$$\frac{dy}{dx} = \frac{d}{dx}[f(u)]$$

$$= \frac{d}{du}[f(u)] \cdot \frac{d}{dx}[g(x)].$$

Or more simply:

$$\frac{dy}{dx} = \frac{dy}{du} \cdot \frac{du}{dx}.$$

To illustrate this rule, let us express the function $f(x) = (4x - 3)^3$ as:

$$y = (4x - 3)^3$$

and

$y = u^3$ **Letting $u = 4x - 3$ and substituting**

$\dfrac{dy}{du} = 3u^2;$ **Differentiating y with respect to u**

also

$\dfrac{du}{dx} = 4$ **Differentiating u with respect to x**

$\dfrac{dy}{dx} = \dfrac{dy}{du} \cdot \dfrac{du}{dx}.$ **Chain rule**

Thus,

$$\frac{dy}{dx} = (\,3u^2\,)(\,4\,)$$ **Substituting**

or

$\dfrac{dy}{dx} = 3(4x - 3)^2(4)$ **Substituting ($u = 4x - 3$)**

$= 12(4x - 3)^2$ **Rearranging**

Now we can see that the chain rule and the power rule give the same result.

EXAMPLE 3   **Application: Pulley System**

A motor-driven lift has a series of three pulleys, as shown in Figure 19.23. The ratios of the radii of the pulleys are $4:1:3$. Determine the rate at which C turns in relation to A's rate.

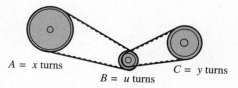

$A = x$ turns

$B = u$ turns

$C = y$ turns

FIGURE 19.23

Solution   If pulley A turns x times, then pulley B turns $u = 4x$ times, and pulley C turns $y = \dfrac{u}{3}x$ times. The instantaneous rate of change, or derivative, is:

$$\frac{dy}{du} = \frac{1}{3},$$

and C turns at $\dfrac{1}{3}$ B's rate. Also,

$$\frac{du}{dx} = 4,$$

and B turns at 4 times A's rate.

Using the chain rule, we can determine the rate of change of y with respect to x, $\dfrac{dy}{dx}$, or the rate at which C turns for each turn of A.

$$\frac{dy}{dx} = \frac{dy}{du} \cdot \frac{du}{dx}$$

$$= \frac{1}{3} \cdot 4$$

$$= \frac{4}{3}$$

The result is that C turns at four-thirds A's rate, or four-thirds of a turn for C for each of A's turns.   ◆

Example 4 illustrates that the chain rule is helpful in determining the derivative of a function that conatins a radical.

EXAMPLE 4   Chain Rule: Radical

Given $f(x) = \sqrt{x^3 + 7x^2 + 5}$, determine $\dfrac{dy}{dx}$.

Solution   We use the chain rule to find the derivative.

$$y = (x^3 + 7x^2 + 5)^{\frac{1}{2}}$$ **Writing in exponential form**

$$y = u^{\frac{1}{2}}$$ **Letting $u = x^3 + 7x^2 + 5$ and substituting**

$$\frac{dy}{du} = \frac{1}{2}u^{-\frac{1}{2}}$$ **Differentiating y with respect to u**

$$\frac{du}{dx} = 3x^2 + 14x$$ **Differentiating u with respect to x**

$$\frac{dy}{dx} = \frac{dy}{du} \cdot \frac{du}{dx}$$ **Chain rule**

$$\frac{dy}{dx} = \left(\frac{1}{2}u^{-\frac{1}{2}} \right)(3x^2 + 14x)$$ **Substituting**

$$= \frac{1}{2}(x^3 + 7x^2 + 5)^{-\frac{1}{2}}(3x^2 + 14x)$$ **Substituting**

$$= \frac{3x^2 + 14x}{2\sqrt{x^3 + 7x^2 + 5}}$$ **Law of exponents and rearranging**   ◆

EXAMPLE 5   Chain Rule: Product

Given $y = (3x + 7)^4(x^2 + 3)$, determine $\dfrac{dy}{dx}$.

Solution   This problem is different in that it is a product of two quantities, but the first quantity is raised to a power. Therefore, first apply the product rule and then apply the chain rule when finding the derivative of the first quantity.

$$\frac{dy}{dx} = \frac{d}{dx}[(3x + 7)^4](x^2 + 3) + (3x + 7)^4 \frac{d}{dx}(x^2 + 3)$$

Product rule

$$\frac{d}{dx}[(3x + 7)^4] = 4(3x + 7)^3(3)$$

Using the chain rule, where
$$u = 3x + 7 \text{ so } \frac{dy}{du} = 4u^3$$
and $\dfrac{du}{dx} = 3$

$$\frac{dy}{dx} = 12(3x + 7)^3 (x^2 + 3) + (3x + 7)^4 (2x)$$

Substituting and differentiating

$$= 12(x^2 + 3)(3x + 7)^3 + 2x(3x + 7)^4$$
$$= 12(x^2 + 3)(3x + 7)^3 + 2x(3x + 7)(3x + 7)^3$$
$$= (3x + 7)^3(18x^2 + 14x + 36). \quad \text{**Factoring**} \quad \blacklozenge$$

EXAMPLE 6   **Chain Rule: Slope of Tangent Line**

Determine the slope of the tangent line to the curve $y = \left(\dfrac{x + 2}{x - 1}\right)^3$ for the following.

a. At any point   **b.** At the point (2, 64)

Solution

a. The slope of the tangent line at any point is found by taking the derivative of the function. However, to find the derivative we must first apply the chain rule and then the quotient rule.

$$\frac{dy}{dx} = 3\left(\frac{x + 2}{x - 1}\right)^2 \frac{d}{dx}\left(\frac{x + 2}{x - 1}\right) \quad \text{**Chain rule**}$$

$$\frac{d}{dx}\left(\frac{x + 2}{x - 1}\right) = \frac{(x - 1)(1) - (x + 2)(1)}{(x - 1)^2} \quad \text{**Quotient rule**}$$

$$= \frac{-3}{(x - 1)^2}$$

$$\frac{dy}{dx} = 3\left(\frac{x + 2}{x - 1}\right)^2\left(\frac{-3}{(x - 1)^2}\right) \quad \text{**Substituting**}$$

$$= \frac{-9(x + 2)^2}{(x - 1)^3} \quad \text{**Multiplying**}$$

Thus, the slope of the tangent line at any point of the curve is:

$$\frac{dy}{dx} = \frac{-9(x + 2)^2}{(x - 1)^3}.$$

b. To determine the slope of the tangent line at the point (2, 64), all we need to do is substitute 2 for x in part **a.**

$$\frac{dy}{dx} = \frac{-9(2 + 2)^2}{(2 - 1)^3}$$

$$= -144$$

Therefore, the slope of the tangent line at the point (2, 64) is -144. $\blacklozenge$

19.7 Exercises

In Exercises 1–14, determine the indicated derivative.

1. $s = 5(7 - t)^4$, $\dfrac{ds}{dt} = ?$

2. $y = 2(x^3 - 1)^7$, $\dfrac{dy}{dx} = ?$

3. $w = 4(x^3 - 4x + 2)^5$, $\dfrac{dw}{dx} = ?$

4. $y = 5(5t - t^2)^3$, $\dfrac{dy}{dx} = ?$

5. $x = -3(4 - 11s^2)^5$, $\dfrac{dx}{ds} = ?$

6. $M = \dfrac{5(3x - 2)^3}{7}$, $\dfrac{dM}{dx} = ?$

7. $y = \dfrac{(4x - x^3)^{11}}{5}$, $\dfrac{dy}{dx} = ?$

8. $y = \sqrt{3x - 11}$, $\dfrac{dy}{dx} = ?$

9. $u = \sqrt[3]{1 - 3t^2}$, $\dfrac{du}{dt} = ?$

10. $u = \sqrt[5]{(1 - t^2)^2}$, $\dfrac{du}{dt} = ?$

11. $s = \dfrac{1}{(3t + 1)^7}$, $\dfrac{ds}{dt} = ?$

12. $s = \dfrac{5}{8(3 - 2t)^6}$, $\dfrac{ds}{dt} = ?$

13. $R = \dfrac{1}{(2x - 1)^8}$, $\dfrac{dR}{dx} = ?$

14. $R = \dfrac{1}{5(4x^2 - 7)^7}$, $\dfrac{dR}{dx} = ?$

In Exercises 15–24, determine the derivative, $f'(x)$.

15. $f(x) = (2x - 5)^3(5x - 7)$

16. $f(x) = (3x^2 + 7)^4(2x^2 - 9)$

17. $f(x) = \dfrac{(x + 2)^2}{x - 1}$

18. $f(x) = \dfrac{2x - 3}{(x - 4)^3}$

19. $f(x) = \left(\dfrac{2x - 5}{x - 4}\right)^4$

20. $f(x) = \left(\dfrac{3x - 8}{x + 9}\right)^7$

21. $f(x) = x\sqrt{2x^2 + 11}$

22. $f(x) = 3x\sqrt[3]{3x + 7}$

23. $f(x) = \dfrac{\sqrt{2x + 11}}{(3x - 8)^2}$

24. $f(x) = (2x - 9)^2\sqrt{3x + 7}$

25. Given the equation $y = (x^2 - 2x)^3$, perform the following.
  a. Determine the slope of the tangent line at $(1, -1)$.
  b. At which point on the curve is the slope of the tangent line zero?

26. A particle moves along a straight line so that its position at time t is given by $x = \sqrt{1 + t^2} - 1$, $t \geq 0$. What is its velocity at the following times?
  a. $t = 2$ s   **b.** $t = 10$ s

27. Given the function $y = (x - 2)^4(x + 3)^4$, perform the following.
  a. Differentiate it as a product of two different composite functions.
  b. Differentiate the equivalent function $y = (x^2 + x - 6)^4$.
  c. Show that both methods give the same result.

28. Given the function $y = \dfrac{1}{x^7 - 3x^4 + 11}$, perform the following.
  a. Differentiate it as a quotient.
  b. Differentiate the equivalent composite function $y = (x^7 - 3x^4 + 11)^{-1}$.
  c. Show that both methods give the same result.

29. When an electrostatic charge of $q = 5.00 \times 10^{-4}$ coulombs is a distance r meters from a stationary charge $Q = 3.00 \times 10^{-6}$ coulombs, the magnitude of the force of repulsion of Q on q is $F = \dfrac{Qq}{4\pi\epsilon_0 r^2}$ N, where $\epsilon_0 = 8.85 \times 10^{-12}$. If q is moved directly away from Q at 2 m/s, how fast is F changing when $r = 2$ m?

30. If R ohms resistance is in series with X ohms reactance, the impedance is $Z = \sqrt{R^2 + X^2}$ ohms. If $R = 6\,\Omega$, determine the instantaneous rate of change of Z with respect to X.

31. Starting at $t = 0$, the charge on a certain capacitor is given by $q(t) = \dfrac{t}{t^2 + 9}$. At what instant is the current to the capacitor equal to zero?

32. The period of a pendulum is directly proportional to the square root of its length. Determine the derivative of the period with respect to the length for a pendulum 1 m long that has a period of $\dfrac{\pi}{35}$.

33. When the volume of a gas is changed suddenly, the pressure P is given by $P = 307\,V^{\frac{3}{2}}$. Determine the change of the pressure with respect to the change in volume for a gas when the initial volume is 1.0 m³. What is the pressure when the volume is 3.0 m³? The result is measured in pressure per unit of volume.

34. The total solar radiation R on a cloudless day during the daylight is $R = \dfrac{390}{(0.87t^6 + 100)^{\frac{1}{2}}}$, where t is measured in hours from 12 P.M. (noon). Hence, 3:00 P.M. has $t = 3$, and 9:00 P.M. has $t = 9$. What is the rate at which R is changing when the time is 5:00 P.M.? What is the rate at which R is changing when the time is 5:30 P.M.?

19.8

Differentiation of Implicit Functions

A function of the form $y = f(x)$ is called an **explicit function;** that is, the dependent variable y is expressed explicitly in terms of the independent variable x. With the rules discussed to this point, we can find the derivatives of explicit functions. However, we can be asked to find the derivatives of the following types of expressions:

$$x + y - 2x^2 + 1 = 0$$
$$3x^2 + y^2 - 4y = 0 \qquad y \geq 2$$
$$xy^3 + 11x^2 + 5xy^2 = 4.$$

These expressions are examples of **implicit functions.** In each case, the expression defines y as an implicit function of x. We easily can solve the first equation for y in terms of x.

$$y = 2x^2 - x - 1$$

We just as easily can determine the derivative.

$$\frac{dy}{dx} = 4x - 1$$

The derivative of the equation $x + y - 2x^2 + 1 = 0$ also can be found by another method, which does not require that we solve for y in terms of x. This method is called **implicit differentiation.** Remember that y is a function of x and that we want to find the rate of change of y with respect to x, $\dfrac{dy}{dx}$.

$$x + y - 2x^2 + 1 = 0$$

$$\frac{d}{dx}(x + y - 2x^2 + 1) = \frac{d}{dx}(0)$$ **Differentiating both sides with respect to x**

$$\frac{d}{dx}(x) + \frac{d}{dx}(y) - \frac{d}{dx}(2x^2) + \frac{d}{dx}(1) = \frac{d}{dx}(0)$$ **Derivative of the sum is the sum of the derivatives**

$$1 \cdot \frac{dx}{dx} + 1 \cdot \frac{dy}{dx} - 4x \cdot \frac{dx}{dx} + \quad 0 \quad = \quad 0$$

$\dfrac{dx}{dx}$ indicates that we have taken the derivative of x with respect to x, and $\dfrac{dx}{dx} = 1$.

$\dfrac{d}{dx}(y) = 1 \cdot \dfrac{dy}{dx}$ indicates we have taken the derivative of y with respect to x. Thus,

$$1 \cdot \boxed{1} + 1 \cdot \boxed{\frac{dy}{dx}} - 4 \cdot \boxed{1} + 0 = 0$$

$$1 + \frac{dy}{dx} - 4x + 0 = 0 \qquad \textbf{Substituting}$$

$$\frac{dy}{dx} = 4x - 1. \quad \textbf{Solving for } \frac{dy}{dx}$$

Therefore, we see that the same result is obtained by both methods.

We also can solve the second equation $3x^2 + y^2 - 4y = 0$ for y in terms of x and obtain the explicit function $y = 2 + \sqrt{4 - 3x^2}$. Differentiating, the result is $\dfrac{dy}{dx} = \dfrac{-3x}{(4 - 3x^2)^{\frac{1}{2}}}$. We can obtain the same result by differentiating implicitly.

$$3x^2 + y^2 - 4y = 0$$

$$\frac{d}{dx}(3x^2 + y^2 - 4y) = \frac{d}{dx}(0)$$ **Differentiating both sides with respect to x**

$$\frac{d}{dx}(3x^2) + \frac{d}{dx}(y^2) - \frac{d}{dx}(4y) = \frac{d}{dx}(0)$$ **Derivative of the sum is the sum of the derivatives**

$$3(2x) \cdot \frac{dx}{dx} + 2y \cdot \frac{dy}{dx} - 4(1) \cdot \frac{dy}{dx} = 0$$

$$6x(\boxed{1}) + 2y\boxed{\frac{dy}{dx}} - 4\boxed{\frac{dy}{dx}} = 0$$

$$6x + (2y - 4)\frac{dy}{dx} = 0 \qquad \textbf{Factoring}$$

$$(2y - 4)\frac{dy}{dx} = -6x$$

$$\frac{dy}{dx} = \frac{-6x}{2y - 4} \qquad \textbf{Solving for } \frac{dy}{dx}$$

Show that when $2 + \sqrt{4 - 3x^2}$ is substituted for y that $\dfrac{dy}{dx} = \dfrac{-3x}{(4 - 3x^2)^{\frac{1}{2}}}$.

Using the pattern from the examples, we can develop a procedure for determining the derivative of an implicit function.

PROCEDURE FOR DETERMINING IMPLICIT DIFFERENTIATION

1. Differentiate both sides of the equation with respect to x.

2. Collect the terms with $\dfrac{dy}{dx}$ on one side of the equal sign.

3. Factor out $\dfrac{dy}{dx}$.

4. Solve for $\dfrac{dy}{dx}$ by dividing.

The third equation is extremely difficult to solve for y. Can we still determine the derivative? Yes, using implicit differentiation. Remember that y is a function of x and that we want to determine the rate of change of y with respect to x, $\left(\dfrac{dy}{dx}\right)$.

$$xy^3 + 11x^2 + 5xy^2 = 4$$

$$\frac{d}{dx}(xy^3 + 11x^2 + 5x^2y) = \frac{d}{dx}(4) \qquad \begin{array}{l}\textbf{Differentiating}\\ \textbf{both sides with}\\ \textbf{respect to } x\end{array}$$

$$\overset{\textcircled{1}}{} \qquad \overset{\textcircled{2}}{} \qquad \overset{\textcircled{3}}{} \qquad \overset{\textcircled{4}}{}$$

$$\frac{d}{dx}(xy^3) + \frac{d}{dx}(11x^2) + \frac{d}{dx}(5x^2y) = \frac{d}{dx}(4) \qquad \begin{array}{l}\textbf{Derivative of}\\ \textbf{sum is the}\\ \textbf{sum of the}\\ \textbf{derivatives}\end{array}$$

Now examining each part (identified by number) separately, we have the following.

$$\textcircled{1}\frac{d}{dx}(xy^3) = \frac{d(x)}{dx}y^3 + x \cdot \frac{d}{dx}(y^3) \qquad \begin{array}{l}\textbf{Derivative of}\\ \textbf{product}\end{array}$$

$$= 1 \cdot \frac{dx}{dx} \cdot y^3 + x \cdot 3y^2 \cdot \frac{dy}{dx}$$

$$= y^3 + 3xy^2 \frac{dy}{dx}$$

966 Chapter 19 The Derivative

$$\text{②} \frac{d(11x^2)}{dx} = 11\frac{d(x^2)}{dx}$$

Derivative of constant times a function is equal to constant times derivative of function and power rule

$$= 11 \cdot 2x \cdot \frac{dx}{dx}$$

$$= 22x$$

$$\text{③} \frac{d}{dx}(5x^2y) = 5\frac{d}{dx}(x^2y)$$

$$= 5\left[\frac{dx}{dx}(x^2)y + x^2\frac{d}{dx}(y)\right]$$

Derivative of product and power rule

$$= 5\left[2x\frac{dx}{dx}y + x^2 \cdot 1 \cdot \frac{dy}{dx}\right]$$

$$= 10xy + 5x^2\frac{dy}{dx}$$

$$\text{④} \frac{d}{dx}(4) = 0. \quad \textbf{Derivative of constant}$$

Putting all the parts together, we have:

$$y^3 + 3xy^2\frac{dy}{dx} + 22x + 10xy + 5x^2\frac{dy}{dx} = 0.$$

Remember we want to find $\frac{dy}{dx}$; collecting all terms on left side containing the factor $\frac{dy}{dx}$, we have:

$$3xy^2\frac{dy}{dx} + 5x^2\frac{dy}{dx} = -y^3 - 22x - 10xy$$

$$(3xy^2 + 5x^2)\frac{dy}{dx} = -y^3 - 22x - 10xy \quad \textbf{Common factor}$$

$$\frac{dy}{dx} = \frac{-y^3 - 22x - 10xy}{3xy^2 + 5x^2}. \quad \begin{array}{l}\textbf{Dividing by}\\ \mathbf{3xy^2 + 5x^2} \textbf{ gives the}\\ \textbf{desired derivative}\end{array}$$

EXAMPLE 1   Tangent Line: Implicit Differentiation

Determine the slope of the tangent line to the ellipse $x^2 + 4y^2 + 4x = 5$ at the point $(1, 0)$.

Solution To determine the slope of the tangent line to the ellipse $x^2 + 4y^2 + 4x = 5$ we must determine $\dfrac{dy}{dx}$. Using implicit differentiation, we have:

$$\frac{d}{dx}(x^2 + 4y^2 + 4x) = \frac{d}{dx}(5) \qquad \text{Differentiating both sides}$$

$$\frac{d}{dx}(x^2) + \frac{d}{dx}(4y^2) + \frac{d}{dx}(4x) = 0 \qquad \text{Derivative of a sum}$$

$$2x\frac{dx}{dx} + 4 \cdot 2y\frac{dy}{dx} + 4 \cdot 1\frac{dx}{dx} = 0 \qquad \text{Constant and power rule}$$

$$2x + 8y\frac{dy}{dx} + 4 = 0$$

$$x + 4y\frac{dy}{dx} + 2 = 0 \qquad \text{Dividing each term by 2}$$

$$\frac{dy}{dx} = \frac{-x - 2}{4y}. \qquad \text{Solving for } \frac{dy}{dx}$$

At the point (1, 0),

$$\frac{dy}{dx} = \frac{-(\boxed{1}) - 2}{4(\boxed{0})}.$$

The slope is undefined; therefore, the tangent line is parallel to the y-axis. ◆

19.8 Exercises

In Exercises 1–24, determine $\dfrac{dy}{dx}$ using implicit differentiation.

1. $3x + 7y = 5$
2. $5x - 4y = 12$
3. $x^2 + y^2 = 25$
4. $x^2 + y^2 = 9$

5. $3x^2 + 5y^2 = 12$
6. $4x^2 + 7y^2 = 28$
7. $x^2 - y^2 = 1$
8. $2x^2 - 3y^2 = 6$

9. $5x^2 + 7y^2 - 4 = 0$
10. $7x^2 - 11y^2 + 15 = 0$
11. $xy = 7$
12. $2xy^2 = 5$

13. $xy + 3x = 7$
14. $xy = 2x + 5$
15. $x^2y + xy^2 - x = 3$
16. $x^2y - xy^2 + y = 4$

17. $3y^2 = \dfrac{x}{x - 2}$
18. $y^2 = \dfrac{2x + 1}{x - 3}$
19. $x^2 + 3y^3 - xy = 0$
20. $x^3 - y^2 + xy = 3$

21. $2y^2 + x^2 - x^2y^3 = 0$
22. $3y^2 + 4x - x^3y = 5$
23. $2x^2y^3 + x^7 - 3y^5 = x^2$

24. $x^3y^2 + y^6 - 3x^4 = x^3$

25. Determine the slope of the tangent line to the hyperbola $3x^2 - 7y^2 + 14y = 27$ at the point $(-3, 0)$.

26. Determine the slope of the tangent line to the ellipse $4x^2 + 9y^2 - 16x - 54y + 61 = 0$ at the point (2, 5).

27. The graph of $x^3y^3 + 4y = 3x^2$ is a curve that passes through the point where $x = 2$ and $y = 1$. What is the slope of the curve at that point?

28. Determine the slope of the tangent line to the graph of the equation $x = 5y^3 - 4y^5$ at the point $(1, 1)$.

19.9

Higher-Order Derivatives

In Section 19.6 we learned that the derivative of distance with respect to time is the velocity $\frac{ds}{dt} = v$ and that the derivative of the velocity with respect to time is the acceleration, $\frac{dv}{dt} = a$. We determined the acceleration by taking the derivative of distance with respect to time; that is,

$$a = \frac{d}{dt}\left(\frac{ds}{dt}\right).$$

Thus, we find the acceleartion by taking the derivative of the distance with respect to time and then taking the derivative of that result with respect to time. This is the second derivative of distance s with respect to time t, and is expressed:

$$a = \frac{d^2s}{dt^2}.$$

In this formula the 2s are not exponents; rather, they indicate the second derivative. If the second derivative is differentiable, we can determine the third derivative, $\frac{d^3s}{dy^3}$. We can continue this process as long as the new expression is differentiable.

EXAMPLE 1 **Fifth Derivative of a Function**

Determine the first five derivatives of $y = 3x^4 + 2x^3 + 7$.

Solution

$$\frac{dy}{dx} = 12x^3 + 6x^2$$

$$\frac{d^2y}{dx^2} = 36x^2 + 12x$$

$$\frac{d^3y}{dx^3} = 72x + 12$$

$$\frac{d^4y}{dx^4} = 72$$

$$\frac{d^5y}{dx^5} = 0 \quad \blacklozenge$$

There are several different ways of indicating higher-order derivatives. They are displayed in the following box.

NOTATION FOR HIGHER-ORDER DERIVATIVES

First derivative: $y', f'(x), \dfrac{dy}{dx}, \dfrac{d}{dx}[f(x)], D_x(y)$

Second derivative: $y'', f''(x), \dfrac{d^2y}{dx^2}, \dfrac{d^2}{dx^2}[f(x)], D_x^2(y)$

Third derivative: $y''', f'''(x), \dfrac{d^3y}{dx^3}, \dfrac{d^3}{dx^3}[f(x)], D_x^3(y)$

Fourth derivative: $y^{(iv)}, f^{(iv)}(x), \dfrac{d^4y}{dx^4}, \dfrac{d^4}{dx^4}[f(x)], D_x^4(y)$

nth derivative: $y^{(n)}, f^{(n)}(x), \dfrac{d^n y}{dx^n}, \dfrac{d^n}{dx^n}[f(x)], D_x^n(y)$

EXAMPLE 2   **First Four Derivatives of a Rational Function**

Determine the first four derivatives of $f(x) = \dfrac{2}{x}$.

Solution   Since $f(x) = 2x^{-1}$, using the power rule, we have:

$$f'(x) = -2x^{-2} = \frac{-2}{x^2}$$

$$f''(x) = -4x^{-3} = \frac{4}{x^3}$$

$$f'''(x) = -12x^{-4} = \frac{-12}{x^4}$$

$$f^{(iv)}(x) = 48x^{-5} = \frac{48}{x^5} \quad \blacklozenge$$

Compare the results in Example 1 and Example 2. In Example 1, the degree of the polynomial is four, and the fifth derivative is zero. In fact, for any polynomial of degree n, the $n + 1$ derivative is zero. The rational function in Example 2, has a fourth derivative with a larger power of x in the denominator than in the original function. In general, regardless of how many times we differentiate a rational function of the form $\dfrac{f(x)}{g(x)}$ (the function is in reduced form, where $g(x)$ is a polynomial in x of degree greater than or equal to 1), the result is never zero.

19.9 Exercises

In Exercises 1–10, determine the indicated higher derivative of the function.

1. $f(x) = 3x - 7, f''(x)$

2. $f(x) = x^2 + 3x + 2, f''(x)$

3. $f(x) = 3x^3 + 4x + 5, f''(x)$

4. $f(x) = \sqrt[3]{x}, f'''(x)$

5. $f(x) = x + \dfrac{1}{x}, f'''(x)$

6. $f(x) = x - \dfrac{1}{x}, f'''(x)$

7. $f(x) = 1 + \dfrac{2}{x} - \dfrac{3}{x^2}, f''(x)$

8. $f(x) = 3x(x + 5)^3, f''(x)$

9. $s(t) = \sqrt{5t + 7}, s''(t)$

10. $s(t) = t\sqrt{3t - 1}, s''(t)$

In Exercises 11–20, determine the indicated higher derivative.

11. $y = x + 5, y''$

12. $y = 3x^5 - 7x^2, y''$

13. $y = 5x^4 - 7x^3 + 13x - 8, y'''$

14. $y = (1 + x)^2, y''$

15. $y = 1 - x^{-1} + x^{-2}, y''$

16. $y = 2x^{2/3} + 5x^{1/4}, y'''$

17. $y = \dfrac{x + 1}{x - 1}, y''$

18. $y = \dfrac{x}{2x + 3}, y''$

19. $y = (x + 3)(x^2 + 7x + 2)\, y''$

20. $y = (2x - 4)(x^2 + 7x + 9), y''$

21. A truck's velocity starting from rest is given by $v = \dfrac{100t}{2t + 20}$, where v is measured in feet per second. Find the acceleration at the following times.
   a. 5 s   **b.** 30 s

22. The position function $s(t) = -8.25t^2 + 66t$ tells the distance, s in feet, that a car traveling at a rate of 66 ft/s travels t seconds after the driver applies the brakes. Determine the following.
   a. $s(1)$ and $s(3)$
   b. $v(1)$ and $v(3)$
   c. $a(1)$ and $a(3)$

23. The formulas $v = L\dfrac{di}{dt}$ and $i = \dfrac{dq}{dt}$ were introduced in Section 19.5. Hence, $v = L\dfrac{d^2q}{dt^2}$. What is the voltage of a circuit whose charge amount is represented by the function $q = 3t^2 + 4t$?

24. The angular velocity of a wheel is given by $\omega = \dfrac{d\theta}{dt}$. The angular acceleration is given by a $\alpha = \dfrac{d\omega}{dt}$.
   a. Express the angular acceleration as a second derivative with respect to θ.
   b. If $\theta = 25t^2 - 3.0t$, evaluate α at 25 s.

Review Exercises

In Exercises 1–12, find the indicated limits.

1. $\lim\limits_{x \to 3} 2x$

2. $\lim\limits_{x \to 0} (11x + 9)$

3. $\lim\limits_{x \to 4} (3x^2 + 7x - 11)$

4. $\lim\limits_{x \to -3} (2x^2 + 9x - 15)$

5. $\lim\limits_{x \to -1} \dfrac{x^2 - 1}{x + 1}$

6. $\lim\limits_{x \to 3} \dfrac{x^2 - 3x}{x + 3}$

7. $\lim\limits_{x \to 0} \dfrac{x^4 + 2x^2 + 3}{x^3 + 1}$

8. $\lim\limits_{x \to 3} \dfrac{x^2 - 9}{x^2 - x - 6}$

9. $\lim\limits_{x \to 8+} \sqrt{x - 8}$

10. $\lim\limits_{x \to 7-} 4x^{x-7}$

11. $\lim\limits_{x \to \infty} \dfrac{4x^2 - 5}{5x^2 - 8x + 3}$

12. $\lim\limits_{x \to \infty} \dfrac{\sqrt{x + 7}}{x + 9}$

In Exercises 13–18, determine whether the function is continuous for the value of x.

13. $f(x) = 3x + 11$, $x = 3$

14. $f(x) = x^2 + 5x$, $x = 2$

15. $f(x) = \dfrac{x + 11}{x - 9}$, $x = 1$

16. $f(x) = \dfrac{x - 11}{x + 9}$, $x = 2$

17. $f(x) = \begin{cases} x^2 + 4, & x < 2 \\ 4x - 1, & x \ge 2 \end{cases}$, $x = 2$

18. $f(x) = \begin{cases} \dfrac{x^2 - 3x - 2}{x - 1}, & x \ne 1 \\ -1, & x = 1 \end{cases}$, $x = 1$

In Exercises 19–52, determine the derivative of the function.

19. $f(x) = 2x + 7$

20. $f(x) = 3x - 10$

21. $f(x) = x^2 + 2x - 7$

22. $f(x) = 3x^2 + 5x - 11$

23. $y = x^3 + 7x^2 - 9x + 5$

24. $y = 5x^3 + 7x^2 - 3x + 1$

25. $y = \dfrac{4}{x}$

26. $y = \dfrac{-5}{x}$

27. $y = 7x^{-9}$

28. $y = x^{\frac{3}{4}}$

29. $y = \sqrt{x^3}$

30. $y = \sqrt[3]{4x^4}$

31. $f(x) = \dfrac{2}{x + 3}$

32. $f(x) = \dfrac{3}{x^2 - 7}$

33. $f(x) = \sqrt{3x^2 + 1}$

34. $f(x) = \sqrt[3]{3x + 4}$

35. $g(x) = \dfrac{1}{7}x^5 + \dfrac{1}{3}x^4 - \dfrac{1}{2}x$

36. $g(x) = \dfrac{3}{5}x^5 - \dfrac{3}{4}x^4 + \dfrac{2}{3}x^3$

37. $y = (x^2 + x)(x - 5)$

38. $y = (x + 11)(x^2 - 2)$

39. $y = (2x^2 - 11x + 9)(x^3 - 11x^2)$

40. $y = (7x^3 - 11x^2 + x)(x^7 - 6x^5)$

41. $f(x) = \dfrac{2x^2 + 3x - 5}{x + 13}$

42. $f(x) = \dfrac{3x^2 - 11x + 4}{x + 11}$

43. $f(x) = \dfrac{x^2 + 11x - 9}{x^2 + 13x - 13}$

44. $f(x) = \dfrac{x^2 + 9x - 11}{2x^2 + 3x + 5}$

45. $y = (3x + 4)^3$

46. $y = (2x^2 + 3)^4$

47. $u = \sqrt[5]{(3 - t^2)^2}$

48. $u = \sqrt[3]{(t^2 - 3)^4}$

49. $u = \dfrac{1}{(3x + 4)^7}$

50. $u = \dfrac{5}{(3x^2 - 9)^5}$

51. $f(x) = \dfrac{(x + 4)^3}{x + 2}$

52. $f(x) = \dfrac{3x - 5}{(x - 4)^7}$

In Exercises 53–57, use implicit differentiation to determine $\dfrac{dy}{dx}$.

53. $2x^2 + 11y^2 = 4$ **54.** $3x^2 - 4y^2 = 7$

55. $2x^2 + 4xy^3 - 7y^2 = 4x$ **56.** $x^2y^3 - 4y^2 + 7xy^2 = 5y$

57. a. If the radius of a circle changes from r to $r + \Delta r$, what is the average rate of change of the area of the circle with respect to the radius?
    b. Find the instantaneous rate of change of the area with respect to the radius.

58. Define the concept of slope of a tangent line to a curve at a point on the curve.

59. Define **a.** average velocity and **b.** instantaneous velocity.

60. Define derivative of a function.

61. Find the acceleration, $a = \dfrac{d^2s}{dt^2}$, if the distance a particle moves in time t is
$s = 180t - 16t^2$. When does the acceleration vanish?

62. State the chain rule for derivatives. Give an example of how the rule is used.

63. Write the equation of the tangent line to the curve $x^2 - 2xy + y^2 + 2x + y - 6 = 0$ at $(2, 2)$.

64. If $y = x\sqrt{3x - 11}$, find $\dfrac{d^2y}{dx^2}$.

65. If $y = ax^3 + bx^2 + cx + d$, find $\dfrac{d^3y}{dx^3}$.

66. Find a function whose derivative is the limit stated.
    a. $f'(x) = \lim\limits_{\Delta x \to 0} \dfrac{(x + \Delta x)^6 - x^6}{\Delta x}$ for all real x
    b. $g'(x) = \lim\limits_{\Delta x \to 0} \dfrac{\sqrt[3]{x + \Delta x} - \sqrt[3]{x}}{\Delta x}$ for $x \neq 0$

67. If the distance (D ft) required for a certain driver to stop a car is a function of the velocity (v ft/s) given by the formula $D = 0.5v + 0.02v^2$, $0 \le v \le 100$, find the rate at which D increased with v when v was 30 ft/s.

68. If a force has a constant vertical component 10 and variable horizontal component h, then the magnitude of the force is given by $F = \sqrt{h^2 + 100}$ for $h \ge 0$. Find the rate at which F changes with respect to h when $h = 6$.

69. For the volume of a cone, $V = \dfrac{1}{3}\pi r^2 h$, where $r > 0$ is the radius of the base and $h > 0$ is the height. If r is changed while V is kept constant, h changes accordingly. Find $\dfrac{dh}{dr}$ by implicit and explicit methods.

✎ Writing About Mathematics

1. Write a brief paper explaining how a graphing calculator can be used to help determine the limit of a function. Will this technique work in every case?

2. Where are tangent lines to a curve found in the physical world? A few examples include optics (light striking a lens), physics (direction of a body's motion along a path), and geometry (angle between two intersecting curves). Do research on one or two applications and discuss the mathematical relation between the curve and the tangent line.

3. Your friend claims that the derivative of $y = \sqrt[3]{x} - 4$ exists for all values of x. Write a paragraph or two explaining why your friend's conclusion is not correct.

4. A friend has difficulty with implicit differentiation. Select an implicit function and write a procedure for determining the derivative of the function.

5. In your own words explain what a derivative is. Include at least one or two applications.

Chapter Test

For Exercises 1–3, determine the limit.

1. $\lim\limits_{x \to 4} (3x^3 - 5x + 1)$

2. $\lim\limits_{x \to 3} \dfrac{x^2 - 9}{x - 3}$

3. $\lim\limits_{x \to \infty} \dfrac{14x^2 - 7}{2x^2 - 5x + 6}$

For Exercises 4 and 5, determine whether the function is continuous at the given value of x. If the function is not continuous, and it is possible, remove the discontinuity.

4. $f(x) = 5x + 7, x = 3$

5. $f(x) = \dfrac{x^2 + x - 12}{x - 3}, x = 3$

For Exercises 6–11, determine $f'(x)$.

6. $f(x) = 6x^3 - 5x^2 + 2$

7. $f(x) = \dfrac{4}{x^3}$

8. $f(x) = (3x^4 - 5)(x^2 - 6x)$

9. $f(x) = \sqrt[5]{2x - 7}$

10. $f(x) = \dfrac{3x^4 - 9}{x + 7}$

11. $f(x) = (5x^4 - 7x)^3$

For Exercises 12 and 13, determine $\dfrac{dy}{dx}$.

12. $x^3 - 4xy + 2y^2 - x + 11 = 0$

13. $y = u - \dfrac{3}{u}; u = 5x^{\frac{3}{5}}$ (*Hint:* Use the chain rule.)

14. Determine $\dfrac{d^3y}{dx^3}$, if $y = 5x^{-4} + 7x^2 + 9$.

15. A company determines that the cost for producing a certain widget is given by $C = 0.22x^3 - 3x^2 + 132$. (x is the number of widgets.) By taking the derivative of C with respect to x, the company can get a handle on the average cost per unit. Determine the projected average cost per unit for each of the following.
   a. The company makes 10 widgets.
   b. The company makes 20 widgets.
   c. The company makes 100 widgets.

Applications of the Derivative

When a fast-moving car takes a curve on an icy road and is no longer constrained by friction, it "breaks loose" in a straight line path. The direction of the skid is a path that is tangent to the curve. In fact, the direction of the velocity of an object is always tangent to its path. If an electron moving along a curve is released, its new path is tangent to the original curve. This same concept was used in biblical days, when David took on Goliath with his sling. Where would the stone have to be released to hit its target? To illustrate the path of a car as it skids out of control on a curve see Example 4 in Section 20.1

20.1

Tangents and Normals

In Chapter 19 we found that the slope of a tangent line to a curve at a given point is the derivative of the function, describing the curve, evaluated at the point of tangency. From analytic geometry we know that the slope of a line and the coordinates of a point on the line provide sufficient information for us to write the equation of the line. Combining these ideas, we can write the equation of a tangent line at a given point on a curve in the slope intercept form. If $y = f(x)$ is a differentiable function and if (x_1, y_1) is a point on the graph, then the slope of the tangent line at the point is $f'(x_1)$. Thus, the equation of the tangent line at that point is $y - y_1 = f'(x_1)(x - x_1)$.

> **Equation of Tangent Line**
> $$y - y_1 = f'(x_1)(x - x_1)$$

EXAMPLE 1 Tangent Line to a Parabola

Determine the equation of the line tangent to the parabola $f(x) = -3x^2 + 5x + 7$ at the point $(2, 5)$.

Solution We are given the function and a point on the graph of the function. To write the equation of the tangent line we determine $f'(2)$ and then substitute the values in the equation.

$$f(x) = -3x^2 + 5x + 7 \qquad \text{Original function}$$
$$f'(x) = -6x + 5$$
$$f'(2) = -6(2) + 5 = -7 \qquad \text{Slope of tangent line}$$
$$y - y_1 = f'(x_1)(x - x_1) \qquad \text{Equation of line}$$
$$y - 5 = -7(x - 2) \qquad \text{Substituting for } f'(x) \text{ and } (x_1, y_1)$$
$$y = -7x + 19$$

Therefore, $y = -7x + 19$ is the equation of the tangent line to the parabola $f(x) = -3x^2 + 5x + 7$ at the point $(2, 5)$. The tangent and the parabola are sketched in Figure 20.1. ◆

By definition a line normal to a curve at a point $P(x_1, y_1)$ is perpendicular to the tangent line at the point P (Figure 20.2). Since the normal and the tangent lines are perpendicular to each other, their slopes are negative reciprocals of each other. That is,

$$m_t = \frac{-1}{m_n}.$$

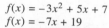

$f(x) = -3x^2 + 5x + 7$
$f(x) = -7x + 19$

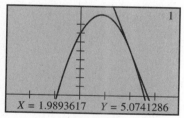

$X = 1.9893617 \qquad Y = 5.0741286$

FIGURE 20.1

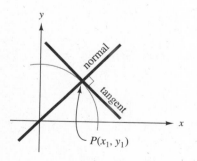

FIGURE 20.2

Using the derivative we can say that the slope to the normal line is the negative reciprocal of the derivative of the function at that point on the curve. That is,

$$m_n = \frac{-1}{f'(x)}$$

at the point P, and the equation of the normal line is $y - y_1 = \frac{-1}{f'(x_1)}(x - x_1)$.

> **Equation of Normal Line**
>
> $$y - y_1 = \frac{-1}{f'(x_1)}(x \rightarrow x_1)$$

When $\frac{dy}{dx}$ is used to indicate the derivative, the notation $\frac{dy}{dx}\Big|_{(x,y)}$ indicates the slope of the tangent at the point (x, y).

EXAMPLE 2   Normal Line to a Hyperbola

Determine the equation of the normal line to the hyperbola $xy = 2$ at the point $(2, 1)$.

Solution   We are given the equation of the curve and a point on the curve. To write the equation of the normal line we determine the slope of the normal $\dfrac{-1}{\dfrac{dy}{dx}}\Big|_{(2,1)}$ and then substitute the values in the equation.

$$y = \frac{2}{x} \qquad \text{Solving for } y$$

$$y = 2x^{-1} \qquad \text{Law of exponents}$$

$$\frac{dy}{dx} = -2x^{-2} \qquad \text{Differentiating}$$

$$\frac{dy}{dx}\Big|_{(2,1)} = -2\left(\frac{1}{2}\right)^2 = -\frac{1}{2} \qquad \text{Slope of tangent line}$$

The slope of the normal line is $\dfrac{-1}{-\dfrac{1}{2}}$ or 2.

$$y - \boxed{1} = \boxed{2}(x - \boxed{2}) \qquad \text{Substituting}$$

$$y = 2x - 3 \qquad \text{Equation of normal line}$$

The equation of the line normal to the hyperbola $xy = 2$ at $(2, 1)$ is $y = 2x - 3$. Figure 20.3 shows the graph of the normal line. ◆

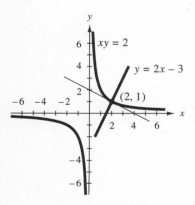

FIGURE 20.3

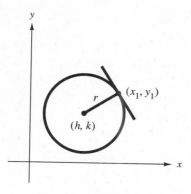

FIGURE 20.4

EXAMPLE 3   Tangent Line to a Circle Is Perpendicular to the Radius

Demonstrate that the tangent line at any point (x_1, y_1) on a circle is perpendicular to the radius of the circle at the point (x_1, y_1). Use Figure 20.4.

Solution   The standard form of the equation of a circle with center (h, k) is $(x - h)^2 + (y - k)^2 = r^2$. Let (x_1, y_1) be any point on the circle. We determine the slope of the tangent to the circle at (x_1, y_1) by differentiating the equation of the circle and evaluating $\dfrac{dy}{dx}$ at (x_1, y_1).

$$(x - h)^2 + (y - k)^2 = r^2 \qquad \text{Equation of circle}$$

$$2(x - h)\frac{dx}{dx} + 2(y - k)\frac{dy}{dx} = 0 \qquad \begin{array}{l}\text{Implicitly differentiating}\\ \text{with respect to } x\end{array}$$

$$\frac{dy}{dx} = -\frac{x - h}{y - k} \qquad \text{Solving for } \frac{dy}{dx}$$

At point (x_1, y_1),

$$m_t = \frac{dy}{dx}\bigg|_{(x_1, y_1)} = -\frac{x_1 - h}{y_1 - k}.$$

Using the center (h, k) and the point (x_1, y_1), the slope of the radius is:

$$m_r = \frac{y_1 - k}{x_1 - h}.$$

The slope of the tangent line to the circle may be rewritten as:

$$m_t = -\frac{1}{\dfrac{y_1 - k}{x_1 - h}} = -\frac{1}{m_r}.$$

This result shows that the slope of the tangent line is the negative reciprocal of the slope of the normal line. Thus, by definition, the two lines are perpendicular. Since (x_1, y_1) could be any point on the circle, the conclusion holds that, at any point on a circle, the radius and the tangent line to the circle at that point are prependicular.   ◆

The direction of a skid when a vehicle moves along a curved path is discussed in Example 4.

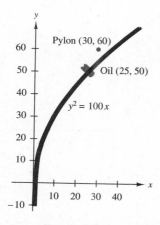

FIGURE 20.5

EXAMPLE 4   Application: Path of a Go-Cart

During a race, a go-cart rounding a curve described by $y^2 = 100x$ (x and y in meters) hits a patch of oil at the point $(25, 50)$ and spins out of control. A pylon marking the course is at $(30, 60)$. Will the go-cart hit the pylon? See Figure 20.5.

Solution The go-cart starts its skid at the point (25, 50). It will tend to slide along a path that is tangent to the curve $y^2 = 100x$. The equation of the line tangent to the curve at the point (25, 50) defines the path of the skid. If the go-cart strikes the pylon, the point (30, 60) must be a point on the tangent line. To write the equation of the tangent line we must first determine the slope at the point (25, 50).

$$y^2 = 100x \qquad \text{Path of cart before skid}$$

$$y = \pm\sqrt{100x} \qquad \text{Solving for } y$$

$$= \sqrt{100x} \qquad \text{Choosing positive root for path in Figure 20.5}$$

$$= (100x)^{\frac{1}{2}} \qquad \text{Exponential notation}$$

$$\frac{dy}{dx} = \frac{1}{2}(100x)^{-\frac{1}{2}}(100) \qquad \text{Taking derivative}$$

$$= \frac{100}{2\sqrt{100x}} \quad \text{or} \quad \frac{5}{\sqrt{x}}$$

Knowing the slope at any point we can now determine the slope at the point (25, 50).

$$\left.\frac{dy}{dx}\right|_{(25,50)} = \frac{5}{\sqrt{25}} = 1 \quad \text{Slope of tangent line}$$

The path of the skid is:

$$y - 50 = 1(x - 25) \qquad \text{Substituting}$$

$$y = x + 25 \qquad \text{Tangent line}$$

Substituting (30, 60) for x and y, we can determine whether the car will strike the pylon.

$$60 = 30 + 25$$

$$60 \neq 55.$$

The location of the pylon does not satisfy the equation of the path of the car, so the car does not strike the pylon. ◆

20.1 Exercises

In Exercises 1–9, perform the following.
a. Determine the equation (in slope intercept form) of the line tangent to the given curve at the stated value(s).
b. Identify the curve (parabola, circle, ellipse, etc.).
c. Sketch the curve and the specific tangent line.

1. $f(x) = x^2 + 2$, $(-2, 6)$
2. $f(x) = 3x^2 - 1$, $(-1, 2)$
3. $f(x) = 4x^2 - 2x + 6$, $x = 3$
4. $x^2 + y^2 = 36$, $y = 2$, $x > 0$
5. $x^2 + y^2 = 16$, $y = 3$, $x > 0$
6. $xy = 12$, $y = 4$

7. $xy = 6, x = 2$

8. $\dfrac{(x-2)^2}{16} + \dfrac{(y-3)^2}{25} = 1, x = 3$ (*Hint:* There are two answers.)

9. $\dfrac{x^2}{16} - \dfrac{y^2 - 4}{9} = 1, x = 4$ (*Hint:* There are two answers.)

In Exercises 10–18, perform the following.
a. Determine the equation (in slope intercept form) of the line normal to the given curve at the stated value(s).
b. Identify the curve (parabola, circle, ellipse, etc.).
c. Sketch the curve and the specific normal line.

10. $x^2 + y - 1 = 0, (2, -3)$ **11.** $x^2 - 2y + 3 = 0, (3, 6)$ **12.** $2x^2 + 2x + 3y - 2 = 0, x = -1$

13. $x^2 - 6y^2 = 25, y = 2, x > 0$ **14.** $x^2 - y^2 = 16, y = 3, x > 0$ **15.** $2xy = 15, x = 2$

16. $3xy = 4, x = 1$ **17.** $\dfrac{y^2 + 3}{4} - \dfrac{x^2 - 1}{9} = 1, x = 1$

18. $\dfrac{(x-4)^2}{9} + \dfrac{y^2}{49} = 1, x = 2$ (*Hint:* There are two answers.)

In Exercises 19 and 20, determine the equation of the tangent lines to the curve at the given points.

19.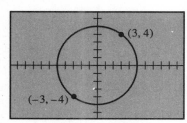

20.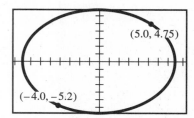

21. **a.** Determine the equation of the tangent line to the circle $(x - 3)^2 + (y + 5)^2 = 29$ at the point $(5, 0)$.
     b. Determine the equation of the normal line to the circle at the point $(5, 0)$.

22. The stone in David's slingshot travels in the circle described by $x^2 + (y - 1.2)^2 = 1.0$. The path of the stone is on a line that is tangent to the circle at the point where the stone leaves the sling. If the stone travels in a straight line when it leaves the sling and if a vital spot on Goliath is at $(3.0, 2.0 \pm 0.1)$ meters, in what range must David release the stone to hit the designated target? Assume that, from David's position, the sling is spinning in a clockwise direction.

23. An automobile is traveling around the curve $(x - 10)^2 + y^2 = 100, 0 \le x \le 10$, $y > 0$, (x and y in meters) when it hits a patch of ice and begins to slide. If it hits a tree at point $(6, 12)$, at what point on the curve is the ice located?

24. An electron is being accelerated along the path $xy = 1, x > 0$ (x and y in meters). When the accelerating field is turned off, the electron is supposed to follow a path at $-30°$ to the positive x-axis. At what point must the accelerating field be turned off?

25. A neutron is accelerated along the path $xy = 1$, $x > 0$ (x and y in meters). If it must strike normal to a flat target placed at $75°$ to the positive x-axis, at what point must the neutron leave the path?

26. In Exercise 25, if the target makes an angle of $50°$ with the positive y-axis, at what point must the neutron leave the path?

27. A suspension cable is attached to supporting columns that are 75 meters apart. The cable sags in the middle by a distance of 10 m, and the resulting curve is parabolic. An engineer needs to find the tension that the cable exerts on the left supporting column by determining the tangent of the curve of the cable at that point.
    a. Write an equation for the parabola. (*Hint:* Choose the origin of the coordinate system at the lowest point on the cable.)
    b. Write the equation of the line along which the tension acts at the left support.

28. Radio waves from a distant star approach a parabolic radio telescope parallel to its central axis (see the sketch). Upon hitting the telescope, the waves are reflected off the surface normally (that is, perpendicularly) and pass through a detector located on the central axis. The equation for the parabola is $y^2 = 8x$. (See the figure.)
    a. If a wave of light hits the telescope at $(4.00, 5.66)$, what is the equation for the line of the reflected wave?
    b. What is the distance between the detector and the point where the light ray hits the telescope?

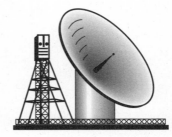

29. When flexible tubing is constrained into a particular shape, the supports must be normal to the tubing. Suppose that a curved shape of tubing follows the equation $y^2 + x^2 = 25$ for $0 < x < 5$ and that supporting cables are located at $x = 1$ and $x = 3$.
    a. Sketch the curved part of the tubing.
    b. For each value of x, how many support points are there?
    c. Write the equation for each of the support lines.

30. A lacrosse player must catch the ball with his stick perpendicular to the flight path of the ball for maximum probability of making a successful catch (See the illustration.) If the ball travels along the path $x^2 + 100y - 300 = 0$ (x and y in meters, $x > 0$), at what optimal angle with the horizontal must the stick be to catch the ball 2 m above the ground?

20.2

Derivatives: Curve Sketching I

In manufacturing, we may be able to write a function that defines maximum production levels. However, we may gain a better understanding of the information provided by the function from a graph of the function and, perhaps, from the graph of its derivative. This happens because derivatives are useful in helping define shapes of the curve and key points.

Let us consider the parabola $f(x) = x^2 - 2x - 8$ as shown in Figure 20.6. By examining the figure we see that as x increases from left to right the functional values $f(x)$ first decrease to the left of the vertex of the parabola at $(1, -9)$ and then increase to the right of the vertex. We also see that $f(x) = x^2 - 2x - 8$ is a decreasing function on the interval $(-\infty, 1)$ and is an increasing function in the interval $(1, \infty)$.

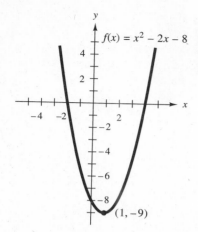

FIGURE 20.6

A function $f(x)$ is **increasing** on an interval if for any $x_2 > x_1$ in the interval $f(x_2) > f(x_1)$. A function $f(x)$ is **decreasing** on an interval if for any $x_2 > x_1$ in the interval $f(x_2) < f(x_1)$.

If we differentiate $f(x) = x^2 - 2x - 8$ we obtain $f'(x) = 2x - 2$. The expression $2x - 2$ is negative for all $x < 1$ and is positive for all $x > 1$. We see from $f'(x) = 2x - 2$ and also from observation of the figure that the slope of the line tangent to the curve is negative when the function is decreasing and that the slope of the line tangent to the curve is positive when the function is increasing. That is, $f'(x) < 0$ for $(-\infty, 1)$ and $f'(x) > 0$ for $(1, \infty)$. At the point on the curve where $f'(x) = 0$, the path of the function changes direction, from increasing to decreasing or decreasing to increasing. The point $(1, -9)$, the point on the graph where $f'(x) = 0$, is the lowest point on the graph.

Sketch the graphs of $f(x) = x^2 - 2x - 8$ and its derivative $f'(x) = 2x - 2$ on the same graph, as shown in Figure 20.7. Where does the line $f'(x) = 2x - 2$ cross the x-axis? At $x = 1$. What else happens when $x = 1$? $f'(x) = 0$ and $(1, -9)$ is the low point on the curve. To the left of $x = 1$ the line $f'(x) = 2x - 2$ is below the x-axis, and the function $f(x) = x^2 - 2x - 8$ is decreasing. To the right of $x = 1$, the line $f'(x) = 2x - 2$ is above the x-axis, and the function is increasing.

If $f'(x) > 0$ for $a < x < b$, then the function is increasing on the interval.
If $f'(x) < 0$ for $c < x < d$, then the function is decreasing on the interval.

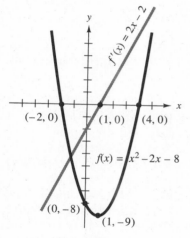

FIGURE 20.7

EXAMPLE 1 **Increasing and Decreasing Intervals of a Function**

On the same graph, sketch the graph of $f(x) = 2x^3 - 3x^2 - 12x + 5$ and $f'(x)$. Determine the intervals where the function is increasing, decreasing, and the points where $f'(x) = 0$.

Solution   We begin by differentiating the function.

$$f(x) = 2x^3 - 3x^2 - 12x + 5$$
$$f'(x) = 6x^2 - 6x - 12$$

Then we set the derivative equal to 0 and factor.

$$x^2 - x - 2 = 0 \qquad (x - 2)(x + 1) = 0$$

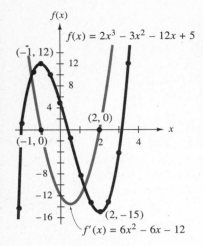

f(x)

$f(x) = 2x^3 - 3x^2 - 12x + 5$

$(-1, 12)$

12

8

4

$(2, 0)$

x

$(-1, 0)$ 2   4

-8

-12

-16 $(2, -15)$

$f'(x) = 6x^2 - 6x - 12$

FIGURE 20.8

We determine that $f'(x) = 0$ when $x = -1$ and $x = 2$. The function and its derivative are graphed in Figure 20.8. The points $(-1, 12)$ and $(2, -15)$ are points on the graph where the curve "flattens out" and $f'(x) = 0$. $f'(x)$ is a parabola that crosses the x-axis at $(-1, 0)$ and $(2, 0)$; it opens up. From the derivative of the function and the graph of the functions we can determine the following information.

Interval or point	Graph of $f'(x)$	$f'(x)$	$f(x)$
$-\infty < x < -1$	Above x-axis	$f'(x) > 0$	Increasing function
$(-1, 0)$	Crosses x-axis	$f'(-1) = 0$	
$(-1, 12)$		$f'(-1) = 0$	High point on curve
$-1 < x < 2$	Below x-axis	$f'(x) < 0$	Decreasing function
$(2, 0)$	Crosses x-axis	$f'(2) = 0$	
$(2, -15)$		$f'(2) = 0$	Low point on curve
$2 < x < \infty$	Above x-axis	$f'(x) > 0$	Increasing function

Thus, the function is increasing in the intervals $-\infty < x < -1$ and $2 < x < \infty$; decreasing in the interval $-1 < x < 2$; and $f'(x) = 0$ at the points $(-1, 12)$ and $(2, -15)$. ◆

From our study of the continuous function $f(x) = 2x^3 - 3x^2 - 12x + 5$ in Example 1, we can see that for the derivative of the function to change from a positive to a negative value, the curve must pass through a point where $f'(x) = 0$. Likewise, for the derivative of the function to change from negative to positive, the curve must pass through a point where $f'(x) = 0$. From the data in Example 1 we see that $f'(x) = 0$ at the points $(-1, 12)$ and $(2, -15)$. Figure 20.8 shows us that the points $(-1, 12)$ and $(2, -15)$ are high and low points on the curve. We say that these points are **relative maximum** and **relative minimum** points, respectively, on the curve. (As can be seen from the figure, the terms relative (local) maximum and relative (local) minimum are used because the functional values may be larger or smaller at other points than they are where the relative maximum or relative minimum points occur on the graph.)

The points $(-1, 12)$ and $(2, -15)$ where the first derivative of the function is equal to zero are special points on the curve called *critical points*.

DEFINITION 20.2

A **critical value** is a number c in the domain of a function f for which $f'(c) = 0$ or for which $f'(c)$ does not exist. The points $(c, f(c))$ are called **critical points**.

> ### Definition 20.3   First-Derivative Test
>
> Given a continuous function $f(x)$ and its derivative $f'(x)$, as x increases through a critical value c and the sign of $f'(x)$ changes then the critical point is a relative maximum or a relative minimum.
>
> 1. If $f'(x)$ changes from positive to negative, the critical point is a **relative maximum.**
> 2. If $f'(x)$ changes from negative to positive, the critical points is a **relative minimum.**

EXAMPLE 2   Critical Value

Sketch the graph of $f(x) = 5x^{\frac{2}{5}}$ for $(-4, 4)$, determine the critical value(s), and use the first-derivative test to determine relative maximum or minimum points.

Solution   The graph of the function is sketched in Figure 20.9. From the graph it appears that the function may have a minimum at $(0, 0)$. To test this we determine $f'(x)$ and then determine the critical value(s).

$$f'(x) = 2x^{-\frac{3}{5}} \quad \text{or} \quad f'(x) = \frac{2}{x^{\frac{3}{5}}}$$

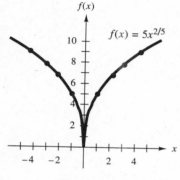

$f(x)$

$f(x) = 5x^{2/5}$

FIGURE 20.9

The derivative is undefined at $x = 0$; the critical value is $x = 0$, and the critical point is $(0, 0)$. To examine the derivative on each side of the critical value, we select any values to the left and right of 0. For ease of calculation, select $x = -1$ and $x = 1$.
When $x = -1$,

$$f'(\boxed{-1}) = \frac{2}{(\boxed{-1})^{\frac{3}{5}}} < 0.$$

When $x = 1$,

$$f'(\boxed{1}) = \frac{2}{(\boxed{1})^{\frac{2}{3}}} > 0.$$

Our results show that the derivative is negative to the left of zero and positive to the right of zero. Thus, by the first-derivative test, the point $(0, 0)$ is a relative minimum.   ◆

EXAMPLE 3   Sketching the Graph

Use the first-derivative test to sketch the graph of $f(x) = \dfrac{x^3}{3} + \dfrac{x^2}{2} - 2x + 1$.

Solution Examining the function $f(x) = \dfrac{x^3}{3} + \dfrac{x^2}{2} - 2x + 1$ we see that when $x = 0, f(0) = 1$. Finding the derivative of $f(x)$ and setting $f'(x)$ equal to zero we can determine the critical values.

$$f(x) = \frac{x^3}{3} + \frac{x^2}{2} - 2x + 1$$
$$f'(x) = x^2 + x - 2.$$

Setting $f'(x) = 0$, we have

$$x^2 + x - 2 = 0.$$

Factoring and solving for x we have $x = -2$ and $x = 1$. This means we have two critical values, $c_1 = -2$ and $c_2 = 1$. Use the first-derivative test when $c_1 = -2$.

$$f'(x) > 0, \text{ for } x < -2 \quad \textbf{For example,} f'(-3) = 4$$
$$f'(x) < 0, \text{ for } x > -2 \quad \textbf{For example,} f'(-1) = -2$$

This tells us that the function has a relative maximum point at $(-2, f(-2))$ or $\left(-2, \dfrac{13}{3}\right)$.

Now use the first derivative test when $c_2 = 1$.

$$f'(x) < 0, \text{ for } x < 1 \quad \textbf{For example,} f'(0) = -2$$
$$f'(x) > 0, \text{ for } x > 1 \quad \textbf{For example,} f'(2) = 4$$

This tells us that the function has a relative minimum point at $(1, f(1))$ or $\left(1, -\dfrac{1}{6}\right)$. There are no points where the derivative is undefined. Therefore, $f(x) = \dfrac{x^3}{3} + \dfrac{x^2}{2} - 2x + 1$ has only two critical points, and is a function continuous for all x. We can use this information to sketch the graph of $f(x)$.

We know that the curve passes through the points $(0, 1)$, $\left(-2, \dfrac{13}{3}\right)$ and $\left(1, -\dfrac{1}{6}\right)$, so these points are plotted in Figure 20.10. We know from our evaluation that the curve is rising in the intervals for $x < -2$ and $x > 1$ and that it is falling for the interval $-2 < x < 1$ because in these intervals the first derivative is positive and negative, respectively. Therefore, we know that the curve crosses the x-axis at three points that lie in the intervals $x < -2$, $-2 < x < 1$, and $x > 1$. ◆

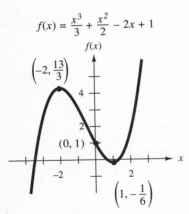

$f(x) = \dfrac{x^3}{3} + \dfrac{x^2}{2} - 2x + 1$

FIGURE 20.10

EXAMPLE 4 Critical Values

Determine the critical values and the relative maximum or minimum points for the function $f(x) = x^4 - 4x^3 + 3$.

Solution Differentiate and set the derivative equal to zero to determine critical values.

$$f(x) = x^4 - 4x^3 + 3$$
$$f'(x) = 4x^3 - 12x^2 \qquad \text{\textbf{Differentiating}}$$
$$4x^3 - 12x^2 = 0$$
$$4x^2(x - 3) = 0 \qquad \text{\textbf{Common factor}}$$
$$x = 0 \quad \text{and} \quad x = 3 \qquad \text{\textbf{Solving for } } x$$

The critical values are $c_1 = 0$ and $c_2 = 3$. The first derivative is a polynomial, and is defined for all values of x. Therefore, there are no critical values for which the first derivative is undefined.

Using the calculator, we sketch the function and the derivative of the function (Figure 20.11). From both the figure of the function and the first derivative of the function, it appears that there may be maximum or minimum points at $(0, 3)$ and $(3, -24)$.

Apply the first-derivative test for $c_1 = 0$.

$$f'(x) < 0, \text{ for } x < 0 \qquad \text{\textbf{For example, } } f'(-1) = -16$$
$$f'(x) < 0, \text{ for } x > 0 \qquad \text{\textbf{For example, } } f'(1) = -8$$

Therefore, the critical point $(0, 3)$ is not a relative maximum or minimum point even though the slope of the line tangent to the function is zero (the tangent line is horizontal). Examining the graph of $f'(x)$ we see that in the interval $(-\infty, 3)$ the curve is never above the x-axis. Thus, the derivative of the function does not change sign on either side of the critical value. Now apply the test for $c_2 = 3$.

$$f'(x) < 0, \text{ for } x < 3 \qquad \text{\textbf{For example, } } f'(2) = -16$$
$$f'(x) > 0, \text{ for } x > 3 \qquad \text{\textbf{For example, } } f'(4) = 64$$

The derivative is negative to the left of 3 and positive to the right of 3; therefore, by the first-derivative test, the critical point $(3, -24)$ is a relative minimum point. We can arrive at the same conclusion by examining the graph of $f'(x)$. To the left of 3, the graph of $f'(x)$ is below the x-axis, and $f'(x) < 0$. To the right of 3, the graph of $f'(x)$ is above the x-axis, and $f'(x) > 0$. Thus, the function has a relative minimum at $(3, -24)$. The point $(0, 3)$ is a flat point on the graph, but it is not a minimum or a maximum point on the graph. ◆

From the examples you can see that the first-derivative test is helpful in sketching the graph of the function. On the other hand, the graph of the function and the graph of the first derivative of the function provides us with information about relative maximum and relative minimum points on the curve.

In Examples 2, 3, and 4 we determined the relative maximum and relative minimum points on the curves. That is, we found the high- and the low-points in a neighborhood, or a section, of the curve. When we restrict the discussion of the functions over a closed interval, we can determine an absolute maximum point and an absolute minimum point on the curve for the closed interval. The **absolute maximum point** is the highest point in the closed interval, and the **absolute minimum point** is the lowest point in the closed interval.

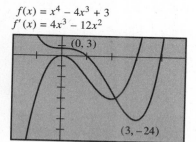

$f(x) = x^4 - 4x^3 + 3$
$f'(x) = 4x^3 - 12x^2$

(0, 3)

(3, −24)

FIGURE 20.11

> **TECHNIQUE FOR DETERMINING ABSOLUTE MAXIMUM AND ABSOLUTE MINIMUM**
>
> 1. Determine the relative maximum and minimum points in the closed interval.
> 2. Evaluate the function at the end points of the closed interval.
> 3. From the results in 1 and 2, select the highest and lowest points. These are the absolute maximum and the absolute minimum points.

For the function in Example 3, determine the absolute maximum and absolute minimum in the closed interval $-4 \le x \le 2$. In Example 3 we found that the relative maximum and relative minimum points are $\left(-2, \frac{13}{3}\right)$ and $\left(1, \frac{-1}{6}\right)$. Now evaluate the functional values at the endpoints of the closed interval.

$$f(\boxed{-4}) = \frac{(\boxed{-4})^3}{3} + \frac{(\boxed{-4})^2}{2} - 2(\boxed{-4}) + 1 = \frac{-13}{3}$$

$$f(\boxed{2}) = \frac{(\boxed{2})^3}{3} + \frac{(\boxed{2})^2}{2} - 2(\boxed{2}) + 1 = \frac{5}{3}$$

The endpoints of the interval are $\left(-4, \frac{-13}{3}\right)$ and $\left(2, \frac{5}{3}\right)$. Thus, the absolute maximum point and and minimum points in the closed interval are $\left(-2, \frac{13}{3}\right)$ and $\left(-4, \frac{-13}{3}\right)$. In this case, the absolute maximum is also a relative maximum point, and the absolute minimum is an endpoint of the interval.

20.2   Exercises

In Exercises 1–18, perform the following.
a. Determine all the critical points for each function.
b. Use the first-derivative test to determine whether the critical points found in part a are maximum or minimum points.
c. Use the results from parts a and b to sketch the graph.

1. $y = 4x^2$

2. $y = \frac{1}{2}x^2$

3. $y = x^2 + 3x + 4$

4. $y = x^2 - 7x + 18$

5. $x^2 = \frac{1}{2}y$

6. $x^2 = 4y$

7. $4x^2 + 16x - y + 17 = 0$

8. $x^2 - 4x - y + 1 = 0$

9. $y = x^3 - x^2 - x$

10. $y = x^3 + x^2 - x$

11. $y = x^3 + 3x^2 - 2$

12. $y = x^3 - 3x^2 + 2$

13. $y = \dfrac{x^3}{3} - \dfrac{5x^2}{2} + 6x - 2$ **14.** $y = \dfrac{x^3}{3} - 9x - 3$ **15.** $y = \dfrac{x^3}{3} - x + 1$

16. $y = x^3 - 2x^2 - 4x + 4$ **17.** $y = x^{\frac{4}{5}}$ **18.** $y = x^{\frac{2}{3}}$

In Exercises 19–24, perform the following.
a. Determine all the critical points for each function.
b. Use the graphics calculator to sketch the graph of the function and the derivative of the function.
c. Use the results from parts a and b and the first-derivative test to determine whether the critical points are maximum or minimum points.

19. $y = x^2 - 192x^{\frac{1}{3}}$ **20.** $y = x^2 - 10x^{\frac{3}{5}}$ **21.** $y = x^3 - 4x^2 + x + 6$

22. $y = x^3 - 9x^2 + 2x + 6$ **23.** $y = \dfrac{x^4}{4} + x^3 - 2x^2 - 12x + 20$ **24.** $y = 3x^4 - 28x^3 + 96x^2 - 144x$

In Exercises 25–28, determine the absolute maximum and absolute minimum points of the function for the closed interval.

25. $f(x) = x^2 - 5x - 4, \ -1 \le x \le 3$ **26.** $f(x) = -4x^2 + 5x + 6, \ -1 \le x \le 2$

27. $f(x) = 2x^3 - 5x^2 - 5, \ -1 \le x \le 3$ **28.** $f(x) = x^3 + 5x + 7, \ -2 \le x \le 2$

20.3

Derivatives: Curve Sketching II

We encounter functions such as $f(x) = x^3 - 3x^2$ when analyzing harmonic systems. These curves have multiple "hills" and "valleys." In Section 20.2 we learned how to determine the relative maximum and minimum points on the curve using the first-derivative test and graphing techniques. In this section we explore how to use the second derivative to determine maximum, minimum, and other important properties of curves.

Let us begin by sketching the graph of $f(x) = x^3 - 3x^2$ and examining the behavior of the function's second derivative.

$$f(x) = x^3 - 3x^2 \quad \textbf{Given function}$$
$$f'(x) = 3x^2 - 6x \quad \textbf{First derivative}$$
$$f''(x) = 6x - 6 \quad \textbf{Second derivative}$$

The first derivative is a polynomial and is defined for all values of x. To determine the critical values, set the first derivative equal to zero and factor.

$$3x^2 - 6x = 0$$
$$3x(x - 2) = 0$$
$$x = 0 \quad \text{or} \quad x = 2$$

There are two critical values, $c_1 = 0$ and $c_2 = 2$. Use the first-derivative test at $c_1 = 0$.

$$f'(x) > 0, \text{ for } x < 0 \quad \textbf{For example, } f'(-1) = 9$$
$$f'(x) < 0, \text{ for } x > 0 \quad \textbf{For example, } f'(1) = -3$$

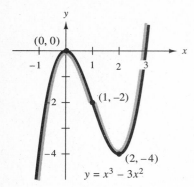

(0, 0)

(1, −2)

(2, −4)

$y = x^3 - 3x^2$

FIGURE 20.12

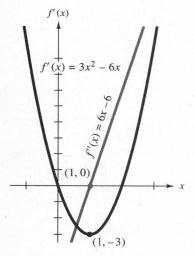

$f'(x)$

$f'(x) = 3x^2 - 6x$

$f''(x) = 6x - 6$

(1, 0)

(1, −3)

FIGURE 20.13

$f(x) = x^3 - 3x^2$
$f'(x) = 3x^2 - 6x$
$f''(x) = 6x - 6$

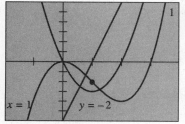

$x = 1$ $y = -2$

FIGURE 20.14

So at $(0, 0)$ we have a relative maximum. Now use the first-derivative test at $c_2 = 2$.

$$f'(x) < 0 \text{ for } x < 2 \quad \text{For example, } f'(1) = -3$$
$$f'(x) > 0 \text{ for } x > 2 \quad \text{For example, } f'(3) = 9$$

So at $(2, -4)$ we have a relative minimum.

Using the information from the first-derivative test, the graph of the function is sketched in Figure 20.12.

Now let us examine the first and second derivative of the function in the same manner that we examined the function and the first derivative. The first derivative $f'(x) = 3x^2 - 6x$ is a quadratic expression, and the graph is a parabola, as shown in Figure 20.13. The second derivative is a linear function $f''(x) = 6x - 6$, and the graph in Figure 20.13 is a straight line. The line $f''(x) = 6x - 6$ crosses the x-axis at $x = 1$. From the second derivative of the function, the graph of the first derivative, and the graph of the second derivative, we can determine the following information.

Interval or Point	Graph of $f''(x)$	$f''(x)$	$f'(x)$
$-\infty < x < 1$	Below x-axis	$f''(x) < 0$	Decreasing function
$(1, 0)$	Crosses x-axis	$f''(1) = 0$	Flat
$(1, -3)$		$f''(1) = 0$	Relative minimum
$1 < x < \infty$	Above x-axis	$f''(x) > 0$	Increasing function

Thus, $f'(x)$ is increasing in the interval $1 < x < \infty$ and decreasing in the interval $-\infty < x < 1$, and $f''(1) = 0$.

Now let us combine these results and look at the function $f(x) = x^3 - 3x^2$, shown in Figure 20.14. To the left of 1, $f'(x)$ is a decreasing function. Also, the graph of $f(x)$ is opening downward, and we say that the curve is *concave down* (spills water). To the right of 1, $f'(x)$ is an increasing function. The graph of $f(x)$ is opening upward, and we say the curve is *concave up*.

DEFINITION 20.4

The graph of a differentiable function $f(x)$ is **concave up** on an interval where $f'(x)$ is increasing and **concave down** on an interval where $f'(x)$ is decreasing.

If the second derivative of the function exists, as is true for $f(x) = x^3 - 3x^2$, we can use the second derivative to determine the concavity. Recall from the discussion and Figure 20.13 that when $f''(x) < 0$, then $f'(x)$ is decreasing, and when $f''(x) > 0$, then $f'(x)$ is increasing. This is summarized into a statement called the second-derivative test for concavity.

> **The Second-Derivative Test for Concavity**
> The graph of $f(x)$ is concave down on any interval where $f''(x) < 0$.
> The graph of $f(x)$ is concave up on any interval where $f''(x) > 0$.

EXAMPLE 1   Concavity Test

Use the second-derivative test to determine the concavity of $f(x) = 2x^2$.

Solution   We must first determine the second derivative of the function.

$$f'(x) = 4x$$
$$f''(x) = 4$$

Since $f''(x) > 0$ for all x, the function $f(x) = 2x^2$ is concave up for all values of x. ◆

For the function $f(x) = x^3 - 3x^2$, the concavity changes form concave down to concave up at the point $(1, -2)$. This point, where the curve changes concavity, is called a *point of inflection*.

> **DEFINITION 20.5**
> A point on the graph of a differentiable function where the concavity changes is called a **point of inflection.**

For $f(x) = x^3 - 3x^2$, the second derivative is negative to the left of 1, positive to the right of 1, and there is a point of inflection at $(1, -2)$. We also found that $f''(1) = 0$. Thus, if there is a point of inflection, the second derivative at the point is zero.

> If a function has a point of inflection at (a, b) and it is possible to differentiate the function twice then $f''(a) = 0$.

EXAMPLE 2   Concavity, Maximum, Minimum, Points of Inflection

For the function $f(x) = \dfrac{x^3}{3} - 2x^2 - 5x + 1$, determine the following.

a.   Relative maximum points
b.   Relative minimum points
c.   Intervals of concavity
d.   Points of inflection

Solution   To determine maximum or minimum points we must determine the critical values.

$$f(x) = \frac{x^3}{3} - 2x^2 - 5x + 1$$

$$f'(x) = x^2 - 4x - 5 \qquad \text{Differentiating}$$

$$0 = (x - 5)(x + 1) \qquad \text{Factoring}$$

$$x = -1 \qquad x = 5 \qquad \text{Solving for } x$$

The critical values are $c_1 = -1$ and $c_2 = 5$.

Apply the first-derivative test for $c_1 = -1$.

$$f'(x) > 0, \text{ for } x < -1 \qquad \text{For example, } f'(-2) = 7$$
$$f'(-1) = 0$$
$$f'(x) < 0, \text{ for } x > -1 \qquad \text{For example, } f'(0) = -5$$

Now apply the test for $c_2 = 5$.

$$f'(x) < 0, \text{ for } x < 5 \qquad \text{For example, } f'(4) = -5$$
$$f'(5) = 0$$
$$f'(x) > 0, \text{ for } x > 5 \qquad \text{For example, } f'(6) = 7$$

a.   Since $f'(x) > 0$, to the left of $c_1 = -1$ and $f'(x) < 0$ to the right of $x = -1$, the first-derivative test tells us there is a relative maximum at $(-1, f(-1))$, or $(-1, 3.7)$.

b.   Since $f'(x) < 0$ to the left of $c_2 = 5$ and $f'(x) > 0$ to the right of $c_2 = 5$, the first-derivative test tells us that there is a relative minimum at $(5, f(5))$, or $(5, -32.3)$.

c.   To determine where the graph is concave up or concave down we determine the intervals where the second derivative is positive or negative.

$$f''(x) = 2x - 4 \qquad \text{Second derivative}$$
$$x = 2 \qquad \text{Second derivative is 0}$$
$$f''(x) < 0, \text{ for } x < 2 \qquad \text{For example, } f''(1) = -2$$
$$f''(2) = 0$$
$$f''(x) > 0, \text{ for } x > 2 \qquad \text{For example, } f''(3) = 2$$

The second derivative test for concavity tells us that the curve is concave down for the interval $(-\infty, 2)$ and concave up for the interval $(2, \infty)$.

d.   From part c we can tell that the concavity changes from concave down to concave up at $(2, -14.3)$ and $f''(2) = 0$. Therefore, the point $(2, -14.3)$ is a point of inflection.   ◆

The graphs of $f(x) = \frac{x^3}{3} - 2x^2 - 5x + 1$, $f'(x) = x^2 - 4x - 5$, and $f''(x) = 2x - 4$ are given in Figure 20.15. The graph is a visual check of the results obtained in Example 2. On the other hand, the calculus is a precise way of determining the desired results in Example 2. Is it possible to determine

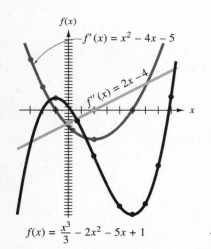

$$f(x) = \frac{x^3}{3} - 2x^2 - 5x + 1$$

FIGURE 20.15

whether we have a picture of a complete graph of a function by using the first and second derivative? We can use the second derivative to determine the relative maximum and relative minimum points of a function.

> **The Second-Derivative Test for Relative Maximum and Minimum**
> If $f'(c) = 0$ and $f''(c) < 0$, then f has a relative maximum at $x = c$.
> If $f'(c) = 0$ and $f''(c) > 0$, then f has a relative minimum at $x = c$.

In Example 2 we found that the relative maximum and relative minimum points on the curve are $(-1, 3.7)$ and $(5, -32.3)$. Use the second-derivative test for relative maximum and minimum.

$$f'(-1) = 0 \quad \text{and} \quad f''(-1) = -6 < 0$$

Therefore, $(-1, 3.7)$ is a relative maximum point.

$$f(5) = 0 \quad \text{and} \quad f''(5) = 6 > 0$$

Therefore, $(5, -32.3)$ is a relative minimum point.

EXAMPLE 3 Applying First- and Second-Derivative Tests

For the function $f(x) = \dfrac{x^3}{3} + \dfrac{x^2}{2} - 2x - 3$, perform the following.

a. Determine the intervals where the curve is increasing.
b. Determine the intervals where the curve is decreasing.
c. Determine all relative maximum points.
d. Determine all relative minimum points.
e. Determine all points of inflection.
f. Determine the intervals where the curve is concave up.
g. Determine the intervals where the curve is concave down.
h. Sketch the graph.

Solution To determine where the curve is increasing or decreasing we must know the critical points.

$$f(x) = \frac{x^3}{3} + \frac{x^2}{2} - 2x - 3$$

$$f'(x) = x^2 + x - 2 \qquad \textbf{Differentiating}$$

$$x^2 + x - 2 = 0$$

$$(x + 2)(x - 1) = 0 \qquad \textbf{Factoring}$$

$$x = -2 \qquad x = 1 \qquad \textbf{Solving for } x$$

Thus, we have critical points at $\left(-2, \frac{1}{3}\right)$ and $\left(1, \frac{-25}{6}\right)$. Now find the second derivative $f''(x) = 2x + 1$.

When $x = -2$ When $x = 1$

$$f''(-2) = 2(-2) + 1 \qquad f''(1) = 2(1) + 1$$
$$= -3 \qquad\qquad\qquad\quad = 3$$

Since the second derivative is negative when $x = -2$, we have a relative maximum at $\left(-2, \frac{1}{3}\right)$. The second derivative is positive when $x = 1$, thus $\left(1, \frac{-25}{6}\right)$ is a relative minimum point.

a. The first derivative is positive for $x < -2$ and for $x > 1$; therefore, the curve is increasing for $x < -2$ and $x > 1$.

b. The first derivative is negative for $-2 < x < 1$; therefore, the curve is decreasing for $-2 < x < 1$.

c. The relative maximum point is at $\left(-2, \frac{1}{3}\right)$.

d. The relative minimum point is at $\left(1, \frac{-25}{6}\right)$.

e. We can determine the points of inflection by setting the second derivative equal to zero.

$$f''(x) = 2x + 1$$
$$2x + 1 = 0$$
$$x = \frac{-1}{2}$$

For $x < \frac{-1}{2}, f'' < 0$, and for $x > \frac{-1}{2}, f'' > 0$. Since there is a change of sign as we move from left to right through the value $x = \frac{-1}{2}$, the point $\left(\frac{-1}{2}, \frac{-23}{12}\right)$ is a point of inflection.

f. We see from part e that $f'' < 0$ for $x < \frac{-1}{2}$; therefore, the curve is concave down for $x < \frac{-1}{2}$.

g. We see from part e that $f'' > 0$ for $x > \frac{-1}{2}$; therefore, the curve is concave up for $x > \frac{-1}{2}$.

h. Using the information from parts a through g and the fact that the y-intercept is the point $(0, -3)$, we can sketch the graph as in Figure 20.16. ◆

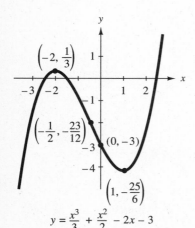

$$y = \frac{x^3}{3} + \frac{x^2}{2} - 2x - 3$$

FIGURE 20.16

20.3 Exercises

In Exercises 1–18, use the second-derivative test to determine the following.
a.   Relative maximum points
b.   Relative minimum points
c.   Points of inflection
d.   Intervals of concavity.

1. $f(x) = x^2 + 2$

2. $f(v) = v^2 - 10$

3. $f(x) = 2x^2 + 3x + 1$

4. $f(n) = n^2 + 4n + 6$

5. $f(x) = 16 - x^2$

6. $f(x) = 25 - x^2$

7. $f(x) = x^3 + 8$

8. $f(x) = x^3 - 27$

9. $f(x) = x^3 + 3x^2 + 1$

10. $f(x) = \dfrac{x^3}{3} - \dfrac{x^2}{2} - 2x + 5$

11. $f(x) = x^3 + 5x^2 - 8x - 5$

12. $f(x) = 2x^3 + 3x + 1$

13. $f(x) = -x^3 + 5x - 1$

14. $f(x) = x^3 + 2x^2 - 8x$

15. $f(x) = (x - 3)^4$

16. $f(x) = \dfrac{2}{x}$

17. $f(x) = x + \dfrac{1}{x}$

18. $f(x) = 3x - \dfrac{1}{3x}$

In Exercises 19–36, perform the following.
a.   Determine the intervals where the curve is increasing.
b.   Determine the intervals where the curve is decreasing.
c.   Determine relative maximum points.
d.   Determine relative minimum points.
e.   Determine points of inflection.
f.   Determine intervals where the curve is concave up.
g.   Determine intervals where the curve is concave down.
h.   Sketch the curve.

19. $f(x) = 2x^2 - 1$

20. $s(t) = 3t^3 - 5$

21. $f(x) = 4x^2 + 5x + 1$

22. $q(r) = r^2 - 4r + 3$

23. $f(x) = x^3 + 27$

24. $f(x) = x^3 - 8$

25. $f(x) = -x^3 + 12x - 3$

26. $f(x) = \dfrac{x^3}{3} - 5x^2 + 9x - 1$

27. $f(x) = 2x^3 - 3x^2 - 36x + 12$

28. $f(x) = x^3 + 6x^2 + 12x - 3$

29. $f(x) = \dfrac{x^3}{3} - 3x^2 + 5x + 1$

30. $f(x) = \dfrac{x^3}{3} - 3x^2 - 7x + 20$

31. $f(x) = \dfrac{x^4}{4} - \dfrac{3}{2}x^2 - 2x + 4$

32. $f(x) = \dfrac{x^4}{4} - \dfrac{x^3}{3} - 2x^2 + 4x$

33. $f(x) = \dfrac{x^4}{4} - \dfrac{x^3}{3} - 4x^2 + 8x$

34. $f(x) = \dfrac{3}{x^3}$

35. $f(x) = \dfrac{-2}{x}$

36. $f(x) = x + \dfrac{1}{x}$

20.4

Applied Maximum and Minimum Problems

The techniques of using the first and second derivatives to test for maximum and minimum values frequently are used in science and technology. At times, then, the physical situation must be expressed in functional form $y = f(x)$ so that we can determine whether the situation is a maximum, a minimum, or neither.

EXAMPLE 1   **Maximum Altitude of a Missile**

A missile is projected into a parabolic path and designed to blow up at maximum altitude. The equation for the path of the missile is given as $y = 100x - 0.01x^2$. What are the coordinates for the expected explosion?

Solution   To determine the maximum height of the missile, we find the critical values and use the second-derivative test.

$$y = 100x - 0.01x^2$$

$$\frac{dy}{dx} = 100 - 0.02x \qquad \text{First derivative}$$

$$\frac{d^2y}{dx^2} = -0.02 \qquad \text{Second derivative}$$

Now we set the first derivative equal to zero to determine the critical value.

$$100 - 0.02x = 0$$

$$x = 5000$$

Since the second derivative is negative for all values of x, the critical value $c = 5000$ is a maximum point on the curve. Substituting the critical value for x in the original function, we can determine the value for y.

$$y = 100(\boxed{5000}) - 0.01(\boxed{5000})^2$$

$$= 250,000$$

Thus, the point on the flight path where the explosion occurs is (5000, 250,000). ◆

EXAMPLE 2   **Application: Low Point of a Cable**

The supporting cable of a pipeline suspension system forms a parabolic arc between the supports, which is described by the equation $y = 0.03125x^2 - 1.25x$. The distances are measured in meters. The origin of the axis system is at the point where the cable attaches to the left support tower. Where is the low point on the cable and how far is it below the attachment point? (See Figure 20.17.)

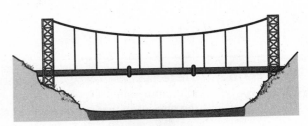

FIGURE 20.17

Solution   To determine the low point of the cable, we determine the critical values and use the second-derivative test.

$$y = 0.03125x^2 - 1.25x$$

$$\frac{dy}{dx} = 0.0625x - 1.25 \qquad \textbf{First derivative}$$

$$\frac{d^2y}{dx^2} = 0.0625 \qquad \textbf{Second derivative}$$

Set the first derivative equal to zero and determine the critical values.

$$0 = 0.0625x - 1.25$$

$$x = 20.0$$

Since the second derivative is positive for all values of x, the critical value $c = 20.0$ is a minimum point on the curve. The low point on the cable occurs 20.0 m to the right of the left tower. Substituting the critical value for x in the original function, we can determine the distance the low point of the cable is below the attachment point.

$$y = 0.03125x^2 - 1.25x$$

$$= 0.03125(\boxed{20})^2 - 1.25(\boxed{20})$$

$$= -12.5 \text{ m}$$

Therefore, the low point of the cable is 20.0 m to the right and 12.5 m below its point of attachment to the support.   ◆

A sketch representing the data is often helpful in developing a function that represents a length, area, or volume that must be maximized or minimized. Example 3 illustrates this.

EXAMPLE 3   **Application: Maximum Area**

The Laws want to install invisible fencing in their front yard to keep their dogs on their property. The design calls for a rectangular area, with the front of the house as one side of the rectangle (see Figure 20.18). With 100 feet of fence, what is the largest area that the Laws can enclose?

Solution   In Figure 20.18 the sides of the rectangle, which are the variable quantities, are labeled. The length and width of the enclosed region are labeled, ℓ and w, respectively. The area of the region is given by the following formula.

$$A = \ell w \tag{1}$$

FIGURE 20.18

The area is to be a maximum, so we want to maximize A. To take the derivative, we must express A in terms of a single variable, either ℓ or w. To do this, we must have a second expression that relates ℓ and w. The amount of fence available is 100 ft, which must be equal to the sum of the three sides of the rectangle. The lengths of the three sides are ℓ, ℓ, and w. Therefore,

$$100 = 2\ell + w.$$

Solve the equation for w.

$$w = 100 - 2\ell \tag{2}$$

Substitute $10 - 2\ell$ for w in Equation (1) and the result is an equation in which A is a function of ℓ.

$$A(\ell) = (\boxed{100 - 2\ell})\ell = 100\ell - 2\ell^2$$

What are realistic values for A and w? A cannot be negative, so let $A(\ell) \geq 0$. When $A(\ell) = 0$,

$$(100 - 2\ell) \geq 0.$$

This inequality is satisfied when ℓ is in $[0, 50]$ and the end points of a closed interval for the function $A(\ell)$ are $\ell = 0$ and $\ell = 50$. To determine the critical values, solve $A'(\ell) = 0$.

$$A'(\ell) = 100 - 4\ell$$
$$0 = 100 - 4\ell$$
$$\ell = 25$$

The critical value is $\ell = 25$. The second derivative $[A''(\ell) = -4]$ is negative for all values of x, and 25 is a relative maximum value. Thus, the three possible absolute maximum values are 25 and the two end points are 0 and 50. Substituting these values in the function $A(\ell) = 100\ell - \ell^2$, the results are:

$$A(25) = 100(\boxed{25}) - 2(\boxed{25})^2 = 1{,}250$$
$$A(0) = 100(\boxed{0}) - 2(\boxed{0})^2 = 0$$
$$A(100) = 100(\boxed{50}) - 2(\boxed{50})^2 = 0$$

With a length of 25, the width is:

$$w = 100 - 2(\boxed{25}) = 50.$$

The largest value is:

$$A(\boxed{25}) = 1{,}250.$$

Therefore, the largest fenced-in area has a length of 25 ft and a width of 50 ft. ◆

Translating written words into equations is often the most difficult part of solving a maximum and minimum problem. The following guidelines may help.

> **GUIDELINES FOR SOLVING MAXIMUM AND MINIMUM PROBLEMS**
>
> 1. Read the problem carefully.
> 2. Identify the quantity that is to be maximized or minimized.
> 3. Draw a sketch and label it, if appropriate.
> 4. Choose variables to represent the unknown quantities.
> 5. Write an equation (or equations) that contain the quantity to be maximized or minimized.
> 6. If necessary, use the equations in Step 5 to write a single equation that contains only the variable to be maximized or minimized.
> 7. Calculate the first derivative and then determine the critical values.
> 8. Use the first-derivative or second-derivative test to determine whether the critical values are the desired maximum or minimum.
> 9. Check your solution.

EXAMPLE 4   **Application: Dimensions of a Cup**

A manufacturer of paper cups is asked to design a cylindrical-shaped cup that will hold 100.0 cc of liquid. Determine the height and radius of the cup that can be made with the least amount of paper.

FIGURE 20.19

Solution   We want to minimize the amount of paper used in making the cup, which is the surface area. The surface area is the sum of the area of the side and the area of the base. (See Figure 20.19.) The formula for the surface area is:

$$SA = 2\pi rh + \pi r^2$$

We must express the surface area in terms of one independent variable, either r or h. Knowing that $V = 100.0$ cc, we write a second expression that relates r and h. With a cylindrical shape, the volume of the cup is:

$$V = \pi r^2 h$$

and

$$100.0 = \pi r^2 h$$

$$h = \frac{100.0}{\pi r^2}. \quad \text{Solving for } h$$

Now we can substitute for h in the equation for surface area.

$$SA = 2\pi r\left(\frac{100.0}{\pi r^2}\right) + \pi r^2$$

$$= \frac{200.0}{r} + \pi r^2$$

$$= 200.0 r^{-1} + \pi r^2$$

We now calculate the first derivative and determine the critical values.

$$\frac{d(SA)}{dr} = -200.0r^{-2} + 2\pi r$$

$$0 = \frac{-200.0}{r^2} + 2\pi r$$

$$0 = \frac{-200.0 + 2\pi r^3}{r^2}$$

$$2\pi r^3 = 200.0 \quad \text{or} \quad r = \sqrt[3]{\frac{100.0}{\pi}}$$

The critical values are $r = 0$ and $r = \sqrt[3]{\dfrac{100.0}{\pi}}$. We know that $r \neq 0$ or there

would be no cup. To test whether $r = \sqrt[3]{\dfrac{100.0}{\pi}}$ is a minimum value, we use the second-derivative test.

$$\frac{d^2(SA)}{dr^2} = 400.0r^{-3} + 2\pi = \frac{400.0}{r^3} + 2\pi$$

Substituting $\sqrt[3]{\dfrac{100.0}{\pi}}$ for r in the second derivative, the result is $\dfrac{d^2(SA)}{dr^2} = 6\pi$, which is always positive. Therefore, r is a minimum. To determine the height, we substitute for r in the equation:

$$h = \frac{100}{\pi r^2} = \frac{100}{\pi \left(\sqrt[3]{\dfrac{100}{\pi}}\right)^2}.$$

The height and radius that minimize the amount of paper used to manufacture the cup is $h = 3.169$ cm and $r = 3.169$ cm. ◆

20.4 Exercises

1. The voltage drop across a capacitance is given by $v = 2t^3 - 6t + 1$. Determine any time $t \geq 0$ when the current is a relative maximum or minimum. What are these voltages?

2. The current through an inductance is given by $i = -2t^3 + 6t^2 - 3$. Determine any time $t \geq 0$, when the current is a relative maximum or minimum. What are these currents?

3. A small factory produces widgets. Its cost is proportional to the square of the number of the widgets, and its gains are related to the number of widgets produced. The company estimates that its profit can be expressed by $P(x) = 10x - 0.01x^2$, where P is the number of dollars per day.
    a. How many widgets should the company make to maximize its profit?
    b. What is the maximum profit per day in dollars?

4. The profit realized by Company Z can be determined with the function $P(x) = \dfrac{-x^3}{3} + x^2 + 8x - 5$, where x is the number of gadgets produced.

    a. Determine the maximum profit.

    b. Determine the number of gadgets the company must manufacture to make the maximum profit.

5. The total cost of manufacturing x gadgets is given by the function $T(x) = x^2 - 2x + 4 + \dfrac{8}{x}$.

    a. Determine the minimum total cost.

    b. Determine the number of gadgets the company can manufacture for the minimum total cost.

6. Psychologists define learning rate $L(x)$ to be the percentage of learning that an individual exhibits in x minutes while solving an intricate puzzle. For a particular puzzle, the average $L(x)$ is given by $L(x) = 100x^2 - 20x^3 + x^4$, for $0 < x < 12$. Determine where $L(x)$ is a minimum.

7. A ball is thrown straight upward so that its height after t seconds is $f(t) = -16t^2 + 32t + 10$, where f is measured in feet.

    a. How high does the ball travel?

    b. How many seconds does it take to reach its maximum height?

8. A company decides to install solar heating devices. The total savings per year (in dollars) S are given by an equation related to the area A of the solar collector: $S(A) = 280A - 0.08A^3$.

    a. What area should be used to maximize savings?

    b. If that area is used, what savings should be realized after one year?

9. A rectangular board for a game requires 81 cm². What should be the dimensions of the rectangle so that the perimeter of the board is a minimum?

10. The base of a box has a rectangular shape and a perimeter of 100 mm. What is the maximum possible area for the base of the box?

11. A square piece of sheet metal 500 mm on a side has squares cut out of its corners so the sides can be folded up to form an open box. What size should the cut-outs be in order to maximize the volume of the box?

12. A bacteria culture undergoes a growth and decline period before it settles into a steady growth period. The number of bacteria in the culture is given by $n = t^3 - 16t^2 + 45t + 210$, with t in hours.

    a. Determine when the early maximum and minimum populations occur.

    b. What are the approximate numbers for these early maximum and minimum populations?

13. The flight of an arrow is described by the equation $y = (-2.00 \times 10^{-3})x^2 + (3.75 \times 10^{-2})x + 1.50$. Assume that the origin of the axis system is at the archer's feet and the distance is measured in meters. What is the greatest height reached by the arrow?

14. A suspension cable has the equation $y = 0.005x^2 - 0.400x + 8.00$. If the origin is in line vertically with the left support, how far away from the support is the low point of the cable? What is the vertical distance between the point of connection at the support and the low point of the cable?

15. The displacement of a cam-operated push-rod is given by $s = 1 + 3t^2 - t^3$, where t is the elapsed time in seconds and s is in inches. What is the maximum displacement of the push-rod?

16. A farmer is planning a rectangular enclosure for his goats. He has 60 meters of fencing and wants to maximize the grazing area.
    a. Write an equation for the area of the enclosure that is dependent upon its length but not on its width.
    b. What length gives a maximum area?

17. A manufacturer wants to make a cost-effective cylindrical drum, which must contain a volume of 10 m³. The cost of the drum is proportional to the surface area of the drum; hence, the manufacturer wants to minimize the surface area.
    a. Let h represent the height of the drum and r represent its radius. Write an equation for the volume and total surface area of the drum.
    b. Combine these equations to express the surface area of the drum exclusively with respect to the radius.
    c. Determine the radius of the drum that minimizes the area.
    d. What must be the height of this special drum?

18. A company that manufacturers aluminum cans receives a bid to produce a can that holds 64π cm³. Determine the height and radius of the can that minimizes the cost of the aluminum per can.

19. A parcel post package is limited by a combined length and girth of 84 in. The girth is the distance around the package in the two smaller dimensions. If the package is to be rectangular with a square cross-section, what is the largest volume that can be accommodated?

20. What is the volume of the largest package according to Exercise 19 if the cross section is to be circular?

21. The dimensions of wood are given with two numbers: the width and the depth. For example, a 2 × 4 is 2 in. wide and 4 in. deep. The strength of the wood is related to the type of wood and the dimension. The strength of a certain type of wood varies in the following way: $S = 300wd^2$. Determine the dimensions of the strongest beam that can be cut from a round log with a diameter of 12 in. (*Hint:* Use the Pythagorean formula to express a relationship between w and d.)

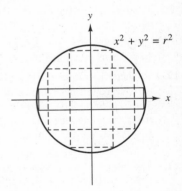

22. A rectangle can be inscribed in a circle $x^2 + y^2 = r^2$ in many ways as shown in the illustration. What lengths of the sides of the rectangle, in terms of r, would cause the area of the rectangle to be a maximum?

23. A satellite has an orbit that can be described in scaled form as $x^2 + 9y^2 - 8x - 36y + 16 = 0$. What are the maximum and/or minimum values of y? (*Hint:* Use implicit differentiation.)

24. A satellite has an orbit that can be described in scaled form as $4x^2 + y^2 - 8x - 10y - 71 = 0$. What are the maximum and/or minimum values of y? (*Hint:* Use implicit differentiation.)

20.5

Related Rates

Many quantities change with time. If a functional relationship exists between two variables it is possible to relate the rate of change with respect to time of one variable to the rate of change with respect to time of the other variable. *This process is called* **related rates** *and is accomplished by differentiating the function with respect to time. The differentials then represent the rates of change of the variables with respect to time.*

For example, in an implicit function of the form $x^2 + y^2 = 25$ if both the x- and the y-variables are dependent upon the variable t, then $x = f(t)$ and $y = f(t)$. If x and y are distances and each is changing with respect to time, then we can determine the change by differentiating x with respect to t and y with respect to t. That is, $\dfrac{dx}{dt}$ is the change of x with respect to t and $\dfrac{dy}{dt}$ is the change of y with respect to t. To take the derivative of each term of the implicit function we need to apply the chain rule. That is, the derivative of x^2 with respect to t is $2x\dfrac{dx}{dt}$, and the derivative of y^2 with respect to t is $2y\dfrac{dy}{dt}$. Example 1 illustrates this procedure.

EXAMPLE 1   Implicit Differentiation

Suppose that x and y are related by the equation $x^2 + y^2 = 64$ and x and y are functions of t.

a.   Determine an equation relating $\dfrac{dy}{dt}$ and $\dfrac{dx}{dt}$.

b.   Determine $\dfrac{dy}{dt}$ when $x = 4$ ft, $y = 5$ ft and $\dfrac{dx}{dt} = 2\dfrac{\text{ft}}{\text{s}}$.

Solution

a.   To determine an equation relating $\dfrac{dy}{dt}$ and $\dfrac{dx}{dt}$, we must think of the equation as an implicit function of t, where both x and y are functions of t. Differentiate each term of the equation with respect to t.

$$\frac{d}{dt}(x^2 + y^2) = \frac{d}{dt}(64)$$

$$\frac{d(x)^2}{dt} + \frac{d(y)^2}{dt} = \frac{d(64)}{dt}$$

$$2x\frac{dx}{dt} + 2y\frac{dy}{dt} = 0 \qquad \textbf{Differentiating with respect to time}$$

$$\frac{dy}{dt} = -\frac{x}{y}\frac{dx}{dt} \qquad \textbf{Solving for } \frac{dy}{dt}$$

b.   To determine the value $\dfrac{dy}{dt}$, we substitute in the result obtained in part a.

$$\frac{dy}{dt} = -\frac{4\text{ ft}}{5\text{ ft}}\,(2\text{ ft/s}) \qquad \textbf{Substituting}$$

$$= -\frac{8}{5}\text{ ft/s}$$

Thus, for the given conditions, the rate of change of the distance y with respect to time t is $-\dfrac{8}{5}$ ft/s.   ◆

In Example 1 we stated that x and y are functions of a third variable t. The result of differentiating the implicit function with respect to t is an equation that shows the relationship between the rate of change of x with respect to t and the rate of change y with respect to t. A problem of this type is called a *related rates problem*. Example 2 illustrates a related rates application.

EXAMPLE 2   Application: Time to Fill a Tank

Waste oil to be reprocessed is pumped into a cylindrical holding tank at a constant rate of 50 ft³/min. If the tank has a radius of 40 ft, is 25 ft high, and the depth of the oil when pumping started was 10 ft, answer the following.

a. At what rate is the height of the oil changing in the tank?
b. How long will it take to fill the tank?

Solution

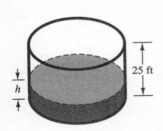

FIGURE 20.20

a. The first thing we do is make a sketch like the one in Figure 20.20. The volume of oil in the tank is a cylindrical mass of height h and radius 40 ft. Using the formula for the volume of a cylinder, $V = \pi r^2 h$, we can write the formula for the volume of oil in the tank,

$$V = \pi(\boxed{40})^2 h.$$

This formula gives us a relationship between the volume of oil and the height of the oil in the tank at any time. We also know that the volume is changing at the constant rate of 50 ft³/min; that is, $\dfrac{dV}{dt} = 50$ ft³/min. We are asked to find the rate at which the height of the oil is changing, that is, $\dfrac{dh}{dt} = ?$ We can determine $\dfrac{dh}{dt}$ by differentiating the formula for volume with respect to time.

$$V = \pi(40)^2 h$$

$$\frac{dV}{dt} = \pi(40)^2 \frac{dh}{dt} \qquad \text{Differentiating with respect to time}$$

$$\boxed{50} = \pi(40)^2 \frac{dh}{dt} \qquad \text{Substituting}$$

$$\frac{dh}{dt} = 9.947 \times 10^{-3} \text{ ft/min} \qquad \text{Solving for } \frac{dh}{dt}$$

b. We know that the remaining distance to the top of the tank is $25 - 10$, or 15 ft, and that the oil is rising at the rate of 9.947×10^{-3} ft/min. Therefore, using the formula distance equals rate times time, we can find the time remaining until the tank is full.

$$d = rt$$

$$h = \frac{dh}{dt}t \qquad \text{Substituting for } d \text{ and } r$$

$$t = \frac{h}{\frac{dh}{dt}} \qquad \text{Solving for } t$$

$$t = \frac{\boxed{15 \text{ ft}}}{\boxed{9.947 \times 10^{-3} \text{ ft/min}}} \qquad \text{Substituting}$$

$$= 1.508 \times 10^3 \text{ min}$$

Therefore, it will take approximately 1510 minutes or 25.1 hours to finish filling the tank. ◆

The following set of guidelines establish a procedure for solving a related rates problem. As you study the examples, refer back to the guidelines.

GUIDELINES FOR SOLVING RELATED RATES PROBLEMS

1. Read the problem carefully.
2. Identify the quantities that are changing with respect to time.
3. Draw a sketch and label it, if possible.
4. Choose variables to represent the unknown quantities.
5. Write an equation (or equations) that relates the quantities that are changing with respect to time.
6. Use implicit differentiation to determine the derivative of the equality with respect to time t. Remember that each variable in the equation is a function of t.
7. Use the results of Step 6 to isolate the quantity for which you are being asked to solve.
8. Substitute the given quantities into the equation found in Step 5 and determine the desired result.
9. Check your solution.

EXAMPLE 3 **Application: Speed of a Canoe**

A canoeist is paddling in still water. Because of paddle slippage through the water, the canoe travels forward during one stroke an amount c that is related to the distance from the starting point of the stroke p by the equation $c = p^{\frac{3}{4}}$. If she pulls at the rate of $1\frac{1}{2}$ m/s through an entire stroke of 1 m, what is the canoe speed at the midpoint of the stroke?

Solution We need to know the speed of the canoe, which would be in meters/second. The given information is in terms of meters and meters/second, and we know an equation that relates distances (meters). We can find an equation relating speeds (meters/second) by differentiating the distance equation with respect to time.

$$c = p^{\frac{3}{4}} \qquad \text{Given}$$

$$\frac{dc}{dt} = \frac{3p^{(\frac{3}{4})-1}}{4}\frac{dp}{dt} \qquad \text{Differentiating with respect to time}$$

$$= \frac{\frac{3}{4}}{p^{\frac{1}{4}}}\frac{dp}{dt}$$

Remember that since c and p have units of meters, $\dfrac{dc}{dt}$ and $\dfrac{dp}{dt}$ have units of meters/second. We know that the stroke length p is 1 m so the midpoint of the stroke comes when $p = \dfrac{1}{2}$ m. The rate of pull at the midpoint of the stroke is $\dfrac{dp}{dt} = 1\dfrac{1}{2}$ m/s. Therefore, the speed of the canoe at the midpoint of the stroke is found by substituting into the equation.

$$\frac{dc}{dt} = \left(\frac{\frac{3}{4}}{p^{\frac{1}{4}}}\right)\left(\frac{dp}{dt}\right)$$

$$= \frac{\frac{3}{4}}{\left(\frac{1}{2}\right)^{\frac{1}{4}}}\left(1\frac{1}{2}\right)$$

$$= 1.338$$

The speed of the canoe at the midpoint of the stroke is 1.34 m/s. ◆

EXAMPLE 4   Application: Slipping Ladder

A painter is on a ladder 7 m long, which is resting against a wall. If the bottom of the ladder starts to slip away from the wall at a constant rate of $\dfrac{1}{4}$ m/s, how fast is the top of the ladder moving down the wall when the bottom of the ladder is each of the following?

a.   2 m from the wall   **b.**   4 m from the wall?

We can represent the ladder resting against the wall as a straight line resting on the x- and y-axes as shown in Figure 20.21. We know the length of the ladder and the distance of its bottom from the wall, both in meters. We know the speed at which the bottom of the ladder is moving away from the wall in meters/second, and we need to know the speed at which the top of the ladder is moving down the wall, which would also have units of meters/second. Therefore, if we can find an equation relating the distances, we can differentiate it with respect

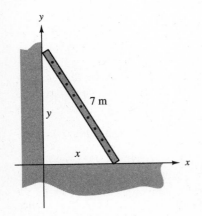

7 m

FIGURE 20.21

to time to obtain an equation relating the speeds. Looking at the right triangle formed by the floor, the wall, and the ladder (Figure 20.21), the Pythagorean theorem seems an appropriate equation.

$$c^2 = a^2 + b^2 \qquad \text{Pythagorean theorem}$$

$$(\boxed{7})^2 = x^2 + y^2 \qquad \text{Substituting}$$

$$y = \sqrt{49 - x^2} \qquad \text{Solving for } y \text{ (we use the positive value of the root because } y \text{ is always positive)}$$

Now differentiate with respect to time.

$$y = (49 - x^2)^{\frac{1}{2}}$$

$$\frac{dy}{dt} = \frac{1}{2}(49 - x^2)^{-\frac{1}{2}}\left(-2x\,\frac{dx}{dt}\right)$$

$$= \frac{-x}{\sqrt{49 - x^2}}\,\frac{dx}{dt}$$

We have obtained an equation that relates the rate of change of the height of the ladder $\left(\dfrac{dy}{dt}\text{ in meters/second}\right)$ to the distance the foot of the ladder is from the wall (x in meters) and the rate of movement of the foot of the ladder $\left(\dfrac{dx}{dt}\text{ in meters/second}\right)$.

a. Substitute for the distance the foot of the ladder is from the wall in the first instance, 2 m, and the rate at which it is moving, $\dfrac{1}{4}$ m/s.

$$\frac{dy}{dt} = \frac{-x}{\sqrt{49 - x^2}}\,\frac{dx}{dt}$$

$$= \frac{-\boxed{2}}{\sqrt{49 - (\boxed{2})^2}}\left(\frac{1}{4}\right) \qquad \text{Substituting for } x \text{ and } \frac{dx}{dt}$$

$$= -0.0745 \text{ m/s}.$$

Since y is measured as positive upward, the negative sign for $\dfrac{dy}{dt}$ means that the top of the ladder is moving downward at the rate of 0.0745 m/s or 74.5 mm/s.

b. When the foot of the ladder is 4 m from the wall

$$\frac{dy}{dt} = \frac{-\boxed{4}}{\sqrt{49 - (\boxed{4})^2}}\left(\frac{1}{4}\right)$$

$$= -0.1741 \text{ m/s}.$$

We see that when the foot of the ladder is 4 m from the wall, the top is sliding downward at 174. mm/s. In this case, the speed at which the top of the ladder moves is dependent upon the position of the foot as well as its speed. ◆

20.5 Exercises

In Exercises 1–8, assuming that x and y are functions of t, perform the following.

a. Differentiate each of the implicit functions with respect to t.

b. Solve for $\dfrac{dy}{dt}$.

c. Determine the value of $\dfrac{dy}{dt}$ when $x = 2$, $y = 1$ and $\dfrac{dx}{dt} = -1$.

1. $x^2 + y^3 = 2$

2. $x^4 - y^2 = 7$

3. $3x^3 - 5x^4 + 8 = 0$

4. $2x^3 - 5y^2 + 3x - y = 5$

5. $x^3 - 7y^2 + 4x - 7y = 2$

6. $4x^2 + 4y^2 + 3x - 5y = 36$

7. $x^2 - 3xy + 3y^2 = 6$

8. $x^2 - y^2 + 5xy - 8 = 0$

9. A satellite moves in a path described by the elliptical equation: $\dfrac{x^2}{28} + \dfrac{y^2}{27} = 1$, where the distance variables x and y are both measured in thousands of miles. When x is 2000 miles ($x = 2$), the x-component, or the speed $\dfrac{dx}{dt}$, is 8000 mph. What is $\dfrac{dy}{dt}$ at this point?

10. When air expands slowly such that there is no change in heat, the type of change is called an adiabatic change. The relationship between pressure p and volume V is $pV^{1.4} = k$, where k is a constant. When the pressure is 20,000 Pa, the volume is 20.0 cc. At that particular pressure, the volume increases at a rate of 1.5 cc/s. What is the rate of change of pressure at that time?

11. When a square steel plate is heated, the length of the side increases at a rate of 0.04 mm/s. How fast is the area increasing when the side of the metal plate is 100 mm long?

12. The area of a circle is increasing at the rate of 24 cm²/min. Determine the rate at which the radius is increasing when the radius is 3 cm.

13. The area of a triangle is increasing at the rate of 12 m²/s. If the height is always twice the base, find the rate at which the base is increasing when the base is 4 m.

14. A spherical balloon is inflated at a constant rate of 0.5612 m³/m. What is the rate at which the radius is increasing $\left(\dfrac{dr}{dt}\right)$ when the radius is 1.0 m?

15. A person's artery has been restricted by fatty deposits. The person is receiving a drug that reduces the fat such that the radius of the opening increases at a rate of 0.010 mm/month. When the artery opens to a radius of 2.0 mm, the patient will be released from the hospital. When patient is released what is the rate at which the cross-sectional opening of the patient's artery is increasing at that radius?

16. A baseball diamond is in the shape of a square, and the bases are 90 ft apart. In the second inning of a game, Adam gets a hit and runs to first base at the rate of 18 ft/s. At the moment Adam starts for first base, Andrew, who has taken a ten-foot-lead off second base, starts running at the same rate for third base. If Adam reaches first base at the same time Andrew reaches third base, how fast is the distance changing between them?

17. A cup-shaped tank in the shape of a hemisphere with radius 12.0 m is being filled at the rate of 500 m³/hr. How fast is the surface level rising when the depth of material is 6.00 m?

18. A tank in the shape of an inverted cone 4 m high and with a 1-m radius opening is being filled at the rate of $\frac{1}{2}$ m³/min. How fast is the surface level rising when the depth of material is 3 m?

19. An electron is being accelerated along the path $xy = 12$ m². If the electron is at (4, 3) and its x-coordinate is increasing at the rate of 1000 m/s, how fast is its y-coordinate changing?

20. The total resistance R of a circuit containing resistors R_1 and R_2 in parallel is $R = \dfrac{R_1 R_2}{R_1 + R_2}$. If R_1 is decreasing at the rate of 0.51 Ω/min and R_2 is decreasing at the rate of 0.83 Ω/min, determine the rate of change of R when R_1 is 5.00 Ω and R_2 is 8.00 Ω.

21. An ice cube is melting such that each surface is decreasing at the same rate. The length of an edge of the ice cube is decreasing at the rate of 0.12 cm/min.

    a. How fast is volume changing when the edge measures 3.1 cm?
    b. How fast is the surface area changing when the edge measures 2.5 cm?

22. A 24-ft ladder leans against a building. If the bottom of the ladder is pulled from the building at the rate of 10.0 ft/s, how fast is the ladder moving down along side the building when the top of the ladder is 5.00 ft from the ground?

23. If $y = (4x + x^3)^{1.5}$ and x is increasing at the rate of 0.50 units/sec, how fast is the slope of the graph changing when $x = 3.0$ units?

24. If $y = (5x - 4^2)^3$ and x is increasing at the rate of 0.40 units/s, how fast is the slope of the graph changing when $x = 3.5$ units?

20.6

Curvilinear Motion

Many of the previous applications were confined to one dimensional motion. Now let us consider the *curved* motions of particles and discuss the applications, which range from electron beams scanning a television set to air molecules moving along a curved-wing section. We begin with some problems dealing with motion in a plane when the motion is expressed in terms of the x- and y-coordinates, locating a point in the plane at any time. (For motion in three dimensions or using different coordinate systems consult texts dealing with kinematics.)

When the motion of a particle is expressed in terms of its x- and y-coordinates, the values of x and y often are given as separate dependent functions of time. The result of writing x and y as functions of t is a set of **parametric equations,** such as

$$x = t + 2 \quad \text{and} \quad y = -t^2 + 12.$$

The independent variable, time, is called the **parameter.** In Example 1 demonstrates a procedure for removing the parameter t and writing the rectangular equation that represents the same graph.

EXAMPLE 1   Path of Parametric Equations

The motion of a point is described by the equations $y = -t^2 + 12$ and $x = t + 2$, where x and y are coordinates of the point at any time t. What kind of path does the point follow?

Solution   To visualize the path of the point we write the rectangular equation, found by eliminating the parameter t.

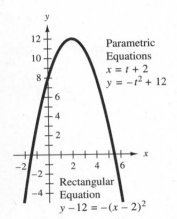

$$x = t + 2 \tag{3}$$
$$y = -t^2 + 12 \tag{4}$$
$$t = x - 2 \qquad \text{Solving (3) for } t \text{ as a function of } x \tag{5}$$
$$y = -(x - 2)^2 + 12 \qquad \text{Substituting (5) for } t \text{ in (4) to determine } y \text{ as a function of } x$$
$$y - 12 = -(x - 2)^2$$

In rectangular form we recognize the path of the parametric equations as a parabola with the vertex at the point $(2, 12)$. The path is illustrated in Figure 20.22. ◆

The equations in Example 1 are specific examples of the parametric equations $x = at + h$ and $y = bt^2 + k$, where $a, b, h,$ and k are constants. If we eliminate the parameter t the result is: $y - k = \dfrac{b}{a^2}(x - h)^2$, which is the general form of a parabola with the vertex at the point (h, k).

To help eliminate the parameter the following guidelines are provided.

Parametric
Equations
$x = t + 2$
$y = -t^2 + 12$

Rectangular
Equation
$y - 12 = -(x - 2)^2$

FIGURE 20.22

GUIDELINES FOR ELIMINATING THE PARAMETER

1. Solve for t in one of the equations.
2. Substitute the value for t found in Step 1 in the second equation.
3. The result is a rectangular equation, which can be written in a standard form.

With an equation in parametric form, we can determine the rate of change of a coordinate with respect to time, as illustrated in Example 2.

EXAMPLE 2   **Change of y-Coordinate with Respect to Time**

If an electron is traveling along a path given by $y = \dfrac{a}{x}$, where x and y are in meters, how fast is its y-coordinate changing with respect to time when x is changing according to $x = t^2$?

Solution

$$y = \frac{a}{x} \qquad \text{Given}$$

$$x = t^2$$

$$y = \frac{a}{t^2} \qquad \text{Substituting for } x$$

To find the rate of change of y, we differentiate with respect to time.

$$y = at^{-2} \qquad \text{Using rule of exponents}$$

$$\frac{dy}{dt} = -2at^{-3} \qquad \text{Differentiating}$$

$$= \frac{-2a}{t^3}$$

Therefore, the rate of change of the y-coordinate of the point is $\dfrac{dy}{dt} = \dfrac{-2a}{t^3}$.

An alternative method of solution would be to differentiate with respect to time first and then substitute.

$$y = \frac{a}{x} \qquad\qquad \text{Given}$$

$$x = t^2$$

$$\frac{dy}{dt} = \frac{-a}{x^2}\frac{dx}{dt} \qquad\qquad \text{Differentiating}$$

$$\frac{dx}{dt} = 2t$$

$$\frac{dy}{dt} = -\frac{a}{(\,t^2\,)^2}(\,2t\,). \qquad \text{Substituting for } x \text{ and } \frac{dx}{dt}$$

Therefore, y is changing with respect to time as $\dfrac{dy}{dt} = -\dfrac{2a}{t^3}$. ◆

It can be shown that the velocity of a particle whose path is known in terms of its x- and y-coordinates is a vector quantity whose components are the speeds

$v_x = \dfrac{dx}{dt}$ and $v_y = \dfrac{dy}{dt}$ of the motion in the x-direction and the y-direction, respectively (Figure 20.23). Therefore, the magnitude of the velocity is the magnitude of the vector whose components are v_x and v_y.

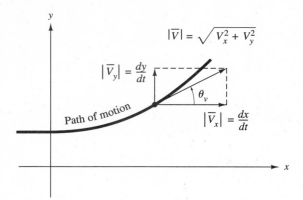

FIGURE 20.23

Magnitude of Velocity

$$ v = \sqrt{v_x^2 + v_y^2} \quad \text{or} \quad v = \sqrt{\left(\frac{dx}{dt}\right)^2 + \left(\frac{dy}{dt}\right)^2} $$

Since the direction of the velocity is always tangent to the path of motion, it can be found from the slope of the path, $\tan \theta_v = \dfrac{|V_y|}{|V_x|}$.

Direction of Velocity

$$ \tan \theta_v = \dfrac{\dfrac{dy}{dt}}{\dfrac{dx}{dt}} $$

EXAMPLE 3   Direction of a Particle

A particle is traveling along the path $y = 2x - \dfrac{1}{x^2}$. What is its direction when $x = -3$?

Solution Since the direction and velocity of a particle are always tangent to its path, we have $\tan \theta_v = \dfrac{dy}{dx}$, where θ_v is the angle the direction of velocity makes with the positive *x*-axis.

$$y = 2x - \frac{1}{x^2} \qquad \text{\textbf{Given path}}$$

$$= 2x - x^{-2}$$

$$\frac{dy}{dx} = 2 - (-2)x^{-3} \qquad \text{\textbf{Differentiating}}$$

$$= 2 + \frac{2}{x^3}$$

When $x = -3$,

$$\frac{dy}{dx} = 2 - \frac{2}{(-3)^3}$$

$$= 2.074$$

$$\tan \theta_v = \frac{dy}{dx}$$

$$= 2.074$$

$$\theta_v = 64.26°.$$

Therefore, the particle is traveling at an angle of 64.3° with respect to the positive *x*-axis. ◆

In Example 4 we determine the velocity of a falling object as it hits the ground.

EXAMPLE 4 **Velocity and Direction of a Falling Particle**

A bird drops a berry that falls along the path $y = -x^2 + 10$ meters above the ground. (The effects of air friction are included.) If $x = \dfrac{1}{2}t$ m, with *t* in seconds, what is the direction and the velocity of the berry when it hits the ground? (See Figure 20.24.)

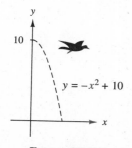

FIGURE 20.24

The graph shows $y = -x^2 + 10$ with the value 10 marked on the *y*-axis.

Solution To determine the magnitude and direction of the berry we need to know $\dfrac{dx}{dt}$ and $\dfrac{dy}{dt}$. Since $x = \dfrac{1}{2}t$, we can differentiate to find $\dfrac{dx}{dt} = \dfrac{1}{2}$. To find $\dfrac{dy}{dt}$, we differentiate $y = -x^2 + 10$ with respect to time.

$$\frac{dy}{dt} = -2x\frac{dx}{dt}$$

We see that to evaluate $\dfrac{dy}{dt}$, we need to know the value of x as well as $\dfrac{dx}{dt}$ at the instant required. When the berry hits the ground $y = 0$, so:

$$y = -x^2 + 10 \quad \textbf{Given}$$
$$0 = -x^2 + 10$$
$$x = \sqrt{10} \text{ m.}$$

Therefore, at the ground $\dfrac{dx}{dt} = \dfrac{1}{2}$. Now substitute.

$$\frac{dy}{dt} = -2(\sqrt{10})\frac{1}{2}$$
$$= -\sqrt{10} \text{ m/s.}$$

The magnitude of the velocity of the berry is:

$$v = \sqrt{\left(\frac{dx}{dt}\right)^2 + \left(\frac{dy}{dt}\right)^2}$$
$$= \sqrt{\left(\frac{1}{2}\right)^2 + (-\sqrt{10})^2}$$
$$= \sqrt{\frac{1}{4} + 10}$$
$$= \sqrt{\frac{41}{4}}$$
$$= \frac{\sqrt{41}}{2} \text{ m/s.}$$

The direction of the velocity of the berry is:

$$\tan\theta_v = \frac{\dfrac{dy}{dt}}{\dfrac{dx}{dt}}$$
$$= \frac{-\sqrt{10}}{\dfrac{1}{2}}$$
$$= -2\sqrt{10}$$
$$\theta_v = -81.02°.$$

Therefore, when it strikes the ground the berry has a speed (magnitude of velocity) of $\dfrac{\sqrt{41}}{2}$ m/s and a downward direction of $81.02°$ with the horizontal.

◆

The acceleration of a particle is also a vector quantity. In rectangular coordinates its magnitude can be expressed as follows.

Acceleration of a Particle

$$a = \sqrt{a_x^2 + a_y^2}, \text{ where } a_x = \frac{d^2x}{dt^2} \text{ and } a_y = \frac{d^2y}{dt^2}$$

Its direction is generally not tangent to the path of motion, but the angle θ_a the acceleration vector makes with the positive x-axis can be found from the following.

Direction of Acceleration Vector

$$\tan \theta_a = \frac{\dfrac{d^2y}{dt^2}}{\dfrac{d^2x}{dt^2}}.$$

Determining the acceleration of a particle is often necessary because according to Newton's law, $\overline{F} = m\overline{a}$, the acceleration of a body is directly proportional to the force required to maintain it in its path.

EXAMPLE 5 **Magnitude and Direction of a Particle**

Using the information from Example 4, what are the magnitude and direction of the acceleration of the berry when it is approximately eye level $\left(1\frac{2}{3} \text{ m above ground} \right)$?

Solution In the previous example we found that $\dfrac{dy}{dt} = -2x\dfrac{dx}{dt}$ and that $\dfrac{dx}{dt} = \dfrac{1}{2}$. Taking the second derivative with respect to time of both expressions, we obtain:

$$\frac{d^2y}{dt^2} = -2\left(\frac{dx}{dt}\right)^2 - 2x\frac{d^2x}{dt^2} \text{ and } \frac{d^2x}{dt^2} = 0.$$

$\left(\text{Don't forget that } x\dfrac{dx}{dt} \text{ is a product.} \right)$ Using the equation $y = -x^2 + 10$ when $y = 1\dfrac{2}{3}$ gives:

$$1\frac{2}{3} = -x^2 + 10$$

$$x^2 = 8\frac{1}{3}.$$

So when $y = 1\dfrac{2}{3}$, $x = \sqrt{8\dfrac{1}{3}}$. Substituting into $\dfrac{d^2y}{dt^2}$, we have:

$$\frac{d^2y}{dt^2} = -2\left(\frac{dx}{dt}\right)^2 - 2x\frac{d^2x}{dt^2}$$

$$= -2\left(\boxed{\frac{1}{2}}\right)^2 - 2\left(\boxed{\sqrt{8\frac{1}{3}}}\right)(\boxed{0})$$

$$= -\frac{1}{2} \text{ m/s}^2.$$

Therefore,

$$a = \sqrt{\frac{d^2x}{dt^2} + \frac{d^2y}{dt^2}}$$

$$a = \sqrt{\boxed{0}^2 + \left(\boxed{-\frac{1}{2}}\right)^2}$$

$$= \frac{1}{2} \text{ m/s}^2$$

and

$$\tan \theta_a = \frac{\dfrac{d^2y}{dt^2}}{\dfrac{d^2x}{dt^2}}$$

$$= \frac{\boxed{-\dfrac{1}{2}}}{\boxed{0}}.$$

Thus, $\tan \theta_a$ is undefined. Therefore, $\theta_a = -90°$, or at eye level, the acceleration is straight down with a magnitude of $\dfrac{1}{2}$ m/s². ◆

20.6 Exercises

In Exercises 1–4, eliminate the parameter.

1. $x = t - 4$
$y = t^2 - 6$

2. $x = 2t - 4$
$y = t^2 - 2t + 8$

3. $x = 5t - 7$
$y = -2t^2 + 15$

4. $x = at + h$
$y = bt^2 + k$

In Exercises 5–14, determine the magnitude and the direction of the velocity vector for the time given.

5. $x = 10, y = 10t, t = 2$
7. $x = -10t, y = t^2, t = 4$

6. $x = 5t, y = 8, t = 1$
8. $x = -2t^2, y = 4t, t = 3$

9. $x = t^2 + 6t + 1, y = -t^2 + 4, t = 0$

10. $x = 2t^2 - 5t + 3, y = -2t^2 + 5, t = 2$

11. $x = t, y = \dfrac{12}{t}, t = 3$

12. $x = \dfrac{t}{36}, y = \dfrac{3}{t}, t = 2$

13. $x = 3t^2 + 2t, y = -t^2 + 1, t = 1$

14. $x = \dfrac{t^4}{4}, y = \dfrac{2}{t^2}, t = 1$

In Exercises 15–24, determine the magnitude and direction of the acceleration for the motions given in Exercises 5–14.

25. A cam follower is tracing a surface described by the equation $x^2 + y^2 = 25$. What is the direction of the velocity of the follower when it is at the point $(3, -4)$ on the surface?

26. The takeoff portion of a ski jump is described by $y = \dfrac{x^2}{16}$ with the lip of the jump located at $(4, 1)$ meters (see the sketch). If a skier has a horizontal velocity component at 8 m/s and a vertical acceleration component of 8 m/s², what are the magnitudes and directions of her velocity and acceleration at takeoff?

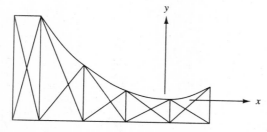

27. A jet plane at an altitude of 1200 m flies parallel to the ground at a speed of 400 m/s. It drops a bomb from its cargo bay. Knowing that the acceleration due to gravity is -9.8 m/s² and using the above information, two equations of motion can be derived:

$$\frac{dx}{dt} = +400 \qquad y = 1200 - \frac{9.8}{2}t^2.$$

a. Determine the magnitude and direction of the velocity one-half minute after the bomb is released.

b. Determine the magnitude and direction of the acceleration one-half minute after the bomb is released.

28. In Example 27, air resistance was neglected. Another way to express the motion in the x-direction is $x = 400t$. When air resistance is taken into account, the equation can be altered to an approximate expression of $x = 400t - 2t^2$. Adopting this change, determine the magnitude and the direction of the velocity and acceleration using the information in Exercise 27.

29. When a car on a circular track speeds up, its equations of motion can be expressed as $x = 200 + 3t^2$ and $y = 200 - 2t^3$. Determine the magnitude and direction of the velocity and acceleration for the car at $t = 3.0$ s.

30. When a space craft is launched, thrust, air resistance, and a gradually diminishing gravity combine to give rather complex equations of motion. As long as the atmosphere remains relatively dense the equations of motion may be approximated as $x = 10(1 + t^4)^{\frac{1}{2}}$ and $y = 35t^{\frac{3}{2}}$.
   a. What are the magnitude and direction of the velocity 20 s after the launch?
   b. What are the magnitude and direction of the velocity 40 s after the launch?

20.7

Differentials (Approximate Increments)

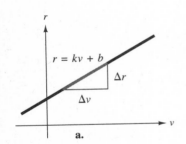

$r = kv + b$

Δr

Δv

a.

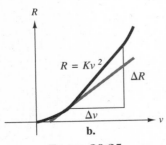

$R = Kv^2$

ΔR

Δv

b.

FIGURE 20.25

If we examine the resistance of an automobile to motion we find that we can approximate its rolling resistance due to friction by $r = kv + b$ and its resistance due to air drag by $R = Kv^2$, where k, b, and K are constant related to a particular model car and $v > 0$. The functions are sketched in Figure 20.25a along with the increments between two points on the curve and a tangent to the curve at one point. In the case of the linear function (Figure 20.25a), the tangent lies along the curve and the resistance depends only on the speed. Hence, an incremental change in v must be accompanied by an incremental change in r. We see that if we change the velocity by Δv that the resistance changes by Δr and that the ratio $\dfrac{\Delta r}{\Delta v}$ is the same as the slope of the function $\dfrac{dr}{dv}$. Therefore,

$\dfrac{\Delta r}{\Delta v} = \dfrac{dr}{dv}$. If we wish to find out how much r changes for given change Δv in the velocity, we can evaluate $\Delta r = \dfrac{dr}{dv} \Delta v$. The quantities dr and dv are called the **differentials** of r and v, respectively. If we compare the slope of the tangent in Figure 20.25b, $\dfrac{dR}{dv}$, we see that it is not equal to the ratio $\dfrac{\Delta R}{\Delta v}$. However, the two quantities approach equality if Δv is small. More specifically, if we pick two points Δv apart on $R = K_v^2$, we have $R_1 = Kv_1^2$ and $R_2 = K(v_1 + \Delta v)^2$. Therefore,

$$\frac{\Delta R}{\Delta v} = \frac{R_2 - R_1}{\Delta v}$$

$$= \frac{K(v_1 + \Delta v)^2 - Kv_1^2}{\Delta v}$$

$$= \frac{K(v_1^2 + 2v_1 \Delta v + \overline{\Delta v}^2) - Kv_1^2}{\Delta v}$$

$$= \frac{K(2v_1 \Delta v + \overline{\Delta v}^2)}{\Delta v}$$

$$= \frac{2Kv_1 \Delta v}{\Delta v} + \frac{K \overline{\Delta v}^2}{\Delta v}$$

$$= 2Kv_1 + K \Delta v.$$

If we differentiate the function $R = Kv^2$ and evaluate it at $v = v_1$, we have $\frac{dR}{dv} = 2Kv$, so $\frac{dR(v_1)}{dv} = 2Kv_1$. Substituting into the expression for $\frac{\Delta R}{\Delta v}$, we find:

$$\frac{\Delta R}{\Delta v} = 2Kv_1 + K \Delta v$$

$$= \frac{dR(v_1)}{dv} + K \Delta v.$$

Therefore, we see that the ratio of the increments of the variables is equal to the derivative of the function evaluated at a point plus a quantity dependent on the increment in the independent variable. If the increment in the independent variable is small we can say that:

$$\frac{\Delta R}{\Delta v} \approx \frac{dR}{dv}.$$

This can be generalized for any function. The ratio of the increments at a point is approximately equal to the ratio of the differentials (the derivative) evaluated at that point.

$$\frac{\Delta y}{\Delta x} \approx \frac{dy}{dx} \quad \text{or} \quad \Delta y \approx \frac{dy}{dx} \Delta x$$

EXAMPLE 1 **Application: Resistance of a Car**

If the rolling resistance of a car in newtons is given by $r = \frac{1}{20}v + 20$ and its air resistance is given by $R = \frac{2}{5}v^2$ (v in m/s), approximately how much does its total resistance change if its speed increases from 20 m/s to 25 m/s (approximately 45 mph to 56 mph)?

Solution The total resistance of the car is $\Re = r + R$, so:

$$\Re = \frac{1}{20}v + 20 + \frac{2}{5}v^2$$

$$\frac{d\Re}{dv} = \frac{1}{20} + \frac{4}{5}v.$$

A speed change from 20 m/s to 25 m/s represents a speed increment of

$$\Delta v = v_2 - v_1$$
$$= 25 \text{ m/s} - 20 \text{ m/s}$$
$$= 5 \text{ m/s.}$$

$$\Delta \Re \approx \frac{d\Re}{dv} \Delta v$$

$$\approx \left(\frac{1}{20} + \frac{4}{5} v \right) \Delta v$$

$$\approx \left[\frac{1}{20} + \frac{4}{5} (\boxed{20}) \right] 5$$

$$\approx 80.25.$$

An increase in speed of 5 m/s increases the resistance approximately 80.3 N at 20 m/s. ◆

EXAMPLE 2 **Application: Change in Volume: Change in Diameter**

A meteorological balloon has a normal diameter at sea level of 2 m. If its volume increases by one half due to air pressure changes as it rises, approximately what is its new diameter?

Solution If the volume of the balloon changes by one half, then:

$$\Delta V = \frac{1}{2} V$$

$$= \frac{1}{2} \left(\frac{4}{3} \pi r^3 \right) \qquad \text{Volume of sphere} = \frac{4}{3} \pi r^3$$

$$= \frac{1}{2} \left[\frac{4}{3} \pi (1)^3 \right]$$

$$= \frac{2\pi}{3}.$$

$$\frac{dV}{dr} = \frac{4}{3} \pi (3) r^2 \qquad \text{Differentiating, } V = \frac{4}{3} \pi r^3$$

$$= 4\pi r^2$$

$$\Delta V \approx \frac{dV}{dr} \Delta r$$

or

$$\Delta r \approx \frac{\Delta V}{4\pi r^2} \qquad \text{Rearranging and substituting for } \frac{dV}{dr}$$

$$\approx \frac{\frac{2\pi}{3}}{4\pi(1)^2} \qquad \text{Substituting for } \Delta v \text{ and } r$$

$$\approx \frac{1}{6} \text{ m.}$$

Since the radius changes by $\frac{1}{6}$ m, the diameter changes by $\frac{1}{3}$ m, and the new diameter is approximately $2\frac{1}{3}$ m. This is a very rough approximation because the volume increment used is relatively large compared with the starting value. ◆

20.7 Exercises

In Exercises 1–16, find the approximate change in the indicated variable.

1. $A = \pi r^2$; $r = 5$, $\Delta r = \frac{1}{2}$, $\Delta A = ?$

2. $V = s^3$; $s = 2$, $\Delta s = 0.01$, $\Delta V = ?$

3. $V = s^3$; $s = 2$, $\Delta V = 0.1$, $\Delta s = ?$

4. $A = \pi r^2$; $r = 5$, $\Delta A = 2$, $\Delta r = ?$

5. $V = \frac{4}{3}\pi r^3$; $r = 4$, $\Delta r = 0.1$, $\Delta V = ?$

6. $S = 2r^2\theta$; $r = 3$, $\Delta\theta = \frac{1}{2}$, $\Delta S = ?$

7. $S = \pi r \sqrt{r^2 + h^2}$; $r = 1$, $h = 4$, $\Delta h = 1$, $\Delta S = ?$

8. $S = \pi r \sqrt{r^2 + h^2}$; $r = 1$; $h = 4$, $\Delta r = 1$, $\Delta S = ?$

9. $T = 2\pi r^2 + 2\pi rh$; $r = 1$; $h = 2$, $\Delta h = \frac{1}{2}$, $\Delta T = ?$

10. $T = 2\pi r^2 + 2\pi rh$; $r = 1$; $h = 2$, $\Delta r = 4$, $\Delta T = ?$

11. $T = \pi r(r + \sqrt{r^2 + h^2})$; $r = 1$, $h = 2$, $\Delta T = 4$, $\Delta r = ?$
12. $T = \pi r(r + \sqrt{r^2 + h^2})$; $r = 1$, $h = 2$, $\Delta T = 4$, $\Delta h = ?$
13. $pV^{\frac{3}{2}} = 12$; $p = 10$, $\Delta V = 1$, $\Delta p = ?$

14. $pV^{\frac{5}{2}} = 100$; $V = 4$; $\Delta p = 1$, $\Delta V = ?$

15. $x = \frac{1}{t} + t^2$; $t = 2$, $\Delta t = 1$, $\Delta x = ?$

16. $y = \frac{1}{t^2} + t$; $t = 2$, $\Delta t = \frac{1}{2}$, $\Delta y = ?$

17. The displacement (loaded weight) of a boat is proportional to the cube of its length. If a 27-ft sailboat displaces 5500 lb, approximately what is the displacement of a similar 30-ft sailboat?

18. The price of a boat is approximately proportional to the cube of its length. If a 5-m runabout costs about $10,000, approximately how much is the price change for a $\frac{1}{4}$-m increase or decrease in length?

19. The price of a metallic object is often proportional to its weight, which in turn is proportional to its volume. If a metallic disk with a 2-m diameter and a volume of 31.4 m³ has its area increased by 0.2 m², approximately how much does its volume increase?

20. The current through a capacitor is given by $i = C\dfrac{dv}{dt}$. If $C = 1\ \mu F$ and $v = t^3 + t - 1$, approximately how much does the current change in one second when $t = 10s$?

21. An oil spill assumes a roughly circular shape in calm waters. Twelve hours after the spill the radius is 1.2 mi and is increasing at a rate of 0.07 mph. What is the rate of increase of the area covered by the oil spill?

22. A spherical balloon increases its radius by 0.002 m with every puff of air. When the radius of the balloon is 0.14 m, what is the incremental increase in each of the following?
    a. The surface area
    b. The volume

23. The force on a particle is given by $F = m\dfrac{dv}{dt}$, where m is the mass. When a 10 kg rocket speeds upward, its equation of motion is $v = 110.2(10 - t) - (9.8/2)t^2$. What is the incremental change in the force for the following?
    a. $t = 2.0$ s
    b. $t = 8.0$ s

24. The height of a projectile is given by $y = 200 + 100t - (9.8/2)t^2$.
    a. Determine the approximate change in the height when the time changes from $t = 9.9$ s to $t = 10.1$ s. Is the object ascending or descending?
    b. Determine the approximate change in the height when the time changes from $t = 19.9$ s to $t = 20.1$ s. Is the object ascending or descending?

Review Exercises

In Exercises 1–6, write the equation (in slope-intercept form) of the line tangent to the given curve at the stated value.

1. $y = x^2 - 2x + 1$; (1, 0)
2. $y = 2x^2 + x - 3$; (−1, −2)
3. $y = 2x^3 + x$; $x = -2$
4. $y = x^3 + 2x^2$; $x = -3$
5. $(x + 2)^2 - xy = 0$; $x = 1$
6. $y^2 - 4xy + 1 = 0$; $y = 2$

In Exercises 7–12, write the equation (in slope-intercept form) of the line normal to the given curve at the stated value.

7. $y = x^4 - 2x^2$; (1, −1)
8. $y = 3x^4 + 4x^3$; (−2, 16)
9. $x^2y = 6$; $x = 2$
10. $y^2x = -2$; $y = -1$
11. $\dfrac{x}{y} = 9$; $y = 2$
12. $\dfrac{y}{x} = 4$; $x = 3$

13. A race car hits an oil slick while going clockwise around a curve given by $y^2 = 400x$, $y > 0$. If it slides safely through a gap between two trees, which is centered at point (4, 50), where on the curve was the oil slick?

14. If the rock from David's sling is traveling along the path $x^2 + 50y - 150 = 0$ and Goliath manages to divert it by placing his shield normal to the path when the rock was 2 m off the ground, what angle does the shield make with the ground?

In Exercises 15–24, use the first derivative test to determine maximum and minimum points on the curve. Sketch the curve.

15. $y = 2x^2$
16. $y = \dfrac{1}{4}x^2$
17. $x^2 = \dfrac{-1}{3}y$
18. $x^2 = 2y$
19. $f(x) = x^4 - 4x^3 - 8x^2$

20. $f(x) = x^4 + \dfrac{4x^3}{3} - 18x^2 - 36x + 4$

21. $y = 5x^{\frac{4}{5}}$

22. $y = 3x^{\frac{2}{3}}$

23. $f(x) = \dfrac{1}{2}x^2 + 3x^{\frac{1}{3}}$

24. $f(x) = \dfrac{1}{2}x^2 - 3x^{\frac{1}{3}}$

In Exercises 25–40, use the second-derivative test to perform the following.
a. Determine maximum points.
b. Determine minimum points.
c. Determine points of inflection.
d. Determine intervals of concavity.
e. Sketch the curve.

25. $y = 3x^2 - 5$

26. $y = -2x^2 + 1$

27. $y = 3x^2 - 17x + 6$

28. $y = 2x^2 + 3x - 9$

29. $y = t^2 - 4t + 7$

30. $y = t^2 + 6t - 5$

31. $y = x^3 + \dfrac{1}{8}$

32. $y = x^3 - \dfrac{1}{8}$

33. $y = \dfrac{x^3}{3} - x^2 - 3x + 9$

34. $y = x^3 - 7x^2 + 7x + 15$

35. $y = 2x^3 - 3x - 6$

36. $y = 4x^3 + x^2 + 5$

37. $y = \dfrac{-5}{x^2}$

38. $y = \dfrac{5}{x^2}$

39. $y = x^2 + \dfrac{1}{x}$

40. $y = x + \dfrac{1}{x^2}$

41. The voltage drop across a capacitance is given by $v = 4t^3 - 12t + 7$. Find any times $t \geq 0$ when the voltage is a relative maximum or minimum. What are those voltages?

42. The current through an inductance is given by $i = -3t^3 - t^2 + 8$. Find any times $t \geq 0$ when the current is a relative maximum or minimum. What are those currents?

43. What is the area of the largest rectangle that can be inscribed in the ellipse $\dfrac{x^2}{36} + \dfrac{y^2}{9} = 1$?

44. What is the volume of the largest right circular cylinder that can be inscribed in a right circular cone with base radius of 400 mm and height of 1600 mm?

45. The impedance in a circuit is given by $Z = \sqrt{R^2 + (X_L - X_C)^2}$. If R and X_C are constants, what value of X_L causes Z to be a minimum?

46. A positron is being accelerated along the path $xy = 36$ m². It is at the point $(9, 4)$ and its y-coordinate is increasing at 6000 m/s. How fast is its x-coordinate changing?

47. The second moment of a plate of unit density is given by $I = \dfrac{1}{12}\,abc(a^2 + b^2)$. If dimensions b and c are held constant, how fast does I change compared to a?

48. A tank in the form of an inverted right circular cone (radius 2 m, height 4 m) is being filled so that when the depth of material is 2 m it is rising at $\dfrac{1}{4}$ m/min. At what rate is material entering the tank?

49. The relationship between the pressure and the volume of a gas is $pV^{\frac{3}{2}} = 100$. What is the relationship between the rate of compression (rate of volume decrease) of the gas and its pressure change?

In Exercises 50–53, what are the magnitudes and directions of the velocity and acceleration for the given motion and time?

50. $x = t^3 + 3t + 2, y = t, t = 4$

51. $x = 4t^2 - 2, y = 6t^3 + 2t, t = 1$

52. $x = \dfrac{4}{t^2}, y = t + 1, t = 3$

53. $x = t^{\frac{3}{2}} + 2t^{\frac{1}{2}}, y = \dfrac{1}{t}, t = 4$

54. A particle is traveling along the path $xy = -12, x < 0$. If it is traveling the generally positive x-direction, where does its velocity make an angle of $30°$ with the x-axis?

In Exercises 55–58, find the approximate change in the indicated variable.

55. $V = \dfrac{4}{3}\pi ab^2; a = 3, b = 2, \Delta b = 1, \Delta V = ?$

56. $V = \dfrac{4}{3}\pi ab^2; a = 3, b = 2, \Delta a = 1, \Delta V = ?$

57. $I = r^2 h\left(r^2 + \dfrac{h^2}{4}\right); r = 2, h = 1, \Delta r = \dfrac{1}{2}, \Delta I = ?$

58. $I = r^2 h(3r^2 + h^2); r = 4, h = 2, \Delta h = 1, \Delta I = ?$

59. The voltage drop across an inductor is given by $v = L\dfrac{di}{dt}$. If $L = 1$ mH and $i = t^4 - 2t^2 + 6$, approximately how much does the voltage change in one second when $t = 12s$?

✍ Writing About Mathematics

1. A friend has purchased 40 ft of fence to enclose an area for her dog. She wishes to enclose a maximum area with the amount of fence purchased. One side of the enclosed area is the back of her garage. Write a design plan for the enclosed pen. In the plan use calculus to support the dimensions to enclose the maximum area.

2. Explain to a friend how to use the graph of a function, the graph of the first derivative of a function, and the graph of the second derivative of a function to determine relative maximum points, relative minimum points, points of inflection, and intervals of concavity. Include examples in your written summary.

3. Write a few paragraphs, including illustrations, showing how related rates are applied in the real world.

4. Write a few paragraphs, including illustrations, showing how parametric equations are applied in the real world.

5. In your own words, explain to a friend how to apply the first- and second-derivative tests to determine maximum and minimum points.

6. In your own words, explain the difference between absolute maximum and minimum and relative maximum and minimum. Include examples in your report.

Chapter Test

In Exercises 1–4, use the function $y = -3x^2 + 5x + 4$ and the point $(1, 6)$ to perform what is requested.

1. Determine the equation (in slope-intercept form) of the line tangent to the given curve at the stated value.

2. Determine the equation (in slope-intercept form) of the line normal to the given curve at the stated value.

3. Identify the curve.

4. Sketch the curve, the tangent line, and the normal line.

In Exercises 5–8, use the function $y = x^3 - 27x + 8$ to perform what is requested.

5. Determine all the critical points for the function.

6. Use the first-derivative test to determine whether the critical points in Exercise 5 are relative maximum or relative minimum points.

7. Use the results from Exercises 5 and 6 to sketch the graph.

8. In the closed interval $-5 \le x \le 5$, determine the absolute maximum and absolute minimum.

In Exercises 9–16, use the function $f(x) = 2x^3 - 24x + 7$ to perform what is requested.

9. Determine the intervals where the curve is increasing.

10. Determine the intervals where the curve is decreasing.

11. Determine the relative maximum points.

12. Determine the relative minimum points.

13. Determine the points of inflection.

14. Determine the intervals where the curve is concave up.

15. Determine the intervals where the curve is concave down.

16. Sketch the curve and label the key points.

Use the following information in Exercises 17–18 to perform what is requested. The total cost of manufacturing x gadgets is given by the function $T(x) = 2x^2 - 2x + 5 + \dfrac{6}{x}$.

17. Determine the minimal cost.

18. Determine the number of gadgets manufactured for the minimum total cost.

19. The area of a circle is increasing at the rate of 36 in.²/min. Determine the rate at which the radius is increasing when the radius is 4 in.

20. Determine the magnitude and the direction of the velocity vector for $x = 2t^2$, $y = 4t$, $t = 3$.

21. In constructing a cube 3 in. on a side, the length of the side may vary ± 0.01 in. What is the approximate change in the volume of the cube?

21

Integration

A session with an exercise machine requires work from the person to stretch the spring. The more the individual stretches the spring, the greater the restraining force (the force against the individual's pull). The force is not constant as the springs are stretched. It changes for each small unit of distance (or increment) that the spring is stretched. The force may be expressed as $F = kx$, where x is the distance stretched and Δx is a small change in x or an increment in x. So what force do we use to determine the stretch? The initial value? The average value? The final value? Actually, we have to analyze this problem over each increment. The amount of work for each increment is $W = F_1 \Delta x$, and the total work is the sum of all the parts. A problem involving the restraining forces is presented in Example 6 of Section 21.4.

21.1

Antiderivative and the Indefinite Integral

In Chapters 19 and 20 we discussed the techniques for finding the derivatives of functions. Many of the applications we studied in *differential calculus* were related rates, velocity, acceleration, and maximum and minimum values.

This chapter introduces another branch of calculus, **integral calculus.** Some applications of integral calculus involve determining the area between curves, the length of curves, centers of mass, and volumes of solids.

Let us begin our exploration of integral calculus by looking at the inverse operation of differential calculus. If we are given the derivative of a function, how do we determine the original function?

For example, if $f'(x) = 5$ or $f'(x) = 4x^3$ or $f'(R) = 6R$, can we find the original function? From our experience with finding derivatives and using good educated guesses we obtain the following functions.

$$\text{Since } f'(x) = 5, f(x) = 5x.$$
$$\text{Since } f'(x) = 4x^3, f(x) = x^4.$$
$$\text{Since } f'(R) = 6R, f(R) = 3R^2.$$

In each case, we found an original function, which we can check by taking the derivative. The process of finding an original function, given the derivative of the function, is called **antidifferentiation** or **integration.**

> **DEFINITION 21.1**
> A function $F(x)$ is called an **antiderivative** of $f(x)$ if $F'(x) = f(x)$ for all x in the domain of $f(x)$.

Suppose, for example, that $F(x) = x^4$. Then $F'(x) = 4x^3$. Now we can work backward to make some sense from all of this. If a function is $4x^3$, then the antiderivative is x^4.

For example, what is $F(x)$ (the original function) such that $F'(x) = 2x$ (the derivative of the original function is $2x$)? Examine the results shown in Table 21.1.

TABLE 21.1

Function		Derivative of the Function
If $F(x) = x^2$	then	$F'(x) = 2x$
If $F(x) = x^2 + 3$	then	$F'(x) = 2x$
If $F(x) = x^2 - 13$	then	$F'(x) = 2x$
If $F(x) = x^2 + \pi$	then	$F'(x) = 2x$

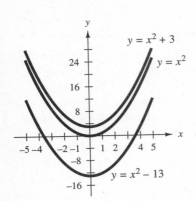

$y = x^2 + 3$

$y = x^2$

$y = x^2 - 13$

FIGURE 21.1

We see from these results that many different functions have the same derivative. Thus, the antiderivative of a derivative is not unique, since each of the functions can have a different constant attached to it. Consequently, we must express the antiderivative in a more general way: $F(x) = x^2 + C$. The letter C represents an arbitrary constant and is part of the notation of the antiderivative of the function $f(x) = 2x$. The function $F(x) = x^2 + C$ represents a family of curves as is suggested by Figure 21.1.

In Example 1 the antiderivative is discussed and in Example 5 how to obtain a specific solution for C is illustrated.

EXAMPLE 1   Antiderivative; Constant C

Determine the antiderivative of $5x^4$.

Solution   To find the antiderivative of $5x^4$ means that we are looking for a function $F(x)$ such that $F'(x) = 5x^4$. Thus, the antiderivative of $5x^4$ is $\dfrac{5x^5}{5} + C$, which we write as $x^5 + C$.

Check this result by taking the derivative of $x^5 + C$. ◆

As mentioned earlier, the process of antidifferentiation is commonly referred to as *integration*, and we denote the process by the symbol ∫, which is called an **integral sign.** The expression $\int f(x)\,dx$ is called the **indefinite integral of $f(x)$.** The symbol indicates the family of curves $F(x) + C$ whose derivative is $f(x)$. That is, if

$$F'(x) = f(x),$$

then

$$\int f(x)\,dx = F(x) + C.$$

In this notation $f(x)$ is the **integrand** of the indefinite integral, the x of dx determines the **variable of integration,** and C is the **constant of integration.** To read the expression $\int f(x)\,dx$ we say either "the antiderivative of the function $f(x)$ with respect to the variable x," or "the integral of $f(x)$ with respect to the variable x."

It is extremely important to recognize that *differentiation is the inverse operation of integration.*

$$\frac{d}{dx}\left[\int f(x)\,dx\right] = f(x), \tag{1}$$

and *that integration is the inverse operation of differentiation:*

$$\int f'(x)\,dx = f(x) + C. \tag{2}$$

The inverse notions (1) and (2) are illustrated by the following examples. Let us begin with (1).

$$\text{Let } f(x) = x^3,$$

then

$$\int x^3\, dx = \frac{x^4}{4} + C;$$

now consider the derivative of the integral:

$$\frac{d}{dx}\left[\frac{x^4}{4} + C\right] = \frac{4x^3}{4} + 0 = x^3.$$

This illustrates that the derivative of the integral of $f(x)$ is equal to $f(x)$. Now look at (2).

$$\text{Let } f(x) = x^3 \quad \text{and} \quad f'(x) = 3x^2,$$

then

$$\int 3x^2\, dx = \frac{3x^3}{3} + C = x^3 + C.$$

This illustrates that the integral of the derivative of $f(x)$ is equal to $f(x)$ plus a constant.

The process of finding the antiderivatives of functions need not be a guessing game, as may have been implied up to this point. To put structure into the process, study the following rules.

Rules of Integration

Constant Rule $\int k\, dx = kx + C$ (k is a constant)

Constant Multiple Rule $\int kf(x)\, dx = k\int f(x)\, dx$

Sum Rule $\int [f(x) + g(x)]\, dx = \int f(x)\, dx + \int g(x)\, dx$

Power Rule $\int x^n\, dx = \frac{x^{n+1}}{n+1} + C,\quad \text{if } n \neq -1$

EXAMPLE 2 **Constant Multiple, Constant, and Sum Rules**

Evaluate $\int (5x - 2)\, dx$.

Solution

$$\int (5x - 2)\, dx = \int 5x\, dx - \int 2\, dx \qquad \text{**Sum rule**}$$

$$= 5 \int x\, dx - 2 \int dx \qquad \begin{array}{l}\text{**Constant multi-**}\\ \text{**ple rule**}\end{array}$$

$$= 5\left[\frac{x^2}{2} + C_1\right] - 2[x + C_2] \qquad \begin{array}{l}\text{**Power rule and**}\\ \text{**constant rule**}\end{array}$$

$$= \frac{5x^2}{2} + 5C_1 - 2x - 2C_2 \qquad \begin{array}{l}\text{**Distributive**}\\ \text{**property**}\end{array}$$

$$= \frac{5x^2}{2} - 2x + 5C_1 - 2C_2.$$

Since $5C_1$ and $2C_2$ are both constants, we combine the two constants. The result is a constant, which we call C. Therefore, $\int (5x - 2)\, dx = \dfrac{5x^2}{2} - 2x + C.$ ◆

EXAMPLE 3 **Power Rule**

Evaluate $\int \sqrt[3]{x}\, dx$.

Solution Before we can integrate, we must rewrite $\sqrt[3]{x}$ as $x^{\frac{1}{3}}$. Now it is in a form to which we can apply the power rule.

$$\int \sqrt[3]{x}\, dx = \int x^{\frac{1}{3}}\, dx$$

$$= \frac{x^{\frac{1}{3}+1}}{\frac{1}{3} + 1} + C \qquad \text{**Power rule**}$$

$$= \frac{x^{\frac{4}{3}}}{\frac{4}{3}} + C$$

$$= \frac{3}{4} x^{\frac{4}{3}} + C. \quad ◆$$

EXAMPLE 4 **Negative Exponent, Power Rule**

Evaluate $\int \dfrac{dx}{x^3}\, dx$.

Solution To evaluate this integral we use the rules of exponents and write the integrand as $x^{-3}\, dx$.

$$\int \frac{dx}{x^3} = \int x^{-3}\, dx \qquad \textbf{Rules of exponents}$$

$$= \frac{x^{-3+1}}{-3+1} + C \qquad \textbf{Power rule}$$

$$= \frac{x^{-2}}{-2} + C$$

$$= \frac{-1}{2x^2} + C \qquad \textbf{Rules of exponents} \quad \blacklozenge$$

EXAMPLE 5 **Value for C, Specific Curve of the Family**

a. Determine $f(x) = \int (4x - x^2)\, dx$.
b. Determine C if $f(3) = 21$.

Solution

a.

$$f(x) = \int (4x - x^2)\, dx$$

$$= \frac{4x^2}{2} - \frac{x^3}{3} + C$$

$$= 2x^2 - \frac{x^3}{3} + C$$

b. We are asked to find a specific value for C. In other words, we must determine a specific curve of the family of curves represented by $f(x) = 2x^2 - \frac{x^3}{3} + C$. To do this, we are told that $f(3) = 21$, so $f(x) = 21$ when $x = 3$.

$$f(x) = 2x^2 - \frac{x^3}{3} + C \qquad \textbf{From part a}$$

$$21 = 2(\,3\,)^2 - \frac{3^3}{3} + C \qquad \textbf{Substituting}$$

$$21 = 18 - 9 + C$$

$$12 = C.$$

Therefore, $f(x) = 2x^2 - \frac{x^3}{3} + 12$ is the equation of the curve satisfying the given conditions. $\blacklozenge$

When the constant of integration is given a specific value, the solution of an indefinite integral is called the **particular solution.** In Example 6 we determine a particular solution.

EXAMPLE 6   **Application: Number of Calories Burned**

Stu takes a stress test on a computerized treadmill. The treadmill starts in a horizontal position at a slow speed; it gradually increases its speed and the slope. Past experience shows that after t minutes a person burns calories at the rate of $2 + 0.86t$ calories per minute. How many calories does Stu burn in a 12-minute stress test?

Solution   Let x be the total number of calories burned after t minutes. The rate of change of x with respect to t is given by $2 + 0.86t$. That is,

$$\frac{dx}{dt} = 2 + 0.86t,$$

which can be written as :

$$dx = (2 + 0.86t)\, dt.$$

We want to find the value of x when $t = 12$. This means that we have to eliminate the incremental notations dx and dt. To do so we integrate both sides of the equation.

$$\int dx = \int (2 + 0.86t)\, dt$$

$$x = 2t + \frac{0.86t^2}{2} + C \quad \textbf{Integrating}$$

$$= 2t + 0.43t^2 + C$$

Before we can determine the number of calories burned after 12 minutes we must calculate the value of C that fits this particular solution. If we assume that when the test starts no calories have been burned, we can say that when $t = 0$, $x = 0$. Substituting these values in the function we have:

$$0 = 2(\,0\,) + 0.43(\,0\,)^2 + C$$
$$C = 0.$$

Thus,

$$x = 2t + 0.43t^2.$$

When $t = 12$,

$$x = 2(\,12\,) + 0.43(\,12\,)^2 \quad \text{or} \quad 85.92.$$

The result tells us that while doing the 12-minute stress test Stu burns approximately 86 calories. ◆

For vertical motion near the earth's surface, Galileo expressed a basic physical principle that states if air resistance is neglected, all objects fall to earth with the same constant acceleration. Specifically, all objects near the surface of the earth are subject to a downward acceleration of approximately 32 ft/s². In symbols, this is expressed:

$$a = -32.$$

The negative sign is necessary because we are assuming that upward motion is inherently positive; the acceleration clearly impels the object downward in the negative direction. From our discussion of the derivative we know that for any given distance $s = f(t)$, $v = \dfrac{ds}{dt}$, and $a = \dfrac{dv}{dt}$. Therefore, reversing the process, we have $dv = a\,dt$ and $ds = v\,dt$, so:

$$v = \int a\,dt \quad \text{and} \quad s = \int v\,dt.$$

EXAMPLE 7   **Formula for Distance of a Falling Body**

An object is dropped from a building 500 ft high. Derive the formula for the distance the object falls in t seconds.

Solution   For a falling body we may take $a = -32$. Thus,

$$v = \int -32\,dt$$

$$= -32t + C_1. \quad \textbf{Integrating}$$

We know that, if the object starts from rest, $v = 0$ when $t = 0$.

$$0 = -32(0) + C_1 \quad \textbf{Substituting}$$
$$C_1 = 0$$

So:
$$v = -32t.$$

Given:

$$s = \int v\,dt$$

$$= \int -32t\,dt$$

$$= -16t^2 + C_2. \quad \textbf{Integrating}$$

When motion started, the object was 500 ft above the ground so when $t = 0$, $s = 500$ ft.

$$500 = -16(0) + C_2$$
$$500 = C_2$$

The position of the object with respect to the ground after t seconds is $s = -16t^2 + 500$. ◆

CAUTION $\int (x^2 + a)^2 \, dx \neq \dfrac{(x^2 + a)^3}{3} + C.$ Example 8 illustrates this point. ∎

EXAMPLE 8   **Integral of a Binomial Squared**

Evaluate $\int (x^3 + 1)^2 \, dx.$

Solution   In order to evaluate this integral with the techniques available we must first expand the binomial.

$$\int (x^3 + 1)^2 \, dx = \int (x^6 + 2x^3 + 1) \, dx \quad \textbf{Expanding binomial}$$

$$= \frac{x^7}{7} + \frac{2x^4}{4} + x + C$$

$$= \frac{x^7}{7} + \frac{x^4}{2} + x + C. \quad \blacklozenge$$

As is illustrated in Example 8, we may need to perform multiplications or divisions before integrating.

21.1   Exercises

In Exercises 1–24, evaluate the indefinite integral and check the result using differentiation.

1. $\displaystyle\int x^6 \, dx$

2. $\displaystyle\int x^7 \, dx$

3. $\displaystyle\int (x^3 + 4) \, dx$

4. $\displaystyle\int (x^2 - 2x + 1) \, dx$

5. $\displaystyle\int (x^4 + 3x^3 - 7) \, dx$

6. $\displaystyle\int (x^5 + 4x^3 - 7x^2) \, dx$

7. $\displaystyle\int (7 - 4y + 11\,y^3) \, dy$

8. $\displaystyle\int (1 - 2y + 7y^4) \, dy$

9. $\displaystyle\int (x^{\frac{3}{2}} + 3x - 7) \, dx$

10. $\displaystyle\int (x^{\frac{2}{3}} + 5x - 4) \, dx$

11. $\displaystyle\int \frac{1}{x^3} \, dx$

12. $\displaystyle\int \left(x + \frac{7}{2\sqrt{x}} \right) dx$

13. $\displaystyle\int \left(x + \frac{3}{\sqrt{x}} \right) dx$

14. $\displaystyle\int \frac{1}{x^2} \, dx$

15. $\displaystyle\int \left(2x^4 + x^{-\frac{2}{3}} - x^{-\frac{5}{3}} \right) dx$

16. $\displaystyle\int \left(3x^2 + x^{-\frac{1}{2}} - x^{\frac{1}{3}} \right) dx$

17. $\displaystyle\int (1 + 3t)t^3 \, dt$

18. $\displaystyle\int (x - 1)(3x - 7) \, dx$

19. $\displaystyle\int (t^2 - 1)^2 \, dt$

20. $\displaystyle\int (y^2 - 2)^2 \, dy$

21. $\displaystyle\int \frac{x^3 + 1}{x^3} \, dx$

22. $\displaystyle\int \left(\frac{y^4 - 2}{y^4} \right) dy$

23. $\displaystyle\int z^2 \sqrt{z} \, dz$

24. $\displaystyle\int (7 - x) \sqrt[3]{z} \, dz$

25. In Example 6, if the speed of the treadmill is $350 + 25t$ ft/min, what is the distance covered by Stu in his 12-minute stress test?

26. a. A ball is dropped from the roof of a building that is 75 meters high. How far does the ball travel in the first 3.0 seconds? In order to properly evaluate C_1, you must know the speed at $t = 0.0$ seconds. What is the speed at $t = 0.0$ seconds? (*Hint:* Remember that in the metric system $a = -9.8$ m/s².)

  b. Suppose the ball is thrown up from the roof of a building with an initial speed of 25 m/s. How far does the ball travel in the first 3.0 seconds? What is the speed at $t = 0.0$ seconds?

  c. Now suppose the ball is thrown down from the roof with an initial speed of 25 m/s. How far does the ball travel in the first 3.0 seconds?

27. Evaporation rate depends on temperature. Suppose a lake loses water at the rate of $\dfrac{T + 6}{3}$ tons per hour, where T is the temperature. Suppose that T varies according to the rule $T = \dfrac{-5}{72}(t - 12)^2 + 30$, where t, the time in hours, varies from 0 to 24. How much water evaporates during the first 3.00 hr?

28. A train runs at a velocity of 55 m/s along a straight track. When the brakes are applied, the deceleration (negative acceleration) is 1.5 m/s². For how long and how far from the station should the brakes be applied so that the train stops at the station?

29. A balloon is rising at the constant rate of 8 ft/s and is 80 ft from the ground at the instant Mike (the aeronaut) drops his binoculars.

  a. How long will it take the binoculars to strike the ground?

  b. At what speed will the binoculars strike the ground?

30. A ball rolls down an incline with an acceleration of 2 ft/s². If the ball is given no initial velocity how far does it roll in t seconds? What initial velocity must be given for the ball to roll 100 ft in 5 s?

31. Find the equation of a curve that has a second derivative equal to $6x - 10$ and passes through the points $(0, 1)$ and $(2, 11)$.

32. Find an equation of the curve whose slope at the point (x, y) is $4x^3 + 6x^2$ if the curve is required to pass through the point $(1, 0)$.

33. Thus far we have considered situations in which the rate of velocity (acceleration) is constant. Depending on the incline and the type of object which is descending, the rate of velocity might vary. Determine the expression for the velocity as a function of time when $v = 3.0$ m/s at $t = 2.0$ s for the following.

  a. $\dfrac{dv}{dt} = \sqrt[3]{t} - 2$

  b. $\dfrac{dv}{dt} = \dfrac{1}{\sqrt[3]{t}} - 2$

34. The current given to a capacitor varies as $i = 3t + 2$. Current is defined as $\dfrac{dq}{dt}$, where dq is the incremental charge moved in the incremental time dt. If the initial charge on the capacitor is 0.034 Coulombs, determine the charge on the capacitor at $t = 2.0$ s.

21.2

The Indefinite Integral

The rules of integration in Section 21.1 provided us with techniques for finding the antiderivatives of relatively simple functions. Is it possible to find the antiderivatives of:

$$6x(3x^2 + 7)^5 \, dx?$$

With the techniques available the answer is yes, if we use the binomial theorem and the distributive property. However, there is a simpler technique. To understand this technique, think how we may have obtained a derivative of this form. That is, what function $f(x)$ has the derivative:

$$f'(x) = 6x(3x^2 + 7)^5?$$

Recalling the power rule for finding the derivative of a function, an original function that would give us this desired result looks like this:

$$\frac{d}{dx}\left[\frac{(3x^2 + 7)^6}{6}\right] = \frac{6(3x^2 + 7)^5(6x)}{6} \quad \text{Note: } \frac{d}{dx}(3x^2 + 7) = 6x$$

$$= 6x(3x^2 + 7)^5.$$

Therefore if we do the inverse operation, the antiderivative, we have the following:

$$\int 6x(3x^2 + 7)^5 \, dx = \frac{(3x^2 + 7)^6}{6} + C.$$

Now let us see if we can relate the power rule of differentiation and the power rule of integration to arrive at a technique for finding the antiderivative. Basically, we are trying to simplify the complex expression into a more manageable one. Let $u = 3x^2 + 7$. By doing this we can replace the complex-looking format with a simpler one:

$$(3x^2 + 7)^5 = u^5.$$

Unfortunately we still have a $6x$ and dx term. We need to express our integral with just one variable, u. To do this, we take the derivative of u with respect to x.

$$\frac{du}{dx} = 6x$$

and

$$du = 6x \, dx \quad \text{Writing as a differential}$$

Using these results we can make the following substitutions:

$$\int (3x^2 + 7)^5(6x \, dx) = \int u^5 \, du \qquad \text{Substituting}$$

$$= \frac{u^6}{6} + C \qquad \text{Power rule}$$

$$= \frac{(3x^2 + 7)^6}{6} + C. \qquad \text{Substituting for } u$$

This illustrates the substitution rule, which provides us with a method for writing an integral in a form to which it may be possible to apply one of the rules of integration.

> **RULE 21.1 Substitution Rule**
>
> If u is a function of x, where $u = g(x)$ and $du = h(x)\, dx$, then:
>
> $$\int [g(x)]^n\, h(x)\, dx = \int u^n\, du$$
>
> $$= \frac{u^{n+1}}{n+1} + C,$$
>
> where $n \neq -1$.

EXAMPLE 1   **Substitution Rule**

Evaluate $\int (3x + 11)^{\frac{1}{3}}\, dx$.

Solution   We cannot expand $(3x + 11)^{\frac{1}{3}}$. Therefore, we must use the substitution rule.

Let $u = 3x + 11$. Then:

$$\frac{du}{dx} = 3$$

$$du = 3\, dx.$$

In the original problem the numerical factor of dx is 1. Therefore, we must divide both sides by 3.

$$\frac{du}{3} = dx$$

$$\int (3x + 11)^{\frac{1}{3}}\, dx = \int u^{\frac{1}{3}}\, \frac{du}{3} \qquad \text{Substituting}$$

$$= \frac{1}{3} \int u^{\frac{1}{3}}\, du \qquad \text{Constant multiple rule}$$

$$= \frac{1}{3} \left[\frac{u^{\frac{1}{3}+1}}{\frac{1}{3}+1} \right] + C \qquad \text{Power rule}$$

$$= \frac{1}{3}\, \frac{u^{\frac{4}{3}}}{\frac{4}{3}} + C$$

$$= \frac{1}{4} u^{\frac{4}{3}} + C \qquad \text{Simplifying}$$

$$= \frac{1}{4} (3x + 11)^{\frac{4}{3}} + C. \qquad \text{Substituting for } u \quad \blacklozenge$$

When the integrand is the product or quotient of two functions, it is generally best to let u be equal to the most complex factor. This is illustrated in Examples 2 and 3.

EXAMPLE 2 **Substitution Rule: Product**

Evaluate $\int (x^3 - 5x + 7)^4(3x^2 - 5x)\, dx$.

Solution Using the substitution rule, we have:

$$u = x^3 - 5x + 7$$

$$\frac{du}{dx} = 3x^2 - 5 \qquad\qquad \text{Differentiating}$$

$$du = (3x^2 - 5)\, dx \qquad\qquad \text{Writing as a differential}$$

$$\int (x^3 - 5x + 7)^4(3x^2 - 5)\, dx = \int u^4\, du \qquad\qquad \text{Substituting}$$

$$= \frac{u^5}{5} + C \qquad\qquad \text{Power rule}$$

$$= \frac{(x^3 - 5x + 7)^5}{5} + C. \qquad \text{Substituting for } u \quad \blacklozenge$$

EXAMPLE 3 **Substitution Rule: Quotient**

Evaluate $\displaystyle\int \frac{2x}{\sqrt{3x^2 + 4}}\, dx$.

Solution Write the integrand in the form $\int (3x^2 + 4)^{-\frac{1}{2}} 2x\, dx$. Now letting:

$$u = 3x^2 + 4$$

$$\frac{du}{dx} = 6x$$

$$\frac{du}{3} = 2x\, dx$$

$$\int (3x^2 + 4)^{-\frac{1}{2}} 2x\, dx = \int u^{-\frac{1}{2}} \frac{du}{3} \qquad\qquad \text{Substituting}$$

$$= \frac{1}{3} \int u^{-\frac{1}{2}}\, du$$

$$= \frac{1}{3} \frac{u^{\frac{1}{2}}}{\frac{1}{2}} + C \qquad\qquad \text{Power rule}$$

$$= \frac{2}{3} u^{\frac{1}{2}} + C \qquad\qquad \text{Simplifying}$$

$$= \frac{2}{3} (3x^2 + 4)^{\frac{1}{2}} + C. \qquad \text{Substituting for } u \quad \blacklozenge$$

EXAMPLE 4 **Substitution Does Not Work**

Evaluate $\int (4x^2 - 7)^2 x^3 \, dx$.

Solution If we let $u = 4x^2 - 7$, then $\dfrac{du}{dx} = 8x$. The factor in the integrand is x^3, not $8x$; therefore we cannot use the substitution rule. We can integrate the expression at this time by expanding the integrand.

$$
\begin{aligned}
\int (4x^2 - 7)^2 x^3 \, dx &= \int (16x^4 - 56x^2 + 49)x^3 \, dx && \text{Squaring } (4x^2 - 7) \\
&= \int (16x^7 - 56x^5 + 49x^3) \, dx && \text{Multiplying by } x^3 \\
&= \frac{16x^8}{8} - \frac{56x^6}{6} + \frac{49x^4}{4} + C && \text{Integrating} \\
&= 2x^8 - \frac{28x^6}{3} + \frac{49x^4}{4} + C && \text{Simplifying} \quad \blacklozenge
\end{aligned}
$$

EXAMPLE 5 **Quotient, Radicals, Direct Method**

Evaluate $\displaystyle\int \frac{1 + \sqrt{x}}{\sqrt{x}} \, dx$.

Solution The integral can be evaluated two different ways. We use the direct method in working this example.

$$
\begin{aligned}
\int \frac{1 + \sqrt{x}}{\sqrt{x}} \, dx &= \int \left(\frac{1}{\sqrt{x}} + \frac{\sqrt{x}}{\sqrt{x}} \right) dx && \text{Rearranging} \\
&= \int (x^{-\frac{1}{2}} + 1) \, dx \\
&= \int x^{-\frac{1}{2}} \, dx + \int dx && \text{Sum rule} \\
&= \frac{x^{\frac{1}{2}}}{\frac{1}{2}} + x + C && \text{Integrating} \\
&= 2x^{\frac{1}{2}} + x + C. && \text{Simplifying}
\end{aligned}
$$

Show that the same result is obtained by using the substitution rule. $\blacklozenge$

21.2 Exercises

In Exercises 1–22, evaluate the indefinite integral.

1. $\displaystyle\int (3x + 4)^8 \, dx$

2. $\displaystyle\int (2x + 7)^5 \, dx$

3. $\displaystyle\int 3x^2(x^3 - 4) \, dx$

4. $\displaystyle\int 10x^4(2x^5 - 4) \, dx$

5. $\displaystyle\int x^4(3x^5 - 8) \, dx$

6. $\displaystyle\int x^3(2x^4 - 11) \, dx$

7. $\displaystyle\int 3x^6(5x^7 - 13) \, dx$

8. $\displaystyle\int 2x^7(3x^8 - 13) \, dx$

9. $\displaystyle\int (3x^2 + 7)(x^3 + 7x)^8 \, dx$

10. $\displaystyle\int (5x^4 - 13)(x^5 - 13x)^{11} \, dx$

11. $\displaystyle\int \frac{3x^2 - 7}{(x^3 - 7x)^4} \, dx$

12. $\displaystyle\int \frac{7x^6 - 3x^2}{(x^7 - x^3)^5} \, dx$

13. $\displaystyle\int \frac{4t}{\sqrt{3t^2 - 7}} \, dt$

14. $\displaystyle\int \frac{5t}{\sqrt{11t^2 - 9}} \, dt$

15. $\displaystyle\int (\sqrt{t} + 1)^2 \, dt$

16. $\displaystyle\int (\sqrt[3]{t} + 1)^3 \, dt$

17. $\displaystyle\int \frac{\sqrt{r^{\frac{1}{3}} + 4}}{r^{\frac{2}{3}}} \, dr$

18. $\displaystyle\int \frac{\sqrt{r^{\frac{2}{3}} - 3}}{r^{\frac{1}{3}}} \, dr$

19. $\displaystyle\int \frac{x + 3x^2}{\sqrt{x}} \, dx$

20. $\displaystyle\int \frac{\sqrt[3]{x} + 1}{\sqrt{x}} \, dx$

21. $\displaystyle\int \frac{x + 1}{(x^2 + 2x + 2)^2} \, dx$

22. $\displaystyle\int \frac{x^2 + 1}{(x^3 + 3x + 7)^3} \, dx$

In Exercises 23–26, determine the equation of the curve having the given slope and passing through the indicated point.

23. $\dfrac{dy}{dx} = 2(3x + 7)^3, \; (-2, 1)$

24. $\dfrac{dy}{dx} = x(x^2 + 1)^4, \; (0, 8)$

25. $\dfrac{dy}{dx} = (2x - 3)(x^2 - 3x)^4, \; (0, -11)$

26. $\dfrac{dy}{dx} = (3x^2 - 4x)(x^3 - 2x^2 + 1)^6, \; (1, -8)$

27. Explain why $\int \sqrt{x^2 + 3x} \, (x^4 \, dx)$ cannot be evaluated using the substitution method.

28. Explain why $\displaystyle\int \frac{x^3}{\sqrt{x^3}} \, dx$ cannot be evaluated using the substitution method.

29. Explain why $\displaystyle\int (3x^5 + 2)^4 \, dx \neq \frac{(3x^5 + 2)^5}{5} + C$.

30. The formula $\dfrac{dw}{dt} = \dfrac{12}{\sqrt{16t + 9}}$ is a method of determining the weight loss w in a log as a function of the number of days drying time t.
   a. Find w as a function of t. (There can be no weight loss before the tree is cut.)
   b. Find the total weight loss after 50 days.

31. The voltage across a capacitor varies with the charge on the capacitor: $v = \dfrac{Q}{C}$. But the charge depends upon the current: $Q = i \, dt$. Suppose the current varies as $i = \sqrt{t + 3}$. If the initial voltage of the capacitor is 6.0 v, what is the voltage when $t = 2.0$ s.

32. A certain curve has a slope given by $x\sqrt{2x^2 - 1}$ that passes through point (3, 3). What is the equation of the curve? $\left(\textit{Hint:} \text{ The slope is } \dfrac{dy}{dx}.\right)$

33. A certain curve has a slope given by $2x^2 - 1$ that passes through the point (3, 3). What is the equation of the curve?

34. A certain curve has a slope given by $x(2x^2 - 1)$ that passes through the point (3, 3). What is the equation of the curve?

35. A certain curve has a slope given by $x(2x^2 - 1)^2$ that passes through the point (3, 3). What is the equation of the curve?

21.3

The Area Under a Curve

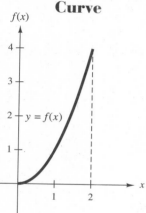

FIGURE 21.2

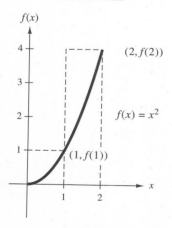

FIGURE 21.3

We can determine the area of a figure whose sides are straight lines relatively easily. We can do so equally easily if the figure is a circle, parabola, or ellipse. For any of these figures we use special formulas to determine the area. For example, to determine the surface area of an elliptical swimming pool, we use the formula $A = \pi ab$, where a and b are the lengths of the semiaxes.

At the beginning of this chapter we learned that one of the applications of integration is determining the area under a curve—a more difficult area to find. In fact, prior to the development of calculus in the 1700s, the area under a curve only could be approximated. However, the process used to approximate the area led to a technique for determining the area.

A technique for approximating the area of a region is to construct rectangles in the region and determine the sum of the areas of the rectangles. Consider the region bounded by the curves $f(x) = x^2$, $x = 0$ (the y-axis), $x = 2$, and $y = 0$ (the x-axis) in Figure 21.2. To approximate the area of the region we divide the region in half and construct two rectangles each of whose upper-right corner is on the curve, as shown in Figure 21.3. The width of each rectangle is 1, and the heights are $f(1) = 1^2 = 1$ and $f(2) = 2^2 = 4$. Thus, the sum of the areas of the rectangles are:

$$(1)f(1) + (1)f(2) = (1)(1) + (1)(4) = 5.$$

However, from Figure 21.3 we can see that the area calculated is larger than the actual area. To gain a better approximation, we can divide the interval, $0 \le x \le 2$, into four equal subintervals.

In general, to approximate the area of a region bounded by the curve $y = f(x)$ (for $f(x) \ge 0$), the x-axis, $x = a$, and $x = b$, divide the interval $[a, b]$ into n equal subintervals so that each subinterval has a width of $\dfrac{(b - a)}{n} = \Delta x$.

Then determine the height of each rectangle by finding the value of $f(x)$ at the upper-right corner of each rectangle. The approximate area is the sum of the areas of the rectangles. This procedure is illustrated in Example 1. The following guidelines should help you understand the procedure.

GUIDELINES FOR APPROXIMATING THE AREA UNDER A CURVE

1. Sketch and label the curve.

2. Determine the width of each rectangle, $\dfrac{(b - a)}{n} = \Delta x$, where a and b are the endpoints of the interval and n is the number of subintervals.

3. $f(x_i)$ is the height of each rectangle (determined by the upper-right corner of each).

4. $A \approx f(x_1)\,\Delta x + f(x_2)\,\Delta x + \cdots + f(x_n)\,\Delta x.$

EXAMPLE 1 **Approximating the Area Under a Curve**

Determine the approximate area of the region bounded by $f(x) = x^2$, $x = 0$, $x = 2$, and the x-axis, using four subintervals.

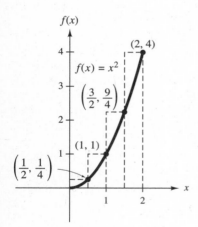

FIGURE 21.4

Solution The curve is sketched in Figure 21.4. The width of each subinterval is:

$$\Delta x = \frac{2 - 0}{4} = \frac{1}{2}.$$

The right end points of each subinterval are $\frac{1}{2}, 1, \frac{3}{2}$, and 2. The approximate area is the sum of the areas of the rectangles.

$$A \approx \Delta x f(x_1) + \Delta x f(x_2) + \Delta x f(x_3) + \Delta x f(x_4)$$

$$\approx \frac{1}{2}f\!\left(\frac{1}{2}\right) + \frac{1}{2}f(1) + \frac{1}{2}f\!\left(\frac{3}{2}\right) + \frac{1}{2}f(2)$$

$$\approx \frac{1}{2} \cdot \frac{1}{4} + \frac{1}{2} \cdot 1 + \frac{1}{2} \cdot \frac{9}{4} + \frac{1}{2} \cdot 4$$

$$\approx \frac{15}{4} = 3\frac{3}{4}$$

Thus, the approximate area of the region is $3\frac{3}{4}$ square units. ◆

The approximate area of $3\frac{3}{4}$ square units is larger than the actual area of the region, which is illustrated by Figure 21.3. To get an even more precise approximation of the area, we increase the number of subintervals and, thus, the number of rectangles. In general, if we keep increasing the number of rectangles, n becomes larger, and the width of each rectangle becomes smaller. Thus, each result is a better approximation of the area of the region. If we increase the number of rectangles to an infinite number, the result is the area of the region, which is $\frac{8}{3}$ square units.

Section 19.2 discussed limits. A limiting process is :

$$A = \lim_{n \to \infty} \quad \textbf{Sum of the areas of } n \textbf{ rectangles}$$

This limiting process is what we mean when we say the area is the **definite integral** of $f(x) = x^2$ from $x = 0$ to $x = 2$. It is written symbolically as:

$$A = \int_{x=0}^{x=2} x^2 \, dx = \frac{8}{3}.$$

We read the symbol as : "the area A equals the integral from $x = 0$ to $x = 2$ of the function $f(x) = x^2$." The 0 is called the **lower limit of integration,** the 2 is called the **upper limit of integration,** the function $f(x) = x^2$ is called the **integrand,** and the dx tells us that we are integrating with respect to the variable x.

This discussion is summarized in Definition 21.2.

DEFINITION 21.2

If $f(x)$ is continuous on the interval $[a, b]$ and $[a, b]$ is divided into n subintervals whose right-hand points are $x_1, x_2, x_3, \ldots, x_n$, then the **definite integral** of $f(x)$ from $x = a$ to $x = b$ is;

$$\int_{x=a}^{x=b} f(x) \, dx = \lim_{n \to \infty} \frac{(b-a)}{n} [f(x_1) + f(x_2) + \cdots + f(x_n)].$$

It is important not to confuse the definite integral with the indefinite integral. The definite integral $\int_{x=a}^{x=b} f(x) \, dx$ is a **real number.** The indefinite integral $\int f(x) \, dx$ is a **set of functions** (a family of curves).

The following formulas are helpful in determining the value of a definite integral when we use Definition 21.2.

$$1 + 2 + 3 + 4 + \cdots + n = \frac{n(n+1)}{2} \tag{3}$$

$$1^2 + 2^2 + 3^2 + 4^2 + \cdots + n^2 = \frac{n(n+1)(2n+1)}{6} \tag{4}$$

In Example 2 we use Definition 21.2 to evaluate a definite integral.

EXAMPLE 2 **Evaluating a Definite Integral**

Determine $\int_{x=0}^{x=3} 4x\,dx$.

Solution The lower- and upper-limits of integration tell us that we are integrating over the closed interval [0, 3]. For n equal subintervals we divide the width of the interval $(3 - 0)$ by n to determine the width of each subinterval.

$$\frac{3-0}{n} = \frac{3}{n}$$

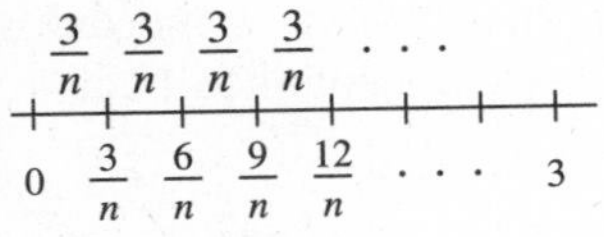

3/n 3/n 3/n 3/n . . .
0 3/n 6/n 9/n 12/n . . . 3

FIGURE 21.5

With the width of each subinterval being $\frac{3}{n}$, the right end points of the subintervals are $\frac{3}{n}, \frac{6}{n}, \frac{9}{n}, \ldots, 3$ (see Figure 21.5). Substituting $a = 0$, $b = 3$, $x_1 = \frac{3}{n}$, $x_2 = \frac{6}{n}, \ldots$, in the definition, we have:

$$\begin{aligned}
\int_{x=0}^{x=3} 4x\,dx &= \lim_{n\to\infty} \frac{(b-a)}{n}[f(x_1) + f(x_2) + \cdots + f(x_n)] \\
&= \lim_{n\to\infty} \frac{(3-0)}{n}\left[f\left(\frac{3}{n}\right) + f\left(\frac{6}{n}\right) + \cdots + f(3)\right] \\
&= \lim_{n\to\infty} \frac{3}{n}\left[\frac{12}{n} + \frac{24}{n} + \cdots + 12\right].
\end{aligned}$$

We now write 12 as $\frac{12n}{n}$ so that we can factor out the common factor $\frac{12}{n}$.

$$\begin{aligned}
&= \lim_{n\to\infty} \frac{3}{n}\left[\frac{12}{n} + \frac{24}{n} + \cdots + \frac{12n}{n}\right] \\
&= \lim_{n\to\infty} \frac{3}{n}\left(\frac{12}{n}\right)[1 + 2 + 3 + \cdots + n] \\
&= \lim_{n\to\infty} \frac{36}{n^2}\,\frac{n(n+1)}{2} && \textbf{Formula (3)} \\
&= \lim_{n\to\infty} \frac{36}{n^2}\left(\frac{n^2}{2} + \frac{n}{2}\right) \\
&= \lim_{n\to\infty} \frac{36}{n^2}\cdot\frac{n^2}{2} + \lim_{n\to\infty} \frac{36}{n^2}\cdot\frac{n}{2} \\
&= 18 + 0 && \textbf{Theorem 19.1 and Definition 19.5} \\
&= 18
\end{aligned}$$

Thus, the $\int_{x=0}^{x=3} 4x\,dx = 18$. ◆

In Definition 21.2 the definite integral is defined in terms of the limiting process. To use this definition to determine the area under a curve we must add one more stipulation, which is $f(x) \geq 0$.

DEFINITION 21.3

If $f(x) \geq 0$ in $[a, b]$, then the area bounded by $f(x)$, the x-axis, $x = a$, and $x = b$ is defined by:

$$\int_{x=a}^{x=b} f(x)\, dx.$$

EXAMPLE 3   Definite Integral: Area of a Region

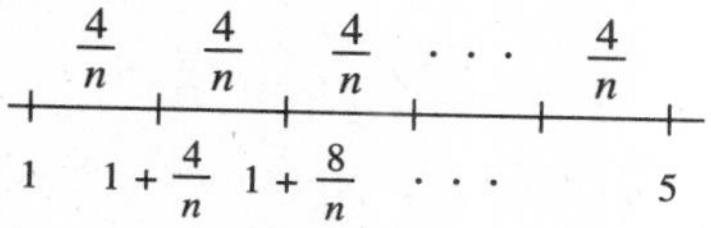

Determine the area of the region bounded by the function $f(x) = 3x + 1$, $x = 1$, $x = 5$, and the x-axis, as shown in Figure 21.6.

Solution We start by dividing the closed interval into n subdivisions, $\dfrac{5-1}{n} = \dfrac{4}{n}$. Since $a = 1$, we must add 1 to $\dfrac{4}{n}$ to determine the first right endpoints. Then we keep adding $\dfrac{4}{n}$ to obtain the other right endpoints. Thus, the right endpoints, of the subintervals area: $\left(1 + \dfrac{4}{n}\right)$, $\left(1 + \dfrac{8}{n}\right)$, . . . , 5, as shown in Figure 21.7.

Substitute $a = 1$, $b = 5$, $x_1 = \left(1 + \dfrac{4}{n}\right)$, $x_2 = \left(1 + \dfrac{8}{n}\right)$, . . . , $x_n = \left(1 + \dfrac{4n}{n}\right)$ in the formula.

$$\int_{x=1}^{x=5} (3x + 1)\, dx = \lim_{n \to \infty} \frac{(b-a)}{n} \left[f(x_1) + f(x_2) + \cdots + f(x_n) \right]$$

$$= \lim_{n \to \infty} \frac{(5-1)}{n} \left[f\left(1 + \frac{4}{n}\right) + f\left(1 + \frac{8}{n}\right) \right.$$
$$\left. + \cdots + f\left(1 + \frac{4n}{n}\right) \right]$$

$$= \lim_{n \to \infty} \frac{4}{n} \left\{ \left[3\left(1 + \frac{4}{n}\right) + 1 \right] + \left[3\left(1 + \frac{8}{n}\right) + 1 \right] \right.$$
$$\left. + \cdots + \left[3\left(1 + \frac{4n}{n}\right) + 1 \right] \right\}$$

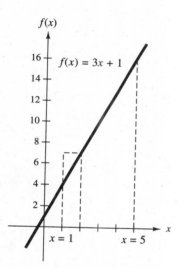

FIGURE 21.6

$$\underset{1}{\quad} \overset{\frac{4}{n}}{\quad} \underset{1 + \frac{4}{n}}{\quad} \overset{\frac{4}{n}}{\quad} \underset{1 + \frac{8}{n}}{\quad} \overset{\frac{4}{n}}{\quad} \cdots \overset{\frac{4}{n}}{\quad} \underset{5}{\quad}$$

FIGURE 21.7

Now multiply by 3 and remove the parentheses inside each set of brackets.

$$= \lim_{x \to \infty} \frac{4}{n} \left\{ \left[4 + \frac{12}{n} \right] + \left[4 + \frac{24}{n} \right] + \cdots + \left[4 + \frac{12n}{n} \right] \right\}$$

$$= \lim_{n \to \infty} \frac{4}{n} \left\{ [4 + 4 + \cdots + 4] + \left[\frac{12}{n} + \frac{24}{n} + \cdots + \frac{12n}{n} \right] \right\}$$

Since there are n subintervals, there are n fours inside the first set of brackets, which can be written as $4n$. Inside the second set of brackets each fraction has a common factor of $\frac{12}{n}$. Rewriting, we have:

$$= \lim_{n \to \infty} \frac{4}{n} \left[4n + \frac{12}{n}(1 + 2 + 3 + \cdots + n) \right]$$

$$= \lim_{n \to \infty} \frac{4}{n} \left[(4n) + \left(\frac{\overset{6}{\cancel{12}}}{\cancel{n}} \right) \frac{\overset{1}{\cancel{n}}(n + 1)}{\underset{1}{\cancel{2}}} \right] \qquad \textbf{Formula (3)}$$

$$= \lim_{n \to \infty} \frac{4}{n} [4n + 6n + 6]$$

$$= \lim_{n \to \infty} \frac{4}{n} [10n + 6]$$

$$= \lim_{n \to \infty} \left[40 + \frac{24}{n} \right]$$

$$= 40 + 0$$

$$= 40.$$

Therefore, the area of the region bounded by the curves is 40 square units. ◆

EXAMPLE 4 **Definite Integral: Area Under a Curve**

Determine the area of the region bounded by the function $f(x) = x^2$, $x = 0$, $x = 1$, and the x-axis, as shown in Figure 21.8.

Solution Dividing the closed interval $[0, 1]$ into n equal subintervals we have $\frac{1 - 0}{n} = \frac{1}{n}$. Thus, the right endpoints of the subintervals are $\frac{1}{n}, \frac{2}{n}, \frac{3}{n}, \ldots, 1$.

Substitute $a = 0$, $b = 1$, $x_1 = \frac{1}{n}$, $x_2 = \frac{2}{n}, \ldots, x_n = \frac{n}{n}$ in the formula.

$$\int_{x=0}^{x=1} x^2 \, dx = \lim_{n \to \infty} \frac{(b - a)}{n} [f(x_1) + f(x_2) + \cdots + f(x_n)]$$

$$= \lim_{n \to \infty} \frac{1}{n} \left[f\left(\frac{1}{n} \right) + f\left(\frac{2}{n} \right) + \cdots + f\left(\frac{n}{n} \right) \right]$$

$$= \lim_{n \to \infty} \frac{1}{n} \left[\left(\frac{1}{n} \right)^2 + \left(\frac{2}{n} \right)^2 + \cdots + \left(\frac{n}{n} \right)^2 \right]$$

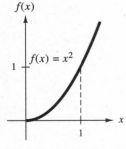

$f(x)$

$f(x) = x^2$

1

1

x

FIGURE 21.8

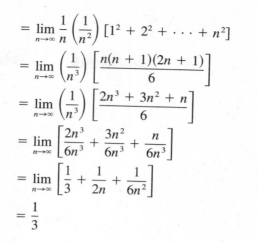

FIGURE 21.9

$$= \lim_{n \to \infty} \frac{1}{n} \left(\frac{1}{n^2} \right) [1^2 + 2^2 + \cdots + n^2]$$

$$= \lim_{n \to \infty} \left(\frac{1}{n^3} \right) \left[\frac{n(n+1)(2n+1)}{6} \right] \qquad \text{Formula (4)}$$

$$= \lim_{n \to \infty} \left(\frac{1}{n^3} \right) \left[\frac{2n^3 + 3n^2 + n}{6} \right]$$

$$= \lim_{n \to \infty} \left[\frac{2n^3}{6n^3} + \frac{3n^2}{6n^3} + \frac{n}{6n^3} \right]$$

$$= \lim_{n \to \infty} \left[\frac{1}{3} + \frac{1}{2n} + \frac{1}{6n^2} \right]$$

$$= \frac{1}{3}$$

Therefore, the area of the region is $\frac{1}{3}$ square units.   ◆

Regions between $y = f(x)$ and the x-axis are interpreted as negative when the region is below the x-axis and as positive when the region is above the x-axis. Then $\int_{x=a}^{x=b} f(x)\, dx$ can be interpreted geometrically as the algebraic sum of these signed areas (see Figure 21.9). This idea is discussed in more detail in Chapter 22.

21.3   Exercises

In Exercises 1–6, determine the approximate area of the region bounded by $f(x) = 2x + 1$, $x = a$, and $x = b$ for n rectangles.

1. $n = 2$, $a = 0$, $b = 2$ **2.** $n = 4$, $a = 0$, $b = 2$ **3.** $n = 8$, $a = 0$, $b = 2$ **4.** $n = 2$, $a = 1$, $b = 5$
5. $n = 4$, $a = 1$, $b = 5$ **6.** $n = 8$, $a = 1$, $b = 5$

In Exercises 7–12, determine the approximate area of the region bounded by $f(x) = x^2 + 1$, $x = a$, and $x = b$, for n rectangles.

7. $n = 2$, $a = 0$, $b = 2$ **8.** $n = 4$, $a = 0$, $b = 2$ **9.** $n = 8$, $a = 0$, $b = 2$ **10.** $n = 2$, $a = 1$, $b = 5$
11. $n = 4$, $a = 1$, $b = 5$ **12.** $n = 8$, $a = 1$, $b = 5$

In Exercises 13–22, determine the exact value of the integral using Definition 21.2.

13. $\displaystyle\int_{x=0}^{x=3} 3x \, dx$ **14.** $\displaystyle\int_{x=0}^{x=2} 5x \, dx$ **15.** $\displaystyle\int_{x=0}^{x=3} (3x + 1) \, dx$ **16.** $\displaystyle\int_{x=0}^{x=2} (5x + 1) \, dx$

17. $\displaystyle\int_{x=0}^{x=3} (2x - 4) \, dx$ **18.** $\displaystyle\int_{x=1}^{x=4} (4x - 3) \, dx$ **19.** $\displaystyle\int_{x=0}^{x=2} x^2 \, dx$ **20.** $\displaystyle\int_{x=2}^{x=3} 2x^2 \, dx$

21. $\displaystyle\int_{x=0}^{x=2} (x^2 + 3) \, dx$ **22.** $\displaystyle\int_{x=2}^{x=3} (x^2 - 4) \, dx$

In Exercises 23–26, determine the area of the region bounded by the function
$y = f(x)$, $x = a$, $x = b$, and the x-axis.

23. $f(x) = 2x + 1$, $a = 1$, $b = 2$ **24.** $f(x) = 3x + 4$, $a = 0$, $b = 2$ **25.** $f(x) = x^2 + 1$, $a = 0$, $b = 2$

26. $f(x) = x^2 - 1$, $a = 0$, $b = 2$

21.4

The Fundamental Theorem of Integral Calculus

The Definite Integral

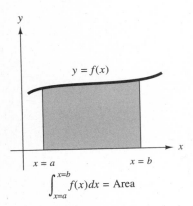

$$\int_{x=a}^{x=b} f(x)\,dx = \text{Area}$$

FIGURE 21.10

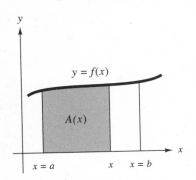

Area from a to x is $A(x)$

FIGURE 21.11

In Section 21.3 we learned that we can determine the area of a region with a definite integral. However, with the tools available to us at this time, evaluating a definite integral using the summation process is rather tedious and time consuming. To provide us with a more efficient method of evaluating the definite integral, we now consider a very important theorem in calculus, the *fundamental theorem of integral calculus*. This discussion will show that the definite integral can be applied in a general manner, and not only to the concept of area (just as we found that the derivative has wider applications than just finding the slope of a line).

To help provide a better understanding of the meaning of the fundamental theorem of integral calculus, let us begin with an intuitive development of the theorem using area as an illustration. (For a precise mathematical development of the theorem see a more theoretical calculus book.)

From our previous discussion we can find the area of a region using the definite integral. Thus,

$$\text{Area} = \int_{x=a}^{x=b} f(x)\,dx,$$

as shown in Figure 21.10.

To develop the theorem we need to introduce a new function called the area function $A(x)$. The function indicates the area of the region under the graph of the function from a to x (Figure 21.11). For $A(x)$ to represent the area, the function must be continuous and non-negative on $[a, b]$.

Now suppose that we increase x by Δx. Then the area under the curve increases by an amount that we call ΔA (Figure 21.12). Using the approximation process, which was introduced in Section 19.3, we can estimate that ΔA is slightly bigger than the area of the inscribed rectangle and slightly smaller than the area of the circumscribed rectangle. In Figure 21.3 the smaller rectangle is inscribed (within the curve) and the larger rectangle is circumscribed.

For the area of the inscribed rectangle, we take the minimum value of $f(x)$ within the closed interval $[x, x + \Delta x]$. We call this minimum value $f(m)$. For the

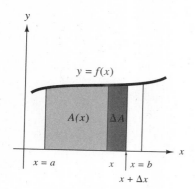

FIGURE 21.12

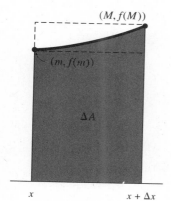

FIGURE 21.13

area of the circumscribed rectangle we take the maximum value within the closed interval $[x, x + \Delta x]$. We refer to this value as $f(M)$. Hence, the minimum area is $f(m) \Delta x$, and the maximum area is $f(M) \Delta x$. See Figure 21.13.

Algebraically we can write:

$$f(m) \Delta x \le \Delta A \le f(M) \Delta x.$$

By dividing each term of the inequality by Δx, we have:

$$f(m) \le \frac{\Delta A}{\Delta x} \le f(M), \qquad \Delta x \ne 0.$$

If we take the limit as $\Delta x \to 0$, then $f(m)$ and $F(M)$ approach the same point on the curve, and both approach $f(x)$. Also,

$$\lim_{\Delta x \to 0} \frac{\Delta A}{\Delta x} = \frac{dA}{dx}.$$

Substituting these results into the inequality, we have:

$$f(x) \le \frac{dA}{dx} \le f(x),$$

which states that:

$$\frac{dA}{dx} = f(x).$$

What we have established is that the area function $A(x)$ is the antiderivative of $f(x)$. This statement says that:

$$A(x) = F(x) + C,$$

where $F(x)$ is the antiderivative of $f(x)$. To determine a real value for $A(x)$ we must solve for C. We know that $A(a) = 0$; thus,

$$0 = F(\,a\,) + C$$

or

$$C = -F(a).$$

Also evaluating $A(b)$, we have:

$$A(b) = F(b) + C$$
$$= F(b) - F(a). \quad \text{\small Substituting}$$

This last equation tells us that if it is possible to find an antiderivative of $f(x)$, we can evaluate the definition integral $\int_{x=a}^{x=b} f(x)\, dx$. All of this is nicely condensed in the fundamental theorem.

> **THEOREM 21.1 Fundamental Theorem of Integral Calculus**
> If a function $f(x)$ is continuous on the closed interval $[a, b]$, then:
>
> $$\int_{x=a}^{x=b} f(x)\ dx = F(x)\Bigg]_{x=a}^{x=b} = F(b) - F(a),$$
>
> where F is any function such that $F'(x) = f(x)$ for all x in $[a, b]$.

It is important to recognize that the fundamental theorem of integral calculus describes a means for evaluating a definite integral. It does not provide us with a technique for finding the antiderivative. To find the antiderivative of a definite integral we use the same techniques we use to find the antiderivative of the indefinite integral. But what, we might ask, happens to the constant C? The constant C drops out as illustrated below.

$$\int_{x=a}^{x=b} f(x)\ dx = (F(x) + C)\Bigg]_{x=a}^{x=b}$$

$$= (F(b) + C) - (F(a) + C) \qquad \textbf{By the fundamental theorem}$$

$$= F(b) - F(a) + C - C \qquad \textbf{Rearranging terms}$$

$$= F(b) - F(a).$$

Before moving on to examples that illustrate the fundamental theorem, examine the list of some useful properties of the definite integral.

> **Properties of Definite Integrals**
> If f and g are continuous functions over the indicated closed intervals and k is a constant, then
>
> **1.** $\displaystyle\int_{x=a}^{x=b} kf(x)\ dx = k \int_{x=a}^{x=b} f(x)\ dx;$
>
> **2.** $\displaystyle\int_{x=a}^{x=b} f(x)\ dx = \int_{x=a}^{x=c} f(x)\ dx + \int_{x=c}^{x=b} f(x)\ dx,$ where $a < c < b;$
>
> **3.** $\displaystyle\int_{x=a}^{x=b} [f(x) + g(x)]\ dx = \int_{x=a}^{x=b} f(x)\ dx + \int_{x=a}^{x=b} g(x)\ dx;$
>
> **4.** $\displaystyle\int_{x=a}^{x=b} f(x)\ dx = -\int_{x=b}^{x=a} f(x)\ dx;$
>
> **5.** $\displaystyle\int_{x=a}^{x=a} f(x)\ dx = 0.$

For the definite integral $\int_{x=a}^{x=b} f(x)\, dx$, the letters a and b are called the **limits of integration**.

EXAMPLE 1 **Definite Integrals: Fundamental Theorem of Calculus**

Evaluate the following definite integrals using the fundamental theorem.

a. $\displaystyle\int_{x=2}^{x=3} x^2\, dx$

b. $\displaystyle\int_{x=1}^{x=4} 3x\, dx$

c. $\displaystyle\int_{x=0}^{x=2} (x^3 - 5x)\, dx$

Solution

a. $\displaystyle\int_{x=2}^{x=3} x^2\, dx = \left.\frac{x^3}{3}\right]_{x=2}^{x=3}$ Antiderivative of x^2

$\qquad\qquad = \dfrac{3^3}{3} - \dfrac{2^3}{3}$ Fundamental theorem

$\qquad\qquad = \dfrac{19}{3}.$

b. $\displaystyle\int_{x=1}^{x=4} 3x\, dx = 3\int_{x=1}^{x=4} x\, dx$ Property 1

$\qquad\qquad = 3\left.\left[\frac{x^2}{2}\right]\right._{x=1}^{x=4}$ Antiderivative of x

$\qquad\qquad = 3\left(\dfrac{4^2}{2} - \dfrac{1^2}{2}\right)$ Fundamental theorem

$\qquad\qquad = 3\left(\dfrac{15}{2}\right)$

$\qquad\qquad = \dfrac{45}{2}.$

c. $\displaystyle\int_{x=0}^{x=2} (x^3 - 5x)\, dx = \int_{x=0}^{x=2} x^3\, dx - 5\int_{x=0}^{x=2} x\, dx$ Properties 1 and 3

$\qquad\qquad = \left.\frac{x^4}{4}\right]_{x=0}^{x=2} - 5\left.\left(\frac{x^2}{2}\right)\right]_{x=0}^{x=2}$ Antiderivative of x^3, x

$\qquad\qquad = \left(\dfrac{2^4}{4} - \dfrac{0^4}{4}\right) - 5\left(\dfrac{2^2}{2} - \dfrac{0^2}{2}\right)$ Fundamental theorem

$\qquad\qquad = -6$ ◆

EXAMPLE 2   **Hypothesis of the Fundamental Theorem**

Evaluate, if possible, the following definite integral using the fundamental theorem.

$$\int_{x=-2}^{x=2} \frac{1}{x^2}\, dx$$

Solution   The function $f(x) = \dfrac{1}{x^2}$ is not continuous over the closed interval

$[2, -2]$. When $x = 0, f(x) = \dfrac{1}{x^2}$ is undefined. From our definition of continuity

we know that the function is discontinuous at $x = 0$. The function $f(x) = \dfrac{1}{x^2}$

does not satisfy the hypothesis of the fundamental theorem; therefore, we cannot apply the theorem.   ◆

It is essential in each case that, before we attempt to integrate, we determine whether the hypothesis of the fundamental theorem is satisfied. That is, is the function continuous over the closed interval? If it is not continuous, we cannot apply the fundamental theorem. And, until we learn other techniques, we cannot evaluate the definite integral.

EXAMPLE 3   **Limits of Integration Are Equal**

Evaluate $\int_{x=3}^{x=3} (x^3 + 7x^2 + 2)\, dx$ using the fundamental theorem.

Solution

$$\int_{x=3}^{x=3} (x^3 + 7x^2 + 2)\, dx = 0 \qquad \textbf{Property 5}$$

We can show that the result is correct by integrating and applying the fundamental theorem.

$$\int_{x=3}^{x=3} (x^3 + 7x^2 + 2)\, dx$$

$$= \left(\frac{x^4}{4} + \frac{7x^3}{3} + 2x \right) \Bigg]_{x=3}^{x=3} \qquad \textbf{integrating}$$

$$= \left[\frac{3^4}{4} + 7\left(\frac{3^3}{3}\right) + 2(3) \right] - \left[\frac{3^4}{4} + 7\left(\frac{3^4}{3}\right) + 2(3) \right] \qquad \begin{matrix}\textbf{Fundamental} \\ \textbf{theorem}\end{matrix}$$

$$= 0. \quad ◆$$

EXAMPLE 4   Substitution Method

Evaluate $\int_{x=-1}^{x=2} x^2(2x^3 - 4)^3 \, dx$ using the fundamental theorem.

Solution   To evaluate $\int_{x=-1}^{x=2} x^2(2x^3 - 4)^3 \, dx$ we must use the method of substitution.

Let $u = 2x^3 - 4$. Then,

$$\frac{du}{dx} = 6x^2,$$

or

$$\frac{du}{6} = x^2 \, dx.$$

We can now write an integral in terms of u.

$$\int_{x=-1}^{x=2} x^2(2x^3 - 4)^3 \, dx = \int u^3 \frac{du}{6}.$$

The original limits of integration are in terms of the variable x. Since we changed the variable of integration to the variable u we may express the limits of integration in terms of the variable u.

We know that $u = 2x^3 - 4$.

When $x = 2$ (upper limit of integration),

$$u = 2(\boxed{2})^3 - 4 \quad \textbf{Substituting}$$
$$= 12.$$

When $x = -1$ (lower limit of integration),

$$u = 2(\boxed{-1})^3 - 4 \quad \textbf{Substituting}$$
$$= -6.$$

Therefore,

$$\int_{x=-1}^{x=2} x^2(2x^3 - 4)^3 \, dx = \int_{u=-6}^{u=12} \frac{u^3}{6} \, du.$$

Substituting for the respective values, the integral in terms of u is:

$$\frac{1}{6}\int_{u=-6}^{u=12} u^3 \, du = \frac{1}{6} \frac{u^4}{4}\bigg]_{u=-6}^{u=12} \qquad \textbf{Integrating}$$

$$= \frac{1}{6}\left[\frac{\boxed{12}^4}{4} - \frac{(\boxed{-6})^4}{4}\right] \qquad \textbf{Fundamental theorem}$$

$$= \frac{1}{6}\left(\frac{20{,}736}{4} - \frac{1296}{4}\right)$$

$$= 810.$$

Therefore, the value of the definite integral $\int_{x=-1}^{x=2} x^2(2x^3 - 4)^3 \, dx$ is 810.

The answer 810 can be obtained by another method. We first can find the antiderivative of:

$$\frac{1}{6} \int u^3 \, du = \frac{1}{6} \frac{u^4}{4} + C \quad \textbf{Integrating}$$

$$= \frac{1}{24} u^4. \quad \textbf{Dropping } C$$

We can drop the constant of integration since it cancels itself when we evaluate the definite integral.

We said earlier that $u = 2x^3 - 4$. Substituting, we have:

$$\frac{1}{24} u^4 = \frac{1}{24} (2x^3 - 4)^4.$$

$$\int_{x=-1}^{x=2} x^2(2x^3 - 4)^3 \, dx = \frac{1}{24} (2x^3 - 4)^4 \Big]_{x=-1}^{x=2}$$

$$= \frac{1}{24} \{[2(\boxed{2})^3 - 4]^4 - [2(\boxed{-1})^3 - 4]^4\}$$

$$\textbf{Fundamental theorem}$$

$$= \frac{1}{24} [(12)^4 - (-6)^4]$$

$$= 810.$$

Thus, we see that we can obtain the same result by both methods. ◆

Example 4 shows that there are two ways of evaluating a definite integral when we use the method of substitution. In one method we change the variable of integration, change the limits of integration, and then integrate and apply the fundamental theorem. Using the other method, we change the variable of integration, integrate the indefinite integral, then substitute back in terms of the original variable and apply the fundamental theorem.

EXAMPLE 5 Area of a Region

Use the fundamental theorem to determine the area of the region bounded by $f(x) = x^2 + 1$, the x-axis, and the lines $x = 0$ and $x = 2$.

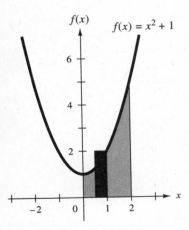

FIGURE 21.14

Solution The first thing we do is sketch the figure (shown in Figure 21.14). From this figure we see the region whose area we want to find. With the help of the indicated sample rectangle we can represent the area with the definite integral:

$$A = \int_{x=0}^{x=2} (x^2 + 1) \, dx.$$

Since the hypothesis of the fundamental theorem is met, we can apply the theorem and find the area.

$$A = \int_{x=0}^{x=2} (x^2 + 1)\, dx.$$

$$= \left(\frac{x^3}{3} + x \right) \Bigg]_{x=0}^{x=2} \qquad \textbf{Integrating}$$

$$= \left(\frac{8}{3} + 2 \right) - (0 + 0) \qquad \textbf{Fundamental theorem}$$

$$= \frac{14}{3} \text{ square units}$$

Thus, the area of the indicated region is $\dfrac{14}{3}$ square units. ◆

Work

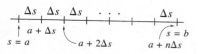

FIGURE 21.15

In physics a general definition of work is the product of the force acting on an object by the distance the object is moved. If s is the distance, F is the force, and W is work, we can express work as:

$$W = Fs.$$

However, we can use this formula only if there is a constant force. If the force varies, then work is determined by a definite integral.

For example, to determine the amount of work needed to move an object from point $s = a$ to point $s = b$, we divide the distance from a to b into many subintervals, as shown in Figure 21.15. We assume that a constant force F is needed to move the object from point a to point $a + \Delta s$. However, the force that is needed to move the object from $a + \Delta s$ to $a + 2\Delta s$, may change. Thus, the amount of work needed to move the object from point a to b is the limit of the sum of all these forces, which is a definite integral. We can represent the work W that is done in moving the object as:

$$W = \int_{s=a}^{s=b} F\, ds.$$

If the force varies, then F is expressed as a function of s; that is, $F = f(s)$.

Hooke's law states that, in stretching a spring, the force F is proportional to the displacement s; that is,

$$F = ks,$$

where k is a constant, called the spring constant. The stiffer the spring, the larger the value of k. The amount of work W done by exerting a force F in stretching a spring from $s = a$ to $s = b$ is:

$$W = \int_{s=a}^{s=b} F\, ds = \int_{s=a}^{s=b} ks\, ds.$$

EXAMPLE 6   **Application: Work in Stretching a Spring**

A spring has a natural length of 1.25 ft. A force of 25.0 lb stretches the spring to a length of 1.75 ft.

a. Determine the spring constant.
b. How much work is required to stretch the spring 2.25 ft beyond its natural length?
c. How far does a 55.0 lb force stretch the spring?

Solution

a. To determine the spring constant we use the formula $F = ks$. We are given that a force of 25.0 lb stretches the spring 0.50 ft; thus, $F = 25.0$ lb and $s = 0.50$ ft. Therefore,

$$25.0 = k(0.50) \quad \text{Substitution}$$
$$k = 50 \text{ ft/lb.}$$

b. Holding the unstressed spring between our hands, the distance $s = 0$. The force required to stretch the spring s ft beyond its natural length is the force required to stretch the spring apart s ft. With $k = 50$ ft/lb, Hooke's law tells us that this force is $F = 50s$. The work required to apply a force that stretches the spring from $s = 0$ to $s = 2.25$ ft is:

$$W = \int_{s=0}^{s=2.25} 50s \, ds$$
$$= 25 \, s^2 \Big]_{s=0}^{s=2.25}$$
$$= 127 \text{ ft/lb.}$$

c. To determine the distance a force of 55.0 lb stretches the spring, substitute $F = 55.0$ in the equation $F = 50s$.

$$55.0 = 50s$$
$$s = 1.10 \text{ ft}$$

Thus, a 55.0 lb force stretches the spring 1.10 ft. ◆

21.4 Exercises

In Exercises 1–26, evaluate, if possible, the definite integrals using the fundamental theorem of integral calculus. Determine whether the hypothesis of the fundamental theorem is met *before* integrating.

1. $\int_{x=1}^{x=2} x \, dx$

2. $\int_{x=1}^{x=3} 2x \, dx$

3. $\int_{x=0}^{x=3} dx$

4. $\int_{x=-1}^{x=2} 3 \, dx$

5. $\int_{x=1}^{x=3} (2x - 1) \, dx$

6. $\int_{x=1}^{x=3} (2x + 1) \, dx$

7. $\int_{x=-1}^{x=1} (3x^2 + 1) \, dx$

8. $\int_{x=-2}^{x=2} (x^2 - 2) \, dx$

9. $\displaystyle\int_{x=-1}^{x=1} (5x^2 - 7x)\, dx$

10. $\displaystyle\int_{x=-2}^{x=2} (6x^2 - 4x)\, dx$

11. $\displaystyle\int_{x=0}^{x=1} 18x^8\, dx$

12. $\displaystyle\int_{x=-1}^{x=1} 14x^6\, dx$

13. $\displaystyle\int_{x=0}^{x=2} \left(\frac{1}{x^2} - 3\right) dx$

14. $\displaystyle\int_{x=0}^{x=3} (3x^{-4} + 5)\, dx$

15. $\displaystyle\int_{x=1}^{x=2} \left(\frac{1}{x^2} - 3\right) dx$

16. $\displaystyle\int_{x=1}^{x=3} (3x^{-4} + 5)\, dx$

17. $\displaystyle\int_{x=1}^{x=4} 2\sqrt{x}\, dx$

18. $\displaystyle\int_{x=4}^{x=9} \frac{2}{\sqrt{x}}\, dx$

19. $\displaystyle\int_{x=1}^{x=3} 5(x^2 - 7)^3\, x\, dx$

20. $\displaystyle\int_{x=1}^{x=2} 3(x^3 + 1)\, x^2\, dx$

21. $\displaystyle\int_{x=2}^{x=4} \sqrt[3]{x-2}\, dx$

22. $\displaystyle\int_{x=-1}^{x=1} \sqrt[5]{2x+3}\, dx$

23. $\displaystyle\int_{x=1}^{x=4} x^2\sqrt{3x^3 - 5}\, dx$

24. $\displaystyle\int_{x=0}^{x=2} x\sqrt[3]{x^2 - 1}\, dx$

25. $\displaystyle\int_{x=1}^{x=2} \frac{x+2}{\sqrt{x^2 + 4x + 1}}\, dx$

26. $\displaystyle\int_{x=0}^{x=1} \frac{x-3}{\sqrt{x^2 - 6x + 1}}\, dx$

In Exercises 27–30, explain why the fundamental theorem cannot be used to evaluate the definite integral.

27. $\displaystyle\int_{x=-3}^{x=2} \frac{dx}{x^3}$

28. $\displaystyle\int_{x=0}^{x=5} \frac{x\, dx}{x^2 - 3}$

29. $\displaystyle\int_{x=-2}^{x=2} \frac{3x\, dx}{x^3 - 1}$

30. $\displaystyle\int_{x=-3}^{x=3} \frac{2x\, dx}{\sqrt{x^2 - 4}}$

In Exercises 31–36, find the area of the region bounded by the x-axis, the function, and the lines. Sketch the graph and indicate a sample rectangle before setting up the integral.

31. $f(x) = 4x^2 - 1$, $x = 1$ and $x = 3$

32. $f(x) = 5x^2 + 1$, $x = 0$ and $x = 2$

33. $f(x) = x^2 - 5x + 6$, $x = 0$ and $x = 2$

34. $f(x) = x^2 - 6x + 8$, $x = 0$ and $x = 2$

35. $f(x) = 5x - x^2$, $x = 1$ and $x = 3$

36. $f(x) = x^2 - 3x + 5$, $x = 1$ and $x = 3$

37. The work in winding up a bucket from a 100 ft well is given by $W = \int F\, ds$, where $F = 4s + 200$. Determine the work done when the distance is evaluated from the bottom of the well to the top of the well.

38. For a spring, the force is given by $F = ks$, where k is called the spring constant. If the spring constant is 150 N/m and the spring is stretched from 0.30 m to 0.60 m, how much work must be done?

39. a. The force on a 100-kg man near the surface of the earth is very nearly a constant value 980 N. Determine the work done to take this man from the surface of the earth to a point 200 m above the earth.

   b. As you move very far away from the earth, the force varies $\dfrac{4 \times 10^{16}}{x^2}$. Determine how much work has to be done to lift a man from the surface of the earth ($R = 6.4 \times 10^6$ m) to a height of 4.6×10^6 m.

21.5

Numerical Integration

In Section 21.3 we used rectangles to approximate the value of a definite integral. Then, in Section 21.4, we learned that if an antiderivative exists for the integral, we can apply the fundamental theorem of integral calculus to determine a value for the definite integral. In this section we study two techniques—the *trapezoidal rule* and *Simpson's rule*—that we can use to evaluate a definite integral for which no antiderivative exists. We also can use a graphics calculator to determine the value of a definite integral. Let us begin with the calculator.

Calculator

The calculator techniques are introduced first so that we can use a calculator to check the results of other problems.

EXAMPLE 1   Numerical Integration Using the Calculator

Use a calculator to determine the value of $\int_{x=1}^{x=3} (x^2 - 4) \, dx$.

Solution   The keystrokes and display for determining the numerical value of the definite integral are given for each model.

fx–7700G		
Keystrokes	**Screen Display**	**Answer**
SHIFT $\int f \, dx$	$\int ($	
(x,Θ,T x^2 – 4	$\int ((x^2 - 4, 1, 3)$	
, 1 , 3)		
EXE		0.666666667

TI–82		
Keystrokes	**Screen Display**	**Answer**
MATH 9	fmInt (	
($x,T,\Theta,$ x^2 – 4	fmInt $((x^2 - 4, x, 1,$	
, x,T,Θ , 1 ,	3)	
3) ENTER		.6666666667

The calculator result for $\int_{x=1}^{x=3} (x^2 - 4) \, dx$, rounded to the nearest thousandths, is 0.667. ◆

Trapezoidal Rule

In Section 21.3 we found the approximate area of a region bounded by a function, the x-axis, and a closed interval by using rectangles. Because a trapezoidal element fits closer to a curve than does a rectangular element, we would expect the use of trapezoidal elements to give a more accurate result than the same number of rectangular elements. How do we approximate the value of $\int_{x=a}^{x=b} f(x) \, dx$ using trapezoids: To apply this method, the function $f(x)$ must be continuous over the closed interval $[a, b]$ and $f(x)$ must be positive over the closed interval. If these conditions are met, the results of applying this technique give us the area of the region bounded by the function $f(x)$, the x-axis, and the lines $x = a$ and $x = b$.

We start by partitioning the interval $[a, b]$ into n subintervals each of width Δx:

$$\Delta x = \frac{b - a}{n},$$

such that $a = x_0 < x_1 < x_2 < \cdots < x_{i=1} < x_i \cdots < x_{n-1} < x_n = b$. In each of the subintervals we construct trapezoids as shown in Figure 21.16. Figure

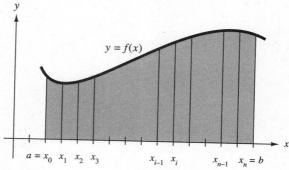

FIGURE 21.16

21.17 is an enlargement of the first trapezoid of Figure 21.16 with the parts labeled. It, like the other trapezoids shown, has vertical bases and horizontal height. The area of a trapezoid with parallel bases is equal to the average of the two bases times the height.

$$A = \frac{(b_1 + b_2)}{2} \cdot h$$

To find the area of the n trapezoids we can find the area of each of the trapezoids and then find the sum of these areas.

$$\text{Area of first trapezoid} = \frac{[f(x_0) + f(x_1)]}{2} \frac{(b - a)}{n}$$

$$\text{Area of second trapezoid} = \frac{[f(x_1) + f(x_2)]}{2} \frac{(b - a)}{n}$$

$$\text{Area of third trapezoid} = \frac{[f(x_2) + f(x_3)]}{2} \frac{(b - a)}{n}$$

$$\text{Area of } n\text{th trapezoid} = \frac{[f(x_{n-1}) + f(x_n)]}{2} \frac{(b - a)}{n}$$

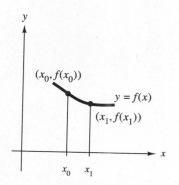

FIGURE 21.17

Since each area contains the factor $\dfrac{(b-a)}{n}$, we can write the sum of the area as:

$$\text{Sum} = \frac{(b-a)}{n}\left[\frac{f(x_0)+f(x_1)}{2} + \frac{f(x_1)+f(x_2)}{2}\right.$$
$$\left. + \frac{f(x_2)+f(x_3)}{2} + \cdots + \frac{f(x_{n-1})+f(x_n)}{2}\right].$$

Factoring out a common factor of $\dfrac{1}{2}$, we have:

$$\text{Sum} = \frac{(b-a)}{2n}[f(x_0)+f(x_1)+f(x_1)+f(x_2)+f(x_2)+\cdots$$
$$+ f(x_{n-1})+f(x_{n-1})+f(x_n)].$$

We see that there are two each of the function values $f(x_1)$ through $f(x_{n-1})$. Thus, we can write:

$$\text{Sum} = \frac{(b-a)}{2n}[f(x_0)+2f(x_1)+2f(x_2)+\cdots+2f(x_{n-1})+f(x_n)].$$

This sum, which is called the trapezoidal rule, is an approximation of the area of the region that can be indicated by the definite integral $\int_{x=a}^{x=b} f(x)\,dx$.

RULE 21.2 Trapezoidal Rule

$$\int_{x=a}^{x=b} f(x)\,dx \approx \frac{(b-a)}{2n}[f(x_0)+2f(x_1)+2f(x_2)+\cdots$$
$$+ 2f(x_{n-1})+f(x_n)]$$

EXAMPLE 2   Trapezoidal Rule

Use the trapezoidal rule to approximate the definite integral $\int_{x=0}^{x=2}(x^2+2)\,dx$. Let $n=4$.

Solution   To find the approximate area of the region bounded by the function $f(x) = x^2 + 2$, the x-axis, and the lines $x = 0$ and $x = 2$ we first need to find Δx.

$$\Delta x = \frac{2-0}{4}$$
$$= \frac{1}{2}$$

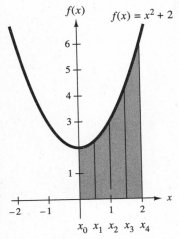

FIGURE 21.18

We next construct the graph of the function showing the four trapezoids as in Figure 21.18. From the figure we see that $x_0 = 0$, $x_1 = \dfrac{1}{2}$, $x_2 = 1$, $x_3 = \dfrac{3}{2}$, and $x_4 = 2$. Now we can calculate each of the functional values.

$$f(x_0) = f(0) = 0^2 + 2 \qquad f(x_1) = f\left(\frac{1}{2}\right) = \left(\frac{1}{2}\right)^2 + 2$$
$$= 2 \qquad\qquad\qquad = \frac{9}{4}$$

$$f(x_2) = f(1) = 1^2 + 2 \qquad f(x_3) = f\left(\frac{3}{2}\right) = \left(\frac{3}{2}\right)^2 + 2$$
$$= 3 \qquad\qquad\qquad = \frac{17}{4}$$

$$f(x_4) = f(2) = 2^2 + 2$$
$$= 6$$

With $n = 4$, the trapezoid rule provides us with the following formula.

$$\int_{x=0}^{x=2} (x^2 + 2)\, dx \approx \frac{(b-a)}{2n}[f(x_0) + 2f(x_1) + 2f(x_2) + 2f(x_3) + f(x_4)]$$

$$\int_{x=0}^{x=2} (x^2 + 2)\, dx \approx \frac{2-0}{2(4)}\left[2 + 2\left(\frac{9}{4}\right) + 2(3) + 2\left(\frac{17}{4}\right) + 6\right] \qquad \textbf{Substituting}$$

$$\approx \frac{2}{8}[27]$$

$$\approx 6\frac{3}{4}$$

Therefore, using the trapezoidal rule, we find that the value of
$$\int_{x=0}^{x=2} (x^2 + 2)\, dx \approx 6\frac{3}{4}.$$

In this case, we could have found the exact value $\int_{x=0}^{x=2} (x^2 + 2)\, dx = 6\frac{2}{3}$.

We see that the trapezoidal rule gives us only an approximation to the value of the integral. Show that by increasing the number of intervals a better approximation is obtained. ◆

The calculations involved in using the trapezoidal rule are performed more easily using a calculator. In fact, the result can be checked by evaluating the definite integral with a calculator.

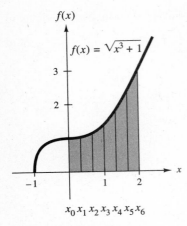

$f(x)$

$f(x) = \sqrt{x^3 + 1}$

$x_0\, x_1\, x_2\, x_3\, x_4\, x_5\, x_6$

FIGURE 21.19

EXAMPLE 3   Trapezoidal Rule

Use the trapezoidal rule to approximate the definite integral $\int_{x=0}^{x=2} \sqrt{x^3 + 1}\; dx$. Let $n = 6$.

Solution We cannot find the antiderivative of the definite integral $\int_{x=0}^{x=2} \sqrt{x^3 + 1}\; dx$ by any technique that we have learned so far. So to approximate the value of the definite integral $\int_{x=0}^{x=2} \sqrt{x^3 + 1}\; dx$, we first find the value of Δx and then sketch the curve as in Figure 21.19.

$$\Delta x = \frac{2 - 0}{6} \quad \text{or} \quad \frac{1}{3}.$$

From the figure we can see that $x_0 = 0$, $x_1 = \frac{1}{3}$, $x_2 = \frac{2}{3}$, $x_3 = 1$, $x_4 = \frac{4}{3}$, $x_5 = \frac{5}{3}$, and $x_6 = 2$. Now, programming our calculator, we find the following functional values.

$$f(x_0) = f(\,0\,) = 1.000$$
$$f(x_1) = f\left(\frac{1}{3}\right) = 1.018$$
$$f(x_2) = f\left(\frac{2}{3}\right) = 1.139$$
$$f(x_3) = f(\,1\,) = 1.414$$
$$f(x_4) = f\left(\frac{4}{3}\right) = 1.836$$
$$f(x_5) = f\left(\frac{5}{3}\right) = 2.373$$
$$f(x_6) = f(\,2\,) = 3$$

With $n = 6$, the trapezoidal rule provides us with the following.

$$\int_{x=0}^{x=2} \sqrt{x^3 + 1}\; dx \approx \frac{(b - a)}{2n}[f(x_0) + 2f(x_1) + 2f(x_2) + 2f(x_3) + 2f(x_4)$$
$$+ 2f(x_5) + f(x_6)]$$

$$\approx \frac{(\,2 - 0\,)}{2(6)}[\,1.000\, + 2(\,1.018\,) + 2(\,1.139\,) + 2(\,1.414\,) + 2(\,1.836\,)$$
$$+ 2(\,2.373\,) + 3\,] \quad \textbf{Substituting}$$

$$\approx 3.26.$$

Thus, the approximate value of the definite integral $\int_{x=0}^{x=2} \sqrt{x^3 + 1}\; dx$ is 3.26 when n is 6. ◆

The trapezoidal rule provides the desired degree of accuracy for the value of a definite integral by increasing the number of trapezoids n. The trapezoidal rule may be considered a linear approximation, since we use a linear function in the procedure.

The second method, Simpson's rule, uses a second-degree polynomial to provide the approximation to the curve. Before discussing Simpson's rule we need to look at a theorem for evaluating integrals of second-degree polynomials.

> ### THEOREM 21.2
>
> If $p(x) = Ax^2 + Bx + C$, and a and b are the end points of the closed interval, then:
>
> $$\int_{x=a}^{x=b} p(x)\, dx = \left(\frac{b-a}{6}\right)\left[p(a) + 4p\left(\frac{a+b}{2}\right) + p(b)\right].$$

Simpson's Rule

Using the trapezoidal rule to approximate the definite integral $\int_{x=a}^{x=b} f(x)\, dx$, we divide the closed interval $[a, b]$ into n subintervals of equal length $\dfrac{b-a}{n}$. When we use Simpson's rule there is a difference— n must be an even number. With n even, there are an equal number of subintervals, and the subintervals can be grouped in pairs.

$$a = \underbrace{x_0 < x_1 < x_2}_{[x_0,\, x_2]} \underbrace{< x_3 < x_4}_{[x_2,\, x_4]} < \cdots < \underbrace{x_{n-2} < x_{n-1} < x_n}_{[x_{n-2},\, x_n]} = b.$$

For the interval $[x_0, x_2]$, we approximate the values of $y = f(x)$ by the second-degree polynomial that passes through the points (x_0, y_0), (x_1, y_1) and (x_2, y_2) as shown in Figure 21.20. We now can use the polynomial $p(x)$ that passes through these points to obtain an approximation to the curve for that subinterval. That is,

$$\int_{x=x_0}^{x=x_2} f(x)\, dx \approx \int_{x=x_0}^{x=x_2} p(x)\, dx$$

$$\int_{x=x_0}^{x=x_2} p(x)\, dx = \left(\frac{x_2 - x_0}{6}\right)\left[p(x_0) + 4p\left(\frac{x_2 + x_0}{2}\right) + p(x_2)\right].$$

Since $\dfrac{b-a}{n}$ is the width of one subinterval, $x_2 - x_0 = 2\dfrac{(b-a)}{n}$, and $x_2 - x_0$ is the width of two subintervals. Also, since the subintervals are of equal width, $\dfrac{x_2 + x_0}{2} = x_1$. Substituting, we have:

$$\int_{x=x_0}^{x=x_2} p(x)\, dx = \left(\frac{2(b-a)}{6n}\right)[p(x_0) + 4p(x_1) + p(x_2)]$$

$$= \left(\frac{b-a}{3n}\right)[p(x_0) + 4p(x_1) + (x_2)].$$

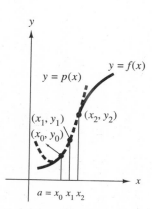

FIGURE 21.20

Since the polynomial passes through the three points on the curve, the best approximation would be the function itself; that is, $P(x_0) = f(x_0)$, $P(x_1) = f(x_1)$, and $P(x_2) = f(x_2)$, Thus,

$$\int_{x=x_0}^{x=x_2} f(x)\, dx \approx \left(\frac{b-a}{3n}\right)[f(x_0) + 4f(x_1) + f(x_2)].$$

If we repeat the process for each subinterval, we find that:

$$\int_{x=x_2}^{x=x_4} f(x)\, dx \approx \left(\frac{b-a}{3n}\right)[f(x_2) + 4f(x_3) + f(x_4)]$$

$$\int_{x=x_{n-2}}^{x=x_n} f(x)\, dx \approx \left(\frac{b-a}{3n}\right)[f(x_{n-2}) + 4f(x_{n-1}) + f(x_n)].$$

Note that $\dfrac{b-a}{3n}$ is a factor of each of the approximations. If we now find the sum of all approximations over the interval, we have an approximation for the definite integral.

$$\int_{x=a}^{x=b} f(x)\, dx \approx \left(\frac{b-a}{3n}\right)\{[f(x_0) + 4f(x_1) + f(x_2)]$$
$$+ [f(x_2) + 4f(x_3) + f(x_4)] + \cdots + [f(x_{n-2}) + 4f(x_{n-1}) + f(x_n)]\}.$$

Grouping the like terms, the approximation is:

$$\int_{x=a}^{x=b} f(x)\, dx \approx \left(\frac{b-a}{3n}\right)[f(x_0) + 4f(x_1) + 2f(x_2)$$
$$+ 4f(x_3) + 2f(x_4) + \cdots + 2f(x_{n-2}) + 4f(x_{n-1}) + f(x_n)].$$

This final result is known as Simpson's rule.

RULE 21.3 Simpson's Rule

$$\int_{x=a}^{x=b} f(x)\, dx \approx \left(\frac{b-a}{3n}\right)[f(x_0) + 4f(x_1) + 2f(x_2) + 4f(x_3)$$
$$+ 2f(x_4) + \cdots + 2f(x_{n-2}) + 4f(x_{n-1}) + f(x_n)],$$

where n is a positive even integer.

Because the parabola usually provides a closer fit to a curve than the straight side of a trapezoid, Simpson's rule is more accurate than the trapezoidal rule for the same value of n. As with the trapezoidal rule, a calculator is helpful in finding all the functional values when using Simpson's rule.

EXAMPLE 4 Simpson's Rule

Use Simpson's rule to approximate the value of the definite integral $\int_{x=0}^{x=2} \sqrt{x^3 + 1}\, dx$. Let $n = 6$. Express the answer to the nearest hundredth.

Solution For $n = 6$, the general formula to obtain an approximation of the definite integral using Simpson's rule is:

$$\int_{x=a}^{x=b} f(x)\ dx \approx \left(\frac{b-a}{3n}\right)[f(x_0) + 4f(x_1) + 2f(x_2) + 4f(x_3)$$
$$+ 2f(x_4) + 4f(x_5) + f(x_6)].$$

Determine the functional values and then substitute.

$$\int_{x=0}^{x=2} \sqrt{x^3+1}\ dx \approx \left(\frac{2-0}{3(6)}\right)\left[\sqrt{0^3+1} + 4\sqrt{\left(\frac{1}{3}\right)^3+1}\right.$$
$$+ 2\sqrt{\left(\frac{2}{3}\right)^3+1} + 4\sqrt{(1)^3+1}$$
$$+ 2\sqrt{\left(\frac{4}{3}\right)^3+1} + 4\sqrt{\left(\frac{5}{3}\right)^3+1} + \left.\sqrt{(2)^3+1}\right]$$
$$\approx \frac{1}{9}[1 + 4(1.018) + 2(1.139) + 4(1.414) + 2(1.836)$$
$$+ 4(2.373) + 3]$$
$$\approx 3.241.$$

Therefore, the approximate value of the definite integral $\int_{x=0}^{x=2} \sqrt{x^3+1}\ dx$, using Simpson's rule, is 3.24 ◆

The approximation of $\int_{x=0}^{x=2} \sqrt{x^3+1}\ dx$ in Example 3 using the trapezoid rule was 3.26. Using the calculator the result is 3.24130. The two methods agree only in the tenths position. Simpson's rule may require more work; but as we see by comparing the result with the calculator result, it provides a better approximation of a definite integral.

EXAMPLE 5 **Simpson's Rule**

Use Simpson's rule to approximate the value of $\int_{x=1}^{x=2} e^x\ dx$, when $n = 8$. Express the answer to the nearest hundredth.

Solution For $n = 8$, the general formula we use to obtain the approximation using Simpson's rule is:

$$\int_{x=a}^{x=b} f(x)\ dx \approx \left(\frac{b-a}{3n}\right)[f(x_0) + 4f(x_1) + 2f(x_2) + 4f(x_3)$$
$$+ 2f(x_4) + 4f(x_5) + 2f(x_6) + 4f(x_7) + f(x_8)].$$

Since $n = 8$, divide the interval from 1 to 2 into 8 equal parts. Also, since the interval starts with 1, $x_0 = 1$ and $x_1 = 1 + \frac{1}{8} = \frac{9}{8}, \ldots, x_n = 2$. Substituting we have:

$$\int_{x=1}^{x=2} e^x \, dx \approx \frac{2 - 1}{3(\boxed{8})} [e^{\boxed{1}} + 4e^{\frac{9}{8}} + 2e^{\frac{5}{4}} + 4e^{\frac{11}{8}} + 2e^{\frac{3}{2}} + 4e^{\frac{13}{8}}$$
$$+ 2e^{\frac{7}{4}} + 4e^{\frac{15}{8}} + e^{\boxed{2}}]$$
$$\approx 4.671.$$

Using Simpson's rule the approximate value of $\int_{x=1}^{x=2} e^x \, dx$ to the nearest hundredth is 4.67. ◆

In Chapter 23 we will learn how to determine the antiderivative of e^x and be able to show that the exact value of $\int_{x=1}^{x=2} e^x \, dx$ is 4.67 to the nearest hundredth.

21.5 Exercises

In Exercises 1–6, use the trapezoidal rule to approximate the value of each definite integral. Round the answer to the nearest hundredth. Compare your results with the exact value of the definite integral.

1. $\displaystyle\int_{x=1}^{x=3} x^2 \, dx, \, n = 4$

2. $\displaystyle\int_{x=2}^{x=4} (x^2 - 4) \, dx, \, n = 4$

3. $\displaystyle\int_{x=0}^{x=1} \left(\frac{x^2}{2} + 1\right) dx, \, n = 4$

4. $\displaystyle\int_{x=1}^{x=3} x^3 \, dx, \, n = 4$

5. $\displaystyle\int_{x=0}^{x=1} 2x(x^2 + 1)^{\frac{3}{2}} \, dx, \, n = 6$

6. $\displaystyle\int_{x=1}^{x=3} \frac{2x}{\sqrt{x^2 + 1}} \, dx, \, n = 6$

In Exercises 7–12, use the trapezoidal rule to approximate the value of each of the definite integrals. Round the answer to the nearest hundredth. To check your results, evaluate the definite integral with a calculator.

7. $\displaystyle\int_{x=1}^{x=3} \frac{dx}{x}, \, n = 6$

8. $\displaystyle\int_{x=0}^{x=1} \sqrt{9 - x^2} \, dx, \, n = 3$

9. $\displaystyle\int_{x=0}^{x=2} \sqrt{1 + x^3} \, dx, \, n = 6$

10. $\displaystyle\int_{x=0}^{x=1} \frac{dx}{\sqrt{1 + x^2}}, \, n = 5$

11. $\displaystyle\int_{x=0}^{x=2} \frac{x \, dx}{\sqrt{1 + x}}, \, n = 6$

12. $\displaystyle\int_{x=0}^{x=1} (x^2 + 1)^3 \, dx, \, n = 6$

In Exercises 13–18, use Simpson's rule to approximate the value of each definite integral. Round the answer to the nearest hundredth. Compare your results with the exact value of the definite integral.

13. $\displaystyle\int_{x=2}^{x=4} x^2 \, dx, \, n = 6$

14. $\displaystyle\int_{x=1}^{x=3} (x^2 - 6) \, dx, \, n = 6$

15. $\displaystyle\int_{x=2}^{x=3} \left(\frac{x^2}{3} - 1\right) dx, \, n = 8$

16. $\displaystyle\int_{x=1}^{x=2} x^3 \, dx, \, n = 8$

17. $\displaystyle\int_{x=0}^{x=1} x(x^2 + 1)^{\frac{5}{2}} \, dx, \, n = 6$

18. $\displaystyle\int_{x=2}^{x=3} \frac{3x^2}{\sqrt{x^3 - 1}} \, dx, \, n = 6$

In Exercises 19–24, use Simpson's rule to approximate the value of each definite integral. Round the answer to the nearest hundredth. To check your results, evaluate the definite integral with a calculator.

19. $\displaystyle\int_{x=1}^{x=3} \frac{dx}{x}, \, n = 6$

20. $\displaystyle\int_{x=0}^{x=2} \sqrt{4 - x^2} \, dx, \, n = 4$

21. $\displaystyle\int_{x=0}^{x=1} e^{2x} \, dx, \, n = 8$

22. $\displaystyle\int_{x=1}^{x=2} \ln x \, dx, \, n = 8$

23. $\displaystyle\int_{x=0}^{x=1} \frac{1}{2 + x + x^2} \, dx, \, n = 4$

24. $\displaystyle\int_{x=0}^{x=1} \frac{x}{x^2 + x + 1} \, dx, \, n = 4$

Review Exercises

In Exercises 1–16, evaluate the indefinite integral and check the result with differentiation.

1. $\int x^8 \, dx$

2. $\int 3x^{11} \, dx$

3. $\int (x^2 + 5x + 6) \, dx$

4. $\int (-2x^2 + 7x - 11) \, dx$

5. $\int (3x^3 - 7x^2 + 5x) \, dx$

6. $\int (11x^4 - 7x^2 + 5x + 2) \, dx$

7. $\int \left(\sqrt[3]{x + 1} + \dfrac{5}{\sqrt{x + 1}} \right) dx$

8. $\int \dfrac{3}{x^5} \, dx$

9. $\int (3x^2 - 2x)^2 \, dx$

10. $\int x^3 \sqrt{x} \, dx$

11. $\int t^2 \sqrt{2t^3 - 7} \, dt$

12. $\int t \sqrt{4t^2 - 13} \, dt$

13. $\int \dfrac{3 - \sqrt{x}}{\sqrt{x}} \, dx$

14. $\int \dfrac{5x}{\sqrt{7x^2 - 2}} \, dx$

15. $\int \dfrac{3x^2 + 3}{(x^3 + 3x + 5)^3} \, dx$

16. $\int \dfrac{6x^2 + 3x + 4}{\left(2x^3 + \dfrac{3}{2}x^2 + 4x \right)^4} \, dx$

In Exercises 17–20, determine the equation of the curve having the given slope and passing through the indicated point.

17. $\dfrac{dy}{dx} = 6x^2 + 10x - 7, (-1, 13)$

18. $\dfrac{dy}{dx} = \dfrac{-1}{(x + 1)^2}, \left(0, \dfrac{1}{2} \right)$

19. $\dfrac{dy}{dx} = \dfrac{1}{2(x + 3)^{\frac{1}{2}}}, (-3, 1)$

20. $\dfrac{dy}{dx} = 9x^2 - 10x, (1, 0)$

21. A balloon is rising at the constant rate of 3.00 m/s and is 30 m from the ground at the instant Sue drops a bag of sand from the balloon.
  a. How long does it take the bag of sand to strike the ground?
  b. At what speed does the bag of sand strike the ground?

22. Find the equation of the curve that has a second derivative equal to $12x$ and passes through the points $(0, 1)$ and $(1, -4)$.

23. Find an equation of the curve whose slope at the point (x, y) is $12x^3 - 6x^2$, if the curve is required to pass through the point $(1, 0)$.

24. A ball rolls down an incline with an acceleration of 1 m/s². If the ball is given no initial velocity, how far does it go in t seconds? What initial velocity must be given for the ball to roll 40 m in 5 s?

In Exercises 25 and 26, determine the following.
  a. The approximate value of the definite integral using rectangles
  b. The exact value using integration

25. $\displaystyle\int_{x=1}^{x=4} (5x - 3) \, dx, \ n = 3$

26. $\displaystyle\int_{x=1}^{x=4} (7x + 6) \, dx, \ n = 6$

In Exercises 27–44, determine the following.
a. Whether the hypothesis of the fundamental theorem is satisfied
b. The value of the definite integral using the fundamental theorem, if the hypothesis
    is satisfied

27. $\int_{x=1}^{x=4} 3x\,dx$

28. $\int_{x=2}^{x=5} 5x\,dx$

29. $\int_{x=2}^{x=6} 4\,dx$

30. $\int_{x=1}^{x=7} -5\,dx$

31. $\int_{x=2}^{x=5} (3x-4)\,dx$

32. $\int_{x=-2}^{x=2} (2x+7)\,dx$

33. $\int_{t=-3}^{t=3} (5t^2-2t+1)\,dt$

34. $\int_{t=0}^{t=4} (3t^2+5t+1)\,dt$

35. $\int_{t=-2}^{t=2} 13t^7\,dt$

36. $\int_{t=1}^{t=2} 9t^{14}\,dt$

37. $\int_{t=0}^{t=4} \sqrt{t^2+1}\,t\,dt$

38. $\int_{t=1}^{t=3} \sqrt[3]{t^2-1}\,t\,dt$

39. $\int_{x=-2}^{x=2} (x^{-2}+4)\,dx$

40. $\int_{x=1}^{x=3} (x^{-2}+4)\,dx$

41. $\int_{x=1}^{x=3} 4(2x^2-7)x\,dx$

42. $\int_{x=0}^{x=4} (x^3-7)x^2\,dx$

43. $\int_{x=-1}^{x=2} \frac{3x^2+8x}{\sqrt{x^3+4x^2+7}}\,dx$

44. $\int_{x=2}^{x=4} \frac{2x^3-5x+1}{\sqrt{x^4-5x^2+2x}}\,dx$

In Exercises 45–48, determine the area of the region bounded by the x-axis, the
function, and the indicated interval. Sketch the graph and indicate a sample rectangle
before setting up the integral.

45. $f(x)=2x^2-5,\ [2,5]$

46. $f(x)=3x^2+2,\ [-1,3]$

47. $f(x)=x^3-x+4,\ [-1,2]$

48. $f(x)=x^3-x^2-2,\ [0,3]$

In Exercises 49–52, use the trapezoidal rule to approximate the value of the definite
integrals to the nearest hundredth.

49. $\int_{x=1}^{x=3} \sqrt{x^2+5}\,dx,\ n=4$

50. $\int_{x=2}^{x=4} \sqrt{x^3-4}\,dx,\ n=4$

51. $\int_{x=1}^{x=3} \frac{dx}{x+1},\ n=6$

52. $\int_{x=1}^{x=3} (x^2-1)^2\,dx,\ n=8$

In Exercises 53–56, use Simpson's rule to approximate the value of each definite
integral to the nearest hundredth.

53. $\int_{x=3}^{x=5} \sqrt{x^2-7}\,dx,\ n=4$

54. $\int_{x=0}^{x=2} \sqrt{2x^3+4}\,dx,\ n=4$

55. $\int_{x=2}^{x=4} \frac{dx}{x-1},\ n=6$

56. $\int_{x=1}^{x=3} (x^2+1)^4\,dx,\ n=8$

57. If $f(x)=px^2+qx+r$ and $f(x)\geq 0$ for all x, prove that the area under the graph
of f from 0 to b is $p\left(\dfrac{b^3}{3}\right)+q\left(\dfrac{b^2}{2}\right)+rb$.

58. Verify that $\displaystyle\int \frac{\sqrt{x^2-a^2}}{x^4}\,dx = \frac{\sqrt{(x-a^2)^3}}{3a^2x^3}$.

59. Let W be the work done in charging a capacitor having (constant) capacitance C with Q coulombs of charge. If E is the voltage drop across the capacitor and I is the current in amperes flowing into the capacitor, then $dW = E\,dQ$, $Q = CE$, and $dQ = I\,dt$, where t denotes time in seconds.

    a. Show that the instantaneous power $\dfrac{dW}{dt}$ required to charge the capacitor is given

       by $\dfrac{dW}{dt} = EI$.

    b. Assuming that $W = 0$ when $E = 0$, prove that $W = \dfrac{1}{2}CE^2$.

✍ Writing About Mathematics

1. Your company is developing an exercise machine that uses a spring. To sell the machine, the company must determine the amount of work required to stretch the spring, which in turn, helps determine the number of calories burned. You tell your boss that you can determine the amount of work using integral calculus. She is skeptical and asks you to write a report explaining why and how integral calculus will work.

2. In this chapter we discovered that integration is the inverse of differentiation. In a similar manner, multiplication is the inverse of division. Consider the function $\dfrac{dx}{dt} = 32t^2$. Note that the dt is connected to the dx by division. Explain, using only words and complete sentences, how you solve for x as a function of t. Your explanation must include two statements about inverses.

3. Your friend has difficulty with the substitution rule. Write a procedure that will help him use the substitution rule when evaluating an integral. Include the following in the procedure: when the rule can be applied, and how someone determines what to let u equal, and what is substituted for dx.

4. Before the formula for determining the area of a circle was discovered, mathematicians constructed regular polygons inside and outside the circle. By determining the areas of these inscribed and circumscribed polygons, they were able to approximate the area of the circle. Write a short paper explaining how this process could have led the mathematicians to approximate the area under a curve.

Chapter Test

In Exercises 1–5, evaluate the indefinite integral and check the result using differentiation.

1. $\displaystyle\int x^5\,dx$ **2.** $\displaystyle\int (x^4 + 3x^7)\,dx$ **3.** $\displaystyle\int \dfrac{1}{x^4}\,dx$ **4.** $\displaystyle\int x^3\sqrt{3x^4 + 1}\,dx$

5. $\displaystyle\int \dfrac{\sqrt{x^{\frac{1}{3}} + 4}}{x^{\frac{2}{3}}}\,dx$

6. A certain curve has a slope given by $x(2x^2 - 1)^3$ that passes through the point $(1, 1)$. What is the equation of the curve?

7. Determine the approximate area of the region bounded by $f(x) = 3x - 1$, $x = 1$, $x = 3$, for n rectangles where $n = 4$.

8. Determine the exact value of $\int_{x=0}^{x=2} (4x + 1) \, dx$ using

$$\int_{x=a}^{x=b} f(x) \, dx = \lim_{n \to \infty} \frac{(b-a)}{n} [f(x_1) + f(x_2) + \; + f(x_n)]$$

In Exercises 9–12, evaluate, if possible, the definite integrals using the fundamental theorem of integral calculus. Determine whether the hypothesis of the fundamental theorem is met before integrating.

9. $\displaystyle\int_{x=1}^{x=3} (x + 3) \, dx$

10. $\displaystyle\int_{x=-1}^{x=3} (3x^2 - 5x) \, dx$

11. $\displaystyle\int_{x=-3}^{x=3} \frac{dx}{x^3} \, dx$

12. $\displaystyle\int_{x=1}^{x=4} \frac{2}{\sqrt{x}} \, dx$

13. A spring has a natural length of 1.35 ft. A force of 35.0 lb stretches the spring to a length of 1.95 ft.
    a. Determine the spring constant.
    b. How much work is required to stretch the spring 3.25 ft beyond its natural length?
    c. How far does a 65.0 lb force stretch the spring?

14. Use the trapezoidal rule to approximate the value of the definite integral $\int_{x=0}^{x=2} \sqrt{2 + x^2} \, dx$, $n = 6$ to the nearest hundredth.

15. Use Simpson's rule to approximate the value of the definite integral $\int_{x=0}^{x=2} \sqrt{2 + x^2} \, dx$, $n = 6$ to the nearest hundredth.

Applications of Integration

A fishing-tackle designer is developing a new style of spinner blade for a lure. She plans to form the blade from a rectangular piece of brass (called a blank), which is stamped from sheet metal. The blank is to be the same size and shape as the area bounded by the parabola $y = -\frac{1}{4}x^2 + 25$ and the lower portion of the circle $x^2 + y^2 = 100$. To balance the blade properly, the designer must determine where the center of gravity of the blade is located. Example 7 in Section 22.4 answers her problem.

22.1

Area Between Two Curves

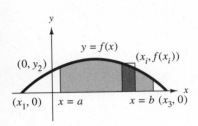

FIGURE 22.1

In Chapter 21 we learned the technique for finding the area of a region under a curve. This was a special case, where $f(x) \geq 0$ for the interval. In this section the technique for finding the area of a region under a curve is extended to finding the area *between* two curves.

A few general guidelines help make our job easier when we find the area of a region. The shaded region in Figure 22.1 is bounded by the curves $y = f(x)$, x-axis ($y = 0$), $x = a$, and $x = b$. The figure also shows a sample rectangle, which we shall call the ith rectangle. We assume that there is a value of x, say x_i that, when substituted in the function, gives a good approximation for the height of the rectangle. This value of x may be at any point on the base of the ith rectangle. Thus, the area of the ith rectangle is:

$$A_i = \text{Height of rectangle} \times \text{Width of the base}$$
$$= f(x_i) \, \Delta x_i.$$

Now, to determine the area of the region, we need to sum the area of all of the rectangles and take the limit of the sum. In the previous section we did this by writing out the terms. This process can be shortened by using the summation symbol Σ (read sigma). Recall that sigma is a Greek letter and is interpreted as the word "add." Examples illustrating the use of the summation symbol are:

$$\sum_{i=1}^{5} i = 1 + 2 + 3 + 4 + 5 = 15$$

$$\sum_{i=1}^{n} i = 1 + 2 + 3 + \cdots + n = \frac{n(n + 1)}{2}$$

$$\sum_{i=1}^{n} x_i = x_1 + x_2 + x_3 + \cdots + x_n$$

The values at the bottom and top of the summation symbol tell us where to start and stop. In the first example, we start with 1, since $i = 1$, substitute 2, 3, and 4, and end with 5, the number on the top of the summation symbol.

Using the summation symbol we can write the area in a more compact form.

$$A = \lim_{n \to \infty} \sum_{i=1}^{n} f(x_i) \, \Delta x_i$$

A set of guidelines that help to determine the area between two curves are given in the following box.

GUIDELINES TO DETERMINE THE AREA BETWEEN TWO CURVES

1. First sketch the figure and label the key parts to identify which region contains the area to be found.
2. Draw a sample rectangle in the region; this helps to set up the correct integral.
3. Determine the area of the sample rectangle using:

$$A_i = f(x_i)\, \Delta x_i.$$

4. The area of the region is the limit of the sum of the areas of the rectangles, which is equal to the definite integral. That is,

$$A = \lim_{n \to \infty} \sum_{i=1}^{n} f(x_i)\, \Delta x_i = \int_{x=a}^{x=b} f(x)\, dx.$$

EXAMPLE 1   **Area of a Region Above the x-Axis**

Determine the area of the region bounded by the curves $f(x) = -x^2 + 5$, $y = 0$, $x = -1$ and $x = 2$.

Solution   The curves are sketched in Figure 22.2, the desired region is shaded, and a sample rectangle is indicated on the figure. To determine the area of the sample rectangle, we multiply the length $f(x_i)$ times the width of the base Δx_i. Thus, the area of the sample rectangle is:

$$A_i = [-(x_i)^2 + 5]\, \Delta x_i.$$

The area is the limit of the sum of the areas of all the rectangles.

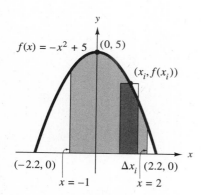

$f(x) = -x^2 + 5$ (0, 5)

$(x_i, f(x_i))$

$(-2.2, 0)$ Δx_i $(2.2, 0)$

$x = -1$ $x = 2$

FIGURE 22.2

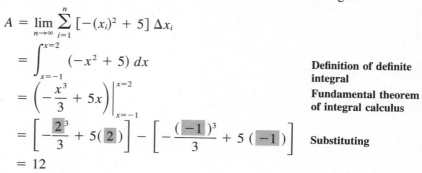

$$A = \lim_{n \to \infty} \sum_{i=1}^{n} [-(x_i)^2 + 5]\, \Delta x_i$$

$$= \int_{x=-1}^{x=2} (-x^2 + 5)\, dx \qquad \text{**Definition of definite integral**}$$

$$= \left(-\frac{x^3}{3} + 5x \right) \Bigg|_{x=-1}^{x=2} \qquad \text{**Fundamental theorem of integral calculus**}$$

$$= \left[-\frac{2^3}{3} + 5(2) \right] - \left[-\frac{(-1)^3}{3} + 5(-1) \right] \qquad \text{**Substituting**}$$

$$= 12$$

The area of the region bounded by the curves $f(x) = -x^2 + 5$, $y = 0$, $x = -1$, and $x = 2$ is 12 square units.   ◆

If we are not careful when determining the area of a region that lies below the x-axis, the result may be negative. Example 2 illustrates how to avoid this pitfall, since area cannot be negative.

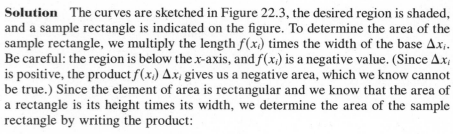

EXAMPLE 2   Area of a Region Below the x-Axis

Determine the area of the region bounded by the curves $f(x) = x^2 - 9$, $y = 0$, $x = -2$ and $x = 3$.

Solution   The curves are sketched in Figure 22.3, the desired region is shaded, and a sample rectangle is indicated on the figure. To determine the area of the sample rectangle, we multiply the length $f(x_i)$ times the width of the base Δx_i. Be careful: the region is below the x-axis, and $f(x_i)$ is a negative value. (Since Δx_i is positive, the product $f(x_i)\, \Delta x_i$ gives us a negative area, which we know cannot be true.) Since the element of area is rectangular and we know that the area of a rectangle is its height times its width, we determine the area of the sample rectangle by writing the product:

$$(y_{\text{top}} - y_{\text{bottom}})(x_{\text{right}} - x_{\text{left}}).$$

In this case, y_{top} is on the line $y = 0$ (x-axis) for every rectangle, and y_{bottom} is on the curve $f(x) = x^2 - 9$. Therefore,

$$y_{\text{top}} - y_{\text{bottom}} = 0 - [(x_i)^2 - 9].$$

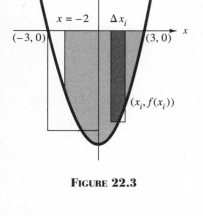

FIGURE 22.3

The width of the rectangle is:

$$x_{\text{right}} - x_{\text{left}} = \Delta x_i$$

$$A_i = \{0 - [(x_i)^2 - 9]\}\, \Delta x_i \qquad \text{Area of sample rectangle}$$

$$A = \lim_{n \to \infty} \sum_{i=1}^{n} -[(x_i)^2 - 9]\, \Delta x_i$$

$$= \int_{x=-2}^{x=3} -(x^2 - 9)\, dx \qquad \text{Definition of definite integral}$$

$$= -\left(\frac{x^3}{3} - 9x\right)\Bigg|_{x=-2}^{x=3} \qquad \text{Fundamental theorem of integral calculus}$$

$$= -\left[\left(\frac{(3)^3}{3} - 9(3)\right) - \left(\frac{(-2)^3}{3} - 9(-2)\right)\right] \qquad \text{Substituting}$$

$$= \frac{100}{3}.$$

The area of the region bounded by the curves $f(x) = x^2 - 9$, $y = 0$, $x = -2$ and $x = 3$ is $\dfrac{100}{3}$ square units.   ◆

In Examples 1 and 2, the entire region for which we determined the area was either above or below the x-axis. Example 3 requires that we determine the area of a region that is partially above and partially below the x-axis.

EXAMPLE 3   Area of Two Regions

Determine the area of the region bounded by the curves $f(x) = x^3 - x^2 - 2x$, $y = 0$, $x = -1$, and $x = 2$.

Solution   The curves are sketched in Figure 22.4. From the sketch we can see that part of the region is above the x-axis and part is below the x-axis. This is a special situation; if we treat the region as a single area, in summing from left to right, the sum of the positive area above the x-axis and the negative area below the x-axis are not equal to the total area. To prevent this from happening, we divide the region into two areas A_1 and A_2. The sum of A_1 and A_2 are equal to the total area A.

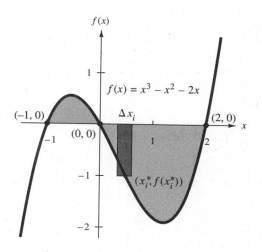

FIGURE 22.4

First we determine the area of A_1 $(-1 \le x \le 0)$.

$$A_i = f(x_i)\,\Delta x_i \qquad \text{Area of sample rectangle}$$

$$A_1 = \lim_{n \to 8} \sum_{i=1}^{n} [(x_i)^3 - (x_i)^2 - 2(x_i)]\,\Delta x_i$$

$$= \int_{x=-1}^{x=0} (x^3 - x^2 - 2x)\,dx \qquad \begin{array}{l}\textbf{Definition of} \\ \textbf{definite integral}\end{array}$$

$$= \left(\frac{x^4}{4} - \frac{x^3}{3} - \frac{2x^2}{2}\right)\Bigg]_{x=-1}^{x=0} \qquad \textbf{Fundamental theorem}$$

$$= 0 - \left[\frac{(\boxed{-1})^4}{4} - \frac{(\boxed{-1})^3}{3} - (\boxed{-1})^2\right] \qquad \textbf{Substituting}$$

$$= \frac{5}{12}$$

Then we determine the area of A_2 ($0 \le x \le 2$). Since the area of A_2 is below the x-axis, we use the same technique as in Example 2.

$$A_i = 0 - f(x_i)\,\Delta x_i \qquad \text{Area of sample rectangle}$$

$$A_2 = \lim_{n\to\infty} \sum_{i=1}^{n} -[(x_i)^3 - (x_i)^2 - 2x_i]\,\Delta x_i$$

$$= \int_{x=0}^{x=2} -(x^3 - x^2 - 2x)\,dx$$

$$= -\left(\frac{x^4}{4} - \frac{x^3}{3} - \frac{2x^2}{2}\right)\Bigg]_{x=0}^{x=2} \qquad \text{Fundamental theorem}$$

$$= -\left[\left(\frac{2^4}{4} - \frac{2^3}{3} - 2^2\right) - 0\right] \qquad \text{Substituting}$$

$$= \frac{8}{3}$$

The total area of the region $A = A_1 + A_2$.

$$A = \frac{5}{12} + \frac{8}{3}$$

$$= \frac{37}{12} \quad \blacklozenge$$

The process of finding the area between two curves is similar to the procedure used in Examples 2 and 3. Let us consider the region bounded by $y = f(x)$, $y = g(x)$, and the lines $x = a$ and $x = b$. The diagrams in Figures 22.5–22.7 depict various cases. Figure 22.5 shows a case where both curves are above the x-axis. Figure 22.6 is a case in which one curve is above and the other is below the x-axis. Figure 22.7 shows a case where both curves are below the x-axis. The sample rectangle is drawn on each figure. The heights of the rectangles are $f(x_i) - g(x_i)$ in each case. Also, in each case, the height of the sample strip is positive. (Therefore, when it is multiplied by the width Δx_i, the product is positive. If we summed these areas and found the limit of the sums for each case, the result is the integral stated under each figure.

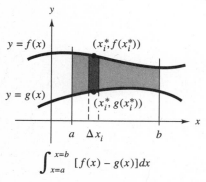

$$\int_{x=a}^{x=b} [f(x) - g(x)]dx$$

FIGURE 22.5

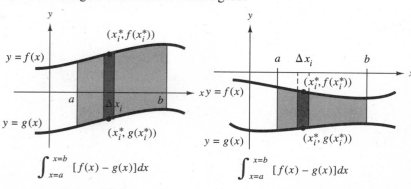

$$\int_{x=a}^{x=b} [f(x) - g(x)]dx$$

FIGURE 22.6

$$\int_{x=a}^{x=b} [f(x) - g(x)]dx$$

FIGURE 22.7

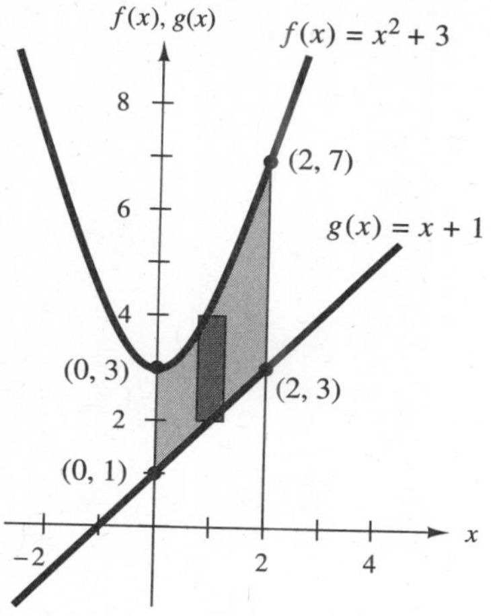

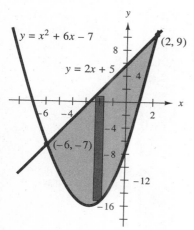

FIGURE 22.8

EXAMPLE 4 **Area of Region Between Two Curves**

Determine the area of the region bounded by $f(x) = x^2 + 3$, $g(x) = x + 1$, $x = 0$, and $x = 2$.

Solution The figure is sketched in Figure 22.8. From the figure we see that $f(x) \geq g(x) > 0$ for all x in the interval $0 \leq x \leq 2$. This is similar to the case shown in Figure 22.5, so we can use the integral stated at the bottom of that figure.

$$\begin{aligned}
A &= \int_{x=a}^{x=b} [f(x) - g(x)]\,dx \\
&= \int_{x=0}^{x=2} [(x^2 + 3) - (x + 1)]\,dx \\
&= \int_{x=0}^{x=2} (x^2 - x + 2)\,dx \\
&= \left(\frac{x^3}{3} - \frac{x^2}{2} + 2x\right)\Bigg]_0^2 \\
&= \left(\frac{2^3}{3} - \frac{2^2}{2} + 2(2)\right) - 0 \quad \text{or} \quad \frac{14}{3}.
\end{aligned}$$

The area of the region between the two curves is $\frac{14}{3}$ square units. ◆

EXAMPLE 5 **Area of Region Between Two Curves**

Determine the area of the region bounded by $y = 2x + 5$ and $y = x^2 + 6x - 7$.

FIGURE 22.9

Solution In order to determine a and b, the limits of integration, we must find the points where the two curves intersect. Since both expressions in terms of x are equal to y we can set them equal to each other and solve for x.

$$\begin{aligned}
2x + 5 &= x^2 + 6x - 7 \\
0 &= x^2 + 4x - 12 \\
0 &= (x + 6)(x - 2) \\
x &= -6 \quad \text{or} \quad 2.
\end{aligned}$$

Substituting in either equation, we find the points of intersection to be $(-6, -7)$ and $(2, 9)$. Thus, the limits of integration are -6 and 2. The curves are sketched in Figure 22.9. Since $2x + 5 \geq x^2 + 6x - 7$ for the $[-6, 2]$, the area can be determined with the integral shown in Figure 22.6.

$$A = \int_{x=a}^{x=b} [f(x) - g(x)] \, dx$$

$$A = \int_{x=-6}^{x=2} [(\,2x + 5\,) - (\,x^2 + 6x - 7\,)] \, dx$$

$$= \int_{x=-6}^{x=2} (-x^2 - 4x + 12) \, dx$$

$$= \left. \left(\frac{-x^3}{3} - \frac{4x^2}{2} + 12x \right) \right]_{x=-6}^{x=2}$$

$$= \left[\frac{-(\,2\,)^3}{3} - \frac{4(\,2\,)^2}{2} + 12(\,2\,) \right] - \left[\frac{-(\,-6\,)^3}{3} - \frac{4(\,-6\,)^2}{2} + 12(\,-6\,) \right]$$

$$= \frac{256}{3} \text{ square units.} \quad \blacklozenge$$

There are situations where it is easier to find the area of the region between two curves if we take a horizontal sample strip rather than a vertical sample strip. With a horizontal sample strip, we sum the rectangle in the direction of the y-axis. A sketch usually suggests whether to sum in the direction of the x-axis or in the direction of the y-axis.

EXAMPLE 6 **Horizontal Sample Strip**

Determine the area of the region bounded by $y^2 = x + 4$ and $y = x - 2$.

Solution Solving the two equations simultaneously we find that the points of intersection are $(5, 3)$ and $(0, -2)$. If we take a sample strip as shown in Figure 22.10 to the left of $x = 0$, the sample strip has both ends on the curve $y^2 = x + 4$. To the right of $x = 0$, the sample strip has the upper end on the curve $y^2 = x + 4$ and the lower end on the curve $y = x - 2$. It is not possible to find the area of the region with a single integral by using vertical strips. We would have to divide the region into two parts and find the area of each part separately.

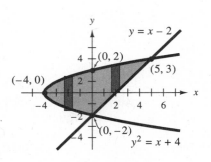

FIGURE 22.10

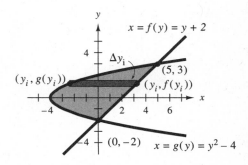

FIGURE 22.11

On the other hand, if we take a horizontal strip as shown in Figure 22.11, the right end of the strip is on the curve $f(y) = y + 2$ and the left end is on the curve $g(y) = y^2 - 4$. This is true regardless of where in the region we take the strip. The area of the strip is

$$A_i = [f(y_i) - g(y_i)]\, \Delta y_i.$$

We then would sum the strips from $y = -2$ to $y = 3$, so these values are the limits of integration.

$$A = \int_{y=-2}^{y=3} [(y + 2) - (y^2 - 4)]\, dy \quad \text{Substituting}$$

$$= \int_{y=-2}^{y=3} (-y^2 + y + 6)\, dy$$

$$= \left(\frac{-y^3}{3} + \frac{y^2}{2} + 6y\right)\Bigg]_{y=-2}^{y=3}$$

$$= \left[\frac{-(3)^3}{3} + \frac{3^2}{2} + 6(3)\right] - \left[\frac{-(-2)^3}{3} + \frac{(-2)^2}{2} + 6(-2)\right]$$

$$\text{Substituting}$$

$$= \frac{125}{6} \text{ square units.} \quad \blacklozenge$$

Now determine the area of the region in Example 6 by dividing the region into two areas and summing vertical strips.

22.1 Exercises

In Exercises 1–22, determine the area of the region bounded by the given curves.

1. $y = -x^2 + 25,\ y = 0$
2. $y = -x^2 + 16,\ y = 0$
3. $y = -x^2 + 3x + 4,\ y = 0$
4. $y = -x^2 - x + 12,\ y = 0$
5. $y = x^2 - 4,\ y = 0$
6. $y = x^2 - 9,\ y = 0$
7. $y = x^2 - 2x - 8,\ y = 0$
8. $y = x^2 + 3x - 10,\ y = 0$
9. $y = x^2 + 4,\ x = -1,\ x = 2,\ y = 0$
10. $y = x^2 + 3,\ x = 0,\ x = 2,\ y = 0$
11. $y = x^3 + x^2 - 2x,\ y = 0$
12. $y = x^3 - x^2 - 6x,\ y = 0$
13. $y = x^3 + x^2 - 20x,\ y = 0$
14. $y = x^3 - 2x^2 - 35x,\ y = 0$
15. $y = x^3 - 25x,\ y = 0$
16. $y = x^3 - 36x,\ y = 0$
17. $y = x + 4,\ y = 1,\ x = -1,\ x = 2$
18. $y = -x + 5,\ y = 2,\ x = 1$
19. $4y = 3x - 12,\ y = 2,\ x = 0$
20. $y = x - 6,\ y = 2,\ x = 1$
21. $y = \sqrt{x},\ y = 0,\ x = 4$
22. $y = \sqrt{3x},\ y = 0,\ x = 3$

23. Determine the area of the regions bounded by the curve $y = x^3$ and the lines $y = 0$, $x = -1$, and $x = 2$.

24. Determine the area of the regions bounded by the curve $y = x^3 + 2$ and the lines $y = 0$, $x = -1$, and $x = 2$.

25. Determine the area of the region bounded between the two curves $y = x^2 - 3$ and $y = -x^2 + 5$.

26. Determine the area of the region bounded between the two curves $y = x^2 - 6$ and $y = -x^2 + 12$.

27. Determine the area of the region bounded between the two curves $y^2 = x + 4$ and $y = x - 2$.

28. Determine the area of the region bounded between the two curves $y^2 = -x + 4$ and $y = -x - 2$.

29. Determine the area of the region bounded by the curves $f(x) = x^2 - 4x + 3$ and $g(x) = -x^2 + 2x + 3$.

30. Determine the area of the regions bounded by the curves $f(x) = (x - 1)^3$ and $g(x) = x - 1$.

31. The total amount of solar radiation on a certain surface is given by a rate equation $\dfrac{dE}{dt} = 3600(12t^2 - 0.90t^3)$, where E is the energy in joules and t is the number of hours the sun shines on the surface.
    a. What is the total amount of energy received in 12 hr?
    b. What is the amount of energy received in the first hour?
    c. What is the amount of energy received during the fourth hour?

32. A parabolic glass window is designed to withstand pressures of 15,000 psi. The length of the base of the window is 1.90 m and the height is 0.90 m, as shown in the figure. The thickness of the glass is 0.30 m.
    a. What is the equation of the parabola?
    b. What is the surface area of the glass in the window?
    c. What is the volume of the glass in the window?
    d. What is the mass of the glass needed for the window if the density of glass is 2.2 g/cm³?

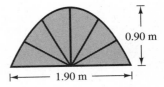

33. A parabolic glass window is designed to withstand pressures of 15,000 psi. The length of the base of the window is 1.20 m and the height is 0.90 m as shown in the figure. The thickness of the glass is 0.20 m.
    a. What is the equation of the parabola?
    b. What is the surface area of the glass in the window?
    c. What is the volume of the glass in the window?
    d. What is the mass of the glass needed for the window if the density of glass is 2.2 g/cm³?

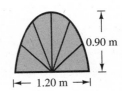

22.2

Volumes of Revolution: Disk and Washer Methods

In Section 22.1 we discussed the techniques of determining the area of a region bounded by two or more curves. In this section, a three-dimensional figure is created by rotating a region about either the x-axis, y-axis, or a line. Having created the figure, we then discuss the techniques of determining the volume of the figure.

Let f be a nonnegative, continuous function on the closed interval $a \le x \le b$. Let R be the region bounded by the function, the x-axis, and the lines $x = a$ and $x = b$ (Figure 22.12). As we rotate the figure about the x-axis, the region R generates a solid (see Figure 22.13). A cross-sectional slice of the figure is a circular disk, whose volume is pi times radius squared times thickness.

$$\text{Volume of disk} = \pi(\text{radius})^2(\text{thickness})$$

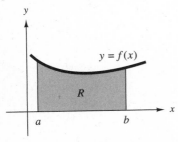

FIGURE 22.12

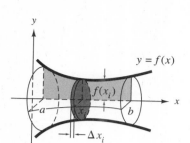

FIGURE 22.13

If we call the thickness of the disk Δx_i and the point where the disk cuts the x-axis x_i, then the radius of the disk is $f(x_i)$, and the volume of the circular disk is:

$$V_i = \pi [f(x_i)]^2 \, \Delta x_i.$$

If we sum all the circular disks from $x = a$ to $x = b$ and take the limit of the sum, the result is the volume of the figure.

> **Disk Method (Rotation About the x-Axis)**
>
> $$V = \pi \int_{x=a}^{x=b} [f(x)]^2 \, dx$$

The formula provides us with a general technique for finding the volume of a solid of revolution. To find the volume of a solid of revolution by the disk method, draw a typical slice, construct a disk, and obtain the desired integral from the disk.

EXAMPLE 1   Volume of a Solid: Disk Method

Determine the volume of a solid generated when the region under the curve $y = \sqrt{x} + 1$ within the interval $0 \le x \le 3$ is revolved about the x-axis.

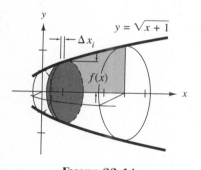

FIGURE 22.14

Solution   The sketch of the figure shown in Figure 22.14 contains a sample disk. The thickness of the ith disk is Δx_i, and the radius of the disk is $f(x_i)$, or $\sqrt{x_i} + 1$. Thus, the volume of the ith disk is:

$$V_i = \pi [\sqrt{x_i} + 1]^2 \, \Delta x_i.$$

Knowing that the figure is bounded in the interval $0 \le x \le 3$, we sum the sample disks from $x = 0$ to $x = 3$. Using this information, we can write the formula for determining the volume of the figure.

$$V = \pi \int_{x=0}^{x=3} (\sqrt{x} + 1)^2 \, dx$$

$$= \pi \int_{x=0}^{x=3} (x + 1) \, dx \qquad (\sqrt{x})^2 = x$$

$$= \pi \left(\frac{x^2}{2} + x \right) \Bigg]_{x=0}^{x=3}$$

$$= \pi \left[\left(\frac{3^2}{2} + 3 \right) - 0 \right] \qquad \textbf{Fundamental theorem}$$

$$= \frac{15}{2} \pi.$$

The volume of the solid of revolution is $\dfrac{15}{2} \pi$ cubic units.   ◆

There are situations when we must rotate the region about the y-axis rather than about the x-axis. The technique of determining the volume of the solid formed by rotating about the y-axis is discussed in Example 2.

EXAMPLE 2   Volume of a Solid: Disk Method

Determine the volume of the solid generated when the right half of the circle $x^2 + y^2 = r^2$ is revolved about the y-axis.

Solution   If we revolve the right half of a circle about the y-axis, the result is a sphere. The equation for the right half of the circle is $x = \sqrt{r^2 - y^2}$. A sample disk, as shown in Figure 22.15, has a radius x and thickness Δy. Thus, the volume of the ith disk is:

$$V_i = \pi[x_i]^2 \, \Delta y_i.$$

Since the radius of the sphere is r, we sum the sample disks from $y = -r$ to $y = r$. The volume can be expressed as:

$$V = \pi \int_{y=-r}^{y=r} (\sqrt{r^2 - y^2})^2 \, dy$$

$$= \pi \int_{y=-r}^{y=r} (r^2 - y^2) \, dy$$

$$= \pi \left(r^2 y - \frac{y^3}{3} \right) \Big]_{y=-r}^{y=r}$$

$$= \pi \left[\left(r^2(r) - \frac{r^3}{3} \right) \cdot \left(r^2(-r) - \frac{(-r)^3}{3} \right) \right]$$

$$= \left(r^3 - \frac{r^3}{3} \right) - \left(-r^3 + \frac{r^3}{3} \right)$$

$$= \frac{4}{3} \pi r^3.$$

The volume of the solid is $\frac{4}{3} \pi r^3$ cubic units. The formula for determining the volume of a sphere with radius r is, $V = \frac{4}{3} \pi r^3$. Thus, our result is correct.   ◆

In Example 2, we can obtain the same result by rotating the upper half of the circle about the x-axis. Try it and show that the same result is obtained.

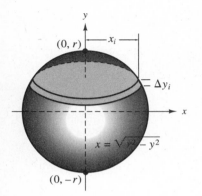

FIGURE 22.15

Disk Method (Rotation About the y-Axis)

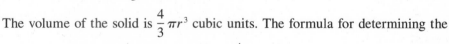

$$V = \pi \int_{y=a}^{y=b} [f(y)]^2 \, dy$$

Let $f(x)$ and $g(x)$ be two continuous curves such that $f(x) \geq g(x)$ for $a \leq x \leq b$. R is the region enclosed by the two curves and the lines $x = a$ and $x = b$ (Figure 22.16). When this region R is revolved about the x-axis, it gener-

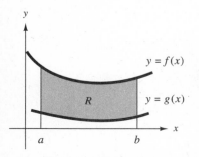

FIGURE 22.16

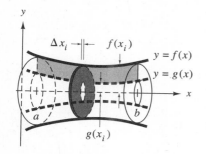

FIGURE 22.17

ates a solid that has a hole in the center. A cross-sectional slice of the figure is a washer with inner radius $g(x_i)$, outer radius $f(x_i)$, and thickness Δx_i as shown in Figure 22.17. The volume of the sample slice is determined by finding the total volume of the slice and then subtracting the volume of the hole in the center. That is,

$$V_i = \pi[f(x_i)]^2 \, \Delta x_i - \pi[g(x_i)]^2 \, \Delta x_i.$$

Factoring the common factors we can write the expression as:

$$y_i = \pi\{[f(x_i)]^2 - [g(x_i)]^2\} \, \Delta x_i.$$

If we sum all the sample slices (washers) from $x = a$ to $x = b$ and take the limit of the sum, we can obtain the volume. This technique of finding the volume is called the washer method.

Washer Method (Rotation About the x-Axis)

$$V = \pi \int_{x=a}^{x=b} \{[f(x)]^2 - [g(x)]^2\} \, dx$$

EXAMPLE 3   Volume of a Solid: Washer Method

Determine the volume of the solid that results when the region bounded by the curves $f(x) = x^2 + 2$, $g(x) = x + 1$, and the lines $x = 0$ and $x = 2$ are revolved about the x-axis.

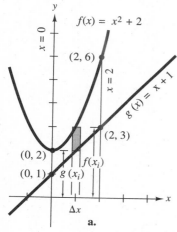

a.

Solution   The curves are sketched in Figure 22.18a with a sample strip indicated. From the sketch we see that when the sample strip is revolved about the x-axis, the result is a washer as shown in Figure 22.18b. The radius of the outer edge of the washer is $f(x) = x^2 + 2$, and the radius of the inner edge of the washer is $g(x) = x + 1$. We sum these sample strips from $x = 0$ to $x = 2$. Substituting in the volume formula, we have:

$$V = \pi \int_{x=a}^{x=b} \{[f(x)]^2 - [g(x)]^2\} \, dx$$

$$= \pi \int_{x=0}^{x=2} \{[x^2 + 2]^2 - [x + 1]^2\} \, dx$$

$$= \pi \int_{x=0}^{x=2} (x^4 + 3x^2 - 2x + 3) \, dx \quad \textbf{Squaring and combining like terms}$$

$$= \pi\left(\frac{x^5}{5} + x^3 - x^2 + 3x\right)\Bigg]_{x=0}^{x=2}$$

$$= \pi\left[\left(\frac{2^5}{5} + 2^3 - 2^2 + 3(2)\right) - 0\right]$$

$$= \frac{82}{5}\pi$$

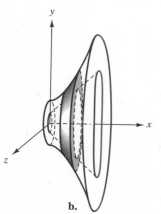

b.

FIGURE 22.18

The volume of the solid formed by revolving the region about the x-axis is $\dfrac{82}{5}\pi$ cubic units. ◆

To determine the volume of a region that is rotated about the y-axis the integral is expressed in terms of the variable y.

> **Washer Method (Rotation About the y-Axis)**
>
> $$V = \pi \int_{y=a}^{y=b} \{[f(y)]^2 - [g(y)]^2\}\, dy$$

Rotation about the y-axis is illustrated in Example 4.

EXAMPLE 4 Volume of a Solid: Washer Method

Determine the volume of the solid that is generated when the region bounded by the curve $y = \sqrt{x}$, and the lines $y = 0$, and $x = 4$ is revolved about the y-axis.

Solution Since the region is revolved about the y-axis, we will draw a sample strip parallel to the x-axis as shown in Figure 22.19a. The solid formed by the sample strip as it revolves about the y-axis is a washer as shown in Figure 22.19b. When the region is rotated about the y-axis, we must express x in terms of y. That is, for $y = \sqrt{x}$, we must square both sides and solve for x.

$$x = y^2$$

The length of the sample strip is $4 - x_i$ and the volume of a typical sample strip is:

$$V_i = \pi[(4)^2 - (x_i)^2]\,\Delta y_i.$$

Summing the sample strips from $y = 0$ to $y = 2$ and finding the limit of the sum, we have:

$$V = \pi \int_{y=0}^{y=2} [4^2 - (y^2)^2]\, dy \qquad \text{Substituting } x = y^2$$

$$= \pi \int_{y=0}^{y=2} (16 - y^4)\, dy$$

$$= \pi \left(16y - \frac{y^5}{5}\right)\Bigg]_{y=0}^{y=2}$$

$$= \pi\left[\left(16(2) - \frac{(2)^5}{5}\right) - 0\right]$$

$$= \frac{128\pi}{5}.$$

The volume of the solid formed by revolving the region about the y-axis is $\dfrac{128\pi}{5}$ cubic units. ◆

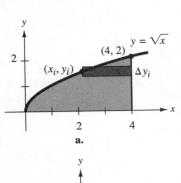

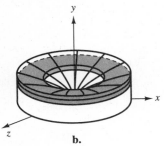

FIGURE 22.19

22.2   Exercises

In Exercises 1–12, determine the volume of the solid formed by revolving the given region about the x-axis.

1. $y = -x + 2$, x-axis, y-axis

2. $y = 9 - x^2$, x-axis, y-axis

3. $y = \sqrt{9 - x^2}$, x-axis, y-axis

4. $y = x^3$, $x = 2$, x-axis

5. $y = x^{\frac{3}{2}}$, x-axis, $x = 4$

6. $y = \sqrt{a^2 - x^2}$, x-axis

7. $y = x^2 + 1$, x-axis, $-2 \le x \le 2$

8. $y = 1 - \dfrac{x^2}{9}$, x-axis,

9. $y = x^2 + 3$, $y = 1$, $x = 0$, $x = 3$

10. $y = x^2$, $y = x^3$, $0 \le x \le 1$

11. $y = \dfrac{1}{x}$, $y = x$, $x = 4$

12. $y = x^{\frac{3}{2}}$, y-axis, $y = 4$, first quadrant

In Exercises 13–20, determine the volume of the solid formed by revolving the given region about the y-axis.

13. $y = x + 2$, y-axis, $y = 3$

14. $y = x^2$, y-axis, $y = 4$, first quadrant

15. $y = \sqrt{9 - x^2}$

16. $y = x^{\frac{1}{3}}$, y-axis, $y = 2$

17. $y^2 = -4x + 4$, y-axis

18. $x = -y^2 + 6y$, y-axis

19. $y = x$, $x + y = 3$, y-axis

20. $x = y^2$, $x + y = 2$

21. Use the disk method to verify that the volume of a right circular cone is $\dfrac{1}{3}\pi r^2 h$, where r is the radius of the base and h is the height.

22. Determine the volume of a prolate spheroid (football) by revolving the upper half of the ellipse $9x^2 + 16y^2 = 144$ about the x-axis.

23. Determine the volume of an oblate spheroid by revolving the right half of the ellipse $9x^2 + 16y^2 = 144$ about the y-axis.

24. A silversmith is making a mold for the base of a candelabra. The shape of the mold is determined by rotating the region determined by the curve $y = x^3 - 5x^2 + 5x + 4$ and the lines $x = 0$, and $x = 4$ about the x-axis. Each unit is one inch.
   a. Determine the volume of the mold.
   b. Determine the number of ounces of silver needed to make a solid candelabra if one ounce of silver weighs 0.9116 troy ounce.
   c. Determine the cost of the candelabra if one ounce of silver costs \$5.43 per troy ounce. (Weight of silver is 0.378 lb/in.³.)

22.3

Volumes of Revolution: Cylindrical-Shell Method

In the previous section, we used the disk and washer methods for finding volumes by constructing the sample strip perpendicular to the axis of rotation. When using these two methods, we may encounter cases when it is difficult to set up an integral and/or impossible to evaluate the integral. An alternative technique to the disk and washer methods is the *cylindrical-shell method*, in which we take the sample slice parallel to the axis of rotation.

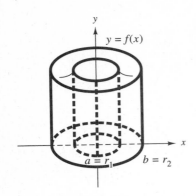

FIGURE 22.20

To develop this technique let $y = f(x)$ be a continuous function in the interval $a \le x \le b$. Now, if we rotate this region about the y-axis, the solid formed is a cylinder as shown in Figure 22.20. The volume of the cylinder is found by multiplying the cross-sectional area (area of the base) times the height.

$$\text{Volume of cylinder} = (\text{Cross-sectional area})(\text{Height})$$

The cross-sectional area is composed of the area between two concentric circles.

$$\text{Cross-sectional area} = \text{Area of outer circle} - \text{Area of inner circle}$$
$$\text{Cross-sectional area} = \pi r_2^2 - \pi r_1^2$$

$$
\begin{aligned}
V &= [\pi r_2^2 - \pi r_1^2]h && \textbf{Substituting}\\
&= \pi(r_2 + r_1)(r_2 - r_1)h && \textbf{Factoring}\\
&= 2\pi\frac{(r_2 + r_1)}{2}\,h(r_2 - r_1) && \textbf{Multiplying by }\left(\frac{2}{2}\right) = 1
\end{aligned}
$$

$$V = 2\pi\,(\text{Average radius})(\text{Height})(\text{Thickness})$$

We can now use this equation to develop the method of cylindrical shells. In Figure 22.21, we have constructed a cross-sectional slice of the cylinder in the first quadrant. If we let x_i^* be the distance from the y-axis to the center of the sample strip, x_i^* is the average radius. Let $f(x_i^*)$ be the height and Δx_i the width of the sample strip. Then the volume of the ith sample strip as it is rotated about the y-axis is:

$$V_i = 2\pi x_i^* f(x_i^*)\,\Delta x_i.$$

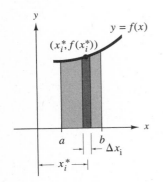

FIGURE 22.21

If we now sum the volumes of all the cylinders formed by the sample strips in $a \le x \le b$ and take the limit of the sum, the result is a definite integral.

Cylindrical-Shell Method (Rotation About the y-Axis)

$$V = 2\pi \int_{x=a}^{x=b} x f(x)\, dx$$

This integral applies when the sample strip is parallel to the y-axis. For cases where the sample strip is parallel to the x-axis, we use a similar method and conclude the following.

Cylindrical-Shell Method (Rotation About x-Axis)

$$V = 2\pi \int_{y=c}^{y=d} y f(y)\, dy$$

In either case, always sketch the figure, including a sample strip, to properly evaluate the problem and determine the direction of rotation (as mentioned in Section 22.2).

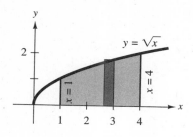

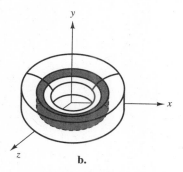

FIGURE 22.22

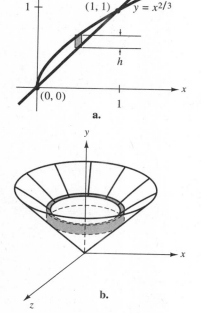

a.

b.

FIGURE 22.23

EXAMPLE 1 Rotation About y-Axis

Use the cylindrical-shell method to determine the volume of the solid generated when the region bounded by $y = \sqrt{x}$, $x = 1$, $x = 4$, and the x-axis is revolved about the y-axis.

Solution The region is sketched in Figure 22.22a, with a sample strip indicated. The height of the curve at any point is $y = f(x) = \sqrt{x}$. Figure 22.22b illustrates the results of the rotation, a three-dimensional figure. We are summing the strips from $x = 1$ to $x = 4$, so these values are the limits of integration.

$$V = 2\pi \int_{x=a}^{x=b} xf(x)\,dx$$

$$= 2\pi \int_{x=1}^{x=4} x(\sqrt{x})\,dx \quad \text{Substituting}$$

$$= 2\pi \int_{x=1}^{x=4} x^{\frac{3}{2}}\,dx$$

$$= 2\pi \left(\frac{x^{\frac{5}{2}}}{\frac{5}{2}}\right)\Bigg]_{x=1}^{x=4}$$

$$= \frac{4}{5}\pi[4^{\frac{5}{2}} - 1^{\frac{5}{2}}]$$

$$= \frac{124\pi}{5}.$$

The volume of the cylinder is $\dfrac{124\pi}{5}$ cubic units. ◆

EXAMPLE 2 Rotation About y-Axis

Use the cylindrical-shell method to determine the volume of the solid generated when the region in the first quadrant is bounded by the curves $y = x$, $y = x^{\frac{2}{3}}$ is revolved about the y-axis.

Solution A sketch of the region is given in Figure 22.23a indicating a sample strip and the points of intersection of the curves. The sample strip generates a cylinder of height $x^{\frac{2}{3}} - x$ and radius x, shown in Figure 22.23b. We sum these strips from $x = 0$ to $x = 1$. Thus, the area of the surface is:

$$2\pi x(x^{\frac{2}{3}} - x),$$

and the volume of the solid is:

$$V = 2\pi \int_{x=0}^{x=1} x(x^{\frac{2}{3}} - x)\, dx$$

$$= 2\pi \int_{x=0}^{x=1} (x^{\frac{5}{3}} - x^2)\, dx$$

$$= 2\pi \left[\frac{3}{8}x^{\frac{8}{3}} - \frac{x^3}{3}\right]_{x=0}^{x=1}$$

$$= 2\pi \left[\left(\frac{3}{8}(\boxed{1})^{\frac{8}{3}} - \frac{\boxed{1}^3}{3}\right) - 0\right]$$

$$= \frac{\pi}{12}.$$

The volume of the cylinder is $\dfrac{\pi}{12}$ cubic units. ◆

EXAMPLE 3   Rotation About the x-Axis

Use the cylindrical-shell method to determine the volume of the solid generated when the region in the first quadrant bounded by the curves $y = -x + 9$, $y = 2x$, and $y = 1$ is revolved about the x-axis.

Solution   A sketch of the region is given in Figure 22.24a, indicating a horizontal sample strip since we are rotating about the x-axis. Solving the equations of the lines for x, which determine the length of the sample strip, we have:

$$y = -x + 9 \quad \text{and} \quad y = 2x,$$

$$x = -y + 9 \quad \text{and} \quad x = \frac{y}{2}.$$

Thus, the sample strip generates a cylinder of height $(-y + 9) - \dfrac{y}{2}$ and an average radius of y (See Figure 22.24b). We sum these strips from $y = 1$ to $y = 6$, and these values are the limits of integration.

$$V = 2\pi \int_{y=c}^{y=d} y f(y)\, dy$$

$$= 2\pi \int_{y=1}^{y=6} y\left[(-y + 9) - \frac{y}{2}\right] dy \qquad \textbf{Substituting}$$

$$= 2\pi \int_{y=1}^{y=6} \left(\frac{-3y^2}{2} + 9y\right) dy \qquad \textbf{Combining like terms}$$

$$= 2\pi \left(\frac{-3y^3}{2 \cdot 3} + \frac{9y^2}{2}\right)\Bigg]_{y=1}^{y=6}$$

$$= 2\pi \left[\left(-\frac{1}{2}(\boxed{6})^3 + \frac{9(\boxed{6})^2}{2}\right) - \left(-\frac{1}{2}(\boxed{1})^3 + \frac{9(\boxed{1})^2}{2}\right)\right]$$

$$= 100\pi.$$

The volume of the cylinder is 100π cubic units. ◆

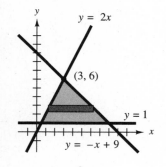

a.

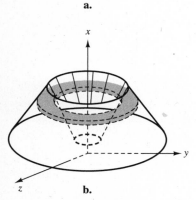

b.

FIGURE 22.24

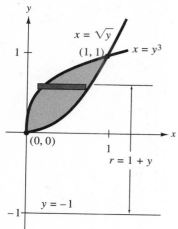

a.

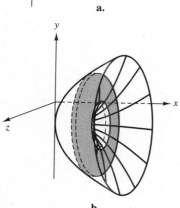

b.

FIGURE 22.25

The axis of rotation does not have to be the x-axis or y-axis; it can be any other line on the plane.

EXAMPLE 4   Rotation About the Line $y = -1$

Use the cylindrical-shell method to determine the volume of the solid generated when the region in the first quadrant bounded by the curves $x = \sqrt{y}$ and $x = y^3$ is revolved about the line $y = -1$.

Solution   A sample strip is shown in Figure 22.25a. Since the axis of rotation is $y = -1$, the radius is $1 + y$, the distance from the sample strip to the axis of rotation, as shown in Figure 22.25b. The height is $\sqrt{y} - y^3$, the length of the sample strip. The sample strips are summed from $y = 0$ to $y = 1$, so these are the limits of integration.

$$V = 2\pi \int_{y=0}^{y=1} (\,1 + y\,)(\,\sqrt{y} - y^3\,)\, dy$$

$$= 2\pi \int_{y=0}^{y=1} (y^{\frac{1}{2}} + y^{\frac{3}{2}} - y^3 - y^4)\, dy$$

$$= 2\pi \left[\frac{2y^{\frac{3}{2}}}{3} + \frac{2y^{\frac{5}{2}}}{5} - \frac{y^4}{4} - \frac{y^5}{5} \right]_{y=0}^{y=1}$$

$$= 2\pi \left[\frac{2(\,1\,)^{\frac{3}{2}}}{3} + \frac{2(\,1\,)^{\frac{5}{2}}}{5} - \frac{(\,1\,)^4}{4} - \frac{(\,1\,)^5}{5} \right]$$

$$= \frac{37\pi}{30}.$$

The volume of the cylinder is $\dfrac{37\pi}{30}$ cubic units.   ◆

22.3   **Exercises**

In Exercises 1–12, use the cylindrical-shell method to determine the solid generated when the region is revolved about the y-axis.

1. $y = x^2, x = 2, y = 0$

2. $y = \sqrt{x}, x = 4, x = 9, y = 0$

3. $y = 3x, y = -x + 8, y = 1$

4. $y = -x + 10, y = x + 2, y = 2$

5. $y = 4x - x^2, x$-axis

6. $y = x^2 - 2x, x$-axis

7. $y = x^2, y = 8 - x^2$

8. $y = x, y = 2x, x = 3$

9. $y = \sqrt{x^2 - 2}, x = 2, x = 5, x$-axis

10. $y = x^2, x = y^2$

11. $y = x^3 + 1, x + 2y = 2, x = 1$

12. $y = 4x - x^3, x$-axis

In Exercises 13–24, use the cylindrical-shell method to determine the volume of the solid generated when the region is revolved about the x-axis.

13. $x = y^2, x$-axis, $x = 4$

14. $x = \sqrt{y}, y$-axis, $y = 9$

15. $x = y, x + 2y = 3, x$-axis

16. $y = x - 2$, $2y = -x + 20$, x-axis   **17.** $y^3 = x$, $y = 3$, y-axis   **18.** $x = 9 - y^2$, y-axis

19. $y = x^2$, $x = 1$, x-axis   **20.** $xy = 4$, $x + y = 5$ **21.** $y = 2x$, $y = 6$, y-axis

22. $2y = x$, $y = 4$, $x = 1$ **23.** $y = 2x^2$, $y^2 = 4x$ **24.** $x = 2y - y^2$, y-axis

25. A round hole of radius $r = 40$ mm is bored through the center of a solid sphere of radius 160 mm. Find the total volume of the sawdust.

26. A round hole of radius 25 mm is bored through the center of a solid sphere radius 100 mm. Find the volume of the remaining part of the sphere.

27. The nose cone of a rocket is a paraboloid. A paraboloid is formed by rotating a parabolic curve about the y-axis. As shown in the diagram, the diameter and height of the paraboloid are 2.0 m and 3.0 m, respectively.
a. What is the equation of the parabola?
b. What is the volume of the paraboloid?

28. A solid lead bullet is designed to combine a paraboloid at the tip (to reduce air resistance) with a cylinder as shown in the diagram. The diameter and height of the cylinder are 0.60 cm and 0.90 cm respectively, and the height of the paraboloid is 1.20 cm.
a. Without using integration, determine the volume of the cylinder.
b. Using integration, determine the volume of the paraboloid.
c. The density of lead is 11 g/cm³. What is the mass of the bullet?

22.4

Centroids

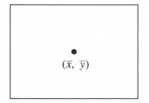

FIGURE 22.26

The **center of gravity** is a point within an object where we say all the mass is concentrated. Determining the placement of the center of gravity is an easy task for symmetrically shaped objects. For example, the center of gravity is in the center of a sphere or cube.

For a shape that has no volume, like a perfectly flat shape, it makes no sense to consider the center of gravity because the shape has no mass. However, the shape still possesses a geometric center called the **centroid.** For a flat piece of metal that has uniform density, the centroid is located at the same point as the center of gravity. The location of the centroid is the point $(\bar{x}, \bar{y})$, as shown in Figure 22.26.

When a body is suspended from its centroid no rotation occurs unless an external twist is applied. Actually, a centroid can be determined by using rotation. When a body rotates without an external twist, two conditions must be specified: the axis of rotation and the distance from the centroid to the axis of

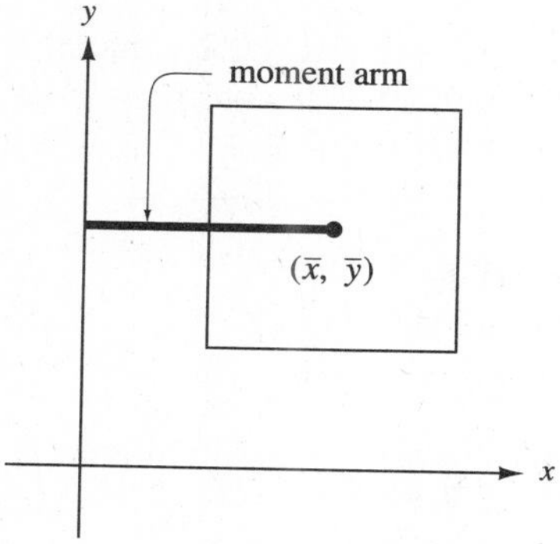

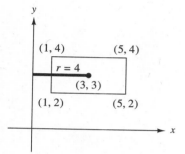

FIGURE 22.27

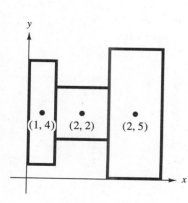

FIGURE 22.28

rotation. The **moment arm** is the perpendicular distance from the centroid of the rotating body to the axis of rotation (Figure 22.27). The **moment of area** is defined as:

$$\text{Moment of area} = (\text{Moment arm})(\text{Area}).$$

Using the definition of moment of area we can determine the centroid of complex shapes. First, however, in Example 1 we explain the procedure for relatively simple shapes.

EXAMPLE 1 Moment of Area

Determine the moment of area for a rectangular shape of uniform density that is rotated about the y-axis. Use the information provided in Figure 22.28.

Solution To determine the moment of area, we must determine the length of the moment arm r and the area of the shape. To determine r we first must know the location of the centroid of the rectangle, that is the point $(\bar{x}, \bar{y})$.

$$\bar{x} = \frac{1+5}{2} = 3$$

$$\bar{y} = \frac{2+4}{2} = 3$$

The centroid of the rectangle is $(3, 3)$. The length of the moment arm for rotation about the y-axis plus the distance from the edge of the rectangle to the centroid (3 units) is the distance from the y-axis to the edge of the rectangle (1 unit). Thus the length of the moment arm is $1 + 3$ or 4. The area of the rectangle is 2×4, or 8 square units.

$$\text{Moment of area} = (\text{Moment arm})(\text{Area})$$
$$= (\,4\,)(\,8\,)$$
$$= 32 \text{ cubic units} \quad \blacklozenge$$

EXAMPLE 2 Centroid of Complex Shape

Determine the centroid of the three rectangular shapes of uniform density that are placed together in the form of an H, as shown in Figure 22.29. The dimensions of the rectangles are 2×8, 4×4, and 4×10 (in centimeters).

Solution The problem is to determine the centroid of the combined shape. To do this, we must determine the total moment about the y-axis as well as the total area of the combined figure.

The total area is equal to the sum of the individual areas:

$$16 \text{ cm}^2 + 16 \text{ cm}^2 + 40 \text{ cm}^2 = 72 \text{ cm}^2.$$

FIGURE 22.29

Since we are dealing with rectangular shapes of uniform mass, the centroid of each is the center of the rectangle. We can see in Figure 22.29 that the centroids are (1, 4), (2, 2), (4, 5) and (2, 5), respectively, and the moment arms are 1, 4 and 8, respectively. Now we can compute the moment of area about the y-axis for each of the shapes.

$$M_{y1} = (\,1\text{ cm}\,)(\,16\text{ cm}^2\,) = 16\text{ cm}^3$$
$$M_{y2} = (\,4\text{ cm}\,)(\,16\text{ cm}^2\,) = 64\text{ cm}^3$$
$$M_{y3} = (\,8\text{ cm}\,)(\,40\text{ cm}^2\,) = 320\text{ cm}^3$$

The total moment of area about the y-axis is the sum of the individual moments.

$$M_y = 16\text{ cm}^3 + 64\text{ cm}^3 + 320\text{ cm}^3 = 400\text{ cm}^3$$

Now, we have already established the equation for moment of area:

$$M = rA,$$

where r is the moment arm and A is the area. So, to determine r for the y-axis, we take the total moment (M_y) and divide it by the total area.

$$r_y = \frac{M_y}{A}$$
$$= \frac{400\text{ cm}^3}{72\text{ cm}^2}$$
$$= 5.56\text{ cm}$$

In a similar manner, we can determine that r_x is 5.00 cm. Therefore, the centroid of the figure is at (5.00 cm, 5.56 cm). ◆

If we choose a differential element of area dA at point $P(x, y)$ in a plane object as in Figure 22.30, **we define its moment with respect to the x-axis as $dM_x = y\,dA$. Its moment with respect to the y-axis is $dM_y = x\,dA$.** In order to obtain the moment of the entire object, we must sum the moments of each part of the object. Therefore, we can say that moments of an area with respect to the x- and y-axes, respectively, are as follows.

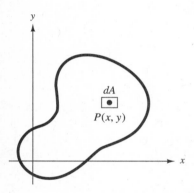

FIGURE 22.30

Moments of Area

$$M_x = \oint_A y\,dA \quad \text{and} \quad M_y = \oint_A x\,dA,$$

where x and y are the coordinates of the infinitesimal element of area dA and $\oint_A$ means the integral across the entire area.

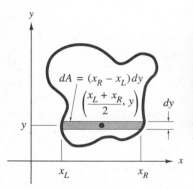

FIGURE 22.31

If the element dA has a finite dimension, then x and y are the coordinates of the centroid of that element. If we choose a rectangular element parallel to the x-axis as shown in Figure 22.31, we can see that the area dA of the element is

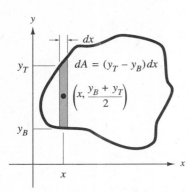

FIGURE 22.32

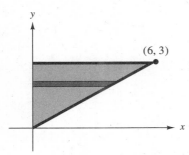

FIGURE 22.33

given by $dA = (x_R - x_L)\,dy$, where x_L and x_R are the x-coordinates of the left and right ends of the element. The centroid of the element is at its middle, which has coordinates $\left(\dfrac{x_L + x_R}{2}, y\right)$. If we instead choose an element parallel to the y-axis as shown in Figure 22.32, its area is given by $dA = (y_T - y_B)\,dx$, and its middle has coordinates $\left(x, \dfrac{y_B + y_T}{2}\right)$. As a result we find that it is often, but not always, easier to calculate $M_x = \oint_A y\,dA$ using an element of area dA parallel to the x-axis and that it is often, but not always, easier to calculate $M_x = \oint_A x\,dA$ using an element of area parallel to the y-axis.

EXAMPLE 3 **Moments for a Triangular Shape**

Determine the moments around the x- and y-axis for the triangular shape shown in Figure 22.33.

Solution Consider the horizontal sample strip that has an area dA, where $dA = x\,dy$. The value of x varies. So in order to integrate this expression, we must express x as a function of y. The line on which the strip ends has the general equation:

$$y = mx + b.$$

The slope of the line is $(3 - 0)/(6 - 0) = 0.5$, and the y-intercept is $(0, 0)$. Hence, the specific equation of the line is:

$$y = \frac{1}{2}x \quad \text{and} \quad x = 2y.$$

Returning to the expression for the area of the strip and substituting, we have:

$$dA = x\,dy = 2y\,dy.$$

Now find M_x, using the definition of moment with respect to the x-axis.

$$\begin{aligned}
M_x &= \oint_A y\,dA \\
&= \oint_A (y)(\,2y\, - \,0\,)\,dy \\
&= \int_0^3 2y^2\,dy \\
&= \left(\frac{2y^3}{3}\right)\Big]_0^3 \\
&= 18 \text{ units.}
\end{aligned}$$

To determine M_y we can use the same sample strip that we used to determine M_x. However, it is imperative to remember that we are imagining a rotation about the y-axis. Clearly, the moment arm is now x. For the particular strip we need the average value, not the whole value, for x. Recall that the moment arm is the distance to the center of the strip. The center of the strip lies between $x = 0$ and $x = 2y$. Therefore, the average is $\dfrac{0 + 2y}{2}$. This is the value of x that we use to find M_y.

$$
\begin{aligned}
M_y &= \oint_A x\, dA \\
&= \oint_A \left(\frac{0 + 2y}{2}\right)(2y - 0)\, dy \\
&= \int_0^3 2y^2\, dy \\
&= \left(\frac{2y^3}{3}\right)\Bigg]_0^3 \\
&= 18 \text{ units.}
\end{aligned}
$$

Therefore, we can find M_x and M_y by using an element either parallel to the designated axis or perpendicular to it. ◆

When we try to determine the centroid of a complex shape, we can rearrange the notions of moments.

If the coordinates of the centroid of an area are $(\bar{x}, \bar{y})$, we can write the moment of the area with respect to the x-axis as either:

$$M_x = \oint_A y\, dA,$$

or

$$M_x = \bar{y}A.$$

Therefore,

$$\bar{y}A = \oint_A y\, dA,$$

or

$$\bar{y} = \frac{\displaystyle\oint_A y\, dA}{A}.$$

Since we can calculate the moment of the area $\oint_A y\, dA$ and we can calculate the area A (using $A = \oint_A dA$ if necessary) we have obtained a method for determining the location of the centroid of an area.

> **DEFINITION 22.1**
> The coordinates of the centroid $(\bar{x}, \bar{y})$, of an area A are found from the following.
>
> $$\bar{x} = \frac{\oint_A x\,dA}{A} \quad \text{and} \quad \bar{y} = \frac{\oint_A y\,dA}{A}$$
>
> $$= \frac{M_y}{A} \qquad\qquad = \frac{M_x}{A}$$

Analogous expressions apply to volumes and weights.

EXAMPLE 4   **Coordinates of the Centroid**

What are the coordinates of the centroid of the triangle shown in Figure 22.34? (We determined the moments of the area of the triangle in Example 3 to be $M_x = 18$ and $M_y = 18$.)

Solution   The coordinates of the centroid of an area are given by the following.

$$\bar{x} = \frac{\oint_A x\,dA}{A} \quad \text{and} \quad \bar{y} = \frac{\oint_A y\,dA}{A}$$

$$= \frac{M_y}{A} \qquad\qquad = \frac{M_x}{A}$$

The given triangle has a base of 6 units and an altitude of 3 units, so its area is:

$$A = \frac{1}{2}\,bh$$

$$= \frac{1}{2}(\boxed{6})(\boxed{3})$$

$$= 9 \text{ units}^2.$$

$M_x = 18$ units^3 and $M_y = 18$ units^3 are given. Substitute these values in the formulas.

$$\bar{x} = \frac{M_y}{A} \qquad\qquad \bar{y} = \frac{M_x}{A}$$

$$= \frac{18 \text{ units}^3}{9 \text{ units}^2} \qquad\qquad = \frac{8 \text{ units}^3}{9 \text{ units}^2}$$

$$= 2 \text{ units} \qquad\qquad = 2 \text{ units}$$

The centroid of the given triangular area lies at coordinates $(\bar{x}, \bar{y}) = (2, 2)$ as shown in Figure 22.35. ◆

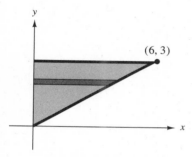

FIGURE 22.34

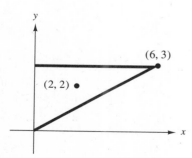

FIGURE 22.35

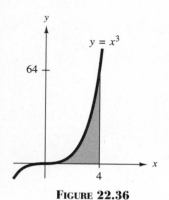

FIGURE 22.36

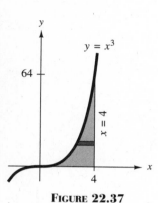

FIGURE 22.37

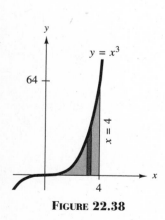

FIGURE 22.38

EXAMPLE 5 **Coordinates of the Centroid**

Determine the coordinates of the centroid of the area shown in Figure 22.36.

Solution To find the centroidal coordinates we need $M_x = \oint_A y\,dA$, $M_y = \oint_A x\,dA$, and the area $A = \oint_A dA$. Using an element parallel to the x-axis, as in Figure 22.37, we have the element of area $dA = (x_R - x_L)\,dy$. The right side of the element is on the line $x = 4$, so $x_R = 4$. The left side of the element is on the curve $y = x^3$, or solving for x as a function of y, $x = y^{\frac{1}{3}}$, so $x_L = y^{\frac{1}{3}}$. Therefore, $dA = (4 - y^{\frac{1}{3}})\,dy$, so:

$$
\begin{aligned}
A &= \oint_A dA \\
&= \int_0^{64} (4 - y^{\frac{1}{3}})\,dy \qquad \text{When } x = 4, y = (4)^3 \text{ or } 64 \\
&= \left(4y - \frac{3y^{\frac{4}{3}}}{4}\right)\Bigg]_0^{64} \\
&= 64.
\end{aligned}
$$

$$
\begin{aligned}
M_x &= \oint_A y\,dA \\
&= \int_0^{64} y(4 - y^{\frac{1}{3}})\,dy \\
&= \int_0^{64} (4y - y^{\frac{4}{3}})\,dy \\
&= \left(\frac{4y^2}{2} - \frac{3y^{\frac{7}{3}}}{7}\right)\Bigg]_0^{64} \\
&= 1170.
\end{aligned}
$$

To find M_y, we can use the same element and take the x-coordinate of the centroid of the element as $\dfrac{x_L + x_R}{2}$ or $\dfrac{y^{\frac{1}{3}} + 4}{2}$, so:

$$
\begin{aligned}
M_y &= \oint_A x\,dA \\
&= \int_0^{64} \frac{(y^{\frac{1}{3}} + 4)}{2}(4 - y^{\frac{1}{3}})\,dy.
\end{aligned}
$$

Alternatively, we can use an element parallel to the y-axis as shown in Figure 22.38. For this element, $dA = (y_T - y_B)\,dx$, so the top of the element $y_T = x^3$. The bottom of the element is on the x-axis, so $y_B = 0$. Therefore, $dA = (x^3 - 0)\,dx$. Since the area is the same no matter what element we use to calculate it, the value of the area is still $A = 64$. Therefore, we need not recalculate A using the new element. We need only to calculate M_y.

$$M_y = \oint_A x\,dA$$

$$= \int_0^4 x(x^3 - 0)\,dx$$

$$= \int_0^4 x^4\,dx$$

$$= \left(\frac{x^5}{5}\right)\Bigg]_0^4$$

$$= 204.8.$$

Therefore,

$$\overline{x} = \frac{M_y}{A} \qquad \overline{y} = \frac{M_x}{A}$$

$$= \frac{204.8}{64} \qquad\quad = \frac{1170}{64}$$

$$= 3.20, \qquad\quad = 18.3.$$

For the area given, the coordinates of the centroid are $(\overline{x}, \overline{y}) = (3.20, 18.3)$. Verify that using an element parallel to the x-axis to calculate M_y produces the same result.   ◆

EXAMPLE 6   Centroid of Volume of Revolution

Determine the centroid of the volume of revolution formed by rotating $y = x^4$ about the y-axis and bounding it by the plane $y = 3$ as shown in Figure 22.39.

Solution   We can see from symmetry that the centroid must lie on the y-axis, so we need only to calculate $\overline{y}$. Taking a circular disk for our element of volume dV, we have $dV = \pi r^2\,dy$, where $r = x$ is the radius of the disk. On the curve $y = x^4$, $x = y^{\frac{1}{4}}$, so:

$$dV = \pi r^2\,dy$$

$$= \pi x^2\,dy$$

$$= \pi(y^{\frac{1}{4}})^2\,dy$$

$$= \pi y^{\frac{1}{2}}\,dy.$$

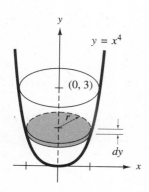

FIGURE 22.39

Therefore,

$$V = \oint_V dV$$

$$= \int_0^3 \pi y^{\frac{1}{2}} \, dy$$

$$= \pi \left(\frac{2 y^{\frac{3}{2}}}{3} \right) \Big]_0^3$$

$$= 10.88.$$

Now we determine the moment with respect to the x-axis.

$$M_x = \oint_V y \, dV$$

$$= \int_0^3 y \left(\pi y^{\frac{1}{2}} \right) dy$$

$$= \pi \int_0^3 y^{\frac{3}{2}} \, dy$$

$$= \pi \left(\frac{2 y^{\frac{5}{2}}}{5} \right) \Big]_0^3$$

$$= 19.59$$

To determine $\overline{y}$ we use the equation $\overline{y} = \dfrac{\oint_V y \, dy}{v}$ which can be written:

$$\overline{y} = \frac{M_x}{V}$$

$$= \frac{19.59}{10.88}$$

$$= 1.80. \quad \blacklozenge$$

EXAMPLE 7   **Application: Centroid of Fishing Lure**

Help the designer mentioned in the chapter opener determine the centroid of the fishing lure to the nearest tenth. Figure 22.40 provides the shape of the lure. All dimensions are in millimeters.

Solution   Since both curves are symmetric about the y-axis, $\overline{x} = 0$. To determine $\overline{y}$, we need M_x and the area of the lure. To determine the area we can divide the lure into two parts: A_1, the area below the x-axis, and A_2, the area above the x-axis. A_1 is the area of a semicircle of radius 10. Thus,

$$A_1 = \frac{\pi (10)^2}{2} = 50\pi \quad \text{or} \quad 157.08$$

$$A_2 = 2 \int_{x=0}^{x=10;} \left(-\frac{1}{4} x^2 + 25 \right) dx = 333.33.$$

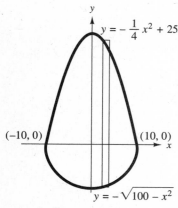

$y = -\frac{1}{4} x^2 + 25$

$(-10, 0)$

$(10, 0)$

$y = -\sqrt{100 - x^2}$

FIGURE 22.40

The total area is:

$$A = A_1 + A_2 = \boxed{157.08} + \boxed{333.33} = 490.41.$$

To determine M_x, we take a vertical strip and again divide the figure into two pieces, we then determine M_{x1}, moment of the part below the x-axis, and M_{x2}, the moment of the part above the x-axis.

$$M_{x1} = \int_{x=-10}^{x=10} \left(\frac{0 + (-\sqrt{100 - x^2})}{2} \right) (-\sqrt{100 - x^2})\, dx$$

$$= \frac{1}{2} \int_{x=-10}^{x=10} (100 - x^2)\, dx$$

$$= 666.67$$

$$M_{x2} = \int_{x=-10}^{x=10} \left(\frac{-0.25x^2 + 25}{2} \right) (-0.25x^2 + 25)\, dx$$

$$= \frac{1}{2} \int_{x=-10}^{x=10} (0.0625x^4 - 12.5x^2 + 625)\, dx$$

$$= 3333.33$$

$$M_x = \boxed{666.67} + \boxed{3333.33} = 4000.00$$

Thus, $\bar{y} = \dfrac{4000.00}{490.41} = 8.16$. To the nearest tenth, the centroid is $(0, 8.2)$ ◆

22.4 Exercises

In Exercises 1–12, determine the following for the given area.
a. M_x and M_y
b. The centroid

1.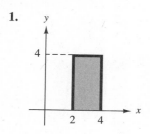
2.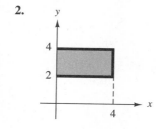
3.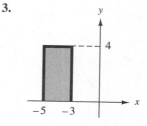
4.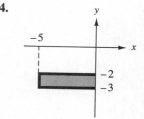
5.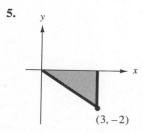
6.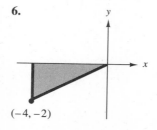
7.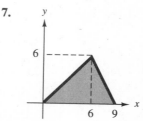
8.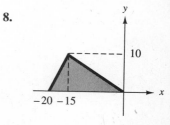

9.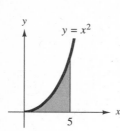

10.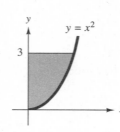

11.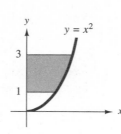

12.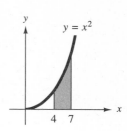

In Exercises 13–16, the letters U, P, G, and O are formed with rectangles. The dimensions of each rectangle are written inside the rectangle. For each rectangle, determine the centroid.

13.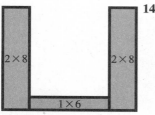

14.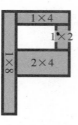

15.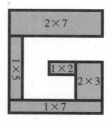

16.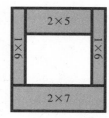

In Exercises 17–20, determine the centroids of the volumes.

17.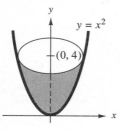

18.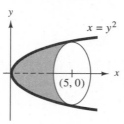

19.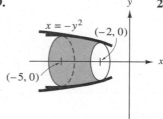

20.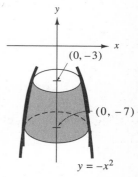

21. A road divider pylon is made of a conical shape with a $\frac{1}{4}$-m diameter base and a 1-m altitude as shown in the sketch. To determine whether it will blow over in a wind, its centroid must be known. Find the centroid. (*Hint:* A cone can be formed by rotating a triangular shape about an axis.)

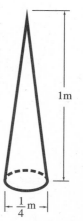

22. A parabolic dish antenna bounded by $y = \frac{x^2}{4}$ and $y = 5$ must have the axis of its supports passing through its centroid. Determine the centroid.

22.5

Moments of Inertia

The ability of an object to rotate is dependent upon its size, shape, and axis of rotation. For example, we find it easier to swing a bat when we "choke up" than when we grip the bat at its end. We find this to be true because a longer bat resists the rotation more and is harder to swing.

This resistance to rotation is called the **moment of inertia,** and is usually represented with the letter I. Experimentally it was discovered that the moment of inertia varies with the square of the moment arm. Hence, the moment of inertia is sometimes referred to as the *second moment,* meaning "to the second power." Physically, the moment of inertia is also dependent upon the mass.

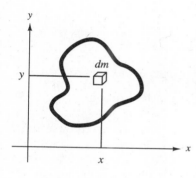

For a plane mass (as shown in the sketch), the moments of inertia with respect to the x- and y-axes respectively are defined as $I_{xx} = \oint_M y^2 \, dm$ and $I_{yy} = \oint_M x^2 \, dm$, where x and y are the coordinates of the centroid of the element of mass dm.

If we are dealing with flat shapes of uniform density, the mass is proportional to the area. Thus, we can study the second moment of the area and obtain comparable results without worrying about the mass distribution. Therefore the equations shown in the box can be modified to:

$$I_{xx} = \oint_A y^2 \, dA \qquad I_{yy} = \oint_A x^2 \, dA.$$

EXAMPLE 1 **Moment of Inertia**

Calculate the second moments of the area shown in Figure 22.41.

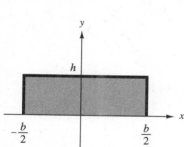

FIGURE 22.41

Solution Taking a horizontal element to calculate I_{xx}, we have $dA = \left(\dfrac{b}{2} - \dfrac{(-b)}{2}\right) dy$ or $dA = b \, dy$.

$$I_{xx} = \oint_A y^2 \, dA \qquad \text{Moment of inertia with respect to the } x\text{-axis}$$

$$= \int_0^b y^2 b \, dy \qquad \text{Substituting}$$

$$= b\left(\frac{y^3}{3}\right)\Big]_0^b \qquad \text{Integrating}$$

$$= \frac{bh^3}{3}$$

Taking a vertical element to calculate I_{yy}, we have $dA = (h - 0) \, dx$ or $dA = h \, dx$.

$$I_{yy} = \oint_A x^2 \, dA \qquad \text{Moment of inertia with respect to the } y\text{-axis}$$

$$= \int_{-\frac{b}{2}}^{\frac{b}{2}} x^2 h \, dx \qquad \text{Substituting}$$

$$= h \int_{-\frac{b}{2}}^{\frac{b}{2}} x^2 \, dx$$

$$= h \left(\frac{x^3}{3} \right) \Bigg]_{-\frac{b}{2}}^{\frac{b}{2}} \qquad \text{Integrating}$$

$$= \frac{hb^3}{12}$$

The second moment of area of a rectangle with respect to an axis along its base is $I_{xx} = \dfrac{bh^3}{3}$; with respect to an axis through its centroid it is $I_{yy} = \dfrac{hb^3}{12}$. ◆

Note that the second moment of an object with respect to an axis through its centroid is not zero, whereas the first moment with respect to the centroid is always zero. This, in part, defends our choice of using the second moment to define moment of inertia. We know that a rectangular shape resists rotation and is hard to rotate. If we had used the first moment, the equation would have suggested that there would be no trouble rotating the shape.

EXAMPLE 2 **Second Moment of a Triangle**

Determine the second moment of area of the triangle shown in Figure 22.42.

Solution The triangle is bordered by the lines $x = 5$, $y = 2$, and $y = \dfrac{-2x}{5} + 2$.
Choosing a horizontal element to calculate I_{xx}, we find:

$$dA = \left\{ 5 - \left[\frac{-5}{2}(y - 2) \right] \right\} dy$$

$$= \frac{5y}{2} \, dy.$$

$$I_{xx} = \oint_A y^2 \, dA$$

$$= \int_0^2 y^2 \left(\frac{5y}{2} \, dy \right)$$

$$= \int_0^2 \frac{5y^3}{2} \, dy$$

$$= \left(\frac{5}{2} \frac{y^4}{4} \right) \Bigg]_0^2$$

$$= 10.$$

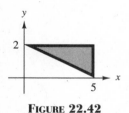

FIGURE 22.42

Choosing a vertical element to calculate I_{yy}, we have:

$$dA = \left[2 - \left(\frac{-2x}{5} + 2\right)\right] dx$$

$$= \frac{2x}{5} \, dx.$$

$$I_{yy} = \oint_A x^2 \, dA$$

$$= \int_0^5 x^2 \left(\frac{2x}{5} \, dx\right)$$

$$= \int_0^5 \frac{2x^3}{5} \, dx$$

$$= \left(\frac{2}{5} \frac{x^4}{4}\right)\Bigg]_0^5$$

$$= 62.5.$$

Therefore, $I_{xx} = 10$, and $I_{yy} = 62.5$. ◆

EXAMPLE 3 I_{xx} and I_{yy} **for an Area**

Determine I_{xx} and I_{yy} for the area shown in Figure 22.43.

Solution   Choosing an element of area parallel to the x-axis, we have $dA = (y^{\frac{1}{3}} - 0) \, dy$. So,

$$I_{xx} = \oint_A y^2 \, dA$$

$$= \int_1^4 y^2 (y^{\frac{1}{3}} \, dy)$$

$$= \int_1^4 y^{\frac{7}{3}} \, dy$$

$$= \left(\frac{3}{10} y^{\frac{10}{3}}\right)\Bigg]_1^4$$

$$= 30.2.$$

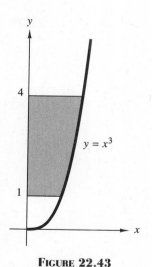

FIGURE 22.43

If we choose an area element parallel to the y-axis, we find that the boundaries of the element are not the same throughout the area. For $0 < x < 1$, the element is bounded by the lines $y = 4$ and $y = 1$. For $1 < x < \sqrt[3]{4}$ the element is bounded by the line $y = 4$ and the curve $y = x^3$. Using Property 2 of definite integrals (page 1048) we can find I_{yy} using the sum of two integrals. Since $I_{yy} = \oint_A y^2 \, dA$, we can write

$$I_{yy} = \int_0^1 y^2 \, dA + \int_1^{\sqrt[3]{4}} y^2 \, dA \qquad \textbf{Property 2 of definite integrals}$$

$$= \int_0^1 x^2(4 - 1) \, dx + \int_1^{\sqrt[3]{4}} x^2(4 - x^3) \, dx$$

$$= \int_0^1 3x^2 \, dx + \int_1^{\sqrt[3]{4}} (4x^2 - x^5) \, dx$$

$$= x^3 \Big]_0^1 + \left(\frac{4x^3}{3} - \frac{x^6}{6} \right) \Big]_1^{\sqrt[3]{4}}$$

$$= (1^3 - 0) + \left[\left(\frac{4(\sqrt[3]{4})^3}{3} - \frac{(\sqrt[3]{4})^6}{6} \right) - \left(\frac{4(1)^3}{3} - \frac{1^6}{6} \right) \right]$$

$$= 2.5$$

Therefore, $I_{xx} = 30.2$, and $I_{yy} = 2.5$. Show that I_{xx} and I_{yy} for an area are in linear units to the fourth power. ◆

22.5 Exercises

In Exercises 1–16, determine I_{xx} and I_{yy} for the areas given.

1.

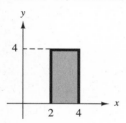

2.

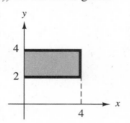

3.

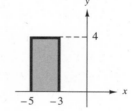

4.

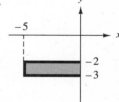

5.

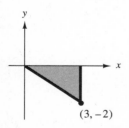

6.

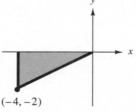

7.

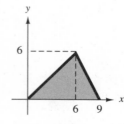

8.

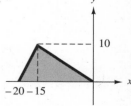

9.

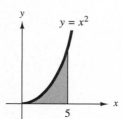

10.

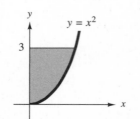

11.

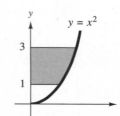

12.

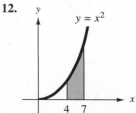

13.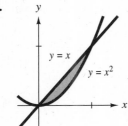

$y = x$
$y = x^2$

14.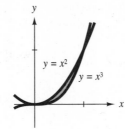

$y = x^2$
$y = x^3$

15.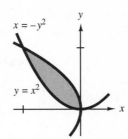

$x = -y^2$
$y = x^2$

16.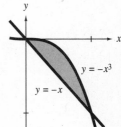

$y = -x^3$
$y = -x$

22.6

Other Applications

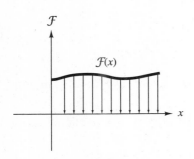

$\mathcal{F}(x)$

FIGURE 22.44

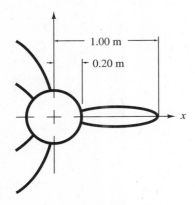

1.00 m
0.20 m

FIGURE 22.45

Integrals can be applied in all areas of technology. This section illustrates a few of these applications.

Often forces are not concentrated at a single point on a structure but are distributed over a finite area. It is easier to analyze these situations if we can replace the forces by an equivalent single force acting at a single point. If the distributed force is unidirectional we can replace it by a single force F equal in magnitude to the area bounded by the loading function $\mathcal{F}(x)$ (Figure 22.44) and passing through the centroid of this area. *A loading function is a function that describes how the force is distributed on a surface.*

EXAMPLE 1 Application: Force on Windmill Blade

The force on a windmill blade (see Figure 22.45) is found to vary as the distance x in meters along the blade according to $\mathcal{F}(x) = 40x^3$ N/m. Determine the equivalent force and locate it on the blade.

Solution   A sketch of the force distribution along the blade is shown in Figure 22.46. The equivalent force is equal in magnitude to the area of the force distribution diagram. Therefore, taking a vertical element of force dF, we have $dF = \mathcal{F}(x)\, dx$.

$$F = \int_{0.20}^{1.00} \mathcal{F}(x)\, dx$$

$$= \int_{0.20}^{1.00} 40x^3\, dx$$

$$= \left(40\frac{x^4}{4} \right)\Bigg]_{0.20}^{1.00}$$

$$= 9.984 \text{ N}$$

Instead of using many different forces acting on different parts of the fan, the net force of 9.984 N may be used as it acts on the centroid. But, where is the centroid? Since the force is perpendicular to the x-axis, we need locate only the x-coordinate of the centroid. Unfortunately, we don't know the equation for the

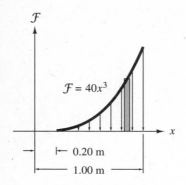

FIGURE 22.46

area of the fan. However, we still can use our notions about moment arms to determine the centroid. The first moment of force is:

$$\bar{x} = \frac{\oint_A x \, dF}{\oint_A dF}$$

$$= \frac{\int_{0.20}^{1.00} x(40x^3 \, dx)}{F} \qquad \oint_A dF = F$$

$$= \frac{40 \int_{0.20}^{1.00} x^4 \, dx}{9.984}$$

$$= \frac{40\left(\dfrac{x^5}{5}\right)\Big]_{0.20}^{1.00}}{9.984} \quad \text{or} \quad 0.8010 \text{ m.}$$

Therefore, an equivalent force of 10.0 N may be placed on the blade 0.80 m from the hub center. ◆

When an object is immersed in a static fluid, the pressure of the fluid (gas or liquid) always acts perpendicular to the surface of the object. The force due to pressure p on an element of area dA is $dF = p \, dA$.

EXAMPLE 2 **Application: Force on a Dam**

The upstream face of a dam has water against it to a depth of 40.0 ft (see Figure 22.47). If the dam is 80.0 ft wide and the pressure variation with respect to its depth is 62.4y, where y denotes feet below the surface, what is the total pressure on the face of the dam?

Solution If we take an element of area on the face of the dam as shown in Figure 22.48, the area $dA = 80.0 \, dy$, and the pressure on it is everywhere the same. Substituting in the equation $dF = p \, dA$, we have $dF = (62.4y)(80.0 \, dy)$, so:

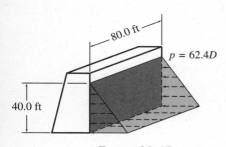

FIGURE 22.47

$$F = \oint_A dF$$

$$= \int_{y=0.0}^{y=40.0} (62.4y)(80.0 \, dy)$$

$$= 4992 \int_{y=0.0}^{y=40.0} y \, dy$$

$$= 4992\left(\frac{y^2}{2}\right)\Big]_{y=0.0}^{y=40.0}$$

$$= 3.99 \times 10^6.$$

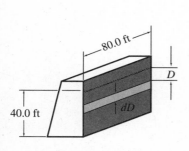

FIGURE 22.48

The total force on the face of the dam is 3.99 million pounds. ◆

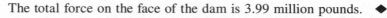

If the force on an object is in the same direction as its motion, we can write the work done by the force during a movement dx as $dW = F(x)\,dx$. Therefore the total work done by the force is $W = \int dW$ or $W = \int_{x_1}^{x_2} F(x)\,dx$.

EXAMPLE 3   **Application: Work in an Electric Field**

The voltage change on a charged particle is a measure of the work required to move the particle from one position to another in an electric field. If an electron has on it a force F in newtons given by $F = -\dfrac{10^{-6}}{x^2}$, with x in meters, what is the work done to move it from $x = 2.00$ m to $x = 2.04$ m?

Solution

$$W = \int_{x_1}^{x_2} F(x)\,dx$$

$$= \int_{2.00}^{2.04} -\frac{10^{-6}}{x^2}\,dx$$

$$= -10^{-6}\int_{2.00}^{2.04} \frac{1}{x^2}\,dx$$

$$= -10^{-6}\left(-\frac{1}{x}\right)\Bigg]_{2.00}^{2.04} \quad \text{or} \quad -9.80 \times 10^{-9} \text{ N m.} \quad \blacklozenge$$

If the current through or the voltage across a resistance R are functions of time, the energy consumed by the resistance in a time interval dt can be written as $dE = R[i(t)]^2\,dt$ or $dE = \dfrac{1}{R}[v(t)]^2\,dt$. When R is in ohms, i is in amperes, v is in volts, and t is in seconds; E is in watt-seconds.

EXAMPLE 4   **Application: Energy Used in a Speaker Coil**

The current in a speaker coil varies as $i(t) = (3t^2 + 2) \times 10^{-3}$ A. If the coil resistance is 1000 Ω, how much energy is used between one and two seconds after start-up?

Solution   Since $dE = R[i(t)]^2\,dt$, we have:

$$E = \int_{t_1}^{t_2} dE$$

$$= \int_1^2 1000[(3t^2 + 2) \times 10^{-3}]^2\,dt$$

$$= 10^{-3}\int_1^2 (9t^4 + 12t^2 + 4)\,dt$$

$$= 10^{-3}\left(9\frac{t^5}{5} + 12\frac{t^3}{3} + 4t\right)\Bigg]_1^2 \quad \text{or} \quad 87.8 \times 10^{-3}.$$

The energy consumed is 87.8 mW-s. $\blacklozenge$

22.6 Exercises

1. What is the total force of fresh water on the face of a dam 600 ft wide and 50 ft high?

2. What is the total force of fresh water on the face of a dam 150 ft wide and 25 ft high?

3. The dam in Exercise 2 has a 2-ft-wide by 1-ft-high floodgate whose top is 21 ft below the surface. What is the total force on the floodgate?

4. The dam in Exercise 1 has a 5-ft-wide by 2-ft-high floodgate 46 ft below the surface. What is the total force on the floodgate?

5. The weight of grain in pounds on a floor joist is as shown in the illustration. What is the total weight of the grain?

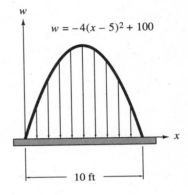

6. Dog food is piled in a bin so that its weight in pounds on any plank is given by the function shown in the illustration. What is the total weight on any plank?

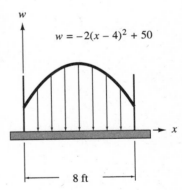

7. Sand piled against a wall assumes the shape shown in the illustration. What is the total weight of sand along a 10-m span of wall? (Weight is in newtons.)

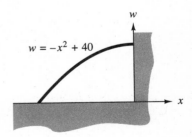

8. Portland cement piled against a wall assumes the shape shown in the illustration. What is the total weight of the cement piled along a 7-m span of wall? (Weight is in newtons.)

9. Where should the equivalent weight in Exercise 7 be located?

10. Where should the equivalent weight in Exercise 8 be located?

11. The force in a spring is given by $F = 40x$ N. How much work is required to stretch the spring 0.2 m starting from its unstretched length?

12. The force in a spring is given by $F = 0.5x$ N. How much work is required to stretch the spring 0.2 m starting from its unstretched length?

13. How much work is required to stretch the spring in Exercise 11 by 0.2 m starting from an initial extension of 0.1 m?

14. How much work is required to stretch the spring in Exercise 12 by 0.2 m starting from an initial extension of 0.3 m?

15. The force on a charged particle is given by $F = \dfrac{10^{-6}}{x^2}$ N. What is the work required to move it from $x = \infty$ to $x = 1$ m?

16. The force on a charged particle is given by $F = \dfrac{-10^{-12}}{x^3}$ N. What is the work required to move it from $x = \infty$ to $x = 2$ m?

In Exercises 17–24, what is the energy consumed by a 1-MΩ resistor under the conditions given?

17. $i(t) = t^2 - 1; 0 \le t \le 4$

18. $i(t) = t - t^2; 0 \le t \le 2$

19. $i(t) = t^3 - t; 2 \le t \le 4$

20. $i(t) = 1 - t^2; 1 \le t \le 3$

21. $v(t) = (t - 1)^2; 0 \le t \le 3$

22. $v(t) = t - 4t^2; 0 \le t \le 2$

23. $v(t) = (t + 1)^3; 0 \le t \le 2$

24. $v(t) = (t - 2t^2)^2; 0 \le t \le 1$

25. A rectangular window is 1.2 m high and 2.0 m long is fit into the dam wall with its center 10 m below water level. The pressure varies in the following fashion: $p = 62.4y$, where y is the distance below the surface of the water. Hence, the water pressure on the top of the window is less than the water pressure on the bottom of the window. What is the total force on the window?

26. A force of 10 N stretches a spring from its relaxed length of 10 cm to a length of 10.6 cm.

   a. The equation of force for a spring can be represented with $F = k\Delta x$, where Δx is how much the spring has been stretched and k is the spring constant. What is the spring constant for this spring?

   b. How much work is required to stretch the spring this distance?

   c. How much work is required to stretch the spring from 12.0 cm to 12.6 cm?

$w = -5x^2 + 25$

Review Exercises

In Exercises 1–10, determine the area of the region bounded by the given curves.

1. $y = 2x - x^2, y = 0$

2. $y = x^2 - x^3$, y-axis

3. $y^2 = x, x = 4$

4. $y = 2x - x^2, y = -3$

5. $y = x^2, y = x$

6. $y = x^4 - 2x^2, y = 2x^2$

7. $\sqrt{x} + \sqrt{y} = 1, x = 0, y = 0$ **8.** $y = x^2 - 3, y = -x^2 + 5$ **9.** $y = x, y = \dfrac{1}{x^2}, x = 2$

10. $y = x + 1, y = 3 - x^2$

11. Determine the volume of the solid generated when the first quadrant region bounded by the arc of $y^2 = 1 - x$ is revolved about each of the following.
   a. x-axis
   b. y-axis

12. Determine the volume of the solid generated by revolving the region bounded by $y = 5, y = 0, x = 0,$ and $x = 7$ about each of the following:
   a. x-axis
   b. y-axis

13. Determine the volume of the solid generated by revolving the region bounded by $y = x, x = 3,$ and $y = 0$ about each of the following.
   a. x-axis
   b. y-axis

14. Determine the volume of the solid generated by revolving the region bounded by $y = 2 - x, x = 0,$ and $y = 0$ about each of the following.
   a. x-axis
   b. y-axis
   c. The line $y = -1$.

15. Determine the volume of the solid generated by revolving the region bounded by $y = x, y = 0, x = 1,$ and $x = 5$ about each of the following.
   a. x-axis
   b. y-axis
   c. The line $x = 5$

In Exercises 16–21, use the cylindrical-shell method to determine the volume generated by revolving the given region about the given line.

16. $y = x, y = 0, y = 2$, about the y-axis   **17.** $y = 1 - x, y = 0, x = 0$, about the y-axis

18. $y = x, y = 0, x = 2$, about the x-axis   **19.** $y = 2 - x, y = 0, x = 0$, about the x-axis

20. $y = x^2 + 4, y = 8, x \geq 0$, about the y-axis   **21.** $y = x^2, y = 4x - x^2$, about the y-axis

22. Determine the volume of the solid generated by revolving the region bounded by $y = x^3, y = 0,$ and $x = 2$ about each of the following.
   a. x-axis
   b. y-axis
   c. Line $x = 4$

23. Determine the volume of the solid generated by revolving the region bounded by $y = -2x^2 + 8x - 6$ and $y = 0$ about the y-axis.

24. A solid is generated by revolving the region bounded by $y = \dfrac{1}{2}x^2$ and $y = 2$ about the y-axis. A hole, centered along the axis of revolution, is drilled through this solid so that $\dfrac{1}{3}$ the volume is removed. Determine the diameter of the hole.

25. Determine the volume of the solid generated by revolving the ellipse $\dfrac{x^2}{a^2} + \dfrac{y^2}{b^2} = 1$ about each of the following.
    a. Its minor axis (oblate spheroid)
    b. Its major axis (prolate spheroid)

26. A gas tank is an oblate spheroid generated by revolving the region bounded by $\dfrac{x^2}{25} + \dfrac{y^2}{16} = 1$ about its minor axis (units are in feet).
    a. How many gallons are contained in the tank? ($1\ \text{ft}^3$ contains 7.481 gal)
    b. Determine the depth of gasoline in the tank when it is filled to $\dfrac{1}{3}$ of its total capacity.

27. Determine the centroid of the region bounded by $y = x^2$ and $y = 2x + 3$.

In Exercises 28–31, determine each of the following using the given figure.
a. the centroid
b. I_{xx}
c. I_{yy}

28.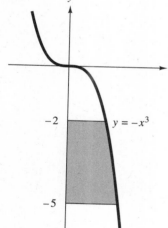

29.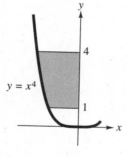

30.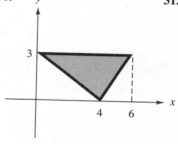

31.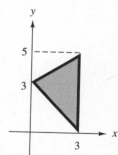

In Exercises 32 and 33, determine the centroid for the given figure.

32.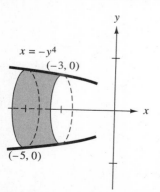

33.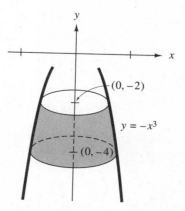

34. Determine and locate the equivalent force in the sketch.

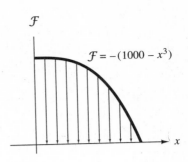

35. The pressure on a plate is a constant 100 N/m². What is the total force on the plate? (Use the sketch).

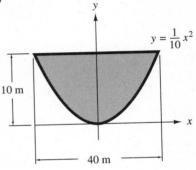

36. The force in a tow bar pulling a vehicle over a hill is given by $F = -\dfrac{1}{100}(x - 100)^2 + 100$ N, where x is measured along the hill. What is the work done towing from $x = 40$ m to $x = 180$ m?

37. What is the energy consumed by a 1000-r resistance if the current passing through it is given by $i(t) = t^2 - 2$? The current is turned on for four seconds.

38. If a spring is 120 mm long, compare the work done in stretching it from 120 mm to 130 mm with the work done in stretching it from 130 mm to 140 mm.

39. a. Sketch the region R in the first quadrant such that the volume of revolution obtained by revolving R about the y-axis is:

$$V = \int_{1}^{5} 2\pi x \sqrt{x - 1}\ dx \text{ (shell method).}$$

  b. Write an equivalent integral for the volume of the solid by using the disk method. Evaluate the integral.

40. Determine the volume of the solid generated by revolving the region bounded by the graphs $y = \sqrt{1 - x^2}$, $y = -\sqrt{1 - x^2}$, and $x = 0$, which lines in the first and fourth quadrants, about the y-axis.

*41. Determine the volume of the solid generated by revolving the right triangle with vertexes at $(a, 0)$, $(b, 0)$, and (a, h) about the y-axis. Assume that $0 < a < b$ and $h > 0$.

*42. A tent is made by stretching canvas from a circular base of radius a to a semicircular rib erected at right angles to the base at the ends of a diameter. Determine the volume enclosed by the tent.

*43. How much work is done by a colony of ants in building a conical anthill with height and diameter both 1 ft, using sand initially at ground level and with density of 150 lb/ft³?

*44. A nose cone for a space re-entry vehicle is designed so that a cross section, taken x feet from the tip and perpendicular to the axis of symmetry, is a circle of radius $\frac{1}{4}x^2$ ft. Determine the volume of the nose cone given that its length is 20 ft.

*45. Locate the centroid and determine the second moments for the area shown in the illustration.

*46. Determine the height of a parabolic segment $y = ax^2$, $0 \leq y \leq h$ so that the centroid and the focus of the parabola are the same point (see the sketch).

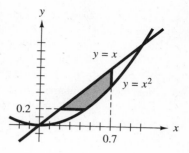

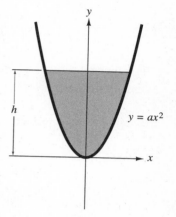

*47. Show that the second moments of an area must always be positive.

✐ Writing About Mathematics

1. You are working in a design studio. A design engineer has designed a bowl whose shape is determined by $y = \dfrac{-5}{8}\sqrt{64 - x^2} + 5$, $y = 0.1x^2 + 1$, and $y = 4$. To determine the cost of manufacturing the bowl, it is necessary to determine the volume of the shell of the bowl. Use one of the techniques in this chapter to determine the volume. Write a report explaining to your supervisor how you arrived at the volume of the shell.

2. Write a brief report explaining in what circumstances you would use the disk method and when you would use the washer method.

3. One of your classmates has difficulty understanding when to use the cylindrical-shell method for determining the volume. Write a few paragraphs explaining in what circumstances the cylindrical-shell method rather than the disk or washer method is used.

4. In the design of a tool the center of gravity must be placed at a key point. Write a report to your supervisor explaining how to determine the center of gravity.

5. The units for the first moment are cm³, and for the second moment units are cm⁴. But the first moment varies with moment arm r, while the second moment varies with moment arm r^2. Should the moments not vary as cm¹ and cm² or cm³ and cm⁶ to show the relationship between the first and second moments? Explain the units in the first moment and second moment.

6. Note that the fundamental dimensions of the cone and the cylinder are the same. Yet each has a different moment of inertia. Why is this so? What is it about the shapes that changes the moment of inertia? Write a few paragraphs to answer these questions.

Chapter Test

1. Use integration to determine the area of the region bounded by $y = x + 5$, $y = 1$, $x = -1$, and $x = 2$.

2. Determine the area of the region bounded between the two curves $y^2 = x + 4$ and $y = -x + 2$.

3. Determine the volume of the solid formed by revolving the region determined by the curves $y = \sqrt{25 - x^2}$ and the x-axis about the x-axis.

4. Determine the volume of the solid formed by revolving the region determined by the curves $y = \dfrac{1}{x}$, $y = x$, and $x = 5$ about the x-axis.

5. Use the cylindrical-shell method to determine the volume of the solid generated when the region determined by the curves $y = 4x$, $y = -x + 5$, and $y = 1$ is revolved about the y-axis.

6. For the area given in the figure, determine each of the following.
    a. M_x and M_y
    b. The centroid

7. Determine I_{xx} and I_{yy} for the area given in the illustration.

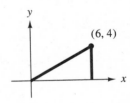

8. The Sells have a swimming pool whose base is formed by two parabolas: $x = y^2$ and $x = -y^2 + 8$.
    a. Determine the area of the base of the pool.
    b. If the pool is 3 ft deep, determine the volume of the pool in cubic feet.

9. A woodworker is turning a candleholder on a lathe. The shape is determined by rotating the region determined by the curves $y = \dfrac{3}{x}$, x-axis, and the lines $x = 1$ and $x = 3$ about the x-axis. Determine the volume of the candleholder.

10. What is the total force of fresh water on the face of a dam 400 ft wide and 30 ft high?

Differentiation and Integration of Transcendental Functions

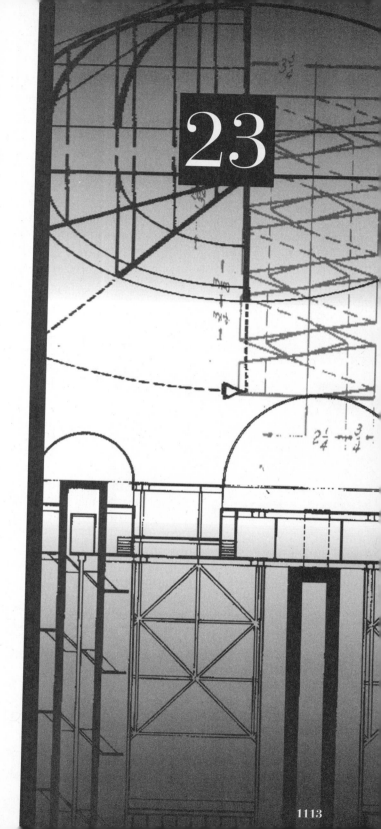

23

Companies often use population studies to predict increases in consumer usage. For example, a pharmaceutical company may want to determine if the rate at which it produces serum for infant inoculations will keep pace with the population growth. It needs to know not just the projected number of infant births, but also the projected change in the birth rate. In other words, assuming that a population grows exponentially, how does the rate of growth increase?

Another application of functions, which we explore in Example 9 of Section 23.2, involves the effects of drug use on the human body. In the ingestion of certain drugs (legal or illegal), as the dosage or rate of consumption increases, the effects on the body increase at an exponential rate.

23.1

Derivatives and Integrals of Logarithmic Functions

Introduction

The study of the operation of differentiation in Chapters 19 and 20 provided us with the tools to differentiate algebraic functions. We recall that an algebraic function is one that can be defined using the algebraic operations of addition, subtraction, multiplication, division, powers, and roots. An interesting note here is that, with all our study of the operation of integration in Chapters 21 and 22, we still cannot integrate all the algebraic functions.

For example, in discussing the power rule of integration:

$$\int x^n \, dx = \frac{x^{n+1}}{n+1} + C,$$

we always add the condition that $n \neq -1$. When $n = -1$ we have what appears to be a simple antiderivative that is algebraic.

$$\int x^{-1} \, dx = \int \frac{dx}{x}.$$

However, the antiderivative of this function turns out to be a logarithmic function. There are many other integration problems where the integrand is an algebraic function and the antiderivative is a logarithmic, trigonometric, or inverse trigonometric function.

In this chapter we learn how to differentiate and integrate the exponential, logarithmic, trigonometric, and inverse trigonometric functions. Since we will use the rules of logarithms and exponents in the development of the formulas for finding the derivatives and the integrals to review the laws of exponents, let us refer back to chapter 13.

LAWS OF EXPONENTS

Where $a > 0$, $b > 0$, and m and n are real numbers.

1. $a^n \cdot a^m = a^{n+m}$
2. $(a^n)^m = a^{n \cdot m}$
3. $(a \cdot b)^n = a^n b^n$
4. $\left(\dfrac{a}{b}\right)^n = \dfrac{a^n}{b^n}$
5. $\dfrac{a^m}{a^n} = a^{m-n}$
6. $a^0 = 1$
7. $\begin{cases} a^{-n} = \dfrac{1}{a^n} \\[2mm] \dfrac{1}{a^{-n}} = a^n \end{cases}$

Recall that while studying logarithms we selected the special irrational number e to be a base. In our discussion of the derivative of logarithmic functions it becomes obvious why we did this. However, before we discuss derivatives, let us define e in a more formal fashion, using the concept of limit.

TABLE 23.1 $f(x) = (1 + x)^{\frac{1}{x}}$

x	$(1 + x)^{\frac{1}{x}}$
-0.2	3.051758
-0.1	2.867972
-0.05	2.789510
-0.01	2.731999
0.01	2.704814
0.05	2.653298
0.1	2.593742
0.2	2.488320

> **DEFINITION 23.1** **The Number e**
>
> $$e = \lim_{n \to \infty}\left(1 + \frac{1}{n}\right)^{n} \quad \text{or} \quad e = \lim_{x \to 0}(1 + x)^{\frac{1}{x}}$$
>
> Therefore,
>
> $$e = 2.718281 \ldots .$$

The actual proof that $\lim_{x \to 0}(1 + x)^{\frac{1}{x}}$ is the irrational number e is perhaps beyond the scope of this textbook. However, using a programmable calculator or a computer program, it can be shown that as we take values of x closer and closer to 0, the value of the function $(1 + x)^{\frac{1}{x}}$ gets closer and closer to e. To illustrate this, Table 23.1 contains a few values of the function close to zero. The values are illustrated in the graph in Figure 23.1.

One more important fact that we must remember is the relationship between the exponential and logarithmic functions.

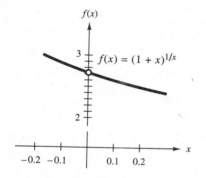

FIGURE 23.1

LAWS OF LOGARITHMS

For any base $b > 0$, where $b \neq 1$.

1. $\log_b(AB) = \log_b A + \log_b B$
2. $\log_b\left(\dfrac{A}{B}\right) = \log_b A - \log_b B$
3. $\log_b(A)^n = n \log_b A$
4. $\log_b 1 = 0$
5. $\log_b b = 1$

In logarithmic notation we write $y = \log_b x$. In exponential notation we write $x = b^y$, or we can write $y = \log_b x$ if, and only if, $x = b^y$, $b > 0$, and $b \neq 1$. This simply means that a logarithm is an exponent. It is the exponent to which b must be raised to get x, as illustrated in Example 1.

EXAMPLE 1 **Solving an Exponential Equation**

Solve the exponential equation $2^y = 32$ for y.

Solution We can write $2^y = 32$ as $2^y = 2^5$; thus $y = 5$. That is, we must raise 2 to the 5th power to get 32. ◆

Derivatives of Logarithmic Functions

We use the rules of exponentials and logarithms to develop the operations of differentiation and integration for these functions. We begin by developing a rule for determining the derivative of the logarithmic function $f(x) = \log_b x$, $b > 0$, $b \neq 1$, $x > 0$. To develop this we must recall the definition of the derivative.

$$f'(x) = \lim_{\Delta x \to 0} \frac{f(x + \Delta x) - f(x)}{\Delta x}$$

$$= \lim_{\Delta x \to 0} \frac{1}{\Delta x} [\log_b(x + \Delta x) - \log_b x]$$

$$= \lim_{\Delta x \to 0} \frac{1}{\Delta x} \left[\log_b \frac{x + \Delta x}{x} \right] \quad \text{Rule 2 of logarithms}$$

$$= \lim_{\Delta x \to 0} \frac{1}{x} \frac{x}{\Delta x} \left[\log\left(1 + \frac{\Delta x}{x}\right) \right] \quad \text{Multiplying by } \frac{x}{x} = 1$$

If we let $p = \dfrac{\Delta x}{x}$ and have x remain constant, then as $\Delta x \to 0$, $p \to 0$. Substituting, we have:

$$= \lim_{p \to 0} \frac{1}{x} \frac{1}{p} \log_b(1 + p) \quad \text{Substituting}$$

$$= \lim_{p \to 0} \frac{1}{x} \log_b(1 + p)^{\frac{1}{p}} \quad \text{Rule 3 of logarithms}$$

$$= \frac{1}{x} \log_b[\lim_{p \to 0}(1 + p)^{\frac{1}{p}}] \quad \text{Properties of limits}$$

$$= \frac{1}{x} \log_b e \quad \text{Definition of } e$$

Recall that $f(x) = \log_b x$; thus, $\dfrac{d}{dx}(\log_b x) = \dfrac{1}{x} \log_b e$. This formula takes on a very simple form for one particular base. We said earlier that it would be obvious why we wanted the irrational number e as a base. Since $\log_e e = 1$ and $\ln x = \log_e x$, we substitute and:

$$\frac{d(\ln x)}{dx} = \frac{d(\log_e x)}{dx}$$

$$= \frac{1}{x} \log_e e$$

$$= \frac{1}{x}.$$

If we let $y = f(u)$, where u is a function of x and apply the chain rule, we can summarize these rules for differentiation.

Differentiation Rules for Logarithmic Functions
For the composite function $u = u(x)$.

$\{1\}$ If $y = \ln u$, then $\dfrac{dy}{dx} = \dfrac{1}{u}\dfrac{du}{dx}$.

$\{2\}$ If $y = \log_b u$, then $\dfrac{dy}{dx} = \dfrac{1}{u}(\log_b e)\dfrac{du}{dx}$.

EXAMPLE 1   Derivative of ln

Determine the derivative of $y = \ln(x^2 + 3x)$

Solution   If we let $u = x^2 + 3x$, then $\dfrac{du}{dx} = 2x + 3$.

$$y = \ln(x^2 + 3x)$$
$$ = \ln u \qquad\qquad \textbf{Substituting}$$
$$\frac{dy}{dx} = \frac{1}{u}\frac{du}{dx} \qquad\qquad \textbf{Differentiation rule \{1\}}$$
$$\phantom{\frac{dy}{dx}} = \frac{2x + 3}{x^2 + 3x} \qquad\qquad \textbf{Substituting}\;\blacklozenge$$

EXAMPLE 2   Derivative of ln to a Power

Determine the derivative of $y = (\ln x)^2$.

Solution   Since $(\ln x)^2$ is a composite function, we let $u = \ln x$, then $\dfrac{du}{dx} = \dfrac{1}{x}$.

$$y = (\ln x)^2$$
$$ = u^2 \qquad\qquad \textbf{Substituting for ln } x$$
$$\frac{dy}{dx} = 2u\frac{du}{dx} \qquad\qquad \textbf{Differentiation rule \{1\}}$$
$$\phantom{\frac{dy}{dx}} = \frac{2}{x}\ln x \qquad\qquad \textbf{Substituting for } u \textbf{ and } \dfrac{du}{dx}\;\blacklozenge$$

EXAMPLE 3   Derivative of $\log_{10}$

Determine the derivative of $y = \log_{10}(x^2 + 5x)$.

Solution To differentiate a logarithmic function in a base other than e we use differentiation rule {2}.

If $y = \log_b u$, then:

$$\frac{dy}{dx} = \frac{1}{u}(\log_b e)\frac{du}{dx}. \quad \text{Differentiation rule \{2\}}$$

If we let $u = x^2 + 5x$, then:

$$\frac{du}{dx} = 2x + 5.$$

$$\frac{dy}{dx} = \frac{1}{(x^2 + 5x)}(\log_{10} e)(2x + 5) \quad \text{Substituting}$$

$$= \frac{2x + 5}{x^2 + 5x}\log_{10} e. \quad \blacklozenge$$

The profit $P(x)$ of a company is determined by subtracting the cost $C(x)$ from the revenue $R(x)$. That is, $P(x) = R(x) - C(x)$. This equation is used in Example 4.

EXAMPLE 4 **Application: Maximum Profit**

The Fun Fare Company has determined that the revenue and cost equations for their new product are projected to be $R(x) = 300 \ln(6x + 12)$ and $C(x) = \frac{x}{6}$.

a. Determine the number of items that the company should sell to maximize its profit.
b. Determine the maximum profit.

Solution

a. We begin by substituting in the profit equation.

$$P(x) = R(x) - C(x)$$

$$= 300 \ln(6x + 12) - \frac{x}{6} \quad \text{Profit equation}$$

To maximize the profit we must determine $P'(x)$ and determine the value of x for which $P(x)$ is a maximum.

$$P'(x) = 300\frac{1}{6x + 12}(6) - \frac{1}{6}$$

$$= \frac{300}{x + 2} - \frac{1}{6}$$

$$= \frac{1800 - (x + 2)}{(x + 2)6}$$

$$= \frac{1798 - x}{6(x + 2)}$$

Setting the fraction equal to zero and solving for x, the critical values are $x = 1798$ and $x = -2$. Using the first derivative test for maximum values, we find that $x = 1798$ is a maximum value. Thus, to maximize the profit, the company should manufacture 1798 units.

b. To determine the maximum profit, we substitute 1798 for x.

$$P(\boxed{1798}) = 300 \ln[6(\boxed{1798}) + 12] - \frac{1798}{6}$$

$$= 2486.52$$

If the company manufactures the maximum amount of 1798 items, it makes a profit of $2486.52. ◆

Integration of Logarithmic Functions

By using the notion of the inverse we can turn around our derivative expression and determine the rules for integration.

$$\int \frac{1}{x} \, dx = \ln x + C, \qquad x > 0$$

Now, at last, we have a method of integrating $f(x) = \frac{1}{x}$. In the event that $x < 0$, we can show that:

$$\int \frac{1}{x} \, dx = \ln |x| + C.$$

Since we may not always be sure whether x is positive or negative, we write the rule with the absolute value sign, which covers both cases ($x < 0$ and $x > 0$).

$$\int \frac{1}{x} \, dx = \ln |x| + C, \qquad x \neq 0$$

Integration Formula for $\dfrac{1}{u}$

For the composite function $u = u(x)$.

$$\{3\} \quad \int \frac{du}{u} = \ln |u| + C$$

The result of using this formula always is a natural logarithmic function.

EXAMPLE 5   **Integration Formula for $\dfrac{1}{u}$**

Evaluate the indefinite integral $\displaystyle\int \frac{3}{x} \, dx$.

Solution

$$\int \frac{3}{x}\, dx = 3\int \frac{dx}{x}$$
$$= 3\ln|x| + C \quad \text{Integration formula \{3\}}$$
$$= \ln|x^3| + C \quad \blacklozenge$$

EXAMPLE 6   Integration Formula for $\frac{1}{u}$, Substitution

Evaluate the indefinite integral $\int \dfrac{3x^2}{x^3 - 5}\, dx$.

Solution   If we let $u = x^3 - 5$, then $\dfrac{du}{dx} = 3x^2$ and $du = 3x^2\, dx$

$$\int \frac{3x^2}{x^3 - 5}\, dx = \int \frac{du}{u} \qquad \text{Substituting}$$
$$= \ln|u| + C \qquad \text{Integration formula \{3\}}$$
$$= \ln|x^3 - 5| + C. \qquad \text{Substituting for } u \quad \blacklozenge$$

EXAMPLE 7   Integration of a Fractional Expression

Evaluate the indefinite integral $\int \dfrac{x^3 + x + 4}{x^2}\, dx$.

Solution   We can rewrite the integral in the form $\int \dfrac{x^3 + x + 4}{x^2}\, dx = \int \left(\dfrac{x^3}{x^2} + \dfrac{x}{x^2} + \dfrac{4}{x^2}\right) dx$. Now, using the sum formula:

$$\int [f(x) + g(x)]dx = \int f(x)dx + \int g(x)\, dx$$
$$\int \left(\frac{x^3}{x^2} + \frac{x}{x^2} + \frac{4}{x^2}\right) dx = \int x\, dx + \int \frac{dx}{x} + 4\int \frac{dx}{x^2}$$
$$= \frac{x^2}{2} + \ln|x| - \frac{4}{x} + C. \qquad \text{Integrating} \quad \blacklozenge$$

EXAMPLE 8   Integration Formula for $\frac{1}{u}$

Evaluate the definite integral $\displaystyle\int_{x=1}^{x=e} \frac{dx}{x}$.

Solution

$$\int_{x=1}^{x=e} \frac{dx}{x} = \ln x \Big]_{x=1}^{x=e} \qquad \text{Integral formula \{3\}}$$

$$= \ln e - \ln 1 \qquad \text{Fundamental theorem}$$
$$\qquad\qquad\qquad\qquad \log_b 1 = 0 \text{ and } \log_b b = 1$$

$$= 1.$$

Here we did not need to use the absolute value sign since e and 1 are both positive. ◆

Recall that when we use the substitution method with a definite integral, we may find it more convenient to express the limits of integration in terms of the new variable. This is demonstrated in Examples 9 and 10.

EXAMPLE 9 **Integration Formula for $\dfrac{1}{u}$, Substitution**

Evaluate the definite integral $\displaystyle\int_{x=2}^{x=3} \frac{dx}{x+1}$.

Solution If we let $u = x + 1$, then $du = dx$. Substituting 2 and 3 for x in the equation $u = x + 1$, the new limits of integration are:

$$u = 2 + 1 = 3 \quad \text{and} \quad u = 3 + 1 = 4.$$

Therefore,

$$\int_{x=2}^{x=3} \frac{dx}{x+1} = \int_{u=3}^{u=4} \frac{du}{u} \qquad \text{Substituting}$$

$$= \ln u \Big]_{u=3}^{u=4} \qquad \text{Integration formula \{3\}}$$

$$= \ln 4 - \ln 3$$

$$= \ln \frac{4}{3}$$

$$= 0.288.$$

We did not need to use the absolute value sign since $u > 0$ in the interval $[3, 4]$ ◆

EXAMPLE 10 **Area Under a Curve**

Determine the area bounded by the curve $y = \dfrac{x}{x^2 + 4}$ and the lines $x = 1$, $x = 4$, and $y = 0$.

Solution A sketch of the figure is given in Figure 23.2. The area of the sample strip is:

$$A_i = \frac{x_i}{x_i^2 + 4}$$

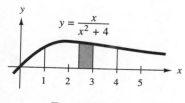

$$y = \frac{x}{x^2 + 4}$$

FIGURE 23.2

Summing the sample strips, the area is represented by the definite integral:

$$A = \int_{x=1}^{x=4} \frac{x}{x^2 + 4}\, dx.$$

To evaluate the integral, let $u = x^2 + 4$, then $\dfrac{du}{dx} = 2x$, or $\dfrac{du}{2} = x\, dx$. When $x = 1$, $u = 5$, and when $x = 4$, $u = 20$.

$$\int_{x=1}^{x=4} \frac{x}{x^2 + 4}\, dx = \int_{u=5}^{u=20} \frac{1}{u}\frac{du}{2} \qquad \text{Substituting}$$

$$= \frac{1}{2}\int_{u=5}^{u=20} \frac{du}{u}$$

$$= \frac{1}{2}\ln u \Big]_{u=5}^{u=20} \qquad \text{Integration formula \{3\}}$$

$$= \frac{1}{2}[\ln 20 - \ln 5\,] \qquad \text{Fundamental theorem}$$

$$= \frac{1}{2}\ln \frac{20}{5} \qquad \text{Rule 2 of logarithms}$$

$$= \frac{1}{2}\ln 4$$

$$= 0.693 \quad \blacklozenge$$

CAUTION

$$\int_{x=1}^{x=4} \frac{x}{x^2 + 4}\, dx \neq \int_{x=1}^{x=4} \frac{1}{u}\frac{du}{2}$$

and

$$\int_{x=1}^{x=4} \frac{x}{x^2 + 4}\, dx \neq \int_{u=1}^{u=4} \frac{1}{u}\frac{du}{2} \quad \blacksquare$$

EXAMPLE 11   Using Laws of Logarithms and Then Differentiating

Determine the derivative of $f(x) = \ln\left[\dfrac{x(x^2 - 3)^2}{(x^2 - 4)^{\frac{1}{2}}}\right]$.

Solution   In examining the function we see that inside the brackets we have a product, a quotient, and a power. Thus, to find the derivative, we would apply the rules for finding the derivative of ln u and then find the derivative of a product, quotient, and power. This all seems rather complicated, but can be simplified if we use the rules of logarithms and write the function in the following manner.

$$f(x) = \ln x + \ln(x^2 - 3)^2 - \ln(x^2 - 4)^{\frac{1}{2}} \quad \textbf{Rules of logarithms}$$

$$= \ln x + 2 \ln(x^2 - 3) - \frac{1}{2} \ln(x^2 - 4)$$

Now the process of finding the derivative is easier since all we have to do is to find the derivative of $\ln u$. Therefore,

$$f'(x) = \frac{1}{x} + 2 \frac{1}{x^2 - 3}(2x) - \frac{1}{2} \frac{1}{x^2 - 4}(2x)$$

$$= \frac{1}{x} + \frac{4x}{x^2 - 3} - \frac{x}{x^2 - 4}$$

$$= \frac{4x^4 - 20x^2 + 12}{x(x^2 - 3)(x^2 - 4)}. \quad \blacklozenge$$

23.1 Exercises

In Exercises 1–20, determine the derivative of the functions.

1. $y = \ln t$

2. $y = \ln z$

3. $y = \ln(x + 2)$

4. $y = \ln(3x - 4)$

5. $y = t \ln t^2$

6. $y = t^2 \ln(t + 1)$

7. $y = \ln(x^2 + 3x + 2)$

8. $y = \ln(x^3 - 7x + 5)$

9. $y = \log_{10}(3x^2 + 7)$

10. $y = \log_{10}(x^3 - 7x + 5)$

11. $y = \dfrac{\ln 5x}{x^8}$

12. $y = \dfrac{\ln(3x + 4)}{x^4}$

13. $y = \ln(1 - 3x^2)^{\frac{1}{3}}$

14. $y = \ln(x^3 - 7x^2)^{\frac{2}{3}}$

15. $y = \log_{10} \sqrt{(x^2 - 7x)} + x^3$

16. $y = \log_{10} \sqrt[3]{(x^3 - 5x)} - x^4$

17. $y = \ln \sqrt{\dfrac{t + 1}{t - 1}}$

18. $y = \ln \sqrt{\dfrac{t - 1}{t + 1}}$

19. $y = \ln\left[\dfrac{x^2 + 5x}{x^{\frac{2}{3}}}\right]$

20. $y = \ln\left[\dfrac{x(x^3 + 1)^4}{x^2 - 1}\right]$

In Exercises 21–40, evaluate the integral.

21. $\displaystyle\int \frac{1}{t + 1}\, dt$

22. $\displaystyle\int \frac{1}{t + 2}\, dt$

23. $\displaystyle\int \frac{1}{2t + 3}\, dt$

24. $\displaystyle\int \frac{1}{3 - 2t}\, dt$

25. $\displaystyle\int \frac{x}{2x^2 + 3}\, dx$

26. $\displaystyle\int \frac{x^2}{4 - x^3}\, dx$

27. $\displaystyle\int \frac{1}{(x + 1)^2}\, dx$

28. $\displaystyle\int \frac{1}{(3x + 1)^2}\, dx$

29. $\displaystyle\int_{x=e}^{x=3} \left(4x^2 + \frac{1}{x}\right) dx$

30. $\displaystyle\int_{z=1}^{z=e} \left(3z + \frac{1}{z}\right) dz$

31. $\displaystyle\int_{z=3}^{z=8} \frac{1}{z - 2}\, dz$

32. $\displaystyle\int_{z=-1}^{z=3} \frac{1}{z + 3}\, dz$

33. $\displaystyle\int \frac{x + 3}{x^2 + 6x + 5}\, dx$

34. $\displaystyle\int \frac{x^2 - 4}{x^3 - 12x + 4}\, dx$

35. $\displaystyle\int \frac{\ln x}{x}\, dx$

36. $\displaystyle\int \frac{\ln x^2}{x}\, dx$

37. $\displaystyle\int \frac{1}{x \ln x}\, dx$

38. $\displaystyle\int \frac{1}{x \ln x^2}\, dx$

39. $\displaystyle\int_{x=1}^{x=e} \frac{(1 + \ln x)^2}{x}\, dx$

40. $\displaystyle\int_{x=0}^{x=1} \frac{x - 1}{x + 1}\, dx$

41. $\displaystyle\int_{x=-3}^{x=4} \frac{dx}{x} = \ln|x| = \ln 4 - \ln 3$

42. $\displaystyle\int_{x=2}^{x=11} \frac{dx}{x-5} = \ln|x-5| = \ln 6 - \ln 3$

43. Evaluate $\displaystyle\frac{d}{dx}\left(\int_{t=1}^{t=x} dt\right), x \geq 0.$

44. Determine the equation of the line tangent to the curve $y = \ln(x^2 - 2x - 2)$ when $x = 3$.

45. Determine the equation of the line tangent to the curve $y = \ln(x^2 - 3)$ when $x = 2$.

46. The loudness L of a sound is related to the intensity I of the vocal message relative to the threshold intensity I_0 of the ear. The loudness is measured in decibels and given by the equation $L = 10 \log\left(\dfrac{I}{I_0}\right)$. Suppose the intensity of a song increases in the following fashion: $I = 5t^2$, where t is the time in seconds. What is the rate of increase in loudness at the crescendo that occurs at the 10-second mark? (*Hint: I_0 may be set equal to 1.*)

47. The Super Bubble Gum Company has determined that the revenue and cost equations for their new product are projected to be $R(x) = 500 \ln(7x + 21)$ and $C(x) = \dfrac{x}{7}$.

    a. Determine the number of items the company should sell to maximize its profit.
    b. Determine the maximum profit.

48. An automobile repair shop determines that if x is the total number of hours worked per week by its mechanics, its revenue is given by the equation $R(x) = 480 \ln(x + 2)$. The corresponding cost function is $C(x) = 2400 + 3x$. Determine the total number of hours mechanics must work to maximize the shop's profit.

49. One type of telegraph cable consists of a conducting circular core surrounded by a circular layer of insulation. If x is the ratio of the radius of the core to the thickness of the insulation, then the speed of the signal is proportional to $x^2 \ln \dfrac{1}{x}$. For what value of x is the speed of the signal a maximum?

50. The true strain e_t of a material at length ℓ is related to its original length ℓ_0 by $e_t = \displaystyle\int_{\ell_0}^{\ell} \frac{d\ell}{\ell}$. What is the true strain at any length ℓ?

51. Determine the area of the region bounded by the curves $y = \dfrac{3x^2}{x^3 + 5}, y = 0$, and the lines $x = 1$, and $x = 3$.

52. Determine an equation of the line tangent to $y = (x + 1)(x + 2)^2(x + 3)^2$ at the point where $x = 0$.

53. In an experiment with bacteria, it is observed that the relative activeness A of the bacteria colony is $A = b \ln\left(\dfrac{T}{a - T} - a\right)$, where a and b are constants and T is the surrounding temperature. $\dfrac{dA}{dT}$ gives an expression that indicates how A changes with

respect to temperature T. Determine $\dfrac{dA}{dT}$.

54. At an inflation rate of r percent a year, prices double in approximately n years, where

$$n = \dfrac{\ln 2}{\ln(1 + r)}.$$

a. Determine $\dfrac{dn}{dr}$.

b. Determine the rate of change when $r = 3\%$.

c. Determine the rate of change when $r = 6\%$.

23.2

Derivatives and Integrals of Exponential Functions

Recall from our earlier discussions that the definition of the exponential function is:

$$y = b^x, \qquad b > 0 \quad \text{and} \quad b \neq 1.$$

To find the derivative of any function, we can use the definition of the derivative. However, with the rules developed in Section 23.2 for logarithms, we can now use these rules and do not have to start with the definition of the derivative.

We start by taking the natural logarithm of both sides of the exponential function.

$$y = b^x$$
$$\ln y = \ln b^x$$
$$\ln y = x \ln b \qquad \textbf{Rule 3 of logarithms}$$

Remember that y is a function of x. In order to find the derivative of y with respect to x $\left(\dfrac{dy}{dx}\right)$ we must use implicit differentiation.

$$\dfrac{d}{dx}(\ln y) = \dfrac{d}{dx}(x \ln b) \qquad \textbf{Implicit differentiation}$$

$$\dfrac{1}{y}\dfrac{dy}{dx} = \dfrac{dx}{dx}\ln b + 0 \qquad \textbf{Derivative of a constant is 0}$$

$$\dfrac{1}{y}\dfrac{dy}{dx} = \ln b$$

$$\dfrac{dy}{dx} = y \ln b \qquad \textbf{Solving for } \dfrac{dy}{dx}$$

$$= \boxed{b^x}\ \ln b \qquad \textbf{Recall } y = b^x \textbf{, substituting}$$

So the derivative of b^x has the same form as b^x with a constant ($\ln b$) thrown in.

If we let $b = e$, we can obtain an expression for finding the derivative of the function $y = e^x$.

$$y = e^x$$

$$\frac{dy}{dx} = e^x \ln e \quad \textbf{Letting } b = e$$

$$= e^x \quad \ln e = 1$$

The easiest function to differentiate is e^x. It does not matter how many times we differentiate e^x, the result is e^x. The simplicity of this function gives us another reason for introducing the irrational number e.

$$\frac{d^n(e^x)}{dx^n} = e^x$$

To find the derivative of e^u or b^u, where u is a function of $x[u = u(x)]$, we use the chain rule and obtain the following results.

$$y = e^u \qquad\qquad y = b^u$$

$$\frac{dy}{dx} = \frac{d}{dx}(e^u) \qquad \frac{dy}{dx} = \frac{d}{dx}(b^u)$$

$$= e^u \frac{du}{dx} \qquad\qquad = b^u \ln b \frac{du}{dx}$$

Differentiation Rules for Exponential Functions

For the composite function $u = u(x)$.

{4} If $y = e^u$, then $\dfrac{dy}{dx} = e^u \dfrac{du}{dx}$.

{5} If $y = b^u$, then $\dfrac{dy}{dx} = b^u \ln b \dfrac{du}{dx}$.

EXAMPLE 1 **Derivative of e^u**

Determine the derivative of $y = e^{3x+1}$.

Solution $y = e^{3x+1}$ is a composite function where $u = u(x) = 3x + 1$.

$$\frac{dy}{dx} = \frac{d(e^{(3x+1)})}{dx}$$

$$= e^{(3x+1)} \frac{d(3x + 1)}{dx} \quad \textbf{Differentiation rule \{4\}}$$

$$= (e^{(3x+1)})(3)$$

$$= 3e^{(3x+1)} \quad \blacklozenge$$

EXAMPLE 2 **Derivative of e^u**

Determine the derivative of $y = e^{(4-x^2)}$.

Solution $y = e^{(4-x^2)}$ is a composite function where $u = u(x) = 4 - x^2$.

$$\frac{dy}{dx} = \frac{d(e^{(4-x^2)})}{dx}$$

$$= e^{(4-x^2)}\frac{d(4-x^2)}{dx} \qquad \text{Differentiation rule \{4\}}$$

$$= e^{(4-x^2)}(-2x)$$

$$= -2xe^{(4-x^2)} \quad \blacklozenge$$

EXAMPLE 3 **Applying the Quotient Rule and the Exponential Rule**

Determine the derivative of $y = e^3 - \dfrac{e^{4x}}{x^4}$.

Solution We note that e^3 is a constant and $\dfrac{e^{4x}}{x^4}$ is a quotient; applying the quotient rule of derivatives, we have:

$$y = e^3 - \frac{e^{4x}}{x^4}$$

$$\frac{dy}{dx} = 0 - \frac{\left[x^4 \dfrac{d(e^{4x})}{dx} - e^{4x}\dfrac{d(x^4)}{dx}\right]}{(x^4)^2} \qquad \text{Quotient rule}$$

$$= -\left[\frac{x^4(\boxed{e^{4x}})(\boxed{4}) - e^{4x}(\boxed{4x^3})}{x^8}\right] \qquad \text{Differentiation rule \{4\}}$$

$$= \frac{-4x^4e^{4x} + 4x^3e^{4x}}{x^8}$$

$$= \frac{-4x^3e^{4x}(x-1)}{x^8} \qquad \text{Factoring}$$

$$= \frac{-4e^{4x}(x-1)}{x^5}. \quad \blacklozenge$$

EXAMPLE 4 **Derivative of b^u, $u = u(x)$**

Determine the derivative of $y = 5^{(3x+2)}$.

Solution $y = 5^{(3x+2)}$ is a composite function of the form b^u, where $b = 5$ and $u = 3x + 2$.

$$\frac{dy}{dx} = (\,\boxed{5^{3x+2}}\,)(\ln 5)\,\frac{d(\,3x+2\,)}{dx} \qquad \textbf{Differentiation rule \{5\}}$$

$$= (5^{3x+2})(\ln 5)(3)$$

$$= 3(5^{3x+2})(\ln 5) \quad \blacklozenge$$

EXAMPLE 5 **Application: Maximum Points, Minimum Points, Points of Inflection**

For the function $f(x) = x^2 e^x$, determine those values of x for which there are maximum points, minimum points, and points of inflection. Sketch the curve indicating the critical points.

Solution The values of x for which the first and second derivatives are equal to zero are the critical values of the function. Knowing the critical values, we then can determine whether we have maximum points, minimum points, or points of inflection.

$$f(x) = x^2 e^x$$
$$f'(x) = 2xe^x + x^2 e^x \qquad \textbf{Derivative}$$
$$f'(x) = xe^x(2 + x) \qquad \textbf{Factoring}$$

Setting both factors equal to zero, we have:

$$xe^x = 0 \qquad \text{or} \qquad 2 + x = 0;$$
$$xe^x = 0, \text{ when } x = 0 \quad \text{and} \quad x + 2 = 0, \text{ when } x = -2.$$

Therefore $x = 0$ and $x = -2$ are critical values. To determine whether we have maximum or minimum values for the function we apply the second-derivative test.

$$f''(x) = 2e^x + 2xe^x + 2xe^x + x^2 e^x$$
$$= 2e^x + 4xe^x + x^2 e^x$$
$$= e^x(2 + 4x + x^2) \qquad \textbf{Factoring}$$

Substituting $x = 0$ and $x = -2$ in $f''(x)$, we find that:

$$f''(0) > 0,$$

and therefore, $(0, 0)$ is a relative minimum point on the curve.

$$f''(-2) < 0,$$

and therefore, $(-2, 0.541)$ is a relative maximum point on the curve.

To determine whether or not there are any points of inflection we set the second derivative equal to zero.

$$e^x(2 + 4x + x^2) = 0$$

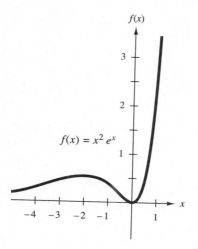

$f(x) = x^2 e^x$

This product can be zero only when $2 + 4x + x^2 = 0$. Using the quadratic formula, we find that the factor equals zero when $x = -2 \pm \sqrt{2}$. Therefore, the points of inflection on the curve are $(-0.586, 0.191)$ and $(-3.41, 0.384)$. Using the maximum point, minimum point, and the points of inflection as guides, we sketch the function in Figure 23.3. ◆

Now that we have rules for finding the derivatives of exponential functions, we easily can determine rules for the integration of exponential functions.

$$\int e^x \, dx = e^x + C \quad \text{and} \quad \int b^x \, dx = \frac{b^x}{\ln b} + C$$

Integration Formulas for Exponential Functions
For composite functions $u = u(x)$.

$$\{6\} \quad \int e^u \, du = e^u + C$$

$$\{7\} \quad \int b^u \, du = \frac{b^u}{\ln b} + C$$

We see from rules 6 and 7 that when e is the base, it greatly simplifies the process of finding the integral of a function. Thus, whenever possible, the exponential functions are expressed with a base e. To reflect this we provide more examples with base e.

CAUTION $\int 5^x \, dx \neq 5^{(x+1)} + C$. Verify that this statement is correct by differentiating the right side of the statement, showing that the derivative is not the same as the integrand. ■

EXAMPLE 6 Integral of e^u

Evaluate the indefinite integral $\int e^{(3x+2)} \, dx$.

Solution $e^{(3x+2)}$ is a composite function of the form e^u. If we let $u = 3x + 2$, then $\dfrac{du}{dx} = 3$, or we can write $\dfrac{du}{3} = dx$.

$$\int e^{(3x+2)} \, dx = \int e^u \, \frac{du}{3} \qquad \text{Substituting}$$

$$= \frac{1}{3} \int e^u \, du$$

$$= \frac{1}{3} e^u + C \qquad \text{Integration formula \{6\}}$$

$$= \frac{1}{3} e^{3x+2} + C \qquad \text{Substituting} \quad ◆$$

EXAMPLE 7   Application: Tangent Line

The slope of the tangent line to a curve is given by $\dfrac{dy}{dx} = 2x^2 e^{x^3}$. Determine the

equation of the curve if it passes through the point $\left(0, \dfrac{2}{3}\right)$.

Solution   The slope of the tangent line was determined by differentiating the function. Thus, to determine the function, we must integrate both sides of the equation.

$$\int dy = \int 2x^2 e^{x^3}\, dx$$

The function e^{x^3} is a composite function of the form e^u, where $u = x^3$. Differentiating with respect to x, we have $\dfrac{du}{dx} = 3x^2$ or $\dfrac{du}{3} = x^2\, dx$.

$$y = \int 2e^u \frac{du}{3} \qquad \textbf{Integrating and substituting}$$

$$= \frac{2}{3}\int e^u\, du$$

$$= \frac{2}{3} e^u + C \qquad \textbf{Integration formula \{6\}}$$

$$= \frac{2}{3} e^{x^3} + C \qquad \textbf{Substituting}$$

To determine the original equation, if the curve passes through the point $\left(0, \dfrac{2}{3}\right)$,

substitute 0 for x and $\dfrac{2}{3}$ for y to determine the value for C.

$$\frac{2}{3} = \frac{2}{3} e^{0^3} + C$$

Therefore, $C = 0$, and the equation of the curve is $y = \dfrac{2}{3} e^{x^3}$. ◆

EXAMPLE 8   Integral of b^u

Evaluate the definite integral $\int_{x=2}^{x=3} 10^x\, dx$.

Solution

$$\int_{x=2}^{x=3} 10^x \, dx = \left. \frac{10^x}{\ln 10} \right]_{x=2}^{x=3} \qquad \text{Integration formula \{7\}}$$

$$= \frac{10^{\boxed{3}}}{\ln 10} - \frac{10^{\boxed{2}}}{\ln 10} \qquad \text{Fundamental theorem}$$

$$= 391 \quad \blacklozenge$$

EXAMPLE 9   Application: Medical

The extensive research on drinking and driving has provided solid data relating the risk $R(\%)$ of having a motor vehicle accident to the blood alcohol level $b(\%)$. As mentioned in the chapter opener, the ingestion of a drug (alcohol) affects our reaction time. The graph in Figure 23.4 shows the effects of a 180 lb person consuming the indicated number of drinks over a two hour period. A drink could be a 12 oz serving of beer, a 5 oz serving of wine, or a 1.5 oz serving of liquor since all three of these volumes have the same alcohol content. The data from the research approximate an exponential curve. The function that seems to fit the data is $\dfrac{dR}{db} = kR$.

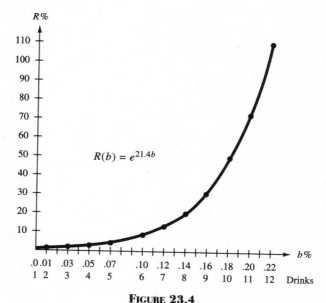

FIGURE 23.4

a. Determine the exponential equation. Assume initial conditions $b = 0\%$ and $R = 1\%$.

b. Determine k using the point (0.14, 20).

c. Write the exponential equation with the value of k determined in part b.

d. Determine at what blood alcohol level the risk of having an accident is greater than 50%.

Solution

a. To determine the exponential equation we write $\dfrac{dR}{db} = kb$ in the form:

$$\frac{dR}{R} = k\, db \quad \text{or} \quad \int \frac{1}{R}\, dR = \int k\, db.$$

$$\ln R = kb + C$$

$$\ln(1) = 0 + C \qquad\qquad\qquad \text{Initial conditions } R_0 = 1,\, b = 0$$

$$0 = C$$

$$\ln R = kb$$

$$e^{\ln R} = e^{kb}$$

$$R = e^{kb} \qquad\qquad\qquad\qquad e^{\ln x} = x$$

b. Substituting the values $b = 0.14$ and $R = 20$, into the equation found in part a, we have:

$$20 = e^{k\,0.14}$$

$$\ln 20 = \ln e^{0.14k} \qquad \text{Taking the ln of both sides}$$
$$\phantom{\ln 20 = \ln e^{0.14k}} \qquad \text{of the equation}$$

$$\ln 20 = 0.14k \qquad\quad \ln e^x = x$$

$$\frac{\ln 20}{0.14} = k$$

$$21.4 = k.$$

c.

$$R = e^{21.4\,b} \qquad \text{Substituting for } k$$

d. To determine at what blood alcohol level the risk of an accident is greater than 50%, we solve the inequality:

$$50 \le e^{21.4b}$$

$$\ln 50 \le \ln e^{21.4b}$$

$$\ln 50 \le 21.4b$$

$$\frac{\ln 50}{21.4} \le b$$

$$0.18 \le b.$$

Thus, when $b \ge 0.18\%$, there is a 50% or greater risk of a motor vehicle accident. The statistics indicate that when a 180 lb person drives after 12 drinks within a two-hour period, that person runs a 50% or greater chance of having an accident. In reality, any amount of alcohol may render a person unable to drive. ◆

23.2 Exercises

In Exercises 1–20, determine the derivative of the functions.

1. $y = e^t$

2. $y = e^x$

3. $y = e^{5x}$

4. $y = e^{-8x}$

5. $y = e^{(3x+5)}$

6. $y = e^{(7x-11)}$

7. $y = e^{t^2}$

8. $y = e^{t^3}$

9. $y = 5^{(x+1)}$

10. $y = 3^{(x-3)}$

11. $y = e^{\sqrt{x}}$

12. $y = e^{-\frac{1}{x^2}}$

13. $y = (e^{-x} + e^x)^2$

14. $y = xe^x - e^{-x}$

15. $y = (e^{3t} - 1)^4$

16. $y = (e^{t^3} + 4)^5$

17. $y = xe^{x \ln x}$

18. $y = \dfrac{e^x - e^{-x}}{e^x + e^{-x}}$

19. $y = 5^{(x^2-x)}$

20. $y = 7^{(4-3x^5)}$

In Exercises 21–38, evaluate the integral.

21. $\displaystyle\int e^t \, dt$

22. $\displaystyle\int e^2 \, dz$

23. $\displaystyle\int 4e^{4t} \, dt$

24. $\displaystyle\int 6e^{6t} \, dt$

25. $\displaystyle\int_{t=0}^{t=3} e^t \, dt$

26. $\displaystyle\int_{t=0}^{t=4} e^t \, dt$

27. $\displaystyle\int \dfrac{e^x - e^{-x}}{4} \, dx$

28. $\displaystyle\int \dfrac{e^x + e^{-x}}{4} \, dx$

29. $\displaystyle\int xe^{(5x^2+1)} \, dx$

30. $\displaystyle\int x^2 e^{(4-3x^3)} \, dx$

31. $\displaystyle\int 5^x \, dx$

32. $\displaystyle\int 7^x \, dx$

33. $\displaystyle\int_{x=0}^{x=2} (x^2 - 2)e^{(x3-6x+4)} \, dx$

34. $\displaystyle\int_{x=1}^{x=\sqrt{2}} (e^{4x} - 5)e^{4x} \, dx$

35. $\displaystyle\int e^x \sqrt{1 - e^x} \, dx$

36. $\displaystyle\int x^e \, dx$

37. $\displaystyle\int x \, 5^{x^2} \, dx$

38. $\displaystyle\int 8^{(7-3x^2)}(-6x \, dx)$

39. Find $\dfrac{d}{dx}\left(\displaystyle\int_{t=1}^{t=x} e^t \, dt \right)$

40. Find $\dfrac{d}{dx}\left(\displaystyle\int_{t=1}^{t=x} e^{(t+1)} \, dt \right)$

41. Determine the equation of the line tangent to the graph of $y = e^x$ when $x = 2$.

42. The equation $T = 50e^{-0.2t}$ is used to determine the temperature difference T between a warm body and its surroundings. The letter t indicates time in minutes.

 a. Determine $\dfrac{dT}{dt}$ when $t = 10$.

 b. Determine t when $\dfrac{dT}{dt} = -5.0°$ per minute.

43. The equilibrium constant k of a balanced chemical reaction changes with absolute temperature T, varying according to the law

$$k = \frac{k_o e^{[-q(T_0-T)]}}{T_0 T},$$

where k_0, q, and T_0 are constants. Determine the rate of change of k with respect to T.

44. For the function $f(x) = x^2e^{-x}$, determine those values of x for which there are maximum points, minimum points, and points of inflection. Sketch the curve and indicate the critical points.

45. The formula for compounding continuously is $A = Pe^{rt}$. If \$1000 is the total amount invested for t years at 6% compounded continuously, the total amount is given by $A(t) = 1000e^{0.06t}$.

   a. Determine $A'(t)$.
   b. Determine $A'(3)$ and interpret the results.
   c. Determine $A'(6)$ and interpret the results.

46. If \$20,000 is the total amount invested for t years at 5.5% compounded continuously, the total amount is given by $A(t) = 20,000e^{0.055t}$.

   a. Determine $A'(t)$.
   b. Determine $A'(2)$ and interpret the results.
   c. Determine $A'(5)$ and interpret the results.

47. The temperature of a heated cup of coffee T is related to the time elapsed t (in minutes) by $T = 150e^{-0.02t} + 65$, $(t \geq 0)$. Is the temperature increasing or decreasing?

48. $i = \left(\dfrac{E}{R}\right)(1 - e^{-\frac{Rt}{L}})$, where $E = 150$ V, $R = 15\ \Omega$, $L = 0.75$ henry, i is in amperes, and t is in seconds.

This is an equation giving the relation between current i and time t in a series circuit containing inductance L and resistance R. A source of 150 V is connected across this series circuit when $t = 0$. After that the current rises in accordance with the equation.

   a. Determine $\dfrac{di}{dt}$ when $t = 0$.
   b. Determine the maximum value of the current. When will the maximum value be reached?
   c. If the current increases at the same rate it was increasing when $t = 0$, how long does it take for the current to reach its maximum value?
   d. Compare the value found in part c with the ratio $\dfrac{L}{R}$, where L is the inductance and R is the resistance. This ratio is called in the time constant of the circuit.
   e. Sketch the graph of the function.

49. Show that $y = e^{-3x}$ satisfies the differential equation $\dfrac{d^2y}{dx^2} + 2\dfrac{dy}{dx} - 3y = 0$.

50. If $y = \dfrac{e^{2x} - 1}{e^{2x} + 1}$ show that $\dfrac{dy}{dx} = 1 - y^2$.

51. Use implicit differentiation to find $\dfrac{dy}{dx}$ for $e^{-x}\ln y + e^y \ln x = 4$.

52. If $\dfrac{dy}{dx} = \dfrac{2}{e^x}$ and $y = 0$ when $x = 5$, find y as a function of x.

53. If the rate of flow of revenue into a firm is given by $R'(t) = \dfrac{50{,}000{,}000}{t + 10}$, where t is measured in years, determine the total revenue obtained during the period $0 \le t \le 5$.

54. If the rate of flow of revenue into a firm is given by $R'(t) = 200t + 20e^{-t}$, where t is measured in years, determine the total revenue received during the time period $1 \le t \le 5$.

55. After a certain experimental drug is injected into an animal, infected cells are reduced at the rate of $D'(t)$ cells per day, where t is the number of days after the drug injection. If $D'(t) = 40e^{4t}$, determine the total decrease in infected cells during the first two days following the injection of the drug.

56. The temperature of a heated metal disk increases at the rate of $\dfrac{dT}{dt} = 3e^{\frac{t}{100}}$ deg/min.

Determine the temperature of the plate 5 min after being heated from its initial temperature of 68°.

57. For the function $y = e^{-x^2}$, perform the following.
  a. Determine the first derivative.
  b. Determine the critical values.
  c. Use the second-derivative test to determine whether the critical values are maximum or minimum values and determine the points on the curve.
  d. Use your graphing calculator to check the result found in part c.

58. The number of bacteria in a culture grows exponentially as $N = 10{,}000\, e^{0.20t}$, where t is measured in hours.
  a. At what rate is the population increasing when $t = 1.0$ hr?
  b. At what rate is the population increasing when $t = 10.0$ hr?
  c. At what rate is the population increasing when $t = 100.0$ hr? What constraints might alter this rate of increase?

59. The current that flows through a 1000 Ω resistor is given by $i = 10e^{\frac{-t}{100}}$. Due to a capacitor storing the charge, the current gradually gets smaller and smaller. The charge stored on the capacitor plates can be found from the equation $i = \dfrac{dg}{dt}$.
  a. What is the current flowing in the circuit after 10 seconds have passed? How much charge has been stored on the capacitor plates at this time?
  b. What is the current flowing in the circuit after 40 seconds have passed? How much charge has been stored on the capacitor plates at this time?

60. When light travels through any medium, some of it is absorbed. The amount of the light that is transmitted I is related to the distance x the light penetrates into the medium. If the initial intensity of the light is I_0 and the amount of absorption is related to α, then $I = I_0 e^{-\alpha x}$.
  a. Sketch the curve.
  b. Determine the first derivative of I with respect to x and explain what this expression means.
  c. Sketch the curve of the derivative.

23.3

Derivatives of the Sine and Cosine Functions

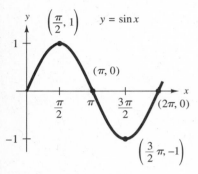

FIGURE 23.5

Now we look at trigonometric functions and their derivatives and integral expressions. Figure 23.5 shows a sine curve. By thinking about the slope of a tangent at various points along the sine curve, we can obtain a rough idea of what the derivative of the sine curve resembles.

Note that the sine curve hits a high at the point $\left(\dfrac{\pi}{2}, 1\right)$ and a low at the point $\left(\dfrac{3\pi}{2}, -1\right)$. At these two points we would expect the first derivative of the function $y = \sin x$ to be zero. At the point $(\pi, 0)$ the curve is falling, and we would expect the derivative to be negative. At the points $(0, 0)$ and $(2\pi, 0)$, the curve is rising, and we would except the derivative to be positive. The derivative at these points are examined in the exercises.

There are a few trigonometric identities that we need to help us in the development of the derivatives of the sine and cosine functions. They are listed in the box.

TRIGONOMETRIC IDENTITIES

1. $\sin (A + B) = \sin A \cos B + \cos A \sin B$

2. $\sin \left(\dfrac{\pi}{2} - A\right) = \cos A$

3. $\cos \left(\dfrac{\pi}{2} - A\right) = \sin A$

You may recall Section 19.1 stated that the concept of limits is an essential part of integral and differential calculus. There are two special limit statements that we need in our development of the rules for differentiating the sine and cosine functions.

Limits of Sine and Cosine

4. $\displaystyle\lim_{\Delta x \to 0} \dfrac{\sin \Delta x}{\Delta x} = 1$

5. $\displaystyle\lim_{\Delta x \to 0} \dfrac{\cos \Delta x - 1}{\Delta x} = 0$

We will not take the time to prove these limit statements. Proofs can be found in more advanced calculus books. We want to show that

$$\frac{d}{dx}(\sin x) = \cos x.$$

Using the definition of the derivative, we have:

$$\frac{d}{dx}(\sin x) = \lim_{\Delta x \to 0} \frac{\sin(x + \Delta x) - \sin x}{\Delta x}.$$

Using Identity 1, we have

$$\sin(x + \Delta x) = \sin x \cos \Delta x + \cos x \sin \Delta x;$$

so

$$\frac{d}{dx}(\sin x)$$

$$= \lim_{\Delta x \to 0} \left[\frac{\sin x \cos \Delta x + \cos x \sin \Delta x - \sin x}{\Delta x} \right] \qquad \textbf{Substituting}$$

$$= \lim_{\Delta x \to 0} \left[\frac{\sin x \cos \Delta x - \sin x + \cos x \sin \Delta x}{\Delta x} \right] \qquad \textbf{Rearranging}$$

$$= \lim_{\Delta x \to 0} \left[\frac{\sin x (\cos \Delta x - 1)}{\Delta x} + \frac{\cos x \sin \Delta x}{\Delta x} \right] \qquad \textbf{Factoring}$$

$$= \sin x \lim_{\Delta x \to 0} \left[\frac{(\cos \Delta x - 1)}{\Delta x} \right] + \cos x \lim_{\Delta x \to 0} \left[\frac{\sin \Delta x}{\Delta x} \right] \qquad \begin{array}{l}\textbf{Properties of}\\\textbf{limits}\end{array}$$

$$= (\sin x)(0) + (\cos x)(1) \qquad \textbf{Identity 1 and 2}$$

$$= \cos x.$$

Therefore, we see that $\frac{d}{dx}(\sin x) = \cos x$.

Can we now use this formula to find the derivative of $y = \sin kx$, where k is a constant? To do this we resort to our substitution technique and the chain rule. Both were discussed in Chapter 19.

To find the derivative of $y = \sin kx$ we let $u = kx$ and:

$$y = \sin u \qquad \textbf{Substituting}$$

$$\frac{dy}{dx} = \frac{d}{dx}(\sin u)\frac{du}{dx}. \qquad \textbf{Chain rule}$$

Since $u = kx$, then $\frac{du}{dx} = k$. Therefore,

$$\frac{dy}{dx} = (\cos u)(k), \qquad \textbf{Substituting}$$

or

$$\frac{dy}{dx} = k \cos kx.$$

> **Differentiation Rule for Sine Function**
> In general, if $u = u(x)$, we can state the following.
>
> $$\{8\} \text{ If } y = \sin u, \text{ then } \frac{dy}{dx} = \cos u \frac{du}{dx}.$$

EXAMPLE 1   Derivative of Sine Function

Determine the derivative of each of the following.

a. $y = \sin 3x$ **b.** $y = \sin (x + 2)$ **c.** $y = \sin x^3$

Solution

a. For $y = \sin 3x$, if we let $u = 3x$, then $\dfrac{du}{dx} = 3$. Substituting in the general formula, we have:

$$\frac{dy}{dx} = \cos u \frac{du}{dx} \quad \textbf{Differentiation rule \{8\}}$$

$$= (\,\text{cox } 3x\,)(\,3\,) \quad \text{or} \quad 3 \cos 3x.$$

b. For $y = \sin (x + 2)$, if we let $u = x + 2$, then $\dfrac{du}{dx} = 1$. Substituting in the general formula, we have:

$$\frac{dy}{dx} = \cos u \frac{du}{dx} \quad \textbf{Differentiation rule \{8\}}$$

$$= [\cos (\,x + 2\,)](\,1\,) \quad \text{or} \quad \cos (x + 2).$$

c. For $y = \sin x^3$, if we let $u = x^3$; then $\dfrac{du}{dx} = 3x^2$. Substituting in the general formula, we have:

$$\frac{dy}{dx} = (\cos x^3) \frac{d}{dx} (x^3) \quad \textbf{Differentiation rule \{8\}}$$

$$= (\,\cos x^3\,)(\,3x^2\,) \quad \text{or} \quad 3x^2 \cos x^3. \quad \blacklozenge$$

To determine the derivative of $y = \cos x$, we use $\dfrac{d}{dx} (\sin x) = \cos x$ and the identity $\cos x = \sin \left(\dfrac{\pi}{2} - x \right)$.

$$\frac{d}{dx}(\cos x) = \frac{d}{dx}\left[\sin\left(\frac{\pi}{2} - x\right)\right] \qquad \textbf{Identity 2}$$

$$= \cos\left(\frac{\pi}{2} - x\right)\frac{d}{dx}\left(\frac{\pi}{2} - x\right) \qquad \textbf{Differentiation rule \{8\}}$$

$$= \cos\left(\frac{\pi}{2} - x\right)(-1)$$

$$= -\sin x. \qquad \textbf{Identity 3}$$

Therefore, $\dfrac{d(\cos x)}{dx} = -\sin x$.

Differentiation Rule for Cosine Function

In general if $u = u(x)$, then we can state the following.

$\{9\}$ If $y = \cos u$, then $\dfrac{dy}{dx} = -\sin u \dfrac{du}{dx}$.

EXAMPLE 2 **Derivative of Cosine Function**

Determine the derivative of the following.

a. $y = \cos(2x - 4)$ **b.** $y = 4 \cos x^4$

Solution

a. For $y = \cos(2x - 4)$, if we let $u = 2x - 4$, then $\dfrac{du}{dx} = 2$. Substituting in the general formula, we have:

$$\frac{dy}{dx} = [-\sin(2x - 4)](2) \qquad \textbf{Differentiation rule \{9\}}$$

$$= -2\sin(2x - 4).$$

b. For $y = 4 \cos x^4$, if we let $u = x^4$, then $\dfrac{du}{dx} = 4x^3$. Substituting in the general formula, we have:

$$\frac{dy}{dx} = 4(-\sin x^4)(4x^3) \qquad \textbf{Differentiation rule \{9\}}$$

$$= -16x^3 \sin x^4. \quad \blacklozenge$$

EXAMPLE 3 **Application: Slope of Tangent Line**

Determine the slope of the graph $y = \sin 2x$ at $\left(\dfrac{\pi}{4}, 1\right)$ and sketch the graph showing the tangent line at that point.

Solution   To determine the slope of the curve $y = \sin 2x$ at $\left(\dfrac{\pi}{4}, 1\right)$ we must determine the derivative of:

$$y = \sin 2x$$

$$\frac{dy}{dx} = 2 \cos 2x.$$

FIGURE 23.6

At $\left(\dfrac{\pi}{4}, 1\right)$,

$$\frac{dy}{dx} = 2 \cos\left[2\left(\frac{\pi}{4}\right)\right]$$

$$= 2 \cos\frac{\pi}{2} \quad \text{or} \quad 0.$$

Since the slope is equal to zero we know that the tangent line is parallel to the x-axis at the point $\left(\dfrac{\pi}{4}, 1\right)$. The graph is shown in Figure 23.6.   ◆

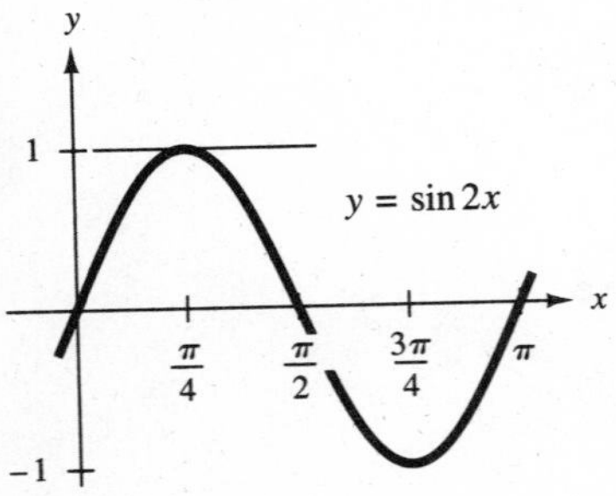

EXAMPLE 4   Derivative of the Product of Sine and Cosine

Determine the derivative of $y = \sin 3x \cos 2x$.

Solution   To determine the derivative of $y = \sin 3x \cos 2x$, we use the product rule.

$$\frac{dy}{dx} = \left[\frac{d}{dx}(\sin 3x)\right](\cos 2x) + (\sin 3x)\frac{d}{dx}(\cos 2x) \quad \textbf{Product rule}$$

$$= (3 \cos 3x)(\cos 2x) + (\sin 3x)(-2 \sin 2x) \quad \textbf{Differentiation rules \{8\}, \{9\}}$$

$$= 3 \cos 3x \cos 2x - 2 \sin 3x \sin 2x. \quad ◆$$

EXAMPLE 5   Applying the Chain Rule

Determine the derivative of $y = 2 \cos^3 (3x^4)$.

Solution   An equivalent way of writing the equation $y = 2 \cos^3 (3x^4)$ is $y = 2[\cos (3x^4)]^3$. Now,

$$\text{let } u = \cos (3x^4), \; y = 2u^3, \quad \text{and} \quad \frac{dy}{du} = 6u^2 ;$$

$$\text{let } v = 3x^4, \; u = \cos v, \quad \text{and} \quad \frac{du}{dv} = -\sin v ;$$

and since $v = 3x^4$,

$$\frac{dv}{dx} = 12x^3 .$$

Applying the chain rule, we have:

$$\frac{dy}{dx} = \frac{dy}{du}\frac{du}{dv}\frac{dv}{dx}$$

$$= (\,6u^2\,)(\,-\sin v\,)(\,12x^3\,) \qquad \textbf{Substituting}$$

$$= 6[\cos (3x^4)]^2\,[-\sin (3x^4)]\,(12x^3) \qquad \textbf{Substituting}$$

$$= -72x^3[\cos 3x^4]^2\,[\sin (3x^4)]. \quad \blacklozenge$$

In Example 5 we substituted twice in order to find the derivative. We could have done the differentiation directly, but the substitution shows each detail, and by doing this, fewer mistakes are made in differentiating.

23.3 Exercises

In Exercises 1–24, determine the derivative.

1. $y = \sin 3x$

2. $y = \sin 4x$

3. $y = \cos 5x$

4. $y = \cos 8x$

5. $y = 4 \sin e^x$

6. $y = 8 \sin e^{2x}$

7. $y = 7 \cos e^{2x}$

8. $y = 13 \cos e^{4x}$

9. $y = \cos \left(x + \dfrac{\pi}{2}\right)$

10. $y = 3 \cos \left(2x + \dfrac{\pi}{6}\right)$

11. $y = 3 \sin (\ln x)$

12. $y = 4 \cos (\ln 3x)$

13. $y = \sin x \cos x$

14. $y = \cos 2x \sin 3x$

15. $y = \ln(\sin 3x)$

16. $y = \ln(\cos 5x)$

17. $y = \dfrac{\sin x}{\cos x}$

18. $y = \dfrac{\cos x}{\sin x}$

19. $y = \sin^3 (\pi x^2)$

20. $y = 4 \cos^4 (x^2 - \pi)$

21. $y = \ln(\sin e^{2x})$

22. $y = \ln(\cos e^{3x})$

23. $y = e^{\sin 2x}$

24. $y = e^{\cos x^2}$

25. For the function $y = \sin x$ in the interval $0 \le x \le 2\pi$, perform the following.

   a. Determine $\dfrac{dy}{dx}$.

   b. Determine critical values.

   c. Determine the relative maximum and minimum points on the curve.

   d. Determine the intervals where the slope is positive.

   e. Determine the intervals where the slope is negative.

   f. Determine the points of inflection.

   g. Sketch and label the curve.

26. For the function $y = \cos x$ in the interval $0 \le x \le 2\pi$, perform the following.

   a. Determine $\dfrac{dy}{dx}$.

   b. Determine critical values.

   c. Determine the relative maximum and minimum points on the curve.

   d. Determine the intervals where the slope is positive.

   e. Determine the intervals where the slope is negative.

   f. Determine the points of inflection.

   g. Sketch and label the curve.

27. Show that $y = 2 \sin x + 3 \cos x$ satisfies the differential equation $\dfrac{d^2y}{dx^2} + y = 0$.

28. Show that $y = \sin 2x - \cos 2x$ satisfies the differential equation $\dfrac{d^2y}{dx^2} + 4y = 0$.

In Exercises 29–34, sketch the function for the interval $[0, 2\pi]$. Determine the slope of the tangent line to the curve at $(0, 0)$. Now compare the value found for the slope with the number of cycles in the interval $[0, 2\pi]$.

29. $y = \sin 2x$

30. $y = \sin x$

31. $y = \sin \dfrac{x}{2}$

32. $y = \sin \dfrac{3x}{2}$

33. $y = \sin \dfrac{x}{3}$

34. $y = \sin \dfrac{x}{4}$

35. Determine the equation of the line tangent to $y = \sin^2 x$ at $x = \dfrac{\pi}{3}$ and sketch the graph showing the tangent line at that point.

36. Determine the slope of the graph $y = \cos 2x$ at $\left(\dfrac{\pi}{4}, 0\right)$ and sketch the graph showing the tangent line at that point.

37. Determine the slope of the graph $y = \cos x \sin 2x$ at $\left(\dfrac{\pi}{2}, 0\right)$ and sketch the graph showing the tangent line at that point.

38. The amount of voltage produced by an electrical generator t seconds after starting is given by $v(t) = 120 \sin 120\pi t$. Determine the rate of change of voltage with respect to time $t = 30$ seconds after starting.

39. Records of the air quality on working days in a certain city show that the amount of air pollution t hours after midnight for a day is given by $P(t) = 2 + 0.6 \sin \dfrac{\pi}{12}(t - 5), 1 \le t \le 24$.

   a. During which hour of the day is there maximum pollution?
   b. During which hour of the day is there minimum pollution?

40. A section of the roller-coaster track at an amusement park is described by $r(x) = 2 \sin x - \cos 2x + 3$ for $0 \le x \le 2\pi$. Determine the relative maximum and minimum heights on this part of the track. Sketch the graph of $r(x)$. (Assume each unit is 10 ft.)

41. A mass is suspended vertically by a spring. When the system is stretched and then released, the mass oscillates up and down. Its position is given by $y = 8.0 \sin 0.50t$. The velocity of the object is the first derivative, or $\dfrac{dy}{dt}$. The acceleration of the object is the second derivative, or $\dfrac{d^2y}{dt^2}$.

   a. Determine the velocity.
   b. Evaluate the velocity of the object at 0.25 s and at 0.50 s.
   c. Determine the acceleration.
   d. Evaluate the acceleration at 0.25 s and 0.50 s.

42. The current in an alternating circuit is given by $i = 5.0 \sin(120\pi t)$, where t is measured in seconds. The voltage across a coil of wire is related to the rate of change of current $\dfrac{di}{dt}$ and given by $v = L\left(\dfrac{di}{dt}\right)$. If $L = 4.0$, what is the instantaneous voltage at $t = 12$?

43. Assuming that there is no air resistance, the range of a projectile is given by $R = \dfrac{v_0{}^2 \sin 2\theta}{g}$, where v_0 is the initial velocity, θ is the angle of the tilt of the cannon, and g is 9.8 m/s^2. Since the initial velocity of a shell is usually constant, the range to a target is governed by θ. For this problem assume v_0 is 400 m/s.
   a. When $\theta = 35°$, what is the range of the projectile?
   b. Determine $\dfrac{dR}{d\theta}$.
   c. What is $\dfrac{dR}{d\theta}$ when $\theta = 25°$?
   d. If the angle of determination is good to within $0.50°$, how much uncertainty is there in the range? (*Hint:* Now we can let $\Delta R \approx \dfrac{dR}{d\theta}\,\Delta\theta$, where ΔR is the uncertainty in the range.)
   e. At what angle(s) does a slight error in the angle alignment cause the most change in R?

44. The shear stress on a plane rotated ϕ degrees with respect to the longitudinal axis of a part with cross-sectional area A and loaded by a longitudinal force F is given by $s = \dfrac{F}{A}\cos^2 \phi$. For what orientation ϕ is the shear stress a maximum? A minimum?

45. To determine the vertical work done by a hydraulic ram it is necessary to differentiate $x^2 = (d \sin \theta - e)^2 + (a - d \cos \theta)^2$, where a, d, and e are constants. What is $\dfrac{dx}{d\theta}$?

46. In geodesic-dome design it is necessary to determine $\dfrac{dt}{da}$, where:

$$t = \left\{ s^2 + 2r_1 r_2 \left[\cos\left(\frac{360°}{n} + a\right) - \cos a \right] \right\}^{\frac{1}{2}},$$

and s^2, r_1, r_2, and n are constants. Determine $\dfrac{dt}{da}$.

23.4

Integration of the Sine and the Cosine Functions

In the previous section we learned that:

$$\frac{d}{dx}(\sin x) = \cos x \quad \text{and} \quad \frac{d}{dx}(\cos x) = -\sin x.$$

Combining these results with the definition of the indefinite integral of a function (Chapter 21), we can write the integral formulas.

Trigonometric Integration Formulas for Indefinite Integrals

$$\{10\} \quad \int \sin x \, dx = -\cos x + C$$

$$\{11\} \quad \int \cos x \, dx = \sin x + C$$

Having formulas for the indefinite integrals we can extend these immediately to the definite integral.

EXAMPLE 1   Area of a Region

Determine the area of the region bounded by $y = \cos x$, $y = 0$, $x = 0$, and $x = \dfrac{\pi}{2}$.

Solution   The curve $y = \cos x$ is sketched in Figure 23.7 showing the region for which we want to determine the area.

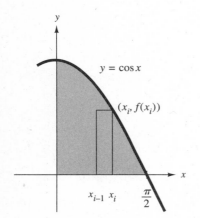

$y = \cos x$

$(x_i, f(x_i))$

$x_{i-1} \; x_i \quad \dfrac{\pi}{2}$

FIGURE 23.7

$$\text{Area} = \int_{x=0}^{x=\frac{\pi}{2}} \cos x \, dx$$

$$= \sin x \Big]_{x=0}^{x=\frac{\pi}{2}} \qquad \textbf{Integrating}$$

$$= \sin \frac{\pi}{2} - \sin 0 \qquad \textbf{Fundamental theorem}$$

$$= 1 \text{ square unit} \quad \blacklozenge$$

From the general differentiation formulas:

$$\frac{d}{dx}(\sin u) = \cos u \frac{du}{dx}$$

and

$$\frac{d}{dx}(\cos u) = -\sin u \frac{du}{dx},$$

we can determine the more general integration formulas.

Trigonometric Integration Formulas
Where $u = u(x)$.

$$\{12\} \quad \int \sin u \, du = -\cos u + C$$

$$\{13\} \quad \int \cos u \, du = \sin u + C$$

EXAMPLE 2   **Integrate Sine Using Substitution**

Evaluate the indefinite integral $\int \pi \sin 3\pi x \, dx$.

Solution   To evaluate the integral, we use the substitution technique. Let

$$u = 3\pi x;$$

then

$$\frac{du}{dx} = 3\pi,$$

or

$$\frac{du}{3} = \pi dx.$$

$$\int (\sin 3\pi x)(\pi \, dx) = \int \sin u \frac{du}{3} \qquad \textbf{Substituting}$$

$$= \frac{1}{3} \int \sin u \, du$$

$$= \frac{1}{3}(-\cos u) + C \qquad \textbf{Integration formula \{12\}}$$

$$= -\frac{1}{3} \cos 3\pi x + C. \qquad \textbf{Substituting}$$

To check the answer we would differentiate the answer, and the derivative should be the same as the original integrand. ◆

EXAMPLE 3   **Integral Cosine Using Substitution**

Evaluate the indefinite integral $\int x \cos x^2 \, dx$.

Solution   Using the substitution method, we let $u = x^2$, then

$$\frac{du}{dx} = 2x,$$

or

$$\frac{du}{2} = x \, dx.$$

$$\int (\cos x^2)(x \, dx) = \int \cos u \frac{du}{2} \qquad \textbf{Substituting}$$

$$= \frac{1}{2} \sin u + C \qquad \textbf{Integration formula \{13\}}$$

$$= \frac{1}{2} \sin x^2 + C \qquad \textbf{Substituting}$$

Check the answer. ◆

EXAMPLE 4 **Indefinite Integral Using Substitution**

Evaluate the indefinite integral $\int \sqrt[3]{\cos x} \sin x \, dx$.

Solution We can write the integral in the form of $u^n \, du$.

$$\int \sqrt[3]{\cos x} \sin x \, dx = \int (\cos x)^{\frac{1}{3}} \sin x \, dx$$

Let:

$$u = \cos x,$$

then:

$$\frac{du}{dx} = -\sin x$$

$$-du = \sin x \, dx.$$

$$\int (\cos x)^{\frac{1}{3}} \sin x \, dx = \int u^{\frac{1}{3}} (-du) \qquad \text{**Substituting**}$$

$$= -\int u^{\frac{1}{3}} \, du$$

$$= -\frac{u^{\frac{1}{3}+1}}{\frac{1}{3}+1} + C \qquad \text{**Integrating**}$$

$$= \frac{-3}{4} u^{\frac{4}{3}} + C$$

$$= -\frac{3}{4} (\cos x)^{\frac{4}{3}} + C \quad \blacklozenge$$

The average value of a periodic wave is given by the formula $v_{av} = \frac{1}{T} \int_0^T v \, dt$. An equivalent formula for a sine wave in terms of the angle θ is:

$$V_{av} = \frac{1}{2\pi} \int_{\theta=0}^{\theta=2\pi} v \, d\theta.$$

The angle θ in this case must be in radians.

EXAMPLE 5 **Application: Average Value**

Determine the average value of the function $v = V \sin \theta$ for the interval $[0, 2\pi]$.

Solution   To find the average value we use the equation:

$$v_{av} = \frac{1}{2\pi} \int_{\theta=0}^{\theta=2\pi} v \, d\theta$$

$$= \frac{1}{2\pi} \int_{\theta=0}^{\theta=2\pi} V \sin \theta \, d\theta \qquad \text{Substitution}$$

$$= \frac{-V}{2\pi} \cos \theta \bigg]_{\theta=0}^{\theta=2\pi} \qquad \text{Integrating}$$

$$= \frac{-V}{2\pi} (\cos 2\pi - \cos 0) \qquad \text{Fundamental theorem}$$

$$= 0.$$

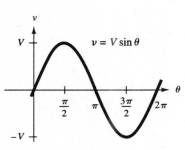

The average value of the sine wave over one complete cycle is zero. This is what we would expect when looking at the graph of the sine function in Figure 23.8 since the positive area cancels out the negative.   ◆

FIGURE 23.8

23.4   Exercises

In Exercises 1–18, evaluate the indefinite integrals and check your results by differentiating.

1. $\displaystyle\int \cos t \, dt$

2. $\displaystyle\int \sin t \, dt$

3. $\displaystyle\int \sin 3t \, dt$

4. $\displaystyle\int \cos 3t \, dt$

5. $\displaystyle\int \cos 5x \, dx$

6. $\displaystyle\int \sin \pi x \, dx$

7. $\displaystyle\int \sin x \cos x \, dx$

8. $\displaystyle\int \cos^5 x \sin x \, dx$

9. $\displaystyle\int \sin^4 x \cos x \, dx$

10. $\displaystyle\int \sqrt[3]{\sin^2 x} \cos x \, dx$

11. $\displaystyle\int x^2 \sin x^3 \, dx$

12. $\displaystyle\int (t + 3) \cos (t + 3)^2 \, dt$

13. $\displaystyle\int e^x \sin e^x \, dx$

14. $\displaystyle\int e^x \cos e^x \, dx$

15. $\displaystyle\int \frac{\cos x \ln(\sin x)}{\sin x} \, dx$

16. $\displaystyle\int \frac{\sin x}{\cos x} \, dx$

17. $\displaystyle\int \frac{\cos x}{\sin x} \, dx$

18. $\displaystyle\int \frac{\sin x \ln(\cos x)}{\cos x} \, dx$

In Exercises 19–22, evaluate the definite integrals and obtain exact results.

19. $\displaystyle\int_{x=\frac{\pi}{4}}^{x=\frac{\pi}{2}} \sin x \, dx$

20. $\displaystyle\int_{x=\frac{\pi}{6}}^{x=\frac{2\pi}{3}} \cos x \, dx$

21. $\displaystyle\int_{x=\frac{\pi}{2}}^{x=\pi} \cos \left(\frac{x}{2} + \pi\right) dx$

22. $\displaystyle\int_{x=\frac{\pi}{3}}^{x=\frac{\pi}{2}} x \sin x^2 \, dx$

In Exercises 23–26, evaluate the definite integral. Use a calculator to determine the result to the nearest thousandth.

23. $\displaystyle\int_{x=0.5}^{x=1.6} \cos x \, dx$

24. $\displaystyle\int_{x=1}^{x=2} \sin x \, dx$

25. $\displaystyle\int_{x=0.5}^{x=3.1} x \sin x^2 \, dx$

26. $\displaystyle\int_{x=0.75}^{x=2.5} x \cos x^2 \, dx$

In Exercises 27–30, sketch the region, indicate a sample rectangle, and determine the area.

27. $y = \sin x$, x-axis, and the lines $x = \dfrac{\pi}{4}$ and $x = \dfrac{3\pi}{4}$.   28. $y = \cos x$, x-axis, and the lines $x = \dfrac{\pi}{6}$ and $x = \dfrac{\pi}{2}$.

29. $y = \cos x$, x-axis, and the lines $x = \dfrac{\pi}{6}$ and $x = \dfrac{\pi}{4}$.   30. $y = \sin x \cos x$, x-axis, and the lines $x = 0$ and $x = \dfrac{\pi}{2}$.

In Exercises 31 and 32, determine the volume of the solid when the indicated region is revolved about the x-axis. Sketch the region and indicate a sample slice in the sketch.

31. $y = 3 \sin 2x$, the x-axis, and the lines $x = 0$ and $x = \dfrac{\pi}{8}$.

32. $y = \sin x$, and $y = \cos x$, and the lines $x = 0$ and $x = \dfrac{\pi}{4}$.

33. The velocity of a particle is given by $v = 3 \cos 2t$. Determine the distance traveled from $t = 0$ to $t = 1$ s.

34. The velocity of a particle is given by $v = 3 \sin 2t \cos 2t$. Determine the distance traveled from $t = 0$ to $t = 2$ s.

35. Power is defined as $\dfrac{dE}{dt}$, where E is the energy in a system at a given time t. The power of an electric circuit is given by $P = 100 \cos (120t)$.
    a. How much energy is delivered between $t = 0.012$ s and $t = 0.013$ s?
    b. How much energy is delivered between $t = 0.013$ s and $t = 0.014$ s?
    c. Since the time intervals in parts a and b are equal in duration, why is the energy output different?

36. What is the average value of the full-wave-rectified sine wave illustrated in the sketch? (*Hint:* See Example 5.)

37. What is the average value of the pulsed sine wave shown in the illustration? (*Hint:* $T = 10\pi$, see Example 5.)

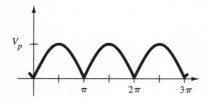

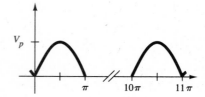

38. The impulse on a spherical satellite due to molecular collisions in a rarefied atmosphere is given by

$$I = KV_s^2 \left[\int_{\theta=0}^{\theta=\frac{\pi}{2}} \sin 2\theta \, d\theta + \frac{1}{2} \int_{\theta=0}^{\theta=\frac{\pi}{2}} \sin 4\theta \, d\theta \right],$$

where V_s is the velocity of the satellite and K is a constant based on the mass and density of the atmospheric molecules. Find I in terms of K and V_s.

23.5

Derivatives and Integrals of the Other Trigonometric Functions

The derivatives of the tangent, cotangent, secant, and cosecant functions can be found by expressing them in terms of the sine and cosine functions. To find the derivative of $y = \tan x$ we use the basic trigonometric identity $\tan x = \dfrac{\sin x}{\cos x}$.

$$y = \tan x$$

$$= \frac{\sin x}{\cos x} \qquad\qquad \textbf{Substituting}$$

$$\frac{dy}{dx} = \frac{\cos x(\cos x) - \sin x(-\sin x)}{(\cos x)^2} \qquad\qquad \textbf{Derivative of quotient}$$

$$= \frac{\cos^2 x + \sin^2 x}{\cos^2 x}$$

$$= \frac{1}{\cos^2 x} \qquad\qquad \textbf{Pythagorean identity}$$

$$= \sec^2 x \qquad\qquad \textbf{sec}\ \textit{x} = \frac{1}{\cos x}$$

Therefore,

$$\frac{d}{dx}(\tan x) = \sec^2 x.$$

Differentiation Rule for Tangent

In general if $u = u(x)$, then we can state the following.

$$\{14\} \quad \text{If } y = \tan u, \text{ then } \frac{dy}{dx} = \sec^2 u\, \frac{du}{dx}.$$

EXAMPLE 1   **Derivative of tan *u***

Determine the derivative of $y = 3 \tan 2x$.

Solution Let $u = 2x$, then $\dfrac{du}{dx} = 2$.

$$y = 3 \tan 2x$$
$$= 3 \tan u \qquad \text{Substituting}$$
$$\frac{dy}{dx} = 3\frac{d}{dx}(\tan u) \qquad \text{Differentiation rule \{14\}}$$
$$= 3 \sec^2 u \frac{du}{dx}$$
$$= 3(\sec^2 2x)(2) \qquad \text{Substituting}$$
$$= 6 \sec^2 2x$$

Thus, the derivative of $y = 3 \tan 2x$ with respect to x is $6 \sec^2 2x$. ◆

To find the derivative of $y = \cot x$, we use the basic trigonometric identity $\cot x = \dfrac{\cos x}{\sin x}$.

$$y = \cot x$$
$$= \frac{\cos x}{\sin x} \qquad \text{Substituting}$$
$$\frac{dy}{dx} = \frac{\sin x(-\sin x) - \cos x(\cos x)}{(\sin x)^2} \qquad \text{Derivative of quotient}$$
$$= \frac{-\sin^2 x - \cos^2 x}{\sin^2 x}$$
$$= \frac{-1(\sin^2 x + \cos^2 x)}{\sin^2 x} \qquad \text{Factoring}$$
$$= \frac{-1}{\sin^2 x} \qquad \text{Pythagorean identity}$$
$$= -\csc^2 x \qquad \csc x = \frac{1}{\sin x}$$

Therefore,

$$\frac{d}{dx}(\cot x) = -\csc^2 x.$$

Differentiation Rule for Cotangent
In general if $u = u(x)$, then we state the following.

$$\{15\} \quad \text{If } y = \cot u, \text{ then } \frac{dy}{dx} = -\csc^2 u \frac{du}{dx}.$$

EXAMPLE 2 **Derivative of cot u**

Determine the derivative of $y = 4 \cot kt$.

Solution Let $u = kt$, then $\dfrac{du}{dt} = k$.

$$y = 4 \cot kt$$

$$= 4 \cot u \qquad \textbf{Substituting}$$

$$\frac{dy}{dt} = 4 \frac{d}{dt}(\cot u) \qquad \textbf{Differentiation rule \{15\}}$$

$$= 4(-\csc^2 u)\frac{du}{dt}$$

$$= -4(\csc^2 kt)(k) \qquad \textbf{Substituting}$$

$$= -4k \csc^2 kt$$

Thus, the derivative of $y = 4 \cot kt$ with respect to t is $-4k \csc^2 kt$. ◆

To find the derivative of $y = \sec x$ we use the basic trigonometric identity $\sec x = \dfrac{1}{\cos x}$.

$$y = \sec x$$

$$= \frac{1}{\cos x} \qquad \textbf{Substituting}$$

$$\frac{dy}{dx} = \frac{\cos x(0) - 1(-\sin x)}{(\cos x)^2} \qquad \textbf{Differentiating}$$

$$= \frac{\sin x}{\cos^2 x}$$

$$= \frac{\sin x}{\cos x}\frac{1}{\cos x}$$

$$= \tan x \sec x \qquad \textbf{Trigonometric substitution}$$

Therefore,

$$\frac{d}{dx}(\sec x) = \tan x \sec x.$$

Differentiation Rule for Secant

In general if $u = u(x)$, then we can state the following.

$$\{16\} \quad \text{If } y = \sec u, \text{ then } \frac{dy}{dx} = \tan u \sec u \frac{du}{dx}.$$

EXAMPLE 3 **Derivative of sec *u***

Determine the derivative of $y = \sec \dfrac{x}{2}$.

Solution   Let $u = \dfrac{x}{2}$, then $\dfrac{du}{dx} = \dfrac{1}{2}$.

$$y = \sec \frac{x}{2}$$

$$= \sec u \qquad\qquad \text{Substituting}$$

$$\frac{dy}{dx} = \tan u \sec u \frac{du}{dx} \qquad\qquad \text{Differentiation rule \{16\}}$$

$$= \left(\tan \frac{x}{2} \sec \frac{x}{2} \right)\left(\frac{1}{2} \right) \qquad \text{Substituting}$$

$$= \frac{1}{2} \tan \frac{x}{2} \sec \frac{x}{2}$$

Therefore, the derivative of $y = \sec \dfrac{x}{2}$ with respect to x is $\dfrac{1}{2} \tan \dfrac{x}{2} \sec \dfrac{x}{2}$.   ◆

To find the derivative of $y = \csc x$ we use the basic trigonometric identity $\csc x = \dfrac{1}{\sin x}$.

$$y = \csc x$$

$$= \frac{1}{\sin x} \qquad\qquad \text{Substituting}$$

$$\frac{dy}{dx} = \frac{(\sin x)(0) - 1(\cos x)}{(\sin x)^2} \qquad \text{Differentiating}$$

$$= \frac{-\cos x}{\sin^2 x}$$

$$= \frac{-\cos x}{\sin x} \frac{1}{\sin x}$$

$$= -\cot x \csc x \qquad\qquad \text{Trigonometric substitution}$$

Therefore,

$$\frac{d}{dx} (\csc x) = -\cot x \csc x.$$

Differentiation Rule for Cosecant

In general if $u = u(x)$, then we can state the following.

$$\{17\} \quad \text{If } y = \csc u, \text{ then } \frac{dy}{dx} = -\cot u \csc u \frac{du}{dx}.$$

EXAMPLE 4   **Derivative of Sum of sin and csc**

Determine the derivative of $y = \sin x + 3 \csc x$.

Solution

$$y = \sin x + 3 \csc x$$

$$\frac{dy}{dx} = \cos x + 3(-\cot x \csc x) \quad \text{**Derivative of a sum**}$$

$$= \cos x - 3 \cot x \csc x$$

Therefore, the derivative of $y = \sin x + 3 \csc x$ with respect to x is $\cos x - 3 \cot x \csc x$. ◆

EXAMPLE 5 **Derivative of Difference of tan and sec**

Determine the derivative of $y = \tan^2 t - \sec^2 t$.

Solution If we let $u = \tan t$ and $v = \sec t$, where u and v are functions of t, $\frac{du}{dt} = \sec^2 t$ and $\frac{dv}{dt} = \tan t \sec t$.

$$y = \tan^2 t - \sec^2 t$$

$$= u^2 - v^2 \quad \text{**Substituting**}$$

$$\frac{dy}{dt} = 2u\frac{du}{dt} - 2v\frac{dv}{dt} \quad \text{**Differentiating**}$$

$$= 2(\,\tan t\,)(\,\sec^2 t\,) - 2(\,\sec t\,)(\,\tan t \sec t\,) \quad \text{**Substituting**}$$

$$= 2 \tan t \sec^2 t - 2 \tan t \sec^2 t$$

$$= 0$$

The result is what we would expect for the answer since by the Pythagorean identity,

$$\tan^2 t - \sec^2 t = 1,$$

and the derivative of a constant is 0. ◆

EXAMPLE 6 **Derivative of Product of cos and cot**

Determine the derivative of $y = \cos 5x \cot x^3$.

Solution To find the derivative of $y = \cos 5x \cot x^3$ we must use the product rule and the formulas for the derivative of cosine and cotangent. If we let $u = 5x$ and $v = x^3$, then $\frac{du}{dx} = 5$ and $\frac{dv}{dx} = 3x^2$.

$$y = \cos 5x \cot x^3$$

$$= \cos u \cot v \quad \text{**Substituting**}$$

$$\frac{dy}{dx} = \left[(-\sin u)\frac{du}{dx}\right](\cot v) + (\cos u)(-\csc^2 v)\frac{dv}{dx} \quad \text{**Product rule**}$$

$$= (-\sin 5x)(5)(\cot x^3) + (\cos 5x)(-\csc^2 x^3)(3x^2) \quad \text{**Substituting**}$$

$$= -5 \sin 5x \cot x^3 - 3x^2 \cos 5x \csc^2 x^3 \quad ◆$$

The following integration formulas can be written as a result of the derivative formulas that we developed in this section.

Trigonometric Integration Formulas

$$\{18\} \quad \int \sec^2 u \; du = \tan u + C$$

$$\{19\} \quad \int \sec u \tan u \; du = \sec u + C$$

$$\{20\} \quad \int \csc^2 u \; du = -\cot u + C$$

$$\{21\} \quad \int \csc u \cot u \; du = -\csc u + C$$

You may now be wondering where the integration formulas for $\tan u$, $\cot u$, $\sec u$, and $\csc u$ are.

We can find an integration formula for $\tan u$ by writing:

$$\int \tan x \; dx = \int \frac{\sin x}{\cos x} \; dx, \; \cos x \neq 0.$$

If we let $u = \cos x$, then $\dfrac{du}{dx} = -\sin x$ and $-du = \sin x dx$,

$$\int \tan x \; dx = -\int \frac{1}{u} \; du$$

$$= -\ln |u| + C.$$

If $\cos x \neq 0$, then $\int \tan x \; dx = -\ln |\cos x| + C.$

Since $-\ln |\cos x| = \ln \left| \dfrac{1}{\cos x} \right| = \ln |\sec x|$, then we can also write:

$$\int \tan x \; dx = \ln |\sec x| + C.$$

Integration Formula for Tangent

In the case where $u = u(x)$ we can write the general form:

$$\{22\} \quad \int \tan u \; du = \ln |\sec u| + C,$$

or

$$\int \tan u \; du = -\ln |\cos u| + C, \; \cos u \neq 0.$$

In the same manner, by using the elementary identity $\cot u = \dfrac{\cos u}{\sin u}$, we can develop an integral formula for $\cot u$.

> ### Integration Formula for Cotangent
> In the case where $u = u(x)$, we can write the general form:
>
> $$\{23\} \quad \int \cot u\, du = \ln |\sin u| + C,\ \sin u \neq 0.$$

The method of developing the formulas for integrating $\sec u$ and $\csc u$ are not as straightforward, so we just list the formulas.

> ### Integration Formulas for Secant and Cosecant
> In the case where $u = u(x)$ we can write the general form:
>
> $$\{24\} \quad \int \sec u\, du = \ln |\sec u + \tan u| + C;$$
>
> $$\{25\} \quad \int \csc u\, du = \ln |\csc u - \cot u| + C.$$

EXAMPLE 7 Integral of tan u

Evaluate the definite integral $\displaystyle\int_{x=0}^{x=\frac{\pi}{4}} \tan \frac{x}{2}\, dx$.

Solution If $u = \dfrac{x}{2}$, then $\dfrac{du}{dx} = \dfrac{1}{2}$, or $dx = 2\,du$. Also when $x = 0, u = 0$, and when $x = \dfrac{\pi}{4}, u = \dfrac{\pi}{8}$.

$$\int_{x=0}^{x=\frac{\pi}{4}} \tan \frac{x}{2}\, dx = \int_{u=0}^{u=\frac{\pi}{8}} \tan u(2\,du) \quad \textbf{Substituting}$$

$$= 2 \int_{u=0}^{u=\frac{\pi}{8}} \tan u\, du$$

$$= 2 \ln(\sec u)\Big]_{u=0}^{u=\frac{\pi}{8}}$$

We do not need to use the absolute value sign, since sec u is positive in the interval $0 \leq u \leq \dfrac{\pi}{8}$.

$$\int_{x=0}^{x=\frac{\pi}{4}} \tan \frac{x}{2}\, dx = 2\left[\ln\left(\sec \frac{\pi}{8} \right) - \ln(\sec 0) \right]$$

$$= 2(\ln 1.082 - \ln 1)$$

$$= 2(\ln 1.082) \quad \text{or} \quad 0.158 \quad \blacklozenge$$

EXAMPLE 8   **Integral of Product of tan and sec**

Evaluate the indefinite $\int \tan 3x \sec 3x\, dx$.

Solution   If we let $u = 3x$, then $\dfrac{du}{dx} = 3$ or $\dfrac{du}{3} = dx$.

$$\int \tan 3x \sec 3x\, dx = \frac{1}{3} \int \tan u \sec u\, du \qquad \text{Substituting}$$

$$= \frac{1}{3} \sec u + C \qquad\qquad \text{Integration formula \{19\}}$$

$$= \frac{1}{3} \sec 3x + C \qquad\qquad \text{Substituting} \quad \blacklozenge$$

EXAMPLE 9   **Indefinite Integral, csc u**

Evaluate the indefinite integral $\int (x + \csc 5x)\, dx$.

Solution   Using the rules of integration, we can write the integral as the sum of two integrals.

$$\int (x + \csc 5x)\, dx = \int x\, dx + \int \csc 5x\, dx.$$

The first integral on the right side is straightforward. To integrate the second integral let $u = 5x$; then $\dfrac{du}{5} = dx$.

$$\int (x + \csc 5x)\, dx = \frac{x^2}{2} + \frac{1}{5} \int \csc u\, du \qquad\qquad \text{Substituting}$$

$$= \frac{x^2}{2} + \frac{1}{5} \ln |\csc u - \cot u| + C \qquad \text{Integration formula \{25\}}$$

$$= \frac{x^2}{2} + \frac{1}{5} \ln |\csc 5x - \cot 5x| + C \qquad \text{Substituting} \quad \blacklozenge$$

EXAMPLE 10   Application: Area Between Curves

Determine the area of the region bounded between the curves $f(x) = \tan x$ and $g(x) = \sin x$ and the lines $x = \dfrac{\pi}{6}$ and $x = \dfrac{\pi}{3}$.

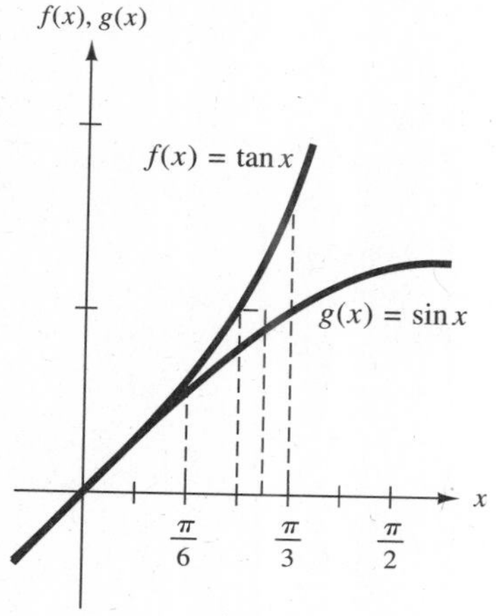

Solution   A sketch of the desired region, including a sample rectangle, is given in Figure 23.9. The height of the sample rectangle is found by taking the difference of the heights of the curves when $x = i$. That is, for the ith rectangle, the height is $f(x_i) - g(x_i)$, and the width is Δx_i.

$$
\begin{aligned}
\text{Area} &= \lim_{n \to \infty} \sum_{i=1}^{n} [f(x_i) - g(x_i)]\, \Delta x_i \\
&= \int_{x=\frac{\pi}{6}}^{x=\frac{\pi}{3}} (\tan x - \sin x)\, dx \qquad \text{\footnotesize \textbf{Definition of definite integral}} \\
&= (\ln|\sec x| + \cos x \Big]_{x=\frac{\pi}{6}}^{x=\frac{\pi}{3}} \\
&= \left(\ln \sec \frac{\pi}{3} + \cos \frac{\pi}{3} \right) - \left(\ln \sec \frac{\pi}{6} + \cos \frac{\pi}{6} \right) \\
&= 0.1833 \text{ square unit}
\end{aligned}
$$

Therefore, the area of the defined region is 0.183 square unit. ◆

$f(x), g(x)$

$f(x) = \tan x$

$g(x) = \sin x$

$\dfrac{\pi}{6}$ $\dfrac{\pi}{3}$ $\dfrac{\pi}{2}$

FIGURE 23.9

23.5   Exercises

In Exercises 1–24, determine the derivative of the functions.

1. $y = \tan 3x$

2. $y = 5 \tan \pi x$

3. $y = 2 \cot 3t$

4. $y = \cot 8x$

5. $y = \sec \pi t$

6. $y = 4 \sec 5x$

7. $y = 4 \csc 2x$

8. $y = \csc t$

9. $y = 2 \tan (x + 3)^2$

10. $y = 4 \tan^2 3x$

11. $y = 4 \cot \sqrt{x^2 - 1}$

12. $y = \cot^3 (x + 1)^2$

13. $y = \sec^2 x^3$

14. $y = \sec^4 (t - 1)$

15. $y = 2 \csc^3 (x + 2)$

16. $y = \dfrac{1}{2} \csc (t^2 + 1)^2$

17. $y = \dfrac{1 + \tan 2x}{\csc 3x}$

18. $y = \dfrac{\sec 2x}{1 - \cot x}$

19. $y = \dfrac{\sin^2 x + 1}{\tan^2 x}$

20. $y = \dfrac{\sec^2 x - 1}{\cos^2 x}$

21. $y = \tan 2x - \cot 2x$

22. $y = \sec 3x - \csc 3x$

23. $y = \dfrac{1}{2} \tan x - \tan^2 x$

24. $y = \dfrac{1}{2} \cot^2 x - \cot x$

25. Determine the slope of the line tangent to the curve $y = 3 \tan 2x$ at $\left(\dfrac{\pi}{8}, 3 \right)$.

26. Determine the slope of the line tangent to the curve $y = 2 \csc \left[3x - \left(\dfrac{\pi}{2} \right) \right]$ at $\left(\dfrac{\pi}{3}, 2 \right)$.

27. Determine the equation of the line tangent to the curve $y = 2 \cot 3x$ at $\left(\dfrac{\pi}{6}, 0\right)$.

28. Determine the equation of the line tangent to the curve $y = \sec\left[2x - \left(\dfrac{\pi}{4}\right)\right]$ at $\left[\dfrac{\pi}{8}, 1\right]$.

In Exercises 29–48, evaluate the integrals.

29. $\displaystyle\int \tan x \, dx$

30. $\displaystyle\int \cot x \, dx$

31. $\displaystyle\int_{\theta=0}^{\theta=\frac{\pi}{4}} \sec^2 \theta \, d\theta$

32. $\displaystyle\int_{x=0}^{x=1} (x - \sec x \tan x) \, dx$

33. $\displaystyle\int \tan 2x \sec 2x \, dx$

34. $\displaystyle\int \csc 4x \cot 4x \, dx$

35. $\displaystyle\int 4 \sec^2 4x \, dx$

36. $\displaystyle\int 3 \csc^2 3x \, dx$

37. $\displaystyle\int_{x=0}^{x=\frac{\pi}{4}} \tan 2\pi x \, dx$

38. $\displaystyle\int_{x=\frac{\pi}{8}}^{x=\frac{\pi}{4}} \cot 2\pi x \, dx$

39. $\displaystyle\int \cot 3x \sin 3x \, dx$

40. $\displaystyle\int x^2 \cot x^3 \csc x^3 \, dx$

41. $\displaystyle\int (\tan 3x + \sec 3x) \, dx$

42. $\displaystyle\int_{x=0}^{x=\frac{\pi}{4}} \sqrt{\tan x} \, \sec^2 x \, dx$

43. $\displaystyle\int \tan x \sec^2 x \, dx$

44. $\displaystyle\int \dfrac{\sec^2 x}{\sqrt{1 + \tan x}} \, dx$

45. $\displaystyle\int (\sin x)e^{\cos x} \, dx$

46. $\displaystyle\int e^x \tan e^x \, dx$

47. $\displaystyle\int \dfrac{\cos^2 x}{\csc x} \, dx$

48. $\displaystyle\int \dfrac{\sin^3 x}{\sec x} \, dx$

49. Determine the area of the region bounded by the curves $y = \sin 2x$ and $y = \tan x$ and the lines $x = -\dfrac{\pi}{4}$ and $x = \dfrac{\pi}{4}$.

50. For the curve $y = \tan x$, determine those values of x for which there are maximum points, minimum points, and points of inflection for the interval $0 \le x \le 2\pi$. Sketch the curve indicating the critical points.

51. In geodesic-dome design it is necessary to determine the derivative of the dip AB with respect to angle α, where

$$AB = \frac{(2 - 2\cos\theta - 2\sin\theta\tan\alpha + \tan^2\alpha)^{\frac{1}{2}}}{2\tan\alpha}.$$

Determine $\dfrac{d(AB)}{d\alpha}$, where θ is a constant.

23.6

Derivatives and Integrals of Inverse Trigonometric Functions

Before we consider the derivatives and integrals of inverse trigonometric functions, let us briefly review the definitions of these functions (Table 23.2).

To determine the derivative of $y = \text{Arcsin } x$, we use the definition of the inverse trigonometric function from Table 23.2. We have no way of differentiating Arcsin x, but we can differentiate the equivalent expression $\sin y = x$ using implicit differentiation.

TABLE 23.2

Function	Domain	Range
$y = \text{Arcsin } x$ if, and only if, $x = \sin y$.	$-1 \leq x \leq 1$	$\dfrac{-\pi}{2} \leq y \leq \dfrac{\pi}{2}$
$y = \text{Arccos } x$ if, and only if, $x = \cos y$.	$-1 \leq x \leq 1$	$0 \leq y \leq \pi$
$y = \text{Arctan } x$ if, and only if, $x = \tan y$.	$-\infty < x < \infty$	$\dfrac{-\pi}{2} < y < \dfrac{\pi}{2}$
$y = \text{Arccot } x$ if, and only if, $x = \cot y$.	$-\infty < x < \infty$	$0 < y < \pi$
$y = \text{Arcsec } x$ if, and only if, $x = \sec y$.	$|x| \geq 1$	$0 \leq y \leq \pi, y \neq \dfrac{\pi}{2}$
$y = \text{Arccsc } x$ if, and only if, $x = \csc y$.	$|x| \geq 1$	$\dfrac{-\pi}{2} \leq y \leq \dfrac{\pi}{2}, \quad y \neq 0$

$$\sin y = x$$

$$\cos y \frac{dy}{dx} = 1 \qquad \textbf{Implicit differentiation}$$

$$\frac{dy}{dx} = \frac{1}{\cos y} \qquad \textbf{Solving for } \frac{dy}{dx}$$

We recall the Pythagorean identity $\cos^2 y + \sin^2 y = 1$ or $\cos y = \pm\sqrt{1 - \sin^2 y}$. For the interval $-\dfrac{\pi}{2} < y < \dfrac{\pi}{2}$, $\cos y$ is positive, and we can write:

$$\cos y = \sqrt{1 - \sin^2 y}.$$

Also, since $\sin y = x$, $\sin^2 y = x^2$,

$$\cos y = \sqrt{1 - x^2}. \qquad \textbf{Substituting}$$

Therefore, since $y = \text{Arcsin } x$, $\dfrac{dy}{dx} = \dfrac{1}{\cos y}$, and $\cos y = \sqrt{1 - x^2}$,

$$\frac{d}{dx}(\text{Arcsin } x) = \frac{1}{\sqrt{1 - x^2}}. \qquad \textbf{Provided } |x| < 1$$

In a more general sense, if $u = u(x)$, where $u(x)$ is a differentiable function and $|u(x)| < 1$, applying the chain rule we have:

$$\frac{d}{dx}(\text{Arcsin } u) = \frac{1}{\sqrt{1 - u^2}}\frac{du}{dx}, |u| < 1.$$

In a similar manner we can develop differentiation rules for the other inverse trigonometric functions. To determine the rule for finding the derivative of $y = \text{Arctan } x$ we write the equivalent function:

$$\tan y = x$$

$$\sec^2 y \frac{dy}{dx} = 1 \qquad \textbf{Implicit differentiation}$$

$$\frac{dy}{dx} = \frac{1}{\sec^2 y}. \qquad \textbf{Solving for } \frac{dy}{dx}$$

Substituting $x = \tan y$ in the Pythagorean identity $\sec^2 y = 1 + \tan^2 y$, we obtain the equation:

$$\sec^2 y = 1 + x^2$$

$$\frac{d}{dx}(\text{Arctan } x) = \frac{1}{1 + x^2}. \qquad \textbf{Substituting}$$

With the use of the chain rule, the preceding rule can be generalized to:

$$\frac{d}{dx}(\text{Arctan } u) = \frac{1}{1 + u^2}\frac{du}{dx}.$$

Differentiation Rules for Inverse Trigonometric Functions

{26} If $y = \text{Arcsin } u$, then $\dfrac{dy}{dx} = \dfrac{1}{\sqrt{1 - u^2}}\dfrac{du}{dx}, |u| < 1.$

{27} If $y = \text{Arccos } u$, then $\dfrac{dy}{dx} = \dfrac{-1}{\sqrt{1 - u^2}}\dfrac{du}{dx}, |u| < 1.$

{28} If $y = \text{Arctan } u$, then $\dfrac{dy}{dx} = \dfrac{1}{1 + u^2}\dfrac{du}{dx}.$

{29} If $y = \text{Arccot } u$, then $\dfrac{dy}{dx} = \dfrac{-1}{1 + u^2}\dfrac{du}{dx}.$

{30} If $y = \text{Arcsec } u$, then $\dfrac{dy}{dx} = \dfrac{1}{u\sqrt{u^2 - 1}}\dfrac{du}{dx}, |u| > 1.$

{31} If $y = \text{Arccsc } u$, then $\dfrac{dy}{dx} = \dfrac{-1}{u\sqrt{u^2 - 1}}\dfrac{du}{dx}, |u| > 1.$

EXAMPLE 1 **Derivative of Arcsin u**

Determine $f'(x)$ if $f(x) = \text{Arcsin } x^4$.

Solution Using rule {26} with $u = x^4$, we have:

$$f'(x) = \frac{1}{\sqrt{1 - (x^4)^2}} \frac{d}{dx}(x^4)$$

$$= \frac{4x^3}{\sqrt{1 - x^8}}. \quad \blacklozenge$$

EXAMPLE 2 **Derivative of Arctan u**

Determine $f'(x)$ if $f(x) = $ Arctan e^{3x}.

Solution Using rule {28} with $u = e^{3x}$, we have:

$$f'(x) = \frac{1}{1 + (e^{3x})^2} \frac{d}{dx}(e^{3x})$$

$$= \frac{3e^{3x}}{1 + e^{6x}}. \quad \blacklozenge$$

EXAMPLE 3 **Derivative of Arcsec u**

Determine $\dfrac{dy}{dx}$ if $y = $ Arcsec x^3.

Solution Using rule {30} with $u = x^3$, we have:

$$\frac{dy}{dx} = \frac{1}{x^3 \sqrt{(x^3)^2 - 1}} \frac{d}{dx}(x^3)$$

$$= \frac{3x^2}{x^3 \sqrt{x^6 - 1}} \quad \text{or} \quad \frac{3}{x \sqrt{x^6 - 1}}. \quad \blacklozenge$$

As we found in previous cases, the differentiation rules provide us with the antidifferentiation formulas or integration formulas. The formulas most commonly used are listed here.

Integration Formulas

$$\{32\} \quad \int \frac{1}{\sqrt{1 - u^2}}\, du = \text{Arcsin } u + C, |u| < 1$$

$$\{33\} \quad \int \frac{1}{1 + u^2}\, du = \text{Arctan } u + C$$

$$\{34\} \quad \int \frac{1}{u \sqrt{u^2 - 1}}\, du = \text{Arcsec } u + C, |u| > 1$$

EXAMPLE 4   **Integral of Expression Equals Arcsin**

Evaluate $\displaystyle\int \frac{e^x}{\sqrt{1 - e^{2x}}}\, dx$

Solution   If we let $u = e^x$, then $\dfrac{du}{dx} = e^x$ or $du = e^x\, dx$. Using formula {32}, we have:

$$\int \frac{e^x}{\sqrt{1 - e^{2x}}}\, dx = \int \frac{1}{\sqrt{1 - u^2}}\, du \qquad \textbf{Substituting}$$

$$= \text{Arcsin } u + C$$

$$= \text{Arcsin } e^x + C. \qquad \textbf{Substituting} \quad \blacklozenge$$

EXAMPLE 5   **Integral of Expression Equals Arctan**

Evaluate $\displaystyle\int \frac{1}{a^2 + x^2}\, dx$, where a is a nonzero constant.

Solution   If we had a 1 in place of the a^2, we could use formula {33} to evaluate the integral. We can get a 1 in place of the a^2 by dividing the numerator and the denominator by a^2.

$$\int \frac{\dfrac{dx}{a^2}}{\dfrac{a^2}{a^2} + \dfrac{x^2}{a^2}} \quad \text{or} \quad \int \frac{\dfrac{dx}{a^2}}{1 + \dfrac{x^2}{a^2}}.$$

Now, if we let $u = \dfrac{x}{a}$, then $\dfrac{du}{dx} = \dfrac{1}{a}$, or $a\, du = dx$.

$$\int \frac{1}{a^2 + x^2}\, dx = \int \frac{\dfrac{a}{a^2}}{1 + u^2}\, du \qquad \textbf{Substituting}$$

$$= \frac{1}{a} \int \frac{1}{1 + u^2}\, du$$

$$= \frac{1}{a} \text{Arctan } u + C$$

$$= \frac{1}{a} \text{Arctan } \frac{x}{a} + C \qquad \textbf{Substituting} \quad \blacklozenge$$

Generalized results of formulas {32}, {33}, and {34} follow.

Integration Formulas

{35} $\displaystyle \int \frac{1}{\sqrt{a^2 - u^2}}\, du = \text{Arcsin } \frac{u}{a} + C,\, a > 0,\, |u| < a$

{36} $\displaystyle \int \frac{1}{a^2 + u^2}\, du = \frac{1}{a}\,\text{Arctan } \frac{u}{a} + C,\, a \neq 0$

{37} $\displaystyle \int \frac{1}{u\sqrt{u^2 - a^2}}\, du = \frac{1}{a}\,\text{Arcsec } \frac{u}{a} + C,\, a > 0,\, |u| > a$

EXAMPLE 6   Application: Largest View of Picture

A picture 3 ft high is hung with its bottom 2 ft above eye level. How far back should a person stand so that the picture appears to be the largest?

Solution   We want the angle the eye makes with the top and the bottom of the picture to be a maximum. We call this angle θ as illustrated in Figure 23.10. We

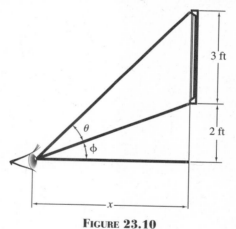

FIGURE 23.10

call the lower angle ϕ, and the distance from the eye to the wall is x ft. Thus we can write:

$$\tan (\theta + \phi) = \frac{5}{x}.$$

We want to maximize θ; therefore, we want to find $\dfrac{d\theta}{dx}$. It is more convenient to find $\dfrac{d\theta}{dx}$ if θ is by itself on one side of the equal sign. We can accomplish this by writing the equation in an equivalent form using the Arctangent function.

$$\theta + \phi = \text{Arctan } \frac{5}{x} \quad \text{and} \quad \theta = \text{Arctan } \frac{5}{x} - \phi$$

From Figure 23.10 we can see that $\tan \phi = \dfrac{2}{x}$, which we can write as $\phi = $ Arctan $\left(\dfrac{2}{x}\right)$.

$$\theta = \text{Arctan } \frac{5}{x} - \text{Arctan } \frac{2}{x} \qquad \textbf{Substituting}$$

$$\frac{d\theta}{dx} = \frac{1}{1 + \left(\dfrac{25}{x^2}\right)}\left(\frac{-5}{x^2}\right) - \frac{1}{1 + \left(\dfrac{4}{x^2}\right)}\left(\frac{-2}{x^2}\right) \qquad \textbf{Differentiating}$$

$$= \frac{-5}{x^2 + 25} + \frac{2}{x^2 + 4}$$

$$= \frac{-3x^2 + 30}{(x^2 + 25)(x^2 + 4)}$$

Solving for the critical value c, we find:

$$c^2 = 10 \quad \text{or} \quad c = \pm\sqrt{10}$$

Since c is a distance, it cannot be negative; thus, $c = \sqrt{10}$. But is this the distance that makes the angle θ a maximum? A quick check using the first-derivative test tells us that for $x < \sqrt{10}$, $\dfrac{d\theta}{dx} > 0$ and for $x > \sqrt{10}$, $\dfrac{d\theta}{dx} < 0$.

Thus, $\sqrt{10}$ is the distance one should be away from the picture for θ to be a maximum. Show that when $c = \sqrt{10}$, $\theta = 25.4$ degrees.   ◆

23.6 Exercises

In Exercises 1–20, determine the derivatives of the functions.

1. $f(x) = \text{Arcsin } 3x$

2. $f(x) = \text{Arccos } 5x$

3. $f(x) = \text{Arccos } (x + 4)$

4. $f(x) = \text{Arctan } (x - 5)$

5. $f(x) = \text{Arctan } 11x$

6. $f(x) = \text{Arcsin } x^2$

7. $f(x) = \text{Arccot } e^x$

8. $f(t) = \text{Arcsec } t^2$

9. $f(t) = \text{Arcsec } e^{2t}$

10. $f(z) = \text{Arccsc } z^3$

11. $f(t) = \text{Arccsc } (t + 3)$

12. $f(t) = \text{Arccot } (2t + 1)$

13. $f(x) = \ln(\text{Arcsin } x)$

14. $f(x) = e^x \text{ Arccos } e^{-x}$

15. $f(x) = \text{Arccos } x - \text{Arcsin } x$

16. $f(x) = \text{Arctan } x + \text{Arccot } x$

17. $f(t) = \ln(\text{Arctan } t^2)$

18. $f(t) = \text{Arcsin } (\ln t)$

19. $f(t) = \text{Arctan } \dfrac{x + 1}{x - 1}$

20. $f(t) = \text{Arcsec } \sqrt{x^3 - 1}$

In Exercises 21–40, evaluate the integrals.

21. $\displaystyle\int_{x=0}^{x=1} \frac{1}{x^2 + 1}\, dx$

22. $\displaystyle\int_{x=0}^{x=1} \frac{1}{\sqrt{1 - x^2}}\, dx$

23. $\displaystyle\int_{x=\sqrt{2}}^{x=2} \frac{1}{x\sqrt{x^2 - 1}}\, dx$

24. $\displaystyle\int_{x=-1}^{x=\sqrt{3}} \frac{1}{\sqrt{1 - \left(\dfrac{x^2}{4}\right)}}\, dx$

25. $\displaystyle\int_{x=2\sqrt{3}}^{x=\frac{9}{2}} \frac{1}{x\sqrt{x^2 - 1}}\, dx$

26. $\displaystyle\int_{x=-\frac{1}{\sqrt{3}}}^{x=1} \frac{1}{x\sqrt{4x^2 - 1}}\, dx$

27. $\displaystyle\int_{x=0}^{x=\frac{9}{2}} \frac{1}{\sqrt{a^2 - x^2}}\, dx,\ a > 0$

28. $\displaystyle\int \frac{1}{x\sqrt{4x^2 - 9}}\, dx$

29. $\displaystyle\int \frac{e^x}{\sqrt{16 - e^{2x}}}\, dx$

30. $\displaystyle\int \frac{\sec x \tan x}{1 + \sec^2 x}\, dx$

31. $\displaystyle\int \frac{1}{\sqrt{x}(1 + x)}\, dx$

32. $\displaystyle\int \frac{1}{x\sqrt{25x^2 - 1}}\, dx$

33. $\displaystyle\int \frac{x}{x^4 + 1}\, dx$

34. $\displaystyle\int \frac{\sec^2 x}{\sqrt{1 - \tan^2 x}}\, dx$

35. $\displaystyle\int \frac{\sin t}{\cos^2 t + 1}\, dt$

36. $\displaystyle\int \frac{1}{x\sqrt{1 - (\ln x)^2}}\, dx$

37. $\displaystyle\int \frac{1}{t\sqrt{t^2 - 7}}\, dt$

38. $\displaystyle\int \frac{e^{2t}}{\sqrt{e^{2t} - 81}}\, dt$

39. $\displaystyle\int \frac{1}{z\sqrt{7z^2 - 5}}\, dz$

40. $\displaystyle\int \frac{1}{\sqrt{13 - 11z^2}}\, dz$

41. Determine an equation of the line tangent to $y = \text{Arcsec } 5x$ at $x = \dfrac{2}{\sqrt{10}}$.

42. Determine the area of the region bounded by $y = \dfrac{1}{1 + x^2}$, the x-axis, and the lines $x = 0$ and $x = 2$.

43. Determine the volume of the solid generated when the region bounded by the graphs of $y = \dfrac{1}{\sqrt{1 + x^2}}$, $x = 1$, $x = 0$, and $y = 0$ is rotated about the x-axis.

44. A plane is flying at an altitude of 4 km at a speed of 130 km/hr away from an airport. How fast is the angle of elevation changing when the plane is above a point that is 16 km away from the airport? (See the sketch.)

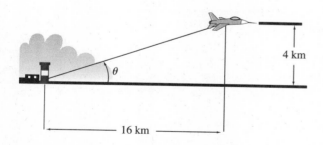

45. A company is placing a billboard that is 5 m high and 3 m above the eye level of a passing motorist along a highway. How far from the highway should the billboard be placed in order to maximize the angle it subtends the motorist's eyes?

46. A steel worker on one side of a building 200 ft wide sees a fellow worker drop a tool. (See the sketch.) As she watches the tool fall downward, she notices that her eyes first move slowly, then faster, then more slowly again. At what angle θ would the tool seem to be moving the most rapidly; that is, when would $\dfrac{d\theta}{dt}$ be a maximum? (*Hint:* Use an inverse trignometric function.)

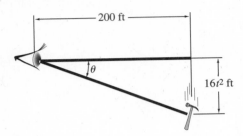

47. As Juan approaches a building, he must raise his head higher and higher to view the top of the building.
 a. Express the angle of elevation θ in terms of the height of the building h and the distance from the building x. For the purposes of this problem, assume that Juan's height is negligible.
 b. How does the angle of view change as Juan walks closer and closer? To discover this, determine $\dfrac{d\theta}{dx}$.

48. A lighthouse is located 3 mi off a straight shore. If the light revolves at 2 rpm, how fast is the beam moving along the coastline at a point 2 mi down the coast?

Review Exercises

In Exercises 1–20, determine the derivative of the function.

1. $y = \ln(x + 1)$
2. $y = \ln(2t + 3)$
3. $y = e^{\ln x}$
4. $y = \ln e^x$

5. $y = \log_{10}(x^2 + 3x + 5)$
6. $y = \log_5(x^3 + 7x)^{\frac{1}{2}}$
7. $y = e^{\sin x^2}$
8. $y = e^{\tan x}$

9. $y = \ln(\cos t^2)$
10. $y = \text{Arctan } e^{2t}$
11. $y = \cot(x^2 + 3)$
12. $y = \sec \dfrac{x + 1}{x + 2}$

13. $y = 5^{(x^2+1)}$
14. $y = 8^{(3x-4)}$
15. $y = \csc(t^2 + 1)$
16. $y = \text{Arcsin } (t^2 - 1)^{\frac{1}{2}}$

17. $y = \text{Arcsec } (t + 5)$
18. $y = e^x \sin x^3$
19. $y = e^{x^2} \tan (x + 2)^2$
20. $y = \sin^2 x \cos^2 x$

In Exercises 21–38, evaluate the integral.

21. $\displaystyle\int_{x=1}^{x=4} \dfrac{1}{x + 2}\, dx$

22. $\displaystyle\int_{t=1}^{t=3} \dfrac{1}{2t + 3}\, dt$

23. $\displaystyle\int x^2 e^{x^3}\, dx$

24. $\displaystyle\int \cos x \, e^{\sin x}\, dx$

25. $\displaystyle\int_{x=0}^{x=\frac{\pi}{6}} \sin 5x\, dx$

26. $\displaystyle\int_{x=0}^{x=\frac{\pi}{4}} \cos (4x + 1)\, dx$

27. $\displaystyle\int \sin^2 3x \cos 3x \, dx$

28. $\displaystyle\int \sec (2t + 1) \tan(2t + 1) \, dt$

29. $\displaystyle\int \sec (3z + 4) \, dz$

30. $\displaystyle\int \csc (11z - 9) \, dz$

31. $\displaystyle\int_{x=0}^{x=\frac{x}{6}} \tan (2x + 4) \, dx$

32. $\displaystyle\int \frac{\cot (\ln x) \, dx}{x}$

33. $\displaystyle\int \frac{1}{1 + 9x^2} \, dx$

34. $\displaystyle\int \cos^3 5x \sin 5x \, dx$

35. $\displaystyle\int \frac{e^x + e^{-x}}{e^x - e^{-x}} \, dx$

36. $\displaystyle\int \sec^2 3x \, dx$

37. $\displaystyle\int \csc^2 5t \, dt$

38. $\displaystyle\int [e^{\ln x} - (\sin^2 x + \cos^2 x)] \, dx$

In Exercises 39–42, first apply the rules of logarithms and then determine $\dfrac{dy}{dx}$.

39. $y = x\sqrt[3]{1 + x^2}$

40. $y = \sqrt[5]{\dfrac{x - 1}{x + 1}}$

41. $y = \dfrac{(x^2 - 8)^{\frac{1}{3}} \sqrt{x^3 + 1}}{x^6 - 7x + 5}$

42. $y = \dfrac{\sin x \cos x \tan^3 x}{\sqrt{x}}$

43. A linear speed of a point on the tread of an automobile tire is given by the formula $v = 32\pi \sin 8\pi t$.

   a. Determine $\dfrac{dv}{dt}$.

   b. Determine $\dfrac{dv}{dt}$ when $t = 2$ s.

44. If inflation makes the dollar worth 5% less each year, the value of $100 in t years is $v = 100e^{-0.054t}$. What is the approximate change in the value during the fourth year?

45. For the curve $y = \sin x$, determine those values of x for which there are maximum points, minimum points, and points of inflection on the curve for the interval $0 \le x \le 2\pi$.

46. A 22-ft ladder leans against a vertical wall. If the bottom of the ladder slides away from the base of the wall at the rate of 4 ft/s, how fast is the angle between the ladder and the wall changing when the top of the ladder is 20 ft above the ground?

47. Find the equation of the line tangent to the curve $y = 2e^{-4x^2}$ when $x = 0$.

48. Find the area of the region bounded by the curves $y = \tan x$, $y = \sec^2 x$, and the lines $x = 0$ and $x = \dfrac{\pi}{4}$.

49. For a person at rest, the velocity v (in liters per second) of air flowing during a respiratory cycle is $v = 0.85 \sin \dfrac{\pi t}{4}$, where t is the time in seconds. Find the volume (in liters) of air inhaled during one cycle by integrating this function over the interval $0 \le t \le 2$.

50. A rocket ship is sitting vertically on a launch pad. The distance from the ground to the tip of the rocket is 40 m. If the rocket blasts off at the rate of 800 m/s, at what rate is the angle of elevation changing when the rocket is 100 m above the ground? (See the diagram.) The observer is 1 km from the launch pad.

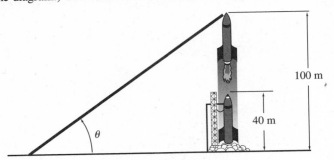

51. Scientists have proved that cooking food changes its nutritional content. If the nutritional content of a certain food is determined by the formula $y = \dfrac{x}{2 + \ln x}$, how long should the food be cooked to maximize nutritional content?

52. When the position of objects that are very far away are measured, angles are often used. Suppose a math text is dropped out of the window of the World Trade Center and a surveyor at ground level watches its fall. The book is dropped from a height of 200 m, and the surveyor is 300 m away. The height of the book changes in the following fashion: $h = 200 - 4.9t^2$.
  a. Write an expression for the angle seem by the surveyor.
  b. How fast is the angle of elevation changing after 1.0 s?
  c. How fast is the angle of elevation changing after 2.0 s?

53. The frictional force on an incline plane varies with the angle of the incline by $f(\theta) = \mu ng \cos \theta$. What is the first derivative of the function? Explain what it means.

54. As an object falls, it encounters air resistance, and its speed is governed by $v = v_t(1 - e^{-kt})$, where v_t is the terminal velocity of the object and k is related to its mass and the viscosity of the fluid through which the object falls. For this problem v_t and k are constants.
  a. Determine the expression for the acceleration of this object. Recall that $a = \dfrac{dv}{dt}$.
  b. Determine the expression for the displacement y of the object. Recall that $v = \dfrac{ds}{dt}$.

*55. For the function $y = \ln x$ show that if $x < 0$, then $\dfrac{d}{dx}(\ln |x|) = \dfrac{1}{x}$ and if $x > 0$, then $\dfrac{d}{dx}(\ln |x|) = \dfrac{1}{x}$.

*56. Solve $(\ln x)^2 - 4 \ln x - 8 = 0$.

*57. Let $f(x) = e^x - 1 - x$. Show that $f'(x) \geq 0$ if $x \geq 0$ and that $f'(x) \leq 0$ if $x \leq 0$. Use this fact to prove that $e^x \geq 1 + x$ and that $e^{-x} \geq 1 - x$.

*58. Let $f(x) = ce^{kx}$. Find a formula for the nth derivative of $f(x)$.

*59. **a.** Evaluate $\int \sin x \cos x \, dx$ by two methods. First let $u = \sin x$, then let $u = \cos x$.
   b. Explain why the two apparently different answers obtained in part a are really equivalent.

*60. Determine the maximum and minimum values of $y = a \sin x + b \cos x$. Show that it has a point of inflection whenever it crosses the x-axis.

*61. The functions $y = \sinh x$ and $y = \cosh x$ are called hyperbolic functions. Both of these functions can be defined in terms of the exponential e^x.

$$\sinh x = \frac{e^x - e^{-x}}{2} \quad \cosh x = \frac{e^x + e^{-x}}{2}.$$

   a. Show that the derivative of $\sinh x = \cosh x$
   b. Show that the derivative of $\cosh x = \sinh x$.
   c. Show that $\cosh^2 x - \sinh x^2 = 1$.

*62. Write a computer program to find an approximate value of

$$\lim_{x \to 0}(1 + x)^{\frac{1}{x}}.$$

Does the value obtained by the computer program approach the value of e?

✍ Writing About Mathematics

1. A classmate has difficulty understanding why it is not possible to apply the rule $\int x^n \, dx$ when determining the solution to $\displaystyle\int \frac{dx}{x}$. Write a few paragraphs explaining why the rule cannot be used and include instructions on the proper method of evaluating the integral.

2. When we evaluate an integral whose solution is a natural logarithmic function sometimes we use the absolute value symbol in the solution and other times we do not. Write a paragraph or two indicating under what conditions both forms of the solution may be correct.

3. You are designing a new roller-coaster ride at an amusement park. The path of the ride is defined by a sinusoidal function. The function may be similar to the function in Exercise 40 in Section 23.3. Explain how you can determine the maximum heights, the minimum heights, the slope of the climb, and the slope of the descent of the ride. Also in your report explain why this information is important in the design.

4. You are asked to determine the area of the region bounded by $y = \cos x$, $y = 0$, $x = 0$, and $x = \pi$. You evaluate the integral $\int_{x=0}^{x=\pi} \cos x \, dx$ and obtain the result 0, which you know is an incorrect answer. Write a short paper explaining why the answer is incorrect and include a correct procedure for determining the correct answer.

5. For the function $y = 10 \sin 2\theta$, determine the first and second derivatives for the interval $0 \le \theta \le 2\pi$. Write an explanation of how a maximum or a minimum can be verified using the first and second derivatives.

6. This is a particularly challenging chapter. Decide which material is unclear to you. Write a letter to the author, citing these areas and explaining why you are confused.

Chapter Test

In Exercises 1–8, determine the derivative of the function.

1. $y = \ln(x^2 + 3)$

2. $y = 6(x + 1)$

3. $y = \dfrac{\sin x}{\cos x}$

4. $y = 7 \tan \pi x$

5. $y = 3 \sec 2x$

6. $y = \csc(3x + 4)$

7. $y = \text{Arcsin}(2x + 3)$

8. $y = \text{Arc} \tan e^x$

In Exercises 9–16, evaluate the integral.

9. $\displaystyle\int xe^{3x^2}\, dx$

10. $\displaystyle\int x\,6^{x^2}\, dx$

11. $\displaystyle\int e^x \cos e^x\, dx$

12. $\displaystyle\int \sin^3 x \cos x\, dx$

13. $\displaystyle\int \sec^2 2x\, dx$

14. $\displaystyle\int \tan 3x \sec 3x\, dx$

15. $\displaystyle\int_{x=0}^{x=2} \dfrac{1}{x^2 + 1}\, dx$

16. $\displaystyle\int_{x=0.50}^{x=1.0} \dfrac{1}{\sqrt{1 - x^2}}\, dx$

In Exercises 17 and 18, sketch the region, indicate a sample strip, and determine the area.

17. $y = \tan 2x$, $y = 0$, $x = 0$, and $x = \dfrac{\pi}{8}$.

18. $y = \cos x$, $y = 0$, $x = \dfrac{\pi}{6}$, and $x = \dfrac{\pi}{2}$.

19. Determine the slope of the graph $y = \sin 2x$ at $\left(\dfrac{\pi}{4}, 1\right)$ and sketch the graph showing the tangent line at that point.

20. If the rate of flow of revenue into a firm is given by $R'(t) = \dfrac{20,000,000}{t + 10}$, where t is measured in years, determine the total revenue during the period $0 \le t \le 5$.

21. For the function $f(x) = 3xe^{-x}$, perform the following.
   a. Determine those values of x for which there are maximum points.
   b. Determine those values of x for which there are minimum points.
   c. Determine those values of x for which there are points of inflection.
   d. Sketch the curve and label key points.

22. A machine is programmed to move an etching tool back and forth according to the equation $x = 2.3 \cos 3.0t$.
   a. What is the equation for the speed of the tool?
   b. What is the acceleration of the tool at $t = 0.20$ s?

Methods of Integration

Alternating currents, which dominate modern society, are easily described by $i = I_0 \cos 120t$. However, things are actually more complex than this equation is. All circuits have a discharging period during which the equation does not work. This discharging period develops exponentially, and the equation is altered to become $i = I_0 e^{-2t} \cos 120t$. If we want to determine how much charge passes through a point as time progresses, we must use the equation $i = \dfrac{dq}{dt}$. Ultimately, then we must rely on integration. But, now we have a complex function. How do we integrate both an exponential function and, at the same time, a trigonometric function? In Example 3 in Section 24.1 we illustrate how to integrate the product of trigonometric and exponential functions.

24.1

Integration by Parts

In Chapters 21 and 23 we learned some of the basic techniques of integration to solve problems like $\int x^2\,dx$ and $\int \sin x\,dx$. But, how do we evaluate an integral whose integrand is the product of two functions such as:

$$\int x \sin x\,dx \quad \text{and} \quad \int xe^x\,dx \quad \text{and} \quad \int x \ln x\,dx?$$

To solve integrals of this type, we introduce a new technique called integration by parts.

To help us develop this technique, recall the product rule for differentiation. Assuming that u and v are both functions of x, then:

$$\frac{d}{dx}(uv) = u\frac{dv}{dx} + v\frac{du}{dx}.$$

Since u and v are both functions of x, we can use a shorthand notation to express the derivative:

$$\frac{dv}{dx} = v' \quad \text{and} \quad \frac{du}{dx} = u'.$$

$$\frac{d}{dx}(uv) = uv' + vu' \qquad \text{Substituting}$$

$$uv' = \frac{d}{dx}(uv) - vu' \qquad \text{Solving for } uv'$$

$$\int uv'\,dx = \int \frac{d}{dx}(uv)\,dx - \int vu'\,dx \qquad \text{Integrating with respect to } x$$

Recall that $\int \frac{d}{dx}[f(x)]\,dx = f(x)$; thus, $\int \frac{d}{dx}(uv)\,dx = uv$. Now, if we let u and v' represent the functions in the integrand, we have a method for integration by parts.

$$\int uv'\,dx = uv - \int vu'\,dx$$

The result obtained uses the product rule to reverse the process of differentiation and gives the formula for integration by parts.

Integration-by-Parts Formula

$$\{1\} \quad \int uv'\,dx = uv - \int vu'\,dx$$

EXAMPLE 1 **Integration by Parts: Polynomial Times an Exponential**

Evaluate the integral $\int xe^x \, dx$.

Solution To evaluate the integral we use the technique of integration by parts. Let $u = x$ and $v' = e^x$.

$$\frac{d(u)}{dx} = \frac{d(x)}{dx} \qquad \int v' dx = \int e^x \, dx$$

$$u' = 1 \qquad v = e^x$$

Substituting in the formula:

$$\int uv' \, dx = uv - \int vu' \, dx, \qquad \textbf{Integration-by-parts formula \{1\}}$$

we have,

$$\int xe^x \, dx = x \, (\, e^x) - \int e^x \, (\, 1) \, dx \qquad \textbf{Substituting}$$

$$= xe^x - e^x + C.$$

Thus, $\int xe^x \, dx = xe^x - e^x + C.$ ◆

The constant of integration appears only in the last step. If we integrate more than once, we can combine the constants as a single constant.

What would have happened in Example 1 if we had let $u = e^x$ and $v' = x$?

$$\frac{d(u)}{dx} = \frac{d}{dx}(e^x) \qquad \int v' \, dx = \int x \, dx$$

$$u' = e^x \qquad v = \frac{x^2}{2}$$

Substituting in the integration-by-parts formula, we have:

$$\int xe^x \, dx = (e^x)\left(\frac{x^2}{2}\right) - \int \left(\frac{x^2}{2}\right)e^x \, dx.$$

This result is a more difficult integral on the right-hand side of the equation than the original integral. The choice of u and v' can make the resulting expression easier or more difficult to evaluate—or impossible to evaluate. Nevertheless, regardless of the choice for u and v', if the expression can be evaluated, the result is the same. A few helpful hints that take some of the guesswork out of which function to choose for u and which function to choose for v' are given in the box.

SUGGESTED PROCEDURES

1. Let v' be equal to the most complicated part of the integrand that we can easily integrate.
2. Let u be equal to that part of the integrand such that u' is a simpler function than u.

These hints must be used with good judgment since they do not work in every case.

EXAMPLE 2   **Integration by Parts: Polynomial Times a Logarithm**

Evaluate the integral $\int x^2 \ln x \, dx$.

Solution   Since it is difficult to integrate $\ln x$, we let $u = \ln x$ and $v' = x^2$.

$$u' = \frac{d(\ln x)}{dx} \qquad v = \int x^2 \, dx$$

$$= \frac{1}{x} \qquad\qquad = \frac{x^3}{3}$$

Therefore,

$$\int x^2 \ln x \, dx = \ln x \left(\frac{x^3}{3}\right) - \int \left(\frac{x^3}{3}\right)\left(\frac{1}{x}\right) dx + C \qquad \text{Integration-by-parts formula \{1\}}$$

$$= \frac{x^3}{3} \ln x - \frac{1}{3} \int x^2 \, dx + C$$

$$= \frac{x^3}{3} \ln x - \frac{1}{3}\frac{x^3}{3} + C \qquad \text{Integrating}$$

$$= \frac{x^3}{3} \ln x - \frac{x^3}{9} + C.$$

Thus, $\int x^2 \ln x \, dx = \dfrac{x^3}{3} \ln x - \dfrac{x^3}{9} + C.$ ◆

EXAMPLE 3   **Integration by Parts: Exponential Times a Trigonometric Function**

Evaluate the integral $\int e^x \sin x \, dx$.

Solution In this problem the hints do not help, so we could let u be either e^x or $\sin x$. We let $u = e^x$ and $v' = \sin x$.

$$u' = \frac{d(e^x)}{dx} \qquad v = \int \sin x \, dx$$

$$= e^x \qquad\qquad = -\cos x$$

Now substitute in the formula.

$$\int e^x \sin x \, dx = e^x \, (\, -\cos x\,) - \int (\,-\cos x\,)\, e^x \, dx \qquad \textbf{Integration-by-parts}$$
$$\textbf{formula \{1\}}$$

$$= -e^x \cos x + \int e^x \cos x \, dx.$$

It appears that we have not made any progress since we cannot evaluate the new integral. However, the form of the new integral prompts us to apply the technique a second time and see what happens.

Let $u = e^x$ and $v' = \cos x$.

$$u' = \frac{d(e^x)}{dx} \qquad v = \int \cos x \, dx$$

$$= e^x \qquad\qquad = \sin x$$

Now substitute in the second formula.

$$\int e^x \sin x \, dx = -e^x \cos x + e^x \, \sin x - \int (\,\sin x\,)(\,e^x\,) \, dx \qquad \textbf{Integration-by-parts}$$
$$\textbf{formula \{1\}}$$

At this stage it looks as though we are back at the original problem. However, if we add $\int e^x \sin x \, dx$ to both sides, we have:

$$2 \int e^x \sin x \, dx = -e^x \cos x + e^x \sin x$$

$$\int e^x \sin x \, dx = \frac{-e^x \cos x + e^x \sin x}{2} + C. \qquad \begin{array}{l}\textbf{Divide both}\\ \textbf{sides by 2}\end{array}$$

Thus, $\int e^x \sin x \, dx = \dfrac{-e^x \cos x + e^x \sin x}{2} + C.$ ◆

If we are integrating a definite integral, the integration-by-parts formula is:

Integration-by-Parts Formula Definite Integral

$$\{2\} \quad \int_{x=a}^{x=b} uv' \, dx = uv \Big]_{x=a}^{x=b} - \int_{x=a}^{x=b} vu' \, dx$$

EXAMPLE 4   Integration by Parts: Definite Integral

Evaluate the integral $\int_{x=1}^{x=2} x(x-1)^6 \, dx$.

Solution   Using our hints we could let $v' = (x-1)^6$ since it is the most complex function and it is easily integrable. Also, letting $u = x$ means that u' is a simpler function than u. So we let $u = x$ and $v' = (x-1)^6$.

$$u' = \frac{d(x)}{dx} \qquad v = \int (x-1)^6 \, dx$$

$$= 1 \qquad\qquad = \frac{(x-1)^7}{7}$$

Therefore,

$$\int_{x=1}^{x=2} x(x-1)^6 \, dx = \left. \frac{x(x-1)^7}{7} \right]_{x=1}^{x=2} - \int_{x=1}^{x=2} \frac{(x-1)^7}{7} (1) \, dx$$

Integration-by-parts formula {2}

$$= \left[\frac{2(2-1)^7}{7} - \frac{1(1-1)^7}{7} \right] - \left. \frac{(x-1)^8}{56} \right]_{x=1}^{x=2}$$

Substituting fundamental theorem.

$$= \frac{2}{7} - \left[\frac{(2-1)^8}{56} - \frac{(1-1)^8}{56} \right]$$

Substituting

$$= \frac{2}{7} - \frac{1}{56} \quad \text{or} \quad \frac{15}{56}.$$

Therefore, $\displaystyle\int_{x=1}^{x=2} x(x-1)^6 \, dx = \frac{15}{56}. \quad \blacklozenge$

24.1   Exercises

In Exercises 1–26, evaluate the integrals. Solve by the easiest technique, some may not require integration by parts.

1. $\displaystyle\int x \, e^{2x} \, dx$

2. $\displaystyle\int x \, e^{3x} \, dx$

3. $\displaystyle\int x \sin x \, dx$

4. $\displaystyle\int x \cos x \, dx$

5. $\displaystyle\int x \sec^2 x \, dx$

6. $\displaystyle\int x \tan^2 x \, dx$

7. $\displaystyle\int x \csc^2 x \, dx$

8. $\displaystyle\int x \cot^2 x \, dx$

9. $\displaystyle\int e^x \sin e^x \, dx$

10. $\displaystyle\int e^x \cos e^x \, dx$

11. $\displaystyle\int x \ln x \, dx$

12. $\displaystyle\int x^3 \ln x \, dx$

13. $\displaystyle\int_{x=1}^{x=3} \ln 3x \, dx$

14. $\displaystyle\int \frac{x}{\sqrt{2x+1}} \, dx, \; x > -\frac{1}{2}$

15. $\displaystyle\int \frac{x}{\sqrt{3x+1}} \, dx, \; x > -\frac{1}{3}$

16. $\displaystyle\int_{x=2}^{x=4} \ln 5x \, dx$

17. $\displaystyle\int \frac{x}{\sqrt{4-x^2}} \, dx, \; x < 2$

18. $\displaystyle\int \frac{x^2 \, dx}{9 + x^6}$

19. $\displaystyle\int \operatorname{Arccos} x \, dx$

20. $\displaystyle\int \operatorname{Arctan} x \, dx$

21. $\displaystyle\int x^2(x-3)^{11} \, dx$

22. $\displaystyle\int x\sqrt[3]{x+1} \, dx$

23. $\displaystyle\int e^x \cos x \, dx$

24. $\displaystyle\int e^{-x} \sin x \, dx$

25. $\displaystyle\int_{x=1}^{x=3} \ln(2x+1) \, dx$

26. $\displaystyle\int_{x=1}^{x=4} \frac{\ln x}{x^3} \, dx$

27. Determine the area of the region bounded by the curve $y = x \ln(x-1)$, the x-axis, and the lines $x = 2$ and $x = 5$.

28. Determine the area of the region bounded by the curve $y = x \, e^{-2x}$, the x-axis, and the lines $x = -1$ and $x = 1$.

29. The slope of a function $y = f(x)$ for any point on its curve is $\dfrac{dy}{dx} = \sin x - x \cos x$. Determine the equation of the function if its curve passes through the origin.

30. The slope of a function $y = f(x)$ for any point on its curve is $\dfrac{dy}{dx} = e^{4x} + 4xe^{4x}$. Determine the equation of the function if its curve passes through the origin.

31. Determine the volume generated by rotating the area bounded by $y = \sqrt{x} \sin x$, $y = 0$, $x = 0$, and $x = 2$ about the x-axis.

32. Determine the volume generated by rotating the area bounded by $y = \tan x$, $y = 0$, $x = 0$, and $x = \dfrac{\pi}{4}$ about the x-axis.

33. Use the cylindrical-shell method to determine the volume generated by revolving the area under the curve $y = \cos x$, where $0 \le x \le \dfrac{\pi}{2}$, around the y-axis.

34. Determine the centroid of the area bounded by $y = \sin x$, the x-axis, $x = 0$, and $x = \pi$.

35. The capacitive nature of any current, even an alternating current, gives rise to an exponential reliance. Suppose a capacitive circuit yields the following expression for current: $i = I_0 e^{-2t} \cos 120t$. Recalling that $i = \dfrac{dq}{dt}$ and assuming that $q_0 = 0$, determine an expression for the amount of charge that passes a point as a function of time.

24.2

Trigonometric Integrals

Recall from Chapter 23 the discussion of the techniques for determining the integrals of the basic trigonometric functions. In this section we will use these techniques to help find the integrals of trigonometric functions to higher powers. These basic identities are needed to simplify the integrands. The identities are listed in the box

---

BASIC TRIGONOMETRIC IDENTITIES

$$\cos^2 x + \sin^2 x = 1$$
$$\tan^2 x + 1 = \sec^2 x$$
$$\cot^2 x + 1 = \csc^2 x$$
$$\cos^2 \frac{x}{2} = \frac{1}{2}(1 + \cos x)$$
$$\sin^2 \frac{x}{2} = \frac{1}{2}(1 - \cos x)$$

The integrals of trigonometric functions to higher powers fit into one of three general forms.

$$\int \sin^n u \cos^m u \, du \qquad [1]$$

$$\int \tan^n u \sec^m u \, du \qquad [2]$$

$$\int \cot^n u \csc^m u \, du \qquad [3]$$

The identity we use to help find the integral of these trigonometric functions is determined by whether the exponents are odd or even positive integers. It is also possible that one of the exponents may be zero.

> To integrate a product of sine and cosine functions as in form [1], where the exponents are not both odd or both even, use the identity $\sin^2 x + \cos^2 x = 1$.

EXAMPLE 1   **The Integrand Is Sine to an Even Power and Cosine to an Odd Power**

Evaluate the integral $\int \sin^2 x \cos^3 x \, dx$.

Solution   To help solve this integral we use the facts that $\cos^3 x = \cos^2 x \cos x$ and $\cos^2 x = 1 - \sin^2 x$.

$$\int \sin^2 x \cos^3 x \, dx = \int \sin^2 x \cos^2 x \cos x \, dx \qquad \textbf{Substituting}$$

$$= \int \sin^2 x \, (\boxed{1 - \sin^2 x}) \cos x \, dx \qquad \textbf{Substituting for cos}^2 \textbf{\textit{x}}$$

$$= \int (\sin^2 x \cos x - \sin^4 x \cos x) \, dx$$

$$= \int \sin^2 x \cos x \, dx - \int \sin^4 x \cos x \, dx$$

If we let $u = \sin x$, then $du = \cos x \, dx$. So:

$$\int \sin^2 x \cos^3 x \, dx = \int u^2 \, du - \int u^4 \, du \qquad \textbf{Substituting}$$

$$= \frac{u^3}{3} - \frac{u^5}{5} + C \qquad \textbf{Integrating}$$

$$= \frac{\sin^3 x}{3} - \frac{\sin^5 x}{5} + C. \qquad \textbf{Substituting}$$

Therefore, $\displaystyle\int \sin^2 x \cos^3 x \, dx = \frac{\sin^3 x}{3} - \frac{\sin^5 x}{5} + C.$ ◆

If the exponent of the sine function n or the exponent of the cosine function m is even, we use a technique similar to that used in Example 1. If the exponents of the sine and cosine functions are even, we use the half-angle formulas $\sin^2\left(\dfrac{x}{2}\right) = \dfrac{1}{2}(1 - \cos x)$ and $\cos^2\left(\dfrac{x}{2}\right) = \dfrac{1}{2}(1 + \cos x)$ to write the integrand in a form that is integrable.

Examples 2 and 3 develop the technique for an integrand to an even power.

EXAMPLE 2 **Integrand: Cosine to an Even Power**

Evaluate the integral $\int \cos^2 3x \, dx$.

Solution We can evaluate the integral $\cos 3x$, but we cannot do so for $\cos^2 3x$ without changing the form. To change the form of the integrand, we use the identity for the half-angle formula.

$$\cos^2 3x = \frac{1}{2}(1 + \cos 6x)$$

$$\int \cos^2 3x \, dx = \int \frac{1}{2}(1 + \cos 6x) \, dx \qquad \textbf{Substituting}$$

$$= \frac{1}{2} \int (1 + \cos 6x) \, dx$$

$$= \frac{1}{2}\left[\int dx + \int \cos 6x \, dx\right].$$

Let $u = 6x$, then $du = 6\,dx$ or $\dfrac{du}{6} = dx$.

$$\int \cos^2 3x\,dx = \frac{1}{2}\left[\int dx + \frac{1}{6}\int \cos u\,du\right] \quad \text{Substituting}$$

$$= \frac{1}{2}\left[x + \frac{1}{6}\sin u\right] + C$$

$$= \frac{1}{2}x + \frac{1}{12}\sin 6x + C \quad \begin{array}{l}\text{Removing}\\ \text{brackets and}\\ \text{substituting for } u\end{array}$$

Therefore, $\displaystyle\int \cos^2 3x\,dx = \frac{1}{2}x + \frac{1}{12}\sin 6x + C.$ ◆

EXAMPLE 3   **Integrand: Even Powers of Both Sine and Cosine**

Evaluate the integral $\int \sin^2 x \cos^2 x\,dx$.

Solution   To evaluate $\int \sin^2 x \cos^2 x\,dx$, we cannot let u be equal to $\sin x$ or $\cos x$ because, in either case, we encounter problems. If we use the half-angle formulas, we can obtain an integrand in terms of the cosine function.

$$\int \sin^2 x \cos^2 x\,dx = \int \left[\frac{1}{2}(1 - \cos 2x)\right]\left[\frac{1}{2}(1 + \cos 2x)\right]dx$$

$$\text{Substituting}$$

$$= \frac{1}{4}\int (1 - \cos^2 2x)\,dx$$

It appears that we have run into a problem with $\cos^2 2x$. Don't panic! Use the half-angle formula again. Let $\cos^2 2x = \frac{1}{2}(1 + \cos 4x)$.

$$\int \sin^2 x \cos^2 x\,dx = \frac{1}{4}\int \left[1 - \frac{1}{2}(1 + \cos 4x)\right]dx \quad \text{Substituting}$$

$$= \frac{1}{4}\int \left(\frac{1}{2} - \frac{1}{2}\cos 4x\right)dx$$

$$= \frac{1}{8}\int (1 - \cos 4x)\,dx \quad \text{Factoring, common factor}$$

$$= \frac{1}{8}\left[\int dx - \int \cos 4x\,dx\right]$$

Let $u = 4x$, then $\dfrac{du}{dx} = 4$ or $\dfrac{du}{4} = dx$.

$$\int \cos 4x \, dx = \int \cos u \, \frac{du}{4} \qquad \textbf{Substituting}$$

$$= \frac{1}{4} \int \cos u \, du$$

$$= \frac{1}{4} \sin u + C_1 \qquad \textbf{Integrating}$$

$$= \frac{1}{4} \sin 4x + C_1 \qquad \textbf{Substituting for } u$$

Now, going back to the original integration, we know that $\int dx = x$, so we can write:

$$\int \sin^2 x \cos^2 x \, dx = \frac{1}{8}\left[x - \frac{1}{4}\sin 4x + C_1 \right] \qquad \textbf{Substituting}$$

$$= \frac{1}{8}x - \frac{1}{32}\sin 4x + \frac{1}{8}C_1. \qquad \textbf{\(\frac{1}{8}C_1 = a\) constant,}$$

$$\textbf{which we call } C$$

Therefore, $\displaystyle\int \sin^2 x \cos^2 x \, dx = \frac{1}{8}x - \frac{1}{32}\sin 4x + C.$ ◆

> To integrate a product of tangent and secant functions as in form [2] when the exponents are both even, we use the identity $\tan^2 x + 1 = \sec^2 x$.

EXAMPLE 4 **Integrand: Even Powers of Tangent and Secant**

Evaluate the integral $\int \tan^2 x \sec^4 x \, dx$.

Solution Let us recall how we would integrate $\int \tan^2 x \sec^2 x \, dx$. If we let $u = \tan x$ then $du = \sec^2 x \, dx$. Substituting, the integral becomes $\int u^2 \, du$, which we can integrate. We can change $\int \tan^2 x \sec^4 x \, dx$ into an integrable form by writing $\sec^4 x$ as $\sec^2 x \sec^2 x$ and then substituting $\tan^2 x + 1$ for $\sec^2 x$.

$$\int \tan^2 x \sec^4 x \, dx = \int \tan^2 x \sec^2 x \sec^2 x \, dx \qquad \textbf{Substituting}$$

$$= \int \tan^2 x \, (\, \boxed{\tan^2 x + 1}\,) \sec^2 x \, dx \qquad \textbf{Substituting}$$

$$= \int (\tan^4 x \sec^2 x + \tan^2 x \sec^2 x) \, dx$$

$$= \int \tan^4 x \sec^2 x \, dx + \int \tan^2 x \sec^2 x \, dx$$

If we let $u = \tan x$, then $du = \sec^2 x \, dx$, so:

$$\int \tan^2 x \sec^4 x \, dx = \int u^4 \, du + \int u^2 \, du \qquad \textbf{Substituting}$$

$$= \frac{u^5}{5} + \frac{u^3}{3} + C \qquad \textbf{Integrating}$$

$$= \frac{\tan^5 x}{5} + \frac{\tan^3 x}{3} + C. \qquad \textbf{Substituting}$$

Therefore, $\displaystyle\int \tan^2 x \sec^4 x \, dx = \frac{\tan^5 x}{5} + \frac{\tan^3 x}{3} + C.$ ◆

A key to evaluating the integral of a product of the tangent and the secant functions is whether or not there is an even power of the secant function. If we have an even power of the secant function and any power of the tangent function, we can use the technique discussed in Example 4.

What happens when the secant function has an odd power and the tangent function has an odd power? Example 5 shows us.

EXAMPLE 5 **Integrand: Odd Powers of Both Tangent and Secant**

Evaluate the integral $\int \tan^3 x \sec^3 x \, dx$.

Solution Since we have an odd power of the secant function, we cannot use the technique developed in Example 4. Recall that if we let $u = \sec x$, then $du = \sec x \tan x \, dx$. Therefore, if we can write the integrand of $\int \tan^3 x \sec^3 x \, dx$ as a product of secant to a power times the product of secant tangent, we can integrate.

$$\int \tan^3 x \sec^3 x \, dx = \int \tan x \, (\tan^2 x) \sec^3 x \, dx$$

$$= \int \tan x \, (\,\sec^2 x - 1\,) \sec^3 x \, dx \qquad \textbf{Substituting}$$

$$= \int (\sec^5 x \tan x - \sec^3 x \tan x) \, dx$$

$$= \int \sec^4 x \sec x \tan x \, dx - \int \sec^2 x \sec x \tan x \, dx$$

Now, if we let $u = \sec x$, then $du = \sec x \tan x \, dx$, and:

$$\int \tan^3 x \sec^3 x \, dx = \int u^4 \, du - \int u^2 \, du \qquad \textbf{Substituting}$$

$$= \frac{u^5}{5} - \frac{u^3}{3} + C \qquad \textbf{Integrating}$$

$$= \frac{\sec^5 x}{5} - \frac{\sec^3 x}{3} + C. \qquad \textbf{Substituting}$$

Therefore, $\displaystyle\int \tan^3 x \sec^3 x \, dx = \frac{\sec^5 x}{5} - \frac{\sec^3 x}{3} + C.$ ◆

To evaluate the case where the tangent function has an even exponent and the secant function an odd exponent, we reduce the entire integrand to powers of secant alone.

To integrate equations of form [3], we use the techniques similar to those for form [2] except we use the identity $\cot^2 x + 1 = \csc^2 x$.

EXAMPLE 6 **Integrand: Even Powers of Cotangent**

Evaluate the integral $\int \cot^4 x \, dx$.

$$\int \cot^4 x \, dx = \int \cot^2 x \cot^2 x \, dx \qquad \textbf{Factoring}$$

$$= \int \cot^2 x \, (\csc^2 x - 1) \, dx \qquad \textbf{Substituting}$$

$$= \int (\cot^2 x \csc^2 x - \cot^2 x) \, dx$$

$$= \int \cot^2 x \csc^2 x \, dx - \int \cot^2 x \, dx$$

$$= \int \cot^2 x \csc^2 x \, dx - \int (\csc^2 x - 1) \, dx \qquad \textbf{Substituting}$$

$$= \int \cot^2 x \csc^2 x \, dx - \int \csc^2 x \, dx + \int dx$$

If we let $u = \cot x$, then $du = -\csc^2 x \, dx$.

$$\int \cot^2 x \csc^2 x \, dx = -\int u^2 \, du$$

$$= -\frac{u^3}{3}$$

$$= -\frac{\cot^3 x}{3}$$

$$\int \csc^2 x \, dx = -\cot x$$

Substituting these individual results into the equation, we have

$$\int \cot^4 x \, dx = -\frac{\cot^3 x}{3} - (-\cot x) + x + C$$

$$= -\frac{\cot^3 x}{3} + \cot x + x + C.$$

Therefore, $\int \cot^4 x \, dx = -\dfrac{\cot^3 x}{3} + \cot x + x + C.$ ◆

The rms (root mean square) value of a waveform tells us the heating value of the waveform. This rms value or heating value is the value of a DC voltage or current that would produce exactly the same amount of heat in a resistor as the time-varying waveform. The formula for the rms value of a voltage wave form is:

$$v_{rms} = \sqrt{\frac{1}{T} \int_0^T v^2 \, dt},$$

where T is the period of the voltage function, t is the time, and v is the voltage as a function of time. We may also calculate an rms value for any periodic function.

EXAMPLE 7   Application: rms Value of a Sine Function

Determine the rms value for the sine wave defined by $v = k \sin \theta$.

Solution   To determine the rms value we use the formula:

$$v_{rms} = \sqrt{\frac{1}{T} \int_0^T v^2 \, d\theta}.$$

The period of the sine function is 2π; therefore, $T = 2\pi$. Also since $v = k \sin \theta$, $v^2 = k^2 \sin^2 \theta$. Substituting, we have:

$$v_{rms} = \sqrt{\frac{1}{2\pi} \int_0^{2\pi} k^2 \sin^2 \theta \, d\theta}$$

$$= \sqrt{\frac{k^2}{2\pi} \int_0^{2\pi} \sin^2 \theta \, d\theta}.$$

To determine the rms value we first evaluate the integral.

$$\int_0^{2\pi} \sin^2 \theta \, d\theta = \int_0^{2\pi} \frac{1}{2} (1 - \cos 2\theta) \, d\theta \qquad \text{Substituting}$$

$$= \frac{1}{2} \left(\int_0^{2\pi} d\theta - \int_0^{2\pi} \cos 2\theta \, d\theta \right)$$

$$= \frac{1}{2} \left(\theta \Big]_0^{2\pi} - \frac{\sin 2\theta}{2} \Big]_0^{2\pi} \right) \qquad \text{Integrating}$$

$$= \frac{1}{2} \left[(2\pi - 0) - \frac{1}{2} (\sin 2(2\pi) - \sin 0) \right] \qquad \text{Substituting}$$

$$= \pi.$$

So,

$$\int_0^{2\pi} \sin^2 \theta \, d\theta = \pi.$$

Substituting back into the original equation, we have:

$$v_{\text{rms}} = \sqrt{\frac{k^2}{2\pi} \int_0^{2\pi} \sin^2 \theta \, d\theta} = \sqrt{\frac{k^2}{2\pi}(\pi)}$$

$$= \frac{k}{\sqrt{2}}$$

$$= 0.707k.$$

Therefore, the rms value of a sine wave over one complete cycle is 0.707 times the amplitude of the sine curve. ◆

24.2   Exercises

In Exercises 1–30, evaluate each of the integrals.

1. $\displaystyle\int \sin^4 x \cos^3 x \, dx$

2. $\displaystyle\int \sin^2 3x \cos^3 3x \, dx$

3. $\displaystyle\int \sin^3 x \cos^2 x \, dx$

4. $\displaystyle\int \sin^3 x \cos^4 x \, dx$

5. $\displaystyle\int \sin^3 x \, dx$

6. $\displaystyle\int \cos^3 x \, dx$

7. $\displaystyle\int \cos^3 2x \, dx$

8. $\sin^3 2x \, dx$

9. $\displaystyle\int \sin^3 x \cos^3 x \, dx$

10. $\displaystyle\int \cos^2 2x \, dx$

11. $\displaystyle\int \sin^2 x \, dx$

12. $\displaystyle\int \sin^4 x \, dx$

13. $\displaystyle\int \sin^4 x \cos^4 x \, dx$

14. $\displaystyle\int \tan^2 x \sec^2 x \, dx$

15. $\displaystyle\int \tan^6 x \sec^2 x \, dx$

16. $\displaystyle\int \tan^5 x \sec^2 x \, dx$

17. $\displaystyle\int \tan^5 x \, dx$

18. $\displaystyle\int \sec^4 x \, dx$

19. $\displaystyle\int \tan^3 2x \sec^3 2x \, dx$

20. $\displaystyle\int \tan^5 x \sec^5 x \, dx$

21. $\displaystyle\int \csc^4 x \, dx$

22. $\displaystyle\int \cot^6 x \, dx$

23. $\displaystyle\int \cot^3 x \csc^3 x \, dx$

24. $\displaystyle\int \csc^2 x \cot^3 x \, dx$

25. $\displaystyle\int_0^{\frac{\pi}{2}} \sin^2 x \cos^2 x \, dx$

26. $\displaystyle\int_0^{\pi} (1 + \cos \theta) \, d\theta$

27. $\displaystyle\int_0^{\frac{\pi}{2}} \sin^2 \theta \, d\theta$

28. $\displaystyle\int_0^{\frac{\pi}{4}} \sin^3 \theta \, d\theta$

29. $\displaystyle\int_0^{\frac{\pi}{6}} \sec^3 \theta \tan \theta \, d\theta$

30. $\displaystyle\int_0^{\frac{\pi}{2}} \tan^5 \frac{\theta}{2} \, d\theta$

31. The region bounded by the x-axis and the arc of $y = \cos^2 x$ from $x = 0$ to $x = \dfrac{\pi}{2}$ is revolved about the x-axis. Find the volume of the resulting solid.

32. The region between the graphs of $y = \tan^2 x$ and $y = 0$ from $x = 0$ to $x = \dfrac{\pi}{4}$ is revolved about the x-axis. Determine the volume of the resulting solid.

33. Suppose that the velocity (at time t) of a point on a coordinate line is $\sin^2 \pi t$ m/s. How far does the point travel in six seconds? $(0 \le t \le 6)$

34. The acceleration (at time t) of a car moving along a straight path is $\sin^2 t \cos t$ m/s^2. At $t = 0$ the car is at the starting point, and its velocity is 10 m/s. Determine the distance moved at time t.

35. Determine the rms value for $v = k \sin \theta$ when $T = \pi$. (*Hint:* See Example 7.)

36. Determine the rms value for $v = k \cos \theta$ when $T = 2\pi$. (*Hint:* See Example 7.)

37. Integrate $\int \sec^2 x \tan x \, dx$ in two different ways and show that the result differs only by a constant.

38. Determine the volume of the solid generated when the region under the curve $y = \sec^2 x$ between $x = 0$ and $x = \dfrac{\pi}{3}$ is rotated about the x-axis.

24.3

Trigonometric Substitutions

In this section we substitute trigonometric functions in place of variables to help integrate algebraic functions. To do this, we use integrals that involve expressions of the form $\sqrt{a^2 - u^2}$, $\sqrt{a^2 + u^2}$, and $\sqrt{u^2 - a^2}$, where a is a positive constant. First of all, let us consider an example, which is followed by more guidelines.

EXAMPLE 1 **Integrand Is of the Form $\sqrt{1 - x^2}$**

Evaluate the integral $\displaystyle\int \frac{1}{\sqrt{1 - x^2}} \, dx$.

Solution Let $x = \sin \theta$, then $dx = \cos \theta \, d\theta$.

$$\int \frac{1}{\sqrt{1 - x^2}} \, dx = \int \frac{\cos \theta}{\sqrt{1 - \sin^2 \theta}} \, d\theta \qquad \textbf{Substituting}$$

$$= \int \frac{\cos \theta}{\sqrt{\cos^2 \theta}} \, d\theta \qquad \textbf{Pythagorean identity}$$

$$= \int d\theta$$

$$= \theta + C \qquad \textbf{Integrating}$$

$$= \text{Arcsin } x + C \qquad \textbf{Since } x = \sin \theta \quad \blacklozenge$$

Therefore, $\displaystyle\int \frac{1}{\sqrt{1 - x^2}} \, dx = \text{Arcsin } x + C.$

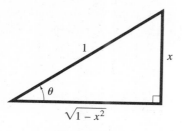

FIGURE 24.1

The result, Arcsin $x + C$, is the result we would expect from our work in Chapter 23. Using this simple example, we can show a geometric relationship by letting $x = (1) \sin \theta$ and labeling a right triangle as in Figure 24.1. We see that the length of the side adjacent to the angle θ is $\sqrt{1 - x^2}$. Therefore, $\cos \theta = \dfrac{\sqrt{1 - x^2}}{1}$, or $\cos \theta = \sqrt{1 - x^2}$. This suggests how we can draw and label a right triangle and do the substitutions directly.

Using right triangle trigonometry, we can relate each of the radical expressions $\sqrt{a^2 - u^2}$, $\sqrt{a^2 + u^2}$, and $\sqrt{u^2 - a^2}$ to a right triangle. With these right triangles, shown in Figure 24.2, we can compile a list of appropriate substitutions for each of the radical expressions.

In Figure 24.2(a), we see that

$$\sin \theta = \frac{u}{a} \quad \text{and} \quad \cos \theta = \frac{\sqrt{a^2 - u^2}}{a}$$
$$a \sin \theta = u \quad\quad a \cos \theta = \sqrt{a^2 - u^2}, a > 0.$$

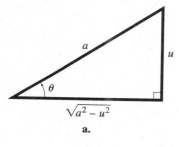

a.

In Figure 24.2(b),

$$\tan \theta = \frac{u}{a} \quad \text{and} \quad \sec \theta = \frac{\sqrt{a^2 + u^2}}{a}$$
$$a \tan \theta = u \quad\quad a \sec \theta = \sqrt{a^2 + u^2}, a > 0.$$

In Figure 24.2(c),

$$\sec \theta = \frac{u}{a} \quad \text{and} \quad \tan \theta = \frac{\sqrt{u^2 - a^2}}{a}$$
$$a \sec \theta = u \quad\quad a \tan \theta = \sqrt{u^2 - a^2}, a > 0.$$

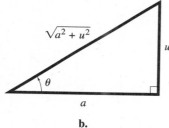

b.

These trigonometric substitutions with restrictions are summarized in the following box.

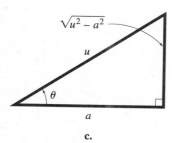

c.

FIGURE 24.2

TRIGONOMETRIC SUBSTITUTIONS

1. If the integrand involves $\sqrt{a^2 - u^2}$, where $0 < u < a$, let $u = a \sin \theta$, where $0 < \theta < \dfrac{\pi}{2}$. Then $du = a \cos \theta\, d\theta$, and $\sqrt{a^2 - u^2} = a \cos \theta$.

2. If the integrand involves $\sqrt{a^2 + u^2}$, where $u > 0$ and $a > 0$, let $u = a \tan \theta$, where $0 < \theta < \dfrac{\pi}{2}$. Then $du = a \sec^2 \theta\, d\theta$, and $\sqrt{a^2 + u^2} = a \sec \theta$.

3. If the integrand involves $\sqrt{u^2 - a^2}$, where $u > a > 0$, let $u = a \sec \theta$, where $0 < \theta < \dfrac{\pi}{2}$. Then $du = a \sec \theta \tan \theta\, d\theta$, and $\sqrt{u^2 - a^2} = a \tan \theta$.

When we use the trigonometric substitutions to evaluate, the integral produces an answer with a different variable. However, we can express the answer in terms of the original variable by referring to the appropriate right triangle. This is illustrated in Example 2.

EXAMPLE 2   Integrand Is of the Form $\sqrt{a^2 - u^2}$

Evaluate the integral $\displaystyle\int \frac{\sqrt{9 - x^2}}{x^2}\,dx$.

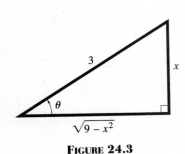

FIGURE 24.3

Solution   The radical is of the form $\sqrt{a^2 - u^2}$. Let $x = 3 \sin \theta$ (see Figure 24.3). Differentiating we have $\dfrac{du}{d\theta} = 3 \cos \theta$, or $du = 3 \cos \theta\, d\theta$. From the right triangle $\left(\text{where } \sin \theta = \dfrac{x}{3}\right)$, we see that $\sqrt{9 - x^2} = 3 \cos \theta$. Thus,

$$\int \frac{\sqrt{9 - x^2}}{x^2}\,dx = \int \frac{(\,3 \cos \theta\,)}{(\,3 \sin \theta\,)^2}(3 \cos \theta)\, d\theta \qquad \textbf{Substituting}$$

$$= \int \frac{\cos^2 \theta}{\sin^2 \theta}\,d\theta$$

$$= \int \frac{1 - \sin^2 \theta}{\sin^2 \theta}\,d\theta \qquad \textbf{Substituting}$$

$$= \int \left(\frac{1}{\sin^2 \theta} - 1\right) d\theta$$

$$= \int (\,\csc^2 \theta\, - 1)\, d\theta \qquad \textbf{Substituting}$$

$$= -\cot \theta - \theta + C. \qquad \textbf{Integrating}$$

Referring again to Figure 24.3, we see that $\cot \theta = \dfrac{\sqrt{9 - x^2}}{x}$ and $\sin \theta = \dfrac{x}{3}$, or $\operatorname{Arcsin} \dfrac{x}{3} = \theta$. Therefore,

$$\int \frac{\sqrt{9 - x^2}}{x^2}\,dx = -\frac{\sqrt{9 - x^2}}{x} - \operatorname{Arcsin} \frac{x}{3} + C. \quad \blacklozenge$$

EXAMPLE 3   Integrand Is of the Form $\sqrt{u^2 + a^2}$

Evaluate the integral $\displaystyle\int \frac{1}{t\sqrt{t^2 + 5}}\,dt$.

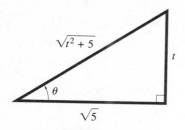

$\sqrt{t^2 + 5}$ t θ $\sqrt{5}$

FIGURE 24.4

Solution The radical is of the form $\sqrt{u^2 + a^2}$. Let $t = \sqrt{5} \tan \theta$ (see Figure 24.4). Differentiating we have $\dfrac{dt}{d\theta} = \sqrt{5} \sec^2 \theta$, or $dt = \sqrt{5} \sec^2 \theta \, d\theta$. From the right triangle we see that $\sqrt{t^2 + 5} = \sqrt{5} \sec \theta$. Thus,

$$\int \frac{1}{t\sqrt{t^2 + 5}} \, dt = \int \frac{\sqrt{5} \sec^2 \theta}{\sqrt{5} \tan \theta \sqrt{5} \sec \theta} \, d\theta \qquad \text{Substituting}$$

$$= \frac{1}{\sqrt{5}} \int \frac{\sec \theta}{\tan \theta} \, d\theta$$

$$= \frac{1}{\sqrt{5}} \int \frac{1}{\sin \theta} \, d\theta \qquad \text{Substituting } \frac{1}{\cos \theta} \cdot \frac{\cos \theta}{\sin \theta}$$

$$= \frac{1}{\sqrt{5}} \int \csc \theta \, d\theta \qquad \text{Substituting}$$

$$= \frac{1}{\sqrt{5}} \ln |\csc \theta - \cot \theta| + C. \qquad \text{Integrating}$$

Referring again to Figure 24.4, we see that $\csc \theta = \dfrac{\sqrt{t^2 + 5}}{t}$ and $\cot \theta = \dfrac{\sqrt{5}}{t}$. Therefore,

$$\int \frac{1}{t\sqrt{t^2 + 5}} \, dt = \frac{1}{\sqrt{5}} \ln \left| \frac{\sqrt{t^2 + 5} - \sqrt{5}}{t} \right| + C. \quad \blacklozenge$$

EXAMPLE 4 **Integrand Is of the Form $\sqrt{u^2 - a^2}$**

Evaluate the integral $\displaystyle\int \frac{x}{\sqrt{x^2 - 9}}$.

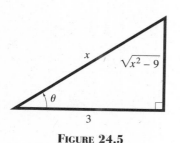

x $\sqrt{x^2 - 9}$ θ 3

FIGURE 24.5

Solution Since the radical is of the form $\sqrt{u^2 - a^2}$, we use the substitution $x = 3 \sec \theta$ (see Figure 24.5). Differentiating we have $\dfrac{dx}{d\theta} = 3 \sec \theta \tan \theta$, or $dx = 3 \sec \theta \tan \theta \, d\theta$. From the right triangle we see that $\sqrt{x^2 - 9} = 3 \tan \theta$. Thus,

$$\int \frac{x}{\sqrt{x^2 - 9}} \, dx = \int \frac{(3 \sec \theta)(3 \sec \theta \tan \theta)}{3 \tan \theta} \, d\theta \qquad \text{Substituting}$$

$$= 3 \int \sec^2 \theta \, d\theta$$

$$= 3 \tan \theta + C. \qquad \text{Integrating}$$

Referring again to Figure 24.5, we see that $\tan \theta = \dfrac{\sqrt{x^2 - 9}}{3}$. Therefore,

$$\int \frac{x}{\sqrt{x^2 - 9}} \, dx = \sqrt{x^2 - 9} + C. \quad \blacklozenge$$

24.3 Exercises

In Exercises 1–26, use an appropriate trigonometric substitution to evaluate each integral.

1. $\displaystyle\int x\sqrt{4 - x^2}\, dx$

2. $\displaystyle\int x^3\sqrt{9 - x^2}\, dx$

3. $\displaystyle\int x^3\sqrt{16 + x^2}\, dx$

4. $\displaystyle\int x\sqrt{25 + x^2}\, dx$

5. $\displaystyle\int x\sqrt{x^2 - 9}\, dx$

6. $\displaystyle\int x^3\sqrt{x^2 - 36}\, dx$

7. $\displaystyle\int \frac{1}{x^2\sqrt{3 - x^2}}\, dx$

8. $\displaystyle\int \frac{1}{x\sqrt{5 - x^2}}\, dx$

9. $\displaystyle\int \frac{x^3}{\sqrt{9 + x^2}}\, dx$

10. $\displaystyle\int \frac{1}{x^2\sqrt{1 + x^2}}\, dx$

11. $\displaystyle\int \frac{1}{x^2\sqrt{x^2 - 25}}\, dx$

12. $\displaystyle\int \frac{1}{x^4\sqrt{x^2 - a}}\, dx$

13. $\displaystyle\int \frac{t^2}{4 - 9t^2}\, dt$

14. $\displaystyle\int \sqrt{9 - 2t^2}\, dt$

15. $\displaystyle\int \frac{1}{t^4\sqrt{4 + t^2}}\, dt$

16. $\displaystyle\int \frac{7t^3}{(4t^2 + 9)^{\frac{3}{2}}}\, dt$

17. $\displaystyle\int \frac{1}{(4u^2 - 9)^{\frac{3}{2}}}\, du$

18. $\displaystyle\int u^3\sqrt{u^2 - 1}\, du$

19. $\displaystyle\int \frac{1}{(25 - x^2)^{\frac{3}{2}}}\, dx$

20. $\displaystyle\int_0^2 x\sqrt{16 - 4x^2}\, dx$

21. $\displaystyle\int_0^1 \frac{x^2}{(4 + x^2)^2}\, dx$

22. $\displaystyle\int_{-1}^1 \frac{1}{(1 + x^2)^3}\, dx$

23. $\displaystyle\int_{\sqrt{2}}^2 \frac{1}{u^2\sqrt{u^2 - 1}}\, du$

24. $\displaystyle\int_{-1/\sqrt{2}}^{1/\sqrt{2}} (1 - 2u^2)^{\frac{3}{2}}\, du$

25. $\displaystyle\int e^x\sqrt{1 - 2e^x}\, dx$

26. $\displaystyle\int \frac{\cos\theta}{\sqrt{2 - \sin^2\theta}}\, d\theta$

27. Find the area under the curve $y = \sqrt{9 + x^2}$ from $x = 0$ to $x = 4$.

28. Find the area under the curve $y = x\sqrt{9 - x^2}$ from $x = 0$ to $x = 2$.

29. The field strength H of a magnet of length 2ℓ on a particle r units from the center of the magnet is given by $H = \dfrac{2m\ell}{(r^2 + \ell^2)^{\frac{3}{2}}}$, where $+m$ and $-m$ are the constant pole strengths of the magnet (see the sketch.) Determine the average field strength as the particle moves from 0 to R units away from the center by evaluating the integral

$$\frac{1}{R}\int_0^R \frac{2m\ell}{(r^2 + \ell^2)^{\frac{3}{2}}}\, dr.$$

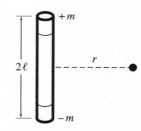

30. Determine the area of the region enclosed by the graph of $9x^2 + y^2 = 25$.

31. Imagine a uniformly charged wire (charge $= q$) that has a length of $2a$. This wire establishes a complex electric potential at various points in space. The voltage just along the perpendicular bisector of the wire is given by $V = kq\displaystyle\int_{x=-a}^{x=a} \frac{dx}{(b^2 + x^2)^{\frac{1}{2}}}$, where k is a constant, x is the distance along the wire, and b is the distance from the center of the wire to some point P along the perpendicular bisector. Evaluate the integral.

24.4

Partial Fractions

With the various techniques of integration that we have discussed to this point, we still do not have a neat way of evaluating the following integral:

$$\int \frac{2x + 1}{x^2 + 5x + 6}\, dx.$$

A technique for integrating rational functions is developed in this section. Recall that a rational function is of the form $R(x) = \dfrac{P(x)}{Q(x)}$, where $P(x)$ and $Q(x)$ are polynomials.

Let us begin by determining the sum of the fractions $\dfrac{-3}{x + 2}$ and $\dfrac{5}{x + 3}$. To determine the sum, we first must determine the least common denominator; that is, $(x + 2) \cdot (x + 3)$.

$$\frac{-3}{x + 2} + \frac{5}{x + 3} = \frac{-3}{x + 2} \cdot \frac{x + 3}{x + 3} + \frac{5}{x + 3} \cdot \frac{x + 2}{x + 2}$$

$$= \frac{2x + 1}{(x + 2)(x + 3)}$$

$$= \frac{2x + 1}{x^2 + 5x + 6}$$

The result is the integrand of the integral in the beginning of this section.

$$\int \frac{2x + 1}{x^2 + 5x + 6}\, dx = \int \left(\frac{-3}{x + 2} + \frac{5}{x + 3}\right) dx$$

With the rational function written as the sum of two rational functions we easily can evaluate the integral. The simple rational functions $\dfrac{-3}{x + 2}$ and $\dfrac{5}{x + 3}$ are called the **partial fractions** of $\dfrac{2x + 1}{x^2 + 5x = 6}$.

Our problem then is how to break up an integrand so that we can integrate, or how to reverse the process of combining fractions over a common denominator. The technique is known as the **method of partial fractions.**

CAUTION   To use the technique of partial fractions, the degree of the numerator must be less than the degree of the denominator. ■

If the degree of the numerator is larger than the degree of the denominator, divide the numerator by the denominator. The result is of the form:

$$R(x) = \text{Quotient} + \frac{\text{Remainder}}{\text{Divisor}} \quad \text{or} \quad R(x) = P(x) + \frac{r(x)}{Q(x)}.$$

The degree of the remainder $r(x)$ must be at least one degree less than the degree of the divisor $Q(x)$. We call the rational function $\dfrac{r(x)}{Q(x)}$ a *proper fraction.*

Distinct Linear Factors

The process of finding the partial fractions of a rational function is easiest to apply when the denominator factors into distinct linear factors. In this situation, we must find a numerator for each factor of the denominator. For example, if we have:

$$\frac{P(x)}{Q(x)} = \frac{P(x)}{(ax + b)(cx + d)} = \frac{A}{ax + b} + \frac{B}{cx + d},$$

where A and B are constants, our challenge is finding A and B. Let us illustrate the process with our original problem.

$$\frac{2x + 1}{x^2 + 5x + 6} = \frac{2x + 1}{(x + 2)(x + 3)} = \frac{A}{x + 2} + \frac{B}{x + 3}.$$

Now we must determine the values for A and B. To do this we find the sum of the two partial fractions.

$$\frac{2x + 1}{(x + 2)(x + 3)} = \frac{A}{x + 2} \cdot \frac{x + 3}{x + 3} + \frac{B}{x + 3} \cdot \frac{x + 2}{x + 2}$$

$$= \frac{Ax + 3A + Bx + 2B}{(x + 2)(x + 3)}$$

$$= \frac{(A + B)x + (3A + 2B)}{(x + 2)(x + 3)}$$

Recall that two fractions are equal if, and only if, their numerators and denominators are equal. We see that the denominators of these fractions are equal. For the numerators to be equal, the coefficients of like-degree terms must be equal. That is, the coefficients of the x-terms must be equal, $A + B = 2$, and the constant terms must be equal, $3A + 2B = 1$. We now have a system of equations in two unknowns.

$$A + B = 2$$
$$3A + 2B = 1$$

Solving the system of equations we find that $A = -3$ and $B = 5$ Therefore,

$$\frac{2x + 1}{(x + 2)(x + 3)} = \frac{-3}{x + 2} + \frac{5}{x + 3}.$$

Integrating the rational function, we have:

$$\int \frac{2x + 1}{(x + 2)(x + 3)}\, dx = \int \frac{-3}{x + 2}\, dx + \int \frac{5}{x + 3}\, dx$$

$$= -3 \ln|x + 2| + 5 \ln|x + 3| + C$$

$$= -\ln|x + 2|^3 + \ln|x + 3|^5 + C$$

$$= \ln\left[\frac{|x + 3|^5}{|x + 2|^3}\right] + C.$$

EXAMPLE 1 **Rational Function: Denominator Has Distinct Factors**

Evaluate the integral $\displaystyle\int \frac{6x^2 - 2x - 2}{x^3 - x^2 - 2x} \, dx$.

Solution The integrand is a proper fraction, so we start by factoring the denominator.

$$x^3 - x^2 - 2x = x(x + 1)(x - 2)$$

Since the denominator factors into three distinct linear factors, we can set the rational function equal to the sum of the three partial fractions.

$$\frac{6x^2 - 2x - 2}{x(x + 1)(x - 2)} = \frac{A}{x} + \frac{B}{x + 1} + \frac{C}{x - 2}.$$

To determine the constants A, B, and C we multiply both sides of the equation by $x(x + 1)(x - 2)$.

$$6x^2 - 2x - 2 = A(x + 1)(x - 2) + Bx(x - 2) + Cx(x + 1)$$

Multiplying and collecting like terms, we have:

$$6x^2 - 2x - 2 = (A + B + C)x^2 + (-A - 2B + C)x + (-2A).$$

The two polynomials will be equal if the coefficients of corresponding powers of x are the same; that is:

$$A + B + C = 6$$
$$-A - 2B + C = -2$$
$$-2A = -2.$$

Solving this system of simultaneous equations, we find that $A = 1$, $B = 2$, and $C = 3$. Having found the values for A, B, and C, we can write the integral as:

$$\int \frac{6x^2 - 2x - 2}{x^3 - x^2 - 2x} \, dx = \int \left(\frac{1}{x} + \frac{2}{x + 1} + \frac{3}{x - 2} \right) dx$$
$$= \ln|x| + 2\ln|x + 1| + 3\ln|x - 2| + C. \text{ Integrating}$$

Using the laws of logarithms, we can write the result in a more compact form. Therefore,

$$\int \frac{6x^2 - 2x - 2}{x^3 - x^2 - 2x} \, dx = \ln|x(x + 1)^2(x - 2)^3| + C. \quad \blacklozenge$$

Repeated Linear Factors

What is the sum of $\dfrac{3}{x + 1} + \dfrac{4}{(x + 1)^2}$?

$$\frac{3}{x + 1} + \frac{4}{(x + 1)^2} = \frac{3(x + 1)}{(x + 1)^2} + \frac{4}{(x + 1)^2}$$
$$= \frac{3x + 7}{(x + 1)^2}$$

Now, how do we go about expressing the fraction $\dfrac{3x + 7}{(x + 1)^2}$ in terms of the sum of partial fractions? Since the denominator is a repeated factor (in fact, it is repeated twice), we have two partial fractions.

$$\frac{3x + 7}{(x + 1)^2} = \frac{A}{x + 1} + \frac{B}{(x + 1)^2}$$

Show that $A = 3$ and $B = 4$. The rational function $\dfrac{P(x)}{Q(x)}$, where the linear factors of $Q(x)$ are not all distinct requires special attention. For example, if

$$Q(x) = (x + 2)(x - 2)(x + 2)(x + 3)(x - 2)(x - 2),$$

then the linear factor $(x + 2)$ appears twice, and the linear factor $(x - 2)$ appears three times. The first thing we do is rewrite the product using exponents. Thus,

$$Q(x) = (x - 2)^3(x + 2)^2(x + 3).$$

With the repeated factors, we do the following to find the partial fractions. For all factors of the form $(ax + b)^k$, where $k > 1$, we must provide k corresponding partial fractions of the form:

$$\frac{A_1}{ax + b} + \frac{A_2}{(ax + b)^2} + \frac{A_3}{(ax + b)^3} + \cdots + \frac{A_k}{(a + b)^k},$$

where $A_1, A_2, A_3, \ldots A_k$ are the constants that must be determined. Nonrepeating factors are handled as before. For example,

$$\frac{x^4 + x^3 - x^2 + x + 1}{(x - 2)^3(x + 2)^2(x + 3)} = \frac{A_1}{x - 2} + \frac{A_2}{(x - 2)^2} + \frac{A_3}{(x - 2)^3} + \frac{B_1}{x + 2}$$
$$+ \frac{B_2}{(x + 2)^2} + \frac{C}{x + 3}.$$

This technique is illustrated in Example 2.

EXAMPLE 2   Rational Function: Denominator Has Repeated Linear Factors

Evaluate the integral $\displaystyle\int \frac{4x^2 - 11x + 12}{(x - 2)^2(x + 1)}\, dx$.

Solution   The integral is a proper fraction, so we begin by trying to find the partial fractions:

$$\frac{4x^2 - 11x + 12}{(x - 2)^2(x + 1)} = \frac{A}{x - 2} + \frac{B}{(x - 2)^2} + \frac{C}{x + 1}.$$

To find the constants A, B, and C, we multiply both sides by $(x - 2)^2(x + 1)$.

$$4x^2 - 11x + 12 = A(x - 2)(x + 1) + B(x + 1) + C(x - 2)^2$$

Multiplying and collecting like terms, we have:

$$4x^2 - 11x + 12 = (A + C)x^2 + (-A + B - 4C)x + (-2A + B + 4C).$$

Equating the coefficients of like powers of x, we have the following system of equations.

$$A \qquad + C = 4 \qquad\qquad (1)$$
$$-A + B - 4C = -11 \qquad\qquad (2)$$
$$-2A + B + 4C = 12 \qquad\qquad (3)$$

Subtracting Equation (3) from Equation (2), the result is:

$$-A + B - 4C = -11$$
$$\underline{-2A + B + 4C = \quad 12}$$
$$A \qquad\quad -8C = -23. \qquad\qquad (4)$$

Now multiplying Equation (1) by -1 and adding the result to Equation (4), the result is:

$$-A - \quad C = -\ 4$$
$$\underline{A - 8C = -23}$$
$$-9C = -27$$
$$C = 3.$$

Knowing that $C = 3$, we see from Equation (1) that $A = 1$, and substituting into Equation (2), we see that $B = 2$. Therefore,

$$\int \frac{4x^2 - 11x + 12}{(x - 2)^2(x + 1)}\, dx = \int \left(\frac{1}{x - 2} + \frac{2}{(x - 2)^2} + \frac{3}{x + 1} \right) dx$$

$$= \int \frac{1}{x - 2}\, dx + 2 \int \frac{1}{(x - 2)^2}\, dx + 3 \int \frac{1}{x + 1}\, dx$$

$$= \ln|x - 2| + 2\left(\frac{-1}{x - 2}\right) + 3 \ln|x + 1| + C.$$

With the rules of logarithms, we can write the result in a more compact form. Therefore,

$$\int \frac{4x^2 - 11x + 12}{(x - 2)^2(x + 1)}\, dx = \ln|(x - 2)(x + 1)^3| - \frac{2}{x - 2} + C. \quad \blacklozenge$$

Quadratic Factors

There are times when the denominator $Q(x)$ of the rational function $\frac{P(x)}{Q(x)}$ cannot be factored completely into linear factors. Leonhard Euler (1707–1783) proved that every polynomial $Q(x)$ whose coefficients are real numbers can be factored completely into a finite number of polynomials, each of which is either linear or quadratic. If a quadratic polynomial is prime, it cannot be factored under the set of real numbers. We easily can check this using the quadratic discriminate, $b^2 - 4ac$. For every quadratic polynomial that is prime $b^2 - 4ac < 0$.

For any rational function $\dfrac{P(x)}{Q(x)}$, where the prime factorization of $Q(x)$ has linear and quadratic factors, we must break $\dfrac{P(x)}{Q(x)}$ into a sum of partial fractions. We know what to do when the factors are linear, so let us look at what to do when the prime factors are quadratic.

For each factor of the form $(ax^2 + bx + c)^k$, the partial fraction decomposition must include the following sum of k fractions:

$$\frac{B_1 x + C_1}{ax^2 + bx + c} + \frac{B_2 x + C_2}{(ax^2 + bx + c)^2} + \cdots + \frac{B_k x + C_k}{(ax^2 + bx + c)^k}.$$

Note that the numerators of the partial fractions are of lesser degree (first degree) than the degree of the denominators (second degree).

EXAMPLE 3   A Quadratic Factor in the Denominator

Evaluate the integral $\displaystyle\int \frac{x^3 - 5x^2 - 3x - 8}{x(x + 2)(x^2 + 1)}\, dx.$

Solution   Since the denominator has both quadratic and linear factors, we write:

$$\frac{x^3 - 5x^2 - 3x - 8}{x(x + 2)(x^2 + 1)} = \frac{A}{x} + \frac{B}{x + 2} + \frac{Cx + D}{x^2 + 1}.$$

We multiply both sides by $x(x + 2)(x^2 + 1)$, and obtain:

$$
\begin{aligned}
x^3 - 5x^2 - 3x - 8 &= A(x + 2)(x^2 + 1) + Bx(x^2 + 1) \\
&\quad + (Cx + D)(x)(x + 2) \\
&= (A + B + C)x^3 + (2A + 2C + D)x^2 \\
&\quad + (A + B + 2D)x + 2A.
\end{aligned}
$$

As we have done in previous examples, we equate coefficients of like powers of x and obtain the system of equations:

$$
\begin{array}{rcll}
A + B + C & = 1 & \qquad (5) \\
2A \qquad + 2C + D & = -5 & \qquad (6) \\
A + B \qquad + 2D & = -3 & \qquad (7) \\
2A \qquad\qquad & = -8. & \qquad (8)
\end{array}
$$

From Equation (8) we can see that $A = -4$. Substituting -4 for A in Equations (5), (6), and (7), we get the new set of equations:

$$
\begin{array}{rcll}
B + C & = 5 & \qquad (9) \\
2C + D & = 3 & \qquad (10) \\
B \qquad + 2D & = 1. & \qquad (11)
\end{array}
$$

Solving Equation (9) for B we have $B = 5 - C$. Substituting in Equation (11), we have a new equation, $-C + 2D = -4$. Multiply the equation by 2, $(-2C + 4D = -8)$ and call it Equation (12). Now add Equations (10) and (12).

$$2C + D = 3 \tag{10}$$
$$\underline{-2C + 4D = -8} \tag{12}$$
$$5D = -5$$
$$D = -1$$

Substituting -1 for D in Equation (10), we see that $C = 2$. Substituting 2 for C in Equation (9), we see that $B = 3$. Thus,

$$\int \frac{x^3 - 5x^2 - 3x - 8}{x(x+2)(x^2+1)} \, dx = \int \frac{-4}{x} + \frac{3}{x+2} + \frac{2x-1}{x^2+1} \, dx$$

$$= -4 \int \frac{1}{x} \, dx + 3 \int \frac{1}{x+2} \, dx + \int \frac{2x}{x^2+1} \, dx - \int \frac{1}{x^2+1} \, dx$$

$$= -4 \ln|x| + 3 \ln|x+2| + \ln|x^2+1| - \text{Arctan } x + C$$

$$= -\ln|x|^4 + \ln|x+2|^3 + \ln|x^2+1| - \text{Arctan } x + C.$$

Using the rules of logarithms, we can write the solution in a more compact form.

$$\int \frac{x^3 - 5x^2 - 3x - 8}{x(x+2)(x^2+1)} \, dx = \ln \frac{(x+2)^3(x^2+1)}{x^4} - \text{Arctan } x + C. \quad \blacklozenge$$

EXAMPLE 4 **A Quadratic Factor in the Denominator Raised to a Power**

Evaluate the integral $\displaystyle\int \frac{2x^3 + x^2 + 2x + 4}{(x^2+1)^2} \, dx$

Solution The denominator has repeated quadratic factors. Therefore we write:

$$\frac{2x^3 + x^2 + 2x + 4}{(x^2+1)^2} = \frac{Ax+B}{x^2+1} + \frac{Cx+D}{(x^2+1)^2}.$$

We multiply both sides by $(x^2+1)^2$ and obtain:

$$2x^3 + x^2 + 2x + 4 = (Ax+B)(x^2+1) + Cx + D$$
$$= Ax^3 + Bx^2 + (A+C)x + (B+D).$$

As we have done in previous examples, we equate coefficients of like powers of x and obtain the system of equations:

$$A = 2$$
$$B = 1$$
$$A + C = 2$$
$$B + D = 4.$$

Since $A = 2$ and $B = 1$, we can easily see that $C = 0$ and $D = 3$. Thus,

$$\int \frac{2x^3 + x^2 + 2x + 4}{(x^2 + 1)^2} \, dx$$

$$= \int \left(\frac{2x + 1}{x^2 + 1} + \frac{3}{(x^2 + 1)^2} \right) dx$$

$$= \int \frac{2x}{x^2 + 1} \, dx + \int \frac{1}{x^2 + 1} \, dx + 3 \int \frac{1}{(x^2 + 1)^2} \, dx.$$

The first two integrals are found by basic definitions:

$$\int \frac{2x}{x^2 + 1} \, dx = \ln |2x + 1|$$

$$\int \frac{1}{x^2 + 1} \, dx = \text{Arctan } x.$$

To integrate:

$$\int \frac{1}{(x^2 + 1)^2} \, dx,$$

we can use trigonometric substitution, and let $x = \tan \theta$ and $dx = \sec^2 \theta \, d\theta$. Then from the right triangle shown in Figure 24.6,

$$\sqrt{x^2 + 1} = \sec \theta$$

$$(\sqrt{x^2 + 1})^4 = \sec^4 \theta$$

$$(x^2 + 1)^2 = \sec^4 \theta$$

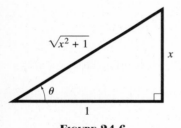

Figure 24.6

$$3 \int \frac{1}{(x^2 + 1)^2} \, dx = 3 \int \frac{\sec^2 \theta}{\sec^4 \theta} \, d\theta \qquad \text{Substituting}$$

$$= 3 \int \frac{1}{\sec^2 \theta} \, d\theta$$

$$= 3 \int \cos^2 \theta \, d\theta \qquad \text{Substituting}$$

$$= 3 \int \frac{1}{2} (1 + \cos 2\theta) \, d\theta \qquad \text{Substituting}$$

$$= 3 \left[\frac{1}{2} \left(\theta + \frac{\sin 2\theta}{2} \right) \right] \qquad \text{Integrating}$$

$$= \frac{3}{2} \left[\theta + \frac{2 \sin \theta \cos \theta}{2} \right]. \qquad \text{Substituting}$$

From Figure 24.6, we see that $\theta = \text{Arctan } x$, $\sin \theta = \dfrac{x}{\sqrt{x^2 + 1}}$, and

$\cos \theta = \dfrac{1}{\sqrt{x^2 + 1}}$.

$$3 \int \frac{1}{(x^2+1)^2}\,dx = \frac{3}{2}\left[\text{Arctan } x + \left(\frac{x}{\sqrt{x^2+1}}\right)\left(\frac{1}{\sqrt{x^2+1}}\right)\right]$$

$$= \frac{3}{2}\text{Arctan } x + \frac{3x}{2(x^2+1)}$$

Therefore,

$$\int \frac{2x^3 + x^2 + 2x + 4}{(x^2+1)^2}\,dx = \ln|2x+1| + \text{Arctan } x + \frac{3}{2}\text{Arctan } x + \frac{3x}{2(x^2+1)} + C$$

$$= \ln|2x+1| + \frac{5}{2}\text{Arctan } x + \frac{3x}{2(x^2+1)} + C. \quad \blacklozenge$$

24.4 Exercises

In Exercises 1–26, evaluate the integrals.

1. $\int \frac{1}{x^2-1}\,dx$

2. $\int \frac{1}{4x^2-9}\,dx$

3. $\int \frac{3x+5}{x^2+2x-3}\,dx$

4. $\int \frac{x-13}{x^2+4x-5}\,dx$

5. $\int \frac{-x-3}{2x^2-x-1}\,dx$

6. $\int \frac{9x-4}{3x^2-5x-2}\,dx$

7. $\int \frac{5x-18}{x^2-9x}\,dx$

8. $\int \frac{3x-4}{x^2-4x}\,dx$

9. $\int \frac{x^2-1}{x^2-2x-15}\,dx$

10. $\int \frac{x^3+3x^2+6x+8}{x^2+3x+2}\,dx$

11. $\int \frac{2x^3-4x^2-15x+5}{x^2-2x-8}\,dx$

12. $\int \frac{4x^2+2x-1}{x^3+x^2}\,dx$

13. $\int \frac{x^2+1}{x^3+2x^2+x}\,dx$

14. $\int \frac{9x^2+18x-24}{(x^2-4)(x+4)}\,dx$

15. $\int \frac{x^2}{(x+1)^3}\,dx$

16. $\int \frac{1}{x^3+x}\,dx$

17. $\int \frac{x}{(x+1)(x^2+1)}\,dx$

18. $\int \frac{x^2+2}{(x^2+1)^2}\,dx$

19. $\int \frac{x^2}{x^4-1}\,dx$

20. $\int \frac{2x}{(x+2)(x^2-1)}\,dx$

21. $\int_2^4 \frac{x}{(x+1)(x+2)}\,dx$

22. $\int_1^2 \frac{5t^2-3t+18}{t(9-t^2)}\,dt$

23. $\int_2^3 \frac{4Z^5-3Z^4-6Z^3+4Z^2+6Z-1}{(Z-1)(Z^2-1)}\,dZ$

24. $\int_3^5 \frac{Z^2-2}{(Z-2)^2}\,dZ$

25. $\int_1^2 \frac{4}{t^3+4t}\,dt$

26. $\int_1^2 \frac{1-t^2}{t(t^2+1)}\,dt$

27. Determine the area of the region in the first quadrant bounded by the curve $(x+1)^2 y = 2 - x$.

28. Determine the volume of the solid generated by revolving the region in Exercise 27 about the x-axis.

29. Determine the partial fractions decomposition of $\dfrac{ax^3 + bx^2 + cx + d}{(x^2+1)^2}$.

30. If the velocity v in meters per second of a particle that is moving along a straight line is expressed by the formula $v = \dfrac{5t^2 - t + 2}{t^3 + t}$, determine the distance s, $s = \int v\,dt$, in feet that the particle travels from $t = 1$ to $t = 3$ seconds.

24.5

Integration by Use of Tables

In our discussion of integration, we have encountered what appear to be many different integration formulas. These, however, are only a few of the hundreds of integration formulas that appear in tables.

Students often wonder why they have to spend all this time studying integration techniques if all they need do is look at a table and read the results. Well, as usual, life is not quite so simple. Consider

$$\int \frac{\cos (\ln x)}{x}\,dx.$$

When we examine an extensive table of integration formulas, this particular one does not appear. However, if we use substitution and let $u = \ln x$, then $\dfrac{du}{dx} = \dfrac{1}{x}$, or $du = \dfrac{dx}{x}$. Then we have:

$$\int \frac{\cos (\ln x)}{x}\,dx = \int \cos u\,du$$
$$= \sin u + C$$
$$= \sin (\ln x) + C.$$

The point here is that many problems can be solved with simple substitution; therefore, they may not be found in a table. Also, in many problems you need to know how to use substitution to *use* the formula in the table.

To solve the remaining problems in this section, we use the integration formulas found in Appendix D.

EXAMPLE 1   Integrating Using Integration Formula

Use the table of integrals to evaluate $\displaystyle\int \frac{x}{3x - 2}\,dx$.

Solution   We can write:

$$\int \frac{x}{3x - 2}\,dx = \int x(3x - 2)^{-1}\,dx. \quad \textbf{Law of exponents}$$

The right-hand side is the same as Formula 8 in the table.

$$\int x(ax + b)^{-1}\,dx = \frac{x}{a} - \frac{b}{a^2}\ln|ax + b| + C \quad \textbf{Integral Formula 8}$$

Let $a = 3$ and $b = -2$; then we can write:

$$\int x(3x - 2)^{-1} \, dx = \frac{x}{3} - \frac{-2}{(3)^2} \ln|3x + (-2)| + C \qquad \text{Substituting}$$

$$= \frac{x}{3} + \frac{2}{9} \ln|3x - 2| + C. \quad \blacklozenge$$

EXAMPLE 2 Integrand Is the Product of an Exponential Function and a Sine Function

Use the table of integrals to evaluate $\int e^{3\theta} \sin 2\theta \, d\theta$.

Solution In examining the forms involving the exponential and the sine functions, we see that Formula 89 can be used to evaluate the integral.

$$\int e^{ax} \sin bx \, dx = \frac{e^{ax}}{a^2 + b^2}(a \sin bx - b \cos bx) + C \qquad \text{Integration Formula 89}$$

If we let $a = 3$, $b = 2$, and $x = \theta$, we have:

$$\int e^{3\theta} \sin 2\theta \, d\theta = \frac{e^{3\theta}}{3^2 + 2^2}(3 \sin 2\theta - 2 \cos 2\theta) + C \qquad \text{Substituting}$$

$$= \frac{e^{3\theta}}{13}(3 \sin 2\theta - 2 \cos 2\theta) + C. \quad \blacklozenge$$

When a polynomial expression is part of the integrand, we may need to complete the square before we can use substitution. This is illustrated in Example 3.

EXAMPLE 3 Integrand Is the Square Root of a Polynomial Expression

Use the table of integrals to evaluate $\int \sqrt{x^2 - 4x + 13} \, dx$.

Solution Studying the list of integral formulas of algebraic functions, there appears to be no formula that we can use. As mentioned earlier, sometimes we need to use substitution to use a formula. By completing the square, we obtain:

$$x^2 - 4x + 13 = x^2 - 4x + 4 + 9$$
$$= (x - 2)^2 + 3^2. \qquad \text{Factoring}$$

Now, if we let $u = x - 2$, we have $(x - 2)^2 + 3^2 = u^2 + 3^2$, and:

$$\int \sqrt{x^2 - 4x + 13} \, dx = \int \sqrt{u^2 + 3^2} \, du.$$

Re-examining the list, we find Formula 20.

$$\int \sqrt{a^2 + x^2}\, dx = \frac{x}{2}\sqrt{a^2 + x^2} + \frac{a^2}{2}\ln(x + \sqrt{x^2 + a^2}) + C$$

Letting $x = u$ and $a = 3$, we can write:

$$\int \sqrt{x^2 - 4x + 13}\, dx = \frac{u}{2}\sqrt{3^2 + u^2} + \frac{3^2}{2}\ln(u + \sqrt{u^2 + 3^2}) + C.$$

Now substitute for u.

$$= \frac{x - 2}{2}\sqrt{(3)^2 + (x - 2)^2} + \frac{3^2}{2}\ln[(x - 2) + \sqrt{(x - 2)^2 + 3^2}] + C$$

$$= \frac{x - 2}{2}\sqrt{x^2 - 4x + 13} + \frac{9}{2}\ln[(x - 2) + \sqrt{x^2 - 4x + 13}] + C. \quad \blacklozenge$$

Example 3 illustrates the importance of understanding the substitution principle.

In the list of integration formulas, a number of formulas have an integral sign on the right-hand side of the equal sign. An example is Formula 19.

$$\int \frac{dx}{(a^2 - x^2)^2} = \frac{x}{2a^2(a^2 - x^2)} + \frac{1}{2a^2}\int \frac{dx}{a^2 - x^2}.$$

A formula of this type is called a **reduction formula.** It carries this name since it reduces the integral to the sum of a function and a simpler integral. The use of reduction formulas is demonstrated in Examples 4 and 5.

EXAMPLE 4 Reduction Formula: Rational Function

Use the table of integrals to evaluate $\displaystyle\int \frac{\sqrt{3t - 5}}{t}\, dt$.

Solution We can use Formula 12.

$$\int \frac{\sqrt{ax + b}}{x}\, dx = 2\sqrt{ax + b} + b\int \frac{dx}{x\sqrt{ax + b}}.$$

Letting $a = 3$, $b = -5$, $dx = dt$, and $x = t$,

$$\int \frac{\sqrt{3t + (-5)}}{t}\, dt = 2\sqrt{3t + (-5)} + (-5)\int \frac{dt}{t\sqrt{3t + (-5)}}$$

$$= 2\sqrt{3t - 5} - 5\int \frac{dt}{t\sqrt{3t - 5}}.$$

Having used a reduction formula, we need to integrate again. To evaluate this new integral on the right-hand side, we can use Formula 13a since $b < 0$.

$$\int \frac{dx}{x\sqrt{ax + b}} = \frac{2}{\sqrt{-b}}\operatorname{Arctan}\sqrt{\frac{ax + b}{-b}} + C. \quad \textbf{Arctan } \tan^{-1} x$$

Letting $a = 3$, $b = -5$, and $x = t$,

$$\int \frac{dt}{t \sqrt{3t + (-5)}} = \frac{2}{\sqrt{-(-5)}} \text{Arctan} \sqrt{\frac{3t + (-5)}{-(-5)}} + C$$

$$= \frac{2}{\sqrt{5}} \text{Arctan} \sqrt{\frac{3t - 5}{5}} + C.$$

Therefore, substituting this result for $\int \dfrac{dt}{t \sqrt{3t - 5}}$

$$\int \frac{\sqrt{3t - 5}}{t} \, dt = 2\sqrt{3t - 5} - 5 \left(\frac{2}{\sqrt{5}} \text{Arctan} \sqrt{\frac{3t - 5}{5}} \right) + C$$

$$= 2\sqrt{3t - 5} - \frac{10}{\sqrt{5}} \text{Arctan} \sqrt{\frac{3t - 5}{5}} + C. \quad \blacklozenge$$

EXAMPLE 5 **Reduction Formula: Trignometric Functions**

Use the table of integrals to evaluate $\int \sin^3 5x \cos^2 5x \, dx$.

Solution We can use Formula 55.

$$\int \sin^n ax \cos^m ax \, dx = \frac{\sin^{n+1} ax \cos^{m-1} ax}{a(m + n)} + \frac{m - 1}{m + n} \int \sin^n ax \cos^{m-2} ax \, dx.$$

Letting $n = 3$, $m = 2$, and $a = 5$,

$$\int \sin^3 5x \cos^2 5x \, dx$$

$$= \frac{\sin^{3+1} 5x \cos^{2-1} 5x}{5(2 + 3)} + \frac{2 - 1}{2 + 3} \int \sin^3 5x \cos^{2-2} 5x \, dx$$

$$= \frac{\sin^4 5x \cos 5x}{25} + \frac{1}{5} \int \sin^3 5x \, dx.$$

We could integrate on the right-hand side of the formula by writing $\sin^3 5x$ as $\sin 5x(1 - \cos^2 5x)$, or we can use Formula 46.

$$\int \sin^n ax \, dx = \frac{-\sin^{n-1} ax \cos ax}{na} + \frac{n - 1}{n} \int \sin^{n-2} ax \, dx.$$

Letting $n = 3$ and $a = 5$,

$$\int \sin^3 5x \, dx = \frac{-\sin^{3-1} 5x \cos 5x}{3(5)} + \frac{3 - 1}{3} \int \sin^{3-2} 5x \, dx$$

$$= \frac{-\sin^2 5x \cos 5x}{15} + \frac{2}{3} \int \sin 5x \, dx.$$

The integral on the right we can integrate directly.

$$\int \sin 5x \, dx = \frac{-1}{5} \cos 5x + C.$$

Putting it all together, we have:

$$\int \sin^3 5x \cos^2 5x \, dx$$

$$= \frac{\sin^4 5x \cos 5x}{25} + \frac{1}{5}\left[\frac{-\sin^2 5x \cos 5x}{15} + \frac{2}{3}\left(-\frac{1}{5} \cos 5x\right)\right] + C.$$

Therefore,

$$\int \sin^3 5x \cos^2 5x \, dx = \frac{\sin^4 5x \cos 5x}{25} - \frac{\sin^2 5x \cos 5x}{75} - \frac{2 \cos 5x}{75} + C. \quad \blacklozenge$$

24.5 Exercises

In Exercises 1–30, use the integration formulas in Appendix D to evaluate the integrals.

1. $\int (3x - 7)^4 \, dx$

2. $\int \frac{dx}{x(-2x + 7)}$

3. $\int \frac{\sqrt{4t - 11}}{t} \, dt$

4. $\int \frac{t}{-5t - 3} \, dt$

5. $\int \frac{dZ}{Z\sqrt{11Z + 13}}$

6. $\int \frac{dZ}{(25 + Z^2)^2}$

7. $\int \frac{dZ}{16 - Z^2}$

8. $\int Z^2\sqrt{16 - Z^2} \, dZ$

9. $\int \frac{dt}{t\sqrt{13 + t^2}}$

10. $\int \sqrt{17 - t^2} \, dt$

11. $\int \frac{dx}{x\sqrt{7 - x^2}}$

12. $\int \frac{\sqrt{x^2 - 11}}{x} \, dx$

13. $\int \sin^4 3x \, dx$

14. $\int \cos^4 3x \, dx$

15. $\int \sin 3\phi \cos 2\phi \, d\phi$

16. $\int \sin 2\theta \sin 3\theta \, d\theta$

17. $\int \sin^4 A \cos^3 A \, dA$

18. $\int \sin^5 A \cos^4 A \, dA$

19. $\int \frac{d\theta}{1 + \sin 3\theta}$

20. $\int \frac{d\theta}{2 + 3 \cos \pi\theta}$

21. $\int x^3 \sin 2x \, dx$

22. $\int x^3 \cos 2x \, dx$

23. $\int \tan^3 2x \, dx$

24. $\int \sec^4 3x \, dx$

25. $\int \text{Arcsin} \frac{x}{3} \, dx$

26. $\int \text{Arctan} \frac{x}{4} \, dx$

27. $\int x^3 e^{7x} \, dx$

28. $\int e^{5x} \sin 7x \, dx$

29. $\int x^5 \ln \frac{x}{5} \, dx$

30. $\int \ln(3a + b)x \, dx$

31. Determine the area of the region bounded by the curve $y = e^{2x} \cos \frac{x}{2}$, the x-axis, and the lines $x = 0$ and $x = 3$.

32. Determine the area of the region bounded by the curve $y = x^2\sqrt{25 - x^2}$, the x-axis, and the lines $x = 3$ and $x = 5$.

33. A chemical company discovers that the time t for a certain material to dissolve is related to the amount of the material still undissolved x by the formula $t = 260 \int \dfrac{dx}{x(x + 4)}$.

  a. If 30 kg are dumped into a vat, how long does it take for half of the material to dissolve? (*Hint:* Evaluate from 30 kg to 15 kg.)

  b. Is the time it takes the material to dissolve independent of the amount of material dissolved? Suppose only 10 kg are dissolved. What is the length of time taken to dissolve 10 kg?

34. The force for a certain complex spring is $F = 2x(2x + 1)^{\frac{1}{2}}$, where x is the distance that the spring is stretched.

  a. Recalling that work is the integral of the force times incremental change in distance, determine the work done if the spring is stretched from 12 cm to 17 cm, a change of 5 cm.

  b. Is the force the same if the spring is stretched from 17 cm to 22 cm (still a total change of 5 cm)?

35. In studying the scattering of small projectiles (like subatomic particles) hitting larger masses (like atoms), angles describe the results better than linear measurements. The amount of scattering depends, in part, upon the number of atoms n in a given unit volume. The equation for n is $n = \int e^{a\cos\theta} \sin\theta \, d\theta$, where a is a constant. Perform the integration and evaluate from $\theta = 0$ to $\theta = \pi$.

Review Exercises

In Exercises 1–30, use the integration formulas developed throughout the text, but *do not* use Appendix D. It is important that you recognize various types of integrals. This set of exercises is designed for that purpose.

1. $\displaystyle\int x \ln x \, dx$

2. $\displaystyle\int x \sin 5x \, dx$

3. $\displaystyle\int \cos^3 \theta \, d\theta$

4. $\displaystyle\int t\sqrt{t^2 - 4} \, dt$

5. $\displaystyle\int \frac{dx}{\sqrt{x^2 + 25}}$

6. $\displaystyle\int e^{2t} \sin t \, dt$

7. $\displaystyle\int (x + 5)e^{2x} \, dx$

8. $\displaystyle\int \sqrt{16 - 25x^2} \, dx$

9. $\displaystyle\int \sin^2\left(x - \frac{\pi}{4}\right) dx$

10. $\displaystyle\int (\sin\theta - \cos\theta)^2 \, d\theta$

11. $\displaystyle\int \sin^3 \frac{\theta}{2} \cos^2 \frac{\theta}{2} \, d\theta$

12. $\displaystyle\int \sqrt{x} \ln 3x \, dx$

13. $\displaystyle\int \frac{x}{(9 - x^2)^2} \, dx$

14. $\displaystyle\int \frac{e^x}{3 + e^x} \, dx$

15. $\displaystyle\int \frac{dt}{\sqrt{64 - t^2}}$

16. $\displaystyle\int \sin^2 \theta \, d\theta$

17. $\displaystyle\int \frac{\ln x}{x} \, dx$

18. $\displaystyle\int \text{Arctan } x \, dx$

19. $\displaystyle\int \frac{5x + 17}{x^2 + 5x - 14} \, dx$

20. $\displaystyle\int \frac{\cos (\tan x)}{\cos^2 x} \, dx$

21. $\displaystyle\int (\sec x - \tan x)^2 \, dx$

22. $\displaystyle\int \frac{dx}{x^2 \sqrt{81 - x^2}}$

23. $\displaystyle\int \frac{dx}{x(1 + \ln^2 x)}$

24. $\displaystyle\int \sec^4 \theta \, d\theta$

25. $\displaystyle\int \sin^2 x \cos^3 x \, dx$

26. $\displaystyle\int \frac{e^x}{6 + e^{2x}} \, dx$

27. $\displaystyle\int \ln^2 x \, dx$

28. $\displaystyle\int \frac{\sin x}{1 - \cos x} \, dx$

29. $\displaystyle\int x e^{3x} \, dx$

30. $\displaystyle\int \frac{x^3 - 12x^2 + 3x - 12}{x(x - 1)(x^2 + 4)} \, dx$

31. Determine the area of the region bounded by the curve $y = x^2 \ln x$, the x-axis, and the lines $x = 1$ and $x = 3$.

32. Determine the centroid of the region bounded by the curve $y = \dfrac{x + 4}{x^2 + 3x + 2}$, the x-axis, and the lines $x = 0$ and $x = 2$.

33. Determine the volume of the solid generated by revolving the region bounded by the curves $y = \dfrac{x^2}{\sqrt{x^2 - 9}}$, the x-axis, and the lines $x = 4$ and $x = 6$ about the x-axis.

34. If the velocity in meters per second of a particle that is moving along a straight line is expressed by the formula $v = \dfrac{2t^2 + 4t + 4}{t^3 + 4t^2 + 4t}$, determine the distance in meters, $s = \int v\, dt$, that the particle travels from $t = 1$ s to $t = 4$ s.

35. Determine the area under the curve $y = x^2 e^{-x}$ from $x = 0$ to $x = 1$.

36. Determine the centroid of the region bounded by the curve $y = \sin^3 x$, the x-axis, and the lines $x = 0$ and $x = \dfrac{\pi}{2}$.

*37. Evaluate the integral $\displaystyle\int \frac{\csc^3 8x}{\cot^4 8x}\, dx$.

*38. Prove that the trigonometric substitution $u = \tan \theta$ converts the integral $\displaystyle\int \frac{du}{(u + 1)^k}$ into the form $\int \cos^n \theta\, d\theta$, where $n = 2(k - 1)$.

*39. Use integration by parts to verify that: $\displaystyle\int \sin^n x\, dx = -\frac{\sin^{n-1} x \cos x}{n} + \frac{n - 1}{n} \int \sin^{n-2} x\, dx$.

*40. Evaluate the integral $\displaystyle\int \frac{e^x}{(e^{2x} + 1)(e^x - 1)}\, dx$ by using the substitution $u = e^x$.

*41. The region bounded by the circle $(x - 4)^2 + y^2 = 1$ is revolved about the y-axis. The resulting doughnut-shaped solid is called a torus. (See the diagram.) Determine the volume of this solid.

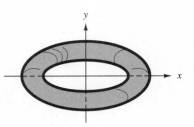

✎ Writing About Mathematics

1. In working with Appendix D we found that understanding the technique of substitution is very helpful. Write a note to your instructor explaining why the technique of substitution is so important when working with a table of integral formulas. Include an example to illustrate your point.

2. A friend, who is a junior in high school, wants to be an engineer and, so, will need to take calculus. Write your friend a letter describing the mathematical skills she needs to succeed in calculus.

3. A classmate is having problems with trigonometric substitution. Write a set of guidelines to help him overcome his problems.

Chapter Test

In Exercises 1–15, evaluate the integrals.

1. $\displaystyle\int xe^{4x}\,dx$

2. $\displaystyle\int x\sin 2x\,dx$

3. $\displaystyle\int e^{2x}\sin 2x\,dx$

4. $\displaystyle\int_{x=1}^{x=3} x^4\ln x\,dx$

5. $\displaystyle\int \sin^6 x\cos^3 x\,dx$

6. $\displaystyle\int \sin^5 x\,dx$

7. $\displaystyle\int \sin^3 x\cos^3 x\,dx$

8. $\displaystyle\int \tan^4 x\sec^2 x\,dx$

9. $\displaystyle\int_{x=0}^{x=3} x\sqrt{9-x^2}\,dx$

10. $\displaystyle\int x^3\sqrt{25+x^2}\,dx$

11. $\displaystyle\int x^3\sqrt{x^2-25}\,dx$

12. $\displaystyle\int \tan^4 x\,dx$

13. $\displaystyle\int \frac{1}{9x^2-16}\,dx$

14. $\displaystyle\int \frac{2x+3}{x^2+4x-5}\,dx$

15. $\displaystyle\int_{x=0}^{x=2\sqrt2} \frac{x^3}{x^2+1}\,dx$

16. Determine the area of the region in the first quadrant bounded by the coordinate axes, the line $x=1$, and the curve $y=\dfrac{3}{x^2-5x+6}$.

In Exercises 17–20, use Appendix D to evaluate the integrals.

17. $\displaystyle\int \frac{3x}{5x-7}\,dx$

18. $\displaystyle\int \sqrt{x^2-6x+13}\,dx$

19. $\displaystyle\int \frac{\sqrt{4x+5}}{x}\,dx$

20. $\displaystyle\int \sin^4 3x\cos^3 3x\,dx$

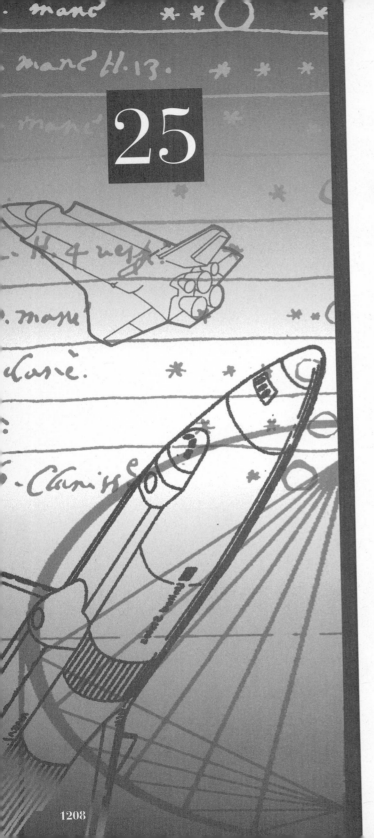

25

First Order Differential Equations

I n Chapters 19, 20, and 23 we examined the derivatives of functions, which in themselves are functions. We call such functions differential equations. In this and the next two chapters our challenge is to learn how to solve differential equations.

The angular velocity of the arm of a rowing machine can be approximated by a differential equation. The rate at which a population P expands is proportional to the population that is present at any time. Thus, one model for population growth is a differential equation. The rate at which bacteria grow can be expressed as a differential equation. Example 3 in Section 25.5 uses a differential equation to determine the growth period of a particular bacteria.

25.1

What Is a Differential Equation?

In many applied areas, a physical reaction is expressed in terms of a rate of change. For example, if an object is dropped from the top of a building, the rate of acceleration of the falling body can be expressed as $a = -9.8$ m/s^2.

Recall that the acceleration can be expressed in terms of velocity or distance; that is,

$$a = \frac{dv}{dt} \quad \text{or} \quad a = \frac{d^2s}{dt^2}.$$

Both of these equations are differential equations.

> **DEFINITION 25.1**
>
> A **differential equation** is an equation that involves derivatives of an unknown function of one or more variables.

If the unknown function depends on only one variable, then the derivative is an ordinary derivative, and the equation is called an **ordinary differential equation.** The following differential equations are examples of ordinary differential equations.

$$\frac{dv}{dt} = -32 \tag{1}$$

$$\frac{d^2s}{dt^2} = -32 \tag{2}$$

$$\frac{d^2y}{dx^2} + 2x\frac{dy}{dx} + y = 3 \tag{3}$$

$$\left(\frac{d^3y}{dx^3}\right)^2 + 2\frac{d^2y}{dx^2} - \frac{dy}{dx} + x^2\left(\frac{dy}{dx}\right)^3 = 0 \tag{4}$$

$$(2x + 3y) + (x - 4y)\frac{dy}{dx} = 0 \tag{5}$$

Before we begin discussing methods of determining solutions for ordinary differential equations, let us look at two other ways of classifying differential equations and define what is meant by a solution to a differential equation.

> **DEFINITION 25.2**
>
> The **order** of a differential equation is the order of the highest derivative occurring in the equation.

EXAMPLE 1   Order of Differential Equation

Determine the order of the five equations given as examples of ordinary differential equations.

Solution   Equations (1) and (5) are first order differential equations. Equations (2) and (3) are second order, and equation (4) is a third order differential equation. Note that the order does not depend upon the exponent of a term of the differential equation. Rather, it depends upon how many times the derivative is taken.   ◆

Another way of classifying an ordinary differential equation is to determine whether it is linear or nonlinear.

DEFINITION 25.3

A differential equation of order n is called a **linear equation** if it can be written in the form:

$$a_n(x)\frac{d^n y}{dx^n} + a_{n-1}(x)\frac{d^{n-1}y}{dx^{n-1}} + \cdots + a_1(x)\frac{dy}{dx} + a_0(x)y = F(x),$$

where $F(x), a_n(x), \ldots, a_1(x), a_0(x)$ are functions of x and $a_n(x)$ is not zero. The equation is said to be linear in y and the derivatives of y.

An ordinary differential equation that does not meet the conditions of Definition 25.3 is called a **nonlinear differential equation.**

EXAMPLE 2   Linear or Nonlinear Differential Equation

Determine whether the five ordinary differential equations are linear or nonlinear.

Solution   We can write equations (1), (2), and (3) in the following form:

$$v' = -32 \tag{1}$$
$$s'' = -32 \tag{2}$$
$$y'' + 2xy' + y = 3. \tag{3}$$

Therefore, by Definition 25.3, these are linear differential equations. If we rewrite equations (4) and (5), we obtain the following results.

$$(y''')^2 + 2y''y' + x^2(y')^3 = 0 \tag{4}$$
$$xy' - 4yy' + 3y = -2x \tag{5}$$

Since the terms $(y''')^2$, $2y''y'$, $x^2(y')^3$, and $4yy'$ are not linear in y, these equations are nonlinear differential equations.   ◆

> **DEFINITION 25.4**
>
> A **solution** of a differential equation is any function $y = f(x)$ or $f(x, y)$ that, when substituted in the differential equation, reduces the equation to an identity; that is, it satisfies the equation.

EXAMPLE 3 **Solution of a Differential Equation**

Show that $y = x^2 + x + C$ is a solution of the differential equation $\dfrac{dy}{dx} = 2x + 1$

Solution

$$y = x^2 + x + C \quad \text{Given function}$$

$$\frac{dy}{dx} = 2x + 1 \quad \text{Differentiating}$$

$$2x + 1 = 2x + 1 \quad \text{Substituting for } \frac{dy}{dx}$$

Thus, $y = x^2 + x + C$ is a solution of the differential equation $\dfrac{dy}{dx} = 2x + 1$. ◆

From our experience in calculus, we know we can obtain a specific solution to a differential equation if we are given a value for the independent variable and a value for the dependent variable. For example, for the function $y = f(x) = x^2 + x + C$ if $f(1) = 3$, then

$$3 = 1 + 1 + C. \quad \text{Substituting for } x \text{ and } f(x)$$

Thus, $1 = C$, and $y = f(x) = x^2 + x + 1$ is a specific solution.

If we are to determine a solution to a differential equation subject to conditions on the unknown function and its derivatives specified for one value of the independent variable, the conditions are called **initial conditions.** Given initial conditions, the problem is called an **initial value problem.** If we are to determine a solution to a differential equation subject to the unknown function specified at two or more values of the independent variable, the conditions are called **boundary conditions.** A problem with boundary conditions is called a **boundary value problem.**

To obtain a specific solution, we must be given initial conditions or boundary conditions. If we have a second order differential equation, we must be given two sets of initial conditions. This is illustrated in Example 4.

EXAMPLE 4   **Specific Solution: Initial Conditions**

Determine a specific solution for the differential equation $\dfrac{d^2s}{dt^2} = -32$ ft/s^2 so that the function satisfies the initial conditions $s = 0$ when $t = 0$ and $\dfrac{ds}{dt} = 2$ ft/s when $t = 0$.

Solution   A specific solution means that we ultimately must end with $s(t)$ equal to some function of time. Note we have a second order differential equation. Therefore, we must integrate twice, and each time we integrate, the solution contains a constant. Thus, we need two initial conditions to solve for the constants and obtain a specific solution.

$$\frac{d^2s}{dt^2} = -32 \qquad\qquad \textbf{Given}$$

$$\frac{ds}{dt} = -32t + C_1 \qquad\qquad \textbf{Integrating both sides}$$

$$2 = -32(\,0\,) + C_1 \qquad\qquad \textbf{Substituting initial conditions}$$
$$\qquad\qquad\qquad\qquad\qquad \frac{ds}{dt} = 2 \text{ and } t = 0$$
$$C_1 = 2$$

$$\frac{ds}{dt} = -32t + 2 \qquad\qquad \textbf{Substituting for } C_1$$

$$s = -16t^2 + 2t + C_2 \qquad\qquad \textbf{Integrating both sides}$$
$$0 = -16(\,0\,) + 2(\,0\,) + C_2 \qquad \textbf{Substituting initial}$$
$$\qquad\qquad\qquad\qquad\qquad \textbf{conditions } s = 0 \text{ and } t = 0$$
$$C_2 = 0$$

$$s = -16t^2 + 2t \qquad\qquad \textbf{Substituting for } C_2 \;\blacklozenge$$

Without initial conditions or boundary values, we obtain a **general solution** to the differential equation. With initial conditions or boundary values, we obtain a **specific solution** to a differential equation.

25.1   Exercises

In Exercises 1–10 determine the following.
a.   Whether the differential equation is linear or nonlinear
b.   The order of each differential equation

1. $y' = 3x^2 + y$

2. $y'' + 3y' + 11y = 3x$

3. $yy'' + 5xy' = 4$

4. $y''' + 2(y')^3 - y = 0$

5. $y^{(vi)} - y'' = 0$

6. $y'' - (y')^2 + xy = 0$

7. $\dfrac{d^3y}{dx^3} - \dfrac{d^2y}{dx^2} + y = 2x$

8. $\dfrac{d^2y}{dx^2} + 5x = \cos y$

9. $(3s + 4t)\,dt + (5s + 6t)\,ds = 0$

10. $x^3\,dy + (xy - e^x)\,dx = 0$

In Exercises 11–16, show that the function is a solution to the differential equation.

11. $y = e^x + e^{2x}$; $y'' - 3y' + 2y = 0$

12. $y = \tan x$; $y' = 1 + y^2$

13. $y = x - x \ln x$; $xy' + x - y = 0$

14. $y = A \sin x + B \cos x$; $\dfrac{d^2y}{dx^2} + y = 0$

15. $y = (x + C) e^{-x}$; $\dfrac{dy}{dx} + y = e^{-x}$

16. $y = C_1 e^{2x} + C_2 e^{-4x} + 2xe^{2x}$; $\dfrac{d^2y}{dx^2} + 2\dfrac{dy}{dx} - 8y = 12e^{2x}$

In Exercises 17–18, use the given initial conditions and the given general solution of the differential equation to determine a specific solution.

17. $xy = C$; $y = 1$ for $x = 2$

18. $\sin(xy) + y = C$; $y = 1$ for $x = \dfrac{\pi}{4}$

In Exercises 19–21, determine a specific solution of the differential equation using the given initial conditions.

19. $y' = \cos x$; $y = 1$ for $x = 0$

20. $y'' = e^x$; $y = 1$ and $y' = 0$ for $x = 1$

21. $y' = \dfrac{1}{x}$; $y = 0$ for $x = 2$

In Exercises 22–24, determine a specific solution of the differential equation using the given boundary conditions.

22. $y'' = 2$; $y' = 1$ for $x = 0$, $y = 3$ for $x = 1$

23. $y'' = \sin x$; $y' = 0$ for $x = \dfrac{\pi}{4}$, $y = 1$ for $x = \dfrac{3\pi}{2}$

24. $y^{(iv)} = 0$; $y = 1$ for $x = 1$ and $x = 3$, $y' = 1$ for $x = 2$, $y'' = 1$ for $x = 1$.

25.2

Separation of Variables

In this and the next two sections we examine techniques for solving first order differential equations. This section discusses a technique called **separation of variables** used for equations in which the variables can be separated in such a way that a solution can be obtained by rewriting the equation and integrating.

The differential equation $\dfrac{dy}{dx} = 2xy$ is an example of an equation that can be solved by this technique. For example, using differential notation, we can rewrite:

$$\frac{dy}{dx} = 2xy$$

as:

$$\frac{dy}{y} = 2x \, dx.$$

We can now integrate both sides of the equation

$$\ln |y| = x^2 + C$$

In general, a first order differential equation can be written in the form:

$$\frac{dy}{dx} = f(x, y).$$

A special case of the general form is:

$$\frac{dy}{dx} = g(x)h(y),$$

which can be written as:

$$\frac{dy}{h(y)} = g(x)\, dx.$$

In this case, the variables are separated and we can obtain the general solution by integrating.

$$\int \frac{dy}{h(y)} = \int g(x)\, dx$$

This technique is illustrated in Example 1.

EXAMPLE 1   Separation of Variables

Determine the general solution of $\sec x \dfrac{dy}{dx} + \tan x = 0$.

Solution   We first rewrite the equation so that all the terms containing a variable x are on one side of the equal sign and those containing the variable y are on the other side.

$$dy = -\frac{\tan x}{\sec x}\, dx \qquad \textbf{sec } x \neq 0$$

$$dy = -\frac{\dfrac{\sin x}{\cos x}}{\dfrac{1}{\cos x}}\, dx \qquad \textbf{Trigonometric identities}$$

$$dy = -\sin x \, dx \qquad \textbf{Simplifying}$$

$$y + C_1 = \cos x + C_2 \qquad \textbf{Integrating}$$

$$y = \cos x + C_2 - C_1 \qquad \textbf{Solving for } y$$

Since $C_2 - C_1$ is a constant, we can write this as a single constant C. Thus, the general solution of the differential equation is $y = \cos x + C$.   ◆

EXAMPLE 2   Separation of Variables

Determine the general solution of $\dfrac{dy}{dx} = 2xy^2$.

Solution

$$\frac{dy}{dx} = 2xy^2 \qquad \textbf{Given}$$

$$y^{-2}\,dy = 2x\,dx \qquad \textbf{Multiplying both sides of the equation by } y^{-2} \textbf{ and } dx$$

$$\int y^{-2}\,dy = \int 2x\,dx \qquad \textbf{Integrating}$$

$$-\frac{1}{y} = x^2 + C$$

$$y = \frac{-1}{x^2 + C} \qquad \textbf{Solving for } y \quad \blacklozenge$$

The general solution of a differential equation can be interpreted as a family, or families, of curves. For example, the general solution of the differential equation $\frac{dy}{dx} = 2x$ is $y = x^2 + C$. For each value of the constant C, the figure is a parabola with the vertex located at $(0, C)$. In this case, the family of curves is a set of parabolas as shown in Figure 25.1.

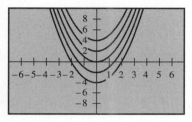

FIGURE 25.1

EXAMPLE 3 Specific Solution

For the general solution (family of curves) obtained in Example 2, determine the solution whose graph passes through the point $(3, -1)$.

Solution   The initial condition is $y(3) = -1$.

$$y = \frac{-1}{x^2 + C} \qquad \textbf{General solution}$$

$$-1 = \frac{-1}{(\,3\,)^2 + C} \qquad \textbf{Substituting initial condition}$$

$$C = -8 \qquad \textbf{Solving for } C$$

$$y = \frac{-1}{x^2 - \boxed{8}} \qquad \textbf{Substituting for } C$$

$$= \frac{1}{8 - x^2} \qquad \textbf{Undefined when } x = \pm\sqrt{8}$$

The function is not defined when x equals $\pm\sqrt{8}$. Thus, the portion of the curve that passes through the point $(3, -1)$ is defined only in the interval $\sqrt{8} < x < \infty$. $\quad \blacklozenge$

We can obtain a visual check of Example 3 by graphing the function, as shown in Figure 25.2.

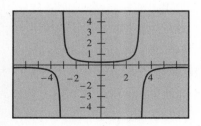

FIGURE 25.2

EXAMPLE 4 **General and Specific Solution**

a. Determine the general solution of $\dfrac{dy}{dx} = \dfrac{x^3 + 1}{y - 3}$.

b. Determine a specific solution for the initial condition $y(2) = 1$.

Solution

a.
$$\frac{dy}{dx} = \frac{x^3 + 1}{y - 3} \qquad \text{Given}$$

$$(y - 3)\, dy = (x^3 + 1)\, dx \qquad \text{Separating variables}$$

$$\frac{y^2}{2} - 3y = \frac{x^4}{4} + x + C \qquad \text{Integrating and combining constants}$$

b. Substituting the initial condition $y(2) = 1$ in the general solution, we have:

$$\frac{(\boxed{1})^2}{2} - 3(\boxed{1}) = \frac{(\boxed{2})^4}{4} + \boxed{2} + C$$

$$C = -8.5.$$

Thus, the specific solution of the differential equation with the given initial condition is $\dfrac{y^2}{2} - 3y = \dfrac{x^4}{4} + x - 8.5$. ◆

Sometimes it is easier to write a solution in terms of the dependent variable, or in a more compact form, if we use the definitions and rules of exponents and logarithms. This is illustrated in Example 5.

EXAMPLE 5 **Solution in Exponential Form**

Determine the general solution of $\dfrac{dy}{dx} = 2xy$.

Solution

$$\frac{dy}{dx} = 2xy \qquad \text{Given}$$

$$\frac{dy}{y} = 2x\, dx \qquad \text{Separating variables}$$

$$\ln |y| = x^2 + C_1 \qquad \text{Integrating}$$

$$y = e^{(x^2 + C_1)} \qquad \text{Recall } y = \log_b x \text{ if } x = b^y$$

$$y = e^{x^2} e^{C_1} \qquad \text{Law of exponents}$$

Since C_1 is an arbitrary constant, we can let $C = e^{C_1}$ and write the solution as $y = Ce^{x^2}$, where C is an arbitrary positive constant. ◆

If all the terms except the constant term of a general solution are in logarithmic form, it is advantageous to write the constant as a logarithm ($\ln C$). This is illustrated in Example 6.

EXAMPLE 6 **Using Logarithms to Simplify the Solution**

a. Determine the general solution of $\dfrac{dy}{dx} = -\dfrac{y}{x}$.

b. Determine a specific solution given the initial condition $y(2) = 4$.

Solution

a.
$$\frac{dy}{dx} = -\frac{y}{x} \quad \textbf{Given}$$

$$\frac{dy}{y} + \frac{dx}{x} = 0 \quad \textbf{Separating variables}$$

$$\ln|y| + \ln|x| = C_1 \quad \textbf{Integrating}$$

Since $\ln C$ is equal to a constant, we can let $C_1 = \ln C$.

$$\ln|y| + \ln|x| = \ln C \quad \textbf{Substituting}$$
$$\ln|y||x| = \ln C \quad \textbf{Law of logarithms}$$
$$yx = C$$

b. To illustrate that the specific solution is not affected by letting $C_1 = \ln C$, we determine the constant term for each case.

For $\ln|y||x| = C_1$	For $yx = C$	
$\ln (4)(2) = C_1$	$(4)(2) = C$	**Substituting**
$C_1 = \ln 8$	$C = 8$	**Solving for C**
$\ln|y||x| = \ln 8$	$yx = 8$	**Substituting for C**
$yx = 8$	$yx = 8$ ◆	

Example 6b illustrates that when we write the constant term as a logarithm, the solution is not affected.

25.2 Exercises

In Exercises 1–20, determine the general solution for each of the differential equations.

1. $x\,dy + y\,dx = 0$

2. $dx + dy = 0$

3. $x^2\,dx + y^2\,dy = 0$

4. $x\,dy + 3y\,dx = 0$

5. $xy' = 5y$

6. $\csc^2 x\,dy + \sec y\,dx = 0$

7. $\sin x\,y' + \cos y = 0$

8. $\cos x \cos y\,dx + \sin x \sin y\,dy = 0$

9. $e^x y' + y^2 = 0$

10. $xy\,dx + (x^2 + 1)\,dy = 0$

11. $(y')^2 = 1 - y^2$

12. $y' + (y - 1)\cos x = 0$

13. $(1 + x^2)\,dy + (1 + y^2)\,dx = 0$

14. $e^x e^y\,dx - e^{-2y}\,dy = 0$

15. $\sqrt{1 - x^2}\,dy = \sqrt{1 - y^2}\,dx$

16. $\dfrac{dV}{dP} = -\dfrac{V}{P}$

17. $e^x y \, dy - (e^{-y} + e^{-(2x+y)}) \, dx = 0$

18. $(e^y + 1)^4 e^{-y} + (e^x + 1)^5 e^{-x} y' = 0$

19. $t^2 y \dfrac{dy}{dt} = t^2 + 1$

20. $\dfrac{ds}{dr} = ks$

In Exercises 21–26, determine a specific solution that satisfies the initial condition.

21. $3\theta \dfrac{dr}{d\theta} - r = \theta \dfrac{dr}{d\theta}$, $\theta = 1$ when $r = 3$

22. $\sin x(e^{-y} + 1) \, dx = (1 + \cos x) \, dy$; $y(0) = 0$

23. $2y \, dx = 3x \, dy$; $y(-2) = 1$

24. $\sin^2 y \, dx + \cos^2 x \, dy = 0$; $y\left(\dfrac{\pi}{4}\right) = \dfrac{\pi}{4}$

25. $2y \cos x \, dx + 3 \sin x \, dy = 0$; $y\left(\dfrac{\pi}{2}\right) = 2$

26. $\sin x \cos y \, dx + \cos x \sin y \, dy = 0$; $y(0) = 0$

27. The slope of a family of curves at the point (x, y) is $\dfrac{y - 1}{1 - x}$. Determine the equation of the curve that passes through the point $(4, -3)$.

28. When a gas expands without gain or loss of heat, the rate of change of pressure with volume varies directly as the pressure and inversely as the volume. Determine the law connecting pressure and volume in this case.

25.3

Special Forms of Differential Equations

In the previous section we learned that if we could separate the variables on different sides of the equal sign, we could determine a solution for the differential equation. Consider now the equation $x \, dy = 4x^3 \, dx - y \, dx$. A quick look at this equation suggests that it defies such a separation. We cannot get the x-terms exclusively on one side and the y-terms exclusively on the other side. So the question arises—are there techniques we can use to change an equation to a form where the variables are separable or to a form where the variables are in integrable combinations? An **integrable combination** is an expression that may result from differentiating a product or a quotient or some other expression that we can recognize. Example 1 illustrates what is meant by an integrable combination.

EXAMPLE 1   **Integrable Combination**

Determine the general solution for $x \, dy = (4x^3 - y) \, dx$.

Solution   If we apply the distributive property on the right side and collect all terms containing ys on the left side, the equation becomes:

$$x \, dy + y \, dx = 4x^3 \, dx.$$

Since the term on the right side of the equal sign does not contain a y, we can integrate the term on this side. The result is x^4. The terms on the left side of the equal sign contain both x- and y-terms; this is a problem. But look, if we differentiate the product xy, the result is:

$$x \, dy + y \, dx,$$

the expression on the left side of the equal sign. Thus the expression $x\,dy + y\,dx$ is an integrable combination. Hence, by integrating both sides, we determine a general solution, which is $xy = x^4 + C$. ◆

Some of the frequently occurring integrable combinations and the functions for which they are exact differentials are shown in the box.

Integrable Combinations

$\{1\}$ $\qquad x\,dy + y\,dx = d\,(xy)$

$\{2\}$ $\qquad \dfrac{x\,dy - y\,dx}{x^2} = d\left(\dfrac{y}{x}\right)$

$\{3\}$ $\qquad \dfrac{y\,dx - x\,dy}{y^2} = d\left(\dfrac{x}{y}\right)$

$\{4\}$ $\qquad \dfrac{2xy\,dy - y^2\,dx}{x^2} = d\left(\dfrac{y^2}{x}\right)$

$\{5\}$ $\qquad \dfrac{2xy\,dx - x^2\,dy}{y^2} = d\left(\dfrac{x^2}{y}\right)$

EXAMPLE 2   **Creating an Integrable Combination**

Determine the general solution of $x\,dy - y\,dx \;=\; x\,dx$.

Solution   In its present form the expression is not an integrable combination. However, if we multiply both sides by $\dfrac{1}{x^2}$, the equation becomes:

$$\frac{x\,dy - y\,dx}{x^2} = \frac{dx}{x},$$

and the left side is of the same form as $\{2\}$. Integrating, we have:

$$\frac{y}{x} = \ln|x| + C. \quad ◆$$

A differential equation such as $\dfrac{x\,dy - y\,dx}{x^2} = \dfrac{dx}{x}$ whose terms are exact differentials or whose terms can be rearranged to form exact differentials is called an **exact differential equation.** This means that the equation is in a form such that a general solution can be found by integrating. To get the equation of Example 2 into an exact form, we multiplied every term by the function $\dfrac{1}{x^2}$, which is called an integrating factor. (For more information on the rules for determining integrating factors, see a textbook on differential equations.) In Section 25.4 we will discuss a special integrating factor.

EXAMPLE 3 **Creating an Integrable Combination**

Determine the general solution of $2xy\,dx - x^2\,dy = xy^2\,dx$.

Solution The left side of this equation is the same as the numerator of the left side of {5}. If we had a y^2 in the denominator, this side would be integrable. Dividing both sides of the equation by y^2, we have:

$$\frac{2xy\,dx - x^2\,dy}{y^2} = x\,dx.$$

Now both sides of the equation are integrable, and the solution is:

$$\frac{x^2}{y} = \frac{x^2}{2} + C. \quad \blacklozenge$$

A differential equation that can be written in the form:

$$\frac{dy}{dx} = f\!\left(\frac{y}{x}\right)$$

is called a **homogeneous differential equation.** A function is homogeneous if the total degree of each term is the same. For example,

$$f(x, y) = 3xy^2 - 7x^2y.$$

The function is homogeneous of degree 3.

Now look at:

$$f(x, y) = xy - 2x^2y.$$

The function is not homogeneous since the degrees of the two terms are different.

Each homogeneous differential equation can be transformed into an equation with variables separated. To achieve this transformation, we let

$$v = \frac{y}{x}, \quad \text{or} \quad y = vx \quad \text{and} \quad \frac{dy}{dx} = v + x\frac{dv}{dx}.$$

This substitution changes the dependent variable from y to v and keeps x as the independent variable. This can be shown by differentiating $y = vx$ with respect to x and then recalling the original function $\frac{dy}{dx} = f\!\left(\frac{y}{x}\right)$.

$$\frac{dy}{dx} = v + x\frac{dv}{x} \qquad \textbf{Differentiating } y = vx$$

$$\frac{dy}{dx} = f\left(\frac{y}{x}\right) \qquad \textbf{Recalling original equation}$$

$$v + x\frac{dv}{dx} = f(\,v\,) \qquad \textbf{Substituting for } \frac{dy}{dx} \textbf{ and } \frac{y}{x}$$

$$x\frac{dv}{dx} = f(v) - v$$

$$\frac{dx}{x} = \frac{dv}{f(v) - v} \qquad \textbf{Separating variables}$$

The variables are separated and we can obtain the solution. This method is illustrated in Example 4.

EXAMPLE 4   **Homogeneous Equation**

Determine the general solution of $xy' = 2x + 3y$.

Solution   Begin by dividing each term by x.

$$\frac{dy}{dx} = 2 + \frac{3y}{x}$$

In this form, we see that the equation is homogeneous. In the previous discussion, we let $y = vx$ or $v = \dfrac{y}{x}$ and $\dfrac{dy}{dx} = v + x\dfrac{dv}{dx}$. Substituting for $\dfrac{dy}{dx}$ and $\dfrac{y}{x}$, we have:

$$v + x\frac{dv}{dx} = 2 + 3v \qquad \textbf{Substituting}$$

$$x\frac{dv}{dx} = 2 + 2v$$

$$\frac{dv}{v + 1} = \frac{2\,dx}{x} \qquad \textbf{Separation of variables}$$

$$\ln|v + 1| = 2\ln|x| + \ln|C| \qquad \textbf{Integrating}$$

$$v + 1 = Cx^2 \qquad \textbf{Law of logarithms}$$

$$\frac{y}{x} + 1 = Cx^2 \qquad \textbf{Substituting for } v$$

$$y = Cx^3 - x. \quad \blacklozenge$$

EXAMPLE 5   Homogeneous Equation

a. Determine the general solution of $y' = \dfrac{y}{x} + \sin \dfrac{y}{x}$.

b. Determine a specific solution for the initial condition $y(2) = \dfrac{\pi}{3}$.

Solution

a. The equation is a homogeneous differential equation; each term is of the same degree.

$$v + x \frac{dv}{dx} = v + \sin v \qquad \text{Substitution}$$

$$x \frac{dv}{dx} = \sin v$$

$$\frac{dv}{\sin v} = \frac{dx}{x} \qquad \text{Separating variables}$$

$$\csc dv = \frac{dx}{x}$$

$$\ln |\csc v - \cot v| = \ln |x| + \ln |C| \qquad \text{Integrating}$$

$$\csc v - \cot v = Cx \qquad \text{Law of logarithms}$$

$$\csc \frac{y}{x} - \cot \frac{y}{x} = Cx \qquad \text{Substituting for } v$$

b. Now use the result of part a to determine the specific solution for the initial condition $y(2) = \dfrac{\pi}{3}$.

$$\csc \left(\frac{\frac{\pi}{3}}{2} \right) - \cot \left(\frac{\frac{\pi}{3}}{2} \right) = 2C \qquad \text{Substituting}$$

$$2 - \sqrt{3} = 2C$$

$$\frac{2 - \sqrt{3}}{2} = C \qquad \text{Solving for } C$$

The equation $\csc \dfrac{y}{x} - \cot \dfrac{y}{x} = \dfrac{2 - \sqrt{3}}{2} x$ is the specific solution of the differential equation for the initial condition $y(2) = \dfrac{\pi}{3}$. ◆

25.3 Exercises

In Exercises 1–20, determine the general solution of the differential equation.

1. $x\,dy + y\,dx = 3y^2\,dy$

2. $y\,dx = -x\,dy + x^2\,dx$

3. $x\,dy - y\,dx = x^4\,dx$

4. $x\,dy = y\,dx + x^2y\,dy$

5. $x\,dy - y\,dx = y^2\,dx$

6. $2xy\,dx - x^2\,dy = x^2y^2\,dx$

7. $2xy\,dy - y^2\,dx = x^3\,dx$

8. $xy\,dy = -\dfrac{y^2}{2}\,dx + x^2\,dx$

9. $2xy\,dx - x^2\,dy = y^2x\,dx$

10. $x^2\,dy = 2xy\,dx + y^4\,dy$

11. $\dfrac{dy}{dx} = 1 + \dfrac{y}{x}$

12. $x\dfrac{dy}{dx} = 2x + y$

13. $xy' = 3x + 4y$

14. $y' = \dfrac{y}{x} + \dfrac{x^2}{y^2}$

15. $y' = \dfrac{y}{x} + \dfrac{y^2}{x^2}$

16. $y' = \dfrac{x - y}{x + y}$

17. $y' = \dfrac{x + y}{x - y}$

18. $(x^2 - y^2)\,dx - 2xy\,dy = 0$

19. $(x + 2y)\,dx + (2x + y)\,dy = 0$

20. $y' = \dfrac{y + x\csc\left(\dfrac{y}{x}\right)}{x}$

21. Given the differential equation $xyy' = 3y^2 + x^2$, determine the equation of the solution curve that passes through the point $(-1, 2)$.

22. Given the differential equation $xe^{\frac{y}{x}}\,dx + y\,dx = x\,dy$, determine the real portion of the solution curve that passes through the point $(1, 0)$.

23. Determine the specific solution $y = f(t)$ for the homogeneous equation $t^2y' = y^2 + 2ty$ with initial condition $y(1) = 2$.

24. Determine the specific solution $y = f(t)$ for the homogeneous equation $2tyy' = 2y^2 - t^2$ with initial condition $y(1) = 2$.

25.4

First Order Differential Equations

A differential equation of the form:

$$\frac{dy}{dx} + P(x)y = Q(x),$$

where P and Q are functions of x is a **linear differential equation of first order in standard form.** The equation is linear since the dependent variable y and its derivative $\dfrac{dy}{dx}$ are of the first degree.

The differential equation can be written as:

$$dy + P(x)y\,dx = Q(x)\,dx. \tag{6}$$

We say that a linear equation of first order in this form is in **exact form.** The right-hand side of the equation is integrable since $Q(x)$ is a function of x. If we could do something to the left side to make it an integrable combination such as we encountered in the previous section, then the entire equation would be integrable.

Let us start by letting $Q(x)$ be equal to zero. Then the equation becomes

$$dy + P(x)y\, dx = 0 \tag{7}$$

In this form, the variables are separable, and we can write the equation as:

$$\frac{dy}{y} + P(x)\, dx = 0$$

$$\ln|y| + \int P(x)\, dx = \ln|C| \quad \textbf{Integrating}$$

$$e^{(\ln|y| + \int P(x)\, dx)} = e^{\ln|C|} \quad \textbf{Writing both sides as an exponential}$$

$$(e^{\ln|y|})(e^{\int P(x)\, dx}) = e^{\ln|C|}$$

$$ye^{\int P(x)dx} = C \qquad e^{\ln|y|} = y \text{ \textbf{Rules of exponents}} \tag{8}$$

If we differentiate both sides with respect to x, we expect the result to be the same as Equation (7).

$$e^{\int P(x)\, dx}\frac{dy}{dx} + P(x)ye^{\int P(x)\, dx} = 0 \quad \textbf{Differentiating (8)}$$

$$e^{\int P(x)\, dx}\, dy + P(x)ye^{\int P(x)\, dx}\, dx = 0 \quad \textbf{Multiplying each term by } dx \tag{9}$$

After differentiating there is a factor $e^{\int P(x)\, dx}$ in Equation (9) that was not present in Equation (7). Note that this factor $e^{\int P(x)\, dx}$ makes the left side an integrable combination. Also note that this factor cannot be equal to zero or the equation does not exist. Therefore, if we divide each term by the factor $e^{\int P(x)\, dx}$ Equation (9) is the same as Equation (7). $e^{\int P(x)\, dx}$ is a special factor called an **integrating factor.** This integrating factor is independent of the variable y. That means that we can multiply each term of Equation (6) by the integrating factor, and the right side is still integrable since it is only a function of x.

The result of all this work is a very important fact. A solution can be found for the first order linear differential equation:

$$\frac{dy}{dx} + P(x)y = Q(x)$$

by determining the integrating factor $e^{\int P(x)\, dx}$.

EXAMPLE 1   Solution Using Integrating Factor

Determine the general solution of $\dfrac{dy}{dx} + 2y = e^x$.

Solution Rewriting the equation, we have:

$$dy + 2y\,dx = e^x\,dx,$$

which is a first order equation in exact form. To solve the equation, we must change the left side to an integrable combination. We can do that by finding an integrating factor. To determine the integrating factor let $P(x) = 2$ and substitute in the equation.

$$e^{\int P(x)\,dx} = e^{\int \boxed{2}\,dx}$$

$$= e^{2x} \qquad \textbf{Integrating factor}$$

$$e^{2x}\,dy + 2e^{2x}y\,dx = e^{2x}e^x\,dx \qquad \textbf{Multiplying each term by the integrating factor } e^{2x}$$

Since $\dfrac{d(ye^{2x})}{dx} = e^{2x}\dfrac{dy}{dx} + 2e^{2x}\,y\,\dfrac{dy}{dx}$, the left side is an integrable combination. Therefore, we can write the following equations.

$$ye^{2x} = \int e^{3x}\,dx + C$$

$$ye^{2x} = \frac{1}{3}e^{3x} + C \qquad \textbf{Integrating}$$

Thus, the general solution of the differential equation is $y = \dfrac{1}{3}\,e^x + Ce^{-2x}$. ◆

In the process of determining integrating factors, we encounter expressions of the form $e^{\ln|u|}$. From the definitions of exponents and logarithms, recall that if $y = e^{\ln|u|}$, then $\ln|y| = \ln|u|$ or $y = u$. Thus, $e^{\ln|u|} = u$. We use this definition in Example 2.

EXAMPLE 2 **Integrating Factor of the Form $e^{\ln|u|}$**

Determine the general solution of $x\dfrac{dy}{dx} + 2y = x^2$.

Solution Writing the equation in exact form, we have:

$$dy + \frac{2}{x}y\,dx = x\,dx. \qquad \textbf{Multiplying each term by } \frac{1}{x} \textbf{ and } dx$$

Thus, $P(x) = \dfrac{2}{x}$ and $Q(x) = x$.

$$e^{\int P(x)\,dx} = e^{\int \frac{2}{x}\,dx} \qquad \textbf{Substituting for } P(x)$$

$$= e^{2\ln|x|} \qquad \textbf{Integrating factor}$$

$$= e^{\ln|x^2|} \qquad \textbf{Rules of logarithms}$$

$$= x^2 \qquad e^{\ln|x|} = x$$

$$x^2\,dy + 2xy\,dx = x^3\,dx \qquad \textbf{Multiplying each term by the integrating factor}$$

The left side is an integrable combination, and the right side is integrable. The result is:

$$x^2 y = \frac{x^4}{4} + C.$$

Thus, the general solution of the differential equation is $y = \frac{x^2}{4} + Cx^{-2}$. ◆

From the examples we see that the solution of the left side of the first order linear differential equation is always of the form $ye^{\int P(x)dx}$. Thus, immediately after finding the integrating factor, we can write the equation as:

$$ye^{\int P(x)\,dx} = \int e^{\int P(x)\,dx}\, Q(x)\,dx + C.$$

EXAMPLE 3   General and Specific Solution

a. Determine the general solution of $\dfrac{dy}{dx} - (\tan x)y = \sin x$.

b. Determine a specific solution that satisfies the initial condition $y\!\left(\dfrac{\pi}{4}\right) = 1$.

Solution   Writing the equation in exact form, we have:

$$dy - (\tan x)y\,dx = \sin x\,dx.$$

a. We see that $P(x) = -\tan x$ and $Q(x) = \sin x$. Thus, the integrating factor is:

$$e^{\int -\tan x\,dx} = e^{-\int \frac{\sin x}{\cos x}\,dx}$$

$$= e^{\ln|\cos x|}$$

$$= \cos x.$$

$$y \cos x = \int \cos x \sin x\,dx + C \qquad \text{\small\textbf{Substituting into general form of the solution}}$$

$$y \cos x = \frac{\sin^2 x}{2} + C. \qquad\qquad \text{\small\textbf{Integrating}}$$

b. To determine the general solution, we let $x = \dfrac{\pi}{4}$ and $y = 1$.

$$(\boxed{1})\cos\left(\frac{\pi}{4}\right) = \frac{1}{2}\sin^2\left(\frac{\pi}{4}\right) + C \qquad \text{\small\textbf{Substituting}}$$

$$C = \frac{1}{4}(2\sqrt{2} - 1) \qquad\qquad \text{\small\textbf{Solving for } C}$$

The general solution can be written as:

$$y = \frac{1}{2}\sin x \tan x - C \sec x,$$

and the specific solution as:

$$y = \frac{1}{2}\sin x \tan x + \frac{1}{4}(2\sqrt{2} - 1)\sec x. \quad ◆$$

EXAMPLE 4 **Integration by Parts**

Determine the solution of $x \dfrac{dy}{dx} - 4y = x^6 e^x$.

Solution Writing the equation in exact form, we have:

$$dy - \frac{4}{x} y \, dx = x^5 e^x \, dx.$$

Thus, $P(x) = \dfrac{-4}{x}$.

$$e^{-\int \frac{4}{x} dx} = e^{-4 \ln |x|} \quad \textbf{Substituting for } P(x)$$
$$= e^{\ln x^{-4}}$$
$$= x^{-4} \qquad \textbf{Integrating factor}$$

Multiplying each term by x^{-4}, we see that the left side is an integrable combination and the right side is:

$$x^{-4} x^5 e^x \quad \text{or} \quad x e^x.$$

Thus,

$$x^{-4} y = \int x e^x \, dx + C$$

$$x^{-4} y = x e^x - e^x + C. \quad \textbf{Integrating by parts}$$

Thus, the general solution is $y = x^5 e^x - x^4 e^x + C x^4$. ◆

25.4 Exercises

In Exercises 1–20, determine the general solution of the differential equation.

1. $\dfrac{dy}{dx} + y = e^{-x}$

2. $\dfrac{dy}{dx} - y = e^x$

3. $xy' + 2y = x^2$

4. $y' - xy = e^{(\frac{1}{2})x^2} \cos x$

5. $y' = y + 3x^2 e^x$

6. $y' + 3y = x + 1$

7. $y' - 2y = \cos 3x$

8. $y' - y = 2e^x$

9. $y' - \dfrac{2}{x} y = 1 - x^2$

10. $y' + \dfrac{1}{x} y = \ln x - 2$

11. $y' + (\cot x)y = \cos x$

12. $\dfrac{dy}{dt} + (\cot t)y = 3 \sin t \cos t$

13. $t \dfrac{dy}{dt} + \dfrac{1}{\ln t} y = 1$

14. $\dfrac{dy}{dx} = \dfrac{3 + xy}{2x^2}$

15. $\dfrac{dy}{dt} = -\dfrac{3}{t} y + t + 4$

16. $\dfrac{dy}{dt} = \dfrac{1}{t} y + 1 + t$

17. $\dfrac{dy}{dt} - y = \sin 2t$

18. $\dfrac{dy}{dt} + 2ty = 4t$

19. $t^2 \dfrac{dy}{dt} + ty + 1 = 0$

20. $\dfrac{dy}{dx} - \dfrac{y}{x} = \dfrac{x - y}{x - 2}$

In Exercises 21–28, determine the general solution and the specific solution for the given initial condition.

21. $\dfrac{dy}{dt} + y = 2t + 5;\ y(0) = 4$

22. $\dfrac{dy}{dt} + 4y = 6\sin 2t;\ y(0) = -\dfrac{3}{5}$

23. $\dfrac{dy}{dt} - 2y + e^{2t} = 0;\ y(0) = 2$

24. $\dfrac{dy}{dt} - 2y = t^2 e^{2t};\ y(0) = 2$

25. $(t^2 + 1)\dfrac{dy}{dt} - 2ty = t^2 + 1;\ y(1) = \pi$

26. $(2y - xy - 3)\,dx + x\,dy = 0;\ y(1) = 1$

27. $x^2\,dy - \sin 2x\,dx + 3xy\,dx = 0;\ y\!\left(\dfrac{\pi}{2}\right) = 0$

28. $\dfrac{dr}{d\theta} = \theta - \dfrac{r}{3\theta},\ r(1) = 1$

25.5

Applications of First Order Differential Equations

In this section we apply the techniques we learned in the previous four sections of this chapter to solve applied problems in the areas of geometry, growth and decay, electronics, motion, and fluids.

EXAMPLE 1   Equation of a Curve

Determine the equation of the curve through the point $(2, -1)$ that has, at each point (x, y) on the curve, a slope of $\dfrac{3y}{x}$.

Solution   The slope of the curve at any point on the curve is equal to the derivative of the function. Thus, we can write the differential equation.

$$\frac{dy}{dx} = \frac{3y}{x}$$

$$\frac{dy}{y} = \frac{3\,dx}{x} \qquad \textbf{Separating variables}$$

$$\ln|y| = 3\ln|x| + \ln C \qquad \textbf{Integrating}$$

$$\ln|y| = \ln x^3 C \qquad \textbf{Rules of logarithms}$$

$$y = Cx^3$$

To determine the curve of the family of curves that passes through the point $(2, -1)$, we substitute the values in the general equation.

$$y = Cx^3$$

$$-1 = C(\,2\,)^3$$

$$\frac{-1}{8} = C$$

$$y = -\frac{1}{8}x^3 \qquad \textbf{Substituting for } C$$

Therefore, the specific equation of the curve is $8y + x^3 = 0$. The curve is illustrated in Figure 25.3. ◆

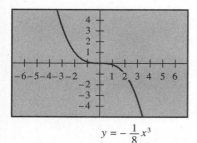

$y = -\dfrac{1}{8}x^3$

FIGURE 25.3

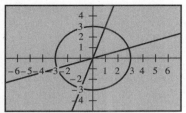

$x^2 + y^2 = 9$, $y = 3x$, $y = \frac{1}{3}x$

FIGURE 25.4

In our study of lines we learned that two lines on a plane are perpendicular if, and only if, the slopes of the lines are negative reciprocals of each other. That is $\ell_1 \perp \ell_2$ if, and only if, $m_1 = -\dfrac{1}{m_2}$. If the tangent lines to the curves at the points of intersection of the curves are perpendicular, as shown in Figure 25.4, then the curves are said to be **orthogonal trajectories** of each other. Thus, if a family of orthogonal trajectories is the solution of the differential equation:

$$P(x, y)\, dx + Q(x, y)\, dy = 0,$$

then the other family of orthogonal trajectories are determined by the solution of the differential equation:

$$Q(x, y)\, dx - P(x, y)\, dy = 0,$$

because the corresponding slopes of the families of curves are negative reciprocals.

$$\frac{dy}{dx} = -\frac{P(x, y)}{Q(x, y)} \quad \text{and} \quad \frac{dy}{dx} = \frac{Q(x, y)}{P(x, y)}$$

This is illustrated in Example 2.

EXAMPLE 2   Orthogonal Trajectories

Determine the orthogonal trajectories of the circles $x^2 + y^2 = C$.

Solution   To determine the orthogonal trajectories of the circles, we must determine the derivative $\dfrac{dy}{dx}$ and replace $\dfrac{dy}{dx}$ with $-\dfrac{dx}{dy}$. Then we solve the new differential equation.

$x^2 + y^2 = C$	**Given**
$2x + 2y\dfrac{dy}{dx} = 0$	**Differentiating**
$\dfrac{dy}{dx} = -\dfrac{x}{y}$	**Solving for $\dfrac{dy}{dx}$**
$\dfrac{dy}{dx} = \dfrac{y}{x}$	**Differential equation of the orthogonal trajectories**
$\dfrac{dy}{y} = \dfrac{dx}{x}$	**Separating variables**
$\ln y = \ln x + \ln C$	
$y = Cx$	**Rules of logarithms**

Thus, the orthogonal trajectories are the family of straight lines that pass through the origin. This is the result we would expect since the radii of a circle are perpendicular to the lines tangent to a circle. ◆

In biology it is often observed that the rate $\dfrac{dy}{dt}$ at which certain microorganisms grow is proportional to the number of microorganisms present at any given time. That is,

$$\frac{dy}{dt} = ky.$$

The solution of this differential equation can be used to predict the number of microorganisms present at different intervals of time.

EXAMPLE 3   Growth Period for Bacteria

A certain bacteria grows at a rate that is proportional to the number present at a particular time. If the number of bacteria initially present is N_0 and at time $t = 1$ hr the number of bacteria is $\dfrac{5}{2} N_0$, determine the time necessary for the number of bacteria to quadruple.

Solution   The equation that represents the rate of growth is:

$$\frac{dN}{dt} = kN,$$

where k is a constant and $N(0) = N_0$. This is a linear equation with separable variables. Therefore, we can write the equation as:

$$\frac{dN}{N} = k\, dt$$

$$\ln |N| = kt + C_1 \quad \textbf{Integrating}$$
$$N(t) = Ce^{kt} \quad \textbf{Solving for } N$$

Substituting the initial condition $N(0) = N_0$, we can determine a value for C.

$$N_0 = Ce^{(k)(0)}$$
$$N_0 = C$$

Substituting for C,

$$N = N_0 e^{kt}.$$

To determine the value of k, we use the condition that $N(1) = \frac{5}{2}N_0$.

$$\frac{5}{2}N_0 = N_0 e^{k(1)}$$

$$\frac{5}{2} = e^k$$

$$k = \ln\left(\frac{5}{2}\right)$$

$$= 0.9163 \quad \textbf{Four decimal places}$$

Thus,

$$N = N_0 e^{0.9163t}$$

To determine the time when the bacteria have quadrupled, we use the fact that $N = 4N_0$. Substituting this value for N, we can determine the time t.

$$4N_0 = N_0 e^{0.9163t}$$

$$4 = e^{0.9163t}$$

$$0.9163t = \ln 4$$

$$t = \frac{\ln 4}{0.9163} \approx 1.51 \text{ hr}$$

Therefore, it takes approximately 1.51 hr for the original number of bacteria to quadruple. ◆

A general formula of the type developed in Example 3 can be used in problems to determine rates of decay, heating and cooling, compound interest, evaporation, mixture, and others.

Newton's second law of motion states, "The time rate of change in momentum of a body is proportional to the net force acting on the body is proportional to the net force on the body and has the same direction as the force." The momentum of an object is defined to be its mass m multiplied by its velocity v.

The rate of change of the momentum with respect to time is $\dfrac{d(mv)}{dt}$. If the force acting on a body is represented by F and k is the constant of proportionality, then the second law can be expressed as:

$$\frac{d(mv)}{dt} = kF$$

$$m\frac{dv}{dt} = kF \qquad \textbf{If } \textit{m} \textbf{ is a constant}$$

$$ma = kF \qquad a = \frac{dv}{dt}\text{, acceleration}$$

$$F = \frac{ma}{k}. \qquad \textbf{Solving for } \textit{F}$$

The value of k depends on the units being used. Using centimeters, grams, and seconds (CGS) in the SI system with $k = 1$, $F = ma$ becomes

$$F = 1 \text{ g} \cdot 1 \text{ cm/s}^2 = 1 \text{ g} \cdot \text{cm/s}^2.$$

This unit of force is called a *dyne*. In the U.S. Customary system (USCS) with $k = 1$, $F = ma$ becomes:

$$F = 1 \text{ slug ft/s}^2,$$

or 1 lb (the mass unit is the slug). Using CGS units in the SI system, near the earth's surface $g = 981$ cm/s^2, while in the USCS $g = 32.2$ ft/s^2 (sometimes taken for simplicity as 32 ft/s^2). Near the earth's surface the weight w is related to the mass m by $w = mg$. In the SI system, with $k = 1$, $F = ma$ becomes:

$$F = 1 \text{ kg m/s}^2,$$

or 1 newton. Also, in SI system, $g = 9.81$ m/s^2 is often given as 9.8 m/s^2.

The concept of force is illustrated in Example 4.

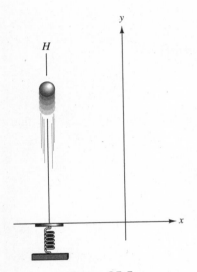

FIGURE 25.5

EXAMPLE 4 Velocity, Distance, and Time of Flight of a Ball

A ball is propelled vertically upward by a spring device with an initial velocity of 64 ft/s. How high does it rise? What is its velocity after 3s? When will it return to the starting position?

Solution We take the y-axis as the vertical axis, with 0 at the starting point (see Figure 25.5). We assume that an upward movement is in the positive direction. Thus, the weight of the ball is the force of gravity acting on the ball, shown in Figure 25.6, which can be expressed as $-mg$ (negative sign signifies downward). Therefore, the differential equation for the motion is $F = ma$, or:

$$-m(32) = m\frac{d^2y}{dt^2}$$

or

$$\frac{d^2y}{dt^2} = -32.$$

mg

Force diagram

FIGURE 25.6

If we let $v = \dfrac{dy}{dt}$, then $\dfrac{dv}{dt} = \dfrac{d^2y}{dt^2}$.

$$\frac{dv}{dt} = -32 \qquad \textbf{Substituting}$$

$$v = -32t + C_1 \qquad \textbf{Integrating}$$

When $t = 0$ and $v = 64$, then $C_1 = 64$.

$$v = -32t + \boxed{64} \qquad \text{Substituting for } C_1$$

$$\boxed{\frac{dy}{dt}} = -32t + 64 \qquad \text{Substituting}$$

$$y = -16t^2 + 64t + C_2 \quad \text{Integrating}$$

We know that $y = 0$ when $t = 0$; therefore, $C_2 = 0$ and:

$$y = -16t^2 + 64t.$$

To determine how high the ball rises, we use the fact that the upward motion ceases when $v = 0$; that is,

$$0 = -32t + 64$$

$$t = \frac{64}{32} = 2.$$

Thus, when $t = 2$ s, the upward motion ceases. Substituting this value in the equation for y, we have:

$$y = -16 \,(\boxed{2})^2 + 64(\boxed{2})$$

$$= 64.$$

Thus, the ball rises to a maximum height of 64 ft.

To determine the velocity after 3 s, we use the formula:

$$v = -32t + 64.$$

Substituting $t = 3$,

$$v = -32(\boxed{3}) + 64 = -32.$$

Thus, the velocity after 3 s is -32 ft/s. The negative sign indicates that the ball is moving down at the rate of 32 ft/s.

The ball returns to the ground when $y = 0$, or:

$$-16t^2 + 64t = 0$$

$$-16(t - 4) = 0.$$

Therefore, $t = 0$ or $t = 4$. Since $t = 0$ is the initial position, the ball returns to the starting position when $t = 4$ s. ◆

In a series circuit containing only a resistor and inductor, Kirchoff's second law states that the sum of the voltage drop across the inductor $L\dfrac{di}{dt}$ and the voltage drop across the resistor iR is the same as the impressed voltage $E(t)$ on the circuit (see Figure 25.7). Using this information, we can write the required differential equation as:

$$L\frac{di}{dt} + Ri = E(t),$$

where L and R are constants known as the inductance and the resistance, respectively. Example 5 illustrates the use of this equation.

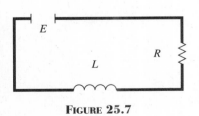

FIGURE 25.7

EXAMPLE 5 **Current in a Simple Circuit**

A 9-volt battery is connected to a simple series circuit in which the inductance is $\frac{1}{2}$ henry and the resistance is 5 ohms. Determine the current if the initial current is zero.

Solution The problem provides us with the following information: $L = \frac{1}{2}$ henry, $R = 5$ ohms, $E(t) = 9$ volt, and $i(0) = 0$.

$$L\frac{di}{dt} + Ri = E(t) \quad \text{Given equation}$$

$$\frac{1}{2}\frac{di}{dt} + 5i = 9 \quad \text{Substituting}$$

Writing the differential equation in exact form, we have:

$$di + 10i\,dt = 18\,dt.$$

The integrating factor for the equation is e^{10t}.

$$ie^{10t} = \int 18e^{10t}\,dt + C$$

$$ie^{10t} = \frac{9}{5}e^{10t} + C \quad \text{Integrating}$$

$$i = \frac{9}{5} + Ce^{-10t} \quad \text{Solving for } i$$

With the initial condition $i(0) = 0$, by substituting, we see that:

$$0 = \frac{9}{5} + C \quad \text{or} \quad C = -\frac{9}{5}.$$

Therefore, the current with initial condition zero is $i(t) = \frac{9}{5} - \frac{9}{5}e^{-10t}$. ◆

25.5 Exercises

1. Determine the equation of the curve through the point $(4, 0)$ that has, at each point (x, y) on the curve, a slope of $\frac{-9x}{16y}$.

2. Determine the equation of the curve through the point $(2, 2)$ that has, at each point (x, y) on the curve, a slope of $\frac{-2y}{x}$.

3. Determine the equation of the curve through the point (4, 1) that has, at each point (x, y) on the curve, a slope of $\dfrac{yx^2 - y^2}{xy}$.

4. The slope of a curve at (x, y) is proportional to $\dfrac{y}{x + 1}$. The curve passes through the points (0, 1) and (1, 16). Determine its equation.

5. Determine the equations of the orthogonal trajectories of the family of curves $y^2 = x^2 + C$.

6. Determine the equations of the orthogonal trajectories of the family of curves $y = Cx^3$.

7. The rate at which mold A grows on bread is proportional to the amount present at a particular time. When first seen, the amount of mold on a loaf of bread is 3 g. In three days the quantity has increased to 24 g. Determine the quantity of mold in five days.

8. The rate at which a radioactive substance loses mass is proportional to the mass present. If m is the mass at any time t and k is the constant of proportionality, perform the following.
   a.   Derive a differential equation to represent the rate of decay.
   b.   Determine m at any time t.

9. Determine an equation of a curve passing through (0, −5) if the slope of the curve at that point is six times the abscissa of the point.

10. A radioactive substance loses mass in accordance with the formula established in Exercise 8.
   a.   Determine m at any time t.
   b.   Determine the half-life if $m = 50$ g when $t = 0$ yr and $m = 45$ g when $t = 20$ yr. Note that the half-life is the time required for the mass to decrease to half its original value.

11. A particle moves along the x-axis so that its velocity at any point is equal to half its abscissa minus three times the time. At time $t = 2$, $x = -4$. Determine the motion along the x-axis.

12. Suppose that the acceleration of a particle moving on the x-axis is given by $\dfrac{d^2x}{dt^2} = 18t - 12$, and when $t = 0$, $\dfrac{dx}{dt} = 5$ and $x = 4$. Determine the motion along the x-axis.

13. A warm or cold object is placed in a container whose temperature stays constant, loses, or gains heat. The container gains heat as measured by its temperature, at a rate approximately proportional to the difference between the object's temperature and the temperature of the container. This is known as Newton's law of cooling, which can be represented by the formula $\dfrac{d\theta}{dt} = -k\theta$, where t is the time and θ is the difference between the temperature of the object and the temperature of the container.
   a.   Determine θ for any time t.
   b.   Determine θ when an object with an initial temperature of 600°F is placed in a container whose temperature is 60°F, and the temperature of the object drops to 90°F 20 min. after being placed in the container.

14. An object whose temperature is initially 5°C is placed in a container with a temperature at 35°C. If the temperature of the object rises 3°C in the first five minutes, how long does it take to reach a temperature of 18°C? (See Exercise 13.)

15. A 10 kg mass is dropped from a hot-air balloon that is $\frac{1}{2}$ km above the ground. Assuming no air resistance, at what time and with what velocity does the weight reach the ground?

16. At $t = 0$, an emf of 30 volts is applied to a circuit consisting of an inductor of 3 henrys in series with a 60-ohm resistor. If the current is zero at $t = 0$, what is it at any time $t \geq 0$?

17. Determine the current for any time $t \geq 0$ in Exercise 16 if the emf is 300 sin 30 t.

18. Determine the orthogonal trajectories of the family of curves for $y^2 = C(1 + x^2)$ and determine the member of the family of curves passing through the point $(-2, 5)$.

19. A certain electrical capacitor holding an initial charge of q_0 coulombs discharges at a rate proportional to the charge present at any time t. Determine q in terms of t.

20. In an RL circuit the inductance is 4 henrys, the resistance is 20 ohms, and E is constantly equal to 100 volts. Given that $i = 0$ when $t = 0$, determine the relationship between i and t for $t \geq 0$.

21. Solve $L\dfrac{di}{dt} + Ri = E$ for the conditions $i = 0$ when $t = 0$ and L, R, and E are positive constraints.

22. If a fluid flows from an orifice at the bottom of a container, the differential equation $\dfrac{dv}{dt} = -4.8\,Ah^{\frac{1}{2}}$ can be used to provide a good approximation of the volume V at any time t. A is the area of the orifice, and h is the height of the liquid in the container.

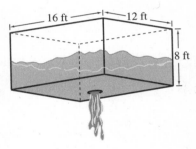

A rectangular tank 8 ft deep with a cross section of 12 ft by 16 ft is initially filled with water that runs out the orifice of radius 6 in. The orifice is located in the bottom of the tank, as shown in the diagram.
a. Determine the equation for the time t required for the tank to empty.
b. Determine the height of the water in the tank 20 min after it starts to drain. (*Hint:* express V as a function of h.)

23. A reservoir in the shape of a cone, as shown in the sketch, has an orifice of 3-in. radius at the vertex of the cone. Determine the time required to empty the reservoir if it is filled with water at time $t = 0$. (See Exercise 22.)

24. Determine the value of k such that the family of parabolas $y = C_1x^2 + k$ are the orthogonal trajectories of the family of ellipses $x^2 + 2y^2 - y = C_2$.

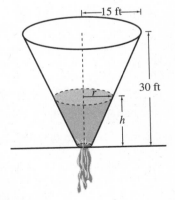

25. A body weighing 16 lb is released from rest and falls toward earth from a great height. As it falls, air resistance acts upon it. Assume that this resistance (in pounds) is numerically equal to $2v$, where v is the velocity of the body in ft/s. Determine the velocity and distance fallen in t seconds.

26. The velocity of a rocket ship leaving the earth is given by $v\dfrac{dv}{dr} = -\dfrac{GM}{r^2}$, where r represents the distance from the ship to the center of the earth, M is the mass of the earth, and G is a constant. If $v = 2.4 \times 10^5$ m/s at the surface of the earth, then solve for v as a function of the radius. (You have to determine the radius of the earth.)

27. A business needs to know how much profit P is realized from only one single change in its operation. This is called the marginal profit function, and it is expressed as $\dfrac{dP}{dx}$, where x is the amount invested into the change.

Suppose a quality control officer is hired for an assembly line. The marginal profit function is assumed to be $\dfrac{dP}{dx} = 2e^{-x} - 3P(x)$. So that a reasonable salary can be set for the quality control officer, determine the total profit P as a function of x. Obtaining an initial condition is rather obvious here. If the company pays nothing ($x = 0$), they get no quality control officer, and thus, no profit is realized ($P = 0$).

28. When a bullet is fired through wood its acceleration $\left(\dfrac{dt}{dv}\right)$ is $-12v^{\frac{1}{2}}$, where v is the velocity in m/s. When will the bullet stop if it enters the wood with a speed of 400 m/s?

Review Exercises

In Exercises 1–6, determine the following.
a. Whether the differential equation is linear or nonlinear
b. The order of the differential equation

1. $y' = -5x^3 + y$

2. $y'' + 2y' = 3y$

3. $y''' - 6y + 3(y')^2 = 0$

4. $(2x + 3y)\,dy + (7x - 4y)\,dx = 0$

5. $(y')^2 + 2xy = x^3$

6. $\left(\dfrac{d^3s}{dt^3}\right)^2 + 2\dfrac{ds}{dt} + t^2 = 0$

In Exercises 7–10, show that the function is a solution to the given differential equation.

7. $y = Cx^{-\frac{2}{3}};\ 2xy^3 + 3x^2y^2\dfrac{dy}{dx} = 0$

8. $x^2 - y^2 = C;\ y\dfrac{dy}{dx} = x$

9. $y = x^2 + Cx;\ x\dfrac{dy}{dx} = x^2 + y$

10. $y = A \sin x + B \cos x;\ \dfrac{d^2y}{dx^2} + y = 0$

In Exercises 11–24, determine the general solution of the differential equations.

11. $x\,dy - y\,dx = 0$

12. $x^2\,dy - y^2\,dx = 0$

13. $\tan x\,dy + \sin y\,dx = 0$

14. $\sec^2 x\,dy + \csc y\,dx = 0$

15. $2xy\,dx + (x^2 - 1)\,dy = 0$

16. $e^{3x}e^{2x}\,dx + e^{-2x}\,dy = 0$

17. $2xy\,dx + x^2dy = 0$

18. $x\,dy + y\,dx = 4x^3\,dx$

19. $\dfrac{2xy' - 2y}{x^2} = 0$

20. $\dfrac{3xy' - 3y}{x^2} = 2y^3$

21. $\dfrac{dy}{x} + y\,dx = 2\,dx$

22. $\sin x\,dy - y\cos x\,dx = \tan x\,dx$

23. $\cos x\,dy - y\,dx = 0$

24. $\dfrac{dy}{dt} = \dfrac{3 + ty}{2t^2}$

In Exercises 25–30, determine the specific solution for the differential equation.

25. $(t + 2s)\,dt + ds = 0,\ s = -1$ if $t = 0$

26. $x^2\,dy - \sin x\,dx + 3xy\,dx = 0,\ y = 0$ if $x = \pi$

27. $dy\,(2x\cot y) = dx$, $y = \dfrac{\pi}{4}$ if $x = 0$

28. $\cos x\sin y\,dy = 2\sin x\cos x\,dx$, $y = 0$ if $x = \pi$

29. $\dfrac{dy}{dx} = e^{(ax+by)}$, $y = 0$ if $x = 0$

30. $\dfrac{dy}{dx} - \dfrac{2y}{x} = x + x^2$, $y = 2$ if $x = 1$

31. Determine the equation of the curve through the point $(8, 6)$ having, at each point (x, y) on the curve, a slope of $\dfrac{y}{3x}$.

32. Determine the equation of the curve through the point $(\pi, 0)$ having, at each point (x, y) on the curve, a slope of $y\sin x - 3\sin x$.

33. Bacteria in a certain culture increase at a rate proportional to the number present. If the number doubles in 1 hr, how long does it take for the original number to quadruple?

34. An object with a temperature of 150°C cools to 140°C in 1 min when placed in a cooler of constant temperature of 100°C. How long would it take for the temperature of the body to drop for each of the following?
a. From 120°C to 110°C
b. 10 degrees from any temperature T

35. A ball is thrown up vertically from a building 100 ft high with an initial velocity of 64 ft/s.
a. How high does the ball go?
b. What is the velocity of the ball as it passes the building on its way down?
c. When does the ball hit the ground and with what velocity?

36. In an RL circuit, the inductance is 20 henrys, the resistance is 30 ohms, and E is constantly equal to 200 volts. Given that $i = 0$ when $t = 0$, determine the relationship between i and t for $t \geq 0$.

*37. A family whose orthogonal trajectories are members of the same family is said to be self-orthogonal. Show that the family $y^2 = 4Cx + C^2$ is self-orthogonal.

*38. Determine a differential equation of the two-parameter family $y = C_1e^{2x} + C_2e^{-2x}$.

✍ Writing About Mathematics

1. The solution to the differential equation $x\,dx + y\,dy = 0$ is the family of concentric circles whose centers are at the origin. Write a brief paper explaining where and how someone might use the differential equation of this family of curves.

2. You are working in a development office and your supervisor requests a report on the rate of change in the teenage population of the city. In the report, you support your argument by using a differential equation. Explain how you do this.

3. You are the detective who must determine the number of hours a homicide victim has been dead. Newton's law of cooling helps you estimate the time of death. For additional help see J. F. Hurley, "An Application of Newton's Law of Cooling," *Mathematics Teacher* 67 (1974): 141–42 and David A. Smith, "The Homicide Problem Revisited," *The Two-Year College Mathematics Journal* 9 (1978): 141–45. Write a brief summary of how differential equations can help you solve the problem.

4. Your classmate has problems recognizing and solving a homogeneous equation. Write a set of guidelines to help your friend with homogeneous equations.

Chapter Test

In Exercises 1–4, without solving, classify each of the equations as either separable, homogeneous, or linear.

1. $\dfrac{dy}{dx} = y - x$

2. $\dfrac{dy}{dx} = \dfrac{y^2 + y}{x^2 + x}$

3. $\dfrac{dy}{dx} = \dfrac{x - y}{x}$

4. $\dfrac{dy}{dx} = 4 + 5y + y^2$

In Exercises 5–6, show that the function is a solution to the differential equation.

5. $y = C(1 + x)$; $(1 + x)\,dy - y\,dx = 0$

6. $y = x + \tan x$; $y' + 2xy = 2 + x^2 + y^2$

In Exercises 7–11, determine the general solution for each differential equation.

7. $(1 + x)\,dy - y\,dx = 0$

8. $dx + dy = 0$

9. $x\,dy + y\,dx = 4y^2\,dy$

10. $(x - y)\,dx + x\,dy = 0$ (homogeneous)

11. $x\dfrac{dy}{dx} + 3y = x^3$

In Exercises 12–14, determine the specific solution for the given initial condition.

12. $\dfrac{dy}{dx} = -\dfrac{x}{y}$; $y(8) = 6$

13. $\dfrac{dy}{dt} + 5y = 4\,e^{2t}$; $y(0) = 1$

14. $x\dfrac{dy}{dx} = 2x + y$, $y(1) = 3$

15. Determine the equation of the curve through the point $\left(\dfrac{\pi}{2}, 2\right)$ that has at each point (x, y) on the curve a slope of $3\cos x + 2\sin x$.

16. Determine the equations of the orthogonal trajectories of the family of curves $y = x^2 + C$.

17. A 20 kg mass is dropped from a hot-air balloon that is 1 km above the ground. Assuming no air resistance, at what time and with what velocity does the weight reach the ground?

26

Higher Order Linear Differential Equations and Laplace Transforms

A higher order differential equation may be used to solve a problem involving a vibrating mechanical system. In a vacuum, under ideal conditions, a spring keeps moving forever, and a pendulum never stops swinging. In the real world, many external forces, such as friction, air currents, gravity, and created retardatives, slow the vibration. These external forces have a *damping effect* on the motion. A damping effect used for automobiles is the shock absorber, which helps reduce the spring vibrations and produces a smoother ride. On storm doors a screw on the door closer provides the proper amount of damping for the door to close without slamming, and it can be adjusted so that the door remains open. An application of a damping effect on a vibrating motion is illustrated in Example 3 of Section 26.4.

26.1

Higher Order Linear Homogeneous Equations

In Chapter 25 we discussed techniques for solving first order differential equations. In this chapter we discuss techniques for solving equations of higher order. For example, the equation:

$$3 \frac{d^3y}{dx^3} - 5 \frac{d^2y}{dx^2} + 2 \frac{dy}{dx} - 6y = \sin 2x$$

is a linear equation of order 3, and:

$$a_n(x) \frac{d^ny}{dx^n} + a_{n-1}(x) \frac{d^{n-1}y}{dx^{n-1}} + \cdots + a_1(x) \frac{dy}{dx} + a_0(x)y = F(x)$$

is a linear equation of order n, according to Definition 25.3. In this section the discussion is restricted to a special form of the nth order linear equation—the case where the coefficients $a_i(x)$ are all constants. In fact, the discussion begins with one additional restriction: $F(x) = 0$.

> **DEFINITION 26.1**
>
> A **homogeneous differential equation** is:
>
> $$a_n(x) \frac{d^ny}{dx^n} + a_{n-1}(x) \frac{d^{n-1}y}{dx^{n-1}} + \cdots + a_1(x) \frac{dy}{dx} + a_0(x)y = 0,$$
>
> with $F(x) = 0$.
>
> The differential equation with $F(x) \neq 0$ is a **nonhomogeneous differential equation.**

EXAMPLE 1   **Types of Equations**

Classify each equation.

a. $\dfrac{d^2y}{dx^2} + 3 \dfrac{dy}{dx} - 5y = e^x$

b. $\dfrac{d^3y}{dx^3} + 5 \dfrac{dy}{dx} + 4y = 0$

Solution

a. It is a nonhomogeneous linear equation of second order.
b. It is a homogeneous linear equation of third order.   ◆

To simplify writing the differential equations, we use D in place of $\dfrac{dy}{dx}$, D^2 in place of $\dfrac{d^2y}{dx^2}$, and D^n in place of $\dfrac{d^ny}{dx^n}$. The symbols $D, D^2, \ldots, D^n$ are called

D operators because they indicate that the operation of differentiation is to be performed. Using this notation, we can write the homogeneous equation in Example 1 as:

$$D^3y + 5Dy + 4y = 0,$$

or more conveniently, as:

$$(D^3 + 5D + 4)y = 0.$$

A third order equation has three solutions, while a second order equation has two solutions, etc. Theorem 26.1 tells us that the linear combination of these solutions is also a solution. A **linear combination** is the sum of the individual solutions. This linear combination of the solutions of the homogeneous equation is called the **complementary solution.**

THEOREM 26.1

If y_1 and y_2 are any two solutions of:

$$(a_2D^2 + a_1D + a_0)y = 0,$$

then $y_c = C_1y_1 + C_2y_2$.
The complementary solution is also a solution. In the equation, C_1 and C_2 are arbitrary constants.

The theorem can be extended for an nth order differential equation; that is,

$$y_c = C_1y_1 + C_2y_2 + \cdots + C_ny_n.$$

Theorem 26.1 is illustrated, but not proven, in Example 2.

EXAMPLE 2   Showing That the Linear Combination Is Also a Solution

The differential equation $(D^2 + 5D + 4)y = 0$ has solutions $y_1 = e^{-x}$ and $y_2 = e^{-4x}$. By Theorem 26.1 $y_c = C_1e^{-x} + C_2e^{-4x}$ is also a solution. Show that y_1, y_2, and y_c are solutions to the differential equation.

Solution   To show that $y_1 = e^{-x}$ is a solution, we substitute e^{-x} for y, and write:

$$(D^2 + 5D + 4)\,e^{-x} = 0$$
$$D^2(e^{-x}) + 5D(e^{-x}) + 4(e^{-x}) = 0 \qquad \text{Recall the } D^2(e^{-x}) \text{ means we differentiate } e^{-x} \text{ twice.}$$
$$e^{-x} - 5e^{-x} + 4e^{-x} = 0 \qquad \text{Differentiating}$$
$$0 = 0.$$

To show that $y_2 = e^{-4x}$ is a solution, we substitute e^{-4x} for y, writing:

$$(D^2 + 5D + 4)\,e^{-4x} = 0$$
$$D^2(e^{-4x}) + 5D(e^{-4x}) + 4(e^{-4x}) = 0 \quad \textbf{Distributive property}$$
$$16e^{-4x} - 20e^{-4x} + 4e^{-4x} = 0 \quad \textbf{Differentiating}$$
$$0 = 0.$$

To show that $y_c = C_1e^{-x} + C_2e^{-4x}$ is a solution, we substitute $C_1e^{-x} + C_2e^{-4x}$ for y, and write:

$$(D^2 + 5D + 4)(\,C_1e^{-x} + C_2e^{-4x}\,) = 0$$
$$D^2(C_1e^{-x}) + D^2(C_2e^{-4x}) + 5D(C_1e^{-x}) + 5D(C_2e^{-4x}) + 4C_1e^{-x} + 4C_2e^{-4x}$$
$$= 0$$
$$\textbf{Distributive property}$$
$$C_1e^{-x} + 16C_2e^{-4x} - 5C_1e^{-x} - 20C_2e^{-4x} + 4C_1e^{-x} + 4C_2e^{-4x} = 0$$
$$\textbf{Differentiating}$$
$$(1 - 5 + 4)C_1e^{-x} + (16 - 20 + 4)C_2e^{-4x} = 0$$
$$\textbf{Factoring}$$
$$0 = 0.$$

We have demonstrated for this differential equation that y_1, y_2, and the linear combination $y_c = y_1 + y_2$ are all solutions. ◆

Theorem 26.1 and Example 2 show that the complementary solution of a second order homogeneous differential equation with two distinct solutions is a linear combination of these solutions. We can also infer from these that the solution of an nth order homogeneous differential equation is a linear combination of the n distinct solutions.

Now let us concentrate on determining solutions to homogeneous differential equations with constant coefficients. Consider a simple equation:

$$\frac{dy}{dx} - 2y = 0.$$

This is a linear equation of first order in standard form. To solve the equation we first must determine the integrating factor. We use the notation $I.F.$ to indicate the integrating factor. Using the D operator, we can write:

$$Dy - 2y = 0$$
$$I.F. = e^{-2x}.$$

Thus, the solution is:

$$ye^{-2x} = C_1 \quad \text{or} \quad y_c = C_1e^{2x}.$$

Now consider the equation:

$$y'' - 3y' - 4y = 0.$$

Written in operator notation, we have:

$$(D^2 - 3D - 4)y = 0.$$

From our limited experience with first order equations in Chapter 25 and Example 2, we suspect that a solution may be $y = e^{mx}$, where m is a constant. Thus, if $y = e^{mx}$ is a solution, it must satisfy the differential equation $(D^2 - 3D - 4)y = 0$.

$$(D^2 - 3D - 4)e^{mx} = 0 \quad \text{Substituting for } y$$

$$m^2 e^{mx} - 3me^{mx} - 4e^{mx} = 0 \quad \text{Differentiating}$$

$$(m^2 - 3m - 4)e^{mx} = 0 \quad \text{Common-term factoring}$$

Since e^{mx} is never equal to zero, the only way we can obtain zero on the left side is if:

$$m^2 - 3m - 4 = 0.$$

Using our knowledge of solving quadratic equations, we find that $m = -1$ and $m = 4$ are roots of the equation. Thus, the two possible solutions of the equation are $y_1 = e^{-x}$ and $y_2 = e^{4x}$. The complementary solution of the equation is the linear combination of y_1 and y_2; that is,

$$y_c = C_1 e^{-x} + C_2 e^{4x}.$$

The quadratic equation $m^2 - 3m - 4 = 0$ is called the *auxiliary* or *characteristic* equation of the differential equation $(D^2 - 3D - 4)y = 0$.

DEFINITION 26.2

$a_n m^n + a_{n-1} m^{n-1} + a_{n-2} m^{n-2} + \cdots + a_0 = 0$ is the **auxiliary equation** of the homogeneous linear differential equation:

$$n \frac{d^n y}{dx^n} + a_{n-1} \frac{d^{n-1}y}{dx^{n-1}} + \cdots + a_1 \frac{dy}{dx} + a_0 = 0,$$

with constant coefficients.

As illustrated above, the roots of the auxiliary equation determine the solution of the differential equation. Auxiliary equations are used to solve the differential equations in Examples 3, 4, and 5.

EXAMPLE 3   **Solution, Using Auxiliary Equation**

Determine the complementary solution of $4\dfrac{d^2 y}{dx^2} - 21\dfrac{dy}{dx} + 5y = 0$.

Solution   The auxiliary equation is:

$$4m^2 - 21m + 5 = 0$$

$$(4m - 1)(m - 5) = 0 \quad \text{Factoring}$$

$$m_1 = \frac{1}{4} \quad \text{or} \quad m_2 = 5 \qquad \textbf{Solving for } \textit{m}$$

$$y_1 = C_1 e^{(\frac{1}{4})x} \quad \text{and} \quad y_2 = C_2 e^{5x}.$$

The complementary solution is $y_c = C_1 e^{(\frac{1}{4})x} + C_2 e^{5x}.$ ◆

EXAMPLE 4   Nonfactorable Auxiliary Equation

Determine the general solution of $(D^2 - 4D - 6)y = 0$.

Solution   The auxiliary equation is:

$$m^2 - 4m - 6 = 0.$$

Since the equation is not easy to factor, we use the quadratic formula to obtain the roots.

$$m = \frac{-(-4) \pm \sqrt{(-4)^2 - 4(1)(-6)}}{2(1)}$$

Thus,

$$m_1 = 2 + \sqrt{10} \quad \text{and} \quad m_2 = 2 - \sqrt{10}.$$

The complementary solution of the differential equation is $y_c = C_1 e^{(2+\sqrt{10})x} + C_2 e^{(2-\sqrt{10})x}$ ◆

EXAMPLE 5   Solution of a Third Order Equation

Determine the complementary solution of $(D^3 - D^2 - 4D + 4)y = 0$.

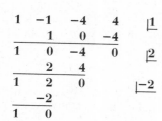

Solution   The auxiliary equation is:

$$m^3 - m^2 - 4m + 4 = 0.$$
$$(m - 1)(m - 2)(m + 2) = 0. \qquad \textbf{By synthetic division}$$
$$m_1 = 1, m_2 = 2, \quad \text{or} \quad m_3 = -2$$

Thus, the complementary solution of the differential equation is $y_c = C_1 e^x + C_2 e^{2x} + C_3 e^{-2x}.$ ◆

In a problem like the one in Example 5, we can estimate the roots of the auxiliary equation by using the graphics calculator.

A method of determining a specific solution given initial values was discussed in Chapter 25. Recall that with a first order linear equation we need only one set of initial conditions. For a second order linear equation, we need two sets of initial conditions. For a third order, we need three sets of initial conditions. Example 6 illustrates a method of determining a specific solution.

EXAMPLE 6   Specific Solution

Given $y(0) = 1$ and $y'(0) = 2$, determine a specific solution for the differential equation in Example 3.

Solution   The complementary solution for $4\dfrac{d^2y}{dx^2} - 21\dfrac{dy}{dx} + 5y = 0$ is:

$$y_c = C_1 e^{(\frac{1}{4})x} + C_2 e^{5x}.$$

Substituting $y(0) = 1$,

$$\boxed{1} = C_1 + C_2. \tag{1}$$

It is obvious at this point that we cannot determine values for C_1 and C_2. So we differentiate the complementary solution and substitute $y'(0) = 2$.

$$y_c' = \frac{1}{4}C_1 e^{(\frac{1}{4})x} + 5C_2 e^{5x}$$

$$\boxed{2} = \frac{1}{4}C_1 + 5C_2 \tag{2}$$

Equations (1) and (2) form a system of equations in two unknowns, C_1 and C_2.

$$1 = C_1 + C_2 \tag{1}$$

$$2 = \frac{1}{4}C_1 + 5C_2 \tag{2}$$

Solving the system, we determine that $C_1 = \dfrac{12}{19}$ and $C_2 = \dfrac{7}{19}$. Substituting these values for C_1 and C_2 in the complementary equation, we have the specific solution $y = \dfrac{\boxed{12}}{19} e^{(\frac{1}{4})x} + \dfrac{\boxed{7}}{19} e^{5x}$. ◆

The solution can be checked by substituting $y(0) = 1$ and $y'(0) = 2$ in the specific solution and the derivative of the specific solution, respectively.

26.1 Exercises

In Exercises 1–22, determine the complementary solution of the differential equation.

1. $y'' + y' = 0$

2. $y'' - 9y = 0$

3. $y'' - 16y = 0$

4. $y'' + 4y' = 0$

5. $y'' - 5y' + 6y = 0$

6. $y'' - 7y' + 6y = 0$

7. $\dfrac{d^2y}{dx^2} + 5\dfrac{dy}{dx} + 4y = 0$

8. $\dfrac{d^2y}{dx^2} - 3\dfrac{dy}{dx} - 4y = 0$

9. $3\dfrac{dy^2}{dx^2} - 2\dfrac{dy}{dx} = 0$

10. $5\dfrac{d^2y}{dx^2} + 3\dfrac{dy}{dx} = 0$

11. $\dfrac{d^2y}{dx^2} - 3\dfrac{dy}{dx} - 10 = 0$

12. $\dfrac{d^2y}{dx^2} + 2\dfrac{dy}{dx} - 15 = 0$

13. $(D^2 + D - 12)y = 0$

14. $(D^2 - 2D - 63)y = 0$

15. $(2D^2 + 11D - 21)y = 0$

16. $(3D^2 + 16D - 12)y = 0$

17. $(3D^2 + 7D - 20)y = 0$

18. $(15D^2 + 14D - 8)y = 0$

19. $(D^3 - 4D^2 - 21D)y = 0$ **20.** $(D^3 - 11D^2 + 28D)y = 0$ **21.** $(D^3 - 3D^2 - 4D + 12)y = 0$

22. $(D + 4)(D^2 - 9)(D^2 - 25)y = 0$

In Exercises 23–26, determine a differential equation in terms of the D operator whose associated auxiliary polynomial has the given roots.

23. $m_1 = 2, m_2 = -3$

24. $m_1 = 4, m_2 = -5$

25. $m_1 = -3, m_2 = 3, m_3 = 0$

26. $m_1 = -4, m_2 = 4, m_3 = 2$

In Exercises 27–30, solve the initial value problems.

27. $y'' - 16y = 0$; $y(0) = 2, y'(0) = 2$

28. $y'' - y = 0$; $y(0) = 1, y'(0) = 1$

29. $y'' + 6y' + 5y = 0$; $y(0) = 0, y'(0) = 3$

30. $y''' + 13y'' + 36y' = 0$; $y(0) = 0, y'(0) = 1, y''(0) = 41$

26.2

Auxiliary Equations with Repeated or Complex Roots

In Chapter 11, we learned that a quadratic equation can have real distinct roots, repeated roots, or complex roots. Section 26.1 used auxiliary equations with real distinct roots; this section discusses equations with repeated roots and complex roots.

Repeated Roots

Let us begin by discussing the differential equation:

$$\frac{d^2y}{dx^2} - 8\frac{dy}{dx} + 16y = 0.$$

The auxiliary equation is:

$$m^2 - 8m + 16 = 0,$$

which we can factor as:

$$(m - 4)^2 = 0.$$

The roots of the auxiliary equation are $m_1 = 4$ and $m_2 = 4$, which are real but not distinct roots. Using the techniques of Section 26.1, we can write the solutions as:

$$y_1 = C_1e^{4x} \quad \text{and} \quad y_2 = C_2e^{4x}.$$

Thus, the complementary solution is:

$$y_c = C_1e^{4x} + C_2e^{4x}.$$

Using common-term factoring, we have:

$$y_c = (C_1 + C_2)e^{4x}.$$

Since C_1 and C_2 are constants, it can be written as:

$$y_c = Ce^{4x}.$$

This clearly is not a complementary solution since a second order equation has two distinct solutions. The complementary solution is a linear combination of two linearly independent solutions, and these two solutions are not linearly independent. Now let us examine how we can arrive at a complementary solution with linearly independendent solutions. Writing the equation in the form:

$$(D - 4)(D - 4)y = 0 \tag{3}$$

let:

$$(D - 4)y = z \tag{4}$$
$$(D - 4)z = 0 \qquad \text{Substituting (4) into (3)} \tag{5}$$
$$z = C_1e^{4x} \qquad \text{Solution of (5)}$$
$$(D - 4)y = C_1e^{4x} \qquad \text{Substituting for } z \text{ in (4)} \tag{6}$$

To solve equation (6), which is a first order linear differential equation, we determine the integrating factor e^{-4x} and write:

$$ye^{-4x} = \int e^{-4x}(C_1e^{4x})\, dx$$
$$ye^{-4x} = C_1x + C_2 \qquad \text{Integrating}$$
$$y = (\,C_1x\, + \,C_2\,)e^{4x}. \qquad \text{Solving for } y$$

Thus, the complementary solution of the equation is:

$$y_c = C_1xe^{4x} + C_2e^{4x}.$$

We now have a complementary solution that is the linear combination of two linearly independent solutions.

For the differential equation:

$$(D - a)(D - a)(D - a)y = 0,$$

the auxiliary equation is:

$$(m - a)(m - a)(m - a) = 0.$$

We see that there are repeated roots, and the root is $m = a$. Using the result of the previous discussion, we can write the complementary solution

$$y_c = (\,C_1\, + \,C_2x\, + \,C_3x^2\,)e^{ax}.$$

Using this pattern, the solution of the differential equation $(D - a)^5y = 0$ is:

$$y_c = (\,C_1\, + \,C_2x\, + \,C_3x^2\, + \,C_4x^3\, + \,C_5x^4\,)e^{ax}.$$

The technique of solving an equation with sets of repeated roots is illustrated in Example 1.

EXAMPLE 1   Auxiliary Equation with Repeated Roots

Determine the complementary solution of $y^{(iv)} + 2y''' + y'' = 0$.

Solution Writing the equation in D operator notation, we have:

$$(D^4 + 2D^3 + D^2)y = 0$$

$$m^4 + 2m^3 + m^2 \quad = 0 \qquad \text{Auxiliary equation}$$

$$m^2(m + 1)(m + 1) = 0 \qquad \text{Factoring}$$

$$m_1 = 0, \ m_2 = 0, \ m_3 = -1, \ m_4 = -1. \qquad \text{Roots of auxiliary equation}$$

There are two sets of repeated roots; therefore, the complementary solution is:

$$y_c = C_1 e^{0x} + C_2 x e^{0x} + C_3 e^{-x} + C_4 x e^{-x},$$

which we would write as:

$$y_c = C_1 + C_2 x + (C_3 + C_4 x)e^{-x}. \quad \blacklozenge$$

Complex Roots

The auxiliary equation of the differential equation:

$$y'' + 4y' + 5y = 0$$

is

$$m^2 + 4m + 5 = 0.$$

Solving the auxiliary equation, the roots are:

$$m_1 = -2 + j \quad \text{and} \quad m_2 = -2 - j.$$

Recall from our study of complex numbers that m_1 is the complex conjugate of m_2. Thus, we have two distinct complex roots for the auxiliary equation. Using the technique of Section 26.1, we can write the complementary solution:

$$y_c = C_1 e^{(-2+j)x} + C_2 e^{(-2-j)x}.$$

Using rules of exponents, we can write this complementary solution as:

$$y_c = e^{-2x}(C_1 e^{jx} + C_2 e^{-jx}).$$

Using a formula developed by Leonhard Euler (1707–1783), we can write this solution with real functions instead of complex exponents.

Euler's Formula

$$e^{j\theta} = \cos \theta + j \sin \theta$$

Substituting for $e^{j\theta}$, the complementary solution becomes:

$$y_c = e^{-2x}\{C_1(\cos x + j \sin x) + C_2[\cos(-x) + j \sin(-x)]\}.$$

Recall that $\cos(-x) = \cos x$ and $\sin(-x) = -\sin x$.

$$y_c = e^{-2x}[C_1(\cos x + j \sin x) + C_2(\cos x - j \sin x)]$$
$$= e^{-2x}[(C_1 + C_2)\cos x + (C_1 j - C_2 j)\sin x].$$

Now let $C_3 = C_1 + C_2$ and $C_4 = C_1 j + C_2 j$, where C_3 and C_4 are arbitrary constants.

$$y_c = e^{-2x}(C_3 \cos x + C_4 \sin x).$$

We can make this substitution since $e^{-2x} \cos x$ and $e^{-2x} \sin x$ are fundamental solutions to the differential equation for all values of x.

In general, if the auxiliary equation has roots:

$$m_1 = a + bj \quad \text{and} \quad m_2 = a - bj,$$

then the complementary solution is:

$$y_c = e^{ax}(C_1 \cos bx + C_2 \sin bx).$$

This technique of determining a solution when the auxiliary equation has complex roots is illustrated in Example 2.

EXAMPLE 2   Complex Roots

Determine the complementary solution of the differential equation $2y'' - 2y' + y = 0$.

Solution   The auxiliary equation of $2y'' - 2y' + y = 0$ is:

$$2m^2 - 2m + 1 = 0.$$

The roots of the equation are:

$$m_1 = \frac{1}{2} + \frac{1}{2}j \qquad m_2 = \frac{1}{2} - \frac{1}{2}j.$$

From our previous discussion we can write the solution as:

$$y_c = e^{\frac{x}{2}}\left(C_1 \cos \frac{x}{2} + C_2 \sin \frac{x}{2} \right). \quad \blacklozenge$$

EXAMPLE 3   Auxiliary Equation with Real and Complex Roots

Determine the complementary solution of the differential equation $y''' - 4y'' + 7y' = 0$.

Solution   The auxiliary equation of:

$$y''' - 4y'' + 7y' = 0$$

is:

$$m^3 - 4m^2 + 7m = 0$$
$$m(m^2 - 4m + 7) = 0$$
$$m_1 = 0, \, m_2 = 2 + j\sqrt{3}, \, m_3 = 2 - j\sqrt{3}$$

Thus, the complementary solution may be written as:

$$y_c = C_1 + e^{2x}(C_2 \cos \sqrt{3} \, x + C_3 \sin \sqrt{3} \, x). \quad \blacklozenge$$

---

1. If the auxiliary equation has real repeating roots where:

$$m = a, a, a, a, \ldots, a,$$

then the general solution of the equation is:

$$y = C_1 e^{ax} + C_2 x e^{ax} + C_3 x^2 e^{ax} + \cdots + C_n x^{n-1} e^{ax}.$$

2. If the auxiliary equation has complex roots:

$$m_1 = a + bj \quad \text{and} \quad m_2 = a - bj,$$

then the complementary solution of the equation is:

$$y_c = e^{ax}(C_1 \cos bx + C_2 \sin bx).$$

26.2 Exercises

In Exercises 1–20, determine the complementary solution of the differential equations.

1. $y'' + 2y' + y = 0$

2. $y'' + 4y' + 4y = 0$

3. $y'' + 14y' + 49y = 0$

4. $y'' + 12y + 36 = 0$

5. $D^4 y = 0$

6. $D^5 y = 0$

7. $\dfrac{d^2y}{dx^2} - 4\dfrac{dy}{dx} + 5y = 0$

8. $\dfrac{d^2y}{dx^2} - 2\dfrac{dy}{dx} + 3y = 0$

9. $\dfrac{d^2y}{dx^2} + y = 0$

10. $\dfrac{d^2y}{dx^2} + k^2 y = 0$

11. $\dfrac{d^2y}{dx^2} - 2\dfrac{dy}{dx} + 4y = 0$

12. $\dfrac{d^2s}{dt^2} - 4\dfrac{ds}{dt} + 13s = 0$

13. $\dfrac{d^2y}{dx^2} - 2\dfrac{dy}{dx} - y = 0$

14. $\dfrac{d^2y}{dx^2} + 3\dfrac{dy}{dx} - y = 0$

15. $36\dfrac{d^2s}{dt^2} + 36\dfrac{ds}{dt} + 13s = 0$

16. $4y''' - 3y' + y = 0$

17. $2y''' + y'' - 4y' - 3y = 0$

18. $y^{(iv)} + 2y''' - 2y' - y = 0$

19. $\dfrac{d^4y}{dx^4} - 4y = 0$

20. $\dfrac{d^4y}{dx^4} - 2\dfrac{d^3y}{dx^3} + \dfrac{d^2y}{dx^2} + 2\dfrac{dy}{dx} - 2y = 0$

21. Determine the particular solution of $y'' + 4y' + 4y = 0$ that satisfies the initial conditions $y(2) = 1$ and $y'(2) = 2$.

22. Determine the particular solution of $y'' + 9y = 0$ that satisfies the initial conditions $y = 5$ and $y' = 6$ when $x = 0$.

23. Determine the complementary solution of $y'' + ky' - \ell y = 0$.

24. Determine a curve having slope 2 at the origin and satisfying the differential equation $y'' + 2y' + y = 0$.

In Exercises 25–28, you are given the solution of a differential equation. For each solution, write a differential equation of the form $(D^2 + aD + b)y = 0$ that represents the solution.

25. $y = C_1 e^{2x} + C_2 e^{-4x}$

26. $y = (C_1 + C_2 x)e^{3x}$

27. $y = C_1 \cos 2x + C_2 \sin 2x$

28. $y = e^{-3x}(C_1 \cos 4x + C_2 \sin 4x)$

26.3

Nonhomogeneous Equations

The general solution of the nonhomogeneous differential equation:

$$a_n \frac{d^n y}{dx^n} + a_{n-1} \frac{d^{n-1} y}{dx^{n-1}} + \cdots + a_1 y' + a_0 y = f(x)$$

is the sum of a particular solution to this equation and the complementary solution to the homogeneous equation:

$$a_n \frac{d^n y}{dx^n} + a_{n-1} \frac{d^{n-1} y}{dx^{n-1}} + \cdots + a_1 y' + a_0 y = 0.$$

Thus, the **general solution** of the nonhomogeneous equation takes the form:

$$Y = y_c + y_p,$$

where y_c is the **complementary solution,** which is the solution of the homogeneous equation, and y_p is the **particular solution.** We learned how to determine y_c in Sections 26.1 and 26.2. In this section we learn how to determine the particular solution of the equation.

EXAMPLE 1   Checking the General Solution

The general solution of the differential equation $(D^2 - 2D - 8)y = 10e^{3x}$ is $Y = C_1 e^{4x} + C_2 e^{-2x} - 2e^{3x}$. Show that Y is the general solution of the differential equation.

Solution   The general solution $Y = C_1 e^{4x} + C_2 e^{-2x} - 2e^{3x}$ can be broken up into two parts.

$$y_c = C_1 e^{4x} + C_2 e^{-2x} \quad \text{and} \quad y_p = -2e^{3x}$$

Thus, to show that Y is the general solution, the following must be true:

$$(D^2 - 2D - 8)y_c = 0 \quad \text{and} \quad (D^2 - 2D - 8)y_p = 10e^{3x}.$$

First we substitute $C_1 e^{4x} + C_2 e^{-2x}$ for y_c to determine whether this linear combination is a complementary solution.

$$(D^2 - 2D - 8)(C_1e^{4x} + C_2e^{-2x}) = 0 \qquad \text{Substituting for } y_c$$

$$D^2(C_1e^{4x}) + D^2(C_2e^{-2x}) - 2D(C_1e^{4x})$$
$$-2D(C_2e^{-2x}) - 8(C_1e^{4x}) - 8(C_2e^{-2x}) = 0 \qquad \text{Distributive property}$$

$$16C_1e^{4x} + 4C_2e^{-2x} - 8C_1e^{4x}$$
$$+ 4C_2e^{-2x} - 8C_1e^{4x} - 8C_2e^{-2x} = 0 \qquad \text{Differentiating}$$

$$0 = 0 \checkmark$$

The result shows that the linear combination is the complementary solution. Now substitute $-2e^{3x}$ for y_p to determine whether it is the particular solution.

$$(D^2 - 2D - 8)(-2e^{3x}) = 10e^{3x} \qquad \text{Substituting for } y_p$$

$$D^2(-2e^{3x}) - 2D(-2e^{3x}) - 8(-2e^{3x}) = 10e^{3x} \qquad \text{Distributive property}$$

$$-18e^{3x} + 12e^{3x} + 16e^{3x} = 10e^{3x} \qquad \text{Differentiating}$$

$$10e^{3x} = 10e^{3x} \checkmark$$

Since both check, Y is the general solution of the differential equation. ◆

The question now is how do we determine y_p? To illustrate this, let us determine y_p for the differential equation $(D^2 - 2D - 8)y = 10e^{3x}$. We saw in Example 1 that if we replace y with y_p and perform the operations on y_p, the result must equal $10e^{3x}$. Thus, y_p must be of the form Ae^{3x}. We can differentiate Ae^{3x} as many times as we wish, and the result still is a constant times e^{3x}. Therefore, with $y_p = Ae^{3x}$, we set up the equality:

$$(D^2 - 2D - 8)Ae^{3x} = 10e^{3x}$$

$$D^2(Ae^{3x}) - 2D(Ae^{3x}) - 8(Ae^{3x}) = 10e^{3x} \qquad \text{Distributive property}$$

$$9Ae^{3x} - 6Ae^{3x} - 8Ae^{3x} = 10e^{3x} \qquad \text{Differentiating}$$

$$-5Ae^{3x} = 10e^{3x}.$$

In order for the expressions to be equal, the coefficients of e^{3x} must be equal; that is,

$$-5A = 10 \quad \text{or} \quad A = -2.$$

Thus, $y_p = -2e^{3x}$. This technique of determining the particular solution is called the **method of undetermined coefficients.**

EXAMPLE 2 **Determining a General Solution**

Determine a general solution of the differential equation $y'' - 5y' + 6y = 6x^2 + 2x - 2$.

Solution Since $f(x) = 6x^2 + 2x - 2$ is a polynomial of degree 2, y_p is a polynomial of degree 2. Therefore, let $y_p = Ax^2 + Bx + C$.

$$(D^2 - 5D + 6)(Ax^2 + Bx + C) = 6x^2 + 2x - 2$$

Substituting y_p for y

$$2A - 5(2Ax + B) + 6(Ax^2 + Bx + C) = 6x^2 + 2x - 2$$

Differentiating

$$6Ax^2 + (-10A + 6B)x + (2A - 5B + 6C) = 6x^2 + 2x - 2$$

$$6A = 6$$

Equating coefficients of like terms

$$-10A + 6B = 2$$

$$2A - 5B + 6C = -2$$

Solving the system of three equations with three unknowns for A, B, and C, we find $A = 1$, $B = 2$, and $C = 1$. Substituting these values in our general form for y_p we have:

$$y_p = x^2 + 2x + 1.$$

The auxiliary equation is:

$$m^2 - 5m + 6 = 0 \quad \text{or} \quad (m - 3)(m - 2) = 0.$$

With $m = 3$ and $m = 2$, we can write the complementary solution:

$$y_c = C_1 e^{2x} + C_2 e^{3x}.$$

Thus the general solution is:

$$Y = C_1 e^{2x} + C_2 e^{3x} + x^2 + 2x + 1.$$ ◆

To help take some of the guesswork out of the process, the discussion in this section is limited to differential equations where $f(x)$ either is a function defined by one of the following.

1. x^n, where n is a positive integer.
2. e^{ax}, where a is a nonzero constant.
3. $\sin bx$, where b is a nonzero constant.
4. $\cos bx$, where b is a nonzero constant.

Or $f(x)$ is a function that is the result of a finite product of two or more of the four types of functions in the list.

EXAMPLE 3   **General Solution Where $f(x) = \sin x$**

Determine a general solution of the differential equation $y'' + 4y' + 4y = \sin x$.

Solution   From the previous section we know that the auxiliary equation is $m^2 + 4m + 4 = 0$, which suggests repeated roots $m_1 = 2$ and $m_2 = 2$. Thus,

$$y_c = C_1 e^{2x} + C_2 x e^{2x}$$

The particular solution must contain $\sin x$. Successively differentiating the sine function, the derivative alternates $\cos x$, $\sin x$, etc. Thus, all possible cases are included, and we select $y_p = A \sin x + B \cos x$. Differentiating and substituting into the equation, we have:

$$y''_p + 4y'_p + 4y_p = -A \sin x - B \cos x + 4A \cos x$$
$$-4B \sin x + 4A \sin x + 4B \cos x$$
$$= (3A - 4B) \sin x + (4A + 3B) \cos x.$$

Thus, $(3A - 4B) \sin x + (4A + 3B) \cos x = \sin x$. Since this last equation is an identity, we can set up the following system of equations.

$$3A - 4B = 1$$
$$4A + 3B = 0$$

Solving for A and B, we find that $A = \dfrac{3}{25}$ and $B = -\dfrac{4}{25}$. Therefore, the general

solution is $Y = C_1 e^{2x} + C_2 x e^{2x} + \dfrac{3}{25} \sin x - \dfrac{4}{25} \cos x$. ◆

GUIDELINE 1 FOR DETERMINING y_p

Form a linear combination of $f(x)$ and all the distinct derivatives of $f(x)$. If none of these, $f(x)$, or the derivatives, are the same as any part of the complementary solution, then this linear combination may be used as the general form of y_p.

In Example 3, $f(x) = \sin x$ and $f'(x) = \cos x$, $f''(x) = -\sin x$. Neither of these are part of the complementary solution. The distinct derivative of $f(x)$ is $\cos x$, and the linear combination we used for the general form of y_p was $y_p = A \sin x + B \cos x$. Example 4 illustrates a case where the exponential in the complementary solution is distinct from the exponential in $f(x)$.

EXAMPLE 4   **General Solution Where $f(x) = xe^{ax}$**

Determine the general solution of the differential equation $y'' + 4y' + 3y = 15xe^{2x}$.

Solution   The auxiliary equation is $m^2 + 4m + 3 = 0$. From this we can write the complementary solution:

$$y_c = C_1 e^{-x} + C_2 e^{-3x}.$$

With $f(x) = 15xe^{2x}$, we find that the distinct forms of $f(x)$ and its derivatives are e^{2x} and xe^{2x}, and none of these are elements of the complementary solution. Knowing this we can write the particular solution as:

$$y_p = Ae^{2x} + Bxe^{2x}.$$

Now substituting y_p for y, we have:

$$y''_p + 4y'_p + 3y_p = 4Ae^{2x} + (4B + 4Bx)e^{2x}$$
$$+ 4[2Ae^{2x} + (B + 2Bx)e^{2x}]$$
$$+ 3Ae^{2x} + 3Bxe^{2x}$$
$$= (15A + 8B)e^{2x} + 15Bxe^{2x} = 15xe^{2x}.$$

Since the last equation is an identity, we can set up the following system of equations.

$$15A + 8B = 0$$
$$15B = 15$$

Solving for A and B, we find that $A = -\dfrac{8}{15}$ and $B = 1$. Therefore, the general

solution of the differential equation is $Y = C_1e^{-x} + C_2e^{-3x} - \dfrac{8}{15}e^{2x} + xe^{2x}$.

$\blacklozenge$

If a function, say e^x or $\sin x$, are part of the complementary solution and the particular solution, then we must use a different technique to determine the particular solution. This technique is illustrated in Example 5.

EXAMPLE 5 e^x Is Part of y_p and y_c

Determine the general solution of the differential equation $y'' + 3y' - 4y = e^x$.

Solution   The auxiliary equation is $m^2 + 3m - 4 = 0$, and:

$$y_c = C_1e^x + C_2e^{-4x}.$$

Using the guidelines for finding y_p, we see that $f(x)$ and its derivatives are all of the form e^x. Since e^x is also part of the complementary solution, we must use another method to find the particular solution. To find the correct form for y_p, we differentiate:

$$y'' + 3y' - 4y = e^x \tag{7}$$

and obtain

$$y''' + 3y'' - 4y' = e^x. \tag{8}$$

Subtracting (7) from (8), we have:

$$y''' + 2y'' - 7y' + 4y = 0.$$

We now have a homogeneous equation and use the rules of Sections 26.1 and 26.2 to obtain a solution. The auxiliary equation is:

$$m^3 + 2m^2 - 7m + 4 = 0$$
$$(m + 4)(m - 1)(m - 1) = 0.$$

With the repeated roots, we obtain a general solution, which we can designate as follows.

$$Y = \underbrace{C_1 e^x + C_2 e^{-4x}}_{y_c} + \underbrace{C_3 x e^x}_{y_p}$$

Our particular solution cannot have an arbitrary constant; thus, we must determine a value for C_3. With this information, we let $y_p = Axe^x$ and substitute for y.

$$y_p'' + 3y_p' - 4y_p = A(2 + x)e^x + 3A(1 + x)e^x - 4xAe^x$$
$$5Ae^x = e^x$$

Solving for A, we find that $A = \dfrac{1}{5}$. Therefore, the general solution is:

$$Y = C_1 e^x + C_2 e^{-4x} + \boxed{\dfrac{1}{5}} xe^x. \qquad \blacklozenge$$

The problem of determining a general form for y_p in Example 5 was not covered in Guideline 1. To cover this situation, we introduce Guideline 2.

GUIDELINE 2 FOR DETERMINING y_p

Form a linear combination of $f(x)$ and all the distinct derivatives of $f(x)$. If any term of this linear combination appears in the complementary solution, then those terms that appear in y_c should be multiplied by the lowest power, of x that eliminates the like terms. This new linear combination becomes y_p.

EXAMPLE 6 **General Solution and Specific Solution**

a. Determine the general solution of the differential equation $y'' + 4y = 8 \sin 2x$.
b. Determine the specific solution for the initial values $y(0) = 6$ and $y'(0) = 8$.

Solution

a. The auxiliary equation is $m^2 + 4 = 0$; solving we have roots $m = \pm 2j$. The complementary solution is $y_c = C_1 \cos 2x + C_2 \sin 2x$. To determine y_p, we differentiate $\sin 2x$, which gives the results $\cos 2x$ and $\sin 2x$. We see that $f(x)$ and its derivatives are in y_c; therefore, we must apply Guideline 2. We write the general form of y_p, as:

$$y_p = Ax \sin 2x + Bx \cos 2x.$$
$$y''_p + 4y_p = (-4Ax \sin 2x - 4B \sin 2x + 4A \cos 2x$$
$$- 4Bx \cos 2x) + (4Ax \sin 2x + 4Bx \cos 2x) \quad \textbf{Differentiating}$$

$$- 4B \sin 2x + 4A \cos 2x = 8 \sin 2x \qquad \text{Substituting}$$
$$-4B = 8 \quad \text{and} \quad 4A \cos 2x = 0$$
$$B = -2 \qquad\qquad A = 0$$

Thus, $y_p = -2x \cos 2x$, and the general solution is $Y = C_1 \cos 2x + C_2 \sin 2x - 2x \cos 2x$.

b. To determine the solution for the initial values we find the derivative of the general solution found in **a.**

$$Y' = -2C_1 \sin 2x + 2C_2 \cos 2x + 4x \sin 2x - 2 \cos 2x$$

Now substitute the initial values in Y and Y'.

$$6 = C_1 \cos(\boxed{0}) + C_2 \sin(\boxed{0}) - 2(\boxed{0}) \cos(\boxed{0})$$
$$8 = -2C_1 \sin(\boxed{0}) + 2C_2 \cos(\boxed{0}) + 4(\boxed{0}) \sin(\boxed{0}) - 2 \cos(\boxed{0})$$
$$C_1 = \boxed{6} \quad \text{and} \quad C_2 = \boxed{5} \quad \text{Solving for } C_1 \text{ and } C_2$$

Substituting these values for C_1 and C_2 in Y, we obtain the specific solution:

$$Y = \boxed{6} \cos 2x + \boxed{5} \sin 2x - 2x \cos 2x. \qquad\qquad \blacklozenge$$

26.3 Exercises

In Exercises 1–6, show that Y is a general solution of the differential equation.

1. $y'' - y = e^x$; $Y = C_1 e^x + C_2 e^{-x} + \dfrac{1}{2} x e^x$

2. $y'' + 4y = e^{2x}$; $Y = C_1 \cos 2x + C_2 \sin 2x + \dfrac{1}{8} e^{2x}$

3. $y'' + y = \dfrac{3}{2} \sin 2x$; $Y = C_1 \cos x + C_2 \sin x - \dfrac{1}{2} \sin 2x$

4. $y'' + y' + y = x^2$; $Y = e^{(-\frac{1}{2}x)}\left(C_1 \cos \dfrac{\sqrt{3}}{2} x + C_2 \sin \dfrac{\sqrt{3}}{2} x \right) + x^2 - 2x$

5. $y'' + 3y' = 6x^2 + x$; $Y = C_1 + C_2 e^{-3x} + \dfrac{2}{3} x^3 - \dfrac{1}{2} x^2 + \dfrac{1}{3} x$

6. $y'' + y' + y = x \sin x$; $Y = e^{(-\frac{x}{2})}\left(C_1 \cos \dfrac{\sqrt{3}}{2} x + C_2 \sin \dfrac{\sqrt{3}}{2} x \right) - x \cos x + \sin x + 2 \cos x$

In Exercises 7–22, determine the general solution of the differential equation.

7. $y'' - 4y' - 5y = 10$

8. $y'' + 3y' + 2y = x^2 + 1$

9. $y'' - 4y' + 4y = 5 \sin x$

10. $y'' - 3y' + 2y = 6e^{3x}$

11. $y'' + 2y' - 3y = 5e^{2x}$

12. $y''' + y' = 4 \cos x$

13. $y'' + y' = 3x^2$

14. $y'' - 4y' + 4y = 12xe^{2x}$

15. $(D^2 + 3D + 2)y = 12$

16. $(D^2 + 4D + 3)y = 3x$

17. $(D^2 - D - 2)y = 8e^{3x}$

18. $(D^2 + 2D)y = 40 \sin 4x$

19. $y'' + y' - 2y = 4 \sin 2x$

20. $y'' + 3y' + 2y = 36xe^x$

21. $(D^2 - 7D + 12)y = 12x^2 + 10x - 11$

22. $(D^2 - 1)y = 5e^x \sin x$

In Exercises 23–30, determine the specific solution for each of the differential equations.

23. $y'' + y = x^2$; $y(0) = 0$, $y'(0) = 1$

24. $y'' - 4y' + 3y = e^{2x}$; $y(0) = 0$, $y'(0) = 0$

25. $y'' + y = 2x$; $y(0) = 1$, $y'(0) = 2$

26. $y'' - y = \sin x$; $y(0) = 1$, $y'(0) = -\dfrac{3}{2}$

27. $(D^2 - 4)y = 25e^{3x}$; $y(0) = 0$, $y'(0) = 5$

28. $(D^2 - 1)y = x^2 - 3x$; $y(0) = 0$, $y'(0) = 2$

29. $(D^2 + 1)y = 8\cos 2x$; $y\left(\dfrac{\pi}{2}\right) = -1$, $y'(\pi/2) = 0$

30. $(D^2 + 4)y = \sin 3x$; $y\left(\dfrac{\pi}{2}\right) = 1$, $y'\left(\dfrac{\pi}{2}\right) = -1$

26.4

Applications

The second order differential equation is used to solve many different types of problems. Some of these involve the vibrations in springs, pendulums, acoustic systems, and mechanical systems, such as air planes and bridges that are acted on by external forces. Also involved are problems in electrical circuits, cardiology, and economics.

The simplest system with which to study vibrating motions is a spring attached to a fixed support, as shown in Figure 26.1a. If a weight W is hung on the spring, the spring stretches a distance s below the original length of the spring (see Figure 26.1b). This position is called the **equilibrium position** ($W = mg = ks$).

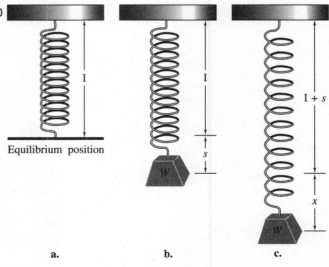

a.   b.   c.

FIGURE 26.1

If the weight is pulled down a distance x and released, the spring vibrates back and forth about the equilibrium position, as indicated in Figure 26.1c. Hooke's law tells us that the spring exerts a force F, called the *resting force* (in lb or Ns), opposite to the direction of the elongation of the spring and proportional to the amount of elongation (in ft or m). That is, $F_{spring} = k(s + x)$, where k is the constant of proportionality.

According to Newton's second law, the net forces of all forces acting is:

$$F_{net} = ma$$

and since $a = \dfrac{d^2x}{dt^2}$,

$$F_{net} = m\frac{d^2x}{dt^2}.$$

If we take a free body diagram of the weight (see Figure 26.2), measuring position downward, so:

$$F_{resultant} = W - F_{spring}$$
$$\frac{d^2x}{dt^2} = \frac{W - k(s + x)}{m}.$$

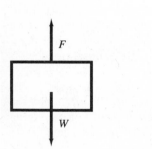

F

W

$+x$

FIGURE 26.2

But, recall that $W = ks$. When the weight was originally placed on the spring and stretched it to the equilibrium position, the force in the spring was the weight W and x was zero. Therefore,

$$W = k(s + 0) = ks$$
$$m\frac{d^2x}{dt^2} = ks - k(s + x)$$
$$= -kx.$$

The mass, m, of an object is its weight divided by the acceleration due to gravity. Example 1 illustrates a spring problem.

EXAMPLE 1   **Application: Spring Problem**

A weight of 5 lb stretches a spring 6 in. and is pulled 3 in. below the equilibrium position and released.

a.   Determine the differential equation and conditions that describe the motion.
b.   Determine the position of the weight as a function of time.
c.   Sketch the graph of the function.
d.   Determine the position, velocity, and acceleration of the weight 1 s after it has been released.

Solution

a. Using Hooke's law, $F = kx$, we can find k since 6 in. $= \dfrac{1}{2}$ ft.

$$5 = k\left(\frac{1}{2}\right) \quad \text{or} \quad k = \boxed{10}$$

The differential equation is:

$$m\frac{d^2x}{dt^2} = -kx$$

$$\frac{\boxed{5}}{\boxed{32}}\frac{d^2x}{dt^2} = -\boxed{10}x \quad m = \frac{w}{g} = \frac{5}{32}$$

$$\frac{d^2x}{dt^2} = -64x$$

$$\frac{d^2x}{dt^2} + 64x = 0.$$

The weight is 3 in. below the equilibrium position, so we have $x = \frac{1}{4}$ ft when $t = 0$. In addition, when the weight is released, $t = 0$ and $\frac{dx}{dt} = 0$ (zero velocity).

b. The auxiliary equation is $m^2 + 64 = 0$, which has roots $m_1 = 8j$ and $m_2 = -8j$. The solution of the differential equation is:

$$x = C_1 \cos \boxed{8}t + C_2 \sin \boxed{8}t.$$

Using the conditions $x = \frac{1}{4}$ ft when $t = 0$, we can determine that $C_1 = \frac{1}{4}$. Thus,

$$x = \frac{1}{4}\cos 8t + C_2 \sin 8t$$

$$\frac{dx}{dt} = -2 \sin 8t + 8C_2 \cos 8t. \quad \textbf{Differentiating}$$

With initial conditions $t = 0$ and $\frac{dx}{dt} = 0$, we determine that $C_2 = 0$. The required equation is:

$$x = \frac{1}{4}\cos 8t.$$

c. The graph of the function is shown in Figure 26.3.

d. To determine the velocity and acceleration, we must differentiate $x = \frac{1}{4}\cos 8t$.

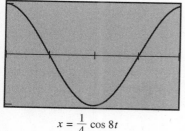

$x = \frac{1}{4}\cos 8t$

FIGURE 26.3

$$v = \frac{dx}{dt} = -2 \sin 8t \quad \text{and} \quad a = \frac{dv}{dt} = -16 \cos 8t$$

Now, substituting $t = 1$ in each of the equations, we find that:

$$x = \frac{1}{4}\cos(8)(\boxed{1}) = -0.036$$

$$v = -2\sin(8)(\boxed{1}) = -1.979$$

$$a = -16\cos(8)(\boxed{1}) = 2.328.$$

The results tell us that after 1 s, the weight is 0.036 ft above the equilibrium position, is traveling upward with velocity of 1.979 ft/s, and has a downward acceleration of 2.328 ft/s². ◆

In the idealized situation in Example 1, the spring would continue vibrating the same distances above and below the equilibrium position forever. However, in real situations, friction and other external effects cause the distance x to become smaller, and eventually, the spring comes to rest. The combined forces that cause the spring to come to rest are referred to as the **damping force.** The magnitude of the damping force is approximately proportional to the instantaneous speed (velocity) of the weight on the spring. The magnitude of the damping force is given by:

$$\beta\left|\frac{dx}{dt}\right|,$$

where β is the constant of proportionality called the **damping coefficient** or **resistance coefficient.** The damping force acts opposite to the motion of the spring; that is, when the weight moves down, the damping force acts up and vice versa. With downward motion positive, the damping force must be negative when $\frac{dx}{dt}$ is positive. With $b > 0$, the damping force must be given both in magnitude and direction by $-\beta\frac{dx}{dt}$. Combining the restoring force and the damping force in Newton's law the differential equation of motion is:

$$m\frac{d^2x}{dt^2} = -\beta\frac{dx}{dy} - kx \quad \text{or} \quad m\frac{d^2x}{dt^2} + \beta\frac{dx}{dt} + kx = 0.$$

Basic Equation of Damped Vibrating Motion

$$m\frac{d^2x}{dt^2} + \beta\frac{dx}{dt} + kx = 0$$

EXAMPLE 2   **Application: Spring with Damping Force**

In addition to the information given in Example 1, assume that a damping force with a coefficient of 2.5 lb acts on the weight.

a. Set up the differential equation and associated conditions.
b. Determine the position of x of the weight as a function of time t.

Solution

a. With a damping force of $-2.5\dfrac{dx}{dt}$, the equation becomes:

$$\frac{5}{32}\frac{d^2x}{dt^2} + 2.5\frac{dx}{dt} + 10x = 0$$

$$\frac{d^2x}{dt^2} + 16\frac{dx}{dt} + 64x = 0.$$

b. The auxiliary equation is $m^2 + 16m + 64 = 0$, and has roots $m_1 = m_2 = -8$. The general solution of the differential equation is:

$$x = e^{-8t}(C_1 + C_2t).$$

Using the initial conditions from Example 1, we can determine values for C_1 and C_2 when $t = 0$ s and $x = \dfrac{1}{4}$ ft.

$$\frac{1}{4} = e^{-8(0)}(C_1 + C_2(0)) \qquad \textbf{Substituting for } x \textbf{ and } t$$

$$\frac{1}{4} = C_1$$

$$x = e^{-8t}\left(\frac{1}{4} + C_2t\right) \qquad \textbf{Substituting for } C_1$$

$$\frac{dx}{dt} = -8e^{-8t}\left(\frac{1}{4} + C_2t\right) + e^{-8t}(C_2) \qquad \textbf{Differentiating}$$

Substituting $t = 0$ and $\dfrac{dx}{dt} = 0$ and solving, we find that $C_2 = 2$. Thus,

$$x = e^{-8t}\left(\frac{1}{4} + 2t\right). \qquad \textbf{Substituting for } C_1 \textbf{ and } C_2$$

The position x of the weight as a function of time t is given by the equation:

$$x = e^{-8t}\left(\frac{1}{4} + 2t\right).$$

A graph of the solution is given in Figure 26.4. ◆

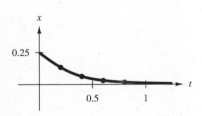

FIGURE 26.4

For the general equation:

$$m\frac{d^2x}{dt^2} + \beta\frac{dx}{dt} + kx = 0,$$

if we divide each term by m, the equation becomes:

$$\frac{d^2x}{dt^2} + \frac{\beta}{m}\frac{dx}{dt} + \frac{k}{m}x = 0.$$

For convenience sake, we let $\frac{\beta}{m} = 2\lambda$ and $\omega = \sqrt{\frac{k}{m}}$; substituting, the equation is:

$$\frac{d^2x}{dt^2} + 2\lambda\frac{dx}{dt} + \omega^2 x = 0.$$

The auxiliary equation is:

$$m^2 + 2\lambda m + \omega^2 = 0.$$

The roots of the auxiliary equation are:

$$m_1 = -\lambda + \sqrt{\lambda^2 - \omega^2} \quad \text{and} \quad m_2 = -\lambda - \sqrt{\lambda^2 - \omega^2}.$$

Examining the discriminate $\lambda^2 - \omega^2$ there are three possible solutions: *overdamped, critically damped,* and *underdamped.* Let us examine each in turn.

a. Overdamped: When $\lambda^2 - \omega^2 > 0$, the solution takes the form: $x = C_1 e^{m_1 t} + C_2 e^{m_2 t}$.
    In this case, shown in Figure 26.5, the system is said to be **overdamped** since the damping coefficient λ is large when compared to the spring constant ω. In this case, there is no vibrating motion.

b. Critically damped: When $\lambda^2 - \omega^2 = 0$, the solution takes the form: $x = e^{m_1 t}(C_1 t + C_2)$.
    In this case, shown in Figure 26.6, the system is said to be **critically damped** since any slight increase in the damping force results in vibrating motion. The roots m_1 and m_2 of the auxiliary equation are real and equal. Example 2 illustrates this case.

c. Underdamped: When $\lambda^2 - \omega^2 < 0$, shown in Figure 26.7, the system is said to be **underdamped** since the damping coefficient is small compared to the spring constant. The roots of the auxiliary equation are:

$$m_1 = -\lambda + \sqrt{\lambda^2 - \omega^2}\, j \quad \text{and} \quad m_2 = -\lambda - \sqrt{\lambda^2 - \omega^2}\, j.$$

The general solution is:

$$x = e^{-\lambda t}(C_1 \cos\sqrt{\omega^2 - \lambda^2} + C_2 \sin\sqrt{\omega^2 - \lambda^2}).$$

The sine and cosine functions indicate vibration is present, but the factor $e^{-\lambda t}$ eventually causes the vibrations to stop—to damp out. Example 3 illustrates this case.

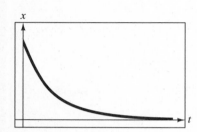

FIGURE 26.5

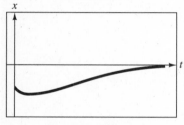

FIGURE 26.6

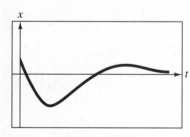

FIGURE 26.7

EXAMPLE 3   Application: Spring Underdamped

A 4-lb weight is attached to a 7-in. spring. At equilibrium, the spring measures 10 in. The weight is pushed 6 in. above the equilibrium position and released. Determine the displacement x if it is further known that the surrounding medium offers a resistance numerically equal to twice the instantaneous velocity.

Solution   The weight stretches the spring 3 in.; therefore, using Hookes's law:

$$F = ks$$

$$4 = k\left(\frac{3}{12}\right)$$

$$k = 16 \text{ lb/ft.}$$

Since $m = \dfrac{w}{g}$, $m = \dfrac{4}{32} = \dfrac{1}{8}$ slug, and the damping constant β is 2. The differential equation of motion is:

$$\frac{1}{8}\frac{d^2x}{dt^2} + 2\frac{dx}{dt} + 16x = 0$$

$$\frac{d^2x}{dt^2} + 16\frac{dx}{dt} + 128x = 0.$$

We solve this equation with the conditions that $x(0) = \dfrac{-6}{12} = -\dfrac{1}{2}$ and $\dfrac{dx}{dt} = 0$ when $t = 0$. The roots of the auxiliary equation $m^2 + 16m + 128 = 0$ are:

$$m_1 = -8 + 8j \quad \text{and} \quad m_2 = -8 - 8j,$$

which tells us the system is underdamped.
Thus, the general solution of the equation is:

$$x = e^{-8t}(C_1 \cos 8t + C_2 \sin 8t).$$

For $x(0) = -\dfrac{1}{2}$, we have:

$$-\frac{1}{2} = e^{-8(0)}(C_1 \cos 8(0) + C_2 \sin 8(0)$$

$$C_1 = -\frac{1}{2}.$$

Therefore,

$$x = e^{-8t}\left(-\frac{1}{2}\cos 8t + C_2 \sin 8t\right)$$

$$\frac{dx}{dt} = -8e^{-8t}(-\frac{1}{2}\cos 8t + C_2 \sin 8t) + e^{-8t}(4\sin 8t + 8 C_2 \cos 8t).$$

Substituting $t = 0$ and $\dfrac{dx}{dt} = 0$ in the equation, we have:

$$0 = -8\left(-\frac{1}{2}\right) + 8C_2$$

$$-\frac{1}{2} = C_2.$$

Thus, the displacement x for anytime t is:

$$x = e^{-8t}\left(-\frac{1}{2} \cos 8t - \frac{1}{2} \sin 8t \right). \quad \blacklozenge$$

In applied situations of the spring problem, there are occasions when, in addition to the restoring and damping forces, other external forces act on the spring. These external forces could come from the movement of the support holding the spring or from driving forces applied directly to the weight. If we denote the external force by $f(t)$, the differential equation for motion of a spring is:

$$m\frac{d^2x}{dt^2} + \beta\frac{dx}{dt} + kx = f(t).$$

This equation is often referred to as the **equation of forced vibrations.** We solve this equation using the technique of undetermined coefficients.

For a single loop series electric circuit containing an inductor, resistor, and capacitor, shown in Figure 26.8, Kirchhoff's second law states that the sum of the voltage drops across each part of the circuit is the same as the impressed voltage $E(t)$. The charge on the capacitor at any time is denoted by $q(t)$, and the current $i(t)$ is given by $i = \dfrac{dq}{dt}$. It is known that the voltage drops across the three passive devices are:

$$v_L = L\frac{di}{dt} \qquad \textbf{Inductor}$$

$$= L\frac{d^2q}{dt^2};$$

$$v_C = \frac{1}{C}q; \qquad \textbf{Capacitor}$$

$$v_R = iR \qquad \textbf{Resistor}$$

$$= R\frac{dq}{dt}.$$

In these formulas, L, C, are R are constants called inductance, capacitance, and resistance, respectively. We can determine $q(t)$ by solving the differential equation:

$$L\frac{d^2q}{dt^2} + R\frac{dq}{dt} + \frac{1}{C}q = e(t).$$

When $e(t) = 0$, the roots of the auxiliary equation tell us whether the solution is sinusoidal. Example 4 illustrates an electrical circuit problem.

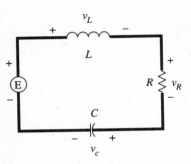

FIGURE 26.8

EXAMPLE 4 Application: Determining Charge and Current

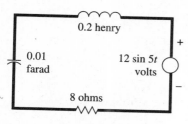

FIGURE 26.9

An inductor of 0.2 henry is connected in series with a resistor of 8 ohms, a capacitor of 0.01 farad, a generator having alternating voltage given by 12 sin 5t, $t \geq 0$, and a switch K, as shown in Figure 26.9.

a. Set up a differential equation for the instantaneous charge on the capacitor.
b. Determine the charge and current at time t if the charge on the capacitor is zero when the switch K is closed at $t = 0$.

Solution

a. The voltage drop across the resistor is 8i; across the inductor it is $0.2\frac{di}{dt}$; and across the capacitor it is $\frac{q}{0.01}$, or 100q. Applying Kirchhoff's law, we can write:

$$8i + 0.2\frac{di}{dt} + 100q = 12 \sin 5t.$$

Since $i = \frac{dq}{dt}$,

$$0.2\frac{d^2q}{dt^2} + 8\frac{dq}{dt} + 100q = 12 \sin 5t \quad \text{Substituting}$$

$$\frac{d^2q}{dt^2} + 40\frac{dq}{dt} + 500q = 60 \sin 5t$$

b. The conditions are $q = 0$ and $i = \frac{dq}{dt} = 0$ at $t = 0$. The auxiliary equation $m^2 + 40m + 500 = 0$ has roots $m_1 = -20 + 10j$ and $m_2 = -20 - 10j$. Thus,

$$q_c = e^{-20t}(A \cos 10t + B \sin 10t).$$

Assuming a particular solution of:

$$q_p = a \sin 5t + b \cos 5t$$

and using the method of undetermined coefficients, we find $a = 0.154$ and $b = -0.045$. Thus, the general solution is:

$$q = e^{-20t}(A \cos 10t + B \sin 10 t) + 0.154 \sin 5t - 0.045 \cos 5t.$$

Using the initial conditions, we find that $A = 0.045$ and $B = 0.037$. Therefore, the required solution is:

$$q = e^{-20t}(0.045 \cos 10t + 0.037 \sin 10t) + 0.154 \sin 5t - 0.045 \cos 5t.$$

The complementary part of the solution in Example 4:

$$[e^{-20t}(0.045 \cos 10t + 0.037 \sin 10t)]$$

is referred to as the **transient solution.** The particular part of the solution:

$$[0.154 \sin 5t - 0.045 \cos 5t]$$

is referred to as the **steady-state solution.** For large values of t, the transient solution goes to zero, and the steady-state solution remains.

26.4 Exercises

In Exercises 1–4, perform each of the following.
a. Determine the differential equation and conditions that describe the motion.
b. Determine the position of the weight as a function of time.
c. Sketch the graph of the function.
d. Determine the position, velocity, and acceleration of the weight for the indicated time after it has been released.

1. A weight of 6 lb stretches a spring 6 in. The weight is pulled 4 in. below the equilibrium position and released. $\left(t = \dfrac{1}{2} \text{ s}\right)$

2. A weight of 12 lb stretches a spring 6 in. The weight is pulled 2 in. below the equilibrium position and released. $\left(t = \dfrac{1}{4} \text{ s}\right)$

3. A weight of 18 lb stretches a spring 12 in. When the equilibrium position is reached, the weight is started downward with a velocity of 3 ft/s. $\left(t = \dfrac{1}{4} \text{ s}\right)$

4. A weight of 18 lb stretches a spring 6 in. When the equilibrium position is reached, the weight is raised 4 in. above the equilibrium position and released. $\left(t = \dfrac{1}{8} \text{ s}\right)$

5. An object weighing 20 pounds, when hung on a helical spring, causes the spring to stretch 3 in. The object is then pulled down 4 in. and released.
   a. Write an equation that defines its motion at any time t.
   b. Determine the period.
   c. Determine the frequency of the motion.

6. An object weighing 10 lb, when hung on a helical spring, causes the spring to stretch 1 in. The object is then pulled down 2 in. and released.
   a. Write an equation that defines its motion at any time t.
   b. Determine the period.
   c. Determine the frequency of the motion.

7. The equation $\dfrac{d^2y}{dt^2} + 100y = 0$ represents a simple harmonic motion.

   a. Determine the general solution of the equation.

   b. Determine the constants of integration if $y = 10$ and $\dfrac{dy}{dt} = 50$ when $t = 0$.

   c. Determine the frequency.
   d. Determine the period.
   e. Determine the amplitude.

8. What must be true of k in the equation $\dfrac{d^2s}{dt^2} + k\dfrac{ds}{dt} + 30s = 0$ if the motion represented is oscillatory?

9. An oscillatory motion is represented by $\dfrac{d^2y}{dt^2} + 6\dfrac{dy}{dt} + 10y = 0$.

   a. Determine the period of oscillation of y.
   b. Determine the damping factor.
   c. Determine the time required for the damping factor to decrease 50 percent.

In Exercises 10–12, name the type of damping associated with each equation.

10. $\dfrac{d^2x}{dt^2} + \dfrac{dx}{dt} + 3x = 0$

11. $\dfrac{d^2x}{dt^2} + 2\dfrac{dx}{dt} + x = 0$

12. $\dfrac{d^2x}{dt^2} + \dfrac{dx}{dt} + \dfrac{1}{5}x = 0$

13. A 32-lb weight is attached to a spring. The spring constant is $k = 36$ lb/ft, and the resistance coefficient is $\beta = 8$. Determine the resulting motion if $x = \dfrac{1}{2}$ ft and $v = 0$ ft/s when $t = 0$ s with impressed force of $f(t) = 72 \cos 6t$.

14. A 32-lb weight is attached to a spring. The spring constant is $k = 16$ lb/ft, and the resistance coefficient is $\beta = 2$. The block is acted on by an external force $f(t) = 96 \cos 4t$. For $x = \dfrac{1}{2}$ ft and $v = 0$ ft/s when $t = 0$ s, what is the displacement and velocity at time t?

15. A block weighing 6 lb is attached to a spring that has a spring constant of 27 lb/ft. In the equilibrium position, when $t = 0$, an external force $f(t) = 12 \cos 20t$ is applied to the system. Determine the resulting displacement as a function of time. Assume that the damping factor is not measurable.

16. A block weighing 10 lb is attached to a spring that has a spring constant of 20 lb/ft. In the equilibrium position, when $t = 0$, an external force $f(t) = 10 \cos 8t$ is applied to the system. A damping force that is five times the instantaneous velocity $\left(5\dfrac{dx}{dt}\right)$ is also applied. Determine the displacement of the weight as a function of the time.

17. A 6-lb block is hung on a coil spring, which stretches the spring 4 in. from the equilibrium position. When $t = 0$ an external force $f(t) = 27 \sin 4t - 3 \cos 4t$ is applied to the system. There is a resistance to the system in pounds that is equivalent to three times the instantaneous velocity that is measured in feet per second. Determine the displacement as a function of the time.

18. An inductor of 0.2 henry is connected in series with a resistor of 10 ohms and electromotive force of 40 V. If the initial current is 0, determine the current at time $t > 0$.

19.   Consider the RLC circuit, shown in the diagram, where there is an initial charge of 2 coulombs on the capacitor and no initial current.

    a.   Set up the initial value problem describing the charge $Q(t)$ on the capacitor for any given emf $E(t)$.

    b.   Determine the charge $Q(t)$ if $E(t) = 0$. What is the current $I(t)$?

    c.   Determine the charge $Q(t)$ if $E(t) = 20$ volts. What is the steady-state term?

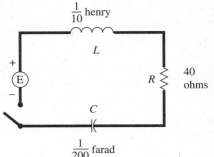

20.   The differential equation of an RLC circuit with $L = \dfrac{1}{7}$ henry, $R = 4$ ohms, and $C = \dfrac{1}{35}$ farad is $\dfrac{1}{7}Q'' + 4Q' + 35Q = E(t)$. Determine the general solution and the steady-state solution for emf $E(t) = 7$.

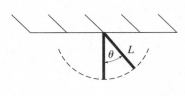

21.   Using the information given in Exercise 20, determine the general solution and the steady-state solution if $E(t) = 7 \cos 7t$.

22.   The motion of the simple pendulum shown in the sketch is described by the differential equation $\dfrac{d^2\theta}{dt^2} = -\dfrac{g}{L}\theta$. ($\theta$ in radians.)

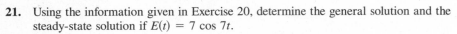

    a.   Determine the general solution for θ as a function of t.

    b.   Show that the period T of the pendulum is given by $T = 2\pi\sqrt{\dfrac{L}{g}}$.

26.5

Laplace Transforms

Oliver Heaviside (1850–1925), an English electrical engineer, devised a method of solving ordinary differential equations that led to the development of the Laplace transforms. Having a direct method of obtaining a particular solution makes the Laplace transform an attractive method of solving differential equations in electric circuits and other areas where particular solutions are required.

> **DEFINITION 26.3 Laplace Transform**
>
> Let $f(t)$ be a real valued function defined for $t \geq 0$. The Laplace transform of f denoted by $\mathcal{L}\{f(t)\}$ is defined by:
>
> $$\mathcal{L}\{f(t)\} = \int_{t=0}^{t=\infty} e^{-st}f(t)\, dt.$$

Since the upper limit of the integral is infinity, we have an improper integral and must rewrite it as:

$$\mathcal{L}\{f(t)\} = \lim_{b\to\infty} \int_{t=0}^{t=b} e^{-st}f(t)\, dt.$$

The result is to be expressed in terms of s, so we write:

$$\mathcal{L}\{f(t)\} = F(s).$$

EXAMPLE 1 **Laplace Transform of $f(t) = 1$**

Determine the Laplace transform of the function $f(t) = 1$.

Solution

$$\mathcal{L}\{1\} = \int_{t=0}^{t=\infty} e^{-st} (1) \, dt$$

$$= \lim_{b \to \infty} \int_{t=0}^{t=b} e^{-st} \, dt$$

$$= \lim_{b \to \infty} \left(\frac{e^{-st}}{-s} \right) \Big]_{t=0}^{t=b}$$

$$= \lim_{b \to \infty} \left(\frac{-e^{-sb}}{s} + \frac{1}{s} \right)$$

$$= \frac{1}{s}, \text{ provided } s > 0. \quad \blacklozenge$$

EXAMPLE 2 **Laplace Transform of $f(t) = t$**

Determine the Laplace transform of the function $f(t) = t$.

Solution

$$\mathcal{L}\{t\} = \int_{t=0}^{t=\infty} e^{-st}(t) \, dt$$

We can determine the Laplace transform using integration by parts. Let $u = t$ and $dv = e^{-st}$ then $du = dt$ and $v = \dfrac{-e^{-st}}{s}$.

$$\mathcal{L}\{t\} = \lim_{b \to \infty} \left\{ \frac{-te^{-st}}{s} \Big]_{t=0}^{t=b} - \int_{t=0}^{t=b} \frac{-e^{-st}}{s} \, dt \right\}$$

$$= \lim_{b \to \infty} \left[\frac{-te^{-st}}{s} - \frac{e^{-st}}{s^2} \right]_{t=0}^{t=b}$$

$$= \lim_{b \to \infty} \left[\left(\frac{-be^{-sb}}{s} - \frac{e^{-sb}}{s^2} \right) - \left(\frac{-1}{s^2} \right) \right]$$

We use the fact that $\lim_{b \to \infty} be^{-sb} = 0$ and that the $\lim_{b \to \infty} e^{-sb} = 0$, and conclude $\mathcal{L}\{t\} = \dfrac{1}{s^2}, s > 0.$ $\quad \blacklozenge$

We could continue finding the Laplace transforms of many functions, but since this becomes boring after a while, those commonly used are listed in Table 26.1.

TABLE 26.1   Laplace Transforms

	$f(t)$	$\mathcal{L}[f(t)]$
1.	1	$\dfrac{1}{s}$
2.	t	$\dfrac{1}{s^2}$
3.	$t^n, \ n = 1, 2, 3, \ldots$	$\dfrac{n!}{s^{n+1}}$
4.	e^{at}	$\dfrac{1}{s-a}$
5.	te^{at}	$\dfrac{1}{(s-a)^2}$
6.	$t^n e^{at}$	$\dfrac{n!}{(s-a)^{n+1}}$
7.	$\sin at$	$\dfrac{a}{s^2+a^2}$
8.	$\cos at$	$\dfrac{s}{s^2+a^2}$
9.	$t \sin at$	$\dfrac{2as}{(s^2+a^2)^2}$
10.	$t \cos at$	$\dfrac{s^2-a^2}{(s^2+a^2)^2}$
11.	$e^{at} \sin bt$	$\dfrac{b}{(s-a)^2+b^2}$
12.	$e^{at} \cos bt$	$\dfrac{s-a}{(s-a)^2+b^2}$
13.	$1 - \cos at$	$\dfrac{a^2}{s(s^2+a^2)}$
14.	$at - \sin at$	$\dfrac{a^3}{s^2(s^2+a^2)}$
15.	$\sin at - at \cos at$	$\dfrac{2a^3}{(s^2+a^2)^2}$
16.	$\sin at + at \cos at$	$\dfrac{2as^2}{(s^2+a^2)^2}$

Recall from integral calculus that

$$\int_{x=a}^{x=b} [\partial f(x) + \beta g(x)]\, dx = \partial \int_{x=a}^{x=b} f(x)\, dx + \beta \int_{x=a}^{x=b} g(x)\, dx.$$

An operation having this property is said to be a **linear operation.** Since the Laplace transform is an integral, we can write:

$$\mathcal{L}\{af(t) + bg(t)\} = a\mathcal{L}\{f(t)\} + b\mathcal{L}\{g(t)\}. \qquad \{1\}$$

Thus, the Laplace transform is a **linear operator.**

Using the results of Table 26.1 and the fact that the Laplace transform is a linear operator, we can determine the Laplace transform of more complex functions.

EXAMPLE 3   **Laplace Transform of the Sum of Two Functions**

Determine the Laplace transform of $f(t) = t^3 + 4 \cos 5t$.

Solution

$$f(t) = t^3 + 4 \cos 5t \qquad \textbf{Given}$$

$$\mathcal{L}\{f(t)\} = \mathcal{L}\{t^3 + 4 \cos 5t\}$$

$$= \mathcal{L}\{t^3\} + 4\mathcal{L}\{\cos 5t\} \quad \textbf{Linear property 1}$$

$$= \frac{3!}{s^{3+1}} + \frac{4s}{s^2 + 5^2} \qquad \textbf{Transforms 3 and 8}$$

$$= \frac{6}{s^4} + \frac{4s}{s^2 + 25} \quad \blacklozenge$$

Remember our goal is to solve differential equations using Laplace transforms. To accomplish this goal we must be able to determine the Laplace transform of derivatives of functions.

$$\mathcal{L}\{f'(t)\} = \lim_{b \to \infty} \int_{t=0}^{t=b} e^{-st}f'(t)\, dt$$

We can determine the Laplace transform using integration by parts. Let $u = e^{-st}$ and $dv = f'(t)\, dt$, then $du = -se^{st}\, dt$ and $v = f(t)$. Thus,

$$\mathcal{L}\{f'(t)\} = \lim_{b \to \infty} e^{-st}f(t) \Big]_{t=0}^{t=b} - \lim_{b \to \infty} \int_{t=0}^{t=b} f(t)(-se^{-st}\, dt)$$

$$= \lim_{b \to \infty} (e^{-sb}f(b) - e^{-s(0)}f(0)) + s \lim_{b \to \infty} \int_{t=0}^{t=b} e^{-st}f(t)\, dt.$$

By definition $\mathcal{L}\{f(t)\} = \lim_{b \to \infty} \int_{t=0}^{t=b} e^{-st}f(t)\, dt$, so substituting, we have:

$$= 0 - f(0) + s\mathcal{L}\{f(t)\}.$$

Therefore,

$$\mathscr{L}\{f'(t)\} = s\mathscr{L}\{f(t)\} - f(0). \qquad \{2\}$$

Using the same technique we can show that:

$$\mathscr{L}\{f''(t)\} = s^2\mathscr{L}\{f(t)\} - sf(0) - f'(0). \qquad \{3\}$$

EXAMPLE 4 **Determining the Laplace Transform of a Differential Equation**

Determine the Laplace transform of the differential equation $f''(t) + 3f'(t) + 2f(t) = 0$, with initial conditions $f(0) = 1$ and $f'(0) = 2$.

Solution

$$\mathscr{L}\{f''(t) + 3f'(t) + 2f(t)\} = 0$$

$$\mathscr{L}\{f''(t)\} + 3\mathscr{L}\{f'(t)\} + 2\mathscr{L}\{f(t)\} = 0 \qquad \text{Linear operation 1}$$

$$[s^2\mathscr{L}\{f(t)\} - sf(0) - f'(0)] + 3[s\mathscr{L}\{f(t)\} - f(0)]$$
$$+ 2\mathscr{L}\{f(t)\} = 0 \qquad \text{Formulas 2 and 3}$$

$$(s^2 + 3s + 2)\mathscr{L}\{f(t)\} - s - 5 = 0$$

Solving for $\mathscr{L}\{f(t)\}$, we have:

$$\mathscr{L}\{f(t)\} = \frac{s + 5}{s^2 + 3s + 2}. \qquad \blacklozenge$$

26.5 Exercises

In Exercises 1–6, determine the Laplace transform of the function using Definition 26.3.

1. $f(t) = t^2$
2. $f(t) = e^{at}$
3. $f(t) = te^{at}$
4. $f(t) = \sin at$
5. $f(t) = \cos at$
6. $f(t) = t \sin at$

In Exercises 7–14, determine the Laplace transform of the function using Table 26.1.

7. $f(t) = t^3$
8. $f(t) = e^{5t}$
9. $f(t) = te^{5t}$
10. $f(t) = \cos 7t$
11. $f(t) = e^{3t} \sin 4t$
12. $f(t) = 1 - \cos 7t$
13. $f(t) = 4t - \sin 5t$
14. $f(t) = \sin 3t - 3t \cos 3t$

In Exercises 15–20, determine the Laplace transform using formulas {2} and {3}.

15. $f''(t) - 3f(t) = 0, f(0) = 1, f'(0) = 2$

16. $f''(t) + 2f'(t) = 0, f(0) = 2, f'(0) = 3$

17. $3f''(t) + 5f(t) = 0, f(0) = -2, f'(0) = 4$

18. $4f''(t) - 3f'(t) + f(t) = 0, \ f(0) = 3, \ f'(0) = -5$

19. $15f''(t) - 7f'(t) - 2f(t) = 0, f(0) = -1, f'(0) = 0$

20. $6f''(t) - 5f'(t) - 2f(t) = 0, f(0) = 1, f'(0) = -1$

26.6

Inverse Laplace Transforms

In the previous section we learned how to take the Laplace transform of a differential equation. There we stated the result of taking the transform in terms of the variable s. Now, in order to express the solution of the equation in terms of the original variables of the differential equation, we must take the inverse Laplace transform. The inverse Laplace is denoted by:

$$\mathcal{L}^{-1}\{F(s)\} = f(t).$$

EXAMPLE 1 Inverse Laplace Transform

Determine the inverse Laplace transform of $\dfrac{1}{(s-3)^2}$.

Solution Examining Table 27.1 we see that the Laplace transform of te^{3t} is $\dfrac{1}{(s-3)^2}$ (transform 5). Thus, the $\mathcal{L}^{-1}\left\{\dfrac{1}{(s-3)^2}\right\} = te^{3t}$. ◆

EXAMPLE 2 Inverse Laplace Transform

Determine the inverse Laplace of $\dfrac{5}{s^2+4}$.

Solution Examining Table 26.1 we see that no transform has the result $\dfrac{5}{s^2+4}$. If we rewrite the transform in the form $\dfrac{5}{2}\dfrac{2}{s^2+4}$, we have a constant times the $\mathcal{L}\{\sin 2t\}$. That is,

$$\mathcal{L}^{-1}\left\{\frac{5}{2}\frac{2}{s^2+4}\right\} = \frac{5}{2}\mathcal{L}^{-1}\left\{\frac{2}{s^2+4}\right\}$$

$$= \frac{5}{2}\sin 2t. \ \blacklozenge$$

Let us determine the $\mathcal{L}^{-1}\left\{\dfrac{4s+7}{s^2+5}\right\}$. In the two previous examples we used Table 26.1 to find the inverse. In this case we use our knowledge of fractions, algebra, and the linear property to write:

$$\mathcal{L}^{-1}\left\{\frac{4s+7}{s^2+5}\right\} = 4\mathcal{L}^{-1}\left\{\frac{s}{s^2+5}\right\} + \frac{7}{\sqrt{5}}\mathcal{L}^{-1}\left\{\frac{\sqrt{5}}{s^2+5}\right\}.$$

With the Laplace transform in this form, we can determine the inverse transform using Table 26.1.

$$\mathcal{L}^{-1}\left\{\frac{4s + 7}{s^2 + 5}\right\} = 4 \cos \sqrt{5}\, t + \frac{7}{\sqrt{5}} \sin \sqrt{5}\, t$$

In order to determine some inverses, we must understand the technique of partial fractions. In Examples 3, 4, and 5, that technique is used to determine a solution.

EXAMPLE 3   Inverse Laplace Transform, Using Partial Fractions

Determine the $\mathcal{L}^{-1}\left\{\dfrac{1}{(s + 1)(s - 2)(s + 3)}\right\}$.

Solution   Examining Table 26.1, we see that there are no Laplace transforms whose denominators are the product of three distinct linear factors. To be able to determine the inverse transform, we must break up the fraction into the sum of fractions whose denominators are $(s + 1)$, $(s - 2)$, and $(s + 3)$. We can do this using the method of partial fractions.

There are constants A, B, C so that

$$\frac{1}{(s + 1)(s - 2)(s + 3)} = \frac{A}{s + 1} + \frac{B}{s - 2} + \frac{C}{s + 3}$$
$$= \frac{A(s - 2)(s + 3) + B(s + 1)(s + 3) + C(s + 1)(s - 2)}{(s + 1)(s - 2)(s + 3)}.$$

For the fractions to be equal, the numerators must be identical; that is,

$$1 = A(s - 2)(s + 3) + B(s + 1)(s + 3) + C(s + 1)(s - 2)$$

or

$$1 = (A + B + C)s^2 + (A + 4B - C)s + (-6A + 3B - 2C).$$

Also the coefficients of like terms must be equal.

$$A + B + C = 0 \quad \text{Coefficients of } s^2 \text{ terms}$$
$$A + 4B - C = 0 \quad \text{Coefficients of } s \text{ terms}$$
$$-6A + 3B - 2C = 1 \quad \text{Coefficients of constant terms}$$

Solving for A, B, and C we have $A = -\dfrac{1}{6}$, $B = \dfrac{1}{15}$, and $C = \dfrac{1}{10}$.

Substituting for A, B, and C:

$$\mathcal{L}^{-1}\left\{\frac{1}{(s + 1)(s - 2)(s + 3)}\right\}$$

$$= \mathcal{L}^{-1}\left\{\frac{\frac{-1}{6}}{s + 1}\right\} + \mathcal{L}^{-1}\left\{\frac{\frac{1}{15}}{s - 2}\right\} \mathcal{L}^{-1}\left\{\frac{\frac{1}{10}}{s + 3}\right\}$$

$$= -\frac{1}{6}\,\mathcal{L}^{-1}\left\{\frac{1}{s+1}\right\} + \frac{1}{15}\,\mathcal{L}^{-1}\left\{\frac{1}{s-2}\right\} + \frac{1}{10}\,\mathcal{L}^{-1}\left\{\frac{1}{s+3}\right\}$$

$$= -\frac{1}{6}\,e^{-t} + \frac{1}{15}\,e^{-2t} + \frac{1}{10}\,e^{-3t}.$$

**Linear property
Table 26.1,
Transform 4** ◆

Example 3 is an illustration of determining an inverse Laplace transform where the denominator contains only distinct linear factors. Example 4 illustrates determining the inverse Laplace transform when the denominator of the fraction has repeated linear factors. In Example 5, the denominator of the fraction has a quadratic factor.

EXAMPLE 4   **Inverse Laplace Transform, Using Partial Fractions**

Determine the $\mathcal{L}^{-1}\left\{\dfrac{s+2}{s^2(s+1)}\right\}$.

Solution   We need to express the fraction in a simpler form, so we use partial fractions. Again, we assume equal fractions.

$$\frac{s+2}{s^2(s+1)^2} = \frac{A}{s} + \frac{B}{s^2} + \frac{C}{s+1} + \frac{D}{(s+1)^2}$$

$$= \frac{As(s+1)^2}{s^2(s+1)^2} + \frac{B(s+1)^2}{s^2(s+1)^2} + \frac{Cs^2(s+1)}{s^2(s+1)^2} + \frac{Ds^2}{s^2(s+1)^2}$$

Setting the numerators equal, we have:

$$s + 2 = As(s+1)^2 + B(s+1)^2 + Cs^2(s+1) + Ds^2$$
$$= (A+C)s^3 + (2A+B+C+D)s^2 + (A+2B)s + B.$$

The system of equations to be solved is:

$$A + C = 0 \qquad \textbf{Coefficients of } s^3 \textbf{ terms}$$
$$2A + B + C + D = 0 \qquad \textbf{Coefficients of } s^2 \textbf{ terms}$$
$$A + 2B = 1 \qquad \textbf{Coefficients of } s \textbf{ terms}$$
$$B = 2. \qquad \textbf{Coefficients of constant terms}$$

Solving for A, B, C, and D we have $A = -3$, $B = 2$, $C = 3$, and $D = 1$. Substituting, the problem becomes:

$$\mathcal{L}^{-1}\left\{\frac{s+2}{s^2(s+1)^2}\right\}$$

$$= \mathcal{L}^{-1}\left\{\frac{-3}{s}\right\} + \mathcal{L}^{-1}\left\{\frac{2}{s^2}\right\} + \mathcal{L}^{-1}\left\{\frac{3}{s+1}\right\} + \mathcal{L}^{-1}\left\{\frac{1}{(s+1)^2}\right\}$$

$$= -3\mathscr{L}^{-1}\left\{\frac{1}{s}\right\} + 2\mathscr{L}^{-1}\left\{\frac{1}{s^2}\right\} + 3\mathscr{L}^{-1}\left\{\frac{1}{s+1}\right\} + \mathscr{L}^{-1}\left\{\frac{1}{(s+1)^2}\right\}$$

$$= -3(1) + 2(t) + 3(e^{-t}) + te^{-t}$$

Table 26.1,
Transforms
1, 2, 4, 5

$$= -3 + 2t + 3e^{-t} + te^{-t}. \quad \blacklozenge$$

EXAMPLE 5   Inverse Laplace Transform, Using Partial Fractions

Determine the $\mathscr{L}^{-1}\left\{\dfrac{3s+2}{s(s^2+4)}\right\}$.

Solution   Again we must break down the fraction into a simpler form, and to do this, we use partial fractions.

$$\frac{3s+2}{s(s^2+4)} = \frac{A}{s} + \frac{Bs+C}{s^2+4}$$

$$= \frac{A(s^2+4)}{s(s^2+4)} + \frac{(Bs+C)s}{s(s^2+4)}$$

Setting the numerators equal, we have:

$$3s + 2 = A(s^2+4) + (Bs+C)s.$$

The system of equations to be solved is $4A = 2$, $C = 3$, $A + B = 0$.

Solving for A, B, and C, we have $A = \dfrac{1}{2}$, $B = -\dfrac{1}{2}$, and $C = 3$. Substituting for A, B and C, we have:

$$\mathscr{L}^{-1}\left\{\frac{s+2}{s(s^2+4)}\right\} = \frac{1}{2}\mathscr{L}^{-1}\left\{\frac{1}{s}\right\} - \frac{1}{2}\mathscr{L}^{-1}\left\{\frac{s}{s^2+4}\right\} + \frac{3}{2}\mathscr{L}^{-1}\left\{\frac{2}{s^2+4}\right\}$$

$$= \frac{1}{2} - \frac{1}{2}\cos 2t + \frac{3}{2}\sin 2t. \quad \text{Transforms 1, 8, 7} \quad \blacklozenge$$

A more detailed explanation of partial fractions is found in Chapter 24.

26.6 Exercises

In Exercises 1–14, determine $f(t)$, the inverse Laplace transform of $F(s)$.

1. $F(s) = \dfrac{3}{s}$

2. $F(s) = \dfrac{4}{s^2}$

3. $F(s) = \dfrac{120}{s^6}$

4. $F(s) = \dfrac{3}{s^5}$

5. $F(s) = \dfrac{7}{(s-5)^4}$

6. $F(s) = \dfrac{4}{s(s^2+4)}$

7. $F(s) = \dfrac{3s+5}{s^2+7}$

8. $F(s) = \dfrac{10s}{(s^2+25)^2}$

9. $F(s) = \dfrac{3s^2 - 12}{(s^2 + 4)^2}$ **10.** $F(s) = \dfrac{1}{4s + 1}$ **11.** $F(s) = \dfrac{2s - 6}{s(s + 2)(s - 3)}$ **12.** $F(s) = \dfrac{3s + 15}{(s + 2)(s^2 - 9)}$

13. $F(s) = \dfrac{s + 1}{s^2(s^2 + 1)}$ **14.** $F(s) = \dfrac{s}{(s - 1)(s^2 + 1)}$

26.7

Solutions of Linear Equations with Laplace Transforms

Now that we understand how to determine Laplace transforms and inverse Laplace transforms, we can use these tools to determine the solutions of differential equations with initial conditions. The procedure of solving a differential equation using Laplace transforms requires the following three basic steps.

PROCEDURE FOR SOLVING DIFFERENTIAL EQUATIONS USING LAPLACE TRANSFORMS

1. Take the Laplace transform of both sides of the differential equation. The result is an algebraic equation involving s and $\mathcal{L}\{y\}$. (Recall that $\mathcal{L}(y) = y(s)$.)
2. Solve this algebraic equation for $y(s)$ in terms of s.
3. Determine the inverse Laplace transform of $y(s)$. Recall that $\mathcal{L}^{-1}\{y(s)\} = Y(t)$. The result is $Y(t)$, the solution to the differential equation.

EXAMPLE 1 **Using Laplace Transforms to Solve a First Order Differential Equation**

Solve the differential equation $y'(t) - 2y(t) = 4$ with $y(0) = 3$ using Laplace transforms.

Solution

$$\mathcal{L}\{y'(t) - 2y(t)\} = \mathcal{L}\{4\}$$ Laplace transform, Step 1

$$\mathcal{L}\{y'(t)\} - 2\mathcal{L}\{y(t)\} = \mathcal{L}\{4\}$$ Linear property

$$[sy(s) - y(0)] - 2y(s) = \frac{4}{s}$$

$$sy(s) - 3 - 2y(s) = \frac{4}{s}$$ Substituting initial conditions

$$(s - 2)y(s) = \frac{4}{s} + 3$$

$$y(s) = \frac{4 + 3s}{s(s - 2)}$$ Solving for $y(s)$; Step 2

$$\mathscr{L}^{-1}\{y(s)\} = \mathscr{L}^{-1}\left\{\frac{4 + 3s}{s(s - 2)}\right\} \qquad \text{Inverse Laplace of both sides; Step 3}$$

$$Y(t) = \mathscr{L}^{-1}\left\{\frac{4}{s(s - 2)}\right\} + 3\mathscr{L}^{-1}\left\{\frac{1}{s - 2}\right\}$$

$$= \mathscr{L}^{-1}\left\{\frac{-2}{s}\right\} + \mathscr{L}^{-1}\left\{\frac{2}{s - 2}\right\} + 3\mathscr{L}^{-1}\left\{\frac{1}{s - 2}\right\}$$

$$\text{Partial fractions}$$

$$= -2 + 2e^{2t} + 3e^{2t} \qquad \text{Transforms 1, 4}$$

$$= -2 + 5e^{2t}$$

The solution $Y(t) = -2 + 5e^{2t}$ can be checked easily by substituting the solution into the differential equation. ◆

EXAMPLE 2   Solution Using Laplace Transforms

Solve the differential equation $y'(t) + y(t) = 5e^{2t}$ with $y(0) = 1$.

Solution

$$\mathscr{L}\{y'(t) + y(t)\} = \mathscr{L}\{5e^{2t}\} \qquad \text{Laplace of both sides}$$

$$\mathscr{L}\{y'(t)\} + \mathscr{L}\{y(t)\} = \mathscr{L}\{5e^{2}\} \qquad \text{Linear property}$$

$$[sy(s) - y(0)] + y(s) = \frac{5}{s - 2}$$

$$sy(s) - 1 + y(s) = \frac{5}{s - 2} \qquad \text{Substituting initial conditions}$$

$$y(s) = \frac{s + 3}{(s + 1)(s - 2)} \qquad \text{Solving for } y(s)$$

$$\mathscr{L}^{-1}\{y(s)\} = \mathscr{L}^{-1}\left\{\frac{s + 3}{(s + 1)(s - 2)}\right\} \qquad \text{Inverse Laplace of both sides}$$

$$Y(t) = \mathscr{L}^{-1}\left\{\frac{-\dfrac{2}{3}}{(s + 1)}\right\} + \mathscr{L}^{-1}\left\{\frac{\dfrac{2}{3}}{s + 1}\right\} \qquad \text{Partial fractions}$$

$$= -\frac{2}{3}e^{-t} + \frac{5}{3}e^{2t} \qquad \text{Transform 4} ◆$$

EXAMPLE 3   Using Laplace Transforms to Solve a Second Order Differential Equation

Solve $y''(t) - 10y'(t) + 25y(t) = t^2e^{5t}$ with $y(0) = 2$ and $y'(0) = 6$.

Solution

$$\mathcal{L}\{y''(t) - 10y'(t) + 25y(t)\} = \mathcal{L}\{t^2e^{5t}\}$$ **Laplace of both sides**

$$\mathcal{L}\{y''(t)\} - 10\mathcal{L}\{y'(t)\} + 25\mathcal{L}\{y(t)\} = \mathcal{L}\{t^2e^{5t}\}$$ **Linear property**

$$[s^2y(s) - sy(0) - y'(0)] - 10[sy(s) - y(0)] + 25y(s) = \frac{2}{(s-5)^3}$$

$$[s^2y(s) - 2s - 6] - 10[sy(s) - 2] + 25y(s) = \frac{2}{(s-5)^3}$$

Substituting initial conditions

$$(s^2 - 10s + 25)y(s) = \frac{2}{(s-5)^3} + 2s - 14$$

$$y(s) = \frac{2}{(s-5)^5} + \frac{2s - 14}{(s-5)^2}$$

$$\mathcal{L}^{-1}\{y(s)\} = \mathcal{L}^{-1}\left\{\frac{2}{(s-5)^5} - \frac{4}{(s-5)^2} + \frac{2}{s-5}\right\}$$

Inverse Laplace of both sides

$$Y(t) = \frac{2}{4!}\mathcal{L}^{-1}\left\{\frac{4!}{(s-5)^5}\right\} - 4\mathcal{L}^{-1}\left\{\frac{1}{(s-5)^2}\right\} + 2\mathcal{L}^{-1}\left\{\frac{1}{s-5}\right\}$$

$$= \frac{1}{12}t^4e^{5t} - 4te^{5t} + 2e^{5t}$$ **Transforms 6, 5, 4**   ◆

In a single loop or series circuit, Kirchhoff's second law states that the sum of the voltage drops across an inductor, resistor, and capacitor is equal to the impressed voltage $E(t)$, as shown in Figure 26.10. This circuit can be represented by the differential equation:

$$Lq''(t) + Rq'(t) + \frac{1}{C}q(t) = E(t).$$

The solution in Example 4 is determined by solving a form of this differential equation.

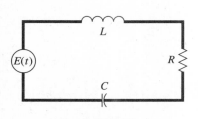

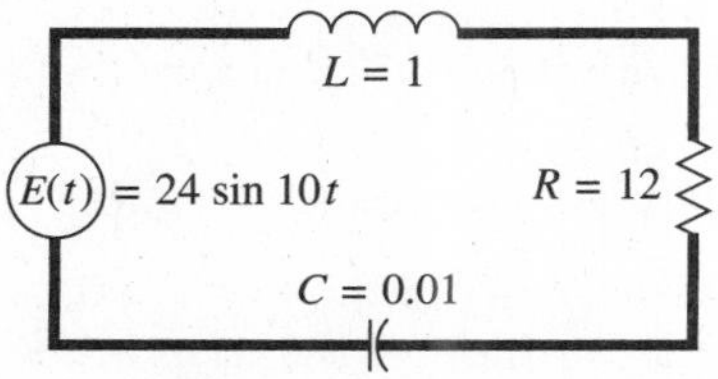

FIGURE 26.10

EXAMPLE 4   Application: Determining the Charge and Current in a Circuit

A circuit consists of an inductor of 1 henry, a resistor of 20 ohms, a capacitor of 0.01 farad, and a generator having voltage given by $E(t) = 24 \sin 10t$, as shown in Figure 26.11. Determine the charge q and the current i at time t if $q(0) = 0$ and $i(0) = q'(0) = 0$.

$L = 1$
$E(t) = 24 \sin 10t$ $R = 12$
$C = 0.01$

FIGURE 26.11

Solution   With the given information we can write the differential equation that we must solve to determine q and i.

$$q''(t) + 20q'(t) + 100q(t) = 24 \sin 10t$$

To solve the differential equation we first take the Laplace transform of the equation.

$$\mathcal{L}\{q''(t) + 20q'(t) + 100q(t)\} = \mathcal{L}\{24 \sin 10t\}$$

Laplace of both sides

$$\mathcal{L}\{q''(t)\} + 20\mathcal{L}\{q'(t)\} + 100\mathcal{L}\{q(t)\} = 24\mathcal{L}\{\sin 10t\}$$

Linear property

$$[s^2q(s) - sq(0) - q'(0)] + 20[sq(s) - q(0)] + 100q(s) = 24\left(\frac{10}{s^2 + 100}\right)$$

$$s^2q(s) + 20\,sq(s) + 100q(s) = \frac{240}{s^2 + 100}$$

$$q(s) = \frac{240}{(s^2 + 100)(s + 10)^2}$$

$$= \frac{-0.12s}{s^2 + 100} + \frac{0.12}{s + 10} + \frac{1.2}{(s + 10)^2}$$

Partial fractions

$$\mathcal{L}^{-1}\{q(s)\} = \mathcal{L}^{-1}\left(\frac{-0.12s}{s^2 + 100} + \frac{0.12}{s + 10} + \frac{1.2}{(s + 10)^2}\right)$$

Inverse Laplace

$$q(t) = -0.12\mathcal{L}^{-1}\left\{\frac{s}{s^2 + 100}\right\} + 0.12\mathcal{L}^{-1}\left\{\frac{1}{s + 10}\right\}$$

$$+ 1.2\mathcal{L}^{-1}\left\{\frac{1}{(s + 10)^2}\right\}$$

$$= -0.12 \cos 10t + 0.12\, e^{-10t} + 1.2te^{-10t}$$

We can determine $i(t)$ by finding $q'(t)$.

$$i(t) = 1.2 \sin 10t - 12te^{-10t} \quad \blacklozenge$$

Example 5 illustrates the concentration of a drug in an organ at any time.

EXAMPLE 5   Application: Concentration of a Drug in an Organ

A fluid carries a drug into an organ, which has volume V cm³, at a rate of a cm³/s, and the drug leaves at a rate of b cm³/s. The concentration of the drug in the entering fluid is c g/cm³.

a.   Write a differential equation for the concentration of the drug in the organ at any time t.

b.   Solve the equation using Laplace transforms.

Solution

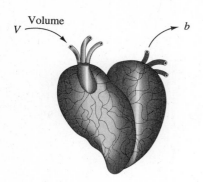

Volume V ↗ ↖ b

FIGURE 26.12

a. The sketch in Figure 26.12 shows an organ of volume V together with intake and outtake. If we let x be the concentration of the drug in the organ (i.e., the number of grams of the drug per cubic centimeter), the amount of drug in the organ at any time t is given by:

$$(V \text{ cm}^3)(x \text{ g/cm}^3) = xV \text{ g.} \tag{9}$$

The number of grams per second entering the organ at time t is given by:

$$(a \text{ cm}^3/\text{s})(c \text{ g/cm}^3) = ac \text{ g/s.} \tag{10}$$

The number of grams per second leaving the organ is given by:

$$(b \text{ cm}^3/\text{s})(x \text{ g/cm}^3) = bx \text{ g/s.} \tag{11}$$

The rate of change of the amount of the drug in the organ is equal to the rate at which the drug enters minus the rate at which it leaves. Using statements (9), (10) and (11), we can write:

$$\frac{d(xV)}{dt} = ac - bx.$$

If we assume that the concentration of the drug in the organ at $t = 0$ is x_0, then $x(0) = x_0$. To simplify our calculations, we also assume that $a, b, c,$ and V are constants. The differential equation is written:

$$V \frac{dx}{dt} = ac - bx.$$

b. Since x is a function of time, the equation can be written as:

$$Vx'(t) = ac - bx(t).$$

Taking the Laplace of both sides and letting $\mathscr{L}\{x(t)\} = u(s)$, we have:

$$V[su(s) - x(0)] = \frac{ac}{s} - bu(s)$$

$$V[su(s) - x_0] = \frac{ac}{s} - bu(s) \qquad \text{\textbf{Substituting initial conditions}}$$

$$u(s) = \frac{Vx_0}{Vs + b} + \frac{ac}{s(Vs + b)} \qquad \text{\textbf{Solving for } } u(s)$$

$$= \frac{x_0}{s + \dfrac{b}{v}} + \frac{ac}{b} \frac{1}{s} - \frac{b}{Vs + b} \qquad \text{\textbf{Partial fractions}}$$

$$\mathscr{L}^{-1}\{u(s)\} = x_0 \mathscr{L}^{-1}\left\{ \frac{1}{s + \dfrac{b}{V}} \right\} + \frac{ac}{b} \mathscr{L}^{-1}\left\{ \frac{1}{s} \right\} - \frac{ac}{b} \mathscr{L}^{-1}\left\{ \frac{1}{s + \dfrac{b}{v}} \right\}$$

$$x(t) = x_0 e^{-\left(\frac{b}{V}\right)t} + \frac{ac}{b} - \frac{ac}{b} e^{-\left(\frac{b}{V}\right)t}. \quad \blacklozenge$$

Let us look at variations of the equation in Example 5. In case 1, $a = b$, and in case 2, $a = b$ and $x_0 = 0$.

Case 1: In this case the drug enters and leaves the organ at the same rate, and the equation becomes:

$$x(t) = c + (x_0 - c)e^{-\left(\frac{b}{V}\right)t}.$$

Case 2: In this case the rate of inflow equals the rate of outflow, and the initial concentration of the drug in the organ is 0. Thus, the equation becomes:

$$x(t) = c(1 - e^{-\left(\frac{b}{V}\right)t}).$$

26.7 Exercises

In Exercises 1–16, solve the differential equation with the given initial conditions using Laplace transforms.

1. $y'(t) - 5y(t) = 0$, $y(0) = 1$
2. $3y'(t) + y(t) = 0$, $y(0) = -1$
3. $y'(t) - 3y(t) = 2$, $y(0) = 1$
4. $2y'(t) - 3y(t) = 6$, $y(0) = 2$
5. $y'(t) - 2y(t) = e^{3t}$, $y(0) = -1$
6. $y'(t) - 2y(t) = 3e^{5t}$, $y(0) = 1$
7. $y'(t) + 3y(t) = te^{2t}$, $y(0) = 2$
8. $y'(t) - y(t) = \sin 3t$, $y(0) = 1$
9. $y''(t) + 4y(t) = 0$, $y(0) = 2$, $y'(0) = 1$
10. $3y''(t) + y(t) = 0$, $y(0) = 1$, $y'(0) = -1$
11. $y''(t) + 9y(t) = 9$, $y(0) = 2$, $y'(0) = 3$
12. $y''(t) + 16y(t) = t$, $y(0) = -1$, $y'(0) = 4$
13. $y''(t) + 4y(t) = 6te^{2t}$, $y(0) = 0$, $y'(0) = 1$
14. $y''(t) + 2y(t) = \cos 3t$, $y(0) = 0$, $y'(0) = 1$
15. $y''(t) + 2y'(t) + y(t) = 6$, $y(0) = 1$, $y'(0) = -1$
16. $y''(t) + 3y'(t) - 4y(t) = t^2$, $y(0) = 2$, $y'(0) = -1$

In Exercises 17 and 18, the problems represent the forced vibration of an undamped or damped spring. Determine the solution using Laplace transforms.

17. $y''(t) + 4y(t) = 3t \cos t$, $y(0) = 0$, $y'(0) = 0$
18. $y''(t) + 4y'(t) + 3y(t) = 3t \sin t$, $y(0) = 0$, $y'(0) = 0$

19. A circuit consists of an inductor of 2 henrys, a resistor of 9 ohms, a capacitor of 0.1 farad, and a generator having voltage given by $E(t) = 25 \cos 10t$. Determine the charge q and the current i at time t if $q(0) = 0$ and $i(0) = 0$.

20. A spring is stretched 6 in. by a 12-lb weight. The 12-lb weight is pulled down 3 in. below the equilibrium point and then released. If there is an impressed force of magnitude $9 \sin 4t$ lb, describe the motion.

For Exercises 21–23, see Example 5.

21. A fluid carries a drug into an organ of volume 400 cm³ at a rate of 8 cm³/s and leaves at the same rate. The concentration of the drug in the entering fluid is 0.08 g/cm³. Assuming that the drug is not present in the organ initially, determine the concentration of the drug in the organ after each of the following.
     a. 30 s.
     b. 120 s.

22. How long would it take for the concentration of the drug in the organ in Exercise 21 to reach each of the following?
   a. 0.03 g/cm³?
   b. 0.05 g/cm³

23. In Exercise 21, change the initial concentration of the drug in the organ to 0.02 g/cm³ and then determine the concentration of the drug in the organ after each of the following.
   a. 30 s.
   b. 120 s.

Review Exercises

In Exercises 1–20, determine the general solution of the differential equation.

1. $y'' + y' - 20 = 0$ **2.** $y'' - 49y = 0$

3. $y'' + 11y' + 28 = 0$ **4.** $y'' - 5y' = 0$

5. $y'' - 6y' + 9y = 0$ **6.** $y'' + 2\sqrt{3}\,y' + 3y = 0$

7. $y''' = 0$ **8.** $y'' + y = 0$

9. $y''' + y' = 0$ **10.** $y'' + 4y + 6 = 0$

11. $y'' - 5y' - 14y = 8$ **12.** $y'' + 4y' + 3y = x + 2$

13. $y'' - 121y = 3 \sin 2x$ **14.** $y'' + 2\sqrt{3}\,y' + 3y = 6 \cos 3x$

15. $y'' + 4y = 5 \cos 5x$ **16.** $y'' + 9y = 3 \sin 4x$

17. $(D^2 - 6D + 9)y = 5e^{2x}$ **18.** $(D^2 - 5D - 24)y = 2e^{3x}$

19. $(D^3 - 9D^2 + 27D - 27)y = x + 1$ **20.** $(D^3 - 3D^2 - 4D)y = xe^{-x}$

In Exercises 21–30, determine the specific solution for each of the differential equations.

21. $y'' - 2y' - 3y = 6 + e^{-t}$; $y(0) = 0$, $y'(0) = 0$

22. $y'' - 4y = te^t$; $y(0) = 1$, $y'(0) = 2$

23. $y'' + y' - 2y = -8 \sin t + 7 \cos t$; $y(0) = 0$; $y'(0) = 1$

24. $y'' + 4y' + 4y = 3e^{-x}$; $y(0) = 3$, $y'(0) = 1$

25. $y'' + 2y' - 3y = 5e^{2x}$; $y(0) = 6$, $y'(0) = 4$

26. $(D - 1)^2(D + 1)y = 10 \cos 2x$; $y(0) = \dfrac{2}{5}$, $y'(0) = -\dfrac{3}{5}$, $y''(0) = -\dfrac{3}{5}$

27. $y'' + y' = 3x^2$; $y(0) = 1$, $y'(0) = 4$

28. $y'' - y' - 2y = 10 \cos x$; $y(0) = -5$, $y'(0) = 3$

29. $y'' + 6y' + 5y = 4e^{-x} \cos 2x$; $y(0) = 1$, $y'(0) = 1$

30. $(D^2 - 3D + 2)y = 2x^3 - 9x^2 + 6x$; $y(0) = 3$, $y'(0) = 4$

31. A circuit has in series an emt of 100 V, a resistor of 10 ohms, and a capacitor of 2×10^{-4} farads. The switch is closed at time $t = 0$, and the charge on the capacitor at this instant is zero. Determine the charge and the current at time $t > 0$.

32. A circuit has in series an emf of 100 sin 200t V, a resistor of 40 ohms, an inductor of 0.25 henry, and a capacitor of 4×10^{-4} farads. If the initial current is zero and the initial charge is 0.01 coulomb, determine the current at any time $t > 0$.

33. A block weighing 4 lb stretches a spring 6 in. The block is pulled down 8 in. from its equilibrium position and released from rest. There is a damping force of 2.5 v lb, where v is in ft/s. Determine a formula for the position at time t. Is the motion overdamped, critically damped, or underdamped?

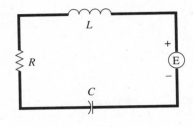

34. Consider the RLC circuit shown in the illustration. Assume $L = 0.5$ henrys, $R = 20$ ohms, $C = 0.1$ farad, $E(t) = 2 \sin 4t$ volts, $Q_0 = 0$ and $i(0) = 0$. Determine the steady state and transient currents for the loop.

In Exercises 35–42, determine the solution of the differential equations using Laplace transforms.

35. $y' + y = e^{-x}$, $y(0) = 1$

36. $y'' + 3y' + 2y = x$, $y(0) = 1$, $y'(0) = -1$

37. $(D^2 + 2D + 2)y = \sin x$, $y(0) = 0$, $y'(0) = -1$

38. $y'(t) + 3y(t) = e^{2t}$, $y(0) = 1$

39. $y'' + 4y' + 5y = 25t$, $y(0) = -5$, $y'(0) = 7$

40. $y' + 2y = \sin \pi t$, $y(0) = 0$

41. The motion of a block undergoing rectilinear motion is governed by the differential equation $y'' + 8y' + 36y = 72 \cos 6t$, $y(0) = \frac{1}{2}$, $y'(0) = 0$. Determine y in terms of t.

42. A circuit consists of an inductor of 3 henrys, a resistor of 20 ohms, a capacitor of 0.3 farad, and a generator having voltage given by $E(t) = 20 \cos 15t$. Find the charge q and the current i at time t if $q(0) = 0$ and $i(0) = 0$.

*43. A 16-lb weight suspended from a linear spring stretches it 2 ft. Let $\mu(t)$ denote the distance of the mass from equilibrium, with positive direction upward, and assume that there is a damping force acting on the spring with damping constant equal to μ lb-s/ft. Determine the differential equation that $y(t)$ must satisfy (assume that $g = 32$ ft/s^2) for each of the following conditions.
   a. For the values $\mu = 2, 4$, and 5 compute $y(t)$ explicitly if the mass is initially at equilibrium and hit upward with velocity of 2 ft/s.
   b. For the values $\mu = 2, 4$, and 5 compute $y(t)$ explicitly if the mass is initially pulled 2 ft below equilibrium and released.

44. For the motion represented by $\dfrac{d^2x}{dt^2} + 4\dfrac{dx}{dt} + kx = 0$, determine the numerically least integral value of k for parts a and c and the greatest integral value for part b.
   a. Critical damping
   b. Overdamping
   c. Underdamped
   d. Sketch the curve of part b.

*45. For the differential equation $y'' + 2y' + \beta y = 0$ determine the general solution in terms of the real number β.

*46. The method of undetermined coefficients requires that the nonhomogeneous term have a finite differential family. The equation $y'' + y = \sec t$, $0 \leq t < \dfrac{\pi}{2}$, does not have a finite family of derivatives. Therefore, we cannot use the method of undetermined coefficients to solve this differential equation. A method that we can use is called variation of parameters, which you can find in a differential equations book. Determine the solution of the equation using the method of variation of parameters.

*47. Determine a second order differential equation having a solution defined by $y = C_1 \sin x + C_2 \cos x + e^x$.

*48. Determine a solution of the differential equation $xy'' - 4y' = x^6$ by letting $u = y'$.

*49. Use Laplace transforms to determine the solution of the differential equation $(D^3 + 1)^3 = e^{-x}$, $y(0) = 0$, $y'(0) = 0$, $y''(0) = 0$

*50. Investigate the improved Euler formula for finding the approximate solution to a differential equation. Use the method to obtain a four decimal approximation for $y' = 2x - 3y + 1$, $y(1) = 5$; let $h = 0.1$. Determine an approximation for $y(1.5)$.

*51. Solve the system of differential equations using Laplace transforms.
$$x'' - 4y = 1, \quad y'' - 4x = 1, \quad x(0) = y(0) = x'(0) = y'(0) = 0$$

✍ Writing About Mathematics

1. You have been introduced to the concept of damping in this chapter. Many examples of damping occur all around us. Select at least three devices where damping is essential and write a paper describing how and why damping is used in the device.

2. Three types of damping are described in this chapter. Write an explanation of the three types and how they differ.

3. In your own words, discuss the difference between a homogeneous and a nonhomogeneous differential equation.

4. Write a procedure for determining the solution to a homogeneous differential equation.

5. All of the literature on drinking and driving tells us that, when alcohol is consumed, it takes a certain period of time before the effect of the alcohol wears off. Is it possible to determine the concentration of alcohol in the blood at any time? Could this be done using differential equations? Write a few short paragraphs answering the questions.

6. Write a step-by-step procedure for solving a differential equation using Laplace transforms. Illustrate your procedure with an example.

Chapter Test

In Exercises 1–8, determine the complementary solution of the differential equation.

1. $y'' - 4y' = 0$
2. $y'' - 4y = 0$
3. $\dfrac{d^2x}{dt^2} - 8\dfrac{dx}{dt} + 7 = 0$
4. $\dfrac{d^2x}{dt^2} - 10\dfrac{dx}{dt} + 25 = 0$
5. $(D + 3)^2(D^2 - 16)y = 0$
6. $(D^3 - 4D^2 - D + 4)y = 0$

7. Determine a differential equation in terms of the D operator whose associated auxiliary polynomial has the roots $m_1 = 3$, $m_2 = -3$, and $m_3 = 0$.

1288   Chapter 26   Higher Order Linear Differential Equations and Laplace Transforms

8. Determine a specific solution to the differential equation $y'' + 6y' + 5y = 0$ given that $y(0) = 1$ and $y'(0) = 1$.

9. Given the solution $y = (C_1 + C_2x)e^{4x}$ to a differential equation, write a differential equation of the form $(D^2 + aD + b)y = 0$ that represents the solution.

In Exercises 10–12, determine the general solution of the differential equation.

10. $y'' + 4y' + 3y = x^2 + 2$ 11. $y'' - 6y' + 9y = 4 \sin x$ 12. $y'' + 3y' - 4y = 7e^{3x}$

13. Determine the specific solution for the differential equation $(D^2 - 1)y = \cos x$;
    $y(0) = 1, y'\left(\dfrac{\pi}{2}\right) = 1$.

In Exercises 14–16, determine the Laplace transform of the function using Table 26.1.

14. $f(t) = 7t^2$ 15. $f(t) = e^{3t} \sin 5t$

16. Determine the Laplace transform of
$$2f''(t) - 5f'(t) + 2f(t) = 0, f(0) = 1, f'(0) = 2.$$

17. Determine $f(t)$, the inverse Laplace transform of $F(s)$, for $F(s) = \dfrac{3s}{s^2 + 5}$.

18. Solve the differential equation with the given initial conditions using Laplace transforms for $y'(t) - 3y(t) = 0, y(0) = 1$.

19. A circuit consists of an inductor of 3 henrys, a resistor of 11 ohms, a capacitor of 0.2 farad, and a generator having voltage given by $E(t) = 22 \cos 5t$. Determine the charge q and the current i at time t if $q(0) = 0$ and $i(0) = 0$.

20. A 2-lb weight suspended from a spring stretches it 1.5 in. If the weight is pulled 3 in. below the equilibrium position and released, determine the following.
    a. Determine the differential equation and conditions that describe the motion.
    b. Determine the position of the weight as a function of time.
    c. Sketch the graph of the function.
    d. Determine the position, velocity, and acceleration of the weight $\dfrac{\pi}{64}$ s after the weight has been released.

Appendix A Measurement

The System International d'Unites (International System of Units or SI) has been adopted as the internationally recognized system of measurement. Currently the United States is the only industrial country not officially on the SI system. Much of science and technology is in the process of converting to this system.

Basic Units

Seven basic independent units are the foundation of the SI system. Each is listed here with a brief informal description. We can find the exact scientific definition in books on the SI system.

Length: The basic unit of length in the SI system is the **meter (m).** The meter is a little longer than a yard. A tall person is about two meters tall.

Mass: The basic unit of mass is the **kilogram (kg).** A kilogram is a little more than two pounds. The mass of an average-size man is about 75 kilograms.

Time: The basic unit of time is the **second (s).** The standard unit for velocity in the metric system is meters per second (m/s).

Electric current: The basic unit of electric current is the **ampere (A),** which is a measure of the rate of flow of electric current. One ampere of current is approximately the amount required for a 100-watt light bulb. Amperes are measured with an instrument called an ammeter.

Temperature: The basic unit for measuring temperature is the **Kelvin (K),** named in honor of Lord Kelvin (William Thompson, 1824–1907). For practical use, the **Celsius** scale (**C**) is used to measure temperature. On the Celsius scale the freezing point of water is at zero and the boiling point is at 100 degrees. (See the diagram on the next page that compares the Kelvin, Celsius, and Farenheit scales.)

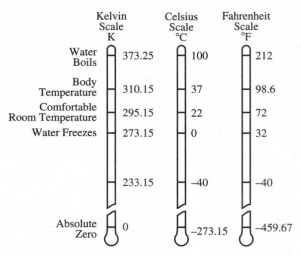

Luminous intensity: The SI unit for luminous intensity is the **candela (cd).** The candela is a unit that measures the amount of light produced by a light source. The luminous intensity emitted from a 40-watt fluorescent bulb is about 4.5 times greater than that of a 40-watt light bulb.

Amount of substance: The **mole (mol)** is the unit that describes the quality of an element or compound. The mole is useful in chemistry for measuring the amounts of chemicals involved in reactions.

Supplementary Units

There are two supplementary units—the plane angle (radian, rad) and the solid angle (steradian, sr).

All other units are derived units. Some of the more common derived units with special names and symbols are described briefly below.

Derived from Length, Mass, and Time

Area: The enclosed area in a square one meter on a side is a **square meter (m²).** (See the sketch.) In the future you may purchase carpet or floor covering by the square meter rather than the square yard.

Volume: The space enclosed in a cube one meter on a side is a **cubic meter (m³).** (See the diagram.) The liter is a measurement of liquid volume and is equal to one cubic decimeter (dm³). The liter is a little larger than a quart. A liter can be divided into 1000 parts each of which is called a milliliter (mℓ). Table A.1 shows the relationship between volume and mass.

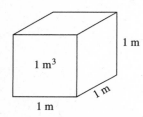

TABLE A.1

Volume in Cubic Units		Volume in Liters		Mass
1 cm³	=	1 mℓ	=	1 g
1 dm³	=	1 ℓ	=	1 kg*
1 m³	=	1 kℓ	=	1 t (1000 kg)

*1 kg water has a volume of 1 ℓ.

Frequency: In an alternating current, the number of complete cycles per second is called the frequency of the current. The SI unit for frequency is the **hertz (Hz);** that is, 1 Hz = 1 cycle/s.

Velocity: The velocity is the distance per unit of time, which may be measured in **meters per second (m/s).**

Acceleration: Acceleration is the velocity per unit of time, which may be measured in **meters per second squared (m/s²).**

Force: Force is mass times acceleration, which may be measured as **kilogram meter per second squared (kg · m/s²).** The measure is called a **newton (N).** One newton applied to a mass of one kilogram gives an acceleration of one meter per second squared.

Pressure (stress): Pressure is the force per unit area, which may be measured as **newton per square meter (N/m²).** The measure is called a **pascal (Pa).** One pascal is the pressure applied by one newton against one square meter.

Energy (work): Energy may be thought of as anything that can be converted to work. Work is equal to a force component times the displacement, which may be measured in **newton meters (N · m).** This measure is called a **joule (J).** When a force of one newton is applied through a distance of one meter, the work done is one joule.

Power: Power is the rate at which work is performed. Power is the energy per unit of time that may be measured in joules per second. The unit of measurement is the **watt (W).** One joule of work done in one second is one watt of power.

Derived Units: From Ampere, Meter, Kilogram, and Second

Electric potential: Electric potential is the power per unit of current, which may be measured in **watt per ampere (W/A).** The unit of measure is called the **volt (V).** The volt is the difference of electric potential between two points of a conductor carrying a constant current of one ampere, when the power between these points is equal to one watt.

Electric resistance: The opposition to the flow of electricity is called electric resistance. The resistance may be measured in **volt per ampere (V/A).** The unit of measure is called the **ohm (Ω).**

Quantity of electricity: The quantity of electricity is equal to current times time, which may be measured in **ampere second (A · s).** The unit of measure is called the **coulomb (C).** The coulomb is the quantity of electricity transported in one second by a current of one ampere.

Electric capacitance: Electric capacitance is the quantity of electricity per unit potential, which may be measured in **coulomb per volt (C/V).** The unit of measurement is called the **farad (F).**

Electric inductance: An inductor or coil is a device in which changing current produces an opposing voltage. The ability of the inductor to induce voltage by current buildup is called its **inductance (L).** The inductance may be measured in **volt seconds per ampere (V · S/A).** The unit of measure is the **henry (H).**

Comparison of the International System and the U.S. Customary System (USCS)

The major difference between the International System and the **U.S. Customary System (USCS)** as used in the United States is that, in SI, mass (kg) is a base unit, and force (N) is a derived unit. In the USCS, force (lb) is a base unit, and mass (slug) is a derived unit. Often, an object incorrectly is said to have a weight given in kilograms. A weight, however, is the force of gravitational attraction exerted on the object, and because it is a force, it should properly be expressed in newtons when using SI units. Weight or mass is understood by comparison to another known object so this does not cause any problems. When weight or mass is to be used in an equation, take care to use proper units.

The relationship between the weight of an object w and its mass m is given by $w = mg$, where g is the local gravitational acceleration. Near the surface of the earth, the standard local gravitational acceleration (to four significant digits) in SI units is taken to be 9.807 m/s², and in USCS units is 32.17 ft/s².

EXAMPLE 1   Weight and Mass

a.   A body is said to weigh 10 kg. Is this correct?
b.   A body weighs 10 lb. What is its mass in USCS?
c.   A body weighs 10 lb. What is its weight in SI?
d.   A body weighs 10 lb. What is its mass in SI?

Solution

a.   It is incorrect to say an object weighs 10 kg (even though it is often done). The kilogram is a mass unit, and in SI, the weight, which is a force, is properly expressed in newtons. Recalling $w = mg$ and assuming that the body is near the surface of the earth, we have:

$$w = 10(9.807) = 98.1 \text{ N}.$$

Properly the object has a mass of 10 kg and a weight of 98.1 N. (The reason that the units used in the equation $w = mg$ do not appear to be consistent is because g is used in two different ways. It represents both the local acceleration of an object in free-fall and the local weight/mass ratio.)

b.   In USCS, the mass of a body that weighs 10 lb near the surface of the earth is found using:

$$w = mg$$
$$10 = m(32.17)$$
$$m = \frac{10}{32.17} \quad \text{or} \quad 0.311 \text{ slug.}$$

c.   Given the weight of the body in USCS units, we can find its weight in SI units by using the conversion factor 1 lb = 4.448 N (see Table A.3). Since we are converting weight to weight and not weight to mass this conversion is valid anywhere in the universe.

$$(10 \text{ lb})\left(\frac{4.448 \text{ N}}{1 \text{ lb}}\right) = 44.5 \text{ N}$$

d. From part c we know that a weight of 10 lb is equivalent to 44.48 N (not rounded). Recalling $w = mg$ and assuming that the body is near the surface of the earth so that we can use $g = 9.807$, in SI units we have:

$$w = mg$$
$$44.48 = m(9.807)$$
$$m = \frac{44.48}{9.807} \quad \text{or} \quad 4.54 \text{ kg.}$$

A body weighing 10 lb near the surface of the earth has a mass of 4.54 kg. ◆

To express quantities in numbers of reasonable size, a system of SI prefixes has been standardized so that the quantities usually can be expressed in terms of numbers between 10^{-3} and 10^3 multiplied by a factor that may be specified by a prefix (Table A.2). When we use numbers in lists, it is good practice to adjust

TABLE A.2   SI Prefixes

Factors	Prefix	Symbol	Pronunciation
$1\ 000\ 000\ 000\ 000\ 000\ 000 = 10^{18}$	exa	E	ĕx′á
$1\ 000\ 000\ 000\ 000\ 000 = 10^{15}$	peta	P	pĕt′á
$1\ 000\ 000\ 000\ 000 = 10^{12}$	tera	T	tĕr′á
$1\ 000\ 000\ 000 = 10^9$	giga	G	jĭ′gá
$1\ 000\ 000 = 10^6$	mega	M	mĕg′á
$1\ 000 = 10^3$	kilo	k	kĭl′ō
$100 = 10^2$	hecto	h	hĕk′tō
$10 = 10^1$	deka	da	dĕk′á
$1 = 10^0$			
$0.1 = 10^{-1}$	deci	d	dĕs′ĭ
$0.01 = 10^{-2}$	centi	c	sĕn′tĭ
$0.001 = 10^{-3}$	milli	m	mĭl′ĭ
$0.000\ 001 = 10^{-6}$	micro	μ	mî′krō
$0.000\ 000\ 001 = 10^{-9}$	nano	n	năn′ō
$0.000\ 000\ 000\ 001 = 10^{-12}$	pico	p	pē′cō
$0.000\ 000\ 000\ 000\ 001 = 10^{-15}$	femto	f	fĕm′tō
$0.000\ 000\ 000\ 000\ 000\ 001 = 10^{-18}$	atto	a	ăt′tō

them so that they all use the same prefix. (For example, list distances of 1.86 km, 2.41 km, and 837 m either as 1.86 km, 2.41 km, and 0.837 km or as 1860 m, 2410 m, and 837 m.) When units are spoken, the prefix is accented. (Kilometer is pronounced with the accent on the first syllable.) In general, the prefixes exa-, peta-, hecto-, deka-, deci-, and centi- are not commonly used. Hecto-, deka-, deci-, and centi- are sometimes used when dealing with area or volume, and centimeter is used when expressing human body dimensions or clothing sizes. Do not use more than one prefix with any dimension. (Don't use megamillimeters.)

EXAMPLE 2   Converting in Metric Units

Convert each of the following.

a. 14.3 meters to millimeters
b. 47.0 μ farads to farads
c. 378. cm^3 to m^3

Solution

a.   From Table A.2, we see that the factor corresponding to the prefix milli- is 10^{-3}. Therefore, 1 mm $= 1 \times 10^{-3}$ m. Using this equality, we can form two conversion factors either by dividing both sides of the equality by 1 mm or by dividing both sides of the equality by 1 m. Using the abbreviations for the units, we have:

$$1 \text{ mm} = 10^{-3} \text{ m} \qquad 1 \text{ mm} = 10^{-3} \text{ m}$$

$$\frac{1 \text{ mm}}{1 \text{ mm}} = \frac{10^{-3} \text{ m}}{1 \text{ mm}} \qquad \frac{1 \text{ mm}}{10^{-3} \text{ m}} = \frac{10^{-3} \text{ m}}{10^{-3} \text{ m}}$$

$$1 = \frac{10^{-3} \text{ m}}{1 \text{ mm}} \qquad \frac{1 \text{ mm}}{10^{-3} \text{ m}} = 1.$$

Both conversion factors are equal to 1, which is the multiplicative identity. Therefore, if we multiply any quantity by either one of the conversion factors, we do not change its value because we are effectively multiplying it by 1. In the given instance, if we multiply 14.3 m by the conversion factor on the left, we obtain

$$(14.3 \text{ m}) \frac{10^{-3} \text{ m}}{1 \text{ mm}} = 14.3 \times 10^{-3} \frac{\text{m}^2}{\text{mm}}.$$

This result is mathematically correct, but the units of the result are not millimeters. If we multiply 14.3 m by the conversion factor on the right we obtain the correct units.

$$(14.3 \text{ m}) \frac{1 \text{ mm}}{10^{-3} \text{ m}} = \frac{14.3}{10^{-3}} \text{ mm}$$

$$= 14.3 \times 10^3 \text{ mm}. \qquad \textbf{Properties of exponents}$$

Therefore, 14.3 m is equivalent to 14.3×10^3 mm or 14,300 mm. We must decide which of the conversion factors to use by checking to see which of them will give us the result in the desired units.

b. To convert 47.0 μ farads to farads, we see from Table A.2 that μ is the symbol for the prefix micro-, which has a factor of 10^{-6}. Therefore, 1 μ farad is 1 microfarad = 10^{-6} farad, and we have conversion factors of:

$$\frac{1\ \mu\ F}{10^{-6}\ F} \quad \text{or} \quad \frac{10^{-6}\ F}{1\ \mu\ F}.$$

By observation we see that, in order to end up in units of farads, we must use the second form of the conversion factor. Therefore,

$$(47.0\ \mu\ F)\left(\frac{10^{-6}\ F}{1\ \mu\ F}\right) = 47.0 \times 10^{-6}\ F$$

$$= 0.000047\ F.$$

c. To convert 378. cm^3 to m^3, we see from Table A.2 that 1 cm = 10^{-2} m, so the available conversion factors are:

$$\frac{1\ cm}{10^{-2}\ m} \quad \text{or} \quad \frac{10^{-2}\ m}{1\ cm}.$$

We see that multiplication by the second conversion factor does not result in units of meters. Rather it gives:

$$(378.\ cm^3)\left(\frac{10^{-2}\ m}{1\ cm}\right) = 378. \times 10^{-2}\ cm^2\ m.$$

However, if we multiply by $\left(\dfrac{10^{-2}\ m}{1\ cm}\right)^3$, we obtain the desired result.

$$(378.\ cm^3)\left(\frac{10^{-2}\ m}{1\ cm}\right)^3 = (378.\ cm^3)\left(\frac{10^{-6}\ m^3}{1\ cm^3}\right) \quad \textbf{Rules of Exponents}$$

$$= 378. \times 10^{-6}\ m^3. \quad \blacklozenge$$

Do not capitalize the names of the units, even if they are derived from the name of a person. Do not capitalize the abbreviations for the units *unless* they are derived from the name of a person. We may use plurals when the name of a unit is spelled out, but do not use them with the abbreviations. Spell prefixes with the unit as one word. Do not use periods with the abbreviations of the units. Avoid using units with prefixes (except kilogram) in a denominator. Leave a space between the quantity referred to and the unit symbol. (Write microfarad, not micro farad; mm not m m.; 41.6 kg, not 41.6kg.)

Because of the two systems of units presently in use in the United States, we sometimes need to convert from one system to the other. The conversion factors given in Table A.3 will aid us in converting between measurement systems. In addition, the miscellaneous constants and factors in Table A.4 occasionally may be useful.

TABLE A.3 Table of Conversion Factors

USCS to SI	Quantity Converted	SI to USCS
$1 \text{ ft}^2 = 9.290 \times 10^{-2} \text{ m}^2$	Area	$1 \text{ m}^2 = 10.76 \text{ ft}^2$
$1 \text{ in.}^2 = 6.452 \text{ cm}^2$		$1 \text{ cm}^2 = 0.1550 \text{ in.}^2$
$1 \text{ yd}^2 = 0.8361 \text{ m}^2$		$1 \text{ m}^2 = 1.196 \text{ yd}^2$
$1 \text{ lb/ft}^3 = 157.1 \text{ N/m}^3$	Density	$1 \text{ N/m}^3 = 6.366 \times 10^{-3} \text{ lb/ft}^3$
$1 \text{ ft lb} = 1.356 \text{ J}$	Energy (work)	$1 \text{ J} = 0.7376 \text{ ft lb}$
$1 \text{ Btu} = 1.055 \times 10^3 \text{ J}$		$1 \text{ J} = 9.480 \times 10^{-4} \text{ Btu}$
$1 \text{ lb} = 4.448 \text{ N}$	Force (weight)	$1 \text{ N} = 0.2248 \text{ lb}$
$1 \text{ ft} = 0.3048 \text{ m}$	Length	$1 \text{ m} = 3.281 \text{ ft}$
$1 \text{ in.} = 25.40 \text{ mm (exact)}$		$1 \text{ mm} = 3.937 \times 10^{-2} \text{ in.}$
$1 \text{ yd} = 0.9144 \text{ m}$		$1 \text{ m} = 1.094 \text{ yd}$
$1 \text{ mi} = 1.609 \text{ km}$		$1 \text{ km} = 0.6214 \text{ mi}$
$1 \text{ slug} = 14.59 \text{ kg}$	Mass	$1 \text{ kg} = 6.853 \times 10^{-2} \text{ slug}$
$1 \text{ hp} = 745.7 \text{ W}$	Power	$1 \text{ W} = 1.341 \times 10^{-3} \text{ hp}$
$1 \text{ psi} = 6.895 \times 10^3 \text{ Pa}$	Pressure (stress)	$1 \text{ Pa} = 1.450 \times 10^{-4} \text{ psi}$
$1 \text{ ft}^3 = 2.832 \times 10^{-2} \text{ m}^3$	Volume	$1 \text{ m}^3 = 35.31 \text{ ft}^3$
$1 \text{ in.}^3 = 16.39 \text{ cm}^3$		$1 \text{ cm}^3 = 6.102 \times 10^{-2} \text{ in.}^3$
$1 \text{ yd}^3 = 0.7646 \text{ m}^3$		$1 \text{ m}^3 = 1.308 \text{ yd}^3$
$1 \text{ gal} = 3.785 \text{ }\ell$		$1 \text{ }\ell = 0.2642 \text{ gal}$

TABLE A.4 Miscellaneous Constants and Conversion Factors

Quantity	Constant or Conversion Factor
Acceleration of gravity (standard)	$1 \text{ g} = 9.807 \text{ m/s}^2$ $1 \text{ g} = 32.17 \text{ ft/s}^2$
Angular measure	$1 \text{ rad} = 57.30°$
Heat	$1 \text{ cal} = 4.187 \text{ J}$
Length	$1 \text{ ft} = 12 \text{ in.}$ $1 \text{ yd} = 3 \text{ ft}$ $1 \text{ mi} = 5280 \text{ ft}$
Mass/Weight (at standard acceleration only)	$1 \text{ N} = 0.1020 \text{ kg}$ $1 \text{ kg} = 9.807 \text{ N}$ $1 \text{ slug} = 32.17 \text{ lb}$ $1 \text{ lb} = 0.4536 \text{ kg}$ $1 \text{ kg} = 2.205 \text{ lb}$
Power	$1 \text{ hp} = 550 \text{ ft lb/s}$
Pressure	$1 \text{ atm} = 14.70 \text{ psi}$
Temperature	$°C = \dfrac{5}{9}(°F - 32)$ $°F = \dfrac{9}{5}°C + 32$
Volume	$1 \text{ }\ell = 1000 \text{ cm}^3$ $1 \text{ gal} = 4 \text{ qt}$ $1 \text{ ft}^3 = 7.481 \text{ gal}$

Appendix A   Measurement   **1297**

EXAMPLE 3   **Conversions: USCS to SI and SI to USCS**

Perform the following conversions.

a. $\dfrac{1}{2}$ hp to watts   **b.** 6 m³ to cubic yards   **c.** 781 cc to quarts

Solution

a. To convert $\dfrac{1}{2}$ hp to watts, we are converting from USCS measurement to SI measurement. From Table A.3, we find 1 hp = 745.7 W, which (as was shown in the previous example) can be expressed equivalently as $1 = \dfrac{745.7\,\text{W}}{1\ \text{hp}}$.

$$\frac{1}{2}\ \text{hp} = \left(\frac{1\ \text{hp}}{2}\right)\left(\frac{745.7\ \text{W}}{1\ \text{hp}}\right)$$

$$= \frac{745.7\ \text{W}}{2}$$

Therefore, $\dfrac{1}{2}$ hp = 373. W.

b. To convert 6 m³ to cubic yards, we are converting from SI measurement to USCS measurement. From Table A.3, we find 1 m³ = 1.308 yd³, which can be expressed equivalently as $1 = \dfrac{1.308\ \text{yd}^3}{1\ \text{m}^3}$.

$$6\ \text{m}^3 = (6\ \text{m}^3)\left(\frac{1.308\ \text{yd}^3}{1\ \text{m}^3}\right)$$

$$= 6(1.308)\ \text{yd}^3$$

Therefore, 6 m³ = 7.85 yd³.

c. To convert from 781 cc to quarts we are converting from SI measurement to USCS measurement. (Note that cc is an abbreviation for cubic centimeters, which is also abbreviated as cm³.) From Table A.3, we see that the conversion of cm³ from SI to USCS is 1 cm³ = 6.102 × 10⁻² in³. However, in Table A.4, we see that 1 ft³ = 7.481 gal and that 1 gal = 4 qt. We also know that 1 ft = 12 in. Using these form conversion factors such that when the original units are multiplied by the conversion factors, we end up with the units we desire.

$$781\ \text{cm}^3 = (781\ \text{cm}^3)\left(\frac{6.102 \times 10^{-2}\ \text{in}^3}{1\ \text{cm}^3}\right)$$

$$\times \left(\frac{1\ \text{ft}}{12\ \text{in.}}\right)^3\left(\frac{7.481\ \text{gal}}{1\ \text{ft}^3}\right)\left(\frac{4\ \text{qt}}{1\ \text{gal}}\right)$$

$$= 781(6.102 \times 10^{-2})\left(\frac{1}{12}\right)^3(7.481)(4)\ \text{qt}$$

Therefore, 781 cm³ = 0.825 qt. ◆

Review Exercises

In Exercises 1–10, match the description to the prefix.

1. deci- _____
2. milli- _____
3. hecto- _____
4. tera- _____
5. mega- _____
6. nano- _____
7. centi- _____
8. micro- _____
9. deka- _____
10. kilo- _____

a. 1000 times base unit
b. $\frac{1}{100}$ of base unit
c. $\frac{1}{1000}$ of base unit
d. 10^{-6} times the base
e. 100 times the base unit
f. 10^{12} times the base unit
g. $\frac{1}{10}$ of the base
h. 10^{6} times the base unit
i. 10^{-9} times the base unit
j. 10 times the base unit

11. In the same length of time, Carlos ran 100 m, while Yolanda ran 100 yd. Who ran faster?

12. Vonda and Melvin are driving the exact route from point A to point B. If both start at the same time and Melvin drives his motorcycle at an average rate of 46.6 mph and Vonda drives her car at an average rate of 75.0 km/hr, which one arrives at point B first?

In Exercises 13–20, fill in the missing values.

13. 2 m = _____ cm
14. 2.4 MΩ = _____ Ω
15. 1.34 mℓ = _____ kℓ
16. 1.378 mg = _____ Mg
17. 2.345 dam = _____ mm
18. 2589 nm = _____ km
19. 14.27 hℓ = _____ ℓ
20. 0.0567 mm = _____ m

In Exercises 21–24, arrange each set of measures in order from smallest to largest.

21. 64 dm, 67 cm, 680 mm
22. 5.6 dam, 0.47 km, 620 cm
23. 4.3 ℓ, 420 cℓ, 0.045 kℓ
24. 2.2 kg, 2 4000 g, 24 3000 dg

25. Cameron cut a piece of lumber 7.8 meters in length into six equal pieces. What was the length of each piece for each of the following?
    a. Meters
    b. Centimeters

26. The distance between each base on a baseball diamond is 27 m. If the batter hits a home run, how far does he run for each of the following?
    a. Meters
    b. Kilometers
    c. Millimeters

27. At 70 cents per meter, what will it cost Karen to buy lace to use as a border around a bedspread that measures 76.0 in. by 45.6 in.?

28. The filter pump on an aquarium circulates 360 mℓ of water every minute. If the aquarium holds 30 ℓ of water, how long will it take to circulate all the water?

29. Dale drove 1200 km and used 187 ℓ of gasoline. What was her average rate of gas used for the trip for the following? In
   a. km/ℓ
   b. mi/gal

30. A mixture of 15 g of salt and 16 g of baking soda is poured into 250 mℓ of water. What is the total mass of the mixture in grams?

31. The following question was selected from a nursing exam. Can you answer it? In caring for a patient after delivery, you are to give 0.2 mg Ergotrute. The ampule is labeled $\frac{1}{300}$ grain/mℓ. How much would you draw and give? (60 mg = 1 grain)
   a. 15 cc   b. 1.0 cc   c. 0.5 cc   d. 0.01 cc

Appendix B Scientific Calculator Introduction

There are two general types of scientific handheld calculators used in technical mathematics courses. These calculators do not have the graphics capability that is discussed in the text. The difference between the two types is the method of entry. A calculator with the algebraic entry has an $\boxed{=}$ key. The machine with Reverse Polish Notation (RPN) has an $\boxed{\text{ENTER}}$ key. Both graphics calculators discussed in the text are algebraic entry.

The two systems of entry have one common basic feature; they use real numbers in decimal form. The machines operate with decimal approximations of all real numbers correct to the capacity of the particular machine. Many scientific calculators do not handle imaginary numbers. For example, try finding the solution to $\sqrt{-4}$ on your calculator. The machine either blinks at you, says "error" or has some other way of saying "no solution on this machine." Another way of discovering the error symbol is to press the keys $\boxed{0}$ and $\boxed{1/x}$. Thus, whenever this error symbol appears, you have performed an unacceptable operation.

Basic Operations

At this time we assume that you are somewhat familiar with the keyboard of your calculator. The first example shows how to add, subtract, multiply, and divide real numbers. The only way to develop an understanding is to turn on your calculator and press the proper keys as you read.

EXAMPLE 1 **Calculator Keystrokes**

Evaluate each of the given expressions using a calculator.

a. $7 + 6$ **b.** $17 - (-15)$ **c.** $(-18)(5)$ **d.** $18 \div (-3)$

Solution   An operation performed by the calculator is indicated with a rectangle drawn around the symbol to form a key.

a.   Algebraic: 7 $\boxed{+}$ 6 $\boxed{=}$ $\boxed{13}$
RPN: 7 $\boxed{\text{ENTER}}$ 6 $\boxed{+}$ $\boxed{13}$

b.   Algebraic: 17 $\boxed{-}$ 15 $\boxed{+/-}$ $\boxed{=}$ $\boxed{32}$
RPN: 17 $\boxed{\text{ENTER}}$ 15 $\boxed{\text{CHS}}$ $\boxed{-}$ $\boxed{32}$

c.   Algebraic: 18 $\boxed{+/-}$ $\boxed{\times}$ 5 $\boxed{=}$ $\boxed{-90}$
RPN: 18 $\boxed{\text{CHS}}$ $\boxed{\text{ENTER}}$ 5 $\boxed{\times}$ $\boxed{-90}$

d.   Algebraic: 18 $\boxed{\div}$ 3 $\boxed{+/-}$ $\boxed{=}$ $\boxed{-6}$
RPN: 18 $\boxed{\text{ENTER}}$ 3 $\boxed{\text{CHS}}$ $\boxed{\div}$ $\boxed{-6}$ ◆

The order of operations for most scientific calculators is such that they automatically perform the operations in the order stated in Section 1.3. Example 2 illustrates the keystrokes used to determine roots and powers.

EXAMPLE 2   **Keystrokes for Powers and Roots**

Perform the indicated operations using a calculator.

a. 12^2 **b.** $3^2 \cdot 5^2$
c. $\sqrt{7}$ **d.** $\sqrt{7} \cdot \sqrt{343}$
e. $\dfrac{\sqrt{2}}{\sqrt{3}}$ **f.** $\sqrt{\dfrac{2}{3}}$

Solution

a.   Algebraic: 12 $\boxed{x^2}$ $\boxed{144}$
RPN*: 12 $\boxed{x^2}$ $\boxed{144}$

b.   Algebraic: 3 $\boxed{x^2}$ $\boxed{\times}$ 5 $\boxed{x^2}$ $\boxed{=}$ $\boxed{225}$
RPN: 3 $\boxed{x^2}$ 5 $\boxed{x^2}$ $\boxed{\times}$ $\boxed{225}$

c.   Algebraic†: 7 $\boxed{\sqrt{x}}$ $\boxed{2.6457513}$
RPN: 7 $\boxed{\sqrt{x}}$ $\boxed{2.6457513}$

d.   Algebraic: 7 $\boxed{\sqrt{x}}$ $\boxed{\times}$ 343 $\boxed{\sqrt{x}}$ $\boxed{=}$ $\boxed{49}$
RPN: 7 $\boxed{\sqrt{x}}$ 343 $\boxed{\sqrt{x}}$ $\boxed{\times}$ $\boxed{49}$

e.   Algebraic: 2 $\boxed{\sqrt{x}}$ $\boxed{\div}$ 3 $\boxed{\sqrt{x}}$ $\boxed{=}$ $\boxed{0.8164966}$
RPN: 2 $\boxed{\text{ENTER}}$ $\boxed{\sqrt{x}}$ 3 $\boxed{\sqrt{x}}$ $\boxed{\div}$ $\boxed{0.8164966}$

f.   Algebraic: $\boxed{(}$ 2 $\boxed{\div}$ 3 $\boxed{)}$ $\boxed{\sqrt{x}}$ $\boxed{=}$ $\boxed{0.8164966}$
RPN: 2 $\boxed{\text{ENTER}}$ 3 $\boxed{\div}$ $\boxed{\sqrt{x}}$ $\boxed{0.8164966}$ ◆

With the algebraic type entry in part f, we may be tempted to do the following:

$$2 \boxed{\div} 3 \boxed{\sqrt{x}} \boxed{=} .$$

*Some RPN calculators require shifts before performing certain operations.

† Some algebraic entry type calculators require pressing $\boxed{\text{2nd}}$ before $\boxed{\sqrt{x}}$.

The result is $2 \div \sqrt{3}$. Due to the logic of the calculator, it finds the $\sqrt{3}$ first, then divides 2 by $\sqrt{3}$. Thus, we must use parentheses.

Understanding the logic, or the order of performance of operations, of your calculator is always important. However, it becomes more important when performing chain operations.

EXAMPLE 3 Keystrokes for Chain Operations

Perform the indicated operations using a calculator.

a. $(-11)(3)(-6)$

b. $\dfrac{(-6)(8)}{12}$

c. $(-6)(4) + 9 \div 3 + 6$

d. $\dfrac{(4)(-8) - 5(-4)}{4 - 7}$

Solution

a. Algebraic: 11 $\boxed{+/-}$ $\boxed{\times}$ 3 $\boxed{\times}$ 6 $\boxed{+/-}$ $\boxed{=}$ 198
   RPN: 11 $\boxed{\text{CHS}}$ $\boxed{\text{ENTER}}$ 3 $\boxed{\times}$ 6 $\boxed{\text{CHS}}$ $\boxed{\times}$ 198

b. Algebraic: 6 $\boxed{+/-}$ $\boxed{\times}$ 8 $\boxed{\div}$ 12 $\boxed{=}$ -4
   RPN: 6 $\boxed{\text{CHS}}$ $\boxed{\text{ENTER}}$ 8 $\boxed{\times}$ 12 $\boxed{\div}$ -4

c. Algebraic: 6 $\boxed{+/-}$ $\boxed{\times}$ 4 $\boxed{+}$ 9 $\boxed{\div}$ 3 $\boxed{+}$ 6 $\boxed{=}$ -15
   RPN: 6 $\boxed{\text{CHS}}$ $\boxed{\text{ENTER}}$ 4 $\boxed{\times}$ 9 $\boxed{\text{ENTER}}$ 3 $\boxed{\div}$ $\boxed{+}$ 6 $\boxed{+}$ -15

The calculator multiplies $-6 \times 4 = -24$, sets this product aside, then divides $9 \div 3 = 3$, recalls the -24 and adds this to 3, which yields -21, and finally adds -21 and 6 for a result of -15.

d. The immediate response may be to work the problem in the following manner:

Algebraic: 4 $\boxed{\times}$ 8 $\boxed{+/-}$ $\boxed{-}$ 5 $\boxed{\times}$ 4 $\boxed{+/-}$ $\boxed{\div}$ 4 $\boxed{-}$ 7 $\boxed{=}$ -34

The answer, −34, is wrong. The calculator multiplies and divides, then it adds and subtracts. To obtain the correct result, we must do the problem in two parts: Obtain the result in the numerator of the fraction and then the result in the denominator. Then find the quotient of the two results. We can accomplish this with the calculator using parentheses or the storage unit as the following examples illustrate.

Algebraic: $\boxed{(}$ 4 $\boxed{\times}$ 8 $\boxed{+/-}$ $\boxed{-}$ 5 $\boxed{\times}$ 4 $\boxed{+/-}$ $\boxed{)}$ $\boxed{\div}$ $\boxed{(}$ 4 $\boxed{-}$ 7 $\boxed{)}$
   $\boxed{=}$ 4

Algebraic: 4 $\boxed{\times}$ 8 $\boxed{+/-}$ $\boxed{-}$ 5 $\boxed{\times}$ 4 $\boxed{+/-}$ $\boxed{=}$
   $\boxed{\text{STO 1}}$ 4 $\boxed{-}$ 7 $\boxed{=}$ -3 $\boxed{1/x}$ $\boxed{\times}$ $\boxed{\text{RCL 1}}$ $\boxed{=}$ 4

RPN: 4 $\boxed{\text{ENTER}}$ 8 $\boxed{\text{CHS}}$ $\boxed{\times}$ 5 $\boxed{\text{ENTER}}$ 4 $\boxed{\text{CHS}}$
   $\boxed{\times}$ $\boxed{-}$ 4 $\boxed{\text{ENTER}}$ 7 $\boxed{-}$ $\boxed{\div}$ 4 ◆

In the remaining discussions, the keystroke explanations generally are limited to the algebraic scientific calculator.

Algebraic calculators have a $\boxed{y^x}$ or $\boxed{x^y}$ key that is used to raise a number to a power other than 2. This is illustrated in Example 4.

EXAMPLE 4 Using the y^x Key

Determine the numerical value of the expressions.

a. $\dfrac{4^{\frac{5}{3}}}{3^3}$ **b.** $(13^3 x^7 y^{\frac{1}{3}})^{\frac{2}{9}}$, $x = 0.5$, $y = 1.3$

Solution

a. $4\;\boxed{y^x}\;\boxed{(}\;5\;\boxed{\div}\;3\;\boxed{)}\;\boxed{\div}\;3\;\boxed{y^x}\;3\;\boxed{=}\;\boxed{0.3733099}$

or $4\;\boxed{y^x}\;\boxed{(}\;5\;\boxed{\div}\;3\;\boxed{)}\;\boxed{\times}\;3\;\boxed{y^x}\;3\;\boxed{+/-}\;\boxed{=}\;\boxed{0.3733099}$

since $\dfrac{4^{\frac{5}{3}}}{3^3} = 4^{\frac{5}{3}} \cdot 3^{-3}$

b. $13\;\boxed{x^2}\;\boxed{\times}\;.5\;\boxed{y^x}\;\boxed{(}\;14\;\boxed{\div}\;3\;\boxed{)}\;\boxed{\times}$

$1.3\;\boxed{y^x}\;\boxed{(}\;2\;\boxed{\div}\;9\;\boxed{)}\;\boxed{=}\;\boxed{7.0534372}$ ◆

We can use the calculator to solve problems in scientific notation. The keys with the symbols $\boxed{EE}$, $\boxed{EXP}$, or $\boxed{EEX}$ are used to enter a number in scientific notation. The number 8×10^6 is entered into the calculator as shown in the box that also indicates the screen display.

Keystroke	Display
8	8
$\boxed{EE}$	8 00
6	8 06

The solution of $8 \times 10^6 \times 7 \times 10^3 \div 2 \times 10^{-4}$ using the calculator is as follows.

Keystrokes	Display
$8\;\boxed{EE}\;6\;\boxed{\times}\;7\;\boxed{EE}\;3\;\boxed{\div}\;2\;\boxed{EE}\;4\;\boxed{+/-}\;\boxed{=}$	2.8 14

The answer is 2.8×10^{14}.

EXAMPLE 5 Scientific Notation

Perform the indicated operations on the calculator using scientific notation.

$$\frac{(5.36 \times 10^{12})^3}{(1.82 \times 10^{-7})(4.63 \times 10^{-9})}$$

Solution The solution using the calculator is as follows

Keystrokes	Display
5.36 [EE] 12 [y^x] 3 [÷] 1.82 [EE] 7 [+/−] [÷] 4.63	
[EE] 9 [+/−] [=]	1.8274352 53

Thus, $\dfrac{(5.36 \times 10^{12})^3}{(1.82 \times 10^{-7})(4.63 \times 10^{-9})} = 1.83 \times 10^{53}$ ◆

Trigonometric Functions

On a scientific calculator, we use trigonometric keys labeled [SIN], [COS], and [TAN] to compute function values for these functions. To find sin 0, enter the 0 value on the keyboard and then press [SIN]. We must instruct the calculator whether the angle measure is in radians or degrees. Many calculators are in **degree mode** when they are first turned on. Most calculators have a key labeled [DRG] (check your instruction book), which we use to change the operation of the calculator to the **radian mode** and back to the degree mode.

EXAMPLE 6   **Calculating Values for Sine**

Compute to three decimal places using a scientific calculator.

a. sin 14.3°   **b.** sin 14.3

Solution

a. The degree symbol next to the 3 in the expression sin 14.3° tells us that we want the sine of 14.3 degrees. Make sure the calculator is in the degree mode, then enter 14.3 and press [SIN]. The calculator will display 0.246999. Round off to 0.247.

b. Since there is no degree symbol next to the 3, we are asked to find the sine of 14.3 rad. Make sure the calculator is in the radian mode and enter 14.3, then press [SIN]. The calculator displays 0.986772. Round off to 0.987. ◆

Note that the trigonometric keys operate only on the value in the display. Therefore, to find the trigonometric value of a result of a calculation, the calculation must be completed and be in the display before pressing the trigonometric key.

EXAMPLE 7   Calculating Values for Cosine and Tangent

Compute to five decimal places.

a. $\cos\left[30° + 2(360°)\right]$ **b.** $\tan\dfrac{3\pi}{7}$

Solution

a. Place the calculator in the degree mode. Compute $[30 + 2(360)]$ and observe the result 750 in the display. Then press COS . The result to five decimal places is 0.86603.

b. Place the calculator in the radian mode. Compute $\dfrac{3\pi}{7}$ and observe the result 1.3463969 in the display. Then press TAN . The result to five decimal places is 4.38129. ◆

Now that we can find the trigonometric functional values of sine, cosine, and tangent, we can perform the inverse operation and determine the angle when given the functional value. Section 5.4 discussed this concept for simple algebraic functions. The procedure for trigonometric functions is easier since we need learn only a few steps on the calculator or learn how to read a table. (A rigorous definition of inverse trigonometric functions is contained in Chapter 15.)

The sequence of keystrokes for finding the angle θ is different on different calculators. The keys on the calculators may be: ARC , INV , SIN⁻¹ , COS⁻¹ , and TAN⁻¹ . In addition, you may have to use a color-coded key or the 2nd key. Consult your handbook for specific details. For example, if the trigonometric function of θ is $\sin\theta = 0.5$, then you can find θ on most calculators by using one of the following sequences:

$$0.5\ \boxed{\text{INV}}\ \boxed{\text{SIN}}\quad\text{or}\quad 0.5\ \boxed{\text{ARC}}\ \boxed{\text{SIN}}\quad\text{or}\quad 0.5\ \boxed{\text{SIN}^{-1}}.$$

If the calculator is in the degree mode, the number 30 appears on the display. If it is in the radian mode, the number 0.5235988 appears.

EXAMPLE 8   Determining the Angle

Use your calculator to verify the following. The measure in degrees is given to the nearest tenth, and the measure in radians is given to the nearest hundredth.

a. If $\sin\theta = 0.45879$, then in degrees, $\theta = 27.309056$, or $\theta = 27.3°$; in radians, $\theta = 0.4766329$, or $\theta = 0.48$.

b. If $\tan\theta = 11.32001$, then in degrees, $\theta = 84.951645$, or $\theta = 85.0°$; in radians, $\theta = 1.4826859$, or $\theta = 1.48$.

We can check each of these results by taking the value obtained for θ (not rounded off) and pressing the function key. For example,

$$27.309056\ \boxed{\text{SIN}}\ 0.45879.\quad◆$$

The discussion up to this point has been limited to the three functions: sine, cosine, and tangent. The discussion of the reciprocal functions cosecant, secant, and cotangent has been delayed because they do not appear on the calculator. In Section 6.2 we discovered that

$$\sin \theta = \frac{1}{\csc \theta}, \quad \sec \theta = \frac{1}{\cos \theta}, \quad \cot \theta = \frac{1}{\tan \theta}.$$

With these relationships, we can use the calculator to calculate the cosecant θ, secant θ, and cotangent θ even though there are no keys with these names.

EXAMPLE 9   Calculating Values for csc θ and sec θ

Compute to five decimal places.

a.   csc 32.7°   **b.**   sec 0.27

Solution

a.   The statement of equality $\csc \theta = \dfrac{1}{\sin \theta}$ provides us with a method for solving the problem. Find the value of $\sin \theta$ and then find the reciprocal of this value by pressing $\boxed{1/x}$. $\left(\text{In other words, csc } 32.7° = \dfrac{1}{\sin 32.7°}.\right)$ To find csc 32.7°, make sure that the calculator is in the degree mode, then press, 32.7 $\boxed{\text{SIN}}$ $\boxed{1/x}$.
The calculator displays 1.851028. Round off to 1.85103. *Warning:* Be sure that you see a display after pressing $\boxed{\text{SIN}}$ and before you press $\boxed{1/x}$. If you press them too rapidly, you may obtain a wrong value.

b.   The statement of equality $\sec \theta = \dfrac{1}{\cos \theta}$ provides us with a method for solving the problem. Find the value of $\cos \theta$ and then find the reciprocal of this value by pressing $\boxed{1/x}$. $\left(\text{In other words sec } 0.27 = \dfrac{1}{\cos 0.27}.\right)$ To find sec 0.27, make sure that the calculator is in the radian mode and press 0.27 $\boxed{\text{COS}}$ $\boxed{1/x}$.

The calculator will display 1.037591, which rounded to five decimal places is 1.03759. ◆

In determining the angle given the functional value, the thing to remember about this procedure is to find the reciprocal first and then find the inverse of the function. By finding the reciprocal first we are changing the functional value of cosecant, secant, or cotangent to sine, cosine, or tangent, respectively, before finding the inverse. This is essential since we have only sine, cosine, and tangent on the calculator.

EXAMPLE 10   Determing the Angle

Use your calculator to verify the following. The measure in degrees is given to the nearest tenth and the measure in radians is given to the nearest hundredth.

a. If cot θ = 2.41421, then in degrees, θ = 22.50003, or θ = 22.5; in radians, θ = 0.3926996, or θ = 0.39.

b. If sec θ = 1.12738, then in degrees, θ = 27.49981, or θ = 27.5°; in radians, θ = 0.4799622, or θ = 0.48.

Solution   We may check each of the results by finding the reciprocal functional value for θ (not rounded off) and then pressing $\boxed{1/x}$.

For example, 22.50003 $\boxed{\text{TAN}}$ $\boxed{1/x}$ 2.41421, which checks part a.   ◆

Logarithms

EXAMPLE 11   Calculating log x

Find the logarithm to the base 10 of **a.** 1.73 and **b.** 17,300 to four decimal places using the scientific calculator.

Solution

a.   Using the calculator to find the log 1.73, we have:

$$1.73 \boxed{\text{LOG}} \boxed{0.2380461}.$$

Therefore, log 1.73 = 0.2380.

b.   Using the calculator to find the log 17,300, we have:

$$17,300 \boxed{\text{LOG}} \boxed{4.2380461}.$$

Therefore, log 17,300 = 4.2380.   ◆

EXAMPLE 12   Calculating log x

Find the logarithm to the base 10 of **a.** 0.173 and **b.** 0.00173 to four decimal places using the scientific calculator.

Solution

a.   Using the calculator to find log 0.173, we obtain:

$$0.173 \boxed{\text{LOG}} \boxed{-0.7619539}$$

Therefore, log 0.173 = −0.7620.

b.   0.00173 $\boxed{\text{LOG}}$ $\boxed{-2.7619539}$

Therefore, log 0.00173 = −2.7620.   ◆

As Chapter 13 explains, the logarithmic system with base e = 2.71828 . . . is frequently encountered in equations describing natural phenomena. Logarithms to base e are called natural logarithms or Naperian logarithms. Because they occur frequently, the symbol ln x is used in place of $\log_e x$. Natural logarithms may be found by transformation from base 10 or by direct use of the calculator.

EXAMPLE 13   Calculating ln x

Find ln 17.3 using the following.

a. Transformation from base 10
b. Direct calculator keystrokes

Solution

a. Using Definition 13.6 and recalling that $e = 2.71828 \ldots$,

$$\ln 17.3 = \log_e 17.3$$

$$= \frac{\log 17.3}{\log 2.71828}$$

$$= \frac{1.238}{0.4343}$$

$$= 2.851 \quad \text{or} \quad 2.85.$$

b. By direct calculator keystrokes,

17.3 | ln x | 2.8507065   or   ln 17.3 = 2.85. ◆

EXAMPLE 14   Antilogarithm

Use the calculator to find the natural antilogarithm of 6.81.

Solution   The natural antilogarithm of 6.81 can be found using an approach like that used to find the antilogarithm of a common logarithm (in Section 13.4). To find antiln 6.81 using the calculator, the keystrokes are:

6.81 | INV | ln x | 906.87081   or   6.81 | e^x | 906.87081

Therefore, antiln 6.81 = 907. ◆

Review Exercises

In Exercises 1–16, evaluate each of the expressions using a scientific calculator.

1. $17^2 \cdot \pi^2$

2. $5 - (\sqrt{7})(5.986)^2$

3. $\dfrac{\sqrt{13}}{\sqrt{23}}$

4. $\sqrt{\dfrac{17}{59}}$

5. $\dfrac{\sqrt{17} - \sqrt{33}}{\sqrt{72}}$

6. $\sqrt{\dfrac{23.875 - 13.543}{47.912}}$

7. $(-2.345)(11.368)(-15.954)$

8. $(\sqrt{5.25})(-3.423)(-13.467)$

9. $\dfrac{(-11.23)(15.897)}{13.456}$

10. $\dfrac{-33.89}{(-27.96)(4.567)}$

11. $(2.34 \times 10^8)(5.987 \times 10^{13})$

12. $(2.345 \times 10^{-15})(7.893 \times 10^{-15})$

13. $(2.456 \times 10^{15})^2(4.987 \times 10^{11})^3$

14. $(5.35 \times 10^9)^4(6.985 \times 10^{-11})^4$

15. $\dfrac{(5.3 \times 10^8)(6.87 \times 10^7)^2}{1.85 \times 10^5}$

16. $125{,}000{,}000{,}000 \times 3{,}690{,}000{,}000{,}000{,}000$

In Exercises 17–32, compute the functional value to five decimal places.

17. $\sin 13°$

18. $\sin 79.3°$

19. $\cos 59°$

20. $\cos 13.2°$

21. $\tan 18°$

22. $\tan 1.34°$

23. $\sec 57°$

24. $\cos \dfrac{3\pi}{4}$

25. $\cot 53°$

26. $\tan \dfrac{5\pi}{6}$

27. $\csc 34°$

28. $\sin \dfrac{3\pi}{4}$

29. $\sin 3.4$

30. $\cos 2.5$

31. $\tan 1.3$

32. $\sin 1.3$

In Exercises 33–42, determine the angle measure to the nearest tenth of a degree.

33. $\sin \theta = 0.54464$

34. $\sin \theta = 0.83867$

35. $\cos \theta = 0.83867$

36. $\cos \theta = 0.54464$

37. $\tan \theta = 0.75355$

38. $\tan \theta = 1.11061$

39. $\csc \theta = 1.83608$

40. $\sec \theta = 1.83867$

41. $\cot \theta = 1.32705$

42. $\cos \theta = 0.50000$

In Exercises 43–48, determine the logarithm to the base 10 of the given number to four decimal places.

43. 1.73

44. 9.86

45. 2.86×10^5

46. 0.2345

47. 0.000777

48. 1.65×10^{-5}

In Exercises 49–54, determine the antilog, base 10, of the given number to three significant digits.

49. 0.7654

50. 0.9863

51. 2.9876

52. 3.9876

53. 4.3452

54. -2.6543

In Exercises 55–60, determine the natural logarithm of the given number to four decimal places.

55. 2.34

56. 8.95

57. 0.234

58. 1.345

59. 2.987

60. 1.34×10^4

Appendix C Areas of a Standard Normal Distribution

The table entries represent the area under the standard normal curve from 0 to the specified value of z. (See the illustration.)

z	.00	.01	.02	.03	.04	.05	.06	.07	.08	.09
0.0	.0000	.0040	.0080	.0120	.0160	.0199	.0239	.0279	.0319	.0359
0.1	.0398	.0438	.0478	.0517	.0557	.0596	.0636	.0675	.0714	.0753
0.2	.0793	.0832	.0871	.0910	.0948	.0987	.1026	.1064	.1103	.1141
0.3	.1179	.1217	.1255	.1293	.1331	.1368	.1406	.1443	.1480	.1517
0.4	.1554	.1591	.1628	.1664	.1700	.1736	.1772	.1808	.1844	.1879
0.5	.1915	.1950	.1985	.2019	.2054	.2088	.2123	.2157	.2190	.2224
0.6	.2257	.2291	.2324	.2357	.2389	.2422	.2454	.2486	.2517	.2549
0.7	.2580	.2611	.2642	.2673	.2704	.2734	.2764	.2794	.2823	.2852
0.8	.2881	.2910	.2939	.2967	.2995	.3023	.3051	.3078	.3106	.3133
0.9	.3159	.3186	.3212	.3238	.3264	.3289	.3315	.3340	.3365	.3389
1.0	.3413	.3438	.3461	.3485	.3508	.3531	.3554	.3577	.3599	.3621
1.1	.3643	.3665	.3686	.3708	.3729	.3749	.3770	.3790	.3810	.3830
1.2	.3849	.3869	.3888	.3907	.3925	.3944	.3962	.3980	.3997	.4015
1.3	.4032	.4049	.4066	.4082	.4099	.4115	.4131	.4147	.4162	.4177
1.4	.4192	.4207	.4222	.4236	.4251	.4265	.4279	.4292	.4306	.4319
1.5	.4332	.4345	.4357	.4370	.4382	.4394	.4406	.4418	.4429	.4441
1.6	.4452	.4463	.4474	.4484	.4495	.4505	.4515	.4525	.4535	.4545
1.7	.4554	.4564	.4573	.4582	.4591	.4599	.4608	.4616	.4625	.4633
1.8	.4641	.4649	.4656	.4664	.4671	.4678	.4686	.4693	.4699	.4706
1.9	.4713	.4719	.4726	.4732	.4738	.4744	.4750	.4756	.4761	.4767
2.0	.4772	.4778	.4783	.4788	.4793	.4798	.4803	.4808	.4812	.4817
2.1	.4821	.4826	.4830	.4834	.4838	.4842	.4846	.4850	.4854	.4857
2.2	.4861	.4864	.4868	.4871	.4875	.4878	.4881	.4884	.4887	.4890
2.3	.4893	.4896	.4898	.4901	.4904	.4906	.4909	.4911	.4913	.4916
2.4	.4918	.4920	.4922	.4925	.4927	.4929	.4931	.4932	.4934	.4936
2.5	.4938	.4940	.4941	.4943	.4945	.4946	.4948	.4949	.4951	.4952
2.6	.4953	.4955	.4956	.4957	.4959	.4960	.4961	.4962	.4963	.4964
2.7	.4965	.4966	.4967	.4968	.4969	.4970	.4971	.4972	.4973	.4974
2.8	.4974	.4975	.4976	.4977	.4977	.4978	.4979	.4979	.4980	.4981
2.9	.4981	.4982	.4982	.4983	.4984	.4984	.4985	.4985	.4986	.4986
3.0	.4987	.4987	.4987	.4988	.4988	.4989	.4989	.4989	.4990	.4990
3.1	.4990	.4991	.4991	.4991	.4992	.4992	.4992	.4992	.4993	.4993
3.2	.4993	.4993	.4994	.4994	.4994	.4994	.4994	.4995	.4995	.4995
3.3	.4995	.4995	.4995	.4996	.4996	.4996	.4996	.4996	.4996	.4997
3.4	.4997	.4997	.4997	.4997	.4997	.4997	.4997	.4997	.4997	.4998
3.5	.4998									
4.0	.49997									
4.5	.499997									
5.0	.4999997									

Appendix D Short List of Integrals

1. $\displaystyle\int u \, dv = uv - \int v \, du$

2. $\displaystyle\int a^u \, du = \frac{a^u}{\ln a} + C, \qquad a \neq 1, \qquad a > 0$

3. $\displaystyle\int \cos u \, du = \sin u + C$

4. $\displaystyle\int \sin u \, du = -\cos u + C$

5. $\displaystyle\int (ax + b)^n \, dx = \frac{(ax + b)^{n+1}}{a(n + 1)} + C, \qquad n \neq -1$

6. $\displaystyle\int (ax + b)^{-1} \, dx = \frac{1}{a} \ln|ax + b| + C$

7. $\displaystyle\int x(ax + b)^n \, dx = \frac{(ax + b)^{n+1}}{a^2}\left[\frac{ax + b}{n + 2} - \frac{b}{n + 1}\right] + C, \quad n \neq -1, -2$

8. $\displaystyle\int x(ax + b)^{-1} \, dx = \frac{x}{a} - \frac{b}{a^2} \ln|ax + b| + C$

9. $\displaystyle\int x(ax + b)^{-2} \, dx = \frac{1}{a^2}\left[\ln|ax + b| + \frac{b}{ax + b}\right] + C$

10. $\displaystyle\int \frac{dx}{x(ax + b)} = \frac{1}{b} \ln\left|\frac{x}{ax + b}\right| + C$

11. $\displaystyle\int (\sqrt{ax + b})^n \, dx = \frac{2}{a}\frac{(\sqrt{ax + b})^{n+2}}{n + 2} + C, \qquad n \neq -2$

12. $\displaystyle\int \frac{\sqrt{ax + b}}{x} \, dx = 2\sqrt{ax + b} + b \int \frac{dx}{x\sqrt{ax + b}}$

13. a. $\displaystyle \int \frac{dx}{x\sqrt{ax+b}} = \frac{2}{\sqrt{-b}} \tan^{-1}\sqrt{\frac{ax+b}{-b}} + C, \qquad \text{if } b < 0$

b. $\displaystyle \int \frac{dx}{x\sqrt{ax+b}} = \frac{1}{\sqrt{b}} \ln\left|\frac{\sqrt{ax+b}-\sqrt{b}}{\sqrt{ax+b}+\sqrt{b}}\right| + C, \qquad \text{if } b > 0$

14. $\displaystyle \int \frac{ax+b}{x^2} dx = -\frac{\sqrt{ax+b}}{x} + \frac{a}{2}\int \frac{dx}{x\sqrt{ax+b}} + C$

15. $\displaystyle \int \frac{dx}{x^2\sqrt{ax+b}} = -\frac{\sqrt{ax+b}}{bx} - \frac{a}{2b}\int \frac{dx}{x\sqrt{ax+b}} + C$

16. $\displaystyle \int \frac{dx}{a^2+x^2} = \frac{1}{a}\tan^{-1}\frac{x}{a} + C$

17. $\displaystyle \int \frac{dx}{(a^2+x^2)^2} = \frac{x}{2a^2(a^2+x^2)} + \frac{1}{2a^3}\tan^{-1}\frac{x}{a} + C$

18. $\displaystyle \int \frac{dx}{a^2-x^2} = \frac{1}{2a}\ln\left|\frac{x+a}{x-a}\right| + C$

19. $\displaystyle \int \frac{dx}{(a^2-x^2)^2} = \frac{x}{2a^2(a^2-x^2)} + \frac{1}{2a^2}\int \frac{dx}{a^2-x^2}$

20. $\displaystyle \int \sqrt{a^2+x^2}\, dx = \frac{x}{2}\sqrt{a^2+x^2} + \frac{a^2}{2}\ln(x+\sqrt{x^2+a^2}) + C$

21. $\displaystyle \int \frac{dx}{x\sqrt{a^2+x^2}} = -\frac{1}{a}\ln\left|\frac{a+\sqrt{a^2+x^2}}{x}\right| + C$

22. $\displaystyle \int \frac{dx}{x^2\sqrt{a^2+x^2}} = -\frac{\sqrt{a^2+x^2}}{a^2x} + C$

23. $\displaystyle \int \sqrt{a^2-x^2}\, dx = \frac{x}{2}\sqrt{a^2-x^2} + \frac{a^2}{2}\sin^{-1}\frac{x}{a} + C$

24. $\displaystyle \int x^2\sqrt{a^2-x^2}\, dx = \frac{a^4}{8}\sin^{-1}\frac{x}{a} - \frac{1}{8}x\sqrt{a^2-x^2}(a^2-2x^2) + C$

25. $\displaystyle \int \frac{\sqrt{a^2-x^2}}{x} dx = \sqrt{a^2-x^2} - a\ln\left|\frac{a+\sqrt{a^2-x^2}}{x}\right| + C$

26. $\displaystyle \int \frac{\sqrt{a^2-x^2}}{x^2} dx = -\sin^{-1}\frac{x}{a} - \frac{\sqrt{a^2-x^2}}{x} + C$

27. $\displaystyle \int \frac{x^2}{\sqrt{a^2-x^2}} dx = \frac{a^2}{2}\sin^{-1}\frac{x}{a} - \frac{1}{2}x\sqrt{a^2-x^2} + C$

28. $\displaystyle \int \frac{dx}{x\sqrt{a^2-x^2}} = -\frac{1}{a}\ln\left|\frac{a+\sqrt{a^2-x^2}}{x}\right| + C$

29. $\displaystyle \int (\sqrt{x^2-a^2})^n\, dx = \frac{x(\sqrt{x^2-a^2})^n}{n+1} - \frac{na^2}{n+1}\int (\sqrt{x^2-a^2})^{n-2}\, dx, \qquad n \neq -1$

30. $\displaystyle\int \frac{dx}{(\sqrt{x^2 - a^2})^n} = \frac{x(\sqrt{x^2 - a^2})^{2-n}}{(2-n)a^2} - \frac{n-3}{(n-2)a^2} \int \frac{dx}{(\sqrt{x^2 - a^2})^{n-2}}$, $n \neq 2$

31. $\displaystyle\int x(\sqrt{x^2 - a^2})^n \, dx = \frac{(\sqrt{x^2 - a^2})^{n+2}}{n+2} + C$, $n \neq -2$

32. $\displaystyle\int \frac{x^2 - a^2}{x} \, dx = \sqrt{x^2 - a^2} - a \sec^{-1}\left|\frac{x}{a}\right| + C$

33. $\displaystyle\int \frac{dx}{x\sqrt{x^2 - a^2}} = \frac{1}{a} \sec^{-1}\left|\frac{x}{a}\right| + C = \frac{1}{a} \cos^{-1}\left|\frac{a}{x}\right| + C$

34. $\displaystyle\int \frac{dx}{x^2\sqrt{x^2 - a^2}} = \frac{\sqrt{x^2 - a^2}}{a^2 x} + C$

35. $\displaystyle\int \frac{dx}{\sqrt{2ax - x^2}} = \sin^{-1}\left(\frac{x-a}{a}\right) + C$

36. $\displaystyle\int \sqrt{2ax - x^2} \, dx = \frac{x-a}{2} \sqrt{2ax - x^2} + \frac{a^2}{2} \sin^{-1}\left(\frac{x-a}{a}\right) + C$

37. $\displaystyle\int (\sqrt{2ax - x^2})^n \, dx = \frac{(x-a)(2ax - x^2)^n}{n+1} + \frac{na^2}{n+1} \int (\sqrt{2ax - x^2})^{n-2} \, dx$

38. $\displaystyle\int \frac{dx}{(\sqrt{2ax - x^2})^n} = \frac{(x-a)(\sqrt{2ax - x^2})^{2-n}}{(n-2)a^2} + \frac{(n-3)}{(n-2)a^2} \int \frac{dx}{(\sqrt{2ax - x^2})^{n-2}}$

39. $\displaystyle\int x\sqrt{2ax - x^2} \, dx = \frac{(x+a)(2x - 3a)\sqrt{2ax - x^2}}{6} + \frac{a^3}{2} \sin^{-1}\frac{x-a}{a} + C$

40. $\displaystyle\int \frac{\sqrt{2ax - x^2}}{x} \, dx = \sqrt{2ax - x^2} + a \sin^{-1}\frac{x-a}{a} + C$

41. $\displaystyle\int \frac{\sqrt{2ax - x^2}}{x^2} \, dx = -2\sqrt{\frac{2a-x}{x}} - \sin^{-1}\left(\frac{x-a}{a}\right) + C$

42. $\displaystyle\int \frac{x \, dx}{\sqrt{2ax - x^2}} = a \sin^{-1}\frac{x-a}{a} - \sqrt{2ax - x^2} + C$

43. $\displaystyle\int \frac{dx}{x\sqrt{2ax - x^2}} = -\frac{1}{a}\sqrt{\frac{2a-x}{x}} + C$

44. a. $\displaystyle\int \sin ax \, dx = -\frac{1}{a} \cos ax + C$

　　b. $\displaystyle\int \cos ax = \frac{1}{a} \sin ax + C$

45. a. $\displaystyle\int \sin^2 ax\, dx = \frac{x}{2} - \frac{\sin 2ax}{4a} + C$

b. $\displaystyle\int \cos^2 ax\, dx = \frac{x}{2} + \frac{\sin 2ax}{4a} + C$

46. $\displaystyle\int \sin^n ax\, dx = \frac{-\sin^{n-1} ax \cos ax}{na} + \frac{n-1}{n}\int \sin^{n-2} ax\, dx$

47. $\displaystyle\int \cos^n ax\, dx = \frac{\cos^{n-1} ax \sin ax}{na} + \frac{n-1}{n}\int \cos^{n-2} ax\, dx$

48. a. $\displaystyle\int \sin ax \cos bx\, dx = -\frac{\cos (a+b)x}{2(a+b)} - \frac{\cos (a-b)x}{2(a-b)} + C, \qquad a^2 \neq b^2$

b. $\displaystyle\int \sin ax \sin bx\, dx = \frac{\sin (a-b)x}{2(a-b)} - \frac{\sin (a+b)x}{2(a+b)}, \qquad a^2 \neq b^2$

c. $\displaystyle\int \cos ax \cos bx\, dx = \frac{\sin (a-b)x}{2(a-b)} + \frac{\sin (a+b)x}{2(a+b)}, \qquad a^2 \neq b^2$

49. $\displaystyle\int \sin ax \cos ax\, dx = -\frac{\cos 2ax}{4a} + C$

50. $\displaystyle\int \sin^n ax \cos ax\, dx = \frac{\sin^{n+1} ax}{(n+1)a} + C, \qquad n \neq -1$

51. $\displaystyle\int \frac{\cos ax}{\sin ax}\, dx = \frac{1}{a} \ln|\sin ax| + C$

52. $\displaystyle\int \cos^n ax \sin ax\, dx = -\frac{\cos^{n+1} ax}{(n+1)a} + C, \qquad n \neq -1$

53. $\displaystyle\int \frac{\sin ax}{\cos ax}\, dx = -\frac{1}{a} \ln|\cos ax| + C$

54. $\displaystyle\int \sin^n ax \cos^m ax\, dx = -\frac{\sin^{n-1} ax \cos^{m+1} ax}{a(m+n)} + \frac{n-1}{m+n}\int \sin^{n-2} ax \cos^m ax\, dx, \qquad n \neq -m$

(If $n = -m$, use No. 68.)

55. $\displaystyle\int \sin^n ax \cos^m ax\, dx = \frac{\sin^{n+1} ax \cos^{m-1} ax}{a(m+n)} + \frac{m-1}{m+n}\int \sin^n ax \cos^{m-2} ax\, dx, \qquad m \neq -n$

(If $m = -n$, use No. 69.)

56. $\displaystyle\int \frac{dx}{b + c \sin ax} = \frac{-2}{a\sqrt{b^2 - c^2}} \tan^{-1}\left[\sqrt{\frac{b-c}{b+c}} \tan\left(\frac{\pi}{4} - \frac{ax}{2}\right)\right] + C, \qquad b^2 > c^2$

57. $\displaystyle \int \frac{dx}{b + c \sin ax} = \frac{-1}{a\sqrt{c^2 - b^2}} \ln \left| \frac{c + b \sin ax + \sqrt{c^2 - b^2} \cos ax}{b + c \sin ax} \right| + C, \qquad b^2 < c^2$

58. $\displaystyle \int \frac{dx}{1 + \sin ax} = -\frac{1}{a} \tan\left(\frac{\pi}{4} - \frac{ax}{2}\right) + C$

59. $\displaystyle \int \frac{dx}{1 - \sin ax} = \frac{1}{a} \tan\left(\frac{\pi}{4} + \frac{ax}{2}\right) + C$

60. $\displaystyle \int \frac{dx}{b + c \cos ax} = \frac{2}{a\sqrt{b^2 - c^2}} \tan^{-1}\left[\sqrt{\frac{b - c}{b + c}} \tan \frac{ax}{2}\right] + C, \quad b^2 > c^2$

61. $\displaystyle \int \frac{dx}{b + c \cos ax} = \frac{1}{a\sqrt{c^2 - b^2}} \ln \left| \frac{c + b \cos ax + \sqrt{c^2 - b^2} \sin ax}{a + c \cos ax} \right| + C, \qquad b^2 < c^2$

62. a. $\displaystyle \int \frac{dx}{1 + \cos ax} = \frac{1}{a} \tan \frac{ax}{2} + C$

b. $\displaystyle \int \frac{dx}{1 - \cos ax} = -\frac{1}{a} \cot \frac{ax}{2} + C$

63. a. $\displaystyle \int x \sin ax \, dx = \frac{1}{a^2} \sin ax - \frac{x}{a} \cos ax + C$

b. $\displaystyle \int x \cos ax \, dx = \frac{1}{a^2} \cos ax + \frac{x}{a} \sin ax + C$

64. $\displaystyle \int x^n \sin ax \, dx = -\frac{x^n}{a} \cos ax + \frac{n}{a} \int x^{n-1} \cos ax \, dx$

65. $\displaystyle \int x^n \cos ax \, dx = \frac{x^n}{a} \sin ax - \frac{n}{a} \int x^{n-1} \sin ax \, dx$

66. a. $\displaystyle \int \tan ax \, dx = -\frac{1}{a} \ln|\cos ax| + C$

b. $\displaystyle \int \cot ax \, dx = \frac{1}{a} \ln|\sin ax| + C$

67. a. $\displaystyle \int \tan^2 ax \, dx = \frac{1}{a} \tan ax - x + C$

b. $\displaystyle \int \cot^2 ax \, dx = -\frac{1}{a} \cot ax - x + C$

68. $\displaystyle \int \tan^n ax \, dx = \frac{\tan^{n-1} ax}{a(n - 1)} - \int \tan^{n-2} ax \, dx, \qquad n \neq 1$

69. $\displaystyle \int \cot^n ax \, dx = -\frac{\cot^{n-1} ax}{a(n - 1)} - \int \cot^{n-2} ax \, dx, \qquad n \neq 1$

70. $\displaystyle \int \sec ax \, dx = \frac{1}{a} \ln|\sec ax + \tan ax| + C$

71. $\displaystyle\int \csc ax\, dx = \frac{1}{a}\ln|\csc ax - \cot ax| + C$

72. $\displaystyle\int \sec^2 ax\, dx = \frac{1}{a}\tan ax + C$

73. $\displaystyle\int \csc^2 ax\, dx = -\frac{1}{a}\cot ax + C$

74. $\displaystyle\int \sec^n ax\, dx = \frac{\sec^{n-2} ax \tan ax}{a(n-1)} + \frac{n-2}{n-1}\int \sec^{n-2} ax\, dx, \qquad n \neq 1$

75. $\displaystyle\int \csc^n ax\, dx = -\frac{\csc^{n-2} ax \cot ax}{a(n-1)} + \frac{n-2}{n-1}\int \csc^{n-2} ax\, dx, \qquad n \neq 1$

76. $\displaystyle\int \sec^n ax \tan ax\, dx = \frac{\sec^n ax}{na} + C, \qquad n \neq 0$

77. $\displaystyle\int \csc^n ax \cot ax\, dx = -\frac{\csc^n ax}{na} + C, \qquad n \neq 0$

78. $\displaystyle\int \sin^{-1} ax\, dx = x \sin^{-1} ax + \frac{1}{a}\sqrt{1 - a^2x^2} + C$

79. $\displaystyle\int \cos^{-1} ax\, dx = x \cos^{-1} ax - \frac{1}{a}\sqrt{1 - a^2x^2} + C$

80. $\displaystyle\int \tan^{-1} ax\, dx = x \tan^{-1} ax - \frac{1}{2a}\ln(1 + a^2x^2) + C$

81. $\displaystyle\int x^n \sin^{-1} ax\, dx = \frac{x^{n+1}}{n+1}\sin^{-1} ax - \frac{a}{n+1}\int \frac{x^{n+1}\, dx}{\sqrt{1 - a^2x^2}}, \qquad n \neq -1$

82. $\displaystyle\int x^n \cos^{-1} ax\, dx = \frac{x^{n+1}}{n+1}\cos^{-1} ax + \frac{a}{n+1}\int \frac{x^{n+1}\, dx}{\sqrt{1 - a^2x^2}}, \qquad n \neq -1$

83. $\displaystyle\int x^n \tan^{-1} ax\, dx = \frac{x^{n+1}}{n+1}\tan^{-1} ax - \frac{a}{n+1}\int \frac{x^{n+1}}{1 + a^2x^2}, \qquad n \neq -1$

84. $\displaystyle\int e^{ax}\, dx = \frac{1}{a}e^{ax} + C$

85. $\displaystyle\int b^{ax}\, dx = \frac{1}{a}\frac{b^{ax}}{\ln b} + C, \qquad b > 0, b \neq 1$

86. $\displaystyle\int xe^{ax}\, dx = \frac{e^{ax}}{a^2}(ax - 1) + C$

87. $\displaystyle\int x^n e^{ax}\, dx = \frac{1}{a}x^n e^{ax} - \frac{n}{a}\int x^{n-1}e^{ax}\, dx$

88. $\displaystyle\int x^n b^{ax}\, dx = \frac{x^n b^{ax}}{a \ln b} - \frac{n}{a \ln b}\int x^{n-1}b^{ax}\, dx, \qquad b > 0, b \neq 1$

89. $\displaystyle\int e^{ax}\sin bx\, dx = \frac{e^{ax}}{a^2 + b^2}(a\sin bx - b\cos bx) + C$

90. $\displaystyle\int e^{ax}\cos bx\, dx = \frac{e^{ax}}{a^2 + b^2}(a\cos bx + b\sin bx) + C$

91. $\displaystyle\int \ln ax \, dx = x \ln ax - x + C$

92. $\displaystyle\int x^n \ln ax \, dx = \frac{x^{n+1}}{n+1} \ln ax - \frac{x^{n+1}}{(n+1)^2} + C, \qquad n \neq -1$

93. $\displaystyle\int x^{-1} \ln ax \, dx = \frac{1}{2}(\ln ax)^2 + C$

94. $\displaystyle\int \frac{dx}{x \ln ax} = \ln|\ln ax| + C$

Answers to Selected Exercises

Chapter 1

Section 1.1 (page 6)

1. Counting **3.** Counting **5.** Neither **7.** Negative integer **9.** Rational **11.** Rational **13.** Irrational **15.** Rational
17. Irrational **19.** Rational **21.** **23.** **25.**

27. **29.** **31.** **33.**

35. Complex **37.** Real **39.** Complex **41.** Pure imaginary **43.** Complex **45.** Pure imaginary **47.** Real
49.

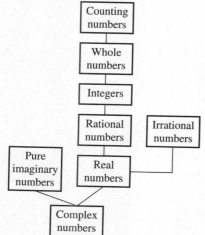

Section 1.2 (page 11)

1. $3 < 14$ **3.** $-\sqrt{3} < -\sqrt{2}$ **5.** $\sqrt{5} < 2.236068\ldots$ **7.** $\dfrac{14}{4} > 3\dfrac{1}{4}$ **9.** $\dfrac{1}{2} + \dfrac{1}{3} = \dfrac{3}{2} - \dfrac{2}{3}$

11. $x = -8, -4, 0$ **13.** $x = -12, -8, -4$ **15.** $x = -4, 0, 4$

$x < 1.4$

$$-12\ -8\ -4\quad 0\quad 4\quad 8\quad 12$$

$x < -2$

$$-12\ -8\ -4\quad 0\quad 4\quad 8\quad 12$$

$x \le 4$

$$-12\ -8\ -4\quad 0\quad 4\quad 8\quad 12$$

17. $x = 3, 4, 8$ **19.** $x = -10, 0, 12$ **21.** $x = -4, 0, 7$

$3 \le x < 11$

$$-12\ -8\ -4\quad 0\quad 4\quad 8\quad 12$$

$-10.4 \le x < 12.2$

$$-12\ -8\ -4\quad 0\quad 4\quad 8\quad 12$$

$x \le 1 \qquad x \ge 7$

$$-12\ -8\ -4\quad 0\quad 4\quad 8\quad 12$$

23. $x < 3$ **25.** $x \le -\sqrt{7}$ **27.** $-5 \le x \le 0$ **29.** $-3 \le x < 3$ **31.** $x \le -2$ or $x > 1$ **33.** 9   **35.** 4
37. 2   **39.** 4   **41.** 4   **43.** 4.361

Section 1.3 (page 20)

1. -6 **3.** 5   **5.** 7   **7.** $\dfrac{1}{3}$ **9.** $9 \cdot 8$ **11.** $(4 \cdot 6) \cdot 7$ **13.** $3 \cdot a + 3 \cdot b$ **15.** $(a + b) + 3$ **17.** $(3 + 14) + 19$

19. Additive identity   **21.** Commutative property of addition   **23.** Distributive property
25. Inverse property of addition   **27.** Distributive property   **29.** 13   **31.** -3 **33.** 6   **35.** -45 **37.** 12   **39.** -12

41. 0.45   **43.** 0.6   **45.** -7 **47. a.** $(5 - 3 = 2) \ne (3 - 5 = -2)$ **b.** $(6 \div 3 = 2) \ne \left(3 \div 6 = \dfrac{1}{2}\right)$ **49.** -44

51. 60   **53.** -336 **55.** 8   **57.** -10 **59.** -10 **61.** 1   **63.** 44   **65.** $55 - (-5) = 60°\text{F}$
67. $95.97 + 444.80 + (-54.75) + (-350.00) = 136.02$
69. The sum of the forces equal 0; the object remains at rest.

Section 1.4 (page 23)

1. 3   **3.** $\sqrt{6}$ **5.** 0   **7.** Undefined   **9.** 1   **11.** Indeterminate   **13.** No, indeterminate if $a = 0$
15. No, indeterminate if $a = b$ **17. a.** Error   **b.** A very large number

Section 1.5 (page 34)

1. -8 **3.** -44 **5.** -84 **7.** 4   **9.** 2   **11.** -1 **13.** 4   **15.** 1667.963144   **17.** 0.4837794468   **19.** -0.4615256084
21. 1845.096403   **23.** -5980.06399 **25.** 0.5694051198   **27.** Undefined   **29.** -12.16237379 **31.** -14.33304173
33. 681.3309245   **35.** 350212.1827 mm^3 **37.** 0.63661977 s   **39.** 0.3575616706
41.   **TI–82**   **fx–7700G**

Window Format	Range
Xmin $= -6$	Xmin: -6
Xmax $= 12$	max: 12
Xscl $= 2$	scl: 2
Ymin $= -3$	Ymin: -3
Ymax $= 8$	max: 8
Yscl $= 2$	scl: 2

43.

TI–82

> **Window Format**
> Xmin $= -2$
> Xmax $= 52$
> Xscl $= 5$
> Ymin $= -2$
> Ymax $= 34$
> Yscl $= 5$

fx–7700G

> Range
> Xmin: -2
> max: 52
> scl: 5
> Ymin: -2
> max: 34
> scl: 5

45. 48   **47.** 108

49.

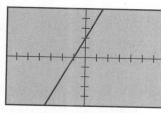

51.

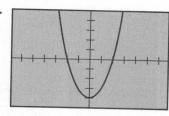

53.

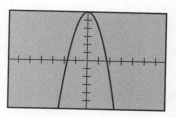

55.

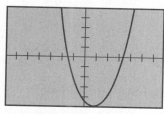

57.

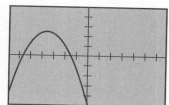

59.

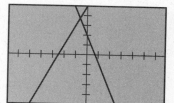

61.

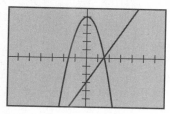

Section 1.6 (page 43)

1. 5.00%   **3.** 3.08%   **5.** 1.08%   **7.** 480   **9.** 2.3   **11.** 87   **13.** 0.0466   **15.** 4600   **17.** 820.3   **19.** 450.6
21. -489.20 **23.** 248.   **25.** 5000   **27. a.** 0.29   **b.** 0.286   **c.** 0.28571   **29.** 425.8 kg   **31.** 44.0 m²   **33.** 26.000 in.
35. 0.7786   **37.** 44.5   **39.** -30. cm

Review Exercises (page 45)

[1.1] *****1.** Rational, real   **3.** Rational, real   **5.** Pure imaginary   **7.** Rational, real   **9.** Complex
[1.2] **11.** $<$ **13.** $=$ **15.** $<$ **17.** $=$
19. $x = 5, 6, 7$ **21.** $x = -6, -5, -4$ **23.** $x = -8, 0, 10$ **25.** $x = -3, 0, 4$

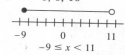

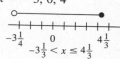

*[1.1] indicates section where material is discussed.

27. 3   **29.** 3   **31.** $\dfrac{5}{12} \cong 0.416667$ [1.3] **33.** Identity property of addition   **35.** Multiplicative inverse

37. Distributive property of multiplication over addition   **39.** -21 **41.** -5 [1.4] **43.** 0   **45.** Indeterminate
[1.5] **47.** 3.6   **49.** -11.577 **51.** -1.7110 [1.4] **53.** Undefined   [1.5] **55.** -11.678 **57.** 0.8   [1.6] **59.** 151.2 lb
61. 47.0 yd^2 **63.** 61000. mm   **65.** 12.1 in.   **67.** 6.58 mm^2 [1.3] **69.** $-17°$F
71. a.
```
10  INPUT X
20  LET A = 5*X 3
30  LET B = 6*SQR(X)
40  LET C = 7*x ∧ (1/4)
50  LET Y = A+B−C+2
60  PRINT X,Y
70  END
```
b. $138.1797\ldots$, $6,664.151\ldots$, $4,705,999.\ldots$, $75,346,183.\ldots$

Chapter Test (page 47)

1. Integer, rational number, real number, complex number   **2.** Rational number, real number, complex number
3. Complex number   **4.** Irrational number, real number, complex number
5. Counting number, whole number, integer, rational number, real number, complex number
6. Rational number, real number, complex number   **7.** $=$ **8.** $<$ **9.** $>$
10.
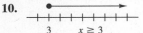
$x \geq 3$
11.
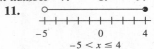
$-5 < x \leq 4$
12.

$-4 \geq x$

13. 7   **14.** $\dfrac{2}{35} \cong 0.0571429$ **15.** Commutative property of addition

16. Distributive property of multiplication over addition   **17.** Additive inverse
18. Associative property of multiplication   **19.** -20 **20.** 0   **21.** undefined   **22.** 0.049   **23.** 2100   **24.** 22,000
25. 626.   **26.** -30.   **27.** 0.952   **28.** 350.

Chapter 2

Section 2.1 (page 53)

1. $6x$ and y **3.** $\dfrac{a + b - c}{3}$ **5.** $-4xy$ and $(6x + 4y)$; $(6x + 4y)$ has terms $6x$ and $4y$ **7.** $8a + 7b$ **9.** $2x + 6y$

11. $6r - 16s$ **13.** $3\pi r - 2\pi r^2$ **15.** $\dfrac{5}{6}x + \dfrac{1}{4}y$ **17.** $\dfrac{7}{8}x^2$ **19.** $-1.36r^2$ **21.** $-2.79r^2$ **23.** 22   **25.** $8x - 7$

27. $3a + 5$ **29.** $15x + 8y$ **31.** $-47a + 5b$ **33.** $6r - s$ **35.** $-12r + 15s$ **37.** $2a + 25b$ **39.** $7x$

41. $-8 + \dfrac{x}{2}$ **43.** $\sqrt{5x} + x^2$ **45.** $(x + y)^3$ **47.** $2xyz + xyz$ **49. a.** $10 - 2.1I_1 - 0.2I_2$ **b.** $11 + 3.5I_1 - 3.7I_2$

c. $1 + 5.6I_1 - 3.5I_2$ **51.** $\dfrac{7\pi}{32}d^2\sigma^2 + \dfrac{\pi}{4}D^2\sigma^2$

Section 2.2 (page 65)

1. x^{17} **3.** $3^{20}x^{55}$ **5.** $\dfrac{1}{x^{15}}$ **7.** 1   **9.** a^3 **11.** $\dfrac{1}{a^4}$ **13.** $\dfrac{7^5}{4^3}$ **15.** $x^{7/12}$ **17.** z^{15} **19.** $\dfrac{3a}{5}$ **21.** $125a^{64}b^5$ **23.** $\dfrac{z^n}{x^n}$ **25.** $\dfrac{c^2}{a^4b^6}$

27. $\dfrac{x^8z^8}{5^43^8y^4}$ or $\dfrac{x^8z^8}{45^4y^4}$ **29.** 81, 81 equal   **31.** 20,736, 20,736 equal   **33.** 1416.   **35.** 1   **37.** 3   **39.** 18.9375   **41.** 0.0005

43. 0.000000069   **45.** 1   **47.** 0.0024   **49.** 28   **51.** 5.0   **53.** 0.2
55. Product rule, add the exponents; power rule, multiply the exponents. $a^3 \cdot a^4 = a^{3+4} = a^7$; $(a^3)^4 = a^{3\times4} = a^{12}$
57. Not true; $7x^{-2} = \dfrac{7}{x^2}$ **59.** $F = \dfrac{F_2 F_1 d(F_1 F_2)}{F_1 d(F_1 F_2) + F_2 d(F_1 F_2) - F_1 F_2}$

Section 2.3 (page 71)

1. $5. \times 10^3$ **3.** $4. \times 10^{-4}$ **5.** 7×10^0 **7.** $3. \times 10^{-1}$ **9.** 5.6×10^7 **11.** 8.77×10^{-7} **13.** 3.937×10^4
15. 1.745×10^{-2} **17.** 3.82×10^5 **19.** 1.61×10^{-19} **21.** 2.838×10^9 **23.** 8500   **25.** 0.000567   **27.** 0.3
29. 9   **31.** 87.500.000.   **33.** 0.0000000034   **35.** 0.0005682   **37.** 750,000,000   **39.** 0.00000000000000000000000000000167
41. 8,649,000   **43.** 0.0000005   **45.** 1×10^2 **47.** 7×10^4 **49.** 3×10^{12} **51.** 6.8×10^4 **53.** 1.3×10^1 hr
55. 0.080 m³/s or 8.0×10^{-2} m³/s   **57.** 1.6×10^7 s   **59.** 1.67248×10^{-15} **61.** $2308.50   **63.** 1.0×10^{16} joules;
1.5×10^{13} pots of coffee   **65.** 7.4×10^9 s; 240 yr   **67.** $\dfrac{36}{(19)^2} \times 10 \approx 1$

Section 2.4 (page 84)

1. $\sqrt{15}$ **3.** $\sqrt[3]{13}$ **5.** $\sqrt[7]{2}$ **7.** $\sqrt{\dfrac{5}{13}}$ **9.** 6   **11.** 11   **13.** 4   **15.** 0.1   **17.** 2   **19.** $13x^2$ **21.** a^2bz^3 **23.** $-4x^3y^2$
25. 4   **27.** $2\sqrt{3} \approx 3.464\ldots$ **29.** $3\sqrt[3]{3} \approx 4.327\ldots$ **31.** $0.731\ldots$ **33.** $x\sqrt{7}$ or $(2.645\ldots)x$
35. $5ab^2\sqrt{2a}$ or $(7.071\ldots)a^{3/2}b^2$ **37.** $x^2\sqrt[3]{3}$ or $(1.442\ldots)x^2$ **39.** $x^2\sqrt[3k-1]{x^2}$ or $x^{\frac{6k}{3k-1}}$ **41.** 3
43. 0, providing $x \neq 0, y \neq 0$ **45.** $5\sqrt{3}$ **47.** $25\sqrt{3}$ **49.** 0   **51.** $(1 + y + y^3)\sqrt{y}$ **53.** $11\sqrt{2}$ **55.** $xy\sqrt{wz}$
57. 5   **59.** xy **61.** $\sqrt{3}$ **63.** $\dfrac{\sqrt{77}}{7}$ **65.** $\sqrt[6]{3}$ or $3^{1/6}$ **67.** $a^{1/6}b^{7/12}$ or $\sqrt[12]{a^2b^7}$ **69.** $\dfrac{4(\sqrt{x}-2)}{x-4}$ **71.** $\dfrac{(\sqrt{x}+2)\sqrt{x-1}}{x-1}$
73. $\dfrac{3x\sqrt{2x-1}}{2x-1}$ **75. a.** 2.49%   **b.** 1.41%   **77.** 45 m/s

Section 2.5 (page 90)

1. $15x^6$ **3.** $28x^3y^4$ **5.** $-18x^{11}y^{13}$ **7.** $-45a^3b^3$ **9.** $96p^5q^5$ **11.** $-4a^4b^2$ **13.** $8a^2 + 32a$ **15.** $5p^4 + 35p^3$
17. $6a^3b^2 - 15a^2b$ **19.** $a^2 + 2ab + b^2$ **21.** $6x^2 + 29x + 35$ **23.** $a^2 - b^2$ **25.** $4x^2 - 9$ **27.** $6x^2 + 13xy - 28y^2$
29. $35a^4b + 14a^2bc + 15a^2c^2 + 6c^3$ **31.** $3x^3 - 2x^2 + 7x + 20$ **33.** $4a^3 + 20a^2 + 37a + 24$
35. $4x^4 + 2x^3 - 22x^2 + 4x$ **37.** $4x^2 + 12xy + 9y^2$ **39.** $18a^2 - 84a^2b + 98ab^2$ **41.** $x^3 + 4x^2 - 17x - 60$
43. $8a^3 - 36a^2b + 54ab^2 - 27b^3$ **45.** $20x^4 - 40x^3 - 55x^2 - 15x$ **47.** $\dfrac{v^2 - v_0^2}{2a}$ or $\dfrac{v^2}{2a} - \dfrac{v_0^2}{2a}$
49. $4000 - 400n - 150n^2$ **51. a.** $3^3 + 4^3 = 91$; $(3+4)^3 = 343$ **c.** $(x+y)^3 = x^3 + 3x^2y + 3xy^2 + y^3 \neq x^3 + y^3$

Section 2.6 (page 94)

1. $\dfrac{2}{3}$ **3.** $-7ab^3$ **5.** $4b^{11}$ **7.** $2 - 4y$ **9.** $2 + 6r$ **11.** $-5xy + 4x - 2y$ **13.** $m + 4$ **15.** $3a + 2b$ **17.** $4x - 7y$
19. $16x^2 - 4x + 1$ **21.** $y^2 - 6y + 11$ **23.** $q^2 + 3q + 4$ **25.** $a - 4 + \dfrac{-10}{a-3}$ **27.** $2x - 1 + \dfrac{10}{3x+5}$
29. $t^2 - 6t + 19 + \dfrac{-34}{t+2}$ **31.** $x + 6 + \dfrac{x+14}{x^2-x-2}$ **33.** $8x^3 - 36x^2y + 54xy^2 - 27y^3$
35. $90 + 80\alpha\gamma - 40\alpha^2\beta\gamma^2$ **37.** $\dfrac{(EI+P)^2}{2} + \dfrac{6}{(EI+P)} - 2$ **39.** $t^2 + t - 5/3 + \dfrac{8/3}{3t-2}$

Section 2.7 (page 100)

1. $3(a + 4b)$ **3.** $7(r + 2s)$ **5.** $2x(1 - 6y)$ **7.** $5x(y - 3xz + 4z)$ **9.** $(3b + 5d)(a + 4c)$ **11.** $(5r - 6k)(3s - 5t)$

13. $5(3R_1 - 7R_2)(I_1 - 2I_2)$ **15.** $A(3AB - 7C)(2D - 5B)$ **17.** $\left(\dfrac{1}{3}A + \dfrac{2}{5}P\right)(B - 2C)$

19. $y(0.2x + 0.3z)(0.3x + 0.4y)$ **21.** $(x + 1)(x + 4)$ **23.** $3(a + 1)^2$ **25.** $(t + 7)(t + 2)$ **27.** $(r + 7)(r - 7)$
29. $(3b + 1)(b - 2)$ **31.** Prime **33.** $(5z - 4)(z + 3)$ **35.** $2(3s - 1)(s + 3)$ **37.** $(17p - 11q)(p + q)$
39. $(2w - 5)^2$ **41.** Prime **43.** $(z + 3)(3z + 1)$ **45.** $(7R - 2)(R - 2)$ **47.** $(w + 7)^2$
49. $(3p + 7)^2$ **51.** $(2s - 3)^2$ **53.** $(8m - 5n)^2$ **55.** $(9w - 4z)^2$ **57.** $(x - 3)(x + 3)$ **59.** $(9r - 4)(9r + 4)$
61. $(xy - 5)(xy + 5)$ **63.** $(a + b - d)(a + b + d)$ **65.** $(p - 4)(p + 4)(q + 3)$ **67.** $(x - 4)(x^2 + 4x + 16)$

69. $-2(2x + 3y)(4x^2 - 6xy + 9y^2)$ **71.** $\dfrac{\pi}{4}(D - d)(D + d)$ **73.** $6x^2 - 3xy - 30y^2$

75. $\pi\left(R - \dfrac{r}{2}\right)\left(R + \dfrac{r}{2}\right)$ or $\dfrac{3\pi r^2}{4}$ **77.** $\dfrac{4}{3}\pi(r_2 - r_1)(r_2^2 + r_2 r_1 + r_1^2)$

Section 2.8 (page 105)

1. $\dfrac{4}{7}$ **3.** pm **5.** $\dfrac{x(x + 3)}{6(x - 2)}$ or $\dfrac{x^2 + 3x}{6x - 12}$ **7.** $\dfrac{2}{3}$ **9.** $\dfrac{4(a - b)}{3}$ **11.** $\dfrac{2(x + 3)}{z(x - 3)}$ **13.** 1 **15.** $-\dfrac{3p}{3 + p}$ **17.** $\dfrac{8}{5}$ **19.** $\dfrac{1}{3ac}$
21. $\dfrac{(a + 3)(a - 6)}{(a + 4)(a + 5)}$ or $\dfrac{a^2 - 3a - 18}{a^2 + 9a + 20}$ **23.** $\dfrac{3}{4}$ **25.** $\dfrac{1}{p - 2}$ **27.** $\dfrac{1}{x - y}$ **29.** 1 **31.** $\dfrac{3x^3}{4y^3}$ **33.** $\dfrac{9z}{2x}$ **35.** $\dfrac{R}{R - r}$ **37.** $\dfrac{F\ell}{A\,\Delta\ell}$

Section 2.9 (page 110)

1. $\dfrac{7}{2x}$ **3.** $\dfrac{p + q + 1}{p + q}$ **5.** $\dfrac{5a - b}{a^2 b^2}$ **7.** $\dfrac{2}{(x - 1)(x + 1)}$ **9.** $\dfrac{-(a + 1)}{(a + 2)(a + 3)}$ **11.** 0 **13.** $\dfrac{-a}{a - 1}$ **15.** $\dfrac{-4m}{(m + 1)(m - 1)}$
17. $\dfrac{3y^2 z - 5x^4}{x^5 y^4 z}$ **19.** $\dfrac{a(a - 1)}{(a - 2)(a + 2)}$ **21.** 0 **23.** $\dfrac{4x^3 - x - 2}{(2x + 1)^2(2x - 1)}$ **25.** $\dfrac{a^2 r^2 + b^2 q^2 - c^2 p^2}{p^2 q^2 r^2}$
27. $\dfrac{x^2 - 3x - 3}{(x - 3)(x + 3)}$ **29.** $\dfrac{2b}{a - b}$ **31.** $\dfrac{-2y + 1}{(x - y)(x + y)}$ **33.** $\dfrac{24(t + 6)}{(t - 5)(t + 5)(t - 7)(t + 7)}$ **35.** $\dfrac{m^2 - 3m - 11}{2 + m}$
37. $\dfrac{a^2 c + ab^2 + bc^2}{8}$ **39.** $\dfrac{-1}{a - b}$ **41.** $\dfrac{5x - 3}{(x - 3)(x + 3)}$ **43.** $\dfrac{2x^2 + 8x - 10}{(x + 3)^2}$ **45.** $-\dfrac{13}{17}$ **47.** $\dfrac{mn + 1}{mn - 1}$ **49.** $\dfrac{IR}{IR - E}$
51. $\dfrac{y}{x}$

Review Exercises (page 111)

[2.2] **1.** a^{16} **3.** $x^{28} y^{21}$ **5.** 1 **7.** $5xy^5$ **9.** $\dfrac{p^3 q^3}{8m^9 n^{12}}$ [2.3] **11.** 3.250×10^3 **13.** $4. \times 10^{-1}$ **15.** 5.79×10^{-6}
17. 7,052,000 **19.** 4.32 **21.** 0.0000462 **23.** 0.035 oz **25.** 1×10^{12} **27.** 8.8×10^{-5} **29.** $8. \times 10^{-13}\,j$
[2.4] **31.** $2\sqrt[3]{4}$ **33.** 4 **35.** y^n **37.** $-4\sqrt{2}$ **39.** $\sqrt{143}$ **41.** $ab\sqrt{cd}$ **43.** $3y\sqrt{x}$ [2.1] **45.** $-8xy^2$
47. $-6\phi\pi + 16RI$ **49.** $\dfrac{58}{105}b^2$ **51.** $-18x + 13$ **53.** $35p - 24q$ [2.5, 2.6, 2.7] **55.** $-143r^5 s^6$
57. $-3pq$ **59.** $\dfrac{(m + 2)(m + 4)}{(m + 3)(m - 5)}$ or $\dfrac{m^2 + 6m + 8}{m^2 - 2m - 15}$ **61.** $\dfrac{1}{3xy^2}$ **63.** $\dfrac{2(x^3 - y^3)}{x^2 y^2}$
65. $15R^3 + 43R^2 - 89R + 36$ **67.** $x^3 + 6x^2 - 7x - 60$ **69.** $3y(a - 2)$
71. $a + 6$ **73.** 1 **75.** 2 **77.** $\dfrac{5RI}{3}$ **79.** $\dfrac{31m + 1}{6(m + 1)(m - 1)}$ **81.** $\dfrac{(x - 1)^2}{(x + 1)(x + 5)}$
[2.7] **83.** $x(3xy - 7z)(2w - 5y)$ **85.** $6(r - s)(r + s)$ **87.** $2(8x + 1)(4x - 1)$
89. $(5a + 6b)(2a - 3b)$ **91.** No; $(10^4)^3 = 10^{12}$, $10^{4^3} = 10^{64}$, $10^{12} \neq 10^{64}$

93. $\sqrt{5 + 2\sqrt{6}} = \sqrt{2} + \sqrt{3}$
$(\sqrt{5 + 2\sqrt{6}})^2 = (\sqrt{2} + \sqrt{3})^2$
$5 + 2\sqrt{6} = 2 + 2\sqrt{2}\sqrt{3} + 3$
$= 5 + 2\sqrt{6}$

95. Average of the squares is greater than the square of the averages.

Chapter Test *(page 114)*

1. $6x^2y^2$ **2.** $\dfrac{1}{r - s}$ **3.** 1.23×10^8 **4.** 2.14×10^{-4} **5.** $3,400,000$ **6.** 0.0000000287 **7.** 3.76×10^5 **8.** 1.2×10^{11}

9. $5x\sqrt{2x}$ **10.** $7x^3y - 3xy + 9$ **11.** $\dfrac{y}{x}$ **12.** $\dfrac{6 + x}{x + 2}$ **13.** $(r + 7t)(2s + 7u)$ **14.** $(3x - 5)(3x + 5)$

15. $x(x^2 + 8x - 32)$ **16.** $(5x + 2y)^2$ **17.** $x - 5 + \dfrac{1}{x + 3}$ **18.** $\dfrac{x - 1}{y}$ **19.** $\dfrac{17x + 13}{(x - 7)(x + 7)}$ **20.** $\dfrac{26r^4}{15s^2}$

Chapter 3

Section 3.1 *(page 125)*

1. $15.3°$ **3.** $104.63°$ **5.** $72.989°$ **7.** $73.9950°$ **9.** $73° \, 20'$ **11.** $105° \, 25'$ **13.** $121° \, 27'20''$ **15.** $27° \, 27' \, 31''$
17. $61° \, 28' \, 46''$ **19.** $a = 133° \, 47'$, $b = 133° \, 47'$, $c = 46° \, 13'$, $d = 46° \, 13'$ **21.** $\angle CBD = 60°$, $\angle CDB = 30°$
23. $50°$, isosceles triangle   **25.** $60°$, equiangular triangle   **27.** $\angle A = \angle B = 15°$, isosceles triangle
29. $\alpha = 48°$, $\beta = 65°$, $\varepsilon = 25°$, $\gamma = 65°$ **31. a.** Yes   **b.** No   **c.** To determine a unique triangle, a rigid figure

Section 3.2 *(page 133)*

1. Not similar, corresponding sides not proportional   **3.** Similar, Definition 3.2   **5.** Definition 3.3, $\alpha = 55°$, $b = 120$ mm
7. Definition 3.2, $\theta = 22° \, 37'$, $x = 28$ in., $y = 12$ in.   **9.** 40.5 ft   **11.** 33 in.   **13.** 20 m   **15.** 56 mm   **17.** 2.2 in.
19. Equiangular triangle, therefore equilateral triangle and $\overline{AB} = \overline{AC} = .275$ km   **21. a.** $\triangle$ are similar, Def. 3.2   **b.** 96 ft
23. ASA $\alpha = 70°$, $\beta = 60°$, $x = 100.0$ m   **25.** 120 yd   **27.** Yes, you can see Lake Ontario.
29. a. $\angle ATR = \angle ABC$, corresponding $\angle S$; $AB \parallel RS \; \angle TRS = \angle ATR$, alternate interior $\angle$; $\angle RTS = \angle BAC$,
corresponding $\angle$; ($\triangle$) are similar, Definition 3.2   **b.** $\angle TSR = \angle ACB$, $\angle TRS = \angle ABC$, $\angle RTS = \angle BAC \; RS : BC$,
$TR : AB$, $TS : AC$

Section 3.3 *(page 143)*

1. 13.2 m, 11 m²   **3.** 12.4 ft, 7.2 ft²   **5.** 90. ft, 140 ft.²   **7.** 110 mm, 570 mm²
9. 150 mm, 1500 mm²   **11. a.** $90°$ **b.** 1.142 in.²   **13. a.** 42,000. ft²   **b.** 65.927. ft²   **c.** 13,383. ft²   **d.** 26,093. ft²
e. 39,961. ft²   **f.** trapezoid, 62,510 ft²   **g.** 249,874 ft²

15. 21.5%   **17.** $1 : 9$ **19.** No, $\sqrt{2}\dfrac{d}{2}$ **21.** 15 m   **23.** 127 ft   **25.** 71 N   **27.** 7850 ft², diameter

Section 3.4 *(page 152)*

1. 4.9230×10^7 mi²   **3.** 120 ft³   **5.** 13 ft   **7.** 6400 in.²   **9.** 245 yd³   **11.** 14.5 lb   **13.** 36.5 ft, 3440 ft²   **15.** 11 gal
17. $144   **19.** 825 spheres   **21.** No, the pool filled with water will weigh 28,700 lb.

Review Exercises *(page 155)*

[3.1] 1. $17.38°$ **3.** $123.58°$ **5.** $254.206°$ **7.** $38.4919°$ **9.** $33° \, 40'$ **11.** $176° \, 8'$ **13.** $1° \, 14' \, 00''$ **15.** $25° \, 25' \, 56''$

[3.1, 3.2] **17.** $\alpha = 61.9°$, $x = 36$ m　**19.** $\angle 1 = 52°$, $\angle 2 = 52°$　**21.** $\alpha = 35°$, $\beta = 55°$, $\phi = 34°$, $\theta = 56°$, $\epsilon = 56°$
[3.3, 3.4] **23.** 12 ft　**25.** 7490 in.^{3}　**27.** Increases the volume by a factor of 3.375　**29.** 383.2 mm^2　**31.** 150 in.^{2}
33. Yes, (68 in.^{3})　**35.** 1.23×10^7 ℓ　**37.** 420 ft^3　**39.** 4.85 yd^3　**41.** 1.95 m, 2.76 m　**43.** 20 ft, 19,000 gal
45. Count the letters in the words.

Chapter Test (page 160)

1. 28.28°　**2.** 29° 20′ 50″　**3.** 54.2°　**4.** 53°　**5.** 46° 35′ 11″　**6.** $\angle B = 76°$ and $\angle C = 76°$
7. $\angle 1 = 145°$; $\angle 2 = 145°$; $\angle 3 = 145°$; and $\angle 4 = 35°$.　**8.** $x = 32$　**9.** $x = 12$　**10.** $P = 16$; $A = 14$
11. $C = 2.35$ in.　**12.** $A = 0.442$ in.^{2}　**13.** $V = 0.0276$ in.^{3}　**14.** $V = 90,000$ mm^3　$T = 20,000$ mm^2
15. Weight $= 23.4$ lb

Chapter 4

Section 4.1 (page 167)

1. $r = 6$　**3.** $n = 2$　**5.** $R = -\dfrac{9}{14}$　**7.** $r = 4$　**9.** $N = 9$　**11.** $x = -\dfrac{5}{3}$　**13.** $s = 6$　**15.** $x = 1.3$　**17.** $a = -2$

19. $S = -\dfrac{4}{9}$　**21.** $x = -0.89$　**23.** $x = \dfrac{207}{58}$ or $x \cong 3.5689655$　**25.** All real numbers

Section 4.2 (page 174)

1. $x = \dfrac{2}{3}$　**3.** $R = \dfrac{-8}{27}$　**5.** $Y = 3$　**7.** $R = 3$　**9.** $t = 7$　**11.** $a = 0$　**13.** No solution　**15.** $x = -26$

17. No solution (extraneous root)　**19.** $x = 16$　**21.** $P = \dfrac{nRT}{V}$　**23.** $h = \dfrac{2A}{b}$　**25.** $w = \dfrac{P}{2} - \ell$　**27.** $F = \dfrac{9}{5}C + 32$

29. $h = \dfrac{3V}{\pi r^2}$　**31.** $m_2 = \dfrac{Fr^2}{Gm_1}$　**33.** $R_2 = \dfrac{R_T R_1}{R_1 - R_T}$　**35.** $P = \dfrac{100A}{St}$　**37. a.** $S_1 = \Delta t\left(-I_1 - \dfrac{I_2}{2} + \dfrac{Q_1}{2} + \dfrac{S_2}{\Delta t} + \dfrac{Q_2}{2}\right)$;

b. $S_2 = \Delta t\left(\dfrac{I_2}{2} - \dfrac{Q_2}{2}\right)$　**39.** $D_i = \dfrac{fD_0}{D_0 - f}$　**41.** $v = 2\sqrt{\dfrac{E}{n}}$

Section 4.3 (page 181)

1. c　**3.** g　**5.** l　**7.** a　**9.** k　**11.** f　**13.** 8500 mm　**15.** 15. mi　**17.** 400 m　**19.** 88 ft/s
21. 1,300,000 milliamperes　**23.** 3.0 m^2　**25.** 5.6×10^b Ω　**27.** 2550 J/s　**29.** 0.0700 mΩ　**31.** 17 km/ℓ
33. $130\overline{,}000$ W　**35.** 22,633 s　**37.** 3.56 ℓ　**39.** 161 hp　**41.** 216 ft^2　**43.** 4.45 m　**45.** 61.0 in.^{3}　**47.** 2.20 lb

Section 4.4 (page 186)

1. 0.46 A　**3.** 65.7 ft　**5.** 7.08 in.　**7.** 423,000 kg-m^2/s^2　**9.** 51,800 kg-m^2/s^2　**11.** 24 m　**13.** 1.6 s　**15.** 45.0 Ω
17. 1800 lb　**19. a.** 41 kg-m/s^2　**b.** 41 kg-m/s^2　**21. a.** 750 slugs/ft-s^2　**b.** 7500 slugs/ft-s^2　**c.** 330,000 slugs/ft-s^2
23. 0.065 m^3; 0.52 m^3; if r is doubled, the new volume is 8 times greater than the old volume: 1.76 m^3

Section 4.5 (page 193)

1. $\dfrac{5}{16}$　**3.** $\dfrac{3}{4}$　**5.** $\dfrac{1}{4}$　**7.** $\dfrac{3}{4}$　**9.** $\dfrac{5}{7}$　**11.** $\dfrac{7}{5}$　**13.** $\dfrac{2y}{3x}$　**15.** $\dfrac{x - 2y}{x + 2y}$　**17.** $\dfrac{4.17}{1}$, greater than　**19.** $\dfrac{3.14}{1}$　**21.** 7.5
23. 10.5　**25.** 2.5　**27.** 3.00　**29.** 370,000 cm　**31.** 33.58 mm　**33.** 87.5 m　**35.** 110. km　**37.** 17. g　**39.** 81. mm
41. 1700 lb　**43.** 2.0 gal　**45. a.** 2.4 : 1　**b.** 25 in.　**c.** $\dfrac{(25.4)^2}{(10.6)^2}$, 5.7 times clearer　**47. a.** $\dfrac{2}{3}$　**b.** $\dfrac{4}{9}$　**c.** $\dfrac{8}{27}$

d. 3 times more heat

Section 4.6 (page 201)

1. $D = kt$ **3.** $C = kd$ **5.** $R = kVW$ **7.** $Z = \dfrac{kD^2}{T}$ **9.** $T = \dfrac{kR\sqrt{x}}{y^2}$ **11.** $L = \dfrac{k}{V}$ **13.** $A = ka^2$ **15.** $T = kd^4n^2$

17. $R = \dfrac{k\ell}{d^2}$ **19.** $t = \dfrac{ky}{X}$ **21. a.** $y = kx^2$ **b.** $k = \dfrac{y}{x^2}, k = \dfrac{3}{16}$ **c.** $y = \dfrac{3}{16}x^2$ **d.** $x = 8$ **23. a.** $E = klw$

b. $k = \dfrac{E}{lw}, k = 1.14 \text{ s/in.}^2$ **c.** $E = 1.14 \text{ s/in.}^2 \, lw$ **d.** 55 s   **25. a.** $P = kt$ **b.** $k = \dfrac{P}{t}, k = 0.05017$

c. $P = 0.05017t$ **d.** $P = 23.7 \text{ lb/in.}^2$ **27. a.** $I = \dfrac{k}{R}$ **b.** $k = IR, k = 30$ **c.** $I = \dfrac{30}{R}$ **d.** 200 Ω

29. a. $s = \dfrac{kwd^2}{\ell}$ **b.** $k = \dfrac{s\ell}{wd^2}, k = 62.5$ **c.** $s = 62.5\dfrac{wd^2}{\ell}$ **d.** 250 lb   **31. a.** $n = \dfrac{k\sqrt{f}}{\ell d}$ **b.** $k = \dfrac{n\ell d}{\sqrt{f}}$ $k = \dfrac{0.008n_1 \, m^2}{5\sqrt{N}}$

c. $n = \dfrac{0.008n_1 \, m^2 \, \sqrt{f}}{5\ell d \, \sqrt{N}}$ **d.** 100 N   **33. a.** $P = \dfrac{k\ell}{d^2}$ **b.** $k = \dfrac{Pd^2}{\ell}, k = 0.567$ **c.** $P = \dfrac{0.567 \text{ N/m } \ell}{d^2}$

d. $P = 490 \text{ N/m}^2$ **35. a.** $M = kr$ **b.** $k = \dfrac{M}{r}, k = 7.84 \times 10^{-28}$ **c.** $M = 7.84 \times 10^{-28} \, r$ **d.** $M = 9.09 \times 10^{-31}$ kg

Review Exercises (page 203)

[4.1, 4.2] **1.** $x = 9$ **3.** $a = 11$ **5.** $z = 1$ **7.** $R = 30$ **9.** $x = \dfrac{-23}{3}$ or 7.67   [4.3] **11.** 11,500 mm   **13.** 121 km

15. 30.1 yd   **17.** 100 km/hr   **19.** 29,840 W   [4.5] **21.** For every three people in the class, there are two girls.
23. 18   **25.** $\dfrac{r - 3s}{r + 3s}$ **27.** $\dfrac{x + 3}{x - 4}$ **29.** $\dfrac{1}{x + 2}$ **31.** $36\dfrac{1}{4}$ mi   **33.** 126 mi   [4.6] **35.** $c = kt$ **37.** $z = \dfrac{kRt^2}{w^2}$
39. $p = kh$ [4.4] **41.** $a = 3.2 \text{ ft/s}^2$ **43.** 3; three significant digits in the area; 60.7 mm   [4.7] **45.** 52,104
47. $110, $100   [4.4] **49.** 4 ft   [4.5] **51.** $z = 12.25$ **53.** $7 : 1$ [4.4] **55.** 7490 in.^{3}
[4.5] **57.** 27 times greater   [4.7] **59.** 3.5 hr   [4.5] **61. a.** lb/ft^3 **b.** $w = \dfrac{E}{8\ell h^2}$ **c.** 20 ft

Chapter Test (page 207)

1. $\dfrac{1}{4}$ or $1 : 4$ **2.** $\dfrac{1}{7}$ or $1 : 7$ **3.** $\dfrac{(x - 4)}{(x - 2)}$ or $(x + 4) : (x - 2)$ **4.** $x = \dfrac{4}{3}$ **5.** $y = 24$ **6.** $x = 4$ **7.** $x = 3$

8. $x = \dfrac{-65}{8}$ or -8.125 **9.** $x = 0.876$ **10.** $h = \dfrac{V}{\pi r^2}$ **11.** $\dfrac{2V}{\pi r^2} - h_1 = h_2$ **12.** 157.1 mm   **13.** 0.5327 m^2

14. $1.85\dfrac{\text{BTU}}{\text{min}}$ **15.** 2010 hp   **16. a.** $d = kg$ **b.** $k = 20\dfrac{\text{mi}}{\text{gal}}$ **c.** 150 mi   **17.** $x = 28$ **18.** $x = 0.6$ **19.** 14 mm
20. $w = 30$ cm.

Chapter 5

Section 5.1 (page 221)

1. i: d; d: P   **3.** i: F; d: c   **5.** i: r, θ; d: s   **7.** i: m, a; d: F   **9. a.** 43   **b.** 15   **c.** $3a^2 + 6ah + 3h^2 + 7a + 7h - 5$
11. a. -7 **b.** 2.8   **c.** -0.19 **13. a.** 35.69   **b.** 103.4   **c.** 34.12   **15. a.** 14.7 in.   **b.** 40.3 m   **c.** 2.856 in.
17. a. 84.4 m^2 **b.** 220 in.^{2} **c.** 1081 mm^2 **19. a.** 7.5   **b.** 18   **21. a.** 4.9   **b.** 97   **23. a.** -4 **b.** 5   **c.** 7
25. a. 2   **b.** 4   **c.** 6   **27.** $D = \{x \mid x \in \mathcal{R}\}, R = \{y \mid y \in \mathcal{R}\}$ **29.** $D = \{t \mid t \in \mathcal{R}\}, R = \{f(t) \mid f(t) \geq 5\}$

31. $D = \{r \mid r \geq 0\}, R = \{f(r) \mid f(r) \geq 0\}$ **33.** $D = \{x \mid x \neq 0\}, R = \{y \mid y \neq 0\}$ **35.** $D = \{x \mid x \neq 1\}$,
$R = \{y \mid y \neq 1\}$ **37.** $D = \{t \mid t \geq -2\}, R = \{s \mid s \geq 0\}$ **39.** $D = \{\ell \mid \ell \geq -3\}, R = \{p \mid p \geq 0\}$
41. $f(h) = 3600h$ **43.** $f(t) = 55t$ **45.** $A(w) = w(20 - w)$ **47.** $f(c) = \$0.15c + \$300$ **49.** $C(x) = \$1.80x + \$400$

Section 5.2 *(page 228)*

1.

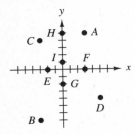

3.

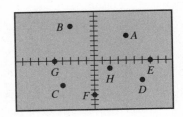

5. $A(2, -4), B(4, -2), C(2, 0), D\ (4, 5), E(0, 3), F(-3, 3), G(-5, 0), H(-4, -2)\ I(-3, -4)$ **7.** 5.39   **9.** 19.1
11. 2.9   **13.** 16.6

15.

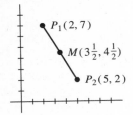

17.

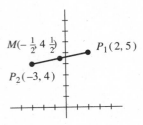

19.

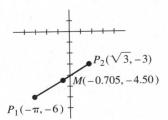

21. a. $\overline{AC} = \overline{BC} = 3.16$ **b.** 9.15   **23.** $\overline{AB} = \overline{AC} = \sqrt{32}, \overline{BC} = 8, \overline{AB}^2 + \overline{AC}^2 = \overline{BC}^2$ **25.** $\overline{AC} = \sqrt{32}, \overline{BC} = \sqrt{18}$,
$\overline{AB} = \sqrt{50}, \overline{AC}^2 + \overline{BC}^2 = \overline{AB}^2$ **27.** If diagonals of rectangle then diagonals are congruent: $\overline{AC} = \overline{BD} = \sqrt{68}$
29. (18, 2), (-4, 2)   **31.** Given points $A(0, 3), B(0, 6), C(4, 3), D(4, 0)$ and $E(2, 3)$ then $\overline{AE} = \overline{EC} = 2$ and
$\overline{BE} = \overline{ED} = \sqrt{13}$ **33.** 10 square units   **35.**

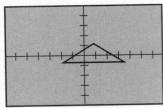

isosceles right triangle

Section 5.3 *(page 237)*

1. a. $x = 1, y = 0$ **b.** $x = 2, y = 0$ **c.** $x = 3, y = 0$ **3.** b   **5.** f   **7.** a   **9.** h   **11.** b

13. $f(x) = x + 3$

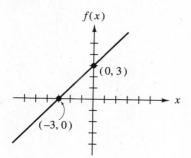

15. $f(s) = \dfrac{1}{s + 2}$

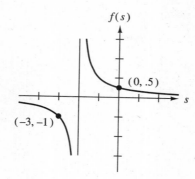

17. $f(t) = \dfrac{t + 5}{t - 3}$

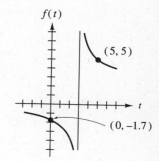

19. $f(x) = \sqrt{2 - x}$

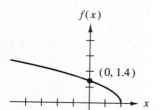

21. $f(x) = x^3 - 1$

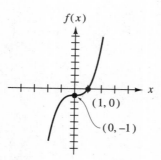

23. $f(x) = 2x - 1$
$D = \{x \mid x \in \mathcal{R}\}$
$R = \{f(x) \mid f(x) \in \mathcal{R}\}$
no asymptotes

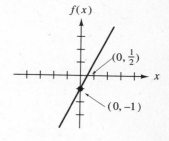

25. $f(x) = 3x^2 - 5x - 7$

$D = \{x \mid x \in \mathcal{R}\}$
$R = \{f(x) \mid f(x) \geq -9.1\}$
no asymptotes

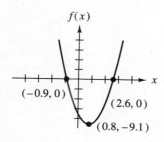

27. $f(s) = \dfrac{s - 3}{s - 4}$

$D = \{s \mid s \neq 4\}$
$R = \{f(s) \mid f(s) \neq 1\}$
asymptotes $s = 4, f(s) = 1$

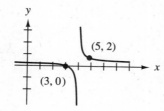

29. $D = \{-2, -1, 0, 1, 2, 3, 4, 5\}$

$R = \{-7, -6, -5, -4, -3, -2, -1, 0\}$

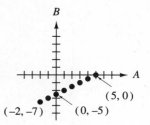

31. $d = 4\left(\dfrac{a}{500}\right)$

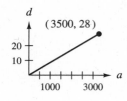

33. a. $V = 6.0I$ **b.** 7.8 v

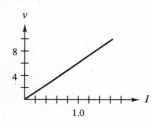

c. $I = \dfrac{120}{R}$ **d.** 6

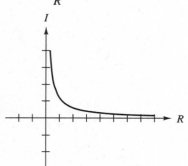

35. a. $F = \dfrac{10^{-9}}{r^2}$ **b.** about 5m

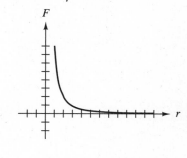

37. $y = 2x + 5$, $y = -2x + 5$

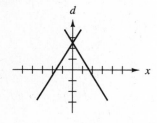

similar: linear functions, same y-intercept $(0, 5)$, same domain $D = \{x \mid x \in \mathcal{R}\}$, same range $R = \{y \mid y \in \mathcal{R}\}$; different: one line increases moving from left to right, the other line decreases; x-intercepts $\left(\dfrac{-5}{2}, 0\right)$, $\left(\dfrac{5}{2}, 0\right)$

39. $y = \sqrt{x - 6}$, $y = \sqrt{x}$, $y = \sqrt{x + 6}$;

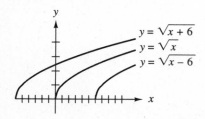

similar: shape, increasing functions, $R = \{y \mid y \geq 0\}$; different: the constant value causes a shift to the left or right of the origin; x =intercepts $(6, 0)$, $(0, 0)$, $(-6, 0)$

41. $y = x^2$, $y = 2x^2$, $y = 4x^2$;

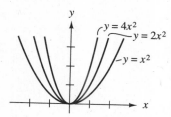

similar: parabolas, all open upward, vertex at origin, y-axis line of symmetry, $D = \{x \mid x \in \mathcal{R}\}$, $R = \{y \mid y \geq 0\}$; different: the larger the constant of the x^2-term, the narrower the curve.

43. $y = x^2 + 3x - 5$, $y = x^2 - 3x - 5$;

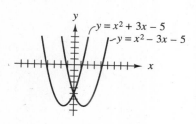

similar: parabolas open upward, same shape, domain and range are the same, $D = \{x \mid x \in \mathcal{R}\}$, $R = \left\{y \mid y \geq \dfrac{-29}{4}\right\}$, line of symmetry for both curves is parallel to y-axis, same y-intercept $(0, -5)$
different: the vertex of one curve is to the left of the y-axis and the other is to the right of the y-axis. By writing the functions in the form $y = x(x - 3) - 5$ and $y = x(x + 3) - 5$, we can see the reason for the shift: the $+3$ and the -3

45. $y = \dfrac{1}{x - 3}$, $y = \dfrac{4}{x - 3}$;

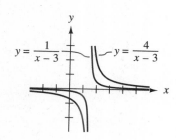

similar: same shape, same asymptotes $x = 3$ and $y = 0$, same domain and range $D = \{x \mid x \neq 3\}$, $R = \{y \mid y \neq 0\}$;
different: y-intercepts $(0, -\frac{1}{3})$ and $(0, -\frac{4}{3})$, constant 4 increases the y-value by a multiple of 4

47. $y = \dfrac{x + 2}{x - 3}$, $y = \dfrac{x - 2}{x - 3}$;

similar: same shape, same horizontal asymptote $y = 1$ and vertical asymptote $x = 3$, same domain $D = \{x \mid x \neq 3\}$ and same range $R = \{y \mid y \neq 1\}$;
different: x-intercepts $(-2, 0)$ and $(2, 0)$, y-intercepts $(0, -\frac{2}{3})$, $(0, \frac{2}{3})$, the $x + 2$ and the $x - 2$ shift the curve to the left and to the right

Section 5.4 (page 244)

1. $f^{-1}(x) = x - 5$ **3.** $f^{-1}(x) = \dfrac{x - 7}{2}$ **5.** $f^{-1}(x) = \dfrac{x}{2} + 4$ **7.** $f^{-1}(x) = 2x - 4$ **9.** $f^{-1}(x) = \dfrac{-3x + 9}{2}$

11. $f^{-1}(x) = 2x - 7$ **13.** $f^{-1}(x) = \dfrac{-2x + 1}{x}$ **15.** $f^{-1}(x) = \dfrac{-x + 2}{x}$ **17.** $f^{-1}(x) = \dfrac{-2x + 5}{3x}$

19. $f(x) = x + 5, f^{-1}(x) = x - 5$

21. $f(x) = 2x + 7, f^{-1}(x) = \dfrac{x - 7}{2}$

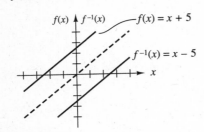

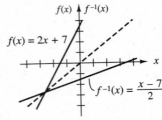

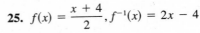

23. $f(x) = 2(x - 4), f^{-1}(x) = \dfrac{x}{2} + 4$

25. $f(x) = \dfrac{x + 4}{2}, f^{-1}(x) = 2x - 4$

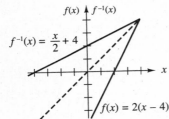

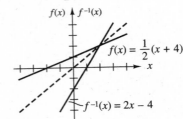

27. $f(x) = \dfrac{9 - 2x}{3}, f^{-1}(x) = \dfrac{-3x + 9}{2}$

29. $f(x) = \dfrac{1}{2}(x + 7), f^{-1}(x) = 2x - 7$

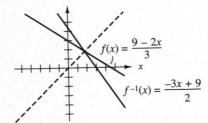

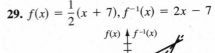

31. $f(x)$: $D = \{x \mid x \neq -2\}$, $R = \{f(x) \mid f(x) \neq 0\}$; $f^{-1}(x)$: $D = \{x \mid x \neq 0\}$, $R = \{f^{-1}(x) \mid f^{-1}(x) \neq -2\}$

33. $f(x)$: $D = \{x \mid x \leq -1\}$, $R = \{f(x) \mid f(x) \neq 0\}$; $f^{-1}(x)$: $D = \{x \mid x \neq 0\}$, $R = \{f^{-1}(x) \mid f^{-1}(x) \neq -1\}$

35. $f(x)$: $D = \left\{x \mid x \neq \dfrac{-2}{3}\right\}$, $R = \{f(x) \mid f(x) \neq 0\}$; $f^{-1}(x)$: $D = \{x \mid x \neq 0\}$, $R = \left\{f^{-1}(x) \mid f^{-1}(x) \neq -\dfrac{2}{3}\right\}$

37. a. $f^{-1}(x) = \dfrac{x - 5}{3}$ **b.** $f(x) = 3x + 5$ **39.** $D = \left\{x \mid x \neq \dfrac{9}{11}\right\}$, $R = \{f(x) \mid f(x) \neq 0\}$, $f(x) = \dfrac{2}{11x} + \dfrac{9}{11}$

41. The function and its inverse are mirror images with respect to the line $y = x$

Review Exercises (page 244)

[5.1] **1. a.** -4 **b.** 11   **c.** $3h + 14$ **3. a.** 13   **b.** 13   **c.** $z^2 + 4z + 8$ **5. a.** 13.0 mm³   **b.** 395 m³   **c.** 0.630 in.³

7. a. $3\dfrac{\text{kg} - \text{m}}{s^2}$ **b.** $1.6 \times 10^7 \dfrac{\text{kg} - \text{m}}{s^2}$ **9. a.** 6   **b.** 39   **c.** 53

11. $D = \{x \mid x \in \mathcal{R}\},\ R = \{y \mid y \in \mathcal{R}\}$ **13.** $D = \{t \mid t \in \mathcal{R}\},\ R = \{f(t) \mid f(t) \geq -5\}$

15. $D = \{x \mid x \neq 1\},\ R = \{f(x) \mid f(x) \neq 0\}$ **17.** $D = \{L \mid L \neq 1\},\ R = \{F(L) \mid F(L) \neq 1\}$

19. $D = \{s \mid s \geq -5\},\ R = \{g(s) \mid g(s) \geq 0\}$ **21.** $\ell(g) = 3.7854g$ **23.** $C(g) = kg$ **25.** $C(d) = \pi d$

27. $A(w) = (3w + 5)w$ [5.2] **29.** 4.24, $\left(\dfrac{11}{2}, \dfrac{7}{2}\right)$ **31.** 10, (0, 0)   **33.** 11, $(-4.8, -1.5)$

[5.3] **35.** $f(x) = x + 7$

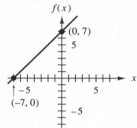

37. $B = 3A^2 - 4A + 5$

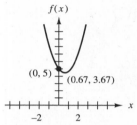

39. $f(t) = \dfrac{t + 3}{t - 2}$

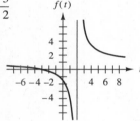

41. $y = 12x - x^3$

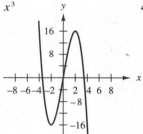

43. $s = \sqrt{5 - t}$

[5.4] **45. a.** $f^{-1}(x) = x + 7$
   b. $f: D = \{x \mid x \in \mathcal{R}\},$
      $R = \{f(x) \mid f(x) \in \mathcal{R}\};$
      $f^{-1}: D = \{x \mid x \in \mathcal{R}\},$
      $R = \{f(x) \mid f(x) \in \mathcal{R}\}$
   c. $f(x) = x - 7,\ f^{-1}(x) = x + 7$

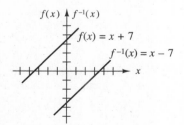

47. a. $f^{-1}(x) = \dfrac{x + 4}{5}$
   b. $f: D = \{x \mid x \in \mathcal{R}\},$
      $R = \{f(x) \mid f(x) \in \mathcal{R}\};$
      $f^{-1}: D = \{x \mid x \in \mathcal{R}\},$
      $R = \{f^{-1}(x) \mid f^{-1}(x) \in \mathcal{R}\}$
   c. $f(x) = 5x - 4;\ f^{-1}(x) = \dfrac{x + 4}{5}$

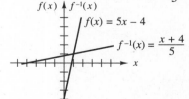

49. a. $f^{-1}(x) = \dfrac{3x - 4}{2}$

    b. $f: D = \{x \mid x \in \mathcal{R}\},$
$R = \{f(x) \mid f(x) \in \mathcal{R}\};$
$f^{-1}: D = \{x \mid x \in \mathcal{R}\},$
$R = \{f^{-1}(x) \mid f^{-1}(x) \in \mathcal{R}\}$

    c. $f(x) = \dfrac{2x + 4}{3},$

$f^{-1}(x) = \dfrac{3x - 4}{2}$

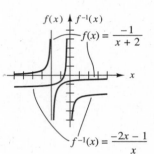

51. a. $f^{-1}(x) = \dfrac{3}{x}$

    b. $f: D = [x \mid x \neq 0\},$
$R = \{y \mid y \neq 0\};$
$f^{-1}: D = [x \mid x \neq 0\},$
$R = \{y \mid y \neq 0\}$

    c. $f(x) = \dfrac{3}{x}, f^{-1}(x) = \dfrac{3}{x}$

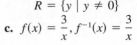
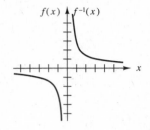

53. a. $f^{-1}(x) = \dfrac{-2x - 1}{x}$

    b. $f: D = [x \mid x \neq -2\},$
$R = \{y \mid y \neq 0\};$
$f^{-1}: D = [x \mid x \neq 0\}$
$R = \{y \mid y \neq -2\}$

    c. $f(x) = \dfrac{-1}{x + 2},$

$f^{-1}(x) = \dfrac{-2x - 1}{x}$

[5.2–5.3] 55. $d = \sqrt{(300t)^2 + \left(\dfrac{600}{5280}\right)^2}$

57. a. $\overline{AC} = \sqrt{13}, \overline{BC} = \sqrt{13},$
$\overline{DC} = \sqrt{13}$

    b. $r = \sqrt{13}$

59. a. $f(x) = 1.15 + 0.25(6x), n \leq x < (n + 1)$
where $n = 0, 1, 2, 3, \ldots$ and n represents
the number of $\frac{1}{6}$ of a mile.

    b.

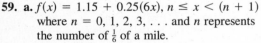

miles: x	$\frac{1}{3}$	1	$1\frac{1}{3}$
cost: $f(x)$	\$1.65	\$2.65	\$3.15

    c.

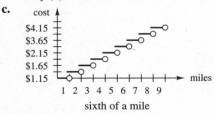

61. Each distance is $\dfrac{1}{2}\sqrt{a^2 + b^2}$.

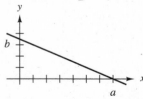

(*Hint:* Let vertices be $(0, 0)(8, 0)$ and $(6, 0)$, then
midpoint of hypotenuse is $(4, 3)$ and all distances are
5 units.)

63. $3x + 2y - 17 = 0$

65.
```
10 PRINT "WHAT ARE THE COORDINATES OF THE FIRST POINT"
20 INPUT X1,Y1
30 PRINT "WHAT ARE THE COORDINATES OF SECOND POINT"
40 INPUT X2,Y2
50 LET D = SQR((X1 - X2)^2+(Y1 - Y2)^2)
60 PRINT "THE DISTANCE IS"; D
70 END
```

Chapter Test (page 247)

1. $l: x, d: y$ **2.** $i: r, d: C$ **3.** 31   **4.** $3a^2 + 6ah + h^2 - 7a - 7h + 5$ **5.** $D = \{x \mid x \in \mathcal{R}\}, R = \{f(x) \mid f(x) \geq 2\}$
6. $D = \{x \mid x \geq -5\}, R = \{f(x) \mid f(x) \geq 0\}$ **7.** 10   **8.** $(1, -1)$
9. No asymptotes for $f(x) = 3x - 4$

10. Asymptotes for $f(x) = \dfrac{1}{x + 1}$: $x = -1, y = 0$

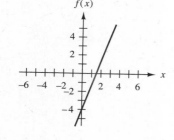

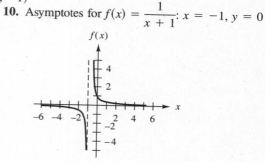

11. $f^{-1}(x) = \dfrac{x + 4}{3}$ **12.** $f^{-1}(x) = \dfrac{-3x + 2}{x}$ **13.** $f(x) = 12x$ **14.** $C(x) = \$125 + \$0.50x$ **15.** $\$285.50$

16. $y = -5x + 6, y = 5x + 6$;

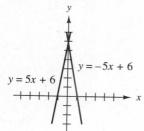

similar: linear functions, same y-intercept $(0, 6)$,
same domain $D = \{x \mid x \in \mathcal{R}\}$,
same range $R = \{y \mid y \in \mathcal{R}\}$;
different: one line increases, moving from left to right; the
other line decreases, x-intercepts $\left(\dfrac{-6}{5}, 0\right), \left(\dfrac{6}{5}, 0\right)$

17. $y = 3x^2 - 4, y = -3x^2 - 4$;

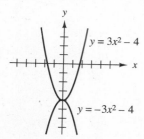

similar: parabolas, same y-intercept $(0, -4)$, line of
symmetry is same for both, y-axis, same vertex $(0, -4)$;

different: one parabola opens up and the other opens
down: parabola that opens up has x-intercepts at
$\left(\dfrac{-2\sqrt{3}}{3}, 0\right)$ and $\left(\dfrac{2\sqrt{3}}{3}, 0\right)$; the other parabola has no
x-intercepts.

Chapter 6

Section 6.1 (page 254)

1. (b)   **3.** (c)   **5.** (b)   **7.** (a)   **9.** (c)   **11.** (a)

13. $\sin \theta = \dfrac{4}{5}$, $\csc \theta = \dfrac{5}{4}$,

$\cos \theta = \dfrac{3}{5}$, $\sec \theta = \dfrac{5}{3}$,

$\tan \theta = \dfrac{4}{3}$, $\cot \theta = \dfrac{3}{4}$,

15. $\sin \theta = \dfrac{1}{\sqrt{2}}$, $\csc \theta = \dfrac{\sqrt{2}}{1}$,

$\cos \theta = \dfrac{1}{\sqrt{2}}$, $\sec \theta = \dfrac{\sqrt{2}}{1}$,

$\tan \theta = 1$, $\cot \theta = 1$

17. $\sin \theta = \dfrac{5}{\sqrt{41}}$, $\csc \theta = \dfrac{\sqrt{41}}{5}$,

$\cos \theta = \dfrac{4}{\sqrt{41}}$, $\sec \theta = \dfrac{\sqrt{41}}{4}$,

$\tan \theta = \dfrac{5}{4}$, $\cot \theta = \dfrac{4}{5}$

19. $\sin \theta = \dfrac{2}{\sqrt{5}}$, $\csc \theta = \dfrac{\sqrt{5}}{2}$,

$\cos \theta = \dfrac{1}{\sqrt{5}}$, $\sec \theta = \dfrac{\sqrt{5}}{1}$,

$\tan \theta = \dfrac{2}{1}$, $\cot \theta = \dfrac{1}{2}$

21. $\sin \theta = \dfrac{1}{2}$, $\csc \theta = \dfrac{2}{1}$,

$\cos \theta = \dfrac{\sqrt{3}}{2}$, $\sec \theta = \dfrac{2}{\sqrt{3}}$,

$\tan \theta = \dfrac{1}{\sqrt{3}}$, $\cot \theta = \dfrac{\sqrt{3}}{1}$

23. $\sin \theta = 0.77$, $\csc \theta = 1.3$

$\cos \theta = 0.64$, $\sec \theta = 1.6$,

$\tan \theta = 1.2$, $\cot \theta = 0.82$

25. $\sin \theta = 0.859$, $\csc \theta = 1.16$,
$\cos \theta = 0.513$, $\sec \theta = 1.95$,
$\tan \theta = 1.67$, $\cot \theta = 0.597$

27. $\sin \theta = 0.829$, $\csc \theta = 1.21$,
$\cos \theta = 0.559$, $\sec \theta = 1.79$,
$\tan \theta = 1.48$, $\cot \theta = 0.674$

29. $\sin \theta = \dfrac{1}{\sqrt{17}}$, $\sec \theta = \dfrac{\sqrt{17}}{4}$

31. $\cos \theta = \dfrac{3}{\sqrt{13}}$, $\csc \theta = \dfrac{\sqrt{13}}{2}$

33. $\sin \theta = \dfrac{2}{5}$, $\tan \theta = \dfrac{2}{\sqrt{21}}$

35. $\csc \theta = \dfrac{2}{\sqrt{2}}$, $\tan \theta = 1$

37. $\sin \theta = 0.93$, $\sec \theta = 2.7$

39. $-325°, 395°$

41. $315°, -405°$

43. $225°, -495°$

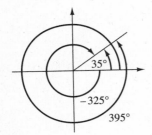

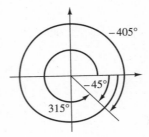

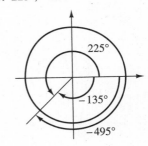

45. $\tan \theta = \dfrac{\sin \theta}{\cos \theta}$

$= \dfrac{y/r}{x/r}$

$= \dfrac{y}{x}$

$= \tan \theta$

47. $\sec \theta = \dfrac{1}{\cos \theta}$

$= \dfrac{1}{x/r}$

$= \dfrac{r}{x}$

$= \sec \theta$

Section 6.2 (page 263)

1. $x = 4.3$, $y = 2.5$ **3.** $r = 16$, $y = 8\sqrt{3}$ **5. a.** 8.8 mi   **b.** 8.8 mi   **7.** 34.82 m   **9.** 3.820×10^5 mi   **11.** 0.980
13. 0.980   **15.** 0.682   **17.** 0.437   **19.** 0.9030   **21.** 5.1×10^{-3} **23.** 0.2899   **25.** 0.2717   **27.** 0.36470   **29.** 0.35982
31. 0.76725   **33.** $\sin 45° = \dfrac{1}{\sqrt{2}}$, $\csc 45° = \dfrac{\sqrt{2}}{1}$, $\cos 45° = \dfrac{1}{\sqrt{2}}$, $\sec 45° = \dfrac{\sqrt{2}}{1}$, $\tan 45° = 1$, $\cot 45° = 1$
35. 0.39   **37.** 0.67   **39.** 0.384   **41.** 13°, 13°   **43.** 13°, 13°   **45.** 30.0°, 30° 00′   **47.** 60.0°, 60° 00′   **49.** 60.0°, 60° 00′
51. 45.00°, 45° 00′   **53.** 4.8   **55.** 4.8   **57.** 2.9   **59.** 1.47   **61.** 1.21   **63.** 1.355   **65.** 3°, 3°   **67.** 0°, 0°   **69.** 2°, 2°
71. 30.00°, 30° 00′   **73.** 29.94°, 29° 56′   **75.** 30.00°, 30° 00′   **77.** 29 km/hr   **79.** 1.9 km/hr   **81. a.** 1.61   **b.** 1.52
83. a. 29 lb   **b.** 7.8 lb   **85.** 18 in., 42 in.

Section 6.3 (page 271)

1. $B = 74.7°$, $C = 90.0°$,   **3.** $A = 25.8°$, $C = 90.0°$,   **5.** $B = 45°$, $C = 90°$, $a = 1.8$ mm
  $a = 4.33$ in., $b = 15.8$ in. $a = 13.7$ mm, b = 28.3 mm $c = 2.6$ mm
7. $A = 71.8°$, $C = 90.0°$,   **9.** $A = 45°$, $C = 90°$, $a = 22$ in.   **11.** $B = 30°$, $C = 90°$, $b = 8.5$ m
  $b = 1.00$ ft, $c = 3.20$ ft $c = 32$ in. $c = 17$ m
13. 10.6 m²   **15.** 731 mm²   **17.** 156 ft   **19.** 53°, 37°   **21.** 1353 ft   **23.** 403.1 ft   **25.** 71 m   **27.** 3660 ft
29. Circle field; would need a glide angle of 22.3° for straight-line approach.
31. $X_L = 11.3$ Ω, 30.8°   **33.** 4.34 VA, 3.98 W   **35.** 9 in.   **37.** 1.4 m   **39.** 49 mm, 100 mm   **41.** $x = 7.0$ ft, $y = 26$ ft
43. $\alpha = 41.4°$, $\beta = 48.6°$

Section 6.4 (page 278)

1. Positive   **3.** Positive   **5.** Negative   **7.** Positive   **9.** Negative   **11.** Positive   **13.** Negative
15. Negative   **17.** Negative   **19.** Positive   **21.** Positive   **23.** Positive   **25.** Positive   **27.** Positive
29. Negative   **31.** Negative   **33.** Positive   **35.** Positive
37. $\sin\theta = \dfrac{4}{5}$, $\csc\theta = \dfrac{5}{4}$, $\cos\theta = -\dfrac{3}{5}$, $\sec\theta = -\dfrac{5}{3}$, $\tan\theta = -\dfrac{4}{3}$, $\cot\theta = -\dfrac{3}{4}$
39. $\sin\theta = -\dfrac{4}{5}$, $\csc\theta = -\dfrac{5}{4}$, $\cos\theta = -\dfrac{3}{5}$, $\sec\theta = -\dfrac{5}{3}$, $\tan\theta = \dfrac{4}{3}$, $\cot\theta = \dfrac{3}{4}$
41. $\sin\theta = -\dfrac{5}{\sqrt{41}}$, $\csc\theta = \dfrac{\sqrt{41}}{5}$, $\cos\theta = \dfrac{4}{\sqrt{41}}$, $\sec\theta = \dfrac{\sqrt{41}}{4}$, $\tan\theta = -\dfrac{5}{4}$, $\cot\theta = -\dfrac{4}{5}$
43. $\sin\theta = \dfrac{5}{\sqrt{29}}$, $\csc\theta = \dfrac{\sqrt{29}}{5}$, $\cos\theta = \dfrac{2}{\sqrt{29}}$, $\sec\theta = \dfrac{\sqrt{29}}{2}$, $\tan\theta = \dfrac{5}{2}$, $\cot\theta = \dfrac{2}{5}$
45. First   **47.** Second   **49.** Third   **51.** First   **53.** Fourth   **55.** Second

Section 6.5 (page 283)

1. d   **3.** a   **5.** b   **7.** c   **9.** h   **11.** −0.63   **13.** −0.895   **15.** 2.41   **17.** 0.09700   **19.** −0.3840   **21.** −15.5
23. 0.665   **25.** 1.0   **27.** −1.16019   **29.** −1.43   **31.** 48°, 132°   **33.** 37°, 217°   **35.** 138.2°, 221.8°   **37.** 33.00°, 147.00°
39. 55.00°, 235.00°   **41.** 120.49°, 239.51°   **43.** 33.3°, 146.7°   **45.** 35.0°, 215.0°   **47.** 123.3°, 236.7°   **49.** −0.80902

51. 0.98481   **53.** −0.70021   **55. a.** $x = \dfrac{0.5\,V_0{}^2}{g}$ **b.** $x = \dfrac{V_0{}^2}{g}$ **c.** $x = \dfrac{0.50\,V_0{}^2}{g}$ **d.** Sine function has a maximum

value at 90°, which occurs when $\alpha = 45°$ **57. a.** The flight time is greater, giving the defenders more time to move

downfield and prevent a run back. When $A = 55°$, $t = \dfrac{0.82V}{g}$, in football this flight time is called hang time   **b.** When

$A = 35°$, this gives the same distance as $A = 55°$, but $t = \dfrac{0.57V}{g}$, and the hang time is less.   **59.** 27,900 m²

Section 6.6 (page 290)

1. $\dfrac{\pi}{12}$

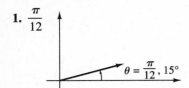

$\theta = \dfrac{\pi}{12},\ 15°$

3. $\dfrac{\pi}{4}$

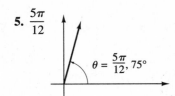

$\theta = \dfrac{\pi}{4},\ 45°$

5. $\dfrac{5\pi}{12}$

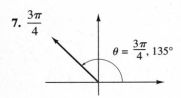

$\theta = \dfrac{5\pi}{12},\ 75°$

7. $\dfrac{3\pi}{4}$

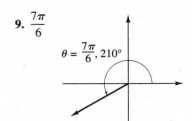

$\theta = \dfrac{3\pi}{4},\ 135°$

9. $\dfrac{7\pi}{6}$

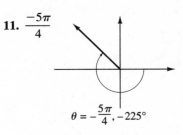

$\theta = \dfrac{7\pi}{6},\ 210°$

11. $\dfrac{-5\pi}{4}$

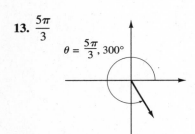

$\theta = -\dfrac{5\pi}{4},\ -225°$

13. $\dfrac{5\pi}{3}$

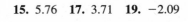

$\theta = \dfrac{5\pi}{3},\ 300°$

15. 5.76 **17.** 3.71 **19.** −2.09

21. 0.68 **23.** 5.06

25. 22.5°

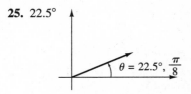

$\theta = 22.5°,\ \dfrac{\pi}{8}$

27. 45°

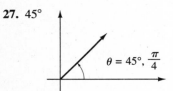

$\theta = 45°,\ \dfrac{\pi}{4}$

29. 75°

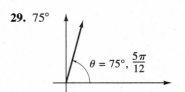

$\theta = 75°,\ \dfrac{5\pi}{12}$

31. 135°

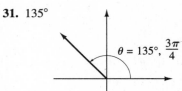

$\theta = 135°,\ \dfrac{3\pi}{4}$

33. 210°

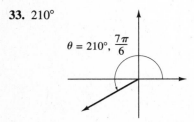

$\theta = 210°,\ \dfrac{7\pi}{6}$

35. 225°

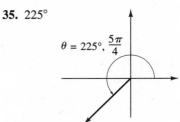

$\theta = 225°,\ \dfrac{5\pi}{4}$

37. 300°

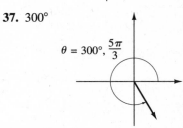

$\theta = 300°,\ \dfrac{5\pi}{3}$

39. 330°

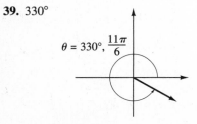

$\theta = 330°,\ \dfrac{11\pi}{6}$

41. 99.7°

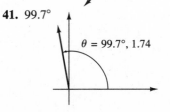

$\theta = 99.7°,\ 1.74$

43. 57.3°

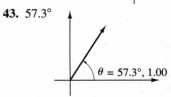

$\theta = 57.3°,\ 1.00$

45. 0.969 **47.** 1.74 **49.** −0.764 **51.** −0.304 **53.** 2.01 **55.** 0.960 **57.** −0.275 **59.** 1.32°, 1.82° **61.** 1.05°, 4.19°
63. 2.44°, 3.84° **65.** 3.45°, 5.97° **67.** 1.11°, 4.25° **69.** 0.284°, 6.00° **71.** 3.42°, 6.00° **73.** 2.095°, 5.236°
75. 1.57, 90° **77.** $(3.58 \times 10^{-4})°$ **79.** 990 m **81.** The wheel that rotates 5.00 rad is closer to the inside of the curve.
83. $628.3 \dfrac{\text{rad}}{\text{s}}$ **85.** 15 revolutions **87.** Yes, more chain passes over the rear sprocket wheel. 11.1 rad

Section 6.7 (page 298)

1. 20.0°, 0.349 rad **3.** 3.0×10^{-5} rad, 0.0017° **5. a.** 4.0×10^{-5} **b.** 8.0×10^{-5} **c.** 2.0×10^{-5} **d.** A good microscope requires high resolving power. This can be accomplished by increasing the distance between the lens and the object or both **7.** 1221 m **9.** $1080 **11. a.** 7.7 m **b.** 4.0 m **13.** $335 \dfrac{\text{mi}}{\text{min}}$ **15.** 4 revolutions **17.** $12,300 \dfrac{\text{mi}}{\text{day}}$
19. $2.990 \dfrac{\text{rev}}{\text{hr}}$ **21.** 52.8 m **23. a.** $1050 \dfrac{\text{in.}}{\text{min}}, 209 \dfrac{\text{rad}}{\text{min}}$ **b.** $630 \dfrac{\text{in.}}{\text{min}}, 209 \dfrac{\text{rad}}{\text{min}}$ **25. a.** $13 \dfrac{\text{rad}}{\text{s}}$ **b.** $380 \dfrac{\text{mm}}{\text{s}}$
c. The linear velocity represents the speed of the belt **d.** $5.0 \dfrac{\text{rad}}{\text{sec}}$ **e.** 48 rpm **27.** $1.5 \times 10^4 \dfrac{\text{in.}}{\text{min}}$ **29. a.** 24°
b. 600 m **c.** 84° **d.** 120° **31. a.** 2400 m **b.** 2700 m

Review Exercises (page 303)

[6.6] 1. $\dfrac{7\pi}{18}$, 1.2 **3.** $\dfrac{11\pi}{9}$, 3.8 **5.** $-\dfrac{4\pi}{9}$, −1.4 **7.** 165° **9.** 115°

[6.1] 11. $\sin \theta = \dfrac{7}{\sqrt{74}}$, $\csc \theta = \dfrac{\sqrt{74}}{7}$, $\cos \theta = \dfrac{5}{\sqrt{74}}$, $\sec \theta = \dfrac{\sqrt{74}}{5}$, $\tan \theta = \dfrac{7}{5}$, $\cot \theta = \dfrac{5}{7}$

13. $\sin \theta = -\dfrac{\sqrt{2}}{2}$, $\csc \theta = \dfrac{2}{-\sqrt{2}}$, $\cos \theta = -\dfrac{\sqrt{2}}{2}$, $\sec \theta = \dfrac{2}{-\sqrt{2}}$, $\tan \theta = 1$, $\cot \theta = 1$

15. $\sin \theta = \dfrac{1}{\sqrt{2}}$, $\csc \theta = \sqrt{2}$, $\cos \theta = \dfrac{1}{\sqrt{2}}$, $\sec \theta = \sqrt{2}$, $\tan \theta = 1$, $\cot \theta = 1$

17. $\cos \theta = \dfrac{2}{\sqrt{7}}$, $\csc \theta = \dfrac{-\sqrt{7}}{3}$ **19.** $\csc \theta = \dfrac{5\sqrt{2}}{7}$, $\cos \theta = \dfrac{1}{5\sqrt{2}}$
[6.2, 6.4, 6.5] 21. 0.319 **23.** −6.65 **25.** −1.04 **27.** −0.953 **29.** 0.541
[6.4, 6.5, 6.6] 31. 13.566°, 166.434°, 0.2368, 2.9048 **33.** 13.500°, 193.500°, 0.23562, 3.3772
35. 105.0000°, 255.0000°, 1.83260, 4.45059 **37.** 40.000°, 320.000°, 0.69813, 5.5850 **39.** 270.0000°, 4.71239
[6.3, 6.6] 41. $B = 51.5°$, $a = 12.1$ mm, $b = 15.2$ mm **43.** $A = 65.3°$, $a = 22.0$ in., $b = 10.1$ in.
45. $B = 0.82$, $b = 16$ m, $c = 21$ m **47. a.** 690 ft **b.** 990 ft **49.** $\theta = 120°$ **51.** $812 \dfrac{\text{rad}}{\text{min}}$ **53.** 59 mm **55.** 100 ft
57. a. 8 cm **b.** −5.1 cm **c.** 7.9 cm **59.** 2.2 cm² **61.** 141°, 2.5 **63.** 148 ft² **65.** 1.2 m² **67. a.** 43° **b.** 140°
69. (40, 0), (12, 38), (−32, 23), (−32, −23), (12, −38)

Chapter Test (page 308)

1. $\dfrac{29\pi}{36}$, 2.53   **2.** $\dfrac{11\pi}{36}$, 0.96   **3.** 120°   **4.** 85.94°

5. $\tan\theta = \dfrac{4}{3}$; $\csc\theta = \dfrac{5}{4}$.   **6.** $\sin\theta = \dfrac{1}{2}$; $\sec\theta = \dfrac{2}{\sqrt{3}}$.   **7.** 0.552   **8.** -0.0192 **9.** -1.3 **10.** 1.7   **11.** 1.671

12. 1.56   **13.** $\theta = 35°$ and $145°$ **14.** $\theta = 140.0°$ and $320.0°$.   **15.** $\theta = 135.01°$ and $224.99°$.   **16.** $A = 51°$

17. $a = 22$ mm   **18.** 45 ft   **19.** 19 m²   **20.** 47,300 mph

Chapter 7

Section 7.1 (page 315)

1. Vector   **3.** Vector   **5.** Scalar   **7.** Scalar   **9.** Vector

11, 13.

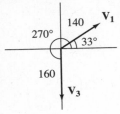

15.

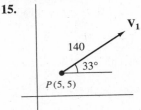

17.

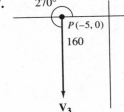

19. $V_H = 6.5\ \dfrac{\text{cm}}{\text{s}}$

$V_V = 75\ \dfrac{\text{cm}}{\text{s}}$

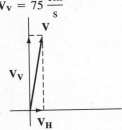

21. $F_H = -3.7$ lb

$F_V = 25$ lb

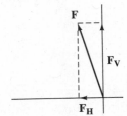

23. $D_H = -37$ km

$D_V = -31$ km

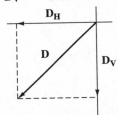

25. $F_H = 110$ dynes

   $F_V = -89$ dynes

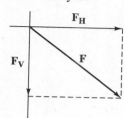

31. $F = 59$ lb $\underline{/53°}$

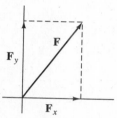

37. $V = 89 \dfrac{cm}{s} \underline{/-118°}$ or

   $V = 89 \dfrac{cm}{s} \underline{/242°}$

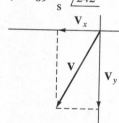

43. 810 lb $\underline{/68°}$ **45. a.** 65 lb   **b.** 55 lb

27. $A_H = 18 \dfrac{cm}{s^2}$

   $A_V = 26 \dfrac{cm}{s^2}$

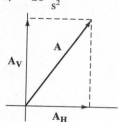

33. $D = 370$ km $\underline{/-21.4°}$ or

   $D = 370$ km $\underline{/338.6°}$

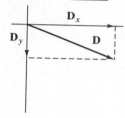

39. $A = 74 \dfrac{ft}{s^2} \underline{/104°}$.

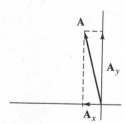

29. $F_H = -0.25$ kg

   $F_V = -0.022$ kg

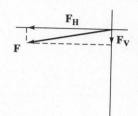

35. $F = 38.0$ dyne $\underline{/117°}$

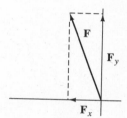

41. $V = 12.6 \dfrac{mi}{hr} \underline{/18.4°}$

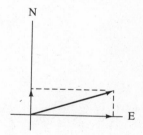

47. 45.5 Ω $\underline{/-56.7°}$ or 45.5 Ω $\underline{/303.3°}$

Section 7.2 (page 324)

1. $A + B = 5$ in. $\underline{/37°}$

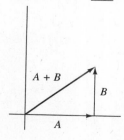

3. $C + D = 8$ in. $\underline{/95°}$

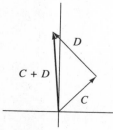

5. $A + C + E = 5$ in. $\underline{/45°}$

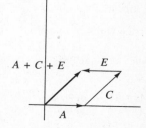

7. $\mathbf{B} + \mathbf{C} + \mathbf{E}$ = 7 in. $/91°$

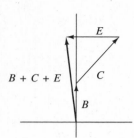

9. $\mathbf{A} - \mathbf{B}$ = 5 in. $/-37°$ or
$\mathbf{A} - \mathbf{B}$ = 5 in. $/323°$

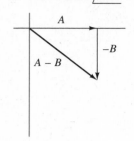

11. $\mathbf{E} - \mathbf{A}$ = 8 in. $/180°$

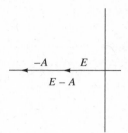

13. $\mathbf{A} + \mathbf{B} + \mathbf{C} + \mathbf{E}$ = 7 in. $/60°$

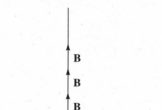

15. $\mathbf{B} + \mathbf{B} + \mathbf{B}$ = 9 in. $/90°$

17. $2\mathbf{A} + 3\mathbf{B}$ = 12 in. $/48°$

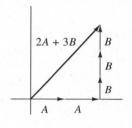

19. $4\mathbf{D} - 5\mathbf{E}$ = 17 in. $/80°$

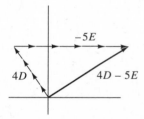

21. 4.41 cm $/38.6°$ **23.** 5.5 ft $/110°$ **25.** 33 m $/136°$ **27.** 0.78 in. $/128°$ **29.** 38 cm $/142°$ **31.** 0.91 in. $/67°$
33. 4.6 m $/1.3°$ **35.** 15 ft $/136°$ **37.** 1.7 mi   **39.** 300 mph east   **41.** 290 lb $/-4°$ **43. a.** 310 lb   **b.** 300 lb
45. $A_y = -1.5 \times 10^4$ kg m/s, $B_y = 1.5 \times 10^4$ kg m/s; the y-component of the resultant vector is the sum of A_y and B_y,
which is zero. Thus, the momentum in the direction of the y-axis is zero   **b.** $A_x = 1.5 \times 10^4$, $B_x = 1.5 \times 10^4$,
$x = 3.0 \times 10^4$ kg m/s   **c.** $V = 7.1$ m/s

Section 7.3 (page 333)

1. 5 km/hr $/30°$ **3.** 174 $\dfrac{\text{km}}{\text{h}}$, 985 $\dfrac{\text{km}}{\text{h}}$ **5.** 14 mph $/96°$ **7. a.** 438 lb   **b.** 204 lb   **c.** 602 lb
9. Brace 280 lb, cable 46. lb   **11.** 11,000 N   **13. a.** $f = 0.210$ **b.** $N = 4400$ N, $f = 620$ N   **15.** 97 lb   **17.** 1790 N

Section 7.4 (page 340)

1. $C = 33°$, $a = 7.6$ cm, $c = 4.4$ cm   **3.** $B = 42°$, $C = 78°$, $c = 10.0$ in.   **5.** $B = 49.1°$, $C = 66.6°$, $c = 5.24$ mm
7. $A = 11.1°$, $B = 138.4°$, $b = 13.7$ in.   **9.** No solution   **11.** $A = 39°50'$, $a = 295$ ft, $c = 214$ ft
13. $A = 90°$, $B = 45°$, $a = 3.5$ in.   **15.** $C = 0.34$, $b = 48$ cm, $c = 18$ cm   **17.** $A = 94°$, $C = 4°$, $a = 2.3 \times 10^5$ m
19. $A = 14°$, $a = 2.2$ ft, $c = 6.7$ ft   **21.** $B = 70°$, $C = 50°$, $c = 64$ mm, $B' = 110°$, $C' = 10°$, $c' = 14$ mm

23. Yes, for $A = 90°$ the ratio $\dfrac{a}{\sin 90°} = \dfrac{a}{1} = a$, $B = 45°$ **25.** 922.5 ft   **27.** P, 1.9 km   **29.** 45 m   **31.** Given one side we can determine the measure of the three angles. **b.** 56 mi

Section 7.5 (page 345)

1. $A = 57°$, $B = 45°$, $c = 14$ **3.** $B = 65.9°$, $C = 54.2°$, $a = 42.7$ **5.** $A = 48.3°$, $C = 27.4°$, $b = 67.5$
7. $A = 44°$ $B = 60°$, $C = 76°$ **9.** $A = 91.8°$, $B = 35.3°$, $C = 52.9°$ **11.** $A = 52.22°$, $B = 33.96°$, $C = 93.82°$
13. 34° 31′   **15.** $c = 38$ cm   **17.** 7221.3 m   **19.** 673 lb   **21.** 14.0 mi   **23.** 34.0°

Review Exercises (page 346)

[7.1] **1.** $\mathbf{V_H} = 18 \dfrac{\text{cm}}{\text{s}}$

$\mathbf{V_V} = 44 \dfrac{\text{cm}}{\text{s}}$

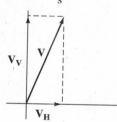

3. $\mathbf{F_H} = -30$ lb

$\mathbf{F_V} = -18$ lb

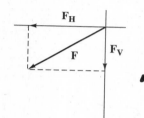

5. $\mathbf{D_H} = 100$ km

$\mathbf{D_V} = -60$ km

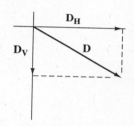

7. $\mathbf{a_H} = -4.79 \dfrac{\text{cm}}{\text{s}^2}$

$\mathbf{a_V} = 31.8 \dfrac{\text{cm}}{\text{s}^2}$

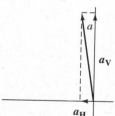

9. $\mathbf{F} = 242.0$ lb $\underline{/4.3°}$

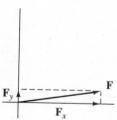

11. $\mathbf{D} = 21$ km $\underline{/133°}$

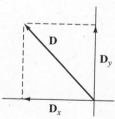

13. $\mathbf{F} = 1.07$ N $\underline{/-116.1°}$ or

$\mathbf{F} = 1.07$ N $\underline{/243.9°}$

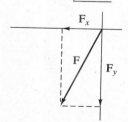

15. $\mathbf{V} = 0.029 \dfrac{\text{m}}{\text{s}} \underline{/-76°}$ or

$\mathbf{V} = 0.029 \dfrac{\text{m}}{\text{s}} \underline{/284°}$

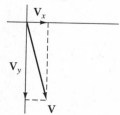

[7.2, 7.3] **17.** 3.6 mph $\underline{/304°}$ (compass heading)

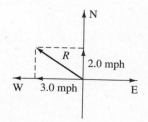

19. 50 Ω $\angle -49°$ or 50 Ω $\angle 311°$ **21. a.** 2400 lb   **b.** 20,000 lb   **23.** 96 cm $\angle 75°$ **25.** 1.4 $\angle 38°$ **27.** 500 lb   **29.** 64 N
[7.4, 7.5]   **31.** $C = 74.7°$, $a = 28.2$, $c = 44.0$ **33.** $B = 30.78°$, $C = 126.22°$, $c = 23.21$ $B = 149.22°$, $C = 7.78°$,
$c = 3.893$ **35.** $A = 55°$, $B = 85°$, $b = 23$ $A = 125°$, $B = 15°$, $b = 5.8$ **37.** $B = 71.6°$, $C = 57.5°$, $a = 36.8$
39. $A = 52.1°$, $B = 57.2°$, $C = 70.7°$ **41.** $A = 21.6°$, $B = 12.1°$, $c = 317$ **43.** 3.7 mi $\angle 31°$ **45.** 20 in.   **47.** 0.626 mi
49. 3.4 km   **51.** 6.3 ft.

Chapter Test *(page 349)*

1.

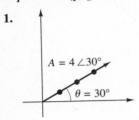

$A = 4 \angle 30°$
$\theta = 30°$

2.

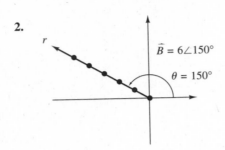

$\vec{B} = 6 \angle 150°$
$\theta = 150°$

3. $A_x = -2.6$ **4.** $A_y = 2.4$ **5.** $\mathbf{A} + \mathbf{B} = 2.2 \angle 192°$ **6.** $\mathbf{B} - \mathbf{A} = 12 \angle 219°$ **7.** $\mathbf{R} = 12 \angle 289°$
8. $C = 78°$; $a = 12$; $c = 17$ **9.** For $A = 50°$, $C = 95°$; $c = 21$. For $A = 130°$, $C = 15°$; $c = 5.4$
10. $a = 20$; $C = 75°$; $B = 30°$ **11.** $A = 25°$; $C = 120°$; $c = 24$ **12.** $c = 20$; $A = 8°$; $B = 12°$
13. $A = 28°$; $B = 81°$; $c = 71°$
14. 567 mph
15. 178 N; 82.9 N   **16.** $B = 55°$; $a = 2.7$ mi; $c = 2.2$ mi
17. 104 ft   **18.** 285 lb.

Chapter 8

Section 8.1 *(page 355)*

1. a. Yes   **b.** 1   **3. a.** No   **5. a.** Yes   **b.** π **7. a.** Yes   **b.** π **9. a.** Yes   **b.** 2π **11. a.** Yes   **b.** 4
13. a. Yes   **b.** 10
15.

θ	$2\pi/3$	$3\pi/4$	$5\pi/6$	π	$7\pi/6$	$5\pi/4$	$4\pi/3$	$3\pi/2$	$5\pi/3$	$7\pi/4$	$11\pi/6$	2π
$\sin\theta$	0.866	0.707	0.500	0	-0.500	-0.707	-0.866	-1	-0.866	-0.707	-0.500	0
$\cos\theta$	-0.500	-0.707	-0.866	-1	-0.866	-0.707	-0.500	0	0.500	0.707	0.866	1

Section 8.2 *(page 368)*

1. e   **3.** f   **5.** d   **7.** Domain set of real numbers; range $-1 \le y \le 1$; y-intercept is 0; θ intercepts are 0, $\pm\pi$,
$\pm 2\pi, \ldots$; $\sin(\pi + \theta) = -\sin\theta$; $\sin(-\theta) = -\sin\theta$; $\sin(\pi - \theta) = \sin\theta$; periodic: $p = 2\pi$.   **9.** Domain reals except

$\theta \ne \pm\dfrac{\pi}{2}, \pm\dfrac{3\pi}{2}, \pm\dfrac{5\pi}{2}, \ldots$; range set of real numbers, y-intercept is 0; θ intercepts are 0, $\pm\pi$, $\pm 2\pi, \ldots$;

$\tan(\pi + \theta) = \tan\theta$; $\tan(-\theta) = -\tan\theta$; $\tan(\pi - \theta) = -\tan\theta$; periodic: $p = \pi$; undefined at $\pm\dfrac{\pi}{2}, \pm\dfrac{3\pi}{2}, \ldots$.

11. Domain reals except $\theta \neq \pm\dfrac{\pi}{2}, \pm\dfrac{3\pi}{2}, \pm\dfrac{5\pi}{2}, \dots$; range reals $y \leq -1$ or $y \geq 1$; y-intercept is 1; θ intercepts: none; $\sec(\pi + \theta) = -\sec\theta$; $\sec(-\theta) = \sec\theta$; $\sec(\pi - \theta) = -\sec\theta$; periodic: $p = 2\pi$; undefined at $\pm\dfrac{\pi}{2}, \pm\dfrac{3\pi}{2}, \pm\dfrac{5\pi}{2}, \dots$.

13.

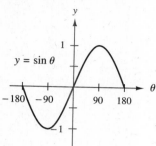

15.

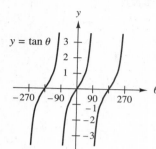

17.

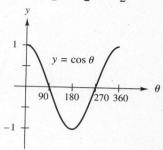

19.

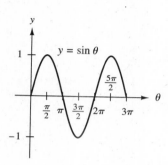

21.

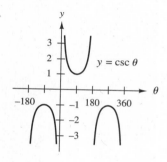

23.

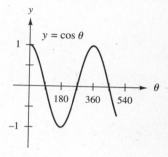

25.

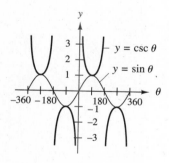

27.

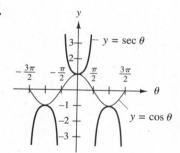

29. Sine, cosine   **31.** Tangent, cotangent   **33.** Tangent, cotangent   **35.** Tangent, secant
37. Secant, cosecant   **39.** Sine, tangent   **41.** Sine, tangent   **43.** Sine, tangent, cosecant, cotangent
45. Sine, cosine, cosecant, secant

Section 8.3 (page 380)

1. a. 3 **b.** 2π **c.** 1 **3. a.** 0.75 **b.** 2π **c.** 1 **5. a.** 1 **b.** $\dfrac{2\pi}{3}$ **c.** 3 **7. a.** 6 **b.** 8π **c.** $\dfrac{1}{4}$

9. a. $\frac{1}{3}$ **b.** 6π **c.** $\frac{1}{3}$ **11. a.** $\frac{1}{3}$ **b.** 6π **c.** $\frac{1}{3}$ **13.** b **15.** a

17. $y = 2 \sin \theta$

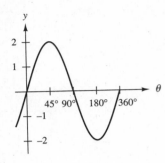

19. $y = \cos \dfrac{\theta}{2}$

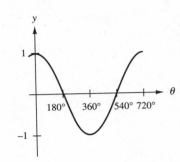

21. $y = 3 \cos 2\theta$

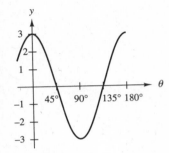

23. $y = 0.5 \sin \dfrac{\pi}{4} \theta$

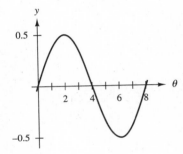

25. $y = \dfrac{1}{2} \sin 2\theta$

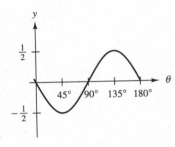

27. $y = -\dfrac{2}{3} \cos \dfrac{2}{3} \theta$

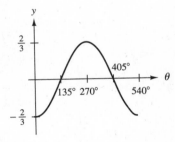

29. $y = 3 \sin \theta$

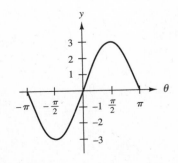

31. $y = \dfrac{1}{2} \cos \theta$

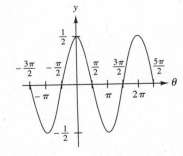

33. $-\dfrac{\pi}{4} \le \theta \le \dfrac{\pi}{4}$, many correct answers possible   **35.** $y = 2 \sin 2\theta$ **37.** $y = \dfrac{1}{3} \sin \dfrac{\theta}{2}$

39. a. $\dfrac{5}{2}, \dfrac{\pi}{4}$

    b. $y = \dfrac{5}{2} \cos 8t$

    c. 2.1

41. a. 2.5 s   **b.** 24

    c. $y = 0.25 \cos 2.5t$

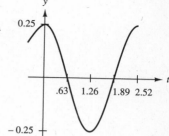

43. a. E note   **b.** $I = 20 \sin (2\pi \cdot 262t)$ **c.** $I = 10 \sin (2\pi \cdot 392t)$

Section 8.4 (page 381)

1. a. 2π **b.** $\dfrac{\pi}{2}$ **c.** 1   **d.** $\dfrac{5\pi}{2}$

    e. $y = \sin\left(\theta - \dfrac{\pi}{2}\right)$

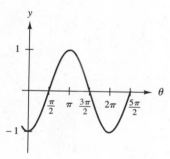

3. a. $360°$ **b.** $30°$ **c.** 1   **d.** $390°$

    e. $y = \sin (\theta - 30°)$

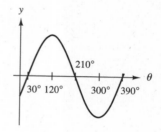

5. a. $\dfrac{2\pi}{3}$ **b.** $\dfrac{\pi}{9}$ **c.** 1   **d.** $\dfrac{7\pi}{9}$

    e. $y = \cos\left(3\theta - \dfrac{\pi}{3}\right)$

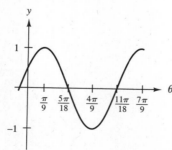

7. a. $72°$ **b.** $-3°$ **c.** 1   **d.** $69°$

    e. $y = \cos (5\theta + 15°)$

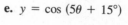

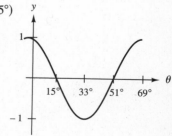

9. a. 2π **b.** $\dfrac{\pi}{2}$ **c.** 3 **d.** $\dfrac{5\pi}{2}$

e. $y = 3 \sin \left(\theta - \dfrac{\pi}{2} \right)$

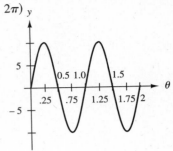

11. a. 1 **b.** 1 **c.** 10 **d.** 2

e. $y = 10 \sin (2\pi\theta - 2\pi)$

13. a. $\dfrac{2}{15}$ **b.** $-\dfrac{1}{15}$ **c.** 30 **d.** $\dfrac{1}{15}$

e. $y = -30 \cos (15\pi\theta + \pi)$

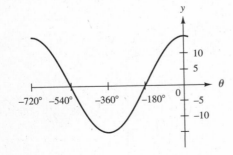

15. a. 120° **b.** 40° **c.** 25° **d.** 160°

e. $y = 25 \sin (3\theta - 120°)$

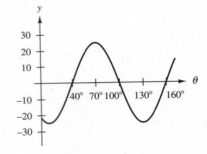

17. a. 720° **b.** −720° **c.** 15° **d.** 0°

e. $y = 15 \cos \left(\dfrac{\theta}{2} + 360° \right)$

19. a. 360° **b.** 720° **c.** 35° **d.** 1080°

e. $y = 35 \cos (\theta - 720°)$

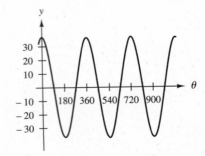

21. $y = 5 \sin\left(\dfrac{2}{3}\theta - \dfrac{\pi}{6}\right)$ **23.** $y = 2 \cos\left(4\theta + \dfrac{2\pi}{3}\right)$ **25.** $y = -2 \sin\theta$ or $y = 2 \sin(\theta + \pi)$

27. $y = 2 \cos\left(\theta + \dfrac{\pi}{2}\right)$

29. a. 8, 12, 2

b. $y = 8 \cos\left(\dfrac{\pi t}{6} - \dfrac{\pi}{3}\right)$

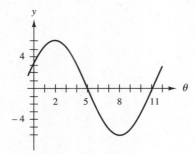

c. $(4, 4)$, $(7, -7)$

31. a. $8, \dfrac{1}{15}, -\dfrac{1}{120}$

b. $y = 8 \cos\left(30\pi t + \dfrac{\pi}{4}\right)$

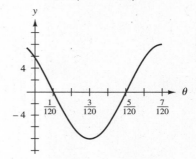

c. $\left(\dfrac{1}{130}, 0.48\right)$

33. $v = 1.5 \cos(3t + 2)$

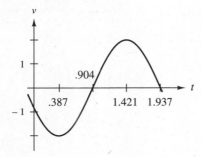

35. a. $i = 750 \sin[(360°)\, 30t + 25°]$
b. $13.3°$
c.

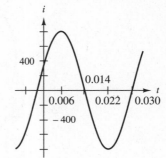

37. a. $v = 150 \sin(360)(700t)$
b. $i = 12 \sin((360)\,(700t) - 30)$
c.

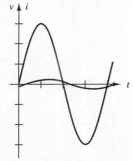

39. 34 ft

Section 8.5 (page 399)

1. h **3.** a **5.** g **7.** e **9.** b

11.

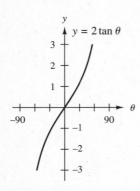

$y = 2 \tan \theta$

13.

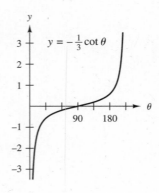

$y = -\frac{1}{3} \cot \theta$

15.

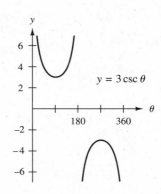

$y = 3 \csc \theta$

17.

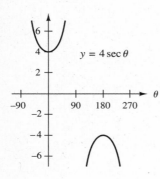

$y = 4 \sec \theta$

19.

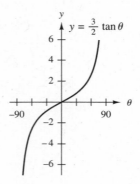

$y = \frac{3}{2} \tan \theta$

21.

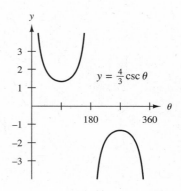

$y = \frac{4}{3} \csc \theta$

23.

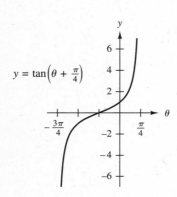

$y = \tan\left(\theta + \frac{\pi}{4}\right)$

25.

$y = \frac{3}{2} \cot\left(2\theta - \frac{2\pi}{3}\right)$

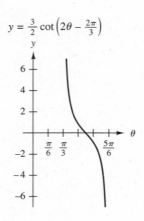

27.

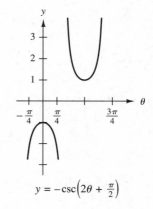

$y = -\csc\left(2\theta + \frac{\pi}{2}\right)$

29.

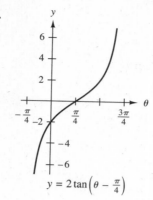

$$y = \sec\left(\theta - \frac{\pi}{6}\right)$$

31.

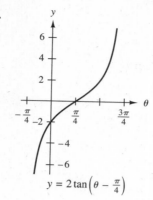

$$y = 2\tan\left(\theta - \frac{\pi}{4}\right)$$

33.

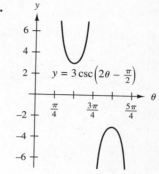

$$y = 3\csc\left(2\theta - \frac{\pi}{2}\right)$$

35.

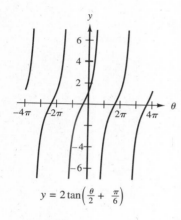

$$y = 2\tan\left(\frac{\theta}{2} + \frac{\pi}{6}\right)$$

37.

$$y = 3\cot(2\theta - 135°)$$

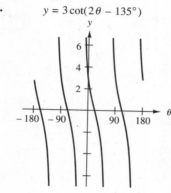

39.

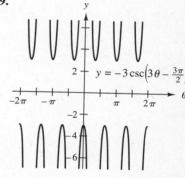

$$y = -3\csc\left(3\theta - \frac{3\pi}{2}\right)$$

41.

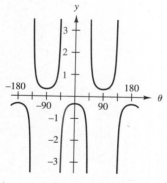

$$y = \frac{1}{3}\sec(2\theta - 180°)$$

43. $y = \dfrac{1}{3}\tan\left(2\theta + \dfrac{\pi}{6}\right)$ **45.** $y = -3\sec\left(\dfrac{1}{2}\theta + 15°\right)$ **47.** $y = 3\csc(2\theta - \pi)$ **49.** $y = -\dfrac{3}{4}\cot\left(3\theta + \dfrac{3\pi}{4}\right)$

Section 8.6 *(page 405)*

1. a. 360° or 2π **b.** 360° or 2π **c.** 1
d.

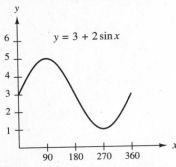

$$y = 3 + 2\sin x$$

3. a. 720° or 4π **b.** 720° or 4π **c.** 1
d.

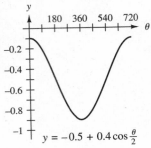

$$y = -0.5 + 0.4\cos\frac{\theta}{2}$$

5. a. Not periodic, 360° or 2π **b.** Not periodic
c. Not applicable
d.

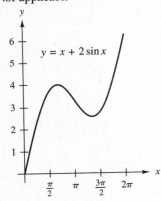

$$y = x + 2\sin x$$

7. a. Not periodic, 360° or 2π
b. Not periodic   **c.** Not applicable
d.

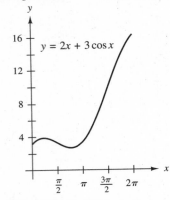

$$y = 2x + 3\cos x$$

9. a. 360°, 360°   **b.** 360°   **c.** 1, 1
d.

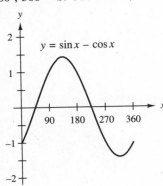

$$y = \sin x - \cos x$$

11. a. 360°, 360°   **b.** 360°   **c.** 1, 1
d.

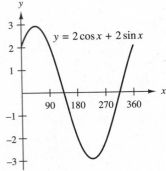

$$y = 2\cos x + 2\sin x$$

13. a. 360°, 360° **b.** 360° **c.** 1, 1
d.

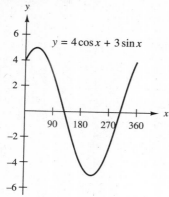

$y = 4\cos x + 3\sin x$

15. a. 180°, 180° **b.** 180° **c.** 1, 1
d.

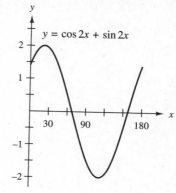

$y = \cos 2x + \sin 2x$

17. a. 120°, 360° **b.** 360° **c.** 3, 1

d. $y = \cos 3x + 2\cos(x + 45°)$

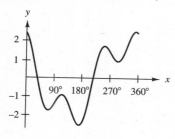

19. a. $\dfrac{2}{3}\pi$, 2π **b.** 2π **c.** 3, 1

d. $y = \sin 3x + \sin\left(x + \dfrac{\pi}{4}\right)$

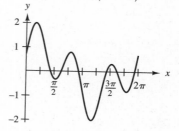

21. $y = \sin \pi x - 2\cos x$

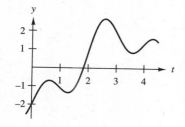

23. $y = 2\sin \dfrac{\pi}{2}x + \cos 3x$

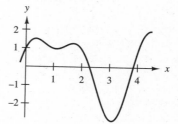

25. $y = \dfrac{3}{x + 3}\cos \pi x$

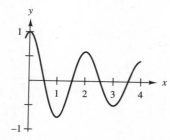

27. $y = \dfrac{6 + x}{5}\sin \dfrac{\pi}{2}x$

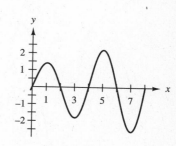

29. $y = 2.3 \cos 1.7t + 2.0 \sin 1.7t$

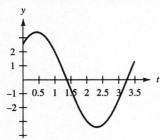

31. $t = 3.14$ s; $d = 20 \cos (t + \pi) + 10 \sin \left(t - \dfrac{\pi}{2}\right)$

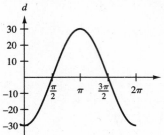

33. $y = 2.5 \cos x + 3 \sin \left(\dfrac{\pi}{2} + x\right)$

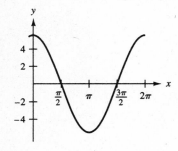

35. $y = 600 \sin^2 120 \,\pi t$

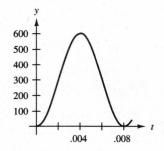

37. a. $y = 15 \sin 7t + 8 \cos 7t$
$y = 17 \sin(17t + 0.490)$

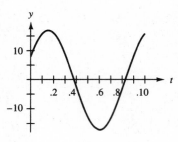 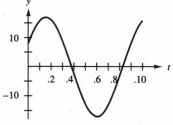

b. Yes   **c.** No

Review Exercises (page 407)

[8.2, 8.3]**1.**

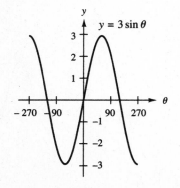

3.

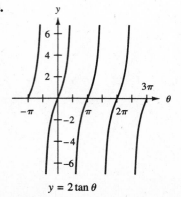

5.

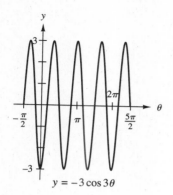

$y = -3\cos 3\theta$

7.

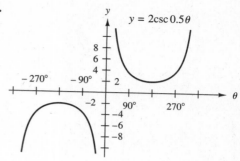

$y = 2\csc 0.5\theta$

9.

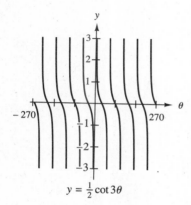

$y = \frac{1}{2}\cot 3\theta$

[8.4]11. a. $\dfrac{2\pi}{3}$ **e.**

b. $-\dfrac{\pi}{12}$

c. 2

d. $\dfrac{7\pi}{12}$

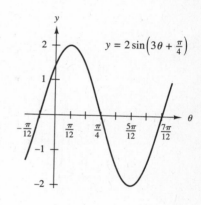

$y = 2\sin\left(3\theta + \frac{\pi}{4}\right)$

13. a. 1080°   **e.**
b. −90°
c. 4
d. 990°

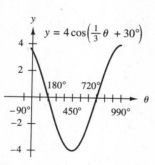

$y = 4\cos\left(\frac{1}{3}\theta + 30°\right)$

15. a. 36,000°   **e.**
b. −1500°
c. 16
d. 34,500°

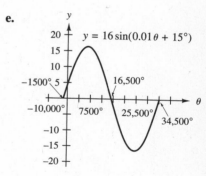

$y = 16\sin(0.01\theta + 15°)$

[8.5] 17.

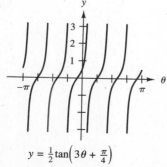

$y = \frac{1}{2}\tan\left(3\theta + \frac{\pi}{4}\right)$

19.

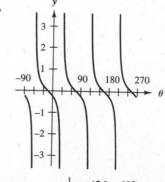

$y = \frac{1}{2}\cot(2\theta - 60°)$

21.

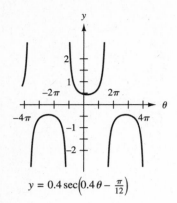

$$y = 0.4\sec\left(0.4\theta - \frac{\pi}{12}\right)$$

[8.6] **23. a.** 720°   **b.** 720°   **c.** 1

d.

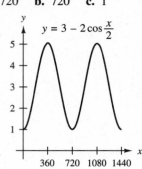

$$y = 3 - 2\cos\frac{x}{2}$$

25. a. 6, 8   **b.** 24   **c.** 4, 3

d.

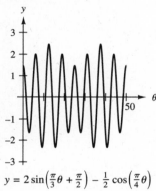

$$y = 2\sin\left(\frac{\pi}{3}\theta + \frac{\pi}{2}\right) - \frac{1}{2}\cos\left(\frac{\pi}{4}\theta\right)$$

27. a. 12, 6   **b.** 12   **c.** 1, 2

d.

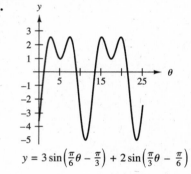

$$y = 3\sin\left(\frac{\pi}{6}\theta - \frac{\pi}{3}\right) + 2\sin\left(\frac{\pi}{3}\theta - \frac{\pi}{6}\right)$$

29. $t = 0.47$

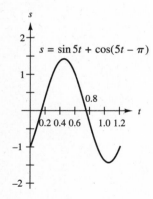

$$s = \sin 5t + \cos(5t - \pi)$$

[5.2–5.6] **31.** $5, \dfrac{1}{30}$

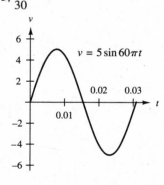

$$v = 5\sin 60\pi t$$

33. $y = 2.5 \sin\left(\dfrac{\pi}{10}x\right)$

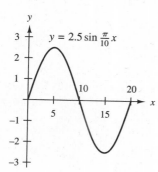

35. a. 3   **b.** π **c.** $y = 3 \sin 2x$

37. $y = 2 \sin\left(2x - \dfrac{\pi}{4}\right)$

39. a. 2   **b.** 0.006

41.

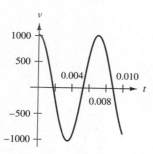

$$v = 8500\left(\tfrac{2\pi}{60}\right)\tfrac{2.35}{2} \sin\left[8500\left(\tfrac{2\pi}{60}\right)t + \tfrac{\pi}{2}\right]$$

43. a. 10 ft   **b.** $y = 35 + 25 \sin\left(20\pi t - \dfrac{3\pi}{10}\right)$

c.

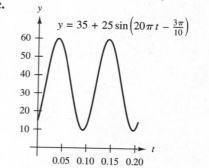

d. (1) 15ft (2) 55 ft   **e.** 0.0682s

45. a. π **b.** $\alpha = 2 \cos 2t$

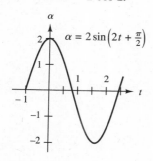

49.

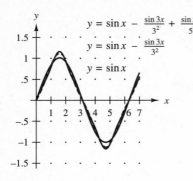

$$y = \sin x - \frac{\sin 3x}{3^2} + \frac{\sin 5x}{5^2}$$

$$y = \sin x - \frac{\sin 3x}{3^2}$$

$$y = \sin x$$

```
 10  PRINT "X IS IN DEGREES"
 20  PRINT
 30  PRINT "ONE TERM"
 40  PRINT "X ","Y "
 50  P=3.14159 / 180
 60  FOR X = 0 TO 360 STEP 15
 70  A = X * P
 80  Y = SIN (A)
 90  PRINT X,Y
100  NEXT X
110  PRINT
120  PRINT "TWO TERMS"
130  PRINT "X ","Y "
140  FOR X = 0 TO 360 STEP 15
150  A = X * P
160  Y =  SIN (A) —  SIN (3 * A) / 9
170  PRINT X,Y
180  NEXT X
190  PRINT
200  PRINT "THREE TERMS"
210  PRINT "X ","Y "
220  FOR X = 0 TO 360 STEP 15
230  A = X * P
240  Y =  SIN (A) —  SIN (3 * A) / 9 +  SIN (5 * A) / 25
250  PRINT X,Y
260  NEXT X
270  END
```

Chapter Test (page 412)

1. Amplitude = 3, period = 360°, phase shift = 0°   **2.** Amplitude = 5, period = 120°, phase shift = 0°
3. Amplitude = 1, period = 180°, phase shift = 22.5°   **4.** Amplitude not applicable, period = 45°, phase shift = −15
5. a. amplitude = 2   **b.** period = 120°
   c. phase shift = 20°   **d.** 140°
   e.

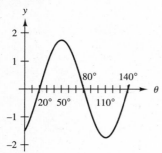

6. a. amplitude = $\dfrac{1}{2}$ **b.** period = 2π

   c. phase shift = $\dfrac{-\pi}{6}$ **d.** $\dfrac{11\pi}{6}$

   e.

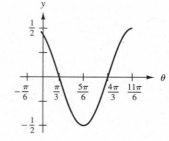

7. a. not applicable   **b.** period $= 180°$
   c. phase shift $= 45°$ **d.** $135°$
   e.

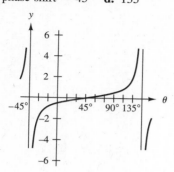

8. a. not applicable   **b.** period $= 360°$
   c. phase shift $= -30°$,   **d.** $330°$
   e.

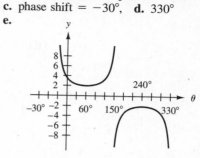

9. a. Period: $y_2 = 360°$ **b.** Period: $360°$ **c.** 1 cycle
   d.

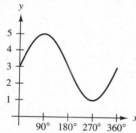

10. a. Period: $y_1 = 360°$, $y_2 = 360°$ **b.** $360°$
   c. y_1: 1 cycle, y_2: 1 cycle; therefore, the combination
   is one cycle
   d.

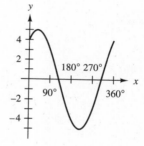

11. a. $y = 2 \sin \dfrac{2\pi x}{15}$
   b.

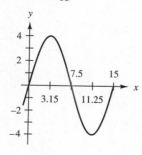

12. a. period $= \dfrac{10}{3}$ s   **b.** $\dfrac{\text{time}}{\text{one period}} = \dfrac{10}{3}$ s

   c. $y = \dfrac{2}{5} \sin \dfrac{3\pi}{5} t$

d. Positive values inhale; negative values
   exhale
e. $y = 0.03$

Chapter 9

Section 9.1 (page 419)

1. $8j$ **3.** $4j\sqrt{2}$ **5.** a^2j **7.** p^3j **9.** j **11.** $-j$ **13.** $-j$ **15.** -1 **17.** 1 **19.** $-j$ **21.** -132 **23.** -48
25. $8j\sqrt{3}$ **27.** 6 **29.** $-24j\sqrt{5}$ **31.** $-44j\sqrt{5}$ **33.** $\dfrac{-j\sqrt{70}}{14}$ **35.** $\sqrt{7}$ **37.** $\dfrac{1}{2}$ **39.** $12j\sqrt{2}$ **41.** $9j\sqrt{5}$
43. $-13 + 17j$ **45.** 0 **47.** 1 **49.** -1 **51.** 1 **53.** $-j$

Section 9.2 (page 425)

1. $x = \dfrac{3}{2}, y = 5$ **3.** $x = -8, y = 4$ **5.** $x = -1, y = -1$ **7.** $11 - j$ **9.** $12 + 17j$ **11.** $5 - 10j$ **13.** $-2 + 24j$
15. $-27 + 36j$ **17.** $12 + j\sqrt{3}$ **19.** 5 **21.** $-\dfrac{1}{5} + \dfrac{8}{5}j$ **23.** $\dfrac{5}{9} - \dfrac{11\sqrt{5}}{45}j$ **25.** $\dfrac{23}{10} + \dfrac{11}{10}j$ **27.** $\dfrac{4}{25} - \dfrac{3}{25}j$
29. $7 + 5j$ **31.** A real number   **33.** $-2.86 + 0.88j$ **35.** $\dfrac{7}{5} - \dfrac{1}{5}j$ **29.** $7 + 5j$ **37.** $\dfrac{320}{281} + \dfrac{100}{201}j$

Section 9.3 (page 433)

1.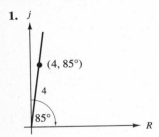

3.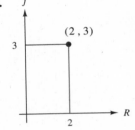

5.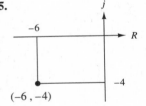

7.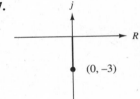

9.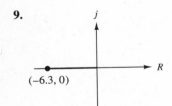

11.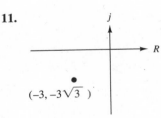

13. $2\sqrt{2}(\cos 45° + j \sin 45°)$ **15.** $2(\cos 330° + j \sin 330°)$ **17.** $12.5(\cos 208.6° + j \sin 208.6°)$
19. $3.4(\cos 1.05 + j \sin 1.05)$ **21.** $7.34(\cos 3.93 + j \sin 3.93)$ **23.** $3.00(\cos 5.50 + j \sin 5.50)$
25. $2 + 2j$ **27.** $-2.28 - 0.542j$ **29.** $-2 + 2j$ **31.** $1 - 2j$ **33.** $-0.69 + 8.7j$ **35.** $5.0 + 0.22j$ **37.** $5e^{1.0j}$
39. $2e^{\pi j}$ **41.** $8.7e^{1.65j}$ **43.** $4e^{5j}$ **45.** $9e^{1j}$ **47.** $4e^{2j}$
49. $0.647e^{5.29j}$, the magnitude of the current is approximately 0.647 A   **51.** Magnitude is 20, 146°, -16, 11

Section 9.4 (page 441)

1. $20\underline{/77°}$ **3.** $2e^{4.538j}$ **5.** $26\underline{/141°}$ **7.** $5 + 5j$ **9.** $80(\cos 191.0° + j \sin 191.0°)$ **11.** $676e^{4.58j}$ **13.** $4\underline{/9°}$
15. $2.0(\cos 51° + j \sin 51°)$ **17.** $0.8241e^{6.093j}$ **19.** $-\dfrac{8}{13} - \dfrac{1}{13}j$ **21.** $0.34\underline{/272.9°}$ **23.** $0.675(\cos 97° + j \sin 97°)$

25. $16\underline{/70°}$ **27.** $4e^{5.062j}$ **29.** $22\underline{/72°}$ **31.** $-3 + 4j$ **33.** $72\,(\cos 171.2° + j\sin 171.2°)$ **35.** $600e^{3.888j}$
37. $125\underline{/126°}$ **39.** $e^{6.021j}$ **41.** $180\underline{/135°}$ **43.** $18 - 26j$ **45.** $830\,(\cos 316.2° + j\sin 316.2°)$ **47.** $21000e^{7.92j}$
49. $4\underline{/16°},\ 4\underline{/136°},\ 4\underline{/256°}$ **51.** $2\underline{/24°},\ 2\underline{/114°},\ 2\underline{/204°},\ 2\underline{/294°}$ **53.** $1\underline{/45°},\ 1\underline{/225°}$

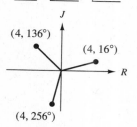

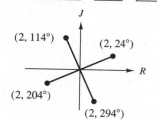

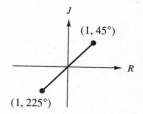

55. $2\underline{/0°},\ 2\underline{/120°},\ 2\underline{/240°}$ **57.** $32\ \text{V}\ \underline{/104°}$ **59.** $2\underline{/65°}$

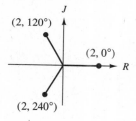

Section 9.5 *(page 446)*

1. $306\underline{/11.6°}$ **3.** $407\cos(251t + 39°)$ **5. a.** $3 + 4j\,\Omega$ **b.** $10\ \text{A}$ **c.** $-53°$ **7. a.** $3 - 4j\,\Omega$ **b.** $10\ \text{A}$ **c.** $53°$
9. a. $1.9 + 1.4j\ \Omega$ **b.** 2.1A **c.** $-37°$ **11. a.** $1.9 - 1.4j\ \Omega$ **b.** 21A **c.** $37°$ **13.** $47.0\ \Omega\ \underline{/45.6°}$
15. $45.8 + 17.5j\,\Omega$ **17.** $34 + 6.2j\,\Omega$ or $3.5\Omega\ \underline{/10°}$ **19.** $26 + 75j\,\Omega$ or $79j\,\Omega\underline{/71°}$ **21.** $28 + 1.9j\,\Omega$
23. $53 - 91j\,\Omega$ **25.** $62\Omega\underline{/-63°}$ **27.** $4.1\sin(\omega t + 14°)$

Review Exercises *(page 448)*

[9.1] **1.** $-j$ **3.** -342 **5.** $286j\sqrt{3}$ **7.** 28 **9.** $-j\sqrt{7}$ **11.** $(21\sqrt{2} + 11\sqrt{3})j$ **13.** $11 + 2j$ **15.** $1396 - 251j\sqrt{6}$
17. $0.332 + 0.0185j$
[9.3] **19.** $13.6\ \underline{/126.0°}$ **21.**   **23.**

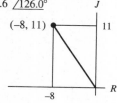

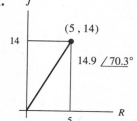

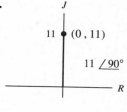

[9.3] **25.** $-1.7 - 2.9j$ **27.** $-0.70 + 1.2j$ **29.** $5.0 + 0.30j$ [9.4] **31.** $75\underline{/115°}$ **33.** $56.25\underline{/166.8°}$
35. $291\underline{/19.6°}$ **37.** $6\underline{/38°},\ 6\underline{/158°},\ 6\underline{/278°}$ **39. a.** $-35.8°$ **b.** $100 + 72j\ \Omega$ **c.** $1.79\ \text{A}$
41. $8.78\underline{/3.9°}\ \Omega$ **43.** $\sqrt[3]{60°} = 3\underline{/60°},\ 3\underline{/180°},\ 3\underline{/300°};\ \sqrt[6]{729} = 3\underline{/0°},\ 3\underline{/60°},\ 3\underline{/120°},\ 3\underline{/180°},\ 3\underline{/240°},\ 3\underline{/330°}$

Chapter Test (page 449)

1. j **2.** -1 **3.** -49 **4.** 20   **5.** $-5 + 12j$ **6.** 58   **7.** $2 - 10j$

8.

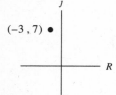

9.

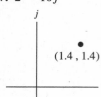

10. $6.4e^{2.25j}$ **11.** $-3.5 + 3.5j$, $5e^{5.5j}$ **12.** $5.5\underline{/-55°}$ or $5.5\underline{/305°}$ **13.** $20e^{3.5j}$ **14.** $625e^{6.0j}$ **15.** $= 7.88\underline{/-6.8°}$

16. $3\underline{/0°}$, $3\underline{/90°}$, $3\underline{/180°}$, $3\underline{/270°}$ **17.** $\dfrac{0.588}{1.96} + \dfrac{4.48}{1.96}j$ **18. a.** $200\underline{/41.3°}\ \Omega$ **b.** 1.7A   **c.** $-41°$

Chapter 10

Section 10.1 (page 461)

1. c   **3.** h   **5.** g   **7.** d

9.

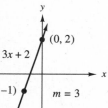

11.

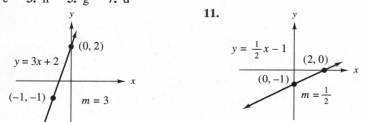

13.

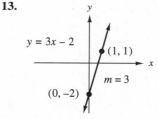

15.

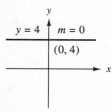

17.

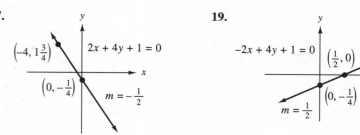

19.

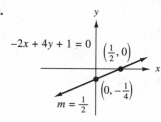

21.

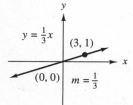

23. Parallel   **25.** Perpendicular   **27.** Neither   **29.** $y = 2x + 2$ **31.** $y = -6x + 3$ **33.** $y = x$

35. $y = 4$ **37.** $y = \dfrac{1}{2}x + 2$ **39.** $y = -\dfrac{3}{2}x + \dfrac{3}{2}$ **41.** $y = 4$ **43.** $y = -3x + 1$ **45.** $y = \dfrac{1}{2}x + \dfrac{1}{2}$

47. $y = 3$ **49.** $y = -x + 7$ **51.** $y = -\dfrac{5}{2}x + 8$ **53.** $y = \dfrac{2}{3}x + 2$ **55.** $y = 3x - 5$ **57.** $y = 2$ **59.** $x = -3$

61. $y = x + 2$ **63.** Yes, it is of the form $y = mx$ **65.** $y = -\dfrac{1}{50}x$

67. a.

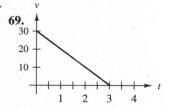

69.

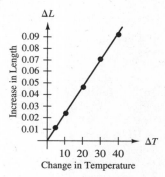

Change in Temperature

b. $v = -9.8t + 30$ **c.** $-9.8\,\dfrac{m}{s^2}$;

This means the tape measure is heading back down.

71. 4.1×10^{-3} ft   **73.** $v(t) = -75t + \$375$

75. a.

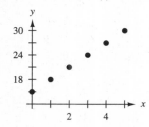

b. $m = 3$ **c.** $y = 3x + 15$

Section 10.2 (page 473)

1. $(-6, 2)$ **3.** $(3, 16)$ **5.** $(4, 27)$ **7.** $(3, 2)$ **9.** $(3, 5)$ **11.** $(-1, -8)$ **13.** $(6, 5)$ **15.** $(6, 0)$ **17.** $(0, -3)$

19. $\left(1\dfrac{1}{2}, 2\right)$ **21.** Inconsistent   **23.** Dependent   **25.** $(0, -2)$ **27.** Inconsistent   **29.** Inconsistent

31. $A = -15,000, B = 13,000$ **33.** $y = -x + 4, y = x - 6$ **35.** $y = x + 3, y = 2x + 4$

37. $y = -2x, y = 3x - 5$ **39.** $y = -3x + 4, y = -3x - 4$ **41. a.** $v = 50 + 3.4t$ **b.** $t = 3.5$ s

43. $m_1 = m_2$ **45.** 14 cc of 10%, 16 cc of 25%   **47.** 80 lb of lettuce,   20 lb of bean sprouts

49. mold = \$300, metal per truck = \$0.25   **51.** $a = 3, b = -4$

53. a. $(5, 15)$ **b.** $y = 30 - 3x, y = 4x - 5$

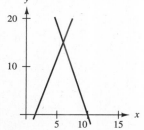

55. a. $\left(\dfrac{12}{11}, -\dfrac{19}{22}\right)$ **b.** $4y + 5x = 2, 2y - 3x = -5$

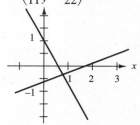

Section 10.3 (page 487)

1. $(1, 3)$ **3.** $(-1, 2)$ **5.** $(-3, -2)$ **7.** $(10, 20)$ **9.** $(1, -6)$ **11.** $\left(-\dfrac{10}{3}, -\dfrac{17}{3}\right)$ **13.** $(1.5, 2.5)$

15. $(17.5, 1.5)$ **17.** $(0.79, 0.73)$ **19.** $(-0.27, -0.46)$ **21.** Inconsistent   **23.** Dependent

25. $S_s = 16{,}000,\ S_A = -32{,}000$ **27.** $R_1 = 1600$ N, $R_2 = 3900$ N   **29.** $i_1 = 0.28,\ i_2 = -0.16$

31. $\left(\dfrac{1}{12}\text{ oz}, \dfrac{2}{3}\text{ oz}\right)$ **33.** $\left(40\text{ m}, -\dfrac{4}{5}\text{ m}\right)$ **35.** \$20,000 at 9% \$15,000 at 11%   **37. a.** 32.5 months   **b.** Model B

39. a. $v = 50 + 3.4t$ **b.** 110 mph

Section 10.4 (page 493)

1. $(1, 0, -1)$ **3.** $(-8, 16, 1)$ **5.** $(0, -2, 1)$ **7.** $(-17.8, 14.5, 16.1)$ **9.** $(-1, -1, -2)$

11. $(-1, 0, -1)$ **13.** $(0, -2, 1)$ **15.** $(0, -1, 0)$ **17.** $(-1, 0, 0)$ **19.** $(0.298, -0.651, -1.25)$ **21.** $(2, 1.4, 1)$

23. $(100, 100, 62.5)$ **25.** $(-2230, 2000, 0)$ **27.** $A = 100$ g, $B = 100$ g, C $= 200$ g

29. nut $= 0.032$ kg, bolt $= 0.008$ kg, washer $= 0.004$ kg

31. $55.6\ \ell$ of 20%, $18.5\ \ell$ of 32%, $25.9\ \ell$ of 50%

33. \$75,000 salaries,   \$200,000 equipment,   \$25,000 overhead

35. \$1.07 hot dog,   \$1.47 cheeseburger,   \$1.77 sandwich

Section 10.5 (page 501)

1. 116   **3.** 153   **5.** -16 **7.** -840 **9.** -74 **11.** 169   **13.** $(-8, 16, 1)$ **15.** $(1, 0, -1)$

17. $(0, -2, 1)$ **19.** $(-17.8, 14.5, 16.1)$ **21.** $(-1, -1, -2)$ **23.** $(-1, 0, -1)$ **25.** $(0, -2, 1)$

27. $T_1 = 417$ kg, $T_2 = 156$ kg, $T_3 = 557$ kg   **29.** $x = 76.9\%,\ y = 19.2\%, z = 3.85\%$ **31.** $(6\text{ m}\ell, 2\text{ m}\ell, 1\text{ m}\ell)$

33. a. 2461 lb   **b.** 70 lb   **c.** 172 lb   **d.** 43 lb

Section 10.6 (page 509)

1. $a = 1, b = 2, c = -1, d = 0$ **3.** $a = 1, x = 1$ **5.** 1×3 **7.** 2×2 **9.** 3×3 **11.** 3×2

13. $\begin{pmatrix} 7 & 6 & -4 \\ 3 & -2 & -1 \\ 6 & 14 & 9 \end{pmatrix}$ **15.** $\begin{pmatrix} -2 & -12 \\ -2 & -3 \\ 4 & 3 \end{pmatrix}$ **17.** $(13, -2, -11)$ **19.** $\begin{pmatrix} -9 & -4 \\ 0 & 1 \end{pmatrix}$

21. a. $2 \times 2, 2 \times 2$ **b.** $\begin{pmatrix} 10 & 48 \\ 5 & 33 \end{pmatrix}$ **c.** $\begin{pmatrix} 16 & 38 \\ 9 & 27 \end{pmatrix}$ **23. a.** $1 \times 3\quad 3 \times 2$ **b.** $(18, 23)$ **c.** Cannot be multiplied

25. a. $2 \times 3, 3 \times 2$ **b.** $\begin{pmatrix} 101 & 38 \\ 13 & 26 \end{pmatrix}$ **c.** $\begin{pmatrix} 16 & -2 & -4 \\ 34 & 48 & 39 \\ 18 & 72 & 63 \end{pmatrix}$ **27. a.** $1 \times 3, 3 \times 3$ **b.** $(19, 54, -7)$ **c.** Cannot be multiplied

29. a. $3 \times 3, 3 \times 1$ **b.** $\begin{pmatrix} x + 2y + 3z \\ -2x + y \\ 5x - 4y + 6z \end{pmatrix}$ **c.** Cannot be multiplied

31. a.

	Tony's	Elaine's	Ruth's
Drama	34.05	35.25	35.05
Band	67.20	68.30	67.00
Rugby	79.20	82.90	83.80

b. Drama club: Tony's,
dance band: Ruth's,
rugby team: Tony's

Section 10.7 (page 522)

1. a. $\begin{pmatrix} 1 & 1 & 1 \\ 2 & 1 & 2 \\ -1 & -2 & -3 \end{pmatrix} \begin{pmatrix} x \\ y \\ z \end{pmatrix} = \begin{pmatrix} -2 \\ 0 \\ 5 \end{pmatrix}$ **b.** $\begin{pmatrix} \frac{1}{2} & \frac{1}{2} & \frac{1}{2} \\ 2 & -1 & 0 \\ 1\frac{1}{2} & \frac{1}{2} & -\frac{1}{2} \end{pmatrix}$ **c.** $(1\frac{1}{2}, -4, \frac{1}{2})$ **3. a.** $\begin{pmatrix} 1 & 1 & 1 \\ 0 & 1 & -1 \\ 2 & -2 & -1 \end{pmatrix} \begin{pmatrix} x \\ y \\ z \end{pmatrix} = \begin{pmatrix} 7 \\ 7 \\ 0 \end{pmatrix}$

b. $\begin{pmatrix} \frac{3}{7} & \frac{1}{7} & \frac{2}{7} \\ \frac{2}{7} & \frac{3}{7} & -\frac{1}{7} \\ \frac{2}{7} & -\frac{4}{7} & -\frac{1}{7} \end{pmatrix}$ **c.** $(4, 5, -2)$ **5.** $(-4, 1, 1)$ **7.** $(-0.286, -0.142, 0.468)$ **9.** $\left(-1\frac{1}{3}, -1\frac{1}{3}, 1\frac{2}{3}\right)$

11. $(1, 0, -1)$ **13.** $(-1, -2, -1)$ **15.** $(-1, -1, 1)$ **17.** $(14, 135, -158)$ **19.** $(0, -2, 1)$ **21.** $(0, 0, 0)$
23. $(1000, -1000, 1000)$ **25.** $A = 50, B = 30, C = 20$ **27.** Original solution 50 gal, add 30 gal of water, total solution
100 gal

Review Exercises (page 523)

[10.1] 1. $-1, 4$ **3.** $\dfrac{1}{2}, 2$ **5.** $0, 3$

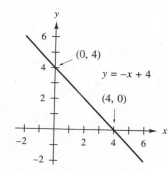

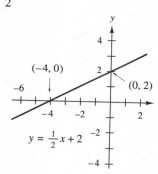

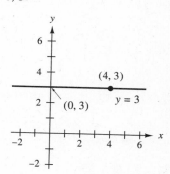

7. $-\dfrac{1}{2}, \dfrac{1}{3}$ **9.** $y = -4x$ **11.** $y = \dfrac{1}{2}x + 3$ **13.** $y = -2$ **15.** $y = 1.32x + 4.17$

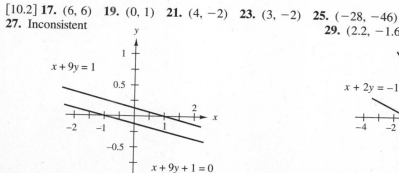

[10.2] 17. $(6, 6)$ **19.** $(0, 1)$ **21.** $(4, -2)$ **23.** $(3, -2)$ **25.** $(-28, -46)$
27. Inconsistent   **29.** $(2.2, -1.6)$

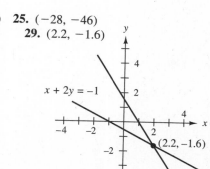

31. Dependent   [10.3] **33.** 51   **35.** −8   **37.** −1   **39.** (3, 9)   **41.** (32, 22)   **43.** (4, 20)   **45.** (3, 2)   **47.** (4, −2)
49. (−4, 1)   [10.4] **51.** (0, −1, −1)   **53.** (2, −1, 0)   **55.** (−11, 1, −11)   **57.** (0, 3, −2)   [10.5] **59.** −24   **61.** 62
63. −5   **65.** −71   **67.** (−1, 3, −2)   **69.** $\left(-\dfrac{1}{2}, 0, \dfrac{1}{2}\right)$ **71.** (1, −1, −1)   **73.** (−1, −2, 1)   [10.1–10.3]
75. $F_1 = 6592$, $F_2 = -561$ **77.** $F_A = 9438$, $F_B = 4718$ **79.** $i_1 = 3$, $i_2 = 2$ **81.** $v_1 = 50$, $v_2 = 40$ **83.** $i_1 = 2.98$,
$t_2 = 2.47$, $t_3 = 0.93$ **85.** Plug = \$1.85, jig = \$0.46, spoon = \$1.85
[10.6] **87.** $\begin{pmatrix} 3 & 2 \\ 1 & 4 \\ 8 & 9 \end{pmatrix}$ **89.** $\begin{pmatrix} 16 & 22 & 6 \\ -2 & -6 & 1 \\ 30 & 12 & 27 \end{pmatrix}$ **91.** $\begin{pmatrix} 5 & 8 & 2 \\ 11 & 5 & 2 \\ -1 & -14 & 3 \end{pmatrix}$ [10.7] **93.** (−1, 2, 2)   **95.** (−4, 1, 1)

[10.1–10.7] **97.** A line that goes up and to the right has a positive slope and makes an angle with the positive x-axis of θ such that $0° \leq \theta < 90°$. For the line going down and left, the angle is $\theta \pm 180°$. Therefore, $m = \tan(\theta \pm 180°)$

$$= \frac{\tan \theta \pm \tan 180°}{1 \mp \tan \theta \tan 180°}$$
$$= \frac{\tan \theta \pm 0}{1 \mp 0}$$
$$= \tan \theta$$

99. $\dfrac{y - 2}{x - 3} = \dfrac{-1 - 2}{5 - 3}$,

$$y = -\frac{3}{2}x + \frac{13}{2},$$

$$\frac{y - y_1}{x - x_1} = \frac{y_2 - y_1}{x_2 - x_1}$$

101. $\begin{vmatrix} a & b \\ c & d \end{vmatrix} = a|d| - b|c|$

$$= ad - bc$$

Chapter Test (page 527)

1. $m = 2$, $(0, -3)$,

$y = 2x - 3$

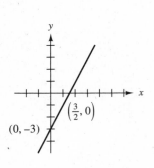

2. $m = -\dfrac{1}{2}$, $\left(0, -\dfrac{1}{2}\right)$,

$2y + x + 1 = 0$

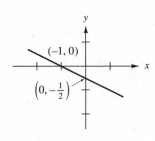

3. $y = \dfrac{2}{3}x + 2$ **4.** $3y = 2x + 11$ **5.** (−3, 5)   **6.** (2, 2, −1)   **7.** (−3, −2)   **8.** (1, 0, −1)

9. $\begin{pmatrix} 19 & 11 \\ 39 & 33 \end{pmatrix}$ **10.** (0, 3, 1)   **11.** $5y = 2x + 26$ **12.** $3y = 2x + 11$ **13.** 8 g of 40%, 12 g of 60%

14. smaller angle = 30°; next larger angle = 40°; and the largest angle = 110°

Chapter 11

Section 11.2 (page 538)

1. 2, 3   **3.** 5, −3   **5.** 5, $-1\dfrac{1}{2}$ **7.** 4, $-\dfrac{2}{5}$ **9.** $-\dfrac{2}{9}$, −1   **11.** 8, −8   **13.** $5\sqrt{5}$, $-5\sqrt{5}$ **15.** −9, −9

17. $-\dfrac{3}{4}, -\dfrac{3}{4}$ **19.** $13j, -13j$ **21.** $0, -13$ **23.** $0, -3$ **25.** $-\dfrac{2}{3}, 3$ **27.** $1, -7$ **29.** $-2, -4$

31. $x^2 + 2x - 15 = 0$ **33.** $5x^2 - 17x + 6 = 0$ **35.** $x^2 - 10x + 25 = 0$ **37.** $6x^2 - 5x + 1 = 0$

39. $x^2 + 2x + 5 = 0$ **41.** $5, -3$ **43.** $-\dfrac{11}{2}, \dfrac{5}{3}$ **45.** $4, 3$ **47.** $-11, -11$ **49.** $3, 3$

51. a. $h^2 - 40h + 300 = 0$ **b.** 10 cm, 30 cm **d.** $h = 10$ cm, $w = 30$ cm

53. a. $x^2 - 8x - 48 = 0$ **b.** $-4, 12$ **d.** 12.00 m **55. b.** $-12, 8$ **d.** 8 mph

57. a. $16 = 4t^2 + 8t$ **b.** $-3.24, 1.24$ **d.** 1.24 s **59. b.** 0.000, 0.089 **d.** Starting point 0.000, hrts x-axis at 0.089.

Section 11.3 (page 544)

1. $14, 2$ **3.** ± 3 **5.** $-5 + \sqrt{15}, -5 - \sqrt{15}$ **7.** $\dfrac{2}{3}, -4$ **9.** $3 + 4j, 3 - 4j$

11. $1, -4$ **13.** $\dfrac{-7 + \sqrt{61}}{2}, \dfrac{-7 - \sqrt{61}}{2}$ **15.** $\dfrac{3}{2}, -3$ **17.** $\dfrac{-1 + \sqrt{21}}{10}, \dfrac{-1 - \sqrt{21}}{10}$

19. $\dfrac{9 + \sqrt{129}}{6}, \dfrac{9 - \sqrt{129}}{6}$ **21.** $3, -1$ **23.** $3, 3$ **25.** $\dfrac{-5 + j\sqrt{3}}{2}, \dfrac{-5 - j\sqrt{3}}{2}$ **27.** $\pm 2\sqrt{5}$

29. $\dfrac{-b + \sqrt{b^2 - 4ac}}{2a}, \dfrac{-b - \sqrt{b^2 - 4ac}}{2a}$ **31.** 70 m **33.** 72 units

Section 11.4 (page 554)

1. $18, -2$ **3.** $-2, -12$ **5.** $2.354, -0.354$ **7.** $-1 \pm 0.707j$ **9.** $3, 0.167$ **11.** $1.5, 1.25$ **13.** $3.73, 0.268$

15. $3.37, 1.63$ **17.** $1.24, -3.24$ **19.** $1 \pm 2j$ **21.** $-2 \pm 3j$ **23.** $4.79, 0.209$ **25.** $5.00, -1.33$

27. $1.12, -7.12$ **29.** $0.433, \pm 0.559j$ **31.** $0.437, -1.87$ **33.** $0.181, -0.283$ **35.** 0.667 s, -4.00 s **37.** 53.4 in.

39. 2.00 **41.** 8 **43. a.** 13 m $\times$ 4 m **b.** 120 m^2 **45.** 90 mm, 150 mm, 30 mm **47.** 2 units **49.** 242 ft $\times$ 242 ft

Section 11.5 (page 559)

1. $+$; real, rational, unequal **3.** $-$; imaginary, unequal **5.** $+$; real, irrational, unequal

7. $+$; real, irrational, unequal **9.** 0; real, rational, equal **11.** $+$; real, rational, unequal

13. $-$; imaginary, unequal **15.** $+$; real, unequal **17.** $+$; imaginary, unequal **19.** 0; imaginary, equal

21. 9 **23.** 25 **25.** $1, -\dfrac{1}{3}$ **27.** Discriminant is negative so roots are imaginary.

29. $2\sqrt{21}, -2\sqrt{21}$

Section 11.6 (page 566)

1. $-5.0, -1.0$

3. $-1.0, -4.0$

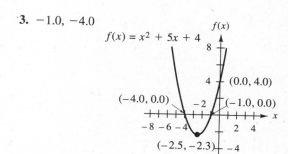

5. 1.7, −1.0

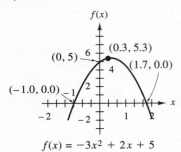

$$f(x) = -3x^2 + 2x + 5$$

7. 2.5, −1.0

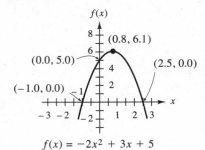

$$f(x) = -2x^2 + 3x + 5$$

9. Imaginary

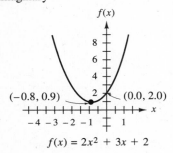

$$f(x) = 2x^2 + 3x + 2$$

11. 4.0, 0.0

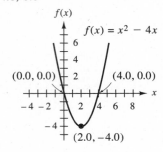

$$f(x) = x^2 - 4x$$

13. 2.5

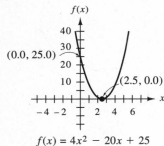

$$f(x) = 4x^2 - 20x + 25$$

15. 0.8, −2.1

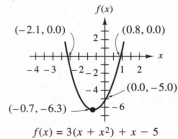

$$f(x) = 3(x + x^2) + x - 5$$

17. $\dfrac{2}{3}$

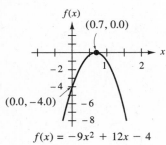

$$f(x) = -9x^2 + 12x - 4$$

19. −0.5, −2.0

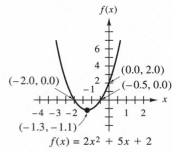

$$f(x) = 2x^2 + 5x + 2$$

21. One   **23.** Two   **25.** None   **27.** Two   **29.** Two   **31. a.** 1   **b.** $m < 1$ **c.** $m > 1$
33. a. ±4   **b.** $m < -4, m > 4$ **c.** $-4 < m < 4$

35. a. $h = 4 + 64t - 16t^2$

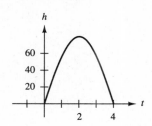

37. 450 ft², 15 ft × 30 ft
39. $w = 4.2$ ft, $h = 2.1$ ft,
   arc = 6.6 ft

41. a. $y = 0.58x - 6.5x^2$

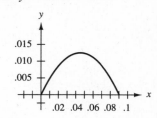

b. 2.0 s   **c.** 68 ft

b. About 0.013   **c.** 0.09   **d.** 0.089

Section 11.7 (page 575)

1. $\pm\sqrt{10}, \pm 1$ **3.** $\pm\sqrt{7}, \pm\sqrt{2}$ **5.** $0, \pm\sqrt{5}, \pm\sqrt{3}$ **7.** $\pm\sqrt{3}, \pm\frac{1}{2}$ **9.** 1   **11.** $-\frac{1}{8}, -1$ **13.** $\pm 1, \pm 0.289$

15. 46656, 1   **17.** $\frac{1}{4}, 4$ **19.** $3, -3$ **21.** $2, \pm 2j$ **23.** -2 **25.** $2.538, -0.788$ **27.** 12   **29.** $-4, 11$ **31.** $\dfrac{1}{5}$

33. $-2, -7$ **35.** 13   **37.** 1   **39.** No solution   **41.** $\dfrac{9}{8}$ **43.** $\dfrac{65}{32}$ **45.** 0   **47.** 43 cm, 17 cm   **49.** 130,000 parcels

51. 12 in. from the brighter lamp (or 8 in. from the 4 candle power lamp)   **53.** $V = \pm 0.011$

Review Exercises (page 577)

[11.4] **1.** $11 -4$ **3.** $2\frac{1}{2}, -2\frac{1}{3}$ **5.** $0.573, -2.907$ **7.** $9, 6$ **9.** $-1.167, \pm 0.553j$

[11.2] **11.** $x^2 - 7x + 12 = 0$ **13.** $x^2 - 8x + 16 = 0$ **15.** $x^2 + 6x + 25 = 0$ [11.3]**17.** $5, -11$ **19.** $\dfrac{4}{5}, -1\dfrac{1}{2}$

[11.5] **21.** +; real, rational, unequal   **23.** −; imaginary, unequal   **25.** +; real, rational, unequal
[11.6] **27.** 9.0, 2.0   **29.** $-0.8, -6.2$

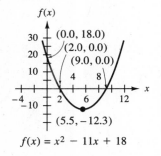

$f(x) = x^2 - 11x + 18$

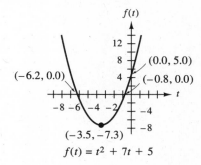

$f(t) = t^2 + 7t + 5$

31. 1.3, −1.5

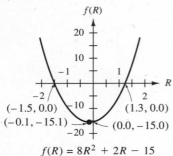

$$f(R) = 8R^2 + 2R - 15$$

33. Imaginary

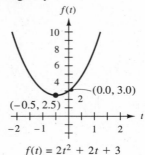

$$f(t) = 2t^2 + 2t + 3$$

35. 1.3, −1.3

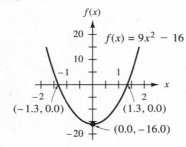

[11.8] **37.** ±2, ±3

39. 1, 531441   **41.** −0.791, 3.791   **43.** No solution   **45.** 0.571, −0.491   **47.** 0   **49.** No solution (0.5 ± 2.179 j)
[11.4] **51.** 1 Ω, 1 Ω   **53.** 144 ft   **55.** 7.6 ft   **57.** 25 cm × 9 cm   **59.** 6 hr programmer 12 hr assistant

61. $q = \dfrac{p}{pd - 1}$ **63.** 50 mph   **65.** 25.99, 6.013   **67.** 12.7 hr, 15.7 hr   **69.** 25   **73.** 239 m²

Chapter Test (page 581)

1. $x = 9$ and $x = 2$ **2.** $x^2 + x - 12 = 0$ **3.** $x^2 - 8x + 16 = 0$ **4.** $b^2 - 4ac > 0$; + ; real, irrational, unequal
5. $b^2 - 4ac < 0$; − ; imaginary, unequal
6.

$$f(x) = x^2 + 2x - 15$$

7. No solution   **8.** $x = \dfrac{-b \pm \sqrt{b^2 - 4ac}}{2a}$ **9.** $w = 5$ and $l = 10$

10. a. 288 ft.   **b.** 8.24 sec.   **c.** 32 ft.

Chapter 12

Section 12.1 (page 589)

1. $6x + 22 + \dfrac{68}{x - 4}$ **3.** $x^2 + x - 4 + \dfrac{4}{x + 1}$ **5.** $-x^2 + 2x - 2 + \dfrac{6}{x + 2}$ **7.** $2x^2 - 2x - 1$

9. $x^3 + x^2 - x + 2$ **11.** $2z^3 - 4z^2 + 19z - 65 + \dfrac{253}{z + 4}$ **13.** $3x^4 + 3x^3 + 3x^2 - 6x$

15. $9R^3 - 12R^2 + 9R + 6 + \dfrac{3}{R - \frac{1}{3}}$ **17.** $2x^3 + 5x^2 - \dfrac{5}{2}x - \dfrac{27}{4} + \dfrac{107/8}{x + \frac{1}{2}}$

19. $y^4 + 2y^3 + 4y^2 + 8y + 16$ **21.** $-Z^6 + 2Z^5 - 4Z^4 + 12.3Z^3 - 24.6Z^2 + 42Z - 84 + \dfrac{168}{Z + 2}$

23. $-80z^3 + 240cz^2 - 721c^2z + 2160c^3 - \dfrac{6480\, c^4}{z + 3c}$ **25.** $2x^3 + x^2 + 1 + \dfrac{x + 12}{(x - 2)(x + 3)}$

Section 12.2 (page 594)

1. 4 **3.** -11 **5.** -4 **7.** 0 **9.** 0 **11.** Yes **13.** No **15.** No **17.** No **19.** Yes **21. a.** $2x^2 + 14x + 15$
b. $2, -1.32, -5.68$ **23. a.** $2x^2 - 7x + 15$ **b.** $7, 1.75 \pm 2.11j$ **25.** $4, -1$ **27.** True **29.** True **31.** True
33. Yes **35.** 2

Section 12.3 (page 606)

1. a. $-3, 2, -5$ **b.** $1, 3, 2$ **3. a.** $-\dfrac{7}{2}, \dfrac{5}{3}, \dfrac{3}{5}$ **b.** $2, 3, 4$ **5 a.** $-5, 7$ **b.** $3, 1$ **7.** $x^2 - 16x + 39 = 0$

9. $x^3 - 6x^2 + 11x - 6 = 0$ **11.** $x^4 - 2x^3 - 3x^2 + 4x + 4 = 0$ **13.** $12x^3 + 32x^2 + 15x - 9 = 0$
15. $x^5 - 7x^4 + 19x^3 - 25x^2 + 16x - 4 = 0$ **17.** $-2 \pm j$

19. a. 1 **b.** 2 or 0 **c.** $\pm 1, \pm 3$ **d.** $-1, -1, 3$
e.

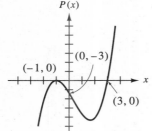

21. a. 1 **b.** 2 or 0 **c.** $\pm 1, \pm 2$ **d.** $-2, -1, 1$
e.

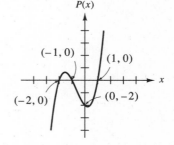

23. a. 1 **b.** 2 or 0 **c.** $\pm \dfrac{1}{2}, \pm \dfrac{3}{2}, \pm 1, \pm 2, \pm 3, \pm 6$

d. $\dfrac{1}{2}, -2, -3$ **e.**

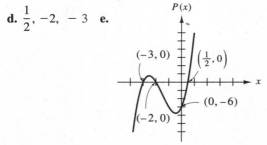

25. a. 1 **b.** 2 or 0 **c.** $\pm 1, \pm 2, \pm 3, \pm 4, \pm 6, \pm 12$

d. No rational roots **e.**

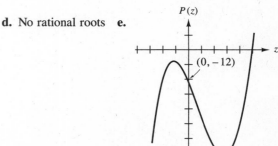

27. a. 3 or 1 **b.** 0 **c.** 1, 2, 4 **d.** 2

e.

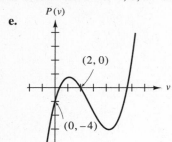

29. a. 2 or 0 **b.** 1

c. $\pm\frac{1}{3}$, $\pm\frac{2}{3}$, $\pm\frac{11}{3}$, $\pm\frac{22}{3}$, ± 1, ± 2, ± 11, ± 22 **d.** $-\frac{2}{3}$

e.

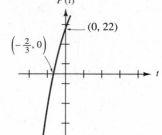

31. a. 4 or 2 or 0 **b.** 0 **c.** 1, 3, 5, 7, 15, 21, 35, 105

d. 1, 3, 5, 7 **e.**

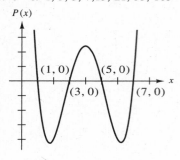

33. a. 1 **b.** 3 or 1

c. $\pm\frac{1}{4}$, $\pm\frac{1}{2}$, $\pm\frac{3}{4}$, $\pm\frac{3}{2}$, $\pm\frac{9}{4}$, $\pm\frac{9}{2}$, ± 1, ± 3, ± 9

d. $-\frac{3}{2}$, $-\frac{3}{2}$ **e.**

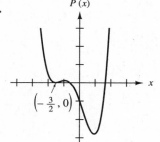

35. a. 5 or 3 or 1 **b.** 0 **c.** $\frac{1}{8}$, $\frac{1}{4}$, $\frac{1}{2}$, 1

d. No rational roots **e.**

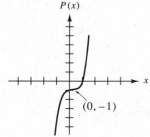

37. a. 2 or 0 **b.** 2 or 0 **c.** $\pm\frac{1}{6}$, $\pm\frac{1}{3}$, $\pm\frac{1}{2}$, ± 1, $\pm\frac{3}{2}$, ± 2, ± 3, ± 6

d. $-\frac{1}{3}$, $\frac{3}{2}$ **e.**

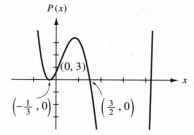

39. a., b.

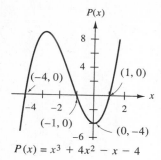

$P(x) = x^3 + 4x^2 - x - 4$

c. Increasing $x < -2.8$ and $x > 0$;
decreasing $-2.8 < x < 0$

d. $x \to \infty$, $P(x) \to \infty$; $x \to -\infty$, $P(x) \to -\infty$

41. a., b.

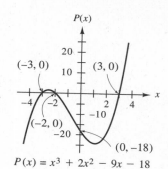

$P(x) = x^3 + 2x^2 - 9x - 18$

c. Increasing $x < -2.5$ and $x > 1.2$;
decreasing $-2.5 < x < 1.2$

d. $x \to \infty$, $P(x) \to \infty$; $x \to -\infty$, $P(x) \to -\infty$

43. a., b.

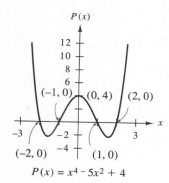

$P(x) = x^4 - 5x^2 + 4$

c. Increasing $-1.6 < x < 0$ and $x > 1.6$; decreasing $x < -1.6$ and $0 < x < 1.6$ **d.** Degree of polynomial is even and $a_n > 0$, therefore as $x \to \infty$, $P(x) \to \infty$ and as $x \to -\infty$, $P(x) \to \infty$

45. Theorem 12.4 states that every polynomial of degree n has exactly n roots. The complex roots occur in pairs, thus for an odd degree polynomial there must be at least one real root.

47. $t = 1$ s, $t = 3$ s   **49.** $x = 2$ ft   **51.** 3 m, 4 m, 5 m   **53.** 1.0   **55.** $1\frac{1}{2}$ s, 2 s, 3 s

Section 12.4 *(page 612)*

1. $P(0) = -1$, $P(1) = 1$ **3.** $P(-0.6) = 1.11$, $P(-0.5) = -0.19$ **5.** $P(0.14) = 0.0227$, $P(0.15) = -0.0466$
7. $P(2.12) = -0.0719$, $P(2.13) = 0.0136$ **9.** $P(1.94) = -0.168$, $P(1.95) = 0.0224$ **11.** 0.2   **13.** 3.7
15. -1.2 **17.** 0.7   **19.** 1.7
21.

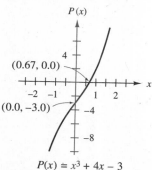

$P(x) = x^3 + 4x - 3$

23.

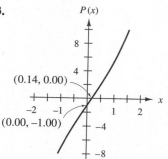

$P(x) = x^3 + 7x - 1$

25.

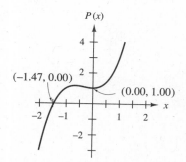

$P(x) = x^3 + x^2 + 1$

27. $P(x) = x^4 + 5x^3 + 6x^2 + 4x + 12$ **29.**

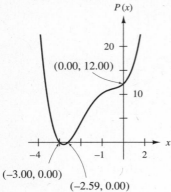

(0.00, 12.00)

(−3.00, 0.00)

(−2.59, 0.00)

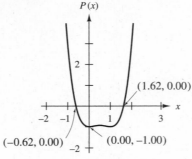

(1.62, 0.00)

(−0.62, 0.00)   (0.00, −1.00)

$P(x) = x^4 - 2x^3 + x^2 - 1$

31. 0.31 m   **33.** 1.63 μF, 0.24 μF   **35.** 16.3 mm   **37. a.** 1   **b.** 1   **c.** 1   **d.** 3   **e.** 1   **f.** 1   **g.** If p and q are integers, $p < 0$ and $|q| < |p|$, then there are three roots. If p and q are integers and $p > 0$, then there is only one solution.

Review Exercises (page 613)
[12.1] **1.** $Q(x) = 2x^2 - 11x - 12$, $R(x) = 0$ **3.** $Q(x) = 2x^2 + x - 3$, $R(x) = -2$

5. $Q(x) = 2x^3 + 25x^2 + 9x - 36$, $R(x) = -2$ **7.** $Q(x) = 6x^3 + 3x^2 - 7x - \dfrac{128}{3}$, $R(x) = -\dfrac{115}{9}$

9. $Q(x) = x^4 - 2x^3 - 19x^2 + 8x + 60$, $R(x) = -3$ [12.3] **11.** $x^3 - 4x^2 - 17x + 60 = 0$
13. $15x^3 + 47x^2 - 50x + 8 = 0$ **15.** $x^4 - 4x^3 + 26x^2 - 100x + 25 = 0$ [12.3]

17. a. 2 or 0   **b.** 1   **c.** ±1, ±2, ±5, ±10   **d.** −2, 1, 5   **19. a.** 1,   **b.** 2 or 0   **c.** ±$\dfrac{1}{2}$, ±$\dfrac{5}{2}$, ±$\dfrac{25}{2}$, ±1, ±5, ±25

e. $P(x) = x^3 - 4x^2 - 7x + 10$ **d.** 0.5   **e.** $P(v) = 2v^3 + 11v^2 + 44v - 25$

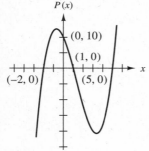

(0, 10)

(1, 0)

(−2, 0)

(5, 0)

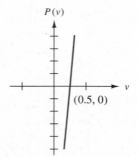

(0.5, 0)

21. a. 3 or 1   **b.** 1   **c.** ±1, ±3, ±9   **d.** 3, 3
e. $P(v) = v^4 - 10v^3 + 32v^2 - 30v - 9$

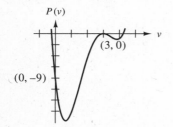

(3, 0)

(0, −9)

23. a. 1 **b.** 3 or 1

c. $\pm 1, \pm 2, \pm 4, \pm 8, \pm 16, \pm \dfrac{1}{3}, \pm \dfrac{2}{3}, \pm \dfrac{4}{3}, \pm \dfrac{8}{3}, \pm \dfrac{16}{3},$

$\pm \dfrac{1}{5}, \pm \dfrac{2}{5}, \pm \dfrac{4}{5}, \pm \dfrac{8}{5}, \pm \dfrac{16}{5}, \pm \dfrac{1}{15}, \pm \dfrac{2}{15}, \pm \dfrac{4}{15}, \pm \dfrac{8}{15}, \pm \dfrac{16}{15}$

d. $\dfrac{2}{5}, -\dfrac{2}{3}$ **e.** $P(y) = 15y^4 + 4y^3 + 56y^2 + 16y - 16$

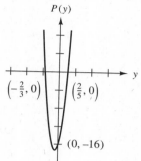

25. a. 3 or 1 **b.** 1

c. $\pm 1, \pm 2, \pm 3, \pm 4, \pm 6, \pm 8, \pm 12, \pm 16, \pm 24, \pm 32,$
$\pm 48, \pm 96$ **d.** $-1, 3$

e. $P(x) = x^4 - 2x^3 + 29x^2 - 64x - 96$

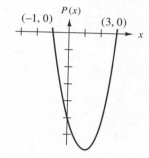

[12.3, 12.4] **27.** $P(x) = x^3 + 2x^2 - 7$

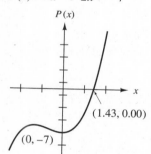

29. $P(x) = x^4 + 6x^3 + 8x^2 - 6x - 24$

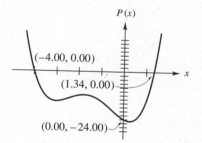

31. $P(R) = 2R^4 - 3R^3 + 6R^2 - 19R + 15$

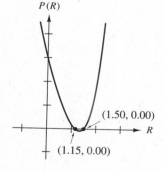

33. $P(z) = 2z^3 + 7z^2 - z - 15$

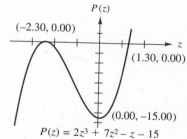

35. $P(t) = t^3 - 5t^2 + 12t + 26$

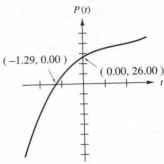

$P(t)$

$(-1.29, 0.00)$

$(0.00, 26.00)$

t

[12.3, 12.4] **37.** 40 mm   **39.** 2   **41.** 1.31 m   **43.** 21 units (To keep $c(x) < \$700$ we do not round)
45. $P(x) = x - a - bj$ **47.** $P(+x)$ will have no sign changes, so according to Descartes' rule of signs, there can be no positive real roots. Also $P(-x)$, since all powers of x are even, will have no sign changes, so there can be no negative roots. Therefore, all the roots must be complex.   **49.** $P(x) = x^4 - 1$ **51.** 0.469, 2.673, 3.519, 4.712, 5.906   **53.** 3.0 cm

Chapter Test (page 616)

1. $(x + 1)(3x^3 - 2x^2 - 5) + 8$ **2.** 2 is not a root   **3.** 2 is a root   **4.** 3 is an upper bound   **5.** There are two changes in sign; thus, there are 2 or 0 positive roots   **6.** There is one change in sign; thus, there is one negative root

7. $\pm\dfrac{1}{2}, \pm\dfrac{3}{2}, \pm 1, \pm 2, \pm 3, \pm 6$ **8.** $-2, 3,$ and $1/2$ **9.** $3j, -3j$ and -2.   **10.** Since the sign of the functional values

changes from negative to positive, $f(x)$ must be zero between $x = -0.8$ and $x = 0.8$.   **11.** To the nearest tenth, the positive roots are 1.0 and 1.8.   **12.** The company can manufacture only 22 units and keep its costs at or below $1600.   **13.** The size of the square is 2 cm by 2 cm.   **14.** 4 products

Chapter 13

Section 13.1 (page 626)

1. 8994.86   **3.** 3.90   **5.** 14.86   **7.** 1236.65   **9.** 109.95   **11. d**   **13. h**   **15. g**   **17. f**

19. $y = 4^x$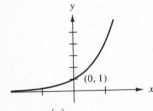

(0, 1)

21. $y = 4^{x+1}$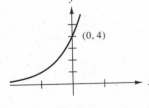

(0, 4)

23. $y = \left(\dfrac{3}{2}\right)^x$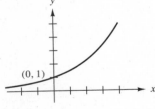

(0, 1)

25. $y = \left(\dfrac{1}{4}\right)^x$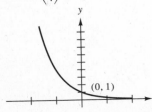

(0, 1)

27. $y = 3^{x+2}$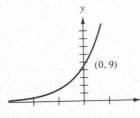

(0, 9)

29. $y = 3^x - 2$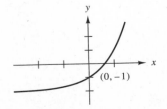

(0, -1)

31. $p = 0.625^t$

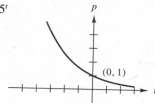

33. $p = 2(3^{-t})$

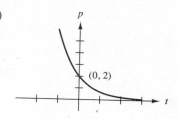

35. 30 s   **37.** 2.5 s   **39.** 0.69 s   **41. b.** 10,000,000   **43.** 13,122 bacteria   **45.** 0.936 kg   **47.** $5.36   **49.** $701.05
51. 10%   **53.** $2720.98   **55.** $2745.57   **57.** $2754.26
59. $f(x) = e^{-x}$

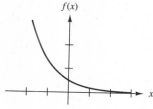

The larger the dose, the smaller fraction of a surviving tumor cell.
61. 13.5%

Section 13.2 (page 636)

1. $\log_5 25 = 2$ **3.** $\log_4 64 = 3$ **5.** $\log_5 y = 2$ **7.** $\log_{10} p = t$ **9.** $\log_n m = x$ **11.** $49 = 7^2$ **13.** $625 = 5^4$
15. $144 = 12^y$ **17.** $123 = 6.3^v$ **19.** $t = m^r$ **21.** 16   **23.** $\dfrac{1}{5}$ **25.** 49   **27.** 5   **29.** 6   **31.** 2   **33.** 2   **35.** 2
37. 5   **39.** -2 **41.** -2 **43.** -2 **45.** 1   **47.** 1   **49.** 0   **51.** 0
53. $y = \log_4 x$ and $y = \log_2 x$

55. $y = \log_4 x$ and $y = \log_{1/2} x$

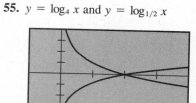

Both functions are increasing functions, they both cross the x-axis at (1, 0). $\log_2 x < \log_4 x$ for $0 < x < 1$ and $\log_2 x > \log_4 x$ for $x > 1$

Both functions cross the x-axis at (1, 0). $y = \log_4 x$ is an increasing function, and $y = \log_{1/2} x$ is a decreasing function

57. $y = \log_6 x$ **59.** $y = \log_6 x$

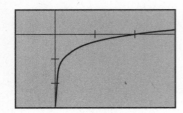   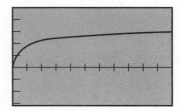

61. 20 hr

Section 13.3 (page 643)

1. $\log_b 7 + \log_b x$ **3.** $\log_b x - \log_b 7$ **5.** $1 + \log_5 x + \log_5 y$ **7.** $\log_5 x - 1 + \log_5 y$

9. $\frac{1}{2}(\log_3 4 + \log_3 x + \log_3 y)$ **11.** $\log_3(x - 4) - \log_3(y + 5)$ **13.** 1   **15.** $2(\log_a x + \log_a y + \log_a 2)$

17. $3 - \log_b 2 - \log_b 3$ **19.** $2 - \log_{10}(x + 1) = \log_{10}\frac{10^2}{x + 1}$ **21.** $\log_2 x^3 y$ **23.** $\log_3 3x$ **25.** $\log_2 \frac{xy}{z}$ **27.** $\log_b \frac{2 - x}{x + 2}$

29. $\log_3(2x - 1)^5(x + 2)^5$ **31.** $\log_2 x^4$ **33.** $\log_{6.3} x^{3.5}y^{2.1}$ **35.** $\log_b \frac{x^4(x + 1)^{13}}{y^5}$ **37.** $\log_b \frac{(y + 3)^5(y + 2)^{1/3}}{\sqrt[4]{y - 5}}$

39. 1.00   **41.** 1.63   **43.** 1.84   **45.** -0.052 **47.** $\log_{10}\left(\frac{I}{I_0}\right)^{10}$

Section 13.4 (page 650)

1. a. 0.21219   **b.** 1.21219   **c.** 2.21219 The mantissa is the same for **a, b,** and **c.** The characteristic increases as the number of digits to the left of the decimal point increase   **3. a.** 6.73239   **b.** 7.73239   **c.** 8.73239 The mantissa is the same for **a, b,** and **c.** With the numbers in scientific notation, the characteristic is the same as the exponent of 10
5. 0.39967   **7.** 3.72263   **9.** -3.14206 **11.** 4.710   **13.** 58.16   **15.** 0.008149   **17.** 6.731×10^{19} **19.** 0.48858
21. 4.06389   **23.** 5.65599   **25.** -0.38126 **27.** -11.35592 **29.** 1.9601   **31.** 3.8863   **33.** 0.12383   **35.** 408,520,000
37. 39.5   **39.** 15,000   **41.** 2200   **43.** 0.437   **45.** 0.891   **47.** 1.4650   **49.** 2.7633   **51.** 2.1534   **53.** 2.3666
55. 2.3869   **57.** -3.1699 **59.** 120 db   **61.** 25.1 db   **63.** 780°K   **65.** 5.50 mol/ℓ **67.** 6.93 units   **69.** 0.29 v
71. 12.8 mph   **73.** 2.93 J   **75.** 3.93 s

Section 13.5 (page 657)

1. 0.774   **3.** 1.16   **5.** 2.72   **7.** $\pm 2j$ **9.** -1 **11.** 0.162   **13.** -5.39 **15.** No solution   **17.** 12.7

19. 0.786 or $\frac{11}{14}$ **21.** 0.414   **23.** 0.414   **25.** No solution   **27.** 1.0001   **29.** 0.414   **31.** 1.04×10^{10} yr

33. 6.3×10^{-6} W/cm²   **35.** 0.020   **37.** $y = \frac{y_1 - y_2 \, e^{z/k}}{1 - e^{z/k}}$ **39.** $V = \sqrt[\gamma]{\frac{ec}{p}}$

Section 13.6 (page 667)

1. $y = 2x^3$ **3.** $y = 4x^2$ **5.** $y = 0.200x^{0.667}$

7.

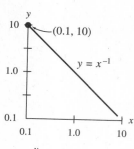

9.

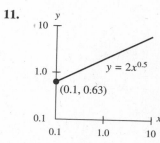

11.

13.

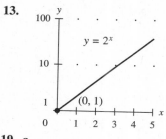

15.

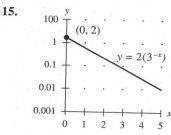

17.

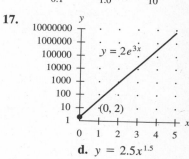

19. a.

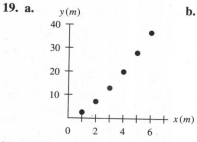

b.

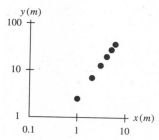

c.

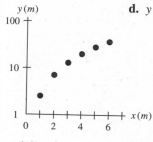

d. $y = 2.5x^{1.5}$

21. a.

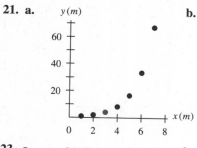

b.

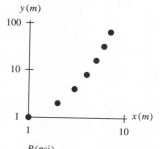

c.

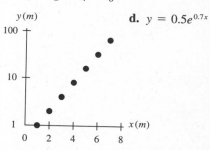

d. $y = 0.5e^{0.7x}$

23. a.

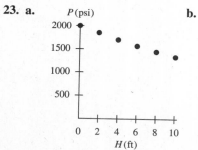

b.

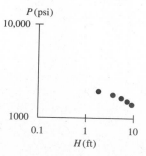

c.

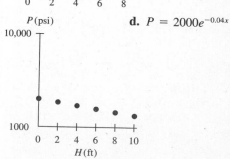

d. $P = 2000e^{-0.04x}$

25. $y = 3.53 \times 10^{-10}x^{3.88}$ **27.** $y = 90.9\ x^{0.0866}$ **29.** $y = 0.952\ e^{-1.16x}$ **31.** $y = 85\ e^{0.058x}$

Review Exercises (page 669)

[13.1] **1.**
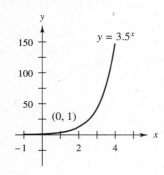

3.
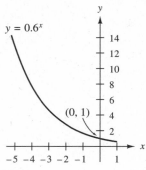

5.
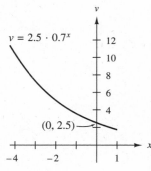

7.
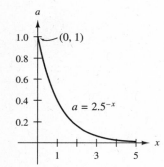

9.
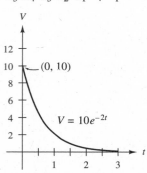

11.
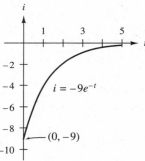

[13.1] **13.** 1.67 s   **15.** 30 s   **17.** $547.36   **19.** 11 bacteria   [13.2] **21.** $\log_7 49 = 2$ **23.** $\log_5 P = t$ **25.** $5^3 = 125$
27. $1.3^s = 416$ **29.** 36   **31.** 0.0313   **33.** 1   **35.** -3 **37.** 0
[13.2] **39.**
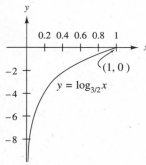

41.
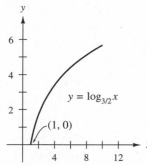

43. 5.64 hr   [13.3] **45.** $2[\log_7 5 + 3 \log_7 x]$ **47.** $\log_2 (x - 2) - \log_2 (x + 2)$ **49.** $\log_{10} (x^2 + y^2) - 3$ **51.** $\log_{10} x$

53. $\log_2 \dfrac{x + 3}{x - 3}$ **55.** 1   **57.** 4.38   **59.** -2.92 **61.** 3.55   **63.** $\dfrac{\log \dfrac{A}{P}}{\log(1 + r)}$ [13.4] **65.** 1.1523; c = 1, m = 0.1523
67. -1.9469; c = -2, m = 0.0531   **69.** 6.1761; c = 6, m = 0.1761   **71.** -1.0915; c = -2, m = 0.9085   **73.** 324.
75. 0.00152   **77.** 43.4   **79.** 501.   **81.** 2.31   **83.** 3.16   **85.** 3   **87.** -3.00 **89.** 1.16   **91.** 1.43   **93.** 0.0794
95. 2.32   **97.** 600.°K   [13.5] **99.** -1.66 **101.** -20.6 **103.** No solution   **105.** No solution   **107.** 672. yr
109. 4.58 rad

[13.6] 111. a.
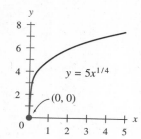

b.
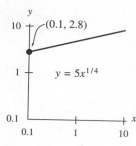

c.
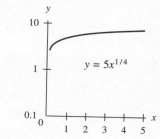

113. a.
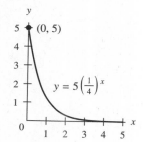

b.
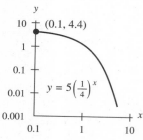

c.
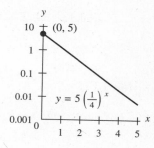

115. a.
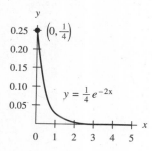

b.
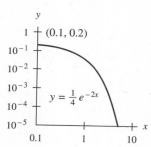

c.
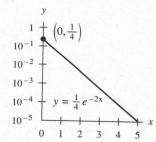

117.
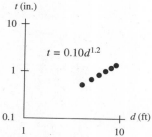

119.
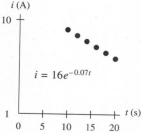

Chapter Test (page 673)

1. 3460   **2.** 31.98   **3.** $y = 3(2^x)$

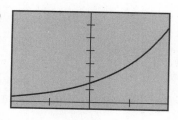

4. $\log_{11} 121 = 2$ **5.** $1.3^z = 234$ **6.** $x = 64$ **7.** $x = \sqrt{11}$. **8.** $x = \log_4 12 - 1$
9. 6.33 **10.** 5.16 **11.** 1.54033 **12.** -6.08 **13.** 27.0 **14.** 0.00340 **15.** 50.3 **16.** 33.2 **17.** 2.14 **18.** 3.16
19. $y = \log(x - 1)$ **20.** $y = \log_5 2x$

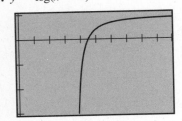
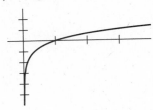

21. $t = -\dfrac{P_0}{R}$ **22. a.** 3000 **b.** 3218 **23.** $y = 5x^3$

Chapter 14

Section 14.1 (page 683)

1.   **3.**   **5.**   **7.**

9.   **11.** $-5 \geq -16$ **13.** $R + 6 > 11$ **15.** $2 < Z \leq 11$ **17.** $6 - x < 0$ **19.** $R - 2 \leq 0$

21. $<$ **23.** $<$ **25.** $>$ **27.** $<$ **29.** $>$ **31.** $y > 0$ **33.** $x < 0, y > 0$ **35.** $x > 4$
37. $x < 3$ **39.** $r \geq 7$ **41.** $R \geq 0$

43. $R \leq -\dfrac{1}{2}$ **45.** $t \leq \dfrac{4}{5}$ **47.** $z \leq \dfrac{-16}{9}$

49. $-1 < y < 5$ **51.** $-9 \leq x \leq 5$ **53.** $-0.45 < x < 1.27$

55. $190 < (110 + 0.20x); x > 400$ mi **57.** $-0.005 \leq d, -0.5 \leq 0.005; 0.495$ mm $\leq d \leq 0.505$ mm

59. $35 < \dfrac{5}{9}(F - 32) < 39; 95° < F < 102.2°$ **61.** $19 < 92 - 32t < 32; 1.95 < t < 2.35$

63. $\dfrac{1}{8} < \dfrac{1}{15} + \dfrac{1}{R_2} < \dfrac{1}{5}; \dfrac{15}{2}\Omega < R_2 < \dfrac{120}{7}$ **65.** $40 < \dfrac{d}{4} < 55; 160$ mi $< d < 220$ mi

67. 24 rad/s $< \omega < 32$ rad/s **69.** 6.49×10^7 yr $< t < 6.51 \times 10^7$ yr

Section 14.2 (page 696)

1. c. $(-\infty, -2)$ or $(3, \infty)$ **3. c.** $[-3, 5]$ **5. c.** $(-\infty, 0]$ or $[3, \infty)$ **7. c.** $(-\infty, -5)$ or $(0, \infty)$ **9. c.** $(-1, 5)$

11. c. $(-\infty, -2]$ or $\left[-\dfrac{1}{2}, \infty\right)$ **13. c.** $[-3.186, -0.314]$ **15. c.** No real x **17. c.** $(-\infty, \infty)$ **19. c.** $(0.279, 2.387)$

21. a. 15   **b.** $10 \le x \le 15$ **23.** $2 \text{ s} \le t \le 8 \text{ s}$ **25.** $m \le 15$ mm   **27.** $b \le 0.5$ m   **29.** $0 \text{ mm} < r < 50$ mm

31. $(-\infty, -2)$ or $(1, 4)$ **33.** $(-7, 3)$ or $(4, \infty)$ **35.** $(-\infty, \infty)$ **37.** $(-5, 0)$ or $\left(\dfrac{3}{4}, \infty\right)$ **39.** $R \ne 0$

41. $(-\infty, 0)$ or $(3, \infty)$ **43.** $(-\infty, -1)$ or $(0, \infty)$ **45.** $y > -3$ **47.** $(-\infty, \infty)$ **49.** $(-\infty, -8]$ or $(-3, 0]$ or $(2, \infty)$

51. $(-\infty, 3]$ or $[5, \infty)$ **53.** $10 \text{ ft} < L < 25$ ft

Section 14.3 (page 706)

1. $-8, 2$ **3.** $1\dfrac{2}{3}, 2\dfrac{1}{3}$ **5.** $-\dfrac{2}{5}, -\dfrac{4}{5}$ **7.** $-4, 7\dfrac{1}{3}$ **9.** $-2\dfrac{1}{3}, 5$ **11.** $\dfrac{3}{35}, \dfrac{18}{35}$ **13.** $-\dfrac{5}{23}, \dfrac{5}{19}$ **15.** $-\dfrac{3}{5}, \dfrac{3}{10}$

17. $[-8, 8]$ **19.** $(-\infty, -13)$ or $(13, \infty)$ **21.** $t \ne 0$ **23.** $[-3, 11]$

25. $(-\infty, 3]$ or $[11, \infty)$ **27.** $\left(-\dfrac{7}{3}, -1\right)$ **29.** $\left(-\infty, -\dfrac{1}{3}\right)$ or $[3, \infty)$ **31.** No solution

33. $\left(-\infty, \dfrac{1}{7}\right)$ or $\left(\dfrac{9}{7}, \infty\right)$ **35.** $(-1, 7)$ **37.** No solution   **39.** $\left(-\dfrac{7}{12}, -\dfrac{1}{12}\right)$

41. $(-0.25, -0.15)$

43. $|x| < 3$ **45.** $|x| > 2$ **47.** $|x - 1| < 2$ **49.** $[13.44 \text{ mm}, 13.46 \text{ mm}]$

51. $[3.97 \text{ mm}, 4.03 \text{ mm}]$ **53. a.** $\left|y - \dfrac{4}{3}\right| < 9$ **b.** $\left(-7\dfrac{2}{3}, 10\dfrac{1}{3}\right)$ **55. a.** $|R + 13| \le 5$ **b.** $[-18, -8]$

57. $40 \text{ in.} < w < 60$ in.   **59.** $\$91.46 < \text{Cost} < \$92.14$

Section 14.4 (page 712)

1.

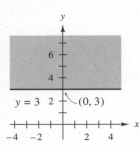

3.

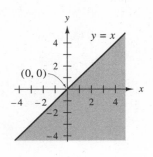

5.

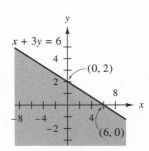

7.

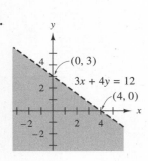

9.

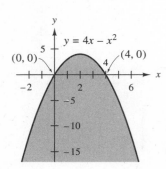

11.

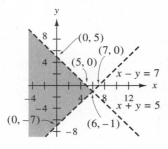

13.

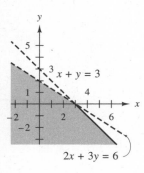

15.

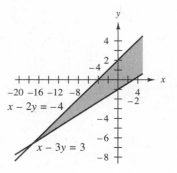

17.

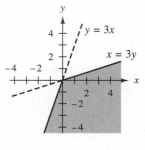

19.

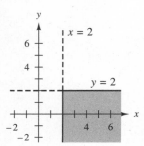

21.

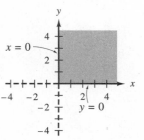

23.

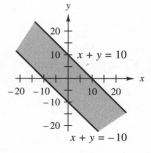

25.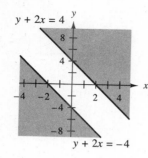

$y + 2x = 4$

$y + 2x = -4$

27.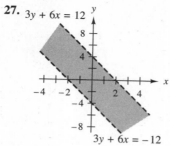

$3y + 6x = 12$

$3y + 6x = -12$

29.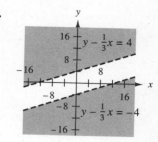

$y - \frac{1}{3}x = 4$

$y - \frac{1}{3}x = -4$

31.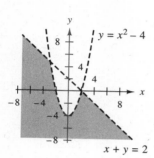

$y = x^2 - 4$

$x + y = 2$

33.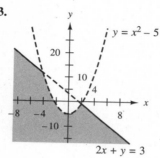

$y = x^2 - 5$

$2x + y = 3$

35.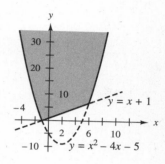

$y = x + 1$

$y = x^2 - 4x - 5$

37.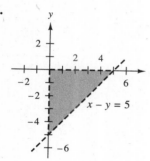

$x - y = 5$

39.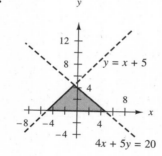

$y = x + 5$

$4x + 5y = 20$

41. $s + l \geq 200$
$1.00s + 1.50l \leq 300$
$s \geq 0 \quad l \geq 0$

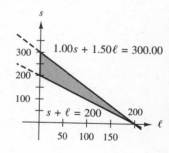

$1.00s + 1.50\ell = 300.00$

$s + \ell = 200$

43. $Q \geq 2H \quad Q + H \geq 10$
$3Q + 4H \leq 80$
$Q \geq 0 \quad H \geq 0$

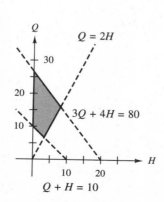

$Q = 2H$

$3Q + 4H = 80$

$Q + H = 10$

Review Exercises (page 713)

[14.1] **1. a.**

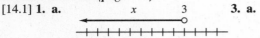

b. $(-\infty, 3)$

3. a.

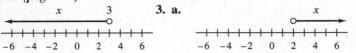

b. $(2, \infty)$

5. a.

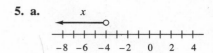

b. $(-\infty, -4)$

7. a.

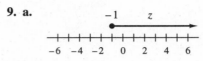

b. $(-\infty, 0]$

9. a.

b. $[-1, \infty)$

11. a.

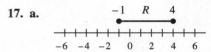

b. $(-\infty, \infty)$

13. a.

b. $\left[-\dfrac{3}{2}, \infty\right)$

15. a.

b. $[1, \infty)$

17. a.

b. $[-1, 4]$

19. a.

b. $(-3, 1]$

[14.3] **21. a.**

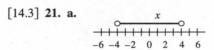

b. $(-4, 4)$

23. a.

b. $(-\infty, -5)$ or $(5, \infty)$

25. a.

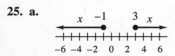

b. $(-\infty, -1]$ or $[3, \infty)$

27. a.

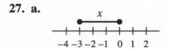

b. $[-3, 0]$

[14.2] **29. a.**

b. $(-3, 0)$

31. a.

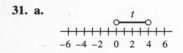

b. $(0, 4)$

33. a.

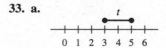

b. $[3, 5]$

35. a.

b. $(-\infty, -2]$ or $[3, \infty)$

37. a.

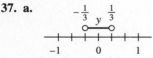

b. $\left(-\dfrac{1}{3}, \dfrac{1}{3}\right)$

39. a.

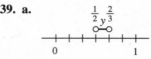

b. $\left(\dfrac{1}{2}, \dfrac{2}{3}\right)$

41. a.

b. $(-3, 5)$ or $(7, \infty)$

43. a.

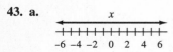

b. $(-\infty, \infty)$

45. a.

b. $(-\infty, -2]$ or $\left[0, \dfrac{1}{6}\right]$

47. a.

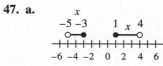

b. $(-5, -3]$ or $[1, 4)$

49. a.

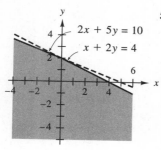

b. $[-3, 3]$

[14.4] **51.**

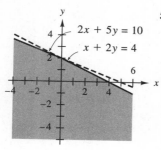

53.

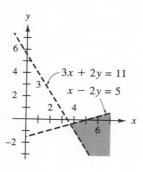

55.

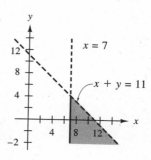

57.

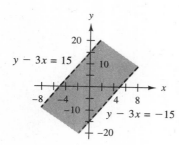

59.

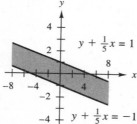

61. (68°F, 86°F)   **63.** $\dfrac{84}{17}\Omega < R_2 < \dfrac{231}{10}\Omega$ **65.** $64 \le \dfrac{r}{5} \le 88$; 320 km $\le r \le$ 440 km

67. $|t - 2| \le 0.04$; 1.96 mm $\le t \le$ 2.04 mm

[all] **69. a.** $x > -3$, concave up; $x < -3$ concave down   **b.** $x < \dfrac{3}{2}$, concave up; $x > \dfrac{3}{2}$, concave down

71. $0 < NT < 0.752$ **75.** $(-6, -4], [-\sqrt{2}, \sqrt{2}], [4, 6)$

Chapter Test (page 000)

1.

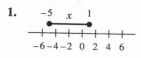

2.

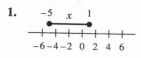

3. $0 \le t < 7$ **4.** $t - 5 \ge 0$

5. $x \le 5$

6. $x > \dfrac{11}{5}$

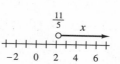

7. $(-\infty, -2)$ or $(0, 3)$ **8.** $\left(3, \dfrac{17}{4}\right)$ **9.** $x = -\dfrac{1}{2}, x = -\dfrac{5}{6}$ **10.** No solution.

11. all real numbers

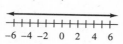

12. $-14 < x < 10$

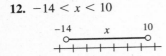

13.

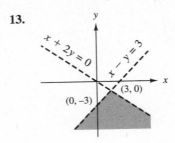

14.

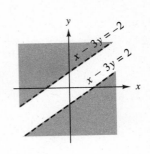

15. $x + y > 200$
$\$2x + \$3y < \$600$
$x \geq 0$
$y \geq 0$

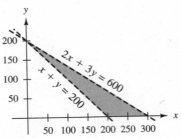

Chapter 15

Section 15.1 (page 723)

17. $\cos A$ **19.** $\tan A$ **21.** $\sin^2 A$ **23.** $\cos \theta$ **25.** $-\sin^3 \theta$ **27.** $\cot \theta$ **29.** $\cot^2 A$ **31.** $\csc^2 B$ **33.** 1
35. 1   **37.** 1   **55.** $1 + \cos \theta$

Section 15.2 (page 730)

1. $\dfrac{1 - \sqrt{3}}{2\sqrt{2}}$ **3.** $\dfrac{1 - \sqrt{3}}{2\sqrt{2}}$ **5.** $\dfrac{1 + \sqrt{3}}{2\sqrt{2}}$ **7.** $-\dfrac{\sqrt{3} + 1}{2\sqrt{2}}$ **9.** $\dfrac{1 + \sqrt{3}}{2\sqrt{2}}$ **11.** $-2 - \sqrt{3}$ **13.** $\dfrac{1}{\sqrt{2}}$ **15.** $\dfrac{1}{\sqrt{2}}$

17. $\dfrac{\sqrt{3}}{2}$ **19.** $\dfrac{1}{2}$ **21.** 1   **23.** $-\sin \theta$ **25.** $-\cos \theta$ **27.** $\sin \theta$ **29.** $\dfrac{\sqrt{3} \sin \theta + \cos \theta}{2}$ **35.** $5\sqrt{3} \cos 20t + 5 \sin 20t$

37. $3 \cos t - 6\sqrt{3} \sin t$ **39.** $1.4 \sin 377t + 1.2 \cos 377t$ **41.** $260 \cos 4t + 200 \sin 4t$

43. $y = -20 \sin \dfrac{\pi t}{4} - 2.8 \cos \dfrac{\pi t}{4}$ **49.** $y = 2 \sin (x + 30°)$ **53.** $W = T_2 \dfrac{\sin (d + \beta)}{\sin \alpha}$

Section 15.3 (page 736)

1. $\dfrac{\sqrt{3}}{2}$ **3.** $-\sqrt{3}$ **5.** $-\dfrac{1}{2}$ **7.** $\dfrac{1}{2}\sqrt{2 - \sqrt{3}}$ **9.** $\dfrac{1}{2 - \sqrt{3}}$ **11.** $\dfrac{1}{2}\sqrt{2 + \sqrt{3}}$ **13.** $\dfrac{1}{2}\sqrt{2 - \sqrt{3}}$ **15.** $\dfrac{1}{2}\sqrt{2 - \sqrt{3}}$

17. $\tan 2A = \dfrac{2 \tan A}{1 - \tan^2 A}$ **25.** $\dfrac{s_1 + s_3}{2} + \dfrac{s_1 - s_3}{2}\cos 2\theta - s_2 \sin 2\theta$ **27.** $x = \dfrac{v_0{}^2 \sin 2\alpha}{g}$ **29. a.** $34.4° < \theta < 90°$
b. $\theta > 17.7°$ **c.** $17.7° < \theta < 34.4°$

Section 15.4 (page 743)

1. $0°, 45°, 180°, 225°$ **3.** $0°, 45°, 135°, 180°$ **5.** $270°$ **7.** $45°, 225°$ **9.** $22\frac{1}{2}°, 112\frac{1}{2}°, 202\frac{1}{2}°, 292\frac{1}{2}°$ **11.** $30°, 150°, 270°$
13. $60°, 300°$ **15.** $60°, 180°, 300°$ **17.** $0°, 90°$ **19.** $90°, 120°, 240°, 270°$ **21.** $0°, 45°, 180°, 225°$ **23.** $0°$
25. $0°, 60°, 300°$ **27.** $30°, 150°, 210°, 330°$ **29.** $30°, 150°$ **31.** $0°$ **33.** $53.6°, 147.5°, 212.5°, 306.4°$ **35.** $90°, 270°$
37. 0.364 **39.** $7.0°, 214.1°$ **41.** $70.1°, 161.5°$ **43.** $130.0°, 320.0°$

Section 15.5 (page 750)

1. a.

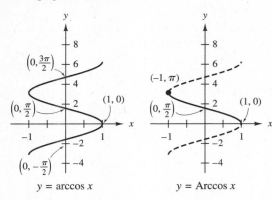

$y = \arccos x$ $y = \text{Arccos } x$

b. $[-1, 1], [-1, 1]$

c. $(-\infty, \infty), [0, \pi]$

5. a.

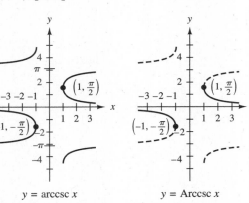

$y = \text{arccsc } x$ $y = \text{Arccsc } x$

3. a.

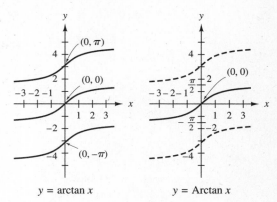

$y = \arctan x$ $y = \text{Arctan } x$

b. $(-\infty, \infty), (-\infty, \infty)$

c. $(-\infty, \infty)\ y \neq \dfrac{\pi}{2} \pm k\pi, \left(-\dfrac{\pi}{2}, \dfrac{\pi}{2}\right)$

b. $(-\infty, -1]$ or $[1, \infty)\ (-\infty, -1]$ or $[1, \infty)$

c. $(-\infty, \infty)\ y \neq \pm k\pi, \left[-\dfrac{\pi}{2}, 0\right)$ or $\left(0, \dfrac{\pi}{2}\right]$

7. $0°$ **9.** $45°$ **11.** $45°$ **13.** $-76.0°$ **15.** $-30°$ **17.** $150°$ **19.** $60°$ **21.** $143.1°$ **23.** $56.2°$ **25.** $-87.3°$ **27.** 0.866

29. 0.707 **31.** 1.333 **33.** 0.750 **35.** 0.621 **37.** -0.280 **39.** Undefined   **41.** -1.152 **43.** 0 **45.** $\dfrac{\sqrt{b^2 - a^2}}{b}$

47. $\dfrac{\sqrt{1 - x^2}}{x}$ **49.** $12.99°$ **51.** $13.26°$ **53.** Because $y = \text{Arctan } x$ is a function whose equivalent function is $x = \tan y$

and $\tan y$ is undefined for $y = -\dfrac{\pi}{2}$ and $\dfrac{\pi}{2}$ **55.** $24°$

Review Exercises (page 751)

[15.1] **5.** $\sec \theta$ **7.** $\cos^2 \theta$ **9.** $2 \csc^2 A$ **11.** $\cos^2 B$ **13.** $\tan^2 A$ **15.** $\cos^2 \theta$ **17.** 1 **19.** 1

[15.2, 15.3] **29.** $\dfrac{\sqrt{3} - 1}{2\sqrt{2}}$ **31.** $-\dfrac{1}{\sqrt{2}}$ **33.** $-(2 + \sqrt{3})$ **35.** -1 **37.** $-2 + \sqrt{3}$ **41.** $\dfrac{2 \tan \theta}{1 - \tan^2 \theta}$

43. $\dfrac{2 \tan A}{1 + \tan^2 A}$ [15.4] **47.** $45°, 225°$ **49.** $90°, 450°$ **51.** $30°, 90°, 150°, 270°$ **53.** $0°, 60°, 300°$

55. $30°, 45°, 135°, 150°, 210°, 225°, 315°, 330°$ **57.** $0°, 120°, 240°$ **59.** $33.7°, 135°, 213.7°, 315°$ **61.** $165°, 345°$

63. 300°   **65.** 65.0°, 217.7°   **67.** 6.8°, 120.1°   [15.5] **69.** 30°   **71.** 36.9°   **73.** 60°   **75.** 30°   **77.** 0.385   **79.** 0.471
81. 0.292   **83.** 0.114   **85.** 0.960   **87.** 0.282   **89.** $\dfrac{x}{\sqrt{1-x^2}}$ [All] **91.** If $r = 0$, θ does not exist
93. $C \cos \omega t - D \sin \omega t$ **95.** No difference
99.
```
   10    PRINT "WHAT IS X?"
   20    INPUT X
   30    LET R = X*3.14159/180
   40    LET S = R−R ∧ 3/6+R∧5/120−R ∧ 7/5040
   50    PRINT "SIN ";X;" = ";S
   60    END
```

Chapter Test *(page 755)*

1. csc x **2.** No answer   **3.** No answer   **4.** sin (180° + 45°)   **5.** sin 2 (30°)   **6.** $x = 45°, 135°, 225°, 315°$
7. $x = 0°, 76°, 180°, 256°$ **8.** $x = 65.5°, 294.5°$ **9.** 45°   **10.** 0.577   **11.** −60°   **12.** No answer
13. No answer   **14.** Not an identity

Chapter 16

Section 16.1 *(page 763)*

1. a. 2, 4, 6, 8, 10, 12
   b. Increasing
   c.
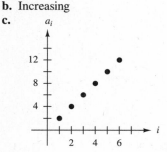

3. a. 4, 5, 6, 7, 8, 9
   b. Increasing
   c.
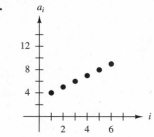

5. a. −2, 1, 4, 7, 10, 13
   b. Increasing
   c.
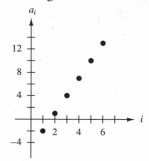

7. a. $-\dfrac{3}{4}, -\dfrac{3}{2}, -\dfrac{9}{4}, -3, -\dfrac{15}{4}, -\dfrac{9}{2}$
   b. Decreasing
   c.
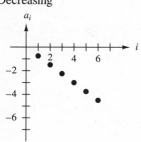

9. a. −2, 0, −2, 0, −2, 0
   b. Neither
   c.
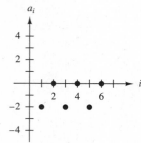

11. a. $-3, 1, -3, 1, -3, 1$
b. Neither
c.

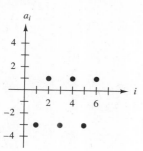

13. a. $0, 3, 8, 15, 24, 35$
b. Increasing
c.

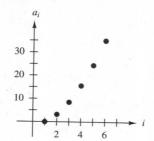

15. a. $\dfrac{3}{4}, \dfrac{4}{5}, \dfrac{5}{6}, \dfrac{6}{7}, \dfrac{7}{8}, \dfrac{8}{9}$
b. Increasing
c.

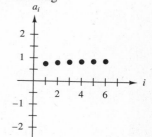

17. a. $0, \dfrac{3}{5}, \dfrac{4}{5}, \dfrac{15}{17}, \dfrac{12}{13}, \dfrac{35}{37}$
b. Increasing
c.

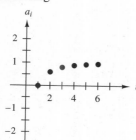

19. a. $1, -\dfrac{1}{2}, \dfrac{1}{3}, -\dfrac{1}{4}, \dfrac{1}{5}, -\dfrac{1}{6}$
b. Neither
c.

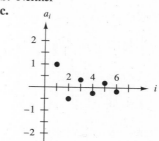

21. $i - 1$ **23.** $2i$ **25.** $2i - 1$ **27.** $4i - 1$ **29.** $-2i$ **31.** $i + 3$ **33.** $4i - 8$ **35.** $\dfrac{1}{i}$ **37.** $\dfrac{1}{i^2}$ **39.** $\dfrac{i+1}{i}$

41. $6x, ix$ **43.** $x + 18y, x + 3iy$ **45.** $x + 25y, x + (5i - 5)y$ **47.** \$3.50, \$3.85, \$4.24, \$4.66, \$5.12, \$5.64

Section 16.2 (page 768)

1. a. 14 **b.** 510 **3. a.** 0.481 or $\dfrac{13}{27}$ **b.** 0.500 or $\dfrac{3280}{6561}$ **5. a.** 21 **b.** 96 **7. a.** -0.0833 or $-\dfrac{1}{12}$ **b.** 2.0673 or

$\dfrac{11461}{5544}$ **9. a.** 2 **b.** -4 **11.** $4 - 4 + \dfrac{16}{3} - 8 + \dfrac{64}{5} = 10.13$ or $\dfrac{152}{15}$ **13.** $9 + \dfrac{27}{2} + 27 + \dfrac{243}{4} = 110.25$

15. $-2 + 2 - \dfrac{8}{3} + 4 - \dfrac{32}{5} + \dfrac{32}{3} = 5.60$ **17.** $\log 3 + \log 4 + \log 5 + \log 6 + \log 7 = 3.40$

19. $\dfrac{3}{7} + \dfrac{1}{2} + \dfrac{5}{9} + \dfrac{3}{5} + \dfrac{7}{11} + \dfrac{2}{3} + \dfrac{9}{13} = 4.08$ **21.** $\sum\limits_{i=1}^{n} 3i$ **23.** $\sum\limits_{i=1}^{n} \dfrac{(-1)^i i}{2^{i-1}}$ **25.** $\sum\limits_{i=1}^{5} \dfrac{1}{i^2}$ **27.** $\sum\limits_{i=1}^{5} \dfrac{(-1)^{i+1}}{i^2}$ **29.** $\sum\limits_{i=1}^{n} \dfrac{i(i+1)}{2}$

31. $\sum\limits_{i=1}^{5} (z + i + 3)$ or $\sum\limits_{i=4}^{8} (z + i)$ **33.** $\sum\limits_{i=1}^{5} \dfrac{z - 2 - i}{z + 2 + i}$ or $\sum\limits_{i=3}^{7} \dfrac{z - i}{z + i}$ **35.** $\sum\limits_{i=1}^{5} z^{2+i}(z + 2 + i)$ **37.** $\dfrac{1}{5} + \dfrac{4}{7} + 1 + \dfrac{16}{11} + \dfrac{25}{13}$

39. $\ln 1 + 2 \ln 2 + 3 \ln 3 + 4 \ln 4$ **41.** $-2 + 0 + 12 + \cdots + n(n + 1)(n - 2)$ **43.** $1 + \dfrac{49}{4} + 64 + \cdots +$

$\left(\dfrac{2n^2 - n + 1}{2}\right)^2$ **45.** $(y^2 + 4y + 2) + (y^2 + 6y + 3) + (y^2 + 8y + 4) + (y^2 + 10y + 5)$ **47. a.** 10 ft **b.** 54 ft

49. a. \$1900.00 **b.**

Year	1	2	3	4	5	6	7	8	9	10
Interest	\$1000	900	800	700	600	500	400	300	200	100

c. \$5500 **d.** $\sum\limits_{i=1}^{10} 0.10[10{,}000 - (i - 1)(1000)]$

Section 16.3 (page 777)

1. Yes, $d = 2$　**3.** No　**5.** Yes, $d = \dfrac{1}{2}$　**7.** No　**9.** No　**11. a.** 13, 18, 23, 28, 33　**b.** 115　**13. a.** 15, 12, 9, 6, 3

b. 45　**15. a.** 11, 10, 9, 8, 7　**b.** 45　**17. a.** $-5\dfrac{1}{2}, -5, -4\dfrac{1}{2}, -4, -3\dfrac{1}{2}$　**b.** $-22\dfrac{1}{2}$　**19. a.** $0, \dfrac{3}{5}, \dfrac{6}{5}, \dfrac{9}{5}, \dfrac{12}{5}$

b. 6　**21.** $a_8 = 15$, $S_8 = 64$　**23.** $n = 7$, $S_7 = 84$　**25.** $d = 3$, $S_8 = 60$　**27.** $d = 0.3$, $a_6 = 1.9$　**29.** $a_1 = 23$, $a_6 = 3$

31. a. 29　**b.** 1218　**33. a.** 51　**b.** 5508　**35.** \$18,600.00　**37. a.** 240 ft　**b.** 1024 ft　**39.** 365 ft, 54 ft　**41.** $6\dfrac{1}{3}, 6\dfrac{2}{3}$,

$7, 7\dfrac{1}{3}, 7\dfrac{2}{3}$　**43.** 7600 bacteria/day　**45.** $n = 12$ s

Section 16.4 (page 785)

1. Yes; $r = 2$, $a_7 = 64$　**3.** No　**5.** Yes; $r = \dfrac{1}{4}$, $a_7 = \dfrac{1}{4096}$　**7.** Yes; $r = \dfrac{1}{7}$, $a_7 = \dfrac{1}{7}$　**9.** Yes; $r = \dfrac{1}{2}$, $a_7 = \dfrac{1}{192}$

11. a. 3, 6, 12, 24, 48　**b.** 93　**13. a.** $2, \dfrac{2}{3}, \dfrac{2}{9}, \dfrac{2}{27}, \dfrac{2}{81}$　**b.** 2.99 or $\dfrac{242}{81}$　**15. a.** $\dfrac{2}{5}, \dfrac{1}{10}, \dfrac{1}{40}, \dfrac{1}{160}, \dfrac{1}{640}$　**b.** 0.533 or $\dfrac{341}{640}$

17. a. $-4, -0.4, -0.04, -0.004, -0.0004$　**b.** -4.44　**19. a.** $4, -8, 16, -32, 64$　**b.** 44　**21.** $a_4 = \dfrac{1}{54}$, $S_4 = \dfrac{22}{7}$ or

0.741　**23.** $r = 2$, $S_6 = 189$　**25.** $a_1 = 3$, $a_6 = 96$　**27.** $a_4 = -9$, $S_n = -6\dfrac{2}{3}$　**29.** $a_1 = \dfrac{3}{4}$, $a_4 = \dfrac{3}{500}$　**31.** 732 N, 3053

N (assuming only one lift per week)　**33.** 69,120 bacteria　**35.** -5 or 4　**37. a.** \$2000　**b.** \$0.01, \$0.02, \$0.04, \$0.08,
\$0.16　**c.** \$10, 485.75　**d.** Second method, it pays \$8,485.75 more than the first method　**39.** 10.1 gal　**41.** 0.472
43. a. 0.25,　**b.** 1.33　**45. a.** 9 billion　**b.** 16 billion

Section 16.5 (page 794)

1. 1　**3.** None　**5.** None　**7.** None　**9.** None　**11.** $\dfrac{32}{3}$　**13.** None　**15.** $\dfrac{5}{2}$　**17.** $\dfrac{40}{7}$ or 5.71　**19.** None　**21.** $\dfrac{1}{3}$

23. $\dfrac{9}{9} = 1$　**25.** $\dfrac{5}{11}$　**27.** $\dfrac{137}{999}$　**29.** $\dfrac{4876}{9999}$　**31.** $\dfrac{1430}{999}$　**33.** $\dfrac{1}{2}$　**35.** $\dfrac{2}{5}$　**37.** $\dfrac{93}{100}$　**39. a.** 2.99 m　**b.** 195 m
41. Forever　**43.** 1.33 A

Section 16.6 (page 803)

1. 40,320　**3.** 42　**5.** 35　**7.** 36　**9.** 1　**11.** 455　**13.** 15　**15.** 78　**17.** 100　**19.** n　**21.** $x^4 + 4x^3 + 6x^2 + 4x + 1$
23. $a^4 + 4a^3b + 6a^2b^2 + 4ab^3 + b^4$　**25.** $243a^{10} + 810a^8b^2 + 1080a^6b^4 + 720a^4b^6 + 240a^2b^8 + 32b^{10}$

27. $16x^8 - 96x^6t^2 + 216x^4t^4 - 216x^2t^6 + 81t^8$　**29.** $4x^4 - \dfrac{8\sqrt{2}}{3}x^3y + \dfrac{4}{3}x^2y^2 - \dfrac{4\sqrt{2}}{27}xy^3 + \dfrac{1}{81}y^4$

31. $a^{18} - 12a^{15}b + 60a^{12}b^2 - 160a^9b^3 + 240a^6b^4 - 192a^3b^5 + 64b^6$
33. $a^{7/2} + 7a^3b^{1/2} + 21a^{5/2}b + 35a^2b^{3/2} + 35a^{3/2}b^2 + 21ab^{5/2} + 7a^{1/2}b^3 + b^{7/2}$
35. $x^{14/3} - 14x^4y^{2/3} + 84x^{10/3}y^{4/3} - 280x^{8/3}y^2 + 560x^2y^{8/3} - 672x^{4/3}y^{10/3} + 448x^{2/3}y^4 - 128y^{14/3}$

37. $1 - 17x^2 + 136x^4 - 680x^6 + \cdots$　**39.** $u^{36} + 9u^{34}v^3 + \dfrac{153}{4}u^{32}v^6 + 102u^{30}v^9 + \cdots$

41. $4,194,304x^{22} - 138,412,032x^{21}y + 2,179,989, 505x^{20}y^2 - 2.1799895 \times 10^{10}x^{19}y^3 + \cdots$

43. $1 + 96y + 4560y^2 + 142{,}880y^3 + \cdots$ **45.** 1.07214 **47.** 4.82681 **49.** 0.913517 **51.** $924a^6x^6$
53. $18{,}564u^{12}v^{24}$ **55.** $673{,}596$ **57.** $\dfrac{7}{18}$

Review Exercises (page 804)

[16.1] **1.** $3i - 2$ **3.** $3 - 2i$ **5.** $3 + (-1)^i$

7. a. $4, 8, 12, 16, 20, 24$
   b. Increasing
   c.

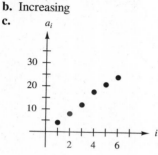

9. a. $-4, -3, -2, -1, 0, 1$
   b. Increasing
   c.

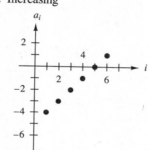

11. a. $-\dfrac{2}{3}, -\dfrac{4}{3}, -2, -\dfrac{8}{3}, -\dfrac{10}{3}, -4$
   b. Decreasing
   c.

13. a. $2, 4, 2, 4, 2, 4$ **b.** Neither
   c.

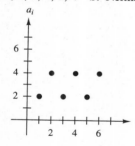

15. a. $-\dfrac{2}{5}, -\dfrac{1}{6}, 0, \dfrac{1}{8}, \dfrac{2}{9}, \dfrac{3}{10}$ **b.** Increasing
   c.

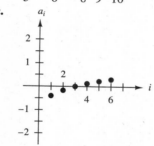

[16.2] **17.** $S_4 = 4 + 16 + 64 + 256,\ S_6 = 4 + 16 + 64 + 256 + 1024 + 4096$

19. $S_4 = 1 + 3 + 5 + 7,\ S_6 = 1 + 3 + 5 + 7 + 9 + 11$ **21.** $S_4 = 0 + \dfrac{1}{4} + \dfrac{2}{5} + \dfrac{3}{6},\ S_6 = 0 + \dfrac{1}{4} + \dfrac{2}{5} + \dfrac{3}{6} + \dfrac{4}{7} + \dfrac{5}{8}$

23. $S_4 = 2 - 3 + 4 - 5,\ S_6 = 2 - 3 + 4 - 5 + 6 - 7$ **25.** $2 - \dfrac{8}{3} + 4 - \dfrac{32}{5} + \dfrac{32}{3} = \dfrac{38}{5}$

27. $4 + 4 + \dfrac{16}{3} + 8 + \dfrac{64}{5} = \dfrac{512}{15}$ **29.** $-2 - 1 - \dfrac{1}{2} - \dfrac{1}{5} + 0 + \dfrac{1}{7} = -\dfrac{249}{70}$ **31.** $S_n = \displaystyle\sum_{i=1}^{n} (4 + i)$

33. $S_n = \displaystyle\sum_{i=1}^{n} \dfrac{(-1)^i(i + 1)}{2^{i-1}}$ **35.** $S_n = \displaystyle\sum_{i=1}^{5} \dfrac{i}{3^{(2i-1)}}$ **37.** $0 + 1 + \dfrac{2}{3} + \dfrac{3}{5} + \dfrac{4}{7} + \dfrac{5}{9} + \dfrac{6}{11}$ **39.** $5 + \dfrac{7}{4} + 1 + \cdots + \dfrac{2n + 3}{n^2}$

41. $4z^2 + 30z - 10$ [16.3] **43.** $11 + 6 + 1 + \cdots + (16 - 5n) + \cdots$ **45.** $9 + 8 + 7 + \cdots + (10 - n) + \cdots$

47. $-6\dfrac{1}{2} - 6 - 5\dfrac{1}{2} + \cdots \left(\dfrac{1}{2}n - 7\right) + \cdots$ **49.** $d = 1,\ S_{11} = 88$ **51.** $d = \dfrac{22}{5},\ a_6 = 19,\ n = 6$

53. $a_1 = 0$, $n = 2$, $S_2 = \dfrac{4}{5}$ [16.4] **55. a.** $\dfrac{1}{3}, \dfrac{2}{3}, \dfrac{4}{3}, \dfrac{8}{3}, \dfrac{16}{3}, \dfrac{32}{3}$ **b.** 21 **57. a.** $2, \dfrac{1}{2}, \dfrac{1}{8}, \dfrac{1}{32}, \dfrac{1}{128}, \dfrac{1}{512}$ **b.** 2.67

59. a. $-2, -0.2, -0.02, -0.002, -0.0002, -0.00002$ **b.** -2.22 **61.** $a_4 = \dfrac{1}{12}$, $S_4 = \dfrac{5}{4}$ **63.** $a_1 = -\dfrac{3}{31}$, $a_5 = -\dfrac{48}{31}$

65. $r = 3$, $S_7 = 364\dfrac{1}{3}$ [16.5] **67.** $\dfrac{1}{4}$ **69.** $\dfrac{50}{49}$ or 1.02 **71.** No sum **73.** $\dfrac{7}{3}$ [16.6] **75.** $\dfrac{8}{9}$ **77.** $\dfrac{412}{333}$

79. $Z^4 + 20Z^3 + 150Z^2 + 500Z + 625$ **81.** $81u^8 - 216u^6w^2 + 216u^4w^4 - 96u^2w^6 + 16w^8$
83. $a^2 + 6a^{5/3}b^{1/3} + 15a^{4/3}b^{2/3} + 20ab + 15a^{2/3}b^{4/3} + 6a^{1/3}b^{5/3} + b^2$
85. $x^{18} + 18x^{17}y + 153x^{16}y^2 + 816x^{15}y^3 + \cdots$ **87.** $a^{16/3} + 16a^5b^{1/3} + 120a^{14/3}b^{2/3} + 560a^{13/3}b + \cdots$
89. $110{,}565x^{11}$ **91.** 248.2 mm **93.** \$5,368,709.12 **95.** 128 **97.** 23.94 in. **99.** \$72.09 **103. a.** Neither a geometric
or arithmetic sequence **b.** $a_{12} = 144$, $a_{13} = 233$

105. $d = \dfrac{1 - r^4}{2(1 - r)} - 2$ **107.**

```
          .
          .
          .
100 PRINT "INPUT A1, D AND N"
110 INPUT A1, D, N
120 S = N*A1 + N*(N - 1)*D
          .
          .
          .
```

109.

```
          .
          .
          .
100 PRINT "INPUT A1 AND R"
110 INPUT A1, R
120 S = A1/(1 - R)
```

Chapter Test *(page 808)*

1. $a_i = 2 - 4(i - 1)$ **2.** $a_1 = \dfrac{2^{i-1}}{5}$

3. 5, 7, 9, 11, 13, increasing

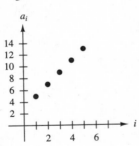

4. $-\dfrac{1}{2}, -1, -\dfrac{3}{2}, -\dfrac{5}{2}$, decreasing

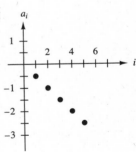

5. $\displaystyle\sum_{i=1}^{5} (3i - 2)$ **6.** $\displaystyle\sum_{i=1}^{5} \left(\dfrac{i - 2}{i + 1}\right)$ **7.** $-\dfrac{1}{3} + 0 + \dfrac{1}{5} + \dfrac{1}{3} + \dfrac{3}{7} = \dfrac{22}{35}$ **8.** $\displaystyle\sum_{i=1}^{n} \left(\dfrac{(-1)^i i}{2^{i-1}}\right)$ **9.** $0 - \dfrac{1}{3} - \dfrac{1}{2} + \cdots + \dfrac{1 - n}{1 + n}$

10. $10 + 8 + 6 + 4 + 2 + \cdots + [10 + (n - 1)(-2) + \cdots]$ **11.** $n = 5$, $a_1 = \dfrac{26}{5}$, $a_5 = -\dfrac{14}{5}$

12. a. $27, 9, 3, 1, \dfrac{1}{3}$ **b.** 40.3 **13.** $a_1 = \dfrac{2}{5}$, $a_n = \dfrac{16}{5}$ **14.** $S = 2$ **15.** $n = \dfrac{254}{99}$

16. $(2a^3)^4 + 4(2a^3)^3(-b^2) + 6(2a^3)^2(-b^2)^2 + 4(2a^3)(-b^2)^3 + (-b^2)^4 = 16a^{12} - 32a^9b^2 + 24a^6b^4 - 8a^3b^6 + b^8$
17. $a^{14} - 21a^{12} + 189a^{10}$ **18.** $10{,}206a^5b^8$ **19.** \$24,500 **20.** 6317 people

Chapter 17

Section 17.1 (page 814)

1. 100 **3.** 14.5–19.5, 19.5–24.5, 24.5–29.5, 29.5–34.5, 34.5–39.5, 39.5–44.5, 44.5–49.5

5. 50–54 **7.**

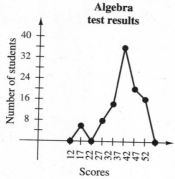

Algebra test results

9. 10 **11.** 24.5, 34.5, 44.5, 54.5, 64.5, 74.5

13.

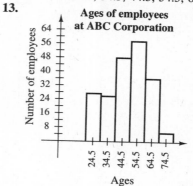

Ages of employees at ABC Corporation

15.

Quiz Scores of Mathematics Class

Scores	Number
1–2	8
3–4	6
5–6	14
7–8	25
9–10	27

17. 1.5, 3.5, 5.5, 7.5, 9.5 **19.**

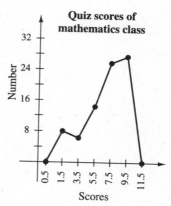

Quiz scores of mathematics class

21. 29.5–30.5, 30.5–31.5, 31.5–32.5, 32.5–33.5, 33.5–34.5, 34.5–35.5, 35.5–36.5, 36.5–37.5, 37.5–38.5

23.

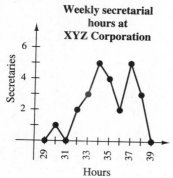

Weekly secretarial hours at XYZ Corporation

25.

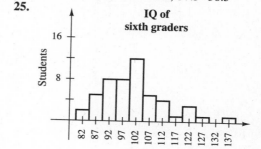

IQ of sixth graders

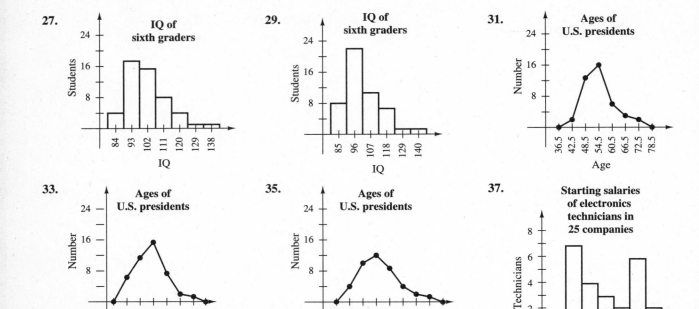

39. 29   **41.** 237   **43.** 5.5–6.5, 6.5–7.5, 7.5–8.5, 8.5–9.5, 9.5–10.5, 10.5–11.5, 11.5–12.5   **45.** 13
47. 87.5–92.5, 92.5–97.5, 97.5–102.5, 102.5–107.5, 107.5–112.5, 112.5–117.5, 117.5–122.5, 122.5–127.5, 127.5–132.5
49. 135,000
51.

Class mark	Frequency (thousands)
1	0
3	10
5	30
7	40
9	35
11	15
13	5
15	0

Section 17.2 (page 824)

1. 7.22, 7, 10   **3.** 72.7, 78, no mode   **5.** 8, 8, no mode   **7.** 72.9, 60, no mode   **9.** 12.7, 13, 14
11. 470.2, 456, no mode   **13.** 77.2, 76, no mode   **15. a.** 5, 4, 4   **b.** a. 5.4, 4, 4   **c.** Mean   **d.** Yes, mean
17. $90.54, $95.055, no mode   **19.** 4.68 m   **21.** 68, 69, 70, 71, 72   **23. a.** At least 33   **b.** No   **c.** At least 22   **d.** At least 82   **25.** $11.70   **27. a.** 34.4   **b.** 34.4   **c.** 31–35   **29. a.** 21.4   **b.** 19.75   **c.** 17–19   **31. a.** 47.75   **b.** 49.3

Section 17.3 (page 834)

1. The first set has a larger range   **3.** All the data have the same value   **5.** 12, 4.51   **7.** 8, 4.04   **9.** 8, 3.19
11. 24, 8.60   **13.** 15, 4.64   **15.** $13.00, $4.92   **17.** 3.9 hrs, 0.283 hr   **19.** 343.8 psi, 7.47 psi

21. a.

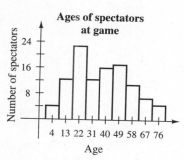

b. 36.9 yr   **c.** 18.5 yr

23. a.

Reaction Time of Water on Steel	
Time (Min)	Frequency
0–1	6
2–3	32
4–5	35
6–7	16
8–9	11

b.

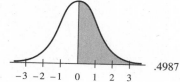

c. 4.38 min   **d.** 2.16 min

Section 17.4 (page 842)

1.

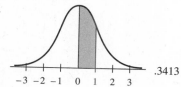

.3413

3.

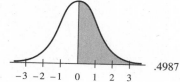

.4987

5.

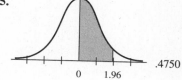

.4750

7.

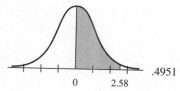

.4951

9.

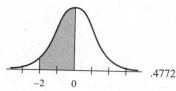

.4772

11.

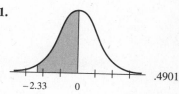

.4901

13.

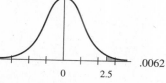

.0062

15.

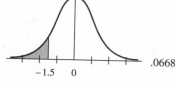

.0668

17.

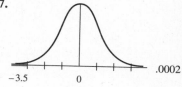

.0002

19.

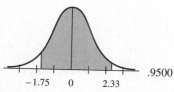

.9500

21.

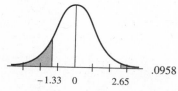

.0958

23. $\mu = 2$ and $\sigma = 3$; the larger the standard deviation, the greater the spread from the mean
25. a. 0.4332 **b.** 0.1151 **c.** 0.8181 **27. a.** 0.4772 **b.** 0.8351 **c.** 0.0013 **29. a.** 0.6915 **b.** 0.6853 **c.** 0.4978
31. a. 0.3472 **b.** 0.0869 **c.** 0.0869 **33. a.** 0.2615 **b.** 0.1949 **c.** 0.0003 **35. a.** 0.0606 **b.** 0.0047 **c.** 0.6915
37. a. 0.2578 **b.** 0.0026 **c.** 0.9305 **39. a.** 0.0808 **b.** 0.2743 **c.** 0.9950

Review Exercises (page 844)

[17.2, 17.3] **1.** 5.2, 4, no mode, 10, 3.90 **3.** 6.125, 6.5, no mode, 7, 2.42 **5.** 25.9, 26, 27, 6, 1.96 **7.** 73.1 **9.** 70–79
[17.1] **11.**

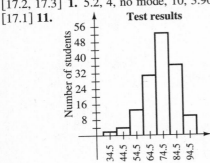

13. 79.2 **15.** 10.9 lb **17.** 5.69 **19.**

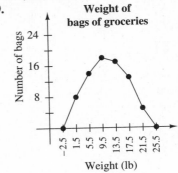

21.

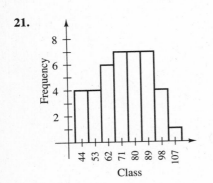

23. 73.5 **25.** 17.2 [17.4] **27.** 0.3085 **29.** 0.9938 **31.** 4.0% **33.** 22.7%
35.
```
10   PRINT "HOW MANY POINTS?"
20   INPUT N
30   PRINT "ENTER EACH POINT SEPARATELY"
40   LET S = 0
50   FOR I = 1 TO N
60   INPUT X
70   LET S = S + X
80   NEXT I
90   PRINT "MEAN = "; S/N
100  END
```

7. & 8.

Chapter Test (page 846)

1. $\bar{X} = 5$ **2.** Median = 4 **3.** Mode = 4 **4.** Range = 7 **5.** $s = 2.45$
6. Class

Limits	x_i	f_i	$x_i f_i$	$x_i^2 f_i$
0–5	2.5	8	20	50
6–11	8.5	8	68	578
12–17	14.5	10	145	2102.5
18–23	20.5	10	205	4202.5
24–29	26.5	9	238.5	6320.25
		45	676.5	13253.5

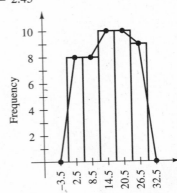

9. $\overline{X} = 15.0$ **10.** $md = 15.4$ **11.** $s = 8.4$ **12.** No modal class
13. 0.2358   **14.** 0.00003   **15.** 0.9544

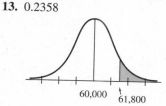

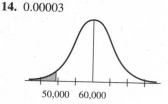

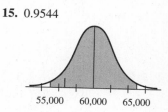

Chapter 18

Section 18.1 (page 854)

1. $x^2 + y^2 = 9$ **3.** $(x - 3)^2 + y^2 = 9$ **5.** $x^2 + y^2 = 25$ **7.** $(x - 6)^2 + (y + 6)^2 = 72$
9. $(x + 9)^2 + (y + 6)^2 = 317$ **11.** $(x + 5)^2 + (y - 4)^2 = 16$ **13.** $x^2 + y^2 = 9$ **15.** $x^2 + y^2 = 25$
17. $(x - 6)^2 + (y + 6)^2 = 100$ **19.** $(x + 9)^2 + (y + 6)^2 = 317$
21.   **23.**   **25.**

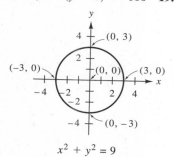

$x^2 + y^2 = 9$

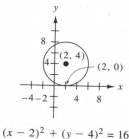

$(x - 2)^2 + (y - 4)^2 = 16$

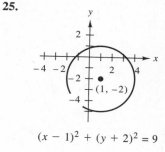

$(x - 1)^2 + (y + 2)^2 = 9$

27. $(x - 2.3)^2 + (y - 0.3)^2 = 5.78$ **29.** $(4, 2), r = 2$ **31.** $\left(0, \dfrac{1}{2}\right) r = \dfrac{3}{2}$ **33.** $(-1, 1) \ r = \sqrt{2}$
35. No, coefficients of x^2 and y^2 are not equal   **37.** No, r^2 must be > 0 **39.** $x = 3$ **41.** No   **43.** 2.31 m
45. $(7.20, -3.10), r = 2.50$
47. a.–c.   **d.**

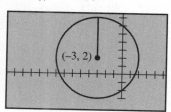

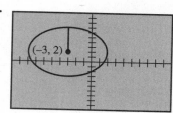

The figure in part a looks more like a circle. The horizontal scale to the vertical scale is 3 : 2. In part d the scale was adjusted, horizontal to vertical to 1:1

49. a. Infinite number   **b.** $y = \dfrac{1}{5}x - 2$

Section 18.2 (page 864)

1.–3.

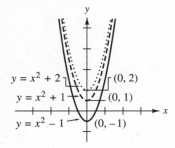

The change in value of the constant causes the curve to shift up or down. Thus the vertices of the parabolas are located at $(-1, 0)$, $(1, 0)$, and $(2, 0)$.

4.–6.

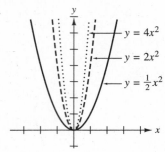

Since the constant is zero, all three curves have the same vertex. The coefficient of the x^2-term changes the shape of the curve. The larger the coefficient, the steeper the slope and the narrower the curve.

7.–9.

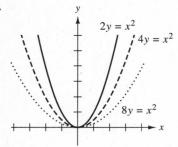

The larger the y-coefficient, the wider the curve.

10.–12.

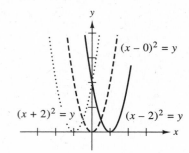

The constant shifts the curve to the left if it is positive and to the right if it is negative.

13.

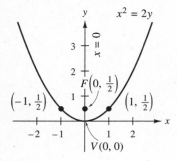

15.

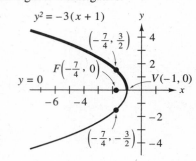

17.

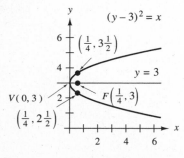

19.

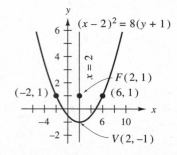

21.

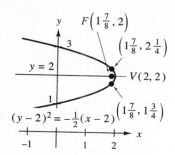

$F\left(1\frac{7}{8}, 2\right)$

$\left(1\frac{7}{8}, 2\frac{1}{4}\right)$

$y = 2$

$V(2, 2)$

$(y - 2)^2 = -\frac{1}{2}(x - 2)$ $\left(1\frac{7}{8}, 1\frac{3}{4}\right)$

23.

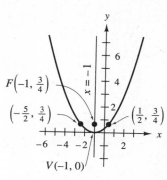

$F\left(-1, \frac{3}{4}\right)$

$\left(-\frac{5}{2}, \frac{3}{4}\right)$ $\left(\frac{1}{2}, \frac{3}{4}\right)$

$x = -1$

$V(-1, 0)$

$x^2 + 2x - 3y + 1 = 0$

25. $x^2 = \dfrac{72}{7}y$ **27.** $y^2 = -\dfrac{49}{11}(x + 3)$ **29.** $(x - 5)^2 = 10(y - 1)$ **31.** $12y = x^2$ **33.** $y^2 = 12x$

35. $(x - 5)^2 = \dfrac{64}{3}(y - 1)$ **37.** $x^2 = -200\left(y - \dfrac{1}{2}\right)$ **39.** $(x - 200)^2 = 800(y + 50)$ **41.** $x^2 = 120(y - 15)$

43. Minimum width is 12 in.

Section 18.3 (page 876)

1. $\dfrac{x^2}{9} + \dfrac{y^2}{4} = 1$

2. $\dfrac{x^2}{4} + \dfrac{y^2}{9} = 1$

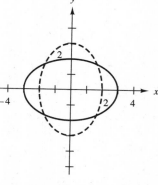

3. $\dfrac{x^2}{16} + \dfrac{y^2}{4} = 1$

4. $\dfrac{x^2}{16} + \dfrac{y^2}{15} = 1$

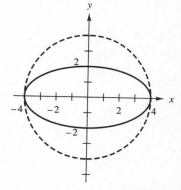

Interchanging the constants 4 and 9 the ellipse is rotated 90°.

When the values of a and b are close to each other, the ellipse becomes more circular in shape.

5.

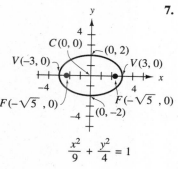

$C(0, 0)$ $(0, 2)$

$V(-3, 0)$ $V(3, 0)$

$F(-\sqrt{5}, 0)$ $F(-\sqrt{5}, 0)$

$(0, -2)$

$\dfrac{x^2}{9} + \dfrac{y^2}{4} = 1$

7.

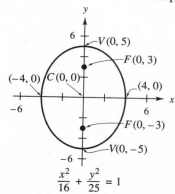

$V(0, 5)$

$F(0, 3)$

$(-4, 0)$ $C(0, 0)$ $(4, 0)$

$F(0, -3)$

$V(0, -5)$

$\dfrac{x^2}{16} + \dfrac{y^2}{25} = 1$

9.

$F(1 - \sqrt{5}, 2)$

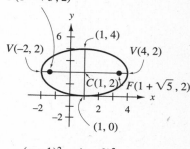

$V(-2, 2)$ $(1, 4)$ $V(4, 2)$

$C(1, 2)$ $F(1 + \sqrt{5}, 2)$

$(1, 0)$

$\dfrac{(x + 1)^2}{9} + \dfrac{(y - 2)^2}{4} = 1$

11.

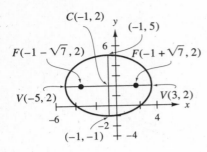

$C(-1, 2)$
$(-1, 5)$
$F(-1 - \sqrt{7}, 2)$ $F(-1 + \sqrt{7}, 2)$
$V(-5, 2)$ $V(3, 2)$
$(-1, -1)$

13. $\dfrac{(x+5)^2}{4} + \dfrac{y^2}{\sqrt{5}} = 1$

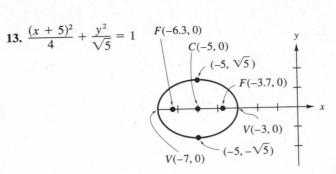

$F(-6.3, 0)$
$C(-5, 0)$
$(-5, \sqrt{5})$
$F(-3.7, 0)$
$V(-3, 0)$
$(-5, -\sqrt{5})$
$V(-7, 0)$

$$\dfrac{(x+1)^2}{16} + \dfrac{(y-2)^2}{9} = 1$$

15. $\dfrac{x^2}{36} + \dfrac{y^2}{9} = 1$ **17.** $\dfrac{(x-2)^2}{9} = \dfrac{y^2}{25} = 1$ **19.** $\dfrac{(x-6)^2}{49} + \dfrac{(y+2)^2}{16} = 1$ **21.** $\dfrac{(x+3)^2}{4} + \dfrac{(y-2)^2}{1} = 1$

23. $\dfrac{(x-8)^2}{16} + \dfrac{(y-2)^2}{4} = 1$ **25.** $\dfrac{(x-1)^2}{25} + \dfrac{(y-3)^2}{9} = 1$

27. ⑮$(-3\sqrt{3}, 0)$ and $(3\sqrt{3}, 0)$ ⑰$(2, -4)$ and $(2, 4)$ ⑲$(6 - \sqrt{33}, -2)$ and $(6 + \sqrt{33}, -2)$

29. $\dfrac{(x-5)^2}{25} + \dfrac{(y-3)^2}{9} = 1$ **31.** $\dfrac{x^2}{0.04} + \dfrac{y^2}{0.36} = 1$ **33. a.** $d = 114$ m **b.** 8.8 laps **35.** $\dfrac{x^2}{25} + \dfrac{y^2}{16} = 1$

37.

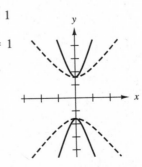

$(-3, 2)$
$V(-8, -1)$ $V(2, -1)$
$F(-7, -1)$ $F(1, -1)$
$C(-3, -1)$ $(-3, -4)$

$$\dfrac{(x+3)^2}{25} + \dfrac{(y+1)^2}{9} = 1$$

39. $\dfrac{x^2}{2.25} + \dfrac{y^2}{7.43} = 1$ **41.** The satellite came within 1980 m of the surface of the earth **43.** $(-5.45, 0)$, $(5.45, 0)$

45. $e = 0.12$

Section 18.4 (page 890)

1.&2. $\dfrac{x^2}{9} - \dfrac{y^2}{4} = 1$

$\dfrac{x^2}{4} - \dfrac{x^2}{9} = 1$

By interchanging a and b, the vertex of the hyperbola is shifted and the opening of the curve is changed.

3.&4. $\dfrac{y^2}{25} - \dfrac{x^2}{4} = 1$

$\dfrac{y^2}{25} - \dfrac{x^2}{24} = 1$

With the same value of a in both equations, the vertex did not change. Increasing the value of b changed the opening of the curve.

5.

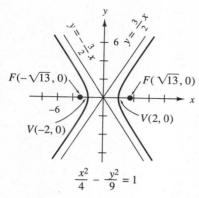

$$\frac{x^2}{4} - \frac{y^2}{9} = 1$$

9. $4x^2 - y^2 = 16$

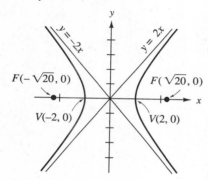

7.

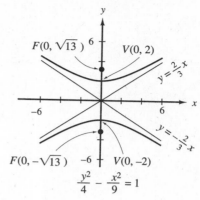

$$\frac{y^2}{4} - \frac{x^2}{9} = 1$$

11. $4y^2 - 25x^2 = 100$

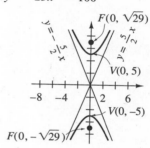

13.

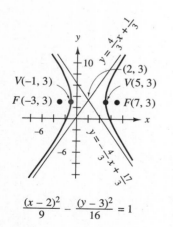

$$\frac{(x-2)^2}{9} - \frac{(y-3)^2}{16} = 1$$

15.

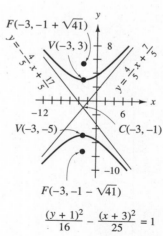

$$\frac{(y+1)^2}{16} - \frac{(x+3)^2}{25} = 1$$

17. $\dfrac{x^2}{16} - \dfrac{y^2}{4} = 1$ **19.** $\dfrac{y^2}{25} - \dfrac{x^2}{16} = 1$ **21.** $\dfrac{(x-4)^2}{16} - \dfrac{4(y-3)^2}{9} = 1$ **23.** $\dfrac{(y+2)^2}{9} - \dfrac{(x+5)^2}{9} = 1$ **25.** $\dfrac{x^2}{16} - \dfrac{y^2}{9} = 1$

27. $\dfrac{(x+1)^2}{9} - \dfrac{(y+2)^2}{3} = 1$

29.

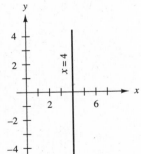

$C(-1, 2)$

$F(-1 - \sqrt{2}, 2)$ $F(-1 + \sqrt{2}, 2)$
$V(-2, 2)$ $V(0, 2)$

$(x + 1)^2 - (y - 2)^2 = 1$

31. $\dfrac{x^2}{5600} - \dfrac{y^2}{34{,}000} = 1$ **33.** $\dfrac{(y - 2.25)^2}{5.0625} - \dfrac{x^2}{22.5} = 1$ **35.** $\dfrac{\left(y + \dfrac{3}{2}\right)^2}{\dfrac{1}{4}} - \dfrac{x^2}{2} = 1$

Section 18.5 (page 897)

1. $y' = x'$ **3.** $v' = 3t'$ **5.** $y' = -2x' - 7$ **7.** $y' = -\dfrac{x'}{2} - \dfrac{9}{2}$ **9.** $(x')^2 + (y')^2 = 16$ **11.** $(x')^2 = -2(y')$, $V(0, 0)$

13. $\dfrac{(x')^2}{25} + \dfrac{(y')^2}{9} = 1$, $V(\pm 5, 0)$, $F(\pm 4, 0)$ **15.** $(y')^2 = \dfrac{1}{2}(x')^2$, $V(0, 0)$ **17.** $(x')^2 = -8(y')$, $V(0, 0)$

19. $\dfrac{(x')^2}{9} + \dfrac{(y')^2}{4} = 1$, $V(\pm 3, 0)$, $F(\pm\sqrt{5}, 0)$ **21.** $(x' - 1)^2 + (y' - 3)^2 = 16$, $C(1, 3)$

23. $(x')^2 = -2(y' - 2)$, $V(0, 2)$ **25.** $\dfrac{(x' + 2)^2}{25} + \dfrac{(y' - 1)^2}{9} = 1$, $C(-2, 1)$

27. $(x' + 5)^2 - (y' - 3)^2 = 1$, $C(-5, 3)$ **29.** $(x' + 200)^2 = 800(y' + 50)$ **31.** $\dfrac{(x' \pm 5280)^2}{(1650)^2} - \dfrac{(y')^2}{(4120)^2} = 1$

Review Exercises (page 898)

[18.1] 1. $(x + 3)^2 + (y - 4)^2 = 64$ **3.** $(x - 4)^2 + (y + 7)^2 = 25$ **5.** $(x - 3)^2 + y^2 = 36$ **[18.2] 7.** $y^2 = \dfrac{25}{6}x$

[18.3] 9. $\dfrac{(x - 6)^2}{25} + \dfrac{(y - 3)^2}{9} = 1$ **[18.4] 11.** $\dfrac{(x - 2)^2}{1} - \dfrac{y^2}{4} = 1$

[18.1] 13.

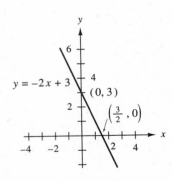

15.

$y = -2x + 3$ $(0, 3)$ $\left(\dfrac{3}{2}, 0\right)$

[18.2] 17. $(x - 1)^2 + (y - 2)^2 = 34$ **19.** 64.5 mm above left corner and 46.5 mm to the right of the lower left corner

21. 4.28 units, $\left(-\dfrac{25}{34}, \dfrac{3}{34}\right)$

23.

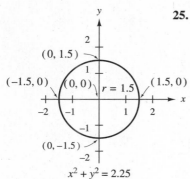

$x^2 + y^2 = 2.25$

25.

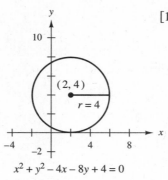

$x^2 + y^2 - 4x - 8y + 4 = 0$

[18.2] 27.

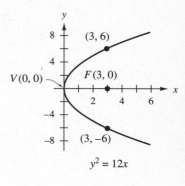

$y^2 = 12x$

29.

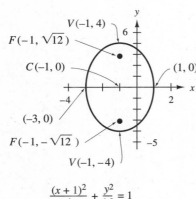

$-3x + y^2 + 2y + 1 = 0$

31. $x^2 = \dfrac{2500}{3}y$ **33.** $y^2 = 80(x + 20)$ [18.3] **35.** $\dfrac{(x + 3)^2}{24} + \dfrac{y^2}{25} = 1$

37. $\dfrac{(x + 3)^2}{9} + \dfrac{(y - 3)^2}{25} = 1$

39.

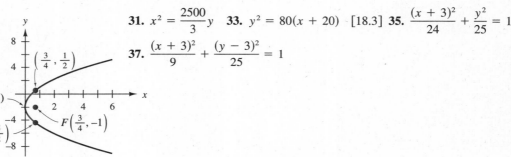

$\dfrac{(x + 1)^2}{4} + \dfrac{y^2}{16} = 1$

41.

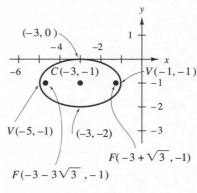

$x^2 + 6x + 8y + 4y^2 + 9 = 0$

43. $\dfrac{4x^2}{1089} + \dfrac{y^2}{144} = 1$ **45.** $\dfrac{(x + 2828)^2}{9 \times 10^6} + \dfrac{y^2}{1 \times 10^6} = 1$ or $\dfrac{(x - 2828)^2}{9 \times 10^6} + \dfrac{y^2}{1 \times 10^6} = 1$

[18.4] **47.**

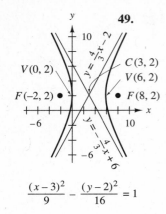

49.

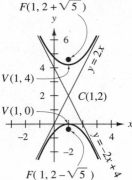

$$\frac{(x-3)^2}{9} - \frac{(y-2)^2}{16} = 1$$

$$-4x^2 + y^2 + 8x - 4y - 4 = 0$$

51. $\dfrac{(y-5)^2}{9} - \dfrac{x^2}{16} = 1$ **53.** $(y')^2 = 12x',\ V(0, 0)$ **55.** $(x')^2 + (y')^2 = 4,\ C(0, 0)$

57.

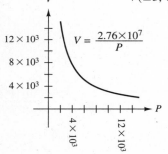

$V(\pm 3, 0)\ F(\pm 5, 0)$

[18.1] **59.** No; since $x^2 + y^2 = \ell x + my + n = 0$ can be written as $\left(x + \dfrac{\ell}{2}\right)^2 + \left(y + \dfrac{m}{2}\right)^2 = \dfrac{\ell^2}{4} + \dfrac{m^2}{4} - n$ if

$\dfrac{\ell^2}{4} + \dfrac{m^2}{4} - n \le 0$, the radius is imaginary or zero and the circle does not exist

[18.5] **61.** $\dfrac{(x')^2}{4} + (y')^2 = 1$, an ellipse

Chapter Test (page 902)

1. $(x + 4)^2 + (y - 3)^2 = 49$ **2. a.** $V(-2, 2)$ **b.** $(-2, 1)$ **c.** $(-4, 1), (0, 1)$ **d.** $x = -2$
e. $(x + 2)^2 = -4(y - 2)$

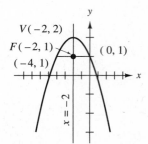

3. a. $(3, 1)$ **b.** $(3 - \sqrt{28}, 1), (3 + \sqrt{28}, 1)$ **c.** $(-5, 1), (11, 1)$ **d.** $(3, -5), (3, 7)$

e. $\dfrac{(x - 3)^2}{64} + \dfrac{(y - 1)^2}{36} = 1$

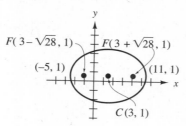

4. a. $(0, 0)$ **b.** $(0, -5), (0, 5)$ **c.** $(0, -4), (0, 4)$ **d.** $y = \pm\dfrac{4}{3}x$

e. $\dfrac{y^2}{16} - \dfrac{x^2}{9} = 1$

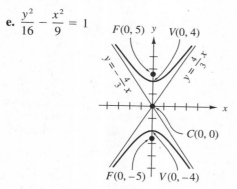

5. $(x')^2 - (y')^2 = 25$ **6.** $(x')^2 = 4(y'), V(0, 0), F(0, 1)$ **7.** $(-1.1, 0), (1.1, 0)$ **8.** $\dfrac{x^2}{2.25} - \dfrac{y^2}{1.75} = 1$

9.&10. both ellipses have the center at the origin. Increasing the value of a has the effect of stretching the ellipse. The vertices change from $(\pm 3, 0)$ to $(\pm 4, 0)$ **11.&12.** By increasing the value of a, the vertex has been shifted, and the second curve appears to get wider as one moves away from the vertex in the positive and the negative direction.

Chapter 19

Section 19.1 (page 917)

1. 1   **3.** 8   **5.** -1 **7.** 0   **9.** 1   **11.** 20   **13.** 10   **15.** $\dfrac{3}{7}$ **17.** 0   **19.** 3   **21.** $\dfrac{4}{3}$ **23.** 0   **25.** 1   **27.** $\dfrac{3}{2}$ **29.** 1

31. True   **33.** True   **35.** True   **37.** True   **39.** True   **41.** True   **43. a.** Does not exist   **b.** 0   **c.** Does not exist
d. 1.5   **e.** 0   **f.** 4   **45.** Function is continuous at $x = 2$ **47.** Function is continuous at $x = 2$

49. Not continuous; discontinuity is removable if the function is defined as: $f(x) = \begin{cases} \dfrac{x^2 - 36}{x - 6} & \text{for } x \neq 6 \\ 12 & \text{for } x = 6 \end{cases}$

51. Not continuous $f(x) \neq \lim\limits_{x \to >3} f(x)$; discontinunity is removable if function is defined as

$$f(x) = \left\{ \begin{array}{ll} \dfrac{x^2 - 9}{x - 3} & \text{for } x \neq 3 \\ 6 & \text{for } x = 3 \end{array} \right\} x = 3$$

53. Not continuous; limit does not exist   **55.** 30,000   **57.** The steady state current is $\dfrac{V}{r}$.

Section 19.2 (page 924)

1. 5   **3.** $\dfrac{1}{3}$ **5.** 17   **7.** -3.6 **9.** 0.25

11.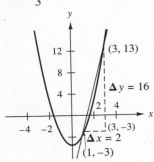

13.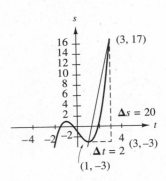

15.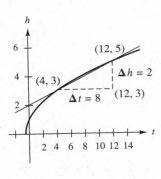

17. a. 2   **b.** 2   **19. a.** $2x + \Delta x - 6$ **b.** 0.1   **21. a.** $4x + 2\Delta x - 5$ **b.** 3.2   **23. a.** $\pi(2r + \Delta r)$ **b.** 4.1π

25. a. $\dfrac{1}{\sqrt{t + \Delta t} + \sqrt{t}}$ **b.** $\dfrac{1}{7}$ **27. a.** 14.4   **b.** 15.84

29. 11.2, 9.2; slope of the secant line through (3.0, 10) and (3.2, 12.24) is greater than the slope of the secant line through (2.0, 0) and (2.2, 1.84)   **31.** 1170; as the number of acres changes from 20 to 50, the profit is increasing at the rate of $1170 per acre.   **33.** 25; the average rate of inflation is $25 per year   **35.** 8.1 m/s; the average change in distance is 8.1 m/s.   **37.** 2.4, 2.8, 2.98, 2.998, 2.9998; limit is 3 cm/s.

Section 19.3 (page 932)

1. 3   **3.** 5   **5.** $2x$ **7.** $-2x$ **9.** $32x - 7$ **11.** $\dfrac{-7}{x^2}$ **13.** $\dfrac{-3}{(x + 3)^2}$ **15. b.** -5 **d.** 3   **e.** $f'(1) = -5; f'(5) = 3$

17. 7   **19.** 42   **21.** -8 **23.** 11

25. a. $y' = -2x + 7$ **b.** 1

27. a. $y' = 6x - 6$ **b.** -6

c.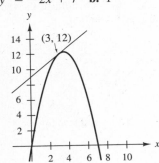

c.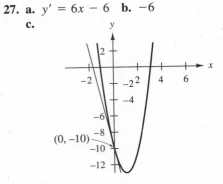

29. $s'(t) = -32t + 120$ **31.** $s'(t) = -32t + 32$ **33.** $D'(x) = 5 - 4x$ **35.** $D'(x) = -14x + 21$
37. a. $f'(x) = 1$ **b.** $f'(x) = 2x$ **c.** $f'(x) = 32x^2$ **d.** $f'(x) = 4x^3$

Section 19.4 (page 939)

1. $f'(x) = 1$ **3.** $f'(x) = 2x + 3$ **5.** $f'(x) = 4x + 3$ **7.** $y' = 3x^2 - 5$ **9.** $y' = 7x^6 + 20x^3 - 9$

11. $y' = 5x^4 + 18x^2 + 5$ **13.** $y' = x^3 - 2x^2$ **15.** $y' = 5\sqrt{3}\,x^4 + \dfrac{2}{3}x - 7$ **17.** $f'(r) = 2\pi$ **19.** $f'(r) = 4\pi r^2$

21. a. $f'(x) = 4x + 5$ **b.** $f'(1) = 9$ **c.**

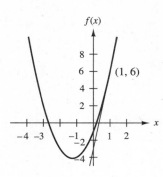

23. $x = 2$ **25.** $x = 3, x = -2$
27. $x = 0, x = -1 \pm \sqrt{2}$

Section 19.5 (page 944)

1. $v = 2$ **3.** $v = 2t + 7$ **5.** $v = 7t^6 + 20t^3 - 10t$ **7.** $a = 6t - 7$ **9.** $a = 6t^2 - 10t + 11$
11. $a = 21t^6 + 25t^4 - 4t$ **13. a.** $v = 12t^2 + 6t + 1$ **b.** 169 mph **15. a.** $a = 30t + 7$

b. 109 m/s² **17. a.** $v = 10 + 10t$ **b.** 40 m/s, 50 m/s **c.** $\dfrac{\Delta s}{\Delta t} = 45$ m/s **19.** $v = (450t - 210)$ mV **21.** $v = 15\ \mu$V

23. a. $i = 3760\ \rho$A or 3.76 nA **b.** 6580 ρA or 6.58 nA **25.** $v = (972t^5 + 1620t^2)$ V **27. a.** $i = 1.2$ A
b. $i = 3.52$ A

Section 19.6 (page 953)

1. $\dfrac{dy}{dx} = 2x + 4$ **3.** $\dfrac{dy}{dx} = 30x + 62$ **5.** $\dfrac{dy}{dx} = 15x^2 + 52x - 38$ **7.** $\dfrac{dy}{dx} = \dfrac{-7}{(x-3)^2}$ **9.** $\dfrac{dy}{dx} = \dfrac{5}{(3x-11)^2}$

11. $\dfrac{dy}{dx} = \dfrac{x^2 - 6x + 8}{(x-3)^2}$ **13.** $f'(x) = 4x^3 + 36x^2 + 88x + 57$ **15.** $f'(x) = 20x^9 + 54x^5 - 104x^3$

17. $f'(x) = 5x^4 + 4x^3 + 21x^2 + 16x + 1 - 14x^{-3} - 3x^{-4}$ **19.** $f'(x) = \dfrac{-2x^2 - 12x - 60}{(x^2 + 11x + 3)^2}$

21. $f'(x) = \dfrac{-4x^9 + 15x^8 + 12x^2 - 24x}{(x^7 + 4)^2}$ **23.** $f'(x) = \dfrac{10x + 25}{(x^2 + 5x + 11)^2}$ **25.** $\dfrac{dy}{dx} = \dfrac{2}{3x^{\frac{1}{3}}}$ **27.** $\dfrac{dy}{dx} = \dfrac{-2}{3x^{\frac{5}{3}}}$

29. $\dfrac{dy}{dx} = \dfrac{6}{x^4}$ **31.** $\dfrac{5\sqrt{3}}{2} \cdot \sqrt{x^3}$ **33.** $\dfrac{-\sqrt{x}}{2x^2}$ **35.** $x = 1.90, x = -1.23$ **37.** $x = 2.58, x = 1.42$ **39.** $x = 2$

41. $y = 18x - 18$ **43. a.** $\dfrac{dv}{dt}_{(t=1)} = -\$2650.00/\text{yr}$ **b.** $\dfrac{dv}{dt}_{(t=2)} = -\$1586.78/\text{yr}$ **45.** $V = -\dfrac{2x^4 - 42x^3}{(x^2 - 7)^2}$

47. a. $y = \dfrac{xP}{x - P}$ **b.** $\dfrac{dy}{dx} = \dfrac{-P^2}{(x-P)^2}$ **49.** $\dfrac{dV}{dt}_{(t=2.0)} = 4.3$ $\dfrac{dV}{dt}_{(t=20)} = 1200$

Section 19.7 (page 962)

1. $\dfrac{ds}{dt} = -20(7 - t)^3$ **3.** $\dfrac{dw}{dx} = 20(x^3 - 4x + 2)^4(3x^2 - 4)$ **5.** $\dfrac{dx}{ds} = 330s(4 - 11s^2)^4$ **7.** $\dfrac{dy}{dx} = \dfrac{-33}{5}x^2(4 - x^3)^{10}$

9. $\dfrac{du}{dt} = \dfrac{-2t}{\sqrt[3]{(1 - 3t^2)^2}}$ **11.** $\dfrac{ds}{dt} = \dfrac{-21}{(3t + 1)^8}$ **13.** $\dfrac{dR}{dt} = \dfrac{-16}{(2x - 1)^9}$ **15.** $f'(x) = (2x - 5)^2(40x - 67)$

17. $f'(x) = \dfrac{(x + 2)(x - 4)}{(x - 1)^2}$ **19.** $f'(x) = \dfrac{-12(2x - 5)^3}{(x - 4)^5}$ **21.** $f'(x) = \dfrac{4x^2 + 11}{(2x^2 + 11)^{\frac{1}{2}}}$ **23.** $f'(x) = \dfrac{-9x - 74}{(3x - 8)^3\sqrt{2x + 11}}$

25. a. $m = 0$ **b.** $(0, 0), (1, -1), (2, 0)$ **27. a.** $\dfrac{dy}{dx} = 4(x - 2)^3(x + 3)^3(2x + 1) = 4[(x - 2)(x + 3)]^3(2x + 1)$

b. $\dfrac{dy}{dx} = 4(x^2 + x - 6)^3(2x + 1) = 4[(x - 2)(x + 3)]^3(2x + 1)$ **29.** $\dfrac{dF}{dr_{(r\,=\,2)}} = -3.37 \times 10^{-8}$ N/m   **31.** $t = 3$

33. $\dfrac{dP}{dV_{(v\,=\,1.0)}} = 461$ P/m³, $\dfrac{dP}{dV_{(v\,=\,3.0)}} = 798$ P/m³

Section 19.8 (page 967)

1. $\dfrac{dy}{dx} = \dfrac{-3}{7}$ **3.** $\dfrac{dy}{dx} = \dfrac{-x}{y}$ **5.** $\dfrac{dy}{dx} = \dfrac{-3x}{5y}$ **7.** $\dfrac{dy}{dx} = \dfrac{x}{y}$ **9.** $\dfrac{dy}{dx} = \dfrac{-5x}{7y}$ **11.** $\dfrac{dy}{dx} = \dfrac{-y}{x}$ **13.** $\dfrac{dy}{dx} = \dfrac{-y - 3}{x}$

15. $\dfrac{dy}{dx} = \dfrac{-2xy - y^2 + 1}{x^2 + 2xy}$ **17.** $\dfrac{dy}{dx} = \dfrac{-1}{3y(x - 2)^2}$ **19.** $\dfrac{dy}{dx} = \dfrac{-2x + y}{9y^2 - x}$ **21.** $\dfrac{dy}{dx} = \dfrac{-2x + 2xy^3}{4y - 3x^2y^2}$

23. $\dfrac{dy}{dx} = \dfrac{-7x^6 + 2x - 4xy^3}{6x^2y^2 - 15y^4}$ **25.** $\dfrac{dy}{dx_{(-3,\,0)}} = \dfrac{9}{7}$ **27.** $\dfrac{dy}{dx_{(2,\,1)}} = 0$

Section 19.9 (page 970)

1. $f''(x) = 0$ **3.** $f''(x) = 18x$ **5.** $f'''(x) = \dfrac{-6}{x^4}$ **7.** $f''(x) = \dfrac{4}{x^3} - \dfrac{18}{x^4}$ **9.** $s''(t) = \dfrac{-25}{4}(5t + 7)^{\frac{-3}{2}}$ **11.** $y'' = 0$

13. $y''' = -120x - 42$ **15.** $y'' = -2x^{-3} + 6x^{-4}$ **17.** $y'' = \dfrac{4}{(x - 1)^3}$ **19.** $y'' = 6x + 20$ **21. a.** $a(5\text{ s}) = 2.2$ ft/s²

b. $a(30\text{ s}) = 0.31$ ft/s²   **23.** $v = 6L$ V

Review Exercises (page 971)

1. 6   **3.** 65   **5.** -2 **7.** 3   **9.** 0   **11.** $\dfrac{4}{5}$ **13.** Continuous   **15.** Continuous   **17.** Not continuous

[19.3, 19.4] **19.** $f'(x) = 2$ **21.** $f'(x) = 2x + 2$ **23.** $\dfrac{dy}{dx} = 3x^2 + 14x - 9$ **25.** $\dfrac{dy}{dx} = \dfrac{-4}{x^2}$ **27.** $\dfrac{dy}{dx} = \dfrac{-63}{x^{10}}$

29. $\dfrac{dy}{dx} = \dfrac{3}{2}\sqrt{x}$ **31.** $f'(x) = \dfrac{-2}{(x + 3)^2}$ **33.** $f'(x) = \dfrac{3x}{(3x^2 + 1)^{\frac{1}{2}}}$ **35.** $g'(x) = \dfrac{5}{7}x^4 + \dfrac{4}{3}x^3 - \dfrac{1}{2}$

[19.6] **37.** $\dfrac{dy}{dx} = 3x^2 - 8x - 5$ **39.** $\dfrac{dy}{dx} = 10x^4 - 132x^3 + 390x^2 - 198x$ **41.** $f'(x) = \dfrac{2x^2 + 52x + 44}{(x + 13)^2}$

43. $f'(x) = \dfrac{2x^2 - 8x - 26}{(x^2 + 13x - 13)^2}$ [19.7] **45.** $\dfrac{dy}{dx} = 9(3x + 4)^2$ **47.** $\dfrac{du}{dt} = \dfrac{-4t}{5(3 - t^2)^{\frac{3}{5}}}$ **49.** $\dfrac{dy}{dx} = \dfrac{-21}{(3x + 4)^8}$

51. $f'(x) = \dfrac{(x + 4)^2(2x + 2)}{(x + 2)^2}$ [19.8] **53.** $\dfrac{dy}{dx} = \dfrac{-2x}{11y}$ **55.** $\dfrac{dy}{dx} = \dfrac{2 - 2x - 2y^3}{6xy^2 - 7y}$

[19.2] **57. a.** $\dfrac{\Delta A}{\Delta r} = \pi(2r + \Delta r)$ **b.** $\dfrac{dA}{dr} = 2\pi r$ **59. a.** $\dfrac{\Delta v}{\Delta t} = \dfrac{f(t + \Delta t) - f(t)}{\Delta t}$ **b.** $v = \lim\limits_{\Delta t \to 0} \dfrac{f(t + \Delta t) - f(t)}{\Delta t}$

[19.5] **61.** $a = -32$; the acceleration never vanishes   [19.8] **63.** $y = -x + 6$ [19.9] **65.** $\dfrac{d^3y}{dx^3} = 6a$

67. $\dfrac{dD}{dv_{(v\,=\,30)}} = 1.7$ ft/s   **69.** $\dfrac{dh}{dr} = \dfrac{-6V}{\pi r^3}$

Chapter Test (page 973)

1. 173   **2.** 6   **3.** 7   **4.** The three conditions are met, and the function is continuous at $x = 3$ **5.** The function is not continuous at $x = 3$. The discontinuity can be removed if the function is defined as:

$$f(x) = \begin{cases} \dfrac{x^2 + x - 12}{x - 3} & x \neq 3 \\ 7 & x = 3 \end{cases} \; x = 3.$$

6. $f'(x) = 18x^2 - 10x$ **7.** $f'(x) = \dfrac{-12x^2}{x^6} = \dfrac{-12}{x^4}$ **8.** $f'(x) = 18x^5 - 90x^4 - 10x + 30$ **9.** $f(x) = \dfrac{2}{5\sqrt[5]{(2x-7)^4}}$

10. $f'(x) = \dfrac{9x^4 + 84x^3 + 9}{(x+7)^2}$ **11.** $f'(x) = 3(5x^4 - 7x)^2(20x^3 - 21)$ **12.** $\dfrac{dy}{dx} = \dfrac{-3x^2 + 4y + 1}{-4x + 4y}$

13. $\dfrac{dy}{dx} = \dfrac{75x^{\frac{6}{5}} + 9}{25x^{\frac{8}{5}}}$ **14.** $\dfrac{d^3y}{dx^3} = -600x^{-7}$ **15.** $\dfrac{dc}{dx} = 0.66x^2 - 6x$ **a.** 6   **b.** 144   **c.** 6000

Chapter 20

Section 20.1 (page 978)

1. a. $y = -4x - 2$ **b.** Parabola
c.

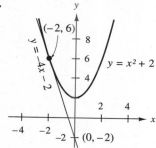

3. a. $y = 22x - 30$ **b.** Parabola
c.

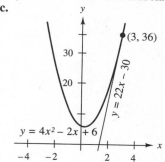

5. a. $y = -\dfrac{\sqrt{7}}{3}x + \dfrac{16}{3}$ **b.** Circle
c.

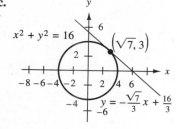

7. a. $y = \dfrac{-3}{2}x + 6$ **b.** Hyperbola
c.

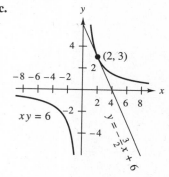

9. a. $y = \dfrac{9}{8}x - \dfrac{5}{2}, y = -\dfrac{9}{8}x + \dfrac{5}{2}$ **b.** Hyperbola
c.

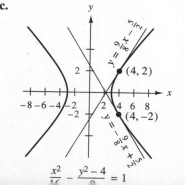

11. a. $y = -\dfrac{1}{3}x + 7$ **b.** Parabola
c.

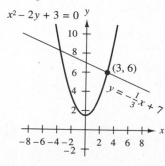

13. a. $y = -\dfrac{12}{7}x + 14$ **b.** Hyperbola

c.

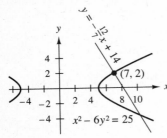

15. a. $y = \dfrac{8}{15}x + \dfrac{161}{60}$ **b.** Hyperbola

c.

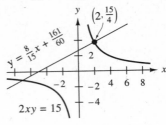

17. a. $y = -\dfrac{9}{4}x + \dfrac{13}{4}, y = \dfrac{9}{4}x - \dfrac{13}{4}$ **b.** Hyperbola

c.

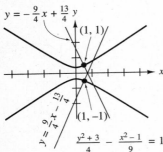

19. $y = -\dfrac{3}{4}x + \dfrac{25}{4}, y = -\dfrac{3}{4}x - \dfrac{25}{4}$ **21. a.** $y = -\dfrac{2}{5}x + 2$ **b.** $y = \dfrac{5}{2}x - \dfrac{25}{2}$ **23.** The ice is located at $(1.69, 5.56)$

25. The neutron must leave the path at $(1.93, 0.52)$ **27. a.** $y = \dfrac{x^2}{140.625}$ **b.** $y = -0.53x - 10$

b. 2   **c.** $y = \pm 4.90x, y = \pm 1.33x$

29. a.

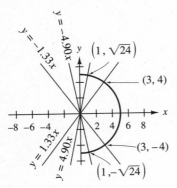

Section 20.2 (page 986)

1. a. $(0, 0)$ **b.** Relative minimum
   c.

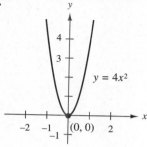

5. a. $(0, 0)$ **b.** Relative minimum
   c.

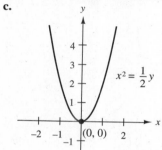

9. a. $\left(\dfrac{-1}{3}, \dfrac{5}{27}\right)$, $(1, -1)$

   b. Relative maximum, relative minimum
   c.

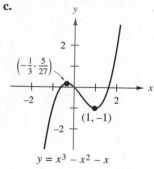

3. a. $\left(\dfrac{-3}{2}, \dfrac{7}{4}\right)$ **b.** Relative minimum
   c.

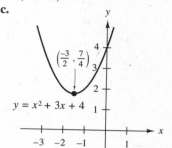

7. a. $(-2, 1)$ **b.** Relative minimum
   c.

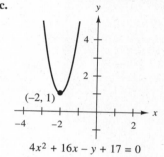

11. a. $(0, -2)$, $(-2, 2)$
    b. Relative minimum, relative maximum
    c.

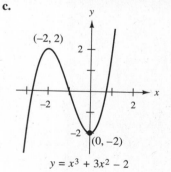

13. a. $\left(3, \dfrac{5}{2}\right), \left(2, \dfrac{8}{3}\right)$

    b. Relative minimum, relative maximum

    c.

$$y = \dfrac{x^3}{3} - \dfrac{5x^2}{2} + 6x - 2$$

15. a. $\left(-1, \dfrac{5}{3}\right), \left(1, \dfrac{1}{3}\right)$

    b. Relative maximum, relative minimum

    c.

$$y = \dfrac{x^3}{3} - x + 1$$

17. a. $(0, 0)$ **b.** Relative minimum

    c.

$$y = x^{4/5}$$

19. a. $(0, 0), (8, -320)$

    b.

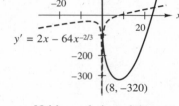

$$y = x^2 - 192x^{1/3}$$
$$y' = 2x - 64x^{-2/3}$$
$$(8, -320)$$

    c. Neither, relative minimum

21. a. $(2.535, -0.879), (0.131, 6.065)$

    b.

$$(0.131, 6.065)$$
$$(2.535, -0.879)$$
$$y' = 3x^2 - 8x + 1$$
$$y = x^3 - 4x^2 + x + 6$$

    c. Relative minimum, relative maximum

23. a. $(-3, 31.25), (-2, 32), (2, 0)$

    b.

$$y = \dfrac{x^4}{4} + x^3 - 2x^2 - 12x + 20$$
$$(-2, 32)$$
$$(-3, 31.25)$$
$$(2, 0)$$
$$y' = x^3 + 3x^2 - 4x - 12$$

    c. Relative minimum, relative maximum, relative minimum

25. $(-1, 2)$ absolute maximum, $(2.5, -10.25)$ absolute minimum

27. $(-1, -12)$ absolute minimum, $(3, 4)$ absolute maximum

Section 20.3 (page 993)

1. a. None **b.** (0, 2) **c.** None **d.** $-\infty < x < \infty$ concave up **3. a.** None **b.** $\left(-\dfrac{3}{4}, -\dfrac{1}{8}\right)$ **c.** None
d. $-\infty < x < \infty$ concave up **5. a.** (0, 16) **b.** None **c.** None **d.** $-\infty < x < \infty$ concave down
7. a. None **b.** None **c.** (0, 8) **d.** $-\infty < x < 0$ concave down, $0 < x < \infty$ concave up **9. a.** $(-2, 5)$ **b.** (0, 1)
c. $(-1, 3)$ **d.** $-\infty < x < -1$ concave down, $-1 < x < \infty$ concave up **11. a.** $(-4, 43)$ **b.** $(0.667, -7.815)$
c. $(-1.667, 17.593)$ **d.** $-\infty < x < -1.667$ concave down, $-1.667 < x < \infty$ concave up

13. a. $\left(\sqrt{\dfrac{5}{3}}, 3.303\right)$ **b.** $\left(-\sqrt{\dfrac{5}{3}}, -5.303\right)$ **c.** $(0, -1)$ **d.** $0 < x < \infty$ concave down, $-\infty < x < 0$ concave up
15. a. None **b.** (3, 0) **c.** None **d.** $-\infty < x < \infty$ concave up **17. a.** $(-1, -2)$ **b.** (1, 2) **c.** (0, undefined)
d. $-\infty < x < 0$ concave down, $0 < x < \infty$ concave up

19. a. $x > 0$ **b.** $x < 0$ **c.** None **d.** $(0, -1)$
e. None **f.** $-\infty < x < \infty$ **g.** None
h.

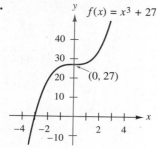

21. a. $x > -\dfrac{5}{8}$ **b.** $x < -\dfrac{5}{8}$ **c.** None **d.** $\left(-\dfrac{5}{8}, -\dfrac{9}{16}\right)$
e. None **f.** $-\infty < x < \infty$ concave up **g.** None
h.

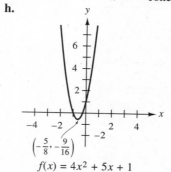

23. a. $-\infty < x < \infty$ **b.** None **c.** None
d. None **e.** (0, 27) **f.** $0 < x < \infty$ **g.** $(-\infty, 0)$
h.

$f(x) = x^3 + 27$

(0, 27)

25. a. $-2 < x < 2$ **b.** $-\infty < x < -2$ and $2 < x < \infty$
c. (2, 13) **d.** $(-2, -19)$ **e.** $(0, -3)$
f. $-\infty < x < 0$ **g.** $0 < x < \infty$
h.

(2, 13)

(0, -3)

$(-2, -19)$

$f(x) = -x^3 + 12x - 3$

27. a. $x < -2$ or $x > 3$ **b.** $-2 < x < 3$ **c.** $(-2, 56)$
d. $(3, -69)$ **e.** $\left(\dfrac{1}{2}, \dfrac{-13}{2}\right)$ **f.** $-\infty < x < \dfrac{1}{2}$

g. $\dfrac{1}{2} < x < \infty$ **h.**

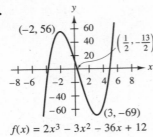

$f(x) = 2x^3 - 3x^2 - 36x + 12$

29. a. $x < 1$ or $x > 5$ **b.** $1 < x < 5$
c. $\left(1, \dfrac{10}{3}\right)$ **d.** $\left(5, -\dfrac{22}{3}\right)$ **e.** $(3, -2)$ **f.** $x > 3$
g. $x < 3$ **h.**

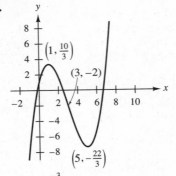

$f(x) = \dfrac{x^3}{3} - 3x^2 + 5x + 1$

31. a. $x > 2$ **b.** $x < 2$ **c.** None   **d.** $(2, -2)$
e. $\left(-1, \dfrac{19}{4}\right), \left(1, \dfrac{3}{4}\right)$ **f.** $x < -1$ or $x > 1$
g. $-1 < x < 1$ **h.**

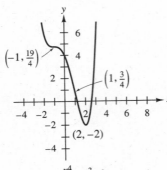

$f(x) = \dfrac{x^4}{4} - \dfrac{3}{2}x^2 - 2x + 4$

33. a. $-2\sqrt{2} < x < 1$ and $x > 2\sqrt{2}$
b. $x < -2\sqrt{2}$ or $1 < x < 2\sqrt{2}$ **c.** $(1, 3.917)$
d. $(-2\sqrt{2}, -31.084), (2\sqrt{2}, -0.915)$
e. $\left(-\dfrac{4}{3}, -16.197\right), \left(2, \dfrac{4}{3}\right)$ **f.** $x < -\dfrac{4}{3}$ or $x > 2$
g. $-\dfrac{4}{3} < x < 2$ **h.**

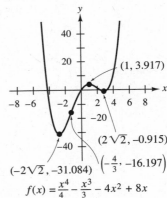

$f(x) = \dfrac{x^4}{4} - \dfrac{x^3}{3} - 4x^2 + 8x$

35. a. $x < 0, x > 0$ **b.** None   **c.** None   **d.** None   **e.** None
f. $x < 0$ **g.** $x > 0$ **h.**

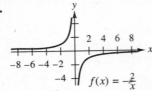

$f(x) = -\dfrac{2}{x}$

Section 20.4 (page 998)

1. $(1, -3)$, minimum   **3. a.** 500 widgets   **b.** \$2500   **5. a.** 8   **b.** 2   **7. a.** 26 ft   **b.** 1 s   **9.** 9 cm $\times$ 9 cm

11. $\dfrac{250}{3}$ mm   **13.** 1.68 m   **15.** 5 in.   **17. a.** $V = \pi r^2 h$, $SA = 2\pi rh + 2\pi r^2$ **b.** $SA = 2\pi r^2 + \dfrac{20}{r}$

   c. $r = \sqrt[3]{\dfrac{5}{\pi}} \approx 1.2$ m   **d.** $h = \dfrac{10}{\pi\left(\dfrac{5}{\pi}\right)^{\frac{2}{3}}} \approx 2.3$ m   **19.** 5488 in.^{3} **21.** $(4\sqrt{3}$ in., $4\sqrt{6}$ in.)

23. $(4, 0)$ minimum, $(4, 4)$ maximum

Section 20.5 Exercises (page 1006)

1. a. $2x\dfrac{dx}{dt} + 3y^2\dfrac{dy}{dt} = 0$ **b.** $\dfrac{dy}{dt} = \dfrac{-2x}{3y^2}\dfrac{dx}{dt}$ **c.** $\dfrac{4}{3}$ **3. a.** $9x^2\dfrac{dx}{dt} - 20y^3\dfrac{dy}{dt} = 0$ **b.** $\dfrac{dy}{dt} = \dfrac{9x^2}{20y^3}\dfrac{dx}{dt}$ **c.** $-\dfrac{9}{5}$

5. a. $3x^2\dfrac{dx}{dt} - 14y\dfrac{dy}{dt} + 4\dfrac{dx}{dt} - 7\dfrac{dy}{dt} = 0$ **b.** $\dfrac{dy}{dt} = \dfrac{3x^2 + 4}{14y + 7}\dfrac{dy}{dt}$ **c.** $-\dfrac{16}{21}$

7. a. $2x\dfrac{dx}{dt} - 3x\dfrac{dy}{dt} - 3y\dfrac{dx}{dt} + 6y\dfrac{dy}{dt} = 0$ **b.** $\dfrac{dy}{dt} = \dfrac{-2x + 3y}{-3x + 6y}\dfrac{dy}{dt}$ **c.** undefined   **9.** $\dfrac{dy}{dt} = \pm 3000$ mph

11. Area increases at the rate of 8 mm^2/s.   **13.** $\dfrac{db}{dt} = \dfrac{3}{2}$ m/s   **15.** $\dfrac{dA}{dt} = 0.13$ mm^2/month   **17.** $\dfrac{dh}{dt} = 1.47$ m/hr

19. $\dfrac{dy}{dt} = -750$ m/s   **21. a.** $\dfrac{dv}{dt} = 3.5$ cm^3/min   **b.** 3.6 cm^2/min   **23.** $\dfrac{dy}{dt} = 150$ units/s

Section 20.6 (page 1015)

1. $y = x^2 + 8x + 10$ **3.** $y = \dfrac{-2}{25}x^2 - \dfrac{28}{25}x + \dfrac{277}{25}$ **5.** $|v| = 10$, $\theta = 90°$ **7.** $|v| = 12.8$, $\theta = 141.3°$

9. $|v| = 6$, $\theta = 0°$ **11.** $|v| = -\dfrac{4}{3}$, $\theta = 306.7°$ **13.** $|v| = 8.25$, $\theta = 346.0°$ **15.** $a = 0$ **17.** $a = 2$, $\theta = 90°$

19. $a = 2.83$, $\theta = 315°$ **21.** $a = \dfrac{8}{9}$, $\theta = 90°$ **23.** $a = 6.32$, $\theta = 341.6°$ **25.** $\theta = 323.1°$

27. a. $|v| = 496$ m/s, $\theta = 323.7°$ with the horizontal   **b.** $a = 9.8$ m/s^2, $\theta = -90°$
29. $v = 57$ units/s, $\theta_v = 288°$, $a = 37$ units/s^2, $\theta_a = 279°$

Section 20.7 (page 1019)

1. 5π **3.** 0.0083   **5.** 6.4π **7.** $\dfrac{4\pi}{\sqrt{17}}$ **9.** π **11.** 0.272   **13.** -13.3 **15.** $\dfrac{15}{4}$ **17.** 7300 lb   **19.** 2 m^3
21. 0.53 m^2/hr   **23. a.** -196 m,   **b.** -784 m

Review Exercises (page 1020)

[20.1] **1.** $y = 0$ **3.** $y = 25x + 32$ **5.** $y = -3x + 12$ **7.** Vertical line $x = 1$ **9.** $y = \dfrac{2}{3}x + \dfrac{1}{6}$
11. $y = -9x + 164$ **13.** $(1, 20)$

[20.2] **15.** Relative minimum $(0, 0)$, relative maximum none

17. Relative maximum $(0, 0)$, relative minimum none

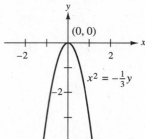

19. Relative maximum $(0, 0)$, relative minimum $(-1, -3)$ and $(4, -128)$

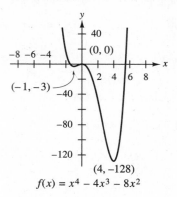

$f(x) = x^4 - 4x^3 - 8x^2$

21. Relative minimum $(0, 0)$

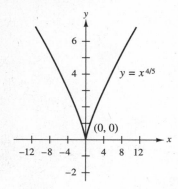

23. Relative minimum $\left(-1, -\dfrac{5}{2}\right)$

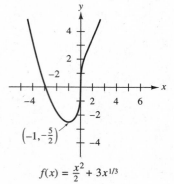

$f(x) = \dfrac{x^2}{2} + 3x^{1/3}$

[20.3] **25. a.** None   **b.** $(0, -5)$ **c.** None
d. Concave up $-\infty < x < \infty$
e.

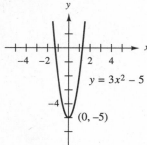

27. a. None   **b.** $\left(\dfrac{17}{6}, \dfrac{-217}{12}\right)$ **c.** None
d. Concave up $-\infty < x < \infty$
e.

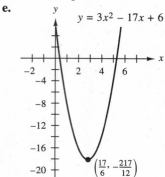

29. a. None **b.** $(2, 3)$ **c.** None
d. Concave up $-\infty < x < \infty$
e.

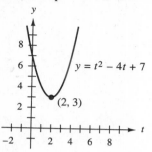

$y = t^2 - 4t + 7$

$(2, 3)$

31. a. None **b.** None **c.** $\left(0, \dfrac{1}{8}\right)$
d. Concave down $x < 0$, concave up $x > 0$
e.

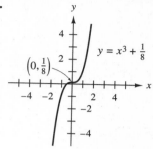

$y = x^3 + \dfrac{1}{8}$

$\left(0, \dfrac{1}{8}\right)$

33. a. $\left(-1, \dfrac{32}{3}\right)$ **b.** $(3, 0)$ **c.** $\left(1, \dfrac{16}{3}\right)$
d. Concave down $x < 1$, concave up $x > 1$
e.

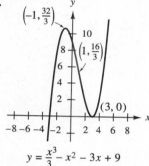

$\left(-1, \dfrac{32}{3}\right)$

$\left(1, \dfrac{16}{3}\right)$

$(3, 0)$

$y = \dfrac{x^3}{3} - x^2 - 3x + 9$

35. a. $(-0.707, -4.596)$ **b.** $(0.707, -7.414)$ **c.** $(0, -6)$
d. Concave down $x < 0$, concave up $x > 0$,
e.

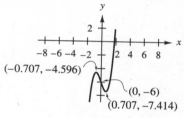

$(-0.707, -4.596)$

$(0, -6)$

$(0.707, -7.414)$

$y = 2x^3 - 3x - 6$

37. a. None **b.** None **c.** None
d. Concave down $x < 0$ and $x > 0$
e.

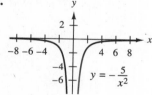

$y = -\dfrac{5}{x^2}$

39. a. None **b.** $\left(\sqrt[3]{\dfrac{1}{2}}, 1.89\right)$ **c.** $(-1, 0)$
d. Concave up $x < -1$ and $x > 0$,
concave down $-1 < x < 0$
e.

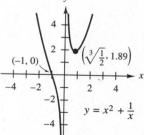

$\left(\sqrt[3]{\dfrac{1}{2}}, 1.89\right)$

$(-1, 0)$

$y = x^2 + \dfrac{1}{x}$

[20.4] **41.** Relative minimum $(1, -1)$ **43.** 36 square units **45.** $X_L = X_C$

[20.5] **47.** $\dfrac{dI}{dt} = \dfrac{1}{12} bc(3a^2 + b^2)\dfrac{da}{dt}$ **49.** $\dfrac{dv}{dt} = -\dfrac{2V}{3p}\dfrac{dp}{dt}$

[20.6] **51.** $|v| = 21.5$, $\theta_v = 68.2°$, $|a| = 36.9$, $\theta_a = 77.5°$ **53.** $|v| = 3.60$, $\theta_v = 359.0°$, $|a| = 0.314$, $\theta_a = 5.7°$

[20.7] **55.** 16π **57.** $\dfrac{33}{2}$ **59.** 1.724 V

Chapter Test *(page 1023)*

1. $y = -x + 7$ **2.** $y = x + 5$ **3.** Parabola, vertex $\left(\dfrac{5}{6}, \dfrac{73}{12}\right)$

4.

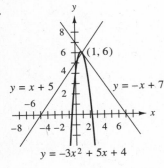

$y = x + 5$ $y = -x + 7$

$y = -3x^2 + 5x + 4$

5. $(3, -46)$, $(-3, 62)$
6. Relative minimum $(3, -46)$, relative maximum $(-3, 62)$

7.

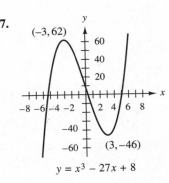

$y = x^3 - 27x + 8$

8. Absolute maximum $(-3, 62)$, absolute minimum $(3, -46)$ **9.** $x < -2$ and $x > 2$ **10.** $-2 < x < 2$
11. $(-2, 39)$ **12.** $(2, -25)$ **13.** $(0, 7)$ **14.** $x > 0$ **15.** $x < 0$ **16.**

17. 15 **18.** 1 gadget **19.** $\dfrac{9}{2\pi}$ in./min
20. $|v| = 4\sqrt{10}$, $\theta_v = 6.3°$ **21.** 0.27 in.^{3}

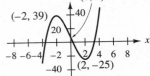

$f(x) = 2x^3 - 24x + 7$

Chapter 21

Section 21.1 *(page 1032)*

1. $\dfrac{x^7}{7} + C$ **3.** $\dfrac{x^4}{4} + 4x + C$ **5.** $\dfrac{x^5}{5} + \dfrac{3x^4}{4} - 7x + C$ **7.** $7y - 2y^2 + \dfrac{11}{4}y^4 + C$ **9.** $\dfrac{2}{5}x^{\frac{5}{2}} + \dfrac{3}{2}x^2 - 7x + C$

11. $\dfrac{-1}{2x^2} + C$ **13.** $\dfrac{x^2}{2} + 6x^{\frac{1}{2}} + C$ **15.** $\dfrac{2}{5}x^5 + 3x^{\frac{1}{3}} + \dfrac{3}{2}x^{\frac{-2}{3}} + C$ **17.** $\dfrac{t^4}{4} + \dfrac{3}{5}t^5 + C$ **19.** $\dfrac{1}{5}t^5 - \dfrac{2}{3}t^3 + t + C$

21. $\dfrac{2x^3 - 1}{2x^2} + C$ **23.** $\dfrac{2}{7}z^{\frac{7}{2}} + C$ **25.** 6000 ft **27.** 28.3 tons **29. a.** 2.5 s **b.** -72 ft/s

31. $y = x^3 - 5x^2 + 11x + 1$ **33. a.** $v = 2t^{\frac{3}{2}} - 2t + 1.3$ **b.** $v = \dfrac{2}{3}t^{\frac{1}{2}} - 2t + 6.1$

Section 21.2 *(page 1038)*

1. $\dfrac{(3x + 4)^9}{27} + C$ **3.** $\dfrac{(x^3 - 4)^2}{2} + C$ **5.** $\dfrac{(3x^5 - 8)^2}{30} + C$ **7.** $\dfrac{3}{70}(5x^7 - 13)^2 + C$ **9.** $\dfrac{(x^3 + 7x)^9}{9} + C$

11. $\dfrac{-1}{3(x^3 - 7x)^3} + C$ **13.** $\dfrac{4}{3}\sqrt{3t^2 - 7} + C$ **15.** $\dfrac{t^2}{2} + \dfrac{4}{3}t^{\frac{3}{2}} + t + C$ **17.** $2(r^{\frac{1}{3}} + 4)^{\frac{3}{2}} + C$

19. $\dfrac{2x^{\frac{3}{2}}}{3} + \dfrac{6x^{\frac{5}{2}}}{5} + C$ **21.** $\dfrac{-1}{2(x^2 + 2x + 2)} + C$ **23.** $y = \dfrac{1}{6}(3x + 7)^4 + \dfrac{5}{6}$ **25.** $y = \dfrac{(x^2 - 3x)^5}{5} - 11$

27. If $u = x^2 + 3$, then $\dfrac{du}{dx} = 2x + 3 \neq kx^4$ **29.** $\dfrac{d}{dx}\dfrac{(3x^5 + 2)^5}{5} = (3x^5 + 2)^4(15x^4) \neq (3x^5 + 2)^4$

31. $v = \dfrac{2}{3C}(5)^{\frac{3}{2}} + 6 - \dfrac{2}{3C}(3)^{\frac{2}{3}}$ **33.** $y = \dfrac{2}{3}x^3 - x - 12$ **35.** $y = \dfrac{1}{12}(12x^2 - 1)^3 - 406.42$

Section 21.3 (page 1045)

1. 8   **3.** 6.5   **5.** 32   **7.** 7   **9.** 5.1875   **11.** 58   **13.** 13.5   **15.** 16.5   **17.** -3 **19.** 2.67   **21.** 8.67   **23.** 4
25. 4.67

Section 21.4 (page 1054)

1. 1.5   **3.** 3   **5.** 6   **7.** 4   **9.** 3.33   **11.** 2   **13.** Function is not continuous on the [0, 2], division by zero is not defined;
$f(0)$ is undefined   **15.** -2.5 **17.** 9.33   **19.** -800 **21.** 1.89   **23.** Function is not continuous on [1, 4]; for $x < \sqrt[3]{\dfrac{5}{3}}$,
$\sqrt{3x^3 - 5}$ is imaginary   **25.** 1.16   **27.** Function is not continuous on the $[-3, 2]$, division by zero is not defined; $f(0)$ is
undefined   **29.** Function is not continuous on the $[-2, 2]$, division by zero is not defined; $f(1)$ is undefined
31. 32.67 sq units   **33.** 4.67 sq units   **35.** 11.33 sq units   **37.** 40,000 ft/lb   **39. a.** 196,000 J   **b.** 2.6×10^9 J

Section 21.5 (page 1064)

1. 8.75   **3.** 1.17   **5.** 1.89   **7.** 1.11   **9.** 3.26   **11.** 1.33   **13.** 18.70   **15.** 1.11   **17.** 1.47
19. 1.10   **21.** 3.19   **23.** 0.37

Review Exercises (page 1065)

[21.1] 1. $\dfrac{x^9}{9} + C$ **3.** $\dfrac{x^3}{3} + \dfrac{5x^2}{2} + 6x + C$ **5.** $\dfrac{3x^4}{4} - \dfrac{7x^3}{3} + \dfrac{5x^2}{2} + C$ **7.** $\dfrac{3}{4}(x + 1)^{\frac{4}{3}} + 10(x + 1)^{\frac{1}{2}} + C$

9. $\dfrac{9}{5}x^5 - 3x^4 + \dfrac{4}{3}x^3 + C$ **[21.2] 11.** $\dfrac{1}{9}(2t^3 - 7)^{\frac{3}{2}} + C$ **13.** $6\sqrt{x} - x + C$ **15.** $\dfrac{-1}{2(x^3 + 3x + 5)^2} + C$

17. $y = 2x^3 + 5x^2 - 7x + 3$ **19.** $y = (x + 3)^{\frac{1}{2}} + 1$ **21. a.** 2.80 s.   **b.** -24.4 m/s.
23. $y = 3x^4 - 2x^3 - 1$ **[21.3] 25. a.** 36   **b.** 28.5   **[21.4] 27. a.** Yes   **b.** 22.5   **29. a.** Yes   **b.** 16
31. a. Yes   **b.** 19.5   **33. a.** Yes   **b.** 96   **35. a.** Yes   **b.** 0   **37. a.** Yes   **b.** 23.0
39. a. No, the function is not continuous on $[-2, 2]$, division by zero is undefined; $f(0)$ is undefined

41. a. Yes   **b.** 48   **43. a.** Yes   **b.** 4.81   **45.** 63 sq units   **47.** $14\dfrac{1}{4}$ sq units   **[21.5] 49.** 6.07   **51.** 0.69

53. 5.90   **55.** 1.10   **57.** $p\left(\dfrac{b^3}{3}\right) + q\left(\dfrac{b^2}{2}\right) + rb$ **59. a.** $dW = E\,dQ,\ dQ = I\,dt,\ dW = EI\,dt,\ \dfrac{dW}{dI} = EI$

b. $Q = CE \rightarrow dQ = C\,dE$ and $dW = E\,dQ \rightarrow dQ = \dfrac{dW}{E}C\,dE = \dfrac{dW}{E} \rightarrow dW = CE\,dE \rightarrow W = \displaystyle\int CE\,dE = C\dfrac{E^2}{2} + C_1;$

since $W = 0$ when $E = 0$, $C_1 = 0$; thus $W = \dfrac{1}{2}CE^2$

Chapter Test (page 1067)

1. $\dfrac{x^6}{6} + C$ **2.** $\dfrac{x^5}{5} + \dfrac{3x^8}{8} + C$ **3.** $\dfrac{-1}{3x^3} + C$ **4.** $\dfrac{1}{18}(3x^4 + 1)^{\frac{3}{2}} + C$ **5.** $2(x^{\frac{1}{3}} + 4)^{\frac{3}{2}} + C$

6. $y = \dfrac{1}{16}(2x^2 - 1)^4 + \dfrac{15}{16}$ **7.** 11.5   **8.** 10 sq units   **9.** 10   **10.** 8   **11.** Hypothesis is not met; function is not defined
at $x = 0$; cannot apply the fundamental theorem   **12.** 4   **13. a.** 58.33 ft/lb   **b.** 308.1 ft-lb   **c.** 1.11
ft   **14.** 3.60   **15.** 3.60

Chapter 22

Section 22.1 (page 1077)

1. $\dfrac{500}{3}$ sq units   **3.** $\dfrac{125}{6}$ sq units   **5.** $\dfrac{32}{3}$ sq units   **7.** 36 sq units   **9.** 15 sq units   **11.** $\dfrac{37}{12}$ sq units

13. $\dfrac{2521}{12}$ sq units   **15.** $\dfrac{625}{2}$ sq units   **17.** $\dfrac{21}{2}$ sq units   **19.** $\dfrac{50}{3}$ sq units   **21.** $\dfrac{16}{3}$ sq units   **23.** $\dfrac{17}{4}$ sq units

25. $\dfrac{64}{3}$ sq units   **27.** $\dfrac{125}{6}$ sq units   **29.** 9 sq units   **31. a.** 8,087,040 j   **b.** 13,590 j   **c.** 391,050 j

33. a. $y = -2.5x^2 + 0.9$ **b.** 0.72 m²   **c.** 0.14 m³   **d.** 308,000 g

Section 22.2 (page 1083)

1. $\dfrac{8}{3}\pi$ cu units   **3.** 36π cu units   **5.** 64π cu units   **7.** 27.47π cu units   **9.** 126.6π cu units   **11.** 20.25π cu units

13. $\dfrac{\pi}{3}$ cu units   **15.** 18π cu units   **17.** $\dfrac{32}{15}\pi$ cu units   **19.** 2.25π cu units   **21.** $\dfrac{1}{3}\pi r^2 h$ cu units   **23.** 64π cu units

Section 22.3 (page 1087)

1. 8π cu units   **3.** 10.7π cu units   **5.** 42.7π cu units   **7.** 16π cu units   **9.** 71.7π cu units   **11.** $\dfrac{11}{30}\pi$ cu units

13. 8π cu units   **15.** π cu units   **17.** 97.2π cu units   **19.** $\dfrac{1}{5}\pi$ cu units   **21.** 72π cu units   **23.** $\dfrac{6}{5}\pi$ cu units

25. 50391.6π cm³   **27. a.** $y = -3x^2 + 3$ **b.** $\dfrac{3}{2}\pi$ cu units

Section 22.4 (page 1097)

1. a. $M_x = 16, M_y = 24$ **b.** (3, 2)   **3. a.** $M_x = 16, M_y = -32$ **b.** (-4, 2)

5. a. $M_x = -2, M_y = 6$ **b.** $\left(2, -\dfrac{2}{3}\right)$ **7. a.** $M_x = 54, M_y = 135$ **b.** (5, 2)

9. a. $M_x = 312.5, M_y = 156.25$ **b.** (3.75, 7.5)   **11. a.** $M_x = 5.84, M_y = 2$ **b.** (0.715, 2.09)

13. (3.4, 5)   **15.** (4.15, 3.53)   **17.** $\left(0, \dfrac{8}{3}\right)$ **19.** (-3.71, 0)   **21.** $\left(0, \dfrac{1}{4}\right)$

Section 22.5 (page 1102)

1. $I_{xx} = 42.7, I_{yy} = 74.7$ **3.** $I_{xx} = \dfrac{128}{3}, I_{yy} = \dfrac{392}{3}$ **5.** $I_{xx} = 2, I_{yy} = \dfrac{27}{2}$ **7.** $I_{xx} = 162, I_{yy} = 769.5$

9. $I_{xx} = 3720.24, I_{yy} = 625$ **11.** $I_{xx} = 13.076, I_{yy} = 1.945$ **13.** $I_{xx} = 0.0357, I_{yy} = 0.05$

15. $I_{xx} = 0.0857, I_{yy} = 0.0857$

Section 22.6 (page 1106)

1. 4.68×10^7 lb   **3.** 2683.2 lb   **5.** 666.67 lb   **7.** 66.67 N   **9.** 7.5 m   **11.** 0.8 N   **13.** 1.6 N   **15.** 10^{-6} N/m

17. 1.66×10^8 **19.** 1.94×10^9 **21.** 6.6×10^{-6} **23.** 3.12×10^{-4} **25.** 1500 lb

Review Exercises (page 1107)

1. $\dfrac{4}{3}$ sq units   3. $\dfrac{32}{3}$ sq units   5. $\dfrac{1}{6}$ sq units   7. $\dfrac{1}{6}$ sq units   9. 1 sq unit   11. a. $\dfrac{1}{2}\pi$ cu units   b. $\dfrac{8}{15}\pi$ cu units

13. a. 9π cu units   b. 18π cu units   15. a. $\dfrac{124}{3}\pi$ cu units   b. $\dfrac{248}{3}\pi$ cu units   c. $\dfrac{112}{3}\pi$ cu units

17. $\dfrac{1}{3}\pi$ cu units   19. $\dfrac{8}{3}\pi$ cu units   21. $\dfrac{16}{3}\pi$ cu units   23. $\dfrac{32}{3}\pi$ cu units

25. a. $\dfrac{4a^2b}{3}\pi$ cu units   b. $\dfrac{4a^2b}{3}\pi$ cu units   27. $(1, 3.4)$

29. a. $(-0.626, 2.58)$ b. $I_{xx} = 27.54$ c. $I_{yy} = 1.96$ 31. a. $\left(2, \dfrac{8}{3}\right)$ b. $I_{xx} = 61.25$ c. $I_{yy} = 33.75$

33. $(0, -3.08)$ 35. $13,333\dfrac{1}{3}$ N/m²   37. $135,466\dfrac{2}{3}$

39. a.

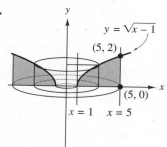

b. $\dfrac{544}{15}\pi$ cu units

41. $\dfrac{\pi h}{3}(b - a)(b + 2a)$ 43. $\dfrac{25\pi}{8}$ ft/lb   45. $(0.501, 0.394)$, $I_{xx} = 0.0155$, $I_{yy} = 0.0242$
47. $I_{xx} = \int y^2\,dA$; since dA is a real area, it is always positive, and y^2 is always positive, so I_{xx} is always positive

Chapter Test (page 1112)

1. 13.5 sq units   2. 20.83 sq units   3. $\dfrac{500}{3}\pi$ cu units   4. $\dfrac{608}{15}\pi$ cu units   5. 19.6875π cu units

6. a. $M_x = 16$, $M_y = 48$ b. $\left(4, \dfrac{4}{3}\right)$ 7. $I_{xx} = 32$, $I_{yy} = 216$ 8. a. $\dfrac{64}{3}$ sq ft   b. 64 cu ft
9. 6π cu units   10. 11,232,000 lb

Chapter 23

Section 23.1 (page 1123)

1. $\dfrac{1}{t}$ 3. $\dfrac{1}{x+2}$ 5. $\ln t^2 + 2$ 7. $\dfrac{2x+3}{x^2+3x+2}$ 9. $\dfrac{6x\log e}{3x^2+7}$ 11. $\dfrac{1-8\ln 5x}{x^9}$ 13. $-\dfrac{2x}{1-3x^2}$

15. $\dfrac{(2x-7)\log e}{2(x^2-7x)} + 3x^2$ 17. $-\dfrac{1}{t^2-1}$ 19. $\dfrac{4x+5}{3x(x+5)}$ 21. $\ln(t+1) + C$ 23. $\dfrac{1}{2}\ln|2t+3| + C$

25. $\dfrac{1}{4}\ln|2x^2+3| + C$ 27. $-\dfrac{1}{x+1} + C$ 29. 9.32   31. 1.79   33. $\dfrac{1}{2}\ln|x^2+6x+5| + C$ 35. $\dfrac{1}{2}\ln^2|x| + C$

37. $\ln|\ln|x|| + C$ **39.** $\dfrac{7}{3}$ **41.** $\dfrac{1}{x}$ discontinuous at $x = 0$ **43.** 1   **45.** $y = 4x - 8$ **47. a.** 3497   **b.** $4553.64

49. 0.6053   **51.** 1.67   **53.** $\dfrac{ab}{(a - T)(T - a^2 + aT)}$

Section 23.2 (page 1133)

1. e^t **3.** $5e^{5x}$ **5.** $3e^{(3x+5)}$ **7.** $2te^{t^2}$ **9.** $5^{(x+1)}\ln 5$ **11.** $\dfrac{e\sqrt{x}}{2\sqrt{x}}$ **13.** $2(e^{2x} - e^{-2x})$ **15.** $12e^{3t}(e^{3t} - 1)$

17. $e^{x\ln}(1 + x + x\ln x)$ **19.** $(2x - 1)(\ln 5)5^{(x^2-x)}$ **21.** $e^t + C$ **23.** $e^{4t} + C$ **25.** 19.1   **27.** $\dfrac{e^x + e^{-x}}{4} + C$

29. $\dfrac{e^{(5x^2+1)}}{10} + C$ **31.** $\dfrac{5^x}{\ln 5} + C$ **33.** -17.9 **35.** $-\dfrac{2(1 - e^x)^{\frac{3}{2}}}{3} + C$ **37.** $\dfrac{5x^2}{2\ln 5} + C$ **39.** e^x

41. $y = xe^2 - e^2$ or $y = 7.4x - 7.4$ **43.** $\dfrac{(qT - 1)k_0 e^{[-q(T_0-T)]}}{T_0 T^2}$

45. a. $60e^{0.0t}$ **b.** $71.83; the increase in the interest earned during the third year over the second year   **c.** $86.00; the increase in the interest earned during the sixth year over the fifth year.   **47.** T is decreasing

51. $\dfrac{y(xe^{-x}\ln y - e^y)}{x(ye^y \ln x + e^{-x})}$ **53.** $20,273,255   **55.** 29,800 decrease in infected cells in first two days.

57. a. $y' = -2xe^{-x^2}$ **b.** $x = 0$ **c.** $(0, 1)$ relative maximum   **59. a.** 9.0 A, 0.95 V   **b.** 6.7 A, 3.3 V

Section 23.3 (page 1141)

1. $3\cos 3x$ **3.** $-5\sin 5x$ **5.** $4e^x \cos e^x$ **7.** $-14e^{2x}\sin e^{2x}$ **9.** $-\sin\left(x + \dfrac{\pi}{2}\right)$ **11.** $\dfrac{3\cos(\ln x)}{x}$

13. $\cos^2 x - \sin^2 x$ **15.** $3\cot 3x$ **17.** $\sec^2 x$ **19.** $6\pi x \sin^2(\pi x^2)\cos(\pi x^2)$ **21.** $2e^{2x}\cot e^{2x}$

23. $2e^{\sin 2x}\cos 2x$ **25. a.** $\cos x$ **b.** $\dfrac{\pi}{2}, \dfrac{3\pi}{2}$ **c.** $\left(\dfrac{\pi}{2}, 1\right)$ relative maximum; $\left(\dfrac{3\pi}{2}, -1\right)$ relative minimum   **d.** Slope is

positive for $0 < x < \dfrac{\pi}{2}$ and $\dfrac{3\pi}{2} < x < 2\pi$ **e.** Slope is negative for $\dfrac{\pi}{2} < x < \dfrac{3\pi}{2}$ **f.** Inflection points at $(0, 0)$, $(\pi, 0)$,

and $(2\pi, 0)$ **g.**

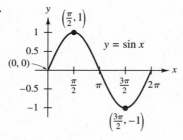

29. $m = 2$, 2 cycles

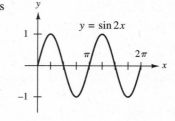

31. $m = \dfrac{1}{2}, \dfrac{1}{2}$ cycle

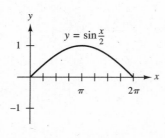

33. $m = \dfrac{1}{3}, \dfrac{1}{3}$ cycle

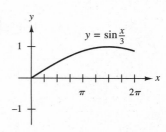

35. $y = 0.866x - 0.157$

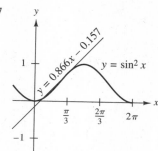

37. $m = 0$

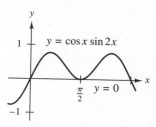

39. a. 11 A.M.   **b.** 11 P.M.   **41. a.** $y' = 4.0 \cos 0.50t$ **b.** 4.0, 3.9   **c.** $y'' = -2.0 \sin 0.5t$ **d.** $-0.25, -0.49$

43. a. 15,000 m   **b.** $\dfrac{dR}{d\theta} = \dfrac{320{,}000 \cos 2\theta}{9.8}$ **c.** 11,000 m   **d.** 16,000 $\cos 2\theta$ **e.** 45°, 135°   **45.** $\dfrac{d(a \sin \theta - e \cos \theta)}{x}$

Section 23.4 (page 1147)

1. $\sin t + C$ **3.** $-\dfrac{\cos 3t}{3} + C$ **5.** $\dfrac{\sin 5x}{5} + C$ **7.** $\dfrac{\sin^2 x}{2} + C$ **9.** $\dfrac{\sin^5 x}{5} + C$ **11.** $-\dfrac{\cos x^3}{3} + C$

13. $-\cos e^x + C$ **15.** $\dfrac{\ln^2 |\sin x|}{2} + C$ **17.** $\ln |\sin x| + C$ **19.** $\dfrac{\sqrt{2}}{2}$ **21.** $-2 + \sqrt{2}$ **23.** 0.520   **25.** 0.976

27. 1.414   **29.** 0.207   **31.** 2.017   **33.** 1.364   **35. a.** 0.007   **b.** -0.005 **c.** Alternating current   **37.** $\dfrac{V_p}{5\pi}$

Section 23.5 (page 1157)

1. $3 \sec^2 3x$ **3.** $-6 \csc^2 3t$ **5.** $\pi \tan \pi t \sec \pi t$ **7.** $-8 \cot 2x \csc 2x$ **9.** $4(x + 3) \sec^2(x + 3)^2$

11. $-\dfrac{4x}{\sqrt{x^2 - 1}} \csc^2 \sqrt{x^2 - 1}$ **13.** $6x^2 \sec^2 x^3 \tan x^3$ **15.** $-6 \csc^3 (x + 2) \cot (x + 2)$

17. $\dfrac{2 \sec^2 2x + 3 \cot 3x(1 + \tan 2x)}{\csc 3x}$ **19.** $\dfrac{2[\sin x \cos x \tan x - (\sin^2 x + 1) \sec^2 x]}{\tan^3 x}$

21. $2(\sec^2 2x + \csc^2 2x)$ **23.** $\sec^2 x\left(\dfrac{1}{2} - 2 \tan x\right)$ **25.** 12   **27.** $y = -6x + \pi$ **29.** $\ln |\sec x| + C$ **31.** 1

33. $\dfrac{1}{2} \sec 2x + C$ **35.** $\tan 4x + C$ **37.** 0.241   **39.** $\dfrac{1}{3} \sin 3x + C$ **41.** $\dfrac{1}{3} (\ln| \sec 3x| + \ln| \sec 3x + \tan 3x|) + C$

43. $\dfrac{1}{2} \tan^2 x + C$ **45.** $-e^{\cos x} + C$ **47.** $-\dfrac{1}{3} \cos^3 x + C$ **49.** 0.31

51. Let $u = 2 - 2 \cos \theta - 2 \sin \theta \tan \alpha + \tan^2 \alpha$; then $\dfrac{d(AB)}{d\alpha} = \dfrac{-2 + 2 \cos \theta + \sin \theta \tan \alpha}{2\mu^{\frac{1}{2}} \sin^2 \alpha}$

Section 23.6 (page 1164)

1. $\dfrac{3}{\sqrt{1 - 9x^2}}$ **3.** $\dfrac{-1}{\sqrt{1 - (x + 4)^2}}$ **5.** $\dfrac{11}{1 + (11x)^2}$ **7.** $-\dfrac{e^x}{1 + e^{2x}}$ **9.** $\dfrac{2}{\sqrt{e^{4t} - 1}}$ **11.** $\dfrac{-1}{(t + 3)\sqrt{(t + 3)^2 - 1}}$

13. $\dfrac{1}{\sqrt{1 - x^2}\,\text{Arcsin}\, x}$ **15.** $\dfrac{-2}{\sqrt{1 - x^2}}$ **17.** $\dfrac{2t}{(1 + t^4)\,\text{Arctan}\, t^2}$ **19.** $\dfrac{-2}{(x - 1)^2 + (x + 1)^2}$ **21.** $\dfrac{\pi}{4}$ **23.** $\dfrac{\pi}{12}$

25. 0.0687   **27.** $\text{Arcsin}\, \dfrac{9}{2a}$ **29.** $\text{Arcsin}\, \dfrac{e^x}{4} + C$ **31.** $2 \,\text{Arctan}\, \sqrt{x} + C$ **33.** $\dfrac{1}{2} \,\text{Arctan}\, x^2 + C$

35. $-\text{Arctan}(\cos t) + C$ **37.** $\dfrac{1}{\sqrt{7}}\text{Arcsec}\dfrac{t}{\sqrt{7}} + C$ **39.** $\dfrac{1}{\sqrt{5}}\text{Arcsec}\dfrac{\sqrt{7}\,z}{\sqrt{5}} + C$ **41.** $y = 0.527x + 0.916$

43. 2.467 cu units   **45.** $2\sqrt{6}$ m   **47. a.** $\theta = \text{Arctan}\left(\dfrac{h}{x}\right)$ **b.** $\dfrac{d\theta}{dx} = \dfrac{-h}{x^2 + h^2}$

Review Exercises (page 1166)

[23.1, 23.2] **1.** $\dfrac{1}{x+1}$ **3.** 1   **5.** $\dfrac{(2x+3)\log e}{x^2 + 3x + 5}$ [23.2, 23.5] **7.** $2x(\cos x^2)e^{\sin x^2}$ [23.1, 23.3]   **9.** $-2t\tan t^2$

[23.5] **11.** $-2x\csc^2(x^2 + 3)$ [23.2] **13.** $2x(\ln 5)5^{(x^2+1)}$ [23.5] **15.** $-2t\cot(t^2 + 1)\csc(t^2 + 1)$

[23.6] **17.** $\dfrac{1}{(t+5)\sqrt{(t+5)^2 - 1}}$ [23.2, 23.5] **19.** $2e^{x^2}[x\tan(x+2)^2 + (x+2)\sec^2(x+2)^2]$ [23.1] **21.** 0.693

[23.2] **23.** $\dfrac{e^{x^3}}{3} + C$ [23.4] **25.** 0.373   **27.** $\dfrac{\sin^3 3x}{9} + C$ [23.5] **29.** $\dfrac{1}{3}\ln|\sec|(3z+4) + \tan(3z+4)| + C$ **31.** 0.344

[23.6] **33.** $\dfrac{1}{3}\text{Arctan}\,3x + C$ [23.2] **35.** $\ln|e^x - e^{-x}| + C$ [23.5] **37.** $\dfrac{-1}{5}\cot 5t + C$ [23.1] **39.** $y\left[\dfrac{1}{x} + \dfrac{2x}{3(1+x)^2}\right]$

41. $y\left[\dfrac{2x}{3(x^2 - 8)} + \dfrac{3x^2}{2(x^3 + 1)} - \dfrac{6x^5 - 7}{x^6 - 7x + 5}\right]$ [23.3] **43. a.** $256\pi^2\cos 8\pi t$ **b.** $256\pi^2$

45. Maximum $\left(\dfrac{\pi}{2}, 1\right)$; minimum $\left(\dfrac{3\pi}{2}, -1\right)$; points of inflection $(0, 0)$, $(\pi, 0)$, $(2\pi, 0)$ [23.2] **47.** $y = 2$

[23.4] **49.** 1.08 ℓ/s   [23.2] **51.** 20.1 min

[23.1] **55.** If $x > 0$, then $|x| = x$ and $\dfrac{d}{dx}(\ln|x|) = \dfrac{d}{dx}(\ln x) = \dfrac{1}{x}$; if $x < 0$, then $|x| = -x$ and $\dfrac{d}{dx}(\ln|x|) = \dfrac{d}{dx}$

$(\ln(-x)) = \dfrac{-1}{-x} = \dfrac{1}{x}$ [23.2] **57.** For $f(x) = e^x - 1 - x$, $f'(x) = e^x - 1$ and $f''(x) = e^x$. Setting $f'(x) = e^x - 1$ equal to

zero, we find $x = 0$; so the point $(0, 0)$ is either relative maximum, minimum, or point of inflection. $f''(x) = e^x$ cannot be zero, so there is no point of inflection. $f''(0) = +1$, so $(0, 0)$ is a relative minimum. Therefore, for $x \geq 0$, $f(x) \geq 0$; so $e^x - 1 - x \geq 0$ and $e^x \geq 1 + x$. For $x \leq 0$, $f(x) \geq 0$, and if we substitute $-x$ (x a positive number) for $x \leq 0$, then $e^{-x} - 1 + x \geq 0$ and $e^{-x} \geq 1 - x$.

[23.4] **59. a.** If $u = \sin x$, $\int \sin x \cos x\, dx = \dfrac{\sin^2 x}{2} + C_1$; if $u = \cos x$, $\int \sin x \cos x\, dx = -\dfrac{\cos^2 x}{2} + C_2$

b. $\dfrac{\sin^2 x}{2} + C_1 = \dfrac{1 - \cos^2 x}{2} + C_1 = \dfrac{1}{2} - \dfrac{\cos^2 x}{2} + C_1$; let $C_2 = \dfrac{1}{2} + C_1$, then $\dfrac{\sin^2 x}{2} + C_1 = -\dfrac{\cos^2 x}{2} + C_2$

[23.2] **61. a.** $\dfrac{d}{dx}(\sinh x) = \dfrac{d}{dx}\left(\dfrac{e^x - e^{-x}}{2}\right)$

$= \dfrac{e^x + e^{-x}}{2}$

$= \cosh x$

b. $\dfrac{d}{dx}(\cosh x) = \dfrac{d}{dx}\left(\dfrac{e^x + e^{-x}}{2}\right)$

$= \dfrac{e^x - e^{-x}}{2}$

$= \sinh x$

c. $\cosh^2 x - \sinh^2 x = \left(\dfrac{e^x + e^{-x}}{2}\right)^2 - \left(\dfrac{e^x - e^{-x}}{2}\right)^2$

$= \dfrac{e^{2x} + 2 + e^{-2x}}{4} - \dfrac{e^{2x} - 2 + e^{-2x}}{4}$

$= 1$

Chapter Test *(page 1170)*

1. $\dfrac{2x}{x^2+3}$ **2.** $6^{(x+1)}\ln 6$ **3.** $\sec^2 x$ **4.** $7\pi \sec^2 \pi x$ **5.** $6\tan 2x \sec 2x$ **6.** $-3\cot(3x+4)\csc(3x+4)$

7. $\dfrac{2}{\sqrt{1-(2x+3)^2}}$ **8.** $\dfrac{e^x}{1+e^{2x}}$ **9.** $\dfrac{1}{6}e^{3x^2}+C$ **10.** $\dfrac{6x^2}{2\ln 6}+C$ **11.** $\sin e^x + C$ **12.** $\dfrac{\sin^4 x}{4}+C$

13. $\dfrac{1}{2}\tan 2x + C$ **14.** $\dfrac{1}{3}\sec 3x + C$ **15.** 1.11   **16.** 1.05   **17.** 0.173   **18.** 0.50

19. $m = 0$

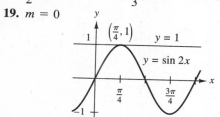

20. \$8,109,302.16

21. a. Relative maximum $\left(1, \dfrac{3}{e}\right)$ **b.** No minimum

   c. $\left(2, \dfrac{6}{e^2}\right)$ **d.**

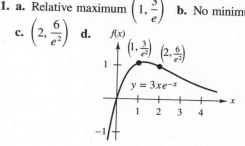

22. a. $\dfrac{dx}{dt} = -6.9\sin 3.0t$ **b.** -17 units/s^2

Chapter 24

Section 24.1 *(page 1176)*

1. $\dfrac{e^{2x}}{4}(2x-1)+C$ **3.** $\sin x - x\cos x + C$ **5.** $x\tan x - \ln|\sec x| + C$ **7.** $-x\cot x + \ln|\sin x| + C$

9. $-\cos e^x + C$ **11.** $\dfrac{x^2}{2}(\ln|x-1|)+C$ **13.** 3.49   **15.** $\dfrac{2x}{3}\sqrt{3x+1}-\dfrac{4}{27}(3x+1)^{\frac{3}{2}}+C$ **17.** $-\sqrt{4-x^2}+C$

19. $x\operatorname{Arccos} x - \sqrt{1-x^2}+C$ **21.** $\dfrac{x^2(x-3)^{12}}{12}-\dfrac{x(x-3)^{13}}{78}+\dfrac{(x-3)^{14}}{84}+C$ **23.** $\dfrac{e^x}{2}(\sin x + \cos x)+C$

25. 3.16   **27.** 9.86   **29.** $y = -2\cos x - x\sin x + 2$ **31.** 5.47   **33.** 3.59

35. $q = \dfrac{I_0 e^{-2t}}{7202}(-\cos 120t + 60\sin 120t)+\dfrac{I_0}{7202}$

Section 24.2 *(page 1185)*

1. $\dfrac{\sin^5 x}{5}-\dfrac{\cos^7 x}{7}+C$ **3.** $\dfrac{\cos^5 x}{5}-\dfrac{\cos^3 x}{3}+C$ **5.** $\dfrac{\cos^3 x}{3}-\cos x + C$ **7.** $\dfrac{\sin 2x}{2}-\dfrac{\sin^3 2x}{6}+C$

9. $\dfrac{\sin^4 x}{4}-\dfrac{\sin^6 x}{6}+C$ **11.** $\dfrac{x}{2}-\dfrac{\sin 2x}{4}+C$ **13.** $\dfrac{3x}{128}-\dfrac{\sin 4x}{128}+\dfrac{\sin 8x}{1024}+C$ **15.** $\dfrac{\tan^7 x}{7}+C$

17. $\dfrac{\sec^4 x}{4}-\sec^2 x + \ln|\sec x| + C$ **19.** $\dfrac{\sec^5 2x}{10}-\dfrac{\sec^3 2x}{6}+C$ **21.** $-\dfrac{\cot^3 x}{3}-\cot x + C$

23. $-\dfrac{\csc^5 x}{5}+\dfrac{\csc^3 2x}{3}+C$ **25.** 0.196   **27.** $\dfrac{\pi}{4}$ **29.** 0.180   **31.** 1.85   **33.** 3   **35.** 0.707 k

37. $\dfrac{\sec^2 x}{2}+C$

Section 24.3 (page 1190)

1. $-\dfrac{1}{3}(4-x^2)^{\frac{3}{2}}+C$ **3.** $\dfrac{(16+x^2)^{\frac{5}{2}}}{5}-\dfrac{4(16+x^2)^{\frac{3}{2}}}{3}+C$ **5.** $\dfrac{1}{3}(x^2-9)^{\frac{3}{2}}+C$ **7.** $-\dfrac{\sqrt{3-x^2}}{3x}+C$

9. $\dfrac{(9+x^2)^{\frac{3}{2}}}{3}-9\sqrt{9+x^2}+C$ **11.** $\dfrac{\sqrt{x^2-25}}{25x}+C$ **13.** $\dfrac{2}{27}\ln\left|\dfrac{2+3t}{\sqrt{4-9t^2}}\right|-\dfrac{t}{9}+C$

15. $-\dfrac{(4+t^2)^{\frac{3}{2}}}{48t^3}+\dfrac{\sqrt{4+t^2}}{16t}+C$ **17.** $-\dfrac{u}{9\sqrt{4u^2-9}}+C$ **19.** $\dfrac{x}{25\sqrt{25-x^2}}+C$ **21.** 0.0159

23. 0.159   **25.** $-\dfrac{(1-2e^x)^{\frac{3}{2}}}{3}+C$ **27.** 14.9   **29.** $\dfrac{2mR}{\ell\sqrt{R^2-\ell^2}}$ **31.** $k_q\ln\left(\dfrac{b^2+2a^2+2a\sqrt{a^2+b^2}}{b^2}\right)$

Section 24.4 (page 1199)

1. $\dfrac{1}{2}\ln\left|\dfrac{x-1}{x+1}\right|+C$ **3.** $\ln|(x+3)(x-1)^2|+C$ **5.** $\ln\left|\dfrac{(2x+1)^{\frac{5}{6}}}{(x-1)^{\frac{4}{3}}}\right|+C$ **7.** $\ln|x^2(x-9)^3|+C$

9. $\ln\left|\dfrac{(x-5)^3}{x+3}\right|+C$ **11.** $x^2+\ln\left|\dfrac{(x-4)^{\frac{3}{2}}}{(x+2)^{\frac{1}{2}}}\right|+C$ **13.** $\dfrac{2}{x+1}+\ln|x|+C$

15. $\dfrac{2}{x+1}-\dfrac{1}{2(x+1)^2}+\ln|(x+1)|+C$ **17.** $\dfrac{1}{2}\operatorname{Arctan}x+\ln\left|\dfrac{(x^2+1)^{\frac{1}{4}}}{(x+1)^{\frac{1}{2}}}\right|+C$

19. $\ln\left(\dfrac{x-1}{x+1}\right)^{\frac{1}{4}}+\dfrac{1}{2}\operatorname{Arctan}x+C$ **21.** 0.300   **23.** 28.2   **25.** 0.458   **27.** 0.901   **29.** $\dfrac{(c-a)x+(d-b)}{(x^2+1)^2}+\dfrac{ax+b}{x^2+1}$

Section 24.5 (page 1204)
(Number before answer indicates integration formula.)

1. 5; $\dfrac{(3x-7)^5}{15}+C$ **3.** 12; $2\sqrt{4t-11}-2\sqrt{11}\operatorname{Arctan}\sqrt{\dfrac{4t-11}{11}}+C$

5. 13b; $\dfrac{1}{\sqrt{13}}\ln\left|\dfrac{\sqrt{11Z+13}-\sqrt{13}}{\sqrt{11Z+13}+\sqrt{13}}\right|+C$ **7.** 18; $\dfrac{1}{8}\ln\left|\dfrac{4x+Z}{4-Z}\right|+C$ **9.** 21; $-\dfrac{1}{\sqrt{13}}\ln\left|\dfrac{\sqrt{13}+\sqrt{13+t^2}}{t}\right|+C$

11. 28; $-\dfrac{1}{\sqrt{7}}\ln\left|\dfrac{\sqrt{7}+\sqrt{7-x^2}}{x}\right|+C$ **13.** 46; $\dfrac{3x}{8}-\dfrac{\sin x}{16}-\dfrac{\sin^3 3x\cos 3x}{12}+C$

15. 48a; $-\dfrac{1}{2}\left(\cos\theta+\dfrac{\cos 5\theta}{5}\right)+C$ **17.** 54; $-\dfrac{\sin^3 A\cos^4 A}{7}-\dfrac{3\sin A\cos^4 A}{35}+\dfrac{3\sin A}{35}-\dfrac{\sin^3 A}{35}+C$

19. 56; $-\dfrac{1}{3}\tan\left(\dfrac{\pi}{4}-\dfrac{3\theta}{2}\right)+C$ **21.** 64; $-\dfrac{x^3}{2}\cos 2x+\dfrac{3x^2}{4}\sin 2x-\dfrac{3}{8}\sin 2x+\dfrac{3x}{4}\cos 2x+C$

23. 68; $\dfrac{1}{4}\tan^2 2x+\dfrac{1}{2}\ln|\cos 2x|+C$ **25.** 78; $x\operatorname{Arcsin}\dfrac{x}{3}+3\sqrt{1-\dfrac{x^2}{9}}+C$

27. 87; $\dfrac{e^{7x}}{2401}(343x^3-147x^2+42x-6)+C$ **29.** 92; $\dfrac{x^6}{6}\left(\ln\left|\dfrac{x}{5}\right|-\dfrac{1}{6}\right)+C$ **31.** 90; 60.3

33. 10;   **a.** 7.23 s   **b.** No; 3.72 s   **35.** 84; $\dfrac{1}{a}(e^a-e^{-a})$

Review Exercises (page 1025)

[24.1–24.4] **1.** $\dfrac{x^2}{2}\ln|x| - \dfrac{x^2}{4} + C$ **3.** $\sin\theta - \dfrac{1}{3}\sin^3\theta + C$ **5.** $\ln|x + \sqrt{x^2 + 25}| + C$ **7.** $\dfrac{e^{2x}}{2}\left(x + \dfrac{9}{2}\right) + C$

9. $\dfrac{1}{2}\left[x - \dfrac{\pi}{4} - \dfrac{1}{2}\sin\left(2x - \dfrac{\pi}{2}\right)\right] + C$ **11.** $\dfrac{2}{5}\cos^5\dfrac{\theta}{2} - \dfrac{2}{3}\cos^3\dfrac{\theta}{2} + C$ **13.** $\dfrac{1}{2(9 - x^2)} + C$

15. $\text{Arcsin}\dfrac{t}{8} + C$ **17.** $\dfrac{(\ln|x|)^2}{2} + C$ **19.** $\ln|(x + 7)^2(x - 2)^3| + C$ **21.** $2\tan x - 2\sec x - x + C$

23. $\text{Arctan}(\ln|x|) + C$ **25.** $\dfrac{\sin^3 x}{3} - \dfrac{\sin^5 x}{5} + C$ **27.** $x\ln^2|x| - 2x\ln|x| + 2x + C$

29. $\dfrac{e^{3x}}{9}(3x - 1) + C$ [24.5] **31.** 7.00   **33.** 252   **35.** 0.1606   **37.** $\dfrac{1}{24}\sec^3 8x + C$ **41.** $8\pi^2 \approx 78.96$ cu units

Chapter Test (page 1027)

1. $\dfrac{1}{4}xe^{4x} - \dfrac{1}{16}e^{4x} + C$ **2.** $-\dfrac{1}{2}x\cos 2x + \dfrac{1}{4}\sin 2x + C$ **3.** $\dfrac{1}{4}e^{2x}(\sin 2x - \cos 2x) + C$ **4.** 43.7

5. $\dfrac{1}{7}\sin^7 x - \dfrac{1}{9}\sin^9 x + C$ **6.** $\dfrac{2}{3}\cos^3 x - \dfrac{1}{5}\cos^5 x - \cos x + C$ **7.** $\dfrac{1}{4}\sin^4 x - \dfrac{1}{6}\sin^6 x + C$ **8.** $\dfrac{1}{5}\tan^5 x + C$

9. 9.00   **10.** $\dfrac{1}{5}(25 + x^2)^{\frac{5}{2}} - \dfrac{25}{3}(25 + x^2)^{\frac{3}{2}} + C$ **11.** $\dfrac{1}{5}(x^2 - 25)^{\frac{5}{2}} + \dfrac{25}{3}(x^2 - 25)^{\frac{3}{2}} + C$

12. $-\dfrac{1}{3}\tan^3 x - \tan x + x + C$ **13.** $\dfrac{1}{8}\ln\left|\dfrac{3x - 4}{3x + 4}\right| + C$ **14.** $\dfrac{5}{6}\ln|x - 1| + \dfrac{7}{6}\ln|x + 5| + C$

15. 2.90   **16.** 0.86

In answers 17–20, the number preceding the answer is the integration formula number.

17. $8; \dfrac{3}{5}x + \dfrac{21}{25}\ln|5x - 7| + C$ **18.** $20; \dfrac{1}{2}(x - 3)\sqrt{x^2 - 6x + 13} + 2\ln|(x - 3) + \sqrt{x^2 - 6x + 13}| + C$

19. $12\ \&\ 13b; 2\sqrt{4x + 5} + \dfrac{5}{\sqrt{5}}\ln\left|\dfrac{\sqrt{4x + 5} - \sqrt{5}}{\sqrt{4x + 5} + \sqrt{5}}\right| + C$ **20.** $55; \dfrac{1}{21}\sin^5 3x\cos^2 3x + \dfrac{2}{105}\sin^5 3x + C$

Chapter 25

Section 25.1 (page 1212)

1. a. Linear   **b.** 1   **3. a.** Nonlinear   **b.** 2   **5. a.** Linear   **b.** 6   **7. a.** Linear   **b.** 3   **9. a.** Nonlinear   **b.** 1

11. $y' = e^x + 2e^{2x}; y'' = e^x + 4e^{2x}: (e^x + 4e^{2x}) - 3(e^{2x} + 2e^{2x}) + 2(e^x + e^{2x}) = 0$

13. $y' = -\ln x, xy' + x - y = x(-\ln x) + x - x + x\ln x = 0$ **15.** $y' = e^{-x} - (x + C)e^{-x}; y' + y = e^{-x}$

17. $xy = 2$ **19.** $y = \sin x + 1$ **21.** $y = \ln\dfrac{|x|}{2}$ **23.** $y = -\sin x + \dfrac{\sqrt{2}}{2}x - \dfrac{3\pi\sqrt{2}}{4}$

Section 25.2 (page 1217)

1. $xy = C$ **3.** $\dfrac{y^3}{3} = -\dfrac{x^3}{3} + C$ **5.** $y = Cx^5$ **7.** $\sec y + \tan y = C(\csc x - \cot x)^{-1}$ **9.** $y = \dfrac{-1}{e^{-x} + C}$

11. $y = \pm\sin(x + C)$ **13.** $y = -\tan(\text{arctan } x + C)$ **15.** $y = \sin(\arcsin x + C)$

17. $(y - 1)e^y = -e^{-x} - \dfrac{1}{3}e^{-3x} + C$ **19.** $y^2 = 2t - \dfrac{2}{t} + C$ **21.** $r^2 = 9\theta$ **23.** $4y^3 = x^2$

25. $y = 2(\sin x)^{-\frac{2}{3}}$ **27.** $y = \dfrac{x - 13}{x - 1}$

Section 25.3 (page 1223)

1. $xy = y^3 + C$ **3.** $\dfrac{x}{y} = \dfrac{x^3}{3} + C$ **5.** $\dfrac{x}{y} = -x + C$ **7.** $\dfrac{y^2}{x} = \dfrac{x^2}{2} + C$ **9.** $\dfrac{x^2}{y} = \dfrac{x^2}{2} + C$ **11.** $y = x \ln|x| + Cx$

13. $y = Cx^4 - x$ **15.** $e^{\frac{x}{y}} = \dfrac{C}{x}$ **17.** $\arctan \dfrac{y}{x} = \ln\left|Cx\left(1 + \dfrac{y^2}{x^2}\right)^{\frac{1}{2}}\right|$ **19.** $-\dfrac{1}{2}\ln\left|\left(\dfrac{y}{x}\right)^2 + 4\left(\dfrac{y}{x}\right) + 1\right| = \ln Cx$

21. $\left(\dfrac{2y^2}{x^2} + 1\right)^{\frac{1}{4}} = 9^{\frac{1}{4}}x$ **23.** $\dfrac{y}{4 + t} = \dfrac{2}{3}t$

Section 25.4 (page 1227)

1. $y = xe^{-x} + Ce^{-x}$ **3.** $y = \dfrac{x^2}{4} + Cx^{-2}$ **5.** $y = x^3 e^x + Ce^x$ **7.** $y = -\dfrac{2}{13}(\cos 3x + \sin 3x) + Ce^{2x}$

9. $y = -x - x^3 + Cx^2$ **11.** $y = \dfrac{\sin x}{2} + \dfrac{C}{\sin x}$ **13.** $y = \dfrac{\ln t}{2} + \dfrac{C}{\ln t}$ **15.** $y = \dfrac{t^2}{5} + t + Ct^{-3}$

17. $y = \dfrac{1}{5}(-\sin 2t - 2\cos 2t) + Ce^t$ **19.** $y = \dfrac{C - \ln t}{t}$

21. General: $y = 2t + 3 + Ce^{-t}$; specific: $y = 2t + 3 + e^{-t}$

23. General: $y = e^{2t}(-t + C)$; specific: $y = e^{2t}(-t + 2)$

25. General: $y = (t^2 + 1)(\arctan t + C)$; specific: $y = \dfrac{1}{4}(t^2 + 1)(4 \arctan t + \pi)$

27. General: $y = \dfrac{4C - 2x \cos 2x + \sin x}{4x^3}$; specific: $y = \dfrac{-\pi + 2x \cos 2x - \sin 2x}{4x^3}$

Section 25.5 (page 1234)

1. $16y^2 + 9x^2 = 144$ **3.** $y = \dfrac{x^2}{3} - \dfrac{52}{3x}$ **5.** $xy = C$ **7.** 96 g **9.** $y = 3x^2 - 5$ **11.** $x = 6t + 12 - 28e^{(\frac{t}{2} - 1)}$

13. a. $\theta(t) = ce^{-kt}$ **b.** $\theta(t) = 60 + 540e^{-0.145t}$ **15.** $t = 10.1$ s; $v = -98.99$ m/s

17. $i(t) = \dfrac{30}{13}e^{-20t} - \dfrac{30}{13}\cos 30t + \dfrac{20}{13}\sin 30t$ **19.** $q = q_0 e^{kt}$ **21.** $i = \dfrac{E}{R} - \dfrac{E}{r}e^{-\frac{(RT)}{L}}$ **23.** $t = 11.4$

25. $v(t) = 16(1 - e^{-4t})$; $s(t) = 16\left(t + \dfrac{1}{4}e^{-4t}\right) - 4$ **27.** $P(x) = \dfrac{1}{4}e^{-4x} - \dfrac{1}{4}e^{-3x}$

Review Exercises (page 1237)

[25.1] **1. a.** Linear **b.** 1 **3. a.** Nonlinear **b.** 3 **5. a.** Nonlinear **b.** 1 **7.** Verifies **9.** Verifies

[25.2] **11.** $y = Cx$ **13.** $-\ln|\csc y - \cot y| = \ln|C \sin x|$ **15.** $y = \dfrac{C}{x^2 - 1}$ **17.** $x^2y = C$

[25.3] **19.** $y = Cx$ **21.** $y = 2 - Ce^{-\frac{x^2}{2}}$ **23.** $y = \dfrac{C\left(\cos \dfrac{x}{2} + \sin \dfrac{x}{2}\right)}{\cos \dfrac{x}{2} - \sin \dfrac{x}{2}}$

[25.4] **25.** $s = \dfrac{1}{4} - \dfrac{1}{2}t - \dfrac{5}{4}e^{-2t}$ **27.** $y = \arcsin(Cx^2)$ **29.** $-\dfrac{1}{b}e^{-by} = \dfrac{1}{a}e^{ax} - \dfrac{a+b}{ab}$

[25.5] **31.** $y = 3x^{\frac{1}{3}}$ **33.** 2 hr   **35. a.** $s = 164$ ft   **b.** $v = -64$ ft/s   **c.** $t = 5.2$ s, $v = -102.4$ ft/s

Chapter Test (page 1239)

1. Linear, homogenous   **2.** Nonlinear, separable   **3.** Linear, homogeneous   **4.** Separable   **5.** Verifies

6. Verifies   **7.** $y = C(1 + x)$ **8.** $y = -x + C$ **9.** $xy = \dfrac{4}{3}y^3 + C$ **10.** $y = -x\ln\left(\dfrac{1}{x}\right) + Cx$

11. $y = \dfrac{x^3}{6} + Cx^{-3}$ **12.** $y^2 + x^2 = 100$ **13.** $y = \dfrac{4}{7}e^{2t} + \dfrac{3}{7}e^{-5t}$ **14.** $y = x\ln x^2 + 3x$

15. $y = 3\sin x - 2\cos x - 1$ **16.** $y = -\dfrac{1}{2}\ln x + C$ **17.** 14.3 s, 140 m/s

Chapter 26

Section 26.1 (page 1246)

1. $y = -C_1 e^{-x} + C_2$ **3.** $y = C_1 e^{4x} + C_2 e^{-4x}$ **5.** $y = C_1 e^{2x} + C_2 e^{3x}$ **7.** $y = C_1 e^{-4x} + C_2 e^{-x}$

9. $y = \dfrac{2}{3}C_1 e^{\frac{2}{3}x} + C_2$ **11.** $y = C_1 e^{5x} + C_2 e^{-2x}$ **13.** $y = C_1 e^{-4x} + C_2 e^{3x}$ **15.** $y = C_1 e^{-7x} + C_2 e^{\frac{3}{2}x}$

17. $y = C_1 e^{-4x} + C_2 e^{\frac{5}{3}x}$ **19.** $y = -\dfrac{1}{3}C_1 e^{-3x} + \dfrac{1}{7}C_2 e^{7x} + C_3$ **21.** $y = C_1 e^{-2x} + C_2 e^{2x} + C_3 e^{3x}$

23. $(D^2 + D - 6)y = 0$ **25.** $(D^3 - 9D)y = 0$ **27.** $y = \dfrac{3}{4}e^{-4x} + \dfrac{5}{4}e^{4x}$ **29.** $y = -\dfrac{3}{4}e^{-5x} + \dfrac{3}{4}e^{-x}$

Section 26.2 (page 1251)

1. $y = e^{-x}(C_1 + C_2 x)$ **3.** $y = e^{-7x}(C_1 + C_2 x)$ **5.** $y = C_4 x^3 + C_3 x^2 C_2 x + C_1$ **7.** $y = e^{2x}(C_1 \cos x + C_2 \sin x)$

9. $y = C_1 \cos x + C_2 \sin x$ **11.** $y = e^x(C_1 \cos \sqrt{3}\,x + C_2 \sin \sqrt{3}\,x)$ **13.** $y = e^x(C_1 e^{-\sqrt{2}x} + C_2 e^{\sqrt{2}x})$

15. $y = e^{-\frac{x}{2}}\left(C_1 \cos \dfrac{x}{3} + C_2 \sin \dfrac{x}{3}\right)$ **17.** $y = e^{-x}(C_1 + C_2 x + C_3 e^{\frac{5x}{2}})$

19. $y = C_1 e^{\sqrt{2}x} + C_2 e^{-\sqrt{2}x} + C_3 \cos \sqrt{2}\,x + C_4 \sin \sqrt{2}x$ **21.** $y = e^{4-2x}(4x - 7)$

23. $y = e^{-\frac{k}{2}x}\left[C_1 \cos\left(\dfrac{\sqrt{-(k^2 + 4\ell)}}{2}x\right) + C_2 \sin\left(\dfrac{\sqrt{-(k^2 + 4\ell)}}{2}x\right)\right]$ **25.** $(D^2 + 2D - 8)y = 0$ **27.** $(D^2 + 4)y = 0$

Section 26.3 (page 1258)

1. Verifies   **3.** Verifies   **5.** Verifies   **7.** $y = C_1 e^{-x} + C_2 e^{5x} - 2$ **9.** $y = C_1 e^{2x} + C_2 x e^{2x} + \dfrac{4}{5}\cos x + \dfrac{3}{5}\sin x$

11. $y = C_1 e^{-3x} + C_2 e^x + e^{2x}$ **13.** $y = C_1 - C_2 e^{-x} - 3x^2 + 6x$ **15.** $y = C_1 e^{-2x} + C_2 e^{-x} + 6$

17. $y = C_1 e^{-x} + C_2 e^{2x} + 2e^{3x}$ **19.** $y = C_1 e^{-2x} + C_2 e^x - \dfrac{1}{5}\cos 2x - \dfrac{3}{5}\sin 2x$

21. $y = C_1 e^{3x} + C_2 e^{4x} + x^2 + 2x + \dfrac{1}{12}$ **23.** $y = 2\cos x + \sin x + x^2 - 2$ **25.** $y = \cos x + 2x$

27. $y = 5e^{2x}(e^x - 1)$ **29.** $y = -\dfrac{8\cos 2x + 11\sin x}{3}$

Section 26.4 (page 1268)

1. a. $\dfrac{d^2x}{dt^2} + 64x = 0,\ x(0) = \dfrac{1}{3},\ x'(0) = 0$ **b.** $x(t) = \dfrac{\cos 8t}{3}$

c.

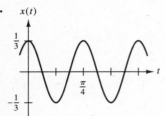

d. $x\left(\dfrac{1}{2}\text{ s}\right) = -0.218$ ft; $x'\left(\dfrac{1}{2}\text{ s}\right) = 2.018$ ft/s; $x''\left(\dfrac{1}{2}\text{ s}\right) = 13.94$ ft/s^2

3. a. $\dfrac{d^2x}{dt^2} + 32x = 0,\ x(0) = 0,\ x'(0) = 3$ ft/s   **b.** $x(t) = \dfrac{3\sqrt{2}}{8}\sin 4\sqrt{2}\,t$

c.

x(t)

0.50

0.56   1.11

0.28   0.83   1.34   t

−0.50

d. $x\left(\dfrac{1}{4}\text{ s}\right) = 0.523$ ft; $x'\left(\dfrac{1}{4}\text{ s}\right) = 0.468$ ft/s; $x''\left(\dfrac{1}{4}\text{ s}\right) = -16.76$ ft/s^2

5. a. $\dfrac{d^2x}{dt^2} + 128x + 0,\ x(0) = \dfrac{1}{3}$ ft, $x'(0) = 0$; $x(t) = \dfrac{\cos 8\sqrt{2}t}{3}$ **b.** $\dfrac{\pi\sqrt{2}}{8}$ s   **c.** $\dfrac{4\sqrt{2}}{\pi}$ cycles/s

7. a. $y(t) = C_1 \cos 10t + C_2 \sin 10t$ **b.** $y(0) = 10,\ y'(0) = 50$: $y(t) = 10 \cos 10t + 5 \sin 10t$ **c.** $\dfrac{5}{\pi}$ **d.** $\dfrac{\pi}{5}$ **e.** 11.2

9. a. 2π **b.** e^{-3t} **c.** 0.231 s   **11.** Critically damped   **13.** $x(t) = e^{-4t}\left(\dfrac{1}{2}\cos 2\sqrt{5}\,t - \dfrac{7\sqrt{5}}{10}\sin 2\sqrt{5}\,t\right) + \dfrac{3}{2}\sin 6t$

15. $x(t) = \dfrac{192}{6319}\cos\dfrac{9t}{4} - \dfrac{192}{6319}\cos 20t$ **17.** $x(t) = e^{-8t}\left(\dfrac{25}{48}\cos 4\sqrt{2}\,t + \dfrac{41\sqrt{2}}{96}\sin 4\sqrt{2}\,t\right) + \dfrac{3}{16}\sin 4t - \dfrac{3}{16}\cos 4t$

19. a. $\dfrac{1}{10}\dfrac{d^2q}{dt^2} + 40\dfrac{dq}{dt} + 200q = e(t),\ q(0) = 2,\ q'(0) = 0$

b. $q_c(t) = \left(1 + \dfrac{10}{\sqrt{95}}\right)e^{20(-10+\sqrt{95})t} + \left(1 - \dfrac{10}{\sqrt{95}}\right)e^{-20(10+\sqrt{95})t}$

$i(t) = \left(1 + \dfrac{10}{\sqrt{95}}\right)(-200 + 20\sqrt{95})e^{20(-10+\sqrt{95})t} + \left(1 - \dfrac{10}{\sqrt{95}}\right)(-200 - 20\sqrt{95})e^{-20(10+\sqrt{95})t}$

c. $q(t) = \left(\dfrac{19}{20} + \dfrac{19}{2\sqrt{95}}\right)e^{20(-10+\sqrt{95})t} + \left(\dfrac{19}{20} + \dfrac{19}{2\sqrt{95}}\right)e^{-20(10+\sqrt{95})t} + \dfrac{1}{100}$

21. General: $Q(t) = e^{-14t}(C_1 \cos 7t + C_2 \sin 7t) + Q_p(t)$; steady-state: $Q_p(t) = \dfrac{1}{56}\cos 7t + \dfrac{1}{56}\sin 7t$

Section 26.5 (page 1274)

1. $\dfrac{2}{s^3}$ **3.** $\dfrac{1}{(s-a)^2}$ **5.** $\dfrac{s}{a^2+s^2}$ **7.** $\dfrac{6}{s^4}$ **9.** $\dfrac{1}{(s-5)^2}$ **11.** $\dfrac{4}{(s-3)^2+16}$ **13.** $\dfrac{4}{s^2}-\dfrac{5}{s^2+25}$ **15.** $\mathscr{L}\{f(t)\}=-\dfrac{s+2}{s^2-3}$

17. $\mathscr{L}\{f(t)\}=\dfrac{3(4-2s)}{3s^2+5}$ **19.** $\mathscr{L}\{f(t)\}=\dfrac{15s-7}{15s^2-7s-2}$

Section 26.6 (page 1278)

1. 3   **3.** t^5 **5.** $\dfrac{7}{6}t^3e^{5t}$ **7.** $3\cos\sqrt{7}t+\dfrac{5\sqrt{7}}{7}\sin\sqrt{7}t$ **9.** $3t\cos 2t$ **11.** $1-e^{-2t}$ **13.** $1+t-\cos t-\sin t$

Section 26.7 (page 1284)

1. $y(t)=e^{5t}$ **3.** $y(t)=-\dfrac{2}{3}+\dfrac{5}{3}e^{3t}$ **5.** $y(t)=2e^{2t}-e^{3t}$ **7.** $y(t)=\dfrac{51}{25}e^{-3t}-\dfrac{1}{25}e^{2t}+\dfrac{1}{5}te^{2t}$

9. $y(t)=2\cos 2t+\dfrac{1}{2}\sin 2t$ **11.** $y(t)=\cos 3t+\sin 3t+1$ **13.** $y(t)=-\dfrac{3}{8}e^{2t}+\dfrac{3}{4}te^{2t}+\dfrac{1}{8}(3\cos 2t+4\sin 2t)$

15. $y(t)=6-5e^{-t}-6te^{-t}$ **17.** $y(t)=t\cos t+\dfrac{2}{3}\sin t-\dfrac{5}{6}\sin 2t$

19. $q(t)=\dfrac{10}{17}e^{-\frac{5t}{2}}-\dfrac{25}{52}e^{-2t}+\dfrac{5}{884}(9\sin 10t-19\cos 10t)$; $i(t)=-\dfrac{25}{17}e^{-\frac{5t}{2}}+\dfrac{25}{26}e^{-2t}+\dfrac{5}{884}(90\cos 10t-190\sin 10t)$

21. a. 0.036 g/cm^3 **b.** 0.073 g/cm^3 **23. a.** 0.047 g/cm^3 **b.** 0.075 g/cm^3

Review Exercises (page 1285)

[26.1, 26.2, 26.3] 1. $C_1e^{4x}+C_2e^{-5x}$ **3.** $C_1e^{-4x}+C_2e^{-7x}$ **5.** $C_1e^{3x}+C_2xe^{3x}$ **7.** $C_1x^2+C_2x+C_3$

9. $C_1\cos x+C_2\sin x+C_3$ **11.** $C_1e^{-2x}+C_2e^{7x}-\dfrac{4}{7}$ **13.** $C_1e^{-11x}+C_2e^{11x}-\dfrac{3}{125}\sin 2x$

15. $C_1\cos 2x+C_2\sin 2x-\dfrac{5}{21}\cos 5x$ **17.** $e^{2x}(C_1e^x+C_2xe^x+5)$ **19.** $C_1e^{3x}+C_2xe^{3x}+C_3x^2e^{3x}-\dfrac{1}{27}(x+2)$

21. $\dfrac{1}{16}e^{-t}(23-4t-32e^t+9e^{4t})$ **23.** $\dfrac{17}{15}e^{-2t}+\dfrac{1}{6}e^t-\dfrac{13}{10}\cos t+\dfrac{31}{10}\sin t$ **25.** $e^{2x}+\dfrac{3}{4}e^{-3x}+\dfrac{17}{4}e^x$

27. $x^3-3x^2+6x+2e^{-x}-1$ **29.** $\dfrac{e^{-5x}}{10}(-3+15e^{4x}-2e^{4x}\cos 2x+4e^{4x}\sin 2x)$

[26.4] 31. $q(t)=\dfrac{1}{50}(1-e^{-500t})$; $i(t)=10e^{-500t}$ **33.** $x(t)=C_1e^{-4t}+C_2e^{-16t}$; overdamped

[26.5, 26.6, 26.7] 35. $y(t)=e^{-t}(t+1)$ **37.** $y(t)=\dfrac{1}{8}e^t-\dfrac{1}{40}e^{-3t}-\left(\dfrac{1}{5}\sin t+\dfrac{1}{10}\cos t\right)$

39. $y(t)=4-\dfrac{19}{2}e^{-t}+\dfrac{1}{2}e^{5t}-5t$ **41.** $y(t)=\dfrac{46}{65}e^{-9t}-\dfrac{27}{130}e^{-4t}+\dfrac{12}{13}\sin 6t$

[All] 43. a. $\mu=2$: $y(t)=-e^{-2t}\left(\dfrac{\sqrt{3}}{6}\sin 2\sqrt{3}t\right)$; $\mu=4$: $y(t)=2te^{-4t}$; $\mu=5$: $y(t)=\dfrac{1}{3}e^{-8t}(e^{6t}-1)$

b. $\mu=2$: $y(t)=-2e^{-2t}(\cos 2\sqrt{3}t)$; $\mu=4$: $y(t)=-2e^{-4t}(4t+1)$; $\mu=5$: $y(t)=\dfrac{2}{3}e^{-8t}(4e^{6t}-1)$

45. $y(t)=C_1e^{(-1+\sqrt{1-\beta})t}+C_2e^{(-1+\sqrt{1+\beta})t}$ **47.** $y''(x)+y(x)=2e^x$ **49.** $y(t)=\dfrac{1}{6}t^3e^{-t}$

51. $x(t)=y(t)=-\dfrac{1}{4}+\dfrac{1}{8}e^{2t}+\dfrac{1}{8}e^{-2t}$

Chapter Test (page 1287)

1. $\dfrac{C_1}{4}e^{4x} + C_2$ **2.** $C_1e^{-2x} + C_2e^{2x}$ **3.** $C_1e^{7t} + C_2e^{t}$ **4.** $e^{5t}(C_1 + C_2t)$ **5.** $C_1e^{-3x} + C_2xe^{-3x} + C_3e^{4x} + C_4e^{-4x}$

6. $C_1e^{-x} + C_2e^{x} + C_3e^{4x}$ **7.** $(D^3 - 9D)y = 0$ **8.** $\dfrac{3}{2}e^{-x} - \dfrac{1}{2}e^{-5x}$ **9.** $(D - 4)^2y = 0$

10. $C_1e^{-3x} + C_2e^{-x} + \dfrac{1}{3}x^2 - \dfrac{8}{9}x + \dfrac{44}{27}$ **11.** $C_1e^{3x} + C_2xe^{3x} + \dfrac{6}{25}\cos x + \dfrac{8}{25}\sin x$ **12.** $C_1e^{-4x} + C_2e^{x} + \dfrac{1}{2}e^{3x}$

13. $\dfrac{e^{x}(3 + e^{\frac{\pi}{2}})}{2(1 + e^{\frac{\pi}{2}})} + \dfrac{e^{\frac{\pi}{2}-x}(3e^{\frac{\pi}{2}} - 1)}{2(e^{\frac{\pi}{2}} + 1)} - \dfrac{\cos x}{2}$ **14.** $\dfrac{14}{s^3}$ **15.** $\dfrac{5}{(s - 3)^2 + 25}$ **16.** $\dfrac{2s - 1}{2s^2 - 5s + 2}$

17. $3\cos\sqrt{5}\,t$ **18.** $y(t) = e^{3t}$

19. $q(t) = \dfrac{44}{29}e^{-2t} - \dfrac{33}{25}e^{-\frac{5t}{3}} - \dfrac{143}{725}\cos 5t + \dfrac{121}{725}\sin 5t;\ i(t) = -\dfrac{88}{29}e^{-2t} + \dfrac{11}{5}e^{-\frac{5t}{3}} + \dfrac{121}{145}\cos 5t + \dfrac{143}{145}\sin 5t$

20. a. $\dfrac{d^2x}{dt^2} + 256x = 0,\ x(0) = \dfrac{1}{4}$ ft, $x'(0) = 0$ **b.** $x(t) = \dfrac{1}{4}\cos 16t$ **c.**

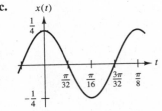

d. $x\!\left(\dfrac{\pi}{64}\right) = \dfrac{\sqrt{2}}{8},\ x'\!\left(\dfrac{\pi}{64}\right) = -2\sqrt{2},\ x''\!\left(\dfrac{\pi}{64}\right) = -32\sqrt{2}$

Index